27th International DAAAM Symposium on Intelligent Manufacturing and Automation 2016

Mostar, Bosnia and Herzegovina
26 – 29 October 2016

Editor:

B. Katalinic

ISBN: 978-1-5108-3300-5

Printed from e-media with permission by:

Curran Associates, Inc.
57 Morehouse Lane
Red Hook, NY 12571

Some format issues inherent in the e-media version may also appear in this print version.

Printed by Curran Associates, Inc. (2016)

For permission requests, please contact DAAAM International Vienna
at the address below.

DAAAM International Vienna
Karlsplatz 13/311, A-1040
Wien, Austria, European Union

Phone: 0043-1-58801-31121
Fax: 0043-1-58801-31199

secretary@daaam.info

Additional copies of this publication are available from:

Curran Associates, Inc.
57 Morehouse Lane
Red Hook, NY 12571 USA
Phone: 845-758-0400
Fax: 845-758-2633
Email: curran@proceedings.com
Web: www.proceedings.com

ISSN 1726-9679

DANUBE ADRIA ASSOCIATION FOR AUTOMATION & MANUFACTURING

DAAAM INTERNATIONAL VIENNA

www.daaam.info

PROCEEDINGS

OF THE 27TH INTERNATIONAL DAAAM SYMPOSIUM
"INTELLIGENT MANUFACTURING & AUTOMATION
26-29TH OCTOBER 2016
MOSTAR, BIH

ORGANIZED BY:
DAAAM INTERNATIONAL VIENNA & UNIVERSITY OF MOSTAR

CO - ORGANIZED BY:
VIENNA UNIVERSITY OF TECHNOLOGY, AUSTRIAN SOCIETY OF ENGINEERS AND ARCHITECTS – ÖIAV 1848, INTERNATIONAL ACADEMY OF ENGINEERING IAE-CEB, UNIVERSITY OF APPLIED SCIENCES – TECHNIKUM, COUNCIL OF MINISTERS OF BOSNIA AND HERZEGOVINA, GOVERNMENT OF FEDERATION OF BOSNIA AND HERZEGOVINA, HERZEGOVINA-NERETVA CANTON GOVERNMENT, CITY OF MOSTAR

UNDER THE AUSPICES OF: THE DANUBE RECTORS' CONFERENCE & RECTORS' HONOR COMMITTEE OF DAAAM INTERNATIONAL

EDITOR: B.[RANKO] KATALINIC

IMPRESSUM
2017

Proceedings of the 27th International DAAAM Symposium "Intelligent Manufacturing & Automation"
26-29th October 2016, Mostar, BiH, Volume 27, No.1,
Abbreviated key title: ***Ann. DAAAM 2017***

ISSN 1726-9679
Ed. B. Katalinic,

Published by DAAAM International, Vienna, Austria, EU, 2016
Layout & Design: B[ranko] Katalinic/ D[amir] Haskovic
Technical Editors: D[amir] Haskovic / I[lija] Zec / R[oberta] Cvetkovska / I[lya] Kukushkin

Publisher:

DAAAM International Vienna
TU Wien, Karlsplatz 13/311, A-1040 Vienna, Austria
E-mails: president@daaam.com / branko.katalinic@tuwien.ac.at
DAAAM homepage: http://www.daaam.info

CONTENTS

AUTHOR INDEX

A

B

P

Q

R

ORGANIZING COMMITTEE FOR 2016

COVIC, DRAGAN – PRESIDENT OF ORGANIZING COMMITTEE
KATALINIC, BRANKO – PRESIDENT OF DAAAM INTERNATIONAL
OSTOJIC, LJERKA – RECTOR OF THE UNIVERSITY OF MOSTAR
GUSEV, BORIS – PRESIDENT OF INTERNATIONAL ACADEMY OF ENGINEERING
UGLESIC, ANTE – COORDINATOR OF ORGANIZING COMMITTEE

BESLIC, LJUBO
BEVANDA, VJEKOSLAV
BOBAN, INJA
BUCHMEISTER, BORUT
COLAK, IVO
COSIC, ILIJA
CVETKOVSKA, ROBERTA
DANKIC, IZABELA
EGOROV, SERGEY
FILARETOV, VLADIMIR
GILJA, MARINKO
HASKOVIC, DAMIR (SECRETARY)
HERCEG, NEVENKO
HOELLER, LIANE
KUKUSHKIN, ILYA
LASIC, DENIS
LEKO, TOMISLAV
LUCIC, MATE
MAJSTOROVIC, VLADO
MILICEVIC, JELKA
NAUMBURGER, JOERG
PANFILOV, PETER
PARK, HONG-SEOK
PLANINIC, IGOR
PRIMORAC, VILIM
PRYANICHNIKOV, VALENTIN
SCHMOELLEBECK, FRITZ
STAZHKOV, SERGEY
STOJKIC, ZELJKO
STOPPER, MARKUS
TEKIC, ANJA
TEKIC, ZELJKO
TOMIC, ZORAN
UDILJAK, TOMA
USHKOVA, NATALIA
ZAVRAZHINA, ALINA
ZEC, ANTE
ZEC, ILIJA
ZEC, LUKA

RECTORS' AND PRESIDENTS' HONOR COMMITTEE

ADAMCZAK, STANISLAW	RECTOR, TECHNICAL UNIVERSITY OF KIELCE
AAVIKSOO, JAAK	RECTOR, TALLINN UNIVERSITY OF TECHNOLOGY
ABRUDAN, IOAN VASILE	RECTOR, TRANSILVANIA UNIVERSITY OF BRASOV
ALEXANDROV, ANATOLIY	RECTOR, BAUMAN MOSCOW STATE TECHNICAL UNIVERSITY
ANDJELINOVIC, SIMUN	RECTOR, UNIVERSITY OF SPLIT
ANTRANIKIAN, GARABED	PRESIDENT, HAMBURG UNIVERSITY OF TECHNOLOGY
AVDISPAHIC, MUHAREM	RECTOR, UNIVERSITY OF SARAJEVO
BAAIJENS, FRANK	RECTOR, EINDHOVEN UNIVERSITY OF TECHNOLOGY
BEIBST, GABRIELE	RECTOR, JENA UNIVERSITY OF APPLIED SCIENCE
BENJAMIN, FLOELLA	CHANCELLOR, THE UNIVERSITY OF EXETER
BORAS, DAMIR	RECTOR, UNIVERSITY OF ZAGREB
BRUNGS, ATTILA	VICE-CHANCELLOR AND PRESIDENT, UNIVERSITY OF TECHNOLOGY, SYDNEY
CHEON, YEON	PRESIDENT, UNIVERSITY OF ULSAN
COSTOIU, MIHNEA	RECTOR, UNIVERSITY POLITEHNICA OF BUCHAREST
DANIELS, MITCHELL	PRESIDENT, PURDUE UNIVERSITY
DEMA, MARJAN	RECTOR, UNIVERSITY OF PRISHTINA
DI SCIASCIO, EUGENIO	RECTOR, POLYTECHNIC OF BARI
FALKOV, VALERIY	RECTOR, TYUMEN STATE UNIVERSITY
FRUNZAVERDE, DOINA	RECTOR, "EFTIMIE MURGU" UNIVERSITY OF RESITA
FURTAK, KAZIMIERZ	RECTOR, CRACOW UNIVERSITY OF TECHNOLOGY
GALIJEV, ILJHAM	PRESIDENT, OMSK STATE TRANSPORT UNIVERSITY

GAZALIYEV, ARSTAN	RECTOR, KARAGANDA STATE TECHNICAL UNIVERSITY
GELENCSER, ANDRAS	RECTOR, UNIVERSITY OF PANNONIA
GILLI, MARCO	RECTOR, POLITECNICO DI TORINO
GOIRIZELAIA, INAKI	RECTOR, UNIVERSITY OF THE BASQUE COUNTRY
GOLIAS, IOANNIS	RECTOR, NATIONAL TECHNICAL UNIVERSITY OF ATHENS
GRIGORYEV, SERGEY	RECTOR, MOSCOW STATE UNIVERSITY OF TECHNOLOGY "STANKIN"
GUDMUNDSON, PETER	PRESIDENT, ROYAL INSTITUTE OF TECHNOLOGY, STOCKHOLM
GUSEV, BORIS V.	PRESIDENT, RUSSIAN ACADEMY OF ENGINEERING, MOSCOW
HAMADA, JUNICHI	PRESIDENT, UNIVERSITY OF TOKYO
HANNULA, MIKA	PRESIDENT, TAMPERE UNIVERSITY OF TECHNOLOGY
HANSELKA, HOLGER	RECTOR, KARLSRUHE INSTITUTE OF TECHNOLOGY
HANSELKA, HOLGER	PRESIDENT, KARLSRUHE INSTITUTE OF TECHNOLOGY
HEEBOLL, KAREN	DIRECTOR, UNIVERSITY OF SOUTHERN DENMARK
HERMIDA ALONSO, JOSE ANGEL	RECTOR, UNIVERSITY OF LEON
HERRMANN, WOLFGANG	PRESIDENT, TECHNICAL UNIVERSITY MUNICH
HOLECEK, MIROSLAV	RECTOR, UNIVERSITY OF WEST BOHEMIA
HORATIU, MIHAI	PRESIDENT, LUCIAN BLAGA UNIVERSITY OF SIBIU
IVANETS, SERGEY V.	PRESIDENT, FAR EASTERN FEDERAL UNIVERSITY
IVANOV, KONSTANTIN	RECTOR, BALTIC STATE TECHNICAL UNIVERSITY
JOZSA, JANOS	RECTOR, BUDAPEST UNIV. OF TECHNOLOGY AND ECONOMICS
KAINZ, HARALD	RECTOR, GRAZ UNIVERSITY OF TECHNOLOGY
KOSIH, ANATOLY	RECTOR, OMSK STATE TECHNICAL UNIVERSITY
LAZARO, CARLOS CONDE	RECTOR, UNIVERSIDAD POLITECNICA DE MADRID
LAZOVSKY, DMITRY	RECTOR, POLOTSK STATE UNIVERSITY
KANOK-NUKULCHAI, WORSAK	PRESIDENT, ASIAN INSTITUTE OF TECHNOLOGY
KARBOWNIK, ANDRZEJ	RECTOR, SILESIAN UNIVERSITY OF TECHNOLOGY, GLIWICE
KMET, STANISLAV	RECTOR, TECHNICAL UNIVERSITY OF KOSICE
KONVALINKA, PETR	RECTOR, CZECH TECHNICAL UNIVERSITY IN PRAGUE
KOSHKIN, VALERIY	RECTOR, SEVASTOPOL STATE UNIVERSITY
KOSIH, ANATOLIY	RECTOR, OMSK STATE TECHNICAL UNIVERSITY
KOSYKH, ANATOLY	PRESIDENT, OMSK STATE TECHNICAL UNIVERSITY
LAZARO, CARLOS CONDE	RECTOR, UNIVERSIDAD POLITECNICA DE MADRID
LAZOVSKY, DMITRY	RECTOR, POLOTSK STATE UNIVERSITY
LODYGOWSK, TOMASZ	RECTOR, POZNAN UNIVERSITY OF TECHNOLOGY
LUCIN, PERO	RECTOR, UNIVERSITY OF RIJEKA
MANZ, CARSTEN	PRESIDENT, KONSTANZ UNIVERSITY OF APPLIED SCIENCES
MAZO, EDUARDO GONZALEZ	RECTOR, UNIVERSITY OF CADIZ
MCDOWELL, JIM	CHANCELLOR, UNIVERSITY OF SOUTH AUSTRALIA
MIHOV, GEORGI	RECTOR, TECHNICAL UNIVERSITY OF SOFIA
MISRA, ASHOK	CHAIRMAN, INDIAN INSTITUTE OF TECHNOLOGY ROORKEE
MUELLER-STEINHAGEN, HANS	RECTOR, DRESDEN UNIVERSITY OF TECHNOLOGY
OSTOJIC, LJERKA	RECTOR, UNIVERSITY OF MOSTAR
OVCHARENKO, SERGEY	RECTOR, OMSK STATE TRANSPORT UNIVERSITY
RADTKE, ULRICH	RECTOR, UNIVERSITY OF DUISBURG-ESSEN
RAIS, KAREL	RECTOR, BRNO UNIVERSITY OF TECHNOLOGY
REDHAMMER, ROBERT	RECTOR, SLOVAK UNIVERSITY OF TECHNOLOGY BRATISLAVA
RESSEL, WOLFRAM	RECTOR, UNIVERSITY OF STUTTGART
RIBICKIS, LEONIDS	RECTOR, RIGA TECHNICAL UNIVERSITY
RITTIDECH, SAMPAN	PRESIDENT, MAHASARAKHAM UNIVERSITY
ROGALEV, NIKOLAI	RECTOR, MOSCOW POWER ENGINEERING INSTITUTE - TU
RT HON LORD MANDELSON	CHANCELLOR, MANCHESTER METROPOLITAN UNIVERSITY
RUDSKOY, ANDREY	RECTOR, PETER THE GREAT ST. PETERSBURG POLYTECHNIC UNIVERSITY
SAHA, PETR	RECTOR, TOMAS BATA UNIVERSITY UNIVERSITY ZLIN
SCHMOELLEBECK, FRITZ	RECTOR, UNIV. OF APPLIED SCIENCES TECHNIKUM, VIENNA
SCHUBERT, ANDREAS	RECTOR, CHEMNITZ UNIVERSITY OF TECHNOLOGY
SEIDLER, SABINE	RECTOR, VIENNA UNIVERSITY OF TECHNOLOGY

Serban, Viorel-Aurel	Rector, Polytechnic University of Timisoara
Smedescu, Ion	President, Romanian American University
Stoll, Wilfried	Managing Partner- Festo
Sung-mo, Kang	President, Korea Advanced Institute for Science and Technology
Szmidt, Jan	Rector, Warsaw University of Technology
Teeri, Tuula	President, Helsinki University of Technology
Thomsen, Christian	President, Berlin Institute of Technology
Ticar, Igor	Rector, University of Maribor
Torfs, Rik	Rector, Catholic University of Leuven
Torma, Andras	Rector, University of Miskolc
Tue, Nguyen Thien	President, Ho Chi Minh city University of Industry
Turkalj, Zeljko	Rector, Strossmayer University of Osijek
Vano, Jean-Philippe	President, Paris Institute of Technology
Vican, Dijana	Rector, University of Zadar
Vlaicu, Aurel	Rector, Technical University of Cluj-Napoca
Vondrak, Ivo	Rector, Technical University of Ostrava
Yuksek, Ismail	President, Yildiz Technical University

International Scientific Committee

Adamczak, Stanislaw (Pol)
Alander, Jarmo Tapani (Fin)
Arai, Tamio (Jpn)
Ares, Enrique (Spa)
Arndt, Guenter (Aus)
Attard, Mario M. (Aus)
Bachynski, Morrel (Can)
Balate, Jaroslav (Cze)
Balic, Joze (Slo)
Beer, Rudolf Julius (Aut)
Belajev, Gennady J. (Blr)
Bilberg, Arne (Den)
Bliedtner, Jens (Deu)
Bohez, Erik L. J. (Tha)
Breido, Iosif (Kaz)
Bright, Glen (Rsa)
Browne, J. (Irl)
Celar, Stipe (Hrv)
Coelho, Reginaldo T. (Bra)
Covic, Dragan (Bih)
Cruz, Fernando M. M. (Por)
Cus, Franci (Slo)
Dar-El, E. (Isr)
De Filippi, A. (Ita)
Devahastin, Sakamon (Tha)
Dincmen, M. (Tur)
Diniz, Anselmo Eduardo (Bra)
Dohda, Kuniaki (Jpn)
Dong, Jianhua (Chn)
Dornfeld, David (Usa)
Dove, Rick (Usa)
Dudas, Illes (Hun)
Egorov, Sergey B. (Rus)
Endter, R. (Deu)
Eversheim, W. (Deu)
Filaretov, Vladimir F. (Rus)
Galantucci, Luigi Maria (Ita)
Gawlik, Jozef (Pol)
Han, Qing Yao (Chn)
Hock, Seo Swee (Sng)
Hofmann, Peter (Deu)
Hryu, Jee-Hwan (Kor)
Ibishi, Ismet Maliq (Kwt)
Iltschev, Velko (Bgr)
Irfan, Mohammad A. (Pak)
Itoh, Kiyoshi (Jap)
Ivan, Nicolae-Valentin(Rom)
Kals, H. J. J. (Ndl)
Katalinic, Branko (Aut)
Kimura, Fumihiko (Jpn)
Kinnander, Anders (Swe)
Knosala, Ryczard (Pol)
Koch, Jan (Pol)
Kondratenko, Yuriy (Ukr)
Kuljanic, Elso (Ita)
Kuosmanen, Petri (Fin)
Kyttner, Rein (Est)
Le, Cung (Vtn)
Lin, Chyi-Yeu Jerry (Twn)
Llorente, J. I. (Esp)
Lombardi, F. (Ita)
Low, Kin Huat (Mal)
Lu Chang, Tao (Chn)
Lu, S. S. (Twn)
Luong, Lee Hong Son (Aus)
Majstorovic, Vlado (Bih)
Marcos, Mariano (Esp)
Meghdari, Ali (Iri)
Mertins, Kai (Deu)
Michelini, R. C. (Ita)
Mikhaylov, Yuri O. (Rus)
Munteanu Radu (Rom)
Nanasi, Jozef (Svk)
Neumann, H. J. (Deu)
Nicoletto, G. (Ita)
Novakovic, Branko (Hrv)
Porter, Brian (Hkg)
Pritschow, Guenter (Deu)
Pryanichnikov, Valentin (Rus)
Przybylski, W. (Pol)
Radermacher, F. J. (Deu)
Radhakrishnan, V. (Ind)
Ramayah T. (Mla)
Rungruang, Channarong (Tha)
Schmid, Steven (Usa)
Segui Llinares, Vicente Jesus (Esp)
Sladek, Jerzy (Pol)
Solomenzev, Juri. M. (Rus)
Stazhkov, Sergey (Rus)
Stefanescu, Doru Michael (Usa)
Tabucanon, Mario T. (Tha)
Takakuwa, Soemon (Jap)
Takala, J. (Fin)
Tikka, H. (Fin)
Topalova, Irina (Bgr)
Torims, Toms (Lva)
Torres-Leza, F. (Esp)
Torvinen, S. (Fin)
Tsai, B. S. Y. (Twn)
Udiljak, Toma (Hrv)
Uglesic, Ante (Cro)
Uhlmann, Eckart (Deu)
Uyar, Erol (Tur)
Van Brussel, Hendrik (Bel)
Van Oudheusden, Dirk L. (Bel)
Varga, Gyula (Hun)
Vavrin, Petr (Cze)
Venkatesh, V. C. (Sgp)
Vetturi, David (Ita)
Vitecek, Antonin (Cze)
Vranjes, Bozo (Hrv)
Waldir, P. O. (Bra)
Wertheim, Rafi (Isr)
Wessely, Emil (Svk)

INTERNATIONAL PROGRAM COMMITTEE

DOLINAY, VILIAM
DOSTAL, PETR
DOSTALEK, PETR
DRAGCEVIC, ZVONKO
DRAGICEVIC, MARIJA
DRSTVENSEK, IGOR
MENZ, PETER
MICHAL, BLIZNAK
MICIETA, BRANISLAV
MICLOSI, VIOREL
MOHORA, CRISTINA
MOK, H.
MORARU, SORIN AUREL
MORIWAKI, T.
MORTOIU, DOINA
MOTOMANCEA, ADRIAN
MRSA, ZORAN
MRVAC, NIKOLA
MUNTEAN, MIHAELA
MUNTEAN, SEBASTIAN
MUNTEANU, ANCA
MUSCA, GAVRIL
MUSTAFIZUL, KARIM
MYSHLYAVTSEV, ALEXANDR
NAGI, MIHAI
NAGLA, K. S.
NAKAMOTO, NOBUYUKI
NEUMANN, PETR
NICA COTET, GABRIELA BEATRICE
NIKOLIC, GOJKO
NKUMBWA, RODGER
NOH, SANG DO
NOVAC, OVIDIU
NOVAKOVIC, BRANKO M.
NOVOSELOVIC, DANIEL
NURUL AMIN A.K.M.
OANCEA, GHEORGHE I.
OBAD, MILENKO
OKAMOTO, J.
OLARU, ADRIAN, DUMITRU
OLARU, ONISIFOR
OLEKSIK, VALENTIN
OLTEANU, CIPRIAN
OMEROVIC, MIDHAT R.
OPRAN, CONSTANTIN
OPREAN, AUREL
OPRUTA, DAN
OSMANI, HYSNI
OSMERA, PAVEL
OTTOSSON, S.
PACAIOVA, HANA
PALCIC, IZTOK
PALOS, RAMONA
PAMINTAS, EUGEN
PANFILOV, PETER
PARASCHIV, NICOLAE
PARK, HONG-SEOK
PARPALA, LIDIA FLORENTINA
PASCUAL, R.COMPAMYS
PASIC, MUGDIM
PASTRAMA, STEFAN DAN
PATACHIA, SILVIA FLORICA CRISTINA
PAULIKOVA, ALENA
PAULOVA, IVETA
PAUNESCU, CARMEN MONICA
PAVIC, IVAN
KRASNIQI, Z.
KREMLJAK, ZVONKO
KRUCINSKA, IZABELLA
KUBALCIK, M.
KUBINGER, WILFRIED
KULENOVIC, MALIK
PECHACEK, FRANTISEK
PELAEZ LOURIDO, GUSTAVO
PENCEA, ION
PERIC, NEDJELJKO
PERJU, DAN
PERNIU, DANA
PERUTKA, KAREL
PETERS, JACQUES
PETKOVIC, DARKO
PETRE, CRISTIAN
PETRE, MARIAN
PETRESCU, VALENTIN
PINCA-BRETOTEAN, CAMELIA
PINTILIE, GHEORGHE
PIRTEA, GABRIEL MARILEN
PIVONKA, PETR
PJERO, ELENICA
PLESA, ANGELA
PODURAEV, JURY
POKORNY, PAVEL
POLJAK, KRUNOSLAV
POMAZAN, VALENTINA
POPA, CICERONE LAURENTIU
POPPEOVA, VIERA
PRODAN, DAN
PROKOP, ROMAN
PROSTREDNIK, DANIEL
PRYANICHNIKOV, VALENTIN
PYTKO, STANISLAW
QUINN, ROGER D.
RADIC, JURE
RAMAJ, VEHBI XH.
RAOS, PERO
RAVAS, RUDOLF
RAY, T. S.
RIZESCU, SABIN VASILE
ROGALE, DUBRAVKO
ROHDE, ULRICH L.
ROMERO, FERNANDO
ROS, OLIMPIA
ROSADO, V.O.G.
RUDAS, IMRE J.
RUGESCU, DRAGOS RADU DAN
RUJA, IOAN
RUSU ANGHEL, NICOLAE
RUSU, LIGIA
RUSU-CASANDRA, AURELIA
RUTEMI, GANI BRAHIM
RUZIC, DRAGO
SALIHU, AVDI HAJDIN
SAMEK, DAVID
SCHREIBER, PETER
SELVAKUMAR, ARUL LAWRENCE
SEONG, POONG HYUN
SERCER, MLADEN
SHIBASAKA, TOSHIROH
SHIH, MING CHANG
SIEMIATKOWSKI, M.
SIHN, WILFRIED
SINAY, JURAJ
MARTINEZ, SUSANA
MASTACAN, LUCIAN
MATUSOVA, MIRIAM
MATUSU, RADEK
MEHTA, NARINDER KUMAR
MEIER, HORST
SISEK, BORIS
SITUM, ZELJKO
SKOCZYNSKI, WACLAW
SKOLUD, BOZENA
SLUGA, ALOJZ
SMIRNOV, ALEXANDER
STEFANIC, NEDELJKO
STOPPER, MARKUS
STRGAR KURECIC, MAJA
SUBRAMANIAN, ARUNKUMAR
SUGAR, PETER
SYROVA, LIVIA
SYSALA, TOMAS
TADAMURA, KATSUMI
TAKALA, JOSU
TAMASILA, MATEI
TAN, KOK KIONG
THIERHEIMER, WALTER WILHELM
TORIMS, TOMS
TURKALJ, GORAN
ULRICH, GUENTHER
VAARIO, JARI
VASEK, VLADIMIR
VELISEK, KAROL
VEZA, IVICA
VISEKRUNA, VOJO
VRANJES, BOZO
VUJASINOVIC, EDITA
WANG, JYHWEN
WEIGANG, WU
WEIGL, K. HEINZ
WESSELY, EMIL
WIENER, PETER
WINKLER, T.
WOZNIAK, ADAM
XING, KE
XUE, D.
YAMAZAKI, K.
YORDANOV, RUMEN STOYANOV
YOSHIDA, SUSUMU
YU, Y.-C.
ZAJAC, JERZY
ZAJAC, JOZEF
ZAPCIU, MIRON
ZAVOIANU, ION
ZEQIRI, HAKIF
ZGODAVOVA, KRISTINA
ZHIRABOK, ALEXEY
ZICHIL, VALENTIN
ZIRRA, DANIELA
ZUEV, ALEXANDER
ZUPERL, UROS

REVIEWING COMMITTEE

27TH DAAAM INTERNATIONAL SYMPOSIUM ON INTELLIGENT MANUFACTURING AND AUTOMATION

Preface

The 27th DAAAM International Symposium on Intelligent Manufacturing and Automation took place in Mostar, BiH, between 26th and 29th October 2016, during the DAAAM International Week. The Symposium was organized by DAAAM International Vienna and University of Mostar and co-organized by Vienna University of Technology, Austrian Society of Engineers and Architects – ÖIAV 1848, International Academy of Engineering IAE-CEB, University of Applied Sciences – Technikum, Council of Ministers of Bosnia and Herzegovina, Government of Federation of Bosnia and Herzegovina, Herzegovina-Neretva Canton Government, City of Mostar and under the auspices of Danube Rectors´ Conference and Rectors´ and Presidents´ Honor Committee of DAAAM for 2016. The Symposium was hosted by University of Mostar. This year symposium aimed at continuing the success of the previous year's focusing on the five-fold traditional objectives of the symposium: the presentation of the most recent high-quality results, support of development of young scientists and researchers, organization of international (summer) doctoral school, inauguration of new members of Central European Branch of International Academy of Engineering and the provision of the necessary setting for stimulating discussions, brainstorming and networking among European and international researchers coming both from the academia government agencies and industry.

DAAAM International Symposium has been established as the leading Central European conference devoted to the cutting-edge advancements in the field of modern production and automation. Symposium provides an international forum for engineers, scientists and industrial managers to present and discuss the current status and impact of advanced manufacturing and automation. The international series of Symposium was initiated by the first meeting in Vienna (1990) and continued in Strbske Pleso (1991), Budapest (1992), Brno (1993), Maribor (1994), Cracow (1995), Vienna (1996), Dubrovnik (1997), Cluj - Napoca (1998), Vienna (1999), Opatija (2000), Jena (2001), Vienna (2002), Sarajevo (2003), Vienna (2004), Opatija (2005), Vienna (2006), Zadar (2007), Trnava (2008), Vienna (2009), Zadar (2010), Vienna (2011), Zadar (2012), Zadar (2013), Vienna (2014), Zadar (2015), Mostar (2016).

Regular organization of annual Symposium is one of the central activities of the DAAAM International association. The association was founded on 5. November 1990 in Vienna on the meeting organised by Professor B. Katalinic in order to celebrate the 175 Years of Vienna University of Technology. The main idea of association was to establish network for new kind of international scientific and academic cooperation in the fields of intelligent manufacturing and automation. Our primer focus is on: all aspects of production of technical products made out of parts and corresponding automation and knowledge, trans-disciplinary application of methods and all kind of education.

DAAAM International activities, during the last twenty-seven years, have been supported by scientists and experts from over fifty countries all over the world. In the course of the last decade, DAAAM International has focused activities on the regional and global long-term co-operation between (technical) universities. The result of this initiative is the founding of the University Network of DAAAM International Vienna and DAAAM International Network for Advanced Technologies which was founded on 2002-10-25 in Vienna.

The main activities of DAAAM are:

- Organisation of annual DAAAM International Symposium
- To maintain and improve the University Network of DAAAM International Vienna as a strategic alliance and permanent international open platform for long-term co-operation and partnership in all fields of technology, research and education.
- Organisation of long-term projects, initially known as Inter-University Doctoral Studies
- Organisation of long-term projects, initially known as the Virtual University.
- Publishing of international journals, books and proceedings
- Promotion of the exchange of people, scientific, and technical information.
- Promotion of international co-operation among scientists and engineers.
- Promotion of R & D, education and industrial application.
- To co-operate with other national and international organisations in matters of common interest.
- Organisation of International Doctoral School as platform for the support of next generation of engineers at the highest level

This proceeding contains 129 refereed papers from 330 authors presented at the 27th DAAAM 2016. In the DAAAM proceedings since first symposium in 1990 we published papers from 8670 authors.

All the papers included in these proceedings are reviewed by the experts in double blind peer review way, and finally revised by the authors for the improvement according to referee's reports and advices.

The papers discuss many aspects of modern manufacturing and automation such as: Algorithms, Artificial Intelligence, CAX, Computer Integration, Control, Cutting Tools, Design, FEM, Invited Lectures, Knowledge, Management, Manufacturing System, Mechatronics, Methodology, Methods, Modelling, Optimization, Robotics, Simulation, Technical Solutions, Technology and Trends.

Appreciation is forwarded to reviewers who contributed to improve the quality of papers reviewing the papers, providing constructive critical comments, improvements and corrections to the authors. We also recognize the support of the reviewers, who were in charge of acceptance/rejection process. We wish to thank the Scientific, Program and Organizing Committees, Session Chairs, student helpers and Administrative assistants, for contributing to the success of this Conference. They did an excellent job.

Finally, we wish to thank all the authors who submitted papers, making this conference possible and the authors of accepted papers for making appropriate modifications and corrections with care and patience and preparing the final manuscript. It is the quality of their presentations and their passion to communicate with the others participants that really makes this conference series a success.

Special thanks to Festo company for support of our activities and support of young researchers and scientists in the form of Festo Prize and Festo scholarships for the participants of the 5th DAAAM International Doctoral School, which was held parallel to the symposium and interacted positively each other.

We would like to thank the Rector of the University of Mostar Professor Dr. Sc. Ljerka Ostojic, her team, management and all those nice colleagues for their excellent support in the organization of 27th DAAAM International Symposium and the 5th DAAAM International Doctoral School.

Vienna, 2016-12-05
Editor
Chair of the 27th DAAAM 2016

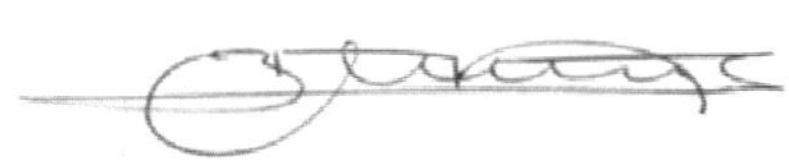

Branko Katalinic

Professor Dr.sc. Dr.mult.h.c.Prof.h.c. Branko Katalinic
President of DAAAM International
TU Wien, IFLT-IMS, Karlsplatz 13/311
1040 Wien, Austria, Europe
Phone:0043-1-58801-31121
Fax#: 0043-1-58801-31199
URL: http://www.daaam.info
E-mail: president@daaam.com
branko.katalinic@tuwien.ac.at

DOI: 10.2507/27th.daaam.proceedings.001

NETWORK–BASED DESIGN OF HETEROGENEOUS MODULAR MOBILE ROBOTIC SYSTEMS

Victor P. Andreev & Yury V. Poduraev

This Publication has to be referred as: Andreev, V[ictor] & Poduraev, Y[uriy] (2016). Network–Based Design of Heterogeneous Modular Mobile Robotic Systems, Proceedings of the 27th DAAAM International Symposium, pp.0004-0009, B. Katalinic (Ed.), Published by DAAAM International, ISBN 978-3-902734-08-2, ISSN 1726-9679, Vienna, Austria
DOI: 10.2507/27th.daaam.proceedings.001

Abstract

The present work deals with a new approach to the construction of hardware and software part of the heterogeneous modular mobile robots. The new approach is to use a network principle for constructing information-measuring and control system of modular mobile robot. The paper discusses a functional-modular principle of construction of a heterogeneous mobile robot. Our concept is based on the functional completeness of each module, which is an electronic and/or mechatronic device with its own control system. The inclusion of any mechatronic devices in the network structure information measurement and control system of the mobile robot is performed via special software –driver.

Keywords: mobile robot, heterogeneous modular robot; local area network; information-measuring and control system

1. Introduction

Robotic systems are widely used in industry, medicine, and everyday life, especially in solving the problems related with emergency situations and defence missions. At present, the application of these systems usually depends on their structure, which leads to a large variety of design and engineering solutions. The main actuator, information and control components of these robotic systems are normally designed to be used in a particular system alone. The electronic and mechanical units of these systems cannot be structurally and functionally separated from their general design. Their software is unique and typically runs on a single powerful computer. All these factors make it difficult or even preclude the use these units in other similar systems. The fact that these robotic systems are expensive restricts the extent of their application.

The modular principle of robotic systems construction makes it possible to substantially scale down the variety of technical solutions, thus reducing the cost of final devices through the unification of hardware and relevant software. The modular architecture has long been used in machine-tool industry and makes it possible to develop flexible and fault-tolerant systems with small mass-dimensional characteristics and low cost of manufacture. Therefore, the use of the modular principle in the design of mobile robotic systems it seems reasonable. In recent years began to appear publications which present the results of current scientific studies on the modular mobile robotics.

2. Modular robotics

The modular principle of robotic systems design was proposed in the Russian State Scientific Center for Robotics and Technical Cybernetics (RTC, St. Petersburg, Russia) and first implemented in the Soviet Union as early as in the 1980s [1]. That time was characterized by the serial production of standard robotic components, such as electromechanical modules, control modules, and modular software systems. The second stage of progress in modular robotics can be attributed to the 1990s, when the accident at the Chernobyl nuclear power plant had occurred. By that time, the Scientific Center for Robotics and Technical Cybernetics prepared a large number of projects for different modules. Thanks to this, a total of 15 mobile robots designed for various purposes were manufactured and put into operation within a couple of months. These robots operated in the disaster area to survey the local radiation situation and clean the buildings and places from radioactive waste.

The experience of operation of these robotic systems resulted in a new research direction in modular robotics, the development of reconfigurable modular robots, i.e., robots with a variable structure. The maximum uncertainty in the conditions of future activities in case of emergency situations, accidents and active opposition makes it impossible to predetermine the composition and configuration of desired robotics. The use of reconfigurable modular robotics allows the robots to be assembled directly on site and adjust their structure in the course of operation.

The modular principle of robotic systems construction is a new research line with its main solutions having been implemented in the CEBOT robot [2] in 1987. The modular robots consist of individual components or nodes (modules), which are combined together to form an integral and independent robotic system. Each module can contain electronic, sensor, and/or control devices and actuators.

By their construction type, the modular robots can be divided into *homogeneous* and *heterogeneous* ones. In *homogeneous* robots, all modules are identical to each other and are separate functional units. These devices can change their configuration by restructuring their components. In *heterogeneous* robots, each module implements a separate function, which makes it possible to change the robot configuration by choosing an appropriate set of modules in line with the intended use of the robot.

By the method of their structure formation, the modular robots can be *reconfigurable* and *self-reconfigurable*. A considerable amount of well-known products in this field addresses self-reconfigurable homogeneous robots, such M-TRAN, ATRON, and SuperBot [3 – 5]. It is believed that in the long-term perspective the self-reconfigurable robots must be heterogeneous because the each module in homogeneous robots has to be additionally supplied with all the necessary components (sensors and/or actuators), which highly complicates the system [6]. Heterogeneous robots are more versatile and flexible [7 – 10] but usually not self-reconfigurable.

3. Modern products of heterogeneous modular robots

Thor (Fig. 1a) [7]. The concept of this robot is partially based on its "ancestor" – the Odin heterogeneous robot [8], the modules of which were morphologically homogeneous. A distinctive feature of Thor is that it is based on the submodular principle of the construction of functional units (a total of eight modules); each module executes only one simple function. In other words, certain function of a single unit of the robot must be divided between several simpler modules. For example, the gripping module (which is composed of the gripper itself, actuator, and batteries) is proposed to be replaced by sub functional modules: actuator, batteries, and a large set of gripping mechanisms, joints and wheels. The main advantage is that the individual submodules can be easily constructed, which allows other submodules to be quickly produced. More expensive submodules of the motor–module type are used only where required. Using different combinations of modules, one can construct mobile robots of different purposes.

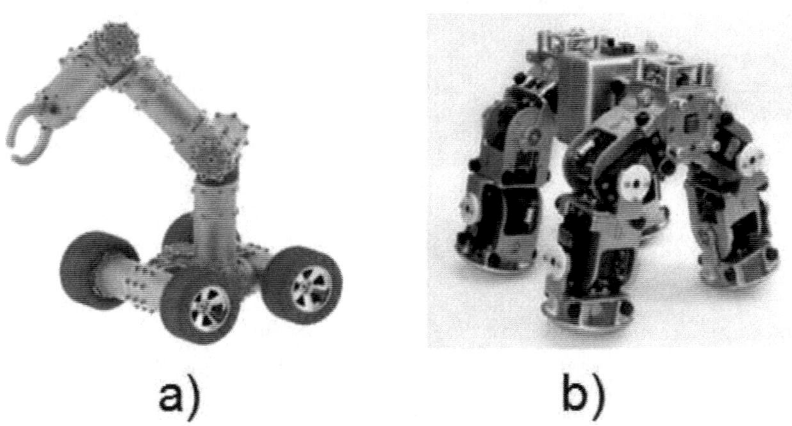

Fig. 1. Computer model of Thor (a); the robot for operations in space (b)

The morphological heterogeneity and subfunctional modularity simplify the robot structure and increase the flexibility of its construction. However, if the modules become extremely simplified, the labour intensity in the construction assembly increases and the corresponding software becomes more complex as an element of the control system.

Robot for operations in space (Fig. 1b) [9]. Originally, the system involved three types of modules: *module-actuator, node module,* and *communication and control module*. The module-actuator is a module with servo-actuator and articulated coupling. The node module is a structure with five pin platforms allowing for different angles of coupling between the modules (including 90° and 120°). Also, the node module provides power transmission and data exchange between coupled modules. The communication and control module allows the robot to operate autonomously or by wire and can be equipped with batteries. Then, "foot"-modules equipped with tactile sensors were developed. In addition, a new module (digital TV-camera) was introduced that can transmit images through a wireless channel to the host computer; this avoids the need for high capacity of the inter-module interface.

The modules communicate with each other and with one or more upper-level computers by using a special-purpose peering network. Any module can send data packets to another module, which ensures a distributed control. The resulting modules were tested in different configurations: a classic "snake-like" manipulator, a robot with multiple manipulators, and a walking robot with three or four walking drivers.

Due to the simple structure of modules, which is focused on rapid production, the use of new functional modules is simple and inexpensive. However, the use of a peer-to-peer network operating on the basis of the equality principle seems to be unjustified because the modules have different functional purposes and therefore are essentially unequal. In this case, the modules can only serve as a source of information (sensors) and/or receivers of simple instructions; the modules have no individual control system.

SMART [10]. This is a reconfigurable heterogeneous modular robotic system consisting of a set of interchangeable modules. The main purpose of this robot is to create modules that preserve the balance between versatility, low production cost, and functionality. The construction of modules implies that they can be quickly developed and produced at minimum economic costs. The system architecture is formed of modules such as *Modules*, *M-Robots*, and *Colonies*.

Modules are the basic components of the system and can be divided into three main types: power/control modules (P/C-modules), joint modules (J), and specialized modules (S). The *M-Robot* is an autonomous system that is based on at least one P/C-module and one or more J and S modules. The *Colony* contains several *M-Robots* operating jointly to accomplish a given task.

The *P/C-module* (power/control) controls over all actuators of other modules, processes the sensor data, and controls the information exchange between other P/C-modules and the control station. The module contains basic electronic devices, power supply, peripheral communications (CAN-bus and Bluetooth), and connection interfaces.

The module control system is based on the master–slave architecture. Accordingly, there are two software types: master software (*Ms*) and slave software (*Ss*). *Ms* controls the external communication via Bluetooth and synchronizes *Ss* within the robot configuration. In the *M-Robot*, *Ms* is responsible for preprocessing of instructions and their distribution among the remaining *Ss* via the CAN-bus. The data transmitted within the *M-Robot* has the form of low-level instructions, such as indicators for feedback control by position, data for synchronization, and data received from sensors and actuators. *Ss* receives messages via the CAN-bus and executes a desired low-level command.

The *Joint module* (*J*) is any active or passive actuator of the rotary, linear (prismatic), or spherical types, and their combinations. The module is controlled by a *P/C-module*, which also supplies it with power. The *J-module* may contain control electronics as well.

The *specialized module* (*S*) is designed for executing specific tasks. The module can be a special sensor that conducts data processing of its own, a specialized tool, or inexpensive constructs of passive "extremities" such as foot, wheel, a hybrid of foot and wheel, flange, and platform as an operating body of the robot.

The control over the configuration of a robot or robotic system (*Colonies*) is implemented through the following two types of communication between modules: Intra-robot and Inter-robot. The Intra-robot communication contains those P/C-modules that are incorporated into the structure of one and the same M-Robot. The Inter-robot connects the M-robot with the control station. The two communication types operate via the CAN-bus and Bluetooth technology, respectively.

The set of modules makes it possible to create different variants of robotic devices. A common variant of the M-Robot is an autonomous device consisting of at least one P/C-module, J-module, and S-module. Different module combinations allow the device to be reconfigured or self-reconfigured.

One of the most important features of the SMART robot is that each module can be designed, on the one hand, as a structurally and constructively simple and inexpensive device and, on the other hand, as a complex and high-performance device with its intellectual component.

The most interesting and promising solution seems to be the SMART robot. However, the use of CAN-bus and Bluetooth as a communication environment restricts the functional capabilities of this robot due to the limited capacity of these communication channels and rules out the possibility of remote and supervisory control over the robot at large distances. In addition, the structure of the information-measuring and control system of MR and the module interaction interfaces do not allow third-party products to be used as modules, which reduces the efficiency of the production of these robots.

The analysis of these examples of different modular heterogeneous robotic systems allows us to conclude that the modular principle of robots construction is of current interest and promising.

4. Objectives and tasks

This study is proposed to be based on the functional-modular principle of the organization of the information-measuring and control system (IMCS) of mobile robots (MR). This is a difficult task because it is necessary to identify the functionally complete parts of the IMCS (modules) and develop universal software and hardware interfaces that allow the modules to be combined into systems of various functional purposes. Then, a new feature appears: depending on the MR purpose, its configuration is changed through an appropriate composition of different modules that are independent functional elements.

The system mobility complicates the solution of this problem since the communication of the mobile part with a control system that is located far away (for variants of remote and supervisory control) is based on a radio channel with all of its shortcomings.

To solve this problem, we propose a new approach: consider the IMCS of the robotic system as a *local area network* (LAN) with mobile nodes and develop appropriate *network software* that makes the system scalable and reconfigurable [11]. Each LAN node is a modular mobile robot with an IMCS that is also a LAN, the nodes of which are functional modules. The final modules produced on the basis of this principle can be directly used in other robotic systems.

The aim of this study is to develop a functional-modular principle of the construction of the software and hardware unit of heterogeneous mobile robotic systems that are composed of functionally independent modules.

The tasks of this study are:

- Develop a MR IMCS structure allowing the robot to be represented as an interaction of individual functionally complete units – modules.
- Determine the minimum set of functionally complete units of the MR IMCS in terms of their implementation as individual mechatronic and/or electronic modules.
- Formulate the functionality of each module and specify the requirements for the module construction as well as communication channels and interfaces (both hardware and software).
- Develop a universal program interface for the interaction of modules, taking into account their upgradability (without changing the functionality) and the possibility of remote dynamic reprogramming of their computing devices.
- Develop a structure of information interaction between MR modules with scalability, reconfigurability, and hardware implementation.
- Develop a user interface for the MR supervisory control system via the Internet and radio channels.
- Develop and produce a laboratory prototype of functional-modular mobile robots, including sensory, execution, control, computing, communication, and program units.
- Conduct an experimental study of the efficiency of the proposed solutions and variants of functional elements.

5. Main concept of the functional–modular principle

The project conception suggests the following main principles for the construction of information-measuring and control systems of a modular MR (functional independence of modules is ensured by):

(1) Each *module* is a functionally complete electronic or mechatronic device.
(2) The robot's IMCS is constructed as a local area network, the nodes of which are *modules*.
(3) The principle of distributed control is used.
(4) The hardware unit of modules is constructed using the principle of distributed stabilized power supply.
(5) Each *module* is supplied with appropriate software (driver).

The functional completeness of modules means a structural and functional independence, which makes it possible to use third-party modules identical to replaced modules by their functionality.

The network principle of ICMS construction makes the robot reconfigurable and scalable. The use of the Ethernet technology provides a capacity of inter-module communication channels of up to 1000 Mbit/s and makes it possible to use the entire family of TCP/IP protocols and other protocols that are widely used on the Internet together with appropriate libraries. In addition, one can use secure digital radio channels (such as those based on the Wi-Fi standard) for remote control.

The principle of distributed control means the following. Each mechatronic or electronic module has its own information-measuring and control system implemented on the basis of a microcontroller and, when necessary, using its own sensory system and actuators. These elements of the module control system ensure its functional completeness. In this case, each module can independently perform a limited set of functions, and the external (general) control system generates a local purpose, determines the sequence of executable functions, and transmits only the necessary instruction parameters to the module, receiving a report on their execution in response: this generates a data loopback [12]. As a result, the whole computing process implementing the general operation of MR can be parallelized.

The principle of distributed stabilized power supply is implemented as follows. The total power of the robot, including the electric actuator, is supplied by a separate module that contains accumulator batteries. The supply voltage of all electrical components, including microcontrollers, is unstable due to load change. Therefore, each module is provided with its own stable voltage supply (SVS), intended for supply only intramodular devices such as microcontrollers. The

SVS input receives voltage directly from accumulator storage batteries. When there are input voltage pulsations caused, for example, by starter currents of electric motors, the SVS prevents the possible interruption of the program code by intramodule computing devices. Besides, this distributed power supply does not impose a limit on the power consumption by each module; it is merely necessary to limit the total power consumed by all modules from accumulator storage batteries. If the total power consumption is exceed, one only needs to replace a single module by a module with more capacious batteries. In addition, the released heat is distributed between modules, rather than concentrated in a single power supply unit, which would require radiators.

"*The concept of drivers*" is that for each mechatronic device a unified network control protocol is created on the basis of existing low-level program interfaces of communication [13]. Then, the producer can relatively easily integrate its device into the IMCS of the modular MR, creating only an additional driver that implements the protocol-based program control instructions and the networking protocols. When a new mechatronic device appears in the network, the software of the central control system automatically puts this device into the network during the interaction with the driver. The fact that each module has a computing device and the network principle of the construction of the IMCS of modular MR make it possible to create an appropriate driver for a given module and thus ensure that third-party modules can be involved in the network without reworking their low-level code.

The functional-modular principle of the construction of mobile robotic systems developed in this project makes it possible to create universal modules, which can be used to construct mobile robotic systems for different purposes [14]. Figure 2 shows possible variants of a heterogeneous modular MR. The wheel robot of vertical design (Fig.2a) can be transformed with the help of a simple modernization into a robot of horizontal design (Fig.2c) or into a raised passableness track-mounted robot (Fig.2b) that is structurally assembled from the same modules.

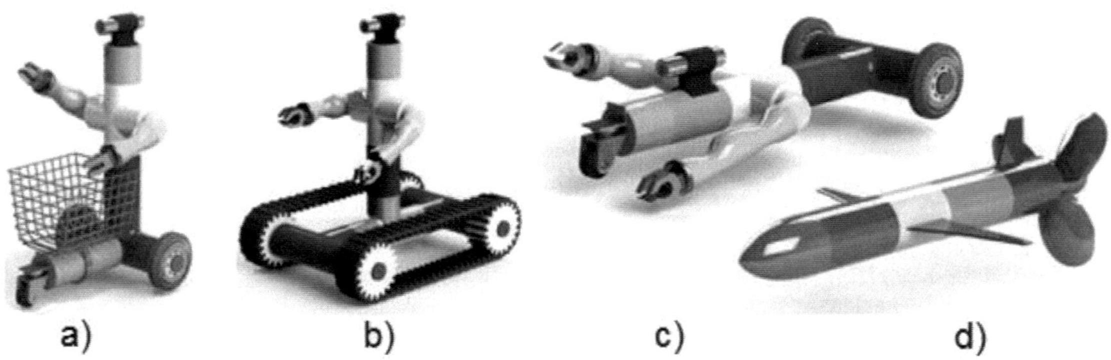

Fig. 2. Variants of heterogeneous modular mobile robots (computer prototypes)

Using functionally similar modules but with another design, one can obtain a variant of the underwater research robot (Fig.2d), but the bodies of its modules are made of a more durable material that can withstand high pressure at large depths.

6. Conclusions

One of the major challenges of modular robotics is to build a inter-module communication interface. It was necessary to offer a solution to this problem, which would provide a relatively simple integration of ICMS of the individual modules into a single ICMS of the robot. Since the structure of a modular MR is formed from the set of special modules, the type, function, and number of which depend on the robot purpose, its structure should be easily changeable and be made up of modules of various designs and different manufacturers. Mentioned problem is solved by constructing ICMS of the modular robot in the form of local area network in which each module represents a node of the LAN. The simplicity of integration ICMS of modules in ICMS of modular robot is ensured by special software – drivers. Manufacturers of modules should develop the drivers based on the rule set using standard network protocols.

Using this approach, we plan to develop a language of inter-module communication and create a modular service mobile robot equipped with a developed sensor system, two hands–manipulators, and a main computer (Fig.2a). The robot is supposed to have a two-tier (short- and long-range) sensor system. The short-range sensor system supplies the motion module with current data on the surrounding space to ensure that the motion module is a fully functional mechatronic system capable of forming the motion trajectory within the analyzed zone and reacting independently to the emergence of dynamic objects in the zone. This module must make self-decisions on the change of motion trajectory or stopping. This main computer uses the long-range sensor system to build the map of the local area, generate local targets for modules, and develop a general strategy of behavior on the basis of given orders.

The proposed principle of modular design of robotic systems can be successfully used in the Bionic Assembly System [15, 16].

5. Acknowledgments

This work was supported by the Russian Foundation for Basic Research, project no. 16-07-00811a.

6. References

[1] A.V.Lopota, E.I.Yurevich (2013), Stages and Developments of Robotic Systems Design Modular Principle, In St. Petersburg State Polytechnical University Journal, 1(164), St. Petersburg (Russia). – pp. 98-103

[2] Fukuda T., S. Nakagawa (1987), A dynamically reconfigurable robotic system (concept of a system and optimal configurations), In Proceedings of IECON '87: 1987 International Conference on Industrial Electronics, Control, and Instrumentation

[3] S. Murata, E. Yoshida, A. Kamimura, H. Kurokawa, K. Tomita, and S. Kokaji (2002), M-TRAN: self-reconfigurable modular robotic system, IEEE/ASME Transactions on Mechatronics, 7(4):432–441

[4] E.H. Østergaard, K. Kassow, R. Beck, and H.H. Lund (2006), Autonomous Robots, Design of the ATRON lattice-based self-reconfigurable robot, 21(2):165–183

[5] Salemi B., M. Moll, and W.-M. Shen (2006), SUPERBOT: A deployable, multi-functional, and modular self-reconfigurable robotic system, In Proceedings, IEEE/RSJ Intl. Conf. on Intelligent Robots and Systems, pages 3636–3641, Beijing, China

[6] Fitch R.C. (2004), Heterogeneous Self-Reconfiguring Robotics, PhD thesis, Dartmouth College, Hanover, New Hampshire, USA

[7] Lyder A.H., Stoy K., Mendoza-Garcia R.-F., Larsen J.C., and Hermansen P. (2013), On sub-modularization and morphological heterogeneity in modular robotics, In Sukhan Lee, Hyungsuck Cho, Kwang-Joon Yoon, and Jangmyung Lee (Eds.), Intelligent Autonomous Systems 12, volume 193 of Advances in Intelligent Systems and Computing, pp. 649-661. Springer Berlin Heidelberg

[8] A. Lyder, R.F.M. Garcia, and K. Stoy (2008), Mechanical design of ODIN, an extendable heterogeneous deformable modular robots, In Proceedings, IEEE/RSJ international conference on intelligent robots and systems, ph.883–888, Nice, France

[9] Hancher M. D., Hornby G. S. (2006), A modular robotic system with applications to space exploration, 2nd IEEE International Conference on Space Mission Challenges for Information Technology (SMC-IT'06), Pasadena, CA: IEEE, pp. 132–140

[10] J. Baca, M. Ferre, and R. Aracil (2012), A heterogeneous modular robotic design for fast response to a diversity of tasks, Robotics and Autonomous Systems, vol. 60, no. 4, pp. 522–531

[11] Pryanichnikov V.E., Andreev V.P. (2012), The Application of Network Technologies to Constructing Group Controlled Systems with Machine Vision for Mobile Robots, In Annals of DAAAM for 2012& Proceedings of the 23th international DAAAM Symposium "Intelligent Manufacturing & Automation" 24-27th October 2012 Zadar, Croatia, ISSN 2304-1382, V.23, No.1. – pp.1167 – 1174

[12] V.Andreev, K.Kirsanov, and V.Pryanichnikov (2016), Geographically distributed "multi-operator" Control for Mechatronic Devices via the Internet, In Proceedings of the 26th DAAAM International Symposium., B. Katalinic (Ed.), Published by DAAAM International, ISBN: 978-1-5108-1839-2, ISSN 1726-9679, Vienna, Austria. 2015. Curran Proceedings, New York,.pp. 0004-0011

[13] Andreev V.P., Kirsanov K.B., Poduraev Yu. V. (2015), Geographically distributed "multi-operator" control of robotized systems using network technologies, In Proceedings of the International Scientific and Technical Conference "Extreme Robotics". – St.Petersburg: "Politechnika-service", pp. 67-71.

[14] Andreev V., and Kim V., Control System and Design of the Motion Module of a Heterogeneous Modular Mobile Robot, Proceedings of the 27th DAAAM International Symposium, pp.xxxx-xxxx, B. Katalinic (Ed.), Published by DAAAM International, ISBN 978-3-902734-08-2, ISSN 1726-9679, Vienna, Austria DOI: 10.2507/27th.daaam.proceedings.xxx

[15] Katalinic B., Kukushkin I., Pryanichnikov V., Haskovic D. (2014), Cloud Communication Concept for Bionic Assembly System. Procedia Engineering, 1877-7058, 69 (2014) pp. 1562–1568 doi:10.1016/j.proeng.2014.03.156.

[16] Katalinic B., Kukushkin I., & Haskovic D. (2014), Bionic Assembly System Cloud: Functions, Information Flow and Behavior. In 9th International Conference of DAAAM Baltic, Industrial Engineering, ISBN 978-9949-23-620-6, ISSN 2346-6138. pp. 103–108, ed. T.Otto, Tallinn, Estonia

DOI: 10.2507/27th.daaam.proceedings.002

EFFECTIVE ALGORITHM FOR DETERMINATION OF THE INITIAL VECTOR FOR THE DEFINITE QUADRATIC PENCIL

Aleksandra Kostic, Zajim Aljicevic, Sefko Sikalo & Melisa Kustura

This Publication has to be referred as: Kostic, A[leksandra]; Aljicevic, Z[ajim]; Sikalo, S[efko] & Kustura, M[elisa] (2016). Effective Algorithm for Determination of the Initial Vector for the Definite Quadratic Pencil, Proceedings of the 27th DAAAM International Symposium, pp.0010-0016, B. Katalinic (Ed.), Published by DAAAM International, ISBN 978-3-902734-08-2, ISSN 1726-9679, Vienna, Austria
DOI: 10.2507/27th.daaam.proceedings.002

Abstract

In this paper we study properties of definite quadratic eigenproblems. Free vibrations of fluid-solid structures are governed by a nonsymmetric eigenvalue problem. This problem can be transformed into a definite quadratic eigenvalue problem. Niendorf and Voss presented an algorithm which determents if quadratic pencil is or is not definite. In their algorithm, the initial vector is taken at random. In this paper we bring a new effective algorithm for determining the initial vector $\mathbf{x}_0$ and thus we accelerate algorithm of Niendorf and Voss. The new algorithm presented in this paper combines earlier results of Kostic and Sikalo with the new results obtained in this study.

Keywords: free vibrations; definite quadratic pencil; eigenvalue; methods for detection; effective algorithm; initial vector.

1. Introduction

In the engineering practice eigenvalue problem is very important because it is related with the mechanical systems and their property of vibration. Vibration is property of every mechanical system. Also it is property of electrical systems in the form of oscillation circuits. Neglecting terms of vibration can lead to resonance. The resonance can have catastrophic consequences (for example the demolition of the bridge), on the other hand it can be used positively. Mathematical description of vibratory conditions leads to a differential equation or a system of differential equations. This problem is further transformed to the eigenvalue problem. This is the motivation for considering the eigenvalue problems. They are divided into linear and nonlinear problems of eigenvalues. The theoretical basics for eigenvalue problems are given in the [7]. In practice, nonlinear eigenproblems commonly arise in dynamic/stability analysis of structures and in fluid mechanics, electronic behavior of semiconductor hetero-structures, vibration of fluid-solid structures, regularization on total least squares problems and stability of delay differential equations.

In practice, most important is the quadratic problem $\mathbf{Q}(\lambda)\mathbf{x} = \mathbf{0}$ where

$$\mathbf{Q}(\lambda) := \lambda^2\mathbf{A} + \lambda\mathbf{B} + \mathbf{C}, \qquad \mathbf{A}, \mathbf{B}, \mathbf{C} \in \mathbb{C}^{n \times n}, \qquad \mathbf{A} \neq \mathbf{0}, \tag{1}$$

More information on the practical application can be found in [12].

Since we, in this paper, deal with the improving of the algorithm to determine whether a quadratic eigenvalue problem is definite or not, let us briefly summarize what has been done so far in the literature up to this point.

For quadratic hyperbolic pencils, Higham, Tisseur and Van Dooren [5] proposed a method for testing hyperbolicity and for constructing a definite linearization. Another method for detecting whether a Hermitian quadratic matrix polynomial is hyperbolic is based on cyclic reduction and was first introduced by Guo and Lancaster [3] and later on, accelerated by Guo, Higham and Tisseur [2].

Niendorf and Voss [10] take advantage of the fact that all eigenvalues of a definite matrix polynomial can be characterized as minmax values of the appropriate Rayleigh functional and that the extreme eigenvalues in each of the intervals $(-\infty, \xi)$, (ξ,μ) and $(\mu,+\infty)$ are the limits of monotonically and quadratically convergent sequences. If a given Hermitian quadratic matrix polynomial is definite, Niendorf and Voss concurrently determine parameters ξ and μ such that the matrices $\mathbf{Q}(\xi) < 0$ and $\mathbf{Q}(\mu) > 0$, which allows for the computation of a definite linearization.

Kostic and Voss [9] applied the Sylvester's law of inertia on the definite quadratic eigenvalue problems.

In paper [8] Kostic and Sikalo gave improvement of the method of Niendorf and Voss [10] for detecting whether a quadratic pencil $\mathbf{Q}(\lambda)$ is definite. They have used some properties of matrices **A, B** and **C** to get improvment of the algorithm. These properties are rank of the matrices **A, B** and **C** and eigenvectors of these matrices which corespond to the egenvalue zero. Kostic and Sikalo in [8] obtained good results when matrix **B** is singular. If matrix **B** is not singular they would have to determine eigenspace of the matrix **A**, to the eigenvalue zero. Also, they would have to determine eigenspace of the matrix **C,** to the eigenvalue zero. This reduces efficiency of their idea.

This paper gives effective algorithm for determination of the initial vector $\mathbf{x}_0$. The algorithm is based on results of the paper [8] and on application of eigenvector $\mathbf{q}_1$ which is corresponding vector of eigenvalue λ_{1A} of matrix **A**, and also on application of eigenvector $\mathbf{f}_n$ which is corresponding vector of eigenvalue λ_{nB} of matrix **B**.

Our paper is organized as follows. In Section 2 we provide definition of the definite eigenvalue problem and results of previous papers based on this topic. Section 3 gives a new effective algorithm for determining of the initial vector $\mathbf{x}_0$. This section also presents new results which lead to our algorithm. A critical review of the proposed algorithm and the future research plan is given in Section 4.

2. Definite quadratic pencils

In this paragraph we give the theoretical basics for the definite quadratic pencil. Since the definite quadratic polynomial is generalization of the hyperbolic quadratic polynomial, let us briefly consider the hyperbolic quadratic polynomial.

A quadratic matrix polynomial

$$\mathbf{Q}(\lambda) := \lambda^2\mathbf{A} + \lambda\mathbf{B} + \mathbf{C}, \qquad \mathbf{A} = \mathbf{A}^H > 0, \qquad \mathbf{B} = \mathbf{B}^H,\ \mathbf{C} = \mathbf{C}^H \tag{2}$$

is hyperbolic if for every $\mathbf{x} \in \mathbb{C}^n, \mathbf{x} \neq \mathbf{0}$ the quadratic polynomial

$$\mathrm{f}(\lambda; \mathbf{x}) := \lambda^2\mathbf{x}^H\mathbf{A}\mathbf{x} + \lambda\mathbf{x}^H\mathbf{B}\mathbf{x} + \mathbf{x}^H\mathbf{C}\mathbf{x} = 0 \tag{3}$$

has two distinct real roots:

$$p_+(\mathbf{x}) := -\frac{\mathbf{x}^H\mathbf{B}\mathbf{x}}{2\mathbf{x}^H\mathbf{A}\mathbf{x}} + \sqrt{\left(\frac{\mathbf{x}^H\mathbf{B}\mathbf{x}}{2\mathbf{x}^H\mathbf{A}\mathbf{x}}\right)^2 - \frac{\mathbf{x}^H\mathbf{C}\mathbf{x}}{\mathbf{x}^H\mathbf{A}\mathbf{x}}}, \qquad p_-(\mathbf{x}) := -\frac{\mathbf{x}^H\mathbf{B}\mathbf{x}}{2\mathbf{x}^H\mathbf{A}\mathbf{x}} - \sqrt{\left(\frac{\mathbf{x}^H\mathbf{B}\mathbf{x}}{2\mathbf{x}^H\mathbf{A}\mathbf{x}}\right)^2 - \frac{\mathbf{x}^H\mathbf{C}\mathbf{x}}{\mathbf{x}^H\mathbf{A}\mathbf{x}}} \tag{4}$$

Functionals in (4) are the Rayleigh functionals of the quadratic matrix polynomial (2). The Rayleigh functionals are the generalization of the Rayleigh quotient.

The hyperbolic problem is very importance because all of its eigenvalues can be variational characterized. More information is contained in [1].

Higham, Mackey and Tisseur [6] generalized the concept of the hyperbolic quadratic polynomial by waiving the positive definiteness of the leading matrix **A**.

Definition 1

A quadratic pencil (1) is definite if:

$\mathbf{A} = \mathbf{A}^H, \mathbf{B} = \mathbf{B}^H,\ \mathbf{C} = \mathbf{C}^H$are Hermitian, there exists $\mu \in \mathbb{R} \cup \{\infty\}$ such that $\mathrm{Q}(\mu)$ is positive definite, and for every fixed $\mathbf{x} \in \mathbb{C}^n, \mathbf{x} \neq \mathbf{0}$ the quadratic polynomial (3) has two distinct roots in $\mathbb{R} \cup \{\infty\}$.

Free vibrations of a fluid-solid structures are governed by a non-symmetric eigenvalue problem [4,8,10,11]. This problem can be transformed into a definite quadratic eigenvalue problem and it is motivation for further consideration of the definite quadratic pencil and eigenvalue problems for these pencils.

The following Theorem has been proved by Duffin [1].

Theorem 1. A Hermitian matrix polynomial $\mathbf{Q}(\lambda)$ is definite if and only if any two (and hence all) of the following properties hold:

- $(\mathbf{x}) := (\mathbf{x}^H\mathbf{B}\mathbf{x})^2 - 4(\mathbf{x}^H\mathbf{A}\mathbf{x})(\mathbf{x}^H\mathbf{C}\mathbf{x}) > 0$ for every $\mathbf{x} \in \mathbb{C}^n, \mathbf{x} \neq \mathbf{0}$
- $\mathbf{Q}(\mu) > 0$ for same $\mu \in \mathbb{R} \cup \{\infty\}$
- $\mathbf{Q}(\xi) < 0$ for same $\xi \in \mathbb{R} \cup \{\infty\}$.

Without loss of generality, we take that $\xi < \mu$.

$f(\mu; \mathbf{x}) = \mathbf{x}^H\mathbf{Q}(\mu)\mathbf{x}$

$f(\xi; \mathbf{x}) = \mathbf{x}^H\mathbf{Q}(\xi)\mathbf{x}$

where $\mathbf{Q}(\xi) < 0 < \mathbf{Q}(\mu)$, $\xi < \mu$.

The Rayleigh functional is explicitly given by

$$p(\mathbf{x}) := \begin{cases} -\dfrac{\mathbf{x}^H\mathbf{B}\mathbf{x}}{2\mathbf{x}^H\mathbf{A}\mathbf{x}} + \sqrt{\left(\dfrac{\mathbf{x}^H\mathbf{B}\mathbf{x}}{2\mathbf{x}^H\mathbf{A}\mathbf{x}}\right)^2 - \dfrac{\mathbf{x}^H\mathbf{C}\mathbf{x}}{\mathbf{x}^H\mathbf{A}\mathbf{x}}} & \text{if } \mathbf{x}^H\mathbf{A}\mathbf{x} > 0 \\ -\dfrac{\mathbf{x}^H\mathbf{C}\mathbf{x}}{\mathbf{x}^H\mathbf{B}\mathbf{x}} \quad \text{if} \quad \mathbf{x}^H\mathbf{A}\mathbf{x} = 0 \\ -\dfrac{\mathbf{x}^H\mathbf{B}\mathbf{x}}{2\mathbf{x}^H\mathbf{A}\mathbf{x}} - \sqrt{\left(\dfrac{\mathbf{x}^H\mathbf{B}\mathbf{x}}{2\mathbf{x}^H\mathbf{A}\mathbf{x}}\right)^2 - \dfrac{\mathbf{x}^H\mathbf{C}\mathbf{x}}{\mathbf{x}^H\mathbf{A}\mathbf{x}}} & \text{if } \mathbf{x}^H\mathbf{A}\mathbf{x} < 0 \end{cases}.$$

The quadratic eigenvalue problem $\mathbf{Q}(\lambda)\mathbf{x} = \mathbf{0}$ has exactly n eigenvalues $\lambda_1 \leq \cdots \leq \lambda_n$ in (ξ, μ) which satisfy a minmax characterization with respect to p, and the safeguarded iterations aiming at λ_1 and λ_n converge globally and monotonically decreasing and increasing, respectively. Niendorf and Voss [10] gave the following algorithm (Fig.1) for detecting definite quadratic matrix polynomial.

Require: initial vector $\mathbf{x}_0 \neq \mathbf{0}$
if $d(\mathbf{x}_0) := (\mathbf{x}_0^H\mathbf{B}\mathbf{x}_0)^2 - 4(\mathbf{x}_0^H\mathbf{A}\mathbf{x}_0)(x_0^H C x_0) < 0$ **then**
STOP: $\mathbf{Q}(\lambda)$ is not definite
end if
determine $\sigma_0 = p(\mathbf{x}_0)$
for $k = 1, 2, \ldots$ **until convergence do**
determine eigenvector x_k of $\mathbf{Q}(\sigma_{k-1})$ corresponding to its largest eigenvalue
if $d(\mathbf{x}_k) := (\mathbf{x}_k^H\mathbf{B}\mathbf{x}_k)^2 - 4(\mathbf{x}_k^H\mathbf{A}\mathbf{x}_k)(\mathbf{x}_k^H\mathbf{C}\mathbf{x}_k) < 0$ **then**
STOP: $\mathbf{Q}(\lambda)$ is not definite
end if
determine $\sigma_k = p(\mathbf{x}_k)$
if $\sigma_k \geq \sigma_{k-1}$ **then**
STOP: $\mathbf{Q}(\lambda)$ is not definite
end if
if $\mathbf{Q}(2\sigma_k - \sigma_{k-1})$ is negative definite then
STOP: $\mathbf{Q}(\lambda)$ is definite
end if
end for

Fig. 1.Algorithm 1.

Now we will briefly explain the Algorithm 1 (Fig.1). If $\mathbf{Q}(\lambda)$ is definite we can determine by Algorithm 1 a parameter ξ such that $\mathbf{Q}(\xi)$ is negative definite. We have to allow for one violation of the monotonicity requirement to incorporate the possible jump of the iterations from one unbounded interval to the other. A second similar sweep aiming at λ_n discovers a parameter μ such that $\mathbf{Q}(\mu) > 0$. Once parameters ξ and μ are found such that $\mathbf{Q}(\xi) < 0 < \mathbf{Q}(\mu)$, the definite linearization can be determined similarly.

Let's analyze the advantages and disadvantages of the Algorithm 1. Its advantages include clarity in the functioning of the algorithm and the ease of programming. The shortcomings of the Algorithm1 are cost of the algorithm which means a large number of flops in every iteration and random selection of the initial vector $\mathbf{x}_0$. Let us look first at the cost of the algorithm.

The k-th step of Algorithm 1 requires $n^3/3$ operations for computing 1 Cholesky factorization, $4n^2$ operations for evaluating $\mathbf{Q}(2\sigma_k - \sigma_{k-1})$, 3 matrix-vector products ($6n^2$ operations), 3 scalar products ($6n$ operations), and the determination of the largest eigenvalue and corresponding eigenvector of a matrix $\mathbf{Q}(\sigma_{k-1})$ which seems to be the most expensive part.

Let a quadratic matrix polynomial

$$\mathbf{Q}(\lambda) := \lambda^2\mathbf{A} + \lambda\mathbf{B} + \mathbf{C}, \qquad \mathbf{A} = \mathbf{A}^H, \qquad \mathbf{B} = \mathbf{B}^H, \ \mathbf{C} = \mathbf{C}^H \tag{5}$$

be definite.

Kostic and Sikalo in [8] developed a strategy for targeting of the initial vector $\mathbf{x}_0$. This leads to reduction in the number of iterations and the algorithm becomes cheaper. Here are the basic theorems on which their strategy is based. Proofs of these theorems are given in [8].

Lemma 1 Problem of the eigenvalues corresponding to the matrix polynomial (5) has 0 as an eigenvalue if and only if **C** is singular.

Without loss of generality, Kostic and Sikalo in [8] took that $\xi < \mu$. They considered rank of matrices **A**, **B** and **C**.

Let $\text{Rank}(\mathbf{A}) = n - p$, $\text{Rank}(\mathbf{B}) = n - q$, $\text{Rank}(\mathbf{C}) = n - r$ where $0 < p < n$, $0 \le q \le n$ and $0 < r < n$ and let $\{\mathbf{y}_1, \mathbf{y}_2, \cdots, \mathbf{y}_n\}$, $\{\mathbf{z}_1, \mathbf{z}_2, \cdots, \mathbf{z}_n\}$ and $\{\mathbf{w}_1, \mathbf{w}_2, \cdots, \mathbf{w}_n\}$ be orthonormal base of vector space $\mathbb{C}^n$ consisting of eigenvectors of the corresponding matrix **A**, **B** and **C** respectively. Without loss of generality, Kostic and Sikalo in [8] took

$$\mathbf{A}\mathbf{y}_i = 0 \ i = 1,2,\cdots,p, \qquad \mathbf{B}\mathbf{z}_j = 0, j = 1,2,\cdots,q, \qquad \mathbf{C}\mathbf{w}_k = 0 \qquad k = 1,2,\cdots,r. \tag{6}$$

Theorem 2 Let $\mathbf{Q}(\xi) < 0 < \mathbf{Q}(\mu)$ and $\xi < \mu$ then stands $\mu > a_1 > c_1 > \xi$ where

$$a_1 := max\left\{-\frac{\mathbf{y}^H\mathbf{C}\mathbf{y}}{\mathbf{y}^H\mathbf{B}\mathbf{y}} : \mathbf{y} \in \{\mathbf{y}_1, \mathbf{y}_2, \cdots, \mathbf{y}_p\}\right\} \text{ and } c_1 := min\left\{-\frac{\mathbf{y}^H\mathbf{C}\mathbf{y}}{\mathbf{y}^H\mathbf{B}\mathbf{y}} : \mathbf{y} \in \{\mathbf{y}_1, \mathbf{y}_2, \cdots, \mathbf{y}_p\}\right\}.$$

Theorem 3 Let $\mathbf{Q}(\xi) < 0 < \mathbf{Q}(\mu)$ and $\xi < \mu$ then stands $\mu > a_2$ and $c_2 > \xi$ where

$$a_2 := \max\left\{-\frac{\mathbf{w}^H\mathbf{B}\mathbf{w}}{2\mathbf{w}^H\mathbf{A}\mathbf{w}} : \mathbf{w} \in \{\mathbf{w}_1, \mathbf{w}_2, \cdots, \mathbf{w}_r\} \text{ and } \mathbf{w}^H\mathbf{A}\mathbf{w} > 0\right\}$$

and

$$c_2 := \min\left\{-\frac{\mathbf{w}^H\mathbf{B}\mathbf{w}}{2\mathbf{w}^H\mathbf{A}\mathbf{w}} : \mathbf{w} \in \{\mathbf{w}_1, \mathbf{w}_2, \cdots, \mathbf{w}_r\} \text{ and } \mathbf{w}^H\mathbf{A}\mathbf{w} < 0\right\}$$

Theorem 4 Let $\mathbf{Q}(\xi) < 0 < \mathbf{Q}(\mu)$, $\xi < \mu$ and $a := \max(a_1, a_2)$ and $c := \min(c_1, c_2)$ then applies $\mu > a$ and $c > \xi$ where a_1, a_2, c_1 and c_2 are defined as in Theorem 2 and Theorem 3.

If we now apply Algorithm 1 to determine the parameter μ for the initial vector $\mathbf{x}_0$ we take the vector through which we specify a. Analogously if we apply Algorithm 1 (Fig. 1) for determining parameter ξ for initial vector $\mathbf{x}_0$ we take the vector through which we have determined c.

There remains to consider what happens in the case of the singularity of the matrix **B**.

Theorem 5 Let $\mathbf{Q}(\xi) < 0 < \mathbf{Q}(\mu)$, $\xi < \mu$, $\text{Rank}(\mathbf{B}) = n - q$ and $q > 0$ than $\mu > 0$ or $\xi < 0$.

Remark: In the proof of the Theorem 5 [8] it is clear that $\mathbf{z}^H\mathbf{A}\mathbf{z}$ has the same sign for all $\mathbf{z} \in \{\mathbf{z}_1, \mathbf{z}_2, \cdots, \mathbf{z}_q\}$. If $\mathbf{z}^H\mathbf{A}\mathbf{z} > 0$ we can multiply problem (5) by -1 and get a new definite eigenvalue problem.

$$\mathbf{Q}(\lambda) := \lambda^2\mathbf{A}_1 + \lambda\mathbf{B}_1 + \mathbf{C}_1, \qquad \mathbf{A}_1 = -\mathbf{A}, \qquad \mathbf{B}_1 = -\mathbf{B}, \qquad \mathbf{C}_1 = -\mathbf{C} \tag{7}$$

Problem (7) has the same eigenvalues as problem (5), but with $\mathbf{z}^H\mathbf{A}_1\mathbf{z} < 0$ and $\xi < 0$. Therefore, in the case of singularity of the matrix **B** without loss of generality we can consider that $\mathbf{z}^H\mathbf{A}\mathbf{z} < 0$ and $\xi < 0$. In the case of singularity of the matrix **B** in order to reduce the costs of determining the initial vector $\mathbf{x}_0$ it is taken $\mathbf{x}_0 = \mathbf{z}_1$.

3. The effective algorithm for determining the initial vector

Kostic and Sikalo in [8] have used some properties of the matrix **A**, **B** and **C** for determination of the initial vector $\mathbf{x}_0$. Those properties are for example ranks of the matrices, eigenvectors belonging to the eigenvalue zero of the **A**, **B** and **C** matrices. For the implementation of this strategy they would have to find eigenspace of the matrices **A, B** and **C**, which are associated with the eigenvalue zero.

The basic idea of this paper is still using nonzero eigenvalues of matrices **A** and **B** to get a better location parameters and better initial vector. The acceleration of the algorithm is also obtained by improvement of the initial vector $\mathbf{x}_0$. Our aim is getting high-quality algorithm for the determination of the initial vector. This algorithm is obtained by using results of the paper [8] and this paper.

In order to accelerate Algorithm 1 we consider a better localization of the parameters μ and ξ which hold $\mathbf{Q}(\mu) > 0 > \mathbf{Q}(\xi)$ and $\mu > \xi$. From this localization we will get the initial vector $\mathbf{x}_0$.

In [8] is noticed that matrices **A** and **C** must be singular. The singularity of matrix **B** can but doesn't have to be satisfied. If matrix **B** is singular than considering Remark for the initial vector we take $\mathbf{x}_0 = \mathbf{z}_1$. If the matrix **B** is not singular theorems 2, 3 and 4 give us possibility of choosing initial vector $\mathbf{x}_0$. However this means we must determine all eigenvectors of the matrix **A** that belong to eigenvalue zero and all eigenvalues of the matrix **C** that belong to eigenvalue zero. Also, this procedure is weighted by finding a large number of products of vectors and matrices and scalar products. This brings additional algorithm costs (a large number of flops and long time computing). To avoid the large number of flops in the case when matrix **B** is not singular, we consider a new strategy.

We assume that the matrix **A** has at least one positive and at least one negative eigenvalue. We denote the largest eigenvalue of matrix **B** with $\lambda_{n\mathbf{B}}$ and the corresponding eigenvector with $\mathbf{q}_1$. We denote the smallest eigenvalue of matrix **A** with $\lambda_{1\mathbf{A}}$ and the corresponding eigenvector with $\mathbf{f}_n$. It is clear that $\lambda_{1\mathbf{A}}<0$.

For every fixed $\mathbf{x} \in \mathbb{C}^n, \mathbf{x} \neq \mathbf{0}$ is

$$\mu^2 \mathbf{x}^H \mathbf{A}\mathbf{x} + \mu \mathbf{x}^H \mathbf{B}\mathbf{x} + \mathbf{x}^H \mathbf{C}\mathbf{x} > 0 \tag{8}$$

$$\xi^2 \mathbf{x}^H \boldsymbol{A}\mathbf{x} + \xi \mathbf{x}^H \mathbf{B}\mathbf{x} + \mathbf{x}^H \mathbf{C}\mathbf{x} < 0 \tag{9}$$

Subtracting the inequality (9) from inequality (8) yields

$$(\mu - \xi)(\mu + \xi)\mathbf{x}^H \mathbf{A}\mathbf{x} + (\mu - \xi)\mathbf{x}^H \mathbf{B}\mathbf{x} > 0 \overset{\xi<\mu}{\Longrightarrow} (\mu + \xi)\mathbf{x}^H \mathbf{A}\mathbf{x} + \mathbf{x}^H \mathbf{B}\mathbf{x} > 0. \tag{10}$$

Further we get

$$(\mu + \xi)\frac{\mathbf{x^H A x}}{\mathbf{x^H x}} + \lambda_{n\mathbf{B}} > (\mu + \xi)\frac{\mathbf{x^H A x}}{\mathbf{x^H x}} + \frac{\mathbf{x^H B x}}{\mathbf{x^H x}} > 0 \tag{11}$$

From (11) we get for every fixed $\mathbf{x} \in \mathbb{C}^n, \mathbf{x} \neq \mathbf{0}$

$$(\mu + \xi)\frac{\mathbf{x^H A x}}{\mathbf{x^H x}} > -\lambda_{n\mathbf{B}}. \tag{12}$$

Specially for $\mathbf{x} = \mathbf{q}_1$ we get

$$(\mu + \xi)\frac{\mathbf{q_1}^{\mathbf{H}}\mathbf{A}\mathbf{q_1}}{\mathbf{q_1}^{\mathbf{H}}\mathbf{q_1}} > -\lambda_{n\mathbf{B}}. \tag{13}$$

Since $\mathbf{q_1}$ is eigenvector of eigenvalue $\lambda_{1\mathbf{A}}<0$ from (13) is obtained

$$(\mu + \xi)\lambda_{1\mathbf{A}} > -\lambda_{n\mathbf{B}}. \tag{14}$$

From (14) and $\lambda_{1\mathbf{A}}<0$ we get

$$\mu + \xi < -\frac{\lambda_{n\mathbf{B}}}{\lambda_{1\mathbf{A}}}. \tag{15}$$

From (14) and $\mu - \xi > 0$ we get

$$\xi < -\frac{\lambda_{n\mathbf{B}}}{2\lambda_{1\mathbf{A}}}. \tag{16}$$

This way we get upper bound for parameter ξ in the case where **B** is not singular. In this case because of the great significance of the matrix **B** for definite eigenvalue problem we will take $\mathbf{x}_0 = 0.9\mathbf{q_1} + 0.1\mathbf{f_n}$ for the initial vector. Now we will give an effective algorithm for determination of the initial vector $\mathbf{x}_0$.

Examine singularity of matrix **B**
If matrix **B** is singular **then** $\mathbf{x}_0 = \mathbf{z}_1$
else
determine smallest eigenvalue of matrix **A** $\lambda_{1\mathbf{A}}$ and corresponding eigenvector $\mathbf{q_1}$
determine largest eigenvalue of matrix **B** $\lambda_{\mathbf{nB}}$ and corresponding eigenvector $\mathbf{f_n}$
$\mathbf{x}_0 = 0.9\mathbf{q_1} + 0.1\mathbf{f_n}$
end if

Fig.2. Algorithm 2

Fig.2. gives the effective algorithm for determining of the initial vector $\mathbf{x}_0$. This algorithm is based on the structure of the definite quadratic eigenvalue problem. At the end we would like to point out the following. To determine whether the matrix **B** is singular we have to decomposition this matrix. From this decomposition we can see all the eigenvalues and the corresponding eigenvectors of matrix **B**. Therefore, if the matrix **B** is not singular, we already have an eigenvector belonging to the largest eigenvalue at no additional cost.

The following two examples numerically show justification for applying Algorithm 2.

Example 1
In the quadratic matrix polynomial $\mathbf{Q}(\lambda) := \lambda^2\mathbf{A} + \lambda\mathbf{B} + \mathbf{C}$ we will take

$$\mathbf{A} = \begin{pmatrix} 1 & 2 & 3 \\ 2 & 5 & 6 \\ 3 & 6 & 8 \end{pmatrix}, \mathbf{B} = \begin{pmatrix} 0 & 0 & 0 \\ 0 & 10^9 & 0 \\ 0 & 0 & -10^{15} \end{pmatrix} \text{ and } \mathbf{C} = \begin{pmatrix} 4 & 0 & 0 \\ 0 & 5 & 9 \\ 0 & 9 & 5 \end{pmatrix}.$$

We tested the speed of determination weather the matrix is or is not definite. First, we used only with Algorithm 1. Numerical tests were performed by using MATLAB. The initial vector is determined by using function randn. We did 100 tests and we obtained that average number of expensive iterations for getting the answer is 4.62. When we use Algorithm 1 and Algorithm 2 we get the answer immediately in the preparatory phase of the Algorithm 1. Therefore, the answer is obtained without expensive iterations.

Example 2
In the quadratic matrix polynomial $\mathbf{Q}(\lambda) := \lambda^2\mathbf{A} + \lambda\mathbf{B} + \mathbf{C}$ we will take

$$\mathbf{A} = \begin{pmatrix} 1 & 2 & 3 \\ 2 & 5 & 6 \\ 3 & 6 & 8 \end{pmatrix}, \mathbf{B} = \begin{pmatrix} -10^9 & 0 & 0 \\ 0 & -10^9 & 0 \\ 0 & 0 & 1 \end{pmatrix} \text{ and } \mathbf{C} = \begin{pmatrix} 4 & 0 & 0 \\ 0 & 5 & 9 \\ 0 & 9 & 5 \end{pmatrix}.$$

Once again we did numerical experiment with MATLAB. The initial vector was determined by using function randn. We did 100 tests and we obtained that the average number of expensive iterations for getting the answer is 4.17. When we used Algorithm 1 and Algorithm 2 we obtained the answer after one expensive iteration.
Example 1 and Example 2 show the importance of using Algorithm 1 for determination of the initial vector. This is especially important when matrix **B** is singular or when there exists great difference between absolute value of the biggest and the absolute value of the smallest eigenvalue as in Example 2.

4. Conclusion

This paper gives the effective algorithm for the determination of the initial vector $\mathbf{x}_0$. If the initial vector $\mathbf{x}_0$ is better determined we get the algorithm acceleration. In previous papers there were attempts of getting a better initial vector $\mathbf{x}_0$ that were based on the exploitation of some properties of the matrices **A**, **B** and **C**. Previously used properties were ranks of matrices and their eigenvectors which belong to eigenvalue zero. In this paper we also give exploitation of the eigenvector $\mathbf{q_1}$ which is corresponding vector of the eigenvalue $\lambda_{1\mathbf{A}}$ of matrix **A**. We also give exploitation of the eigenvector $\mathbf{f_n}$ which is corresponding vector of the eigenvalue $\lambda_{\mathbf{nB}}$ of matrix **B**. In Example 1 and Example 2 we have concretely shown the justification for applying the Algorithm 2.
Therefore, in this paper we sum up results from previous papers with the results of this paper. This way we get a high-quality algorithm for determining initial vector $\mathbf{x}_0$. This algorithm enables significant acceleration of the Algorithm 1. The further research will go in the direction of the use of features of the matrix **A**, **B** and **C**, which determine square pencil that is used for improvement of localization of the parameters ξ and μ.

References

[1] Duffin, R. J. (1955). A minimax theory for overdamped networks, J. Rat. Mech. Anal., 4 (1955) 221–233
[2] Guo, C.-H.; Higham, N.J. & Tisseur, F. (2007). Detecting and solving hyperbolic quadratic eigenvalue problems, Technical Report 117, The Manchester Institute of Mathematical Sciences,The University of Manchester, (2007)
[3] Guo, C.-H. & Lancaster P. (2005). Algorithms for hyperbolic quadratic eigenvalue problems, Math.Comp., 74 (2005) 1777–1791, ISSN 0025-5718
[4] Harris, C. & Piersol, A. (2002). Harris Shock and Vibration Handbook, 5 ed., McGraw-Hill, NewYork, 2002
[5] Higham, N. J.; Tisseur, F.& Van Dooren, P.M. (2002). Detecting a definite Hermitian pair anda hyperbolic or elliptic quadratic eigenvalue problem, and associated nearness problems, Linear Algebra Appl., (2002) 351–352 455–474, ISSN 0024-3795
[6] Higham, N.; Mackey, D. & Tisseur, F. (2009). Definite matrix polynomials and their linearization by definite pencils, SIAM J. Matrix Anal. Appl. 33 (2009) 478–502, ISSN 0895-4798
[7] Kostic, A. (2016). Eigenvalue Problems In: Applied linear algebra in action, Katsikis, N. (Ed.),57-83, InTech ISBN 978-953-51-2419-1, Rijeka
[8] Kostic, A. &. Sikalo, S. (2015). Definite Quadratic Eigenvalue Problems. Procedia Engineering 100 (2015) 56 – 63, ISSN 1877-7058
[9] Kostic, A. & Voss, H. (2013). On Sylvester's law of inertia for nonlinear eigenvalue problems, Electr. Trans. Numer. Anal., 40 (2013) 82–93, ISSN 1068-9613
[10] Niendorf, V. & Voss, H. (2010). Detecting hyperbolic and definite matrix polynomials, Linear Algebra Appl., 432 (2010) 1017–1035, ISSN 0024-3795
[11] Stammberger, M. & Voss, H. (2010). On an unsymmetric eigenvalue problem governing free vibrations of fluid-solid structures, Electr. Trans. Numer. Anal., 36 (2010), 113–125, ISSN 1068-9613
[12] Tisseur, F. & Meerbergen, K. (2001). The quadratic eigenvalue problem. SIAM Review, 43 (2001) 235–286, ISSN 0036-1445

DOI: 10.2507/27th.daaam.proceedings.003

ENERGY AND MATERIAL SAVING TECHNOLOGIES IN SLOVENIAN MANUFACTURING FIRMS

Iztok Palcic & Borut Buchmeister

This Publication has to be referred as: Palcic, I[ztok] & Buchmeister, B[orut] (2016). Energy and Material Saving Technologies in Slovenian Manufacturing Firms, Proceedings of the 27th DAAAM International Symposium, pp.0017-0022, B. Katalinic (Ed.), Published by DAAAM International, ISBN 978-3-902734-08-2, ISSN 1726-9679, Vienna, Austria
DOI: 10.2507/27th.daaam.proceedings.003

Abstract

The main objective of this paper is to map the adoption of technologies for energy reduction and resources consumption in production. The aim is also to contribute to the identification and understanding of the characteristics of the manufacturing firms that use this kind of energy and material saving technologies. Our research is based on the data from the largest European manufacturing survey and it includes data from Slovenia. The results show that the use of specific energy saving technologies and material saving technologies in Slovenian manufacturing firms is still modest. Dividing manufacturing firms based on technology intensity sectors and based on the relative energy efficiency we have concluded that firms in high technology industries focus less on energy efficiency than low technology firms. We also found out that firms with more implemented energy and material saving technologies are more likely to be more energy efficient.

Keywords: energy efficiency; manufacturing firm; survey; technology

1. Introduction

Manufacturing is defined as the transformation of materials and information into goods for the satisfaction of human needs. Turning raw materials into consumer products is also a major source of generating environmental pollution. Waste coming out from manufacturing activities is an environmental threat originating from several regions around the world [1]. Therefore, in recent years, mostly in response to increasing pressure from environmental regulations, many manufacturing firms have made significant efforts in cleaner production [2-3].

This paper is based on an empirical study that tries to contribute to world literature in the field of energy and material efficient technologies. The objective of this paper is firstly to map the adoption of the technologies for reduction of energy and resources consumption in production and, second to contribute to the identification and understanding of the characteristics of the manufacturing firms that use this kind of innovative technologies.

The paper is organized as follows. Introductory background and literature review about the energy efficiency in production is recalled. Next, we present our research methodology and methods used to analyse the characteristics of energy and material saving technologies' adoption and their adopters. Results and findings are presented for the manufacturing firms with the use of descriptive statistics. In the end we discuss our results and present some implications.

2. Literature review

In recent years, firms have faced strong pressure from their stakeholders to implement environmental management, especially since the Rio Declaration in 1992 and the Kyoto Protocol in 1997 [4]. Manufacturing firms have been affected particularly strongly because through their production activities they are reputedly primary polluters [5]. Climate change, insecure energy supplies and rising energy prices are topics of increasing importance in today's society [6]. New energy efficient technologies can contribute to higher energy efficiency of enterprises, but there is a trade-off between a firm's environmental and economic performance [7].

Industrial energy efficiency plays a central role as the manufacturing industry accounts for about 75% of the world's yearly coal consumption, 44% of the world's natural gas consumption, and 20% of global oil consumption. In addition, these manufacturing firms also use 42% of all the electricity generated [8]. Although renewable energy technologies, such as photovoltaic technology, might be a long-term solution, more efficient energy use can make the highest and most economic contribution towards solving these problems in the short run. Using the available energy more efficiently is an effective countermeasure to rising energy needs and unsecure energy supplies [9, 10]. Bunse et al. [6] argue that examples in the literature and in the world of practice show that although the manufacturing sector has made continuous improvement in energy efficiency, economically beneficial energy efficiency potential is not yet exploited [6, 11].

Improving energy efficiency is regarded as one of the most important options to reduce the emissions of greenhouse gases and the dependency of countries on energy imports [12]. Measuring energy efficiency is the basis for controlling energy consumption in the production processes, for deciding about improvement measures and for tracking changes and improvements in energy efficiency [6]. Studies on energy consumption of manufacturing processes have provided fundamental information for improving energy efficiency and build a comprehensive foundation towards reducing the energy consumption of manufacturing processes [13]. There is also an on-going debate on the reasons why profitable investments to reduce energy consumption are not realized in firms [14, 15]. There are several barriers to implementing energy efficiency improvement measures in firms, e. g.: payback periods, limited capital, a low priority given to energy efficiency by the management, lack of information, or "difficult-to-measure components" of energy investments [6, 16].

Bunse et al. [6] define energy efficiency as "the ratio of energy services out to energy input [meaning] getting the most out of every energy unit you buy". Increased energy efficiency may be accomplished by more efficient technology, energy recovery in the same process or further use of energy waste in different processes, increased energy conversion efficiency or optimized operational practices. Usually, energy efficiency indicators are ratios describing the relationship between an activity and the required energy. In the industrial sector, activities such as the production process of a product can be described in either economic or physical terms resulting in either economic or physical indicators. Economic indicators are useful at an aggregated level, such as for comparing different sectors; however, to gain insight into particular manufacturing processes, physical indicators are more suitable [17]. Examples of physical indicators are specific energy consumption [17-19], final energy efficiency improvement [20] etc. There is no single energy efficiency indicator that can be applied in every situation, but the appropriate indicators have to be defined depending on the decision to be made or decision tool to be applied [19].

Although there are many studies on energy efficiency analysing different aspects of the researched field, we have found no studies that analyse the use of specific energy and material saving technologies in manufacturing environment. Our study contributes to this area in a specific geographical area.

3. Methodology

We used data from European Manufacturing Survey (EMS) from 2009 for our research. The EMS is the largest European survey on manufacturing activities, coordinated by the Fraunhofer Institute for Systems and Innovation Research – ISI, Germany. The survey collects data on manufacturing strategies, the application of innovative organisational and technological concepts in production, personnel deployment and qualification, the production off-shoring and back-sourcing activities, cooperation patterns etc. In addition, data on firm characteristics and performance indicators (R&D expenses, productivity, returns on sales,) is collected. This paper uses data from Slovenian sub-sample, which had 64 responses. The survey was performed in manufacturing firms (NACE codes from 17 to 35) with at least 20 employees.

In recent years, only a few surveys have been launched in the world that analyse energy efficiency in manufacturing firms and their energy saving technologies (EST) and material saving technologies (MST) use. These existing surveys cover only some industrial sectors monitoring very specific technologies or cover only American and Asian countries. None of them include the European countries covered by EMS that also encompasses majority of manufacturing industries. Therefore, our survey added several questions relating environmental and energy issues. In that sense, EMS defines 10 general groups of technologies, 8 for energy efficiency and 2 for material consumption saving. These wide groups allow classifying any specific technology into one of them obtaining a global map of their use and level of implementation.

EST included were:
T1 control system for shut down of machines in off-peak periods;
T2 electric motors with rotation speed regulation;
T3 compressed air contracting;
T4 highly efficient pumps;
T5 low-temperature joining processes;
T6 retrieval of kinetic and process energy;
T7 combined cold, heat and power – Bi-/Tri-generation and
T8 waste material for in-house energy generation.

We included two MST:
T9 utilisation of recycled material in product manufacturing and
T10 product recovery after product life cycle.

EST and MST are characterized in terms of use and also in terms of usage levels (extent of use) through a descriptive and a frequency analysis. Extent of actual use is referred to comparing the actual use of the technology in the firm to the most reasonable potential use. There are three levels: Extent of utilised potential "low" for an initial attempt to utilise, "medium" for partly utilized and "high" for extensive use.

We have analysed the characteristics of EST and MST adopters according to the OECD's taxonomy of manufacturing industries classified by their technological intensity [21]. We have formed two groups: "Low and Medium-Low technology" (LMT) with firms from NACE 17-19, 20-22, 23, 25, 26, 27, 28; and "Medium-High and High technology" (MHT) with firms from NACE 29, 30, 31, 32, 34 and 35. We created a discrete variable to group this classification into both categories: "Low and Medium-Low technology" – value 1, and "Medium-High and High technology" – value 2.

Next, we have classified technology adopters in three groups that represent the relative energy and materials consumption efficiency in production. These groups have been created from the responses of the question regarding the perception of their production efficiency in terms of actual material and energy consumption in comparison with other factories in their industry. Energy efficiency is therefore measured on a relative scale with values from 1 to 5. The scale ranges from 1 meaning considerably much less efficient to value 5 considerably much more efficient. The value 2 indicates rather less efficient, 3 indicates equally efficient and 4 indicates rather more efficient. In the analyses, three groups have been created from this variable: "Less efficient" (including firms rated with values 1 or 2), "Equally efficient" (including firms rated with value 3) and "More efficient" (including firms rated with values 4 or 5).

4. Results and findings

Fig. 1 depicts the use of EST and MST for all manufacturing sectors presented. It is shown that "Speed control" is the most used technology with a 54% of affirmative responses. The second position and quite far from the first one was EST "Control system for shut down of machines in off-peak periods" (37%). The third and fourth technology in the use ranking were MST, namely "Recycled material in production" with a 34%, and "Product recovery" with a 21%. The high share of "Speed control" technology and its distance to other technologies could be misleading. The understanding of this technology could be misunderstood or widely interpreted. The term "Electric motors with rotation speed regulation" could be understood in the sense that almost each machine that produces any kind of motion or rotation with a common speed regulation system over the engine, have implemented this technology. For most machines this is not an option, but an intrinsic characteristic. A doubt arises to what extent "Speed control", presented as it is, should be considered an EST.

The graph in Fig. 2 presents a distribution of technologies used according their implementation degree and ranked by the highest implementation level to the lowest. This ranking compared to the simple use has changed. The "Speed regulation" technology was the most widely used technology, but only 22% of firms acknowledge high use of this technology – rank 7. This fact could be again related to the possible misunderstanding of the term "Speed regulation". Control system for shut down of machines in off-peak periods is highly used in just 16% of manufacturing firms that have this EST implemented. "Waste material for in-house energy generation" is the only EST highly used in at least 50% of all implementations. "Compressed air contracting" and "Retrieval of kinetic and process energy" have no implementations with high use. Nevertheless, a reduction in the dispersion of the per cents of the highly implemented technologies is perceived compared with the per cents of the simple use of these technologies. This fact is more evident for EST, less for MST as both MST technologies are more widely used and both have relatively high extensive share of use.

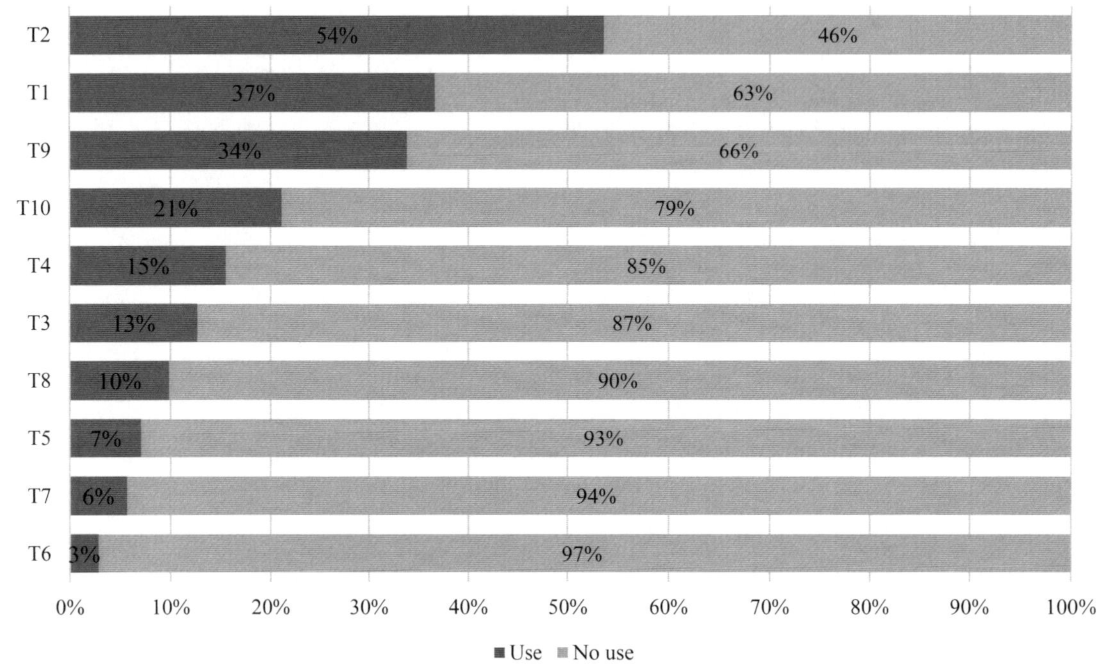

Fig. 1. Use of EST and MST for all manufacturing sectors

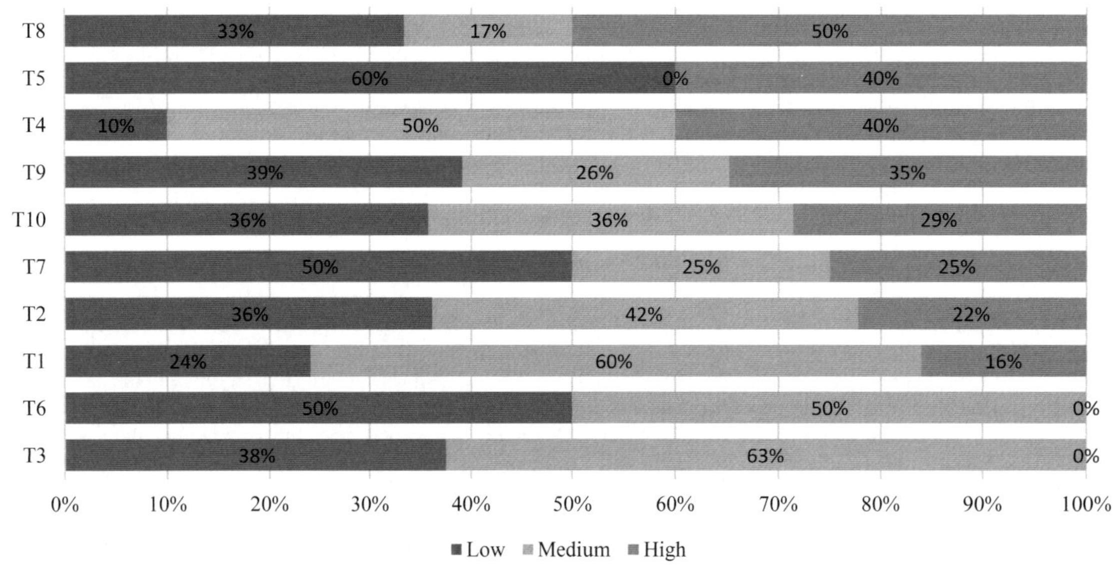

Fig. 2. Implementation degree of EST and MST for all manufacturing sectors

Fig. 3 presents EST and MST in accordance to three technological intensity groups. The technologies are ranked based on the share of use in "Medium-High and High technology" group (from the highest to the lowest share). "Compressed air contracting" is the technology with the highest per cent of use in "Medium-High and High technology" group with 78%, followed by "Low-temperature joining processes" with 60%. "Highly efficient pumps ", "Speed regulation" and "Bi-/Tri-generation" are the other three EST used in at least 50% of firms within "Medium-High and High technology" group. "Control system for shut down of machines in off-peak periods" technology is used in slightly less than 50% of "Medium-High and High technology" group. Both MST technologies are more often used in "Low and Low-Medium technology" group (around 60%). "Retrieval of kinetic and process energy" is never used in "Medium-High and High technology" group. "Waste material for in-house energy generation" is also predominately used in "Low and Low-Medium technology" group.

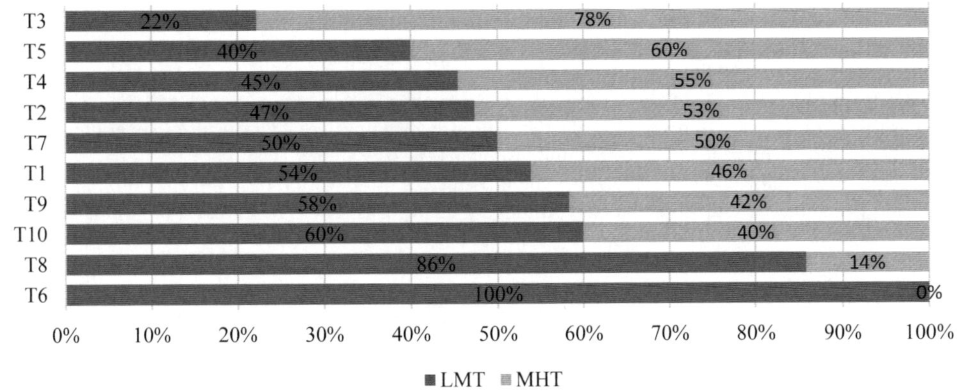

Fig. 3. Implementation per cent of EST and MST by technological sector

Fig. 4 presents EST and MST in accordance to three groups that represent the relative energy and materials consumption efficiency in production. The technologies are ranked based on the share of use in the "More effective" group (from the highest to the lowest share). In general, the most predominant group is "More effective" group, since 8 of 10 technologies are implemented in most of the firms from this group. Only "Highly efficient pumps" and "Compressed air contracting" are most often used in "Equally efficient" group. "Low-temperature joining processes" is the technology with the highest share of "More efficient" group with a 80%, followed by "Product recovery" and "Waste material for in-house energy generation" technologies. It is very obvious that EST and MST are hardly used in "Less efficient" group with very low shares of use. The only exception is "Bi-/Tri-generation" technology, but its implementation in general is very low. 5 out of 10 EST and MST are not implemented in any of "Less efficient" group.

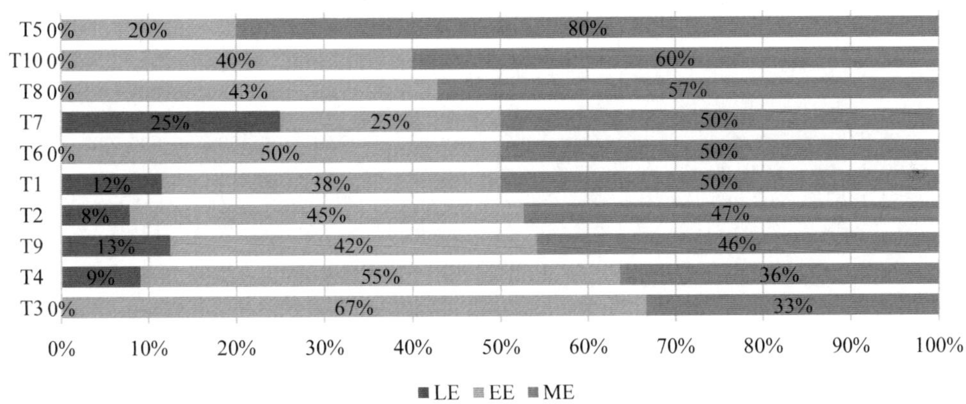

Fig. 4. Implementation per cent of EST and MST by level of efficiency relative to the sector

5. Discussion

Based on our analysis several conclusions can be drawn. General observation on the use of EST and MST is that the use of these technologies in manufacturing firms is still relatively low (from 3% to 37%). The only exception is "Speed control" technology with 54%. The first conclusion is the fact that analysing energy efficiency groups we have observed that there is a slight decrease of the technological intensity values from "Less efficient" group to "More efficient" group. The average technological intensity in "Less efficient" group was 1,71, in "Equally efficient" group 1,44 and "More efficient" group 1,40. On the other hand, "Low and Low-Medium technology" group has a slightly higher average of material and energy efficiency in production than "Medium-High and High technology" group (3,39 vs. 3,13). Both these facts could reveal a possible negative relationship between energy efficiency in production and technological intensity of firms, at least on average. This could lead to a conclusion that firms in high technology industries focus less on energy efficiency than low technology firms. Secondly, both MST are ranked third and fourth in general use. But it is interesting to see that they are mostly used in low and medium technology sector, not in high technology one. Thirdly, only 10% of all manufacturing firms claims to be less energy efficient than firms from their sector, 35% believes they are more energy efficient than others. We calculated that MST and EST are used on average in 50% of firms within "More efficient" group, 43% within "Equally efficient" group and 7% within "Less efficient" group. Based on this fact we could assume that manufacturing firms are more efficient if they use at least one EST or MST.

Our research has several limitations. The first is that only descriptive statistics was used to map the characteristics of energy efficient technologies and their adopters. To draw further conclusions in the future several advanced statistical methods will be used (e. g. linear regression for quantitative independent variables and ordinal logistic regression). We will further explore the relationship between the implementation of energy efficient technologies and environmental performance of manufacturing firms. Our limitation is also the narrow geographical coverage and the fact that no similar previous data exists to compare our findings. We will expand our research with data from other countries from EMS consortium. Despite these shortcomings, our contribution explains the use of energy and material efficient technologies, the characteristics of their adopters and indicates a possible influence of these technologies on environmental performance of the manufacturing firms, thus giving some business implications to managers to decide to invest in these technologies.

6. References

[1] Marland, G., Boden, T. A. & Andres, R. J. (2007). Global, Regional, and National CO2 Emissions, In: Trends: A Compendium of Data on Global Change, Carbon Dioxide Information Analysis Center, Oak Ridge National Laboratory, US Department of Energy, Oak Ridge, Tenn., USA

[2] Tseng, M. L., Lin, Y. H. & Chiu, A. S. F. (2009). Fuzzy AHP-based study of cleaner production implementation in Taiwan PWB manufacturer, Journal of Cleaner Production, Vol. 17, No. 14, pp. 1249-1256

[3] Kliopova, I. & Staniskis, J. K. (2006). The evaluation of cleaner production performance in Lithuanian industries, Journal of Cleaner Production, Vol. 14, pp. 1561-1575

[4] Nishitani, K., Kaneko, S., Hidemichi, F. & Komatsu, S. (2011). Effects of the reduction of pollution emissions on the economic performance of firms: an empirical analysis focusing on demand and productivity, Journal of Cleaner Production, Vol. 19, pp. 1956-1964

[5] Dessus, S. & Bussolo, M. (1998). Is there a trade-off between trade liberalization and pollution abatement?, Journal of Policy Modeling, Vol. 20, pp. 11-31

[6] Bunse, K., Vodicka, M., Schönsleben, P., Brülhart, M. & Ernst, F. O. (2011). Integrating energy efficiency performance in production management – gap analysis between industrial needs and scientific literature, Journal of Cleaner Production, Vol. 19, pp. 667-679

[7] Micieta, B., Zavodska, L., Rakyta, M. & Binasova, M. (2015). Sustainable concept for green logistics and energy efficiency in manufacturing, In: DAAAM Scientific Book 2015, B. Katalinic (Ed.), pp.391-400, DAAAM International, ISBN 978-3-902734-05-1, Vienna, Austria

[8] Thollander, P., Danestig, M. & Rohdin, P. (2007). Energy policies for increased industrial energy efficiency: Evaluation of a local energy programme for manufacturing SMEs, Energy Policy, Vol. 35, pp. 5774-5783

[9] Tanaka, K. (2008). Assessment of energy efficiency performance measures in industry and their application for policy, Energy Policy, Vol. 36, pp. 2887-2902

[10] International Energy Agency. (2008). Assessing Measures of Energy Efficiency Performance and Their Application in Industry. Available from: http://www.iea.org/papers/2008/JPRG_Info_Paper.pdf Accessed: 2016-07-01

[11] International Energy Agency. (2009). Implementing Energy Efficiency Policies 2009. Are IEA Member Countries on Track? Available from: http://www.gbv.de/dms/zbw/613955536.pdf Accessed: 2016-07-01

[12] Neelis, M., Ramirez-Ramirez, A., Patel, M., Farla, J., Boonekamp, P. & Blok, K. (2007). Energy efficiency developments in the Dutch energy-intensive manufacturing industry, 1980–2003, Energy Policy, Vol. 35, pp. 6112-6131

[13] Li, W., Winter, M., Kara, S. & Herrmann, C. (2012). Eco-efficiency of manufacturing processes: A grinding case. CIRP Annals - Manufacturing Technology, Vol. 61, pp. 59-62

[14] De Groot, H. L. F., Verhoef, E. T. & Nijkamp, P. (2001). Energy saving by firms: decision making, barriers and policies, Energy Economics, Vol. 23, pp. 717-740

[15] Paton, B. (2001). Efficiency gains within firms under voluntary environmental initiatives, Journal of Cleaner Production, Vol. 9, pp. 167-178

[16] Sardianou, E. (2008). Barriers to industrial energy efficiency investments in Greece, Journal of Cleaner Production, Vol. 16, pp. 1416-1423

[17] Phylipsen, D., Blok, K., Worrell, E. & de Beer, J. (2002). Benchmarking the energy efficiency of Dutch industry: an assessment of the expected effect on energy consumption and CO2 emissions, Energy Policy, Vol. 30, pp. 663-679

[18] Farla, J., Blok, K. & Schipper, L. (1997). Energy efficiency developments in the pulp and paper industry: a cross-country comparison using physical production data, Energy Policy, Vol. 25, pp. 745-758

[19] International Energy Agency. (2007). Tracking Industrial, Energy Efficiency and CO2 Emissions. Available from: http://www.iea.org/textbase/nppdf/free/2007/tracking_emissions.pdf Accessed: 2016-07-01

[20] Irrek, W. & Thomas, S. (2006). Der Energie Spar Fonds für Deutschland (Energy Saving Funds for Germany), Hans-Böckler-Stiftung, Düsseldorf, Germany

[21] OECD. (2005) Directorate for Science, Technology and Industry, stan indicators (2005 edition): 1980–2003. Organisation for Economic Co-operation and Development, Paris, France. Available from: http://www.oecd.org/industry/industryandglobalisation/40230754.pdf Accessed: 2016-07-01.

DOI: 10.2507/27th.daaam.proceedings.004

NEW ELECTRONIC DIDACTIC TOOL FOR NONLINEAR SYSTEMS LABORATORY

Karel Perutka

Faculty of Applied Informatics, Tomas Bata University in Zlin
kperutka@fai.utb.cz

This Publication has to be referred as: Perutka, K[arel] (2016). New Electronic Didactic Tool for Nonlinear Systems Laboratory, Proceedings of the 27th DAAAM International Symposium, pp.0023-0033, B. Katalinic (Ed.), Published by DAAAM International, ISBN 978-3-902734-08-2, ISSN 1726-9679, Vienna, Austria
DOI: 10.2507/27th.daaam.proceedings.004

Abstract

The paper presents new electronic didactic tool that was created for the laboratory where nonlinear systems are taught and practiced by simulations. The tool in the laboratory works with several control strategies and models of nonlinear systems including Hammerstein and Wiener models and it was created in MATLAB and SIMULINK environment. Students had worked with it during last semester and positively evaluated it which is summarized at the end of the paper.

Keywords: Didactic tool; Hammerstein model; MATLAB; Nonlinear control; SIMULINK; Wiener model.

1. Introduction

Most of the physical, biological, chemical, economical systems are nonlinear in the real world. In the engineering practice, the nonlinear systems are almost everywhere. There is often possible to control most of the nonlinear systems using the methods for linear control, but only in the close area of steady states. Nonlinear control enlarges the area of control of such systems. Some systems have nonlinearities which are hardly linearly approximated and therefore, there is necessary or useful to employ the nonlinear control [28]. Nonlinear control is of big interest during last years which is proved by the huge amount of publications.

There is always necessary to have the theoretical background for practical applications. Methods of nonlinear control are developing very fast and in huge amount of them can be found in the literature, such as stability analysis of nonlinear systems [3], optimal nonlinear control of continuous systems by adaptive dynamic programming [7], optimal control of nonlinear time delay systems [8], robust decentralized control methodology for a class of interconnected nonlinear systems [9], fuzzy dynamic surface control for uncertain nonlinear systems under input saturation [10], adaptive control for a class of switched nonlinear systems with modelled dynamics [11], targets tracking algorithm for a class of nonlinear systems with feedforward compensations [12], block oriented feedforward control [13], fuzzy control for uncertain non-affine nonlinear systems with dead-zone inputs [15], over-coming time-domain performance limitation by nonlinear control [17], and, finally, the adaptive self-structuring fuzzy control for nonlinear uncertain systems [32].

Modelling of many nonlinear systems is based on Hammerstein and Wiener models, as can be also proven by many publications during last years. Some publications are just focused on Hammerstein models and the other ones for Wiener models. But there are also papers dealing with of these models, for example the paper describing the recovering the Wiener-Hammerstein nonlinear state-space models using linear algebra [21], or control, such as the robust model predictive control of nonlinear processes represented by Wiener or Hammerstein models [29], and nonlinear predictive control for Hammerstein-Wiener systems [27].

Hammerstein models are mentioned for example in applications to pH process control, where is used the pole-placement self-tuning control of nonlinear Hammerstein system [30], or in servo systems, where is employed the nonlinear modelling and predictive functional control of Hammerstein system [31], or like a recursive parametric estimation algorithm of multivariable systems described by Hammerstein models [26].

Wiener models are also used, for example as a robust nonlinear internal model control of stable Wiener systems [20], or block-oriented feedforward control with demonstration to Wiener modelling [25], or modelling of memristor-based chaotic systems using nonlinear Wiener adaptive filters [24]. Identification of such systems is solve for example by the recursive least squares algorithm [22].

Nonlinear systems and their control can be found in many areas, for example using nonlinear model predictive control in automotive industry [6], receding horizon control of vehicle formation [16], stochastic nonlinear model predictive control in film deposition [2], nonlinear PI control of wind turbine [14], neural network for model predictive control of continuous reactor [23], nonlinear model predictive control of robot manipulator [18], steam valve control of multi-machine power systems [1], or controlling galloping vibrations [4], nonlinear control of boost converter [19], adaptive control of nonlinear finance system [5], to name some. During last years, nice implementation of new creative educational technologies for inter-university network was realized [33].

Most of the papers mentioned hereinbefore are so nice that they will definitely be used in future to enlarge tasks portfolio in the laboratory of automation where several systems and methods are verified by simulations in MATLAB.

After introductory part, the created models are presented. These models were used in the new electronic didactic tool, which described after that. Finally, there was done the evaluation of the created didactic tool in the form of the questionnaire. The results are presented in the graphs. There were no more studies about this topic at my university and therefore the theme and its realisation is the unique one.

2. Created Models

For the laboratory, there were created and used following models with nonlinearities which are based on the previous work of my student [28].

2.1. Spherical Fluid Reservoir

The balance equation is used to create the mathematical model of the reservoir

$$input = output + accumulation \quad (1)$$

$$q_v = q + \frac{dV}{ct} \quad (2)$$

$$q = a\sqrt{h} \quad (3)$$

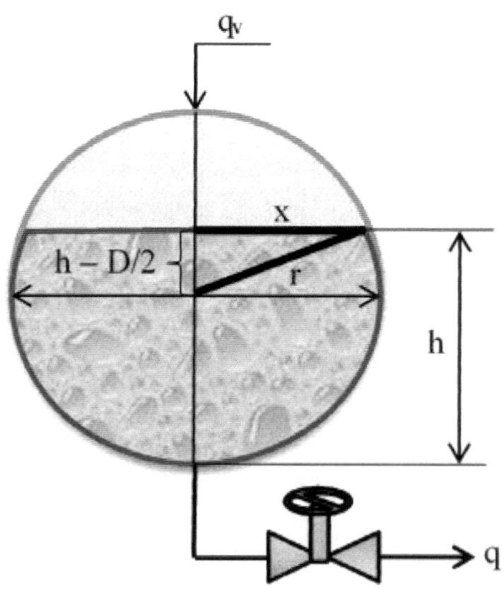

Fig. 1. Scheme of the spherical fluid reservoir

and for the accumulation it holds

$$\frac{dV}{dt} = S\frac{dh}{dt} \tag{4}$$

$$S = S(h) = \pi x^2 \tag{5}$$

$$x^2 = \left(\frac{D}{2}\right)^2 - \left(h - \frac{D}{2}\right)^2 = hD - h^2 = h(D-h) \tag{6}$$

Then we can easily derive that

$$q_v - q = S\frac{dh}{dt} \tag{7}$$

$$\frac{dh}{dt} = \frac{1}{\pi h(D-h)}\left(q_v - a\sqrt{h}\right), h(0) = h^S \tag{8}$$

where q_v ($m^3.s^{-1}$) is the volumetric in-flow of the liquid, S (m^2) is the cross-section of the reservoir, D (m) is the reservoir diameter, h (m) is the liquid height in the reservoir, a ($m.s^{-1}$) is the known constant of the valve.

It is clear from the equations that it is single input single output system with nonlinearities - division and root. Now we select the state space variables as $x = h$ and $u = q_v$ and we obtain the nonlinear system

$$\dot{x} = \frac{1}{\pi x(D-x)}\left(u - a\sqrt{x}\right) \tag{9}$$

$$f = \frac{-a\sqrt{x}}{\pi x(D-x)} \tag{10}$$

$$G = \frac{1}{\pi x(D-x)} \tag{11}$$

In case that volumetric flow of liquid q_v is random, we suppose it as unmeasurable disturbance v, i.e. $q_v = v$. Mathematical model of the reservoir is than

$$\dot{x} = \frac{1}{\pi x(D-x)}(u - v) \tag{12}$$

$$f = \frac{-v}{\pi x(D-x)} \tag{13}$$

$$G = \frac{1}{\pi x(D-x)} \tag{14}$$

Next two pictures display the scheme of these 2 models in MATLAB-SIMULINK software.

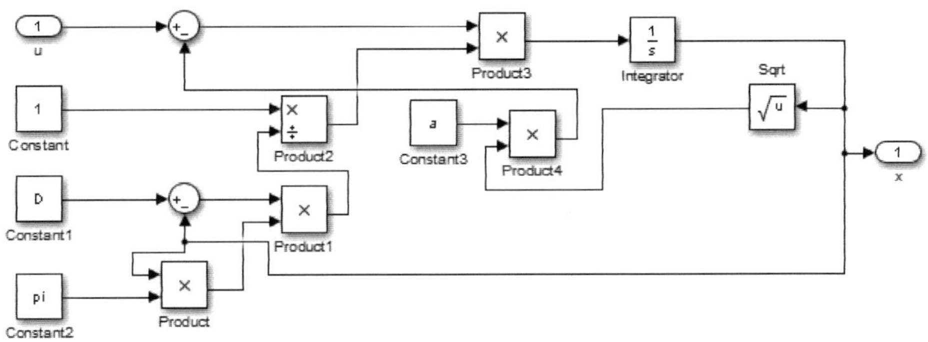

Fig. 2. Model of the spherical fluid reservoir in MATLAB-SIMULINK environment

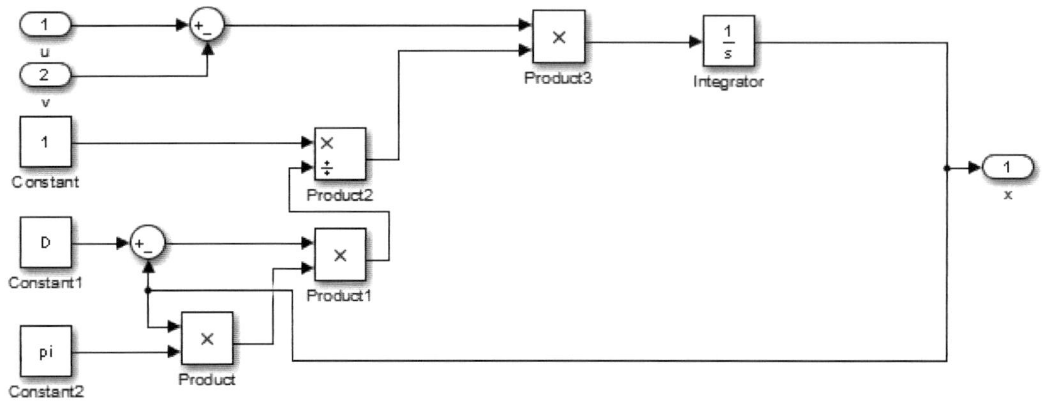

Fig. 3. Model of the spherical fluid reservoir with disturbance in MATLAB-SIMULINK environment

2.2. Inverse Conical Fluid Reservoir with Non-Constant Cross Section

We use equation 1 as the balance equation to get the mathematical model of the reservoir and for the accumulation in the reservoir it holds

$$\frac{dV}{dt} = S\frac{dh}{dt} \qquad (15)$$

$$S = S(h) \qquad (16)$$

$$tg\,\alpha = \frac{x}{h} = \frac{d/2}{H} \qquad (17)$$

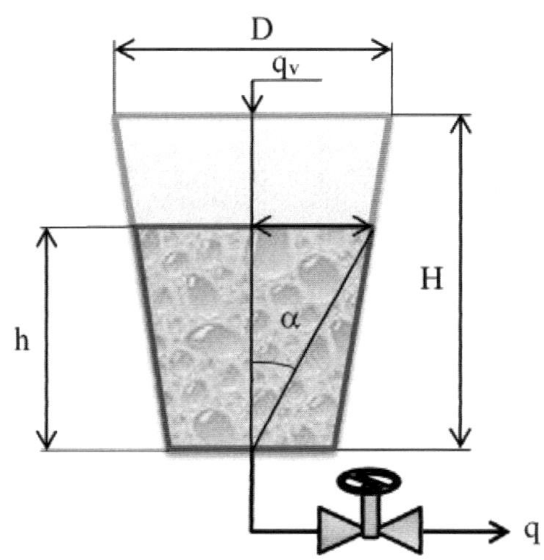

Fig. 4. Scheme of the inverse conical fluid reservoir with non-constant cross section

$$S = S(h) = \pi x^2 \qquad (18)$$

$$\frac{dV}{dt} = \pi\frac{D}{4H^2}h^2\frac{dh}{dt} \qquad (19)$$

Than we can derive that

$$q_v - q = S\frac{dh}{dt} \qquad (20)$$

$$\frac{dh}{dt} = \frac{4H^2}{\pi D^2 h^2}\left(q_v - a\sqrt{h}\right), h(0) = h^S \qquad (21)$$

where q_v ($m^3.s^{-1}$) is the volumetric in-flow of the liquid, S (m^2) is the cross-section of the reservoir, D (m) is the reservoir diameter, h (m) is the liquid height in the reservoir, H (m) is the reservoir height, a ($m.s^{-1}$) is the known constant of the valve.

It is clear from the equations that it is again the single input single output system with nonlinearities - division and root. Now we select the state space variables as $x = h$ and $u = q_v$ and we obtain the nonlinear system

$$\dot{x} = \frac{4H^2}{\pi D^2 h^2}(u - a\sqrt{x}) \tag{22}$$

$$f(x) = -\frac{4H^2}{\pi D^2 h^2} a\sqrt{x} \tag{23}$$

$$G(x) = \frac{4H^2}{\pi D^2 h^2} \tag{24}$$

In case that the volumetric flow of the liquid q_v is random, we suppose it as unmeasurable disturbance v, i.e. $q_v = v$. Mathematical model of the reservoir is than

$$\dot{x} = \frac{4H^2}{\pi D^2 h^2}(u - v) \tag{25}$$

$$f(x) = -\frac{4H^2}{\pi D^2 h^2} v \tag{26}$$

$$G(x) = \frac{4H^2}{\pi D^2 h^2} \tag{27}$$

Next two pictures display the scheme of these 2 models in MATLAB-SIMULINK software.

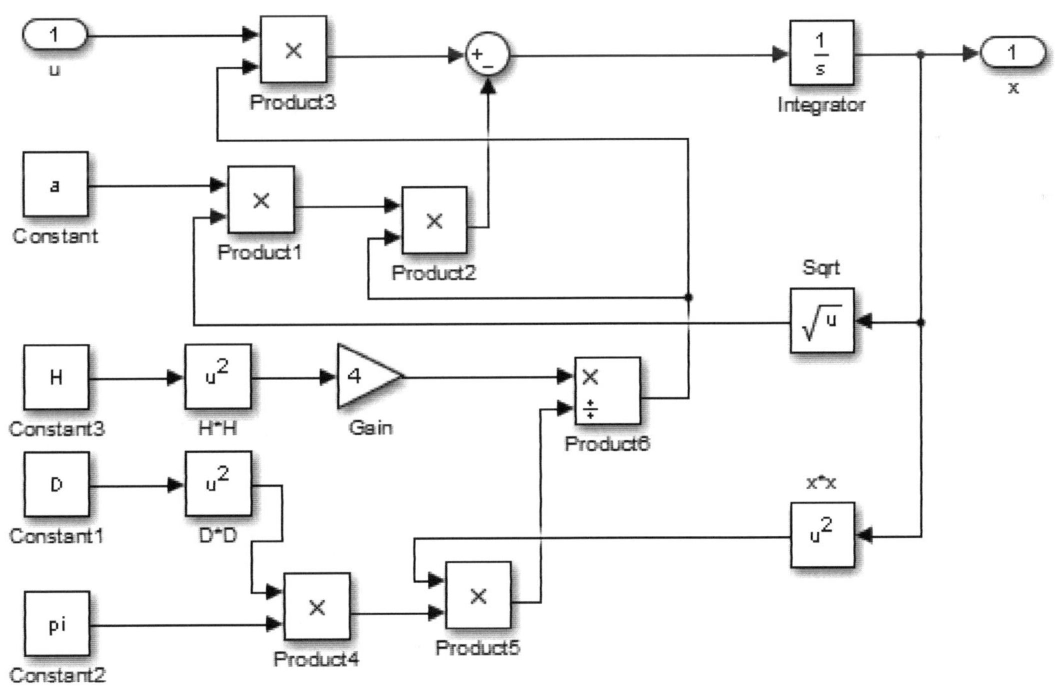

Fig. 5. Model of the inverse conical fluid reservoir with non-constant cross section in MATLAB-SIMULINK environment

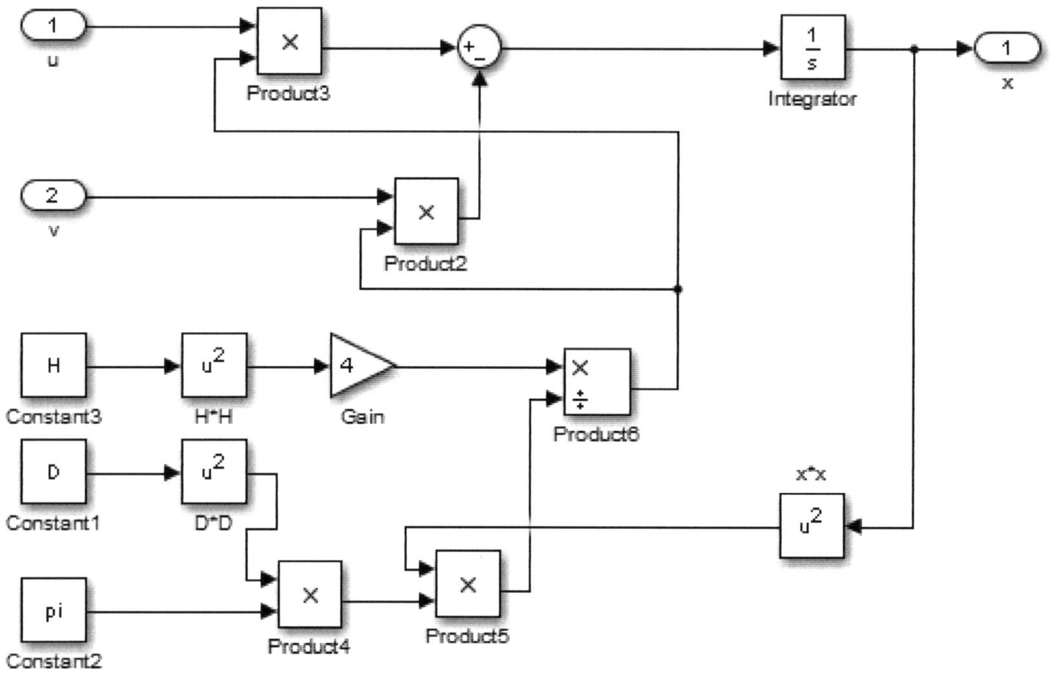

Fig. 6. Model of the inverse conical fluid reservoir with non-constant cross section with disturbance in MATLAB-SIMULINK environment

2.2. Cylindrical Flow Heat Exchanger with Stirring

We can also use the balance equation to get the mathematical model as

$$\frac{dT}{dt} = -\frac{S\alpha}{V\rho c_p}T + \frac{S\alpha}{V\rho c_p}T_c + \frac{T_v - T}{V}q \tag{28}$$

$$\frac{dT_c}{dt} = -\frac{S\alpha}{V_c\rho_c c_{pc}}T + \frac{S\alpha}{V_c\rho_c c_{pc}}T_c + \frac{T_{cv} - T_c}{V_c}q_c \tag{29}$$

$$q \geq 0, q_c \geq 0 \tag{30}$$

where q and q_c ($m^3.s^{-1}$) are the volumetric flows of the liquid, S (m^2) is the heat transfer surface of the heat exchanger, T_v and T_{cv} (°C) are the input temperatures of the liquid in-flows in the heat exchanger, T and T_c (°C) are the temperatures of the warmer and cooler liquid in the heat exchanger, c_p and c_{pc} ($kJ.kg^{-1}.K^{-1}$) are the specific heat capacities of the warmer and cooler liquid, V and V_c (m^3) are the volumes of the warmer and cooler liquid in the heat exchanger, ρ and ρ_c ($kg.m^3$) are the densities of the warmer and cooler liquid and α ($kJ.m^{-2}.K^{-1}.s^{-1}$) is the heat transfer coefficient.

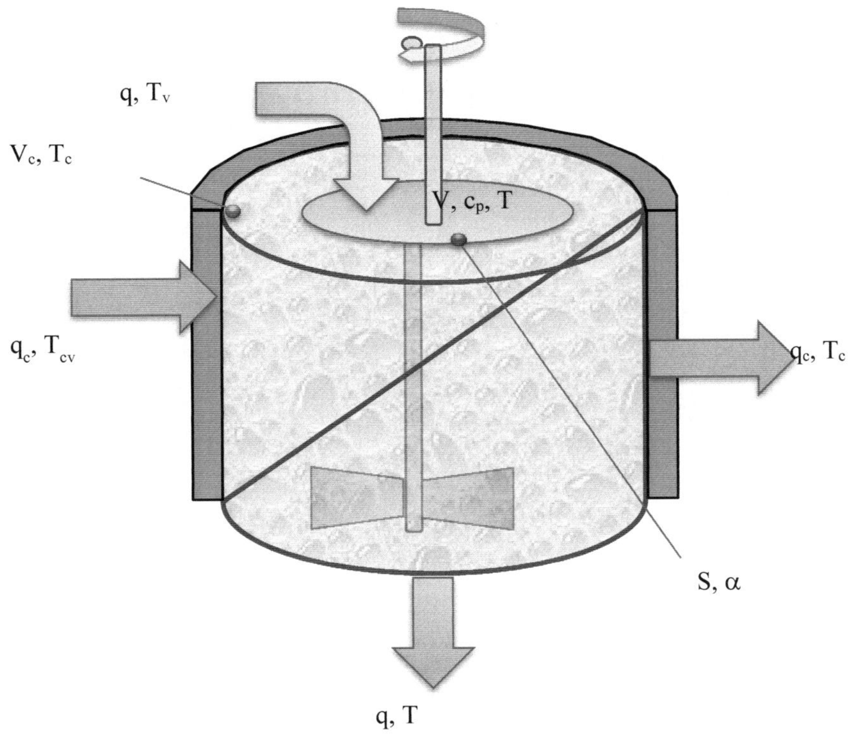

Fig. 7. Scheme of the cylindrical flow heat exchanger with stirring

It is clear from the equations that it is the system with multiple inputs multiple outputs, namely system with two inputs and two outputs. Now we select the state space variables as $x_1 = T$, $x_2 = T_c$, $u_1 = q$ and $u_2 = q_c$ and we obtain the nonlinear system as

$$\dot{x_1} = -\frac{S\alpha}{V\rho c_p}x_1 + \frac{S\alpha}{V\rho c_p}x_2 + \frac{T_v - T}{V}u_1 \tag{31}$$

$$\dot{x_2} = -\frac{S\alpha}{V_c\rho_c c_{pc}}x_1 + \frac{S\alpha}{V_c\rho_c c_{pc}}x_2 + \frac{T_{cv} - T_c}{V_c}u_1 \tag{32}$$

$$\mathbf{f} = \begin{bmatrix} -\frac{S\alpha}{V\rho c_p}x_1 + \frac{S\alpha}{V\rho c_p}x_2 \\ \frac{S\alpha}{V_c\rho_c c_{pc}}x_1 - \frac{S\alpha}{V_c\rho_c c_{pc}}x_2 \end{bmatrix} \tag{33}$$

$$\mathbf{G} = \begin{bmatrix} \frac{T_v - T}{V} & 0 \\ 0 & \frac{T_{cv} - T_c}{V_c} \end{bmatrix} \tag{34}$$

Next picture displays the scheme of this model created in MATLAB-SIMULINK software.

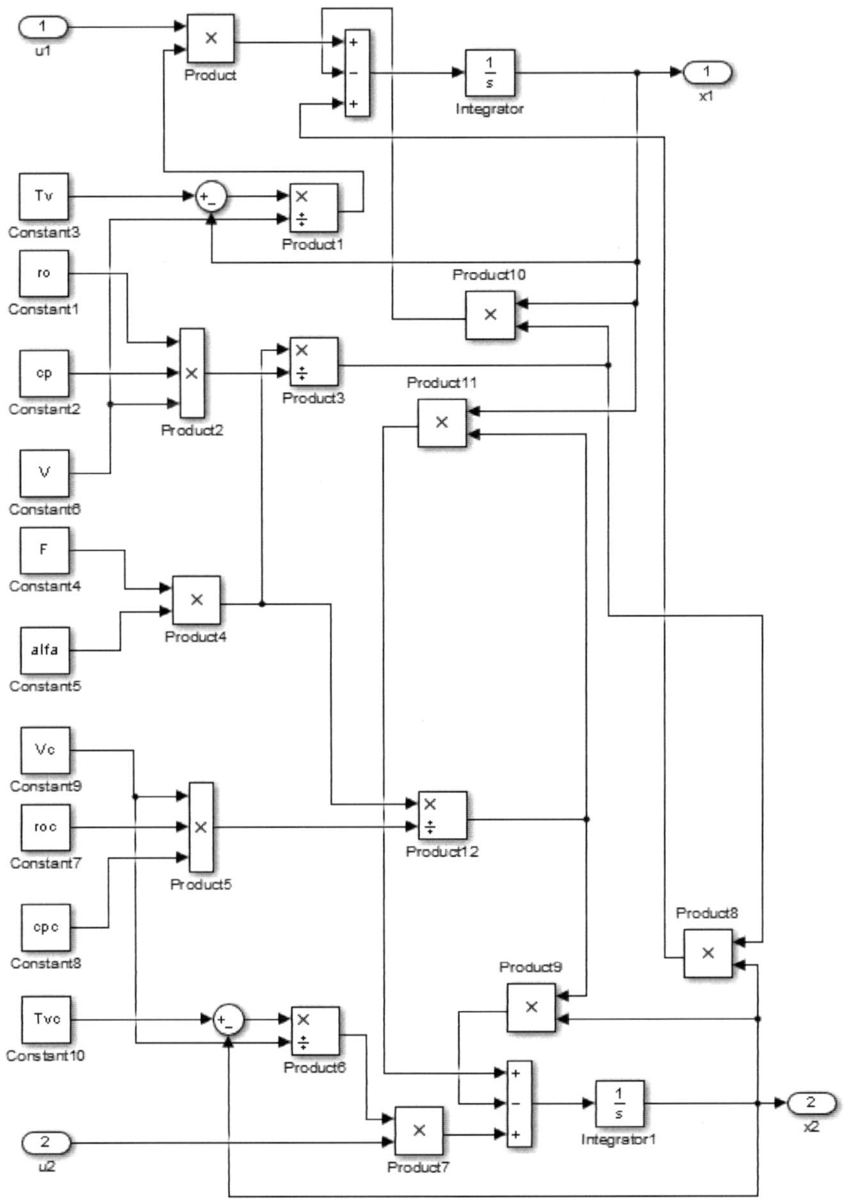

Fig. 8. Model of the cylindrical flow heat exchanger with stirring in MATLAB-SIMULINK environment

3. Description of Created Didactic Tool

The models described before were used as a “core” of the didactic tool. There was created the GUI in MATLAB that offers to select the nonlinear model and set the parameters and realize the response of the model on the one of the possible input signals for each input. There is also possible to set the parameters of the nonlinear models in GUI. Plus it is also possible to set the range of the parameters and visualize it all in one graph with the color change. That is one part of the tool.

The set of the videos describing the model derivation and its realization in MATLAB-SIMULINK with explanation what each block means in the derived nonlinear model is the other part of the tool. This part is positively evaluated mostly by students of distance curriculum who did not work with MATLAB in practice.

The other part of didactic tool consists of web pages explaining the basis of nonlinear systems, their modelling and control including for example types of nonlinearities, examples of Hammerstein and Wiener models, and some methods of nonlinear control of nonlinear systems. All materials are in the Czech language only.

4. Evaluation of Didactic Tool

The didactic tool was created for undergraduate students of Bachelor curricula. They have subjects about control theory and practice only marginally because their curricula are mostly oriented on programming. The didactic tool was offered to students during last academic year in both semesters, in regular and distance curriculum.

There was created the following questionnaire:

1) How often you have used the didactic tool? (Always/ Often/ Rarely/ Never)
2) What would you like to change? (Enlarge/ Models from other areas/ Version for smartphones)

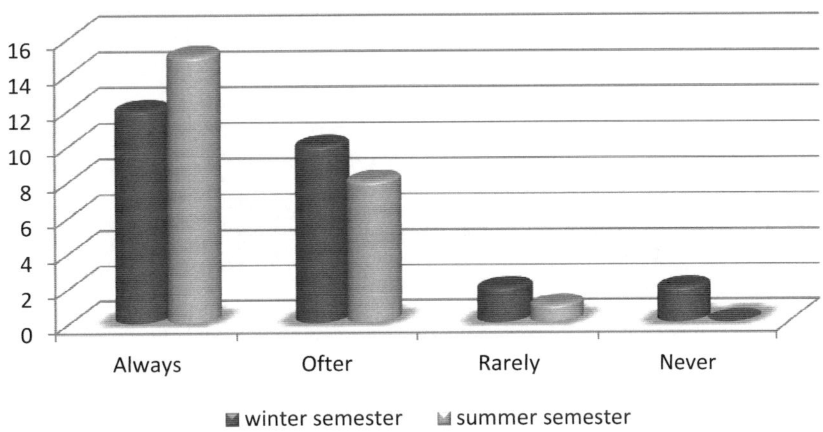

Fig. 9. Evaluation Questionnaire – Question 1 (Regular Curriculum)

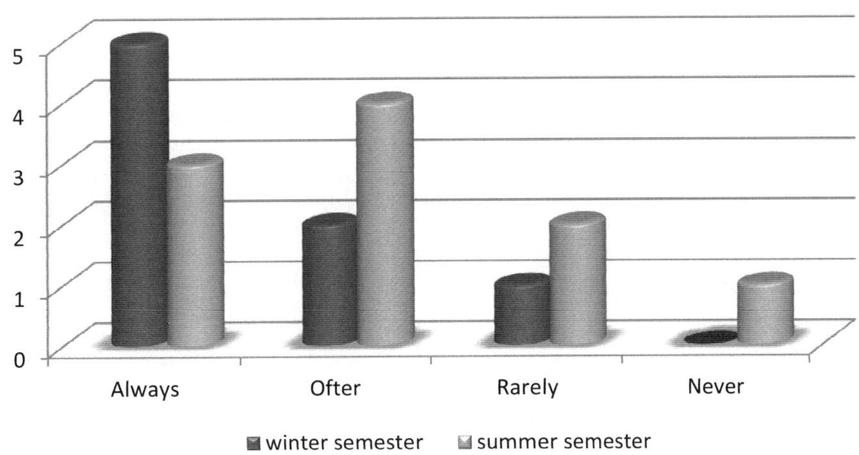

Fig. 10. Evaluation Questionnaire – Question 1 (Distance Curriculum)

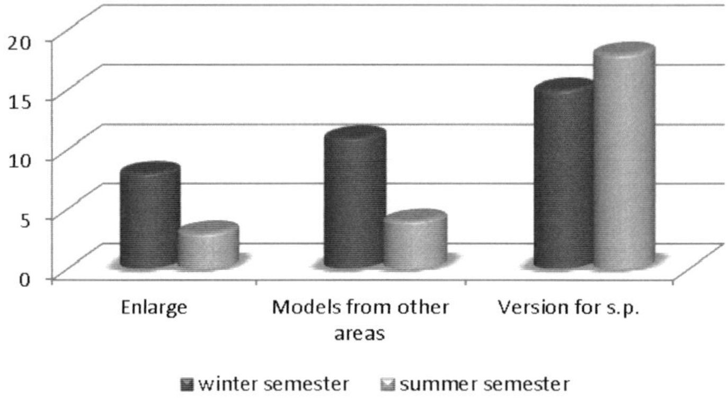

Fig. 11. Evaluation Questionnaire – Question 2 (Regular Curriculum)

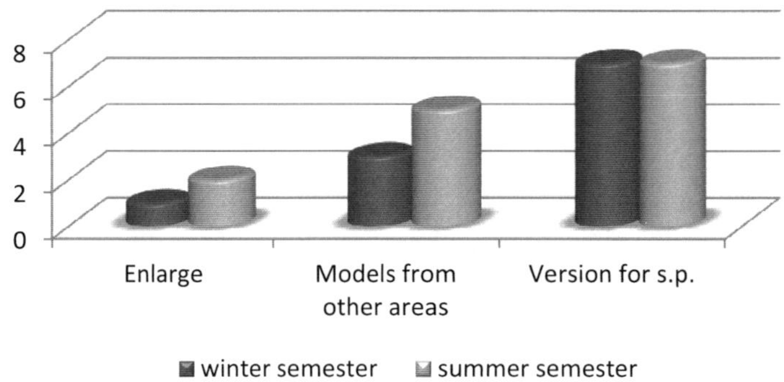

Fig. 12. Evaluation Questionnaire – Question 2 (Distance Curriculum)

5. Conclusion

The paper presented the created didactic tool for students of curricula where also control theory is taught. There were also presented the created and used models and evaluation by students. The main problem was to make the teaching process more attractive. This problem was solved by modern way and the solution was inspired by the experience of the author from the teaching the other subjects using electronic didactic tools, the created new didactic tool is the main result of this work.

Students mostly positively evaluated the created tool and wanted also the new version for smartphones. Therefore, future work will be focused on new version for smartphones. However, it will be only the smaller one than the one for PC. Other enlargements of the PC version will include the programs in MATLAB-SIMULINK realizing the control of nonlinear systems, videos describing the approach of controller design plus describing the realization of control in MATLAB-SIMULINK.

6. References

[1] A. M. Fombu, G. Kenné, J. D. Nguimfack-Ndongmo, R. Kuate-Fochie, "Decentralized nonlinear coordinated excitation and steam valve adaptive control for multi-machine power systems," Electrical Power and Energy Systems, vol. 75, pp. 117-126, 2016.

[2] S. Rasoulian, L. A. Ricardez-Sandoval, "Stochastic nonlinear model predictive control applied to a thin film deposition process under uncertainty," Chemical Engineering Science, vol. 40, pp. 90-13, 2016.

[3] H. Omran, L. Hetel, M. Petreczky, J.-P. Richard, F. Lamnabhi-Lagarrigue, "Stability analysis of some classes of input-affine nonlinear systems with aperiodic sampled-data control," Automatica, vol. 70, pp. 266-274, 2016.

[4] H. L. Dai, A. Abdelkefi, L. Wang., "Usefulness of passive non-linear energy sinks in controlling galloping vibrations," International Journal of Non-Linear Mechanics, vol. 81, pp. 83-94, 2016.

[5] O. I. Tacha, Ch. K. Volos, I. M. Kyprianidis, I. N. Stouboulos, S. Vaidyanathan, V.-T. Pham, "Analysis, adaptive control and circuit simulation of a novel nonlinear finance system," Applied Mathematics and Computation, vol. 276, pp. 200-217, 2016.

[6] P. Chen, J. Wang, "Estimation and adaptive nonlinear model predictive control of selective catalytic reduction systems in automotive applications," Journal of Process Control, vol. 40, pp. 78-92, 2016.

[7] J. Zhang, H. Liang, T. Feng, "Optimal control of nonlinear continuous systems by adaptive dynamic programming based on fuzzy basis functions," Applied Mathematical Modelling, article in press.

[8] L. Rodríguez-Guerrero, O. Santos-Sánchez, S. Mondié, "A constructive approach for an optimal control applied to a class of nonlinear time delay systems," Journal of Process Control, vol. 40, pp. 35-49, 2016.

[9] C. Zhang, H. Li, "A generalized robust decentralized control methodology for a class of interconnected nonlinear systems subject to uncertainties and disturbances," Nonlinear Analysis: Hybrid Systems, vol. 22, pp. 55-71, 2016.

[10] S. Gao, B. Ning, H. Dong, "Fuzzy dynamic surface control for uncertain nonlinear systems under input saturation via truncated adaptation approach," Fuzzy Sets and Systems, vol. 290, pp. 100-117, 2016.

[11] J. Mao. Z. Xiang, S. Huang, "Adaptive finite-time tracking control for a class of switched nonlinear systems with unmodeled dynamics," Neurocomputing, vol. 196, pp. 42-52, 2016.

[12] P. Jiang, "Multiple-targets tracking algorithm for a class of nonlinear systems with feedforward compensations," Neurocomputing, vol. 196, pp. 210-213, 2016.

[13] D. K. Rollins, Y. Mei, S. D. Loveland, N. Bhandari, "Block-oriented feedforward control," Chemical Engineering Research and Design, vol. 109, pp. 397-404, 2016.

[14] Y. Ren, L. Li, J. Bridley, L. Jiang, "Nonlinear PI control for variable pitch wind turbine," Control Engineering Practice, vol. 50, pp. 84-94, 2016.

[15] L.-B. Wu, G.-H. Yang, "Adaptive fuzzy tracking control for a class of uncertain nonaffine nonlinear systems with dead-zone inputs," Fuzzy Sets and Systems, vol. 290, pp. 1-21, 2016.

[16] H. Li, Y. Shi, W. Yan, "Distributed receding horizon control of constrained nonlinear vehicle formations with guaranteed γ-gain stability," Automatica, vol. 68, pp. 148-154, 2016.

[17] B. G. B. Hunnekens, N. v. d. Wouw, D. Nešić, "Overcomming a fundamental time-domain performance limitation by nonlinear control," Automatica, vol. 67, pp. 277-281, 2016.

[18] J. Wilson, M. Charest, R. Dubay, "Non-linear model predictive control schemes with application on 2 link vertical robot manipulator," Robotics and Computer-Integrated Manufacturing, vol. 41, pp. 23-30, 2016.

[19] D. J. Pradeep, M. M. Noel, N. Arun, "Nonlinear control of a boost converter using a robust regression based reinforcement learning algorithm," Engineering Applications of Artificial Intelligence, vol. 52, pp. 1-9, 2016.

[20] K.-K. K. Kim, E. Ríos-Patrón, R. D. Braatz, "Robust nonlinear internal model control of stable Wiener systems," Journal of Process Control, vl. 22, pp. 1468-1477, 2012.

[21] P. Dreesen, M. Ishteva, J. Schoukens, "Recovering Wiener-Hammerstein nonlinear state-space models using linear algebra," IFAC-PapersOnLine, vol. 48, pp. 951-956, 2015.

[22] F. Ding, X. Liu, M. Liu, "The recursive least squares identification algorithm for a class of Wiener nonlinear systems," Journal of Franklin Institute, vol. 353, pp. 1518-1526, 2016.

[23] S. Li, Y. Li, "Model predictive control of an intensified continuous reactor using a neural network model," Neurocomputing, vol. 185, pp. 93-104, 2016.

[24] Y. Zhao, Y. Jiang, J. Feng, L. Wu, "Modeling of memristor-based chaotic systems using nonlinear Wiener adaptive filters based on backlash operator," Chaos, Solitons and Fractals, vol. 87, pp. 12-16, 2016.

[25] D. K. Rollins, Y. Mei, S. D. Loveland, N. Bhandari, "Block-oriented feedforward control with demonstration to nonlinear parametrized Wiener modeling," Chemical Engineering Research and Design, vol. 109, pp. 397-404, 2016.

[26] H. Salhi, S. Kamoun, "A recursive parametric estimation algorithm of multivariable systems described by Hammerstein mathematical models," Applied Mathematical Modelling, vol. 39, pp. 4951-4962, 2015.

[27] M. Lawryńczuk, "Nonlinear predictive control for Hammerstein-Wiener systems," ISA Transactions, vol. 55, pp. 49-62, 2015.

[28] M. Bubenikova, "Control of SISO systems with given type of nonlinearity", thesis, Tomas Bata University in Zlin, 89 p., 2007.

[29] F. Khani, M. Haeri, "Robust model predictive control of nonlinear processes represented by Wiener or Hammerstein models," Chemical Engineering Science, vol. 129, pp. 223-231, 2015.

[30] Z. Zhou, D. Zhao, X. Liu, Y. Guo, Ch. Guan, W. Feng, N. Guo, "Pole-placement self-tuning control of nonlinear Hammerstein system and its application to pH process control," Chinese Journal of Chemical Engineering, vol. 23, pp. 1346-1368, 2016.

[31] Q. Zhang, Q. Wang, G. Li, "Nonlinear modeling and predictive functional control of Hammerstein system with application to the turnable servo system," Mechanical Systems and Signal Processing, vol. 72-73, pp. 383-394, 2016.

[32] N. Wang, J.-Ch. Sun, Y.-Ch. Liu, "Direct adaptive self-structuring fuzzy control with interpretable fuzzy rules for a class of nonlinear uncertain systems," Neurocomputing, vol. 173, pp. 1640-1645, 2016.

[33] V.E. Pryanichnikov, B. Katalinic, A.A. Kirilchenko, R.V. Khelemendik, S.V. Kuvshinov, D. Vician, A. Uglesic, "New Creative Educational Technologies for Inter-university Network," Procedia Engineering, vol. 100, pp. 259-268, 2015.

DOI: 10.2507/27th.daaam.proceedings.005

DEVELOPMENT OF THE PARTICULAR VESSEL MATHEMATICAL MODELS

Vadim Kramar

This Publication has to be referred as: Kramar, V[adim] (2016). Development of the Particular Vessel Mathematical Models, Proceedings of the 27th DAAAM International Symposium, pp.0034-0039, B. Katalinic (Ed.), Published by DAAAM International, ISBN 978-3-902734-08-2, ISSN 1726-9679, Vienna, Austria
DOI: 10.2507/27th.daaam.proceedings.005

Abstract

The technique of constructing a particular mathematical model of a given point vessel automatic deduction, taking into account the mutual influence ofmovement different type, as well as the impact of external factors, such as current, regular wave. The equations describe the motion of the vessel allows for different modes of operations performed.The particular vessel mathematical model was received that takes into account the operations performed on it at the time of its dynamic positioning.

Keywords: mathematical model; vessel;external factors;dynamic positioning control.

1. Introduction

Ships of various design and deadweight are used in today's worldwide practice for marine drilling aimed at probing or extracting minerals. There are automatic keeping systems over a drilling site in some of them. Some problems, such as, principles of automatic keeping systems, mathematical description of a drilling vessel as a control object of the automatic deduction contained require consideration in respect of theory as well as applying to technical implementation.

Development of ships mathematical models, taking into account the specificity of vessel design, equipment, including drilling, if we consider drilling vessel and process is an integral part of building a system of automatic deduction of a given point. When constructing mathematical models need to be sure to obtain the calculated ratios for the calculation of external forces and moments acting on the ship when it is spatial displacement, taking into account wind, currents, sea stat.

Contained in the various literature, for example [1-5], the output of a mathematical model of the vessel often overlapping and contains no presuppositions, which makes it difficult to carry out analytical studies and the formation of the algorithm for generating navigate a vessel when holding at a given point. Also, the information about the mathematical description of the vessel as a control object of the automatic deduction do not contain analytical relation vessels hydroaerodynamics characteristics and do not reveal the principles of such systems. Also, it is necessary to study the dynamics of the spatial movement of the ship hold. Obviously, at a specified vessel operation vessel may have different from each other dynamic models describing various process periods of operation. In this article, we will get particular mathematical models of the ship as well as the mathematical model of the vessel based on the current.

Obtained in this paper a mathematical model of the vessel, taking into account such external influence as a trend, which is one of the major external factors hindering [5-7] the ship is at a given point during a given time allowed to become the basis of the algorithm for generating dynamic positioning control.

2. Mathematical model of a vessel

In [8] we obtain a generalized model ofthe interrelatedmovementsof the vessel on a regularwaves thatfor the studyof hisspeakersneed to be supplementedby equationsfor forcedcomponentsof vertical andhorizontaltransverse motion ofthe shipon a regular waves.In [8] we consideration the systems of coordinate, which characterize spatial position of the vessel and which is analog as [9].

Where all coefficients in (1) are obtain of the following arguments [8].

$$\begin{aligned}
&(m+\lambda_{11})\dot{\upsilon}_x+\dot{\lambda}_{11}\upsilon_x+mz_c\ddot{\psi}+[m\dot{z}_c+(m+\lambda_{33})\upsilon_z+\mu_{33}\varsigma_b]\dot{\psi}-\\
&\qquad-[(m+\lambda_{22})\upsilon_y+\mu_{22}\eta_b\cos\varphi]\dot{\varphi}=F_x,\\
&(m+\lambda_{22})\dot{\upsilon}_y+\dot{\lambda}_{22}\upsilon_y-(mz_c-\lambda_{24})\ddot{\theta}+[\mu_{22}\varsigma_b+\mu_{24}-m\dot{z}_c+\dot{\lambda}_{24}-(m+\lambda_{33})\upsilon_z-\mu_{33}\varsigma_b]\dot{\theta}+\\
&+(\dot{\mu}_{22}\varsigma_b+\mu_{22}\dot{\varsigma}_b+\dot{\mu}_{24})\theta+\lambda_{26}\ddot{\varphi}+[\dot{\lambda}_{26}-\mu_{22}\eta_b\sin\varphi+(m+\lambda_{11})\upsilon_x]\dot{\varphi}+(\dot{\mu}_{22}\eta_b+\mu_{22}\dot{\eta}_b)\cos\varphi=F_y,\\
&(m+\lambda_{33})\dot{\upsilon}_z+\dot{\lambda}_{33}\upsilon_z+\lambda_{35}\ddot{\psi}+[\dot{\lambda}_{35}+\mu_{33}\eta_b\sin\varphi+\mu_{35}-(m+\lambda_{11})\upsilon_x]\dot{\psi}+\\
&+(\dot{\mu}_{33}\eta_b\sin\varphi+\mu_{33}\dot{\eta}_b\sin\varphi+\dot{\mu}_{35})\psi+[(m+\lambda_{22})\upsilon_y+(\mu_{22}-\mu_{33})\eta_b\cos\varphi]\dot{\theta}-\\
&-(\dot{\mu}_{33}\eta_b\cos\varphi+\mu_{33}\dot{\eta}_b\cos\varphi)\theta+\mu_{35}\dot{\varsigma}_b+\dot{\mu}_{35}\varsigma_b=F_z,\\
&(J_x+\lambda_{44})\ddot{\theta}+[J_x+\dot{\lambda}_{44}+\mu_{24}\varsigma_b+\mu_{44}+(mz_c-\lambda_{24})\upsilon_z-\dot{m}z_c^2]\dot{\theta}+\\
&+[\dot{\mu}_{24}\varsigma_b+\mu_{24}\dot{\varsigma}_b+\dot{\mu}_{44}-\mu_{33}\eta_b\cos\varphi\upsilon_y-(\mu_{24}+\mu_{22}\varsigma_b)\upsilon_z]\theta-(mz_c-\lambda_{24})\dot{\upsilon}_y-\\
&-[m\dot{z}_c-\lambda_{24}-(\mu_{33}\eta_b\sin\varphi+\mu_{35})\psi-(\lambda_{33}-\lambda_{22})\upsilon_z-(\lambda_{35}+\lambda_{26})\dot{\psi}-\mu_{33}\varsigma_b]\upsilon_y-\\
&-[\mu_{24}\eta_b\sin\varphi+(\lambda_{35}+\lambda_{26})\upsilon_z+\mu_{35}\varsigma_b+mz_c\upsilon_x]\dot{\varphi}+(\mu_{24}\dot{\eta}_b+\dot{\mu}_{24}\eta_b-\mu_{22}\eta_b\upsilon_z)\cos\varphi=M_x,\\
&(J_y+\lambda_{55})\ddot{\psi}+(\dot{J}_y+\dot{\lambda}_{55}+\mu_{55}+\mu_{35}\eta_b\sin\varphi+mz_c\upsilon_z-\lambda_{35}\upsilon_x-\dot{m}z_c^2)\dot{\psi}+\\
&+(\dot{\mu}_{55}+\dot{\mu}_{35}\eta_b\sin\varphi+\mu_{35}\dot{\eta}_b\sin\varphi-\mu_{33}\eta_b\sin\varphi\upsilon_x-\mu_{35}\upsilon_x)\psi+mz_c\ddot{\upsilon}_x+[m\dot{z}_c+\\
&+(\lambda_{11}-\lambda_{33})\upsilon_z+\mu_{33}(\eta_b\cos\varphi\cdot\theta-\varsigma_b)]\upsilon_x+\lambda_{35}\dot{\upsilon}_z+\dot{\lambda}_{35}\upsilon_z-[\mu_{35}\eta_b\cos\varphi+\lambda_{26}\upsilon_y+\\
&+\mu_{26}\eta_b\cos\varphi)\dot{\theta}-(\dot{\mu}_{35}\eta_b\cos\varphi+\mu_{35}\dot{\eta}_b\cos\varphi)\theta+[\mu_{24}\eta_b\cos\varphi-(mz_c-\lambda_{24})\upsilon_y]\dot{\varphi}+\\
&+\mu_{35}\dot{\varsigma}_b+\dot{\mu}_{35}\varsigma_b=M_y,\\
&(J_z+\lambda_{66})\ddot{\varphi}+(\dot{J}_z+\dot{\lambda}_{66}-\mu_{26}\eta_b\sin\varphi+\lambda_{26}\upsilon_x)\dot{\varphi}+(\dot{\mu}_{26}\eta_b+\mu_{26}\dot{\eta}_b-\mu_{24}\eta_b\dot{\psi}+\\
&+\mu_{22}\eta_b\upsilon_x)\cos\varphi+\lambda_{26}\dot{\upsilon}_y+[\dot{\lambda}_{26}-\lambda_{24}\dot{\psi}+(\lambda_{22}-\lambda_{11})\upsilon_x]\upsilon_y+(\mu_{26}\varsigma_b+\dot{\lambda}_{35}\upsilon_z+\\
&+\mu_{35}\varsigma_b+\lambda_{24}\upsilon_x)\dot{\theta}+[\dot{\mu}_{26}\varsigma_b+\mu_{26}\dot{\varsigma}_b+(\mu_{22}\varsigma_b+\mu_{24})\upsilon_x]\theta=M_z.
\end{aligned}\tag{1}$$

The movement of the vessel may be represented as consisting oftranslational motion with the starting point G of the coordinate system, closely related to the vessel, and rotation around the point G. Vessel movement equations comprisemovement equations of the point G dynamics, which coincides with the vessel gravity center at the moment, equations of rotation movement towards the point G and equations ofthe gravity center dislocation in the course of the drilling operations.

In [8] we examine the movement of the vessel on the free infinite liquid surface and disregard liquid elasticity as a system "body+liquid in the vacuum" as it is carried out in [9,10].

With the help of relations for $\vec{M}^*$ – the motion quantity instant vector of the particles and $\vec{F}^*$ – the motion quantity vectorof the particles, thrown off the body surface within a unit of time;

$$\vec{M}^*=\sum_{\nu=1}^{n}\vec{r}_\nu\times\frac{dm_\nu}{dt}\vec{u}_\nu,\qquad \vec{F}^*=\sum_{\nu=1}^{n}\frac{dm_\nu}{dt}\vec{u}_\nu,\tag{2}$$

where $\vec{u}_\nu$ is the absolute rate of the particles thrown offby a point having the mass m_ν, $\vec{r}_\nu$ is the position vector of the point m_ν with respect to the static coordinate system, the plane of whichcoincides with theundisturbed freesurface of the liquid, and one axis is directed downward along the bore hole.

In problems about balancing on the water surface the following congruence may occur

$$\lambda_{jk} - \frac{i}{\sigma_k}\mu_{jk} = -\rho\iint_{\Omega}\varphi_j\frac{\partial\varphi_k}{\partial n}d\Omega \quad (j,k=1,2,\ldots,6) \tag{3}$$

where σ_k is pitching frequency; λ_{jk} and μ_{jk} are hydrodynamic coefficients which determine inertia and damping forces of wave nature, the functions $\varphi_i(x,y)$ depend on oscillation frequency and rate of sailing
On denoting $\upsilon_{gx}=\upsilon_1, \upsilon_{gy}=\upsilon_2, \upsilon_{gz}=\upsilon_3, \omega_x=\upsilon_4, \omega_y=\upsilon_5, \omega_z=\upsilon_6$, we obtain

$$\Phi = \sum_{i=1}^{6}\upsilon_i\varphi_i \tag{4}$$

where Φ is the potential offended progressive waves velocities.

To go further we shall discuss a series of assumptions. Let us assume that: 1) The angles θ and ψ are small, therefore $\sin\theta=\theta$, $\sin\psi=\psi$, $\cos\theta=\cos\psi=1$; 2) movement through the points ψ and θ are forced and cyclic with oscillation frequency σ_k, depending on the regular waves frequency σ and the vessel velocity v: $\sigma_k=\sigma-vk\cos\chi$, where k is the module of sea wave shape, χ – the angle between the direction of the vessel and that of wave advance; 3) movementalong the verticaland transversehorizontalcoordinates in thefixed coordinate system consists ofthe cyclic component(frequency σ_k) and the noncyclic one; 4) we disregard damping fluid properties towards the axis $O\xi$; 5) the elements containing products of the values $\psi,\dot\psi,\theta,\dot\theta,\dot\varphi,\ddot\varphi$ are small and can be disregarded.

In accordance with the assumptions accepted we can formulate:

Having small values $\theta,\psi,\dot\varphi$ for angular and linear rates in the related standard of rest we obtain

$$\begin{aligned}
&\omega_x=\dot\theta;\ \ \omega_y=\dot\psi;\ \ \omega_z=\dot\varphi;\\
&\upsilon_x=a_1\dot\xi_g+a_2\dot\eta_g+a_3\dot\varsigma_g=\dot\xi_g\cos\varphi+\dot\eta_c\sin\varphi+i\sigma_k(\eta_b\sin\varphi+\sigma_k^{-2}\dot\psi\dot\varsigma_g);\\
&\upsilon_y=b_1\dot\xi_g+b_2\dot\eta_g+b_3\dot\varsigma_g=-\dot\xi_g\sin\varphi+\dot\eta_c\cos\varphi+i\sigma_k(\eta_b\cos\varphi+\varsigma_b\theta)+\theta\dot\varsigma_c;\\
&\upsilon_z=c_1\dot\xi_g+c_2\dot\eta_g+c_3\dot\varsigma_g=(\psi\cos\varphi+\theta\sin\varphi)\dot\xi_g+(\psi\sin\varphi-\theta\cos\varphi)\dot\eta_g+\dot\varsigma_c+\\
&\qquad+i\sigma_k[\eta_b(\psi\sin\varphi-\theta\cos\varphi)+\varsigma_b].
\end{aligned} \tag{5}$$

Writing at small values θ, ψ, $\dot\varphi$ the expressions for the angular and linear velocity related coordinate system, substituting them into the expression for $R_x, R_y, R_z, N_x, N_y, N_z$ and taking into account only thereall partsget

$$\begin{aligned}
&R_x=(m+\lambda_{11})\upsilon_x+mz_c\dot\psi;\\
&R_y=(m+\lambda_{22})\upsilon_y-(mz_c-\lambda_{24})\dot\theta+(\mu_{22}\zeta_b+\mu_{24})\theta+\lambda_{26}\dot\varphi+\mu_{22}\eta_b\cos\varphi;\\
&R_z=(m+\lambda_{33})\upsilon_z+\lambda_{35}\dot\psi+(\mu_{33}\eta_b\sin\varphi+\mu_{35})\psi-\mu_{33}\eta_b\theta\cos\varphi+\mu_{33}\zeta_b;\\
&N_x=(J_x+\lambda_{44})\dot\theta-(mz_c-\lambda_{24})\upsilon_y+\mu_{24}(\eta_b\cos\varphi+\theta\zeta_b)+\mu_{44}\theta;\\
&N_y=(J_y+\lambda_{55})\dot\psi+(\mu_{55}+\mu_{35}\eta_b\sin\varphi)\psi+mz_c\upsilon_x+\lambda_{35}\upsilon_z-\mu_{35}\eta_b\theta\cos\varphi+\mu_{35}\xi_b;\\
&N_z=(J_z+\lambda_{66})\dot\varphi+\lambda_{26}\upsilon_y+\mu_{26}(\eta_b\cos\varphi+\zeta_b\theta).
\end{aligned} \tag{6}$$

Substituting (6) inequation of the equation for projections of motion quantity andmotion quantity instants of the particles,we obtain [8]a generalized model (1) ofthe interrelatedmovementsof the vesselon a regularwaves thatfor the studyof hisspeakersneed to be supplementedby equationsfor forcedcomponentsof vertical andhorizontaltransverse motion ofthe shipon a regular waves.

The system of equations(1) should be supplemented by equations for forced components ς_b,η_b of vertical and horizontal transverse motion of the ship on a regular excitement

$$\dot\eta_b=-\sigma_k\eta_0(r,\sigma,\chi)\sin\sigma_k t, \qquad \dot\varsigma_b=-\psi\upsilon_x+\theta\upsilon_y+\upsilon_z-\dot z_c. \tag{7}$$

Here χ – the angle between the direction of wave propagation and the longitudinal axis of the vessel. To determine the amplitude $\eta_0(r,\sigma,\chi)$ as a function of the parameters of the sea waves, can be applied dependence [3]

$$\eta_0(r,\sigma,\chi) = \aleph_n \eta_1 \tag{8}$$

where $\aleph_n$ – reduction factor, which is calculated based on the geometric dimensions of the ship and sea waves parameters is performed according to the method proposed in [5]; η_1 - the amplitude of the wave-particle oscillations in the direction of wave propagation.

The system of integrated equations of the drilling vessel changes, obtained in [8], is not an analogue of a real vessel, as the latter, depending on the type of operations performed, may have one of the four dynamic analogues corresponding to the stages(initial, preliminary, drilling, ultimate) of drilling activity.

So, on assuming the dimension $m, \lambda_{ij}, \mu_{ij}, J_x, J_y, J_z, z_c$ from (1)as constant and regarding the equation

$$\dot{\xi}_g = a_1\upsilon_x + b_1\upsilon_y + c_1\upsilon_z, \qquad \dot{\eta}_g = a_2\upsilon_x + b_2\upsilon_y + c_2\upsilon_z, \qquad \dot{\zeta}_g = a_3\upsilon_x + b_3\upsilon_y + c_3\upsilon_z, \tag{9}$$

we obtain the equation of the vessel changes at the initial stage.

To obtain the equations reflecting vessel movement at the preliminary period, if we consider drilling vessel [8] we write down:

$$\begin{aligned} &\psi = \psi_0 e^{j\sigma_k t}, && \theta = \theta_0 e^{j\sigma_k t}, \\ &\zeta_g = \zeta_0 e^{j\sigma_k t}, && \eta_g = \eta_0 e^{j\sigma_k t}. \end{aligned} \tag{10}$$

On assuming the dimensions m, λ_{ij}, μ_{ij} $(i, j = 1, 2, \ldots, 6)$ J_x, J_y, J_z, z_c as constant, on carrying out calculations similar to (8)- (10) we obtain equations of the preliminary period.

$$\begin{aligned}
&(m+\lambda_{11})\dot{\upsilon}_x + mz_c\ddot{\psi} + [(m+\lambda_{33})\upsilon_z + \mu_{33}\varsigma_b]\dot{\psi} - \\
&\qquad -[(m+\lambda_{22})\upsilon_y + \mu_{22}\eta_b \cos\varphi]\dot{\varphi} = F_x, \\
&(m+\lambda_{22})\dot{\upsilon}_y + (\lambda_{24} - mz_c - (\mu_{22}/\sigma_k^2)\upsilon_z)\ddot{\theta} + [\mu_{22}\varsigma_g - (\mu_{22}/\sigma_k^2)\dot{\upsilon}_z + \\
&+\mu_{24} - (m+\lambda_{33})\upsilon_z - \mu_{33}\varsigma_b]\dot{\theta} + \mu_{22}\upsilon_z\theta + \lambda_{26}\ddot{\varphi} + \lambda_{26}\ddot{\varphi} + \\
&+[(m+\lambda_{11})\upsilon_x - \mu_{22}\eta_g \sin\varphi]\dot{\varphi} + \mu_{22}\dot{\eta}_b \cos\varphi = F_y, \\
&(m+\lambda_{33})\dot{\upsilon}_z + \mu_{33}\upsilon_z + (\lambda_{35} - (\mu_{33}/\sigma_k^2)\upsilon_x)\ddot{\psi} + [\mu_{35} - (\mu_{33}/\sigma_k^2)\dot{\upsilon}_x - (m+\lambda_{11})\upsilon_x]\dot{\psi} + \\
&+(\mu_{33}/\sigma_k^2)\upsilon_y\ddot{\theta} + [(\mu_{33}/\sigma_k^2)\dot{\upsilon}_y + (m+\lambda_{22})\upsilon_y + \mu_{22}\eta_g \cos\varphi]\dot{\theta} = F_z, \\
&(J_x + \lambda_{44} - (\mu_{24}/\sigma_k^2)\upsilon_z)\ddot{\theta} + [\mu_{44} + \mu_{24}\zeta_g - (\mu_{24}/\sigma_k^2)\dot{\upsilon}_z + (mz_c - \lambda_{24})\upsilon_z + (\mu_{33}/\sigma_k^2)\upsilon_y^2]\dot{\theta} - \\
&-\mu_{22}\varsigma_g\upsilon_z\theta + [(\lambda_{26} + \lambda_{35} - (\mu_{33}/\sigma_k^2)\upsilon_x)\upsilon_y + \mu_{26}\eta_g \cos\varphi]\dot{\psi} + \mu_{35}\upsilon_y\psi - \\
&-[\mu_{24}\eta_g \sin\varphi + mz_c\upsilon_x + (\lambda_{35} + \lambda_{26})\upsilon_z + \mu_{35}\varsigma_g]\dot{\varphi} + (\lambda_{24} - mz_c)\dot{\upsilon}_y + \mu_{33}\varsigma_g\upsilon_y + \\
&+[(\lambda_{33} - \lambda_{22})\upsilon_y - \mu_{22}\eta_g \cos\varphi]\upsilon_z + \mu_{24}\dot{\eta}_g \cos\varphi = M_x \\
&(J_y + \lambda_{55} + (\mu_{35}/\sigma_k^2)\upsilon_x)\ddot{\psi} + [m_z\upsilon_z - (\lambda_{35} - (\mu_{35}/\sigma_k^2)\upsilon_x)\upsilon_x + \mu_{55} - (\mu_{35}/\sigma_k^2)\dot{\upsilon}_x]\dot{\psi} - \\
&-\mu_{35}\upsilon_x\psi + (\mu_{35}/\sigma_k^2)\upsilon_y\ddot{\theta} + [(\mu_{35}/\sigma_k^2)\dot{\upsilon}_y - \lambda_{26}\upsilon_y - \mu_{26}\eta_g \cos\varphi - (\mu_{35}/\sigma_k^2)\upsilon_x\upsilon_y]\dot{\theta} + \\
&+[(-mz_c + \lambda_{24})\upsilon_y + \mu_{24}\eta_g \cos\varphi]\dot{\varphi} + mz_c\dot{\upsilon}_x + [(\lambda_{11} - \lambda_{33})\upsilon_z - \mu_{33}\zeta_g]\upsilon_x + \lambda_{35}\dot{\upsilon}_z + \mu_{35}\upsilon_z = M_y, \\
&(J_z + \lambda_{66})\ddot{\varphi} + (\lambda_{26}\upsilon_x - \mu_{26}\eta_g \sin\varphi)\dot{\varphi} + (\mu_{26}/\sigma_k^2)\upsilon_z\ddot{\theta} + [\mu_{26}\eta_g - (\mu_{26}/\sigma_k^2)\dot{\upsilon}_z + \lambda_{35}\upsilon_z + \mu_{24}\eta_b\dot{\psi} + \\
&+\mu_{35}\varsigma_g - (\mu_{26}/\sigma_k^2)\upsilon_z\upsilon_x + \lambda_{24}\upsilon_x)\dot{\theta} + [\mu_{26}\upsilon_z + (\mu_{22}\varsigma_g + \mu_{24})\upsilon_x]\theta - (\lambda_{24}\upsilon_y + \mu_{24}\eta_g \cos\varphi)\dot{\psi} + \\
&+\lambda_{26}\dot{\upsilon}_y + \mu_{26}\eta_g \cos\varphi + [(\lambda_{22} - \lambda_{11})\upsilon_y + \mu_{22}\eta_g \cos\varphi]\upsilon_x = M_z. \\
&\dot{\xi}_g = \upsilon_x \cos\varphi - \upsilon_y \sin\varphi + (\psi\cos\varphi + \theta\sin\varphi)\upsilon_z, \\
&\dot{\eta}_g = \upsilon_x \sin\varphi + \upsilon_y \cos\varphi + (\psi\sin\varphi - \theta\cos\varphi)\upsilon_z, \\
&\dot{\zeta}_g = -\psi\upsilon_x + \theta\upsilon_y + \upsilon_z.
\end{aligned} \tag{11}$$

Now, let us assume that current is possible only in the horizontal plane and it is characterized by velocity υ_T and direction (the angle α_T between the velocity vector $\bar{\upsilon}_T$ and the axis $O\xi$) that is its components:

$$\dot{\xi}_T = \upsilon_T \cos\alpha_T, \quad \dot{\eta}_T = \upsilon_T \sin\alpha_T, \quad \dot{\zeta}_T = \zeta_T = 0. \tag{12}$$

In the projections upon dynamic axes of reference $Gxyz$ we obtain:

$$\begin{aligned} \upsilon_{xT} &= a_1\dot{\xi}_T + a_2\dot{\eta}_T + a_3\dot{\zeta}_T = \upsilon_T \cos(\alpha_T - \varphi), \\ \upsilon_{yT} &= b_1\dot{\xi}_T + b_2\dot{\eta}_T + b_3\dot{\zeta}_T = \upsilon_T \sin(\alpha_T - \varphi), \\ \upsilon_{zT} &= c_1\dot{\xi}_T + c_2\dot{\eta}_T + c_3\dot{\zeta}_T = -\upsilon_T\theta\sin(\alpha_T - \varphi) + \upsilon_T\psi\cos(\alpha_T - \varphi). \end{aligned} \tag{13}$$

Here a_i, b_i, c_i, $(i = 1, 2, 3)$ are factors determined through the above-mentioned anglesby the equation

$$\begin{aligned} & a_1 = \cos\varphi\cos\psi, a_2 = \sin\varphi\cos\psi, a_3 = \sin\psi, \\ & b_1 = \sin\theta\cos\varphi\sin\psi - \cos\theta\sin\varphi, b_2 = \cos\theta\cos\varphi + \sin\theta\sin\varphi\sin\psi, b_3 = \sin\theta\cos\varphi, \\ & c_1 = \cos\theta\cos\varphi\sin\psi + \sin\theta\sin\varphi, c_2 = \cos\theta\sin\varphi\sin\psi - \sin\theta\cos\varphi, c_3 = \cos\theta\cos\psi. \end{aligned} \tag{14}$$

The resultant velocity of the vessel will be composed of vessel velocity $\vec{\upsilon}$ regardless of the current and current velocity $\vec{\upsilon}_T$. The components of the resultant vessel velocity in the dynamic standard of rest are determined through the equation:

$$\begin{aligned} \upsilon_{x1} &= \upsilon_x + \upsilon_{xT} = \upsilon_x + \upsilon_T \cos(\alpha_T - \varphi), \\ \upsilon_{y1} &= \upsilon_y + \upsilon_{yT} = \upsilon_y + \upsilon_T \sin(\alpha_T - \varphi), \\ \upsilon_{z1} &= \upsilon_z + \upsilon_{zT} = \upsilon_z - \upsilon_T \sin(\alpha_T - \varphi) + \upsilon_T\psi\cos(\alpha_T - \varphi). \end{aligned} \tag{15}$$

On plugging the formulations (15) into the equation (9), we obtainequations ofvessel gravity center displacement in the static standard of rest $O\xi\eta\zeta$ regarding the current:

$$\begin{aligned} \dot{\xi}_g &= \upsilon_x \cos\varphi - \upsilon_y \sin\varphi + (\psi\cos\varphi + \theta\sin\varphi)\upsilon_z + \upsilon_T \cos\alpha_T, \\ \dot{\eta}_g &= \upsilon_x \sin\varphi + \upsilon_y \cos\varphi + (\psi\sin\varphi - \theta\cos\varphi)\upsilon_z + \upsilon_T \sin\alpha_T, \\ \dot{\zeta}_g &= -\psi\upsilon_x + \theta\upsilon_y + \upsilon_z. \end{aligned} \tag{16}$$

3. Conclusion

In this article, we received a particular vessel mathematical model that takes into account the operations performed on it at the time of its dynamic positioning. For example, if we consider drilling vessel, depending on the type of work undertaken vessel can have one of four dynamic models corresponding periods (initial, preparatory drilling, end) drilling. Also, when constructing a mathematical model of vessel movement is obviously necessary to consider external factors. This article discusses the current account and the regular excitement in the construction of a mathematical model. To build a complete model of the need to integrate marine equipment used during operations on the parameters of the equations of dynamics of the vessel.

4. References

[1] Fossen, Thor I.,Guidance and Control of Ocean Marine Vehicles. John Wiley and Sons Ltd.New York, 1994.

[2] Lewis, E.V., Principles of Naval Architecture vol III: Motions in Waves and Controllability. Society of NavalArchitecture and Marine Engineers, New York,1988.

[3] Jensen, Andreas G. Fluid dynamic derivatives: marine and wind engineering approaches. Journal of Wind Eng.and Industrial Aerodynamics 69 (1997) 777–793.

[4] Abkowitz, M.A. Lectures on ship hydrodynamics- steering and manoeuvrability. Technical Report Rep.No.Hy-5.Hydro ogAerodynamiskLaboratorium, Lyngby, Denmark, 1975.

[5] Ruoshi Cha, Decheng Wan, Numerical Investigation of Motion Response of Two Model Ships in Regular Waves. Procedia Engineering, Volume 116, 2015, pp. 20-31. DOI:10.1016/j.proeng.2015.08.260

[6] Ivanova O., Dushko V., Rodkina A. Experimental Researches Automation of Spatial Oscillations of the Floating Ocean Engineering Systems in the Wave Basin. Proceedings of the 26th DAAAM International Symposium, pp.1059-1067,B. Katalinic (Ed.), Published by DAAAM International, ISBN 978-3-902734-07-5, ISSN 1726-9679, Vienna, AustriaDOI:10.2507/26th.daaam.proceedings.149.

[7] Nesin Danilo Y., Dushko Veronica R. Numerical model of the large carrying capacity crane ship with the fullyrevolving topside. Procedia Engineering (25th DAAAM International Symposium on Intelligent Manufacturing and Automation, DAAAM 2014), Volume 100, 2015, Pages 1082–1091.

[8] Kramar V. Investigation of Changes and Development of Mathematical Model for a Drilling Vessel. Proceedings of the 26th DAAAM International Symposium, pp.0049-0055,B. Katalinic (Ed.), Published by DAAAM International, ISBN 978-3-902734-07-5, ISSN 1726-9679, Vienna, AustriaDOI:10.2507/26th.daaam.proceedings.007

[9] T. Perez, A. J. Sørensen, and M. Blanke.Marine Vessel Models in Changing Operational Conditions – A Tutorial.In Proceedings of the 16th IFAC Symposium on System Identi_cation, Newcastle, Australia, 2006.

[10] Blagoveshensky S.N. Handbook of statics and dynamics of the ship. Shipbuilding Publishers, St. Petersburg,Russia,1976.

DOI: 10.2507/27th.daaam.proceedings.006

ANALYSIS OF IMAGE ENHANCEMENT METHODS FOR OUTDOOR-APPLICATIONS

Philipp Gujon, Wilfried Kubinger

This Publication has to be referred as: Gujon, P[hilipp] & Kubinger, W[ilfried] (2016). Analysis of Image Enhancement Methods for Outdoor-Applications, Proceedings of the 27th DAAAM International Symposium, pp.0040-0043, B. Katalinic (Ed.), Published by DAAAM International, ISBN 978-3-902734-08-2, ISSN 1726-9679, Vienna, Austria
DOI: 10.2507/27th.daaam.proceedings.006

Abstract

For autonomous robots and driver assistance systems it is necessary to react adequately to sudden changes in the surrounding area. Camera-systems are often affected by environmental influences like sun light, shades or sun-reflecting surfaces which makes it hard to obtain well exposed images for further utilization. In this paper, two algorithms for image enhancement are compared in several test scenes to examine their performance with regard to contrast enhancement and computing time. In comparison one off-the-shelf high dynamic range camera is tested, too. The results from this study show that histogram equalization works significantly faster than the successive mean quantization transform. Both algorithms are appropriate to be used for autonomous robots and driver assistance systems if the camera is incapable to high dynamic range images.

Keywords: High dynamic range imaging; successive mean quantization transform; histogram equalization; autonomous robots; driver assistance systems

1. Introduction

Regarding to the steady growth of autonomous machines in industry and agriculture it always becomes more inevitably to be able to react on many environmental influences. To solve this setting of tasks camera-systems are frequently used [1-2].

However, especially in outdoor applications it is hard for camera-systems to work adequately because of a huge variation in the objects appearance due to changing weather and light conditions. Performing the task automatically without human intervention is particularly hard in image processing. Many methods have been proposed to handle this challenge.

According to [3] these methods to achieve a contrast enhancement can be divided into two groups. One way to solve the problem is to take minimum 3 pictures of the same scene with different exposures and combine them to one well exposured picture with enhanced contrast. Different approaches and techniques have been suggested to solve this problem [4]-[6].

But in fast changing environments it is possible, that some fragments are not shown on the right place. This phenomenon is called "Ghost Effect".

To prevent pictures with not trustworthy information a lot of algorithms for digital image processing and contrast enhancement have been proposed [7]-[11]. Especially automotive suppliers are very interested in video cameras which can cope with varying lightning conditions. For advanced driver assistance systems (ADAS) it is mandatory to be robust against direct sunlight or dark areas – e.g., when driving in or out of a tunnel [12]-[17].

2. Contrast Enhancement

Every digital image consists out of single pixels. In grey scale pictures each of these image points corresponds to a brightness value. Nowadays, there are different kind of sensors available. The difference between them is how many steps are between the brightest and the darkest point. An 8 bit sensor, for example, can distinguish the luminance of an image into 256 steps. According to these insights, a picture can be interpreted as a matrix of values in the area [0,255]. The graphical representation of the distribution of the image points with the prevailing intensities is called histogram.

This is the basis of the histogram equalization procedure. This method causes a contrast enhancement by an equipartition of the brightness-value distribution. The calculation occurs by means of accumulated sum of the pixels. This entails that those areas of the histogram with higher appearing grey values are stretched and those with smaller amount are compressed.

With this method it is possible to make the full use of the whole range of intensities in a picture. The result is a contrast enhancement. The histogram equalization automatically flattens and stretches the histogram and therefore, the dynamic range of the image.

However, the successive mean quantization transform perform an automatic structural breakdown of information and these characteristic make the transform interesting for automatic enhancement of any images [9]. The smallest unit of the arithmetic process is the Mean Quantization Unit (MQU). The SMQT function can be described by a binary tree of MQU's as it is shown in Figure 1.

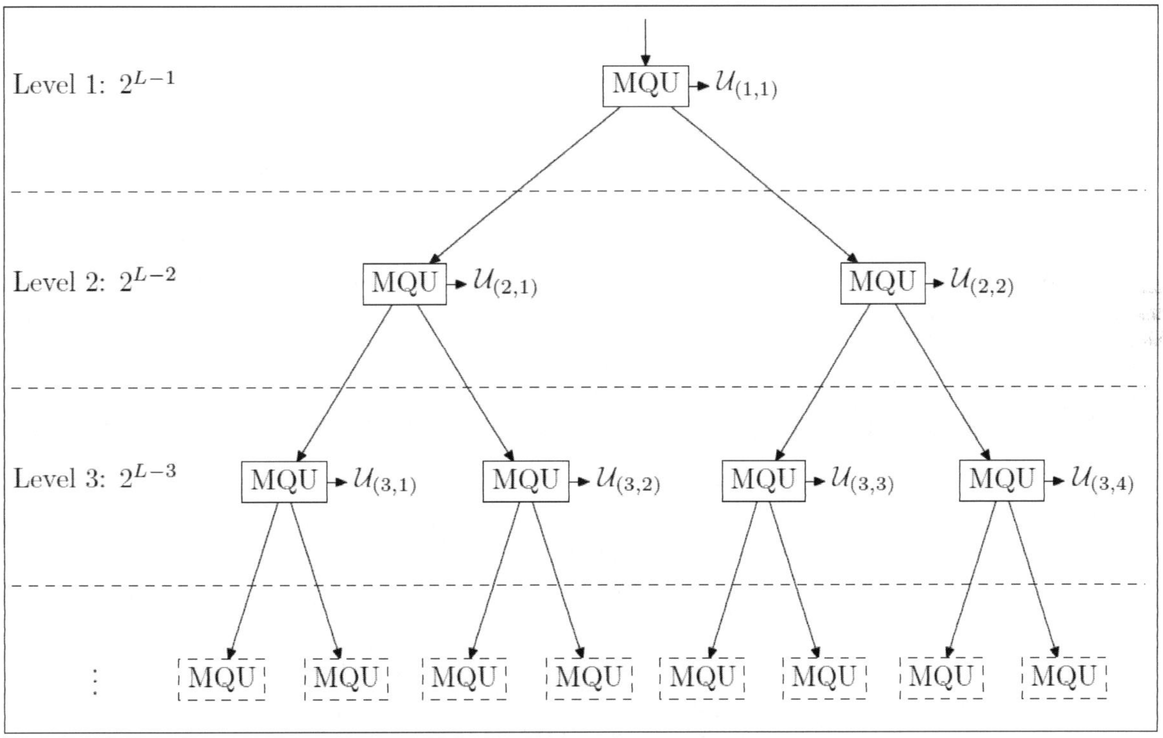

Fig. 1. The SMQT as a binary tree of MQU's (source: [7])

Histogram equalization, which was developed for image contrast enhancement, is similar to the successive mean quantization transform. However, both algorithms are used to enhance the luminance of low dynamic range images to extend the dynamic range.

3. Experimental Results

In fact that the histogram equalization and successive mean quantization transform algorithms calculate all pixels new in the picture, the computing time is given in 10^{-9} sec/pixel (ns/pixel).

This allows a further objective comparison to different sized pictures. As time measurement the tic-toc function in MatLab on a computer with an i7-4700MQ 2,4GHz processor was used.

In Table 1 the arithmetic average of 20 procedures with 100 image calculations are listed. The computing time and the maximum possible picture rate with a picture format of 1024x768 pixels is shown, too.

Method	Computing time [ns/pixel]	Max. frame rate [fps]
SMQT grey level	90,56	14
SMQT colour	528,81	2
HE grey level	2,2	574
HE colour	50,66	25

Table 1. Computing time and maximum frame rate of SMQT and HE.

The results of this comparison shows that the HE method is faster to compute than the SMQT method and therefore, it is possible to get higher frame rates which is necessary for fast changing environments.

Regarding to the fact that in grey-level pictures every pixel represents only the value of saturation, the computing time is much faster than for colour pictures. The tested algorithms were applied in images with different dynamism extent of the motive. The test scenarios contained under- and over- exposed pictures as well as backlight conditions to simulate any environmental factors like in real-world outdoor applications.

In Figure 2 the results of the image enhancement of a picture with poor contrast and for comparison an image of the same scene with a high dynamic range camera is shown. The differences are evident between images take with a customary industrial camera and a camera with raised dynamism extent as well as the contrast enhancement with the algorithms proposed on top.

Both methods for contrast enhancement deliver satisfactory results, nevertheless, the quality of the images is clearly worse than those of the HDR camera [18]. In closer consideration it is apparent that as a consequence of the contrast enhancement the white noise increases in the dark regions of the picture.

Fig. 2. (A) Low dynamic range image with back-light conditions. (B) The enhanced image with HE. (C) The enhanced image using the SMQT method. (D) High dynamic range image taken with an HDR-camera DFK AFU130-L53 from The Imaging Source [18].

5. Conclusion

In this paper the histogram equalization and the successive mean quantization transform methods for contrast enhancement of a low dynamic range image were evaluated and a comparison with an HDR- camera was shown. Both algorithms for image enhancement were compared in several test scenes to examine their performance with regard to computing time and level of contrast enhancement. For comparison, the results of a high dynamic range camera were demonstrated, too.

We show, that the histogram equalization works significantly faster than the successive mean quantization transform. Both algorithms deliver satisfying results in contrast enhancement but the SMQT shows more focus on details.

However, HDR-cameras provide excellent images even in dynamic environments.

6. Acknowledgments

This project has been funded in part by MA23 - City of Vienna within the Project Call 16-02 "Photonics: Foundations and industrial applications" and by the Austrian Security Research Programme KIRAS – an initiative of the Austrian Federal Ministry for Transport, Innovation and Technology (bmvit).

7. References

[1] Woeber, W.; Kefer, M.; Kubinger, W. & Szuegyi, D. (2012). Evaluation of Daylight and Thermal Infra-Red based Detection for Platooning Vehicles, Annals of DAAAM for 2012 & Proceedings of the 23rd International DAAAM Symposium, ISBN 978-3-901509-91-9, ISSN 2304-1382, pp 0719 - 0722, Editor B[ranko] Katalinic, Published by DAAAM International, Vienna, Austria.

[2] Spulak, D.; Otrebski, R & Kubinger, W. (2015). Evaluation of PCA, LDA and Fisherfaces in Appearance-based Object Detection in Thermal Infra-red Images with Incomplete Data. Procedia Engineering, Volume 100, 2015, pp. 1167-1173.

[3] Myszkowski, K.; Mantjuk, R. & Krawczyk, G. (2008). High dynamic range video. Lectures on Computer Graphics and Animation, Morgan & Claypool Publishers, 2008, 1, 1-158.

[4] Li, S.; Kang, X. & Hu, J. (2013). Image fusion with guided filtering. Image Processing, IEEE Transactions on, 2013, 22, 2864-2875.

[5] Matsuoka, R.; Ishibashi, H.; Baba, T. & Okuda, M. (2015). Multiple exposure integration for restoring all in-focus images. Image Processing (ICIP), 2015 IEEE International Conference, 2015, 4165-4169.

[6] Park, S. & Kim, J. H. (2006). Wide dynamic range vision for robot eyes based on double capturing scheme sensors. IEEE Conference, 2006, 5, 1473-1476.

[7] Nilsson, M.; Dahl, M. & Claesson, I. (2005). Gray-scale image enhancement using the SMQT. IEEE International Conference on Image Processing (ICIP).

[8] Im, J.; Jeon, J.; Hayes, M. & Paik, J. (2011). Single image-based ghost-free high dynamic range imaging using local histogram stretching and spatially-adaptive denoising. Consumer Electronics, IEEE Transactions, 2011, 57, 1478-1484.

[9] Nilsson, M.; Dahl, M. & Claesson, I. (2005). The Successive Mean Quantization Transform. IEEE Internation Conference on Acoustics, Speech and Signal Processing (ICASSP).

[10] Sahani, M.; Rout, S. K.; Mohan Satpathy, L. & Patra, A. (2015). Design of an embedded system with modified contrast limited adaptive histogram equalization technique for real-time enhancement. International Conference on Communications and Signal Processing (ICCSP), 2015, 332-335.

[11] Li, C.-C.; Hsu, S.-W.; Shen, P.-C. & Guo, J.-I. (2014). A single-camera high dynamic range technique by using contrast enhancement and exposure control. Asia-Pacific Signal and Information Processing Association (APSIPA).

[12] Prinet, V.; Wang, J., Lee, J., & Wettergreen, D. (2016). 3D road curb extraction from image sequence for automobile parking assist system. In: Image Processing (ICIP), 2016 IEEE International Conference on. IEEE, pp. 3847-3851.

[13] Ziebinski, A.; Cupek, R., Erdogan, H., & Waechter, S. (2016). A Survey of ADAS Technologies for the Future Perspective of Sensor Fusion. In: International Conference on Computational Collective Intelligence. Springer International Publishing, pp. 135-146.

[14] Song, Y. & Liao, C. (2016). Analysis and review of state-of-the-art automatic parking assist system. In: Vehicular Electronics and Safety (ICVES), 2016 IEEE International Conference on. IEEE, pp. 1-6.

[15] Hertel, D. W. & Chang, E. (2007). Image quality standards in automotive vision applications. In Proc. IEEE Intelligent Vehicles Symposium, pp. 404-409.

[18] Knoll, P. M. (2007). HDR vision for driver assistance. In High-Dynamic-Range (HDR) Vision, Springer Berlin Heidelberg, pp. 123-136.

[17] Hoefflinger, B. (2016). HDR-and 3D-Vision SensorsVision Sensors. In CHIPS 2020 Vol. 2, Springer International Publishing, pp. 201-209.

[18] http://www.theimagingsource.com, (2016). The Imaging Source, DFK AFU130-L53 Color Camera data sheet, Accessed on: 2016-09-26

DOI: 10.2507/27th.daaam.proceedings.007

OPTIMIZATION OF IMPACT VIBRATION ISOLATION

Davorka Šaravanja, Marko Grbešić & Mario Dragičević

This Publication has to be referred as: Saravanja, D[avorka]; Grbesic, M[arko] & Dragicevic, M[ario] (2016). Optimization of Impact Vibration Isolation, Proceedings of the 27th DAAAM International Symposium, pp.0044-0050, B. Katalinic (Ed.), Published by DAAAM International, ISBN 978-3-902734-08-2, ISSN 1726-9679, Vienna, Austria DOI: 10.2507/27th.daaam.proceedings.007

Abstract

Impact machines are designed to generate high dynamic forces with each impact hammer. Conventionally designed bases of these machines require relatively large reinforced concrete blocks to transfer the impact force to the ground. Size basic block is dependent on a loading force produced a hammer and the characteristics of the soil. The vibrations generated by blows of hammers also be transmitted to the ground. These vibrations can have large impact and consequences on equipment, employees and even to the building structure. Recently, large investments in research and development of new technologies vibration isolation for impact machinery using the steel coil springs and viscous fluid dampers. The old solutions of elastic foundations are required large inertial mass under the hammer. With constant research and development to come up with new solutions that provide direct support to the impact machines using springs, without using a inertial mass.

Keywords: impact machines; vibration isolation; hammer, isolator.

1. Introduction

In recent years, increasingly used vibration isolators with coil springs and elastic elements, because they showed better qualities compared to conventional anti-vibration stand, because they are designed to be low-level and they are very effective for low frequency vibrations. The combination of coil springs and elastic elements enables greater efficiency for higher frequency range making it even more effective in the control of vibration. [1]

Their quality to conventional anti-vibration pads in the specific surface that acts as a sound barrier and increases the effect of isolation at high frequencies, which are transmitted through the spring.

In today's conditions often exceed the levels permitted by environmental noise due to the impact of industrial plants or craft workshops. In addition to noise, heavy machinery used in industrial production generate very strong vibrations. The reasons may be different from non-quality spatial planning, housing industrial facilities and craft workshops, inappropriate places in the area (near residential buildings) or poor isolation. Overdrafts level can be several times higher than permitted, and in some cases the level of vibration cannot be reduced to the appropriate technical measurement frameworks.

Sound isolation plant using silencers is the core problem, but no less important is not reducing vibrations that create impact and very heavy machinery. As much as it takes care of the design, balancing and installation (assembly), it is inevitable that the work of these machines follows the emergence of cyclical vibration phenomena of varying intensity, which run the entire mass. The problem consists in the isolation of the machine from the substrate structure to prevent vibration can be transmitted. Therefore, between the machine and the structure of the substrate should be set element that absorbs vibration energy in its two alternative components (usually vertical). This element must be flexible in order to stop the direction of the forces, which means should be installed anti-vibration elements-insulators. Isolating element should be resistant to the weight of the machine, ie. should be chosen as a function of its weight and the natural frequency of vibration of the machine. [2]

This problem is particularly acute in the area located on the rocky soil that is subject to the transmission of vibrations to the environment of the installed plant of this type.

2. Designing optimal procedure vibration isolation

Technical methods for solving these problems are in the optimization process of vibration isolation on mechanical machines that generate unacceptable levels of vibration and which not only threaten the characteristics of the machine, but also affect the threat to equipment in their environment. The starting point for the selection of such an isolator to form an optimal relationship between quantities such as ω_n-natural frequency (Hz); ω-excitation frequency (Hz); T-transmissibility function (%) and the d-static deflection (mm).

Thus, when designing the optimal procedure vibration isolation external excitation forces must follow the basic principles for calculating transfer functions, depending on the modification of the vibration characteristics (frequency, stiffness and damping coefficients).

The vibration isolation is a way of reducing the dynamic forces that are transmitted from the vibrating system to another system that should protect the construction or equipment nearby.

Thus, the isolation is a process in which isolator receives vibrational energy of the base system (in return use force on the system that is equal and opposite to vibrational excitation), ie. preventing vibration motion, and the energy it receives, gradually dissipates, so it is advisable and specific damping the isolation. Otherwise, the two resonances that occur in isolation, would have infinite value, and dampers isolators have just role up maximum resonance value.

Optimal vibration isolator with its frequency corresponds to a very short field frequency relationship, because only in this area allows a certain balance. Terms of designing optimal isolator, can be realized in several ways:

- Reducing the excitation force;
- The change of activity and the frequency of the excitation force;
- By changing the stiffness of the system in order to avoid adverse cases resonance;
- Based on the elastic energy absorbing the excitation forces and so prevent the passage of energy to the sensitive parts of the system;
- Proper selection of the machine which influences the acceptable dynamic response of a structure;
- Elimination of local vibration increasing rigidity of vibrating elements, which can be achieved by installing stiffeners.

In large impact machines with very high dynamic loads due to the rapid movement of workpieces or machine parts with very high requirements, improve vibration isolation can be achieved by adding additional mass in the system. Figure 1. shows the measured values obtained from the analysis of vibration system with and without isolators. [2]

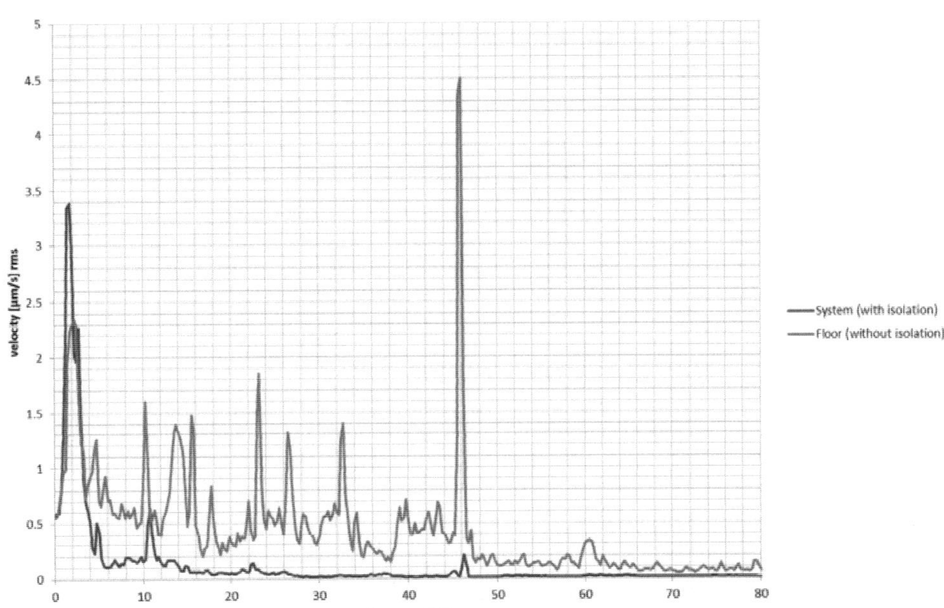

Fig. 1. Vibration analysis with and without vibration isolation

2.1. Vibration isolation heavy impact machines-impact hammers and forging machines

When performing the isolation of heavy impact machines the same should be placed on vibrating isolators and thus isolate them from the dynamic communication aimed at these machines. In practice, there are so called single layered, double and sometimes a triple layer type of vibration isolation. In the single-type vibration isolating used different variations of suspension by vibration isolators on the structure that is to be isolated. Between the machine and the vibration isolators is usually by the relevant plates of certain characteristics customized design suspension (steel, aluminum or even steel-concrete). Today, optimization impact vibration isolators realized in the form of special supports, so they can be:

- Spring vibration isolators, whose main operating element of one or several steel coil springs (cylindrical coil springs, cylindrical or conical);
- Rubber or rubber-metal vibration isolators whose main working body of the rubber complex forms;
- Vibration isolators are made of thin pressed steel parts (known as "metal rubber");
- In the form of an elastic material placed between the layers and foundations;
- In the form of a floor on an elastic basis which is commonly used in two-layer mode of vibration isolation.

Hammers are machines dynamic action, in which the kinetic energy of the moving mass is converted into work required for execution the process of deformation – forging (Fig. 2). According to the basic design characteristics hammers are divided into: hammers with anvil and without anvil, ie. anti-impact hammers.

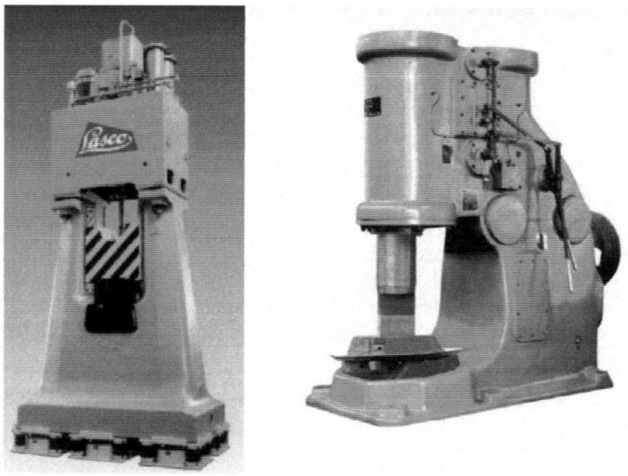

Fig. 2. Impact hammer and press forging-examples

Construction of heavy impact machines (impact hammers and forging machines) makes: the support structure, propulsion system, working part of the management system, lubrication system, auxiliary systems and security systems.

Hammers are selected on the basis of data obtained by designing technological process of forging. Basic specifications of hammers: nominal (available) energy, mass hair, no hair strokes per unit time, the maximum stroke of hair, forehead hair surface, the surface of the table, face the pressure of the working fluid, the working fluid consumption in a certain period of time, the weight of an anvil, dimensions the hammer and the total mass of the hammer. [2]

Starting the pneumatic hammer mallet is carried out with compressed air from the compressor which is a constituent part of the hammer. While the piston compressor down is done raising the pile, and to move the piston compressor up, accelerating the hammer down. Stroke rate of hair corresponds to the number of strokes of the piston compressor unit time.

Foundations hammer is the basis on which is placed a hammer and anvil. When operating the hammer foundation receives strong impact loads causing his movement and vibration. Since the foundation structure largely depends on the proper functioning of the hammer, as well as the noise level in operation.

Foundations can be a support suitable for reception of strong impact and dynamic loads. Impact is based on the share based rigid structures and vibroisolation based.

The support foundation - is a reinforced concrete block to which is attached a hammer. It is used in anti-shock hammer (hammer with movable anvil) where there is no strong shocks and vibrations. On the basis of the casted anchors for binding hammer. This method is based mainly used in areas where it is not hard court. [4]

Impact based applied for admission shock loads in all hammer with anvil. Impact foundation of reinforced concrete block certain mass and the supporting surface, which is determined by examining the technical documentation for the machine that is being prepared.

Being built as a one-piece or multi-piece. If the drive there is more hammer is based can not be separated or connected in a linear configuration. Anvil hammer is placed on the rubber basis in order to mitigate impact load. The diagram (Fig. 3) shows vibration values of hammers with and without isolation system. Vibrations are reduced by more than 80% compared to rigid foundation. [5]

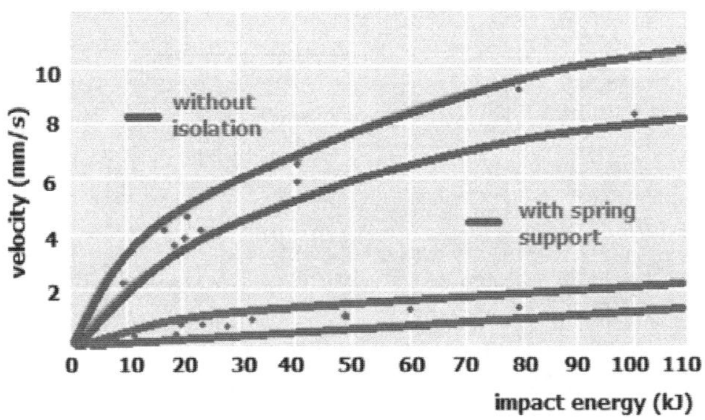

Fig. 3. Measured vibrations velocities of impact machine (in a distance of 10 m)

Vibration isolation of impact machines is a process which is very difficult to do well because of the large mass of the machine, and the enormous force of the blows produced by the machine.

2.2. Characteristics of vibration isolators

There are two types vibroisolators: isolators with series soil springs with viscous fluid dampers and elastomeric isolation systems.

Proper selection and optimization of structural parameters of these types of isolators leads to better characteristics vibration isolation impact machines.

Both types have proven to be very effective in isolating hammer and impact forces. Isolators with springs and viscous dampers provide better isolation characteristics, but they are quite expensive, require expensive foundations and tend to maintenance and repair. Viscous dampers in isolators with springs are difficult to protect, especially if you have a hole for the foundation prone to flooding, as in this case require the replacement of the fluid in the shock absorbers. [6] [8]

Some shock absorbers/dampers are designed to have cracks to leak into the walls of the foundation, which prevents their replacement in case of need and repair. For these reasons, these types of isolators are becoming the standard in the isolation of vibration and impacts (Fig. 4).

They offer the best isolating properties, with no major inertial mass, in case of flooding and no major maintenance costs related to the isolators with coil springs. They are very easy to install and maintain and are very long-lasting.
This work seeks to contribute to solving the technical problems involved in the process of isolation of forging machines. When it comes to protective coating process, forging machine with impact hammer must be analyzed while: a bump in the lowering phase, while performing an action on an object and when it is in the stage of raising.

The ability of the impact machine is generally determined by the level of energy that can be obtained from the dropping weight and the upper parts of the machine.

Most impact machine is designed so that their weight starts to drop the initial zero and reaches a speed of 6-7 m/s when striking the workpiece. Some also hammers accelerating fall by a piston driven by steam, pneumatic or hydraulic pressure accumulator. It is important that the isolating system applies a sufficient amount of damping so that there is no movement when there is going to blow.

If the system is moving down when he gets the next blow, the blow will increase the amplitude of movement down more than the previous impact, which affects an overloaded system isolation and leads to instability of the whole structure. For softer installed isolating systems, when it comes to speeding up work hammer, the hammer can be loaded isolating system which leads to instability.

Usually, the hammers have enough weight so that their recoil is less of a problem than hitting the object.

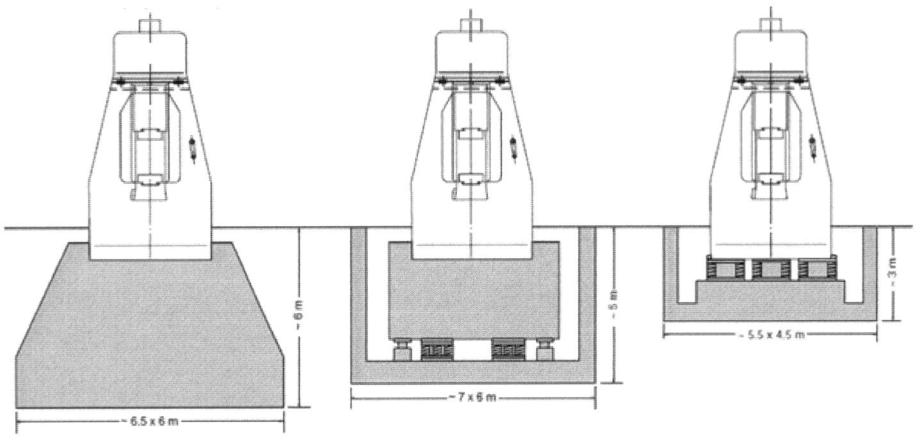

Fig. 4. Examples of vibroisolation foundations in section [5]

For a short time from when it happens contact with the workpiece and its deformation is the most important time in the working oparation of impact hammer. There is a wide range of forge work that can be done with a hammer, so that the magnitude of the induction force and duration of the force can vary significantly.

Increased heat workpiece allows a lower magnitude of the impact force and shortening the duration of the force on the workpiece. The final blows in forging blows the hardest. Analysis of the reaction anvil shot is actually simplified by the fact that many more massive anvil of tools for impact and that the impact takes a very short time. Tool for shock moving down while the speed of the anvil increases to equal the speed of the tool, and rebounds. Engineers have to understand that the cause maximum force on hammer, so that the anvil must be much more massive than him.

The theoretical force induced on the hammer relative to the extremely massive anvil is 100 times the hammer. It has been reported that up to 10 times the mass of the anvil increased very little changes attained the size of the force.

Vertical movement forgings depends on several parameters, the most important are the mass of the anvil and rigid systems, ie. changes in the system of hammers decreases when the mass anvil increases and movement increases with less rigid systems of lower natural frequency.

If applied generally accepted limit of several mm size of the movement, it is clear that the coil springs and elastic systems there is a need for such weight anvil which would maintain the natural frequency and characteristics of isolation performance, as shown in Figure 5. [5]

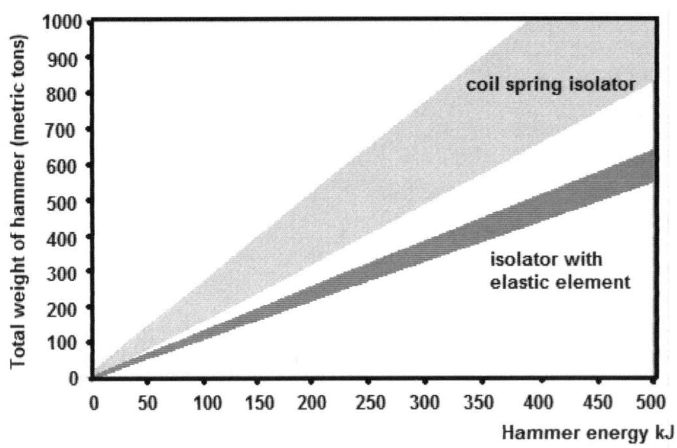

Fig. 5. Required hammer weight to achieve a 7 mm peak maximum response

Because the cost of adding concrete or steel inertial mass, isolator with elastic elements are more economical while keeping the same quality of isolation. Steel inertial mass are more economical, because the steel plate is denser and requires less space, thereby reducing the space and the size of the foundation. [7]

Studies have shown that the steel inertial masses more durable the same load. The force transmitted to the foundation is the product of isolation system dynamic stiffness and movement of anvil.

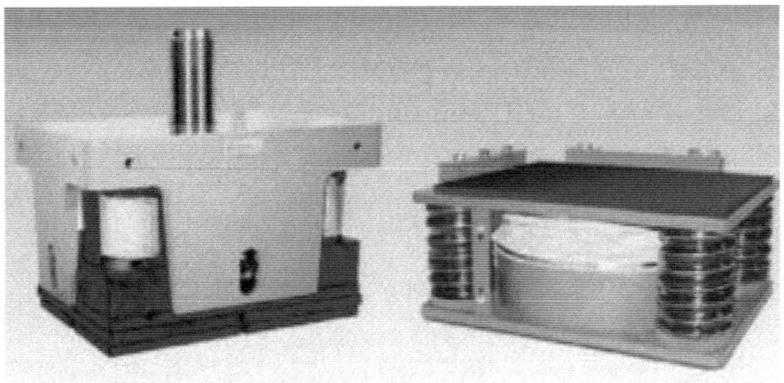

Fig.6. Isolator with elastic elements and steel coil spring type isolator [4]

For coil spring isolators, the addition mechanism with viscous damper adds a small amount of force to the force transmitted by support the springs. [9] For isolators with elastic element, the damping of the material is included in the real dynamic stiffness. The forces generated in the prime area of the enormous (Figure 6).

The exact size and duration of of the operation is usually not known, because measuring force is not possible. However, experienced workers on machines with impact hammers can be easily noted a significant reduction in fatigue of the components of the structure in relation to the traditional machines with impact hammers without isolators of this type. The correct application of optimal isolation system will result a significant reduction in the enormous influence of impact hammers.

Isolation systems with springs or elastic elements affect drastically reducing shock loads of high amplitude. Optimization of isolation of impact vibration to reduce health problems, reduce maintenance costs of the construction of the facility in which these machines are installed, reducing the formation of noise at work, and reduces the cost of the foundation. [5]

3. Conclusion

The main conclusion is that shown vibration isolators, ie. elastic systems with special configurations and proper selection of its parameters, have a decisive role in increasing the degree of isolation of the base structure.Vibration isolation of impact machines require a special consideration because they transmit dynamic loads to soil in addition to static loads due to weight of foundation, machine and accessories. The dynamic load due to operation of the machines is generally small compared to the static weight of machines and the supporting foundation.

The paper presents results of research on the impact machinery proving optimal design vibration isolators with coil springs and elastic elements in relation to the vibrational isolators with inertial masses. Studies have shown significant differences in the amount of vibrational amplitude of the transmitted dynamic force of impact. For greater impact energy that is transmitted to the vibration isolator with coil springs and elastic elements, the speed of vibration is reduced by up to ten times.

On the basis of impact machines occurs dynamic load during a long period of time, but its small size. Therefore perform optimization vibration insulation foundations to allow better flexibility because, otherwise, the deformation becoming greater with each load cycle which can become quite unacceptable.

Isolation systems with coil springs or elastic elements have proven to be very effective in isolation shock vibration on machines with impact hammers. Compared with traditional machines with striking hammer and their installation methods, such as wood and rubber flooring, isolation systems with coil springs and elastic elements are superior for faster and easier installation and assembly of the separate building. [10]

The amplitude of the vibration of the impact machine with its operating frequency, the coefficient of stiffness and damping vibration isolators are most important parameters that define in optimizing vibration isolation, with the nature frequency of the foundation structure of the machine. The two systems of isolation are cost effective and extremely durable.

In the last decade, research in this area have developed very good technology for the isolation of the impact of vibration on machines with impact hammers. But, recommendations for future research would be to prove even better quality combined vibration isolators. This is evidenced by the chart shows the relationship of the total weight of the impact of the machine (hammer) vs. the energy hammer that is transmitted to one or another type of isolators.

5. References

[1] http://antivibration-systems.com/wp-content/uploads/2013/03/Catalog

[2] http://www.vibrodynamics.com/english/hammers.html Vibration and Shock Isolation Systems for Forging Machinery, Vibro/dynamics, Technical bulletin M/L- 697

[3] Schmidt, J. H.; Granchi, M.; Barth, T.M.: Optimal Vibration Isolation for Shock-Sensitive Machines, Bilz Vibration tehnology, Leonberg, Germany

[4] Chowdhury, I.: Dasgupta, S. P. (2009). Dynamics of Structure and Foundation, Taylor&Francis Group

[5] https://www.gerb.com

[6] Fulan, P. (1988). Theory and Experiment of Hammer Foundation Vibration, Proceedings: Second International Conference on Case Histories in Geotechnical Engineering, St. Louis, Mo., Paper No. 4.05

[7] Prakash, S.; Puri, V. K. (2006). Foundations for Vibrating Machines, Special Issue, of the Journal of Structural Engineering, SERC, Madras, India.

[8] Broch, J. T. (1984). Mechanical Vibration and Shock Measurements, Brüel & Kjær, ISBN 8787355345

[9] Dragičević, M.; Šaravanja, D. (2012). Vibration Analysis Within Dynamic Testing of the Diesel Generator HE Mostarsko Blato, Zbornik radova Fakulteta strojarstva i računarstva Sveučilišta u Mostaru, ISSN 1986-5236, Mostar

[10] Leopa, A; Nastac, S.; Debeleac, C. (2012). Researches on damage identification in passive vibro-isolation devices, Shock and Vibration 19, 803–809, DOI 10.3233/SAV-2012-0689 IOS Press

DOI: 10.2507/27th.daaam.proceedings.008

RISK EVALUATION OF PRODUCTION AND IMPLEMENTATION OF THE PROJECT

Bozica Rados & Ante Rados

This Publication has to be referred as: Rados, B[ozica] & Rados, A[nte] (2016). Risk Evaluation of Production and Implementation of the Project, Proceedings of the 27th DAAAM International Symposium, pp.0051-0058, B. Katalinic (Ed.), Published by DAAAM International, ISBN 978-3-902734-08-2, ISSN 1726-9679, Vienna, Austria
DOI: 10.2507/27th.daaam.proceedings.008

Abstract

One of the key parts during the project cycle of the technical projects is also a risk evaluation of production and implementation of the project itself in all phases of the production. In this paper, we will examine the correlation of one of the risks (i.a. political risk) that are taken in consideration during risk analysis production and which are in direct correlation with the current status of country's development status in which the projects are conducted, specifically in this case of countries which are new members of EU. Critical risk in that case is the policy and institutional aspects, including existing economic policies and development plans, organisation and management of services to be provided/developed by the project, as well as capacity and quality of the institutions involved.

Keywords: risk evaluation; political risk; technical projects; approach from EU projects; Monte Carlo method

1. Introduction

A major issue that occurs during preparation of large long term infrastructure projects is always a preparation of project documentation. If we are planning a long-term infrastructure project all risks have to be foreseen. In a country with a marginal economic growth all bigger infrastructure projects have to be partially funded by the EU. In those countries one of bigger risks is the possibility of changes in a strategical approach towards those projects due to inexistent and incoherent national long-term development strategy. Standard approach in projects funded by the EU consists of two elements. First element is based on methodology of risk analysis which is described in a CBA guide 2014, issued by the EU. Second element is an international project experience of the Consultant in a project cycle. CBA is an analytical tool to be used to appraise an investment decision in order to assess the welfare change attributable to it and, in so doing, the contribution to EU cohesion policy objectives. The purpose of CBA is to facilitate a more efficient allocation of resources, demonstrating the convenience for society of a particular intervention rather than possible alternatives. Risk analysis within EU is defined as an assessment of probability leading to changes in variables or tasks that may occur [1]. By assigning an appropriate distribution of probability to the critical variables, distribution of probability for performance indicators can be estimated if applicable. The risk analysis is done in order to assess potential impacts of risks on the project respectively the planed measures. In general, the standard approach is applied as follows:

Risk analysis is done in 5 steps.

Step 1: A list of risks will be elaborated. That list will be prepared including most relevant risks for projects focusing on implementation.

Step 2: For each risk estimate the probability for occurrence of the risk (high – medium – low) and the description of the potential events leading to it will be assessed and described. This should be done by a panel of experts knowing the project and market conditions.

Step 3: For each risk estimate the potential impact on project performance (significant – medium – insignificant) and the description in which way the risk would affect the project performance will be assessed and described (panel assessment). In fact, the impact on delay and cost will be mentioned in order to use that for an overall impact the cost estimation and project schedule. This should be done by a panel of experts knowing the project and market conditions.

Step 4: The next point includes the preparation of a risk matrix and identification of critical risks (= medium or high probability for occurrence AND medium or significant impact).

	Low risk			
	Important risk, risk should be monitored	**Probability**		
	Critical risks, mitigation measures needed	Low	Medium	High
Impact	**Insignificant**			
	Medium			
	Significant			

Table 1: Impact/probability table

Step 5: The final step includes the quantitative risk analysis (Monte Carlo simulation) of only critical risks (if applicable respectively assessment of the risk and data availability), subject to data availability. Panel estimated probability for risk occurrence includes possible ranges of risks (expected, minimum and maximum possible values) which mean triangle probability distribution is applied.

Finally, a mitigation measures will be described and recommend for selected critical risk but also for the determined important risks in order to handle the probability and impact depending on the overall results [2].

2. Risk analysis preparation

The risk preparation deals with the description of the risks, the assessment and final prioritization of the risks [3].

List of risks

The main risks that are relevant for the further analysis using the partition as follows:

- Political, legal and administrative risk
- Technical (including design and construction) and operational risks
- Financial risks
- Social and cultural risk
- Environmental risk

In this paper authors will emphasize the relation of political, legal and administrative risks on technical projects.

Political, legal and administrative risks

No.	Risk	Description
1.1	Missing the target - implementing of project main goals	Delay in implementing of project main goals leads to impacts on the connected project in EU. To be expressed in month of delays.
1.2	Missing the target - implementing of project main tasks	Delay in implementing project main tasks leads to impacts on the connected project in EU. To be expressed in losses of revenues.

1.3	Missing the target- Increase of competitiveness of project	Delay in implementing project main goals and tasks leads to impacts on the connected project in EU. To be expressed in losses of revenues.
1.4	Change of strategy of project	Can cause changes in design phases and losses of already invested funds. To be expressed in monetary terms and months of delays.
1.5	Main project supplier's contract	Negotiation and signing of the contract (selection of supplier) can cause changes in project start and roll out, selection of supplier contract negotiation. To be expressed in in months of delays.
1.6	Contractor's bankruptcy (supplier)	Can cause new tendering procedures and negotiation of contracts. To be expressed in monetary terms and months of delays.
1.7	Technical allocation	Approval by the national applicable law authority. To be expressed in monetary terms months of delays.
1.8	Delays related to project goal implementation	Implementation of the technologies requires information and coordination procedures with operators of neighbouring states in depending on their technologies. To be expressed in months of delays.
1.9	Service Level Agreement (SLA) with the supplier (not signed)	SLA (defined service level) must be agreed in order to avoid additional costs within operation. To be expressed in monetary terms.
1.10	Spare part management / contract	Regulations according to the spare part handling (supplier) has to be defined and agreed (contract, negotiation, specification). To be expressed in monetary terms.
1.11	Contracts for project implementers	Permission according to communication in the neighbouring states is necessary for starting operations. To be expressed in months of delays.
1.12.	Analysis after project implementation	Future development may be different than forecasted. To be expressed in % foreseen.

Table 2: Risk description - political, legal and administrative risks

Risk matrix

The risk matrix, as shown below, illustrates the assessment of the risks that have already been described. The assessment includes the probability, the impact and the type of impact.

	Risk	**Probability (high-medium-low)**	**Impact (significant-medium-insignificant)**	**Type of impact (costs-delay)**	**Description**
1	**Political, legal and administrative risk**				
1.1	Missing the target - implementing of project main goals	Low	Significant	Delay	The planned measures and implementation on the project main goals change due political decisions as soon as the project implementation reaches the major target.
1.2	Missing the target - implementing of project main tasks	Low	Significant	Revenues	The planned measures and implementation on the project main goals change due political decisions as soon as the project implementation reaches the major target.
1.3	Missing the target- Increase of competiveness of project	Low	Significant	Revenues	The planned increase of competiveness on the project main tasks change due political decision as soon as the project implementation reaches the major target.
1.4	**Change of strategy of project**	**High**	**Significant**	**Delay & Costs**	**Current strategy of project is defined due previously defined main goals and main tasks. The implementation strategy has not yet been finalized and approved.**
1.5	Main project supplier's contract	Low	Medium	Delay	Study phase is ongoing and further planning phases will detail the specification, tender process and contract negotiation

	Risk	Probability (high-medium-low)	Impact (significant-medium-insignificant)	Type of impact (costs-delay)	Description
1.6	Contractor's bankruptcy (supplier)	Low	Medium	Delay & Costs	Economical (strong) suppliers are mostly responsible for systems. The tender procedure should reflect on economic and technical capability of the supplier.
1.7	Technical allocation	Low	Significant	Delay & Costs	The application (inquiry) should be started immediately in order to consider times of the authorities. In general, (EU), it should be no problem. In some cases, costs could accrue. If that could not be handled the infrastructure and equipment could not be commissioned.
1.8	Delays related to project goal implementation	Low	Medium	Delay	Information and coordination procedure must be implemented with neighboring countries. (EU benchmarks are available).
1.9	Service Level Agreement (SLA) with the supplier (not signed)	Low	Significant	Delay	The task should be considered before operation and awareness must be taken because other supplier will not solve problems with the systems. Agreement could be prepared on the basis of experiences. SLA could be negotiated and contracted (standard).
1.10	Spare part management / contract	Low	Medium	Delay	This is part of the tender documents and must be clarified before in order to keep the system running during operation (decrease in availability).
1.11	contracts for project implementers	Low	Insignificant	Delay	The coordination with neighbouring states needs to be done (EU standards, and experiences are available).
1.12	Analysis after project implementation	Medium	Insignificant	%foreseen results	Moderate growth rates forecasted and the foreseen result is partly declining. The project implementation focuses on international projects.

Table 3: Risk matrix

The risk analysis will only consider the highlighted measures of the highlighted option. The qualitative risk analysis (as initial preparation) is reflecting the stage of the project development cycle at which project is currently in: study preparation is ongoing, design has not yet been started, and works are not yet tendered. That analysis could also be done for further project phases as design or tendering [4]. The risk analysis excludes the current projects that under preparation (different phases of realization) and focuses on the overall preparation [5].

Prioritization

The prioritization can be separated as follows considering the two categories:

On the basis of the risk matrix and their assessment the prioritization is prepared as shown below [6], [7]. The prioritization includes the two categories "critical risk" and "important risk".

- Critical risks, mitigation measures needed
- Important risk, risk should be monitored and

The critical risks are summarized as follows:

- Change of strategy due political changes
- Inadequate construction cost estimate
- Lack of national finance

The risks that need to be monitored are summarized as follows:

- Missing the target - Improvement foreseen by project main tasks and goals
- Service Level Agreement (SLA) is not signed
- Inadequate site surveys and investigation
- Change in requirements
- Operational migration
- Failure existing technology
- Lack of information to technical requirements in tender documentation
- Lack of EU finance (funding)
- Inadequate supervision cost estimates
- Cost overruns during construction
- Risk prevention and mitigation

The further on report deals with risk prevention and mitigation according to he defined critical risks as well as risk that should be monitored [8].

Critical risks

Risk	Prevention and / or mitigation	Timing (short, medium, long)	Residual risk (in comparison to risk matrix)	
			Probability	Impact
Change of strategy	The project strategy implementation must be defined. Function in charge: state	Short	Medium	Significant
Lack of national finance	The national co-financing (EU funding plus national funding) must be clarified and agreed. A strategy for using (innovative) alternative financial instruments should be developed (e.g. EIB loan). Function in charge: investor, ministry	Medium	Medium	Significant
Inadequate construction cost estimation	The planning phase and further required investigations must be started on the basis of the agreed strategy. Function in charge: investor	Medium	Medium	Significant

Table 4: Risk prevention / mitigation - critical risks

Risks that need to be monitored

Risk	Prevention and / or mitigation	Timing	Residual risk (in comparison to risk matrix)	
		(short, medium, long)	Probability	Impact
Missing the target - implementing of project main goals/tasks	The project must be implemented (promote) as fast as possible for reaching the target focus. The design phase and needed investigation must be set up. Function in charge: state, investor	Short	Low	Medium
Service Level Agreement (SLA) with the supplier (not signed)	The task should be considered / negotiated before operation. SLA could be negotiated and contracted (standard). Project experiences are available in other countries Function in charge: investor			
		Short	Low	Medium
Inadequate site surveys and investigation	The planning phases have to be started and supported by further required surveys and investigations. Function in charge: investor	Long	Low	Medium

Risk	Prevention and / or mitigation	Timing	Residual risk (in comparison to risk matrix)	
		(short, medium, long)	Probability	Impact
Chance in requirements	The requirements must be detailed as soon as possible. Firstly, the general conditions e.g. technical level and step by step detailed due to the ongoing planning process. The requirements are based on the decision according to the strategy. Function in charge: investor			
Operational migration	The required interfaces must be investigated and defined (survey, study). Function in charge: investor	Medium	Low	Medium
Failure existing technology	The interfaces and the existing technology must be monitored and maintained (preventive maintenance). Function in charge: investor	Short	Low	Medium
Lack of information to technical requirements in tender documentation	The technical documentations (design) need to be on a required level and the tender document must be written by experienced experts / on the basis of the already prepared projects. The know-how according to the technology has to be ensured and tender specification defined. Function in charge: investor	Short	Low	Medium
		Long	Low	Medium
Lack of EU finance (funding)	The preparation of the project application must be started early. EU financial technical assistance must be involved early in the project cycle to reduce time for project approval. Function in charge: state, ministry, investor	Long	Low	Insignificant
Inadequate supervision cost estimates	The supervision costs should be extended on a basis of project experiences (national / international). Function in charge: state, ministry, investor			
Cost overruns during construction	The preparation phase must be detailed with surveys, investigation and planning's. Experienced (local / international) experts must supervise the projects during and the contract with the suppliers must be under consideration. Investment cost estimates should be compared well with costs experienced with similar projects implemented in the EU in the last years. Consultations with plant and equipment manufacturers were carried out to crosscheck estimates with current market conditions. Function in charge: state, ministry, investor	Short	Low	Medium
		Long	Low	Medium
		Long	Low	Medium

Table 5: Risk prevention / mitigation - monitored risks

3. Results of the risk analysis (critical risks)

The risk prevention / mitigation in the last chapter show measures for the critical risks which leads changes in their assessment. Anyhow, the risks are further critical and must be considered in a quantitative analysis. The overall risk assessment is finished here by discussing the contingencies in relation to the probability of overrun / underrun the cost estimate for the implementation of project strategy [9], [10] (without design costs).

Firstly, the probability distributions of the selected critical risks have to be determined. Afterwards, the Monte Carlo method will be applied for analysing the overall cost estimate. Finally, the potential amount of contingency will be discussed by applying and illustrating the statistical results.
In general, the risk analysis including the Monte-Carlo method should give an understanding of overrunning (or underrunning) of the cost estimate (point estimate) by using a probability distribution for critical variables and illustrating the probability distribution of the overall project costs as well. In making so, the accuracy of the cost estimate can be illustrated by using a certain statistical interval of confidence.

Probability distribution

The probability distribution of the selected variables will firstly be determined. As already mentioned the distribution based on expert discussions and apply a triangle distribution for each variable, due to the missing data from the past. So, the minimum and the maximum of the variables have been determined in discussions with experts [11].
The following table presents the variables and the established distribution. After specifying the distribution of the variables, the Monte-Carlo method have been run (n = 4,000).

Risk		**Minimum**	**Expected**	**Maximum**
Category				
Political, legal and administrative risk	Chance of strategy [month]	0	6	12
Financial risks	Lack of national finance [month]	0	6	12
	Inadequate construction cost estimate [mEUR]	90	100	130

Table 6: Quantitative risk analysis - probability distribution (triangle distribution)

One month delay amounts to 150,000 EUR in the consideration (assumption by the Consultant on the basis of project experience).

The figure below**Error! Reference source not found.** presents the results in form of a probability distribution and a cumulative distribution of the cost estimate after the Monte-Carlo analysis. This first result doesn't include a number of contingencies.

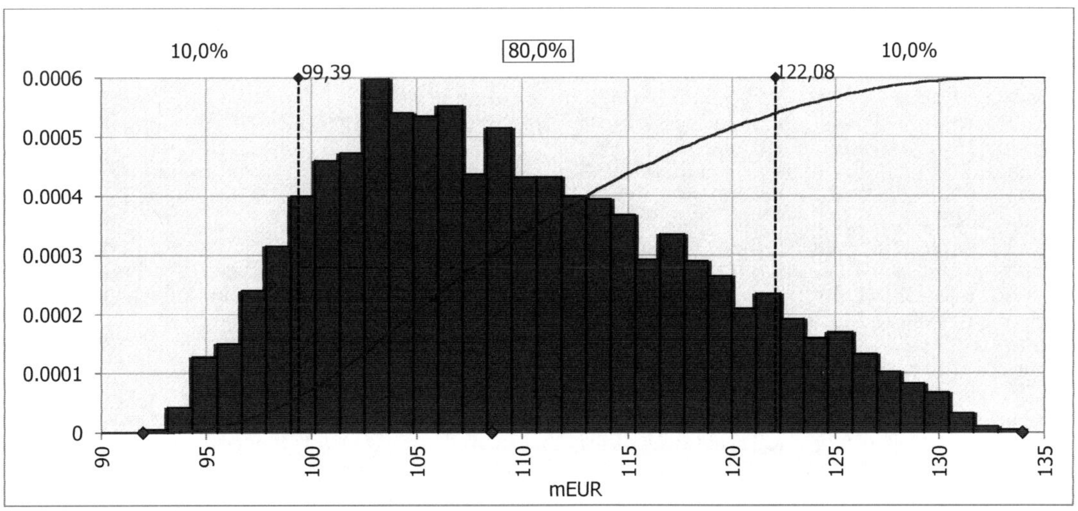

Figure 1: Quantitative analysis - probability of cost estimate (technical project) and cumulative probability (without contingencies), 80 % confidence

The value represents the point estimate of the consideration and amounts to 100 m EUR. In the case of 80 % confidence level the estimate ranges between 99 m EUR and 122 m EUR. This implies, the accuracy range (without contingencies) amounts to -1 % to 22 %. After describing the case without contingencies, the consideration will be extended to the case including contingencies. The recommended level of the contingencies depends on the assumed probability of overrun as well as underrun of the point estimate among other things.**Error! Reference source not found.** The table below presents a comparison of different estimate accuracies for several probabilities of overrun as 50 %, 30 %, 20 % as well as 10 %.

Probability of overrun	Base Cost m EUR	Contingency		Total Cost m EUR	Accuracy (80% confidence)
		Amount MEUR	%		
50 %	100	8	8	108	-9 % to 13 %
30 %	100	11	11	114	-15 % to 7 %
20 %	100	18	18	118	-22 % to 3 %
10 %	100	22	22	122	-23 % to 0 %

Table 7: Quantitative analysis - Determination of the contingencies and the accuracy in depending on the probability of overrun

The chosen contingency during the estimation of the CAPEX amounts to 25 %. The current analysis has shown that in the case of a probability of overrunning of 10 % (80 % confidence) the contingency is recommended with an amount of 22 %. This implies, the selected contingency covers the uncertainties considering the assumptions and explanations.

4. Conclusion

The presented analysis shows correlations and, between other risks, the risk of changing political situations in correlation with financial risks. In cases of technical projects, especially those whom are related to long-term infrastructural investments, the risk of changing the main goals of the project due to unstable and incoherent national strategical development strategies is humongous. For that reason, this type of risk evaluation has a great impact on foreseen results. In phases of project preparation risk evaluation is the most important step in order to have a clear view of a situation. In countries, which are new members of the EU the planning of big long-term technical infrastructure projects is very delicate due to above described risks. It is extremely important to do a competent risk evaluation before closing the financial part of the project funded by the EU. [12]. Cost Benefit Analysis (CBA) is explicitly required, among other elements, as a basis for decision making on the co-financing of major projects included in operational programmes (OPs) of the European Regional Development Fund (ERDF) and the Cohesion Fund. Those funds are critical for financing long-term infrastructure projects. In conclusion, the analysis described in this article allows for the selection of projects not only on the basis of the best estimation, but also based on the risk associated with it, simply by weighing the performance with the risk [13].The expected performance, and not the modal one, is the value that should be reported in the application form for major projects requiring EU assistance, whenever a probabilistic risk analysis is carried out. In order to evaluate the result, one very important aspect is the compromise to be made between high-risk projects with high social benefits on the one hand, and low-risk projects with low social benefits on the other.

5. References

[1] Buchmeister, B.; Pandza, K.; Kremljak, Z. & Polajnar, A. (2004). Possibilities of practical use of risk management, Proceedings of 3rd International scientific conference Business systems management, University of Mostar

[2] Kremljak, Z. & Buchmeister, B. (2006). Uncertainty and development of capabilities, DAAAM International Publishing, Vienna

[3] http://ec.europa.eu/regional_policy/sources/docgener/studies/pdf/cba_guide.pdf(2016). European commission, Guide for CBA. Accessed on: 2016-09-13

[4] Kremljak, Z.(2011). Risk Management Methods, Chapter 10 in DAAAM International Scientific Book 2011, pp. 119-132, B. Katalinic (Ed.), Published by DAAAM International, ISBN 978-3901509-84-1, ISSN 1726-9687, Vienna, Austria

[5] Dvorak, J.: (2011) Risk Management of Technical Products in Their Life Cycle, Sbornik prispevku - Soutezni prehlidka studentskych a doktorskych praci FST 2011. Eds.: P. Zitek, ISBN 978-80-7043-995-1, Pilsen

[6] Cechova, L;Simon, M.(2011). Annals of DAAAM for 2011 & Proceedings of the 22nd International DAAAM Symposium, Volume 22, No. 1, ISSN 1726-9679 ISBN 978-3-901509-83-4

[7] Kafol, C. (2016). Project Risk Management Methodology in Practice, Proceedings of the 26th DAAAM International Symposium, pp.0445-0452, B. Katalinic (Ed.), Published by DAAAM International, ISBN 978-3-902734-07-5, ISSN 1726-9679, Vienna, Austria

[8] Kostelac, D.; Matrljan, D. & Dobovicek, S. (2011): Relationship between processes and project management, Annals of DAAAM for 2011 & Proceedings of the 22nd International DAAAM Symposium, Volume 22, No. 1, ISSN 1726-9679, ISBN 978-3-901509-83-4, Editor B. Katalinic, Published by DAAAM International, Vienna, Austria, EU, 2011

[9] Avram, E.L., Savu, L.D., Avram, C. Investment decision making with economic and mathematic model. 2010. Annals of DAAAM and Proceedings of the International DAAAM Symposium, pp. 343-344.

[10] Spicar, M.; Januska, M.(2014).Use of Monte Carlo Modified Markov Chains in Capacity Planning. 25th DAAAM International Symposium on Intelligent Manufacturing and Automation, DAAAM 2014

[11] Hromada, M., The European Critical Infrastructure Operator Duties, In: Security Magazín, Číslo 95, 2010, ISBN–12108723

[12] JASPERS (2013), Staff Working Papers – Project Preparation and CBA of RDI Infrastructure Projects, JASPERS Knowledge Economy and Energy Division. Accessed on: 2016-09-13; Available at: http://www.jaspersnetwork.org/plugins/servlet/documentRepository/ displayDocumentDetails?documentId=184

[13] ESFRI (2011), ESFRI Evaluation Report 2011. Accessed on: 2016-09-13; Available at: http://ec.europa.eu/research/infrastructures/pdf/esfri_evaluation_ report_2011.pdf

DOI: 10.2507/27th.daaam.proceedings.009

LIGHTING QUALITY, PRODUCTIVITY AND HUMAN HEALTH

Ruzena Kralikova & Emil Wessely

This Publication has to be referred as: Kralikova, R[uzena] & Wessely, E[mil] (2016). Lighting Quality, Productivity and Human Health, Proceedings of the 27th DAAAM International Symposium, pp.0059-0065, B. Katalinic (Ed.), Published by DAAAM International, ISBN 978-3-902734-08-2, ISSN 1726-9679, Vienna, Austria
DOI: 10.2507/27th.daaam.proceedings.009

Abstract

Measuring the increase in productivity due to the influence of lighting is not a simple task. In a production plant, information about productivity is an important indicator of performance. This paper describes an interdisciplinary approach of experts from the fields of psychology, ergonomics, medicine and lighting technology jointly applying a set of evaluation instruments with goal to describe and analyse the effects on well-being and stress, motivation and sleeping quality, the capacity to regenerate after work, as well as the quality of the lighting situation could be useful.

Keywords: lighting quality; illuminance; productivity; working environment; visual conditions.

1. Introduction

Good lighting in the workplace with well-lit task areas is essential for optimising visual performance, visual comfort and ambience, especially with an ageing workforce. And the beneficial effects of good lighting extend much further than it was originally thought. In fact, over the last two decades, medical science has consistently shown that light has a positive influence on health and wellbeing. With better lighting, performance can be improved by speeding up tasks and reducing failure rates and when added to the energy saving aspects. Sustainable lighting can be a powerful tool to improve business also [1,5].

Studies have shown that the body's hormone levels rise and fall during the day in response to the availability of natural light [1,3,6,7]. Increasing light levels make people feel energised and active but as daylight intensity diminishes, the opposite is true. Light has a profound effect on our physical and emotional health [1,17]. Good lighting solutions delivers the right level of light at different times of the day and enhances people's sense of wellbeing, improving concentration, motivation and performance. Suitable lighting at the workplace, with appropriate lighting of places of work tasks, is a prerequisite for optimizing visual performance, visual comfort and comfort of working environment, especially in the context of an aging workforce and the positive effects of adequate lighting can be extended even further [24,25].

In a production plant, information about productivity is an important indicator of performance. Measuring the increase in productivity due to the influence of lighting is not a simple task. Where an increase is observed, this may often be due to a variety of influences. In the sphere of lighting, there are many possibilities to create good visual conditions, even through static lighting meeting standard requirements as a minimum. For dynamic brightness sequences, it is possible to modify the brightness or colour temperature of the lighting. The effect on the workers may be assessed through various interviews, questionnaire surveys and measurements [2, 7].

2. Lighting quality defining

Lighting quality is depending on several factors. It depends largely on people's expectations and past experiences of electric lighting. Those who experience elementary electric lighting for the first time, for example in remote villages in developing countries, have different expectations and attitudes towards lighting from office workers in industrialized countries. There are also large individual differences in what is considered comfortable lighting, as well as cultural differences between different regions. Lighting quality is much more than just providing an appropriate quantity of light. Other factors that are potential contributors to lighting quality include e.g. illuminance uniformity, luminance distributions, light colour characteristics and glare.

There are many physical and physiological factors that can influence the perception of lighting quality. Lighting quality cannot be expressed simply in terms of photometric measures nor can there be a single universally applicable recipe for good quality lighting [8, 9]. Light quality can be judged according to the level of visual comfort and performance required for our activities. This is the visual aspect. It can also be assessed on the basis of the pleasantness of the visual environment and its adaptation to the type of room and activity. This is the psychological aspect. There are also long term effects of light on our health, which are related either to the strain on our eyes caused by poor lighting (again, this is a visual aspect), or to non-visual aspects related to the effects of light on the human circadian system [4].

A number of different approaches have been suggested to define lighting quality. The definition that seems most generally applicable is that lighting quality is given by the extent to which the installation meets the objectives and constraints set by the client and the designer. In this way lighting quality is related to objectives like enhancing performance of relevant tasks, creating specific impressions, generating desired pattern of behaviour and ensuring visual comfort [10].

The constraints may be set by the available financial budgets and resources, set time-lines for completing the project and possible predetermined practices and design approaches that need to be followed. Lighting quality means achieving an optimum balance among *human needs* (visibility, task performance, visual comfort, health, safety, well-being, social communication, aesthetic...), *architectural considerations* (daylighting, standards, safety and security, composition, style) and *economics, energy efficiency, environment* (installation, energy saving, operation, maintenance), Figure 1 [25].

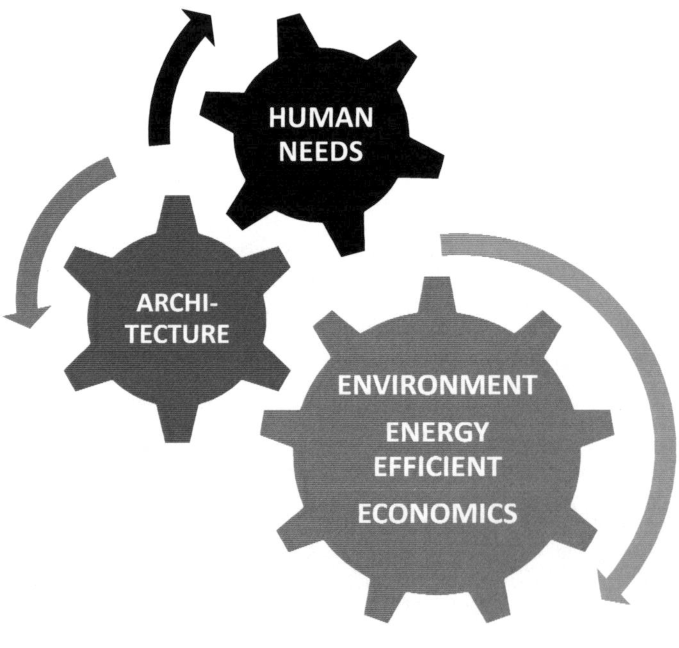

Fig.1. Lighting quality

3. Luminous environment and human performance

Light has effects that are fully or partly separated from the visual system and biological effects of light, called also the non-visual and they influence on the well-being, safety work, productivity, health, social communication and aesthetic judgement.

3.1 Visual aspects of the lighting

The influence of artificial lighting on the psychophysiological well-being and the productivity of workers in a production plant depends on kind of lighting source that with respect to[4]:

- brightness - illuminance,
- light colour - spectrum and
- light distribution - luminance distribution in the room.

The effects of artificial lighting on people are more pronounced, that natural influence of daylight inside working hours is very low especially during winter season. One of the major aspects of the lighting practice and recommendations is to provide adequate lighting for people to carry out their visual tasks. Ensuring adequate and appropriate light levels - quantity of light is only an elementary step in creating comfortable and good-quality luminous and visual environments. Lighting that is adequate for visual tasks and does not cause visual discomfort is not necessarily good-quality lighting. Depending on the specific application and case, both insufficient lighting or too much light can lead to bad-quality lighting. There are a number of lighting-related factors that may cause visual discomfort [13]. The current indoor lighting recommendations give ranges of illuminance values for different types of rooms and activities [10].

In addition, guidelines on light distribution in a space, the limitation of glare, and the light colour characteristics are given. The colour characteristics of light in space are determined by the spectral power distribution (SPD) of the light source and the reflectance properties of the surfaces in the room [16]. The colour of light sources is usually described by two properties, namely the correlated colour temperature (CCT) and general colour rendering index (CRI) [21]. The colour appearance of a light source is evaluated by its correlated colour temperature (CCT). For example, incandescent lamps with CCT of 2700 K have a yellowish colour appearance and their light is described as warm. Certain type of fluorescent lamps or white LEDs have CCT of around 6000 K with bluish appearance and light described as cool. The CRI of the CIE measures how well a given light source renders a set of test colours relative to a reference source of the same correlated colour temperature as the light source in question of International Commission on Illumination (CIE). The general CRI is calculated as the average of special CRIs for eight test colours. The reference light source is Planckian radiator (incandescent type source) for light sources with CCT below 5000 K and a form of a daylight source for light sources with CCT above 5000 K. The higher the general CRI, the better is the colour rendering if a light source, the maximum value being 100. The Light source spectrum, i.e. radiant power distribution over the visible wavelengths, determines the light colour characteristics. Examples of spectra of the lamps are shown on Figure 2 - 4 [24].

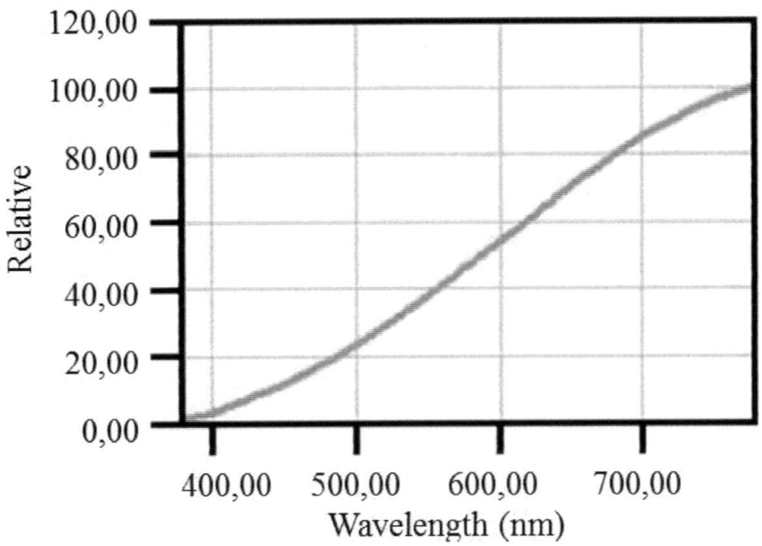

Fig. 2. Light source spectrum - Example of spectra of an incandescent lamp (CCT= 2690 K, CRI= 99)

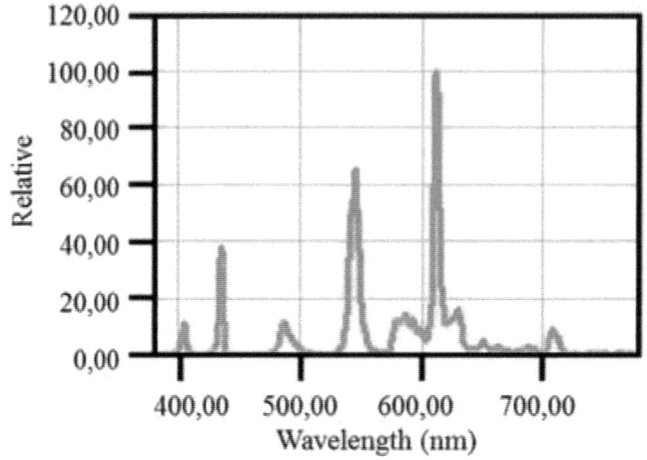

Fig. 3. Light source spectrum - example of spectra of a compact fluorescent lamp (CCT= 2780 K, CRI = 83)

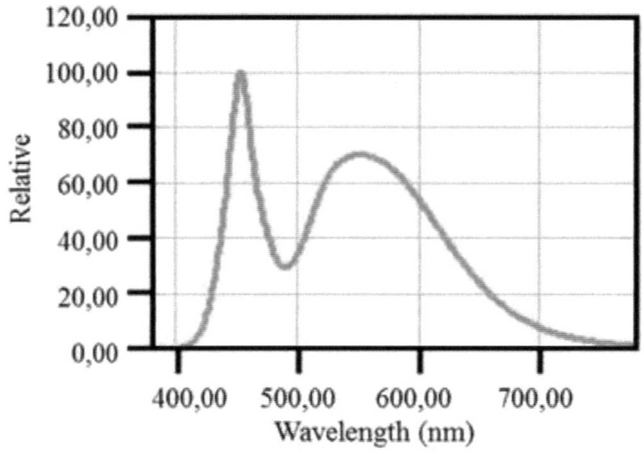

Fig. 4. Light source spectrum - example of spectra of a white LED lamp (CCT= 6010 K, CRI = 78)

3.2 Lighting standards

A completely uniform space is usually undesirable whereas too nonuniform lighting may cause distraction and discomfort. Lighting standards and codes usually provide recommended illuminance ratios between the task area and its surroundings according EN 12 464-1 [21]. Most indoor lighting design is based on providing levels of illuminances while the visual system deals with light reflected from surfaces i.e. luminances. For lighting there are recommended luminance ratios between the task and its immediate surroundings. Room surface reflectances are an important part of a lighting system and affect both the uniformity and energy usage of lighting. Compared to a conventional uniform lighting installation with fluorescent lamps, LEDs provide opportunities to concentrate light more on actual working areas and to have light where it is actually needed. This provides opportunities to increase the energy efficiency of lighting in the practise [19].

3.3 Psychological aspects of space lighting

Several researchers have investigated psychological aspects of space lighting [3,17]. The luminous environment can be perceived in many ways e.g. as more or less agreeable, more or less attractive, more or less appropriate to the function of the space, more or less highlighting the company image. Variations of luminances and colours can strengthen attractiveness, trigger emotions, and affect worker's mood, the impact of lighting depends much on the individuals and their state of mind. Unacceptable lighting conditions may impact on task performance and thus productivity through motivation. A lighting installation that does not meet the user's expectations can be considered unacceptable even if it provides the conditions for adequate visual performance. People perceive their luminous environment through their eyes [17].

3.4 Biological effects of light

Light has also effects that are fully or partly separated from the visual system. Biological effects of light are called also the non-visual effects of light, related to the human circadian photoreception. The biological effects of light and their effects on human performance are not yet very well known. Research work is needed to generate an improved understanding of the interaction of the effects of different aspects of lighting on behavioural visual tasks and cortical responses and on how the biological effects of lighting could be related to these responses [17].

The biological effects of light and their effects on human performance are not yet very well known. Lighting should be designed to provide people with the right visual conditions that help them to perform visual tasks efficiently, safely and comfortably. The luminous environment acts through a chain of mechanisms on human physiological and psychological factors, which further influence human performance and productivity, Fig.5 [6].

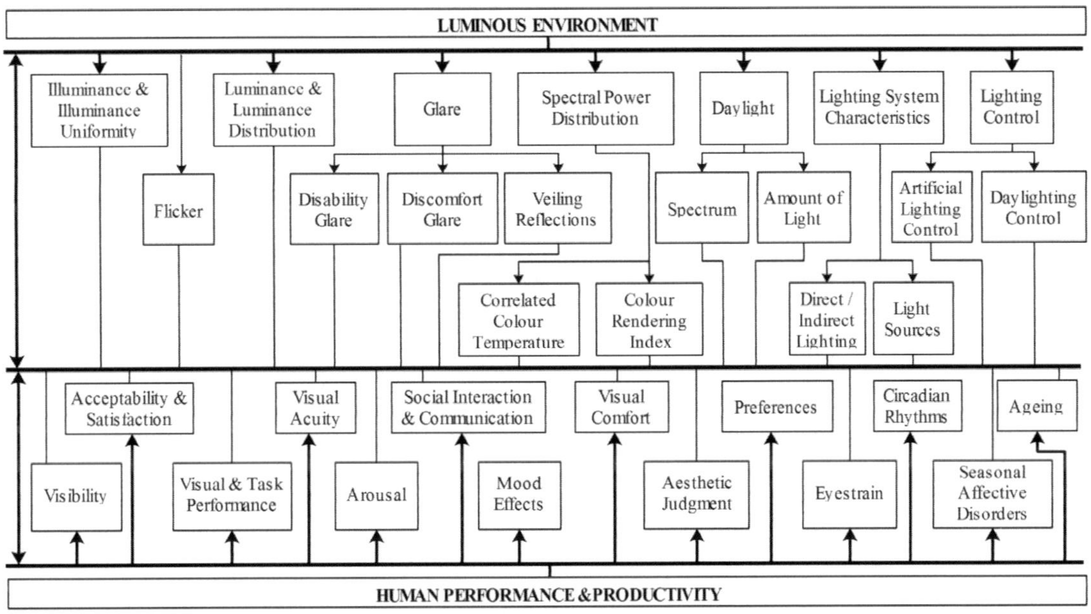

Fig. 5. Luminous environment and human performance

4. The effect of lighting on productivity

Lighting should be designed to provide people with the right visual conditions that help them to perform visual tasks efficiently, safely and comfortably [8]. The luminous environment acts through a chain of mechanisms on human physiological and psychological factors, which further influence human performance and productivity [6].

The effect of lighting on productivity is ambiguous. The difficulty in finding the relations between lighting and productivity is that there are several other factors that simultaneously affect human performance. These factors include motivation, relationships between workers and the management and the degree of having personal control to the working conditions [11,22]. With appropriate lighting the ability to perform visual tasks can be improved and visual discomfort can be avoided. This can provide conditions for better visual and task performance and, ultimately, productivity.

The difficulty of field studies in working environments is the degree of experimental control required. Several studies have investigated the effect of increase in illuminance on task performance. However, illuminance is only one of the many aspects in the lighting conditions. In making changes to lighting, which lighting aspects are changed (e.g. illuminance, spectrum, and luminance distribution) and whether there are other factors that are simultaneously changed in the working conditions (e.g. working arrangements, people, supervision of work) need to be controlled and analysed. Recently, several studies are investigating the effects of light spectrum on human performance and the possibilities to use blue-enriched light to improve human performance through the non-visual effects of light. Poor lighting conditions can easily result in losses in productivity of employees and the resulting production costs of the employer can be much higher than the annual ownership cost of lighting.

5. Discussion

The effects of lighting conditions on productivity have been published in the several studies the earliest studies indicated that lighting conditions can improve performance by providing adequate illuminance for the visual tasks. Their results are sometimes contradictory. For example, a study in office work indicated that an increase in illuminance from 500 lx to 1500 lx could increase the performance of office workers by 9%, while another study showed that lower illuminance levels (150 lx) tended to improve performance of a complex word categorisation task as compared to a higher level (1500 lx) [1, 5, 11].

A field study in industrial environment measured direct productivity increases in the range from 0 to 9 % due to changes in lighting. The literature shows some examples of null results than clear-cut effects of illuminance on task performance, over a wide range of illuminance levels and for a variety of complex and simple tasks in working environment [5].

6. Conclusion

Observing the effect of light on human beings and health is a topic that is addressed by various sciences such as psychology, medicine, ergonomics and lighting technology. The interdisciplinary application of methods shows that measuring the effect of lighting on human beings is complex. If no results are obtained, that does not mean that there is no effect, but that the instruments of observation and measurement have not been optimally used. Further efforts are required to improve the measuring methods, instruments and evaluations.

Here are a few guidelines how can to create a setting that increases the satisfaction and productivity:

- Take the biological aspect of lighting into account. Light synchronises the circadian rhythm and it influences to sleep-wake cycle and this has a big influence on people health and productivity.
- Adapt the lighting to the daytime and season and to the individual needs of the user.
- Provide adequate lighting for people to carry out their visual tasks, create a sense of space and depth gives a visual comfort and adds a pleasant atmosphere to the working environment.

From research studies Good lighting creates pleasant working conditions in a functional and atmospheric way. Both daylight and artificial light play a key role in our health and wellbeing. Satisfaction and efficiency will improve if will enough daylight intake and a well-lit working space. Artificial lighting that does not just help you see, but also meets your biological lighting needs, is the future of work space lighting. Productivity, visual comfort and employees health will be positively affected by well-designed illumination.

7. Acknowledgments

The research in this paper was supported by a grant „ Research of influence of chosen parameters of working environment on working power and productivity" VEGA – No 1/0537/2015 supported by Slovak Research and Development Agency (SRDA).

8. References:

[1] Boyce, P. R. (2004). Lighting research for interiors: the beginning of the end or the end of the beginning. In: Lighting Research and Technology.Vol. 36,No.4, 2004), pp.283-293.

[2] Chin Kim Gan; Ahmad Farid Sapar; Yik Chee Mun & Kuan Eng Chong (2013). Techno-Economic Analysis of LED Lighting (2013): A Case Study in UTeM's Faculty Building, Procedia Engineering, Vol. 53, 2013, pp. 208-216, ISSN 1877-7058.

[3] Durak, A. ; Olguntürk, N. C.; Yener, C.; Güvenç, D. & Gürçınar, Y. (2007). Impact of lighting arrangements and illuminances on different impressions of a room. Building and Environment, 42(10), 3476-3482.

[4] Frumkin, H. (2003). Healthy places: exploring the evidence. In: American journal of public health , Vol. 93,No.9, 2003, pp.1451-1456.

[5] Gligor, A.; Grif, H.; & Oltean, S. (2006). Considerations on an intelligent buildings management system for an optimized energy consumption. In: Automation, Quality and Testing, Robotics, IEEE International Conference on. IEEE, 2006, pp. 280-284.

[6] Gligor V. (2004). Luminous Environment and Productivity at Workplaces. Thesis, 2004. Espoo:Helsinki University of Technology.

[7] Hughes, P. C. & McNelis, J. F. (1978). Lighting, productivity, and the work environment. Lighting Design and Application, Vol. 8, No.12, 1978, pp.32–38.

[8] Kadir Amasyali & Nora El-Gohary (2016). Building Lighting Energy Consumption Prediction for Supporting Energy Data Analytics, Procedia Engineering, Vol. 145, 2016, pp. 511-517, ISSN 1877-7058.

[9] Kõrbe Kaare,K. & Tauno O.(2015). Smart Health Care Monitoring Technologies to Improve Employee Performance in Manufacturing, Procedia Engineering vol.100.2015, pp. 826 – 833.

[10] Leppiman, A.; Kotka, T.; Kõrbe Kaare, K. & Koppel, O.(2015). Decision-making framework for industrial-size datacenters. Proceedings of the International Conference of DAAAM Baltic "Industrial Engineering", 2015-January, pp. 237-242.

[11] Manlig, F.; Šlaichová, E.; Koblasa, F. & Vavruška, J. (2014). Innovation of business processes by means of computer-aided simulation Applied Mechanics and Materials, 474, pp. 67-72.

[12] Micieta, B. & Binasova, V. (2013). Defining requirements for energy efficiency in manufacturing. DAAAM International Scientific Book 2013, pp. 887-894. ISSN 1726-9687.

[13] Micieta, B.; Markovic, J. & Binasova, V. (2016). Advances in sustainable energy efficient manufacturing system, MM Science Journal, 2016-June, pp. 918-926.

[14] Petek, J.; Glavič, P. & Kostevšek, A. (2016). Comprehensive approach to increase energy efficiency based on versatile industrial practices, Journal of Cleaner Production, Vol. 112, 2016, pp. 2813-2821.

[15] Perlova,E.; Platonova,M.; Gorshkov, A. & Rakova, X. (2015). Concept Project of Zero Energy Building, Procedia Engineering. Vol. 100. 2015. pp. 1505 – 1514.

[16] Varga, G.; Szigeti, F. & Desző, G. (2015). Examination of surface roughness of burnished workpieces In: Scientific Bulletin: Mechanics, Tribology, Machine Manufacturing Technology. Vol. 29, 2015, pp. 94-98, ISSN: 1224-3264.

[17] Veitch, J.A. (2001). Psychological processes influencing lighting quality. In: Journal of the Illuminating Engineering Society, Vol. 30, No. 1, 2001, pp.124-140.

[18] Murgul V.; Vatin N. & Zayats, I. (2015). The Role of the Solar Light Quantity in the Architectural Forming of Buildings, Procedia Engineering, Vol. 117, 2015, pp. 819-824, ISSN 1877-7058.

[19] Vuorio, J.; Nikkilä, V.; Teivastenaho, V.; Peltola, J.; Partanen, J.; Kiviluoma, P. & Kuosmanen P. (2014). Additive Manufacturing with UV Light Cured Resin, 9th International DAAAM Baltic Conference "INDUSTRIAL ENGINEERING" 24-26 April 2014, Tallinn, Estonia, pp. 317-322.

[20] Wei Li, Jinzhong Zhu & Zhimin Zhu (2012). The Energy-saving Benefit Evaluation Methods of the Grid Construction Project Based on Life Cycle Cost Theory, Energy Procedia. Vol.17(A), 2012. pp. 227–232.

[21] EN 12464-1, Light and lighting – Lighting at work places – Part I: Indoor Work Places, European Standard.

[22] European Directive 2006/25/EC, 2006. On the minimum health and safety requirements regarding the exposure of workers to risks arising from physical agents (artificial optical radiation). Official Journal of the European Union, L 114, 38-59

[23] http://www.aivc.org/sites/default/files/members_area/medias/pdf/Conf/2002/109_076%20Halonen.pdf

[24] The Lighting Handbook, available: http://www.zumtobel.com/PDB/teaser/EN/lichthandbuch.pdf

[25] https://www.iald.org/Advocacy/Advocacy/Quality-of-Light

DOI: 10.2507/27th.daaam.proceedings.010

METHODICAL PROCEDURE FOR DESIGNING OF EFFICIENT LIGHTING SYSTEMS

Ruzena Kralikova & Emil Wessely

This Publication has to be referred as: Kralikova, R[uzena] & Wessely, E[mil] (2016). Methodical Procedure for Designing of Efficient Lighting Systems, Proceedings of the 27th DAAAM International Symposium, pp.0066-0073, B. Katalinic (Ed.), Published by DAAAM International, ISBN 978-3-902734-08-2, ISSN 1726-9679, Vienna, Austria
DOI: 10.2507/27th.daaam.proceedings.010

Abstract

Based on the lighting systems efficient utilization strategies analysis of lighting systems efficient utilization and through its synthesis with the options of intelligent control were for a design and calculation of lighting systems analyzes of the reference shop energy-saving lighting system designs were carried out. The results of the rationalization procedures and measures that have been made for the purpose of a reduction of energy demand of industrial operations lighting systems, were the basis for a development of a methodical procedure of designing of energy-efficient lighting systems in the sphere of engineering shops. The methodology presented in this paper is drawn up on the basis of valid legislation of the assessment and evaluation of lighting systems in a working environment with regard to ensuring of a visual comfort and creation of requested conditions for a visual activity, taking into account its energy efficiency.

Keywords: Methodology; energy; illumination; evaluation; design.

1. Introduction

Lighting is the way to a greener future. Sustainability meeting our needs today without compromising the resources future generations will need touches everything we do as a company. It's all about reducing environmental impact, a commitment that goes beyond our products to include how the products are used, how they are manufactured and how they are distributed. It even extends to the design of "greener" buildings. "Green building" incorporates environmentally responsible practices into the design of new construction and the renovation of existing buildings. Green facilities aren't just healthier for people and the planet. Using only as much light as is needed when it is needed is key to reducing energy consumption [7].

Characteristics of industry are large and high spaces, fixed working positions, long work periods, adverse conditions, various levels of illuminance, high bay lighting, central lighting control, restricted maintenance opportunities etc. It is one of the reason, why lighting accounts on average for about 15 % of total electricity used in industry [11].

Recent developments in lighting technology combined with planned lighting control strategies can result in very significant cost savings, typically in the range of a third to a half of the electricity traditionally used for lighting [14,17]. Up to 75 % of a conventional lighting cost is wasted cost - money spent on light that simply isn't required. An intelligent lighting system applies the correct level of light when needed to help meet carbon emissions targets, legislation requirements and dramatically reduce expenditure, whilst improving efficiency, comfort and safety [2], [16]. Energy-saving lighting for industry can reduce the impact of CO_2 emissions on the environment, and save energy costs, as well as total costs and can ultimately contribute to improving of the quality and productivity of work [13].

Lighting can have other significance than just to light a dark room. It can improve the functions and the design of environment, improve occupational safety and create flexible spaces that are adapting themselves to the tasks according to the type of the carried out activity. Good lighting solutions delivers the right level of light at different times of the day and enhances people's sense of wellbeing, improving concentration, motivation and performance. Nowadays, with rising energy prices and tightening of legislation in the field of environment, energy-saving lighting can help businesses to achieve the objectives of sustainability [12].

2. Methodical Procedure

The methodology for designing of efficient lighting systems is drawn up on the basis of valid legislation of the assessment and evaluation of lighting systems in a working environment with regard to ensuring of a visual comfort and creation of requested conditions for a visual activity, taking into account its energy efficiency. The main parts of this methodology are divided into individual branches, by putting together of which, a sequence of steps and conclusions arise, which the methodology was developed for it [3].

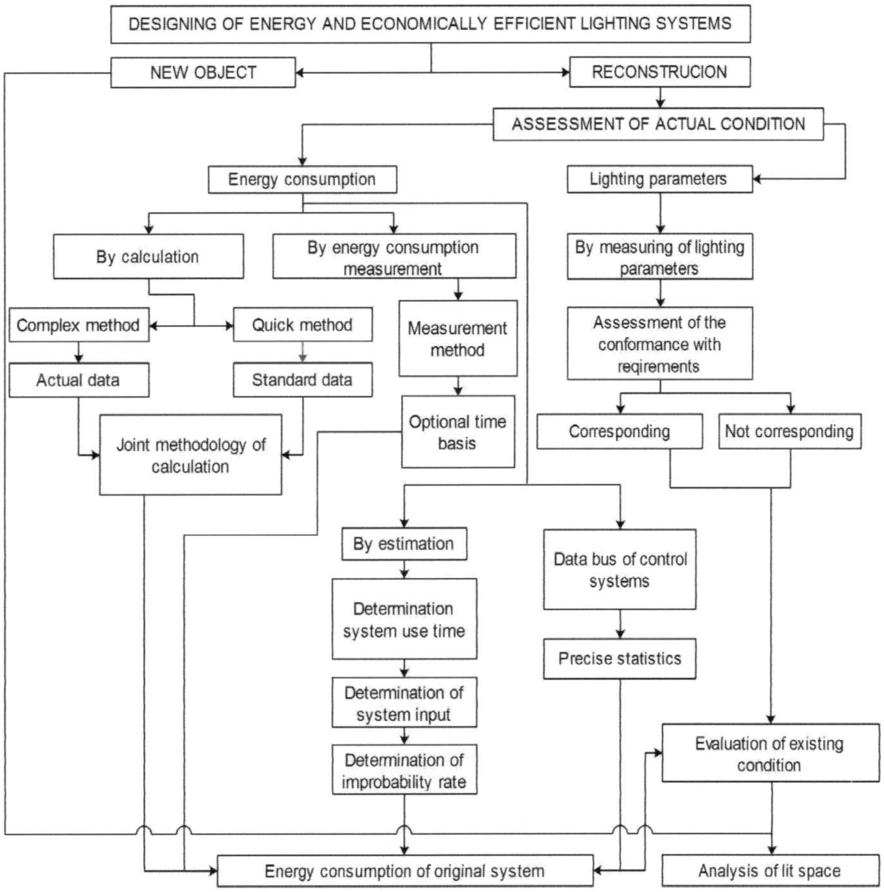

Fig.1. Diagram of assessing the current state of the objects it illuminates

The first step (Fig. 1) in the procedure of designing of energy and economically efficient lighting systems, should take into account the information if it is a newly implemented project or a reconstruction of a lighting system. In the case of a reconstruction, it is necessary to assess the actual condition of lighting in operation, from the point of view of energy as well as quality. It is needed to assess, whether the original lighting system meet legislative and normative requirements for the given space type and activity performed.

For the assessment of energy demanding character, several methods can be applied. Either the methods of calculation established by STN EN 15 193:2008 Energy performances of buildings [20]. Energy requirements for lighting, or it will be determined by energy consumption measurement or alternatively, by an estimation, where, however, it is needed to define the improbabilities of a result. The most accurate manner, how to determine the lighting energy consumption in industrial installations, is the utilization of collected data implemented in the systems of a lighting control [10].

2.1 Analyse of the evaluation illuminated space

Upon the current status evaluation, the next step in the process of an ecologisation of lighting systems is the analysis of lit space (Fig. 2), where in addition to dimensional and construction characteristics of a room, a classification of the assessed space into a correct workplace type will be made based on legal and normative regulations [3,4]. The correct classification of workplace by a type of operations performed plays an important role in the development of designs of energy-saving lighting systems, since the illuminance values, which the lighting system is to be dimensioned to, are based just on these documents [8].

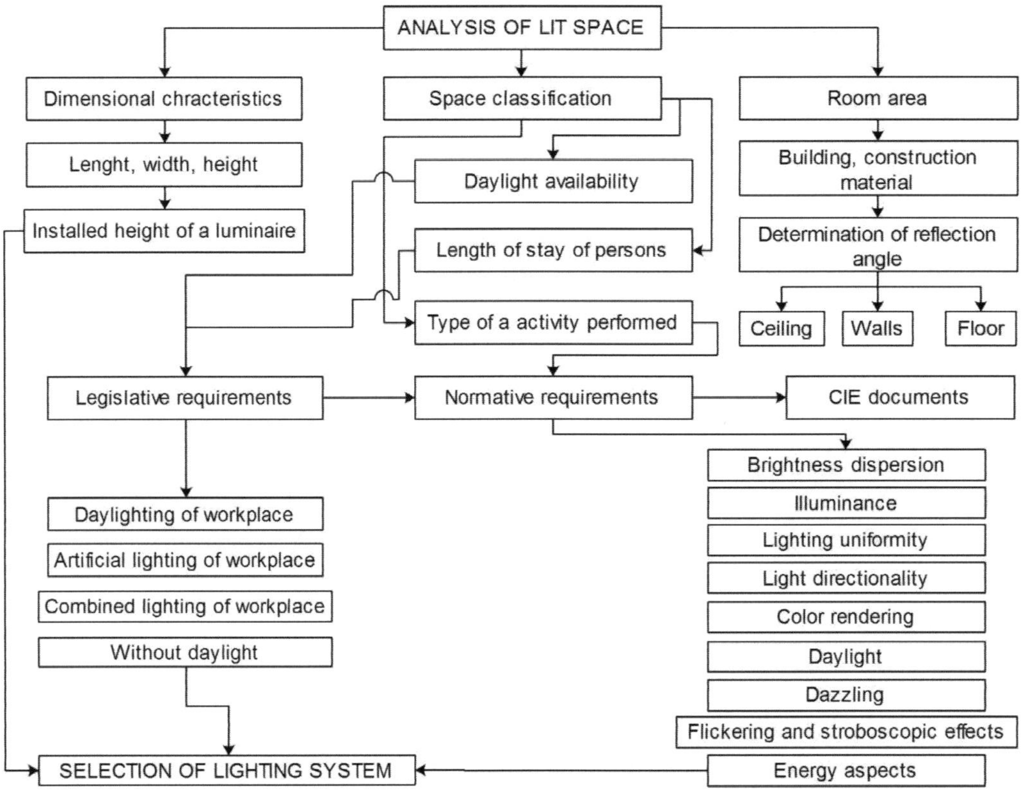

Fig. 2. Diagram of analyses evaluation illuminated space

2.2 Choice of luminaire, technical devices and accessories

The performance of a luminaire should be considered just as carefully as its cost. In the long term a well designed, well constructed luminaire will be cheaper than a poor quality unit; and the salient features of a good quality luminaire are [5]:

- Adequate screening of high luminance lamps to minimise discomfort and glare
- High light output ratio with the appropriate light distribution
- Ease of installation, cleaning and maintenance

By gradual defining of requirements related to a lighting system, taking occupational safety and health protection of employees into account, the selection of a lighting system was proposed as the next step, within the framework of the methodology (Fig. 3). Within this point, a careful analysis of parameters of technical devices of the lighting system (lamps, luminaires, ballasts, etc.) is needed, since just a suitable selection of technical devices and based on their parameters, it is possible to achieve the considerable energy savings in the process of the lighting systems ecologisation, while ensuring a maximum occupational safety and health protection of employees. A due attention is to be paid just to this step in the whole process of designing of energy-saving lighting systems.

The resulting material for further development of the lighting ecologisation project contains optional solutions of the technical devices the parameters of which were thoroughly compared and the most efficient of them were then selected.

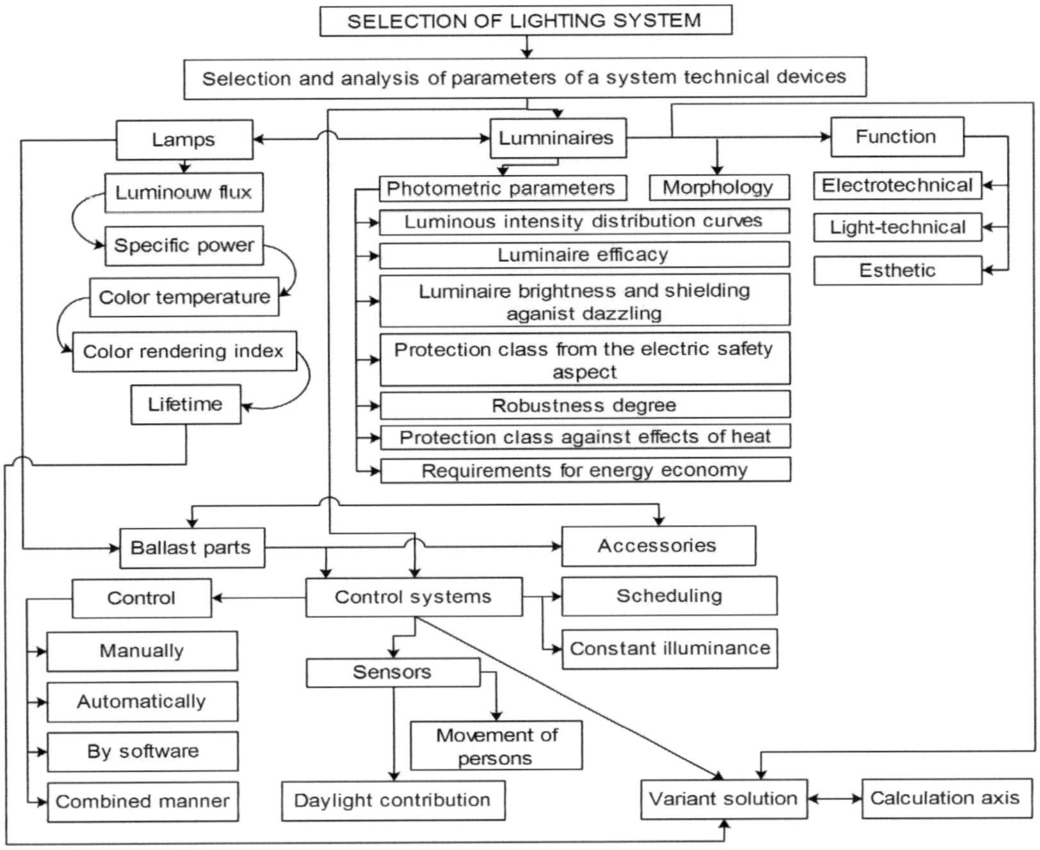

Fig. 3. Diagram choosing the media lighting system

Once the system of lighting has been selected, the most appropriate light source can then be chosen followed by the luminaire.

The following attributes should be studied when choosing the light source [19,21]:
- Light output (lumens)
- Total input wattage
- Efficacy (lumens per Watt)
- Lifetime
- Physical size
- Surface brightness/glare
- Colour characteristics
- Electrical characteristics
- Requirement for control gear
- Compatibility with existing electrical system
- Suitability for the operating environment

A number of factors also affect luminaire choice:
- Characteristics of the light source and control gear
- Luminaire efficiency (% lamp light output transmitted out of the fixture)
- Light distribution
- Glare control
- Finish and appearance
- Size
- Accessibility of components for maintenance
- Ability to handle adverse operating conditions
- Aesthetics
- Thermal management.

2.3 The calculation of lighting system

Upon defining of the characteristics of the system technical devices, for further advance in the ecologisation process, it is needed to choose an appropriate tool for a calculation of lighting system (Fig. 4). It is possible to carry out the calculation by various calculation methods or by designing and calculating computer tools intended for the elaboration of the lighting systems design and calculation. In this step, it is important to choose an appropriate type of the calculation and to take all circumstances related to the lighting system into account [1]. Very important parameter that is to be taken into consideration, is the maintaining factor value that is directly connected with a proposal of the maintenance schedule of the lighting system. The determination of a correct value of the maintenance factor has a large influence on total economy of the lighting system during the period of its utilization [8]. The applications of correct calculations and proposal results, in this stage of designing, in elaborated optional solutions which comply with the occupational and health protection requirements and, at the same time, are energy-efficient and economically profitable.

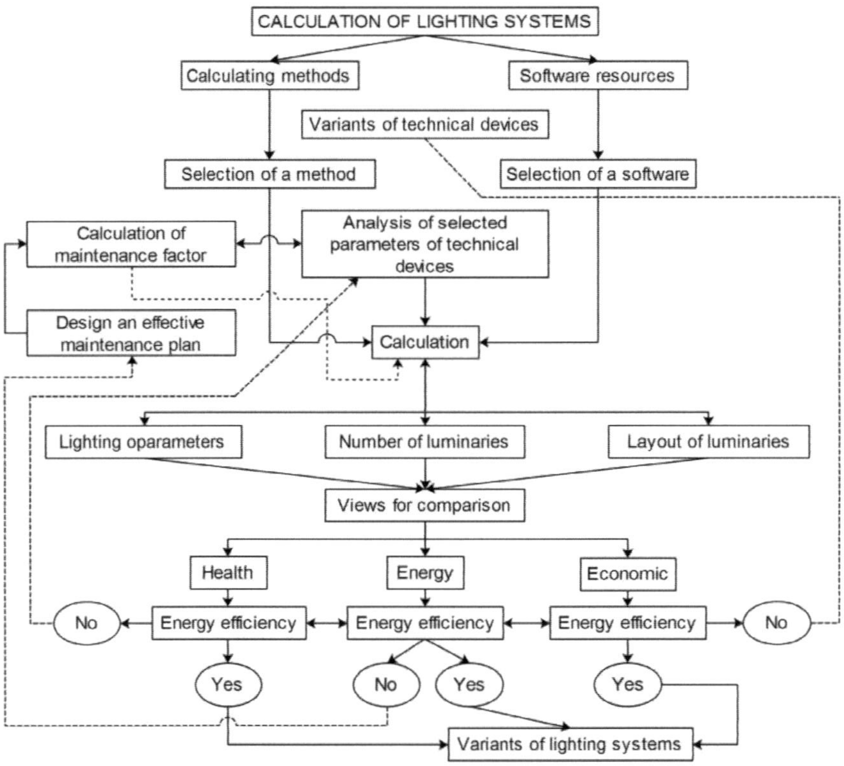

Fig.4. Diagram calculations of lighting systems

Lighting calculation methods fall into three broad categories:

1. Manual calculation methods
2. Three dimensional modelling
3. Visualisation

The advent of computer modelling enables a more flexible approach to lighting design and significantly increasing the information available to the designers. There are a wide range of computation methods for the calculation of different lighting aspects. These include complex methods for calculating the illuminance from a wide variety of shapes of luminous objects. The majority of these have now been superseded by computer programs. Photometric data for light sources and luminaires is commercially available to contribute to these calculations.

It is important in using a lighting calculation program that the output records the type of luminaire used, the location of the luminaires, the assumed lumen output of the lamp, the light loss factor and the aiming points. If this is not recorded you have only a picture of the installation and no way of making it a reality.

For visualisation exist more programs that create a perspective rendering of the space in levels of detail that vary from a block representation of the space, to photographic quality renderings, depending on the sophistication of the program and the level of detail of the interior are entered. The programs fall into two basic types:

- Flux transfer or radiosity calculations
- Ray tracing calculations

Visualisation programs are a useful tool in the presentation of a design, as a tool for the designer to check that the design is consistent with his own visualisation of the space, and to model specific lighting solutions. The programs are mainly calculation tools and can show the designer how a specific design will look to assess the acceptability of a design. The output should include:

- Installation information – the type and location of all luminaires and the aiming information. The lamp details should be included, the specific catalogue number of photometric file etc.
- Light technical parameters – the illuminance, uniformity and other parameters that have been calculated to achieve the design.
- Verification information – adequate details to enable the lighting calculation to be verified. This should include the luminaire type, the photometric file, surface reflectances that were assumed, light loss factors, lumen output of lamps and mounting and aiming locations.

2.4 The procedure for selecting the optimal alternative solutions

In the last stage of the lighting systems designing process (Fig. 5), the elements of intelligent control systems are implemented into the optional proposals, which offer considerable energy sources savings based on the implementation of sensors and daylight contribution, human motion sensors, time regimes introduced or by a creation of the luminaire groups and therefore it is important, considering the given type of the workplace, to assess all possibilities of the lighting control and regulation in the installation [11].

In the final stage of the lighting systems ecologisation, particular options are compared from the point of view of energy efficiency and total investment and operational costs. In the case of a lighting system reconstruction, a comparison of the initial status and proposed option will be elaborated, taking energy savings and related financial savings, as well as attained lighting parameters of the system into account.

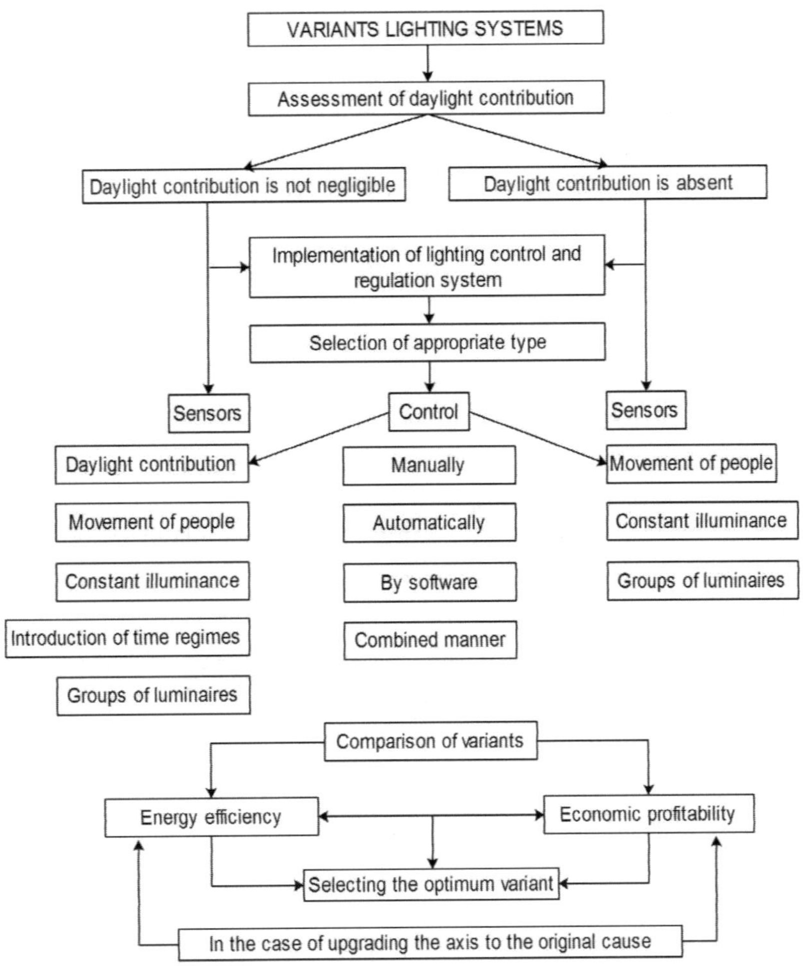

Fig. 5. Diagram selected optimum variants

The effectiveness and efficiency of any lighting installation is affected as much by the control system as by the light sources and fixtures chosen [12]:

- Providing multiple switches to control the number of lights that come on at any one time. Using one switch to turn on all the lights in a large room is very inefficient.
- Placing switches at the exits from rooms and using two-way switching to encourage lights to be turned off when leaving the room.
- Using 'smart' light switches and fittings which use movement sensors to turn lights on and off automatically. These are useful in rooms used infrequently where lights may be left on by mistake, or for the elderly and disabled.

3. Discussion and conclusion

The presented contribution deals with actual issues, the search for rationalization measures in the energy consumption for lighting. Lighting systems is no small part of the share of total energy consumption, so it is necessary to look for energy savings opportunities in this area. In Slovakia, especially in industrial plants, a large amount of lighting systems that do not meet the requirements of energy efficiency, but what is worse, employees do not ensure an nor the minimum lighting requirements at work, which is the primary function of lighting workplaces. For that reason, is important to take design of lighting systems increased attention. However, not only after aspects of energy efficiency, which will ultimately brings also financial savings for consumers, but mainly with a view to ensuring a safe working environment for employees.

Nowadays 30% of energy used in building is wasted. On a personal, corporate or national level, lighting plays an important role in any energy saving strategy. Therefore, if we would to lower energy consumption we will reduce environmental footprint also. There are many good reasons to use energy-efficient lighting systems:

- Lower operating costs. The long-term savings can quickly repay the capital investment. Additionally, incentives may be available to owners to help offset the financial investment in new energy-efficient systems. These include allowing owners to use the money saved on energy bills to pay for the new system, utility incentive programs and government tax incentive programs.
- Increased property value - upgrading a new or existing facility with energy-efficient systems increases its fair value.
- Energy Attractiveness - energy efficient systems help to lowering operating costs.
- Improved workplace conditions. Many of today's lighting systems deliver more comfortable illumination and provide employees with the ability to select light levels that are suitable and comfortable for specific tasks. In many cases, employees chose lower light levels, which translate into greater energy savings.
- Installing energy-efficient systems provides an opportunity to identify the company as an environmentally conscious member of the community.

Based on a synthesis of theoretical assumptions and practical verification options to streamline lighting systems engineering operations is presented in the paper designing methodical process economic and energy-efficient lighting systems. Processed methodology served as the basis for drawing up. Within this methodology were conducted measurements of parameters of lighting in production areas of engineering establishments, which were found unsatisfactory. After analysing quantitative and qualitative parameters and variants with respect to options to achieve energy savings have been processed variants linked with the steering system and regulation lighting. For the draft to arrive at a solution that provided a reduction of energy consumption by half at double the amount of illuminance values in operation, which is how we assume sufficient results in search of rationalization measures with regard to the reducing energy consumption.

4. Acknowledgments

The research in this paper was supported by a grant „ Research of influence of chosen parameters of working environment on working power and productivity" VEGA – No 1/0537/2015 supported by Slovak Research and Development Agency (SRDA).

5. References

[1] Chin Kim Gan; Ahmad Farid Sapar; Yik Chee Mun & Kuan Eng Chong (2013). Techno-Economic Analysis of LED Lighting (2013). A Case Study in UTeM's Faculty Building, Procedia Engineering, Vol. 53, 2013, pp. 208-216, ISSN 1877-7058.

[2] Csapo, A. & Baranyi, P. (2012). A unified terminology for the structure and semantics of CogInfoCom channels In: Acta Polytechnica Hungarica Vol. 9, No. 1, 2012, pp. 85-105, ISSN: 1785-8860.

[3] Despotovic-Zrakic, M. ; Barac, D.; Bogdanovi Z.; Jovanic B.; Radenkovic B. (2014). Software environment for learning continuous system simulation In: Acta Polytechnica Hungarica Vol. 11, No. 2, 2014, pp. 187-202, ISSN: 1785-8860.

[4] Dulina,L. & Bartanusova, M. (2014). CAVE design using in digital factory, Proceedings of Procedia Engineering: international symposium on intelligent manufacturing and automation, Katalinic, B. (Ed.), DAAAM 2014, Vol. 100, pp. 291-298, ISSN 1877-7058, Vienna

[5] Durak, A.; Olguntürk, N. C.; Yener, C.; Güvenç, D. & Gürçınar, Y. (2007). Impact of lighting arrangements and illuminances on different impressions of a room. Building and Environment, 42(10), 3476-3482.

[6] Gastermann, B. C.; Stopper, M. & Katalinic, B. (2012). Adapting CONWIP Characteristics for Conventional Production Planning. In DAAAM International Scientific Book 2012, pp. 553-564, Editor B. Katalinic, Publisher by DAAAM International, ISBN 978-3-901509-86-5, Vienna, Austria

[7] Kovacic Lukman R.; Glavic, P.; Carpenter A. & Virtič, P. (2016): Sustainable consumption and production – Research, experience, and development – The Europe we want. In: Journal of Cleaner Production . August 2016, DOI: 10.1016/j.jclepro.2016.08.049

[8] Kadir Amasyali & Nora El-Gohary (2016). Building Lighting Energy Consumption Prediction for Supporting Energy Data Analytics, Procedia Engineering, Vol. 145, 2016, pp. 511-517, ISSN 1877-7058.

[9] Leppiman, A.; Kotka, T.; Kõrbe Kaare, K. & Koppel, O.(2015). Decision-making framework for industrial-size datacenters. Proceedings of the International Conference of DAAAM Baltic "Industrial Engineering", 2015-January, pp. 237-242.

[10] Manlig, F.; Šlaichová, E.; Koblasa, F. & Vavruška, J. (2014). Innovation of business processes by means of computer-aided simulation Applied Mechanics and Materials, 474, pp. 67-72.

[11] Micieta, B., Binasova, V. & Kubinec, L.(2015): Information and Communication Technology Enabled Energy Efficiency In: DAAAM International Scientific Book 2015: Chapter 08 P. 087-096 ISBN: 978-3-902734-05-1.

[12] Micieta, B.; Zavodska, L.; Rakyta, M. & Binasova, V. (2015). Sustainable concept for green logistics and energy efficiency in manufacturing. In: DAAAM International Scientific Book 2015. Chapter 33 p. 391-400 ISSN: 1726-9687 ISBN: 978-3-902734-05-1.

[13] Perlova,E.; Platonova,M.; Gorshkov, A. & Rakova, X. (2015). Concept Project of Zero Energy Building, Procedia Engineering. Vol. 100. 2015. pp. 1505 – 1514.

[14] Petek, J.; Glavič, P. & Kostevšek, A. (2016). Comprehensive approach to increase energy efficiency based on versatile industrial practices In: Journal of Cleaner Production Vol. 112, 2016 , pp. 2813-2821, ISSN: 0959-6526.

[15] Sapi, J. ;Drexler, D. A. ; Harmati, I.; Szeles, A.; Kiss, B.; Sápi, Z. & Kovács, L. (2013). Tumor growth model identification and analysis in case of C38 colon adenocarcinoma and B16 melanoma In: SACI 2013: 8th IEEE International Symposium on Applied Computational Intelligence and Informatics, Romania, pp 303-308, ISBN: 978-1-4672-6400-3 .

[16] Satwiko, P. (2011). Pemakaian perangkat lunak Dialux sebagai alat bantu proses belajar tata cahaya In: Jurnal Arsitektur KOMPOSISI Vol. 9, No. 2, 2011, pp. 142-153.

[17] Sarkans, M.; Pikner, H.; Sell, R. & Sonk, K. (2014). Energy efficiency monitoring system for technology mapping driven by FoF concept. Proceedings of the International Conference of DAAAM Baltic "Industrial Engineering", 2014-January, pp. 187-192.

[18] Vuorio, J.; Nikkilä, V.; Teivastenaho, V.; Peltola, J.; Partanen, J.; Kiviluoma, P. & Kuosmanen P. (2014). Additive Manufacturing with UV Light Cured Resin, 9th International DAAAM Baltic Conference "INDUSTRIAL ENGINEERING" 24-26 April 2014, Tallinn, Estonia, pp. 317-322.

[19] EN 12464-1, Light and lighting – Lighting at work places – Part I: Indoor Work Places, European Standard.

[20] EN 15 193:2008 Energy performances of buildings. European Standard.

[21] European Directive 2006/25/EC, 2006. On the minimum health and safety requirements regarding the exposure of workers to risks arising from physical agents (artificial optical radiation). Official Journal of the European Union, L 114, 38-59

DOI: 10.2507/27th.daaam.proceedings.011

RISK ANALYSIS OF SPECIFIC PROJECT PROBLEMS

Zvonko Kremljak

This Publication has to be referred as: Kremljak, Z[vonko] (2016). Risk Analysis of Specific Project Problems, Proceedings of the 27th DAAAM International Symposium, pp.0074-0081, B. Katalinic (Ed.), Published by DAAAM International, ISBN 978-3-902734-08-2, ISSN 1726-9679, Vienna, Austria
DOI: 10.2507/27th.daaam.proceedings.011

Abstract

Risk management is one of the key project management processes performed in the initial project planning. The risk management process includes many steps, described in the paper.

Claims management is one of the most important challenges in large projects and normally linked with many financial losses. Effects of probabilistic claims clearly state a need for a comprehensive model to analyse claim risks in projects. In the present research, list of frequent claims are identified in a piping project. Then based on risk types, probabilistic claims are evaluated by questionnaire aimed at calculation of probability percentage and also probability of success. In order to determine reliability of questionnaire, statistical test of internal consistency coefficient was applied.

Result of the present study is to quantify claims in implementation phase of piping projects and finally by analysing cost effects. Difference between probability of final project cost before and after application of claim indicates a more accurate estimation of project cost given portions of claims for project; impact of claim on project success rate in this project is 8 %.

Keywords: Risk management; Project; Claims; Statistical distribution; Cost.

1. Introduction

Risk management is the process of identifying, analysing, responding to, and controlling project risks. Risk management is a proactive project management process, since you are trying to deal with potential future events before they occur. Risk management is performed in the initial project planning. The process includes the following steps [1]:

- create a Risk Management Plan (RMP),
- identify all potential risks,
- analyse the risks using qualitative techniques,
- utilize quantitative analysis for all high-level risks (optional),
- create a response plan for each high-level risk,
- move the activities associated with the risk response plans to the project schedule.

The remaining, follow-up steps are:

- monitor the risk response plan,
- periodically re-evaluate risks [1].

Risk management is not a one-time process. Project manager should always identify risks at the beginning of the project during the up-front planning process, but he should also periodically look at remaining work to identify any new risks.

Since risks are generally perceived as bad, it makes sense that the first instinct of a project manager would be to eliminate them. Mitigation is a risk response that tries to minimize the probability of a risk occurring or to minimize the impact of the risk on your project. However, there are a number of other options for responding to a risk, including: leave it, monitor the risk, avoid the risk or move the risk [2]. None of these options include ignoring the risk. Even the option to leave the risk is the result of a conscious decision.

When identifying risks on the project, the first place to look is the inherent risks, which are based on the general characteristics of a project rather than the specific circumstances of the project. Since they are general in nature, they can be identified and placed in a checklist for all project managers to review (tight deadlines, project size, new vendors, new technologies etc.). This should be the starting point for the risk identification process. On the other hand, risks are future conditions or circumstances outside the control of the project team that will have an adverse impact on the project should they occur.

This paper is focused to claims in construction industry as the potential problems related to risk and its control.

The management of claims is the greatest challenge facing with contractors in today's challenging business environments. Nowadays, construction projects become increasingly at risk due to a variety of factors which may cause to time extension and cost recovery [3].

Claim management is somehow similar to risk management procedure and consists of the following four processes: claim identification, claim quantification, claim prevention, claim resolution [4].

Reliable prediction of construction duration and consequently budget control is consolidated in decision making process and is an essential part of a successful management [5].

2. Overview of theory

2.1. Risk management

The risk management loop according to Fig. 1 is a guideline for establishment of a risk management system.

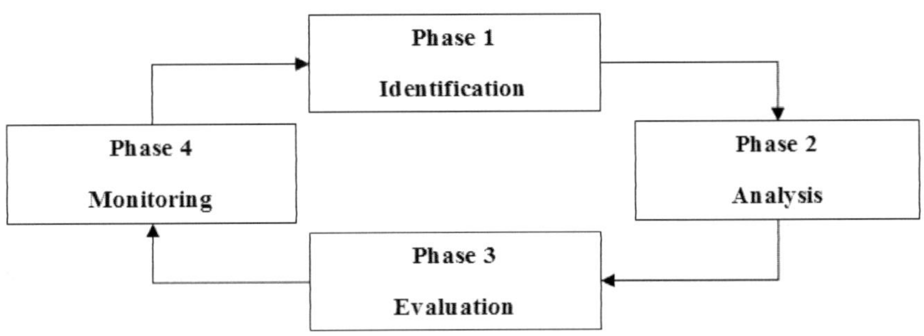

Fig. 1. Schematic Risk Management Circle according to [6]

When risk identification and registration are completed, a probability of occurrence and consequence are assigned to each task. This assignment can be done through a quantitative assessment or a qualitative one or a combination of both [7].

2.2. Claim management

Claim was initially proposed in 1978 as a risk and probability and consequences of the various claims considered during a project life cycle [8].

While there is no a unique definition of this subject in the literature, a claim can be defined as "right given to the party who deserves a request for compensation and damages incurred by the other party" [9]. It may be also described as investigation of consideration or change by one of the parties involved in the project process. A claim happens when one party to the contract has suffered a detriment by the other party [10].

Claim management describes the processes required for elimination, prevention or reduction of construction claims when they are expected to occur [4].

2.3. Construction claims

No project can be considered isolated from a potential claim. Also much kind of claims can lead to financial damages [11]. Construction claims are observed by many participants as an unpleasant event in a project [12].

In general, claims are a common part in construction projects and may happen as a result of several reasons which can contribute delaying a project and/or increasing its relevant costs. On time accomplishing of a project is a difficult task to accomplish in the uncertain, complex, and dynamic environment of construction projects [10]. Claims may arise on a construction project due to number of reasons. Some well-known ones include as given below [13]:

- creep in scope of work (changes, extras and errors),
- inadequate bid / tender information,
- faulty and/or lately supply of equipment and materials by owner,
- low quality of drawings and/or specifications,
- insufficient time through biding analysis,
- interruptions through proceeding of the operations due to lack of coordination, design information or equipment,
- blocked work,
- re-schedule of works ordered by owner, etc.

The most influencing factors of claims are unclear documentations, weak instructions, variations initiated by the employer/engineer, measurement related issues, weather conditions, change and time extension assessment [14].

3. Claims of piping projects

Deterministic claims are not in our research focus. Probabilistic claims in piping construction projects may include [15]: material control, organizational communications, documents and piping plans.

Main approach of this research is to identify and study potential claims which are discussed in piping construction projects and by prioritization and ranking of the claims in a specialized form.

Two factors entitled "probability" and "impact" had been evaluated by questionnaire and required data were collected. To find out the importance or magnitude of a cause from a set of causes for occurrence of claims, two factors entitled "probability of occurrence" and "chance of claim occurrence" were examined and evaluated the most probabilistic causes of claims.

Thus, a questionnaire including 15 questions related to probabilistic claims in piping construction were prepared.

According to 5-fold Likert scale (very low, low, moderate, high, very high) rate of each index for each claim was rolled up and scored [16]. The statistical population's size was selected based on comments made by project managers, piping managers, planning and project control managers, contract managers with high executive experiences. Selection of people was performed based on communications and their involvements in projects. About 25 answered questionnaires were collected for claim analysis and monitoring stage.

3.1. Verification and validation

In the present study, to check the validation for questions of questionnaires, surface and content of questionnaire were examined by independent interviews and finally by comparison among conditions of the present and past projects, potential claims of the projects were evaluated. The Cronbach's alpha coefficient for questions about probability was 0.77 and for questions was 0.71 which are acceptable measurements.

First stage in project planning is to include claim through deterministic planning.

In a project plan along with risk analysis, different outcomes and ways were examined for the given project and impacts of claim risks on the project success were determined.

3.2. Risk analysis for cost and time of project

Through running risk analysis, three optimistic (minimum), the most probabilistic, and pessimistic (maximum) scenarios have to be considered instead of incorporating a deterministic cost and time analysis. By applying project claim analysis, more realistic scheduling is obtained. Before a cost is allocated with a risk, a resource should be defined so that the cost could be allocated. Cost allocation to an activity is the best cost estimation approach for the activity definition. In piping construction projects, once resource for an activity is defined, planned cost for any activity is determined and then triangular distribution has been used to estimate minimum and maximum costs for any individual activity (90 % and 110 % of the estimated cost), see an example in Fig. 2.

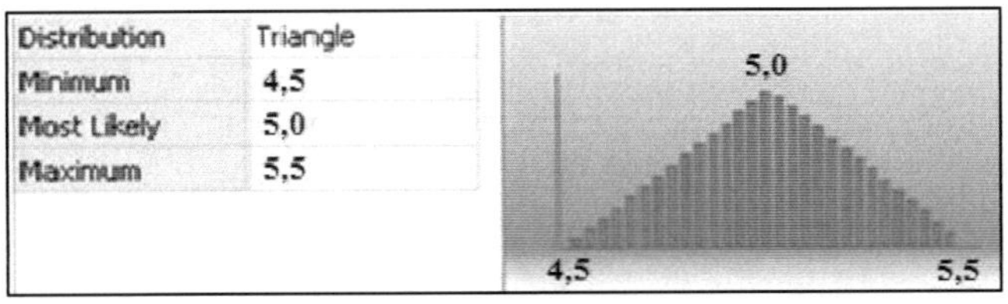

Fig. 2. Triangular distribution of activity cost

3.3. Claims identification and screening process

A check-list of potential claims were prepared for running a comparison between identified potential claims in current and in the projects completed in the past. Outcome of this stage was a list of potential claims for the present project with a certain range of influencing factors. To find the most important claims, two factors "probability of occurrence" and "claim's chance of success" (intensity of claim) were examined and evaluated by the questionnaire. Variables used in scoring include:

- Claim's probability of occurrence (P),
- Claim's impact or chance of occurrence (I),
- Claim's degree of risk (C).

Description	Numerical score
Very low (VL)	1
Low (L)	3
Medium (M)	5
High (H)	7
Very high (VH)	9

Table 1. Impact and probability scoring factors

To rank and classify claim risks, once average probability and impact of claim's risk is calculated, by probability of occurrence (P) multiplied by impact (I), claim's degree of risk (C) is obtained:

$$C = P \times I \quad (1)$$

According to prioritization of claims, 6 important claims that are likely to have the highest effect on the project were investigated.

According to the data entry of quantitative values for probability and impact of claims, those could be scored and finally ranked. Since it is possible that available resources in organization are not enough in order to deal with all claims presented in the project, ranking the risks based on the obtained scores could be very helpful to identify important claims of the project.

A risk could have positive or negative effects on the project [17]. If it has a positive effect on the project, it is called an opportunity and if it has a negative effect on the project, it is called a threat.

Claims could be mapped to an individual activity in the project to run an integrate analysis.

Once questionnaires collected, risk ranking matrix may specify position of each claim based on probability and its effect on Probability–Impact matrix (PI matrix), see Table 2.

	Impact				
Probability	Very low	Low	Medium	High	Very high
Very high [%]	9	27	45	63	81
High [%]	7	21	35	49	63
Medium [%]	5	15	25	35	45
Low [%]	3	9	15	21	27
Very low [%]	1	3	5	7	9

Table 2. Claims PI matrix

By ranking of associated risks, we could specify the most important and effective claims. Since scores of claims are determined in terms of probability and risk impact, the ranking indicates the importance of risks as compared to each other properly. The most important claims are given in Table 3 by preference.

Rank	**Claim event**	**Factor**	**Average score**	**Score**
				(P×I) × 100
1	Contractors' claims, in case of issuance of material issue voucher (MIV) and delay in delivery of material or Non-Issue (NIS)	Probability	7.6	55.48
		Impact	7.3	
2	Requirements for piping operations to be hold or not met	Probability	7.2	51.12
		Impact	7.1	
3	Client delays in delivering isometric drawings	Probability	6.6	42.90
		Impact	6.5	
4	Discrepancies and shortcomings in contract terms	Probability	6.7	42.88
		Impact	6.4	
5	TQ (Technical Query) due to failure, inadequacy and deficiencies information plan	Probability	6.2	36.58
		Impact	5.9	
6	Delay in piping invoices payment	Probability	5.8	34.22
		Impact	5.9	

Table 3. Most important piping construction claims

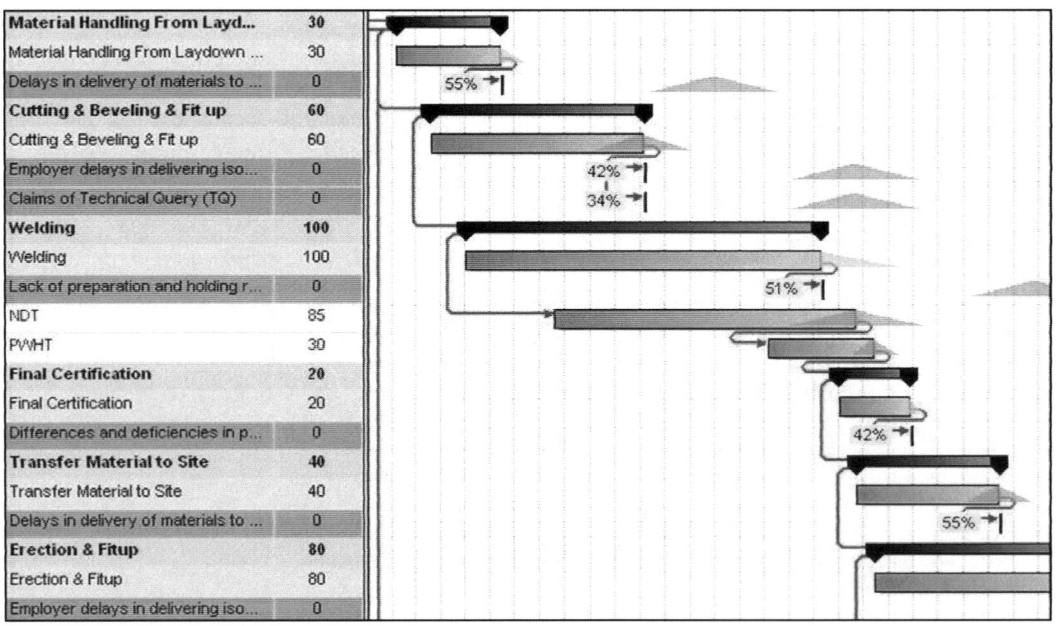

Fig. 3. Project schedule with included claims (segment)

In this project triangular distribution is used to schedule activities (Fig. 3). Then triangular distribution is selected and applied for costs of resources. By doing risk analysis of graphs before applying claim, cost of the project can be registered.

3.4. Cost distributions

After completion of analysis, cost for any repetition (using random values from the triangular distributions) is displayed on a graph [18] which illustrates distribution of project cost (Fig. 4).

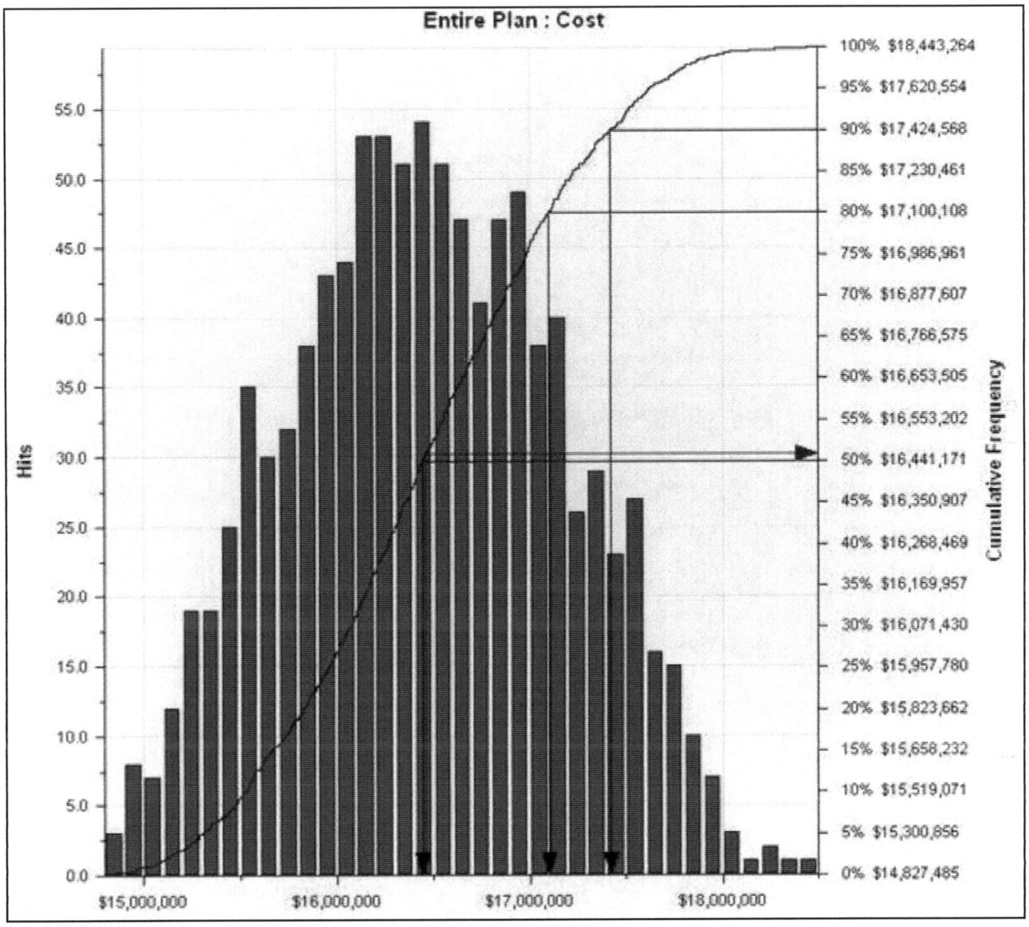

Fig. 4. Cost distribution of piping construction project in uncertain conditions

As can be seen from the Fig. 4, with a 90 % confidence interval, total cost of piping construction project is 17,424,568 USD (see the cumulative frequency) and the estimated probability for completing the estimated cost equals to 16,460,200 USD is only 51 %.

Implementing claim risk analysis by application of claims after taking a preventive (post-mitigation) action, next distribution graph is registered.

As could be seen in Fig. 5, there is 43 % probability the project is completed with the initial estimation 16,460,200 USD. Also, this figure shows with an 80 % confidence the final cost of project could be considered 17,185,148 USD.

Handling claims efficiently and accurately is essential to keeping projects within acceptable time and cost limits and improving your overall performance. Decision management for claims enables the organization to combine predictive analytics with industry best practices and existing business processes. As a result, claims adjustors and others with in-depth project knowledge can quickly and easily define how risk should be assessed and automate many routine decisions while retaining full control of the claims handling process.

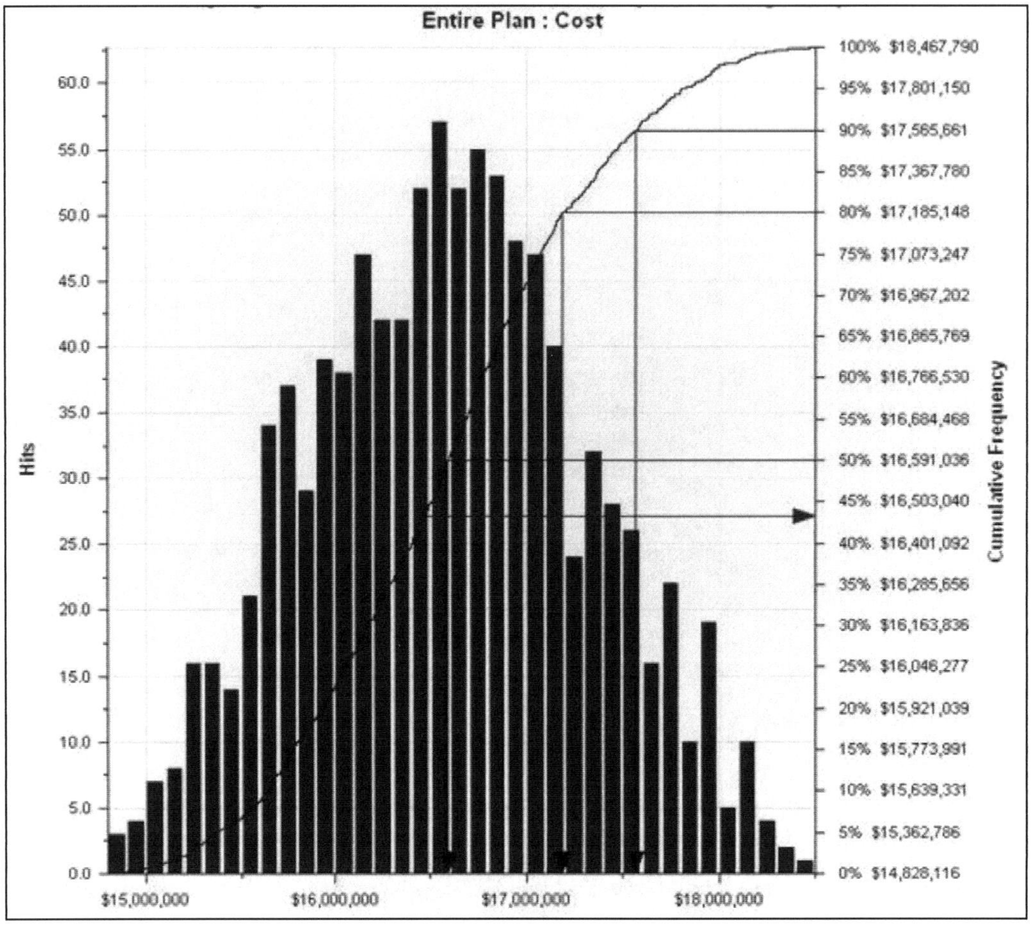

Fig. 5. Cost distribution of piping construction project after application of claims

Table 4 shows statistical values including minimum, maximum, mean, 90 % confidence level, probability of occurrence, and project costs for both states before and after application of claims.

		Cost [USD]			
	Deterministic	Minimum	Mean	Maximum	90 % confidence level
Project cost with uncertainty taken into account before application of claims	51 %	14,827,485	16,465,689	18,443,264	17,424,568
Project cost with uncertainty taken into account after application of claims	43 %	14,828,116	16,583,807	18,467,790	17,565,661

Table 4. Comparison of costs before and after application of claims

4. Conclusion

Analyses of claim risks in projects are very important. This paper, identifying common claims in piping construction contract was performed by questionnaire distributed among senior experienced managers and experts in this industry. Once the most important claims were determined by quantifying the claims project success rates before and after applying the claim were found and difference between the obtained values indicated impact percentage of claim risk in piping construction projects.

This study has important achievements for piping construction projects and the important results include:

1. Identification of actionable claims in piping construction projects and also ranking and prioritization of the most important claims.
2. Claims: impact on project success rate is 8 % for project cost overrun.
3. Understanding cost sensitivity indicates impact of cost of an activity on the total cost.
Using this information, the project could be estimated and evaluated more accurately. Claims affecting aims of project are identified and documented according to expert judgments.

Project risk is one of those exciting topics that everyone (executives, functional managers, project managers or engineers) has an opinion about. Any good project has plenty of risk. The claim strategy depends on our answers to the questions about our relationship to the contractual partner, the policies of other involved units, project goals, business goals etc.

Future research work will be oriented to other types of industries too. It will include some improvements of the model, for example: more parameters for observation, related to specific branch, application of fuzzy modelling through project claim management, inclusion of decision support system, adopted methods of analysis etc.

5. References

[1] Mochal, T. & Mochal, J. (2011). Lessons in Project Management, 2nd edition, Apress, Springer-Verlag, ISBN 978-1-4302-3834-8, New York
[2] Kremljak, Z. & Kafol, C. (2014). Types of risk in a system engineering environment and software tools for fisk analysis, Proceedings of the 24th DAAAM International Symposium on Intelligent Manufacturing and Automation, Katalinic, B. (Ed.), Procedia Engineering, ISSN 1877-7058, Vol. 69, pp. 177-183, Elsevier, DOI:10.1016/j.proeng.2014.02.218
[3] Kululanga, G.; Kuotcha, W.; McCaffer, R. & Edum-Fotwe, F. (2001). Construction contractors' claim process framework, Journal of Construction Engineering and Management, Vol. 127, No. 4, pp. 309-314, ISSN 0733-9364, DOI:10.1061/(ASCE)0733-9364(2001)127:4(309)
[4] PMI. (2013). A Guide to the Project Management Body of Knowledge (PMBOK® Guide), 5th edition, Project Management Institute, ISBN 978-1-9355-8967-9, Newtown Square
[5] Al-Momani, A. H. (2000). Construction delay: a quantitative analysis, International Journal of Project Management, Vol. 18, No. 1, pp. 51-59, ISSN 0263-7863, DOI: 10.1016/S0263-7863(98)00060-X
[6] Kremljak, Z. (2004). Decision making under risk, DAAAM Publishing series – Management Science, ISBN 3-901509-44-5, DAAAM International, Vienna
[7] Smith, C. A. (2011). Integrated scenario-based methodology for project risk management, PhD Thesis, University of Maryland, College Park
[8] Bosch, R. V. (1978). Identifying construction claims, Transactions of the American Association of Cost Engineers, pp. 320-329
[9] Simon, M. S. (1979). Construction Contracts and Claims, McGraw-Hill, ISBN 978-00705-7433-5 New York
[10] Kartam, S. (1999). Generic methodology for analyzing delay claims, Journal of Construction Engineering and Management, Vol. 125, No. 6, pp. 409-419, ISSN 0733-9364, DOI:10.1061/(ASCE)0733-9364(1999)125:6(409)
[11] Enshassi, A.; Mohamed, S. & El-Ghandour, S. (2009). Problems associated with the process of claim management in Palestine: Contractors' perspective, Engineering, Construction and Architectural Management, Vol. 16, No. 1, pp. 61-72, ISSN 0969-9988, DOI:10.1108/09699980910927895
[12] Ho, S. P. & Liu, L. Y. (2004). Analytical model for analyzing construction claims and opportunistic bidding, Journal of Construction Engineering and Management, Vol. 130, No. 1, pp. 94-104, ISSN 0733-9364, DOI:10.1061/(ASCE)0733-9364(2004)130:1(94)
[13] Jergeas, G. F. & Hartman, F. T. (1994). Contractors' construction-claims avoidance, Journal of Construction Engineering and Management, Vol. 120, No. 3, pp. 553-560, ISSN 0733-9364, DOI:10.1061/(ASCE)0733-9364(1994)120:3(553)
[14] Yogeswaran, K.; Kumaraswamy, M. M. & Miller, D. R. A. (1998). Claims for extensions of time in civil engineering projects, Construction Management & Economics, Vol. 16, No. 3, pp. 283-293, ISSN 0144-6193, DOI:10.1080/014461998372312
[15] Yi, H. Y. & Guo, P. (2015). Modelling risk coordination of supply chains with put option contracts and selective return policies, International Journal of Simulation Modelling, Vol. 14, No. 3, pp. 551-562, ISSN 1726-4529, DOI:10.2507/IJSIMM14(3)CO15
[16] Hwang, C.-L. & Yoon, K. (1981). Multiple Attribute Decision Making: Methods and Applications – A State-of-the-Art Survey, Springer-Verlag, ISBN 978-3-540-10558-9, Berlin
[17] Ocampo, L. A.; Clark, E. E.; Tanudtanud, K. V. G.; Ocampo, C. O. V.; Impas Sr., C. G.; Vergara, V. G.; Pastoril, J.; Tordillo, J. A. S. (2015). An integrated sustainable manufacturing strategy framework using fuzzy analytic network process, Advances in Production Engineering & Management, Vol. 10, No. 3, pp. 125-139, ISSN 1854-6250, DOI:10.14743/apem2015.3.197
[18] Palisade Corp. (2013). @RISK for Project (ver. 6), Palisade, Middlesex

DOI: 10.2507/27th.daaam.proceedings.012

OPTIMAL DESIGN AND ANALYSIS OF QUARTER VEHICLE SUSPENSION SYSTEM BY USING MATLAB

Likaj, R. Bruqi, M. Shala, A. & Bajrami, Xhevahir

This Publication has to be referred as: Likaj, R[ame]; Bruqi, M[irlind]; Shala, A[hmet] & Bajrami, X[hevahir] (2016). Optimal Design and Analysis of Quarter Vehicle Suspension System by Using Matlab, Proceedings of the 27th DAAAM International Symposium, pp.0082-0090, B. Katalinic (Ed.), Published by DAAAM International, ISBN 978-3-902734-08-2, ISSN 1726-9679, Vienna, Austria
DOI: 10.2507/27th.daaam.proceedings.012

Abstract

The paper deals with the optimal design and analysis of quarter car vehicle suspension system. For optimal designs are used the optimal parameters which have been derived by comparison of two optimisation algorithms: Sequential Quadratic Program (SQP) and Genetic Algorithms (GA's), for a five chosen design parameters. The goal function is chosen to provide the possibility to emphasize three main objectives of vehicle suspensions; ride comfort, suspension travel and road holding. Fuzzy Logic Control (FLC) is considered to control active suspension for the optimal parameters derived by GA's, and the rule base can be tuned to improve each of the above objectives, while the main focus is to minimise the vertical vehicle body acceleration. It also deals with parametric analysis, state space modelling, Laplace Transform, Transfer Function, Stability, Controllability, Observability and many other important attributes to analyse quarter car vehicle model.

Keywords: Optimisation; parametric analysis; optimal control; GA; FLC.

1. Introduction

Nowadays the focus of vehicle suspension design has switched from pure numerical analysis to the application of different algorithms which are designed based on optimization methods. The development of optimizing methods is connected with the development of an automatic control of systems, which is the case in our study.

Algorithms designed based on optimizing methods enables us to fasten the search process of finding optimal solution expressed by an objective function under the fulfillment of constraint conditions, which evidently has a great influence on the development of the whole group of modern control methods that for the calculation of the control effect utilize a mathematical model of the process determined based on the measured process data (MPC methods and algorithms). At present, a great success has been achieved by agent and colonic control methods as well as methods and genetic algorithms used in the system of off-line techniques applied to the searching for a global extreme of a goal function. The applications of the mentioned methods and their algorithms to nonlinear processes modeling and control, systems, significantly contribute to the improvement of approximation properties of modeling processes as well as to the improvement of a control quality of selected types of a control [3]. Genetic algorithms (GAs) have been used in various applications such as function optimization, system identification and control systems. GAs is general-purpose stochastic optimization methods for solving search problems to seek a global optimum. However, GAs is characterized by a large number of function evaluations [4].

On the other hand, traditional methods, such as sequential quadratic programming (SQP), are well known to exploit all local information in an efficient way, provided that certain conditions are met and the function to be minimized is 'well-conditioned' in the neighbourhood of a unique optimum. These methods require adequate local information to be known (such as the gradient and Hessian matrix). If the basic requirements are not satisfied, the reliability of the SQP method is greatly jeopardized.

The aim of this study is, for quarter car model to find optimal design parameters, by minimising a nonlinear objective function subjected to a set of constraints with two different optimisation algorithms; SQP and GA in order to improve ride quality. By means of the ride quality analysis in the frequency domain, the vertical vehicle body acceleration (VBA), suspension working space (SWS) and dynamic tire load (DTL) can be obtained [4].

In this design optimization, the main objective is to minimize the VBA acceleration. In the meantime, the SWS and DTL are constrained. If the SWS is too small, the sprung mass will strike the un-sprung mass and this may lead to damage of the vehicle. If the DTL is greater than the static tire load, the vehicle's tires will bounce off the road [5] and this will result in unstable modes of vehicle motion. Therefore, we aimed to present the development of a method for obtaining poles, eigenvectors, natural frequencies and mode shapes, in order to perform a vibration analysis for a quarter car model by using three different approaches [5], [6], [7] and comparison of the natural frequencies which have been obtained for the optimised model with the natural frequencies defined by the laws of the vibration theory. The parametric analysis gave information on how the natural frequency is affected when one of the parameters is perturbed. Parametric studies of dynamical systems are an important task in all major engineering design. This is especially true in the design of control systems. It is important to know when one parameter is changed, how that affects the system; if the system becomes unstable, do the natural frequencies change, etc. In addition, we presented a task of state space modelling, Laplace Transformation, Transfer Function, Stability, Controllability, Observability as one of most important attributes to analyse quarter car vehicle model.

Another aim was to develop a Fuzzy Logic Controller (FLC) for a quarter car model with optimised parameters, with a rule base [8], [5], which can be tuned to improve each of three objectives; ride comfort, suspension working travel and handling.

In addition, in this study we have applied in integrative way software programs MATLAB and SIMULINK for optimisation, analysis, design and control of a quarter car model.

2. Dynamical model of quarter car vehicle model

The model of the quarter-car active suspension system used in this paper with two degree of freedom is shown in Fig. 1. The model represents a single wheel of a car in which the wheel is connected to the quarter portion of the car body through a hydro pneumatic suspension. The dynamics of quarter car model in Fig. 1 are described by:

$$\begin{aligned} m_1\ddot{z}_k &= f_a - k_1(z_k - z_p) - c_s(\dot{z}_p - \dot{z}_k) \\ m_2\ddot{z}_p &= -f_a + k_1(z_k - z_p) + c_s(\dot{z}_k - \dot{z}_p) - k_2(z_p - z_r) \end{aligned} \tag{1}$$

The model parameters and their respective units are summarized in Tab. 1.

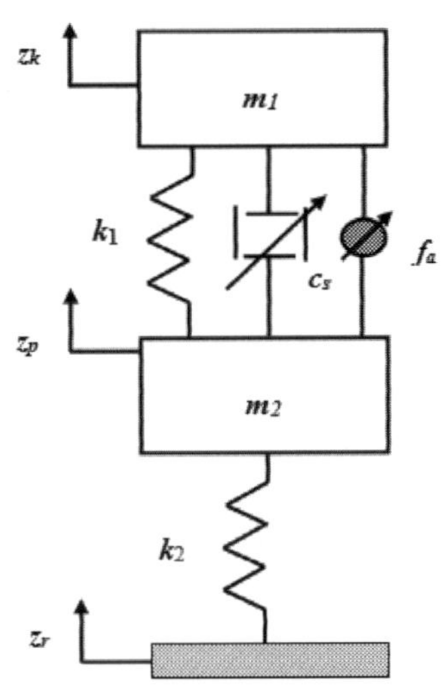

Fig. 1. Quarter car suspension model of 2 DOF

Symbol	Description	Value
m_1	Body mass	200 kg
m_2	Wheel mass	33 kg
c_s	Damping ratio	1600 Ns/m
k_1	Spring constant	9000 N/m
k_2	Spring constant	200000 N/m

Table 1. Quarter car vehicle model parameters

Equations (1) can be represented in matrix form as:

$$M \cdot \ddot{z} + C \cdot \dot{z} + K \cdot z = f \tag{2}$$

where M, C, and K represent the mass, damping, and stiffness matrices described by:

$$M = diagonal\,\left[m_k, m_p\right] \tag{3}$$

$$C = \begin{bmatrix} c_s & -c_s \\ -c_s & c_s \end{bmatrix} \tag{4}$$

$$K = \begin{bmatrix} k_1 & -k_1 \\ -k_1 & k_1 + k_2 \end{bmatrix} \tag{5}$$

Standard second order matrix form for the passive quarter car suspension system is shown in the following:

$$\begin{bmatrix} m_1 & 0 \\ 0 & m_2 \end{bmatrix} \cdot \begin{bmatrix} \ddot{z}_k \\ \ddot{z}_p \end{bmatrix} + \begin{bmatrix} c_s & -c_s \\ -c_s & c_s \end{bmatrix} \cdot \begin{bmatrix} \dot{z}_k \\ \dot{z}_p \end{bmatrix} + \begin{bmatrix} k_1 & -k_1 \\ -k_1 & k_1 + k_2 \end{bmatrix} \cdot \begin{bmatrix} z_k \\ z_p \end{bmatrix} - \begin{bmatrix} 0 \\ k_2 \end{bmatrix} \cdot z_r = 0 \tag{6}$$

For free vibration, the general form leads to:

$$A \cdot z - B \cdot z = 0 \tag{7}$$

The response of this system takes the form:

$$z(t) = v \cdot e^{\lambda \cdot t} \tag{8}$$

where $\boldsymbol{v}$ is a constant vector.

The model represents a single wheel of a car in which the wheel is connected to the quarter portion of the car body through a hydropneumatic suspension.

Substituting (8) in (7) leads to eigenvalue problem;

$$\left(A - \lambda \cdot B\right) \cdot v = 0 \tag{9}$$

Vector $\boldsymbol{v}$ is called the eigenvector and λ represents eigenvalues.

3. Natural frequencies and mode shapes simulation in Matlab

Parametric studies of dynamical systems are an important task in all major engineering design. This is especially true in the design of control systems, when the focus is for low sensitivity to parameter variations [6]. It is important to know when one parameter is changed, how that affects the system; if the system becomes unstable, do the natural frequencies change, etc.

Since we have obtained matrices M, C, and K, we can calculate natural frequencies and mode shapes in MATLAB; Natural Frequencies are:

w =

6.5612 0 and 0 79.5945

Mode shapes are:
u=
-0.0707 -0.0012
-0.0031 0.1741

Through simulation are obtained the values from V1 to V4 of a complex conjugate pairs of vectors using MATLAB with three different programming approaches. The simulation results are shown in Table 2.

First approach:
```
a=[m1 cs k1];
% poles of the system
r =
 -24.3582 +73.2172i
 -24.3582 -73.2172i
  -3.8842 + 5.5424i
  -3.8842 - 5.5424i
```

Second approach,
```
r =
  -7.1146 -15.0873i
  -7.1146 +15.0873i
   0.0009 + 0.0011i
   0.0009 - 0.0011i
p =
 -24.3582 +73.2172i
 -24.3582 -73.2172i
  -3.8842 + 5.5424i
  -3.8842 - 5.5424i
k =
0.0050 -0.3127
```

Third approach: Frequency response and poles of the system
```
v =
  -0.0011 + 0.0008i  -0.0011 - 0.0008i  -0.0991 - 0.0395i  -0.0991 + 0.0395i
  -0.0042 - 0.0120i  -0.0042 + 0.0120i   0.0002 - 0.0049i   0.0002 + 0.0049i
  -0.0317 - 0.0984i  -0.0317 + 0.0984i   0.6039 - 0.3961i   0.6039 + 0.3961i
   0.9834 - 0.0166i   0.9834 + 0.0166i   0.0264 + 0.0200i   0.0264 - 0.0200i
s =
 -24.3582 +73.2172i       0                  0                  0
 0        -24.3582 -73.2172i       0                  0
       0            0          -3.8842 + 5.5424i        0
       0            0                0           -3.8842 - 5.5424i
```

As we can see for all cases similar results have been obtained.

Poles	Value
s11	-24.3582 +73.2172i
s22	-24.3582 -73.2172i
s33	-3.8842 + 5.5424i
s44	-3.8842 - 5.5424i

Table 2. Poles of the system

4. Stability of the model by Lyapunov theorem in Matlab

The Lapunov criterion will be used to check whether or not the given system is stable. Starting point for stability analysis is state-space form of system [1], [5]. Matrix ***A*** must be definite, x is the state vector containing n state variables. The quadratic form of Lyapunov function is:

$$V(x) = x^T(t) \cdot Q \cdot x(t) \tag{10}$$

where $\boldsymbol{Q}$ must be a positive definite symmetric matrix $n \times n$. General form of Lyapunov matrix equation is:

$$A \cdot Q + Q \cdot A^T = -C \tag{11}$$

where $\boldsymbol{C}$ is a positive definite symmetric matrix. Usually, matrix $\boldsymbol{Q}$ is taken $\boldsymbol{Q=I}$.

From MATLAB, for closed loop is:
eig(A-B*K)

```
ans =
-24.3567 +73.2176i
 -24.3567 -73.2176i
  -3.8842 + 5.5424i
  -3.8842 - 5.5424i
```

5. Full Controllability

If any modes of the open-loop system are unstable, they can be stabilized with a feedback using linear quadratic control. A necessary condition for solving the controller problem is that the system is controllable. This means that all the modes can be excited or controlled by the input u. In terms of the state space matrices, the condition of controllability can be shown by constructing the controllability matrix $\boldsymbol{Qc}$ defined by:

$$Q_c = \left[B, AB, ..., A^{m-1}B\right] \tag{12}$$

Controllability is ensured when the rank of Q_C equals m, where m is the order of the system.

6. State Controllability

A system is said to be "(state) controllable" if for any t_0 and any initial state $x(t_0) = x_0$ and any final state x_f, there exists a finite time $t_1 > t_0$ and control $u(t)$, such that $x(t_1) = x_f$.

A LTI system with the state equation $\dot{x} = Ax + Bu$ is controllable if the controllability matrix $Qc = \left[B, AB, A^2B, ..., A^{m-1}B\right]$ has rank m.

```
Controllability
rank(Qc)
  1.0e+08 *
   0.0000  -0.0003  -0.0163   2.5186
        0   0.0003  -0.0185  -1.0447
        0        0   0.0105  -0.5923
        0        0        0   0.1678
ans = 4
```

Since rank of Qc is 4 equals with size of matrix A, system is controllable.

7. State Observability

A system is said to be "(state) observable" if for any t_0 and any initial state $x(t_0) = x_0$ there exists a finite time $t_1 > t_0$ such that knowledge of $u(t)$ and $y(t)$ for $t_0 \leq t \leq t_1$ suffices to determine x_0
A LTI system with the state space model

$$\dot{x} = Ax + Bu$$
$$y = Cx \tag{13}$$

Only if observability matrix $Q_0 = \left[C, CA, CA^2, \ ... \ , CA^{n-1}\right]^T$ has rank n. Functions in MATLAB

```
Observability
rank(Q0)
Q0 =
  1.0e+07 *
        0        0   0.0000   0.0000
```

```
        0   0.0013   0.0000        0
   0.0852   0.0000        0        0
  -4.8140  -8.4945  -2.0177  -0.7093
ans = 4
```

Since rank of Qo is 4 equals with size of matrix A, system is observable.

8. Stochastic Road Modeling and PSD Response

Road irregularity or unevenness represents the main disturbing source for either the rider or vehicle structure itself. The road profile elevation is usually expressed in terms of the power spectral density (PSD). The PSD of the road profile elevation is expressed as:

$$G_q(n) = G_q(n_0)(n / n_0)^{-w} \tag{14}$$

For the purposes of design optimization, according to James' principle, the root mean square (RMS) of the sprung mass acceleration $\ddot{z}_2$ can be expressed as:

$$\sigma_{\ddot{z}} = \left\{ \pi RV \left[\frac{k_2 c_s}{2 m_1^{3/2} k_1^{1/2}} + \frac{(m_1 + m_2) k_1^2}{2 c_s m_1^2} \right] \right\}^{1/2} \tag{15}$$

The RMS of the suspension working space f_d is:

$$\sigma_{fd} = \left\{ \pi RV \left[\frac{(m_1 + m_2)(m_1 k_1)^{1/2}}{2 m_1 c_s} \right] \right\}^{1/2} \tag{16}$$

The RMS of the relative dynamic tire load can be calculated as:

$$\sigma_{Fd/G} = \left\{ \pi RV \left[\frac{k_2^2 m_2}{2 c_s (m_1 + m_2)^2} + \frac{(m_1 + m_2) k_1^2}{2 m_1^2 c_s} - \frac{k_2 k_1 m_2}{c_s m_1 (m_1 + m_2)} + \frac{c_s k_1}{2 m_1 m_2} \right] \right\}^{1/2} \tag{17}$$

9. Design Optimisation

In this section, the sprung mass vertical acceleration is minimized, while the design constraints on the suspension working space and dynamic tire load should be satisfied. To implement the design optimization, the two optimization algorithms, i.e, SQP and GA, will be applied, respectively.

Optimization based on SQP algorithm and GA

The SQP algorithm is a non-linear programming technique that is used for the purpose of minimizing a smooth non-linear function subjected to a set of constraints with upper and lower bounds. The objective function and the constraint functions are assumed to be at least twice continuously differentiable. This algorithm is a gradient-based search method [4],[10]. This algorithm is well-suited for constrained design optimizations. The reliability for finding the optimum decreases with the increase of number of design variables when using SQP method. In contrast, whether the number of design variables increase the GA can still reliably find the optimum. This can be explained by the fact that GA works on a population of variables in parallel, not on a unique point. GAs are a global search methods that are based on the Darwin's principle of natural selection and genetic modification and have higher reliability for the global optimum with minimum number of computational operations.

The RMS of the acceleration of a sprung mass $\sigma_{\ddot{z}_k}$ is frequently used to evaluate the riding quality of a vehicle. A rider's comfort improves as the acceleration decreases. Ride comfort is chosen to be the design criterion. The suspension working space and dynamic tire load σ_{fd} are selected as the design constraints. The design variables are m_1, m_2, k_1, k_2 and c_s, respectively.

Thus, the design optimization problem can be obtained by minimisation of sprung mass acceleration [4], [9] given in (18) subject to constraints (19).

$$\sigma_{\dot{z}_k}(m_1,m_2,k_2,k_1,c_s)=\left\{\pi\ R\ V\left[\frac{k_2c_s}{2m_1^{3/2}k_1^{1/2}}+\frac{(m_1+m_2)k_1^{2}}{2c_sm_1^2}\right]\right\}^{1/2} \tag{18}$$

Subject to:

$$\begin{cases}\sigma_{Fd/G}(m_1,m_2,k_1,k_2,c_s)\le a=0.5\\ \sigma_{fd}=(m_1,m_2,k_1,k_2,c_s)\le b=0.05\\ 150\le m_1\le 200\\ 32\le m_2\le 40\\ 200000\le k_2\le 220000\\ 8500\le k_1\le 91000\\ 1580\le c_s\le 1861\end{cases} \tag{19}$$

In this sub-section, the optimization results are derived for a vehicle travelling at the speed of 40 m/s on the road with an irregularity coefficient of power spectrum taking the value of 6.5x10^{-6} m^3.

	Original values	**SQP method**	**GA**
Initial values		[10 10 10 10 10]	[10 10 10 10 10]
m_1 [kg]	200	200.0	199.918612498565
m_2 [kg]	33	32.0	32.04226032330942
k_2 [N/m]	200000	200000.0	200305.81925474654
k_1 [N/m]	9000	9100.0	9099.82776078753
cs [Ns/m]	1600	1580.0	1591.7473971815848
$\sigma_{\dot{z}_1}$		1.2673254442031	1.26976709377
		Optimum found: 52 iterations	Optimum found: 51 generations

Table 3. Optimal design variables based on the SQP and GA for minimizing the sprung mass vertical acceleration, vehicle speed 40 m/s

10. Fuzzy Logic Controller

Fuzzy Logic Control has accelerated in recent years in many areas, including feedback control. By using empirical rules according to the designers knowledge and experience, which are represented linguistically with the conditional statements and resulting assertion.

A fuzzy rule base has a very significant effect on the control strategy in FLC, in other words it defines the strategy of the controller. To the active suspension system there are at least three main objectives, namely ride comfort, suspension travel and handling. The rule base can be tuned to improve each of the above objectives.

The fuzzy logic controller used in the active suspension has three inputs: body acceleration $\ddot{z}_b$, body velocity $\dot{z}_b$, body deflection velocity $\dot{z}_b-\dot{z}_t$ and one output: desired actuator force f_a, shown in Fig. 2.

$$f_a=\frac{\int_{f_a} f^*\mu_D(f)df}{\int_{f_a}\mu_D(f)df} \tag{20}$$

The output of the fuzzy controller μ_D is corresponding membership function.

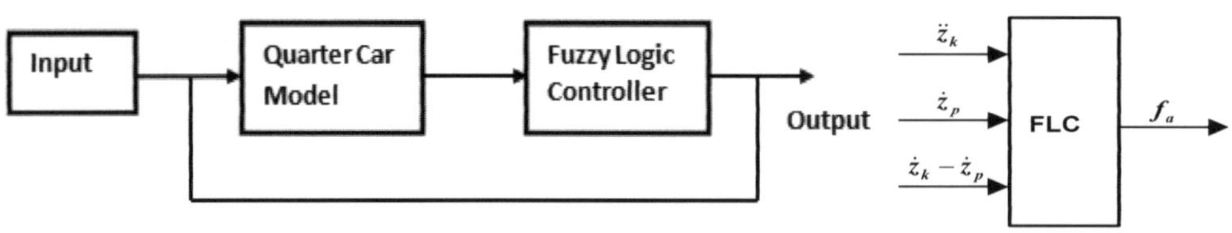

Fig. 2. Control scheme of quarter car model

The control system itself consists of three steps: fuzzification, fuzzy inference machine and deffuzification. During the fuzzification process the real numbers (crisp) inputs will be converted into fuzzy values, where after fuzzy interference machine processes the input data and computes in cope with the rule base and database. The obtained outputs (fuzzy values) are converted into real numbers by the defuzzification step. Membership functions are chosen for the inputs and the output variables: NV-negative very big, NB-negative big, NM-negative medium, NS-negative small, N-negative, ZE-zero, P-positive, PS-positive small, PM-positive medium, PB-positive big, PV-positive very big.

The fuzzy rule based system modeled by designer's knowledge and experience is shown in Fig. 3. [7], [8], [5]. $\dot{z}_b - \dot{z}_t$, $\dot{z}_b$, $\ddot{z}_b$ and f_a, which are characterised by their membership functions: $-1 \le z_b - z_t \le 1$; $-1 \le \dot{z}_b \le 1$; $-4 \le \ddot{z}_b \le 4$; $-4000 \le f_a \le 4000$. Simulation results for acceleration of unsprunged mass for passive quarter car model are shown in Fig. 4. Accelerations of the active quarter car model are shown in Fig. 5.

$\ddot{z}_b$	$\dot{z}_b$	$\dot{z}_b - \dot{z}_t$	f_a
ZE	P	P	NS
ZE	P	ZE	NM
ZE	P	N	NB
ZE	ZE	P	PS
ZE	ZE	ZE	ZE
ZE	ZE	N	NS
ZE	N	P	PB
ZE	N	ZE	PM
ZE	N	N	PS
P or N	P	P	NM
P or N	P	ZE	NB
P or N	P	N	NV
P or N	ZE	P	PM
P or N	ZE	ZE	ZE
P or N	ZE	N	NM
P or N	N	P	PV
P or N	N	ZE	PB

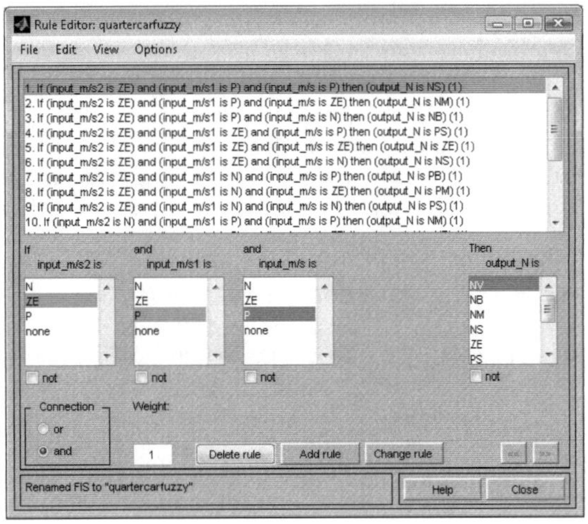

Fig. 3. Fuzzy rules

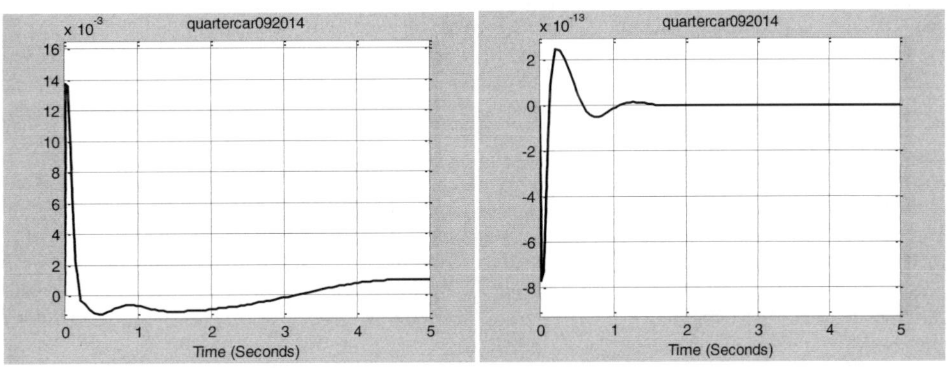

Fig. 4. Acceleration of unsprunged mass for passive and active quarter car model

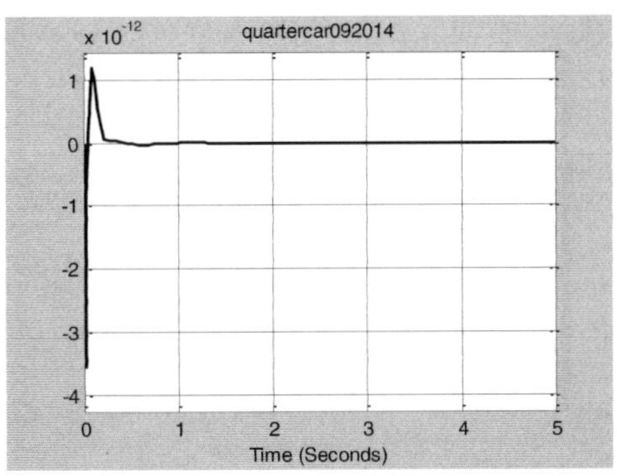

Fig. 5. Acceleration of sprunged mass for active quarter car model

11. Conclusions

A comparative study of two optimization algorithms (genetic algorithms, GAs and sequential quadratic programming, SQP), has been conducted through minimizing the vertical sprung mass acceleration subjected to a suspension working space and dynamic tire load.

By optimizing the design parameters compared with the original design, the sprung mass (body) acceleration decreases. The suspension working space and the dynamic tire load satisfy the specified design constraints. Based on the simulation results the optimum found by GAs at 19 generations, while by using the SQP the optimum is found after 51 iterations. As shown from the numerical simulation results the max amplitude of body displacement using optimized design variables is reduced for 9%, while maximal amplitude of body acceleration is reduced around 22%.

Numerical experiments reveal the fact that to improve vehicle ride quality and satisfy the specified suspension working space and relative dynamic tire load, different vehicle speed and road irregularity have different requirements on the design variables, in particular, the un-sprung mass.

From the parametric analysis it can be concluded that mass is a very sensitive parameter, and must be handled with special attention, while the results of the natural frequencies obtained in both ways shows that the values are almost the same.

Active suspension systems using FL Controllers can reduce vertical accelerations, as shown in Fig. 5. This means, that the main properties; ride comfort and road holding were achieved for the quarter car model. For the design of an FLC, an accurate vehicle model is not needed, but it's a very difficult task to express the knowledge and experience in terms of fuzzy logic.

From the results of this study it became evident to us that the combination of MATLAB and SIMULINK represent a powerful tool on research of parametric analysis, optimisation and optimal control of vehicle suspension system.

Since the quarter car model is one the bests and represents the simplest model for the analysis of ride comfort, the same approach will be applied in our future research to analyse half car and full car vehicle suspension system

12. References

[1] Jadlovska, A.; Katalinic, B.; Hrubina, K., & Wessely, E. (2013). *On Stability of Nonlinear Systems and Application to Apm Modeling*, Chapter 11 in DAAAM International Scientific Book 2013, pp. 257-276, B. Katalinic & Z. Tekic (Eds.), Published by DAAAM International, ISBN 978-3-901509-94-0, ISSN 1726-9687, Vienna, Austria , DOI: 10.2507/daaam.scibook.2013.11

[2] Jadlovska, A.; Katalinic, B.;Hrubina; K.; Macurova, A.& Wessely, E. (2012). *Solution to the Problem Control of a Distributed Parameter Process*, Chapter 15 in DAAAM International Scientific Book 2012, pp. 169-186, B. Katalinic (Ed.), Published by DAAAM International, ISBN 978-3-901509-86-5, ISSN 1726-9687, Vienna, Austria DOI:10.2507/daaam.scibook.2012.15

[3] Jadlovska, A.; Katalinic, B.;Hrubina, K.; Macurova, A. & Wessely, E. (2011). Optimal Control of Nonlinear Systems with Constraints, Chapter 22 in DAAAM International Scientific Book 2011, pp. 265-282, B. Katalinic (Ed.), Published by DAAAM International, ISBN 978-3-901509-84-1, ISSN 1726-9687, Vienna, Austria DOI:10.2507/daaam.scibook.2011.22

[4] Likaj R.; Shala A.; Bruqi M.; Qelaj M. (2009). Optimal Design of Quarter Car Vehicle Suspension System, Trends in the Development of Machinery and Associated Technology – ISSN 1840-4944; pp. 417-420; Livorno, Italy

[5] Likaj R.(2005). Fuzzy Logic Control on Nonlinear Vehicle Suspension System, Phd thesis, Prishtina, Kosovo

[6] Likaj, R.; Geca, A.; Pajaziti, A.; Osmanaj S. (2005). Parametric and sensitivity analysis of four DOFs vehicle model, 4th DAAAM International Conference on Advanced Technologies for developing Countries, September 21-24, 2005, Slavonski Brod, Croatia

[7] Likaj R.; Pajaziti A., (2002). Nonlinear and fuzzy Logic Control of Vehicle Suspensions, "Intelligent Manufacturing & Automation: Learning from Nature" 13th International DAAAM Symposium; Annals Proceeding, ISBN 3-901509-29

[8] Cai,B.; Konik,D. (1993). Intelligent Vehicle Active Suspension Control using Fuzzy Logic, IFFAC World Congres, Vol.2, pp. 231-236..

[9] Zhongzhe Ch., Yuping H., Natarer G.: Design Optimization of suspension with a Quarter- Vehicle Model, CSME/ de la SCGM, Vol.32, No .2, pp 297-312, 2008.

[10] Yuping H: Design Optimization of a Quarter Vehicle suspension system using a Genetic Algorithm and SQP, Department of Systems Design Engineering, University of Waterloo, 1999

DOI: 10.2507/27th.daaam.proceedings.013

KEY ELEMENTS ANALYSIS OF PROJECTS' ENVIRONMENTAL IMPACT ASSESSMENT

Vlado Majstorovic, Maja Medic & Marijana Bandic Glavas

This Publication has to be referred as: Majstorovic, V[lado]; Medic, M[aja] & Bandic Glavas, M[arijana] (2016). Key Elements Analysis of Projects' Environmental Impact Assessment, Proceedings of the 27th DAAAM International Symposium, pp.0091-0094, B. Katalinic (Ed.), Published by DAAAM International, ISBN 978-3-902734-08-2, ISSN 1726-9679, Vienna, Austria
DOI: 10.2507/27th.daaam.proceedings.013

Abstract

In all areas of human activity, the greatest achievements have been achieved by projects. Beside all the positive impacts, some projects impact on the environment negatively. This is why the developed countries of the European Union and the world in their legislature began to regulate environmental protection, bringing numerous legal acts which, among other things, are project related. Guided by this fact, the paper highlights the key elements in the projects' environmental impact assessment and gives a short review of the EU legal framework in this area. Emphasis is placed on the directives on the assessment of the impacts of certain projects on the environment and their amendments. In order to reduce the negative impact on the environment, a substantial number of important elements in the procedures of environmental impact assessment and decision making have been introduced, and the legislation instruments have continuously been reviewed and improved.

Keywords: key elements; impact assessment; project; environment; directive.

1. Introduction

The 'project' term is used to describe activities that are not carried out continuously but occasionally and optionally. Generally speaking, project can be defined as a set of mutually and logically connected activities of limited duration, aimed to achieve goals being set. Thus, a project can be defined as any process that achieves goal or set of goals. [8] [9] Projects as an economic or social activity are damaging environmental stability, biodiversity, or in any other way (usually negatively) impact on the environment. [11] Therefore, the EU and other world countries have begun creating the laws in order to regulate environmental protection by bring a number of legal acts related to the environmental impact assessment of certain projects.

Environmental impact assessment is the process of evaluating the acceptability of the planned intervention (project) with respect to the environment and determination of necessary environment protection measures to be implemented in the context of preparation of the planned intervention. Usually those are the projects for construction of infrastructural facilities (transport, energy, water), manufacturing and sports facilities, waste management facilities, tourism and catering, shopping centers, interventions that exploit mineral resources, etc. [11]

The aim of environmental impact assessment is to determine the potential environmental, social and health impacts of implementation of a specific project. The goal is to estimate the physical, biological and socioeconomic impacts in a form that would allow a rational decision-making. [4]

Guided by this fact, this paper provides a short review about the necessity for the legal regulation and a short key elements analysis of projects' environmental impact assessment.

2. The emergence of the need for legal regulation of the projects' environmental impact assessment

The first significant legislation in this area - the Law on State policy of the US Environmental Protection was signed in 1970. The adoption of this act triggered the interest of other countries, and from 1973, this process has become a part of the environmental legislation in Canada, from 1975 in future EU countries, namely in Germany, in 1976 in France and Italy, in Luxemburg in 1978, in 1986 in the Netherlands and the United Kingdom, followed by other countries: Norway, Sweden, Finland, Japan, Australia, New Zealand. [7]

The fact that the impact assessment is a necessary instrument for the environmental care and prevention of distortion and degradation of the environment was confirmed by the European Economic Commission in its second action plan for environmental protection in 1977. There, the important role was played by Directive 85/377/EEC from 1985 on the analysis of the effects of certain public and private projects on the environment, which was aimed to harmonize the laws in the EU countries and to contribute to better achievement the EU objectives in the field of environmental protection and life quality. [7]

3. EU legal framework in the field of environmental protection

EU legislation in the field of environmental protection (Environmental acquis communautaire) makes about 300 legal acts (directives, regulations and decisions), divided into eleven subsectors that represent the various spheres of environmental management [5]: (1) horizontal legislation; (2) air quality; (3) waste management; (4) water quality; (5) environmental protection; (6) control of industrial pollution and risk management; (7) chemicals and genetically modified organisms; (8) noise; (9) forestry; (10) climate change and (11) civil protection.

Horizontal legislation includes environmental legislation on various issues that are included in different areas of the environment, opposed to the legislation related to every single sector (e.g. air or water). It includes the integration of environmental protection in all economic sectors, strategic assessment of plans and programs, assessment of impact of projects on the environment, access to information and reporting on the implementation of directives about the environment. [4]

The following are directives of horizontal legislation on the assessment of effects of certain projects on the environment and their amendments [2]:

- Directive on the assessment of the effects of projects on the environment 85/377/EEC applies to those projects which may have significant effects on the environment. In this sense, the project means the execution of construction works or other installations or projects in the natural environment and landscape. [1] The EIA Directive 85/377/EEC of 1985 has been amended three times, in 1997, 2003 and 2009:
- Directive 97/11/EC brought the Directive in line with the Espoo Convention on EIA in a Transboundary Context. The Directive of 1997 widened the scope of the EIA Directive by increasing the types of projects covered, and the number of projects requiring mandatory environmental impact assessment (Annex I). It also provided for new screening arrangements, including new screening criteria (at Annex III) for Annex II projects, and established minimum information requirements.
- Directive 2003/35/EC was seeking to align the provisions on public participation with the Aarhus Convention on public participation in decision-making and access to justice in environmental matters.
- Directive 2009/31/EC amended the Annexes I and II of the EIA Directive, by adding projects related to the transport, capture and storage of carbon dioxide (CO2).
- Directive 2011/92/EU of the European Parliament and the Council of 13 December 2011 on the assessment of the effects of certain public and private projects on the environment, as amended, known as the "EIA" (environmental impact assessment) Directive, requires that an environmental assessment to be carried out by the competent national authority for certain projects which are likely to have significant effects on the environment by virtue, inter alia, of their nature, size or location, before development consent is given. The projects may be proposed by a public or private person.
- Directive 2014/52/EU of the European Parliament and the Council from April 16th 2014 amending Directive 2011/92/EU on the assessment of the effects of certain public and private projects on the environment. In order to strengthen the quality of the process of environment impact assessment, to harmonize this process with the principles of quality regulation and improve connections and synergies with other legislation and policies of the Union as well as strategies and policies that have been developed by Member States in the areas of national jurisdiction, it was necessary to amend the previous Directive from 2011.

Besides the above mentioned directives, the legal framework for environmental impact assessment includes international conventions because of which the directives have been amended: the Espoo, Aarhus and Kiev.

In 1991, in Espoo, Finland, the Convention on Environmental Impact Assessment in a Transboundary Context was adopted. The Convention sets out obligation of the parties to assess the impact of certain activities on the environment at an early stage of planning. It also obliges states to notify and consult each other in all major projects being under consideration which may have significant environmental impact across national borders. By convention, the neighboring countries have the right to participate in the environmental impact assessment if the planned project in one country can have an impact on the environment across its borders. [5]

Convention on Access to Information, Public Participation in Decision-making and Access to Justice in Environmental Matters was adopted in Aarhus, Denmark, in 1998. The Aarhus Convention is an international agreement which defines common standards the public's right to information, participation and access to justice. The aim was to contribute to protection of the rights of every person in present and future generations to live in the environment suitable for their health and well-being. [5]

Protocol on Strategic Environmental Assessment with the Convention on Environmental Impact Assessment in a Transboundary Context, was adopted in Kiev 2003. It aims to provide a high level of environmental protection, including health. Parties are required to provide a procedure that includes public participation. [5]

4. Key Elements Analysis Of Projects' Environmental Impact Assessment

When we look in which direction the legislation on environmental impact assessment has been developing, we can see that each new modification of the directive has included a new element to which attention should be payed. Thus, after adoption of the original Directive 85/377/EEC on the assessment of the effects of certain public and private projects on the environment, in 1997 the need for its modification due to compliance with the provisions of the Espoo Convention has been shown. Then the area of operation of the directive was extended to the environmental impact assessment across national borders. The next step was adoption of common standards on the public's right to information, participation and access to justice with the adoption of the Aarhus Convention, and therefore the directive was amended again in 2003. With the third amendment in 2009 the projects related to transportation, capture and storage of carbon dioxide were added into the directive.

After three amendments of the original Directive, by Directive 2011/92/EU in 2011, the EU legislation harmonized the principles for assessing the impact of projects on the environment by introducing minimum requirements with regard to the type of projects that are estimated, the most important obligations of the project coordinators, content of evaluation and participation of competent authorities and the public. It contributes to a high level of environmental protection and human health.

In order to strengthen the quality of the process of environmental impact assessment, to harmonize it with the principles of quality regulation and improve connections and synergies with other legislation and policies of the Union as well as with strategies and policies that have been developed by Member States in areas of national jurisdiction, it was necessary to amend the previous Directive from 2011, and in 2014, Directive 2014/52/EU which amends Directive 2011/92/EU on the assessment of effects of certain public and private projects on the environment was adopted. [6]

Furthermore, trends in environmental impact assessment are increasingly emphasizing the need to include some other elements in assessment procedures and decision-making process. In addition to the earlier inclusion of health and social impacts in environmental assessment, some even more important elements that have been emphasized in recent decade are: resource efficiency and sustainability; conservation of biological diversity; climate changes; risks of accidents and disasters; impact on the land; protection of marine environment; protection and promotion of cultural heritage and others.

More extensive studies of effectiveness and sustainability of resources have been included in the review of Directive 2011/92/EU.

There is also a need to assess significant adverse impacts of projects on biodiversity, in order to avoid or reduce their environmental impact to a minimum. [6]

Climate changes will continue to cause harm to the environment and endanger economic development. In this respect, it is appropriate to assess the impact of projects on climate (e.g. emissions of greenhouse gases) and their susceptibility to climate changes. [6]

In order to ensure a high level of environmental protection, it is necessary to take precautions in specific projects that, due to their susceptibility to major accidents and/or natural disasters (such as floods, sea level rise, earth quakes), are likely to have significant adverse impacts on the environment. For such projects it is important to consider their vulnerability (exposure and resistance) to major accidents and/or disasters, the risk of such accidents or disasters and indicators for the likelihood of significant adverse impacts on the environment. [6]

For public and private projects should be considered to limit their impact on the land, especially with regard to land use, and on the ground, including organic matter, erosion, compaction and closure ground. [6]

In order to ensure a high level of protection of marine environment, particularly species and habitats, should be taken into account features of such projects taking into account the used technologies. [6]

For better preservation of historical and cultural heritage and landscape, it is important to deal with the visual impact of projects. [6]

We can see that an environmental impact assessment has a significant effect on numerous projects or their modifications. In addition to positive effects such as the identification of key environmental issues, higher standards of mitigation of adverse consequences, making right decisions, it has its adverse effects. EIA Directive applies only to actions, without setting new environmental standards and it does not include stricter financial requirements. Application of environmental impact assessment depends on implementation of its principles into national law. In addition, the environmental impact assessment of individual projects is not enough, there is no alternative locations or methods, the dynamics of the environment and development in general is not processed, and the cumulative effects are ignored as well. [4]

Thus, due to the complexity of some influence and many synergistic effects is not enough to analyze individual object or only certain element but it is necessary make simultaneous analysis of many wider area and all the elements (air, water, soil, biota, etc.). In this respect, the significant is occurrence of strategic environmental impact assessment, which is not related to a specific project but to a much wider system, ultimately the relevant plans and programs which contain and coordinate the many individual projects. [10] It is in a certain way a complement to the environmental impact assessment and it is always current.

Strategic environmental assessment is a procedure that evaluates possible environmental impacts made by strategies, plans and programs, and their alternatives, proposed measures necessary to combat negative impacts, as well as the program of monitoring the impact. Primary purpose is application of process of ensuring sustainable development at a higher level of planning. In the world, it has been applied in the last ten years, and the EU member states, in accordance with the provisions of the Strategic Environmental Assessment (SEA) Directive 2001/42/EC from 2001, have been obliged to apply it since 2004. [3]

SEA Directive is an extension of EIA directive that raises individual projects on the level of plan and programme. The difference between these two directives is that SEA Directive is applied only in public programs which make political decision makers directly responsible for a negative impact on the environment. [4]

Therefore we can say that the EIA is an administrative instrument that applies to the particular project and represents a lower level of decision-making, and SEA is an advisory instrument of higher level of decision-making. [5]

In both cases of the environmental impact assessment, it is about procedure to ensure that in the decision on the acceptance of an intervention, plan, program or project, their possible impacts on the environment should be considered.

5. Conclusion

The EU Directive 85/377/EEC played an important role in the analysis of a certain public and private projects' environmental impact assessments. Its goal was to mutually harmonize national laws of the EU countries and to improve EU goal reach in the area of environmental protection and life quality.

After three amendments to the original Directive, the Directive from 2011 harmonized principles for assessing the impact of projects on the environment. In order to strengthen the quality of the projects' environmental impact assessment, then to synchronize the principles of quality regulation and to improve the coherence and synergies with other legislation and policies of the Union as well as with the strategies and policies developed by the Member States in the areas of national jurisdiction, it was necessary to modify the previous directive.

Every modification is driven by the necessity for the inclusion of the new elements in projects' environmental impact assessment. With the increased number of important elements in the procedures of environmental impact assessment and decision making and continuous review and improving instruments of legislation, aims at reducing the negative impact on the environment.

In order to reduce the negative impact on the environment, it is necessary to introduce a substantial number of important elements in the procedures of environmental impact assessment and decision making, as well as it is to continuously review and improve the legislation instruments.

6. References

[1] Access to the legislation of the European Union, Council Directive 85/337/EEC, Available from: http://eur-lex.europa.eu/LexUriServ/LexUriServ.do?uri=OJ:L:1985:175:0040:0048:EN:PDF, Accessed: 2016-06-17

[2] European Commission (2013), Assessing the impact of projects on the environment, Available from: http://ec.europa.eu/environment/eia/pdf/eia_case_law.pdf, Accessed: 2016-06-10

[3] CARDS2003 (2003), Assessment of Development Strategies on Environment, Strategic Environmental Assessment, Available from: http://www.zavod.pgz.hr/docs/zzpuHR/documents/139/Original.pdf, Accessed: 2016-06-20

[4] Črnjar, M. & Črnjar, K. (2009), Management of Sustainable Development (economy - ecology – environmental protection), University of Rijeka, Rijeka

[5] Herceg, N. (2013), Environment and Sustainable Development, University of Mostar, Mostar

[6] Official Journal of the European Union (2014), Directive 2014/52/EU of the European Parliament and of the Council, Available from: http://eur-lex.europa.eu/legal-content/EN/TXT/?uri=uriserv:OJ.L_.2014.124.01.0001.01.ENG, Accessed:2016-07-10

[7] Hodolić, J., Badida M., Majernik M. & Šebo D. (2005), Mechanical Engineering in Environmental Protection, Faculty of Technical Sciences in Novi Sad, Novi Sad

[8] Majstorović, V. (2010), Project Management, University of Mostar, Mostar

[9] Majstorović V., (2001), Upravljanje proizvodnjom i projektima (Production and Project Management), DAAAM International Vienna and University of Mostar

[10] Malbaša, N. & Jelavić V. (2013), History and current problems of environmental impact assessment in Croatia, First Regional Conference On Environmental Impact Assessment, Zadar

[11] Rajković, D. (2011) Cost-benefit analysis of the environmental impact assessment. University of Zagreb, Zagreb

DOI: 10.2507/27th.daaam.proceedings.014

INVERSE DYNAMIC ANALYSIS OF HOBBY ROBOT UARM BY MATLAB/SIMULINK

Vjekoslav Damic, Maida Cohodar & Marko Tvrtkovic

This Publication has to be referred as: Damic, V[jekoslav]; Cohodar, M[aida] & Tvrtkovic, M[arko] (2016). Inverse Dynamic Analysis of Hobby Robot uArm by Matlab/Simulink, Proceedings of the 27th DAAAM International Symposium, pp.0095-0101, B. Katalinic (Ed.), Published by DAAAM International, ISBN 978-3-902734-08-2, ISSN 1726-9679, Vienna, Austria
DOI: 10.2507/27th.daaam.proceedings.014

Abstract

Inverse dynamic analysis of robot manipulator with closed kinematic structure is performed in the paper. Dynamic model is developed using second generation blocks from Simulink/Simscape/SimMechanics library. Presented approach is applied on hobby robot uArm. Manipulator's links are created using 3D CAD models in form of step files. The mass and inertia properties of links are automatically calculated based on geometry defined by step files and link density, defined by user. The actuator torques acting at joints needed to acheive imposed joint motion are computed using inverse dynamics. Validity of the dynamic model is confirmed by comparison of coordinates of TCP obtained during simulation with analytical solution of direct kinematics.

Keywords: closed kinematic chain; inverse dynamic model; robot manipulator

1. Introduction

Investigation of industrial robot performances gains attention of many researchers and goes in several directions. Many papers are devoted to dynamic analysis of robots. Dynamic model of industrial robot IRB 4400 by bond graphs is subject of research in [1]. In this paper two models are developed: dynamic and visual one, using two different software packages BondSim and BondSimVisual, respectively [2], [3]. Object oriented environment of software BondSim provides systematically development of more complex systems. Between two models two-way communication is obtained providing exchange necessary information. In other words, two models behave as one. Dynamic analysis of systems with closed structure is subject of investigation in [1], [4] where bond graph are also used as suitable technique [1].

Calibration technique and algorithms of the robot is in focus of extend investigations. A geometrical approach for axis orientation parameters calibration of industrial robots using laser tracker is proposed in [5]. Noncontact methods for calibration of an industrial robot and identifying certain parameters of the robot model using a street vision system is subjects of research in [6]. Development of intelligent CAD/CAM systems is presented in [7]. A procedure for the dynamical parameters identification of a 6-DOF industrial robot has been presented in [8]. An efficient algorithm for identification of robot parameters including drive characteristics is applied on 4-DOF SCARA manipulator in [9].

With improvement of electronic component performances and reducing their manufacturing costs, the use of robots becomes significantly important in small companies, households, hospitals, as educational, hobby robots etc. This papers deals with small hobby robot uArm, found as open source project in [10]. Structure of robot, dynamic analysis and programing is explained in [11]. It is developed according to design of ABB IRB460. Actuator torques needed to achieve desired motion are computed using inverse dynamics.

Simulation model is developed using second generation blocks of Simscape/SimMechanics library [12]. Robot parts are imported as step files. An internal visualization window opens during simulation. It is very useful especially in the case of dynamic model development of a complex system analysis.

2. Kinematic model of robot uArm

The main subsystems of robot uArm are: mechanical part of the manipulator with the vacuum gripper, four servomotors and a microcontroller (Arduino), Fig.1. Manipulator has four degrees of freedom. Joints J1, J2 and J3 are actuated by servomotors, placed on the drive unit and provide positioning of the end-effector. Joint J13 is driven by the fourth servomotor to orient the vacuum gripper.

The manipulator consists of two mechanisms. Mechanism from the right side, Fig.1a, services to obtain the main movement of the arm. Parallelogram geometry between the shoulder and the elbow of an anthropomorphic arm creates closed kinematic chain. On the left side of uArm (Fig.1b), the other mechanism is integrated to manipulator. Because it holds $\overline{J2J7} = \overline{J8J9}$J=$\overline{J8J10} = \overline{J11J12}$, orientation of the Link 2 does not change during the motion. On this way, orientation of the vacuum gripper, attached to the Link 2 by joint J13, only changes by angle rotation about its axis provided by the fourth servomotor.

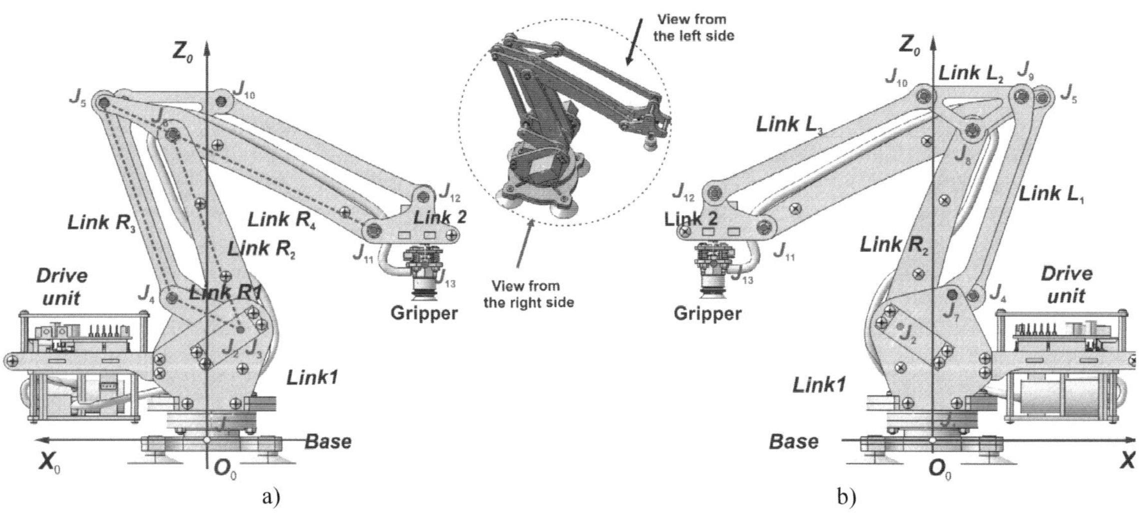

Fig. 1. The uArm manipulator mechanism: a) the right side; b) the left side

To develop kinematic model we used DH convention and followed approach presented in [13]. Joint J6 is adopted as cut joint. The uArm manipulator with attached coordinate frames is given in Fig.2. DH parameters are presented in Table 1.

Link	a_i	α_i	d_i	θ_i
Link 1	$-L_1$	90^0	L_0	θ_1
Link R1	L_2	0	0	$\theta_{2a}= \theta_2+55^0$
Link R3	L_3	0	0	θ_4
Link R4 (part O_3O_4)	L_2	0	0	θ_5
Link R2	L_3	0	0	$\theta_{3a}= \theta_3+55^0$
Link R4 (part O_4O_5)	L_4	0	0	0

Table 1. DH parameters for uArm manipulator

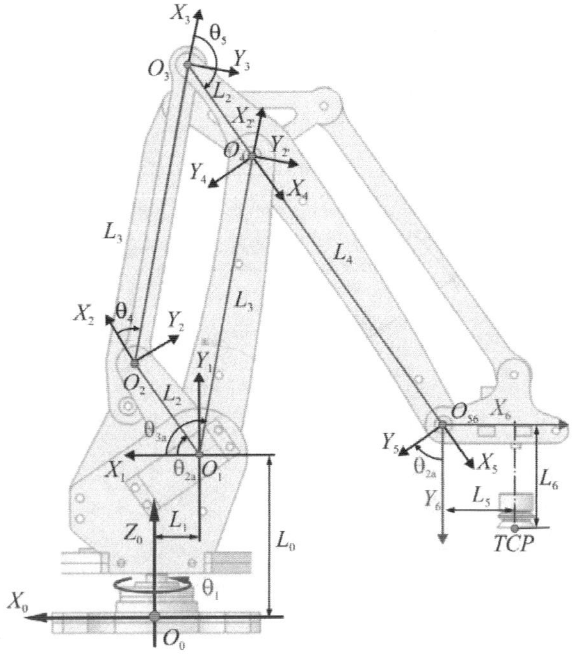

L_0 = 78.767 mm
L_1 = 20.728 mm
L_2 = 54 mm
L_3 = 148 mm
L_4 = 160 mm
L_5 = 35.450 mm
L_6 = 49.963 mm
$\theta_{2a} = \theta_2 + 55^0$
$\theta_{3a} = \theta_3 + 100^0$

Fig. 2. The uArm manipulator in its initial position (θ_i= 0, i=1,2,3) with attached coordinate frames

Matrices of homogenous transformations, according to defined DH parameters are given by:

$$\boldsymbol{A}_1^0 = \begin{bmatrix} c\theta_1 & 0 & s\theta_1 & -L_1 c\theta_1 \\ s\theta_1 & 0 & -c\theta_1 & -L_1 s\theta_1 \\ 0 & 1 & 0 & L_0 \\ 0 & 0 & 0 & 1 \end{bmatrix},$$

$$\boldsymbol{A}_4^1 = \boldsymbol{A}_2^1\boldsymbol{A}_3^2\boldsymbol{A}_4^3 = \begin{bmatrix} c\theta_{2a}\theta_4\theta_5 & -s\theta_{2a}\theta_4\theta_5 & 0 & L_2c\theta_{2a} + L_3c\theta_{2a}\theta_4 + L_2c\theta_{2a}\theta_4\theta_5 \\ s\theta_{2a}\theta_4\theta_5 & c\theta_{2a}\theta_4\theta_5 & 0 & L_2s\theta_{2a} + L_3s\theta_{2a}\theta_4 + L_2s\theta_{2a}\theta_4\theta_5 \\ 0 & 0 & 1 & 0 \\ 0 & 0 & 0 & 1 \end{bmatrix}, \tag{1}$$

$$\boldsymbol{A}_5^4 = \begin{bmatrix} 1 & 0 & 0 & L_4 \\ 0 & 1 & 0 & 0 \\ 0 & 0 & 1 & 0 \\ 0 & 0 & 0 & 1 \end{bmatrix}.$$

The frame $O_6X_6Y_6Z_6$ is attached to the link 2 whose orientation does not change during the motion due to action of the left mechanism (Fig.1b). Its orientation is defined by angle θ_{2a} regarding to frame $O_5X_5Y_6Z_5$ according to:

$$\boldsymbol{A}_6^5 = \begin{bmatrix} c\theta_{2a} & s\theta_{2a} & 0 & 0 \\ -s\theta_{2a} & c\theta_{2a} & 0 & 0 \\ 0 & 0 & 1 & 0 \\ 0 & 0 & 0 & 1 \end{bmatrix}. \tag{2}$$

For closed structure of manipulator (Fig.2), following conditions have to be satisfied:

$$\begin{aligned} &L_2[\cos(\theta_{2a}) + \cos(\theta_{2a} + \theta_4 + \theta_5)] + L_3[\cos(\theta_{2a} + \theta_4) - \cos(\theta_{3a})] = 0, \\ &L_2[\sin(\theta_{2a}) + \sin(\theta_{2a} + \theta_4 + \theta_5)] + L_3[\sin(\theta_{2a} + \theta_4) - \sin(\theta_{3a})] = 0. \end{aligned} \tag{3}$$

That implies:

$$\theta_4 = \theta_{3a} - \theta_{2a},\ \theta_5 = \pi - \theta_{3a} + \theta_{2a}, \tag{4}$$

where

$$\theta_{2a} = \theta_2 + 55^0, \theta_{3a} = \theta_3 + 100^0. \tag{5}$$

Using (3) and (4) matrix $\boldsymbol{A}_6^5$ can be reorganized and we obtain:

$$\boldsymbol{A}_5^0 = \boldsymbol{A}_1^0 \boldsymbol{A}_4^1 \boldsymbol{A}_5^4 = \begin{bmatrix} -c\theta_{2a} & s\theta_{2a} & 0 & L_3 c\theta_{3a} - L_4 c\theta_{2a} \\ -s\theta_{2a} & -c\theta_{2a} & 0 & L_3 s\theta_{3a} - L_4 s\theta_{2a} \\ 0 & 0 & 1 & 0 \\ 0 & 0 & 0 & 1 \end{bmatrix}. \quad (6)$$

Coordinates of TCP regarding to the frame $O_6X_6Y_6Z_6$ do not change during motion and can be defined by (in [mm]):

$$\boldsymbol{r}_{TCP}^6 = \begin{bmatrix} 35.450 \\ 49.963 \\ 2.500 \end{bmatrix}. \quad (7)$$

They can be expressed in the reference frame according to:

$$\mathbf{r}_{\mathrm{TCP}}^0 = \mathrm{A}_6^0 \cdot \mathbf{r}_{\mathrm{TCP}}^6, \quad (8)$$

and given by (in [mm]):

$$\begin{aligned} X_{TCP}^0 &= 2.5 \cdot \sin(\theta_1) - \cos(\theta_1) \cdot [160 \cdot \cos(\theta_2 + 0.9599) - 148 \cdot \cos(\theta_3 + 1.7453) + 56.178] \\ Y_{TCP}^0 &= -2.5 \cdot \cos(\theta_1) - \sin(\theta_1) \cdot [160 \cdot \cos(\theta_2 + 0.9599) - 148 \cdot \cos(\theta_3 + 1.7453) + 56.178] \\ Z_{TCP}^0 &= -160 \cdot \sin(\theta_2 + 0.9599) + 148 \cdot \sin(\theta_3 + 1.7453) + 28.804. \end{aligned} \quad (9)$$

3. Dynamic model of manipulator uArm

Dynamic model is developed using the second generation blocks of Simscape/SimMechanics of software Matlab/Simulink, Fig.3.

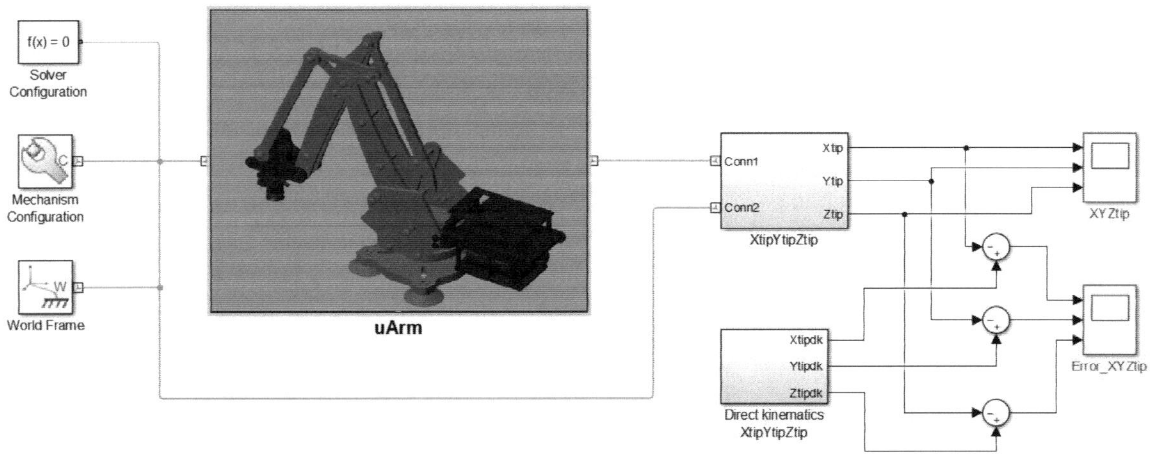

Fig. 3. Dynamic model of the uArm manipulator by SimMechanics

Dynamic model of uArm is organized on several levels by subsystems. Robot manipulator is presented by subsystem uArm on the first level. Its structure is depicted in Fig.4. Links are considered as rigid bodies. They are represented by 'solid' blocks taken from the library Simscape/SimMechanics/Second generation (We used Matlab v.2015b; this library is called Multibody in Matlab/Simulink, v.2016). The block 'solid' provides representation of the rigid body geometry by some basic shapes, but also importing 3D CAD models in form of step or stl files. In this paper we imported the manipulator links in form of step files. To create step file of each link, we used 3D CAD models of robot parts from [10]. They are imported in CATIA in which robot assembly has developed. Manipulator consists of many parts: screws, nuts, levers, bearings etc. represented by theirs 3D CAD models. They are grouped in order to create links. Each link is exported from CATIA in form of step files.

Block 'solid' enables importing step files and the mass and inertial properties of the links, based on its geometry and specified density by user, are automatically calculated.

Revolute joints are represented by blocks 'Revolute joint', which permits rotation about its *Z*-axis. Because of this, we need rigid transformation before and after rotation that is realized by blocks 'Rigid transformation' as depicted in Fig.5a. Joints J1, J2 and J3 are actuated. Joint actuator is represented by subsystem 'Joint *i* actuated'.

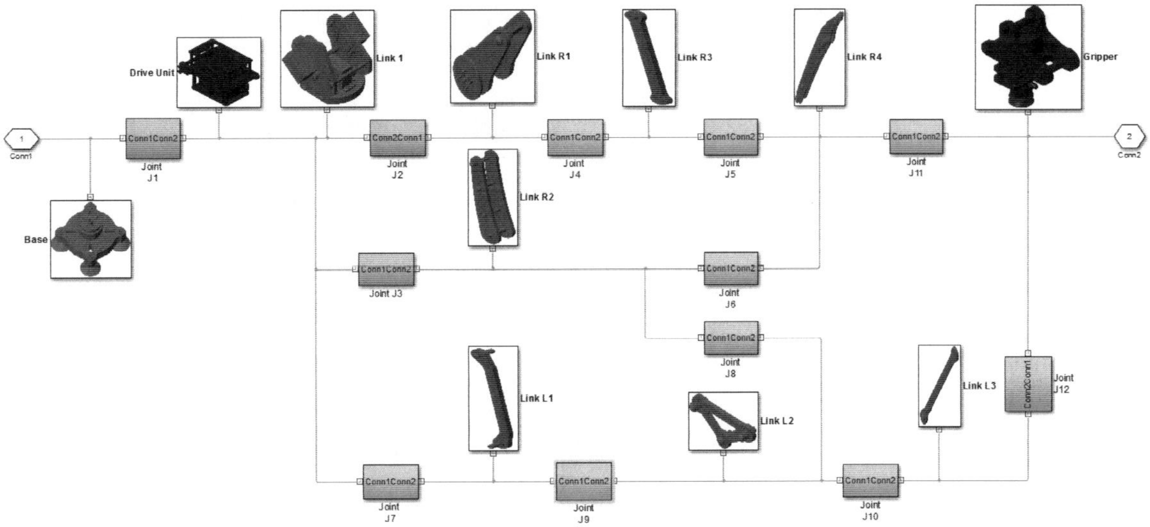

Fig. 4. Structure of the uArm manipulator model

Structure of subsystem 'Direct kinematics' from the first level of decomposition (Fig.3) is shown in Fig.5b. This provides calculating coordinates of TCP, according to (9). Subsystem denoted by XtipYtipZtip consists of sensor to sense coordinate of TCP point.

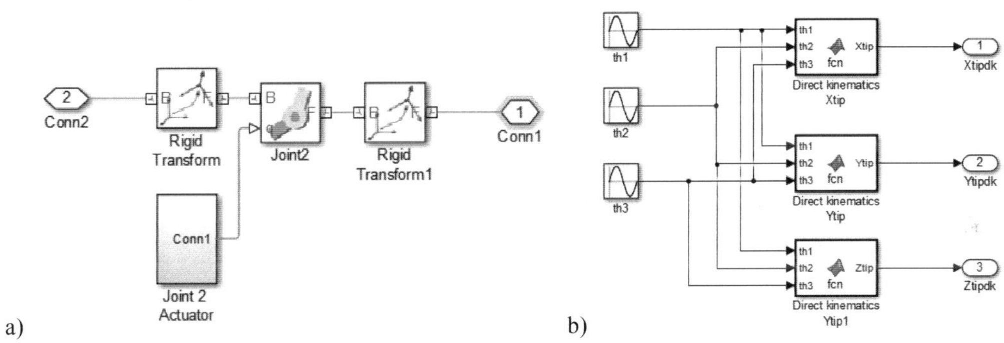

Fig. 5. Structure of subsystem: a) Revolute joint; b) Direct kinematic model

Joints J1, J2 and J3 are actuated according to laws (in [deg]):

$$\theta_1 = 150 \cdot \sin(0.25t),\ \theta_2 = 15 \cdot \sin(t);\ \theta_3 = 20 \cdot \sin(t). \tag{10}$$

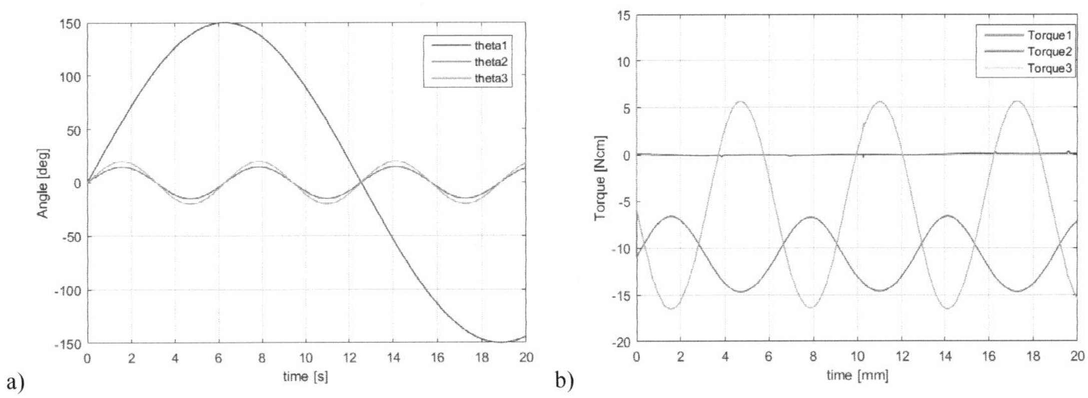

Fig. 6. Simulation results: a) Change joint angles with time; b) Change actuator torques with time

Simulation time was 20 s. During the simulation solver ode45 (variable step) is automatically chosen. It is the Runge-Kutta (4,5) solver of a fifth-order methods to a fourth-order estimate of the error [12]. Simulation results are presented in Figs.6 and 7. Imposed motions in joints and actuator torques needed to obtain desired motion are shown in Fig.7a and 7b. Changes of coordinates of TCP with time is presented in Fig.7a. Values of TCP coordinates are compared with ones obtained analytically, by (9). Errors in TCP coordinates, as shown in Fig.7b are order of 1e-7 m that confirms validity of simulation model.

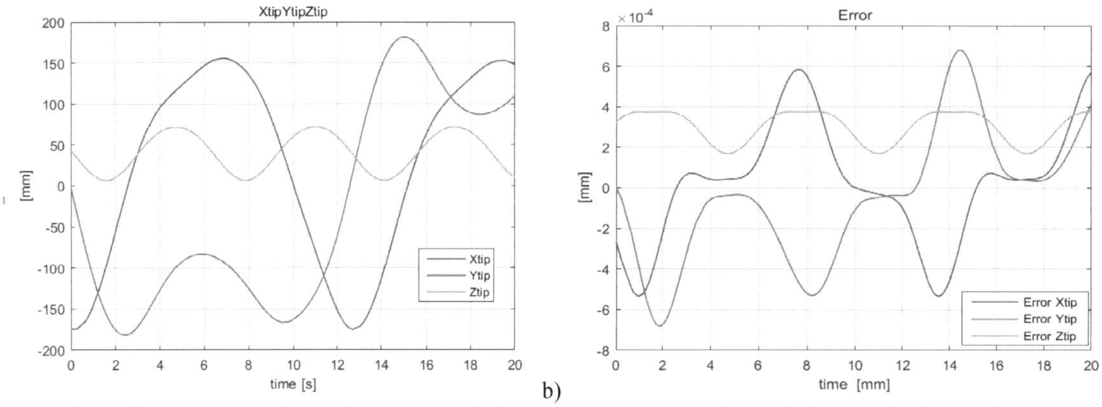

Fig. 7. Simulation results: a) Coordinates of TCP point in the global frame; b) Error in TCP coordinates

Animation screen opens during simulation providing visualization of robot motion (Fig.8). This is grateful aid in building of model.

Fig. 8. Internal animation screen during simulation

Dynamic model, based on an external 3D CAD files, can be automatically generated using Simscape Multibody importer. Firstly, we developed assembly of robot uArm in SolidWorks taking 3D CAD parts taken from [10]. Robot assembly is exported from SolidWorks and imported to Simulink. Model is generated using blocks of second generation of Simscape/SimMechanics on the one level, as shown in Fig.9a. It can be modified as any other Simscape models. Obtained simulation results are the same as previously reported. Animation screen is depicted in Fig.9b.

4. Conclusion

Inverse dynamic analysis of hobby robot manipulator uArm is considered in this paper. The paper explains the procedure for development of robot dynamic model based on 3D CAD models of robot links. The model is developed using blocks from second generation of Simscape/SimMechanics Matlab/Simulink library. Links are presented in form of step files providing automatically calculation of mass and geometric properties. To compute the mass properties, the link density is specified by user. During simulation the first three joints are actuated according to sine law. Actuator torques needed to obtain imposed motion in actuated joints are computed using inverse dynamics. To verify developed model, computed *X*-,*Y*- and *Z*-coordinates of TCP in the global frame are compared with analytical solution of the direct kinematic model. Comparison shows very good agreement of obtained results.

In the future investigation the dynamic model will be extended on the whole robot including servo-motors. Simulation results will be compared with experimentally obtained results.

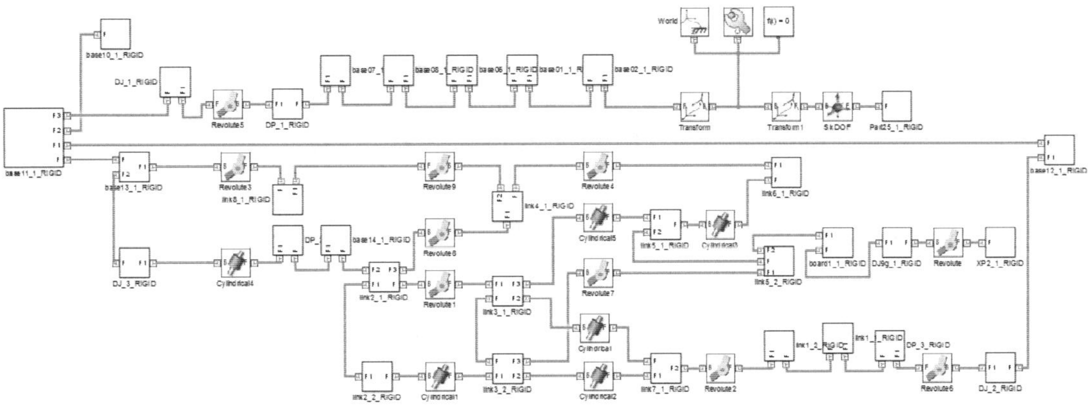

a)

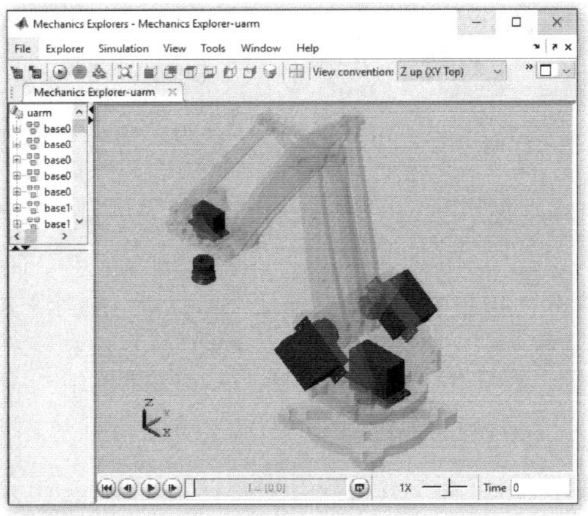

b)

Fig. 9. a) Automatically generated dynamic model; b) Animation screen

5. References

[1] Damic, V. &. Cohodar, M. (2015). Dynamic analysis and visualization of spatial manipulators with closed structure, Proceedings of the 26th DAAAM International Symposium, pp.0109-0117, B. Katalinic (Ed.), Published by DAAAM International, ISBN 978-3-902734-07-5, ISSN 1726-9679, Vienna, Austria DOI: 10.2507/26th.daaam.proceedings.016

[2] Damic, V. &. Montgomery, J. (2015). *Mechatronics by Bond Graphs, An object oriented Approach to Modelling and Simulation*, Springer-Verlag, ISBN 978-3-662-49002-0, Berlin Heidelberg

[3] www.bondsimulation.com, Accessed: 2016-09-01

[4] Damic, V. & Cohodar, M. (2015). Dynamic analysis of Stewart platform by bond graphs, *Procedia Engineering,* Vol.100, 2015, pp.226-233, doi: 10.1016/j.proeng.2015.01.362.

[5] Vorotnikov, A; Bashevskaya, O; Ilyukhin, Y; Romash, E; Isaev, A. V. & Poduraev, Y. (2016). Geometrical Approach for Industrial Robot Axis Calibration Using Laser Tracker, Proceedings of the 26th DAAAM International Symposium, pp.0897-0904, B. Katalinic (Ed.), Published by DAAAM International, ISBN 978-3-902734-07-5, ISSN 1726-9679, Vienna, Austria DOI: 10.2507/26th.daaam.proceedings.125

[6] Svaco, M; Sekoranja, B; Suligoj, F. & Jerbic, B. (2014). Calibration of an Industrial Robot using a stereo vision system, *Procedia Engineering,* Vol.69, 2014, pp.459-463, doi: 10.1016/j.proeng.2014.03.012.

[7] Klancnik, S; Brezocnik, M; Balic, J. (2016) Intelligent CAD/CAM system for programming of CNC machine tools. *International journal of simulation modelling*, ISSN 1726-4529, 2016, vol. 15, no. 1, pp. 109-120

[8] Ding, L; Wu, H; Yao, Y. & Yang, Y.(2015). Dynamic model Identification for 6-DOF Industrial Robots, *Journal of Robotics*, Vol.2015, 2015, https://www.hindawi.com/journals/jr/2015/471478/, http://dx.doi.org/10.1155/2015/471478

[9] Chan, S.P. An Efficient Algorithm for Identification of Robot Parameters Including Drive Characteristics, Journal of Intelligent and Robotic Systems (2001) 32: 291. doi:10.1023/A:1013918927148

[10] www.ufactory.cc/ Accessed: 2016-09-01

[11] Tvrtkovic, M. (2016). Modeling and dynamic analysis of robot manipulator, MSc thesis, Faculty of Mechanical Engineering University of Sarajevo, B&H

[12] Matlab tutorials, www.mathworks.com, Accessed: 2016-09-01

[13] Siciliano, B.; Sciavicco, L.; Villani, L. & Oriolo G. (2009). *Robotics, Modelling, Planning and Control,* Springer-Verlag, London

DOI: 10.2507/27th.daaam.proceedings.015

LIGHTWEIGHT EDUCATIONAL SCARA ROBOT FOR PALLETIZING OF USB STICKS

Kemajl Stuja, Mirlind Bruqi, Erich Markl & Mohamed Aburaia

This Publication has to be referred as: Stuja, K[emajl]; Bruqi, M[irlind]; Markl, E[rich] & Aburaia, M[ohamed] (2016). Lightweight Educational Scara Robot for Palletizing of USB Sticks, Proceedings of the 27th DAAAM International Symposium, pp.0102-0108, B. Katalinic (Ed.), Published by DAAAM International, ISBN 978-3-902734-08-2, ISSN 1726-9679, Vienna, Austria
DOI: 10.2507/27th.daaam.proceedings.015

Abstract

This paper deals with design of customized lightweight SCARA- manipulator for educational purposes. In addition to this task shows this paper one concept solution for palletizing of small pieces like USB- Sticks using lightweight Scara. Design and optimization of the mechanical structure was accomplished using the powerful CAE tool -SolidWorks from Dassault Systèmes. Robot parts are generated using additive manufacturing technology- laser sintering of polyamide. The core part of robot controller consist of Arduino- card. Programming will be done using simple editor, which translates a moving and logical instructions of robot into Arduino real time instructions. The vision system mounted at the stand will be responsible for recognition of work-pieces, sending a key information's about the position and orientation of unsorted parts on the incoming conveyor. The Robot pick up the parts and places in ongoing trays into second conveyor. The results of this problem are shown and concisely discussed in this paper. Further work will be focused on developing HMI-human machine interface for using small devices, such are mobile phones and tablets, to jog and programme the robot targets.

Keywords: Scara Robot; CAD; Simulation; Motion Control and Palletizing.

1. Introduction

The modern production systems are inclined for custom-made solutions. In the past customized solutions are accompanied with very high engineering costs and very long setup time. Nowadays with the support of a powerful computer design software and advanced manufacturing systems it is possible to manufacture and implement components into production systems. Additive manufactured manipulator controlled with low cost controller card besides the educational purpose will be enough accurate to palletize small parts like a USB-Stick. The main task of the manipulator is to pick up the unsorted USB- Sticks from an input conveyor and to place this into trays - output conveyor. The distance between the two conveyors is about 350 mm at the same level.

This work shows stage by stage a design of the robot system as well as its integration to the palletizing cell. As mentioned before for this stage of the research a CAE tool SolidWorks was used. The results of this work confirmed that the designed and manufactured robot system under certain circumstances can be used not only in the education but as far for accomplishing a simple manufacturing tasks. The main advantage of this robot is the light weight and the small moment of inertia. So the fast acceleration and short retardation of the robot joints are very significant compared to the heavy casting robot. The energy consumption is the fundamental feature of this robot system. The main disadvantage of this robot is the rigidity compared to heavy casting robot.

In the previous work [1] a small 4 axis serial kinematic polyamide robot for palletizing handy covers, was presented. The main disadvantage of this robot- caused by serial kinematic- was vibrations and an elongated time for stabilisation of manipulator, when a high speed was required. In the second work [2] was designed a serial Research Robot for Didactical Development with 6 DoF. The main advantages of this robot can be described as:

- Low- Cost
- Lightweight
- Open hardware and software architecture
- Modular Structure
- Safe use
- Easy to disassemble, in order to explain the internal Structure of the Robot to students etc.

Open hardware and software architecture enables the possibility for mechatronic students to research and develop the future modular structures and the motion control algorithms. In the figure 1 a) was shown the CAD Model of palletizing robot. Figure 1 b) shows the 6 axis didactical vertically knick arm robot.

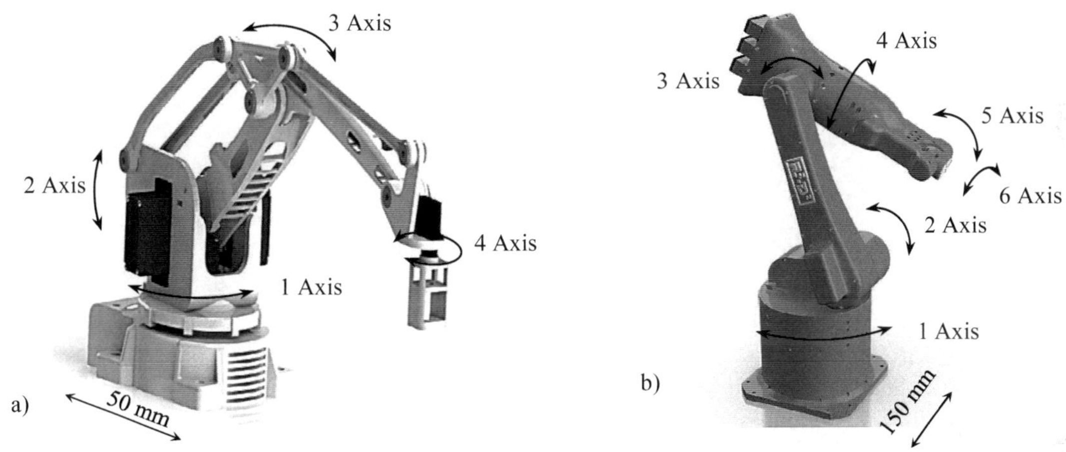

Fig. 1. Education Robots a) 4 DoF Palletizing Robot and b) 6DoF Knickarm Robot

2. Design of Scara Robot

Robot links are made by using of the plastic laser sintering system EOS FORMIGA P 110. This Machine uses laser to sinter powdered material, binding it together to create a solid structure of the parts. The slice unit of this platform is about 0.1 mm (layer thicknesses). The used powder material is PA 2200 which is a non-filled powder on basis of PA 12 with the following characteristics [3]:

- High strength /stiffness.
- good chemical resistance,
- excellent long-term constant behavior,
- high selectivity and detail resolution,
- Various finishing possibilities etc.

2.1. Mechanical Design of the prior 3D Draft

The first step, which is necessary to be done, is a drafting of links and Joints of the Scara. Traditionally the procedure starts with the calculation and selection of the motors and gears. Whereas industrial design starts with 3D modelling in order to calculate the motors and the gear. This paper shows the second procedure. Since the robot will be manufactured using additive technology, it is very imported to take care of the feasibility of the 3D printing. The Links which are very long and outside of work area of the 3D Printer, will be fragmented into more parts. This approach is known in the literature as "Designing for manufacturing". The software tool used for 3D modelling was SolidWorks from Dassault Systèmes. SolidWorks covers a wide range in the engineering area such as Mechanical design and optimization, static simulation and dynamic motion analysis. The biggest motor Nema23 of the first joint and second motor Nema 17 of the second joint are placed collinear and bolded to the static part of the robot- to base. The last two motors Nema 11 concerns rotation and translation of the last link of the robot. The picture 2 b) shows the kinematic structure of the 4 axis Scara – RRR T (R- Rotational joint, T- Translational joint) and the main dimensions.

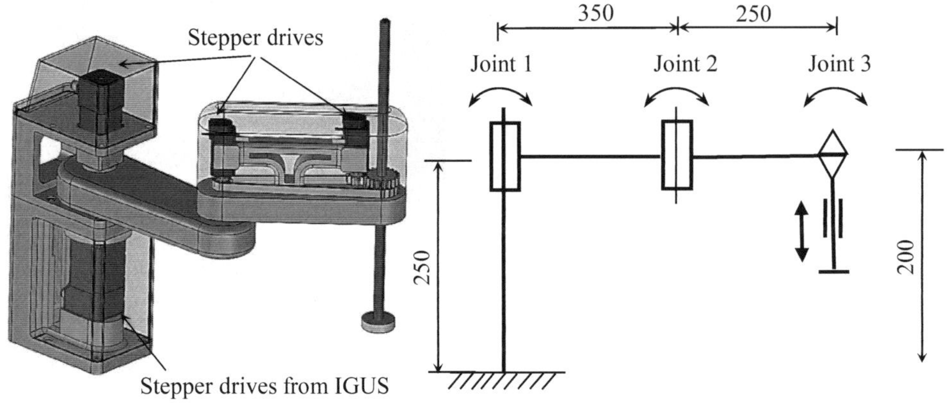

Fig. 2. Scara robot a) CAD design and position of drives, b) kinematic structure of the Scara

2.2. *Drive calculation for the cycle time*

Traditional drive calculations procedure starts with moving profile for the cycle time. In order to achieve the cycle time all drives must ensure the required moment of torque. As shown in the figure below every joint of the robot has three moving stages: acceleration, constant speed and retardation. Following this when the angular velocity is constant the Torque is Zero. In fact for the acceleration and retardation it's very important to calculate the moment of inertia [3].

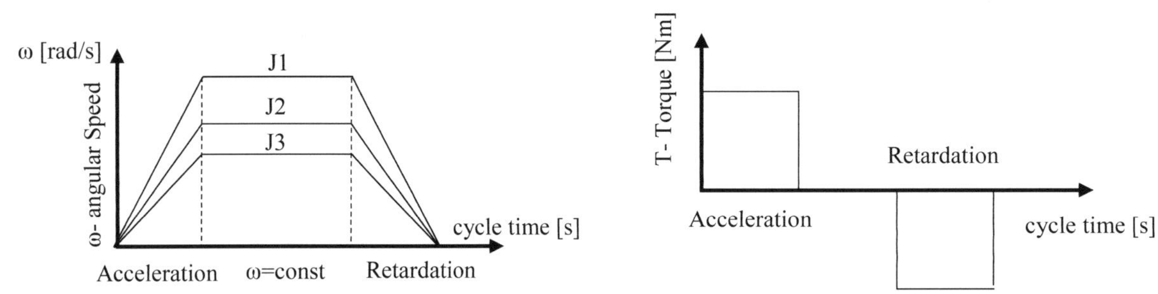

Fig. 3. Movement profile a) The first three joints of SCARA, b) torque

In order to select appropriate motor-gear system or drive system a torque must to be considered:

$$M = I * \varepsilon \quad (1)$$

Where: M –Torque [Nm], I - inertia [kgm^2] and ε angular acceleration [rad/sec^2].

The second argument is angular acceleration. Cycle time required in this work for pick and place operation was two seconds. That means angular acceleration can be calculated easily from moving profile. A fist argument is mass moment of inertia. Calculation takes very long time, if no CAD system is being used. The more complex the geometry is, the more complex is the calculation. Using CAD system is only a matter of a few clicks. Mass moment of inertia is defined like:

$$I = \sum_{i=1}^{n} r_i^2 \, dm_i \quad (2)$$

r [m] is the distance from the centre of gravity and rotation axis is usually expressed in meter and m is the mass expressed in kilogram. It is apparent from the (2) formula, that not only the mass of robot parts is a critical parameter for the calculation of the system. A very importing parameter is the distance from the rotation axis. For this reason the second motor was placed to the fixed base in order to reduce the mass of the system and to displace the centre of gravity near to the first link. Secondly, using the PA 2200 material allows to reduce the mass of robot and to take care of energy consumptions. The figure 4 a) is a given a snapshot using 3D CAD system SolidWorks for calculation of the centre of gravity and moment of inertia tensor. In figure 4 b) is the simulation of the first joint in SolidWorks is shown. For the acceleration of the rest of the system together with end-effector and work piece was needed approximately about 2 [Nm].Meaning that the drive system must be selected for this quantity in order to fulfil cycle time requirements from the research demands, for the worst case scenario situation.

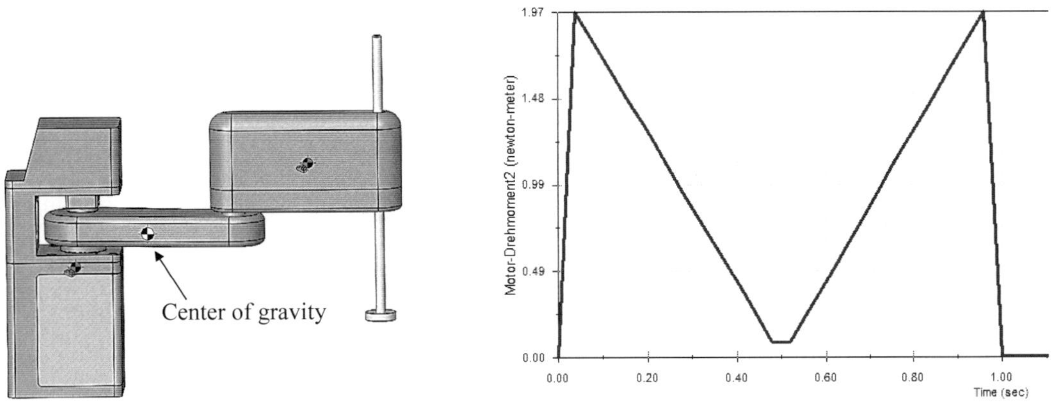

Fig. 4. Scara robot a) CAD design, b) kinematic structure of the Scara

2.3. Robot kinematics

Primarily the inverse kinematics was solved. Inverse kinematics deals with the problem of finding the required joint angles to produce a certain desired position and orientation of the end-effector. Finding the inverse kinematics solution for a general manipulator can be a very tricky task. Generally they are non-linear equations. Close-form solutions may not be possible and multiple, infinity, or impossible solutions can arise. Nevertheless, special cases have a closed-form solution and can be solved [3].

For solving the robot kinematics, a very simple approach has been followed. By simplifying the 3 dimension problematic into plan problem (x, y coordinates) and using Pythagoras and Law of cosines to find the position for the first and the second axis. After that the shift in the third demining (z-coordinate)

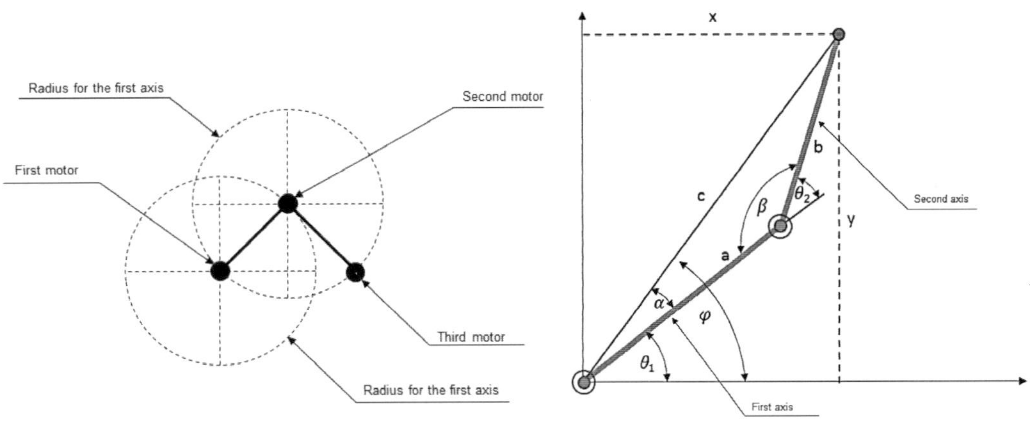

Fig. 6. Solving the inverse kinematics for the first and second axis using Pythagoras and Law of cosines

Finding θ_2- angle for the second joint of robot:

$$c^2 = x^2 + y^2 \quad (3)$$

$$c^2 = a^2 + b^2 - 2abcos(\beta) \quad (4)$$

$$\beta = 180 - \theta_2 \rightarrow \cos(\beta) = \cos(180 - \theta_2) = -\cos(\theta_2) \quad (5)$$

$$\theta_2 = \arccos\left(\frac{-a^2 - b^2 + c^2}{2ac}\right) \tag{6}$$

$$why\ \ c = \sqrt{x^2 + y^2}, \text{ then} \tag{7}$$

$$\theta_2 = \arccos\left(\frac{-a^2 - b^2 + x^2 + y^2}{2a\sqrt{x^2 + y^2}}\right) \tag{8}$$

Finding θ_1- angle for the first joint of robot:

$$b^2 = a^2 + c^2 - 2accos(\alpha) \tag{9}$$

$$\alpha = arcos(\frac{a^2 - b^2 + c^2}{2ac}) \tag{10}$$

$$\alpha = arcos(\frac{a^2 - b^2 + x^2 + y^2}{2a\sqrt{x^2 + y^2}}) \tag{11}$$

$$\theta_1 = \varphi - \alpha \tag{12}$$

$$\varphi = \arctan\left(\frac{y}{x}\right) \tag{13}$$

$$\theta_1 = (\arctan\left(\frac{y}{x}\right) - \arccos(\frac{a^2 - b^2 + x^2 + y^2}{2a\sqrt{x^2 + y^2}})) \tag{14}$$

2.4. Robot Control System

Stepper motors have found their way into many different areas of automation and control systems. The wide popularity of these motors can be attributed in part to the various ways the motor can be driven and because of its compatibility with digital systems. In particular, stepper motors are ideal for control systems that require discrete, easily repeatable movements at moderate to low frequencies. The figure below shows an example block diagram of a system with microcontroller, stepper motor, and encoder- feedback. In this research, stepper motors with a feedback provided for precision assistance will be used [4].

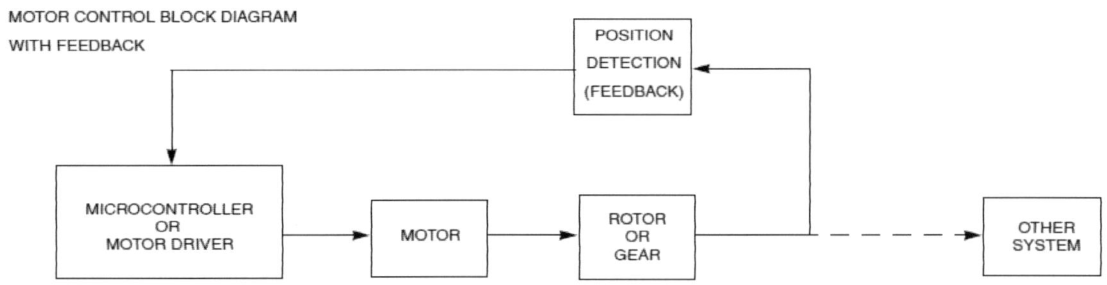

Fig. 6. Example System with a Microcontroller, Stepper Motor, and Feedback [4]

For controlling the robot system, the open Open-source electronic prototyping platform Arduino will be used. Simulink (the add-on product to MATLAB) provides an interactive, graphical environment for modelling, simulating, and analysing of dynamic systems. It enables rapid construction of virtual prototypes to explore design concepts at any level of detail with minimal effort [5]. Simulink has also a support package for Arduino hardware, which allows users to develop and simulate algorithms that run standalone on the Arduino and configuring and accessing the sensors and actuators on Arduino [7]. The following figure shows how to use the Stepper Motor Driver and Stepper Motor blocks together to implement a controlled permanent magnet stepper motor with Simulink [6].

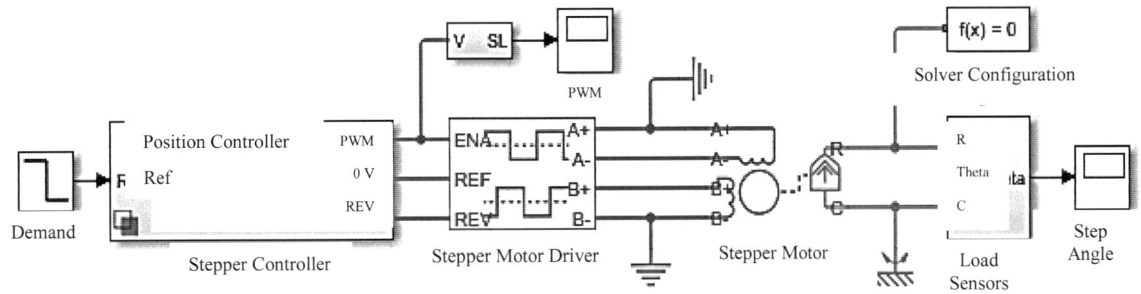

Fig. 7. Controlling a stepper motor with Simulink [6]

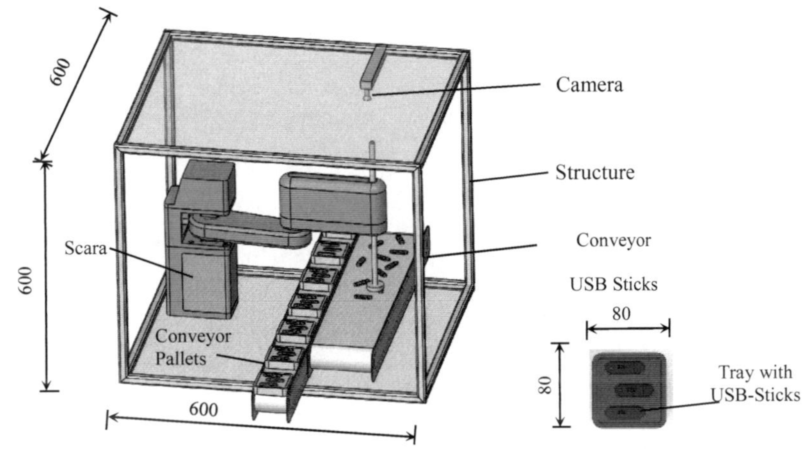

Fig. 8. Layout of robot cell for palletizing of USB- Sticks

5. Conclusion

The goal of this research was to design and manufacturing robot system for educational purposes and to improve a performance of this system for by simple manufacturing task, such is palletizing of USB- Sticks (see the Figure 8). The design stage was done using the powerful CAD tool- SolidWorks. This powerful tool reduced radically the engineering effort and provided a possibility for the optimization of robot mechanical structure. In this perspective was reduced the weight of manipulator converting them into lightweight robot class. Lightweight ability of robot allows the fast acceleration as well as short the breaking time. The energy consumption per cycle time was far below comparing him with industrial robot. The outcomes of this research confirmed our statement, which for the simple manufacturing tasks and moderate accuracy, additive manufactured robots provides passable solution.

However this work left some opened issues and future works to be done. The future works are focused in optimizing the structure of robots and human machine interface. The heavy step motors will be replaced by lightweight servo motors in order to reduce the mass of the robot as well for better positioning / path accuracy. By using of high resolution of encoders it will be possible to capture the vibrations in order to counteract with control logic. The second task will be focused on developing HMI- human machine interface. It will be possible to use the tablets and small devices like mobile phones to jog and programme the robot targets.

6. Acknowledgments

We would like to express our special thanks of gratitude to IGUS- Company- Austria for sponsoring the key components of the Scara robot. Without this support it would not be possible to realise this research inside the planed period. IGUS produces long life industrial components such as cables, connectors, bearings and joints-links for small robots.

7. References

[1] Aburaia, B.; Markl, E. & Stuja, K. (2014). New concept for design and control of 4 axis robot using the additive manufacturing technology, 25th DAAAM International Symposium on Intelligent Manufacturing and Automation, Elsevier, ISSN 1877-7058, Katalinic, B. (Ed.), pp. 1364-1369, Vienna.

[2] Schweidler, R.; Aburaia, M. & Engelhardt-Nowitzki, C. (2016). Design of an Industrial Robot with Six Degrees of Freedom for Educational Purposes, OAGM & ARW joint workshop on "Computer Vision and Robotics" Kurt Niel, Peter M. Roth, Markus Vincze. (Ed.), pp. 155-161, Wels, Austria

[3] Pires, J. N. (2007), *Robot Manipulators and Control Systems*, ISBN: 978-0-387-23325-3, Publisher: Springer US, New York, DOI 10.1007/b101252.

[4] http://pro.sculpteo.com (2014), PA2200 Datasheet, Fine Polyamide PA 2200 for EOSINT P, Available at: http://pro.sculpteo.com/media/data/faq/pa2200, Accessed: 2014-09-20.

[5] Matthew Grant, (2005). Quick Start for Beginners to Drive a Stepper Motor, Available from: http://www.nxp.com/files/microcontrollers/doc/app_note/AN2974.pdf Accessed: 2016-09-20.

[6] https://ewh.ieee.org (2016), IEEE- Institute of Electrical and Electronics Engineers, New York, Available from: https://ewh.ieee.org/r1/ct/sps/PDF/MATLAB/chapter8.pdf, Accessed: 2016-09-20.

[7] http://de.mathworks.com (2016), MathWorks, Natick, Massachusetts, USA Available from: http://de.mathworks.com/ help/physmod/elec/ examples/controlled-stepper-motor.html, Accessed: 2016-09-20.

[8] http://de.mathworks.com (2016), MathWorks, Natick, Massachusetts, USA Available from: http://de.mathworks .com/hardware-support/arduino-simulink.html?requestedDomain=de.mathworks.com, Accessed: 2016-09-20.

DOI: 10.2507/27th.daaam.proceedings.016

PROGRAMMING OF ROBOT SYNERGISM IN MULTI-AGENT SIMULATORS

Valentin Pryanichnikov, Denis Davydov, Kirill Kirsanov & Stanislav Eprikov

This Publication has to be referred as: Pryanichnikov, V[alentin] E[.]; Davydov, D[enis]; Kirsanov, K[irill] & Eprikov, S[tanislav] (2016). Programming of Robot Synergism in Multi-Agent Simulators, Proceedings of the 27th DAAAM International Symposium, pp.0109-0115, B. Katalinic (Ed.), Published by DAAAM International, ISBN 978-3-902734-08-2, ISSN 1726-9679, Vienna, Austria
DOI: 10.2507/27th.daaam.proceedings.016

Abstract

This article reviews the problem of building software systems for control of distributed hardware sets of mechatronic systems, mobile robots and their virtual models. For this purpose, application-dependent software ("middleware") was built, which allows performing operations on asynchronous data buses with correct network input/output of various controlled devices along with combination of work of real robots and their virtual images. One of the well-known approaches to building adequate mathematic models of mobile robots is the use of V-REP robotic simulator, its abilities for creating new solid models out of the known components and modules. However, such an approach allows us to solve the design problem of separately functioning robotic systems. This study describes an approach to create a distributed robotic simulator, which allows programming the control of mobile service robots by generating the so-called "synergism" – generic templates of more elementary (or previously constructed) actions.

Keywords: mobile service robots with distributed control; robotic simulator; simulated environment; V-REP.

1. Introduction

On the basis of the conducted research of various options of technical implementation of the project for creating a network of Internet laboratories called "Intellectlnaya Robotronika" and building supervisory control via Internet by mobile training robots AMUR (*АМУР*) (International Laboratory "Sensorika") and Robotino (Festo) with organization of video monitoring of command execution, a new program architecture has been proposed for the distributed data measuring and control system (DMCS). This approach to building DMCS has shown stability against network delays and allowed to overcome the main disadvantages of the existing approaches by means of constructing three principal components (Fig.1):

- Worker – This component implements a parallel process, which performs communication between the mechatronic device and other system components.
- Connector – This component implements a network protocol for access to mechatronic components.
- Locator / Combinator – This component enables the search of mechatronic components within the network, accumulates, and provides access to, data buses, provides dispatching of messages and aggregation service.

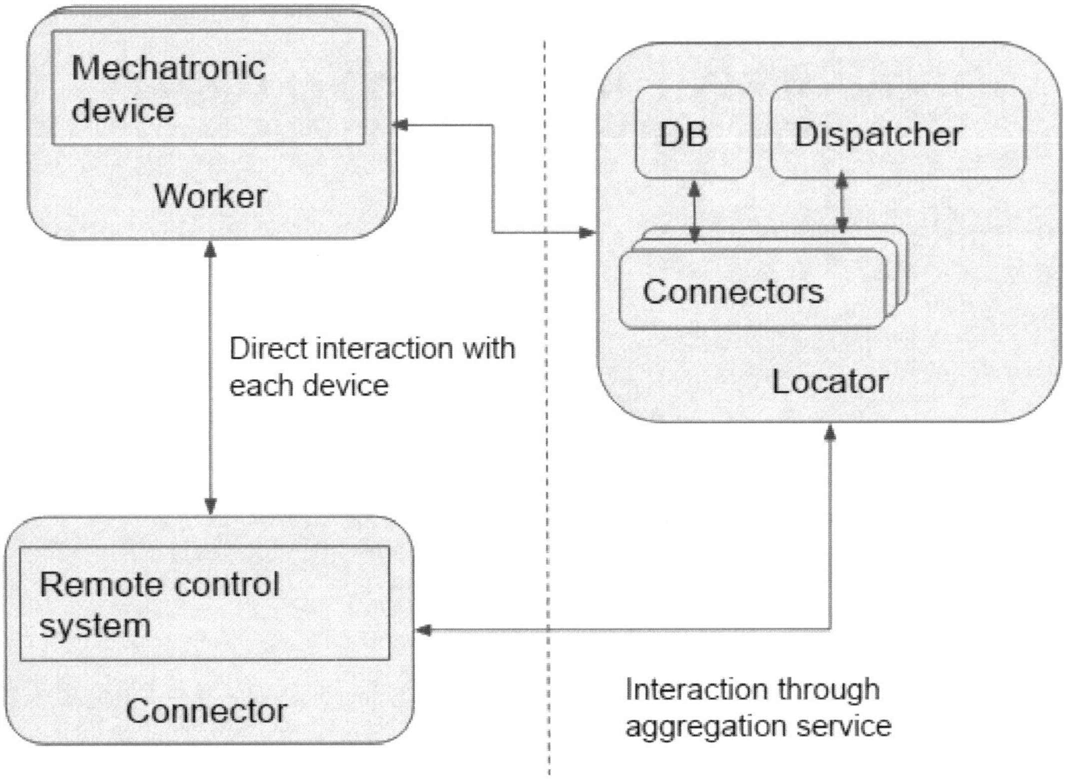

Fig. 1. The scheme of interaction of software components: Onboard devices interact directly with each other. Remote control system gains access to aggregating service. It allows for direct communication with remote control system for debugging purposes

In consequence of adding a database into the structure of controlling software, there appears an opportunity to analyze and compare the measurements recorded and commands either in real time mode, or after the event. DB data are stored in "rough form", i.e. without additional conversion, and for this reason one needs to know their structure in order to work with them.

This information is contained in a manifest file, which is a part of the driver. It contains information on the transmitted data, commands and their SI-system values as well as on operation frequencies of the main loops. This metadata allows organizing the interoperation of mechatronic devices on a higher level by means of intelligent interpolation and extrapolation, synchronization of loops with different frequencies.

Also, the use of DB makes it possible to simplify the indication process of key synergies of the mechatronic system (Fig.2). For this task, a prototype of the editing tool having access to both reference measurements and their interpolated/extrapolated forms.

It provides the possibility to:

1. View the measurements and commands at various scales;
2. Automatically calculate the basic statistic values – mathematical expectation, dispersion, correlation with other measurements or commands;
3. Provide displacement and scaling of measurements or commands in order to model a new behavior of the mechatronic system. However, this requires certain models to be implemented, for example, with the help of virtual simulators or similar software tools called "physics engines". These programs often reflect the real mechanical processes, are not based on dynamic equations for mechanical systems, i.e. they are phenomenological models.

2. The use of V-Rep simulator for robotic systems modeling

In the tasks of robotic systems design, it's often necessary to preliminarily test the algorithms for control and construction of the robot's virtual model. Today, there are a set of simulation environments is available for these purposes, for example, Gazebo, Microsoft Robotics Developer Studio as well as V-REP.

Fig. 2. An example of ribbon cutting synergism by Robotino robot's manipulator

V-REP is a simulation environment for various types of robots, at that, the user does not have to have physical access to the real robot, which saves time and costs. Its most distinctive features are: its free-of-charge basis for non-commercial use, various methods of the simulated robot programming and a vast range of robot types, which can be modeled simultaneously.

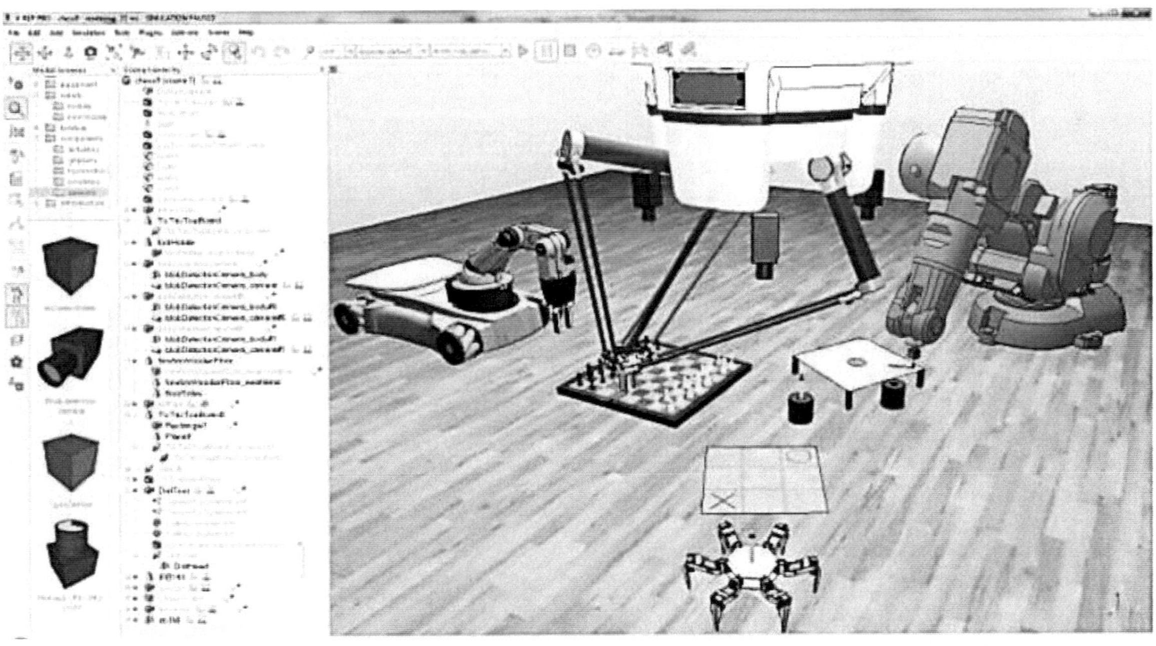

Fig.3. An example of simulation cycle in V-REP simulator

V-REP environment provides a convenient interface for visualization of the robot's actions in 3D virtual space mycg earlier than the real robot prototype will be created (Fig. 3-4).
V-REP is an ideal tool for: fast prototyping and verification; remote monitoring; quick algorithm development; education robotics and factory automation modeling systems.

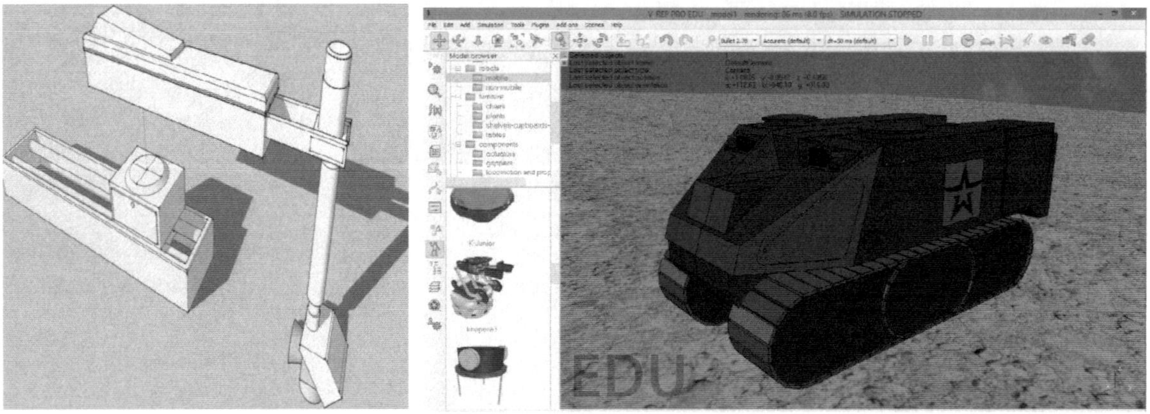

Fig. 4. Building new objects in V-REP simulator

In the simulator, there are already prepared models of such industrial and mobile robots as Kuka LBR4, ABB IRB 140, Pioneer p3dx, etc. On top of that, there is a possibility for a user to create his/her own models out of the existing components.

These components include:
1. Connections. This component may interlink several objects and work in rotation / sliding mode as an engine/actuator.
2. Contactless sensors. They calculate the distances to figures (objects), which are within the range of sensor's detection.
3. Visual sensors. These sensors allow extracting complex images from the simulation scene and work as cameras.
4. Force sensors. They allow writing the applied forces and rotation moments into files and can break if the preset threshold is exceeded.
5. Molds. They are used to simulate the volumes of solid bodies.
6. Rotary tools. They are used for simulation of rotary cutting, laser-beam cutting, etc.
7. Charts. They are able to perform recording of various data, which may vary during the simulation process. The saved data can be displayed on the chart or written to a file.

The simulator supports several programming methods, for example, inline scripts. With this method, it is possible to write a control program right in the simulator without installing any external software and explicit compilation. Scripts are written in a built-in editing program in Lua script language. A script can be opened within or out of stream implementation.

Remote clients. Remote interface provides an opportunity to interact with scene components by means of sockets. On the client side, any software environment can be applied if it has the option to connect dynamic-link libraries. The simulator comes with libraries for remote operation of V-REP for some of the most popular programming languages (C/C++, Python, Java, Matlab, Octave, Lua, Urbi). The advantages of such method are the use of familiar programming environment and integration with the existing solutions.

Despite different approaches to programming of the simulated object, the software interface is almost identical for both client and server.

5. Development of network-type robotic simulator

The disadvantage of V-REP simulator reviewed above is the set of limitation for its use in networks, especially in case of parallel operation of several operators with one robot.

In V-REP simulator, control is performed by means of connection to particular components of the simulated robot. In case of multiple connections, the simulator doesn't perform user authorization and regularization of components available to a user based on privileges. Because of that, situations may arise when the developer gets access to the components of someone else's robot.

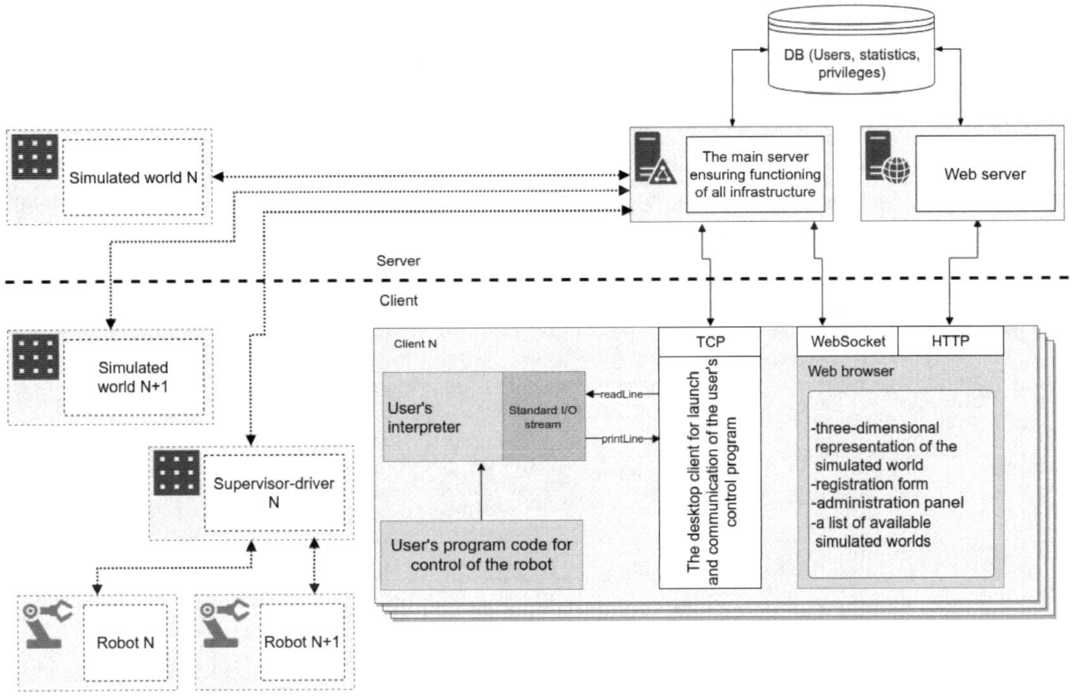

Fig. 5. The structure of the developed network-type robotic simulator

Besides robot programming, it is necessary to set the behavior of the scene's surrounding objects. Such behavior is set by the scene creator and must not be available to be changed by the user. For this purpose, V-REP supports inline scripts in a special scripting language, but it is limited by the simulator's program interface and doesn't allow changing low-level settings of the physics engine.

Cumulatively, the above-named reasons make it more difficult to use V-REP simulator in a multi-user environment. In order to solve these problems, a prototype was developed.

The developed simulator prototype is going to be used for simulation of one or more robots, which can perform competitive or cooperative tasks. A control program can be developed separately for each robot and started up on separate machines.

The simulator's architecture, which has actor model as the basis, allows to scale each component of the system on physically separate computers, and also to design the system with account of high fault tolerance.

Main advantages of the proposed architecture:

- Distributed server. Simulated world (physics engine plus a set of rules) can be started on the client side, which reduces the load on central server;
- Robot control client program works on the client machine, which allows the client to use any software environment, any programming languages, libraries, etc. This reduces the load on central server.
- Robot client control program communicates with the robot by means of standard input/output stream, i.e. only 2 standard functions (printLine, readLine) are needed, which can be found in any programming language. This reduces the simulator entry threshold, allows starting to write the control program immediately without configuring the environment and libraries. Since the system is centralized (started up on the central server), this allows:
    - several robots to take part in one simulated world;
    - gathering the statistics on tasks performed;
    - maintaining the administration of simulated worlds.
- Visual feedback for the developer takes place in a web-browser. In order to work with the simulator, it is only required to download a special launcher program for starting of controlling software.

Such an approach also has a number of disadvantages:

- Security problems. Physics engine works on the client side, and it is potentially vulnerable to external intervention from the client side;

- Client software works on a client, and therefore it becomes possible to receive information from external sources, for example, through manual control of the robot;
- If the physics engine is started on the client, the values pass through the central before they get to the console, and only then the client receives them. This creates or may create a significant delay in feedback for controlling software.

Methods of controlling the disadvantages specified above make it more difficult to use the simulator in real time mode:

- Naturally, the increase in delay between discrete control steps to ~5 seconds when using visual feedback won't allow an opportunity for a person to efficiently control the robot manually, but leaves a possibility to analyze the control afterwards.
- The use of discrete modeling (lockstep model). In this case, all clients are polled and the answer from all clients is expected (with a certain limited time-out). Only after that, the results of using the implemented robot control will be calculated with the use of the physics engine. After all calculations within the system, simulator switches to the next discrete step in time. This allows avoid synchronization of the robot's world model.

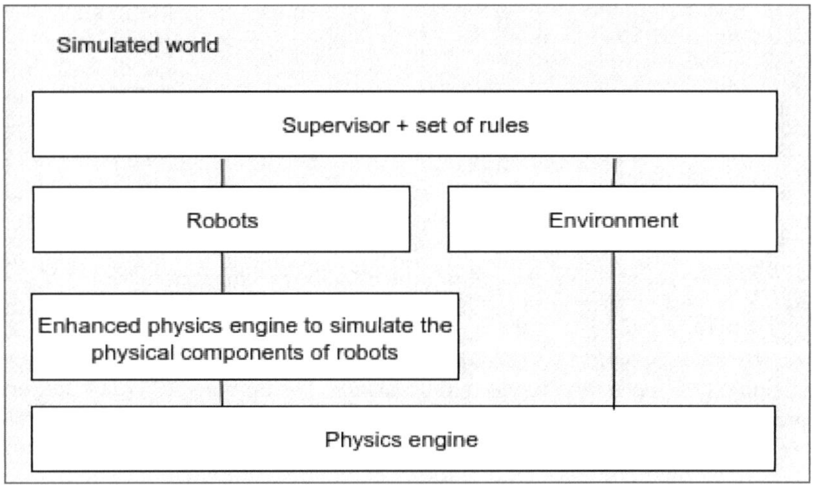

Fig. 6. The structure of the simulated world

In the current implementation of the simulated world, there's no possibility to configure the scene environment and its behaviors through the graphical interface for a user having no programming skills. Any changes in the scene require recompilation of the world. Because of that, V-REP simulator can serve as the simulated world, since it provides simpler scene setting and robot construction processes for people who are not qualified as software developers.

The developed simulator can partially replaces the functionality of other simulators and can be used as an add-in, which expands the functionality of the existing solutions allowing them to operate together. Since any robot control is performed through a centralized server, the client gets a uniform robot control interface. This is achieved by means of manifest files.

The protocol for data transmission between the client and server is generated on the basis of manifest files. They define the robot's structure, its components and possible operations on them. Such robot description in a formal language allows for automatic generation of software interfaces for different programming languages and software environments. Also, such manifests play the role of documentation, which the client-developer can use to review the methods of working with one or another robot. As the result, the user gets a generated software interface, which is connected to the interpreter/compiler. The user can start to work with the robot immediately without thinking about the details of implementation and data transmission method.

6. Conclusion

The developed application-dependent software («middleware») has allowed building a system, which makes it possible to combine the work of real robots and their virtual mathematical models. The main distinctive feature of the developed tools is operation in network access mode to all subsystems of the developed simulator and to real mobile robots (Amur, Robotino) as well. The created technology has allowed programing and reprogramming the mobile robots by generating synergies in network access and even during their functioning. It also allows the use of virtual models.

7. References

[1] Emmerich, W. The impact of research on the development of middleware technology / Wolfgang Emmerich, Mikio Aoyama, Joe Sventek // ACM Transactions on Software Engineering and Methodology. – N. Y.: ACM – 2008. Vol.17, № 4. – pp.19-48.

[2] Pryanichnikov V., Andreev V. The Application of Network Technologies to Constructing Group Controlled Systems with Machine Vision for Mobile Robots // In Proceedings of the 23th DAAAM Intern. Symp. "Intelligent Manufacturing & Automation" 24-27th October 2012 Zadar, Croatia, 2012. – V.23, No.1, – pp.1167 – 1174.

[3] Andreev V.P. Technology Supervisory Control for Mechatronic Devices via the Internet / V.P.Andreev, K.B.Kirsanov, P.F.Pletenev, Yu.V.Poduraev, V.E.Pryanichnikov, E.A.Prysev // 25th DAAAM Intern. Symp. "Intelligent Manufacturing & Automation". Procedia Engineering, 2015. – V.100, – pp.33 – 40.

[4] Andreev V. Education on the basis of virtual learning robotics laboratory and group-controlled robots / Andreev V., Pryanichnikov V., Poduraev Y., Kuvshinov S. // 24th DAAAM Intern. Symp. on Intelligent Manufacturing and Automation, DAAAM 2013. Procedia Engineering, 2014. – V.69. – pp.35 – 40.

[5] Carl Hewitt, et al. Actor Induction and Meta-evaluation // Conference Record of ACM Symposium on Principles of Programming Languages, January 1974.

[6] V-REP simulator documentation URL [electronic resource]. – Access mode: http://www.coppeliarobotics.com/helpFiles/

[7] Rohmer Eric. Singh and Marc Freese. V-REP: a Versatile and Scalable Robot Simulation Framework / Eric Rohmer, P.N. Surya. – 2013.

[8] Pryanichnikov V.E., Andreev V.P., Prysev E.A. Group control of mobile robots, based on the net-technologies // Proceedings of the Int. conference on Robotics for Security & Space (Planet & Earth rovers). St. Peterburg: Poliectnika service. 2010. P. 279–283.

[9] Katalinic B., Pryanichnikov V.E., Ueda K., Kukushkin I., Cesarec P., Kettler R. Bionic Assembly System: hybrid control structure, working scenario and scheduling // Proceedings of 9th National Congress on Theoretical and Applied Mechanics, Brussels. 2012. P. 111–118.

[10] Pyanichnikov V.E. Algorithmic base for remote sensors of mobile robots // Mechatronics, Automation, Control. 2008. № 10(91). P. 10–21.

[11] Kirsanov K., Levinsky B., Pryanichnikov V. Integrating software for intelligent robots // Informational-Measuring and Control Systems, Radiotechnika. 2009. V. 7. № 6. P. 35–43.

[12] Kirillchenko A.A., Pryanichnikov V.E., Rogozin K.V. Limits of validity and reliability of proofs. Scepticism in mathematics, functions, and traditions // Information-Measuring and Control Systems. 2013. V. 11. № 4. P. 57–65.

DOI: 10.2507/27th.daaam.proceedings.017

EXPERT SYSTEM OF GENARATIVE CAPP SYSTEM IN FUNCTION CUTTING SPEED

Senad Rahimic & Vojo Visekruna

This Publication has to be referred as: Rahimic, S[enad] & Visekruna, V[ojo] (2016). Expert System of Genarative Capp System in Function Cutting Speed, Proceedings of the 27th DAAAM International Symposium, pp.0116-0121, B. Katalinic (Ed.), Published by DAAAM International, ISBN 978-3-902734-08-2, ISSN 1726-9679, Vienna, Austria
DOI: 10.2507/27th.daaam.proceedings.017

Abstract

In this paper will be show model for expert system process planning with minimum cost per piece in function to cutting speed. The numbers variants depends of resource by manufactured system, machine labour and overhead rate and tools. The optimization model will be use in generative CAPP system for rotation parts. The Mathematically model devolpment for minimum cost per piece, which have all parameters of process planning. For determining machining and manufacturing tims importing for maximum profit, what is leading aim all manufacturing of industry. The optimal of cutting speed determining which to receive maximum profit. The maximum profit introducing combination of criterion minimum cost per piece variant process planning and criterion optimal of cutting speed.

Keywords: Expert system; generative CAPP; process planning.

1. Introduction

The manufacturing industry has been pushed to adopt more effective and efficient production strategies to meet the challenge of shorter life cycle, higher quality, lower cost, wider variety of customer demands [1].
This increased emphasis on achieving highly adaptive manufacturing to reduce manufacturing costs and to better utilize manufacturing capacity has led to a critical focus on agile manufacturing as a strategy to achieve these goals. In manufacturing, process planning is the task that transforms the design information into the manufacturing processes and determines the operation sequence.

CAPP considered a crucial link between Computer -Aided Design (CAD) and Computer -Aided Manufacturing (CAM) [2]. Research of over 30 years in CAPP has resulted in a wealth of knowledge on CAPP and many experimental and commercial CAP'P systems have been developed as a result. Computer-assisted process planning was originated in the 1960s (Niebel 1965) and since has been a very active area of research and development. During the late 1970s, the science of computer-aided process planning (CAPP) evolved into two basic approaches: variant or generative process planning (Chang 1998). Modern approaches toward CAPP include using case-based reasoning. [3]

In this paper proposed of methodology show on figure 1, which is start by geometric information about part using basic features. Target make all alternative variants process planning, as organized by sequence phase, scheduled and operations, this technological to steps different.

The module developed for searching basic mechanical features (groove revolution, slot, thread, plane, cylinder cone outer, cylinder cone inner, etc.) must has two conditions, the first condition is all features to be simple, and make together complex features defined rotation part. The second condition had defined features by every simple features to be individual.

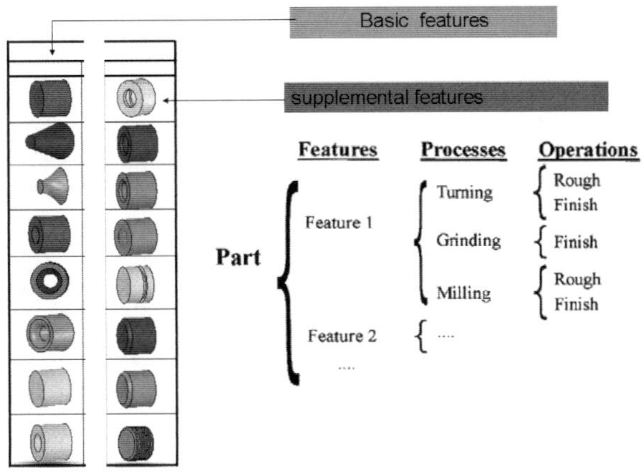

Fig. 1. Assignment of alterative processes and operations

2. An object- oriented knowledge representation scheme

Whole part considerationed, searching from plans simple features which made complex rotation part, with their individual informations possibled define process and commission technology for resumption destitute operations for to do any assignment basic features. stage enactive all features, all process and operations which can to be applicable for make with acceptable declaration quality.

We propose an appropriate knowledge representation scheme which attems to address various facets of the process planning task. We recommend object-oriented common attributes and/or their values from higher level objects. Taxonomies of features, processes, machines and tools exist in the semantic network. Inheritance is used to infer attributes and behaviour of more specific object. (Hang-Wai, L.; & Hon-Yuen, T.) a number of represent manufacturing capabilities and reasoning in process planning. Now we will explain some of these taxonomies and relations in more detail.

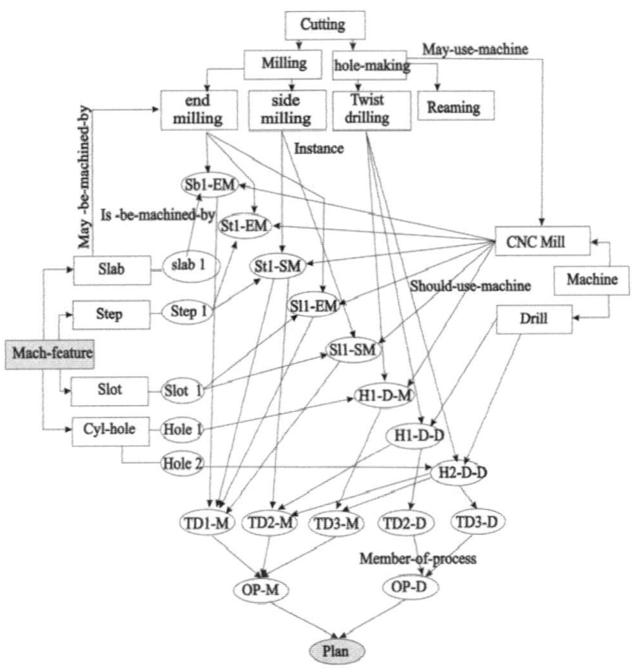

Fig. 2. The process planning semantic network

3. Minimum cost per piece

The average cost per piece to produce a work piece consists of the following costs: nonproductive cost per piece, machining time cost per piece, tool changing cost per piece and tooling cost per piece. Mathematically, this can be expressed as equality (1).

$$c_1 = c_o t_1 + c_o t_c + c_o t_a \left(\frac{t_{ac}}{T}\right) + c_a \cdot \left(\frac{t_{ac}}{T}\right) \tag{1}$$

Upon partially differentiating c1 with respect to v, equating to zero, and solving, we obtain the minimum unit cost cutting speed (v_{min}) as follows equality (2).

$$\frac{\partial c_u}{\partial v} = 0 \tag{2}$$

Upon partially differentiating c1 with respect to v, equation to zero, equating to zero, and solving for v, we obtain showed on equality (3).

$$v_{coj} = \sum_{j=1}^{i_p} \frac{C}{\left[\left(\frac{1}{n}-1\right)\left(\frac{c_o \cdot t_{aj} + c_a}{c_o}\right)\right]^n} \tag{3}$$

Optimization mathematically systems, this can be expressed as systems of equality (4).

$$\begin{aligned}
c_{1f1} &= c_o\left[\frac{t_{ps}}{N_0} + t_{ss} + \left(t_p + \frac{L}{1000 v_{ph}}\right) i_p\right] + \sum_{j=1}^{i_p}\left[c_o\left(\frac{\pi D_j l}{1000 v_{cj} f}\right) + c_o\left(\frac{\pi D_j l}{1000 v_{cj} f}\right)\left(\frac{v_{cj}}{C}\right)^{1/n} t_a + c_a\left(\frac{\pi D_j l}{1000 v_{cj} f}\right)\left(\frac{v_{cj}}{C}\right)^{1/n}\right] \\
c_{1f2} &= c_o\left[\frac{t_{ps}}{N_0} + t_{ss} + \left(t_p + \frac{L}{1000 v_{ph}}\right) i_p\right] + \sum_{j=1}^{i_p}\left[c_o\left(\frac{\pi D_j l}{1000 v_{cj} f}\right) + c_o\left(\frac{\pi D_j l}{1000 v_{cj} f}\right)\left(\frac{v_{cj}}{C}\right)^{1/n} t_a + c_a\left(\frac{\pi D_j l}{1000 v_{cj} f}\right)\left(\frac{v_{cj}}{C}\right)^{1/n}\right] \\
c_{1fn} &= c_o\left[\frac{t_{ps}}{N_0} + t_{ss} + \left(t_p + \frac{L}{1000 v_{ph}}\right) i_p\right] + \sum_{j=1}^{i_p}\left[c_o\left(\frac{\pi D_j l}{1000 v_{cj} f}\right) + c_o\left(\frac{\pi D_j l}{1000 v_{cj} f}\right)\left(\frac{v_{cj}}{C}\right)^{1/n} t_a + c_a\left(\frac{\pi D_j l}{1000 v_{cj} f}\right)\left(\frac{v_{cj}}{C}\right)^{1/n}\right]
\end{aligned} \tag{4}$$

The minimum cost per piece for one operation in process planning are representation sums all phases consisted in operation as inform of equality (5)

$$c_{1o} = \sum_{i=1}^{f} c^i{}_{1f} \tag{5}$$

4. Example of part

Showed on figure 3 testing rotation part for given mathematically of module. They correspond to the following types:

- Groove revolution: GRO1, GRO2.
- Slot: SLO1, SLO2, SLO3.
- Thread: THR1.
- Plane: OD1, OD3, OD5, OD7, OD10.
- Cylinder cone outer: OD2, OD4, OD6, OD8, OD9.
- Cylinder cone inner: HO1.

The production media for which the process plan is developed consisting of a group of machines–tool and fixtures that provide support for conventional machining processes. The capacities established for these manufacturing systems are the usual ones for standard quality.

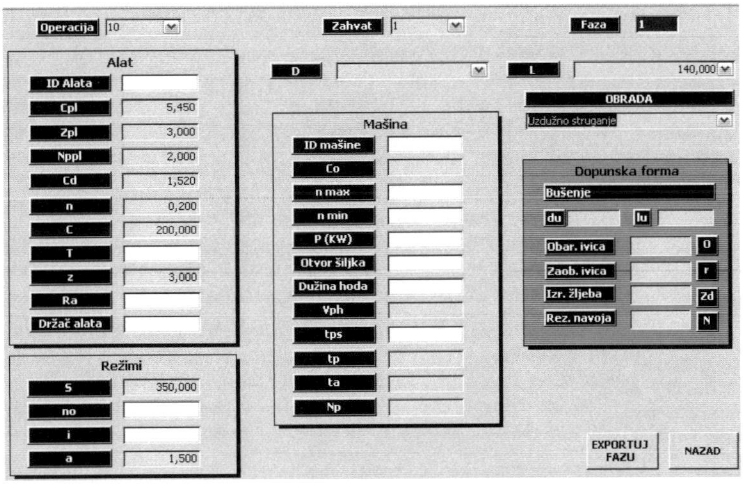

Fig. 3. Example Expert System for generative CAPP system

Figure 3. shows the expert system for generative CAPP systems for rotating parts, using the method of object oriented programming.

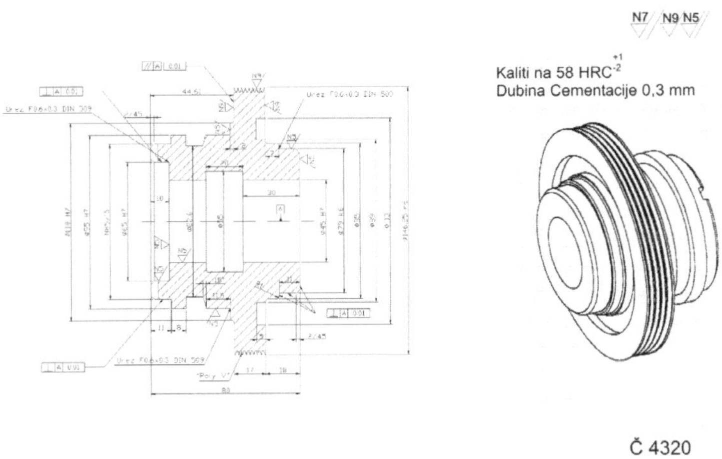

Fig. 4. *Drawing model with features*

Using CAD software in this paper is Solid Works as an example of the rotation parts on figure 3 [5]. Graph of function for minimum cost per piece for all possible variants process planning shown in figure 4.

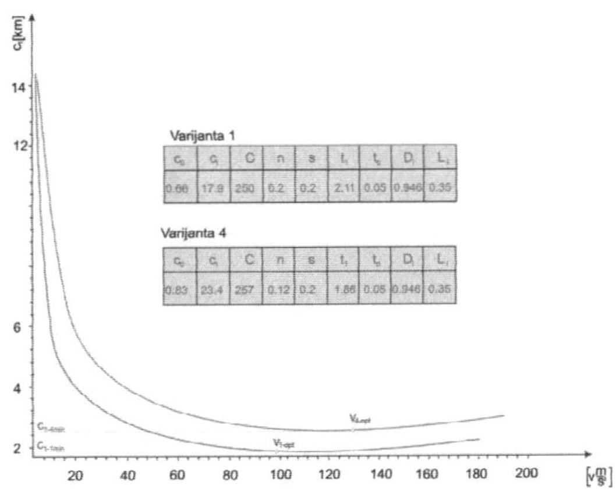

Fig. 5. Graph of function for minimum cost per piece for cutting

Showed on figure 5 testing for testing rotation part, we get for all possible variants for process planning in function cutting speed and minimum cost per piece and number workpieces.

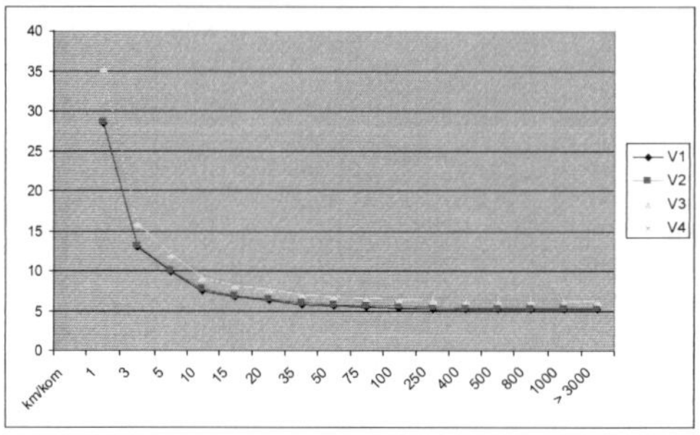

Fig. 6. Shown all possible variants

On graph showed function for minimum cost per piece for different machining regime possible variants process planning shown in figure 6 [7].

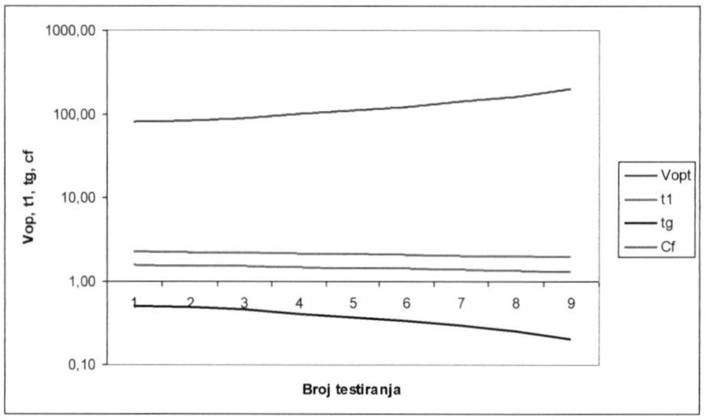

Fig. 7. The considered variant 1-4.

On the table I: showed minimum cost per piece for variants with different machining regime in function of number workpieces and given coefficient rate of dependent [7].

number workpieces	V1	V4	k
	28,83	34,27	0,16
3	12,95	16,17	0,20
5	9,84	12,28	0,20
10	7,51	9,37	0,20
15	6,73	8,4	0,20
20	6,34	7,91	0,20
35	5,84	7,29	0,20
50	5,64	7,04	0,20
75	5,49	6,84	0,20
100	5,41	6,75	0,20
250	5,27	6,57	0,20
400	5,24	6,53	0,20
500	5,22	6,51	0,20
800	5,21	6,49	0,20
1000	5,2	6,48	0,20
> 3000	5,18	6,46	0,20

Table 1. Achieving coefficient rate of dependent

On figure 7 rate variant 1 and variant 4, in function on numbers workpieces, we can conclusion to there mathematically dependence which is declare through coefficient showed on equality (6).

$$k = \frac{c_1 - c_4}{c_4} \tag{6}$$

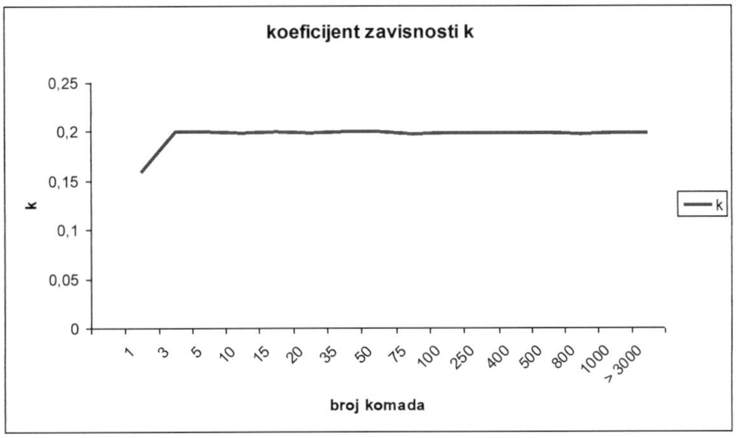

Fig. 8. The coefficient dependence between variants.

5. Conclusion

In this paper proposed an expert system suitable for generative the process plan of machining parts. It is a proposal in which the peripheral processes needed for these parts are taken into account and they enable the development of alternative process plans.

The process plans offered by the system constitute all of the alternatives for the sequence of phases, guaranteed by a high degree of optimization with regard to cost and number of phases. These alternatives explicitly include feasible alternatives for machines and, tools for operations (depending on the type of machine for each phase).

These characteristics are largely based on the methodology proposed, the functional structure, the use of general information models and the general functional procedures. All these factors working together give the system its qualities. The development of the CAPP expert system based on this proposal has demonstrated the system's feasibility and its optimal qualities.

The target is connecting CAD and CAM system to realization flow over information's by design of product then to its manufacturing. The development of the CAPP system based on this proposal has demonstrated the system's feasibility and efficient.

The process plans offered in CAPP system alternative variants, which are composed of sequence and phases. The optimization of the process variant depends of the minimal cost per peace and optimal speed of cutting.

5. References

[1] Ben-Arieh, D.; Gutin, G.; Penn, M.; & Zverovitch, Y.; Process planning for rotational parts using the generalized travelling salesman problem, int. j. prod. res., 2003, vol. 41, no. 11, 2581–2596

[2] Gonzalez, F.; & Rosado, P.; General and flexible methodology and architecture for CAPP: GF-CAPP system, int. j. prod. res., 2003, vol. 41, no. 12, 2643–2662

[3] Kusiak, A.; Decomposition in Data Minning: An Industrial Case Study, IEEE Transactions on electronics packagin manufacturing, 2000

[4] Hang-Wai, L.; & Hon-Yuen, T.; Object-oriented analysis and design of computer aided process planning system, Int. Journal CIM – vol 13, 2000

[5] Rahimić, S. & Višekruna, Model minimalnog jedinčnog troška varijante tehnološkog procesa, CIMOSov dan raziskav 2006 – CRD '06, Koper, 17. november 2006

[6] Rahimić, S., Višekruna V & Balič J., Optimization of variant process palnning for genarative CAPP system, 11th International Scientific Conference on Production Engineering –CIM2007, Croatian Association of Production Engineering, Zagreb, 2007

DOI: 10.2507/27th.daaam.proceedings.018

METHODS FOR THE REDUCTION OF CO_2 EMISSIONS IN THE CEMENT INDUSTRY

Jerzy Duda, Mariusz Kolosowski & Jacek Tomasiak

This Publication has to be referred as: Duda, J[erzy]; Kolosowski, M[ariusz] & Tomasiak, J[acek] (2016). Methods for the Reduction of CO2 Emissions in the Cement Industry, Proceedings of the 27th DAAAM International Symposium, pp.0122-0130, B. Katalinic (Ed.), Published by DAAAM International, ISBN 978-3-902734-08-2, ISSN 1726-9679, Vienna, Austria
DOI: 10.2507/27th.daaam.proceedings.018

Abstract

Cement burning process of high temperature is counted among industries particularly disruptive for the environment. This is especially true of carbon dioxide emissions, which is the sum of emissions from the combustion of fuel and emissions resulting from the technological process. The issue of carbon dioxide emissions reduction has been presented on the example of cement industry. Existing activities in this area resolve mainly to the reduction of heat consumption in the process of clinker burning and limiting its content in cement. Those activities have almost completely exhausted the possibility of further CO2 reduction. New, even greater reductions, which the European Commission plans to implement after 2020, require a different approach to this problem. The authors conduct research on possibilities of the reduction of greenhouse gases emission in the cement industry, among others through the use of the ORC (the organic Rankine cycle) and CCS (carbon capture and storage) technologies.

Keywords: cement production; rotary kiln; greenhouse gases; waste; ORC.

1. Introduction

Despite the significant reduction in the energy intensity of production process and various measures being implemented to reduce the emissions of greenhouse gases (GHGs), recent years witnessed a progressive growth of atmospheric CO_2 concentrations and the resulting increase of temperatures on Earth. According to the forecasts by the International Energy Agency, the civilizational and economic growth will cause continued increase in the consumption of electricity, with non-renewable (fossil) fuels having a significant share (ca. 80%) in the production thereof. As a consequence, the greenhouse gas emissions will increase leading to adverse climate changes. The changes are presented in the Fig. 1.

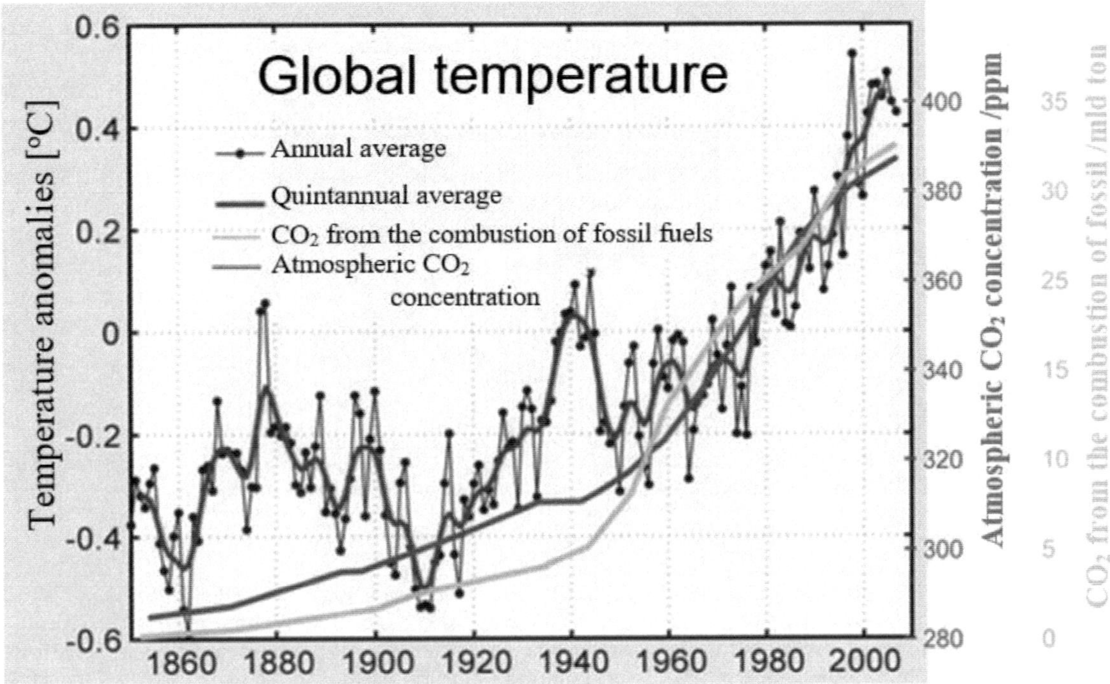

Fig. 1. Changes in Earth temperatures [1] and atmospheric CO_2 levels

The research conducted by the Intergovernmental Panel on Climate Change demonstrated that atmospheric concentration of carbon dioxide increased by 1.5 to 3 ppm each year starting from mid-19th century to reach ca. 380 ppm in the baseline year (2005). Already in 1992, the United Nations Framework Convention on Climate Change (UNFCCC) was adopted during the "Earth Summit" Conference in Rio de Janeiro, with basic objective to "achieve (...) stabilization of greenhouse gas concentrations in the atmosphere at a level that would prevent dangerous anthropogenic interference with the climate system". Consequently, an international agreement, referred to as the Kyoto Protocol, was developed in 1997 with the aim to reduce anthropogenic emissions of greenhouse gases (as expressed in carbon dioxide equivalents) in the reference period of 2008-2012 in signatory countries. Considering the above, the countermeasures for climate changes were considered one of the priorities of the new energy policy of the European Union. The target scenario of the European Commission assumed that the Earth temperature increase would be restricted to 2°C. According to expert opinions, a condition to meet this restriction would be to reduce the worldwide emissions of CO_2 by 50% as compared to 2005. Such a great reduction in carbon dioxide emissions requires a drastic reduction in the use of fossil fuels and increase in the use of renewable energy sources (RES). This is particularly difficult to achieve in Poland where coal is the main industrial fuel due to its cost and availability. Similar to the energy production sector, the cement industry consumes significant quantities of coal. Therefore, the quest to reduce the energy intensity and thus the production costs as well as to reduce the harmful environmental impact of cement is still valid.

According to the Integrated Pollution Prevention Control (IPCC) directive, acceptable emission levels should be determined on the basis of BREFs (BAT Reference Documents, i.e. reference documents for best available techniques not causing excessive increase in costs and preventing or reducing air pollution). The technological changes in worldwide industry resulting from the strive to reduce energy intensities and to meet the BREF-defined environmental protection standards pertain mainly to energy intensive processes such as cement clinker sintering technologies and raw material and cement milling techniques [2, 3]. Recent years witnessed a significant development of novel cement production technologies which were successfully implemented during modernization of the cement industry in Poland. The Fig. 2. presents the impact of the modernization of the cement industry in Poland on the energy consumption and ecological parameters.

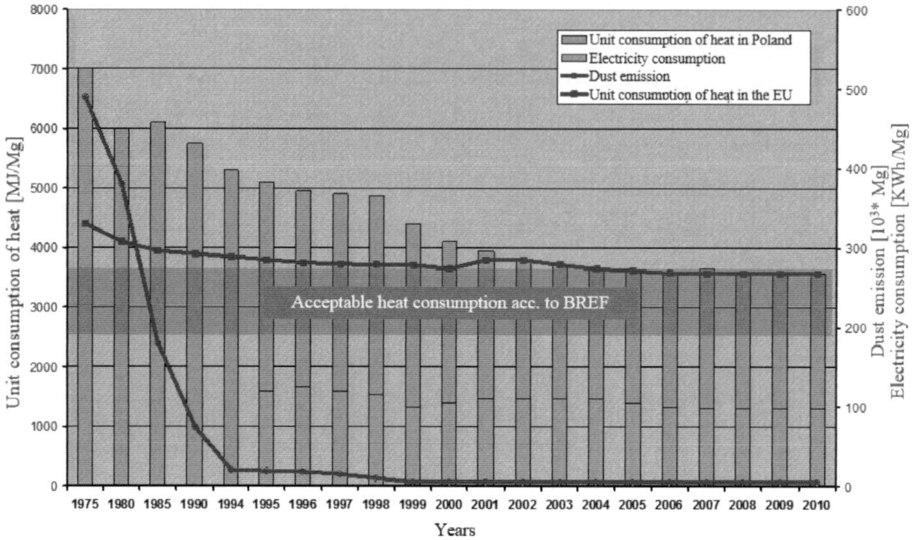

Fig. 2. The impact of the modernization of the cement industry on energy intensity and dust emissions

As shown by the presented data, implementation of innovative technologies in Polish cement industry led to a significant (50%) reduction in energy intensity and dust emissions. To all purposes, our industry meets the BAT standards and is one of the best cement industries in Europe. The problem currently faced by the cement industry and requiring a quick solution consists in the reduction of greenhouse gas emissions, particularly due to the announced new, significantly lower limits of acceptable emissions to be implemented in the future.

2. Cement production vs. CO_2 emission

As part of the UNFCCC (Kyoto Protocol of 1997), Poland has committed itself to reduce the emissions of GHGs by 6% as compared to the level of 1988 (the baseline year). The cement industry in Poland, which had undergone a radical modernization in years 1990-2000, had virtually no problems with meeting the new limits. The first problems occurred when the European Commission, having considered the positive changes resulting from the restructuring of Polish industry (and the generated CO_2 emission reserves), made a decision to reduce the limits for Poland in 2005. In recent years, a progressive increase in restrictions on acceptable GHG emissions can be observed. Whereas the requirement to reduce the acceptable CO_2 emission by 20% until 2020 (according to the 3x20 climate initiative) can be achieved, the expected further reductions to be entered in force after 2020 pose a problem for the cement industry in Poland. In light of the novel assumptions regarding the restrictions in CO_2 emissions after 2020, published by the European Commission in the Green Book in 2013, novel aims are set to be achieved by 2030, including:

- CO_2 emission being reduced by 40% ,
- RES contribution to reach 27% of overall energy production.

The assumed new acceptable emission limit is particularly unfavorable for the cement industry where the main part of emissions does not originate from combustion; instead, emissions are generated in the technological process of the decarbonation of limestones used as the main raw material in the production of cement clinker. The compliance with further restrictions of GHG emissions expected after 2020 will require novel, innovative techniques and technologies of cement production.

Similar to the production of electricity, the cement production process is characterized by high emission of carbon dioxide being not only the result of fuel (carbon) combustion processes. The main part of CO_2 emissions in cement production processes originate from decarbonation of calcium and magnesium carbonates being the basic components of kiln feed. Decarbonation proceeds in line with the following equations:

$$CaCO_3 + Q = CaO + CO_2 \uparrow \tag{1}$$

$$MgCO_3 + Q = MgO + CO_2 \uparrow \tag{2}$$

Carbon dioxide present in the rotary kiln waste gas originates from the combustion of fuel (fuel-related CO_2^P) and from the decomposition of carbonates (material-related CO_2^S):

$$\sum CO_2 = CO_2^P + CO_2^S \tag{3}$$

The production of 1 kg of cement clinker (the main component of cement) in the rotary kiln emits ca. 0.86 kg of CO_2 . Of these 0.56 kg of CO_2 (ca. 60%) is ascribed to emission resulting from decarbonation of the kiln feed whereas the remaining 0.3 kg of CO_2 results from the combustion of fuel in the clinker sintering process.

The extent of fuel-related emission (CO_2^P) depends on the type of fuel and energy intensity of the clinker sintering process, whereas the extent of material-related emission (CO_2^S) depends on the chemical composition of material feed (carbonate and non-carbonate carbon content). The Table 1 presents the emissions of carbon dioxide depending on the clinker sintering technology in use.

Quantity	Unit	Dry method[1]	Dry method[2]	Wet method
Heat consumption	GJ/Mg cl.	3.4	4.2	6.8
Fuel-related emission benchmark*	kg CO_2/GJ	94.5	94.5	94.5
Fuel-related emission	kg CO_2/Mg cl.	321	397	643
Raw material-related emission*	kg CO_2/Mg cl.	535	535	535
Total emission	kg CO_2/Mg cl.	**856**	**932**	**1178**

* Fuel- and raw material-related rates as used for the first level of accuracy in CO_2 emission monitoring [4],
[1] short kiln with four-stage cyclone heat exchanger;
[2] long dry method kiln.

Table 1. The impact of the clinker sintering method on CO_2 unit emission rates

As shown by the presented data, emission is largely dependent on the sintering technology and its energy intensity (unit heat consumption). As the result of recent modernization of the cement industry in Poland, consisting mainly in the elimination of the energy-intensive wet method and implementation of novel, dry sintering methods in kilns with multistage cyclone heat exchangers and preliminary calcination systems. The modernization brought about a significant reduction in CO_2^P emission from fuel combustion processes (ca. 300 kg CO_2/Mg cl.).

Currently, when virtually all technological possibilities to reduce the energy intensity were used up, the search is ongoing for alternative ways to further reduce the emission of carbon dioxide.

3. The impact of waste fuels and materials on the reduction of CO_2 emission

The deficit in natural raw materials due to the mass production of cement combined with the high costs of fuels and the requirements to reduce the emission of greenhouse gases are the driving forces behind the industry's search for novel technologies for utilization of waste fuels and materials. The Fig. 3 presents the possible uses of waste fuels and materials in cement production processes.

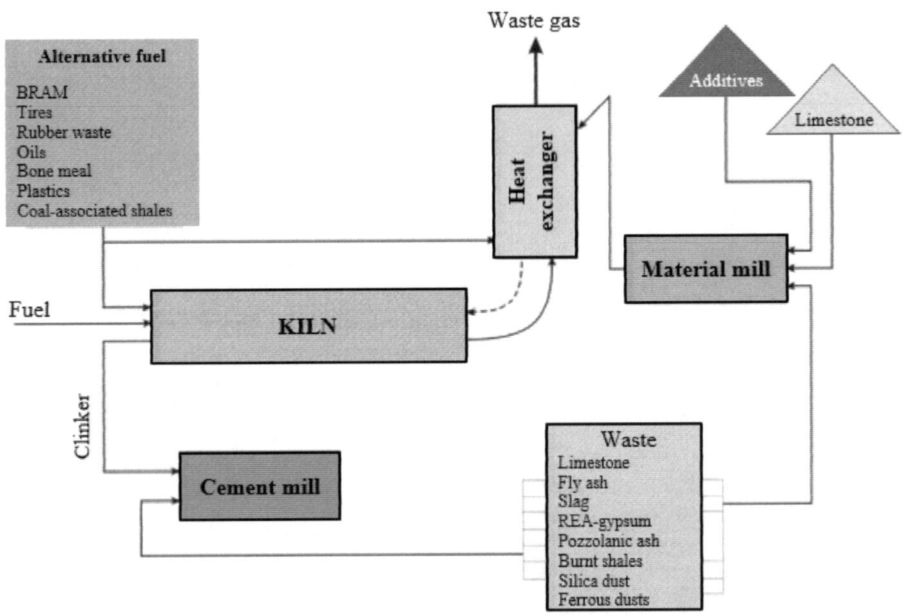

Fig. 3. Possible uses of waste fuels and materials in cement production

A large part of waste materials such as fly ash from power plants, blast-furnace slag, silica dust or gypsum from flue gas desulphurization plants are already commonly used in cement mills. Similarly, utilization of waste fuels are common. Novel technologies of cement sintering in kilns featuring cyclone heat exchangers and preliminary decarbonation systems (precalcinators) facilitate combustion of various types of low-calorific fuels or combustible waste [4]. The temperature conditions within rotary kilns as well as the possibility to load alternative fuel into different sections of the kiln (as shown in the Fig. 3) facilitate the use of fuels of various physical forms (liquids, chunks, and dusts) [5].

The major issue in thermal utilization of waste material consists in the emission of hazardous gases - polychlorinated dibenzodioxins and dibenzofurans (PCDD and PCDF). Due to high temperatures of gas streams (> 2000°C) and the sintered materials (ca. 1430°C) as well as to the turbulences and flow times within the high temperature zone (> 1100°C) co-combustion of alternative fuels with carbon dust in rotary cement kilns meet all the requirements of the Ordination of the Minister of Economy dated 21 March 2002 on the conditions of thermal waste transformation processes (Journal of Laws 2002.37.339). As seen from the data presented in the diagram of a typical modern rotary kiln (Fig. 4), gases dwell in the kiln at temperatures of > 1100°C over a period of ca. 7 s which secures the requirements associated with the destruction of toxic PCDD and PCDF compounds.

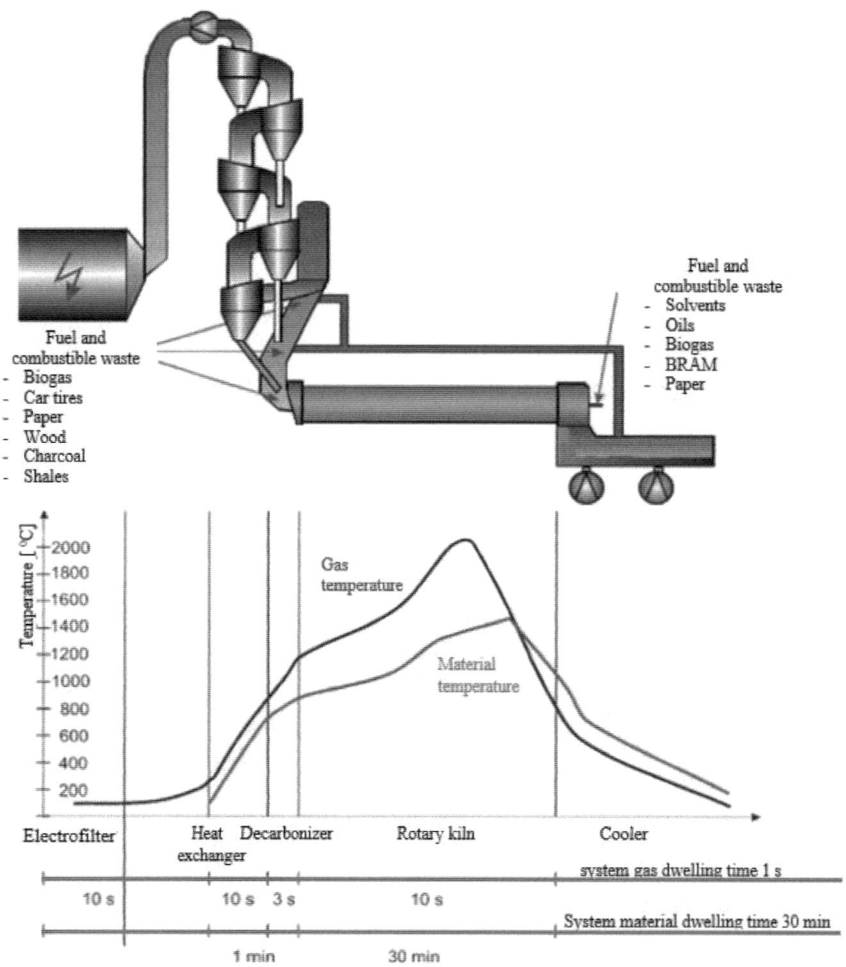

Fig. 4. A diagram of the rotary kiln with distribution of gas and sintered material temperatures

The use of waste fuels as substitutes for natural fuels (carbon) in the cement production process is in line with the principles of sustained growth and the objectives of the 3x20 climate initiative. Besides the reduction in the consumption of non-renewable (fossil) fuels, an important effect of the use of waste fuels consists in the reduction of the emission of hazardous gases (CO_2 and NO_x). The reduction in the emission of carbon dioxide stems from the reduced consumption of hard coal which is characterized by high benchmark value (ca.95 kg CO_2/GJ) as compared to the emission from the combustion of alternative-derived fuels, commonly containing biomass with a significantly lower benchmark value (ca.55 kg CO_2/GJ). Thus, CO_2 emission may be reduced by ca. 40kg CO_2/GJ compared to hard coal combustion. Waste utilization is an important ecological effect of the use of alternative-derived fuels. Compared to other methods of thermal utilization , rotary cement kilns feature a number of advantages that ensure waste-free, environmentally safe combustion of alternative fuels and thus the combustion of waste fuels in rotary cement kilns is increasingly considered to be one of pro-ecological means for utilization of hazardous waste materials [6]. Despite the significant percentage share (ca. 40%) of alternative fuels in the clinker sintering process and the resulting environmental benefits, these measures are insufficient to achieve the required levels of CO_2 emissions (reduction by 20%). This shows that no further reduction in CO_2 emission is possible for the conventional clinker sintering technology. The remaining possibility to reduce the emissions from the sintering process consists in modification of the phase composition of clinker (alite reduction). The energy-related effects of this modification include the reduction in the temperature and energy intensity of the clinker production process. However, the applicability of this modification are limited due to the required properties of the cement product. A new method for reducing hazardous gas (CO_2 and NO_x) emissions, hitherto not implemented in cement industry, is the combustion of fuels in oxygen-enriched air (zero-emission oxy-fuel combustion). It is assumed that implementation of oxy-fuel technology would allow to reduce the CO_2 emission by ca. 4%. To date, the technology can't make its way to being applied in cement industry as well as in power production mainly due to the protection of oxygen and the costs thereof.

As shown in the aforementioned assessment of the possibilities to reduce CO_2 emission in clinker sintering processed, virtually all reserves have already been used up. Therefore, further reduction in emissions must be sought for outside the sintering process.

A method that allows to achieve significant indirect reduction in CO_2 emission consists in the reducing the quantities of clinker used in the cement and replacing it by additives with hydraulic properties such as blast furnace slag, fly ash from power plants, silica dust or limestone. New cement standards permit the use of various cement additives. The Table 2 presents examples of the effects of cement additives on the reduction in the emission of carbon dioxide from cement production processes [7].

Cement type	Emission [kg CO_2/kg cem.]	Additive percentage share [%]	CO_2 reduction degree in [%]
CEM I 32.5	0.81	0	0
CEM III/A 32.5	0.52	45	35
CEM II/A-V 32.5	0.6	20	25

Table 2. The effects of cement additives on the reduction of CO_2 emission

A novel method to reduce the energy intensity and the CO_2 emissions from cement production process, hitherto not implemented in Poland, is the production of hydraulic binders from mineral waste materials. One of such methods consists in a production of clinker substitute (belite) with a lower lime saturation factor, characterized by lower sintering temperature of ca. 1200°C, and the resulting lower emission of CO_2. The Fig. 5 presents a diagram of such technology (referred to as K-Tech) based on low-carbonate materials and waste fuels.

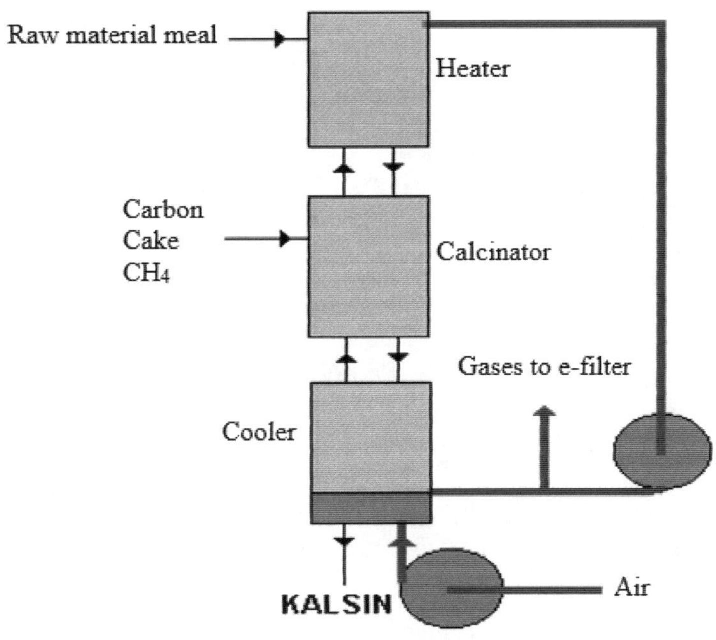

Fig. 5. The diagram of the K-Tech system

The sinter produced in the process (known as Kalsin) is a replacement (in the amounts of up to 50%) of conventional clinker. When used in cement, it facilitates the reduction of its energy intensity by ca. along with a significant reduction in CO_2 emission compared to conventional cement with the same properties. Compared to conventional cement characterized by the emission level of ca 0.85 kg CO_2/kg cl., the emission of cement prepared with the addition of Kalsin is about 0.3 kg CO_2/kg cl. [2]. K-Tech is one of the new promising technologies for the production of binding materials that meets the requirements of the 3x20 initiative as well as the future limits of greenhouse gas emissions.

Research is also conducted on the modification of the geometry of the cyclone systems used in the cement industry. Cyclones are among the most popular mechanical dust separators used to separate solid particles from carrier gas in industrial processes. They are used for air pollution control and for conducting technological processes. The main parameters determining their performance is a constant separation efficiency and pressure drop [8].

Another indirect method for the reduction of CO_2 emission, already used in the cement industry and continuing to grow rapidly, is the production of electricity in waste heat recovery (WHR) systems. Electricity produced in these systems by means of utilization of waste enthalpy from technological processes is not burdened by any emission; it is the so-called "clean" (zero-emission) energy. Therefore, every kilowatt hour of electricity produced in the WHR systems indirectly reduces the emission (ca. 0.8 kg CO_2/kWh) that would otherwise be generated in production of the same amount of energy in a conventional combustion plant.

In cement mills, electricity may be generated from the rotary kiln waste heat by the following methods:

- the classical steam cycle (Clausius-Rankine cycle);
- the organic Rankine cycle (ORC);
- the Kalina cycle.

Modern WHR systems used in cement industry provide as much as 30% of the total electricity demand of cement mills. The Fig. 6 presents the values of electric power that can be generated in WHR systems and the resulting reductions in CO_2, emission depending on kiln efficiency and the moisture content of materials.

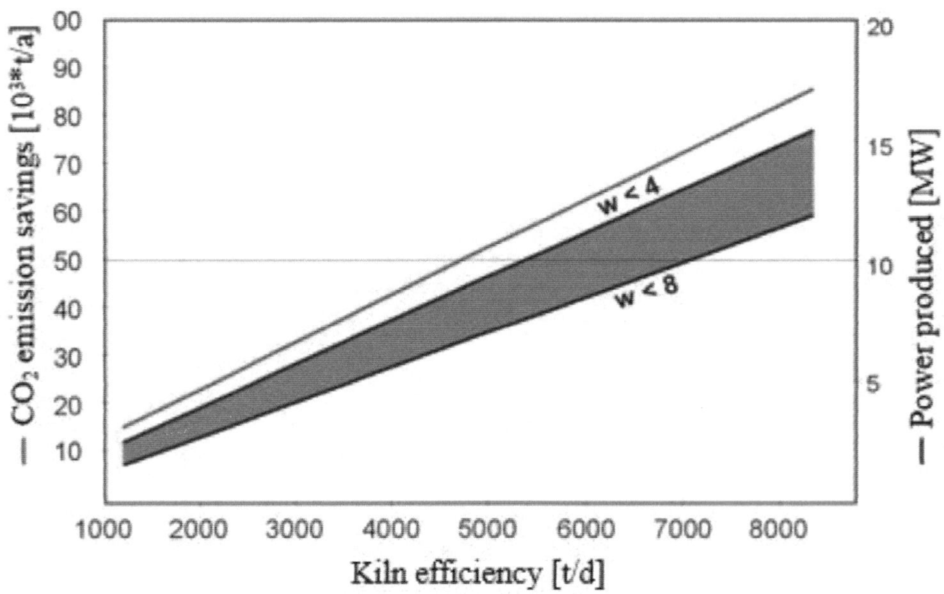

Fig. 6. The electric power as a function of kiln efficiency and materials' moisture content

The mean energy recovery from WHR systems in cement mills is about 32-45 kWh/ton. Therefore, in Polish conditions cement mills would be capable of producing 4-16MW of clean energy and thus indirectly reducing CO_2 emissions by 10,000 to 70,000 Mg CO_2/year [7].

No WHR systems have been installed in Polish cement mills to date. WHR systems may be a rescue solution for cement mills when all other means to reduce the greenhouse gas emissions are used up. Because of progressive increase in electricity prices usually based systems Rankine cycle are used in heat power engineering [9].

4. New technologies for CO_2 reduction

As demonstrated by the methods for the reduction of CO_2 emission and the related efficacy data presented above, the problem of CO_2 emission in cement production remains unsolved. The effects of the reduction measures are insufficient to meet the 3x20 initiative's limits of permissible emission of greenhouse gases. The situation is even worse when new CO_2 emission levels proposed by the European Commission to be in force from the year 2020 are considered. The current deficit of CO_2 emission, destined to increase significantly after new limits are introduced in 2020, may result in the need to reduce the production down to the allocated CO_2 emission levels or to purchase emission permits. Further economic development and the expansion of road infrastructure and residential areas in Poland will require the production to be increased rather than reduced. Also, the option of purchasing the remaining CO_2 emission is infeasible as it would lead to an increase in the price of cement produced in Poland and to possible imports of cheaper cement from another country. Therefore, other method must be sought by the industry to reduce CO_2 emission. The solutions should also take into account the future scenarios regarding emission reductions coming into force after the year 2030. The Blue Map scenario assumes that the emission required to produce 1 ton of cement would be reduced from the current value of 680 kg CO_2/Mg cem. to 406 kg CO_2/Mg cem (the minimum variant) or even down to 338 kg CO_2/Mg cem. (the optimistic variant). At current state of the art and with carbon being used as the technological fuel, such reduction of greenhouse gases could be obtained by implementation of carbon capture and storage (CCS) systems (pursuant to the Directive 2009/31/EC of the European Parliament and of the Council on the geological storage of carbon dioxide. However, this method is not only expensive in terms of the required investment and maintenance costs, but it would also cause significant logistical problems [10].

Due to the current threats for conventional cement production technologies resulting from the carbon dioxide emission reduction requirements, the search continues for new binders characterized by properties similar to those of cement with lower energy intensity and CO_2 emissions. Examples of such studies include the development of a novel binder referred to as geopolymer cement. The name, originating from a mineral polymer formed in geochemical processes, was introduced in 1978 by prof. Josef Davidovits [11]. Geopolymer cement is characterized by energy intensity that is 2-3 times lower and CO_2 emission that is over 4 times lower than that of the conventional cement. Due to this fact, it was nicknamed the "green cement". Despite its ecological as well as quality-related advantages, geopolymer cement has not found wider use to date.

Similar characteristics are presented by another green cement, known as Celitement, which was developed at the Karlsruhe Institute of Technology in Germany. Similarly to geopolymer cement, Celitement is characterized by high strength parameters, energy intensity being nearly half of that of conventional concrete, as well as significantly (ca. 50%) lower CO_2 emissions. These are the technologies of the future which require further research and thus may not yet be considered valid methods for the reduction of CO_2 emissions [12].

5. Conclusion

Due to the high-temperature process of clinker sintering and to the emission of CO_2 due to decarbonation of raw materials, cement production is a high environmental impact process that emits large quantities of greenhouse gases. The ever-stricter limits for acceptable GHG emission require that novel production methods that would allow to meed these limits are being sought for. In relation to the new, significant reduction of CO_2 emissions (40%) scheduled after the year 2020, the cement industry must search for the ways to reduce these emissions. New technologies and equipment implemented in recent years in the industry and meeting the recent BREF requirements of 2013 exhaust the technological possibilities to reduce CO_2 emissions. Further reduction in emissions must be achieved outside the sintering process. A large chance for reducing the acceptable reduction limits in cement mills consists in the production of clean energy in WHR systems based on waste heat generated in the clinker sintering process. Considering the continued increase in worldwide cement production and the resulting increase in carbon dioxide emission as well as the already limited technological possibilities to reduce the emission, the only efficient solution available at the current state of art would consist, similar as in the energy production sector, is implementation of CSS technologies. The authors hope that the effective implementation of mentioned above technologies in the cement industry will lead to the reduction of CO_2 emissions and thereby improve the efficiency of the manufacturing process of cement.

6. References

[1] The Report „Poland to the Convention on Climate Change" (in Polish: Raport „Polska wobec postanowień Konwencji Klimatycznej), Instytut na rzecz Ekorozwoju, December 2000.

[2] Duda, J. (2004). Energy-saving and environment-friendly technology for cement clinker burning (in Polish: Energooszczędne i proekologiczne techniki wypalania klinkieru cementowego), IMMB Papers (in Polish: Prace IMMB), Opole.

[3] BREF 2013, Cement - Lime and Magnesium Oxide Manufacturing Industries, May 2013, European Commission, www.cembureau.be.

[4] Polish Minister of Environment of 12 September 2008 on the method of monitoring the size of emissions covered by the EU system of emissions trading (in Polish: Rozporządzenie Ministra Środowiska z dnia 12 września 2008 r. w sprawie sposobu monitorowania wielkości emisji substancji objętych wspólnotowym systemem handlu uprawnieniami do emisji, Dz. U. Nr 183/2008 poz. 1142).

[5] Duczkowska, A. & Duda, J. (2014). Municipal and industrial waste as alternative raw materials and fuels in the cement production process (in Polish: Odpady komunalne i przemysłowe alternatywnymi surowcami i paliwami w procesie produkcji cementu), ICiMB Papers (in Polish: Prace ICiMB), No. 18, 2014, pp.172-187.

[6] Jõgi, G.; Bashkite, V. & Karaulova, T. (2010). Analysis of first waste-to-energy plant production line at Kunda Nordic Cement AS, Proceedings of 7th International Conference of DAAAM Baltic Industrial Engineering: 7th International Conference of DAAAM Baltic Industrial Engineering, Tallinn, 22-24 April 2010. Küttner, R. (Ed.), Tallinn University of Technology Press, pp. 358–363.

[7] Tomasiak, J. & Duda, J. (2015). The impact of innovative techniques WHR to improve the energy efficiency of technological processes (in Polish: Wpływ innowacyjnych technik WHR na poprawę efektywności energetycznej procesów technologicznych), In: Innovations in Management and Production Engineering (in Polish: Innowacje w zarządzaniu i inżynierii produkcji), Knosala, R. (Ed.), Polish Association for Production Management, Opole, 2015, pp. 510-519.

[8] Wasilewski, M. & Duda, J. (2016). Multicriteria optimisation of first-stage cyclones in the clinker burning system by means of numerical modelling and experimental research, ELSEVIER, Powder Technology, Vol. 289, February 2016, pp. 143-158.

[9] Apostol, V.; Pop, H.; Dobrovicescu, A.; Tudor Prisecaru, T; Alexandru, A. & Prisecaru, M. (2015). Thermodynamic Analysis of ORC Configurations Used For WHR from a Turbocharged diesel engine, 25th DAAAM International Symposium on Intelligent Manufacturing and Automation, ELSEVIER, Procedia Engineering, Vol. 100, 2015, pp. 549-558.

[10] Duda, J; Kolosowski, M. & Tomasiak, J. (2015). CCS technology in the cement plant (in Polish: Technologia CCS w zakładzie cementowym), Logistics (in Polish: Logistyka), No. 3, 2015, pp. 1040-1050.

[11] Stemmermann, P.; Schweike, U.; Garberr, K. & Beuchle G. (2010). Celitement – a sustainable prospect for the cement industry, Cement International, No. 5, 2010, pp.53-65.

[12] Król, M. & Błaszczyński, T. (2013). Geopolymers in construction (in Polish: Geopolimery w budownictwie), Insulations (in Polish: Izolacje), No. 5, 2013, pp. 38–44.

DOI: 10.2507/27th.daaam.proceedings.019

EFFECT OF PROCESSING OF WOOD PLANNING ON STATIC AND DYNAMIC WETTING ANGLE ADHESIVES

Murco Obucina, Seid Hajdarevic & Almir Akvic

This Publication has to be referred as: Obucina, M[urco]; Hajdarevic, S[eid] & Akvic, A[lmir] (2016). Effect of Processing of Wood Planning on Static and Dynamic Wetting Angle Adhesives, Proceedings of the 27th DAAAM International Symposium, pp.0131-0135, B. Katalinic (Ed.), Published by DAAAM International, ISBN 978-3-902734-08-2, ISSN 1726-9679, Vienna, Austria
DOI: 10.2507/27th.daaam.proceedings.019

Abstract

The objective of this research was to study the effect of planning processing operations on the surface wettability of beech wood. For surface preparation were used two different machine types, Unimat 500 and Unimat Super 4. Quality parameters were monitored by surface roughness and the size of the wettability measured by estimated contact angle within the method of stationary droplet (sessile drop method) and dynamic wetting. Gluing was implemented with water-resistant PVAc adhesive, typically used for gluing massive wood. The results were statistically analyzed and compared with each other using the Student's t-Test.
There were no significant differences in surface roughness obtained by conventional planning and fig machine (sawing), although the processing of these two machines results in different types of surfaces. Slightly higher roughness was determined for planning processing. Established wetting angle is a significantly different, depending to the wood surface roughness and type of processing.

Keywords: Technology wood gluing; Surface roughness; Contact angle; Static wetting; Dynamic wetting

1. Introduction

Quality of solid wood panels gluing highly depends on the surface preparation. Planning, sawing (fig) machine are the two most common types of wood processing as preparation for applying the adhesive. Surface sanding may also be used, but it is uncommon type of technology for applying of glue on wood surface. *Planning machine* is used very often in technology of solid wood panels gluing, with its main disadvantage resulting in cycloid surface formed due to rotation of tool and simultaneous linear movement of work piece. Other processing type is machine with special circular saws tool often named *fig machine*.

Wood gluing takes a very important place in the wood technological processes. It significantly affects the market value and quality of wood products. Gluing is a complex process, during which several complex physicochemical processes takes place, due to the properties of wood and glue, their interactions and gluing technology. The main goal of gluing process is to achieve a combination of high strength and long-lasting strength, which in principle should not be less than the strength of the wood [2], [3], [5].

The strength of the bonded joint depends on several factors, the most on the adhesion, in addition to the properties of the adhesive. The adhesion of the glue-wood primarily depends on the wetting surface, glue penetration, porosity, pH value, moisture, surface tension, surface condition and anatomical directions of wood [4], [6]. Technological process of bonding and surface preparation of wood affects the quality of the shear strength, especially on the consistent quality of glued joint [5].

Bonding cannot be achieved without wetting surfaces to be bonded. Spreading of glue on the wood surface is the most important prerequisite for efficient and successful bonding. Wetting is in relation to surface tension and surface free energy of the system: hardwood body (wood)-glue-air [1], [7].

The wetting angle, as very important parameter, can be used to determine surface energy and adhesion. It has been found if adhesive angle is less than 90 degrees wood surface has good wetting – the lower wetting angle leads to better wetting. The optimal wetting is for surface where angle of wetting is close to zero degrees and in that case glue tends to be completely spread [4]. Based on research, it is recommended that the contact angle should be measured for volume greater than 0,1 ml [8].

In this paper beech samples (Fagus sylvatica) were used for research of static and dynamic wetting angle, as well as roughness, by processing them on two different types of machines – planning and fig (sawing) machine.

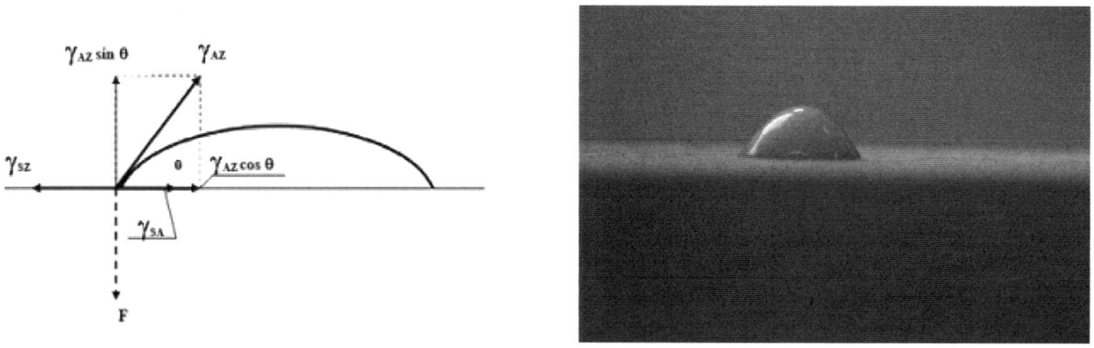

Fig. 1. Wetting angle of adhesive

2. Methodology

The first part of this research was to determine wood surface quality in "Laboratory of metal cutting and tools design" at Faculty of Mechanical Engineering Sarajevo University. Device "Mitutoyo SJ-201P Portable Surface Roughness Tester" was used for measuring of roughness, where the main point was on determining average length between the peaks and valleys and the deviation from the mean line on the entire surface within the sampling length (Ra) and vertical distance from the highest peak to the lowest valley (Rz).

For measuring of static wetting angle simple dispenser was used. Through dispenser, droplets of the same size were discharged on wood surface. After a contact between droplet and wood surface had been made, the droplet started its spreading. The whole process was recorded with special PCO camera. Two hundred photos were saved between the moments of droplet discharge from dispenser until the moment of completely spreading. This process required patience and precision, as well as analysing those photos in software "Image J" with "drop analysis" plug-in, which allows precise measurement of the wetting angle. The purpose of this measuring was to compare wetting angles for beech samples, processed by conventional planning machine and fig machine, as well as to determine if there was significant difference between angles.

Fig. 2. PCO camera and wetting angle measuring by "ImageJ" software

Dynamic wetting angle was testing by simulation of gluing in real production conditions. Practically, two glue droplets were discharged on wood surface sample by dispenser. There was space between two droplets, at the beginning of simulation, but droplets were close enough to become one after the end of process. That means the other wood sample was moving parallel to first wood piece and this was the simulation of real conditions of solid wood panels gluing. At the first phase, wood sample was moving down making partial contact with two droplets. Through phases 2 and 3, wood sample was getting closer to the other sample. The final stage resulted in a full contact of wood samples.

Simulation of solid wood panels gluing was done by lowering one piece of wood sample to another by slim aluminium 1 mm thick pieces. Those aluminium sheets were made using the contact calibrator for metal. At the stage 1, there were 6 pieces on both sides of droplets, which resulted in 6 mm distance between wood samples. They have been pulling out from both sides simultaneously and the whole experiment was recorded by PCO camera that made 500 pictures in a second this time.

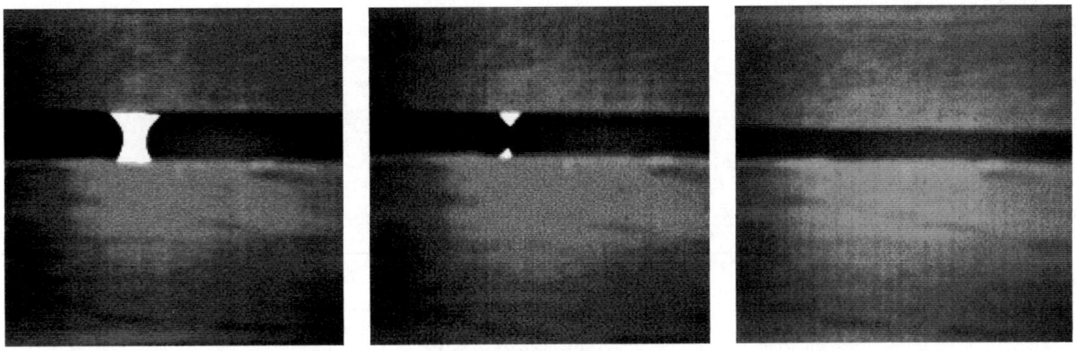

Fig. 3. PCO camera photos of simulation for measuring dynamic wetting angles
Phases 2, 3 and 4

3. Results

By the end of roughness testing, all results have been processed and analysed. Statistical methods have been used, including Student's t-Tests as a part of data analysis of Microsoft Excel. The final results led to the conclusion there was no significant difference for wood samples earlier processed by planning machine, as well as fig (sawing) machine. Surface roughness still had higher value for samples processed by conventional planning machine (*figure 4*).

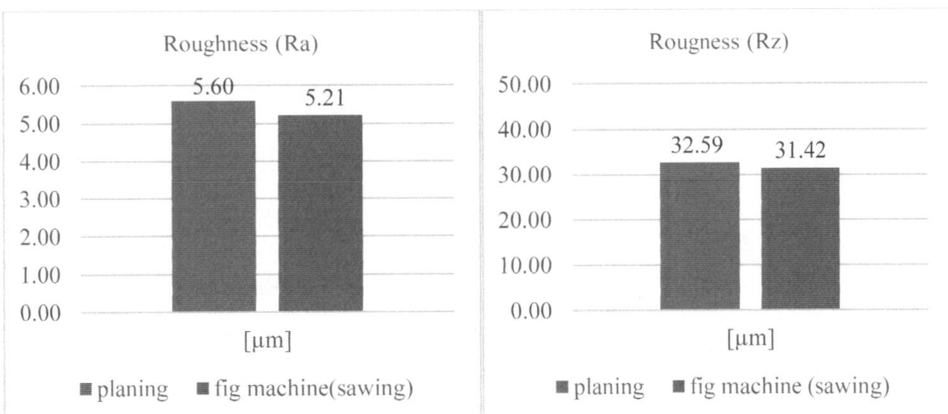

Fig. 4. Roughness measured for two types of processing

By comparing surface roughness for two types of wood processing, there was no significant difference and null hypothesis came to be true. Software "Image J" was used for editing of photos and measuring of wetting angle. Angle was measured from both sides of droplets, and arithmetic mean was used as the resulting angle. Data is presented in *tables1 and 2* and on *figure 5*.

As it is shown in charts above, static angle of wetting resulted in significant difference between variables. Wood samples processed by conventional planning method had significantly higher static angle value, in compare to wood samples processed by fig (sawing) machine. Null hypothesis was rejected, and alternative hypothesis came true, using the standard $\alpha = 0.05$ cutoff.

No.	PLANING MACHINE			FIG (SAWING) MACHINE		
	Left angle [°]	Right angle [°]	Mean [°]	Left angle [°]	Right angle [°]	Mean [°]
1	117	94	105,5	91	92	91,5
2	103	98	100,5	82	85	83,5
3	86	90	88	84	87	85,5
4	87	92	89,5	87	87	87
5	90	101	95,5	84	85	84,5
6	85	59	72	88	83	85,5
7	94	81	87,5	87	83	85
8	94	96	95	84	94	89
9	87	92	89,5	96	64	80
10	97	100	98,5	80	87	83,5

Table 1. Static angle full data

	PLANING MACHINE			FIG (SAWING) MACHINE		
	Left angle [°]	Right angle [°]	Mean [°]	Left angle [°]	Right angle [°]	Mean [°]
μ [°]	94,00	90,30	92,15	86,30	84,70	85,50
St.dev. [°]	9,88	12,41	90,67	4,64	8,10	84,83
K.var. [%]	10,51	13,74	89,74	5,38	9,56	85,06
Min. [°]	85,00	59,00	90,04	80,00	64,00	84,98
Max.[°]	117,00	101,00	90,06	96,00	94,00	84,77

Table 2. Static angle data

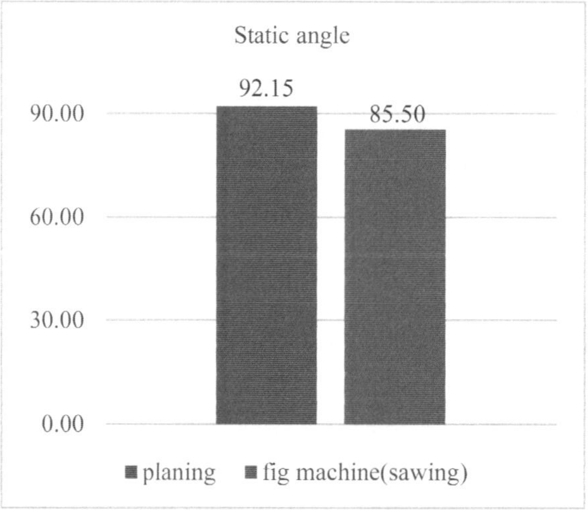

Fig. 5. Static angle measured for two types of processing

As the chart above shows it, static angle of wetting resulted in significant difference between variables. Wood samples processed by classic planing had significantly higher static angle value, in compare to wood samples processed by fig (sawing) machine. Null hypothesis was rejected and alternative hypothesis came true, using the standard $\alpha = 0.05$ cut off.

Dynamic wetting angle made this research more interesting by simulating real gluing conditions of solid wood. Results for dynamic angle measuring are given in *table 3* and *charts 6.a. and 6.b.*

Phase No	Mean value α [°]	Mean value ß [°]	Mean value γ [°]	Mean value δ [°]
1	126.5	128.3	148	146
2	131	131.6	152.5	156.5
3	127	125.2	149	145
4	134	134.4	122	137.5

Table 3. Dynamic angle data

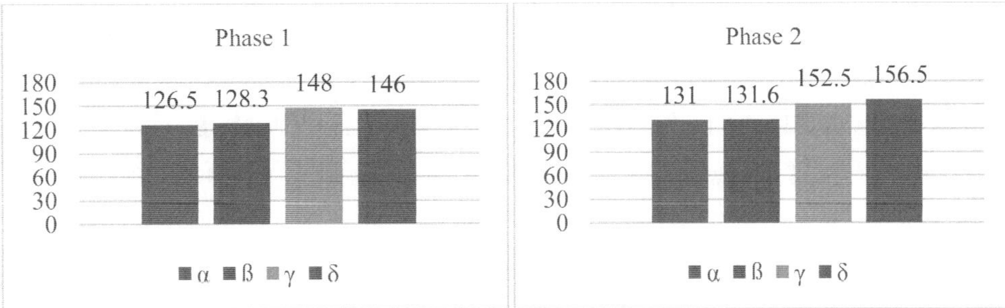

Fig. 6.a. Dynamic angle data measured in phases 1 and 2

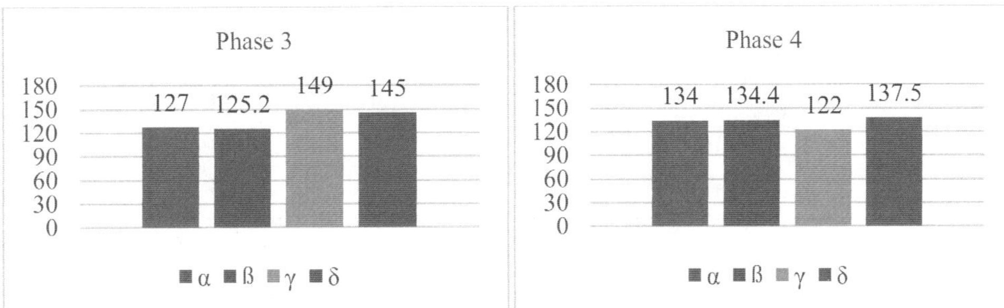

Fig. 6.b. Dynamic angle data measured in phases 3 and 4

In the future researches, wetting angles and both side adhesive penetration should be considered, as well as quality of gluing depending on these two factors.

4. Conclusion

Although processing on these machines resulted in different surface types and surface roughness, there was no significant difference by comparing surface roughness values for samples processed on conventional planning machine and fig (sawing) machine. For samples processed on planning machine surface roughness resulted in slightly higher values. In this paper static and dynamic angle were measured. By comparing static angle values for samples processed on planning and fig (sawing) machine, result sled to conclusion there was significant, but not highly significant difference. It is important to mention the coefficient of variation that resulted in lower values for samples processed on fig (sawing) machine. The conclusion for static angle of surface wetting, resulted in higher values for samples processed by conventional planning process, due to cycloid surface. Despite this conclusion, difference was not significant. Fig (sawing) machine processing sawing tools are parallel and that fact ensure, not only precise processing, but also optimally prepared surfaces for gluing. This is the reason why fig (sawing) machine is widely used in solid wood panel productions.

Dynamic wetting angle should be measured by proper mechanical simulator of gluing and that method could improve new knowledge about change of adhesive angles during solid wood panels gluing.

5. References

[1] Bogner, A.; Grbac, I.; Despot, R. (2002).Adhesion and Optimum Surface Tension of Adhesives, Wood Research 47

[2] Gindl, M.; Tschegg, S. (2002). Significance of the acidity of wood to the surface free energy components of different wood species, Langmuir, 18, 3209–3212

[3] Marra, A.A. (1992). Technology of wood bonding: principles in practice, Van Nostrand Reinhold, New York

[4] Obucina, M.; Gondzic, E.; Manso, E. The Influence of Adhesion Temperature to the Shear Strength of Width Glued Wooden Elements ". 25th International Symposium on Intelligent Manufacturing and Automation, DAAAM 2014. Procedia Engineering 100 (2015) 321–327

[5] Obucina M.; Akvic, A."Influence of gluing technology on the shear strength in the gluing of solid wood",Proceedings of the 26th DAAAM International Symposium,pp.0182-0188, B. Katalinic (Ed.), Published by DAAAM International, ISBN 978-3-902734-07-5, ISSN 1726-9679, Vienna, Austria DOI:10.2507/26th.daaam.proceedings.026

[6] I. Rowell; R, M. Chemical Modification of Wood for Improved adhesion in Composites, Wood Adhesives, Forest Products Society, Madison, (1996) WI., 69-78

[7] Sernek, M.; Resnik, J.; Kamke, F.A. (1999). Penetration of liquid urea-formaldehyde adhesive into beech wood, Wood and Fiber Sci. 31(1), 41–48

[8] McGuire, J; Yang, J. (1991). The effect of drop volume on contact angle, J. Food Protection, 54, 3, 232–235

DOI: 10.2507/27th.daaam.proceedings.020

THE POSSIBILITIES OF EXPLOITATION OF E-LEARNING IN TECHNOLOGY-ORIENTED COURSES

Jana Moravčíková

This Publication has to be referred as: Moravcikova, J[ana] (2016). The Possibilities of Exploitation of E-Learning in Technology-Oriented Courses, Proceedings of the 27th DAAAM International Symposium, pp.0136-0141, B. Katalinic (Ed.), Published by DAAAM International, ISBN 978-3-902734-08-2, ISSN 1726-9679, Vienna, Austria
DOI: 10.2507/27th.daaam.proceedings.020

Abstract

Developing trends in area of modernization of E-learning are mainly based on the technical possibilities of information and communication technologies. These trends supplement the distance learning implemented through E-learning about new possibilities of formation, presentation and distribution of curriculum. Information and communication technologies allow develop a new way of teaching for institutions, streamlining and modernizing the possibilities of study and become an important means to support teaching. E-learning modules serve as a complement to the comprehensive study. The paper is focused on the possibility of creating and using E-learning to support teaching of technologically-oriented courses at Slovak Technical University. Contribution describes the possibility of creation of E-learning and study support for technological course "Fundamentals of production technologies II" in an environment Academic Information System (AIS) STU. Contribution aims to determine the outline of the course, instructions on how to set up E-learning and subsequently prepare E-learning support, its individual parts and describes connection of support with the course and students. It describes in detail the creation of E-learning library and its E-objects that are part of E-learning. E-learning enables active use of video, images and tests.

Keywords: E-learning; E-object; E-learning support; E-learning project; technology.

1. Introduction

E-learning can be understood as electronic media supported of learning process. The E-learning system, which is considered to be a new didactic method, enables to provide education not only in a distant form, however there are new possibilities for other forms of education, as well [1]. American founder of the E-learning industry Eliott Masie defines the E-learning as a tool that uses network technology to create, distribute, selection, administration and continuous updating of educational materials [1]. The company "Urdan and Weggen" used in their study the term of E-learning as providing information, educational materials and educational content by using of various forms of electronic media - internet, intranet, extranet, CD-ROM, satellite broadcast, interactive TV [2]. E-learning is used as synonym for education supported by technologies TBT - Technology based training. Other companies, such as "Block and Dobell," consider e-learning as a subset of distance education, and this process is defined as the interaction between the students and the source of information who is physically placed in a remote location [4]. The way of using ICT resources and the availability of information materials depends primarily on educational goals and content, the nature of the education process, needs and possibilities for all the actors of the education process [5]. The issue of increasing the quality of the learning process using different methods of E-learning is currently is currently the object of interest of many authors [6-19]. The problem is the absence of development of AIS STU E-learning in methodic. The aim of E-learning is to have the skills and knowledge to be useful in the job or can helping individuals to achieve personal learning goals [3]. Contribution describes the possibility of creation of E-learning and study support for technological course "Fundamentals of production technologies II" on bachelor study degree in an environment Academic Information System (AIS) STU.

2. The process of creation and implementation of E-learning in the AIS STU

2.1. Scheme of creation of E-learning in AIS environment

Schematic design of process making E-learning is shown in Fig. 1. It is a graphical representation of each step in the creation of E-learning courses in the AIS environment. Schema contains guide on how to establish E-learning and subsequently prepare E-learning support. The library is designed to store of particular file used in E-learning project. Uploaded files are, for example, different texts (e. g. PDF), animated images, videos and sound records.

The project distinguishes three types of libraries:

- The personal library of E-learning objects - objects that can be used for example in tests, only for personal use.
- Project library of E-learning objects - objects related to specific projects and objects may only be used by persons who are directly connected to the project.
- Public library of E-learning objects - objects who are defined as public, so they are available to the general public

After entering to the folder can be added to the library various "sub-folders" containing individual objects and therefore can be categorized each other.

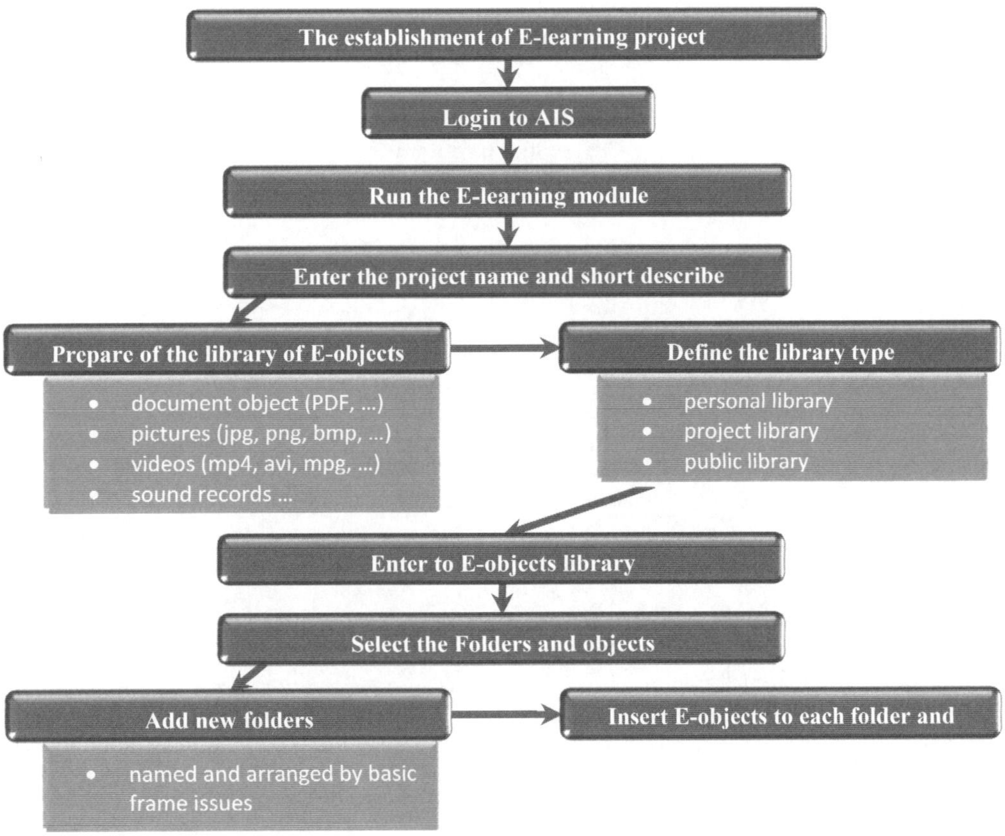

Fig. 1. Scheme of creation of E-learning in AIS environment

2.2 Structure determination methods of machining

At the beginning of the creation of E-learning it is necessary to create an outline - chapters of the course to be found in E-learning. Subfolders in the E-learning library will have the name of each chapter in E-learning support. Machining technology, for example, for grinding, it may be characterized by the following points:

- Process characterisation
- Kinematics of grinding
- Dynamics of grinding
- Basic methods
- Formation of chips
- Grinding tools
- Toolholder
- Workpiece clamping
- Machine tools
- Surface quality
- Wear and dressing of grinding tools

3. Course and e-learning condition, course structure of grinding

3.1 Create of E-learning project

The E-learning in AIS can create the registered user only. Created project can be made available to the whole university or only for a specific course. The project can be establish using the application eLearning projects (Fig. 2). Basic information about the created project is shown in Fig. 3. The project is prepared to teach in the Slovak language.

Fig. 2. Menu for establish of E-learning project

Project title:	E-Learning of selected studying themes - Fundamentals of production technologies II
Administrator:	Ing. Jana Moravčíková, PhD.
Project description:	Study support of the course Fundamentals of production technologies II
Language:	Slovak
Department:	Institute of Production Technologies
Protected by password:	yes
Project status:	upcoming
Project start date :	14. 02. 2016
Number of workers in the project:	4
Number of tasks in the project:	0/0
Number of documents in the project:	0/0
Number of projects in the projects library:	101
Sample tests designed for the project:	0
Number of questions in test base:	30
Supports designed for the project:	9

Fig. 3 Basics information about created project

For E-learning was created the library of E-objects. The structure for technology characterisation was described in chapter 2.2. Into created sub-folders can be loaded various objects and files which will complement E-learning project. Fig. 4 documented the example of created subfolders for grinding technology.

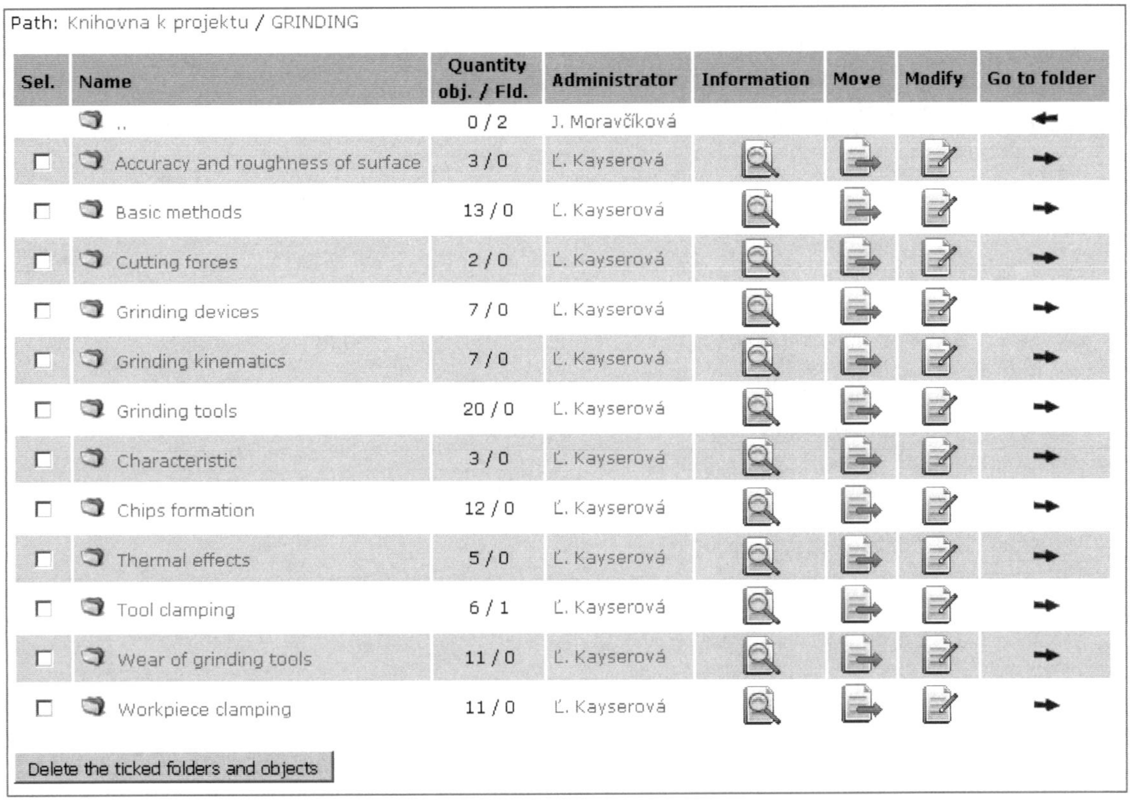
Path: Knihovna k projektu / GRINDING

Sel.	Name	Quantity obj. / Fld.	Administrator	Information	Move	Modify	Go to folder
	..	0 / 2	J. Moravčíková				
	Accuracy and roughness of surface	3 / 0	Ľ. Kayserová				
	Basic methods	13 / 0	Ľ. Kayserová				
	Cutting forces	2 / 0	Ľ. Kayserová				
	Grinding devices	7 / 0	Ľ. Kayserová				
	Grinding kinematics	7 / 0	Ľ. Kayserová				
	Grinding tools	20 / 0	Ľ. Kayserová				
	Characteristic	3 / 0	Ľ. Kayserová				
	Chips formation	12 / 0	Ľ. Kayserová				
	Thermal effects	5 / 0	Ľ. Kayserová				
	Tool clamping	6 / 1	Ľ. Kayserová				
	Wear of grinding tools	11 / 0	Ľ. Kayserová				
	Workpiece clamping	11 / 0	Ľ. Kayserová				

Delete the ticked folders and objects

Fig. 4 The list of created subfolders for grinding

After entering to subfolder is possible to add the required E-object. For example, to subfolder "*Basic methods*" were continually added 11 E-objects (pictures) which name defines the content of each E-object (Fig. 5). Next were prepared and added 5 videos, which documents the basic grinding methods.

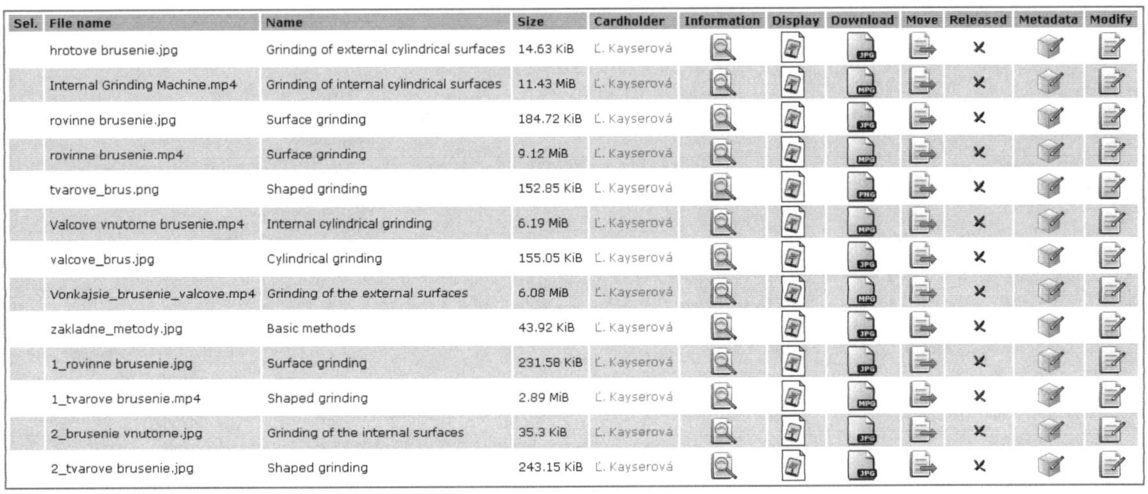

Sel.	File name	Name	Size	Cardholder	Information	Display	Download	Move	Released	Metadata	Modify
	hrotove brusenie.jpg	Grinding of external cylindrical surfaces	14.63 KiB	Ľ. Kayserová					×		
	Internal Grinding Machine.mp4	Grinding of internal cylindrical surfaces	11.43 MiB	Ľ. Kayserová					×		
	rovinne brusenie.jpg	Surface grinding	184.72 KiB	Ľ. Kayserová					×		
	rovinne brusenie.mp4	Surface grinding	9.12 MiB	Ľ. Kayserová					×		
	tvarove_brus.png	Shaped grinding	152.85 KiB	Ľ. Kayserová					×		
	Valcove vnutorne brusenie.mp4	Internal cylindrical grinding	6.19 MiB	Ľ. Kayserová					×		
	valcove_brus.jpg	Cylindrical grinding	155.05 KiB	Ľ. Kayserová					×		
	Vonkajsie_brusenie_valcove.mp4	Grinding of the external surfaces	6.08 MiB	Ľ. Kayserová					×		
	zakladne_metody.jpg	Basic methods	43.92 KiB	Ľ. Kayserová					×		
	1_rovinne brusenie.jpg	Surface grinding	231.58 KiB	Ľ. Kayserová					×		
	1_tvarove brusenie.mp4	Shaped grinding	2.89 MiB	Ľ. Kayserová					×		
	2_brusenie vnutorne.jpg	Grinding of the internal surfaces	35.3 KiB	Ľ. Kayserová					×		
	2_tvarove brusenie.jpg	Shaped grinding	243.15 KiB	Ľ. Kayserová					×		

Fig. 5 Example of added objects in subfolder "Basic methods"

After adding all required E-objects, user can show the preview of object number in subfolders by "Tree of folders" (Fig. 6) which can be used for quick and simple orientation of created project. The "Folders tree" shown the all subfolders of project and number of inserted E-objects in individual subfolders.

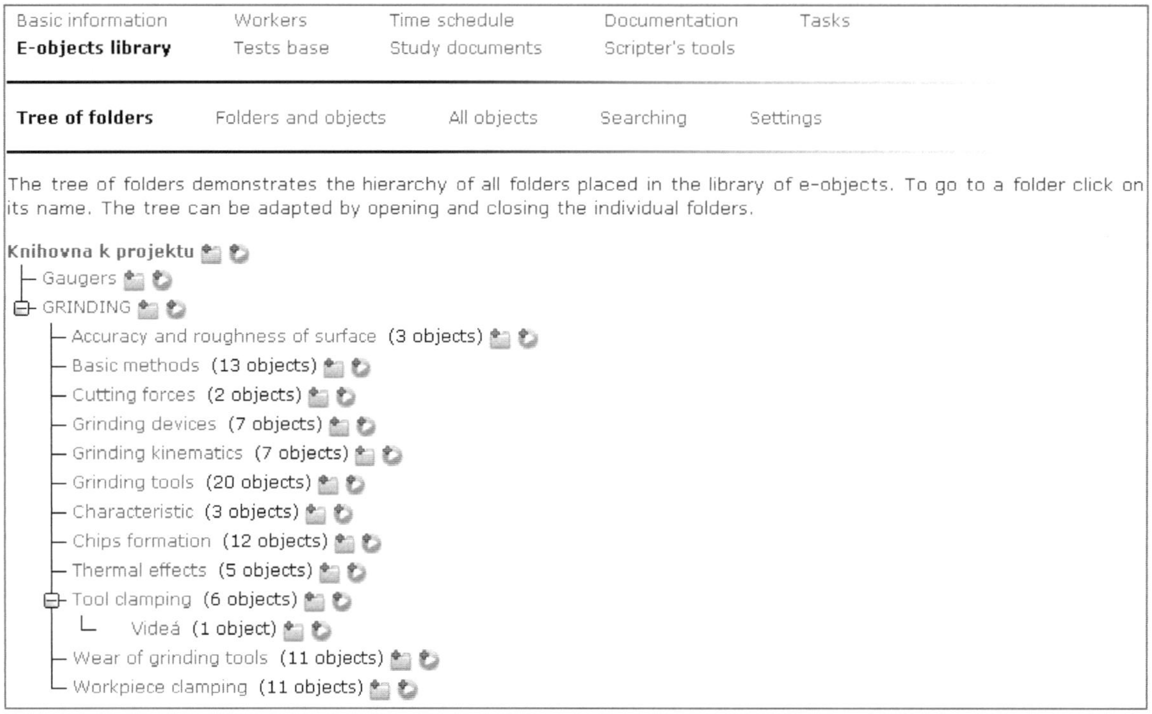

Fig. 6 Subfolders tree in E-objects library

3.2. Development of study support

The study support is possible create only after inserting all of required E-objects which is the E-support created. The study support contain interactive parts, such as glossary of terms, animated pictures and videos. Sample of prepared study support of part "workpiece clamping during grinding" is shown on Fig. 7. From the created E-learning support can be accessible progressively formed individual parts of the topics with are actually teaches or set up automatic publishing of the topics on the specified date.

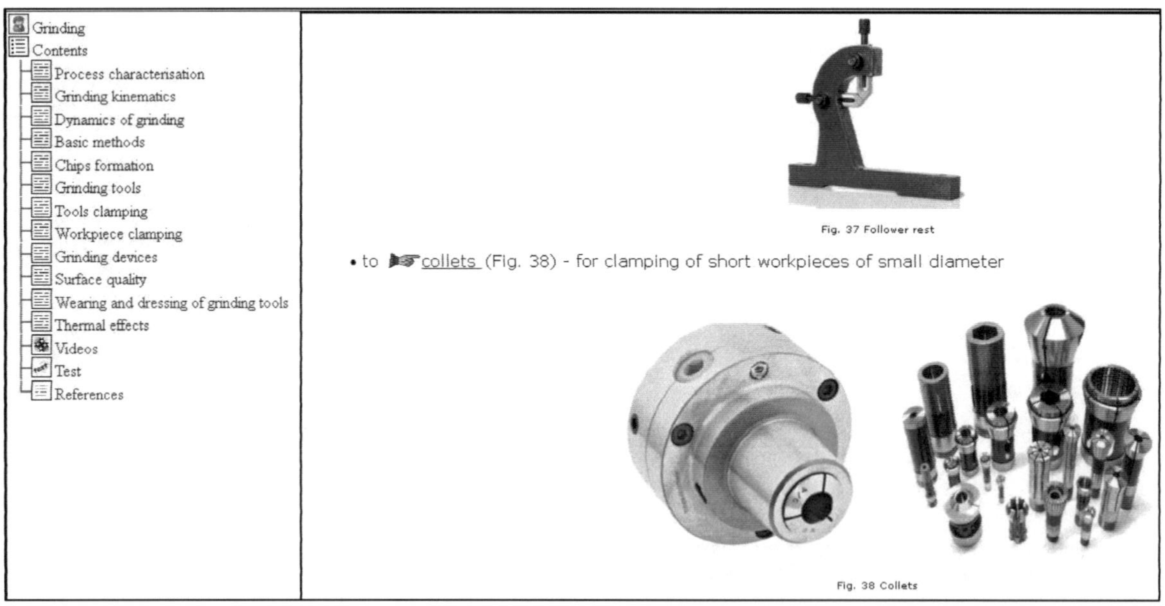

Fig. 7 Sample of study support of „Workpiece clamping"

From the created E-learning support can be accessible progressively formed individual parts of the topics with are actually teaches or set up automatic publishing of the topics on the specified date.

4. Conclusion

The aim was to create an E-learning project for the course Fundamentals of Production Technologies II. For the processing of support was used and processing using the machining many wedges cutting tool with undefined geometry of cutting wedge - grinding. Part of the course support was created using the E-learning in Academic information system of STU, in which was incremental added E-learning materials for students of the faculty which have teaching and testing character. Prepared project is ready to process and characterize other machining technologies, which are content of the course Fundamentals of production technologies II. To the E-learning for the E-learning project "Fundamentals of production technology II" were added 13 sub-sections - chapters related to grinding, one of which contained only links to videos from Youtube.com website and another test questions. Within the support, it is appropriate to establish testing test base too, that allows continuous verification of the acquired knowledge. Prepared support is implemented into education of the course. As a result, it was observed improvement in exam results from acquired knowledge.

5. Acknowledgements

This contribution is a part of the KEGA project of Ministry of Education, Science, Research and Sport of the Slovak Republic, No. 032STU-/2014 “Blended Learning principles implementation into teaching of programming of CNC machine tools with advanced kinematic structure”.

6. References

[1] Lepiš, F. (2015). E-learning – Tomorrow Communication, Available from: http://www.pulib.sk/elpub2/FHPV/Pavelka1/8.pdf Accessed: 2015-11-18
[2] http://teachersupport.info/blog/19-01-2015/introduction-elearning#.VmKg5Pnqiko, (2015). Education support, Accessed on: 2015-11-26
[3] Clark, R.; Mayer R. E. (2011). E-learningand the Science of Instruction, John Wiley &Sons
[4] Baresova, A. (2011). E-learning ve vzdělávání dospělých, page 29, 197 s. ISBN 978-808-7480-007, Praha Vox
[5] Zounek, J. (2009). E-learning - jedna z podob učení v moderní společnosti, Masarykova univerzita, Brno, 2009, page 37-38
[6] Jordan, A; Carlile, O; Stack, A. (2008). Approaches to Learning: A Guide for teachers. 2008 Publisher: McGraw Hill, Glasgow
[7] Ally, M.; Fahy, A. (2004). Foundations of educational Theory for online learning. Edited by: Anderson, T.; Elloumi, F. Theory and practice of online learning, pages: 3-31, Publisher: Athabasca University, Athabasca, Canada
[8] Ambrose, S.A.; Bridges, M.W.; Di Pietro, M.; et al. (2010). How Learning Works: 7 Research-Based Principles for Smart Teaching. Publisher: John Wiley & Sons, Jossey-Bass, San Francisco, CA
[9] Clark, R. C.; Mayer, R. E. (2003). E-Learning and the science of instruction. Publisher: Wiley., New York, NY
[10] MacDonald, J. (2008). Blended learning and online tutoring: A good practice guide. 2008. Publisher: Gower, Aldershot, UK
[11] Moore, Joi L.; Dickson-Deane, C.; Galyen, K. (2011). E-Learning, online learning, and distance learning environments: Are they the same? Internet and higher education, Volume: 14 Issue: 2 Special Issue: SI, pages: 129-135, Published: MAR
[12] Bates, A. (2015). Teaching in a digital age Published: URL: http://opentextbc.ca/teachinginadigitalage/
[13] McVey, M. E. (2016). E-learning and education for sustainability. International review of education, Volume: 62, Issue: 1, Pages: 117-121
[14] Graham, C. R. (2006). Blended learning systems. CJ Bonk & CR Graham, The handbook of blended learning: Global perspectives, local designs, Publisher: Pfeiffer, Available from: ttp://www.click4it.org/images/a/a8/Graham.pdf
[15] Fryer, L.K., Bovee, H. N. (2016). Supporting students' motivation for e-learning: Teachers matter on and offline. Internet and higher education, Volume: 30, Pages: 21-29
[16] Keller, John M. (2008). First principles of motivation to learn and E-learning. Distance Education, Volume: 29, Issue: 2 Pages: 175-185
[17] Esteban, A. M.; Zapata R. M. (2016). Learning Strategies and E-Learning. Some Notes on the Foundations of Learning Design in Virtual Learning Environments. Red-Revista de Educacion Adistancia, Issue: 50, Article Number: 15
[18] Makraiova, J., Pokorna, E. (2013). Person-organisation fit in the context of cultural learning. 24TH DAAAM International Symposium on Intelligent Manufacturing and Automation, Book Series: Procedia Engineering 69 (2014) 712 – 719.
[19] Andreev, Kuvshinov, S., Pryanichnikov, V., Poduraev, Y. (2013). Education on the basis of virtual learning robotics laboratory and group-controlled robots. 24TH DAAAM International Symposium on Intelligent Manufacturing and Automation, Book Series: Procedia Engineering 69 (2014) 35 – 40.

DOI: 10.2507/27th.daaam.proceedings.021

INVESTIGATION OF ROBUST STABILITY FOR DISCRETE-TIME POLYNOMIALS WITH NONLINEAR UNCERTAINTY STRUCTURE

Radek Matušů

This Publication has to be referred as: Matusu, R[adek] (2016). Investigation of Robust Stability for Discrete-Time Polynomials With Nonlinear Uncertainty Structure, Proceedings of the 27th DAAAM International Symposium, pp.0142-0146, B. Katalinic (Ed.), Published by DAAAM International, ISBN 978-3-902734-08-2, ISSN 1726-9679, Vienna, Austria
DOI: 10.2507/27th.daaam.proceedings.021

Abstract

The main objective of this contribution is to present the application of a graphical approach to robust stability analysis of discrete-time uncertain polynomials with special emphasis to polynomials with a complicated nonlinear uncertainty structure. The generalization of the value set concept and the zero exclusion condition under robust *D*-stability framework is used for this purpose. The relative simplicity of the applied technique is shown by means of the illustrative examples.

Keywords: Robust stability; discrete-time polynomial; parametric uncertainty; nonlinear uncertainty structure; value set; zero exclusion condition.

1. Introduction

Stability is a critical property of the control loops. In the case of uncertain systems, frequently described by the models with parametric uncertainty, stability must be ensured for all possible members from the assumed family, i.e. so-called robust stability is investigated. Naturally, robust stability represents attractive and deeply studied discipline of the last decades, especially in the continuous-time domain [1], [2], [3], [4], [5], [6]. A number of interesting results have been developed also for the discrete-time systems [7], [8], [9], [10], [11], but many of them are applicable only for simple uncertainty structures (e.g. interval polynomials) and suffer from the lack of generality as they require some restrictions and pre-conditions. And even if some results are more general, they are (in author's opinion), relatively complicated to use.

This contribution is intended to describe the application of the value set concept in combination with the zero exclusion condition under robust *D*-stability framework adopted from [1] for analyzing robust stability of discrete-time polynomials with nonlinear uncertainty structure. The similar ideas as in this work have been already published in the conference paper [12] and its extended version [13]. Moreover, robust stability of continuous-time polynomials with nonlinear uncertainty structure was studied in [14] and robust stability of discrete-time polynomials with a simple uncertainty structure was presented in [15] (the example within the mentioned paper is focused only on discrete-time interval polynomial).

2. Problem formulation

The problem of stability of systems can be supposed as the problem of stability of their characteristic polynomials. The discrete-time polynomial with the uncertain parameters can be written in the form:

$$p(z,q) = \sum_{i=0}^{n} \rho_i(q) z^i \tag{1}$$

where z is the complex variable, q is the vector of uncertainty and ρ_i are coefficient functions. Then, the family of polynomials is defined by [1]:

$$P = \{p(\cdot,q) : q \in Q\} \tag{2}$$

where Q is the uncertainty bounding set (represented by a multidimensional box in this work). Generally, the family of polynomials (2) is robustly stable if and only if $p(\cdot,q)$ is stable for all $q \in Q$, i.e. all roots of $p(z,q)$ must lie within the unit circle. The direct calculation of roots is a possible solution, but it can be very impractical due to potentially extremely long computation times. On that account, the more efficient techniques have to be used.

Among a number of existing methods which were developed and described e.g. in [7], [8], [9], [10], [11], the graphical approach based on combination of the value set concept and the zero exclusion condition under robust *D*-stability framework represents very universal, elegant and powerful tool. Its universality is indispensable especially for the most complicated (nonlinear) uncertainty structures, which are studied in this contribution, because the universal analytical tools practically do not exist for this kind of structures.

3. Value set concept and zero exclusion condition for robust *D*-stability

The basic continuous-time version of the value set concept and the zero exclusion condition can be found in [1] and subsequently e.g. in [6]. This contribution employs its improved version, described hereafter, which is extended and generalized to so-called robust *D*-stability framework [1]. The main idea remains the same, but it allows investigating robust stability for an arbitrary stability region *D*, e.g. for a unit circle in case of discrete-time polynomials with parametric uncertainty.

Suppose a family of polynomials (2). The value set at any evaluation point $x \in \mathbf{C}$ is determined by [1]:

$$p(x,Q) = \{p(x,q) : q \in Q\} \tag{3}$$

In other words, $p(x,Q)$ is the image of Q under $p(x,\cdot)$. For instance, in discrete-time case, substitute z for a point at the unit circle in a family $P = \{p(z,q) : q \in Q\}$ and let the vector of uncertain parameters q range over the set Q.

In this contribution, the value sets for the families with complicated nonlinear uncertainty structure are plotted by using a suitable sampling (gridding) of the uncertain parameters and direct calculation of related partial points of the value sets for a supposed frequency range. The relative simplicity and applicability for the complex uncertainty structures represents the main advantage of this approach. On the other hand, a long computational time for a high number of uncertain parameters is the weakness.

The zero exclusion condition formulated in [1] says: Let *D* be an open subset of the complex plane and assume that (2) is a family of polynomials with invariant degree and uncertainty bounding set Q which is pathwise connected. Moreover, suppose that the coefficient functions $\rho_i(q)$ are continuous and that (2) has at least one *D*-stable member $p(\cdot,q_0)$. Then (2) is robustly *D*-stable if and only if:

$$0 \notin p(x,Q) \tag{4}$$

for all $x \in \partial D$, where ∂D denotes the boundary of *D*.

Moreover, under the assumption that *D* is bounded (such as unit circle for the case of discrete-time polynomials), the pre-condition of degree invariance is no longer needed.

4. Illustrative examples

First, assume the family of discrete-time polynomials with nonlinear (or more specifically polynomial) uncertainty structure:

$$p(s,q) = 7z^5 + \left(q_1^2 + q_2^2 + 5\right)z^4 + \left(q_1^4 - q_2^2 + 5\right)z^3 + \left(q_1^2 + q_1 q_2^2 + 3\right)z^2 + \left(2q_1^3 + 3q_2^3 + 2\right)z + \left(3q_1^2 q_2^2 + 1\right)$$
$$q_1, q_2 \in [-0.7, 0.7] \qquad (5)$$

An arbitrary member of the family is taken (e.g. for $q_1 = q_2 = 0$) and its stability is verified. Then, the value sets for 100 points on the unit circle and for sampling both uncertain parameters with the step 0.01 are shown in Fig. 1. As can be seen, the origin of the complex plane is excluded from the value sets and thus the family (5) is robustly stable.

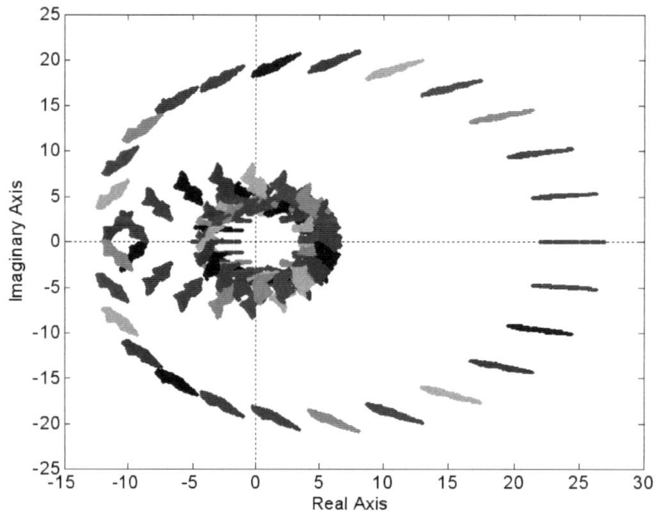

Fig. 1. The value sets of the family (5)

In the next step, the possible intervals for the uncertain parameters in the family (5) are supposed to be wider while the structure itself remains the same, i.e. the following family is considered:

$$p(s,q) = 7z^5 + \left(q_1^2 + q_2^2 + 5\right)z^4 + \left(q_1^4 - q_2^2 + 5\right)z^3 + \left(q_1^2 + q_1 q_2^2 + 3\right)z^2 + \left(2q_1^3 + 3q_2^3 + 2\right)z + \left(3q_1^2 q_2^2 + 1\right)$$
$$q_1, q_2 \in [-1, 1] \qquad (6)$$

The value sets (for the same sampling as in the previous Fig. 1) are depicted in Fig. 2 (full view) and Fig. 3 (zoomed view for better sight near the origin of the complex plane). The figures clearly prove the robust instability of the family (6) because the complex plane origin is included in the value sets.

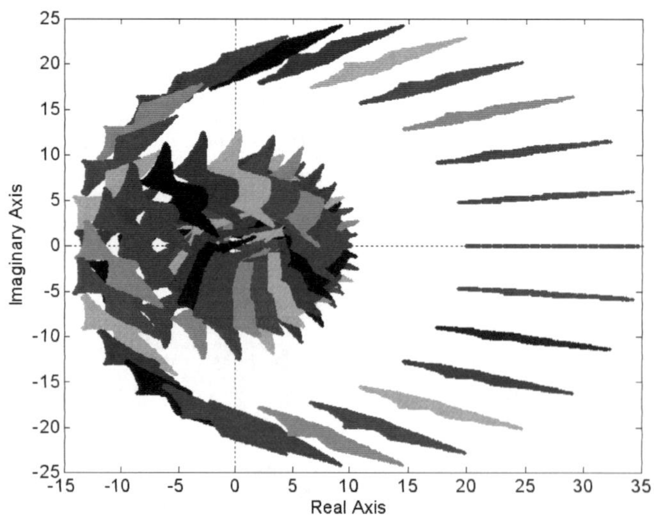

Fig. 2. The value sets of the family (6) – the full view

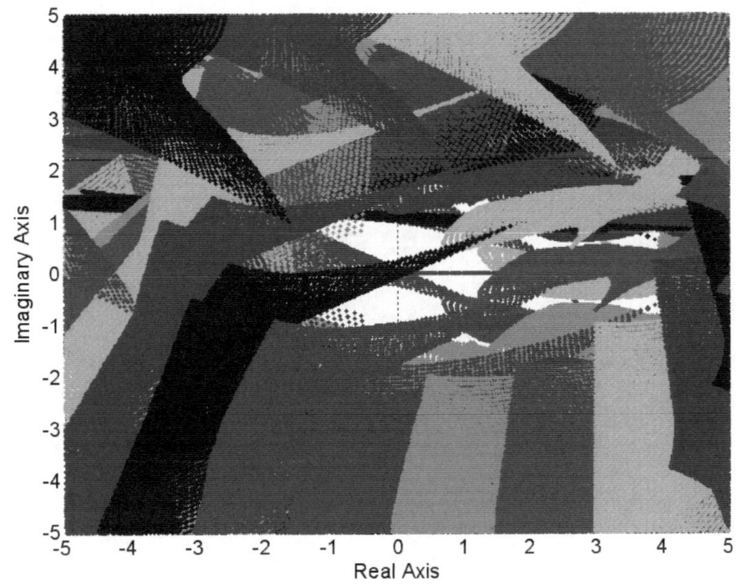

Fig. 3. The value sets of the family (6) – a zoomed view

Finally, assume the family of discrete-time polynomials with nonlinear (now completely general) uncertainty structure:

$$\begin{aligned} p(s,q) &= 7z^3 + [\cos(q_1 q_2)+3]z^2 + \left[5\sqrt{|q_1|} - \sin q_2 - \cos(q_1 q_2) + 2\right]z + \left[-3\sqrt{|q_2|} + 4\sin q_1 + 2\cos(q_1 q_2) + 1\right] \\ q_1, q_2 &\in [-0.3, 0.3] \end{aligned} \tag{7}$$

As in the previous example, an arbitrary member of the family is taken (e.g. again for $q_1 = q_2 = 0$) and its stability is checked. Subsequently, the value sets for 100 points on the unit circle and for sampling both uncertain parameters with the step 0.005 are shown in Fig. 4. Obviously, the family (7) is stable due to the exclusion of the zero point from the value sets.

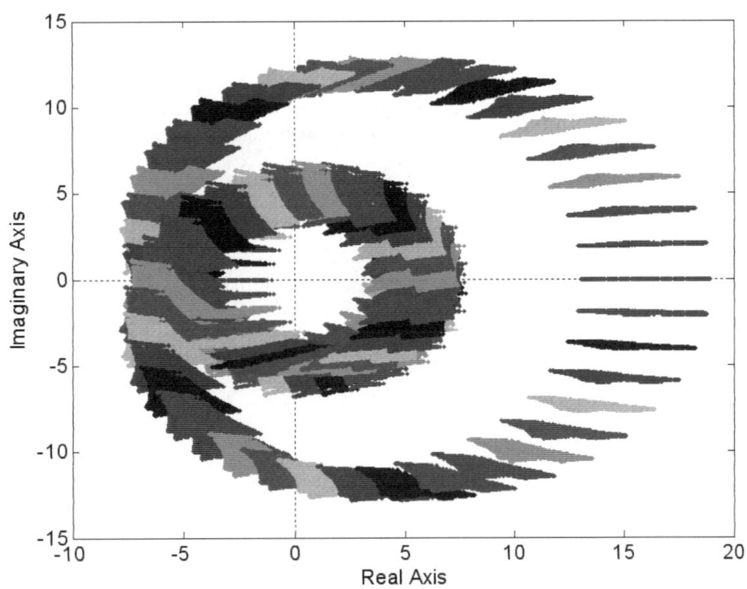

Fig. 4. The value sets of the family (7)

5. Conclusion

This contribution was focused on application of a graphical approach to robust stability analysis of families of discrete-time polynomials with special emphasis to families with a complex nonlinear uncertainty structure. The employed technique is based on the generalization of the value set concept and the zero exclusion condition under robust D-stability framework. The relative simplicity of the method is demonstrated via the set of the illustrative examples.

6. Acknowledgments

This work was supported by the Ministry of Education, Youth and Sports of the Czech Republic within the National Sustainability Programme project No. LO1303 (MSMT-7778/2014).

7. References

[1] Barmish, B. R. (1994). *New Tools for Robustness of Linear Systems*, Macmillan, New York, USA

[2] Bhattacharyya, S. P.; Chapellat, H. & Keel, L. H. (1995). *Robust Control: The Parametric Approach*, Prentice Hall, Englewood Cliffs, New Jersey, USA

[3] Ackermann, J.; Bartlett, A.; Kaesbauer, D.; Sienel, W. & Steinhauser, R. (1993). *Robust Control – Systems with Uncertain Physical Parameters*, Springer-Verlag, London, Great Britain

[4] Dullerud, G. E. & Paganini, F. (2000). *A Course in Robust Control Theory. A Convex Approach*, Springer, New York, USA

[5] Sánchez-Peña, R. S. & Sznaier, M. (1998). *Robust Systems: Theory and Applications*, John Wiley & Sons, New York, USA

[6] Matušů, R. & Prokop, R. (2011). Graphical analysis of robust stability for systems with parametric uncertainty: an overview. *Transactions of the Institute of Measurement and Control*, Vol. 33, No. 2, pp. 274-290

[7] Kraus, F. J.; Anderson, B. D. O. & Mansour, M. (1988). Robust Schur polynomial stability and Kharitonov's theorem. *International Journal of Control*, Vol. 47, No. 5, pp. 1213-1225

[8] Mansour, M.; Kraus, F. J. & Anderson, B. D. O. (1988). Strong Kharitonov theorem for discrete systems, Proceedings of the 27th Conference on Decision and Control, Austin, Texas, USA

[9] Barmish, B. R. (1989). An Extreme Point Result for Robust Stability of Discrete-Time Interval Polynomials, Proceedings of the 28th Conference on Decision and Control, Tampa, Florida, USA

[10] Kraus, F., Mansour, M. & Jury, E. I. (1992). Robust Schur Stability of Interval Polynomials. *IEEE Transactions on Automatic Control*, Vol. 37, No. 1, pp. 141-143

[11] Jetto, L. & Orsini, V. (2008). Some remarks on the Schur stability of interval polynomials, Proceedings of the 16th Mediterranean Conference on Control and Automation, Ajaccio, France

[12] Matušů, R. & Prokop, R. (2013). Graphical Approach to Robust Stability Analysis for Discrete-Time Interval Polynomials, Proceeding of the 15th WSEAS International Conference on Mathematical and Computational Methods in Science and Engineering, Kuala Lumpur, Malaysia, pp. 34-37

[13] Matušů, R. (2014). Robust Stability Analysis of Discrete-Time Systems with Parametric Uncertainty: A Graphical Approach. *International Journal of Mathematical Models and Methods in Applied Sciences*, Vol. 8, pp. 95-102

[14] Matušů, R. & Prokop, R. (2009). Systems with Nonlinear Uncertainty Structure: Robust Stability Analysis, Annals of DAAAM for 2009 & Proceedings of 20th International DAAAM Symposium, Vienna, Austria, pp. 1373-1374

[15] Matušů, R. (2010). A Graphical Approach to Robust Stability Analysis of Discrete-Time Systems with Parametric Uncertainty, Annals of DAAAM for 2010 & Proceedings of 21st International DAAAM Symposium, Zadar, Croatia, pp. 485-486

DOI: 10.2507/27th.daaam.proceedings.022

ROBUST STABILIZATION OF A CONTINUOUS STIRRED TANK REACTOR MODEL

Radek Matušů

This Publication has to be referred as: Matusu, R[adek] (2016). Robust Stabilization of a Continuous Stirred Tank Reactor Model, Proceedings of the 27th DAAAM International Symposium, pp.0147-0150, B. Katalinic (Ed.), Published by DAAAM International, ISBN 978-3-902734-08-2, ISSN 1726-9679, Vienna, Austria
DOI: 10.2507/27th.daaam.proceedings.022

Abstract

The main goal of this contribution is to outline an approach to robust stabilization of a continuous stirred tank reactor (CSTR) by means of PI controller. The linearized mathematical model of the CSTR with an exothermic reaction is supposed in the form of the fourth order interval system and its robust stabilization is based on combination of a method for calculating/plotting the stability boundary, the sixteen plant theorem and an algebraic approach to control design. The validity of the technique is verified on a simulative example of CSTR control.

Keywords: Robust stability; continuous stirred tank reactor; CSTR; interval system; PI controller; stability boundary; stability region.

1. Introduction

Chemical reactors belong among the most critical and interesting processes in all chemical engineering [1], [2]. They are frequently affected by complex behaviour and bounded with potential safety problems. From the viewpoint of control engineering, they represent the systems suitable for application of robust control [3], [4]. A common type of reactor is known as a continuous stirred tank reactor (CSTR) [1], [2]. The mathematical model of CSTR, robustly stabilized in this contribution, was constructed in [5]. Furthermore, the same work [5] presented stabilization of the CSTR by using the technique from [6], [7] and the pole-placement method. Besides, robust static output feedback control was applied to this CSTR in [8]. The robust stabilization in the paper [9] (which published the same results as can be found here) was based on the similar idea as in the contribution [5], but it used the alternative methods, namely the combination of the Kronecker summation method [10] and the algebraic approach to control design under the ring of proper and stable rational functions (R_{PS}) [11], [12], [13], [14], [15]. The same combination and its application to a nonlinear electronic model described by the interval system have been already studied in [16], [17].

This contribution is focused on brief description of design of robustly stabilizing PI controllers for a CSTR in which an exothermic reaction occurs. The controlled CSTR is modelled as a fourth order interval system. The PI controller is designed by means of combination of a method for calculating/plotting the stability boundary, sixteen plant theorem and an algebraic approach. The robust stabilization is verified on a simulative example. The same results as in this work have been already published in the journal paper [9].

2. Key results

The controlled process is represented by a CSTR for hydrolysis of propylene oxide to propylene glycol. The detailed description of this CSTR can be found e.g. in [5], [8], [9]. Some physical parameters (reaction enthalpy, pre-exponential factor, and overall heat transfer coefficient) of the CSTR are not known exactly but they are assumed to vary within the intervals. This contribution adopts the mathematical model of the CSTR introduced in [5], where it was obtained in the linearized form of the fourth order interval transfer function:

$$G(s)=\frac{[-0.0291,-0.0245]s^2+[-0.0199,-0.0127]s+[-0.000574,-0.0003549]}{s^4+[0.5801,0.903]s^3+[0.1002,0.2299]s^2+[0.0062,0.0142]s+[0.0001094,0.0002412]} \quad (1)$$

where the input is the volumetric flow rate of the coolant, and the output is the temperature of the reacting mixture.

In the first step, the aim is to determine all possible robustly stabilizing PI controllers by means of plotting the stability boundary. It can be done e.g. via the Tan's (et al.) method [6], [7] or through the Kronecker summation method [10] and their combination with the sixteen plant theorem [3], [18]. A brief overview of these techniques can be found also in [19], [20]. The Fig. 1 shows the open areas of stabilizing or non-stabilizing pairs of proportional (k_P) and integral (k_I) gains of PI controller for the first of the sixteen Kharitonov plants [3], [18]:

$$G_{1,1}(s)=\frac{-0.0245s^2-0.0199s-0.000574}{s^4+0.903s^3+0.2299s^2+0.0062s+0.0001094} \quad (2)$$

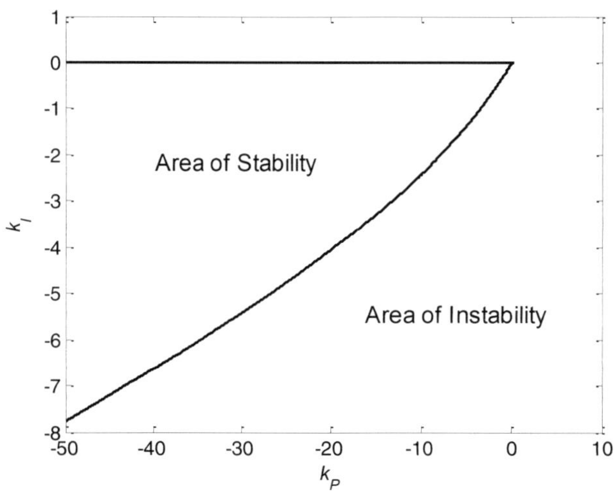

Fig. 1. Areas of stability/instability for the first Kharitonov plant (2)

The stability areas for all sixteen Kharitonov plants [3], [18] are plotted in Fig. 2. The intersection of these stability areas defines the region of stabilizing (k_P, k_I) pairs for all sixteen Kharitonov plants together, i.e. it gives the robust stability region for the original interval plant (1). This region is zoomed and highlighted in Fig. 3.

The next step is to select a convenient PI controller from the obtained robust stability region. This contribution uses an algebraic approach based on general solutions of Diophantine equations in R_{PS}, Youla-Kučera parameterization of controllers, conditions of divisibility in R_{PS}, and additional single parameter tuning. The method is grounded in the ideas from [11], [12] and its details can be found e.g. in [13], [14], [15]. The PI controller designed for a nominal system obtained from the interval plant (1) is:

$$C(s)=\frac{k_P s+k_I}{s}=\frac{-1.238s-0.029706}{s} \quad (3)$$

Obviously, the controller (3) lies inside the robust stability region from Fig. 3. The full design is described in [9].

Finally, the robust stability is confirmed and the control performance demonstrated in Fig. 4 which shows the set of output signals of the feedback control loop with designed PI controller (3) and 128 selected members of the interval family (1) with all variations of minimum or maximum values of each interval parameter. The level of reference signal was chosen just for the simulation purposes and it does not correspond to any real life values. As can be seen, the only one off-line tuned feedback PI controller with fixed coefficients can be successfully applied to robustly stabilize the fourth order interval model (1) of the CSTR.

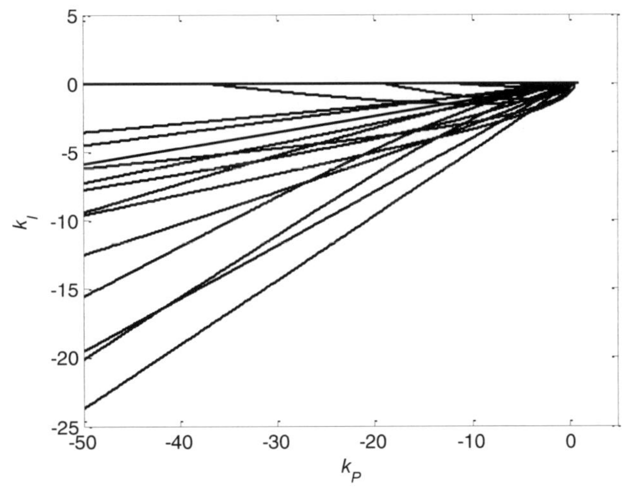

Fig. 2. Areas of stability for sixteen Kharitonov plants

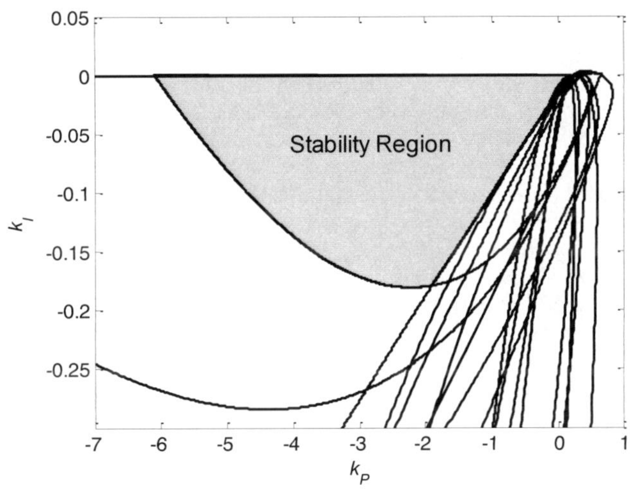

Fig. 3. Robust stability region for the interval plant (1)

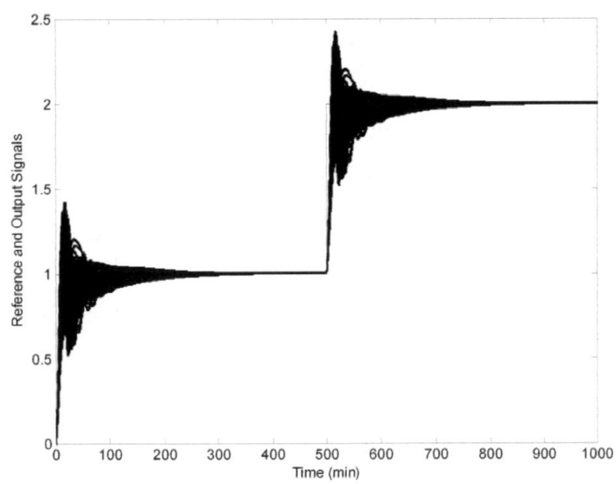

Fig. 4. Output (controlled) signals of 128 selected members of the interval family (1)

3. Conclusion

This contribution was focused on brief description of an approach to robust stabilization of CSTR by means of PI controller. The CSTR with an exothermic reaction was considered as the fourth order interval plant and its robust stabilization was based on combination of a technique for calculating/plotting the stability boundary, the sixteen plant theorem and an algebraic approach to PI controller design. The validity of the method was verified through a simulative example with a set of CSTR control results.

4. Acknowledgments

This work was supported by the Ministry of Education, Youth and Sports of the Czech Republic within the National Sustainability Programme project No. LO1303 (MSMT-7778/2014).

5. References

[1] Luyben, W. L. (2007). Chemical Reactor Design and Control, John Wiley & Sons, New Jersey, USA

[2] Antonelli, R. & Astolfi, A. (2003). Continuous stirred tank reactors: easy to stabilise? *Automatica*, Vol. 39, No. 10, pp. 1817-1827

[3] Barmish, B. R. (1994). *New Tools for Robustness of Linear Systems*, Macmillan, New York, USA

[4] Vasičkaninová, A.; Bakošová, M.; Čirka, Ľ. & Kalúz, M. (2015). Comparison of Robust Control Techniques for Use in Continuous Stirred Tank Reactor Control, *IFAC-PapersOnLine* (8th IFAC Symposium on Robust Control Design ROCOND 2015), Vol. 48, No. 14, pp. 284-289

[5] Závacká, J.; Bakošová, M. & Vaneková, K. (2010). Design of robust PI controllers for control of an exothermic chemical reactor, Proceedings of the 14th WSEAS International Conference on Systems, Corfu, Greece

[6] Tan, N. & Kaya I. (2003). Computation of stabilizing PI controllers for interval systems, Proceedings of the 11th Mediterranean Conference on Control and Automation, Rhodes, Greece

[7] Tan, N.; Kaya, I.; Yeroglu, C. & Atherton D. P. (2006). Computation of stabilizing PI and PID controllers using the stability boundary locus. *Energy Conversion and Management*, Vol. 47, No. 18-19, pp. 3045-3058

[8] Bakošová, M.; Puna, D.; Dostál, P. & Závacká, J. (2009). Robust stabilization of a chemical reactor. *Chemical Papers*, Vol. 63, No. 5, pp. 527-536

[9] Matušů, R; Závacká, J.; Prokop, R. & Bakošová, M. (2011). The Kronecker Summation Method for Robust Stabilization Applied to a Chemical Reactor. *Journal of Control Science and Engineering*, Vol. 2011, 7 p.

[10] Fang, J.; Zheng, D. & Ren, Z. (2009). Computation of stabilizing PI and PID controllers by using Kronecker summation method. *Energy Conversion and Management*, Vol. 50, No. 7, pp. 1821-1827

[11] Vidyasagar, M. (1985). *Control system synthesis: a factorization approach*, MIT Press, Cambridge, Massachusetts, USA

[12] Kučera, V. (1993). Diophantine equations in control - A survey. *Automatica*, Vol. 29, No. 6, pp. 1361-1375

[13] Prokop, R. & Corriou, J. P. (1997). Design and analysis of simple robust controllers. *International Journal of Control*, Vol. 66, No. 6, pp. 905-921

[14] Matušů, R. & Prokop, R. (2011). Single-parameter tuning of PI controllers: Theory and Application. *Journal of the Franklin Institute*, Vol. 348, No. 8, pp. 2059-2071

[15] Matušů, R.; Vaneková, K.; Prokop, R. & Bakošová, M. (2010). Design of robust PI controllers and their application to a nonlinear electronic system. *Journal of Electrical Engineering*, Vol. 61, No. 1, pp. 44-51

[16] Matušů, R.; Prokop, R.; Matejičková, K. & Bakošová, M. (2010). Application of Kronecker Summation Method in Computation of Robustly Stabilizing PI Controllers for Interval Plants, Proceedings of the 14th WSEAS International Conference on Systems, Corfu, Greece

[17] Matušů, R.; Prokop, R.; Matejičková, K. & Bakošová, M. (2010). Robust Stabilization of Interval Plants using Kronecker Summation Method. *WSEAS Transactions on Systems*, Vol. 9, No. 9, pp. 917-926

[18] Barmish, B. R.; Hollot, C. V.; Kraus, F. J. & Tempo, R. (1992). Extreme point results for robust stabilization of interval plants with first order compensators. *IEEE Transactions on Automatic Control*, Vol. 37, No. 6, pp. 707-714.

[19] Matušů, R. (2008). Robust Stabilization of Interval Plants Using PI Controllers, Annals of DAAAM for 2008 & Proceedings of the 19th International DAAAM Symposium, Trnava, Slovakia, pp. 831-832

[20] Matušů, R. & Prokop, R. (2010). Two Approaches to Computation of Stabilizing PI Controllers, Annals of DAAAM for 2010 & Proceedings of the 21st International DAAAM Symposium, Zadar, Croatia, pp. 487-488

DOI: 10.2507/27th.daaam.proceedings.023

INCREASE OF WEAR AND FRETTING RESISTANCE OF MINING MACHINERY PARTS WITH REGULAR ROUGHNESS PATTERNS

Viktor Krasnyy, Viacheslav Maksarov & Jüri Olt

This Publication has to be referred as: Krasnyy, V[iktor]; Maksarov, V[iacheslav] & Olt, J[ueri] (2016). Increase of Wear and Fretting Resistance of Mining Machinery Parts With Regular Roughness Patterns, Proceedings of the 27th DAAAM International Symposium, pp.0151-0156, B. Katalinic (Ed.), Published by DAAAM International, ISBN 978-3-902734-08-2, ISSN 1726-9679, Vienna, Austria
DOI: 10.2507/27th.daaam.proceedings.023

Abstract

The paper discusses the advantages of surfaces with regular roughness patterns, machined using the method of vibration burnishing, over those machined with the use of conventional techniques, with regard to improvement in the related parts' performance. The results of the investigation disclosed show a change in the part performance specifications as a function of microgroove profile variations, both with and without lubrication. The diagrams are included, showing surface roughness parameters versus machining conditions. The curves can be used for optimization of the machining parameters, in order to achieve the preselected roughness pattern, thus providing the needed part performance specifications. Vibration burnishing is a simple and cost-efficient machining method, which may be recommended as a substitute for a variety of conventional, environmentally harmful techniques. Examples of items, including mining machines, are given, which can be recommended for the proposed method of processing.

Keywords: vibration burnishing; regular roughness; wear resistance; friction pair.

1. Introduction

Any of the part performance specifications, primarily its reliability and life time, are greatly dependent on the part's surface roughness. For every combination of part operation conditions an optimum roughness can be specified to provide the designed life time [1; 3; 6; 7; 12; 16].

This problem is relevant for a number of friction pairs of mining machinery parts working at high loads and vibrations. Currently, various types of mining, transport and enrichment equipment is used in mines and careers, and in recent years the equipment used has been very powerful, sometimes even heavy-duty. The pressure their use causes is oftentimes the reason behind the considerable wear of crucial parts, which in many cases leads to extreme cases of wear, such as scuffing, galling, jamming and fretting corrosion damage [8].

In many cases, various covers for parts' work surfaces are used to prevent this type of damage. At the same time, positive results can often be obtained through implementing special surface finishing methods that are based on the plastic surface deformation of the surface layers of the part, such as surface vibration burnishing with the creation of a regular microgroove pattern.

Most of the conventional finishing methods (turning, grinding, lapping, honing) offer few possibilities for roughness optimization, since they will typically yield roughness patterns, which can be described as irregular, if not chaotic [8; 9].

The considerable improvement on conventional techniques would be the implementing of a finishing method that provides a regular surface microgroove pattern, in particular, the vibration burnishing process [1; 4]. Among the advantages, involved in a regular roughness pattern, are the following:

1) all the rolled microgrooves are identical in shape, size and relative position;
2) microgroove ridge and bottom radii can be increased by 2 to 3 orders;
3) the areas with such a roughness pattern gain higher strength and surface hardness;
4) the possibility appears to separately control the depth, pitch and relative location of grooves, which can be precalculated as functions of needed part performance specifications;
5) on the other hand, the said values can be derived from a combination of the vibration burnishing process parameters, thus enabling the latter to be calculated as a function of the part's desirable performance.

Upon using this method (alongside the effect caused by the plastic surface deformation finishing that guarantees the hardening of surface layers) the presence of grooves caused by vibration burnishing allows for the creation of “oil reservoirs” on the surface, thus facilitating the work of the friction pair in the case of boundary lubrication or its absence, avoiding the scuffing and galling of surfaces.

2. Experimental

Vibration burnishing involves, unlike the cutting process, the fine plastic deformation of the metal surface layers by means of a hard tool (hardened ball or rounded diamond tip), which is forced into the surface and moves relative to the latter as shown in Fig. 1. The surface roughness pattern can be fully or partially regular (non-contiguous grooves, contiguous grooves, intersecting grooves). The pattern can be varied and finely adjusted across a wide range of values, by changing the speed/frequency of the part rotation and/or the tool feed.

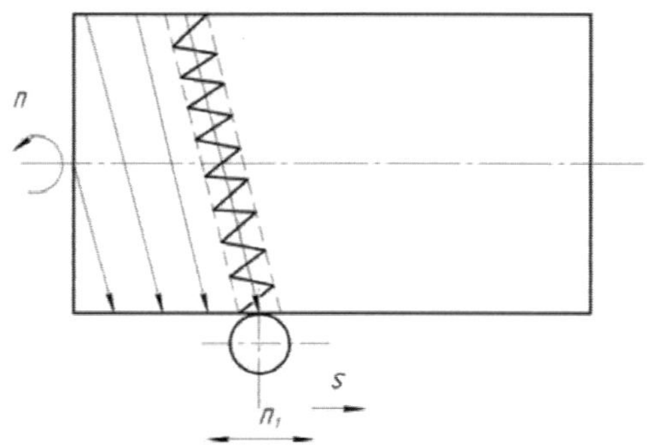

Fig.1. Vibration burnishing kinematics.

In addition to cylindrical ones, surfaces with a regular microgroove pattern can be achieved in edge, flat, spherical, profile, screw and involute surfaces [10].

The research carried out for many years has resulted in defining the optimum regular groove patterns for a number of typical groups of friction wearing parts, which can substantially contribute to the improvement in the part performance characteristics, namely:

1) life time and reliability;
2) wear resistance, run-in period, friction-related losses;
3) ability to stem fretting, scoring and scuffing;
4) contact rigidity and fatigue strength;
5) corrosion resistance etc. [10].

This paper presents some results of the research staged with a view to reveal a correlation between the regular groove pattern parameters and the wear resistance characteristics of the mining machined friction pairs.

3. Results and Discussion

One of the key factors influencing the friction and wear processes is the lubrication of the surfaces. The regular groove pattern contributes to retaining the lubricant and facilitating its spreading over the surface, especially during the run-in [10]. Fig. 2 shows the run-in duration and wear for various machining techniques turning, polishing, lapping, roller burnishing, vibration burnishing. The materials of the tested friction pair parts were bronze and

C50E, with pressure feed lubrication. Both the wear rate and run-in period substantially decrease as the groove peak and valley radii increase, or as the groove depth dispersion decreases. Improvement in the wear resistance during run-in can be, to a certain extent, accounted for the surface layer hardening, but the simultaneous reduction in the run-in duration can be only explained by the dominant influence of the improved microgeometry and the higher regularity of the groove shapes and sizes [13].

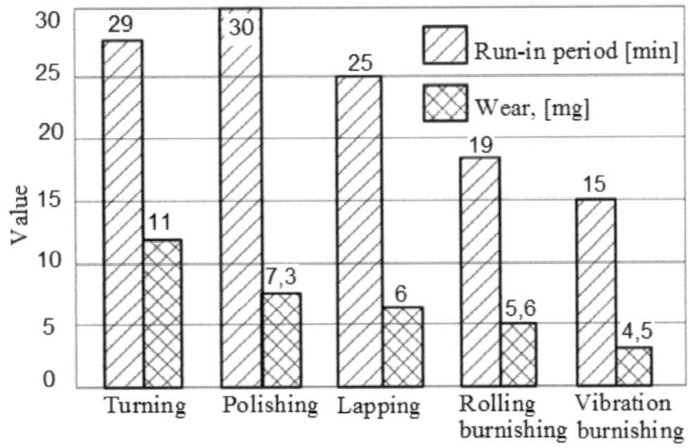

Fig.2. Wear and run-in period after machining with varying methods: 1 – turning; 2 – polishing; 3 – lapping; 4 – roller burnishing; 5 – vibration burnishing.

It should be noted that the curve, which represents the relationship between the scuffing resistance and the machining parameters that influence the resulting regular roughness pattern, has an apparent optimum point. This can be seen from Fig. 3, which shows a diagram of scuffing resistance vs. vibration burnishing speed, with (1) and without (2) lubrication.

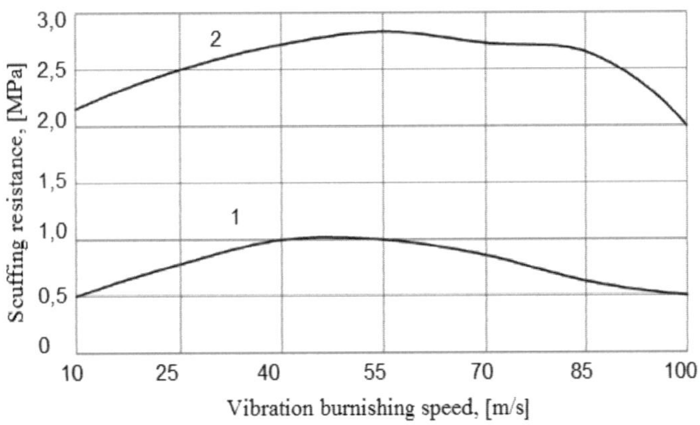

Fig.3. Scuffing resistance vs vibration burnishing speed: 1 – with lubricant; 2 – with no lubricant.

Sharp groove edges, particularly those on non-regular microgeometry surfaces, act as stress concentrators. Huge instantaneous stress values are developed in them, accompanied with excessive heat release. This results in the emergence of scuffing nuclei, proliferating as an avalanche. On the other hand, contact of too smooth surfaces with low oil retaining capacity results in the adhesion becoming the key friction factor. Both effects are considerably moderated by vibration burnishing.

This can be illustrated by the comparative run-in test results for some automotive engine parts. In particular, vibration burnished cylinder liners have shown a run-in wear rate 1.3 to 1.6 times lower than the honed ones, the run-in period being 2 to 3 times shorter.

During run-in, parts are most vulnerable to such abnormal forms of friction wear as galling and scuffing. As a result of the next series of tests, friction surface temperature vs. load curves have been plotted (Fig. 4) to characterize the scuffing resistance after: 1 – polishing, 2 – vibration burnishing. Curve 1 indicates the scuffing point for a polished surface at a load of as little as 4.9 MPa.

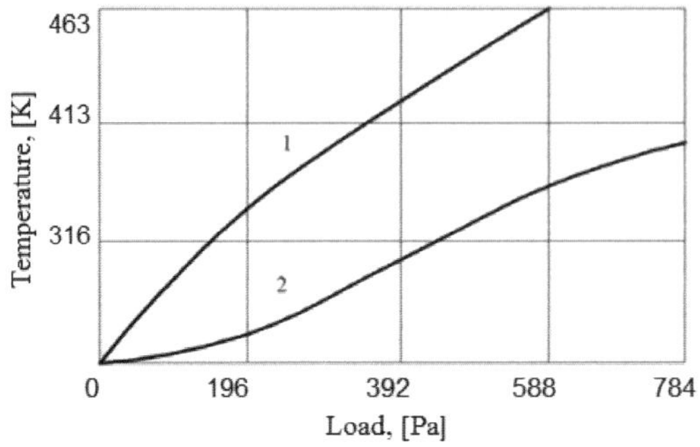

Fig. 4. Surface temperature vs. load curves for parts: 1 – polished; 2 – vibration burnished.

All the test conditions being equal (pressure, speed, lubrication), the vibration burnished samples have a surface temperature 52°C lower than the polished ones, and develop no scuffing.

Most of the known experiments have only revealed the relationship between the friction coefficient and the microgroove height of the contacting surfaces. But it should be noted that the friction coefficient is proportional to the effective contact area F_k, which, in its turn, depends on the ridge rounding radius divided by the (biggest) groove height.

The friction coefficients without lubrication were defined for pairs of samples, each consisting of a grey cast iron bushing and a disc of normalized Steel 45. The discs were machined by turning, R_a = 0.7–1.25 µm. The three bushing samples were lathe bored, reamed and vibration burnished correspondingly. The friction coefficient of the latter was found to be 25–30% lower, than in case of the pairs machined by standard cutting methods.

Fretting wear was investigated on vibration burnished and polished samples. A sample bushing was put in rotational reciprocation movement, while pressed against the immobile counterpart at variable vibration amplitude and load. Pairs of parts from Steel 20 were tested at a load of 20 MPa, a vibration amplitude of 100 µm and an oscillation frequency of 900 Hz. Pairs of grey cast iron/C22E were tested at a load of 87 MPa, a vibration amplitude of 50 µm and an oscillation frequency of 500 Hz. The results (mass and volume wear) after 500000 cycles are shown in Fig. 5 (1, 2 - steel/steel pair, 3, 4 – cast iron/steel pair; 1, 3 – polished surfaces, 2, 4 – vibration burnished surfaces). The vibration burnished samples showed wear rates 3 to 4 times lower than the polished ones.

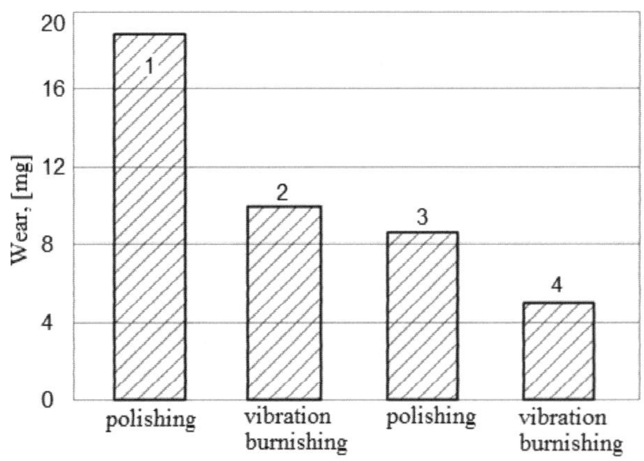

Fig. 5. Fretting wear for pairs: 1, 2 – steel part vs. steel counterpart; 3, 4 – cast iron part vs. steel counterpart.

The difference between the two types remained also after electroplating the surfaces or the application of polymer coatings. Coated vibration burnished surfaces demonstrated wear rates 30–35% lower, than the polished ones, the difference under lubrication being 25–30%.

Friction surfaces of polished samples are populated with cavities filled with oxidized debris, typical of fretting wear. There are also spots, covered with particles of several micrometers in diameter without vestiges of oxidation, where typical brittle fracture cracks are clearly visible (Fig. 6, a). Friction tracks on vibration burnished samples are split into separate areas (domains), commensurate in size with the fretting dither magnitude (Fig. 6, b). Such a structure of the friction track may possibly be accounted for the process of splitting the stress and strain waves on surfaces with regular microgeometry, which may substantially contribute to the improvement of fretting resistance.

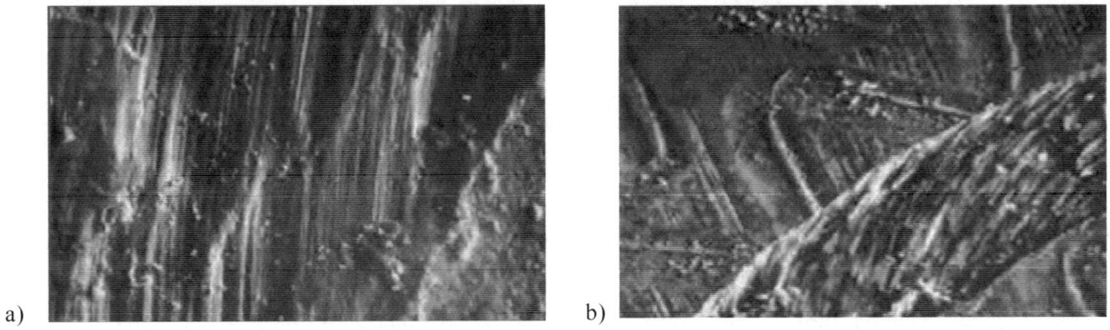

Fig. 6. Friction track on fretted steel surface [2]: a – polished; b – vibration burnished.

Increase of the resistance to fretting as a result of implementing regular surface roughness patterns is also noted in [11; 12; 13; 14].

The conducted experiments [2, 5] allow to make the conclusion that the wear of steel and cast iron samples after vibration burnishing with the formation of a regular microgroove pattern in fretting conditions is comparable with the wear of samples with galvanic and sprayed surfaces, friction-mechanical brass plating, laser strengthening, and in a number of cases it looks more preferable. Fig. 7 shows the results of the experiments made on a steel sample (Fig. 7,a) and cast iron counterpart (Fig. 7,b): polished without a surface cover (1), as well as with various surface covers: galvanic copper (2, 3) and bronze (4) of different thicknesses, friction-mechanical brass plating (5), molybdenum spraying (6), with laser treatment in hard phase (8), in comparison to vibration burnished samples without covering (7).

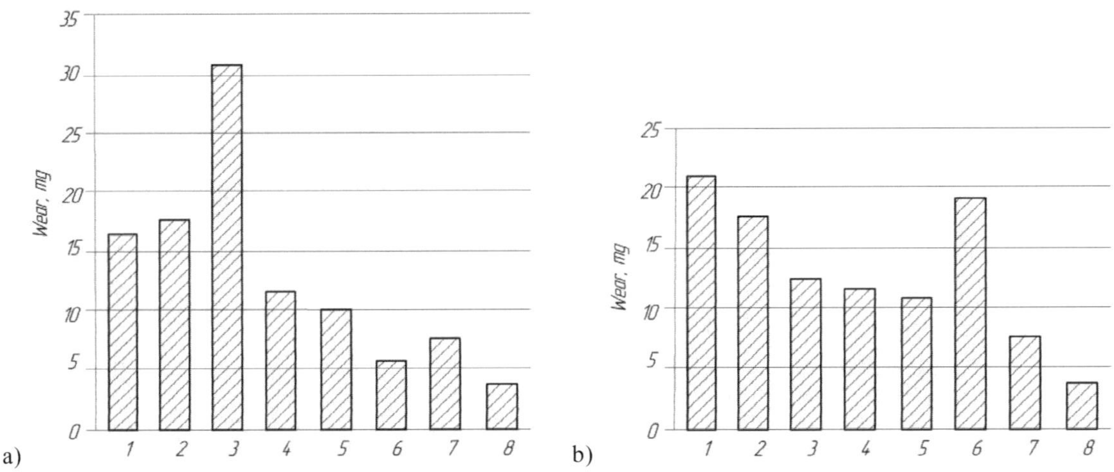

Fig. 7. The results of the experiments for fretting wear: a – sample wear, mg; b – counterpart wear, mg. Samples: 1 – no surface cover; 2 – galvanic copper 5 –8 μm; 3 – galvanic copper 30 μm; 4 – electrolytic bronze 15 μm; 5 – friction mechanical brass plating 3 –5 μm; 6 – molybdenum spraying 0.2 mm; 7 – vibration burnished with the creation of regular microgrooves; 8 – with laser treatment in hard phase.

The experiment conditions were as follows: normal pressure 25 MPa, amplitude 100 μm, frequency 900 cycles per minute, imitating real conditions that arise during the use of the loaded junctions in mining machines. The number of experiment cycles was 5x105. The wear of vibration-burnished steel samples during the experiments proved to be smaller than of samples employing surface covers, and was insignificantly higher compared to the wear of the samples that had been laser-treated and sprayed with molybdenum, whereas the wear of cast iron counterparts was insignificantly lower only in the case of laser treatment.

It can be concluded based on the results that in a number of cases the use of vibration burnishing with the formation of a regular microgroove pattern may successfully compete with the more expensive possibilities of special wear-proof surface covers.

The combination of vibration burnishing and thin surface covers seems especially promising, as this method preserves the loads that had been planned during the construction of the junction.

4. Conclusion

1. Vibration burnishing as a method of providing regular surface roughness patterns offers vast possibilities to improve a number of part performance specifications, primarily – the wear and fretting resistance. The vibration burnished surface is highly hardened. Its roughness pattern is formed as a set of regular microgrooves, uniform in shape, depth and location, acting as lubricant accumulators. The method is simple and reliable.

2. Use of a controlled roughness pattern improves the supporting capacity of the contact surface. Regularly arranged grooves contribute to the improvement of lubricant circular leakage conditions, the removal of wear products from the contact zone, the splitting of stress and strain waves, thus effectively preventing fret development.

3. In the conditions of fretting corrosion, the vibration burnishing method that creates a regular microgroove pattern is able to compete with the traditional methods of surface covering (galvanic, sprayed and other). A rather promising solution seems to be the combination of vibration burnishing and fine surface covers capable of reducing fretting wear without disrupting the loads that were planned during construction.

4. The paper shows the positive effect of regular microgeometry obtained on vibration burnished surfaces on the wear and fretting resistance of friction pairs. The vibration burnishing method can be offered as an advantageous substitute for such labour-consuming and environmentally harmful technologies as abrasive polishing, scraping and other cutting-related processes.

5. References

[1] Bougharriou A., Bouzid W., Saï K. (2014). Analytical modelling of surface profile in turning and burnishing, International Journal of Advanced Manufacturing Technology, 75, 547–558, ISSN: 0268-3768.

[2] Krasnyy, V., Maksarov, V., Olt, J. (2016). Improving fretting resistance of heavily loaded friction machine parts using a modified polymer composition. Agronomy Research 14 (Special Issue 1), 1023-1033; ISSN:1406-894X.

[3] Kreines L., Halperin G., Etsion I., Varenberg M., Hoffman A., Akhveldiani R. (2004). Fretting wear oft thin diamond films deposited on steel substrates. Diamond and Related Materials, 13, 9, 1731–1739, ISSN: 0925-9635.

[4] Li F, Xia W., Zhou Z., Zhoa J., (2010). Tang Z. Dynamic analysis of ball vibration assisted burnishing, Advanced Materials Research, 139–141, 925–928, ISSN: 1022-6680.

[5] Maksarov V.V., Krasnyy V.A. (2015). The mechanisms of friction of thin-layer nano-coatings under conditions of fretting. St. Petersburg Polytechnic University Journal of Engineering Sciences and Technology, 226, 3, 111–120.

[6] Marinescu I.D., Rowe W.B., Dimitrov B., Ohmori H. (2013). Tribology of Abrasive Machining Process, 2-ed., Elsevier, ISBN: 978-1-4377-3467-6.

[7] Nakano M., Ando Y. Recent studies on the application of microfabrication technologies for improving tribological properties. Lubrication Science, 23, 3, 2011, 99–117, ISSN: 0954-0075.

[8] Ostrovskiy M.S. (2011). Fretting as a cause of reduced reliability of mining machines. Mining informational and analytical bulletin, 12, 315–331, ISSN

[9] Rostamy N., Bergstom D.J., Sumner D., Bugg J.D. The effect of surface roughness on the turbulence structure of a plane wall jet, Phys. Fluids, 23, 2011, 1–10, ISSN: 1070-6631.

[10] Stępień P. 2009. Regular surface texture generated by special grinding process, Journal of Manufacturing Science and Engineering. Transactions of the American Society of Mechanical Engineers, 131, 1, 151–157, ISSN: 1087-1357.

[11] Tian Z.F., Inthavong K., Tu J., Yeoh G. Numerical investigation into the effects of wall roughness on a gas-particle flow in a 90 degrees bend, Int. J. Heat Mass Transf., 51, 2008, 1238–1250, ISBN: 0017-9310.

[12] Travieso_Rodrigues, J.A., Gras, G.G., Peiro, J.J., Carrillo, F., Dessein, G., Alexis, J., Rojas H.G. (2015). Experimental study on the mechanical effects of the vibration-assisted ball-burnishing process. Materials and Manufacturing Processes Vol 30, Issue 12, 1490-1497: ISSN: 1042-6914.

[13] Vadiraj A., Kamaraj M. Effect of surface treatments on fretting fatigue damage of biomedical titanium alloys, Tribology International, 40, 2007, 82–88, ISSN: 0361-679.

[14] Vadiraj A., Kamaraj M., Kamachi Mudali U., Raj B. Fretting fatigue of biomaterials, Transactions of the Indian Institute of Metals, 58, 5, 2005, 911–929, ISSN: 0972-2815.

[15] Varenberg M., Halperin G., Etsion I. Different aspects of the role of wear debris in fretting wear. Wear, 11–12, 252, 2002, 902–910, ISSN: 0043-1648.

[16] Wang Z., Wu J., Chen M., Zheng H., Zheng X.A review of surface texture of tribological interfaces, Applied Mechanics and Materials, 37–38, 2010, 41–45, ISSN: 1660-9336.

DOI: 10.2507/27th.daaam.proceedings.024

OPTIMIZATION OF PROCESS PARAMETERS FOR CUT QUALITY IN CO_2 LASER CUTTING USING TAGUCHI METHOD

Derzija Begic-Hajdarevic, Mugdim Pasic, Ahmet Cekic & Muhamed Mehmedovic

This Publication has to be referred as: Begic-Hajdarevic, D[erzija]; Pasic, M[ugdim]; Cekic, A[hmet] & Mehmedovic, M[uhamed] (2016). Optimization of Process Parameters for Cut Quality in CO2 Laser Cutting Using Taguchi Method, Proceedings of the 27th DAAAM International Symposium, pp.0157-0164, B. Katalinic (Ed.), Published by DAAAM International, ISBN 978-3-902734-08-2, ISSN 1726-9679, Vienna, Austria
DOI: 10.2507/27th.daaam.proceedings.024

Abstract

Laser cutting is a popular manufacturing process utilized to cut various types of materials. There are many non-linear interaction factors responsible for the performance of the laser cutting process. Identification of dominant factors that significantly affect cut quality is important. In this paper, the effect of the process parameters is analysed on the laser cut quality of an uncommon alloy, the tungsten alloy sheet with thickness of 1 mm. The experiments are conducted at three levels of the selected process parameters such as assist gas type (oxygen, air and nitrogen) and assist gas pressure. Taguchi design of experiments is used to obtain the optimum conditions through the study of signal to noise ratio for three quality characteristics such as kerf width, heat affected zone and surface roughness. Results of this study indicate that the optimal assist gas pressure is entirely different for kerf width, heat affected zone and surface roughness. The nitrogen as assist gas is suggested as the optimal assist gas for all of three quality characteristics.

Keywords: CO_2 laser cutting; cut quality; process parameters; Taguchi method; tungsten alloy.

1. Introduction

Laser cutting is one of the important applications of lasers in industry, especially for machining materials that are difficult to cut. Compared with other conventional machining processes, laser cutting removes little material, involves highly localized heat input to the work piece, minimizes distortion, and offers no tool wear [1]. Particular interests of manufacturers using laser cutting are the productivity and the quality of components made by the laser cutting process. Both aspects are managed by selection of appropriate laser process parameters, which are unique for each material and thickness. Most work reviewed in the literature considers only one or two characteristic properties of the laser cut surface to describe quality [2]. Kerf width, surface roughness and size of heat affected zone are often used to describe laser cut quality. Some researchers [3] show that the size of heat affected zone increases with increase of the laser power and decreases with increase of the cutting speed and gas pressure. Oxygen gas as assist gas produces better surface roughness compared to air and nitrogen during laser cutting of tungsten composite materials using pulsed Nd: YAG [4]. They also observe that the nitrogen assist gas develops an oxide free surface and a low discoloration, while the oxygen assist gas surface is strongly oxidized and discoloured. The effect of laser power, cutting speed and oxygen assist gas pressure on the cut quality in laser cutting of tungsten alloy is analysed in [5]. They define optimal cutting parameters in laser cutting examined alloy by using oxygen as assist gas.

There are various aspects of laser beam machining that can be modelled with different methods in order to predict quality characteristics which are essential for any manufacturing process [6]. In [7] technique to predict surface roughness in the laser cutting process is developed for the first time by analysing the dynamic phenomenon that happens within the cutting front. The quality characteristics (such as kerf width, surface roughness and cut edge slope) were observed for the various cutting parameters such as laser power, cutting speed and assist gas pressure during pulsed CO_2 laser cutting of Al6061/SiCp/Al_2O_3 composite [8]. Hybrid approach of grey based response surface methodology for predicting the optimal combination of laser cutting parameters is used. A grey relational analysis is used to determine a single optimized set of cutting parameters in precision laser cutting of three different thermoplastics [9]. It is found that the laser power has dominant effect on heat affected zone for all thermoplastics. The relation between process parameters and quality characteristics in CO_2 laser cutting of tungsten alloy was modelled with artificial neural network [10]. It is shown that the proposed model could be useful tool for surface roughness and kerf width prediction. In [11] comparison of surface roughness during CO_2 laser cutting of tungsten alloy plate using oxygen as assist gas, based on control charts made by statistical process control (SPC) approach is reported. For further analysis, it is especially interesting to analyse uncommon materials and alloys where the common knowledge is not applicable [12].

Comprehensive review of the literature shows that research work in the area of laser cutting of tungsten alloy is limited. Hence, this paper aims to investigate the effect of laser cutting process parameters on the cut quality features and then to obtain the optimal cutting conditions by using Taguchi design of experiments through the study of signal to noise ratio that would lead to the desired quality features. The process parameters such as assist gas type (oxygen, nitrogen and air) and assist gas pressure are considered as the experimental variables. The experimental work is carried out on tungsten alloy sheet with thickness of 1 mm in CO_2 laser cutting process.

2. Experimental procedure

The experiments are carried out on a Rofin CO_2 laser system with a nominal output power of 2000 W in CW mode at a wavelength of 10.6 μm with a high quality beam, as shown in figure 1. The experimental investigations are conducted at the University of Applied Science Jena in Germany. Tungsten alloy sheet (92.5 % pure tungsten in a matrix of nickel and iron) with thickness of 1 mm is used for this experiment. The laser beam is focused using a 127 mm focal length lens. Assist gases are used coaxially with the laser beam via a 2 mm exit diameter nozzle. Focus position, nozzle stand of distance (stand-off), laser power and cutting speed are kept constant throughout the main experimentation. The constant process parameters are shown in table 1.

Assist gas type	Laser power	Cutting speed	Stand-off	Focus position
Oxygen	2000 W	4000 mm/min	1.00 mm	-0.50 mm
Nitrogen	2000 W	1750 mm/min	0.75 mm	-0.50 mm
Air	2000 W	1500 mm/min	1.00 mm	- 1.00 mm

Table 1. Constant process parameters and their values

Assist gas type and pressure are selected as variable process parameters. Process parameters with their units and values at different levels are listed in table 2. Testing the effect of one parameter on the cut quality requires the variation of one parameter while keeping the other parameters at the pre-selected values.

Parameter	Unit	Level 1	Level 2	Level 3
Assist gas type	---	Oxygen	Air	Nitrogen
Assist gas pressure	bar	10	12.50	15

Table 2. Process parameters and their different levels

Observed quality characteristics are top kerf width, heat affected zone (HAZ), surface roughness and dross formed at bottom side. Visual inspections of each cut are carried out to ensure that no pitting and burrs are present within the cut areas. Measurements and sample geometry are shown in figure 2.

Surface roughness of the cut edge is measured in terms of the average roughness R_a, using a Taylor-Hobson stylus instrument. Roughness is measured along the length of a cut approximately in the middle of the thickness. Five consistent surface roughness values of each sample are measured and an average value is calculated for each sample.

The top kerf width and size of heat affected zone are measured using Stemi optical microscope fitted with a video camera and a zoom lens. Kerf width and heat affected zone are measured in the microscopic images of each sample at five different places and an average value is calculated for each sample and each controlled parameters. The dross formed at bottom side is observed using a TesaVisio microscope.

Fig. 1. CO_2 laser system

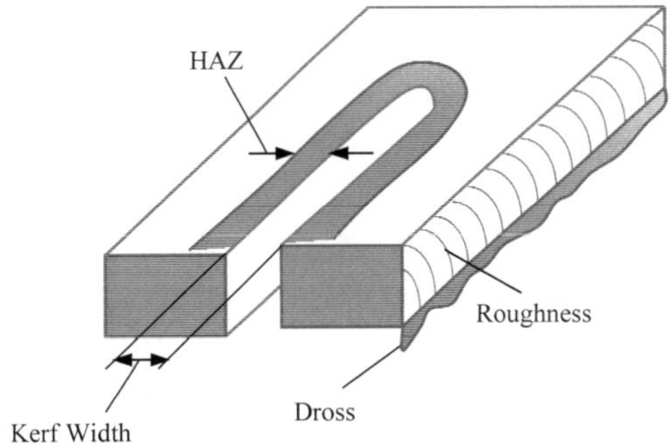

Fig. 2. The measured characteristics of cut quality

3. Experiment design

Taguchi method is used widely in engineering analysis to optimize performance characteristics by the means of settings of design parameters [13]. Taguchi methodology for robust parameter design is an off-line statistical quality control technique in which the level of controllable factors or input process parameters are chosen in a way to nullify the variation in responses due to uncontrollable factors such as humidity, vibration, noise etc. The objective of Taguchi approach is to determine the optimum setting of process parameters or control factors, thereby making the process insensitive to the sources of variations due to uncontrollable factors [14]. Taguchi has tabulated 18 basic types of designed experimental matrixes known as orthogonal arrays. The selection of orthogonal arrays is based on the number of controllable factors and their levels and interactions [15]. In this method main process parameters or control factors which influence process results are set as input parameters and the experiment is performed per specifically designed orthogonal array. More than one test per trail can be used to conduct the experiments. It increases the sensitivity of experiments to detect small changes in averages of responses. Economic consideration too can be performed for conducting the repeated experiments with the same experimental run.

In this study L_9 Taguchi orthogonal array design is selected. Measured experimental values obtained for different quality characteristics in each experimental run are given in table 3.

Exp. no	**Process parameter level**		**Measured parameters**		
	Gas Type	Gas Pressure	Kerf width, mm	HAZ, mm	*Ra*, µm
1	1	1	0.282	1.386	5.805
2	1	2	0.306	1.322	7.532
3	1	3	0.310	1.326	7.303
4	2	1	0.240	1.477	6.150
5	2	2	0.244	1.395	7.026
6	2	3	0.247	1.357	7.073
7	3	1	0.200	1.457	5.167
8	3	2	0.174	1.322	5.917
9	3	3	0.175	1.196	6.215

Table 3. Kerf width, heat affected zone and surface roughness obtained from experiments

Taguchi method uses the S/N ratio to measure characteristic deviating from the desired value. The S/N ratio is the ratio of the mean to the standard deviation. Taguchi method suggests that the signal to noise ratio (S/N) can be used as a quantitative analysis tool. Since smaller kerf width, heat affected zone and surface roughness values are desired in this experiment the signal-to-noise ratio is chosen as:

$$S/N = -10\log\left(\frac{1}{n}\sum_{i=1}^{n} y_i^2\right) \quad (1)$$

where: y_i is the observed data of quality characteristic at the i-th trial and n is the number of repetitions at the same trial. The S/N ratio represents the desired part/ undesired part and the aim is always to maximize the S/N ratio whatever nature of quality characteristics are. Based on the experimental results S/N ratio for each controlled parameters is obtained as shown in table 4.

Exp. no	Process parameter level		S/N ratio		
	Gas Type	Gas Pressure	Kerf Width	HAZ	*Ra*
1	1	1	10.99	-2.83	-15.27
2	1	2	10.28	-2.42	-17.54
3	1	3	10.17	-2.45	-17.27
4	2	1	12.39	-3.39	-15.78
5	2	2	12.25	-2.89	-16.93
6	2	3	12.14	-2.65	-16.99
7	3	1	13.98	-3.27	-14.26
8	3	2	15.19	-2.42	-15.44
9	3	3	15.14	-1.55	-15.87

Table 4. S/N ratio calculated for kerf width, heat affected zone and surface roughness

4. Results and discussion

4.1. The effect process parameters on top kerf width

The analysis of the experimental results revealed the assist gas type as the most important parameter determining the kerf width (see figure 3). The kerf width slightly changes as the assist gas pressure increases.

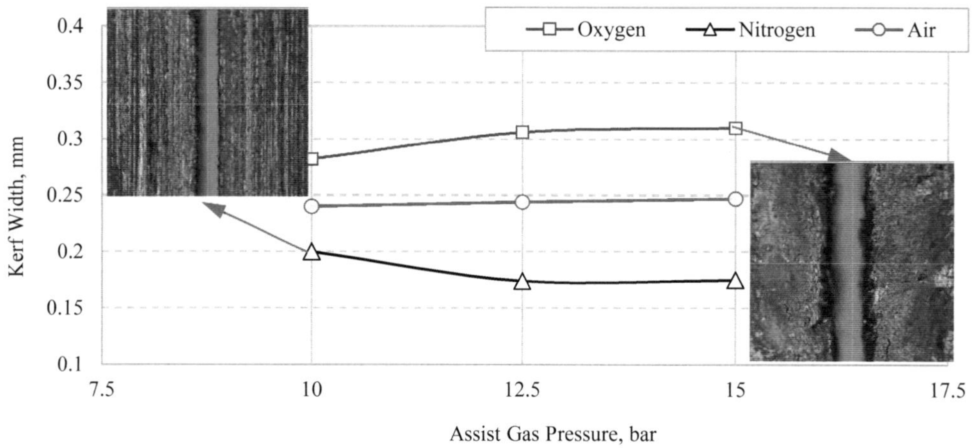

Fig. 3. Experimental observation for kerf width

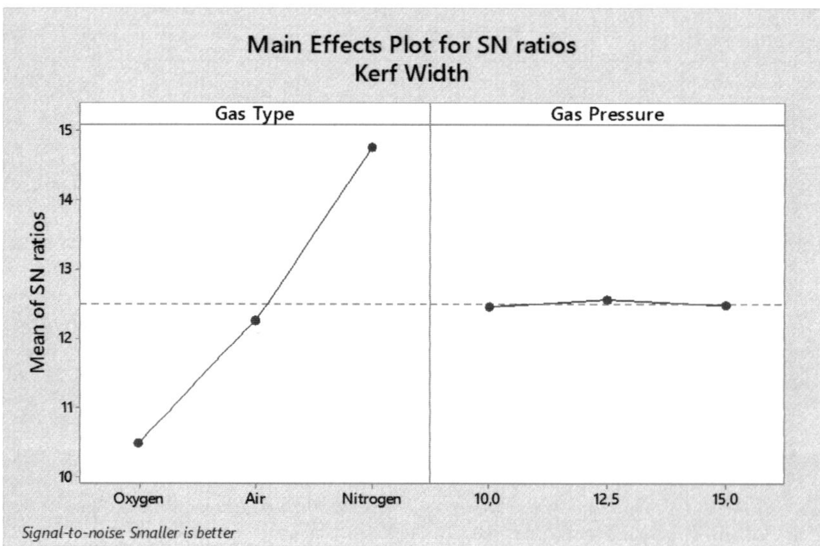

Fig. 4. The effect of process parameters on kerf width for S/N ratio

From S/N ratio analysis shown in figure 4 optimal process parameters for kerf width are determined as assist gas pressure 12.5 bar and nitrogen as assist gas. Also it can be observed form figure 5 that kerf width slightly increases as assist gas pressure increases. Narrowest kerf width is obtained in laser cutting when used nitrogen as assist gas for all of three levels assist gas pressure. While the largest kerf width is made during CO_2 laser cutting when used oxygen as assist gas.

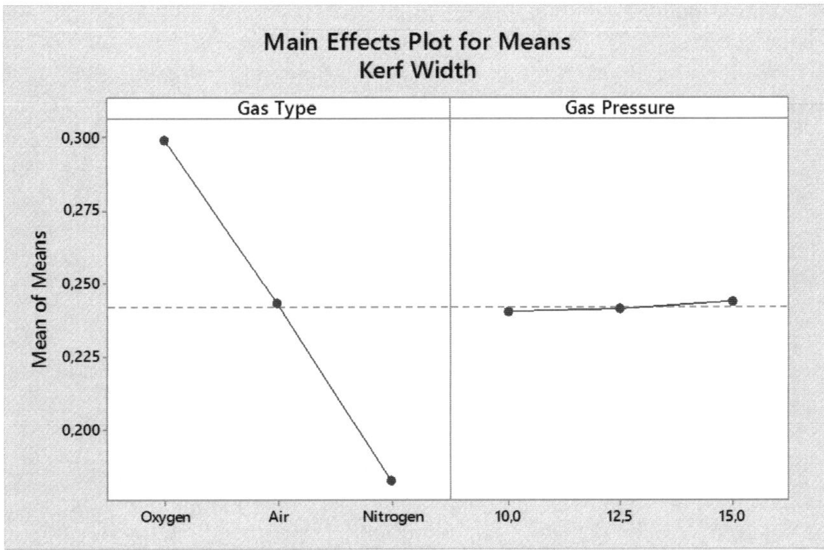

Fig. 5. The effect of process parameters on kerf width for the means

4.2. The effect process parameters on heat affected zone

Measured experimental values obtained for the size of heat affected zone are shown in figure 6. It can be observed that the largest heat affected zone is obtained in CO_2 laser cutting when used air as assist gas. The reason for this may be that the lowest cutting speed is selected when air is used as an assist gas in comparison to two other assist gasses that are used in the experiments.

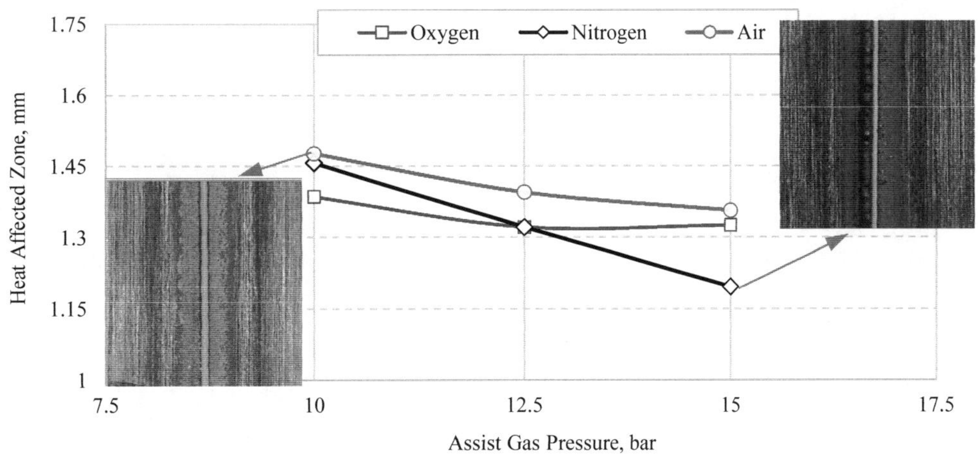

Fig. 6. Experimental observation for heat affected zone

The effect of process parameters on the size of heat affected zone for the S/N ratio and for the means are shown in figures 7 and 8, respectively. From the S/N ratio analysis shown in figure 7 optimal process parameters for the heat affected zone are determined as assist gas pressure 15 bar and nitrogen as assist gas. From figure 8 it can be observed that the size of heat affected zone decreases as the assist gas pressure increases. Smallest size of heat affected zone is obtained when nitrogen as assist gas is used.

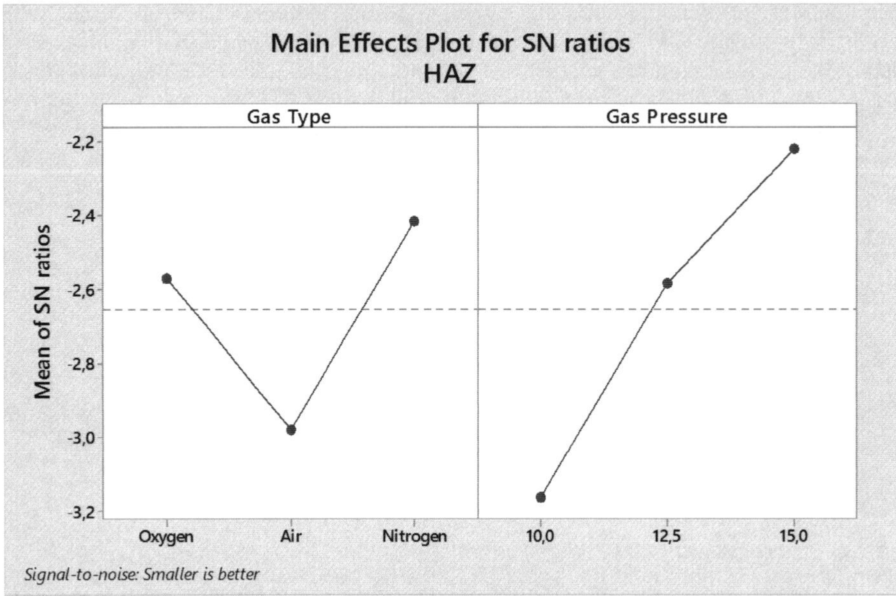

Fig. 7. The effect of process parameters on heat affected zone for S/N ratio

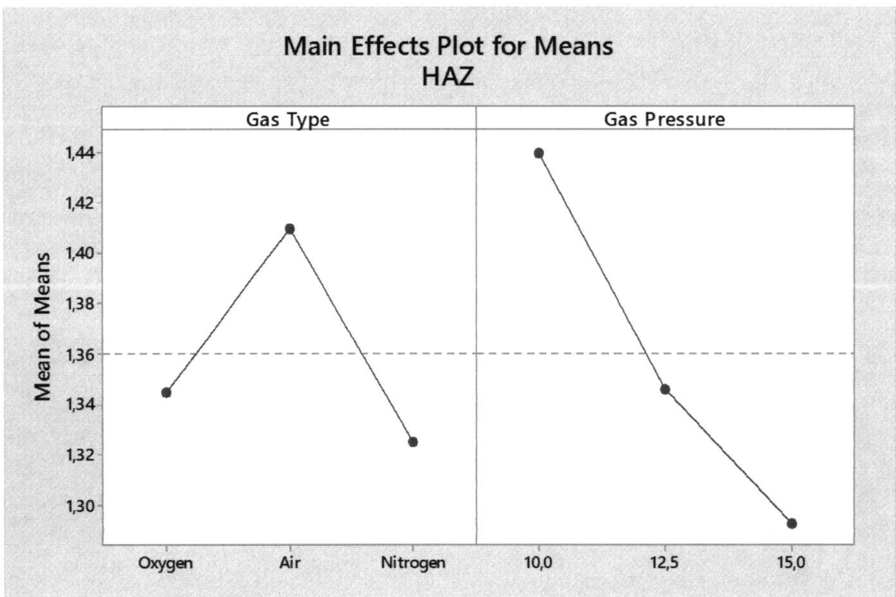

Fig. 8. The effect of process parameters on heat affected zone for the means

4.3. The effect process parameters on surface roughness

The measured experimental values obtained for surface roughness are shown in figure 9. It can be observed that the smallest surface roughness is obtained when nitrogen as assist gas is used. It can be seen that surface roughness increases as the assist gas pressure increases for all of three assist gasses which are used in the experiments. On the microscopic images shown in figure 9 it can be seen that the dross is formed on the bottom side of sample along the entire length of a cut at nitrogen pressure of 10 bar, while this is not the case at a nitrogen pressure of 12.5 bar.

From the S/N ratio analysis shown in figure 10 optimal process parameters for surface roughness are determined as assist gas pressure 10 bar and nitrogen as assist gas. Also it can be observed form figure 11 that surface roughness slightly increases as assist gas pressure increases over 12.5 bar. It can be concluded that the measured experimental values of surface roughness which are obtained for all of three levels assist gas pressure belonged to the same class of roughness.

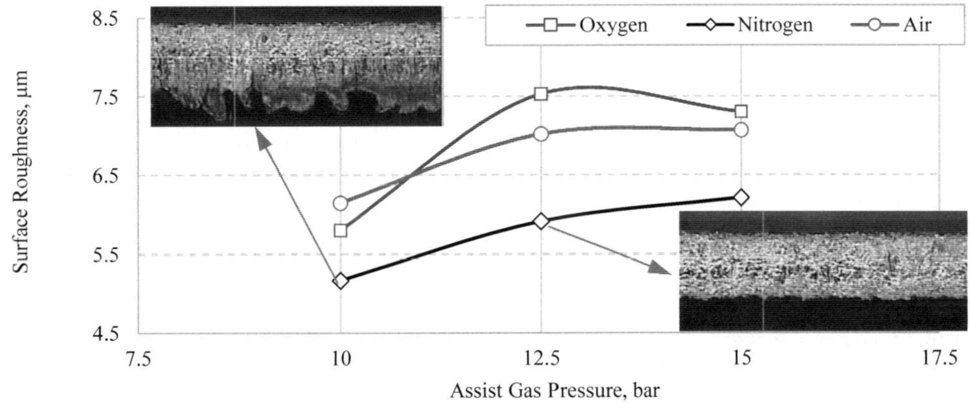

Fig. 9. Experimental observation for surface roughness

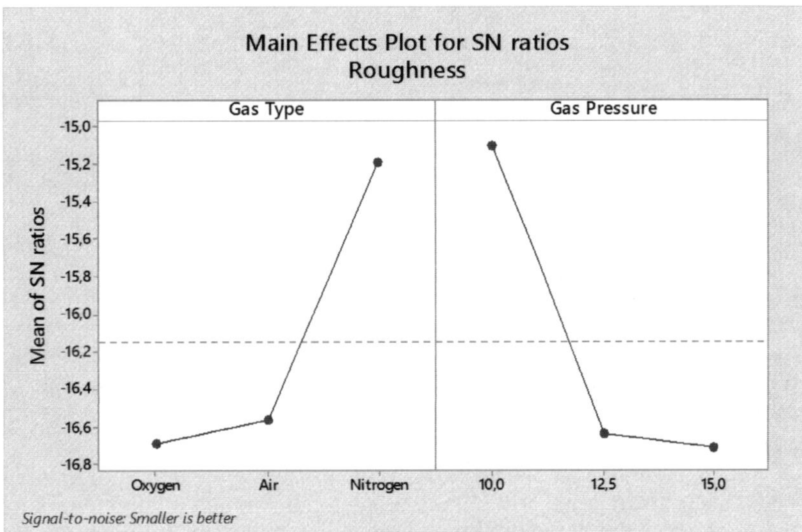

Fig. 10. The effect of process parameters on surface roughness for S/N ratio

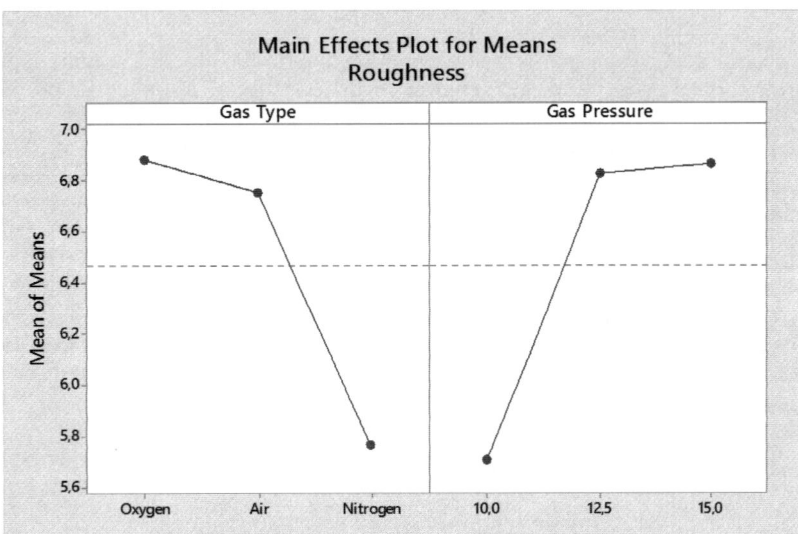

Fig. 11. The effect of process parameters on surface roughness for the means

5. Conclusion

In this paper the effect of the process parameters on the quality characteristics such as top kerf width, heat affected zone and surface roughness in CO_2 laser cutting of tungsten alloy with thickness of 1 mm is analysed using Taguchi orthogonal arrays method. Type of assist gas has the most significant effect on the cut quality in CO_2 laser cutting process.

Medium level of assist gas pressure and nitrogen as assist gas as the optimal process parameters for desired cut quality in CO_2 laser cutting of tungsten alloy are recommended.

Future work should include application of Taguchi orthogonal arrays method in laser cutting process, including the effect of different process parameters such as laser power, cutting speed, focus position and other parameters.

5. Acknowledgments

The authors would like to thank the Federal Ministry of Education and Science, Bosnia and Herzegovina for financial support of this study through the project entitled: "Non-conventional machining processes: laser and plasma cutting". Also, thanks to the Department of Laser and Opto-Technologies at the University of Applied Science Jena, Germany where experimental work was performed.

6. References

[1] Ready, J.F. (1997). Industrial Applications of Lasers, Academic Press, ISBN: 978-0-12-583961-7, United States of America.

[2] Avanish, K.D. & Vinod, Y. (2008). Laser beam machining – A review, International Journal of Machine Tools & Manufacture, Vol. 48, Issue 6, (May 2008) pp. 609-628, ISSN: 0890-6955.

[3] Hanadi, G.S.; Mohy, S.M.; Yehya, B. & Wafaa, A.A. (2008). CW Nd:YAG laser cutting of ultra-low carbon steel thin sheets using O_2 assist gas, Journal of Materials Processing Technology, Vol. 196, (January 2008) pp. 64-72, ISSN: 0924-0136.

[4] Uebel, M.; Buerger, W.; Schoele, H.; Stoerzner, F.; Meudtner, A. & Stibritz, G. (2008). Requirements to precision laser cutting processing of refractory metals, Annals of DAAAM for 2008 & Proceedings of the 19th International DAAAM Symposium, 22-25th October 2008, Trnava, ISSN 1726-9679, ISBN 978-3-901509-68-1, Katalinic, B. (Ed.), pp. 1419-1420, Published by DAAAM International, Vienna.

[5] Begic, D.; Kulenovic, M.; Cekic, A. & Bliedtner, J. (2009). CW CO_2 laser cutting of tungsten alloy using O_2 assist gas, Annals of DAAAM for 2009 & Proceedings of the 20th International DAAAM Symposium, 25-28th November 2009, Vienna, ISSN 1726-9679, ISBN 978-3-901509-70-4, Katalinic, B. (Ed.), pp. 1345-1347, Published by DAAAM International, Vienna.

[6] Schaaf, P. (2010). Laser processing of materials, Springer Series in Materials Science, 139, Springer-Verlag, ISBN: 978-3-642-13280-3, Berlin.

[7] Pietro, P.D. & Yao Y. (1995). A new technique to characterize and predict laser cut striations, International Journal of Machine Tools and Manufacture, Vol. 35, Issue 7, (July 1995) pp. 993-1002, ISSN: 0890-6955.

[8] Adalarasan, R.; Santhanakumar, M. & Rajmohan, M. (2015). Optimization of laser cutting parameters for Al6061/SiCp/Al2O3 composite using grey based response surface methodology (GRSM), Measurement, Vol. 73, (September 2015) pp. 596-606, ISSN: 0263-2241.

[9] Tamrin, K.F.; Nukman, Y.; Choudhury, I.A. & Shirley, S. (2015). Multiple-objective optimization in precision laser cutting of different thermoplastics, Optics and Lasers in Engineering, Vol. 67, (April 2015) pp. 57-65, ISSN: 0143-8166.

[10] Klancnik, S.; Begic-Hajdarevic, D.; Paulic, M.; Ficko, M.; Cekic, A. & Cohodar Husic, M. (2015). Prediction of Laser Cut Quality for Tungsten Alloy Using the Neural Network Method, Strojniški vestnik-Journal of Mechanical Engineering, Vol. 61, No. 12, (December 2015) pp. 714-720, ISSN: 0039-2480.

[11] Begic-Hajdarevic, D.; Pasic, M.; Vucijak, B. & Cekic, A. (2016). Statistical Process Control of Surface Roughness during CO_2 Laser Cutting using Oxygen as Assist Gas, Proceedings of the 26th DAAAM International Symposium, 21st-24th October 2015, Zadar, ISSN 1726-9679, ISBN 978-3-902734-07-5, Katalinic, B. (Ed.), pp. 0247-0255, Published by DAAAM International, Vienna, DOI:10.2507/26th.daaam.proceedings.034.

[12] El-Labban, H.F.; Mahmoud, E.R.I. & Al-Wadai, H. (2014). Laser cladding of Ti-6Al-4V alloy with vanadium carbide particles, Advances in Production Engineering & Management, Vol. 9, No. 4, (December 2014) pp. 159-167, ISSN: 1854-6250.

[13] Taguchi, G. (1962). Tables of orthogonal arrays and linear graphs, Maruzen, Tokyo.

[14] Phadke, M.S. (1989). Quality engineering using robust design, Prentice-Hall, ISBN-10: 0137451679, New Jersey.

[15] Ross, P.J. (1988). Taguchi techniques for quality engineering, McGraw Hill, ISBN-10: 0070539588, New York.

DOI: 10.2507/27th.daaam.proceedings.025

HYDRAULIC BALANCING SYSTEMS WITH LOW POWER CONSUMPTION AND CONSTANT FLOW PUMPS

Dan Prodan, Tiberiu Dobrescu, Anca Bucuresteanu & Adrian Motomancea

This Publication has to be referred as: Prodan, D[an]; Dobrescu, T[iberiu]; Bucuresteanu, A[nca] M[onica] & Motomancea, A[drian] (2016). Hydraulic Balancing Systems With Low Power Consumption and Constant Flow Pumps, Proceedings of the 27th DAAAM International Symposium, pp.0165-0174, B. Katalinic (Ed.), Published by DAAAM International, ISBN 978-3-902734-08-2, ISSN 1726-9679, Vienna, Austria
DOI: 10.2507/27th.daaam.proceedings.025

Abstract

In this paper, authors show a cost-saving method of replacing variable flow pumps with pressure regulator within hydraulic balancing systems for heavy machine-tools, with simple, constant flow pumps. The system proposed is relevant for heavy machine-tools where large loads need balancing: cross beams, slides, enclosures, etc. The basic hydraulic schematics, calculation method and results of certain simulations are displayed here. A characteristic of these systems is the use of routine hydraulic equipment, thus providing low costs. Hydraulic equipment used in such balancing systems may successfully replace complex and expensive equipment used in feeding / positioning kinematic chains within CNC heavy machine-tools, like: Horizontal Boring and Milling Machines (HBM), Vertical Lathes (VL) and Gantry Machines.

Keywords: heavy machine-tools; feeding / positioning kinematic chain; hydraulic balancing; low power consumption.

1. Introduction

In case of heavy machine-tools, with vertically displaced heavy elements, hydraulic systems are used for discharging the feeding / positioning kinematic chains [1, 2, 3].

Hydraulic balancing is achieved with hydraulic cylinders. Figure 1 shows the hydraulic balancing of a cross beam within a portal milling machine [1, 4].

In figure 1 the following legend is used: 1 - balanced mobile cross beam, 2 - dual-output electric motor, 3 - reducing gears, 4 - feed screws, 5 - milling head(s), 6 - fixed cross beam, 7 - props, 8 - balancing cylinders.

The mobile cross beam 1 together with milling heads 5 have a large weight (possibly exceeding 15 t) which, in the absence of a balancing system, is displaced on the guides of props 7 by the motor 2 acting through reducing gears 3 and feed screws 4. The portal is closed with fixed cross beam 6. Cylinders 8 are used to accommodate fully or partially the displaced weight.

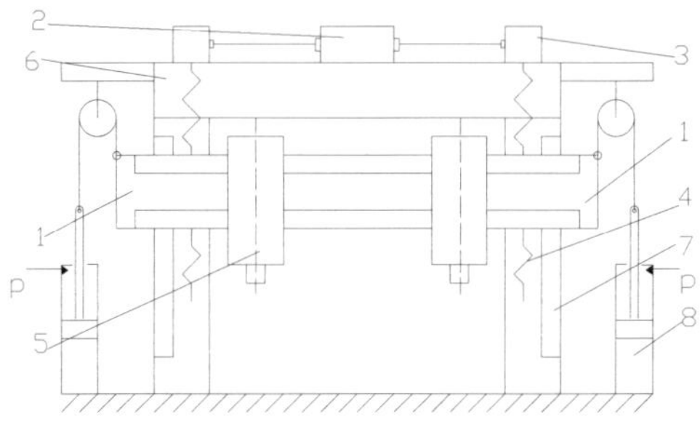

Fig. 1. Hydraulic balancing of mobile cross beams

Figure 2 shows the drive schematics for enclosure balancing.

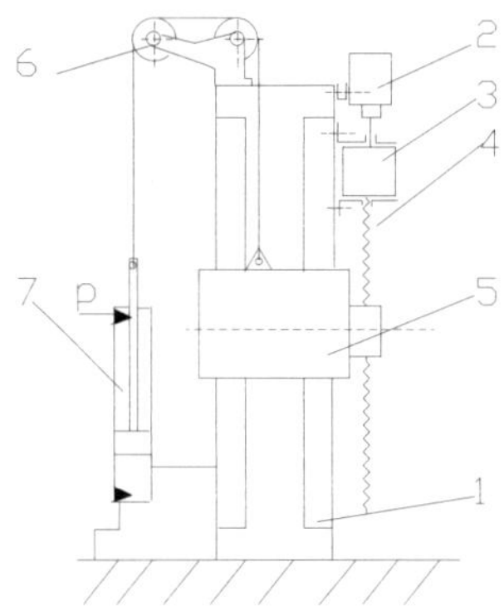

Fig. 2. Vertical displacement of hydraulically balanced loads

In figure 2 the following legend is used: 1 - prop with vertical guides, 3 - electric motor that drives the feed kinematic chain, 3 - feed box (reducing gears), 4 - feed screw, 5 - hydraulically balanced enclosure, 6 - sheaves and cables system, 7 - hydraulic cylinder. Enclosure 5 is displaced on guides of prop 1 by the feeding / positioning kinematic chain consisting of motor 2, reducing gears 3 and feed screw 4. For unloading the feed kinematic chain [5], enclosure 5 is balanced using the sheaves and cables systems 6, and cylinder 7.

The stationary force developed by each cylinder has the following expression (if losses are omitted):

$$F = p \cdot \frac{\pi \cdot (D^2 - d^2)}{4} \tag{1}$$

In equation above the following symbols are used: D - piston diameter, d - piston rod diameter, p - pressure in hydraulic system. Depending on each machine, the force developed by the cylinder may be lower than resisting forces (under-balancing) or higher (over-balancing).

In stationary conditions, the travel speed v dictated by the feed kinematic chain has the following expression:

$$v = n_{EM} \cdot i \cdot p_{BS} = \frac{4 \cdot Q}{\pi \cdot (D^2 - d^2)} \tag{2}$$

In (2) the following symbols are used: i - reducing gear's (feed box) transfer ratio, p_{BS} - lead of feed balls screw, Q - flowrate provided by the hydraulic balancing system.

2. Hydraulic balancing systems

Normally, for vertical feed kinematic chains, in case of enclosures, slides and even cross beams, if operating axes of CNC machine tools are involved [6], balancing systems with reducing valves, pressure regulating pumps or closed loop systems with accumulators are used [7]. All these systems involve expensive equipment and, given the operating schedule (full time), they raise equipment heating problems. Operating axes, controlled by the command system, must be actuated even in STOP stages, their position being maintained by elevation control read by transducer on the ruler. The hydraulic balancing installations used in such cases are under permanent stress, which requires special measures to mitigate heating.

These measures include [5, 7]:

- use of reducing valves, usually at lower loads;
- use of pressure-regulating variable flow pumps, an option that yields high quality results for affordable prices at pressures up to 100-150 bar for impeller pumps and up to 400 bar for axial piston pumps [8, 9];
- use of pneumo-hydraulic accumulator batteries, in which case the price is quite high and balancing is achieved with variable forces, depending on the position of displaced skid;
- use of complex, electro-hydraulic systems which involve the use of proportional hydraulic devices and/or electric power recovery systems. These systems are the most expensive and they are fit for very large machines.

In case of kinematic chains used exclusively for positioning, hydraulic balancing is not required to operate permanently. It is activated before each positioning and becomes usable only after the cross beam is unblocked and de-indexed [1].

3. Hydraulic balancing with variable flow pump and pressure regulator

The system in figure 3 is destined to feed the balancing cylinder of an existing machine-tool.

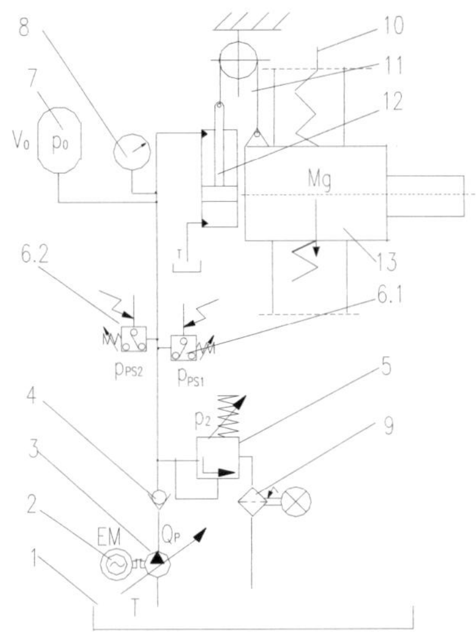

Fig. 3. Balancing system with variable flow pump

In figure 3 the following legend is used: 1 - tank, 2 - electric motor, 3 - variable flow pump, 4 - check valve, 5 - pressure valve, 6.1, 6.2 - pressure switches, 7 - accumulator, 8 - pressure gauge, 9 - filter, 10 - feed kinematic chain, 11 - sheaves system, 12 - hydraulic cylinder, 13 - mass balancing element M.

The variable flow pump 3, driven by the electric motor 2, sends oil to the balancing equipment from tank 1. Downstream from check valve 4, a pressure valve 5 is installed, set to a pressure higher that the pressure set on pump's regulator. Minimum and maximum balancing pressures are confirmed by the pressure switches 6.1 and 6.2. The accumulator 7, having the volume V_0 charged at pressure p_0, provides the flow peaks needed during startup phase [5]. Pressure gauge 8 allows reading the pressure. Return filter 9 provides oil purity. Cylinder 12, through the sheaves/cables system 11, discharges the feed kinematic chain 10 which displaces the enclosure 13 with mass M.

Operating features of the system that needs to provide the three stages, ascending, descending and STOP, are shown in figure 4.

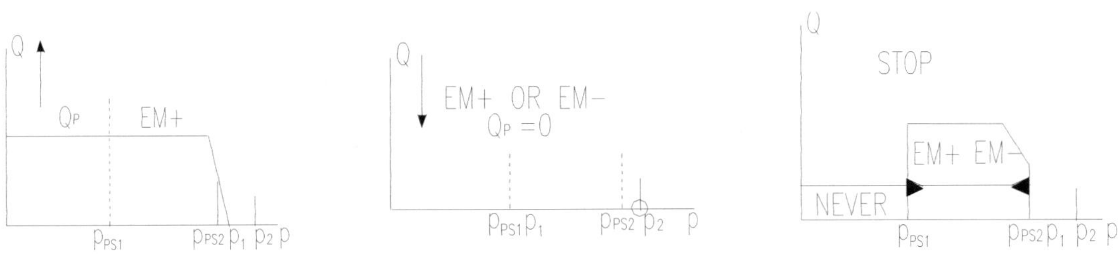

a b c

Fig. 4. Operating features of balancing systems with variable flow pumps

In figure 4 the following legend is used: p - pressure, p_{PS1} - minimum balancing pressure, p_{PS2} - maximum balancing pressure, p_1 - pressure set on pump's regulator, p_2 - pressure set on pressure valve, Q - flowrate provided, EM - electric motor. In ascending stages, the maximum useful flowrate equals the maximum pump flow. In descending stages, check valve 4 closes and the pump supplies a very low flowrate, theoretically zero. In this case pressure is p_2. In STOP stages, in a correctly executed, zero-losses system, pressure is p_1 and the flowrate supplied by the pump is zero. The two pressure switches may actuate and stop the pump's motor when pressure drops. In this case the pump shall cover any possible losses. A pressure drop below p_{PS1} is a warning of a possible failure.

The system executed as per schematics in figure 3 is shown in figure 5, where the legend notations are maintained.

Fig. 5. Hydraulic balancing system with variable flow pump

Maximum flowrate of pump is $Q_P = 20$ l/min. Set pressures have the following values: $p_1 = 80$ bar, $p_2 = 90$ bar, $p_{PS1} = 70$ bar, $p_{PS2} = 75$bar. The remaining elements in figure 5 are destined for performance of other hydraulic functions.

These installations are very frequently used in heavy machine tools [5]. Usually the pumps used are impeller pumps (up to 160 bar) or axial piston pumps (up to 400 bar). These are expensive pumps, produced by specialized manufacturers. In order to replace these pumps with simpler, gear pumps, which are also much less expensive, specific simulation software packages were used to analyze the dynamic behavior of balancing system shown in figure 3 [10]. In ascending stages, depending on the speed dictated by the feed kinematic chain, the system may supply flowrates between 0 and 20 l/min. Normally the maximum travel speed is limited to a value lower than the value corresponding to maximum flowrate.

Next, a supposition is made as per (2), that the maximum achievable speed is of v = 10 m/min. In theory, at zero pressure, the flowrate supplied equals the maximum flowrate, like in figure 6a. This flowrate is usable up until pressure p_1 is reached, as set on the pump's regulator. Figures 6b and 6c show the operating features at 26 bar and at 55 bar.

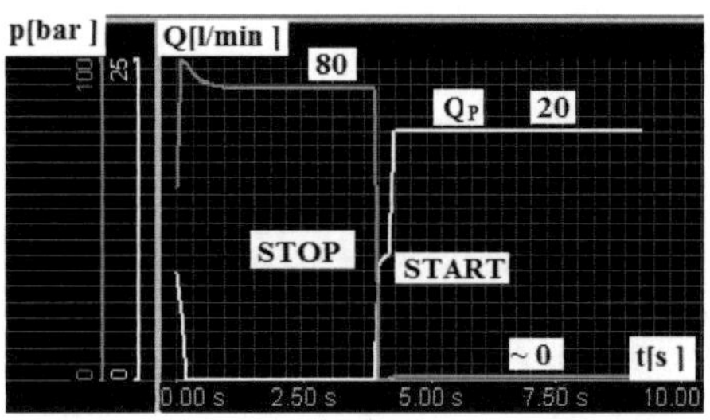

a. $p_1 = 80$ bar, $Q_P = 20$ l/min, p = 0, v = 10m/min

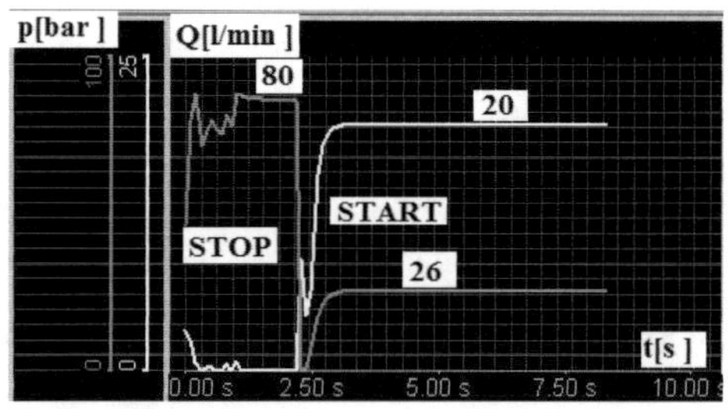

b. $p_1 = 80$ bar, $Q_P = 20$ l/min, p = 26 bar, v = 10 m/min

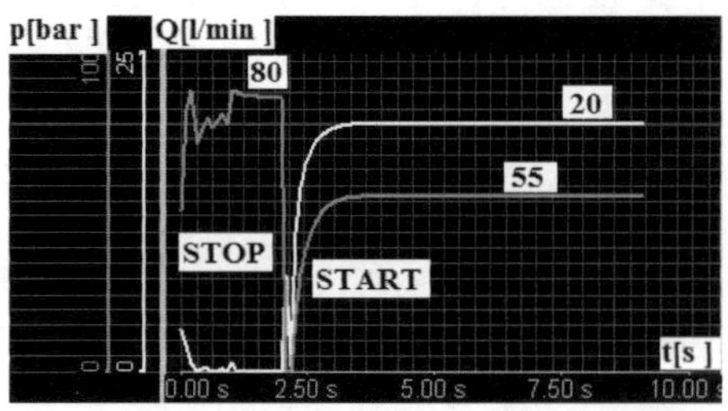

c. $p_1 = 80$ bar, $Q_P = 20$ l/min, p = 55 bar, v = 10 m/min

Fig. 6. Simulation of operational conditions of balancing systems with variable flow pump. Possible quick displacements

When the pressure set on regulator (p_1) is reached, the system reaches the operating phase of ascending stages. Figure 7 show the operating features for actual speeds dictated by the feed kinematic chain.

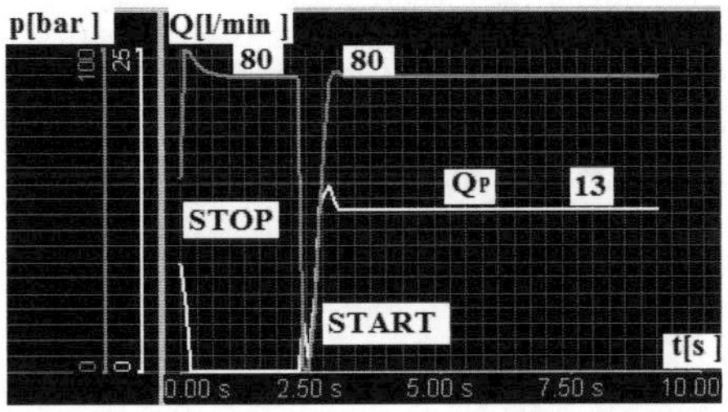

a. p_1 = 80 bar, Q = 13 l/min, p = 80 bar, v = 6500 mm/min

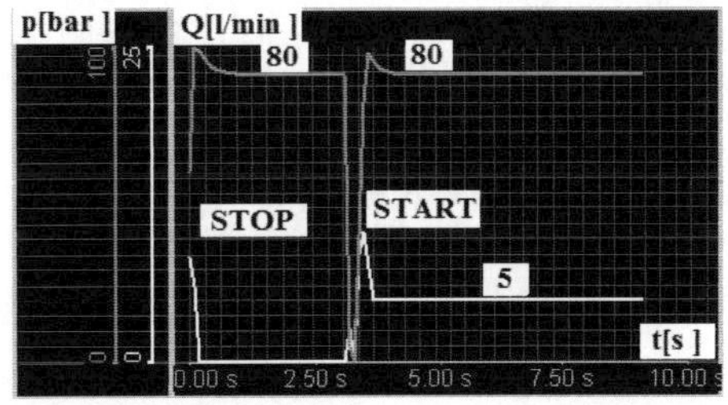

b. p_1 = 80 bar, Q = 5 l/min, p = 80 bar, v = 2500 mm/min

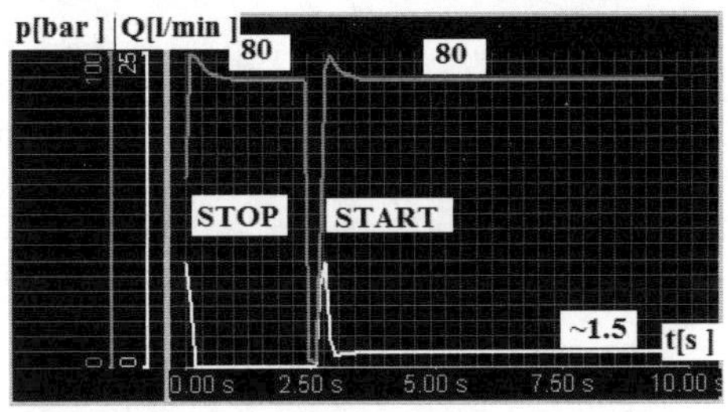

c. p_1 = 80 bar, Q = 1.5 l/min, p = 80 bar, v = 750 mm/min

Fig. 7. Simulation of operational conditions of balancing systems with variable flow pump. Quick displacements and ascending feed

The operating features of the system in descending and STOP stages are shown in figure 8.

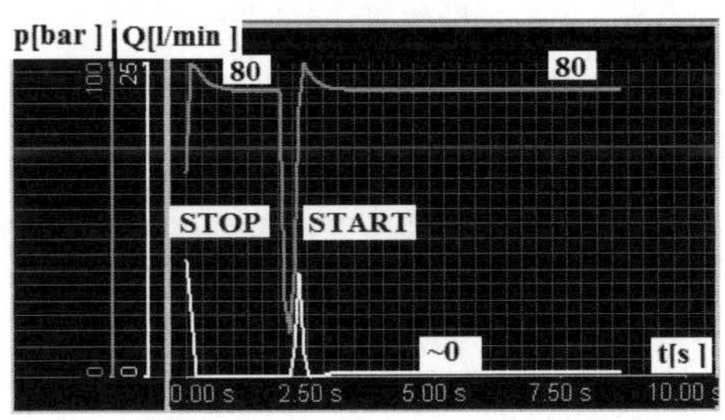

p_1 = 80 bar, Q = 0 l/min, p = 80 bar

Fig. 8. Simulation of operational conditions of balancing systems with variable flow pump. Quick displacements and descending feed and the STOP stage

Based on simulations made, the flowrate / pressure function was drawn as shown in figure 9.

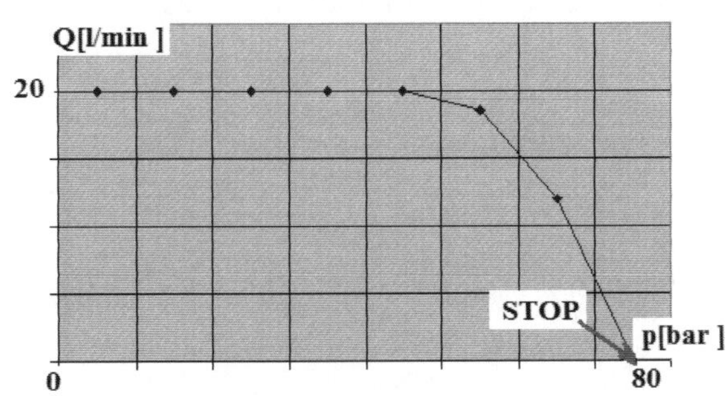

Fig. 9. Flowrate / pressure function of the balancing system using variable flow pump and pressure regulator.

4. Hydraulic balancing using constant flow pumps

The hydraulic system described below shall supply the same cylinder like the previous system, but it uses two constant flow pumps, which are normally gear pumps [4, 7, 9]. Hydraulic schematics is shown in figure 10.

In figure 10 the following legend is used: 1 - tank, 2 - suction filters, 3, 4 - constant flow pumps, 5 - electric driving motor, 6 - check valve, 7.1, 7.2 - pressure switches, 8 - hydraulic cylinder, 9 - accumulator, 10.1, 10.2 - pressure valves, 11 - hydraulic distributor, 12 - return filter, 13 - pressure gauge, 14 - sheaves system, 15 - feed kinematic chain.

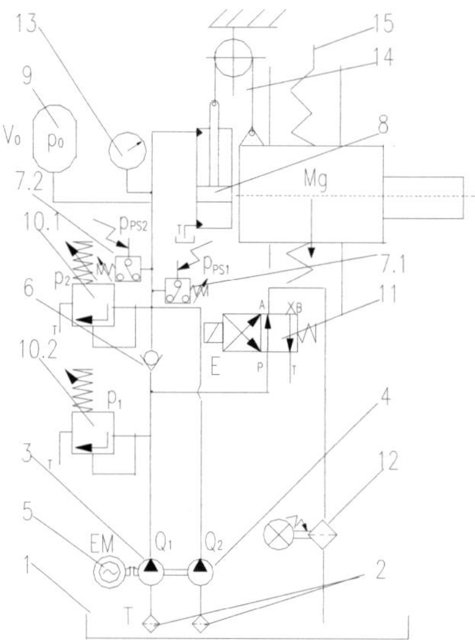

Fig. 10. Hydraulic schematics of balancing system using constant flow pumps

Pumps 3 (with flowrate Q_1) and 4 (with flowrate Q_2), driven by motor 5, take the oil from tank 1 and send it into the system through strainers 2. The high-rate pump 3 feeds the balancing circuit through check valve 6. The low-rate pump 4 feeds the balancing circuit directly. The pressure in balancing circuit is confirmed by the pressure switches 7.1 and 7.2. Maximum values of balancing pressure are adjusted using the pressure valves 10.1 and 10.2. It is recommended that p_2-$p_1 \geq 10$ bar. The flow of larger pump 3 may be discharged back into the tank, without feeding the balancing circuit, through distributor 11. Maintenance of in-circuit pressure after shutdown of pumps is provided, in certain conditions [5, 7], by accumulator 9, which is supplied by the balancing circuit together with cylinder 8. When pump 3 discharges directly into the tank, oil is passed through the return filter 12. Pressure in the balancing circuit is read on pressure gauge 13. Cylinder 8 performs balancing of mass M using the sheaves system 14. Vertical travel is provided by feed kinematic chain 15.

Three operating phases may be noticed in the schematics above:

1. ASCENDING. The electric motor starts and, after relay 7.1 confirms that set pressure is achieved, voltage is applied to electromagnet E of distributor 11. The following condition must be checked:

$$Q_U = \pi \cdot \frac{D^2 - d^4}{4} \cdot v < Q_1 + Q_2 \tag{3}$$

In (3) the following legend was used: v - travel speed of cross beam, Q_U - necessary ascending flow.

The flow discharged through pressure valve 10.2 Q_{PV} is:

$$Q_{PV} = Q_1 + Q_2 - Q_U \tag{4}$$

The power loss in the ascending phase through the balancing system is:

$$P_{LU} = p \cdot Q_{PV} \tag{5}$$

2. STOP after ascension. In a first stage, voltage is removed from electromagnet E and then, if no positioning is performed next, motor 5 may be shut down. If pressure drops below the value set on relay 7.1, the motor is restarted until pressure set on pressure relay 7.2 is confirmed. Power loss in this phase is zero.

3. DESCENDING. To perform this phase, electric motor 5 is started and the system waits for pressure to be confirmed by pressure switch 7.2. In this stage electromagnet E is not actuated. If the travel speed is still v, the following may be considered:

$$Q_D = \pi \cdot \frac{D^2 - d^4}{4} \cdot v \tag{6}$$

$$Q_{PV} = Q_D + Q_2 \tag{7}$$

$$P_{LD} = p \cdot Q_{PV} \tag{8}$$

In equations above the following symbols are used: Q_D - flow necessary for descending, P_{LD} - power lost as heat. In this power only the element ΔP comes from the hydraulic system, where ΔP is:

$$\Delta P = p \cdot Q_2 \tag{9}$$

4. STOP after descending. If the required position is achieved, the positioning kinematic chain is shut down. The electric motor may be shut down.

In all these phases, accumulator 9 provides the peak flows for start-ups and maintenance of a certain volume of oil under pressure [7].

Operating features of the system that needs to provide the three stages, ascending, descending and STOP, are shown in figure 11.

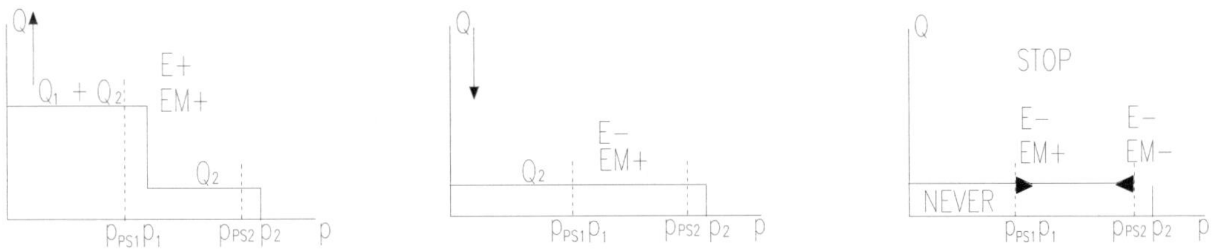

a b c

Fig. 11. Operating features of balancing systems with constant flow pumps

In figure 11 the following legend is used: p - pressure, p_{PS1} - minimum balancing pressure, p_{PS2} - maximum balancing pressure, p_1 - pressure set on pump's regulator, p_2 - pressure set on pressure valve, Q_1, Q_2 - flowrates supplied, E - distributor's electromagnet, EM - pump driving electric motor

The system executed as per schematics in figure 10 is shown in figure 12, where the legend notations are maintained.

Fig. 12. Hydraulic balancing system with constant flow pumps

Maximum flowrate of the large pump is $Q_1 = 14.5$ l/min and of the small pump is $Q_2 = 6$ l/min. The set pressures have the following values: $p_1 = 70$ bar, $p_2 = 80$ bar, $p_{PS1} = 65$ bar, $p_{PS2} = 75$ bar.

In ascending stages, depending on the speed dictated by the feed kinematic chain, the system may supply flowrates between 0 and ~20 l/min delivered by pump 4 or both pumps.

Next, a supposition is made as per equation (2), that the maximum achievable speed is of v = 10 m/min. It is considered that ascension is ordered after descending to 80 bar.

In theory, at zero pressure, the flowrate supplied equals the maximum flowrate, like in figure 13a. This flowrate is usable up until pressure p_1 is reached, as set on pressure valve 10.2.

Figures 13b and 13c show the operating features at 10 bar and at 55 bar.

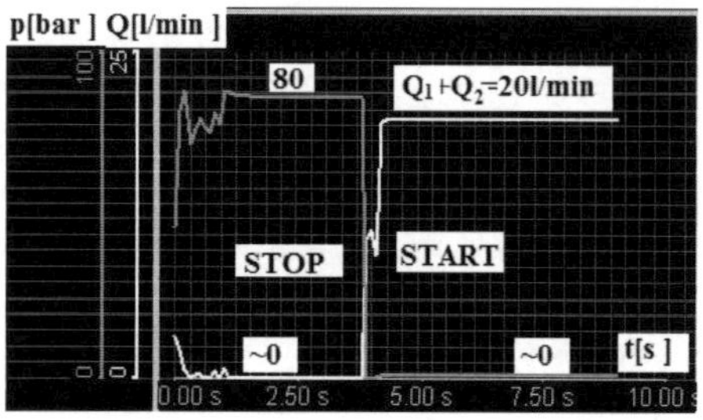

a. p_1=70 bar, p_2=80bar, Q_1+Q_2=20 l/min, p=0 bar, v=10 m/min

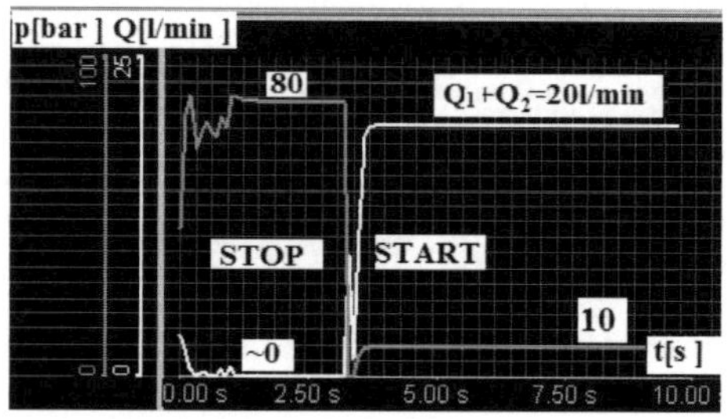

b. p_1=70 bar, p_2=80 bar, Q_1+Q_2=20 l/min, p=10 bar, v=10 m/min

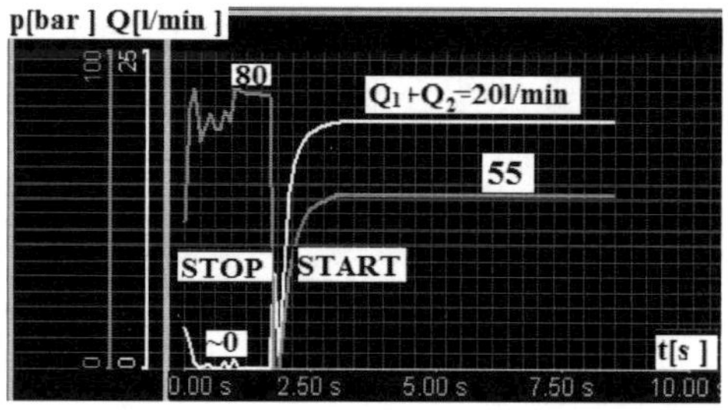

c. p_1=70 bar, p2=80 bar, Q_1+Q_2=20 l/min, p=55 bar, v=10 m/min

Fig. 13. Simulation of operational conditions of balancing systems with constant flow pump. Possible quick displacements in ascending phase

A correctly adjusted balancing system will operate in versions shown in figure 13 only for quick ascending displacements. When the pressure set on valve 10.2 (p_1) is reached, the system reaches the operating phase of ascending feed stages. Figure 14 show the operating features for actual speeds dictated by the feed kinematic chain.

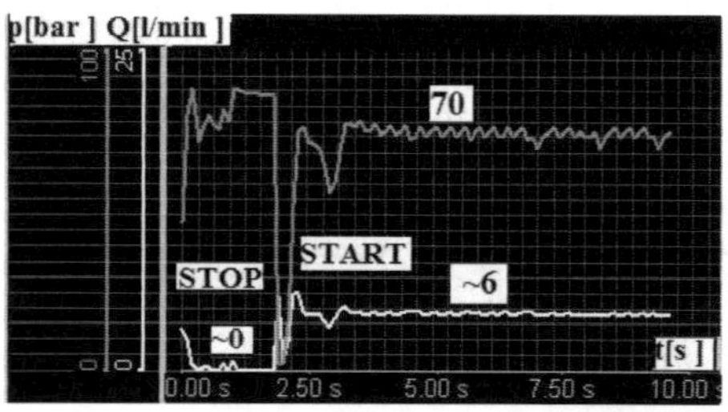

a. p_1=70 bar, p_2=80bar, Q_2=6 l/min, p>p_1=70 bar, v=3000 mm/min

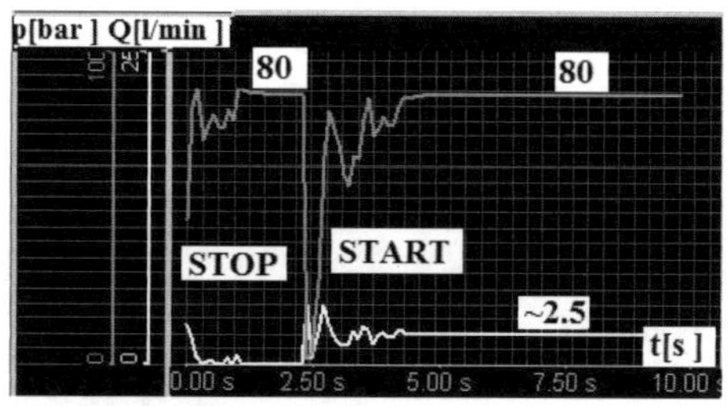

b. p_1=70 bar, p_2=80 bar, Q=2.5 l/min<Q_2, p=p_2=80 bar, v=1250mm/min

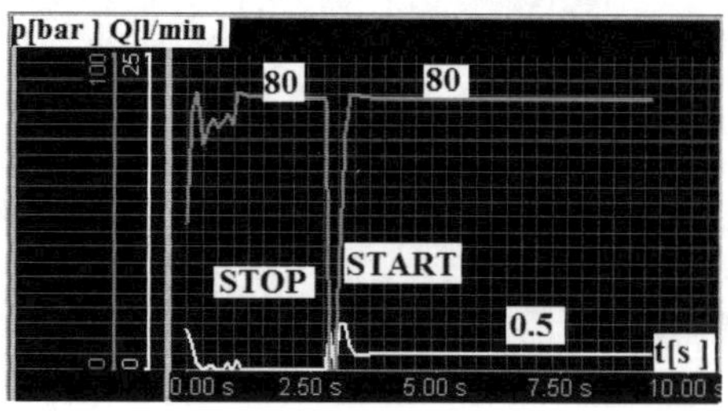

c. p_1=70 bar, p_2=80 bar, Q=0.5 l/min<Q_2, p= p_2=80 bar, v=250 mm/min.

Fig. 14. Simulation of operational conditions of balancing systems with constant flow pump. Quick displacements and ascending feed

The operating features of the system in descending and STOP stages are shown in figure 15.

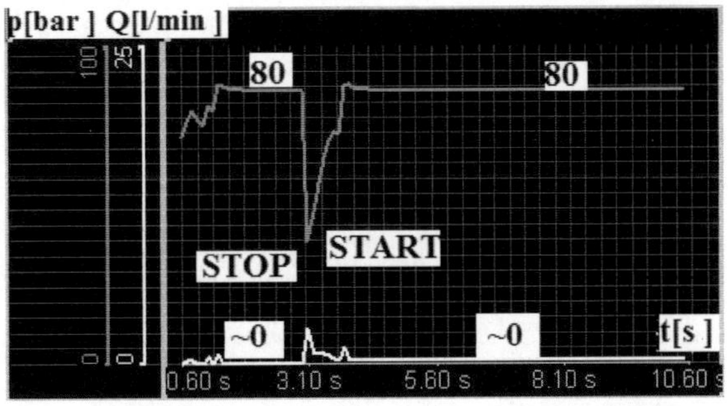

p_1=70 bar, p_2=80 bar, Q=0, p=p_2=80 bar, v=0

Fig. 15. Simulation of operational conditions of balancing systems with constant flow pump. Quick displacements and descending feed and the STOP stage

Based on simulations made, the flowrate / pressure function was drawn as shown in figure 16.

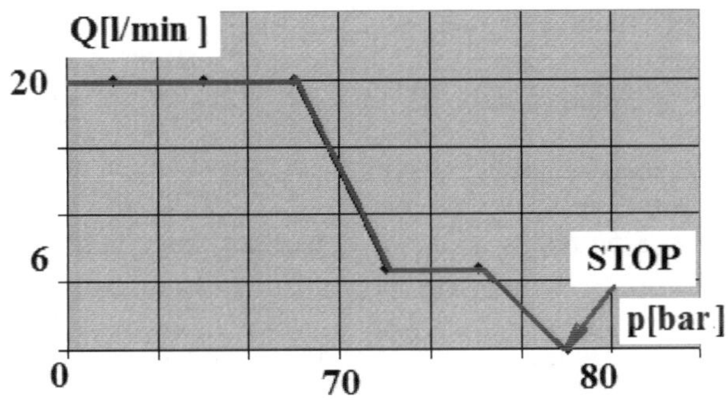

Fig. 16. Flowrate / pressure function of the balancing system using constant flow pump

In order to compare the two versions, it is considered that quick displacement of load has v = 8 m/min and feed displacement speed is of v = 100 mm/min, the balancing piston having a useful area of S = 20 cm^2. Comparative data is shown in tables 1 and 2.

	p [bar]	**Balanced load Ascending [daN]**	**Consumed power Ascending [kW]**	**Balanced load Descending [daN]**	**Consumed power Descending [kW]**	**Balanced load STOP [daN]**	**Consumed power STOP [kW]**
Variable flow pump	80	1600	2.4	1800	~0	1800	~0
Constant flow pump	65	1300	2.6	1600	max. 1	1600	0

Table 1. v = 8 m/min

	p [bar]	**Balanced load Ascending [daN]**	**Consumed power Ascending [kW]**	**Balanced load Descending [daN]**	**Consumed power Descending [kW]**	**Balanced load STOP [daN]**	**Consumed power STOP [kW]**
Variable flow pump	80	1600	0.1	1800	~0	1800	~0
Constant flow pump	80	1600	1.2	1600	max. 1	1600	0

Table 2. v = 100 mm/min

5. Conclusions

In case of positioning of high loads, in heavy machine tools, simple, inexpensive hydraulic systems may be installed which, if used infrequently, may replace the traditional balancing systems specific to feed kinematic chains.

Variable flow pumps are more expensive compared to constant flow pumps, but their use provides, among others, the following benefits: low noise levels, no other instruments necessary (valves, distributors, etc.).

The use of constant flow pumps increases system's power consumption, but provides, for slow feed speeds, a constant balancing force.

Opportunities for replacement of hydraulic system's components are better when using constant flow pumps, since the number of manufacturers is quite large.

In case the vertical displacement of load is performed only for positioning purpose and only rarely, like in the case of mobile cross beams of vertical lathes and heavy gantry machines, the use of constant flow pumps is preferred. In such case the load must be blocked after positioning. In this case, after blocking, the hydraulic system may perform other functions. The use of pneumo-hydraulic accumulators is recommended in this case too, as they improve system's viability.

The units shall be calculated in static but also dynamic conditions, by means of specialized simulation programs.

6. References

[1] Prodan, D.; Bucuresteanu, A. & Balan, E. (2010). Heavy machine-tools. feed-positioning systems for crossrails, The 21st INTERNATIONAL DAAAM SYMPOSIUM" Intelligent Manufacturing & Automation: Theory, Practice & Education"20-23rd OCTOBER, Zadar, Croatia, ISSN 1726-9679, ISBN 978-3-901509-73-5, Katalinic, B. (Ed.), pp. 0273-0274, Published by DAAAM International, Vienna, Austria, DOI: 10.2507.

[2] Perovic, B. (2006). *Machine tools handbook (Handbuch Werkzeug-Maschinen)*, Carl Hanser Verlag, ISBN 10:3-446-40602, Munchen.

[3] Joshi, P. H. (2007). *Machine tools handbook*, McGraw-Hill, ISBN 978-0-07-14943, New Delhi.

[4] Prodan, D.; Bălan, E.; Bucureşteanu, A. & Constantin, G. (2014). Solutions for hydraulic balancing of the positioning and feed kinematic chains in portal milling machines, Proceedings in Manufacturing Systems, Volume 9, Issue 1, pag, 25–30, ISSN 2067-9238, Bucharest.

[5] Prodan, D. (2010). *Heavy machine tools. Mechanical and Hydraulic Systems (Masini-unelte grele. Sisteme mecanice si hidraulice)*, Printech Publishing, ISBN 978-606-521-474-3, Bucharest.

[6] Kief, H. B. & Roschiwal, H. A. (2012). *NC Handbook*, McGraw-Hill Education, ISBN 9780071799485, New Delhi.

[7] Deaconescu, T. (2007). *Hydraulic Drives (Actionari hidraulice)*, Publishing House of Transilvania University, ISBN 978-973-598-121-1, Brasov, Romania.

[8] *** Bosch Rexroth, Berarma, Parker, Atos, Aron catalogs.

[9] Esnault, F & Beneteau, P. (1997). *Hydrostatic 1. Power transmission. (Hydrostatique 1. Transmission de puissance)*, Ellipses /Edition Marketing S.A., ISBN 2-7298-4777-4, Paris.

[10] *** Software package: Automation Studio by Famic Technologies Inc.

DOI: 10.2507/27th.daaam.proceedings.026

AUTOMATED VISUAL INSPECTION SYSTEM IN PRALINE INDUSTRY

Irina Topalova

This Publication has to be referred as: Topalova, I[rina] (2016). Automated Visual Inspection System in Praline Industry, Proceedings of the 27th DAAAM International Symposium, pp.0175-0179, B. Katalinic (Ed.), Published by DAAAM International, ISBN 978-3-902734-08-2, ISSN 1726-9679, Vienna, Austria
DOI: 10.2507/27th.daaam.proceedings.026

Abstract

There are many industrial applications of sophisticated food packaging systems developed nowadays. An essential part of these systems is the visual inspection of the package quality. One of the most important tasks in the automated food packaging is the determination of missing elements in the package. The major drawback of these systems is the implementation of too complex techniques, hardware and software methods which raises the cost of the automated system. This paper represents a hardware, software method and algorithm solution for determining missing pralines when automatically loading them into the trays. The main objectives of the research are focused on developing a simple visual method and algorithm using standard hardware components and communication interfaces, easily compatible with different control systems. An image histogram correlation determination is applied on the off-line programing stage. Based on the results, the system is trained with a simple parametrical vector including different histogram parameters to define the number of missing pralines in the tray. The system was tested with different blurred tray images to simulate the movement of the conveyer belt. The results of the obtained high classification accuracy and fast response are represented and discussed. The opportunities for further development of the system are also presented.

Keywords: Automated system; Visual inspection; Histogram; Praline packaging.

1. Introduction

There are many industrial applications of automated food packaging systems developed nowadays, that implement visual inspection of the package quality. But the major disadvantage of these systems is the implementation of too complex techniques, hardware and software methods which makes the automated system too expensive. One of the most important tasks in the automated food packaging is the determination of missing parts in the package. There are many automated packaging systems in praline industry, but they are too complex or not developed for checking of missing pralines. For example, *Mecmesin Corporation* [1] produces comprehensive range of standard package testing machines but each user defined case, demands development and adjustment of specific equipment. They apply an autonomous mechanical system also for checking the presence of pills in a blister. *Gerhard Schubert GmbH* [2] offers an excellent praline packaging automated line with a visual system for recognition of praline type or damaged exemplars, before filling the trays. But the system does not include checking for missing pralines, after automated arranging the tray. *AMP Rose Chocolate* investor [3] supplies chocolate processing machinery lines to suit customer requirements with no package control. Some systems developed by *Mettler Toledo* [4] offer detection of missing food parts applying *X-Rey inspection*, which acquires additional equipment or sometimes causes food spoilage. The authors in [5] use support vector machines to analyze the texture feature vector of fruits in order to classify them in groups. Analyzing the hue plane of an image, every pixel is described with the help of its feature vector by analysis a small neighborhood of 7×7 pixels, with the current pixels as a center. For each pixel a 6-value-vector is constructed, including a mean value, standard deviation, smoothness, third moment, uniformity and entropy. In [6] the authors calculate the area, perimeter, shape, roundness und apply a fuzzy logic method for automatic analysis of X-ray images of industrial products for defect detection. Both methods use too much complex preprocessing calculations, constructing unnecessary detailed feature vector. This causes increase in computational time. To overcome these disadvantages a simple method, algorithm and visual inspection system is proposed. It uses standard hardware components and communication interfaces, easily compatible with different control systems. An image histogram correlation determination is applied on the off-line programing stage. Based on the results, the system is trained with a simple parametrical vector including different histogram parameters to define the number of missing pralines in the tray. The system was tested with different blurred tray images to simulate the movement of the conveyer belt. The results of the obtained high classification accuracy and fast response are represented and discussed. The opportunities for further development of the system are also presented.

2. Proposed visual inspection system

The proposed visual inspection system consists of a conventional smart camera with triggered infrared lighting and a Vision Builder software package for development of user defined visual inspection algorithms. Different integrated standard interface drivers give good opportunity for further connection with any real-time working control system. The developed algorithm consists of a pre-processing stage for image histogram extraction, correlation analysis between images with different number of missing pralines, histogram reduction and extraction of histogram parameters. The extracted reduced histogram parameters form train parametrical vectors for a minimum mean distance classifier. Different blurred tray images were used for testing purpose to simulate the image acquisition by different conveyer belt velocities. The achieved high classification accuracy, fast response and standard interface compatibility turn it into a reliable and easy to use automated system.

2.1 Proposed algorithm

The initial step of the algorithm includes image acquisition and histogram extraction. Four classes were investigated

Fig. 1. Exemplars of the four investigated classes - *full*, *miss-2*, *miss-3*, *miss-5*

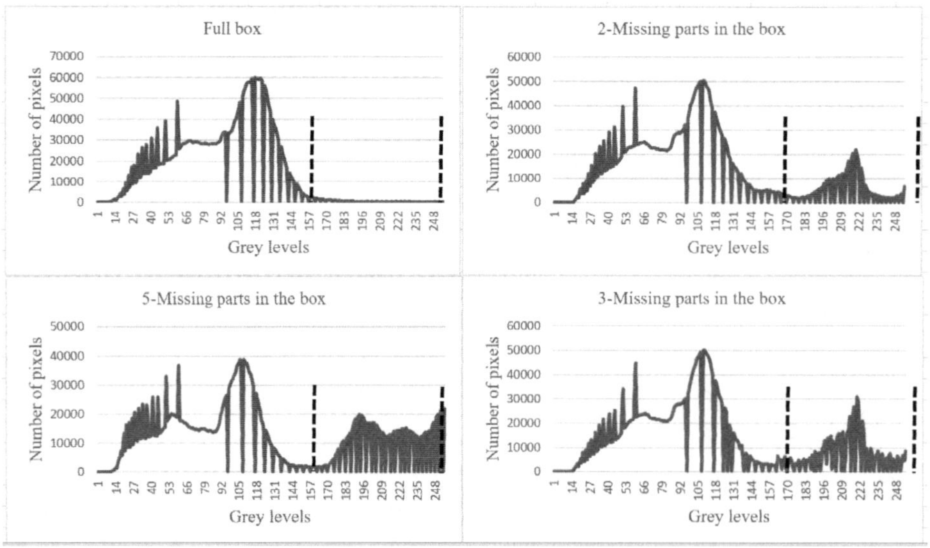

Fig. 2. Grey level histograms of the four classes - *full*, *miss-2*, *miss-3*, *miss-5*

with a full tray, with missing 2, 3 or 5 pralines having corresponding names *full*, *miss-2*, *miss-3*, *miss-5*, shown in Fig.1.

- Next the grey level histograms were calculated on the basis of calculated mean RGB value. The histograms are calculated only over Region of Interest (ROI), fixed on the praline nests in the tray. To obtain invariance against rotation and translation of the tray, the ROIs are related to a coordinate system, which is fixed to a predefined pattern. This pattern is found after scanning the image and searching the highest value of the cross-correlation coefficient $C(i,j)$ between the chosen pattern and the image

$$C(i,j) = \sum_{x=0}^{L-1} \sum_{y=0}^{K-1} t(x,y) f(x+i, y+j), \tag{1}$$

where $i = 0,1, \dots N-1,\ \ j = 0,1, \dots M-1$, where $t(x, y)$ is the pattern, M and N are the image dimensions, L and K are pattern dimensions. The partial cross-correlation [7] between each two histograms moving the sliding window with a step of five grey levels is calculated. The highest value of the cross-correlation coefficient varies between 0.78 and 0.89 for the range between 0 and 160 grey level and the lowest values between -0.09 and 0.02 for the range between 0 and 160 grey levels. This means the lower grey level values of all histograms are high overlapping, while the range between 160 and 255 grey level values are much more distinguishable.

- As next step all histograms are reduced taking only the low correlated ranges between 160 and 255 grey levels. Only four parameters of the reduced histograms are chosen to form the parametrical vector, used to train the classifier. These parameters are [8]: mean histogram value (*mean*), standard deviation (*stdev*), median value (*median*) and histogram skew (*skew*). For each class j a mean parametrical vector $\bar{\boldsymbol{M}}_j$ is calculated over 10 exemplars.

$$\bar{\boldsymbol{M}}_j = \{\overline{mean^j}, \overline{stdev^j}, \overline{median^j}, \overline{skew^j}\} \tag{2}$$

- The calculated parametrical vector for each current image $\boldsymbol{X}_i = \{X_1^i, X_2^i, \dots, X_{n_i}^i\}$, (where in our case $n_i = 4$), is classified applying minimum mean distance classifier. It means $\boldsymbol{X}_i$ belongs to class Cj if (3) is true.

$$\boldsymbol{X}_i \epsilon\ Class\ C_j, if\ d(\boldsymbol{X}, \bar{\boldsymbol{M}}_j) = \min d(\boldsymbol{X}, \bar{\boldsymbol{M}}_j) \tag{3}$$

where for d Manhattan distance is chosen.

- In the last step, the current calculated vector parameters and the recognized class are visualized. The results are prepared for export through standard interfaces.

2.2 System development

The experiments are implemented using a Smart Camera NI 1742 with triggered infrared lighting and software Vision Builder AI 2014. The constructed vision inspection system is shown in Fig.3. To simulate the blur noise [8] added to the images because of the conveyer belt moving, the calculated blur in number of pixels is added to each image. For a typical conveyer belt velocity of 25m/min 417 mm/sec and image resolution of 300 dpi 118,11 dp(cm) 11,81 dp(mm), the calculated conveyer belt velocity measured in Pixel per second is Vp= 417x11,8 = 4920 Pix/sec. The resulting Motion Blur = (Vp) x (Exposure time) = 4920 x 1/125 40 Pix Motion Blur. In our case, a short value of Exposure time = 1/500 is chosen which corresponds to 9 Pix Motion Blur.

Fig. 3. The constructed vision inspection system

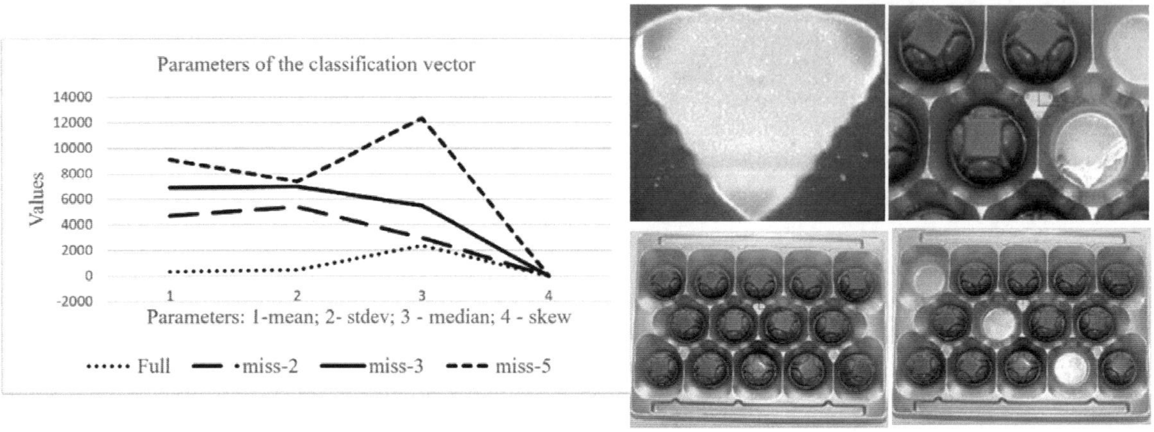

a b

Fig. 4. a/ Parameters of the classification vector $\bar{\boldsymbol{M}}_j$ for classes *full*, *miss-2*, *miss-3*, *miss-5*
b/ Pattern determination, fixed coordinate system and ROIs

This value is added to some of the test images [9]. Fig.4 a/ shows the obtained parameters of the classification vector $\bar{\boldsymbol{M}}_j$ for classes *full*, *miss-2*, *miss-3*, *miss-5* calculated with mean values of each parameter calculated over 10 different images. It is used as support/reference vector for the calculations in (3). Fig.4 b/ represents the determination of chosen Pattern, fixed with the pattern coordinate system and the ROIs used for histogram calculations. The given images represent slightly rotated trays. The corresponding steps in the algorithm are named as follows: Image Acquisition, Image Improvement with histogram calculation and reduction, Match Pattern, Set Coordinate System, Parameter calculation and Display of the results, data preparation for export through standard interfaces.

3. Experimental results

The system functionality was tested with 20 exemplars/images of each class. To some of them Motion Blur of 9Pixels, of 15Pixels and even of 25 Pixels was added as in [10]. Fig.5 shows the distribution of parameters *median*, *mean* and *stdev* for 20 exemplars of classes: *miss-2*, *miss-3* and *miss-5* respectively. Obviously there are some exemplars overlapping in the parametrical feature space. The proposed method and algorithm separates them very good at the system output, because the obtained classification accuracy is 100%. The accuracy is calculated as [(number of all exemplars - number of exemplars with bad line fitting)/ number of all exemplars].100 [%]. Fig. 6 shows the output of the developed user program display when recognizing class *miss-2.* Table 1 represents State/Step elapsed times in *ns* based on 100 inspection run(s). The very short export time of 1.1186 *ms* ensures the possibility of using all available standard export interfaces in Vision Builder AI 2014. They are Serial I/O, Pulse I/O, Modbus, TCP/I/O, Ethernet/IP [11]. The obtained short Total execution time gives opportunity for application in real-time control systems.

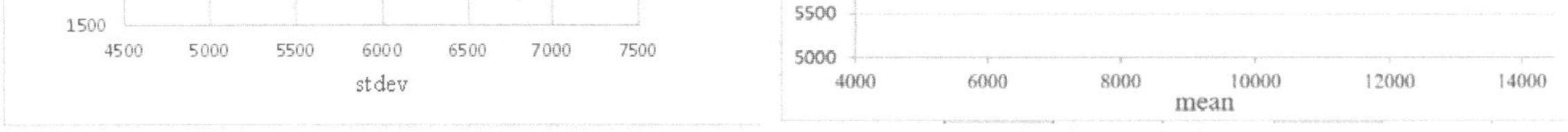

a b

Fig. 5. a/ Distribution of parameters *median* and *stdev* for 20 exemplars of classes *miss-2* and *miss-3*; b/ Distribution of parameters *mean* and *stdev* for 20 exemplars of classes *miss-3* and *miss-5*

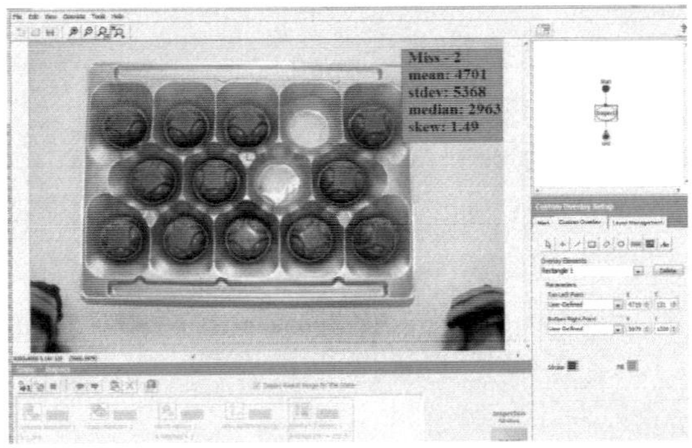

Fig. 6 Output of the developed user program display when recognizing class *miss-2*

Step Name	Iterations	Total Time	Average	Std-Dev	Shortest	Longest
Inspect	100	188066.277	1880.663	25.604	1840.41	2041.688
Image Acquisition	100	46291	462.91	4.557	443	473
Image Improvement	100	19549	195.49	2.335	189	204
Match Pattern	100	62413.682	624.137	20.595	591.062	760.59
Set Coordinate System	100	35.562	0.356	0.044	0.252	0.453
Parameter calculation	100	58486.004	584.86	9.589	566.288	620.191
Display/data export	100	1185.75	11.857	3.614	9.681	40.702

Table 1. State/Step elapsed times in *ns* based on 100 inspection run(s).

4. Conclusion

The developed inspection system offers high classification accuracy, beside simple technical equipment. Applying the proposed calculations only over the reduced histograms and only for the defined ROIs contributes to fast computational response. Using a relevant coordinate system, fixed to a chosen pattern enables a recognition invariant against image rotation and translation. The system is easy-coupled to various control devices because of data export to different interfaces. The achieved high classification accuracy, fast response and standard interface compatibility turn it into a reliable and easy to use automated system. In the future a connection to a National Instruments real-time working controll system is forseen. Also as future work different tests with normalized histograms is planed, intending to make the system invariant against variations in the image intensity.

5. References

[1] http://www.mecmesin.com/flexible-packaging-welded-seal-strength-cs6, Accessed on: 2016-08-15
[2] http://www.gerhard-schubert.de/de/verpackungsmaschinen-aus-crailsheim, Accessed on: 2016-08-10
[3] http://www.amp-rose.com/used-machines/chocolate/, Accessed on: 2016-08-08
[4] http://www.mt.com/int/en/home/applications/Product-Inspection_2/PI-ProductInte.html, Accessed on: 2016-08-11
[5] Rakun, J.; Berk, P. & Lakota, M. (2015). Fruit Sorting Based on Texture Analysis and Support Vector Machine Classification, Chapter 19 in DAAAM International Scientific Book 2015, B. Katalinic (Ed.), Published by DAAAM International, pp. 209-218, ISBN 978-3-902734-05-1, ISSN 1726-9687, Vienna, Austria, DOI: 10.2507/daaam.scibook.2015.19.
[6] Amza, C. G. & Cicic. D. T. (2014) Industrial Image Processing Using Fuzzy-Logic, Proceedings of 25th DAAAM International Symposium on Intelligent Manufacturing and Automation, Vol. 100-2015, ISSN 1877-DAAAM 2014 7058, Katalinic, B. (Ed.), pp. 492 – 498.
[7] Leydesdorff, L. & Vaughan, L. (2006). Co-occurrence matrices and their applications in information science: extending ACA to the Web environment. Journal of the American Society for Information Science and Technology 57(12), 1616-1628.
[8] Gonzalez, R., C. & Woods, R., E. (2002) Digital image processing, 2nd ed. Boston, MA, USA: Prentice-Hall of India, 2002.
[9] B. G. Batchelor, B., G. (2012) Machine Vision Handbook, Springer, ISBN 9781849961684, 2012.
[10] Topalova, I., (2015) Method for Parameter Adjustment for Automated Visual Inspection of Bottled Liquids, IARIA/ ICAS'2015 – 9th International Conference on Autonomic and Autonomous Systems, May, pp. 16-19, ISBN: 9781-61208-405-3, Rome, Italy, 2015.
[11] National Instruments, (2009) NI Vision Builder for Automated Inspection User Manual, 373379C-0. National Instruments, 2009.

DOI: 10.2507/27th.daaam.proceedings.027

USING VIRTUAL MODELLING FOR AS/RS MATERIAL FLOW MANAGEMENT

Constantin Adrian Popescu, Cicerone Laurentiu Popa,
Costel Emil Cotet & George Enciu

This Publication has to be referred as: Popescu, C[onstantin] -[Adrian]; Popa, C[icerone] L[aurentiu]; Cotet, C[ostel] E[mil] & Enciu, G[eorge] (2016). Using Virtual Modelling for AS/RS Material Flow Management, Proceedings of the 27th DAAAM International Symposium, pp.0180-0186, B. Katalinic (Ed.), Published by DAAAM International, ISBN 978-3-902734-08-2, ISSN 1726-9679, Vienna, Austria
DOI: 10.2507/27th.daaam.proceedings.027

Abstract

In today's industries, the storage of workpieces, components and final products has become an issue due to lack of storage spaces and to high storage costs. Companies are trying to find new solutions for material flow optimization in warehouses in order to obtain the best storage position of products. The main goal of the paper is to present how mathematical and virtual modelling can be applied in this particular area of industrial logistics. The first part of the paper describes a mathematical model based on the Markov chain theory used for industrial logistics applications. The second part of the paper contains case studies which demonstrate how Markov chain theory can be used for virtual material flow modelling of AS/RS (Automated storage/Retrieval system) systems in order to obtain the best storage position of products considering: available position, estimated storage time, product type, weight etc.

Keywords: Markov chain; virtual modelling; AS/RS system; material flow management.

1. Introduction

In this paper we present some of the research results obtained in the industrial logistics laboratory from our faculty where multiple logistic systems and platforms have been made from scratch in the last four years for educational and research purposes. For each developed system and platform a virtual model was made using dedicated software applications (CATIA, SolidWorks, Inventor) and for each system and platform a material flow simulation was made to achieve a preliminary diagnosis and to validate the identified optimization solution for the chosen architecture configuration. The material flow modelling, simulation and optimization for the storage systems represent some of the major concerns, considering the demands coming from our industry partners. Due to lack of storage spaces and to high storage costs, the storage of workpieces, components and final products has become an issue. Companies are trying to find new solutions for material flow optimization in warehouses in order to obtain the best storage position of products considering: available position, product type, estimated storage time etc.

In one of our previous papers we presented how CAD (Computer Aided Design) and the simulation results of a virtual model were used to develop and to optimize the performances for an educational platform considering various manufacturing scenarios [1]. In the present paper we extend our research by looking for the best storage solution using the Markov chain model. In the second part of the paper we will present a case study which demonstrate how Markov chain theory can be used for virtual material flow modelling of AS/RS (Automated storage/Retrieval system) systems in order to obtain the best storage position of products considering: available position, estimated storage time, product type, weight etc.

2. Literature Review

The function of a material storage system is to store materials for a period of time and to permit access to those materials when required [2]. An automated storage/retrieval system (AS/RS) represents a combination of equipment and controls, which handles, stores and retrieves materials with precision, accuracy and speed under a defined degree of automation [3]. Automated Storage and Retrieval Systems (AS/RSs) are widely used in the industry in order to store and retrieve unit-loads without interference of an operator. An AS/RS is used for raw material, work-in-process and finished goods. We can find AS/RS systems in many domains such as: warehousing and distribution, manufacturing, automotive, pharmaceutical etc. The main advantages of AS/RSs are savings in labour costs and floor space, increased reliability, and reduced error rates [4].

Elements from Markov chain theory are used in many domains, among which: physics, medicine, chemistry, economics, sociology, IT&C, data storage etc. Markov chain models are widely used in engineering and manufacturing in detecting machine failures in a production system [5]. Literature review shows many approaches regarding the use of Markov chain in the manufacturing and industrial logistics domains. For example, a Markovian model is used to determine optimal equipment adjustment in multi-stage production systems taking variable cost into consideration. The goal is to maximize the expected profit per item of a multi-stage production system by determining best adjustment points of the equipment used based on technical product specifications defined by designer [6]. A similar study regarding the analysis of a Markov chain model of a multistage manufacturing system with inspection, rejection and rework was conducted by Yu and Briker (1990) [7]. Atali and Ozer (2012) show how to better manage multi-item production systems with production smoothing constraints at each stage when the demand environment fluctuates. Also, in their paper a model is presented which provides a cost upper bound for production systems that allow a manager to keep inventory of intermediate products. [8].

Markov chain models have been used by many authors to capture the influence on demand by factors such as weather, product age, economic conditions, and price competition. In manufacturing, random production yields are influenced by factors such as breakdowns, repairs, maintenance. [9]. Pillai and Chandrasekharan present a Markov model which describes a material flow in a production system under uncertainties due to scrapping and reworking. Their analysis shows clearly the interaction between design and control decisions, and it provides an opportunity for the management to analyse quality-related problems. [10]

Virtual modelling represents an important step during the development of a logistic system. However, in the studied papers the virtual modelling of the storage system or manufacturing system was not made. In our paper, in addition to the Markov chain model we will present the virtual model of the storage system (AS/RS storage system). Virtual modelling reduces the design time, can be used to test structures and constructions for strengths before they are physically built, can be very easily and quickly modified at any time. Also, by using virtual modelling, mistakes in design or production can be avoided, the structural elements can be used in the material flow simulation etc.

3. Mathematical and Virtual Modelling

Due to the probabilistic stages through which the physical systems generally go through, it can be admitted that their evolution is described by a random process, defined by a family of random variables. Knowing the stages prior to the moment n has a contribution in knowing the stage at the moment n. Similarly, the functioning of a logistic subsystem is characterized through a succession of stages in which the subsystem can be at a certain moment, and depending on this, decisions can be taken regarding its evolution.

The algorithm for determining the possible configurations for the passing from the initial stage to the present stage for a storage structure specific for AS/RS systems with n storage positions is the following:

- the matrix of the current stage corresponding to the AS/RS storage system is selected;
- all the combinations that have as basis the matrix for the current stage and that can be obtained through the adding of a single element in the matrix structure on one of the free positions (marked with 0) are identified.
- the two stages are repeated for all elements equal with 0, until the matrix of the current stage has all the elements equal with 1;
- for each passing the relation between the initial stage and the current stage is identified, so that no identical stages will exist corresponding for the same number k;
- at the end, for every passing from an initial stage and the current stage, p possibilities for passing where $p = \mathrm{n} - k$, and n represents the number of storage positions and k represents the parts stored at a given moment.

3.1 The Mathematical Model

The matrix associated to AS/RS system will have the following structure:

$$A = \begin{bmatrix} a_{n1} & a_{n2} & \cdots & a_{nj} & \cdots & a_{nm} \\ \vdots & \vdots & & \vdots & & \vdots \\ a_{i1} & a_{i2} & \cdots & a_{ij} & \cdots & a_{im} \\ \vdots & \vdots & & \vdots & & \vdots \\ a_{21} & a_{22} & \cdots & a_{2j} & \cdots & a_{2m} \\ a_{11} & a_{12} & \cdots & a_{1j} & \cdots & a_{1m} \end{bmatrix} \quad (1)$$

where $1 \leq i \leq m$ şi $1 \leq j \leq n$

and $a_{ij} = \begin{cases} 0, & \text{for free position} \\ 1, & \text{for occupied position} \end{cases}$

The general shape of the passing matrix results:

$$P = \begin{bmatrix} \frac{n_{11}}{x_1} & \frac{n_{12}}{x_1} & \frac{n_{13}}{x_1} & \cdots & \frac{n_{1c}}{x_1} \\ \frac{n_{21}}{x_2} & \frac{n_{22}}{x_2} & \frac{n_{23}}{x_2} & \cdots & \frac{n_{2c}}{x_2} \\ \vdots & & & & \\ \frac{n_{c1}}{x_c} & \frac{n_{c2}}{x_c} & \frac{n_{c3}}{x_c} & \cdots & \frac{n_{c}}{x_c} \end{bmatrix} \quad (2)$$

Elements of the passing matrix can be:

$$\begin{cases} 0 \ - \text{if no transition can't be made from the } s_{i_T} \text{ stage to the } s_{i_T} \text{ stage} \\ 0 < \frac{n_{i_T j_T}}{x_{j_T}} < 1 - \ \text{if transition can be made from the } s_{i_T} \text{ stage to the } s_{i_T} \text{stage} \end{cases}$$

$i_T \nrightarrow j_T$ if $p_{i_T j_T} = 0$
$i_T \rightarrow j_T$ if $p_{i_T j_T} \neq 0$

Determining the passing matrix for the stages in which the storage structure of the AS/RS system can be found is based on the following algorithm:

- for all possible combinations, in each matrix the most favourable storage position is identified, taken into account the storage time (the time necessary for the transport of the workpiece from the in/out position to the storage place) and the occupation matrix. The storage time is determined depending on the transportation speed specific for the transfer/transport system that varies depending on the weight of the workpiece that needs to be stored;
- the vector corresponding to the stages in which the AS/RS specific storage structure can be found is written;
- based on the algorithm for establishing the possible configurations for passing from stage S_{i_T} to the present stage S_{j_T} the passing matrix is completed with the values calculated for each position corresponding the storage spaces;
- the $\sum n_{i_T j_T} = x_{j_T}$ formula is applied for each of the lines of the resulted matrix;
- The matrix is determined in the following way: each element of the passing matrix that differs from 0 will be:

$$p_{i_T j_T} = \frac{n_{i_T j_T}}{\sum_{j_T=1}^{c} n_{i_T j_T}} \quad (3)$$

3.2 The Virtual Model of the AS/RS System

The complete functioning cycle of the transfer-transport system specific to the AS/RS system, at the loading on the storage structure is:

- Positioning the transport/transfer system in front of the entry conveyor;
- Extending the mobile segments specific to the transfer subsystem for picking up the part – x axis;
- The vertical movement of the transfer subsystem on a short distance – z axis;
- The retreat of the mobile segments specific to the transfer subsystem in the 0 position – the x axis;

- The movement of the transfer/transport system on the two axes (y and z) for the positioning in front of the specific storage position where the part will be placed;
- The extension on the mobile segments specific to the transfer subsystem for placing the stored part on the x axis;
- The vertical movement of the transfer subsystem on a short distance – the z axis, in order to place the part in the specific storage position;
- The withdrawal of the mobile segments specific to the transfer subsystem in the 0 position, on the x axis.

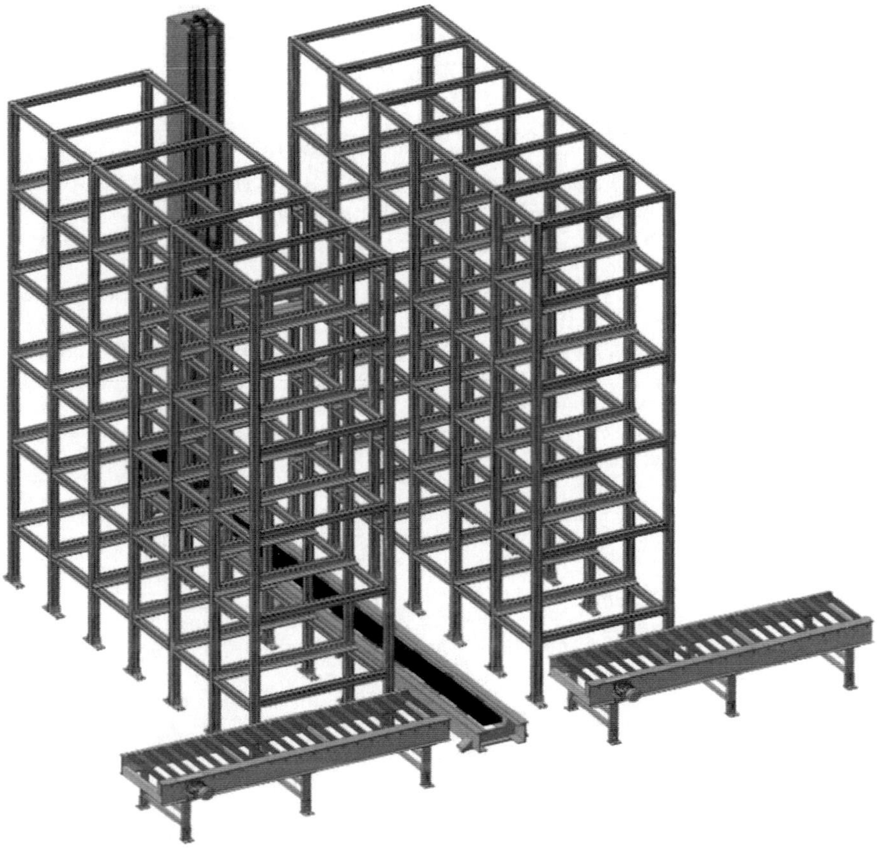

Fig. 1. The virtual model of the AS/RS system

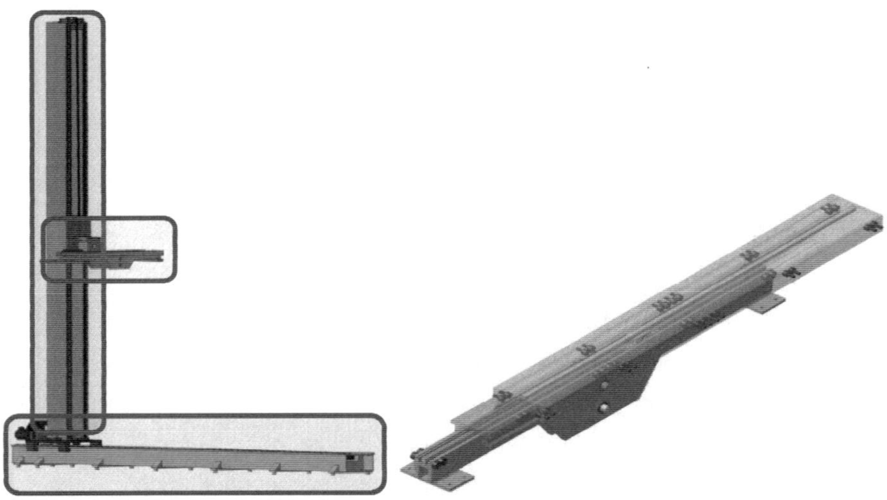

Fig. 2. Main transport/transfer components of AS/RS system

4. Case Study

The main objective is to reduce the total functioning time of the specific transport-transfer AS-RS system for the presented assembly. The maximum travelled distance for the same type and number of parts will be underlined. For the simplification of the analysis, 9 storage places were considered corresponding to a single rack.

Entry data:

- Nr.ord - The order in which the parts will be removed from the rack;
- L - The length of a storage position;
- H - The height of a storage position;

The general entry data specific for the application: L: 0.77 m; H: 1.10 m

For running the application two classes are used:

- The Element.cs class is used to store the initial vector and it has the following proprieties: stringRepresentation (ex: 000100111), OneCount (ex. "1" from StringReprezentation), DifferencePosition, DifferenceRowPosition, DifferenceColumnPosition, DijValue (Dij value for current matrix), TijValue (Tij value for current matrix).
- The Representation class permits the running of all steps necessary for obtaining the final result and it has the following proprieties: MinValue, MinValuePosition, FinalVector, LowerElement.

The program based on the mathematical model, personalized for 9 storage positions, is formed from a single "MainWindow" view that contains a button for launching the program and textboxes for showing the results.

```
public bool Compare(string value, int lines, int columns)
{
int diferences = 0;
if (this.StringReprezentation.Length != value.Length)
    {
        return false;
    }
    for (int i = 0; i < this.StringReprezentation.Length; i++)
    {
        if (this.StringReprezentation[i] != value[i])
        {
            diferences++;
            this.DifferencePosition = i;
            if (diferences > 1)
            {
                this.DifferencePosition = -1;
                return false;
            }
        }
    }
this.DifferenceRowPosition = (int)(this.DifferencePosition / columns);
this.DifferenceColumnPosition = (int)(this.DifferencePosition % columns);
    return true;
}
```

The virtual model of AS/RS system	The matrix
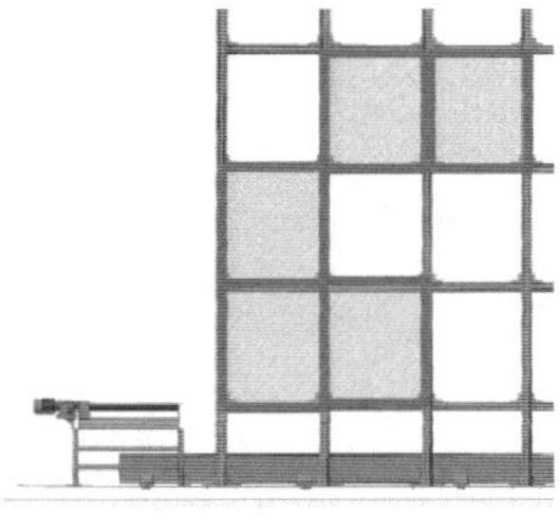	$\begin{matrix} a31 & a32 & a33 \\ a21 & a22 & a23 \\ a11 & a12 & a13 \end{matrix} \quad \begin{matrix} 0 & 1 & 1 \\ 1 & 0 & 0 \\ 1 & 1 & 0 \end{matrix} \quad \begin{matrix} 0 & R4 & R5 \\ R2 & 0 & 0 \\ R1 & R3 & 0 \end{matrix}$

Table 1. The configuration from which we start in our case study

After obtaining the passing matrix, it is multiplied with the vectors of the current stage, obtaining a vector for which the minimal value will be determined. After running the program, the data is shown on the screen with the method "Show results()". For a complete run on the program (meaning from the start of the initial matrix = 0) the program assures a running of maximum 3x3. Because the program generates all possible stages, the passing stage will contain 29 lines and 29 columns, a longer execution time being necessary in this case. The configuration from which we start is presented in table 1.

Case 1. The loading of the storage structure with parts depending on the minimal distance which the transfer – transport system travels. Storage time is considered as being constant.

Nr. crt.	Action	Storage position	Part Type	Nr. ord.	Distance [m]
1	Loading	a[1][3]	T5	-	2.31
2	Loading	a[2][2]	T4	-	1.89
3	Unloading	a[1][1]	T1	-	0.77
4	Loading	a[1][1]	T5	-	0.77
5	Unloading	a[1][2]	T2	-	1.54
6	Loading	a[1][2]	T4	-	1.54
7	Unloading	a[2][1]	T3	-	1.34
8	Unloading	a[3][2]	T4	-	2.68
9	Unloading	a[3][3]	T5	-	3.19
10	Loading	a[2][1]	T1	-	1.34
11	Loading	a[3][1]	T2	-	2.33
12	Unloading	a[1][3]	T1	-	2.31
13	Unloading	a[3][1]	T2	-	2.33
14	Loading	a[1][3]	T2	-	2.31
15	Unloading	a[2][2]	T4	-	1.89
16	Loading	a[2][2]	T1	-	1.89
17	Unloading	a[1][3]	T5	-	2.31
18	Loading	a[1][3]	T2	-	2.31
The total distance travelled					35.05

Table 2. Results for case 1

Case 2. The loading of the storage structure with parts, depending on the minimal distance which the transfer – transport system travels and the time the parts will remain stored (the order number – Nr. ord.).

Nr. crt.	Action	Storage position	Part Type	Nr. ord.	Distance [m]
1	Loading	a[2][3]	T5	5_1	2.55
2	Loading	a[3][1]	T4	4_1	2.33
3	Unloading	a[1][1]	T1	1	0.77
4	Loading	a[1][1]	T5	5_2	0.77
5	Unloading	a[1][2]	T2	2	1.54
6	Loading	a[2][2]	T4	4_2	1.89
7	Unloading	a[2][1]	T3	3	1.34
8	Unloading	a[3][2]	T4	4	2.68
9	Unloading	a[3][3]	T5	5	3.19
10	Loading	a[1][2]	T1	1	1.54
11	Loading	a[2][1]	T2	2	1.34
12	Unloading	a[1][2]	T1	1	1.54
13	Unloading	a[2][1]	T2	2	1.34
14	Loading	a[1][2]	T2	2_1	1.54
15	Unloading	a[3][1]	T4	4_1	2.33
16	Loading	a[2][1]	T1	1	1.34
17	Unloading	a[2][3]	T5	5_1	2.55
18	Loading	a[1][3]	T2	2_2	2.31
The total distance travelled					32.89

Table 3. Results for case 2

The results for the two cases are presented in figure 3 (Case 1- blue, Case 2 – orange)

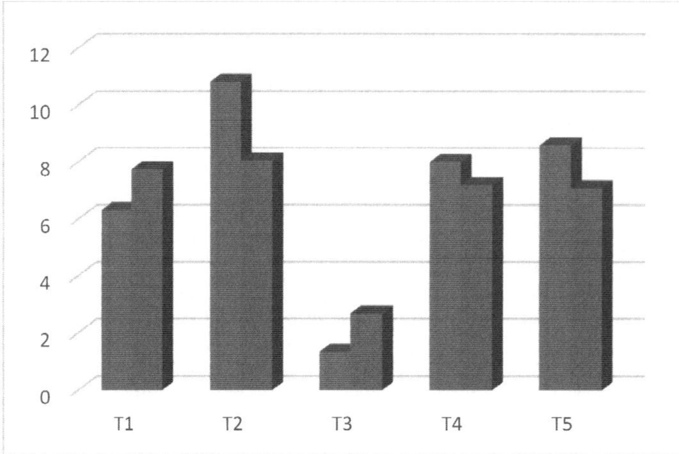

Fig. 3. The specific distance for every type of part for the two cases

5. Conclusion

This paper is centred on the use of mathematical and virtual modelling in optimizing the functioning of various systems in industrial logistics. The case study illustrating the proposed optimization algorithms was performed in the industrial logistics laboratory from our faculty. The application developed from the mathematical model based on the Markov chain theory allows for the reduction of the distance travelled by the transfer-transport system and of its functioning time at the moment of loading and unloading of parts. Because the passing matrix which allows the identification of the present stage of the system – i, from the previous stage – i-1, has a very large number of components, the case study was made for a reduced number of storage positions. Also, the passing matrix was obtained only by the quantification of the distance specific for every storage position, in relation with the in/out position and with the time period in which the part will remain stored. In the future we plan to use a different algorithm for obtaining the passing matrix and a higher number of characteristics specific to the stored part (weight of the part, the storage frequency for a certain type of part etc.).

6. References

[1] Cotet, C. E.; Popa, C. L.; Enciu, G.; Popescu, A.; Dobrescu T. (2016). Using CAD and Flow Simulation for Educational Platform Design and Optimization. International Journal of Simulation Modelling, Vol. 15, No. 1, 5-15, ISSN 1726-4529

[2] Kumar, S. A. & Suresh N. (2009). Production and Operations Management. New Age International Publishers, ISBN 978-8122421774, New Delhi

[3] Material Handling Industry, http://www.mhi.org

[4] Lerher, T.; Ekren, Y. B.; Sari, Z.; Rosi, B. (2015). Simulation Analysis of Shuttle Based Storage and Retrieval Systems. International Journal of Simulation Modelling, Vol. 14, No. 1, 48-59, ISSN 1726-4529

[5] Allen H. Tai; Wai-Ki Ching; Chan L. Y. (2009). Detection of Machine Failure: Hidden Markov Model Approach. Computers & Industrial Engineering, Volume 57, Issue 2, 608–619, ISSN 0360-8352

[6] Fallah Nezhad, M. S.; Niaki, S. T. A.; Shahin, E. (2013). Markov Model to Determine Optimal Equipment Adjustment in Multi-stage Production Systems Considering Variable Cost. Iranian Journal of Operations Research, Vol. 4, No. 2, 146-160, ISSN 2008-1189

[7] Yu, K. Y. & Bricker, D. L. Analysis of a Markov Chain Model of a Multistage Manufacturing System with Inspection, Rejection, and Rework. ISSN 0360-8352 available from: http://user.engineering.uiowa.edu/~dbricker/or_sample_quizzes/papers/MC_Mfg.pdf, Accessed: 2016-06-16

[8] Atali, A. & Ozer, O. (2012). Stochastic Multi-item Inventory Systems with Markov-Modulated Demands and Production Quantity Requirements. Probability in the Engineering and Informational Sciences, Volume 26, Issue 2, 263-293, ISSN 0269-9648

[9] Gallego, G. & Hu, H. (2004). Optimal Policies for Production/Inventory Systems with Finite Capacity and Markov-Modulated Demand and Supply Processes. Annals of Operations Research, Vol. 126, Issue 1, 21–41, ISSN 0254-5330

[10] Pillai, V. M., Chandrasekharan, M. P. (2008). An Absorbing Markov Chain Model for Production Systems with Rework and Scrapping. Computers & Industrial Engineering, Volume 55, Issue 3, 695–706, ISSN 0360-8352

DOI: 10.2507/27th.daaam.proceedings.028

MATHEMATICAL MODELLING OF SURFACE ROUGHNESS THROUGH MACHINING PARAMETERS AND MACHINING TIME DURING THE DRY MILLING PROCESS

Nexhat Qehaja, Ilir Doçi*, Mirlind Bruçi, Fitore Abdullahu, Kaltrinë Jakupi & Fatlume Zhujani

This Publication has to be referred as: Qehaja, N[exhat]; Doci, I[lir]; Bruqi, M[irlind]; Abdullahu, F[itore]; Jakupi, K[altrina] & Zhujani, F[atlume] (2016). Mathematical Modelling of Surface Roughness Through Machining Parameters and Machining Time During the Dry Milling Process, Proceedings of the 27th DAAAM International Symposium, pp.0187-0194, B. Katalinic (Ed.), Published by DAAAM International, ISBN 978-3-902734-08-2, ISSN 1726-9679, Vienna, Austria
DOI: 10.2507/27th.daaam.proceedings.028

Abstract
This paper investigates the effects of milling parameters on the surface roughness and surface texture by applying RSM. The most important measures of surface quality during the machining process is the average surface roughness (Ra), and it is mostly caused by many machining parameters, such as true rake angle and side cutting edge angle, cutting speed, feed rate, depth of cut, nose radius, machining time etc.
In this work, an experimental investigation through mathematical modelling was carried out to study the effect of cutting parameters such as; cutting speed, feed rate and machining time on surface roughness during the dry end milling process of cold rolled steel C62D. The experiment is executed on the basis of a three level factorial design. The influences of all machining parameters on surface roughness have been analyzed based on the developed mathematical model. The developed prediction equation shows that the most significant parameter is cutting speed followed by feed rate and lastly machining time. The result from this research is useful to be implemented in industry to reduce time and cost in surface roughness prediction.

Keywords: Machining; cutting process; milling; roughness; tool; machining time

1. Introduction

Machinability is a fundamental technological feature of the machined metal and is an economic feature of the cutting operations. Machinability of metals and the tool cutting capacities are mutually related terms and are defined by the same method [11]

Metal cutting is one of the most significant manufacturing processes in the area of material removal. Black [2] defined metal cutting as the removal of metal chips from a workpiece in order to obtain a finished product with desired attributes of size, shape, and surface roughness [1]. One important parameter in the qualification of cut surfaces is their roughness, and its indexes. The roughness has great significance primarily at mating, sliding surfaces. This has been one more reason for the researchers' increased interest for a long time to predict these indexes for a given process within the specified cutting conditions. Several modeling procedures and techniques were worked-out, which essentially can be classified into four groups: l) analytical models, 2) experimental methods, 3) DoE (Design of Experiment)-based methods and 4) AI (Artificial Intelligence)-based methods [2, 3].

In order to establish an adequate functional relationship between the responses (such as surface roughness, cutting force, tool life/wear) and the cutting parameters (cutting speed, feed, and depth of cut), a large number of tests are needed, requiring a separate set of tests for each and every combination of cutting tool and work piece material. This increases the total number of tests and as a result the experimentation cost also increases. As a group of mathematical and statistical techniques, response surface methodology (RSM) is useful for modeling the relationship between the input parameters (cutting conditions) and the output variables. RSM saves cost
and time by reducing number of experiments required [4].

Surface roughness has received serious attentions for many years. It has formulated an important design feature in many situations such as parts subject to fatigue loads, precision fits, fastener holes and esthetic requirements. In additions to tolerances, surface roughness imposes the most critical constraints for selection of machines and cutting parameters in process planning [5].

The surface finish in milling is found to be influenced in varying amounts by a number of factors, such as cutting speed, feed rate, depth of cut, material characteristics, tool geometry, workpiece deflection, stability and stiffness of the machine tool - cutting tool - workpiece system, built-up edge, cutting fluid, etc. [6].

There are various parameters used to evaluate surface roughness. In the present research for surface finish characterization in turning operations, the average surface roughness (R_a) is selected. It is the most widely used surface finish parameter in industry. Many authors suggested linear and exponential empirical models for surface roughness as functions of machining parameters by the following.

The progress in the development of predictive models, based on cutting theory, has not yet met the objective; the most essential cutting performance measures, such as, tool life, cutting force, roughness of the machined surface, energy consumption, … etc., should be defined using experimental studies. Therefore, further improvement and optimization for the technological and economic performance of machining operations depend on a wellbased experimental methodology. Unfortunately, there is a lack of information dealing with test methodology and data evaluation in metal cutting experiments [7].

Various methodologies and practices are being employed for the prediction of surface roughness, such as machining theory, classical experimental design, the Taguchi method and artificial intelligence or soft computing techniques [6].

The aim of this research is to develop the model for predict the in-process surface roughness in ball-end milling process which can be used in practice. The in-process surface roughness models are developed under various cutting conditions by employing the exponential function with the aid of the multiple regression analysis and the use of the least square method[17].

2. Nomenclature

R_a the average surface roughness

P power

n RPM

f feed rate

d_{max} workpiece diametër

L tail stock

N factorial design

K number of factors

N_0 number of additional tests

v cutting speed

T cutting time

c_0, c_1, c_2,c_3 constants

y logarithmic value of the measured surface roughness

β_0, β_1, β_2, β_3 regression coefficients

x_0 unit vector

x_1, x_2, x_3 logarithmic values of cutting speed, feed rate, cut of depth

3. Experimental conditions

Machine tool: Universal milling machine *GKA-3* motor power:8 kW, spindle speed range: 40 - 2000 rpm, feed rate range: 0,02 - 2,0 mm/rev, working are: X/Y/Z 400x2000x300 mm, max. tool diameter: 80 mm, max. tool length: 200 mm.

Workpiece was made of cold rolled steel C62D. Its chemical composition is as follows: (0.62-0.65)% C; (0.56-0.78)% Mn; 0.22% Si; 0.032% P, 0.03% S, and 98,28% Fe. Tensile strength is 230-247 N/mm2, and hardness 236-245 N/mm2. The workpiece dimensions are: the length 300 mm, the diameter 70 mm, and it is machined under dry turning conditions. Cutting tool: HSS-E Co10, with data in table 1. To guarantee the initial conditions of each test, a new tool is used in each experiment.

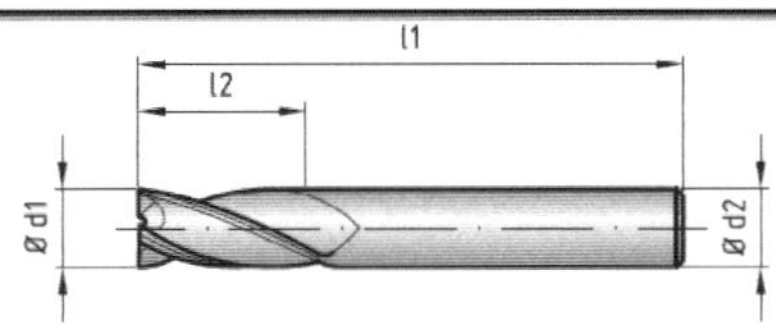

DIN 844	d_1 [mm]	d_2 [mm]	l_1 [mm]	l_2 [mm]	z
HSS-E Co10	20	20	104	38	3

Table 1. Cutting tool data

Roughness measuring equipment: HADRON, SRT-6210. Spectrometer Metorex Arcmet 930, Hardness meter Krautkramermic.10.DL.

4. Experimental setup

Experimental design. This work is an experimental study focused on the effect of cutting parameters and machining time on surface roughness, developing a correlation between them. The experimental design involves variation of three factors at three levels (low, medium and high), including cutting speed (v), feed rate (f) and machining time (T) as indicated in Table 2.

A design matrix was constructed on the basis of the selected factors and factor levels as shown on the table 2. The selected design matrix was a full factorial design N=2k+N0 (k=3 - number of factors, N0=4 – number of additional tests for three factors) consisting of 12 rows of coded/natural factors, corresponding to the number of trials. This design provides a uniform distribution of experimental points within the selected experimental hyper-space and the experiment with high resolution [7, 8].

The factor ranges were chosen with different criteria for each factor, aiming at the widest possible range of values, in order to have a better utilization of the proposed models. At the same time, the possibility of the mechanical system and manufacturer's recommendations are taken into account.

Machining conditions used in the experiment also are shown in Table 2 [9, 10, 11]. All of the trials have been conducted on the same machine tool, with the same tool type and the same cutting conditions.

Measured values of surface roughness, as the results of testing twelve experimental points defined by experiment plan matrix, are shown in Table 3. The mentioned values of surface roughness are input data for mathematical modeling of results, which was made by multiple regression analysis.

Cutting factors and their levels					
No.	Factors	Code	High level	Middle level	Low level
		level	1	0	-1
1	v, m/min	X_1	94.20	72.96	56.52
2	f, mm/rev	X_2	0.3	0.18	0.1
3	T, s	X_3	1200	1049	900

Table 2. Experimental setup at three level factors.

a. Regression based modeling

The main task for regression analysis is to show relationship between the roughness and machining independent variables. Many authors suggested linear and exponential empirical models for surface roughness as functions of machining parameters [13, 14, 15, 16], by the following:

$$R_a = c_0 \cdot v^{c_2} \cdot f^{c_1} \cdot T^{c_3} \quad (1)$$

Coded factors	Performance measures

Test No.	X_0	X_1	X_2	X_3	Ra	Y = ln R a
1	+1	+1	-1	+1	2.783	1.023529483
2	+1	-1	+1	+1	2.985	1.093599747
3	+1	+1	+1	-1	1.982	0.684106436
4	+1	-1	-1	-1	1.524	0.421338457
5	+1	0	0	0	1.963	0.674473915
6	+1	0	0	0	1.882	0.632335041
7	+1	-1	-1	+1	1.479	0.391366184
8	+1	+1	-1	-1	1.256	0.227932068
9	+1	+1	+1	+1	1.145	0.135404637
10	+1	-1	+1	-1	3.182	1.15750993
11	+1	0	0	0	1.714	0.53882982
12	+1	0	0	0	1.825	0.601579987

Table 3. Experimental results.

Three parameters: cutting speed (v), feed rate (f), and cutting time (T), were selected for this study, which are based on experimental results of tool life in earlier stage for the same cutting conditions [12]. Ra is the surface roughness in μm, f - feed rate in mm/rev, r-nose radius in mm, T-cutting time in sec., and respectively c_0, c_1, c_2, and c_3 are constants.

Multiple linear regression models for surface roughness can be obtained by applying a logarithmic transformation that converts non-linear form of eq. (1) into following linear mathematical form:

$$ln\ R_a = ln\ c_0 + c_1\ ln\ f + c_2\ ln\ r + c_3\ ln\ T \quad (2)$$

The linear model of eq. (3) in term of the estimated response can be written as:

$$y = \beta_0 + \beta_1 x_1 + \beta_2 x_2 + \beta_3 x + \varepsilon \quad (3)$$

where y is the logarithmic value of the measured surface roughness, β_0, β_1, β_2, β_3 are regression coefficients to be estimated, x_0 is the unit vector, x_1, x_2, x_3 are the logarithmic values of cutting speed, feed rate, cut of depth and ε is the random error.

The above equation in matrix form becomes:

$$y = X\beta + \varepsilon \quad (4)$$

Thus, the least squares estimator of β is

$$\beta = (X'X)^{-1}X'y \quad (5)$$

The fitted regression model is

$$\acute{Y} = X\beta \quad (6)$$

The difference between the experimentally measured and the fitted values of response is:

$$e = y - \dot{y} \quad (7)$$

The regression analysis technique using least squares estimation was applied to compute the coefficients of exponential model. The following empirical exponential model for surface roughness was determined and is given, respectively:

$$R_a = 1.942\ v^{-05426} f^{0.250}\ T^{0.215} \quad (8)$$

5. Results and discussion

Table 3 presents experimental results of surface roughness criteria R_a for various combinations of cutting speed, feed rate and machining time to full factorial design. Minimal value of surface roughness criteria R_a was obtained at V = 94.20 m/min, f = 0,1 mm/rev, T=900 s, (test No. 8). That means increasing of cutting speed with the lowest feed rate and machining time lead to decreasing of surface roughness.

It is found that cutting speed has the most significant effect on surface roughness, followed by feed rate and machining time.

Maximal value of surface roughness criteria R_a was registered at V = 56.52 m/min, f = 0.3 mm/rev, T =1200 s, (test No. 2). In order to achieve better surface finish, the highest level of cutting speed, and the lowest level of feed rate and machining time, should be recommended.

The analysis of obtained mathematical models and cutting data influence at surface roughness has been made using diagrams shown in Figures 4, 5, 6.

Fig. 1 which highlights the main factor plots for Ra appears to be an almost linear decreasing function of cutting speed and an increasing function of feed rate (f) and cutting time (T).

Figs. 2 a, b, c and d illustrate 3D surface plots of Ra according to the predictive exponential empirical model (9).

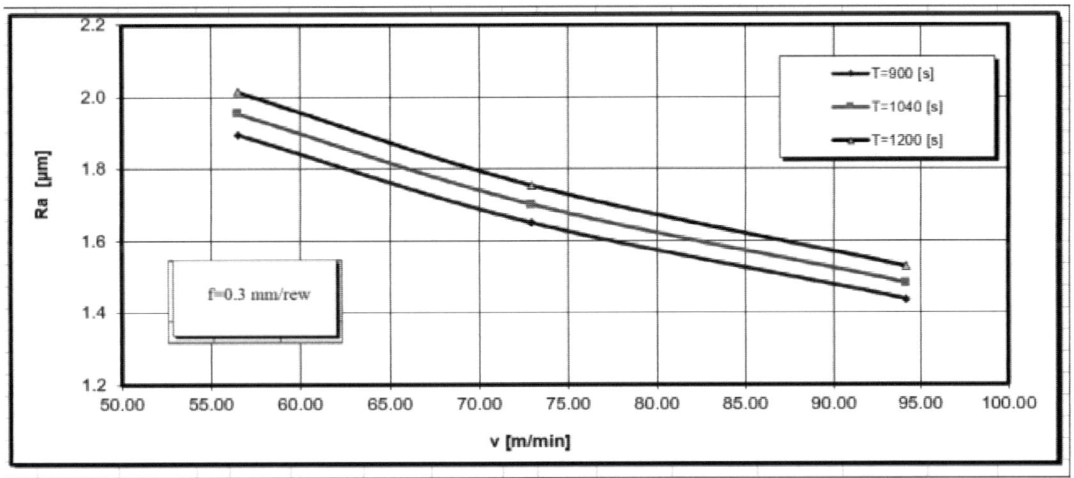

a)

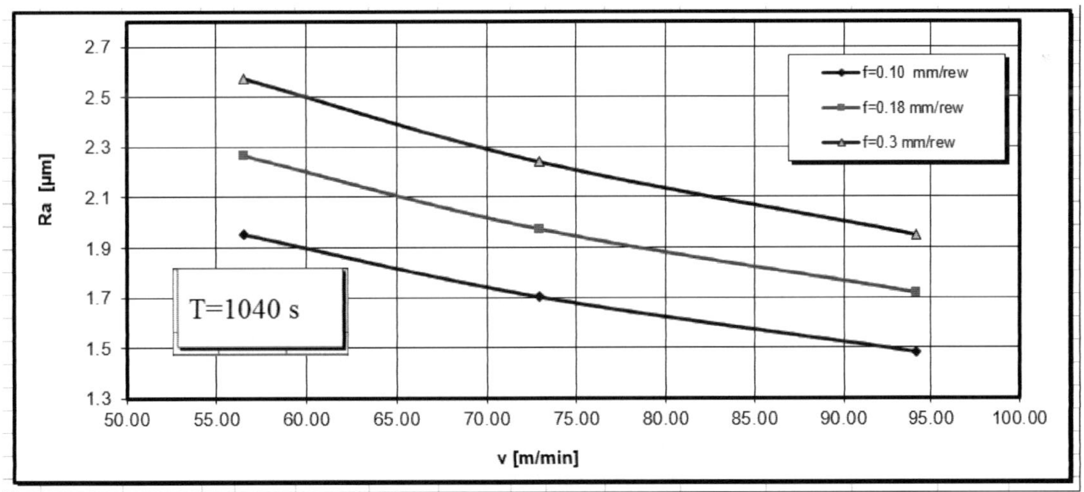

b)

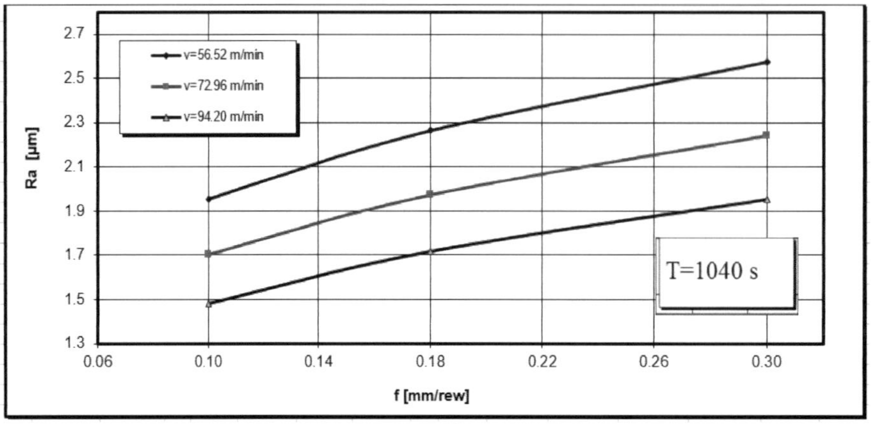

c)

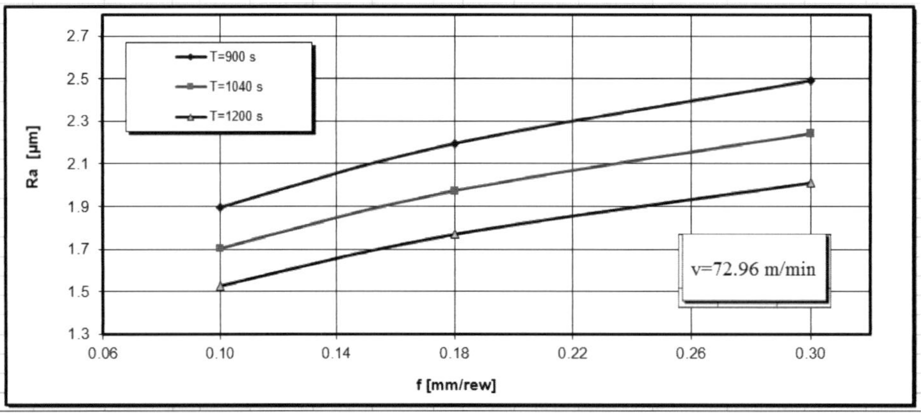

d)

Fig. 1. The dependence of surface roughness on: a) cutting speed and various values of cutting time, b) cutting speed and various values of feed rate, c) feed rate and various values of cutting speed, d) feed rate and various values of cutting time

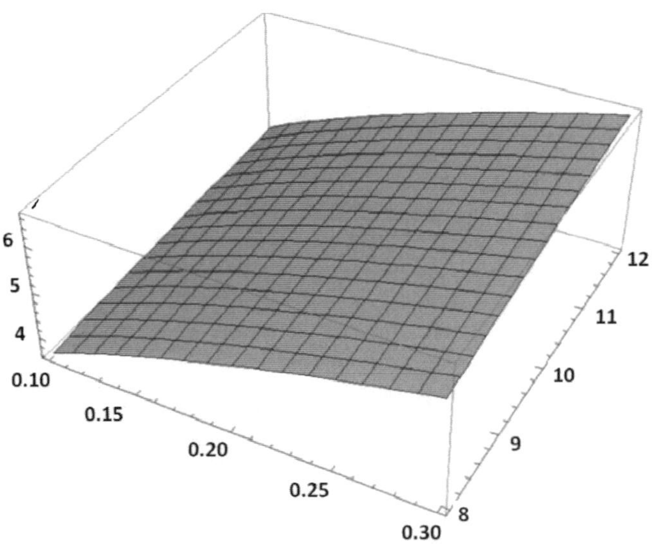

a)

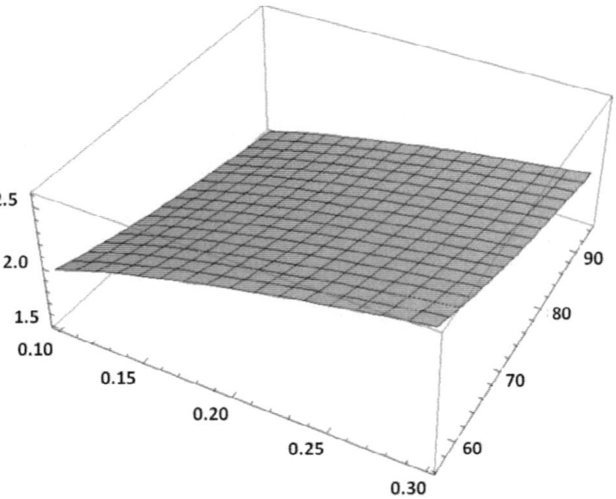

b)

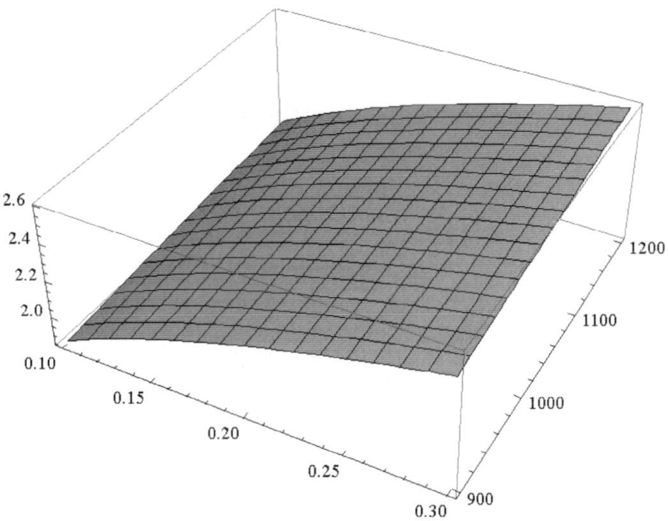

c)

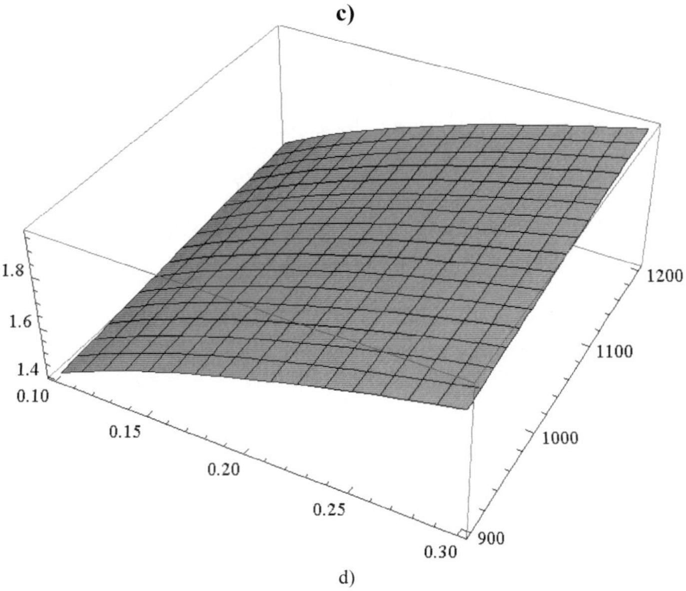

d)

Fig.2. 3D surface plots: Plot3D [6.795*(56.52^0.542)*(f^0.250)*(T^0.215),{f,0.1,0.3},{T,900,1200}]

6. Conclusions

This paper presents research of various cutting parameters affecting the surface roughness in dry milling of carbon steel using HSS cutters.The investigations of this study indicate that the cutting parameters like cutting speed and feed rate are the primary influencing factors, which affect surface roughness.

Statistical models deduction defined the degree of influence of each cutting regime element on surface roughness criteria. The results revealed that cutting speed seems to influence surface roughness (0.542) more significantly than feed rate (0.250). However, machining time is less significant (0.215). With the regression equation generated, the best combination of design independent variables for achieving the optimization of cutting processes.

The relations of the surface roughness, the cutting speed, the feed rate, and the cutting time are investigated to develop the prediction models of surface roughness. The exponential function is employed to represent the relation of the arithmetic average surface roughness, the cutting force ratio, and the cutting parameters. The multiple regression analysis has been utilized to calculate the regression coefficients of the in-process prediction of surface roughness model by using the least square method.

7. References

[1] M. T. Hayajneh, M. S. Tahat, J. Bluhm, A Study of the Effects of Machining Parameters on the Surface Roughness in the End-Milling Process, Jordan Journal of Mechanical and Industrial Engineering, Volume 1, Number 1, Sep. 2007, ISSN 1995-6665, Pages 1 – 5.

[2] D. K. Badek, H. S. Kim, Optimization of Feed Rate in a Face Milling operation Using a Surface Roughness Made, International Journal of Machine Tools Manufacture. 41:451-462, 2001

[3] T.L. Ginta, A.K.M.N. Amin, A.N.M. Karim, A.U. Patwari, M.A. Lajis, Modeling and Optimization of Tool Life and Surface Roughness for End Milling Titanium Alloy Ti–6Al–4V Using Uncoated WC-Co Inserts, CUTSE International Conference 2008, 24-27 November 2008, Miri, Sarawak, Malaysia

[4] S. Thamizhmanii, S. Saparudin and S. Hasan; Analyses of Surface Roughness by Turning Process Using Taguchi Method, Journal of Achievements in Materials and Manufacturing Engineering, Volume 20 Issues 1-2 January-February 2007

[5] V. P. Astakhov: Metal Cutting Mechanics, CRC press, 1999.

[6] G.C. Onwubolu, A note on Surface Roughness Prediction Model in Machining of Carbon Steel by PVD Coated cutting Tool, American Journal of Applied Sciences 2 (6): 1109-1112, 2005, ISSN 15469239

[7] N. Qehaja, H. Zeqiri, A. Salihu, H. Osmani, F. Zeqiri, Machinability of Matals, Methods, and Practical Application, Annals of DAAAM & Proceedings 2012, The 23RD World Symposium, 24-27th October 2012, University of Zadar, Croatia, URL: http:\\www.daaam.com, Vienna University of Technology, Karlsplatz 13/311, A-1040 Wien, Austria, Europe

[8] A. Salihu, H. Zeqiri, A. Bunjaku, N. Qehaja, F. Zeqiri, Research of Surface Roughness Average of Steel C45 During Turning, Anals of DAAAM for 2010 & Procedings of the 21st International DAAAM symposium, Volume 21, PP. 0185÷0186, ISBN 978-3-901509-73-5. No 1. ISSN 1726-9679. Vienna University of Technology, Karlsplatz 13/311, A-1040 Wien, Austria, Europe

[9] L. Huang and C. Chen, A Multiple Regression Model to Predict In-process Surface Roughness in Turning, Journal of Industrial Technology • Volume 17, Nr. 2 • February 2001 to April 2001

[10] N. Qehaja, A. Bunjaku, J. Kacani. A.Salihu, H. Zeqiri, H. Osmani, Mathematical models of flank wear curve during the turning of steel C60E4, The 20thInternational DAAAM SYMPOSIUM "Intelligent Manufacturing & Automation: Focus on Next Generation of Intelligent Systems and Solutions" 22-25th October Vienna, Austria,2009

[11] N. Qehaja, A. Salihu, H. Zeqiri, H. Osmani & F. Zeqiri, Machinability of metals, methods and practical application, Annals of DAAAM for 2012 & Proceedings of the 23rd International DAAAM Symposium, Volume 23, No.1, ISSN 2304-1382 ISBN 978-3-901509-91-9, CDROM version, Ed. B. Katalinic, Published by DAAAM International, Vienna, Austria, EU, 2012

[12] H. Zeqiri, N. Qehaja, F. Zeqiri, A. Salihu, H. Osmani, Research of Flank Wear in Turning of Steel Ck 60, Annals of DAAAM & Proceedings 2012, The 23rd DAAAM World Symposium, 20012

[13] A. Salihu, H. Zeqiri, A. Bunjaku, N. Qehaja, H. Osmani, A. Kyçyky, Research of the horizontal parameters of the roughness machined surface by turning, The 20th International DAAAM Symposium "Intelligent Manufacturing & Automation: Focus on Next Generation of Intelligent Systems and Solution", 2009

[14] M. Durairaja, S. Gowri, Parametric Optimization for Improved Tool Life and Surface Finish in Micro Turning using Genetic Algorithm, International Conference On DESIGN AND MANUFACTURING, IConDM 2013, Procedia Engineering 64 (2013) 878 – 887

[15] S. Chinchanikar, A.V. Salve, P. Netake, A. More, S. Kendre and R. Kumar, Comparative evalutions of surface roughness during hard turning under dry and with water- based and vegetable- oil-based cutting fluids, International Conference on Advances in Manufacturing and Materials Enginneering AMNE 2014, Procedia Materials Science 5 (2014) 1966-1975

[16] N. Qehaja, K. Jakupi, A. Bunjaku, M. Bruçi, H. Osmani, Effect of machining parameters and machining time on surface roughness in dry turning process, 25th DAAAM International Symposium on Intelligent Manufacturing and Automation, DAAAM 2014, Procedia Eng 100:135–140, 2015

[17] S. Tangjitsitcharoen, A. Senjuntichai, In- Process monitoring and prediction of surface roughness in ball-end milling process, Annals of DAAAM for 2010 & Proceedings of the 21st International DAAAM Symposium, Volume 21, No. 1, ISSN 1726-9679 ISBN 978-3-901509-73-5, Editor B. Katalinic, Published by DAAAM International, Vienna, Austria, EU, 2010

DOI: 10.2507/27th.daaam.proceedings.029

MATHEMATICAL MODELLING OF SURFACE ROUGHNESS FOR EVALUATING THE EFFECTS OF CUTTING PARAMETERS IN DRILLING PROCESS

Fatlume Zhujani, Nexhat Qehaja, Fitore Abdullahu* & Mirlind Bruçi

This Publication has to be referred as: Zhujani, F[atlume]; Qehaja, N[exhat]; Abdullahu, F[itore] & Bruqi, M[irlind] (2016). Mathematical Modelling of Surface Roughness for Evaluating the Effects of Cutting Parameters in Drilling Process, Proceedings of the 27th DAAAM International Symposium, pp.0195-0202, B. Katalinic (Ed.), Published by DAAAM International, ISBN 978-3-902734-08-2, ISSN 1726-9679, Vienna, Austria
DOI: 10.2507/27th.daaam.proceedings.029

Abstract

The optimization of surface integrity in drilling process using response surface method (RSM) is presented. This paper investigates the machining parameters affecting the roughness of surfaces produced in dry drilling process. Three parameters were selected for study: cutting speed, feed rate and drill diameter. In this study developed a model of surface roughness based on the response surface method, logarithmic linearized approach for determining the processing parameters in drilling process of steel EN 1.0038, using HSS coated TiN drill tools. The experiment has been designed and carried out on the basis of a three level factorial design. Obtained results are in good accordance with the experimentally obtained data, confirming the effectiveness of regression analysis in modeling of surface roughness in the dry drilling process. The established predictive model shows that the surface roughness increases with the increase of feed rate and depth of cut but decreases with cutting speed increase.

Keywords: Machining; cutting process; drilling; roughness; tool

1. Introduction

Drilling is one of the most important machining processes. Approximately 75% of all metal cutting process involves drilling operation [1]. In automotive engine production, costing of drilling hole is among the highest [2].

Therefore, surface integrity is an important parameter in manufacturing engineering. It is because, surface integrity can influence the performance of final parts and it's quality [3]. In order to study the relationship between drilling process parameters and surface integrity, a systematically approach, the design of experiments (DOE); response surface method (RSM) can be used effectively [4].

In the manufacturing industries, various machining processes (turning, milling, drilling etc.) are adopted for removing the material from the workpiece to obtain finished product. Among the various metal removing processes, drilling is the one of most important metal removing process as compared to other traditional machining processes. Drilling is use for making the hole in the workpieces. Hole making is a most important machining process in manufacturing. During the drilling, the drill rotates and feed into the work Different drilling tools and hole making methods are used for drilling. The selection of different tools and methods depends on the type of workpiece, size of the hole, the quantity of holes, and the quantity of the holes in given time periods.[5].

The surface quality is one of the most specified customer requirements and the major indicator of surface quality on machined parts is surface roughness. The surface roughness is mainly a result of various controllable or uncontrollable process parameters and it is harder to attain and track than physical dimensions are. A considerable number of studies have researched the effects of the cutting speed, feed, depth of cut, nose radius and other factors on the surface roughness. In recent studies the effects of some factors on surface roughness has been evaluated and models has been developed.

A central task in science and engineering practice is to develop models that give a satisfactory description of physical systems being observed. The goal of this study is to obtain a mathematical model that relates the surface roughness to three cutting parameters in face milling, precisely to the cutting speed, feed rate and depth of cut. [17],[12].

Although metal cutting methods have improved in the manufacturing industry, conventional drilling process still remains one of the most common processes. Drilling can be applied to various workpiece materials. The surface quality is important for the functional behavior of the mechanical parts [6],[. The most obvious factors influencing the accuracy of drilled holes are cutting speed and feed rate [9]. Cutting speed and feed rate significantly affect the surface roughness of the machined surface whereby high cutting speed and low feed rate resulted in the better surface finish [3]. In material removal processes, an improper selection of cutting conditions will result in rough surfaces and dimensional errors. Therefore, it is necessary to understand the relationship among the various controllable parameters and to identify the important parameters that influence the quality of drilling.[10]

Metal cutting is one of the most significant manufacturing processes in the area of material removal. Black [14] defined metal cutting as the removal of metal chips from a workpiece in order to obtain a finished product with desired attributes of size, shape, and surface roughness [15]. One important parameter in the qualification of cut surfaces is their roughness, and its indexes. The roughness has great significance primarily at mating, sliding surfaces. This has been one more reason for the researchers' increased interest for a long time to predict these indexes for a given process within the specified cutting conditions. Several modeling procedures and techniques were worked-out, which essentially can be classified into four groups: l) analytical models, 2) experimental methods, 3) DoE (Design of Experiment)-based methods and 4) AI (Artificial Intelligence)-based methods.

Experiment aim is to define adequate mathematical model that is used to determine the influence of independent factors and cutting data, at surface roughness. The following independent factors are selected: cutting speed (v), drill diameter (d), and feed rate (f). For a concrete case, that means determination of criteria:

- surface roughness: $R_a = f(v,f,d)$

2. Nomenclature

R_a the average surface roughness
P power
n RPM
f feed rate
d tool drill diameter
L lenth
B width
H hight
N factorial design
K number of factors
N_0 number of additional tests
v cutting speed
c_0, c_1, c_2,c_3 constants
y logarithmic value of the measured surface roughness
β_0, β_1, β_2, β_3 regression coefficients
x_0 unit vector
x_1, x_2, x_3 logarithmic values of cutting speed, feed rate, cut of depth

3. Experimental conditions

Machine tool: Universal milling machine *GKA-3,* 8 kw spindle power, spindle speed range: 250 - 1400 rpm, feed rate range: 0,02 - 2,0 mm/rev, working are: X/Y/Z 400x2000x300 mm, max. tool diameter: 80 mm, max. tool length: 200 mm. Workpiece was made of forgings steel EN 1.0038 (EN 10250-2), vith dimensions; BxLxH=100x100x10 mm. Its chemical composition is as follows: (0.16-0.178)% C; (1.35-1.45)% Mn; 0.032% S; 0.034% P, (0.52-0.56) Cu, (0,34-0,36)CE. Tensile strength is (380-510) MPa, and hardness 240-250 N/mm2.. Cutting tool: HSS-TiN coated jobber drills DIN 338, A1211TiN, with data in table 1. To guarantee the initial conditions of each test, a new drill tool is used in each experiment.

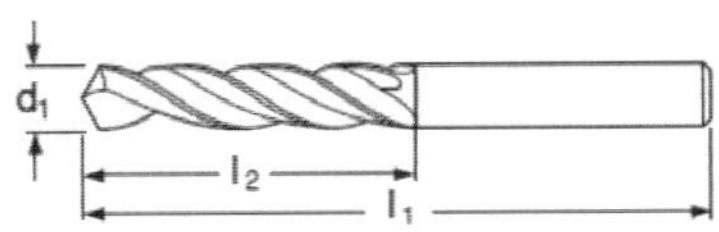

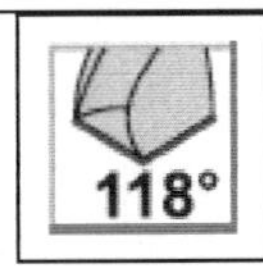

DIN 338	d_1 [mm] h8	l_1[mm]	l_2 [mm]
1	8	117	75
2	10	133	87
3	12	151	101

Table 1. Cutting tool data

The measurement of surface roughness was made by; HADRON, SRT-6210. The measurement of chemical composition was made by; Spectrometer Metorex Arcmet 930 and hardness by; meter Krautkramermic.10.DL.

4. Experimental setup

Experimental design: This work is an experimental study focused on the effect of cutting and geometrical parameters on surface roughness, developing a correlation between them. The experimental design involves variation of three factors at three levels (low, medium and high), including cutting speed (v), feed rate (f) and drill tool diameter (d) as indicated in Table 2.

A design matrix was constructed on the basis of the selected factors and factor levels as shown on the table 2. The selected design matrix was a full factorial design N=2k+N0 (k=3 - number of factors, N0=4 – number of additional tests for three factors) consisting of 12 rows of coded/natural factors, corresponding to the number of trials. This design provides a uniform distribution of experimental points within the selected experimental hyper-space and the experiment with high resolution [7[, [8].

The factor ranges were chosen with different criteria for each factor, aiming at the widest possible range of values, in order to have a better utilization of the proposed models. At the same time, the possibility of the mechanical system and manufacturer's recommendations are taken into account.

Machining conditions used in the experiment also are shown in Table 2 [9], [10], [11]. All of the trials have been conducted on the same machine tool, with the same tool type and the same cutting conditions.

Measured values of surface roughness, as the results of testing twelve experimental points defined by experiment plan matrix, are shown in Table 3. The mentioned values of surface roughness are input data for mathematical modeling of results, which was made by multiple regression analysis.

Input factors	Coded values of input factors		
	Low level	*Midlle level*	*High level*
	-1	0	+1
$x_1 = v$ [m/min]	15	20	25
$x_2 = f$[mm/rew]	0.1	0.175	0.3
$x_3 =$ d [mm]	8	10	12

Table 2. Experimental setup at three level factors.

a. *Regression based modeling*

The main task for regression analysis is to show relationship between the roughness and machining independent variables. Many authors suggested linear and exponential empirical models for surface roughness as functions of machining parameters [13], [14], [15], [16], by the following:

$$R_a = c_0 \cdot v^{c_2} \cdot f^{c_1} \cdot d^{c_3} \tag{1}$$

Three parameters: cutting speed (v),feed rate (f), and drill tool diameter (d). Ra is the surface roughness in µm, f - feed rate in mm/rev, d-drill tool diameter, and respectively c_0, c_1, c_2, and c_3 are constants.

Test No.	Coded factors				Performance measures	
	X_0	X_1	X_2	X_3	Ra	Y = ln R a
1	+1	+1	-1	+1	2.425	0.885831524
2	+1	-1	+1	+1	4.232	1.442674695
3	+1	+1	+1	-1	3.241	1.175881924

4	+1	-1	-1	-1	3.564	1.27088351
5	+1	0	0	0	3.582	1.275921304
6	+1	0	0	0	3.322	1.200567011
7	+1	-1	-1	+1	5.826	1.762330659
8	+1	+1	-1	-1	2.542	0.932951173
9	+1	+1	+1	+1	7.321	1.990746931
10	+1	-1	+1	-1	8.824	2.177475282
11	+1	0	0	0	3.501	1.253048642
12	+1	0	0	0	3.614	1.284815192

Table 3. Experimental results.

Multiple linear regression models for surface roughness can be obtained by applying a logarithmic transformation that converts non-linear form of eq. (1) into following linear mathematical form:

$$ln\ R_a = ln\ c_0 + c_1\ ln\ f + c_2\ ln\ f + c_3\ ln\ d \quad (2)$$

The linear model of eq. (3) in term of the estimated response can be written as:

$$y = \beta_0 + \beta_1 x_1 + \beta_2 x_2 + \beta_3 x + \varepsilon \quad (3)$$

where y is the logarithmic value of the measured surface roughness, β_0, β_1, β_2, β_3 are regression coefficients to be estimated, x_0 is the unit vector, x_1, x_2, x_3 are the logarithmic values of cutting speed, feed rate, cut of depth and ε is the random error.

The above equation in matrix form becomes:

$$y = X\beta + \varepsilon \quad (4)$$

Thus, the least squares estimator of β is

$$\beta = (X'X)^{-1}X'y \quad (5)$$

The fitted regression model is

$$\acute{Y} = X\beta \quad (6)$$

The difference between the experimentally measured and the fitted values of response is:

$$e = y - \dot{y} \quad (7)$$

The regression analysis technique using least squares estimation was applied to compute the coefficients of exponential model. The following empirical exponential model for surface roughness was determined and is given, respectively: Final mathematical model of surface roughness R_a is obtained:

$$R_a = 46.566 \cdot v^{-0.816} \cdot f^{0.44} \cdot d^{0.323} \quad (8)$$

5. Results and discussion

Table 3 presents experimental results of surface roughness criteria R_a for various combinations of cutting speed, feed rate and drill tool diameter to full factorial design. Minimal value of surface roughness criteria R_a was obtained at V=25 m/min, f = 0,1 mm/rev, d=12 mm, (test No. 1). That means increasing of cutting speed and drill tool diameter with the lowest feed rate lead to decreasing of surface roughness.

It is found that cutting speed has the most significant effect on surface roughness, followed by feed rate and drill tool diameter.

Maximal value of surface roughness criteria R_a was registered at V = 15 m/min, f = 0.3 mm/rev, d=8 mm, (test No. 10). In order to achieve better surface finish, the highest level of cutting speed, and the lowest level of feed rate and drill tool diameter, should be recommended.

Fig. 1 which highlights the main factor plots for Ra appears to be an almost linear decreasing function of cutting speed and an increasing function of feed rate (f) and drill tool diameter (d). Figs. 2 a, b, c and d illustrate 3D surface plots of Ra according to the predictive exponential empirical model (9).

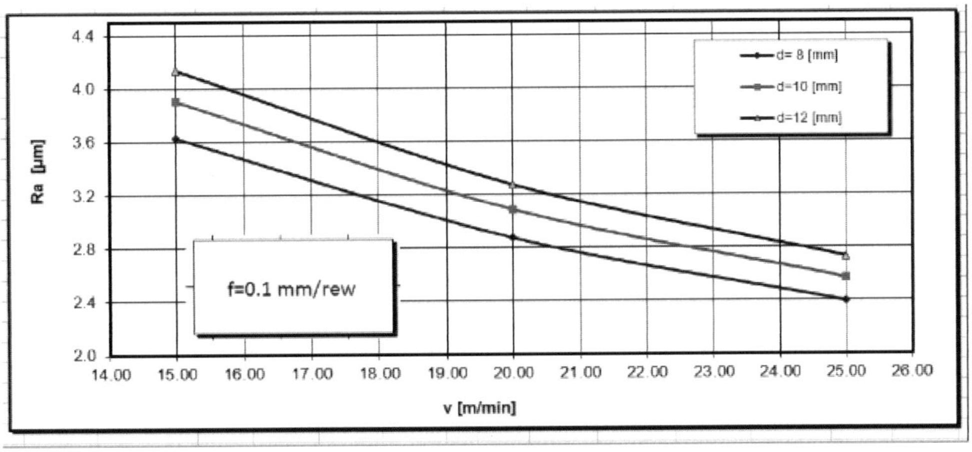

a)

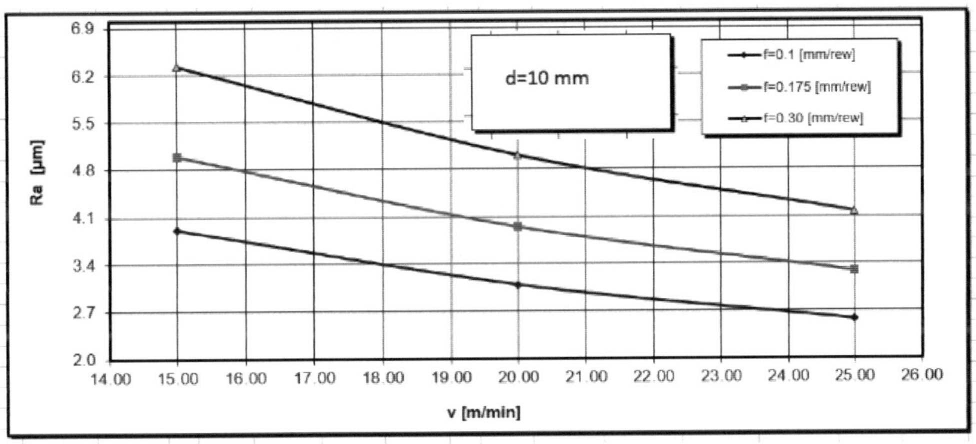

b)

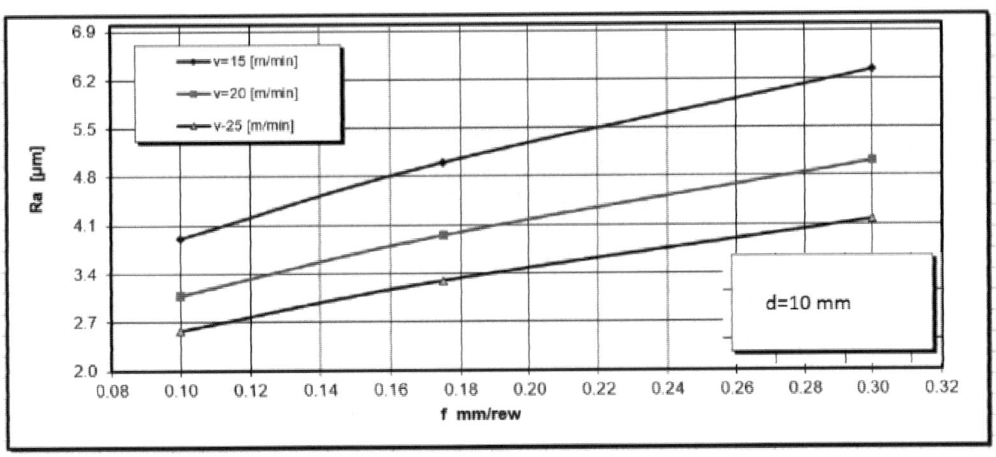

c)

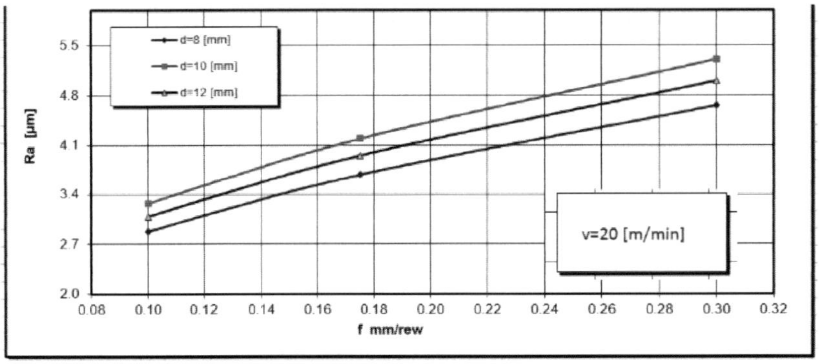

d)

Fig. 1. The dependence of surface roughness on: a) cutting speed and various values of tool diameter, b) cutting speed and various values of feed rate, c) feed rate and various values of cutting speed, d) feed rate and various values of tool diameter

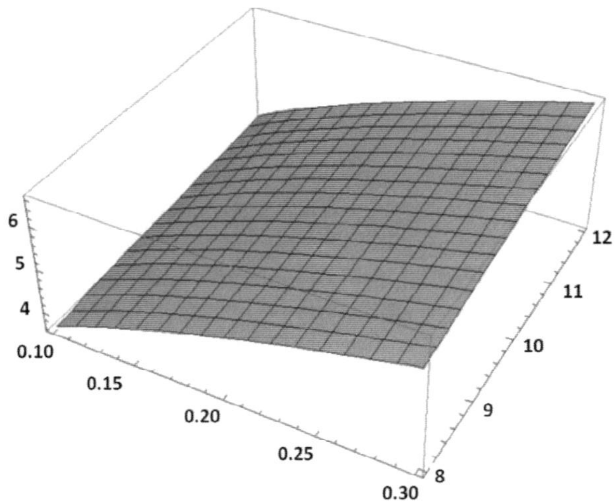

a)

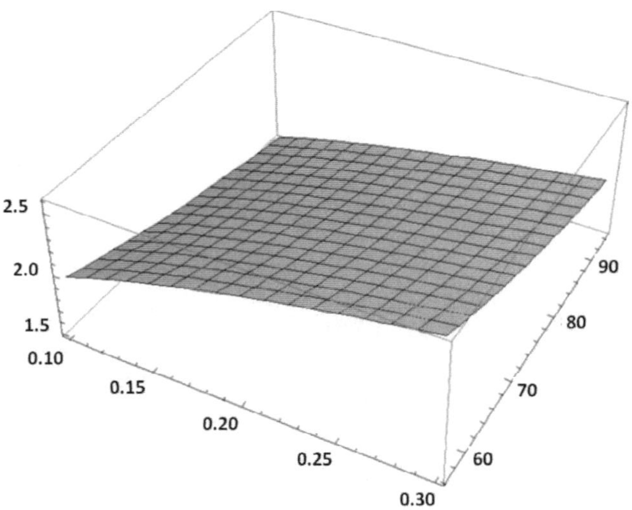

b)

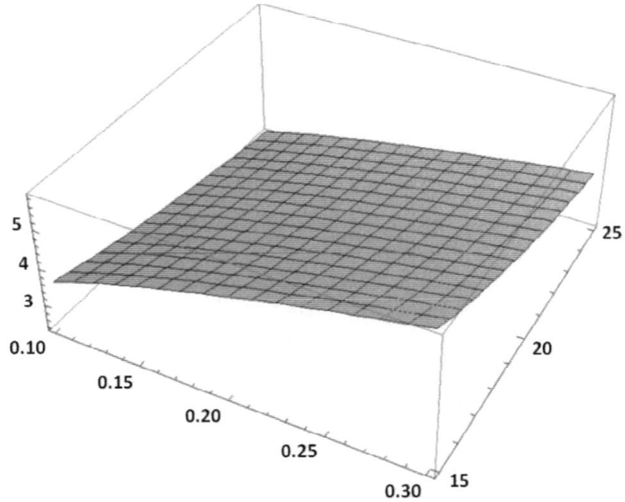

c)

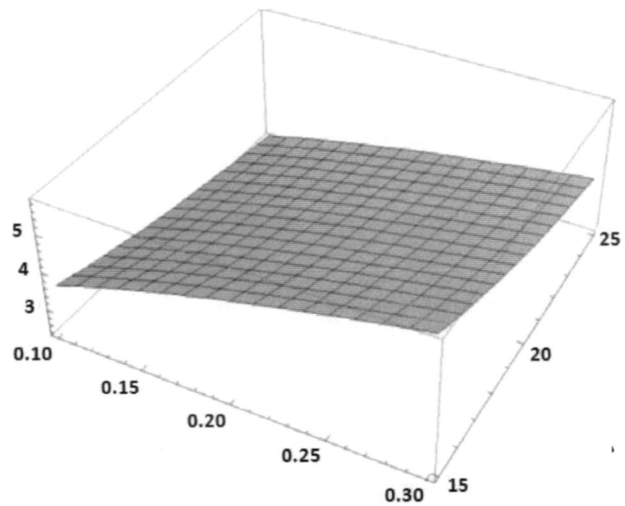

d)

Fig.2. 3D surface plots: **3PlotD[46.567 * (12^0.32) * (f^0.44) * (v^ − 0.816),{f,0.1,0.3},{v,15,25}]**

6. Conclusions

This paper presents research of various cutting parameters affecting the surface roughness in dry drilling of forgings steel EN 1.0038 using HSS coated TiN drill tools.
The investigations of this study indicate that the cutting parameters like cutting speed and feed rate are the primary influencing factors, which affect surface roughness.
Statistical models deduction defined the degree of influence of each cutting parameter element on surface roughness criteria.
The results revealed that cutting speed seems to influence surface roughness (0.816) more significantly than feed rate (0.44). However, drill tool diameter is less significant (0.323).
With the regression equation generated, the best combination of design independent variables for achieving the optimization of cutting processes.
The predicted developed mathematical model can be used by the manufacturer while selecting the machining parameters.
The predicted results are in good agreement with the measured ones. These relationships are applicable within the range of tested parameters.

7. References

[1] M. A. Amrana,*, S. Salmah, N.I.S. Hussein, R. Izamshah, M. Hadzley, Sivaraos, M.S. Kasimb, M.A. Sulaiman, Effects of machine parameters on surface roughness using response surface method in drilling process, The Malaysian International Tribology Conference 2013, MITC2013, Procedia Engineering 68 (2013) 24 – 29, Available online at www.sciencedirect.com

[2] Kalpakjian, S., Chmid, S., 2006. Manufacturing Engineering and Technology, Singapore, Pearson

[3] S. Garg, R. k. Goyal, A Study of Surface Roughness in Drilling of AISI H11 Die Steel using Face Centered Design , IJIRST –International Journal for Innovative Research in Science & Technology| Volume 1 | Issue 12 | May 2015 ISSN (online): 2349-6010 All rights reserved by www.ijirst.org 464

[4] S. Thamizhmanii, S. Saparudin and S. Hasan; Analyses of Surface Roughness by Turning Process Using Taguchi Method, Journal of Achievements in Materials and Manufacturing Engineering, Volume 20 Issues 1-2 January-February 2007

[5] J. Jurko, Study of Changes in a Machined Stainless Steels Surface during Drilling, Annals of DAAAM for 2010 & Proceedings of the 21st International DAAAM Symposium, 20-23rd October 2010, Zadar, Croatia, ISSN 1726-9679, ISBN 978-3-901509-73-5, Katalinic, B. (Ed.), pp. 0051-0052, Published by DAAAM International Vienna, Vienna

[6] A. Z. Sultana, Safian Sharifb,*, Denni Kurniawanb, Effect of Machining Parameters on Tool Wear and Hole Quality of AISI 316L Stainless Steel in Conventional Drilling, Procedia Manufacturing 2 (2015) 202 – 207 Available online www.sciencedirect.com, 2nd International Materials, Industrial, and Manufacturing Engineering Conference, MIMEC2015,4-6 February 2015, Bali Indonesia.

[7] N. Qehaja, H. Zeqiri, A. Salihu, H. Osmani, F. Zeqiri, Machinability of Matals, Methods, and Practical Application, Annals of DAAAM & Proceedings 2012, The 23RD World Symposium, 24-27th October 2012, University of Zadar, Croatia, URL: http:\\www.daaam.com, Vienna University of Technology, Karlsplatz 13/311, A-1040 Wien, Austria, Europe

[8] A. Salihu, H. Zeqiri, A. Bunjaku, N. Qehaja, F. Zeqiri, Research of Surface Roughness Average of Steel C45 During Turning, Anals of DAAAM for 2010 & Procedings of the 21st International DAAAM symposium, Volume 21, PP. 0185÷0186, ISBN 978-3-901509-73-5. No 1. ISSN 1726-9679. Vienna University of Technology, Karlsplatz 13/311, A-1040 Wien, Austria, Europe

[9] L. Huang and C. Chen, A Multiple Regression Model to Predict In-process Surface Roughness in Turning, Journal of Industrial Technology • Volume 17, Nr. 2 • February 2001 to April 2001.

[10] N. Qehaja, A. Bunjaku, J. Kacani. A.Salihu, H. Zeqiri, H. Osmani, Mathematical models of flank wear curve during the turning of steel C60E4, The 20thInternational DAAAM SYMPOSIUM "Intelligent Manufacturing & Automation: Focus on Next Generation of Intelligent Systems and Solutions" 22-25th October Vienna, Austria,2009

[11] N. Qehaja, M. Bruçi, H. Zeqiri, A. Salihu and A Gjelaj; Tool Life Prediction of Tungsten Carbide Cutting Tools in Turning Process, International Conference on Innovative Technologies In-Tech2014, p.271-275, Budapest, Hungary

[12] H. Zeqiri, N. Qehaja, F. Zeqiri, A. Salihu, H. Osmani, Research of Flank Wear in Turning of Steel Ck 60, Annals of DAAAM & Proceedings 2012, The 23rd DAAAM World Symposium, 2012

[13] A. Salihu, H. Zeqiri, A. Bunjaku, N. Qehaja, H. Osmani, A. Kyçyky, Research of the horizontal parameters of the roughness machined surface by turning, The 20th International DAAAM Symposium "Intelligent Manufacturing & Automation: Focus on Next Generation of Intelligent Systems and Solution", 2009

[14] M. Durairaja, S. Gowri, Parametric Optimization for Improved Tool Life and Surface Finish in Micro Turning using Genetic Algorithm, International Conference On DESIGN AND MANUFACTURING, IConDM 2013, Procedia Engineering 64 (2013) 878 – 887

[15] C.G. Onwubolo, S. Kumar, Response Surface Methodology Based Approach To CNC Drilling Operation, Journal of Materials rocessing Technology 171, pp. 41-47. 2006.

[16] N. Qehaja, K. Jakupi, A. Bunjaku, M. Bruçi, H. Osmani, Effect of machining parameters and machining time on surface roughness in dry turning process. Procedia Eng 100:135–140, 2015

[17] S. Tangjitsitcharoen, A. Senjuntichai, In- Process monitoring and prediction of surface roughness in ball-end milling process, Annals of DAAAM for 2010 & Proceedings of the 21st International DAAAM Symposium, Volume 21, No. 1, ISSN 1726-9679 ISBN 978-3-901509-73-5, Editor B. Katalinic, Published by DAAAM International, Vienna, Austria, EU, 2010

DOI: 10.2507/27th.daaam.proceedings.030

CALCULATION COOLING COMPRESSORS

Halima Hadziahmetovic

This Publication has to be referred as: Hadziahmetovic, H[alima] (2016). Calculation Cooling Compressors, Proceedings of the 27th DAAAM International Symposium, pp.0203-0212, B. Katalinic (Ed.), Published by DAAAM International, ISBN 978-3-902734-08-2, ISSN 1726-9679, Vienna, Austria
DOI: 10.2507/27th.daaam.proceedings.030

Abstract

In many systems of compressed air, besides the proper selection of the compressor, there are significant opportunities to save energy, including energy recovery, pressure reduction, reducing leakage losses and optimization of systems and control systems. When planning investments in the construction of the compressed air system, it is necessary to examine and evaluate them. In addition, importance must be given to possible future requirements that the system must meet, as well as all other possible influences on the compressed air system. Typical examples of the mentioned influences are the environmental conditions requirements, requirements for energy savings, and increased quality requirements in production and possible future investment in the growth of production. In this paper is explained calculation cooling compressors. Cooling of compressors is going to be done in two cycles. In first cycle technological water circulates by pumps of first cycle. In second cycle "demi" water will circulate and cooled compressors and coolers of compressed air and also will be drived by pumps.

Keywords: pneumatic conveying; cooling; compressor; compressed air.

1. Introduction

Recently, much attention is paid to preserving the environment, and build systems that prevent pollution of the environment. Thermal power plants are one of the biggest environmental polluters. In the process of electricity production as coal combustion products occur large amounts of waste material. Transportation and disposal of fly ash and bottom ash are among the vital technological systems of each power plant that burns coal. The system of transport of fly ash and bottom ash in the thermal power plant can be: transportation by belt conveyors, transportation with dumper-trucks, hydraulic transport and pneumatic conveying [5,9].

The pneumatic transport involves the transportation powder, granular and piece of material and is based on the phenomenon that at the appropriate velocity of air in the pipeline, the solid particles are brought in the desired direction. Selection and efficiency of pneumatic transport depends of physical and chemical characteristic of the fly ash [2,11].

Compressors are mechanical devices used for raising the pressure of gas or vapour either by lowering its volume (as in the case of positive displacement machines) or by imparting to it a high kinetic energy which is converted into pressure in a diffuser (as in the case of centrifugal machines). The selection of compressors for different applications is a crucial issue in the process industry. It is usually the most expensive piece of equipment and has dominant influence on cycle efficiency. The common types of compressors used in industry are reciprocating, twin screw, single screw, centrifugal, scroll and rotary vane. In this paper is explained calculation cooling compressors. For properly system sizing ability, it is important to know the system function condition, pipe network, equipment position, and conveying medium characteristics (in this case compressed air) [3,6,10].

In this paper is presented a system of preparation and distribution of compressed instrument air to the consumer in the system of internal pneumatic transport of fly ash under the electrofilter, flues and steam air heaters and calculation cooling compressors. System of compressed air for electrofilter is an integral part of a whole, the pneumatic transport of fly ash for the silo. The quantity of fly ash shown in table 1 and list of design input data are shown in tables 2 and 3.

Input data	**Value**
Fly ash:	165 – 232,2 t/h
Ash from boiler hoppers and air preheater hopper: (this ash is included in the sum for fly ash)	14,0 – 17,8 t/h

Table 1. The quantity of ash produced by one unit

Input data	**Value**
Bulk density of fly ash	0,74 t/m³
Specific mass density	2,22 g/cm³

Table 2. Design data

Input data	**Value**
Long distance fly ash pneumatic conveying	260 t/h per unit (including15% reserve)

Table 3. Capacity data of the system

2. Technical description

2.1. Compressed air distribution

Compressed air is distributed by seamless steel galvanized pipelines. Distribution pipeline is oriented towards consumers, apropos to all valves on pneumatic drive which are located under electrofilter (EPS) of unit 1 and unit 2, and to the consumers in pump stations 1 and 2. In function of assigned pressure, complete pipeline is dimensioned, apropos diameters of certain sections are specified.

2.2. Compressed air consumers

Compressed air is used for manipulation and control in pneumatic transport process as well as for manipulation of pneumatic valves which are within automatic control of pneumatic transport. Locations which are supplied with compressed air: pressure vessels for pneumatic fly ash transport, air transport pipeline, armatures of ash receiver (plate gate valves), ash transport pipelines and pneumatic valves in bager station. Consumers, apropos places which are needed to be supplied with compressed air, altogether there are 226 pieces. Function of these valves is in program control of fly ash transport process which is managed by PLC. Safe operation of installation, apropos maintaining pressure in installation within limitations of minimal and maximal permissible pressure is supplied with regulation system which is located in the compressor. Compressor is in mutual automatic operation with reservoir of instrumental air where pressure should be between 6,3 – 7,5 bars. When pressure in reservoir drops under 7,5 bars, and reaches value of minimum permissible operation pressure p_{min} = 6,3 bars, automatically one compressor activates until maximal wanted pressure is not reached [4].

3. Selection of compressor for compressed air under electrofilter

- Air consumption by consumer for transmition in block 1 : Q_1= 53 m³/h.
- Air consumption by consumer for transmition in block 2: Q_2= 53 m³/h.
- Air consumption by consumer for transmition in pump stations 1 and 2 like consumers on air preheaters and under channel of flue gases: Q_3= 15 m³/h.
- Total air consumption increased for 20% : $Q= 1,2 \cdot (Q_1 + Q_2 + Q_3) = 1,2 \cdot 121 = 145,2$ m³/h.

On the basis of calculated consumption of 145,2 m^3/h compressor of firm Atlas Copco is selected:
Type: GA15$^+$, Capacity: 160 m^3/h , Operation pressure: 7,5 bara , Power of electromotor: 15 kW, Noise level: 64 dB, Dimensions 1400 x 1225 x 650 [1,4].

Fig 1. Compressor

On the base of selected compressor and according to Atlas Copco recommendation absorption dryers type CD 44 (2 piece) are chosen (Fig. 1). And on the base of experience in designing system of compressed air, instrumental air reservoir is selected, volume V = 2 m^3.

4. Calculation cooling compressors

Compressors (Fig. 2) of compressed air are water cooled, while compressors of control air are air cooled. Water for cooling of compressed air and compressors is used in secondary circuit which is closed. A water tank will be used to supply additional water should any losses of water occur during circulation. Water from primary circuit is supplied from unit 1 and flows to bager station. Circulation pump, heat exchanger and pipeline were selected in accordance with calculation.

Fig 2. Compressors

4.1. Calculation of quantity of water for cooling of compressor and compressed air of secondary circuit

Calculation of quantity of water for cooling of compressor and compressed air of secondary circuit be calculated according to the following:

$$Q_{uk} = 5Q_{k1} + 2Q_{k2} \tag{1}$$

where $Q_{K1} = 11{,}5 l/s = 41{,}4 m^3/h$ is quantity of water for cooling of compressor and compressed air (Q=5600m^3/h) and $Q_{K2} = 4{,}66 l/s = 16{,}7 m^3/h$ is quantity of water for cooling of compressor and compressed air (Q=940m^3/h).

$$Q_{uk} = 5 \cdot 41{,}4 + 2 \cdot 16{,}7 = 240 m^3/h$$

Reserve of water for one compressor is applied in above (Q=5600 m^3/h).

4.1.1.Determining dimensions of secondary circuit pipeline of cooling water

Determining dimensions of secondary circuit pipeline of cooling water is given by:

$$d = \sqrt{\frac{4 \cdot Q_{uk}}{\pi \cdot v}}, \tag{2}$$

where v= 2 m/s is assumed water velocity in cooling pipeline.

$$d = \sqrt{\frac{4 \cdot 240}{\pi \cdot 2 \cdot 3600}} = 206 mm$$

Nominal diameter DN 200 (Ø219,1 X 5,9) is selected.

4.1.2. Calculation of actual water velocity in pipeline

Actual water velocity in pipeline be calculated according to the following:

$$v = \frac{4 \cdot Q_{uk}}{\pi \cdot d_p^2} = 1{,}97 m/s, \tag{3}$$

where is d_p= 219,1 – 2 x 5,9 = 207,3 mm = 0,2073 m.

Water velocity in cooling pipeline:

$$v = \frac{4 \cdot 240}{\pi \cdot 0{,}2073^2 \, 3600} = 1{,}976 m/s.$$

4.1.3. Pipeline features and local losses

Length of pipeline is L= 60m. List of pipeline features and local losses is shown in table 4.

Pipeline features	**Local losses**
10 elbows 90° R= 3D	10 * 0,5 = 5
10 elbows 45° R= 3D	10 * 0,5 = 5
20 valves	20 * 0,6 = 12
6 T- pieces	6 * 0,6 = 3,6
	Σ ξ = 25,6

Table 4. Pipeline features and local losses

4.1.4. Calculation of pressure drop in secondary cooling circuit

Pressure drop in secondary cooling circuit can be determined by:

$$\Delta p_u = \Delta p + \Delta p_k + \Delta p_i , \tag{4}$$

where $\Delta p = \Delta p_1 + \Delta p_2$ is pressure drop in pipeline (Δp_1 - pressure drop in straight pipe and Δp_2 - pressure drop due to local losses), $\Delta p_k = 0{,}55\ bar$ is pressure drop through compressors and $\Delta p_i = 0{,}45\ bar$ is pressure drop through heat exchanger.

Pressure drop in straight pipe should be calculated according to the following:

$$\Delta p_1 = \lambda \frac{L}{d_p} \cdot \frac{v^2}{2g} , \tag{5}$$

where λ=0,0144+0,00947/$\sqrt{v}$ =0,0211 is coefficient of air friction using Weissbach method.

$$\Delta p_1 = 0{,}12 bar.$$

Pressure drop due to local losses should be calculated according to the following:

$$\Delta p_2 = \frac{v^2}{2g} \cdot \xi , \tag{6}$$

where $\xi = 25{,}6$ is local losses and pressure drop in pipeline is given by:

$$\Delta p_2 = 0{,}528\, bar$$

and pressure drop in pipeline

$$\Delta p = 0{,}12 + 0{,}528 = 0{,}648\ bar.$$

Pressure drop in secondary cooling circuit can be determined by (4):

$$\Delta p_u = 0{,}648 + 0{,}55 + 0{,}45 = 1{,}65\ bar.$$

4.1.5. Selection of circulation pump in secondary circuit

Pump power is given by:

$$N = \frac{Q_{uk} \cdot H_p \cdot \rho \cdot g}{367 \cdot \eta} , \tag{7}$$

where $\rho = 1000\ kg/m^3$ is water density in secondary circulit , $\eta = 0{,}7$ is coefficient of device usage, Q_{uk} - total water flow in secondary circuit and H_p - pump head.

Total water flow in secondary circuit can be determined by (1):

$$Q_{uk} = 5 \cdot 41{,}4 + 2 \cdot 16{,}7 = 240 m^3 / h ,$$

where $Q_{K1} = 11{,}5 l/s = 41{,}4 m^3/h$ is quantity of water for cooling of compressor and compressed air (Q=5600m^3/h) and $Q_{K2} = 4{,}66 l/s = 16{,}7 m^3/h$ is quantity of water for cooling of compressor and compressed air (Q=940m^3/h).

Pump head should be calculated according to the following:

$$H_p = \Delta p_i + \Delta p_k + \Delta p, \tag{8}$$

where $\Delta p_i = 0{,}45 bar$ is pressure drop in heat excanger, $\Delta p_k = 0{,}55 bar$ is pressure drop in compressor and $\Delta p = 0{,}65 bar$ pressure drop local losses.

$$H_p = 0{,}45 + 0{,}55 + 0{,}65 = 1{,}65 bar$$

Pump power can be determined by (7):

$$N = \frac{240 \cdot 1{,}65 \cdot 1 \cdot 9{,}81}{367 \cdot 0{,}7} = 15{,}1 kW.$$

$$N_m = \frac{N}{\mu}, \tag{9}$$

where μ= 0,9 is coefficient of device safety,

$$N_m = 16{,}8 kW.$$

Standard pump motor with N= 18,5 kW will be used.

The characteristics of the pump:

Firm: WILO, Type: IL150/260-18,5/4, Flow range: 216 m³/h, Motor power: 18,5 kW and Weight 309 kg [7].

4.2. Selection of circulation pump in primary circuit

Total water flow in primary circuit can be determined by:

$$\dot{m}_{tv} = \frac{i_2^{ul} - i_2^{iz}}{i_1^{iz} - i_1^{ul}} \cdot \dot{m}_{dv}, \tag{10}$$

where are:

$\dot{m}_{dv} = 60{,}02 l/s = 216 m^3/h$ - total water flow in secondary circuit,
$i_2^{ul} = 209{,}34\ kJ/kg \quad (t = 50°C)$ - enthalpy secundary circulit water at outlet heat excanger,
$i_2^{iz} = 167{,}53\ kJ/kg \quad (t = 40°C)$ - enthalpy secondary circulit water at outlet heat excanger,
$i_1^{ul} = 117{,}37\ kJ/kg \quad (t = 28°C, p = 3{,}5 bar)$ - enthalpy primary circulit water at inlet heat excanger,
$i_1^{iz} = 142{,}25\ kJ/kg \quad (t = 35°C, p = 3{,}5 bar)$ - enthalpy primary circulit water at outlet heat excanger.

$$\dot{m}_{tv} = \frac{209{,}34 - 167{,}53}{142{,}25 - 117{,}37} \cdot 60{,}02 = 100 l/s = 360 m^3/h$$

4.2.1. Determining dimensions of primary circuit pipeline of cooling water

Determining dimensions of primary circuit pipeline of cooling water is given by:

$$d = \sqrt{\frac{4 \cdot \dot{m}_{tv}}{\pi \cdot v}}, \tag{11}$$

where v= 2 m/s is assumed water velocity in cooling pipeline.

$$d = \sqrt{\frac{4 \cdot 360}{\pi \cdot 2 \cdot 3600}} = 252{,}3 mm$$

Nominal diameter DN 250 (Ø273 X 6,3) is selected.

4.2.2. Calculation of actual water velocity in pipeline

Actual water velocity in pipeline be calculated according to the following:

$$v = \frac{4 \cdot \dot{m}_{tv}}{\pi \cdot d_p^2}, \tag{12}$$

where is d_p= 273 – 2 x 6,3 = 260,4 mm = 0,2604 m.

$$v = \frac{4 \cdot 360}{\pi \cdot 0,2604^2 \cdot 3600} = 1,88 m/s.$$

4.2.3. Pipeline features and local losses

Length of pipeline is L= 700m. List of pipeline features and local losses is shown in table 5.

Pipeline features	Local losses
15 elbows 90° R= 3D	15 * 0,5 = 7,5
2 elbows 45° R= 3D	2 * 0,5 = 1
2 elbows 30° R= 3D	2 * 0,5 = 1
6 valves	6 * 0,6 = 3,6
6 T- pieces	6 * 0,6 = 3,6
	Σ ξ = 16,7

Table 5. Pipeline features and local losses

4.2.4. Calculation of pressure drop in primary cooling circuit

Pressure drop in straight pipe should be calculated according to the following (5):

$$\Delta p_1 = 1,02 bar$$

where λ=0,0144+0,00947/$\sqrt{v}$ =0,0211 is coefficient of air friction using Weissbach method.

Pressure drop due to local losses should be calculated according to the following (6):

$$\Delta p_2 = 0,3 bar,$$

where ξ = 16,7 is local losses and pressure drop in pipeline should be calculated according to the following:

$$\Delta p = 1,02 + 0,3 = 1,32\ bar.$$

Pressure drop in secondary cooling circuit can be determined by (4):

$$\Delta p_u = 1,32 + 0,55 + 0,45 = 2,32\ bar,$$

where $\Delta p_k = 0,55\ bar$ is pressure drop through compressors and $\Delta p_i = 0,45\ bar$ is pressure drop through heat exchanger.

4.2.5. Selection of circulation pump in primary circuit

Pump power is given by:

$$N = \frac{\dot{m}_{tv} \cdot H_p \cdot \rho \cdot g}{367 \cdot \eta}, \tag{13}$$

where $\rho = 1000 kg/m^3$ is water density in secondary circuit, $\eta = 0,7$ is coefficient of device usage, H_p - pump head and $\dot{m}_{tv}$ - total water flow in primary circuit ($99 l/s = 360 m^3/h$).

Pump head should be calculated according to the following (8):

$$H_p = 0,45 + 0,55 + 1,32 = 2,32 bar,$$

where $\Delta p_i = 0,45 bar$ is pressure drop in heat exchanger, $\Delta p_k = 0,55 bar$ is pressure drop in compressor and $\Delta p = 1,32 bar$ pressure drop local losses.

Pump power can be determined by (13):

$$N = \frac{360 \cdot 2{,}32 \cdot 1 \cdot 9{,}81}{367 \cdot 0{,}7} = 31kW,$$

and

$$N_m = \frac{31}{0{,}9} = 35kW,$$

where $\mu = 0{,}9$ is coefficient of device safety.

Standard pump motor with $N = 45$ kW will be used.

The characteristics of the pump: Firm: WILO, Type: IL200/320-45/4, Flow range: 360 m³/h, Motor power: 45 kW and Weight: 512 kg [7].

4.3. Calculation of area heat exchange

4.3.1. Characteristics of water in secondary circuit

$p_1 = 100kPa$ - water pressure in secondary circuit,
$t_2^{ul} = 50°C$ - inlet water temperature in secondary circuit exchanger,
$t_2^{iz} = 40°C$ - outlet water temperature in secondary circuit exchanger and
$\dot{m}_{tv} = 66l/s$ – mass flow of water.

4.3.2. Water performance in primary circulit

$p_2 = 3{,}5\,bar = 350\,kPa$ - pressure water of primary circulit,
$t_1^{ul} = 28°C$ - inlet temperature water in heat exchanger primary circulit and
$t_1^{iz} = 35°C$ - outlet temperature water in heat exchanger primary circuit.

4.3.3. Heat quantity which recive primary circulit water

Heat quantity which give in secondary circulit water be calculated according to the following:

$$Q_d = \dot{m}_{tv}\left(i_2^{ul} - i_2^{iz}\right) \qquad (14)$$

where are:

$i_2^{ul} = 209{,}34\ kJ/kg \quad (t = 50°C)$ - enthalpy secundary circulit water at outlet heat excanger,
$i_2^{iz} = 167{,}53\ kJ/kg \quad (t = 40°C)$ - enthalpy secondary circulit water at outlet heat excanger and

$$Q_d = 66\ (209{,}34 - 167{,}53) = 2759{,}46\,kW\cdot$$

Heat quantity which recive primary circulit water be calculated according to the following:

$$Q_{tv} = \eta \cdot Q_d, \qquad (15)$$

where $\eta = 0{,}985$ is coefficient of used heating.

$$Q_{tv} = 2759{,}46 \cdot 0{,}985 = 2718kW$$

4.3.4. Calculation of area heat exchange

Flow of technological wather through heat exchanger can be determined by:

$$\dot{m}_{tv} = \frac{Q_{tv}}{\left(i_1^{iz} - i_1^{ul}\right)}, \tag{16}$$

$$\dot{m}_{tv} = \frac{2718}{(142{,}45 - 117{,}37)} = 108 l/s.$$

Area of heat exchange is given by:

$$A_{tv} = \frac{Q_{tv} \cdot K}{\Delta T_{\log}}, \tag{17}$$

where Q_{tv} is heat quantity which recive primary circulit water (tehnical water), $K = 1/R$ is coefficient of resistance of heat passage, $R = 4{,}5\ \frac{Wm^2}{K}$ is resistance of heat passage and $\Delta T_{\log}$ is middle logarithmic temperature.
Middle logarithmic temperature be calculated according to the following:

$$\Delta T_{\log} = \frac{\Delta T_v - \Delta T_m}{l_n \frac{\Delta T_v}{\Delta T_m}}, \tag{18}$$

$$\Delta T_{\log} = \frac{\left(t_2^{ul} - t_1^{iz}\right) - \left(t_2^{iz} - t_1^{ul}\right)}{l_n \frac{t_2^{ul} - t_1^{iz}}{t_2^{iz} - t_1^{ul}}}, \tag{19}$$

$$\Delta T_{\log} = \frac{(50-35)-(40-28)}{l_n \frac{50-35}{40-28}} = \frac{3}{l_n \frac{15}{12}} = 13{,}44.$$

Area of heat exchange should be calculated according to the following (17):

$$A_{tv} = \frac{2718 \cdot \frac{1}{4{,}5}}{13{,}44} = 44{,}94 m^2$$

The characteristics of the plate heat exchanger: Firm: Alfa Laval, Test pressure: 13 bar, Design pressure: 10 bar, Min. temperature: 0 °C, Netweight: 941 kg and Operating weight: 1117 kg [8].

5. Conclusion

The selection of compressors for different applications is a crucial issue in the process industry. It is usually the most expensive piece of equipment and has dominant influence on cycle efficiency. In this paper is presented a system of preparation and distribution of compressed instrument air to the consumer in the system of internal pneumatic transport of fly ash under the electrofilter, flues and steam air heaters and calculation cooling compressors. System of compressed air for electrofilter is an integral part of a whole, the pneumatic transport of fly ash for the silo.

Compressed air is distributed by seamless steel galvanized pipelines. Distribution pipeline is oriented towards consumers, apropos to all valves on pneumatic drive which are located under electrofilter (EPS) of unit 1 and unit 2, and to the consumers in pump stations 1 and 2. In function of assigned pressure, complete pipeline is dimensioned, apropos diameters of certain sections are specified.

Compressed air is used for manipulation and control in pneumatic transport process as well as for manipulation of pneumatic valves which are within automatic control of pneumatic transport. Locations which are supplied with compressed air: pressure vessels for pneumatic fly ash transport, air transport pipeline, armatures of ash receiver (plate gate valves), ash transport pipelines and pneumatic valves in bager station. Consumers, apropos places which are needed to be supplied with compressed air, altogether there are 226 pieces. Function of these valves is in program control of fly ash transport process which is managed by PLC. Safe operation of installation, apropos maintaining pressure in installation within limitations of minimal and maximal permissible pressure is supplied with regulation system which is located in the compressor. Compressor is in mutual automatic operation with reservoir of instrumental air where pressure should be between 6,3 – 7,5 bars. When pressure in reservoir drops under 7,5 bars, and reaches value of minimum permissible operation pressure p_{min} = 6,3 bars, automatically one compressor activates until maximal wanted pressure is not reached.

Selection of compressor for compressed air under electrofilter: air consumption by consumer for transmition in block 1 : Q_1= 53 m³/h, air consumption by consumer for transmition in block 2: Q_2= 53 m³/h, air consumption by consumer for transmition in pump stations 1 and 2 like consumers on air preheaters and under channel of flue gases: Q_3= 15 m³/h and total air consumption increased for 20% : $Q= 1{,}2 \cdot (Q_1 + Q_2 + Q_3) = 1{,}2 \cdot 121 = 145{,}2$ m³/h.

On the basis of calculated consumption of 145,2 m³/h compressor of firm Atlas Copco is selected: Type: GA15⁺, Capacity: 160 m³/h , Operation pressure: 7,5 bara , Power of electromotor: 15 kW, Noise level: 64 dB, Dimensions 1400 x 1225 x 650. On the base of selected compressor and accrding to Atlas Copco recommendation absorption dryers type CD 44 (2 piece) are chosen. And on the base of experience in designing system of compressed air, instrumental air reservoir is selected, volume V = 2 m³.

Compressors of compressed air are water cooled, while compressors of control air are air cooled. Water for cooling of compresed air and compressors is used in secondary circuit which is closed. A water tank will be used to supply additional water should any losses of water occur during circulation. Water from primary circuit is supplied from unit 1 and flows to bager station. Circulation pump, heat exchanger and pipeline were selected in accordance with calculation.

In this paper is explained calculation cooling compressors. Cooling of compressors is going to be done in two cycles. In first cycle technological water circulates by pumps of first cycle. In second cycle "demi" water will circulate and cooled compressors and coolers of compressed air and also will be drived by pumps. Possible water lost through cooling process will be refunded from additional water tank which is connect with supplying water pipe line for cooling. Selection of circulation pump in secondary circuit (the characteristics of the pump): Firm: WILO, Type: IL150/260-18,5/4, Flow range: 216 m³/h, Motor power: 18,5 kW and Weight 309 kg and selection of circulation pump in primary circuit: Firm: WILO, Type: IL200/320-45/4, Flow range: 360 m³/h, Motor power: 45 kW and Weight: 512 kg. The characteristics of the plate heat exchanger: Firm: Alfa Laval, Test pressure: 13 bar, Design pressure: 10 bar, Min. temperature: 0 °C, Netweight: 941 kg and Operating weight: 1117 kg. Optimized operation of the compressor is becoming increasingly important, especially for larger systems of compressed air depending on the industry. As the production rate in plants grow with plant development, the operating conditions of the compressor will change. It is therefore important that the compressed air supply system is based both on the current needs, and the needs in the future.

6. References

[1] Atlas Copco Airpower NV, (2015), Compressed Air Manual 8th edition, Belgium, ISBN 9789081535809

[2] Bloch, H. P., (2006), Compressors and Modern Process Applications, John Wiley & Sons, Hoboken, New Jersey, ISBN 0-471-72792-X

[3] Bloch, H. P., (2006), A Practical Guide to Compressor Technology, 2nd Ed., John Wiley & Sons, Hoboken, New Jersey, ISBN 0-471-727930-8

[4] Hadziahmetovic, H., Dzaferovic, E., Hodzic, N., (2012), System of compressed air for electrofilrer in thermal power plant, Annals of DAAAM for 2012 & Proceedings of the 23rd International DAAAM Symposium, Volume 23, No.1, ISSN 2304-1382, ISBN 978-3-901509-91-9, Ed. B. Katalinic, Published by DAAAM International, Vienna, Austria

[5] Mills D., (2004), Pneumatic Conveying Design Guide, Second Edition, Butterworth-Heinemann, ISBN 0750654716, UK.

[6] Seshaiah, N.,(2006), Experimental and Computational Studies on Oil Injected Twin-Screw Compressor, Doctoral Thesis submitted to National Institute of Technology, Rourkela

[7] Wilo Product Catalog (2014), Pumps and systems for Building Services, Water Management, and Groundwater, Dortmund, Germany

[8] http://doc.texnikoi.gr/ylikadata/rtfs/laval4.pdf (2016), Alfa Laval plate heat exchangers, A product catalogue for comfort heating and cooling

[9] Hadziahmetovic, H. & Dzaferovic, E., (2009), Ash pneumatic conveying from existing silos no. 4 to two new silos and ash loading in autocisterns - The 20th INTERNATIONAL DAAAM SYMPOSIUM "Intelligent Manufacturing & Automation: Theory, Practice & Education", 25-28th November 2009, Vienna, Austria

[10] Bogović, I-N., Barišić, B., Katalinić, B., Kršulja, M., Car, Z.,(2011), Digitizing system ATOS - measuring turbo compressor housing, Annals of DAAAM for 2011 & PROCEEDINGS of the 22nd International DAAAM Symposium / Katalinić, Branko (ur.). - Vienna : DAAAM International Vienna , 2011. 1367-1368 (ISBN: 978-3-901509-83-4)

[11] Bibire L., Ghenadi A., Topliceanu L., (2011), Maintenance -Reliability Duality For A 40 Bars Compressor From A Pet Bottling Line, The 22nd DAAAM World Symposium "Intelligent Manufacturing & Automation: Power of Knowledge and Creativity", Vienna, 23-26th November 2011, ISBN 978-3-901509-83-4, ISSN 1726-9679, pag. 0029-0030

DOI: 10.2507/27th.daaam.proceedings.031

RFID READER'S CONSTRUCTION IN TERMS OF ELECTROMAGNETIC SUSCEPTIBILITY

Martin Pospisilik[a], Zdenek Korytak[a], Peter Janku[a],
Rui Miguel Soares Silva[b] & Petr Neumann[a]

[a]Faculty of Applied Informatics, Tomas Bata University, Nad Stranemi 4511, Zlin, Czech Republic
[b]Polytechnic Institute of Beja, Laboratory UbiNET, Rua Pedro Soares, Beja, Portugal

This Publication has to be referred as: Pospisilik, M[artin]; Korytak, Z[denek]; Janku, P[eter]; Silva, R[ui] M[iguel Soares] & Neumann, P[etr] (2016). RFID Reader's Construction in Terms of Electromagnetic Susceptibility, Proceedings of the 27th DAAAM International Symposium, pp.0213-0218, B. Katalinic (Ed.), Published by DAAAM International, ISBN 978-3-902734-08-2, ISSN 1726-9679, Vienna, Austria
DOI: 10.2507/27th.daaam.proceedings.031

Abstract

RFID technology has established itself in many sectors of human activity, including security and access systems. Just at this context, the demands on the hardware design in terms of its electromagnetic compatibility become critical. This paper provides a description of an immunity test against the electrostatic discharge according to the standard EN 61000-4-2 that was applied to an RFID reader which was primarily intended to be applied in access systems, employing the on-board recognition of the RFID tags. It was shown that such device can be endangered by the external electrostatic pulse; the findings described in this paper may be beneficial for the developers of the hardware for the RFID devices.

Keywords: Electrostatic discharge; RFID reader; Immunity test; ESD generator; Radiation pattern

1. Introduction

The issues on electrostatic discharges have become important deep in the nineteenth century when several paper mills had faced the accidents caused by ignition of the paper dust by spark discharges. Even earlier, the effects on spark discharges on gunpowder have also been known [3]. However, the problems of the electrostatic discharges have been comprehensively tackled since the first field-effect semiconductors begun to be employed. Nowadays the electrostatic discharge (ESD) protection is one of the most important issues at the electronic devices` development as the current semiconductors exhibit high sensitivity to the energy released during the discharge.

At present, testing of devices on the immunity to electrostatic discharges falls within the tests performed within the framework of electromagnetic compatibility (EMC) testing. In European Union, the set of immunity tests is prescribed by the standard EN 61000-4-1 and the conditions and criteria of testing on the ESD immunity are described by the standard EN 61000-4-2. This standard prescribes the criteria that must be fulfilled when the tested device is in operation as well [6] [7]. The device should be enclosed in its cover.

1.1 Electrostatic discharge

The local electrostatic discharge occurs between two surfaces provided there is a significant difference between their charges that are determined by the number of accumulated electrons. Lightning is probably the most dramatic effect of ESD, but in practice, even in much smaller discharges high powers are dissipated, which can result in unpleasant consequences. Although the typical energy released at one ESD reaches the order of mJ, due to very short discharge times (nanoseconds) the levels of voltage and current are destructive to most of the semiconductor devices. According to [2] the charge on the surface of a human body can reach up to 15 kV when walking or rubbing on clothing. A typical current waveform when ESD occurs is depicted in Fig. 1.

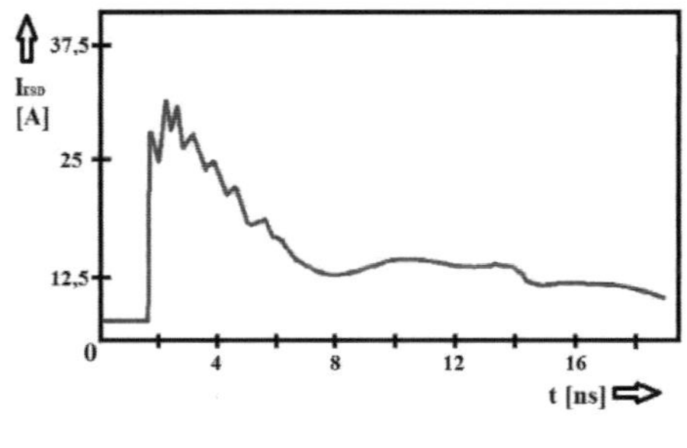

Fig. 1. Typical current waveform during the electrostatic discharge [2]

1.2 Standardization

As mentioned above, testing of any electronic device`s immunity against ESD falls among the issues on electromagnetic susceptibility. The generic standard EN 61000-6-1 defines functional criteria according to which the performance of the tested device can be objectified.

The ESD immunity test is defined by the standard EN 61000-4-2. By means of this standard, the voltage levels, pulse waveforms and the configuration of the experiment are prescribed as well as the steps of the experiment. As the source of energy, ESD simulators are used. According to the appropriate standards, the amount of energy released during the discharge is specified by the level of voltage on a capacitor, the capacitance of which is also defined by the standards. Therefore, the discharge voltage (4, 8 kV etc.) is specified instead of the amount of the electric charge. In addition, the dimensions and construction of the replaceable tips of the ESD generator are also specified by the appropriate standards.

The following test types can be applied:

- Discharge through the air gap
- Discharge through the direct contact
- Discharge through the coupling plane

2. Description of the experiment

The experiment was performed on the RFID reader that has been developed at the Tomas Bata University in Zlin. The goal of the experiment was to verify the level of the immunity of the RFID reader to the external electrostatic discharge. Moreover, the RFID´s reader antenna radiation pattern has been obtained by means of the near field probe. The details are provided in subchapters below.

2.1 Description of the reader

The block diagram of the RFID reader that was tested within this experiment is provided in Fig. 2. The construction is based on STM32F family microcontroller that has a direct connection to the RFID reader chip MFRC523. This chip provides a complete RFID interface that allows reading the tags of the appropriate access cards. The detailed description of the chip is provided in [4].

There are several ports available to establish the data transfer between the RFID reader's module and the connected data processing unit, because the RFID reader was intended to operate as a universal unit.

These ports are as follows:

- CAN bus based on the engine MCP 2550
- USB implemented directly on the processor's chip
- Direct parallel port driven by the microprocessor's I/O pins.

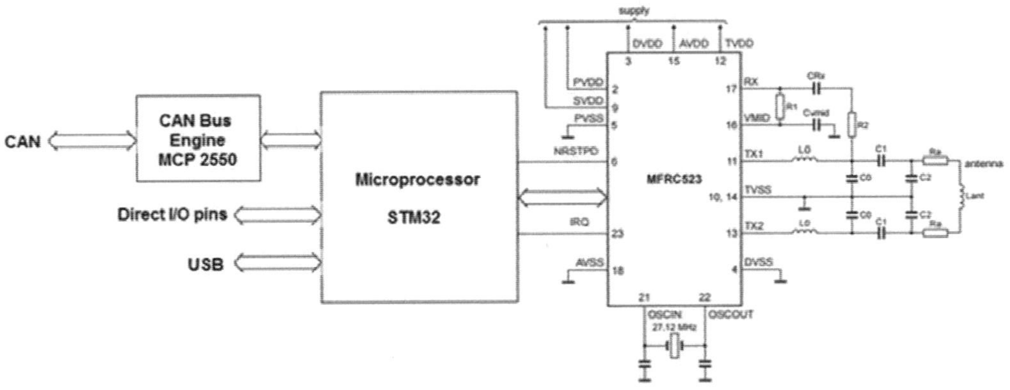

Fig. 2. Tested RFID reader's block diagram

For the purposes of the test, the following algorithm has been implemented to the reader's microcontroller: The RFID chip MFRC523 is periodically requested to detect and read the tag of the prospective RFID tag. Once the tag is detected and read, its unique number is compared to all numbers enlisted in the microcontroller's memory. When the tag is recognized and approved, a set of logical bits is send to the parallel port of the reader. This enables direct control of the connected device. For example, the door lock cen be opened directly by the RFID reader's unit. Moreover, there is a LED connected to one of the output pins to indicate the logical level at the appropriate output. Methods described in [8] have been applied in the unit's design.

2.2 Details of the experiment

The experiment has been performed inside the Laboratory of Electromagnetic compatibility of Tomas Bata University in Zlin in two steps. In the first step, the immunity of the reader's unit to the electrostatic discharge has been tested by means of the ESD generator ONYX 16 according to the standards EN 61000-6-1 and EN 61000-4-2. The test voltages were increased from the basic test level of 1 kV up to the levels prescribed by the abovementioned standards:

- ± 4 kV for a contact discharge
- ± 8 kV for a discharge through the air gap

All three test types, as described in the chapter 1.2 have been applied. The points of the RFID reader's unit that were hit by the ESD discharge were as follows:

- Input and output pins
- Status LED
- RFID antenna

Afterwards, another functional sample of the reader has been employed and its transmitting radiation pattern has been visualized by means of a close electrical field probe and the measuring receiver Rohde & Schwarz ESU 8. In this case, the intensity of the carrier frequency transmitted by the reader's antenna (13.56 MHz) has been mapped in the horizontal plane of the antenna in 70 different points and the results, processed by Maple software, has been visualised by means of the 3D graph. In both cases the tested reader was supplied from a laboratory power source.

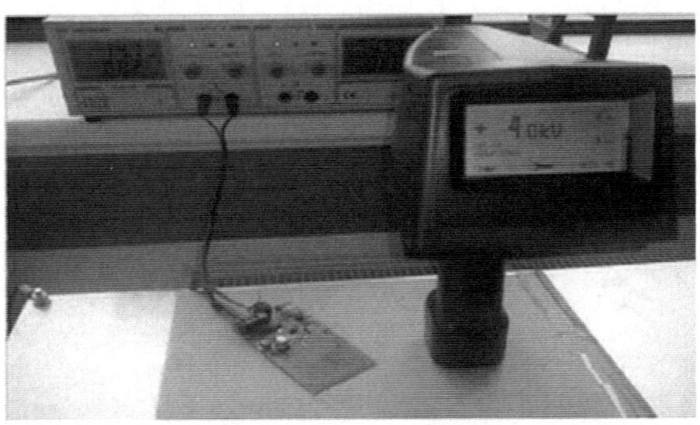

Fig. 3. The functional sample of the RFID reader and the ESD generator ONYX 16 [5]

Fig. 4. Measurement of the RFID reader's antenna radiation pattern

3. Results

The immunity test results are enlisted in Table 1. Different discharge types and voltage levels has been applied.

Point of ESD application	Charge voltage	Discharge type	Polarity	Result
I/O pins	4 kV	Contact	(+)	B
			(-)	B
LED (output state indicator)	4 kV	Contact	(+)	B
			(-)	B
RFID antenna	4 kV	Contact	(+)	C
			(-)	B
The whole reader through a vertical coupling plane	4 kV	Vertical coupling plane	(+)	A
			(-)	B
I/O pins	8 kV	Air gap	(+)	B
			(-)	B
LED (output state indicator)	8 kV	Air gap	(+)	B
			(-)	B
RFID antenna	8 kV	Air gap	(+)	B
			(-)	B

Table 3. Results of the immunity test

To fulfil all the requirements of the standard EN 61000-4-2, the worst achieved result should not be worse than B. As obvious from the Table 3, in one case this requirement has not been met. However, it must be considered, that the tested device was operated without the proper cover and the contact discharge was targeted directly to the receiver's antenna. This problem can be solved by enclosing the reader into a suitable cover.

After processing of the test the device has been checked for its proper operation and it has been found that no permanent damage occurred. Unfortunately, the test showed another problem that is not covered by the requirements of the standard. When the device was influenced by the ESD, several logical 1 states occurred at the direct parallel bus. That means that there exists a real risk if the RFID reader is used directly to drive the actuators (for example the door lock), consisting in the fact that the attacker would bypass the authentication by giving the ESD shock to the RFID reader. On the other hand, this risk can be eliminated by using the modulated and/or encoded output peripherals as the CAN bus or USB. In this case, the RFID reader only gives the information on reading of the tag to another device that manages the actuators driving and the existence of accidental pulses on the communication bus cause no harm to the safety of the authentication system.

The results of the antenna`s radiation pattern are depicted in Fig. 5. From this figure it is obvious, that the antenna`s radiation pattern is symmetrical in terms of the space perpendicular to its z-axis. This determines the most critical direction of the interfering energy that could be sent by the aggressor in order to overload the RFID reader`s input.

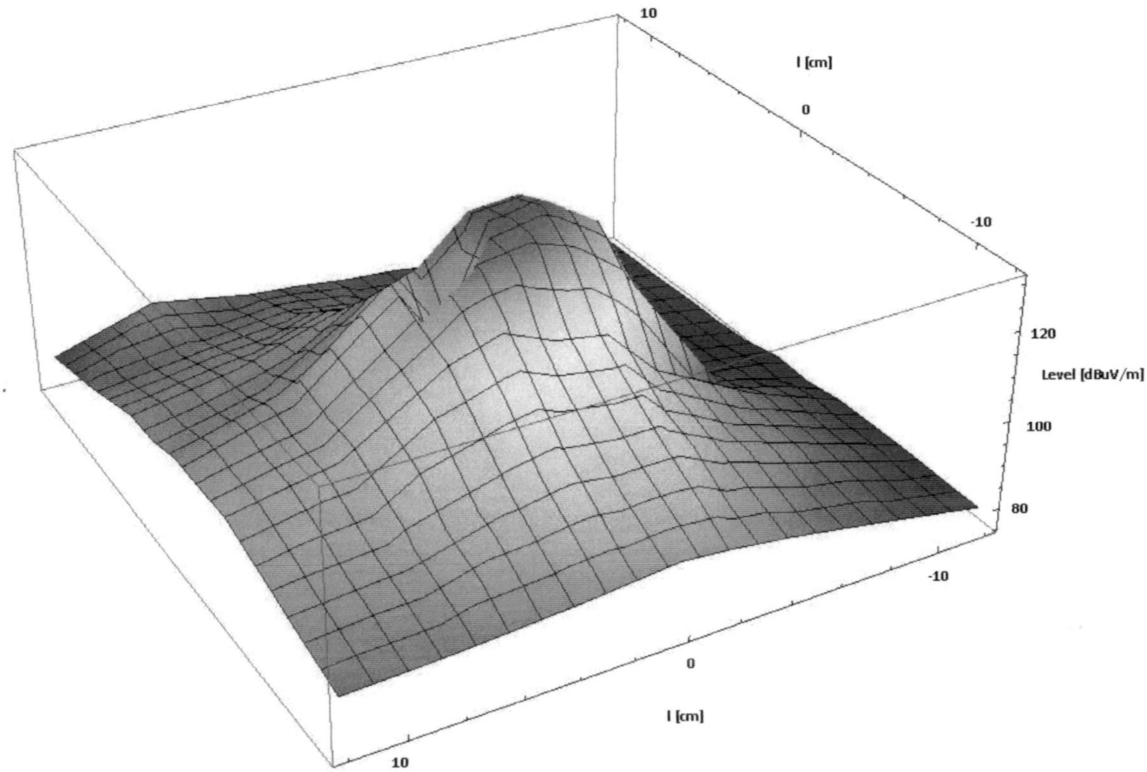

Fig. 5. RFID reader's antenna radiation pattern [5] (intensity vs. position)

4. Conclusions

In this paper a description of the ESD immunity test on a typical construction of the RFID reader is provided. The results of the test indicate, that if the RFID reader is intended to be used in a safety critical application, for example as a controller of the door lock, the phenomenon of the electrostatic discharge cannot be omitted, as the ESD can be generated deliberately as a kind of attack to the system. In this test a possibility of generating false positive signals at the output of the reader was confirmed. Therefore we recommend not to use the RFID reader directly to recognize the tags, but to process the information on the tags in the neighbourhood of the reader`s antenna by the remote device.

In addition, the radiation pattern of the RFID reader`s antenna has been obtained in order to check the antenna`s design and determine the direction of the external energy that will probably have the most harmful effect on the reader`s operation.

The hereby described experience with the tested device helped its manufacturer to improve its design in order to reach safer operation.

5. Acknowledgments

This work was supported by the Ministry of Education, Youth and Sports of the Czech Republic within The National Sustainability Programme Project No. LO1303 (MSMT-7778/2014) and also by The European Regional Development Fund under the project CEBIA-Tech No. CZ.1.05/2.1.00/03.0089.

6. References

[1] Svacina, J.. (2001). *Electromagnetic compatibility*. VUT Brno, Czech Republic.

[2] Horsky, P. (2014). Electrostatic discharge and testing of its inuence on integrated circuits. Online. URL: http://design.georgius.cz/sites/infocube/eem/horsky_eem081.pdf

[3] Poole, I. (2014). ESD Protection. Online. URL: https://radio-electronics.com

[4] MFRC523 (2016). Datasheet

[5] Korytak, Z. (2015). *Testing the Electromagnetic Susceptibility of Electrical Appliances*. Tomas Bata University in Zlin. Czech Republic

[6] Paul, C. R. (2006). *Introduction to electromagnetic kompatibility*. Wiley-Interscience. Hoboken.

[7] Vaculikova, P.; Vaculik, E. (1998). *Electromagnetic compatibility of electrotechnical systems*. BEN. Czech Republic

[8] Shah, R[ima]; Park, H[ong-Seok] & Lee, G[yu] B[ong] (2016). Design For Assembly: An approach to increase Design Efficiency of Electronics Home Appliance, Proceedings of the 26th DAAAM International Symposium, pp.0877-0882, B. Katalinic (Ed.), Published by DAAAM International, ISBN 978-3-902734- 07-5, ISSN 1726-9679, Vienna, Austria. DOI: 10.2507/26th.daaam.proceedings.122

DOI: 10.2507/27th.daaam.proceedings.032

USE OF WATERJET IN MANUFACTURING TEST BARS OF HIGH-STRENGTH STEELS

Stepan Jenicek, Kotesovec Vratislav, Tomas Kalina & Bohuslav Masek

This Publication has to be referred as: Jenicek, S[tepan]; Kotesovec, V[ratislav]; Kalina, T[omas] & Masek, B[ohuslav] (2016). Use of Waterjet in Manufacturing Test Bars of High-Strength Steels, Proceedings of the 27th DAAAM International Symposium, pp.0219-0224, B. Katalinic (Ed.), Published by DAAAM International, ISBN 978-3-902734-08-2, ISSN 1726-9679, Vienna, Austria
DOI: 10.2507/27th.daaam.proceedings.032

Abstract

This article introduces a procedure which has been used for developing and optimizing a process of making tensile test specimens of ultra-high-strength sheet metal samples. Thermomechanically-processed materials of this kind are characterized by high hardness, strength, and toughness. Making specimens of these materials by classical chip cutting is therefore difficult, time-consuming, and costly. For the sake of comparison, various methods are listed here which are potential candidates for making such specimens. In this case, an alternative process was chosen: waterjet cutting. From the production viewpoint, the design of the workholding fixture for this process is of major importance. The fixture has an indirect but very significant impact on the manufacturing process and, ultimately, on the quality of the test specimens produced. In this experimental programme, the suitability of the workholding fixture, the process chosen, and the cutting conditions were examined

Keywords: Abrasive waterjet cutting; Machining; AWJ; Workholding fixture; AHSS; Unconventional technologies; Test bars.

1. Introduction

In order to produce high-quality test specimens from high-strength metal sheets, a new procedure that could deliver the desired quality and effectiveness had to be developed and put in place. Thermomechanically-processed materials of this kind are characterized by high hardness, strength, and toughness [10],[11]. For this reason, making specimens of these materials by classical chip cutting is difficult and costly. Cutting plate is a difficult and time consuming task, especially when the job calls for a tight tolerance or an unusual shape. In those cases, can choose abrasive waterjet cutting [9],[4]. Abrasive waterjet cutting is an attractive process with many advantages. The process is suitable for cutting steel of small thickness where it provides a smooth and very precise cut. In terms of precision, waterjet cutting can even outperform laser cutting, as the resulting edges are smoother and free from melting, burning and thermal distortion [1]. Waterjet is now the most cost effective method to cut steels [9]. Therefore, an alternative machining process of abrasive waterjet cutting was employed.

2. Available methods

Where materials are to be parted or cut, a decision must be made as to the process to be used. Basically, one can choose from four primary cutting processes: oxy-fuel cutting, plasma-arc cutting, laser cutting, waterjet technologies or perhaps wire electrical discharge machining (WEDM). Tensile test specimens must not develop any heat-affected zone during manufacture because in this case the properties imparted by thermomechanical treatment are to be measured. This is the reason why all high-temperature cutting processes [2],[3] have been ruled out, i.e. those which have severe heat effects on the cut area. Hence, the only processes that are still available are waterjet cutting and wire electrical discharge machining.

2.1 *Abrasive waterjet cutting*

Abrasive waterjet is an appropriate choice for cutting any dense materials, ranging from very thin foils to parts of 300 mm or even greater thickness. Cuts of the highest precision can be made with the aid of draft angle compensation in materials of up to 50–100 mm thickness, depending on the machine design. Rough cuts can be completed in materials much thicker than 100 mm. [4] The process is suitable for cutting steel of small thickness where it provides a smooth and very precise cut. In terms of precision, waterjet cutting can even outperform laser cutting, as the resulting edges are smoother and free from melting, burning and thermal distortion. Waterjet is not constrained by thickness, unlike laser and plasma technologies. Higher costs, on the other hand, are a disadvantage. The investment cost of the equipment exceeds that of plasma cutting equipment while being lower than in laser equipment. By contrast, the operating costs of waterjet cutting are much higher than laser cutting, which is mainly due to the price of the abrasive. [1]

2.2 *Wire electrical discharge machining (WEDM).*

Wire electrical discharge machining can only be used for electrically conductive materials. Sculptured products with a thickness up to 500 mm can be machined [6]. The greatest advantage of the process is its ability to machine electrically conductive materials regardless of their hardness. The process does not generate any conventional cutting forces. During cutting, only electric erosion forces occur. The cut must be made through thickness of the material. The method is well-suited for applications that required shape and dimensional accuracy. With the introduction of NC systems, this technology became one of the leading machining processes in terms of precision. The contour accuracy in state-of-the-art machines is several micrometres and the quality of the cut surface matches the parameters of ground surfaces. The most common diameter of brass wire used for WEDM is 0.25 mm. Today's top-quality machines can use wires with diameters down to 0.02 mm. Electrical discharge machining does not introduce adverse stresses into the material. It can also produce parts with very thin walls [5]. Nevertheless, the operating costs of WEDM are much higher than those of waterjet, mainly because of its slow machining speed. The investment costs of the equipment are approximately the same as those of waterjet cutting equipment, provided that the cutting space in the machine is considerably smaller.

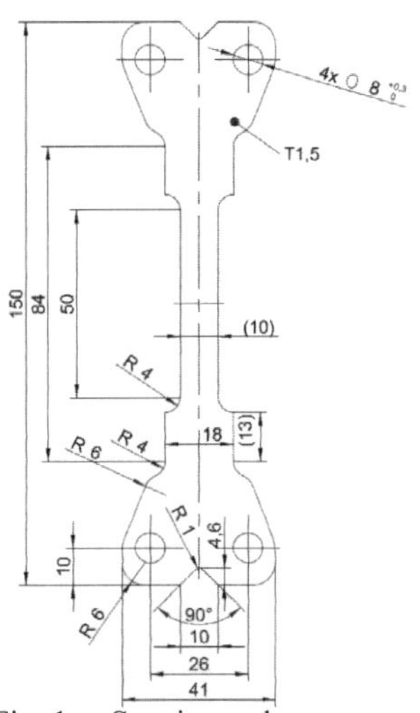

Fig. 1. – Specimen shape

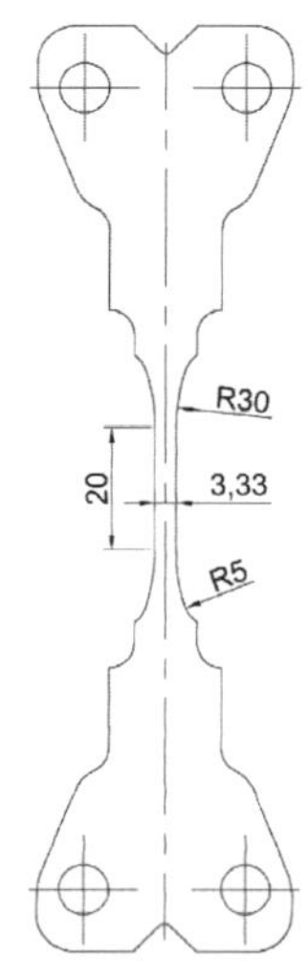

Fig. 2 – Modified specimen shape

2.3 *Process selection*

Optimized test specimens (Fig. 1) are thermomechanically treated in a thermomechanical simulator [7]. They must be converted to standard-shape specimens (Fig. 1) for tensile testing. For reasons related to economy and speed of specimen preparation, abrasive waterjet cutting was chosen. For this process, a workholding fixture for accurate and firm clamping of the specimens with respect to the tool had to be designed.

2.4 *Process selection*

The fixture is a very important item, as it has a major impact on the test bar quality. The main reason for using a fixture is the need for quality and dimensional, shape and geometric accuracy.

The requirements for the workholding fixture for waterjet cutting were as follows:

- secure positioning
- firm clamping of the part
- clamping the part in a defined position with respect to the tool

The design (3 and 4) was developed using the SolidEdge ST8 software.

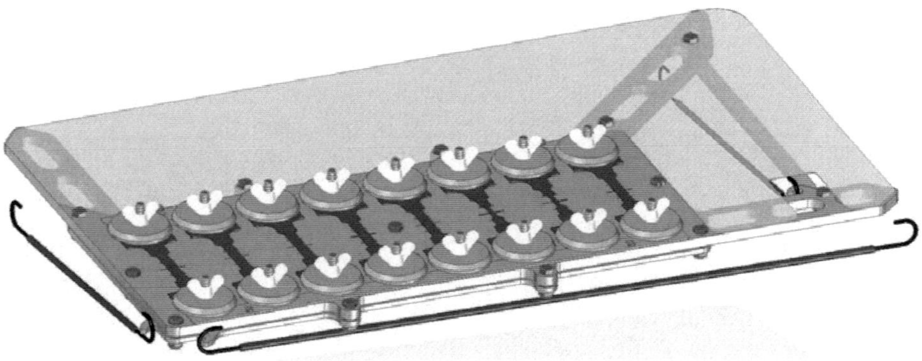

Fig. 3: Visualized fixture with eight test specimens

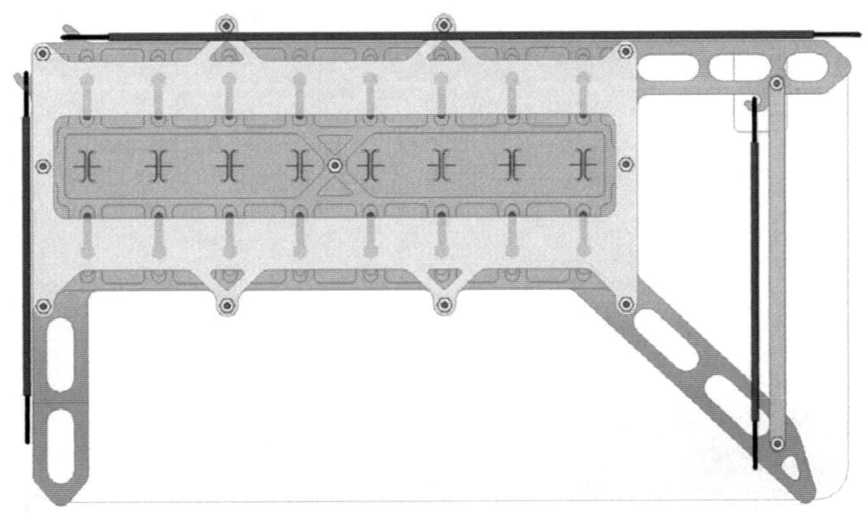

Fig. 4 – Plan view from bottom

The design of the fixture allows up to 8 specimens to be cut out in a single series. This solution is motivated by the economy of and time savings in production. The test specimen is centred in the fixture with V-grooves. Mating cylindrical surfaces are guided accurately in the fixture. At the top, they are provided with threads for holding the specimens in place. The transverse plane of the fixture is provided with marks in the form of channels. These are used for centring the specimens before tightening.

To verify the design and function of the fixture (Fig. 5), it was manufactured of S235 steel. This material is not the optimal choice for this fixture because it lacks corrosion resistance but it is fully sufficient for verifying its function. Semi-finished products were also made by waterjet cutting. The fixture was designed to be producible from sheet metal.

3. Experimental

The process was tested in the WJ 2830-2Z-Cobra-PJ60 machine. The experiment compared the alignment of cutouts from the original, and the final shape. A total of 23 specimens were tested (Tab 1). The measurement was carried out using the CMM Carl Zeiss Prismo 7 Navigator 3D measuring machine. A total of four parameters were measured (Fig. 5): specimen width – dimension A, the distance between the right edge of the cutout from the initial contour – dimension B, the distance between the left edge of the cutout and the initial contour – dimension C, and the angle between the centreline of the initial specimen and the centreline of the newly-cut specimen.

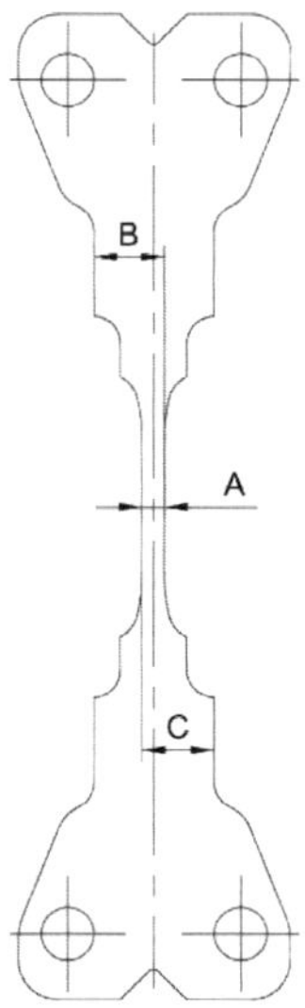

Fig. 5 – Specimen dimensions of interest

Test arrangement:

- PTV JETS 3.8/60-COMPACT high pressure pump (max. operating pressure 415 MPa, max. flow 3.8 l/min)
- X-Y cutting table PTV WJ 2830-2Z (working space of 3000×2000 mm)
- Continuously variable cutting speed 0–20 m/min

Experimental conditions

- Operating pressure [MPa]: 400
- Abrasive type: GMA Australian garnet
- Abrasive particle size [MESH]: 200
- Mass flow of abrasive – [g/min]: 30
- Waterjet nozzle diameter [mm]: 0.10
- Abrasive nozzle diameter [mm]: 0.508
- Standoff distance [mm]: 4

Workpiece material

- Material – EN10131 DC01
- Thickness [mm] 1.5

As evidenced by statistical characteristics of the set of results, the fixture meets the requirements for clamping precision (Tab. 1). Since the cutting error is caused by random factors, the two sigma rule can be applied. According to this rule, an interval of two standard deviations around the mean value will contain 95.4 % of the actual measured values according to Gauss distribution. In this case, an accuracy of +/- 0.03 mm from the mean value is achieved with 95.4 % confidence. As a consequence of inaccurate setting of kerf width compensation, the deviation of the mean width of the test specimen from the required value was 0.15 mm. In a subsequent run, this deviation was eliminated by precise measurement of the first piece and by resetting the kerf width compensation function. The difference of 0.4 mm between the B and C values was due to inaccurate movement to the zero reference point. This error was also eliminated by an improved fixture positioning procedure in the machine. In no case did the misalignment between the initial and the final contour exceed the critical limit of 1°.

Specimen number	Dimension A [mm]	Dimension B [mm]	Dimension C [mm]	Angle [°]
1	3.181	10.335	10.697	0.77
2	3.174	10.385	10.696	3.27
3	3.182	10.336	10.678	0.23
4	3.177	10.351	10.674	1.10
5	3.186	10.269	10.788	0.45
6	3.163	10.325	10.721	7.22
7	3.178	10.322	10.767	3.19
8	3.168	10.335	10.732	5.38
9	3.199	10.360	10.740	4.33
10	3.142	10.412	10.640	0.52
11	3.202	10.252	10.840	0.03
12	3.180	10.364	10.690	1.42
13	3.168	10.228	10.856	4.97
14	3.182	10.370	10.745	2.05
15	3.186	10.318	10.787	0.55
16	3.190	10.312	10.734	4.43
17	3.196	10.303	10.763	2.35
18	3.175	10.412	10.664	0.45
19	3.207	10.286	10.865	3.27
20	3.193	10.407	10.683	4.85
21	3.189	10.255	10.773	1.97
22	3.209	10.376	10.677	3.88
24	3.186	10.354	10.734	7.83
Mean value	3.183	10.333	10.737	2.805
Standard deviation	0.015	0.050	0.060	2.218

Tab 1. – Results of measurement

4. Conclusion

As many roads lead to the final destination, each product can be manufactured in various ways. When producing of the product, the goal was to produce the product for cutting tensile tests specimens in such way, which will fullfill not only all reguirements of drawing documentation, but also will be produced in the cheapest way. Another aim was to demonstrate practical experience of making the product with the technology of waterjet cutting.

Waterjet cutting was used for making a fixture for preparing test specimens for tensile testing. As the materials in question were AHSS steels, the specimens were impossible to be prepared in required quality, quantity and at the required cost by classical machining routes. The fixture for making eight test specimens was tested. It was found that all requirements for the test specimen quality have been met.

In the future, we will focus on optimizing cutting conditions and their influence on surface integrity.

5. Acknowledgements

The present contribution has been prepared under the project LO1502 'Development of the Regional Technological Institute under the auspices of the National Sustainability Programme I of the Ministry of Education of the Czech Republic aimed to support research, experimental development and innovation.

5. References

[1] http://www.esabna.com/us/en/education/blog/what-is-the-best-way-to-cut-steel-plate.cfm, (2013). What Is The Best Way To Cut Steel Plate, ESAB KNOWLEDGE CENTER, Accessed on: 2016-08-12

[2] Kašpar J. (2011). Thermal oxidative cutting, Tepelné dělení materiálu kyslíkem, *MM Průmyslové spectrum*, Vol. 2011, No. 24, pp 25, ISSN: 1212-2572

[3] Roubíček M., (2016). Criteria for the choice of method and trends in materials thermal cutting. Kriteria volby metody a trendy tepelného dělení materiálu, Available from: http://www.airliquide.cz/file/otherelement/pj/roubicek49122.pdf, Accessed: 2016-08-10

[4] http://www.flowwaterjet.com/Resources/Ask-Dr-Hashish-a-Question, (2016). Ask Dr. Hashish - Dr. Mohamed Hashish is the inventor of the abrasive waterjet and an expert in high pressure design and tribology., Accessed on: 2016-08-12

[5] Zeman Z. (2013). What is the secret of electrocorrosion. V čem spočívá tajemství elektroeroze, *Technický týdeník*, Vol. 2013, No. 24, pp 25, ISSN: 0040-1064

[6] Dvoráková, J.; Dvorák, J. (2007). Milling and Engraving With Unconventional Technology AWJ, Frézování a gravírování nekonvenční technologií AWJ, *MM Průmyslové spectrum*, Vol. 2007, No. 4, pp 56, ISSN: 1212-2572

[7] Vorel, I., Vancura, F., Pilecek, V., Jirkova, H., Masek, B. (2015). Material-technological Modelling of C45 Steel die Forgings. In *25th DAAAM International Symposium Intelligent Manufacturing & Automation -2014,* Vienna, ISSN: 1877-7058 , Katalinic, B. (Ed.), pp. 714-721, Published by Elsevier Ltd, doi:10.1016/j.proeng.2015.01.424

[8] Martinec, P.; Sitek, L. (2015). Abrasives and possibilities of increase in efficiency of abrasive waterjets. In: *Vodní paprsek 2015 - výzkum, vývoj, aplikace*, Sitek, L., Klichová, D. (Ed.), pp 157-163, Ústav geoniky AV ČR, ISBN 978-80-86407-56-2, Ostrava

[9] Radocovic M. (2007) Abrasive waterjet cutting. *Modelling And Optimization In The Machines Building Field.* Volume III , No. 13, 2007, ISSN 1224-7480

[10] Aišman, D., Staňková, H., Skálová , L., Mašek, B. (2008). Testing of the parameters of the Q-P process in high strength low-alloyed steel. In *Danube Adria Association for Automation and Manufacturing*. Vienna Austria: DAAAM International, Vienna Austria, 2008. ISBN: 978-3-901509-68-1, Vienna, Austria

[11] Kana, J; Vorel, I & Ronesova, A. (2016). Simulator of Thermomechanical Treatment of Metals, Proceedings of the 26th DAAAM International Symposium, pp.0513-0518, B.Katalinic (Ed.), Published by DAAAM International, ISBN 978-3-902734-07-5, ISSN 1726-9679, Vienna, Austria

DOI: 10.2507/27th.daaam.proceedings.033

INTELLIGENT ROBOTRONICS: HARDWARE-SOFTWARE COMPLEXES OF ROBOTARIUMS

V. E. Pryanichnikov, A. Ya. Ksenzenko, S. V. Kuvshinov,
Yu. V. Poduraev, E. A. Prysev, R. V. Helemendik & S. R. Eprikov

This Publication has to be referred as: Pryanichnikov, V[alentin]; Ksenzenko, A[lexander]; Kuvshinov, S[ergey]; Poduraev, Y[uriy]; Prysev, E[vgeny] & Khelemendik, R[oman] (2016). Intelligent Robotronics: Hardware-Software Complexes of Robotariums, Proceedings of the 27th DAAAM International Symposium, pp.0225-0229, B. Katalinic (Ed.), Published by DAAAM International, ISBN 978-3-902734-08-2, ISSN 1726-9679, Vienna, Austria
DOI: 10.2507/27th.daaam.proceedings.033

Abstract

The article describes the technology of development of advanced mobile service robots and basic hardware and software elements as well as their use for implementation of the network of associated laboratories-robotariums within the framework of the project called "Intellectual Robotronics" organized by the Central European Branch of the International Academy of Engineering in conjunction with KIAM of the Russian Academy of Sciences, the International Institute of New Education Technologies (IINET) of the Russian State University for the Humanities, the universities of Russia, Croatia and Austria. Within the framework of this network, the access is provided to the entire range of the developed mechatronic devices and mobile robots having systems for remote charging and video stream exchange, and also to 3D-prototyping tools placed in the participating organizations, and to their guidance papers. Mobile robots and their simulation models repeat the basic functionality and appearance of significantly more difficult and expensive robotic complexes (RTC) of the Ministry of Emergency Situations and other agencies, which allows for the effective conducting of training of the operators of such RTCs and the development of integrating software ("middleware"), creating the knowledge representation technology and automatic generation of action plans.
This paper was partially supported by the Russian Foundation for Basic Research, Project No. 16-07-01264a, 16-07-00935a, 16-07-00811a.

Keywords: robotarium network; hardware and software simulators of robotic complexes, knowledge representation and pentalogic, intellectual robotronics.

1. Introduction

This article describes the key directions of development of advanced intellectual robotic complexes. The first problem is the need to unify the procedure of embedding various mechatronic devices by creating the respective integrating software. This allows for significant simplification the transfer from the inevitable and expensive building of low-level sensory control systems to the development of the top-level of control systems (CS). The second issue is insufficient development of deductive mechanisms and knowledge representation methods for CS intellectualization.

Solving these problems is difficult due to absence of systematic training of specialists, so-called "knowledge engineers". We have created application-dependent software and hardware complexes – space-distributed robotic laboratories (Fig.1) providing full access to all mechatronic devices included into them remotely through the Internet. This has allowed developing an effective education technology on the basis of the interuniversity network created by the International Institute of New Education Technologies of the Russian State University for the Humanities together with the Institute of Applied Mathematics named after Keldysh of the Russian Academy of Sciences (IAM) and the International Laboratory "Sensorika". As a part of the proposed approach, 10 AMUR mobile robots have been developed and produced with three integrated Robotino mobile robots (including elephant-trunk manipulator made by Festo) and integrated industrial manipulators. We have solved the problems of synchronization, delay time tracking and access rights differentiation. Also, mesh networks with moving repeaters are being built. Implementing the system has allowed to solve education-related problems during laboratory practicums closely related to conducting of timely scientific research studies [1-7].

2. Developing software and hardware framework and education programs

The problems we are solving have appeared due to the necessity of remote online reprogramming sensory control systems developed and installed on Brokk-110D and Brokk-330 mobile robots. These robots have been rebuilt to execute operations in dangerous areas with limited access for humans. The customers needed to have possibility to change it during mission execution without stopping robots, to integrate new mechatronic devices and third-party sensors. For this reason, the corresponding remote access tools and multi-camera vision systems have been developed [8, 9].

These results have found the use in AMUR educational and scientific robotic centers [10]. A technology for distribution of control processes has been created on the basis of integrating software ("middleware"), which allows combining hundreds of mechatronic devices with provision of remote access to them, and at the same time, providing dynamic change of system components configuration without stopping robots and automatic switching in case of faults, including auto-configuration. This software supports various operating systems, provides for integrating new drivers for sensors and drives, which can be installed even in field conditions [11]. Data streams were organized as special data structures ("buses"), and the dissection of control tasks into separate processes was the key to success for online collective (parallel) programming and fast prototyping. Teaching students to use this method helps them to avoid one of the problems of modern education related to "tape-record" acquirement of lectures and absence of creativity. The proposed solution of this problem is to motivate students to take part in scientific research studies as "knowledge engineers", to teach them how to build the required algorithms and effective formalizations of tasks [12]. As a base for such training research studies, AMUR mobile robots have been developed, each consisting of a body, track or two wheels (differential drive), or walking mechanisms, TV cameras (from 1 to 6), video-servers, ultrasonic locators (measurable ranges from 50 to 3000 mm), odometers, on-board controller, accumulators and charging devices, four analog channels and 16 on/off channels, power keys with PWM and PID controllers 2-10A (12...32V), on-board micro-PC and a remote PC connected through Wi-Fi with maximum range up to 50 meters or 1 kilometer when using additional antennas, and also a TV module for external supervision through WiFi-"satellite" with PTZ (Pan, Tilt, Zoom) and >10x optical zoom. AMUR mobile robots work on Linux in Python and C++ with the use of a distributed technology for programming of all types of sensory and video data processing, with logical filters on the remote PC and/or on board. All data are recorded in real time mode by buses (DTF), which reflect all the details of operation for the last week/month of mobile robots' functioning. DTF buses provide not just recording of data into files, but also synchronization, extrapolation and interpolation of data, which allows avoiding problems of data transmission delays and gaps in data streams, thanks to the respective fast logical filtering.

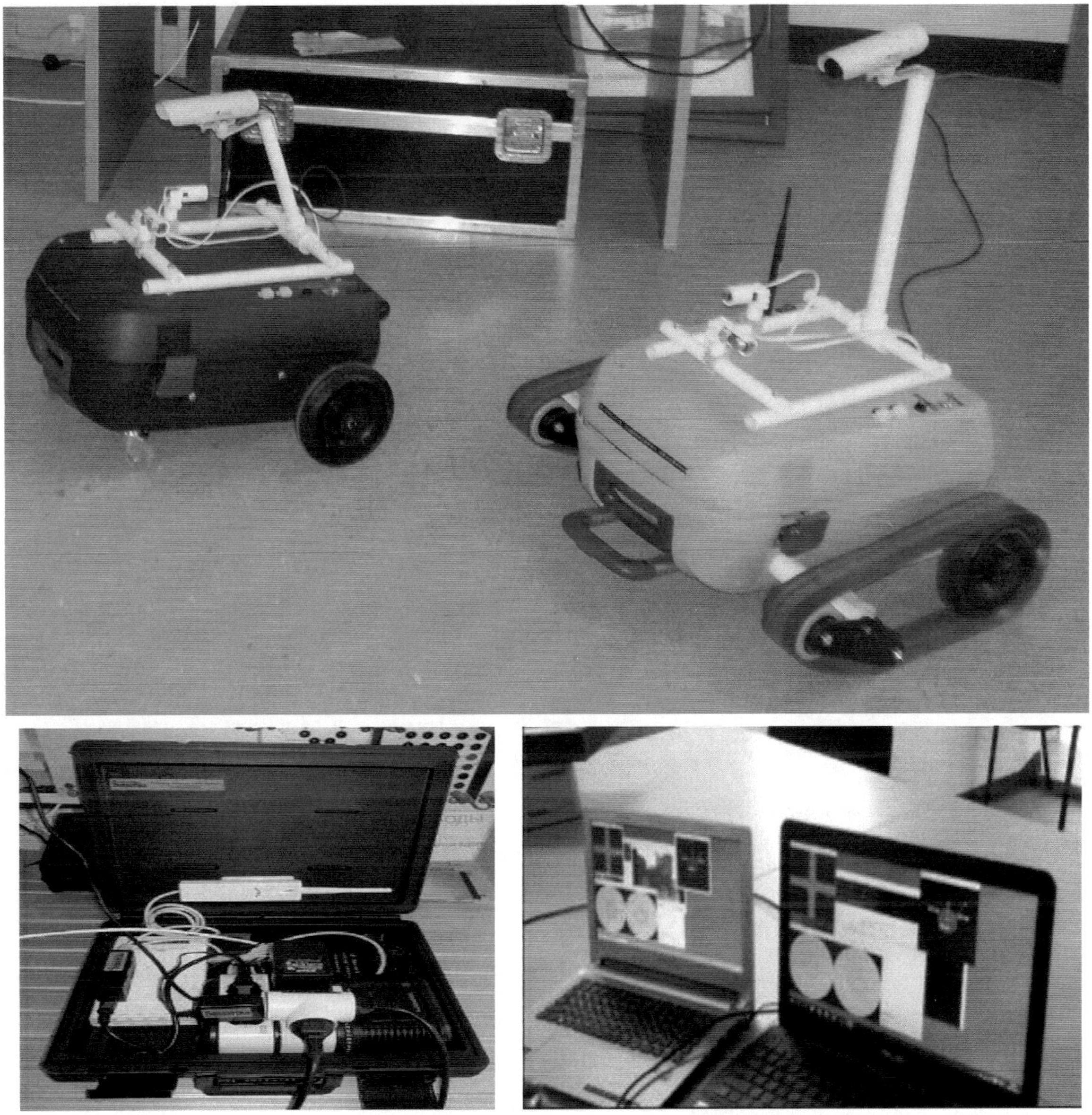

Fig. 1. Complex of AMUR mobile robots. Network forming kit – bottom left. Robot consoles – bottom right

Weight and dimensions of the implemented mobile robots AMUR-5, AMUR-6, AMUR-7, AMUR-105 and AMUR-107 are as follows: weight – 17-23 kg, dimensions – L580xB510xH340 mm. Produced velocity – 10…750 mm/s, battery life – 1-2 hours.

The main advantage of our approach is the use of our in-house developed simulator or building mathematical models on the basis of V-REP simulator in combination with hinging additional body elements imitating other (more expensive) complexes on the real AMUR mobile robots. This allows performing full-scale training of robot operators with significantly lower expenses by using both the existing models and virtual reality options (Fig. 2).

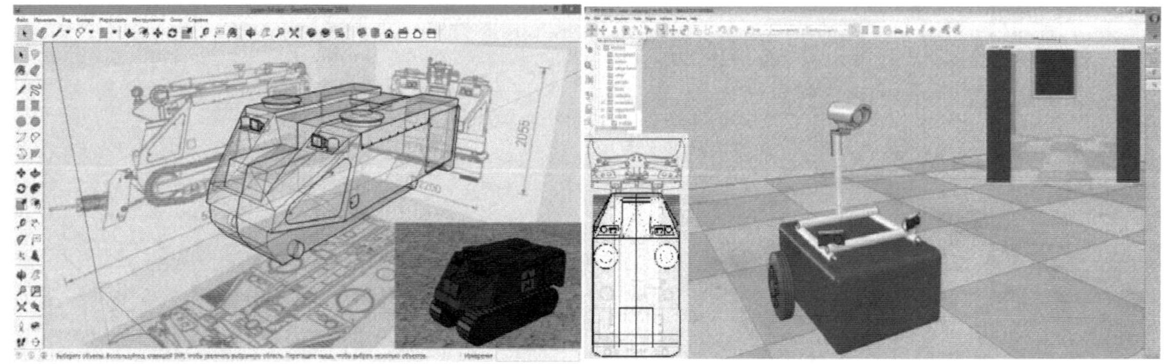

Fig. 2. Virtual mathematical model based on simulation V-Rep

3. Hardware solutions for contactless power supply

Contactless power supply system of autonomous robots. One of the discovered solutions, which were used during our research, was the power supply system designed for real-time automatic or semi-automatic recharging of accumulator batteries of mobile robotic devices without using direct connection of electrical contacts (for terrestrial and underwater operating environments.

Applicability of this system as a part of underwater robotic complexes is due to the possibility of performing long-continuing deep-water operations without lifting robots to the surface for recharging or replacement of on-board accumulator batteries. At that, a set of time-consuming and laborious actions is eliminated: surfacing, unsealing, recharging or battery replacement, subsequent sealing, check of functioning, submersion, and search of the work area. Excluding the necessity to periodically lift the autonomous robot improves the efficiency of underwater works significantly, especially depending on depth (Table 1). Along with recharging, high-speed transmission of data and multi-channel video is provided.

Functioning principle of the contactless power supply system is the use of electrodynamic induction. The system consists of two parts: the radiating unit and the receiving unit. The radiating unit includes: converter of electrical energy (generally, direct current) into high-power high-frequency current, and the converter of this current into electromagnetic field radiated in the required direction. The radiating unit is installed on a static or slow-moving part of the robotic complex and has stable guaranteed power supply (Fig.3). The receiving device consists of a receiving antenna, which converts the received electromagnetic oscillation into high-frequency current, and an impedance-matching device, which converts the received energy into normalized direct current with parameters acceptable for normal charging mode of the on-board accumulator of the autonomous robotic device. The receiving device is installed on a fully autonomous robot to be provided with contactless power supply. The main problem in developing such devices is the provision of electromagnetic compatibility of all RTC systems considering the effect of significant electromagnetic radiation, and also the provision of independence and concurrency of processes of on-board battery charging and high-speed data transmission (up to 10 Mbit/s).

Characteristic	Underwater robot	Terrestrial robot
Distance of efficient energy transmission	0 – 20 mm	10 – 30 mm
Received power	20 W	300 W
Efficiency of system	25 – 45%	20 – 35%
Input voltage of transmission device	+48 V	~220 V
Weight of receiver	450 g	1800 g

Table. 1. Specifications of the contactless power supply system of underwater and terrestrial autonomous robots

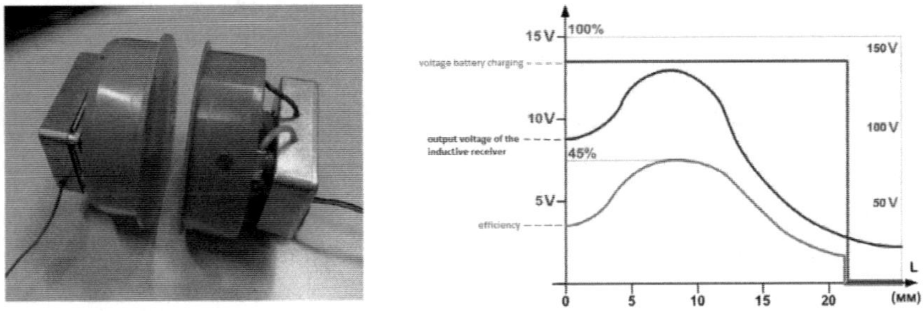

Fig. 3. AMUR-205 mobile robot, contactless power supply and data transmission diagram.

4. Conclusion

This article describes the results of studies, which have led to the development of hardware and software that can be used for solving a wide range of mechatronic device control tasks as well as for training tasks. Along with this, if some other solutions are possible for single robots, the proposed approach has almost no alternative as regards to large distributed mobile robotic complexes. The obtained results are being used for optimization of knowledge representation technologies, development of new heuristic methods for intellectualization of robots with advanced abilities, and for building new creative education technologies.

5. Acknowledgments

This paper was partially supported by the Russian Foundation for Basic Research, Project No. 16-07-01264a, 16-07-00935a, 16-07-00811a.

6. References

[1] Pryanichnikov V.E., Biyelich T., Vitsan D., Katalinich B., Kirsanov K.B., Kuvshinov S.V., Marzanov Y.S., Podurayev Y.V., Khelemendik R.V., Prysev E.A., Ugleshik A., Kharin K.V. Development of education technologies and the network of associated laboratories-robotariums. Data Measuring and Control Systems, M.: "Radiotekhnika". 2015, т.13, N7 – ISSN 2070-0814, C.7-25

[2] Pryanichnikov V.E., Andreev V.P. Intellectulazation of special mobile robots, including return algorithm to a zone of stable RC // Proceedings of the XXI Int. Conference on Extreme Robotics. St. Peterburg: Poliectnika-service. 2010. P. 46–49

[3] Pryanichnikov V.E., Andreev V.P., Prysev E.A. Group control of mobile robots, based on the net-technologies // Proceedings of the Int. conference on Robotics for Security & Space (Planet & Earth rovers). St. Peterburg: Poliectnika service. 2010. P. 279–283

[4] Katalinic B., Pryanichnikov V.E., Ueda K., Kukushkin I., Cesarec P., Kettler R. Bionic Assembly System: hybrid control structure, working scenario and scheduling // Proceedings of 9th National Congress on Theoretical and Applied Mechanics, Brussels. 2012. P. 111–118

[5] Pyanichnikov V.E. Algorithmic base for remote sensors of mobile robots // Mechatronics, Automation, Control. 2008. № 10(91). P. 10–21

[6] Kirsanov K., Levinsky B., Pryanichnikov V. Integrating software for intelligent robots // Informational-Measuring and Control Systems, Radiotechnika. 2009. V. 7. № 6. P. 35–43

[7] Kirillchenko A.A., Pryanichnikov V.E., Rogozin K.V. Limits of validity and reliability of proofs. Scepticism in mathematics, functions, and traditions // Information-Measuring and Control Systems. 2013. V. 11. № 4. P. 57–65.

[8] Klyuev V.V., Kondrat'ev Yu.A., Okhotsimsky D.E., Popov E.P., Pryanichnikov V.E., et al. Sensor Systems and Adaptive Industrial Robots // monograph: Mashinostroenie. 1985. Ch. 3–5. P. 56–172

[9] Okhotsimsky D.E., Veselov V.A., Platonov A.K., Pryanichnikov V.E. Educational Modular transport robot with PC control // Proceedings of the IV Int. conf. on Flexible production systems. Moscow: Int. R&D Institute on the Control Problems. 1987. P. 62–64

[10] Andreev V., Kuvshinov S., Pryanichnikov V., Poduraev Yu. Education on the basis of virtual learning robotics laboratory and group-controlled robots // 24th DAAAM International Symposiumon Intelligent Manufacturing and Automation. 2013. Procedia Engineering. 2014. V. 69. P. 35–40

[11] Pryanichnikov V., Andreev V., Bobrov P., Biryukova E., Frolov A., Kharin K., Kirsanov K., Kostin A., Kuvshinov S., Marzanov Y., Prysev E. Scientific-educational distributed laboratory - software and hardware means // Annals of DAAAM for 2012 & Proceedings of the 23rd International DAAAM Symposium. 2012. V. 23. № 1. DAAAM International. Vienna. Austria. EU

[12] Pryanichnikov V.E., Bielic T., Katalinic B., Kirilchenko A.A., Kirsanov K.B., Khelemendik R.V., Kuvshinov S.V., Marzanov Ju.S., Prysev E.A., Vican D., Uglesic A. Implementation of the main elements of the net of associated laboratories and robotariums - project «Intelligent robotronics»: Sc. Tech. Report#. 2015-01. Zadar: University of Zadar. Moscow: KIAM Russian Academy of Sciences. 2015. 21p

DOI: 10.2507/27th.daaam.proceedings.034

STIFFNESS CONTROL OF ORTHOPAEDIC MATTRESS

Radek Votrubec

This Publication has to be referred as: Votrubec, R[adek] (2016). Stiffness Control of Orthopaedic Mattress, Proceedings of the 27th DAAAM International Symposium, pp.0230-0234, B. Katalinic (Ed.), Published by DAAAM International, ISBN 978-3-902734-08-2, ISSN 1726-9679, Vienna, Austria
DOI: 10.2507/27th.daaam.proceedings.034

Abstract

The paper deals with the stiffness control of enhanced orthopaedic mattress. There are several pneumatic elements into body of the mattress. Pressures in the elements are measured by the pressure sensors. Input and output flows of air are controlled by set of pneumatic valves. All components including compressor are implemented in the bed. First control algorithm was designed in LabVIEW. The final solution is implemented by a microprocessor Arduino. It is possible to set desired stiffness on two zones of mattress, in place of shoulders and in place of hips. The system has two modes. Desired values can be adjusted manually according to achieving equal position of the spine. In the second case, the stiffness can be changed dynamically and may act against bedsore.

Keywords: vibroisolation; orthopaedics; sleep quality; Arduino; PUR foam

1. Introduction

Getting enough sleep is important for all people, because the body regenerates during it. We spend sleeping approximately one third of our lives. Choice of beds and mattresses is important for the quality of sleep. There is a large amount of mattresses on the market. Some of them are adjustable. Adjustability mostly consists in the possibility to replace inserts with a different hardness. Usually two or three options are only possible, hard-soft or hard-medium-soft. It is described in [1]. The goal was to develop user friendly pneumatic system for fluent control stiffness of mattress in areas of shoulders and hips. It enables to set the correct position of the spine, fig.1

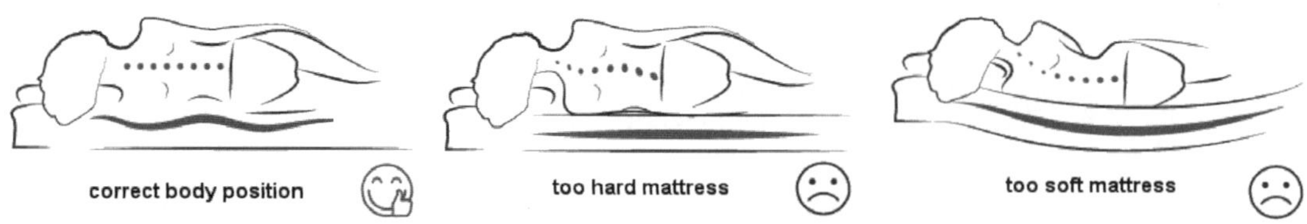

Fig. 1. Correct and wrong body position

The innovation of mattresses is part of project of computer-aided simulation tools [2]. The most critical factor is the speed of implementation of the particular innovation. High demands on the quality of process control require the use of supporting tools that enable us to get complex and detailed analysis of the process. We have dealt with similar systems in our department at our university. It is especially controlled stiffness of car seat or other pneumatic systems [3, 4]. This topic is related to work on other vibroisolation systems such as active ambulance stretcher [5] or platform stabilized by means of gyroscopes [6].

2. Description of mattress

Lamellar orthopaedic mattress was used for our system, fig.2. The basis of the mattress is lamella grid. Three beams made of PUR foam supports the grid. Replaceable inserts with different hardness were replaced by pneumatic elements, fig.3. The force acting from the bottom on the lamella grid depends on the pressure in the pneumatic element. PUR foam is inserted into pneumatic elements. It guarantees minimal force during zero pressure. We have discussed how many zones is optimal. Then we chose two zones only.

Fig. 2. Lamellar orthopaedic mattress

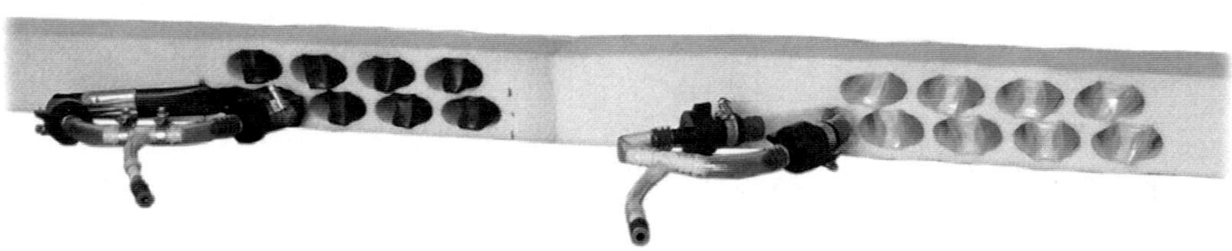

Fig. 3. Pneumatic elements

Articles [7,8] deal with measurement and modelling of damping force of compressed polyurethane foam. Force is described as function of displacement and velocity (1), where coefficients α and b_α are difficult to determine.

$$F_d(x,\dot{x}) = b_\alpha x^\alpha \dot{x} \tag{1}$$

This is a hysteresis curve, which can be identified by measuring. Special tool was constructed for this purpose, fig 4. Measured force in case of various pressures is in fig. 5. Achieved hardness values of mattress with pneumatic elements exceed the range of hardness of standard original mattresses. Interaction between body and parts of some subject is described in [9]. Used methods can be useful for us in the future.

Fig. 4. Tool for measurement of PUR foam force

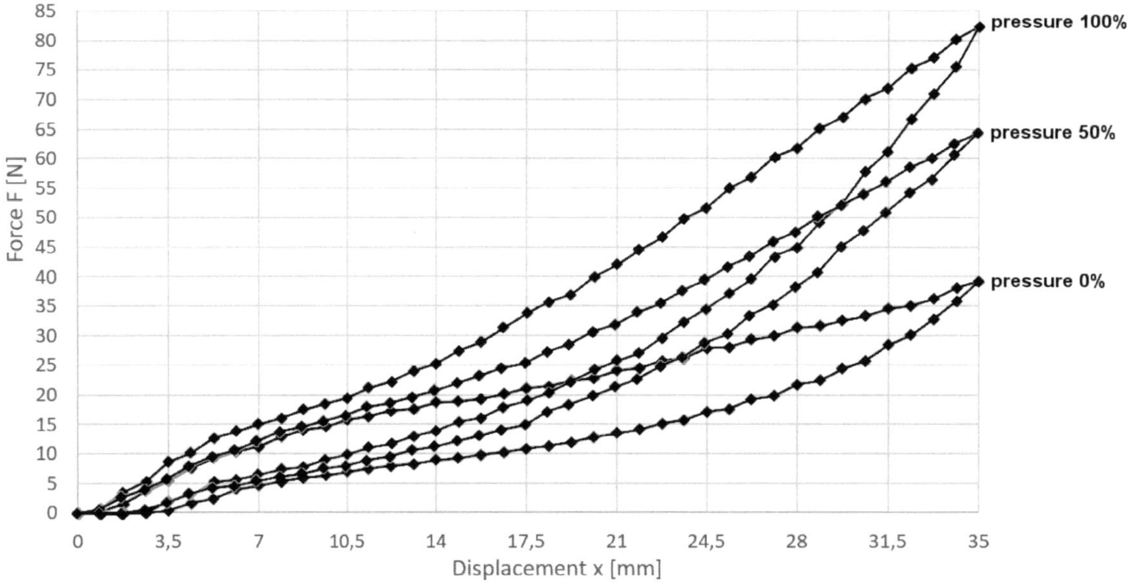

Fig. 5. Measurement of PUR foam force

3. Electric and pneumatic equipment

The scheme of control circuit with actuators, sensors and accessories is in fig. 6. Compressor K supplies pressure air to reservoir R. Sensor PK measures pressure in reservoir. Two state on/off control is used to controlling the compressor. Couple of discrete valves is used to control pressure in each zone. Pressures in pneumatic elements are measured with sensors P1 and P2. All components are connected to Arduino. Maximal current of Arduino digital output is 20mA only. We designed contactless relays for digital outputs, fig. 6 on the right side. They consist of optical isolation circuit and power MOSFET transistor.

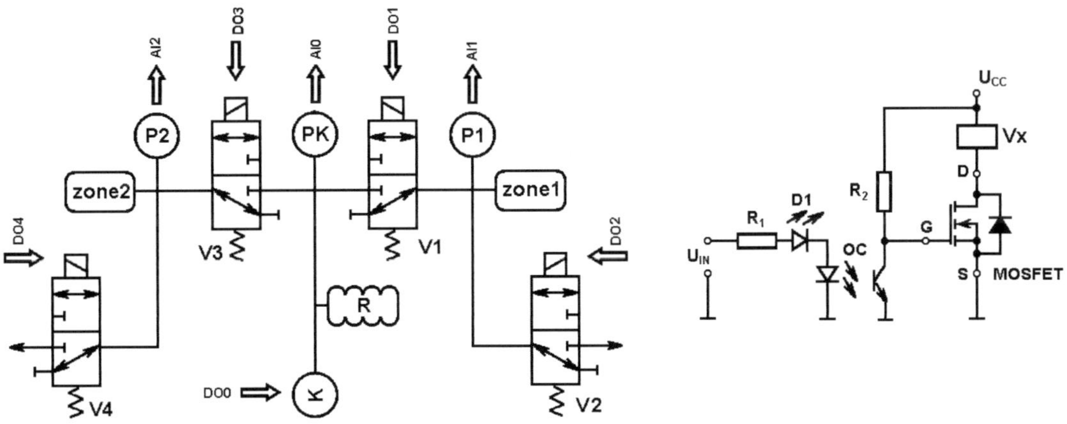

Fig. 6. Pneumatic and electric scheme

The pressure in both zones can be independently adjusted using potentiometers. They are connected to analog inputs of Arduino. Control algorithm is finally realized in Arduino. Proper adjustment pressures in both zones to achieve the correct position of the spine and hips while sleeping, fig. 7. There are two situations in fig.7, wrong position of spine in upper picture and correct position of spine in lower picture. In this case In this case the hip is at right angles to ground.

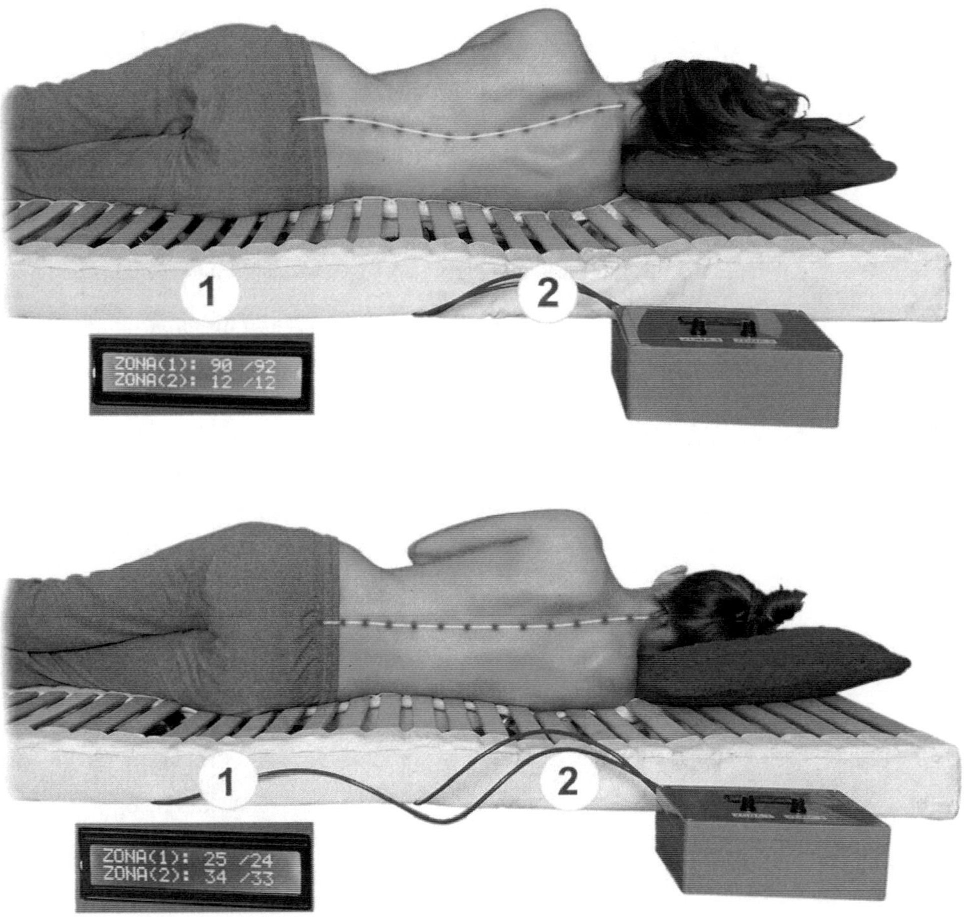

Fig. 7 Wrong and correct position of spine and hips while sleeping

4. Conclusion

The problem was to design user friendly pneumatic system for control stiffness of mattress in areas of shoulders and hips that enables to set the correct position of the spine during sleeping. Five solutions were designed and the best one was selected. The prototype of adjustable orthopaedic mattress with variable stiffness was created. Its control system was designed and realized. Control application with control algorithms was programmed in Arduino. Satisfactory results of control pressure in pneumatic element were obtained. Currently, a group of people testing the mattress. They evaluate the quality of sleep. We are going to offer our solution to furniture manufacturers in Czech Republic. Next it is possible to change standalone microprocessor board instead of Arduino. We are going to find cheaper components suitable for series production. In the future we are going to design some equipment to measure the position of spine. It will enable good setting of pressure in both zones.

5. Acknowledgments

The research has been supported by project SGS 21001-115 "Research and development of control structures of pneumatic, hydraulic and electrical components".

6. References

[1] http://www.jelinek.eu, (2006). The Furniture Jelinek, Accessed on: 2016-09-30

[2] Manlig, F., Slaichova, E., Koblasa, F., Vavruska, J.. Innovation of Business Processes by Means of Computer-Aided Simulation.Applied Mechanics and Materials. 2014, vol. 474, p. 67-72. DOI: 10.4028/www.scientific.net/AMM.474.67. ISSN 1660-9336

[3] Votrubec, R. Control system of the seat with variable stiffness. In: Proceedings of the 16th International Conference on Mechatronics - Mechatronika 2014. IEEE, 2014, p. 314-317. ISBN 978-80-214-4816-2

[4] Votrubec, R. & Vavrousek, M. Control system of a rotary pneumatic motor. In: Proceedings of the 16th International Conference on Mechatronics - Mechatronika 2014. IEEE, 2014, p. 588-593. ISBN 978-80-214-4816-2

[5] Votrubec, R. Control of active sanitary coach, In: Annals of DAAAM for 2011 & Proceedings of the 22th International DAAAM Symposium, s. 617-618, DAAAM,International, Vienna, Austria, 2011, ISSN: 1726-9679, ISBN: 9783901509834

[6] Votrubec, R. Stabilization of platform using gyroscopes, In: Annals of DAAAM for 2013 & Proceedings of the 24rd International DAAAM Symposium "Intelligent Manufacturing & Automation", Editor B.[ranko] Katalinic, ISSN 1877-7058, 23-26th October 2013, Zadar, Croatia

[7] Cirkl,D. Modelling of damping force of dynamically loaded specimen of polyurethane foam in dependence on its density. Shock and Vibration 17 (2010), ISSN: 1070-9622, 373-381

[8] Jiga, G, Stamin, S., Dinu, G., Dobrescu, T, Popovici, D. Comparative Studies on the Impact Behavior of Two Sandwich Structures. Procedia Engineering [online]. 2015, 100, 418-427 [cit. 2016-10-03]. DOI: 10.1016/j.proeng.2015.01.386. ISSN 18777058

[9] Harih G., Borovinsek M., Ren Z., Dolsak B. (2015). Optimal Products' Hand-Handle Interface Parameter Identification. Int. Journal of Simulation Modelling, Vol. 14, No. 3, p. 404-415

DOI: 10.2507/27th.daaam.proceedings.035

COMPARISION OF EDGE DETECTION METHODS FOR OBSTACLES DETECTION IN A MOBILE ROBOT ENVIRONMENT

Boris Crnokić, Snježana Rezić, Slaven Pehar

University of Mostar, Faculty of Mechanical Engineering and Computing, Matice hrvatske bb, 88000, Mostar, BiH

This Publication has to be referred as: Crnokic, B[oris]; Rezic, S[njezana] & Pehar, S[laven] (2016). Comparision of Edge Detection Methods for Obstacles Detection in a Mobile Robot Environment, Proceedings of the 27th DAAAM International Symposium, pp.0235-0244, B. Katalinic (Ed.), Published by DAAAM International, ISBN 978-3-902734-08-2, ISSN 1726-9679, Vienna, Austria
DOI: 10.2507/27th.daaam.proceedings.035

Abstract

This paper presents application of different edge detection methods for detecting of obstacles in the environment of mobile robot. Autonomous mobile robots operating in an unknown environment must be able to cope with hazards and obstacles, and at the same time to navigate through the environment performing required tasks. Detection and avoidance of obstacles, collisions and hazardous situations is a very important part of reliable and secure navigation of mobile robots. Vision sensor (camera) provides large amount of data about obstacles in the environment, however it is necessary to extract only the most important details. This paper will present a comparison of the four most common edge detection methods on examples of obstacles detection in different conditions. Methods used in this study are: Sobel, Prewitt, Roberts and Canny.

Keywords: Mobile Robot; Robot Vision; Obstacle Recognition; Edge Detection.

1. Introduction

In the last two decades mobile robots are attracting more attention of scientists because their range of use is increasing in various fields such as space research, underwater exploration, automotive, military, medical, service robots, robots for serving and entertainment, etc. Mobile robots have to possess some autonomy, intelligence and ability to navigate in an unknown surrounding, which is a fundamental problem for mobile robots. Robot navigation requires guidance of mobile robot through desired path to the desired goal avoiding obstacles and hazards encountered in an unknown environment.[1] Detection and avoidance of obstacles, collisions and hazardous situations is in the first place, however, path planning and arrival at the desired goal is also a very important part of reliable and secure navigation of mobile robots. Planning the optimal path requires optimization of specific performance of navigation, such as minimum time until robot reaches desired goal with a minimum of control, but also requires to comply with certain restrictions in robot motion, such as avoiding obstacles with the maximum speed of the robot.

Generally, all obstacles or hazards in the urban surrounding of the robot, can be classified as [2]:

- Drop offs (sidewalk curbs, downward stairs, steps);
- Static obstacles (walls, furniture);
- Dynamic obstacles (people, doors);
- Invisible obstacles (glass doors and glass walls);
- Overhangs (table tops, railings, tree branches);
- Inclines (wheelchair ramps, curb cuts);
- Rough surfaces (travel paths, grass beds).
- Narrow regions (doorways, elevators).

To avoid collision with obstacles in the environment, robots should be equipped with sensor system that will enable detection of these obstacles. Informations obtained from sensors are used in obstacle avoidance strategies, or in strategies for finding best solution in case of emergency stop of robot before collision.[1] Obstacle detection can be classified into two types [3]:

- *Ranged-based obstacle detection.*
- *Appearance-based obstacle detection.*

Obstacle criterion is important difference between these two obstacle detection systems. In range-based systems, obstacles are objects that protrude a minimum distance from the ground. In appearance-based systems, obstacles are objects that differ in appearance from the ground.[4] In case of range-based obstacle detection, sensors scan the area and detect any obstacle within the range, also sensors calculate the range between the robot and the obstacle. In appearance based obstacle detection, the physical appearance of the obstacle is detected from the environment, usually by image processing.[3] Vision sensors provide large quantity of data from the environment that are useful for obstacles detection. However, this large quantity of data also represents great computational complexity for processing circuitry of the mobile robot, so it is necessary to extract only the most important details. One way to extract only the most important details are methods for detecting the edges of obstacles placed in the robot's surrounding. In this paper we will present comparison of the four most commonly used methods for edge detection: *Sobel, Prewitt, Roberts* and *Canny.*

These edge detection methods were used in various systems for edge detection in different environments and scenarios. Each method has strengths and weaknesses and its particularities. Sobel, Prewitt and Roberts methods uses derivatives on an intensity map to calculate the maximum change in the gradient at an edge [5]. Canny algorithm has the ability to achieve a low error-rate by eliminating almost all non-edges and improving the localization of all identified edges [6]. Various authors have compared these methods on a variety of examples in the application of detected edges to perform different tasks. On the example of edge detection for mobile robot on lunar surface and surroundings [5] it is noticed that Canny edge algorithm results in highly detailed edges but it is not desirable in a defined textured scene, good result provided Sobel operator for detecting high concentration edges. Comparison of these methods was done through a navigation system of a mobile robot using Pan Tilt Zoom network camera mounted on the robot as the primary sensor for robot navigation [7]. Application and comparison of these methods is presented trough edge detection algorithm for vision guided robotics which showed that Canny's edge detection algorithm gives better result and minimum error [8]. The edge detection techniques Canny, Prewitt, Roberts and Sobel are compared in paper to detect the obstacles with the precision of free Google Maps API [9]. In the papers [10]–[12] through various experiments, comparisons of these four methods have shown that Canny algorithm is computationally more expensive compared to Sobel, Prewitt and Robert methods. However, the Canny algorithm performs better in different environments and scenarios, also under noisy conditions and indicates better performance.

This paper presents comparison of Sobel, Prewitt, Roberts and Canny methods to select best edge detection method that will be used for mobile robot indoor navigation. Future work is based on the development of mobile robot navigation system with a fusion of infrared sensors and camera, where most suitable of these edge detecting methods will be used.

2. METHOD USED

2.1. Hardware and Software Used

Hardware used in this paper includes a mobile robot Robotino 2 (Fig. 1.) and a laptop, while the application algorithms were implemented in MATLAB/Simulink programming environment. Robotino is a fully functional, high quality mobile robotic system with omnidirectional wheel drive. The system can work independently but also linked to an external computer system (Wi-Fi connection). Robotino possesses different types of sensors. Some of these sensors are built-in as standard equipment, such as: VGA camera, infrared sensors, incremental encoder and bumper. Other sensors come as an upgrade, such as: analog inductive distance sensor, gyroscope or a navigation system. Robotino controller unit consists of 3 components [13], [14]:

- PC 104 processor, 300 MHz, and Linux operating system with real-time kernel, SDRAM 128 MB
- Compact flash card with C++ API for controlling Robotino
- Wireless LAN access point

Controller unit is equipped with the following interfaces: Ethernet, 2 ea. USB and VGA. Power is supplied by two rechargeable 12 V batteries. Robotino is equipped with a camera system. Its height and inclination can be adjusted, and connects to Robotino via USB port. The camera makes it possible to display live images using different programming interfaces, including MATLAB/Simulink. Technical specifications of the camera are shown in Table 1.

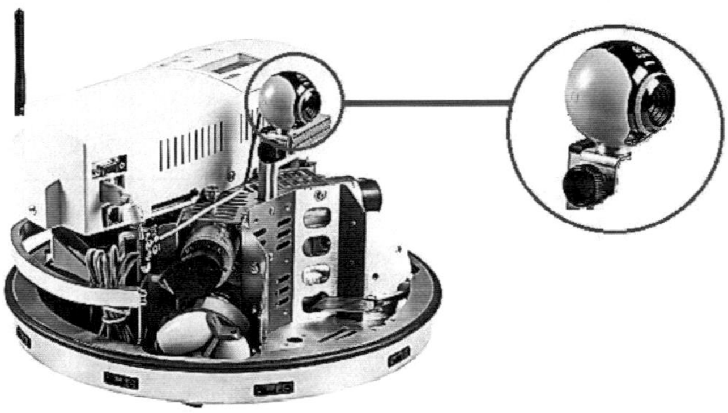

Fig. 1. Mobile robot Robotino 2 and VGA camera [13]

Technical Specifications	
Image Sensor	*Colour VGA CMOS*
Colour Depth	*24 Bit True Colour*
PC-connection	*USB 2.0 (compatible with USB 1.1)*
Video resolutions	*320 x 240, (QVGA), 30 images/second* *640 x 480, (VGA), 30 images/second* *Robotino View can process this resolution* *800 x 600(SVGA, software-interpolated), 30 images/second*
Still image resolutions	*640 x 480 (VGA)* *800 x 600(SVGA)* *1024 x 768 (SXGA, software-interpolated)*
Still capture format	*BMP, JPG*

Table 1. Technical specificatios of VGA camera [13]

2.2. Edge Detection Methods

Edge detection is one of the most important parts in the image processing systems of mobile robots navigation. It allows the extraction and display of features such as curves, lines and angles for the purposes of identifying images from the environment. Images represented by edges are widely used to simplify the process of identifying and interpreting images to the navigation systems of mobile robots which use camera as primary sensor.[15]

Main idea in edge detection process is based on an abrupt change in the intensity of pixels between adjacent pixels in the image. The edge represents a location that forms the border between the pixels of high and low intensity. Abrupt transition in the intensity of pixels requires the application of derivative operators to create a gradient images. The edge detector can be defined as a mathematical operator that corresponds to spatial variations and discontinuities in the grayscale pixel set in the image. The edge is indicated with a rapid changes within the picture that shows the characteristic features and can therefore be characterized as a set of pixels whose intensity of environmental follows continuous variation.

The quality of edge detection largely depends on light conditions, the presence of objects of similar intensity, density of edges in the scene and noises. Different edge detectors work better under different conditions, and therefore there is no uniform algorithm that can be adapted to all conditions in the environment. An "optimal" edge detector means [16]:

- *Good detection* – the algorithm should mark real edges in the image as possible.
- *Good localization* – Marked edges should be as close as possible to the edge in the real image.
- *Minimal response* – a given edge in the image should only be noticeable once, and where possible, image noise should not generate false edges.

Ideal would be to use an algorithm that simultaneously uses different detectors applying each of them when conditions in environment are ideal for that particular method. This paper will present comparison of four methods for detecting edges of obstacles that mobile robot encounters. These methods are: *Sobel, Prewitt, Roberts and Canny.*

Sobel Edge Detector

Sobel operator uses two 3×3 kernels which are convolved with the original image to calculate approximations of the derivatives, one for horizontal changes, and one for vertical changes. To produce separate measurements of the gradient component in each orientation, kernels can be applied separately to the input image (*Gx* - in direction of the X-axis; *Gy* – in direction of Y-axis), as follows: [18]

$$G_x = \begin{bmatrix} -1 & 0 & +1 \\ -2 & 0 & +2 \\ -1 & 0 & +1 \end{bmatrix} \quad and \quad G_y = \begin{bmatrix} -1 & -2 & -1 \\ 0 & 0 & 0 \\ +1 & +2 & +1 \end{bmatrix} \tag{1}$$

The x-coordinate is defined as increasing in the "right"-direction, and the y-coordinate is defined as increasing in the "down"- direction. These can then be combined together to find the resulting gradient approximations at each point in the image, to give the gradient magnitude:

$$G = \sqrt{G_x^2 + G_y^2} \tag{2}$$

Then, it is possible to calculate the gradient's direction:

$$\theta = arctan\left(\frac{G_y}{G_x}\right) \tag{3}$$

Prewitt edge detector

Prewitt operator is very similar to the Sobel operator, and also utilizes two masks which enable calculation of pixel intensity gradient in the vertical and horizontal directions. The overall gradient and edge orientation is calculated in the same way [17], [18]:

$$G = \sqrt{G_x^2 + G_y^2} \quad and \quad \theta = arctan\left(\frac{G_y}{G_x}\right) \tag{4}$$

Horizontal (*Gx*) and vertical (*Gy*) derivative approximations are computed as:

$$G_x = \begin{bmatrix} -1 & 0 & +1 \\ -1 & 0 & +1 \\ -1 & 0 & +1 \end{bmatrix} \quad and \quad G_y = \begin{bmatrix} -1 & -1 & -1 \\ 0 & 0 & 0 \\ +1 & +1 & +1 \end{bmatrix} \tag{5}$$

Roberts edge detector

The Roberts detector performs a simple, quick to compute, 2-D spatial gradient measurement on an image. Pixel values at each point in the output represent the estimated absolute magnitude of the spatial gradient of the input image at that point. The operator consists of a pair of 2×2 convolution kernels. These kernels are designed to respond maximally to edges running at 45° to the pixel grid, one kernel for each of the two perpendicular orientations. The kernels can be applied separately to the input image, to produce separate measurements of the gradient component in each orientation (*Gx* and *Gy*).This is very similar to the Sobel operator. In order to perform edge detection with the

Roberts operator initially it is necessary to convolve the original image, with the following two kernels [11], [17]:

$$G_x = \begin{bmatrix} +1 & 0 \\ 0 & -1 \end{bmatrix} \quad and \quad G_x = \begin{bmatrix} 0 & +1 \\ -1 & 0 \end{bmatrix} \tag{6}$$

These can then be combined together to find the absolute magnitude of the gradient at each point and the orientation of that gradient:

$$G = \sqrt{G_x^2 + G_y^2} \tag{7}$$

The direction of the gradient can also be defined as follows:

$$\theta = arctan\left(\frac{G_y}{G_x}\right) - \frac{3\pi}{4} \tag{8}$$

Canny Edge Detector

The Canny edge detector is known as the optimal edge detector. John F. Canny in 1986 introduced a multi-stage algorithm for detecting a wide range of edges. He discussed a mathematical problem of derivation of optimal smoothing filter with respect to the criteria of detection, localization and minimization of multiple responses of one edge.[19] Canny saw the edge detection problem as a signal processing optimization problem, so he developed an objective function to be optimized. The solution to this problem was a rather complex exponential function, but Canny found several ways to approximate and optimize the edge-searching problem. Steps in the canny edge detector are as follows [17], [19]:

1. *Smoothing*: Blurring of the image to remove noise.
2. *Finding gradients*: The edges should be marked where the gradients of the image has large magnitudes.
3. *Non-maximum suppression*: Only local maxima should be marked as edges.
4. *Double thresholding*: Potential edges are determined by thresholding.
5. *Edge tracking by hysteresis*: Final edges are determined by suppressing all edges that are not connected to a very certain (strong) edge.

The block diagram of Canny edge detection algorithm is shown in Figure 2.

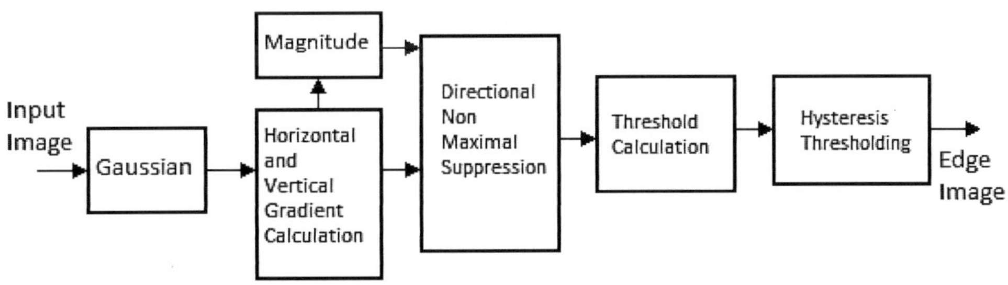

Fig. 2. Block diagram of Canny edge detector [19]

3. Implementation

Sobel, Prewitt, Roberts and Canny edge detection algorithms are implemented using MATLAB Simulink models by using Video and Image Processing Blocksets from Simulink of MATLAB 7.10 (2010a). Manufacturer (Festo) of mobile robot Robotino has provided an upgrade for MATLAB/Simulink. This also includes Simulink Blockset with blocks for connecting to Robotino and to control all its components, including the camera, motors, sensors, etc.

Exactly the same blocks and the same structure is used for Sobel, Prewitt and Roberts edge detector Simulink models (Fig. 3.).

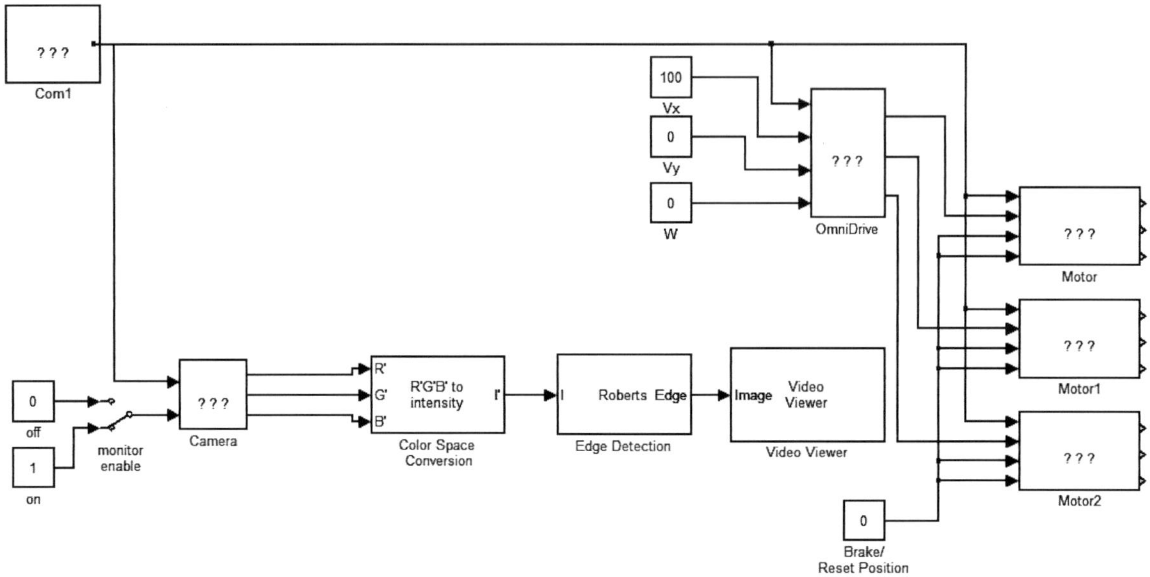

Fig. 3. Sobel edge detection Simulink model for Robotino

The only thing that was changed for each method is the selection of exactly preferred method in the "Function Block Parameters" of "Edge Detection" block (Fig. 4.). For Sobel, Prewitt, and Roberts, the "Edge Detection" block finds the edges in an input image by approximating the gradient magnitude of the image. The block convolves the input matrix with the Sobel, Prewitt, or Roberts kernel, and outputs two gradient components of the image, which are the result of this convolution operation. Alternatively, the block can perform a thresholding operation on the gradient magnitudes and output a binary image, which is a matrix of Boolean values. If a pixel value is 1, it is an edge. [16], [17]

After starting the simulation, Robotino is moving and observing with a camera various obstacles in the environment. Image from the camera passes through the "Edge Detection" block and appears as an image with edges in the "Video Viewer" block. For comparison, on other monitor is possible to see real image of the obstacle.

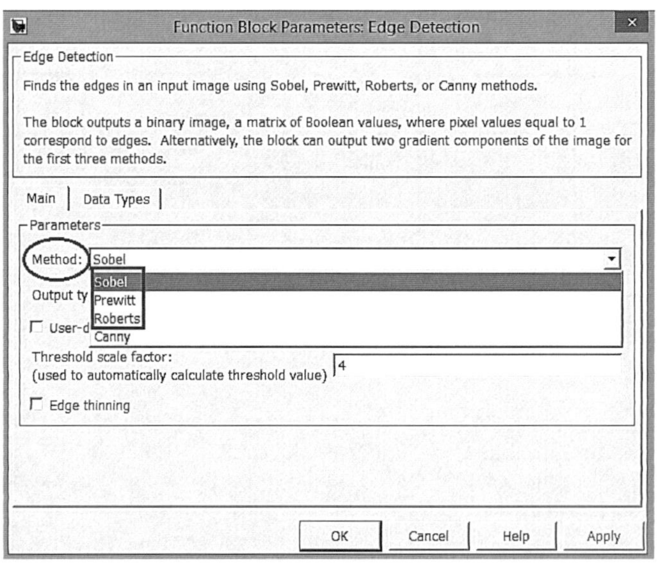

Fig. 4. Selection of methods for edge detection

Canny edge detector Simulink model is created with the same blocks as in the previous case and with some new blocks (Fig. 5.). [20] Fixed-point input is not supported for Canny edge method in Simulink. "Gaussian Noise Generator" and "Median Filter" blocks are used. It is necessary to add noise to the image using a Gaussian noise generator in the order to test the model in noisy environment. "Median Filter" is used to reduce the effect on noise present in the image.[21] For Canny method, the "Edge Detection" block finds edges by looking for the local maxima of the gradient of the input image. It calculates the gradient using the derivative of the Gaussian filter. The Canny method uses two thresholds to detect strong and weak edges. It includes the weak edges in the output only if they are connected to strong edges. As a result, the method is more robust to noise, and more likely to detect true weak edges.[17], [21], [22]

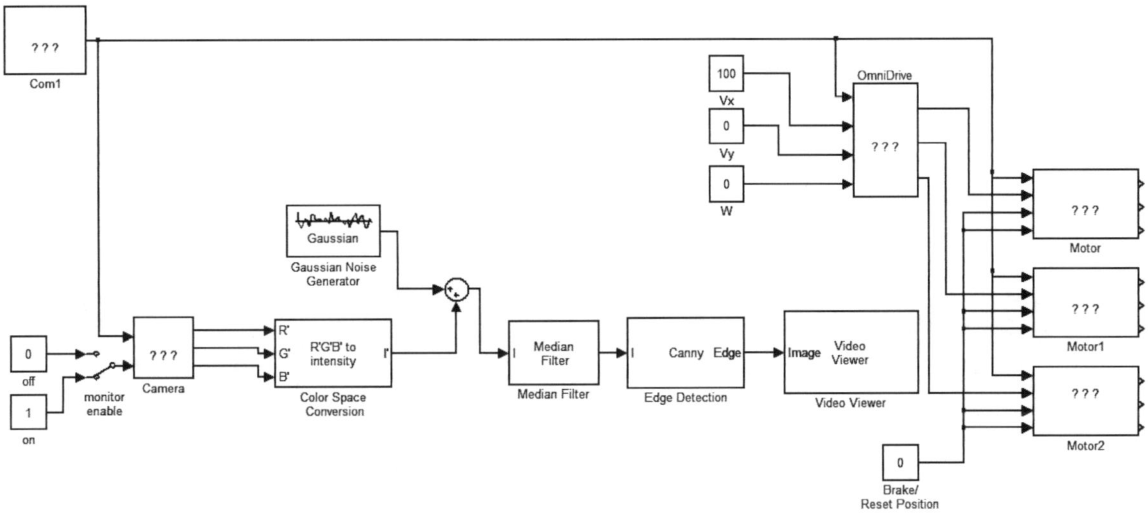

Fig. 5. Canny edge detection Simulink model for Robotino

4. Results

Results of edge detection of some characteristic obstacles are presented in this section. First of all, the Simulink models of Sobel, Prewitt, Roberts and Canny edge detector were tested with different parameters, under different lighting conditions, and thereafter edge detection was performed with the best configuration for each model. Sobel, Prewitt and Roberts Simulink model were tested with different values of "*Threshold scale factor*" (***TSF***), which is used to automatically calculate threshold value. Threshold scale factor was combined with the option "*Edge thinning*" (***ET***). Values of the Threshold scale factor were *from 1 to 4*. Canny model was tested with different values of "*Approximate percentage of weak edge and nonedge pixels*" (***APP***), which is used to automatically calculate threshold value. Values of Approximate percentage of weak edge and nonedge pixels were *from 70 to 99*. Value of "*Standard deviation of Gaussian filter*" whole time of the testing was set to 1, because in the previous use of these tools it has been proved that this is the most appropriate value. Fig. 6. shows testing of all edge detecting methods on the example of detecting edges while the robot passes through the narrow passage between the wall and the desk.

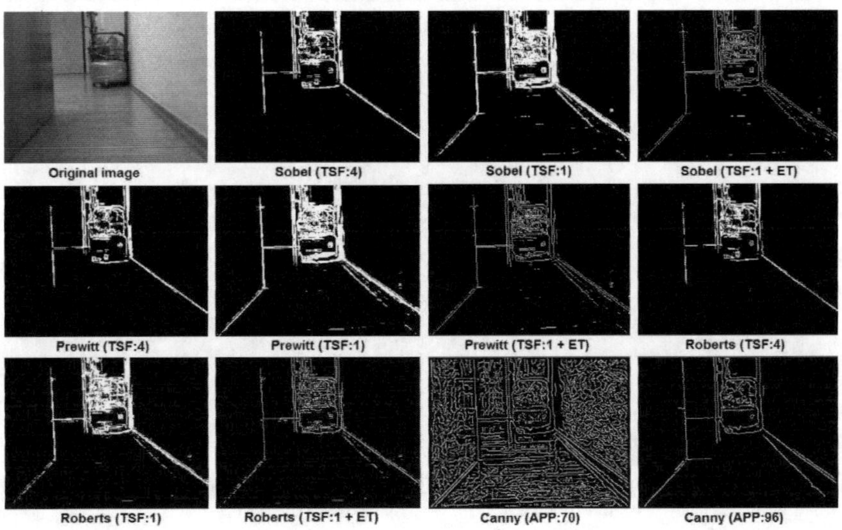

Fig. 6. Testing of edge detecting methods – Obstacle: passage between the wall and the desk; Room: poorly illuminated

After testing on different scenes, it is possible to notice that the following parameters are most suitable:

- Sobel, Prewitt and Roberts: ***TSF**: **1 or 4** plus **ET***
- Canny: ***APP**: **95-97***

Following images show detection of edges in the accordance with previously adopted parameters. These settings provide the best edge detection of the obstacles that Robotino encounters.

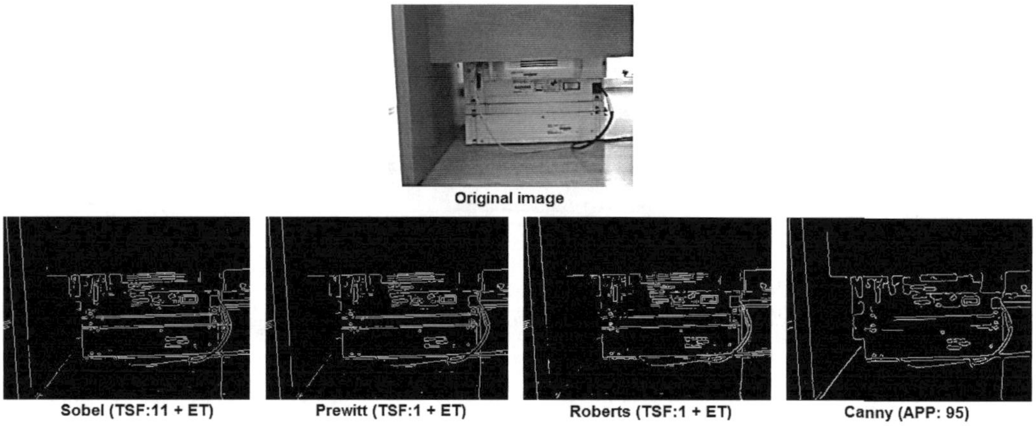

Fig. 7. Edge detection - Obstacles: table and crosspiece; Room: moderately illuminated

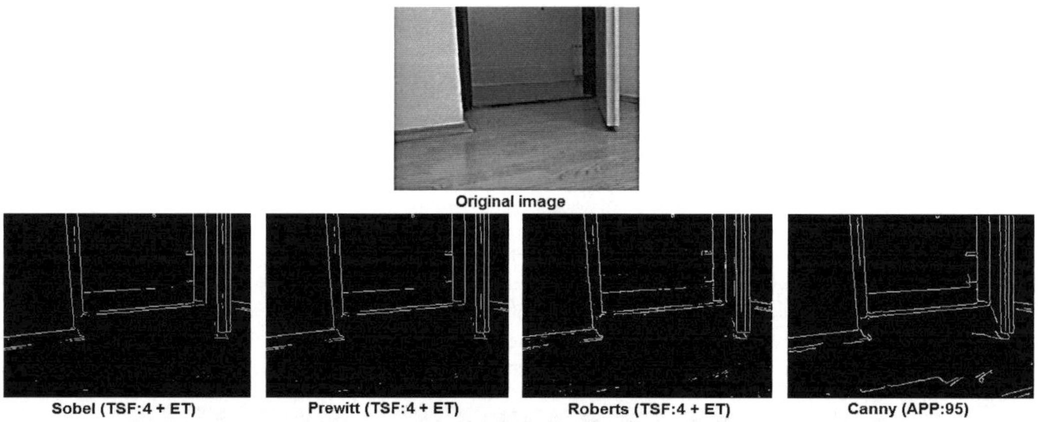

Fig. 8. Edge detection - Obstacles: door, wall, doorstep; Room: very illuminated

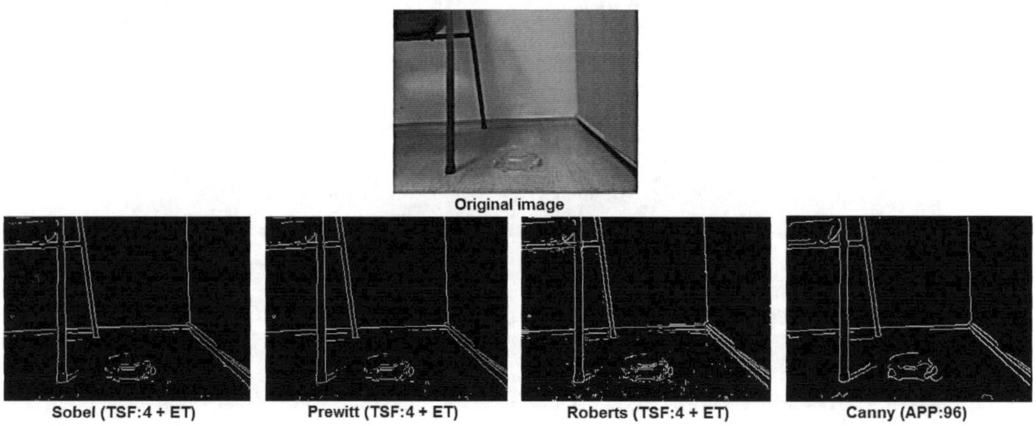

Fig. 9. Edge detection - Obstacles: chair, closet, glass ashtray; Room: very illuminated

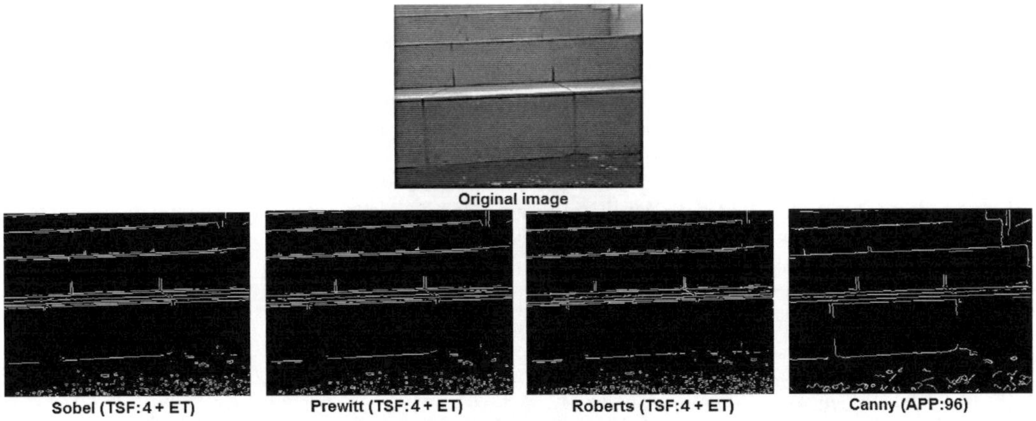

Fig. 10. Edge detection - Obstacles: upward stairs; Room: very illuminated

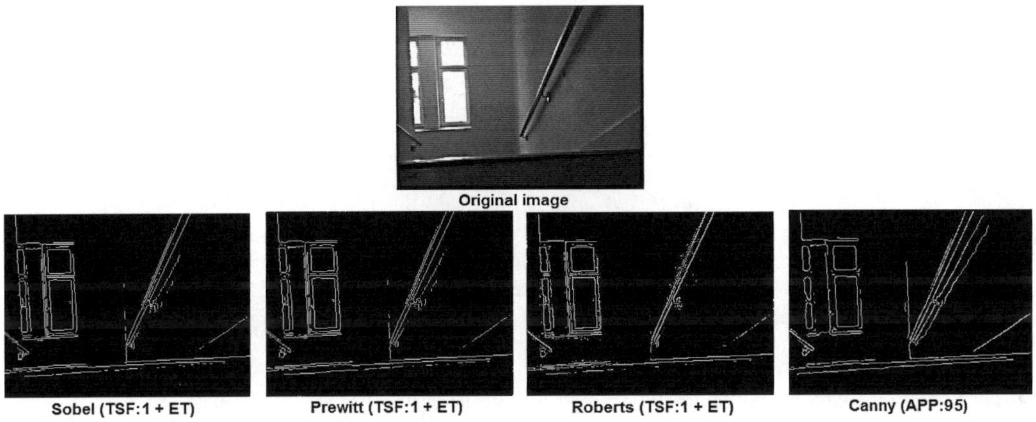

Fig. 11. Edge detection - Obstacles: upward stairs; Room: moderately illuminated

Analysing examples of the previous images it can be concluded that Sobel, Prewitt and Roberts produced thicker edges. Sobel, Prewitt and Roberts have shown simplicity in edge detection and their orientations, but Sobel gives some false edges in case when images have more details. However, they are quite sensitive to noises in the detection of the edges and their orientations, and have decreased accuracy as the gradient magnitude of the edges decreases. Canny method gave the best results and provides sharpest and clearest obstacle edges, lines of edges are continuous and clear, and this is a very important feature in the application of edge detection method in the mobile robot navigation system. Canny gives better results in case of noise conditions, and provides a thin lines using non-maximal suppression. Use of the high threshold eliminates almost all the false edges. Canny edge detector algorithm is computationally more expensive compared to Sobel, Prewitt and Roberts. Also, main disadvantage is time consumption because of complex computation, and it is difficult to reach real time response. To improve real time response, these disadvantages will be compensated by using more powerful computer processing unit for mobile robot navigation system that will be implemented through this project. Finally, as confirmed in comparison with other studies referenced in this paper, Canny edge detector has the best results, Sobel and Prewitt mutually have very similar characteristics of the edges, Roberts has weakest results with the highest number of false edges.

It has been shown that the best settings for edge detection, compared to environmental conditions, are as follows:

- In the case of poor illumination, for Sobel, Prewitt and Roberts method, Threshold scale factor should have a lower value (in this simulation it is set to 1).
- When it comes to increased illumination, for the same methods, Threshold scale factor should have a higher value (in this simulation it is set to 4).
- Canny edge detector showed consistent results in both cases, whether it is a poor or increased illumination. The results are very similar, compared to lighting conditions, when the approximate percentage of weak edge and nonedge pixels is in the range 95-97.

5. Conclusion

This paper presents a Simulink models of Sobel, Prewitt, Roberts and the Canny edge detectors, used to compare the quality of edges applied on detection of obstacle edges in navigation system of mobile robot. For mobile robot navigation and detection of obstacles, it is very important that obstacles are visible, respectively, lines of edges must be clear without false edges. Since edge detection is the initial step in obstacles detection and recognition, it is very important to choose a method that will provide greatest number of correct data with least computational complexity. As shown through the examples in this paper, Canny edge detector has the best results and it is able to provide sharpest and clearest obstacle edges, with continuous and clear lines. Disadvantage of Canny method is computational complexities, so it is necessary that robot has a powerful computer system in order to detect obstacles in the real time. Future work will be focused on development of mobile robot navigation algorithm, which will be based on Canny edge detector.

6. References

[1] B. Crnokić and M. Grubišić, "Comparison of Solutions for Simultaneous Localization and Mapping for Mobile Robot," Proc. Fac. Mech. Eng. Comput. Univ. Most., p. 8, 2014
[2] A. Muraka, M. Sridahan, and B. Kuipers, "Detecting Obstacles and Drop-offs using Stereo and Motion Cues for Safe Local Motion," Intell. Robot. Syst. 2008. IROS 2008. IEEE/RSJ Int. Conf., pp. 702–708, 2008
[3] F. Shahdib and H. Mahmud, "Obstacle Detection and Object Size Measurement for Autonomous Mobile Robot using Sensor," Int. J. Comput. Appl., vol. 66, no. March 2013, pp. 28–33, 2013
[4] I. Ulrich and I. Nourbakhsh, "Appearance-Based Obstacle Detection with Monocular Color Vision," Artif. Intell., no. August, pp. 866–871, 2000
[5] M. Iqbal, "Edge Detection for Mobile Robots on Lunar Surface and Surroundings," BRAC University; Dhaka, Bangladesh, 2012
[6] C. Gentsos, C. L. Sotiropoulou, S. Nikolaidis, and N. Vassiliadis, "Real-time canny edge detection parallel implementation for FPGAs," 2010 IEEE Int. Conf. Electron. Circuits, Syst. ICECS 2010 - Proc., pp. 499–502, 2010
[7] M. S. Guzel, "Mobile Robot Navigation using a Vision Based Approach," Computer (Long. Beach. Calif)., 2003
[8] B. K. Balabantaray, P. Jha, and B. B. Biswal, "Application of edge detection algorithm for vision guided robotics assembly system," Sixth Int. Conf. Mach. Vis., no. November, pp. 1–7, 2013
[9] A. Q. Mahdi, G. A. QasMarrogy, and M. M. Ihassan, "Comparison of the Edge Detection Methods to Detect , Identify and Locate the Obstacles for Agricultural Robotic Vehicles," Int. J. Res. Comput. Appl. Robot., vol. 2, no. January, pp. 5–17, 2014
[10] P. S. Deokar, "Review on Distributed Canny Edge Detector using FPGA," vol. 1, no. 4, pp. 37–45, 2014.
[11] R. Maini and H. Aggarwal, "Study and comparison of various image edge detection techniques," Int. J. Image Process. ..., vol. 147002, no. 3, pp. 1–12, 2009
[12] B. Bansal, J. S. Saini, V. Bansal, and G. Kaur, "Comparison of Various Edge Detection Techniques," J. Inf. Oper. Manag., vol. 3, no. 1, pp. 103–106, 2012
[13] Festo Didactic, "Robotino-Workbook With CD-ROM Festo Didactic 544307 en." Denkendorf, p. 124, 2012
[14] I. Breido, R. Markvardt, and A. Zarnitcin, "Modernization of Communication Interface Between Mobile Robot Robotino and Matlab Software," Ann. DAAAM 2012 Proc. 23rd Int. DAAAM Symp., vol. 23, no. 1, pp. 1107–1110, 2012
[15] W. Lee, D. Kim, and I. Kweon, "Automatic edge detection method for the mobile robot application," Intell. Robot. Syst. 2003. (IROS 2003). Proceedings. 2003 IEEE/RSJ Int. Conf., vol. 3, pp. 2730–2735, 2003.
[16] P.H. Pawar R.P. Patil, "Image Edge Detection Techniques using MATLAB Simulink," Int. J. Eng. Res. Technol., vol. 3, no. 6, pp. 2149–2153, 2014
[17] M.R. Rakesh, "Design and Simulation of Matlab / Simulink Model for Edge Detection Techniques in Image Segmentation," Int. J. Adv. Res. Electr. Electron. Instrum. Eng., vol. Vol. 2, no. Issue 12, pp. 5828–5834, 2013.
[18] N. H. Seng, "A comparison of various edge detection methods," Seberang Perai, 2013
[19] N. M. Chaithra and K. V. R. Reddy, "Implementation of Canny Edge Detection Algorithm on FPGA and displaying Image through VGA Interface," no. 6, pp. 243–247, 2013
[20] B. Crnokić and S. Rezić, "Edge Detection for Mobile Robot using Canny method," in 3rd International Conference „NEW TECHNOLOGIES NT-2016 „Development and Application, 2016, pp. 106–111
[21] The MathWorks Inc, "Edge Detection," Find edges of objects in images using Sobel, Prewitt, Roberts, or Canny method. [Online]. Available: https://www.mathworks.com/help/vision/ref/edgedetection.html. [Accessed: 08-Mar-2015]
[22] J. K. Sahu, P. R. Mishra, P. K. Sethy, and S. Behera, "Edge detection of a still image using simulation and real-time implementation using TMS320C6713DSK," UGC Spons. Natl. Semin. Emerg. Trends Inf. Technol. Dev., no. March, 2014

DOI: 10.2507/27th.daaam.proceedings.036

NUMERICAL SOLUTION OF POISSON'S EQUATION IN AN ARBITRARY DOMAIN BY USING MESHLESS R-FUNCTION METHOD

Vedrana Kozulic & Blaz Gotovac

This Publication has to be referred as: Kozulic, V[edrana] & Gotovac, B[laz] (2016). Numerical Solution of Poisson's Equation in an Arbitrary Domain by Using Meshless R-Function Method, Proceedings of the 27th DAAAM International Symposium, pp.0245-0254, B. Katalinic (Ed.), Published by DAAAM International, ISBN 978-3-902734-08-2, ISSN 1726-9679, Vienna, Austria
DOI: 10.2507/27th.daaam.proceedings.036

Abstract

This paper describes a numerical procedure that uses solution structure method, atomic basis functions and a collocation technique. Solution structure method is based on the theory of R-functions. The solution of a boundary value problem is expressed in the form of formulae called solution structure which depends on three components: the first component describes the geometry of the domain exactly in analytical form, the second describes all boundary conditions exactly, while the third component is called differential component because it contains information about governing equation. Unknown differential component of the solution structure is represented by a linear combination of basis functions. Here, we propose to use atomic basis functions because of their good approximation properties. To determine the coefficients of linear combination in the solution structure, a collocation technique is used. Combination of atomic basis functions and solution structure method gives the meshfree method that can be applied for solving boundary value problems in domains of arbitrarily complex geometry with complex boundary conditions. This paper summarizes the main principles of the proposed method and presents its application to solution of the torsion problem.

Keywords: meshless method; solution structure; collocation; boundary conditions; atomic basis functions.

1. Introduction

Widely used mesh-based numerical methods such as finite element, finite difference, and finite volume methods, introduce a finite number of nodes to specify boundary conditions and perform numerical computations, and use spatial grids to approximate the geometric shape of a model. However, in modelling problems with complex geometry, difficulties often appear in creating good spatial grid that conforms to the shape of the model. To overcome this obstacle, a new class of numerical methods has been developed called meshfree or meshless methods. These methods may still use spatial grids to construct the basis functions and perform numerical computations, but such grids do not necessarily have to conform to the geometric model. To date, many different meshfree methods have been developed. Their detailed review and comparison can be found in many references [1], [2], [3], [4], [5]. Many meshfree methods use radial basis functions to represent solutions of engineering problems [6]. Using meshfree methods significantly simplified the meshing process, but, at the same time, it made the treatment of boundary conditions as demanding task [7].

Here, a numerical method for solving engineering problems that enables exact treatment of all prescribed boundary conditions at all boundary points and does not require numerical integration is presented. It combines meshfree method known as solution structure method, atomic basis functions (ABFs) and a collocation technique.

In the solution structure method, a solution is sought in the form of formulae called solution structure. The original idea is due to Kantorovich [8]. He proposed that the homogeneous Dirichlet conditions may be satisfied exactly by representing the solution as the product of two functions: (1) an real-valued function that takes on zero values on the boundary points; and (2) an unknown function that allows to satisfy (exactly or approximately) the differential equation of the problem. Such a solution structure was used by Kantorovich and his students to solve boundary value problems on geometrically simple domains.

Rvachev [9] suggested using R-functions — the real valued functions that behave as continuous analogy of logical Boolean functions. R-functions allow construction of a set of functions vanishing on the boundary that possess desired differential properties and may be assembled into a solution structure. Based on the theory of R-functions, the R-Function Method (RFM), also known as solution structure method, is developed which can be applied to problems with arbitrarily complex domains and boundary conditions. Over the last several decades, the theory of R-functions have been applied to numerous scientific and engineering problems by Rvachev and his students [10], [11]. The RFM has been applied to problems of thermo conduction, elasticity, magneto-hydrodynamics, various problems in inhomogeneous media, and many other areas [12].

This paper presents the use of atomic basis functions to approximate unknown differential component of the solution structure. They are infinitely-differentiable functions with compact support [13], [14], [15]. Rvachev and Rvachev [13], in their pioneering work, called these basis functions "atomic" because they span the vector spaces of all three fundamental functions in mathematics: algebraic, exponential and trigonometric polynomials. In numerical modelling, we applied Fup basis functions that belong to the atomic functions of algebraic type [16], [17], [18]. All derivatives of atomic Fup basis functions required by differential operators in the solution structure can be used directly in the numerical procedure. This fact allows to use procedures based on strong formulation. To determine the coefficients of linear combination in the solution structure, a collocation technique is used.

2. Solution Structure Method: Basic principle

The original idea of the solution structure method [8] is to express the solution of two-dimensional boundary value problem with homogeneous Dirichlet boundary conditions

$$u|_{\Gamma} = 0 \quad (1)$$

by formula called solution structure in the form of the product of two functions:

$$u = \omega \Phi \quad (2)$$

where $\omega : R^n \to R$ is a known function that takes on zero values on the boundary of the domain Γ and is positive in the interior of the domain Ω, and Φ is some unknown function that allows to satisfy (exactly or approximately) the differential equation of the problem.

In most practical situations, unknown Φ is represented by a linear combination of basis functions

$$\Phi = \sum_{i=1}^{n} C_i F_i \quad (3)$$

where C_i are scalar coefficients and F_i are some basis functions. The solution structure does not place any constraints on the choice of basis functions. Numerical values of the coefficients C_i can be obtained by using different numerical methods.

For complex domains, Rvachev [9] set the theory of R-functions which was the basis for the development of the R-function method.

2.1. Theory of R-functions

The question is how to generate functions that will simply describe the given domain and satisfy different boundary conditions on any part of the boundary? There is an elegant way by using ω functions.

Generating ω functions over the complex area is proposed by the Ukrainian scientist V. L. Rvachev. Rvachev [9] came up with the idea that logic operations of Boolean algebra are applied to the functions. In this way he created so called semi-algebra. Basic R-operations are shown in Fig. 1.

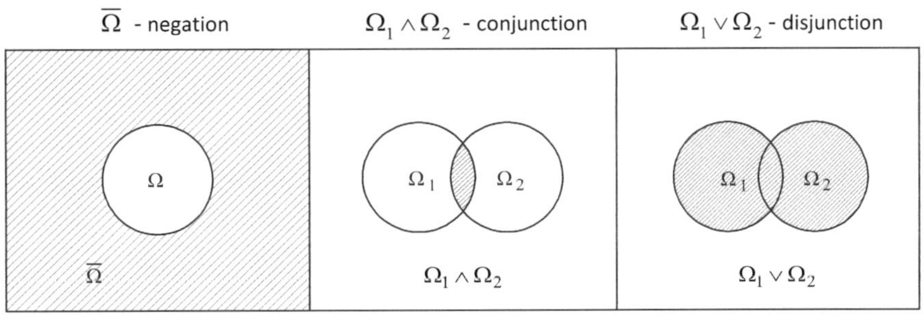

Fig. 1. Basic R-operations

R – conjunction (section) ($f_1 \wedge f_2$):

Boolean function is logical "and" ($\wedge$). It is defined in the form: $F_1(f_1,f_2) = f_1 \wedge f_2 \equiv f_1 + f_2 - \sqrt{f_1^2 + f_2^2}$

R – disjunction (union) ($f_1 \vee f_2$):

Boolean function is logical "or" ($\vee$). It is defined in the form: $F_2(f_1,f_2) = f_1 \vee f_2 \equiv f_1 + f_2 + \sqrt{f_1^2 + f_2^2}$

R – negation ($\neg f = -f$). The logical negation of the function is the change of sign of that function: $F_3(f) = \overline{f} \equiv -f$.

Using these operations, we can determine the function ω over the very complex domains. Then such functions are called R-functions.

3. General solution structures

Once we constructed the solution structure $u = \omega\,\Phi$ for the boundary value problems with homogeneous Dirichlet boundary conditions, it is easy to obtain the solution structure for nonhomogeneous conditions

$$u\big|_{\Gamma} = \varphi_0 \tag{4}$$

Let φ be the extension of the φ_0 inside the domain Ω. Then the solution structure

$$u = \omega\,\Phi + \varphi \tag{5}$$

satisfies the prescribed boundary conditions exactly. In practice, the function φ_0 may be specified in a piecewise fashion, with a different value φ_i prescribed on each portion of the boundary Γ_i. Such individual boundary conditions may be combined into a single global function φ [9], [12]:

$$\varphi = \frac{\sum\limits_{i=1}^{n} \varphi_i \prod\limits_{j=1, j\neq i}^{n} \omega_j}{\sum\limits_{i=1}^{n} \prod\limits_{j=1, j\neq i}^{n} \omega_j} \tag{6}$$

Detailed and systematic derivations of solution structures for more general boundary conditions can be found for example, in references [9] and [12]. General solution structure for the second-order boundary value problem with mixed boundary conditions:

$$u\big|_{\Gamma_1} = \varphi_0 \quad ; \quad \left(\frac{\partial u}{\partial n} + hu\right)\bigg|_{\Gamma_2} = \psi_0 \tag{7}$$

can be written in the form that interpolates boundary conditions on Γ_1 and Γ_2 according to [9]:

$$u = \omega_1\Phi - \omega\cdot\left(D_1^{\omega_2}(\omega_1\Phi) - h\,\omega_1\Phi + D_1^{\omega_2}(\varphi) + \psi - h\,\varphi - \omega_2\Phi\right) + \varphi \tag{8}$$

where h is a part of the Robin boundary condition. ω_1 is part of the primary function ω that belongs to the part of the boundary Γ_1 while ω_2 is part of the primary function ω that belongs to the part of the boundary Γ_2. For example, for the domain in Fig. 2 can be written:

$$\Gamma_1 = L_1 \cup L_3 \cup L_4 \cup L_6 \cup L_8 \;,\; \Gamma_2 = L_2 \cup L_5 \cup L_7 \cup L_9 \;,\; \Gamma_1 \cup \Gamma_2 = \Gamma \;;\; \omega_1\big|_{\Gamma_1} = 0 \;,\; \omega_2\big|_{\Gamma_2} = 0 .$$

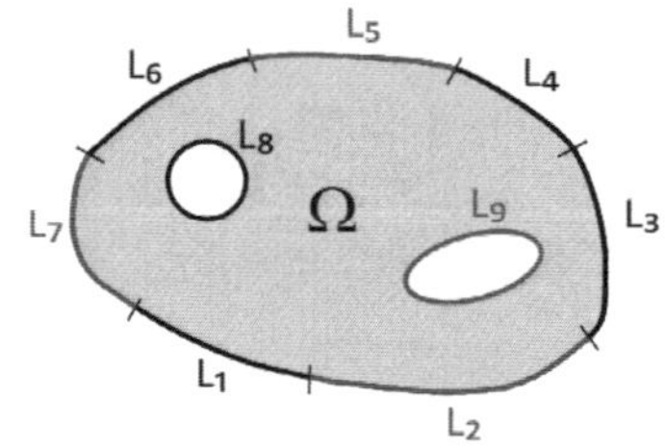

Fig. 2. Two-dimensional domain of arbitrary shape with mixed boundary conditions

Differential operator of the first order $D_1^{\omega_2}(\bullet)$ is:

$$D_1^{\omega_2}(\bullet) = \frac{\partial \omega_2}{\partial x} \cdot \frac{\partial (\bullet)}{\partial x} + \frac{\partial \omega_2}{\partial y} \cdot \frac{\partial (\bullet)}{\partial y} \tag{9}$$

Differential operators in the solution structures transmit information about the dynamic boundary conditions from the boundary to the domain. To this extension be consistent, the primary function ω must be normalized function [11]. From the general solution structure, one can derive solutions for all cases of boundary conditions. For h=0, Robin boundary condition becomes Neumann boundary condition:

$$\left.\frac{\partial u}{\partial n}\right|_{\Gamma_2} = \psi_0 \tag{10}$$

For these mixed boundary conditions, the solution structure is:

$$u = \omega_1 \Phi - \omega \cdot (D_1^{\omega_2}(\omega_1 \Phi) + D_1^{\omega_2}(\varphi) + \psi - \omega_2 \Phi) + \varphi \tag{11}$$

For only Neumann boundary conditions, with substitutions $\omega_1 = 1$, $\omega_2 = \omega$, the solution structure can be written in this form:

$$u = \Phi - \omega D_1^{\omega}(\Phi) - \omega \psi + \omega^2 \Phi \tag{12}$$

where $D_1^{\omega}(\Phi) = \nabla\omega \cdot \nabla(\Phi)$ is a differential operator in the direction of the internal normal to the boundary Γ. The resulting function ψ interpolates the individual values ψ_i :

$$\psi = \frac{\sum_{i=1}^{n} \psi_i \prod_{j=1, j\neq i}^{n} \omega_j}{\sum_{i=1}^{n} \prod_{j=1, j\neq i}^{n} \omega_j} \tag{13}$$

For only Dirichlet boundary conditions $u|_{\Gamma} = \varphi_0$, with substitutions $\omega_1 = \omega$, $\omega_2 = 0$, the solution structure can be written in the previously mentioned form $u = \omega \Phi + \varphi$.

4. Combining solution structure method with atomic basis functions and collocation method

A physical field being modelled is represented by a solution structure that can be written in a general form:

$$u = u(\omega, \Psi, \Phi) \tag{14}$$

The solution component ω exactly introduces all information on the domain geometry. So, ω is called the primary function of the solution. The second component Ψ accurately introduces all information about boundary conditions. The third or differential component Φ can be determined as a linear combination of chosen basis functions in the form

$\Phi = \sum_{i=1}^{n} C_i\, F_i$. Approximation properties of the solution structure to a large extent are determined by selection of basis functions $\{F_i\}_{i=1}^{n}$. Here, we propose to use *Fup$_n$(x)* basis functions because they are well conditioned, they have compact support and good approximation properties [19]. Index *n* denotes the highest degree of the polynomial that can be expressed exactly in the form of linear combination of *Fup$_n$(x)* basis functions. The length of the Fup function support is determined according to expression $[-(n+2)2^{-n-1};(n+2)2^{-n-1}]$.

For the Fup basis functions, a criterion of choice of collocation points exists [20]. It is optimal to perform collocation in natural knots of basis functions, i.e. vertices of basis functions situated in a closed domain as shown in Fig. 3. This selection of collocation points provides the simplest numerical procedure, banded collocation matrix is obtained, which is diagonally dominant and thus well conditioned. This selection also implies uniformly distributed nodes set.

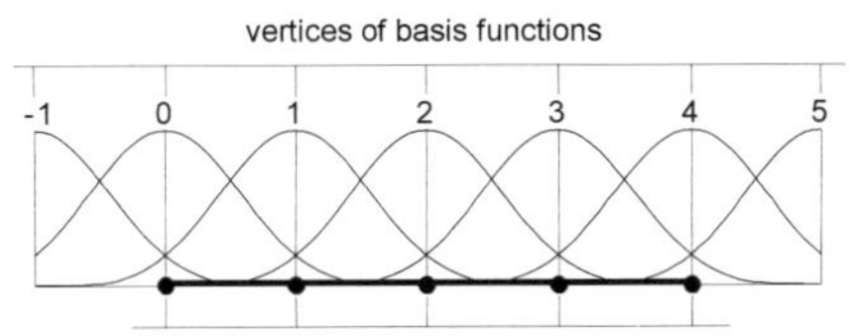

Fig. 3. Layout of the Fup basis functions and positions of collocation points

To basis functions set can be complete, we must keep all basis functions with vertices outside the domain that have values inside the domain different from zero, and we have to write one conditional equation for each of them.

In the case of 1D problems, if *Fup$_4$(x)* basis functions are selected, an unknown component of the solution is:

$$\Phi(x) = \sum_{i=-\infty}^{\infty} C_i \cdot Fup_4\left(\frac{x}{2^4 \cdot \Delta x} - \frac{i}{2^4}\right) \tag{15}$$

Then the function $\Phi(x)$ is defined on the whole real axis. This linear combination can accurately represent an arbitrary algebraic polynomial of the fourth degree *P$_4$(x)*. The R-function method satisfies exactly all boundary conditions using the solution structure, so it is only necessary to satisfy the differential equation in collocation points inside the domain [*a*, *b*].

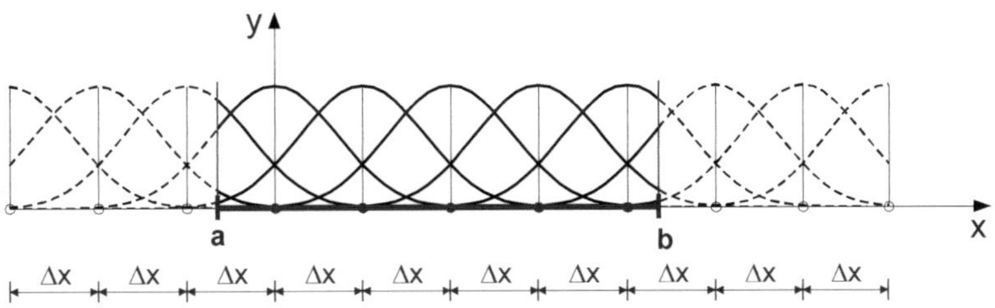

Fig. 4. Layout of Fup$_4$ basis functions in relation to the boundaries of the one-dimensional domain

For the coefficients of basis functions with vertices outside the domain (see Fig. 4), the following recursive formulas that represent connections between external and internal coefficients are used:

$$\begin{aligned} C_k &= 5\,C_{k+1} - 10\,C_{k+2} + 10\,C_{k+3} - 5\,C_{k+4} + C_{k+5} \quad &\textit{(for negative direction of the x axis)} \\ C_k &= 5\,C_{k-1} - 10\,C_{k-2} + 10\,C_{k-3} - 5\,C_{k-4} + C_{k-5} \quad &\textit{(for positive direction of the x axis)} \end{aligned} \tag{16}$$

These recursive formulas are obtained from the condition that the fifth derivative in the middle of the collocation points is equal zero ($\Phi^V(x)=0$ in $x=a+\Delta x/2$ and $x=b-\Delta x/2$, see Fig. 4). In this way, an arbitrary solution function outside of the domain is naturally extended by corresponding polynomial of the fourth degree *P$_4$(x)*.

The basis function for numerical analyses of two-dimensional problems is obtained from the Cartesian product of two one-dimensional *Fup* functions defined for each direction:

$$Fup_n(x,y) = Fup_n(x) \cdot Fup_n(y) \tag{17}$$

Calculation of all required derivatives of function *Fup$_n$(x,y)* can be written in an analogue form [16].

5. Numerical examples

In this section, we will investigate numerical properties of the proposed approach by solving boundary value problems for Poisson equation with homogeneous Dirichlet boundary conditions and nonhomogeneous Neumann boundary conditions and compare the obtained results with analytic solutions. To perform this comparison, we will choose a benchmark problem with known analytic (exact) solutions.

5.1. Example of homogeneous Dirichlet problem: torsion problem

We now apply R-function method to the torsion problem for a bar with the cross section shown in Fig. 5a). The elastic torsion of a bar is a classical problem in the theory of elasticity [21], [22]. This problem may be reduced to the boundary value problem with Poisson equation and homogeneous Dirichlet boundary conditions:

$$\frac{\partial^2 u(x,y)}{\partial x^2}+\frac{\partial^2 u(x,y)}{\partial y^2}=-2G\vartheta \quad ; \quad u(x,y)\big|_\Gamma=0 \tag{18}$$

where $u(x,y)$ is the stress function, G is the shear modulus, while ϑ is the angle of twist per unit length of a bar. Shear stress components are determined according to the following expressions:

$$\tau_{xz}=\partial u/\partial y \quad ; \quad \tau_{yz}=-\partial u/\partial x \tag{19}$$

Exact solution for this domain has been derived by algebraic polynomials in [23] and will be used here for comparison. An analytic solution for the stress function and maximum shear stress are expressed in term of a parameter a:

$$u(x,y)=\frac{G\vartheta a^2}{2}\cdot\left[\left(\frac{x}{a}\right)^3-\left(\frac{x}{a}\right)^2-3\left(\frac{x}{a}\right)\left(\frac{y}{a}\right)^2-\left(\frac{y}{a}\right)^2+\frac{4}{27}\right]; \quad \tau_{max}=\frac{G\vartheta a}{2}. \tag{20}$$

For $G=1.0$, $\vartheta=1.0$ and $a=12.0$ an exact values are $u_{\mathrm{A\,exact}}=10.66\dot{6}$, $\tau_{\mathrm{max\,exact}}=6.0$.

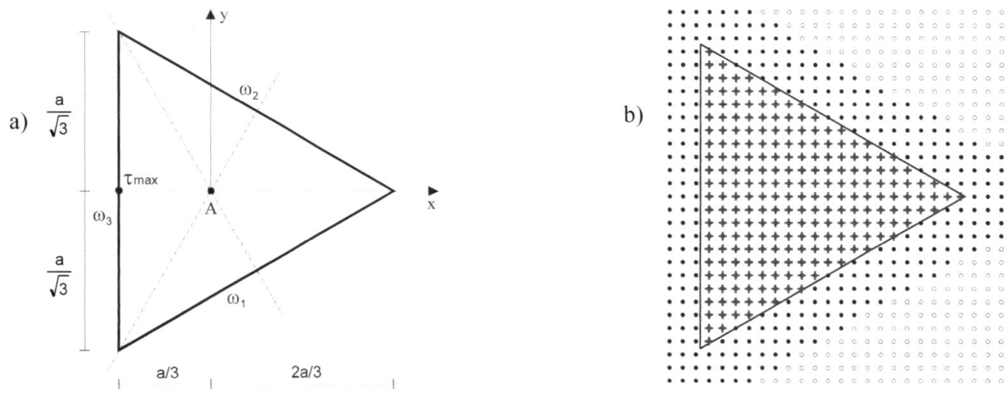

Fig. 5. a) Triangular cross section; b) Uniform grid of points that covers the domain

We will represent approximate solution in the form (2), with function ω defined by (21). For the undetermined function Φ we choose a linear combination of $Fup_4(x,y)$ basis functions on a uniform 26×29 grid shown in Fig. 5b). Only points within the domain are collocation points; for points outside of the domain (black dots in Fig. 5b) recursive formulas are written according to (16) while the remaining points (white dots in Fig. 5b) are not included in the procedure. So, the number of equations to be solved is less than 26×29. Numerical solution obtained with this grid is $u_A=10.663$, $\tau_{max}=6.003$.

$$\omega=\omega_1\wedge\omega_2\wedge\omega_3$$

$$\omega_1(x,y)=-\frac{x}{2}+\frac{\sqrt{3}}{2}y+\frac{a}{3} \quad ; \quad \omega_2(x,y)=-\frac{x}{2}-\frac{\sqrt{3}}{2}y+\frac{a}{3} \quad ; \quad \omega_3(x,y)=x+\frac{a}{3} \tag{21}$$

Practically good enough numerical solution is obtained on a grid that has a minimum number of collocation points within the domain that allows the implementation of the procedure of solution structure method with the selected basis functions. The minimum number of collocation points means that within a domain there is a core of 5×5 points which allows writing recursive equations (16). It defines uniform 14×15 grid with 38 collocation points and a total of 176 equations. This grid gives numerical solution $u_A = 10.646$ which is 98.81% of the exact solution. Such accuracy is expected because the exact solution (20) is a cubic polynomial and linear combination of *Fup₄(x,y)* basis functions can expressed exactly an algebraic polynomial of the fourth degree.

Figures 6a) and 6b) show the computed shearing stresses τ_{xz} and τ_{yz} for geometric domain with $a = 12.0$.

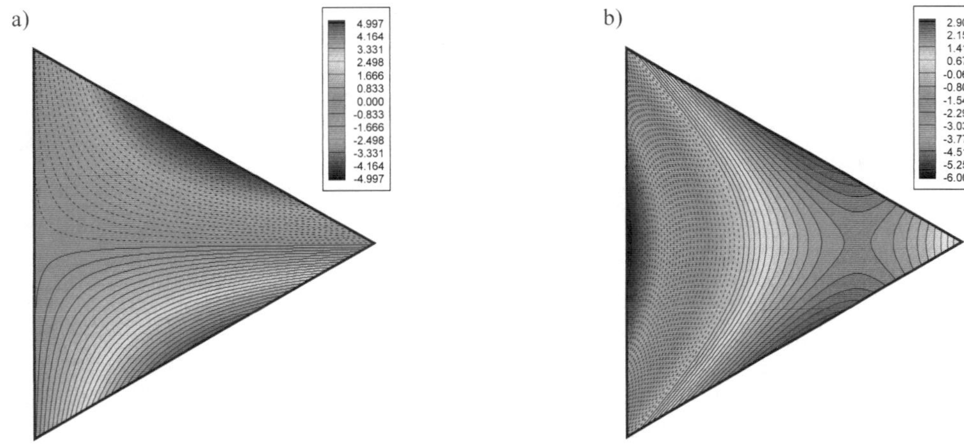

Fig. 6. Stresses τ_{xz} (a) and τ_{yz} (b) computed by solution structure method

5.2. Dirichlet and Neumann boundary value problems

Let consider a bar with square cross-section length of sides $2d = 10\ \text{cm}$, $G = 1.0\ \text{kN/cm}^2$, $\vartheta = 1$, Fig. 7a).

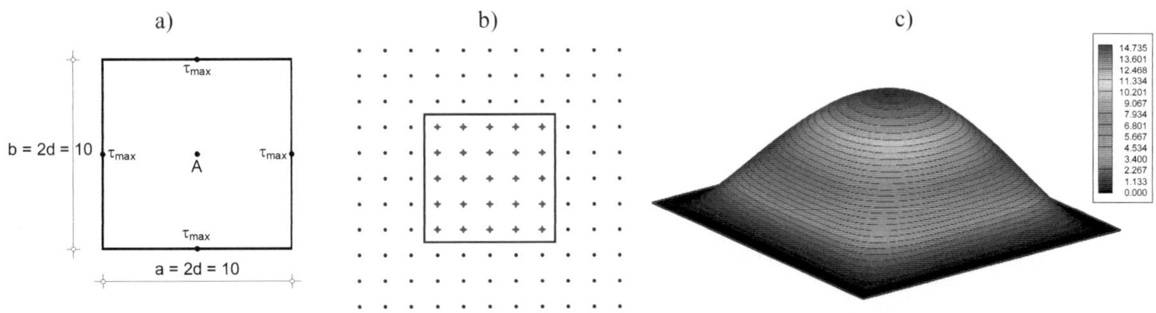

Fig. 7. a) Square cross section; b) Uniform grid with the minimum number of collocation points; c) Stress function computed by 55×55 grid

For this cross-sectional shape, arbitrarily exact solution can be found using the development of the stress function into infinite trigonometric Fourier series. According to the analytical expressions for maximum value of the stress function and maximum shear stress [21], the following values are obtained:

$$u_{A\,exact} = 0.1473427 \cdot G \cdot \vartheta \cdot (2d)^2 \quad ; \quad \tau_{max\,exact} = 0.6753145 \cdot (2G) \cdot \vartheta \cdot d \qquad (22)$$

First, the problem (18) was analyzed by proposed method using solution structure (2) with a different density of uniform collocation points which make up a grid that covers the given two-dimensional domain but does not conform to it. Fig. 7c) presents stress function over the domain. The number of collocation points is varied from a minimum required number 25 (Fig. 7b) to maximum number 2401. Numerical solution converges to the exact solution with the increase in the number of points N×N for N=5,7,13,25,49 as shown in Fig. 8.

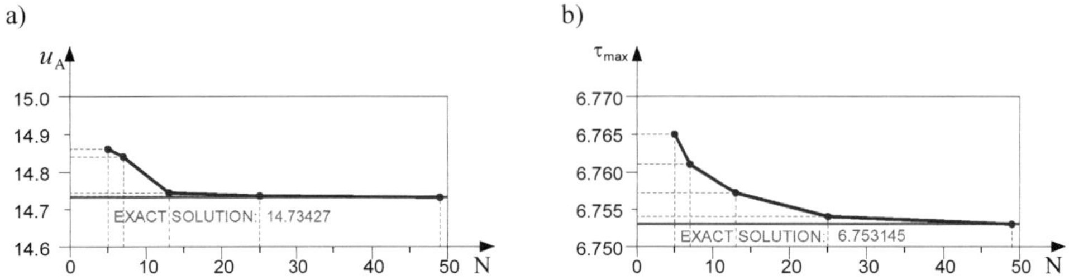

Fig. 8. Convergence diagram for: a) a value of the stress function in the point A; b) value of the maximum shear stress

The elastic torsion of a bar can be described also in formulation of warping function. This problem may be reduced to the boundary value problem with Laplace equation and nonhomogeneous Neumann boundary conditions:

$$\frac{\partial^2 w}{\partial x^2}+\frac{\partial^2 w}{\partial y^2}=0 \quad ; \quad \left.\frac{dw}{dn}\right|_{\Gamma}=y\cdot n_1-x\cdot n_2=\psi(s), \quad s\subseteq\Gamma \tag{23}$$

where w is the warping function, n is the outer normal to the boundary Γ, n_1 and n_2 denote components of the outer normal in x and y directions, respectively. Shear stress components are determined according to the following expressions:

$$\tau_{xz}=G\vartheta(\partial w/\partial x-y) \quad ; \quad \tau_{yz}=G\vartheta(\partial w/\partial y+x)\,. \tag{24}$$

Now, the problem (23) is analyzed by proposed method using solution structure (12). The resulting function ψ that interpolates the individual values ψ_i according to (13) transmit information about the dynamic boundary conditions from the boundary to the domain. Fig. 9 presents warping function obtained by 31×31 grid with 625 collocation points.

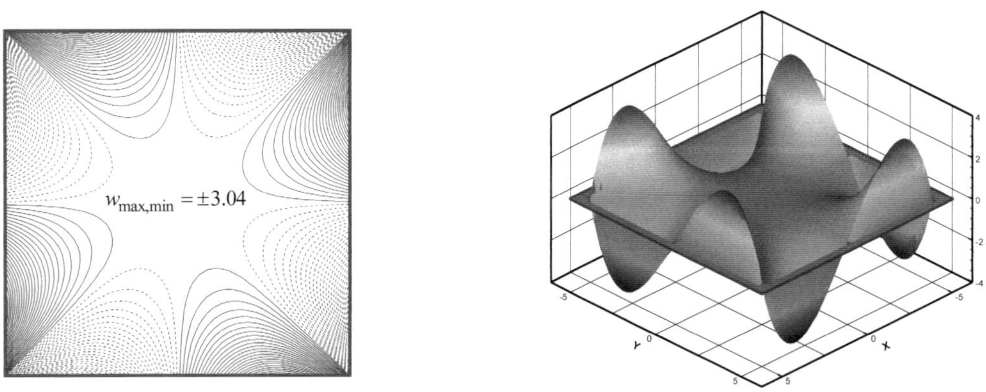

Fig. 9. Contour and the shape of the warping function w obtained by RFM

Numerical value for maximum shear stress converges to the exact solution with the increase in the number of points N×N for N=7,9,13,19,41,49 as shown in Fig. 10. Thus, we can obtain the numerical solution of the torsion problem in two ways, i.e. by means of the formulation using the stress function and by means of the formulation using the warping function. It is very useful that the exact solution is securely between these two numerical values, as can be seen in Fig. 8b) and Fig. 10.

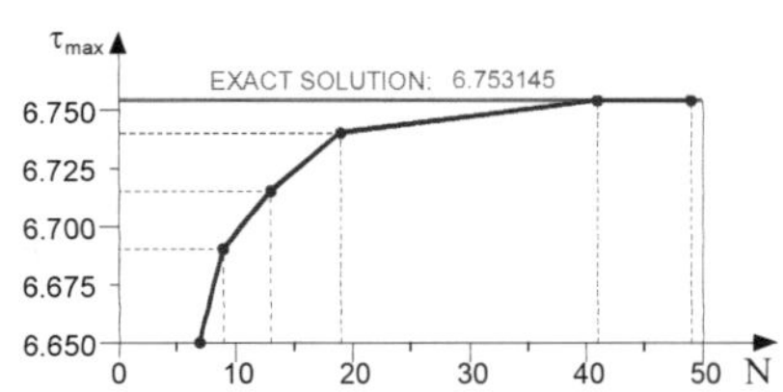

Fig. 10. Convergence diagram for the value of the maximum shear stress

5.3. Example of complex cross section

To demonstrate the applicability of the proposed method combined of solution structure method, atomic basis functions and a collocation technique, we analyzed practical engineering torsion problem for a rod with the cross section shown in Fig. 11a). This is a textbook problem [21] with many good approximations already known. For example, an approximate analytic expression for torque in terms of parameters *r*, *a*, and *b* has been derived for the same domain in [24] and will be used here for comparison.

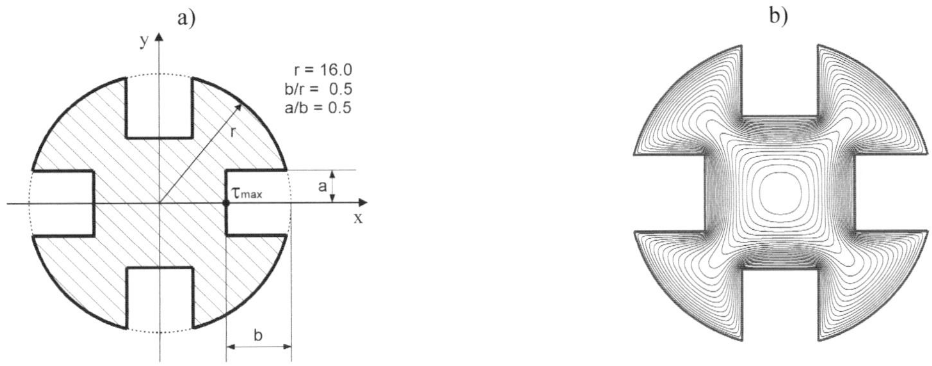

Fig. 11. a) Geometry of the domain; b) Contour of the function ω

The primary function of the solution ω which exactly introduces all information on the domain geometry is constructed using R-operations described in the part 2.1. Figure 11.b) shows the contour of the function ω which represents a surface "inflated" over the domain. The numerical solution of the Poisson equation (18) was obtained by using uniform 55×55 grid (1217 collocation points).

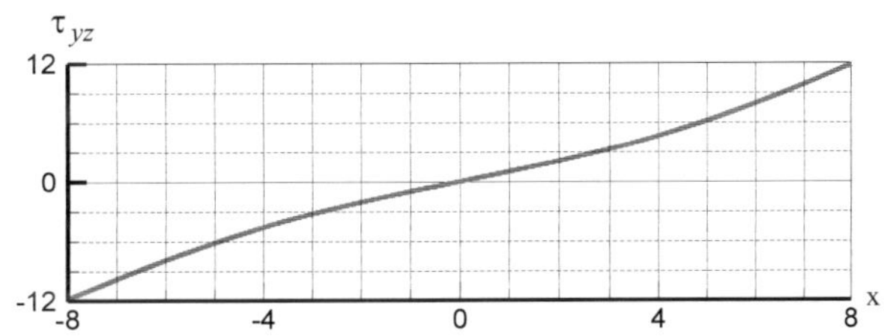

Fig. 12. Computed shearing stresses τ_{yz} along the horizontal axis of symmetry

By using approximate analytic expression for this cross-sectional shape [24], we verified our numerical model. For the fixed values *r*, *a* and *b* (Fig. 11a), the computed maximum value of shear stress component τ_{yz} is 11.819 while the value predicted by the closed-form expression in [24] is 11.956. Figure 12 shows the diagram of the computed shearing stresses τ_{yz} along the horizontal axis of symmetry.

6. Conclusion

This paper presents numerical method created to solve the modelling of problems with complex geometry in a way that spatial grids, that are needed to construct the basis functions and perform numerical computations, do not have to conform to the geometric shape of the model and at the same time boundary conditions can be well treated.

The basic principle of R-function method is described. We have combined RFM with atomic basis functions and get a method that connects the advantages of solution structure method, collocation technique and good approximation properties of atomic Fup basis functions.

Numerical examples considered here have shown that properly constructed solution structures are complete in the sense that they converge to the exact solution of a problem. Proposed method is unique amongst meshfree methods because RFM solution structures can be constructed to satisfy all boundary conditions exactly. This property allows that the solution procedure does not require mapping the geometry of the domain. Combination of atomic basis functions and solution structure method gives numerical solutions that have characteristics of analytical solutions because solution structure method enables exact treatment of boundary conditions while ABFs ensure numerical solutions with desired level of accuracy.

RFM has one significant disadvantage when compared to the usual spatial discretization and other meshfree methods: the solution structure is constructed, differentiated, and integrated at runtime. Thus, a fully automatic implementation of RFM requires automatic construction of implicit functions and structures, given a geometric model and prescribed boundary conditions, automatic differentiation of the constructed functions and structures at various points of the domain, constructing and solving the resulting linear system for values of the coefficients C_i in the undetermined functional component Φ. In our examples, Φ is a uniform set of atomic Fup basis functions, so the resulting sparse banded system can be solved in linear time using standard techniques. All these tasks are feasible today with existing technology, so the proposed method is promising numerical method for solving technical problems.

Our future research will be directed towards the development of numerical models for analysis 2D heat transfer problems, groundwater flow modelling in karst aquifers, nonlinear problems, plate bending problems, and towards automation of the numerical process with development of adaptive techniques.

7. Acknowledgments

This paper is a result of research in the frame of national project "Groundwater flow modelling in karst aquifers" funded by Croatian Science Foundation.

8. References

[1] Atluri, S. N. (2005). Methods of Computer Modeling in Engineering & the Sciences, Volume I, Tech Science Press, ISBN 0-9657001-9-4, University of California, Irvine

[2] Griebel, M. & Schweitzer, M. A. (Eds.) (2003). Meshfree Methods for Partial Differential Equations, Vol. 26, Springer-Verlag, ISBN 3-540-43891-2, Berlin

[3] Liu, G. R. (2003). Mesh Free Methods: Moving Beyond the Finite Element Method, CRC Press LLC, ISBN 0-8493-1238-8, Boca Raton

[4] Belytschko, T.; Krongauz, Y.; Organ, D.; Fleming, M. & Krysl, P. (1996). Meshless methods: An overview and recent developments, Comput. Methods Appl. Mech. Eng., Vol. 139, No. 1–2, 1996, pp. 3–47

[5] Fries, T. P. & Matthies, H. G. (2004). Classification and Overview of Meshfree Methods, Institute of Scientic Computing Technical University Braunschweig, Brunswick

[6] Chen, W.; Fu, Z. J. & Chen, C. S. (2014). Recent Advances in Radial Basis Function Collocation Methods, Springer Briefs in Applied Sciences and Technology, ISBN 978-3-642-39571-0, Springer

[7] Gunter, F. C. & Liu, W. K. (1998). Implementation of boundary conditions for meshless methods, Comput. Methods Appl. Mech. Eng., Vol. 163, 1998, pp. 205–230

[8] Kantorovich, L. V. & Krylov, V. I. (1958). Approximate Methods of Higher Analysis, Interscience Publishers, New York

[9] Rvachev, V. L. (1982). Theory of R-functions and Some Applications, Naukova Dumka, Kiev. In Russian

[10] Rvachev, V. L.; Sheiko, T. I.; Shapiro, V. & Tsukanov, I. (2000). On completeness of RFM solution structures, Comput Mech, Vol. 25, 2000, pp. 305–317

[11] Rvachev, V. L. & Sinekop, N. S. (1990). R-functions Method in Problems of the Elasticity and Plasticity Theory, Naukova dumka, Kiev. In Russian

[12] Rvachev, V. L. & Sheiko, T. I. (1995). R-functions in boundary value problems in mechanics, Applied Mechanics Reviews, Vol. 48, No. 4, 1995, pp. 151–188

[13] Rvachev, V. L. & Rvachev, V. A. (1971). On a finite function, Dokl. Akad. Nauk Ukrainian SSR, ser. A, 1971, pp. 705

[14] Rvachev, V. L. & Rvachev, V. A. (1979). Non-classical Methods for Approximate Solution of Boundary-Value Problems, Naukova dumka, Kiev. In Russian

[15] Kravchenko, V. F. (2003). Lectures on the Theory of Atomic Functions and Their Some Applications, Radiotechnika, Moscow

[16] Kozulic, V.; Gotovac, B. & Colak, I. (2006). Multilevel mesh free method for the torsion problem, In: DAAAM International Scientific Book 2006, Katalinic, B., (Ed.), Chapter 29, pp. 365-384, DAAAM International Vienna, ISBN 3-901509-47-X, Vienna

[17] Kozulic, V.; Gotovac, B. & Sesartic, R. (2008). Mesh free modeling of the curved beam structures, In: DAAAM International Scientific Book 2008, Katalinic, B., (Ed.), Chapter 34, pp. 395-408, DAAAM International Vienna, ISBN 978-3-901509-66-7, Vienna

[18] Kozulic, V. & Gotovac, B. (2015). Computational modeling of structural problems using atomic basis functions, In: Advanced Structured Materials, Vol. 70: Mechanical and Materials Engineering of Modern Structure and Component Design, Ochsner, A.; Altenbach, H., (Eds.), Chapter 17, pp. 207-230, Springer

[19] Gotovac, B. & Kozulic, V. (1999). On a selection of basis functions in numerical analyses of engineering problems, Int. J. Eng. Modelling, Vol. 12, No. 1-4, 1999, pp. 25-41

[20] Prenter, P. M. (1989). Splines and Variational Methods, John Wiley & Sons, ISBN 0-471-50402-5, New York

[21] Timoshenko, S. P. & Goodier, J. N. (1961). Theory of Elasticity, McGraw-Hill, New York

[22] Lurie, A. I. (1970). Theory of Elasticity, Nauka, Moskva

[23] Kostrencic, Z. (1982). Theory of Elasticity, School Book, Zagreb

[24] Pilkey, W. D. (2005). Formulas for Stress, Strain, and Structural Matrices, John Wiley & Sons

DOI: 10.2507/27th.daaam.proceedings.037

MASS TRANSFER COEFFICIENT IN THE CASE OF BRAKE-UP OF A CONTINUOUS FALLING FILM

Džana Kadrić, Mirela Alispahić, Berina Delalić

This Publication has to be referred as: Kadric, D[zana]; Alispahic, M[irela] & Delalic, B[erina] (2016). Mass Transfer Coefficient in the Case of Brake-Up of a Continuous Falling Film, Proceedings of the 27th DAAAM International Symposium, pp.0255-0260, B. Katalinic (Ed.), Published by DAAAM International, ISBN 978-3-902734-08-2, ISSN 1726-9679, Vienna, Austria
DOI: 10.2507/27th.daaam.proceedings.037

Abstract

Two-phase flow of the thin, vertical liquid film and the counter-current air flow are studied experimentally in conditions in which liquid film brakes up. Parameters related to the two-phase flow are shown as a function of the ratio between the liquid flow rate and air flow rate and film Reynolds number. All relevant parameters change in time, so their values are shown for two time periods: when break-up starts and after a two and a half hours when final value of the degree of the wetted area is reached. Mass transfer coefficient is a function of the film Reynolds number and time, and it increases as film Reynolds number increases but decreases in time as surface de-wetting occurs. Proportion of wetted surface area as well as value of evaporated water vapour, changes as process of falling film break-up occurs.

Keywords: falling film; counter-current flow; instability; mass transfer coefficient.

1. Introduction

A thin, falling liquid films are encountered in many industrial applications (e.g. distillation facilities, cooling towers, vertical tube evaporators, falling film chemical reactors, food processing facilities, etc.), due to their high efficiency in heat and mass transfer. Critical conditions occur during the facility start-up, since significant amount of liquid is needed to completely cover a dry surface with continuous liquid film. When continuous liquid film is formed, it is very important to maintain the fluid minimum mass flow rate above its critical value to prevent surface to dry out. Liquid-flow rate that corresponds to these regimes is called a minimum wetting rate and the corresponding film thickness is called a minimum film thickness [1].

Theoretical models derived for prediction of the minimum wetting rate and corresponding film thickness are mostly based on two criteria: the force balance in the stagnation point and the minimum total energy of rivulets [1–5].

Breakup of the isothermal, distilled water falling film on a vertical plate in three regimes (wetting of the dry plate, rewetting and break-up of the continuous falling film) is examined in [6] and [7]. Influence of contact angle (static and dynamic) on the minimum wetting rate is confirmed as well as the influence of the plate inclination angle [7]. In the processes where air flows in counter-current or co-current to the film flow, problem is stated as a two-phase flow, and studied in many experimental and theoretical studies [8].

Influence of the air flow on the film brake-up process and relevant parameters has been studied theoretically in [9].

In presented experiment, influence of the counter-current air flow on the parameters related to the two-phase flow, precisely on the mass transfer coefficient, in the case of the film brake-up, is examined. Minimum flow rate for the case of the film brake-up without the air flow was very low, since measured receding contact angle is zero which is an indication of high adhesion between the plate material and the liquid. Therefore, falling film remains stable even for the very small flow rate (liquid Reynolds number smaller than 1). Results of experimental investigation for this liquid-plate material combination showed that minimum wetting rate, area of dry surface and mass transfer coefficient changes abruptly in the case of the brake-up while the counter-current air flow is introduced.

2. Experiment setup

The experimental setup used in the present examination is illustrated in Fig. 1. It consists of two lines: air line and the liquid line as it is shown schematically in Figure 1. These two lines were combined in test section where these two fluids in direct contact exchange mass in counter-current flow. Air was supplied from compressor.

Following equipment was installed on the air line: air cleaner, pressure regulator, flow rate regulators, and in-line flow meters. Air cleaner was installed to prevent the fouling of equipment. Pressure regulator was used to decrease pressure from 7 bars, supplied from the compressor, to the desired value. Flow rate regulators are in line with flow meters to control the air flow rate at the inlet of test section. The water feed line leads from the liquid reservoir, through the distributor towards the target surface. A laboratory pump drives the working fluid into the distributor.

In order to obtain a better spreading of the liquid over the plate, a narrow strip of filter paper was placed on the higher edge of the plate. The narrow target plate was placed vertically or inclined onto support. The stainless steel plate dimensions are 500×100 with wetted length of 410 mm. The plate was sandblasted, with resulting average roughness given in Table 1. Working fluid was distilled water; water temperature measured at the water inlet section was 12°C.

Material	Plate thickness, mm	Average roughness, μm
Stainless steel, X5CrNi 18–10	1,5	1,3

Table 1. Plate characteristics

Experiments were carried out on a vertical upward air–water flow in a closed vertical rectangular channel. A schematic diagram of the loop is shown in Fig. 1. Experimental procedure was unified for all measurements. The plate was previously fully wetted and then the air was introduced into the channel. The water flow was set on a constant value and the air flow rate was decreased gradually in order to form a dry zone on the plate. The degree of wetted plate surface was observed for two and half hours, as it takes for one complete measuring procedure. Mass transfer takes place in direct contact while the fluids flow in countercurrent.

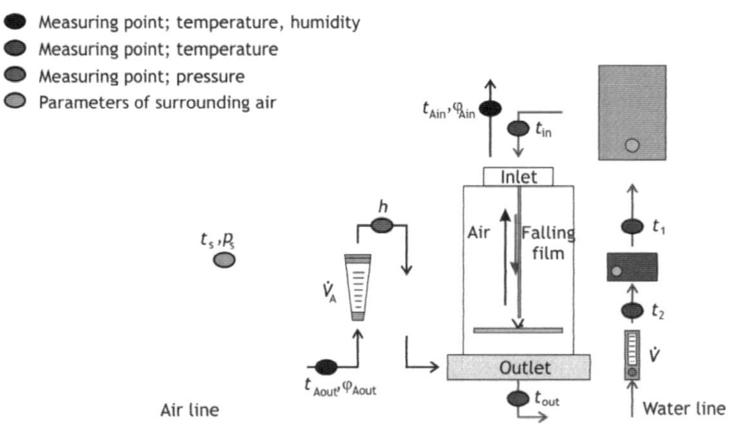

Fig. 1. Experimental setup

3. Flow hydrodynamics (observation)

Developed experiment procedure ensured analysis of the impact of fluid flow and air flow reduced to standard conditions on wetted surface, defined as the ratio of wetted to the overall plate surface.

The flow corrected to standard conditions is calculated from the flow measured on the flowmeter, as:

$$\dot{V}_{A,s} = \dot{V}_A \sqrt{\frac{p_{A,out} + p_{at}}{1{,}013} \frac{288{,}15}{173{,}15 + t_{A,out}}} \tag{1}$$

where $\dot{V}$ is the volumetric air flow rate measured on the flowmeter, $p_{A,out}$ is the pressure at the inlet of the test section, p_{at} is the atmospheric pressure and $t_{A,out}$ is the temperature at the inlet of the flowmeter, as marked in Fig. 1. Pressure p at the exit of the flowmeter is measured using the water manometers (*U*-tube).

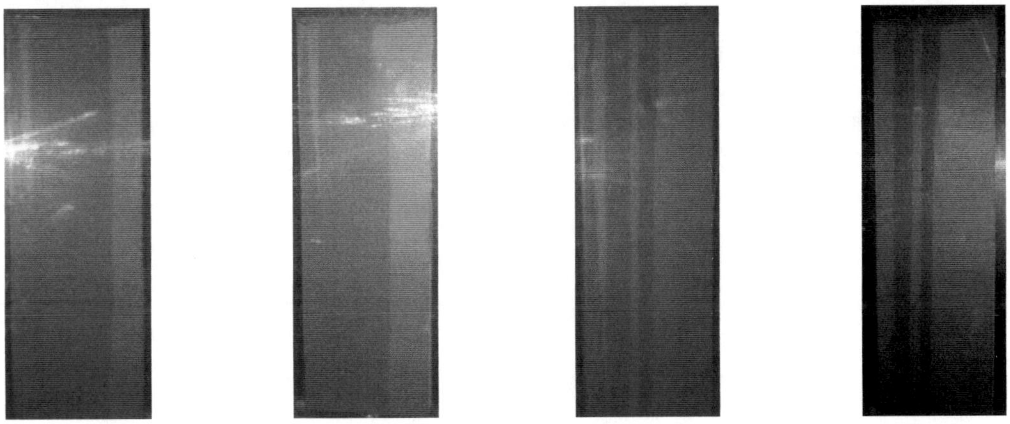

a) *b)* *c)* *d)*

Fig. 2. Plate dry out over time: water flow rate 1 ml/min and mean air velocity u_A= 0,566 m/s (duration of the experiment τ= 2,5 h)

In Fig. 2. a set of measurements is showed. Liquid flow rate was constant 1 ml/min, which represents flow rate above minimum wetting rate for conditions without air flow rate. Duration of measurement was two and a half hours. When air is supplied in to a test section plate, de-wetting occurs rapidly (Fig. 2a.). De-wetted area increases in time which can be noticed in Fig. 2 (b-c)., and finally two narrow rivulets are formed as it is shown in Fig. 2.d.

The degree of the wetted surface is a function of the time needed to carry out the complete measuring procedure, as in Fig. 2. In order to ensure stable state and parameters it was concluded that the time could not be less than two and half hours.

Liquid flow rates: a) 1 ml/min b) 1,7 ml/min c) 2,4 ml/min

Fig. 3. Wettability of the plate for different volumetric liquid flow rates and mean air velocity u_A=0,552 m/s

Fig. 3. shows significant impact of the liquid flow rate on the degree of the wetted surface. It is noticeable that with increase of the liquid flow rates the degree of the wetted surface is increasing for the same air flow rate. For smaller water flow rates, only about 30% of the plate was wetted. This small percentage of the wetted surfaces for small flow rates confirms previous observations on liquid microfilm covering blasted hydrophilic surface of the plate, which dries out as soon as the air flow starts. Mean air velocity in the test section ranged from 0,548 m/s to 0,736 m/s.

4. Analytical considerations

For a two-phase flow consideration, dimensionless numbers are used: Sherwood number as a representative of the mass transfer and Reynolds number as a representative of the flow hydrodynamics. Similar experiments conducted on this experimental line showed that sensible heat transfer is significantly smaller than latent heat transfer which resulted in a following relationship [10]:

$$\mathrm{Sh} = \frac{\beta D_m}{Mc\boldsymbol{D}} = 0{,}66\mathrm{Re}_{\mathrm{A}}^{0{,}69}\mathrm{Re}^{0{,}27} \tag{2}$$

where β is the mass transfer coefficient, D_{m} is the characteristic mass transfer diameter, M is the molecular weight for the water vapor, c is the molar concentration and $\boldsymbol{D}$ is the diffusion coefficient which is calculated as:

$$\boldsymbol{D} = 1{,}97 \cdot 10^{-5}\frac{p_0}{p}\left(\frac{T}{T_0}\right)^{1{,}685}, \tag{3}$$

which is suitable for $273<T<373$ K, where p_0=1 atm and T_0=256 K.

Reynolds numbers for air and falling film are defined as:

$$\mathrm{Re}_{\mathrm{A}} = \frac{u_{\mathrm{A}} D_{\mathrm{h}}}{v_{\mathrm{A}}}, \mathrm{Re} = \frac{4\gamma}{\mu} \tag{4}$$

respectively, where D_{h} is the hydraulic diameter, O_{h} is the wetted perimeter, u_{A} is the mean air velocity, ν_{A} is the kinematic viscosity and γ is liquid flow rate per unit perimeter.

5. Results and discussion

In Fig. 4. degree of the wetted surface is shown as a function of the ratio of liquid to air flow rates. It is noticeable that increase of liquid to air flow ratio results in an increase of the degree of the wetted surface. It is noticeable that, for ratio higher than 2,8, portion of wetted surface remains very high. For smaller ratios, portion of the wetted surface is small which certainly adverse process of the heat and mass transfer.

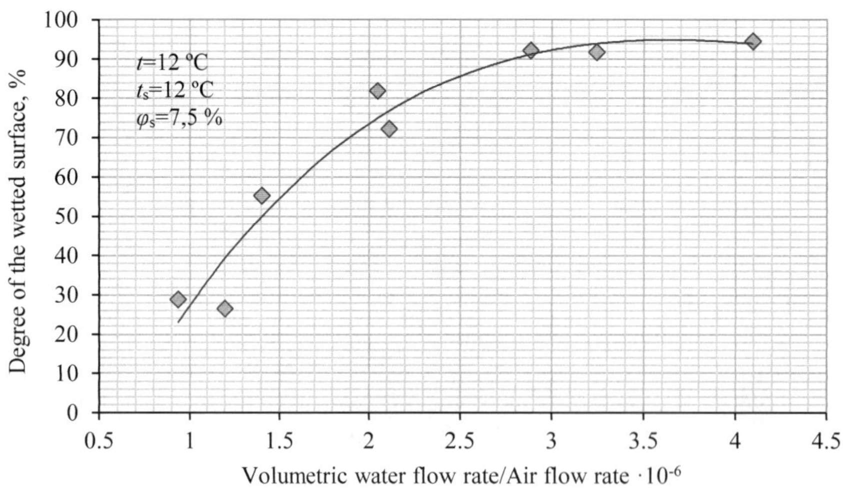

Fig. 4. The degree of the wetted surface in terms of the ratio of liquid to air flow rates

Values of mass transfer coefficient, calculated from Eq. (2) changes in time, as it is shown in Fig. 5. When measurements start, mass transfer coefficient trend is practically constant, with small deviations, as film Reynolds number changes. Contact area between air flow and liquid film is approximately equal for all starting measurements since whole plate is covered with thin liquid film. While plate de-wetting occurs and air wetted perimeter decreases, mass transfer coefficient decreases, especially for small film Reynolds numbers. For film Reynolds number higher than 1,5 mass transfer coefficient approaches initial value and shows that it is not a function of time.

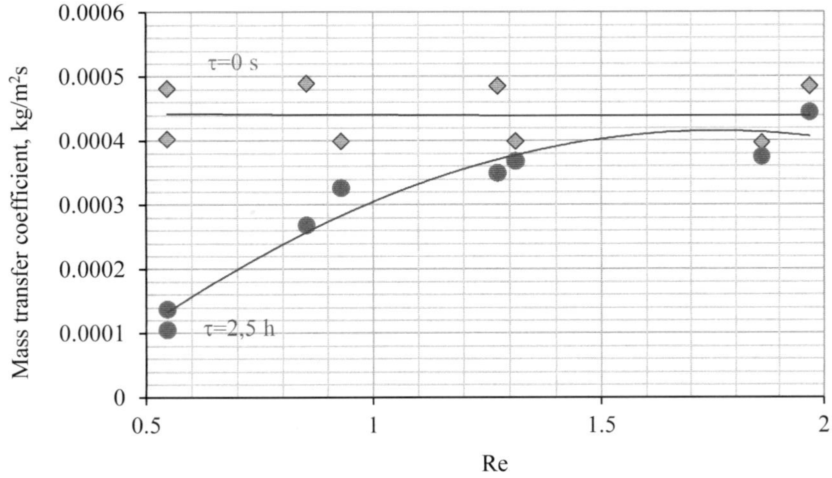

Fig. 5. Mass transfer coefficient in terms of the liquid film Reynolds number

Using the values of mass transfer coefficient, percentage of evaporated liquid in the net liquid flow is calculated and shown in terms of ratio of liquid to air flow rates in Fig. 6. For small ratio when measurements start, evaporated liquid ratio is very high and it decreases in time, since large portion of plate is de-wetted and liquid-air interface decrease abruptly. For higher ratio of liquid to air flow rates, percentage of evaporated liquid remains constant in time and has the value of about 30 %.

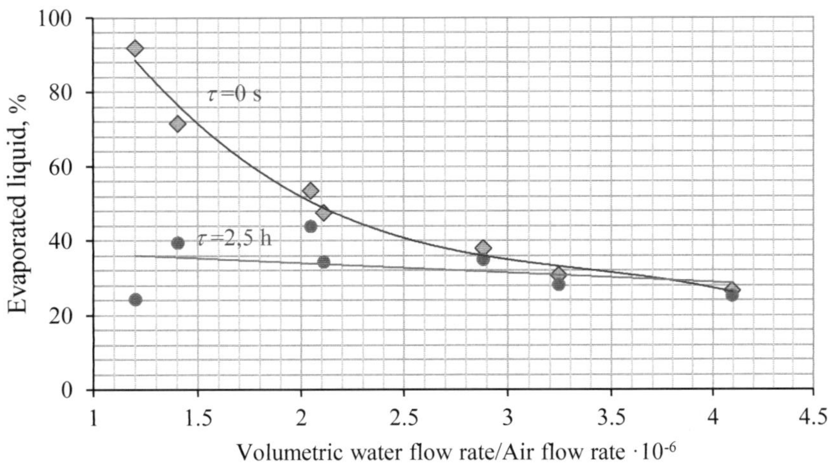

Fig. 6. Percentage of the evaporated liquid in net liquid flow in terms of the ratio between the volumetric flow rates of liquid and air.

6. Conclusions

Results of experimental investigation of two-phase flow parameters for conditions where film brake-up occurs, are presented. Thin liquid film is formed on the vertical plate. Minimum flow rate for the case of the film brake up without the air flow was very low, since measured receding contact angle is zero which is an indication of high adhesion between the plate material and the liquid. Therefore, falling film remains stable even for the very small flow rate (liquid Reynolds number smaller than 1).

Results for tested liquid-plate material combination showed that area of dry surface and mass transfer coefficient changes abruptly when counter-current air flow is introduced. All relevant parameters changes in time, so their values are shown for two time periods: when break-up starts and after a two and a half hour when final value of the degree of the wetted area is reached. It is shown that ratio of liquid to air flow rate can be included in to a consideration, so important two-phase flows parameters can be shown in term of this ratio. Since, ratio of liquid flow rate to air flow rate vas in the range of 1,2 to $4{,}2\cdot10^{-6}$, air humidity and temperature change within the test section could not be measured. Therefore, relation for Sherwood number and air and falling film Reynolds number is taken from the similar experiments conducted on the same experimental line. This ensures that mass transfer coefficient could be evaluated and presented as a function of the film Reynolds number and time. Results shows that it increases as film Reynolds number increases but decreases in time as surface de-wetting occurs. Proportion of wetted surface area as well as value of evaporated water vapour, changes as process of falling film break-up occurs.

It is shown that for ratio of liquid to air flow rates higher than 2,8, portion of wetted surface remains very high and portion of evaporated liquid remains constant in time. For smaller ratios, portion of the wetted surface that is interface between liquid film and air flow is small which adverse process of the heat and mass transfer. Also, while plate de-wetting occurs and air wetted perimeter decreases, mass transfer coefficient decreases, especially for small film Reynolds number. For film Reynolds number higher than 1,5 mass transfer coefficient does not change in time significantly.

Presented experiments were conducted in a specific condition; plate-liquid material with resulting zero receding contact angle and high adhesion forces. For different plate surface treatments, other contact angles and liquid flow rates for film brake-up are measured. It would be very useful to conduct similar experiment on plates from other materials and surface treatments, to investigate the mass transfer coefficient and liquid evaporate rate in presence of counter-current air flow.

Nomenclature			
c	Molar concentration [mol m^{-3}]	$\boldsymbol{D}$	Diffusion coefficient [m^2 s^{-1}]
D_h	Hydraulic diameter [m]	D_m	Characteristic mass transfer diameter [m]
M	Molecular weight for the water vapor [kg mol^{-1}]	p	Pressure [bar]
O_h	Wetted perimeter [m]	Re	Reynolds number
t T	Temperature [°C, K]	Sh	Sherwood number
$\dot{V}$	Volumetric flow rate [m^3 s^{-1}]	u	Velocity [m s^{-1}]
γ	Mass flow rate per unit perimeter [kg m^{-1} s^{-1}]	β	Mass transfer coefficient [kg m^{-1} s^{-2}]
ν	Kinematic viscosity [m^2/s]	μ	Dynamic viscosity [N s m^{-2}]
φ	Relative humidity [%]	τ	Time [s]
Abbreviations			
A	air	at	atmospheric
in	inlet	out	outlet
s	surrounding		

Table 2. Nomenclature and abbreviations

7. References

[1] M.S. El-Genk and H.H. Saber(2001). Minimum thickness of a flowing down on a vertical surface, Int. J. Heat Mass Transfer, 44:2809-2825

[2] D.E. Hartely, W. Murgatroyd(1964). Criteria for the break–up of thin liquid layers flowing isothermally over solid surface, Int. J. Heat Mass Transfer, 7: 1003-1015

[3] T. Hobler (1964). Minimum surface wetting, Chem. Stosow, 2B:145-159

[4] S.G. Bankof (1971). Minimum thickness of a draining liquid film, Int. J. Heat Mass Transfer, 14:2143-2146

[5] J.Mikielewicz and J.R. Moszynski(1976). Minimum thickness of a liquid film flowing vertically down a solid surface, Int. J. Heat Mass Transfer, 19:771-776

[6] D. Moalem–Maron, N. Brauner (1983). Transport characteristics of wavy thin films on inclined surfaces, Advance in Two-Phase flow and Heat Transfer, Vol. 1:155–181

[7] Dž. Kadrić, M. Alispahić, Š. Šikalo, S. Jakirlić and E. N. Ganić (2015), E181, Proceedings of the Eight International Symposium on Turbulence, Heat and Mass Transfer, K. Hanjalić, T. Miyauchi, D. Borello. M. Hadžiabdić, P. Venturini (Editors); ISBN: 978-1-56700-428-8, Begell House Inc.

[8] Šikalo, Šefko; Berberović, Edin (2014). Analysis of Droplet Deposition in a Vertical Air-Water Dispersed Flow, Proceedings of the 25th DAAAM International Symposium, pp.xxxx-xxxx, B. Katalinic (Ed.), Published by DAAAM International, ISBN 978-3-902734-08-2, ISSN 1726-9679, Vienna, Austria

[9] M.S. El–Genk, H.H. Saber(2004). On the break-up of thin liquid film subject to interfacial shear, J. Fluid Mech. 501:113–133

[10] Kadrić, Dž., Šikalo, Š., Delalić, N. and Ganić, E.N. (2007). Experimental investigation of a laminar falling film, 6th International Conference on Multiphase Flow,ISBN 978-3-86010-913-7

DOI: 10.2507/27th.daaam.proceedings.038

INFLUENCE OF TEMPERATURE ON A FALLING FILM INSTABILITY

Džana Kadrić, Berina Delalić & Mirela Alispahić

This Publication has to be referred as: Kadric, D[zana]; Delalic, B[erina] & Alispahic, M[irela] (2016). Influence of Temperature on a Falling Film Instability, Proceedings of the 27th DAAAM International Symposium, pp.0261-0265, B. Katalinic (Ed.), Published by DAAAM International, ISBN 978-3-902734-08-2, ISSN 1726-9679, Vienna, Austria
DOI: 10.2507/27th.daaam.proceedings.038

Abstract

Results of experimental research of influence of a film temperature on a falling film brake-up are shown for case of the continuous film brake-up and wetting of the previously dry plate, which represents very important regimes for heat transfer equipment such as some models of heat exchangers. Experiments are conducted on a vertical plate and distilled water as a working fluid. For case of the brake-up of the continuous falling film, which was observed for Reynolds numbers smaller than 1, and inlet liquid film temperature has no influence on the minimum wetting rate. During the facility start-up high flow rates are needed to completely cover the plates. This regime was investigated for elevated film temperature and two material surface treatments. It is shown that sandblasted plates provide better ratio of wetted width than polished plates, with smaller advancing contact angles while rivulets hydrodynamics was driven by the higher surface tension forces.

Keywords: experimental work; heat transfer equipment; film temperature; film instability.

1. Introduction

Thin, falling liquid films are encountered in many industrial applications (e.g. distillation facilities, cooling towers, vertical tube evaporators, falling film chemical reactors, food processing facilities, etc.), due to their high efficiency in heat and mass transfer. Three different cases are recognized as critical conditions; wetting of the previously dry plate during the facility start-up, break-up of the continuous liquid film and rewetting of plate after a short stop. Liquid-flow rate that corresponds to these regimes is called a minimum wetting rate and the corresponding film thickness is called a minimum film thickness [1]. Theoretical models derived for prediction of the minimum wetting rate and corresponding film thickness are mostly based on two criteria: the force balance in the stagnation point and the minimum total energy of rivulets [1–5]. Hysteresis of minimum wetting rate and minimum film thickness for three regimes is recorded and experimentally confirmed [6], [7]. Influence of contact angle (receding, advancing and equilibrium) on the minimum wetting rate is confirmed as well as the influence of the plate inclination angle [7]. Many researchers have shown that increase of fluid temperature results in a decrease of the minimum wetting rate [6], [8]. This was studied experimentally for combination distilled water/sandblasted copper plate for which receding contact angle is equal to zero therefore adhesion forces are very strong. Due to the strong adhesion forces, continuous falling film remain stable for very small

Reynolds numbers, smaller than 1. Hysteresis of flow rate for wetting of the previously dry plate compared to break-up of continuous falling film is recorded, where for previous film regime the Reynolds number is approximately 150.

2. Experimental setup

Experimental setup used in presented examination is illustrated in Fig. 1. Liquid line consists of liquid reservoir with temperature regulation and laboratory pump which drives the working fluid through distributor towards the target surface. The narrow stainless steel (X5CrNi 18-10) plate was placed vertically onto support. The plate dimensions are 500×100 with wetted length of 410 mm. Average roughness of stainless steel plate is R_a=1,15 μm (R_z=2,04 μm), while sandblasted plate had resulting average roughness of R_a=1,3 μm (R_z=11,3μm). During the experiment, liquid temperature was ranged from 20 °C to 52 °C, while surrounding temperature was 19 °C.

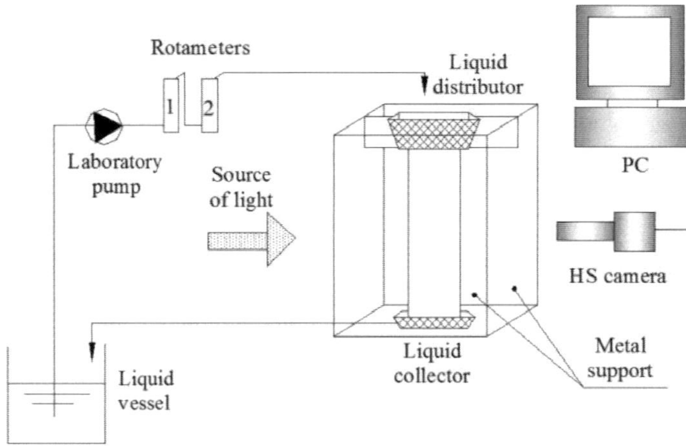

Fig. 1. Experimental setup

3. Experimental results

Two sets of experiments were carried out: measurement of brake-up for various inlet liquid temperature on the sandblasted vertical plate and measurements of wetting of the previously dry plate for various liquid temperature performed on the sandblasted and on a polished plate, both from stainless steel. Contact angle for tested liquid and plate are recorded, following the general experimental procedure given in detail in [7].

3.1. Brake-up of the continuous liquid film

Experiments performed on isothermal liquid film and sandblasted surface showed that continuous liquid film remains stable even for very small liquid flow rates, with equivalent Reynolds numbers lower than 1 and fluid velocity at the entrance zone lower than 0,002 m/s.

Fig. 2a. Receding contact angle between liquid and plate

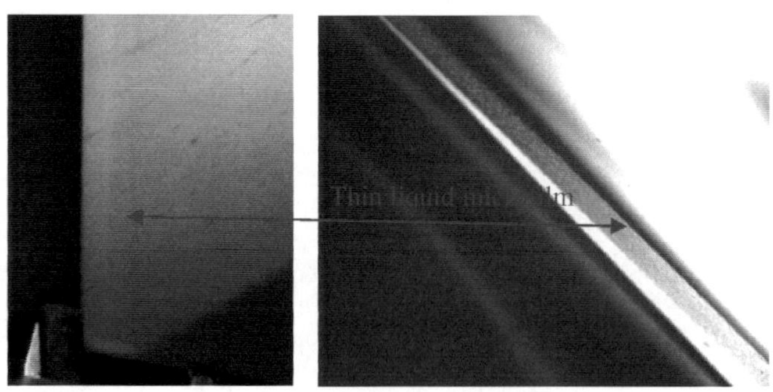

Fig. 2b. Microfilm formed on the plate for Re=0,4 (photograph and laboratory camera view)

Adhesive forces between liquid and plate material are very strong, which is indicated by zero receding contact angle, as shown in Fig. 2a. Very thin layer of microfilm is observed while liquid flow rate decreases, which also confirms that high adhesive forces are presented (Fig. 2b.).

Due to the very small flow rates, the increase of liquid temperature shows no impact on the film brake-up, even for conditions with temperature of liquidup to 50°C, which was significantly higher than the ambience temperature.

Thermography has proven to be useful tool for various engineering applications [9], [10] so it has been used during the measurement in order to show liquid temperature in a qualitatively manner. In order to obtain exact temperature field some parameters must be precisely known (such as emissivity and temperature of ambient) however, thermographic images provide useful information. In Fig.3. temperature field of liquid film is shown for inlet liquid temperature of 29 °C for two liquid flow rates. It is noticeable how temperature of the liquid film decrease as the flow rate decreases, for flow rates higher than brake-up flow rates. Since minimum wetting flow rate was approximately 0,02 l/m·min, (wetted perimeter equals to the plate width), temperature of the film decreases and process parameters became like in the isothermal process. This was recorded for every tested temperature.

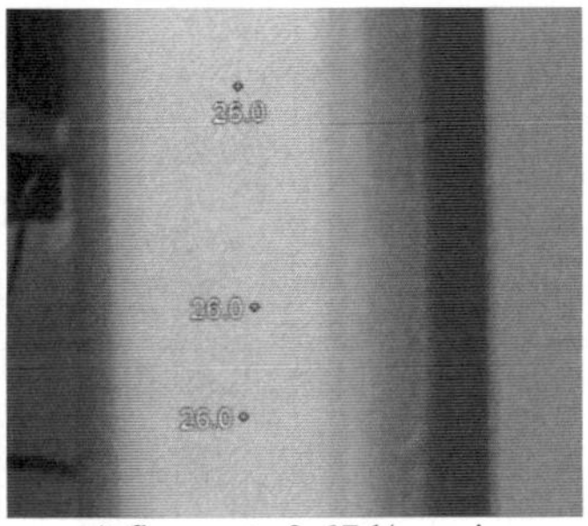

a) flow rate 2,67 l/m·min

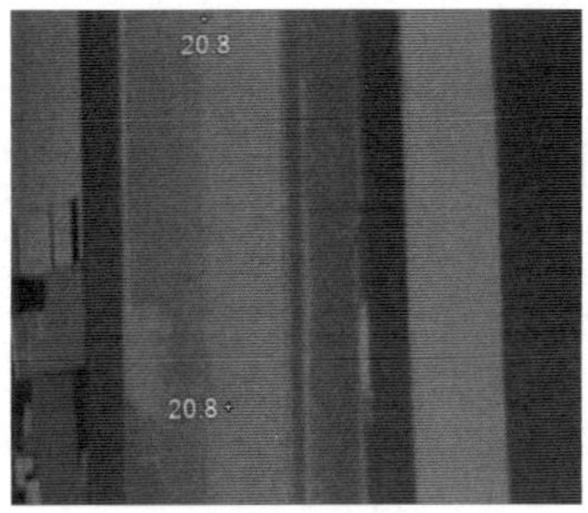

b) flow rate 0,2 l/m·min

Fig. 3. Temperature field of liquid film for decrease of flow rate (liquid temperature of 29 °C)

3.2. Wetting of the previously dry plate

Hysteresis of flow rates has been confirmed experimentally [7], according to which the flow rates necessary to form a continuous liquid film on the previously dry plate are significantly higher that flow rates detected when continuous film brakes-up.

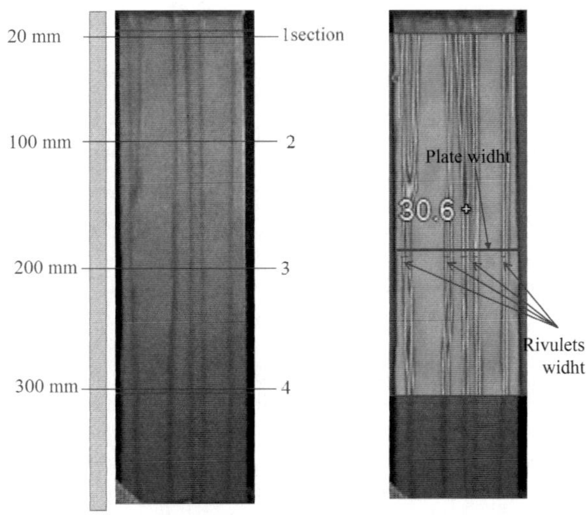

Fig. 4. Liquid rivulets recorded for liquid flow rate 0,24 l/min

Contact angle is in correlation with the minimum wetting rate as follows: a receding contact angle for the breakup of a continuous liquid film and advancing contact angle for wetting of a previously dry surface. Measured advancing contact angle for distilled water and sandblasted stainless steel has value of 75,5 °, which is significantly higher than receding contact angle. Minimum flow rate needed to form a continuous liquid film on a dry plate is in average 1 l/min in case of isothermal film.

For tested temperatures, resulting flow rate was the same as for the isothermal film, but flow hydrodynamics was in some aspects different. This was shown through parameter called „ratio of wetted width'' determined for several sections along the plate, as shown in Fig. 4. Ratio of wetted width presets ratio of rivulets width to plate width (total wetted width).

Dependence of the parameter ratio of wetted surface on volume flow rate is shown in Fig. 5. for two different cases of liquid inlet temperatures and flow rates ranging from 0,15l/min to 1 l/min. It is noticeable that, for lower volume flow rate, the ratio of wetted width is similar for both cases of inlet temperature and it increases while volume flow rate increases. For colder fluid it increases gradually, while for higher liquid temperature it increases steeply. Ratio of wetted width increases gradually since kinetic energy force dominates over the surface forces, until a dry patch disappears. For heated liquid flow, the rivulets are thin and they are getting cooled in direct contact with cold plate. Therefore, the wetting mechanism is similar as in case of cold fluid. As volume flow rate increases, average rivulets and film temperature also increases therefore surface tension forces decreases. Meanwhile, kinetic energy of rivulets increases since volume flow rate increase which all together results in a sudden increase of ratio of wetted width. After that the force balance again and establishes a somewhat milder trend of ratio of wetted width as volume flow rate increases, until the final formation of the front liquid film.

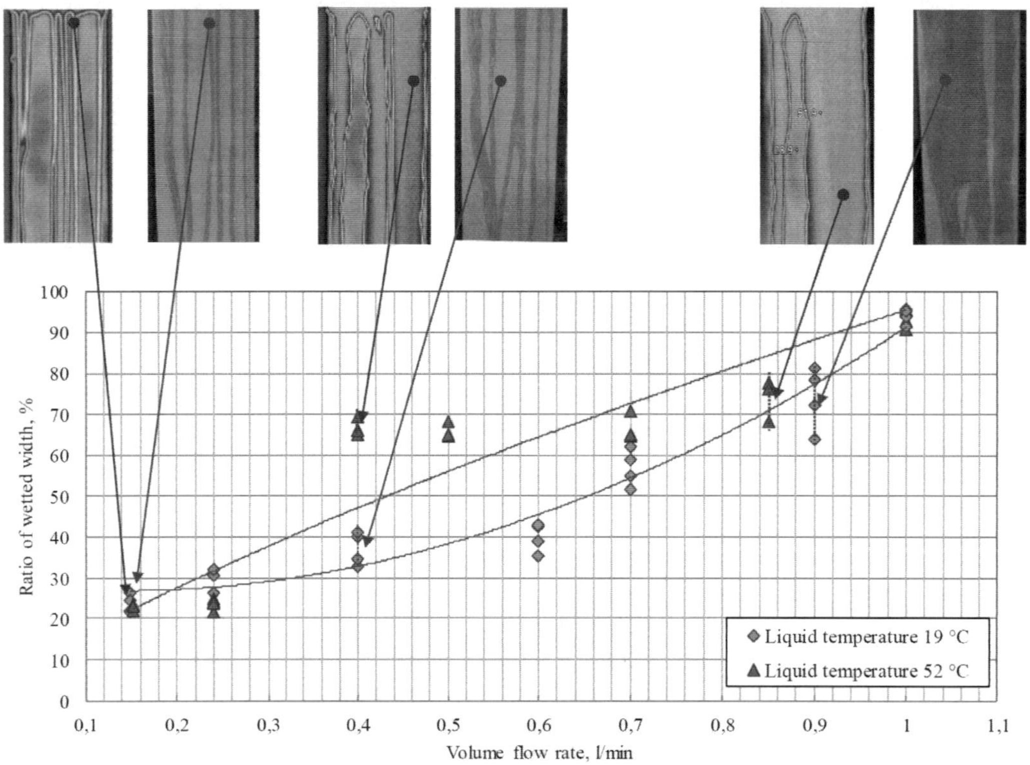

Fig. 5. Ratio of wetted width as a function of volume flow rate for two cases of liquid temperature

3.3. Wetting of the previously dry plate with different surface treatment

In order to test an influence of the surface treatment on a formation of the continuous liquid film, wetting of dry stainless steel plate subjected to the sandblasting and to the polishing is recorded. Liquid temperature was 52 °C. Starting liquid flow rate was 0,15 l/min and is was gradually increased up to 1 l/min.

Values of the ratio of the wetted width for various liquid flow rates are shown in Fig. 6.

Comparison of ratio of wetted width for two cases of surface treatment shows that sandblasted surface presents surface with higher wetting potential. For polished surface, even for higher liquid flow rate, ratio of wetted width slightly increases. This means that thickness and velocity of rivulets increases while liquid flow rate increases but ratio of wetted width remains smaller than 50 % in average.

For smaller flow rates, meandering rivulets are formed and they merge into a narrow liquid front. For sandblasted surface, rivulets flow in parallel and gradually combine as flow rate increases. Plate roughness is smaller for polished surface, which explains difference in minimum wetting flow rates. Also, even after a detail cleaning of the polished surface, it is reasonable to conclude that remains of polishing pasta are present on a plate surface, which results in a hydrophobic surface.

Advancing contact angle for polished surface is 80 °, which is higher than for the sandblasted surface. All these facts are working in favor with the recorded results.

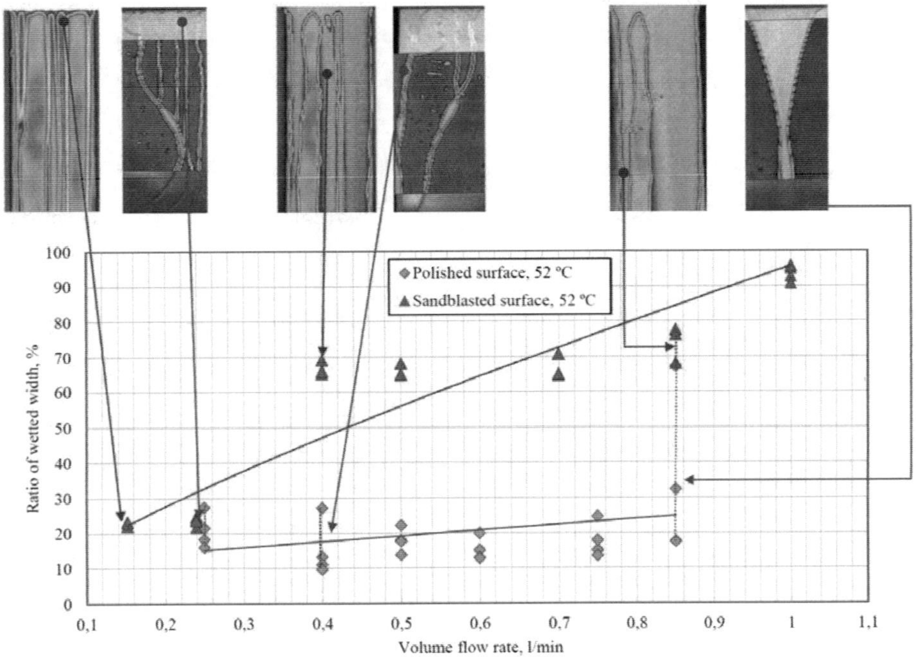

Fig. 6. Ratio of wetted width as a function of volume flow rate for two cases of surface treatment

4. Conclusion

Results of experiments conducted on a vertical plate and distilled water as a working fluid for the case of the brake-up of the continuous falling film, observed in region of Reynolds numbers smaller than 1, showed that liquid film temperature has no influence on the minimum wetting rate. It is shown that sandblasted plates provide better ratio of wetted width than polished plates, with smaller advancing contact angles while rivulets hydrodynamics was driven by the higher surface tension forces. Elevated liquid temperature has no effect on minimum wetting rate on the previously dry surface, but experimental results reveals how process is driven by the balance of the surface tension forces and kinetic energy forces.

5. References

[1] M.S. El-Genk and H.H. Saber(2001).Minimum thickness of a flowing down on a vertical surface, Int. J. Heat Mass Transfer, 44:2809-2825

[2] D.E. Hartely, W. Murgatroyd(1964). Criteria for the break–up of thin liquid layers flowing isothermally over solid surface, Int. J. Heat Mass Transfer, 7: 1003-1015

[3] T. Hobler (1964). Minimum surface wetting, Chem. Stosow, 2B:145-159

[4] S.G. Bankof (1971). Minimum thickness of a draining liquid film, Int. J. Heat Mass Transfer, 14:2143-2146

[5] J.Mikielewicz and J.R. Moszynski(1976). Minimum thickness of a liquid film flowing vertically down a solid surface, Int. J. Heat Mass Transfer, 19:771-776

[6] D. Moalem–Maron, N. Brauner (1983). Transport characteristics of wavy thin films on inclined surfaces, Advance in Two-Phase flow and Heat Transfer, Vol. 1:155–181

[7] Džana, Kadrić; Mirela, Alispahić; Šefko, Šikalo; Suad, Jakirlić and Ejup N. Ganić (2015), Proceedings of the Eight International Symposium on Turbulence, Heat and Mass Transfer, K. Hanjalić, T. Miyauchi, D. Borello, M. Hadžiabdić, P. Venturini (Editors), Begell House, Inc. New York, Wallingford (UK). ISBN 2377-2816 (Online); ISBN 2377-4169 (CD-ROM); Extended abstract p. 221–224

[8] Moalem–Maron, D. and Brauner, N. (1983). Transport characteristics of wavy thin films on inclined surfaces, Advance in Two-Phase flow and Heat Transfer, Vol. 1, 155–181

[9] Stefan, Franko; Eva, Babusova; Miroslav, Badida; (2011). Annals of DAAAM for 2011 & Proceedings of the 22nd International DAAAM Symposium, ISSN 1726-9679 ISBN 978-3-901509-83-4, Editor B. Katalinic, Published by DAAAM International, Vienna, Austria, Volume 22, No. 1, 1233-1234

[10] Džana, Kadrić; (2011). Annals of DAAAM for 2011 & Proceedings of the 22nd International DAAAM Symposium, ISSN 1726-9679 ISBN 978-3-901509-83-4, Editor B. Katalinic, Published by DAAAM International, Vienna, Austria, Volume 22, No. 1, 1063-1065

DOI: 10.2507/27th.daaam.proceedings.039

T-PHANTOM: A NEW PHANTOM DESIGN FOR NEUROSURGICAL ROBOTICS

Marko Švaco, Bojan Jerbić, Ivan Stiperski, Domagoj Dlaka,
Josip Vidaković, Bojan Šekoranja & Filip Šuligoj

This Publication has to be referred as: Svaco, M[arko]; Jerbic, B[ojan]; Stiperski, I[van]; Dlaka, D[omagoj]; Vidakovic, J[osip] & Sekoranja, B[ojan] (2016). T-Phantom: a New Phantom Design for Neurosurgical Robotics, Proceedings of the 27th DAAAM International Symposium, pp.0266-0270, B. Katalinic (Ed.), Published by DAAAM International, ISBN 978-3-902734-08-2, ISSN 1726-9679, Vienna, Austria
DOI: 10.2507/27th.daaam.proceedings.039

Abstract

In this paper we propose a novel phantom design for measuring application accuracy of neurosurgical robotic systems and stereotactic frames. We develop a novel phantom (T-Phantom) which enables simultaneous localization of translational displacements in entry and target points. The phantom consists of multiple trajectories positioned around a localizer feature simulating approach trajectories in neurosurgical procedures on the intracranial space. Each trajectory consists of two parallel and coaxial hollow cylinders printed in selective laser sintering technology. We apply a stereo vision measuring method for precise measurements of translational displacements in target and entry positions. The paper further provides a systematic comparison of phantom designs originating from stereotactic frames and neurosurgical robotic systems. To the author's knowledge, the developed T-Phantom is the first stereotactic phantom which enables simultaneous measurements both in deviation from target and entry positions and angular deviation from the planned trajectory.

Keywords: Robotics; Neurosurgery: Stereotactic phantom: Stereotactic localization.

1. Introduction

The use of robotics in neurosurgery is experiencing constant growth and novel robotic systems are continuously being developed to assist in complex neurosurgery tasks. In particular, neurosurgery operations are extremely lengthy and tedious and the application of robots is expected to provide the surgeon and patient with multiple benefits [1], [2]. Neurosurgical procedures performed by a robotic system yield better and faster performance, are less invasive and enable faster recovery of the patient [3]. By this means the utilization of hospital operational resources is considerably improved. Prior to performing in vivo procedures, it is necessary to make a diverse range of testing on phantoms [1], [4] for testing the accuracy and consistency of robotic systems. The phantom representing a patient head (more specifically the intracranial space) does not necessarily need to be anthropomorphic. Phantoms developed for stereotactic procedures need to provide an accurate replica of targeted trajectories in stereotactic procedures. Phantom designs can also provide functionalities [5] for drilling procedures and tissue simulations with different characteristics of the human head (skin, bone, brain). This research discusses only the former problem, that of translational and angular displacements in trajectories. Each set of translational displacements can be further expressed as Euclidian distance errors. In the operating room these errors are accounted from CT or MRI scans, errors in registration procedures, mechanical errors in robotic systems or stereotactic frames and other.

In this paper we have developed a novel phantom called the T-Phantom (Trajectory Phantom) which is to the author's knowledge the first stereotactic phantom which enables simultaneous measurements both in deviation from target and entry positions and angular deviation from the planned trajectory. In the first part of the paper we give a systematic overview of various phantom designs for both robotic neurosurgical systems and for stereotactic frames. The problem of phantom design in neurosurgery has been tackled by many researchers. In each phantom design and associated accuracy measurement method we identify measurable variables and give a critical overview. Finally we discuss the design of the T-Phantom and apply a previously developed stereo vision measurement method [6] for objective accuracy measurement.

2. Neurosurgical phantom designs

The Neuromate (Renishaw) anthropomorphic phantom described in [7] (Fig. 1a) uses five implantable frameless markers (Fischer-Leibinger, Freiburg, Germany), randomly distributed on the surface, which are used for the registration in the infrared region. A ZD (Fischer-Leibinger, Freiburg, Germany) stereotactic frame is fitted onto the phantom. A different phantom design for the Neuromate robot [8] uses a technique of milling foam blocks (Fig. 1.b) to a predefined shape. Each of the foam blocks is measured after the milling procedure to acquire volumetric errors.

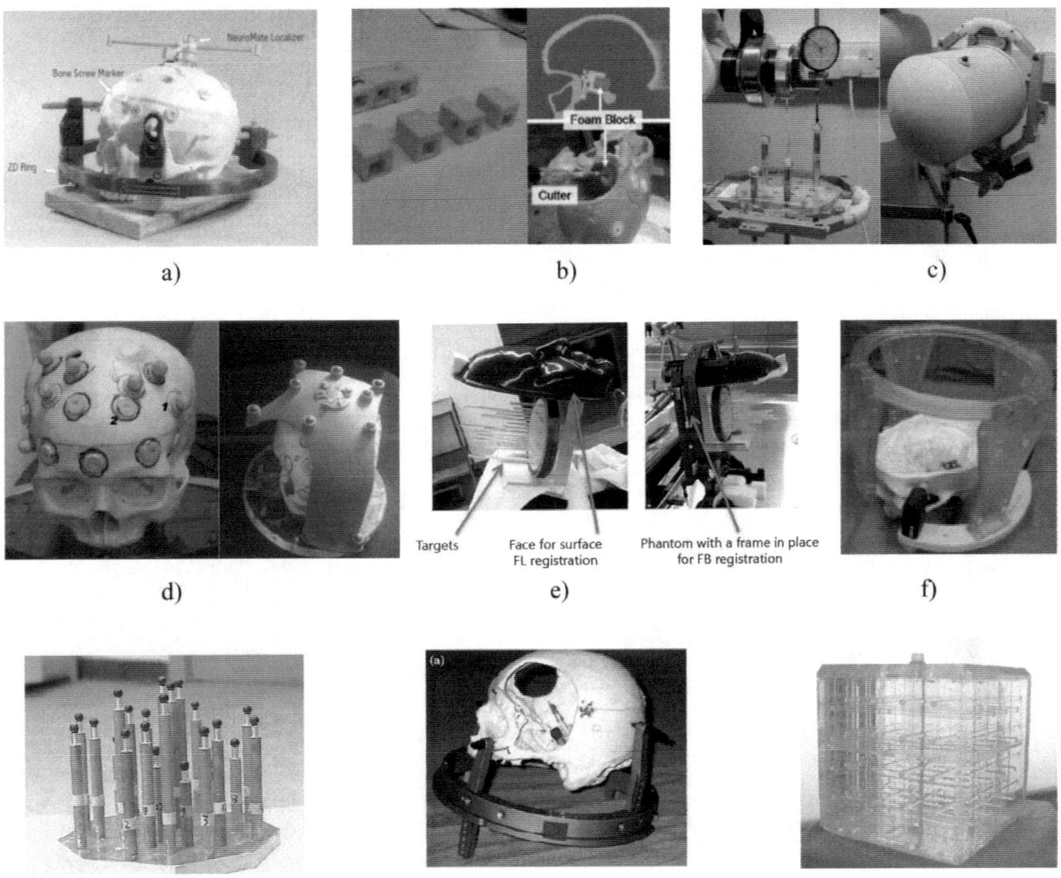

Fig. 1. Diverse neurosurgical phantom designs: a) Neuromate phantom [7] b) Neuromate phantom [8] c) Pathfinder phantom [9] d) Pathfinder phantom [4] e) ROSA phantom [10] f) MARS phantom [11] g) NeuroMaster phantom [12] h) CRW frame phantom [13] i) Leksell frame phantom [14]

The phantom used to check the robot system Pathfinder [9] consists of spherical targets (Fig. 1c) and detachable cylindrical surface serving as a simulation of skin to which the markers can be attached for the registration procedure. Spherical targets are 10 mm in diameter and are located at positions that simulate the most common target depths during neurosurgical operation procedures. A depth gauge is used to measure the application accuracy. Another phantom design for the Pathfinder robot [4] employs an anthropomorphic phantom. Onto a replica of the human skull registration markers and ten target surface points are mounted (Fig. 1d). The interior of the phantom is equipped with nine (9) depth target points. The measurements were made using vernier calipers.

The NeuroMaster [12] phantom (Fig. 1g) has a similar design and is made of cylinders with plastic spheres. Four spheres are used as markers for the registration of phantom while the other are used as target points.

The phantom [10] for the robotic system ROSA (Medtech innovative surgical technology) is shown in Fig. 1e. The phantom is made of a base with hollow polymer cylinders representing targets at different depths. A removable water tank gives a possibility of recording the phantom on MRI. The phantom has a hollow plastic face that can be filled with a contrast agent to make it suitable for scanning on MRI scanners.

The MARS stereotactic robot [11] uses commercially available phantoms made as replicas of the human skull (Fig. 1f). Inside the skull cylindrical conical cylinders represent target points. The measurement is performed with two vertically mounted cameras which measures the position of the probe tip deviation at the target point.

Phantom designs for stereotactic frames such as Leksell frame [14] and CRW frame [13] are shown in Fig. 1i and Fig. 1h. Other phantom designs include gel based phantoms [15] dedicated for MRI scanning, special phantoms with divots [16], hollow cylindrical phantoms [17] with different target materials and resin based phantoms [18] with implantable titanium screws.

3. T-Phantom

For the purpose of measuring the accuracy of robotic System RONNA [1] a phantom design called the T-Phantom is proposed. The T-Phantom (Fig. 2) consists of a Plexiglas construction and a localization plate (RONNAmarker) with three (or four) spherical markers. The spherical markers are used to define the phantom coordinate system. The phantom has hollow cylinders printed in selective laser sintering technology (SLS) which simulate operation trajectories.. The phantom is designed to simulate trajectories in neurosurgical applications under four tilt angles (45 °, 35 °, 25 °, 15 °), and four trajectories perpendicular to the phantom base. The maximum spatial angle between two trajectories is 70°. Trajectories are selected in accordance with common trajectory angles in actual operations. The top of the T-Phantom has a radius simulating the top of the human head For attaching the localization plate (RONNAmarker). The phantom has machined grooves on each of the tilted trajectories so that the distance from entry to target point can be adjusted. Each trajectory consists of two parallel rectangular prisms with coaxial cylindrical bores. Three rings made from selective laser sintering (SLS) are located in each of the bores. The outer white cylinders are 2 mm thick and the inner black cylinder is 6 mm thick as shown in Fig. 3.

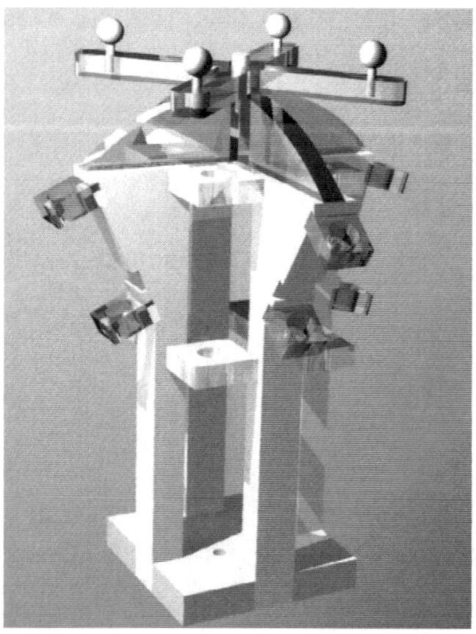

Fig. 2. a) T-phantom CAD prototype b)Actual T-Phantom

The measurement procedure is as follows. The T-Phantom is scanned on a CT scanner (512x512 with 0,75mm slice thickness, no gantry). The trajectories are planned in the operation planning software. The phantom is positioned in a Mayfield clamp after which the RONNA robotic system localizes the phantom and positions the tool guide to the planned trajectory. The surgical tool (probe) is calibrated to a predefined depth as the robot tool center point. The insertion of the probe is done manually to the predefined depth. At this step the measurement is carried out on the entry and target point for each trajectory as shown in from Fig. 3.

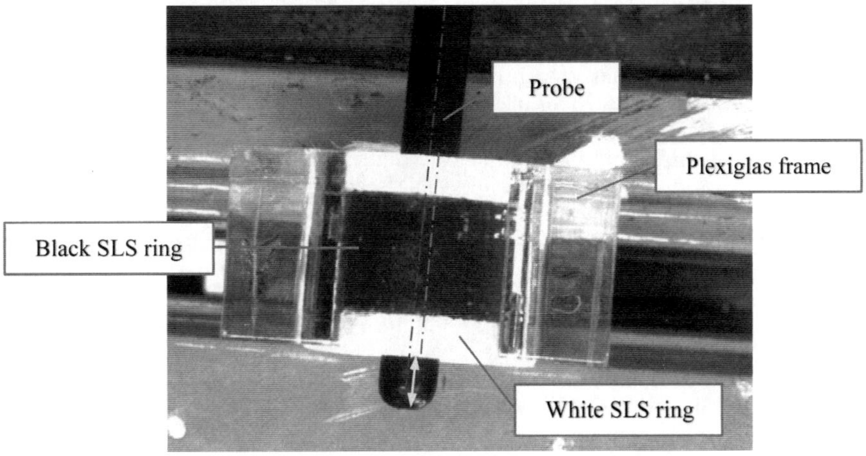

Fig. 3. a) Vision system measurement of the target point

In Fig. 3 the main measurement concept is depicted through a picture taken by our stereo vision system [6]. The picture shows a target point in which the depth error in the Z direction is depicted by the yellow arrow. The red dotted line represents the rotational axis of the probe while the blue dotted line represents the rotational axis of the cylindrical SLS inserts. By measuring the position of the probe relative to the phantom in two perpendicular camera directions the total Euclidian error can be computed for both the target and entry point. By measuring the XY translational error at the target point and the XY translational error of the Entry point the angular error is calculated. The angular error reflects the total (spatial) angular displacement error of the probe (red dotted line) with respect to the ideal trajectory (blue dotted line).

4. Conclusion and future work

Table 1. shows a detailed comparison of thirteen phantom designs and the new T-Phantom design. Both neurosurgical robotic phantoms and phantoms intended for stereotactic frames are compared (CRW Precision™ Arc system, ZD stereotactic system, Leksell Stereotactic System®). The CT scan column (Computed Tomography) depicts if the phantom design is suitable for CT technology. MRI scan (Magnetic Resonance Imaging) depicts if the phantom design is suitable for MRI technology. Target point and entry point error define if it is possible to localize positioning errors of neurosurgical tools (probes, drills, etc.) with respect to planned operation points in a given phantom design. The angular error further takes into considerations errors derived from entry and target points. The last column in Table 1. differentiates anthropomorphic[5] phantoms from non-anthropomorphic phantom designs.

Phantom (system)	CT scan	MRI scan	Target point error	Entry point error	Angular error	Anthropomorphic phantom
Neuromate® [7]	yes	no	yes	no	no	yes
Neuromate® [8]	yes	no	yes	no	no	yes
Pathfinder [9]	yes	no	yes	no	no	no
Pathfinder [4]	yes	yes	yes	no	no	yes
ROSA™ [10]	yes	yes	yes	yes	no	no
MARS [11]	yes	yes	yes	no	no	yes
NeuroMaster [12]	yes	no	yes	no	no	no
CRW™ frame [13]	no	yes	yes	no	no	yes
Leksell® frame [14]	yes	yes	yes	no	no	no
Gel based phantom [15]	no	yes	yes	no	no	no
Phantom with divots [16]	yes	yes	yes	no	no	no
Cylindrical phantom [17]	yes	yes	yes	no	no	no
Resin based phantom [18]	yes	no	yes	no	no	yes
T-Phantom	yes	no	yes	yes	yes	no

Table 1. A systematic comparison of 13 phantom designs with the developed T-Phantom

It can be observed that only the ROSA™ phantom design enables simultaneous measurements of both errors in target and entry points. The ROSA phantom design with hollow cylinders does not enable objective quantitative measurements but only measurements of type "better than". The main advantage of the novel T-Phantom design is that it enables simultaneous objective measurements of translational errors in positioning of neurosurgical instruments at target and entry points. Our design also provides a possibility of angular displacement calculations and deviations from planned trajectories. The T-Phantom can be also used or adapted to other fields and applications where there is a need for measurements of deviations from planned translational (linear) trajectories. The T-Phantom design is scalable so that it can be used ranging from applications dedicated for microsurgery to applications where distances from target to entry point considerably exceed 120mm (longest trajectories in neurosurgery).

In future research we plan to optimize the phantom design so that it can be machined from polymer components with higher accuracy. Also a new redesign should be done to enable scanning of the phantom using MRI technology. Furthermore, we plan to make a detailed accuracy analysis of the measurement method by comparing the stereo vision measurements with ground truth measurements made with either a coordinate measuring machine or a high accuracy vision system [19].

5. Acknowledgments

Authors would like to acknowledge the Croatian Scientific Foundation through the research project ACRON - A new concept of Applied Cognitive Robotics in clinical Neuroscience.

6. References

[1] B. Jerbić, G. Nikolić, D. Chudy, M. Švaco, and B. Šekoranja, "Robotic application in neurosurgery using intelligent visual and haptic interaction," International Journal of Simulation Modelling, vol. 14, no. 1, pp. 71–84, 2015

[2] G. R. Sutherland, S. Wolfsberger, S. Lama, and K. Zarei-nia, "The evolution of neuroArm," Neurosurgery, vol. 72 Suppl 1, pp. 27–32, doi: .1227/NEU.0b013e318270da19 2013

[3] M. Hoeckelmann, I. J. Rudas, P. Fiorini, F. Kirchner, and T. Haidegger, "Current Capabilities and Development Potential in Surgical Robotics," International Journal of Advanced Robotic Systems, p. 1, 2015

[4] M. S. Eljamel, "Validation of the PathFinder™ neurosurgical robot using a phantom," The International Journal of Medical Robotics and Computer Assisted Surgery, vol. 3, no. 4, pp. 372–377, doi: .1002/rcs.153 2007

[5] A. Müns, J. Meixensberger, and D. Lindner, "Evaluation of a novel phantom-based neurosurgical training system," Surgical Neurology International, vol. 5, no. 1, p. 173, 2014

[6] M. Švaco, B. Šekoranja, F. Šuligoj, and B. Jerbić, "Calibration of an Industrial Robot Using a Stereo Vision System," in Procedia Engineering, 2014, vol. 69, pp. 459–463

[7] Q. H. Li, L. Zamorano, A. Pandya, R. Perez, J. Gong, and F. Diaz, "The application accuracy of the NeuroMate robot--A quantitative comparison with frameless and frame-based surgical localization systems," Comput. Aided Surg., vol. 7, no. 2, pp. 90–98, doi: .1002/igs.10035 2002

[8] T. Haidegger, "Improving the Accuracy and Safety of a Robotic System for Neurosurgery," 2008

[9] P. S. Morgan, T. Carter, S. Davis, A. Sepehri, J. Punt, P. Byrne, A. Moody, and P. Finlay, "The application accuracy of the Pathfinder neurosurgical robot," International Congress Series, vol. 1256, pp. 561–567, Jun. 2003

[10] M. Lefranc, C. Capel, A. S. Pruvot, A. Fichten, C. Desenclos, P. Toussaint, D. Le Gars, and J. Peltier, "The Impact of the Reference Imaging Modality, Registration Method and Intraoperative Flat-Panel Computed Tomography on the Accuracy of the ROSA® Stereotactic Robot," Stereotactic and Functional Neurosurgery, vol. 92, no. 4, pp. 242–250, 2014

[11] M. Heinig, "Design and Evaluation of the Motor Assisted Robotic Stereotaxy System MARS," Lubeck, 2012

[12] J. Liu, Y. Zhang, and Z. Li, "The application accuracy of neuromaster: a robot system for stereotactic neurosurgery," in Mechatronic and Embedded Systems and Applications, Proceedings of the 2nd IEEE/ASME International Conference on, 2006, pp. 1–5

[13] A. Quiñones-Hinojosa, M. L. Ware, N. Sanai, and M. W. McDermott, "Assessment of Image Guided Accuracy in a Skull Model: Comparison of Frameless Stereotaxy Techniques vs. Frame-Based Localization," Journal of Neuro-Oncology, vol. 76, no. 1, pp. 65–70, Jan. 2006

[14] C. Yu, "An image fusion study of the geometric accuracy of magnetic resonance imaging with the Leksell stereotactic localization system," Journal of Applied Clinical Medical Physics, vol. 2, no. 1, p. 42, Jan. 2001

[15] A. D. Squires, Y. Gao, S. F. Taylor, M. Kent, and Z. T. H. Tse, "A Simple and Inexpensive Stereotactic Guidance Frame for MRI-Guided Brain Biopsy in Canines," Journal of Medical Engineering, vol. 2014, pp. 1–7, 2014

[16] D. Á. Nagy, T. Haidegger, and Z. Yaniv, "A Framework for Semi-Automatic Fiducial Localization in Volumetric Images," in Augmented Environments for Computer-Assisted Interventions, Springer, 2014, pp. 138–148

[17] S. Poggi, S. Pallotta, S. Russo, P. Gallina, A. Torresin, and M. Bucciolini, "Neuronavigation accuracy dependence on CT and MR imaging parameters: a phantom-based study," Physics in medicine and biology, vol. 48, no. 14, p. 2199, 2003

[18] G. Eggers and J. Muhling, "Template-based registration for image-guided skull base surgery," Otolaryngology - Head and Neck Surgery, vol. 136, no. 6, pp. 907–913, Jun. 2007

[19] F. Suligoj, B. Jerbic, M. Švaco, B. Sekoranja, D. Mihalinec, and J. Vidakovic, "Medical applicability of a low-cost industrial robot arm guided with an optical tracking system," 2015, pp. 3785–3790

DOI: 10.2507/27th.daaam.proceedings.040

MATHEMATICAL MODEL OF HYDROLYSIS REACTION FOR THE COLLAGEN HYDROLYZATE PRODUCTION FROM LEATHER SHAVINGS

Hana Vaskova & Vladimir Vasek

This Publication has to be referred as: Vaskova, H[ana] & Vasek, V[ladimir] (2016). Mathematical Model of Hydrolysis Reaction for the Collagen Hydrolyzate Production From Leather Shavings, Proceedings of the 27th DAAAM International Symposium, pp.0271-0274, B. Katalinic (Ed.), Published by DAAAM International, ISBN 978-3-902734-08-2, ISSN 1726-9679, Vienna, Austria
DOI: 10.2507/27th.daaam.proceedings.040

Abstract

An integral part of the automated control implementation is mathematical description of the basic processes and mathematical modeling. In this paper, the mathematical model for alkaline hydrolysis of shavings, a leather waste material, is presented. Leather industry produces annually a large amount of wastes. For the environment-friendly solution it is efficient to process the certain part of wastes arising from the leather industry. Collagen protein contained in e.g. leather shavings can be further utilized. Hydrolysis processes seems to be reasonable solution. The description of the hydrolysis process is based on the linearized state model. The mathematical-physical model is built on the mass balance of the input substances and the resulting hydrolyzate protein, moreover on an enthalpy balance for the reaction mixture and saturated steam serving in reactor as the heat transfer medium.

Keywords: Mathematical modeling; hydrolysis; leather; waste; automatic control.

1. Introduction

The topic of waste management and waste processing is widely discussed topic nowadays. A significant amount of waste is in the leather industry. The leather tanning industry is historically one of the first industries occupying an important function in terms of by-products of other industry - meat industry. However, it is also perceived as a resources consumer and pollutants producer [1]. Tanning industry transforms raw hides of slaughter and game animals through physical and chemical processes to a stable, technically usable product - the leather. On one hand it uses meat industry waste as a feedstock on the other hand this industry produces another hazardous waste.

Production of 200 kg of leather requires one ton of raw hides processing, i.e. feedstock usage is only 20% efficient. From this one ton of raw hides about 250 kg of non-tanned waste, 200 kg of tanned wastes containing chromium is also produced. The remaining share comprises modified solid waste and sludge. In the production of 200 kg of leather about 50,000 kg of waste water containing about 5 kg of chromium is generated. Approximately 10 million tons of raw hide is processed annually [2].

Another topic comprising toxic hexavalent chromium and the ways of its transformation from and into trivalent form is closely related with the production of leather and with the area of tannery waste disposal. This topic should not be neglected in terms of human health, the environment [3] and the detection of harmless substances [4].

Mathematical modeling gains a great importance for the analysis and prediction of the behaviour of processes and phenomena in areas of natural, technical, industrial, but also economic and social in recent years. It offers the possibility to transfer the problem from the field of application to be processed mathematically. This allows to penetrate through the analysis into its nature and to obtain useful information about investigated problem. Using mathematical means and information technology, procedures leading to deeper and more thorough understanding can be created.

The literature search shows, that the automatic control in leather/hide processing respectively leather waste processing is not very widespread. The most literary sources talks rather about the use of automated and robotic devices in secondary manufacturing industries. These sectors include especially footwear industry, which process over the half of the produced leather, textile industry and extensive use is also for upholstery and automotive industry.

The level of measurement and control systems applications is low due to the lack of quantitative models allowing the use of some modern control methods. However, in recent years, efforts are being made to implement the automation into the leather processing to increase productivity, precision of operations, time efficiency and last but not least to reduce physically demanding work, restrict acts of the health hazards of employees and contribute to the work environment improvement.

The problems of processing the leather industry wastes including proposals for technological devices equipped with automatic control are solved in last decades at our workplace, at the Faculty of Applied Informatics at Tomas Bata University in Zlín, Czech Republic [5], [6] and [7].

2. Leather waste processing

2.1. Waste treatment process

The complex processing of leather shavings can be, in simplified way, divided into three stages as is shown in Fig. 1 (left). Collagen protein is liquefied and separated from the chromium sludge in the first stage. Chromium sludge can be subjected to revitalization for obtaining chromium in a form of salts for further leather tanning [8]. One part of resulting gelatinous protein, a high-quality gelatine finds cost-effective usage applications such as in pharmacy, food industry, or cosmetics. The rest of gelatinous protein is used for further processing. Enzyme activity leads in alkaline conditions to molar mass decrease of gelatinous hydrolyzate. Further splitting of protein chain can continue by the act of acid. The hydrolyzate with lower molar mass is acquired subsequently and can find application in agriculture.

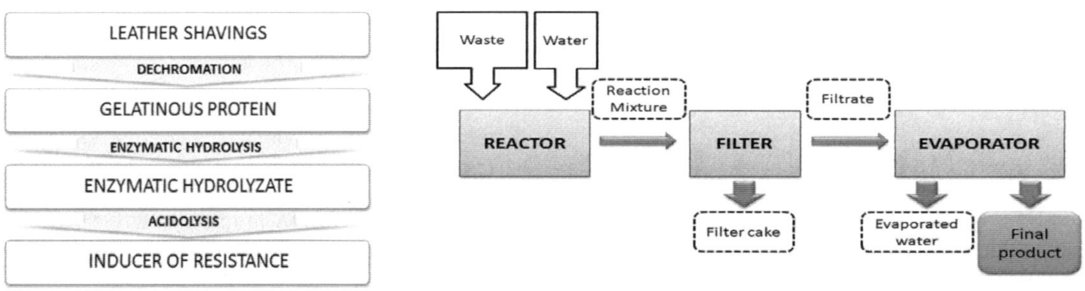

Fig. 1. The structure of collagen protein hydrolysis (left). The diagram of the production system (right)

2.2. *Production system*

The production system for producing the hydrolyzate is composed of three basic elements: the reactor, the filter and the vacuum evaporator. The simplified diagram is shown in Fig. 1 (right). Feedstocks in the reactor are leather waste, water and an alkali, creating a reaction mixture. When the desired degree of hydrolysis is reached, hot heterogeneous mixture is filtered and the resulting filtrate is further transferred to an evaporator, where the concentration of hydrolysed substance is increased.

3. Mathematical model

For the process description the linearized state mathematical model including input, output variables and inner-state variables is used. Linearized model is obtained by introducing deviations of state and input variables from their stationary states and then linearized using Taylor series. The detailed mathematical treatments are presented in [3] Continuous linear system is described by the state equation and output status

$$\Delta \dot{\mathbf{X}} = \mathbf{A}\Delta\mathbf{X} + \mathbf{B}\Delta\mathbf{U} \tag{1}$$

$$\Delta\mathbf{Y} = \mathbf{C}\Delta\mathbf{X} + \mathbf{D}\Delta\mathbf{U} \tag{2}$$

where A is state matrix, dim A = $n \times n$, B is the input matrix, dim B = $n \times r$, C is the output matrix, dim C = $m \times n$, D is the zero matrix, dim D = $\mathrm{m} \times r$.

This system as a real system fulfils the strong physical condition of the feasibility, i.e. the outputs are functions only of the state variables. This means that for

$$\Delta\mathbf{Y} = \Delta\mathbf{X} \tag{3}$$

the matrix C is the identity matrix and D is the zero matrix.

Mathematical-physical model for alkaline hydrolysis of shavings is based on the mass balance of the input substance, the resulting hydrolyzate protein, enthalpy balance for the reaction mixture and the heat transfer medium. As a heat transfer medium the saturated steam is considered. The hydrolysis process takes place in a flow reactor with a stirrer ensuring the state variable independence on the position in the reactor. It is considered as system with lumped parameters. The mixing the heat transfer medium is also presumed. The mathematical model is given by the balance equations (4) – (7). The used symbols are specified in Table 1.

Symbol	Physical quantity	Unit
m_R	Mass of reaction mixture (RS)	kg
m_P	Mas of water	kg
$\dot{m}_{RS}$	Mass flow of reaction mixture	kg/s
$\dot{m}_P$	Mass flow of heating stem, respectively water	kg/s
a_P	Initial mass fraction of protein in RS	1
a_B	Mass fraction of decomposed protein in the reactor	1
a_E	Mass fraction of undecomposed protein in the reactor	1
T_{RS}	Temperature of RS	K
T_0	Temperature of stem, the output	K
T	Temperature of the reactor shell = temperature of water, the output	K
c_{RS}	Specific heat of RS	J/kgK
c_P	Specific heat of water	J/kgK
K	Heat transfer coefficient	W/m^{2}K
S	Area of reactor heated by steam	m^2
H	Specific latent heat of vaporization	J/kg

Table 1. Variables used in the model

Mass balance for protein

$$\dot{m}_{RS}\, a_P = \dot{m}_{RS}\, a_B + k m_R a_B + m_R \frac{\mathrm{d}a_B}{\mathrm{d}t} \tag{4}$$

Mass balance for hydrolyzate

$$0 = \dot{m}_{RS}\, a_E - k m_R a_B + m_R \frac{\mathrm{d}a_E}{\mathrm{d}t} \tag{5}$$

Enthalpy balance for reaction mixture

$$\dot{m}_{RS}\, c_{RS} T_{RS} + KS(T - T_R) = \dot{m}_{RS}\, c_{RS} T_R + m_R c_{RS} \frac{\mathrm{d}T_R}{\mathrm{d}t} \tag{6}$$

Enthalpy balance for heat transfer fluid

$$\dot{m}_P H + \dot{m}_P c_P T_0 = \dot{m}_P c_P T + KS(T - T_R) + m_P c_P \frac{\mathrm{d}T}{\mathrm{d}t} \tag{7}$$

Taking the Laplace transforms of (1) and (2) together with (3) we get

$$G(s) = (sI - A)^{-1} B \tag{8}$$

Expression of matrices A and B is then based on the mathematic model (4) – (7). The transfer matrix is calculated from (8). Transferred to the dimensionless form and expressed using values of physical quantities for a real reactor we got a matrix G* depicted in equation (9).

$$G^* = \begin{bmatrix} \frac{1}{4{,}09+s} & \frac{0{,}27}{4{,}09+s} & 0 & 0 & 0 \\ \frac{3{,}09}{4{,}09+5{,}09+s^2} & -\frac{0{,}27+0{,}27s}{4{,}09+5{,}09+s^2} & 0 & 0 & 0 \\ 0 & -\frac{50(22{,}75+s)}{343(28{,}13+29{,}13s+s^2)} & \frac{22{,}75+s}{28{,}13+29{,}13s+s^2} & \frac{3{,}44}{28{,}13+29{,}13s+s^2} & \frac{2{,}17}{28{,}13+29{,}13s+s^2} \\ 0 & -\frac{227250}{343(28{,}13+29{,}13s+s^2)} & \frac{4545}{28{,}13+29{,}13s+s^2} & \frac{4{,}09+0{,}64s}{28{,}13+29{,}13s+s^2} & \frac{2{,}58+0{,}41s}{28{,}13+29{,}13s+s^2} \end{bmatrix} \tag{9}$$

Simulations based on the state description concerning matrixes A and B and the calculated transfer matrix G were performed in Matlab Simulink. These simulations are not described within this paper.

4. Conclusion

Mathematical-physical model for the hydrolysis of collagen protein from tannery wastes is presented in the paper. This way of treatment offers safe, environmentally friendly method for the reduction of waste gained from leather manufacturing sector. Simulations of the model serves for further laboratory and pilot scale experiments and contribute to automated control of the production process. Further steps leads to the optimization of the main operating costs for the production of collagen hydrolyzate what brings another benefit in terms of economical effectivity.

5. Acknowledgement

This work was supported by the Ministry of Education, Youth and Sports of the Czech Republic within the National Sustainability Program project No. LO1303 (MSMT-7778/2014).

6. References

[1] Mwinyihija, Mwinyikione. (2010). Main Pollutants and Environmental Impacts of the Tanning Industry. Ecotoxicological Diagnosis in the Tanning Industry. New York, 978-1-4419-6266-9, NY: Springer New York

[2] http://www.fao.org/fileadmin/templates/est/MARKETS_MONITORING/HidesSkins.pdf World Statistical Compendium for raw hides and skins, leather and leather footwear 1993-2012, Accessed on: 2015-06-11

[3] Vaskova, H. (2015) Modeling of chemical reactor for leather waste dechromation. Ph.D. Dissertation, Tomas Bata Univerzity in Zlin, Czech Republic

[4] Vašková, H. (2010) Raman spectroscopy: a modern technique for material properties identification. Proceedings of the 21st International DAAAM Symposium 2010 Zadar, Croatia. Vienna: DAAAM International Vienna, pp. 1321-1322. ISBN 978-3-901509-73-5

[5] Vašek, V., Dolinay, J., Dostálek, P., Janáčová, D., Kolomazník K. (2009). Microcontrollers in Tannery Waste Recycling Control. 1th International Leather Engineering Symposium "Leather Industry, Environment and Progressive Technologies". Izmir, Age University

[6] Kolomazník, K., Vašek, V., Janáčová, D. (2009). Control recycling technology of tannery chromium wastes. Proceedings of the 13th WSEAS International Conference on Computers. Rhodes, WSEAS Press, pp. 550-554

[7] Charvátová, H., Vašek, V., Drga, R., Křenek, J. (2011). Software application for solving of nonstationary temperature fields in twolayer plate. Annals of DAAAM and Proceedings of the International DAAAM Symposium Zadar, Croatia, pp. 359-360

[8] Kolomaznik, K,. Mladek, M., Langmaier, F., Janacova, D., Taylor, M., M.(2000). Experience in Industrial Practice of Enzymatic Dechromation of Chrome Shavings. JALCA, Vol. 94, pp. 55- 6. ISBN: 0002-9726

DOI: 10.2507/27th.daaam.proceedings.041

REAMING OF VERY PRECISE AND DEEP HOLES WITH CERMET REAMER

Jaroslava Fulemova & Jan Rehor

This Publication has to be referred as: Fulemova, J[aroslava] & Rehor, J[an] (2016). Reaming of Very Precise and Deep Holes With Cermet Reamer, Proceedings of the 27th DAAAM International Symposium, pp.0275-0282, B. Katalinic (Ed.), Published by DAAAM International, ISBN 978-3-902734-08-2, ISSN 1726-9679, Vienna, Austria
DOI: 10.2507/27th.daaam.proceedings.041

Abstract

Reaming is a technology which is ranked among finishing machining. Reamers are usually tools with one or more teeth. Teeth can be made from different cutting materials. The most common cutting materials for reamers are HSS and Sintered Carbide (SC) with or without coated thin layer. Less frequently used cutting materials are Cermet, CBN and PKD. During reaming, the basic criterion is quality of reamed hole with regard to ensure the optimal level of technology – economy criteria of machining process. It means tool life, productivity of machining process, cost of reaming a hole, etc. Selecting of cutting material has to match the requirements which are mentioned above. This article deals with application cermet reamers in comparison with SC reamers during reaming a hole which is 8D deep and is produced in accuracy class H6. This hole can be found on VDI tool holder. Further, article describes problems which were caused by SC reamers and successful solution with cermet reamers.

Keywords: reaming; cermet; sintered carbide; tool life; VDI tool holder.

1. Introduction

As was mentioned above reaming is finishing technology. In most cases it is last technological operation. Nevertheless there are situations, when there is demand for more precise holes, where it is necessary to use the other technological operations like grinding, rolling, honing, etc. Drilling and boring precedence reaming and they are operations which influence final accuracy parameters of reamed hole. If there is something wrong like machining parameters, technology, geometry of cutting tools, etc. the reamer is not able to improve geometry and accuracy of produced hole. Drilling is roughing technology, boring and reaming improves size, roughness and geometry.

1.1. Problem statement

This article is based on diploma thesis [1], which was solved in cooperation with a Czech company. The company is producer of engineering components and also produces its own tools. A part of their production is VDI tool holders. Production process consist a lot of technologies and one of them is very difficult. Namely it is reaming a hole ø8H6 and deep 69mmm. The main problems are:

- tool life of the cutting edge is between 50 ÷ 300 pieces per a reamer
- excess of criteria value Ra which is set up on 0.8μm
- excess of rotational accuracy criteria value which is set up on 0.05mm

These problems cause:

- low reproducibility of production
- low reliability of cutting process
- higher costs on tool
- waste of the time
- a lot of nonconforming products

Fig. 1. VDI tool holder [2]

Specifications of process parameters, which are realized in the company, are in the Table 1.

Machine tool	**Type**		**Name**	**Fabrication year**	**Number of axis**	**Power of spindle [kW]**	**Spindle speed [min^{-1}]**
	Vertical machining centre		AXA-DBZ	2005	3	13	30 ÷ 6000
Tool	**Type**	**Cutting material**	**Coating**	**Number of teeth**	**Cooling**	**Cutting speed [m/min]**	**Feed rate [mm.rev^{-1}]**
	Solid	Sintered Carbide	TiAlN	6	Inner, 30bar, 8% concentration	80	0.4
Product	**Material**	**Heat treatment**	*Measured parameters*	**Diameter [mm]**	**Depth of the hole [mm]**	**Surface roughness**	**Rotational accuracy [mm]**
	C45	Soft annealing		8H6	69	Ra = 0.8	0.05

Table 1. Specified parameters for reaming the hole

1.2. Literature research of specified problems

Based on problems, which were specified in the chapter 1.1., and on 3 questions was did literature research to find out a solution.

Question:	Solution:
Why is the tool life so low?	Check the type of tool wear. Change cutting conditions. Choose different type of grade. Choose different type of cutting tool material. Change pressure of inner cooling.
Why is excessed value Ra?	Check the type of tool wear. Change cutting conditions. Change geometry of tool.
What can cause higher value of rotational accuracy?	Check the type of tool wear. Change tool holder. Choose tool holder with possibility to regulate eccentricity.

Table 2. Questions which can help solve the problem [3] and [4]

Tool life is parameter expressed in minutes, metres or number of pieces. If we measure tool life we have to check type of tool wear. During machining soft or stainless steels the main problem is formation BUE. BUE is type of adhesion when small pieces of workpiece stick on a cutting edge. Sometimes we talk about micro-welding. BUE has some advantages but a lot of disadvantages. Protection of cutting edge is ranged among advantages. On the other hand BUE creates new cutting edge which can scratch machined surface. Sometimes the BUE can be broken away and this broken can damage cutting edge. When reaming we can recognize BUE on machined surface, because there can be seen spiral. If we monitor so scratched machined surface we have to measure higher values of surface roughness.[4] and [5] Sintered carbides are very sensitive to BUE. Rehor et al.[6] observed low tool life and higher value of surface roughness of machined surface during reaming cast iron. Reamer was made from sintered carbide and also coated with TiAlN. Photo of the cutting edge with BUE plus differential analysis of the same cutting edge can be seen in the Fig. 2. In the same figure, there is also a photo of machined surface. Surface was created with BUE on the cutting edge of sintered carbide. Coating protected cutting edge only for 2 metres of machining after that basic substrate was detected and BUE started to grow up. [6]

Fig. 2. BUE on the cutting edge of a reamer and picture of reamed hole with BUE on the cutting edge of sintered carbide [6]

So the answer for first and second question is tool wear, to be specific formation of BUE on the cutting edge. World-wide tool producers recommend following:

- Increase cutting speed
- Increase feed rate
- Use coated tool with higher toughness
- Use positive geometry
- When tool life is too low, use huge amount of cooling liquid or not to cool if there is not enough cooling liquid
- Use different kind of cutting material [7] and [8]

Only one possibility, how to improve tool life, was to use different kind of cutting material. Based on literature research was chosen ceremet. Cermet is sintered carbide which is consisted very hard particles of titanium. Cermet is combination of words Ceramic and METal. Cermets of third generation have very high cutting power thanks to increasing content of nitrogen or nitride. Mechanical properties of Ti (C, N) cermets are influenced very difficult microstructure. Their microstructure is much more difficult than microstructure of sintered carbides. Microstructure is influenced with original powder, size of particles, lay-out of particles and sintering atmosphere (vacuum, atmosphere N2 or Ar). Cermets are formed by two phases. First one is very hard phase (TiNC) and the second one is metal binder (Ni, Co, or their combination). Cermet is suitable to use for finishing machining. Feed rate can be the same like for cutting ceramic and cutting speed like for sintered carbide with coatings or even higher. Ideal cutting conditions are stable machining process and uninterrupted cut. Cutting edge stays sharp for long time. On the other hand it is very brittle. Cermet is a queen of smooth surfaces. [7], [8] and [9]

Next part of literature research was focused on reducing of rotational inaccuracy. Nowadays, there is necessary to reduce radial eccentricity of rotational tools. There are used more powerful cutting tools, higher cutting speeds and there are also demands for higher quality of surface roughness. These can be influenced with more precise clamping systems and possibility to reduce eccentricity. Ideal clamping system must be solid, stiff, precise and reliable. Choosing of right clamping system means better dimensional accuracy, better geometric tolerations, smoother surface roughness, higher tool life and their economical usage. Classic tool holders like Weldon and whistle notch mountings, ER collet chucks, hydro-plastic tool holder, Tribos system, heat shrinking tool holder, etc. we can call "passive". The second group we can call "active". Active clamping systems are based on passive system, but there is possibility to set the quantity of eccentricity of tool tip up to micrometers. Tool producers of passive systems write to the catalogue information about maximal eccentricity (in micrometers), but this value is not quantified in the field of cutting edge. This value is expressed like x-times tool diameter which is measured from the front part of the tool holder. The values for each tool holder are written in the Tab. 3. The longer tool the higher radial eccentricity. For that kind of tools is good choice to use tool holders with possibility to regulate eccentricity. [10]

Type of tool holder	Weldon	ER collet	Hydro-plastic	Heat shrinking	Tribos
Radial eccentricity	15-20 μm	10-20 μm	≤ 3 μm*	≤ 3 μm**	≤ 3 μm*

* measured in the distance 2.5D from the front part of tool holder
** measured on the front part of tool holder

Table 3. Radial eccentricity of each tool holder [10]

System for regulation eccentricity was firstly patented by Robert E. Blades in 1956. This system is called Compensation tool holder and was solving regulation of tool eccentricity which was clamped in three-jaw chuck. Systems of active tool centring offer tool producers like Mapal, Dihart, Komet, Ham-final, etc. These companies are also specialized in reaming technology and reamers. [10]

2. Experiment

Based on literature research and specification of problems were set up the experiment. Experiment was done in two steps. First one was in laboratory conditions and second one was in company on real semi-product. All experiments were done on turn-mill complete machining centre CTX beta 1250 TC 4A. As a workpiece was used block of material C45 (soft annealed) which was clamped into three-jaw chuck. The dimensions of workpice were: 160mm diameter and 70mm high. There were 93 holes in one workpiece (thickness of wall between holes were 6mm). The machine tool and workpiece with reamer are in the Fig. 3. Chemical composition with microstructure is in Tab. 4.

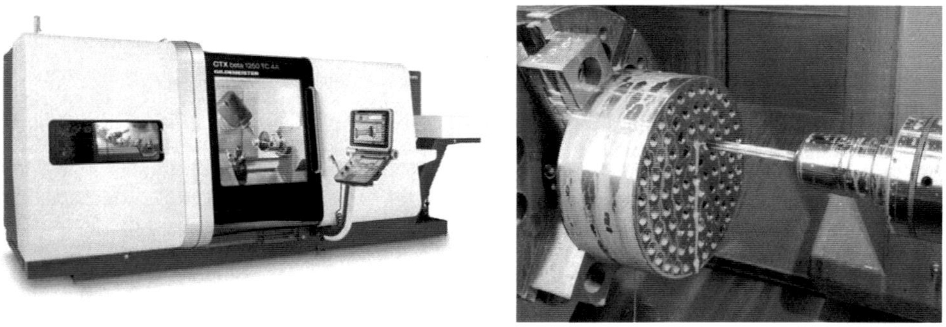

Fig. 3. Machining centre CTX beta 1250 TC 4A and workpiece with reamer

C45 (soft annealed)	Chemical element	C	Mn	Si	P	S	Cr	Ni	Cu
	% representation	0.491	0.675	0.248	0.0106	0.0147	0.0524	0.0702	0.179
	Chemical composition	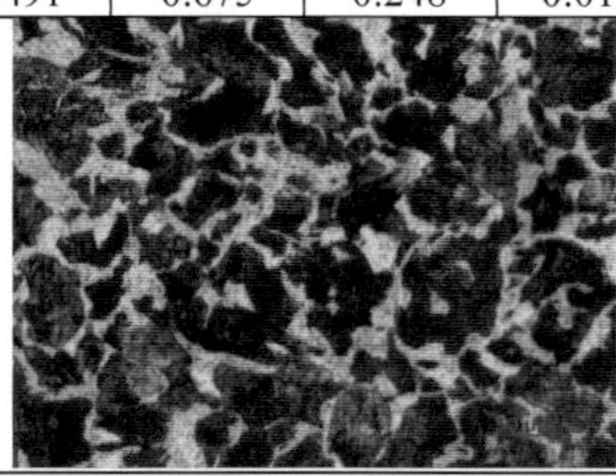				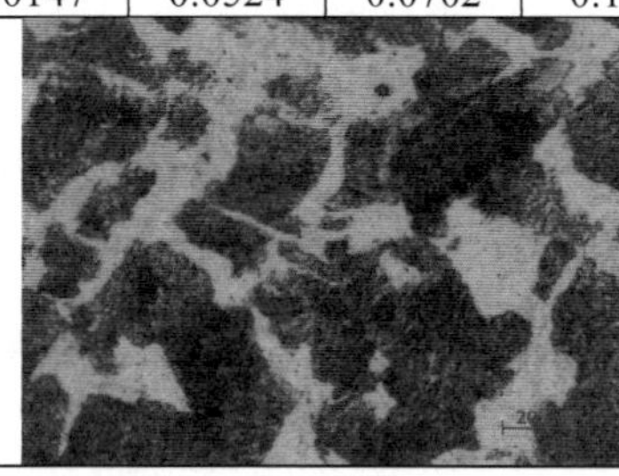			

Table 4. Chemical composition and microstructure of steel C45

Further, next very important part is production process of the experiment. In VDI tool holder there is reamed blind hole. Before reaming it is necessary to make few technologies, it means produce hole for reaming. Production process is in the Tab. 5. Because reamed hole is deep, the first step is called spot-drilling. That technology is realized with end mill cutter and size of prepared hole is ø7.8 x 6.5mm in the accuracy class ISO 2768 m. Second operation is drilling of the hole ø7.8 x 62mm. This hole will be reamed and the working allowance for reamer is 0.2mm. Third step is drilling of the hole ø7.1 x 69mm. Deeper but smaller hole enables reaming up to depth 61mm (generally there should be place between bottom of the hole and lower position of the reamer). Further, there will be other channels for inlet of cooling liquid. Next to last step is chamfering. This technological operation should not be omitted! Last operation was the subject of the experimental research. Plan of the experiment (further specification of the technology nr. 5) is in the Table 6.

Nr. of technology	1.	2.	3.	4.	5.
Technology operation	Spot-drilling	Drilling	Drilling	Chamfering	Reaming
Dimension [mm]	ø7.8 x 6.5	ø7.8 x 62	ø7.1 x 69	60° x 1.5	ø8H6 x 61
Cutting tool	End mill cutter	Twist drill	Twist drill	Conical cutter	Reamer
Cutting tool material	SC	SC with TiAlN	SC with TiAlN	SC	*
Clamping system	ER collet	ER collet	ER collet	ER collet	*
Nr. of teeth	2	2	2	4	*
Cutting speed [m/min]	80	50	50	95	*
Feed rate [mm/rev.]	0.1	0.1	0.1	0.02	*
Pressure of internal cooling liquid [bar]	-	30	30	-	*

*parameters which are subject of the experiment

Table 5. Production process of experiment

Based on literature research and on requirements of the project sponsor the only one possibility how to remove problems with tool life and higher values of surface roughness and rotational accuracy was to find a new cutting tool. Because machined material is very "soft" there was chosen cermet as cutting material. Cermet has better properties than sintered carbide, because it has better tool life and machined surface is smoother (sometimes we says that cermet is a queen of machined surfaces). In addition there cannot be observed BUE on cermet cutting edges.

Nr.	Cutting tool	Clamping system	Pressure of internal cooling liquid [bar]	Radial eccentricity on the tip of cutting edge [µm]	Cutting speed [m/min]	Feed rate [mm/rev.]	Repetition
1.	Cermet reamer (size: ø8.009)	RC tool holder	30	2	140	0.3	2x
2.	Cermet reamer (size: ø8.007)	RC tool holder	100	2	140	0.3	-
3.	Cermet reamer (size: ø8.007)	Hydro-plastic	30	12	140	0.3	-
4.	Sintered carbide (size: ø8.005)	RC tool holder	30	2	80	0.4	1x

Table 6. Plan of the experiment

RC tool holder is tool holder which enables to Regulate Centricity/eccentricity. Clamping system is hydro-plastic and it is mounted on the system of deformation elements. Centricity is regulated with thrust and puller screw.

3. Experimental results

Draw the experimental plan (see in the Table 6) was set up to test all possible variants. Number sequence of experiments was randomized. If the tool was clamped in RC tool holder, the radial eccentricity was 2µm. The tool clamped in the hydro-plastic tool holder had radial eccentricity 12µm. Results were compared in terms of real diameter (see in the Fig. 5) and surface roughness (see in the Fig. 7). Real diameter of the hole was measured with pneumatic measuring system Millimar S1840. Measured places were at the beginning of and on the bottom of the hole. Final value, which can be found in the Fig. 4 is an average value. The red line represents the border (accuracy class H6 means that lower size of the machined hole is 8.000m and upper size of machined hole is 8.009mm). Everything what is above these borders is wrong and we call it as nonconformity products. Cermet cutting tool with diameter 8.009mm produced wrong pieces. So it is not suitable to use the tool with the same dimension like maximal upper size of the hole. What is more, if we look at the real diameter of the hole at the beginning and at the end of machining, we can see that the difference is only 1µm. So it means if we produce cermet tool with lower diameter, for example ø8.007 ÷ ø8.005, the cermet tool will work properly. Cermet tool, clamped in RC tool holder with pressure of cooling liquid 100bar and sintered carbide tool exceeded upper border only at the beginning of machining. Strictly speaking 7 and 20pieces of VDI tool holders was wrong. The reamer, which produces almost 190 pieces of VDI tool holders was cermet tool clamped in hydro-plastic tool holder and with pressure of cooling liquid 30bar. How is it possible that the reamer with the same diameter, but clamped in different tool holders, was able to produce products with different results? The pressure of cooling liquid is the answer. To explain this problem in more detail, we need to know other information. For example the way of tool wear. Look at the figure 6, there are 3D models and differential analysis of cutting edges after 11meters of machining. If we compare these tools (Cermet_8.007_30bar_hydro and Cermet_8.007_100bar_RC) we can find out that tool cooled with 1000bar is extremely worn. A part of cutting edge is missing, so we talk about extremely chipped cutting edge. What has happened? High pressure of cooling liquid in the blind holes caused vibration of the tool. Vibration damaged cutting edge and influenced the quality of machined holes.

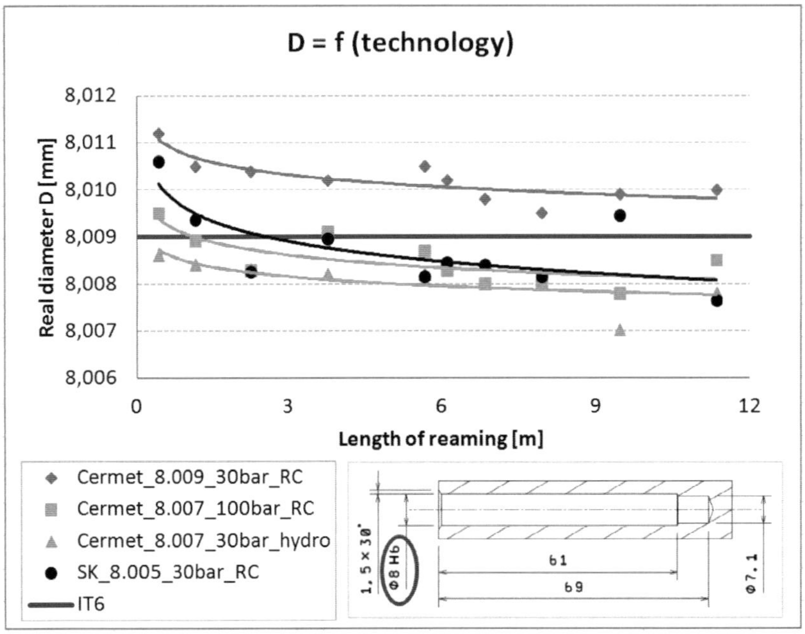

Fig. 5. Influence of technology on the real diameter of machined hole

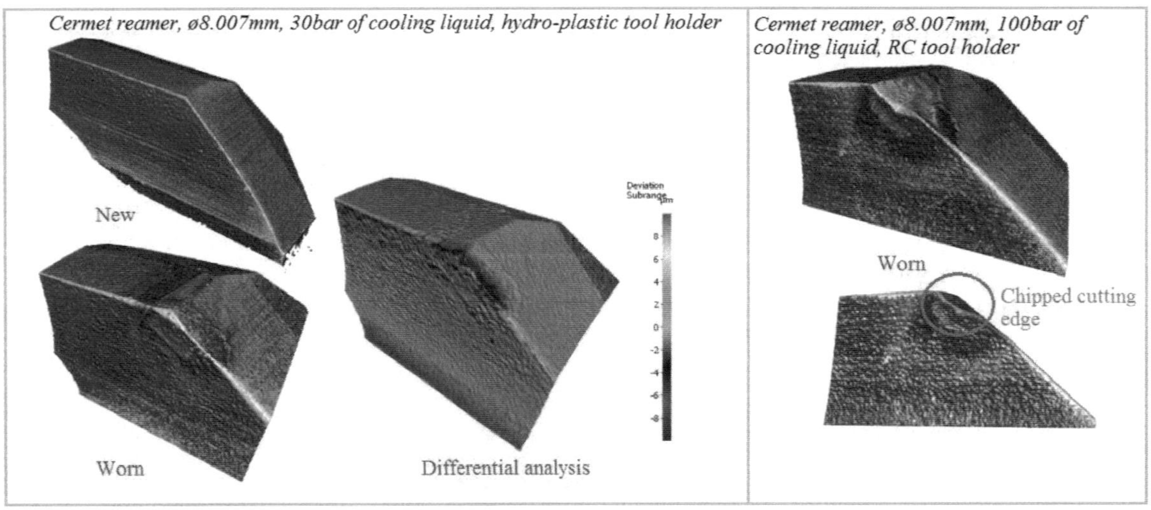

Fig. 6. Comparison of two reamers

The next part of experimental study is in the Fig. 7. There was investigated the influence of technology (it means different cutting tool, clamping system and pressure of inner cooling) on the surface roughness of reamed hole. The worst results were done with sintered carbide reamer. The criteria value of surface roughness (Ra = 0.8μm) was exceeded after 6.5m (112 pieces of VDI tool holder). Surface roughness was influenced with cutting edge condition. There was BUE on the cutting edge and this statement is confirmed with pictures from IFM G4 Alicona microscope and optical microscope Multicheck PC 500, see in the Fig. 8. BUE on the cutting edges produces very rough surface, surface looks like scratched, there is usually possible to see helix mark. On the other hand, surface roughness after machining with cermet reamer is very low; the machined surface is very smooth. There are also some differences between results of cermet reamer. The smoother surface roughness was achieved with cermet_8.009_30bar_RC. Results are better than the experiment with the same cutting tool (cermet_8.007_30bar_hydro) which is clamped in the hydro-plastic tool holder (we can neglect the influence of different tool diameter). Why are results different? The answer to this question is connected with the system of clamping. Radial eccentricity of cutting tools is different, to be specific one of them has 2μm and the next one has 12μm. Radial eccentricity influenced the tool wear of the cutting edges, see in the Fig. 9. The tool wear on the main and second flank face for Cermet_8.007_30bar_hydro is bigger than for Cermet_8.009_30bar_RC.

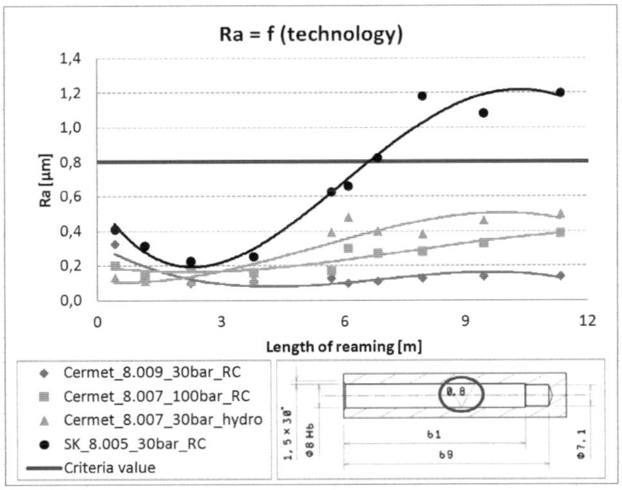

Fig. 7. Influence of the technology on the surface roughness of the hole

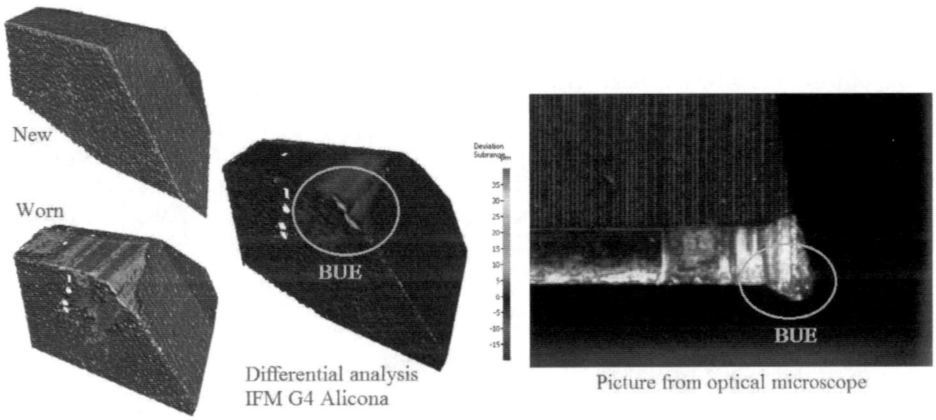

Fig. 8. BUE on the reamer made from sintered carbide

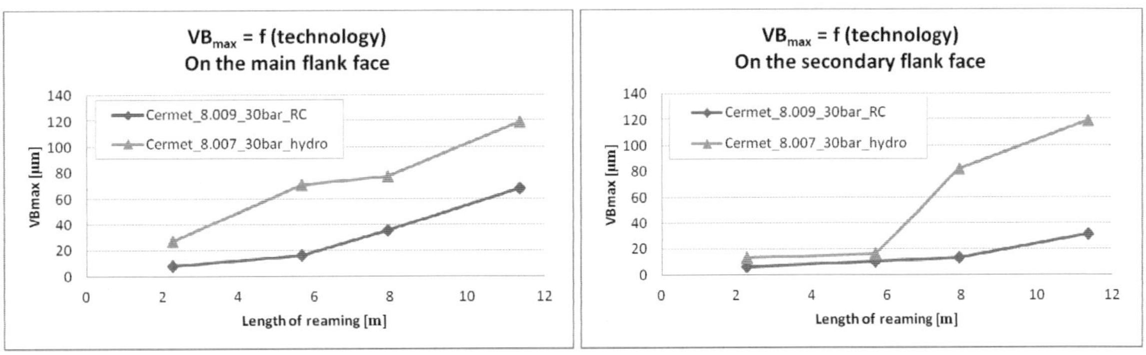

Fig. 9. Tool wear on the main and second flank face for two different clamping systems

4. Conclusion

This article was divided into two parts. The first one was focused on literature research and the second one on the solution of problems in real production process. The engineering company is manufacturing VDI tool holders. The tool holder consists a very precise hole and the company solves how to produce it well. Actual production problem is very poor tool life, because they are able to machine about 50 ÷ 300 pieces per a reamer. So it means that reliability of production process is very low. The main reason of poor tool life is overrun of criteria value Ra or maximal size of produced hole. The company does not want to machine VDI tool holders on a new machine tool or cooperate with different company. At the beginning there was done literature background research to find out what we can do or change. The table of important questions was written down (Table 2). Only two possible solutions could be changed, namely cutting tool material and better clamping system. Sintered carbide reamer was substituted for cermet reamer and there was done a few of experiments. Plan of the experiment can be seen in Table 6. The findings are following:

- Pressure of internal cooling liquid 100bar has negative influence on tool life. High pressure of cooling liquid in the blind holes caused vibration of the tool. Vibration damaged cutting edge and influenced the quality of machined holes. We can see extremely chipped cutting edge in the Fig.6. High pressure of cooling liquid also helped to change the shape and size of a chip, from long cohesive chip to tiny split chip. In the first case, the long cohesive chip stayed at the bottom of the hole (to be specific, it was cram-full in the diameter 6.1mm). In the second case, the tiny split chip was flushed out. The small pieces of the chip could also help to damage the cutting edges. The conclusion is following. Pressure 100bar of internal cooling is not suitable for reaming of this hole type. It is better to use 30bar.
- RC tool holder helped to improve the quality of machined surface (Fig. 7) and reduced tool wear (Fig. 9). RC tool holder regulates the radial eccentricity on the tip of the cutting edge and it has good influence on the cutting process and its results. On the other hand it is suitable to use it in a mass production or there where we need very precise holes. The main disadvantage of RC tool holder is necessity to adjust it right on the machine tool. So it is connected with idle time. For right setting of cermet reamer in RC tool holder is necessary to buy/use a dial indicator with low thrust ($\leq$0.3N) and scale division 0.002mm or lower. RC tool holder improved results of surface roughness and tool life, but if the production charge is low, like in our case (maximal number of workpieces in production charge is 300), it is necessary to count economic return of investments into RC tool holder and accessories. Results of reamer, clamped in hydro-plastic tool holder, are also sufficient.
- Sintered carbide reamer is not suitable for reaming so "soft" workpiece material, because of its inclination to BUE on cutting edge. BUE has negative influence on topography of final surface. Usually we can measure higher values of surface roughness and we can see spiral on machined surface. BUE also influence tool life. There are some general advices how to avoid formation of BUE on cutting edge, but all of them are not possible to use during reaming. In addition, cermet is cutting material which is recommended for finishing technologies and theory says that formation of BUE is less common in comparison with sintered carbide. So the next logical step was to replace sintered carbide as a cermet. Cermet reamer helped to remove all problems during reaming.

The results of the experiments were presented in the company. Also there was done short exhibition of machining with cermet reamer. The company was satisfied and they are going to exchange sintered carbide reamer to cermet.

This real machining problem brightly demonstrated, there are not so many information about influence of all factors (definable and indefinable) on the results of machining very precise holes. So University of West Bohemia in Pilsen, Regional Technological Institute, Laboratory of Metal Cutting decided to focus on this process. So the future steps will be about basic research, it means detailed understanding of reaming process and after that writing a handbook for all who are interested in machining precise holes.

5. Acknowledgments

This paper was created due to the project GA ZCU v Plzni: **SGS-2016-005** "Research and development for innovation in field of Manufacturing processes – Technology of metal cutting II".

6. References

[1] Halaburdova, J. (2016). Improvement of reaming efficiency of a 8D hole it IT6. Plzen. Diploma thesis. ZCU v Plzni

[2] Drzak vyvrtavaci tyce forma E2. MK GROUP CZ s.r.o. [online]. 2014 [cit. 2016-10-01]. Dostupné z: http://mkgroupcz.cz/eshop/kemmler/vdi/drzak-vyvrtavaci-tyce-forma-e2.html

[3] Moravcikova, J. (2015) Cutting Material Influence on the Quality of the Machined Surface. Procedia Engineering [online]. 100, 328-333. DOI: 10.1016/j.proeng.2015.01.375. ISSN 18777058

[4] Piska, M. & Polzer, A. (2012). On the advanced PVD coatings for threading in austenitic steel. 23rd DAAAM International Symposium on Intelligent Manufacturing and Automation 2012, 2, pp. 831-834

[5] Pilny, L.; Chiffre, de L.;, Piska, M. & Villumsen, M.F. (2012). Hole quality and burr reduction in drilling aluminium sheets. CIRP Journal of Manufacturing Science and Technology [online]. 5(2), 102-107. DOI: 10.1016/j.cirpj.2012.03.005. ISSN 17555817.

[6] Rehor, J.; Fulemova, J.; Rut, D.; Trisková, V.; Kutlwaser, J. & Kouril, K. (2015). Reaming of very precise holes in hydrostatic component. Manufacturing Technology, 15 (3), pp. 409-415

[7] Zeqiri, H.M.; Qehaja, N.E.; Zeqiri, F.; Salihu, A.H. & Osmani, H. (2012). Research of flank wear in turning of steel CK60. 23rd DAAAM International Symposium on Intelligent Manufacturing and Automation 2012, 1, pp. 33-36

[8] Identifying and addressing eight common insert failure modes. Seco Tools [online]. (2016) Http: https://www.secotools.com/en/Global/presscentre/Left/Archive-2014/Identifying-and-addressing-eight-common-insert-failure-modes

[9] Tsuda, K. (2016) History of development of cemented carbides and cermet. SEI Technical Review, (82), pp. 16-20

[10] Matous, P. (2015). Technical illustration and animation of HAM-FINAL RC tool system for practical use. Plzen, Diploma thesis, ZCU v Plzni [11] Schmid, J., Jones, B. & Braun, J. (2008). The title of the journal paper", Journal Name, Vol. 24, No. 11, 2008, pp. 1872-1880., ISSN, DOI

DOI: 10.2507/27th.daaam.proceedings.042

THE MANAGEMENT TOOLS AND METHODS APPLICATION OF THE ORGANIZATION TO ENVIRONMENTAL PROTECTION

Miroslav Rusko, Ruzena Kralikova & Monika Mikulova

This Publication has to be referred as: Rusko, M[iroslav]; Kralikova, R[uzena] & Mikulova, M[onika] (2016). The Management Tools and Methods Application of the Organization to Environmental Protection, Proceedings of the 27th DAAAM International Symposium, pp.0283-0288, B. Katalinic (Ed.), Published by DAAAM International, ISBN 978-3-902734-08-2, ISSN 1726-9679, Vienna, Austria
DOI: 10.2507/27th.daaam.proceedings.042

Abstract

This paper describes the approaches development of organizations to environmental protection through several stages from complete passivity, over so-called the reactive approach to the proactive principles. The passive approach relies on the power of nature to dispose of waste by own assimilative capacity, the reactive approach is based on a strategy of control and management especially in relation to the legislative measures and the proactive principle, applied in developed countries since the nineties of the twentieth century is based on the fact of prevention of pollution occurrence being cheaper and more effective than to subsequently remove it, or bear the consequences. A proactive approach focuses on eliminating problems before they have a chance to appear and a reactive approach is based on responding to events after they have happened. The difference between these two approaches is the perspective each one provides in assessing influences and their impacts. The aim of environmentally oriented management based on the proactive principles is use the special management tools and methods application are the subject of this contribution.

Keywords: environment; management; protection; organisation; policy.

1. Introduction

Approach of organizations to environmental protection issues is important because it determines the strategy of environmental protection that will be implemented into action at an organization.

Enterprises in the past responded to a variety of environmental challenges. First they did not respond at all, then respond only when a crisis occurs, and eventually integrating environmental management started into overall enterprise management through defined environmental management system in order to ensure sustainable development of society. It is a gradual transition from egocentric approach, allowing for the fact that nature will deal with any activity, towards the eco-centric one, which takes into account environmental considerations in economic activities of the company [5].

For the business enterprise, sustainable development means adopting business strategies and activities that meet the needs of the enterprise and its stakeholders today while protecting, sustaining and enhancing the human and natural resources that will be needed in the future. If sustainable development is to achieve its potential, it must be integrated into the planning and measurement and management systems of business enterprises [1]. Considering the rising costs of natural resources, the promulgation of environmental enforcement, new ways of thinking, patterns of future liability (e.g. Polluter Pays) and the increasingly competitive commercial market, it is indisputable that the application of voluntary tools in industry is a "win-win" situation, where production costs are reduced, energy is conserved, waste and emissions are reduced and relationships with stakeholders strengthened through increased transparency [10].

This paper considers the potential role for voluntary initiatives and environmental regulations in the transition toward more sustainable industrial systems. Several pervasive problems have hindered evaluation of their environmental effectiveness so far including poorly specified objectives, inadequate data on results and poorly specified baselines for comparison. Economic results indicate that in different circumstances, voluntary approaches may increase or decrease economic efficiency [8]. These considerations suggest that while voluntary initiatives can help us move toward more sustainable industrial systems, considerable advances in the design and analysis of voluntary initiatives will be required to harness their full potential.

2. Modern pro-environmental oriented management approaches

Modern pro-environmental oriented management approaches (as of BS 7750 standards, EMS, EMAS etc.) emphasizes the principle of continuous improvement, based on the Deming Cycle, Fig.1 [11].

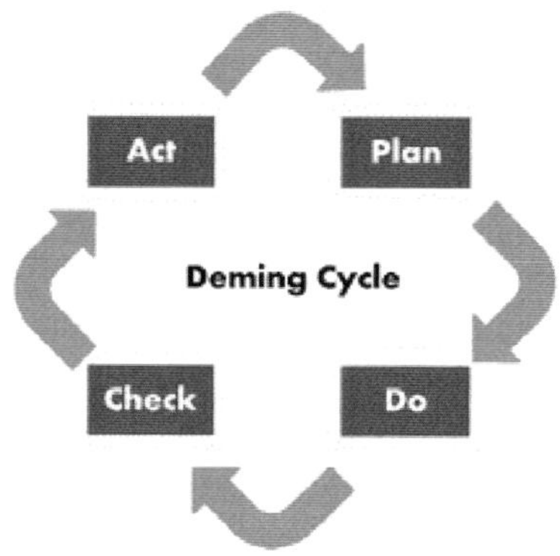

Figure 1. Deming cycle

The aim of environmentally oriented management is to minimize negative impacts on the environment, optimize the use of raw materials and energy resources, minimize waste and create conditions for sustainable development. Sustainable development is of an upmost priority of global society. We are talking of synchronized platform for:

- Integration, activity leaning towards sustainable development, monitoring and assessment.
- Support of domestic technological development, research in innovations including the need of suitable political environment for the development and besides other also industrial diversification.
- Modernization of infrastructure and renovation of industry in order to make them sustainable.
- Higher effectivity use of resources and better acceptance of clean and environmentally sound technologies and industrial processes.
- Achieving sustainable management and effective use of natural resources.

An important role has played and continues to play a shift in the threshold of general public and professional audience on the issue of environmental pollution. Based on consumer pressure there has been a shift towards pro-environmental activities in research, production and management organizations. This was followed by informing the consumer sector of such development e.g. the environmentally-friendly products [2]. Followed by received feedback, which is consumers have become a driving force, for those who make decisions on the market:

- Primary, which product is acceptable for them in terms of environmental protection,
- Secondary, purchasing environmentally sound products (or making provision for the purchase of a product weather the producer has implemented EMS (EMAS), decide which organization is successful in the market.

Such an important role can be successfully proven for consumers who are pro-environmentally oriented and environmental protection is at their value-among the priorities. In terms of implementation of environmental policy in practice it is not a priority to persuade once convinced, but to provide an appropriate form of information (such as environmental labeling type I) to persuade undecided, respectively consumerist-oriented consumers to become pro-environmentally oriented driving force in the market [3], [4].

Because an organization with an established EMS requires that subcontractors gradually introduce EMS and so gradually widened the application of voluntary environmental policy instruments. Although it is voluntary system, organization with an established EMS prefer subcontractors and partner organizations with established EMS. This creates pressure on the corporate sector, should an organization want to succeed in the market, it has to "voluntarily" implement EMS. Organization is not forced by legislation and sanction based environmental policy instruments, but is voluntarily moved into pro-environmental oriented path where the results often go beyond the measures of legislative and penalty mechanisms of the state. A similar approach and orientation can be seen with environmental product labelling type I. This way, we find a gradual solution to tackle a serious environmental problems and their elimination. The relevant threshold shift of an organization is applied in addition to the environmental sensitivity and moment of competition and selection. Competition among organizations to achieve success, also due to their pro-environmental orientation within its own economic policies are the driving force of the market. Thus such oriented organizations in selection of their partners in the context of their business prefer organizations that are similarly pro-environmentally oriented respectively they tolerate to a certain time (given impulse), while partner organization (subcontractor) starts in its economic activities to implement optional pro-environmental tools. In this way, more and more organizations involved in voluntary pro-environmental oriented activities.

	Type of behavior		
Intensity of organization involvement	Social and environmental duty	Social and environmental response	Social and environmental sensitivity
Emphasis on:	Economic and environmental legislative responsibility of an organization	Economic and environmental legislative responsibility of an organization	Economic, environmental, legislative, social and public responsibility of an organization

Table 1. Subject behaviour within the its own environmental responsibility

BASIC APPROACHES OF AN ORGANIZATION TOWARDS ENVIRONMENTAL PROTECTION				
Passive approach ▲		The organization practically does not care about the issue of environmental protection. (Note: the state organizations may be constrained towards pro-environmental oriented activities e.g. the availability of funds from a central source, respectively disinterest of competent bodies)		
▼ **Reactive approach** ▲		• Organization responds to the development of environmental legislation. • Organization addresses issues of environmental protection solely on the ground to bring it into line with regulatory measures (legislative and sanctions) however itself it does not proactively seeks ways to minimize their negative effects on the environment.		
VOLUNTARY APPROACH	▼ **Preventive approach** ▲	**Based on principles**		
		Prevention	**Wariness**	**Integration**
		Prevention of pollution of the environment (it is cheaper and more effective than additionally remove it, resp. bears its consequences).	Prevention of secondary unwanted objects, i.e. not only dangerous and toxic products	Complex evaluation process, ie the process in question is considered integrally in all stages and effects of all environmental impacts that are considered inseparable
	▼ **Voluntary proactive approach** ▲	Organization itself is actively seeking ways in which the impact of its activities on the environment can be reduced	One-sided commitments as environmental polluters	
			Public voluntary programs, i.e. voluntary environmental solutions initiated by public institutions	
			Negotiated agreements, i.e. voluntary environmental agreements	
			The organization builds environmentally oriented management approach (EMS/EMAS, resp. IMS) at all levels of organization management. This is based on the principles of continuous improvement, in terms of own active environmental policy, with possible application of additional / other environmental oriented tools. The aim is to be constantly improving the environmental efficiency and environmental performance of an organization.	
	▼ **Voluntary proactive and sustainable management approach**	Organization itself is actively seeking ways to align its activities with sustainable development	Organization builds environmentally oriented management approach at all levels of management, which is based on the principles of continuous improvement, in terms of its active environmental policy. Using EST is improving environmental efficiency and environmental performance of an organization in accordance with the paradigm of sustainability.	

Explanatory note: ▲▼ - no strict boundaries exist between individual approaches, transition stages of development.

Table 2. Systemization of basic approaches of and organization towards environmental protection.

The result should be that organizations, which proactively address the potential, respectively existing environmental problems, reduce the negative impact of their business on the environment [9]. The measures implemented by the

company in the context of environmental management can have a diverse nature - from minor organizational measures to technical and technological changes that require major investments. In its ultimately lead to more efficient use of resources, environmentally respectful processes, reduced-emission and impacts etc. [5].
The organization must realize that environmental responsibility lies within themselves. It is not enough to rely solely on meeting the standards set by the legislative-sanction instruments of state policies and thus transfer the responsibility of the environment to the state [6]. Types of subject (organization) behavior as part of own environmental responsibility can be viewed in Table. 1. Systemization of organization approach towards environmental protection is possible from several viewpoints are list in Table 2.

The voluntary instruments include activities (methods, procedures, patterns of behavior etc.), which are enumerated or standardized at the international level and which are implemented by businesses on voluntary basis, outside the scope of legislation [4]. The examples of management tools and methods application within the organization towards different approaches to environmental protection are showed on Table 3.

<table>
<tr><th colspan="4">MANAGEMENT TOOLS AND METHODS APPLICATION</th></tr>
<tr><td colspan="2">Passive approach</td><td colspan="2">• From historical point
• Currently it is more issue of legislation breach and ignorance of legislative regulations</td></tr>
<tr><td colspan="2">Reactive approach</td><td colspan="2">• Dilution of pollutants prior their discharges into the environment
• End technologies („end of pipe“ technology), i.e. capture of pollutants before they enter the environment
• Concentration control of pollutants prior their release into the environment
• Execution of the so-called No Waste (low-waste) technology
• Recycling technologies</td></tr>
<tr><td rowspan="6">VOLUNTARY APPROACH</td><td>Preventive approach</td><td colspan="2">• Various forms of cleaner strategies of production - Cleaner Production
• Waste Reduction
• Waste Minimization
• Waste Reduction at the Source
• Reducing Pollution at the Source
• Pollution Prevention
• Eco-efficiency</td></tr>
<tr><td rowspan="4">Voluntary proactive approach</td><td>One-sided commitments as environmental of polluters</td><td>• Program Responsible Care</td></tr>
<tr><td>Public voluntary programs, i.e. voluntary environmental solutions initiated by public institutions</td><td>• EMAS (Eco-Management and Audit Scheme); within EÚ
• Environmental evaluation and labelling of products, national and multinational programs</td></tr>
<tr><td>Negotiated agreements, i.e. voluntary environmental agreements</td><td>• Conventions („covenants “)
• Individual public environ. agreements
• Private environ. agreements</td></tr>
<tr><td>The organization builds environ-mentally oriented management: EMS/EMAS, IMS.
Continuous improvement in terms of own active environmental policy, with application of additional/other environmental oriented tools.
Improving the environmental efficiency and environmental performance of an organization.</td><td>• EMS (Environmental Management Systems)
• EMAS (Eco-Management and Audit Scheme)
• EA - Environmental Audit
• ER - Environmental Reporting
• Environmental evaluation and labelling of products: EL - Eco-labelling
• LCA - Life Cycle Assessment; Eco-balance
• Eco-design
• EPE - Environmental Performance Evaluation</td></tr>
<tr><td>Voluntary proactive and sustainable management approach</td><td colspan="2">Environmental tool mix based on application of generic approaches EMS/EMAS (resp. IMS), and other environmentally oriented approaches aimed at application of EST especially with the possible use of EL, LCA, etc..</td></tr>
</table>

Table 3. Examples of management tools and methods application

3. Case study from the enterprise

Case study relates to a company which is engaged in manufacturing of automotive components. The company has implemented environmental management system according to ISO standards 14000. This standard of environmental management system helps enterprises decrease its ecological intensity, and reduce the pollution and waste. The company opted for implementation of the environmental management system according to standard ISO 14001, because it is a global company located in the United States. The implementation of environmental management system according to EMAS II relates to the countries of the European Union. Application of pro-environmental voluntary instruments at an organization as EMS takes its environmental responsibilities beyond the standards set by the legislative-sanction instruments of state policy.

The company set itself the aim to reduce its energy intensity by 30 percent by year 2018 and have achieved a 11.7 percent reduction through the end of year 2012. Other environmental goals by year 2018 are to reduce waste intensity by 20 percent, to reduce water intensity by 10 percent and to reduce greenhouse gas emissions intensity by 30 percent. The company its goals gradually the meets and everything publishes using the environmental reporting. Sustainable development also focus extends to their supply partners. It cooperates with more than 100,000 direct and indirect suppliers throughout the world and expects them to conduct their operations in an environmentally and socially sustainable manner. They use a variety of tools to assess suppliers, including background checks, self-assessment surveys, site visits and audits. Of course, companies should be deal with how to mitigate its impact on the environment, but it is also important to give priority to cooperation with other companies that are also environmentally responsible. Therefore, these companies have an advantage in the competitive struggle [7]. Pro-environmental application of voluntary instruments has become not only part of the competitive struggle among them but also company's image.

5. Conclusion

While most previous studies focused on an index of emissions toxicity, this study examines three areas of impacts, none of which have been explored in the literature: natural resource use, solid waste generation, and wastewater effluent. The study is also unique in that the effectiveness of standards ISO 14000 are considered in relation to environmental regulations. Our findings are summarized as follows. First, both ISO 14001 and report publication help reduce all three impacts; the former appears more effective in all areas except wastewater. Second, environmental regulations do not weaken the effect of ISO 14001. Third, assistance programs offered by local governments a voluntary approach promote facilities' adoption of ISO 14001. These findings suggest that governments can use command-and-control and voluntary approaches concurrently. The prospect of such an approach should be sustainable society development. Formalized environmental management systems are currently being implemented a worldwide. EMS, either according standards ISO 14000 or EMAS, are among the most famous and the best tools for all types of large enterprises allowing to integrate environmental aims into their activities especially. It is a voluntary proactive tool giving effect to the more integrated and more flexible measures to reduce the negative impacts of the organization and its activities on the environment.

The main goal of this paper is to show on the potential role for voluntary initiatives and environmental regulations in the transition toward more sustainable industrial systems, in light of these criteria. The application of environmental regulations being instituted in most countries and using of voluntary management tools to environmental protection may encourage proper environmental planning for future industrial development. In addition, voluntary initiatives can solve some problems that are not covered by environmental legislation and may lead to cost-effective solutions. Future research could also examine whether positive the results from implementation of environmentally oriented approaches are consistent for small and medium enterprises (SME) and across countries.

6. Acknowledgments

This paper was written in frame of the work on the projects VEGA 1/0990/15 „The readiness of industrial enterprises to implement the requirements of standards for quality management systems ISO 9001:2015 and environmental management systems ISO 14001:2014"and project VEGA 1/0537/15 „Research of influence of chosen parameters of working environment on working power and productivity ".

7. References

[1] Dvorakova, L. & Zborkova, J. (2014). Integration of Sustainable Development at Enterprice Level In: Procedia Engineering Vol. 24, no. 69, p. 686-695 ISSN: 1877-7058

[2] Hurt. U.; Otto. T.; Korbe. K. & Koppel. O. (2015) New Approach to Knowledge Transfer Environment Development. Procedia Engineering, Volume 69, 2014, pp. 273-281

[3] Lee, G. B. & Ko, M. (2012) Assessment of Environmental Impacts of the Post-Drill Cleaning Process for PCB using LCA, Annals of DAAAM for 2012 & Proceedings of the 23rd International DAAAM Symposium, Volume 23, No.1, pp. 49-52. ISSN 2304 -1382

[4] Kralikova, R., Rusko, M. & Badida, M.(2013) Six sigma method applying within environmental management, Advanced Materials Research, Vol. 739, pp. 700-705. ISSN: 10226680

[5] Micieta, V. et al. Sustainable concept for green logistics and energy efficiency in manufacturing In: DAAAM International Scientific Book 2015 : Chapter 33 P. 391-400 ISSN: 1726-9687 ISBN: 978-3-902734-05-1

[6] Petek, J., Glavic, P., Kostevsek, A. et al. (2016) Comprehensive approach to increase energy efficiency based on versatile industrial practices In: Journal of Cleaner Production Vol. 112. pp. 2813-2821 ISSN: 0959-6526

[7] Polankova, M. ; Manlig, F. & Kralikova, R. (2015). Environmental reporting in the enterprise and related issues MM Science Journal, 2015 (October), pp. 691-695. ISSN: 18031269

[8] Paton, B.(2000) Voluntary environmental initiatives and sustainable industry.In: Business Strategy and the Environment; Vol.9 Sep/Oct 2000; 9, 5; ProQuest Central, p. 328

[9] Rusko, M.; Sablik,J.; Markova, P.; Lach, M. & Friedrich, S. (2014) Sustainable Development, Quality Management System and Environmental Management System in Slovak Republic, 24th DAAAM. Procedia Engineering , Volume 69, 2014, pp. 486-491.

[10] UNEP (2004) Voluntary Environmental Initiatives For Sustainable Industrial Development. United Nations Publication. 55 p. ISBN 92-807-2480-0

[11] Environmental management: The ISO 14000 Family of International Standards [online]. ISO Central Secretariat. Switzerland, 2009 ISBN 978-92-67-10500-0.
Available from <http://www.iso.org/iso/theiso14000family_2009.pdf> [cit. 2016-04-14]

DOI: 10.2507/27th.daaam.proceedings.043

DESIGN OF ASSEMBLY SYSTEM – MIXED REALITY MODELLING

Peter Malega & Jozef Kovac

This Publication has to be referred as: Malega, P[eter] & Kovac, J[ozef] (2016). Design of Assembly System – Mixed Reality Modelling, Proceedings of the 27th DAAAM International Symposium, pp.0289-0297, B. Katalinic (Ed.), Published by DAAAM International, ISBN 978-3-902734-08-2, ISSN 1726-9679, Vienna, Austria
DOI: 10.2507/27th.daaam.proceedings.043

Abstract

Within production system designing are important especially these activities: philosophic and conceptual aspects of technological innovation, modelling, existing solutions certified in praxis, financial and material security of innovation changes and multi-purpose and purpose innovation projects. The aim of every project is optimal reduction of designing processes by simultaneous observance of required quality and costs minimisation. Nevertheless it is important to use innovation methods, tools and procedures within solving the various designing problems. This article is oriented on the designing of assembly system through the interconnection of virtual reality and physical reality. The result of this connection is mixed reality, which can be used as a useful tool in the 21st century.

Keywords: design; assembly system; mixed reality; 2D and 3D modelling.

1. Introduction

In the present time, innovation methods, practices, tools as well as modern technical and software equipment significantly influence the area of assembly systems design. Contemporary development in this area is affected mainly by progress, which is achieved in modelling of simulation, information, communication and computer technologies [1, 2, 18, 22].

Production systems are converted into finished production processes and the most important representative of these systems is assembly [3, 4, 38, 39].

Design of assembly system despite the achievement progress requires another integrated development and enhancement about new principles and solutions.

Within assembly system design are important especially these activities: philosophic and conceptual aspects of technological innovation, modelling, existing solutions certified in praxis, financial and material security of innovation changes and multi-purpose and purpose innovation projects [5, 9,13, 32].

Verification and tested software CAx systems will allow creation of digital models of assembly system and their simulation verification in virtual reality, which allows to optimize designing processes and to reduce risks of non-optimized proposals [6, 10, 16, 41].

New algorithmic procedures of assembly system design on the base of mixed reality modelling supported with CAx technologies originate primarily as the consequence of experimental verification implication [15, 22, 23, 42, 43].

2. Theoretical support of innovative design

The philosophy of innovative, alternative, interactive and participatory design is based on the assumption that the solution of the assembly system has three stages [8, 10, 16, 24, 40]:

1. It is proposed the initial set of assembly system of variant solutions in line with created goals. It can be used various methods for determining the variants of solution (analytic, morphological, hierarchical generation, usage of reference solutions, etc.). In the case of less complex tasks it can be used the intuitive concepts of generating the variants on the base of the reference type models.

2. It is defined acceptable solutions and excluded the inappropriate variants of solution for assembly system either on the base of the intuitive approach, respectively on the base of usage of logical and mathematical methods.

3. It is selected the optimal variant of assembly system solution as the final stage. There are applied also the decision-making methods, respectively optimization criteria. Optimal variant can be obtained on the basis of comparative methods (reference projects), on the basis of simulation activity, etc.

Illustration of the approach to the design of assembly systems in accordance with this philosophy is shown in Fig.1 [14, 15, 18, 26, 33, 39].

Innovative design of assembly systems currently uses modelling in all stages of solution and it is supported with the software products.

The importance of assembly system modelling is important for the development of new approaches in designing and optimizing their solutions. Current development trends in the production companies that lead to quickly "adapt to market" structures require their adaptability to all hierarchical levels of design and implementation [17, 29, 32, 35].

Used hierarchical decomposition on the individual levels of solution enables optimization of project assembly structure. This is done in the related detailed activities of the projected process. As the basic construction units are considered autonomous assembly modules. Assembly systems of higher rank are created from these modules [12, 18, 31, 33, 37].

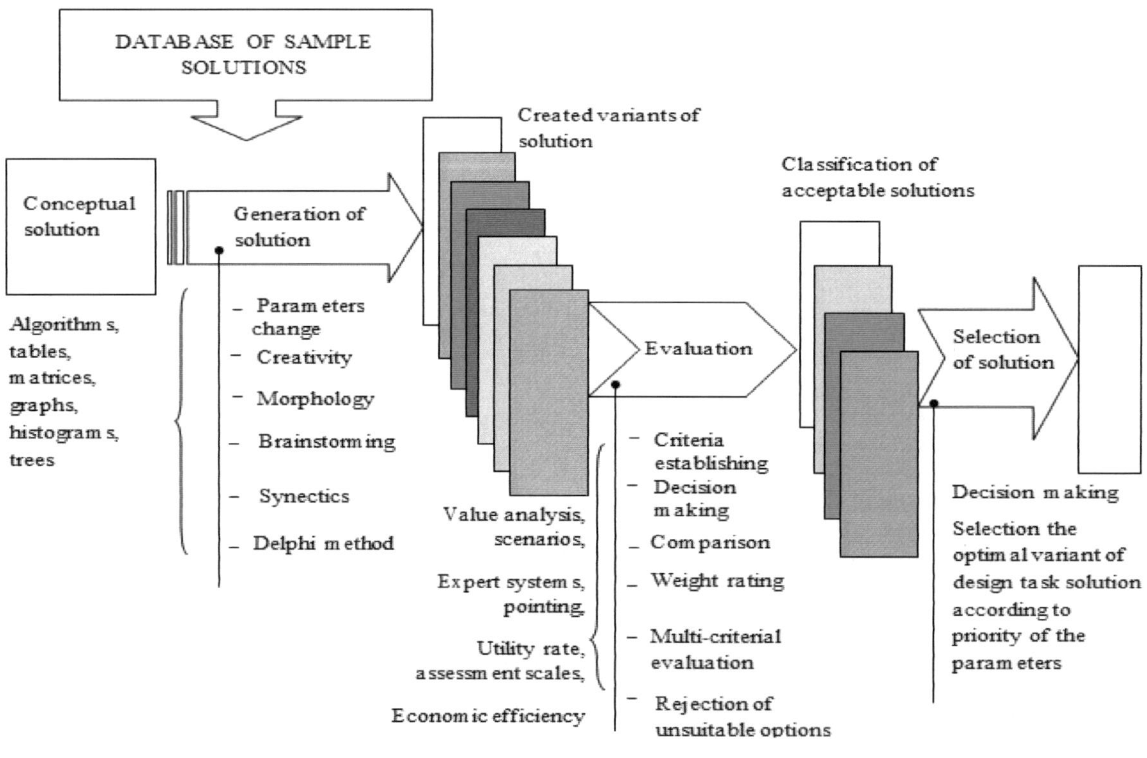

Fig. 1. Illustration of the approach to creating design solutions of assembly systems

3. Methodical procedure of assembly process modelling

The methodological procedure algorithm of model creation of assembly system is shown in Fig. 2 [32, 34, 38].

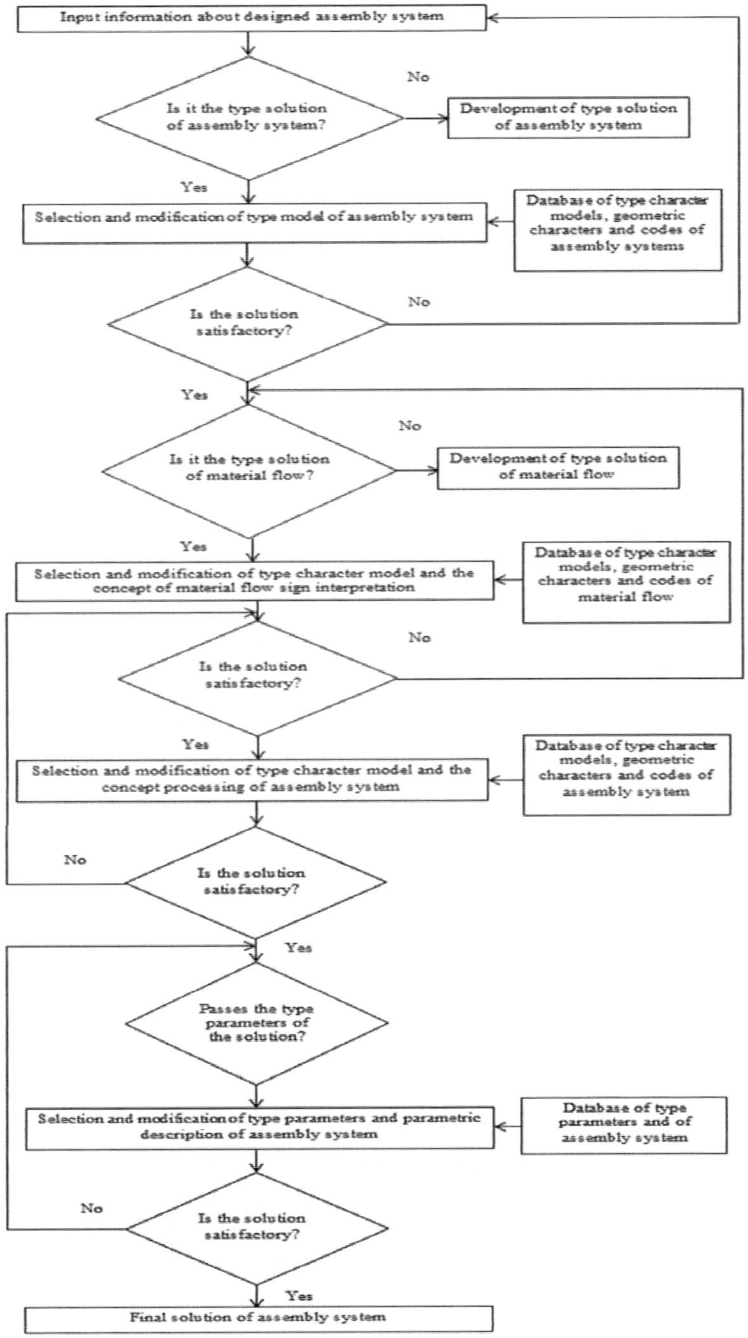

Fig. 2. The methodological procedure algorithm of model creation of assembly system

Methodical procedure of assembly processes modelling and their implementation systems can be divided into four levels [21, 27, 37, 40]:

- At the first level, it is created a conceptual model for assembly process solving. The conceptual model expresses the principled scheme of assembly process and determines the composition and sequence of implemented stages. At this level of modelling it is appropriate to create several variants of symbolic functional structures of assembly process, especially if the optimal sequence sub-processes implementation is a precondition for the effectiveness of the entire solution. Graphic presentation of the conceptual model in this case shows the distribution of assembly workplaces and systems, which implements partial or complete sections of assembly process and material flows. On a similar principle, especially in the design of automated production, are modelled information flows. As information tool for generating variants of conceptual symbolic models may be used reference solutions and project types.

- At the second level are created models of assembly operations (processes) and models of the functional activity of production workplaces and systems. Models are oriented on composition and sequence of operations (functions) implemented in all operations and operational activities at the workplace, respectively in the system. If the model is oriented not only on the functional activity, but also on the technical conception, there are specified technical units and tools. Decomposition of the functional activity can be implemented into the micro level corresponding, especially to the movements of elements or units. The result of this stage of modelling is again a number of variants from which is then selected optimal variant within the selected decision-making and optimization method.
- At the third level it is implemented parametric description of models. At this level of solution, are quantitative parameters assigned to individual functions (time characteristics, ranges, speed of movements, etc.). The necessary condition is that these parameters are specified in preceding stages of the design solution, or in the reference patterns.
- At the fourth level are elaborated spatial and time charts. These charts reflect the spatial and motion functions performed with the implementation tools, as well as corresponding time dependences.

4. Creating of virtual reality in laboratory conditions

In our laboratories were realized several kinds of laboratory experiments with appropriate laboratory equipment [3, 4, 5, 6, 26, 31].

a. Integrated laboratory of managerial and project system

This laboratory includes:

Horizontal design table (work with project table – Fig. 3) designed for creating 3D physical models of the assembly system concepts and plants, including planning and managing their activities. It enables teamwork of different specialists. It uses the 3D models of construction equipment (machines, robots, vehicles, etc.) for generating variants of production systems in appropriate scales.

Digital camera system is used for capturing and digitizing the projected assembly system (scenes) for the CA technologies treatment.

2D presentation system displays 2D conceptual computer generated scenes on project table.

PC system with CA technologies is used for the assembly of digital and virtual projects at conceptual level and detailed level including their simulation verification (DELMIA software modules for the design of digital factories, respectively other CA systems).

3D (2D) is used as presentation system for displaying virtual project scenes.

Virtual technical equipment (data glove, helmet display, 3D projector, etc.) is used for generating of the work in virtual reality, e.g. assembly.

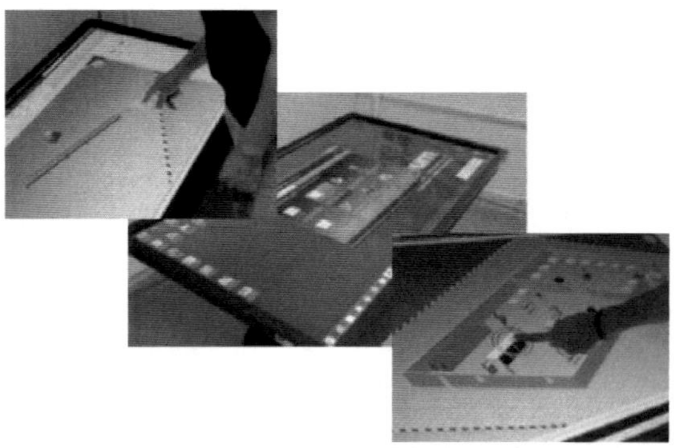

Fig. 3. Work with project table

It enables the new form of interaction between the computer and the designer. Multitouch technology combines display technology with sensors that are able to watch more entry points. The technology lets users the interaction with computers.

The device is used for visualization and manipulation of objects in virtual reality. For the implementation and sensing of Multi-Touch points is used the applications CCV (multitouch open-source framework).

b. Department of virtual reality

It consists from modern hardware and software equipment that supports work in virtual reality.

Data gloves CyberGlove II (right and left) sense the spatial movement with 18 high-precision measurement sensors. They use patented resistive technology for accurate transformation of hand and finger moves in real-time data [23, 24, 25].

Wireless glove CyberGlove (Fig. 4) from Immersion Corporation is equipped with 18 sensors that detect movement of the fingers and hand. It uses a precise and sensitive motion technology to fingers and hands in real space and transmits them in to the virtual space.

Fig. 4. Cyber Glove II – physical movements of data glove in the virtual reality

The basic system includes a glove, 2 batteries, battery charger and USB wireless adapter with drivers [30].

Sensing runs under magnetic method. It uses sensors that are clamped to the glove. They respond to low frequency magnetic field generated by the source. The computer then analyses the given location and movement in a magnetic field and this motion is transferred into the 3D environment in which the device operates. Sensing of the hand position in space provides "track" facility. Ascension Flock of Birds is a device that captures and transmits the behaviour of the motion sensor to the computer. Flock consists of the device itself, transmitter and motion sensor, which is mounted on the wrist. Connection is made via the serial RS-232C port.

c. Computer network

Computer network with hardware and software for virtual reality modelling is equipped with 2 workstations, 5 - specialized PC (CA), 10 classic PC stations with the accessories, CAD and other specialized software. Software modules of the DELMIA allow virtual 3D productions. Development and validation of production processes is in the connection to the product in all its development stages. Academic license includes the whole portfolio of solutions DELMIA.

DELMIA PLM Express is used in the laboratory conditions primarily for digital plant solution in the small and medium enterprises. Human Work Analysis is used for simulation and ergonomic analysis of the human body in the production environment and the factors that act on it. It includes modules Workplace Simulation (HSU), Human Catalogues (HCU) and Human Work Analysis (HAU).

For the valuation of human physical activities in non-automated work activities are used the function RULA. Setting the parameters and analysis results are displayed in the dialog boxes and tables. In the score section of the RULA analysis dialog window shows partial results, respectively burden assessment of body parts and the resulting evaluation score. [19, 20] 3D model with ergonomic analysis and 3D animation of selected workplace are shown in Fig. 5 [16, 17].

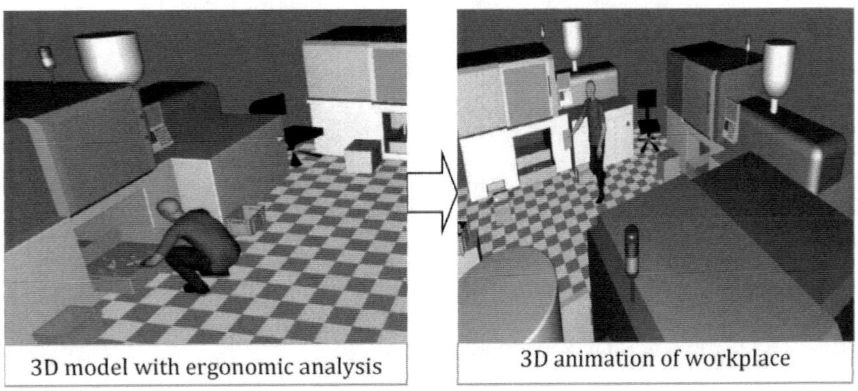

Fig. 5. 3D model with ergonomic analysis and 3D animation of workplace

On this workplace were realized RULA analyses.

The final score can have a value from 1 to 7 and is also expressed in colour. The meaning of several values is following:

- 1 and 2 (green) – indicates, that the posture can be accepted under the assumption, that it is not the same for a long time or it is not often repeated for a long time,
- 3 a 4 (yellow) – indicates the need of further detailed research and then, maybe be necessary to change the postures,
- 5 a 6 (orange) – indicates, that the research and the changes are necessary in very short time,
- 7 (red) – indicates, that is necessary immediate to research and to change of the postures.

This final score is accounted as a middle value from the partial scores of the particular body parts. For every body part exist a separate evaluating scale (see Table 1).

Segment – part of body	Score	1	2	3	4	5	6
Upper arm	from 1 to 6	Green	Green	Yellow	Yellow	Red	Red
Forearm	from 1to 3	Green	Yellow	Red			
Wrist	from 1to 4	Green	Yellow	Orange	Red		
Wrist twist	from 1to 2	Green					
Neck	from 1to 6	Green	Green	Yellow	Yellow	Red	Red
Trunk	from 1to 6	Green	Green	Yellow	Yellow	Red	Red

Table 1. Realized RULA analysis

5. Creating of physical reality in laboratory conditions

On the basis of virtual reality, we made experiments also in physical reality. This research was made according to results from virtual reality and on this research it was used the physical equipment of our laboratories. View of the experimental work carried out by the data glove is shown in Fig. 6 [33, 39].

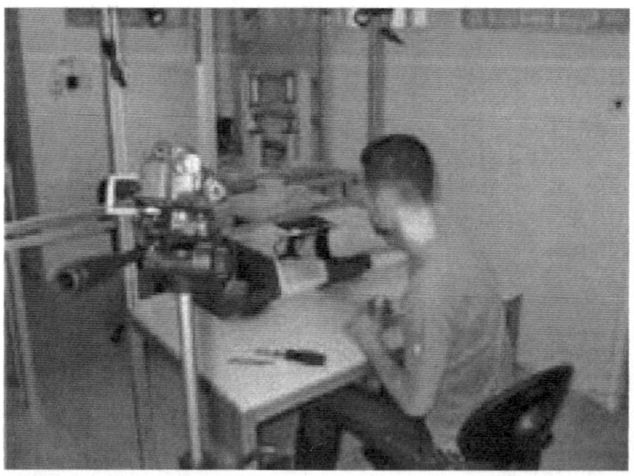

Fig. 6. Physical realization of work by data glove

During the realization of work (e.g. assembly) are sensed the hand movement trajectories with the data glove. Implemented movement is recorded and displayed in virtual reality on a computer terminal screen of workstation. Each trajectory corresponds with the motion of data glove sensor [35, 36].

6. Creating of mixed reality in laboratory conditions

In terms of the assembly system conception and its spatial structure are innovative practices based on the mixed reality modelling of assembly systems particularly effective in the development of assembly clusters of higher rank [7, 9, 28, 40, 41, 42, 43].

Manipulation with physical model through virtual hand is shown in Fig. 7 [11, 25, 31].

This is the result of our research, where we found interconnection between virtual and physical reality in laboratory conditions and we have shown and also we believe, that mixed reality is the future of the innovative design of assembly systems. This mixed reality has advantages from two areas (physical and virtual reality) and it puts them into one bigger reality, named mixed reality.

Our results can be also interpreted as an experiment, in which we tried to find the basis for the future research in the area of innovative design of assembly systems, because in the mechanical engineering area is still the place for improvements, which will people use in various areas of their life as a useful tool in the 21st century.

So, we can state, that our hypothesis about mixed reality was correct and we found the connection between virtual and physical reality and we tested it in real conditions and results show the exploitability of this tool in the 21st century.

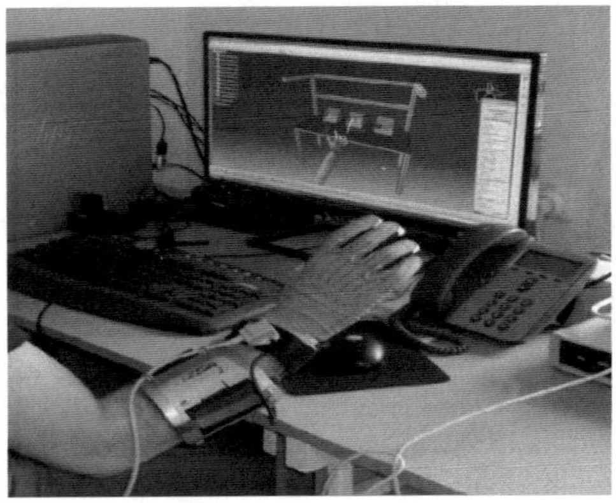

Fig. 7. Transformation of physical movements of data glove into the virtual reality

7. Conclusion

Assembly system conceptions projected as the new generation systems represents complex integrated solutions created on the basis of exploitation modern software, information and communication technologies [1, 18, 24, 40].

Technical and software tools contribute to increasing of productivity, projection activities, quality, products and services value and decreasing of assembly time [6, 14].

The problem was to create innovative design of assembly systems in the area of mixed reality, which is in today´s competitive environment one of the biggest challenge for the industrial companies. We proposed the methodological procedure algorithm of model creation of assembly system, which we can use as a scheme for our research today and also in the future. Then we put all inputs in the virtual reality, in which we realized our research. After modelling in the virtual reality, we focused on the creating of physical reality and in the final stage we focused on the transformation of virtual and physical reality into the mixed reality and in this area we modelled the assembly systems. So, in the laboratory conditions, we found the interconnection between virtual and physical reality and the creation of mixed reality as the result of our research will help us in the future research, which we have to focus on the modelling of assembly systems, but not only in the laboratory conditions. Our vision is to help small and medium industrial enterprises to see their future possibilities in the area of assembly systems modelling.

8. Acknowledgments

This article was created by implementation of the grant project VEGA 1/0853/16 "New project technologies for the creation and implementation of future factories".

9. References

[1] Abele,T. (2006) Verfahren fúr das Technologie-Roadmapping zur Unterstútzung des strategischen Technologiemanagements. JostJetterVerlag, Heimsheim

[2] Bandic Glavas, M. & Majstorovic, V.: Future Virtual Project Management Development Trends, Annals of DAAAM for 2012 & Proceedings of the 23rd International DAAAM Symposium, ISBN 978-3-901509-91-9, ISSN 2304-1382, pp 1159 - 1162, Editor Katalinic, B., Published by DAAAM International, Vienna, Austria

[3] Erler, M. & Essers, M. (2012) Approach for Generation of Complete Virtual Machine Tools based on a Unified Model Description, Annals of DAAAM for 2012 & Proceedings of the 23rd International DAAAM Symposium, ISBN 978-3-901509-91-9, ISSN 2304-1382, pp 0779 - 0782, Editor Katalinic, B., Published by DAAAM International, Vienna, Austria

[4] Cheng, H. et al. A virtual reality-based system for hand function analysis. // Computer Methods and Programs in Biomedicine. Vo. 69, No. 3 (2002), pp. 189-196. ISSN 0169-2607

[5] Cheng, Y., Huang, R. (2012) Using virtual reality environment to improve joint attention associated with pervasive developmental disorder. // *Research in Developmental Disabilities.* Vol. 33, No. 6 (2012), pp. 2141-2152. ISSN 0891-4222

[6] Choi, S.H., Chan, A. M. (2004) A virtual prototyping system for rapid product development. // *Computer-Aided Desing*. No.. 36, Vol. 5 (2004), pp. 401-412. ISSN 0010-4485

[7] http://www.daaam.info/Downloads/Pdfs/proceedings/proceedings_2015/039.pdf, DAAAM International Symposium 2015, Accessed on: 2015-11-13

[8] http://www.daaam.info/Downloads/Pdfs/proceedings/proceedings_2015/107.pdf, DAAAM International Symposium 2015, Accessed on: 2015-11-13

[9] http://www.daaam.info/Downloads/Pdfs/science_books_pdfs/2012/Sc_Book_2012-019.pdf, DAAAM International Symposium 2012, Accessed on: 2012-11-18

[10] http://www.sciencedirect.com/science/article/pii/S1877705814002021, DAAAM International Symposium 2014, Accessed on: 2014-12-09

[11] http://www.sciencedirect.com/science/article/pii/S1877705815004464, DAAAM International Symposium 2014, Accessed on: 2014-12-09

[12] http://www.sciencedirect.com/science/article/pii/S187770581500449X, DAAAM International Symposium 2014, Accessed on: 2014-12-09

[13] Kim, Y., Park, J. (2014) Study on interaction-induced symptoms with respect to virtual grasping and manipulation. *International Journal of Human-Computer Studies*. Vol. 72, No. 2 (2014), pp. 141-153. ISSN 1071-5819

[14] Kováč, J. et al (2011) Integrated Design manufacturing systems based on physical and virtual modelling. *Acta Mechanica Slovaca*. Vol. 14, No. 2 (2011), pp. 41-47. - ISSN 1335-2393

[15] Kováč, J., Knežo, D., Ižaríková, D. (2011) The model of topological relations-relation between two constructed equipment. *Acta Mechanica Slovaca*. Vol. 15, No. 2 (2011), pp. 60-68. - ISSN 1335-2393

[16] Kováč, J., Rudy, V., Kováč, Ju. (2008) Creation of 3D models of production systems and clusters. // *Transfer inovácií* Vol. 10, No. 11 (2008), pp. 261-264. The web version <http://web.tuke.sk/sjf-icav/stranky /transfer/8-2005/pdf/261-264.pdf>

[17] Kováč, J., Szombathyová, E. (2006) Ergonomics principles of assembly systems projection. // CO-MAT-TECH 2006. Bratislava : STU in Bratislava, 2006, pp. 593-597. ISBN 80-227-2472-6

[18] Kováč, M., Kováč, J. (2011) Innovative designing of production processes and systems. Košice : TU-SjF, 2011. ISBN 978-80-553-0805-0

[19] Kwon, O. et al. (2009) Determination of key dimensions for a glove sizing system by analysing the relationships between hand dimensions. *Applied Ergonomics*. Vol. 40, No. 4 (2009), 2008, pp. 762-766. ISSN 0003-6870

[20] Lee, K.-S., J, M.-Ch. (2015) Ergonomic Evaluation of Biomechanical Hand Function. *Safety and Health at Work*. Vol. 6, No. 1 (2015), pp. 9-17. ISSN 2093-7911

[21] Madarász, L., et al. (2002) Implementation of symbolic modelling in generation of production processes and systems for customer oriented production. RAAD 2002. Budapest : Polytechnic, 2002 pp. 391-395. - ISBN 963-715-409-4

[22] Malega, P., Mihok, J. (2006) Downtimes and losses versus effectiveness of production // *Transfer inovácií* Vol. 8, No. 9 (2006), pp. 97-100. The web version <http://web.tuke.sk/sjf-icav/stranky/transfer/9-2006/PDF/97-100.pdf>.

[23] Malloy, M. K., Milling, S. L. (2010) The effectiveness of virtual reality distraction for pain reduction: A systematic review. *Clinical Psychology Review*. Vol. 30, No. 8 (2010), pp. 1011-1018. ISSN 0272-7358

[24] Mareš, A., et al. Virtual reality – application of data glove in the solution of selected tasks. *Success: productivity and innovation in context*. No. 4 (2010), pp. 10-12. ISSN 1803-5183

[25] Mareš, A., et al. Data glove - intuitive tool of handling objects in CAD. *IT CAD*. Vol. 19, No. 4 (2009), pp. 30-31. - ISSN 1802-0011 The web version http://www.cad.cz

[26] Mujber, T. S. et al. Virtual reality applications in manufacturing process simulation. *Journal of Materials Processing Technology*. Vol. 155-156, 2004, pp. 1834-1838. ISSN 0924-0136

[27] Novak - Marcincin, J.; Barna, J.; Fecova, V. & Novakova - Marcincinova, L. (2012) Augmented Reality Applications in Manufacturing Engineering, Annals of DAAAM for 2012 & Proceedings of the 23rd International DAAAM Symposium, ISBN 978-3-901509-91-9, ISSN 2304-1382, pp 0065 - 0068, Editor Katalinic, B., Published by DAAAM International, Vienna, Austria, 2012

[28] Nomura, J., Sawada, K. (2001) Virtual reality technology and its industrial applications. *Annual Reviews in Control*. Vol. 25, 2001, pp. 99-109. ISSN 1367-5788

[29] Ostertagová, E., Kováč, J., Malega, P. (2011) Application of morphological analysis by selection of material flow structure and storage allocation to production system. 1 electronic optic disk (CD-ROM). MMaMS 2011 : Modelling of Mechanical and Mechatronical Systems : proceedings of the 4th international conference : Herľany, Slovakia, 20. - 22. 9. 2011. Košice : Technical University, 2011 pp. 374-379. ISBN 978-80-553-0731-2

[30] Pilacidi, G. et al. (2013) Overall design and implementation of the virtual glove. *Computers in Biology and Medicine*. Vol. 43, No. 11 (2013), pp. 1927-1940. ISSN 0010-4825

[31] Robinson, G. et al. (2009) Automated design process modelling and analysis using immersive virtual reality. *Computer-Aided Design*. Vol. 41, No. 12 (2009), pp. 1082-1094. ISSN 0010-4485

[32] Rudas, I., J. (1999) CA - system for preparation of production base modernization. Proceedings of the jubilee international conference. Budapest., 1999. pp. 173-176. ISBN 9-637-154-03-5

[33] Saggio, G. (2014) A novel array of flex sensors for a goniometric glove. *Sensors and Actuators A: Physical*. Vol. 205, 2014, pp. 109-125. ISSN 0924-4247

[34] Siddique, Z., Rosen, D. (1997) A virtual prototyping approach to product disassembly reasoning. *Computer-Aided Desing*. Vol. 29, No. 12 (1997), pp. 847-860. ISSN 0010-4485

[35] Spear, B. (2002) Virtual reality: patent review. *World Patent Information*. Vol. 24, No. 2 (2002), pp. 103-109. ISSN 0172-2190

[36] Steffan, R. - Schull, U., Kuhlen, T. (1998) Integration of virtual reality based assembly simulation into CAD/CAM environments. Industrial Electronics Society, Aachen, Germany, 1998. IECON '98. Proceedings of the 24th Annual Conference of the IEEE, pp. 2535-2537 Vol.4, ISBN 0-7803-4503-7

[37] Svoboda, M., Senderská, K., Kováč, J. (2004) Assembled product analysis as a part of assembly systems design procedure. *Acta Mechanica Slovaca*. Vol. 8, No. 1 (2004), pp. 91-96. - ISSN 1335-2393

[38] Straka, Ľ. (2005) New Trends in Technology System Operation. Proceedings of the 7th conference with international participation, Presov, pp. 385, 2005

[39] Šutaj-Eštok, A., Liberko, I., Sirková, M. (2012) Process management in relation to the systems thinking. In: Management 2012 : research management and business in the light of practical needs. - Prešov : Bookman, 2012. p. 214-218. ISBN 978-80-89568-38-3

[40] Wang, Q.-H., Li, J.-R. (2006) Interactive visualization of complex dynamic virtual environments for industrial assemblies. *Computers in Industry*. Vol. 57, No. 4 (2006), pp. 366-377. ISSN 0166-3615

[41] Weidlich, D. et al. (2007) Virtual Reality Approaches for Immersive Design. *CIRP Annals – Manufacturing Technology*. Vol. 56, No. 1 (2007), pp. 139-142. ISSN 0007-8506

[42] Yang, R. D. et al. (2007) Virtual assembly Technologies based on constraint and DOF analysis. *Robotics and Computer-Integrated Manufacturing*. Vol. 23, No. 4 (2007), pp. 447-456. ISSN 0736-5845

[43] Zachmann, G. et al. (1999) Virtual reality as tool for verification of assembly and maintenance processes. *Computers & Graphics*. Vol. 23, No. 3 (1999), pp. 389-403. ISSN 0097-8493

DOI: 10.2507/27th.daaam.proceedings.044

STATE-OF-THE-ART OF MESSAGING FOR DISTRIBUTED COMPUTING SYSTEMS

Stipe Celar, Eugen Mudnic & Zeljko Seremet

This Publication has to be referred as: Celar, S[tipe]; Mudnic, E[ugen] & Seremet, Z[eljko] (2016). State-Of-The-Art of Messaging for Distributed Computing Systems, Proceedings of the 27th DAAAM International Symposium, pp.0298-0307, B. Katalinic (Ed.), Published by DAAAM International, ISBN 978-3-902734-08-2, ISSN 1726-9679, Vienna, Austria
DOI: 10.2507/27th.daaam.proceedings.044

Abstract

Modern software applications rarely live in isolation and nowadays it is common practice to rely on services or consume information provided by remote entities. In such a distributed architecture, integration is key. Messaging, for more than a decade, is the reference solution to tackle challenges of a distributed nature, such as network unreliability, strong-coupling of producers and consumers and the heterogeneity of applications. Thanks to a strong community and a common effort towards standards and consolidation, message brokers are today the transport layer building blocks in many projects and services, both within the physics community and outside. Moreover, in recent years, a new generation of messaging services has appeared, with a focus on low-latency and high-performance use cases, pushing the boundaries of messaging applications. This paper will present messaging solutions for distributed applications going through an overview of the main concepts, technologies and services.

Keywords: messaging; message-oriented middleware; MQ; message queuing; distributed systems.

1. Introduction

This paper presents an overview of messaging concepts, functionalities and modern technologies. It starts with an introduction of messaging for distributed communication and system integration. A review of the main messaging features is then provided, followed by an overview of the major technologies for messaging, from broker to broker-less systems. In conclusion, a list of successful CERN's stories concerning the use of messaging for solving the communication problem of distributed applications is presented.

2. Message-oriented middleware

To cope with increasing demands on scalability, flexibility, and reliability, a message-oriented middleware (MOM) is an infrastructure for loosely coupled interprocess communication in an enterprise service bus or clouds [1] [2]. Particularly in clouds, loose coupling allows to rapidly scale message producers and consumers. A message with respect to MOM is an autonomous, self-contained entity that models an event and separates into a header and a body or payload. The middleware provides technical means of exchange, so a peer can exchange messages with other connected peers. A central concept in MOM is the notion of a message queue (or channel) for storing, transforming, and forwarding messages. Message queues enable asynchronous interaction, and a simple form is a First-In-First-Out (FIFO) queue. There are two different approaches to MOM using message queues as shown in Fig. 1.:

- Peer-to-peer messaging. A unified middleware component in every peer coordinates discovery and interaction between peers.
- Broker-based messaging. The middleware acts as a broke to provide a messaging infrastructure between the heterogeneous peers.

Peers can participate as client, service, or both [1]. A broker reduces the communication complexity between a numbers of peers but can incur delays in real-time applications because an additional store-and-forward procedure is necessary. [3]

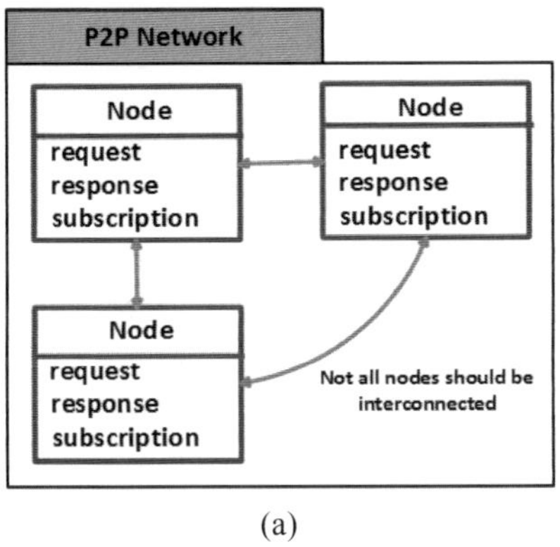

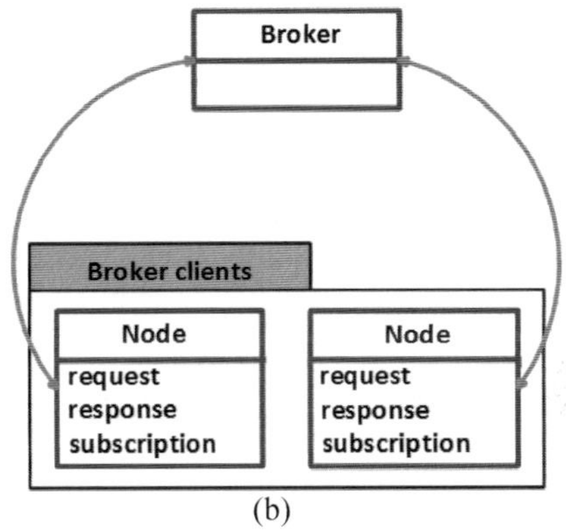

Fig. 1. A message-oriented middleware abstracts communication between heterogeneous peers: (a) Peer-to-peer messaging; (b) Broker-based messaging

In terms of interaction patterns, a trivial message queue allows bilateral Send and Receive, for asynchronous messaging, and multilateral One-to-Many Send, e.g., publish-subscribe. Using message queues in a broker architecture allows to implement sophisticated routing patterns. In general, a MOM is characterized by Curry [1]:

- *Messaging specification*. A MOM needs to specify th format of messages and transport mechanisms. Interconnectin proprietary MOM systems is achieved through adapters or bridges.
- *Message filtering*. A core functionality of a MOM is filtering for message delivery. Curry [1] distinguishes:
  - A channel-based system offers predefined groups of events as channels, where clients can subscribe to.
  - Messages in a subject-based system carry metadata in the message header, e.g., a subject. A client subscribes messages, where the metadata matches some given pattern.
  - In a content-based system, a client subscribes messages, where the message body satisfies a set of properties expressed in a query language.
  - Composite events functionality extends a content based filtering with property matching across sets or sequences of messages.
- *Message transformation*. Messages can originate from various heterogeneous sources and consequently carry all kinds of content types as payload. A MOM can offer APIs to modify messages, e.g., XML transformations.
- *Integrity, reliability, and availability*. A MOM can have properties to increase the overall Quality-of-Service:
  - Transactions and Atomic Multicast Notification;
  - Reliable message delivery: at-least-once, exactly-once, or at-most-once;
  - Guaranteed message delivery by acknowledgments;
  - Prioritization of messages;
  - Load balancing over several brokers or queues; and
  - Message broker clustering for fault tolerance.

A MOM is typically accessed through an API to abstract the technical details of message exchange. Due to the transport-agnostic design of SOAP/WS-* services, a MOM can also serve as a transport mechanism for SOAP messages. [3]

2.1. Java Message Service

The general purpose API named Java Message Service (JMS) [4] is maintained in a Java community process for MOM support. JMS defines a number of operations for creating, sending, receiving, and reading messages. It is transport-agnostic to abstract messaging from MOM implementations and therefore relaxes vendor lock-in. JMS is a universal interface for interacting with heterogeneous messaging systems [1]. A message body is dynamically typed according to the content type information stored in the header.

Some examples for JMS-enabled software implementations are the JMS reference implementation OpenMQ [5], IBM Websphere MQ [6], or TIBCO Enterprise Message Service [7]. [3]

2.2. RESTful Messaging Service

The motivation for RESTful Messaging Service (RestMS) [8] is Web-compatible messaging by using HTTP as transport mechanism and REST principles to describe locations, i.e., URLs, where messages can be posted to and received from. RestMS is an API specification, where XML-based messages are sent and received using HTTP methods. With respect to the REST service, resource locations are distinguished into feeds for incoming and pipes for outgoing messages. Feeds are joined with pipes on the service-side for message distribution. Message types in RestMS refer to XML, JSON, and a set of MIME content types for dynamically typing data. The specification also includes profiles to connect to other messaging infrastructures, e.g., AMQP. [3]

2.3. Open Middleware Agnostic Messaging API

Due to the diversity in middleware standards and wire formats, the Open Middleware Agnostic Messaging API (OpenMAMA) [9] initiative is an attempt to provide a single API for developing applications spanning across multiple MOMs. For correct translation messages and operations, a MOM has to provide a so-called OpenMAMA bridge implementation.

OpenMAMA is available as open-source library. It offers a built-in bridge for AMQP-enabled Apache Qpid and supports several bridges for proprietary messaging infrastructures in the finance sector. [3]

2.4. Proprietary messaging solutions

MSMQ [10] is a MOM for standalone integration or as a transport mechanism in Microsoft's WCF, next to Web services and COM+. It offers guaranteed message delivery, message routing, transactions, prioritization, and a simple type system for message body types. When used as a transport in WCF, a message body is either XML, binary, or ActiveX format. Beside its proprietary protocols, messages can also be transmitted over COM+. In terms of security, MSMQ allows authentication and encryption of messages. There is no broker in MSMQ; similar to Fig. 18a, a queue is hosted locally on a peer, and processes can store and retrieve messages. In terms of service interaction patterns, MSMQ is bilateral Send and Receive. MSMQ can exploit IP multicast to replicate a message for addressing multiple queues. A Microsoft alternative with brokerage support is SQL Server Service Broker [11].

Other proprietary MOM software products are the brokerless TIBCO Rendezvous [12], which uses direct connections between peers similar to MSMQ, Oracle Tuxedo Message Queue [13] as part of the Oracle Tuxedo application server for cloud middleware, and Terracotta Universal Messaging [14]. [3]

2.5. Advanced Message Queuing Protocol

Historically, MOM solutions have relied on proprietary protocols, and JMS is an attempt to agree on a compatible interface. Interoperability between varying MOM solutions is still difficult; costly JMS adapters or bridges are necessary to connect different transport mechanisms. AMQP [15] unifies messaging through an agreed-on wire format and has a similar role like HTTP in Web applications. While the OASIS AMQP 1.0 standard is restricted to the transport model for interoperability over the Internet, messaging architectures are specified by the AMQP working group [16].

The AMQP specification distinguishes a transport model and a queuing model [17]. The semantic queuing model defines terms like message, queue, exchange, and binding with respect to AMQP. Messages always end up in queues which are analogous to postal mailboxes. A queue stores messages and offers functionality for searching, reordering, or transaction participation. If a client wants to send a message, it chooses a broker-like exchange which is responsible for delivering messages to queues. An exchange can be offered as a service, and there exists an individual URI scheme (amqp: or amqps:) [18] to locate an exchange. A binding is a set of queue-specific arguments for an exchange. As shown in Fig. 2., there are different exchange types with respect to message filtering capabilities [19]:

- In a direct exchange, a message has a routing key and is sent to the queue, whose binding is equivalent to the routing key. In case of multiple queues with identical bindings, multiple message copies are delivered, i.e., a channel-based system.
- A topic exchange forwards copies of a message to all client queues, where the message routing key matches a queue's binding pattern, i.e., a subject-based system for publish-subscribe delivery.
- In a fan-out exchange, messages are forwarded to a set of queues without a specified binding, i.e., channel-based system.
- A headers exchange matches the headers of a message against predicate arguments of client queues beyond the routing key, i.e., a content-based system.

Messages are finally fetched from queues by consumer processes. AMQP provides guaranteed delivery, authentication, wire-level encryption, and transaction-based messaging for reliability. In terms of patterns, an exchange applies pattern Send in case of direct delivery or One-to-Many Send in other cases. Due to the self-contained type system and self describing message content, messages are dynamically typed in AMQP.

Examples for JMS-compatible broker implementations are OpenAMQ [20], JORAM [21], WSO2 Message Broker [22], SwiftMQ [23], Apache Qpid [24], and Red Hat Enterprise MRG [25].

AMQP defines four types of exchanges. A producer creates a message and sends it to an exchange. Depending on the exchange type and bindings, the message is delivered to queues, where consumers can fetch it from (a) direct exchange; (b) topic exchange; (c) fan-out exchange; (d) headers exchange. [3]

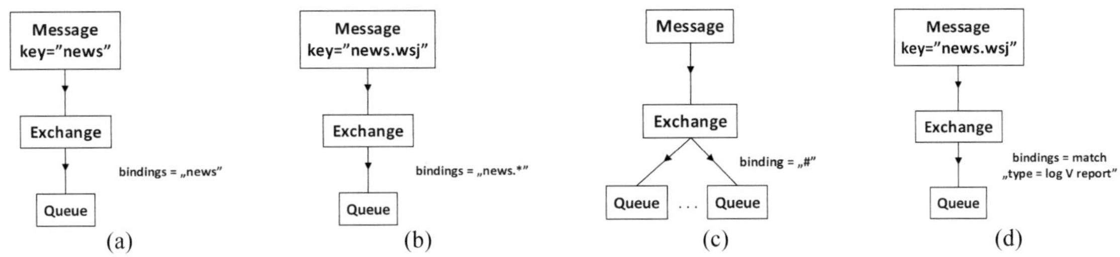

Fig. 2. AMQP types of message exchange

2.6. *Extensible Messaging and Presence Protocol*

While the XMPP [26][27][28] has been intended as an open standard for instant messaging, presence information, and contact list maintenance in chat applications, it also has middleware properties. In its base specification, XMPP exchanges messages as XML stanzas in client-to-service and service-to-service communication for federated services. An XMPP service therefore takes the role of a broker.

XMPP is particular attractive for MOM scenarios, where Web agents are involved because it supports HTTP as transport mechanism and most Web browsers and JavaScript runtime environments are capable of processing XML stanzas. Furthermore, XMPP is also considered as a suitable messaging protocol for Internet of Things applications [29]. The protocol is extensible, and extensions are specified in a community process. MOM-specific extensions are:

- Transfer of Base64-encoded binary content with an assigned MIME media type [30];
- RPC over XMPP [31];
- Service discovery [32];
- Publish-subscribe [33] for broker scenarios, extended addressing [34] for message routing, and event notification extensions [35][36];
- Reliable message transport [37]; and
- SSL/TLS protected transport mechanism and S/MIME [38] for end-to-end message encryption.

By default, messages in XMPP are XML stanzas and bodies are restricted to text only; there exists a notion of message type, but it is limited to instant messaging applications. Therefore, out-of-band signaling or a custom protocol, e.g., XMPP bits of binary [39], is required to discover message content types in a middleware scenario.

XMPP can also serve as a messaging infrastructure for SOAP/WS-* Web services [40][41]. Beside instant messaging, XMPP has been successfully deployed in the VIRTUS middleware for Internet of Things applications [42] using the real-time collaboration server software OpenFire [43]. Another software that offers XMPP messaging over Web-Socket is the Kaazing WebSocket Gateway [44]. [3]

2.7. *Streaming Text Oriented Messaging Protocol*

The simple text-based wire protocol STOMP [45] is for asynchronous message exchange between a client and a service or broker with simplicity and interoperability in mind. In the open standard of STOMP, a client and a service establish a session and asynchronously exchange frames of type Message, Receipt, or Error; a frame is partitioned into a command, header fields for metadata, and content of a certain MIME type. Messages are therefore dynamically typed. The protocol supports transactions and acknowledgments for reliable message delivery.

STOMP supports either bilateral messaging, i.e., Send and Receive, or broker-based publish-subscribe for One-to-Many Send interaction. Two notable service implementations are CoilMQ [46] and, for the latest protocol version 1.2, Stampy [47]. [3]

2.8. Message Queue Telemetry Transport

MQTT [48] originates from IBM and is now an open OASIS standard [49] for lightweight machine-to-machine messaging and Internet of Things applications, where bandwidth is limited. MQTT is intended for broker-based publish-subscribe architectures, One-to-Many Send interaction. An MQTT message can encapsulate binary payload up to 256 megabytes, but there is no notion of content type. The participating parties therefore have to agree on allowed formats out-of-band. For reliability, the protocol offers acknowledgments and retransmissions, but there is no transaction functionality.

Two notable MQTT broker software implementations are HiveMQ [50] and Mosquitto [51]. Both support Web clients using WebSocket. Another application that relies on MQTT messaging is Facebook Messenger [52].

2.9. Data Distribution Service for real-time systems

The open standard DDS [53] specifies a machine-to-machine MOM for publish-subscribe message distribution, real-time message delivery, scalability, and high throughput. Fields of application include the finance and defense sector, industry, aerospace, Internet of Things, and mobile devices [54].

Contrary to MQTT, DDS facilitates a data-centric, peer-to-peer interaction in the spirit of Fig. 1. (a). A *domain* partitions entities such as publisher, subscriber, and topic. A topic in a domain has a unique name and a strong datatype for publishing; these types are specified in an IDL, and messages are therefore statically typed. Subscribers in the domain request data via the topic, and publishers in the domain are responsible for message distribution [55].

DDS supports rich Quality-of-Service policies for data transmission. Interoperability between software implementations is achieved by the RTPS [56] wire protocol. To locate endpoints of peers, DDS provides dynamic discovery of publishers, subscribers, topics, and datatypes with respect to topics [54]. Reliable message delivery is achieved by negative acknowledgment when data is missing [57]. Security extensions for DDS, e.g., encrypted transport, are still in a beta state at time of writing [58].

Notable software implementations are OpenDDS [59], RTI Connext DDS [60], PrismTech OpenSlice DDS [61], and Twin Oaks CoreDX DDS [62]. [3]

2.10. Apache Kafka

Developed by LinkedIn, Apache Kafka [63] is a message broker specification and implementation for high-throughput publish-subscribe messaging, i.e., One-to-Many Send interaction. Kafka has an individual binary wire format protocol on top of TCP, and for fault tolerance, it supports clustering of brokers, persistent storage, and replication of messages.

On a conceptual level, Kafka distinguishes between topics for messages, producers that publish messages, and consumers that subscribe to topics. For every topic, a Kafka cluster maintains a partitioned log, where every partition stores an ordered sequence of published messages. The messages are kept for a configurable timespan, and partitions are replicated and distributed over servers in the Kafka cluster for fault tolerance and performance. The distributed log in Kafka guarantees the ordering of published and consumed messages in a certain topic. For a subscribed topic, a consumer maintains an offset in the message sequence to keep track of already processed ones. Through this offset, a consumer can also access older messages if they are still available on the cluster.

A message body is a byte sequence of a certain length and has no notion of type. Content type information therefore needs to be agreed out-of-band or by using a custom protocol. An interface for Web clients to subscribe to Kafka over WebSockets is already in an experimental state [64].

2.11. Polyglot message brokers

A natural approach for interconnecting several MOM standards is polyglot message brokerage. Three notable JMS-compliant software implementations in this area are Apache ActiveMQ [65], RabbitMQ [66], and JBoss HornetQ [67].

Beside features for scaling and clustering, the messaging core of Apache ActiveMQ, referred to as Apollo [68], uses the OpenWire [69] wire format, but also supports standards like AMQP, MQTT, and STOMP over WebSockets. ActiveMQ furthermore provides a proprietary HTTP-based RESTful API for Web clients.

RabbitMQ supports AMQP, STOMP, MQTT, and also HTTP as transport. Messages over HTTP can be sent in three ways: a native Web management API, STOMP over WebSockets, and JSON-RPC for Web browser integration.

HornetQ [67] is a MOM that originates from the JBoss application server. It supports AMQP, has an HTTP-based RESTful Web interface, and provides STOMP over Web-Sockets for Web clients. [3]

2.12. Message queuing as a service

Message brokerage has become an attractive cloud service. A broker is a critical component in a MOM architecture and needs fault tolerance, regular maintenance, and scalability; a message queue cloud service can eventually reduce cost. Amazon Web Services offers Simple Queue Services (SQS) [70] for transporting untyped text-based messages up to 256 kilobytes. SQS operates on a SOAP/WS-* Web service stack accessible through HTTP and HTTPS bindings.

Google's App Engine offers Pull Queues [71] and Push Queues [72] for messaging and App Engine task distribution. Both queue types are accessible through a RESTful API and use JSON format for messages. While Pull Queues need to be polled, Push Queues rely on webhooks for HTTP-based message delivery. Google has also announced Cloud Pub/Sub [73], a broker-based publish-subscribe messaging service for the App Engine, cloud apps, and Web clients. Using a RESTful API, Cloud Pub/Sub distributes JSON based messages according to topics. Subscribers can either poll for new messages or register a webhook for notification. The service supports guaranteed message delivery by maintaining a queue for every subscriber, and messages are removed from the queue, when the client acknowledges the message.

Microsoft also offers two cloud-based messaging solutions: Azure Queues and Service Bus Queues [74]. Azure Queues provide direct messaging between cloud services, and they are accessible through a RESTful interface. Messages are sequences of bytes and therefore not typed similar to Microsoft SQS. Service Bus Queues offer advanced architectures such as publish-subscribe and routing patterns. Windows applications and peers can access a service bus through WCF or directly by HTTP. A Brokered Message in a Service Bus Queue explicitly refers to a user-specified message body content. Service Bus Queues also offer an AMQP interface [75].

Two cloud services that offer AMQP brokerage as a service are StormMQ [76] and IronMQ [77]. CloudAMQP specifically offers the polyglot broker RabbitMQ as a Service [78]. CloudMQTT [79] is another pay-per-use broker for MQTT messaging, e.g., for complex event processing in Internet of Things environments. Rackspace Cloud Queues [80] supports publish-subscribe architectures by a HTTP based RESTful API in the spirit of RestMS. [3]

2.13. ZeroMQ and Nanomsg

Although there are many comprehensive messaging systems available, matching with the application requirements could be very difficult and without required performance characteristics. Consequently, in recent years, a new generation of low level messaging services has appeared such as ZeroMQ [81] and Nanomsg [82].

The intelligent socket library ZeroMQ aims for more flexible connectivity between peers. ZeroMQ offers several network transports, including TCP, UDP, and IP multicast, and a number of sockets types for architectural patterns. Messages are delivered to a thread- or process-local queue and made available through a socket. The specification defines the following socket types:

- REQ and REP for bilateral Send–Receive;
- DEALER and ROUTER for routing patterns;
- PUB and SUB for publish-subscribe One-to-Many Send;
- PUSH and PULL for workload distribution through One-to-Many Send and One-from-Many Receive;
- PAIR for asynchronous Send or Receive between two sockets.

ZeroMQ has no notion of broker because it is a socket abstraction. However, a MOM broker could be implemented using ZeroMQ. Messages are sequences of bytes and do not have a specified content type. The content type needs to be agreed on out-of-band or requires a custom protocol.

An attempt to provide ZeroMQ access in Web environments is NullMQ [83]. The JavaScript library uses Web-Sockets and a modified version of STOMP to bridge ZeroMQ messages into Web browsers. ZeroRPC [84] integrates RPC on top of ZeroMQ. Information is serialized as JSON-based MessagePack format and forwarded over ZeroMQ connections. A service interface is dynamically typed, and an ZeroRPC has been used in the dotCloud PaaS.

Nanomsg, however, is a reimagining of ZeroMQ—a complete rewrite in C. It builds upon ZeroMQ's rock-solid performance characteristics while providing several vital improvements, both internal and external. It also attempts to address many of the strange behaviors that ZeroMQ can often exhibit.

3. USE CASE - CERN

This section presents several implementations where messaging-based communication has been successfully adopted to solve the problem of exchange information in distributed system. [85]

3.1. CERN Beam Control middleware

The Beam Control department at the CERN laboratory is using messaging for highly reliable control/monitoring/alarm applications for the Large Hadron Collider (LHC). Since 2005, a cluster of ActiveMQ brokers, in a store and forward configuration, is used to collect the critical data generated by the safety systems (e.g. 30 producers, 2MB/s, 4.5K msg/s) and to forward it to many consumers (e.g. monitoring tool, dashboards). Being safety data mission critical, the store and forward configuration allow to completely decouple data production from consumption, preventing misbehaving clients to affect data collection and archiving [86]. Moreover, the LHC Control framework has been recently migrated from CORBA to ZeroMQ as communication layer [87]. [85]

3.2. DAQ Online Monitoring

Messaging has been also extensively used in several monitoring tools for Data Acquisition (DAQ) systems, which are responsible to filter and collect data from detectors (e.g. high energy physic experiments) to storage facilities. [85]

3.2.1. The ATLAS TDAQ shifter assistant project

It relies on messaging to distribute operational alarms from private TDAQ network to GPN to a number of heterogeneous consumers. An ActiveMQ cluster is used in a master/slave configuration in order to minimize the impact on the required firewall configuration to a single outbound connection. [88]

3.2.2. The STAR Online framework

It relies on an AMQP-based system for flexible, loosely coupled distribution of detector metadata, using messaging as unified transport layer for processing, storage and monitoring. Moreover, investigation has been done to re-write the control framework over MQTT, profiting from the protocol flexibility and interoperability [89]. [85]

3.3. WLCG Messaging Service

Messaging has been also successfully used on large-scale geographically distributed infrastructure. The WLCG (Worldwide LHC Computing Grid) messaging service is the backbone transport layer used for monitoring WLCG sites and services around the world, with more than 50000 clients and an average message rate of 100 KHz. The monitoring infrastructure is based on STOMP with JSON payload. Thanks to the interoperability of the STOMP protocol across several broker flavours, heterogeneous message-broker clusters (ActiveMQ, Apollo or RabbitMQ) are used in a scenario where client applications produce to any and consume to all [90]. [85]

4. Conclusion

Messaging is pragmatic reaction to the problem of communication in distributed systems. It allows loosely coupled communication acting as intermediate layer between producer and consumer. It brings many benefits in distributed applications flexibility and scalability, with implications in application and infrastructure complexity. Messaging systems are still evolving technology with the AMQP standardization effort pointing in the good direction, but still with partial adoption. Message brokers are solid and reliable technology used as transport layer building blocks in many projects and services, both within the physics community and outside. In the recent years, a new generation of systems is promoting messaging for low-latency / high-throughput / data-intensive communication, like ZeroMQ, narrowing use cases and relaxing assumptions, but pushing the boundaries of messaging applications towards new domains and successfull implementations for demanding CERN applications.

5. References

[1] Curry, E. (2005). Message-oriented middleware. In: *Mahmoud QH (ed) Middleware for communications*. Wiley, Chichester

[2] Celar, S.; Seremet, Z. & Turic, M. (2011). Cloud computing: definition, characteristics, services and models, Annals of DAAAM for 2011 & Proceedings of the 22nd International DAAAM Symposium, Volume 22, No. 1, ISSN 1726 – 9679 ISBN 978-3-901509-83-4, Editor B. Katalinic, Published by DAAAM International, Vienna, Austria, EU, 2011

[3] Lampesberger, H. (2016). Technologies for Web and cloud service interaction: a survey, *Service Oriented Computing and Applications*, June 2016, Volume 10, Issue 2, pp. 71–110

[4] Oracle: Java Message Service (2014). Available from: http://www.oracle.com/technetwork/java/jms/index.html Accessed: 2016-02-19

[5] GlassFish Project: Open Message Queue (2014). Available from: https://mq.java.net/ Accessed: 2016-02-19

[6] IBM: WebSphereMQ (2014). Available from: http://www-03.ibm.com/software/products/en/websphere-mq. Accessed: 2016-08-22

[7] TIBCO: Enterprise Message Service (2014). Available from: http://www.tibco.com/products/automation/enterprise-messaging/enterprisemessage-service/default.jsp. Accessed: 2016-07-21

[8] Wikidot.com: RestMS (2008). Available from: http://www.restms.org/. Accessed: 2016–02-21

[9] OpenMAMA: Introduction to OpenMAMA (2014). Available from: http://www.openmama.org/what-is-openmama/introduction-to-openmama Accessed: 2016–07–09

[10] Microsoft Developer Network: Message Queuing (MSMQ) (2014). Available from: http://msdn.microsoft.com/en-us/library/ms711472.aspx Accessed: 2016-02-20

[11] Microsoft Developer Network: SQL Server Service Broker (2014). Available from: http://msdn.microsoft.com/en-us/library/bb522893.aspx Accessed: 2016-02-23

[12] TIBCO: Rendezvous Messaging Middleware (2014). Available from: http://www.tibco.com/products/automation/enterprise-messaging/rendezvous/default.jsp Accessed: 2016-07-22

[13] Oracle: Oracle Tuxedo Message Queue Product Overview (2013). Available from: http://docs.oracle.com/cd/E35855_01/otmq/docs12c/overview/overview.html. Accessed: 2016-04-28

[14] Software AG: Terracotta Universal Messaging (2014). Available from: https://www.softwareag.com/corporate/images/SAG_Terracotta_Universal_Messaging_FS_Jun14_Web_tcm16-114090.pdf Accessed: 2016–06–04

[15] OASIS: Advanced Message Queuing Protocol v1.0 (2011). Available from: http://docs.oasis-open.org/amqp/core/v1.0/os/amqp-core-overviewv1.0-os.html Accessed: 2016-02-19

[16] OASIS: AMQP Working Group 0–10 (2014). Available from: http://www.amqp.org/specification/0-10/amqp-org-download Accessed 2016–08–06

[17] O'Hara, J. (2007). Toward a commodity enterprise middleware. Queue 5(4), 48–55. Available from: http://queue.acm.org/detail.cfm?id=1255424 Accessed: 2016-07-24

[18] Pivotal Software Inc: AMQP URI Specification (2014). Available from: http://www.rabbitmq.com/uri-spec.html Accessed: 2016-07-27

[19] O'Hara, J. (2007). Toward a commodity enterprise middleware. Queue 5(4), 48–55. Available from: http://queue.acm.org/detail.cfm?id=1255424 Accessed: 2016-07-21

[20] iMatix Corporation: OpenAMQ (2009). Available from: http://www.openamq.org/ Accessed: 2016-02-21

[21] JBoss Community: JBoss Web Services (2014). Available from: https://www.jboss.org/jbossws Accessed: 2016-03-28

[22] WSO2: WSO2 Message Broker (2014). Available from: http://wso2.com/products/message-broker/ Accessed 2016–07-23

[23] SwiftMQ: Enterprise messaging platform (2014). Available from: http://www.swiftmq.com/ Accessed: 2016-07-23

[24] Apache Software Foundation: Apache Qpid (2013). Available from: https://qpid.apache.org/ Accessed: 2016-07-21

[25] Red Hat Enterprise: MRG - Messaging, Realtime, Grid (2009). Available from: http://www.redhat.com/f/pdf/MRG_brochure_web.pdf Accessed: 2016-07-21

[26] Saint-Andre, P. (2011). Extensible messaging and presence protocol (XMPP): Address Format. RFC 6122 (Proposed Standard), Available from: http://www.ietf.org/rfc/rfc6122.txt Accessed: 2016-07-21

[27] Saint-Andre, P. (2011). Extensible Messaging and Presence Protocol (XMPP): Core. RFC 6120 (Proposed Standard), Available from: http://www.ietf.org/rfc/rfc6120.txt Accessed: 2016-07-21

[28] Saint-Andre, P. (2011). Extensible messaging and presence protocol (XMPP): Instant messaging and presence. RFC 6121 (Proposed Standard), Available from: http://www.ietf.org/rfc/rfc6121.txt Accessed: 2016-07-21

[29] XMPP: IoT Systems (2013). Available from: http://wiki.xmpp.org/web/Tech_pages/IoT_systems Accessed: 2016–07-24

[30] Saint-Andre, P. & Simerda, P. (2008). XEP-0231: Bits of Binary, Available from: http://xmpp.org/extensions/xep-0231.html Accessed: 2016-07-25

[31] Adams, D. (2011). XEP-0030: Service Discovery, Available from: http://xmpp.org/extensions/xep-0009.html. Accessed: 2016-07-23

[32] Hildebrand, J.; Millard, P.; Eatmon, R. & Saint-Andre, P. (2008). XEP-0009: Jabber-RPC, Available from: http://xmpp.org/extensions/xep-0030.html Accessed: 2016-07-23

[33] Millard, P., Saint-Andre, P. & Meijer, R. (2010). XEP-0060: Publish-Subscribe, Available from: http://xmpp.org/extensions/xep-0060.html Accessed: 2016-07-23

[34] Hildebrand, J. & Saint-Andre, P. (2004). XEP-0033: Extended stanza addressing, Available from: http://xmpp.org/extensions/xep-0033.html Accessed: 2016-07-23

[35] Saint-Andre, P. & Smith, K. (2010). XEP-0163: PersonalEventing Protocol, Available from: http://xmpp.org/extensions/xep-0163.html Accessed: 2016-07-23

[36] Waher, P. (2014). XEP-0337: Event Logging over XMPP, Available from: http://xmpp.org/extensions/xep-0337.html Accessed: 2016–07-23

[37] Miller, M & Saint-Andre, P. (2005). XEP-0079: Advanced message processing, Available from: http://xmpp.org/extensions/xep-0079.html Accessed: 2016-07-23
[38] Ramsdell, B. & Turner, S. (2010). Secure/Multipurpose Internet Mail Extensions (S/MIME) Version 3.2 Message Specification. RFC 5751 (Proposed Standard), Available from: http://www.ietf.org/rfc/rfc5751.txt Accessed: 2016-07-22
[39] Saint-Andre, P. & Simerda, P. (2008). XEP-0231: Bits of Binary. Available from: http://xmpp.org/extensions/xep-0231.html Accessed: 2016-07-25
[40] Forno, F. & Saint-Andre, P. (2005). XEP-0072: SOAP Over XMPP. Available from: http://xmpp.org/extensions/xep-0072.html Accessed: 2016-07-23
[41] Mansour Fallah, S. (2016). Multi Agent based Control Architectures, Proceedings of the 26th DAAAM International Symposium, pp.1166-1170, B. Katalinic (Ed.), Published by DAAAM International, ISBN 978-3-902734-07-5, ISSN 1726-9679, Vienna, Austria
[42] Conzon, D.; Bolognesi, T; Brizzi, P.; Lotito, A; Tomasi, R. & Spirito, M. (2012). The virtus middleware: an xmpp based architecture for secure iot communications. In: *21st international conference on computer communications and networks*, ICCCN'12, pp. 1–6
[43] ignite realtime: Openfire (2014). Available from: http://www.igniterealtime.org/projects/openfire/ Accessed: 2016-07-23
[44] Kaazing: WebSocket Gateway - XMPP (2014). Available from: http://kaazing.com/products/editions/kaazing-websocket-gateway-xmpp/ Accessed: 2016-07-23
[45] STOMP: The Simple Text Oriented Messaging Protocol v1.2 (2012). Available from: http://stomp.github.io/stomp-specification-1.2.html Accessed: 2016-02-19
[46] CoilMQ: Lightweight Python STOMP message broker (2012). Available from: https://github.com/hozn/coilmq/ Accessed: 2016-07-24
[47] Stampy: Java implementation of the STOMP 1.2 specification (2013). Available from: http://mrstampy.github.io/Stampy/ Accessed: 2016-07-24
[48] IBM Developer Networks: MQ Telemetry Transport (MQTT) V3.1 Protocol Specification (2010). Available from: http://www.ibm.com/developerworks/webservices/library/ws-mqtt/ Accessed: 2016-02-20
[49] OASIS: MQTT Version 3.1.1 (2014). Available from: http://docs.oasis-open.org/mqtt/mqtt/v3.1.1/mqtt-v3.1.1.html Accessed: 2016-07-21
[50] HiveMQ: Enterprise Grade MQTT Broker (2014). Available from: http://www.hivemq.com/ Accessed: 2016–07-22
[51] Mosquitto: an open source MQTT v3.1/v3.1.1 Broker (2014). Available from: http://mosquitto.org/ Accessed: 2016-07-24
[52] Zhang L (2011). Building facebook messenger. Available from: https://www.facebook.com/notes/facebook-engineering/building-facebookmessenger/10150259350998920. Accessed: 2016–06-13
[53] Object Management Group: Documents Associated With Data Distribution Services, V1.2 (2007). Available from: http://www.omg.org/spec/DDS/1.2/ Accessed: 2016-02-20
[54] Twin Oaks Computing Inc: What can DDS do for You? (2013). Available from: http://www.omg.org/hot-topics/documents/dds/CoreDX_DDS_Why_Use_DDS.pdf Accessed: 2016-07-24
[55] Object Computing Inc.: OpenDDS Developer's Guide (2014). Available from: http://download.ociweb.com/OpenDDS/OpenDDS-latest.pdf Accessed: 2016-07-27
[56] Object Management Group: Documents Associated With The Real-Time Publish-Subscribe Wire Protocol DDS Interoperability Wire Protocol Specification (DDSI), V2.1 (2009). Available from: http://www.omg.org/spec/DDSI/2.1/ Accessed: 2016-02-20
[57] Object Computing Inc.: OpenDDS Developer's Guide (2014). Available from: http://download.ociweb.com/OpenDDS/OpenDDS-latest.pdf Accessed: 2016-07-27
[58] Object Management Group: Documents Associated With DDS Security (DDS-Security) 1.0 - Beta 1 (2014). Available from: http://www.omg.org/spec/DDS-SECURITY/1.0/Beta1/ Accessed: 2016-07-25
[59] Object Computing Inc.: Welcome to OpenDDS! (2013). Available from: http://www.opendds.org Accessed: 2016-07-24
[60] RTI: Connext DDS Software (2014). Available from: http://www.rti.com/products/index.html Accessed: 2016-07-27
[61] PrismTech: OpenSplice DDS Community (2014). Available from: http://www.prismtech.com/opensplice/opensplice-dds-community Accessed: 2016-07-24

[62] Twin Oaks Computing Inc: CoreDXDDS data distribution service middleware (2014). Available from: http://www.twinoakscomputing.com/coredx Accessed: 2016-07-24

[63] Apache Software Foundation: Kafka (2014). Available from: http://kafka.apache.org/ Accessed: 2016-07-15

[64] Black, B. (2014). kafka-websocket. Available from: https://github.com/b/kafkawebsocket Accessed: 2016-08-03

[65] Apache Software Foundation: Apache ActiveMQ (2011). Available from: http://activemq.apache.org/ Accessed: 2016-02-21

[66] Pivotal Software Inc: RabbitMQ (2014). Available from: https://www.rabbitmq.com/ Accessed: 2016-02-19

[67] HornetQ: what is HornetQ? (2013). Available from: http://hornetq.jboss.org/ Accessed: 2016-07-24

[68] Apache Software Foundation: Apollo - ActiveMQ's next generation of messaging (2014). Available from: http://activemq.apache.org/apollo/ Accessed: 2016-07-24

[69] Apache Software Foundation: OpenWire version 2 specification (2011). Available from: http://activemq.apache.org/openwire-version-2-specification.html Accessed: 2016-02-20

[70] AmazonWeb Services: Amazon Simple Queue Service (Amazon SQS) (2013). Available from: http://aws.amazon.com/sqs/ Accessed: 2016-02-21

[71] Google Developers: Using Pull Queues in Java (2014). Available from: https://developers.google.com/appengine/docs/java/taskqueue/overview-pull Accessed: 2016–08–07

[72] Google Developers: Using Push Queues in Java (2014). Available from: https://developers.google.com/appengine/docs/java/taskqueue/overview-push Accessed: 2016–08–07

[73] Google Developers: Google Cloud Pub/Sub (2014). Available from: https://developers.google.com/pubsub/overview Accessed: 2016–08–07

[74] Microsoft Azure: Azure queues and service bus Queues—Compared and contrasted (2014). Available from: http://msdn.microsoft.com/library/azure/Hh767287.aspx Accessed: 2016-08-21

[75] Microsoft Azure: Windows Azure AMQP 1.0 Support in Service Bus (2014). Available from: http://www.windowsazure.com/en-us/documentation/articles/service-bus-amqp-overview/ Accessed: 2016-02-21

[76] stormmq Limited: Message queues as a service in the cloud (2013). Available from: http://stormmq.com/ Accessed: 2016-02-21

[77] Iron.io: IronMQ (2014). Available from: http://www.iron.io/mq Accessed: 2016–02–20

[78] CloudAMQP: RabbitMQ as a Service (2014). Available from: http://www.cloudamqp.com Accessed: 2016-07-21

[79] CloudMQTT: Hosted broker for the Internet of Things (2014). Available from: http://www.cloudmqtt.com Accessed: 2016-07-25

[80] Rackspace: Cloud Queues (2014). Available from: http://docs.rackspace.com/queues/api/v1.0/cq-gettingstarted/content/DB_Overview.html Accessed: 2016-07-21

[81] iMatix Corporation: ∅MQ (2013). Available from: http://www.zeromq.org/ Accessed: 2016-02-19

[82] nanomsg (2016). Available from: http://nanomsg.org/ Accessed: 2016-02-19

[83] Lindsay, J.; Shakirzyanov, B. (2014). NullMQ. Available from: https://github.com/progrium/nullmq Accessed: 2016-03-10

[84] dotCloud: ZeroRPC (2013). Available from: http://zerorpc.dotcloud.com/ Accessed: 2016-02-19

[85] Magnoni, L. (2015). Modern Messaging for Distributed Sytems, Journal of Physics: *Conference Series 608* (2015) 012038

[86] Ehm, F. (2011). Running a Reliable Messaging Infrastructure for CERN's Control System. Proceedings of ICALEPCS2011 (Grenoble, FRANCE)

[87] Dworak, A.; Ehm, F.; Sliwinski, W. & Sobczak, M. (2011). Middleware Trends and Market Leaders 2011. Proceedings of ICALEPCS2011 (Grenoble, FRANCE)

[88] Kazarov, A; Miotto, G. L. & Magnoni, L. (2012). The AAL project: automated monitoring and intelligent analysis for the ATLAS data taking infrastructure. Journal of Physics: Conference Series, Volume 368

[89] Arkhipkin, D.; Lauret, J. & Betts, W. (2011). A message-queuing framework for STARs online monitoring and metadata collection. Journal of Physics: Conference Series, Volume 331

[90] Cons, L. & Paladin, M. (2011). The WLCG Messaging Service and its Future. Journal of Physics: Conference Series, Volume 396

DOI: 10.2507/27th.daaam.proceedings.045

IMPACT CRACK INITIATION AND PROPAGATION ENERGY OF PROLONGED EXPLOITED HEAT RESISTANT STEEL

Damir Hodzic & Ismar Hajro

This Publication has to be referred as: Hodzic, D[amir] & Hajro, I[smar] (2016). Impact Crack Initiation and Propagation Energy of Prolonged Exploited Heat Resistant Steel, Proceedings of the 27th DAAAM International Symposium, pp.0308-0311, B. Katalinic (Ed.), Published by DAAAM International, ISBN 978-3-902734-08-2, ISSN 1726-9679, Vienna, Austria
DOI: 10.2507/27th.daaam.proceedings.045

Abstract

In this paper heat resistant steel 14MoV6-3 in conditions as a virgin and exploited material is investigated. This steel was used for the main steamline of thermal power plant for a prolonged life time period under service conditions. As a consequence of microstructure evolution through long period of exploitation at elevated temperature there was a significant decrease of mechanical properties. In order to investigate decrease of impact properties of steel 14MoV6-3, following temperatures were selected for impact testing: 20 °C, 150 °C, 400 °C and 540 °C (service temperature). This was done by testing and comparison of impact crack initiation and propagation energy of virgin and serviced material 14MoV6-3 after 194.207 hours of exploitation. Irrespective to the fact that impact tests cannot be used for assessment of the further steamline safe service time, it is indispensable for assessing the materials deformability and its capability to carry the load in its further service process.

Keywords: Steel 14MoV6-3; Impact testing; Crack initiation and propagation energy; Lifetime assessment.

1. Introduction

Nevertheless to the very long history of low-alloyed steel 14MoV6-3, this material is still built in the numerous power plants, particularly in the boilers and its belonging high-temperature components. According to the German Codes VGB-R 509L and TRD 508 the start or extended material inspection is required after about 70.000 hours for steel 14MoV6-3 and about 100.000 hours of exploitation for the other heat resistant steels [1]. Power plants, which were originally intended to provide the base load, are frequently shut down and powered up. Variations in the steam temperature accompanying the power changes induce thermo-mechanical stresses in components, which lead to material degradation and consequently can cause failure [2]. The effort to extend designed lifetime of industrial plants operating for a long time at elevated temperatures requires the knowledge of residual lifetime of the critical components. Residual lifetime assessment is unthinkable without the knowledge of mechanical properties of materials prior to operation and mechanical properties after actual time of operation (actual mechanical properties), because the material properties can be reduced throughout the service life [3]. Decrease of mechanical properties of exploited material is caused by changes in the steel microstructure due to long-lasting service operation. Reduction of impact properties caused by long-term exploitation of the steel at elevated temperature depends to a large extent on the initial steel microstructure. Some published investigations reported that decrease of impact toughness caused by long-lasting operation is the least in the case of tempered bainite structure [4]. In addition to previous, metallographic examinations of various steel grades after long-term service at elevated temperature revealed that transformations of carbides and morphological changes of phases have the most significant effect on service properties degradation [5].

2. Chemical composition and microstructure

In order to investigate decrease of impact properties of material, low-alloyed steel 14MoV6-3 exposed to 194.207 hours of exploitation, has been compared with same, but virgin material. Investigated material is taken from the Unit 5 main steamline (ø245×28mm) that operated at temperature 540 °C and pressure 13,5MPa in thermal power plant Kakanj, Bosnia and Herzegovina. Chemical composition of investigated material 14MoV6-3 (virgin and exploited) was accomplished in order to confirm that all delivered specimens of steamline are made from the same material, so the results of predicted investigation on virgin and exploited material could be comparable. Method for determination of chemical composition was spectral analysis. Chemical composition of material 14MoV6-3 according to normative DIN 17175/79 is presented in Table 1.

Grade	C, %	Si, %	Mn, %	P, %	S, %	Cr, %	Mo, %	V, %
14MoV6-3	0,10-0,18	0,10-0,35	0,40-0,70	Max.0,035	Max.0,035	0,30-0,60	0,50-0,70	0,22-0,32

Table 1. Chemical composition of steel 14MoV6-3 according to DIN 17175

From the results of chemical composition analysis it was obvious that investigated steamline specimens (virgin and exploited) are made of the same material 14MoV6-3. Slightly less content of Molybdenum, comparing with chemical composition according to DIN 17175, is probably error of measuring without influence on further investigation, because content of Molybdenum is almost the same for virgin and exploited steamline material. Chemical composition of investigated virgin and exploited material 14MoV6-3 is presented in Table 2 and Table 3, respectively.

Grade	C, %	Si, %	Mn, %	P, %	S, %	Cr, %	Mo, %	V, %
14MoV6-3	0,149	0,30	0,57	0,013	0,015	0,59	0,475	0,28

Table 2. Chemical composition of virgin steel 14MoV6-3

Grade	C, %	Si, %	Mn, %	P, %	S, %	Cr, %	Mo, %	V, %
14MoV6-3	0,139	0,32	0,56	0,013	0,013	0,50	0,47	0,28

Table 3. Chemical composition of exploited steel 14MoV6-3

In order to investigate microstructure of exploited steamline material, metallographic testing was accomplished. This was done in laboratory at IWS Institute TU Graz (Institute for materials and welding at Technical University Graz), Austria, by testing and comparison of microstructure of virgin material and exploited material 14MoV6-3 after 194.207 hours of exploitation. One specimen per material condition (virgin and exploited) was used and metallographic testing of material microstructure by optical microscope was done with different magnifications of 200×, 500× and 1000×. Microstructure of investigated material 14MoV6-3 at transversal cross section of steamline pipe with magnification 200× is shown in Fig. 1.

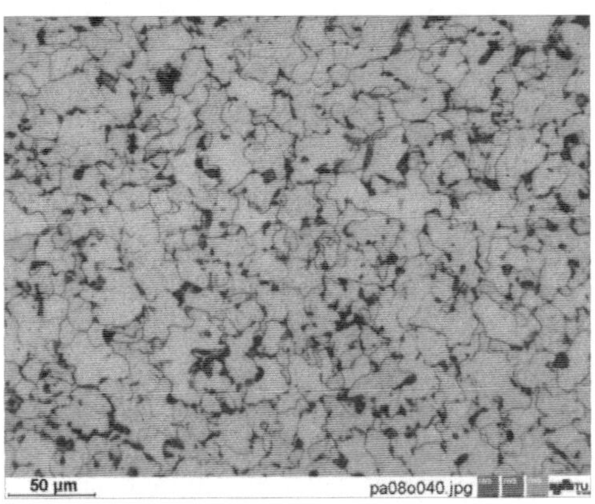

Fig. 1. Microstructure of virgin and exploited steel 14MoV6-3 [6]

According to previous investigations of this material with similar service conditions, that are published in scientific journals, the initial microstructure of the 14MoV6-3 low-alloyed steel features the mixture of bainite with ferrite, sometimes with a small amount of pearlite. Occurrences of the significant amount of the M_3C carbides and numerous, very fine MC type ones, are identified in such material. The final structure image after prolonged exploitation is ferrite with rather homogeneously distributed precipitations inside grains and chains of the significant amount of precipitations on their boundaries, [5]. Results of microstructure investigation that are presented in this paper mainly can confirm previous facts, but in addition to mentioned microstructure evolution, there is also a significant growth of ferrite grain size after long-term operation at elevated temperature.

3. Impact testing of crack initiation and propagation energy

In general, notch toughness is measured in terms of the absorbed impact energy needed to cause fracturing of the specimen. The change in potential energy of the impacting head (from before impact to after fracture) is determined with a calibrated dial that measures the total energy absorbed in breaking the specimen. Other quantitative parameters, such as fracture appearance and degree of ductility/deformation, are also often measured in addition to the fracture energy. Impact tests may also be instrumented to obtain load data as a function of time during the fracture event. The Charpy V-notch test continues to be the most utilized and accepted impact test in use in the industry, [7]. In order to investigate decrease of impact properties of exploited steamline material 14MoV6-3, following temperatures were selected for impact testing: 20 °C, 150 °C, 400 °C and 540 °C (service temperature). This was done by testing and comparison of crack initiation and propagation energy values of virgin material and exploited material 14MoV6-3 after 194.207 hours of exploitation. For every testing temperature 3 Charpy V-notch specimens were used. Results of average (3 specimens) impact crack initiation energy values per testing temperature for virgin and exploited material 14MoV6-3 are presented in Fig. 2.

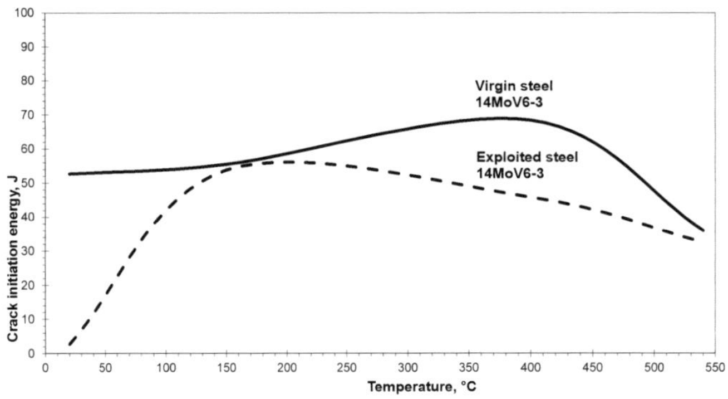

Fig. 2. Impact crack initiation energy

Results of average (3 specimens) impact crack propagation energy values per testing temperature for virgin and exploited material 14MoV6-3 are presented in Fig. 3.

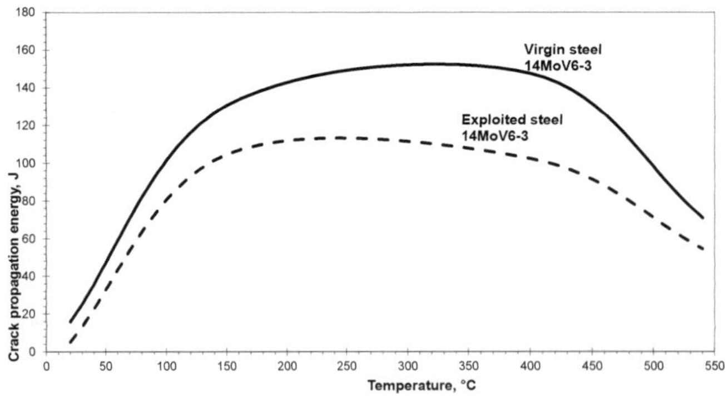

Fig. 3. Impact crack propagation energy

From the results of impact testing it is notable that the impact crack initiation and propagation energy increases slightly up to 400 °C for virgin material 14MoV6-3 and up to 150 °C for exploited material 14MoV6-3. It is reduces significantly, but not drastically above 400 °C for virgin material and above 150 °C for exploited material, so that its values are still more than sufficient at steamline service temperature 540 °C. The most important result of exploited steamline material impact testing is extremely low value of impact crack initiation energy at room temperature 20°C, which is significantly beneath the allowed value. Impact properties of steamline material 14MoV6-3 depends mostly on development of the precipitation processes and also on development of the microstructure changes and structure discontinuities, as well as grain growth, originated during the long period of exploitation at elevated temperature.

4. Lifetime assessment

Components of power plant boiler are exposed to elevated temperatures, aggressive environment, creep, fatigue, and other damage mechanisms that can cause degradation, deformation or cracking of components. Boiler components and belonging steamlines of thermal power plants have exceeded most often significantly their design service time being most often 100.000 hours long. They require forecasting their further safe service for the conditions in exploitation. Evaluation of microstructural evolution in exposed to service materials is a key tool for a correct evaluation of material status and allowable service extension. The knowledge of the structure and degree of damage could be essential for the assessment of residual service life and damage analysis respectively. It should however be pointed out that the above mentioned knowledge alone does not allow a prediction of the residual service life. A reliable life assessment should be made not only by means of microstructural inspection but it is preferable that together with other inspections the same is included. A large number of failures in engineering components occur due to preexisting defects, nonmetallic inclusions or other imperfections (casting, welding defects, etc.) [8]. However, it is of engineering interest to know how and why particular component has failed.

5. Conclusion

Main problem with impact testing is that it cannot be used for the final assessment of the further steamline safe service time and it is not quite useful for the residual life assessment and for determining of the exhaustion extent. Very low impact toughness values of investigated exploited steel 14MoV6-3 could cause crack initiation and propagation by extremely low impact load level in service at room temperature. Results of this investigation show that impact testing should be obviously included together with the other diagnostic methods as a very important indicator. However, it is indispensable for assessing the material's deformability and its capability to carry the load connected with the pressure tests, as well as in limiting the number of banking and setting to work the steamline in its further service process. In the future, not only impact toughness but also the other parameters of fracture mechanics such as fracture toughness could be very interesting for investigation of prolonged exploited heat resistant steels used in power plants.

6. Acknowledgement

This investigation was partly supported by IWS Institute at Technical University Graz and OEAD Austrian Agency for International Cooperation in Education and Research.

7. References

[1] ECCC Recommendations – Volume 9, Part II: High Temperature Component Analysis Overview of Assessment & Design Procedures, ECCC European Creep Collaborative Committee, 2005

[2] Linn, S., Scholz A. (2013). Creep-fatigue Lifetime Assessment with Phenomological and Constitutive Material Laws. Procedia Engineering (6th International Conference on Creep, Fatigue and Creep-Fatigue Interaction CF-6), Vol. 55, 2013, pp. 607-611, ISSN1877-7058, DOI 10.106/j.proeng.2013.03.302

[3] Matocha K. (2014). The Use of Small Punch Test for Determination of Fracture Behaviour of Ferritic Steels. Procedia Engineering (1st International Conference on Structural Integrity ICONS-2014), Vol. 86, 2014, pp. 885-891, ISSN1877-7058, DOI 10.106/j.proeng.2014.11.110

[4] Golanski G., Stachura S., Gajda-Kucharska B., Kupczyk J. (2007). Optimisation of Regenerative Heat Treatment Parameters of G21CrMoV4-6 Cast Steel. Archives of Materials Science and Engineering, Vol. 28, No. 6, (June 2007) page numbers (341-344), ISSN 1897-2764

[5] Dobrzanski J., Zielinski A., Krtzon H. (2007). Mechanical Properties and Structure of the Cr-Mo-V Low-alloyed Steel after Long-term Service in Creep Condition. Journal of Achievements in Materials and Manufacturing Engineering, Vol. 23, No. 1, (July 2007) page numbers (39-42), ISSN 1734-8412

[6] Hodzic D., Hajro I. (2012). Microstructure Degradation after Prolonged Exploitation of Heat Resistant Steel 14MoV6-3. Proceedings of 16th International Research/Expert Conference "Trends in the Development of Machinery and Associated Technology", 10-12 September 2012, Dubai UAE, ISSN 1840-4944, Ekinović S., Yalcin S., Calvet J.V. (Ed.), pp. 155-158, Faculty of Mechanical Engineering in Zenica, Zenica

[7] ASM Handbook Vol. 8 – Mechanical testing and evaluation, Impact toughness testing, ASM International, 2000

[8] Brnic J., Brcic M. (2015). Comparison of Mechanical Properties and Resistance to Creep of 20MnCr5 Steel and X10CrAlSi25. Procedia Engineering (25th DAAAM International Symposium on Intelligent Manufacturing and Automation, 2014), Vol. 100, 2015, pp. 84-89, ISSN1877-7058, DOI 10.106/j.proeng.2015.01.345

DOI: 10.2507/27th.daaam.proceedings.046

METHOD OF FAULT IDENTIFICATION IN MECHATRONIC SYSTEMS

Vladimir Filaretov, Alexey Zhirabok, Alexander Zuev, Alexander Protsenko

This Publication has to be referred as: Filaretov, V[ladimir]; Zhirabok, A[lexey]; Zuev, A[lexander] & Protsenko, A[lexander] (2016). Method of Fault Identification in Mechatronic Systems, Proceedings of the 27th DAAAM International Symposium, pp.0312-0318, B. Katalinic (Ed.), Published by DAAAM International, ISBN 978-3-902734-08-2, ISSN 1726-9679, Vienna, Austria
DOI: 10.2507/27th.daaam.proceedings.046

Abstract

This paper considers a method of synthesis of identification system for faults appearing in the various elements of complex mechatronic systems described by nonlinear differential equations. To solve this problem logic-dynamic approach and special feedback by residual signal is suggested. This approach consists of three main steps: replacing the initial nonlinear system by certain linear logic-dynamic system, obtaining the bank of linear logic-dynamic observers, and transforming these observes into the nonlinear ones. Logic-dynamic approach allows detecting and isolating faults in nonlinear dynamic systems by using the linear methods for diagnosis in nonlinear mechatronic systems. Feedback of special form introduced in the observers after the detection and isolation of faults. It is allowing to determine the value of faults in nonlinear systems with incomplete measurability of state vector. Efficiency and high performance of the proposed method is shown on example of identification of faults arising in electric drives of multilink manipulator.

Keywords: Fault identification; diagnostic; observers; mechatronic systems; nonlinear systems.

1. Introduction

Operative detection and localization of faults that may occur in elements and subsystems in the process of operation it is one of the perspective ways to improve the efficiency of operation of mechatronic systems(MS) in the event of minor defects in the sensor elements and actuators. Timely detection, localization and identification of the values of these faults allow the use of methods of fault-tolerant control and accommodation for the formation of specific control signals that allow to keep the most important characteristics of the system in the presence of faults in it providing guaranteed performance of specified operations and missions.

Currently, there are several different approaches to the synthesis of the system of identification of faults [1-9]. All these approaches are based on the procedures for the real-time identification and localization of faults via diagnostic observers (DO). Several methods for synthesis of diagnostic observers for nonlinear dynamical systems are known, including those based on the differential geometric approach [5], the algebra of functions [5], logic-dynamic approach [1-3, 9] and others [4, 6 - 8]. The first two allow obtaining the optimal solution with a minimum dimension of observer, however, have a rather complicated process of finding a solution, because of complex analytical calculations. Logic-dynamic approach does not guarantee a minimum dimension, but the process of finding a solution on its basis is quite simple because of using of linear methods. The task of identifying the values of the faults on the basis of logic-dynamic approach has been solved in [2] and [3], by introducing into the diagnostic observers the signal of feedback by the residual. However, this approach is valid only for the case when the state vector of the diagnosing object fully measurable and diagnostic observers has the first order.

This paper considers solution to the problem of developing a method of synthesis of precise identification system for faults in a variety of complex MS, which use the DO, described by differential equations with the order above the first. In method a special feedback by the signal of residual is introduced in the diagnostic observers obtained with the logic-dynamic approach for real-time identification of values of faults.

2.Model of diagnosed object

In general, non-linear multidimensional diagnosed object (DO) can be described by a model of the form:

$$\dot{x}(t) = Fx(t) + B(x(t), u(t)) + Gu(t) + Ld(t),$$
$$y(t) = Hx(t), \tag{1}$$

where $x \in R^n$ is state vector, n is dimensionality of DO; $y \in R^m$ is output vector; $u \in R^l$ is input vector, p-dimensional vector $d(t)$ describes the errors occurring in the system because of the appearance of faults in it (in the absence of faults all the elements of the vector are zero; in the presence of a fault corresponding element of the vector becomes unknown function of time); $F \in R^{n\times n}$ is matrix of system dynamics; $G \in R^{n\times l}$ is matrix that determines the effect of the input to the system; $H \in R^{m\times n}$ is matrix that determines the relationship of output with the state vector; $B(x(t),u(t))$ is a vector which determines a non-linear part of the system. In this paper a case where each of elements of vector $d(t)$ included only in one of the equations of (1) is considered, which means that there is only one nonzero element in each of the columns of matrix $L \in R^{n\times p}$. The task of detecting faults and determining their values are invited to decide on the basis of logic-dynamic approach (LDA) [1], which allows the construction of diagnostic observers using only linear methods. After synthesis of the diagnostic observers, the special feedback must be introduced in them to ensure stability and to determine the value of the faults occurring in the system.

The developed method provides a set of nonlinear diagnostic observers, each of which is sensitive to one of the faults and invariant to the others. The output signals of observers are the values of the corresponding faults.

As a result of the well-known procedure of synthesis of diagnostic observers based on the LDA [1] k-dimensional observer sensitive to one of the faults $d_i(t)$, $i = \overline{1, p}$, can be built in the following form:

$$\dot{x}^*(t) = F^*x^*(t) + B^*(x^*(t), u(t)) + G^*u(t) + Jy(t),$$
$$y^*(t) = H^*x^*(t), \tag{2}$$

where $x \in R^k$ is state vector of observer; y^* is output signal of observer; $F^* \in R^{k\times k}$, $G^* \in R^{k\times l}$, $H^* \in R^{1\times n}$ are matrixes corresponding to matrices of system (1), $J \in R^{k\times m}$ is matrix, which defines the use of DO output in observer. All these matrices must be determined during the synthesis of the observer. It is known [1] that F^* и H^* can be chosen as:

$$F^* = \begin{bmatrix} 0 & 1 & 0 & \dots & 0 \\ 0 & 0 & 1 & \dots & 0 \\ \dots & \dots & \dots & \dots & \dots \\ 0 & 0 & 0 & \dots & 0 \end{bmatrix}, \; H^* = \begin{bmatrix} 1 & 0 & 0 & \dots & 0 \end{bmatrix}. \tag{3}$$

On the basis of the outputs of DO and observer following residual is generated:

$$r(t) = \mathrm{C}y(t) - \mathrm{y}^*(t), \tag{4}$$

where C is m-dimensional vector, defining use of DO output in residual generation.

The state vectors of DO and observer are associated by the matrix $\Phi \in R^{k\times n}$, and, in the absence of a mismatch between the DO and the observer (if $r = 0$), satisfy the following equations[1]:

$$x^* = \Phi x, \; G^* = \Phi G, \; CH = H^*\Phi, \; \Phi F = F^*\Phi + JH. \tag{5}$$

In case of any fault the residual (4) becomes different from zero, and the equations (5) is no longer fulfilled. To solve the problem of identification of faults, consider how residual will change in the event of faults. After differentiating the expression (4) taking into account (1) and (2) it can be obtained:

$$\dot{r} = CH(Fx + Gu + B(x,u) + Ld) - H^*(F^*x^* + G^*u + B^*(x^*,u) + Jy).$$

In view of this and (5) it can be obtained:

$$\dot{r} = H^*(F^*(\Phi x - x^*) + \Phi B(x,u) - B^*(x^*,u) + \Phi Ld).$$

Denote the value of $\Phi x - x^*$ as a vector $e \in R^k$ of mismatch of states of DO and the observer: $e = \Phi x - x^*$. Considering introduced vector, it can be obtained:

$$\dot{r} = H^*(F^*e + \Phi B(x,u) - B^*(x^*,u) + \Phi Ld)\,. \tag{6}$$

Consider the relationship of the vector of mismatch e with the residual r, in view of (4):

$$r = Cy - y^* = CHx - H^*x^* = H^*\Phi x - H^*x^* = H^*e. \tag{7}$$

Thus, taking into account the form of vector H^*, the residual r is the first element of mismatch vector e:

$$r = e_1.$$

Next, consider the impact of a fault not only on residual, but on the entire vector e. Then, taking into account (5) and (7) it can be obtained:

$$\dot{e} = F^*e + \Phi B(x(t),u(t)) - B^*(x^*(t),u(t)) + \Phi Ld. \tag{8}$$

From (8) it follows that the change of the mismatch vector depends on the element $\Phi B(x(t),u(t)) - B^*(x^*(t),u(t))$, which characterizes the non-linearity of DO and observer. In the case of incomplete measurability of DO, state vector x and the value of the element $\Phi B(\mathrm{x}(t),\mathrm{u}(t))$ can be unknown. It is making difficult identifying of value of faults. In LDA observer constructed in way to compensate the effect of non-linearity in the residual signal and $\Phi B - B^* = 0$. However, this condition is fulfilled only when there is no mismatch between the state vectors of DO and observer and $e = 0$. To determine the value of fault d_i, which we are interested, it is necessary to introduce feedback by residual signal formed in such way to provide mismatch elimination even in the case of faults.

3. Introduction of feedback

After the introduction of feedback by residual signal in the observer, its model assumes the form:

$$\begin{aligned} \dot{x}^*(t) &= F^*x^*(t) + B^*(x^*(t),u(t)) + G^*u(t) + w(r) + Jy(t), \\ y^*(t) &= H^*x^*(t), \end{aligned} \tag{9}$$

where $w(r) \in R^k$ is a vector specifying feedback by residual signal.
Taking into account the introduced feedback, change of the mismatch vector will be determined by the equation:

$$\dot{e} = F^*e + \Phi B(x,u) - B^*(x^*,u) + \Phi Ld - w(r)\,. \tag{10}$$

If formed feedback ensures that the equalities (4) and $B^*(x(t),u(t)) = \Phi B(x^*(t),u(t))$ are performed, nonlinearity in the process of identification of faults will be compensated.

To ensure the sensitivity of observer to the fault d_i and the invariance of the rest, matrix Φ is constructed so that all the columns of matrix ΦL except ith were zero, so observer will be affected only by one element of fault vector d. Next, consider common case, when the equation (11) can be represented as:

$$\begin{aligned}
&\dot{e}_1 = e_2 - w_1(e_1),\\
&\dot{e}_2 = e_3 - w_2(e_1),\\
&\dot{e}_3 = e_4 - w_3(e_1),\\
&\ldots\\
&\dot{e}_v = e_{v+1} + f_v d_i + a_v - w_v(e_1),\\
&\ldots\\
&\dot{e}_k = -w_k(e_1).
\end{aligned} \tag{11}$$

where $a = \Phi B(x(t), u(t)) - B^*(x^*(t), u(t))$, f_v is element of the ith column of the matrix ΦL.

In this case, the fault and uncompensated nonlinear component a_v are present in only one of the equations (11) In the case of $w_{v+1}(e_1) = 0, \ldots, w_k(e_1) = 0$ and the consistency of the initial conditions of DO and observer, $e_{v+1}, \ldots, e_k$ are zero regardless of the occurrence of the fault. Then further, it is possible to consider only a part of the system of equations (11):

$$\begin{aligned}
&\dot{e}_1 = e_2 - w_1(e_1),\\
&\dot{e}_2 = e_3 - w_2(e_1),\\
&\dot{e}_3 = e_4 - w_3(e_1),\\
&\ldots\\
&\dot{e}_v = f_v d_i + a_v - w_v(e_1).
\end{aligned} \tag{12}$$

Writing the system of equations (13) in the form of a differential equation we get:

$$f_v d_i + a_v = \dot{e}_v + w_v = \ddot{e}_{v-1} + \dot{w}_{v-1} + w_v = \dddot{e}_{v-2} + \ddot{w}_{v-2} + \dot{w}_{v-1} + w_v = e_1^{(v)} + w_1^{(v-1)} + \ldots + w^{(1)}{}_{v-1} + w_v.$$

Accepting elements of the vector of feedback, except the vth, proportional to the residual $w_j = \frac{T_i}{T_0} e_1$ $(j = \overline{1, v-1})$ and $w_v = \frac{T_v}{T_0} e_1 + \frac{1}{T_0} z(e_1)$, it can be obtained:

$$f_v d_i + a_v = e_1^{(v)} + \frac{T_1}{T_0} e_1^{(v-1)} + \ldots + \frac{T_{v-1}}{T_0} e_1^{(1)} + \frac{T_v}{T_0} e_1 + \frac{1}{T_0} z(e_1). \tag{13}$$

If value of $f_v d_i + a_v$ presented as input signal, and e_1 as output, equation (14) can be represented as a transfer function:

$$W = \frac{T_0}{T_0 s^v + \ldots + T_v + Z(s)}.$$

Suppose that input $f_v d_i + a_v$ can be represented as a polynomial of degree q: $f_v d_i + a_v = b_0 + b_1 t + \ldots + b_q t^q$. The Laplace transform of such input is $Q = \frac{b_0}{s} + \frac{b_1}{s^2} + \ldots + \frac{b_q}{s^{q+1}}$.

Then the steady value of e_1 can be found as: $e_1 = \lim\limits_{s \to 0} W Q s = \lim\limits_{s \to 0} \frac{T_0 (b_0 s^{q-1} + \ldots + b_q) s}{(T_0 s^v + \ldots + T_v) s^{q+1} + Z(s) s^{q+1}}$.

Assuming $Z(s) = \frac{T_{v+1}}{s} + \frac{T_{v+2}}{s^2} + \ldots + \frac{T_{v+q}}{s^q} + \frac{T_{v+q+1}}{s^{q+1}}$ we get: $e_1 = \lim\limits_{s \to 0} \frac{T_0 (b_0 s^{q-1} + \ldots + b_q) s}{(T_0 s^v + \ldots + T_v) s^{q+1} + T_{v+1} s^q + \ldots + T_{v+q+1}} = 0.$

Thus introducing a feedback of indicated type result in a steady value of $e_1 = 0$. This also results in resetting of all elements of vector e. For many real systems value of the fault is constant or slowly changing and it is sufficient to introduce one integral of residual in feedback to achieve the required accuracy of the identification of faults:

$$z(e_1) = \int e_1 dt. \tag{14}$$

Thus, after the introduction of this kind of feedback, residual e_1 will tend to zero, after the completion of the transition process, including in the event of faults. This will ensure the synchronization of state vectors of DO and observer.

The next step of solving the problem is estimation of value of the desired fault. Since the introduction of the feedback was able to arrange that $e = 0$, taking into account (12) it can be written:

$$d_i = \frac{w_v(e_1)}{f_v} \quad \text{or} \quad d_i = \frac{1}{f_v T_0} \int e_1 dt. \tag{15}$$

Thus the output of the system of identification of faults after the of transient processes will be sought value d_i. The required system performance can be achieved by selection of $T_0 ... T_v$ coefficients.

4. Example

Let us to consider multilink manipulator(MM) with n degrees of freedom which is driven by n actuators with DC motors. Each actuator has sensors of the position, speed and current. It is assumed that the following types of faults are possible in the actuators of MM: d_1– fault caused by an increase of moment of viscous friction in the link of manipulator; d_2– fault caused by an increase of moment of viscous friction in the actuator; d_3 – fault caused by the change of active resistance of rotor circuit of DC motors (for example, in case of significant changing of temperature). The presence of these faults significantly reduces the quality of the actuators and accuracy of performance of specific technological operations.

In the presence of these faults, each actuator of MM without feedback can be described by the nonlinear differential equations in form (1) where:

$$F = \begin{bmatrix} 0 & 1 & 0 & 0 & 0 \\ -ci^2/H & -(k_{vn}+h)/H & ci/H & 0 & 0 \\ 0 & 0 & 0 & 1 & 0 \\ ci/J & 0 & -c/J & -k_{vd}/J & k_m/J \\ 0 & 0 & 0 & -k_\omega/L & -R/L \end{bmatrix}, B = \begin{bmatrix} 0 \\ (-M_E - M_{stn0}sign(x_2))/H \\ 0 \\ -M_{std0}sign(x_4))/J \\ 0 \end{bmatrix}, x = \begin{bmatrix} x_1 \\ x_2 \\ x_3 \\ x_4 \\ x_5 \end{bmatrix}, G = \begin{bmatrix} 0 \\ 0 \\ 0 \\ 0 \\ k_u/L \end{bmatrix},$$

$$L = \begin{bmatrix} 0 & 0 & 0 \\ 1 & 0 & 0 \\ 0 & 0 & 0 \\ 0 & 1 & 0 \\ 0 & 0 & 1 \end{bmatrix}, H = \begin{bmatrix} 1 & 0 & 0 & 0 & 0 \\ 0 & 0 & 0 & 1 & 0 \\ 0 & 0 & 0 & 0 & 1 \end{bmatrix} d = \begin{bmatrix} d_1 \\ d_2 \\ d_3 \end{bmatrix},$$

where R, L, I are resistance, inductance and current of DC motor rotor circuits, accordingly; k_ω is a counter-EMF coefficient; k_u is an amplifier coefficients; k_m is a moment coefficient; J is a moment of inertia of DC motor shaft and rotating parts of reducers; c is rigidity of mechanical transmission; x_1 is a turn angle of link; i is a reducing ratio of the reducer; h, H, M_E are the variable components of torque influences on actuators, including the interactions between all the degrees of freedom of the manipulator; u is a input voltage of amplifiers.

In this case, the dimension of the output vector of the system less than dimension of state vector Thus, only the value of x_1, x_4 and x_5 are measured, and x_2 and x_3- no.

Next, consider the synthesis of the identification system for fault d_1. Diagnostic observer of the third order, which is sensitive to this fault and insensitive to other faults, will be described by system of equations (10) and the following matrix:

$$F^* = \begin{bmatrix} 0 & 1 & 0 \\ 0 & 0 & 1 \\ 0 & 0 & 0 \end{bmatrix}, G = \begin{bmatrix} 0 \\ 0 \\ 0 \end{bmatrix}, J = \begin{bmatrix} -(k_{vn}+h)/H & 0 & 0 \\ -ci^2/H & 0 & 0 \\ 0 & ci/H & 0 \end{bmatrix}, C = \begin{bmatrix} 1 & 0 & 0 \end{bmatrix}, B = \begin{bmatrix} 0 \\ -\dfrac{M_E + M_{stn0}sign(x_2)}{H} \\ 0 \end{bmatrix}.$$

Since $\Phi L = \begin{bmatrix} 0 & 0 & 0 \\ 1 & 0 & 0 \\ 0 & 0 & 0 \end{bmatrix}$, the fault d_1 is affecting only second equation of observer, and feedback vector have the form: $w = \begin{bmatrix} w_1 & w_2 & 0 \end{bmatrix}^T$.

Since the value of x_2, a member of the vector B is not measured by sensors, it must be defined from (4):

$$x_2 = x_2^* - \frac{k_{vn} + h}{H} x_1^* .$$

In the synthesis of the third order observer is advisable to select feedback coefficients as follows:

$$T_0 = (6T_c)^3, \ T_1 = 3(6T_c)^2, \ T_2 = 18T_c \ ,$$

where T_c is the time during which the output of the system enters the 5% area of the value of the desired fault. Accepting $T_c = 1s$ we get:

$$T_0 = 216, \ T_1 = 108, \ T_2 = 18.$$

Feedback vector in this case will be:

$$w = \begin{bmatrix} \frac{T_1}{T_0} e_1 \\ \frac{T_2}{T_0} e_1 + \int\limits_0 \frac{1}{T_0} e_1 dt \\ 0 \end{bmatrix}.$$

To verify the functionality and effectiveness of the proposed method, the modelling of the synthesized faults accommodation system for the model of the actuator of the second degree of freedom (q_2) of PUMA-type manipulator (see Fig.1) was carried out.

At the synthesis of faults accommodation system were used the following parameters of actuator with DC motor: J=0.0001 kg·m·s^2, k_ω=0.02 V·s, k_u=100, R=0.4Ω, L=0.004Hr, k_m=0.02 N·m/A, i=100, c=2Nm/rad, k_{vn}=10^{-2} N·m·s/rad, kvd=10^{-5} N·m·s/rad, M_{stn}=0.06 N·m, M_{std}=0.0006 N·m.

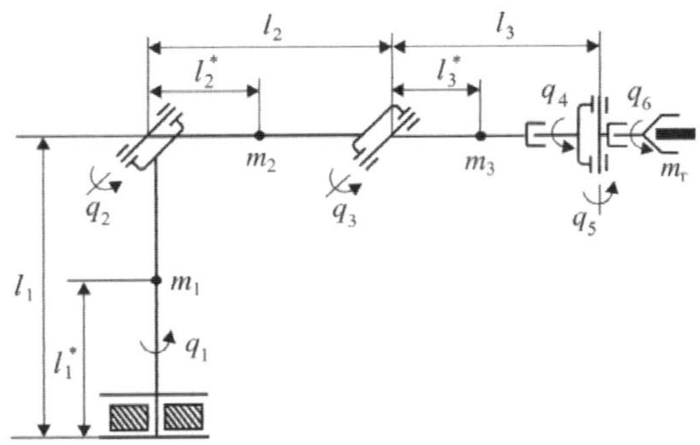

Fig. 1. Kinematical scheme of the manipulator

For the second degree of freedom mutual influence effects in the model are described by the following equations [10]:

$$H_2 = J_{n2} + J_{n3} + m_2 l_2^{*2} + m_3 l_3^{*2} + (m_3 + m_g) l_2^2 + m_g l_3^2 + 2 l_2 (m_3 l_3^* + m_g l_3) \cos q_3 ,$$

$$h_2 = -2 l_2 (m_3 l_3^* + m_g l_3) \dot{q}_3 \sin q_3 ,$$

$$M_{E2} = [J_{n3} + m_3 l_3^{*2} + m_g l_3^2 + (m_3 l_3^* + m_g l_3) l_2 \cos q_3] \ddot{q}_3 - l_2 (m_3 l_3^* + m_g l_3) \dot{q}_3^2 \sin q_3 + g[m_2 l_2^* + (m_3 + m_g) l_2] \sin q_2 +$$

$$+ g(m_3 l_3^* + m_g l_3) \sin(q_2 + q_3) - \frac{1}{2}[(J_{n2} - J_{s2} + m_2 l_2^{*2} + (m_3 + m_g) l_2^2) \sin 2q_2 + (J_{n3} - J_{s3} + m_3 l_3^{*2} + m_g l_3^2) \sin 2(q_2 + q_3) +$$

$$+ 2 l_2 (m_3 l_3^* + m_g l_3) \sin(2q_2 + q_3)] \dot{q}_1^2 ,$$

where J_{si} is the moment of inertia of link i about its longitudinal axis; J_{ni} is the moment of inertia of link i of a transverse axis passing through its centre of mass; g is acceleration of gravity.

In modelling the following values of parameters of MM's links were used: m_1=35.8kg, m_2=3.5kg, m_3=3kg, l_1=0.2m, l_2=0.5m, l_3=0.4m, l^*_1=0.1m, l^*_2=0.12m, l^*_3=0.1m, J_{s2}=0.0047kg·m^2, J_{n2}=0.12kg·m^2, J_{n2}=0.067kg·m^2, m_g=0.5kg.

Faults were simulated by introducing error signals d_1=0.01 from time t_1=10 s. to t_3=20 s. and d_2=d_3=0.1, from t_2=15 s. Output of observer is shown at Fig. 2.

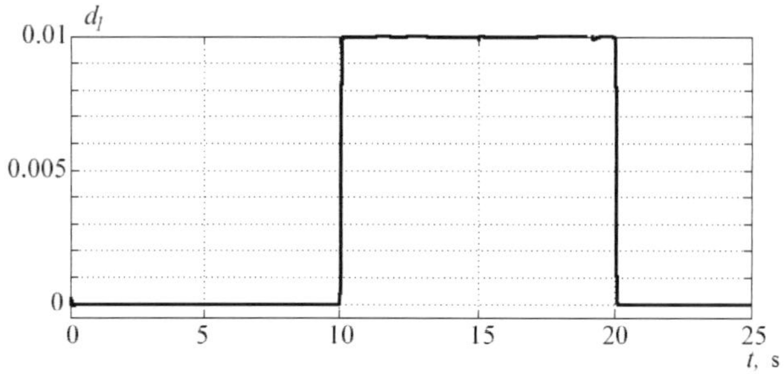

Fig. 2. Output of observer at constant fault value.

As can be seen the proposed system accurately determines the presence or absence of a fault d_1 and its value. In addition, as required, faults d_2 and d_3 does not affect the determination of d_1.

Conclusion

In this paper method of synthesis of accurate identification systems for occurring faults in a variety of complex MS, which use the DO described by differential equations with the order above the first was developed. This method consists of applying of logic-dynamic approach for synthesis of diagnostic observers, guaranteeing the independence of the detection and localization of possible faults and introducing of special feedback for diagnostic observers, allowing identifying values of faults. Efficiency of the proposed method of synthesis of diagnostic observers for identification of faults in nonlinear dynamic systems was confirmed by the results of mathematical modelling. The subject of further research will be application of the proposed method for synthesis of the fault accommodation and failsafe control systems for such objects as industrial manipulators and underwater vehicles.

Acknowledgments

This research was supported by Ministry of Education and Science of Russia (Agreement 14.613.21.0018 from 22.10.2014)

References

[1] Zhirabok, A. & Usoltsev, S. (2001). Linear methods for fault diagnosis in nonlinear systems, Proc. European Control Conf. ECC'01, Porto, Portugal, pp. 1363-1366.

[2] Filaretov, V.F.; Zirabok, A.N.; Zuev A.V. & Protsenko, A.A. (2012). The Development of the Faults Accommodation System for Actuators of Multilink Manipulator, Proc. of the 23rd International DAAAM Symposium, Zadar, Croatia, pp.575-578.

[3] Filaretov, V.F.; Zhirabok, A.N.; Zuev, A.V. & Procenko, A.A. (2013). Synthesis Method of Fault Tolerant Control System for Manipulators, Journal of Advanced Material Researches. Vol. 717. pp. 551-556.

[4] Isermann, R. (2006). Fault-Diagnosis Systems: An Introduction from Fault Detection to Fault Tolerance, Springer-Verlag: Berlin, pp. 475.

[5] Shumsky, A.Ye. & Zhirabok, A.N. (2006). Nonlinear diagnostic filter design: Algebraic and geo-metric points of view, Int. J. Applied Mathematics and Computer Science, vol. 16, no . 1, pp. 115-127.

[6] Frank, P. (1990). Fault diagnosis in dynamic systems using analytical and knowledge-based redundancy – A survey and some new results, Automatica, vol. 26, pp. 459-474.

[7] Staroswiecki, M.; Yang, H. & Jiang, B. (2006). Progressive accommodation of aircraft actuator faults, Proc. IFAC Symp. Safeprocess'2006, Beijing. China, pp. 877–882.

[8] Siqueira, A. & Terra, A. (2004). Fault Tolerant Manipulator Robot Based on H2, H∞ and Mixed H2/H∞ Markovian Controls, Proc. of the IEEE International Conference on Control Applications, Taipei, Taiwan, pp. 309-314.

[9] Filaretov, V.; Zhirabok, A. & Kucher, D. (2008). New approach to robust observer design, Proc. 19-th Intern. DAAAM Symposium, Slovakia, pp.491-492.

[10] Filaretov ,V.F. (2007). Synthesis of Adaptive Control Systems for Electric Servo Actuators of Manipulators, Proc. of the 18th DAAAM Int. Symp. «Intelligent Manufacturing & Automation», Zadar, Croatia, pp. 277- 279.

DOI: 10.2507/27th.daaam.proceedings.047

DEVELOPMENT OF ERP AND OTHER LARGE BUSINESS SYSTEMS IN THE CONTEXT OF NEW TRENDS AND TECHNOLOGIES

Sabrina Suman & Ivan Pogarcic

This Publication has to be referred as: Suman, S[abrina] & Pogarcic, I[van] (2016). Development of ERP and Other Large Business Systems in the Context of New Trends and Technologies, Proceedings of the 27th DAAAM International Symposium, pp.0319-0327, B. Katalinic (Ed.), Published by DAAAM International, ISBN 978-3-902734-08-2, ISSN 1726-9679, Vienna, Austria
DOI: 10.2507/27th.daaam.proceedings.047

Abstract

This paper presents an overview of terms, concepts, trends and technologies that are relevant to today's business. It describes the basics of data and information integration and flow in a company through a central ERP system with concepts of CRM and SCM. The emergence of *big data* as a tributary of a huge number of often unstructured data from different sources can become a central problem or opportunity for advancement and achievement of competitive advantages of a company. Ignorance of key figures and/or the non-acceptance of new business conditions, new technologies and possible deployment solutions are the main reasons for non-productivity and poor business performance. To demonstrate the dynamics of appearance and popularity of terms, concepts, trends and technologies this paper offers a tabular overview of the frequencies based on the data from 3 global databases. Meta analysis shows the expected future development of analytical trends and technologies. This paper is intended for those who lead, run and participate in projects of implementation of large software systems, dealing with quality management of business, or want to understand the complexity of this area and the future directions of development.

Keywords: ERP; BI; DSS; SCM; CRM; big data.

1. Introduction

Getting specific information for specific job-situation-task, means that data from different sources must continuously be collected, stored and analyzed. Care and use of so much data is impossible without a well-structured database and/or data warehouse, and the analysis of this data is impossible without a series of BI (Business Intelligence) tools. Information therefore needs to be within easy reach and analyzed in order to give effect to the business objectives [1], and information systems should be changed, integrated and introduced precisely in order to realize this goal. [2]

Companies whose executives understand data as a core asset and resources of business are six times more likely to be successful than those whose top management does not perceive data in this way (31% vs. 5%) and have a lot more success with BI projects. [3].

Technological progress and the information needs of companies today create a continuous cycle of growing needs for ever more effective ways of coverage and analysis of data in and around the company. Currently, the society is faced with a growing, large and various sources of data, which is often referred to as big data, and is used for decision support. *Big data* are complex, layered, large amounts of data and finding i.e. getting the right information in such a large amount of data is like searching for a needle in a haystack, and the greater the amount of data it is more difficult to find the true value of the data. [4].

Many observers, including [5], [6], [7] and [8] argue that the potential of using big data to improve the personal life or to help companies to compete is unlimited. According to [6], "better access to information and technology for the management and analysis of data is changing the world." In a way, big data will lead to better health, better teachers, improved education, and better decision-making. Gartner hype curve predicts that the big data for two to five years [9] will do much to transform and influence the business.

The goal is not the production and use of vast amounts of data, but the way the data is being analyzed and presented, which techniques are being used to get valuable information and the right support and a base for decision making, and in reality it is about ***big analytics*** not just about ***big data*** [10] [11]. [12] argues that the management and analysis of these data is precisely the most difficult part in the field of big data. Availability and timeliness of the results of quality analysis in the moments when you have to make decisions is a major challenge. Such phenomena of large amounts of data, and the need for high-quality analysis also require new expertise and experts in many organizations [13].

After describing the perceived problems caused by the appearance of massive amounts of data from various sources, we can see the importance of choosing and using various software tools, technologies and techniques and the efficient management of all data concerning the business seems like a prerequisite for the survival of the various activities of the company.

In consideration of the successful management of a company interaction of large-scale integrated enterprise systems such as ERP, CRM, and SCM with the DSS and BIS is also important, where analytics of business intelligence based on integrated data of the entire company can directly affect a more quality, timely and flexible decision support, and decisions will again directly affect the effectiveness of CRM and SCM. [14].

Setting an ERP for the origin of this systematic review is also reflected in the main target for the implementation of an ERP - access to data in a controlled manner at the level of the entire company and information sharing across business processes [15]. [16] suggest that a suitable choice of an ERP system provides results such as increased productivity, punctual delivery, decreased implementation time, and reduced price of a product while a bad choice of an ERP system leads to project failure or performance degradation of a company.

Furthermore, a review links together concepts of BI, DSS - Decision Support Systems, SCM - Supply Chain Management, CRM - Customer Relationship Management, BPM - Business Performance Management with the involvement of the big data sphere. Also shown is the dynamic of representation of concepts from the paper through publications of 3 global scientific-technical databases from 1991-2016 in phases of 5 years. To view the future development of new trends and technologies related to retrieving, storing and analyzing information, generated was a graphical representation obtained by meta analysis of research [17].

Objective of the paper is to review and synthesize the inter-related areas that need to be considered in today's business.

Purpose of the work is to indicate the complexity of the successful implementation of maintenance and future development of ERP and other business systems through a systematic review of interrelationships of large systems, new technologies and trends.

2. Overview of connection between the ERP and business-critical systems

This section will connect the following terms and concepts: ERP (Enterprise Resource Planning), DSS (Decision support system), BI (Business Intelligence), CRM (Customer Relationship Managemenet), SCM-(Supply Chain Managemenet), BPM (Business performance Management), DW (Data Warehouse, Data Warehousing), KM (Knowledge management) and MEC (Multi Enterprise Collaboration). The objectives of this chapter are:

- To stress the importance, function, purpose and potential benefits of ERP systems,
- To point out the known problems and bad practices observed in the planning and/or after the implementation of ERP in the company,
- To include the actuality of today's business conditions, which requires inter-organizational collaboration - MEC (Multi-Enterprise Collaboration) in decision-making and technological implications and possible solutions to such business.
- To show links of the ERP systems with other large systems, the importance of connection between these systems and the consequent implications of these connections on the criteria in the selection of software modules and ways of implementation
- List some guidance on the selection of software packages related to big business systems and the type of BI analytics

2.1. ERP

ERP software automates and integrates business processes and enables the sharing of information and data in a variety of business functions. ERP software enhances the functionality and efficiency of business processes that take place in the departments of finance, human resources, operations and logistics, and sales and markets [18]. Since it affects the entire business, or at least a large part of the business, and its purchase and introduction is a large organizational cost, selection of an ERP system is a particularly complex and delicate task. It is estimated that 40-70% of ERP implementations have experienced some type of failure [19]. Determination of the potential benefits of ERP implementation is a challenging task, because most of the advantages do not come from the ERP system, but from different ways in which the system can be implemented and used. Although this is true for any type of information system (IS), it is a special burning question for the ERP systems, because of their decisive influence on almost all aspects of the organization. This is even more important in the context of ERP implementation through multi-organization, a subject of work by Eckartz et al. [20].

Companies that decide to introduce the ERP systems are often inspired by imitation of successful practices in similar companies. Such a conviction management board can lead to so-called homogenization of companies which weakens the differentiation, i.e. the uniqueness or specificity by which a company and its products are recognized and stand out in the market. The loss of differentiation and uniqueness can weaken the company's potential and performance that delivers business value. [21] emphasizes the importance of detailed and elaborate evaluation protocols and practices to choose the solution that best suits the company instead of imitating the competitors. This argues the need to inform about all the influential factors of the optimal selection of ERP solutions and overall quality management of performance and operations of the company.

Complexity of approach in the selection of an ERP and the number of factors that affect the success of the implementation was discussed gradually in the further part of the review where the function, purpose, benefits and success of ERP are viewed in relation to the relevant terms of the wider area of managing the performance of a company.

2.2. *ERP and decision making support*

Davenport [22] suggests that the main reason and the greatest potential of ERP is making quality and timely decisions, and Palaniswamy and Frank [23] concluded that the ERP is a prerequisite for decision making support. Part of the research on the effects of ERP to support the decision making proved that ERP systems offer significant benefits in the area of decision making support [24].

Although ERP systems integrate knowledge and provide reporting tools for users to analyze data, decision making support is not their primary purpose. This is supported by a multitude of BI software solutions which the company uses to implement a decision support system, which can not be fully developed only with an ERP solution. Therefore, it can be said that the quality of implementation of ERP systems is closely linked to the quality of the decision making system [25].

2.3. *ERP and Business Intelligence*

Business intelligence is a huge opportunity for any company to collect valuable insights from all the data covered by their ERP and those of other systems. Li [26], referring to the need for the ERP systems to link internal and external data, identifies the need to create an effective business intelligence as the primary objective of an ERP system.

BI software enables companies' quality and quick decision-making because of the availability of data and information in an easily customizable form. The standard functionality of the BI software allows DSS to use the data from the data warehouse, shaped to measure. This warehouse provides management with BI analytical tools, the ability to create ad hoc reports, graphics, tables, use graphical dashboards that offer information through financial statements, scorecards and KPIs (key performance indicators).

BI software is also available as a stand-alone package and as modules in ERP solutions. In recent years, ERP vendors have included BI products and sell them in their systems and have thus made way for potentially highly effective solutions that put on the business-user the new burdens of selection, costs of consultants, education etc. In addition to the BI module, many ERP vendors include the very ERP functionality of managing client interactions and promotional campaigns, but such applications are sold as separate software, such as SAP CRM and Oracle's Siebel CRM applications that have additional management capabilities with client activities.

2.4. *ERP and multi-organizational collaboration*

Tools for decision making support and application should be open and flexible enough, not only to provide support within their own organization, but also across organizational boundaries. It stresses the need for research of multi-organizational decision-making and the development of IT which will support and optimize the production [27] and in other areas of business. These includes heterogeneity in terms of the breadth and depth of data, complexity in terms of models, algorithms for solution and data visualization / process, distribution to the scope and range, versatility of domains and paradigms, flexibility, ability to reuse and extensibility [28].

More recently, vendors, including Microsoft, SAP and Oracle have introduced tools and systems that support decision-making aspect of the MEC (MEC, multi-enterprise collaborative). The shift of the market and scientific research towards this field of research is still in its early stages and new perspectives and insights are constantly being acquired as new products are being released.

2.5. *ERP, data warehouse and knowledge management*

[29] point out that the knowledge of the employee is the most important resource of the company, and point out that although the repositories of knowledge in companies often exist, they are not organized in a way that they could be effectively used. The authors [30] observed the importance of integrating the processes of knowledge management (KM) within the Decision Support System so that the decision-makers can combine different types of knowledge (explicit and implicit) and data (internal and external) available in various forms. Improved creation and sharing of knowledge should increase the flexibility and innovation [31]. Literature dealing with ERP success suggests careful knowledge management in order to maximize its potential. Results of the Ifinedo research (Ifinedo, 2006) [32] have shown that knowledge management, including the creation of knowledge, is a strong predictor of ERP success. The more implicit and explicit knowledge the organization has, the more likely that the ERP will be successful [33].

Knowledge exists in the heads of corporate staff and management. Knowledge also exists in the environment outside of the company. The needs of business are also to formalize, store, access and use the entire knowledge of the company.

Data warehousing is not just a problem-solving process, but also a concrete architecture. Different companies have different data warehouses. According to the features of their ERP systems, many companies have designed the architecture of their data warehouses. Data warehouse is responsible for providing the information needed to support decision-making at different levels [34].

Existing data warehouse of a company can be extended to create a knowledge warehouse (KW) [35]. KW will not only facilitate the retrieval and creation of knowledge, but will also improve the retrieval and exchange of knowledge through the organization, providing at the decision-maker's disposal an intelligent platform for analysis. Knowledge warehouses will be continuously upgraded, with the aim of structuring knowledge from previous situations of decision-making and optimize future decisions. Shafiei et al., also mention the concept of a knowledge warehouse (KW) which is "above" the analysis of data from the ERP system via a data warehouse. [28]. [36] is also in the same vein, that with the advancement of technology expects that the existing business system will provide current intelligence information and not just the numbers, which can be used for making business decisions and strategies. This trend and the need are applicable today through various methods of artificial intelligence, especially in areas of text processing and natural language.

2.6. *ERP and CRM*

With the goal of providing "a single face to the customer," the basic principle behind CRM is that every employee in contact with the client must have access to information on the latest customer interactions with the company **[37]**. While CRM tools within an ERP, such as the SAP ERP system, when used properly, can help in the management of customer relationships, companies that opt for the concept of CRM often use a separate CRM system that communicates with the ERP system. The advantage of this approach is that the planning and analysis conducted in the CRM system does not interfere with the performance of the ERP system, which primarily handles large volumes of business transactions.

When compared with ERP, for CRM can be said that it, just like the ERP, offers ways to automate the process and conduct business more efficiently. However, these two systems are designed to streamline the various functions. While CRM is used to manage contacts, accounts, opportunities, activities, marketing, etc., ERP is designed to manage the operations and business functions, such as product planning, purchasing, inventory, customer service, order tracking and other back-end business processes. However, after the ERP vendors included the CRM features in their software, and CRM vendors included ERP capabilities into their tenders, the differences between them began to disappear.

2.7. *ERP and SCM*

The study [38] has empirically proven theoretical assumptions in the existing literature on the impact of ERP benefits on SCM competencies. A typical configuration of business software in the manufacturing enterprise includes at least three large systems ERP, SCM and CRM. All are constructed of one or more database management systems (DBMS) - which usually use the same logical integrated database. ERP, SCM and CRM systems are usually standard software that can adapt to the requirements of individual organizations. Today, these three types of systems tend to be integrated: SCM module, for example, will have access to information available in the ERP system directly or through a common database [39].

Supply Chain Management (SCM) is directed across boundaries, taking into account that companies are increasingly concentrating on their core competencies, leaving the other activities to the partners who have more knowledge. With the growing dependence on partners, effective supply chain has become equally important to the success of the company as the efficiency of internal business processes. Information systems that support supply chain management (SCM systems) are developed either by ERP vendors or software companies specializing in logistics. They have either expanded their ERP systems with additional SCM functions or developed new SCM systems that collaborate with their ERP systems.

Software companies are developing dedicated SCM systems and providing interfaces to standard ERP systems. The reason for this is that the SCM without ERP is hardly possible. An interesting trend is that some specialized SCM vendors were bought by large ERP vendors. In this way, ERP vendors are now able to offer supply chain management as part of their business portfolio. A typical company today uses a large number of information systems. These systems tend to be integrated, so that they can work together [39].

2.8. ERP and BPM

Over time, it turned out that, despite the fact that ERP systems hold most of the organization's data, they can provide insight into valuable information necessary for strategic decision making. ERP vendors are aware of this and offer BI functionality with their software to help "liberate" the data from these complex systems. At its core, BI is designed to consolidate data that extends across different business areas and systems, and help managers to make better decisions faster. In the new paradigm, the ERP system acts as an important source of data, but not necessarily the only source. This opens the door to choice to combine ERP with BI and/or Performance Management Suites and tools that meet the needs of the organization. This may or may not be from the same vendor.

In order to differentiate concise terms BI and BPM it is stated that BI discloses a technology used to access, analyze and inform about relevant business information. It covers a wide range of software solutions, including ad hoc queries, OLAP, dashboards, scorecards, today usually as modules for various BI suite. BPM is characterized as BI + strategic planning, which means that the convergence of BPM and BI planning is on a unified platform, cycle of plan-monitoring-analysis. Processes that include BPM are not new: they exist in every medium to a larger organization, and BPM provides a framework for the integration of processes, methodologies and metrics of other systems into a single solution. Term BPM refers to the processes, methodologies, metrics and technologies used by the company to scale, monitor and manage business performance. BPM is a continuous set of processes which, if properly carried out affects the entire organization from top to bottom. For BPM it is important to synchronize the entire company: it helps users to achieve goals that help the execution of the strategy and the adoption of value for all the stakeholders [40].

3. Emergence of new technologies and trends - *big data* and embedded analytics

The reasons for considering the development and potentials of ERP systems and the entire BPM system and big data are reflected in the new capabilities-functionalities of a broader business software system and exploiting new trends and technologies, as outlined below. A large amount of big data is unstructured, such as audio, video, text, and developed techniques of storage, processing and analysis of data types have evolved, there is a possibility that ERP is stored in and managed by a new system, the "knowledge hub" where the knowledge from inside and surrounding environment of the company will be stored and formalized integrally [41].

Taking into account the big data sphere allows analysis of data and transactions that are created outside the company. In this way, all stakeholders, customers, partners, suppliers are connected and the development of applications which will integrate data, models from different business perspectives is expected. Therefore, big data provides the ERP system with additional features and characteristics in relation to other systems, such as CRM and SCM, and thus adds value to the business.

Using data from big data sphere will also initiate some issues of privacy and data security and protection, and the movement of data will need to be constantly monitored and take into account the latest regulations. Also, the quality of the data in the ERP system can be disrupted after the integration of big data, so it is necessary to set some filters when downloading raw data.

These considerations relating to the question of the position of the ERP system and the expansion of the boundaries of the data relevant for the company, with the development of new technologies certainly point to new opportunities for additional functionality of ERP systems and consequently the entire BPM, whose potential can be achieved if new issues which may occur by using big data types are also considered. Problems and changes are mostly related to the processing, storage, analysis and quality of big data. Also, the integration of ERP systems and big data will require some changes in the activities of the stages in the life cycle of ERP systems as a change of requirements to be met by the ERP [41].

As the pace of business accelerates steadily, the company realizes that it is not enough just to analyze the data but the activities that are being imposed on the basis of the results of the data analysis must also be operationalized as soon as possible. This means that the aim is to exploit and share not only data, but also the results of the analysis, and exchange them through the business processes and applications, in order to shorten the time of decision making.

Operationalizing and embedding analytics consists of gaining insight on the necessary actions within the business processes that can be automated or provide support for decision-making. Analytics of different functionality and complexity are usually built into the dashboards, applications, devices, systems, databases and the like. Operational analytics are in the background of logistics applications, some CRM modules, authorizations and detection of faults, the system recommended content and products for each customer and the like. Embedded analytics, while not entirely a new technology today is especially being taken into account due to the occurrence of big data, and the possibility of operationalization of decisions through programs is very effectively. Also, sharing analytics through a variety of applications allows them to be received and potentially exploited by a large number of stakeholders, which is certainly an added value [17].

4. Research Methodology

In order to show the development of individual concepts through four five-year periods chosen were two science databases (Web of Science, Core Collection, Science Direct), and the database Google Scholar, which shows a large number of scientific and professional publications regardless of the quality to compare dynamics of incidence and popularity. The last columns for each database show the Tendency, which results in a number 1 and the up arrow if constant growth was recorded, and otherwise 0 and arrow down. Attached you can find Table 3 obtained by meta analysis of research (for research details see under [17] which shows the current representation of the use of data types, analytical platforms and types of analytics and prediction for the next three years based on the responses of 300 respondents). The table was sorted by three categories from top to bottom by the intensity of shades: the first category represents concepts that are anticipated to decline in use over the next three years, and the remaining two categories indicate concepts that will be moderately more and much more used in the next designated period.

5. Research results and discussion

Table 1 is the representation of individual frequencies. In the first two databases the decline is shown only by the concept of Business Process Reengineering, while in the Google Scholar database more than one category shows a decline in participation in publications. Each topic was searched with quotation marks.

Table 2 shows the representation of publications that at the same time mention all the concepts related to the operator and a base WoS Core Collection is this time omitted due to the low number of publications for the first three five-year periods. Table 2 shows that in the Science Direct and Google Scholar databases the representation of the publications in which the selected concepts appear together is constantly growing to this day. This also shows the connection between ERP and other large systems with selected concepts, but in the last five years there has also been a noticeable expansion of considerations related to big data and embedded analytics.

	Web of Science Core Collection - Search by Topic						Science Direct- Search by Title,abstract,keywords						Google Scholar- Search all excluded citations and patents					
Topic	1991-1995	1996-2000	2001-2005	2006-2010	2011-present	Tendency	1991-1995	1996-2000	2001-2005	2006-2010	2011-present	Tendency	1991-1995	1996-2000	2001-2005	2006-2010	2011-present	Tendency
Decision Support	1980	2857	3823	6505	11303	↑ 1	1022	1137	1456	2471	4545	↑ 1	18600	33200	92000	224000	154000	↓ 0
Enterprise Resource Planning	2	54	264	350	380	↑ 1	0	11	122	137	217	↑ 1	132	3160	14900	20400	17600	↓ 0
Business Performance Management	0	0	2	6	9	↑ 1	0	0	1	0	7	↑ 1	30	109	860	2410	3590	↑ 1
Business Process Management	6	31	91	172	309	↑ 1	0	7	32	65	186	↑ 1	266	1270	7310	17000	17700	↑ 1
Business Process Reengineering	73	159	90	60	54	↓ 0	25	104	57	29	44	↓ 0	1130	6350	9840	13100	14100	↑ 1
Supply Chain Management	22	243	1028	2337	3592	↑ 1	4	78	446	876	1458	↑ 1	1480	9340	26400	57900	57500	↓ 0
Customer Relationship Management	0	19	224	381	520	↑ 1	0	7	104	137	223	↑ 1	468	2840	16400	20100	26200	↑ 1
Business Intelligence	6	26	104	184	462	↑ 1	2	5	52	68	338	↑ 1	614	2150	12700	16600	23400	↑ 1
Data Warehouse Or Data Warehousing	7	247	503	524	682	↑ 1	32	481	1565	2176	3963	↑ 1	47	2460	5920	9250	10400	↑ 1
Data Mining	27	944	3924	5751	8962	↑ 1	10	190	962	1889	3772	↑ 1	4010	19300	169000	463000	238000	↓ 0
Text Mining	0	33	330	777	1687	↑ 1	0	2	66	233	644	↑ 1	63	827	8670	19300	20900	↑ 1
Web Mining	0	14	155	173	185	↑ 1	0	1	29	64	82	↑ 1	16	461	5850	12300	16300	↑ 1
Business Analytics	0	0	3	13	99	↑ 1	0	0	1	5	50	↑ 1	11	32	423	1590	10400	↑ 1
Knowledge Management	42	447	1783	2470	3111	↑ 1	28	149	541	648	1107	↑ 1	3820	16000	127000	205000	102000	↓ 0
Knowledge Discovery	48	441	880	897	1183	↑ 1	6	108	208	310	454	↑ 1	1430	10400	25800	50900	53600	↑ 1
Big Data	0	0	0	26	3946	↑ 1	0	0	0	4	1689	↑ 1	0	0	721	2060	32200	↑ 1

Table 1. The frequency of appearance of selected topics from 1991-2016

	Science Direct-Search by Title,abstract,keywords						Google Scholar- Search all excluded citations and patents					
Combination of topics	1991-1995	1996-2000	2001-2005	2006-2010	2011-present	Tendency	1991-1995	1996-2000	2001-2005	2006-2010	2011-2016	Tendency
"ERP" and "Decision support"	0	4	37	67	89	↑ 1	124	999	5690	10400	15300	↑ 1
"ERP" and "Business performance management"	-	-	-	-	-	-	2	9	171	639	806	↑ 1
"ERP" and "Business process management"	0	0	0	0	2	↑ 1	24	189	2330	6750	10200	↑ 1
"ERP" and "Business process reengineering"	0	0	8	9	52	↑ 1	14	544	2710	3870	4170	↑ 1
"ERP" and "Supply chain management"	0	3	45	49	63	↑ 1	95	1450	9780	15200	16800	↑ 1
„ERP“ and „Customer relationship management“	0	4	66	71	106	↑ 1	33	590	6130	10000	13200	↑ 1
"ERP" and "Business intelligence"	0	1	33	39	68	↑ 1	22	310	3220	6400	11300	↑ 1
"ERP" and "Business intelligence" and "Decision support" and "Supply chain management" and "Customer relationship management"	0	0	16	21	52	↑ 1	1	18	328	585	965	↑ 1
"ERP" and "Business performance management" and "Business intelligence" and "Decision support" and "Supply chain management" and "Customer relationship management"	0	0	0		1	↑ 1	0	0	10	65	86	↑ 1
"ERP" and "business process reengineering" and "business intelligence" and "decision support" and "supply chain management" and "customer relationship management"	-	-	-	-	-	-	0	7	70	110	140	↑ 1
„ERP“ and „big data“	-	-	-	-	-	-	0	70	129	226	5180	↑ 1
"ERP" + "embedded analytics"	0	0	0		13	↑ 1	0	0	6	34	50	↑ 1

Table 2. The frequency of simultaneous appearances of combination of topics from 1991-2016

The following text includes Table 3 obtained by meta analysis of research (for details see research in Halper, 2015) [17], sorted by categories from the top down: the first category represents concepts that are anticipated to decline in use over the next three years, and the remaining two categories indicate concepts that will be moderately more and much more used in the next designated period. This view of emerging trends and technologies is clearly essential for reflection and selection of a combination of business software suite, BI modules and types of analytics, platforms and ways of implementation in order to exploit the potential of each and the total potential of all of them in a given company.

Data Types Used for Operationalizing Analytics	Statistics of today's use	Will use for 3 years from now
Structured data(tales,records)	90%	8%
App logs	24%	36%
Real time messages	30%	46%
Web logs and clickstreams	21%	40%
Machine generated data	20%	39%
Data from graph databases	15%	32%
Social media data	22%	47%
Streaming, real time continous data	16%	46%
Unstructered data(audio video, text)	15%	48%
Analytics Used for Operationalizing and Embedding	Statistics of today's use	Will use for 3 years from now
Reporting	89%	9%
visualisation	69%	26%
Optimisation	28%	51%
Predictive analytics	28%	59%
Geospatial analytics	17%	37%
What-if simulations	22%	57%
Social media analytics	17%	46%
Prescriptive analytics	18%	51%
Text analytics	15%	45%

Platforms Used for Operationalized Analytics	Statistics of today's use	Will use within3 years
Data warehouse	74%	19%
Other databases or data marts	73%	14%
BI and analytics platforms	73%	24%
Transaction systems	59%	16%
Operational data stores	53%	23%
In-memory analytics platforms	31%	38%
Open source platforms	22%	25%
Data appliances	19%	32%
Public cloud platforms	19%	39%
Native mobile platforms	15%	29%
Hadoop	15%	39%
In-memory grids	13%	27%
Event stream processing engines/	12%	29%

Table 3. Data Types, Platforms and Analytics Used for Operationalizing Analytics

6. Conclusion

This paper presents an overview of the field with the aim of systematically linking important areas, approaches and philosophies whose understanding is necessary for successful implementation of ERP and/or SCM and/or CRM system and the establishment of effective systems for decision making support. During operation described are some issues related to the non-exploitation of potentials and non-integration of large systems, sub-optimal use of analytical tools, lack of new technological solutions, lack of understanding the phenomena big data and the like. The paper therefore proposes a set of guidelines and tips that to bypassed or eliminate these problems.

Chapter which systematically presents the concept of the affected area from selected literature has been upgraded in content around the central concept of the ERP systems. Gradually upgrading with other terms and concepts is done through their description and interpretation of the function and connecting them in relation to the ERP system. Each concept has been given a position, purpose and relationship with other areas in a continuous system of managing business performance. In doing so some guidance in terms of using new technology storage, retrieval and analysis of data has been given.

Also presented was the frequency of appearance which may also be used to monitor the development and character of processed concepts, and thus demonstrating the need to also consider their mutual relations and not just the concepts individually. Meta analysis has produced graphs that describe the anticipated development of certain technologies, types of data, analytical tools, and highlights those that are anticipated to decline in popularity and the ones that will be more and more used in the next few years. This concise view of future trends is also important in the assessment criteria for the selection of software, tools and types of implementation. This paper is intended to gain insight on the complexity of the field of business management and exploitation of the full potential of software packages that are offered today in the market in order to find the best combination for a given company. The further course of research would go into setting up a comprehensive framework which sets the basic structure and configuration of the business suites and after that to create a support system in the selection of such a configuration.

7. Acknowledgments

The research has been conducted under the project "Extending the information system development methodology with artificial intelligence methods" (reference number 13.13.1.2.01.) supported by the University of Rijeka (Croatia).

8. References

[1] Avram,C.D., ERP inside Large Organizations, Informatica Economică, vol. 14, no. 4,2010

[2] Gligora Marković, M.; Pogarčić, I.; Šuman, S.,Managements' role in development of Information system // Annals of DAAAM for 2012. & Proceedings of the 23rd International DAAAM Symposium / Katalinić, Branko (ur.).Vienna : DAAAM International, 2012. 737-742

[3] Eckerson,W.,W.,, Performance Dashboards, John Wiley & Sons, Inc. 2011

[4] Butler, B., Cloud Cronicles, 2015., http://www.networkworld.com/article/2973963/big-data-business-intelligence/5-problems-with-big-data.html , Accessed on: 05.01.2016

[5] Aziza, B., Big Data 'A-Ha' Moment? Forbes CIO Central , February 25, 2013 at URL http://www.forbes.com/sites/ciocentral/2013/02/25/big-data-a-ha-moment/, Accessed on: 12.5.2015

[6] Ehrenberg, R. (2012, January 19). What's the big deal about Big Data? InformationArbitrage.com blog post. At URL http://informationarbitrage.com/post/16121669634/whats-the-big-dealabout-big-data (12.5.2015)

[7] Morris, J. (2012, July 16), Top 10 categories for Big Data sources and mining technologies. ZDNet. At URL http://www.zdnet.com/article/top-10-categories-for-big-data-sources-and-mining-technologies/ (12.5.2015.)
[8] Mayer-Schönberger, V., & Cukier, K. (2013). Big data: A revolution that will transform how we live, work, and think . New York: Houghton Mifflin Harcourt, 2013
[9] Heudecker, N, "Hype Cycle for Big Data, 2013," July 31., 2013, (https://www.gartner.com/doc/2574616).
[10] Devlin, B. (2013a, February 5). B-eye-Network blog . At URL http://www.b-eye-network.com/blogs/devlin/archives/2013/02/, Accessed on: 12.5.2015
[11] Devlin, B. (2013b, March 4). Big data—Please, drive a stake through its heart!. B-eye-Network blog. At URL http://www.b-eye-network.com/blogs/devlin/archives/2013/03/, Accessed on: 12.5.2015
[12] Power, D. J.,Creating a Data-Driven Global Society , Springer International Publishing Switzerland 2015 13 L.S. Iyer, D.J. Power (eds.), Reshaping Society through Analytics, Collaboration, and Decision Support, Annals of Information Systems , Volume 18,2015,pp 13-28
[13] Dyche 2013, Dyche, J. (2013, March 13). Big data's three-legged stool. Information Management . At URL http://www.information-management.com/news/big-data-three-legged-stool-10024077-1.html , Accessed on: 12.5.2015., registration needed
[14] Šuman, S., Gligora Marković, M., Jadro, B., Decision's support and the business intelligence – what needs to be learnt?, Zbornik Veleučilišta u Rijeci, Vol. 2, (2014.), in press, No. 1
[15] Benders, J., Batenburg,R., Van der Blonk, H., Sticking to standards; technical and other isomorphic pressures in deploying ERP systems, Information & Management 43,2006, 194–203
[16] Liao, X., Li, Y., Lu, B. , A model for selecting an ERP system based on linguistic information processing. Information Systems, 32, 2007, 1005–1017
[17] Halper, F., Operationalizing and Embedding Analytics for Action, Best practices report 2015 by TDWI
[18] Hallikainen,P., Kimpimaki, H., Kivijarvi, H., Supporting the module sequencing decision in the ERP implementation process. In Proceedings of the 39th Hawaii international conference on system sciences, 2006.
[19] Frano, J., "ERP System Acquisition Project Planning," in ERP vol. 2010, I. Toolkit, Ed., ed: IT Toolkit, 2008
[20] Eckartz, S., et al., A Conceptual Framework for ERP Benefit Classification: A Literature Review , Technical Report,2009
[21] Lai ,V.S. et al., What influences ERP beliefs — Logical evaluation or imitation? / Decision Support Systems 50 (2010) 203–212
[22] Davenport, T. H., "Putting the Enterprise into the Enterprise System," Harvard Business Review, 1998,Vol. 76, pp. 121-131
[23] Palaniswamy, R., Frank, T., Enhancing manufacturing performance with ERP systems, Information Systems Management (2000 Summer) 43–55
[24] Holsapple ,C.W., Sena, M.P. ,ERP plans and decision-support benefits Decision Support Systems 38 , 2005, 575–590
[25] Tsai M.T. et al., Beyond ERP implementation: The moderating effect of knowledge management on business performance, Total Quality Management Vol. 22, No. 2, February, 2011, 131–144
[26] Li, C., ERP packages: what's next? Information Systems Management (Summer ,1999) 31–35
[27] Riives, J.; Karjust, K.; Küttner, R.; Lemmik, R.; Koov, K.; Lavin, J., Software development platform for integrated manufacturing engineering system , 8th International DAAAM Baltic Conference "INDUSTRIAL ENGINEERING" 19-21 April 2012, Tallinn, Estonia
[28] Shafiei, F. et al. , Multi-enterprise collaborative decision support systemExpert Systems with Applications 39 (2012) 7637–7651
[29] Lemmik, R.; Otto, T.; Küttner, R., Knowledge Management Systems For Service Desk Environment , 9th International DAAAM Baltic Conference "INDUSTRIAL ENGINEERING" 24-26 April 2014, Tallinn, Estonia
[30] Bolloju, N., Khalifa, M.,Turban, E., Integrating knowledge management into enterprise environments for the next generation decision support, Decision Support Systems, Vol. 33, pp.163–176, 2002
[31] Krogh, G., Nonaka, I. and Aben, M.,"Making the most of your company's knowledge: A strategic framework," Long Range Planning, 2001, Vol. 34, No. 4, pp. 421-439
[32] Ifinedo, P. , "Extending the Gable et al. enterprise systems success measurement model: A premilinary study," Journal of Information Technology Management, Vol. 17, No. 1, pp.14-33.,2006
[33] Jeng ,D.J.F. , Dunk, N., Knowledge Management Enablers and Knowledge Creation in ERP System Success, International Journal of Electronic Business Management, 54 Vol. 11, No. 1 ,2013
[34] Zhang ,H., Liang, Y., A Knowledge Warehouse System for Enterprise Resource Planning Systems, Systems Research and Behavioral Science Syst. Res.23,169-176 (2006)
[35] Nemati H, et al.,. Knowledge warehouse: an architectural integration of knowledge management, decision support, artificial intelligence and data warehousing. Decision Support Systems 33(2) ,2002: 143–161
[36] Robb ,D., Analytics Playing Greater Role in ERP, at URL http://www.enterpriseappstoday.com/erp/analytics-playing-greater-role-in-erp.html , Accessed on: 12.5.2015
[37] Monk ,E., F., Wagner B., J., Concepts in Enterprise Resource Planning, Fourth Edition, 2013 Course Technology, Cengage Learning
[38] Su, Y.F., Yang, C., Why are enterprise resource planning systems indispensable to supply chain management?/ European Journal of Operational Research 203 (2010) 81–94
[39] Kurbel ,K., E. Enterprise Resource Planning and Supply Chain Management Functions, Business Processes and Software for Manufacturing Companies, Springer-Verlag Berlin Heidelberg 2013
[40] Turban, E., Sharda, R., Delen, D., Decision Support and Business Intelligence Systems, Pearson , 2010
[41] Elragal,A., ERP and Big Data: The Inept CoupleProcedia Technology 16 (2014) 242 – 249)

DOI: 10.2507/27th.daaam.proceedings.048

A FRAMEWORK FOR OPTIMAL QUALITY OF PARTS IN REVERSE ENGINEERING

Nebojša Rašović & Milenko Obad

This Publication has to be referred as: Rasovic, N[ebojsa] & Obad, M[ilenko] (2016). A Framework for Optimal Quality of Parts in Reverse Engineering, Proceedings of the 27th DAAAM International Symposium, pp.0328-0337, B. Katalinic (Ed.), Published by DAAAM International, ISBN 978-3-902734-08-2, ISSN 1726-9679, Vienna, Austria
DOI: 10.2507/27th.daaam.proceedings.048

Abstract

This paper presents a methodology of reverse engineering process in order to obtain an optimal quality of the scanned surfaces. These surfaces are using for model reconstruction in process which analyses curved geometry. This kind of design process has a key role in modern engineering, because the goal is to get electronic input, in case where a physic model is missing from some reason or is not available. By corresponding software support, we are able to control surfaces and their vector definition. Using advanced techniques, it is possible to set a procedure for measuring complex parts dimensions and transforming them on behalf of design intent. For the paper purpose, it is used physic model of the screw to create the point cloud. Therefore, the crucial phase of reverse engineering process is developing of adequate method for the best manufacturing conditions. Case study has been carried out by NextEngine 3D Scanner ULTRA HD.

Keywords: Reverse engineering; Feature recognition; Surface recognition.

1. Introduction

Today, product development has a tendency for rapidly positioning at the marketplace by various ways. One of these ways for product development is certainly Computer Aided Reverse Engineering (CARE). The CARE process has a few steps before getting a final result, such as usable electronic input. First of all, it is necessary to make measurements at points on the part surfaces. These points are defined by own x, y, z values of the space coordinate system. The CARE system forms collection of these points named point cloud. Point structure helps in feature construction with respect to the components or solids geometry. This approach allows multiple inspections of the real components or of the existing CAD models. For visualization inspection, geometric model has to be discretized into a mesh composition of a finite number of primitive elements [1]. High quality inspection is a real challenge in all activities of modern engineering. Some researchers are proposing structured methodology for the screening and comparison based on programs that are suitable for quality inspection [2].

Reverse engineering is modern technology which provides a lot of benefits and possibilities in the CAD/CAM domain [3]. There are a large number of reasons for using reverse engineering. Some of them are including design reconstruction with impacts on shape analysis and prototyping [4, 5], quality control and inspection on certain parts [6], request for making clone of the original model [7], remanufacture of existing parts [8], and more. Very important fact about reverse engineering process is based on effort to eliminate inaccuracy in the proposed scan algorithm [9]. The generic process has three main phases. The first one is scanning selected part, second one belongs to the point cloud processing in certain 3D software modeler, and the least one is application on specific area of geometric model development (Fig. 1).

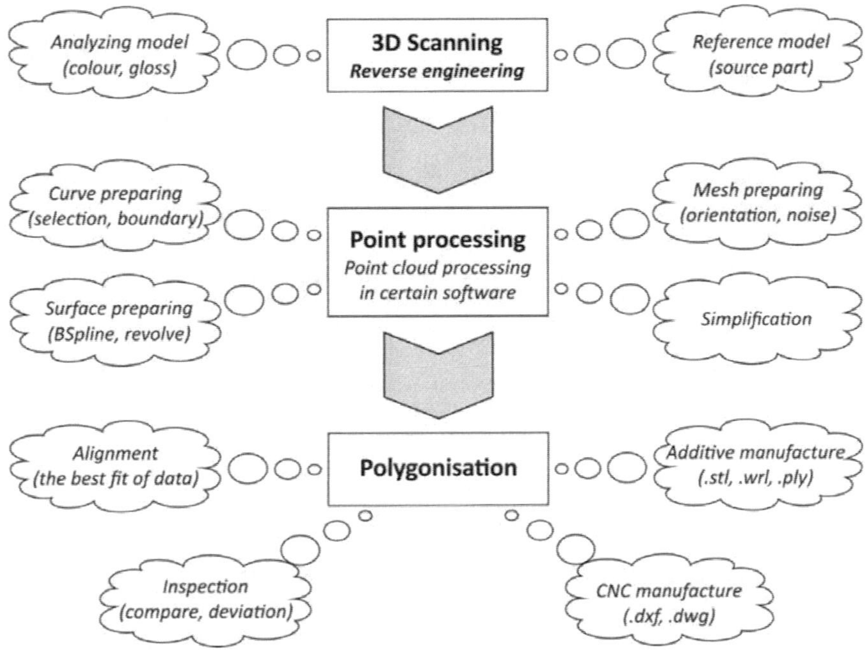

Fig. 1. Generic Algorithm of reverse engineering process

The initial phase involves preparatory activities. First of all, it performs analyses of model appearance in order to solve light fracture issue. This statement occurs at glossy models, where the light is reflected from the surface. According to the above mentioned, it must be adopted a strategy for selected adequate parameters which are available in scanner system options.

The second phase is important from a model of geometric standpoint. Scan data are using for presentation of the point cloud data. The point cloud data sets of data points in a coordinate system, where the points are defined by x, y, z coordinates, which are a surface of some object. 3D scanner creates the point cloud and measures a large number of points of a surface. The results of this phase are introducing into the next phase of reverse engineering process (Fig. 2).

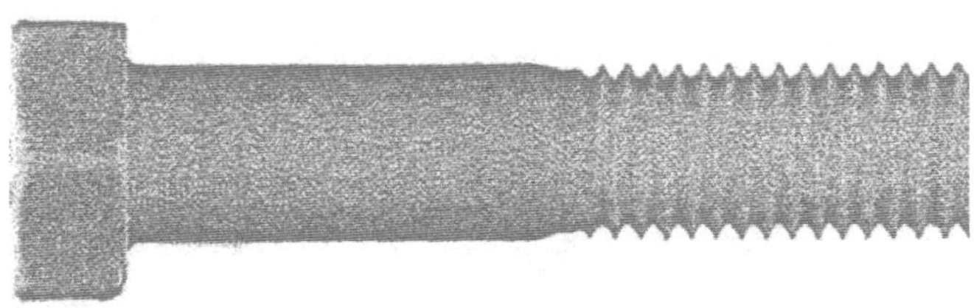

Fig. 2. A point cloud of the screw

The third phase includes a set of activities which provide different direction of application. The point clouds are converted to a polygon or a triangle mesh, or CAD formats through a process referred as a surface reconstruction. The point cloud of a model can be aligned to a CAD model or even another point cloud and compared to check for deviations. These deviations are displayed such as colour maps that give a visual overview of the deviation between the manufactured part and the CAD model. Geometric dimensions and tolerances can also be extracted directly from the point cloud. Getting a usable CAD model presents the hardest thing in the scanning process. The output of this step is a final product of the model geometry written by one of the standard formats in CAD/CAM systems such as IGES, STEP, STL, VRML, PLY, DXF, DEG etc.

2. 3D Scanning process

For the paper purpose, scan data and following processes are carried out by NextEngine 3D Scanner ULTRA HD, which represents a high resolution device with quickly captures millions of points. The equipment consists of scanner, multidrive, part gripper and platter (Fig. 3). In this way, there is no need for calculating distance in order to have cable connection between scanner and multidrive.

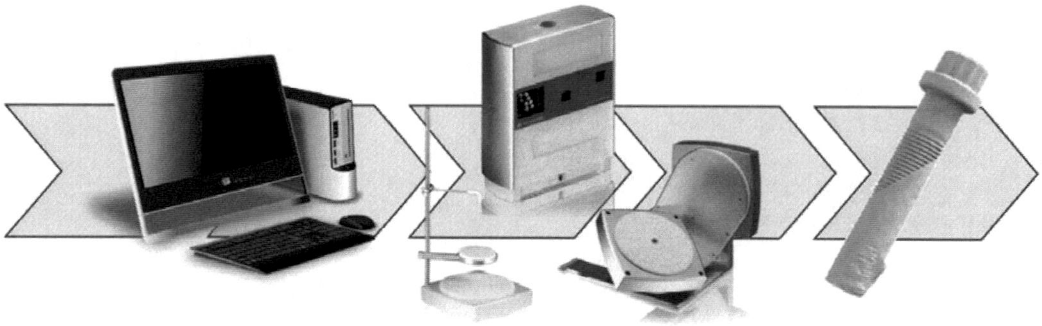

Fig. 3. Apparatuses of the scanning process

Scanning process is carried out in ScanStudio v2.0.2 NextEngine Inc. Before the first scanning, there is a possibility for calibration process of multidrive in order to align a model for well capture reasons. This step is highly recommended, but it is always possible to jump for another step. Models have to be analysed from skin point of view since the laser beam may be reflected in undesired directions from shiny or glossy surfaces. According to that, there is an opportunity for powder coating. For this case of the screw, model has grey matt surfaces all over the volume, so that gives a well capture results of pictures. In the scan options, a selection is made by both, the HD Speed for the greatest resolution and the highest colour accuracy under texture capture mode. An user can control predefines parameters of regenerating points from raw scan data. This applies a simplification of the model before starting of the process, which helps to remove unwanted areas and to decrease complications.

All above mentioned is an important framework condition at the beginning of the scan process. Except that, it is possible to manipulate with five families for scan positioning related to the divisions, start angle, tilt angle, points per inch square, and target. All of them are predefined parameters, but it is always possible for manual setup. After the scan process, the scan data are presented on the Fig. 4.

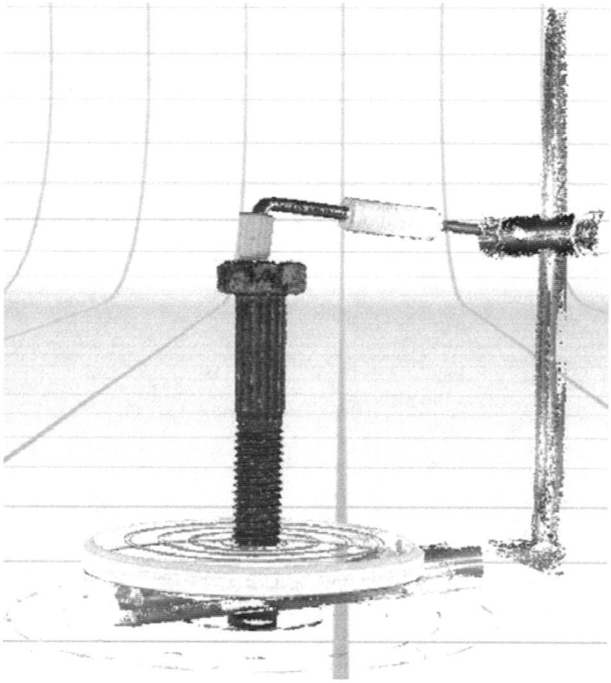

Fig. 4. Raw scan data of the screw

The conditions have been created for applying the most popular common tools, which are typical for the CAD environment. It is obvious, that a part gripper and a platter are over plus and need to be removed by trim tool. Overlapping data have to be trimmed in order to improve fusing, which is the next step in this process (Fig. 5). The selected areas have been marked by red colour, and those are overlap data. This is useful in direction of getting smaller file data and it becomes easily for handling. It is always calculated to reduce file data in order to disburden computer hardware for better results in data processing.

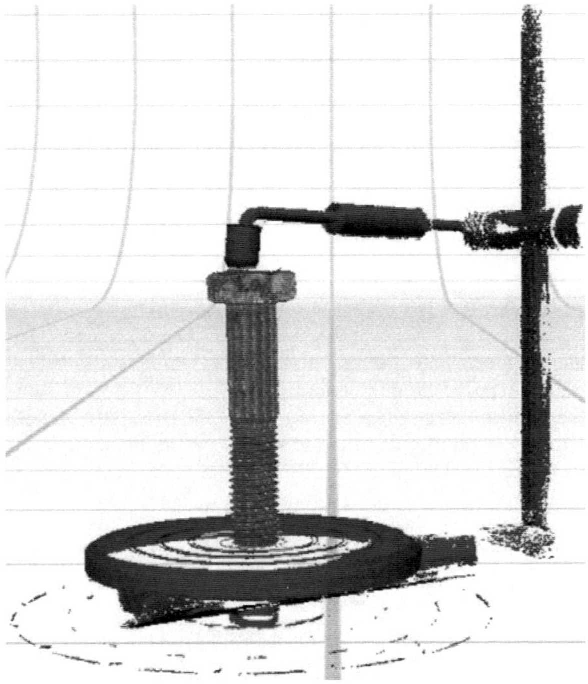

Fig. 5. Highlighted surfaces have to be trimmed

The fuse tool is intended for merges, re-meshes, holes fill, and simplifies scan data. It would be desirable to achieve mesh simplification by entering a deviation tolerance. That makes the file size smaller. It is smart simplification, which means that keeps more points in detailed areas and less points in other areas. Also, the model needs to be polished to get smoothness surfaces. It is possible to polish entire model or some areas, especially holes or threads. In that case, a selector must add a surface for starting process of filling (Fig. 6). The entire model has to be buffed.

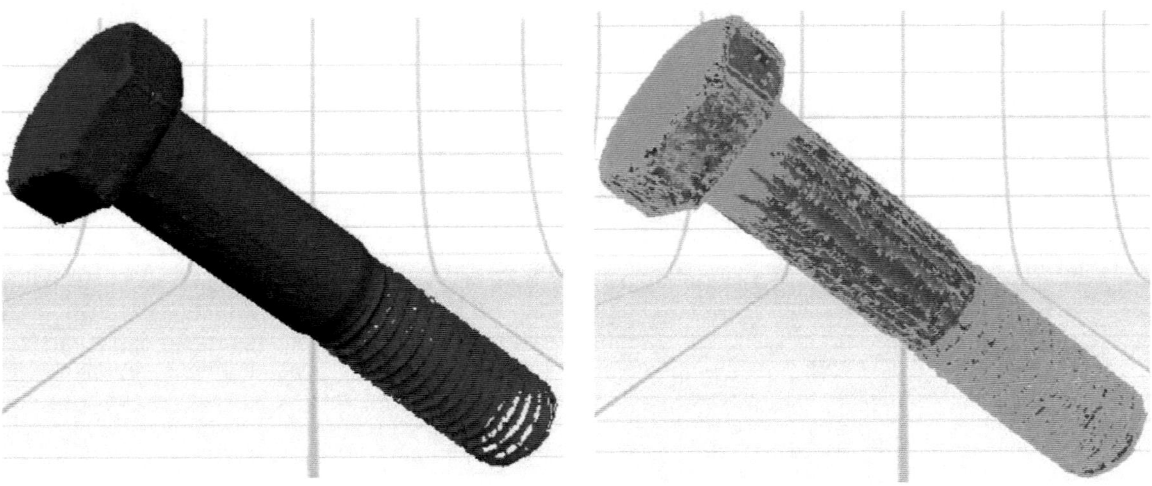

Fig. 6. Filling model areas using available fill tools by software

3. Point cloud processing

After forming a point cloud data, it has been created conditions for importing geometry in SolidWorks, by Dassault Systemes. First of all, file data are importing such as .xyz extension. An imported point cloud looks like a dense mesh with a lot of numbers of active points (Fig. 7).

Fig. 7. A point cloud imported in SolidWorks

A point cloud needs to be analyzed over mesh preparation wizard. On the start phase, a file data has 268353 points. This value has to be much smaller. A point cloud orientation uses the existing orientation for the mesh obtained by importing. A lot of errors are always introduced into scan data where the points are placed in unwanted areas or regions. For the point cloud data, noise removal tool divides the cloud into partitions and removes points that are outside the average distribution, resulting in the removal of noise points. It is necessary to remove separate mesh patches that are small in an area. As a possibility, wizard is offered to delete extraneous point data, but it has already done in ScanStudio v2.0.2 NextEngine, Inc. Simplification reduces the number of vertices for mesh features, or reduces the number of points for point cloud features, resulting in a simpler, smaller file size in additional. Now, we simplify the entire mesh. The percentage value for reduction is about 70%, and resulting by 71155 target points size (Fig. 8).

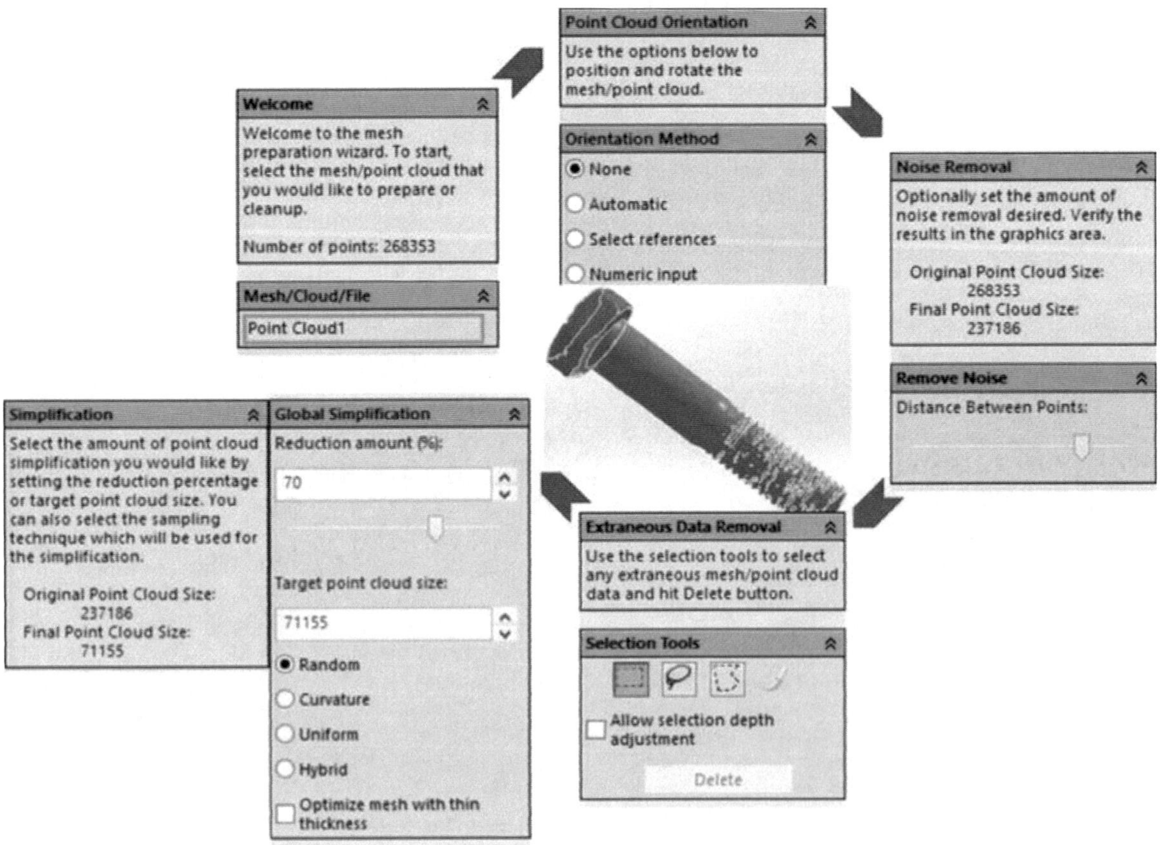

Fig. 8. Mesh preparation process

Level of smoothness can be adjusted by global smoothing slider to set the overall value and for individual regions as well (Fig. 9, left). Also, all holes are automatically detected to be filled. You can control which holes need to be removed, or filled up (Fig. 9, right).

Fig. 9. Process of surfaces smoothing (left) and filling holes (right)

Upon getting the mesh, it is necessary to analyze accuracy of the curved geometry. A curve wizard offers to create boundary and section curves in point cloud and mesh files. The key of this methodology is the fact that the curves are created in 3D sketch and are fully editable. That's obvious, it can be controlled the surface accuracy by parametric curve. Also, an user can add points for more level of control. In this case, section plane is top plane, and will serve for creating set of parallel reference offset planes on the certain distance. Before that, it has to select a pick point on the model, to set active point from that place for distance measuring. For this model, number of faces for setting up the parameters and editing of the curve´s points and selection have to be reduced at the minimum. A start point belongs to the top plane of the model, and that place is at the head screw. Total number of the reference planes are 19 and distance between them are 3.73288117mm (Fig. 10).

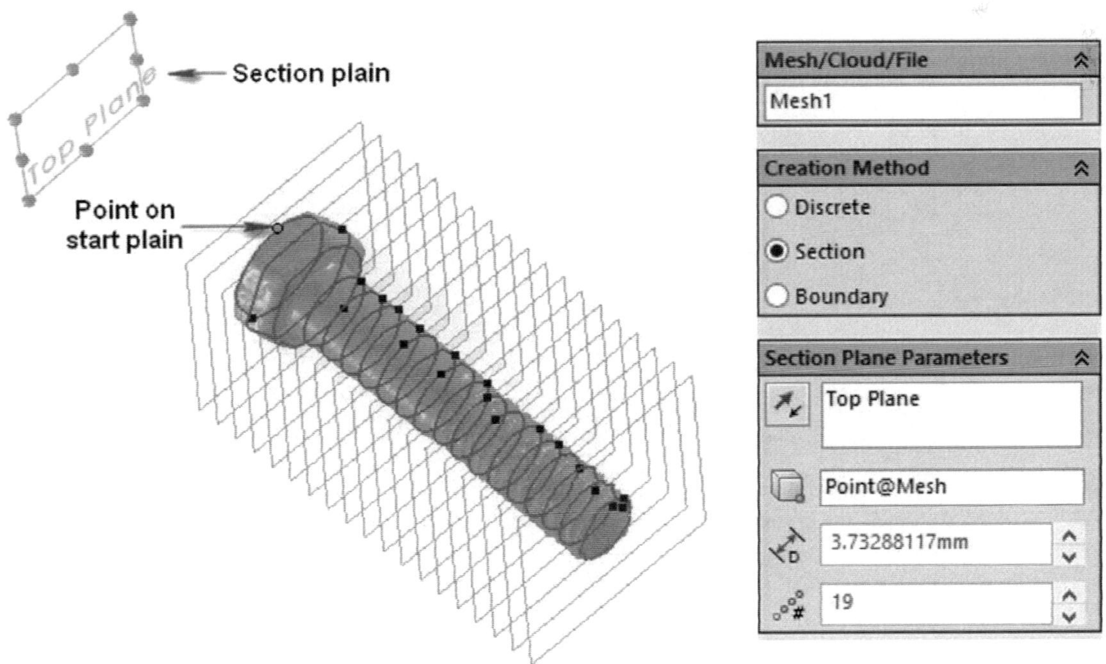

Fig. 10. Adding reference planes parallel with the top plane

Each curve is a sketch in certain plane and contains editable points for accuracy control. In the parametric form, each point on a curve is expressed as a function of a parameter t by equations: $x = x(t), y = y(t), z = z(t)$. Equations in this form are parametric equations for x, y, and z. Equation (1) is the parametric equation for a 3D curve:

$$L(t) = [x, y, z]^T = [x(t), y(t), z(t)]^T; \; where \; is: \; t_{min} < t < t_{max} \quad (1)$$

First of all, it should be selected a curve in editing curve circumstances, and should be adjusted or modified the curve's shape using edit tools such as a traditional CAD user interface. In the other word, if a curve is selected from the offered list, then there is a situation to adjust the curve fit.

A curve fitting should be adjusted by controlling value of the tolerance. If the slider is on the left side of scale, sample more vertices or points, making the curve fit more tightly to the mesh or moves it to the right to loosen the curve fit. If it adds more points, that situation refines the curve position. You can create a curve which more accurately represents the designer's intent by removing points or vertices related to noise in the scanned data (Fig. 11).

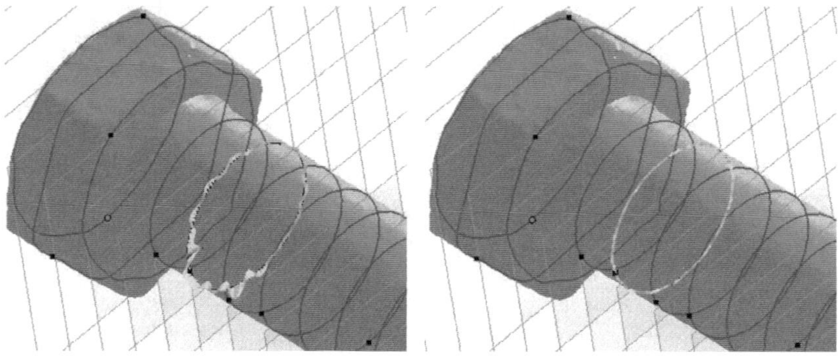

Fig. 11. Adjusting curve fit

In order to control all surfaces and convert them to the solid body, it is necessary to running surfaces wizard analyzed on the existing mesh. Accordance with previous, you need select individual regions in the graphic area or automatically extract all offered faces which are recognized by 3D modeler (Fig. 12).

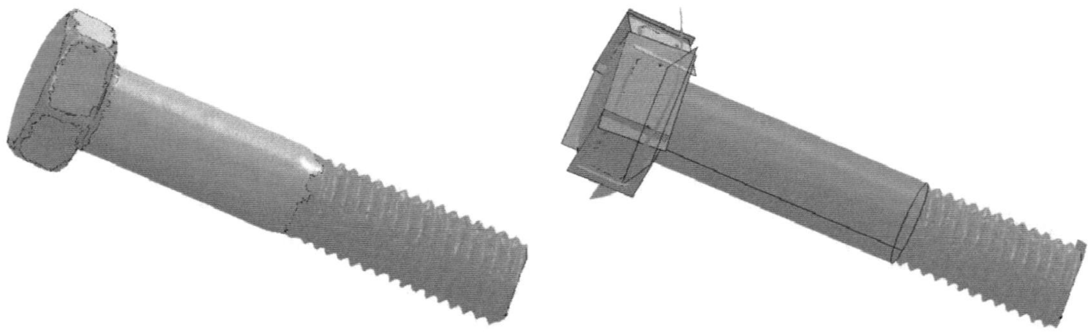

Fig. 12. Extracting a surfaces

A surface wizard helps in simplification of individual surface. It recognizes error on some surfaces and offers to remove them from configuration. Also, it is possible to relax some geometric entities in order to get accurate object, which is a crucial phase in next phase called inspection with original CAD model. After job realisation, the wizard highlighting with a message that the surface model is completed and recommended to inspect the model. A final model layout looks like as shown on the Fig. 13.

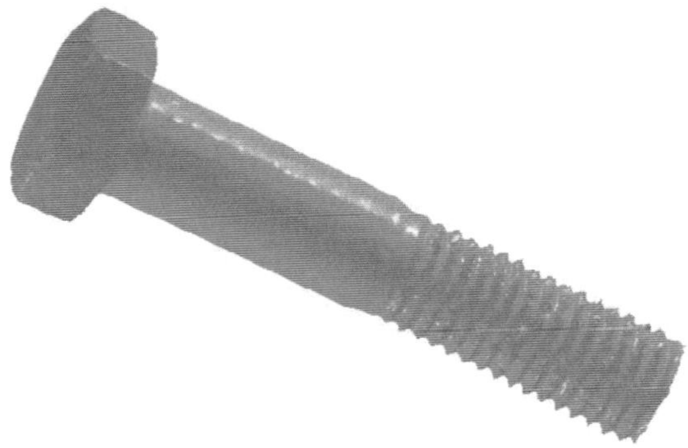

Fig. 13. Final look of the model

4. Polygonisation

A surface´s shape consists of the set of faces. These faces involve a series of geometric entities which has played a key role in forming single entity. Surface quality is important factor in approach of the products design process [10, 11]. The ability to provide the geometry from its original and making better transition from vectors formats to physical model are major challenge in engineering. In order to verify produced shape and its fit and function within assembly, it is implemented the measuring technique by 3D scanner. Scanner is supplied by geometric data from the existing physical object. Thus obtained geometric data are used by accuracy analyse of the object surfaces on standard screw. For the best results, all measurements are automatically merged into one to achieve a common global coordinate system. To perform a comparison of scanned physical and CAD source data, the scanned data are converted into CAD data coordinate position. For comparison purposes, the system requires from user to define the nominal point first at the CAD model, but then to select actual point at the scanned model and repeat that three times. These align operations have been realized in GOM Inspect software. Based on defined reference points, the system calculates the best fit for the actual coordinates and suggests a possible alignment (Fig. 14).

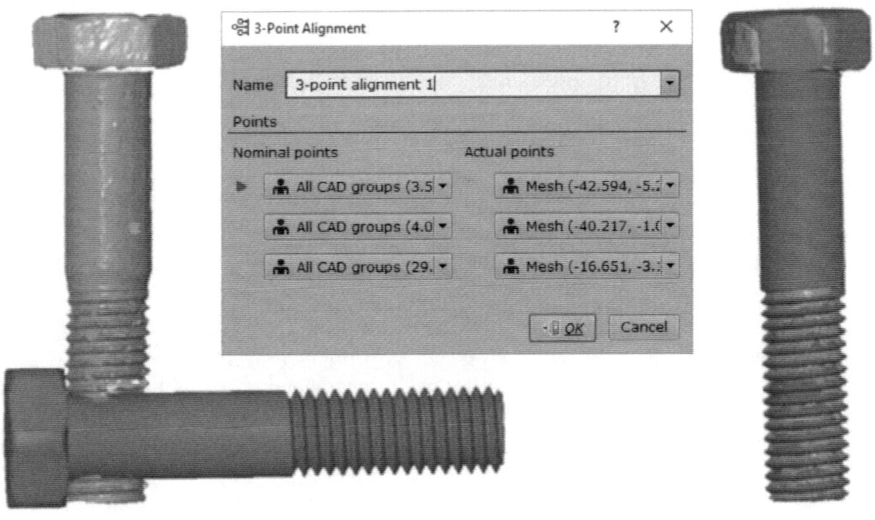

Fig. 14. Models fitting by 3-Point Alignment

In order to visualize surfaces of the models, colors are added to each particular part of the objects. It has been created a color map for easier interpretation and transferring dimensional precision of the produced model, which are compared by the CAD model. The process has done by deviation inspection. It automatically displays colorful deviation scale, which is based on the precisely calculated deviation values. Reddish areas on the map indicate that the scanned surfaces of the test pieces are over measured dimensions in relate to the surfaces of the CAD model. In contrast, the blue areas on the map indicate that the scanned surfaces of the test pieces are measured according to the surfaces of the CAD model. The green areas are the closest to the real dimensions, while the rest of the model is coincident with real dimensions. Deviation scale can also be finely adjusted or adapted by the user (Fig. 15).

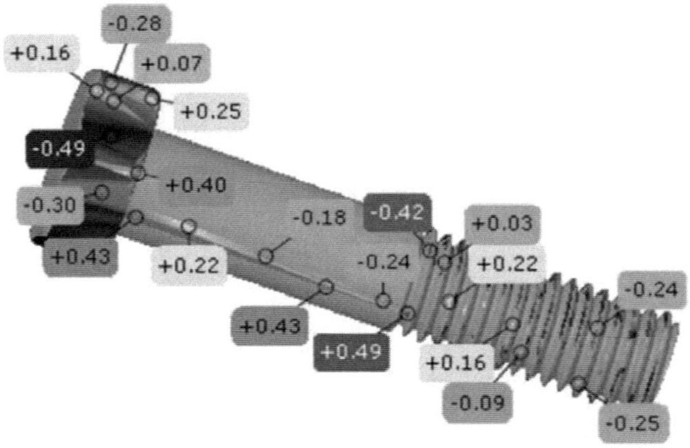

Fig. 15. A Surfaces deviation of the scanned data compared by the CAD model

Based on comparing approach and deviation inspection, it can see clear differences in surfaces precision. The real condition is that the comparing model is approximation of the original part. The main objective has been improved the quality and precision of the polygon model of the screw. On the first view, it is obviously clear that deviation is in the range from +0.5mm to -0.5mm, because the colour map highlights the extremely red colour, which gives a clear picture of the deviation of the part. Cylindrical face is recognized in 3D modeller and confirmed like that. In accordance with that, cylindrical face has zero deviation value on the over 90% of the body (Fig. 15).

The reason that attracts attention is certainly thread part of the volume. From the other hand, thread has more complicated geometry and deviation level is varying. Thread will be more accurate if reference planes have been smaller distance between each other. In that case, there is a solution for creating various distance along the entire body. One distance value for the cylindrical face and another one distance value for the thread. This situation shows a real state of the thread because configuration need to be uniform for better fit. That case brings a smaller number of faces and contributions to model simplification. This note of uniform and variable model configuration will be one of the subjects of the further researches in this area. The extension of this file is prepared for manufacturing by the certain method, additive or CNC.

5. Conclusion

In order to define the appropriate form, designers use various methods of formulation of the product. In any case, the final result is a form, which is precisely defined in terms of proportions and surface classification. While the proportions being relatively easily identifiable and defined, the definition of areas are great challenge of the whole process. In the most cases, designers still produce physical prototypes from design clay. The prototypes are then digitized using a 3D scanner. The result of the scan is a point cloud in space over which surfaces are laid using various techniques.

This paper has presented the framework for optimal quality of parts in reverse engineering process. Based on that, a solution is obtained for acquiring scanned data and their post-processing in some procedures. The overview of the scan process has included parameters which define conditions for better point cloud approximation. The research goal was to coincidence all measurements at actual model compared by CAD model. Polygonisation showed certain amount of surfaces deviation in some segments of the screw. But on the other side, it is getting fully usable 3D model for potential production by additive manufacturing methods.

The goal of the future work is to define an advanced framework of mathematical characteristics of model surfaces, which can be used to describe parametric surfaces. This process shall be conditioned by the skills and experience of an engineer who performs parameterization. The end surfaces definitions are, in spite of the mathematical notation, usually complex. Production processes are not limited any more to conventional technology. New layered manufacturing technologies enforce rapidly. Regarding to the manufacturing process, design recommendations for the design of products will be improved in the future.

6. References

[1] Bénière, R., Subsol, G., Gesquière, G., Le Breton, F. & Puech, W., (2013) A comprehensive process of reverse engineering from 3D meshes to CAD models. Computer-Aided Design. **45**(11): p. 1382-1393

[2] Minetola, P., Iuliano, L. & Calignano, F., (2015) A customer oriented methodology for reverse engineering software selection in the computer aided inspection scenario. Computers in Industry. **67**: p. 54-71

[3] Vinesh Raja & Fernandes, K.J., (2008) Reverse Engineering, Springer London

[4] Wang, J., Gu, D., Yu, Z., Tan, C. & Zhou, L., (2012) A framework for 3D model reconstruction in reverse engineering. Computers & Industrial Engineering. **63**(4): p. 1189-1200

[5] Min, Z., (2011) A New Approach of Composite Surface Reconstruction Based on Reverse Engineering. Procedia Engineering. **23**: p. 594-599

[6] Marciniec, A., Budzik, G., Dziubek, T. & Grzelka, M.a., (2012) Quality control and inspection of bevel gears of the aircraft gearbox utilizing the Atos 3d Scanner. Journal of KONES. Powertrain and Transport Journal of KONES. Powertrain and Transport. **19**(1): p. 261-266

[7] Jiang, Q., Feng, X., Gong, Y., Song, L., Ran, S. & Cui, J., (2016) Reverse modelling of natural rock joints using 3D scanning and 3D printing. Computers and Geotechnics. **73**: p. 210-220

[8] Paulic, M., Irgolic, T., Balic, J., Cus, F., Cupar, A., Brajlih, T. & Drstvensek, I., (2014) Reverse Engineering of Parts with Optical Scanning and Additive Manufacturing. Procedia Engineering. **69**: p. 795-803

[9] Kovács, I., Várady, T. & Salvi, P., (2015) Applying geometric constraints for perfecting CAD models in reverse engineering. Graphical Models. **82**: p. 44-57

[10] Cekic, A., Rasovic, N., Obad, M., Kaljun, J., Dolsak, B. & Begic-Hajdarevic, D., (2016) Production of Optimized Layered Products using Intelligent Support. Proceedings of the 26th DAAAM International Symposium, DAAAM International, Vienna, Austria, pp. 0271-0279. DOI: 10.2507/26th.daaam.proceedings.037

[11] Rašović, N., (2014) Intelligent decision support in design process of products to be produced by layered manufacturing technologies. Ph.D. Thesis, University of Mostar: Mostar, BiH

DOI: 10.2507/27th.daaam.proceedings.049

THE USE OF THREE-DIMENSIONAL VISUALIZATION ON THE BASIS OF THE MOTIONPARALLAX3D TECHNOLOGY IN THE DEVELOPMENT OF COMPLEX TECHNICAL DEVICES AND TRAINING OF ENGINEERS

Konstantin Kharin, Sergey Kuvshinov & Valentin Pryanichnikov

This Publication has to be referred as: Kharin, K[onstantin]; Kuvshinov, S[ergey] & Pryanichnikov, V[alentin] (2016). The Use of Three-Dimensional Visualization on the Basis of the Motionparallax3D Technology in the Development of Complex Technical Devices and Training of Engineers, Proceedings of the 27th DAAAM International Symposium, pp.0338-0343, B. Katalinic (Ed.), Published by DAAAM International, ISBN 978-3-902734-08-2, ISSN 1726-9679, Vienna, Austria
DOI: 10.2507/27th.daaam.proceedings.049

Abstract

We consider some problems in the digitalization of modern engineering education and the need for improved clarity regarding the development of complex technical devices. A stereo display based on the MotionParallax3D technology is proposed for three-dimensional visualization of objects. The actual use of this technology in engineering education processes allows us to conclude that the students can rather quickly develop "three-dimensional" thinking, which increases the creativity of the entire educational process.

Keywords: engineering education; 3D modeling; three-dimensional visualization; MotionParallax3D technology.

1. Introduction

Engineering education is the basis for the development of modern society in the era of transition to a knowledge based society [1]. We realized that there is almost no means to develop three-dimensional thinking and growth the culture of work with three-dimensional models for future engineers. Three-dimensional visual technologies have become a foundation for building the structure of a new digital educational environment and arranging a new type of design, teaching and research, and, most importantly, production space [2]. The implementation of 3D technologies and digital personal production processes in educational institutions provides a unique opportunity for teachers to find new ways of handling with educational and scientific resources using the technology of case studies, particularly in a specially installed place (digital lab).

Due to the digital organization, these laboratories can efficiently use network technologies to create a distributed environment [3] for research and educational purposes [4]. It should also be noted that network technologies can be used to create educational and situational centers for these distributed laboratories with the help of modern three-dimensional visualization tools, such as 3D virtual studios [5]. However, it is modeling and prototyping that form the basis of the majority of advanced technologies and processes for the development of complex technical devices. The article considers the problem of determining the optimal composition of technical means to ensure the development of complete mathematical models of designed objects and the implementation of the respective learning network technologies.

An important, if not the most important, methodological principle of modern design is thus the fact that the "designer" operates with simulating models and model systems, rather than the real processes and phenomena. In addition, the essence of real processes is judged from models. Probably, it can be stated that human reasoning is generally based on model images. Although universal models can hardly be found, the principle of modeling is obviously universal in any activity. With some reservations, a real end product of engineering can also be called a model.

In operating with models, the developers almost always need to visualize the model and analyze it as a virtual structure. In this context, the widespread use of digital visualization tools brings the development process to a new level. We demonstrate this on the example of the *MotionParrallax3D* technology.

One should especially note the fundamentally changed (in recent decades) approach to engineering and designing activities in general due to the widespread use of computer technologies, the focusing on digital production processes, and the significant speedup of design from the initial conception and sketching to the creation of first prototypes and their comprehensive analysis. An expert engineer should possess the following main skills:

- creative and critical thinking;
- analytical culture with knowledge from different fields;
- cooperation in international, intercultural, and interdisciplinary teams;
- independent learning in the age of rapid innovation and improvement of production processes and technologies as well as emergence of new materials;
- development of new ideas; advanced skills to handle with novel high-tech equipment and information systems under regular software updates, modification of regional and international standards for design, etc.

The modern workplace of design engineers, designers and developers involves a graphic workstation with professional software (CAD), which acts as both a "drawing stand" and a tool for the automation of computing associated with the stability and reliability of mechanisms and structures, and their coupling with other existing or projected units and elements of the finished product. The diversity of the equipment designed for working primarily with often three-dimensional objects rather than with classical planar drawings puts forward new requirements for the visualization system in terms of efficient and most realistic visualization of these objects from different angles.

The technological chain of the project development from the model creation to the production of its prototype consists of three interconnected stages (Fig. 1). Here, the second stage (virtual prototyping) makes it possible to save time, materials, and equipment resources in the early stages of design.

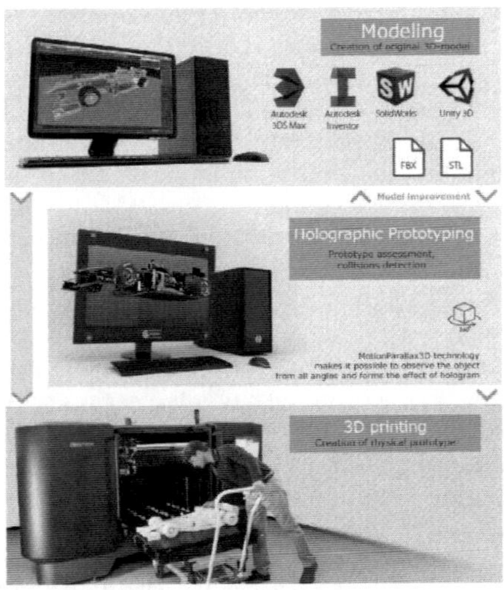

Fig. 1. Technological chain of modeling and prototyping. (1) Creation of original 3D-model using the known CAD and simulation packages. (2) Virtual prototyping. *MotionParallax3D* makes it possible to observe the object from all angles for a comprehensive assessment of the prototype, detection of collisions, design errors, reworking opportunities, etc. (3) Creating a physical prototype by 3D-printing techniques

2. The role and place of modeling and prototyping in the development of complex technical devices

We must distinguish between two fundamentally different types of models — mathematical (reproducing and explaining the real mechanisms of object behavior) and imitation (describing only the basic interaction of an object as a "black box" with the outside world).

Despite the fact that the professional slang involves specific terms such as *experimental*, *base*, *serial*, etc., the above seems inappropriate in the design that is oriented to the end material product, where the product value is incommensurable to the values of the models that are auxiliary by definition. In the function-oriented design, the value (but not the cost) of all models is comparable, and this seems to be obvious. A possible example is the following fact of the history of the first-generation supersonic passenger aircraft created in the 1960s. American engineers actually triggered the development of *Tupolev-144* and *Concorde* aircraft by the Russian and European aviation industries, respectively, but simultaneously they limit themselves to research on virtual models, which gave them a great experience and made it possible to save huge resources. If there arises a need in the near future for second-generation supersonic passenger aircraft, the United States will not have even potential competitors.

The study of models as an independent entity, unlike their practical use, has a relatively recent history; however, their typology is rather simple. The criteria may involve the language of description (presentation): from verbal to one that is implemented to actual product, the degree (depth) of engineering: from rough (unfinished) to detailed (finished), the functional purpose: from demonstration use to mass production, etc. Here, multiple criteria can be used for a single model.

A model is a universal tool for both representing and solving any problem. All design problems can be conditionally described (modeled) by a form defining the relations between entities under consideration, attributes, properties, and behavior (characteristics) in some external environment. The problem has one unknown parameter (entity, environment, or characteristics), which must be determined with the help of the other two components. Depending on the chosen unknown parameter, all problems can be divided into the following three main classes. The first class involves problems of analysis, when the entities are given and one analyzes the characteristics (unknown) that they have in a given external environment. The second class involves problems of synthesis, when the required characteristics are created and it is necessary to determine the entities that ensure these characteristics in a given environment. The third class involves problems of the external environment assessment, when the entities and their characteristics are given and it is necessary to obtain an environment (yet unknown), such that the given characteristics are manifested by the entities.

There is no formal method that would allow one to construct the most appropriate (correct) model for a given problem. During the solution, the problem of model construction and subsequent correction is imposed entirely on humans.

Simulation is an extremely complex, multi-faceted, and responsible process. It is particularly important that the unfinished model be correctly constructed at the stage of conceptual design; therefore, it makes sense to emphasize again that the construction of models in engineering design is to choose a way starting from some unfinished model and reaching a correct finished model.

Innovations in three-dimensional prototyping and improved digital production machines have made it possible to significantly reduce their cost and simplify the setup and overall activities. At present, the digital production laboratory is equipped with a set of computer-controlled universal tools with a possibility to operationally reconstruct the designed product. This outfit allows the laboratory to compete with mass production. In Europe and the United States, the laboratory has already proved to be capable of highly facilitating the development of single high-tech devices for specific purposes. The digital laboratory, where the product passes from its conception to working models, involves a system of audiovisual equipment: 3D video projectors, interactive visualization control tools (such as interactive panels, tables, and boards), three-dimensional sound systems; equipment for remote communication and high-quality multipoint video conferencing; digital production equipment: laser engraving machines, plasma cutters, knife cutters for cutting sheet materials, three-axis CNC machines, computer-controlled subtractive milling and turning machines [6]; rapid prototyping tools: 3D-scanners for a rapid creation of very accurate three-dimensional models of existing components, objects, artifacts, their reengineering and decomposition and 3D-printers for a rapid creation of functional prototypes from the models using different technologies for various materials, such as plastic, photopolymers, metals, and paper.

These digital labs allow students to go from the conception to implementation of their ideas very quickly and, most importantly, obtain the result in a short time, rather than being fixed on presentations and computer models of products. The laboratory generates a new type of project thinking for young people. However, this does not occur immediately. The development of stereoscopic technologies allowing one to observe the real or virtual world in a more natural (three-dimensional) form is only the first step. The technologies of 3D computer graphics that make it possible to simulate virtual objects (spaces), create three-dimensional models, and print them on 3D printers are the next step in the transition to the "three-dimensional" thinking.

The three-dimensional project perception has already been used in most activities, such as construction and design, simulation, and data analysis. Many elements of three-dimensional thinking have already been used in engineering, education, science, medicine, and culture. Relevant educational activities cover the fields of simulation, prototyping, and design. The students learn to think three-dimensionally (systemically) and create three-dimensional material objects. The process of three-dimensional thinking involves three-dimensional perception, three-dimensional simulation, 3D stereo training, interactive 3D stereo applications, development of intrinsic three-dimensional virtual objects, "materialization" of virtual reality, 3D printing of models, and, as a consequence, development of creative skills.

3. The MotionParallax3D technology for technical objects modeling and prototyping

An extension of the capabilities of currently widespread stereo monitors in terms of displaying three-dimensional models with a depth effect is the use of visualizing devices (with some conditionality, these can be called "holographic displays") in various stages; for example, we can mention the devices developed by the Russian company *Nettle* [7]. Alternatively, these devices can be called *MotionParallax3D* displays in view of the technology used [8].

We use the term "holographic" to suggest that the proposed visualization technologies make it possible to correctly and realistically (for humans) transmit the images of three-dimensional objects from different angles by analogy with holography, which has already become traditional for experts and where the images are recorded and displayed with the help of a highly stable laser light. In some sense, the virtual imaging on *MotionParallax3D* displays are close to multiplex holograms proposed by Lloyd Cross [9] but with a much more adequate demonstration of objects, which allows them to be regarded as full-fledged holograms in the direct sense of the word: as a complete optical information about the object.

Finally, the process of designing and prototyping can be characterized by the following key features and advantages:

- a high degree of detailing of models;
- demonstration of the details of any size and scaling;
- object decomposition, assembly/disassembly into elements;
- animation of the operation of units, possibility to observe the mechanism in the course of its operation;
- extended possibility for display: transfer of color and visual properties of materials (metal, plastic, etc.);
- presentation possibilities: the output model can be used in exhibition activities;
- distributed interaction between participants of the process: 3D-models can be sent electronically to remote areas of the world for evaluation;
- saving of time, materials, and equipment resources necessary for 3D-printing [10].

A multi-angle three-dimensional image makes an illusion of a solid object observable from all sides yet formed on a flat screen. In this case, the user is not separated from the real world: he can see both real and virtual objects with no difference in their behavior; the human brain fully believes in this illusion. In this sense, the proposed *MotionParallax3D* displays are superior to virtual reality helmets, which completely isolate the user from the real world, whereas the *MotionParallax3D* displays allow the user to be guided in the surrounding space to a greater or lesser degree. On the other hand, this imposes some restrictions on them: because the user can observe both real and virtual objects, their behavior should be identical. The projections of virtual objects are calculated so that the image observed by the user is fully identical to the image that would be observed if the virtual object was real and located at the appropriate point of the real space. The construction and mapping of correct projections of virtual objects requires that the virtual reality system be provided with the actual coordinates for observing the virtual world (the user's eye position).

Specialized 3D shutter-glasses with active trackers are used to allow the system to track the user's position in space with high accuracy. Based on eye position data, the system calculates the projection of the virtual world on the screen plane. Unlike stereo displays, which use only binocular vision, the *MotionParallax3D* displays use the three-dimensional perception mechanism of parallax motion, i.e., displacement of image parts relative to each other with an angular velocity that is proportional to the difference between them and the observer as their mutual location varies. This perception mechanism is activated by a permanent restructuring of the image according to the actual coordinates of the user's eyes. Due to this, the virtual objects are displaced relative to each other and relative to observable real objects according to the same laws and principles as the real-world objects. This allows the brain to build up a complete picture of the world containing both real and virtual objects with a visually indistinguishable behavior. However, due to the real-time projection realignment in this case, the objects are perceived as having some shape, volume and distance from the user's eyes, which, thus, are realistic from all angles.

Here, the image position and scale on the screen can be controlled both using a traditional computer "mouse" and in more intuitive ways: with a wireless touchpad, tablets and smartphones working under different mobile operating systems, gaming joysticks of different manufacturers or multi-axis 3D-manipulators.

The main focus in these devices is on the geometric correctness of the projection as the most significant indicator for the quality of the information mapping. The geometric correctness of projection is affected by the accuracy of user position tracking and the interval from the time when the user position is determined and the time when the image appears on the screen (the delay time). The tracking accuracy directly affects the correctness of the virtual object projection and can be determined from the general geometric error (depending on the architecture and geometry of tracking tools and the calibration quality) as well as the integral index of noise-induced error.

The tracking system uses unique Russian cameras with a scanning frequency of up to 900 frames per second in the normal mode. The four-camera system guarantees a one-millimeter accuracy and operational stability even in harsh illumination conditions. The working area of the tracking system with respect to vertical and horizontal (in the screen plane) viewing angles exceeds 170°. In this case, the optical tracking system can do without calibration during the operation, even after transportation of the equipment.

The interval from the time when the user position is determined and the time when the image appears on the screen (the delay time) is the main source of geometric incorrectness of the 3D-scene projection in *MotionParallax3D* systems. The delay is caused by the fact that all operations that determine the user's position as well as construct and display the projection require some time for their completion.

A characteristic feature of the technology of *MotionParallax3D* displays is the distributed processing of tracking data: each of the high-speed cameras has a built-in microcontroller with its own software based on *Embedded Linux*, which processes the data and transmits the marker position data that are ready for centralized processing. Also, to improve the quality of perception of virtual three-dimensional objects and significantly reduce the probability of above-mentioned symptoms for the user, the technology of user position prediction is used, which makes it possible to partially compensate for the delay; however, the prediction accuracy and horizon largely depend on many external factors.

4. The use of 3D visualization systems for educational purposes

The visualization systems described above can be efficiently used in education through a ready-made content (primarily, in the disciplines of natural sciences) fully adapted to *MotionParallax3D* displays. Custom content can be created by a built-in development kit. The implementation of this technology in educational processes requires a special approach to the installation of classrooms. This full-featured computer class may include several workplaces with a full set of equipment and standard software as well as *MotionParallax3D* monitors for the interactive study from all angles, construction of models, and practical and creative work. The central workplace of the instructor makes it possible to control the models on the working systems of students. A 3D-projector and a projection screen or an interactive board allow individual angles of the chosen model to be broadcasted in the 2D or stereo (for viewing with glasses) modes to the entire audience (Fig. 2). The concept of this computer class is successfully integrated with the requirements for electronic simulators based on high-speed local networks for robotic laboratories [11].

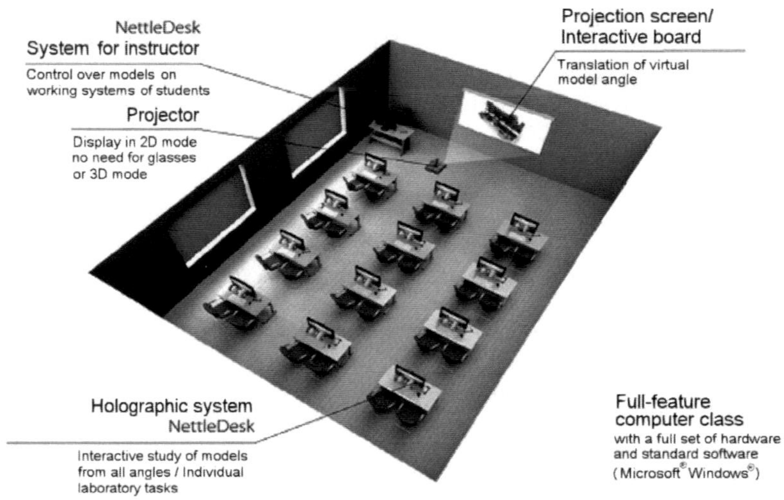

Fig. 2. Integrated solution for hardware of a holographic class

The ready-made educational content includes animated models of objects and phenomena from different fields, mostly causing problems in traditional teaching due to visualization complexities in conventional 2D graphics; for example, an internal combustion engine, the magnetic field of coil with current, different types of radiation, the heliocentric model of the world, the structure of gases, liquids and solids, and the principle of motor operation (in physics); different types of chemical bonds and the structure and form of electron clouds for electrons with different sets of quantum numbers (in chemistry); the structure of animal and plant cells, the process of photosynthesis, the structure of viruses and bacteria, and the structure of human eye (in biology).

Among the prospects of the proposed technologies, one can mention the potential increase in the number of users by increasing the frequency of stereo monitor frames for a correct separation of the video streams intended for each eye or by projecting the image on each eye individually with the help of special glasses. Also, it is important to simplify the integration of visualization systems with software for 3D-modeling or the creation of three-dimensional content in different fields that are not directly associated with digital technologies (including culture and the arts). It is of particular interest to develop devices for navigation and manipulation of virtual objects up to the user's full "immersion" into the resulting virtual world in educational and even entertainment purposes (Fig. 3).

5. Conclusion

The paper shows the way of selecting technical means for the implementation of efficient technologies for prototyping and modeling. For example, to intensify the process of training of future specialists, primarily in engineering, it is quite prospective and already widely used personal production technologies on the basis of laser cutting machines, 3D printers, plotters, CNC machines, etc. The objects have traditionally been visualized using 3D video projectors, stereoscopic monitors; however, a more qualitative display of a realistic three-dimensional picture of the object and a "presence effect" can be achieved by a successful application of the *MotionParallax3D* technology. The actual use of this technology in educational processes makes it possible to conclude that the students can rather quickly develop "three-dimensional" thinking, which sharply reduces the number of designing errors and inaccuracies. The new technologies of three-dimensional mapping catalyze the transition of education from a reproductive level to a creative one.

Fig. 3. Interactive manipulator for editing of virtual objects on a *MotionParallax3D* display

6. References

[1] Katalinic, B. (2010). Engineers for Knowledge Based Society, Annals of DAAAM for 2010 & Proceedings of the 21st International DAAAM Symposium, 20-23rd October 2010, Zadar, Croatia, ISSN 1726-9679, ISBN 978-3-901509-73-5, Katalinic, B. (Ed.), pp. 0001-0002, Published by DAAAM International Vienna, Vienna

[2] Kuvshinov, S.V. (2014). A technology of three-dimensional visualization for teaching in the humanities, in Recording and Viewing 3D Images in Motion Pictures and Other Fields: Proceedings of the VI International Conference, Moscow, April 17–18, 2014, Moscow: VGIK, 2014, pp. 239–245

[3] Pryanichnikov, V., Andreev, V., Bobrov, P., Biryukova, E., Frolov, A., Kharin, K., Kirsanov, K., Kostin, A., Kuvshinov, S., Marzanov, Y., and Prysev, E. (2012). Scientific-educational distributed laboratory — software and hardware means, in Annals of DAAAM for 2012 & Proceedings of the 23th international DAAAM Symposium "Intelligent Manufacturing & Automation" 24–27th October 2012 Zadar, Croatia, 2012, vol. 23, no. 1, pp. 1175–1178

[4] Andreev, V., Pryanichnikov, V., Poduraev, Y., and Kuvshinov, S. (2014). Education on the basis of virtual learning robotics laboratory and group-controlled robots, in 24th DAAAM Int. Symp. on Intelligent Manufacturing and Automation, 2013, Procedia Engineering, 2014, vol. 69, pp. 35–40

[5] Andreev V., Karbanov V., Kharin K., Kuvshinov S., Poduraev Y., and Pryanichnikov V. (2015). Training Situation Center Based on Three-Dimensional Virtual Studio for Distributed Mobile Robotics Laboratory, in Proceedings of the 26th International DAAAM Symposium "Intelligent Manufacturing & Automation" 21-24th October 2015, Zadar, Croatia, EU, Ed. by B. Katalinic, 2015, Vol. 26, No. 1

[6] Kuvshinov, S.V. (2014). EduBrication: An innovative trend in European education, in Innovation Technologies in Motion Pictures and Education: Proceedings of Scientific and Practical Conference, Moscow, October 29–31, 2014, Moscow: VGIK, 2014, pp. 178–184

[7] http://www.rbc.ru/newspaper/2014/05/14/56beb80f9a7947299f72d0cc. (2014). Holographic millions, Accessed on: 2016-09-08

[8] https://en.wikipedia.org/wiki/MotionParallax3D. (2016). MotionParallax3D. Accessed on: 2016-08-30

[9] https://en.wikipedia.org/wiki/Holography. (2016). Holography. Accessed on: 2016-08-25

[10] Kuvshinov, S.V. and Uskov, G.N. (2013). The use of technologies of virtual reality and integrated stereoscopic 3D-systems in educational processes, Mezhdunar. Nauchn. Zh., 2013, no. 4, pp. 57–64

[11] Andreev, V.P., Kirsanov, K.B., Levinsky, B.M., Pryanichnikov, V.E., and Travushkin, A.S. (2009). Development of a classroom of electronic simulation training systems on the basis of high-speed LAN, Information-Measuring and Control Systems, 2009, vol. 7, no. 6, pp. 82–92

DOI: 10.2507/27th.daaam.proceedings.050

EFFECT OF HEAT LOAD ON A MECHANICAL FORGING PRESS

Zdenek Chval & Karel Raz

This Publication has to be referred as: Chval, Z[denek] & Raz, K[arel] (2016). Effect of Heat Load on a Mechanical Forging Press, Proceedings of the 27th DAAAM International Symposium, pp.0344-0348, B. Katalinic (Ed.), Published by DAAAM International, ISBN 978-3-902734-08-2, ISSN 1726-9679, Vienna, Austria
DOI: 10.2507/27th.daaam.proceedings.050

Abstract

This article deals with effect of thermal load to the mechanical forging press. Main topic is assessment of displacement on each part of mechanical forging press caused by thermal load. Accuracy of forged part and durability of press is highly influenced by this displacement. Correct settings of thermal simulations lead to proper setting of clearances in structure of mechanical forging press, specially clearance between lower and upper die and in guidance of ram. This can be done generally for all size of press.

Keywords: thermal simulation; forging press; heat load, FEM simulation.

1. Introduction

Incorrect setting of clearance between upper and lower die can during forging lead to not finished forging or to increased loading of forging machine. Mechanical forging press is during hot forging process working with pre-heated die to temperature 300°C. This heat (thermal load) is then transmitted to other parts of forging machine. It means to the frame, ram, crankshaft and to the connecting rod. This heat is causing dilatation. Main aim of research is proper description of heating process by using modern numerical methods of modelling and elimination of its influence.

Heat sources occurring on forming machines can be divided into 3 groups:

- Heat from forming technology (hot preform, preheated die) – the subject of this article
- Heat generated in the forming machine (friction between moving members - guiding the moving parts, expansion of the working fluid, friction from clutches and brakes, etc.)
- Heat that is transferred from the surroundings

The working area of the forging presses is mostly influenced by the preheated die. This area has the greatest impact on the accuracy of products. Each size of forging presses is specific, and therefore thermal analysis must be carried out separately for each of them. This article is focused on a mechanical forging press with nominal force 25 MN. [1, 2, 3, 4]

Heat affected area, see Fig 1:

- part of frame
- ram
- guides
- press table

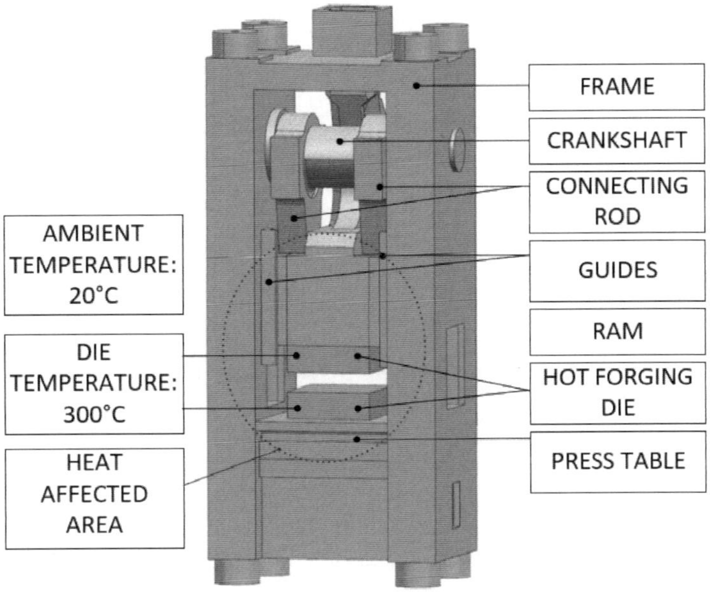

Fig. 1 Description of mechanical forging press and boundary condition of thermal calculation

2. Boundary conditions of thermal analysis

When the boundary conditions were determined, efforts were made to capture the least favourable conditions of the real state. The correct determination of boundary conditions is a necessary prerequisite for obtaining relevant results. Boundary conditions must be specified for each machine separately with regard to the operating technologies.

Temperatures of both parts of the die were set to 300 °C, which corresponds to normal preheating. This temperature is kept constant throughout the calculation time. This boundary condition is based on previous research. There was determined, that thermal load caused by forged part is not increasing temperature of preheated die.

Ambient temperature of the press was set to 20 °C. Radiation and convection to environment was during simulation considered. Each part of the press was connected by thermal contacts [1, 2, 5, 7, 8, 9]

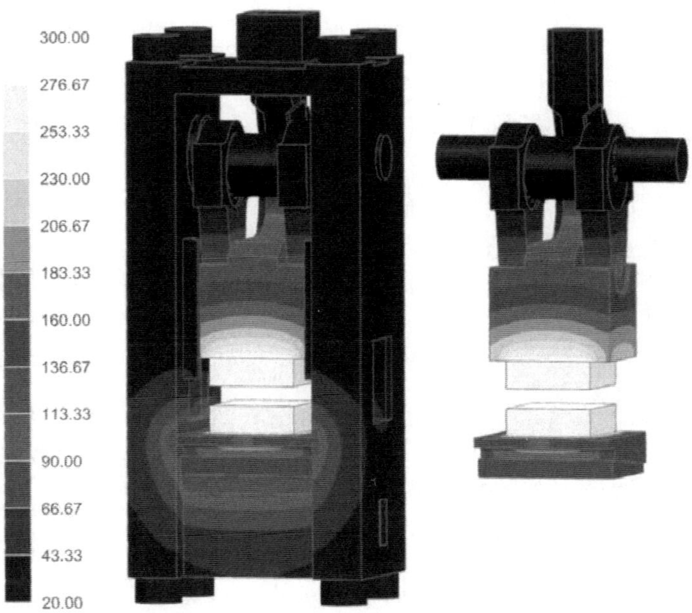

Fig. 2 Temperature load of mechanical forging press [°C]

3. The results of thermal analysis

The results of the calculation represent the thermal steady state of the structure (see Fig 2), for boundary condition defined in previous text.

The calculation shows that in this case cannot be ignored influence of heat load on the machine. Especially, thermal influence of ram and press frame near the press table is significant.

The maximum temperature on frame, near the press table is 121°C. The maximum temperature reached inside the ram guides is 116°C. We can say that the temperature will have a significant influence on defining the clearance between hot forging dies and between ram and guides.

4. Boundary conditions for structural analysis

Results from thermal analysis is used as input for following structural analysis. All considered parts of the press (crankshaft, ram, connecting rods, ram, press table) were connected using contacts. Clearance of the guidance was set to 0.1 mm, clearance in the bearings to 0.2 mm. [1, 6]

5. The results of structural analysis

Another necessary step is the structural calculation for obtaining the results of displacement of the structure. Especially is mentioned comparison between frame elongation (see Fig 3 - ΔL – FRAME) and whole inner system. (see Fig 3 – ΔL – ram, connecting rod crankshaft). This has major impact on clearance between dies.

Following results shows structural loading caused by thermal load from thermal simulation. Displacement results show elongation of frame in area between table and bearing of crankshaft. Value of this elongation is 2.47 mm. Elongation of whole inner system is 4.97 mm. It shows that not considering of thermal influence can lead to machine overloading. It means in this case necessity of elongation 2.5 mm elimination.

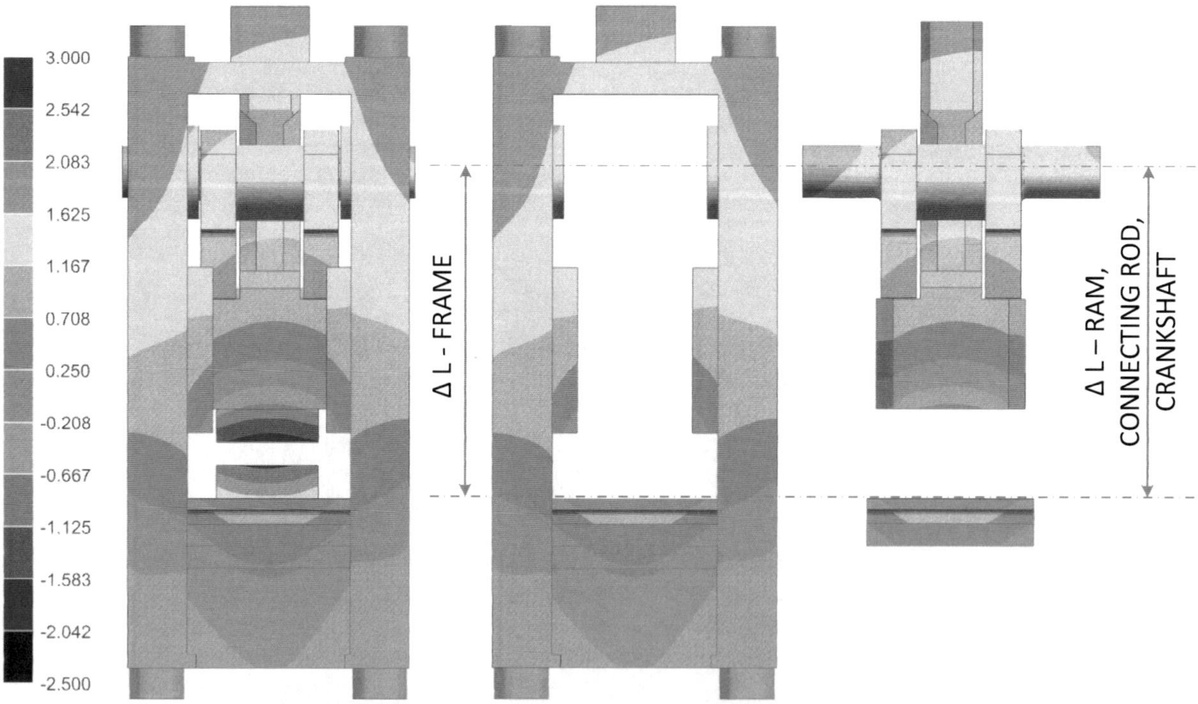

Fig. 3 Displacement of forging press in vertically direction [mm]

Figure 4 shows results of horizontal displacement. Elongation of frame (see Fig 4 - ΔL – FRAME) and ram (see Fig 4 – ΔL – ram) has significant influence on clearances setting in ram guides. From results is obvious that elongation of frame in area of guides is 0.92 mm and ram elongation is 1.2 mm. According these results is necessary to set minimal clearance 0.29 mm for elimination of thermal expansion.

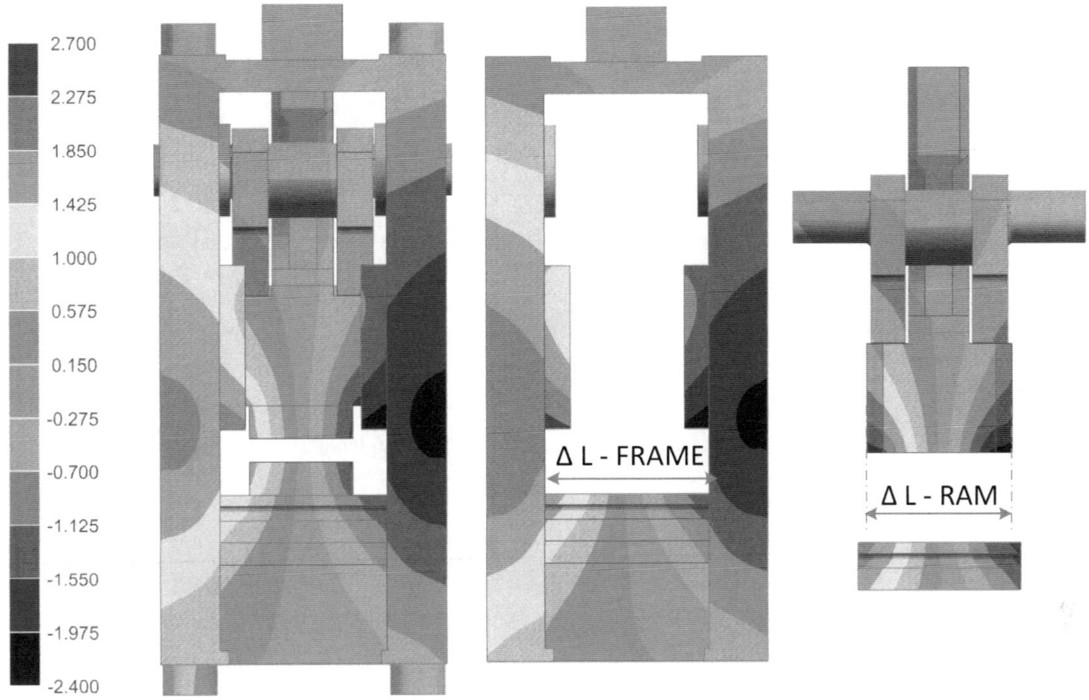

Fig. 4 Displacement of forging press in horizontally direction [mm]

Figure 5 (left) shows stress on ram caused by thermal load. It possible to notice, that clearance in simulation was small and contact between ram and guides occurred. In Figure 5 are also visible areas with higher stress in corners of ram. This are contact locations caused by thermal loading. This situation in praxis means stopping of press. Figure 5 (right) shows stress on ram with changed clearance which allows movement of ram.

Correct settings of clearances have significant influence on forgings quality. When clearances are too big, tilting of ram can occur and this can lead to non-uniform forging. On the other hand, small clearances can lead to press overloading.

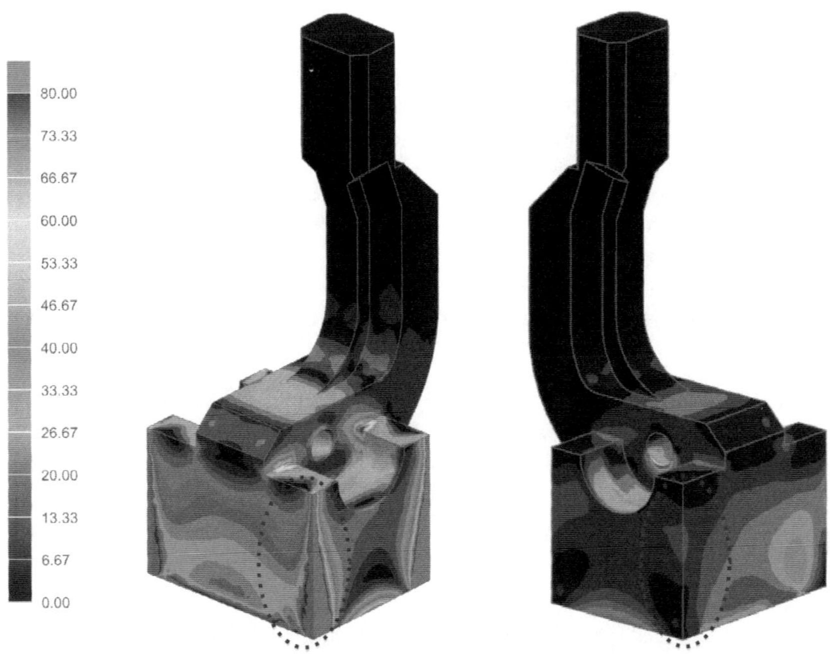

Fig. 5 Stress of ram caused by thermal load with small clearance (left) and sufficient clearance (right) [MPa]

6. Conclusion

By this analysis and research of mechanical forging press is confirmed that influence of thermal load is significant. It is especially visible during clearances investigation. Main influence is on forgings accuracy and press loading.

This research will continue and focused will be on improvement of boundary conditions during virtual simulations. Main aim is including thermal loading directly in structural simulation without additional thermal analysis. Next step is including of die preheating in simulation process.

Next research will be focused also on proper settings of clearance in ram guides with respect to thermal load and also on considering of different types of guides and their influence on accuracy.

Because of energy demands is also important to think about possibilities of energy loos reductions. This loses are caused by heat transfer to surrounding environment.

7. Acknowledgments

The article has been prepared under project LO1502 'Development of the Regional Technological Institute under the auspices of the National Sustainability Programme I of the Ministry of Education of the Czech Republic aimed to support research, experimental development and innovation.

5. References

[1] Chval, Z. (2013). Effect of Heat Load to the Forming Machines, MM (Modern Machinery) Science Journal, October 2013, ISSN 1805-0476, pp. 418-422

[2] Chval, Z. (2013) Effect of Heat Load on a Forging Press, Annals of DAAAM for 2013 & Proceedings of the 24st International DAAAM Symposium, Zadar; Croatia; 23 October 2013 through 26 October 2013, ISSN 1877-7058, ISBN 978-3-901509-73-5, Katalinic, B. (Ed.), pp. 0209-0210, Published by DAAAM International Vienna, Vienna

[3] M. Cechura, J. Housa, Energy Analysis of Forming Machines and Further Proposals for Decreasing of Energy Consumption, research report, V-11-037, VSCVTT, Prague, 2011

[4] F. Tikal, M. Duchek, FEM Analyses of the radiation in heating forging furnace, Annals of DAAAM for 2011 & Proceedings of the 22nd International DAAAM Symposium, ISSN 1726-9679, ISBN 978-3-901509-83-4, Vienna, Austria, EU, 2011

[5] Z. Chval, J. Stanek, Temperature Effect of Molded Piece to Forming Machine, ISBN 978-80-227-3135-5, Bratislava, 2009

[6] M. Cechura, Z. Chval, Convectional Versus Multiple Operating Press, Kovarenstvi, May 2013/17, 67-70, ISSN 1213-9289, Brno, 2013

[7] Selimaj, R. A.; Osmanaj, S. & Sahiti, N. (2010). Modeling of Temperature and Non-Stationary Heat Flux Through a Flat Wall, Annals of DAAAM for 2010 & Proceedings of the 21st International DAAAM Symposium, 20-23rd October 2010, Zadar, Croatia, ISSN 1726-9679, ISBN 978-3-901509-73-5, Katalinic, B. (Ed.), pp. 0209-0210, Published by DAAAM International Vienna, Vienna

[8] Cernecky, J. & Koniar, J. (2010). Research of Local Values of Heat Transfer Coefficients in the Area of Heated Curved Wall, Annals of DAAAM for 2010 & Proceedings of the 21st International DAAAM Symposium, 20-23rd October 2010, Zadar, Croatia, ISSN 1726-9679, ISBN 978-3-901509-73-5, Katalinic, B. (Ed.), pp. 0315-0316, Published by DAAAM International Vienna, Vienna

[9] Gerlich, V. & Zalesak, M. (2010). Experimental Validation of Heat Transfer Model, Annals of DAAAM for 2010 & Proceedings of the 21st International DAAAM Symposium, 20-23rd October 2010, Zadar, Croatia, ISSN 1726-9679, ISBN 978-3-901509-73-5, Katalinic, B. (Ed.), pp. 0969-0970, Published by DAAAM International Vienna, Vienna

[10] Terziqi, A. K.; Haxhiaj, A. B.; Kamberaj, N. S. & Bajraktari, B. V. (2010). Thermal Load Impact on Steam Parameters for Cooling the Shaft Furnace, Annals of DAAAM for 2010 & Proceedings of the 21st International DAAAM Symposium, 20-23rd October 2010, Zadar, Croatia, ISSN 1726-9679, ISBN 978-3-901509-73-5, Katalinic, B. (Ed.), pp. 0173-0174, Published by DAAAM International Vienna, Vienna

DOI: 10.2507/27th.daaam.proceedings.051

USING FISHER'S EXACT TEST IN THE ENGINEERING INDUSTRY

Katerina Bicova &Martin Melichar

This Publication has to be referred as: Bicova, K[aterina] & Melichar, M[artin] (2016). Using Fisher's Exact Test in the Engineering Industry, Proceedings of the 27th DAAAM International Symposium, pp.0349-0352, B. Katalinic (Ed.), Published by DAAAM International, ISBN 978-3-902734-08-2, ISSN 1726-9679, Vienna, Austria
DOI: 10.2507/27th.daaam.proceedings.051

Abstract

This contribution solves the problem of monitoring precision manufacturing. Because, despite the high precision of production, there are fluctuations in precision during the day, the collected data is analyzed. The proposed analysis is to find out whether there is a dependence on the time of day precision. Next, using F-test to determine, whether the fluctuations in precision are significant. Another reason for this analysis are the requirements for continuous improvement, which is based on standard TS 16949 quality management system in the automotive field.

Keywords: Automotive industry; precise manufacturing; F-test; Statistical evaluation; Process.

1. Introduction

Automotive industry belongs to the most competitive branch of mechanical engineering. Companies fight for reach and prove their good names. Progressive technologies, product properties and services for customers are constantly developing.

This contribution describes an issue of precise production of machine parts evaluation and analysis. It works with data that came from serial production in automotive industry, where high precision and small amount of rejected parts play a big role. In our case, in spite of high precision of production, are there still significant fluctuations of precision during the day. From this reason, we did a detailed analysis of provided data form measurement in production line for giving suggestions, which should lead to production process improvement. The requirements for continual improvement of the process came from TS 16949 standard that works with quality management system in automotive industry. Top management shall review the product realization processes and the support processes to assure their effectiveness and efficiency. Statistical studies shall by conducted to analyse the variation present in the results of each type of measuring and test equipment system.[10]

2. Problem description

The subject of analysis is data from the output control produced machine parts. Controlled is a critical part dimensions. Measurement is carried out part of the innovative measuring technology Renishaw Equator (view Fig. 1). This innovative measuring system enables 100% control required because it is a very precise production of critical components in the automotive field.

Why Renishaw Equator? When selecting, assessing and current possibilities of CMM on the market, any of them offer flexibility, used friendliness, performance measurement per unit of time. But only Equator has ability to work in automatic mode, for which the device is very well prepared. It devalues competitors in 3D measuring with the unique architecture, based on the parallel kinematics.

Automatic mode allows 100% control of the entire volume of parts of production. After comparison with the standard it can evaluate whether the measured piece is good or bad and display outputs on the control screen. The basic principle of the measurement process is based on comparing inspection = comparison measured part to the "golden" or etalon. "Golden" part is the one that is made as a model for mastering = initial setup of parameters before or during the measurement, temperature change, achieving limit number of checked items, or the expiry of the time limit, the setting, etc. [2].

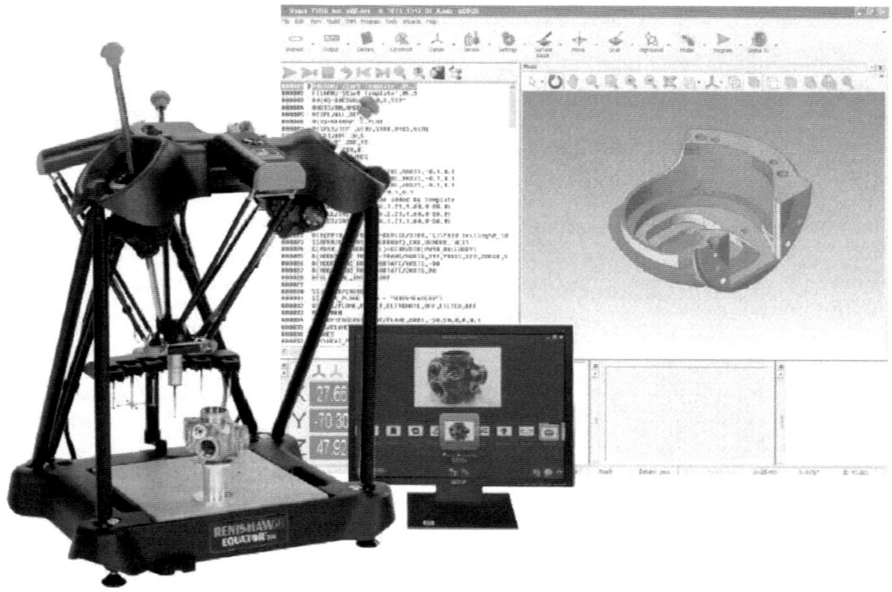

Fig. 1. Measuring device Equator[3]:

Production of components is carried out during the 24-hour standard (on the 3 shifts). People take turns regularly shifts. It is a stable process with high capability 1ppm. Despite stable process fluctuations precision during the day.[1,9]

The proposed analysis is to find out whether there is a dependence on the time of day precision in production. Simultaneously, whether fluctuations in precision are significant. The measured data are accumulated over a period of one month. Of them is then formed an average curve precision during the day.

For analysis of the significance of fluctuations in precision was designed to compare the morning (6:00 to 14:00) and afternoon shifts (14:00 - 22:00). Is selected, 8am and 4pm. Based on the consideration that after coming to work one moment must be incorporated.

The performance value of each person's is variable, dependent on many factors. There are not only significant individual differences between people, but the performance also changes throughout life and even during the week and every working day.[8]

To compare the precision of production during the day could be contemplated Fisher test. This test is used to compare two measuring methods, devices or precision of production on two shifts. I.e. Testing the hypothesis that one shift is lower production precision than the second shift. Or for example, if we are testing a new process, we may only be interested in knowing if the new process is less variable than the old process. [7]

3. Analysis of data - F-test

U Statistical hypothesis testing belongs together with the methods of estimation theory to practice the most important statistical inference. To use parametric tests must be fulfilled the assumption of data normality of monitored variables. Between the parametric tests include primarily Student t-test for testing difference between the two mean values and F-test (Fisher test) for testing the difference of two variances.[4]

Assuming that during the measurement causes no systematic error, measurement precision is defined by the parameter σ. Therefore, when comparing two measurement methods A and B with a value σ_A and σ_B, true that the smaller the value of σ is more precise. If these are not known, they may be used to compare of the standard deviation. A random influences cause variation of standard deviations s_A and s_B around the actual value σ_A and σ_B. [5]

Test criteria assuming hypothesis: H0: $\sigma_A^2 = \sigma_B^2$ is equal to (Eq.1):

$$F = \frac{s_A^2 \cdot \frac{n_1}{n_1 - 1}}{s_B^2 \cdot \frac{n_2}{n_2 - 1}} \tag{1}$$

and simultaneously applies (Eq.2)

$$F_{1-\alpha/2}(f_1, f_2) = \frac{1}{F_{1-\alpha/2}(f_2, f_1)} \tag{2}$$

In this case, testing the hypothesis that one shift is lower production precision than the second shift. I.e. Test for independent series from two normal distributions, that on the morning shift (around 8 hours) is lower than the production precision on the afternoon shift (about 16 hours). Measured values are in following tab.1.

Measured values during the morning and afternoon shifts							
8 am	28.063	28.054	28.055	28.044	28.093	28.041	28.036
	28.031	28.055	28.051	28.036	28.055	28.083	28.066
	28.061	28.042	28.042	28.028	28.045	28.073	28.074
	28.047	28.053	28.049	28.065	28.048	28.051	28.049
16 pm	28.038	28.048	28.047	28.035	28.034	28.028	28.028
	28.05	28.036	28.045	28.049	28.032	28.039	28.062
	28.036	28.054	28.03	28.043	28.064	28.048	28.059
	28.056	28.035	28.041	28.04	28.036	28.034	28.035

Table 1. Measured values during the morning and afternoon shifts

Fisher test of equality of variances		
	morning	*afternoon*
Variances	0.000228	0.000103
Observation	28	28
Reliability	95%	
F	**2.2112**	
F crit	**1.904823**	
Significance value p	**0.04369**	

Table 1. Hypothesis - Fisher test of equality of variance

According to tab.2 is critical field $W_{0,05}$: $F \geq F_{crit}(27,27)$, tedy $F \geq 1{,}904823$ [6]. The value of the test criteria $F = 2.2112$ falls into the critical field because the verdict $2.2112 \geq 1.904823$ applies, the hypothesis of equality of variances therefore the significance level of 0.05 in favor of rejecting the alternative hypothesis is representative of the fact that production is on the afternoon shift more precise than on the morning shift.
Result significance F-test: $p = 0.04369$ represents the probability of the null hypothesis of compliance variances of the two files (the probability of errors). Because this $p < 0.05$, it means that the difference between the variances is statistically significant. So the precision differs significantly.
More precise production on the afternoon shift is confirmed by the following chart, which shows the variation of precision during the day (Fig.2).

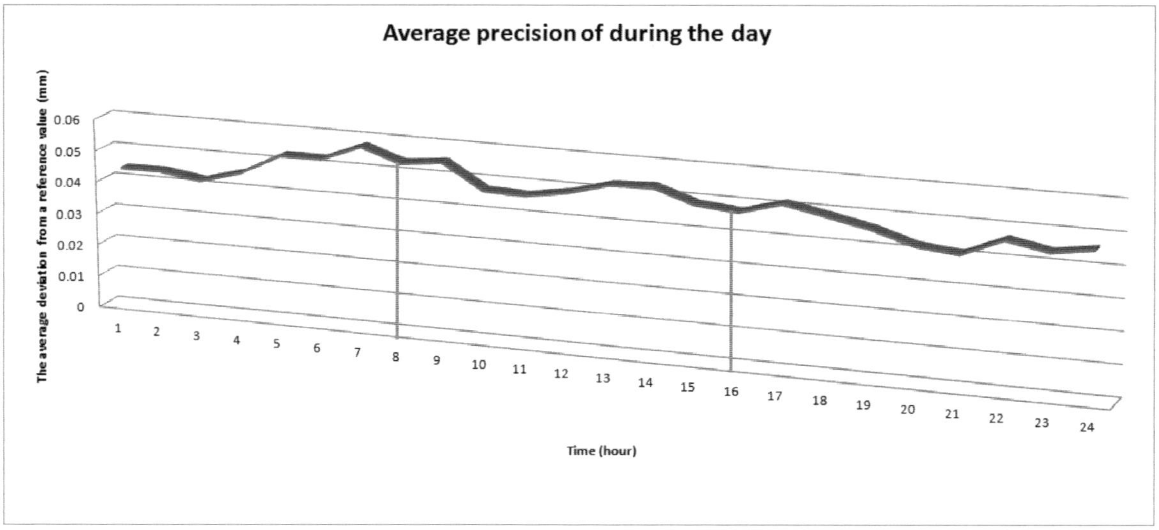

Fig. 2. Views average of manufacturing precision

4. Conclusion

Automotive sector has very strict production accuracy requirement. In business environment the time is important value. Automotive manufacturers are under customer pressure to a keep maximum of 5 ppm, which is very difficult to achieve and maintain.

The aim of the analysis was to determine whether there is a dependence on the daytime of manufacturing process and precision. Dependence on the time of day manufacturing precision is evident from the assembled curve average accuracy during the day (see Fig.2).

The analysis further shows that differences in accuracy between morning and afternoon shifts are significant. It would be appropriate to apply an additional tool for monitoring the process in order to identify the causes of these fluctuations. It is necessary to think that the key to any manufacturing or control process is the man with all his physiological aspects.

The organization shall continually improve the effectiveness of the quality management system through the use of quality objectives, audit results, analysis of data and management review.

5. Acknowledgments

This paper includes results created within the project SGS-2016-005.

6. References

[1] MSA (Measurement System Analysis) Praha : ČSJ, 2011. ISBN: 978-80-02-02323-5

[2] Martin Melichar, Dana Kubátová: Processing Data from Automatic Measurement Device, Procedia Engineering, Volume 100, 2015, Pages 899-906

[3] www.renishaw.cz/cs/equator-univerzalni-merici-system--13465 .[online].[Cited: 20.8.2016]

[4] Lectures - Safety and Quality. [Online] [Cited: 12. 08 2016.] http://cit.vfu.cz/statpotr/POTR/Teorie/Predn3/Ftest.htm.

[5] Kožíšek, J., Stieberová, B. Statistic in example. Praha : Verlag Dashöfer, 2012. ISBN: 978-80-86897-48-6.

[6] Tabs - critical value [Online] [Cited: 9. 5 2016.] http://www.kmt.zcu.cz/person/Kohout/info_soubory/letnisem/tabulky.htm

[7] Engineering Statistics Handbook. [Online] [Cited: 29. 07 2016.] http://www.itl.nist.gov/div898/handbook/eda/section3/eda359.htm

[8] Bicova, K.; Melichar, M.:Influence on the Shift the Product Precision of Machining Process and Eligibility, Proceedings of the 26th DAAAM International Symposium, pp.0508-0512, ISBN 978-3-902734-07-5, ISSN 1726-9679, Vienna, Austria

[9] Liviu Dorin Pop, Nagy Elod: Improving Product Quality by Implementing ISO / TS 16949, Procedia Technology, Volume 19, 2015, Pages 1004–1011

[10] ČSN ISO/TS 16949:2009, Quality management system for automotive industry

DOI: 10.2507/27th.daaam.proceedings.052

ISSUE OF HIGH PRECISION MANUFACTURING ANALYSIS IN AUTOMOTIVE INDUSTRY

Katerina Bicova, Jan Kutlwaser & Josef Sklenicka

This Publication has to be referred as: Bicova, K[aterina]; Kutlwaser, J[an] & Sklenicka, J[osef] (2016). Issue of High Precision Manufacturing Analysis in Automotive Industry, Proceedings of the 27th DAAAM International Symposium, pp.0353-0357, B. Katalinic (Ed.), Published by DAAAM International, ISBN 978-3-902734-08-2, ISSN 1726-9679, Vienna, Austria
DOI: 10.2507/27th.daaam.proceedings.052

Abstract

The contribution deals with issue of evaluation of precise production of machine parts. It works with detailed analysis of data from serial production from automotive industry, where the precise production plays the main role. Authors used a Pareto analysis for evaluation of data, which was collected during production of one batch of parts and analyzed those data. The batch is re-measured in laboratory conditions, new analysis is conducted and several suggestions for production process improvement are given.

Keywords: Automotive industry; precise manufacturing; evaluation; Pareto analysis; roundness.

1. Introduction

Automotive industry belongs to the most competitive branch of mechanical engineering. Companies fight for reach and prove their good names. Progressive technologies, product properties and services for customers are constantly developing.

This contribution describes an issue of precise production of machine parts evaluation and analysis. It works with data that came from serial production in automotive industry, where high precision and small amount of rejected parts play a big role. In our case, in spite of high precision of production, is there still significant number of rejections. From this reason, we did a detailed analysis of provided data form measurement in production line for giving suggestions, which should lead to production process improvement. For the detailed analysis authors choose the Pareto analysis, which belongs to the seven basic tools of quality management system. The requirements for constant improvement of the process came from TS 16949 standard that works with quality management system in automotive industry. [4, 5]

Evolution of market expectations leads to new solutions in quality control. Nowadays, in most cases control of geometry in industry is carried out using devices that apply coordinate measuring technique [8].

2. Problem description

Our input was a series of measurement data that came from 100% part control from a serial production of a part for automotive industry (Fig.1b). All parts at the end of production process were measured by optical comparator KEYENCE IM series. [6] Identified rejected parts are eliminated. In accordance to technical specification, 15 dimensions are measured and evaluated for each part. An example of one position of the part during its evaluation is in (Fig.1a).

The Pareto analysis of tolerance outlier dimensions occurrence frequency was done. Pareto analysis belongs to the seven basic tool of quality control. The main idea of Pareto analysis is that most of effect (80%) comes from minimum cause (20%).The main aim is to graphically display this asymmetrical distribution. The diagram shows causes of a problem, which are divided in accordance to their importance and are expressed, in most cases, by their cumulative frequency.

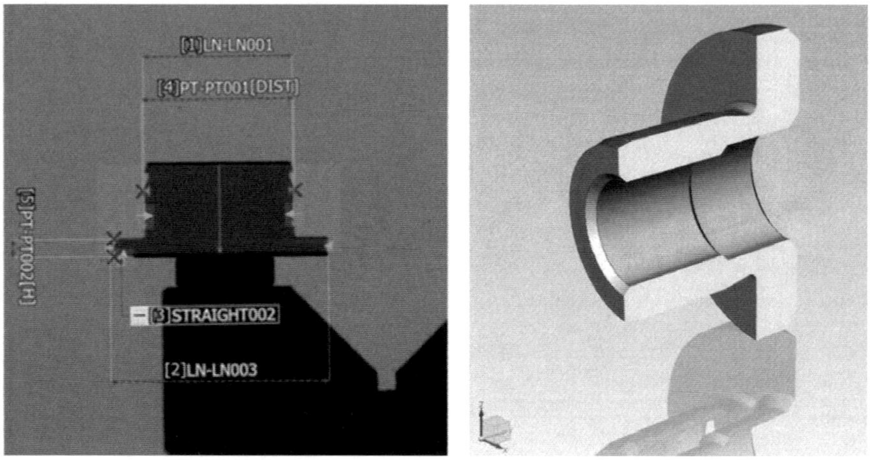

Fig. 1. Manufactured part during optical evaluation (a) and 3-D model of it (b)

In our case are measured dimensions chosen instead of errors and we are expressing cumulative frequency of dimensions out of tolerance. The data was sorted downwards from maximal count of dimensions that lie out of tolerance to the minimal. Then the data was depicted in the histogram. After that, cumulative frequencies were highlighted and Lorentz cumulative curve was created. We can draw a vertical line from the intersection of 80% horizontal line and Lorentz cumulative curve. This vertical line divides our histogram into two parts. The first part shows crucial minority of errors or, in our case, dimensions, which represent the majority of errors or rejected parts occurrence [1].

From this analysis we can say which of the controlled dimension makes the biggest "problems", so it is often out of the tolerance limits. Now we have dimensions that we must take a closer look to.

Controlled dimensions from our case are showed in Table 1 and they are sorted in accordance to number of outliers. The Table 1 also contains relative frequency and cumulative frequency.

Controlled dimension	**Frequency of outliers**	**Relative frequency of outliers**	**Cumulative frequency of outliers**
8	80	47%	47%
9	43	25%	72%
7	18	11%	82%
11	10	6%	88%
6	8	5%	93%
4	7	4%	97%
14	3	2%	99%
5	2	1%	100%
1	0	0%	100%
2	0	0%	100%
3	0	0%	100%
10	0	0%	100%
12	0	0%	100%
13	0	0%	100%
15	0	0%	100%

Table 1. Frequency of outliers

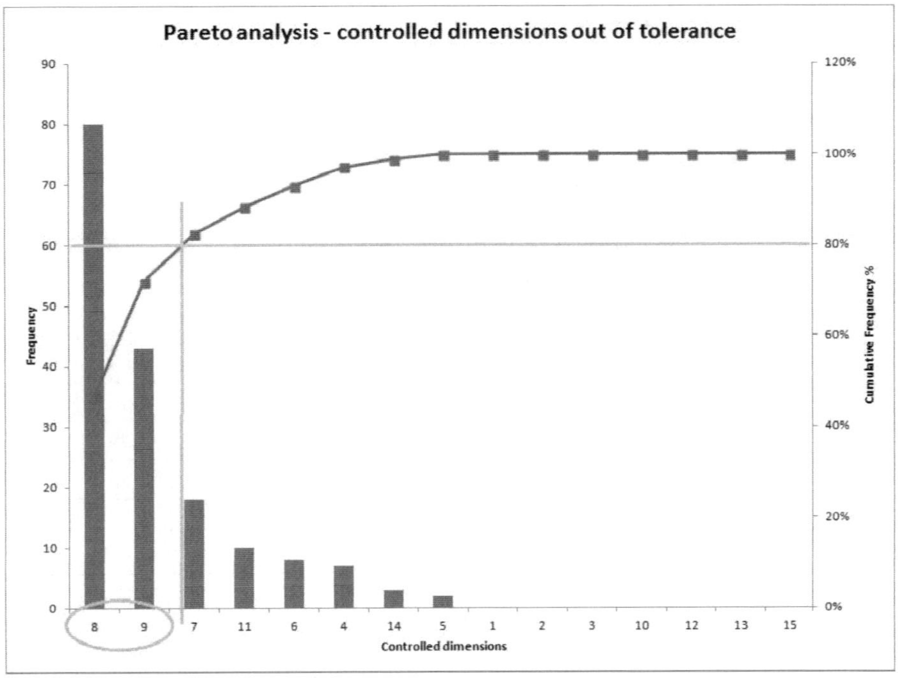

Fig. 2. Pareto chart

From previous analysis we can say which dimensions cause the most of problems. That means they are often out of tolerance. In our case, we can see in the Figure 2 that we must have a closer look at dimensions number 8 and 9. If we eliminate the errors occurrence in those dimensions production, we can also eliminate the most of rejected parts. In both cases, the dimensions represent roundness of the center hole.

3. Analysis evaluation

We took a one batch from the production to measure them using our high precision laboratory equipment. The aim of this new measurement is to prove our idea about using the Pareto analysis for data from serial production analysis and for verification of results the optical measurement system, which should be not sufficient for evaluation of roundness measured at such part as we have. Because the optical measurement system works only with 2-D image that is, in case of the inner hole, build from a projection of the whole cylinder. However, in many cases it is necessary to measure an element with a incomplete contour.[7] International standard [2] says that the produced hole must fulfill the described tolerance in any cross-section that is perpendicular to the axis of analyzed hole.

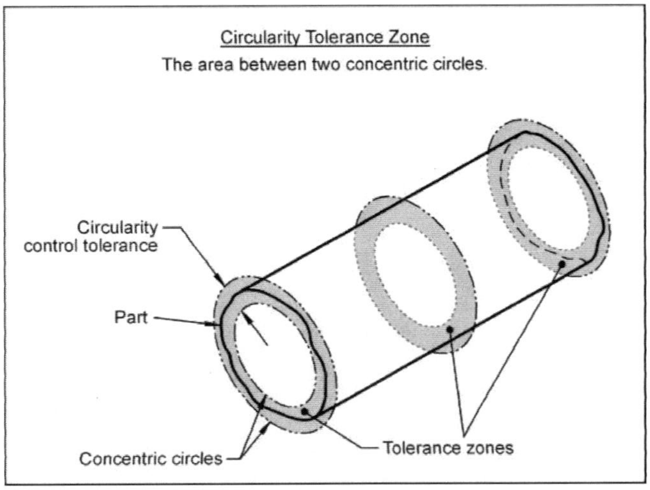

Fig. 3. Usual measurement scheme for roundness (the beginning, the end and the middle of the part) [3]

Usually, the roundness is measured at the beginning, at the end and in the middle of the part (Figure 3). For better image about the produced hole, authors decided to take 5 measurements upon height of the part. The measurement of roundness was carried out using Taylor Hobbson Talyrond 585 high precise laboratory roundness instrument. The rest of dimensions were measured by ZEISS Prismo Navigator 7 CMM. Both measurement instruments work under laboratory conditions and produce very precise results of measurement. The uncertainty of measurement is 0.1μm in case of roundness instrument and 0.9μm in case of CMM.
The example of measured data taken from roundness instrument is shown in Table 2.

Roundness in cross-section – dimension no. 8					
Nr. part	**A**	**B**	**C**	**D**	**E**
1	1.80	1.75	1.35	1.09	**2.32**
2	1.16	0.96	0.68	0.84	1.02
3	1.65	0.94	0.63	1.04	1.78
4	1.47	1.06	0.77	0.75	1.44
5	1.42	1.02	1.10	0.80	1.91
6	1.13	1.07	**2.96**	1.45	1.78

Table 2. An example of measured data

4. Conclusion

An example of measured data (dimension no. 8) is shown in Table 2. We can see that in case of part number 6, the roundness value in the middle cross-section is out of tolerance. That can prove that evaluating the roundness only with optical measurement system is not enough.

From the database we picked up only the maximal values and we repeated the calculation that gives a base for Pareto analysis on our new data and it is showed in Table 3.

Controlled dimension	**Frequency of outliers**	**Relative frequency of outliers**	**Cumulative frequency of outliers**
8	84	47%	47%
9	45	25%	73%
7	18	10%	83%
11	10	6%	89%
6	8	5%	93%
4	7	4%	97%
14	3	2%	99%
5	2	1%	100%
1	0	0%	100%
2	0	0%	100%
3	0	0%	100%
10	0	0%	100%
12	0	0%	100%
13	0	0%	100%
15	0	0%	100%

Table 3. Frequency of outliers – new measurement

From the new measurement, which was carried out with contact method, we can see that the number of faulty parts is slightly higher (84 instead of 80 at number 8 and 45 instead of 40 at number 9), than the optical measurement method can reveal. It means in this case, we prove that the optical measurement method is not able to catch all faulty parts (for example conical hole) that could be rejected and such parts could cause the parts reclamation from the customer side. The other dimensions did not show differences.

Because of high requirements for precise production and production process stability in automotive industry, we have for the company two main suggestions:

- Have a look at the manufacturing technology of the holes. Nowadays the production is done by drilling and then internal turning. We suggest to test other strategies of manufacturing:
  1) Keep the same technology with change of cutting conditions or tool's geometry.
  2) Use new strategy. Drilling and reaming.
- Further statistic evaluation of the production process with evaluation of stability and capability of the process.

Authors used a Pareto analysis for evaluation of measured data. Pareto analysis is usually used for analysis of rejected products and analysis of rejections. From the article we can see that the Pareto analysis could be also used for evaluation of measured data in mass production.

5. Acknowledgments

This paper includes results created within the project SGS-2016-005.

6. References

[1] Andersen, B., Fagerhaug, T.: Analysis of the root causes - Simplified tools and methods, ČSJ, Praha, 2011, ISBN: 978-80-02-02356-2

[2] ISO 1101:2012, pages 40-41

[3] Roundness evaluation, figure available on line at: http://www.engineeringessentials.com/gdt/circularity/circularity.htm, date cited: 08-09-2016

[4] Liviu Dorin Pop, Nagy Elod: Improving Product Quality by Implementing ISO / TS 16949, Procedia Technology, Volume 19, 2015, Pages 1004–1011

[5] ČSN ISO/TS 16949:2009, Quality management system for automotive industry

[6] Martin Melichar, Dana Kubátová: Processing Data from Automatic Measurement Device, Procedia Engineering, Volume 100, 2015, Pages 899-906

[7] Bartosz Gapinski, Michal Wieczorowski: Measurement of Diameter and Roundness on Incomplete Outline of Element with Three-lobbing Deviation, Procedia Engineering, Volume 69, 2014, Pages 247-254

[8] Bartosz Gapinski, Michał Wieczorowski, Lidia Marciniak-Podsadna, Bogdan Dybala, Grzegorz Ziolkowski: Comparison of Different Method of Measurement Geometry Using CMM, Optical Scanner and Computed Tomography 3D, Procedia Engineering, Volume 69, 2014, Pages 255-262

DOI: 10.2507/27th.daaam.proceedings.053

REDUCTION OF PANTOGRAPH INFLUENCE ON ENERGY CONSUMPTION

Karel Raz, Martin Zahalka & Zdenek Chval

This Publication has to be referred as: Raz, K[arel]; Zahalka, M[artin] & Chval, Z[denek] (2016). Reduction of Pantograph Influence on Energy Consumption, Proceedings of the 27th DAAAM International Symposium, pp.0358-0362, B. Katalinic (Ed.), Published by DAAAM International, ISBN 978-3-902734-08-2, ISSN 1726-9679, Vienna, Austria
DOI: 10.2507/27th.daaam.proceedings.053

Abstract

This article deals with energy consumption in train industry. It is specially focused on pantographs mounted on top of train. Simulations of commonly used pantograph were performed during research in order to obtain resistance (drag force). Than was created design of two different modifications in order to improve air flow around pantograph. Simulation of pantograph with additional ramp and pantograph with basement mounted in hole on train´s roof was performed. This results were compared with default design in terms of drag forces and uniformity of flow. For simulation were used modern CFD (computational fluid dynamics) solvers and postprocessor. Results can be applied on different design of pantograph or on pantograph in different position. Results shows how important, comparing to whole train, is during movement energy consumption of pantograph.

Keywords: flow simulation; pantograph; train; drag force; speed.

1. Introduction

Pantograph is a device that is mounted on the train's roof and it is used to collect the energy from the catenary. This electrical energy is than used for train movement. The human being always needed to move from one place to another, is it about move some quantity of objects or materials, or to move himself, and the way it has been done is evolving all the time.

This energy must come from somewhere, and need to be collect it. The most common to do that is a combination of pantograph-catenary. All producers of trains are using different designs and positions of pantograph. Our aim is to get simple answer, what is more suitable.

Aerodynamic resistance of all parts has to be decreased for lower energy consumption. Our research is focused on body of pantograph. Resistance, which is main evaluating criterion, can be called as a drag force. This force is highly speed-dependent (it is on second power, obvious from (1)).

For simulation was used CFD module of advanced solver NX Nastran. Pre-processing of this task was done in software Siemens NX 10. Beginning of this kind of simulations is preparation of simulation model. This procedure is simply described on Fig.1. Parts, which are small are during this process neglected. Than is this simulation model subtracted from the volume of air and is created fine FEM mesh.

It is necessary to use special approach called BLM (boundary layer mesh) to ensure sufficient number of elements at the surface of moving body. During our simulations was considered constant movement of train with speed 100km/h. From definition of aerodynamic force is obvious, that higher speed results in higher energy consumption (1).

$$F = \frac{1}{2}\rho v^2 cA \tag{1}$$

Where: F…Aerodynamic force
ρ… Fluid viscosity
v… Velocity of the body
c… Aerodynamic coefficient
A… Transversal area of the body

For simulation was used the most common design of pantograph. Results from this simulation can be used on different design with respect to their geometry.

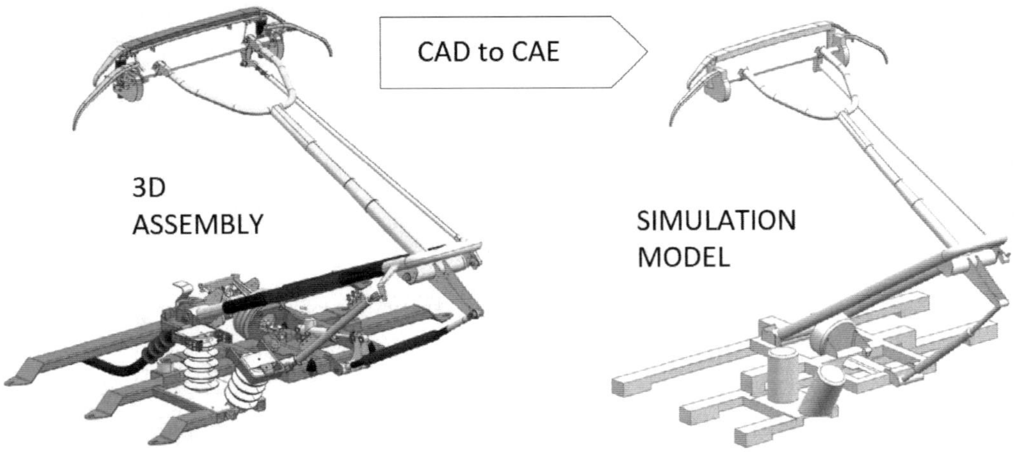

Fig. 1. Initial and simulation model of pantograph

2. Comparing of Variants

Detailed design of pantograph was considered for simulation from computational point of view, but train body was modelled as simplified. Additional devices mounted on top of train were neglected. Reason is because all producers are designing and mounting these devices (such as horns, electric connections…) in different positions. [1,2,3]

During CFD analysis were compared three designs:

- Default design without any change of geometry
- Design with added geometry of ramp in front of pantograph
- Design with change of connection between train and pantograph

Comparing was performed at two different levels. Drag force caused on pantograph body was compared during first step and then was compared uniformity of air flow around pantograph. Simply, it is possible to say that the more uniform flow is, the more suitable is for train. Uniform flow leads also to minimization of vibrations and possible mechanical defects.

For simulation was used mathematical model of flow called SST- Shear Stress Transport. It is combination of K-Epsilon and K-Omega models. Far from walls, the SST — Shear Stress Transport model behaves like the K-Epsilon model. Near the walls, the SST — Shear Stress Transport model behaves like a K-Omega model. The SST — Shear Stress Transport model displays better performance for unconfined flows and flows with strong adverse pressure gradients. [4,5,9]

Boundary conditions were same for all simulations, and are described as:

- Input velocity of 100 km/h
- Pressure of 0.101325 MPa
- Gravity of 9.81 m/s^2
- Boundary layer mesh with approximately 2 million elements

2.1. Default Design

Default design is obvious in Fig. 2. From CFD simulation is on the same picture clearly to see blue area behind pantograph. This area has lower speed of air flow comparing to the rest. Main aim of optimization should be decreasing of this blue area and trying to get in most places the same speed of flow [6,7]. Default design causes resistance force 442,46 N.

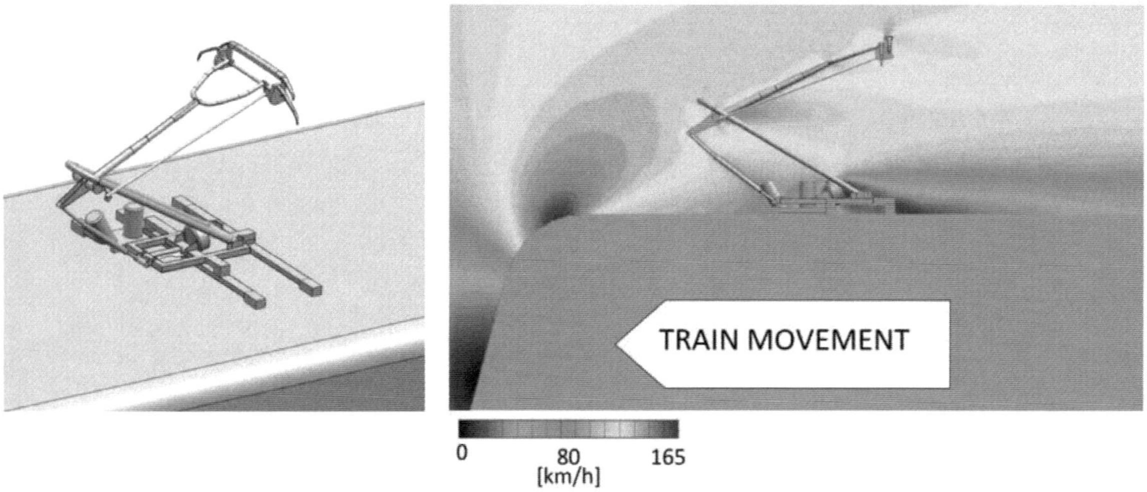

Fig. 2. Air flow around default position

2.2. Design with added geometry of ramp in front of pantograph

Our first try was a small ramp, trying to force the air flow go through up to the base. This concept was designed with the objective to be simple, and from this getting next improvement. In following picture is obvious zone behind the pantograph with slower air flow (Compared to default on Fig.2). This area with lower speed is bigger than in default design. It is resulting in increasing of the drag force and from this point of view is not suitable of using during real traffic. Increasing of resistance force was on level 484,19 N.

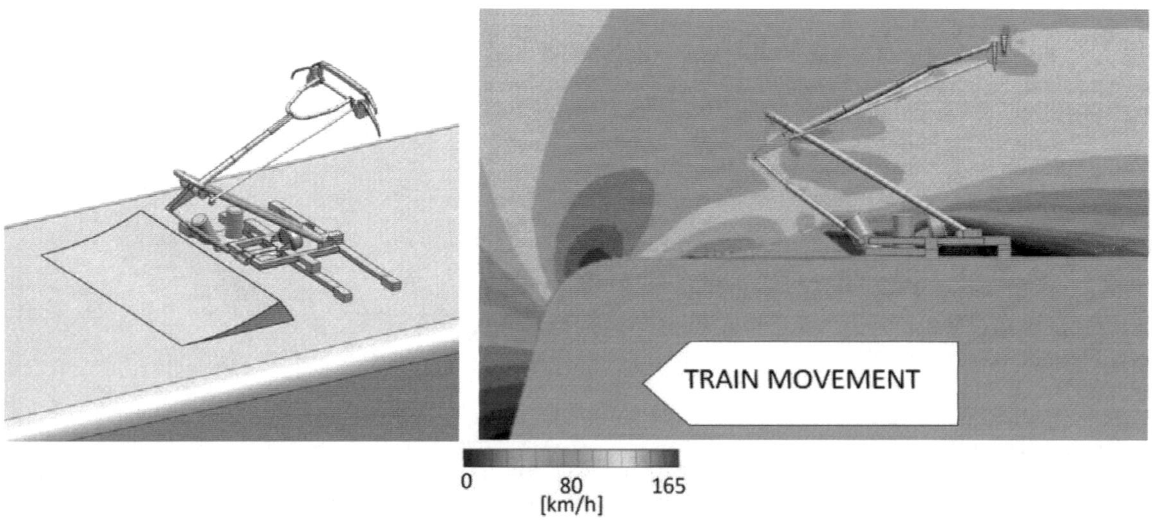

Fig. 3. Air flow around pantograph with added device

2.3. Design with change of connection between train and pantograph

After we have analysed the previous results, we tried to put the pantograph in a lower position on the train's roof. In this configuration, basement of pantograph is attached to a hole on roof. It helps to flow make it more uniform. Nowadays should be approach to put all parts which are not necessary exposed to air flow inside to the train body. Main structure of train is among all producers designed to be aerodynamic and all parts connected from the outside to this body are increasing consumption of energy [8].

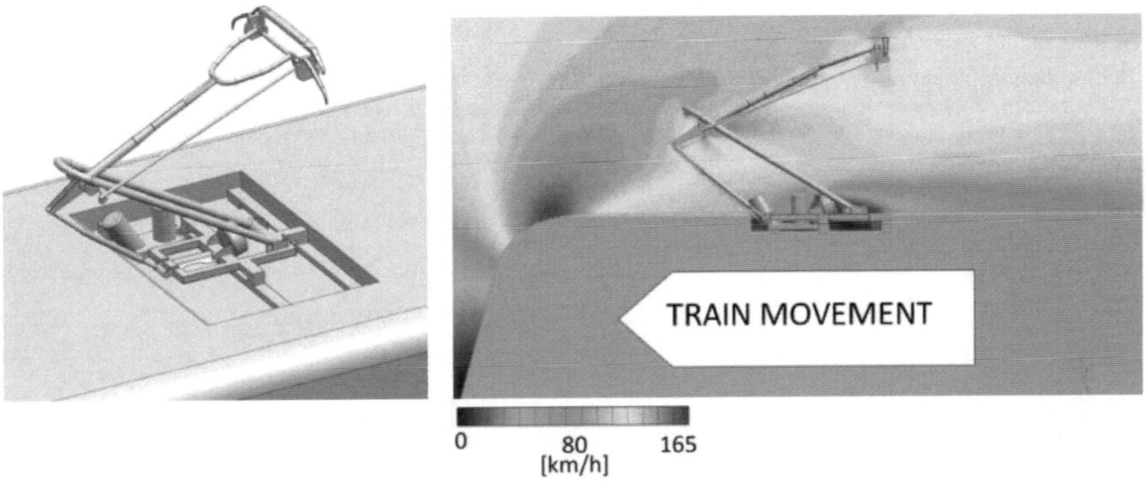

Fig. 4. Air flow around pantograph with improved basement mounted inside roof

This design results in decreasing of resistance force on final level 369,85N. Uniformity of air flow is visible from Fig.4. Blue area behind pantograph is significantly smaller comparing previous designs. This design results in some vorticities. They are visible on Fig.5. There are some streamlines around body of train and pantograph. In hole below pantograph are visible some vortexes which are caused by this design.

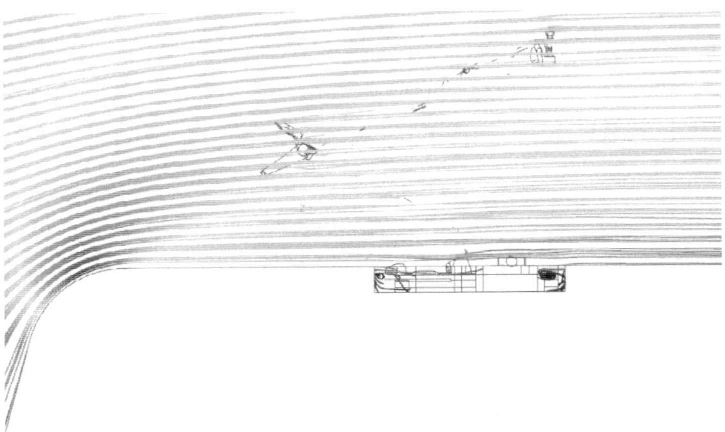

Fig. 5. Streamlines around pantograph and train

3. Conclusion

Results are described in following table. Significant is difference comparing to default design. Is obvious, that adding additional ramp is not resulting in improvement and has negative influence. Good approach is hiding some parts (basement) of pantograph under level of roof. This leads to decreasing of drag force (it means resistance force) by 16%. For clear understanding is added also lift force of each design. This force is acting in vertical direction and is 10% of drag force. Because of that was for comparing used only drag force.

	Lift (N)	Drag (N)	Drag Difference
Default design	49	442,46	0%
Design with ramp in front of pantograph	35,4	484,19	9%
Design with change of connection between train and pantograph	25,36	369,85	-16%

Table 1. Results summary

Before the conclusion, we must be warned that a lot of parts weren't take in account during the simulations (can create interference in the flow nearby the surface) and the pantograph's height can variate during the travel. In these simulations were used the average height, and the higher the pantograph is, the greater the force acting on it. It is caused by the transversal area changes (1).

Further steps should be simulation of air flow inside air tunnel and validating of previous simulations. This is possible to done by comparing streamlines from virtual (CFD) simulations and from real experiment in tunnel.

Nowadays should all products be designed with respect to energy consumption and its minimization. Our research shows, that all parts (even the smaller ones) have influences and it is necessary to optimize them in order to get final products with better properties.

4. Acknowledgments

The article has been prepared under project LO1502 'Development of the Regional Technological Institute' under the auspices of the National Sustainability Programme I of the Ministry of Education of the Czech Republic aimed to support research, experimental development and innovation.

5. References

[1] Ambrósio, J., Pombo, J. (2012). A computational procedure for the dynamic analysis of the catenary-pantograph interaction in high-speed trains, ISSN 1429-2955, Journal of Theoretical and Applied Mechanics, pp. 681-699

[2] Wang, Y., Liu, Z.-G., Huang, K., Gao, S.-B. (2014). Pantograph-catenary surface heat flow analysis and calculations based on mechanical and electrical characteristics , Journal of the China Railway Society, ISSN 1001-8360 pp. 36-43

[3] Bautista, A., Montesinos, J., Pintado, P. (2016). Dynamic interaction between pantograph and rigid overhead lines using a coupled FEM - Multibody procedure, Mechanism and Machine Theory, pp. 100-111, doi:10.1016/j.mechmachtheory.2015.10.009

[4] Arsene, S., Sebesan, I., Popa, G. (2015). The Influence of Wind on the Pantograph Placed on the Railway Electric Vehicles Bodywork, Proceedings of 5th World Conference on Learning, Teaching and Educational Leadership, pp. 1087-1094, doi:10.1016/j.sbspro.2015.04.185

[5] Popa, C. L.; Parpala, R. C. & Cotet, C. E.(2012). Contact Identification for Assembly/Disassembly Process in CAD-CAE Software, Annals of DAAAM for 2012 & Proceedings of the 23rd International DAAAM Symposium, ISBN 978-3-901509-91-9, ISSN 2304-1382, pp 0613 - 0616, Editor B[ranko] Katalinic, Published by DAAAM International, Vienna, Austria

[6] Ahmed, H. & Chacko, S.(2012). Computational Optimization of Vehicle Aerodynamics, Annals of DAAAM for 2012 & Proceedings of the 23rd International DAAAM Symposium, ISBN 978-3-901509-91-9, ISSN 2304-1382, pp 0313 - 0318, Editor B[ranko] Katalinic, Published by DAAAM International, Vienna, Austria

[7] Ramaj, V.; Dhori, A.; Ramaj, V.; Dhoska, K.; Koleci, A. & Konjusha, E. (2010). CFD Code Turbulence Models Validation for Turbulent Flows Over a Wavy Surface, Annals of DAAAM for 2010 & Proceedings of the 21st International DAAAM Symposium, 20-23rd October 2010, Zadar, Croatia, ISSN 1726-9679, ISBN 978-3-901509-73-5, Katalinic, B. (Ed.), pp. 0425-0426, Published by DAAAM International Vienna, Vienna

[8] Chwastyk, P. & Kolosowski, M.(2012). Integration CAD/CAPP/CAM Systems in Design Process of Innovative Products, Annals of DAAAM for 2012 & Proceedings of the 23rd International DAAAM Symposium, ISBN 978-3-901509-91-9, ISSN 2304-1382, pp 0397 - 0400, Editor B[ranko] Katalinic, Published by DAAAM International, Vienna, Austria

[9] Piscan, I.; Janssens, T. & Predincea, N. (2011). Experimental Validation of FEM for Frictional Contacts, Annals of DAAAM for 2011 & Proceedings of the 22nd International DAAAM Symposium, 23-26th November 2011, Vienna, Austria, Volume 22, No. 1, ISSN 1726-9679, ISBN 978-3-901509-83-4, Katalinic, B. (Ed.), pp. 0741-0742, Published by DAAAM International Vienna, Vienna

DOI: 10.2507/27th.daaam.proceedings.054

MODELLING OF EMC BROADBAND ANTENNAS WITH COMPLEX GEOMETRY ABOVE 1 GHZ

Tereza Novakova, Mikulas Bittera, Viktor Smiesko & Eva Kralikova

This Publication has to be referred as: Novakova, T[ereza]; Bittera, M[ikulas]; Smiesko, V[iktor] & Kralikova, E[va] (2016). Modelling of Emc Broadband Antennas With Complex Geometry Above 1 GHz, Proceedings of the 27th DAAAM International Symposium, pp.0363-0370, B. Katalinic (Ed.), Published by DAAAM International, ISBN 978-3-902734-08-2, ISSN 1726-9679, Vienna, Austria
DOI: 10.2507/27th.daaam.proceedings.054

Abstract

Radiofrequency measurements of electromagnetic compatibility are negatively influenced by uncertainty of the measuring antenna. To make quantification of errors caused by the antenna faster, it is useful to create an antenna model and simulate the circumstance. A change in parameter values caused by a change of position of the antenna or a change of surroundings can be determined using a numerical electromagnetic simulation software. Modelling a precise copy of antenna, however, can cause unnecessarily high simulation requirements, therefore it is essential to know how model structure simplification affects the characteristics of the model. For this purpose, broadband antennas commonly used for measurements of electromagnetic compatibility above 1 GHz were taken into account: stacked logarithmic-periodic antenna and double-ridged horn antenna. Multiple models were created with various extent of simplification to achieve reduction in computational time and memory requirements. The level of correspondence of the model to the real antenna was inspected, using chosen electromagnetic simulation software, by obtaining characteristic parameters of the models, i.e. antenna factor, gain, and radiation pattern, and by comparison of results to measured values. Considering the results and computational effort of the simulation, the most suitable model was chosen for each antenna.

Keywords: Broadband antenna; Antenna model; Feature selective validation; Stacked logarithmic-periodic dipole array; Double-ridged horn antenna.

1. Introduction

Numerical simulators have become widely used to solve problems in electromagnetism, since it is very difficult to solve the problems analytically for unsymmetrical and inhomogeneous structures [1,2,3]. When simulating complex antennas, one can come to a point, when the simulation requirements are higher than the computational power of the computers available for computation. In such case, it is very important to know the right approach to modelling, to be able to create a simulation model that copies the real problem at full, but simultaneously does not demand excessively on the computational power and simulation time [2].

This paper is focused on the choice of the optimal model of broadband antennas with complex geometry that are commonly used in measurements of electromagnetic compatibility for frequency range above 1 GHz. To analyse the models' characteristic parameters, e.g. antenna factor, gain, and radiation pattern, numerical simulator FEKO was used.

Characteristic parameters of each model were then compared to measured values and their agreement was validated as a comparison of two data sets using feature selective validation method.

2. Problem analysis

Solving difficult problems in electromagnetism can be done experimentally, analytically, or numerically. Numerical methods, as the most practical way of solving these problems, can be divided into three main categories: frequency domain, time domain and eigenmode or modal solvers, while the most suitable methods are solvers in frequency domain [4]. They generally solve Maxwell's equations in their integral form [5]:

$$\iiint_V \nabla X \, dx \, dy \, dz = Y \tag{1}$$

where Y is the source, X is the unknown function, V is the volume of the analysed area, and ∇ is the Hamiltonian operator. Frequency domain solvers discretize the solution domain, build a matrix, invert the matrix and use iteration to find the solution X [5]:

$$X = [f_n]\,[l_{mn}^{-1}][g_m] \tag{2}$$

where f_n are the basis functions, g_m are the exciting functions, and l_{mn} are the linear operators. The matrix building and its solution must be repeated for each frequency.

Numerical method has to be chosen in regard to the type of application for which it is used and requires some level of compromising. The method used in this work, Method of Moments (MoM), was chosen and tested in previous work and is described in [6]. The simulations were carried out in electromagnetic fields simulator FEKO, which uses MoM in combination with other auxiliary methods, such as Finite Element Method, Finite Difference Domain Method, etc. [7]. Model surface can be discretised into mesh using simple wire segments or triangles.

After retrieving the simulation results, validation is an important step to quantify the comparison of result data sets. For this purpose, Feature Selective Validation (FSV) has been used, which quantifies the comparison by mirroring engineers' visual perception and allows for automated comparisons of large complex data [8]. The data comparison is based on decomposing the original data into two component measures and subsequent recombination of the results to provide a global goodness of fit measure [9]. FSV offers three means of the data comparison [10]:

- Amplitude difference measurement (ADM) – comparison of the amplitudes and 'trends' of the two data sets
- Feature difference measurement (FDM) – comparison of the rapidly changing features as a function of independent variable
- Global difference measurement (GDM) – combination of the previous two

GDM can be expressed as:

$$GDM = \sqrt{ADM^2 + FDM^2} \tag{3}$$

The resulting values can be divided into six natural language descriptor categories, ranging from Excellent to Very Poor. In [11], FSV validation has been presented for computational electromagnetics modelling and simulation.

Previous works that analysed and validated the effect of model accuracy on simulation results for logarithmic-periodic dipole array and hybrid biconical-logarithmic-periodic dipole array are in [4] and [12]. The accuracy of the model dimensions and shape has considerable impact on simulation results. To analyse the extent of this impact further, two geometrically different broadband antennas were taken into account and models were created.

2.1. Modelled antennas

First analysed, stacked logarithmic-periodic array, further denoted as STLP, consists of four identical arms. Two pairs of arms are placed to form a V-shape in order to obtain a higher degree of directivity than observed in standard logarithmic-periodic antennas [13]. The STLP antenna consists of a 1 mm thick metal plate with 2×20 dipole pairs, with the shortest and longest dipole element length being 8 mm and 95 mm respectively. Length of the antenna in the direction of the main lobe is 244 mm. The scale factor is 0.87 and spacing factor 0.21.

Another analysed, a double-ridged horn antenna, further denoted as DRGH, has an aperture dimensions of 243 mm x 155 mm and the length in the direction of the main lobe is 210 mm. In general, horn antennas are strongly influenced by imperfections in manufacturing process. Even a slight deviation from intended dimensions and shape, mainly in the feeding structure of the antenna, can lead to significant change in characteristic parameters of the antenna [14]. As a consequence, also the simulation model has to be very precise, to obtain satisfactory results.

Both STLP and DRGH antenna are used for applications in frequency range 1 GHz – 18 GHz and are widely used in electromagnetic compatibility testing of devices.

2.2. Simulation models

To create most of the models, CAD module of FEKO simulator was used. Two different models of STLP antenna were created, which are shown in Fig. 1:

- Surface model – omitting the thickness of the material, with expectation of shorter simulation time and lower computational requirements (Fig. 1.a);
- Volume model – respecting 1 mm thickness of metallic material, with expectation of more accurate results (Fig. 1.b).

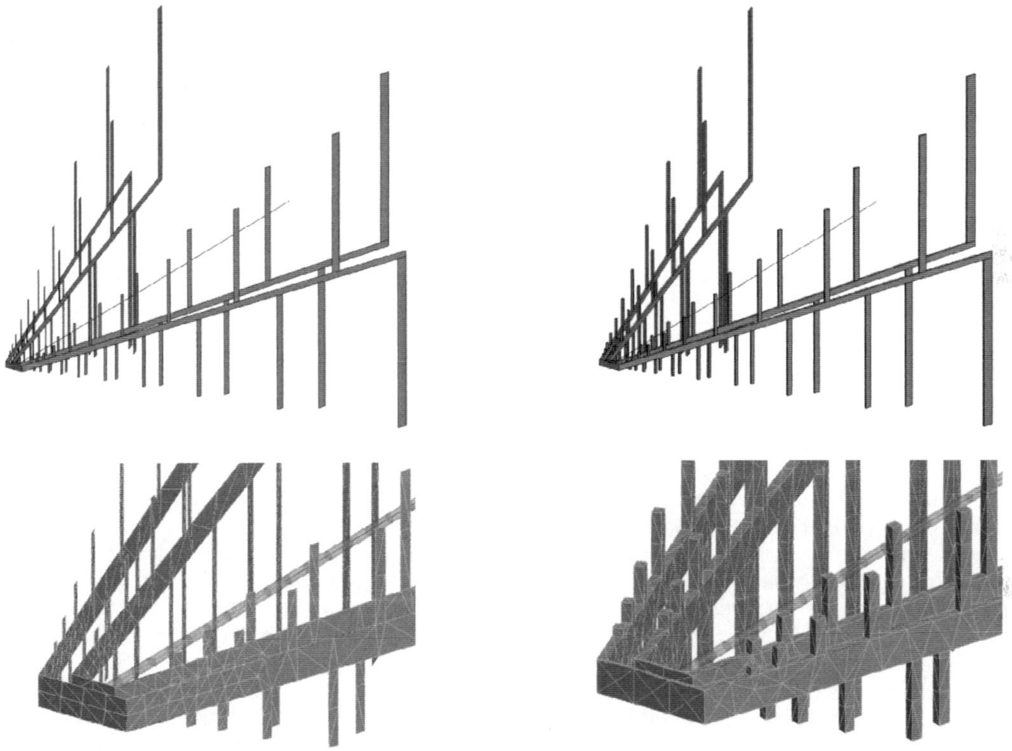

a) b)

Fig. 1. STLP antenna models: (a) surface and (b) volume, without mesh (up) and detailed with mesh (down)

For DRGH antenna, three different models were created, as seen in Fig. 2:

- Surface model – omitting thickness of the horn, omitting perforated side walls, resonator measured and modelled manually, with expectation of shorter simulation time and lower computational requirements (Fig. 2.a)
- Scan model – a precise model obtained by 3D scanning, with expectation of more accurate results (Fig. 2.b)
- Combined model – a combination of previous two - precise scanned resonator, omitted perforated side walls and material thickness, expected reduction in simulation time and computational requirements, while sustaining satisfactory results (Fig. 2.c)

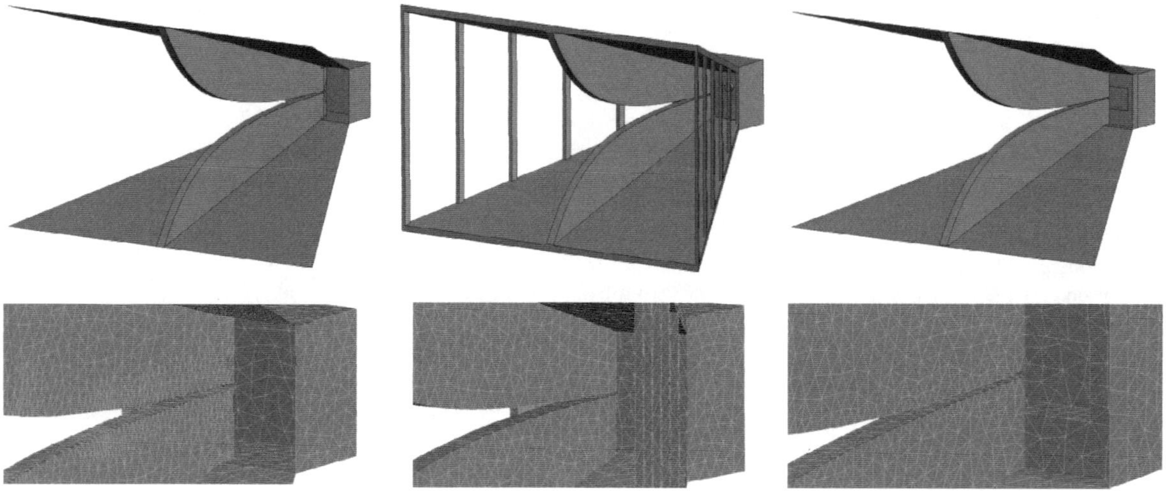

a) b) c)

Fig. 2. DRGH antenna models: (a) surface; (b) scan; (c) combined, without mesh (up) and detailed with mesh (down)

Scan model of DRGH antenna does not contain inner structure behind resonators, which was inaccessible for the optical 3D scanner and was also omitted in surface model for the matter of comparison. All three models of DRGH do not contain feeding part of the antenna, as it was inaccessible without dismantling, which would mean a risk of permanent damage to the antenna.

3. Results

For both antennas, three characteristic parameters were obtained, analysed, and validated: antenna factor, gain and radiation pattern. For each model, the three parameters were obtained in the antennas' intended frequency range, 1 GHz – 18 GHz, and compared to the measured values.

Antenna factor is roughly described as the extent, at which an antenna can transform an electric field it is placed in to voltage. As the most important parameter of an antenna, antenna factor was obtained using known values of electric field radiated from a source at a known distance and the induced antenna voltage, then computed using near-field analysis in FEKO. Measured values of antenna factor were obtained by three-antenna method following the international standard [15].

Gain, a parameter described as radiated field strength in specific direction, which may be expressed relative to radiation of a theoretical isotropic radiator or dipole in the same direction, was obtained using far-field analysis. Gain value is achieved by pattern alternation (directivity), but antenna losses are subtracted. Only the value in the direction of maximum radiation was considered. Measurement of gain was performed also by three-antenna method at the measuring distance of 3 m.

Radiation pattern, described as radiated electric field strength as a function of azimuth and elevation, was also computed using far-field analysis. Measurement of radiation pattern was performed in anechoic chamber, which has to be fully anechoic to avoid the influence of the surroundings on the results.

3.1. STLP antenna

As seen in Tab. 1, volume model of STLP antenna shows more satisfactory results, reaching GDM value of 0.144 for antenna factor and 0.593 for gain, while surface model shows GDM values of 0.179 for antenna factor and 0.717 for gain. Comparison of models' results to measured values is shown in Fig. 3.a for antenna factor and Fig. 3.b for gain.

	Antenna factor			Gain		
Model type	**ADM**	**FDM**	**GDM**	**ADM**	**FDM**	**GDM**
Surface	0.080	0.153	0.179	0.370	0.522	0.717
Volume	0.073	0.110	0.144	0.343	0.409	0.593

Table 1. STLP antenna models: ADM, FDM, and GDM values of antenna factor and gain

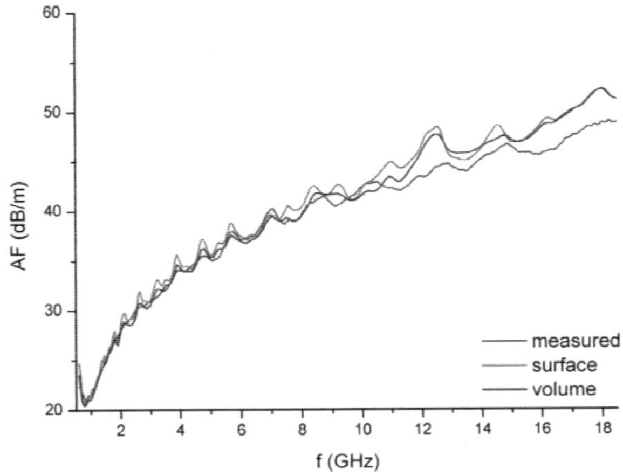

a)

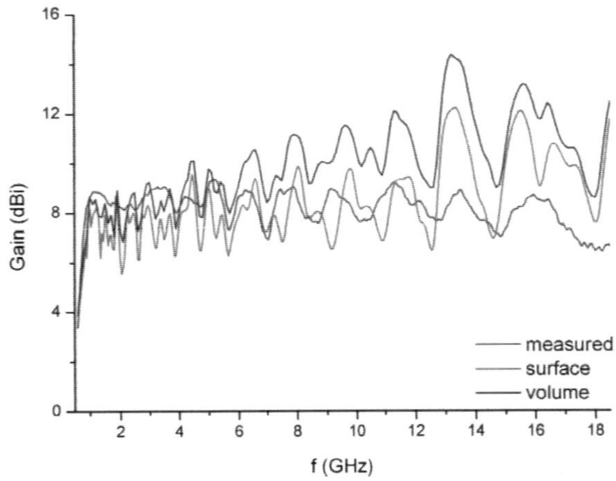

b)

Fig. 3. STLP antenna: Comparison of (a) antenna factor and (b) gain of analysed models to measured values

Both surface and volume model exhibit relatively good agreement of radiation pattern trace in the main lobe, as seen in Fig. 4 for 14 GHz in E-plane and H-plane.

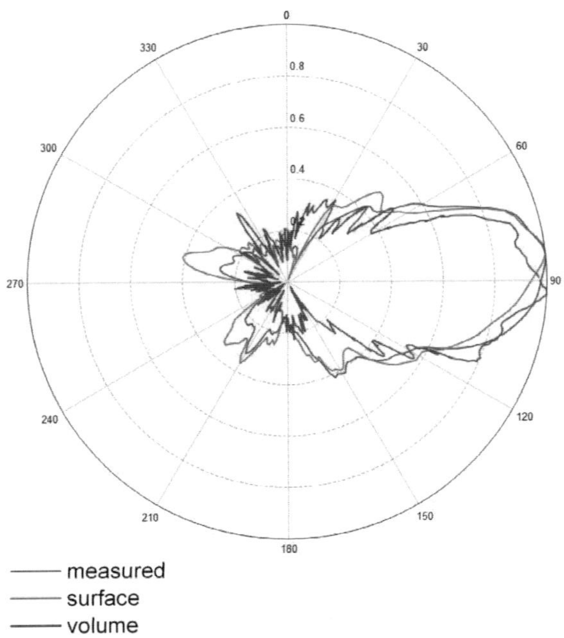

a)

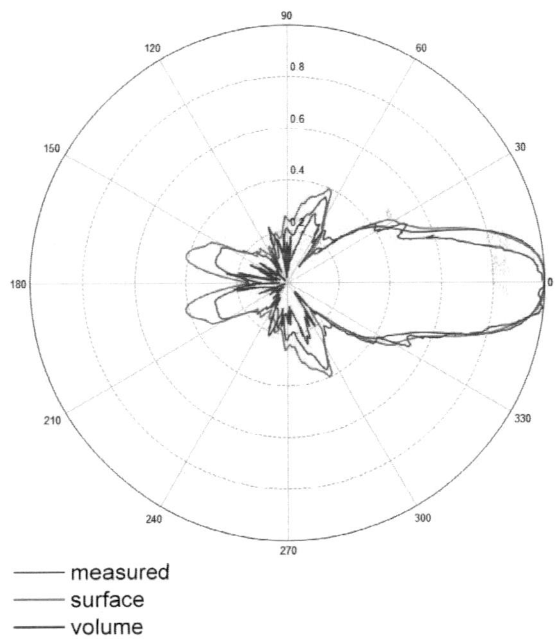

b)

Fig. 4. STLP antenna: Comparison of radiation pattern of different models to measured values in (a) E-plane and (b) H-plane at 14 GHz

To select the most suitable model, not only the simulation results are important, but also the memory size required by computation process and the overall simulation time have to be taken into account. Tab. 2 shows the difference in simulation parameters and the increase of required time and memory caused by different structure of the model.

Model type	Simulation time	Number of triangular segments	Number of wire segments	Required memory RAM	Evaluation matrix size
Surface	813 s	1 220	55	16.97 MB	1 390 x 1 390
Volume	14 882 s	4 788	55	426.39 MB	7 399 x 7 399

Table 2. STLP antenna models: Simulation parameters

3.2. DRGH antenna

Validation results of DRGH antenna are generally less satisfying than those of STLP antenna, both surface and scan model show relatively high values of GDM for antenna factor, that is 0.372 and 0.310 respectively, whereas the values of GDM for gain are as high as 1.260 and 1.350. For combined model, the GDM values are improved to 0.137 for antenna factor and 0.651 for gain. For surface and scan model, the resulting antenna factor and gain trace, shown in Fig. 5.a and Fig. 5.b, suffer from significant deviation seen as undesired peaks at several frequencies.

This may be due to imperfections in the model geometry, primarily imprecise resonator curve in surface model and missing inner part behind the resonators of both models.

This phenomenon is overcome in combined model, consisting of the precise resonator curve and the inner structure. This model, however, maintains visible deviation in both antenna factor and gain trace increasing at higher frequencies.

	Antenna factor			Gain		
Model type	ADM	FDM	GDM	ADM	FDM	GDM
Surface	0.212	0.372	0.450	0.687	0.990	1.260
Scan	0.186	0.310	0.377	0.671	1.050	1.356
Combined	0.077	0.109	0.137	0.445	0.454	0.651

Table 3. DRGH antenna models: ADM, FDM, and GDM values of antenna factor and gain

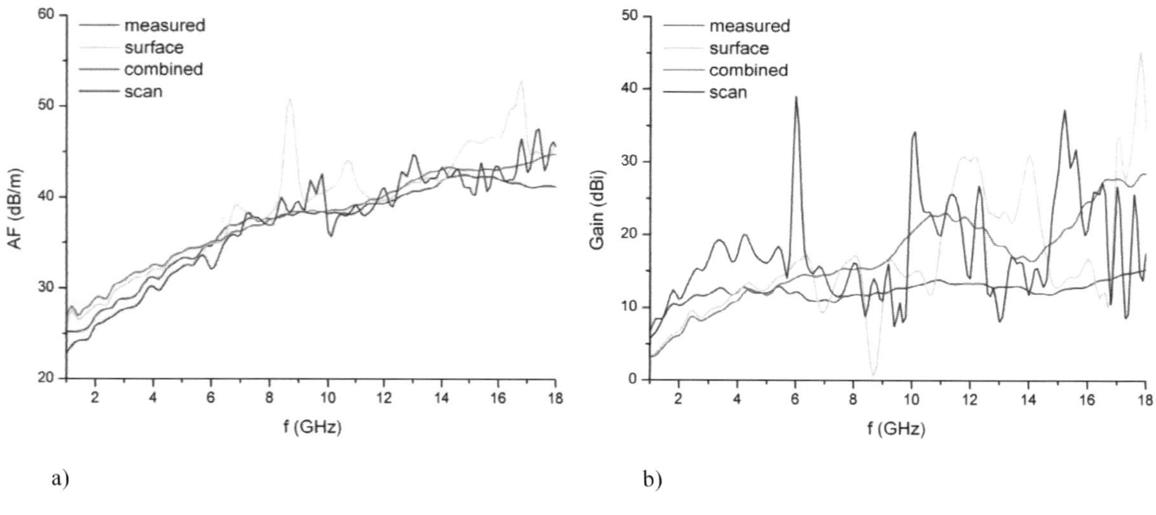

Fig. 5. HORN antenna: Comparison of (a) antenna factor and (b) gain of analysed models to measured values

Models show fair to good agreement in radiation pattern in the main lobe, as seen in Fig. 6 for 14 GHz. Similarly as for gain trace, both surface and scan model fail at several frequencies showing poor accuracy of radiation pattern compared to measured values.

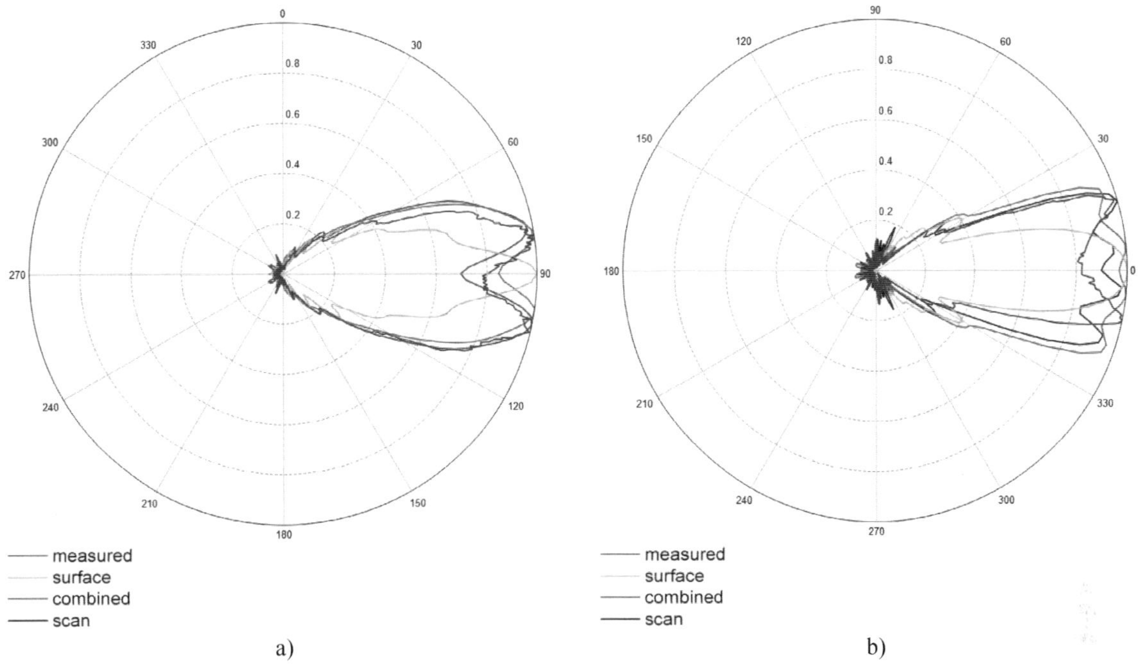

Fig. 6. DRGH antenna: Comparison of radiation pattern of different models to measured values in (a) E-plane and (b) H-plane at 14 GHz

The final simulation parameters for DRGH antenna are given in Tab. 4. In the case of DRGH antenna, not only the results, but also the simulation parameters are markedly improved in combined model, which is caused by simplification of resonator curve surface and replacement of bulk walls of the horn with simple surface walls.

Model type	**Simulation time**	**Number of triangular segments**	**Number of wire segments**	**Required memory RAM**	**Evaluation matrix size**
Surface	33 334 s	9 000	13	1.37 GB	13 476 x 13 476
Scan	80 165 s	13 092	13	2.89 GB	19 629 x 19 629
Combined	29 245 s	8 286	13	1.16 GB	12 411 x 12 411

Table 4. DRGH antenna models: Simulation parameters

4. Conclusion

In practical applications, there are times when duration of the simulation and memory requirements of an antenna model exceed tolerable extent. With simplification of the model structure however, the accuracy of the model parameters decreases. When choosing the right type of a model, it is important to find a compromise.

For STLP antenna, both models, surface and volume, show relatively similar results, although the simulation parameters are much more favourable for surface model, where simulation time and the memory requirements are almost 20-times lower. Therefore, the surface model omitting thickness of the metallic material may be more suitable for use in practical applications.

As for DRGH antenna, the only model showing good resemblance of characteristic parameter values compared to measured values is combined model. This confirms the expected need of high precision of model dimensions in the resonator and inner structure part of the antenna, while at the same time, the model keeps lowest simulation requirements at as much as 1/3 of the time and memory required by scan model. Replacing the horn walls with simple surface and completely omitting perforated side walls has no significant effect on model parameters.

However, omitting detailed feeding part of the antenna might be the reason for the significant deviation, which is the field of interest for further study to acquire even better results while preserving low simulation and memory requirements.

5. Acknowledgements

The work presented in this paper was supported by the Slovak Grant Agencies of Ministry of Education, Science, Research and Sport of the Slovak Republic VEGA, No. 1/0431/15 and Slovak Research and Development Agency APVV, No. APVV-15-0062.

6. References

[1] Cicchetti, R., Testa, O. & Caratelli, D. (2012). A numerical procedure for the analysis of EMC/EMI problems in radio communication systems operating in complex environment, In: IEEE Transactions on Electromagnetic Compatibility, vol. 54, no. 6 (December 2012), pp. 1269-1280.

[2] Ndip, I., et. al. (2012). Modelling the shape, length and radiation characteristics of bond wire antennas, In: IET Microwaves, Antennas & Propagation, vol. 6, no. 10 (July 2012), pp. 1187-1194.

[3] Matyas, J., Vlcek, K. (2011). Experimental antenna design based on natural construction. Proceedings of 22th DAAAM international symposium on Intelligent manufacturing and automation, Zadar, Croatia, vol. 22 (2011), ISSN 1726-9679, Katalinic, B. (Ed.), pp. 1435-1436, DAAAM International, Vienna

[4] Bittera, M. (2013). Modeling broadband wire antennas with complex geometry. Proceedings of 24th DAAAM international symposium on Intelligent manufacturing and automation, Procedia engineering, Zadar, Croatia, vol. 69 (2014), ISSN 1877-7058, Katalinic, B. (Ed.), pp. 1082-1087, Elsevier, doi:10.1016/j.proeng.2014.03.094

[5] Duffy, A., et. al. (2005). The feature selective validation (FSV) method, In: IEEE International Symposium on EMC, Chicago, (August 2005), ISSN 2158-1118, pp. 272-277

[6] Harrington, R.F. (1993) Field computation by moment method, Wiley-IEEE Press, ISBN 0-7803-1014-4, New York

[7] http://www.altairuniversity.com/wp-content/uploads/2015/03/UserManual.pdf, (2015). FEKO 7.0 User's Manual, EM Software and Systems-S-A, Accessed on: 2016-05-02.

[8] http://ing.univaq.it/uaqemc/FSV_4_0_3L/1DFSV_Manual.pdf, (2007). 1D FSV Tool, User's Guide, Version: 4, Accessed on: 2016-07-02

[9] Hartansky, R. & Hallon, J. (2016). Sensors for electromagnetic field measurements, STU, ISBN 978-80-227-4565-9, Bratislava

[10] IEEE 1579.1 (2009). Standard for validation of computational electromagnetics computer modeling and simulation, IEEE Standard

[11] Harrington, R. F. (1993). Field computation by moment methods, first ed., Wiley, ISBN: 978-0-7803-1014-8, New York

[12] Bittera, M., Stih, M. (2009). Modelling of wire antennas, In: Mechatronika 2009: Proceedings of 12th International Conference on Mechatronics, Trencianske Teplice, Slovak Republic, 3-5.6.2009, Trencin: Trencianska univerzita Andreja Dubceka v Trencine, 2009, ISBN 978-80-8075-392-4, pp. 73-76

[13] Kraus, D. P. (2001). Antennas, Third ed., ISBN 0-07-123201-X McGraw-Hill Education, Singapore

[14] Jacobs, B., Odendaal, J. W. & Joubert, J. (2009). Modelling manufacturing tolerances in 1-18 GHz double-ridged horn antennas, In: 39th European Microwave Conference, Rome, 2009, ISBN 978-1-4244-4748-0, pp. 1484-1487

[15] ANSI C63.5 (2006). American national standard for calibration of antennas used for radiated emission measurements into electromagnetic interference, American National Standard

DOI: 10.2507/27th.daaam.proceedings.055

CMM MEASURING CYCLE AND HUMAN FACTOR

Martin Melichar, Dana Kubátová & Jan Kutlwašer

This Publication has to be referred as: Melichar, M[artin]; Kubatova, D[ana] & Kutlwaser, J[an] (2016). CMM Measuring Cycle and Human Factor, Proceedings of the 27th DAAAM International Symposium, pp.0371-0376, B. Katalinic (Ed.), Published by DAAAM International, ISBN 978-3-902734-08-2, ISSN 1726-9679, Vienna, Austria
DOI: 10.2507/27th.daaam.proceedings.055

Abstract

This paper deals with topic of evaluation of precise metrology application. Current state of measuring problems in the field od automotive industry often suppose that inspection of product guarantee the quality of this product. This idea is even stronger if the inspection is done by CMM. Research team tried to confirm or disprove importance of human factor for automatic measuring cycle of CMM. Using real parts with tolerances 11x bigger, that declared possibilities of CMM team of researchers done MSA analysis. Result of the experiment confirmed that the importance of human factor starts way above precision of CMM and it is necessary to focus on.

Keywords: MSA; CMM; measuring; metrology; measuring cycle.

1. Introduction

After being overlooked for a long time, metrology is gaining importance in today's world of technology, all the more so because of current standards in purchasing. In general, customer requirements do not end with products meeting their specifications, and with final inspection reports being presented as evidence of this conformity any more. An ever more frequent requirement concerns evaluation of supplier's measurement system to demonstrate its readiness, repeatability, reproducibility and suitability for the application in question. Final control of product is often implemented on a CMM. Here the question arises - how it affects the human factor in automatic mode CMM

2. Theory of MSA

There is a well-proven and customer-recognized methodology for demonstrating the qualification of a measurement system for a particular application: the MSA (Measurement System Analysis) quality tool. It uses simple experiments to find whether a particular approach chosen by the inspection personnel for a particular issue is suitable or at least adequate. One typical example involves verification of the required repeatability and reproducibility of results on pliable products, such as plastics, obtained by inspection personnel with a manual gauge that exhibits first-order error.

The outcome of the MSA analysis is the GR&R parameter. [3] In practice, it is the percentage of the initial tolerance specified by the customer which the supplier "uses up" owing to ambiguous measurement. If GR&R exceeds 30 %, the measurement system becomes unjustifiably expensive for the supplier (as it drives up the costs of manufacture which therefore must operate within a much narrower tolerance band), and unacceptable for the customer. [7]

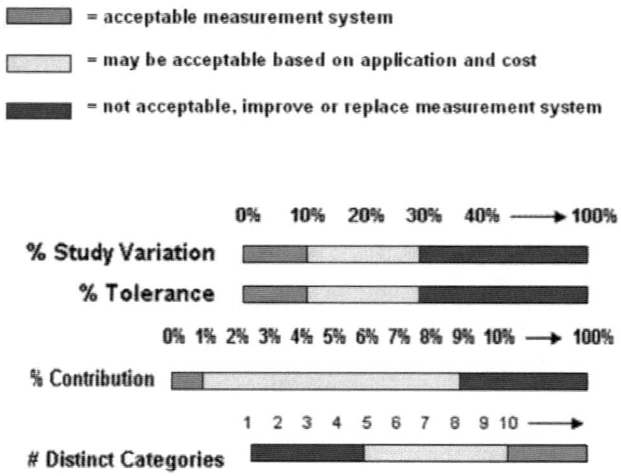

Fig. 1. MSA principle [2]

Manual measurement operations – to which MSA is traditionally applied – are subject to a number of factors linked to the human operator who possesses unique features including:

- Physical abilities
- Experience
- Attention
- Care
- Others.[6]

Since today's metrology laboratories are forced to rely on CMM systems, the effect of the operator can be expected to be minimized, provided that unified methodology and measurement programme are used. It order to test this hypothesis, the metrology laboratories of the Regional Technological Institute have conducted an experiment. [1]

3. Experimental

The measurement system analysis presented here involved CMM measurement of components from a single lot in large-series production. In the framework of MSA, the most widely used method was selected: assessment of average values and spread in a group of ten randomly chosen products from a single lot. The purpose of this experiment was to find whether MSA should be applied to the automatic mode of CMM operation as well or whether CMM can be relied on as a means of eliminating the operator influence. [4]

Input parameters of the experiment:

- On purpose, critical parts for automotive industry were selected whose tolerance was on the order of 0.01 mm (Fig.3)
- A suitable parameter – diameter of a hole – was chosen
- The measurement was performed on the Carl Zeiss Prismo 7 Navigator CMM (Fig.2) which itself guarantees high repeatability (approx. 1.5 micrometre using the index able head)
- For this measurement, a special program was developed for the part type
- All data collection took place under consistent controlled conditions t= 20±1°C, 50 % humidity
- Three operators each of whom possessed a different level of experience with CMM and who did not know each other were selected in order to avoid undesirable interference
- Using a measurement cycle, each operator measured 10 parts three times (in a random order)

Fig. 2. Zeiss Prismo7 Navigator

Fig. 3. Touch the measurement

The only theoretical potential complications affecting repeatability and repeatability are as follows:

- Machine's repeatability error
- Machine vibration – e.g. due to the speed of traversing mechanisms or due to the environment
- System points inadequately recorded by the operator
- Part inadequately clamped by the operator
- Incorrectly chosen stylus tip – e.g. its diameter
- Program error [5]

Operator -- Measurement series		Part number									
		1	2	3	4	5	6	7	8	9	10
A	1	15,430	15,428	15,428	15,437	15,426	15,433	15,438	15,436	15,438	15,428
	2	15,428	15,428	15,429	15,437	15,431	15,432	15,439	15,436	15,437	15,428
	3	15,430	15,425	15,428	15,435	15,426	15,432	15,440	15,436	15,436	15,427
Average		15,429	15,427	15,428	15,436	15,428	15,432	15,439	15,436	15,437	15,428
Spread		0,002	0,003	0,001	0,002	0,005	0,001	0,002	0,000	0,002	0,001
B	1	15,427	15,428	15,427	15,437	15,430	15,434	15,441	15,435	15,436	15,428
	2	15,429	15,427	15,427	15,436	15,431	15,437	15,440	15,435	15,437	15,428
	3	15,430	15,428	15,429	15,437	15,430	15,432	15,442	15,435	15,435	15,429
Average		15,429	15,428	15,428	15,437	15,430	15,434	15,441	15,435	15,436	15,428
Spread		0,003	0,001	0,002	0,001	0,001	0,005	0,002	0,000	0,002	0,001
C	1	15,430	15,429	15,430	15,436	15,430	15,434	15,440	15,435	15,438	15,427
	2	15,426	15,426	15,429	15,438	15,429	15,434	15,440	15,437	15,438	15,427
	3	15,424	15,426	15,429	15,437	15,430	15,432	15,440	15,437	15,436	15,429
Average		15,427	15,427	15,429	15,437	15,430	15,433	15,440	15,436	15,437	15,428
Spread		0,006	0,003	0,001	0,002	0,001	0,002	0,000	0,002	0,002	0,002

Table 1. Table of measured values for the three operators in 10 repetitions

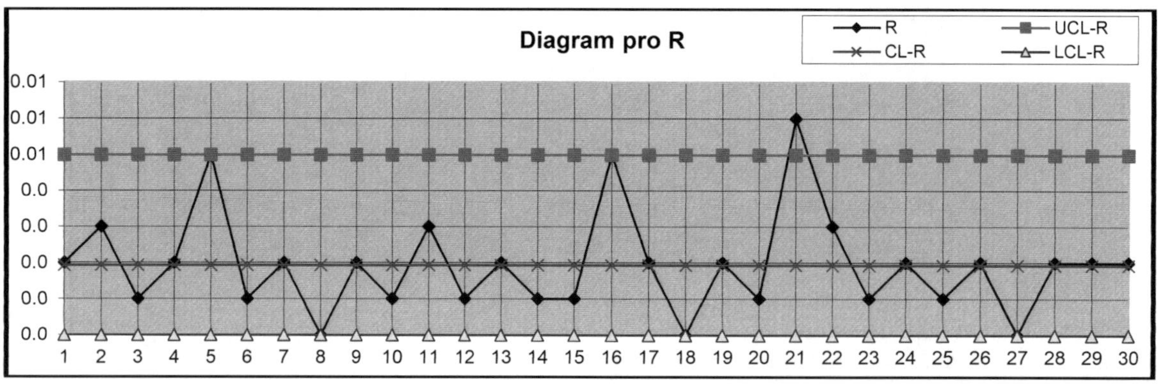

Fig. 1. Graph of scattering of the measured values

Analysis measuring unit	Total variance (TV)
- **Repeatability – scattering instrument (EV)** $EV = \bar{\bar{R}} \times K_1$ $\underline{EV = 0{,}001142213}$	%EV = 100[EV/TV]
- **Reproducibility – scattering operator** $AV = \sqrt{[(\bar{X}_{DIFF} \times K_2)^2 - (EV^2/nr)]}$ $\underline{AV = 0{,}00016}$ n = number of components r = number of selections	%AV = 100[AV/TV] %AV = 0,16
- **Repeatability and reproducibility (R&R)** $R\&R = \sqrt{(EV^2 + AV^2)}$ $\underline{R\&R = 0{,}00115}$	**<u>%GR&R = 1,15</u>**
- **Variance components (PV)** $PV = \sqrt{(TV)^2 - (R\&R)^2}$ $\underline{PV = 0{,}1001}$	%PV=100,0066
- **The total variance (TV)** $TV = (USL - LSL)/6$ $\underline{TV = 0{,}1}$	ndc=1,41*(PV/ R&R) 122,2903

Selections	3
K_1	0,5908

operator	2	3
K2	0,7071	0,5231

Part.	2	3	4	5	6	7	8	9	10
K_3	0,7071	0,5231	0,4467	0,403	0,3742	0,3534	0,3375	0,3249	0,3146

Table 2. Analysis measuring unit [7]

4. Conclusion

Using the experimental method was verified human factor influence on automatic CMM measurements. Applied methodology followed three independent operators, who controlled 10 parts. The experiment suggests that although the GR&R value was found to be deep below the limit of 10 percent. For standard MSA analysis it is the inadmissible system and our measurement seems to be okay, but the results must be analysed extensively.

For instance, if it became necessary to measure components with much better accuracy than in our experimental arrangement, our analysis of our system could soon encounter unacceptable boundary values. Once the required component accuracy drops to approximately 0.005 mm, the entire measurement process would have to be automated to such extent that at least the component clamping process and the initial identification of system points would have to be independent of human intervention.

There is one more important fact that must be taken into account – material component. A majority of those components where a request for MSA could be considered come from the field of processing of soft plastics where the potential effect of the operator can be even higher due to the force on the stylus. Inconsistencies speed of touch while operating by operator can be a significant source of uncertainty.

It is therefore certain that monitoring repeatability and reproducibility of an automatic CMM in a supply chain of critical components is more than advisable. It can help reveal weak spots of the process which can ultimately lead to cost savings in production because a smaller portion of the available tolerance will be used up by measurement.

The investigation of effects of individual factors on the overall outcome of measurement is part of the project no. LO1502 B16.

5. Acknowledgments

This paper was supported by the project SGS-2016-005.

6. References

[1] Antonio Piratelli-Filho, Benedito Di Giacomo, CMM uncertainty analysis with factorial design, Precision Engineering, Volume 27, Issue 3, July 2003, Pages 283-288, ISSN 0141-6359, http://dx.doi.org/10.1016/S0141-6359(03)00035-7

[2] http://www.six-sigma-material.com/MSA.html [online]. [cit. 2016-09-08]

[3] Https://www.moresteam.com/toolbox/measurement-system-analysis.cfm: Measurement System Analysis (MSA) [online]. [cit. 2016-09-08]

[4] Matthias Asplund, Jing Lin, Evaluating the capability of measuring system for measuring the profile of the wheels using GR & amp; R, vol. 92, October 2016, page 19-27, ISSN 0263-2241, 2016.05.090

[5] Burdick, Borror, Montgomery: A review of methods for measurement systems capability analysis (2003) Journal of Quality Technology, 35 (4), pp. 342-354. Cited 133 times. https://www.scopus.com/inward/record.uri?eid=2-s2.0-1442324285&partnerID=40&md5=b85b8c9799a460da650f855106a13f11

[6] Wheeler: Problems with gauge R&R studies(1992) Annual Quality Congress Transactions, 46, pp. 179-185. https://www.scopus.com/inward/record.uri?eid=2-s2.0-0026987231&partnerID=40&md5=3cbac4ee9a86a24e46f38da58903721c

[7] http://www.rubymetrology.com/add_help_doc/MSA_Reference_Manual_4th_Edition.pdf [online]. [cit. 2016-10-06]

[8] Rewilak, J. MSA Planning - A proposition of a method (2015) Key Engineering Materials, 637, pp. 45-56. https://www.scopus.com/inward/record.uri?eid=2-s2.0-84924777858&partnerID=40&md5=a60d05220800743a43ac119db4560745

DOI: 10.2507/27th.daaam.proceedings.056

AUTOMATED GENERATION OF QUESTIONS FOR BASIC ELECTRICAL ENGINEERING EDUCATION

Frano Skopljanac-Macina, Bruno Blaskovic & Damir Pintar

This Publication has to be referred as: Skopljanac-Macina, F[rano]; Blaskovic, B[runo] & Pintar, D[amir] (2016). Automated Generation of Questions for Basic Electrical Engineering Education, Proceedings of the 27th DAAAM International Symposium, pp.0377-0385, B. Katalinic (Ed.), Published by DAAAM International, ISBN 978-3-902734-08-2, ISSN 1726-9679, Vienna, Austria
DOI: 10.2507/27th.daaam.proceedings.056

Abstract

This paper deals with the problem of automated generation of questions for basic electrical engineering education in e-learning systems. One of the main challenges in developing e-learning systems is to create an effective knowledge testing module. In most of e-learning systems teachers need to manually compose test questions, prepare the correct and wrong answers, and added feedback information. The process is time-consuming and prone to errors. We have focused on problems in basic electrical engineering education, specifically series – parallel DC circuits, and we propose a solution for automated generation of template based questions of various degrees of similarity and complexity. Our prototype Web application dynamically generates text of the question and its circuit diagram image, solves the problem, and as a feedback provides correct results and a step-by-step solution. The interface is bilingual, available in Croatian and English. We have used standard Web technologies, such as *JavaScript*, *jQuery* and SVG, together with *MathJax* for mathematical notation. We will recommend our students on the course Fundamentals of Electrical Engineering to use this Web Application for self-assessment and practice. In future work we plan to advance our Web application and to fully integrate it into our developing e-learning system.

Keywords: Automated question generation; automated feedback generation; e-learning systems; random circuit diagram generation.

1. Introduction

E-learning systems use computer technology and Internet in the field of education. E-learning is primarily used in formal education to support teaching and learning process. Today, e-learning systems are also used in corporate environment as a tool for training and testing current and future employees. E-learning is also used semi-formally for distance education in MOOCs (*Massive online open courses*), and informally in collaborative and life-long education, e.g. online (video) tutorials, social network groups and forums [1].

Today, e-learning systems are getting more complex, combining advantages of Intelligent Tutoring Systems (ITS) – its individual approach to each student and student–tutor two way communication, with those of Adaptive Hypermedia Systems (AHS) – adaptable presentation of learning materials and choice of different learning paths. Such advanced e-learning systems are known as Adaptive Web-based Educational Systems [2]. In universities they are usually based on the *Moodle* learning management system [3]. These systems facilitate learning process by providing students with individual adaptive learning environment, but also they can be efficiently used in training of teachers and administrative staff [4]. Online educational technologies can be used to create inter-university networks to connect, share knowledge and to remotely train specialists in different fields [5].

Nevertheless, teachers and instructors must invest a lot of their time and resources in creating and maintaining teaching and learning online materials, basic learning scenarios and learning paths. Finally, the process of devising test questions and preparing testing materials is very important, but usually delicate and time-consuming. For each new question instructors must also give the correct and the wrong answers, sometimes also draw a figure, and preferably add useful hints.

Recently, there is growing research interest in the field of question generation. It deals with the dynamic creation of questions for a specific knowledge domain. Some of the research is based on Natural language processing and Natural language generation, and the goal is to automatically create textual questions based on a given set of factual information. This information can be (semi-)structured (concept maps or OWL ontologies) or even unstructured (free text from articles or handbooks) [6].

There are also interesting research directions in question and problem generation for STEM field education (*Science, Technology, Engineering and Mathematics*). Usually, questions in these fields contain not only text, but also figures and/or formulae. Gulwani et al. in [7], [8], [9] and [10] researched how to dynamically generate tests and questions, as well as useful feedback information in the STEM field education. In [8] the authors propose a model for automatic generation of rich feedback information when a student submits wrong solution to a simple programming assignment on the course Introduction to programming in Python. However, teacher must first prepare an example of a correct implementation for each assignment, as well as typical coding errors. The system then tries to automatically correct student's wrong solution. If it succeeds, the system gives student instant feedback with short explanations together with the list of coding errors. Authors report that for 64% of wrong solutions the system was able to correct them, and provide useful feedback to students. In [9] the authors present an interesting synthesis method of geometric proofs based on a given figure (110 figures scanned from Indian and American high school mathematics handbooks) and a defined set of axioms. In [10] the authors propose a method for automatic generation of new variants of algebra problems, based on original problems from mathematical handbooks. Zitko et al. in [11] presented their ITS system for automatic test generation for evaluating knowledge contained in an OWL ontology. Test provides dynamically generated questions based on a predefined set of generic question templates.

In this paper we will focus on calculation questions in the field of basic electrical engineering. These questions contain text and figures with electrical circuit schema. We asked ourselves is there an effective way to easily generate dynamic questions of this type? That would benefit the assessment process – it would ensure that the most students get different questions of similar complexity.

The rest of the paper is structured as follows: Section 2 contains information about our e-learning system, and the motivation for this research; in Section 3, first we set prerequisites and outcomes of proposed automated question generation method, and then describe its implementation; in Section 4 we discuss the main results of presented research; and Section 5 contains conclusion and an overview of future work.

2. Background

We have developed our own e-learning system *WebOE,* which is primarily used on the freshman year course (over 650 students) Fundamentals of Electrical Engineering at University of Zagreb (Croatia), Faculty of Electrical Engineering and Computing [12]. The professors can use it to check on students' progress during the semester, upload teaching materials, give homework and quizzes, prepare written and oral exams, and analyze their results. And vice versa, students use it keep track of their success and obligations during the course, access various teaching/learning materials, solve homework and online quizzes, and check timetables for laboratory assignments. The *WebOE* system questions database currently contains 2987 questions as it is shown in Table 1.

Question type	Static/dynamic content	Number of questions
multiple choice (1 correct answer)	static	2143
multiple choice (2 or more correct answers)	static	166
fill in	static	32
yes/no	static	33
general questions (oral and written exam)	static	171
homework (calculation)	dynamic	264
variable multiple choice questions (1 correct answer, multiple variations of each question)	dynamic	178
Total		**2987**

Table 1. Question database in *WebOE* e-learning system

From the Table 1 we can see that our question database is very rich, but nevertheless it is obvious that the majority of questions are static (2545) – question text, list of answers and an image (if present) are fixed. But, even with these static questions we can generate tests dynamically, by randomly selecting questions or with permutations of defined question lists. Also, the answer list in multiple choice questions can be permuted. Homework calculation questions are somewhat dynamic, because every student gets a unique combination of numeric inputs and different numeric solutions. Each homework calculation question must have its solution program for automatic checking and grading students' answers. But, the question text and the image are fixed so it is possible that more students get basically the same questions. Variable multiple choice questions are very dynamic, and we primarily use them for Laboratory quizzes. Each of those questions usually has 4 variations: question text can be substantially different in every variation, and different figures and answer lists can be used for each variation. Still, all of these questions have been prepared, checked and entered into the database manually.

Year	Type of questions	Number of students	Success rate
2007	static multiple choice	794	49.11%
2008	static multiple choice	718	76.27%
2009	static multiple choice	832	75.94%
2010	static multiple choice	764	75.86%
2011	static multiple choice	821	81.53%
2012	mixed (static & variable multiple choice)	771	71.21%
2013	variable multiple choice	622	56.24%
2014	variable multiple choice	563	53.20%
2015	variable multiple choice	667	58.40%
2016	variable multiple choice	656	56.96%

Table 2. Laboratory quizzes success rate (third laboratory test from 2007 to 2016)

As expected, the success rate of unchanged static questions rises with every new generation of students because of their collaborative work in gathering old questions and remembering correct answers (e.g. compare 2008 to 2007 success rate in Table 2). The success rate of old homework calculation questions also slowly rises every new academic year. Each year a number of old static questions and homework assignments have to be deactivated and manually replaced with their new variations. The variable questions have proved to be more resilient – since 2013 their success rate is more constant, oscillating below 60% success rate as can be seen in Table 2. That is much more representative than before 2012 when we used only static questions in laboratory quizzes (success rate was over 75%). Different variations, sometimes very subtle, successfully confuse those students who did not bother to learn, but instead tried to memorize correct answers.

Rich instant feedback can be very valuable to students, and can motivate them to continue learning and practicing online, especially in distance learning [13]. Our system offers basic feedback (is the answer correct or not, and some statistical data). Additionally, some questions have useful rich explanations, but their number is rather small (241 of 2987 questions). Recently, we have researched ways to provide more useful feedback information to students. In [14] and [15] we have explored ways to use method of Formal Concept Analysis to give students meaningful visual representation (concept lattice digraph) of their performance on the course, and also quantitative measures (attribute coverage and formal concept similarity). In [16] we used model checking and mutation testing methods to propose a model for checking homework assignments consistency, and a model for generating richer feedback information for failed homework assignments.

In this research our motivation was to combine the approach of our homework calculation questions (dynamic inputs) with that of variable multiple choice questions (variable question text, answers and different images) with automated enriched feedback generation.

3. Prototype Web application for automated question generation

Most of the questions and assignments on our course Fundamentals of Electrical Engineering are based around a DC or AC circuit diagrams. We have also observed that in our e-learning system all of our circuit figures are static and had to be manually created, which is not preferable in an automated system. Furthermore, we know that only a small percentage of available questions in our system have rich (manually written) explanations. Therefore, we wanted to create a system which will automate question generation (including text and figures), as well as question solving and feedback information generation.

We decided to create a Web application for automated generation and solving of template based DC circuit questions. Therefore, our proposed Web application should meet following prerequisites:

- ability to generate simpler and more complex question variants form a basic question template:

  - by changing the number of input parameters
  - by increasing/decreasing complexity of the circuit
  - by changing the number of requested results
- automatic generation of random circuit figures:
  - series – parallel DC circuits
  - one voltage or current source
- automatic solving of the given problem
- automatic generation of step-by-step solution to the given problem using basic circuit laws (Ohm's Law and the principles of Kirchhoff's Laws) as an additional feedback information for the students

1.3. System overview

Our Web application for automated generation of series – parallel DC circuit questions and solutions is available online at http://osnove.tel.fer.hr/questions/DCcircuits.asp. The application is bilingual – it is available in Croatian and English. Its main page is shown in the Fig. 1. Our automated application can create questions of different complexity based on the number of input parameters. It can generate a simple problem (e.g. one resistor connected directly on a voltage or a current source) requiring only one result (e.g. power dissipated on the resistor), and also very complex series – parallel circuits requiring 10 different results (total resistance, unknown resistances, voltage drop across resistors or the source terminals, current through resistors or the source and power dissipated by resistors or generated by the source). In its present setup students can choose between easy questions (1 – 4 resistors), normal questions (5 – 9 resistors) and hard questions (10 – 14 resistors). Generating questions with more than 14 resistors is, of course, possible but the circuit image would not fit usual screen sizes. Every question has its own randomly generated circuit diagram figure. Also, during the automated solving process the circuit diagram will be randomly simplified step-by-step, by calculating equivalent resistances. The whole problem solving process is based on basic circuit laws – Ohm's Law and the principles of Kirchhoff's Laws.

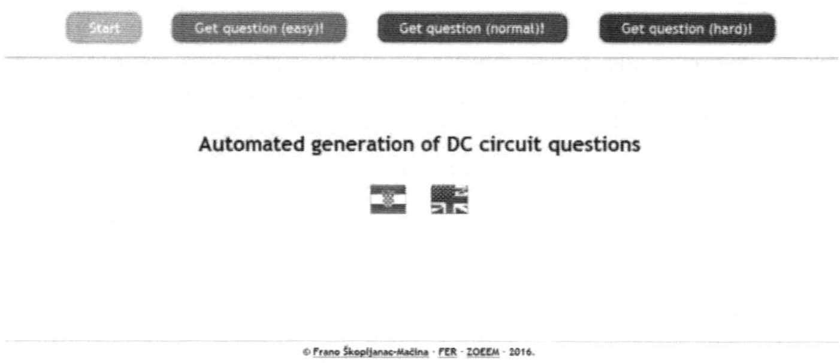

Fig. 1. Web application for automated question generation
http://osnove.tel.fer.hr/questions/DCcircuits.asp

3.2. Technologies

We used standard Web technologies, such as HTML, *JavaScript* and *jQuery* [17]. Circuit diagrams are automatically drawn as a SVG [18] inline image. Mathematical notation is provided using *JavaScript* library *MathJax* [19] using standard LaTeX syntax. *Microsoft ASP* server-side scripting language is also used, but only for the Web application homepage. The core of our prototype application is client-based. Circuit diagram is drawn in SVG inline mode, which means that the image is not server-based, but generated directly on the client's browser from the generated SVG inline code. Still, we must acknowledge one downside of using SVG inline code – it does not work in *Internet Explorer 8* and older *Internet Explorer* browsers.

3.3 Question generation

Question is generated by calling *JavaScript* function **generateProblem (level)**. Therefore, first we need to decide on the difficulty level of the question (easy: up to 4 resistors; normal: form 5 to 9 resistors; hard: more than 9 resistors). Afterwards, the function randomly chooses the number of resistors (between defined lower and upper bounds), and then begins to form schematics (series – parallel circuit) as a nested brackets string of basic components **s(*,*)** – two elements series and **p(*,*)** – two elements parallel. Schematic builder combines those two components by inserting (in place of a random star symbol) one into another and vice versa. It checks not to chain two same components one in another. Also it is possible to randomly extend number of elements between the brackets: **s(*,*)** can become **s(*,*,*,*)**. Next, after the schematic string is complete the stars symbols are iteratively replaced by numbered resistors – e.g. **s(p(r1,r2),r3,p(r4,r5))**.

In the following step the functions decides randomly if this resistor combination will be connected to a voltage source or a current source. Afterwards, all the resistors are given a random value (between predetermined value ranges), as well as only one of the following: source voltage, source current or source power.

Next, the question is completed by deciding which of the unknown values need to be calculated, with random selection from the list of missing source parameters, and also from the list of missing resistors' voltages/currents/powers. Afterwards, the application generates the full text of the question. The text is created simply by appending the unknown values and their description to the simple template ("For the circuit shown in the figure, calculate..."), and by typing the list of known input data. Finally, the question is shown on the screen, together with the generated circuit diagram image (Fig. 2.). After solving the problem *on paper* students can check the results and step-by-step solution.

Currently, our system supports only this basic question type, but we plan to add different new question types by choosing different set of known values, e.g. by providing a current through one of the resistors, all resistor values, but no information about the voltage or current source parameters.

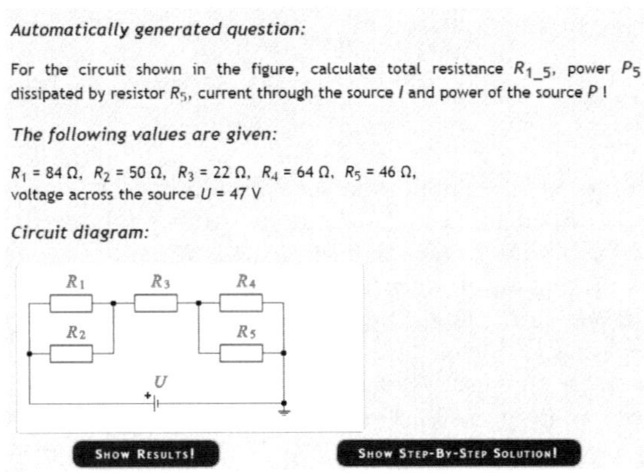

Fig. 2. Automatically generated question for schematic **s(p(r1,r2),r3,p(r4,r5))**

3.4. Circuit diagram drawing

Web application converts schematic into a SVG inline image of a circuit diagram using prepared SVG symbols definition library shown on Fig. 3. Resistor, voltage source and current source are all 100 px wide and 20 px high so it is easier to combine them into a series – parallel circuit. Dot symbol is placed on all exact places where more than two lines meet. Ground symbol is always placed in the bottom right corner of the circuit diagram – on the sources negative terminal. Circuit diagram drawing functions are one of the more complicated parts of our application. Main function for circuit drawing is **drawSchema(str)**. It is a recursive function which runs through the given schematic, by dividing it on smaller sub-schematics, and ensures that all resistors and line segments connect into their place.

After the series-parallel schematic is created it must be connected to the source. That is done in the final steps of the circuit drawing. Another function **highlightedSchematics(find,type)** can highlight in different colours various parts of the circuit diagram. This function is used extensively in feedback generation – to point out the set of resistors in series or in parallel which can be replaced by a single equivalent resistor.

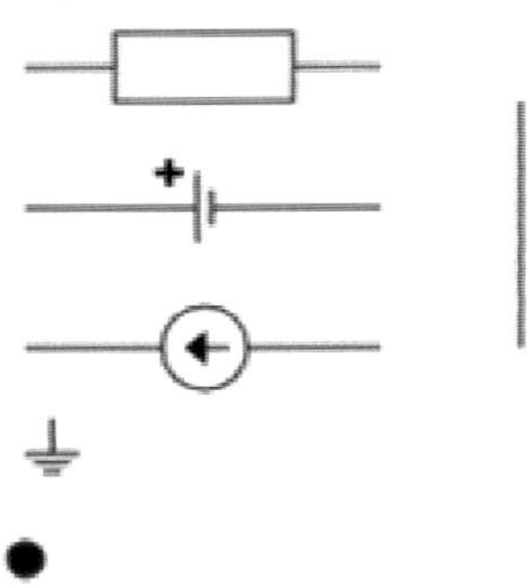

Fig. 3. Custom SVG definition library for DC circuits

3.5. Problem solving and feedback generation

After the question has been set up we can request numeric solutions to the problem by pressing button *Show Results*, or we can get entire step-by-step solution of the problem by pressing the button *Show Step-By-Step Solution*. We have defined a single strategy for solving questions in Section 3.4 (we know all the resistor values and one parameter of the voltage/current source). Function **solveProblem(type)** starts by iteratively simplifying starting circuit diagram, replacing random series or parallel of resistors with their equivalent resistances. After the circuit is reduced to one equivalent resistor, the function can calculate, using Ohm's Law and formulas for power, all the missing parameters of the source. After the total resistance is known, together with the source voltage, current and power the problem solving function moves in the opposite direction – replacing equivalent resistors with their predecessors. At the same time function calculates missing voltage, current and power of every resistor. Finally, the circuit diagram returns to its original state, and in that last step the functions calculates all the remaining unknowns.

Step-by-step solution provides a rich feedback for the students, with little hints, messages and specially highlighted circuit diagrams. Each step of the problem solving function is also typed out in detail using *MathJax* mathematical notation and calculated values.

When we add different types of questions, as hinted at the end of Section 3.4., we should also prepare different adapted strategies for problem solving. But, in essence they would also use only Ohm's Law and the main principles of Kirchhoff's Laws.

4. Results and discussion

Our method and resulting prototype Web application has shown promising results in automated generation of DC circuit questions. In Fig. 4 we can see another generated question with a circuit diagram containing 5 resistors. If we compare it to the question shown in Fig. 2, we can observe that we got two quite different questions of similar complexity. In both questions we got a voltage source, but the schema configuration is slightly different, and, besides total resistance, we need to calculate completely different unknown values. Of course, both questions have different results, and different step-by-step solutions. This illustrates that we can generate a list of different questions of the same complexity – which can be ideal for testing many students simultaneously. We could say that virtually all students in a normal sized group (80 students) would get unique – personalized questions of same difficulty. This approach also sounds promising in generating a large number of personalized homework assignments. But, in this prototype stage we will offer this application to our students mainly for self-assessment, practice and training for the written and oral exams.

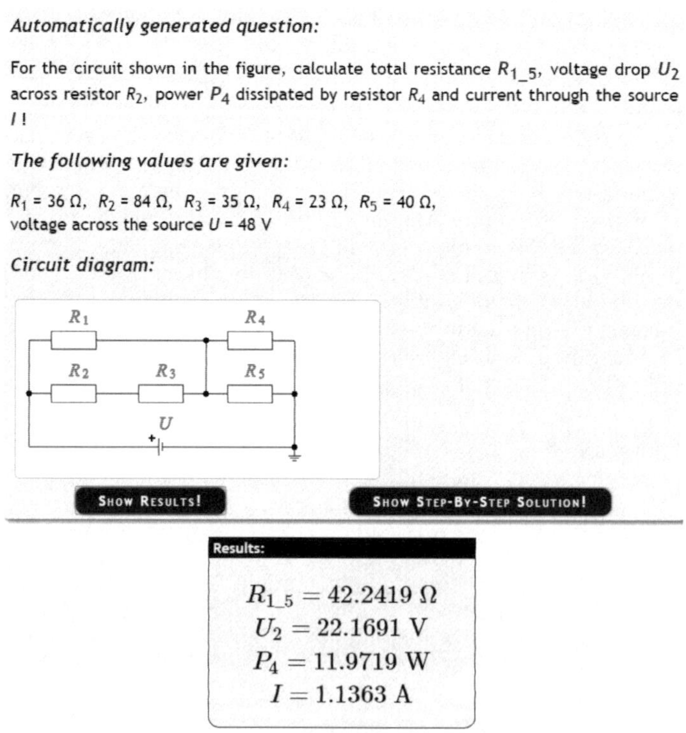

Fig. 4. Randomly generated question and its solutions (difficulty - normal)

Another example of our application's versatility can be seen in Fig. 5 where we have 4 different randomly generated DC circuit diagrams, each consisting of 7 resistors in different configurations. Of course, each circuit diagram is accompanied by a randomly generated text, with different combination of know values and requested unknown values. Also, for each resistor combination we can change the source type (voltage source to current source and vice versa). This increases the number of possible variations of questions with exactly 7 resistors. And by adding new question types, different sets of known and unknown data as pointed out in Section 3.4., the number of variations would further increase.

We can also see in Fig. 5 that all four different random circuit diagrams are not equally complex. Circuit diagrams in Fig. 5 c) and b) are less complex, as they require only one and two simplifications respectively (substitutions of parallel resistors with their equivalent resistor) to become a simple series resistor circuit. Circuit diagram in Fig. 5 a) is more complex, because it requires 3 simplification steps before it transforms in a simple parallel circuit (substitution of R_3 and R_4 resistor parallel with an equivalent resistor, and then substitute resistor series in both parallel branches with their equivalent resistors). Finally, circuit diagram in Fig. 5 d) appears to be most complex, because it needs 5 simplification steps to transform it into an equivalent simple parallel circuit. We start with the simplification of the R_3 and R_4 resistor parallel, then we substitute the resulting equivalent resistor R_{34} and R_2 series with another equivalent resistor R_{234}, afterwards we calculate the R_{234} and R_5 resistor parallel, and finally substitute resistor series in both parallel branches with their equivalent resistors R_{12345} and R_{67}. Now, we have an equivalent simple parallel circuit of resistors R_{12345} and R_{67}.

Therefore, it is necessary to describe these random circuit diagrams not only with the number of resistors, but also with an additional objective complexity measure. One example could be the number of nested parallel and series connections in the circuit diagram. This would correspond with the described number of simplification steps. Using that measure we could state that the circuit in Fig. 5 c) and Fig. 5 b) have circuit complexity 1 and 2 respectively. Circuit in Fig. 5 a) has circuit complexity 3, and the most complex circuit in Fig. 5 d) has circuit complexity 5. By applying this calculated circuit complexity measure the system could ensure that all student get different questions of the same circuit complexity. This measure is especially required in automatic student evaluation, because then we would avoid situations where one student gets a simple circuit diagram (Fig. 5 b)), and another gets a more complex circuit (Fig. 5 d)). The same measure could be also useful in self-assessment and tutoring, because the system could adapt to the current knowledge level of the student, by gradually increasing or decreasing circuit complexity.

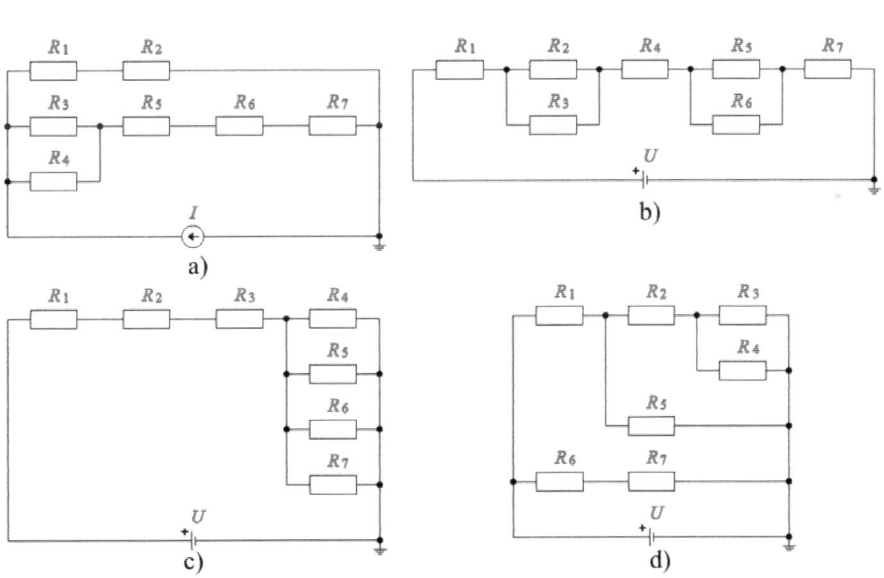

Fig. 5. Four different random circuit diagrams with 7 resistors each

As we have described in Section 3.6., our Web application can calculate and display correct results, as shown in Fig. 4, but also it can provide students with detailed feedback in form of step-by-step solution. The system displays all steps of the calculations using *MathJax* for mathematical notation, and additional explanations and hints. One example is presented in Fig. 6, where we can see a step 4 of a longer calculation process. In this step we identify simplification candidates, randomly choose one of the simple resistor parallels or series and calculate their equivalent resistance. By randomly simplifying the original circuit diagram the system also ensures that the same question can be solved in a number of different ways.

At this moment our web application works as a prototype, but it will show its full potential when we fully integrate it into our e-learning system. Also, we plan to add more question types, and different strategies for problem solving, namely using current divider rule and voltage divider rule. Additionally, it is possible to assign potentials to nodes in circuit diagrams, and that would further increase the number of question variations. In the next phase of development of our automated question generator we could include questions about Thévenin and Norton theorems and also maximum power theorem. Furthermore, the same principals described here could be applied to DC Capacitor networks and later, possibly even to AC circuits.

Before using our application in student evaluation (e.g. laboratory quizzes or homework assignments) we should conduct a proper verification and validation of our system using formal methods such as model checking. Also, we should test the outputs of our application by comparing them to results of a simulation of the same DC circuit in professional electronic circuit simulator (e.g. SPICE). Currently, the application was tested only manually by experts comparing system outputs to correct calculated results.

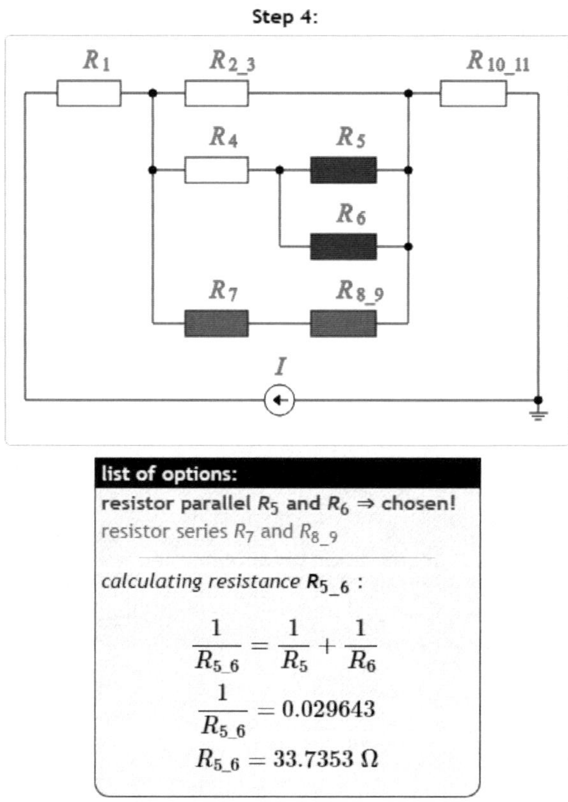

Fig. 6. Excerpt from step-by-step solution (difficulty - hard)

5. Conclusion and future work

In this paper we have presented our method and prototype Web application for automatically generating DC circuit questions. This method has the potential to help solve one of our main problems, and that is static nature of many teaching and learning materials. One of the main contributions of this paper is the circuit diagram random generator using only standard Web technologies. The other is the rich feedback generator offering step-by-step solution to randomly created DC circuit problem.

Currently, the presented Web application supports only one basic type of questions, and has one defined strategy for solving the generated problem. We plan to add other different types of questions and provide adaptive solving strategies.

In future work, we intend to improve our application prototype, making it more interactive so that the students can try different options in step-by-step solutions, and submit and check their answers automatically. Also, we intend to fully integrate presented application into our e-learning system, as a part of a developing adaptive module for student tutoring and knowledge evaluation.

6. Acknowledgments

The work has been fully supported by Croatian Science Foundation under the project (UIP-2014-09-2051 eduMINE – Leveraging data mining methods and open technologies for enhancement of the e-learning infrastructure).

7. References

[1] Lin, X. & Hu, Q. (2015). Trends in e-learning research from 2002-2013: A co-citation analysis, 15th IEEE International Conference on Advanced Learning Technologies, ICALT 2015, Hualien, Taiwan, July 6-9, 2015, G. Sampson D.; Huang R.: Hwang G.; Liu T.; Chen N.; Kinshuk & Tsai C. (Eds.), pp. 483–485, IEEE

[2] Grubisic, A.; Stankov, S. & Zitko, B. (2015) Adaptive courseware: A literature review, J. UCS, vol. 21, no. 9, pp. 1168–1209

[3] https://moodle.org/, (2016). Moodle learning platform, Accessed on 2016-10-13

[4] Rakic, K.; Bandic Glavas, M. & Majstorovic, V. (2013). Development of the Intelligent System for the use of University Information System, 24th DAAAM International Symposium on Intelligent Manufacturing and Automation, Zadar, Procedia Engineering 69 (2014) pp. 402-409

[5] Pryanichnikov, V.E.; Katalinic, B.; Kirilchenko, A.A.; Khelemendik, R.V.; Kuvshinov, S.V.; Vician, D. & Uglesic, A. (2014). New Creative Educational Technologies for Inter-University Network, 25th DAAAM International Symposium on Intelligent Manufacturing and Automation, Vienna, Procedia Engineering 100 (2015) pp. 259-268

[6] Chali, Y & Hasan, S. A. (2012). Towards Automatic Topical Question Generation, COLING 2012, Mumbai, Proceedings of COLING 2012: Technical Papers (2012) pp. 475-492

[7] Gulwani, S., (2014). Example-based learning in computer-aided STEM education, Commun. ACM, vol. 57, no. 8, pp. 70-80

[8] Singh, R.; Gulwani, S. & Solar-Lezama, A. (2013). Automated feedback generation for introductory programming assignments, in ACM SIGPLAN Conference on Programming Language Design and Implementation, PLDI '13, Seattle, WA, USA, June 16-19, 2013 Boehm H. & Flanagan C. (Eds.), pp. 15–26, ACM, 2013.

[9] Alvin, C.; Gulwani, S.: Majumdar R. & Mukhopadhyay S. (2014). Synthesis of geometry proof problems, Proceedings of the TwentyEighth AAAI Conference on Artificial Intelligence, July 27 -31, 2014, Québec City, Québec, Canada., Brodley, C. E. & Stone P. (Eds.), pp. 245–252, AAAI Press

[10] Singh R.; Gulwani S. & Rajamani S. K. (2012). Automatically generating algebra problems, Proceedings of the Twenty-Sixth AAAI Conference on Artificial Intelligence, July 22-26, 2012, Toronto, Ontario, Canada. Hoffmann, J. & Selman, B. (Eds.), AAAI Press

[11] Zitko, B.; Stankov, S.; Rosic, M. & Grubisic A. (2009). Dynamic test generation over ontology-based knowledge representation in authoring shell, Expert Syst. Appl., vol. 36, no. 4, pp. 8185–8196

[12] Pavić, A. & Škopljanac-Mačina, F. (2012). Computer in testing the Fundamentals of electrical engineering at the University of Zagreb Faculty of electrical engineering and computing, MIPRO 2012 - 35th International Convention on Information and Communication Technology, Electronics and Microelectronics, Opatija

[13] Pardanjac, M.; Eleven, E. & Kaurovic, D. (2014). Increase of User Motivation in Teaching Realized Through Distance Learning, Chapter 10 in DAAAM International Scientific Book 2014, pp.131-144, B. Katalinic (Ed.), Published by DAAAM International, ISBN 978-3-901509-98-8, ISSN 1726-9687, Vienna, Austria DOI: 10.2507/daaam.scibook.2014.10

[14] Škopljanac-Mačina, F.; Blašković, B. (2013). Formal Concept Analysis – Overview and Applications, 24th DAAAM International Symposium on Intelligent Manufacturing and Automation, Zadar, Procedia Engineering 69 (2014) pp. 1258-1267

[15] Škopljanac-Mačina, F.; Blašković, B. & Skočir, Z. (2014). Using Formal Concept Analysis for student assessment, Proceedings of ELMAR-2014 56th International Symposium, Zadar, Tralić, D.; Muštra, M. & Zovko-Cihlar, B. (Eds.), pp. 285-288 Zagreb, LotusGRAF, Zagreb

[16] Škopljanac-Mačina, F.; Zakarija, I.; Blašković, B. (2015). Exam questions consistency checking, Proceedings of 38th International Convention on Information and Communication Technology, Electronics and Microelectronics MIPRO 2015, Biljanović, P. (Ed.), pp. 921-924, GRAFIK, Rijeka

[17] https://jquery.com/, (2016). jQuery, Accessed on 2016-10-03

[18] https://www.w3.org/Graphics/SVG/ (2016). SVG, Accessed on 2016-10-03

[19] https://www.mathjax.org/ (2016). MathJax, Accessed on 2016-10-03

DOI: 10.2507/27th.daaam.proceedings.057

WORLD CLASS MANUFACTURING (WCM) MODEL AS A TOOL FOR COMPANY MANAGEMENT

Peter Poor, Marek Kocisko & Radoslav Krhel

This Publication has to be referred as: Poor, P[eter]; Kocisko, M[arek] & Krhel, R[adoslav] (2016). World Class Manufacturing (WCM) Model as a Tool for Company Management, Proceedings of the 27th DAAAM International Symposium, pp.0386-0390, B. Katalinic (Ed.), Published by DAAAM International, ISBN 978-3-902734-08-2, ISSN 1726-9679, Vienna, Austria
DOI: 10.2507/27th.daaam.proceedings.057

Abstract

Rapid pace of world economy development is forcing businesses to adapt to conditions of global competition and find ways of successful application in an environment of constant change. In current business conditions in predominance of demand and open market the question of gain and apply a competitive advantage is critical for establishment and operation of all types of enterprises. Basic requirements of competitiveness are consistent customer orientation, reliability, and quality management. World Class Manufacturing [WCM] is collective term for most effective methodologies and techniques to realize products of consistent high quality, delivery On Time In Full amount ordered and products at lowest possible cost level. Companies engaging in World Class Manufacturing strategies focus on improving operations, strive to eliminate waste and create lean organizations. This often results in higher productivity. But these companies also focus on speed of total throughput from order capture through delivery setting new standards for delivery without heavy dependence on inventory (Muda, Hendry, 2002). This paper deals with use of WCM (World Class Manufacturing) model and its use in company management.

Keywords: WCM; World Class Manufacturing; company; management; industry.

1. Introduction

World Class Manufacturing system is primarily based on a comprehensive Japanese management system called Kaizen. These method improvements in production, increasing quality and flexibility are described in book by Mr. Masaaki Imai. KAIZEN is moving forwards by focusing on production process, unlike western thinking oriented to innovation and results. Before oil crisis in 1980s, world industry development of new technologies and innovations has paid off, but after this event in new situation, where prices for manufacturing resources, quality demands also demands for faster introduction of new products increased, and thus western market was forced to pressure competition resort to Japanese strategy (Lind, 2001).

World Class Manufacturing is an integrated system which improves processes and quality, reduces cost and with increased flexibility fulfils customer expectations. This system is based on 10 pillars of activities and 10 managerial pillars. It is a very convenient program in terms of high sophistication. Globalization and competition coming from low-cost countries brings great challenge for industrial production. Only the strongest and most flexible companies can survive. To become the best, we must constantly improve performance and results, we need to strengthen relationships with suppliers and partners. WCM system is for now a great challenge for the future.

The aim is to reach world-class level under conditions of setting these ambitious goals:

- Quality
- Cost
- Delivery

To achieve the highest quality, it is necessary to achieve zero defects, and becoming the largest competition in the world. Costs must be reduced by at least 30% during the next 3 years. And to fulfil the third objective running time according to the needs of customers must be to shortened. All these goals can be met only through the involvement of people, in this case, company employees. It is necessary to ensure their development and standardization. Only World Class Manufacturing is the proper key to success. The main aspects of all employees, loss and downtime for completeness and also standards methods.

If structure of the system is divided into parts, then we can specify the possibility of reducing incremental costs. First such element is the workplace organization using methods of industrial engineering at a concentration on productivity in order to achieve zero losses. Second element is achieving quality using method of total quality control focus on improving quality in order achieving zero defects. Third important element is fulfilment of maintenance methods concentrating on technical efficiency with the result of achieving zero downtime. The last element is self logistics using "just in time", focus is on level of services with the aim of zero inventory. When properly connected all components of employee involvement, then company creates value and achieves high Customer Satisfaction.

1.1. Main principles of WCM according to (Flynn, Schroeder, Sakakibara, and Bates, 1997):

- World Class Safety is essential to achieving World Class Performance
- WCM Leaders encourage and require creation of and adherence to standards
- World Class Companies are becoming the voice of customer "heart / engine" race
- WCM will not accept any loss (goal is always zero: injuries, errors Supply & quality stock, downtime)
- Consistent use of WCM methodology guarantees the elimination of losses
- World Class race are any abnormalities immediately visible
- WCM takes place directly in workplace, in the process, not in the office
- WCM learn best through team's use of techniques / methods
- WCM strength comes from involvement of people
- World Class companies collect energy from its own successful solutions problems, crises

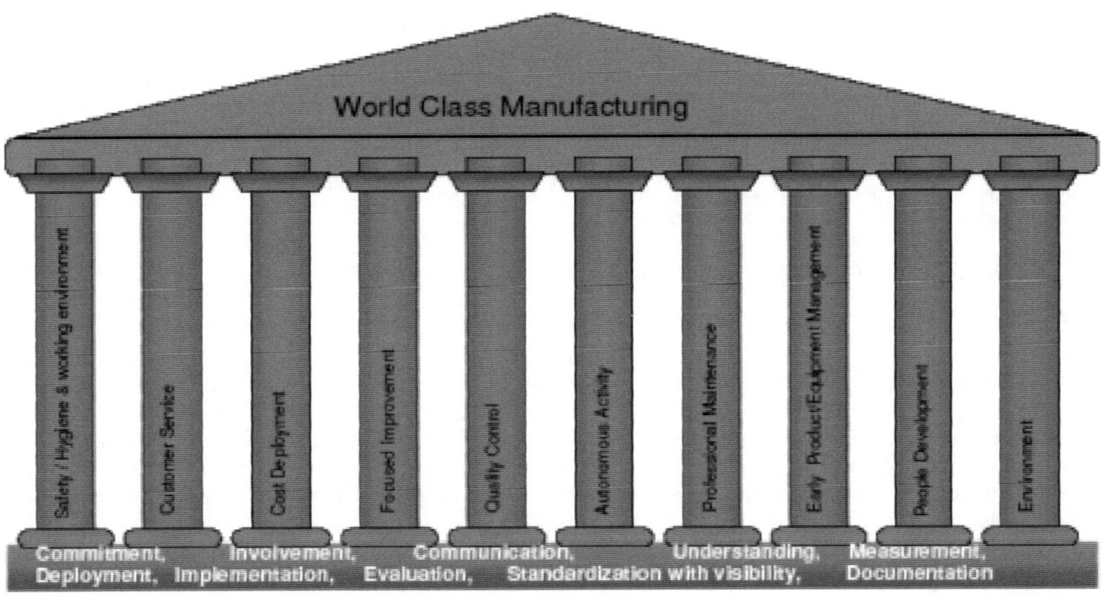

Fig. 1. WCM pillars

1.2. 7 WCM basic tools

1) Prioritization - Organization of importance, identify important issues. They include deployment cost analysis, Pareto diagrams, stratification, ABC Classification for targeted distribution of machines according to their importance in process, QA matrix to identify losses, safety matrix to identify work-related injuries, value stream maps for cost distribution in the process and point mapping of machine failures
2) Systematization and logic - is simply identifying where one problem
3) A description of the problem with sketches - helps to understand the problem
4) 5W + 1H with 5G principles - simple questions for understanding problems, (What, When, Where, Who, Which, How) regarding the resulting accident 5G principles are then used to address possible remedies of an accident
5) Root cause analysis - identifying the cause of a graph causes and action or method 5Why), which is considered sufficient quantity questions to reveal the real causes
6) Visualization - description of phenomena using images and detail, determines what it is hidden behind the problem
7) The way to teach people - "a way to teach people to" search for hidden problems, in a way fault-finding in humans, determining whether he knows correctly process works or how to react in case of a problem.

2. Research goals on WCM implementation into manufacturing company

2.1. Research Objectives

Research objectives are focused on management organization WCM model:

- Implementation of WCM model in the organization,
- Set the entire management system of values and rules of the organization (vision, goals ...)
- Developing an organizational structure,
- Resource management, processes and performance

WCM management in organization includes in particular the overall system configuration management, values and rules of the organization, organizational structure adjustment, resource management, processes and performance. It is a cross-sectional area, are used in her methods of strategic management methods in the area of quality and management efficiency.(Oliver N, Delbridge R, Jones D, Lowe J., 1994)

New management concepts are based on process approach, which is a fundamental principle of continuous improvement of business processes and applied in enterprises as process management. Process management is a form of governance that systematically identifies business processes and their attributes, creating conditions for their efficient operation, coordination, measurement and continuous improvement, resulting in a quality product integrating customer requirements and business strategy.

The object of research of this article is finding representation of individual roles in teams in manufacturing companies in middle and top management. The aim of research is to describe the various staff positions in selected teams.

2.2. Research plan

Research Plan will be divided and implemented in two steps:

I. Finding teamwork, which take the role of workers in work teams,
II. Rating the performance of individual work teams

3. Detailed research on WCM model use as a tool for company management

3.1. Diagnostic tools used in the research:

- Questionnaire teamwork (Belbin): For research purposes we will be using a diagnostic tool developed by R. M. Belbin and collective. Methodology - the test consists of seven sections - questions and will be completed by respondents. Each issue will may have eight different answers (claims) and 10 points. Respondents will be selected from eight answers those most characteristic of them and describe their behavior in the team. These points can be assigned to several descriptions. In extreme cases points may be divided among all the claims, or may be assigned any of 10 points to the only claim in the part. It is recommended to avoid this extreme. Points are marked in the box next to sentence.
    - Evaluation - points will be entered in the evaluation table.
    - Interpretation - The highest score is a team role, where you can most benefit from your team. The second highest score indicates Team roles that can perform well when the smaller group needs your preferred role. Two lowest score indicates role group, you do not like to refer to your weaker points. Instead, however, to

work towards improving them, it is preferable to seek together with my colleagues opposite tendency of behaviour in the team, with a tendency to hold also the roles that you are complementary. Ideally, the team includes representatives of all these roles, mutually recognizing and utilizing their strengths teamwork.

- Questionnaire Rating performance: This research group will be used questionnaire for rating performance. The second questionnaire aims to team members when their supervisor (Pillar leader) evaluates. The questionnaire contains five performance factors:
    - Flexibility
    - Reliability
    - Theoretical knowledge
    - Quality (work passing)
    - Quality (performance dates) of work with eight team roles (Implementer Coordinator, strategy, ideas Seeker, Seeker resources critic evaluator team builder, Finisher).

This questionnaire evaluates performance of subordinates in terms of five factors. The answer is indicated by circling the appropriate number on the scale, where 1 is maximum negative and positive assessment than 9 employee in these factors. Numbers 2-8 are intermediate stages, reflecting the greater or lesser inclination to one of the extreme points.

4. Target group and sampling

First task is to evaluate the results of the questionnaire Belbin team roles according to the research group teams. The tendency of individual employees to team roles as well as representation roles within the team (pillars). The second task is to measure the performance of team members of five factors.(Vollman, Berry, Whybark, 1997) The target group is aimed at members of middle and senior management.

4.1. Test Conditions

Through the approval from organizational structure of project organization and WCM Director of the company personnel department issued an internal notice - Appointment of teams. The questionnaires were distributed by mail from Human Resources department:

- Questionnaire dealing with teamwork (Belbin) was sent to team members and team leader,
- Questionnaire dealing with rating the performance was sent to managers and leaders of individual pillars.

Questionnaires were coded by numbers to guarantee the anonymity of respondents.

5. Conclusion

The aim of this paper was to introduce WCM model as a tool for company management. Role within team can rarely be identified by characteristic features, or outward appearance. This was one of the reasons we chose a powerful diagnostic tool for investigation of various team roles - Belbin teamwork questionnaire. It was also necessary to track the performance of a work team. Therefore questionnaire rating the performance was used. Also we needed to specify the target group according to selected criteria. This article is only the beginning, we will continue in the research and you can read the results. in our future articles, thatswhy the results may not be expected soon. We are working on this research and we will publish our results in our next articles.

6. Acknowledgments

This paper was created with the subsidy of the project n° K-14-008-00.

7. References

[1] Oliver N, Delbridge R, Jones D, Lowe J. World class manufacturing: Further evidence in the Lean production debate1. British Journal of Management. 1994 Dec 1;5(s1):S53-63.

[2] Flynn, B.B., Schroeder, R.G., Flynn, E.J., Sakakibara, S. and Bates, K.A., 1997. World-class manufacturing project: overview and selected results. International Journal of Operations & Production Management, 17(7), pp.671-685.

[3] Lind, J., 2001. Control in world class manufacturing—A longitudinal case study. Management Accounting Research, 12(1), pp.41-74.

[4] Lau, R.S., 1996. Strategic flexibility: a new reality for world-class manufacturing. SAM Advanced Management Journal, 61(2), p.11.

[5] Thomas E.. Vollmann, William L.. Berry and Whybark, D.C., 1997. Manufacturing planning and control systems. Irwin/McGraw-Hill.

[6] Voss, C. and Blackmon, K., 1996. The impact of national and parent company origin on world-class manufacturing: findings from Britain and Germany. International Journal of Operations & Production Management, 16(11), pp.98-115.

[7] Muda, S. and Hendry, L., 2002. Proposing a world-class manufacturing concept for the make-to-order sector. International Journal of Production Research, 40(2), pp.353-373.

[8] Al Falah, K., Zairi, M. and Moneim Ahmed, A., 2003. The role of supply-chain management in world-class manufacturing: An empirical study in the Saudi context. International Journal of Physical Distribution & Logistics Management, 33(5), pp.396-407.

[9] DeFilippo, J.S., 1997. World-class manufacturing in Chengdu: a case study on China's first aviation joint venture. International Journal of Technology Management, 13(5-6), pp.681-694.

[10] Kodali, R., Sangwan, K.S. and Sunnapwar, V.K., 2004. Performance value analysis for the justification of world-class manufacturing systems. Journal of Advanced Manufacturing Systems, 3(01), pp.85-102.

[11] Subramanian, A. & Sun, H.: A Model for Value Chain Integration in a Vertically Integrated Design and Manufacturing Firm

[12] Rakyta, M.; Fusko, M.; Haluska, M. & Grznar, P.: Maintenance Support System for Reconfigurable Manufacturing Systems

[13] Radzevich, S. P., & Krehel', R. (2011). Application priority mathematical model of operating parameters in advanced manufacturing technology. International Journal of Advanced Manufacturing Technology,56(5-8), 835-840.

[14] Novotný, L., Abreu, H.,F.,G., Miranda, H.,C., M. Beres, M.: Simulations in multipass welds using low transformation temperature filler material. Science and Technology of Welding and Joining, Vol. 21, no. 8, pp 680 - 687, 2016

[15] Hendry, L. C. (1998). Applying world class manufacturing to make-to-order companies: Problems and solutions. International Journal of Operations & Production Management, 18(11), 1086-1100

[16] Kucerova, M.; Mlkva, M.; Paulova, I.: APPLYING PRINCIPLE OF CUSTOMER FOCUS IN BUSINESS PRACTICE. Annals of DAAAM & Proceedings; 2010, p695

DOI: 10.2507/27th.daaam.proceedings.058

SIMULATION MODELING ENVIRONMENT FOR COMPLEX DATA DRIVEN DISTRIBUTED SYSTEMS DESIGN

Peter Panfilov, Sergey Salibekyan & Yury Leokhin

This Publication has to be referred as: Panfilov, P[eter]; Salibekyan, S[ergey] & Leokhin, Y[ury] (2016). Simulation Modeling Environment for Complex Data Driven Distributed Systems Design, Proceedings of the 27th DAAAM International Symposium, pp.0391-0399, B. Katalinic (Ed.), Published by DAAAM International, ISBN 978-3-902734-08-2, ISSN 1726-9679, Vienna, Austria
DOI: 10.2507/27th.daaam.proceedings.058

Abstract

The methodology is presented for the data-driven distributed application development which is based on the proprietary object-attribute model of the computational process and data organization, which allows for modeling the parallel processes according to the dataflow principle and to synthesize and modify the data structures directly in the simulation process. We propose a new formalization approach for specification and analysis of dataflow applications that combines the proprietary OA-approach to describe the data synthesis and modifications, the KPN to describe the parallel dataflow computational process, and the theory of automatons to describe the operation of a separate FUs in application. The proposed methodology is realized as an object-attribute programming and simulation environment which is based on and build upon a proprietary OA-programming language.

Keywords: distributed systems; dataflow; object-attribute model; programming and simulation environment.

1. Introduction

In this paper, we discuss new approach to simulation modeling of distributed dataflow computer systems and applications targeted at processing complex structured data such as those found in computer-aided design/manufacturing systems (CAD/CAM), expert and decision support systems (DSS), artificial intelligence systems (AI), and search engines. The contemporary computer systems do not cope well with this problem for different reasons. First, the von Neumann architecture which is based on control flow paradigm that dominates the computer world these days is not flexible enough to construct the complex distributed computational processes. Second, there is a lack of adequate formal models for specifying and performing different kinds of analyses for such systems. We here present a survey of basic models which can pretend to play a significant role in this field.

The automaton model (Final State Machine, FSM) is more suitable for modeling sequential processes. But there are attempts to adapt it to concurrent processes and parallel computations by creating a system of parallel operating automatons which exchange the state numbers or react to events generated by other automatons contained in the computer system [1]. But synchronization of such a system is very complicated because the computer designer must do this manually. The state of affairs with the complexly structured data processing is even worse.

The Petri nets are mainly intended for searching the deadlock conditions in the modeled system (also in parallel systems) but is completely unsuitable for modeling the complexly structured data processing. Moreover, they are inconvenient for modeling the parallel processes because, according to the opinion of Carl Hewitt, Professor of Massachusetts Technology Institute, USA, they have the following significant drawbacks [2]:

- the Petri nets are restrictive because they model the data flow control rather than the data flow itself
- they are too complex to describe simultaneous actions in the computational process
- the "physical" interpretation of transitions in the Petri nets is rather dubious

Different process algebras (π-calculus, Communicating Sequential Processes (CSP), Calculus of Communicating Systems (CCS), etc.) are suitable for modeling the parallel computations (they describe the computational threads and the data exchange between them), but they cannot model the complex data processing.

The frame (or object-oriented) model [3] is well suited for simulation of complexly structured data. This model permits describing highly abstract complex structures and combines the data and the program (the object behavior) in a single information structure. But the information structures are rather restrictive, namely, such models can operate only with a tree-like structure of the data. The OO-paradigm meets difficulties in the simulation of parallel process, because it cannot be used these purposes. The drawbacks of the object-oriented approach are described in [4].

The agent model where the system is represented as a set of agents, i.e., autonomous mutually interacting objects with their own behavior, is ideal for modeling the parallel processes. The agents can be organized in structures (for example, as graphs of agents). It should be also noted that the agent system does not require any centralized control and is very flexible, which permits modeling the self-organizing systems. But several authors (for example, [5]) note that nowadays there are no efficient agent models available for understanding and practical use of architectures. The development of general methodological approach for constructing such a class of models including all stages of development (statement of the problem, its formalization, the simulation program realization, and experiments) is still in progress. There are still no conventional formal models in this field.

A very perspective research area in the simulation modeling is the dataflow (control of computations by using the data flow), because it is ideal for describing the parallel processes. The processes are then synchronized according to the data (i.e., the synchronization occurs as if by itself), which significantly simplifies the modeling of parallel computer systems. The system does not require any centralized control and hence permits modeling self-organizing systems. The dataflow system is a graph whose nodes are executive units exchanging messages which form a token along the graph arcs [6]. A token is a set of data together with some ordering information (the message receiver address, data type, etc.). The process in such a system is an avalanching process, i.e., the messages arrive at executive units; after an executive unit obtains all necessary data, it activates and begins to process them; then the obtained data formed as a token arrive at other executive units; etc.

At present, a method for processing the complexly structured data, which is called the i-structure, is most frequently used in the dataflow-paradigm [6]. The i-structure is a data structure where each field has a data flag. The data in the field is written only once, and hence no modification of the data, not to mention the data structure, can be there.

A commonly accepted formal model in the field of dataflow is the Kahn process network (KPN) which is a set of calculators (or handlers) exchanging tokens through data channels equipped with the processing queue (if an executive unit is processing the preceding token, then the arriving token in placed in the queue until the unit is free). The KPN formalism is ideal for modeling parallel systems, but it does not permit modeling a unique data structure processed by several simultaneously operating parallel executive units.

The discrete event simulation (DES) [7] describes only methods for modeling parallel processes and does not pay attention to the technology of simulation of the data structure processing.

Therefore, the goal in our investigation is to construct (or choose) the simulation method, to develop the mathematical model, and to create the simulation environment for such a class of systems.

2. Simulation modeling methodology in OA-basis

An object-attribute (OA) approach for organizing the computational process and data structures [8, 9] belongs to the dataflow class [6] and underlies the developed methodology because it permits modeling both the parallel processes and any modifications of complex data structures.

Comprehensive facilities for modeling the complex data structures are provided by the OA-graph, which allows one to describe structures of any complexity and topology and to modify them. In fact, the OA-graph is a semantic net whose nodes can denote both physical and abstract objects; the nodes are described by information capsules (IC) which are sets (or sequences) of information pairs (IP). The OA-graph can easily be synthesized or modified by adding or deleting an IP to IC, i.e., the OA-graph is similar to a toy construction set, and any data modification does not violate the data structure integrity. The OA-graph is processed by functional units (FU) which exchange information in the computational process; the information is organized as tokens; the tokens have the simplest form in the OA-system, i.e., they consist of two fields, the attribute (data identifier) and the load (data or reference-pointer).

The object models described by OA-graphs can be endowed with behavior (similarly to the object-oriented approach or OOA), because the OA-graph nodes can also be subroutines (similarly to the OOA methods). An OA-program is a sequence (or a set) of IP (or millicommands) transferred to FU; it can be placed in IC and built in the OA-graph. At any moment, an IP sequence from such an IC can be transferred to the corresponding FU to be processed. The IP loads can also contain the FU pointer to the corresponding OA-program to be performed.

Moreover, the FU themselves can also be structured, because their context (the set of internal registers determining the FU state) can contain pointers to the contexts of other FU (the context is the memory region containing the content of the FU internal registers). For example, to organize the FU as a grid structure, it is necessary to supplement the context of each FU with four registers containing references to the contexts of the neighboring FU (on the right, on the left, above, and below). The FU then use these references to exchange data with their neighbors. Thus, for example, the physical processes can be modeled by the net approximation method, where the considered physical region is divided into several regions by using a grid and the physical region characteristics are computed at the nodes of this grid (each node is associated with its own FU). In such computations, the parameters of the current point on the grid depend on the parameters of the neighboring nodes. The FU structures can be reconstructed in the process of modeling; for this, it is necessary to change the addresses of the computation result receivers in the FU contexts. Thus, both the structure of connections between the FU and the data structure can easily be modified in the model constructed according to the OA-principles.

It should be noted that the proposed approach is versatile. First, it ensures the simulation of structures of any topology, namely, the OA-graph structure can be regular (for example, hypercube, tree) or irregular (net). Second, it is possible to model the self-organizing systems and systems synthesizing and modifying their structures dynamically, because there is no centralized control in the model (the computational process is controlled by a set of independently operating FU). For example, this holds for the semantic analysis of a natural language in the case where the ontological base describing the text meaning is constructed in the analysis of the text; or for the collective behavior of some intelligent objects (for example, a collaborated mobile robots); or for Bionic Assembly System [10].

3. Formalization basics of an OA-model

It was required to formalize the developed OA-approach. The formal model must equally effectively describe the parallel dataflow computational process and the process of data synthesis and modification. To satisfy these requirements, we combined the OA-approach (the notions of IP, IC, and OA-graph) to describe the data synthesis and modifications, the KPN to describe the parallel dataflow computational process, and the theory of automatons to describe the operation of a separate FU.

3.1. Description of data structure

The data in the OA-system are described by IP, IC, and OA-graphs. We now formalize these notions and construct the so-called OA-algebra which allows us to describe such constructions.

We begin with the notion of information pair (IP). An IP is the dyad $\{a,l\}$, where $a \in A$ is the set of attributes and $l \in L$ is the set of loads. The attribute ($A \subseteq$ N, where N is the set of positive integers) is a universal identifier of the load used by the FU to recognize it. The load $l \in L = \{\Re \cup nil \cup \Omega \cup \Theta \cup S \cup A\}$, where L is the data set in the load, $\Re$ is the set of rational numbers, the symbol "nil" denotes the empty load, Ω is the set of global memory addresses (numbers), Θ is the set of FU addresses (numbers), and $S \in \Sigma^*$ is the set of symbol chains in the alphabet Σ.

All IP can be divided into two classes, namely, the information IP (used to describe some objects) and the millicommands (used to exchange information between OA-automatons).

We define IC by the definition of "IP chain". We shall say that η is an **IP chain**, if the following conditions are satisfied:

1. if ω is an empty IP chain (does not contain any IP), then ω is an IP chain;
2. if γ is an IP chain and c is an IP, then γc (or $\gamma \cup c$) is an IP chain (where $\cup$ is the symbol of union of IP chains);
3. η is an IP chain if and only if it is conditions (1) and (2) are satisfied.

We use $|\eta|$ to denote the cardinality (number of IP) of the IP chain η.

To modify IP, we introduce the operation of **concatenation (union) of IP chains** in the OA-algebra**:** if α and β are IP chains, then the IP chain $\alpha\beta$ ($\alpha \cup \beta$) is called the concatenation (chaining, union) of IP chains α and β: $\alpha\omega = \omega\alpha = \alpha$, where ω is an empty IP chain. We also introduce the operation of deleing the last or the first IP from a chain, of deleting an IP with a certain index, of introducing an IP chain in another chain after an IP with a certain index, and some other operation modifying IP chains.

To operate with several IP chains in the model, we introduce the notion of **bunch of IP chains** (W) which is an interrelated data structure consisting of IP chains. The cells of the bunch W can contain IP chains $W = \{w_1, w_2, \ldots w_{|W|}\}$, where $|W|$ is the length of the bunch of symbol chains, $w_1, \ldots, w_{|W|}$. To index a specific IP in the chain of IP chains, we can use the double index W_{ij} denoting the IP contained in the j-th IP on the i-th IP chain in W. In our model, a bunch of IP chains can emulate the common main memory of the computer system, where the data to be processed are stored; this memory can be accessed by several computing devices at once.

The OA-graph in our model is a set of IP chains united by references located in IP loads (the references belong to the set of addresses of global memory Ω) (Fig. 1). For convenience, the chains are united in a bunch of chains so as to operate with certain chains according to their indices.

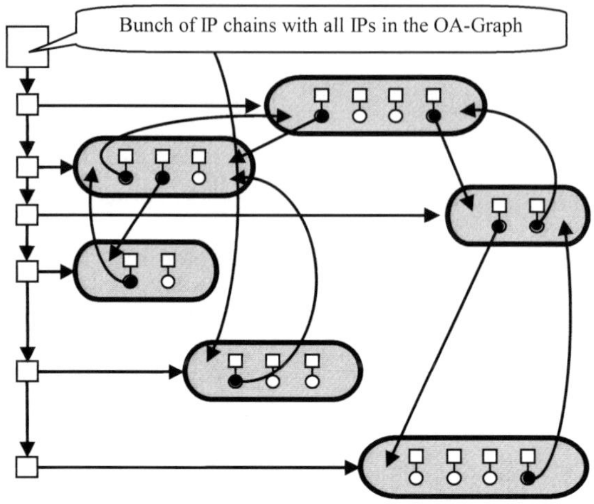

Fig 1. An object-attribute graph

3.2. Modelling the computational process

The model used to describe the process is based on the KPN, where the messages transferred between the nodes are symbols of a certain alphabet. A message in our model is an IP (millicommand). The FU uses the attribute to identify the obtained data stored in the millicommand load and appropriately processes them. If the FU cannot process the obtained millicommand, it is placed in the queue.

The FU operation can be described by an "improved" automaton model. The improvement means that the automaton deals not with the input flow of symbols but with the flow of millicommands. In certain cases, the use of OA-automata with discrete states is insufficient, because the OA-automata must store the data arriving in the millicommand loads (recall that $L = \{\Re \cup nil \cup S \cup \Omega \cup \Theta\}$ and the current state can also be described as a continuous multidimensional space). Therefore, we had to use the formal apparatus of hybrid systems which describe the automata with continuous space of states. So we first give the formal definition of OA-automaton.

The **OA-automaton context** is a set of abstract entities of any nature (scalar or vector variables, countable or continuous sets, etc.). If each entity contained in the OA-automaton context is represented as a measurement in a multidimensional space, then the automaton state K is a point in this space. The OA-automaton context must necessarily contain input and output queues of millicommands which are used to exchange information between OA-automata. A variation in the OA-automaton state is called its activation. An activation may occur only if one of the input millicommand chains contains a millicommand (i.e., there is a millicommand to be processed). Such a transition is determined by a function F contained in the OA-automaton context $K'=F(K)$, where F is the function of the OA-automaton transition from the current state K to a new state K'.

For example, the function F of an OA-automaton with discrete states can be defined by the transition predicate vector $\overline{P}$ and the transition function vector ($\overline{F}$). In this case, the vector dimensions must coincide, i.e., $|\overline{P}| = |\overline{F}|$. The arguments of the predicate P_i and the function of transition to a new state F_i, where i=1,2,…,$|\overline{P}|$, are all the parameters contained in the context K. Here F_i is a vector of functions, where each function returns the value of one of the context parameters $\overline{F}_i=[F_{i1},F_{i2},\dots,F_{i|K|}]$, $K_{j'}=F_j(K_j)$, where $j=1\dots|K|$.

The OA-automaton transition from state K to a new state K' occurs as follows: the predicates in the predicate vector are considered in succession and if $\exists\, i$: $Pi(K)=true$, then $K'=F_i(K)$ (i.e., $K_j=F_{ij}(K)$, where i=1,2,..,$|\overline{F}|$, j=1,2,…, $|K|$) and the search is terminated. In the case where none of the predicates Pi is true, the OA-automaton context remains unchanged. The function F is contained in the OA-automaton context and can be changed after the OA-automaton transition to a new state. Thus, the automaton operation algorithm can be changed directly in the computational process.

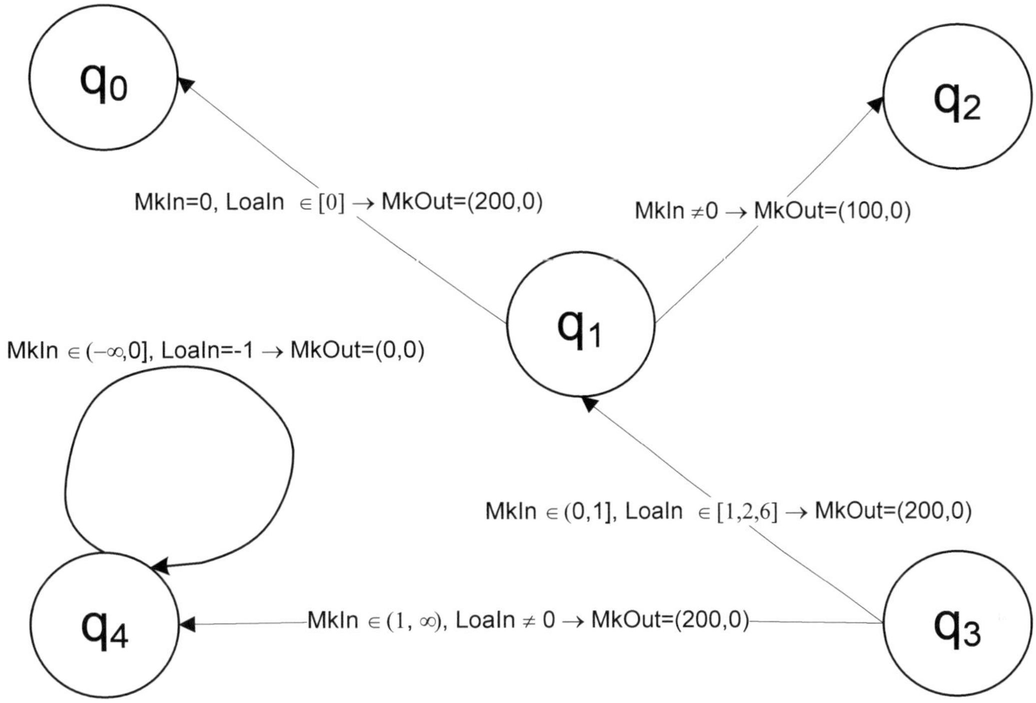

Fig 2. Graphic representation of the OA-automaton context change function

The **OA-automata** can be divided into several types each of which realizes a certain functional set and has its own individual context. But the context of an OA-automaton of any type necessarily contains the input and output queues of millicommands. So the context variation function *F* can be activated only if there is a millicommand at least in one of the input queues of millicommands (after activation of the function *F*, the millicommand at the head of the millicommand queue is deleted). When changing its context, an OA-automaton can place the information at the end of its output millicommand queues and thus transfer the data to other OA-automata for which this queue is an input queue. In this case, one millicommand queue can be contained in the context of one OA-automaton as an output queue and in the context of another OA-automaton as an input queue. The context of an OA-automaton of any type necessarily contains the common main memory W (an IP chain) which stores the OA-graphs for the common processing by OA-automata. Thus, the OA-automata can operate with the data located in this memory (this ensures the OA-automata operation with the common memory). All OA-automata are designed on the basis of the base automaton which has the minimal set of context fields to ensure the operation of the OA-system (OA-network):

AB={A, L, MkIn, MkOut, W, F},

where *A* is the alphabet of attributes, *L* is the data set in the IP load, *MkIn* is the queue of input millicommands, *MkOut* is the queue of output millicommands (an OA-automaton can add an Mk chain to the end of the queue), W is a global memory (bunch of IP chains), *F* is the function of the OA-automaton context change (each type of OA-automaton is associated with its own context change rules).

F enters the context and hence can be changed in the transition of the OA-automaton to another state. The possibility of dynamic reconfiguration (reprogramming) of an OA-automaton in the course of the computational process is thus realized.

A new type of automaton can be obtained by adding new elements to the context of the base automaton.

The set of OA-automata forms an **OA-automaton network**. Some context elements can simultaneously be contained in several OA-automata. For example, one millicommand queue can be an input millicommand chain for one automaton and an output chain for other automata. So, one automaton can place a millicommand in a queue where it is taken to be processed by another OA-automaton. In Fig. 3, we outline the OA-network where the data exchanged is realized through a special FU called «Bus». In such a system, all input millicommand queues are input queues for the FU Bus, and the FU output millicommand queues are input queues for the FU Bus. The FU Bus obtains a millicommand into its output queue from FUs, processes it, determines the millicommand target, and places it in the input queue of the corresponding OA-automaton-target.

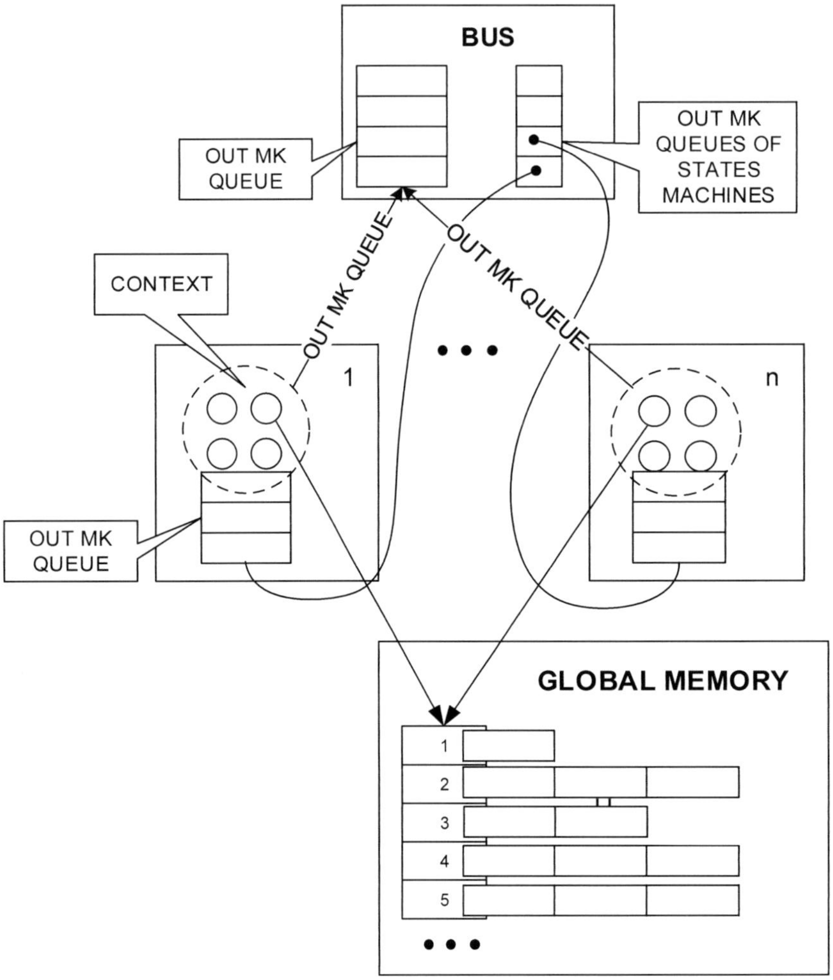

Fig. 3. An OA-Network structure

A sequence of OA-automaton activations in the OA-network is considered as an implementation of the OA-network. Only one OA-automaton can be activated at a moment of the model time. The initial data for the OA-network are formed as millicommands and are placed in the input millicommand chains of the corresponding OA-automata. The OA-network implementation is finished when all input millicommand queues are empty or when a different scenario is used. The states of OA-automata and the millicommands in the output millicommand queues form the simulation result.

4. Language for the OA-model description

A special programming language (OA-language) was developed for practical applications of the proposed OA-modeling and simulation methodology [8]. This language allows one to create virtual FU (or OA-automata), initialize them, input the initial data in the OA-model, and output the simulation results. The OA-language is also used to construct OA-graphs which are then modified by a set of OA-automata (FU).

4.1. OA-language syntax

The Object-Attribute language has a very simple syntax (see Table 1). An IP is denoted by the symbol «=», the attribute mnemonics is placed to the left of this symbol (each index of the attribute is associated with a mnemonics), and the load notation is placed to the right. The load can be a constant (integer or fractional number, Boolean value or row), a variable, or a reference. A capsule is denoted by curly brackets. If an IC is given as an IP load, then it can be written in curly brackets immediately after the symbol «=». A new FU is created by the millicommand NewFU, and an IC with the FU parameters is indicated in its load, namely, the FU type and its mnemonics. In the environment, each FU has its universal name (mnemonics) which is used by the programmer to address it. The mnemonics of a millicommand transferred by an FU is written after the FU mnemonics and is separated by point. An OA-graph in the «list» topology is denoted by the symbol «>».

Name	Notation	Example
InformationPair (IP)	Attribute=Load	FUType=FUIntALU
Millicommand	FUName.DataAttribute=Load	**ALU**.Set=0
Capsule	[CaplsuleName]{IP1 IP2 ...}	{**ALU**.Set=0 ALU.Add=5}
Capsule reference	{Attribute=[CaplsuleName]{IP1 IP2}}	**ALU**.EqProgSet= {Console.LnOut="Zero"}
Declaration of a New FU	NewFU{Mnemo=FUName FUType=FUTYPE}	NewFU{Mnemo="ALU" FUType=FUIntALU}
Comments	\\ * ... *\	\\ to the end of the row * comments *\
Declaraction of Variable-Load	**FUName**.DataAttribute= VariableName (Load)	**ALU_Fib2.**Set=temp(0)
Load with Reference to Variable	**FUName**.DataAttribute = VariableName	**ALU_Fib**.Pop=temp **ALU_Fib2**.Set=temp(0)!
List	>[CaplsuleName]**{...}** >[CaplsuleName]**{...}**	

Table 1. The Main constructions of the OA-programming language

4.2. Programming in OA-language

For example, in the OA-language an FU is constructed, initialized, and controlled as follows:

```
NewFU={Mnemo="ALU" FUType=FUIntAlu} // construct an FU with name «ALU»
ALU.Set=1 // Set the accumulator value 1
ALU.Mul=10 // Multiply the accumulator by 10
ALU.PopMk=Console.LnOut // Output the obtained value to the console.
// Console is an FU that outputs the result to the console.
```

An example of the OA-graph description (in this case, we describe the part of the dictionary which is used in the system of semantic analysis of the natural language in the NLP application):

```
Vocabulary.Set=
>{Mnemo="coach"
  Fork=
    >{PartOfSpeach=Noun SemProp={ Contiguity =Sport Contiguity=Instructor}}
    >{PartOfSpeach=Noun SemProp={ Contiguity =Transport Contiguity=Railway}}
    >{PartOfSpeach=Noun SemProp={ Contiguity =Transport Contiguity=Horse}}
    >{PartOfSpeach=Verb  SemProp={ Contiguity =Transport Contiguity= Horse Contiguity=Moving}}
  }
```

In this case, we describe the multivalued word «coach» which can mean, first the trainer of a sport team; second, a railway car; third, a horse-drawn carriage; fourth, the verb «go by a carriage». The attribute Fork precisely means that the IP load contains a reference to the list of meanings of a single word. SemProp is the attribute of the pointer to the semantic characteristics (properties) of the word. Contiguity is an associative relation (each word has individual associative relations). PartOfSpeach denotes the part of speech to which this word belongs. A special FU is designed to process OA-graphs of type «list»; in our example, such an FU has name Vocabulary, and the FU millicommand Set is used to transfer the reference to the OA-graph.

5. Programming and simulation environment for the OA-applications

We have developed and tested the methodology of simulation modeling of data-driven applications that follow and implement the dataflow paradigm of computing. This methodology is realized as an object-attribute programming and simulation environment which is based on and build upon a proprietary OA-programming language as it presented in previous Section. The OA-environment consists of the following parts.

- Software modules realizing the FU operation logic (OA-platform). Each FU type is a subprogram in the imperative programming language with standard interface. Such an approach is very similar to the actor programming, but the actors have different interfaces which does not improve the unification of the modeling system. The interface of the FU subprogram has three fields: the reference to the memory region containing the FU context; the millicommand index; and the reference to the millicommand load.
- Compiler of the OA-language of programming. It should be noted that the compiler is realized on the basis of the OA-architecture and is written in the OA-language (we used the technology of compiler untwining).
- Console of OA-program input and the console of the program operation result output. It is also possible to create additional input and output consoles. There also exist program modules responsible for the output of the OA-graph and of different graphs.
- Tool panel (on the left in Fig. 5), simplifies the process of programming (list of mnemonics of attributes, variables, pointers, list of FUs created in the environment).
- Tools ensuring the simulation modeling according to the Discrete Events System (DES) principle [7].

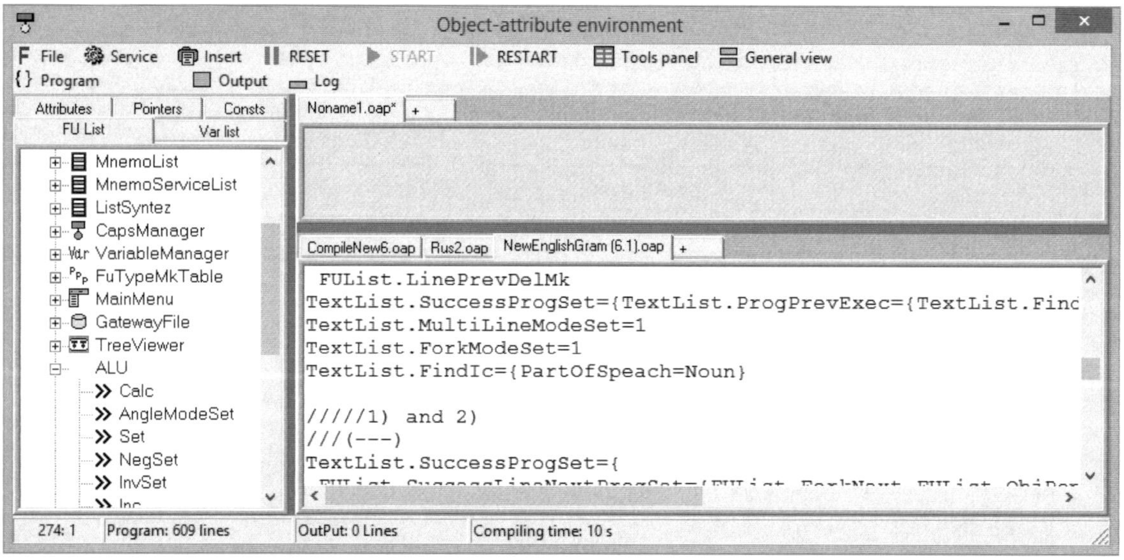

Fig. 4. Interface of OA-environment for data-driven application development and simulation

Our OA-programming and simulation environment is in development phase. At present, more than 80 types of FU have been realized in the OA-environment. There are several types of FU ensuring the OA-language compilation, information processing, input and output of information, and OA-graph processing, and there are FU designed to solve some application-specific problems. Because the OA-approach for organizing the computational process is very flexible, FU of new types can be added to the OA-environment without violation of the entire environment integrity. Any modification of the program implementing the FU operation logic does not also violate the environment integrity, i.e., the FU modification consists in adding new millicommands to the program while the old millicommands preserve their functionality. The refurbishment (which is practically inevitable when rather complicated OO-programs are designed) can be avoided because of the universal interface of the program implementing the FU operation logic. It allows one, without any aftereffects, to change the number of arguments of the commands realized by FU and to supplement FU with new operation modes without apprehensions to violate the system integrity.

6. Conclusion

The methodology of simulation modeling and its formal model were developed. The methodology is based on the OA-principle of the computational process organization and data organization, which allows us to model the parallel processes according to the dataflow principle and to synthesize and modify the data structures directly in the simulation process. To develop the formal model, we analyzed the existing methods and proposed a model based on a combination of the following three methods: the OA-approach for organizing the computational process and data structures, the hybrid automaton model, and the KPN. As a result of our investigations, we constructed the environment of OA-programming and simulation which permits modeling systems with complex dynamically variable structure. The proposed approach allows one to realize models of any level of detail, from conceptual (ESL) to one-to-one models.

The developed methodology permits simulation on multiprocessor and multicomputer complexes. The point is that the FU are autonomous devices «communicating» with other virtual FU (VFU) only through messages, i.e., the VFU are not in the universal address space and can therefore be set in operation on different computational devices united by communication lines (the messages between the VFU are transmitted through these channels). The CS where the simulation is implemented can also be heterogeneous (i.e., can consist of computational nodes of different architecture). Such possibilities can significantly accelerate the simulation process and make the CS much cheaper (the CS consists of many low-power computational nodes, and this is much cheaper than one high-performance CS).

The OA-model design does not require highly qualified programmers. For example, the program can be realized in the imperative language, and one need not insert special language structures in the text of the program (as in the case of the actor programming). The OA-programming language used to describe the FU interaction has extremely simple syntax and minimal time is required to get acquainted with it. Because of the dataflow-paradigm, the programmer need not take care of synchronization of the modeled parallel processes, i.e., the entire synchronization is automatic over the data.

The proposed OA-programming and simulation environment can also be used for modeling and study the system of semantic analysis of a natural language (Natural Language Processing, NLP). The simulation model of such a system can synthesize a semantic net describing a certain ontology from a text in a natural language. The model also takes the polysemanticism (multiple meaning) of lexemes and syntax structures into account.

The process of semantic net synthesis begins with formation of a list of lexeme interpretations. Then this list is transformed into a semantic net in several stages. In the English language, more than 20 stages have been separated (but it may happen that the number of stages will increase as a result of further studies). At each stage, two or more lexemes are united in a single semantic structure and so, step by step, the list of lexemes becomes a semantic net. At present, an FU specializing in transformation of the list of initial lexemes in the semantic net (OA-graph) is integrated in the OA-environment and some experimental bases of knowledge necessary to analyze the Russian and English languages are realized. The NLP-model operates with a restricted dictionary and processes a certain set of rules describing the identified language. The principle of lexeme semantic matching was developed to delete the unnecessary lexeme interpretations and text fragments, which permits omitting the lexeme interpretations inconsistent with the context. For example, the obtained semantic net can be used to seek information, namely, the user request in the natural language is transformed in the semantic graph-request, and it is further determined whether the graph-request is a subgraph of the text graph. If the answer is affirmative, then the search result is positive; otherwise, the result is negative.

The developed environment, which was used to model a special system of matrix multiplication, a system for modeling the computational processes by the net approximation method, and a system for modeling the semantic analysis of a natural language, demonstrated its efficiency.

7. References

[1] Alfredo Milani, Stefano Marcugini, Marco Baioletti. A Planning Model for Concurrent Asynchronous Automata. IJ ITA VOL.6, NO.1, 1999. Available at: http://cswww.essex.ac.uk/conferences/ukpssig/open-univ-19/milani.pdf.

[2] Carl Hewitt. A historical perspective on developing foundations for iInfo™ information systems: iConsult™ and iEntertain™ apps using iInfo™ information integration for iOrgs™ information systems. Available at: http://arxiv.org/ftp/arxiv/papers/0901/0901.4934.pdf

[3] Meyer Bertrand. The Conceptual Perspective. Computer (IEEE). 1996. Vol. 29, No 1. pp 86–88.

[4] Gabriel, R. Objects Have Failed: Notes for a Debate. Available at: http://www.dreamsongs.com/Files/ObjectsHaveFailed.pdf

[5] Gorodetskii, V. I. Self-organization and multi-agent systems. I. Models of many-agent self-organization. Izv. Ross. Akad. Nauk. Teoriya i Sistemy Upravleniya. 2012. № 2. pp. 92–120

[6] Jurij Silk, Borut Robic and Theo Ungerer. Asynchrony in parallel computing: From dataflow to multithreading. Institut Jozef Stefan, Technical Report CDS-97-4, September 1997

[7] J. Banks, J.S. Carson II, B.L. Nelson, D.M. Nicol. Discrete-Event System Simulation. Fifth Edition. Prentice Hall, 2010 – 662 pages

[8] S.M. Salibekyan, P.B. Panfilov. Object-attribute architecture for design and modeling of distributed automation systems. Automation and remote control. Volume 73, 2012, Number 3, pp. 587-595, DOI: 10.1134/S0005117912030174

[9] P.B. Panfilov, S.M. Salibekyan. Dataflow Computing and Its Impact on Automation Applications. Procedia Engineering. Volume 69 (2014), pp. 1286–1295. Available at: http://www.sciencedirect.com/science/article/pii/S1877705814003671

[10] Kukushkin, I. K.; Katalinic, B.; Cesarec, P.; Zdyb, D. & Kettler, R. Modeling of mobile robot behavior for line-less Bionic Assembly System, Annals of DAAAM for 2012 & Proceedings of the 23rd International DAAAM Symposium, ISBN 978-3-901509-91-9, ISSN 2304-1382, pp. 0865 - 0870, Editor B[ranko] Katalinic, Published by DAAAM International, Vienna, Austria, 2012

DOI: 10.2507/27th.daaam.proceedings.059

ELECTROMAGNETIC EMISSION MEASUREMENT OF DEVICES WITH GSM INTERFACE

Jozef Hallon, Mikuláš Bittera, René Harťanský & Karol Kováč

This Publication has to be referred as: Hallon, J[ozef]; Bittera, M[ikulas]; Hartansky, R[ene] & Kovac, K[arol] (2016). Electromagnetic Emission Measurement of Devices With GSM Interface, Proceedings of the 27th DAAAM International Symposium, pp.0400-0404, B. Katalinic (Ed.), Published by DAAAM International, ISBN 978-3-902734-08-2, ISSN 1726-9679, Vienna, Austria
DOI: 10.2507/27th.daaam.proceedings.059

Abstract

The paper deals with the methodology of measurements of conducted and radiated electromagnetic emissions of devices equipped with a GSM interface whereas their full functionality has to be maintained during the electromagnetic compatibility (EMC) measurement. At first it briefly describes a standard measurement method and an auxiliary system securing the GSM signal in a shielded semi-anechoic chamber, where the measurement takes place. Then it picks up the most important factor, which causes the rise of radiated emission level during GSM communication and it demonstrates the level of this emission increase. At the end the paper outlines to ways to suppress this phenomenon and analyses the circumstances by other wireless communication technologies.

Keywords: Electromagnetic compatibility; radiated emission; conducted emission; shielded chamber; GSM; GPRS.

1. Introduction

Wireless networks are increasingly used for transfer of information among technological devices: GSM network, serving primarily for voice services and its part GPRS, or its modified version - EDGE (has different modulation and coding) serving as mobile data networks, gained great popularity. GSM/GPRS interface became part of not only telecommunication devices, but also various telemetric, controlling and monitoring systems, e.g. [1], [2], [3], [4]. It is relatively outdated 2.5G technology, but it is still sufficient for many measurement systems.

Devices placed on the EU market must meet the requirement, that the electromagnetic (EM) interference they produce does not exceed the level, when radio and telecommunication devices or other devices cannot operate with the intended purpose [5]. Special measurement procedures standardized in relevant technical standards [6], [7] are used for detecting the levels of interference of electronic equipment. The requirement of the measurement procedure at measuring the EM emissions is that the device is to be measured in the mode, when its EM radiation is at the highest level [6]. Devices using GSM/GPRS (hereinafter only GSM) interface reach highest level of EM radiation when the GSM modem is in the broadcast mode. This is because the broadcast mode makes high energetic demands on the power supply system generally associated with higher levels of EM emissions.

The measurement of EM emissions is often carried out in shielded chambers specially adapted for this purpose. However the principle of the shielded chamber used for measurement of radiated EM emissions itself prevents the creation of a dataless connection. This requires modification of the measurement procedure and equipping the anechoic shielded chamber, where the measurements of EM emissions are carried out, with new equipment, allowing a wireless connection to outside network of assessed device data interface [8], [9].

Considering the mention facts we describe the standard emission (radiated as well as conducted) measurement setup with additional GSM communication equipment in the second part of the paper. In part 3 we analyse the main reason of emission increase during GSM communication, which is the rise of current supplying the electronic device containing GSM module. On a real device prototype we show the circuit elements, what may help to lower emissions. Then we present the comparison of emission measurement results of two operational modes of the device. Both emission spectra are shown to see the difference. The results are presented both for radiated as well as conducted emissions. The variances are described in the text.

2. Measurement Setup

The standard radiated emission measurement test setup requires that tested device (EUT – equipment under test) is placed on a supporting table or on the floor of the shielded chamber. The chamber is in a semi-anechoic configuration – conducting metallic floor creates reflective plane. The measuring distance is L = 3 m or 10 m [6]. The antenna and the EUT form the two foci of imaginary ellipse, creating an obstruction-free area, which defines a minimum area free from scatterers of EM field. Technical equipment in the chamber must enable EUT to rotate around its axis and antenna to adjust the height in the range of 1 ÷ 4 m above a conductive floor of the chamber (Fig. 1a). The automated measuring system measures radiated emissions (antenna - preamplifier - measuring receiver, which is connected to the PC) at various positions of the EUT and the antenna. Then the system compares the measured emissions with the limit values given in the relevant standard [10].

GSM communication of the EUT requires creating a connection, which provides bidirectional data transmission between the EUT and auxiliary equipment situated outside the shielded chamber. GSM connection to the EUT needs a communication system (outside shielded chamber) consisting of GSM antenna and repeater that receives and amplifies signal of nearby base stations. This signal is transferred via coaxial cables and feed-through to the GSM antenna inside the chamber. The communication system is shown in Fig. 1b. Location of the GSM antenna inside the chamber and GSM signal level have to be chosen to enable tested device a reliable communication, while the measuring antenna should receive GSM signal of sufficiently low level (ideally less than relevant limit level, but definitely cannot exceed maximal input power of a measuring preamplifier). High level of signal can cause a malfunction of the measuring preamplifier and then a degradation of the measurement [9].

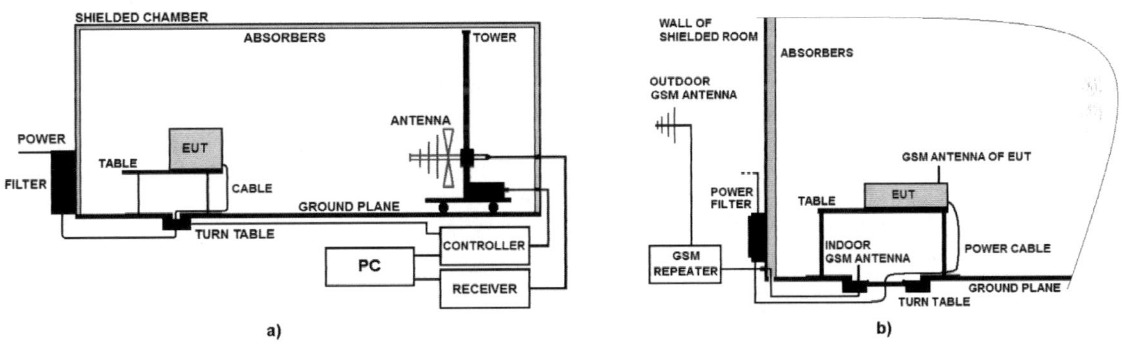

Fig. 1. Test setup for measurement of radiated emissions (a) and GSM communication system (b)

In case of measuring a conducted interference (interfering terminal voltage) the situation is simpler. The measurements can be carried out with an open chamber, allowing normal GSM communication without the mentioned GSM communication system. Only the reference metal floor of the chamber and the supply system, filtered in order to prevent the interfering voltage from penetrating into the shielded chamber via the power supply system, is used for this measurement. Moreover the measurement of radiated emissions of the tested device is performed each time, the measurement of conducted emissions are performed only in case the EUT is connected to an external common AC or DC power distribution or to telecommunication ports. The measurement of emitted disturbance voltages on the power terminals is performed using the LISN (Line Impedance Stabilizing Network). LISN is connected to the measuring receiver (more details are in [6]).

3. Analysis of Cause of Emission Increase during GSM Communication

During operation the current consumed by the GSM modem may differ by several orders of magnitude. During the broadcasting interval the peak current may reach up to 3 A. The amount of this current is proportional to radiated power. The power is proportional to the distance from the base station, dimming of transmission path (e.g. a building) and depends also on properties of the antenna connected to the modem. The Fig. 2 shows the time dependence of the modem current consumption (in GPRS mode) for typical operating modes for modem MC75i [11], [12]. Shown maximal impulse current due to impedances in power circuits ((impedance/resistance on connectors, in the power source and signal tracks) causes the pulse fluctuations of power voltage and transitory phenomenon on pulse edges (Fig. 2). Decreases of power voltage may reach up to several 100 mV (for typical modem power voltage 3.2 V – 4.5 V). This makes high demands on the construction of modem power circuits. In case mistakes are made in the design and realization of power circuits, it has an adverse consequence on the EMC properties of the whole device.

We have more alternatives when powering the device containing GSM modem. If it is a portable device, batteries or rechargeable batteries are used. When the external voltage is small, e.g. 5 V, low drop linear stabilizers are used. Switching regulators are used for higher power voltages, where the linear regulators would be inefficient from the energetic point of view. In terms of EMC this alternative is the most interesting one, because the switching regulator itself produces relatively high broad spectrum interference further augmented by the impulse load of GSM modem.

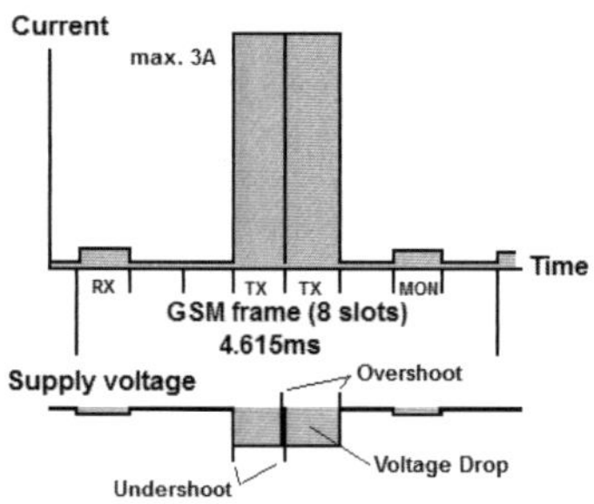

Conditions	I [mA]
TALK mode	
GSM900	460
GSM1800	275
DATA mode; GPRS class 8	
GSM900	495
GSM1800	300
DATA mode; GPRS class 10	
GSM900	680
GSM1800	500
DATA mode; GPRS class 12	
GSM900	730
GSM1800	530

Fig. 2. Time dependence of supply current, voltage and maximal average supply current of modem MC75i.

The capacitor with high capacity, which must cover the power supply of the modem at the peak current during transmission slot (slots) within the GSM frame, has a vital role in the power supply chain. It must be a low ESR (Equivalent Series Resistance) type with values of the order of 10 mΩ and sufficient capacity. The required capacity changes depending on the type of stabilization circuits and ranges from 100μF – 10 mF [11]. Chosen capacitor must be accompanied with a parallel combination of several ceramic condensers (e.g. 100 nF + 10 nF + 100 pF) to reduce the EMI (electromagnetic interference) in the wide frequency range. These elements must be located in the immediate vicinity of the GSM modem; otherwise their effect will not be as expected [13].

As already mentioned the use of switching DC/DC converter is the most difficult solution in terms of emissions minimization, however in terms of efficiency it is often necessary (power from 12 V or 24 V distribution systems). There is a large number of converters manufactured by various companies available. It is necessary to follow wiring diagram prescribed by the manufacturer and ensure an efficient EMI input filter on the supply side. The Fig. 3 shows a block diagram of power circuits mentioned above.

It is very important to pay great attention to the layout of individual components and PCB design for power circuits and respect recommendations of the manufacturer of the DC/DC converter, GSM modem and general construction rules known from the EMC literature [13].

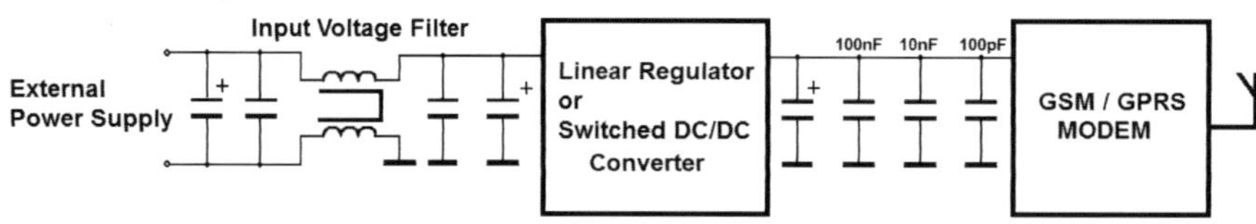

Fig. 3. Scheme of power circuits of GSM modem.

4. Example of Emission Measurement

We performed the measurement on the prototype system sample of the data logger for remote data collection equipped by the GSM modem. The sample had deliberately undersized filter measures as well as the value of the condenser with low ESR for coverage of energy consumption during transmission slots, so that the difference in emission level between the state of the GSM modem in standby mode and operational mode (the modem was broadcasting a data packet with rate of 500 Bytes/s, or was alternately sending a SMS continuously) was significant. The device consisted of a GSM modem MC75i, powered by the step-down DC/DC converter from a 12 V battery. The device measured voltage values using an A/D converter controlled by a microcontroller and concurrently the task of the converter was to format measured data and send it to the modem. The device was situated on one circuit board and it was located in a plastic box. The analog input cable was shielded, it was 1 m long and it ended with typical impedance. The modem output cable was shielded as well, it was 0.5 m long and a GSM antenna (short monopole) was connected to the end of the cable. Power cable (wire pair) leading to the battery was 30 cm long. In order to force the modem work under more difficult conditions, i.e. at a low level of signal from the base station, we connected a 20 dB attenuator to the GSM repeater output. This way the modem worked at the edge of its capabilities, in the worst possible situation.

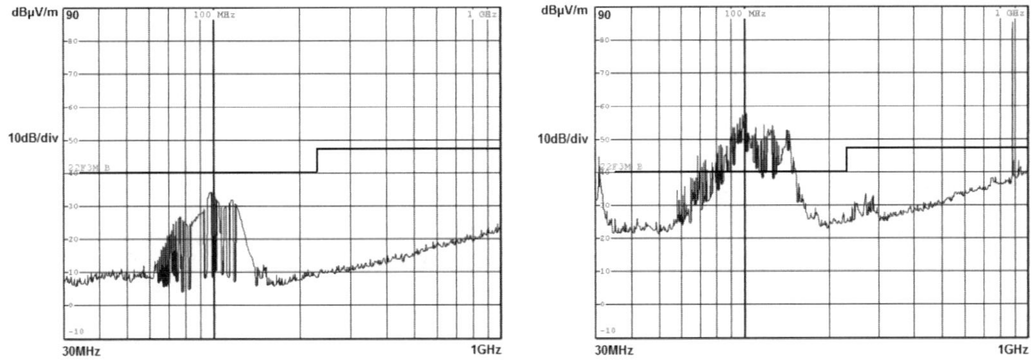

Fig. 4. Radiated emission of remote measuring system with GSM modem (without the GPRS transmission and with GPRS transmission)

Radiated emissions were measured in the frequency range 30 MHz – 1 GHz, with the residual bandwidth (RBW) of 120 kHz, as prescribed in technical standards, which deal with measurement of EM emissions of devices information technology. The limit was chosen for the emission class B (for measuring distance of 3 m) [6].
A high peak, corresponding to the modem working frequency appeared near 900 MHz in case of active operation of the GSM modem. The radiated power was relatively high; therefore we needed to attenuate the input signal of the measuring receiver by 20 dB (to avoid overloading of electronic circuits of the measuring receiver). This was reflected in rising of the apparatus noise level. As shown in Fig. 4 on the right, in case the modem was operational, emissions increased by approximately 25 dB and moreover in the range from 80 MHz – 150 MHz significantly exceeded the limit. Even though the spectrum was measured using an peak detector and the limit is determined for measurement by a quasi-peak (QP) detector, the limit would be exceeded even in the case of measurements by individual peaks with a QP detector (it would be unbearably time consuming to measure the whole spectrum using a QP detector). It is unpleasant even due to the fact that VHF bandwidth for radio is located in this frequency range and higher level of emissions in this area may interfere with the radio reception.

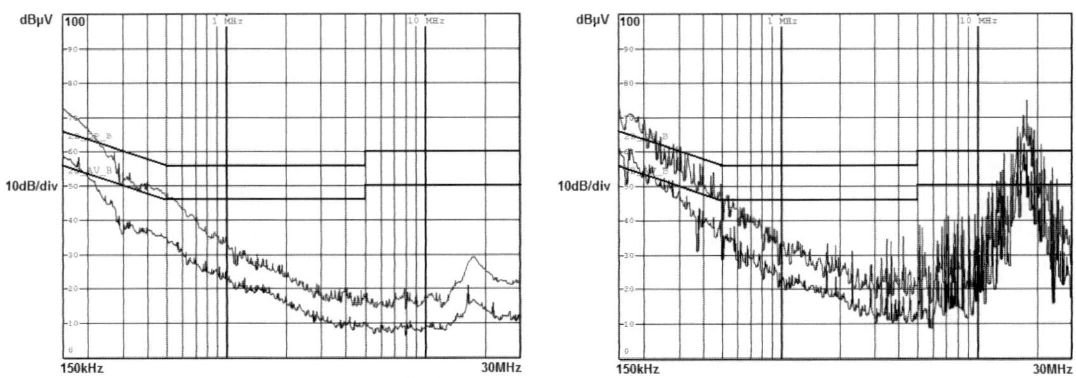

Fig. 5. Conducted emission remote measuring system with GSM modem (without using the GPRS activities and with enabled GPRS activities)

Conducted emissions from power supply were measured in the frequency range 150 kHz – 30MHz, at intermediate frequency bandwidth of 9 kHz, as prescribed in technical standards dealing with measurements of emissions of devices of information technology. The limit was again chosen for the emission class B [6]. Even in this case we can see a significant difference in the emission spectrum at active modem operation (Fig. 5 on the right). The most significant interference appears in range 15 MHz – 25 MHz, where the values exceed relevant limits by approx. 10 dB. Even in this case there will be a problem with measurement using QP and also AV detector, because they would exceed the relevant limits.

5. Conclusion

We showed in the paper the significance of emissions levels increase when changing the mode of operation of electronic device with GSM communication module from quiet to transmit operation. For our example of such device this rising achieved +25 dB for radiated and even more than +40 dB for conducted emissions. It is necessary to mention that studied device was not perfectly designed as our aim was to demonstrate the significance of the GSM transmission operation influence upon the level of emissions of both types. On the basis of this reality one must conclude that within EMC compliance verification procedure it is necessary to measure each electronic device with GSM communication possibilities also during GSM communication operation.

The measurement of radiated emissions of mentioned devices itself requires additional equipping of semi-anechoic shielded chamber with a system ensuring the optional GSM operation. Such system is also needed for immunity to EM field testing. Many EMC laboratories do not have such facilities, so they are not able to ensure GSM data transfer during EMC measurements, and therefore in many cases the measurement could not be carried out in transmitting mode (typical operation mode).

As was demonstrated in the paper the higher level of emissions is caused by higher power supply currents consumed by powerful GSM transmitter output amplifier. It is because this type of communication required up to 2 W of high frequency power. Other wireless data communication systems, e.g. WIFI (the maximum output of 100 mW is allowed in Europe for bandwidth of 2.4 GHz) and Bluetooth (by classes 1 mW – 100 mW) work with the order of less output power. So one may predict, that unintended radiation of these devices would be under limits of corresponding EMC standards.

6. Acknowledgments

Work presented in this paper was supported by the Slovak Grant Agencies of Ministry of Education, Science, Research and Sport of the Slovak Republic VEGA No. 1/0431/15 and KEGA No. 016STU-4/2014, and Slovak Research and Development Agency APVV, No. APVV-15-0062.

7. References

[1] Angelides, K. (2013). Controlling diabetes with a cellular GPRS linked glucometer pedometer, Patent US8568309 B2, 29. 10. 2013

[2] Jayakumar, J.; Queen, J. H. J.; James, T.; Hemalatha, G. & Lonappan, N. (2013). Distribution Transformer Monitoring Using GPRS. International Journal of Scientific & Engineering Research, Volume 4, Issue 6, June-2013, ISSN 2229-5518, pp. 1199-1204

[3] Korhonen, J.; Aalto, O.; Gurtov, A. & Lamanen, H. (2001). Measured performance of GSM, HSCSD and GPRS, Proceedings of Communications ICC 2001, IEEE International Conference on, Helsinki, 11 Jun 2001-14 Jun 2001, ISBN: 0-7803-7097-1, pp. 1330-1334

[4] Bordea, N.; Sintea, S.; Moraru, S.A. & Kristaly, D.M. (2008). Serial Data Communication Connection Over GSM Data Channel, Annals of DAAAM for 2008 & Proceedings of the 19th International DAAAM Symposium, 22-25th October 2008, Trnava, Slovakia, ISSN 1726-9679, ISBN 978-3-901509-68-1, Katalinic, B. (Ed.), pp. 0127-0128, Published by DAAAM International Vienna, Vienna

[5] Directive 2014/53/EU of the European Parliament and ot the Council of 16 April 2014 on the harmonisation of the laws of the Member States relating to the making available on the market of radio equipment, Official Journal of the EU, L 153/62-105, 22. 5. 2014, Available from: http://eur-lex.europa.eu/legal-content/HR/ALL/?uri=CELEX:32014L0053

[6] European standard EN55022:2010 Information technology equipment – Radio disturbance characteristics – Limits and methods of measurement, CENELEC 2010

[7] European standard EN55032:2015 Electromagnetic compatibility of multimedia equipment – Emission requirements, CENELEC 2015

[8] Hallon, J.; Harťanský, R. & Bittera M. (2015). Testing EMC properties of electrical devices equipped wireless communication, Proceedings of the 10th International Conference on Measurement. Smolenice, Slovakia, 25.-28. 5. 2015, ISBN: 978-80-969672-9-2, pp. 249-252

[9] Hallon, J.; Bittera, M. & Kováč, K. (2016). The interface set for the EMC testing shielded chamber. Przeglad elektrotechniczny, R. 92 NR 2/2016, ISSN: 0033-2097, pp. 31-33

[10] Cicakova, O.; Szolik, I. & Krammer, A. (2009). Complex system for EMC immunity tests, Proceedings of Applied Electronics 2009, Pilsen, Czech Republic, 9.-10. 9. 2009, ISSN: 1803-7232, ISBN: 978-80-7043-781-0, pp. 87-90

[11] Power Supply for Wireless Applications, Appl. Note 26, WM_AN26_PwrSupply_v06 2013-12-13, Available from: https://developer.gemalto.com/sites/default/files/wm_an_26_pwr_supply_v06.pdf

[12] Quad-band GSM/GPRS data & voice modules System Integration Manual, 22. 8. 2014, Available from: https://www.u-blox.com/sites/default/files/products/documents/LEON-G1_SIM_(UBX-13004888).pdf

[13] *Goedbloed*, J. (1992). *Electromagnetic Compatibility*, Prentice Hall, ISBN: 0-13249293-8, New Jersey

DOI: 10.2507/27th.daaam.proceedings.060

CONTROL MOBILE ROBOT WITH HYBRID DRIVE

Ilyukhin Yu. V., Tatarintseva A. V.

This Publication has to be referred as: Ilyukhin, Y[ury] & Tatarintseva, A[lexandra] (2016). Control Mobile Robot With Hybrid Drive, Proceedings of the 27th DAAAM International Symposium, pp.0405-0416, B. Katalinic (Ed.), Published by DAAAM International, ISBN 978-3-902734-08-2, ISSN 1726-9679, Vienna, Austria
DOI: 10.2507/27th.daaam.proceedings.060

Abstract

In the paper we obtained the mathematical model and the organization of automatic traffic control of transport mobile robots with hybrid drives. The hybrid drive is a mechatronic system which consists of an internal combustion engine, an electric torque drive and robot control device. When you build a algorithms and structures of control system of robot with hybrid drive, is taken into account the non-linearity of the characteristics of the hybrid drive, processes of energy recuperation, requirements to dynamic properties of the robot and the efficient using of energy resources. The results of theoretical research supported by the results of computer simulation, which confirmed the effectiveness of the structure and the control system algorithms for the robot motion on different types of path segments.

Keywords: mobile robot; motion control; mechatronic system; hybrid drive; internal combustion engine; electric torque drive; computer simulation.

1. Introduction

A lot of research is dedicated to the development of autonomous mobile robots, which are necessary to replace human under the difficult operating conditions [1]. Therefore, automatically functioning transport robots are allocated among the mobile robots. They are intended for the delivery of goods over long distances on the condition of their movement, including over rough terrain with the speed varying over a wide range depending on the prevailing situation controlled by robot. Therefore, modern mobile robots must meet the requirements of high rough-terrain performance and intense dynamics and have a great resource of motion with cost-effective use of energy resources. Currently the most common mobile robots have electric drive and controlled motion by a human operator, as a rule, these types of robots don't have the required resource of motions and not able to independently decide on the choice of the speed and operators have to quickly make decisions to control the mobile robot in a changing environment. [2]. Autonomous mobile robots [3-4] equipped with hybrid drives are largely comply the above-formulated requirements. The hybrid drives of mobile robots are mechatronic systems and consist of internal combustion engine (ICE), torque electric drive (TED) and computer control device [5]. It is necessary to develop structures and algorithms for effective use of a hybrid drive for controlling the movement of the robot. It is important to achieve coordinated work of ICE and TED automatically taking into account the non-linearity characteristics of the hybrid drive components and features of energy recuperation processes. Therefore, the task of motion control of mobile robots, equipped with mechatronic hybrid drives is one of the urgent tasks in the field of robotics.

To a certain extent, this problem is similar to the problem that has been solving in the automotive industry, where everything is increasingly used hybrid propulsion. One of the typical examples of the successful application of the hybrid drive is the car Toyota Prius [6-8]. The difficulty of solving the problem, typical of mobile robotics is due to the need for an autonomous operation of the robot, where the control system on their own, without the involvement of a human operator makes a decision about the most rational way to use features of the hybrid drive for the realization of the desired mode of movement of transport mobile robot.

The paper presents solution to the problem of the control movement of mobile robot with a hybrid drive, which includes a definition of robot structure with the considered type of drive, the formation of its mathematical model, the development of rules and automatic control algorithms in different driving modes, including taking into account energy recuperation processes.

2. The structure of mobile robot with hybrid drive

We consider the mobile robot with four wheels, and two front wheels are leading. These wheels are driven by a parallel hybrid type [9]. The mobile robot with hybrid drive has a structure (fig.1) that built in accordance with the principles and methods of mechatronics [10-12]. The main components of the robot are a robot control device, mechatronic power unit and the moving parts of the robot (the robot mechanism) as a control object.

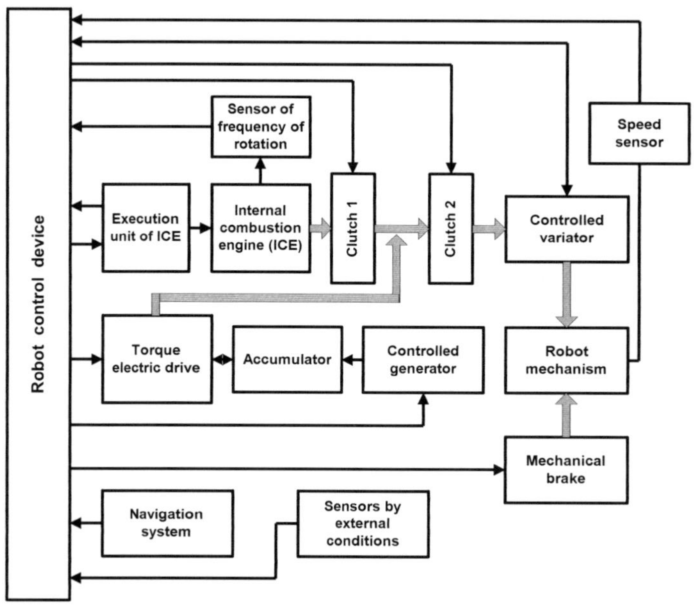

Fig. 1 The structural scheme of mobile robot with hybrid drive

The composition of mechatronic power unit [5] consists of the petrol ICE with executive unit to control ICE, TED, two controllable clutches, controlled gearbox (controlled variator), accumulator, sensor of frequency of rotation ICE, the mechanical brake and controlled generator. Torque electric drive (TED) is built on the basis of the electric motor. So TED is high-speed servo system that is capable of high fidelity reproduction of the torque on the motor shaft.

All of the processes occurring in the robot and its hybrid drive, obey the signals issued by the robot control device. It makes decisions based on data from the speed sensor of the robot, a navigation system and sensors by external conditions. Mobile robot with mechatronic hybrid drive can operate in different modes, and its mechatronic hybrid drive should have a high flexibility in implementing the required torque of the forces. Robot control system can use opportunities the ICE only or separately operated the TED, and can implement a joint coordinated operation of the ICE and the TED.

In this case it is possible to achieve improvement of dynamics during acceleration or deceleration of the robot, as well as improving the energy performance of drive unit, which has a positive effect on the duration of work the robot to the next refueling. Possibility of the ICE and the TED work both separately and jointly provided by using in the hybrid drive two controllable clutches. The main modes of operation of the autonomous mobile robot with a hybrid drive are the movement of the maximum dynamics, the movement at a low speed and energy recuperation mode. In the mode with maximum dynamics, for example, when driving on the rise, mechatronic hybrid drive should cultivate a big torque and create a big acceleration value. In this case, as a rule, the TED helps the ICE, increasing to the required level of the total torque. The intensity of deceleration is also increased, because the TED can create a significant braking torque. In the driving mode with a low speed of the ICE can't operate stably because desired engine speed can be less than the idle speed. Therefore, the robot movement is carried out at the expense of TED. The feature of this mode is that the propulsion system of the robot does not create excessive noise.

A limited number of energy resources makes attentive to the choice of control algorithms. Availability of the TED allows effectively replenish the loss of energy in the accumulator. In control recuperation mode, the TED generates a braking torque overcome by the ICE. This ensures replenishment of electric energy in the accumulator. It should be noted, if the robot moves along the terrain, it can be alternation phases summation the ICE and the TED torques. When robot moves on the rise the accumulator electric energy is consumed by the electric drive. Charge of the accumulator is as a result of recuperation energy of the robot on the descent. This increases the energy efficiency of mobile robot and increases its resource of motion. However, in some modes controlled generator can be used for forcible charging. In this case control algorithms should provide control over the degree of charge of the accumulator and its current limit charge.

3. Mathematical model of mobile robot with hybrid drive as a control object

Mathematical model of the mobile robot with a hybrid drive, which is the object of control, includes descriptions of interacting components, such as the mechanical part mobile robot (robot mechanism), the ICE and the TED. Torques produced by the ICE and the TED constitute resultant torque M_Σ. Transmission of the robot consists of variator with gear ratio k_{VAR} and the main transmission ratio k_{MT}. Translational motion of the mechanical part of mobile robot with a hybrid drive can be described by the following differential equations:

$$\frac{dV}{dt} = m_\Sigma^{-1}(F_{FHD} - F_{MRF}),\qquad \frac{dx}{dt} = V, \tag{1}$$

where V and x – speed and movement of the robot accordingly; F_{FHD} – force produced by a hybrid drive; F_{MRF} – motion resistance force; m_Σ – the equivalent weight of the moving parts of the robot, when

$$m_\Sigma = m_{ROB} + R_W^{-2}(4J_W + J_{FF}) + R_W^{-2}(k_{MT}k_{VAR})^2 J_{RT};\quad F_{FHD} = M_\Sigma k_{MT} k_{VAR} R_W^{-1},\quad F_{MRF} = F_{MRF.ICE} + F_{RF} + F_{FF} + F_{AR} + F_{BR}, \tag{2}$$

where m_{ROB} – robot weight; R_W – wheel radius; J_W – moment of inertia of the wheel; J_{FF} – moment of inertia of the transmission relative to the axis of rotation; J_{RT} – the total moment of inertia of the rotating parts the ICE and the TED; $F_{MRF.ICE}$ – the force, due to the moment of resistance forces in the ICE; F_{RF} – the rolling force, which depends upon the weight and the surface inclination angle at which the robot is located, in relation to the horizontal plane; F_{FF} – the resulting friction force; F_{AR} – the force of aerodynamic resistance of the robot motion; F_{BR} – resistance force caused by brake operation, if they are involved.

$$F_{MRF.ICE} = M_{MR.ICE} n k_{MT} k_{VAR} R_W^{-1}, \tag{3}$$

where $M_{MR.ICE}$ – resistance moment of the ICE; n – control parameter equal to 1 if the ICE is connected via the clutch, and equal to 0 if the ICE is disconnected via the clutch.

$$F_{CK} = m_{POБ} g \sin\alpha, \tag{4}$$

where g – acceleration of gravity; α – angle of inclination of the surface on which robot is in relation to a horizontal plane.

The required value of the variator transmission ratio k_{VAR} is set as a function of its desired value $k_{VAR.D}$ by variator regulation subsystem. This subsystem is dynamic, and its properties are characterized by the equation

$$T_{VAR}\frac{dk_{VAR}}{dt} = k_{VAR.D} - k_{VAR}, \tag{5}$$

where T_{VAR} – the time constant of the variator drive model (this model is given in the form of an aperiodic link).

Mathematical model of the ICE is characterized by nonlinear dependence, which in most cases detected in the analysis of experimental data [13-15]. As a result of analysis of experimental data for the type of internal combustion engine ZMZ-

40524, 98 kW [16] using the least squares method, we've obtained an analytical dependence torque on the shaft of the internal combustion engine of the average in one revolution of the crankshaft M_{ICE} and the angular velocity of the crankshaft Ω_{ICE} and the control action h (the degree of opening of the valve feeding the working mixture). This dependence has the form

$$M_{ICE} = M_0 f_1(\Omega_{ICE}) f_2(\Omega_{ICE}, h) + M_1, \tag{6}$$

where h can change from 0 to 1;

$$f_1(\Omega_{ICE}) = \lg(k_1\Omega_{ICE} + \Omega_0);$$
$$f_2(\Omega_{ICE}, h) = 1 - k_2 \exp[(h_0 - h)b_1(k_3\Omega_{ICE} + b_2)^{-1}]; \tag{7}$$

and M_0 = 37,75 Nm; M_1 = 100 Nm; k_1 = 1,25 s/rad; Ω_0 = 0,0473 rad/s; k_2 = 2,79; h_0 = 0,015; b_1 = 44,9; k_3 = 0,0131 s/rad; b_2 = 1,07. The control action h is formed by the control device of ICE depending on its input control signal U_C. It is described by an aperiodic link with the gain k_{CD} and time constant T_{CD}.

The ratio of the angular speed Ω_{ICE} of the crankshaft to robot speed of the longitudinal movement is given by the equation

$$\Omega_{ICE} = Vk_{MT}k_{VAR}R_W^{-1}. \tag{8}$$

The analysis showed that the proposed form of the dependence of the torque of the ICE from Ω_{ICE} and h is valid for any type of ICE. Therefore, it can be used to describe a hybrid drive of mobile robot. The comparison data obtained via the reduced formulas with experimental data determined that the relative error calculation torque of the ICE is in the range of 5 ... 20% for different types of engines. This demonstrates the suitability of the obtained dependence for the analysis of the properties of the mechatronic hybrid drive of mobile robot.

We consider the torque electric drive (TED) built on the basis of a DC motor. The basis of the TED forms a dynamic drive system, closed current loop the current PI controller and configured to the technical optimum. In considering such a drive system in a first approximation, we can neglect the influence of the motor velocity to the motor current. Then the mathematical model of the system can be described by equation

$$\frac{d^2 i_{TED}}{dt^2} + a_2 \frac{di_{TED}}{dt} = a_1(i_{TED.R} - i_{TED}), \tag{9}$$

where i_{TED} and $i_{TED.R}$ – actual and required current of the TED accordingly; a_1 and a_2 – coefficients, values when tuned to technical optimum depends on cutoff frequency ω_{CT} of open loop system of the TED and determined by the formulas $a_1 = 2\omega_{CT}^2$, $a_2 = 2\omega_{CT}$.

It should be noted that the TED is better to build as closed torque loop servo system. Her straight chain includes series-connected controller and the drive system considered above. Considering that this drive system is static, it is proposed to use an integrated controller to increase the dynamic precision torque control of the TED.

4. The approach to the construction of motion control systems of robots with hybrid drives

The control system of movement mobile robot (fig.2) is designed to automatically control the speed of the mobile robot. The actual speed V is determined at the input signal U_V carrying information about the desired speed V_R. This system is dynamic and closed by a longitudinal movement of the robot speed. The desired speed V_R of movement of the robot is formed by a system of a higher rank in the planning of the route of movement and during the subsequent operational adjustments taking into account the detected obstacles, the state of the road and a number of other factors that affect the speed of traffic on the selected route. Established impact U_V undergoes an additional adjustment to the calculation portion of the desired speed and turns into a signal V_C applied to the input of the closed speed loop of control system.

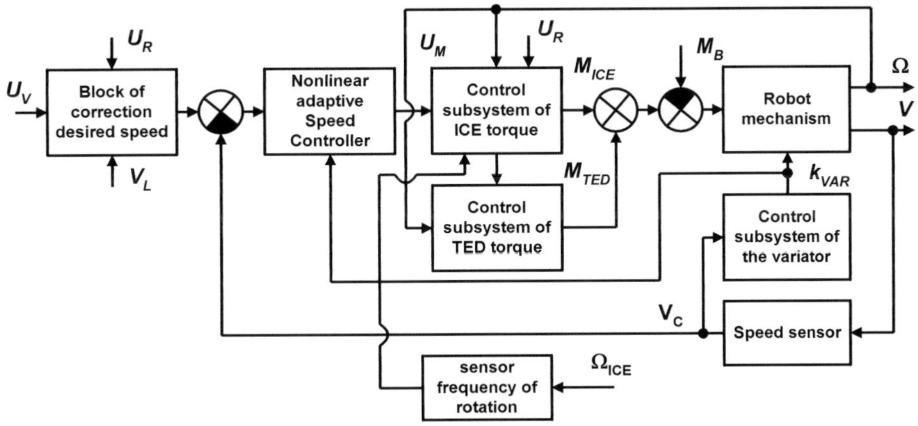

Fig. 2. The structure of the robot control system with hybrid drive

In order to improve the accuracy of the implementation of the desired speed in the main control loop we introduced non-linear adaptive speed control, built on the basis of PI-controller. The values of its adjustable parameters depend on the current value of the transmission ratio of the variator k_{VAR}. The output signal of this regulator U_M includes information on a desired torque of the hybrid drive. He comes to the input to the control subsystems torque of the ICE and affect the value of torques M_{ICE} and M_{TED} produced the engine and TED, respectively. In this case control subsystem of TED torque receives input action generated by converting the signal U_M to the control unit torque distribution, located inside the control subsystem of ICE torque. Variable transmission ratio which converts control system the movement of mobile robot to a non-linear system, is one of the causes of control problems. There are a various solutions of this problem. Proposed idea of variator control is as follows. In the appointment $k_{VAR.D}$ we will strive to ensure that, for any admissible value of the speed of movement of the robot V the velocity of ICE Ω_{ICE} was close to the optimum value $\Omega_{ICE.OPT}$ corresponding to the mode of formation of the maximum torque the ICE. In this case, the robot will be the most dynamic. Other things being equal, he will more quickly accelerate. Choice of the permissible torque of the ICE under conditions of in durability this case is not considered.

Variable transmission ratio which converts control system the movement of mobile robot to a non-linear system, is one of the causes of control problems. There are a various solutions of this problem. Proposed idea of variator control is as follows. In the appointment $k_{VAR.D}$ we will strive to ensure that, for any admissible value of the speed of movement of the robot V the velocity of ICE Ω_{ICE} was close to the optimum value $\Omega_{ICE.OPT}$ corresponding to the mode of formation of the maximum torque the ICE. In this case, the robot will be the most dynamic. Other things being equal, he will more quickly accelerate. Choice of the permissible torque of the ICE under conditions of in durability this case is not considered.

Thus, when variator is controlled, stabilization principle of chosen nominal velocity of ICE is realized. The calculated value of the variator $k_{VAR.C}$ ratio is computed by the formula

$$k_{VAR.C} = \Omega_{ICE.OPT} R_K (V k_{MT})^{-1}, \tag{10}$$

where R_K – drive wheel radius. An analysis of the characteristics of the consideration ICE shows that the range of values of the engine velocity Ω_{ICE}, in which the ICE torque differs by a maximum up 20%, is observed in the selection $\Omega_{ICE.OPT}$ = 350 rad / s.

The desired variator ratio $k_{VAR.D}$ is calculated taking into account limitations by the formula

$$k_{VAR.D} = \begin{cases} k_{VAR.C} & when \quad k_{VAR.MIN} < k_{VAR.C} < k_{VAR.MAX}, \\ k_{VAR.MAX} & when \quad k_{VAR.C} \geq k_{VAR.MAX}, \\ k_{VAR.MIN} & when \quad k_{VAR.C} \leq k_{VAR.MIN}, \end{cases} \tag{11}$$

where $k_{VAR.MAX}$ and $k_{VAR.MIN}$ – the minimum and maximum possible values of the transmission ratio of the variator respectively.

Speed controller provides the adaptation of the control system to change the variator transmission ratio k_{VAR} and the total moment of inertia J_Σ of the moving parts of the robot, relation to the shaft engine. The algorithm of the controller is described by the following equations:

$$\delta_V = V_C - V,$$
$$\frac{du_{VP}}{dt} = \delta_V,$$
$$u_V = \begin{cases} u_{VP}, & if \ |u_{VP}| \le u_{V.MAX}, \\ u_{V.MAX} sign(u_{VP}), & if \ |u_{VP}| > u_{V.MAX}, \end{cases} \tag{12}$$
$$k_{PC} = k_{PI0} k_{VAR} J_\Sigma,$$
$$U_{MP} = k_{PC}(\delta_V + k_{IC} u_V),$$
$$U_M = \begin{cases} U_{MP}, & if \ |U_{MP}| \le U_{M.MAX}, \\ U_{M.MAX} sign U_{MP}, & if \ |U_{MP}| > U_{M.MAX}, \end{cases}$$

where k_{PC} – the gain of the proportional component PI-controller of the speed; k_{PI0} – «basic» gain of the PI-controller; k_{IC} – the gain of the integral component PI-controller; $u_{V.MAX}$ and $U_{M.MAX}$ – levels of limitation signals of speed control;

One of the important challenges facing the control system is to ensure the number of constraints imposed on the variables that characterize the functioning of the processes mobile robot with hybrid drive. First of all, at an acceptable level V_{ACC} the linear speed V the mobile robot should be limited to V_{ACC}. Value V_{ACC} comes to the input the considered of the control system from higher rank control system and is determined taking into account road conditions, the presence of obstacles, the road inclination in relation to the horizon and other factors. This is achieved by limiting the input impact V_C of the block of correction desired speed. To prevent of proslipping of robot driving wheels when a sudden change in the speed of the control action V_C is formed taking into account the desired value of the speed in such a way to limit the acceleration to an acceptable level. In this case the calculation of the impact of the rate V_C of change is made in accordance with an algorithm

$$\frac{dV_C}{dt} = \begin{cases} k_a |V_D - V_C| sign(V_D - V_C) & when \ k_a |V_D - V_C| < k_{LIM} k_{VAR}, \\ k_{LIM} k_{VAR} sign(V_D - V_C) & when \ k_a |V_D - V_C| \ge k_{LIM} k_{VAR}, \end{cases} \tag{13}$$

where k_a and k_{LIM} – proportionality factors affecting the rate of change of the control action outside the limitation and action limitation, respectively.

At the second, to prevent of proslipping (slipping wheels) it should be limited to the total torque M_Σ produced by the hybrid drive in relation to the rotation axis of driving wheels. This operation is performed in the module limitation the total torque, which is part of the adaptive speed controller. At rise or descent maximum value $M_{\Sigma.MAX}$ of total torque M_Σ of hybrid drive should be chosen from the condition.

$$M_{\Sigma.MAX} \le F_{FF.MAX} R_W, \tag{14}$$

where $F_{FF.MAX}$ – the maximum value of the friction force on the of driving wheels of the road surface. This friction force is calculated according to the formula

$$F_{FF.MAX} = k_{FF} mg\left(\frac{1}{2}\cos\alpha - \frac{L}{B}\sin\alpha\right), \tag{15}$$

where k_{FF} – the coefficient of static friction between the drive wheels to the road surface; m – robot weight; g – acceleration of gravity; L – the distance between the axes of the front and rear wheels of the robot; B – the height of the robot's center of gravity relative to the road surface; α – angle of inclination of the surface (road) relative to horizontal surface, относительно горизонтальной плоскости, a positive value is counted counter-clockwise.

The resulting torque, which leads to move the robot, is determined by the formula

$$M_{\Sigma.} = M_{\Sigma} - M_W ,\tag{16}$$

where M_W – torque due to the influence of the weight of the robot, and

$$M_W = R_W m_{ROB} g \sin\alpha .\tag{17}$$

Limitations resulting torque of the hybrid drive and the speed movement of mobile robot become important in the energy recuperation mode. In this case, the rules of torque correction and the desired speed correction determined by the command U_R generated by mobile robot controller.

5. Torque control of hybrid drive

The task of control torque of hybrid drive is subordinated to the task controlling the speed of the mobile robot. The output signal U_M of speed controller, containing information about the desired torque of the hybrid drive is divided into two components U_{ICE} and U_{TED} in the distributor of torque (fig. 3). The component U_{TED} comes to input of the TED and leads to the formation torque M_{TED} of the TED. The second component U_{ICE} is worked by control system of ICE torque and formed the torque M_{ICE}. The sum of M_{ICE} and M_{TED} forms the hybrid drive resultant torque U_{Σ} and allows the mobile robot to move.

Control subsystem of ICE torque constructed as automatic control system, closed loop by torque the internal combustion engine indirectly measured. Control loop of ICE torque consists of the ICE torque controller, limiter of control action, the control device of ICE, sensor control action, the ICE torque calculator. Required to regulation evaluation of ICE torque is based mathematical model of the ICE on the above and the angular velocity Ω_B measured by the sensor of frequency rotation. Subsystem of ICE torque control attaches astatic property to improve the accuracy of control M_{TED}. For this purpose in the control loop an integrated controller is proposed to use.

The results of theoretical analysis and computer simulations have shown that, to ensure the stability of the ICE control processes it required limiter control action h at the level h_{LIM} depends on the velocity of the ICE crankshaft. It is found that it is possible to take $h_{LIM} = k_{h\Omega}\Omega_{ICE}$, where $k_{h\Omega} = 1{,}45\cdot 10^{-3}$.

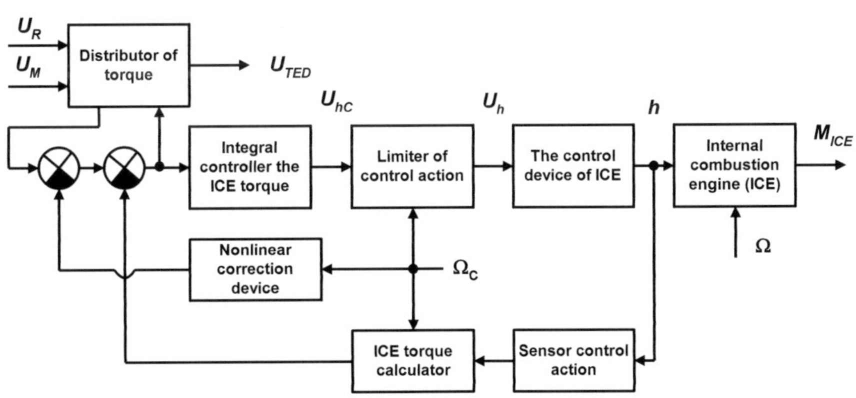

Fig.3 The structure of control subsystem of ICE torque

Additionally, the results indicate advisability of introducing a non-linear correction of the angular velocity of the crankshaft rotation. The dependence Ω_B on the correction signal U_{KORR} generated by the nonlinear correction device in the form of a link with the dead zone, which has the threshold value of the angular velocity Ω_{B0} and the corrective feedback factor k_{KORR}.

One of the problems of control hybrid drive torque is to determine the rule of rational distribution of the desired torque between the ICE and the TED. It is proposed to carry out such distribution, depending on the mode of the hybrid drive of mobile robot. For this purpose the robot control system U_R generates a command that comes to distributor of torque. We will distinguish between normal mode ($U_R = 0$) and energy recuperation mode ($U_R = 1$).

In normal mode, it implements the principle of the priority of the ICE. Distributor of torque passes signal, equal to U_{ICE}, to input of control subsystem of ICE torque, that corresponding to the desired torque $U_{\Sigma.D}$ of the hybrid drive.

As a result, ICE seeks to produce all the necessary torque of the hybrid drive. At the input of the TED signal transmitted, which proportional of the error of control subsystem of ICE torque and calculated in the distributor of torque in accordance with the equation

$$M_{TED.D} = k_{TED}(M_{\Sigma.D} - M_{ICE}), \quad (18)$$

where k_{TED} – the coefficient of proportionality.

Thus, in the normal mode the TED plays a subordinate role and serves to increase the total torque of the hybrid drive in those situations, when the ICE alone can't cope with the task. For example, this situation can occur if the ICE doesn't produce enough torque for acceleration or with extremely intensive robot moving on the rise. The TED is a highly dynamic system and effectively manages with the task of rapid formation of an additional component of the total torque of the hybrid drive, providing high values of acceleration and deceleration during the robot motion.

In the energy recuperation mode desired torque $U_{\Sigma.D}$ is divided into two parts $M_{ICE.D}$ and $M_{TED.D}$ which are defined by the following rules:

$$M_{ICE.D} = k_{ICE.R} M_{\Sigma.D}, \quad M_{TED.D} = k_{TED.R} M_{\Sigma.D}. \quad (19)$$

Moreover, the coefficients $k_{ICE.R}$ and $k_{TED.R}$ the distribution rule satisfy $k_{ICE.R} + k_{TED.R} = 1$. The following values of these coefficients can be recommended on the basis of the research results: $k_{ICE.R} = 0,5$; $k_{TED.R} = 0,5$.

6. Features of the mobile robot control in the energy recuperation mode

Transformation mechanical energy into electrical energy and replenish the accumulator is possible only in modes, when the robot commits on descent or moves along the horizontal surface by the ICE. In this case electric motor, in which current is regulated using the PWM, acts as brake and its torque counteracts external torque or ICE torque [5]. Average for the PWM period current I_A ,flowing through the accumulator, can be found by the formula

$$I_A = \frac{k_E}{k_M U} M_{TED}(M_{TED}\frac{R_A}{k_M k_E} + \Omega_{ICE}), \quad (20)$$

where k_M and k_E – torque and speed coefficients of the motor, respectively, R_A – resistance circuit consisting of the motor circuit, a frequency converter and accumulator; U – voltage of electric drive. Energy recuperation occurs when $I_A < 0$. During $0 < t < \gamma T_{PWM}$ from the beginning of PWM period under the influence of self-induction EMF current flows into the accumulator. Then, after switching transistor pairs for the time $\gamma T_{PWM} < t < (1-\gamma)T_{PWM}$ current is consumed by accumulator and passes through another pair of transistors, which in this period of time are translated into a state of saturation. When robot moves down the slope energy recuperation is possible if speed V is larger than a certain minimum value $V_{\min}$:

$$V > V_{\min} = k_{TED.R}\frac{R_A R^2}{(k_{MT}k_{VAR})^2 k_M k_E} mg\sin\alpha. \quad (21)$$

In this case current of charge I_A must be limited to an acceptable level $I_{A.MAX}$. Otherwise, accumulator is overheating, the electrolyte begins to boil and the accumulator can be damaged. Therefore, the robot movement speed should be limited to the top

$$V \le V_{\min} + \Delta V, \quad (22)$$

besides

$$\Delta V = \frac{I_{A.MAX} k_M U}{k_{TED.R} k_E mg\sin(\alpha + \Delta\alpha)}, \quad (23)$$

where $\Delta\alpha$ – small auxiliary quantity, preventing division by zero at movement on the horizontal segment of the path.

Accumulator charge increases with speed of movement increase. Therefore, in the energy recovery mode advantageously selected speed equal to

$$V_{MAX} = V_{\min} + \Delta V \,. \tag{24}$$

On the other hand, the desired speed V_D of robot motion must be less than allowed speed V_{ACC}, which is determined by taking into account the angle of inclination of the surface and other characteristics of the track. The correction is made in the block correction of desired speed. To fill the accumulator when the robot moves in the horizontal section of the trajectory, the control system of hybrid drive changes the TED to the braking mode. The TED produces the braking torque M_{TED}. In this case mechanical energy produced by the ICE is converted into electrical energy. Torque M_{TED} selected from the condition of existence of the recuperation process

$$0 > M_{TED} > -\Omega_{ICE} k_M k_E R_A^{-1} \tag{25}$$

taking into account the speed of robot movement and the necessity limit the accumulator current to the formula

$$M_{TED} = -b - \sqrt{b^2 + \frac{I_{A.MAX} k_M^2 U}{R_A}} \,, \tag{26}$$

where

$$b = -V \frac{k_{MT} k_{VAR} k_M k_E}{2RR_A} \,. \tag{27}$$

7. The results of computer modeling control processes of speed mobile robot with hybrid drive

In order to determine the main features of the functioning of the mobile robot with the proposed structures and algorithms of control systems, we have carried out computer modeling of its movement in the horizontal section of the track, on the ascent and the descent. In this case we can determine the character of the change speed of robot movement, torque of the hybrid drive and the current accumulator charge in the energy recuperation. We consider the robot with the following values: robot weight is 1000 kg; R_W = 0,3 m; k_{MT} = 4; k_{VAR} = 1 … 4; J_W = 0,225 kg*m^2; J_{TPH} = 0,1 kg*m^2; J_{RT} = 0,66 kg*m^2; B = 0,6 m; L = 3 m; k_{FF} = 0,5; T_{VAR} = 0,05 sec, T_{YY} = 0,02 sec. Electric motor in the TED has following parameters: nominal torque is 50 Nm; nominal angular velocity is 600 rad/sec; U = 250 V; k_M = 0,333 Nm/A; k_E = 0,333 V*sec/рад; R_A = 0,333 Ohm; ω_{CT} = 330 sec^{-1}.

"Basic" of the speed controller gain k_{PI0} is assumed to be 200, and the gain the integral component k_{IS} = 5 sec^{-1}. Gain factors of integral controller of the ICE and the TED have the same values, equal to 30 sec^{-1}. Maximum accumulator current of charge is $I_{A.MAX}$ = 10 A. Fig. 5-10 shows the results of computer simulation of mobile robot operation during acceleration. Movement of the robot begins on the horizontal section, and after a time equal to 20 seconds, starts the rise on angle 10 degrees. The process of changing speed movement of the robot during acceleration from speed 2 m/sec up to 5 m/sec is shown in Fig. 4. The process of forming the resultant torque of the hybrid drive is shown in Fig. 5. In this situation, speed is relatively low. Therefore, the variator ratio is the highest constant value. Robot speed changes smoothly without overshoot and the acceleration process completed in approximately 6 seconds. Robot control system doesn't involve the TED, because the ICE has sufficient capacity to bring the robot to move to the desired rule.

Process robot acceleration from speed 2 m/sec to a speed of 10 m/sec looks somewhat different (fig. 6). It can be seen that the acceleration is more intense and there is a small oscillation in the final stage. The higher intensity of acceleration is due to the automatic adjustment of the variator ratio. With the speed of increasing value k_{VAR} is reduced if the speed increases more rapidly. The process of changing torque of the hybrid drive is shown in Fig. 7. Same as in the previous experiment, it is completely ensured by the ICE.

The process of increase speed during acceleration 2 m/sec to 25 m/sec is shown in Fig. 8. Obviously, the duration of the process of speed control is greater than in the previous cases, because a considerable range of speed changes is set and the torque limit of the drive is shown. Clearly it expressed "zonality" of the process. At a small robot speed the variator ratio is great, then it decreases rapidly and reaches a minimum value by the end of the process. Reducing the value k_{VAR} leads to an increase equivalent coefficient in the transmission of the executive part of the system. Stabilization of dynamic features of control system is provided by adaptive speed controller.

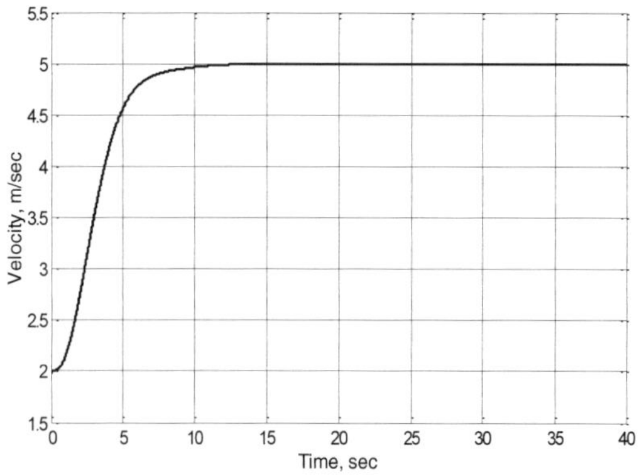

Fig. 4 Graph of speed changes during acceleration from 2 m/sec to 5 m/sec

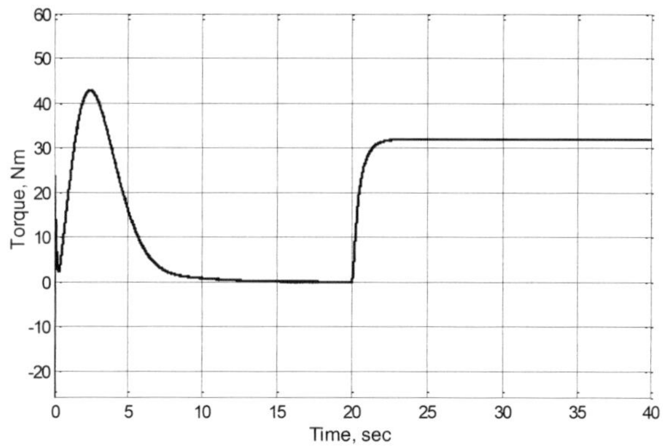

Fig. 5 Graph of torque changes of hybrid drive during acceleration from 2 m/sec to 5 m/sec

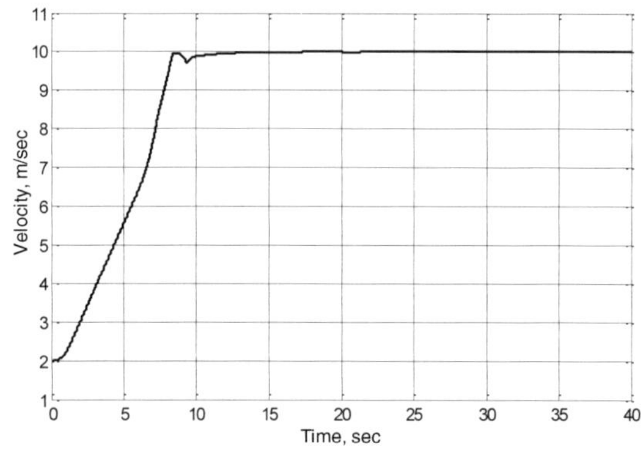

Fig. 6 Graph of robot speed changes during acceleration from 2 m/sec to 10 m/sec

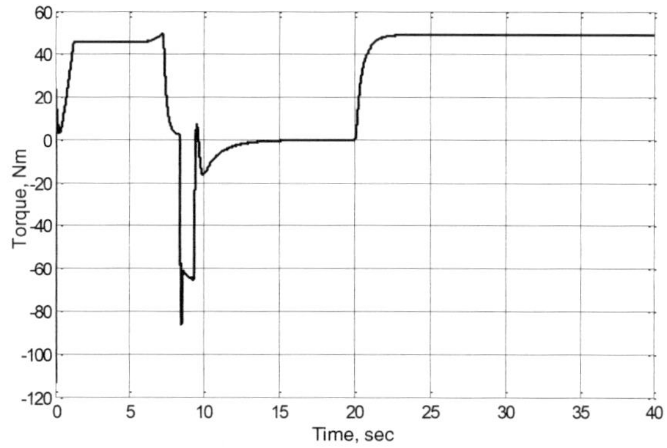

Fig. 7 Graph of torque changes of hybrid drive during acceleration from 2 m/sec to 10 m/sec

The resulting torque of the hybrid drive is also undergoing significant changes (fig. 9). Phases of acceleration and deceleration are observed, when the torque is negative. Negative torque is provided mainly by TED, having a quick performance. The final value of torque is substantially greater than the initial, because robot climbs up the hill, and the value $k_{V_{AR}}$ is minimal.

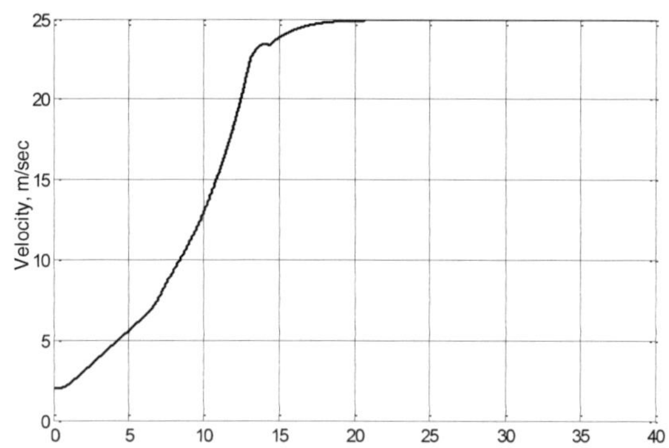

Fig. 8 Graph of robot speed changes during acceleration from 2 m/sec to 25 m/sec

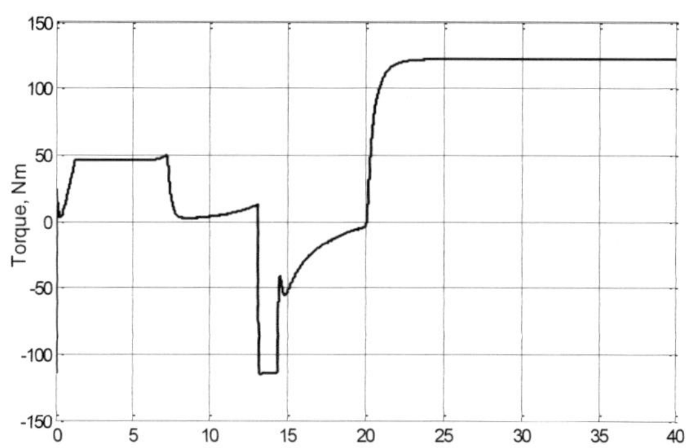

Fig. 9. Graph of torque changes of hybrid drive on descent and acceleration from 2 m/sec to 10 m/sec

The simulation results describing movement of the robot during the descent, are shown in Fig. 10-11. The process begins on the horizontal section, and at time equal 20 seconds, the descent takes place at an angle of -10 degrees. The process of changing the speed of robot movement during its acceleration from speed 2 m/sec up to a speed 10 m/sec is shown in Fig. 10. In this case control system automatically adjusts the speed of movement taking into account the inclination angle, reducing it to a safe level by the TED, which works as a brake. Thus, there is the current change in the

TED (fig. 11). As the result of braking energy recuperation process provides accumulator charging. After the transient processes current reaches the desired level of security equal -10A.

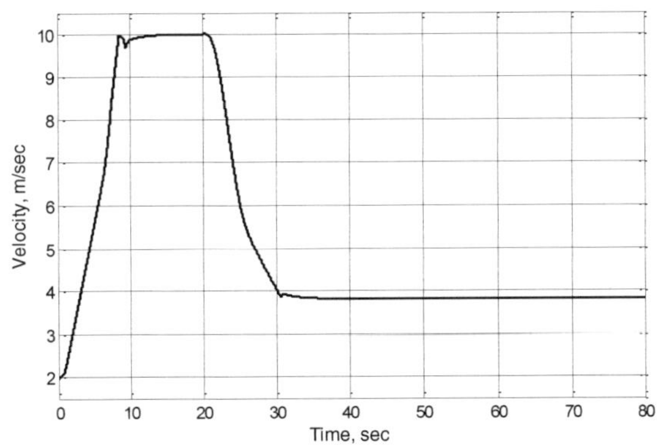

Fig. 10 Graph of robot speed changes on descent and acceleration from 2 m/sec to 10 m/sec

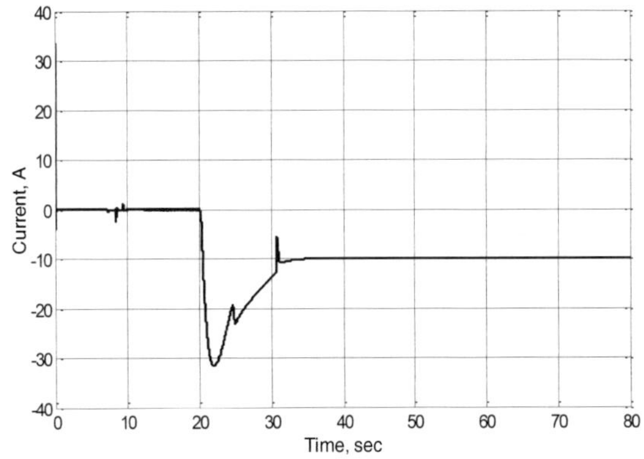

Fig. 11 The process of charging accumulator current when the robot moves on descent

Fig. 12 shows the process of changing the speed of movement of the robot moving the horizontal part of the track. This process begins with acceleration from speed 2 m/sec to speed of 15 m/sec. It is starting at time equal to 20 sec, it is converted into an energy recuperation mode. The TED produces a torque of resistance to movement of the robot. The electric motor mainly operates as a generator and the accumulator charge occurs (fig.13).

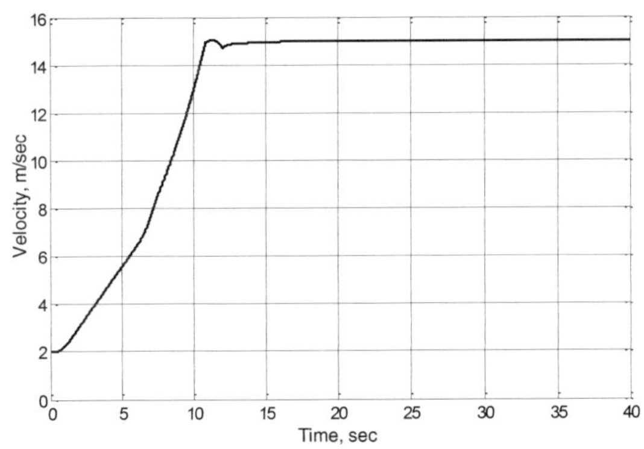

Fig. 12 Graph of robot speed changes on the horizontal section of the path and acceleration from 2 m/sec to 15 m/sec

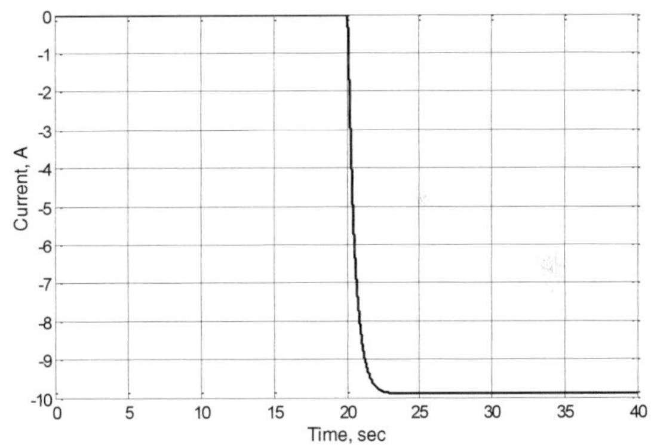

Fig.13 The process of accumulator current control in the recuperation mode, when the robot is moving along the horizontal section

As it follows from Fig. 13, the control system limits the charging current to an acceptable level equal to -10 A.

5. Conclusion

The paper presents solution to the problem of the control movement of mobile robot with a hybrid drive. This problem was solved by analyzing the characteristics of the components that form the hybrid drive, this required the development of new structures and motion control algorithms for robots with such drives. First of all, it should be noted that we obtained a closed system of automatic speed robot motion control with subordinate control subsystem torque of the hybrid drive. In its turn this subsystem consists of two interacting elements that ensure regulation of torques of the ICE and the TED. It was found reasonable to use the adaptive speed controller. It allowed stabilize the dynamic properties of the robot when changing the transmission ratio. The peculiarity of the proposed solutions is also related to the need to introduce a series of non-linear elements in the control of the system. We proposed to use the distributor of torque and block of correction of the desired speed. It is crucial that both the control system components have varying structures and their operation algorithms depend on the selected operating mode.

We solved the problems of definition robot structure with the considered type of drive, the formation of its mathematical model, the development of rules and automatic control algorithms in different driving modes, including taking into account energy recuperation processes. The given results of theoretical studies and computer simulations have confirmed the effectiveness of the proposed solutions. The author's future plans will be aimed at improving control algorithms of robot with hybrid drives, solving questions planning trajectories and profiles of speeds of movement these robots on a given route.

6. Acknowledgments

The present paper was carried out under the financial assistance of the Ministry of Education and Science of the Russian Federation in the framework of the state assignment in the field of scientific activities (assignment № 2014/105, project № 1678).

7. References

[1] Mailah A. H. Ali, M., Tang H. H., Rahim M. A. A., "Autonomous Mobile Robot Platform for On-the-Road Navigation", Advanced Materials Research, Vol. 845, pp. 750-754, 2014

[2] Katsurin, A[lexey] (2016). Planning Trajectory of the Mobile Robot with a Camera, Proceedings of the 26th DAAAM International Symposium, pp.0407-0416, B. Katalinic (Ed.), Published by DAAAM International, ISBN 978-3-902734-07-5, ISSN 1726-9679, Vienna, Austria

[3] Ionescu, G. & Boiangiu, C. A.: Mobile Platform for Environment Analysis, Annals of DAAAM for 2012 & Proceedings of the 23rd International DAAAM Symposium, ISBN 978-3-901509-91-9, ISSN 2304-1382, pp 0587 - 0590, Editor B[ranko] Katalinic, Published by DAAAM International, Vienna, Austria, 2012.

[4] Andreev V.P., Pryanichnikov V.E. Operation Environment of Mobile Robots with Supervision Control, Annals of DAAAM for 2011 & Proceedings of the 22nd International DAAAM Symposium, Volume 22, No. 1, Vienna, Austria, 2011, pp. 0021-0022.

[5] Ilyukhin Y.V., Tatarintseva A.V. Strategy motion control of mobile robots with hybrid drives according to energy recuperation. //Advanced Materials Research Vols.1079-1080 pp895-898 (2015) Trans Tech Publications, Switzerland (2015)

[6] Sun Y. T., Zhang J. Z., Wang Y. L., Fan D. H., Zhu R. F., "Analysis of Typical Driving Condition Test and Control Strategies Based on Toyota Prius", Advanced Materials Research, Vol. 977, pp. 321-325, 2014

[7] Sun Y. T., Zhang H. T., Wang Y. L., "Toyota Prius Control Strategy Experiment Analysis", Advanced Materials Research, Vols. 591-593, pp. 1762-1765, 2012

[8] I. F. Fileru, "Toyota Prius – A Successful Pioneering in Hybrid Vehicle World", Applied Mechanics and Materials, Vols. 809-810, pp. 1139-1144, 2015

[9] Liu Z. T., He H. W., Li W. Q., "Optimization of Economy Shift Schedule for Automated Mechanical Transmission in a Parallel Hybrid Electric Vehicle", Applied Mechanics and Materials, Vols. 260-261, pp. 331-336, 2013

[10] Damic V., Cohodar M., Kulenovic M. Physical modeling and simulation of mechatronics systems by acausal bond graphs. Annals of DAAAM for 2011 & Proceedings of the 22nd International DAAAM Symposium, Volume 22, No. 1, ISSN 1726-9679, ISBN 978-3-901509-83-4, Editor B. Katalinic, Published by DAAAM International, Vienna, Austria, EU, 2011

[11] Ilyukhin Yu.V., Computer control of mechatronic systems., MSTU "STANKIN", Moscow, 2014.

[12] Ilyukhin Yu.V., Poduraev Yu.V., Tatarintseva A.V., Nonlinear Adaptive Correction of Continuous Path Speed of the Tool for High Efficiency Robotic Machining. Volume 100, 2015, Pages 994–1002. (25th DAAAM International Symposium on Intelligent Manufacturing and Automation, 2014)

[13] Ilyukhin Yu.V , Tatarintseva A.V., Mathematical model and features of the mobile robot control with mechatronic hybrid. Journal of the Samara State Aerospace University named after Academician S. Korolev (National Research University) №2 (33). Samara, p. 222-227, 2012

[14] Witwit A. R. H., Yasin A., Gitano H., Yadav T. K., "New Model for Knock Factors Optimization in Internal Combustion Engine (SI)", Advanced Materials Research, Vol. 980, pp. 219-224, 2014

[15] Niu Q. Y., Fan C., Wang X. C., Zhao Y. W., Dong Y. C., "Research on the Parameter Calibration of the Internal-Combustion Engine Work Process Simulation Model", Advanced Materials Research, Vols. 308-310, pp. 953-961, 2011

[16] http://www.zmz.ru/files/PP40524(02).pdf The internal combustion engine ZMZ-40524.10. Instructions for use, maintenance and repair. OAO «Zavolzhsky Engine Plant», 2008.

DOI: 10.2507/27th.daaam.proceedings.061

EXPERIMENTAL RESEARCHES IN THE WAVE BASIN USING NATIONAL INSTRUMENTS TECHNOLOGY

Stanislav Sholar, Olga Ivanova, Veronika Dushko, Vadim Kramar & Anna Rodkina

This Publication has to be referred as: Sholar, S[tanislav]; Ivanova, O[lga]; Dushko, V[eronika]; Kramar, V[adim] & Rodkina, A[nna] (2016). Experimental Researches in the Wave Basin Using National Instruments Technology, Proceedings of the 27th DAAAM International Symposium, pp.0417-0422, B. Katalinic (Ed.), Published by DAAAM International, ISBN 978-3-902734-08-2, ISSN 1726-9679, Vienna, Austria
DOI: 10.2507/27th.daaam.proceedings.061

Abstract

The article is devoted to the possibility of using National Instrument analog-to-digital converter NI USB 6008 in experimental researches in the Sevastopol State University wave basin. The analog-to-digital converter using in experimental researches in experimental basin allows you to record all relevant data, such as the profile of the wave, the force of wave's impact, wave characteristics and their interaction with the test physical models of offshore constructions.

Keywords: wind and wave loads; offshore constructions; wave basin; analog-to-digital converter (ADC); Sevastopol State University (SevSU); National Instruments (NI) technology.

1. Introduction

At present, the world economy is on the threshold of industrial development of the continental shelf (about 22% of the world ocean shelf).

Oil and gas offshore exploitation is released as a priority activity that contribute to the development of the oil and gas industry and the economy as a whole. Shelf exploitation will require the development of a whole range of new technological solutions for the exploitation of hydrocarbons in extreme conditions. In particular, it requires the development and implementation of specialized underwater equipment for the successful development of offshore fields, are characterized by complex and severe ice conditions (in the Arctic); development of exploitation of a wide range of modern technical means for the development of offshore oil and gas resources, including the offshore rigs of various types (offshore ice-resistant drilling rig, semi-submersible drilling rig, offshore ice-resistant drilling rig, drilling vessels, etc.), gas carrier vessels, mining technology platforms (ice-resistant stationary, ship type, SPAR, TLP, etc.).

For example, in the Russian Federation under the program "Development of shipbuilding and equipment for the development of offshore fields in 2015 – 2030 years" is planned by 2030 to increase oil exploitation on the shelf is 5 times – from the current 13 million to 66.2 million tons; gas exploitation is planned to increase by 4 times – from 57 billion to 230 billion m^3. The solution to these massive problems associated with the conduct of research, which include the present work aimed primarily at improving the reliability of these facilities offshore and reducing the risk of man-made disasters, which, unfortunately, occur at this time.

In the investigation of wind and wave loads impact on the ocean engineering systems constructions, as well as in the investigation of the accidents causes, such as corrosion-mechanical and fatigue destruction, the dynamics of soil scouring under template fixed offshore platform or other types of platforms, are used both theoretical and experimental studying methods [1], [2], [3].

In general, theoretical methods in the form of mathematical models allow to describe and explain the relations of the explored system or object elements in a relatively wide range of the variable values. However, the construction of theoretical models inevitably introducing any limitations, assumptions, hypotheses, etc. Therefore, the problem of evaluating the reliability of the resulting model to the real process is arises. To do this, we conducted experimental verification of developed theoretical models. In most cases, the experimental research results provide an impetus to the theoretical generalization of the explored phenomenon. Research works in wave (experimental) basins are held in many areas of applied hydrodynamics, shipbuilding, marine technologies.

Scientists research works in the field of sea drilling platforms creation and use cover the period from 50th years of last century. The works devoted to questions of constructional design, stability and strength of such constructions are executed by A.A. Aliseychik, etc. [4], T. Douson [5], V.R. Dushko [6], A.V. Lukoshkov [7], A.P. Pronkin and S.S. Hvorostovskij [8], V.P. Shostak [9], V.G. Franchuk, etc. [10], [11]. Hydrodynamics researches of such installations can be found in M.F. Barshtejn works, etc. [12], V.R. Dushko, etc. [6], V.M. Kushnir, etc. [13], I.Sh. Halfin [14], N. Srinivasan [15], Z. Demirbilek [16] and other authors. Some questions of installation deduction forces calculations on a soil are considered and systematized by T. Douson [5], as well as in papers [17].

Research works in wave (experimental) basins are held in many areas of applied hydrodynamics, shipbuilding, marine technology. theoretical and experimental research in the directions: nonlinear dynamics ocean engineering systems; dynamic positioning of vessels; the protection of vessels and structures from corrosion and mechanical damage; the spatial distribution of the velocity of the wave and the steady flow over objects ocean technology; influence of soil movement (bottom sediment and dumping) on the stability of subsea pipelines are carried out at the Ocean Technology Research Laboratory of SevSU Graduate Chair of Ocean Technology and Shipbuilding.

The scheme of experimental research in the SevSU Ocean Technology Research Laboratory wave basin is shown in Figure 1. The main parameters of the basin: length L_b = 10 m, breadth B_b = 3 m, basin wall height 1.5 m, depth (liquid level in the basin) h_b = 1.3 m. Waves on the water surface are driven by pneumatically wave generator 7, which consists of an air compressor and a rotary plug pipe socket. The plug rotation period is specified in the range of T = 0.5…2 s. Pipe socket output directed to the water surface, the pressure oscillation amplitude at the output is constant. Wash area which propagate along a basin as free waves is generated as a result of the variable pressure on the water surface. Sloping bottom which deviation angle from the horizontal plane is set in the range of from 5 to 20° is arranged at the opposite end of the basin inclined bottom.

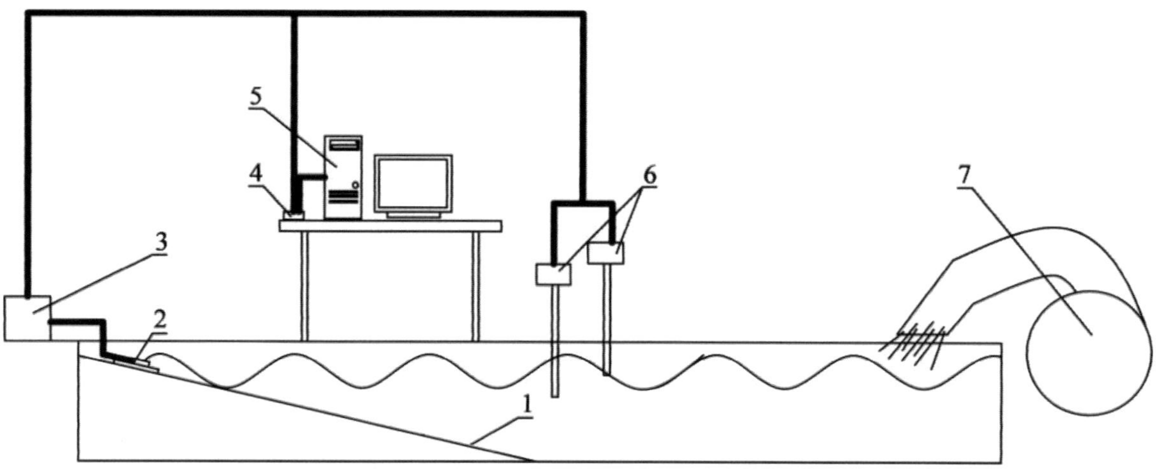

Fig. 1. The ocean technology laboratory test installation design. 1 – the sloping bottom; 2 – the aneroid box; 3 – the active filter box; 4 – the analog-to-digital converter (ADC) NI USB-6008 5 – the personal computer with the installed program LabView 8.5; 6 – the wave-recording gauges; 7 – the wave generator

For the purpose of experimental researches automation in the experimental basin, a complex of digital equipment has been created [18], [19], [20], allow for the measurement of the shock wave loads on the sloping bottom. To convert the electrical signals from the sensors to digital, complex chipset managed software product matter developed for each sensor separately is used [21].

For equipment switching, for converting analog signals to digital and then writing the data to the computer used ADC NI USB-6008 developed by National Instrument. Graphical programming language LabVIEW (Laboratory Virtual Instrumentation Engineering Workbench) are designed by NI to work with its devices. This language significantly changed and simplified a variety issues solving process, it allowed to solve issues without resorting to the professional programmers help in many cases. LabVIEW is a powerful graphical programming, a flexible software environment and user-friendly interface that is used for measurement and analysis.

2. Special aspects of national instruments

Using of National Instruments technology greatly simplifies the measurement process. The advantages of working with ADC NI USB-6008:

- increases accuracy;
- simplifies the implementation of different measurement methods in the certain experiment;
- time reduces the measuring channel settings;
- allows to organize experimental data processing algorithms.

Some experimental measurements can be carried out by the Internet remote access mode (for example, if you install the equipment on the beach).

Among the benefits of National Instruments technology highlighted next factors: the visibility of the code, easy to program and easy to use functional interface, and ease of matching hardware and software of the unified information-measuring system.

Technical aspects of NI USB 6008/6009:

- eight analog input channels, the definition of 12 or 14 bit, digitization rate up to 48 kHz;
- screw terminals for connecting the sensors;
- quick connection to the computer;
- drivers for operating systems Windows, Mac OS X and Linux;
- multifunction input/output for the collection and preservation of data;
- USB bus powering;
- free software to collect and store data.

3. The application of NI technology in the ocean technology laboratory

National Instrument equipment is successfully used in the SevSU experimental basin during the experimental research of wave impacts and shock loads on the physical models of ocean engineering structures. Analog-to-digital converter ADC NI USB 6008 is used to record values of the pneumatic type wave-recording gauge [22] on the basis of differential sensors firm "Motorola" MPX2010DP pressure type with a dynamic range and sensitivity of 10 kPa 2.5 mV / kPa [23]. The first inputs of the sensors are attached to the rigid measuring tube length of 1 m and a diameter of 1 cm, the ends of which are lowered to the desired depth z, where they are exposed to the aqueous medium pressure. The second inputs of sensor output to open air, which provides compensation for atmospheric pressure changes in the output. The main features of the wave recorder include the ability to obtain data on the "inside" the structure of the wave positioning receiving ends of the tubes at different depths, and the relative simplicity and reliability, combined with high sensitivity and stability characteristics. The frequency of synchronous measurements of parameters of waves is equal to 50 Hz, which is the range of possible wave disturbances periods (0.5...2) provides at least 25 samples per period of oscillation.

For treatment of common sets of measured data on selected rows of 10000 samples, the correlation functions of length was 250 samples, provided that the spectral window in the frequency range from 0.1 to 25 Hz. The number of waves received is equal to compute spectra from 100 to 400, respectively, for the most low-frequency and high-frequency disturbances.

Representative examples of wave disturbance spectra are shown in Figure 2. The number of freedom degrees is equal to their estimates 80 that the distribution of Pearson (χ^2) to sample the spectrum estimates corresponding to a 95% probability inequality $0{,}87 \leq (S_D/S) \leq 1{,}13$, where S_D – the actual value, S – the selective evaluation of the spectrum. In Figure 2, the range of the specified sampling variability spectra shown straight line segments estimates.

As already noted, in the experiments using two identical wave recorder installed along the direction of wave propagation at a fixed distance. The phase shift between the readings of these sensors is used to determine the apparent wavelength and parameters of direct and reflected waves [24].

Figure 3 shows the change in amplitude of the wave measured in real time. Figure 3a shows two wave recorder connected to the laptop via the NI USB-6008. The first wave recorder (Figure 3b – lower curve) at a depth of 5 cm, the second (Figure 3b – the upper curve). At a depth of 10 cm, the distance between the wave recorder wave recorder installed 20 cm along the direction of wave propagation at a fixed distance. The phase shift between the readings wave recorder allows mathematically determine the characteristics of waves, such as wavelength λ = 0.8 m, amplitude a_w = 0.03 m, the wave period T = 0,7 c.

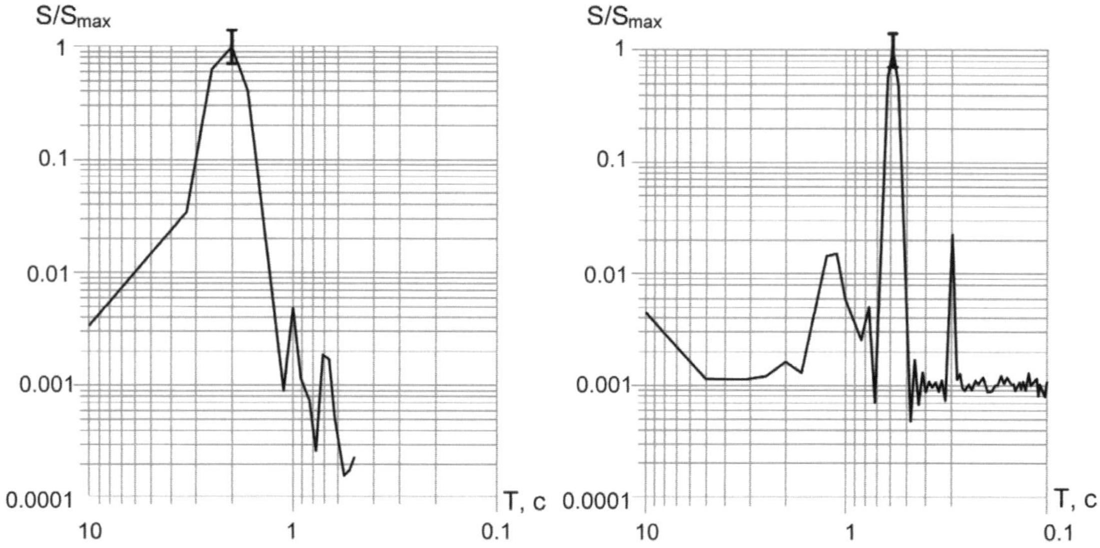

Fig. 2. Characteristic normalized spectra of wave disturbances in the experimental tank:
left at the wave generator frequency of 0.5 Hz, right – 1.7 Hz.
The vertical lines show the range of sampling variability of estimates with 95% probability

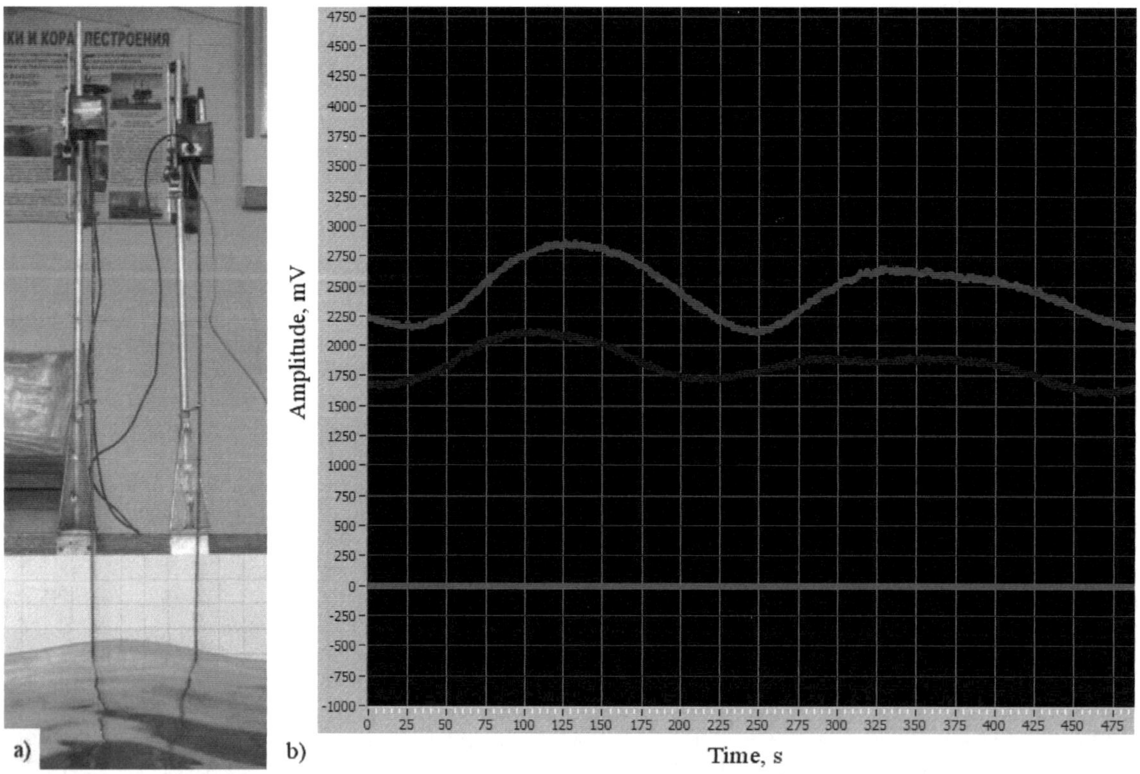

Fig. 3. Wave amplitudes recording in real time:
a – 2 two wave-recording gauges mounted on the wave basin wall; b – screenshot

Figure 4 shows the organization of data records in a text file; recording frequency and the number of points can be adjusted. The signal obtained by ADC can be software strengthened just before recording. As a rule, frequency of 50 Hz. Recording duration is unlimited, but for the statistical programs of laboratory experiments data computing in experimental basins is most applicable to the recording duration of 200 s [21].

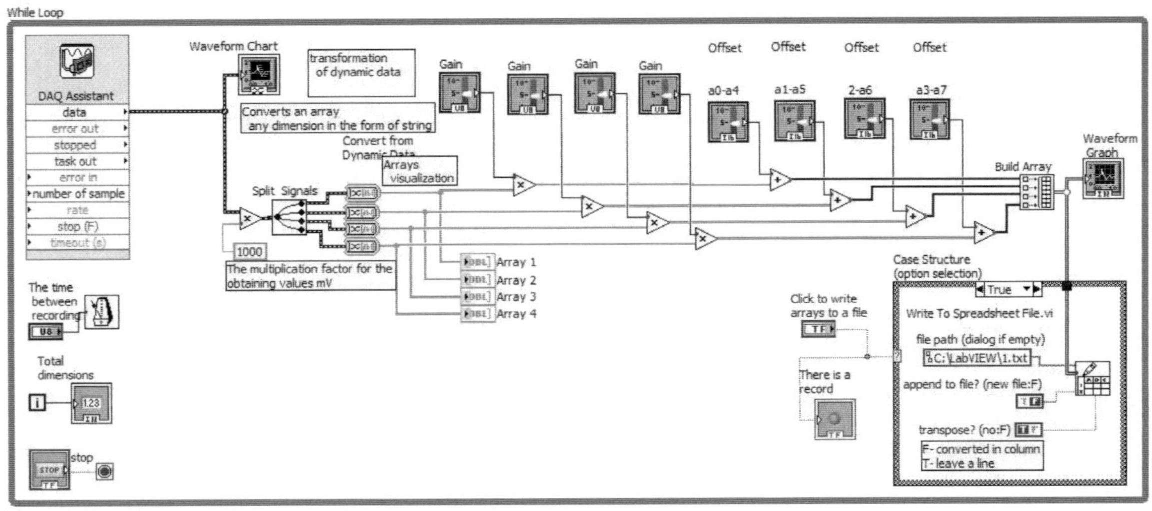

Fig. 4. Screenshot of the program written in a LabView language

Analog-to-digital converter used in experimental researches on the basic part of the governmental target No. 2014/702 «Extreme wind and wave loads on ocean engineering systems and their relation to the risk of technogenic accidents and environmental disasters» and in the research project RFBR No. 14-41-01556/14 «Fundamental research of problems of nonlinear dynamics of surface waves and their impact on ocean engineering installation in the coastal zone of the Azov – Black Sea». The essence of the works lies in the physical modeling of wave breaking processes in the coastal zone and comparing existing mathematical models with numerical results obtained in the experiment. Wave-recording gauges and membrane sensors are connected with the help of ADC. More details about the results of an experiment to investigation the shock wave loads on sloping bottom is written in [25]. Equipment connected by the ADC was calibrated. Wave-recording gauges were calibrated lowering at different depths and comparing the computer recorded code with hydrostatic pressure. The correlation between the pressure and the recorded code was $R^2 = 0{,}99$. Calibration of the sensor membranes was performed by lowering the wave-recording gauge measuring tube [22] at different depths in a bundle with a membrane sensor. The correlation between the pressure and the recorded code was $R^2 = 0{,}945$.

4. Conclusion

Any limitations are inevitably introducing in the theoretical models construction of wave impacts and shock loads on the physical models of ocean engineering structures. Therefore, the problem of evaluating the reliability of the resulting model to the real process is arises, for example, by the experimental method.

In the current investigation, this problem is solved with the use of National Instruments technologies in the Ocean Technology Research Laboratory, which allowed: to increase accuracy; to simplify the implementation of different measurement methods in the certain experiment of wave characteristics determining; to reduce measuring time of the channel settings; to organize experimental data processing algorithms.

The analog-to-digital converter using in experimental researches in experimental basin allows you to record all relevant data, such as the profile of the wave, the force of wave's impact, wave characteristics and their interaction with the test physical models of offshore constructions.

Future plans are using National Instruments Technology:

- for experimental and modeling researches in the wave basin associated with the estimates of the impact of storm surges on coastal, deep water ocean engineering and hydrotechnical constructions;
- in the investigation of the problem of changes in the structure of nonlinear surface waves arising from the severe storms and the formation of impact loading at their destruction;
- for modelling and estimation of extreme wind conditions and parameters of surface waves, for the most promising areas for offshore development in the Black Sea, Caspian Sea, Sea of Okhotsk, the Barents and Kara Seas.

5. Acknowledgments

The reported study was funded by The Ministry of Education and Science of the Russian Federation on the basic part of the governmental target No. 2014/702.

6. References

[1] Kushnir, V., Dushko, V., Kramar, V. (2013). Impact of surface gravity waves on coastal ocean engineering facilities. Eastern-European Journal of Enterprise Technologies. 6/5(66). 36–41.

[2] Kramar, V., Dushko, V., Rodkina, A., Zaiets, A. (2015). Influence of Stress-Corrosion Fractures on Potential of Ship-Building Metals in the Sea Water. Energy Procedia. 100, 1068-1074. http://dx.doi.org/10.1016/j.proeng.2015.01.468

[3] Kramar, V., Dushko, V., Rodkina, A., Alchakov, V. (2016). Neural Network Modeling the Potential of Steels with Different Chrome Content. Proceedings Engineering of the 26th International DAAAM Symposium "Intelligent Manufacturing & Automation" 21-24th October 2015, Zadar, Croatia, EU, Volume 26, No.1, ISSN 1726-9679, Ed. B. Katalinic, Published by DAAAM International. 0641–0647. DOI:10.2507/26th.daaam.proceedings.087

[4] Aliseichik, A.A., Galakhov, I.N., Lithonia, O.E. (1981). Plavuchie burovye platformy. Konstrukciya i prochnost' (Floating drilling platforms. The design and strength). Leningrad: Sudostroenie.

[5] Douson, T. (1986). Proektirovanie sooruzhenij morskogo shelfa (Designing of sea shelf facilites). Leningrad: Sudostroenie.

[6] Dushko, V.R. (2007). Ustojchivost morskih burovih platform na myagkih i podvizhnyh gruntah (Stability of offshore drilling platforms on soft soils and moving). PhD dissertation. Nikolaev.

[7] Lukoshkov, A.V. (1984). Tekhnika Issledovaniya Morskogo Dna (Engineering researches of the seabed). Leningrad: Sudostroenie.

[8] Pronkin, A.P., Hvorostovskij, S.S. (1999). Prognozirovanie napravlenij razvitiya razvedochnogo bureniya na shelfe (Forecasting trends in the development of exploration drilling on the continental shelf). Moscow: Nedra-biznescentr.

[9] Shostak, V.P. (2002). Ehffektivnost tekhniki osvoeniya okeana (The efficiency of ocean exploration technology). Kiev: Naukova dumka.

[10] Franchuk, V.G., Chuprun, V.P., Yaremijchuk, Ya.S., Shvachenko, I.I, Strunevich O.V. (2010). Samopodemnye burovye ustanovki: analiticheskij obzor (Self-lifting drilling rig: analytical review). Oil & gas. 1, 28–39.

[11] Nyrkov, A., Chernyi, S.; Zhilenkov, A. & Sokolov, S. (2016). The use of Fuzzy Neural Structures to Increase the Reliability of Drilling Platforms. Proceedings Engineering of the 26th International DAAAM Symposium "Intelligent Manufacturing & Automation" 21-24th October 2015, Zadar, Croatia, EU, Volume 26, No.1, ISSN 1726-9679, Ed. B. Katalinic, Published by DAAAM International. 0672–0677. DOI:10.2507/26th.daaam.proceedings.091

[12] Barshtejn, M.F., Zubkov, A.N., Maslov, B.E. (1974). Eksperimentalnoe izuchenie kolebanij neftepromyslovyh sooruzhenij pri neregulyarnom volnenii (Experimental study of the fluctuations of oil field facilities with irregular waves). 34. Moscow: Strojizdat.

[13] Kushnir, V.M., Dusko V.R., Fedorov S.V. (2009). Influence of the water environment on systems for shelf resourses development. Sevastopol: SevNTU Publishing.

[14] Halfin, I.Sh. (1990). Vozdejstvie voln na morskie neftegazopromyslovye sooruzheniya (The impact of waves on offshore oil and gas installations). St.Petersburg: Nedra.

[15] Srinivasan, N. (1995). Tension base TLP can support development in 4000 ft depths. Offshore. 48–65.

[16] Demirbilek, Z.Ed. (1989). Tension Leg Platform: a state of the art review. New York: American Society of Civil Engineers.

[17] Tertyshnikova, A.S., Blagovidova, I.L., Kushnir, V.M. (2010). Parameters of the Position System for the Deep-water Drilling Platform. Vestnik SevNTU. 106. 164–167.

[18] Ivanova, O.A. (2014). Automation of leadthrough of experimental researches is in experimental tank. Vestnik SevNTU. 146. 164–167.

[19] Ivanova, O., Dushko, V., Rodkina, A. (2016). Experimental Researches Automation of Spatial Oscillations of the Floating Ocean Engineering Systems in the Wave Basin. Proceedings Engineering of the 26th International DAAAM Symposium "Intelligent Manufacturing & Automation" 21-24th October 2015, Zadar, Croatia, EU, Volume 26, No.1, ISSN 1726-9679, Ed. B. Katalinic, Published by DAAAM International. 1059–1067. DOI:10.2507/26th.daaam.proceedings.149

[20] Kushnir, V.M., Sholar, S.A., Dushko, V.R. (2014). System for measuring impact pressure on destruction of the surface gravity waves in the near-shore. 153. 7–12.

[21] Kushnir, V.M., Dushko, V.R., Moreva, I.N., Ivanova, O.A., Fedorov, S.V. (2012). Digital system for researches of the ocean engineering systems in waves basins. Proceedings of the Odessa National Maritime University. 34. 50–65.

[22] Grekov, A.N., Grekov, N.A., Stroganov, A.S. (2007). Automated hydrodynamic wave recorder gauge. Monitoring systems of environment. 52–57.

[23] Sholar, S.A., Ivanova, O.A., Dushko V.R., Kramar V.A. (2016). The sensor for measuring the pressure of the shock wave basis of the piezoelectric. Monitoring systems of environment. 3(23). 32–37.

[24] Kushnir, V.M., Ivanova, O.A., Sholar, S.A. (2015). Transformation, destruction and shots of marine waves on sloping seabed in off-shore zone. Saarbrücken: LAP LAMBERT Academic Publishing.

[25] Kushnir, V.M., Sholar, S.A., Dushko, V.R. (2016). Experimental research of dynamic loads formed upon breaking of a gravitational wave on a sloping bottom. Instruments and Experimental Techniques. 59. 451–457.

DOI: 10.2507/27th.daaam.proceedings.062

TEMPERATURE INVERSION MEASUREMENTS IN SARAJEVO VALLEY USING UNMANNED AERIAL VEHICLES

Adnan Mašić, Rajfa Musemić & Emina Džaferović-Mašić

This Publication has to be referred as: Masic, A[dnan]; Musemic, R[ajfa] & Dzaferovic-Masic, E[mina] (2016). Temperature Inversion Measurements in Sarajevo Valley Using Unmanned Aerial Vehicles, Proceedings of the 27th DAAAM International Symposium, pp.0423-0427, B. Katalinic (Ed.), Published by DAAAM International, ISBN 978-3-902734-08-2, ISSN 1726-9679, Vienna, Austria
DOI: 10.2507/27th.daaam.proceedings.062

Abstract

Temperature inversions represent an important element of air pollution in cities situated in natural valleys, such as the Sarajevo Valley. Due to the strong thermal stratification, which acts as a natural lid over the city, air circulation is blocked and pollutants accumulate and stay for much longer time. As a consequence, air pollution in the residential area of the city increases dramatically. Investigation of the temperature inversions in Sarajevo Valley was, for the first time, performed using unmanned aerial vehicles.

Keywords: Temperature inversion; Sarajevo Valley; Unmanned Aerial Vehicles.

1. Introduction

Air pollution is becoming a major environmental issue. Cities situated in the natural valleys have particular problem of thermal stratification during the winter [1]. These stratifications sometimes stay for weeks, and during that period concentrations of pollutants rise dramatically. The city of Sarajevo is clear example (Figure 1).

Fig. 1. Sarajevo Valley

Thermal stratification in the Sarajevo Valley was previously investigated by means of computer simulation and turbulence modelling [2]. The turbulent flow of air was simulated using Large Eddy Simulation with subscale model [3]. However, no experimental studies of the temperature inversion layer in the Sarajevo Valley were reported, up to now.

The purpose of this study is to evaluate the possibility of using unmanned aircrafts equipped with data acquisition systems to investigate the phenomenon of the temperature inversion in the Sarajevo Valley. Accurate temperature profiles are required for any successful modelling (analytical or computer based) of this phenomenon. Our hypothesis is: our custom data acquisition system is suitable for installation on the unmanned aerial vehicle for measurements and modelling of the temperature profiles in the inversion layer.

Unmanned Aerial Vehicles (UAVs), often called "drones" in the mainstream, are increasingly popular nowadays. They are being used for aerial imaging, research and commercial purposes. We have developed a research platform which includes quad-rotor in "X" configuration (Figure 2) and custom made data acquisition system (DAS) [4].

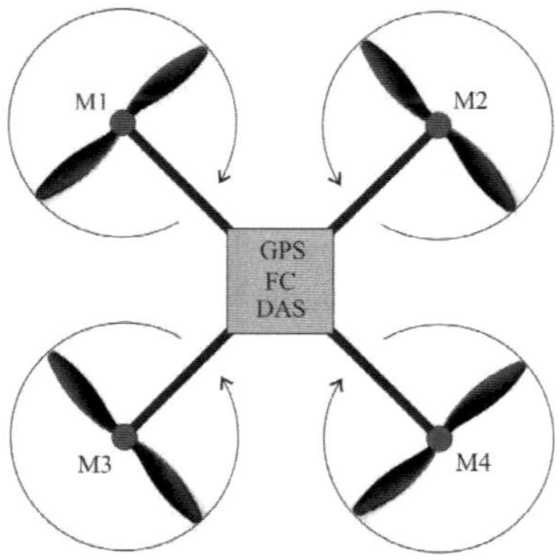

Fig. 2. Schematic view of quadcopter

The flight dynamics [5], [6], [7] and [8] is maintained by different amount of power sent to motors M1 to M4, which is calculated by a computer called flight controller (FC). The hardware of flight controller includes various sensors:

- MPU6050 6-axis gyroscope and accelerometer with Motion Processing Unit
- HMC5883L 3-axis digital magnetometer
- MS5611-01BA03 high precision altimeter
- Ublox NEO-M8N GPS/GLONASS receiver (external).

The CPU of the flight controller, ATMega2560 in our case, monitors all data from the abovementioned sensors, and calculates the required throttle levels for each motor in order to preserve the desired trajectory of the aircraft and to keep it stable in the air. The software in the flight controller can be proprietary (in which case user can only change few adjustable parameters) or open-source. Our UAV is based on an open-source flight controller, which is essential. This enables us to change (when necessary) all flight parameters and routines.

The flight controller uses a PID loop [9], [10] for control of the motors. The stability of PID controller [11] is essential, and the tuning of the propulsion system requires a lot of adjustments which were carefully performed in our laboratories.

Once calculated by the FC, throttle levels are applied to individual motors using the electronic speed controllers (ESC), one for each motor. The ESCs use pulse-width modulation (PWM) [12] to control (brushless) motors.

The flying platform can be controlled manually via radio control unit, or it can fly autonomously (GPS augmented). For this function, extremely capable GPS receiver is necessary. Ublox GPS receiver of the eighth generation uses both GPS and GLONASS satellites. The typical number of acquired satellites is about 20, with HDOP (horizontal dilution of precision) of 2.0 or less.

Battery for aircraft contains lithium-polymer cells, with standard voltage of 3.7 V (per cell). The modular platform of our UAV allows installation of batteries of various capacities, depending on the desired mission parameters. For example, smaller battery can be used if the payload is heavy, and required time of flight is short enough.

Together with the development of the UAV, we have been working on data acquisition system. For this particular purpose, DAS should have fulfilled the following requirements:

- small dimensions
- low weight
- low power consumption
- availability of required sensors and software platform.

The microcontroller based on Arduino [13] fitted perfectly. Some aspects of similar microcontrollers are analysed in [14].

2. Measurements and Results

After successful tests in the laboratory, we connected DHT22 sensor [15] to our DAS and performed full flights. Digital sensor DHT22 measures temperature and relative humidity, and atmospheric pressure is measured by Bosch BMP085 (digital) sensor [16]. Some measurements were performed using proprietary DAS, Pasco Xplorer. We will present two characteristic measurements here: one during the day without any temperature inversion, and other when the thermal stratification was strong. Location of both measurements is indicated by cursor in Figure 1.

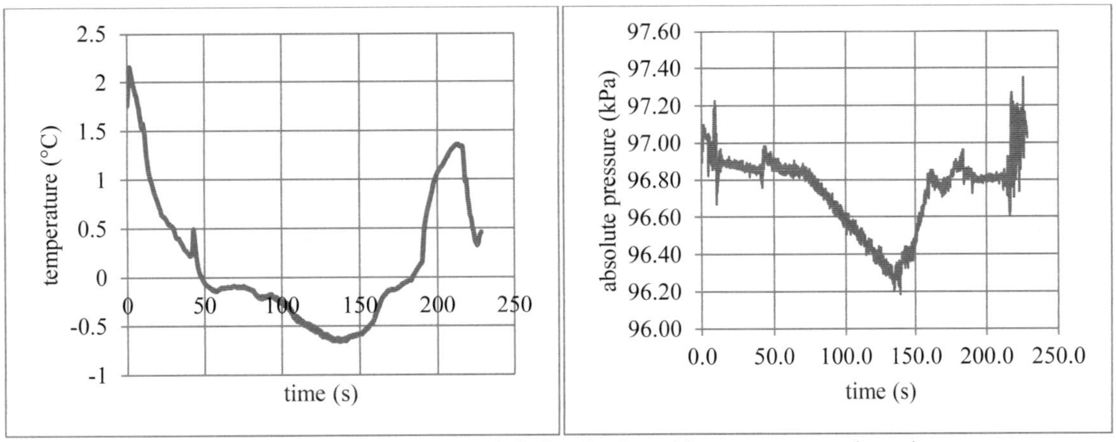

Fig. 4. Pressure and temperature graph during the day without temperature inversions.

Figure 4 shows temperature and pressure readings (Pasco Xplorer DAS was used) during the clear winter day, without temperature inversion. We can see that the temperature decreases with altitude (temperature and pressure measurements are synchronized and pressure indicates the altitude), just as expected. Turbulent nature of air flow (due to propellers) can be seen on the pressure graph. Measurements of the temperature inversions are presented in Figure 5. Sensors DHT22 and BMP085 were used. Altitude above ground was calculated using the barometric pressure readings from BMP085 sensor.

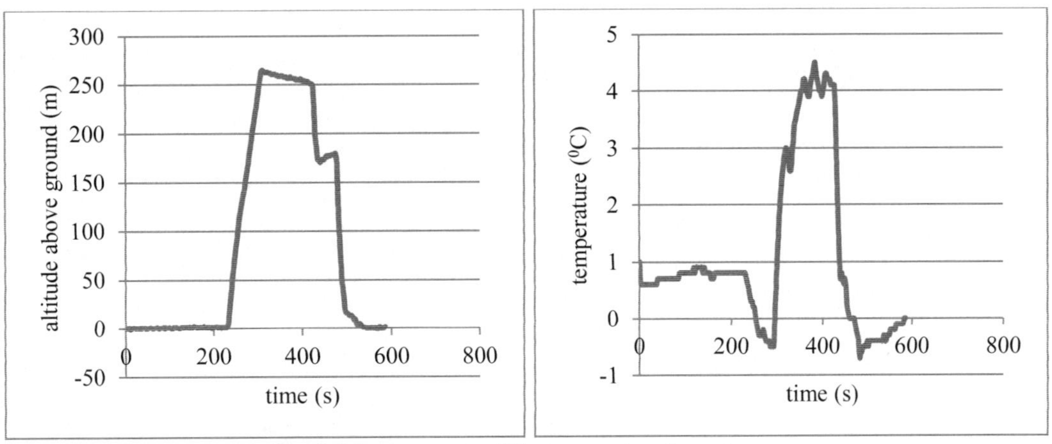

Fig. 5. Measurements of the temperature inversion

The presented graphs have time as independent variable (as they were logged in DAS). It is not difficult to produce temperature-height graphs, by eliminating time. However, due to non-negligible thermal mass of the DHT22 sensor, there is some inertia in the process of measurement of the temperature, which is significant during the fast climbs and descents. This effect produces hysteresis in the temperature-height graph. A convenient way to overcome this problem is to take two temperature values for every height, and calculate the average value of these two temperatures. Furthermore, the difference between values of temperature, during ascend and descent, on the same height gives an indication of the uncertainty of the temperature measurement. Figure 6 shows the temperature-height graph using this processing technique, together with uncertainties. Polynomial fit of third order is added as well. The fit is not perfect, but is reasonably good analytical model of the temperature profile in the inversion layer. This means that our hypothesis is fully confirmed.

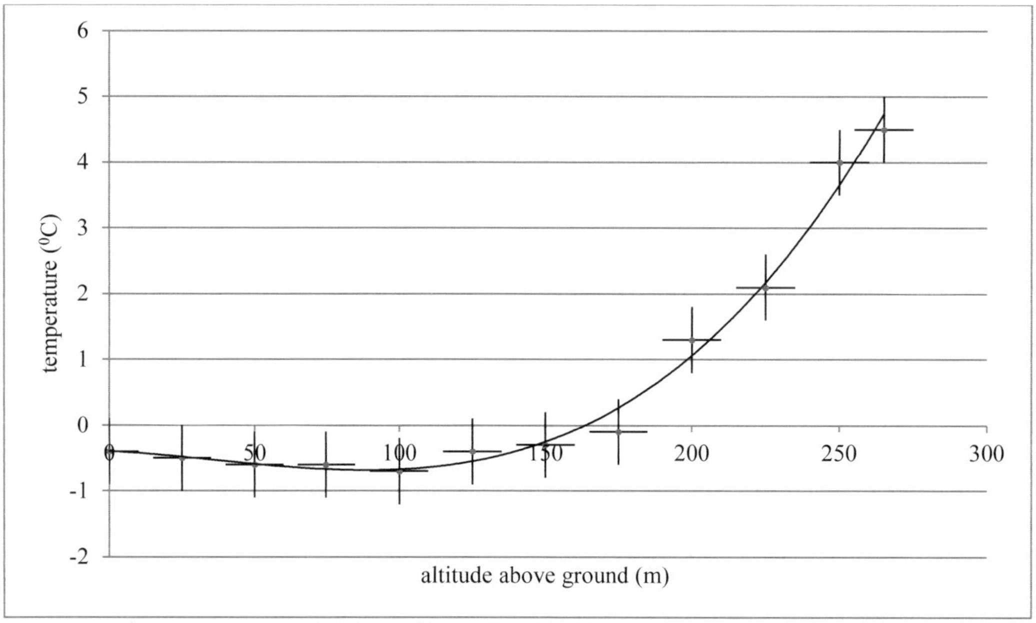

Fig. 6. Temperature–height profile during the day with strong temperature inversion

In the Figure 6 we can see that during the first 100 meters temperature barely changes and even decreases with the altitude. However, from 150 m to 250 m, temperature rises by about 5 degrees! This is inversion of 1 degree Celsius every 20 meters, which is really strong stratification. Figure 7 illustrates the consequence: nothing can escape this natural lid, and the city is completely covered by huge amount of (dirty) air. Some 250 meters above the city, the inversion layer disappears.

Fig. 7. Pollutants above the city of Sarajevo as a consequence of temperature inversion

3. Conclusion

Unmanned aerial vehicle with data acquisition capabilities was built and used for the investigation of temperature inversions above the city of Sarajevo. Open-source flight controller was used and in-house developed data acquisition system. Temperature, pressure and humidity values were collected. This platform was successfully launched during several clear days and those with strong temperature inversion. Temperature profiles exposed that some inversions were as strong as 1 degree Celsius every 20 meters. Such temperature profiles can be approximated by third order polynomial.

In the future, we plan to build larger UAV with capability to carry more instruments and with longer flight time. Obtained data can be combined with computer simulations, for complete understanding of the phenomenon of temperature inversion and its consequences. Another possibility to expand this research is to use equipment for monitoring of certain air pollutants (such as particulate matter or volatile organic compounds) together with temperature profiles. That could be used as input for computer simulations of multiphase flows, and completely new level of modelling. Environmental strategies should also take this data into considerations. Currently, we act after the air pollution parameters reach critical values. Using this method, we can predict the situation and act before we reach the critical level of air pollution.

4. Acknowledgments

This research is supported by a grant of the Government of Sarajevo Canton, Ministry of Education and Science, which is greatly acknowledged.

5. References

[1] Kenjereš, S. & Hanjalić, K. (2002). Combined effects of terrain orography and thermal stratification on pollutant dispersion in a town valley: a T-RANS simulation. Journal of Turbulence, 1468-5248, 3(26): 1-21.

[2] Musemić, R. & Krstović, G. (2005). Numerical Simulation of Actual Heat Flux Effects on Air Ventilation in the Mountainous Town Valley of the City of Sarajevo. 9th International Research/Expert Conference "Trends in the Development of Machinery and Associated Technology" TMT 2005, Antalya, Turkey, 26-30 September, 2005.

[3] Musemić, R. & Krstović, G. (2006). Numerical Simulation of Turbulent Air Pollution Dispersion in a Typical Mountainous Urban Valley. 4th International Conference on Computer Aided Design and Manufacturing, CADAM 2006.

[4] Mašić, A. (2015).Unmanned Aerial Vehicle as Data Acquisition System. Journal of Trends in the Development of Machinery and Associated Technology. Vol. 19, No. 1, 2015, ISSN 2303-4009 (online), p.p. 181-184.

[5] Wiesel,W. E. (2010). Spaceflight dynamics, Aphelion Press, ISBN: 978-1-4528795-9-8, Beavercreek, Ohio, USA.

[6] Etkin, B. & Reid, L. D. (2010). Dynamics of Flight: Stability and Control, John Wiley & Sons, ISBN: 978-8126528912.

[7] Isidori, A. (1999). Nonlinear Control Systems II (Communications and Control Engineering), Springer-Verlag, ISBN: 978-1447111603, London, GB.

[8] Sira-Ramirez, H. & Agrawal, S.K. (2004). Differentially flat systems, Marcel Dekker, ISBN: 978-0824754709, New York, USA.

[9] Kril, S.; Fedoryshyn, R.; Kril, O. & Pistun, Y. (2014). Investigation of Functional Diagrams of Step PID Controllers for Electric Actuators. 25th DAAAM International Symposium on Intelligent Manufacturing and Automation, DAAAM 2014.

[10] Pillay, N. & Govender, P. (2013). A Data Driven Approach to Performance Assessment of PID Controllers for Setpoint Tracking. 24th DAAAM International Symposium on Intelligent Manufacturing and Automation, 2013.

[11] Matusu, R. (2011). Regions of Stability for PID Controllers. Annals of DAAAM for 2011 & Proceedings of the 22nd International DAAAM Symposium, Volume 22, No. 1, ISSN 1726-9679 ISBN 978-3-901509-83-4, Editor B. Katalinic, Published by DAAAM International, Vienna, Austria, EU, 2011.

[12] Vasca, F. & Ianelli, L. (2012). Dynamics and Control of Switched Electronic Systems. Springer-Verlag, ISBN: 978-1447128847, London, GB.

[13] Kushner, D. (2011). The Making of Arduino. IEEE Spectrum.

[14] Dolinay, J.; Dostalek, P. & Vasek, V. (2010). Simple and Cheap Microcontroller Kit for Students, Annals of DAAAM for 2010 & Proceedings of the 21st International DAAAM Symposium, 20-23rd October 2010, Zadar, Croatia, ISSN 1726-9679, ISBN 978-3-901509-73-5, Katalinic, B. (Ed.), pp. 0517-0518, Published by DAAAM International Vienna.

[15] https://www.sparkfun.com/datasheets/Sensors/Temperature/DHT22.pdf (2016). Digital-output relative humidity & temperature sensor/module, Accessed on: 2016-10-17.

[16] https://www.sparkfun.com/datasheets/Components/General/BST-BMP085-DS000-05.pdf (2016). BMP085 Digital Pressure Sensor Data Sheet, Accessed on: 2016-10-17.

DOI: 10.2507/27th.daaam.proceedings.063

TRIBOLOGICAL TESTING OF POLYMERIC MATERIAL

Dana Kubátová, Martin Melichar & Alejandro Pereira

This Publication has to be referred as: Kubatova, D[ana]; Melichar, M[artin] & Pereira, A[lejandro] (2016). Tribological Testing of Polymeric Material, Proceedings of the 27th DAAAM International Symposium, pp.0428-0434, B. Katalinic (Ed.), Published by DAAAM International, ISBN 978-3-902734-08-2, ISSN 1726-9679, Vienna, Austria DOI: 10.2507/27th.daaam.proceedings.063

Abstract

The article deals with testing the properties of specific materials - polymers. The basic idea was to confirm or refute the theory that materials with a higher modulus of elasticity will experimentally have lower wear. A precision tribometer at the experimental laboratories of the University of Vigo (where the international cooperation on the experiments took place) was used to test and find the origin of the wear. For the relevant documentation and independent verification, the map of wear was verified on a highly accurate profile/roughness measurement device at the metrology laboratories in the RTI at UWB in Pilsen.

Keywords: ABS polymers; PA6,6+15% glass fibres; Tribotest Pin-on-Disk; Friction coefficient; Weight lost; Wear map

1. Introduction

In the technical, social and natural fields science there are many unresolved issues and problems that scientists are looking for the answers to, either in the form of answers at the theoretical level or in practical application. After a period of massive development of R&D during the last century, when energy efficiency and environmental protection was not often considered, over the last few decades, and in the third millennium, there has come a sobering awareness of the need to save all kinds of energy. This is occurring against the background of expansion and development of many disciplines relating to the quality of human life, human needs and the desire to discover.

To fulfil these goals today machine and system builders should be equipped with fairly deep knowledge of many disciplines. This information enables them to solve problems in many areas. [1]

Situation in the field of durability and wear of machine parts exposed to friction and wear is completely different. In many cases the designers do not have necessary knowledge and technical documentation about this. The problem stems primarily from the complexity interaction of the materials on the functional surface machine components, the variability of the operating conditions, which are determined by the dominant factors that are friction and wear. Particularly when is contact between moving parts metal on metal or a metal of an alternative material, such as e.g.: polymer.

2. Polymeric material

Polymeric materials are basically divided according to their properties for elastomers, thermoplastics and thermosets. For this test were selected samples from polymeric materials in the field of thermoplastics.

- AKRYLONITRILBUTADIENSTYREN next only ABS
- POLYAMID 6.6 with 15% GLASS FIBRE next only PA6.6+15%
- POLYAMID 6.6 next PA 6.6
- POLYAMID 6.6 with 30% GLASS FIEBRES

Note: bold materials were tested

Why polymeric materials? The answer is simple, because since the time they were discovered, the man is trying various uses. Replace them other materials, which have either limited or a polymer with its specific characteristics surpass. Properties that polymeric materials over conventional materials are numerous. Among the most interesting, which make the polymers used are:

- Lightweight
- Electrical insulation properties
- Low thermal conductivity
- Corrosion Resistance
- Resistance to certain chemicals
- Flexibility / elasticity
- High fatigue resistance
- Resistance to aging

3. Sample

3.1. Preparation sample

For preparation samples was choosen method of injection moulding. Injection was performed on machine Engel Victory 28 that is in industrial laboratories of the University of Vigo.

The samples are in the form of a disc that have diameter Ø 60 mm and a height of 3 mm. Sample size was chosen with regard to the area to carry out all the tests, such as the repeated test tribometer, roughness measurements and other tests.

Parameters for injection were set according to the parameters provided by the manufacturer of the machine. The parameters used for injection are seeing in Table 1 and 2. As a variable parameter was chosen injection pressure in Table 2 marked yellow. These conditions were the same for both materials tested.

Furthermore, the material PA6.6 + 15% GF was dryed before injecting material. Drying was carried out at 80° C for 10 hours.

Set max. Temperature in the cylinders /°C/	265	255	240	220	40
The actual temperature in the cylinder /°C/	249,7	249,5	239,7	220	45,7

Table 1. Temperature during injection

Injection volume	17 cm3	17 cm3
Max. pressure	160 bar	160 bar
Piston diameter	30 mm	30 mm
Time injection	2 s	2 s
Cooling time in the form of	20 s	20 s
Injection pressure	100 bar	150 bar
Volume of the mould	11 cm3	11 m3

Table 2. Injection parameters

3.2. Testing of sample

Own tests selected plastic materials were conducted in cooperation of Metrology lab of University of West Bohemia in Pilsen and Laboratories University of Vigo.

This article partly describes preliminary tests. However, in this article we have already come into field of repeated tests. Because it was funded interesting information during the design of maps wear.

3.3. Tribological tests

The variability of this test allows exploring the different possibilities of test materials. Tribotest setting options are almost limitless.

The task of tribology is description of materials behaviour in a relative motion. Tribology tested two or more materials in contact, or interaction of the material with the environment there is a loss of material from the surface, a process known as wear. The conditions used in the tests are shown in Table 3.

Material	Speed test /rpm/	Radiu test /mm/	Load /N/	Time test /m/	Radius test bodies /mm/
ABS-1-a	200	9	20	200	4
ABS-1-b					
ABS-2-a					
ABS-2-b	300		30		
PA-1-a					
PA-2-a					

Table 3. Testing parameters

4. Friction coefficient

4.1. ABS material

This parameter is the most important value for this test. On the basis of information about the value of the friction coefficient, one is able to predict how the material will behave under load. Thus, both will be wearing, how will heat up or how it will deform...

The parameter is measured during continuous tribological testing and then this value continues to operate. This test measured the coefficient of friction for long term testing since the objective was to determine the mean coefficient of friction.

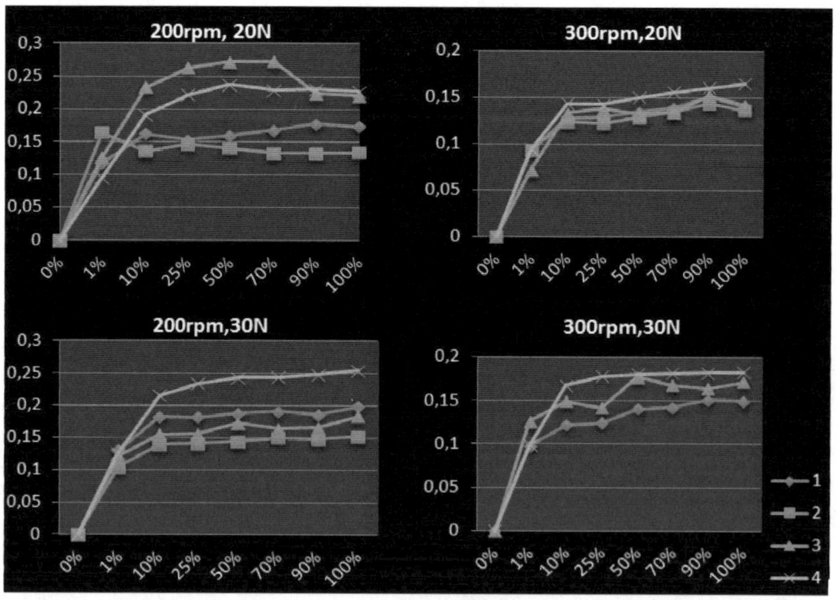

Fig. 1 Friction coefficient during testing for ABS material - according to the test parameters

Label groups:

- 1-150 bar and roughness poor
- 2-100 bar and roughness poor
- 3-100 bar and roughness good
- 4-150 bar and a roughness good

The upward trend of the coefficient of friction is due to the fact that they still rises contact area when pike test specimen and the test sample.

In Fig. 1, graph 200rpm and 20N in Category 3 (gray line) shows a decreasing tendency from the value of 70% and is probably due to a problem in the structure of the sample.

Furthermore, this test confirmed that the friction coefficient depends on the value of surface roughness, as shown Fig. 1 in the following figures and example. At 200 rpm and a load of 30 N it has the best surface roughness of the sample group which has poor roughness.

For better orientation, the effect of the load value, and scanning speed (time) is shown in the following Fig. 2 and 3..

It is easy to see how changes the value of the coefficient of friction changes for each load and speed testing. It is seen that the coefficient decreases with increasing load and speed combination testing.

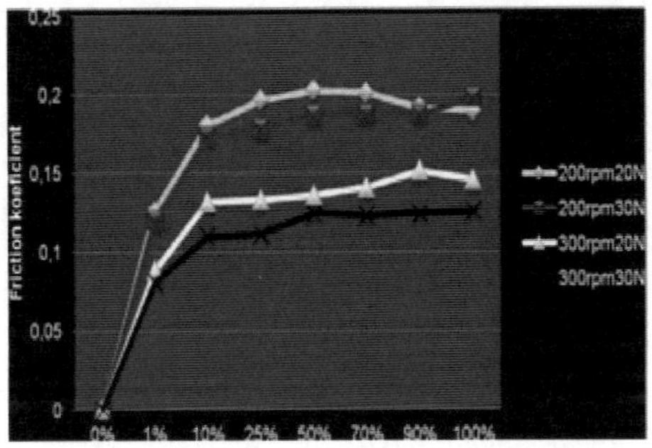

Fig. 2 Friction coefficient – ABS material summarization

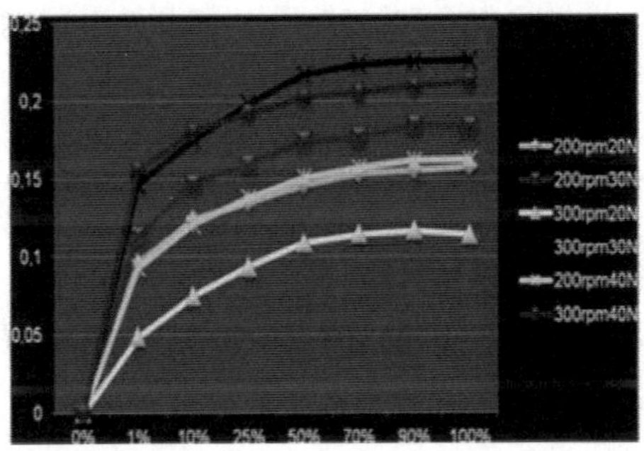

Fig. 3 Friction coefficient – ABS material summarization

4.2. Polyamide material

Even for this material, comparisons were made between the loads for the groups as in the previous tests. However, for these comparisons nothing unusual was found, therefore only the resulting comparison is given.

This comparison is interesting in that the greatest friction coefficient was achieved at maximum load. This can be anything from previous tests to explain why it is necessary to carry out further tests with those conditions and that both the load and speed of rotation but also the same moisture conditions. Moisture is needed in this case the guard due to the fact that PA6.6 + 15% GF is a highly hygroscopic material.

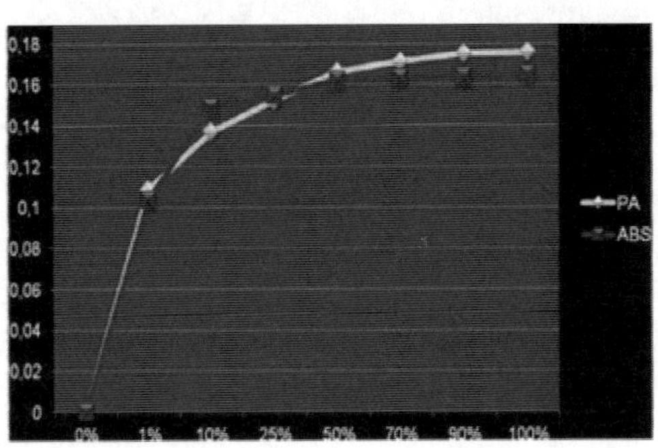

Fig. 4 Friction coefficient – comparison ABS and PA6,6+15% GF

5. Change of weight

The changing weight of the sample describes how much the research sample wears under the influence of higher pressure and cyclic effect (rotary) wear.

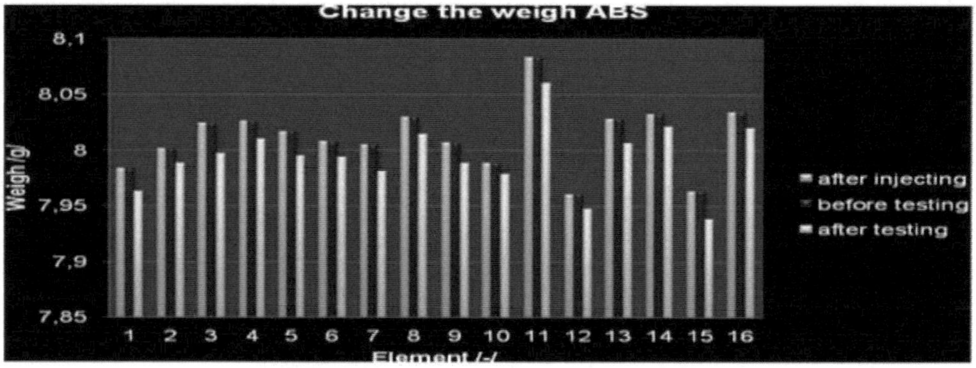

Fig. 5 Differences between weights for ABS material

Here, in the graph for PA6.6 + 15% GF it is important to note that we changed the mass immediately after injection of the sample and before the test. This is due to the very high hygroscopicity material.

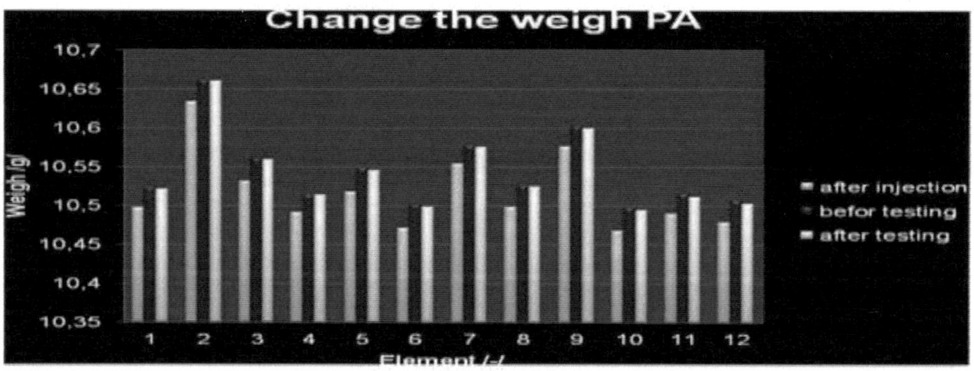

Fig. 6 Differences between weights for PA material

From Chart 3 it is already possible to say how the material will behave during further testing and what results can be expected from additional tests, and that material PA 6.6 + 15%GF will better withstand potential stresses than ABS. In numbers, this is expressed by ABS during the test, lost 0.017ga PA 6.6 + 15%GF lost 0.003 g.

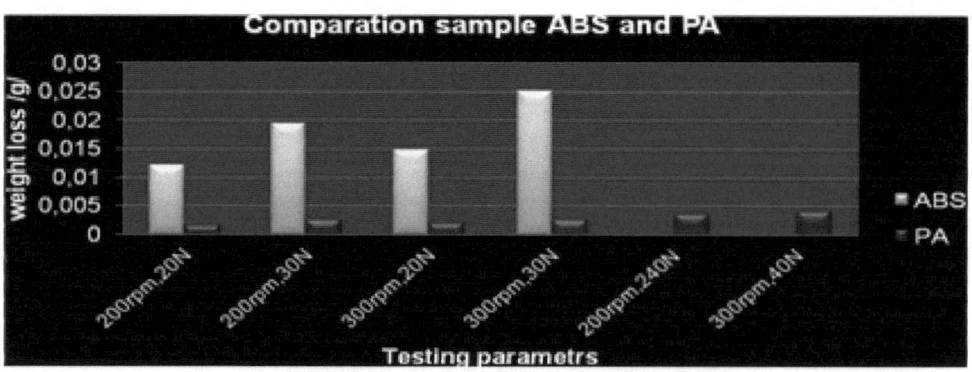

Fig. 7 Comparison ABS and PA material – weights lost

6. Wear map

Wear maps describes how material will behave in the area of tested conditions. Maps are used to get quickly information about the material. Maps are use not only for getting behaviour but the maps are used for comparing two or more of the tested materials and on how they are against each other position of their mechanical properties. However, all this can be expected only if the tests are carried out under the same test conditions - the same testing body, our case was used steel ball diametre 4 mm and test conditions see Table 3.

For construct a wear maps is necessary to know the width, depth and average diameter traces left at sample after the test tribometer.

To determine the actual width tracks and medium diameter traces were used a microscope connected to a PC in the laboratories of the University of Vigo. Images from this measurement can be seen in Fig. 8. The microscope was set lens 10x zoom 1x.

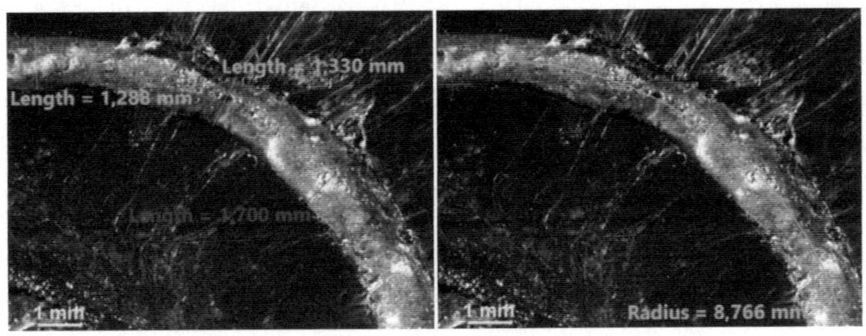

Fig. 8 Measuring stops after testing – width and radius

As an additional parameter is the depth of left traces. This was measured in Laboratories University of West Bohemia device to Etamic Hommel T8000. Device enables to measure both the profile and roughness of components. There was used the profile measurement. The output from the measuring is seeing on Fig. 9. The red box in Fig. 9 is marked path area after tribotest.

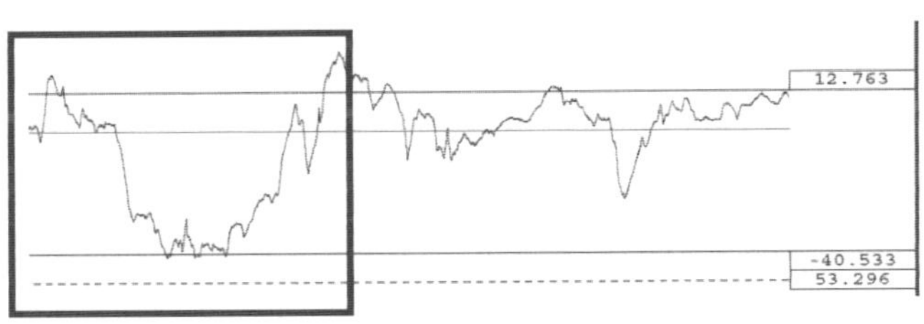

Fig. 9 Measuring stops after testing – deep

After obtaining all the parameters necessary for calculating the wear map has been calculated to obtain data for constructing and plotted into a graph that can be seen in Fig. 10.

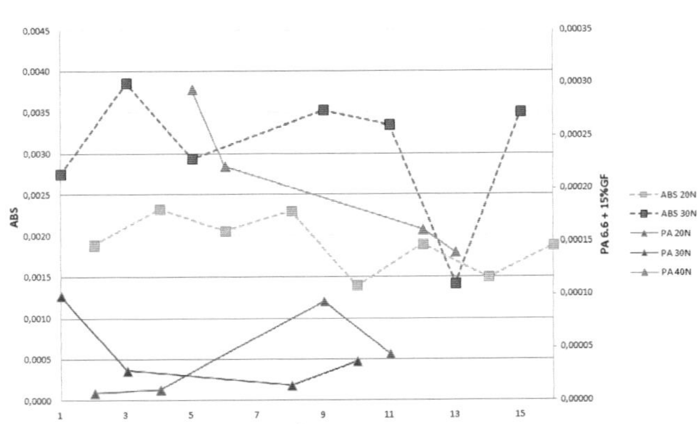

Fig. 10 Wear map for two materials

Usually wear maps are constructed only of individual points. However, when is given two or more materials into one wear maps is difficult to orientation in this map. Therefore in this case is map completed by connecting lines, which enable easier orientation in the chart.

7. Conclusion

ABS and PA are polymers and are thermoplastics. Therefore, they should have roughly the same properties and characteristics as some specific parameters for these materials.

Even in terms of comfort for the test were as arm Vibration noise or better based material PA6.6 + 15% GF.

In the article non-traditional is map designed for wear two materials together. Here the article being wear value related to the ratio of modulus of elasticity. The modulus of elasticity was chosen because it clearly defines the internal structure of the material. Assumption was that for the same test parameters (conditions tribotestu) should result in less wear on the material with higher modulus of elasticity. This hypothesis was confirmed, but not the value of the wear. The wear values could according modulus should differ by about 70%. While the wear values differ by more than 90%.

If the resultant product is used as an exposed part, according to the test the clear choice is ABS. This is covered by acquired better visual structures and surface roughness, and after averaging all values amino slightly smaller friction coefficient than PA66 + 15% GF.

However, when used as a loaded component, this material is very difficult to use. Large wear loss of material occurs when loading, which is adhered in the form of fine dust on all around visibility

8. Anknowligement

The present contribution has been prepared under project LO1502 ‘Development of the Regional Technological Institute‘ under the auspices of the National Sustainability Programme I of the Ministry of Education of the Czech Republic aimed to support research, experimental development and innovation.

9. References

[1] M. Renouf, F. Massi, N. Fillot, A. Saulot, Numerical tribology of a dry contact, Tribology International, Volume 44, Issues 7–8, July 2011, Pages 834-844, ISSN 0301-679X, http://dx.doi.org/10.1016/j.triboint.2011.02.008

[2] T. Mathia, P. Pawlus y M. Wieczorowski, «Recent trends in surface metrology,» Wear, p. 494–508, 2011

[3] D. Kubátová, M. Melichar, A. Pereira, Comparative Study of Tribological Behaviour of Polyamide and Abs Polymers After Processing by Injection

[4] J. Sudeepan, K. Kumar, T.K. Barman, P. Sahoo, Study of Friction and Wear of ABS/Zno Polymer Composite Using Taguchi Technique, Procedia Materials Science, Volume 6, 2014, Pages 391-400, ISSN 2211-8128, http://dx.doi.org/10.1016/j.mspro.2014.07.050

[5] ASM handbook. 10th editon. Materials Park, Ohio: ASM International, 2016. ISBN 1615038272

DOI: 10.2507/27th.daaam.proceedings.064

SIMULATION OF ROBOT FOR USE IN RESCUE OPERATION GROUPS

Ivan Vasilyev, Alla Kashourina, Maxim Krasheninnikov & Ekatherina Smirnova

This Publication has to be referred as: Vasilyev, I[van]; Kashourina, A[lla]; Krasheninnikov, M[axim] & Smirnova, E[katerina] (2016). Simulation of Robot for Use in Rescue Operation Groups, Proceedings of the 27th DAAAM International Symposium, pp.0435-0440, B. Katalinic (Ed.), Published by DAAAM International, ISBN 978-3-902734-08-2, ISSN 1726-9679, Vienna, Austria
DOI: 10.2507/27th.daaam.proceedings.064

Abstract

The rescue of people in disaster through autonomous means of evacuation in some cases is the only way to save their lives. Rescue Mission often takes place in remote locations. The paper studies modeling of a universal rescue vehicle (URV) with a rotary-screw propeller. URV s provided with a pivotally fastened screws. This method of attachment allows to overcome significant obstacles. There is a problem of management and simulation of such screws. This article describes mathematically the kinematic model of the URV.

Keywords: robot; rescue mission; vehicle.

1. Introduction

Universal rescue vehicle (URV) is a passenger all-terrain vehicle of amphibious type, which is provided with four screw propellers, two on each side of the vehicle (Fig. 1).

Fig. 1. URV

Figure 1 illustrates the reason for using "breaking" screws, which is the increase of cross-country motion. Another reason for the use of screw propellers is connected with providing floatation on the water, because screws are hollow and their volume provides additional positive floatability of the whole vehicle. Each screw has an additional actuator for changing its tilt.

Thus, a universal rescue vehicle can move both on hard road and non-cohesive supporting base, including water floatation. And also to overcome the obstacles of considerable height (up to 2 m). Let's consider the kinematic scheme of the Universal rescue vehicle (Fig. 2).

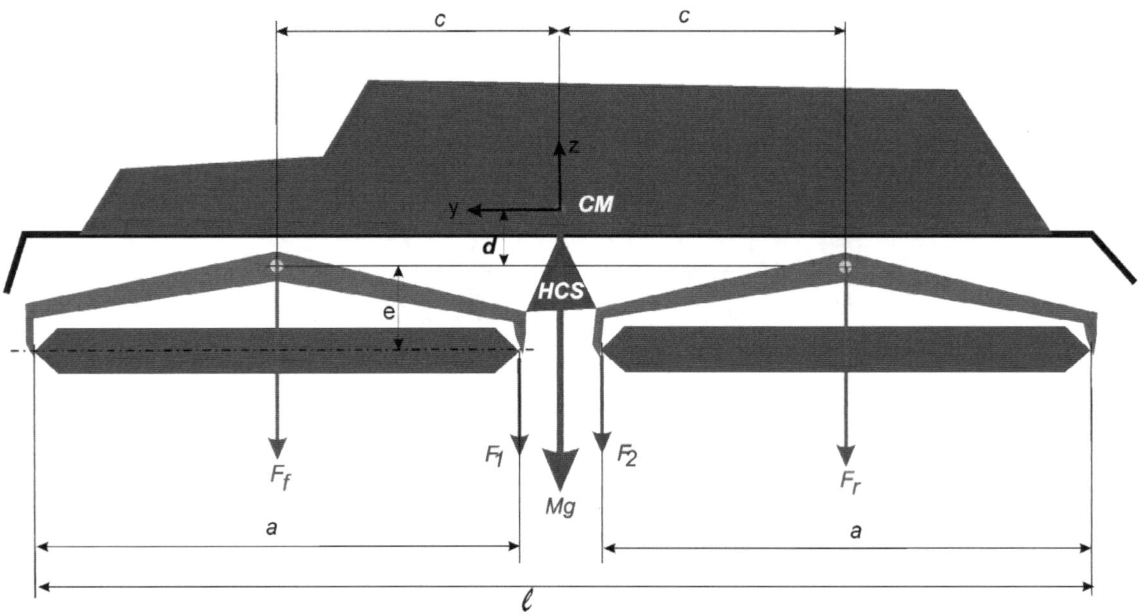

Fig. 2 The kinematic scheme of the URV

In this scheme we denote: CM - center of mass. The fact that CM is exactly halfway between the axes of the mounting screws is quite logical - for the same load of the screws when moving horizontally. In a real situation, of course, it can be (and it is) not. There are many possible reasons for it: from different amount of residual fuel in the tanks to improper seating of passengers in the compartment. But here we consider the situation when everything is perfect. Then we denote: Mg – gravity of the USV center of mass, axes y and z denote the base coordinate system (BCS) of the unit (it‘s more logical to place it in the center of mass), FП - is a component of the USV force of gravity, attributable to the front screw, F3 - a component of the URV gravity attributable to the rear screw.

For simplicity, we assume that the URV is provided with only two screws - front and rear. We "forget" that the URV has a third dimension - width and, accordingly, four screws. For an ideal situation, which is shown in Figure 2, front and rear powers will be equal and will be exactly half of the USV gravity. Abbreviation SHC denotes the system of hydraulic cylinders, providing rotation of the screws. Accordingly, the strengths of these cylinders are designated as F1 and F2. In the described ideal situation, these two forces are equal to zero.

2. Model of the cylinder in water

We assume the distance from the BSC to the axes of rotation of the screws in a horizontal plane is denoted as *c* and a vertical shift as *d*.

Let's try to calculate the depth of immersion of a hollow cylinder of known weight and dimensions. For simplicity, we assume that we just have a cylinder without cone "caps" at the ends. Let the ratio of the weight of the cylinder and the weight of the displaced water be denoted by ξ.

The first case is the cylinder stands in water vertically. This option does not seem the most obvious, but it is the easiest, so we will consider it at first. Let's consider Figure 3, which shows the cylinder in water.

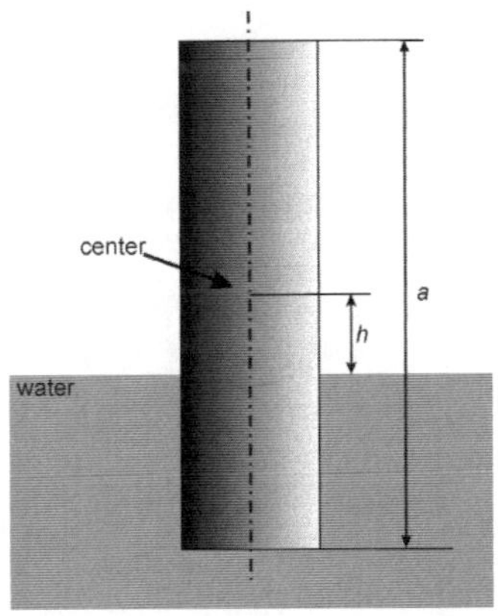

Fig. 3. The cylinder stands vertically in water

Here, the required distance h is easy to find:

$$h = a\,(\frac{1}{2} - \xi),$$

here all symbols are denoted above.
The second case - the cylinder lies horizontally in water. Here the situation is more complicated, let's consider Figure 4.

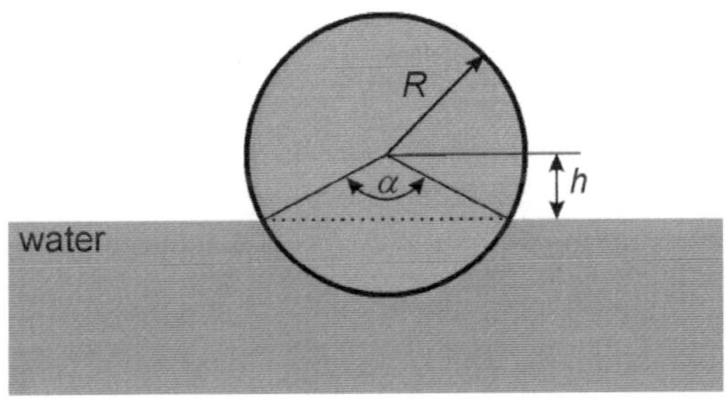

Fig. 4. The cylinder lies horizontally in water. The end view.

Obviously, the ratios of volumes ξ are equal to the ratio of areas of a circle segment below the dotted line and the area of the circle. Therefore we can three-dimensional problem bring to a flat one, shown in Figure 4.

It is easy to obtain a formula for the ratio of the volumes:

$$\frac{Arccos\,\frac{h}{R}}{2\pi} - \frac{h}{\pi\,R^2}\sqrt{R^2 - h^2} = \xi$$

This shows that the equation for the desired height h is strongly nonlinear, which is bad. We substitute $Arccos\,\frac{h}{R}$ with $\frac{\pi}{2} - 1.25\frac{h}{R}$ that is justified because graphics for modeling coincide - Figure 5

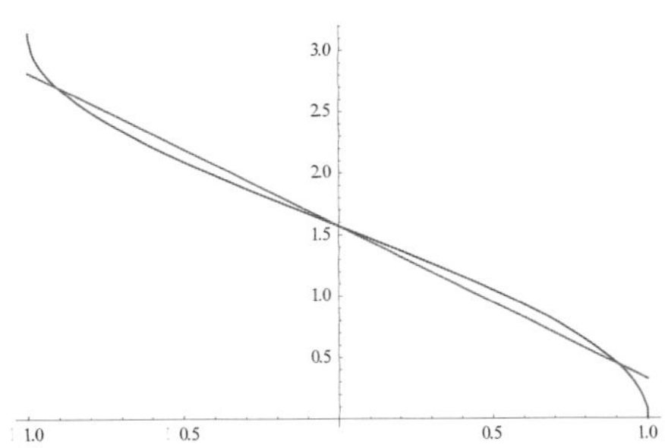

Fig. 5. Graphics Arccos x (blue curve) and π/2-1.25x (red curve)

In the result we get equation of the 4th degree, the solution of which is expressed by a "long" formula, which occupies an entire page. No need to bring it here.
Again, for modeling purposes it is sufficient to substitute the circle by the square of the same area - see Fig. 6.

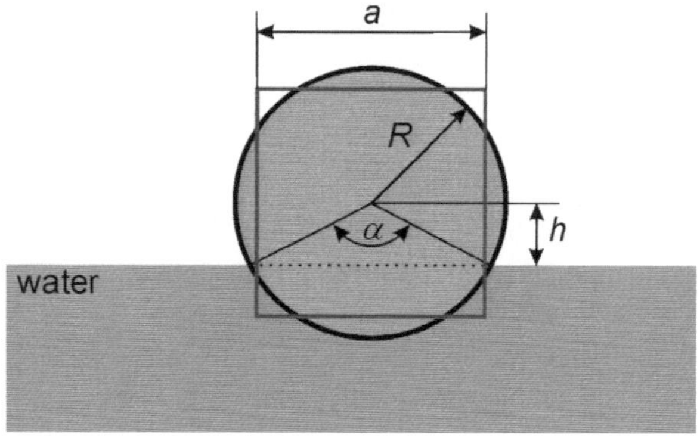

Fig. 6. Replacement of the circle with the square

Here a - is the size of the side of the square. In order to have the same area with the circle the square must have sides equal to:

$$a = \sqrt{\pi}R$$

Now, we get the familiar expression for the height h:

$$h = a\left(\frac{1}{2} - \xi\right)$$

Third, the most common case - when the cylinder is at an angle α to the water surface - see Fig. 7.

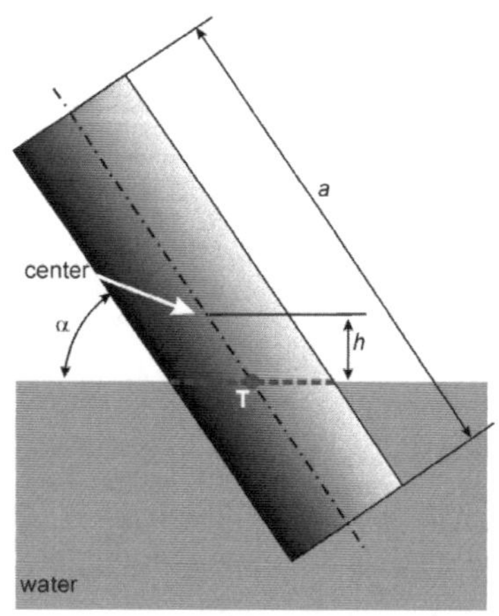

Fig. 7. The cylinder is tilted

Here, the red dotted line shows the level of the water surface. When tilting the cylinder it can be assumed that the cylinder is rotated around the point T, highlighted in Figure 7 as a green dot. Therefore, the value of h is calculated by a simple formula:

$$h = a\left(\xi - \frac{1}{2}\right)\cos\alpha.$$

Modeling of driving of the universal rescue vehicle from the water to the ice. Based on these formulas, we can calculate the position of the URV at its various positions, both in water and when driving over hard surfaces (shore or ice). Let's consider the efforts that arise in such movements - see Fig. 8.

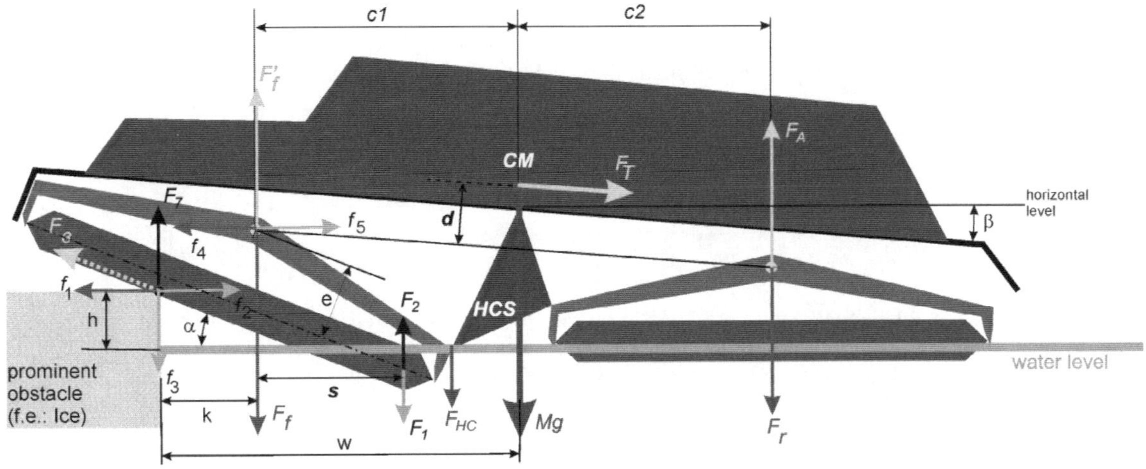

Fig. 8. Demonstration of driving on an obstacle

The following forces are denoted here:

- Mg – weight of the URV;
- F_A и $F'_{\text{п}}$ - reaction in the suspension point of the screws;
- $F_{\text{з}}$ и $F_{\text{п}}$ - load at the point of suspension of the screws;
- $F_{\text{гц}}$ – the force of the hydraulic cylinder by the tilted screw;
- F_1 и F_2 – part of the gravity attributable to the front end of the screw and the corresponding buoyancy force;
- F_7, f_1, f_2 и f_3 –forces arising from the support of the front screw on the obstacle;
- $F_{\text{в}}$ – the strength of the screw screwed into an obstacle;
- F_m – tractive force required for the movement of the unit.

To raise an obstacle the screwing force $F_{\text{в}}$ is not enough because this force depends on the pressure force f_3, and it is may be inadequate at the initial moment because it depends on the values of w и F_1:

$$f_3 = Mg\frac{c_2}{w} - F_1$$

Traction force of screw in water is negligible, so the URV must be equipped with additional propulsion – a water jet.

Knowing the linking coefficient of the prominent obstacle, which is similar to the coefficient of friction, it is possible to calculate the force required for the movement of the Universal rescue vehicle:

$$F_m = F_{\text{в}}\frac{\cos\alpha}{\cos\beta} + F_T$$

where

α – the angle of the screw to the horizon;
β – the angle of URV body to the horizon угол;
F_T - the water jet traction force.

Let's calculate the required tractive force of the water jet. When moving the unit the main force of resistance is the force of gravity, as the URV is being lifted onto the obstacle.

Therefore, the force f_3 to be projected onto the axis of the force $F_{\text{в}}$

$$f_3 \sin\alpha = F_T\frac{\cos\beta}{\cos\alpha} + F_{\text{в}} = F_T\frac{\cos\beta}{\cos\alpha} + \vartheta\, f_3 \cos\alpha$$

Where ϑ – the linking coefficient of the screw on the material obstacles. So the required water jet tractive force:

$$F_T = \frac{\cos\alpha}{\cos\beta} f_3(\sin\alpha - \vartheta\cos\alpha)$$

5. Conclusion

In the article the problem of management and simulation URV. It demonstrated mathematically the kinematic model of URV.

Work on modeling and the creation of a universal rescue vehicle is only at the very beginning. Nevertheless, it is clear that the crossing possibility of the URV in almost any conditions is almost limitless. At present the prototype of the vehicle is being manufactured it will begin to be tested this year.

6. References

[1] Briskin E.S, Chernyshev V.V, Maloletov A.V., Sharonov N.G. Comparative analysis of wheeled, tracked and walking machines. // Scientific and Technical Journal «Robotics and Technical Cybernetics», №1, 2013

[2] Screw Rotary rover. Electronic resource: https://en.wikipedia.org/wiki/Screw-propelled_vehicle

[3] Naumov V.N., Mashkov K.Y., Byakov K.E., Modeling of linear motion of transport and technological unit with rotary-screw propeller. // Journal «Engineering», №12-2013

[4] Naumov V.N., Byakov K.E., Change of towing speed characteristics of transport and technological unit with rotary-screw propulsion by varying the lifting angle of helical blade in the process of motion. // Vestnik MSTU. NE Bauman: electronic edition. 2013

[5] Vasilyev, A. Kashourina, M. Krasheninnikov, E. Smirnova. Use mobile robots groups for rescue missions in extreme climatic conditions // Annals of 25th DAAAM International Symposium on Intelligent Manufacturing and Automation, DAAAM 2014

[6] Smirnova, E.; Stepanov, D. & Goryunov, V.: A Technique of Natural Visual Landmarks Detection and Description for Mobile Robot Cognitive Navigation // Annals of 26th DAAAM International Symposium on Intelligent Manufacturing and Automation, DAAAM 2015

DOI: 10.2507/27th.daaam.proceedings.065

MACHINING OF INCONEL 718 USING UNCOATED CUTTING TOOLS WITH DIFFERENT CUTTING EDGE QUALITY

Tomas Baksa, Vaclav Schornik, Pavel Adamek & Miroslav Zetek

This Publication has to be referred as: Baksa, T[omas]; Schornik, V[aclav]; Adamek, P[avel] & Zetek, M[iroslav] (2016). Machining of Inconel 718 Using Uncoated Cutting Tools With Different Cutting Edge Quality, Proceedings of the 27th DAAAM International Symposium, pp.0441-0446, B. Katalinic (Ed.), Published by DAAAM International, ISBN 978-3-902734-08-2, ISSN 1726-9679, Vienna, Austria
DOI: 10.2507/27th.daaam.proceedings.065

Abstract

This article deals with testing cemented carbide cutting tools during the experimental cutting of Inconel 718 super-alloy. Several carbide end-mills with different cutting edge qualities obtained by different grinding conditions and strategies were used for machining Inconel 718. The main aim is to compare all end-mills in terms of their cutting tool life during cutting. The machining was performed with cutting tools right after grinding without any coating or cutting edge preparation. Cutting forces were measured during machining to find out the influence of the cutting edge quality on the cutting force Fy during the first impact of the tool. Cutting force at the end of the tool life was also measured. The strategy of side milling was used for the cutting tests of tools during machining of Inconel 718. This material is used for highly stressed components in the nuclear industry, such as combustion turbines. The material is very problematic for machining due to its rapid hardening and high heat generation. The influence of the cutting edge after grinding on the cutting process is assessed by means of flank wear and cutting force.

Keywords: Cutting edge; Tool life; Inconel 718; Side milling; Cutting force.

1. Introduction

Solid carbide cutting tools are of great importance in manufacturing technology. The combination of hard carbide particles and a tough metallic binder creates a perfect cutting material with high wear resistance and fractural toughness. Carbide cutting tools can be used for roughing and finishing operations in the machining of metal alloys. The required geometry of a cutting tool is obtained using the grinding process. The quality of its cutting edge is also affected by grinding. The grinding process and the resulting surface are affected by the grain size of the cemented carbide. It was found that cemented carbides with submicron grain size have a higher surface strength after ultra-precision grinding than carbides with micron grain size. This is due to the higher compressive residual stress of the ground submicron-structured carbides [1].

High-strength material such as Inconel 718 is characterized by excellent mechanical properties at high temperatures and is used for heat stressed components in the aerospace, automotive and nuclear industries, and others. However, it is also characterized by its poor machinability. A cutting tool is negatively affected by the high resistance of the material and deformation hardening during the cutting process [2, 3]. Therefore, it is necessary to develop new cutting tools for effective machining of Inconel 718. Cutting efficiency and cutting tool life are affected by the cutting tool material, geometry, micro-geometry, coating and the cutting process [10].

Machining of Inconel 718 has been the subject of many studies. In [4], the important role played by the cutting temperature was determined during high speed machining of Inconel 718. A comparative study of the influence of the type of coating on tool wear in the micro milling of Inconel 718 is described in [3], where several coating materials were used. The flank wear during the cutting process mostly occurred on the cutting edge. Some local fractures on the cutting edge occurred due to fatigue and built-up edge formation [3]. Uncoated cutting tools can achieve better results at lower cutting speeds than PVD coated tools. This was investigated in [7], where the uncoated tool performed better than coated tools at a cutting speed of 25 m/min. The chipping resistance of the uncoated tool was increased by the radius on the cutting edge which gave additional mechanical strength. However, increasing the cutting speed up to 50 m/min brings worse performance of the uncoated tool than PVD-coated tools. This was due to high wear resistance and the low thermal conductivity of the TiN coating [7]. Influence of different thin layers on the machining of Inconel 718 is described in [9].

Chipping and breakage of the cutting edge are the most common causes of cutting tool failure. It was found that increasing the cutting speed leads to reduction of the cutting force due to softening of the work material. This improves the cutting tool life. However, it depends on the type of milling operation. For example, when the cutting speed is further increased during slot milling, chips are welded on the sides of the slot and this slows the chip flow. On the other hand, cutting speed plays a less important role in side milling, where the chip flow is better [4]. Research [5] shows optimization of the cutting process by changing the cutting edge radius. An appropriate choice of carbide grade with a combination of cutting edge preparation also has an effect on the resulting cutting tool life [2]. The different wear behaviour of up and down milling is described in [6]. It was found that tool flank wear propagation in the up-milling operation was more rapid than for the down-milling operation [6]. The chip formation was also different for up and down milling, where up-milling produced a segmented chip and down-milling produced a discontinuous serrated chip [6].

This paper is focused on the experimental cutting testing of several cutting tools with different cutting edge qualities. The different cutting tool qualities were obtained by different grinding conditions and strategies during the grinding of the primary radial relief. The cutting edges were analysed on an electron microscope after grinding to detect the cutting edge quality. The importance of correct measurement is described in [8], where three measuring methods are suggested. The machining experiment was performed using Inconel 718. Cutting tools were tested without any cutting edge preparation and without coatings in the cutting process of Inconel 718. Flank wear was measured during cutting tests of end-mills on an optical microscope. Cutting forces were measured and analysed during milling of Inconel 718. The influence of the cutting edge quality on the cutting force during the impact was determined.

2. The experiment and discussion

Several carbide end-mills were produced using grinding with different grinding conditions. Table 2 shows all the cutting tools and grinding conditions. The same carbide grade was used for all the cutting tools. This carbide grade is produced by standard suppliers and is suitable for the production of precision cutting tools for turning, drilling and milling. The carbide grade is specified in Table 1. Cutting edges were analysed before the cutting test using an electron microscope.

It was found that conditions and the strategy during grinding of the primary radial relief affects the quality of the cutting edge. A high quality cutting edge was achieved using lower feed rates v_{fg}. A little chipping of the cutting edge occurred when the feed rate increased up to 260 mm/min, but there was no significant change. However, a significant change occurred when the spin direction of the grinding wheel was changed from CCW (counter clockwise) to CW (clockwise). Chipping of the cutting edge occurred even at a low feed rate and it was more significant as the feed rate increased (Fig. 1).

Grain size	WC [%]	Co [%]	Density [g/cm^3]	Hardness HRA	Hardness HV30	TRS [MPa]
Sub-micron	90	10	14.35	91.8	1580	3800

Table 1. Specification of the carbide grade

Tool	Wheel speed v_{cg} = 20 m/s	
	Feed rate v_{fg} [mm/min]	Wheel spin direction
T1	20	CCW
T2	50	CCW
T3	80	CCW
T4	110	CCW
T5	140	CCW
T6	170	CCW
T7	200	CCW
T8	230	CCW
T9	260	CCW
T10	20	CW
T11	80	CW
T12	170	CW
T13	260	CW

Table 2. Cutting tools and grinding conditions

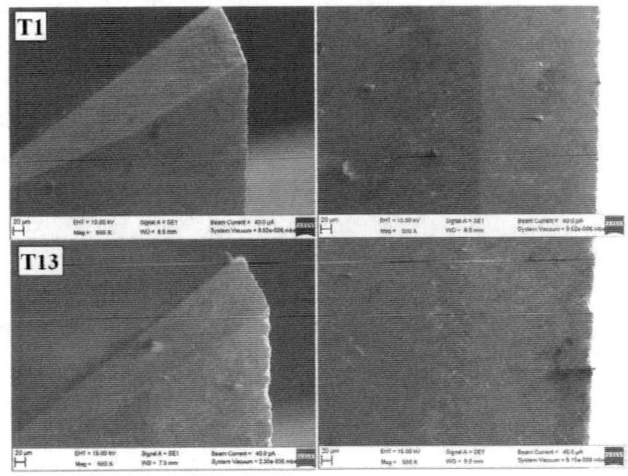

Fig. 1. Cutting edge after grinding

All the milling tools were used in the milling test under constant cutting conditions (Table 3). These cutting conditions were determined by previous experiments of cutting Inconel 718 alloy. Because of the high heat generation during the machining of Inconel 718, external coolant was used. Cutting tools were used right after grinding, so there was no edge preparation or coating of the tools. The aim of the experiment was to determine the influence of the cutting edge quality obtained by grinding on cutting tool life. Machining was carried out on Inconel 718 which was clamped onto a KISTLER 9225A dynamometer.

The strategy of side milling was used for the milling test. Down milling was used to increase the stability of the cutting process. Cutting forces were measured during the milling process at the beginning and at the end of the cutting tool life of each tool. The machining strategy and the directions of the components of the cutting forces are shown in Fig. 2a). The cutting tool life was determined based on the critical flank wear VBcrit on the cutting edge. The flank wear was measured on each cutting edge of each tool and the critical value of the flank wear VBcrit was set to 150 µm. A Multicheck PC500 optical microscope was used for measurements. The measurement site was 2 mm from the cutting tool face as seen in Fig. 2b).

Cutting speed vc [m/min]	Feed speed vf [mm/min]	Axial depth of cut ap [mm]	Radial depth of cut ae [mm]	Cutting environment
35	111	3	0.5	External cooling

Table 3. Cutting conditions

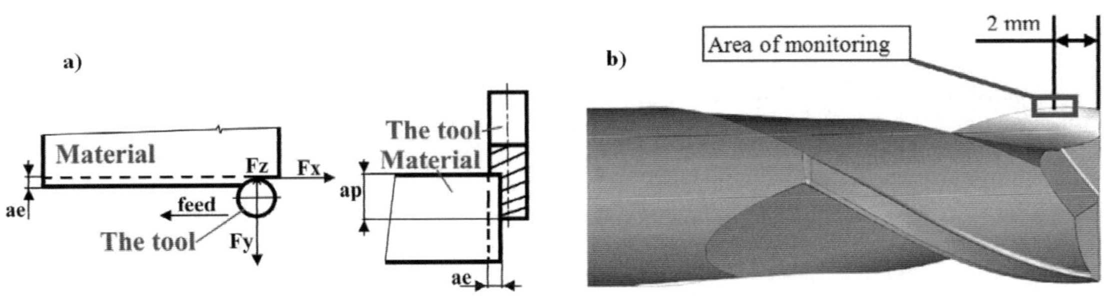

Fig. 2. Cutting strategy (a) and measurement area (b)

In Fig. 4 the results of the cutting test in terms of the cutting tool life are shown. It can be seen that cutting tools ground using CCW strategy T2, T8 and T9 achieve the best results with a durability of 11.7 min. The same durability was achieved by tool T13 which was ground using the CW strategy. The lowest tool life was reached by tools T5 (CCW strategy), T10 and T11 (CW strategy) with a durability of 9.1 min. It seems that the difference in cutting tool life is not significant for tools which were ground at feed rates ranging from 20 to 200 mm/min. However, increasing the grinding feed rate to above 200 mm/min shows better results than at lower grinding feed rates. The strategy of the clockwise grinding (CW) of radial relief at lower feed rate (20 to 80 mm/min) achieved a worse tool life in the cutting test. As in the case of CCW grinding, it seems that the durability of tools ground by the CW strategy increases with grinding feed rate.

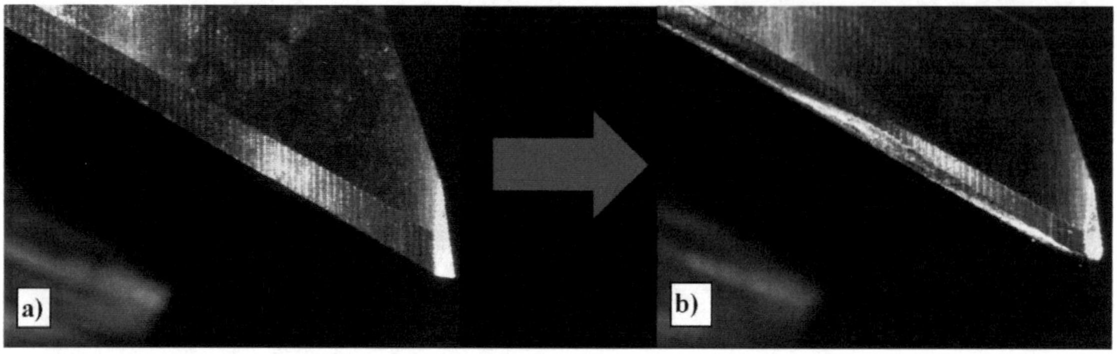

Fig. 3. Sharp cutting edge (a) and worn cutting edge (b)

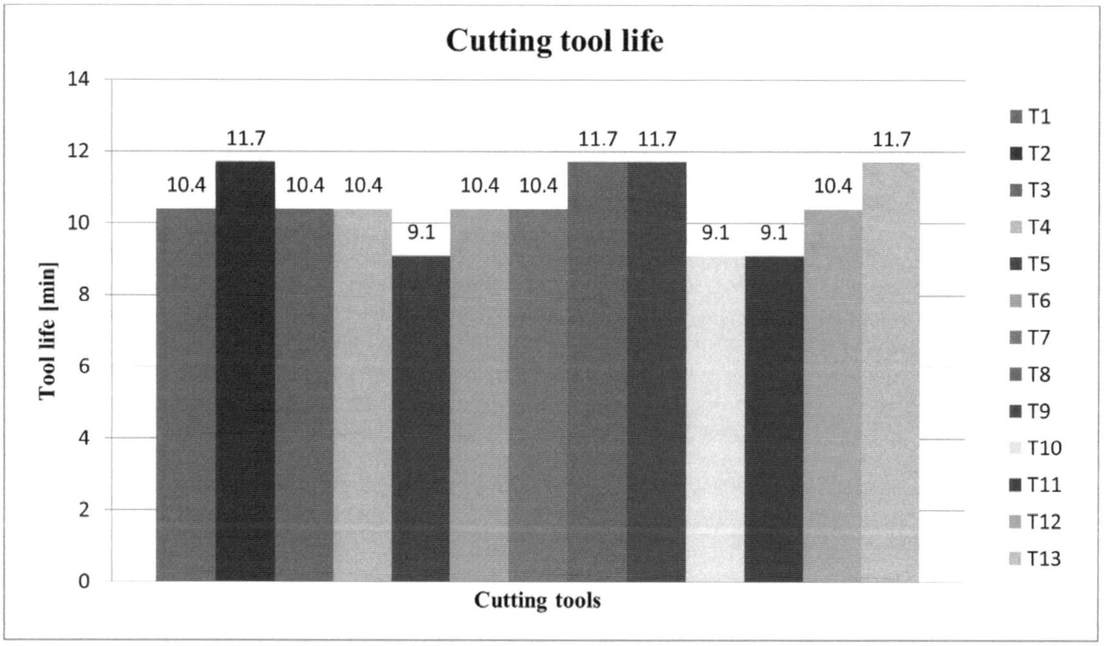

Fig. 4. Cutting tool life

Cutting forces were measured while machining Inconel 718. Measuring was carried out always during the first cut when the wear mechanism started and during the last cut when the flank wear reached the critical value VB_{crit}. The dominant component of the cutting force was the y-component F_y. Fig. 5 shows a summary of cutting forces F_y during the first (Fy00) and the last (Fy150) cut of each tool. All the cutting tools show a very similar Fy150 of about 350 N when reaching VB_{crit}. This is caused by the changing of the cutting edge geometry due to the similar wear mechanism for all the tools. However, some differences between CW and CCW relief grinding strategies can be seen during the first cut. The cutting tools where the CW strategy was used reached higher values of Fy00 than cutting tools with a CCW strategy (Fig. 5).

There is also seen a difference in the force during the first impact into the material (Fig. 6). It is assumed that this impact force is due to the different quality of the cutting edge after grinding. Cutting tools ground using a CCW strategy are characterized by a sharp cutting edge which can more easily penetrate the material due to less cutting resistance. Cutting tools ground using a CW strategy are characterized by the chipping formed on the cutting edge which increases the cutting resistance. The chipping is more significant when the grinding feed rate increases.

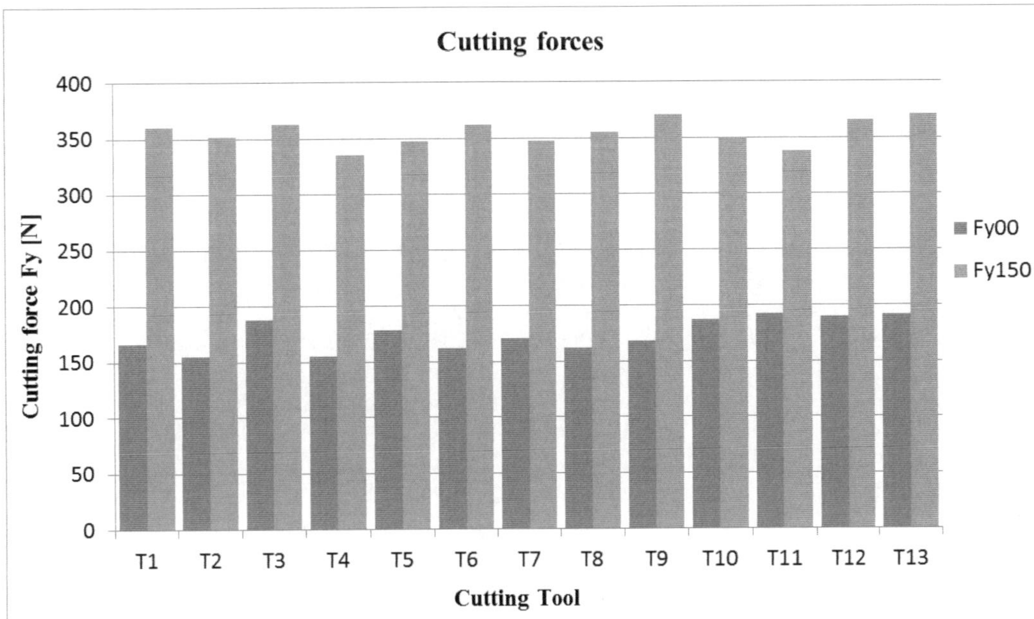

Fig. 5. Cutting force F_y at the beginning and at the end of the tool life

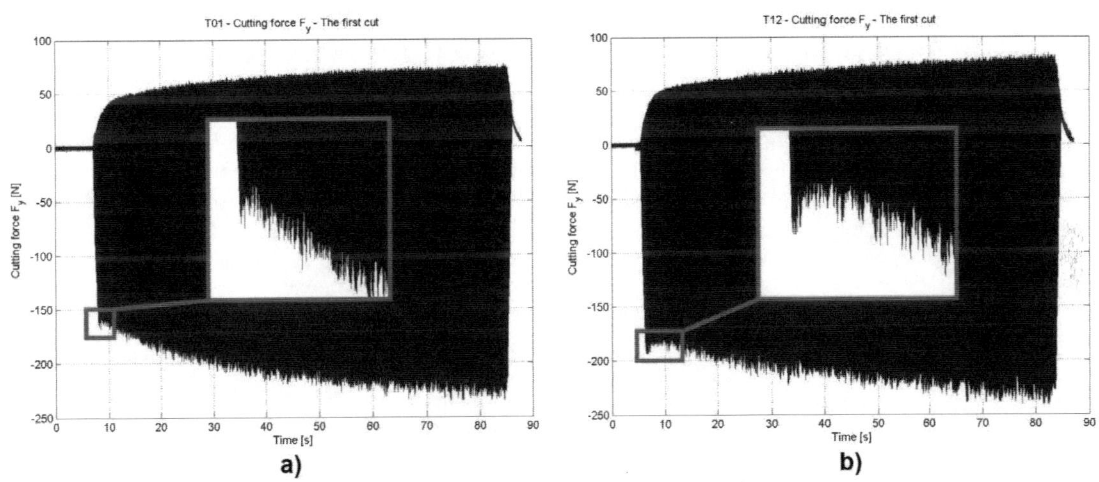

Fig. 6. The impact force F_y of tool T1 (a) and T12 (b)

3. Conclusions

Experimental milling of Inconel 718 was carried out with all cutting tools. The main aim was to determine the influence of the cutting edge quality obtained by grinding on the cutting force and cutting tool life. All cutting tools were without coating or any after-grinding edge preparation. It was found that cutting tools grinded by CCW strategy at the higher feed rate (230 to 260 mm/min) achieve slightly higher durability despite the worse quality of the cutting edge. The same effect occurred for tools grinded by CW strategy. Difference in the impact cutting force was found between CCW and CW strategy. Tools grinded by CW strategy achieved higher force Fy00 during the first impact into the material. The best results in terms of cutting tool life achieved tools T2, T8, T9, T13 with durability of 11.7 min. The worst results were achieved by tools T5, T10, T11 whit durability 9.1 min. This experiment showed that different quality of the cutting edge after grinding has a little effect on the cutting tool life and that the worse quality of the cutting edge can achieve the same or the better results when the cutting tool without coating is used. However the worse cutting edge quality showed the higher impact force which affects the stability of the cutting process. The worse quality of the cutting edge would also affect the adhesion of the thin layer adversely when the coating is used. The results of this experiment will be used for further study where this influence on the coating will be investigated.

4. Acknowledgments

The present contribution has been prepared under project LO1502 'Development of the Regional Technological Institute' under the auspices of the National Sustainability Programme I of the Ministry of Education of the Czech Republic aimed at supporting research, experimental development and innovation.

5. References

[1] Yin, L.; Spowage, A. C.; Ramesh, K.; Huang, H..; Pickering, J. P. & Vancoille, E. Y. J. (2004). Influence of microstructure on ultraprecision grinding of cemented carbides. International Journal of Machine Tools & Manufacture, Vol. 44, pp. 533-543, ISSN: 0890-6955, Elsevier.

[2] Baksa, T.; Moravek, M. & Zetek, M. (2016). Influence of the coated carbide grade on the cutting tool life when machining Inconel 718. Key Engineering Materials, Vol. 674, pp. 271 – 276, ISSN: 1662-9795, TTP.

[3] Ucun, I.; Aslantas, K. & Bedir, F. (2013). An experimental investigation of the effect of coating material on tool wear in micro milling of Inconel 718 super alloy. Wear, Vol. 300, pp. 8-19, ISSN: 0043-1648, Elsevier.

[4] Liao, Y. S.; Lin, H. M. & Wang, J. H. (2008). Behaviors of end milling Inconel 718 superalloy by cemented carbide tools. Journal of Material Processing Technology, Vol. 201, pp. 460-465, ISSN: 0924-0136, Elsevier.

[5] Zetek, M.; Cesakova, I. & Svarc, V. (2014). Increasing Cutting Tool Life when Machining Inconel 718. Procedia Engineering, Vol. 69, pp 1115-1124, ISSN: 1877-7058, Elsevier.

[6] Hadi, M. A.; Ghani, J. A.; Che Haron, C. H. & Kasim, M. S. (2013). Comparison between up-milling and down-milling operations on tool wear in milling Inconel 718. Procedia Engineering, Vol. 68, pp. 647-653, ISSN: 1877-7058, Elsevier.

[7] Jawaid, A.; Koksal, S. & Sharif, S. (2001). Cutting performance and wear characteristics of PVD coated and uncoated carbide tools in face milling Inconel 718 aerospace alloy. Journal of Material Processing Technology, Vol. 116, pp. 2-9, ISSN: 0924-0136, Elsevier.

[8] Cesakova, I.; Zetek, M. & Svarc, V. (2014). Evaluation of Cutting Tool Parameters. Procedia Engineering, Vol. 69, pp 1105-1114, ISSN: 1877-7058, Elsevier.

[9] Schornik, V.; Zetek, M. & Baksa, T. (2015). Durability of the cutting tool with different thin layers when Inconel 718 is machined. Proceedings of the 26th DAAAM International Symposium, pp.0678-0682, B. Katalinic (Ed.), Published by DAAAM International, ISBN 978-3-902734-07-5, ISSN 1726-9679, Vienna, Austria.

[10] Schornik, V.; Zetek, M. & Dana, M. (2014). The influence of working environment and cutting conditions on milling nickel–based super alloys with carbide tools. Procedia Engineering, Vol. 100, pp 1262-1269, ISSN: 1877-7058, Elsevier.

DOI: 10.2507/27th.daaam.proceedings.066

ACTIVE CAMERA POSITIONAL TRACKING FOR AUGMENTED REALITY APPLICATIONS

Jiri Polcar, Sergo Martirosov, Pavel Kopecek

This Publication has to be referred as: Polcar, J[iri]; Martirosov, S[ergo] & Kopecek, P[avel] (2016). Active Camera Positional Tracking for Augmented Reality Applications, Proceedings of the 27th DAAAM International Symposium, pp.0447-0451, B. Katalinic (Ed.), Published by DAAAM International, ISBN 978-3-902734-08-2, ISSN 1726-9679, Vienna, Austria
DOI: 10.2507/27th.daaam.proceedings.066

Abstract

This paper presents our approach for making an active camera tracking application for augmented reality. We have taken into consideration the use of such applications in industry and conditions in production systems., Unity3D with ARToolkit for Unity was chosen as the development platform due to its compatibility and availability, although we needed to conduct a slight modification to the ARToolkit for Unity which we present in the paper. For testing the modified algorithm, we placed fixed markers around a room and made a LiDAR scan of it. We evaluated the quality of registration of the actual view and the LiDAR scan. The results are yet to be further evaluated and the algorithm must be tested in conditions that will be as near as possible to an actual shop floor.

Keywords: ARToolkit for Unity; Augmented Reality; Marker Tracking; Positional Tracking, Unity3D

1. Introduction

According to Gartner's Hype Cycle and DHL, augmented reality (AR) is a very promising emerging technology suitable for many industrial applications that should take about 5 years to enter the consumer market. The combination of a real view for the user and virtual enhancements, either 2D or 3D, is a feature of this technology which is very much in demand. AR users can visualize arrows pointing to an object of interest, be navigated to a certain place or certain object, visualize inside the contents of a box, etc.

In their promo video, SAP SE foresee the use of AR smartglasses for navigating around a production system and warehouses, displaying a warehouse operator the correct box to pick and correct stations to supply. BMW AG also has a promo video that shows their vision of using smartglasses to guide a car mechanic through the replacement of faulty parts.

Although these visions show the great potential of such technology, there are many problems for augmented reality to overcome. The overall quality of current smartglasses can discourage potential buyers from using them for various reasons. They are not very beautiful or comfortable, and the display quality is low, the field of view is too narrow and the displays lack brightness. There are not yet many studies showing the health effects of wearing a display in such a short distance in front of the eyes or a source of various electromagnetic and heat radiation very near to the brain.

Another problem with current smartglasses is their positional tracking in space. Their rotation in space can be tracked autonomously using a gyroscopic sensor, but getting the position is not an easy task. In exterior applications, the GPS signal will not always provide the required accuracy. Indoors, GPS cannot be used and the AR device needs to acquire its position differently. Unfortunately, an accelerometer is not accurate enough and is basically the only simple way to get the position autonomously. There are various ways to achieve more or less precise tracking, including WPS (Wi-Fi position system), magnetic or ultrasonic tracking, self-learning 3D scanning of the room, etc. The method this article will focus on is known as position marker tracking.

There are many AR applications that focus on marker tracking using computer vision and image recognition algorithms. For example, a hybrid method was presented in [1]. It relies on marker recognition to view a 3D model placed in the position of the marker. The marker recognition process can fail due to camera refocusing or when a part of the marker is obstructed. In such cases the position is computed from an InterSense InterMax2 device, an inertial tracking device, to supplement the loss of tracking. The disadvantage of this approach is the price of the InterSense device.

Another study deals with tracking objects using binocular cameras – a dual camera rig [2]. The tracked device, a measuring pen, has three near-IR beacon markers. Using sub pixel recognition algorithms, the authors achieved an accuracy of about 1 millimetre, but the camera rig was quite large and this method was intended to be used at close distances. Another enhancement for video based marker tracking is presented in [3], using QR codes in a marker contrast border as the tracked image. This allows for a basically unlimited number of markers, but with increased demands for marker recognition, because the QR code features are smaller.

The algorithms mentioned above are mainly used for tracking the markers in the camera's field of view. In this case, an inverse algorithm is needed, but the main difference is small: The markers are in fixed, known positions, whereas the camera moves. This would require the markers to be placed and their positions stored in a database. Then, the camera's position is known when the camera can see these markers. This approach can be enhanced by an inertial sensor as mentioned in [1] for cases when visual tracking is lost. For higher accuracy, the approach from [2] can be used with IR beacons around the space, but there would be a problem if there were a lot of them – the patterns must be unique. For tracking in big areas like production halls, QR code markers (mentioned in [3]) might not be very suitable because of their small details.

To get active camera tracking, the markers must be placed at positions that are visible from as wide an area as possible, for example on ceiling support pillars or at aisle crossings. There are even markerless approaches for camera tracking, like the one in [4], which first learn the space by recognition of natural features like edges, but there are many edges in a production hall, making such algorithms prone to errors and instability. Recovery of lost tracking can take a small amount of time or even needs the camera to be placed camera somewhere, where the tracking is guaranteed to recover [5]. Paper [6] deals with making AR tracking more robust and deals with accuracy in different tracking conditions. [7] presents an approach where the ceiling is covered with a lot of markers, one next to another. The user wears a HMD with a front facing camera for tracking real environment and markers inside it. Another camera points directly up to see the markers on the ceiling, providing positional tracking. This can be applied for precision tracking inside one single room, but in production halls, the ceiling is too high and the surface is too big to cover it all with markers.

Combining vision based tracking for idle states with InterSense IS-600 for tracking while moving results in more robust tracking with less jitter and lower latencies was presented by [8]. Unfortunately, the InterSense tracking device requires a static and relatively small tracking space. [9] presents a combination of GPS and vision based tracking based on known marker positions for use in outdoor environments and a prepared indoor environment, all worn in a backpack. The quality of augmented reality applications and tracking can be reduced by errors occurring during the image recognition, registration, etc. Several types of error sources may alter the quality of the tracking of augmented reality tasks. An overview of potential error sources is given in [10].

2. Implementation

From our previous projects, we made augmented assembly manuals requiring positional tracking of supply boxes to be picked by the operator. This was done in Metaio Unifeye Design which allowed placing of virtual objects over markers. The assembly manual was made in two versions: one only showed animated assembly instruction, the other showed arrows over the boxes with correct parts. This implementation required the markers to be on each of the boxes.

We needed an AR tracking solution for the virtual assemblies in our warehouse navigation system, so we decided to go with the ARToolkit for Unity. The Unity3D platform allowed us to make any virtual content and bind it with the ARToolkit. Unfortunately, for active camera tracking, the ARToolkit for Unity was not suitable right out of the box. There was no solution for camera tracking in an environment with fixed markers with known positions. We decided to make our own branch of ARToolkit for Unity with two targets:

- Tracking of virtual objects in camera's field of view.
- Tracking of the camera based on fixed markers with known positions.

The default ARToolkit algorithm is to consider the first found marker as a so-called base marker. The scene camera is placed relative to this marker. When other markers are found in the image, the virtual objects are then placed relative to the base marker. This approach is not convenient for our intended use. We want to have the camera tracked using fixed markers and display virtual models over other markers that are not marked as fixed. We changed the marker tracking algorithm so that all virtual models are placed relative to the camera without considering any markers as base markers. See Fig. 1 displaying the change to the algorithm.

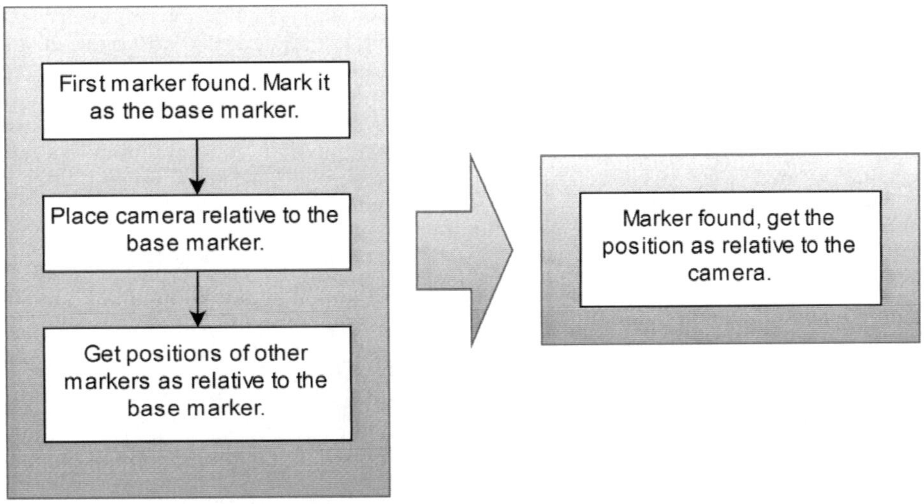

Fig. 1. Marker tracking algorithm modification.

Compared to the original algorithm, our algorithm does not cause intensive jittering of the camera by changing its position relative to the markers all the time. Between the frames, it is not consistent which of the markers will be considered as the base marker. This would not be such a problem, but applications depending on the positions of objects would be affected, because these objects were basically jumping in the scene. The modified algorithm makes the objects jitter only because of tracking accuracy and it allows us to combine it with the next modified algorithm.

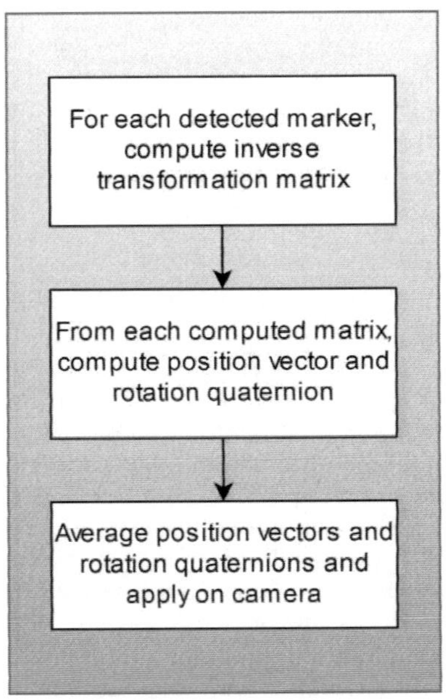

Fig. 2. Camera tracking algorithm.

The other algorithm is for active camera tracking. Considering having markers as fixed and with known positions, we can simply place the camera relative to them and compute an average of positions and rotations. The ARToolkit gets the transformation matrix of the marker in the camera's coordinate system. Simply by inverting this matrix, we have the relative position of the camera in the marker's coordinate system. Then, we need to transform this matrix to the world coordinate system. From the resulting matrix, the rotation and position of the camera can be computed. If more markers are seen, the location vectors and rotation quaternions are averaged. See Fig. 2 for the scheme of the algorithm.

3. Testing, Discussion and Further Work

For testing the modified algorithm, we used a Logitech C910 webcam capable of delivering a FullHD video at 30 frames per second. Then, we placed the markers around our laboratory and captured their exact positions using a Leica ScanStation C5 LiDAR scanning device which we have at our disposal. We imported the resulting point cloud in Unity in order to place objects representing the markers and to check the registration of the point cloud with the actual camera's field of view.

The resulting image jittered slightly, with the accuracy being around 10 cm. This is enough for navigational tasks, but when using this solution for smartglasses tracking, it would be more than enough to cause cyber sickness and unpleasant feelings. The jittering is caused by the ARToolkit algorithm and also by lighting conditions, marker print quality, marker size, camera framerate and resolution. For an example of the result, see Fig. 3.

Fig. 3. Example of tracking results: Registration of a point cloud with a real view.

Although the functional prototype for tracking objects in the visual field (field of view) and cameras in the area was ready, it is still necessary to develop the program further. First it is necessary to find out the precision of the tracking method and then the dependence of its accuracy in different conditions. The size of the markers, the distance to them, lighting, camera resolution and the ink that was used to print the markers, all must be taken in mind when measuring accuracy. In addition to that we must take into account the speed of tracking as well. Tracking speed is very important when using head-mounted displays.

In our further research, we are going to rig a webcam with an InterSense IS-900 tracking device for better accuracy reading. The InterSense IS-900 has a tracking positional accuracy of 2 to 3 millimetres, allowing us to evaluate the accuracy of the tracking more precisely. We will conduct experiments with different cameras, marker print ink and quality, light conditions and other conditions that will be as near to actual shop floors as possible.

4. Intended Use

The modified ARToolkit can be used as a platform for different projects. Below you can see some examples of its possible applications in industry:

- Finding the position of the warehouse operator, who is wearing smartglasses in the warehouse. It is possible to accurately navigate him through the warehouse, though connection to an information system is required. It is also necessary to know the exact position of the markers after their deployment in the warehouse.
- Visualization of a model of a future product in the office. The office should be equipped with marker tracking. The set CAD model is then projected in the same position, for example - on the floor. Placing of the model is done using a modelling tool in a PC (the object then would not require a marker) or with the help of markers. Static objects are resistant to errors in marker tracking and have less risk of displacement or disappearance of their markers in the shade, which is an advantage compared to the classical approach.
- Determining the position of a user's smartglasses in the production system. If suddenly an information system alerted that one of the machines requires maintenance, adjustment or any other service, an arrow and a message would appear on the user's smartglasses leading a worker to the machine location. How it can be used: when passing the production system, it shows the location pointers. When finally reached, it displays the instructions required for fixing the problem.

5. Conclusion

The modified algorithm for object tracking allowed us to tailor the functionality of ARToolkit for Unity for easier and more intuitive development of augmented reality applications. The advantages of ARToolkit are the wide possibilities of the tracking modes, both marker and markerless. It was possible to implement active camera marker based tracking, assuming the positions of the markers are known and they are registered with their virtual counterparts. We now have a suitable platform to develop applications for navigation in interiors with other tracked markers and objects attached to them. The accuracy of the algorithm needs to be improved, as the camera jitters a bit. The algorithm will be improved by using the visual tracking only as positional tracking. The rotation tracking should be more accurate when using gyroscopes. The tracking platform will then be suitable for visualisation of production machine state or for navigation in warehouses.

6. Acknowledgement

This paper is based upon work sponsored by project LO1502 Development of Regional Technological Institute.

7. References

[1] Yang, P.; Wu, W. & Moniri, M. (2007). A Hybrid Marker-Based Camera Tracking Approach in Augmented Reality, Proceedings of 2007 IEEE International Conference on Networking, Sensing and Control, pp. 622-627, London, DOI: 10.1109/ICNSC.2007.372851

[2] Liao G.; Hu Y. & Li Z. (2011). Research on marker tracking method for augmented reality, Fuzzy Systems and Knowledge Discovery (FSKD), Proceedings of 2011 Eighth International Conference on, pp. 2386-2390, Shanghai, DOI: 10.1109/FSKD.2011.6019986

[3] Agusta, G. M.; K. Hulliyah; Arini & Bahaweres R. B. (2012). QR Code Augmented Reality tracking with merging on conventional marker based Backpropagation neural network, Advanced Computer Science and Information Systems (ICACSIS), Proceedings of 2012 International Conference on, pp. 245-248, Depok, DOI: 10.1109/WMSVM.2010.52.

[4] Chen, L.; Peng, X.; Yao, J.; Qiguan, H.; Chen, C. & Ma, Y. (2016). Research on the augmented reality system without identification markers for home exhibition, Proceedings of 2016 11th International Conference on Computer Science & Education (ICCSE), pp. 524-528, Nagoya, Japan. DOI: 10.1109/ICCSE.2016.7581635

[5] Coffin, C.; Lee, C. & Höllerer, T. (2011). Evaluating the impact of recovery density on augmented reality tracking, Mixed and Augmented Reality (ISMAR), Proceedings of 2011 10th IEEE International Symposium on, 2011, pp. 93-101, Basel, DOI:10.1109/ISMAR.2011.6092374

[6] Hirzer, M. (2008). " Marker Detection for Augmented Reality Applications", Inst. For Computer Graphics and Vision Graz University of Technology Austria, Available from: http://studierstube.icg.tugraz.at/thesis/marker_detection.pdf Accessed: 2016-10-05

[7] Jun, J.; Yue, Q. & Qing Z. (2010). An Extended Marker-Based Tracking System for Augmented Reality, Modeling, Simulation and Visualization Methods (WMSVM), Proceedings of 2010 Second International Conference on, pp. 94-97, Sanya, DOI: 10.1109/WMSVM.2010.52.

[8] Yong Liu; Storring, M.; Moeslund, T. B.; Madsen, C. B. & Granum, E. (2003). Computer vision based head tracking from re-configurable 2D markers for AR, Mixed and Augmented Reality, Proceedings of The Second IEEE and ACM International Symposium on, 2003, pp. 264-267. DOI: 10.1109/ISMAR.2003.1240712.

[9] Piekarski, W.; Avery B.; Thomas, B. H. & Malbezin, P. (2004). Integrated head and hand tracking for indoor and outdoor augmented reality, Proceedings of IEEE, pp. 11-276. DOI: 10.1109/VR.2004.1310050

[10] Egnal, G.; Mintz, M. & Wildes, R. (2004) .A stereo confidence metric using single view imagery with comparison to five alternative approaches, Proceedings of Image and Vision computing, pp. 943-957, DOI: 10.1016/j.imavis.2004.03.018.

DOI: 10.2507/27th.daaam.proceedings.067

DIRECT DISPLACEMENT PROBLEM OF THE 6-DOF GOUGH-STEWART PLATFORM

Gebel Elena

This Publication has to be referred as: Gebel, E[lena] (2016). Direct Displacement Problem of the 6-DOFGough-Stewart Platform, Proceedings of the 27th DAAAM International Symposium, pp.0452-0456, B. Katalinic (Ed.), Published by DAAAM International, ISBN 978-3-902734-08-2, ISSN 1726-9679, Vienna, Austria
DOI: 10.2507/27th.daaam.proceedings.067

Abstract

In order to obtain direct displacement solutions of parallel manipulator without divergence in real time, a Newton-Raphson algorithm was proposed for direct displacement analysis of six-degree-of-freedom (DOF) Gough-Stewart platform. Based on geometrical frame of parallel mechanism, the highly nonlinear equations of kinematics were arranged into the homogeneous transformation matrix. The procedure of Newton-Raphson algorithm was programmed in Matlab/Simulink. The performance of the proposed algorithm for the 6-PSS parallel mechanism was analysed and confirmed. Applying Newton-Raphson algorithm, the real generalized pose of movable base is solved by using the given position of actuators.

Keywords: parallel mechanism; direct displacement problem; Newton-Raphson algorithm.

1. Introduction

With development of the manufacturing technology, robots are used in industrial applications where high accuracy, repeatability and stability of operations are required. Various parallel robots with six degree of freedom have been widely used in many fields including flight simulators, rehabilitation robotics and coordinate measuring devices [1]. Compared to other kinds of robots, the parallel robots have special kinematic characteristics because of their limited actuator displacements. As we usually assume that the number of bars of the parallel robot is equal to its degree of freedom, and only one actuator is needed for each bar [2]. Then based on the Chebyshev criterion, each bar of the 6-DOF parallel robot should be a 6-DOF serial kinematic chain. Once one prismatic joint and two spherical joints are applied in one bar, the only chain structure for 6-DOF is P-S-S (Fig. 1).

The workspace of 6-DOF parallel robot is hard to compute and express [3 – 7], because it is a complex six dimensional volume with non-linear boundaries. Direct kinematics is referred to the calculation technique of the movable base's position from the preset values of actuator joints. For the most 6- DOF parallel robot, the nonlinear direct kinematic equations can be solved by the algorisms using Newton method [4, 11], homotopy method [5], neural network [6] and algebraic elimination [7]. The singularity of 6-DOF parallel robot is a special mechanism pose where resultant displacement from actuators acted upon the movable base generates into a dimension less than six. The singularities usually lead the direct kinematic algorithm of 6-DOF parallel robot to converge into wrong solutions if improper initial value is chosen.

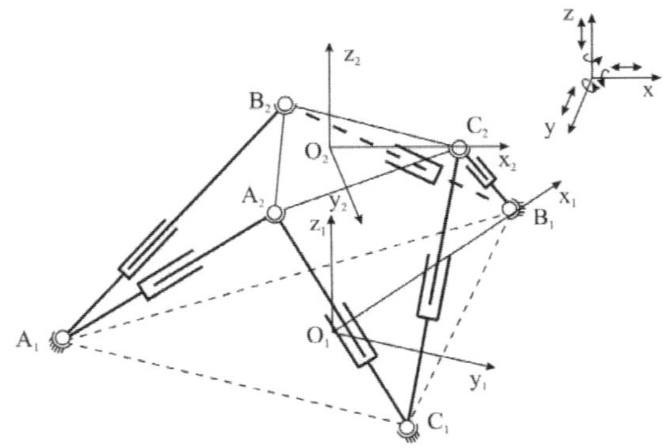

Fig. 1. Kinematic scheme of the 6-DOF parallel mechanism

This paper aims to obtain the common algorithm for the direct displacement analysis based on the Newton method and then the common kinematic constraints of 6-PSS parallel robot for a proper motion domain determination. Compared to the complex six-dimensional workspace, a proper motion range for the actuators is more convenient for application purpose.

2. Direct displacement problem

There are two kinds of the situation: firstly, at the beginning of work (calibration of the robot) and secondly, the real-time control of the parallel mechanism, when it is needed to solve the direct displacement problem. Mentioned above the task for a parallel mechanism is more complex than the inverse kinematics because the corresponding mathematical model can have a lot of possible solutions.

The direct displacement problem is referred to as the expression of the functional dependence $P = f(L)$ describing the ratio between the upper base position and given values of actuator joints. In common way the mathematical model of the direct displacement problem of 6-DOF Gough-Stewart platform can contain six (three angles and the coordinates of the end-effector) of nine (three coordinates of three points on the movable base) unknown variables.

Based on the principals of analytical algebra the equations of the holonomic stationary ideal link between the spherical joints of the upper and lower bases are expressed in the following way:

$$\begin{cases} l_4^2 = (x_{A1} - x_{A2})^2 + (y_{A1} - y_{A2})^2 + (z_{A1} - z_{A2})^2; \\ l_5^2 = (x_{A1} - x_{B2})^2 + (y_{A1} - y_{B2})^2 + (z_{A1} - z_{B2})^2; \\ l_6^2 = (x_{B1} - x_{B2})^2 + (y_{B1} - y_{B2})^2 + (z_{B1} - z_{B2})^2 \\ l_1^2 = (x_{B1} - x_{C2})^2 + (y_{B1} - y_{C2})^2 + (z_{B1} - z_{C2})^2; \\ l_2^2 = (x_{C1} - x_{C2})^2 + (y_{C1} - y_{C2})^2 + (z_{C1} - z_{C2})^2; \\ l_3^2 = (x_{C1} - x_{A2})^2 + (y_{C1} - y_{A2})^2 + (z_{C1} - z_{A2})^2. \end{cases} \quad (1)$$

Where distance l_i is equal to the magnitude of the co-directional vector B_1C_2 with the same bar of the parallel mechanism so $l_2 = \left|\overline{C_1C_2}\right|$, $l_3 = \left|\overline{C_1A_2}\right|$, $l_4 = \left|\overline{A_1A_2}\right|$, $l_5 = \left|\overline{A_1B_2}\right|$, $l_6 = \left|\overline{B_1B_2}\right|$.

Failing dependences for the calculation the movable base's position are found out by using the apparatus of analytical geometry in space:

$$\begin{cases} (x_{A2} - x_{B2})^2 + (y_{A2} - y_{B2})^2 + (z_{A2} - z_{B2})^2 = 3R_2^2; \\ (x_{A2} - x_{C2})^2 + (y_{A2} - y_{C2})^2 + (z_{A2} - z_{C2})^2 = 3R_2^2 \\ (x_{C2} - x_{B2})^2 + (y_{C2} - y_{B2})^2 + (z_{C2} - z_{B2})^2 = 3R_2^2. \end{cases} \quad (2)$$

The bases of the parallel mechanism are equilateral triangles $A_1B_1C_1$ and $A_2B_2C_2$ rotated to each other at the angle 60°. The vertices of the upper and lower basis are specified using by the magnitudes of the circle radii R_1 and R_2 which has been set before.

The resultant mathematical model is combined the formulae (1) and (2) and present the non-linear equation system which can be solved by well-known numerical methods.

For the more convenient mode of control of the studied mechanism, the results of the direct positional problem should be rearranged to the homogeneous matrix that is offered by J. Denavit and R. Hartenberg. The equation of the plane passing through three points namely vertexes of movable base $A_2B_2C_2$ are derived by:

$$\begin{vmatrix} x - x_{A2} & x_{B2} - x_{A2} & x_{C2} - x_{A2} \\ y - y_{A2} & y_{B2} - y_{A2} & y_{C2} - y_{A2} \\ z - z_{A2} & z_{B2} - z_{A2} & z_{C2} - z_{A2} \end{vmatrix} = 0 \tag{3}$$

Rearrange the expression (3) to the equation of the straight line in a space which is passed through the given point:

$$(x - x_{A2}) \cdot d_1 - (y - y_{A2}) \cdot d_2 + (z - z_{A2}) \cdot d_3 = 0 \tag{4}$$

Where variables d_i (i=1,...,3) are the coordinates of the normal vector and equal to the determinants of square matrices:

$$\begin{aligned} d_1 &= (y_{B2} - y_{A2}) \cdot (z_{C2} - z_{A2}) - (y_{C2} - y_{A2}) \cdot (z_{B2} - z_{A2}); \\ d_2 &= (x_{B2} - x_{A2}) \cdot (z_{C2} - z_{A2}) - (x_{C2} - x_{A2}) \cdot (z_{B2} - z_{A2}); \\ d_1 &= (x_{B2} - x_{A2}) \cdot (y_{C2} - y_{A2}) - (x_{C2} - x_{A2}) \cdot (y_{B2} - y_{A2}). \end{aligned} \tag{5}$$

Transfer vector $(x_{O2}\ y_{O2}\ z_{O2})^T$ in the matrix of the homogeneous transformation is found out by using the following formulae:

$$(x_{O2}\, y_{O2}\, z_{O2})^T = \left(\frac{x_{A2} + x_{B2} + x_{C2}}{3}, \frac{y_{A2} + y_{B2} + y_{C2}}{3}, \frac{z_{A2} + z_{B2} + z_{C2}}{3} \right)^T \tag{6}$$

The directional cosines of the normal vector $n = (d_1; -d_2; d_3)$ to the plane that is described through (4) is calculated as follows:

$$\begin{aligned} a_x &= \cos\alpha_{O2z} = \frac{d_1}{\sqrt{d_1^2 + d_2^2 + d_3^2}}; \\ a_y &= \cos\beta_{O2z} = \frac{-d_2}{\sqrt{d_1^2 + d_2^2 + d_3^2}}; \\ a_z &= \cos\gamma_{O2z} = \frac{d_3}{\sqrt{d_1^2 + d_2^2 + d_3^2}}. \end{aligned} \tag{7}$$

Elements s_x, s_y, s_z of the transformation matrix which are characterized the rotation around the axis O_2y are equal to the angle between the unit vector of O_2z axes and vector $\overrightarrow{O_2C_2}$

$$\begin{aligned} s_x &= \cos\alpha_{O2y} = \frac{x_{C2} - x_{O2}}{\sqrt{(x_{C2} - x_{O2})^2 + (y_{C2} - y_{O2})^2 + (z_{C2} - z_{O2})^2}}; \\ s_y &= \cos\beta_{O2y} = \frac{y_{C2} - y_{O2}}{\sqrt{(x_{C2} - x_{O2})^2 + (y_{C2} - y_{O2})^2 + (z_{C2} - z_{O2})^2}}; \\ s_z &= \cos\gamma_{O2y} = \frac{z_{C2} - z_{O2}}{\sqrt{(x_{C2} - x_{O2})^2 + (y_{C2} - y_{O2})^2 + (z_{C2} - z_{O2})^2}}. \end{aligned} \tag{8}$$

Constituents of the rotation angle around the O_2x axes are computed through the vector product of the unit – vectors of the O_2x- and O_2z- axes:

$$\begin{aligned} n_x &= \cos\alpha_{O2x} = s_y \cdot a_z - s_z \cdot a_y; \\ n_y &= \cos\beta_{O2x} = s_z \cdot a_x - s_x \cdot a_z; \\ n_z &= \cos\gamma_{O2x} = s_x \cdot a_y - s_y \cdot a_x; \end{aligned} \tag{9}$$

Combining (6) – (9), the matrix of homogeneous transformation is formed to rearrange the position of the end-effector from relative coordinate system $O_2x_2y_2z_2$ to absolute fixed system $O_1x_1y_1z_1$:

$$T = \begin{bmatrix} n_x & s_x & a_x & x_{O2} \\ n_y & s_y & a_y & y_{O2} \\ n_z & s_z & a_z & z_{O2} \\ 0 & 0 & 0 & 1 \end{bmatrix} \quad (10)$$

Thus, the mathematical model of the direct displacement problem is presented in form of homogeneous matrix (10). The solution of this task is carried out by using the different methods, for examples: the exceptions [7], the polynomial continuation [8], Grobner bases [9], interval analysis [10] etc. The first mention above the method gives no sustainable solution in numerical form since there are solution omissions and side roots. The second method is more stable than first. Using the last two methods excludes the problems mentioned above and a permit providing the given accuracy, though the time taken for its usage is more than it is needed for real-time control of a mechanism.

If priory information (initial approximation) for Gough-Stewart platform is known, the iterative Newton-Raphson or Gauss-Newton methods [4, 11] are more appropriate. For the studied mechanism the direct displacement algorithm is given as follows:

1. Input the desire actuators displacement l_i $(i = 1 \div 6)$, and calculate the corresponding T_i for the Gaugh-Stewart platform, choose the initial position:

$$L_0 = f(X_0). \quad (11)$$

2. Calculate a position correction through the above-mention nonlinear direct displacement algorithm:

$$X_1 = X_0 + A(L_1 - f(X_0)). \quad (12)$$

Where matrix A is obtained as $A = \left(\frac{\partial f}{\partial l_i}\right)^{-1}\Big|_{X_0}$.

3. If $\|f(p, \Delta p)\| \le \varepsilon_p$, then $p = p + \Delta p$; else back to step 2) for new Δp. Where $\| \ \|$ denotes the norms in corresponding parameters space.
4. If $\|f(p, \Delta p)\| \le \varepsilon_C$, then output the present p as the result; else back to step 2) for a new iteration.

3. Simulation of Gough -Stewart platform

Fig. 1 shows the 6-DOF parallel mechanism we used in this research. Based on the mechanism structure parameters and values of radii R_1 =100mm and R_2 = 25mm, the terminal coordinates in the corresponding coordinates system are presented in Table. 1.

	G_1 – Homogeneous coordinates of joints on the lower base in $O_1x_1y_1z_1$ frame, mm	**G_2 – Homogeneous coordinates of joints on the upper base in $O_2x_2y_2z_2$ frame / mm**
Chain 1. B_1C_2	(100, 0, 0, 0)	(-12.5, -21.651,0,0)
Chain 2. C_1C_2	(-50, -86.603, 0, 0)	(-12.5, -21.651,0,0)
Chain 3. C_1A_2	(-50, -86.603, 0, 0)	(-12.5, 21.651,0,0)
Chain 4. A_1A_2	(-50, 86.603, 0, 0)	(-12.5, 21.651,0,0)
Chain 5. A_1B_2	(-50, 86.603, 0, 0)	(25, 0, 0, 0,)
Chain 6. B_1B_2	(100, 0, 0, 0)	(25, 0, 0, 0,)

Table 1. Terminal coordinates for the 6-DOF parallel mechanism

The uniqueness problem for direct kinematic calculation is more complex, which can be transformed into an objective relation proof. In this research, we proposed a robust direct displacement algorithm to determine the uniqueness of direct kinematic solution by numerical verification.

Normally, the numerical direct displacement analysis methods like Newton-Raphson method [4, 11], Jacobi method [12] and Powell method [13] can achieve high accuracy in the parallel robot analysis when the initial pose is not far from the final result. All of these methods may fail to refresh the value of p when $\det(A) = 0$. Then the iteration can be trapped by a local minimum solution. Even the methods without Jacobian matrix (like Powell method) may be affected by the bifurcation or singularities.

Here presented Newton-Raphson method to obtain the direct displacement solution of the 6-PSS parallel mechanism. For researched Gough-Stewart platform, once the actuators output L is known, the whole kinematic information of G_1 is determined. Newton-Raphson method searches the solution from actuator workspace directly, which means all the temporary status appearing during the iteration can be realized by real 6-PSS parallel mechanism.

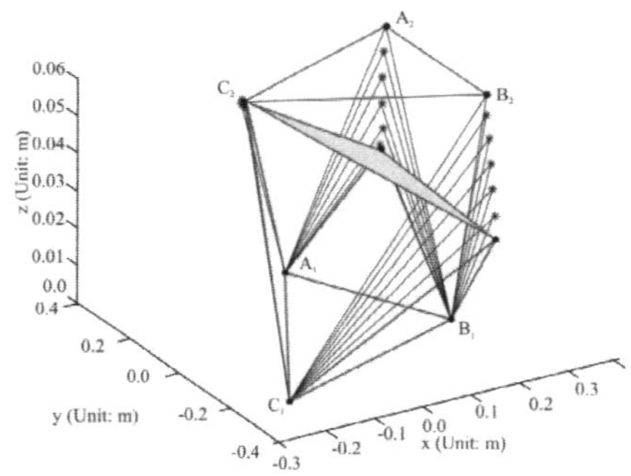

Fig .2. Procedure of direct displacement method

Fig. 2 shows the procedures of the direct kinematic method. The blue sketches are the initial status of the PSS parallel mechanism, the black sketches are the final status for the direct kinematic solutions. In Fig. 2, the numerical iteration steps are limited by penalty coefficient, and there is a maximum step limit for the modification of l_i

5. Conclusion

In this paper, a through direct displacement analysis of 6-PSS parallel mechanism has been experienced. The method for direct displacement problem for the 6-DOF Gough-Stewart platform is proposed respectively. The simulation results validate the effectiveness of the proposed method.

In the future work, the proposed method will be applied for available workspace determination, singularity analysis and bifurcation avoidance.

6. References

[1] Shilong Zeng, et al. (2015) "Kinematics Analysis and Verification on the Novel Reconfigurable Ankle Rehabilitation Robot Based on Parallel Mechanism"(Series: New Tends in Mechanism and Machine Science). Vol. 24, Springer (2015), pp.195 – 202.

[2] Rui Zeng, et al. (2015) "Determination of the proper motion range for the rotary actuators of 6-RSS parallel robot" CCToMM Mechanism, Machines and Mechatronics (M3) Symposium.

[3] Ndrioaia, D., et al. (2012) "Determining the Workspace in Case of the Robots with Parallel Structure Delta 3DOF" In "Proc. of the 23rd International DAAAM Symposium", Vol. 23, No.1, pp. 1123-1126

[4] Parrish, R., Dieudonné, E. and Bardusch, R. (1972) "An actuator extension transformation for a motion simulator and an inverse transformation applying Newton-Raphson's method". NASA Technical report D-7067.

[5] Mu, Z.L. and Kazerounian, K. (2002) "A real parameter continuation method for complete solution of forward position analysis of the general Stewart." Journal of Mechanical Design, Vol. 124, No. 2, pp. 236–244.

[6] Ramesh Kumar, P. and Bandyopadhyay, B. (2013) "The forward kinematic modelling of a Stewart platform using NLARX model with wavelet network" In "Proc. Of 2013 11th IEEE International Conference on Industrial Informatics", pp. 343 – 348, IEEE, Bochum, Germany.

[7] Huang, X.G. et al. (2007) "Forward kinematics of the 6-6 Stewart platform with planar base using algebraic elimination" In "Proc. Of 2007 IEEE International Conference of Automation and Logistics", pp. 2655 – 2659, IEEE, China.

[8] Sommese, A.J. (2005) "Numerical solution of polynomial systems arising in engineering and science", Singapore: WorldScience, p. 401.

[9] Gan, D. et al. (2009) "Forward displacement analysis of a new 1CCC-5SPS parallel mechanism using Grobner theory", Journal of Mechanical engineering science, Vol. 223, No. 5, pp.1233-1241.

[10] Merlet, J.-P. (2004) "Solving the Forward kinematics of a Gough-Type Parallel manipulator with interval analysis', International Journal of Robotics Research, pp. 221 – 235.

[11] Cent, J. (2010) "Forward kinematic analysis of parallel manipulator using global Newton-Raphson method" (Springer, South Univ. Technol.), Vol. 17., pp. 1264-1270.

[12] Wilkinson, J.H., Wilkinson, J.H. and Wilkinson, J.H. (1965) "The algebraic eigenvalue problem", Vol. 87. Clarendon Press Oxford.

[13] Powell, M.J. (1970) "A hybrid method for nonlinear equations." Numerical methods for nonlinear algebraic equations, Vol. 7, pp. 87–114.

DOI: 10.2507/27th.daaam.proceedings.068

IMPLEMENTATION OF CLOUD SERVICES FOR ADVANCE MANAGEMENT OF STEEL TRANSPORT FOR CONTINUOUS CASTING PRODUCTION

Nikita Kudriashov, Ivan Protasov, Sergey Markov, Vyacheslav Potekhin, Yulia Yadgarova, & Victor Taratukhin

This Publication has to be referred as: Kudriashov, N[ikita]; Protasov, I[van]; Markov, S[ergei]; Potekhin, V[yacheslav]; Yadgarova, Y[ulia] & Taratukhin, V[ictor] (2016). Implementation of Cloud Services for Advance Management of Steel Transport for Continuous Casting Production, Proceedings of the 27th DAAAM International Symposium, pp.0457-0462, B. Katalinic (Ed.), Published by DAAAM International, ISBN 978-3-902734-08-2, ISSN 1726-9679, Vienna, Austria
DOI: 10.2507/27th.daaam.proceedings.068

Abstract

Nowadays, manufacturing is on the threshold of the fourth industrial revolution. It means the increase of enterprise productivity requirements. To improve the productivity, factories have to implement advanced analytical algorithms. It means the increasing demand for big data technology and cloud services for modern manufacturing systems. Therefore, different technologies operating information from Web as well as from field level and top level manufacture information systems should offer high degree of interoperability. For instance, the implementation of Industry 4.0 concepts technologies will be presented on the example of steel transportation for continuous casting production, to access benefits unavailable before.
This paper refers to the problems of Cloud Services and Big Data usage in order to enhance productivity of steel transportation system on metallurgical plant.

Keywords: Industry 4.0; Manufacture Execution system; Intelligent control systems; Automation; Cloud services; Casting production

1. Introduction

Due to unprecedented development of IT technologies manufactures new opportunities of increasing production efficiency are now possible. Traditional approach to manufacture management is struggling to keep expected outcome. Contemporary realities require production with higher efficiency, quality, and output rates at lower prices in the shortest time possible. Otherwise, there is a high risk of losing orders due to the high-level competition in the market.

Metallurgical industry is one of the most important part of economy [1], therefore it cannot stand apart from the trend which was mentioned above. The question is how can new technologies be implemented for better productivity and resource efficiency for steel transportation system in particular?

Steel transportation for continuous casting production is a system of cranes and rail transporters, which deliver melted steel in ladles for next preparations and casting. A huge machine works following rigid technological process algorithm.

Involvement to the well-established technological process is unnecessary because of large-scale production, low line flexibility and high level of safety is required. Nevertheless, numerous equipment, which is involved in technological processes, opens great opportunities for maintenance optimisation. Regulations of maintenance, which are able to be more flexible and adaptive, depends on current state on production lines and experience of exploitation in such condition.

To realize the idea of increasing production efficiency by reduction of equipment downtime, we can implement Industry 4.0 concept and IT world technologies[2]. By using Industrial Internet of Things (IIoT) services, we can collect and store data from any devices from production line that provides comprehensive information about production state. The increased data flow can be stored and quickly analysed by Big Data cloud services for statistics management and prognosis of faults. Moreover, implementation of Information Technologies such as machine vision, pattern recognition makes new information about technological process available that was not accessible before. In more detail, about this technologies we will talk further.

2. Critical overview of works on this field

First, it is necessary to look at the set of work on this field. As steel mill ladles have relatively short life cycle (on the average, one lining has to be repaired any 50-70 pouring), it is important to analyze its condition. Otherwise, if do not do any necessary repairing procedures, metal breakout will occur. It entails such problems as equipment destructions and, as the result, huge financial losses. The described problem opens a number of possibilities for engineers to make their researches and projects.

As an example, there is a set of works, connected with steel mill ladles lining analysis using IR cameras. All of them has many common features. This approach for ladle monitoring is to place its IR video cameras in robust housings, and locate them in fixed positions where they have a clear view of a filled ladle as it passes by on the gantry crane. Those cameras capture video images of the ladle in real time, and calculate temperatures over the ladle's surface. A few cameras can cover the entire surface of the ladle, but typically, a few relevant areas are predefined for analysis. [3]

Often these solutions differs in the number of IR cameras and hardware configuration. On the Figure 1 the descriptions of three implementations are shown [4]

Another state is that none of this systems store data about previous conditions of ladles. It means that algorithm only analyze the current state of the ladle and can only make the alarm go off, if metal critical overheating has been detected. On the other hand, it shows the Areas on the ladle, where the temperature indicate excessive wear or a defect are highlighted with alarm colours in the on-screen display. This way, the operator gets the location of a potential problem with the refractory [3]. However, we need an experienced operator to analyse the current state of ladle.

Often, this recourse is enough for metal breakout prevention. However, it would be better if we could analyse state of all ladles in any number of pouring. It could implement not only prevention of alarm situation, but plan the usage of current ladle, according to collected data about its specifications.

Big companies work on improving manufacturing processes, considering ideas of Industry 4.0. At Hannover Messe 2014 implementation of smart services of machine maintenance was demonstrated. It was demonstrated in context of the pump unit interfaces connection to Cloud software, which is enabling early prognosis of system errors [6]. In this paper we expand the machine data to cloud connectivity idea for steel transportation system for continuous casting production. It is a complex task to consolidate specific data from IR-cameras and other sensors with cloud services and develop algorithms for reducing amount of unexpected transport system downtimes as well as lowering the maintenance cost. To achieve this aim we have to operate the stored big data quickly which is possible by using cloud technologies.

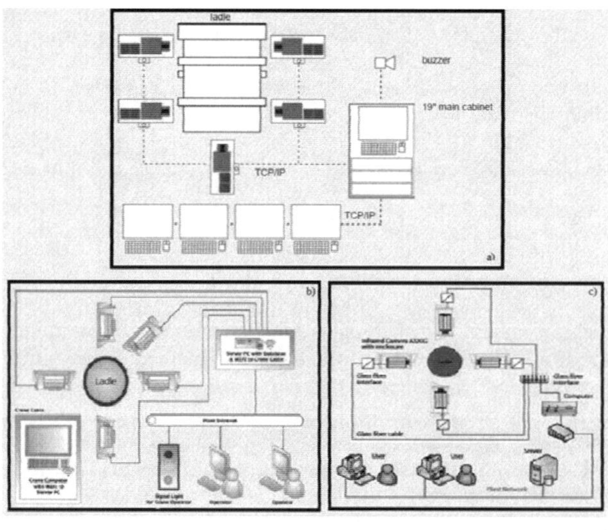

Fig. 1. a – IntraTec [5] b - FLIR solution [3] c - LadlChak [4]

3. System description

Developing system inherits some of the main principals of previous works. However, following issues has to be improved:

- Selection of optimal configuration parameters (number of IR cameras, hardware, connections)
- Software improvement, according to new configuration of the system
- Secondary appointment of the system - multifunctionality, for covering more areas of interest (for example we can reduce speed-measuring sensor by implying machine vision algorithms on IR thermogram).
- Inculcation of Industry 4.0 concepts to implement the system, which can make the decision, based on its experience (Big Data, Neural networks, machine learning algorithms, Cloud services)

Firstly, let us start from the beginning, first question is, how many cameras to chose. There are systems with four cameras (one for each side), and with five cameras (each side and bottom.). To reduce the price of the system, we can reduce one of the cameras, Figure 2.

There are several problems to consider:

- Reducing of one camera entails requirements for its higher resolution. Would it be cheaper?
- Will the quality be good enough to make algorithm work correctly? (On Figure 2 red areas – is the areas of intersection of cameras observation, so the reducing of excess data is needed)
- How accurate would be our system? (Figure 3 – an example of accuracy evaluation for Optris IR Camera)

Therefore, the first step the optimisation task have to be solved. It depends on a huge set of parameters connected with not only camera features, but with specification of current ladle. The optimization goal is to find the cheapest configuration with high value of accuracy and algorithm working time.

Second step of this work is to implement an algorithms, according to chosen configuration of the system. The main requirements to these algorithms:

- Reduce excess information about ladle surface (areas "a" on Figure 2). This goal can be implemented by using such machine vision algorithms as pattern matching, variational methods and contour matching algorithms. In the end, it has to one scan of whole ladle surface.
- Centralized storing of processed ladle surface images. The amount of storing data will expend during the iterations technical process. Big Data technologies, is the only way to solve this problem.
- Decision making algorithms. There are several variants of algorithms for such system. The first one is neural network, which can make a decision, according to its experience,obtained by teaching algorithm. Another way is to use the combination of simple optimization algorithms.

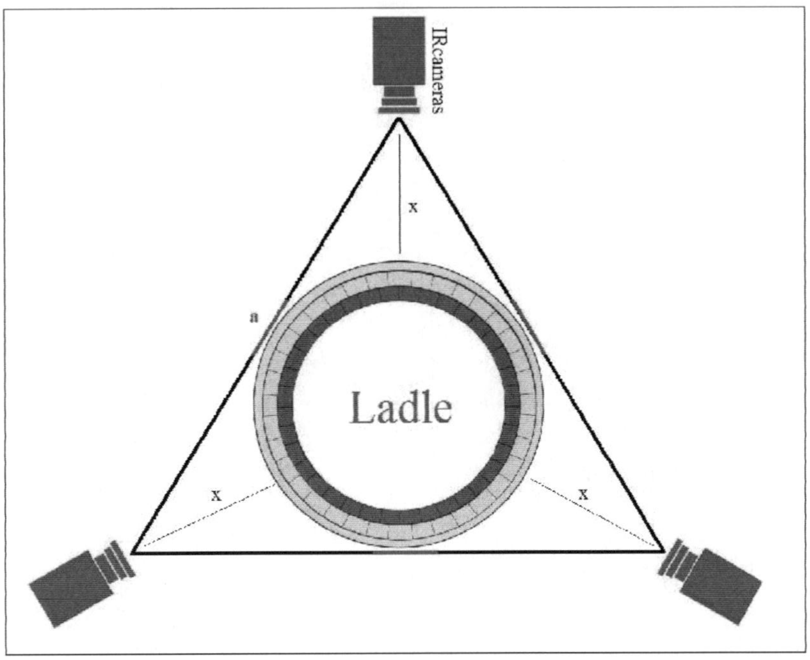

Fig. 2. Three IR camaras configuration

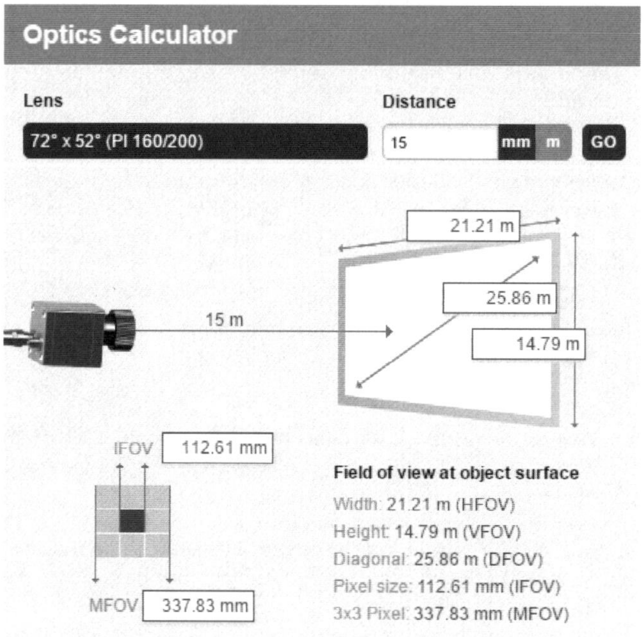

Fig. 3. Parameters evaluation for Optris IR Camera

Thirdly, as it has been said, secondary appointment of the system is multifunctionality. To control steel transport for continuous casting production such parameter as ladle speed has to be measured. Simple contact methods has many limitations. Usage IR Camera can reduce this trivial methods and, by using machine vision algorithms, measure ladle speed according to thermogram[7]. Speed can be measured by using simple tracking algorithms, which takes the most distinctive parts of ladle, and tracks their position during the period of time, then speed values averages, Figure 4.

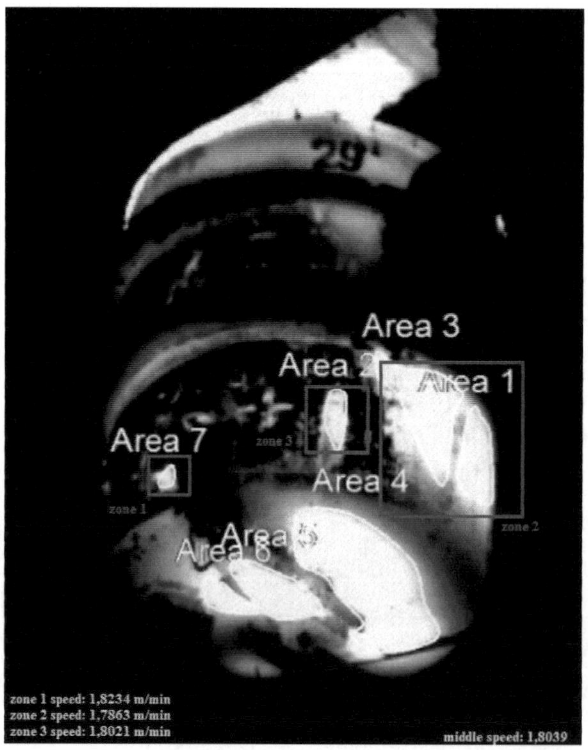

Fig. 4. Tracking of the most distinctive areas of ladle

Finally, implementation of the IT technologies development achievements can significantly expand functionality of the steel transport control system for continuous casting production. Previous system can only answer the question: is this steel ladle have a latent defect or not? The answer is not provide sufficient information for organizing timely supported maintenance schedule that is extremely important in large-scale production of steel. Resource of the ladles depends on differences parameters, such as construction of the ladle, type of produced steel, technological process, composition of slag and it's properties and many others. If we keep statics of the technological process parameters, we can analyze it and make some prognosis about resource of the ladle. Furthermore, the more statistics data we obtain the more accurate prognosis of the ladle resource we have. Based on this analysis we can plan of maintenance more optimal by ladles resource parameter. Implementation of the proposed idea is reflected in the architecture shown in Figure 5.

The necessary data collected from the IR-cameras analysis system and other shop floor sensors by Manufacture Execution System. Then, MES send the data to the distributed cloud storage. After storing and processing of the received data, the Cloud send to MES optimal plan of equipment maintenance. Using of the cloud storage has the following advantages[8]:

- Availability of relevant information from any plant of metallurgical company,
- Special hardware and software architecture for high performance data operations,
- Using powerful cloud Big Data analytical tools.

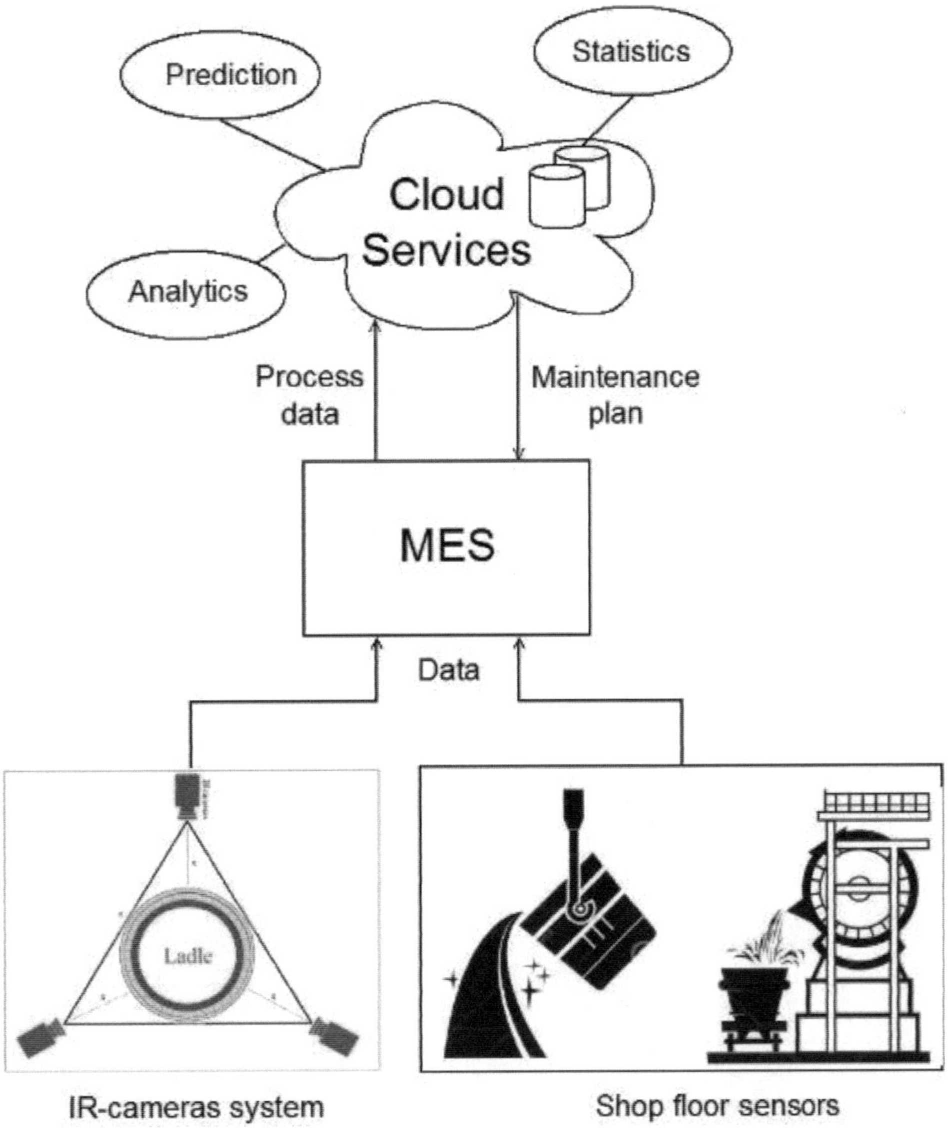

Fig. 5. Architecture of the proposed system

4. Conclusion and outlook

The aim of this work is to propose the implementation of the cloud technologies for steel production costs optimization. According the research, we suggest to reduce cost by optimization of maintenance processes using an example of steel transportation for continuous casting production. As a result we suggest an architecture of equipment state control system which is based on implementation of cloud technologies. Additionally, we consider using of machine vision with IR-cameras for comprehensive analysis of ladle working resource and opening issue of camera quality and quantity, and composition. Furthermore, by using of IR-cameras we can measure ladle speed more precisely by tracking algorithms that will increase functionality of the system. Taking implementation of the system into consideration we have derived main requirements for ladle state recognition algorithm. Through research, we observe the clear trend of convergence of shop floor data and Cloud Services [9]. In this paper we demonstrate how to implement modern trends and technologies in order to improve the efficiency of enterprises.

Beyond the scope of this research there is outstanding issues of IR-cameras control system configuration, implementation of machine vision algorithm, development of analytical algorithms and generation of maintenance schedule, elaboration of data collection. Our future research shall focus on issues mentioned above for further architecture development of factory of future.

5. References

[1] http://tpprf.ru/download.php?GET=6LPAY%2F81BmxbyHP1wg7B0g%3D%3D, Chamber of Commerce and Industry of the Russian Federation, Analytical Report of the State of the Steel Iindustry, Accessed on: 2016-08-30

[2] Yerofeyev S.A., Ipatov O.A., Markov S.A., Potekhin V.V., Sulerova A.S. & Shkodyrev V.P. (2015) Adaptive Intelligent Manufacturing Control System, Proceedings; DAAAM International Symposium of Intelligent Manufacturing and Automation, DAAAM 2015; Vienna; Austria; Volume 26, No. 1, ISSN 2304-1382

[3] http://www.flir.com/uploadedFiles/CS_EMEA/Application_Stories/Media/Downloads/Steel_EN.pdf. Mill Ladle Refractory Monitoring, FLIR solution series, Accessed on: 2016-08-30

[4] LadleCheck. Thermal Imaging System for a Reliable. Condition Monitoring of the Refractory Lining, Automation Technology

[5] Ladle Hot Spot Detection. Thermography-based Ladle Monitoring system. InfraTec.

[6] https://news.sap.com/industry-4-0-two-examples-future-factory/, Industry 4.0: Two Examples for the Factory of the Future, Accessed on: 2016-08-30

[7] Kudriashov N.S., Markov S.A. (2016) Multifunctional diagnostics and process control systems based on Infrared thermography methods and machine vision algorithms, Automated System and Technologies, unpublished.

[8] Gastermann, B., Stopper, M., Kossik, A., Katalinic, B. Secure Implementation of an On-Premises Cloud Storage Service for Small and Medium-sized Enterprises, Proceedings of DAAAM International Symposium on Intelligent Manufacturing and Automation, DAAAM 2014; Vienna; Austria; 26 November 2014 through 29 November 2014, Volume 100, Issue C, 2015, Pages 574-583.

[9] Protasov I.I., Yadgarova Y.V., Potekhin V.V. & Taratukhin V. (2016) Implementation of Cloud Services for Smart Manufacture, Automated System and Technologies, unpublished

DOI: 10.2507/27th.daaam.proceedings.069

CONDITIONAL CONSTRUCTIONS IN NON-PROCEDURAL LANGUAGES OF ROBOTIC SYSTEMS

Ipatov Oleg, Vasilevsky Alexander, Rudskoy Andrey & Yamschikov Yury

This Publication has to be referred as: Ipatov, O[leg]; Vasilevsky, A[lexander]; Rudskoy, A[ndrei] & Yamschikov, Y[ury] (2016). Conditional Constructions in Non-Procedural Languages of Robotic Systems, Proceedings of the 27th DAAAM International Symposium, pp.0463-0466, B. Katalinic (Ed.), Published by DAAAM International, ISBN 978-3-902734-08-2, ISSN 1726-9679, Vienna, Austria
DOI: 10.2507/27th.daaam.proceedings.069

Abstract

The article researched and developed a universal theoretical concept of conditional constructions in non-procedural languages of different robotic systems. A multivariate classification of conditional constructions of non-procedural languages of robotic systems is offered. It is shown that the presented concept allows conditional constructions to be described in a form which allows to present them explicitly as they are common, especially their semantic properties and their typological primarily formal differences. A well-known thesis is discussed in the paper. It says that the conditional non-procedural design language fairly accurately reflects the logical operation of implication, and the notion of material implication is the source for the semantic definition of conditional constructions. A methodological basis of semantic research, pragmatic and communication properties of conditionals in the non-procedural languages of robotic systems is offered. A distinction of conditional constructions in non-procedural languages robotic systems into even and odd patterns is also researched.

Keywords: conditional constructions; non-procedural languages; robotic systems; cognitive processes; knowledge; judgment; methodology; concept; analysis.

1. Introduction

The interest to the problem of conditional constructions in robotic systems can be explained by the following circumstances.

At first. Taking into account the attention to conditional constructions, it is possible to conclude that they are still not enough studied from the universal typological position [6]. There are no any studies based on a correct theoretical information providing a detailed description of the conditional constructions in non-procedural languages of the robotic systems of different structures.

Secondly. Until now the initial setting of research was a well-known thesis that the conditional constructions reflect the logical operation of implication quite accurately. The notion of the material implication is the reference for the semantic definition of conditional constructions. The preliminary results show that the reliance on the logical tradition in the description and interpretation of conventional constructions requires substantial correction [2].

The implicative judgments tested by the logic, constitute only a part of structures that are called conventional. The content of the propositions in the statement as well as the presence / absence of the meaningful communication between these propositions is irrelevant for the logic [7]. The robotic system requires a semantic relationship between propositions, denoted in the statements. To establish how to use implicative judgments on the basis of new knowledge, it is necessary to identify and describe the conditional constructs like an object, with its own meaning and formal specificity.
Thirdly. Until now the main object of the study were bi-predicative constructions in the field of robotic systems. Even a rigorous mathematical proof is much shorter than the reasoning of orders formalized in predicate style [12]. In this regard, it is necessary to have the effective methods of investigation in relation to bi-predicative constructions.

2. The conditional structures in procedural languages of robotic systems

A necessary condition for the creation of robotic systems that would be able to replace a human in hazardous conditions is developing means of intelligent control and autonomous navigation providing the possibility of making independent decisions [11]. So it is very important to develop theoretical concepts, which allow to describe the conditional constructions of non-procedural languages of robotic systems. This provides the ability to represent both their common, first and foremost, semantic properties and their typological primarily formal observed differences explicitly. It demands the creation of the multi-dimensional classification of the calculated conditional constructions and exploring their prototypical and peripheral types, as well as the development of a universal metric of the conditional constructions combining different sets of issues and their characteristics.

The robot acquires information from the environment through its sensors, which provides the input signals to the controller [14]. The understanding of the conceptual and behavioural organization of intellectual capacity to design and interpret conventional constructions ensures comprehension of cognitive processes, linguistic competence and outlet strategies of the robotic systems. There are some studies which cover the classification of prototypical contingent of relevant constructions, their formal and semantic features. The contingent constructions, which in formal terms differ from the prototypical, yet remain unremarkable for the researchers. According to the other popular studies, it should be noted that the problem of synchro-diachronic semantic and syntactic relations of the conditional constructions with other language constructions, especially with time, concessive or causal, still exist [10]. The conditional constructions, as well as the causal, concessive constructions, and the constructions of value and purpose have the total value of the conditionality. The conditional structures are closely related to the relative, interrogative, and some other structures. Formal differences between conditional structures and the others can be quite shaky, whereas the standard measures of conditional values in certain contexts express different meanings. There are also situations where the construction is in principle ambiguous. The construction can allow both the conditional and temporary reading, the difference between them is a little.

3. Some problems of the universal typological analysis of conventional constructions

During the universal typological analysis of conventional constructions there can appear a number of significant problems:

- It is necessary to establish whether the concept of "default value" is undetectable, or it is possible to offer a universal definition that can reliably restrict the empirical data of various non-procedural languages of the robotic systems corresponding to this concept.
- Those substantive parameters that are necessary and sufficient for the construction of a universal pragmatic expedient enumerative classification of the conditional constructions should be identified.
- It is necessary to build a calculated classification of the conditional structures and determine which of the logical possibilities provided by the calculation are implemented in the language of robotic systems. In addition, it is necessary to find the causes of the established distribution of logical possibilities.
- It is necessary to establish and explore the semantic, pragmatic and communicative features of the conventional constructions, determining the preferred use of certain calculation models in the construction of discourse.
- It is necessary to determine which formal indicators and their aggregates, are directly or indirectly used in the non-procedural languages of robotic systems for the labeling of conventional structures and their different semantic types.
- It is necessary to determine which concepts are semantically close to the concept of "default value". It can determine the possibility of using the same formal indicators for the labeling of all semantically close concepts. As a result, it will be possible to determine the certain place of the conditional constructions in the the system of other constructions which are close to each other semantically and formally.
- There should be a metric which will be used for a description of conditional constructions in different languages of therobotic systems.

The methodological concept of solving the problems mentioned above is available in the following content.
The initial setting of semantic research, pragmatic and communicative properties of the conventional constructions is based on the well-known thesis that the conditional construction of the non-procedural language fairly accurately reflects the logical operation of implication. The notion of material implication is the source for the semantic definition of the conditional constructions. The semantics of conditionality is as close as possible to the content of logical and grammatical category of a functional compliance or a function.

In the study of formal syntactic types of conditional constructions it is necessary:

- to determine which empirical facts oft he non-procedural languages of the robotic systems correspond to the concept of "default value" semantically;
- to analyze the formal-syntactic heterogeneity oft he proposals, including the conditional construction;
- to develop metrics for the parameters oft he classification for such proposals.

Among these parameters area formal separability / non-separability of the unextractable parts of conventional structures; the presence / absence of a formal connection between the parts of the conditional structure; a status of the indicator of the communication [4].
The conventional concept must be seen as a universal semantic primitive. All attempts to define a conditional concept containing in the canonical conditional sentences, turned out to be unsuccessful because the definitions have such conceptions as the hypothetical conclusions. The possible worlds that are more complex comparing with the conventional concept need to have the independent definition either.

It should be noted that in the logic of the concept of connection does not appear in the analysis of the implications, which appears in the non-procedural languages of the conventional structures. Some experts in the field of robotic systems proved that the table of material implication loses communication idea, which is present in the case of if-then [1]. It is necessary to examine whether the relationship between the proposition-condition and the proposition-the result is casual, random or conditional. It is necessary to prove that the concept of "notional connection" is fundamental. It can not be reduced to any other concepts as it is a conceptual entity itself.

The conventional ctructures are a complex object, possessing a variety of specific, substantive and formal features. They can be subjected to a number of independent parameters. Some of these parameters are relevant for both conventional structures and other types of structures. Some of them - only for conventional structures. The number of small variations with respect to the basic decision determines the size of the search space [13]. Therefore, each of the independent features can be considered to be a basis for the classification of conventional constructions. Thus, it is possible to offer as many classifications as classification parameters [9].

When it comes to multi-dimensional classification enumerative it is important to pay attention on the selection criteria for this classification parameters, ie, how many and which parameters should include this classification. It seems that there is only one common requirement, which must be demanded from the typologically oriented enumerative classifications. All parameters of such classifications should be semantic. This requirement requires no explanation because only semantics but not a form can be a common component of a wide variety of languages. Only after the building of the universal semantic models of the conditional constructions it is possible to move on to the question concerning the formal means of these models and how they are translated into specific non-procedural languages, and how to classify the languages, depending on what kind of formal tools they use to express the certain universal semantic models of conventional constructions.

From the systemic point of view, all the researched models are logically equivalent for its semantic features, its pragmatic preferences and limitations, and their representation in the texts. These models are noticeably different from each other. It ultimately depends on the specific combinations of the classification parameter values that define the personality of each model [3].

The most important is the distinction between the real conditional constructions (odd models) and unreal conditional constructions (even models) [5]. However, for the realization of the value of both real and unreal conditional constructions the important role is played by this or that specific conditions and time localizations (at least) of the investigation. According to this, all realized models should be divided into three groups: the first group includes those models, in which the condition is localized in the future; the second group consists of those models, in which the condition is localized in the past; the third group consists of those models, in which the condition is localized in the present. The relationships between the cause and the effect influence the representation of the models in the. [8] Thus implemented models can be divided into two groups: the first group includes models in which the condition precedes the effect; the second group consists of models in which the condition ist he same as the investigation. So it is necessary first of all to get semantic-pragmatic characteristics of real conditionals.

4. Conclusion

The importance of researching the problem of conditional constructions in non-procedural languages of the robotic systems is clear. These constructions are the key, giving the opportunity to explore and describe the thought processes of robotic systems. Their specificity is proved by a direct reflection of the cognitive ability to reflect the alternative judgments and draw conclusions, based on incomplete and insufficient information to suggest possible correlations between situations and to understand how the world would could change if the correlations were different.

5. References

[1] Kasevich VB The semantics of the conventions and convention semantics. // Proceedings of the international conference. Vol.1, M. 1995.

[2] Apresyan YD Lexical Semantics. // St. Petersburg, publishing house "Nauka", 1998.

[3] G.T. Stump. The Semantic Variability of Absolute Constructions // Reidel: Dordrecht, 2001.

[4] Fedulaev FG, Karpechenko PA, OA Nikolaev, Jirov VS Inference rules of predicate logic. St. Petersburg, publishing house SPGU 2002.

[5] Bondarenko VK, Fedulaev FG Knowledge representation rules and the conclusion. Abstracts of the second All-Russia conference "Representation and use of knowledge", p. 35, Kazan, 2003.

[6] Nikoltsev VA Korzhavin GA, Podoplekin YF, Vasilevsky AS Calculated classification conditional constructions declarative languages like. Scientific and practical journal "Information and Control Systems". Electronics industry. St. Petersburg, 2006, p. 30-32.

[7] AS Vasilevsky, Podoplekin YF, Lapshin KV Methodology problems of conditional constructions in the systems of presentation and use of knowledge. // Proceedings of the Tenth National Conference on Artificial.

[8] G.F. Luger. Artificial Intelligence: Structures and Strategies for Complex Problem-Solving (6th Edition) // Addison Wesley, 2008.

[9] Alexander Vasilevsky, Lapshin KV Multidimensional classification calculated conditional constructions in non-procedural languages intelligent systems // Proceedings of the Twelfth National Conference on Artificial Intelligence with international participation KII 2010, Volume 1, M., FIZMATLIT 2010.

[10] VV Gorelov, Lapshin KV Research sintatiko morphological problems of conditional constructions in non-procedural languages, knowledge processing systems. // NA-tehn. Sat. FSPC JSC "Concern" Granit-Electron "," Ship and aircraft multi information management systems. " // Vol. №16, p. 11-17, 2011.

[11] Smirnova, E. Stepanov, D. & Goryunov, V. (2016). A Technique of Natural Visual Landmarks Detection and Description for Mobile Robot Cognitive Navigation, Proceedings of the 26th DAAAM International Symposium, pp.0905-0911, B. Katalinic (Ed.), Published by DAAAM International, ISBN 978-3-902734-07-5, ISSN 1726-9679, Vienna, Austria, DOI: 10.2507/26th.daaam.proceedings.126.

[12] Kuznetsov OP Cognitive semantics and artificial intelligence // Artificial intelligence and decision-making, the Federal Research Centre "Information and Management" of the Russian Academy of Sciences, 2012. № 4. S. 32-42.

[13] Diveev AI Shmalko EY Numerical synthesis of group control system of robots by symbolic regression. // Proceedings of the SFU. Engineering, number 10 (171). Pp 29-45, 2015.

[14] Nehmzow U., Akanyeti O., Billings S.A. Towards modelling complex robot training tasks through system identification // Robotics and Autonomous Systems, V. 58, №3. PP. 265-275, 2010.

DOI: 10.2507/27th.daaam.proceedings.070

MICROSTRUCTURE OF MS1 MARAGING STEEL IN 3D-PRINTED PRODUCTS AFTER SEMI-SOLID PROCESSING

Kateřina Rubešová, Štěpán Jeníček, Josef Káňa & Ivana Zetková

This Publication has to be referred as: Rubesova, K[aterina]; Jenicek, S[tepan]; Kana, J[osef] & Zetkova, I[vana] (2016). Microstructure of MS1 Maraging Steel in 3D-Printed Products After Semi-Solid Processing, Proceedings of the 27th DAAAM International Symposium, pp.0467-0472, B. Katalinic (Ed.), Published by DAAAM International, ISBN 978-3-902734-08-2, ISSN 1726-9679, Vienna, Austria
DOI: 10.2507/27th.daaam.proceedings.070

Abstract

The importance of 3D printing in today's industry is continuously increasing. One of the materials which are frequently used in this process is the precipitation-hardenable MS1 Maraging steel whose properties after conventional treatment have been thoroughly mapped. Maraging steels belong to the strongest homogeneous steel materials ever. Their alloying additions include cobalt, molybdenum, titanium and aluminium, and their unusual properties make them candidates for heavy-duty parts operating predominantly under dynamic loading. Semi-solid processing of this particular steel enables near-net shape products to be made using a single production step. Furthermore, unconventional microstructure and physical, chemical and mechanical properties can be obtained by this process. The semi-solid processing route and the mechanical properties achieved by it can expand the potential uses of this material. Semi-solid processing of small volumes of material has already been mastered with high-alloy steels which offer a wide range of forming temperatures. However, as this process is technically demanding, its use for medium and low-alloy ultra-high-strength steels has not been explored in greater detail yet. In the present experiment, MS1 Maraging steel was studied. The experimental feedstock was made by 3D printing. The initial set of semi-solid processing parameters was proposed on the basis of calculation and then tested experimentally. In this article, description of the microstructures of this steel before and after several semi-solid processing routes is given, based on examinations using optical and electron microscopy.

Keywords: Key words: semi-solid processing; rapid solidification; near net shape; maraging steel; 3D printing

1. Introduction

Although the principle of thixoforming has been known since the early 1970s, the process is still ranked among unconventional metalworking methods. It relies on the thixotropic behaviour which occurs in materials heated into the semi-solid state, i.e. to a temperature between their solidus and liquidus. Thixoforming belongs to precision casting and forming methods. During the forming process, shear forces act on the partially melted material and cause its viscosity to decrease significantly. This effect is known as thixotropy [1]. The liquid fraction in the workpiece being formed is normally between 10 and 60 % [2]. The technique was initially developed mainly for low-melting materials: aluminium, magnesium and tin alloys [3]. Thanks to intensive research, commercial machines for the thixoforming of low-melting alloys are already (though scarcely) in use. Yet, when it comes to the thixoforming of steels, basic research is lagging behind due to technological complexity of the issue. It is unfortunate, considering the vast potential of this technology. Nevertheless, semi-solid forming enables difficult-to-form materials to be worked by means of a single forming operation. In addition, more complex and intricate shapes can be produced by this technique than

by conventional forming routes. In addition, unconventional microstructures can be produced by thixoforming even in ordinary materials.

The research into the use of steels has made less progress than with low-melting alloys because it faced substantial technological difficulties. Among them are the high process temperatures, narrow temperature intervals, and the ensuing need for strict control of the temperature field [4]. The only detailed description of this process available so far in literature concerns high-alloy steels. Up to now, no comprehensive research into the behaviour of low-alloy medium-carbon and low-carbon steels has been undertaken – which is the focus of the present experiment.

Its objective was to explore the capabilities of mini-thixoforming of 3D-printed steel workpieces, and obtain the desired type of microstructure by the process. Mini-thixoforming enables near-net-shape products to be obtained. These are geometrically accurate and require only minimal subsequent finishing for the final surface quality and geometric accuracy, unlike the conventional formed products. Mini-thixoforming therefore offers material, energy, and financial savings.

2. Experimental programme

For this investigation, the low-carbon MS1 tool steel in the form of 3D-printed feedstock was chosen as the experimental material (Tab. 1). It is one of the popular materials for 3D printing. It is a precipitation-hardenable steel whose properties after conventional treatment have been thoroughly mapped. Maraging steels belong to the strongest homogeneous steel materials ever. They are alloyed with cobalt, molybdenum, titanium, and aluminium. Their unusual properties make them candidates for heavy-duty parts operating predominantly under dynamic loading. Their semi-solid processing parameters window is very narrow due to their chemical composition. This makes the choice of suitable thixoforming temperature all the more difficult.

Ni	Co	Mo	Ti	Al	Cr, Cu	C	Mn, Si	P, S	Fe
17-19	8.5-9.5	4.5-5.2	0.6-0.8	0.05 - 0.015	≤ 0.5	≤ 0.3	≤ 0.1	≤ 0.01	balance

Table 1. Chemical composition of the MS1 Maraging steel measured by chemical analysis

In order to identify suitable thixoforming parameters, the physical and chemical properties of the material and their profiles throughout the entire heating process must be known. One of the key characteristics relevant to semi-solid processing is the width of the interval between the solidus and liquidus. The interval recommended specifically for mini-thixoforming is 10–30 % [5, 6]. Finding these parameters experimentally is time-consuming. In this case, the first step involved calculations which were completed using the JMatPro software. As the chemistry of this steel is extremely complex, the results should be considered with caution, not to mention the fact that the actual process is very rapid. The calculated theoretical onset of melting is at 1395°C. The material becomes completely melted at 1445 °C (Fig. 1). Hence, the freezing range width is approximately 50°C. According to the calculation, the theoretical interval that is usable for semi-solid forming is 1408–1420°C. The advantage of this material is that its melting curve is smooth without any major curvature. This is favourable for the temperature field control. The narrow processing parameters window and the high temperature pose major difficulties for the process.

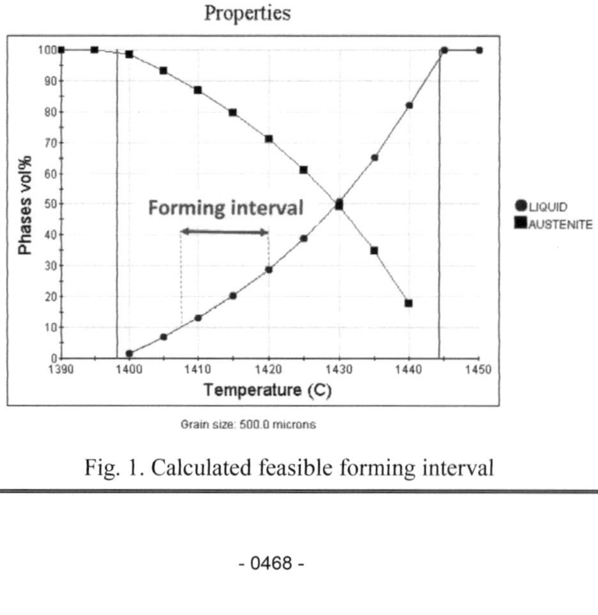

Fig. 1. Calculated feasible forming interval

The experiments were carried out in special mini-thixoforming equipment which had been designed for handling and processing a very small amount of material in partially melted state (Fig. 2). It relies on high-frequency electrical resistance heating for precise temperature control and generation of the required temperature fields in the workpiece [7]. The actual mini-thixoforming operation consisted in lateral extrusion into a mould (Fig. 3).

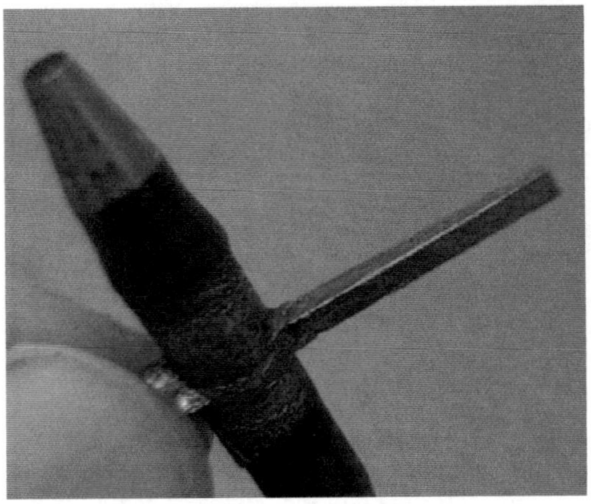

Fig. 2. Typical product manufactured by Mini-thixoforming and the tool in forming machine

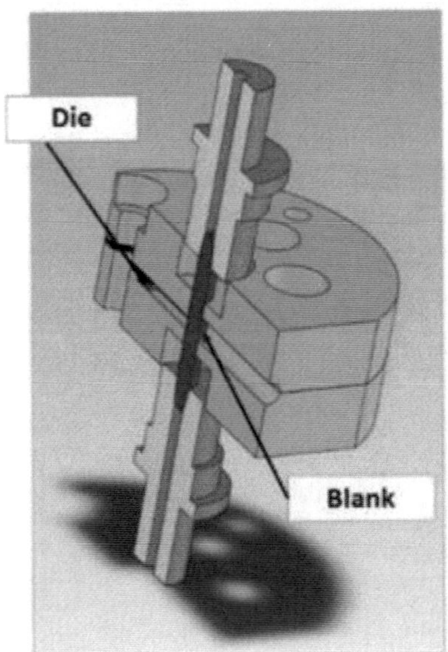

Fig. 3. – Principle of Process

The blanks for the experiment were made of the MS1 Maraging Steel by 3D printing. The total length of the blank was 48 mm, of which the length of the effectively worked part was 38 mm. The diameter was 6 mm. Both ends of the blank were conical with the apex angle of 28°. Their purpose was to centre the blank in the die and provide contact surfaces for energy transfer for heating.

The blanks and the resulting semi-solid processed workpieces were examined under a microscope in both longitudinal and transverse directions. Since the input material for printing was powder, the entire technology chain that governed the microstructural evolution was documented by metallographic techniques.

3. Discusion of results

The input material for making the blanks was the commercial powder supplied by the manufacturer of the 3D printing equipment. Its particle size was 5–50 μm (Fig. 4).

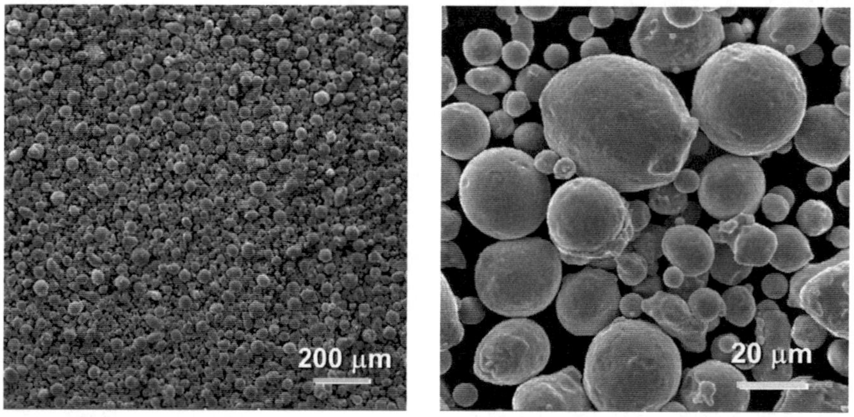

Fig. 4 – Maraging steel powder before printing REM

The 3D printing process was based on laser melting and provided a dense structure of parallel buildup layers with shallow penetration between individual laser paths. In order to acquire excellent properties, maraging steels must be subsequently precipitation-hardened by means of heat treatment. The heat treatment of the buildup therefore provides mainly precipitation of particles of intermediate phases. A major difficulty in the metallographic analysis of these materials is the difference in the etch response of their various states. Nital etching effectively reveals individual laser beam passes and the sharp boundaries between layers of melted metal. In addition, it clearly shows the widths of individual melted regions and the directions of laser passes (Fig. 5a,b). The drawback of this etching reagent is that it inadequately reveals the actual microstructure of the solidified metal. For this reason, an etchant of a mixture of HCl and H_3PO_4 acids (Etchant 2) was used. It effectively reveals the martensitic microstructure of this steel in its non-hardened condition (Fig. 6a,b). At this stage of treatment, the hardness of the workpiece was 360 HV10.

a) Optical microscopy, Nital etch

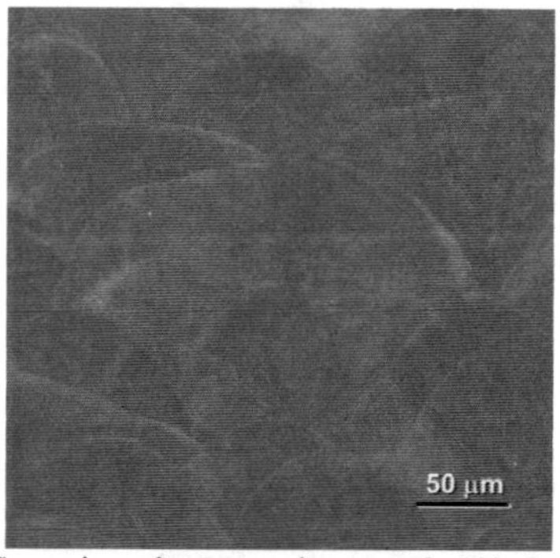

b) Scanning electron micrograph, Nital etch

Fig. 5 – Laser melted microstructure after D3 printing

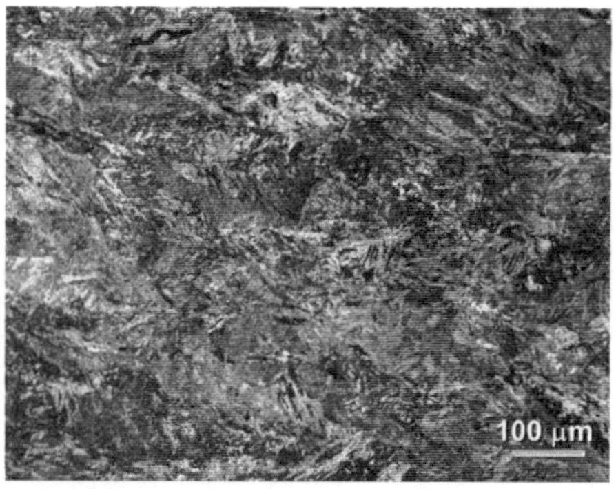

Fig. 6a – Optical micrograph, etched with a mixture of HCl and H_3PO_4 acids

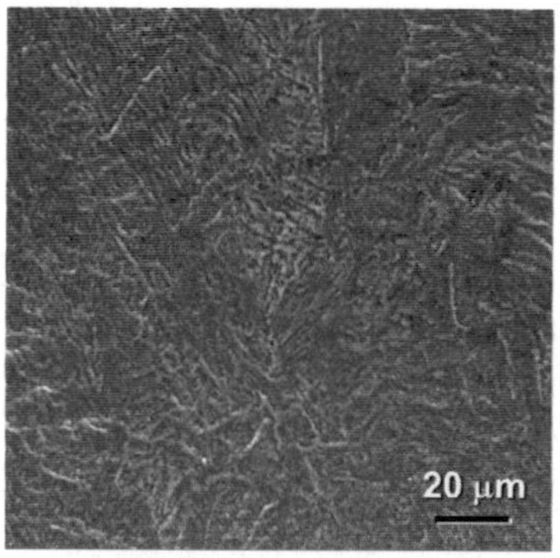

Fig. 6b – REM micrograph, etched with a mixture of HCl and H_3PO_4 acids

The next operation was stress relieving. As expected, the workpieces showed no considerable microstructural changes, apart from the desired reduction in stress. Considerable changes occurred after the hardening step: uniformly-distributed fine precipitates of intermediate phases have formed in the martensitic structure (Fig. 7a,b).

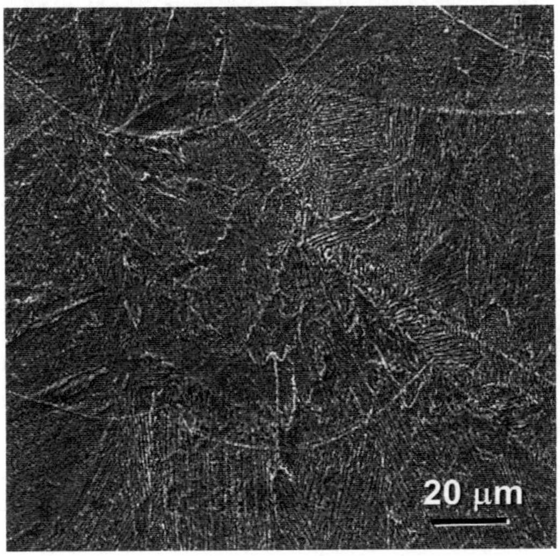

a) melted regions after laser passes, REM

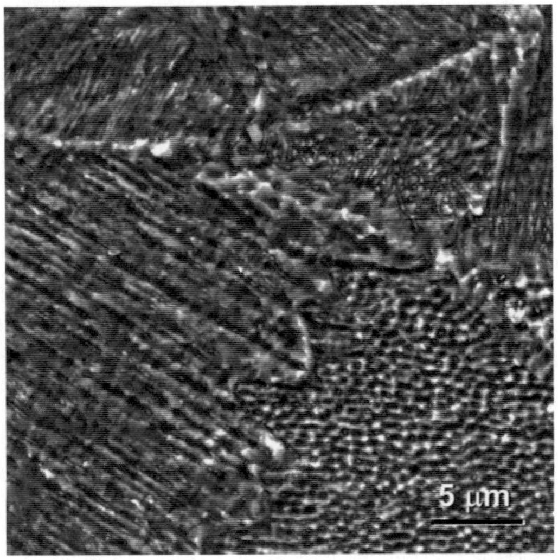

b) precipitates of intermediate phases after hardening detail

Fig. 7 – Microstructure after hardening

The final step of the process chain was mini-thixoforming. It comprised heating to 1440°C, holding for approximately 10 seconds, and lateral extrusion of the workpiece material into the mould held at room temperature. The heating time was 68 seconds. The forming stroke finished in 0.3 seconds and was followed by rapid solidification thanks to the contact with the surface of the metal die cavity.

An analysis of the product has shown that when appropriate process conditions are used, maraging steel can develop a microstructure that is characteristic of products made by semi-solid processing. Such a structure consists of polyhedral islands of 20–40 μm size embedded in a uniform closed network. The fact that it evolves from semi-solid state is confirmed by a characteristic defect which was found by accident in the melted and solidified material between polyhedral islands (Fig. 8). As the region of the semi-solid state available for processing this maraging steel is very narrow, melting also occurred within some of the polyhedral islands. The growth of such melted spots had no immediate constraints which is why they are more or less spherical in shape and have no contact with the network which separates the polyhedral islands. During this process, the thixotropic behaviour was not manifested strongly. Therefore, this material can be considered to be difficult to minithixoform. The next research efforts will focus on the redistribution of chemical elements and on the evolution of microstructural constituents in this process.

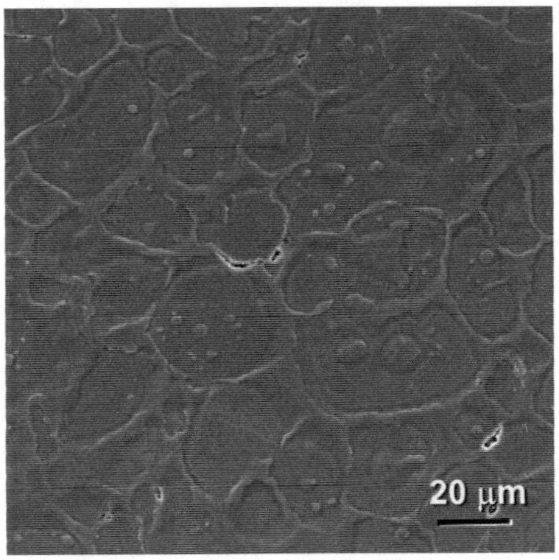

Fig. 8. Micrograph of maraging steel after semi-solid processing

4. Conclusion

A maraging steel was chosen for this mini-thixoforming process. The objective was to test a combination of 3D printing and semi-solid processing. A simple experiment involving lateral extrusion provided information on complex microstructural evolution in the technology chain considered. Suitable and achievable processing parameters have been found which can provide the desired structural state upon semi-solid processing, rapid solidification, and cooling. Given the high processing temperature and the narrow temperature interval, the process became a great technological challenge. Nevertheless, it was completed successfully. Despite that, current findings suggest that at these parameters, the thixotropic property of this material is not favourable. Unless more suitable parameters are found, this material cannot be used in this 3D printing-thixoforming process chain.

5. Acnowledgements

This paper presents results achieved under the project LO1502 Development of Regional Technological Institute. The project is subsidised by the Ministry of Education of the Czech Republic from specific resources of the state budget for research and development.

6. References

[1] Fleming M.C. Behavior of Metal Alloys in the Semisolid State (1991). Metall Trans A, 22: 957-981.

[2] Thixoforming: Semi-solid Metal Processing. Edited by HIRT G. and KOPP R., pp 1 – 27, Copyright 2009 WILEY-VCH Verlag GmbH & Co. KGaA, Weinheim ISBN: 978-3-527-3220 4-6.

[3] Püttgen W., Bleck W., Hirt G., Shimahara H., Thixoforming of Steels – A Status Report, Advanced Engineering Materiale 9, (2007), No. 4 234 – 245.

[4] Vančura F., Vorel I., Pileček V., Mašek B. Material Technological Modeling of Thermomechanical Processing of Die Forging of Microalloyed Steel, Kovárenství, 2015, ISSN 1213-9289.

[5] W. H. Bauer and E.A.Collins, in F.R. Eirich (Ed.),Rheology: Theory and Applications, Vol.4, (1967) Academic Press, New York, , ch. 8 .

[6] Wolf, A., Baur, J., Gullo, G.C. Thixoforging, (2012), Available from: http://www.cct-bw.de/veroeffentlichung_pdf/WoBa%20Massiv01%20english.pdf.

[7] Ronešová, A., Mašek, B., Staňková, H., Štádler, C. Patent CZ c. 299758 - Method of handling and its shaping at a temperature between solid and liquid, (3. 10. 2008).

DOI: 10.2507/27th.daaam.proceedings.071

MACHINING OF DIFFICULT-TO-CUT MATERIALS

Friedrich Bleicher, Daniel Finkeldei & Anton Siller

This Publication has to be referred as: Bleicher, F[riedrich]; Finkeldei, D[aniel] & Siller, A[nton] (2016). Machining of Difficult-To-Cut Materials, Proceedings of the 27th DAAAM International Symposium, pp.0473-0479, B. Katalinic (Ed.), Published by DAAAM International, ISBN 978-3-902734-08-2, ISSN 1726-9679, Vienna, Austria
DOI: 10.2507/27th.daaam.proceedings.071

Abstract

Austenitic steels are used in different areas of application where high strength and corrosion resistance are necessary at low and intermediate temperature levels. Machining these materials therefore induces high mechanical and thermal loading and reduces tool life and the overall process performance. A well-known approach to raise tool life is to dissipate the heat from the cutting edge. This paper presents some findings of a combined externally and internally cooled cutting insert compared to a solely externally cooled cutting insert.

Keywords: longitudinal turning; austenitic stainless steel, internally cooled insert, tool wear, tool lifetime

1. Introduction

Austenitic stainless steels are a group of corrosion resistant materials mainly used for components of industrial manufacturing [1]. With an increasing amount of nickel (up to 50%) as its alloying element, austenitic steels become nickel alloys [2].

Fig. *1* summarizes the material properties of two most widespread austenitic steels and the nickel based alloy Inconel 718. Latter is more heat-resistant and shows higher hardness and strength. However, its elongation and thermal conductivity are significantly lower.

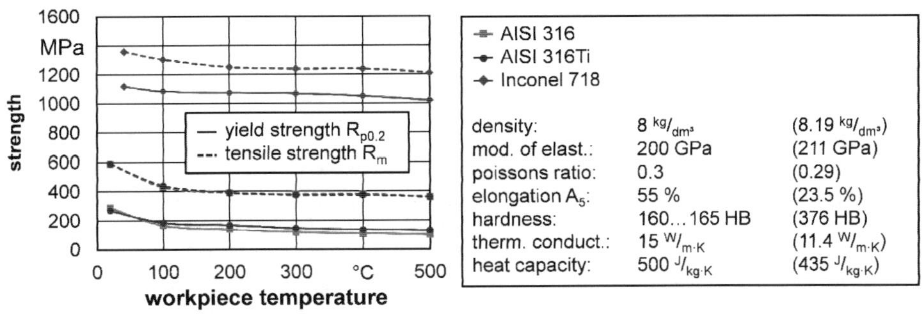

Fig. 1. Material properties of typical Cr-Ni-Mo stainless steels and Inconel 718 (values written in braces) [3, 4]

The applications of austenitic stainless steels in the field of the process industry are in "pulp and paper, textile, food and beverages, pharmaceutical (and) medical (...) processing equipment" [3]. In nuclear power stations these alloys are used for instruments in the reactor, in marine applications e.g. as components of submarines or off-shore wind power plants [1]. Another major field of application is coming from the automotive industry in exhaust gas systems [1]. Nevertheless, main application of these materials is for pipes and fittings for the water supply of apartments and houses [5, 6]. Some petrochemical applications use the austenitic steel AISI 317L or the relatively new P750 for clads, vessels, tubes, valves and other components in oil refineries [7].

The six important alloying elements in austenitic stainless steels are chromium, molybdenum, nickel, manganese, carbon and nitrogen. Nickel and nitrogen are austenite stabilizers who increase the resistance against stress-corrosion-cracking [8] while nitrogen "forms an interstitial solid solution which disrupts dislocation motion and strengthens the steel" [9]. Manganese is a substitution element for nickel and builds up a protective layer of manganese sulphite in sulphurous atmosphere [8]. Chromium supports the corrosion resistance while molybdenum reduces the tendency for pitting and crevice corrosion [8].

Their high strength at room temperature as well as elevated temperatures combined with their low thermal conductivity makes stainless steels difficult to machine. With increasing cutting speed, the temperature at the cutting tool will raise proportionally [10]. The cutting temperature (at the cutting tool tip) is proportional to the cutting speed and below a critical feed rate chatter could result [10]. In addition, because of the ductile material behaviour and the formation of a big, instable Built-Up-Edge (BUE), high surface roughness is observed at higher cutting speeds [11]. For the end-milling process, the use of a system with ultrasonic-assistance resulted in a slightly worse roughness than in conventional milling [12]. Another technology was investigated by Shi and Attia [13]. They analysed cutting forces, workpiece temperature and flow stress during laser-assisted machining of an austenitic steel with a FEM simulation model [13]. Anderson and Shin [9] conducted laser-assisted machining experiments with the austenitic stainless steel P550. With increasing workpiece temperature, tool wear, specific cutting energy and surface roughness decreased. However, depositions of Cr_2N and σ-phase in the microstructure of the material were observed [9]. The following investigations focused on the machining behaviour under dry cutting conditions and under variation of the coating material of the cemented carbides. Kulkarni et al. [14] e.g. conducted high-speed-machining experiments with coated cemented carbide tools using the HPPMS coating process with AlTiCrN [14]. In comparison with Korkut et al. and Ciftci [15, 16], who used coated cemented carbide tools with multi-layers of TiCN-TiC-Al_2O_3 and TiC-TiCN-TiN, the AlTiCrN-coated tools led to a decrease of surface roughness by a factor of 2 to 3 and, because of the smoother tool surface, cutting and feed forces were reduced as well [14]. Selvaraj et al. [17] performed a turning experiment in dry cutting of austenitic stainless steel AISI 304. Using the Taguchi design of experiments method, they optimized the cutting parameters to obtain a minimum surface roughness [17].

Cooling lubricants have a significant influence on the machining process to raise tool life and improve surface finish. In contrast to conventional coolants, Jayadas et al. [18] and Xavior et al. [11] conducted research on coconut oil as a substitution medium [18] with positive results with respect to tool wear and surface roughness [11]. In the past, the use of minimum quantity lubrication (MQL) resulted in improved cutting behaviour for steels and other materials. Different studies discuss the advantages of MQL in machining [19-22]. Dureja et al. [23] did a parameter study based on the Taguchi orthogonal array robust design method. Using this lubricating technique, very low surface roughness was achieved [23]. Another study of Maruda et al. [24] was concerned with the ratio of chip compression and the influence of the shear angle and friction coefficient. They varied cutting speed and feed rate and studied the effects of MQL and dry cutting process [24]. Zhang et al. [25] applied an approach of the specific energy consumption to describe the surface roughness for increased cutting parameters [25]. Other researcher used cryogenic cooling for machining of stainless steels. Mayer et al. [26] studied different cooling methods supplying the process and the tool with liquid nitrogen (LN_2) and carbon dioxide snow (CO_2). Besides external cooling, they used internal coolant supply for local cooling of the tool tip. The tool froze due to the intense heat removal, leading to thermal shrinkage of the tool [26]. A further investigation focused on the turning process while cooling with cryogenic carbon dioxide snow (CO_2). For differently prepared cutting inserts, changes in the microstructure of α'-martensite as well as the surface roughness were observed [27]. Bleicher et al. [28, 29] designed an innovative cooling approach in which cutting inserts are cooled externally or internally with conventional coolant supplied by the machinery. Compared to cryogenic cooling, the cooling temperature is marginally higher and the heat dissipates directly from the tool tip without any thermal shrinking of the tool as observed by Mayer et al. [26]. This cooling strategy shows excellent results for the machining of aluminium and cast iron hybrid components [28, 29]. For this reason, using the same cooling approach is expected to show promising results also for machining of austenitic stainless steels.

2. Experimental setup

The used workpiece material is P750, an austenitic chromium-nickel-nitride stainless steel developed for oil field industry applications. The material properties are comparable to those of conventional austenitic steels (see

Fig. *1*). Its yield and tensile strength are 900 MPa and 1035 MPa, respectively. Its hardness is 32...43 HRC and its elongation of 15% is approximately 40% lower than that of conventional austenitic stainless steels. The tool material is a TiAlN-coated cemented carbide with a triangular shape form as of ISO definition TCGT 16 T3 08-UM. The clearance angle α is 7°, the rake angle γ 9°, the corner angle ε 60° and the cutting edge radius $r_ε$ 6 μm. The experimental setup is shown in Fig.2.

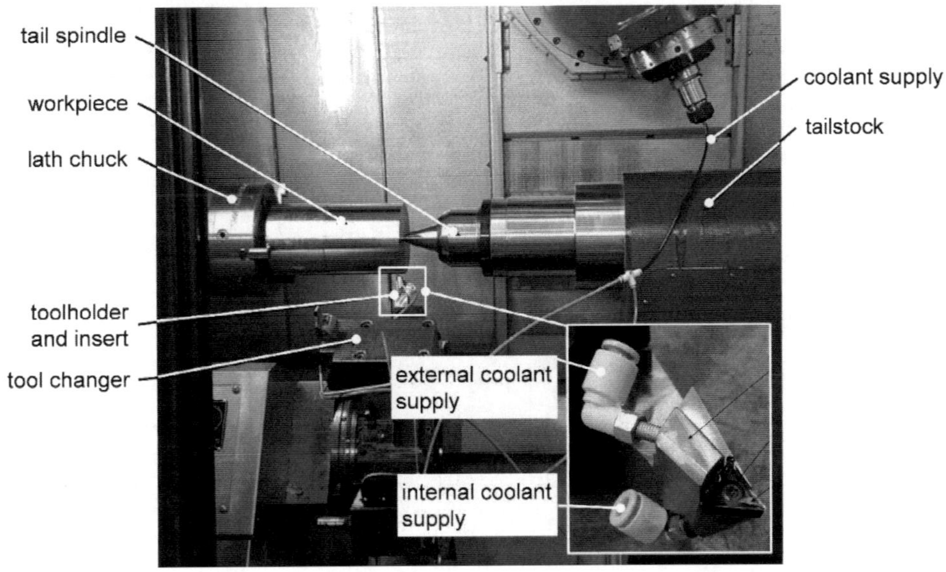

Fig. 2. Experimental setup of the turning process

In order to examine the effect of the coolant supply strategy, three options were considered. At first, the coolant was supplied externally over the rake face, secondly only internally through the tool insert whereas in the third option both external and internal cooling was used. For every internal supply strategy the flow volume of the coolant through the toolholder was set to 0.6 $^{l}/_{min}$. The design of experiment is shown in Table 1. To validate the behaviour of the tools under these experimental variations, the tool wear was measured with a Keyence VW600C microscope with a magnification factor of up to 200. Three dimensional images of the tool tip were additionally obtained on an Alicona InfiniteFocus G4 measurement system.

Experimental design	Setup 1
cutting speed v_c	100 $^{m}/_{min}$
feed rate f	0.1 $^{mm}/_{rev}$
depth of cut a_p	0.5 mm
tool coating	TiAlN-coated
cooling/lubrication strategies	external cooling internal cooling external + internal cooling
flow volume through the toolholder	0.6 $^{l}/_{min}$

Table 1. Design of Experiment

The machining process was stopped as soon as one of the following tool wear criteria was exceeded:

1. Tool wear land VB_{max} exceeds 250 µm
2. Tool notch wear VB_{notch} exceeds 200 µm
3. Crumbling of the cutting edge larger than 150 µm
4. Process time larger than 90 min

3. Experimental results

For all of the three cooling strategies the development of flank wear reached the linear wear progression zone after a cutting time of 6 min (see Fig. 3). While tool wear on the flank and rake face was uncritical, there was a tribo-chemical reaction on the flank face. For the externally and combined cooled insert, the tribo-chemical reaction was more profound but localized on a certain part of the flank face (shown in the middle and right upper image of Fig. 4). On the other hand, the tribo-chemical reaction under the internal cooling condition was more transparent and distributed over the whole tool contact area.

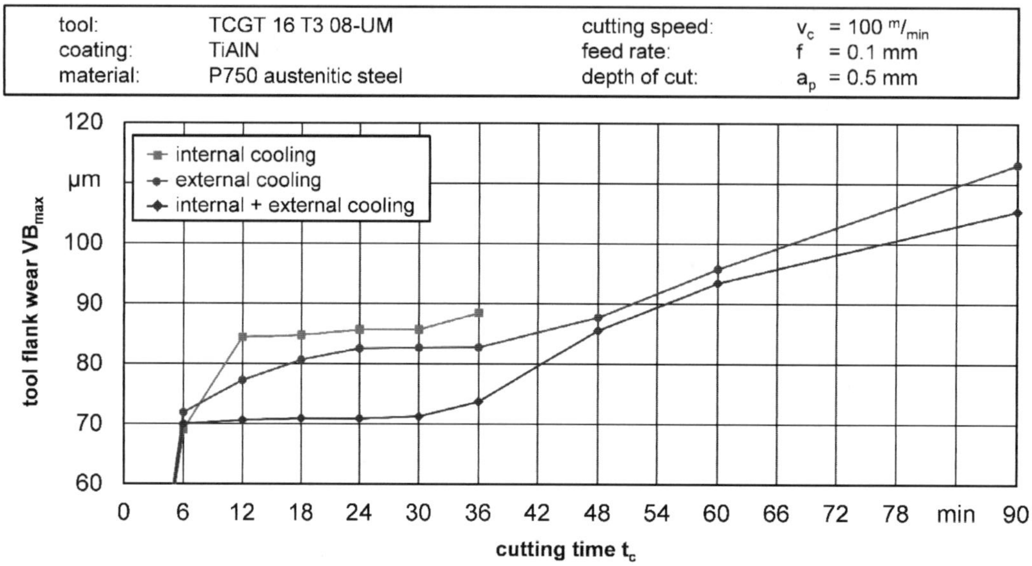

Fig. 3. Visualisation of tool wear over cutting time

After a cutting time of 36 min, significant changes in tool wear occurred (see Fig. 5). The tribo-chemical reaction (described in

Fig. 6) under the external and combined cooling method increases slightly. The flank face with the internally cooled insert had a strong scaling, as observable by the discoloured shape of tribo-chemical surface (seen on the left upper image of Fig. 5). Furthermore, notch wear can be observed on the cutting edge along the flank and rake face. The lack of lubrication fluid is a logical explanation for this wear behaviour. For the machining process, high-strength austenitic stainless steels require some lubrication to reduce friction and, as a consequence of this, the abrasive tool-wear.

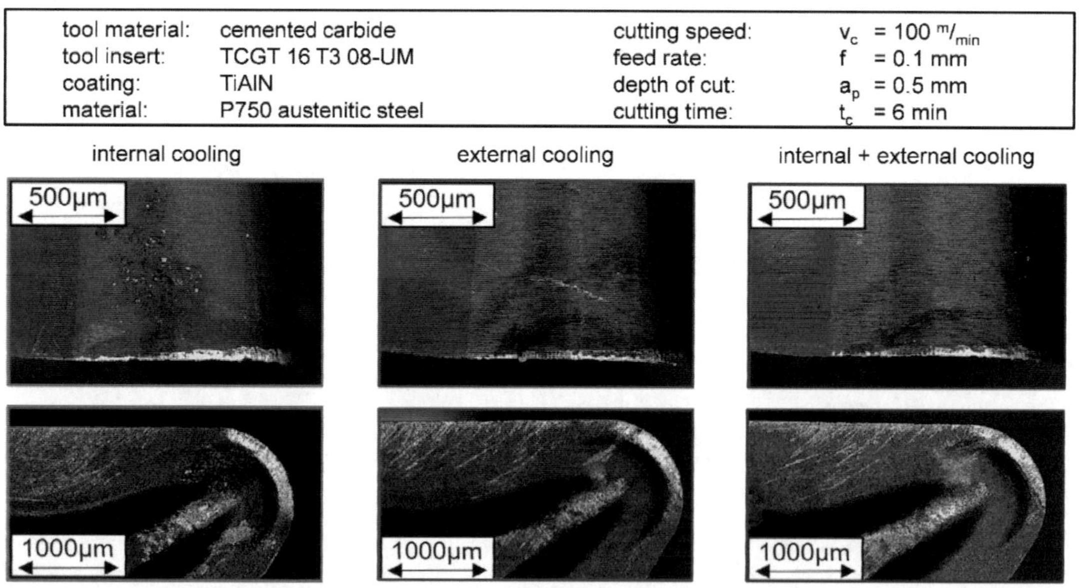

Fig. 4. Tool wear of the rake and flank face after a cutting time of 6 min

In comparison to the two other cooling strategies, the rapidly increasing tribo-chemical wear leads to a stop of the machining process with the internally cooled insert without reaching the tool wear criteria. Additionally, the machining process under the external and combined cooling performed up to a cutting time of 90 min (seen in

Fig. 6). In this situation, notch and flank wear increased with both strategies (seen in the red circles of

Fig. 6). The tribo-chemical reaction is more distinct on the tool flank face for the external cooling strategy.

tool material:	cemented carbide	cutting speed:	v_c = 100 $^m/_{min}$
tool insert:	TCGT 16 T3 08-UM	feed rate:	f = 0.1 mm
coating:	TiAlN	depth of cut:	a_p = 0.5 mm
material:	P750 austenitic steel	cutting time:	t_c = 36 min

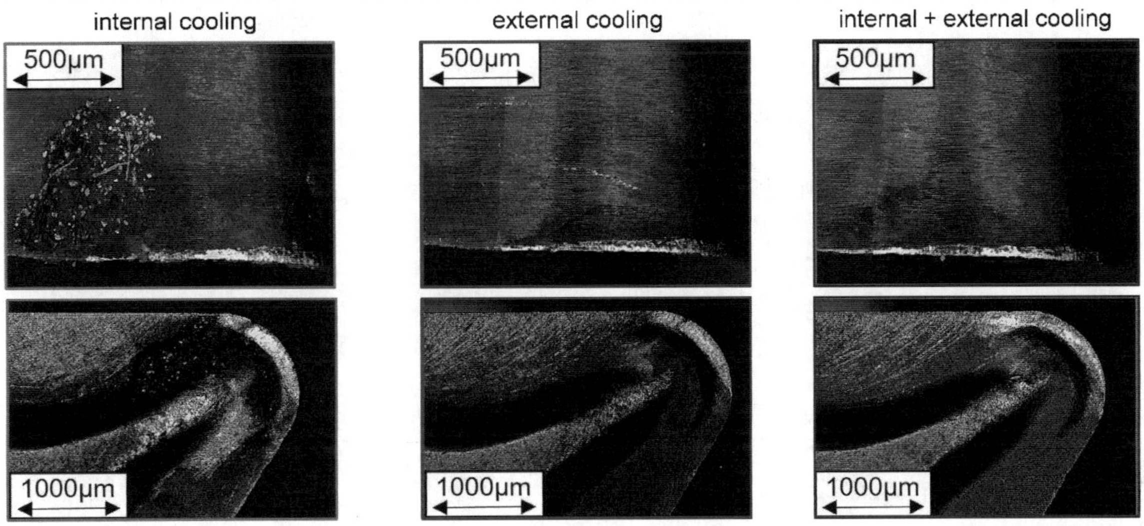

Fig. 5. Tool wear of the rake and flank face after a cutting time of 36 min

Fig. 7 shows three dimensional images of cutting edges using both cooling strategies. The tool notch wear is visible on the right side of the picture. Using the external cooling technique, a shattered surface of the cutting tool along the rake and flank face occurs during machining. The combined cooling strategy stabilized the resistance of the cutting tool edge against tool wear.

tool material:	cemented carbide	cutting speed:	v_c = 100 $^m/_{min}$
tool insert:	TCGT 16 T3 08-UM	feed rate:	f = 0.1 mm
coating:	TiAlN	depth of cut:	a_p = 0.5 mm
material:	P750 austenitic steel	cutting time:	t_c = 90 min

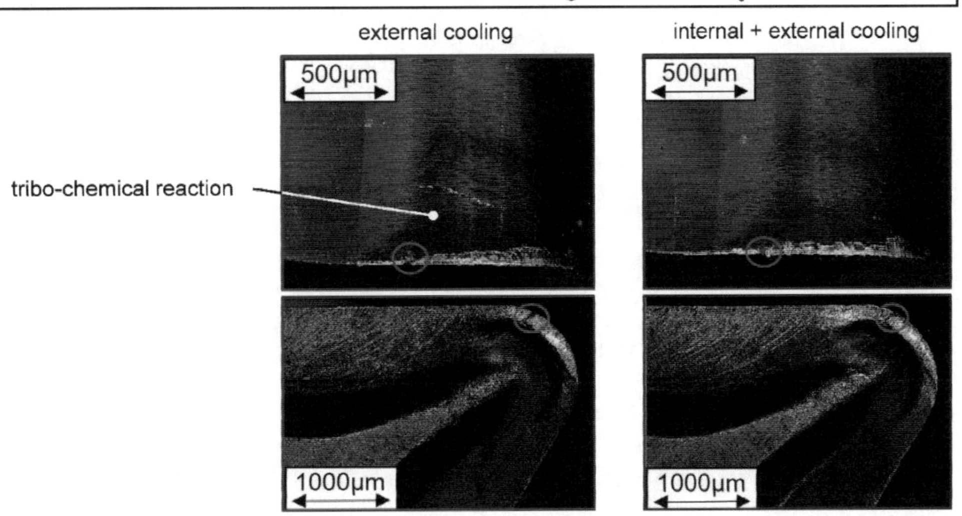

Fig. 6. Tool wear of the rake and flank face after a cutting time of 90 min

Over the whole machining time of 90 min, the combined cooling technique reduced tool wear as it is observable by smaller wear land along the cutting edge. However, between cutting times of 36 min and 60 min the wear progression curves of the external and combined cooling strategies converged (see Fig. 3). Because of the large step range of 6 min, the degressive character of the starting phase of the tool wear progression could not be shown. However, the focus of this work was to investigate the tool wear behaviour at intermediate cutting times.

tool material:	cemented carbide	cutting speed:	v_c = 100 $^{m}/_{min}$
tool insert:	TCGT 16 T3 08-UM	feed rate:	f = 0.1 mm
coating:	TiAIN	depth of cut:	a_p = 0.5 mm
material:	P750 austenitic steel	cutting time:	t_c = 90 min

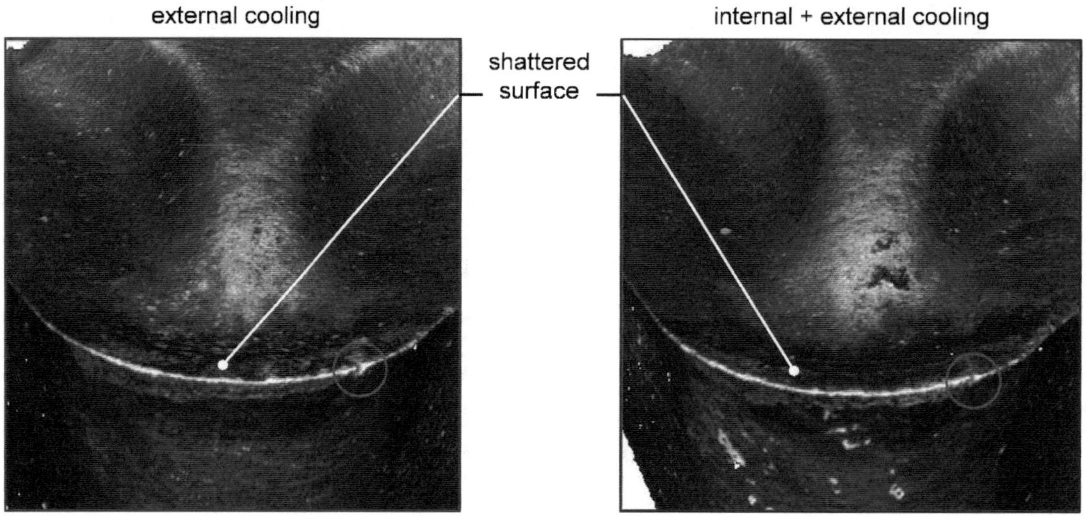

Fig. 7. Three dimensional image of worn tools

5. Conclusion

The purpose of this study was to investigate the tool wear behaviour in machining the high-strength austenitic stainless steel P750 with an internal cooling strategy. When machining this material i.a. high thermal and mechanical tool load emerged near the cutting edge and increase the temperature inside the tool and hence tool wear. Based on a study of Bleicher et al. [28, 29], conventional coolant was led through the cutting insert to dissipate heat directly near the cutting edge. The internal cooling strategy combined with an external coolant supply resulted in a reduction of tool wear when machining P750. In order to reduce tool wear, lubrication is a necessity and an important factor of the cutting process. Low cutting speeds are chosen in this study. To validate the discussed results and to observe the machining behaviour under the applied cooling strategies, higher cutting speeds could be used in a further investigation. Moreover, subsequent studies could focus on an investigation of machining with the internal cooling technique based on minimum quantity lubrication. Besides tool life, the surface finish could be examined as well in order to draw conclusions on the workpiece quality.

6. References

[1] Montanstahl (2016): Datasheet about special steel profiles, http://www.montanstahl.at/de/1.4401, Stand: 07.08.2016

[2] Heubner, U. (1998): Nickel alloys and high-alloy special stainless steels – materials summary and metallurgical principles. In: Nickel Alloys, edited by Heubner, U., pp. 1-43

[3] Outokumpu (2006): Datasheet of Standard Cr-Ni-Mo stainless steels, https://www.outokumpu.com/SiteCollectionDocuments/Standard_Cr-Ni-Mo_Stainless_Steels_Datasheet.pdf

[4] Bibus Metals AG (2016): Datasheet of INCONEL Alloy 718, https://www.bibusmetals.ch

[5] Montanstahl (2016): Datasheet about special steel profiles, http://www.montanstahl.at/de/1.4571, Stand: 07.08.2016

[6] Avery, RE; Lamb, S.; Powel, C.A.; Tuthill, A.H. (2016): Stainless steel for potable water treatment plants (PWTP). In: Nickel Development Institute, https://www.nickelinstitute.org/~/media/Files/TechnicalLiterature/10087_StainlessSteelForPotableWaterTreatmentPlants.ashx

[7] Farneze, H.N.; Ravares, S.S.M.; Pardal, J.M. (2015): Effects of Thermal Aging on Microstructure and Corrosion Resistance of AISI 317L Steel Weld Metal in the FSW Process. In: Materials Research, 18(Suppl. 2), pp. 98-103

[8] Jones, R.H. (2005): Environmental Effects in Engineered Materials. New York: Marcel Dekker, Inc.

[9] Anderson, M.C.; Shin, Y.C. (2006): Laser-assisted machining of an austenitic stainless steel: P550. In: Journal of the Engineering Manufacture, 220, pp. 2055-2067

[10] Lin, W.S. (2008): The study of high speed fine turning of austenitic stainless steel. In: Journal of Achievements in Materials and Manufacturing Engineering, 27, pp. 191-194

[11] Xavior, M.A.; Adithan, M. (2009): Determining the influence of cutting fluids on tool wear and surface roughness during turning of AISI 204 austenitic steel. In: Journal of Materials Processing Technology, 209, pp. 900-909
[12] Maurotto, A.; Wickramarachi, C.T. (2016): Experimental investigations on effects of frequency in ultrasonically-assisted end-milling of AISI 316L: A feasibility study. In: Ultrasonics, 65, pp. 113-120
[13] Shi, B.; Attia, H. (2013): Integrated Process of Laser-Assusted Machining and Laser Surface Heat Treatment. In: Journal of Manufacturing Science and Engineering, 135
[14] Kulkarni, A.P.; Joshi, G.G.; Sargade, V.G. (2013): Dry turning of AISI 304 austenitic stainless steel using AlTiCrN coated insert produced by HPPMS technique. In: Procedia Engineering, 64, pp. 737-746
[15] Korkut, I.; Kasap, M.; Ciftci, I.; Seker, U. (2004): Determination of optimum cutting parameters during machining of AISI 304 austenitic stainless steel. In: Material and Design, 25, pp. 303-305
[16] Ciftci, I. (2006): Machining of austenitic stainless steels using CVD multi-layer coated cemented carbide tools. In: Tribology Internation, 39, pp. 565-569
[17] Selvaraj, D.; Chandramohan, P. (2010): Optimization of surface roughness of AISI 304 austenitic stainless steel in dry turning operation using taguchi design method. In: Journal of Engineering Science and Technology, 5 no. 3, pp. 293-301
[18] Jayadas, N.H.; Prabhakaran Nair, K. (2006): Coconut oil as base oil for industrial lubricants – evaluation and modification of thermal, oxidative and low temperature properties. In: Tribology International, 39, pp. 873-878
[19] Weinert, K.; Insaki, I.; Sutherland, J.W.; Wakabayashi, T. (2004): Dry machining and minimum quantity lubrication. In: Annals of the CIRP, 53, pp. 511-537
[20] Attanasio, A.; Gelfi, M.; Giardini, C.; Remino, C. (2006): Minimal quantity lubrication in turning: Effect on tool wear. In: Wear, 260, pp. 333-338
[21] Liao, Y.S.; Lin, H.M. (2007): Mechanism of minimum quantity lubrication in high-speed milling of hardened steel. In: International Journal of Machine Tools & Manufacture, 47, pp. 1660-1666
[22] Zhang, S.; Li, J.F.; Wang, Y.W. (2012): Tool life and cutting forces in end milling Inconel 718 under dry and minimum quantity cooling lubrication cutting conditions. In: Journal of Cleaner Production, 32, pp. 81-87
[23] Dureja, J.S.; Singh, R.; Singh, T.; Singh, P.; Dogra, M.; Bhatti, M.S. (2015): Performance Evaluation of Coated Carbide Tool in Machining of Stainless Steel (AISI 202) under Minimum Quantity Lubrication (MQL). In: International Journal of Precision Engineering and Manufacturing-Green Technology, 2, pp. 123-129
[24] Maruda, R.W.; Krolczy, G.M.; Nieslony, P.; Krolczyk, J.B.; Legutko, S. (2016): Chip formation zone analysis during the turning of austenitic stainless steel 316L under MQCL cooling condition. In: Procedia Engineering, 149, pp. 297-304
[25] Zhang, Y.; Zou, P.; Li, B.; Liang, S. (2015): Study on optimized principles of process parameters for environmentally friendly machining austenitic stainless steel with high efficiency and little energy consumption. In: International Journal of Advanced Manufacturing Technology, 79, pp. 89-99
[26] Mayer, P.; Skorupskim, R.; Smaga, M.; Eifler, D.; Aurich, J. (2014): Deformation induced surface hardening when turning metastable austenitic steel AISI 347 with different cryogenic cooling strategies. In: Procedia CIRP, 14, pp. 101-106
[27] Mayer, P.; Kirsch, B.; Müller, R.; Becker, S.; von Harbou, E.; Aurich, J. (2016): Influence of cutting edge geometry on deformation induced hardening when cryogenic turning of metastable austenitic stainless steel AISI 347. In: Procedia CIRP, 45, pp. 59-62
[28] Bleicher, F.; Brier, J.; Siller, A. (2016): Simultaneous machining of a material combination with an internally and externally cooled cutting insert. In: Procedia CIRP, 46, pp. 15-18
[29] Bleicher, F.; Pollak, C.; Brier, J.; Siller, A. (2016): Reduction of built-up edge formation in machining Al- and cast iron hybrid components by internal cooling of cutting inserts. In: Annals of the CIRP – Manufacturing Technology, 65, pp. 97-100

DOI: 10.2507/27th.daaam.proceedings.072

UNCERTAINTY ESTIMATION OF AXES DIRECTION DETERMINATION OF INDUSTRIAL ROBOT USING AN ELLIPSOID CONCENTRATION MODEL

Andrei Vorotnikov, Elena Romash, Alexander Isaev, Olga Bashevskaya,
Giacomo Bianchi & Yuri Poduraev

This Publication has to be referred as: Vorotnikov, A[ndrei]; Romash, E[lena]; Isaev, A[lexander]; Bashevskaya, O[lga]; Bianchi, G[iacomo] & Poduraev, Y[uri] (2016). Uncertainty Estimation of Axes Direction Determination of Industrial Robot Using an Ellipsoid Concentration Model, Proceedings of the 27th DAAAM International Symposium, pp.0480-0486, B. Katalinic (Ed.), Published by DAAAM International, ISBN 978-3-902734-08-2, ISSN 1726-9679, Vienna, Austria
DOI: 10.2507/27th.daaam.proceedings.072

Abstract

There is uncertainty while measure different geometrical elements of industrial robots, multi-axis machines or other complex mechanisms during their calibration using CMM. This uncertainty has a large number of components. It is extremely difficult to estimate this uncertainty using ISO [12]. This article represents approach for uncertainty estimation of axes direction determination of industrial robot.

We revealed the concentration ellipsoid – a geometrical view of the spatial spread of error after repeated axis determination of an industrial robot. The mathematical model of the concentration ellipsoid was developed. In this paper we have analysed the big axis value of the concentration ellipsoid using a 3D surface. An iterative uncertainty search algorithm for the industrial robot's axis and its exit criterion was developed. Uncertainty of each axis of the industrial robot was estimated. The standard deviation of the iterative algorithm of uncertainty was calculated. The dependence of uncertainty from increasing the rotation radius of the measuring device reflector was experimentally revealed.

Keywords: concentration ellipsoid; uncertainty estimation; laser tracker; industrial robot calibration.

1. Introduction

Nowadays many studies in the field of mechatronics are related to the use of portable coordinate measuring machines (CMM). CMM are usually used for increasing accuracy of mechatronic devices. That promotes evolution of mechatronics [1] and confirms the importance of the direction of this article [13-15]. Laser tracker LTD 800 that was applied in this research is used in mechatronics for axis position determination of different mechanisms in the implementation of the following movements: linear, rotary or complex. For example the laser tracker provides calibration of industrial robots [2, 3], to determine the spindle axis position of the tool [4-6] or setup angular orientation of optical schemes for high-speed-shooting mechatronic systems of fast moving targets [7,8].

In our research the KUKA KR16 – industrial robot with 6 joints was used as the measuring object. Each joint has its own axis of rotation. There is task when calibrate the industrial robot – to determine of all axes position in laser tracker reference frame. For determining the axis position of the industrial robot it is necessary to determine a point on the axis (joint center of rotation) and the direction (orientation) of this axis. Estimation of error in determining the joint center of rotation is represented in [2]. This article is devoted to uncertainty estimation of axes direction determination of an industrial robot.

CPA method was applied for measuring axis direction of an industrial robot. The description of CPA is represented in [2]. There is an uncertainty during measurement of the axis direction of an industrial robot. This uncertainty has a lot of components. During the research three main factors of uncertainty were revealed, the cumulative effect of which brings the greatest error in the result of uncertainty estimation:

- First, it is the error of the data processing algorithm of the coordinate points that were received on geometrical figures with small dimensions. The well-known example is fitting the data points that lie on a circle. When the data points form a small arc the determination of coordinates of the circle centre is not accurate [9]. In our research the similar problem is found in the approximation of coordinates of points on a plane: the ability to accurately determine the direction of the axis of the robot is lost on small arcs.
- Second, is the error of industrial robot caused by design features of the mechanism: flexibility; splits in kinematic pairs; temperature deformations in links of the robot [10, 11] and an input effect error of the servo drives [18, 19].
- Third, is the repeatability of the CMM that is rising during increasing of the distance between the laser tracker and the measuring object, and errors caused by features of the scanning mode.

So, there are three factors with many components that influence the estimation of uncertainty: error of the data processing algorithm; error anisotropy that is typical for industrial robots and also repeatability of the CMM. The calculation of uncertainty estimation represented in [12] has many components and is extremely difficult. Because of that, this article shows an iterative algorithm for searching uncertainty with the concentration ellipsoid model [16, 17]. Using of this algorithm helps to estimate the uncertainty when determining the axis of an industrial robot. The calculated uncertainty value can be used for analyzing and modeling. Also it should be noted that in this research the systematic error of a laser tracker was not taken into account for uncertainty. Only the dispersion in the conditions of repeatability was taken into account, since it is assumed that the systematic error was eliminated by other methods.

2. Experiment about spatial spread of error

First of all, we determine the geometrical view of spatial spread of errors during multiple determination of direction of the same axis of the industrial robot. It gives the concentration ellipsoid illustrated on figure 1. Based on the concentration ellipsoid an iterative algorithm for searching uncertainty was developed. This algorithm takes all factors with many components of uncertainties described above into account. For illustration of the concentration ellipsoid during multiple determination of the direction of the 5th axis $(\overline{\mathbf{n}})$ of KUKA KR 16 next experiment was carried out: during changing of the joint at an angle $\pi\backslash 3$, the flange with mounted reflector of the laser tracker rotated around the 5th axis while the laser tracker was scanning the trajectory (x_{ij}, y_{ij}, z_{ij}) of that rotation. It is evident that the trajectory of the reflector rotation is a circle that lies at plane α . For collecting the statistical data, 51 rotations $(j = 1..M, M = 51)$ in one direction were made. Each rotation contains 500 scan points $(i = 1..N, N = 500)$.

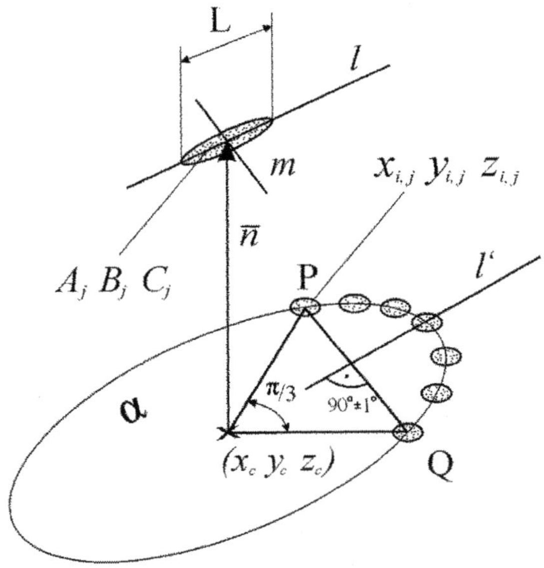

Fig. 1. Concentration ellipsoid determination.

Each coordinates selection (j) of motion trajectory for determining of the 5th axis parameters was fitted obtaining plane parameters (A_j', B_j', D_j') using the next formula:

$$\begin{pmatrix} A'_j \\ B'_j \\ D'_j \end{pmatrix} = \begin{pmatrix} \sum_{i=1}^{N}(x_{i,j}+\delta x_{ij})^2 & \sum_{i=1}^{N}(x_{i,j}+\delta x_{ij})\cdot y_{i,j} & \sum_{i=1}^{N}(x_{i,j}+\delta x_{ij}) \\ \sum_{i=1}^{N}(x_{i,j}+\delta x_{ij})\cdot(y_{i,j}+\delta y_{ij}) & \sum_{i=1}^{N}(y_{i,j}+\delta y_{ij})^2 & \sum_{i=1}^{N}(y_{i,j}+\delta y_{ij}) \\ \sum_{i=1}^{N}(x_{i,j}+\delta x_{ij}) & \sum_{i=1}^{N}(y_{i,j}+\delta y_{ij}) & N \end{pmatrix}^{-1} \cdot \begin{pmatrix} \sum_{i=1}^{N}(x_{i,j}+\delta x_{ij})\cdot(z_{i,j}+\delta z_{ij}) \\ \sum_{i=1}^{N}(y_{i,j}+\delta y_{ij})\cdot(z_{i,j}+\delta z_{ij}) \\ \sum_{i=1}^{N}(z_{i,j}+\delta z_{ij}) \end{pmatrix} \quad (1)$$

Where $(\delta x_{ij}, \delta y_{ij}, \delta z_{ij})$ - the uncertainty when measuring each coordinate (it is unknown in the experimental stage). Changing to parameters of axis direction (a_j, b_j, c_j) with formula:

$$\begin{pmatrix} a_j \\ b_j \\ c_j \end{pmatrix} = \cdot \begin{pmatrix} A'_j / \mathrm{D}'_j \\ B'_j / \mathrm{D}'_j \\ 1/\mathrm{D}'_j \end{pmatrix} \quad (2)$$

Translation them to unit vector using Euclidean norm:

$$\begin{pmatrix} A_j \\ B_j \\ C_j \end{pmatrix} = \cdot \begin{pmatrix} a_j / \sqrt{a_j^2 + b_j^2 + c_j^2} \\ b_j / \sqrt{a_j^2 + b_j^2 + c_j^2} \\ c_j / \sqrt{a_j^2 + b_j^2 + c_j^2} \end{pmatrix} \quad (3)$$

It is necessary to avoid error of the data processing algorithm of (1-3) at fitting when the trajectory points are very close to Cartesian axes (OX, OY, OZ) of the laser tracker coordinate system. We use special technique: two rotations of the laser tracker coordinate system according to the expression:

$$\begin{pmatrix} \mathrm{x}'_{i,j} \\ y'_{i,j} \\ z'_{i,j} \end{pmatrix} = \begin{pmatrix} \cos(\alpha_k) & 0 & -\sin(\alpha_k) \\ 0 & 1 & 0 \\ \sin(\alpha_k) & 0 & \cos(\alpha_k) \end{pmatrix} \cdot \begin{pmatrix} 1 & 0 & 0 \\ 0 & \cos(\alpha_k) & -\sin(\alpha_k) \\ 0 & \sin(\alpha_k) & \cos(\alpha_k) \end{pmatrix} \cdot \begin{pmatrix} \cos(\alpha_k) & -\sin(\alpha_k) & 0 \\ -\sin(\alpha_k) & \cos(\alpha_k) & 0 \\ 0 & 0 & 1 \end{pmatrix} \cdot \begin{pmatrix} \mathrm{x}_{i,j} \\ y_{i,j} \\ z_{i,j} \end{pmatrix} \quad (4)$$

Where the angle of rotation is α_k, $(k = 1..2)$.The first rotation is 30 degrees, the second is 45 degrees. After each rotation the unit vector of the fitted plane (j=1) is determined. Given unit vectors after fitting are turned back. Than three vectors: two after rotation and fitting, and one just fitting. Than three vectors are comparing among themselves. If a significant discrepancy between the three fitted vectors is detected, they are being compared, one of the three is excluded from the maximum mathematical error, and the coordinates are selected from any rotation of the remaining two. It should be noted that this problem shows only when the fitted axis is very close to Cartesian axes (OX, OY, OZ). The received set of unit vectors has a geometrical form of an ellipsoid with two axes (l (big axis) and m (small axis)), as it is shown on figure 1. The third axis of the ellipsoid is located along $(\overline{\mathbf{n}})$ and is negligible.

3. Mathematical model of the concentration ellipsoid

After receiving the geometrical view of the spatial spread of errors during multiple determination of direction of the same axis of the industrial robot we made the decision to develop a mathematical model of the concentration ellipsoid. The mathematical model of the concentration ellipsoid is needed for determining the dimension of L on axis l that depends on the number of points (N) on the arc and also depends on radius of reflector rotation (R). There is an assumption in the basis of model: each i-point on the arc $((\mathrm{x}_i,)_j, (y_i,)_j, (z_i,)_j)$ has normal distribution with equal mean square deviations in Cartesian coordinates $(\sigma / \sqrt{3} = \sigma_x = \sigma_y = \sigma_z)$ because the geometrical view of spatial spread of errors in each i-point is unknown:

$$\sigma = \sqrt{\sigma_x^2 + \sigma_y^2 + \sigma_z^2}$$

The rotation of reflector is imitated by next formula:

$$\begin{pmatrix} x_{mi,j} \\ y_{mi,j} \\ z_{mi,j} \end{pmatrix} = \begin{pmatrix} x_c + R \cdot \cos(\frac{\pi \cdot i}{v \cdot N}) + \sigma_{xi} \\ y_c + R \cdot \cos(\frac{\pi \cdot i}{v \cdot N}) + \sigma_{yi} \\ z_c + \sigma_{zi} \end{pmatrix}_j \quad (5)$$

Where $(x_{mi,j}, y_{mi,j}, z_{mi,j})$ - are the coordinates of model of reflector motion trajectory ($i = 1..N$ (number of points on arc), $j = 1..M$ (number of arcs)), (x_c, y_c, z_c) - coordinates of center rotation, R – radius of reflector rotation, v - number, corresponding to the value of arc. It is necessary to rotate the coordinate frame according to (4) for obtaining coordinates of model of reflector motion trajectory. For determining the unit vectors (A_{mj}, B_{mj}, C_{mj}) of the concentration ellipsoid model it is necessary to use (1-3).

Using the ellipsoid concentration model it is possible to receive ellipsoids with different L that depend on changing parameters of the model. It is reasonable to characterize several dependences: if the arc is more than π, the geometrical form of the ellipsoid changes to a disk, if the arc less than π, the ellipsoid is formed. Very small arc length significantly increases the dimensions of L of the ellipsoid. If the mean square deviation (uncertainty) changes, is almost proportionally changes the ellipsoid dimensions. The main advantage of the received concentration ellipsoid (model) is a knowledge about the projection of the ellipsoid big axis l' on plane α. It is always perpendicular (with deviation about ± 1 degree) to the vector PQ, as illustrated on figure 1. This is a key feature for robotics and mechatronics. Movement of the measuring area (the measured arc) allows to change the direction of the spatial spread of error l.

Applying the mathematical model represented by (1) – (5) we constructed the 3D surface illustrated on figure 2, which shows the L dimension dependence of the concentration ellipsoid from number of points (N) on arc and from radius of reflector rotation (R). The graph of a 3D surface shows the size L considerably increasing at small values of the radius of reflector rotation and small amount of points on an arc. This 3D surface also shows behavior of the mathematical model of concentration ellipsoid with mean square deviation of 0.0375 mm on a π/3 arc.

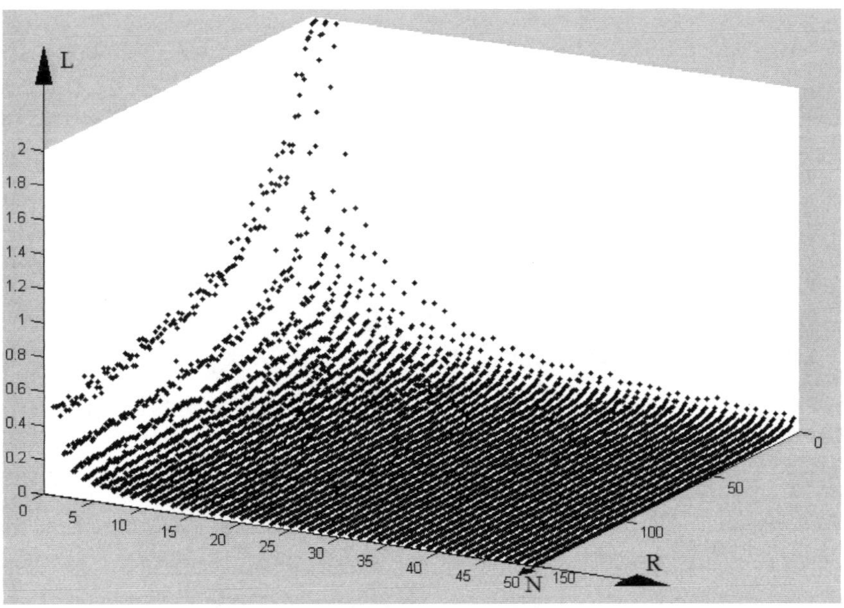

Fig. 2. 3D surface that shows the behaviour of big axis (L) of concentration ellipsoid.

4. Iterative uncertainty search algorithm

Iterative algorithm for searching uncertainty is shown on figure 3. The algorithm is based on the mathematical model of the concentration ellipsoid (1) – (5) described above. The initial value of uncertainty is chosen as 0.001 mm. The increase in value of uncertainty occurs with some step the amount of which is chosen empirically, as a reference serves the step number and the final operating time of the algorithm. At small steps the operating time of an algorithm increases considerably, at big steps the accuracy of searching the uncertainty decreases.

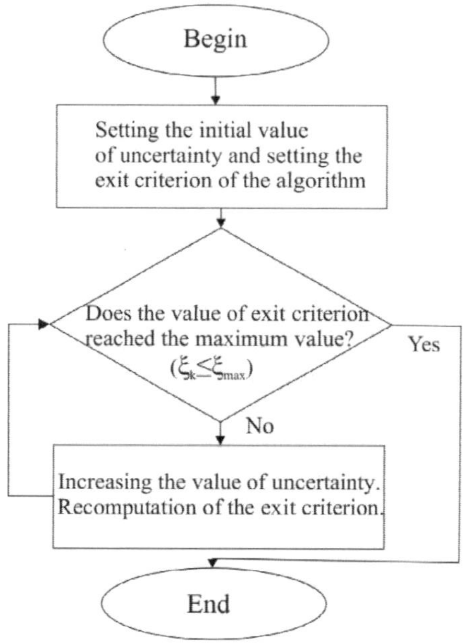

Fig. 3. Iterative algorithm for searching uncertainty.

The exit criterion of iterative algorithm is the mean square deviation ξ_k, where k is the number of algorithm iteration:

$$\xi_k = \sqrt{\frac{1}{M^2} \cdot \sum_{p=1}^{M^2} (\mathrm{d}_p - \mathrm{d}_{mid})^2} \tag{6}$$

Where d_p - are distances between the received points (A_j, B_j, C_j) (or (A_{mj}, B_{mj}, C_{mj})), that create an ellipsoid and characterize the uncertainty $(p = 1..M \cdot M)$. d_{mid} - mean value of samples of d_p. ξ_{max} determines after measuring and receiving the concentration ellipsoid. Recalculation of the exit criterion (mean square deviation ξ_k) occurs according to the formulas (1) – (6) by means of creation of the concentration ellipsoid.

The result of the algorithm work is presented in table 1. The row mathematical model of concentration ellipsoid (1) – (5) shows ellipsoid model parameters. The values which are the result of work of the algorithm are represented in row value of σ_A. Also, in the corresponding row, standard deviations from σ_A are estimated.

Mathematical model of concentration ellipsoid	R=20, mm N=150 M=100 σ=0.01, mm	R=50, mm N=500 M=15 σ=0.1, mm	R=160, mm N=300 M=25 σ=0.035, mm	R=300, mm N=500 M=200 σ=0.25, mm	R=700, mm N=300 M=50 σ=0.05, mm
Value of σ_A	0.011	0.07	0.035	0.245	0.0495
Standard deviations from σ_A	±0.0005	±0.01	±0.0025	±0.03	±0.0005

Table 1. Results of work of iterative algorithm for searching uncertainty using data of concentration ellipsoid model.

Fluctuations and systematic errors of values σ_A from model uncertainty σ are caused by the choice of exit criterion of the algorithm. Different exit criterion give different results from the algorithm. The best results are obtained by applying the algorithm at a value $M = 25;50$. For estimation of real values of uncertainty the experiment with the industrial robot KUKA KR 16 was made. For all joints individually all motion trajectories of the reflector fixed on a robot flange were scanned. A set of coordinates for each circle consisted of 300 points. For receiving the concentration ellipsoid of uncertainty each circle was received 25 times. Results of the experiment are represented in table 2.

Number of joint (axis of industrial robot)	1	2	3	4	5	6
Radius of reflector rotation R	1115.780	1071.275	854.973	24.704	184.542	24.797
Uncertainty σ	0.030	0.016	0.007	0.053	0.005	0.051

Table 2. Results of work of the iterative algorithm for searching uncertainty using experimental data.

It is possible to do the following conclusions using experimental data: uncertainties have big values on small radiuses, therefore it is reasonable to increase rotation radius for finding of the 4th and 6th axes of industrial robots by additional equipment. The development of a special calibration device is a next step of our research. The tendency of increasing uncertainty on big circles was found. That could be also taken into account for further mathematical modelling of the concentration ellipsoid. Because of insufficient quantity of experimental data, the dependence of uncertainty from the radius of reflector rotation was fitted with a considerable error, but has the functional dependence close to:

$$\sigma_A = \frac{f}{R} + g \cdot R^2 + h \cdot R + e$$

Where f, g, h, e - are the parameters of dependence of uncertainty from the radius of the reflector rotation.

5. Conclusion

There is uncertainty when measuring different geometrical elements of industrial robots, multi-axis machines or other complex mechanisms for their calibration using CMM. This uncertainty has a large number of components. It is extremely difficult to estimate this uncertainty using ISO [12]. This article represents approach for uncertainty estimation of axes direction determination of industrial robot.

We revealed the concentration ellipsoid – a geometrical view of the spatial spread of error after repeated axis determination of an industrial robot. The mathematical model of the concentration ellipsoid was developed. In this paper using a 3D surface we have analysed the dimension of the big axis of the concentration ellipsoid that depends on number of points on arc and also on radius of reflector rotation. An iterative uncertainty search algorithm for the industrial robot axis and its exit criterion was developed. Uncertainty of each axis of the industrial robot was estimated. The standard deviation of the iterative algorithm was calculated. The dependence of uncertainty from increasing the rotation radius of the measuring device reflector was experimentally revealed. This dependence has influence on optimal measurement area. By results of the experiments and by results of the work of the algorithm it is recommended to carry out the uncertainty estimation when calibrating the industrial robot or a mechatronic device since values of uncertainty for various axes will be different. It is also reasonable to develop and use the specialized equipment for decreasing the uncertainty in case of determination of axes of industrial robots. The development of specialized equipment is the purpose of further research. Also, it is reasonable to carry out similar estimation of uncertainty for manufacturers of the CMM to provide better performance of their equipment.

6. Acknowledgments

The reported study was partially supported by RFBR, research project №15-58-78024 Итал_а

7. References

[1] Poduraev, Yu. V. (2006). Mechatronics: fundamentals, methods, instrumentation, M.: Engineering, ISBN 5-217-03355-X, Moscow.

[2] Vorotnikov, A. A., Poduraev, Yu. V., Romash E.V. (2015) Estimation of error in determining the centers of rotation of links in a kinematic chain for industrial robot calibration techniques, Measurement Techniques, Vol. 58, No. 8, November, 2015, pp. 864-871. DOI: 10.1007/s11018-015-0809-9

[3] Vorotnikov, A[ndrei]; Bashevskaya, O[lga]; Ilyukhin, Y[uri]; Romash, E[lena]; Isaev, A[.] V[.] & Poduraev, Y[uri] (2016). Geometrical Approach for Industrial Robot Axis Calibration Using Laser Tracker, Proceedings of the 26th DAAAM International Symposium, pp.0897-0904, B. Katalinic (Ed.), Published by DAAAM International, ISBN 978-3-902734-07-5, ISSN 1726-9679, Vienna, Austria. DOI: 10.2507/26th.daaam.proceedings.125

[4] Grechishnikov, V. A., Petukhov, Yu. E., Pivkin, P. M., A. V. Isaev, V. B. Romanov, and P. V. Domnin. Lathe Turning of Complex Shaped Parts Providing Desired Surface Microrelief / Russian Engineering Research, 2016, Vol. 36, No. 3, pp. 229–231. ISSN 1068-798X DOI: 10.3103/S1068798X16030059

[5] Grechishnikov, V. A., Isaev, A. V., Ilyukhin, Yu. V., Pivkin, P. M., Vorotnikov, A. A., Kharchenko, A. N., Bianchi, G., Leonesio, M., Pedrocchi, N., Molinari, Robotic cells for machinning and system of their instrumental providing. M.: FGBOU IN MSTU OF STANKIN, 2015, ISBN 978-5-7028-0683-9, Moscow.

[6] Grechishnikov, V. A., Isaev, A. V., Ilyukhin, Yu. V., Pivkin, P. M., Vorotnikov, A. A., Kharchenko, A. N., Bianchi, G., Leonesio, M., Pedrocchi, N., Molinari, T. L., Building principles of robotic machining systems and their cutting tool system, Vestnik MSTU «STANKIN», №4 (35), 2015. pp. 46-51, ISSN 2072-3172.

[7] D. Klimov, Y. Poduraev, Design and Experimental Testing of a Robotic System for High Speed Recording, in: Annals of DAAAM for 2014 & Proceedings of the 25th International DAAAM Symposium, ISSN 1877-7058, ISBN 978-3-901509-97-1, Editor B[ranko] Katalinic, Published by DAAAM International, Vienna, Austria, 2014, pp. 1332-1337.

[8] Klimov, D[aniil] & Poduraev, Y[ury] (2016). Robotic Flight Follower Systemfor High Speed Recording, Proceedings of the 26th DAAAM International Symposium, pp.0720-0726, B. Katalinic (Ed.),Published by DAAAM International, ISBN 978-3-902734-07-5, ISSN 1726-9679, Vienna, Austria

[9] Chernov, N., (2011). Circular and linear regression fitting circles and lines by least squares. CRC Press Taylor & Francis Group, ISBN 978-1-4398-3590-6, USA.

[10] Bashevskaya O.S., Gulieva R.M., Romash E.V., Melnichenco E.A. Research of temperature deformations of end bar. Vestnik MSTU «STANKIN», №4. pp. 36–40.ISSN 2072-3172.

[11] Bashevskaya, O. S. Bushuev, S. V., Poduraev, Yu. V., Koval'skii, M. G., Kainer, G. B., Romash E. V. An Investigation of the Influence of Temperature Deformations on the Precision of Linear Measurements, Measurement Techniques, Vol. 56, December, 2013, Issue 9, pp 1016–1020, DOI: 10.1007/s11018-013-0322-y

[12] ISO/IEC Guide 98-3:2008 - Uncertainty of measurement

[13] Nubiola, A., Bonev, I. A.. Absolute calibration of an ABB IRB 1600 robot using a laser tracker. Robotics and Computer-Integrated Manufacturing, Vol. 29 (1), February, 2013, pp. 236-245, DOI: 10.1016/j.rcim.2012.06.004

[14] Yahui Gan, Xianzhong Dai, and Donghui Dong. Robot Calibration for Cooperative Process under Typical Installation, Available from: http://dx.doi.org/10.1155/2014/576420

[15] Klimchik A., Pashkevich A., Chablat D.,Hovland G. Compliance error compensation technique for parallel robots composed of non-perfect serial chains, Available from: http://dx.doi.org/10.1016/j.rcim.2012.09.008

[16] Kramer, H. (1975). Mathematical methods in statistics, M.:Inostranaya literature, Moscow.

[17] Belyaev, Yu.K., Chepurin, E.E., (1982) The basis of statistics, M.: Moscow University Publisher, Moscow.

[18] Ilyukhin, Y[uri] & Kolesnichenko, R[uslan] (2016). Impedance Control of High-Precision Geared Servo Drives with Two Motors for Technological Robots, Proceedings of the 26th DAAAM International Symposium, pp.0599-0607, B. Katalinic (Ed.), Published by DAAAM International, ISBN 978-3-902734-07-5, ISSN 1726-9679, Vienna, Austria, DOI: 10.2507/26th.daaam.proceedings.081

[19] Ilyukhin Yu. V. Poduraev Yu. V. Improving the Precision of Mechatronic Robot Drives. // ISSN 1068_798X, Russian Engineering Research, 2016, Vol. 36, No. 4, pp. 328–334. © Allerton Press, Inc., 2016. Original Russian Text © Yu.V. Ilyukhin, Yu.V. Poduraev, 2015, published in STIN, 2015, No. 9, pp. 30–37.

DOI: 10.2507/27th.daaam.proceedings.073

USING 80/20 PRINCIPLE TO IMPROVE DECISION MAKING AT PFMEA

Nikola Banduka, Ivan Mačužić, Željko Stojkić, Igor Bošnjak & Ivan Peronja

This Publication has to be referred as: Banduka, N[ikola]; Macuzic, I[van]; Stojkic, Z[eljko]; Bosnjak, I[gor] & Peronja, I[van] (2016). Using 80/20 Principle to Improve Decision Making at PFMEA, Proceedings of the 27th DAAAM International Symposium, pp.0487-0492, B. Katalinic (Ed.), Published by DAAAM International, ISBN 978-3-902734-08-2, ISSN 1726-9679, Vienna, Austria
DOI: 10.2507/27th.daaam.proceedings.073

Abstract

Turbulent market of today is forcing companies to respond on customer demands very fast, with right quality and acceptable price. Producing in quality is often hard and time-consuming job, with many risks. One of those risks can be defect (failure) appearance during the production process. One of quite often used analysis for this purpose is failure mode and effect analysis. This method is relatively accurate, but it is a very time-consuming method, and companies need to respond fast to customer demands with right quality. Another problem in this method is that all products and failures are ranked with the same value, which is an ideal case. However, more often case is that less percentage of products give most of the company profit. So, in this paper, the principle 80/20 (old Pareto principle) was used to help in decision making due to PFMEA realization.

Keywords: Production process; FMEA; PFMEA; Principle 80/20.

1. Introduction

Turbulent market of today is pressuring companies to respond on customer demands very fast, with right quality and acceptable price. These three factors (Time, quality, and price) are in an interaction and interdependence, so ideal balance between them should be found with the aim to achieve customer satisfactory [1]. Producing in quality is often hard and time-consuming job, with many risks. One of those risks can be defect (failure) appearance during production process. This problem can be controlled with various tools, techniques, and methods, and one very famous is Failure mode and effect analysis wider known as FMEA.

In general FMEA is a systematic method for identifying and preventing product and process problems before they occur [2]. FMEA can be used for specific needs, design, process, system, machine, service implementation phase, operations, maintenance, etc [1, 3]. But the most often it is used for product design or production process. This types are also known as DFMEA and PFMEA. This paper is focused on PFMEA.

Modern PFMEA dates back to 1990s, when three leading American companies Chrysler LLC, Ford Motor Company, and General Motors Corporation forced all their suppliers to use PFMEA as preventing tool [1, 2]. That created problems and confusion at suppliers. So Automotive industry agency group (AIAG) integrated different FMEA regulations in one unity and in 1993, a guide "Potential FMEA" was published (4th edition from 2008 is active today) [4]. In short, these three companies made PFMEA famous and important. Since then PFMEA has been used in various industries and for different purposes. This method also has many shortcomings, and various authors have tried to improve it by modifying, or combining and integrating it with some other methods, but in practise it is still mostly used traditional approach to PFMEA.

Study conducted with 150 automotive companies shows that the highest number of surveyed companies see PFMEA as additional administrative work which wastes a lot of time, and gives back less benefits [5]. The biggest problem lays in this. It should be different, companies should use less time to get more benefits.

Solution might be found in philosophy of the 80/20 principle – the secret to achieving more with less [6]. 80/20 Principle is old Pareto's principle in modern era rehabilitated and improved by Richard Koh. Main idea of this principle is that 80% of total benefits can be achieved with 20% of total effort. And that is exactly what is needed for PFMEA. So, in this paper a combination of 80/20 Principle with PFMEA in a practical example is presented. The idea was to make a model which will calculate PFMEA indexes with additional coefficient achieved from 80/20 principle. The purpose of 80/20 principle in this paper is segregation of products which give the highest percentage of total profit, and giving them higher priority. As result, a model with better risk priority index which facilitates decisions making at PFMEA was achieved.

2. PFMEA

PFMEA is a team based method, and for this purpose, a classical brainstorming technique is usually used. It is deductive type of method with structural top-down approach [1, 7, 8]. Failure problems should be solved according to the highest priority defined with Risk priority number index (RPN). Traditional RPN is calculated by multiplication of S (Severity), O (occurrence) and D (detection) index (see equation 1). And corrections should be done mandatory if RPN exceeds 100, or if one of S, O, or D indexes has 9 or 10 value of failure.

First FMEA was founded for the USA military needs under the name "Military procedure MIL-P-1629" [9]. The Principle that was used at that time for FMEA report fulfilment as well as concept of that analysis is not in use anymore. First formal use of FMEA analysis was in aerospace industry.

Today FMEA is in wide use all over the world and it has become standard practice in Asian, American, and European companies [10]. The most common practice of use of FMEA is in automotive and aerospace industry, and in addition to this it is also used in military industry, industry for manufacturing of electrical components, nuclear industry, and in medicine [1, 7-14]. Addition to this, there is also use of FMEA in retail manufacturing, power plants, construction, telecommunication, mechanical and service industry, companies for computer systems embedding, companies for hardware and software making, information systems, food production companies, plastic injection companies, etc. [1, 2, 8-10, 14-19].

$$RPN = S * O * D \tag{1}$$

3. 80/20 Principle

80/20 Principle is showing how more benefits can be achieved with less investment (see fig. 1). But this does not mean that proportion must be 80/20. In most of cases it is, but not in all. The number even does not need to have 100 in sum.

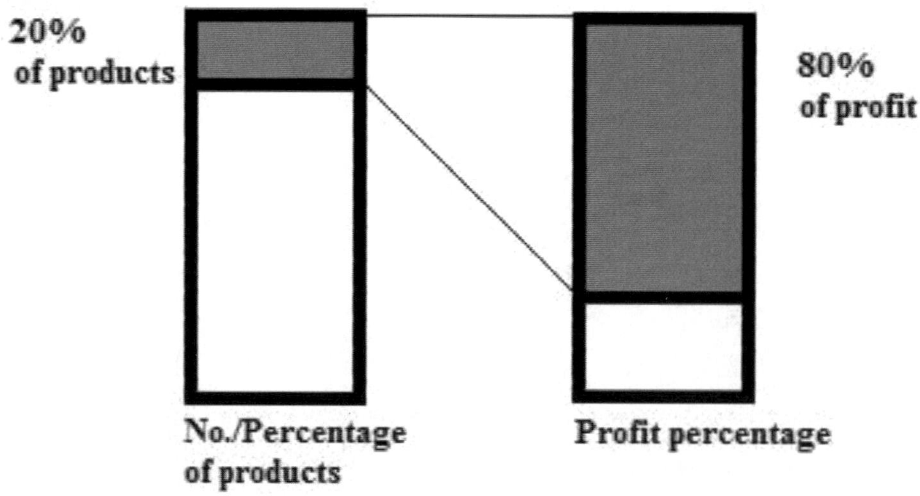

Fig. 1. The ratio of 80/20 principle [6].

The precursor to 80/20 principle was discovered by Italian economist Vilfredo Pareto in 1897 [6]. From that time this principle has had a lot of names, and some of them are: the Pareto principle, the Pareto rule, the rule 80/20, etc. After Pareto's death this principle was forgotten until World War II. Many years later Harvard philology professor Zipf, G. K. [20] started to use Pareto rule again and he named it "Principle of least effort". It is evident that this researcher, among other fields, applied this principle in industry. Another and maybe even more important scientist for this kind of research interests, Juran, J. M. [21] also used this principle just two years later. Juran was responsible for quality revolution from 1950s to 1990s. Opposite to Zipf, Juran was employed in industry as industrial engineer. He used 80/20 principle as a principle for quality improvement, and called it "rule of important minority".

Today, this principle is very often used in everyday life. So it can be used by any intelligent person, organization, social group, company, etc. It is evident that 80/20 principle has brought great discoveries in business and economy. But, it also brought many benefits to various companies such as IBM, Apple, Lotus, Microsoft, etc [6].

4. Using 80/20 principle to improve PFMEA decision making

There are two known applications of Pareto and FMEA which are similar to this approach, but very different in purpose and structure [1, 22, 23]. The idea of using 80/20 principle to improve PFMEA is in focusing on failures which occur at products which are more valuable for company than others. During PFMEA realization, a team is looking on products and failures as they are all the same, and priority are those with the highest RPN coefficient. That is an ideal case. But in practise, it is usually different. Value of failure also depends on product value, and the product value depends on the amount of contribution or profit which it brings to a company. So, in ranking of RPN, value of the products should be included.

Why is this so important? In industry, it is not so simple to invest in everything that is needed. Top management is usually focused on direct return from investments. When failure is occurring, it is normally to made prevention and correction which is time-consuming and requires some additional costs. So, 80/20 principle should point out in which process and failures it is more profitable to invest, and which are less important according to product profitability priority. For this purpose, a coefficient K_{VP} is invented which is included in traditional RPN (see equation 2).

K_{VP} is coefficient of product value, defined by 80/20 principle. K_{VP} may have two numerical values (1 or 2). Less percentage of total products which gives bigger amount of profit gets value 2, while higher percentage which gives less percentage of profit gets value 1. The principle of K_{VP} defining will be shown in the next chapter in a practical example.

$$RPN = S * O * D * K_{VP} \quad (2)$$

5. Example

In order to show improvements and difference between traditional PFMEA and improved PFMEA approach with using of 80/20 principle, 10 different arbitrary failures which respectively occurs on 10 different products were taken. Traditional PFMEA calculation is shown in Table 1.

The highest priority of this 10 failures was defined by classical RPN index. Failures were classified in table 1 according to highest value of RPN respectively from the highest to the lowest. So failures calculated on traditional way are ready for comparation.

Failure	S	O	D	RPN	The highest priority
F1	6	4	3	72	F8
F2	3	7	4	84	F10
F3	5	5	2	50	F2
F4	3	2	3	18	F9
F5	2	8	1	16	F1
F6	3	5	2	30	F3
F7	5	3	8	40	F7
F8	7	6	5	210	F6
F9	4	4	5	80	F4
F10	6	3	7	126	F5

Table 1. Example of traditional PFMEA calculation.

Now is shown an example of improved PFMEA with application of 80/20 principle, with K_{VP} coefficient influence included. Table 2 shows 10 arbitrary used products (with profit data included) which contribute to a company. Software extension for MS Excel – QI Macros 2016 [24] was used for 80/20 principle calculation. In this case all failures belongs to different products respectively. But in some other cases it is possible to have more than one failures for one product. 80/20 chart (see Fig. 2) is based on products on one side and profit data on the other side from Table 2.

Product	Profit (€)	Profit [%]
P1	5 780	1.307
P2	32 400	7.328
P3	37 000	8.369
P4	20 000	4.524
P5	6 950	1.572
P6	13 600	3.076
P7	68 900	15.584
P8	160 000	36.188
P9	19 000	4.297
P10	78 500	17.755
Total	442 130	100

Table 2. Data needed for application of 80/20 principle in determining K_{VP}.

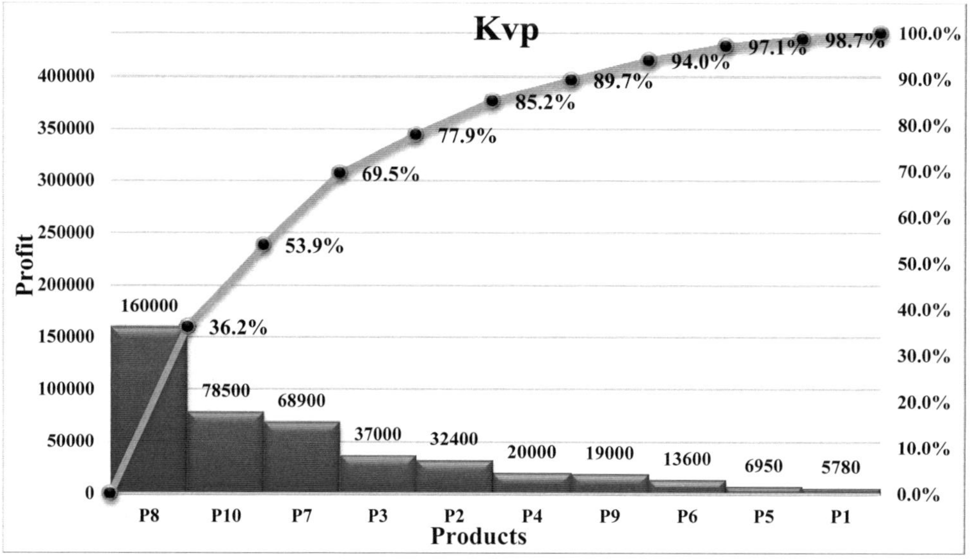

Fig. 2. Using QI Macros 2016 for Defining of K_{VP} coefficient.

As it can be seen in the chart, almost 70% of total profit of the company is achieved from 30% of total products. Therefore, in this case, 70/30 principle is active. Value 2 for K_{VP} coefficient is assigned to products with 30% (P8, P10, and P7) of less, and for the other products (P3, P2, P4, P9, P6, P5, and P1) of more the value of K_{VP} is 1.

After defining K_{VP} coefficient, everything needed for a new approach calculation is ready. Table 3 shows calculation of new RPN with K_{VP} coefficient included.

Failure	S	O	D	K_{VP}	RPN	The highest priority
F1	6	4	3	2	144	F8
F2	3	7	4	1	168	F3
F3	5	5	2	2	200	F2
F4	3	2	3	1	18	F1
F5	2	8	1	1	36	F10
F6	3	5	2	1	60	F9
F7	5	1	8	1	40	F6
F8	7	6	5	2	840	F7
F9	4	4	5	1	80	F5
F10	6	3	7	1	126	F4

Table 3. Calculation of RPN with K_{VP} coefficient included.

Achieved results (see Table 1 and Table 3) show that a new approach sets a priority to the failures different from traditional approach. The difference is seen in 8 failures, while two are in the same place of priority. So this improved method shows that it is more efficient then traditional method.

6. Conclusion

In this paper a new approach with new coefficient included (K_{VP}) is shown which is defining products priority. One of very important things to deffine is which products mostly contribute to the company. In PFMEA all products are the same, as failures. Failures priorities are defined according to the highest RPN number, so importance of products is neglected. In this paper, 80/20 Principle helped in defining the highest priority products, which mostly contribute to the company. This combination of approaches proved bigger accuracy and it is easy for use.

One of shortcomings in this new approach might be separation of failures. Because this approach requires centralized base of failures. But it depends on a point of view. If it is observed from the side of time and cost it is an advantage, because it saves a lot of time and it is oriented to products which contribute the most to the company. But if it is observed from the side of company which wants to solve all problems and have time for all of it, then it can be complicated for use, because this method is primarily decision making oriented. Also, possible solution is that all products give the same value or 50/50 ratio, but that is rare or ideal case.

Future research should be cost oriented failure, because that is one more factor which influences PFMEA, but in combination with a new approach. And it is highly recommended to use it in combination with product priority coefficient K_{VP}.

7. Acknowledgments

This work has been fully supported by Croatian Science Foundation under the project Innovative Smart Enterprise – INSENT (1353).

8. References

[1] Stamatis, D. H. (2003). Failure mode and effect analysis: FMEA from theory to execution. ASQ Quality Press.

[2] Mikulak, R. J., McDermott, R., & Beauregard, M. (2008). The basics of FMEA. CRC Press.

[3] Liu, H. C. (2016). FMEA Using Uncertainty Theories and MCDM Methods. Springer, Singapore.

[4] Chrysler LLC, Ford Motor Company, General Motors Corporation (2008). POTENTIAL FAILURE MODE AND EFFECT ANALYSIS (FMEA), Available from www.engmatl.com/home/finish/20-engineering.../160-fmea-manual, Accessed: 2016-08-29

[5] Johnson, K. G., & Khan, M. K. (2003). A study into the use of the process failure mode and effects analysis (PFMEA) in the automotive industry in the UK. Journal of Materials Processing Technology, Vol.139., No.1., August 2003, pp.348-356, ISSN: 0924-0136

[6] Koch, R. (1999). The 80/20 principle: the secret to achieving more with less, Crown Business, ISBN:0385491743, Available from http://leadershipcoachingblog.com/wp-content/uploads/2012/03/the-80-20-principle-to-achieve-more-with-less-effort1.pdf, Accessed: 2016-08-29

[7] Singh, V., Pungotra, H., Singh, S., & Gill, S. S. (2013). Prioritization of Failure Modes in Process FMEA using Fuzzy Logic. International Journal Of Enhanced Research In Science Technology & Engineering, Vol. 2, No.2, February 2013, ISSN: 2319-7463

[8] Dinmohammadi, F., & Shafiee, M. (2013). A fuzzy-FMEA risk assessment approach for offshore wind turbines. International Journal of Prognostics and Health Management, Vol. 4, No.13, pp.59-68, 2013, ISSN:2153-2648

[9] Santos, F. R. S. D., & Cabral, S. (2008). FMEA and PMBOK applied to project risk management. JISTEM-Journal of Information Systems and Technology Management, Vol.5, No.2, 347-364, 2008, ISSN:1807-1775

[10] Pentti, H., & Atte, H. (2002). Failure mode and effects analysis of software-based automation systems, STUK - Radiation and Nuclear Safety Authority. Helsinki: August 2002, p.37. Available at http://www.fmeainfocentre.com/handbooks/softwarefmea.pdf, accessed June 13, 2016.

[11] Breiing, A. J., & Kunz, A. M. (2002). Critical Consideration and Improvement of the FMEA, Proceedings of the TMCE, May 7 -11, 2012,Karlsruhe, Germany, pp. 519-530

[12] Arabian-Hoseynabadi, H., Oraee, H., & Tavner, P. J. (2010). Failure modes and effects analysis (FMEA) for wind turbines. International Journal of Electrical Power & Energy Systems, Vol. 32, No.7, pp.817-824., 2010, ISSN: 0142-0615

[13] Guimarães, A. C. F., & Lapa, C. M. F. (2007). Fuzzy inference to risk assessment on nuclear engineering systems. Applied Soft Computing, Vol. 7, No.1, pp.17-28., 2007, ISSN: 1568-4946

[14] Deng, W. J., Chiu, C. C., & Tsai, C. H. (2007). The Failure Mode and Effects Analysis Implementation for Laser Marking Process Improvement: A Case Study. Asian Journal on Quality, Vol. 8, No. 1, pp.137-153, 2007, ISSN: 1598-2688

[15] Khasha, R., Sepehri, M. M., & Khatibi, T. (2013). A Fuzzy FMEA Approach to Prioritizing Surgical Cancellation Factors. International Journal of Hospital Research, Vol.2, No.1, pp.17-24, 2013, ISSN:2322-2085

[16] Reiling, J. G., Knutzen, B. L., & Stoecklein, M. (2003). FMEA-the cure for medical errors. Quality progress, Vol. 36, No. 8, pp.67, 2003, ISSN:0033524X

[17] Stålhane, T., & Wedde, K. J. (1997). Modification of safety critical systems: an assessment of three approaches. In: Reliability, Quality and Safety of Software-Intensive Systems, pp.134-147, Springer US, ISBN: 978-0-412-80280-5

[18] Goddard, P. L. (2000). Software FMEA techniques Proceedings of Reliability and Maintainability Symposium, February 2000, Los Angeles, California USA, ISSN:0149-144X ISBN: 0-7803-5849-X, pp.118-123, IEEE

[19] Scipioni, A., Saccarola, G., Centazzo, A., & Arena, F. (2002). FMEA methodology design, implementation and integration with HACCP system in a food company. Food control, Vol. 13, No. 8, pp.495-501, 2002

[20] Zipf, G. K. (1950). Human Behaviour and the Principle of Least Effort. The Economic Journal, Vol. 60, No. 240, pp. 808-810, December 1950, ISSN: 00130133

[21] Juran, J. M. (1954). Universals in management planning and controlling.Management Review, Vol. 43, No. 11., pp.748-761, 1954

[22] Zasadzień, M. (2014). Using the Pareto diagram and FMEA (Failure Mode and Effects Analysis) to identify key defects in a product. Management Systems in Production Engineering, Vol. 4, No. 16, pp.153-156, 2014, ISSN: 2299-0461

[23] Arvanitoyannis, I. S., & Varzakas, T. H. (2007). Application of failure mode and effect analysis (FMEA), cause and effect analysis and Pareto diagram in conjunction with HACCP to a potato chips manufacturing plant. International journal of food science & technology, Vol. 42, No. 12, pp.1424-1442, 2007, SSN: 0022-1155

[24] https://www.qimacros.com/, (2016). QI Macros 2016, KnowWare International, Inc., Denver, CO 80222 USA, Accessed on: 2016-08-29

DOI: 10.2507/27th.daaam.proceedings.074

DESIGN OF A COMPOSITE LEAF SPRING FOR RAILWAY VEHICLES

Frantisek Sedlacek, Petr Bernardin, Vaclava Lasova

This Publication has to be referred as: Sedlacek, F[rantisek]; Bernardin, P[etr] & Lasova, V[aclava] (2016). Design of a Composite Leaf Spring for Railway Vehicles, Proceedings of the 27th DAAAM International Symposium, pp.0493-0500, B. Katalinic (Ed.), Published by DAAAM International, ISBN 978-3-902734-08-2, ISSN 1726-9679, Vienna, Austria

DOI: 10.2507/27th.daaam.proceedings.074

Abstract

The paper deals with the design/optimization of a leaf spring for railway vehicles using composite materials. The optimization was carried out based on the main mechanical parameters of the current design solution for steel leaf springs. A numerical simulation was created to obtain the data, then the simulation was verified using an experimental test. Glass fibre reinforcement with epoxy resin was chosen as the material for the composite leaf spring. The main shape of the composite leaf spring was created using geometrical optimization. An advanced finite element model which modelled individual plies of the laminate was created to check the strength of the composite spring. A maximum stress strength criterion and transverse shear interlaminar criterion were used to evaluate the failure index.

Keywords: composite materials; leaf spring; numerical simulation; geometric optimization; maximum stress failure criterion

1. Introduction

The use of composite materials as structural material is no longer a privilege of elite design in the aerospace industry or motorsport, but it is increasingly applied to private cars and railway vehicles. One of the best applications is for vehicle springs. Using composite materials in the construction of springs not only ensures a great reduction in weight, but also has a whole range of other advantages: the possibility of reducing vibration, noise and ride harshness due to high damping factors; the absence of corrosion (lower maintenance cost); lower tooling cost (lower manufacturing cost) [1].

The design of composite springs has already been dealt with in several works: Beardmore [2] or Morris [3] were dealt with a design composite leaf spring for private cars; Daugherty [4] was dealt with using a composite leaf spring for trucks; Krall [5] was dealt with investigation of the dynamic behaviour of the composite leaf springs and Mahdi [6] and Talib [7] were dealt with very similar solution with elliptic springs for vehicle suspension.

However, this article deals with the optimization of leaf springs for heavy vehicles; specifically, for rail tram vehicles. The basic shape of the composite leaf spring is sought with the help of a geometrical optimization.

2. The current solution

The current solution is a conventional steel leaf spring, which is composed from six plates. The individual plates of the spring are connected using a central housing with a connecting pin. The mounting of the leaf spring is ensured by two eyes which are a part of the lower leaf spring. The current design solution for a steel leaf spring is shown in Fig. 1.

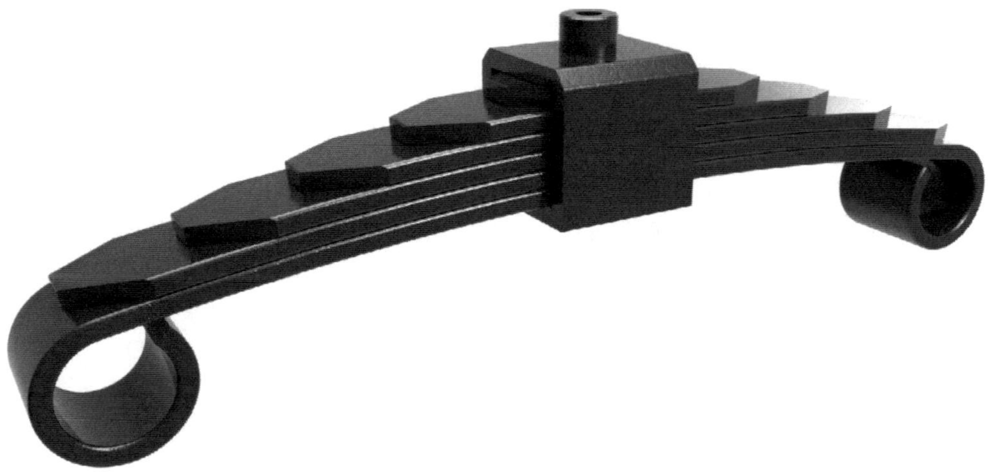

Fig. 1. Current solution - steel leaf spring

3. Determination of the basic parameters

A numerical simulation of the current design solution of the steel leaf spring was carried out to obtain the basic parameters of the spring (in particular, the stiffness of the spring). The five steel leaf spring was experimentally tested to validate the numerical simulation.

3.1. Numerical simulation of the current solution (steel leaf spring)

In the first phase, a finite element model of the assembly of the leaf spring was created. A 3D finite element mesh (combination of the elements; CQUAD8 and CTRIA4) was applied to the model of the spring (see Fig. 2.). Steel 1.7106 (56SiCr7 according to EN10089) was used. This material is one of the most common types used for steel leaf springs. Its mechanical properties are given in Table 1. The physical properties of the FE mesh were defined as nonlinear by using a bilinear engineering stress-strain curve.

E [GPa]	207	*Young's modulus*
$R_{p0.2}$ [MPa]	1350	*Yield strength*
R_m [MPa]	1520	*Tensile strength*
ν [-]	0.3	*Poisson's ratio*
ρ [kg/m^{-3}]	7850	*Density*
A [%]	6	*Elongation*

Table 1. Mechanical and physical properties of spring steel 1.7106 (56SiCr7 - EN 10089)

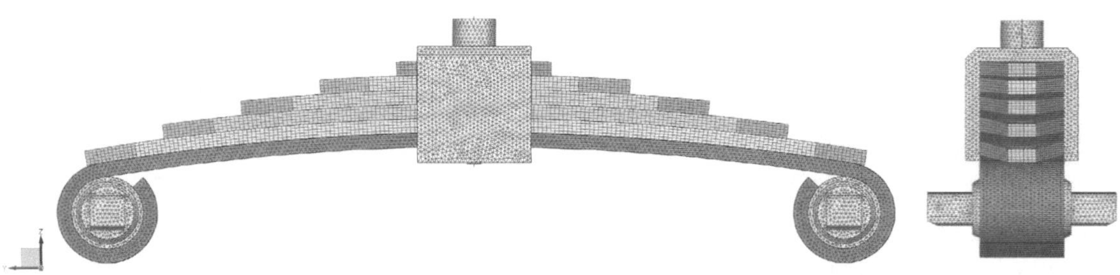

Fig. 2. Finite element model of a steel leaf spring

The numerical simulation was solved as nonlinear by using the NX Nastran 11 solver in Siemens NX 11 software. The nonlinear solver was used because there was a large displacement in the assembly. Boundary conditions of the type 'surface to surface contact' were used between individual parts of the assembly of the leaf spring. The relevant degrees of freedom were removed in the fastening eyes and on the central pivot. Three loading conditions were solved; load from an empty vehicle; load from a fully loaded vehicle; and a dynamically increasing load from a fully loaded vehicle. The results of the individual loads are given in Fig. 3.

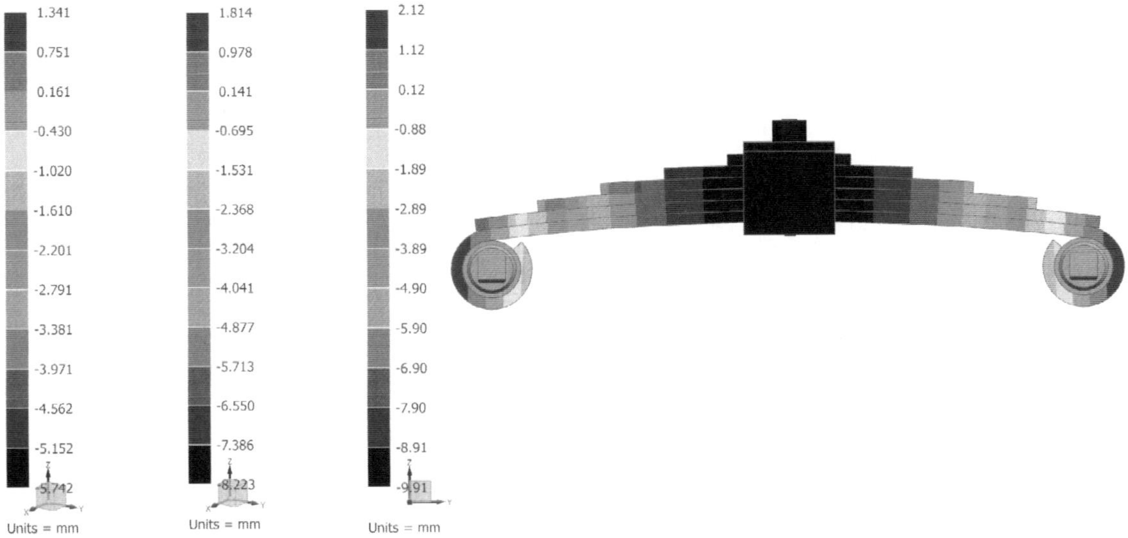

(a) load from empty vehicle **(b) load from fully loaded vehicle** **(c) load from fully loaded vehicle with dynamic increase**

Fig. 3. Displacement in vertical axis (direction 'z') for different load conditions on the steel leaf spring (mm)

3.1. Experimental measurement of the stiffness of the existing steel leaf springs

Experimental testing of the stiffness of an existing steel leaf spring was carried out to verify the numerical simulation. The measurement was done on an Inova FU-O-250 hydraulic testing machine. The spring was attached by using a special jig that allows free motion of the spring in its longitudinal axis (direction 'y'). The testing machine with the jig and the spring attached are shown in Fig. 4.

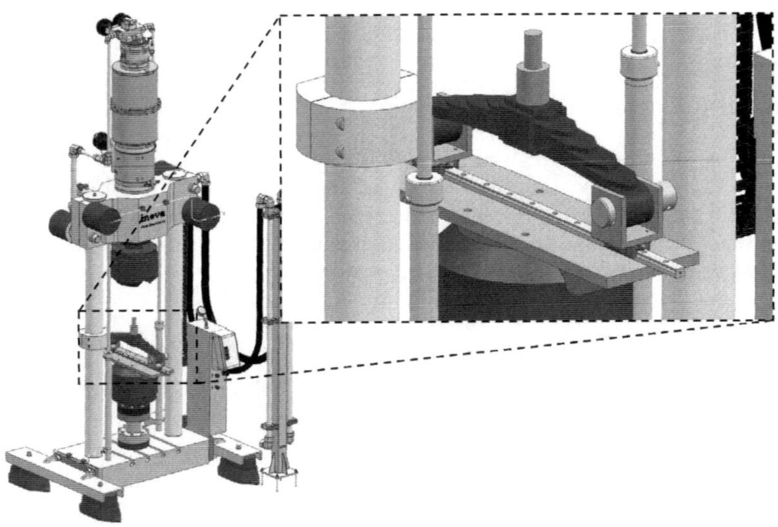

Fig. 4. Attachment of the leaf spring in the Inova F250 testing machine

The stiffness of the first spring is given for individual load states (the load from the empty vehicle F_I = 10.6 (kN); the load from the fully loaded vehicle F_{II} = 16.1 (kN); and the load from the fully loaded vehicle with dynamic increase F_{III} = 19.7 (kN)) in Fig. 5. However, these values indicate the displacement of the spring, including the deflection of the jig in which it is mounted. This 'parasitic' displacement was measured using a horizontal dial test indicator and this was subsequently deducted for the individual states [8]. The final results of the individual load conditions are given in Table 2.

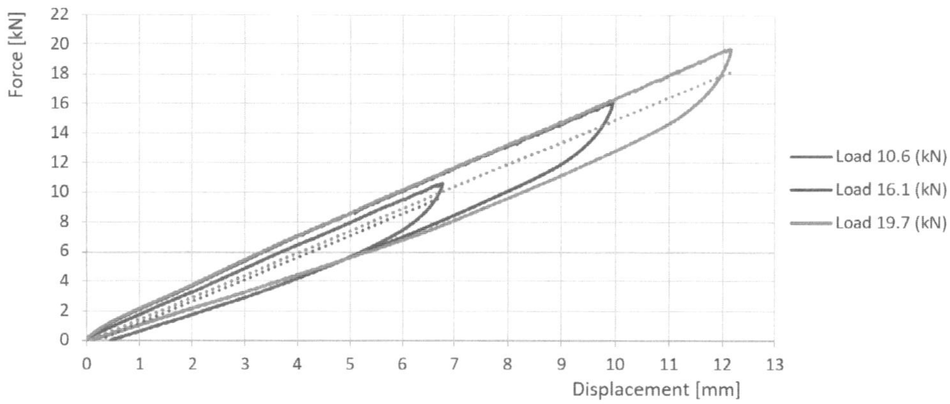

Fig. 5. The results of the experimental tests; stiffness of the steel leaf spring

3.1. Comparison of the numerical simulations and experimental measurements of the steel leaf spring

The data from the numerical simulation and experimental tests were compared. The final results are given in Table 2. The maximum difference between the results is 18% (in the third load condition 'fully loaded vehicle with dynamic increase'). This value is more than satisfactory, because the numerical simulations did not take into account several influencing parameters, such as a slightly simplified geometry of the model, the values of the coefficients of friction between the contact faces of the leaves of the spring, clearance between the leaves and attachments of the spring, and many other things.

The loading condition (kN)		**Numerical simulation (mm)**	**Experimental test (mm)**			**Difference (%)**
			Displacement of the leaf spring and preparation	Parasitic displacement of the preparation	Displacement of the leaf spring	
empty vehicle	**F_I = 10.6**	5.742	6.76	0.218	6.542	15
fully loaded vehicle	**F_{II} = 16.1**	8.223	9.94	0.342	9.598	17
fully loaded vehicle with dynamic increase	**F_{III} = 19.7**	9.91	12.14	0.419	11.721	18

Table 2. Results of the comparison of the numerical simulations and experimental measurements of the steel leaf spring

4. Design of the composite leaf spring

Several designs of the composite leaf spring were created. The choice was between either an integral or a mounted spring design. The 'mounted spring design' of the leaf spring means here the attachment of the spring to the chassis of the tram. This attachment can be created by using integrated or shaped joints. These joints in composite leaf springs are the keystone. This issue has already been dealt with in other works [9], see Fig. 6.

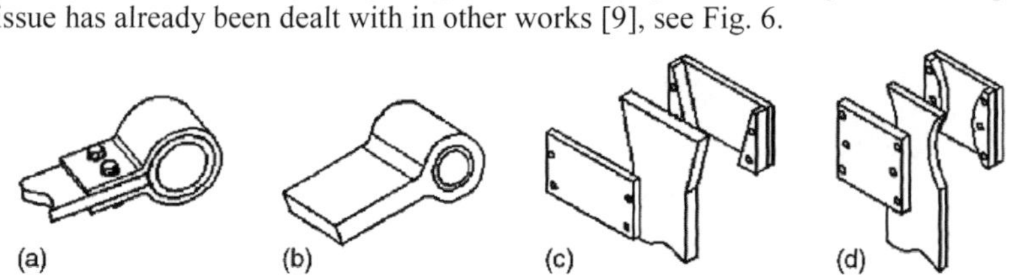

Fig. 6. Different types of joints for attaching the composite leaf spring to the vehicle body; (a) consists of a steel eye that can be bolted or pinned to the spring; (b) the eye and spring are manufactured simultaneously from the same material; (c) the spring end has a conical or (d) concave width profile [10]

The integral design of the leaf spring was chosen as the best one. The attachment part of the spring in the shape of the unclosed eye was selected, because a closed eye creates many problems during design and manufacturing, increasing the production costs. The unclosed eyes of the spring were chosen because the leaf spring is attached using pins with a relatively large diameter, thereby leading to better distribution of the load. Disadvantages of this solution are that it is necessary to increase the stiffness of the eyes of the spring, there are lower damping properties of the spring, and there is higher pressure/stress on the contact surfaces of the spring and pin.

4.1. Selection of the material for the composite leaf spring

A unidirectional E-Glass fibre in combination with high toughness epoxy resin was chosen, after considering the cost/weight/mechanical properties and availability of the material. This type of material in the form of prepreg was chosen due to the large number of plies in the laminate and the relative complexity of the shape. The physical parameters of the material are given in Table 3.

V_f (%)	58	*Fibre Volume Fraction*
V_m (%)	42	*Matrix Volume Fraction*
t (mm)	0.78	*Thickness of the ply*
ρ (kg/m³)	1870	*Density*
E1 (GPa)	43	*Young's Modulus 0°*
E2 (GPa)	7	*Young's Modulus 90°*
G12 (GPa)	4.1	*In-plane Shear Modulus*
ν12 (-)	0.28	*Poisson's Ratio in plane 12*

Table 3. The physical properties of UD E-Glass prepreg 800 gsm

4.1. Finding the geometrical parameters of the composite leaf spring

A geometrical optimization was created to find the main shape (the main geometric parameters) of the spring. A minimization of the volume/weight of the composite body of the spring was chosen as the objective function of the geometric optimization. Individual displacement of the leaf spring in the vertical axis (from numerical simulation of the steel leaf spring) was chosen as the constrain of the geometry optimization. Next, the limits of the geometric parameters of the spring were specified. The range of these values was limited by the mounting space and individual points of the attachment to ensure compatibility with the currently used steel leaf spring. The geometric parameters of the spring are given in Fig. 7.

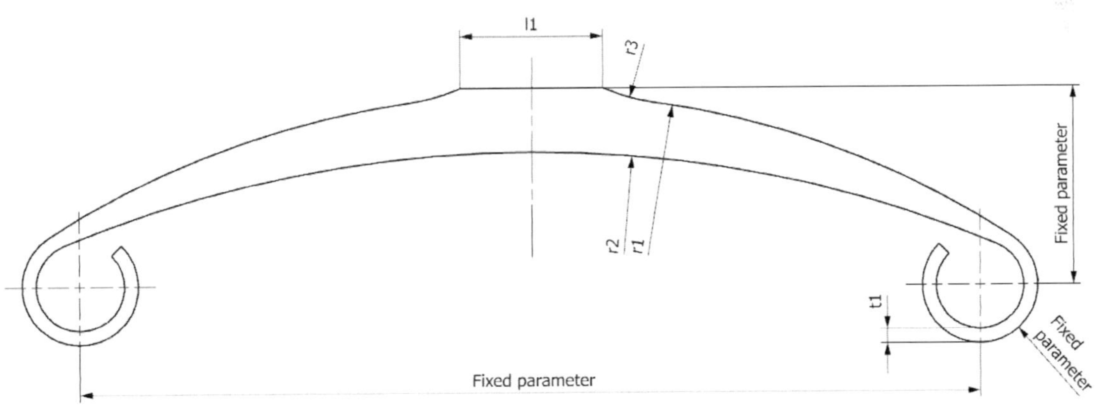

Figure 7. The geometric parameters of the geometric optimization of the composite leaf spring

4.2. Determination of the strength of the composite leaf spring

An advanced numerical model of the composite leaf spring was created to determine the strength of the spring. A 2D FE model of the spring was created from CQUAD4 elements. Then, the required number of the plies of the laminate and their physical, mechanical and geometrical parameters were defined in different zones of the 2D model. This model was transformed into a 3D FE model by using the special function 'extrude laminate' in the laminate module. This 3D FE model allows consideration of the third main direction, resin drops and interlaminar properties of the laminate; see Fig. 8.

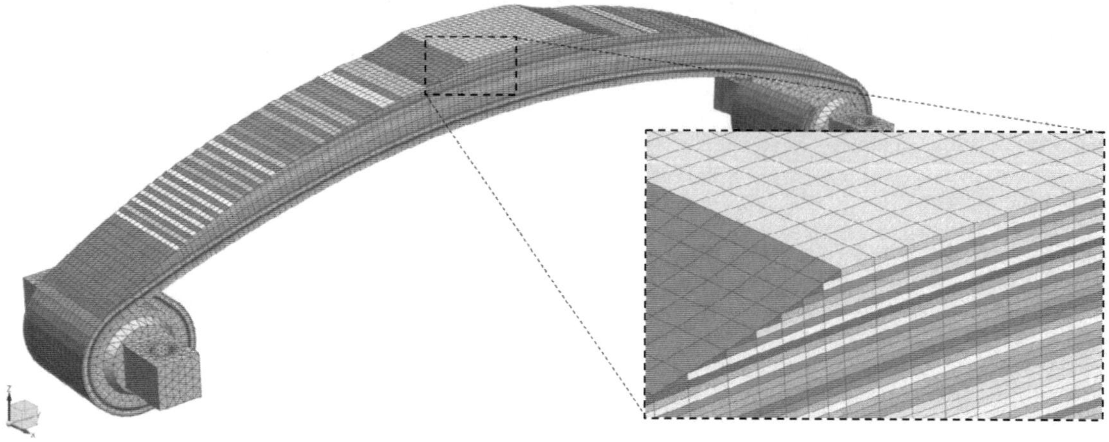

Fig. 8. FE model of the composite leaf spring

The structural analysis was carried out using non-linear solver 601 in the NX Nastran 11 solver. The maximum stress failure criterion was used to evaluate the strength of the spring. According to this theory, failure occurs when any stress component reaches the ultimate strength of the material [11].

Failure indexes	Maximum stress criterion
F_{11}	σ_1/X^T if $\sigma_1 > 0$; σ_1/X^C if $\sigma_1 < 0$
F_{22}	σ_2/Y^T if $\sigma_2 > 0$; σ_2/Y^C if $\sigma_2 < 0$
F_{33}	σ_3/Z^T if $\sigma_3 > 0$; σ_3/Z^C if $\sigma_3 < 0$
F_{12}	$\lvert\tau_{12} / S_{12}\rvert$
F_{23}	$\lvert\tau_{23}/S_{23}\rvert$
F_{13}	$\lvert\tau_{13}/S_{13}\rvert$

Table 4. Failure indexes of Maximum stress criterion (for 3D laminates)

Transverse shear failure criterion for 3D laminates (from orthotropic material) was used as the interlaminar failure criterion; see Table 5.

Failure index	Transverse shear failure criterion (3D laminates)
F_{IL}	$\sqrt{\left(\frac{\tau_{13}}{S_{13}}\right)^2 + \left(\frac{\tau_{23}}{S_{23}}\right)^2}$

Table 5. Interlaminar failure criterion transverse shear (for 3D laminates)

Where σ1 (σ2, σ3) is normal stress in the longitudinal (transverse) direction; XT (YT, ZT) is the longitudinal (transverse) tensile strength in the direction; XC (YC, ZC) is compressive strength in the longitudinal (transverse) direction; τ12 (τ23, τ13) are shear stresses in plane 12 (23 or 13), and S12 (S23, S13) is shear strength in plane 12 (23 or 13) [12].
The strength properties of UD E-Glass prepreg are given in Table 6.

X^T (MPa)	900	*Tensile strength (0°)*
Y^T (MPa)	30	*Tensile strength (90°)*
Z^T (MPa)	30	*Tensile strength (N90°)*
X^C (MPa)	800	*Compression strength (0°)*
Y^C (MPa)	150	*Compression strength (90°)*
Z^C (MPa)	150	*Compression strength (N90°)*
S_{12} (MPa)	55	*In-Plane Shear Strength*
S_{ILSS} (MPa)	60	*In-Lam. Shear Strength*

Table 6. Strength properties in stress of unidirectional E-Glass prepreg

4.3 Results of the numerical simulation of the composite leaf spring

The difference in the results for the displacement of the final design of the composite leaf spring and the steel leaf spring is a maximum of 5% (1% for the first and second load conditions, and 5% for the third load condition). The results of the displacement of the composite leaf spring are given in Fig. 9.

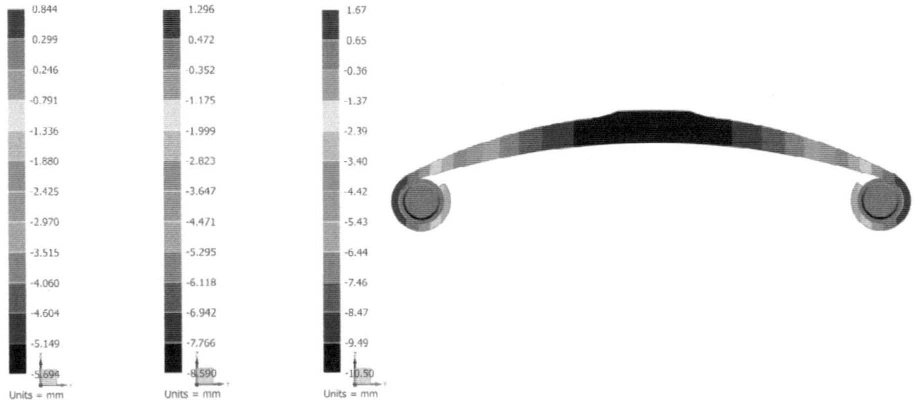

(a) load from the empty vehicle **(b) load from the fully loaded vehicle** **(c) load from the fully loaded vehicle with dynamic increase**

Fig. 9. Displacement of the composite leaf spring in the vertical axis ('z') for different load conditions (mm)

The maximum normal stress in the leaf spring occurs in the first ply of the laminate in directions 11 and 22. The results of the normal stress are shown in Fig. 10. for the whole model and the first ply of the composite leaf spring.

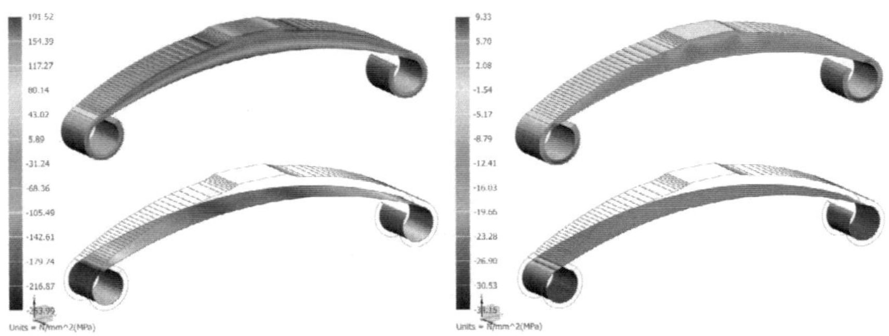

Fig. 10. Maximum normal stress in the composite leaf spring; (a) in direction 11 – whole model and first ply; (b) in direction 33 – whole model and first ply (MPa)

The maximum shear stress in the leaf spring in direction 12 occurs in the first ply of the laminate and occurs in direction 31 in ply 41. The results of the shear stress are shown in Fig. 11.

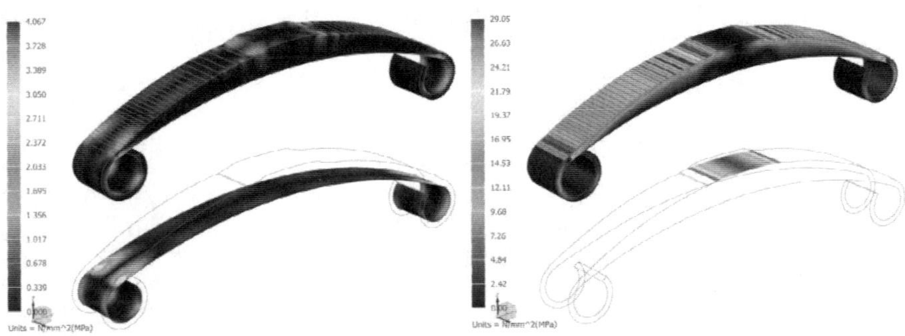

Fig. 11. Maximum shear stress in the composite leaf spring; (a) in direction 12 – whole model and first ply; (b) in direction 31 – whole model and ply 41 (MPa, Absolute value)

Results of the normal and shear stresses were checked by using the 'maximum stress' failure criterion and the 'transverse shear' interlaminar criterion; see Fig. 12. The critical area of the failure index is on the edges of plies number 39-45, especially in shear in direction 31. In terms of the interlaminar failure index, the critical area is above the attachment pin in plies number 1-7.

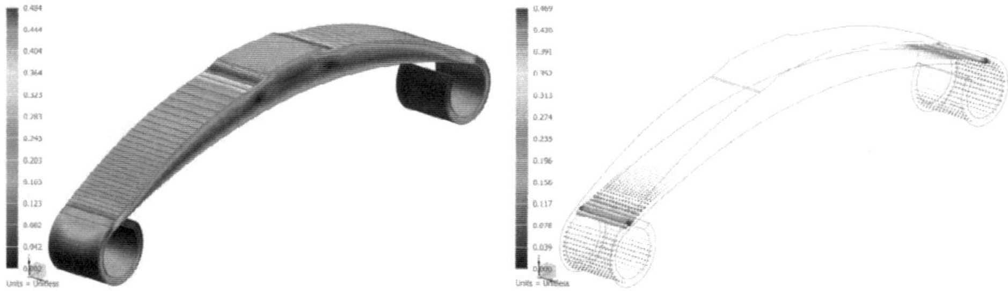

Fig. 12. Failure index of the third load condition of the leaf spring; (a) 'maximum stress' failure criterion; (b) 'transverse shear' interlaminar failure criterion– ply number 6

5. Conclusion

This paper deals with the optimization of a conventional steel leaf spring for a tram vehicle by using composite materials. Numerical simulations and experimental tests were carried out to obtain the main parameters. E-Glass fibre in combination with a high toughness epoxy matrix was chosen as the material. A geometric optimization based on these parameters was created to find the appropriate basic shape for the composite leaf spring. An advanced FE model which modelled the individual layers of the lamina was created. The failure indexes were evaluated by using three dimensional 'maximum stress' failure criterions and the 'transverse shear' interlaminar failure criterion. The weight of the leaf spring was reduced by using this design from 12.6 (kg) to only 1.9 (kg), meaning a weight reduction of more than 80%.

We are currently working on manufacturing several composite leaf springs for experimental testing and verification of the numerical simulations.

6. Acknowledgments

This work has been prepared under the project LO1502 'Development of the Regional Technological Institute' under the auspices of the National Sustainability Programme I of the Ministry of Education of the Czech Republic aimed at supporting research, experimental development and innovation.

7. References

[1] Mallick, P. K. Composites Engineering Handbook. New York: M. Dekker, 1997. Print.

[2] Morris, C.J. Composite integrated rear suspension, Composite Structures, Volume 5, Issue 3, 1986, Pages 233-242, ISSN 0263-8223.

[3] Beardmore, P.; Johnson, C.F. The potential for composites in structural automotive applications, Composites Science and Technology, Volume 26, Issue 4, 1986, Pages 251-281, ISSN 0266-3538.

[4] Daugherty, R. L. Composite Leaf Springs in Heavy Truck Applications. K. Kawata, T.Akasaka (Eds). Composite Materials Proceedings of Japan-US Conference Tokyo, 1981.

[5] Krall, S.; Zemann, R. Investigation of the Dynamic Behaviour of CFRP Leaf Springs, Procedia Engineering, Volume 100, 2015, Pages 646-655, ISSN 1877-7058

[6] Mahdi, E.; Alkoles, O.M.S.; Hamouda, A.M.S.; Sahari, B.B.; Yonus, R.; Goudah, G. Light composite elliptic springs for vehicle suspension, Composite Structures, Volume 75, September 2006, Pages 24-28, ISSN 0263-8223.

[7] Talib A.R.A., Aidy A., Goudah G., Nur Azida Che Lah, Golestaneh A.F., Developing a composite based elliptic spring for automotive applications, Materials & Design, Volume 31, Issue 1, January 2010, Pages 475-484, ISSN 0261-3069.

[8] Wahl, A. M. Mechanical Springs. New York: McGraw-Hill, 1963. Print.

[9] J.P. Hou, J.Y. Cherruault, I. Nairne, G. Jeronimidis, R.M. Mayer, Evolution of the eye-end design of a composite leaf spring for heavy axle loads, Composite Structures, Volume 78, May 2007, Pages 351-358, ISSN 0263-8223.

[10] Mahmood M Shokrieh, Davood Rezaei, Analysis and optimization of a composite leaf spring, Composite Structures, Volume 60, Issue 3, May–June 2003, Pages 317-325, ISSN 0263-8223.

[11] Vasiliev Valery V., Morozov Evgeny V.: Advanced Mechanics of Composite Materials and Structural elements, 3rd Edition. ISBN: 978-0-08-098231-1, Elsevier, Oxford, 2013.

[12] Sedláček, F., Lašová, V., Kottner, R., Bernardin, P. Comparison of Numerical Simulation and Experiment of a Flexible Composite Connecting Rod. In Industrial Engineering. 2015. s. 1-6. ISBN: 978-9949-23-804-0, ISSN: 2346-6138

DOI: 10.2507/27th.daaam.proceedings.075

INJECTION MOLDING SIMULATIONS OF HARDLY PRODUCIBLE PARTS FROM PBT

Karel Raz, Martin Zahalka & Roman Polak

This Publication has to be referred as: Raz, K[arel]; Zahalka, M[artin] & Polak, R[oman] (2016). Injection Molding Simulations of Hardly Producible Parts from PBT, Proceedings of the 27th DAAAM International Symposium, pp.0501-0505, B. Katalinic (Ed.), Published by DAAAM International, ISBN 978-3-902734-08-2, ISSN 1726-9679, Vienna, Austria
DOI: 10.2507/27th.daaam.proceedings.075

Abstract

This article is focused on possibilities of advanced injection molding simulations. Parts made from plastic materials are in almost all technical systems. A lot of various hardly producible plastic parts are used nowadays. Examples are components of electronical systems, mainly various connectors from material polybutylene terephthalate (PBT). These connectors are long in one direction and their geometry is with high number of various small holes. This leads to problems during injection molding. Movement of melted plastic material is slow and also number of weld lines is high. Main aim of this article is comparing of different variants of injection molding process and demonstration of CAE advanced methods usage. This approach leads to optimized production of plastic connectors. During simulation is used accurate material model of PBT which is clearly described.
Results leads to mathematical description of dependence between all injection molding parameters and results from this technological process. Mainly are described temperatures and pressures in various moments of injection process. Results of research describe all influences on quality of product. This is done by evaluation of geometrical, optical and mechanical properties of final connector. As an example is used connector with 54 independent contacts. Validation of simulation is performed by comparing with real experiment.

Keywords: injection molding; PBT; virtual simulation; surface quality; connector.

1. Introduction

Nowadays are really often used parts made from plastic materials. Their market share is rapidly increasing. In all areas of engineering are still produced parts with more complex shapes. It is necessary to shorten time when new product is started to be produced. It is suitable to used modern mold-flow analysis. This kind of analysis is able to show weak points of process and it is possible to optimize it virtually without stopping production line. Connector for electronic industry is chosen as an example (according Fig. 1). The filling of this part is only from one side. This leads to problematic mold filling. This part is also suitable for optimization. Final product should be with better properties and more reliable molding process. [8,9,10]

Our research shows how can be injection molding process optimized with mold flow analysis and how can be described material for this kind of analysis.

2. Simulation model

Simulation model is clearly described in Fig.1. For simulation is critical parameter diameter of entrance gate, which is here 1.3mm. Design of mold is also obvious in Fig.1 (right). There is visible shape of cooling channels. For this kind of virtual simulation is not necessary to model each body of mold cores. We are interested here in cavity of mold. Whole simulation was performed in system Moldex3D. This software is used for advanced simulation of injection molding. 3D design of CAD parts was performed in Siemens NX10 software with usage of exporting data into Moldex 3D. [4,5]

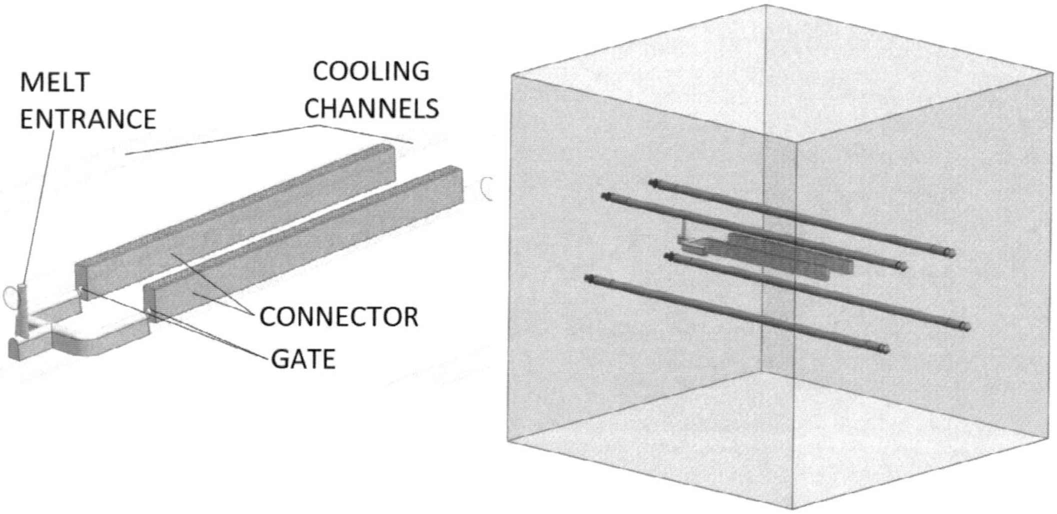

Fig. 1. Detail of cavity filled by plastic material with two connectors and mold design

3. Material model for virtual simulation

Material CELANEX 3316 was used for simulation. It is PBT (polybutylene terephthalate) with 30% of glass fibres. These properties are exactly according material list given by producer Celanese. All these properties are essential for correct results and have to be used as one of inputs for finite element solver [6,7].

One of the most important parameters for molding analysis is viscosity. This parameter is highly dependent on temperature and shear rate. In following graph (Fig.2 left) is visible, how is during simulation viscosity changed comparing to temperature and shear rate. In case that during injection molding is problem with filling of cavity, this problem can be solved by higher viscosity of material. As is visible from following graph, viscosity is changed more with shear rate than with temperature (in comparable ranges). Influence of temperature and pressure on specific volume of part has to be also considered in simulation. In general, this relation can be described in Fig.2 right.

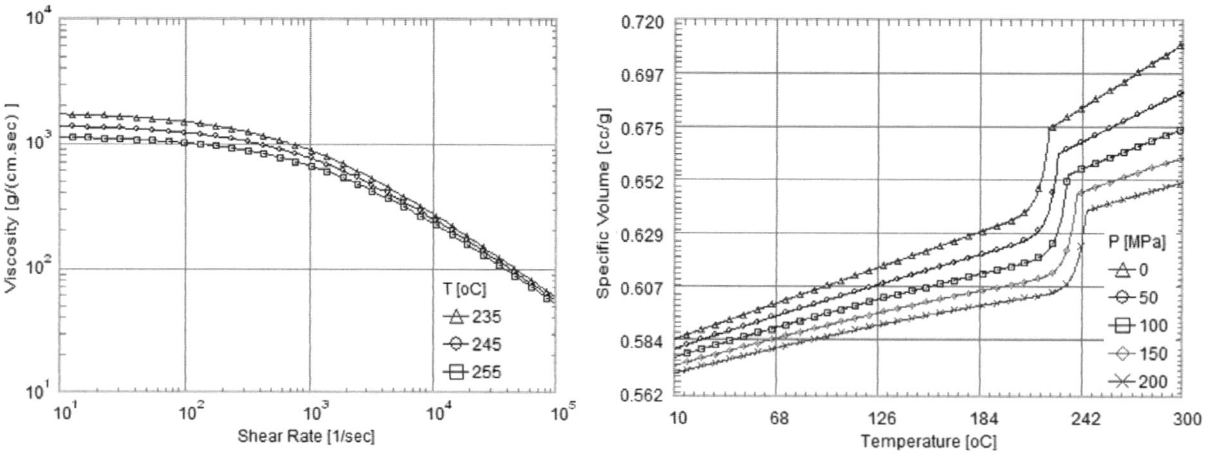

Fig. 2. Description of viscosity (left) and specific volume (right)

4. Simulation of injection molding

First set of parameters is all the time hard to predict. Designer is able to make some iterations with using of virtual simulation for getting more suitable parameters. As first entering parameters were chosen following:

Maximal injection pressure:	197 MPa
Filling time:	0,9 s
Maximal packing pressure:	140 MPa
Time of packing:	2,4 s
Packing levels:	140/80/55 MPa after 0,1/1,6/0,7 s
Temperature of injected material:	299°C
Mold temperature:	85°C
Cooling time:	10 s
Material:	Celanex 3316

First problem is obvious from selected parameters. Because is chosen part with filling only from one side, it is necessary to have maximal temperature of molded material. It was chosen temperature 299°C, which is still material able to survive. Whole process has to be controlled in terms of maximal shear stress (not above 50000 1/s) because higher stress can lead to increasing of temperature in entrance gates and to destroying of material. [3,4]

4.1. Filling results

Filling time is achieved and parts are fully filled as is obvious in Fig.3. The maximal permissible injection pressure wasn't achieved during process. So as a result from filling is visible, that time may be shorter, because maximal injection pressure is not limiting actual pressure.

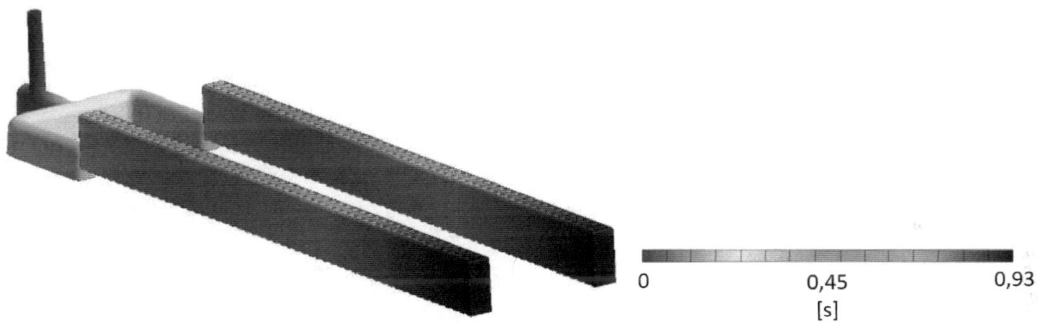

Fig. 3. Filling time

Temperature at the end of filling stage is sufficient close to the gates as is visible in Fig.4. But is obvious that in the middle of part is temperature under 200°C and material is there in solid state. This happened in areas close to shaped cores for holes and generally for areas which are in contact with material of mold.

As far from gate is temperature evaluated as lower is. This leads to decreasing of viscosity and problems with filling.

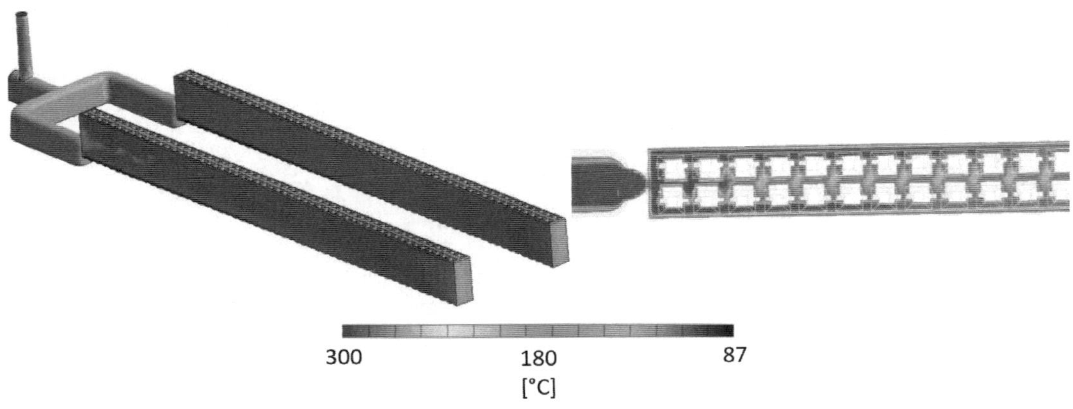

Fig. 4. Temperature at the end of filling- general view and cut

4.2. Packing results

Temperature at the end of packing is in Fig.5. It is visible that material is almost everywhere in solid state and can be ejected from mold.

Fig. 5. Temperature at the end of packing stage

It is possible to show shape of non-solid material as interesting result. Product is in solid state at the end of packing really and can be ejected. It is visible in Fig.6.

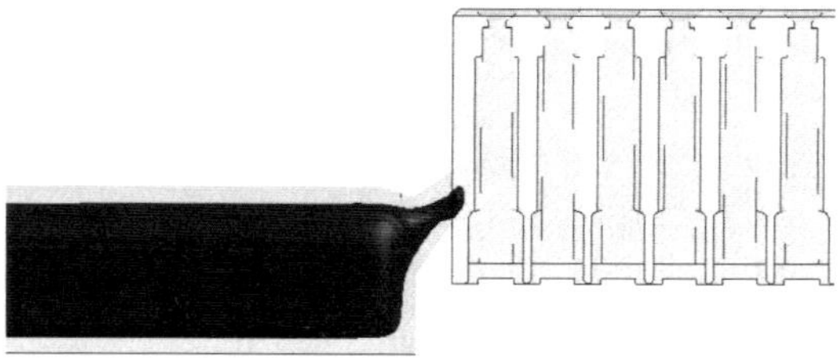

Fig. 6. Non- solid volume of plastic material

5. Findings from simulation and next steps- optimization

From performed simulation is obvious that selected parameters of molding process (times and pressures) were chosen quite good. This process can be still optimized. We were focused on maximization of temperature at the end of filling stage. If temperature will be higher, viscosity will be better, part will be filled easier and visual properties will be on higher level.

Technological process was optimized firstly in terms of shortening time of filling stage. The time was decreased from initial 0.9s to final 0.25s. Decreasing of filling time leads to higher production rate in terms of pieces per unit of time.

As is visible in Fig.7, the temperature compared with first process, is on higher level. Generally, it is up to 10% higher. [1,2]

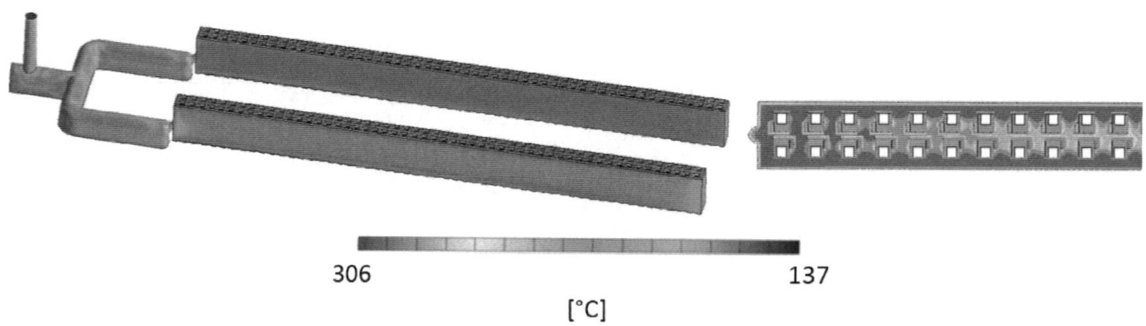

Fig. 7. Temperature at the end of filling after optimization- general view and cut

6. Discussion

These optimized parameters were used on real injection molding machine and final product was with all parameters according performed virtual simulation. This approach was used for virtual simulation validation. Comparing of initial setup and optimized setup of molding parameters was performed and temperature is more uniform after optimization. This leads to better optical properties of product and uniform shrinkage.

Approach of using virtual simulations in iterations leads to optimized production on minimizing of waste times on machines.

7. Conclusion

This article shows possibilities and advantages of using virtual mold-flow analysis. Process of determining parameters for injection molding as is shown here can rapidly decrease time, when production line is stopped because of testing new molding process.

Results from this kind of analysis are highly material-dependent, so it is obligatory to set up simulation material according material list of real material. Material model and input data, which are necessary to put into simulation are shown in chapter material model. Parameters from this simulation were used for real experiment on machine and results were comparable with simulation.

8. Acknowledgments

The article has been prepared under project LO1502 'Development of the Regional Technological Institute' under the auspices of the National Sustainability Programme I of the Ministry of Education of the Czech Republic aimed to support research, experimental development and innovation.

9. References

[1] Popa, C. L.; Parpala, R. C. & Cotet, C. E.(2012). Contact Identification for Assembly/Disassembly Process in CAD-CAE Software, Annals of DAAAM for 2012 & Proceedings of the 23rd International DAAAM Symposium, ISBN 978-3-901509-91-9, ISSN 2304-1382, pp 0613 - 0616, Editor B[ranko] Katalinic, Published by DAAAM International, Vienna, Austria

[2] Chwastyk, P. & Kolosowski, M.(2012). Integration CAD/CAPP/CAM Systems in Design Process of Innovative Products, Annals of DAAAM for 2012 & Proceedings of the 23rd International DAAAM Symposium, ISBN 978-3-901509-91-9, ISSN 2304-1382, pp 0397 - 0400, Editor B[ranko] Katalinic, Published by DAAAM International, Vienna, Austria

[3] Piscan, I.; Janssens, T. & Predincea, N. (2011). Experimental Validation of FEM for Frictional Contacts, Annals of DAAAM for 2011 & Proceedings of the 22nd International DAAAM Symposium, 23-26th November 2011, Vienna, Austria, Volume 22, No. 1, ISSN 1726-9679, ISBN 978-3-901509-83-4, Katalinic, B. (Ed.), pp. 0741-0742, Published by DAAAM International Vienna, Vienna

[4] Opran, C.; Balota, D. A.; Teodorescu, D. F. (2016). Modelling Assisted Injection Molding with Magnetic Field for Complex Polymeric Products, Proceedings of the 26th DAAAM International Symposium, pp.0301-0310, B. Katalinic (Ed.), Published by DAAAM International, ISBN 978-3-902734-07-5, ISSN 1726-9679, Vienna, Austria DOI:10.2507/26th.daaam.proceedings.041

[5] Sabau, A.; Barhalescu, M. L. & Oanta, E.: Modeling of High-Pressure Fuel Injection Systems, Annals of DAAAM for 2012 & Proceedings of the 23rd International DAAAM Symposium, ISBN 978-3-901509-91-9, ISSN 2304-1382, pp 1019 - 1022, Editor B[ranko] Katalinic, Published by DAAAM International, Vienna, Austria, 2012

[6] Stanek, M.; Manas, M.; Manas, D.; Kyas, K.; Navratil, J. & Senkerik, V. (2011). Injection Molding Process and its Optimization, Annals of DAAAM for 2011 & Proceedings of the 22nd International DAAAM Symposium, 23-26th November 2011, Vienna, Austria, Volume 22, No. 1, ISSN 1726-9679, ISBN 978-3-901509-83-4, Katalinic, B. (Ed.), pp. 0155-0156, Published by DAAAM International Vienna, Vienna

[7] Kyas, K.; Stanek, M.; Manas, M.; Manas, D.; Navratil, J. & Senkerik, V. (2011). Influence of Runner System Design on Molding Process, Annals of DAAAM for 2011 & Proceedings of the 22nd International DAAAM Symposium, 23-26th November 2011, Vienna, Austria, Volume 22, No. 1, ISSN 1726-9679, ISBN 978-3-901509-83-4, Katalinic, B. (Ed.), pp. 0733-0734, Published by DAAAM International Vienna, Vienna

[8] Ausperger, A. (2011). Simulation of Deformation and Compression of Fabric During the Back Injection Moulding Process, Annals of DAAAM for 2011 & Proceedings of the 22nd International DAAAM Symposium, 23-26th November 2011, Vienna, Austria, Volume 22, No. 1, ISSN 1726-9679, ISBN 978-3-901509-83-4, Katalinic, B. (Ed.), pp. 1261-1262, Published by DAAAM International Vienna, Vienna

[9] Kyas, K.; Stanek, M.; Manas, M.; Manas, D.; Krumal, M. & Cerny, J. (2010). Polymer Injection Molding Process Simulation, Annals of DAAAM for 2010 & Proceedings of the 21st International DAAAM Symposium, 20-23rd October 2010, Zadar, Croatia, ISSN 1726-9679, ISBN 978-3-901509-73-5, Katalinic, B. (Ed.), pp. 1081-1082, Published by DAAAM International Vienna, Vienna

[10] Rotar, I. D.; Cosma, C.; Tut, V. & Serban, I. (2010). Algoritm for Realization of a die used in Micro Injection Molding, Annals of DAAAM for 2010 & Proceedings of the 21st International DAAAM Symposium, 20-23rd October 2010, Zadar, Croatia, ISSN 1726-9679, ISBN 978-3-901509-73-5, Katalinic, B. (Ed.), pp. 1275-1276, Published by DAAAM International Vienna, Vienna

DOI: 10.2507/27th.daaam.proceedings.076

INCREASE THE SMART CITIES DEVELOPMENT BY USING AN INNOVATIVE DESIGN FOR VERTICAL AXIS WIND TURBINE

Ion Mălăel, Bogdan George Gherman & Ionuţ Porumbel

This Publication has to be referred as: Malael, I[on]; Gherman, G[eorge] B[ogdan] & Porumbel, I[onut] (2016). Increase the Smart Cities Development by Using an Innovative Design for Vertical Axis Wind Turbine, Proceedings of the 27th DAAAM International Symposium, pp.0506-0513, B. Katalinic (Ed.), Published by DAAAM International, ISBN 978-3-902734-08-2, ISSN 1726-9679, Vienna, Austria
DOI: 10.2507/27th.daaam.proceedings.076

Abstract

In the actual urban development context, the smart city concept is frequently used. This concept has not an unique definition, but is always defined as of positive impact. At the base of this concept, is placed the challenge to reduce the environmental impact and the carbon footprint. Smart city is a system with people, financing, services, materials and also energy development. To create a sustainable smart environment is crucial to integrate different renewable energy, transport and data flows in city planning and management. However, the implementation of new technologies must be fitted with the existing structures. In this paper, we study a counter rotating vertical axis wind turbine. To investigate the flow in this new concept we used the CFD methods. For solving the Navier-Stokes system equation the commercial CFD ANSYS Fluent was used. The vorticity magnitude and the momentum and power coefficient variation by blade position resulting from the numerical simulations are presented.

Keywords: Smart city; wind energy; VAWT; CFD; power coefficient.

1. Introduction

The biggest challenge of our times is the development of cities into intelligent and sustainable environments. In the EU's energy policy the smarter energy systems play a key role. The targets of EU for 2020 are to consume 20 % less energy, to increase the share of renewable energy to 20 % and to reduce greenhouse gas emissions by 20 % [1]. The problems associated with the energy sector and climate changes are the top priorities for the EU agenda. The ever diminishing oil resources, combined with ecological concerns have led to a focused attention on renewable energy, which should be inexhaustible and pollutant free [2]. Without the use of fossil fuels, without the change of natural course of rivers and without emitting any green house gases, wind energy has proven itself as a viable solution to the energetic independency of EU member countries. Therefore, the development of wind farms continues to be a successful endeavor, particularly in the EU [3] where multiple member states have already joined the forefront of wind turbine engineering.

The main objective of the smart energy system is to mitigate the environmental impact (climate change) due to the population increase [4]. It became more and more important to optimize all the processes of energy generation and consumption.

The smart energy system allows to integrate energy from unpredictable and intermittent renewable sources and to distribute power much more efficiently [5]. A smart system will deliver electricity more cost-effectively and with lower greenhouse gas emissions (Fig. 1).

Fig. 1 Schematic smart energy system [6]

It was estimated that in the Earth's wind the available energy is around 10 million MW [7]. Wind turbines are of two types, Horizontal Axis Wind Turbine (HAWT) and Vertical Axis Wind Turbine (VAWT). VAWT is mostly viable for places with low wind speed regimes where HAWT is highly uneconomical [8]. The combination of two vertical axis wind turbines was studied by many researchers [9-11]. The common combination is between Darrieus and Savonius [12-14]. This configuration, Darrieus-Savonius was done to help Darrieus wind turbine at the starting moment. It's well known [15] that Savonius rotor creates high torque and is self-starting even at low wind speeds, but is relatively low in efficiency rating. The Darrieus rotor is not a self starting rotor [16], but has much higher efficiency than the Savonius rotor. The combination of rotors increases the total power of the turbine in lower wind speeds [17]. The advantages of VAWT are based on the wind direction independency, but also on the simple design and low cost.

A new counter rotating (CR) VAWT concept was developed at COMOTI. The concept is a combination between two VAWT Darrieus H-type wind turbines, placed concentricallyas shown in Fig. 3, and rotating in opposite directions. A Permanent Magnet Generator (PMG - Fig. 4) is placed on the common rotation axis of the two turbines, and consists in a set of permanent mangnets rotating together with inner wind turbine, and a set of inductance coils, rotating together with the outer wind turbine. The relative movement achieved between the magnets and the coils produces the electrical power generated by the assembly. One of main the advantages of the new concept are related to the fact that it contains no statinary parts, as the two armatures of the PMG (one with permanent magnets and one with coils) will rotate, but in opposite directions. Also, the fact that the two armatures are both rotating increases the relative speed between them, and, consequently, the magnetic field variation rate through the coils increases, leading to the production of more electrical power.

A small model of CR-VAWT with three straight blades will be designed, manufactured and studied in the NRDI COMOTI's wind tunnel. The blades airfoil used is NACA 0021 [18]. In this study, a series of numerical analyses on was carried out in order to validate a new concept. Usually, wind turbine can generate electric power only when it reaches certain speed [19], so the tests will be carried out at 3 wind speeds. The maximum current velocity of COMOTI's wind tunnel is 20 m/s. Also, to capture the streamlines and the interaction between this two VAWT the PIV installation will be used [20].

2. Theoretical approach

The vertical axis wind turbine Darrieus type is based on the lift force [21]. The Darrieus design consists of two or more blades attached to a vertical shaft. In Fig. 2, the forces and the three velocity vectors acting on the Darrieus VAWT are presented [22]. The orientation and magnitude of the drag (F_D) and lift (F_L) forces and the resultant force (F_R) are also represented. The F_R can also be decomposed into a normal force (F_N) and a tangential force (F_T).

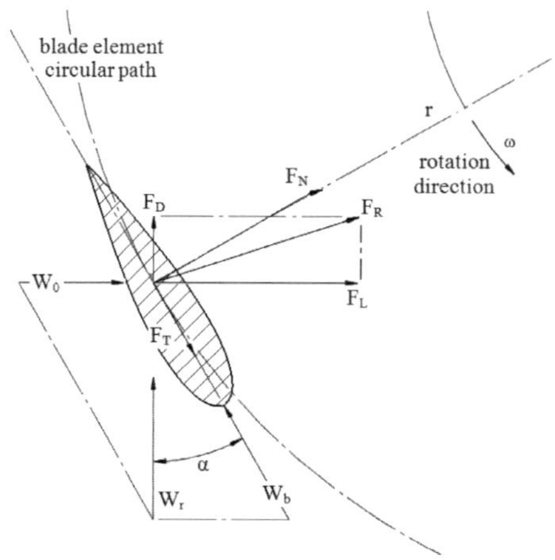

Fig 2. Force components and velocity and for VAWT Darrieus type

For a VAWT rotor with N blades, the average lift and drag forces can be defined as [23]

$$F_L = \frac{Nc}{2\pi}\int_{z=0}^{H}\int_{\theta=0}^{2\pi} q(-C_N sin\theta_b + C_T cos\theta_b)d\theta_b dz \tag{1}$$

$$F_D = \frac{Nc}{2\pi}\int_{z=0}^{H}\int_{\theta=0}^{2\pi} q(C_N cos\theta_b + C_T sin\theta_b)d\theta_b dz \tag{2}$$

where:

$$C_N = C_L cos\varphi + C_D sin\varphi \tag{3}$$

$$C_T = C_L sin\varphi - C_D cos\varphi \tag{4}$$

and C_L and C_D are the lift and drag coefficients.

The torque varies is function to azimuthal angle and the rotor height [24]. For a rotor with N blades, we have:

$$T_R = \frac{c\omega}{2\pi}\int_{z=0}^{H}\int_{0}^{2\pi} qC_T r d\theta_b\, dz \tag{5}$$

The average power is defined as [25]:

$$P = \frac{Nc\omega}{2\pi}\int_{z=0}^{H}\int_{0}^{2\pi} qC_T r d\theta_b dz \tag{6}$$

So, the power coefficient can be obtained as [26]:

$$C_P = \frac{P}{P_{max}} = \frac{81}{64}\frac{1}{2\pi}\frac{Nc\omega}{\rho V_\infty^3 RH}\int_{z=0}^{H}\int_{0}^{2\pi} qC_T r d\theta_b dz \tag{7}$$

3. Geometry definition

In this section, we present details about the geometry definition of the new vertical axis, Darrieus type wind turbinedeveloped at COMOTI. Fig. 3 presents the 3D virtual model of the wind turbine, developed using the SolidWorks software.

This turbine is designed to generate up to 2.5 kW power at a wind speed of 12 m/s. Two variants of the wind turbine, of different diameter and blade chords were studies. Table 1 shows the geometric parameters of the entire system in the two cases. Using the theoretical approach [27], the dimensions of the wind turbine rotors were defined. In this way, the characteristic dimensions, such as height and rotor diameter, were determined. Given that it is desired to have high turbine efficiency at values of TSR's (Tip speed ratio) as small as possible, a value of 0.5 was used for the first turbine solidity and 0.35 for the second one. Thus, using $\sigma = \frac{Nc}{D}$ we determined the aerodynamic chord length of the vertical axis wind turbine blade.

	Turbine 1	Turbine 2
Length blade [m]	3.2	3.2
Blade chord [m]	0.35	0.18
Airfoil	NACA 0021	NACA 0021
Solidity	0.5	0.5
Diameter	2.12	1.06
No. of blades	3	3

Table 1. Geometric parameters

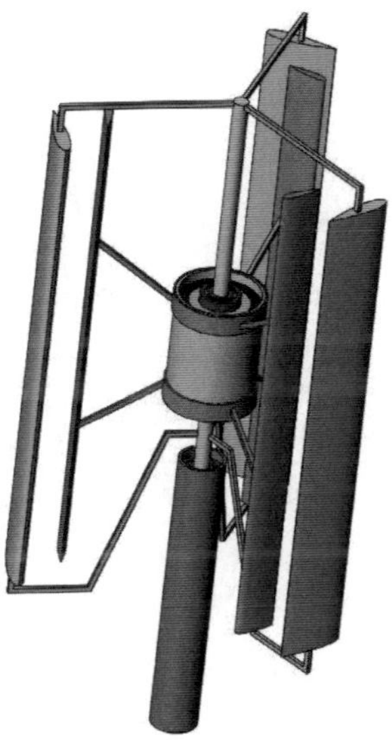

Fig. 3 CR-VAWT design

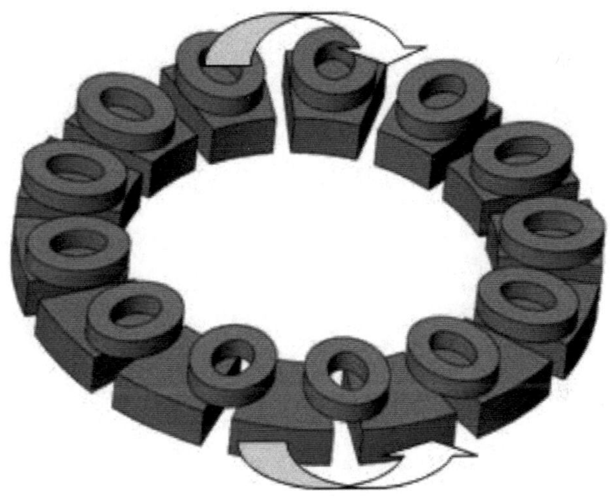

Fig. 4 Generator components and rotation directions

4. CFD methods

Starting from the operating parameters in table 1, we used the CFD (Computational Fluid Dynamics) approach to estimate the efficiency of this new CR(Counter-Rotating) -VAWT Darrieus type. In this case, we used the commercial code ANSYS-Fluent [28]. The SIMPLEC type pressure based solver with a second order upwind discretization scheme was used.

This presented numerical simulations use the URANS (Unsteady Reynolds Average Navier-Stokes) approach, with the k-ω SST turbulence model [29]. This model solves for two additional transport equations for the turbulent kinetic energy (k) and the specific dissipation rate (ω) and combines two standard turbulence models, k-ε and k-ω. The first full turbulence model imagined was the k-ω model. This model was introduced in 1941 by Kolmogorov [30].

The transport equation for the turbulent kinetic energy, k is written as [31]:

$$\frac{\partial k}{\partial t} + \bar{u}\frac{\partial k}{\partial x} + \bar{v}\frac{\partial k}{\partial y} + \bar{w}\frac{\partial k}{\partial z} = \Gamma_k\left(\frac{\partial^2 k}{\partial x^2} + \frac{\partial^2 k}{\partial y^2} + \frac{\partial^2 k}{\partial z^2}\right) + \tilde{G}_k - Y_k \tag{8}$$

Also, the transport equation for the specific turbulent kinetic energy dissipation rate, ω has the form [32]:

$$\frac{\partial \omega}{\partial t} + \bar{u}\frac{\partial \omega}{\partial x} + \bar{v}\frac{\partial \omega}{\partial y} + \bar{w}\frac{\partial \omega}{\partial z} = \Gamma_\omega \left(\frac{\partial^2 \omega}{\partial x^2} + \frac{\partial^2 \omega}{\partial y^2} + \frac{\partial^2 \omega}{\partial z^2}\right) + G_\omega - Y_\omega + D_\omega \quad (9)$$

where: Γ_k and Γ_ω are the effective diffusivities for k and ω; D_ω^+ is the positive portion of the cross-diffusion term; $\tilde{G}_k$ represents the production of turbulence kinetic energy; and G_ω represents the production of ω.

This numerical study was conducted on a combination of two vertical axis wind turbines in a 2D domain. The computing domain consists of three fields with an interface at the boundary between them (Fig. 5). The rotor1 domain has a radius R = 1.6 m, and the stator domain is 6 times bigger. Also, in the Fig. 2 the mesh and the value of y+ for all the blades are presented.

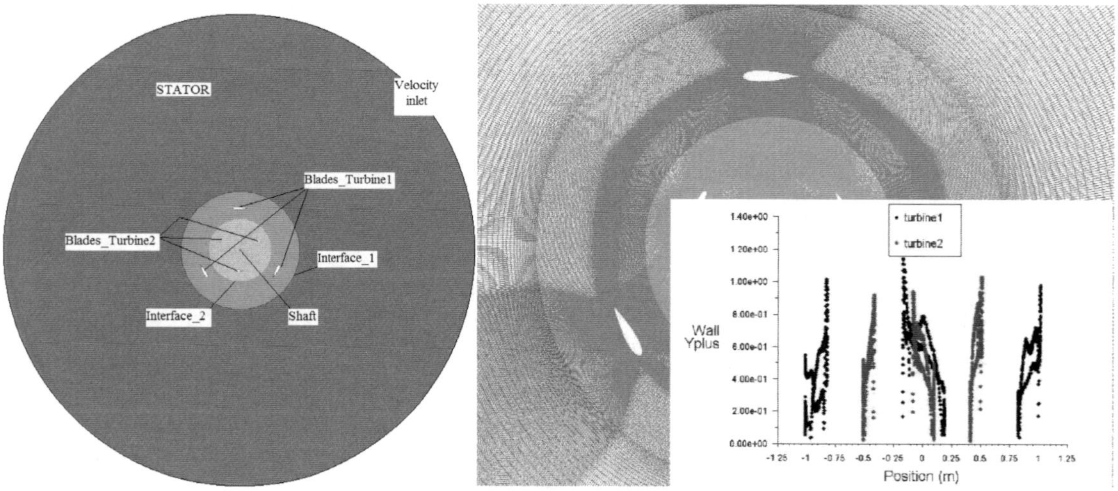

Fig. 5. Computational domain with y+ value for all blades

For the grid generation, the blocking method included in the ICEM CFD meshes software was used. To solve the problems that could occur in the boundary layer, the value of y + was maintained at values less than 1, and the ratio of adjacent elements sizes was kept below 1.05.

The boundary conditions were imposed so that the flow to be close to the real one. On the inlet of the stator domain a 12 m / s velocity was imposed, normal to the boundary surface, and solid, no-slip wall boundary conditions were imposed on the blades and the shaft. Six cases were studied, with TSR values were set between 0.5 and 3.

5. Results

Figs. 6 and 7 present, respectively, the vorticity magnitude for each TSR value and the power coefficient variation around the each of the two wind turbine: the outer turbine is denoted Turine 1, and the inner turbine is Turbine 2.

As the TSR increases, the vortical pattern resulting from the CFD simulation can be observed to change significantly. Thus, at low TSR, large vortices are detaching from each blade. The vortex trail oscillates downstream of each profiles due to the interaction between the inner and the outer blade networks. As the TSR increases, the interaction intensity increases, and the vortex trail tends to become attached to the trailing edges of both the inner and the outer blades. The vorticity intensity decreases, and the volume of the wake increases. At the highest TSR, the waves extend to cover almost the entire circumference of the circles on whch the blades are rotating, and they are deflected either inside, or outside, in an oscillatory motion, by the presence of the leading edge of the adjacent blade.

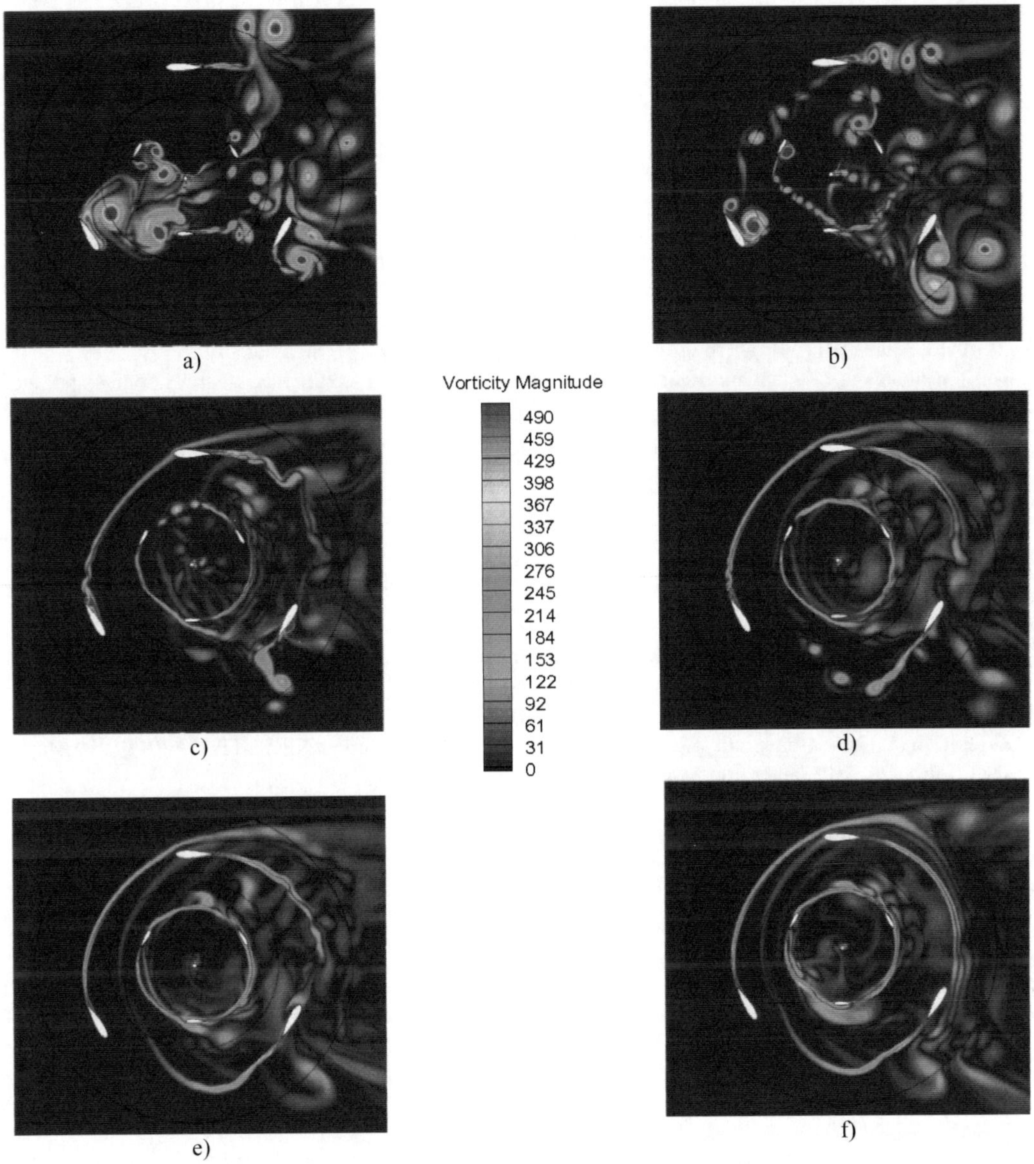

Fig. 6. Vorticity magnitude for: a) TSR=0.5; b) TSR=1; c) TSR=1.5; d) TSR=2; e) TSR=2.5; f) TSR=3;

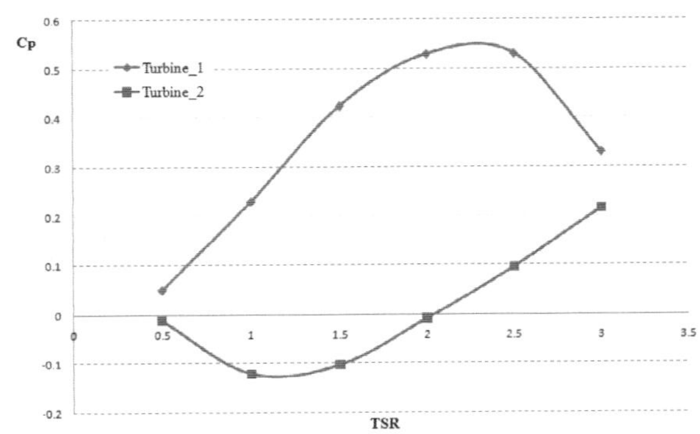

Fig. 7. Cp variations for each wind turbine

Fig. 7 indicates an optimal TSR can be found with respect to the TSR value. Thus, the outer turbine power coefficient is maximum at about TSR = 2.4.

6. Conclusion

In this paper work the flow field around new concept of Darrieus-type (lift-based) vertical axis wind turbine has been predicted numerically using CFD method, namely URANS. Both the wind turbines flow has been analised. The study has been conducted to determine the influence of the inner turbine on the global turbine efficiency. It is found that this influence has a positive impact on the overall turbine performance, due to the interaction of the wakes generated by the two blade networks rotating in opposite directions. An optimium value is found at TSR = 2.4.

The simulations presented here use standard NACA 0021 airfoils, but optimization methods, such as Genetic Algorithms (GA) [33], or ANN [34] can be used to increase also the efficiency of the assembly.

This paper contributes to a better understanding of the flow field around vertical axis wind turbines and suggests strongly that accurate modeling of the unsteadiness effects is an essential condition to high fidelity prediction of the aerodynamics performance of such devices.

7. References

[1] "Cities, towns & Renewable energy" © OECD/IEA, 2009 International Energy Agency

[2] Renewable energy world, jan 1999

[3] European Wind Energy Association, EU energy policy after 2020, www.ewea.org,2011

[4] José Miguel Guzmán, George Martine, Gordon McGranahan, Daniel Schensul, Cecilia Tacoli, "Population Dynamics and Climate Change", Copyright © UNFPA and IIED, 2009 ISBN: 978-0-89714-919-8

[5] Bethany Speer, Mackay Miller, Walter Shaffer, Leyla Gueran, Albrecht Reuter, Bonnie Jang, Karin Widegren, „The Role of Smart Grid in Integrating Renewable Energy", Technical Report NREL/TP -6A20- 63919 May 2015

[6] http://www.citi.io/2015/12/29/the-top-3-smart-city-trends-to-expect-in-2016/

[7] Søren Krohn, Poul-Erik Morthorst, Shimon Awerbuch, "The Economics of Wind Energy" - By the European Wind Energy Association, March 2009

[8] Malael, I[on]; Dragan, V[aleriu] & Gherman, G[eorge] B[ogdan] (2016). Turbulence Intensity Effects on the Vertical Axis Wind Turbine Starting Efficiency, Proceedings of the 26th DAAAM International Symposium, pp.0974-0979, B. Katalinic (Ed.), Published by DAAAM International, ISBN 978-3-902734- 07-5, ISSN 1726-9679, Vienna, Austria DOI: 10.2507/26th.daaam.proceedings.137

[9] Lucia-Andreea Mitulet*, Gabriela Oprina, Rares-Andrei Chihaia, Sergiu Nicolaie, Adrian Nedelcu, Mihail Popescu (2014). Wind Tunnel Testing for a New Experimental Model of Counter-Rotating Wind Turbine, 25th DAAAM International Symposium on Intelligent Manufacturing and Automation, DAAAM 2014

[10] P. Kacor, S. Misak, L.Prokop, Modification of construction design of vertical axis wind turbine, Annals of DAAAM for 2011 & Proceedings of the 22nd International DAAAM Symposium, Volume 22, No. 1, ISSN 1726-9679, ISBN 978-3-901509-83-4, Published by DAAAM International, Vienna, Austria, EU, 2011

[11] Anurag Kumar , et al." Hybrid Kinetic Turbine Rotors: A Review", [IJESAT] [International Journal of Engineering Science & Advanced Technology, Volume-4, Issue 6, 453-463, ISSN: 2250 - 3676

[12] Ali M.E. Estimation of the Performance of the Darrieus - Savonius combined machine. In: Ecologic Vehicles and Renewable Energies, Monaco, Germany, March 26 - 29 2009

[13] Gupta R., Das, R. and Sharma, K.K. Experimental study of a Savonius-Darrieus wind machine. In: International Conference on Renewable Energy for Developing Countries, University of Columbia, Washington DC, 2006

[14] K.K. Sharma, A.Biswas, R. Gupta, "Performance Measurement of a Three-Bladed Combined Darrieus-Savonius Rotor", INTERNATIONAL JOURNAL of RENEWABLE ENERGY RESEARCH A.Biswas et al., Vol.3, No.4, 2013

[15] Abdullah Al-Faruk,Ahmad Sharifian, "Geometrical optimization of a swirling Savonius wind turbine using an open jet wind tunnel", Alexandria Engineering Journal, Available online 30 July 2016, http://dx.doi.org/10.1016/j.aej.2016.07.005

[16] A.R. Senguptaa, A. Biswasa,R. Guptab, "Studies of some high solidity symmetrical and unsymmetrical blade H-Darrieus rotors with respect to starting characteristics, dynamic performances and flow physics in low wind streams", Renewable Energy Volume 93, August 2016, Pages 536–547, http://dx.doi.org/10.1016/j.renene.2016.03.029

[17] Saha U.K., Rajkumar M. J. (2005), on the performance analysis of Savonius rotor with twisted blades, Renewable Energy, 31, pp. 1776-1788

[18] MĂLĂEL, Ion; DRĂGAN, Valeriu; VIZITIU, Georgel, „The vertical axis wind turbine efficiency evaluation by using the CFD methods", Applied Mechanics & Materials . 2015, Vol. 772, p90-95. 6p
[19] Kirke, Brian Kinloch, 1998. "Evaluation of Self-Starting Vertical Axis Wind Turbines for Stand-Alone Applications". Griffith University, Australia. Retrieved from http://www4.gu.edu.au:8080/adt-root/public/adt-QGU20050916.120408/ on November 1, 2005
[20] Hofemann, C., Sim˜ao Ferreira, C., Dixon, K., van Bussel, G., van Kuik, G., and Scarano, F. (2008). 3d stereo piv study of tip vortex evolution on a vawt. In EWEC 2008 - European Wind Energy Conference - Brussels
[21] Paraschivoiu, I. (2002). Wind turbine design with emphasis on Darrieus Concept. Polythecnic International Press.
[22] Pierre Tchakoua, René Wamkeue, Mohand Ouhrouche, Tommy Andy Tameghe, Gabriel Ekemb "A New Approach for Modeling Darrieus-Type Vertical Axis Wind Turbine Rotors Using Electrical Equivalent Circuit Analogy: Basis of Theoretical Formulations and Model Development", Energies 2015, 8, 10684-10717; doi:10.3390/en81010684, ISSN 1996-1073
[23] I. Paraschivoiu, "Wind turbine design: with emphasis on darrieus concept", 2002, Montréal, Québec, Canada:Polytechnic International Press
[24] Paraschivoiu, I. and Delclaux, F. (1983). Double multiple streamtube model with recent improvements. Journal of Energy, 7:250 - 255
[25] M. Raciti Castelli, A. Englaro, and E. Benini, "The Darrieus Wind Turbine: Proposal for a New Performance Prediction Model based on CFD", Energy 36, 2011, pp. 4919-4934
[26] H. Dumitrescu, Al.Dumitrache, Fl. Frunzulica,A. Pal, V. Turbatu,*TORNADO* concept and realisation of a rotor for small VAWTs, INCAS BULLETIN, Volume 5, Issue 3/ 2013, pp. 69 – 75 ISSN 2066 – 8201
[27] K. Suffera, R.Usubamatovb, G. Quadir, K.Ismaild ,: Modeling and Numerical Simulation of a Vertical Axis Wind Turbine Having Cavity Vanes, 5th International Conference on Intelligent Systems, Modelling and Simulation, 2014
[28] I. Malael, L. Moutet, V. Dragan, "Numerical simulation of a vertical axis wind turbine for urban use", Applied Mechanics and Materials Vol. 811(2015) pp 333-338, Trans Tech Publications doi:10.4028 /www.scientific.net/AMM.811.333
[29] F. R. Menter, M. Kuntz, R. Langtry "Ten Years of Industrial Experience with the SST Turbulence Model", turbulence,Heat and Mass transfer 4, 2003 Begell House, Inc
[30] Frisch U. Turbulence The legacy of A.N. Kolmogorov.Cambridge Univ. Press, 1995
[31] S.Ferreira, H.Bijl, G. Van Bussel,G van Kuik , "Simulating dynamic stall in a 2D VAWT", Journal of Physics ,Conferance Series 75,2007
[32] T. Chitsomboon, C. Thamthae, „Adjustment of k −ω SST turbulence model for an improved prediction of stalls on wind turbine blades", World Renewable Energy Congress, 2011-Sweden
[33] C.F. Cuciumita, V.A. Vilag, V. Silivestru, I. Porumbel "Genetic Algorithm for Gas Turbine Blading Design", GT2011-46171, Proceedings of the ASME Turbo Expo 2011, Vancouver, Canada
[34] F.G. Florean, I. Porumbel, "Novel Algorithm for Geometrical Optimization of Flow Channels", Termotehnica, vol. 2, ISSN-L 1222-4057, 2015

DOI: 10.2507/27th.daaam.proceedings.077

COMPACT STEP DOWN VOLTAGE CONVERTER CONSTRUCTED IN TERMS OF EMC

Peter Janku, Martin Pospisilik & Tomas Dulik

[a]Faculty of applied informatics, Tomas Bata University in Zlin, Nad Stranemi 4511, Zlin 760 05, Czech Republic

This Publication has to be referred as: Janku, P[eter]; Pospisilik, M[artin] & Dulik, T[omas] (2016). Compact Step Down Voltage Converter Constructed in Terms of EMC, Proceedings of the 27th DAAAM International Symposium, pp.0514-0523, B. Katalinic (Ed.), Published by DAAAM International, ISBN 978-3-902734-08-2, ISSN 1726-9679, Vienna, Austria
DOI: 10.2507/27th.daaam.proceedings.077

Abstract

Designing of compact step down voltage converter must be provided according to the variety of standards and other requirements. Moreover, the constructed converter usually have to accomplish a set of opposite requirements defined by a target project or a target user (small size vs. high maximum level of provided current, etc.). Therefore the designing of converter could be really tricky. This paper tries to show the possibilities of a really low-end DC/DC converter based on controller AOZ 1284. Moreover, the constructed device is a subject for set of measurement. These measurements are performed in order to confirm that the device meet all requirements defined by standard EN 61000-6-3. All measurement results together with some necessary measures are presented at the end of this paper.

Keywords: EMC; step down converter; power supply.

1. Introduction

In modern electronic designs, the low price and compact size are two basic requirements defined by customers. At the other side, modern electronic devices have to meet really strict limitations defined in local or global EMC standards. These two opposite requirements, in common with wide variety of custom integrated drivers available on market, makes the correct step-down converters design really tricky.

This paper follows the previously based study [1] in which two different step down voltage converters were described. In following sections the step down converter based on AOZ1284 is discussed. The main advantage of this controller is its wide range of working frequency. Thanks to this, one of the most voluminous parts - inductor, can be chosen with relatively low value. The second, but not least advantage is its price - selected controller due to its parameter is one of the cheapest available on the market. [6,12]

The problematic of electromagnetic compatibility is wide discuss theme in scientific publication. Although that EMC can be divided into two separate parts: EMI - electromagnetic interferences and EMS - electromagnetic susceptibility,

EMI is more problematic part for step-down converters. Measurement of radiated field and measurement of interferences on input wires are presented in this paper.[13]

The measurements of real parameters of step down controller are presented in this paper. Moreover, the main disadvantage of this converter – EMC interference level is discussed common with design of measures. All measurements and measures were performed to meet the requirements placed by working conditions and the standard EN 61000-6-3.[2-4,12]

2. Description of the tested converter

The converter, which parameters and performance are discussed in this paper, was constructed according the specifications defined in section 2.1. These specifications were defined on a real embedded project requirements. In this project, new special secured access system will be developed in cooperation with specific company.

2.1. Operating requirements

As was told before, operating requirements were defined on real project base. Whole required electric characteristics are listed in Table 1. In addition, there were special requirement on size of whole DC/DC converter and finally the price had to be as low as possible.

Parameter	Value	
	Min.	Max.
Input voltage	12 V	24 V
Output voltage	5 V	5 V
Output current – continuous	150 mA	300 mA
Output current - peak	500 mA	800 mA
Power efficiency	80 %	-

Table 1. Electrical operating requirements

2.2. Converter circuit description

The converter, which parameters and performance are discussed, is based on AOZ1284 controller. The whole design was performed with respect to the manufacturer's notes provided within the datasheet of the circuit [5]. The circuit diagram of this converter is depicted in Fig.1.

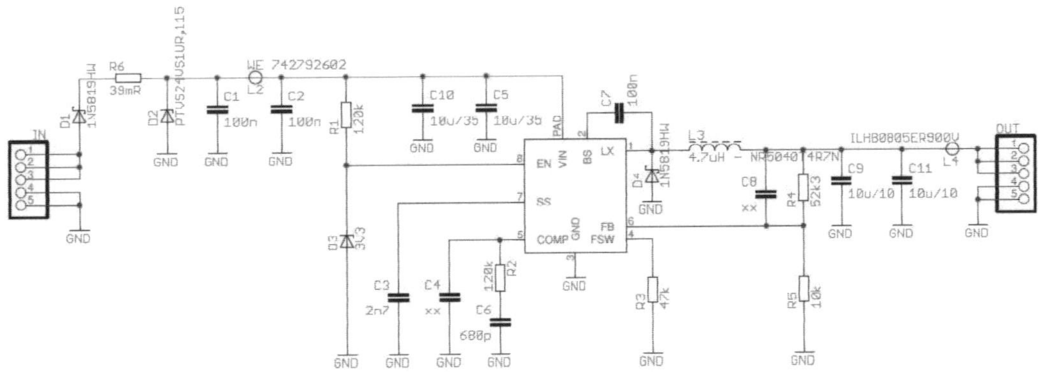

Fig. 1. Circuit diagram of a voltage converter based on AOZ 1284

As could be recognized from provided schematic, the switching transistor is integrated on the chip. The circuit operation is based on Continuous-Conduction Mode (CCM) in fixed frequency.

The inbuilt transistor is N-MOSFET switch. The inductor current is sensed by amplifying the voltage drop across the drain to source of this MOSFET. Output voltage is divided down by external voltage divider (defined by R5 and R4), it's internally compared with internal reference voltage source and amplified by the internal transconductance error amplifier. The error voltage is compared against the current signal, which is sum of inductor current signal and ramp compensation signal, at PWM input. If the current signal is less than the error voltage, the internal high-side switch is on. The inductor current flows from the input through the inductor to the output. When the current signal exceeds the error voltage, the high-side switch is off. The inductor current is freewheeling through the Schottky diode to output.[5,6]

As it is described above, the whole converter works on a fixed frequency defined by the internal oscillator. It can be programmed by external resistor connected to the FSW pin. In this specific circuit the operating frequency is set by R3 to value 961 kHz.

The circuit is equipped with filters (C1, L2 and C2) and with set of two protective diodes. The D1 take care about right input voltage polarization and the D2 works as overvoltage protection. Moreover, there is a Zener diode connected to the enable pin of converter, which is able to disable converter, when the input voltage is under the limit defined by manufacturer. [4,6,10,11]

3. Measurements and results

The described circuit was constructed as a functional sample and tested for achieving of the required parameters. Consequently a set of test was performed in order to gain data set needed for implementation of the DC converter into the final product. As in the previous research and in context of the requirements, the standard EN 61000-6-3 had been chosen, since the target use of the circuits was not specified.

The tests were as follows:

- output voltage stability versus output load,
- power efficiency at different output load and input voltages,
- maximum output power according to the cooling capability of the components,
- electromagnetic interference without a cover according to EN 61000-6-3,
- interference currents on the input cables measured by a current clamp according to EN 61 000-6-3.

3.1. Measurement configurations

Three basic measurement network configurations were used in order to obtain the above described data:

- measurement of DC static parameters,
- measurement of radiated electromagnetic field,
- measurement of interferences on input cables.

3.1.1. Measurement configurations

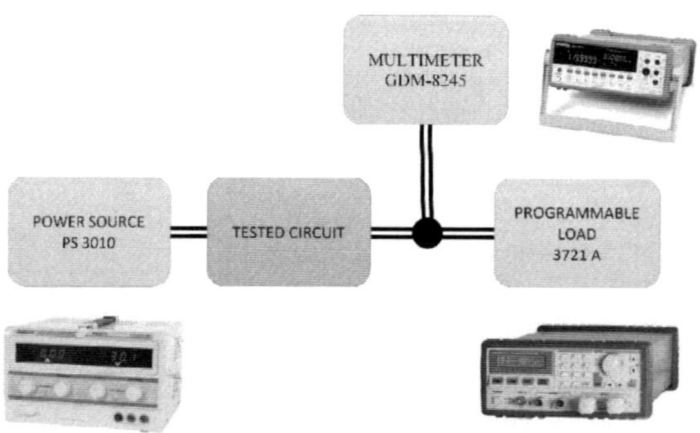

Fig. 2. Configuration of measurement network for validating of converters static DC parameters. [1]

The DC parameters were measured by using stabilized laboratory power source Velleman PS3010 as converters voltage input. The converter output was stretched by programmable electronic load Array 3721A operating in constant current mode. The input voltage and current were measured by set of two RMS multimeters. The output parameters – voltage and current were measured directly by the electronic load. Its accuracy was randomly checked with another RMS multimeter.

3.1.2. *Radiated electromagnetic field*

The intensity of electromagnetic interference was measured in a semi anechoic chamber in order to avoid other signals ordinary presented in environment. The measurement network is visualised in Figure 3. As can be seen, it is composed of a bilogarthmical antenna Teseq Bilog CBL 6112 connected to the Rohde\&Schwartz ESU 8 test receiver and spectral analyser. The measured controller was loaded by the electronic load Array 371A, that sunk a current of 0.2A and as power supply the lead battery was used. The potential interferences caused by the electronic load were excluded by additional measurement during which the converter was bypassed. All obtained data were collected and processed by means of EMC 32 software. [7-9]

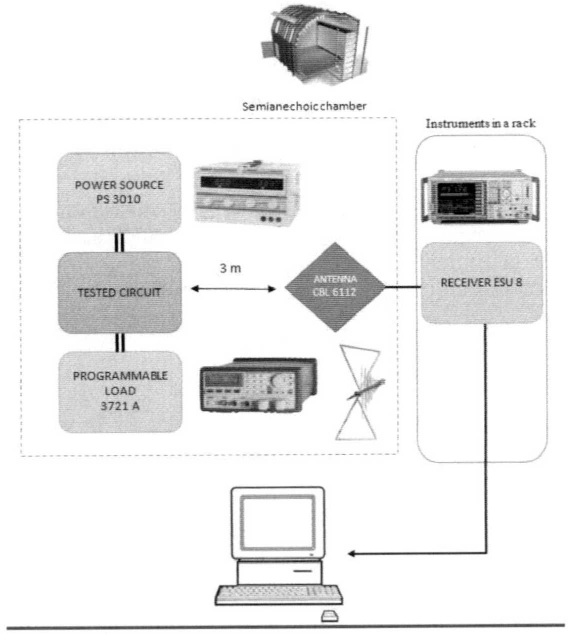

Fig. 3. Configuration of the experiment consisting in measurement of electromagnetic field radiated by the tested voltage converters. [1]

3.1.3. *Interferences on input cables*

The third measurement configuration was built in two stages. In the first one the current probe F-52 was used as the sensor connected to the ESU 8 receiver. This configuration was used in previous research [1]. In the second stage, the current probe was replaced by a LISN Rohde\&Schwarz ENV216 which can meet the requirements of CISPR 16-1-2, EN 55016-1-2. Both stages were compared with positive results – the same values were gained by both sensors. Therefore the second stage measurement network was selected as the one used for whole measuring of interferences on input cables.

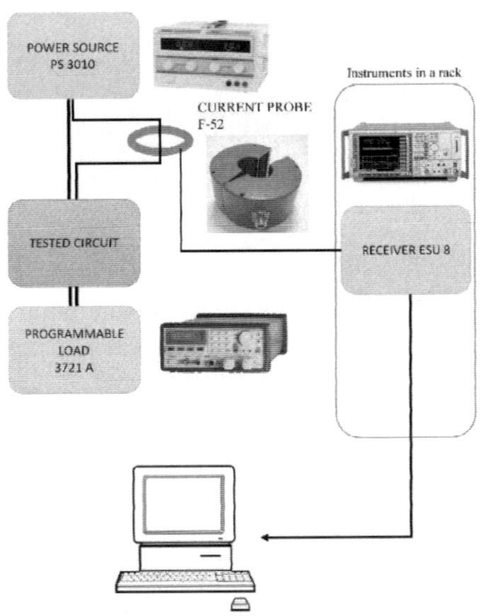

Fig. 4. Configuration of measurement network used for measure on input wires by using current probe. [1]

The obtained data were collected in EMC 32 software and post processed in MS Excel. Tested circuit was powered by laboratory power source PS3010 and the output current was consumed by programmable load Array 3721A. Final experiment configuration is described in picture Fig. 4.

3.2. *Obtained results*

By using measurement networks and devices described in previous parts of this paper, a large set of results was obtained. The most interesting and important results are presented in this section.

3.2.1. *DC parameters*

The first measurement was performed in order to verify required parameters of constructed converter. After static verification of output voltage, the load characteristic was measured. As can be seen in Fig. 5, the output voltage naturally goes down together with rising output current load. Moreover, significant output voltage drop down was measured on 24 V input and 400 mA output current. Despite this fact, maximal voltage drop down achieved 44 mV (in case of 15 V input), which is absolutely acceptable level.

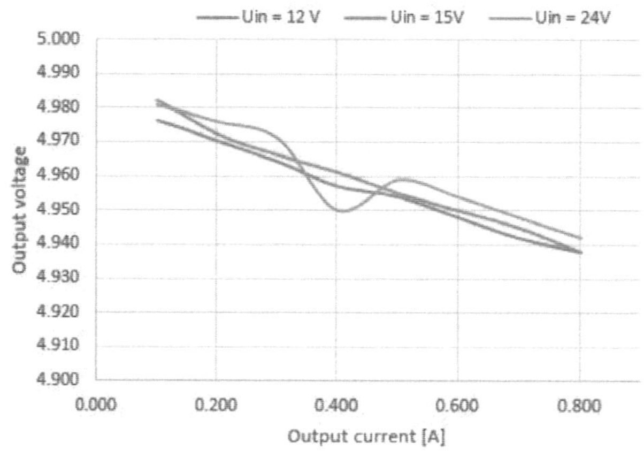

Fig. 5. Output voltage dependence on the output current and input voltage

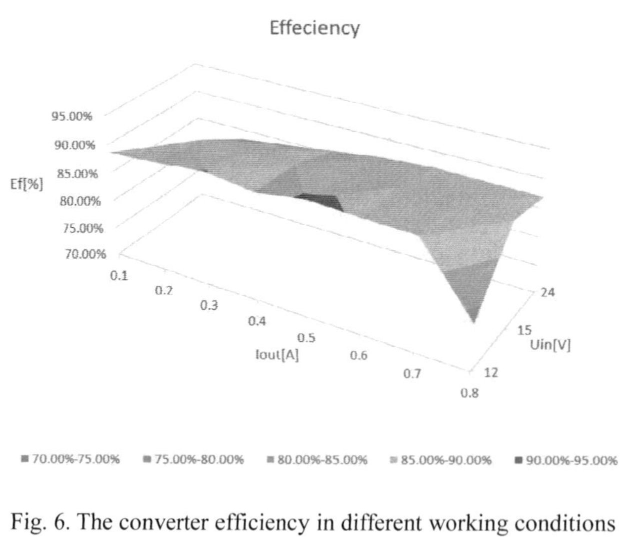

Fig. 6. The converter efficiency in different working conditions

The converter's power efficiency was calculated from the measured input and output power. It is depicted by Fig. 6. As can be noticed, the average value of power efficiency is about 87\%. The highest value was measured at the input level of 12 voltage and the output current of 500 mA (90.33\%). At the other side, the lowest efficiency was observed at the input level of 12 voltage and the output current of 800 mA and at the input leve of 24 voltage and the output voltage of 100 mA.

3.2.2. *Electromagnetic interferences*

As it was described in previous sections, the electromagnetic interferences were measured according to the requirements of the standard EN 61000-6-3. The base band defined by this standard was from 30 MHz to 1 GHz and the measurement was performed on working prototype without any box or another shielding.

The converter mostly passed the test with small exceptions. In the Fig. 8 the worst test case result is shown. As can be seen, the only one problem is at the frequency of about 40 MHz, where the measured level exceed the limit over 1 dBuV/m. Nonetheless, the converter is expected to pass the test in real device with proper case.

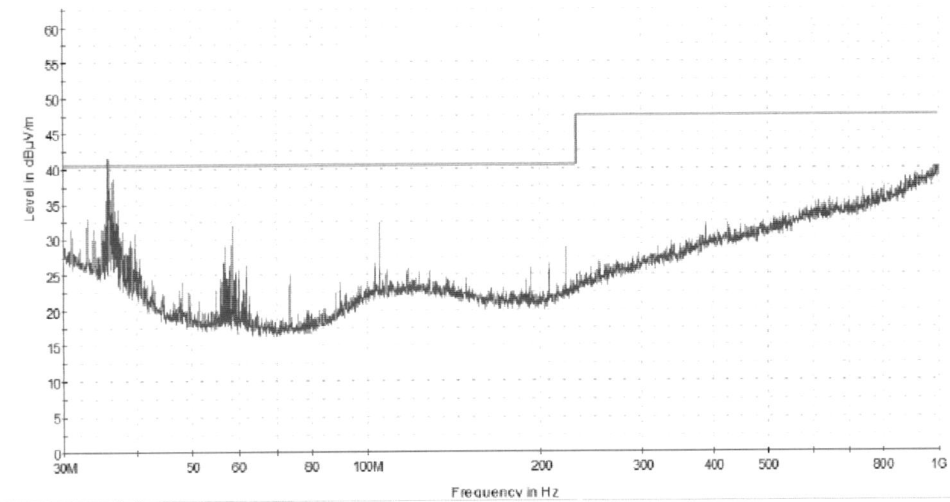

Fig. 7. The electromagnetic interference spectrum

It has to be mentioned, that together with measured sample, there were other devices in chamber (power source and programmable load. Theoretically, these two devices could produce some level of electromagnetic interferences. For this reason, the power source was replaced by a leaded battery. Moreover, the initial measurement was performed with bypassed converter in order to gain electromagnetic background in chamber. Because of significantly low level of electromagnetic interferences in chamber with bypassed converter, it could be claimed, that the measurement is correct and results are suitable.

3.2.3. *Interferences on input wires*

According to the principle of operation of the converters the current ripples and/or voltage peaks are expected to occur at its input and output cables. Within the framework of the standard EN 61000-6-3, a large set of measurements is expected to be processed. Due to limited size of this paper, only data gained on positive power supply wire are discussed here. The measurement configuration was described in previous section. In additional, as was described, the measurement was performed with ENV 2016 LSN network and verified by using a current probe. Both results were the same; therefore, only data measured by ENV 216 LSN are presented.

In the Fig. 8 and Fig. 9 the dependency of the measured voltage peaks spectrums on the output current (at a constant input voltage) is shown. It could be noticed from these pictures, that there are peaks starting from working frequency (about 961 kHz) and continuing on its harmonic frequencies up to 30 MHz. Moreover, most of these peaks break the limits defined by the standard. The most visible exceeded is noticeable at 24 voltage input and low output current. In this stage, the converter is working in a discontinuous mode and it is producing highest level of interferences.

Based on previous findings, it can be expected, that this converter will not pass the standard requiremetns of interferences on the input wires anymore. Therefore, a new input LC (inductor – capacitor) filter has been designed, built and, in the form of a functional sample, connected to the voltage convertor. After that all previous measurements on input wires were repeated.

The second set of input ripple voltage spectrum measurements (after the filter was applied) is shown in pictures Fig. 10 and Fig. 11. It is clear, that the measured results are significantly better after the filter application then before. The whole spectrum is lowered down; especially the peaks on working frequency and another harmonics frequencies. In this consequence, it could be expected that the designed voltage converter with appropriate input filter will be able to pass the tests defined standard EN 61000-6-3.

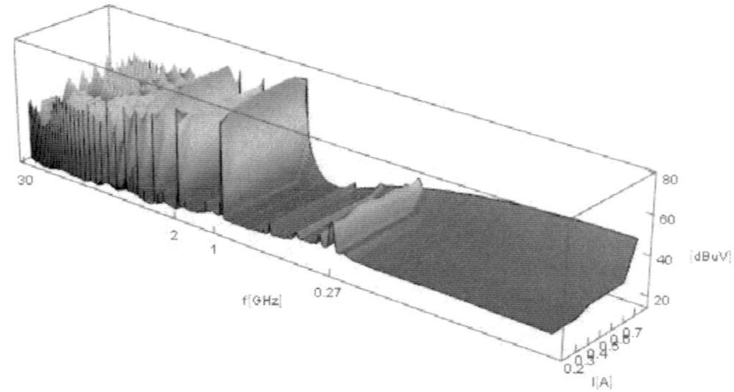

Fig. 8. Ripple current spectrum on the input wire (constant input voltage 12V), no filter applied

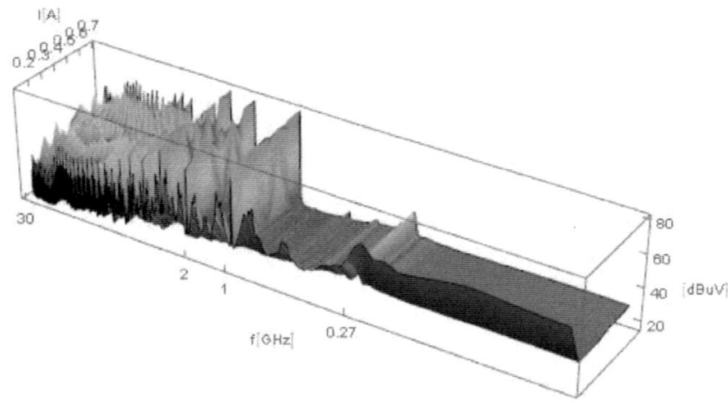

Fig. 9. Ripple current spectrum on the input wire (constant input voltage 24V), no filter applied

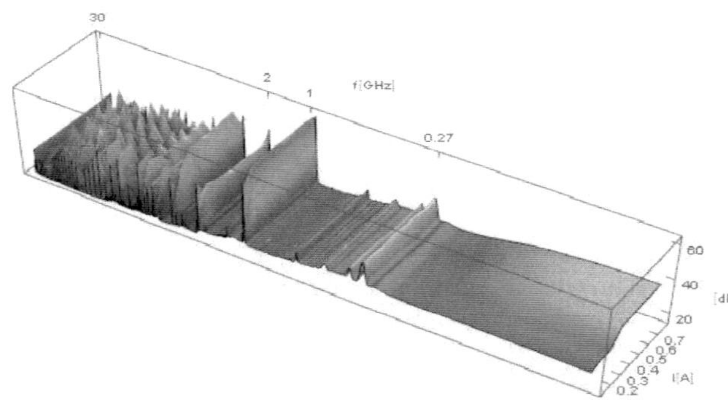

Fig. 10. Ripple current spectrum on the input wire (constant input voltage 12V), designed filter connected

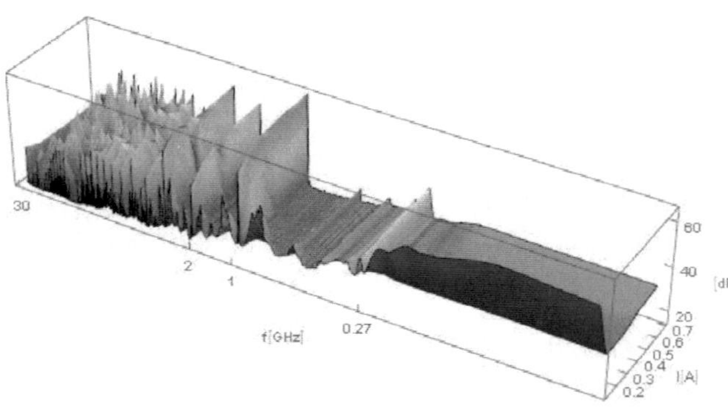

Fig. 11. Ripple current spectrum on the input wire (constant input voltage 24V), designed filter connected

The most significant improvement can be seen if we made a comparison two ripple current spectrums measured at one output current level. As can be found in Fig. 12 and Fig. 13 the improvement gained by the instalation of filter enormous. Moreover, It does not matter if the spectrum contains only peak (Fig. 14 - continuous mode) and/or if the spectrum contains wide range of frequencies (Fig. 15 - discontinuous mode). In both cases the filter is able to reduce maximal measured levels.

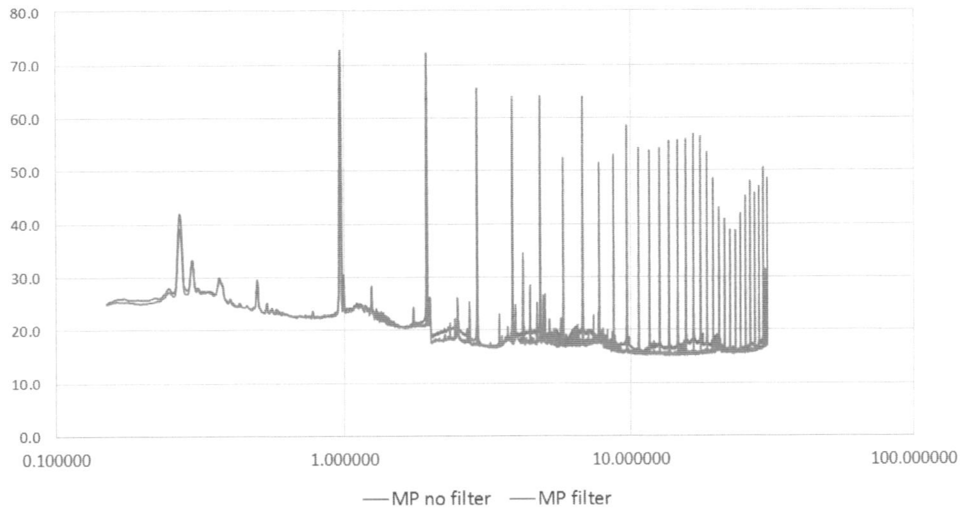

Fig. 12. Comparison of ripple current spectrum on the input wire (constant input voltage 12V and output current 100mA)

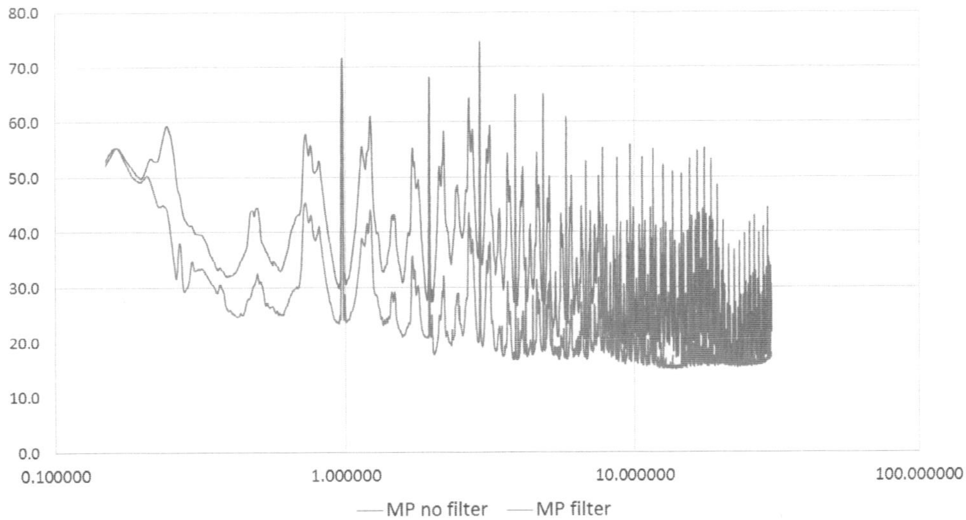

Fig. 13. Comparison of ripple current spectrum on the input wire (constant input voltage 24V and output current 100mA)

All previous measurements were performed by using MaxPeak detector. It means, that the maximum level, which was achieved in spectrum subrange, was taken as a result. But the standard define the second possible detector called QuasiPeak. This detector does not take only one maximum value in subrange, but tries to weigh signals according to their rate in measured subrange. The result obtained by using QuasiPeak detector can be seen in Fig. 14, Fig. 15, Fig. 16 and Fig. 17.

As is shown in these figures, the situation is similar to previous measurement - the use of a proper filters allows meet the defined levels of the standards

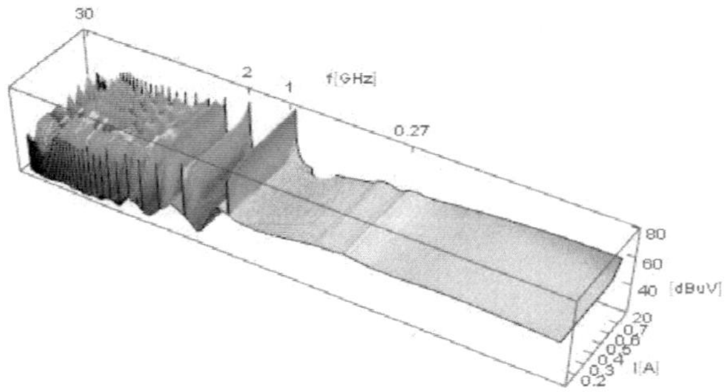

Fig. 14. Ripple current spectrum on the input wire (constant input voltage 12V), no filter applied. QuasiPeak detector used

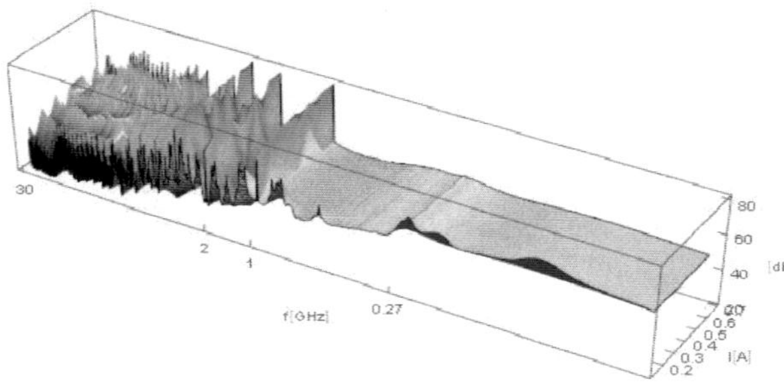

Fig. 15. Ripple current spectrum on the input wire (constant input voltage 24V), no filter applied. QuasiPeak detector used

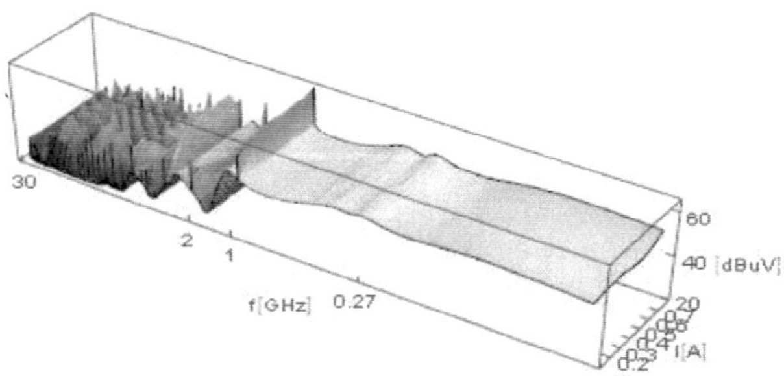

Fig. 16. Ripple current spectrum on the input wire (constant input voltage 12V), designed filter connected. QuasiPeak detector used

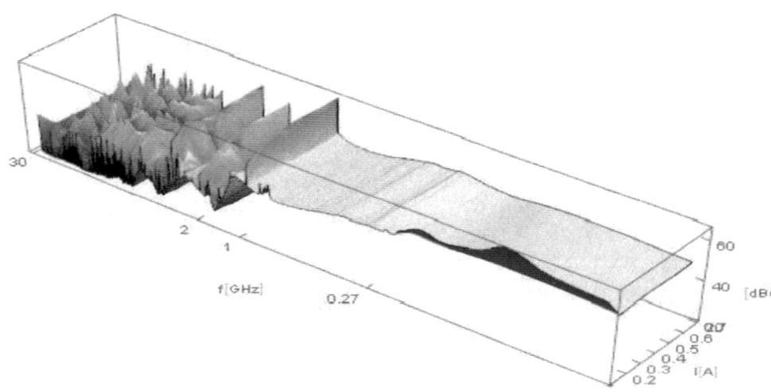

Fig. 17. Ripple current spectrum on the input wire (constant input voltage 24V), designed filter connected. QuasiPeak detector used

4. Conclusion

This paper provides an overview about designing of a small cheap voltage converter, which can be useful for implementation into small embedded devices. As the basic driver, the small and simple controller the AOZ 1284 has been chosen. Based on the datasheets and manufacturers' recommendations, the DC/DC converter circuit was designed and the applicable sample was constructed. After that the static DC characteristics were verified, the set of control measurement of interferences was performed according to the requirements of the standard EN 61000-6-3.

The static DC measurements agree that all requirements defined in the start of this paper were fulfilled. The radiated electromagnetic field measurements shown, that the designed circuit is likely to meet the requirements defined by the standard. At the other hand, the measurement of interferences on input cables revealed a significant problem with the level of these interferences.

According to this information, the simple LC filter was designed and all necessary measurements were repeated. After the application of the filter, the level of interferences has been lowered.

Because of the level of the interferences produced by this converter also depends on the connected load and the PCB design, we did not continued in tuning of the input filter and converter into details.

According to the data we obtained it can be claimed, that the designed converter will be suitable for its designation; moreover, it can be expected, that it will be able to pass all requirements defined in standard EN61000-6-3 after small tuning of its parts. Together with the price of this converter (about 0.4 USD) and final size of whole design (2 cm x 3 cm), this converter can be used as a replacement for frequently used MC34063 drivers and it could be sufficient competition for products of other manufacturers.

We would like to focus on other comparable integrated controllers of step-down converters in further research. Its real features and real measurements on test samples instead of simulations haven't been published well. Moreover, the input power filters and their impact on interferences on input wires will be interesting theme for further research.

5. Acknowledgement

This paper is supported by the Ministry of Education, Youth and Sports of the Czech Republic within the National Sustainability Programme project No. LO1303 (MSMT-7778/2014) and also by the European Regional Development Fund under the project CEBIA-Tech No. CZ.1.05/2.1.00/03.0089 and the Internal Grant Agency at TBU in Zlin, Project No. IGA/FAI/2016/036 .

6. References

[1] M. Pospisilik, M. Adamek, R. M. S. Silva, Diferent constructions of step down voltage converters in termsof EMC, International Journal of Circuits, Systems and Signal Processing, Volume 10, (2016), p. 190-199, ISSN: 1998-4464

[2] J. Svacina, Electromagnetic Compatibility [Elektromagneticka kompatibilita], Brno: VUT Brno, (2002)

[3] C. R. Paul, Introduction to Electromagnetic Compatibility, New York: John Wiley, (1992)

[4] P. A. Chatterton, M. A. Houlden, EMC Electromagnetic Theory to Practical Design, New York: John Wiley, (1991).

[5] ALPHA & OMEGA Semicondutor (2012, March), AOZ1284 Datasheet[online]. Available: http://aosmd.com/res/data_sheets/AOZ1284PI.pdf

[6] M. Pospisilik., T. Dulik, International Journal of Circuits, Systems and Signal Processing, Volume 8, (2014), pp 518 - 528

[7] M. Mann, B. Gutheil, J. Zastrau, P. Weiss, Electromagnetic _eld measurements Means of veri_cation, In Proc. of the 5th WSEAS/IASME Int. Conf. on Electric Power Systems, High Voltages, Electric Machines, Tenerife, Spain, December 16-18, (2005), pp 591-595

[8] M. Pospisilik, J. Soldan, M. Adamek, Inuence of the Properties of a Real Semi Anechoic Chamber on an Internal Electromagnetic Field Distribution, WSEAS Transactions on Systems, Volume 14, (2015), pp 174 186, E-ISSN: 2224-2678

[9] J. A. Cartrysse, Measuring method for the characterization of shielding gaskets, in 8th International Conference on Electromagnetic Compatibility, pp 251 253, Heriot-Watt University, Edinburgh, UK, September (1992)

[10] Z. Trnka, Theory of Electrical Engineering [Teoretick elektrotechnika]. Bratislava: SNTL Alfa, (1972)

[11] V. Kus, E_ects of Converters with Semiconductors to the Power System [Vliv polovodiovch menicu na napjec soustavu], Prague: BEN-Technicka literature, (2002), ISBN: 80-730- 0062-8

[12] Shah, R[ima]; Park, H[ong-Seok] & Lee, G[yu] B[ong] (2016). Design For Assembly: An approach to increase Design Efficiency of Electronics Home Appliance, Proceedings of the 26th DAAAM International Symposium, pp.0877-0882, B. Katalinic (Ed.), Published by DAAAM International, ISBN 978-3-902734-07-5, ISSN 1726-9679, Vienna, AustriaDOI: 10.2507/26th.daaam.proceedings.122

[13] Ismaeel, J.M., Jayanthy, T., Sugadev, M., Kumar, M.A., 2008. Analysis of electromagnetic interface filters for power systems and telephone lines, in: 10th International Conference on Electromagnetic Interference Compatibility, 2008. INCEMIC 2008. Presented at the 10th International Conference on Electromagnetic Interference Compatibility, 2008. INCEMIC 2008, pp. 433–437

DOI: 10.2507/27th.daaam.proceedings.078

THE INFLUENCE OF THE FINISHING STRATEGY ON THE QUALITY OF THE SURFACE

Lubos Kroft

[a]*Faculty of Mechanical Engineering, University of West Bohemia, Pilsen, Czech Republic*

This Publication has to be referred as: Kroft, L[ubos] (2016). The Influence of the Finishing Strategy on the Quality of the Surface, Proceedings of the 27th DAAAM International Symposium, pp.0524-0533, B. Katalinic (Ed.), Published by DAAAM International, ISBN 978-3-902734-08-2, ISSN 1726-9679, Vienna, Austria
DOI: 10.2507/27th.daaam.proceedings.078

Abstract

This article explores the effects of cutting forces in a finishing strategy on the quality, and geometric and dimensional accuracy of the finished surface. It describes another part of the running research into the effectiveness of finishing strategies for sculptured surfaces. The finishing stage is very important to the effectiveness of the entire machining process. This applies to both qualitative and economic aspects. In order to select the appropriate strategy, the process engineer needs to know its impact on the quality of the resulting finished surface. It is, however, nearly impossible to monitor the changes in all parameters of the process. Fortunately, each of these changes is reflected in the magnitude and profile of the cutting force. Therefore, it is desirable to describe how the machining strategy affects the finished surface quality and the cutting force profile. Describing such dependence is the ambition of future research which will draw on the theoretical findings of the present study.

Keywords: Quality of surface; finishing; strategy; cutting forces; tool wear.

1. Introduction

In today's engineering production, there is a strong pressure on productivity and economy. Strict requirements make every aspect of the entire process very important. The goal is to manufacture parts to sufficient quality at reasonable costs. In the process, the resulting utility properties are dictated by finishing operations.

Finishing is therefore of major importance in the production of any part. Owing to the complexity of shaped surfaces, the development of control programs for production machines takes place exclusively in CAM software. CAM systems must therefore include advanced machining strategies for finishing, such as HSC, HFC, and others.

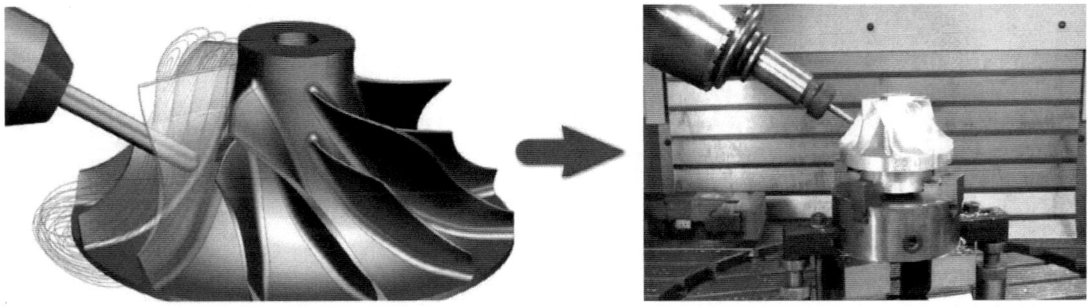

Fig. 0. Simulation of machining in CAM software and the actual process [9], [10]

Several scientific studies dealt with improving the effectiveness of finishing strategies. One of them was [14]. It compared productivity in finishing strategies in 5-axis operations. However, it did not mention any relationship between strategy and the resulting quality of the finished surface. Article [15] deals with the machined surface accuracy but only in terms of interpolation of the surface and not as a result of a particular strategy.

Nevertheless, there are studies which are closer to the issue of our interest. Moodleah and Makhanov [16] worked on improving the efficiency of tool paths through vector field-based tool path generation. Within this vector field, the directions are sought for which paths with maximum feed rate can be generated. Unfortunately, in this study only the efficiency of tool paths was sought, rather than the relation to finishing and to the resulting surface quality.

A completely different method of improving the efficiency, predominantly for finishing processes, has been presented by Imania, Sadeghib and Elbestawi [17], and Takata, Tsai, Inui and Sata. They attempted to predict cutting forces and compare the calculations with measured data. Their studies could lay foundation for future development of finishing operations.

All above-mentioned publications deal with the issue of our interest but none of them links the machining strategy and the cutting forces to the quality and dimensional and geometric accuracy of the finished surface. The objective of the present effort is to provide a theoretical basis for further research into the effectiveness of finishing sculptured surfaces.

table example:

NC	Numerical Control
CNC	Computer Numerical Control
CAD	Computer Aided Design
X, Y, Z	Axes in the cartesian coordinate system
SW	Software
Fx	Components of cutting force parallel to the X axis [N]
Fy	Components of cutting force parallel to the Y axis [N]
Fz	Components of cutting force parallel to the Z axis [N]
F	Total cutting force [N]
$\sum_{i=1}^{n} X_i$	Sum of X elements in the interval between X_1 and X_n
R_a	Arithmetic mean deviation of the profile [μm]
R_z	Maximum height of the profile [μm]
f_z	Feed per tooth [mm]
r_n	Tool nose radius [mm]
z	Number of teeth of the tool
Z	Deviation of the profile [μm]

Table 1. List of abbreviations

2. Quality of machined surface

Finishing operations are carried out to impart the required quality to a surface. Effective finishing therefore means achieving uniform quality and geometric accuracy of the machined surface.

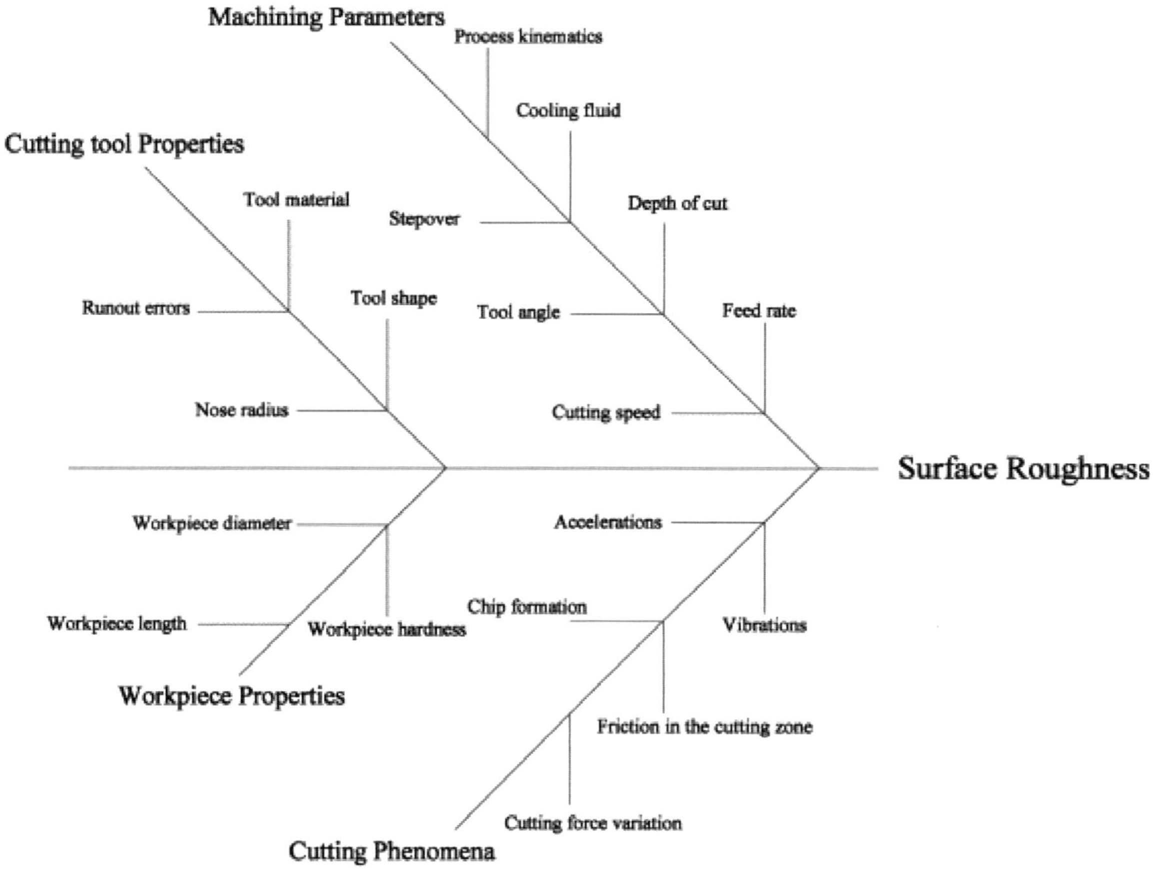

Fig. 0. Parameters that affect surface quality [4]

It follows from what has been said above that the magnitude of and fluctuations in the cutting force, and the proportions of its components have a major impact on the resulting quality of the finished surface. Yet, not every change in the cutting force must reflect substantially in the surface quality. All factors which can play a role in the surface quality are listed in Fig. 2. Whenever a cutting parameter changes, e.g. the cutting speed, stepover, tool diameter or others, the effect can be traced through a change in the cutting force.

The directions of components of the cutting force on a ball-end tool. More details are given in Fig. 3 and 4. The first one illustrates the effect of the inclination of the surface and the stepover on the radial component of the cutting force. [8]

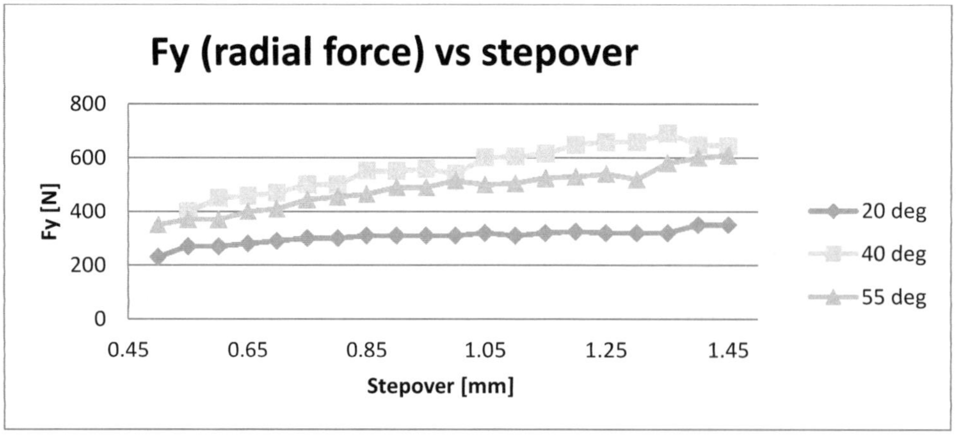

Fig. 3. Radial component of cutting force on the stepover size [9]

The second graph shows the dependence of cutting force on the cutting speed. Although the scatter is substantial, the trend indicated by the polynomial fit is in agreement with the assumption.

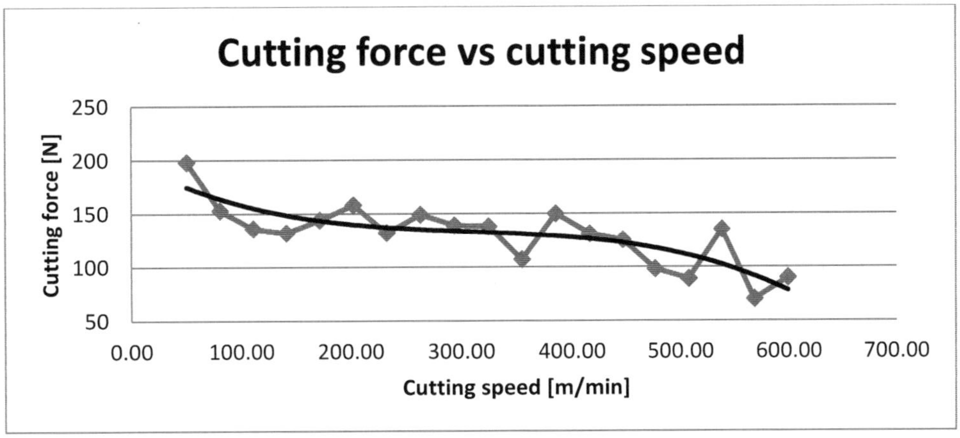

Fig. 4. Plot of cutting force vs cutting speed [9]

It is desirable to use these results to identify the relationship between cutting force variation and the quality of the finished surface. Then, the resulting quality of surface could be predicted on the basis of changes in the cutting force. In order to define such relationship, the cutting force profile must be examined in greater detail. In the following, cutting force will be studied as an indication of the potential quality and dimensional accuracy of the finished surface.

2.1 Effect of cutting force profile on machined surface quality

When dealing with the cutting force, it is important to consider both overall profile, variation, and maximum levels within intervals of up to tens of seconds, as well as the micro-scale variation within several tenths of a second. The quality of the finished surface depends in some way on both these scales.

The fundamental one is the macro-scale, where an entire pass or even multiple passes are considered. In the example in Fig. 5, the force magnitude in the first five seconds into the process is relatively small but it increases with time. The exact values shown in this graph are not essential for gaining understanding of the process Nevertheless, one can assume that the surface quality at the beginning will be different than at the end of the process. Between 5th and 10th second, there is fluctuation of the cutting force which could also affect the machined surface quality.

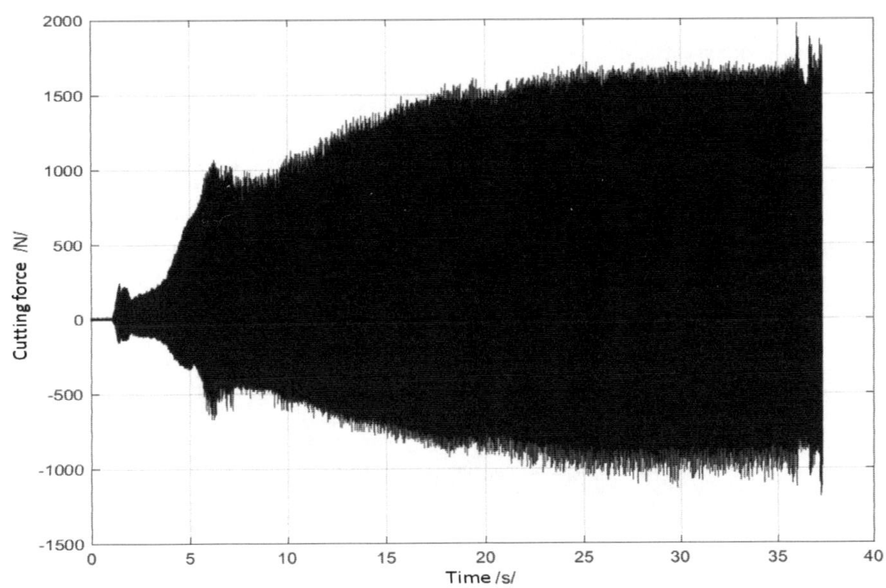

Fig. 5. Cutting force profile – macro-scale

Gradual increase in the force and its variation can arise from varying inclination of the workpiece surface, variation in the feed rate, the stock allowance, as well as from the cutting edge becoming dull or other factors, such as inhomogeneities in the material, and others. The important conclusion is that such variations can impact on the quality of the finished surface. It is therefore desirable to develop a strategy which keeps the cutting force constant, and therefore leads to quality surface even in a sculptured workpiece.

As said above, the cutting force profile can be examined on various scales. One of them is the micro-scale represented by a force plot within an interval of several tenths of a second. Such a plot is shown in Fig. 6. It covers 0.35 seconds of the machining process which means that it includes several revolutions of the tool. The curve indicates shock loads which are typical of the milling process. The force varies from -600 N to approximately 1200 N, causing relatively severe shocks which may affect the quality of the machined surface. These shocks are due to individual teeth engaging the workpiece or more precisely entering and leaving the cut area. This variation as such is unavoidable but its magnitude can be reduced. The shocks become less severe when more teeth are engaged at the same time. In addition, it is possible to adjust the geometry or microgeometry of the cutting edge, and other parameters.

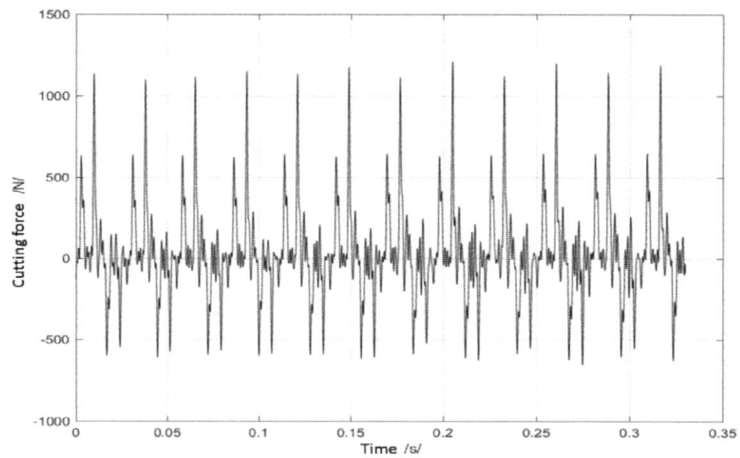

Fig. 6. Cutting force profile – micro-scale

In the finishing of sculptured surfaces, the macro-scale view is more important than the micro-scale one. The impact of the micro-scale profile on the quality of the finishing process arises from the inherent features of the milling process and is virtually independent of the strategy used for tool path generation. On the other hand, the macro-scale profile is substantially dictated by the strategy chosen and therefore becomes the key to the process.

2.2 Evaluation of finished surface quality

Suitable indicators for quality evaluation include surface roughness, macro-scale surface profile, and, understandably, findings from visual inspection.

The number of surface characteristics available today is very large. Practically, the most widely used ones include Ra and Rz values. Thanks to optical scanning devices, areal roughness is being used more frequently as well. Its information value is not to be overstated but the method is suitable for sculptured surfaces thanks to contactless measurement.

One appropriate indicator might be the arithmetic mean deviation of the profile calculated according to:

$$R_a = \frac{1}{n} \times \int_1^n |Z_{(x)}| dx \ [\mu m] \ [11] \tag{1}$$

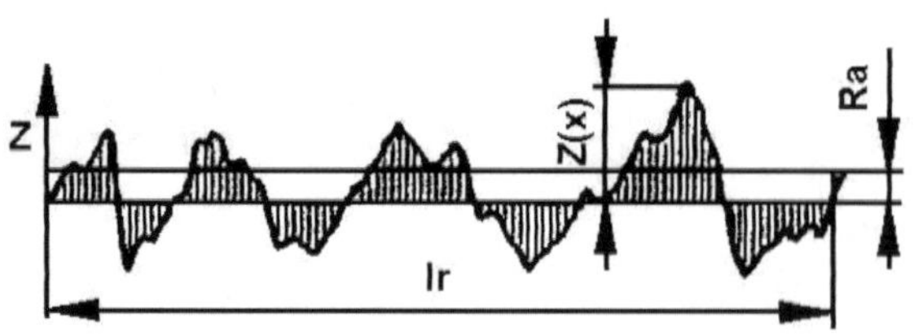

Fig. 7. Illustration of the arithmetic mean deviation of the profile Ra [11]

It is advisable to verify the Ra value by theoretical calculation based on the specified feed rate for the milling process:

$$R_a = \frac{{f_z}^2}{n^2 \times z^2 \times 31{,}2 \times r_n} \ [\mu m] \ [12] \quad (2)$$

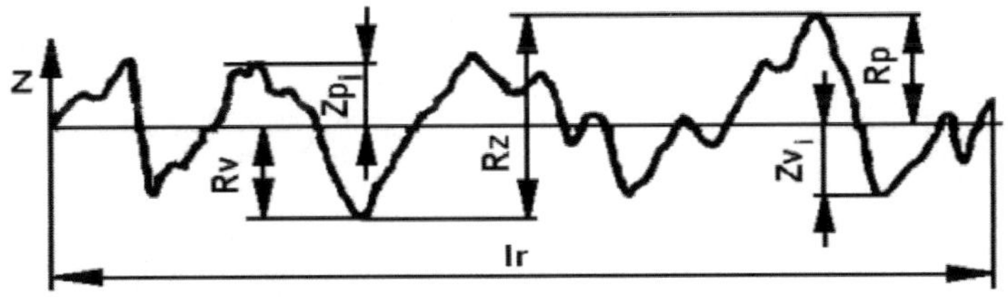

Fig. 8. Other roughness parameters, including Rz [11]

One can also use the arithmetic mean spacing between five peaks and valleys calculated as:

$$R_z = \frac{\sum_{i=1}^{5}|y_{pmi}| + \sum_{i=1}^{5}|y_{rmi}|}{5} \ [\mu m] \ [11] \quad (3)$$

Dimensional analysis and evaluation of individual dimensions of a sculptured surface tend to be difficult. A convenient approach uses measurement by a scanning device and an evaluation of the deviations from the 3D model.

It would be useful to compare the values of cutting forces, evaluate their profiles throughout the measurement run, and their envelopes. In addition, it is possible to evaluate the total cutting force using the following formula:

$$F = \sqrt{{F_x}^2 + {F_y}^2 + {F_z}^2} \ [N] \quad (4)$$

In order to compare the forces that arise in individual strategies, one can calculate the statistical mean value according to:

$$\bar{F} = \frac{1}{n} \times \sum_{i=1}^{n} F_i \ [N] \quad (5)$$

Correlation of qualitative parameters and dimensional values with the values and profiles of cutting force in individual strategies will reveal the dependences, on the basis of which optimized finishing strategies can be developed.

2.3 Summary

From the viewpoint of finishing operations which impart the final quality and dimensional accuracy to the machined surface, the macro-scale profile of cutting force is the key factor. It is because it is affected by the machining strategy which is the topic of the next section. In order to solve this problem, one needs to gain a comprehensive understanding of the cutting process at the finishing stage, and of the impact of individual force components on the quality and dimensional accuracy of the machined surface, and find correlations between these factors.

3. Tool path strategy

Fundamental differences in the tool behavior can be seen between climb milling and conventional milling. The distinction between these two techniques is well known. When dealing exclusively with finishing strategies, the choice of the technique does not have as strong effect as in other stages. The reason is that the load on the tool is relatively low and does not reach critical levels. Therefore, both techniques can be used without any substantial qualitative differences, and time savings can be achieved by eliminating non-cutting traversing movement.

Much greater impact is related to the feed and pick feed directions with respect to the inclination of the surface. All variants are illustrated in Fig. 10.

The first option involves horizontal feed and vertical downward pick feed. The above strategy is referred to as linear machining and is optimal for surfaces with steeper inclination angles, approximately 75°, as suggested in [3]. Here, this strategy leads to the lowest cutting force and to the most favorable variation in the cutting force. This strategy involves downward tool movement and typically leads to abrasive wear on the tool flank. The flank wear tends to be uniform, as shown in Fig. 9 a).

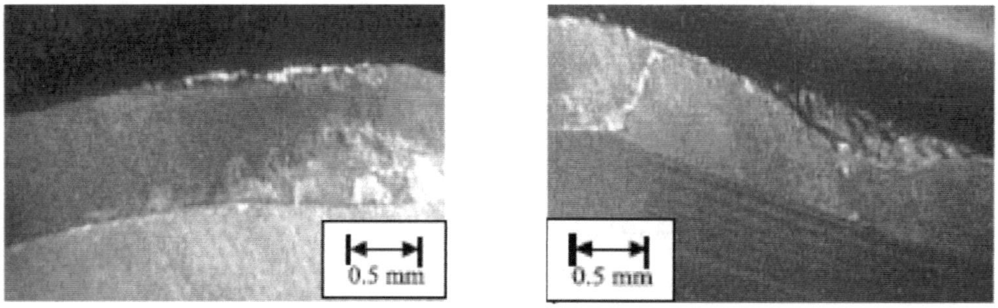

Fig. 9. Tool wear upon a horizontal feed process: a) downward, b) upward [2]

If the strategy involved the same horizontal feed but upward pick feed direction, these forces would increase. The variation would be from 60 to 200 %, predominantly in the F_y component. This force acts in the normal direction with respect to the workpiece surface and has a major impact on the resulting quality and dimensional and geometric accuracy. The mechanical load on the tool is higher as well. The force profile is similar to the first case. Cutting edge chipping may occur, as illustrated in Fig. 9 b). [6]

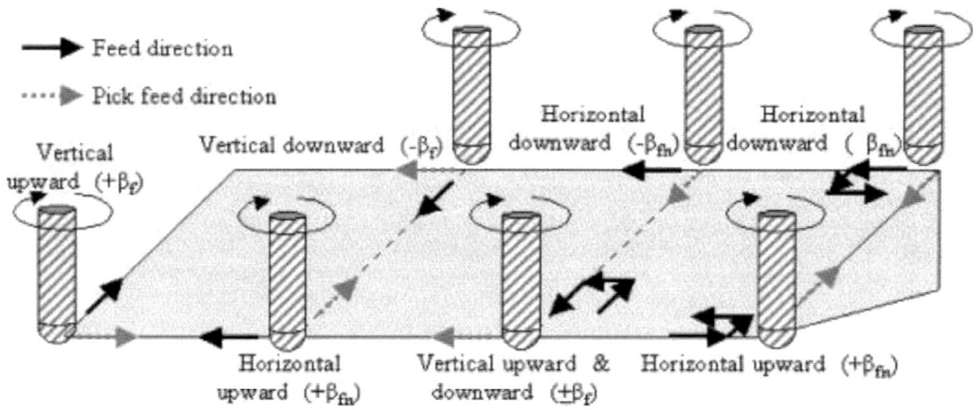

Fig. 10. Tool path strategy [7]

Another option of those listed in Fig. 10 involves vertical feed direction, and starting the cutting process at the top. The pick feed direction is horizontal. In this strategy, the material is being pushed in front the tool. As a result, the tool may deflect toward the workpiece, predominantly due to the Fy component of the cutting force. The variation in the cutting force might range from 30 to 300 %. This strategy is appropriate for surfaces with mild inclination, less than 25° [3, 7].

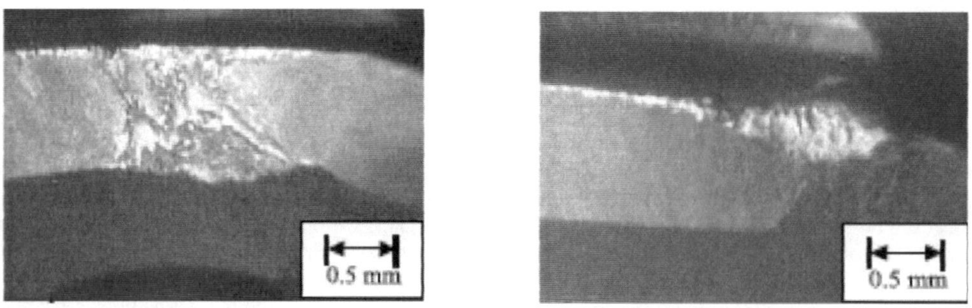

Fig. 11. Tool wear upon vertical feed: a) downward, b) upward [2]

No major change will be registered when machining starts at the bottom and ends at the top of planar surfaces with mild inclination angles.

Near-planar surfaces with inclination angles under 15° are better produced using form-duplicating strategies, according to [13], but the wear is more severe due to lower cutting speed. On surfaces with inclination angles around 45° none of the strategies leads to markedly better results.

These conclusions are based on the literature sources cited, as well as on experiments carried out at University of West Bohemia [6], [8].

3.1 Effect of strategy on dimensional accuracy

As mentioned above, an important aspect of the result of machining is the quality of the machined surface. However, an equally important factor is its dimensional accuracy. Figure 12 a) shows an example of two machining strategies on a shape surface: linear machining and form-duplicating machining.

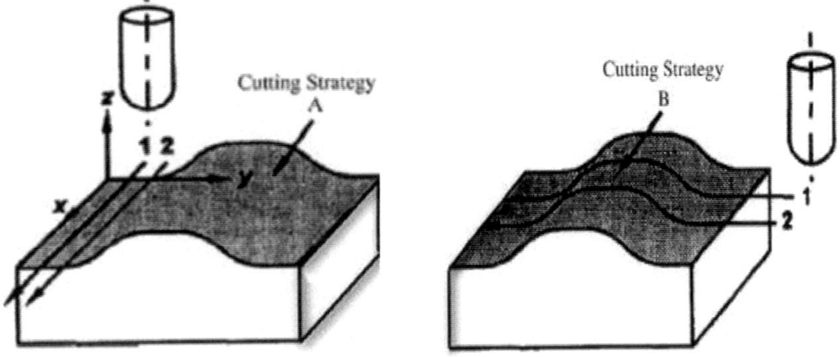

Fig. 12. Schematic representation of tool paths for a) linear machining strategy, b) form-duplicating machining strategy

Strategy A is a linear machining strategy with an upward vertical pick feed direction. We will not explore the suitability of this strategy for a particular shape surface. Instead, we will focus on the dimensional accuracy of the finished surface shown in Fig. 13. We will find that it is not uniform. The largest deviations from the nominal values are produced as the tool descends. While the tool is ascending, dimensional errors arise as well but their magnitude is smaller. The main reason is the variation in the cutting force in various parts of the shape. In particular, this applies to the component that is normal to the surface.

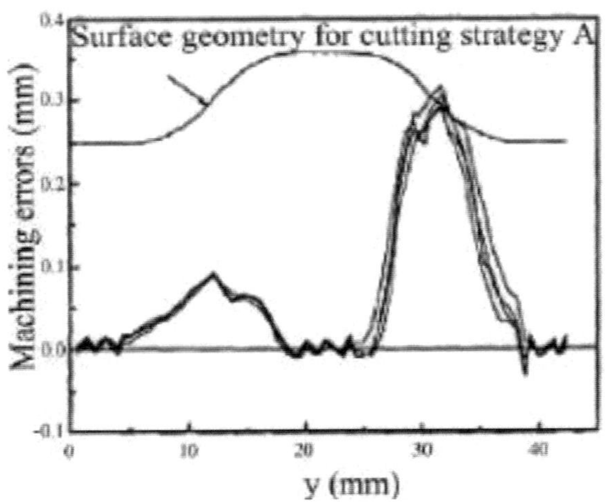

Fig. 13. Dimensional accuracy upon machining with strategy A [1]

By contrast, strategy B shown in Fig. 12 b) traces the shape of the workpiece in the horizontal direction and the pick feed is perpendicular to the feed direction. The texture of the finished surface is completely different from that produced by strategy A. As the tool descends, incomplete cutting occurs. The reason is, again, the normal component of the cutting force. However, the amount of incomplete cut is less because the tool pushes the work in front. Conversely, as the tool ascends, it overcuts because it is being pulled into the work. The results are illustrated in Fig. 14.

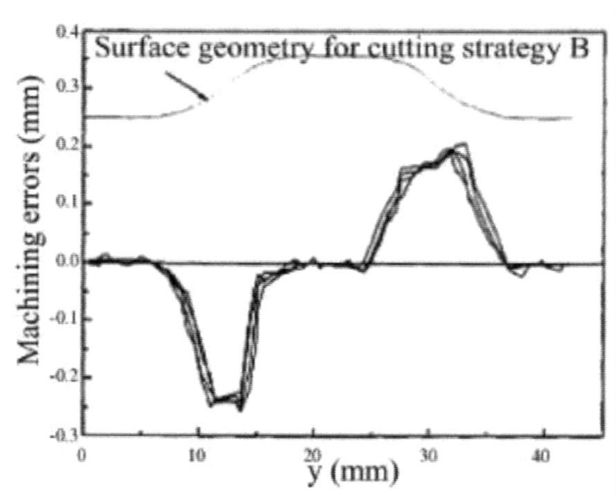

Fig. 14. Dimensional accuracy upon machining with strategy B [1]

These data suggest that the finishing strategy has a major impact on the accuracy of the finished surface. The same can be expected to hold for the quality of the finished surface, although this is not certain as no data on the surface quality are available in this respect.

3.2 Summary

The above-named strategies represent the fundamental available approaches to calculating tool paths for general shape surfaces. In many regions of such surfaces which are, in one way or another, approaching some limit, these strategies begin to fail. In these regions, the pre-set parameters undergo changes – which either impairs the quality of the shape surface or increases the machining time prohibitively.

Besides the machine, the tool, and the properties of the work, the shape itself also dictates the quality of the finished surface. The reason is the changes in the cutting force or more precisely in some component of the cutting force. Such changes affect the other parameters, such as cutting speed, chip volume, stock allowance for finishing, and others. In order to be able to propose a finishing strategy which, at least in part, improves the state of the art, these aspects must be investigated in greater depth. This will be the objective of the next part of the research

4. Conclusion

This study explores modern effective finishing of sculptured surfaces, i.e. one part of the field, which concerns the effect of the tool path strategy on the quality and geometric and dimensional accuracy of the finished surface. Despite today's advanced computing hardware and CAM software tools, finishing operations remain a challenge.

Algorithms that improve the effectiveness of tool paths in finishing are available. Unfortunately, their implementation in automation and CAM systems is so complicated that even today none of the above-named algorithms is part of any commercial manufacturing support software. Consequently, finishing paths have inadequate dimensions, e.g. too small stepover, or combine multiple operations, which is very time-consuming and inefficient because of the overlap of paths.

All these drawbacks have mostly negative effects on geometric accuracy of the finished surface, and lead to non-uniform quality. Together with finishing time, the quality and geometric accuracy are the fundamental indicators of effectiveness of any particular process. The reason is that these indicators show whether the condition of the surface is adequate or whether reworking is necessary, which involves additional machining costs.

In order to be able to prove the effect of cutting parameters in a particular strategy, quality and dimensional accuracy must be linked to a parameter which reflects virtually all changes that occur during machining. This parameter is the cutting force. It varies with almost all changes in the fundamental parameters. Cutting force, specifically its components, is a suitable parameter for evaluating finishing operations. One reason is that its amplitude gives a clear indication of changes in the process, and the other is that its overall profile can be used for tracking the effects of the strategy, finding correlations with the surface quality, and with the geometric accuracy of the finished surface.

Such conclusions can be drawn on the basis of the above-described facts. Further research will aim at finding correlations of the quality and geometric and dimensional errors of the machined surface with variation in the cutting force, and at improving the effectiveness of finishing sculptured surfaces, and thus developing the entire discipline of machining.

5. Acknowledgments

This paper includes results created within the project SGS-2016-005.

6. References

[1] E.M. Lim, H.Y. Feng, C.H. Menq, Z.H. Lin The prediction of dimensional error for sculptured surface productions using the ball end milling process. Part 1. Chip geometry analysis and cutting force prediction Int. J. Mach. Tools Manufact, 35 (8) (1995), pp. 1149–1169

[2] Toh, C. K.: A study of efects of cutter path strategies and orientations in milling, Journal of materials procesing technology 113(2004) 346-356

[3] Poulachon, Gérard, Patric GHIDOSSI, José Outeiro and David Prat. Procedes et moyens de fabrication: Support de cours- TPS. Cluny, 2013

[4] Czerech, Łukasz. Selection of Optimal Machining Strategy in the Manufacture of Elements Bounded by Curvilinear Surfaces. acta mechanica et automatica, 2013, 7.1: 5-10. http://www.degruyter.com/view/j/ama.2013.7.issue-1/ama-2013-0001/ama-2013-0001.xml

[5] Sova, František. Technologie obrábění a montáže. 2nd edition. Plzeň: University of West Bohemia, Faculty of Mechanical Engineering, 1998, 273 pages. ISBN 80-708-2449-2

[6] Řehoř, Jan. Literature search for the project Studium vysokorychlostního obrábění (HSC) ocelí vysoké pevnosti a tvrdosti, Plzeň: University of West Bohemia, Faculty of Mechanical Engineering, 2001

[7] Surface topography analysis in high speed finish milling inclined hardened steel [online]. School of Engineering (Mechanical), University of Birmingham, 2003 [Accessed 2016-01-04]. http://www.sciencedirect.com/science/article/pii/S0022460X03012598

[8] Hnatik, Jan, Lubos Kroft and Jan Kutlwaser. Constant chip volume machining. Elsevier Ltd. [online]. 2015, 7 [Accessed 2016-08-24]

[9] CAD.cz [online]. [Accessed 2015-09-09]. http://www.cad.cz/pdmplm/7-2007/1377-petiose-obrabeni-v-praxi.html

[10] CAD-FEM.CZ, A Blog Of The Company Pemavako S.R.O. [online]. [Accessed. 2015-09-09]. http://cad-fem.cz/component/content

[11] GAJDOŠÍK, D. Měření drsnosti povrchů se zaměřením na měřící techniku Mytutoyo SJ 210. České Budějovice, 2013. Bachelor thesis. University of South Bohemia in České Budějovice. Supervisor PaedDr. Bedřich Veselý, Ph.D.

[12] A.B. Abdullah, L.Y. Chia and Z. Samad, 2008. The Effect of Feed Rate and Cutting Speed to Surface Roughness. Asian Journal of Scientific Research, 1: 12-21

[13] Vopát Tomas, Peterka Jozef, Kováč Mario, Buranský Ivan, The Wear Measurement Process of Ball Nose end Mill in the Copy Milling Operations, Procedia Engineering, Volume 69, 2014, Pages 1038–1047, 24th DAAAM International Symposium on Intelligent Manufacturing and Automation, 2013

[14] Tunç, Lütfi Taner; Ozkirimli, Omer Mehmet; Budak, Erhan. Machining strategy development and parameter selection in 5axis milling based on process simulations. The International Journal of Advanced Manufacturing Technology, 2015, 118. http://link.springer.com/article/10.1007%2Fs0017001580016

[15] Nittler, Kevin M.; Farouki, Rida T. A realtime surface interpolator methodology for precision CNC machining of swept surfaces. The International Journal of Advanced Manufacturing Technology, 2016, 83.14: 561574. http://link.springer.com/article/10.1007%2Fs001700157552x

[16] Moodleah, S.; Makhanov, Stanislav S. 5axis machining using a curvilinear tool path aligned with the direction of the maximum removal rate. The International Journal of Advanced Manufacturing Technology, 2015, 80.14: 6590. http://link.springer.com/article/10.1007%2Fs0017001569589

[17] Imania, B.M., M.H. Sadeghib A M.A. Elbestawia. An improved process simulation system for ball-end milling of sculptured surfaces. International Journal of Machine Tools and Manufacture [online]. 1998, (38) [Accessed 2016-09-06]. http://www.sciencedirect.com/science/article/pii/S0890695597000746

DOI: 10.2507/27th.daaam.proceedings.079

CONCEPTUAL THEORETICAL MODEL FOR LIFE CYCLE ENERGY ANALYSIS OF PHOTOVOLTAIC MODULES

Milovan Medojevic, Ilija Cosic, Nemanja Sremcev, Milovan Lazarevic

This Publication has to be referred as: Medojevic, M[ilovan]; Cosic, I[lija]; Sremcev, N[emanja] & Lazarevic, M[ilovan] (2016). Conceptual Theoretical Model for Life Cycle Energy Analysis of Photovoltaic Modules, Proceedings of the 27th DAAAM International Symposium, pp.0534-0543, B. Katalinic (Ed.), Published by DAAAM International, ISBN 978-3-902734-08-2, ISSN 1726-9679, Vienna, Austria
DOI: 10.2507/27th.daaam.proceedings.079

Abstract

Photovoltaic (PV) systems, as well as other renewable energy systems, strive to be both energy efficient and cost competitive in order to emerge as a prominent mode of electricity production. However, conversion efficiency, defined as the percentage of solar insolation converted to electricity, has been one of the primary performance metrics for evaluating alternative PV technologies. Unfortunately, conversion efficiency only addresses the operational energy efficiency of a PV device. A more comprehensive energy analysis includes the total life-cycle of the PV system, encompassing raw material production, manufacturing, use, maintenance, and end-of-life management. Having this in mind, the aim of this paper is to provide a closer insight on the energy consumption intensity of the relevant processes from resource production processes to the end of life management of PV modules. In this paper a theoretical, conceptual and above all holistic model for life cycle energy analysis has been introduced and analysed, while the crucial process points, relevant in terms of energy recovery are identified and presented.

Keywords: PV module; Life cycle energy analysis; Energy consumption.

1. Introduction

With average annual growth rates in excess of 40% over the past decade [1,2], the success of the PV industry can largely be attributed to the steadfast growth of wafer-based multi-crystalline and mono-crystalline silicon. This growth has been sustained through a powerful combination of three critical competitive advantages: (1) industry-leading full module area sunlight power conversion efficiencies (to date, monocrystalline silicon continues to provide the highest power conversion efficiency among all commercially demonstrated single junction PV modules [3]); (2) product 'bankability' from the appropriately qualified suppliers (with warranties for 80% of original performance after 25 years of service now being standard [4,5]); and (3) a consistent ability to offer competitively priced modules, which has been enabled through an ability to realize cost reductions throughout the c-Si module supply chain. In addition, PV technology offers some unique benefits that are not realized by other renewable energy technologies, such as the fact that it represents a silent energy source which in most cases does not require moving parts [6]. Furthermore, PV's have a long system lifetime with low maintenance costs with experienced substantial reduction in upfront cost over the past two decades. Being a decentralized technology, PV systems increase the resilience of the energy infrastructure, while simultaneously improving national security of supply. This ascertainment constitutes support and points to the importance of research in this area.

Having in mind the aforementioned, determination of production system energy performance and efficiency could be identified by implementation of the life cycle energy analysis (LCEA). Conventionally, life cycle analysis (LCA) takes into account the direct and indirect impacts throughout the entire life cycle of the product, including material sourcing, manufacturing, operation, transportation, disposal, etc. Moreover, as given by many authors, such as Phylipsen [7], Alsema [8], Knapp [9], Phent [10], Meijer [11], LCA is recognized as an invaluable tool to assess the energy and environmental profiles of a PV systems in practice for many years. Likewise, LCEA is an approach to energy management system implementation involving the quantitative evaluation of a product's overall energy performance and impact. Energy requirements throughout the whole life cycle of the product are estimated in order to enable such evaluation and provide results that can be used for related energy analysis and assessment. On the other hand, since life cycle is related to a broad range of variables and is complicated, it is difficult to comprehend the exact significance of the results, where, accordingly, it is highly important to define a purpose for the evaluation.

However, a majority of studies dealing with this issue, lacks a holistic approach in determination of energy payback time and intensity of energy consumption of a typical PV system. In the most studies, the system boundary is defined as the pre-manufacturing, manufacturing, installation and use phases, while recycling and disposal phases are excluded [12,13]. Furthermore, in some cases boundary conditions for analysis only take into account energy consumption of specific quartz sand-to-PV module manufacturing processes [14,15]. Likewise, a shortage of studies that takes into account inter process/operation transportation energy consumption, as well as energy consumed during distribution and logistics phase (before putting the system in operation) has been identified.

Therefore, in this paper, LCEA principles are suggested as a mean for overall efficiency identification regarding crystalline-based PV technology, where a more comprehensive energy analysis includes the total life-cycle of the PV system, encompassing raw material production, manufacturing, use, maintenance, inter process/operation energy consumption, and end-of-life management. This has been concretely and in detail given as a conceptual theoretical model, while it is strongly believed that application of this model ensures systematic monitoring of a system process defined by implied system boundaries.

2. Conceptual model for PV module life cycle energy analysis

Having in mind that the objective of this study was to implement the LCEA approach for identifying and guiding the PV modules life cycle, a detailed description of the LCEA methodology and its application is presented as conceptual theoretical model. Even though the analysis could be deeper and more complex it stimulates logical identification for potential optimization spots, system effectiveness and process energy efficiency.

2.1 Defining the system boundaries

By respecting that the system boundaries should be defined with careful consideration, the scope of the analysis is to evaluate the total energy requirements for the observed system scope level (OSSL), where OSSL considers the complexity and how detailed the system should be analysed.

The manufacturing of solar cells, to produce electrical power, is quite complex and involves many steps starting from the initial grains of sand to the finished solar cell module or panel, it utilization and end of life treatment. (Fig. 1).

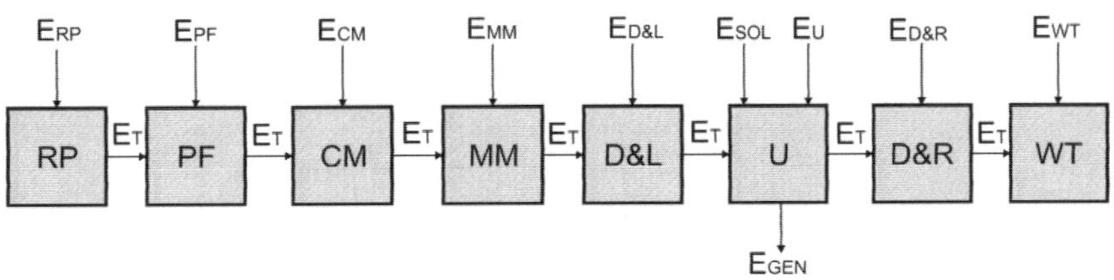

Fig. 1. Observed system scope level and overall process flow

In this study, the manufacturing process is roughly divided into sub categories where some of them are consisted of specific actions and often represent separate processes that in some cases can be outsourced. This also applies to the product end of life treatment processes respectively. However, it is of high importance to establish the important physical components or subsystems for both energy and production systems that play a key role in the supply and demand side of overall process operation. When energy and process flowcharts are put together, valuable information is provided on where, why and what type of energy is used. This represents the basis for decisions on setting up energy cost centres (ECC), segments (i.e. units of equipment or single equipment) where activities or production volume are quantifiable and where a significant amount of energy is used [16]. Given the aforementioned, this model identifies key processes in a PV module life cycle as an ECC's upon which general energy balance equation could be generated and derived depending on research complexity. The Fig. 1 illustrates the OSSL, where:

RP - Resource production process
PF – Part fabrication process
CM – Cell manufacturing process
MM – Module manufacturing process
D&L – Distribution and logistics
U – Utilization
D&R – Disassembly and Recycling
WT – Waste treatment
E_T – Inter-process transportation energy consumption
E_{RP} – RP Energy consumption
E_{PF} – PF Energy consumption
E_{CM} – CM Energy consumption
E_{MM} – MM Energy consumption
$E_{D\&L}$ – D&L Energy consumption
E_{SOL} – Solar gains (Insolation)
E_U – U Energy consumption
E_{GEN} – Generated energy
$E_{D\&R}$ – D&R Energy consumption
E_{WT} – WT Energy consumption

In order to determine overall system energy consumption (E_{sys}), energy inputs for each stage of the life-cycle in the lifetime period are stated in Equation (1):

$$E_{sys} = E_{RP} + E_{PF} + E_{CM} + E_{MM} + E_{D\&L} + E_U + E_{D\&R} + E_{WT} + \sum E_T \quad (1)$$

In addition, energy payback time (EPBT), a metric adopted by several analysts to characterize the energy sustainability of various technologies could be introduced to identify overall energy profitability of observed system. It is the energy analogy to financial payback, defined as the time necessary for a photovoltaic panel to generate the energy equivalent to that used to produce it. In this case EPBT can be determined as given in the Equation (2):

$$EPBT = \frac{E_{sys}}{E_{GEN}} \quad (2)$$

Considering all previously mentioned, the same approach is applied to identified subcategories based on defined research boundaries given in the Fig. 1.

2.2 Resource production

The basic, starting material for solar cells manufacturing is silicon. Silicon is an important resource and semiconductor which is extremely brittle. Resource production process (RP) represents its manufacturing (Fig. 2) by heating quartz sand with the coke at a high temperature in an electric furnace and then processed to metallurgical grade silicon with a purity of about 99%. The silicon is then refined in a complex purification step involving the intermediate trichlorosilane. Polycrystalline silicon ingots are then moulded from the crystalline silicon granules (solar-grade silicon).

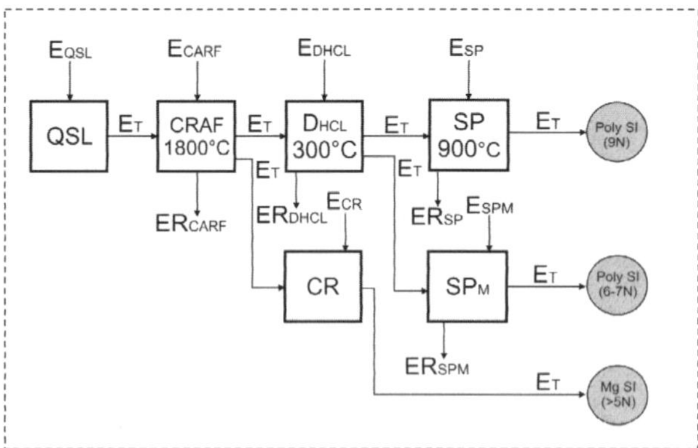

Fig. 2. Resource production process flow

In the Fig. 2 it is possible to distinct three main categories of silicon manufacturing processes, resulting in different purity levels: (1) Electronic-grade silicon (9N), (2) Medium-grade silicon (6-7N) and (3) Metallurgical-grade silicon (>5N). At the beginning, quartz sand is imported via quartz sand loader (QSL), for which, it is need to ensure adequate amount of energy (E_{QSL}). Then, quartz sand is transported to the next operation which indicates that transportation energy consumption (ET) should be also taken into account. The next step is coke reduction (CRAF) where metallurgical-grade silicon with 98.5% purity is produced from quartz sand in an arc furnace at very high temperatures. Here the high temperature (1800°C), necessary for process execution, is provided by electric arc furnace, which indicates significant power draw of the furnace and thereby energy consumption (E_{CARF}). Having in mind the height of temperature which must be reached, potential of energy recovery should be considered. In this case, system modification and upgrade with heat exchangers could ensure that the certain amount of energy could be recovered (ER_{CARF}) and used in the next process step or for heating, hot water preparation, as well as for other heat-demanding processes. After coke reduction, the metallurgical grade silicon powder is dissolved in hydrogen chloride (D_{HCL}) and subsequently distilled to form a saline gas. In most instances, this is the trichlorosilane, but could be others. This process is 6 times less energy intensive in compared to CRAF since it requires temperature of 300°C. This thermal energy (E_{DHCL}) could be largely provided as (ER_{CARF}), if it does not disturb the functioning of the system. Next in a row is so-called Siemens Process (SP) where the polycrystalline silicon is grown at quite high temperatures (900°C). It requires hydrogen and produces more hydrogen-chloride as a by-product. Similarly to previously mentioned CARF, significant power draw occur indicating energy consumption (E_{SP}), while at the same time potential for energy recovery becomes obvious (ER_{SP}). The Siemens Process is the last step in production of electronic-grade silicon (9N). In addition, the big drawback of the standard process as described above is that a Siemens reactor is very expensive and the Siemens process itself requires a lot of energy. A number of new proprietary processes reduce the energy consumption and the capital costs for silicon production, though they are still similar to the traditional Siemens process. These are so called modified Siemens processes (SP_M) where two main alternatives are imposed. The first one is Fluidized Bed Reactor which operates at much lower temperatures and does not produce by-products, while the second one is Vapour to liquid deposition which is similar to Siemens, but ensures faster extraction. In both cases medium-grade silicon (6-7N) is produced. However, SP_M also consume energy (E_{SPM}) and therefore it is equally important to analyse potential for energy recovery (ER_{SPM}). Lastly, in an altogether different process, metallurgical-grade silicon is chemically refined (CR). By blowing gasses through, the silicon melts, the boron and phosphorous impurities are removed, followed by directional solidification. Energy consumption of CR is defined as (E_{CR}), while taking this production route metallurgical-grade silicon (>5N) is generated. However, it could be concluded that by avoiding high purification, manufacturing costs could be reduced significantly [17].

In order to determine RP consumption (E_{RP}), energy inputs (E_i) for each process stage are summed up while energy recovery potential (ER) which could be reused or reapplied is subtracted from the inputs sum. This has been given below by Equations (3) and (4):

$$E_{RP} = \sum E_i - \sum ER \tag{3}$$

$$E_{RP} = E_{QSL} + E_{CARF} + E_{DHCL} + E_{CR} + E_{SP} + E_{SPM} + \sum E_T - (ER_{CARF} + ER_{DHCL} + ER_{SP} + ER_{SPM}) \tag{4}$$

Here the Equation (4) could be derived more, which is not necessary to understand this model in theory, while model practical application generate more precise results by increasing research complexity level.

2.3 Part fabrication

Part fabrication process (PF) or "wafering" is where the ingots are firstly cut to small bricks and then to thin quadratic slices, so called wafers (Fig. 3). The cutting of silicon wafers is one of the most difficult parts of the entire machining process.

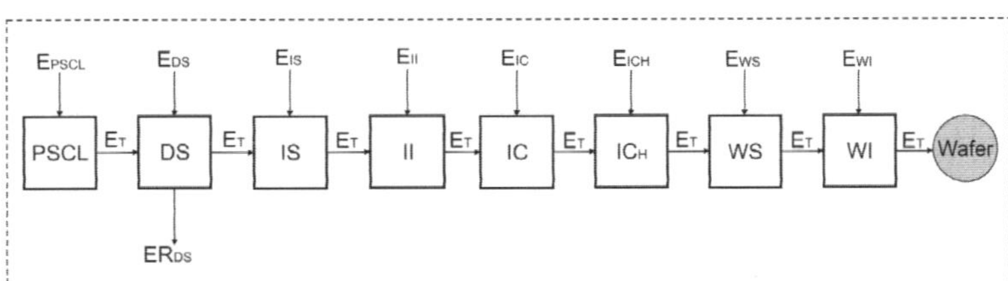

Fig. 3. Part fabrication process flow

Polycrystalline silicon is a material that consists of multiple small silicon crystals. The manufacture method is to solidify the melting silicon (to get the polycrystalline silicon ingots, the melting silicon will be casted via slowly cooling from bottom to top in the squared crucible, then the ingots are being cut in bricks and finally sliced into wafers). Process starts by importing Poly Si chunk via Poly Si chunk loader (PSCL) and consumes energy (E_{PSCL}). Then, in the next step directional solidification occur (DS) where the melting silicon will be casted via slowly cooling from bottom to top in the squared crucible in order to generate silicon ingot. Having in mind the energy intensity of melting silicon (E_{DS}), during the cooling process a certain amount of energy could be recovered (ER_{DS}). Furthermore, generated ingots are being squared (IS) by engaging required energy (E_{IS}), and inspected (II) in order to identify possible fractures and irregularities. Inspection of ingots is not as energy intensive process as previously presented ones but still requires electricity (E_{II}) for its operation. Inspected ingots that meet the desired quality are forwarded to cropping (IC) and then to chamfering (IC_H), while the (E_{IC}) and (E_{ICH}) represent the energy consumption of these processes respectively. Lastly, final-shaped ingots are cut in bricks and sliced into wafers (WS) after which wafers are inspected to meet the required quality (WI). Here energy consumption of wafer slicing operation is presented as (E_{WS}), while energy consumption of final wafer inspection is (E_{WI}).

According to the same principle adopted to determine energy consumption of RP given in the Equation (4), PF energy consumption is given in the Equation (5):

$$E_{PF} = E_{PSCL} + E_{DS} + E_{IS} + E_{II} + E_{IC} + E_{ICH} + E_{WS} + E_{WI} + \sum E_T - ER_{DS} \quad (5)$$

As already mentioned regarding the Equation (4), Equation (5) could be derived more as well, depending on the research complexity.

2.4 Cell manufacturing

In the cell manufacturing process (CM), a functional solar cell is produced from the wafer. Cell production is a multistep physical-chemical process which includes variety of important steps such as doping, the anti reflex coating and electrical contacts soldering (Fig. 4).

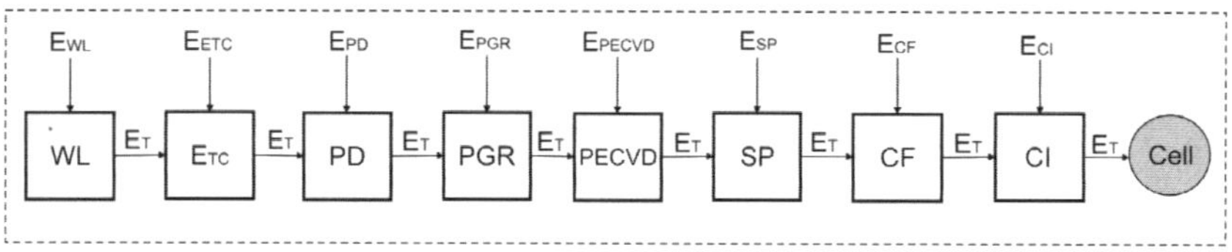

Fig. 4. Cell manufacturing process flow

Cell manufacturing process starts by wafer loading (WL) and cleaning by using acid or alkali liquid to do saw damage removal, and in order to perform work, it will consume certain amount of energy (E_{WL}). The next step is so called etching or texture forming (E_{TC}), where acid or alkali liquid is used to form texture and consumes (E_{ETC}) amount of energy for operation. Coming up next, Phosphorus Diffusion (PD) occur under high temperature vacuum tube process with gas ($POCl_3$) to define positive/negative electric junctions, where (E_{PD}) represent PD energy consumption. Upon completion of PD, phosphorus glass removal (PGR) process, also known as Wet Edge Isolation, starts by using acid liquid in order to remove SiO_2 layer from wafer surface and isolate wafer edge. Next stage in the process flow is the Plasma Enhanced Chemical Vapour Deposition (PECVD) where anti-reflection layer (SiNx) is built on wafer surface. After that, screen printing operation (placing the electrode pattern on both side with metal component paste) starts (SP) followed by operation called co-firing (CF), where high temperature heater is used to make cross-link between electrode and junction for electric current output. After CF, solar cells are inspected (CI) by surface visual inspection with AOI (automated optical inspection) and electricity with IV-tester [18]. All of the previously mentioned operations (PGR, PECVD, SP, CF and CI) are accompanied by specific energy consumptions (E_{PGR}, E_{PECVD}, E_{SP}, E_{CF}, E_{CI}) respectively, where the cell manufacturing process energy consumption (E_{CM}) represents the sum of energy inputs at this process stage as given by Equation (6):

$$E_{CM} = E_{WL} + E_{ETC} + E_{PD} + E_{PGR} + E_{PECVD} + E_{SP} + E_{CF} + E_{CI} + \sum E_T \quad (6)$$

2.5 Module manufacturing

In the module manufacturing process (MM), the solar cell has its characteristic appearance and ability to convert the sunlight into electrical power. Likewise, CM in MM, the solar cells are assembled to ready-for-use modules. Several solar cells are connected together to form solar modules in order to generate a useful amount of electrical power. The MM process flow is shown in the Fig. 5.

The process of module manufacturing starts by loading solar cells in the solar cell loader (SCL), where the (E_{SCL}) represents the amount of energy needed to perform the operation. Next operation considers tabbing and stringing (TS), after which cells are being laid up by automatic cell lay-up machine (AL). This is followed by energy consumption (E_{TS}) and (E_{AL}) respectively. In addition, the laid up cells are forwarded to bus bar soldering (BBS) and electrical inspection (EI), while analogously to previous processes (E_{BBS}) and (E_{EI}) represent energy consumption of these operations. Furthermore, after EI, cells enter the laminator (L). Lamination is one of the most critical processes in the solar module manufacturing line which ensures the quality and durability of the PV module. The lamination main function is module encapsulation by applying the right pressure and temperature to laminate the various components. This is energy intensive process (E_L), after which cells are being cooled down, which indicates the potential for energy recovery (ER_L).

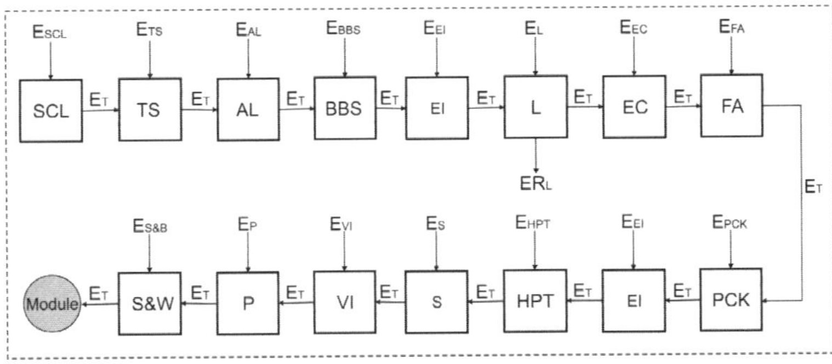

Fig. 5. Module manufacturing process flow

Next operations are edge cutting (EC) and frame assembly (FA), followed by so called punching (PCK) where corner key is being placed. These operations are followed by (E_{EC}), (E_{FA}) and (E_{PCK}) energy consumption in order to perform desired tasks. After PCK, EI is conducted again, and cells are forwarded to Hi-pot testing (HPT). Traditionally, HPT is a term given to a class of electrical safety testing instruments used to verify electrical insulation in finished appliances, other wired assemblies, printed circuit boards, etc. HPT is low energy intensity process (E_{HPT}), but for the sake of consistency it should be taken into account [19]. The following operations, simulation (S), visual inspection (VI), packiging (P), as well as storage and warehousing (S&W) represent the final operations in the module manufacturing process while their energy consumption by individual operation are given as (E_S), (E_{VI}), (E_P), ($E_{S\&W}$) respectively.

MM process energy consumption (E_{MM}) could be determined as given in Equation (7):

$$E_{CM} = E_{SCL} + E_{TS} + E_{AL} + E_{BBS} + 2E_{EI} + E_L + E_{EC} + E_{FA} + E_{PCK} + E_{HPT} + E_S + E_{VI} + E_P + E_{S\&W} + \sum E_T - ER_L \quad (7)$$

2.6 Logistics and distribution

Here, logistics is considered as the finalized solar modules flow management between the point of origin and the point of destination in order to meet arranged requirements of customers or corporations. Especially the core logistics task, transportation of the goods, can reduce costs and energy consumption through efficient energy management [20,21]. The relevant factors are the choice of means of transportation, duration and length of transportation and cooperation with logistics service providers. Roughly, the logistics causes more than 14% percent of CO_2 emissions worldwide [22]. In addition, an often case during distribution is warehousing where the products are being stored before being finally distributed to customers. Having in mind that logistic processes can be quite complex, as well as that variety of logistic models exists today, it was not analysed in detail as previous process in this study. However, in order to determine energy consumption, a specific distribution and logistics model should be identified upon which overall energy consumption ($E_{D\&L}$) could be determined. This process has been briefly shown in the Fig. 1.

2.7 Utilization process

The utilization of PV modules in order to generate electricity is the most diverse process in the LCEA. This observation follows from the fact that this process primarly generates energy in compared to the all aformentioned ones which were more or less intensive energy consumers. Although here energy consumption is also present, it is insignificant compared to the amount of generated energy during operation period. The utiliziation process (U) of PV modules is illustratively given in the Fig. 6.

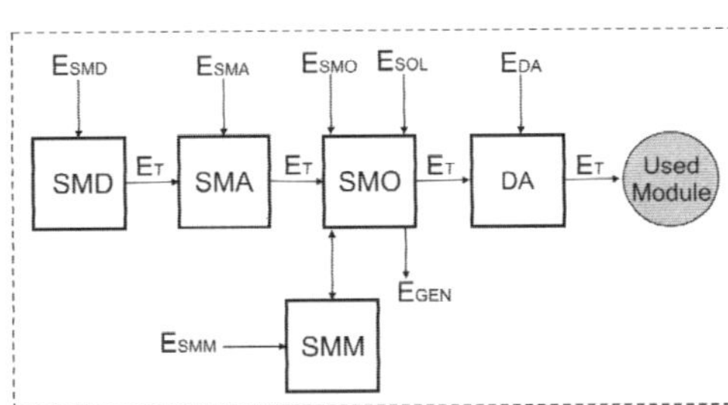

Fig. 6. Utilization process flow

The process starts by distributing solar modules (SMD) to the assembly location where the modules are being assembled (SMA) and prepared for operation phase. In order to complete the operations, a certain energy had to be consumed. These are presenteted as (E_{SMD}) and (E_{SMA}) respectively in the Fig. 6. Next phase is PV module operating phase. This phase represents the crucial phase for the LCEA as given by Equation (2), where EPBT is in inverse proportion to the (E_{GEN}). The general method to estimate the (E_{GEN}) in output of a PV system is given in Equation (8):

$$E_{GEN} = A \cdot \eta \cdot E_{SOL} \cdot P_R \cdot \tau \text{ [kWh/a]} \quad (8)$$

The (E_{GEN}) represents a function of total solar panel area (A, [m^2]), solar panel yield or efficiency (η, [%]), annual average solar radiation on tilted panels (E_{SOL}, [kWh/m^2/a]) and performance ratio or losses coefficient integration (P_R, [%]) over lifecycle operation time (τ, [h]). In addition, η is the yield of the solar module given by the ratio of electrical power (in kWp) of one module divided by its area. For example, the solar panel yield of a PV module of 250 Wp with an area of 1.6 m^2 is 15.6%. It is important to mention that this nominal ratio is given for standard test conditions (insolation: 1000 W/m^2, cell temperature: 25 °C, wind speed: 1 m/s, Reference Solar Spectral Irradiance at Air Mass: 1.5). E_{SOL} is the annual average solar radiation on tilted panels. E_{SOL} is highly determined by PV system location and varies between 200 (kWh/m^2/a) in Norway and 2600 (kWh/m^2/a) in Saudi Arabia. Lastly, PR (Performance Ratio) is a very important value to evaluate the quality of a photovoltaic installation because it gives the performance of the installation independently of the orientation, inclination of the panel, etc. PR value depends on the site location, technology, sizing of the system, etc., while it function is to include overall system losses such as Inverter losses (4% to 10%), Temperature losses (5% to 18%), DC cables losses (1 to 3%), AC cables losses (1 to 3%), Shadings (0% to 80% specific to each site), Losses at weak radiation (3% to 7%), Losses due to dust and snow coverage ($\leq$ 2%), as well as other possible losses (?%) [23].

An example of typical PV on-grid system components with energy transformation flow is given in the Fig. 7. It consists of 15 modules attached into 2 strings (7 and 8, 250 Wp, modules per string), 3.6 kWp Inverter, connected via 6 and 4 mm^2 PV solar cables. The effective capacity of this single phase PV solar system ammounts 3.75 kWp [24].

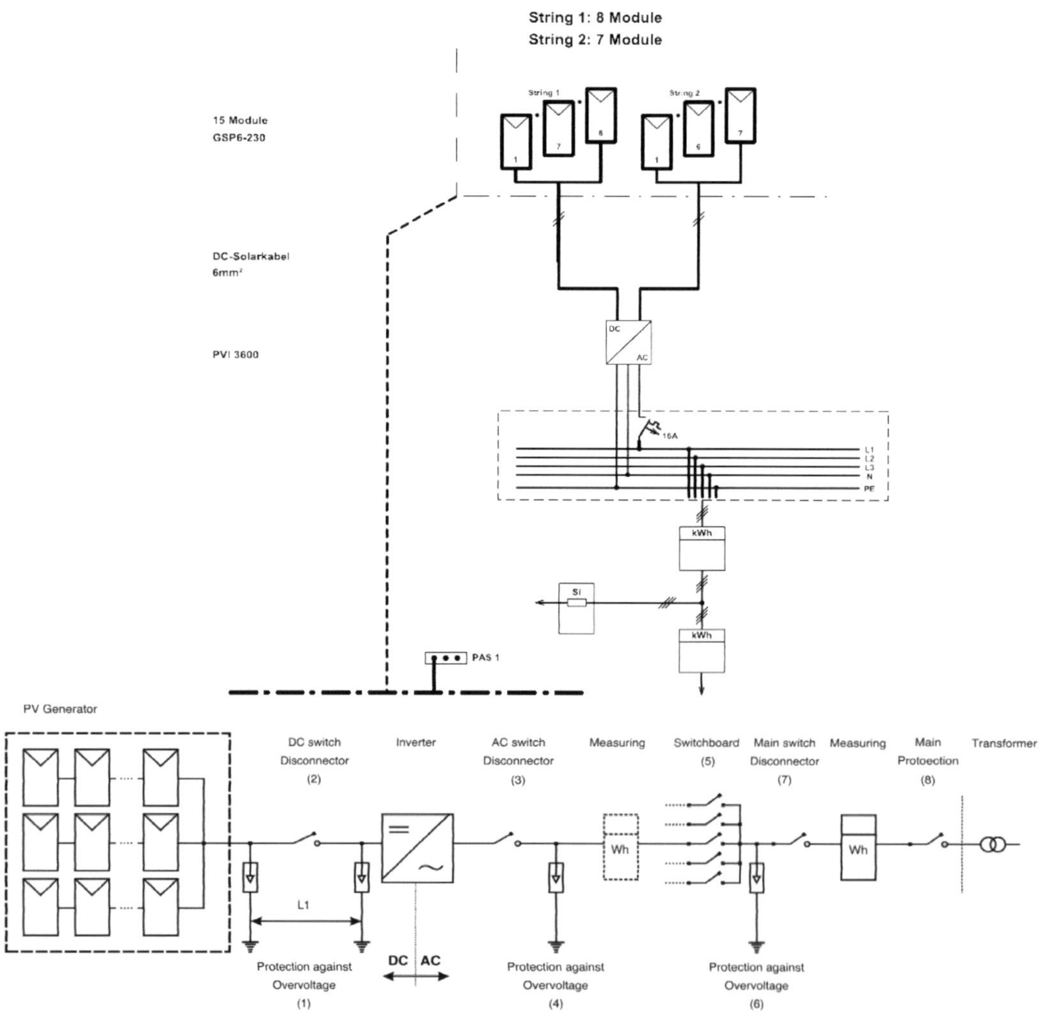

Fig. 7. Typical PV on-grid system components and energy transformation flow [24]

Having in mind the process flow given in the Fig. 8. usefull or better said accountable (E_{GEN}) is amount of energy handed over to the grid.

If the system operation life time is 25 years, overall energy that has been generated (E_{GO}) by the PV system could be determined as given in the Equation (9):

$$E_{GO} = \sum_{a=1}^{a=25} E_{GEN} = \sum_{a=1}^{a=25} A \cdot \eta \cdot E_{SOL} \cdot P_R \cdot \tau \tag{9}$$

During the operation period all the maintenance time occurred due to the system failure or disconnection, should be subtracted from (E_{GO}), because system or part of the system had not been productive during that period. Solar module maintenance (SMM) is accompanied by maintenance energy consumption (E_{SMM}), which in case of large PV power plants can be pretty intensive. After operation life cycle ends, solar modules are being disassembled (DA) and transported to the storage, from which they are being forwarded to disassembly and recycling processes. Transportation energy consumption (E_T), as well as disassembly energy consumption (E_{DA}), are illustratively shown in the Fig. 6.

According to the all previously mentioned, energy consumption of (U) is given in the Equation (10):

$$E_U = E_{SMD} + E_{SMA} + E_{SMO} + E_{SMM} + E_{SDA} + \sum E_T \tag{10}$$

Likewise, Equation (11) stands for overall energy production (E_P) of the (U) process individually:

$$E_P = E_{GO} - E_U \tag{11}$$

2.7 Disassembly, Recycle and Waste Treatment process

An example of PV disassembling and recycling technology that enables the automatic separation consists of three main processes such as aluminium frame removal, backsheet removal, EVA (ethylene-vinyl-acetate) resin burning. This has been illustratevly given in the Fig. 8.

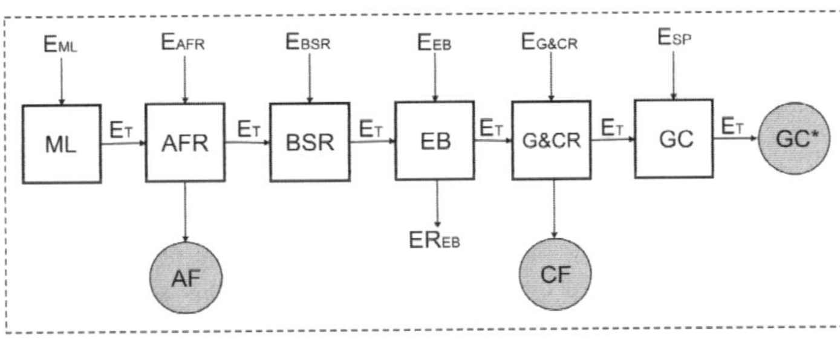

Fig. 8. Disassembly and Recycle process flow

The process starts by loading modules by module loader (ML) after which modules are directed to the operation of automatic frame removal (AFR), where aluminum frames (AF) are being separated and stored. Here, (E_{ML}) and (E_{AFR}) represent the energy consumption of mentioned operations respectively. Next operation is back sheet removal (BSR) followed by EVA film burning (EB). Unlike the BSR, which is only characterized by its energy consumption (E_{BSR}), EB has the potential to partly recover (ER_{EB}) energy consumed (E_{EB}) for the incineration of EVA film burning. Next in a row is glass and cell recovery (G&CR) where cell fragments are separated, while the glass is forwarded to the glass crusher (GC) in order to produce glass cullet (GC*) [25]. These are being followed by their specific energy consumption ($E_{G\&CR}$) and (E_{GC}) respectively.

Conclusively, all components of PV modules are being separated as much as possible in order to recover and recycle valuable materials, so they can be handed out to companies which refine metals and recycle secondary metals. Likewise, glass that can be separated and retain high purity is recycled as glass cullet. Lastly, materials difficult or unable to separate, recover and recycle are being sent to landfill. Depending on subject to regulation and classification of hazardous content, appropriate waste treatment process (WT) is applied where process energy consumption (E_{WT}), as well as the possibility of energy recovery should be taken into account or considered. According to the all previously mentioned, energy consumption of (D&R) is given in the Equation (12):

$$E_{R\&D} = E_{ML} + E_{AFR} + E_{BSR} + E_{EB} + E_{G\&CR} + E_{GC} + E_{WT} \sum E_T - (ER_{EB} + ER_{WT}) \tag{12}$$

3. Discussion

Having in mind that LCEA is a fundamental tool for evaluating and guiding the development of PV technology, comprehensive modelling of a PV system is essential for assessing its full potential as a sustainable energy technology. The model equations presented here, provide a means for evaluating the life-cycle energy performance of a PV system. On the other hand, this paper emphasises a conceptual theoretical model development in order to generate a holistic approach for LCEA of PV systems. As previously mentioned, even though the analysis could be more complex the purpose of this study is to stimulate logical identification of potential optimization spots, system effectiveness and process energy efficiency by respecting carefully defined system boundaries. It is important to mention that the boundaries of the system under investigation are variable depending on the specific scope and objectives of each analysis.

Moreover, carefully conducted LCEA could be considered as major input for Energy payback time determination. This metric assesses the time period necessary for a PV project to become profitable from an energy perspective.

Furthermore, electricity production efficiency should be the metric of choice when comparing competing electricity-generating systems. The maximum electricity production efficiency of a PV system is a function of its useful life, so this metric can evaluate the lifetime energy performance of an electricity-generating system. Energy payback time, in contrast, does not address ultimate efficiency of the system being measured.

Likewise, life-cycle conversion efficiency is a useful metric for directly comparing alternative solar technologies because it measures how efficiently a system converts sunlight into net energy in the form of electricity. Life-cycle conversion efficiency allows another meaningful comparison between various PV devices, such as crystalline and amorphous silicon modules.

By applying this conceptual model in practice, systematic approach to problem solving could be ensured, while the observation complexity could be derived to the desired or needed level.

A more refined model would enable the evaluation of metrics based on cases where energy input requirements are met by PV electricity-generating systems or other renewable sources. Such a model would require material production energy, manufacturing energy and other energy inputs to be disaggregated into specific renewable and non-renewable categories. Substitution of electricity for other energy carriers, such as natural gas and fuel oil, could also be considered. This scenario may result in a situation where the substitution leads to more primary energy consumption than the conventional fuels case. For example, an electrically heated boiler is less efficient on a primary energy basis than a natural gas boiler, resulting in higher total energy consumption.

4. Conclusion

The increasing demand for renewable energy challenges society to find out sustainable and renewable energy source. In this quest, LCEA is imposed as a tool which can be used effectively in assessing the sustainability of renewable energy sources, such as PV systems, while, on the other hand, the collection of actual data for such study is a quite challenging task.

This paper provides a holistic approach regarding PV system energy consumption and generation over the useful life time. In addition, potential energy recovery spots have been identified, but these still represent the subject of change in terms of technology differentiation. Practical implementation of this model combined with multi-objective analysis of environmental, performance, cost and regulatory/policy issues over the life-cycle of a PV system could ultimately provide the most complete basis for design, planning and implementation. This full set of information and data offers a more powerful means for promoting PV technology as an effective source of sustainable energy.

Recognizing this goal, LCEA remains one of the most fundamental components of a multi-objective analysis of an energy-generating system.

5. References

[1] Frankl, P., Nowak, S., Gutschner, M., Gnos, S., Rinke, T. (2010). International Energy Agency Technology Roadmap: Solar Photovoltaic Energy, Available Online at: http://www.iea.org/papers/2010/pv_roadmap.pdfS.

[2] Mints, P. (2011). Photovoltaic Manufacturer Shipments, Capacity and Competitive Analysis 2010/2011. Navigant Consulting Photovoltaic Service Program: Report NPS-Supply 6, Palo Alto, CA.

[3] Smith, D.D., Cousins, P.J., Masad, A., Westerberg, S., Johnson, M., Tu, X., Dennis, T., Harley, G., Solomon, G., Rim, S., Shepherd, M., Herrington, S., Defensor, M., Leygo, A., Tomada, P., Wu, J. (2012). Sunpower's generation III high-efficiency lower-cost technology: transition to full-scale manufacturing, in: Proceedings of the Thirty eighth IEEEP VSC,2012.

[4] Trina Solar Linear Performance Warranty (3.5% in the first year, thereafter 0.68% per year, ending with 80.18% in the 25th year after the Warranty Start Date). Available online at: http://www.trinasolar.com/eu/products/product-resources/product-warrantyS.

[5] Goodrich, A., Hacke, P., Wang, Q., Sopori, B., Margolis, R., James, T.L., Woodhouse, M. (2013). A wafer-based monocrystalline silicon photovoltaics road map: Utilizing known technology improvement opportunities for further reductions in manufacturing costs. Solar Energy Materials & Solar Cells, 114, pp. 110–135.

[6] Bhandari, K.P., Collier, J.M., Ellingson, R.J., Apul D.S. (2015). Energy payback time (EPBT) and energy return on energy invested (EROI) of solar photovoltaic systems: A systematic review and meta-analysis. Renewable and Sustainable Energy Reviews, 47, Jul 2015., pp. 133–141, ISSN: 1364-0321.

[7] Phylipsen, G.J.M., Alsema, E.A. (1995). Environmental life cycle assessment of multicrystalline silicon solar cell modules. Report, 95057, Department of Science Technology and Society of Utrecht University (STS-UU), Netherlands. Available online at: www.chem.uu.nl/nws/www/publica/95057.htm.

[8] Alsema, E.A. (2000). Energy pay-back time and CO_2 emissions of PV systems. Progress in Photovoltaics: Research and Applications, 8, pp. 17-25.

[9] Knapp, K.E., Jester, T.L. (2000) Energy balances for photovoltaic modules: status and prospects. in IEEE Photovoltaics specialists conference. Available online at: www.ieeepvsc.nrel.gov.

[10] Pehnt, M., Bubenzer, A., Räuber, A. (2002) Life-Cycle Assessment of Photovoltaic Systems - Trying to fight Deep-seated Prejudices. Photovoltaics Guidebook for Decision Makers, Springer: Berlin, DE.

[11] Meijer, A., Huijbregts, M.A.J., Schermer, J.J., Reijnders, L. (2003). Life-cycle Assessment of Photovoltaic Modules: Comparison of mc-Si, InGaP and InGaP/mc-Si Solar Modules. Progress in Photovoltaics: Research and Applications, 11, pp. 275-287. DOI: 10.1002/pip.489.

[12] Hyoungseok, K., Kyounghoon, C., Fthenakis, V.M., Parikhit, S., Hur, T. (2014). Life cycle assessment of cadmium telluride photovoltaic (CdTe PV) systems. Solar Energy, 103, pp. 78–88.

[13] Peng, J., Lu, L., Yang, H. (2013). Review on life cycle assessment of energy payback and greenhouse gas emission of solar photovoltaic systems. Renewable and Sustainable Energy Reviews, 19, pp. 255–274.

[14] Sherwani, A.F., Usmani, J.A., Varun, C. (2010). Life cycle assessment of solar PV based electricity generation systems: A review. Renewable and Sustainable Energy Reviews, 14, pp. 540–544.

[15] Stoppato, A. (2008). Life cycle assessment of photovoltaic electricity generation. Energy, 33, pp. 224–232.

[16] Morvay, Z.K.; Gvozdenac, D.D. (2008). *Applied Industrial Energy and Environmental Management*, John Wiley and Sons, Ltd, Publication, ISBN 978-0-470-69742-9, Wiltshire, Great Britain.

[17] Green Rhino Energy, Manufacturing Silicon, Available online: http://www.greenrhinoenergy.com/solar/technologies/pv_manufacturing.php

[18] Tsec, Solar Module/Cell Manufacturing Process Introduction, Available online at: http://www.tsecpv.com/en-global/solar_knowledge/index/zero_house_02

[19] Gintung Energy Corporation, Module manufacturing process, Available online at: http://www.asectw.com/asec_Product_FlowProcess.html

[20] Andrejić, M.; Kilibarda, M. (2013). The problems of measuring efficiency in logistics, Proceedings of 1st Logistics International Conference, Belgrade, Serbia , 28-30 November 2013. Available online at: http://logic.sf.bg.ac.rs/wp-content/uploads/Papers/ID-40.pdf

[21] Wang, W. Comparing the Energy Consumption of Logistics Chains in Traditional Retailing and E-commerce for Popular Products in the Context of China, Available online at: http://press-files.anu.edu.au/downloads/press/n1673/pdf/Wenjia_Wang.pdf

[22] Micieta, B.; Zavodska, L.; Rakyta, M.; Binasova, V. (2015). Sustainable concept for green logistics and energy efficiency in manufacturing, Chapter 33 in DAAAM International Scientific Book 2015, pp.391-400, B. Katalinic (Ed.), Published by DAAAM International, ISBN 978-3-902734-05-1, ISSN 1726-9687, Vienna, Austria, DOI: 10.2507/daaam.scibook.2015.33

[23] Chen, C. J. (2011). *Physics of Solar Energy*, John Wiley and Sons, Ltd, Publication, ISBN 978-0-470-64780-6, Hoboken, New Jersey.

[24] PV solar system diagrams, PV On-grid systems. Available online at: http://pvshop.eu/diagrams.html

[25] IRENA. (2016). End-of-life Management: Solar Photovoltaic Panels, ISBN 978-92-95111-99-8 (PDF, IRENA) Available online at: http://www.irena.org/DocumentDownloads/Publications/IRENA_IEAPVPS_End-of-Life_Solar_PV_Panels_2016.pdf

DOI: 10.2507/27th.daaam.proceedings.080

LASER COMMUNICATIONS PROTOTYPING FOR THE GROUP OF MOBILE ROBOTS

Valentin Pryanichnikov & Daria Stepanova

This Publication has to be referred as: Pryanichnikov, V[alentin] & Stepanova, D[aria] (2016). Laser Communications Prototyping for the Group of Mobile Robots, Proceedings of the 27th DAAAM International Symposium, pp.0544-0550, B. Katalinic (Ed.), Published by DAAAM International, ISBN 978-3-902734-08-2, ISSN 1726-9679, Vienna, Austria DOI: 10.2507/27th.daaam.proceedings.080

Abstract

Laser communications in case of a group of mobile robots communication in extreme conditions can be considered as an alternative or supplement to the radio. Thus, this work aims to develop a laser communication terminal for a mobile robot and with the particular focus on building the pointing, acquisition and tracking (PAT) system. During the present work, the idea of the laser terminal design was proposed based on the analysis of requirements and the model of data transmission between two mobile robots. The mathematical model of the PAT control system and stability analysis showed the feasibility of the design. A separate part of the work was also dedicated to the development of the robot orientation determination system and PAT stabilization system. The prototype and all necessary algorithms were implemented and the hardware and software were created. At this point, the laser terminal has been tested and has achieved acceptable performance of 5 degrees of accuracy under existing constraints. The paper was partially supported by the Russian Foundation for Basic Research, project No. 16-07-01264a, 16-07-00935a, 16-07-00811a.

Keywords: Mobile robots optical communications; guidance system; pointing and tracking.

1. Introduction

The performance of radio frequency (RF) communications is noticeably weaker than the one of optical in a number of applications [1]. Comparing to RF communication, optical links provide larger modulation bandwidth, therefore higher communication speed and higher security level due to less interference between the optical signals. [2] A higher optical antenna directivity and thus, a higher concentration of energy in the laser beam, makes the gain of optical antennas higher than that of the RF with the same efficiency. RF antennas can possibly achieve the same order of directivity, however, the aperture will increase proportionally with the gain affecting the weight of antenna as well [3], becoming impossible to be used on small robots. The advantage of high directivity in optical antennas brings the challenge of accurate pointing for the optical beam [4]. The importance of pointing, acquisition and tracking (PAT) system considerably increases due to the sensitivity of communication speed on pointing accuracy: the lower the accuracy the more link losses will happen during the communication session, less information will be transmitted [5]. Important parameters such as beam divergence and transmitter power depend on pointing accuracy as well and thus, influence the laser terminal design. The lower the pointing accuracy, the higher the beam divergence shall be, the more transmit power shall be used to provide the same communication speed. The aim of this work is to develop a laser terminal and PAT systems for the experimental laser communication testing between two or more mobile robots [6-18].

2. Possible configurations for the group of mobile robots

During present work, several possible configurations of mobile robots operating in a hazardous area and connections between them were considered. Each configuration is based on the following assumptions: a group of mobile robots collects information in a hazardous area and transmits the data about the zone and telemetry through an unmanned aerial vehicle (UAV) working as a relay to a station controlled by people and located outside the danger zone.

Thus, the optical communication link may be established between:
1 - mobile robots; 2 - mobile robots and relay; 3 relay and a repeater station.

Each scenario assumes the development of optical terminal that meets certain specific requirements in each case.

3. System Design

Requirements definition for the design of complex optical and PAT systems is the key to successful performance of the whole optical terminal. The following requirements for the terminal design were derived during present work.

The data transfer rate. To determine the requirements for data rates the amount of data produced by the mobile robot have been evaluated. In most cases, in hazardous locations where people prefer mobile robots to perform all necessary operations, people shall be aware of the place where the robot operates. Therefore, the main type of transmitted information is a real time video; thus, the upper bound of data rate was chosen to be 150 Mbs.

The distance between mobile robots. Within all possible configurations in which the mobile robot can operate in extreme conditions, the distance between robots, robots to the repeater and the repeater to the station can vary from 10m to several kilometres. Thus, the upper bound of the distance was chosen to be 3 km.

The size of the optical terminal. The size of an optical terminal, which is located on the mobile robot, is a critical parameter. Thus, during the present work the compromise between the size, weight and terminal requirements in terms of "endurance" was chosen. One of the main factors limiting the miniaturization of the optical terminal is the telescope, which was chosen to be the Newton telescope.

Other requirements for the laser are shown in Table 1.

Parameter	Value	Parameter	Value
Power at the transmitter	0.5 W	Mirror mass	165 g.
Laser wavelength	1550 nm	Mirror size	200x85mm
Distance between mobile robots	3 km	Maximum beam divergence	3 deg
Terminal mass	< 3 kg	PAT system mass	400 g.

Table 1. Parameters and requirements for laser terminal design

Requirements definition based on the mathematical model. In order to determine the requirements and the parameters for the optical communication terminal and to choose necessary technical equipment for its implementation a mathematical model of the motion of mobile robots and optical channel model were developed. Due to the fact that the mutual movement of mobile robots is arbitrary, several scenarios of their relative motion were considered. Figures below show the trajectories of the oncoming traffic two robots on a path in the form of a square (Fig.1) and on a collision trajectory (Fig.2) together with the change of relative angle (second plot) and the corresponding power changes (third plot).

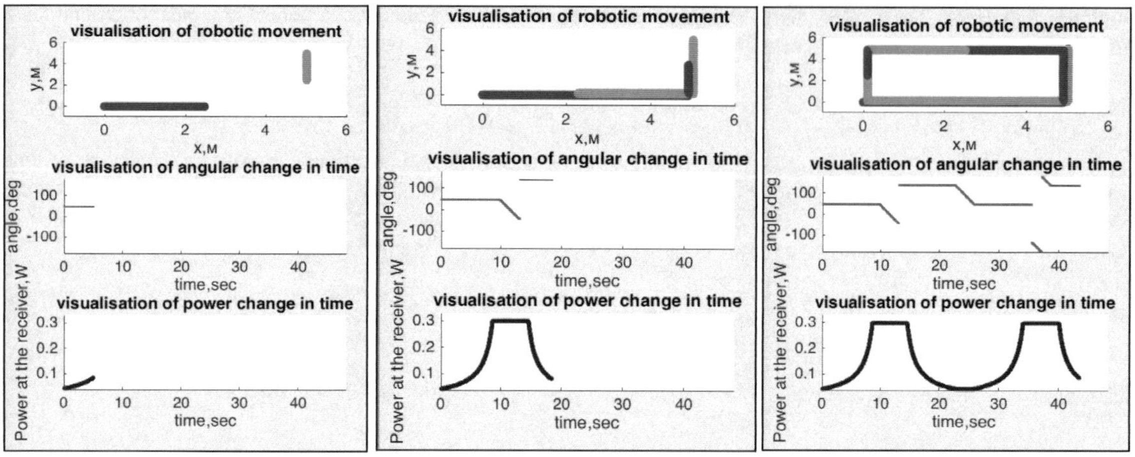

Fig.1. Mobile robots simulation of motion in opposite directions for 3 different points at time for the trajectory "line"

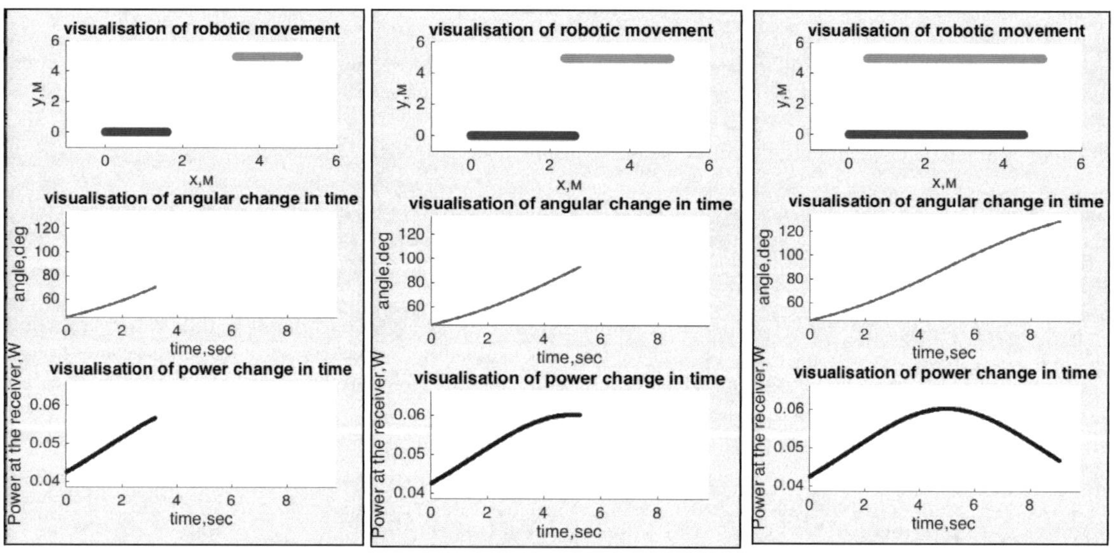

Fig.2. Mobile robots simulation of motion in opposite directions for 3 different points at time for the trajectory "rectangle"

Following parameters varied during the simulation: the width of the optical beam, tracking accuracy, transmitter power, the distance between mobile robots. As a result, critical angular rates, delays in the tracking algorithm were estimated. Some variations of parameters in different scenarios are listed in Table 2.

Parameter	1 Scenario	2 Scenario	3 Scenario
Trajectory type	Robots move towards each other on two lines with 5m difference between them.	Robots move towards each other on two lines with 50m difference between them.	Robots move towards each other on two lines with 500m difference between them.
Optical beam divergence, deg	1	2	3
Power at the receiver, W	0.01	0.1	0.54
Angular rates, deg/sec	5.7 - 11.4	0.5 - 1.1	0.068 – 0.057
Max algorithm delay, sec	0.086	1.71	15.67

Table 2. Parameter comparison depending on scenario

4. Tracking algorithm development

After the connection establishment, the question of its retention and communication quality improvement by compensating for pointing errors appears. For that purpose, the tracking algorithm was developed based on the image processing of the laser spot captured by the CMOS matrix with the size of 460x640 pixels. At this stage of project, the image-processing algorithm is performed on a PC using Matlab program. The developed algorithm allows the detection of an optical spot with other objects on the background, including radiant, determination of the distance between robots based on the spot size and calculation the pointing vector for the control system. Fig. 3 shows examples of the algorithm results.

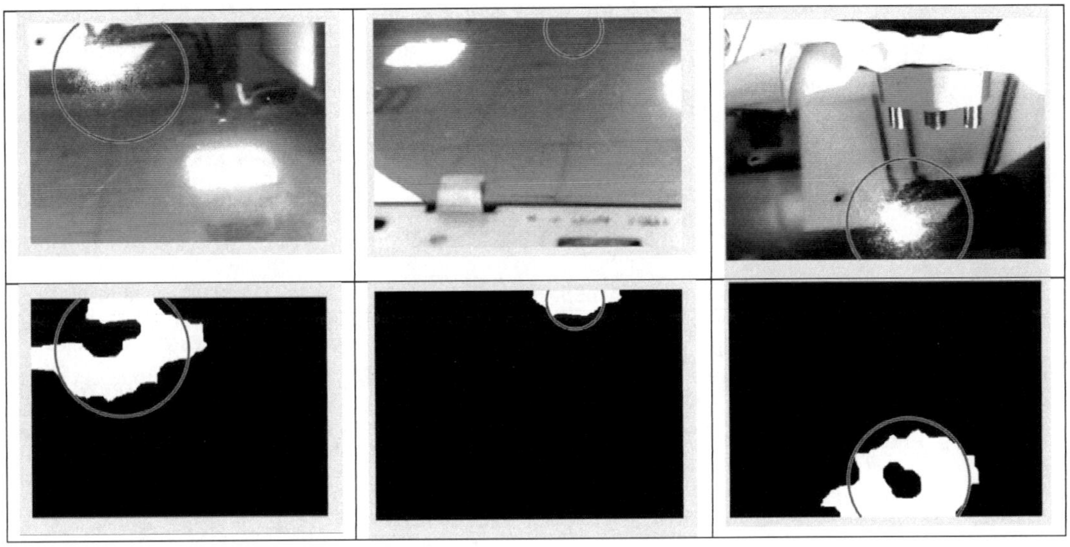

Fig.3 Tracking algorithm development

The algorithm processing time is 0.13 seconds with 7 frames per second, which is not enough for some configurations of mobile robots. However, in future it is planned to optimize the algorithm and implement it on the FPGA, thereby reducing the calculation time to 0.002 seconds.

5. The laser terminal prototype development

After the requirements analysis and optical terminal mathematical model development, the laser terminal was designed with an emphasis on the low-cost component selection. The terminal consists of the following subsystems: optical, electrical, PAT system.

The optical subsystem is represented by the telescope, PAT mirror, lenses, and splitter, which are connected to redirect the optical beam to the receiver and CCD matrix. The laser beam way in the optical system is following: on the receiving side the plane optical wave falls on a mirror of the PAT system. The mirror redirects the optical signal to the main telescope. Here, the optical beam is divided into two by the splitter. Most of the beam falls on an avalanche photodiode, and the rest focuses on the CCD matrix to collect information about the pointing error. This information is used as feedback to the PAT system. The transmitting part consists of a photodiode, the lens, which is necessary for the calibration of the beam divergence, and is mounted on the PAT system.

Electronics and software. The electronic subsystem is divided into the analog signal recovery, which strengthens and purifies raw optical signals, and a digital processing of packages. The redundant communication channel is also present on the platform and is based on the Xbee long-range transceivers. The basic system software is implemented on a microcomputer BeagleBone Black. It runs the basic software: the position forecast, laser and radio transmitter, communication control I2C, part of the PAT software.

The aim of PAT system is to provide the best possible pointing accuracy of the optical beam. The approach for the PAT system design is following: the laser diode and the main mirror of the optical system, which redirects the optical beam in the telescope, rotate in two planes - pitch and yaw. The mirror and laser are mounted on a horizontal axis of rotation; frame is mounted on a mobile platform. PAT system includes a sensor subsystem that is distributed across the platform and motor subsystem. Brushless DC motors perform rotation of the mirror. Arduino Due board with built-in 32-bit ARM core microcontroller Cortex-M3 CPU was chosen to be PAT computer and controls motors, collects data from sensors, calculates mirror position and provides the communication with the host computer. (Fig.4)

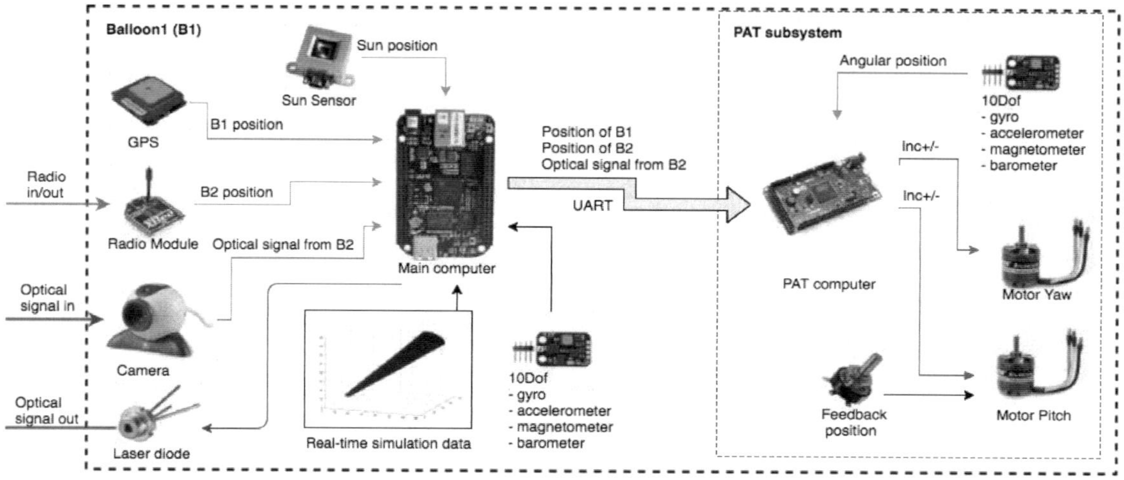

Fig.4. PAT system architecture

The exact definition of the orientation of the mobile robot is a critical task in the PAT system development. Accelerometer, gyroscope and magnetometer are installed on the platform in order to determine the orientation of it. The magnetometer is located as far as possible from motors and the metal parts of the telescope in order to minimize electromagnetic noise, affecting its performance. Another sensor with an accelerometer and a gyroscope is mounted on the mirror to determine its position. The potentiometer is fixed on the motor shaft to determine the yaw position of the mirror and the rotation angle of the motor. PAT computer performs data collection and processing from the sensors located on the PAT system (Fig.5).

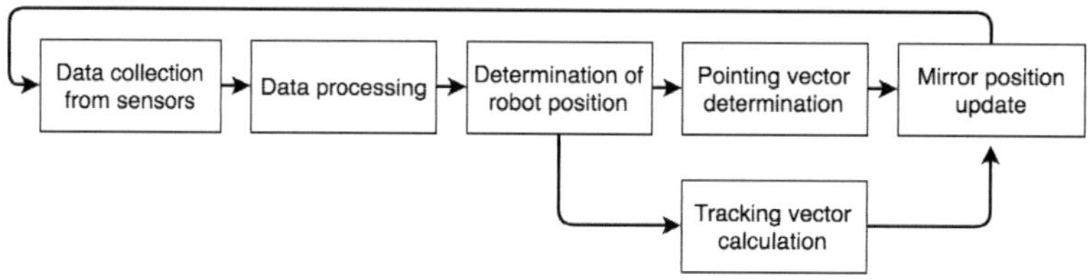

Fig. 5. PAT system representation as finite state machine

The developed control system consists of host computer, PID controller, and load (Fig.6). Pitch and yaw angle change in each cycle. The load consists of a mirror, the laser holder and 10dof sensor, which is placed on the mirror. The orientation of the mirror and the absolute rotation based on potentiometer data are fed to a microcontroller, which generates a sine wave PWM signal and controls the motor via the driver, as a feedback.

During the present work, the structure and drivers of the PAT systems have been developed. The structure consists of the mirror holder and frame. Mirror holder is fixed on the one side to the motor shaft and on the other - to laser metal holder. The frame, which is designed to the mirror holder, is fixed on the yaw motor shaft. 10dof sensor is attached to the mirror holder at the intersection of the two axes of rotation. Yaw motor is placed on the platform. The PAT system center of mass is located at the intersection of the axes of rotation of the engine in order to prevent vibrations.

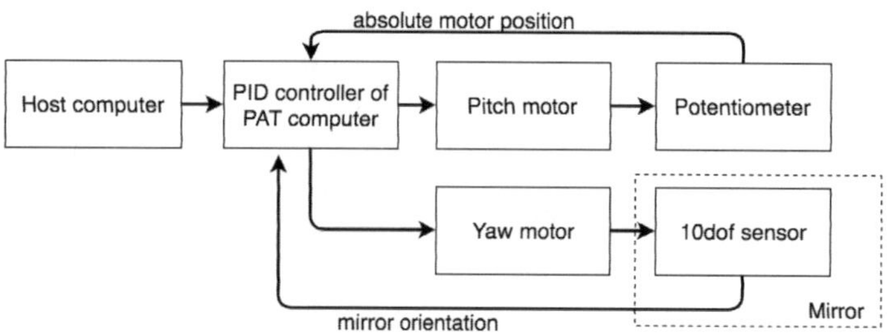

Fig. 6. PAT system algorithm

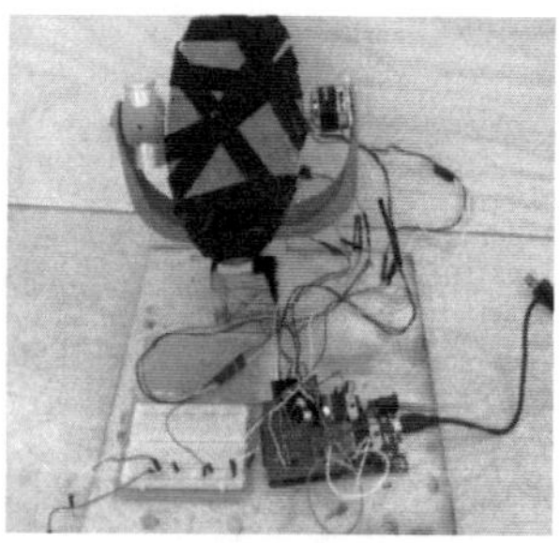

Fig. 7. PAT system design

6. Experimental prototype

On this stage of the work, PAT subsystems, such as stabilization and attitude determination, the optical subsystem have been tested separately. The stabilization system was tested in the laboratory conditions using two PAT prototypes. The experimental scheme is shown in Fig.8. During the stabilization system test platforms were deflected at angles between 0 to 30 degrees and the accuracy of the system was determined.

Due to 10dof sensor measurement noise, mechanical misalignments in mirror mounting, imperfections of 3D printing, a center of mass offset the definition of the pitch angle (relative to the yaw angle) was less accurate. Thus, the error in the pitch angle determination was 4.3 degrees, and the yaw angle - 1.2 deg. Fig. 9 shows the changes in pitch and yaw angles, the control response of the system and for angular deviation of 5, 15, 20 and 30 degrees respectively. Pitch angle deviation included fast (40deg/s) and moderate (10 deg/s) rotation around z-axis. The rotational speed was determined by using a gyroscope. Blue dots indicate the time when pointing error is less than 7 degrees, and pink dots indicate the time when pointing error is less than 4 degrees. Pointing accuracy value determined during the experiment was equal to 5.4 degrees at an angular displacement of 16 deg. Thus, within 40 seconds of the experiment, the number of connections between the two laser terminals was 6 and 20 with the beam divergence equal to 4 and 7 degrees respectively. The performance of the system was sufficient to ensure a data rate of 50Mbs.

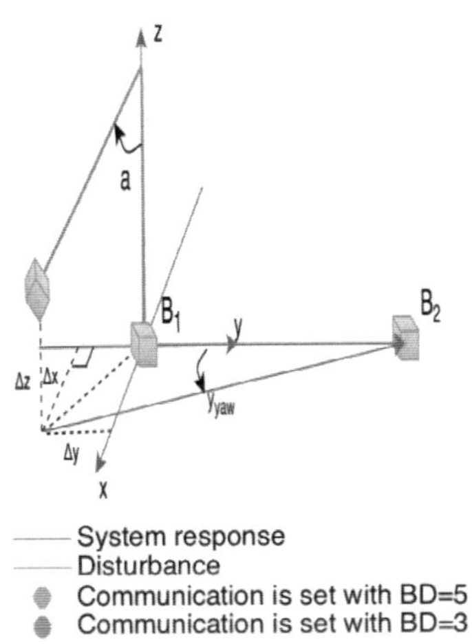

Fig.8. The experimental setup for the stabilization system testing

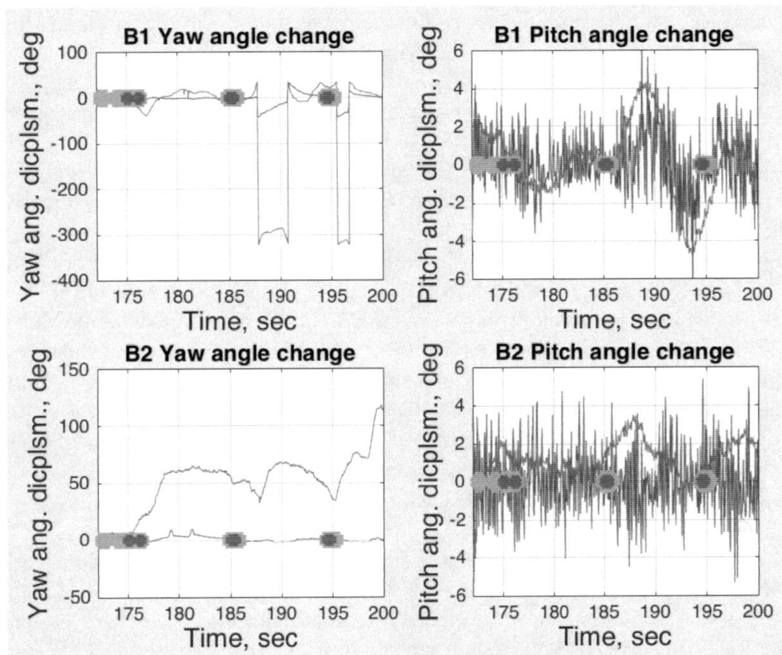

Fig.9. Visualization of PAT system disturbances rejection capabilities at different angles of deflection in two planes and the allocation of time periods in which it is possible to establish optical communication.

7. Conclusions

In the case of mobile robots utilisation in extreme conditions, there is a significant need for the laser secure communication for the transmission of video streams between the mobile robots, robots and the repeater. Thus, this works' aim was to develop a laser terminal in order to provide the optical link between mobile robots. The work started with the literature review, which showed the incoming need in the development of laser communications for mobile robots operating in extreme conditions. Based on the derived system requirements and modelling of a group of mobile robots dynamics, the laser terminal concept, the orientation determination algorithm, pointing, acquisition and tracking (PAT) system, high-speed image processing algorithms have been developed. For the experimental laser terminal design proof of concept, the PAT system design was proposed, the control system and such subsystems, as the orientation determination and stabilisation, were developed. The low-level sensorless control of brushless motor was chosen and implemented together with PID controller; for the orientation determination the Magwick filter-Triad algorithm fusion was proposed. The necessary hardware and software (structural design, orientation algorithm, sensors processing, communication protocols, motor control algorithm, PID controller) were developed and the system was tested "on the table" through several iterations. As a result, the PAT system accuracy achieved 5 degrees, thus providing a data transfer rate close to 50Mbs. In the next stages of work ways to assess and absorb vibrations shall be examined as well as its impact on the pointing accuracy and the final version of the hardware and software of the system shall be developed.

8. References

[1] Chan, Vincent WS. "Free-space optical communications." Journal of Lightwave Technology 24.12 (2006): 4750-4762.

[2] Link: http://www.parabolicarc.com/2015/04/26/digital-solid-state-propulsion/

[3] NASA, "Small Spacecraft Technology - State of the art," Ames Research Center, Moffett Field, California, Mission Design Division, 2015.

[4] Nakagawa, Keizo, and Akio Yamamoto. "Preliminary design of laser utilizing communications equipment (LUCE) installed on optical inter-orbit communications engineering test satellite (OICETS)." Photonics West'95. International Society for Optics and Photonics, 1995.

[5] Alluru, Seshupriya, and Janise McNair. "An optical payload for cubesats." (2010).

[6] Pryanichnikov VE, Bielich T. Vitsa D., B. Katalinich, Kirsanov KB, Jars SV Marzanov YS, Yu Poduraev, Helemendik RV Prysev E .A., Ugleshik A., Harin KV Development of educational technology and the associated laboratory-robotarium. Information-measuring and operating systems, M .: Radio engineering. 2015, t.13, N7 - ISSN 2070-0814, S.7-25 (Пряничников В.Е., Биелич Т., Вицан Д., Каталинич Б., Кирсанов К.Б., Кувшинов С.В., Марзанов Ю.С., Подураев Ю.В., Хелемендик Р.В., Прысев Е.А., Углешик А., Харин К.В. Разработка образовательных технологий и сети ассоциированных лабораторий-роботариумов. Информационно-измерительные и управляющие системы, М.: Радиотехника. 2015, т.13, N7 – ISSN 2070-0814, С.7-25)

[7] Pryanichnikov, V. E., and V. P. Andreev. "Intellectulazation of special mobile robots, including return algorithm to a zone of stable RC." Proceedings of the XXI Int. Conference on Extreme Robotics, St. Peterburg: Poliectnika-service. 2010.

[8] Pryanichnikov, V. E., V. P. Andreev, and E. A. Prysev. "Group control of mobile robots, based on the net-technologies." Proceedings of the Int. conference on Robotics for Security & Space (Planet & Earth rovers), St. Peterburg: Poliectnika service. 2010.

[9] Katalinic, Branko, et al. "Bionic Assembly System: hybrid control structure, working scenario and scheduling." Proceedings of 9th National Congress on Theoretical and Applied Mechanics, Brussels. 2012.

[10] Valentin, Pryanichnikov, and Shipovalov Egor. "Terrain Modeling for Mobile Robot Control Using Cross-Section Curves in Nested Multi-Resolution Grids." Annals of DAAAM & Proceedings26.1 (2015).

[11] Kirsanov K., Levinsky B., Pryanichnikov V. Integrating software for intelligent robots // Informational-Measuring and Control Systems, Radiotechnika. 2009. V. 7. № 6. P. 35–43.

[12] Kirillchenko A.A., Pryanichnikov V.E., Rogozin K.V. Limits of validity and reliability of proofs. Scepticism in mathematics, functions, and traditions // Information-Measuring and Control Systems. 2013. V. 11. № 4. P. 57–65.

[13] Katsurin, Alexey. "Planning Trajectory of the Mobile Robot with a Camera." Annals of DAAAM & Proceedings 26.1 (2015).

[14] Andreev, Victor, et al. "Training Situation Center Based On Three-Dimensional Virtual Studio for Distributed Mobile Robotics Laboratory." Annals of DAAAM & Proceedings 26.1 (2015).

[15] Isaenko, Sergey, et al. "Simulation of Manipulator as a Part of Space Robotic System Weightlessness Device." Annals of DAAAM & Proceedings 26.1 (2015).

[16] Ilyukhin, Yury, and Ruslan Kolesnichenko. "Impedance Control of High-Precision Geared Servo Drives With Two Motors for Technological Robots." Annals of DAAAM & Proceedings 26.1 (2015).

[17] Andreeva, Victor, Pavel Pleteneva, and Yury Poduraeva. "The Method of Network Control of Robotic Systems of Different Models and Manufacturers."

[18] Filaretov, Vladimir, Dmitry Yukhimets, and Eduard Mursalimov. "The Universal Onboard Information-Control System for Mobile Robots." Procedia Engineering 100 (2015): 737-745.

DOI: 10.2507/27th.daaam.proceedings.081

A STUDY OF MATERIAL-TECHNOLOGICAL MODELLING FOR CHOOSING THE IDEAL COOLING RATE FOR DESIGNING PRODUCTION OF CLOSED DIE FORGINGS USING 30MNVS6 STEEL

Khodr Ibrahim, Ivan Vorel, Štěpán Jeníček, Josef Káňa, Kateřina Rubešová, Kateřina Opatová, Vratislav Kotěšovec

This Publication has to be referred as: Khodr, I[brahim]; Vorel, I[van]; Jenicek, S[tepan]; Kana, J[osef]; Rubesova, K[aterina] & Opatova, K[aterina] (2016). A Study of Material-Technological Modelling for Choosing the Ideal Cooling Rate for Designing Production of Closed Die Forgings Using 30MnVS6 Steel, Proceedings of the 27th DAAAM International Symposium, pp.0551-0555, B. Katalinic (Ed.), Published by DAAAM International, ISBN 978-3-902734-08-2, ISSN 1726-9679, Vienna, Austria
DOI: 10.2507/27th.daaam.proceedings.081

Abstract

Production of forgings contains complex and demanding processes that consist of a number of forging operations which can also include post-forge heat treatment. An optimized manufacturing line is a prerequisite for obtaining prime-quality products which in turn are essential for profitable operation of a forging company. Problems may, however, arise from modifications to the manufacturing route due to changing customer needs. To prevent the production line from being temporarily suspended for changeover and optimization. The use of material-technological modelling with the required modifications can be tested and optimized under laboratory conditions outside the plant without disrupting the production. Thanks to material-technological modelling, the process parameters can be well controlled and varied rapidly in response to changes in market requirements. Outcomes of the modelling runs include ideal parameters for the manufacturing process of the forging part and results of microstructure analysis. This study describes the use of material-technological modelling for exploring and studying the impact of the maximum amount of deformation and its effect on the critical deformed area correlated with different rates of cooling of a particular forged part from the finish-forging temperature on its microstructure.

Keywords: Material-technological modelling; 30MnVS6; True strain; Closed die forgings

1. Introduction

A large number of parameters are playing the main roles in forging processes. The key ones include temperature, strain magnitude and strain rate. The combination of their values together with cooling conditions impacts the final resulting properties of forged parts. It is often impossible to map their effects in actual operation, mainly due to financial constraints associated with production. The best available solution is the use of material-technological modelling. This method involves simulation of the real-world forging process in a laboratory in order to study the effects of process parameters without interrupting or slowing down production in the forge shop [1, 2]. As a result, countless trials can be run which – if carried out in the factory – would lead to a result of a major financial burden.

The product of material-technological modelling is a specimen with a processing history identical to the actual manufacturing route for the forged part [3, 4]. This approach enables microstructure characterization, analyzation and mechanical testing unlike FEM simulations. The designing of this modelling process requires specific data collected from the real-world production [5]. This data include temperatures, intervals between operations, and cooling rates, as well as results from the FEM simulation of the forming process which include deformation values, and are used for developing material-technological modelling specifications [6, 7, 8].

Fig. 1. General view of the forged part, a cross-section with the point selected for modelling, and its total true strain level

2. Experimental

This material-technological modelling study was conducted to correlate the true strain level of the critical deformed area which has the minimum thickness and to study the effect of the different cooling rates which follow the deformation process directly from the finish-forging temperature with resulting microstructure of a particular forged part. Unlike the real life forging part which is made by two separate processes, forging and normalizing heat treatment. This forging is used for making part of the driving wheel system of a truck (Fig. 1). Its final microstructure was specified as a mixture of ferrite and pearlite.

Material-technological models were constructed for the critical point on the forging's cross section. For this point, FEM simulations using the DEFORM software reported the following true strain levels: $\varphi = 2.8$. Cooling curves starting at the finish-forging temperature were plotted for these point (Fig. 2). The curves, i.e. the associated cooling rates, were designed to enable a particular limit cooling rate to be identified: the one at which no bainite forms in the material within the critical cooling interval of 950/300 °C.

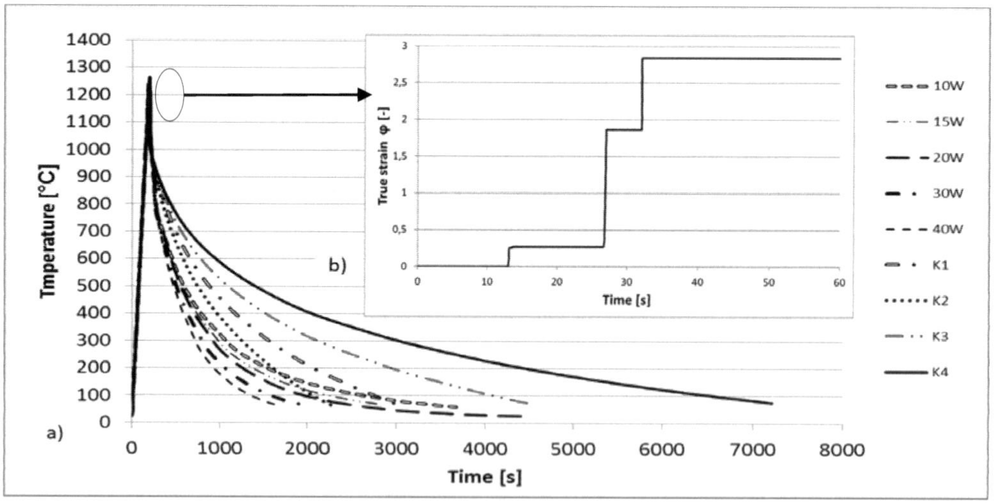

Fig. 2. a) Cooling rates from the finish-forging temperature in the material-technological models of the chosen point, b) Profile of true strain introduced during forging operation at the finish-forging temperature

Modelling was carried out in a thermomechanical simulator provided with an electrical induction-resistance heating system which offers heating rates of up to 100 °C/sec with an accurate cooling rates control. The specimens obtained by material-technological modelling were sectioned to prepare metallographic sections. The experimental material was 30MnVS6 steel (Fig. 3).

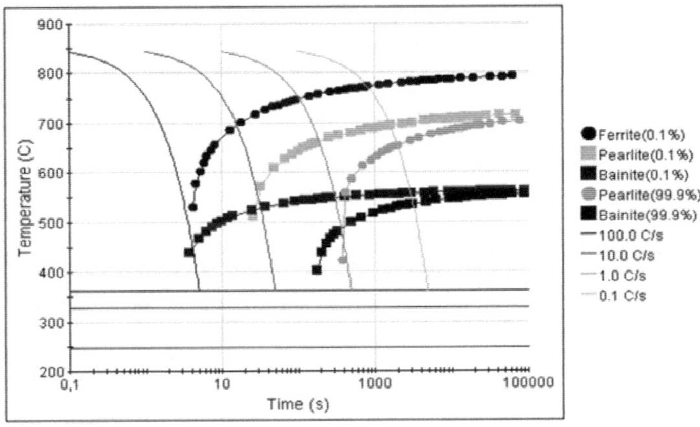

Fig. 3. CCT diagram for the 30MnVS6 steel as computed and constructed using the JMAtPro program

3. Metallographic examination

The metallographic examination was carried out to study the critical area with the highest deformation and thinnest wall of the cross section, in this case, depending on the results from this examination we can choose whether the microstructure will contain a combination of bainite, ferrite and perlite or just a ferritic-pearlitic microstructure. The microstructure of the specimens was revealed with 3 % nital and examined under optical microscope.

All specimens which had cooled according to K1 curve consisted of ferrite, pearlite, and a small amount of bainite (Fig. 4). Bainite was also found in specimens cooling according to K2 curve (Fig. 5). K3–K4 curves led to ferritic-pearlitic microstructures (Fig. 6, 7).

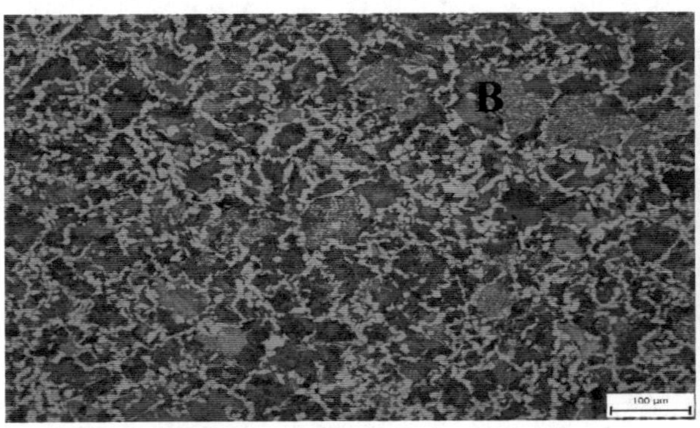

Fig. 4. Micrograph of specimen cooling according to K1 curve

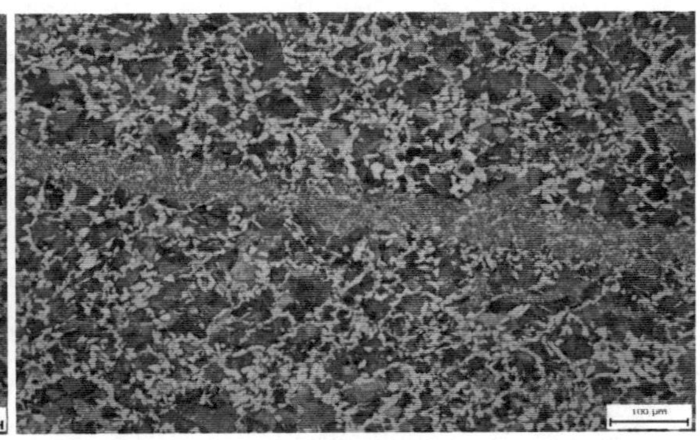

Fig. 5. Micrograph of specimen cooling according to K2 curve

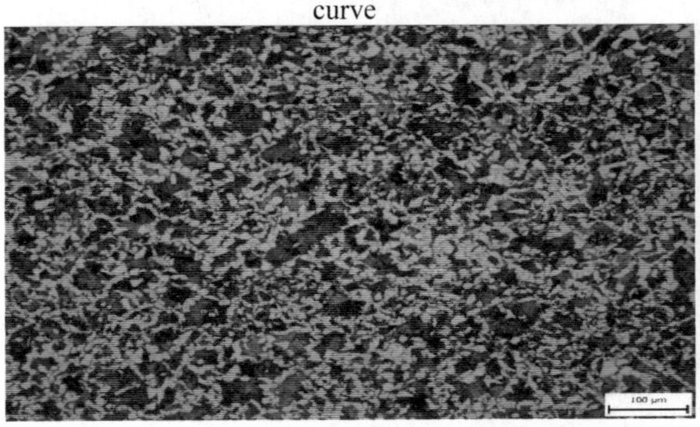

Fig. 6. Micrograph of specimen cooling according to K3 curve

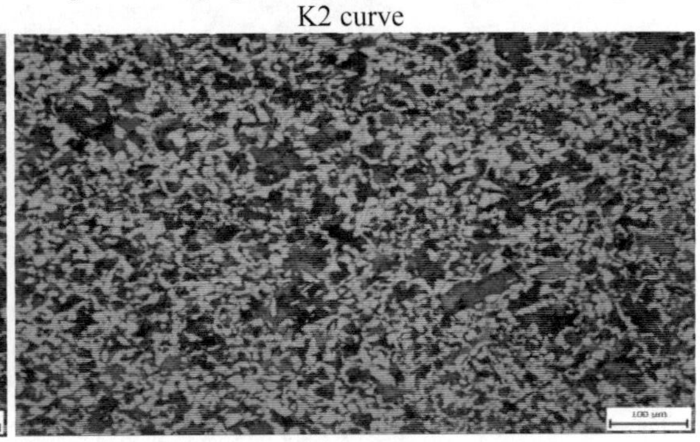

Fig. 7. Micrograph of specimen cooling according to K4 curve

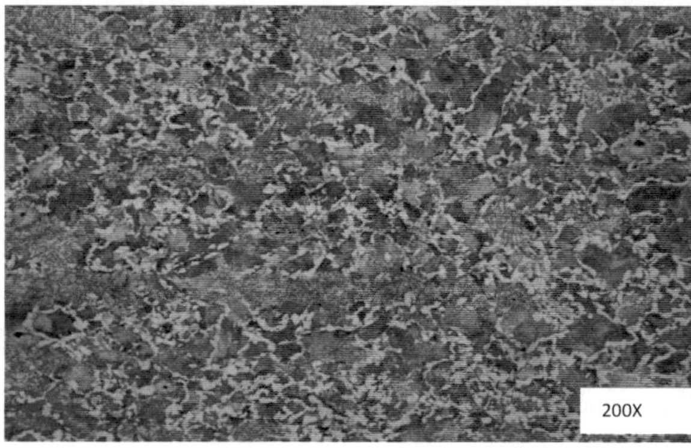

Fig. 8. Micrograph of specimen cooling according to 10W curve

Fig. 9. Micrograph of specimen cooling according to 15W curve

The content of bainite was found also in the specimens according to 10W curve which has faster cooling rate than the previous cooling rates (Fig. 8). The rest of samples which were treated with faster cooling rates contained bainite in their microstructure with a combination with ferrite and perlite as shown in (Fig. 9, 10, 11).

The highest content of bainite was found in the specimens with the cooling curve 40W which has the fastest cooling rate as shown in (Fig. 12).

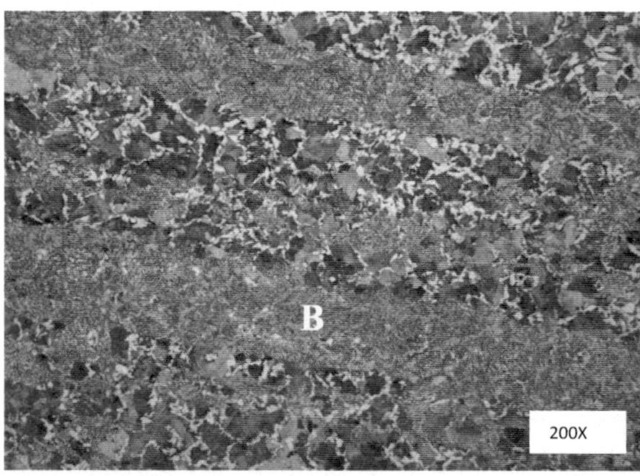

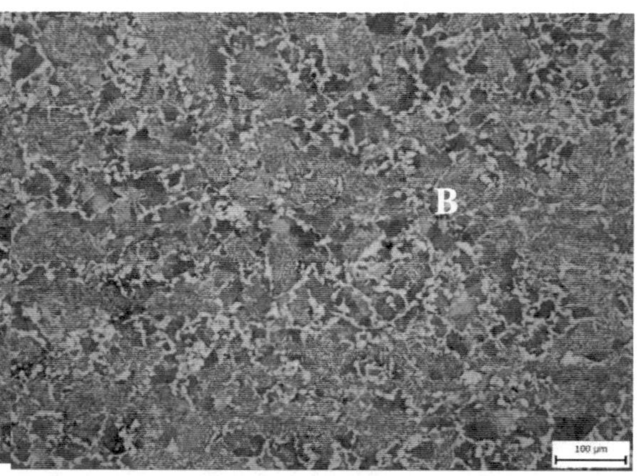

Fig. 10. Micrograph of specimen cooling according to 20W curve

Fig. 11. Micrograph of specimen cooling according to 30W curve

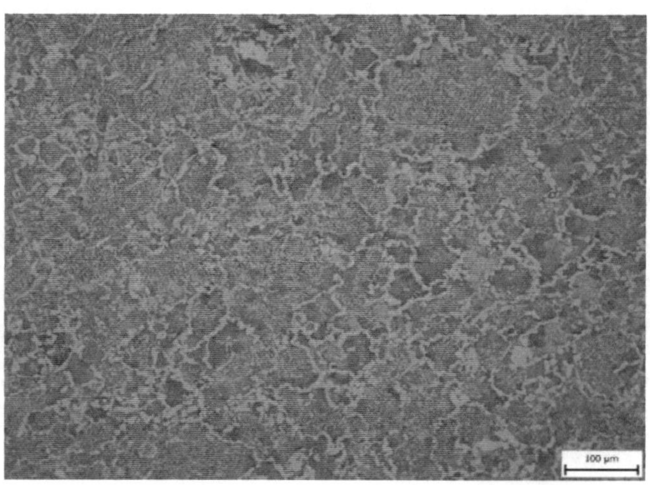

Fig. 12. Micrograph of specimen cooling according to 40W curve

4. Conclusion

Using material-technological modelling, correlation between true strain level, the rate of cooling of a particular forged part from the finish-forging temperature and the resulting microstructure was examined. FEM simulations using the DEFORM software reported the following true strain for the critical deformed area in the cross section: $\varphi=2.8$. Material-technological models were constructed for this critical point to describe the forging process in thermophysical terms. One boundary condition for this study required the final specimens to contain exclusively ferritic-pearlitic microstructure. The critical cross section area was chosen because of the smallest thickness, which means fastest cooling rate compared with thicker areas from the same part. This can insure for us that when the critical area doesn't contain bainite, the whole specimen will not contain it.

Important conclusions were drawn from the experiment and verified. The prescribed ferritic-pearlitic microstructure can be obtained by cooling through the 950/300 °C interval longer than 1600 seconds which is presented in the cooling curve K3 and K4.

For future study, we will try to find out the best cooling process and media to achieve the results obtained from the material-technological modelling in the real life production, so that the forging process will be directly followed by the controlled cooling process.

5. Acknowledgments

This paper includes results created within the project LO1502 Development of Regional Technological Institute. The project is subsidised by the Ministry of Education, Youth and Sports from specific resources of the state budget for research and development.

6. References

[1] Mašek, B.; Jirková, H.; Malina, J.; Skálová, L.; Meyer, L. W. (2007). Physical Modelling of Microstructure Development During Technological Processes with Intensive Incremental Deformation. Key Engineering Materials, Vol. 345-346, No. 1-2, pp. 934-946, ISSN 1013-9826.

[2] Pileček, V.; Vančura, F.; Jirková, H.; Mašek, B. (2014) Material-Technological Modelling of Die Forging of 42CrMoS4 Steel. Materiali in technologije, Vol. 48, Issue 6, pp. 869-873, ISSN 1580-2949.

[3] Vorel, I.; Pileček, V.; Vančura, F.; Jirková, H.; Mašek, B. Material-Technological Modelling of C45 Steel Die Forgings. Procedia Engineering. (DAAAM 2014), 2015, Vol. 100, Issue C, pp. 714-721, ISSN 1876-6102.

[4] Vorel, I.; Vančura, F.; Mašek, B. Material-Technological Modelling of Controlled Cooling of Closed die Forgings from Finish Forging Temperature. In METAL 2015 24th International Conference on Metallurgy and Materials. Ostrava: 2015 TANGER Ltd., 2015. pp. 202-208. ISBN: 978-80-87294-62-8

[5] Mašek, B.; Jirková H.; Kučerová, L.; Ronešová, A.; Malina, J. Material-Technological Modelling of Real Thin Sheet Rolling Process. METAL 2011. 20th Anniversary International Conference on Metallurgy and Materials, 2011, pp. 216-220. Edit. TANGER Ltd., ASM Int, Mat Informat Soc; CSNMT; VSB-TU. ISBN 978-80-87294-24-6.

[6] Mašek, B.; Vančura, F.; Aišman, D.; Jirková, H.; Wagner, M. F. Effect of Input Structure of Blank on Development of Final Structure when Processing at Temperatures between Solidus and Liquidus. In DAAAM 2014. Vienna: Procedia Engineering, 2015. s. 722-729. ISBN: neuvedeno , ISSN: 1877-7058

[7] Simulator of Thermomechanical Treatment of Metals. Káňa, J.; Vorel, I.; Ronešová, A. In: 26th DAAAM International Symposium on Intelligent Manufacturing and Automation, DAAAM International Vienna, 21.-24. 10. 2015, Vienna, Austria, pp. 1-7, ISSN 2304-1382, ISBN 978-3-902734-06- 8

[8] Material-Technological Modelling of Controlled Cooling of Closed die Forgings from Finish Forging Temperature. Vorel, I.; Vančura, F.; Mašek, B. In: 24th International Conference on Metallurgy and Materials, TANGER spol. s r.o., 3.-5.6.2015, Ostrava, Czech Republic, pp. 1-7, ISBN 978-80- 87294-58-1

DOI: 10.2507/27th.daaam.proceedings.082

ADVANCED AXIAL PISTON SWASH PLATE PUMP PARAMETERS RECOMMENDATIONS

Anton Kuzmin, Valeriy Popov, Sergey Stazhkov

This Publication has to be referred as: Kuzmin, A[nton]; Popov, V[alery] & Stazhkov, S[ergey] (2016). Advanced Axial Piston Swash Plate Pump Parameters Recommendations, Proceedings of the 27th DAAAM International Symposium, pp.0556-0561, B. Katalinic (Ed.), Published by DAAAM International, ISBN 978-3-902734-08-2, ISSN 1726-9679, Vienna, Austria

DOI: 10.2507/27th.daaam.proceedings.082

Abstract

Mechatronics is one of rapidly developing spheres of world machinery and science. In many cases its quality depends on drive unit quality. Sensitivity and efficiency of a hydraulic drive unit are mostly determined by hydraulic machines parameters. For the reasons set out an axial-piston swash plate pump with reduced friction forces was designed.

Present work describes the method of computation of the basic parameters of piston – cylinder bore mechanism that provide operability and improved characteristics of the developed axial-piston swash plate pump, and an experimental proof of this computation. The research object is an axial-piston swash plate pump with full hydrostatic slipper bearing. As a result of the preliminary tests of the piston – cylinder bore mechanism of the swash plate pump operational and structural restrictions, which influence the interfaces between pump's elements, were found. Thereby an additional force analysis was performed, applied relations were obtained, which lead to detection of the developed pump swash plate angle range that provides its operability and improved dynamic and energy characteristics.

The conditions of operability of the developed pump were found. An appliance of the present hydraulic machine allows reducing friction forces and significantly improving characteristics of the whole mechatronic system.

Keywords: swash plate pump; machinery; hydraulics; mechatronics; drive unit

1. Introduction

In mechatronic systems, such characteristics as efficiency and sensitivity are crucial and represent the quality of the whole mechanism. From this point, characteristics of each component of the system become important. Drive unit – and prime mover in particular – is a source of significant power losses.

In hydraulic mechatronic systems, hydraulic machines act as prime movers. To reduce power losses caused by friction, there should be fluid friction mode in every kinematic pair. [1]

An axial piston swash plate pump with improved energy and dynamic characteristics was designed [2]. The present hydraulic machine has reduced energy consumption compared to analogues. As a rule, in that kind of pumps a ball and socket joint that connects a piston body to a slipper, is coaxial with the piston body. In that case transverse force, that is a component of a swash plate reaction force, inclines a piston axis with respect to a cylindrical bore axis, which leads to high contact stresses where the piston contacts edges of the cylindrical bore and, therefore, to higher friction forces and reducing efficiency of the hydraulic machine [3].

2. Literature review

There is an axial piston swash plate hydraulic machine, contains a housing including a swash plate and a cylinder block with pistons in its cylindrical bores. Pistons and bores form displacement chambers, and each piston has a ball and socket joint that links it with a slipper. The piston ball's center lies on an axis, directed along the normal to the plane of the swash plate and through the bearing surface of the cylindrical bore at every rotation angle of the machine shaft. In this way, there is a displacement of the piston ball, which is provided by a separator, mounted in a special guide [4].

The shown solution for reduction of the friction force between pistons and cylindrical bores increases contact area that also means an alignment of the piston and the bore. However, a design of the hydraulic machine becomes more complicated. In case of high load, a chance of seizure of the separator can cause the machine failure. Moreover, the present structure does not provide fluid friction mode for the whole operation cycle of hydraulic motor and pump.

To reduce piston transverse forces in the swash plate axial piston machines of fixed displacement type Mohamed Elashmawy proposed circumferential cam contour design. In this machine the piston slipper is replaced by a ball that is rotatably mounted within a ball socket formed at the piston end. The ball runs on a circumferential contour groove formed on the swash plate surface. The sliding friction between swash plate and the slipper is replaced by a rolling friction between ball and circumferential runway groove [5].

Although theoretical computations show positive results and obvious advantages of this development, the essence of replacement of the slipper by the ball in the groove is close to the invention of H. Molly, where reaction of a metal body prevents the piston from inclination. Problems of friction and seizure become even more probable as there is no hydrostatic effect to prevent metal surfaces contacting. Less mobility of the piston outer end leads to the same troubles as in the case of H. Molly's design.

To improve fluid film between a piston and a cylinder load-carrying ability Pelosi [3] designed a barreled shaped piston, which also helped to reduce piston micro-motions. However, profits of this solution were proved only with high shaft speed, when hydrodynamic effects have big influence, and there is no confirmation of advantages of this piston on a whole operating range.

The present hydraulic machine contains piston flat hinge with a slipper axis, directed along the normal to the plane of the swash plate. On the piston surface there are grooves around the piston body lying in a plane parallel to the plane of the swash plate, and one of the grooves has a lubrication orifice, providing operating fluid under pressure from the displacement chamber. An intersection point of the axes of the piston body and the slipper is located between these two grooves. In addition, these grooves stay inside the cylindrical bore during the whole operating cycle.

Preliminary theoretical research of the piston – cylindrical bore mechanism of the swash plate hydraulic machine, such as kinematic and force analysis, allowed to predict significant reduction of the friction force due to compensation of the torque caused by the reaction of the swash plate. This result is achieved by the displaced piston – slipper joint, and, therefore, the point of the swash plate reaction application is displaced, in that way, vectors of reaction in the cylindrical bore become coaxial as the total vector of reaction in the cylindrical bore counterbalances the transverse component of the reaction of the swash plate. The flat hinge determines the piston position concerning the swash plate tilt angle.

3. Preliminary experiment

A special test rig was designed to prove the points of the theoretical substantiation of the developed hydraulic machine. The pistons work in antiphase – in pump and motor mode alternately. The cyclogrammes on Figure 1 show a lengthwise component of the friction force in piston – cylindrical bore mechanism, written at different pressure values.

An operating cycle starts from a position matching with the beginning of the pump mode and lasts at low speed in pump and motor mode. This kind of operating is the most problematic as stability of the slipper increases with its speed [6] and the lift generated by the slipper also grows with the rotational speed of the machine shaft [7]. At the swash plate tilt angle of 15° in motor mode, varying pressure full operating pressure range and assuming the piston position from 0° to 135° concerning the starting position, capsize of the slipper was not recorded, though the friction force increases significantly at 270°, which agrees with the decreasing pressing force of the hydrostatic bearing [8].

At the turn of the piston – cylindrical bore mechanism of 135 ° to 180°, which matches the motor mode, the capsize of the slipper was recorded. That fact causes violation of the structure and kinematics of the developed hydraulic machine.

The goal of the present work is to describe this process and to find the lower limit of the swash plate tilt angle providing the developed axial piston hydraulic machine operability and experimental proof of theoretical computations.

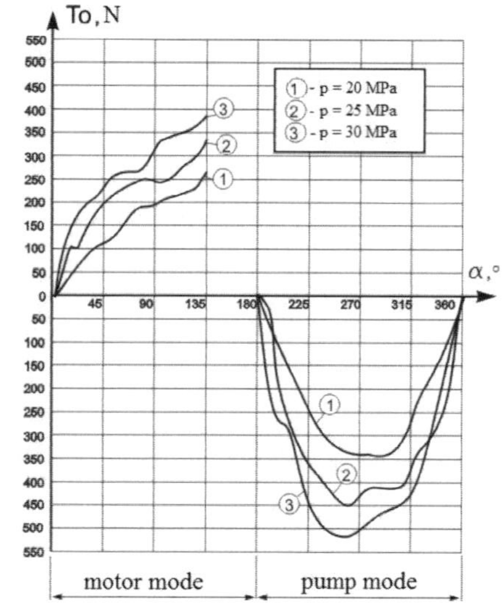

Figure 1. Friction in the cylindrical bore cyclogramme with 15° tilt angle.

4. Theoretical study

The point of the joint between the slipper and the swash plate disclosure is in developed piston design features. For piston axis incline regarding to cylindrical bore axis elimination piston flat hinge has to be displaced to generate torque, opposite to the torque caused by the swash plate tilt angle, in that way, the piston does not rotate around its own axis. Such kinematics causes piston rotation around the cylindrical bore axis and slipper sliding over the swash plate. In case of prevalence of friction forces influence in these kinematic pairs over the reaction of the swash plate effects, that keeps the piston in the operating position, the piston and its slipper are starting to rotate with the cylindrical bore, disclosing the slipper – swash plate joint.

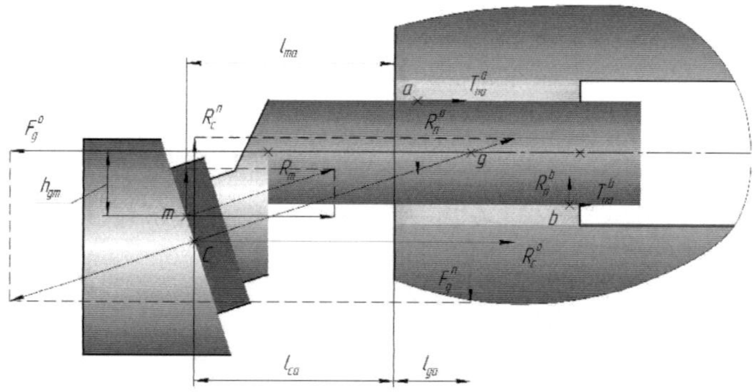

Figure 2. Force model of the piston-cylindrical bore mechanism.

$$R_n^b = \frac{\left[R_m^n l_{ma} - R_m^o\left(h_{gm}+0.5d_n\right)+F_g^n l_{ga}\right]}{l_{ab}} + \frac{\left[0.5F_g^o d_n + R_c^n l_{ca} - R_c^o\left(h_{gc}+0.5d_n\right) - T_{no}^b\right]}{l_{ab}} \tag{1}$$

$$R_n^b = \frac{\left[R_m^n\left(l_{ma}+l_{ab}\right) - R_m^o\left(h_{gm}-0.5d_n\right) - F_g^n\left(l_{ab}-l_{ga}\right)\right]}{l_{ab}} + \frac{\left[-0.5F_g^o d_n + R_c^n\left(l_{ca}+l_{ab}\right) - R_c^o\left(h_{gc}-0.5d_n\right)+T_{no}^a\right]}{l_{ab}} \tag{2}$$

while $T_{no}^a = R_n^a f d_n$, $T_{no}^b = R_n^b f d_n$,
R_n^a – reaction of the external edge of the cylindrical bore,
R_n^b – reaction of the internal edge of the cylindrical bore,
R_c^o – lengthwise component of the reaction of the swash plate,
R_c^n – transverse component of the reaction of the swash plate,
m – the point of contact between the swash plate and the slipper,
c – bearing center of the piston mechanism,

g – the point of intersection of the axis, lying along the normal of the swash plate and through the c point, and the axis of the piston
F_g^n – transverse component of the force, acting from the piston to the swash plate, applied at the g point,
F_g^o – lengthwise component of the force, acting from the piston to the swash plate, applied at the g point,
R_m – reaction of the direct contact between the slipper and the swash plate,
R_m^o – lengthwise component of the swash plate reaction,
R_m^n – transverse component of the swash plate reaction,
a – the point of R_n^a application,
b – the point of R_n^b application,
l_{ab} – axial distance between a and b points,
l_{ma} – axial distance between a and m points,
l_{ca} – axial distance between a and c points,
h_{gm} – transverse distance between g and m points,
h_{gc} – transverse distance between g и c points,
d_n – the piston diameter,
l_{ga} – axial distance between a and g points,
T_{no}^a – lengthwise component of the piston – cylindrical bore friction force in a point,
T_{no}^b – lengthwise component of the piston – cylindrical bore friction force in b point,
f – coefficient of sliding friction.

As a result a relationship for total reaction in piston – cylindrical bore was obtained:

$$R_n^{\Sigma} = R_n^a + R_n^b \tag{3}$$

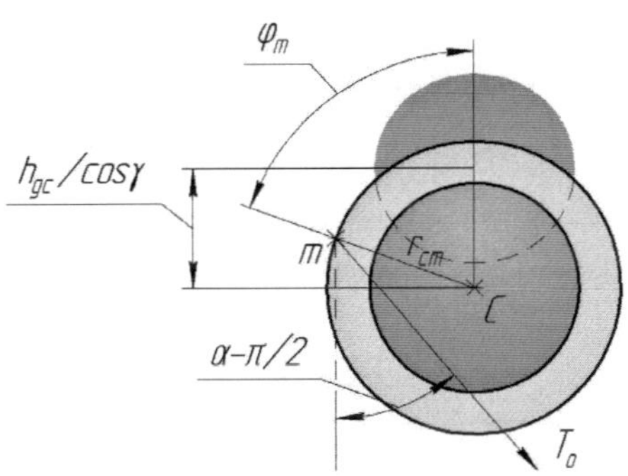

Figure 3. Force model from the swash plate surface.

Present version of the piston mechanism kinematic condition is necessity of the piston rotation relatively to the cylindrical bore with frequency equal to the frequency of the hydraulic machine shaft rotation. This condition is true only if the torque caused by the transverse component of the swash plate reaction in the point of direct contact with the slipper exceeds torques caused by friction forces in kinematic pairs of the piston – cylindrical bore mechanism.

$$M_{R_m} > M_{T_o} + M_{T_n} \tag{4}$$

where $M_{R_m} = \Psi \frac{\pi d_n^2}{4} p \tan\gamma \, r_{cm} \sin(\varphi_m)$,
$M_{T_o} = f\Psi \frac{\pi d_n^2}{4} p \, sin\left(\alpha - \frac{\pi}{2}\right) \cdot (h_{gc} - r_{cm}\cos(\varphi_m) \cdot cos\gamma) + f\Psi \frac{\pi d_n^2}{4} p \, cos\left(\alpha - \frac{\pi}{2}\right) \cos\gamma \cdot r_{cm} \, sin(\varphi_m)$ – the friction force between the slipper and the swash plate torque,
M_{R_m} – the torque caused by the reaction R_m,
Ψ – slipper pressing to the swash plate surface coefficient,
p – operating liquid pressure,
γ – the swash plate tilt angle,
r_{cm} – m point relatively to c point position radius,
φ_m – m point angular coordinate,
α – cylinder rotation angular coordinate,
$M_{T_n} = R_n^{\Sigma} f \frac{d_n}{2}$ – the friction forces between the piston and the cylindrical bore torque.

For this case, considering the fact that the friction forces counteract movements of the slipper, the slipper – swash plate joint operation condition can be calculated as:

$$\tan\gamma > f\left(\frac{R_n^{\Sigma}}{F_p\Psi}\cdot\frac{d_n}{2r_{cm}} + |cos\alpha|\frac{h_{gc}}{r_{cm}} + |sin\alpha|\cdot\cos\gamma\right) \quad (5)$$

In this way, the relationship for the minimum swash plate tilt angle, which provides the operability for the developed hydraulic machine, was obtained. The calculation according to this formula showed, that the tilt angle, which saves the slipper from the capsize, must be not less than 16°.

5. Experimental

To verify technical implementation of the present constructive option of the piston mechanism the bearing between the slipper and the swash plate, providing the tilt angle of 18°, was made. Wherein, although the vector of the swash plate reaction went beyond the cylindrical bore, the operability of the piston mechanism was provided throughout the whole operation cycle, the friction force between the piston and the cylindrical bore was significantly reduced.

Figure 3 shows the cyclogramme of the lengthwise component in the piston – cylindrical bore interface with different shaft rotation frequency, while the piston mechanism with the displaced flat hinge.

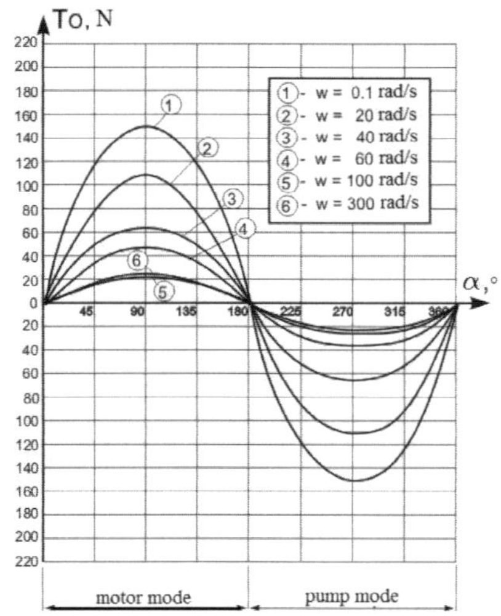

Figure 3. Friction in the cylindrical bore cyclogramme with 18° tilt angle

The developed axial piston swash plate hydraulic machine friction forces in piston – cylindrical bore interface reduced in more than 10 times comparing to the standard design. Better difference was obtained at high pressure values (30-45 Mpa).

6. Results

Firstly of all it was found, that the developed piston – cylindrical bore mechanism has restrictions of its operability. Secondly, boundaries of mechanism operability were defined. The third point is that a method of the operability boundaries computation was formed. The next item is that the experimental verification of the operability range, calculated with the use of the present method, was conducted. The fifth point is that friction forces in the developed swash plate hydraulic machine were significantly reduced.

7. Conclusion and further work

In summary, the present work experimentally substantiates the existence of the limit swash plate tilt angle, which provides the operability of the advanced piston mechanism, and describes the way to find certain values of this limit. This information makes application of the developed piston mechanism possible. Preliminary results of friction forces reduction were reported and discussed in an article by Larchikov, Stazhkov, Yurov [10].

Results of this work allow to design a working sample of the advanced axial piston swash plate hydraulic machine. With this hydraulic machine a hydraulic mechatronic system based on delivery control with improved sensitivity and efficiency can be created.

It is planned to conduct complementary research for elaboration of the permissible swash plate tilt angle. Further work also includes research of the flat hinge as its constructive design may influence its capacity [9].

8. References

[1] Hamrock, B., Schmid, S. and Jacobson, B. (1994) Fundamentals of Fluid Film Lubrication. Marcel Dekker, New York,

[2] Stazhkov,S[ergey] (2013), Development Of An Axial-Piston Hydraulic Machine Of A Drive System, Chapter 12 in DAAAM International Scientific Book 2013, pp. 277-296, B. Katalinic & Z. Tekic (Eds.), Published by DAAAM International, ISBN 978-3-901509-94-0, ISSN 1726-9687, Vienna, Austria.

[3] Matteo Pelosi (2012). An Investigation On The Fluid-Structure Interaction Of Piston/Cylinder Interface, Ph.D. Dissertation, Purdue University, West Lafayette, Indiana, USA.

[4] US Patent office N 3424097. Hydrostatic axial piston unit oft he slipping shoe design / H.Molly, 1969, ser. N 569256

[5] Mohamed Elashmawy (2014). Design of Swashplate Axial Piston Machines Having Low Piston Transverse Forces. International Journal of Mechanical Engineering and Applications. Special Issue: Advanced Fluid Power Sciences and Technology. Vol. 3, No. 1-2, 2015, pp. 17-23. doi: 10.11648/j.ijmea.s.2015030102.13.

[6] Mr Kishan Choudhuri & Dr Prasun Chakraborti (2012), A Computational Approach on the position of Load Centre of a Slipper Bearing, Global Journal of Researches in Engineering Mechanical and Mechanics Engineering Volume 12 Issue 5 Version 1.0, USA, Online ISSN: 2249-4596 Print ISSN:0975-5861.

[7] Jonathan Mark Haynes (2007). Axial Piston Pump Leakage Modelling and Measurement, Ph.D. Dissertation, The University of Cardiff, Cardiff, Wales, United Kingdom.

[8] Yi Sun, Jihai-Jiang (2011), Study On Oil Film Characteristics Of Slipper Within Axial Piston Pump Under Different Working Condition. Proceedings of the 8th JFPS International Symposium on Fluid Power, OKINAWA, pp 534-539, ISBN 4-931070-08-6.

[9] Manring N.D., Wray C.L., Dong Z. (2004). Experimental Studies on the Performance of Slipper Bearings Within Axial-Piston Pumps. Journal of Tribology, Vol. 126, pp.511-518, DOI: 10.1115/1.1698936.

[10] Larchikov, I. ; Stazhkov, S & Yurov, A (2012) The Stusdy Of Hydromechanical Processes In Hydromachines Of Power-Intensive Drives. Annals of DAAAM for 2012 & Proceedings of the 23rd International DAAAM Symposium, Volume 23, No.1, ISSN 2304-1382 ISBN 978-3-901509-91-9, CDROM version, Ed. B. Katalinic, Published by DAAAM International, Vienna, Austria, EU, 2012

DOI: 10.2507/27th.daaam.proceedings.083

SIMULATION FOR ROBOTIC STEREOTACTIC NEUROSURGERY

Josip Vidaković, Bojan Jerbić, Filip Šuligoj, Marko Švaco, Bojan Šekoranja

This Publication has to be referred as: Vidakovic, J[osip]; Jerbic, B[ojan]; Suligoj, F[ilip]; Svaco, M[arko] & Sekoranja, B[ojan] (2016). Simulation for Robotic Stereotactic Neurosurgery, Proceedings of the 27th DAAAM International Symposium, pp.0562-0568, B. Katalinic (Ed.), Published by DAAAM International, ISBN 978-3-902734-08-2, ISSN 1726-9679, Vienna, Austria
DOI: 10.2507/27th.daaam.proceedings.083

Abstract

Through the development of different research and commercial systems, robotic neurosurgery slowly becomes more standard in the medical field. Beyond other advantages, the possibility of frameless neurosurgery is an important reason for their consideration. However, the use of robots in these applications brings problems regarding robot-patient positioning and possible influence of kinematic restrictions during the execution of a surgical plan. To cope with these, a simulation concept for the control of critical positioning and kinematic parameters is presented in this paper. An optical tracking system, simulation software and stereotactic planning mechanism are integrated into one functional unit. It is a simulation tool which is intended to be used shortly after the operation scene is set in the operating room to ensure seamless operation flow. The system is successfully implemented and evaluated in real clinical use.

Keywords: Robotic surgery; Operation simulation; Operation planning; Stereotactic surgery

1. Introduction

Applications of robots in the medical field require a degree of absolute certainty when performing operation tasks. In neurosurgery, this represents a crucial factor for considering the use of modern robotic technology instead or in combination with standard neurosurgical equipment [1]. Generally, there are three main types of neurosurgical systems according to user interaction: automated, telesurgical and systems with shared control [2]. Because of their large dimensions and heavy built, telesurgical systems like the DaVinci system are not flexible in terms of spatial positioning. This results in their ability to operate just in a relatively small predetermined workspace in the operating room. On the other hand, automated systems which are often implemented as light-weight robot arms mounted on mobile platforms, are much more flexible. They can be configured to work in any workspace in the operating room and allow variable robot placement in respect to the patient. This makes them particularly suitable for surgical applications where the position and orientation of the patient can't be identical for every operation case.

In order to satisfy safety requirements in the operating room, technical reliability is one of the main aspects that robotical systems have to guarantee. The other one is functionality, which represents successful utilization of such equipment in individual operation cases. Robotic neurosurgical systems and robotic surgical systems in general have to deal with a high level of potential functional problems. They occur due to the unique parameters of every individual operation case (operation plan, patient position and orientation, collision with the environment). This is especially pronounced with flexible automated systems. Therefore, different planning strategies are developed for such systems with the purpose of solving or minimizing these problems. Here, some of the approaches will be mentioned. In [3], a positioning strategy is shown, which makes use of a workspace analysis for setting an optimization problem for every laparoscopic procedure performed by the laparoscopic robot individually. Another method for planning minimally invasive surgery that optimizes the relative position between the robot, patient and the operating room surrounding is presented in [4]. Other approaches are concerned with the dexterity of the robot actuator. Optimal robot workspaces are determined with the use of a dexterity criterion [5]. This way, a high operation completion rate can be guaranteed for all procedures that are carried out in a certain dexterous workspace.

A typical automated application in neurosurgery is frameless stereotactic surgery [6]. Advantages of robot kinematics are used for precise navigation of surgical instruments in the intracranial space of the patient [7]. The prerequisite for this kind of operation is the creation of a surgical plan. In the planning phase, the surgeon specifies points of interest as entry and target points in a CT or MR representation of the patient. These points are then transformed into the referent coordinate system for the robot. In the intraoperative phase, the robot localizes some referent coordinate system and performs positioning according to previously defined entry and target points which form unique trajectories. The position and orientation of trajectories are defined for every operation procedure according to the needs and specific demands of the patient's case.

The main challenge for the robot is to perform any of the defined trajectories regarding its relative position to the patient. The robot's ability to execute specific planned procedures is determined by the main three parameters: operation trajectories, patient position and robot base position. Here, robot-patient position and robot dexterity are the most important factors for successful operation completion.

Regardless which operation planning strategy is used, successful operation completion has to be guaranteed. For this, a concept for intraoperative procedure validation will be presented. The presented approach is not concerned with providing a strategy for optimal robot placement but giving a tool for final position validation of a stereotactic robot, which can guarantee successful operation completion. It is used at the very beginning of the operation, immediately after the patient and robot positioning is performed. The research is conducted in the scope of the project RONNA (Robotic Neuronavigation) [8].

The system originally consist of a dual arm robot configuration where one robot is responsible for the precise guidance of surgical tools and the second for assisting in different operation tasks (e.g. surgical drilling, biopsy needle/DBS electrode insertion). In the basic version, which will be in the scope of the application presented in this paper, the system is used for stereotactic procedures. This setup of the system consists of one general purpose industrial robot arm and two stereovision systems. One of the vision systems has a relatively wide working volume used for the global localization of the patient. The other one is a custom designed stereovision system mounted on the robot flange which uses a virtual TCP for the final, more accurate localization [9].

2. Materials and methods

The following system for intraoperative setup control is developed for the purposes of the RONNA project. It is composed of a NDI Polaris Spectra optical tracking system (OTS) and central simulation software. The OTS tracks passive retroreflective markers in its workspace volume and is able to give their precise position and orientation information. The system allows tracking of custom markers and is very commonly used in medical applications. In the simulation software, a simplified virtual setup of the operation environment is modelled using static CAD models of objects and instruments present in the operating room. These will be later used for collision detection. Also, a 3D model of the used robot with its corresponding kinematics is incorporated in the environment.

The system is designed to work in five steps:

1. Operation plan is loaded into the simulation software.
2. The OTS registers the position of the robot and the patient.
3. Corresponding virtual scene is set in the simulation environment.
4. The operation is simulated with exact robot movements.
5. Simulation results are evaluated and displayed to the operator.

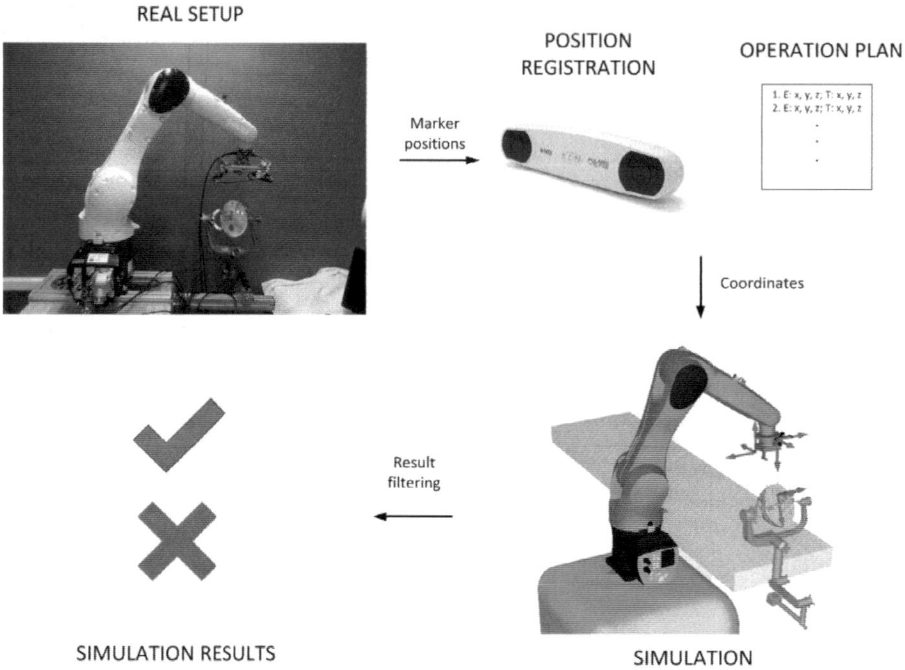

Fig. 1. Intraoperative simulation concept

2.1. Position registration

For the registration of the patient's position and orientation in the robot base frame, the system uses a fix passive retroreflective marker attached to the robot flange and a removable marker attached to a bone screw previously mounted on the patient's skull.

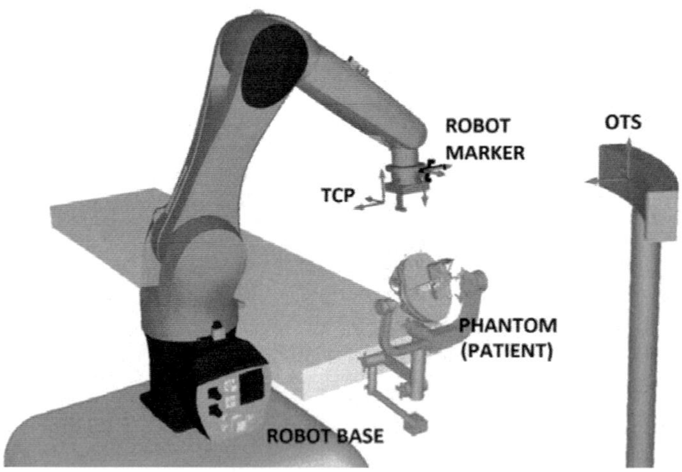

Fig. 2. Simulation setup

Using tracking data from the OTS, the system calculates the position of the patient in respect to the robot base coordinate system with the following homogenous transformations:

$$T_{PATIENT}^{ROBOT\_BASE} = T_{TCP}^{ROBOT\_BASE} \times T_{PATIENT}^{TCP} \tag{1}$$

$$T_{PATIENT}^{TCP} = T_{ROBOT\_MARKER}^{TCP} \times T_{PATIENT}^{ROBOT\_MARKER} \tag{2}$$

The result of these transformations is a transformation matrix which contains the position and orientation of the patient marker. This is than sent to the simulation software via a TCP socket interface.

2.2. Operation plan

The operation plan is created in medical imaging diagnostics software. Points of interest are marked by the surgeon in the form of entry and target points which form spatial trajectories. The trajectories are specifically chosen not to interfere with vital brain structures. Depending on this and the patient's diagnostics, every operation demands specific operation plans. Those points are transformed into the coordinate system of the patient and saved in a form suitable which will later serve as input for the robot in the operation procedure.

2.3. Simulation execution and result filtering

Based on the registered patient and the operation plan, the specific operation case can be simulated. The simulated robot performs the movement in two operational phases: localization and operation. In the first one, the robot moves its TCP into the center of every of three spheres, that the marker attached to the patient consists of, while maintaining a constant orientation. In the operation step, the robot positions its tool frame into the target of every planned trajectory with the orientation defined by the two points that make the trajectory (line in Cartesian coordinate system).

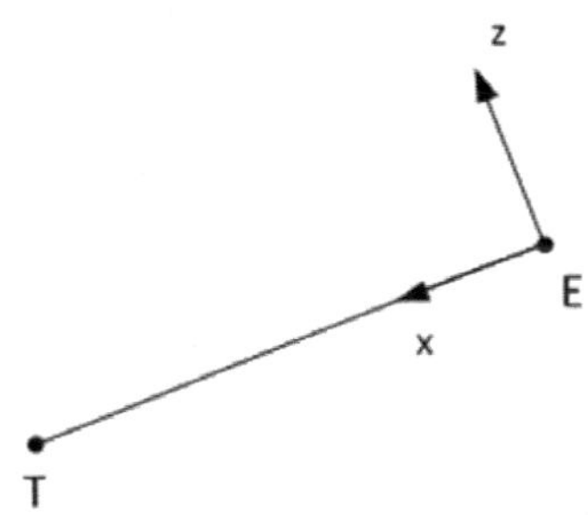

Fig. 3. Trajectory coordinate system orientation in respect to entry and target points

Orientations of trajectory frames are calculated from the two points (entry, target) and one supplementary point which is always vertical in respect to the origin of the coordinate system which is the entry point. The calculation is conducted as shown in (3).

$$\begin{aligned}
&\vec{O} = [X_E \quad Y_E \quad Z_E] \rightarrow \vec{O} = \vec{O} - \vec{O} \\
&\vec{X} = [X_T \quad Y_T \quad Z_T] \rightarrow \vec{X} = \vec{X} - \vec{O} \\
&\overrightarrow{Pxz} = [X_E \quad Y_E \quad Z_E + 100] \rightarrow \overrightarrow{Pxz} = \overrightarrow{Pxz} - \vec{O} \\
&\vec{X} = \frac{\vec{X}}{\| \vec{X} \|}; \quad \overrightarrow{Pxz} = \frac{\overrightarrow{Pxz}}{\| \vec{X} \|}; \quad \vec{Y} = \frac{\vec{X} \times \overrightarrow{Pxz}}{\| \vec{X} \times \overrightarrow{Pxz} \|} \\
&\vec{Z} = \vec{Y} \times \vec{X}
\end{aligned} \tag{3}$$

The homogenous rotation matrix is constructed as follows: $R = \begin{bmatrix} X[0] & Y[0] & Z[0] \\ X[1] & Y[1] & Z[1] \\ X[2] & Y[2] & Z[2] \end{bmatrix}$.

Also, for the simulation of a real stereotactic procedure, the robot is then moved from the patient along the trajectory while maintaining the same orientation. In the simulation software, the robot's inverse kinematic problem is solved for every point on the trajectory *(p, g)*. One of the requirements to this operation is that the robot must not change its configuration in any step of the movement.

$$J_{IK(p,g)} = [J1 \quad J2 \quad J3 \quad J4 \quad J5 \quad J6] \in e.g.front, nonflip, elbowup \tag{4}$$

During linear movement the solution is also tested for singularities.

$$\det(Jacobian(p, g)) = 0 \tag{5}$$

Collision detection during all of these movements is performed for the simplified case of the operation environment. This gives the information about every trajectory and potential problem regarding its execution.

If any of the parameters is not satisfied in the simulation, setup parameters i.e. position parameters of the operation have to be changed in order to retry the simulation and successfully complete the operation from the system point of view. This means the need for robot position change when the simulation does not complete successfully.

3. Experiments and Results

In order to validate the described simulation algorithm, the obtained results are compared with a corresponding experimental setup for stereotactic surgery. The simulation setup validity is here considered in terms of robot positioning accuracy. Spatial position coordinates of trajectory points (entry and target) have been obtained in the robot base coordinate system of the simulated robot and compared with the coordinates read from the real robot. The test is conducted by placing a real phantom for stereotactic testing in different regions of the robot workspace and executing the stereotactic procedure. The obtained results are presented in Table 1 and interpreted by the Euclidian distance as the measure of difference between the simulation and experiment.

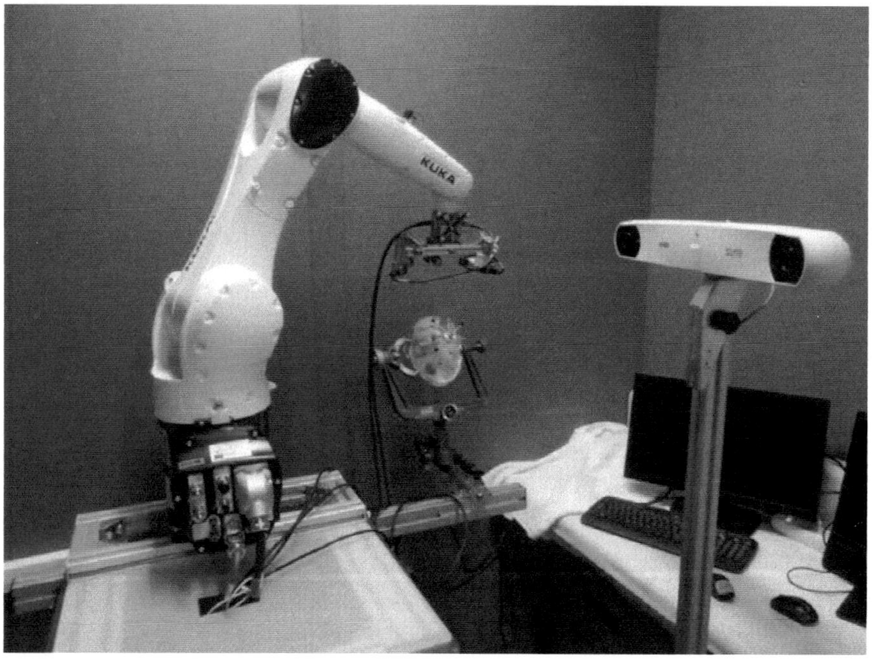

Fig. 4. Laboratory setup

Robot base	avg. error (mm)	min. error (mm)	max. error (mm)
X axis	3.06	2.22	4.81
Y axis	1.84	0.20	3.31
Z axis	4.12	3.05	5.68
Euclidian distance	5.45	3.77	8.14

Table 1. Comparison of results

It is shown that the difference between the simulation and the real setup positioning data is in the range of 4 to 7 mm, mostly depending on the position and orientation of the phantom i.e. patient marker. The error is present mainly due to inaccuracies of the OTS. These consist of general tracking errors and calibration inaccuracies of the OTS tool center point (TCP) obtained by a pivoting method. Robot calibration and positioning inaccuracies are here negligible because they are typically ranging under 1 mm. However, for the relatively low number of 10 compared trajectories shown in the scope of this paper, and further 40-50 trajectories tested informal, the results indicate that this error does not rule out the simulation in terms of trustworthiness of the procedure completion success. In other words, the setup can reliably be used for intraoperative simulation as the last step and final control of the planning strategy, providing good insight for eventual problems in terms of kinematic restrictions and collision.

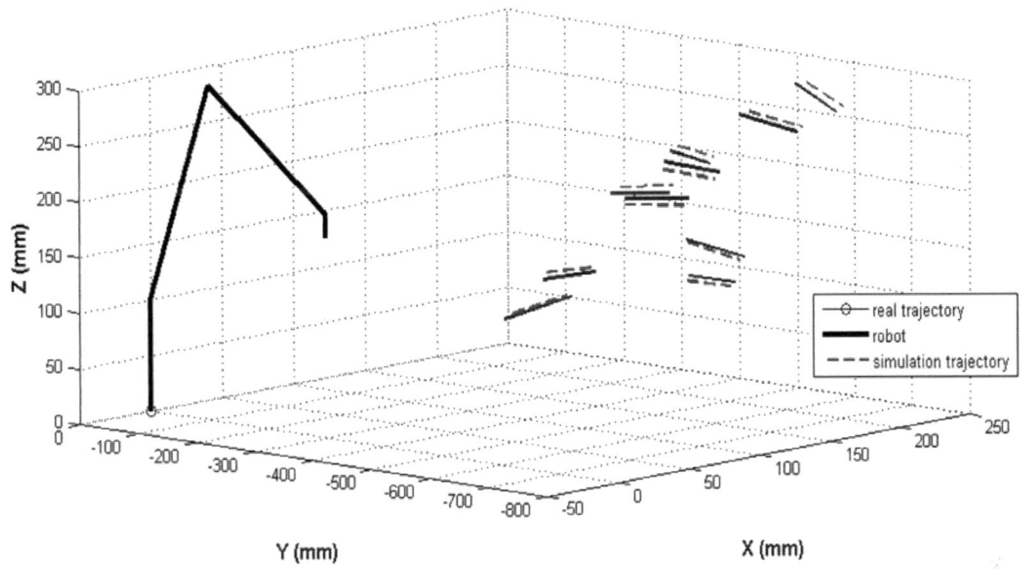

Fig. 5. Spatial comparison of 10 real and simulated trajectories

5. Conclusion

A simulation concept for the intraoperative setup control of a robotic system for neurosugical applications is presented in this paper. The simulation concept represents a simple and straightforward tool for introperative planning compatible with data and equipment already present in the standard medical environment (trajectories, CT models, Polaris OTS).

Robot kinematic constraints represent a crucial boundary when trying to satisfy the needs of spatial positioning of surgical tools in a surgical procedure. The presented approach eliminates eventual problems of this kind during the operation phase. This is done by simulating the procedure with real input parameters in its preparation phase, almost emediately after the patient is positioned in the operating room. The simulation can give all participants the insight into the operating procedure in a step-by-step way which can confirm the validity of the operational setup or emphasise problems regarding robot positioning or collisions. The approach is validated with the comparison of the real and simulated setup for different operation cases (plans).

This concept can be extended for surgical operations with a dual arm robot configuration that will be carried out in scope of the RONNA project. Also, it can be used for intraoperative control of various rmedical robotic applications which whose performans depends on spatial relations to the patient.

Future development will be directed toward developing a fully functional and automated preoprational planning strategy which will guarantee optimal setup configuration for the specific surgical operation plan before entering the surgical operating room. This could minimize the need for intraoperative simulation of this kind or be used along with it to satisfy strict medical safety standards.

6. Acknowledgments

Authors would like to acknowledge the support of the European Regional Development Fund through the project RONNA. Authors would also like to acknowledge the Croatian Scientific Foundation through the research project ACRON - A new concept of Applied Cognitive RObotics in clinical Neuroscience.

7. References

[1] J. D. Geiger and R. B. Hirschl. (2015). Innovation in surgical technology and techniques: Challenges and ethical issues, Semin. Pediatr. Surg., vol. 24, no. 3, pp. 115–121, Jun..

[2] N. Nathoo, M. C. Cavusoglu, M. A. Vogelbaum, and G. H. Barnett. (2005). In Touch with Robotics: Neurosurgery for the Future:, Neurosurgery, vol. 56, no. 3, pp. 421–433, Mar.

[3] B. Pan, Y. Fu, G. Niu, and D. Xu. (2013) Preoperative position planning strategy for laparoscopic robot, in Information and Automation (ICIA), IEEE International Conference on, pp. 1200–1205.

[4] L. Adhami and E. Coste-Manirei. (2003). Optimal planning for minimally invasive surgical robots, IEEE Trans. Robot. Autom., vol. 19, no. 5, pp. 854–863, Oct.

[5] J. Liu, Y. Zhang, T. Wang, H. Xing, and Z. Tian. (2004). Neuromaster: a robot system for neurosurgery, in Robotics and Automation, 2004. Proceedings. ICRA'04. 2004 IEEE International Conference on, vol. 1, pp. 824–828.

[6] C. Faria, W. Erlhagen, M. Rito, E. De Momi, G. Ferrigno, and E. Bicho. (2015). Review of Robotic Technology for Stereotactic Neurosurgery, IEEE Rev. Biomed. Eng., vol. 8, pp. 125–137.

[7] T. A. Mattei, A. H. Rodriguez, D. Sambhara, and E. Mendel. (2014). Current state-of-the-art and future perspectives of robotic technology in neurosurgery, Neurosurg. Rev., vol. 37, no. 3, pp. 357–366, Jul.

[8] B. Jerbic, G. Nikolic, D. Chudy, M. Svaco, and B. Sekoranja. (2015). Robotic application in neurosurgery using intelligent visual and haptic interaction, Int. J. Simul. Model. IJSIMM, vol. 14, no. 1.

[9] M. Švaco, B. Šekoranja, F. Šuligoj, and B. Jerbić. (2014). Calibration of an Industrial Robot Using a Stereo Vision System, Procedia Eng., vol. 69, pp. 459–463.

DOI: 10.2507/27th.daaam.proceedings.084

QUALITY ANALYSIS OF HYDRAULIC SYSTEMS IN FUNCTION OF RELIABILITY THEORY

Orošnjak Marko, Jocanović Mitar & Karanović Velibor,

University of Novi Sad, Faculty of Technical Sciences, Trg Dositeja Obradovića 6, Serbia

This Publication has to be referred as: Orosnjak, M[arko]; Jocanovic, M[itar] & Karanovic, V[elibor] (2016). Quality Analysis of Hydraulic Systems in Function of Reliability Theory, Proceedings of the 27th DAAAM International Symposium, pp.0569-0577, B. Katalinic (Ed.), Published by DAAAM International, ISBN 978-3-902734-08-2, ISSN 1726-9679, Vienna, Austria
DOI: 10.2507/27th.daaam.proceedings.084

Abstract

Through design of a hydraulic systems various methods can be used for increasing system efficiency. Increasing efficiency while at the same time maintaining reliability and energy consumption is far from an easy task. In this article author have used mathematical and empirical evidence while trying to accomplish aforementioned. Design of hydraulic system is based on reliability theory and results are compared with the data collected from empirical sources and articles from previous research in the field. Based on simulation it is possible to increase efficiency of a hydraulic system either at design stage or during exploitation. Results from statistical modeling and simulation show that it is possible to improve system efficiency and therefore achieve higher overall system performance.

Keywords: Hydraulic system performance; efficiency; reliability modelling; maintenance; simulation

1. Introduction

Reliability theory is a science that studies the probability of failures in technical systems and their components (subsystems, set, subset, parts)[1]. Notion of reliability, generally, should be comprehend as the property of a system to operate without a failure in different environmental conditions and modes in a specific period (t). Since authors are investigating design of a system based on reliability, much needed data is required. Therefore, it is a very complex task since, beside requiring large amount of statistical data on components, it also depends on other factors that influence operating conditions of a hydraulic system. Hence, reliability is driven by probability and statistics.

Hydraulic system is a core of control and power transmission. It is very imporant to design system properly or it can backfire with great losses to system overall. Therefore, it is very important to evaluate reliability of complex hydraulic system[2]. Many [2][3][4][5] have recognized the importance of reliability theory in hydraulic maintenance and design in order to improve overall system performance. Analysis of hydraulic system's operation indicates that reliability of a system and components depends on a large number of various factors, which are often interrelated in a complex manner. This factors can include: mechanical properties of components, material fatigue, loads, time in exploatation, oil degradation, contamination, maintenance, etc. Operating conditions have a significant role concerning reliability of a hydraulic component, which have been discussed by various researchers [4][6][7].

Beside aforementioned, similiar investigation concerning reliability of the hydraulic system design have been done in terms of impact that vibration has on a hydraulic pressure pipeline systems. It is claimed by the same authors that the research in terms of reliability of hydraulic components is still incomplete[8].

Some have used different tools and methods to investigate reliability of a failure for specific component in hydraulic system, such as FMEA. For example, Prado and de Souza [9] used Failure Modes and Effects Analysis (FMEA) for hydraulic system (hydraulic press) while aiming to define the critical components and the effects of their failures on the press performance.

Jocanović[7] studies the impact of changes of working parameters in hydraulic system (values of pressure, temperature and viscosity of hydraulic fluid to the number and size of solid particles that pass through the gap in the hydraulic component) and their impact on the reliability of components in the system. Karanović[6] studies the influence of microscale solid particles on the critical operational performances of the most common type of tribo-couples in power hydraulic components – directional valve. From results of experimental analysis by Karanović[6] it can be concluded that contamination has a great impact on system reliability and hydraulic component lifecycle.

Basic equations are used from [1][10] in order to calculate hydraulic system reliability in writing of this paper. Moreover, reliability of a system can be increased exclusively during the designing stage, while at every other lifecycle stage the reliability tends declining. The presented hydraulic model is used from [4] as a guideline example. While redesigning the system and tending to higher the overall mean operational time and reliability, results are far from panaceas but are slightly higher then previously designed model. Another thing to point out is that this example of the system is only used to higher the performance in terms of reliability.

2. Hydraulic system with reliability data of components

2.1 Hydraulic system design

Hydraulic system shown in figure 1. consist of a large group of components, which reliability (i.e. probabilty of failure) of components are given in table 1. Working parameters of a system had been conducted within optimum range of 40°C to 60°C. Reliability-wise all components have been connected serially, so in the case of a failure of any component the whole system stops. System is divided into three sub-systems: accumulation, extrusion and filling sub-system.

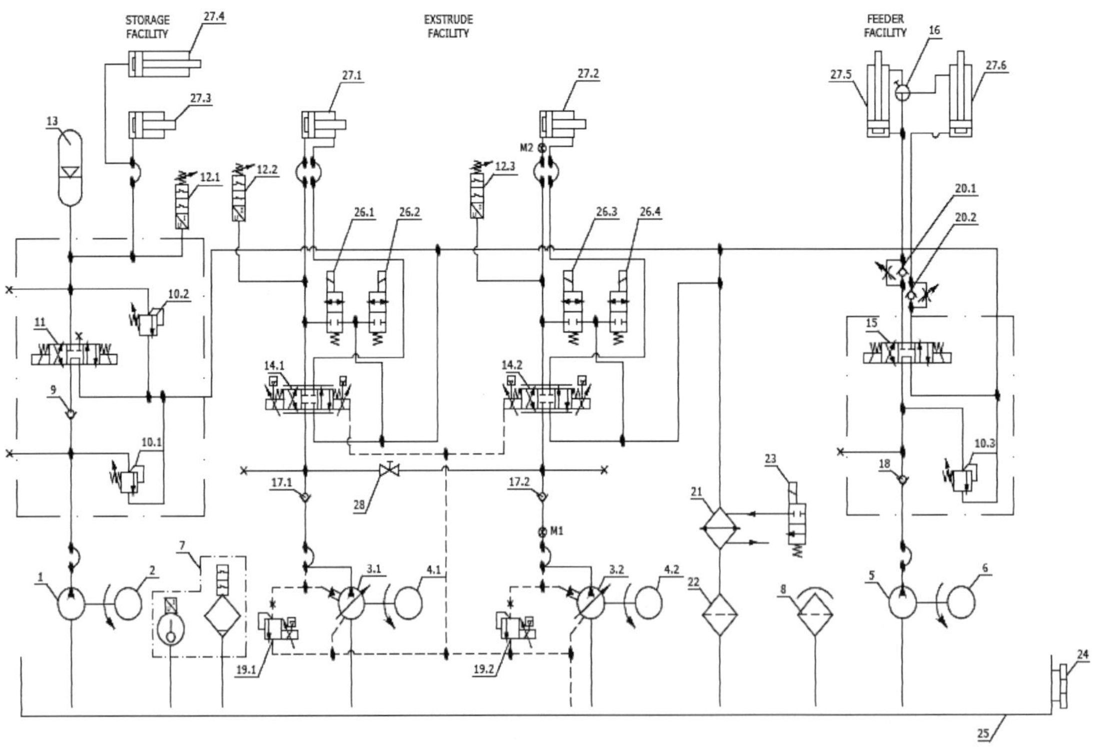

Fig. 1. Block diagram of the hydraulic system used for reliability-analysis-based component design[4]

All data about component failures are based on data given by Fitch & Hong[11]. In their investigation and results, they show the influence of different factors on the reliability of hydraulic components and they explicitly show this in their book. Based on their experiments they have defined the expected failure rates for each hydraulic system component. Therefore, based on the data given by Fitch & Hong[11] it is possible to define the expected reliability each of the sub-system and also to calculate mean operating time and compare it with previous design in terms of performance. The failure rate (range and mean) values of all components in the system from figure 1. are given in the table 1.

Position	Hydraulic component (Failure Rate – marks)	Failure rate (x10^{-6})	
		Range of Value	Mean
(1)	(2)	(3)	(4)
(1), (5)	Gear pump (λ_1, λ_5)		13
(1), (4.1), (4.2), (6)	Electrical motor (λ_2, λ_4, λ_6)	2 to 18	10
(3.1), (3.2)	Axial piston pump (λ_3)	6 to 13	9
(9), (17.1), (17.2), (18)	Check valve (λ_9, λ_{17}, λ_{18})	0.112 to 10.2	6.5
(10.1), (10.2), (10.3)	Relief valve (λ_{10})	1.41 to 8.13	5.88
(11), (15), (23), (26.1), (26.2), (26.3), (26.4)	Electro – hydraulic valve (λ_{11}, λ_{15}, λ_{23}, λ_{26})	2.27 to 19.7	11
(12.1), (12.2), (12.3)	Pressure switch (λ_{12})		10
(13)	Gas charged accumulator (λ_{13})	0.35 to 7.5	6.8
(14.1), (14.2)	Proportional valve (λ_{14})	16.8 to 56	30
(16), (28)	Ball valve ($\lambda_{16,}$ λ_{28})	1.11 to 7.6	4.6
(19.1), (19.2)	Pressure compensated relief valve (λ_{19})		20
(20.1), (20.2)	Check valve – flow restrictor (λ_{20})	0.112 to 10.2	6.5
(21)	Radiator for cooling (λ_{21})		1
(22)	Filter (λ_{22})	0.01 to 1.62	0.79
(25)	Hydraulic reservoir (λ_{28})	0.48 to 2.52	1.5
(27.1), (27.2), (27.3), (27.4), (27.5), (27.6)	Hydraulic cilinders (λ_{27})		0.008
Redesigned hydraulic system			
(29.1) (29.2)	Gear pump with frequency regulation (λ_{27})		13
(30.1)	Relief valve – with relieve of load valve (λ_{30})	0.224 to 14.1	5.7
(31.1)(31.2)	Electro – hydraulic valve (directional control)	1.81 to 7.22	4.6

Table 1. Reliability of constituent components within the considered hydraulic system [4][11]

2.2 Reliabiliy theory used as method in research

Reliability is a tool that can be used in development and analysis of a hydraulic systems, from the selection of components to the planning of efficient maintenance routines[9]. Reliability is also characteristic of an item, generally designated by R(*t*), and expressed by the probability P(*t*) that the item will perform it's required function under given conditions for a stated time interval (*t*). Considered qualitatively, reliability can also be defined as the ability of an item to remain functional. Hence, reliability specifies the probability that no operational interruptions will occur during a stated time interval (*t*) [10]. This does not exclusively reflect only the item or part of the system, but on a system as whole. Basic reason lies in the fact that complex system unites smaller or/and bigger parts of constituent elements (sub-system, assemblies, sub-assemblies, parts) and it can not be judged if one doesnt analyze and analytically cover reliability of each element respectively[1]. Another thing to point out is that in reliability bond diagrams (RBD) components and their bonds are specifically analyzed from reliability standpoint.

Ways to connect parts (components) into system are divided into:

- serial
- parallel
- pasive parallel
- partial parallel
- special (quazy-serial, quazy-parallel)
- complex (combinaton of these bonds elements)

Analysis of a system's reliability has been conducted under assumption that is working under period of random failures. Reliability of a system can be described with exponential distribution law and it is expressed with equation:

$$R(t) = e^{-\lambda \cdot t} \tag{1}$$

where R is system reliability, λ is system failure rate (for the exponential distribution function, failure rate is constant or nearly constant value), t is system operational time.
Reliability of a system arranged in serial order:

$$R(t) = R_1(t) \cdot R_2(t) \cdots R_{n-1}(t) \cdot R_n(t) = \prod_{i=1}^{n} R_i(t) \tag{2}$$

Substituting equation (1) into equation (2):

$$R(t) = e^{-\lambda_1 \cdot t} \cdot e^{-\lambda_2 \cdot t} \ldots e^{-\lambda_{n-1} \cdot t} \cdot e^{-\lambda_n \cdot t} = e^{-(\sum_i^n = 0 \lambda_i) \cdot t} \tag{3}$$

While calculating the reliability using eq.(3) of the sub-systems in time interval taken from range of values: 1, 2, 3, 6, 9, 12, 18, 24, 30, 36 months, Jocanović et al. [4] came with results of each sub-system from one month: a) Filling sub-system *R(730)=0.957;* b) Extrusion sub-system *R(730)=0.781;* c) Accumulation sub-system *R(730)=0.976.* Total reliability of the system had been calculated and charted (figure 2) and results clearly showed that extrusion sub-system was the least reliable within the system and components with failure rates λ_3, λ_{14}, λ_{19}, i.e. axial piston pump, proportional valve and pressure compensated relief valve were replaced. Calculated mean operational time was:

$$T_{op} = \frac{1}{\lambda} = 2268.96\ h \tag{4}$$

which was pretty low comparing to a final redesigned system.

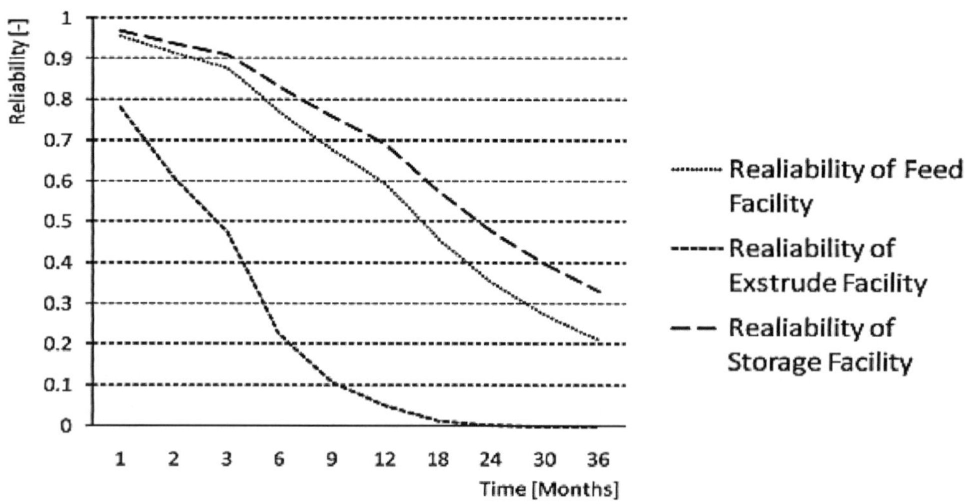

Fig. 2. Total reliability of the three sub-systems of considered hydraulic system

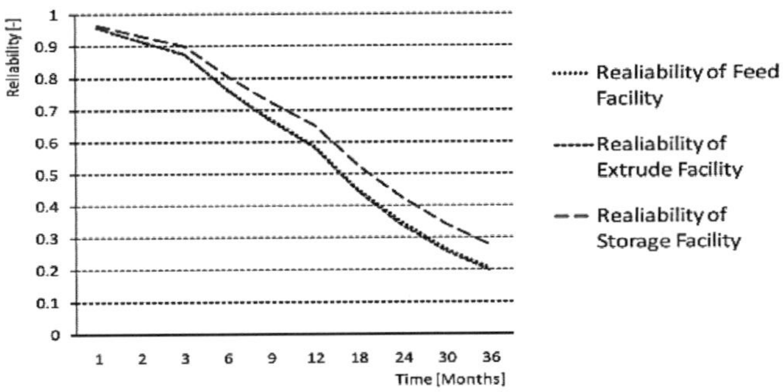

Fig. 3. Total reliability of redesigned hydraulic system in [4]

Even so, system was still failing every 2 hours due to overheating. While monitoring the system it was noticed that the biggest problem was oscilations in pressure and flow and they were the main cause of temperature rise and overheating. Investigation showed that it was problem in hydraulic system design. Axial piston pump is replaced with gear pump with frequency regulation, proportional valve is replaced with internal pressure relief valve, while proportional distributing valve was substituted with electro-hydraulic valve. Newly designed system compared to original system showed 17,43% reliability gain, while mean operational time for the system was T_{op} =*5811,99 h*, i.e. increase of 156%.

The authors of this paper investigated further and came with a conclusion that it is possible to higher the effectivness of the system even more but at the higher cost and energy consumption. This can be done by proposing new design of a system but insted of two main pumps in extruder facility only one pump with higher capacity and power will be used in the system as a main and the other will be there as a passive element, i.e. if main one fails.

3. Newly proposed hydraulic system with higher reliability

Basic idea and goal was to higher the effectiveness of the hydraulic system without adding new elements which can affect reliability overall. New model is represented in figure 4. Based on the model given in figure 4. it is obvious that some components are replaced and others separated from the system. These components were mainly in extrusion sub-system: the axial piston pumps (3.1)(3.2) have been replaced with gear pumps (29.1)(29.2) with frequency regulator. Proportional valves (14.1)(14.2) had been replaced with electro-hydraulic distribution valve (31.1)(31.2), while electro-hydraulic valves (26) had been removed from the system. Notice that extruder facility is now driven by main gear pump (29.1) instead of both working separately. The idea was to use gear pump (29.1) as a main pump while the other gear pump (29.2) will work as a passive component to higher the reliability rate of the system as a whole, since the effectiveness of the extruder sub-system was the lowest.

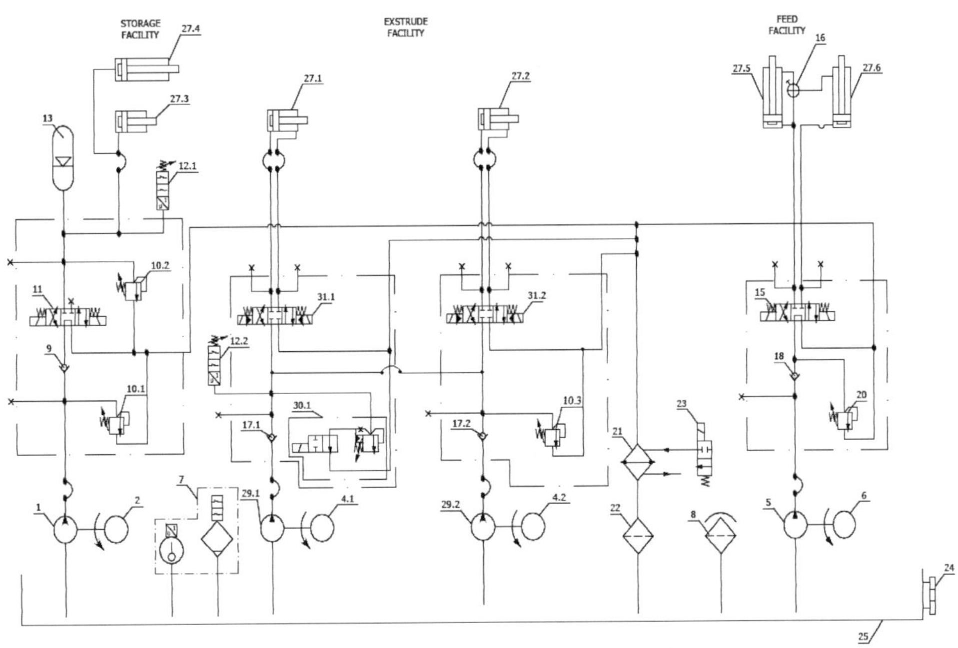

Fig. 4. New block diagram concept of improved hydraulic system

Hydraulic system presented in figure 1 is used for reliability analysis of components. Since, the basic calculations are done by the exponential distribution of serially connected components in the system, the reliability was: a) Filling sub-system *R(730)=0.957;* b) Extrusion sub-system *R(730)=0.781;* c) Accumulation sub-system *R(730)=0.976.* Newly proposed hydraulic system for reliability based analysis showed following results: a) Filling sub-system *R(730)=0.951;* b) Extrusion sub-system *R(730)=0.977;* c) Accumulation sub-system *R(730)=0.954.* Changes in reliability were twofold. Failure rates (λ) in this paper were used by only mean values to calcuate reliability and mean operational time. Second, redesigning system, replacing and removing some components and also using parallel passive connection between components extruder sub-system showed significant improvement in reliability, as system itself.

One of the main reasons to replace axial piston pumps lies in the fact that system doesnt require pressure and flow regulations but rather a constant flow to actuators. Even so, replacing axial piston pump with gear pump is possible with decently high overall performance of gear pump. Lately, gear pumps are constructed and designed with quiet operation and low pulsations even at high speeds and resemble with high volumetric efficiency. This is done by avoiding entrapment between the teeth by using redesigned inclined teeth[12].

Another thing to point out is that proportional valve in opposite to electro-hydraulic valve require higher level of fluid cleanness. Particulate contaminant in the system containing directional control valve can be most cripling substance[13] and in the case of proportional valve demands are even higher. Contaminats in the fluid (especially solid particles) can cause large amount of damage to proportional valves which are very sensitive[14]. Proportional valve has a first stage, a hydraulic amplifier (converts an electrical input to a hydraulic control pressure), and a second stage (which responds to the control pressure and meters an accurate output flow). Hence, several failure modes can occur in the first stage such as ferrous particles that are attracted to magnetic field and cause a large hysteresis. Contaminant lock is the major failure in second stage, since any contaminant can lodge in the clearance and silt up the spool[13]. Therefore the biggest threat to proportional valves by particle contamination is sliding wear and erosion[15] which directly lead to higher system energy consumption, cost, maintenance, time and lower reliability most of all. In order to cope with the obstacle different types of methods can be used[6][16]. Usually using different types of filters and places of installation can help control contamination, but this leads to pressure drops, costs, etc[17]. The higher the fluid cleanness is required the higher are the expences of maintenance overall[18]. Since it was mentioned before that system suffered from overheating, next to particle contamination, heat is the second most common cause of failure of hydraulic components[19][20]. In order to overcome that it needs to be controlled, elseway oil will degrade if working temperature gets to high and therefore viscosity will get lower. Hence, getting viscosity critically low it no longer maintains a boundary lubrication. Therefore, it can lead to rise in component surface temperatures, and subsequent mechanical failures of components. For example, number of helicopter accidents have been caused due to loss of lubrication[21].

Redesigning hydraulic system will be accomplished by the use of reliability bonds. As said, the goal is to increase reliability of the extruder subsystem. This can be done by installation of a passive gear pump which turns on if the main one fails. It is obvious that this leads to higher start up costs for redesigning but it will also increase mean operational time. Therefore, to justify costs of investment it would be interesting to conduct a research of cost-effectivness in this type of solution. Hydraulic system can be represented in block diagrams of reliability (RBD)[10]. For (a) filling sub-system and (c) accumulation sub-system the blocks are serial connected and calculated by eq.(3) and in this case they were not illustrated. While (b) extrusion sub-system is represented in figure 5. for better apprehension.

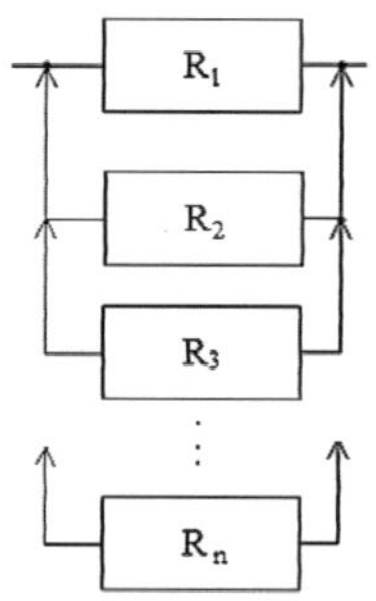

Fig. 5. Passive parallel connection

The RBD in figure 5. does not represent simply parallel connection but a passive parallel connection. Under the passive parallel connection or passive reserve, means that the installation of a passive component (R_2, R_3...R_n) replaces main component, so that component inclusion in the work is achieved only if an active (main) component fails or needs to be supported. Therefore, the idea was to use the gear pump (29.1) as a main component while gear pump (29.2) acts as a passive element. Calculation for passive parallel connection can be done by Poisson distribution:

$$R(t) = \sum_{k=0}^{n} \frac{(\lambda t)^k}{k!} e^{-\lambda \cdot t} = \left| \frac{(\lambda t)^0}{0!} + \frac{(\lambda t)^1}{1!} \frac{(\lambda t)^2}{2!} \cdots \frac{(\lambda t)^n}{n!} \right| e^{-\lambda \cdot t} \quad (5)$$

where is:

n – number of reserve components (elements) in the system;
k – 0 ÷ n – variable – number of elements in passive connection;
λ – const. – intensity of failure exponential law of distribution;
t – time variable.

Hence, only R_1 (fig 2.) presents active (main) component and the others (R_2, R_3... R_n) are there as passive component(s) to replace active component only if it fails, and in this particular case main goal is to include the auxiliary gear pump if the main gear pump fails. Therefore, particular hydraulic extruder sub-system is modelled as follows:

$$R_{active}(t) = e^{-\lambda_{29}\cdot t}\cdot e^{-\lambda_{4}\cdot t}\cdot e^{-\lambda_{17}\cdot t}\cdot e^{-\lambda_{30}\cdot t}\cdot e^{-\lambda_{12}\cdot t}\cdot e^{-\lambda_{31.1}\cdot t}\cdot e^{-\lambda_{27.1}\cdot t}\cdot e^{-\lambda_{31.2}\cdot t}\cdot e^{-\lambda_{27.2}\cdot t} \quad (6)$$

$$R_{pasive}(t) = e^{-\lambda_{29}\cdot t}\cdot e^{-\lambda_{4}\cdot t}\cdot e^{-\lambda_{17}\cdot t}\cdot e^{-\lambda_{10}\cdot t} \quad (7)$$

Including eq. (7) and (6) into eq. (5):

$$R_{ex-sys}(t) = \sum_{k=1}^{1}\frac{(\lambda t)^1}{1!}e^{-\lambda\cdot t} = \left[\frac{(\lambda t)^0}{0!}+\frac{(\lambda t)^1}{1!}\right]e^{-\lambda\cdot t} \quad (8)$$

Calculations show the results of reliability for time $t=730\ h$ (1 month) for 1 passive element as follows:

$$R_{ex-sys}(t) = \sum_{k=1}^{1}\frac{(\lambda t)^1}{1!}e^{-\lambda_i\cdot t} \quad (9)$$

$$R_{ex-sys}(730) = \left[1+\frac{(\lambda_{29}+\lambda_4+\lambda_{17}+\lambda_{10})\cdot t^1}{1!}\right]e^{-(\lambda_{29}+\lambda_4+\lambda_{17}+\lambda_{30}+\lambda_{12}+\lambda_{31}+\lambda_{27}+\lambda_{31}+\lambda_{27}+\lambda_{21}+\lambda_{22}+\lambda_{23})\cdot t} \quad (10)$$

The results are: a) Filling sub-system *R(730) = 0,951*, b) Extrusion sub-system *R(730) = 0,977*, c) Accumulation sub-system *R(730) = 0,954.*

From research based on reliability theory on this particular hydraulic system it has been shown that most unreliable sub-system was extruder, i.e. least reliable assembly of the hydraulic system. For calculations of reliability of all sub-systems, mean failure rates (λ) were used. Results are presented in figure 6. From originally designed hydraulic system reliability gain of 21,51% claims a significant improvement. Replaced and discarded components have led to increased reliability of the whole system. Another thing to point out is mean operational time. The results are as follows: T_{op}'= *6052,995* hours which also led to increase of ~3800 hours from T_{op}=*2268,96* hours originially designed and even more than 200 hours comparing to redesigned system in [4]. This indicates the increase of 167% and 4,15% from originally designed and redesigned by [4], respectivelly.

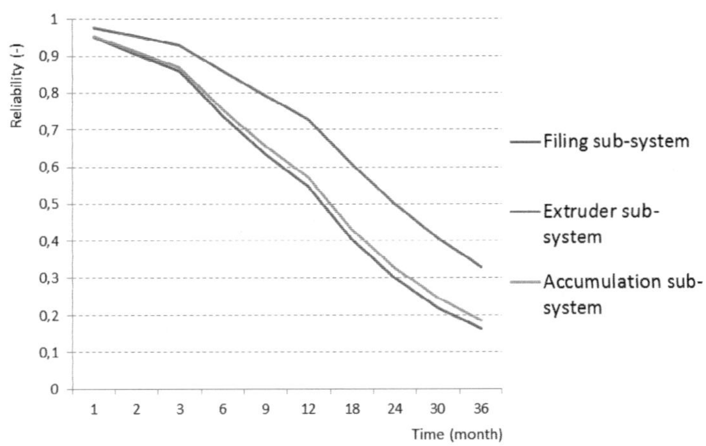

Fig. 6. Reliability of redesigned system

4. Conclusion and further research

From investigation in the field of hydraulic system effectiveness very little research have been done concerning reliability analysis in conjuction with hydraulic systems. During research authors found considerable amount of reliability data in terms of technical systems[22] but most of them lack the research concerning hydraulic systems. Therefore scientific community should consider this as a potential field that lack the research data especially entering the new era of Industry 4.0 where reliability of automatic technical systems and components can be crucial.

Constant failures of a hydraulic system leads to high downtimes, expences, component failures, etc. Main objective was resolving the problem of low mean operational time of hydraulic system using reliability data of components and redesigning the system. From this article it has been shown that it is possible to improve sub-system and system performance overall. The biggest problem was extruder sub-system with axial piston pump. It was realized that system works in a contaminated environment and gear pumps can handle contamination better than axial piston pump with targeted ISO cleanless level[23]. Another thing to point out is that system doesnt require pressure or flow regulations but rather a constant flow to actuators, therefore some of the components, i.e. proportional valve, is replaced with electro-hydraulic valve.

It has been noticed that the extruder facility which is driven by main gear pump (29.1 at figure 4) now leads to twofold improvements. First, failure rates indicates higher mean operational time. Secondly, removing or replacing some components of extruder sub-system showed significant imrpovement in reliability. This could only be done by adding a passive parallel connection with a passive gear pump but also leads to higher investments and costs in system redesign. It can be improved even further but it needs to be conducted cost effective analysis while adding or replacing low/high reliable components. From discussion it can be realized that applying reliability on improving hydraulic system performanse can help identify down-time causes, weakest component/s, value of operating parameters, mean operational time, etc. It can improve maintenance by proactively monitoring reliability of a system in real-time, which can help predict and react to a potential failure.

Another interesting keymark that can be further researched are influence of solid particles, air and water influence on the reliability of a hydraulic system. Therefore, it would be interesting to see how does contaminants in the hydraulic oil influence effectiveness of a system which can be expressed in system reliability and even develop a model. Hence, investigation could go separate ways concerning system reliability in terms of hydraulic design based on the reliability of components and reliability modeling in terms of influence of contaminated oil. It must not be left out that highest reliability-performance can only be done at the system design so that must be duly noted. As for improving reliability based on contamination control it can surely contribute to a better system performance, understanding the failure, determining critical period of a contamination, critical points in reliability hysteresis, etc. Therefore, it represents a potential open field for discussion to scientific community.

5. References

[1] Ivanovic, G., Stanivukovic, D. & Beker, I. (2010). *Reliabilty of Technical Systems*. FTN publish, Novi Sad. (written in Serbian). ISBN 978-86-7892-247-3

[2] Chen, D., Yao. C. & Feng. Z. (2013). Reliability Prediction Method of Hydraulic System by Fuzzy Theory. 6th IFAC Symposium on Mechatronic Systems. The international federation on Automatic Control. April 10-12. Hangzhou, China. Published by Elsevier Ltd. All rights reserved. pp 457-462. DOI: 10.3182/20130410-3-CN-2034.90001.

[3] Avontuur, G.C. & van der Werff, K. (2002). System reliability analysis of mechanical and hydraulic drive systems. *Reliability engineering and System safety*. Vol. 77. Issue 2. pp.121-130. ISSN: 0951-8320

[4] Jocanović, M., Šević, D., Karanović, V., Beker, I. & Dudić, S. (2012). Increased efficiency of Hydraulic Systems Through Reliability Theory and Monitoring of System Operating Parameters. *Strojniški vestnik – Journal of mechanical engineering.* Vol.58, Issue 4. pp.281-288. ISSN:0039-2480. DOI: 10.5545/sv-jme.2011.084

[5] Rahimdel, M. J., Ataei. M., Khalokakaei, R. & Hoseini, S.H. (2013). Reliability based maintenance scheduling of hydraulic system of rotary drilling machines. *International Journal of Mining Science and Technology*. pp. 771-775. ISSN: 2095-2686. DOI: http://dx.doi.org/10.1016/j.ijmst.2013.08.023

[6] Karanović, V. (2015). Development of a solid particle influence model on performance of pistoncylinder contacting pairs for hydraulic components. Ph.D. Thesis. Department of Industrial Engineering and Management. Faculty of Technical Sciences, University of Novi Sad, Serbia, Novi Sad. Available at: http://nardus.mpn.gov.rs/bitstream/handle/123456789/1856/Disertacija.pdf?sequence=1&isAllowed=y

[7] Jocanović, M. (2010). Approach to research and define the model for the calculation of flow of solid particles with a mass of mineral oil through the gaps in a function of the constructive operating parameters of hydraulic components. Ph.D. Thesis. Departman of Industrial Engineering and Management, Faculty of Technical Sciences, University of Novi Sad, Serbia, Novi Sad.

[8] Zhang, T. & Liu, X. (2013). Reliability design for impact vibration of hydraulic pressure pipeline systems. *Chinese Journal of Mechanical Engineering.* Vol. 26. Issue 5. pp. 1050-1055. Available at: http://www.cjmenet.com/EN/Y2013/V26/I5/1

[9] Prado, A. C. & de Souza, G.F.M. (2003). Reliability analysis of hydraulic systems employed in presses: Application to deep drawing press for 250 ton. Proceedings of COBEM 2003. 17th International Congress of Mechanical Engineering. November 10-14. Sao Paolo. Brasil. Available at: http://www.abcm.org.br/anais/cobem/2003/html/pdf/COB03-0185.pdf

[10] Birolini, A. (2010). *Reliability Engineering – Theory and Practice*, 6th ed., Springer, Berlin.

[11] Fitch, E.C. & Hong, I.T. (2004). *Hydraulic system design for service assurance*, BarDyne, Inc., Stillwater.

[12] Opran, C., Pricop, M & Teodoru, C. (2014). Embedded engineering for low noise gear pumps. *DAAAM international scientific book 2014.* Chapter 13. pp.163-170. ISSN 1726-9687. DOI: 10.2507/daaam.scibook.2014.13

[13] Fitch, E.C. (1988). *Fluid Contamination Control*. Oklahoma state University, Stillwater, Oklahoma, USA.

[14] Sasaki, A. (2001). Hydraulic valve problems caused by oil oxidation products. Hydraulic Failure Analysis: Fluids, Components, and System Effects. Authors: Totten, G.E., Wills, D.K., and Feldmann, D.G. ASTM West Conshohocken, USA. pp.474-488. DOI: 10.1520/STP38296S

[15] Stecki, J. (2009). Contamination control and failure analysis. *Lubrication and Maintenance of Industrial Machinery: Best practicies and reliability*. Editors: Gresham, R.M. and Totten, G.E. by Taylor and Francis Group, LLC. United States. pp IV-33. ISBN: 978-1-4200-8936-3 DOI: 10.1201/9781420089363.ch4

[16] Orošnjak, M. (2015). Indicator development for maintaining hydraulic systems based on tribological parameteres. UNS FTN – Proceedings Faculty of Technical Sciences, Novi Sad. Vol. 16. Issue 12. pp. 3025-3029. ISSN 0350-428X. Available at: http://www.ftn.uns.ac.rs/2041336908/broj-16

[17] Savić, V. (1995). Energy consumption of different construction solution of hydraulic systems. Proceedings of HEP. Hydralics in Industry. Issue 1. May.1995.

[18] Savić, V. (2012). Oil Hydraulics 4 part 1: examples of hydraulic system calculations. IKOS, Novi Sad. Written in Serbian. ISBN 978-86-915541-0-1

[19] Savić, V. (1995). *Tribology and Lubrication.* FELJTON. Novi Sad. (written in Serbo-Croatian)

[20] Crosford, J. (2015). Industrial Reliability in Hydraulics. Fluidpower world. Available from: http://www.fluidpowerworld.com/industrial-reliability-in-hydraulics/ Accessed: 05.07.2016

[21] Rashid, H.S.J., Place, C.S., Mba, D., Keong, R.L.C., Healey, A., Kleine-Beek, W., & Romano, M. (2015). Reliability model for helicopter main gearbox lubrication system using influence diagrams. *Realiability Engineering & System Safety.* vol. 139. pp.50-57. ISSN: 0951-8320 DOI: http://dx.doi.org/10.1016/j.ress.2015.01.021

[22] Straka, L.; Rimar, M.; Corny, I. & Mihalcova, J. (2011). Increasing of Operational Reliability of Technical System, Annals of DAAAM for 2011 & Proceedings of the 22nd International DAAAM Symposium, 23-26th November 2011, Vienna, Austria, Volume 22, No. 1, ISSN 1726-9679, ISBN 978-3-901509-83-4, Katalinic, B. (Ed.), pp. 1089-1090, Published by DAAAM International Vienna, Vienna. Available at: http://www.daaam.info/Downloads/Pdfs/proceedings/proceedings_2011/1089_Straka.pdf

[23] Understanding ISO codes (n.d.). Available at: https://www.hyprofiltration.com/clientuploads/directory/Knowledge/PDFs/ISO%20Intro.pdf Accessed: 09.09.2015

DOI: 10.2507/27th.daaam.proceedings.085

BASIC RESEARCH OF THERMAL TRANSFER SIMULATIONS

Václav Marek

This Publication has to be referred as: Marek, V[aclav] (2016). Basic Research of Thermal Transfer Simulations, Proceedings of the 27th DAAAM International Symposium, pp.0578-0585, B. Katalinic (Ed.), Published by DAAAM International, ISBN 978-3-902734-08-2, ISSN 1726-9679, Vienna, Austria
DOI: 10.2507/27th.daaam.proceedings.085

Abstract

The article summarizes basic research into thermal transfer simulations. Problems of thermal influence in mechanical systems are solved there. The first steps focus on matching simple heat-transfer samples with CAE software. Simple cases are performed in a real environment. Thermal values are measured. Cases are also solved using CAE software tools. Solutions are compared. CAE solutions are matched to real values. CAE results are verified or refuted. There are many differences between the options in the solvers, there are steady states and transient-run possibilities, etc. Software tools like Nastran, etc. need many coefficients to solve the problem. This procedure is able to identify specific conditions, fits the solver to the specific sample and performs CAE simulations to get real, verified results. For example, passive radiators heated by an induction heater are used for real tests. Temperature fields are measured by thermal camera and structural deformations by measuring displacement. These values are used in simulations and solved by finite elements method. Simulations are performed in Siemens NX10 software, supported by solvers Nastran, MAYA, and NX Multiphysics. Results are compared and matched in the simulation to acquire a more precise solution in the following steps All these steps are processed to get characteristics of thermal transfer simulation which will be useful in difficult examples of simulation machines, machine tools etc.

Keywords: heat load; thermal flow simulation; FEM

1. Verifying of thermal load

A simple simulation of the heat load was performed in the first part of the research. A small iron cylinder was heated by a high frequency induction heater. The cylinder was heated from 25°C to nearly 250°C as you can see on the Fig.1. The heat load was turned on for 60 seconds, and then it was cooled by radiation and natural convection. The temperature of the cylinder was logged by a thermocouple. The temperature was monitored in relation to time. Thermal fields were monitored by a thermal camera. The thermal camera was used to verify the optimum homogeneous warming. Four measurements were made. Three measurements of a cylinder with diameter 25 mm and 30 mm high are performed. Effective emissivity was guaranteed by paint with guaranteed emissivity (e = 0.95).
This effect was simulated. Derivation of the measured curve provided the loading characteristic. The same loading characteristic is used in the simulation. That obtains two views, measured and simulated, as you can see in the Fig. 1. The results of the experiments are shown in the graphs Figs. 2.
Heat load was measured in the experiment. Measuring determined heat load 219 W. For equivalent simulation was estimated heat load 220 W. Then was computed convection heat exchange. Measured exchange was 16.3 W and simulated heat exchange less, about 14 W. Cooling losses were calculated by derivation of curves in Fig. 2.

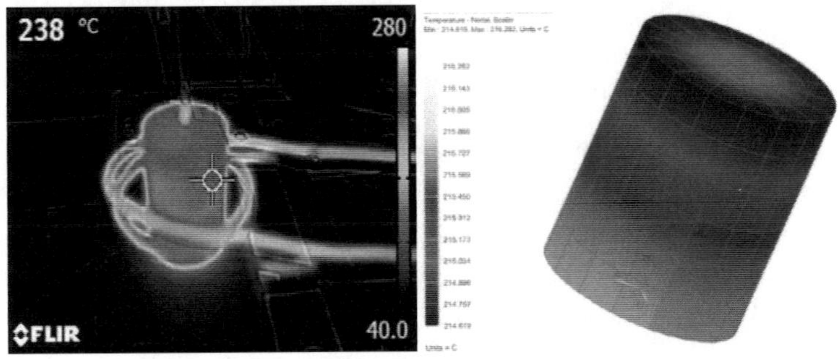

Fig. 1 Temperature fields, heat loading – measured(left) and simulated(right)

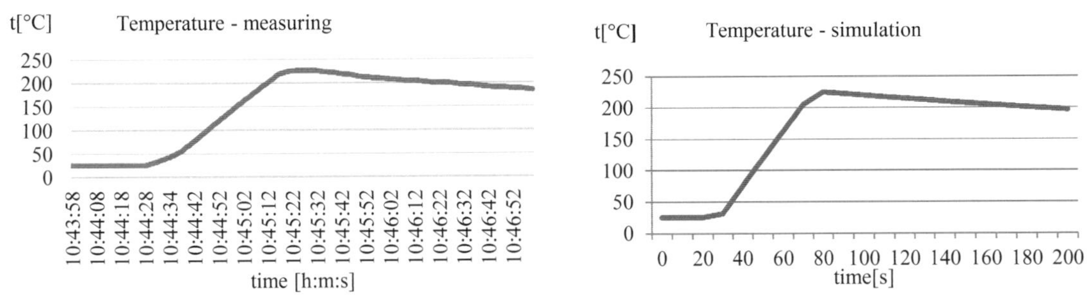

Fig. 2 Heat loading graph – measured (left) and simulated (right)

2. Verification of thermal cooling capacity of passive cooler

The experiment with a passive cooler takes a more precise look at the thermal effects. Thermal effects like free convection and radiation are very difficult to simulate precisely. Results are particularly sensitive to the mesh.[6] Results are also sensitive to the other coefficients. Therefore, it is necessary to perform many experiments to obtain the real coefficients and to acquire information about the basic thermal activity.
The heat transfer is generally expressed by conduction and convection. The major balance equation which describes this problem is represented below (1), where ρ means density, c_p heat capacity, T temperature, t time, k heat conductivity, Q heat source. [10]

$$\rho \cdot c_p \cdot \frac{\partial T}{\partial t} + \nabla(-k \cdot \nabla T) = Q \tag{1}$$

The heat flux is computed by general convection equation, the convective heat flux is given by following equation (2). Where q means heat flux, h heat transfer coefficient, T_2 outer temperature, T_1 temperature of material [10]:

$$q = h \cdot (T_1 - T_2) \tag{2}$$

The essence of the topic is the harmonization of thermal simulation and real measurement.
A high frequency induction-heating device heats the iron test samples (S235). The main disadvantage of this source is that we do not know the precise thermal input into the sample. It is crucial to find the heating parameters.

2.1. Experiment

An experiment was performed for verification. Experiment was consisted of this setup:

- heat load from high frequency heating device(1)
- cooling by free convection (2)
- thermocouple measuring(3)
- thermal camera monitoring(4)

The experimental setup is presented in the Fig.3.

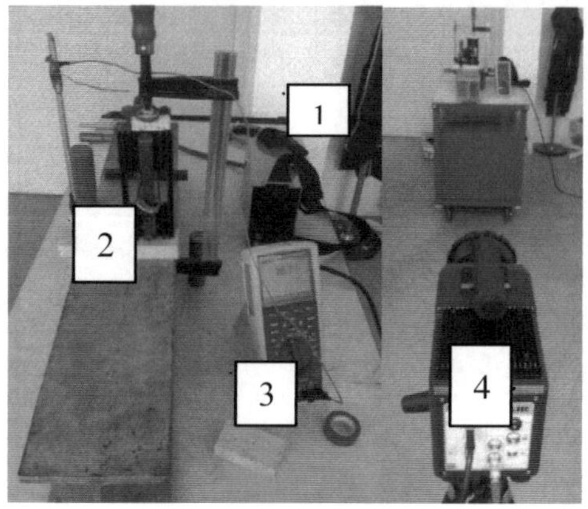

Fig. 3 Experimental measuring, thermocouple measuring(left), thermal camera (right)

2.2. *Heat loading*

Heat load has to be determined and the result is used for the heating of the passive cooler in thermal simulation. This choice is based on the integration of the thermal field in time dependent states. Heating cycles can be seen in Table. 1.

Cycle	Heat load [s]	Cooling [s]	Total time [s]
1	120	180	300
2	90	210	300
3	90	510	600

Table 1. Heating cycles

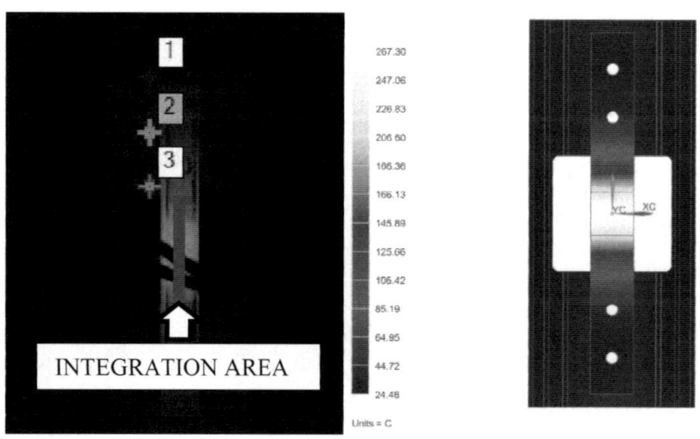

Fig. 4 heat load of passive cooler – measuring by thermal camera and simulation

2.3. *Integration area*

Figure 4 shows the integration area. Temperatures were measured in this area. Energy states are computed by their integration. The results are the energy states in three periods. It takes the heat load by their differences. Graphs of the temperatures are shown in Fig 5. Values, which are represented by graph, are approximated values, because heat conductivity is neglected. It is neglected because of the fast process during 20 s. These results are the input values for the simulation and they will be refined by iterations. Figure 4 also represents graphic result of temperature fields of the first iteration. In Fig.4 can be compared thermal influence of heat source computed by FEM and measured.

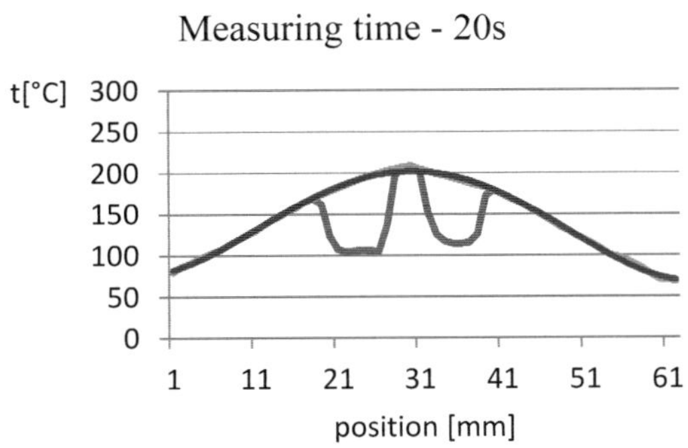

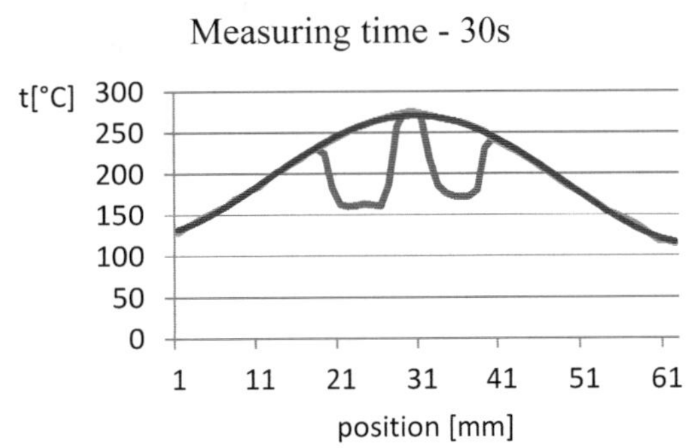

Fig. 5 Measured temperatures in Integration Area, Tempreratures dependent on position, three graphs for three times

Average heat load was estimated to be 347 W by difference of energetic states defined by graphs in Fig5. In these graph can be seen energetic states of measured samples. It is represented in two measuring times 20s and 30s. Integration was performed in excel file by discretization of a steel part.

2.4. *Simulation of heat loading and verification*

Simulations are performed in Siemens NX10 with a palette of solvers. Three options of heat loading states were used there. Simulations and measurements present the states of thermal behaviour during heating and cooling time. States are solved by a pallet of the provided solvers. Solved states are shown in Table 2.

2.5. *DATA for comparison*

All data for procession are given bellow. Data are represented by four sets:

A) NX Thermal NASTRAN

The simplest way is to use NX Nastran solver. Environment provides easy option of boundary condition.

B) Solver NX Thermal/Flow – Thermal solution

More sophisticated solver is Thermal/Flow solver included in NX. It supports more options to specify real states. However, in this case can be used only thermal solver. Convection can be defined merely by coefficients.

C) Solver NX Thermal/Flow – Coupled Thermal/Flow solution

Last performed option is full thermal/flow solution. This option respects many properties of heat and flow effects. [5] You can see result in the Fig. 10.

D) Measuring

Measuring provides experimental data, which are used to verification of computed results.

Four points on the body were verified. Three points are situated on the aluminum cooler (see Fig. 4) and one is situated on the heated iron sample. The evaluation is presented below. For heating in the first iteration was used heating schema as shows Table 2.

Cycle	Heat load [W]	Heat load timing [s]	Cooling [s]	Total time [s]
1	350	120	180	300
2	350	90	210	300
3	350	90	510	600

Table. 1 Heat load cycle

2.6. *Values before solver settings (default set)*

Solver	C1- max temp [°C]	C2- max temp [°C]	C3- max temp [°C]
Nastran	109.1	82.06	82.6
Thermal-Flow: Thermal	130.4	105.1	105.2
Thermal-Flow: Coupled	127	98.2	98.05
Measuring	154.17	97.5	104.7

Table. 2 Maximal temperatures

Solver	C1- finish temp [°C]	C2- finish temp [°C]	C3- finish temp [°C]
Nastran	88.03	79.22	66.37
Thermal-Flow: Thermal	113	90.1	66.4
Thermal-Flow: Coupled	106.8	76.6	49.38
Measuring	151.01	95.6	73.9

Table. 3 Final temperatures

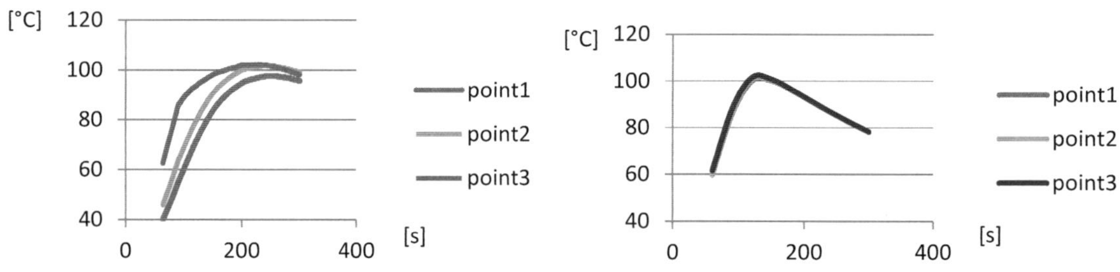

Fig.6 Measured(left) and simulated(left default set) temperatures dependent on time

2.7- Comparison

Tables 3 and 4 present measured temperatures. As can be seen in Fig. 6 , differences in the default simulation with perfect contact are substantial. The large difference in maximum temperature is obviously caused by a more powerful heat load. A useful parameter is the temperature of point 4 on the iron sample. Points 1, 2 and 3 show the effects of face contact between the iron sample and the aluminium cooler. The heat transfer coefficient is a characteristic quantity of convection [4]. Values of point 4 before solver settings are given in the Tables 5 and 6.
Measured points can be seen in the Fig. 4 made by thermal camera. Simulated temperature fields are presented in the Fig. 4 on the right side.

2.8. Values before solver settings (default set) POINT 4

Solver	C1- max temp [°C]	C2- max temp [°C]	C3- max temp [°C]
Nastran	265.9	222	222
Thermal-Flow: Thermal	252.7	219.3	219.7
Thermal-Flow: Coupled	248.8	214.7	176.9
Measuring	-------------	307	325.1

Table. 4 Maximal temperatures

Solver	C1- finish temp [°C]	C2- finish temp [°C]	C3- finish temp [°C]
Nastran	117.39	86.4	86.4
Thermal-Flow: Thermal	116.4	92.15	67.5
Thermal-Flow: Coupled	109.5	78.85	50.46
Measuring	-------------	131	84.4

Table. 5 Final temperature

3. Solution match

Solvers provide a lot of options for improving the results of the simulations. As can be seen, the differences are more than 20%. Generally, good accuracy of thermal simulations is about 10–15%.

Main factors in thermal simulations are below [6]:

- face contact between objects (thermal resistance)
- effective emissivity – radiation
- roughness of the walls
- mesh quality
- flow effects – turbulent/laminar flow, etc.

3.1- Match of solvers

Values provided by three solution setups were verified by the measuring. Solution setups/solvers are given below:

- NX Thermal NASTRAN;
- NX Thermal/Flow – Thermal solution;
- NX Thermal/Flow coupled – This is a coupled thermal and flow solution.

Computation was processed with these boundary conditions:

- contact resistance 0.75 C/W [3]; this temperature change is known as the thermal contact resistance [1, 9, 8]
- heat load 375 W.

Table 7 shows experiment input values in three cycles. Figure 7 shows graphic comparison of cycle 2. Graph shows measured and simulated point 4. Measuring was performed by thermocouple. Graph in the Fig.7 represents temperatures dependent on time and influence of contact resistance. Table 8 and table 9 show results of the second iteration of simulation.

3.2. Heat load – Iteration 2

Cycle	Heat load [W]	Heat load time [s]	Cooling [s]	Total time [s]
1	375	120	180	300
2	375	90	210	300
3	375	90	510	600

Table. 6 Heat load - iteration 2

3.3. Correlated values of point 1

Solver	C1- max temp [°C]	C2- max temp [°C]	C3- max temp [°C]
Thermal-Flow: Thermal	120.78	101.8	109
Thermal-Flow: Coupled	118	91.75	95.9
Measuring	154.17	97.5	104.7

Table. 7 Corellated values

Solver	C1- finish temp [°C]	C2- finish temp [°C]	C3- finish temp [°C]
Thermal-Flow: Thermal	117.18	97.67	76.7
Thermal-Flow: Coupled	112	81.35	56.2
Measuring	151.01	95.6	73.9

Table. 8 Correlated values

3.4. Graphic comparison of cycle 2, point 4 – simulation, measuring – tmc (thermocouple)

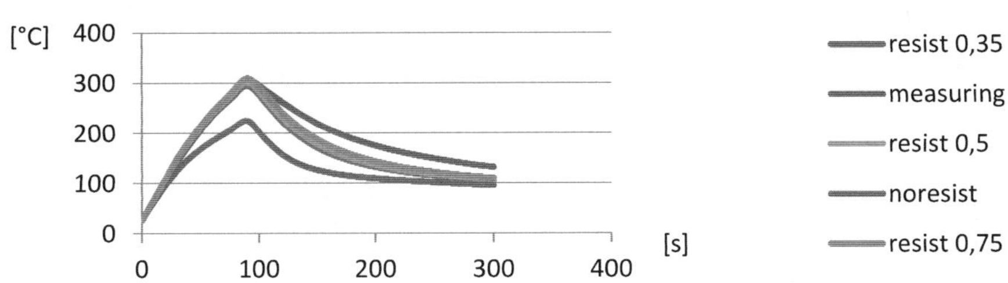

Fig.7 Results for variation of contact resistance [C/W]

3.5. Achieved result

Achieved results can be seen below. You can see influence of a face contact and temperature fields. Figure 9 represents matched solution of the simulation and thermal camera measuring, which verify the simulation. Figure 8 shows achieved solution of thermal simulation in graph. There can be seen differences of temperatures caused by surface contact. Graph shows temperature of point dependent on time. Measured temperature fields are shown in Fig. 9 on the left side. Similar temperature fields are obtained by simulation in the Fig. 9 on the right side. Temperature fields are similar, small differences could be neglected.

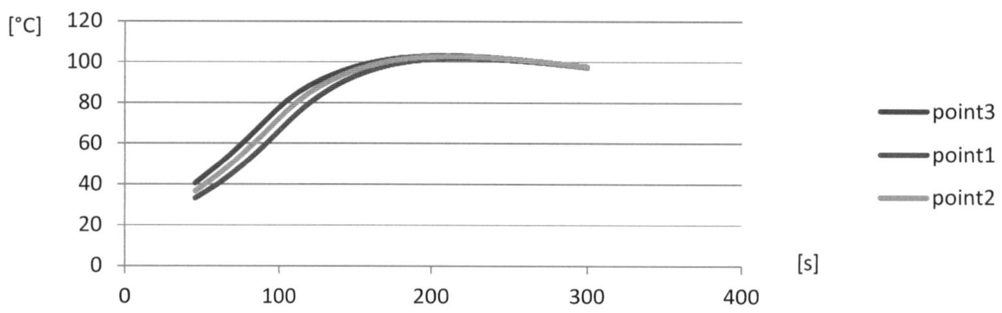

Fig. 8 Achieved solution, heat resistance of face contact 0.75C/W

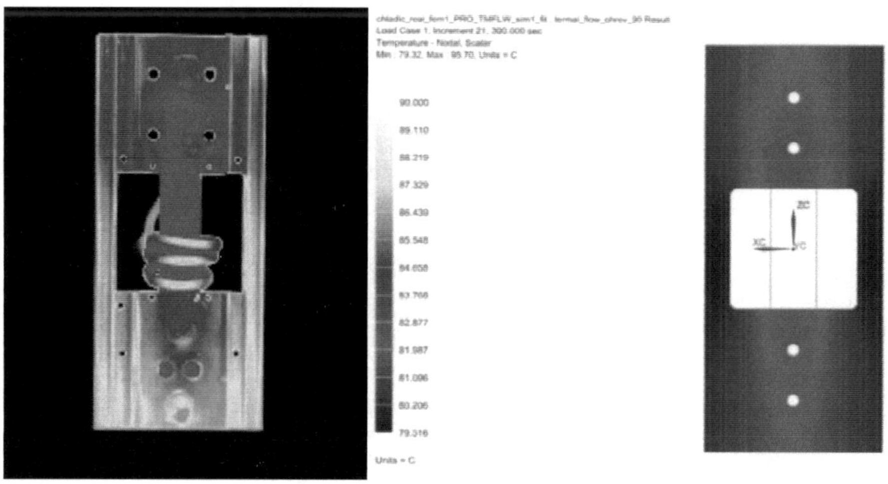

Fig. 9 Measuring(left) and simulation(right) of temperature fields in time 300s

4. Cooling capacity simulation

This part of the article shows how to use the acquired data. Two methods for computing the theoretical cooling capacity were performed. Heat loading process was designed with a heat load of 20 W and 50 W. The cooling process was designed with a free and forced convection cooling. Final computation is processed in two ways. The first computation was processed with Nastran simple thermal simulation. Second computation was processed in Thermal/Flow environment for two conditions - free and force convection. If the fluid is made to move by the action of a pump, a fan, or a blower, we have a case of forced convection. [2, 7]

Results of a flow simulation are represented in Fig. 10 and table 9. Results of thermal/flow environment are more detailed. Computation is able to analyse fluid temperatures, flow directions and temperature of parts.

4.1. Cooling capacity

Solver	Heat load [W]	Heat load timing [s]	Max temperature [°C]
Thermal-flow - Steady state	20	Until steady state	86.33
Thermal-flow - Transient	50	600	143.8
Thermal-Flow Forced convection – Steady state	20	Until steady state	74.81
Thermal-Flow Forced convection – Transient	50	600	143.85

Table. 9 Cooling capacity

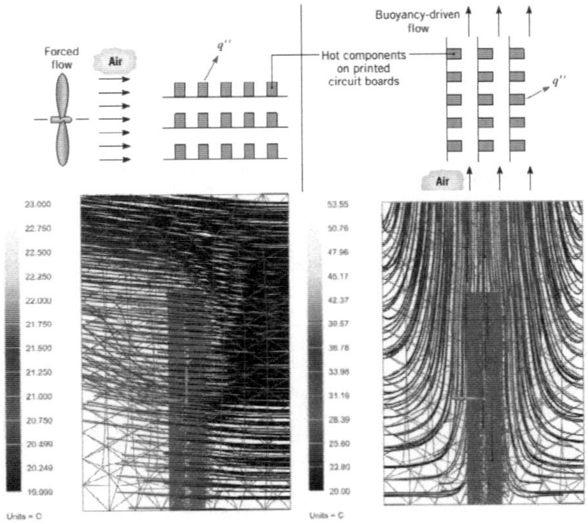

Fig. 10 Simulation of convection – air flow temperatures: forced(left), free(right) [1]

5. Conclusion

Vital issues in complex thermal transfer simulations are the unknown thermal transfer coefficients and the unknown precise definition of the environment condition. These facts cause that the results are not very precise. This problem is solved by a couple of measuring and simulation. Difficult transfer cases are supported by simulation and simply measured cases, which can help us to find the boundary conditions and to identify the coefficients. This article shows how to identify conditions and refine the results. Simply case of thermal transfer was simulated and measured. Results are refined by the iterations. Refined simulation provided information about transfer coefficients in the case. These coefficients are used in complex simulation, which is proofed. The goal of this paper is verified coefficients and also verified simulation in part 3. Low deviation of results of different solvers was also proved. It is possible to get results of simulations with low deviation as can be seen in part 4. Next research will show how to simulate thermal transfer in bearings, transmissions or other important nodes in machines.

6. Acknowledgements

This contribution has been prepared as part of the project LO1502 'Development of the Regional Technological Institute under the auspices of the National Sustainability Program I of the Ministry of Education of the Czech Republic aimed to support research, experimental development and innovation.

7. References

[1] Bergman, T. L. & Frank P. Incropera. (2011). Fundamentals of heat and mass transfer, John Wiley & sons, ISBN 9780470501979, Hoboken

[2] Baskharone, E. A.(2012) Thermal science: essentials of thermodynamics, fluid mechanics, and heat transfer, McGraw-Hill, ISBN 978-0-07-177234-1, New York

[3] Brown, M. E. (1988) Introduction to thermal analysis: techniques and applications, Chapman and Hall, London

[4] Balmer, R. T.(2011) Thermodynamic tables to accompany modern engineering thermodynamics, Elsevier, ISBN 978-0-12-385038-6, Amsterdam

[5] Anaratone ,D. (2010) Engineering heat transfer, Springer, ISBN9783642039324, New York

[6] Goncharov, P. , Artamonov, I. , Khalitov, T.(2014) Engineering Analysis With NX Advanced Simulation. Lulu Publishing Services, ISBN9781483417325

[7] Stephan, P. (2006) VDI-Wärmeatlas, Berechnungsunterlagen für Druckverlust, Wärme- und Stoffübertragung. Springer-Verlag, ISBN 3540255044, Berlin, Heidelberg

[8] Alexandru, T.G., Mantea T.A., Pupaza C., Velicu S, Heat transfer simulation for thermal management of electronic components, Proceedings in Manufacturing Systems, Vol. 11, No. 1 (2016), pp. 15–26

[9] Yovanovich, M. (2005). Four Decades of Research on Thermal Contact, Gap, and Joint Resistance in Microelectronics, IEEE Transactions on components and packaging technologies, vol. 28, NO. 2, June 2005, pp. 182-206, 10.1109/TCAPT.2005.848483

[10] Gerlich, V., Zálešák, M.(2010) Experimental validation of heat transfer model, Annals of DAAAM and Proceedings of the International DAAAM Symposium, 2010, pp. 969-970. ISSN 1726-9679, Zadar

DOI: 10.2507/27th.daaam.proceedings.086

CONTROL SYSTEM AND DESIGN OF THE MOTION MODULE OF A HETEROGENEOUS MODULAR MOBILE ROBOT

Victor Andreev & Valerii Kim

This Publication has to be referred as: Andreev, V[ictor] & Kim, V[alerii] (2016). Control System and Design of the Motion Module of a Heterogeneous Modular Mobile Robot, Proceedings of the 27th DAAAM International Symposium, pp.0586-0594, B. Katalinic (Ed.), Published by DAAAM International, ISBN 978-3-902734-08-2, ISSN 1726-9679, Vienna, Austria
DOI: 10.2507/27th.daaam.proceedings.086

Abstract

We consider the modular principle for the construction of functional units of mobile robots. The main ideas of this concept and the general structure of the heterogeneous robot are presented. Basic functions of the robot are used to determine the minimum set of functional modules. For one of the simple configurations of the modular robot (the service mobile robot), we develop a motion module that is one of the main modules of the mobile robot. We present the design of this motion module and the result of a mathematical model of its control system. The control system combines the tactical level (for the construction of motion sub-targets and intermediate trajectories) and the executive level. The executive level of control system is based on kinematic and dynamic models of the modular robot. Working efficiency of the modular robot's model was tested on different motion tasks.

Keywords: mobile robot; modular robot; heterogeneous modular robot; mechanics of mobile robots; motion along trajectory; path planning; subordinate control loop; control system; servo drive; feedback control loop

1. Introduction

Normally, the design of robotic systems (RS) is accompanied by information about the purpose of the device. Accordingly, all the constructive and program decisions made during the design focus on the achievement of the specific purpose by the final device. Because of this, when the scope of application of the RS is changed, the designer has to revise in a varying degree the device construct ensuring its full compliance with the given purpose. One of the modern approaches to robotic design (and to machinery in general) is to divide the device structure into individual functional components. Each component is responsible for some part of the robot functionality. This unit or module normally has a simpler structure because it is responsible for a single function only. As one of many components of the robot, the module can be easily replaced by a similar or more advanced module. Then, there is no need to redesign the robot: it will suffice merely to replace it by particular functional units. Therefore, the modular robots can provide higher economic efficiency.

2. The modular principle in robotics

The modular principle of mobile robot construction was introduced for the first time in the research of Yurevich E. I. [1]. Also the modular principle of design for industrial robots was implemented in work [2]. This principle became a separate research filed in robotics. One of the early work in this field is the CEBOT [3] robot described by Fukuda and Nakagawa. CEBOT consisted of similar modules–drives and its main purpose was to demonstrate that its configuration can be autonomously modified. Later, the CEBOT-type robots were said to be homogeneous robots, i.e., consisting of similar modules. In this paper, we consider another class of modular robots: heterogeneous modular robots. In heterogeneous robots, each module is responsible for a single function (or a pair of functions); accordingly, the modules of a single robot constructively may be very different. Heterogeneous robots have the advantage that they are more adapted to performing different tasks because the functionality of modules corresponds to the robot purpose and have tools to perform these tasks. However, at this stage, the majority of studies are devoted to homogeneous robots such as M-TRAN, ATRON, and SuperBot [4–6]. In heterogeneous robots, the reconfiguration task takes a back seat, and the versatility, flexibility, and interchangeability of modules becomes important. This fact was well reflected in Thor [7] and SMART [8] robots.

The purpose of this study is to develop a major functional unit of a modular mobile robot (MR) in the framework of the concept of modular system construction. We consider some specific features of this robot as well as what is meant by reconfiguration here.

The project concept is based on the following main principles of modular MR construction [9]:

1. Principle of constructive and functional independence of each module.
2. Principle of universal hardware and software communication interface and interaction between the modules.
3. Principle of distributed control.
4. Network principle of information-measuring and control system.
5. Each module is involved in the network structure with the help of special-purpose software (drivers).

The first principle means that each module should have a design and software that allow it to operate independently of other modules except power supply and most common instructions. *This means that the design and software of a single module do not affect the operation of another module.* Therefore each module of the modular MR is functional complete electronic or mechatronic device.

The second principle requires that the inter-module data communication be strictly standardized and independent of the type of the construction and control system (CS) of modules. This implies the following key feature of the given heterogeneous robot: *the functional reconfigurability*. This means that the inter-module interface should allow modules to execute their *basic functions* as a part of the robot that can be used to solving *different* kinds of tasks under *different* conditions of working environment. The final purpose of the robot determines only the design of modules and, in some cases, their control system, rather than the hardware and software interface of their interaction. For example, a service robot (Fig. 1a) can be converted into a track-mounted robot (Fig. 1b) relatively easily, without a significant change in the design of modules. The configuration of a go-anywhere modular MR (Fig. 1e) requires a change in the design and CS of only the motion module. In the case of transition to the underwater configuration (Fig. 1c), the designs of modules and their control systems should be significantly modified. In all the examples, the data interaction between modules will always be *the same*.

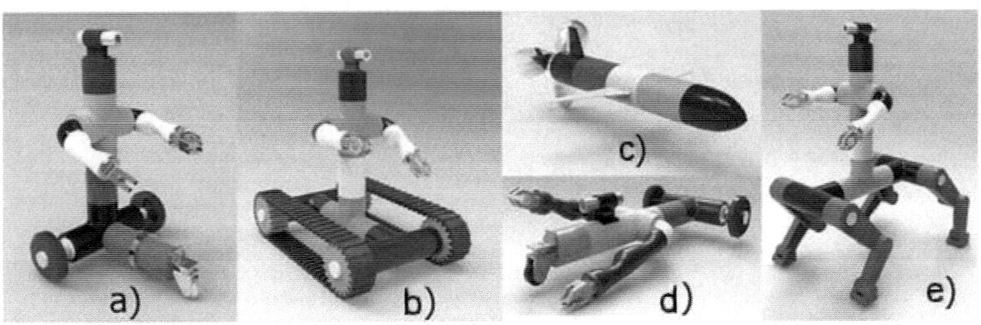

Fig. 1. Different configurations of the heterogeneous modular mobile robot.

The third principle determines the hierarchy of interaction between modules. The control of the operation of modules as a unified system is concerted by a special *supervisory control* system, which includes a *control module* as its core. This module forms objective functions for the remaining modules in the framework of the robot purpose. This is reached by dividing this task into common control instructions and their *distribution* between modules. Accordingly, this control instruction corresponds to the functionality of a given module.

The fourth principle is based on the notion that each module is a unit of the local area network (LAN) [10]. As a result, the modular robot is a network structure with modules exchanging information with each other via standard network protocols (in this case, Ethernet).

The fifth principle, which is based on the network structure information-measuring and control system (IMCS) of the robot, serves as a major hardware and software interface between modules. The "principle of drivers" (its application for distributed control is considered in [11]) enables modules to be automatically involved in the common structure of the modular MR through special software (driver). In the future, not only the modules but also other devices manufactured by third parties can be included in the structure of modular MR.

Main aims of the study:

1. Develop a common structure of heterogeneous modular MR where each module is represented as an independent functional unit.
2. Determine the functionality of each module of the modular MR.
3. Develop the design and control system of the motion module as functional complete mechatronic device on the basis of the concept of modular MR and information about the robot purpose.

Based on the above-mentioned principles of the construction of heterogeneous modular MR, we have developed a common structure of the modular robot system (Fig. 2).

In this structure, each module corresponds to a certain function. The network control module (NCM) is the main control module responsible for *supervisory* control and *distribution* of common instructions between modules. The NCM interacts with each of the modules via the network hardware and software interface. Figure 2 shows that this system has four interfaces. As noted earlier, the design and CS of each module are determined by the robot purpose; however, the NCM structure is least dependent of this parameter.

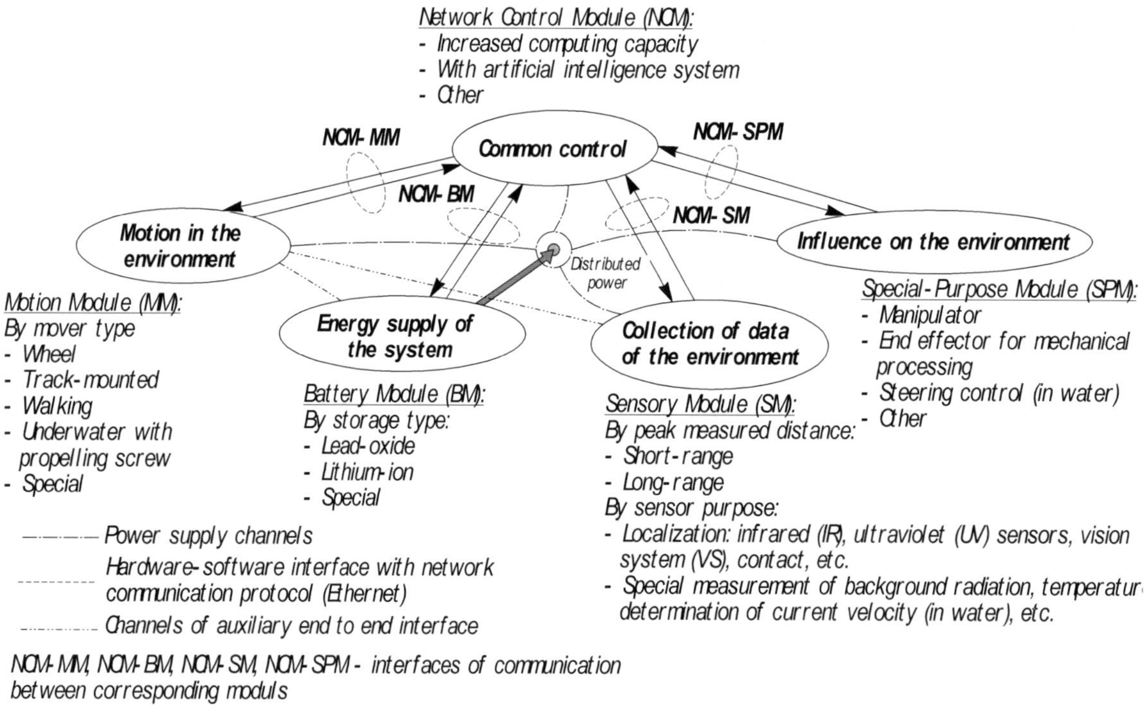

Fig. 2. General structure of heterogeneous modular mobile robot.

In this project, the service modular MR is taken as the original configuration (see Fig. 1a) since this is the simplest assembly of the above-mentioned variants that is not subjected to special requirements.

Like other configurations, the service modular MR (according to the structure shown in Fig. 2) consists of the following modules (Fig. 3): 1 – Motion Module (MM), 2 – Battery Module (BM), 3 – Network Control Module (NCM), 4 – Special-Purpose Module (SPM), 5 – Sensory Module (SM). For this variant of the MR, the SPM is a module with manipulators.

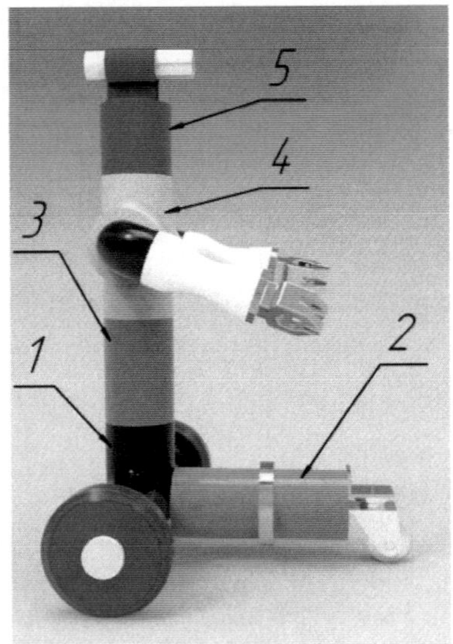

Fig. 3. Service modular mobile robot

3. Design of the motion module

The motion module (MM) is one of the main robot modules responsible for the function of motion in the environment (see Fig. 2). Since the service modular MR operates predominantly in indoor areas, the motion module should provide movement on flat surfaces.

The control system of the motion module combines both tactical and executive control levels. The tactical control level directly interacts with the NCM via the NCM-MM interface and receives from it necessary data to execute motion instructions: information about the coordinates and orientation of the robot in initial and final positions, data on the local map (if available), etc. In turn, the NCM receives actual data on the course of the instruction execution along the MM feedback chain.

The tactical level solves the problem of path planning and tracking along the trajectory. In this study, the tactical control level is implemented by the algorithm described in [12]. A similar approach is used for the mobile robot control [13]. Based on these studies, we developed a mathematical model of the CS of the tactical level using *Simulink* (MATLAB) to calculate the angular velocities of MM wheels aimed at its passing along a given trajectory.

The executive level of the CS consists of kinematic and dynamic models of the modular robot. Analysis of the motion module's control system is made along with considering battery module's dynamic parameters, because the battery module is presented in all configurations of the modular robot. The basis of the dynamic model are two servo drives with DC motors. Each servo drive is close looped with angular speed of the wheel of the motion module. Figure 4 shows the block diagram of the servo drive.

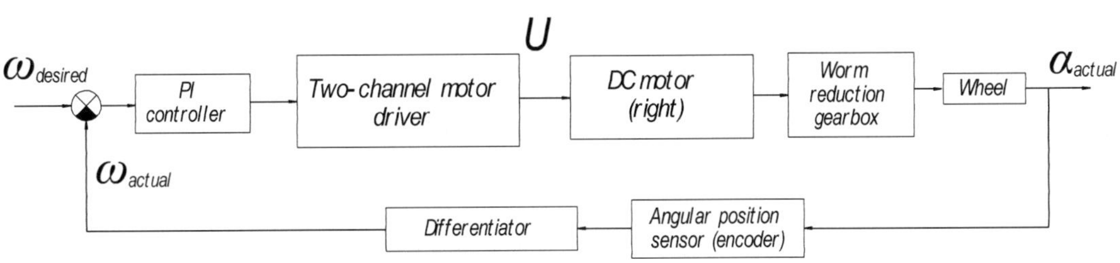

Fig. 4. Structural diagram of the servo drive

The servo drive's controller consists of the PI speed controller with the following transfer function:

$$W_{PI}(s) = K_{PC}\frac{1 + T_{PI}s}{T_{PI}s} \tag{1}$$

where T_{PI} – time constant of the PI speed controller, K_{pc} – PI controller's gain coefficient.

The PI controller is configured on a technical optimum [14]. In accord with this adjustment PI controller's time constant is equal to the time constant of the most inertial aperiodic link. Transfer function of an invariant part of the DC motor has aperiodic link with the electromechanical time constant T_M. It's well known that T_M is much greater than electromagnetic time constant T_e and motor driver's time constant T_{PWM}. Then PI controller's time constant is given by:

$$T_{PI} = T_M = \frac{J_\Sigma R_a}{k_m k_e}, \tag{2}$$

where J_Σ - summary reduced mass moment of inertia, $kg \cdot m^2$, R_a – armature's winding's resistance, Ohm, k_m – torque factor of the DC motor, $N \cdot m/A$, k_e - coefficient of EMF of the DC motor, $V \cdot s/rad$. Summary mass moment of inertia is given by (one wheel of considered drive rotates, another is fixed):

$$J_\Sigma = \frac{J_w}{i^2} + J_m + M \cdot r^2 + J_z \frac{r^2}{l^2}, \tag{3}$$

where J_w – mass moment of inertia of the wheel relatively of it's rotation axis, $kg \cdot m^2$, i – reduction ratio, J_m – mass moment of a rotor's inertia, $kg \cdot m^2$, M – mass of the motion module and the battery module, kg, r – wheel's radius, m, J_z – mass moment of modular robot's inertia relatively of it's vertical rotation axis, $kg \cdot m^2$, l – distance between wheels, m. Then PI controller time constant is:

$$T_{PI} = \frac{R_a}{k_m k_e}\left(\frac{J_w}{i^2} + J_m + M \cdot r^2 + J_z \frac{r^2}{l^2}\right). \tag{4}$$

PI controller's gain coefficient K_{pc} proportional to Q factor of the speed control loop and estimated with consideration of the mass moment of inerta J_Σ.

Consideration of dynamic properties of the modular robot during evaluation of the PI controller's parameters allow to achieve acceptable quality of the servo drive control. Figure 5a shows transient speed process in servo drive and figure 5b shows transient process with backlash in reduction gearbox (about 3°).

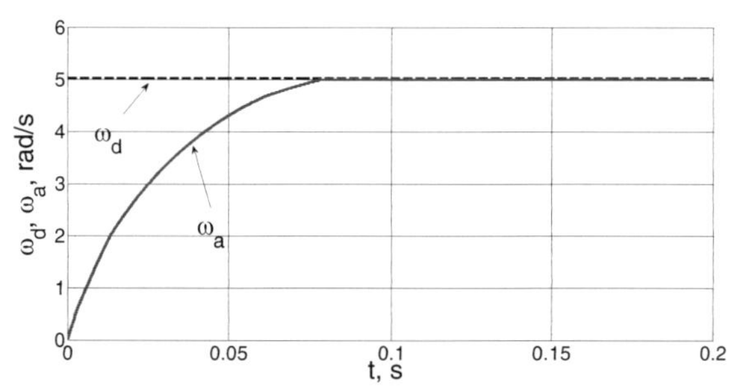

Fig. 5a. Transient speed process

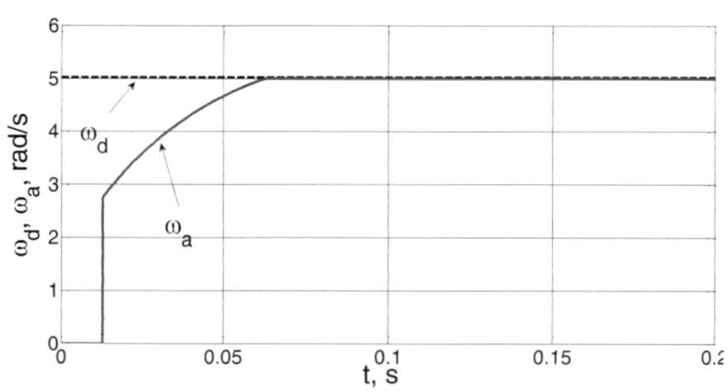

Fig. 5b. Transient speed process with gearbox' backlash

Analysis of graphs shows that overshoot is equal to zero, transient period is less than $0.07\ s$ and stationary error is less than 0.4%. Backlash in reduction gearbox causes speed's jump after short time delay. This can cause negative influence on motor driver and DC motor. It's necessary to use special correction algorithm for smooth speed control.

The computer model of the modular robot was developed. This model includes only *executive* level of the motion module's control system and sensor module. The sensor module is equipped with three laser rangefinders: one measures distance ahead of the robot, two others measure distance at rear sides of the robot. Figure 6 shows this model's structure developed in Simulink:

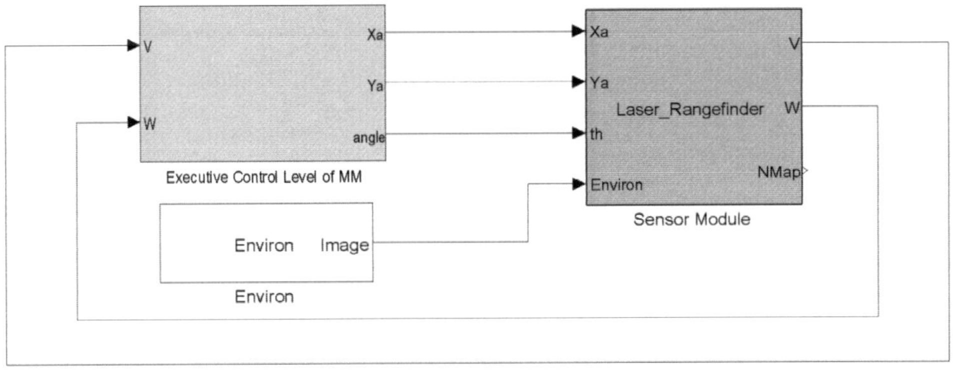

Fig. 6. Structure of modular robot's computer model

This modular robot's model is used for demonstration of the motion module capabilities to perform motion tasks in environment' model.

The executive control level of the motion module takes linear V and angular W velocities as input and actual position (coordinates Xa and Y_a) with orientation (angle) as output. The actual position and orientation of the robot is determined by the sensor module's SLAM algorithm (not considered in this work). The sensor module takes position and orientation of the modular robot and environment's model as input. Environment's model is represented here as image. Since in this model the executive control level of motion module can take as input only linear and angular velocities, the sensor module must executes its algorithms and gives V and W as output for feedback loop.

The modular robot's model was used for task of collision avoidance in room's model. For this task only one laser rangefinder in front of the modular robot was used in the sensor model. If the laser rangefinder detects an obstacle ahead of the robot on defined minimal distance then robot turn to left with angular speed W. Figure 7 shows the trajectory of the robot in the room with obstacles.

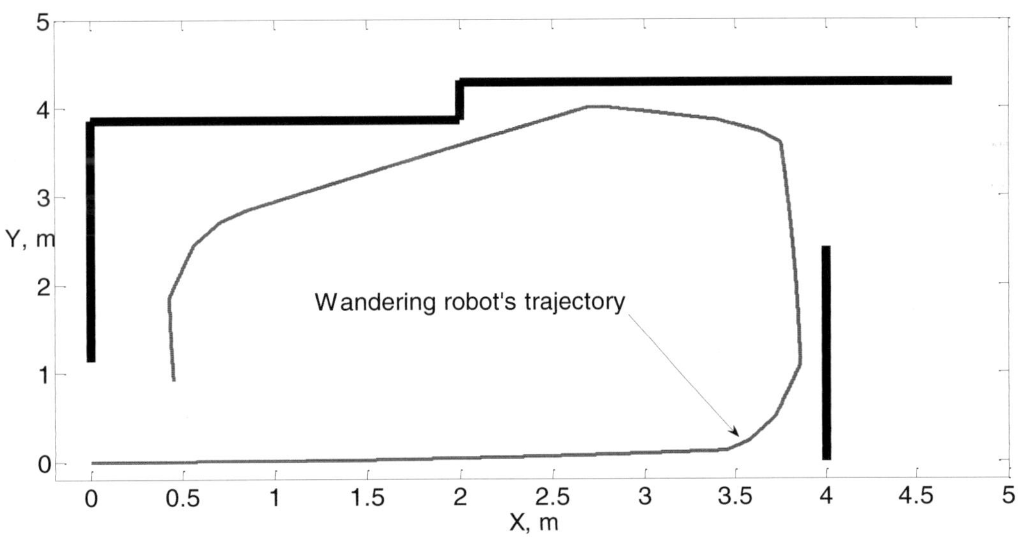

Fig. 7. The robot's trajectory in room with obstacles

From fig. 7 it's clear that the robot avoids obstacles located in front of the robot, but without laser rangefinders located at rear sides of the robot a collision with obstacles from right or left side may occur. This model experiment in only needed to demonstrate that the execution control level of the motion module can perform basic robot's movements.

The task of going through a labyrinth was considered. To perform this task the sensor module's computer model must have three laser rangefinders. Two laser rangefinders at rear sides of the modular robot directly used for the sensor module's algorithm. This algorithm is based on difference of distances between robot and side walls in labyrinth. If a distance to the right wall is greater than distance to the left wall then robot must turn on the right with angular speed W and vice versa. Figure 8 shows the robot's trajectory in the labyrinth.

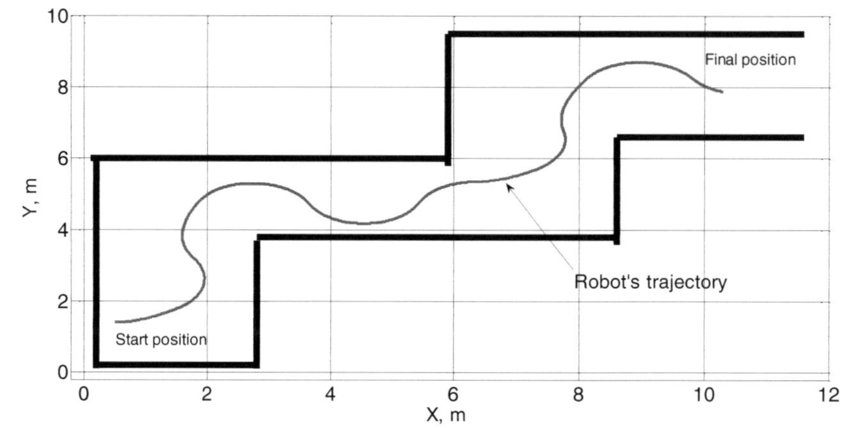

Fig. 8. Robot's trajectory in the labyrinth

From fig. 8 it is clear to see that the robot goes through labyrinth without collisions with walls. But the trajectory isn't optimal because the modular robot doesn't move straight where it would be more effective. This happens due to control algorithm that works without smooth correction of linear and angular velocities. Figure 9 shows graphs of laser rangefinder's measurements that were made during robot's motion through labyrinth.

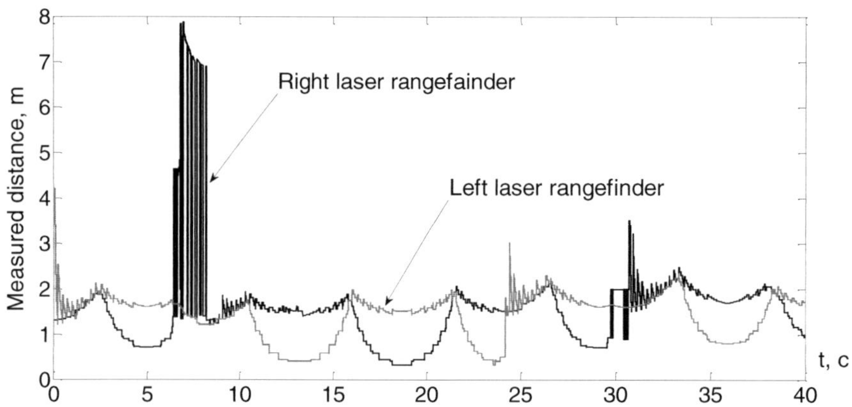

Fig. 9. Laser rangefinder's measurements

The jump in measured distance made by right laser rangefinder is due to detection one of the distant side walls in beginning of the motion.

As said before the tactical control level of motion module is implemented by the algorithm described in [12]. The computer model that was considered above was complemented by tactical level control block. Figure 10 shows tactical and executive control levels and sensor module.

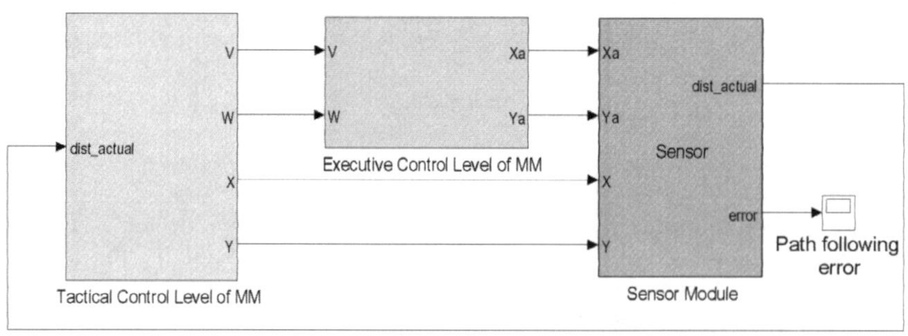

Fig. 10. Tactical and executive control levels of motion module with the sensor module

Tactical control level evaluates intermediate aim positions and creates desirable trajectory of the modular robot. It takes actual distance of traversed path from the sensor module as input and desirable velocities (V and W) and robot's position as output. Executive control level of motion module takes desirable velocities as input and actual position as output. The sensor module takes actual and desirable positions as input and evaluates path following error and actual distance of traversed path. In this model we assume that the tactical control level already took coordinates of final position and desirable robot's orientation in this final position as input and created desirable trajectory. The modular mobile robot must move along a given trajectory with a least deviation from it. The modular robot can move from point to point along trajectory due to position control loop that organized by the sensor module. Figure 11 shows robot's actual and desirable mechanical trajectories after model simulation.

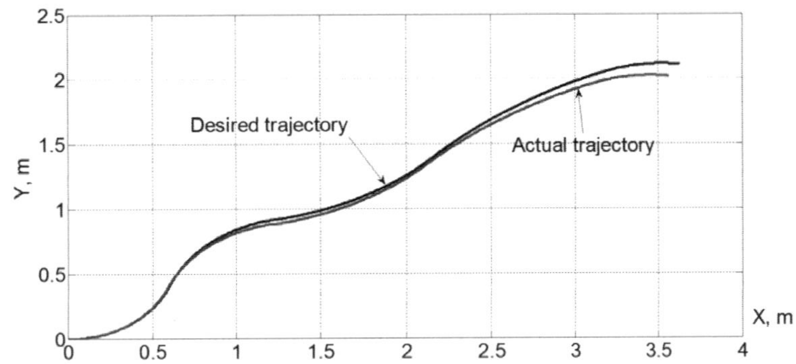

Fig. 11. Desired and actual modular robot's trajectories

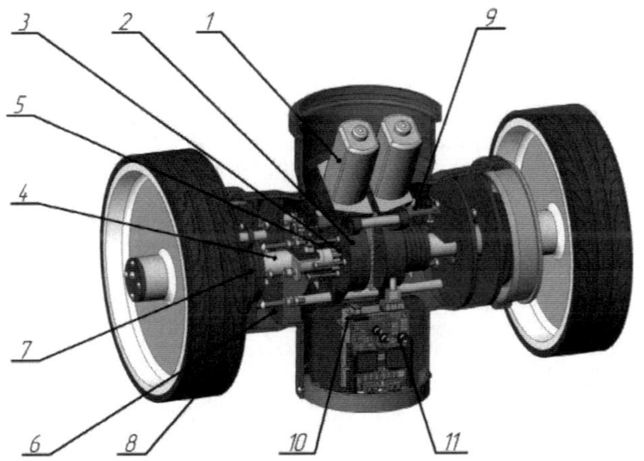

Fig. 12. Motion module.

Fig. 13. Wheel–module.

The plot shows that the deviation of the actual motion from the desired one is sufficiently small (up to 35 mm). We got similar results as in [12] but with greater path following error. The main source of a path following error is the inertia of the modular mobile robot and nonlinear elements in the dynamic robot's model. Development of correction algorithms for robot's position control loop is the objective for further research.Figure 12 shows the isometric section of the MM.

Let us consider some specific features of the MM design. The device consists of individual parts allowing us to relatively easily assemble the module (Fig. 12). The MM design contains two components connected to the drives and an electronic control unit. The MM drive consists of electric DC motor (1) and worm reduction gearbox (2). Gear (3) is connected to shaft (4) at the gearbox output through multilayer clutch (5). The shaft, clutch, frame unit (6), multilayer bearing sleeve (7), wheel (8), and encoder (9) constitute a single unit - *wheel-submodule* (Fig. 13.). The electronic control unit includes microcontroller-based hardware and software platform (10) and electric motor driver (11). Similar design decisions (usage of submodules for example) can be found in the modular robot Thor [7].

The described modular design of the mobile robot can be successfully used in Bionic Assembly System that presented in [15, 16].

4. Conclusions

The main problem of this research was to create the functional complete motion module of the modular mobile robot. This problem was solved by designing special structure of the modular robot's system. At first stage we distributed basic functions of the robot between corresponding modules. Basic function of each module determines its hardware and software components. Principles of the modular mobile robot construction were considered during development of the motion module. Then control system of the module was designed: it's consists of both tactical and execution control levels. The control system receives data with simple structure (desirable position and orientation) and all complex calculations are performed by the motion module itself. At the same time the network control module must provide only actual data about desirable goal position and its software's structure is no matter for motion module. Working capacity of the control system's module was verified by computer modeling (the task of going through a labyrinth). In this model it's clear that interaction between the sensor module and the motion module is also realized by exchange of data with simple structure and with low dimension. In this paper we describe that the motion module's construction is constructively independent. This property is achieved by the simple mechanical interface and special construction arrangement of the module. Analysis of the control system and construction of the motion module allow concluding that this module is functional complete mechatronic device.

Development of the network control module and the sensor module, realization of natural experiments are objectives of further research. Design of the special-purpose module (a module with manipulators) is also objective of further research.

5. Acknowledgments

This work was supported by the Russian Foundation for Basic Research, project no. 16-07-00811a

6. References

[1]. A.V.Lopota, E.I.Yurevich, Stages and Developments of Robotic Systems Design Modular Principle, In St. Petersburg State Polytechnical University Journal, 1(164), St. Petersburg (Russia), 2013. – p.98-103.

[2]. Vorobiev, E.; Kozirev, Yu & Tsarenko, V. (1988). Promishlennie roboti agregatno-modulnogo tipa (Industrial robots of unit-modular type) / Edited by E. P. Popov, Mashinostroenie, ISBN 5-217-00166-6, Moscow (Russia).

[3]. Fukuda T., S. Nakagawa, A dynamically reconfigurable robotic system (concept of a system and optimal configurations), In Proceedings of IECON '87: 1987 International Conference on Industrial Electronics, Control, and Instrumentation, 1987.

[4]. S. Murata, E. Yoshida, A. Kamimura, H. Kurokawa, K. Tomita, and S. Kokaji, M-TRAN: self-reconfigurable modular robotic system, IEEE/ASME Transactions on Mechatronics, 7(4):432–441, 2002.

[5]. E.H. Østergaard, K. Kassow, R. Beck, and H.H. Lund, Autonomous Robots, Design of the ATRON lattice-based self-reconfigurable robot, 21(2):165–183, 2006.

[6]. Salemi B., M. Moll, and W.-M. Shen, SUPERBOT: A deployable, multi-functional, and modular self-reconfigurable robotic system, In Proceedings, IEEE/RSJ Intl. Conf. on Intelligent Robots and Systems, pages 3636–3641, Beijing, China, 2006.

[7]. Lyder A.H., Stoy K., Mendoza-Garcia R.-F., Larsen J.C., and Hermansen P., On sub-modularization and morphological heterogeneity in modular robotics, In Sukhan Lee, Hyungsuck Cho, Kwang-Joon Yoon, and Jangmyung Lee (Eds.), Intelligent Autonomous Systems 12, volume 193 of Advances in Intelligent Systems and Computing, pp. 649-661. Springer Berlin Heidelberg, 2013.

[8]. J. Baca, M. Ferre, and R. Aracil, A heterogeneous modular robotic design for fast response to a diversity of tasks, Robotics and Autonomous Systems, vol. 60, no. 4, pp. 522–531, 2012.

[9]. Andreev, V[ictor], & Poduraev, Y[ury] (2016). Network–based design of heterogeneous modular mobile robotic systems, Proceedings of the 27th DAAAM International Symposium, pp.xxxx-xxxx, B. Katalinic (Ed.), Published by DAAAM International, ISBN 978-3-902734-08-2, ISSN 1726-9679, Vienna, Austria DOI: 10.2507/27th.daaam.proceedings.xxx.

[10]. Pryanichnikov V.E., Andreev V.P., The Application of Network Technologies to Constructing Group Controlled Systems with Machine Vision for Mobile Robots, In Annals of DAAAM for 2012& Proceedings of the 23th international DAAAM Symposium "Intelligent Manufacturing & Automation" 24-27th October 2012 Zadar, Croatia, ISSN 2304-1382, 2012. – V.23, No.1. – pp.1167 – 1174.

[11]. Andreev V.P., Kirsanov K.B., Poduraev Yu. V., Geographically distributed "multi-operator" control of robotized systems using network technologies, In Proceedings of the International Scientific and Technical Conference "Extreme Robotics". – St.Petersburg: "Politechnika-service", 2015. – P.67-71.

[12]. Gulati S., Kuipers B., High performance control for graceful motion of an intelligent wheelchair, In Proc. 2008 IEEE Int. Conf. on Robotics and Automation (ICRA, May 2008), pp. 3932–3938.

[13]. Davydov O.I., Platonov A.K., A control algorithm for a differential drive of RB-2 mobile robot, Preprint Keldysh Institute of Applied Mathematics, RAS. Moscow. 2015. No. 25. - pp.4 - 16.

[14]. Kluchev V. (2001) Teoriya electroprivoda (Theory of electric servo drive), Energoatomizdat, ISBN 5-283-00642-5, Moscow (Russia).

[15]. Katalinic B., Kukushkin I., Pryanichnikov V., Haskovic D. Cloud Communication Concept for Bionic Assembly System. Procedia Engineering, 1877-7058, 69 (2014) (2014), pp. 1562–1568 doi:10.1016/j.proeng.2014.03.156.

[16]. Katalinic B., Kukushkin I., & Haskovic D. Bionic Assembly System Cloud: Functions, Information Flow and Behavior. In 9th International Conference of DAAAM Baltic, Industrial Engineering, ISBN 978-9949-23-620-6, ISSN 2346-6138. pp. 103–108, ed. T.Otto, Tallinn, Estonia, 2014.

DOI: 10.2507/27th.daaam.proceedings.087

DEVELOPING PRINCIPLES OF BUILDING THE SIMULATOR FOR ENERGY WORKER AND TECHNOLOGY FOR COAL MINE ELECTRICAL SERVICE STAFF TRAINING

Parshina Galina, Feshin Boris & Breido Josif

This Publication has to be referred as: Parshina, G[alina]; Feshin, B[oris] & Breido, J[osif] (2016). Developing Principles of Building the Simulator for Energy Worker and Technology for Coal Mine Electrical Service Staff Training, Proceedings of the 27th DAAAM International Symposium, pp.0595-0599, B. Katalinic (Ed.), Published by DAAAM International, ISBN 978-3-902734-08-2, ISSN 1726-9679, Vienna, Austria
DOI: 10.2507/27th.daaam.proceedings.087

Abstract

The article discusses the specifics of developing a virtual computer simulator of energy worker and a technology for training personnel of electrical services of coal mines. The paper describes the scope and main principles of simulator systems. The proposed energy worker's simulator model is described as a dual system: "Knowledge Base subsystem + Knowledge Control subsystem". In the work, we propose to build a training center for energy workers on the basis of the simulator, automated calculation of diagrams for calculating production units' power supply, as well as instruction and knowledge assessment system. The article presents a functional diagram of a training center equipped with a virtual simulator and considers two approaches in distributing software between the server and training center clients analyzing the advantages of each approach. Moreover, the work describes the technology for training coal mine electrical services employees and proposes an algorithm for using the simulator in training and controlling knowledge of this staff focusing on the advantages of using the simulator.

Keywords: simulator; power supply system; coal mine; electrical service, automated.

1. Introduction

Due to IT development, simulators for acquiring professional skills by technological production staff had become widespread over the last few years, especially for production presenting threats to life and health. Virtual simulators are often used as simulators of abnormal situations and emergency. This is justified technically and financially during the operation of technological facilities in highly explosive and fire hazardous industries [1-3].

One of those industries that are dangerous to life and health is production units of coal mines [4, 5]. Quality work which is achieved by many factors, among which not the last place is occupied by such as providing of reliability, security, functionality, necessary and acceptable performance of the mining electromechanical equipment. Calculations of power supply systems of electrical facilities of a mine, tunneling areas and producing units are carried out by electrical service personnel and approved by a chief power engineer of a coal mine. The consequences of errors in such calculations associated with insufficient qualification of the personnel can be disastrous.The works [6, 7] proved that the efficiency and reliability of mining electromechanical equipment and safety of relevant staff greatly depend on the quality of calculation and design of diagrams of production units' power supply.

The article describes the principles of building the simulator and technology for training coal mine electrical services employees to obtain practical skills of calculation, design and operation of power supply systems of production units of coal mines.

2. Basic principles of building the simulator

The proposed simulator model is based on the three principles:

1. The basis for the creating the simulator is a simulator model. In turn, a simulator model is built on the basis of a simulated model of the production process (SM PP). SM PP virtually displays the operation of a technological object in predetermined industrial conditions (including normal, emergency, catastrophic). The present work considers an automated system for calculating power supply of coal mine production unit (ASCPS) as a simulator model [6]. Many industrial conditions are displayed as a multitude of ASCPS conditions which ensures the identity of real production process and simulator models for specific analyzed conditions of a production unit and available electromechanical equipment.
2. Virtual operator interface of a simulator reproduces real production networks of a technological object for specific industrial conditions. When calculating and designing power supply of production units, sequence of automated calculation and choice of instruments is taken similar to the algorithm used in manual calculation and construction (design) of power supply diagrams by a chief power engineers of coal mine.
3. The simulator is a tool of transferring knowledge from the experienced professional to apprentice. Experienced professionals may be the expert group with extensive experience in developing power supply diagrams [8]. Knowledge of the expert group are used to generate lists and content of training tasks, standard problem solutions both for the learning system and knowledge control system of the simulator for a variety of work situations and at different stages of training: beginner, intermediate and final assessment [6].

3. Model of a training center with the simulator

The model and computer simulator of energy worker itself is built on the basis of the subsystem of calculating power supply of workstation of electrical service employee. This ensures the identity of the simulator and the workplace of electrical worker.

The simulator, automated system for calculating power supply of coal mine production unit and training and knowledge control system (TKCS) may be on the same electronic computing machine at the training center which is organized in the form of training places in a format "client-server" connected in a common local/global cloud network.

The main link of the training center is server. The server has the database and simulator-related software installed. Software distribution between server and clients depends on the adopted approach:

1. at thick client technology, minimum software is installed on the server, and the simulator itself and all its accompanying software is located on workstations;
2. at thin client technology, switching software functions in the client, while the simulator and accompanying software is located on the server.

Choice of the first or second approach is determined as a function of bandwidth. At low bandwidth capacity, thick client approach is used with minimizing network traffic. At high bandwidth capacity, thin client approach is used which allows to unload a user's computer and simplify software administration process. Figure 1 is a functional diagram of a training center with a virtual simulator of energy worker.

4. Methods of training

Let us present simulator of energy worker as a dual system "Knowledge Base subsystem + Knowledge Control subsystem" [9, 10]. Apparent effectiveness of the simulator will be achieved if certain standard tested and/or trained employees score the necessary amount of points that characterize positive solution of control tasks over a fixed time and for the entire multitude of industrial mine conditions [8, 11-14]. Suppose that experts who develop training tasks and tests can heuristically evaluate the quality of the simulator work. Evaluation of employee's competence is conducted at all stages of training and control: beginner, intermediate, and final.

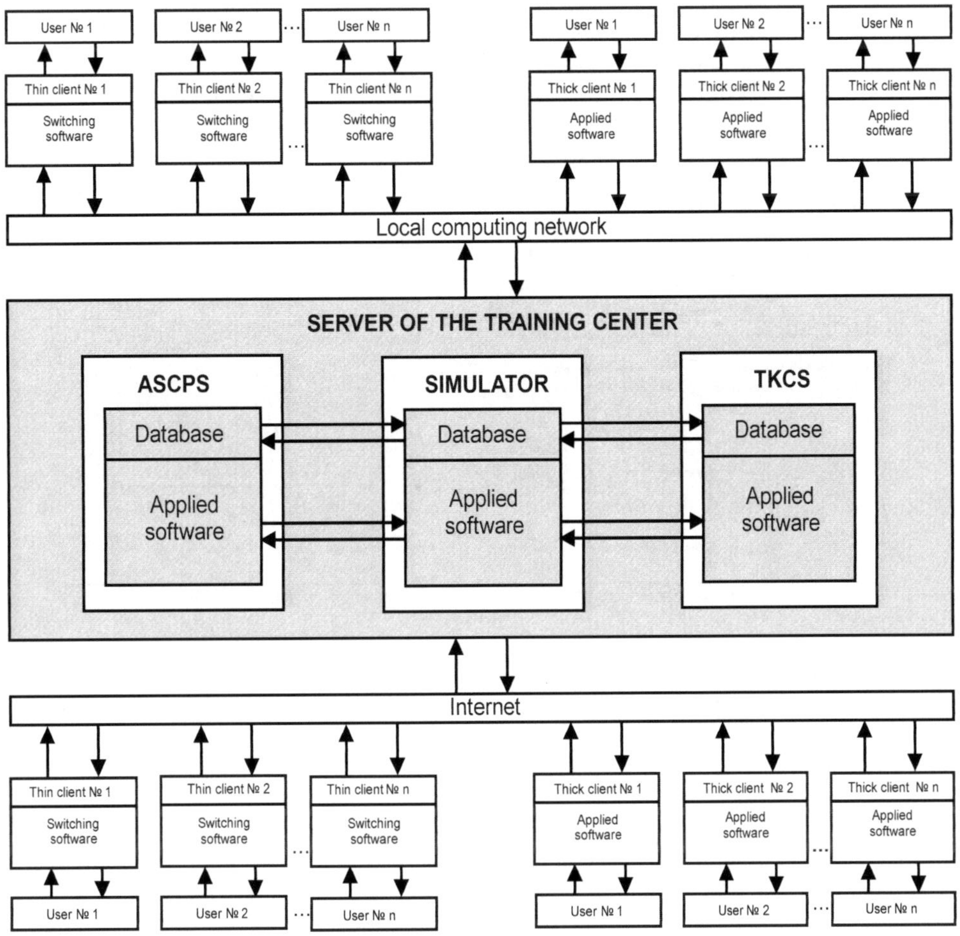

Fig.1. Functional diagram of a training centre with a virtual simulator of energy worker

The process of training and knowledge control by simulator of energy worker is a phased implementation of strict algorithmic operations by a person in split and real-time [6, 7, 11-14] format, including:

1. for evaluating initial knowledge of the tested: generation of the initial data; calculation and design of production unit power supply diagram by the employee through the tools of automated workstation;
2. assessment of the calculation results according to step 1. In case of a negative result, go to step 3. In case of a positive result, go to step 5;
3. training through the simulator tools;
4. assessment of learning outcomes. In case of a negative result, go to step 3. In case of a positive result, go to step 5;
5. training in a variety of control tasks;
6. assessment of training by step 5. In the case of a negative result, go to step 3. In case of a positive result, go to step 7;
7. multiple training of the employee;
8. assessment of learning outcomes by the technology of multiple training according to step 7. In case of a positive result, go to step 10. In case of negative result, go to step 9;
9. administrative and psychological assessment of knowledge and capabilities of the employee, recommendations for re-training and/or change of the working place of the employee.
10. issuance of a certificate and work admission of the mine electrical service employee.

In the training and knowledge control system, the simulator is used at all stages, learning, training and knowledge control.

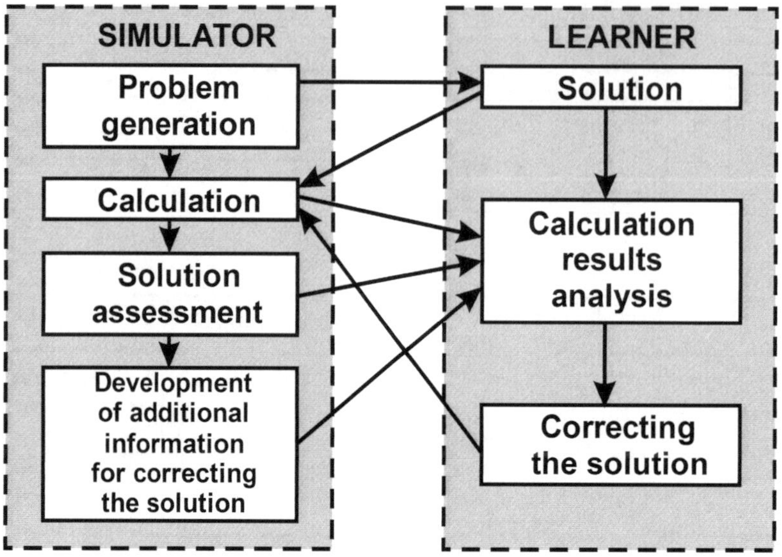

Fig. 2. Diagram of operational scenario of energy worker's simulator

In learning mode, the simulator has a function that allows a learner to compare his actions with the right ones while the training tasks are supplemented with explanation of standard solutions and provide links to sections of the training system on specific issues of decision-making. The algorithm of the simulator work in the learning mode includes the following steps:

1. generation of the initial data of the problem;
2. solution of the problem;
3. comparison of the problem solution with standard answers and assessment of the solution by a 100-point scale;
4. in the case of non-complete compliance of the problem solution results with standard answers, a set of additional data is offered; it is available to form a new control task and then go to step 2. Upon acceptable compliance of the problem solution with the standard answers, transition to step 5 is necessary;
5. end of the simulator work.

5. Conclusion

The offered simulator model built on the basis of the basic principles of creating such systems. Simulator model is described as a dual system: "Knowledge Base subsystem + Knowledge Control subsystem". Functional diagram of a training center is based on two main approaches in distributing software between the server and training center. Technology training and control of knowledge includes the algorithm to use the simulator at all stages of preparation and control.

Creation and implementation of a system consisting of subsystems of training and knowledge control, simulator, and automated calculation and design of power supply systems of coal mine production units provides for the following split and real-time opportunities:

- getting an estimate, a rating of coal mine electrical services personnel for the admission to operation of electrical equipment;
- obtaining the predictive assessment of a failure-free level of electrotechnical facilities of a mine production unit;
- increasing the rating of electrical services personnel by means of targeted training and knowledge control;
- automating the calculation of power supply of production units of coal mines;
- reducing the complexity of calculating the mine power supply diagrams;
- improving the calculation efficiency and quality by automating basic procedures;
- reducing the probability of errors during the calculation of power supply diagrams;
- reducing the risk contributed by individuals when making design decisions concerning power supply systems of coal mines' production units.

6. References

[1] Victor Hreniuc, Ghiorghe Batrinca, A Pleading for Ship Manned Models as a "Physical" Simulator in the Ship Handling Training Process. Procedia Engineering, Volume 69, 2014, pp. 1410-1419, ISSN 1877-7058 doi:10.1016/j.proeng.2014.03.136, Editor B. Katalinic, Published by DAAAM International, Vienna, Austria, 2014.

[2] G.Ya. Pyatibratov, O.A. Kravchenko, A.M. Kivo, Design Principles and Implementation of Advanced Simulators for Training Astronauts to Work in Zero or Low Gravity Conditions. Procedia Engineering, Volume 150, 2016, pp. 1410-1414, ISSN 1877-7058, http://dx.doi.org/10.1016/j.proeng.2016.07.337, Editor B. Katalinic, Published by DAAAM International, Vienna, Austria, 2016.

[3] Przemysław Mądrzycki, Dariusz Karczmarz, Piotr Golański, Usage of Diagnostic Simulator for Ground Engineering Crew Training. Procedia Computer Science, Volume 22, 2013, pp. 395-400, ISSN 1877-0509, doi:10.1016/j.procs.2013.09.117.

[4] Rules for industrial safety in hazardous industrial facilities of coal mines. Approved by order of the Minister for Investment and Development of the Republic of Kazakhstan of December 30, 2014. No. 351. – 225 pp.

[5] Safety code for operation of consumer electric installations. Approved by the RK Government Resolution of 24 October, 2012. No. 1353. Astana. – 97 pp.

[6] Parshina G.I. & Feshin B.N., The automated system of calculation of power supply systems of stope face of coal mines. Proceedings of the International Scientific and Practical Conference "Informatization of Engineering Education", ISBN 978-5-383-00939-0, Moscow, Publishing House MEI, April 12-13, 2016. pp. 390-393.

[7] Parshina G.I. & Feshin B.N., Developing the Structure of a Database for Automated Calculation of Coal Mines Producing Units' Power Supply. University Proceedings, ISSN 1609-1825, Karaganda, KSTU publisher, Issue No. 4(61), 2015, pp. 76-81.

[8] Feshin B., Parshin G., System of Providing of Effective Using of Electrotechnical Complexes of Mining Department of Mines. 22nd World Mining Congress & EXPO 11-16 September Istanbul-2011, volume-III, 2011. ISBN: 978-605-01-0146-1 (Tk.), 978-605-01-0149-2 (III. Vol.), pp. 441-448. Editor Dr. Şinasi Eskikaya, Published by Aydoğdu Ofset, Ankara-Turkey, İSTANBUL, 2011

[9] Ulrika Hurt, Tauno Otto, Kati Kõrbe Kaare, Ott Koppel, New Approach to Knowledge Transfer Environment Development. Procedia Engineering Volume 69, 2014, pp. 273–281, ISSN 1877-7058 doi:10.1016/j.proeng.2014.02.232, Editor B. Katalinic, Published by DAAAM International, Vienna, Austria, 2014.

[10] Rakic Kresimir, Bandic Glavas Marijana, Majstorovic Vlado, Development of the Intelligent System for the Use of University Information System. Procedia Engineering, Volume 69, 2014, pp. 402-409, ISSN 1877-7058, doi:10.1016/j.proeng.2014.03.005, Editor B. Katalinic, Published by DAAAM International, Vienna, Austria, 2014.

[11] Parshina G.I. & Feshin B.N. Expert Knowledge Quality Evaluation Systems of the Mining Electrotechnical Units' Personnel. International workshop "System Analysis, Management and Information Processing" (27th – 30th September, 2010) — Rostov-on-Don: DGTU publisher, 2010.

[12] Parshina G.I., Automated Systems of Production Units' Power Supply Calculation as a Basis of Coal Mines' Operation Safety. University Proceedings, ISSN 1609-1825, Karaganda, KSTU publisher, Issue No. 3, 2010, pp. 89-93.

[13] G.I. Parshina, High Performance System for the Operation of the Electrotecnical Units of the Coal Mines' Production Departments. Automation and Informatics, ISSN 1560-7305, Karaganda, KSTU publisher, Issue No. 1-2 (26- 27), 2010, pp. 32-35.

[14] Parshina G.I. Development of Automated System of Training and Control Knowledge Staff Electrical Services. Proceedings of the International Scientific and Practical Conference "Informatization of Engineering Education", ISBN 978-5-7046-1535-4, Moscow, Publishing House MEI, April 15-16, 2014. pp. 451-454.

DOI: 10.2507/27th.daaam.proceedings.088

DETERMINING HEATING CAPACITY AND COEFFICIENT OF ENERGY TRANSFORMATION OF HYDRODYNAMIC SET

Breido Josif, Kalinin Alexey & Kuchin Valery

This Publication has to be referred as: Breido, J[osif]; Kalinin, A[lexey] & Kuchin, V[alery] (2016). Determining Heating Capacity and Coefficient of Energy Transformation of Hydrodynamic Set, Proceedings of the 27th DAAAM International Symposium, pp.0600-0606, B. Katalinic (Ed.), Published by DAAAM International, ISBN 978-3-902734-08-2, ISSN 1726-9679, Vienna, Austria
DOI: 10.2507/27th.daaam.proceedings.088

Abstract

In the work the developed constructive and technological diagram of the bench based on the frequency controlled electric drive of the pump set and SCADA system for experimental defining power characteristics of the hydrodynamic heaters (HDH) of liquid environments is described.
Based on methods of planning a multiple-factor experiment studies have been carried pit in various HDH operating modes. Preliminary theoretical calculations of the parameters of the HDH hydraulic operating mode have been carried out in the environment of the information-graphic system, at the set of factors similar to that of the pilot studies.
The mathematical dependences establishing interrelations of heating capacity and coefficient of energy transformation with the HDH constructive and operational parameters based on the pipe type reactor have been found.
The results of the studies and tests have been used when developing specifications for designing and producing the HDH prototypes.

Keywords: hydrodynamic heater, bench, model, studies, tests, methods of planning a multifactor experiment

1. Introduction

One of the trends of increasing the efficiency of power systems is obtaining thermal energy from internal energy of liquid environments by means of activation with external sources (for example, an electric drive). This trend is connected with designing, studying and producing hydrodynamic heaters (HDH). Their internal energy is taken from the liquid environment (the so-called "working liquid") in the course of activation by means of electromechanical impact.

In 1993 in the USSR and the USA the inventions of Yu.S. Potapov [1] and James Griggs [2] were patented which could change the technological role of water in power technologies. Water can be used not only as an effective energy carrier but also as an available and cheap energy resource.

The analysis of these experimental tests and technical characteristics of HDH permits to find a set of unresolved problems which slow down a wide use of vortex power effects in power technologies and sets. These problems are caused by the absence of high-quality and reliable bench studies and certification tests of the finished samples of hydrodynamic converters of energy and theoretical generalizations accompanying them. There is no design methodology without which designing and producing a number of standard sizes HDH is impossible [3-6]. At present the processes of transforming electric energy into mechanical energy with the subsequent transforming mechanical one into thermal one are not studied completely.

Today the development of methods of HDH designing and determining its optimum parameters providing the maximum heating capacity and coefficient of transformation of energy is topical.

2. Test bench for studying HDH power characteristics

The authors have developed a HDH constructive and technological diagram. The test bench for studying the HDH power characteristics has been developed [7]. The bench contains an automated system of scientific studies for detecting the regularity of transforming electric energy into mechanical one with the subsequent transforming it to thermal energy [8]. The task has been set: on the basis of multifactor experiment to find the HDH optimum parameters providing the maximum heating capacity and coefficient of energy transformation. Then to obtain mathematical dependences establishing interrelations of heating capacity and coefficient of energy transformation with the HDH constructive and operational parameters.

The HDH constructional units were to provide the working liquid circulation in the closed heat-generating contour. At this in the pipe reactor (Figure 1, [9]) with the liquid vortex motion the effect of cavitation is developed. As a result the allocation of internal energy from the working liquid in the form of heat takes place.

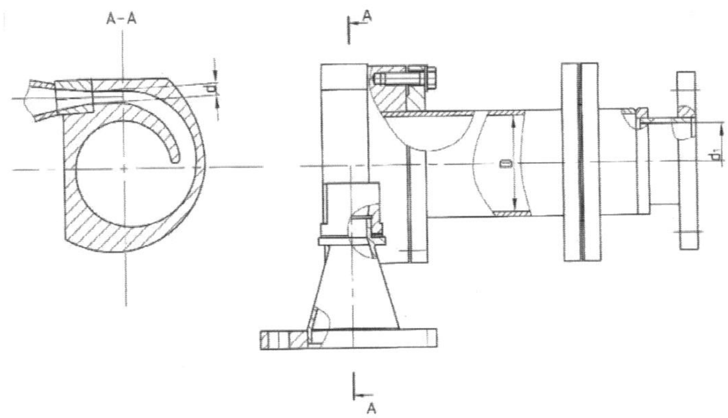

Fig. 1. Hydrodynamic heater sketch

The technological diagram of the test bench of the HDH system with a pipe reactor for studying power characteristics of the experimental series of the hydrodynamic heater is given in Figure 2. The bench permitted to carry out a complex of tasks of planning, collecting and processing measurements of the electric, hydrodynamic and heat-physical processes which were taking place in the HDH elements. In the diagram (Figure 2) there are two contours of the working liquid circulation: a heat-generating contour and a load contour. In the heat-generating contour where from the working liquid the internal energy releases, there is a circulation pump (pos. 1), a hydrodynamic heater (pos. 2) and a heat accumulator (pos. 3) representing a tank store. The load contour serves for accumulation or consumption of the released thermal energy. It contains a circulation pump (pos. 1), a heat accumulator (pos. 3) and an air-heating unit which part is a heater (pos. 4) and a fan (pos. 5). The shutoff valves were used for selecting a joint or a separate operating mode of the contours. At the set points of the technological diagram sensors for digital measurements of the working liquid parameters in both contours, as well as operating modes of the electric motor of the circulation pump are mounted.

The circulation pump unit (pos. 1) with the frequency controlled drive provides the working liquid movement both through the HDH and through the load contour. Rated capacity of the electric motor of the pump unit is spent for the vortex motion of the working liquid combined with the continuous cavitation and overcoming hydrodynamic resistance of all the units of the heat-generating and load contours.

The frequency controlled electric drive of the pump is needed for supporting in the course of the experiment of the working liquid pressure at the intake branch pipe of the HDH at the constant level. The working liquid pressure determines the value of the water stream specific energy therefore its values are to be selected from the pump characteristic "Pressure head – Flow rate" at the maximum frequency of rotation in the vicinity of the working zone where the efficiency and power of the pump have the maximum value. At this the water extreme mechanical activation in the HDH with emission of excess heat is provided. For controlling the course of the experiment it is expedient to use a system of dispatching control and data collection. It is not only to accumulate the measured values, to watch the experiment course in the parameters set by the experimenter, but also to set and support the frequency of circulation pump engine shaft rotation.

The HDH bench is equipped with measuring inserts for digital measuring the working liquid temperature, pressure and consumption. Measuring inserts for measuring the flow rate (pos. 1g and 2g) represent a straight section of the pipeline without local hydraulic resistance on which the flow rate sensor is mounted. Measuring inserts for measuring pressure and temperature of the working liquid represent selected devices for mounting sensors of pressure (pos. 1p, 2p, 3p) and temperature sensors (pos. 1t, 2t, 3t, 4t, 5t) in the set sections of the heat-generating and load contours.

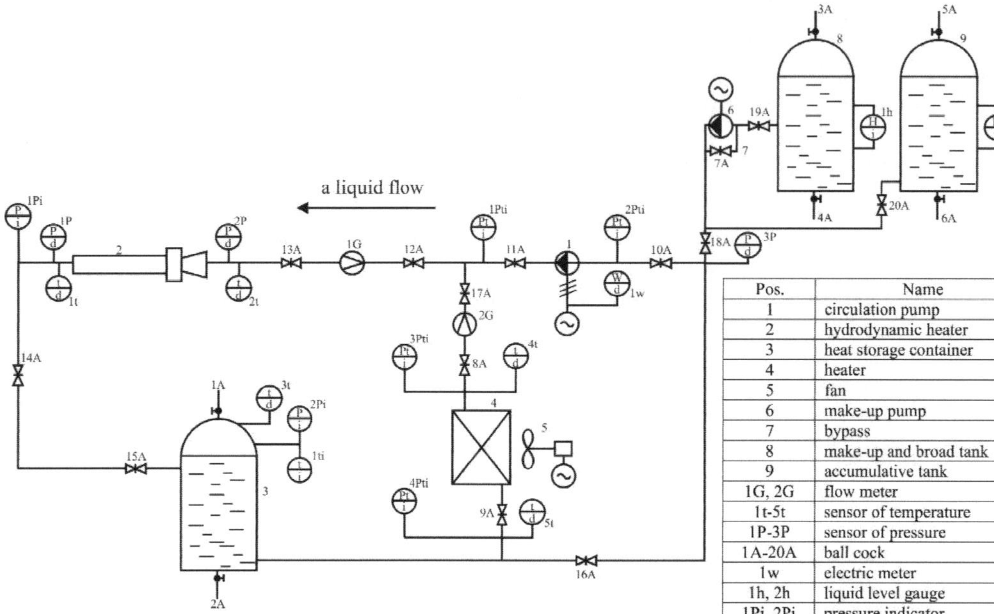

Pos.	Name
1	circulation pump
2	hydrodynamic heater
3	heat storage container
4	heater
5	fan
6	make-up pump
7	bypass
8	make-up and broad tank
9	accumulative tank
1G, 2G	flow meter
1t-5t	sensor of temperature
1P-3P	sensor of pressure
1A-20A	ball cock
1w	electric meter
1h, 2h	liquid level gauge
1Pi, 2Pi	pressure indicator
1ti	thermometer indicator
1Pti- 4Pti	thermomanometer indicator

Fig. 2. Technological network of the bench

The system of controlling and collecting experimental data has been selected the SCADA-system Genesis32 permitting to measure and accumulate in the database in discrete intervals of real time variable conditions of the turbulent movement of the working liquid.

3. Methodology of studying

The task of determining heating capacity and coefficient of energy transformation and their maxima as a function of the HDH power and geometrical parameters has been set.

The criterion function for studying the mechanism of the HDH action is as follows:

$$Q = F(X; P), \text{ at } P = \text{const}, \qquad (1)$$

where $X = (X_1, X_2, \ldots, X_k)$ is the vector of variable factors;

$P = (P_1, P_2, P_3, \ldots, P_m)$ is the vector of fixed parameters.

The pipe type reactor HDH constructive and mode parameters have been selected as variable factors:

1) X_1 – the pipe reactor length, mm;
2) X_2 – internal diameter of the pipe reactor, mm;
3) X_3 – the working liquid pressure at the spiral supply intake, MPa;
4) X_4 – the time of the circulation pump operation, h.

The HDH qualitative and quantitative parameters have been defined as fixed factors:

1) P_1 – the brake unit type;
2) P_2 – the direction of the whirlwind (right rotary, left rotary);
3) P_3 – physical-and-chemical properties of the working liquid;
4) P_4 – the circulation pump type;
5) P_5 – the circulation electric engine power, kW;
6) P_6 – the working liquid starting temperature, °C;
7) P_7 – the working liquid mass in the volume of the storage vessel, kg;
8) P_8 – the spiral supply intake area diameter, mm;
9) P_9 – the cavitator exit area diameter, mm.

For experimental obtaining a mathematical model connecting the HDH criterion function of GDN, the variable factors and the state parameters the plan of experiments based on the Latin squares and the multifactor equations of Protodyakonov-Teder was used [10].

The methodological advantage of such an approach consists in the absence of initial connection between the experiment plan and the form of a possible mathematical model. It gives an opportunity to the experimenter to consider more completely the features of the studied phenomenon on the basis of the creative judgment of both the objective and the obtained results of the experiments. At this the correctness of selecting the models of private functions and the generalizing dependence is estimated by statistical criteria [11].

The pragmatic advantage of such an approach consists, first of all, in the sharp reduction of the needed number of experiments. Another advantage of this methodology is the possibility of expanding the limits of applicability of the generalizing mathematical model independent on physical limits of changing the studied factors [12].

Such a plan of studies which resolution is provided at the minimum number of experiments has been chosen. In the four-factorial studying of the phenomena and the processes such an opportunity belongs to the plan at three levels in which it is necessary to carry out only 9 experiments, every time with a new combination of the factors values. The type of such a plan with code designations of the factors values is given in Table 1.

Experiment No.	**Variable factors**				**Criterion function**
	X1	X2	X3	X4	Q
1	1	1	3	1	Q1
2	2	2	2	1	Q2
3	3	3	1	1	Q3
4	1	2	1	2	Q4
5	2	3	3	2	Q5
6	3	1	2	2	Q6
7	1	3	2	3	Q7
8	2	1	1	3	Q8
9	3	2	3	3	Q9

Table 1. Plan of four-factorial experiment at three levels in the abstract space of variables (Latin square)

Each line of this plan represents specific conditions of the experiment. In the same line the result is registered. From the results the samples for three-point private dependences of heating capacity on each of the factors are formed: $q_1(X_1)$, $q_2(X_2)$, $q_3(X_3)$, $q_4(X_4)$. As to the zero value of private dependences for physical reasons there can correspond a zero value of the generalizing multifactor dependence, it should be formed as the product of private dependences $q_i(Xi)$ in the form:

$$Q_g = \frac{\prod_{i=1}^{k} q_i(X_i)}{q_{av}^{k-1}}, \tag{2}$$

where Q_g is a multifactor function of Protodyakonov;

$q_i(Xi)$ is a private function of one argument for each of the studied factors;

k is the number of private functions (factors);

q_{cp} is the average value of all the considered results of the experiment where the private functions normalizing is executed.

For the indirect measuring of the HDH heat capacity in the i-th experiment the following expression has been used:

$$Q = k \cdot G \cdot c_p \left(\frac{dt}{d\tau}\right)_i, \tag{3}$$

where Q is HDH heat capacity, kW;

k = 1.163 is the coefficient the thermal stream dimension transformation from kcal/h into W;

G is the working liquid mass in the volume of the circulation contour, kg;

c_p is the working liquid specific heat capacity, kcal/kg°C;

$(\frac{dt}{d\tau})_i$ is the working liquid heating temp in the HDH reactor in the i-th experiment.

To express the HDH heat capacity in the form of the criterion P-function Q(L,d,P,t) the following four private functions of one argument heat capacity have been selected:

$q_1(L)$ is the dependence of heat capacity on the pipe reactor length;

$q_2(d)$ is the dependence of heat capacity on the reactor internal diameter;

$q_3(P)$ is the dependence of heat capacity on the pressure at the spiral supply intake;

$q_4(t)$ is the dependence of heat capacity on the pump operation time.

To check the adequacy of the dependence the coefficient of nonlinear multiple correlation has been used [13]:

$$R = \sqrt{1 - \frac{(n-1)\sum_{i=1}^{n}(q_{ei}-q_{ti})^2}{(n-k-1)\sum_{i=1}^{n}(q_{ei}-q_{av})^2}}, \tag{4}$$

where n is the number of described points 9for the accepted experiment plan n = 3);

k is the number of acting factors: for private approximating functions k = 1, and for generalizing function k = 4);

$q_{эi}$ is the experimental value of the result, i.e. heat capacity;

q_{ti} is the theoretical (calculated) value of the result;

q_{cp} is the average experimental value.

Theoretical calculations have been carried out in the environment of the information and graphic GID-99w system on the same set of factors as in the HDH pilot studies. The GID-99w has been developed at Karaganda State Technical University and intended for imitating modeling and the spatial technological analysis of the operational and emergency hydraulic operation of the heat-supplying systems of a variable technological structure and an arbitrary set of the standard heat-mechanical equipment [14-16].

The importance of the coefficient of correlation and of the checked approximating function for the 95% level of reliability is defined by the inequality [13]:

$$t_r = \frac{R\sqrt{n-k-1}}{1-R^2} > 2. \tag{5}$$

4. Experimental results processing

The adequacy of private functions of heating capacity was carried out by means of coefficient R of nonlinear multiple correlations by the criterion of importance t_R (5) with the 95% level of reliability. The results of the adequacy estimates are given in Table 2.

q_i	R	Importance by R	t_R ($t_R > 2$)	Importance by $t_R > 2$
$q_1(L)$	0.7299	Important	1.5622	Unimportant
$q_2(d)$	1	Important	∞	Important
$q_3(P)$	0.9999	Important	4018579.401	Important
$q_4(t)$	0.9999	Important	210614020.5	Important

Table 2. Heat capacity q_i of private functions estimates

The resultant approximating function of the pipe type reactor HDH heating capacity formed by the structure of Protodyakonov multifactor function has obtained the following expression:

$$Q(L,d,P,t) = (5\cdot 10^{-6}L^2 - 0.0114L + 12.035)\cdot(-0.0002d^2 + 0.0304d + 4.9627)\cdot (-67.94P^2 + 87.376P - 21.631)\cdot(-2.7269t^2 + 6.8838t + 2.1474)/5.883^3, \tag{6}$$

At this the function appeared important (R=0.9974: t_R = 197.9978>>2) that permitted to use it in the physically expedient range of parameters (L,d,P,t), that is 714 mm ≤ L ≤ 1500 mm; 51 mm ≤ d ≤ 150 mm; 0.6 Mpa ≤ P ≤ 0.8 MPa; 1 h ≤ t ≤ 2 h.

The experimental and calculated values of the HDH heating capacity obtained by (7) and used for calculation of the criterion of importance are given in Table 3. Numerical studies for the function extremum (7) by the method of complete enumeration of the parameters (L, d, P, t) space with a small step of change has permitted to establish the optimum values of the variable factors and extreme value of the HDH heating capacity given in Table 4.

The obtained extreme value of heating capacity of the experimental series has made 8.118 kW or 6980.22 kcal/h. At this thermal losses through the thermal isolation of the circulating contour have made 204.319 kcal/h. Then the HDH complete heating capacity taking into account thermal losses through the isolation of the circulating contour has made 7184.542 kcal/h.

By the similar technique the resultant approximating function of coefficient of energy transformation has been obtained. The coefficient of energy transformation represents the HDH power efficiency and is defined from the expression:

$$Kn_{ei} = Q_i/P_i\cdot\eta_i, \tag{7}$$

where Q_i, P_i, η_i are respectively the HDH heat capacity, power consumed by the pump aggregate and the pump aggregate efficiency in the mode of the i-th parallel experiment on the current line of the matrix of planning; the efficiency established on the physical prototype of the HDH system of the pump aggregate with 7.5 kW power in the nominal mode (at the maximum efficiency) is equal η = 0.63.

$$Kn_m(L,d,P,t) = (-2\cdot 10^{-7}L^2 + 0.0004L + 1.0773)\cdot(3\cdot 10^{-6}d^2 - 0.0033d + 1.5368)\cdot (-5.6123P^2 + 6.9644P - 0.8747)\cdot(-0.3069t^2 + 0.8619t + 0.6949)/1.246144^3, \tag{8}$$

Experiment No.	L, mm	d, mm	P, MPa	t, h	Qe, kW	Qt, kW	Relative error, %
1	714	51	0.70	1	7.3101	7.4634	2.0541
2	1200	100	0.60	1	6.5444	6.5416	0.0434
3	1500	150	0.50	1	5.0583	4.8784	3.6867
4	714	100	0.50	1.5	6.3165	6.1068	3.4334
5	1200	150	0.70	1.5	5.7280	5.4203	5.6773
6	1500	51	0.60	1.5	6.9680	7.3099	4.6780
7	714	150	0.60	2	5.4955	5.0436	8.9585
8	1200	51	0.50	2	3.8404	4.1524	7.5139
9	1500	100	0.70	2	5.6861	5.6987	0.2224
Average value, x_{cp}					5.883	5.846	
Maximum value, x_{max}					7.310	7.463	
Minimum value, x_{min}					3.840	4.152	
Variation range, $R=x_{max}-x_{min}$					3.469	3.311	
Mean linear deviation, $a = \frac{1}{n}\sum_{i=1}^{n}\lvert x_i - \bar{x}\rvert$					0.8015	0.8972	
Mean-square deviation, $\sigma = \sqrt{\frac{1}{n}\sum_{i=1}^{n}(x_i - \bar{x})^2}$					0.9436	0.9981	
Coefficient of correlation					0.9697		

Table 3. Experimental Qe and theoretical Qt values of the HDH heat capacity predicted by the Q(L,d,P,t) function

Q, kW	L, mm	d, mm	P, MPa	t, h
8.118	714	76	0.65	1.25

Table 4. Extreme values of heat capacity parameters

In Table 5 there are presented experimental and calculated values of the coefficient of energy transformation obtained by (8) and used for calculating the criterion of importance.

Experiment No.	L, mm	d, mm	P, MPa	t, h	$K\pi_e$	$K\pi_t$	Relative error, %
1	714	51	0.70	1	1.3906	1.4016	0.7844
2	1200	100	0.60	1	1.2985	1.3015	0.2281
3	1500	150	0.50	1	1.0608	1.0591	0.1537
4	714	100	0.50	1.5	1.2558	1.2592	0.2674
5	1200	150	0.70	1.5	1.1907	1.1802	0.8872
6	1500	51	0.60	1.5	1.4454	1.4534	0.5489
7	714	150	0.60	2	1.1068	1.1051	0.1560
8	1200	51	0.50	2	1.2968	1.2951	0.1351
9	1500	100	0.70	2	1.1699	1.1682	0.1489
Average value, x_{cp}					1.246	1.247	
Maximum value, x_{max}					1.445	1.453	
Minimum value, x_{min}					1.061	1.059	
Variation range, $R=x_{max}-x_{min}$					0.385	0.394	
Mean linear deviation, $a = \frac{1}{n}\sum_{i=1}^{n}\lvert x_i - \bar{x}\rvert$					0.1014	0.1057	
Mean-square deviation, $\sigma = \sqrt{\frac{1}{n}\sum_{i=1}^{n}(x_i - \bar{x})^2}$					0.1134	0.1175	
Coefficient of correlation					0.9994		

Table 5. Experimental $K\pi_e$ and theoretical $K\pi_t$ values of the coefficient of energy transformation predicted by the $K\pi_t$(L,d,P,t) function

The HDH parameters extreme values for the function of the coefficient of transformation are presented in Table 6.

$K\pi$	L, mm	d, mm	P, MPa	t, h
1.5159	1000	51	0.6	1.4

Table 6. Extreme values of the coefficient of energy transformation parameters

5. Conclusion

As a result of the carried out studies the following results have been obtained:

- the HDH constructive and technological diagram and the experimental bench have been developed for studying the HDH power characteristics of liquid environments with an automated system of the HDH scientific studies;
- the system of automatic control of the bench electric equipment with the frequency-controlled electric drive for detecting the regularity of transforming electric energy into mechanical one with the subsequent transforming it to thermal energy has been developed;
- there have been carried out based on the methods of planning a multifactor experiment studies in the HDH various operating modes [17-20];
- the processes of transforming electric energy into mechanical one and mechanical energy into thermal one with finding the extreme value of heating capacity of the experimental series have been investigated;
- the methods of establishing the HDH optimum parameters providing the maximum heating capacity and coefficient of energy transformation on the basis of multifactor experiment have been defined;
- the mathematical dependences establishing interrelations of heating capacity and coefficient of energy transformation with the constructive and operational parameters of the HDH with the pipe type reactor have been found.

6. References

[1] RF patent No. 2045715, MPC F25B29/00. The heat generator for liquid heating [Text] /the patent holder Potapov Yu.S. (RU). No. 93021742/06; appl. 26.04.1993; publ. 10.10.1995.

[2] U.S. Pat. No. 5188090 A. Apparatus for heating fluids / James L. Griggs - application 08.04.1991, № US 07/682,003; published by 23.02.1993.

[3] Merkulov, A. (1997).Vortex effect and its application in equipment. Samara: Optima, 348 p.

[4] Yong Li & Mian Lin (2010). Hydrodynamic coefficients induced by waves and currents for submerged circular cylinder, Procedia Engineering, Volume 4, 2010, pp. 253–261.

[5] Santhosh Kumar Gugulothu, P.R. Vijaya Kumar, P. Deekshith (2012). Exploring Cavitating Phenomenon With and Without Ultrasonic Transducer, Procedia Engineering, Volume 38, 2012, pp. 154–164

[6] Furmakov, E. (2004). Can hydrodynamic heat generators work superefficiently? Coll. "Problems of studying the Universe": St-Pb. 21 p.

[7] Breido, I., Kalinin, A., & Tomilova, N. Field and simulation study of the hydrodynamic heater for fluid environments, IEEE Xplore, in press.

[8] Kalinin, A. (2014). Automated system of scientific studying hydrodynamic heaters. Proc. of the international scientific and methodical conference Informatization of Engineering Education, ISBN 978-5-7046-1535-4, April 10-11, 2014 NIU MEI, Moscow – M.: MEI publishing house, pp. 359-362.

[9] RK patent No. 6900 PK KZ B. Hydrodynamic heater [Text] / Karasyov N.I., Kuchin V.N., Okrut I.I. Bull.No. 10, 2002.

[10] Protodyakonov, M. & Teder, R.(1970). Methodology of rational planning experiment M.: Nauka, 1970.

[11] Zedginidze, I. (1976). Planning experiment for studying multicomponent systems. M.: Nauka, 390 p.

[12] Brodsky, V. (1976). Introduction to factorial planning of experiment – M.: Nauka, 224 p.

[13] Rumshinsky, L. (1971). Mathematical processing of experimental results – The main editorial office of physical-mathematical science of publ.house “Nauka”, 192 p.

[14] Karasyov, N. (2002). Solver of problems of imitating modeling the modes of large heat supplying systems//Power, telecommunications and higher education in modern conditions: 3rd International scientific and practical conference, ISBN 9965-9206-0-5, Almaty: APEC, pp. 24-28.

[15] Karasyov, N. & Kalinin, A. (2010). Modeling stationaryh flow distribution in the IGS GID-99w environemtn. Proc. of int.symp. «Information-and-communication technologies in industry, education and science», ISBN 9965-04-599-2, October 28–29, 2010 in 2 parts. Part 1. RK Ministry of Education and Science, KSTU, pp.163-165.

[16] Tomilova, N., Kalinin, A., & Tsok, A. (2012). Mathematical models of stationary hydraulic modes of heat supplying systems in terms of regime situations//University Proc., Iss. 1 / KSTU. Karaganda, pp. 92-95, ISSN 1609-1825

[17] RK patent No. 30705. MPC F24H 1/10, F24J 3/00, F24H 7/00, F24D 3/02. Mechanoactivated heat generator [Text] / Kuchin B.H. (KZ); Bekturganov N. S. (KZ); Isaev V. L. (KZ); Kalinin A.A. (KZ); Yurchenko V. V. (KZ) – No. 2014/1278.1; appl. 03.10.2014; publ. 15.12.2015, Bull. No. 12(I), 2015.

[18] Kuchin, V., Yurchenko, V., & Kalinin, A. (2015). Pilot studies of power characteristics of trial installations of environmentally friendly hydrodynamic heaters of liquid environments. Monograph, ISBN 978-601-296-932-0, Karaganda: Publishing house of KSTU, 78 p.

[19] Kuchin, V., Yurchenko, V., Kalinin A., & Mergenov, A. (2015). Pilot studies of power characteristics of hydrodynamic heaters of liquid environments. Bull. of the Kazakhstan National Academy of Natural Sciences No. 1/2015, pp. 32-34, ISSN 2222-3851.

[20] Kuchin, V., Yurchenko, V., Kalinin, A., & Bekturganov, N. (2015). Studying cavitation process of hydrofynamic heater. Bull. of Kazakhstan National Academy of Natural Science. No. 2/2015, pp. 55-58, ISSN 2222-3851.

DOI: 10.2507/27th.daaam.proceedings.089

VACUUM GAUGE PERFORMANCE VERIFICATION SYSTEM

Antons Stekleins, Eriks Gerins & Artis Kromanis

This Publication has to be referred as: Stekleins, A[ntons]; Gerins, E[riks] & Kromanis, A[rtis] (2016). Vacuum Gauge Performance Verification System, Proceedings of the 27th DAAAM International Symposium, pp.0607-0614, B. Katalinic (Ed.), Published by DAAAM International, ISBN 978-3-902734-08-2, ISSN 1726-9679, Vienna, Austria DOI: 10.2507/27th.daaam.proceedings.089

Abstract

The paper analyzes a pressure deviation issues in a vacuum technological process, which occurs due to vacuum gauge inaccuracies during its use. For this purpose a unique vacuum system was designed and used for research. Developed system allows verification of vacuum gauges in manufacturing conditions using various experimental settings for obtaining necessary measuring data and allowing to perform an evaluation about its accuracy, repeatability and the need of calibration. In the following research several experiments with ionization vacuum gauges were carried out to obtain performance data. Presented results and vacuum system allowed to develop a new methodology of vacuum gauge verification under manufacturing conditions.

Keywords: vacuum gauge; vacuum system; verification; pressure measurement

1. Introduction

One of the vacuum system main requirements is to achieve vacuum in the system and maintain it permanently unchanged to a certain level during the technological process, especially during a coating process. Typically, there are two reasons for providing a vacuum: firstly, to control the reactivity of the deposition process and, secondly, to reduce the number of gas particles collisions thus enabling higher energy particles to be used as part of the deposition process [2].

Vacuum gauge calibration is commonly used when pressure measurement accuracy starts to deviate and it is impossible to secure technological process stability. Some studies analyze vacuum gauge accuracy and calibration period, but there are no clear approaches for vacuum gauge verification under manufacturing conditions before calibration operation. [1,5]. Common to all measuring devices, a pressure-measuring instrumentation needs periodic calibration, primarily to monitor changes in performance [1]. Reliable calibration of gauges is impossible without establishing a widely recognized, true models of accuracy, to which gauges can be verified, also known as "primary vacuum standards" [5]. Therefore we are at the opinion that calibration period must be determined individually by applying specific methodology which includes multifunctional vacuum gauge verification using our developed vacuum system.

In this paper we analyze vacuum gauge unreliable pressure measurements during technological coating process at Sidrabe Inc. and present developed vacuum system designed for vacuum gauge verification under manufacturing conditions. The system provides fast, reliable and multifunctional vacuum gauge verification. Ionization vacuum gauges were tested by comparison method using custom made vacuum system.

Main purpose of the paper is to propose a new vacuum verification system which can be used in any vacuum coating company to verify vacuum gauges at any moment, even daily to obtain vacuum gauge pressure deviation data.

2. Vacuum gauge pressure deviation analyses

At present, there have been identified two core problems associated with the technological coating processes: firstly, a vacuum gauges pressure measurement deviation and, secondly, frequency and the necessity of the calibration.

During manufacturing calibrated vacuum gauges are being used to ensure all necessary product characteristics and quality. In general, the vacuum system contains minimum of two different types of vacuum gauges, which are used to cover several ranges of vacuum, but nevertheless a comparison between two gauges indicated pressure should be performed. Prior art discusses a fact that vacuum gauge accuracy can be influenced by several factors such as: operating conditions (vibration, contamination, and temperature), connection, mounting and corrosive gases [5, 10, 9]. Main problem we have encountered during a daily coating process at Sidrabe Inc. is that the vacuum gauges indicated pressure deviates after vacuum chamber evacuation. The deviation is illustrated in Fig.1.

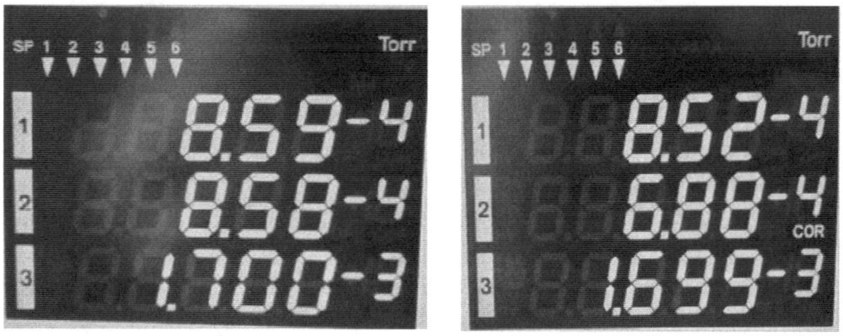

a) b)

Fig.1. Indicated pressure (a) without argon correction; (b) with argon correction.

Mainly two types of vacuum gauges were used: INFICON® diaphragm vacuum gauge under number three at Fig.1 (a, b) and ionization gauge HPG 400 under number two at Fig.1 (a,b) [8]. Indicated pressure measurements by two vacuum gauges differ and it makes it even harder to rely on any gauge with 100% confidence. As it is already known that the diaphragm vacuum gauge is not affected by gas, therefore it is more reliable than ionization gauge [4, 8]. In this experiment as illustrated in Fig.1 (a) an argon is fed into the chamber in volume of 100 sccm or 14,4%. [8]. In order to reach a pressure measurement without correction and change it from 8,58 up to $6{,}88^{-4}$ torr, argon must be reduced to 11,2 % or 78 sccm [8].

If diaphragm vacuum gauge indicated pressure would be considered as true value due to it's resistance to gas, then difference between a true value and ionization gauge indicated pressure with correction turned on would be 59,5%. Current experiment shows used vacuum gauge indicated pressure deviations influence on argon flow amount needed for the process in order to reach required pressure. One of the main problems is that after some time vacuum gauges accuracy start to deviate and it becomes even harder to determine which vacuum gauge indicated correct measurements. Usage of unreliable and inaccurate vacuum gauge may result in following consequences: poor product quality, loss of material, resources and production time for repeating a process (defects). Even minor deviations may leave a significant impact on product characteristics and therefore it's crucial to have reliable vacuum gauge with high accuracy and repeatability. If vacuum system has one vacuum gauge connected, then a determination of gauge pressure precision becomes even more problematic due to lack of comparison operation when two vacuum gauges are connected.

Another issue is related to vacuum gauge calibration necessity and frequency. Prior art discusses that this issue can be addressed with certainty [1]. For majority of instruments the recalibration interval is approximately a year, but if no historical data are available, which relates to the device when used in a manufacturing environment and in a way that is similar to the way it is normally used, then it is not possible to answer the question [1]. Main problem is that during a daily coating process it is possible to reach a point when product characteristics fall outside the permissible limits due to vacuum gauge accuracy failure. Since vacuum gauge operating time, environment inside a vacuum system and other factors are up to the specific coating process, then for the manufacturer it becomes even harder to specify calibration period. If manufacturer suspects that vacuum gauge is defective and unreliable, then the suspected gauge must be disconnected from the vacuum system and sent to the laboratory where it will be calibrated. Usually the manufacturer cannot afford to waste time and resources for calibration operation due to manufacturing intensity, therefore it's necessary to come up with the methodology of vacuum gauge performance verification under production conditions. Using developed methodology would give a possibility to verify a vacuum gauge measurement accuracy, detect deviations in measurements and perform test operations to ensure its accuracy. Since developed system allows all necessary operations to be performed under manufacturing conditions it will facilitate the decision-making about calibration necessity.

3. Custom made vacuum verification system

The aim of the experiment was to build a small size vacuum system in order to reach high vacuum and perform all manipulations needed for vacuum gauge verification using comparison method, gas flow, correction factor and to obtain the necessary data. Custom made vacuum verification system at Sidrabe Inc. can be seen in Fig.2.

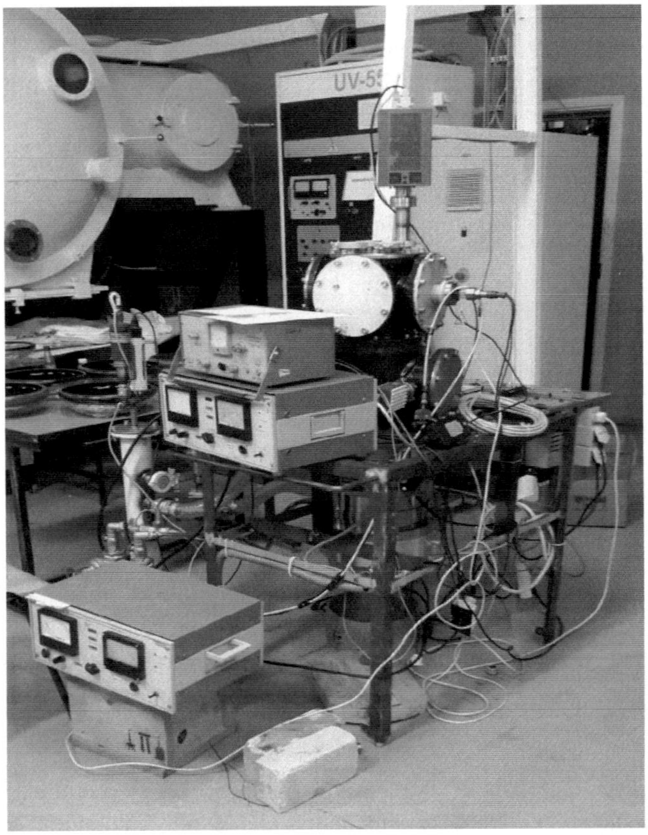

Fig.2. A custom made vacuum verification system.

The vacuum system was design, manufactured and adjusted in two years meeting all selected criteria for fast system evacuation with steady flow of the gas and specific positioning of the vacuum gauges. Vacuum system consists of following components: a chamber, a gate valve, a mechanical pump and a turbo pump, solenoid valves, a manual leak valve, a mass flow controller, flanges with connection ports for vacuum gauges, and a system control unit. Schematic representation of the system is showed in Fig.3.

The system is being controlled manually from control panel and evacuation process is secured by following list of incremental steps. Vacuum system evacuation begins when forevacuum pump receives an input start signal from system control panel and after valves opening provides base pressure of nominally $2x10^{-2}$ Torr. As second step a turbomolecular pump must be enabled, when it accelerates and reaches necessary rotational speed. In the meantime the gate valve should be opened. After the gate is opened pumping provides pressure inside vacuum chamber $2x10^{-6}$ Torr. Gas system consists of two separate lines for argon and oxygen flow with manual valves and mass flow controllers in order to control gas flow during experiments. Gas flow can be operated in range from 1 sccm to 10 sccm.

One of the most important parameters of vacuum system is the design of vacuum chamber. Manufactured vacuum system is capable of providing the lowest pressure of $8x10^{-7}$ Torr. Given pressure can be achieved in short period of time comparing to the coating vacuum systems with high volume for manufacturing purpose. Prior art discusses same idea of vacuum calibration system design, where the easiest way to obtain a constant pressure and maintain gas purity in the high vacuum range is to pass a steady flow of the calibration gas in to the vacuum system and out through the high vacuum pump [7]. Given the importance of uniform pressures and the number of variables that can influence this, it is important to test if the reference gauges and test gauges are sensing the same pressure [7]. Design and the size of verification chamber determine how man vacuum gauges can be mounted simultaneously. Small size vacuum chamber provides faster system evacuation, facilitates to maintain unchanged pressure and significantly reduces the required volume of gas needed.

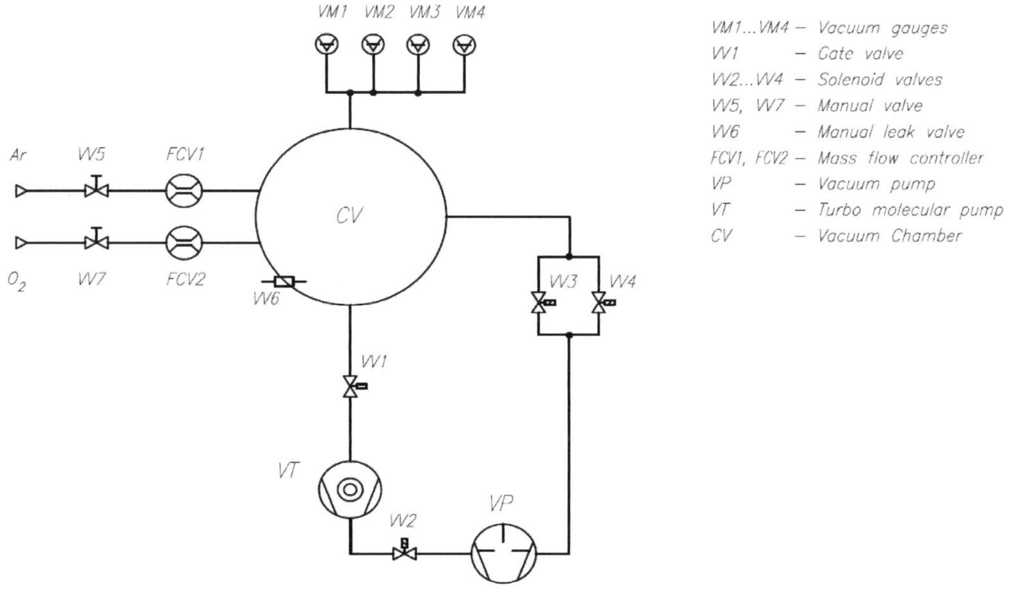

Fig.3. Block scheme of verification vacuum system.

Therefore, several connection ports are necessary to perform vacuum gauges pressure measurement comparison actions, adjustment and calibration if reference standard is available. In order to reach selected criteria the vacuum chamber was designed and manufactured using mechanical machining and blacksmithing work. All the vacuum system components are showed at fig. 4.

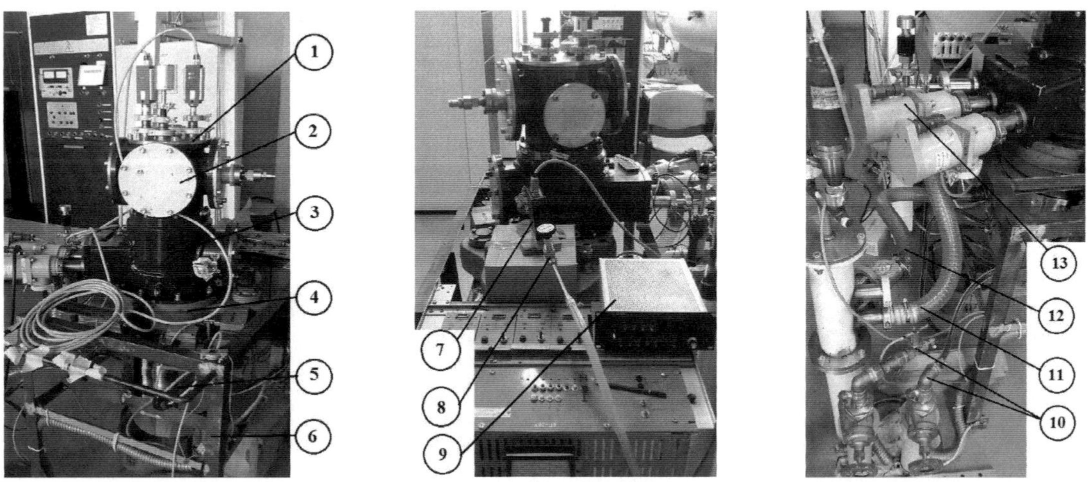

Fig. 4. Vacuum verification system with its components.

As illustarted there are four side covers and one cover at the top of the chamber for vacuum gauges connections. Main idea was to design relatively large connection ports with covers, place them symmetrically to each other. Such design allows to connect up to four vacuum gauges on one cover, thus a distance from pumping orifice and gas inlet port is equal, which provides reliable vacuum gauge pressure measurements. This way both vacuum gauges will sense identical pressure in vacuum chamber. For the experiments there were produced several covers, without connections ports and with different connection ports location. Unique design of vacuum chamber lets switch vacuum gauges positions, connect several gauges simultaneously, change position of the gas inlet system. All the manipulations made with listed elements can provide useful information about vacuum gauge performance and pressure measurement accuracy. After preliminary analysis of collected data, it is possible to find relationships between pressure measurements readings, their positioning and gas flow impact.

The system uses turbomolecular pump VARIAN Turbo-V 3K-T (5) connected to a vacuum chamber (1) though a gate valve (4) and having a pumping speed up to 1900 l/s for argon without inlet screen. This pump consists of high frequency motor driving a turbine fitted with 10 bladed stages with nominal rotational speed up to 31800 rmp, thus a water-cooling system (10) is essential. Pumping line consists of turbomolecular pump (5) connected to a pipe (12) through solenoid valve and forevacuum pump connected to vacuum chamber through a pipe (11) and through a solenoid valve (13) to vacuum chamber as can be seen at Fig. 4.

Gas system consists of several connected tubes, a manual valve (8), a mass flow controller (7) and a four channel mass flow controller power supply readout (9). Using power supply readout (9), a gas flow ratio can be changed from 1 up to 10 sccm. Manual valve is a backup tool for sealing the gas inlet line.

Mainly two types of INFICON vacuum gauges were used in vacuum coating system and for the experiments: a diaphragm vacuum gauge CDG 025D and ionization gauges HPG 400 and BPG 402. Diaphragm vacuum gauge is used as most reliable and stable pressure measuring instrument due to its gas independence. It combines low relative measurement uncertainty with large dynamic range and high stability [6]. Accuracy of gauge also plays significant role and therefore reading uncertainty must be the lowest, thus the diaphragm gauge could meet selected criteria with measurement uncertainty of 0,50% from reading.

High vacuum must be reached in order to implement vacuum gauge performance verification in broad vacuum range including its upper and lower pressure range limits where relative measurement uncertainty reaches the highest value. Other important factor is that high pressure provides an opportunity for vacuum gauge adjustment. Sometimes the word "calibration" is misused to describe the process of altering the performance of an instrument to ensure that the values it indicates are correct within specified limits [1]. This is adjustment, but not calibration [1]. Formally, calibration is said to be "a set of operations the establish, under specified conditions, the relationship between the values of quantities indicated by a measuring instrument and the corresponding values realized by standards" [1]. Measurement standards can be divided in two different types: primary and secondary [3]. A primary standard is one "designed or widely acknowledged as having the highest metrological qualities and whose value is accepted without reference to other standards of the same quality" [1]. Secondary or transfer standards are calibrated against a primary standard of the same unit and depend on a predictable stability to maintain their accuracy [3]. Vacuum gauge calibration in manufacturing conditions cannot be made due to several reasons including the reference standard unavailability; therefore, a vacuum gauge adjustment can be performed. Vacuum gauge operating manual describes that it's recommended to perform a zero adjustment, when gauge is operated for the first time and also due to long time operation or contamination, a zero drift could occur and zero adjustment may become necessary [11]. For adjusting the zero a gauge must be operated under the same constant ambient conditions and in the same mounting orientation as normally [11].

Developed vacuum system was created as a necessary equipment for the vacuum gauges verification under manufacturing conditions in order to test their performance using comparison method, generate adjustment, obtain and save all the data from experiments, make analysis of relationships between several factors and vacuum gauge pressure measurements accuracy.

4. Experiments

In research of vacuum gauge performance, several experiments were designed and conducted using developed verification system. As the first step a vacuum gauge pressure measurements values can be checked by comparison method. For the experiment it was necessary to select minimum of two identical vacuum gauges and third one as a reference gauge. In this particular experiment two INFICON ionization gauges HPG 400 were selected as verifiable gauges and third gauges BPG 402 was selected as a reference.

No.	HPG 400 (SN 566)	BPG 402	HPG 400 (SN 549)
1	2,50E-03 (Torr)	1,90E-03 (Torr)	8,00E-04 (Torr)
2	1,00E-03 (Torr)	8,00E-04 (Torr)	3,20E-04 (Torr)
3	2,10E-04 (Torr)	1,70E-04 (Torr)	6,50E-05 (Torr)
4	5,90E-05 (Torr)	5,30E-05 (Torr)	2,20E-05 (Torr)
5	3,20E-05 (Torr)	3,00E-05 (Torr)	1,40E-05 (Torr)
6	1,30E-05 (Torr)	1,40E-05 (Torr)	7,50E-06 (Torr)
7	8,20E-06 (Torr)	1,00E-05 (Torr)	5,90E-06 (Torr)
8	4,60E-06 (Torr)	7,10E-06 (Torr)	4,50E-06 (Torr)
9	1,80E-06 (Torr)	4,00E-06 (Torr)	2,70E-06 (Torr)

Table 1. HPG and BPG vacuum gauges measured pressures values.

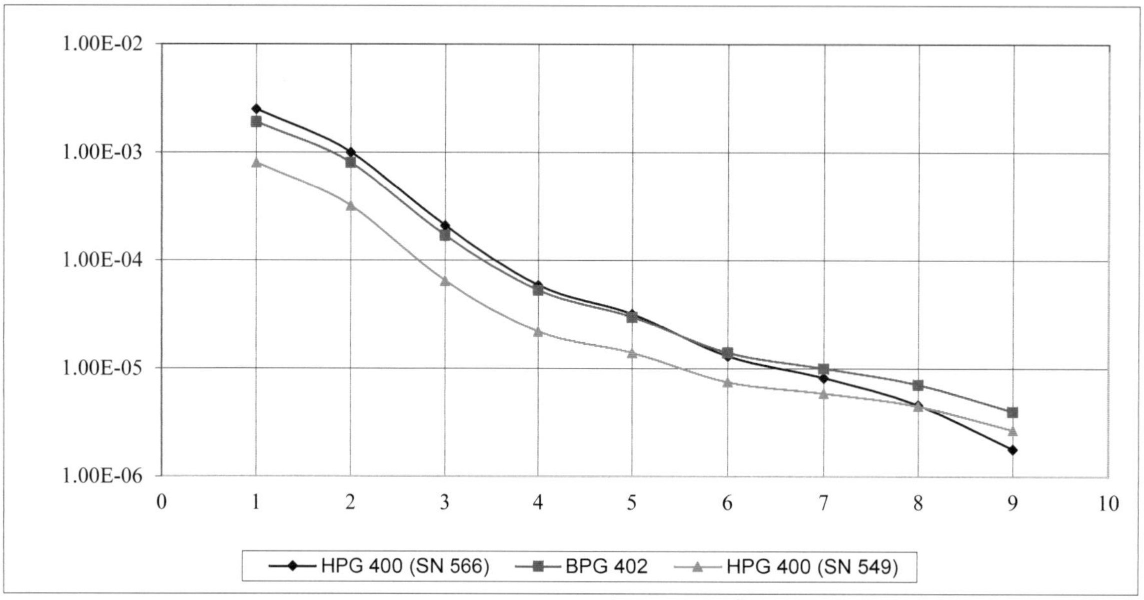

Fig.5. HPG 400 and BPG 402 vacuum gauge performance comparison.

In INFICON HPG 400 ionization vacuum gauge data sheets stated that the accuracy is +/- 15% of reading in range from 10E-5 to 1 mbar [12]. Same accuracy provided by the BPG 402 gauges in the range from 10E-8 to 10E-2 mbar [13]. After preliminary analysis of the results it was possible to conclude that ionization gauge HPG 400 (SN 549) lacks accuracy due to significant difference in pressure measurements comparing to reference gauge. As an example, the first measuring point was compared for HPG 400 (SN549) and BPG 402 gauges. Difference between indicated pressures was 58% if we accept that reference gauge pressure is a true value. Prior art discusses that in order to be able to allocate the largest possible measuring ranges to the individual types of vacuum gauge, one has to accept the fact that the measurement uncertainty rises very rapidly, in some cases up to the greater extent, particularly at the upper and lower limit [5]. First three measuring points do not exceed upper limit for HPG 400 ionization gauges, thus accuracy at specified pressure range complies lowest possible uncertainty. BPG type vacuum gauge accuracy at all measuring points remains the highest due to its relative measurement uncertainty in the middle of range limits.

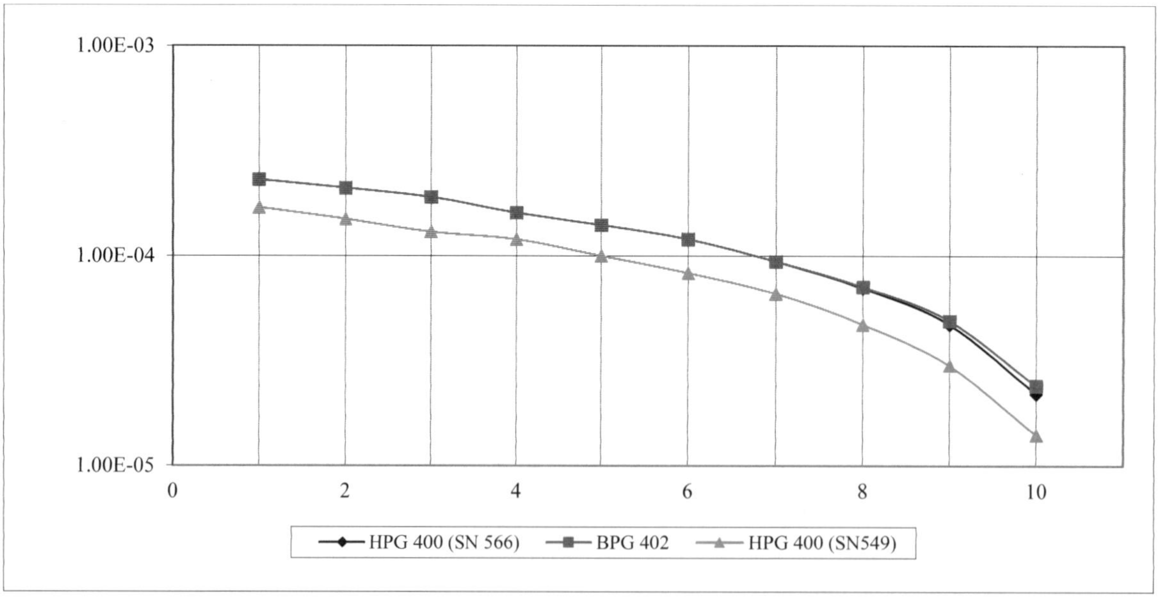

Fig.6. HPG 400 and BPG 402 vacuum gauge performance comparison with argon gas.

No.	Argon flow (sccm)	HPG 400 (SN 566)	BPG 402	HPG 400 (SN549)
1	10sccm	2,30E-04	2,30E-04	1,70E-04
2	9sccm	2,10E-04	2,10E-04	1,50E-04
3	8sccm	1,90E-04	1,90E-04	1,30E-04
4	7sccm	1,60E-04	1,60E-04	1,20E-04
5	6sccm	1,40E-04	1,40E-04	1,00E-04
6	5sccm	1,20E-04	1,20E-04	8,30E-05
7	4sccm	9,40E-05	9,40E-05	6,60E-05
8	3sccm	7,00E-05	7,10E-05	4,70E-05
9	2sccm	4,70E-05	4,90E-05	3,00E-05
10	1sccm	2,20E-05	2,40E-05	1,40E-05

Table 2. HPG and BPG vacuum gauges indicated pressures with argon correction.

In the second experiment a comparison of ionization gauges with turned on argon correction was performed. During all experiments an argon flow was controlled and gradually decreased. This step is necessary to obtain a data of vacuum gauges performance with argon correction factor. Another idea for further research is to verify, how vacuum gauge corresponds to gas flow reduction. Ionization gauges HPG 400 (SN566) and BPG 402 graphic lines are identical and indicated pressure values from Table 2 are identical up to eight measurement point, which is upper range limit of HPG 400 (SN 566) where relative measurement uncertainty rises.

Character of pressure measurements changes should be identical in both experiments. From results it was possible to conclude the fact that vacuum gauge performances are identical with and without gas correction factor, but accuracy of ionization vacuum gauge HPG 400 (SN 549) exceeds the permissible limits. As an example the sixth measuring point can be compared for HPG 400 (SN549) and BPG 402 gauges. Difference between reference and tested gauge at sixth point indicated 31% of deviation. HPG 400 gauges consists of two subgauges, where Pirani gauge filament is exposed to vacuum, thus during opening of a chamber and during technological process, the particles are deposited on the filament. Contamination of vacuum gauge leads to lack of accuracy and performance characteristics.

5. Conclusion

In this paper we analysed vacuum gauge pressure deviation problem. The vacuum system offered in this paper allows performing vacuum gauge verification under manufacturing conditions, adjustment of vacuum in the system, as well as allowing to decide about necessity of calibration period, obtaining and retaining the data. Custom made system has specific design, where vacuum gauges can be tested using various amount of ports and locations, position of gas inlets can be switched, thus small size chamber provides fast system evacuation reaching maximum pressure of 8x10-7 Torr.

Ionization vacuum gauges were tested by comparison method using custom made vacuum system. After preliminary analysis of experimental data it was concluded that ionization vacuum gauge HPG 400 (SN549) lacks accuracy, deviations in measurements are significant compared to reference gauge BPG 402.

This vacuum system can be practically used in any vacuum coating manufacturing company to verify its vacuum gauges, to adjust them, to obtain and retain the data and also make a decision of calibration necessity. This system has unique design which allows further development of various approaches and methods for vacuum gauge verification.

Further research should focus on various approaches and experimental methods which could be implemented for vacuum gauge verification using custom made vacuum system. In result of which a methodology of vacuum gauge verification under manufacturing conditions could be created.

6. References

[1] http://www.npl.co.uk, (1998). National Physical Laboratory. Guide to the Measurement of Pressure and Vacuum, Accessed on: 25.09.2016

[2] Charles A. Bishop. (2007). Vacuum deposition onto Webs, films, and foils, William Andrew Publishing, ISBN-13: 978-0-8155-1535-7 (978-0-8155), 13 Eaton Avenue, Norwich, NY 13815

[3] http://www.glb.nist.gov, 1992. National Institute of Standards and Techonology. Pressure and Vacuum Measurements in Physical Methods of Chemistry, Accessed on 25.09.2016

[4] https://www.leybold.com, 2016. Leybold. Fundamentals of Vacuum Technology , Accessed on 25.09.2016

[5] Akram, H. M., Fasih, A. (2012). Selection criterion of gauges for vacuum measurements of systems with diverse ranges, Physics Procedia, Vol 32., pp. 503 – 512, ISSN: 1875-3892

[6] Barthélémy Daudéa, Hadj Elandaloussia, b, Christof Janssena, b. (2014). On the gas dependence of thermal transpiration and a critical appraisal of correction methods for capacitive diaphragm gauges, Vacuum, Vol. 104, pp. 77 – 87, ISSN: 0042-207X

[7] Ilknur Kocas, Gokce Sevim Sariyerli. 2012. The characterization of new VGMS (vacuum gauge metrology system) by the traceability of multi stage static expansion system, Measurement, Vol. 45, pp., 2430-2433, ISSN: 0263 - 2241

[8] Stekleins A., Gerins E., Kromanis A. (2016). Analysis of selection criteria for vacuum gauges and its accuracy, unpublished

[9] http://www.lesker.com, 2016. Kurt J.Lesker, 9th edition Global Vacuum Product Guide, Accesses on: 25.09.2016

[10] http://www.noshok.com 2016, Measurement solutions, Process Conditions That Affect Pressure Guage Accuracy ^ Performance, Accessed on: 25.092016

[11] http://inficon.com., 2016, Vacuum instruments, CDG 025D Data Sheet, Accessed on: 26.09.2015

[12] http://inficon.com, 2016, Vacuum instruments, HPG 400 Data Sheet, Accessed on: 26.09.2015

[13] http://inficon.com, 2016, Vacuum instruments, BPG 402 Data Sheet, Accessed on: 26.09.2015

DOI: 10.2507/27th.daaam.proceedings.090

INFLUENCE OF GEOMETRICAL MAGNIFICATION ON COMPUTED TOMOGRAPHY DIMENSIONAL MEASUREMENTS

Amalija Horvatić Novak, Biserka Runje, Danijel Butković

This Publication has to be referred as: Horvatic, A[malija]; Runje, B[iserka] & Butkovic, D[anijel] (2016). Influence of Geometrical Magnification on Computed Tomography Dimensional Measurements, Proceedings of the 27th DAAAM International Symposium, pp.0615-0622, B. Katalinic (Ed.), Published by DAAAM International, ISBN 978-3-902734-08-2, ISSN 1726-9679, Vienna, Austria
DOI: 10.2507/27th.daaam.proceedings.090

Abstract

Computed tomography is a widely used method in the field of dimensional measurements. It is a non-contact, non-destructive measurement method that enables insight into both external and internal geometry of measured part which allows measurement of characteristics otherwise unreachable with tactile measurement methods. The biggest problem with computed tomography measurements is lack of metrological traceability. Because of the fact that a lot of parameters influence the whole measurement system, measurement uncertainty is still not evaluated. In this paper, influence of one of the influencing parameters, geometrical magnification, on results of dimensional and geometrical measurement characteristics was observed. Besides experimental research, simulations of computed tomography scanning were done. Measurement results obtained from both scanned and simulated models were compared and observed with regard to reference values. Results obtained from scanned model showed predictable behaviour compared to results obtained from simulated model.

Keywords: Computed tomography; dimensional measurements; influence parameters; geometrical magnification.

1. Introduction

Computed tomography – CT is a new method used in the field of dimensional measurements. It is a non-contact, non-destructive measurement method that enables insight into both external and internal geometry of measured part which allows measurement of characteristics otherwise unreachable with tactile measurement methods. Computed tomography uses X-ray for obtaining large number of 2D scans that arise during rotation of part for 360 degrees, which are later used for rendering real 3D model of measured part. Concerning the fact that this is a new dimensional measurement method with a large number of influencing parameters, measurement uncertainty is still unknown and metrological traceability is still not achieved [1-5]. With aim to assure metrological traceability, identification of all influencing parameters and their contribution to measurement uncertainty is necessary. In this paper influence of geometrical magnification on measurement results has been analyzed. Also, comparison of measured results obtained from real scanned model and simulated model was done. Experimental scanning was performed on CT device for dimensional measurements produced by Nikon, model XT H 225, while simulation of X-ray scanning were performed in software package 'aRTist – *trial version*' (analytical Radiographic Testing inspection simulation tool) by BAM (Federal Institute for Materials Research and Testing) [6-8].

Selected object for this analysis was cylinder made from aluminium, and the following dimensional characteristics were observed: inner and outer diameter and cylinder length, and geometrical characteristics: coaxiality and parallelism.

2. Usage of computed tomography in dimensional measurements

Computed tomography method is known for over 40 years when it was first implemented in field of medicine, and later from 1980s in field of material analysis [9]. Idea for applying computed tomography for dimensional measurements emerged at the beginning of 1990s when first dimensional measurement was done. Accuracy of obtained results was about 0.1 mm [10]. After that event, significant development of CT devices suitable for dimensional measurements has begun. First CT device for dimensional measurement was presented in 2005 on Control Fair in Sinsheim, Germany by producer Werth Messtechnik GmbH [11]. Today, computed tomography method is widely used in field of dimensional measurement, but a problem with lack of measurement uncertainty and connected to that, lack of metrological traceability is still present [12-14]. In order to achieve metrological traceability, it is necessary to assess measurement uncertainty of measured results. Parameters influencing the whole CT measurement process contribute to measurement uncertainty, so the first step in addressing measurement uncertainty is identification of all influencing parameters. Considering the fact that process of dimensional measurement with computed tomography consists from three separate sub-processes, influencing parameters can be divided into three sub-classes: parameters influencing scanning process, parameters influencing modelling process and parameters influencing measurement process. On the other hand, some authors [15-17] classified influencing parameters as: environmental parameters, hardware parameters, software parameters, object related parameters and influence of operator. Here, classification of parameters considers hardware parameters, software parameters and other parameters and it is given in Table 1.

Hardware influencing parameters	**Software influencing parameters**	**Other influencing parameters**
- X-ray source - Rotational table - X-ray sensor (detector)	- 3D reconstruction - Surface determination - Software correction of beam hardening, noise reduction and scattering	- Influence of measured part (dimensions, geometry, surface characteristics – roughness) - Environmental parameters (temperature, humidity, vibrations) - Influence of operator skills and measurement approach (choose of input parameters, object position, number of projection images...)

Table 1. Classification of CT system influencing parameters

One of the parameters related to operator is position of the measured part, and connected with that the parameter of geometrical magnification. Geometrical magnification (1) is defined as ratio between source-to-detector distance (*SD*) and source-to-object (*SO*) distance.

$$Geometrical\ magnification = \frac{SD}{SO} \quad (1)$$

By positioning the object closer to X-ray source, larger resolution can be achieved, but at the same time, picture of the scanned object results in less sharp edge projection with appearance of so called penumbra effect [15]. In standard CT systems, source-to-detector distance is constant, so geometrical magnification depends only on source-to-object distance. With increase of source-to-object distance, geometrical magnification decreases. Lower geometrical magnification means lower resolution, but enables scanning of the whole part, especially when large size objects are investigated. Influence of geometrical magnification on CT measurement results is significant, where deviations in results increase with increase of geometrical magnification [18-20]. In this paper object with simple geometry, aluminium cylinder has been investigated in order to obtain behaviour of results depending on geometrical magnification and define deviations in results obtained by scanning and simulations. Observed were three dimensional characteristics: outer diameter D, inner diameter d and length l and two geometrical characteristics: parallelism of top and bottom cylinder planes and coaxiality between outer and inner cylinders.

3. Experimental research

Experimental researches consist of:

- performance of CT scanning real part, data processing and measurement of reconstructed 3D model,
- performance of CT simulation, where the same scanning settings as in real scanning process were chosen.

3.1. CT measurements

CT scanning was performed on industrial CT device by Nikon, model XT H 225. Chosen scanning parameters are given in Table 2 and Table 3. With the fact that in this research influence of one parameter, geometrical magnification was investigated, other scanning parameters were kept constant. Table 2 presents scanning parameters set for all five cases while in Table 3 are given source-to-object distances and amounts of geometrical magnification according to (1).

Parameter	Amount
Voltage, kV	130
Current, μA	30
No.of projections	1000
Detector size, pixels	3192 x 2296
Pixel size, μm	127 x 127
X-ray spot size, μm	3.9

Table 2. Scanning parameters

Parameter	Experiment No.				
	1	2	3	4	5
Source-to-object distance, mm	200	300	400	500	600
Geometrical magnification	4.921	3.281	2.461	1.968	1.640

Table 3. Source-to-object distances

CT models of scanned parts were generated with usage of software package CT-Pro, and measurements were performed in software VGStudio Max 2.2. Chosen measurement strategy involved fitting simple geometry objects where both outer and inner diameters were fitted with cylinders, using Gaussian approach; length was measured as a distance between two planes; parallelism was observed as a geometrical measurand of two planes, and coaxiality was observed between outer and inner cylinders (Table 4).

Measurand	Strategy
Outer diameter, D, mm	Cylinder
Inner diameter, d, mm	Cylinder
Lenght, l, mm	Plane - Plane
Parallelism	Plane - Plane
Coaxiality	Cylinder - Cylinder

Table 4. Measuring strategy for CT measurements

3.2. CT simulations

CT simulations were conducted in software for radiographic testing named 'aRTist' (analytical Radiographic Testing inspection simulation tool), developed by *Bundesanstalt für Materialforschung und –prüfung*, BAM, Germany. For simulation of the CT process, modules aRTist image view and Ct scan were used. Program setups were the same as real setups and limitations of Nikon XT H 225 device. The same input scanning parameters as the one used for CT scanning of the real part, stated in Tables 2 and 3 were chosen. CAD model of cylinder made according to the actual reference values was simulated. Figure 1 presents drawing of simulated cylinder.

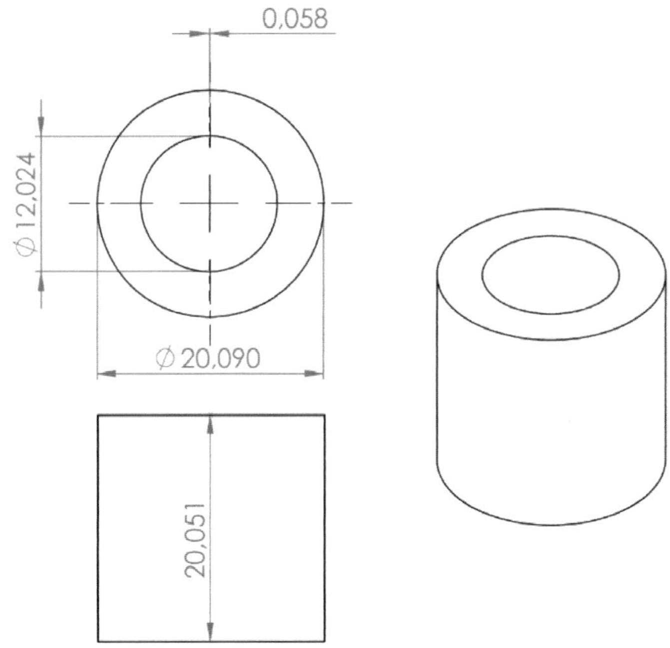

Figure 1. Drawing of cylinder with dimensions and model of cylinder

Models generated with usage of aRTist were also analyzed with software packages for measurements and data analysis VGStudio Max 2.2. The same measurement method was used for measurement performance on simulated models (Table 4.).

4. Results

Results are given graphically as deviations of experimentally obtained results of CT scanning from reference values and as deviations of results from simulation obtained scanned model and measured results from reference values. Deviations are given on the same graph.

Reference measurements were performed on coordinate machine Ferranti Merlin and measurements were done in software MODUS. The reproducibility was obtained by measuring the cylinder three times in different days during the period of two weeks. Table 5 presents measured results and related measurement uncertainties of dimensional and geometrical characteristics.

Measurand	Measured results	Expanded measurement uncertainty U, $k = 2$, $P = 95$ % , µm
Outer diameter, D, mm	20.098	4
Inner diameter, d, mm	12.017	4
Length, l, mm	20.063	4
Parallelism, mm	0.055	1
Coaxiality, mm	0.019	1

Table 5. Reference values

Results obtained from CT measurements for outer diameter presented in Figure 1 show almost linear fall in amount with increase of source-to-object distance. Deviations of measured results obtained in CT simulations show decline in value of cylinder outer diameter with increase of source-to-object distance until the amount of source-to-object distance reaches 400 mm. After that, the value of outer diameter starts to grow significantly. The best agreement between results obtained from CT measurements and CT simulations are in the source-to-object distance range 300-400 mm, which equals geometrical magnification range between 3.281 and 2.461.

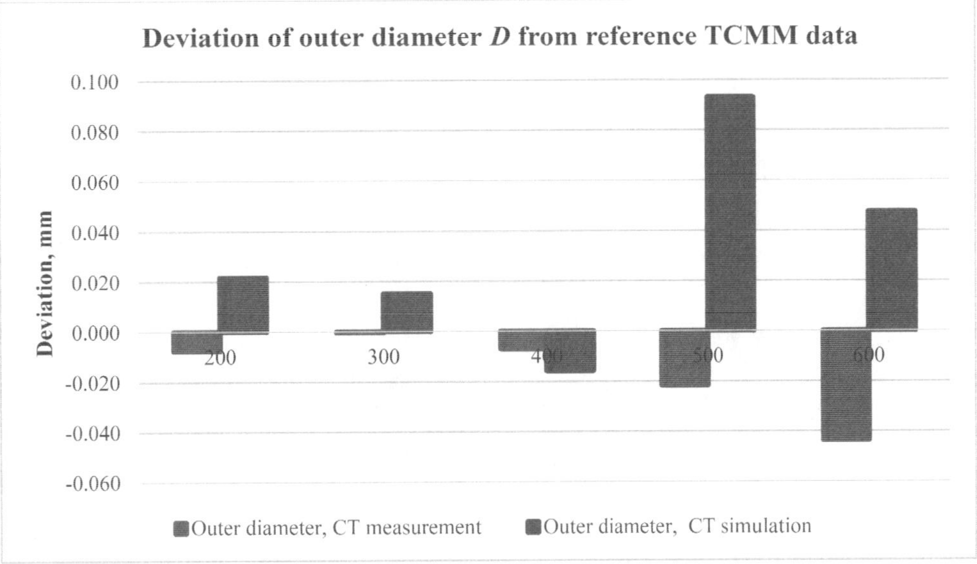

Figure 2. Deviation of outer diameter from reference values

Figure 3 shows deviations of cylinders inner diameter from reference values in dependence of source-to-object distance. Results obtained in CT measurements show increase of deviation in inner diameter with increasing source-to-object distance. Behaviour of simulated results shows significant decrease of deviation in inner diameter until the amount of source-to-object distance equals 400 mm and from 400 mm to 600 mm slight decline in results is observed.

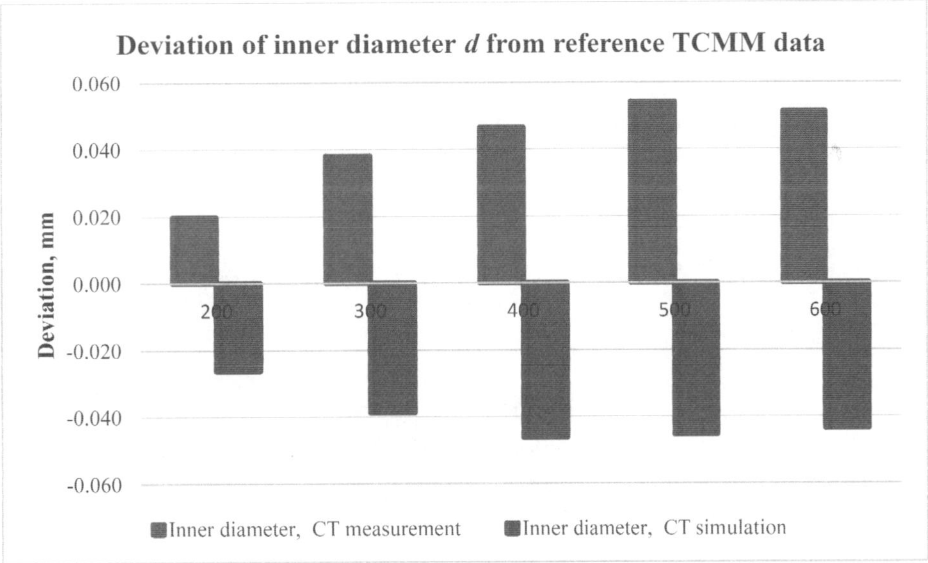

Figure 3. Deviation of inner diameter from reference values

In the Figure 4 an increase in deviation of cylinder length with increase of source-to-object distance from 200 mm to 400 mm distances, for results obtained by CT measurements, is visible. Maximal deviation between cylinder length and reference value is noted for distances from X-ray source in the range of 300 mm to 400 mm, while the best results are obtained when the object is as close as possible to X-ray source, but also in the case when object is the nearest to the detector. Simulated results on the other hand show the biggest deviation from reference value in case where the object is near the X-ray source. Measured and simulated results in dependence of geometrical magnification showed the biggest discrepancy in case of length measurements.

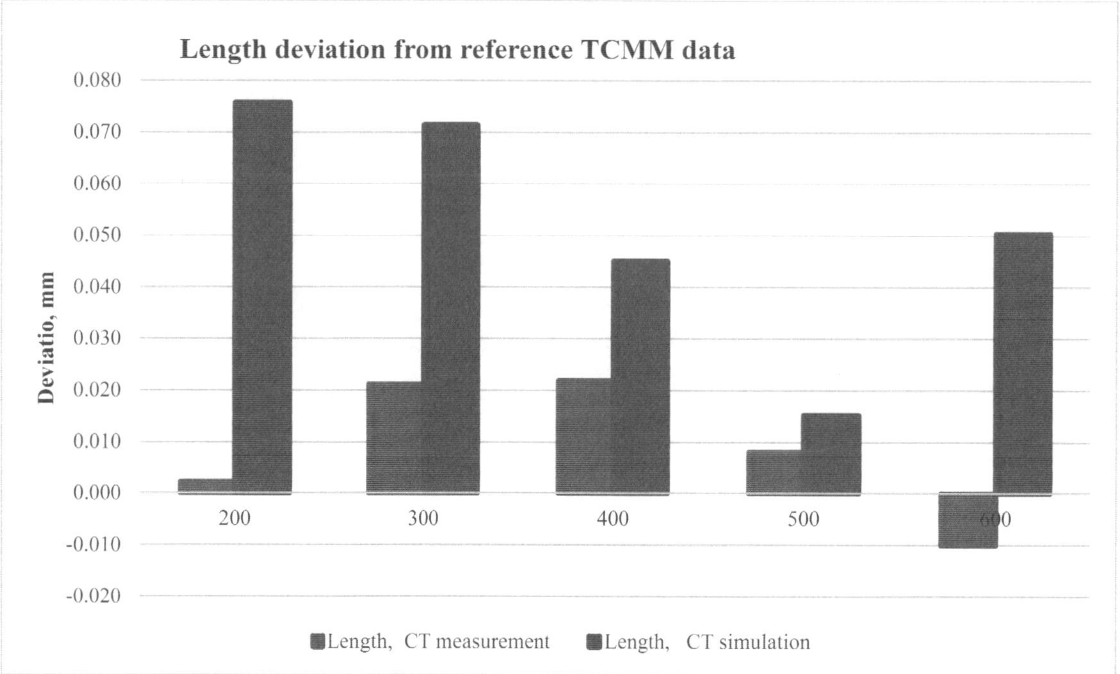

Figure 4. Length deviation from reference values

From geometrical characteristics, parallelism and coaxiality were observed. Results of parallelism obtained from CT measurements and the one obtained from CT simulations show equal behaviour with increase of source-to-object distance (Figure 5).

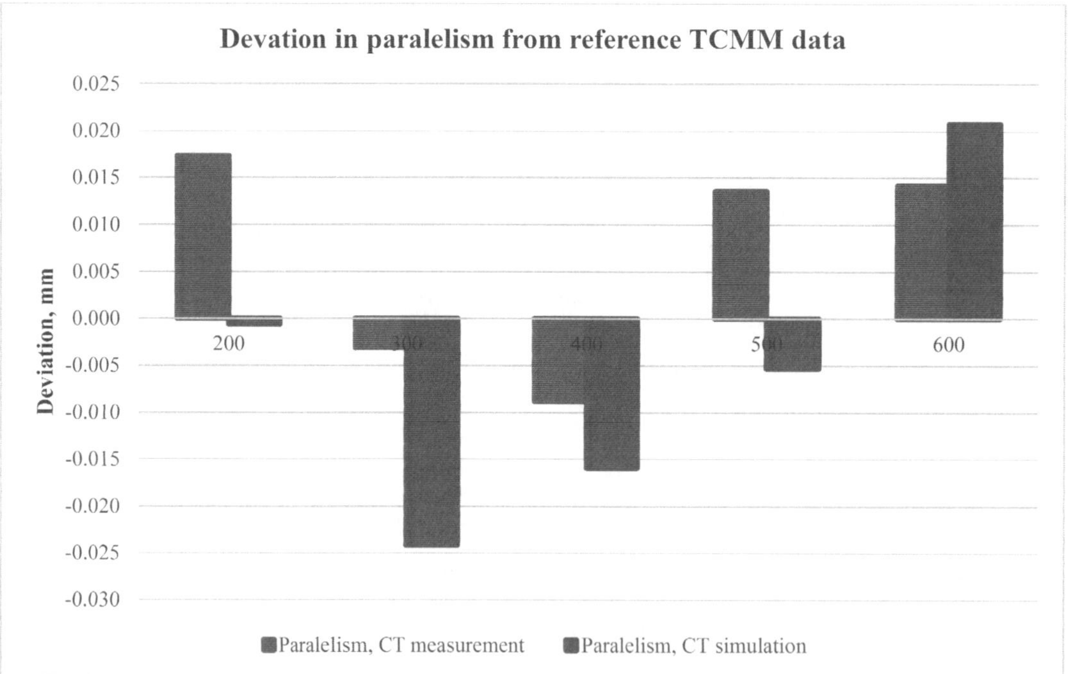

Figure 5. Deviation in parallelism from reference values

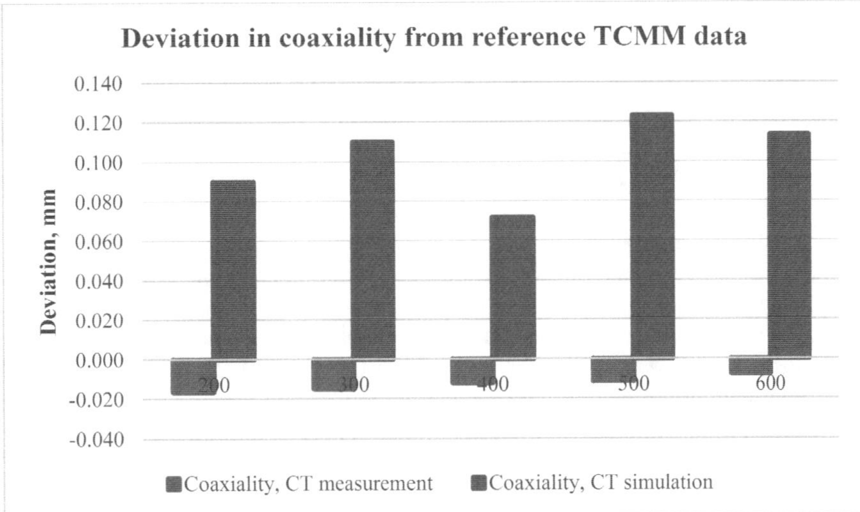

Figure 6. Deviation in coaxiality from reference values

Results of coaxiality (Figure 6.) obtained from CT measurement show invariance with increase of source-to-object distance. On the other hand, results obtained from CT simulations behave totally unpredictable.

5. Conclusion

In this research measurements of dimensional and geometrical characteristics were monitored. Dimensional characteristic of outer and inner diameter, as well as length of the aluminium cylinder were observed. Research included observation of influence parameter geometrical magnification on measurement results obtained with usage of computed tomography. Two approaches were conducted, first one included performance of experiment on CT device Nikon X TH 225 while second approach was simulation of CT scanning process by using simulation software for radiographic testing 'aRTist' (analytical Radiographic Testing inspection simulation tool), developed by *Bundesanstalt für Materialforschung und –prüfung*, BAM, Germany. Observed were deviations of measured characteristics from reference values obtained on coordinate tactile measurement machine. Obtained results showed significant deviations from reference values. General conclusion is that results obtained by CT scanning behave in predictable way, where results of outer diameter fall with decrease of geometrical magnification, while results of inner diameter rise with decrease of geometrical magnification parameter. Results obtained by CT simulation in case of outer diameter showed good agreement with reference values for source-to-object distance from 200 mm to 400 mm, while results in case of higher source-to-object distance showed unpredictable leap. The same unpredictable leap in results obtained by CT scanning occurs in length measurement for source-to-object distance from 500 mm and over. Observation of geometrical characteristics results depending on geometrical magnification showed, in case of parallelism similarities within results obtained by scanning and simulations. Obtained results deviate from reference values approximately 0.020 mm in absolute amount. On the other hand results of coaxiality obtained by CT scanning showed invariance on geometrical magnification change, while results obtained by CT simulation behave totally unpredictable and deviate from reference values for about 0.100 mm.

Further step in research in a field of computed tomography dimensional measurement and achievement of metrological traceability should include evaluation of components of measurement uncertainty with further aim to asses measurement uncertainty of results obtained with CT measurement method which is basis for achievement of metrological traceability.

6. Acknowledgments

The authors would like to thank BAM (Federal Institute for Materials Research and Testing) for providing the simulation tool 'aRTist' - trial version in duration of one month.

7. References

[1] Carmignato, S.; Pierobon, A.; Rampazzo, P.; Parisatto, M. & Savio, E. (2012). CT for Industrial Metrology – Accuracy and Structural Resolution of CT Dimensional Measurements, Proceedings of ICT, Conference on Industrial Computed Tomography, Wels, Austria

[2] Müller, P.; Hiller, J.; Cantatore, A. & De Chiffre, L. (2011). Investigation of measuring strategies in computed tomography, Proceedings of the International conference on Advanced Manufacturing Engineering, Brno, Czech Republic 2011, pp. 31-42

[3] Müller, P.; Hiller; J., Dai; Y., Andreasen; J.L., Hansen H.N. & De Chiffre, L. (2014). Estimation of measurement uncertainties in X-ray computed tomography metrology using substitution method, CIRP Journal of Manufacturing Science and Technology Vol. 7, No. 3, 2014, pp. 222-232, doi:10.1016/j.cirpj.2014.04.002

[4] Müller, P; Cantatore, A; Andreasen, J.L.; Hiller, J. & De Chiffre, L. (2013). *Computed tomography as a tool for tolerance verification of industrial parts*. Proceedings of 12th CIRP Conference on Computer Aided Tolerancing. Doi: 10.1016/j.procir.2013.08.022

[5] Carmignato, S. (2012). *Computed tomography as a promising solution for industrial quality control and inspection of castings.* Metallurgical Science and Technology Vol. 30-1, 2012, pp.5-14

[6] Bellon, C & Jaenisch, G.-R. (2007). *aRTist – Analytical RT Inspection Simulation Tool.* Proceedings of DIR 2007 - International Symposium on Digital industrial Radiology and Computed Tomography, June 25-27, Lyon, France.

[7] Jaenisch, G.-R.; Bellon, C & Ewert, U. (2008). *aRTist – Analytical RT Inspection Simulation Tool for Industrial Application.* Proceedings of 17th World Conference on Nondestructive Testing, October 25-28, Shanghai, China.

[8] Bellon, C; Deresch, A; Gollwitzer, C & Jaenisch, G.-R. (2012). *Radiographic Simulator aRTist: Version 2.* Proceedings of 18th World Conference on Nondestructive Testing, April 16-20, Durban, South Africa.

[9] Gapinski, B.; Wieczorowski, M.; Marciniak-Podsadna, L.; Dybala, B. & Ziolkowski, G. (2013). Comparison of Different Method of Measurement Geometry using CMM, Optical Scanner and Computed Tomography 3D, Procedia Engineering 69, 24th DAAAM International Symposium on Intelligent Manufacturing and Automation, Zadar, Katalinic, B. (Ed.), pp.255-262, Elsevier, DOI:10.1016/j.proeng.2014.02.230

[10] Kruth, J.P.; Bartscher, M.; Carmignato, S.; Schmidt, R.; De Chiffre, L. & Weckenmann A. (2011). Computed tomography for dimensional metrology, CIRP Annals - Manufacturing Technology Vol. 60, No. 2, 2011, pp.821-842, ISSN 0007-8506; DOI 10.1016/j.cirp.2011.05.006

[11] http://www.control-messe.de/en/nc/exhibitors/exhibitors-information/trade-fair-profile/?cHash=0844b798ad4d75440af67c0013350c7e&sword_list%5B0%5D=2005, Accessed on: 2016-03-04

[12] Carmignato, S. *Accuracy of industrial computed tomography measurements:Experimental results from an international* comparison. CIRP Annals - Manufacturing Technology Vol. 61, 2012, pp.491-494, ISSN 0007-8506; DOI: /10.1016/j.cirp.2012.03.021

[13] Hiller, J. & Reindl, L.M. (2012). *A computer simulation platform for the estimation of measurement uncertainties in dimensional X-ray computed tomography.* Measurement Vol. 45, 2012, pp.2166-2182. DOI: 10.1016/j.measurement.2012.05.030

[14] Affenzeller, C; Gusenbauer, C; Reiter, M & Kastner, J. (2015). *Measurement uncertainty evaluation of an X-ray computed tomography system.* Proceedings of Digital Industrial Radiology and Computed Tomography (DIR 2015), June 22-25, Ghent, Belgium.

[15] Welkenhuyzen, F.; Kiekens, K.; Pierlet, M.; Dewulf, W.; Bleys, P.; Kruth, J.-P. & Voet A. (2009). Industrial Computer Tomography for Dimensional Metrology: Overview of Influence Factors and Improvement Strategies, OPTIMESS2009 proceedigsbook, 4th International Conference on Optical Measurement Techniques for Structures and Systems, Antwerp, Buytaert, J & Dirck J.J. (Eds.), ISBN 978-90-423-0366-9

[16] Cantatore, A. & Müller, P. (2011). *Introduction to computed tomography*, Kgs.Lyngby: DTU Mechanical Engineering

[17] Sun, W.; Brown, S.B. & Leach, R.K. (2012) *An overview of industrial X-ray computed tomography*, NPL Report, Teddington

[18] Müller, P.; Hiller, J.; Cantatore, A.; Bartscher, M. & De Chiffre L. (2012). Investigation on the influence of image quality in X-ray CT metrology, Paper presented at 4th Conference on Industrial Computed Tomography (iCT), Wels, Austria

[19] Weckenmann, A & Krämer, P. (2009). *Assesment of measurement uncertainty caused in the preparation of measurements using computed tomography.* Proceedings of XIX IMEKO World Congress Fundamental and Applied Metrology, September 6–11, 2009, Lisbon, Portugal.

[20] Weckenmann, A & Krämer. (2010). Chapter 17 in *Product Lifecycle Management: Geometric Variations*. Eds. Max Giordano, Luc Mathieu and François Villeneuve, ISTE Ltd. Published 2010 by ISTE Ltd.

DOI: 10.2507/27th.daaam.proceedings.091

OPTIMIZING HIGH-PERFORMANCE CUDA DSP FILTER FOR ECG SIGNALS

Ervin Domazet, Marjan Gusev & Sasko Ristov

This Publication has to be referred as: Domazet, E[rvin]; Gusev, M[arjan] & Ristov, S[asko] (2016). Optimizing High-Performance CUDA DSP Filter for ECG Signals, Proceedings of the 27th DAAAM International Symposium, pp.0623-0632, B. Katalinic (Ed.), Published by DAAAM International, ISBN 978-3-902734-08-2, ISSN 1726-9679, Vienna, Austria
DOI: 10.2507/27th.daaam.proceedings.091

Abstract

Digital signal processing is usually used in biomedical engineering research area. ECG digital signal filtering serves as a basis for further data analysis, with the intention to eliminate noise. Noise can stem from electrical switching power, radio waves, physical movement and breathing muscle artifacts. Feature extraction and further analysis are successful on noise-free ECG signal.

In many critical times, real-time processing of ECG signal can save lives. In case thousands of connected sensors transfer ECG signals to the Cloud Data Center, processing needs to be fast in order real-time analysis to be done. In this paper, we introduce several optimization approaches to improve the parallel CUDA algorithm for digital signal filtering on GPU cores. Our approach utilizes the shared memory for signal segments and for the impulse response coefficients. Additionally, we consider using constant memory for the impulse response coefficients. Other optimization approaches, including loop unrolling, smaller calculation precision and combined effects were tested and the best solution is discussed. The provided analysis indicates that using shared memory for signal segments optimized the code for 13-78% proportional to kernel length. Considering shared memory for the impulse response coefficients increased the performance on average for 10%, whereas constant memory for 15%. There is a 1-5% performance gain when loop unrolling is considered. Using smaller calculation precision speeds up the code by a factor of 1. The best combination of the proposed optimizations yields nearly 6 times faster code compared to the naïve parallel code.

Keywords: DSP; ECG; Heart Signal; Optimization; Parallelization; CUDA; GPGPU.

1. Introduction

Digital Signal Processing (DSP) area has introduced a powerful set of tools to deal with digital signals, and some of them can be successfully applied to the analysis of ECG signals. Being one of the most important tools of the DSP area, filtering can be used for noise elimination and further on for extraction of signal features. Thus, main features of ECG can be detected and extracted from the hidden information, such as alternating changes of the wave amplitude and subtle deviation of the heart rhythm.

Potential heart risks can be detected before some time making the real-time processing a critical task. This, in turn, can save lives [3]. Sequential algorithms are insufficient of processing ECG signals real time, especially in the case of a data center intended for real-time processing and analyzing thousands of connected wearable ECG sensors. Our motivation is, therefore, to optimize the high-performance solution for signal processing.

There are alternatives for parallelizing the convolution, such as shared memory OpenMP parallelism, distributed MPI parallelism, Maxeler dataflow parallelism, General Purpose GPU CUDA parallelism and etc.

In this research, we focus on optimizing the parallel version of the filtering algorithm on graphics processing unit (GPU) cores. The goal is to find an optimized solution that speed ups the parallel CUDA solution. The hypothesis set in this paper is to confirm whether the utilization of shared and constant memories on GPU can yield faster execution times on ECG signal filtering. To test the hypothesis we will measure the execution times of naive GPU solution over the optimized solution. We are also interested in determining whether loop unrolling and precision has an effect on the speedup. Moreover, we think it would be worthy to investigate the performance of the element version.

We are aware of the limitations, especially that there is a limit on CUDA-enabled GPU's, on the number of threads per blocks due to the internal synchronization inside a block. Considering the advancements on the GPU arena, we hope that this limit will increase. Yet in ECG signal filtering, current limits on the filter lengths are sufficient.

The paper is organized as follows. The background about ECG signals, DSP processing and CUDA obstacles for the high-performance computing are presented in Section 2. Section 3 presents the optimization methods, while Section 4 the experimental methodology. Obtained results of measured processing time and performance analysis of the optimized code are presented in Section 5. Section 6 discusses the obtained results and compares to other approaches. Finally, the paper is concluded and future work is elaborated in Section 7.

2. Background

This section presents a brief description of ECG signals, DSP processing, and CUDA obstacles in order to better understand the paper.

2.1 ECG signal essentials and DSP processing

Electrocardiogram (ECG) signal contains information about the complex heart condition. It is a collection of the electrical activity of the beating heart muscle over a period of time. The electrical activities are captured over time via attached electrodes, measuring the electrical potential reaching to the human skin. These are detected as tiny electrical signals and then amplified and represented on a specialized paper format.

Since the level of ECG signals is very low, there is a considerable amount of accompanied noise, mostly due to amplification of tiny electrical signals. Radio waves, internal human breathing and physical movement are considered as secondary sources of noise. For precise data analysis and interpretation, data pre-processing is necessary.

ECG signal is the representation of impulses generated by beating heart muscle over a period of time (Figure 1) and generally consists 5 features, marked as P, Q, R, S, and T.

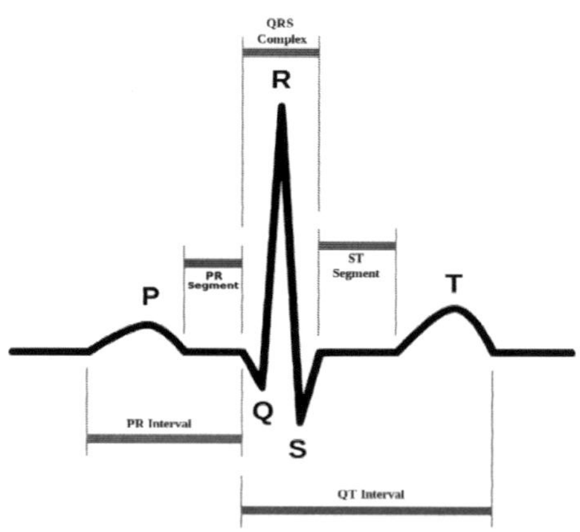

Fig. 1. A general representation of an ECG signal [6]

Characteristic signal of an ECG holds representative and significant information regarding the complex cardiovascular condition of the heart. The general methodology for processing and analyzing ECG signals consist of data pre-processing, feature space reduction and feature extraction steps [3,10]. In this research, we explicitly target the data pre-processing phase.

The power of DSP field rapidly increased during last decades [5]. It is possible to underline that today's most complex systems are built on top of DSP tools. In this research, we are focused on data pre-processing phase of ECG signal analysis.

Convolution, as a mathematical operation, combines the *input data stream* $x(i)$ and *impulse response coefficients* $h(i)$ to generate a new *output data stream* $y(i)$, for $i = 0, \ldots$ by the calculation expressed in Equation (1).

$$y(i) = \sum_{j=0}^{M-1} h(j)\, x(i-j) \tag{1}$$

Noise elimination on ECG signal is achieved by using three classic filters: low-pass, high-pass and band-pass filters. The main functionality of low-pass filters is to weaken all the frequencies above the cut-off frequency, known as a *stopband* and pass all frequencies below the *passband* [5]. On the other side, high-pass filters are the exact opposite of low-pass. The combination of low-pass and high-pass filters is considered as a band-pass filter, which generally passes a predefined region of frequencies and eliminates other frequencies.

Figure 2 shows the effect of applying all these filters on the input signal. For example, a low-pass filter with a 30 Hz cut-off frequency eliminates the 50Hz noise caused by the electricity. A high-pass filter eliminates the baseline drift, which is caused by breathing and physical movement. A band-pass filter by its definition is the combination of low and high-pass filters and combines both the eliminations of high-frequency noise and baseline drift.

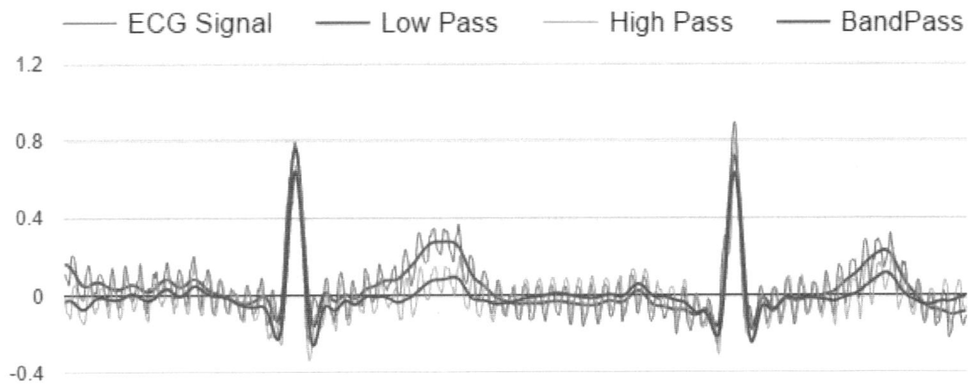

Fig. 2. ECG signal segment with two QRS complexes, filtered with a low-pass filter of 30Hz, a high-pass filter of 0.5Hz and a band pass filter between 0.5Hz and 30Hz.

2.2 CUDA Background

Compute Unified Device Architecture (CUDA) is an application programming interface (API) model developed by NVIDIA. This technology ensures a significant increase in the software performance by using the massive computation power of GPU.

CUDA enables massively parallel hardware designed to run generic or non-graphic code on a general purpose GPU. The goal of CUDA is to write code that can run on compatible massively parallel SIMD architectures. A GPU consists of thousands of low-frequency cores, intended for efficiently handling multiple similar tasks simultaneously.

2.3 Identifying CUDA GPU obstacles for high-performance convolution

CUDA allows massive parallelism, which should be utilized in an efficient way. In order to design an optimized code, it is important to optimize memory alignment and thus accesses. Generally, most important performance consideration is the coalescing of the memory accesses. Figure 3 illustrates the coalescing concept.

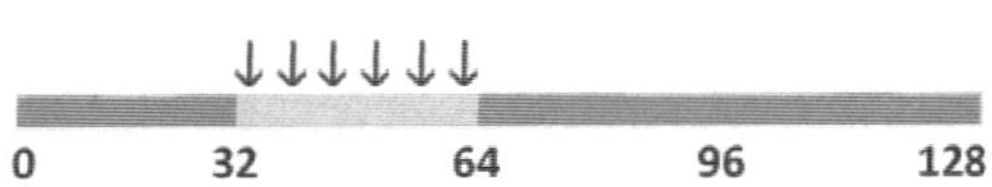

Fig. 3. Coalesced access to memory - all threads access one cached line.

Moreover, utilization of shared memory is another performance improvement. Shared memory is located on-chip and generally provides much higher bandwidth and lower latency than the global memory [4]. This is valid especially when there are no bank conflicts. When multiple addresses of a memory request values of the same memory bank, the accesses are serialized, which degrades the performance significantly.

Constant memory is another CUDA improvement that enables a fast access to data. Particularly, consecutive reads of the same address do not incur any additional memory traffic. A single read from constant memory can broadcast to other nearby threads. This will ensure that no bank conflict will occur in the case of access to constant memory.

A high-performance solution will include a combination of the shared memory and constant memory. Since the filter kernel coefficients are not changing over the execution, a natural solution will be to use the shared memory for storing the input signal and constant memory for the filter kernel, as presented in Figure 4.

Another obstacle is the bank conflict that may occur while accessing the data in the shared memory. An example of 2-way bank conflict is illustrated in Figure 5.

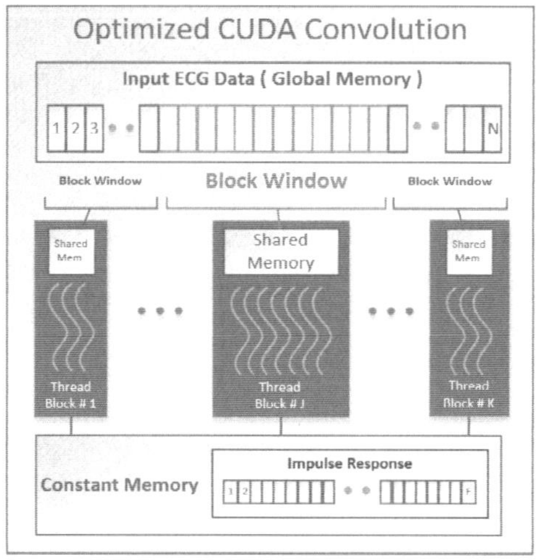

Fig. 4. Data flow in a solution that uses both the shared and constant memory.

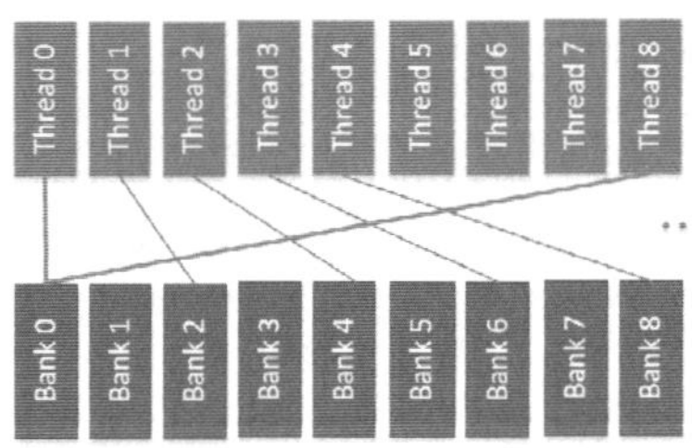

Fig. 5. A two-level bank conflict.

The solution needs a highly accurate filter and fast processing. Therefore the FIR filter is the best candidate for the filtering. A problem will occur if the filter length is bigger than the maximum number of threads per block to eliminate propagation of partial results and introducing internal synchronization inside a block. Currently, NVIDIA devices have 1024 threads per block. In our case, to process an ECG signal, a FIR filter length of 1000 is acceptable, since it generates 30*db* attenuation with a relatively small ripple passband of 0.1db.

3. Optimization approaches

A sequential algorithm for performing convolution is an iterative procedure that repeats the kernel item for each new data element. The complexity of the algorithm $O(nM)$ depends on the input n and the kernel stream length M. The flow on CPU is sequential, where inner loop length depends on the kernel size.

In our previous research [2], we have used a naïve parallel CUDA version of the DSP filtering algorithm running on a GPU. Since it did not include any optimization, we have faced lower performance due to lots of concurrent memory reads. In this research, we target all identified bottleneck problems by optimizing the concurrent memory reads, aligning memory accesses and reorganization of computations.

The following optimization approaches will be tested:

- **O1** – using shared memory for the input signal,
- **O2a** – using shared memory for the filter kernel,
- **O2b** – using constant memory for the filter kernel,
- **O3** – using loop unrolling,
- **O4** – smaller calculation precision, and
- **O5** – element version.

3.1 *Utilizing Shared Memory*

The O1 optimization approach utilizes the shared memory for storing relevant segment from the actual ECG signal, although the complete input is expected to be stored on the Global Memory.

Let any thread block consist of *TPB* threads, and the filter length is *F*. Each block will eventually require an input segment of length *TPB* + *F* from the original signal. The first *TPB* threads (within a given thread block) initialize the relevant shared memory. Before each convolution operation, the block threads are synchronized by the intrinsic CUDA synchronization primitive *__syncthreads()* in order to ensure consistent data when initializing the shared memory.

Similarly, the O2a optimization approach utilizes the shared memory for storing corresponding filter kernel coefficients. In this case, O1 and O2a will both occupy the shared memory and therefore, the capacity limitation will decrease the level of performance gain.

3.2 *Utilizing Constant Memory*

The O2b optimization approach, which is complementary to the O2a approach, uses CUDA constant memory to store the read-only filter kernel (weight) coefficients. In this manner, consecutive reads of the same address do not generate any additional memory traffic. It is important to note that no bank conflict happens when using constant memory.

A combination of the O1 and O2b optimization approaches will use shared memory for the input signal segment and constant memory for impulse response coefficients. This avoids the bank conflicts that occur when utilizing only shared memory. Figure 4 presents the idea of this version of the parallel CUDA algorithm.

3.3 *Loop unrolling*

The O3 optimization approach is based on unrolling the loop in the thread. Eventually, this will eliminate instructions that increment the loop index and test if the limit is reached and combine two or more loop bodies into one loop body. The technique decreases the number of realized operations per thread.

Figure 6 presents a simplified code segment that loops 100 times. Figure 6a shows the basic method, and Figure 6b a code that uses twice unrolled loop. The procedure decreases the number of overall processed instructions, since instead of 100 checks for the looping condition and 100 instructions to increase x, the program is now performing those instructions only 50 times.

```
for (x=0; x<100; x++)
{
        process(x);
}
```

a) Normal loop

```
for (x=0; x<100; x+=2)
{
        process(x);
        process(x+1);
}
```

b) Unrolling by a factor of 2

Fig. 6.A loop unrolling example.

3.4 *Precision decrease*

The naive parallel CUDA code used double precision as a base type. On devices of computing capability 2.x, each bank has a bandwidth of 32 bits every two clock cycles, and successive 32- bit words are assigned to successive banks [8]. The warp size is 32 threads accessing 32 locations in a memory bank, each with a 32-bit precision. When double precision is used, in the case of accessing 64-bit numbers, the total number of memory locations is 16. Thus, a simultaneous access of 32 threads is not possible in a warp to 32 words of 64-bit length will cause, and only a half warp can be executed.

In order to increase the performance, the O3 approach decreases the precision to 32-bit single precision mode. In such case bank capacity limitation will be eliminated since 32 threads in a warp will simultaneously access 32 words of 32-bit length. Devices with computing capability 3.x allow a bandwidth of 64 bits every clock cycle [4]. Thus, bank capacity limitation arising from using double precision will be eliminated for devices with computing capability 3.x or higher. In addition, processing a single precision instruction is faster than the double precision.

3.5 *Element version*

Moreover, the element version of convolution is designed and implemented for ECG signal. In this version, instead of rows, a thread computes multiplication and summation of elements. Threads are defined for each combination of input and kernel items. Summation, on the other hand, is performed in logarithmic time. However, this approach leads to use of a lot of synchronization, which can cause a performance decrease.

4. Experimental Methodology

This section describes the conducted experiments.

4.1 Testing Environment

The sequential and parallelized codes are tested on an Amazon EC2 - G2 2xlarge instance. It consists of an 8-core Intel(R) Xeon(R) CPU E5- 2670 2.60GHz 64-bit system with a 15GB of memory. Moreover, the parallelized code is tested on a NVIDIA GRID GPU (Kepler GK104) device, having 1,536 CUDA cores with an 800Mhz system clock and 4GB RAM. The total amount of constant memory is 64KB where each block has 48KB of shared memory. This device has a computing capability 3.x, and input data is 64-bit double precision.

4.2 Experiments and test cases

The experiments were defined by testing the each previously described optimization approach independently and in a combination with the other approaches. Each experiment had several test runs for various sizes of the input data stream that consist of 10.000 up to 500.000 samples of an ECG signal with incremental steps of 10.000 samples.

Since the sampling frequency is 500Hz, the input data stream present an ECG signal from 2 seconds up to 1000 seconds. The filter kernels tested consisted of M=100, 250, 500, 750 and 1000 elements.

Each test run for the experiments was tested at least five times and an average value of measured times was calculated and used for further processing. A functional verification was conducted for each test run to verify if the functional characteristics of the naïve and optimization parallel algorithm executions obtain identical results.

4.3 Test Data

The measured parameters in the experiments are T_{pn} as time required by executing the naïve parallel algorithm using n cores, and T_{on} as time required by the given experiment with an appropriate combination of optimization approaches. The speed-up is calculated by Equation (2) as a ratio of the measured times for execution of the sequential and parallel algorithms.

$$S_n = \frac{T_{pn}}{T_{on}} \tag{2}$$

5. Performance analysis

This section describes the conducted experiments and presents the obtained results.

5.1. Shared Memory

Figure 7 shows the speedup of the O1 optimization using the shared memory for the input signal. The optimized version is more efficient, and the speedup increases with the filter kernel size. An average value of 13% increase is obtained as speedup for the filter length of 100, and 78% for filter length of 1000. We observe that for each filter size, the speedup $S_n > 1$ and the greater speedup is achieved when increasing the filter length, which proves the scalability of the improvement.

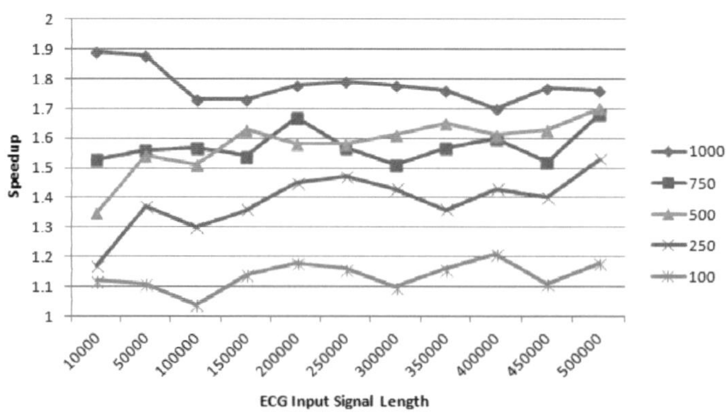

Fig. 7. Speedup of the O1 optimization using shared memory for the input signal in the case of different kernel sizes.

Additionally, Figure 8 shows the speedup of O2a optimization only using the shared memory for filter kernel. It can be noted code is optimized by roughly 10% and that speedup increases with the increasing filter kernel length. The higher the filter kernel is, the better performance gain is obtained. Better results are obtained for the usage of shared memory for storing the input signals, instead of the filter coefficients.

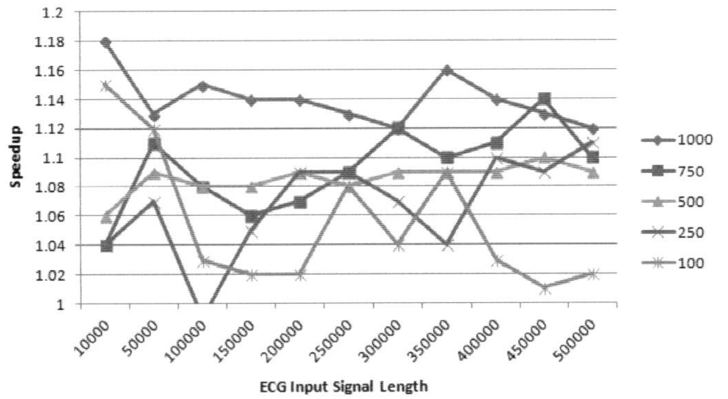

Fig. 8. Speedup of the O2a optimization using shared memory for different kernel sizes

5.2. Constant Memory

Figure 9 presents the analysis made on the constant memory defined by the O2b optimization approach. One can conclude that using a constant memory for the kernel will increase the performance by average 15%. We have compared the O2a and O2b approaches, as presented in Figure 10. The solution using constant memory instead of shared memory for storing the filter kernels obtains a higher performance by an average speedup of 4%.

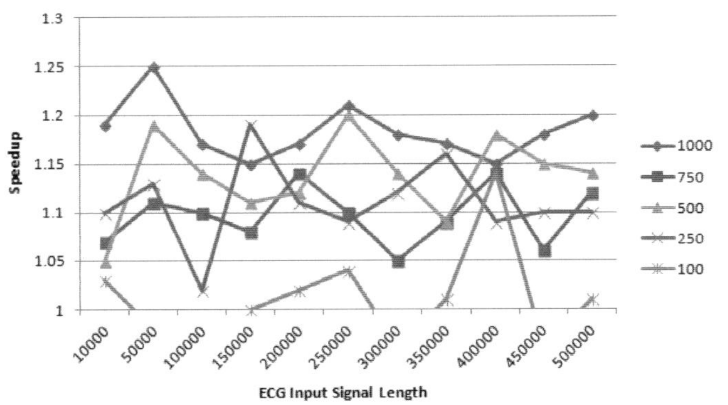

Fig. 9. Speedup of the O2b optimization approach using Constant Memory for the filter kernel.

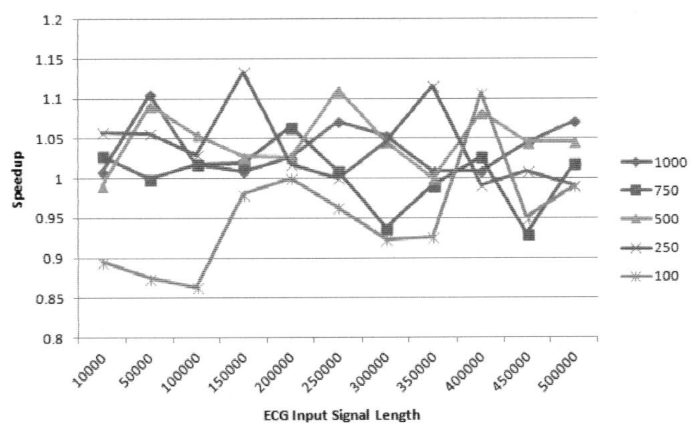

Fig. 10. Speedup of the O2b optimization approach compared to the O2a.

5.3. Loop Unrolling

The speedup obtained by loop unrolling O3 optimization approach is presented in Figure 11. On average, the loop unrolling optimization technique increases the performance for 1 to 5 percent. The best performance gain was obtained for the unrolling length of 64.

5.4. Decreasing the double to single precision

The conducted experiments generally use a 64-bit double precision for storing and calculating the convolution. However, a GPU double precision calculation is a costly operation [7]. We have tested the performance by decreasing the precision to 32-bit, since the functional testing showed difference in both approaches expressed in order of 10^{-5}. Figure 12 presents the effect of the decreased precision on the performance. As input size increases, the single precision version is approximately 40 to 60 percent faster than the pure parallel code.

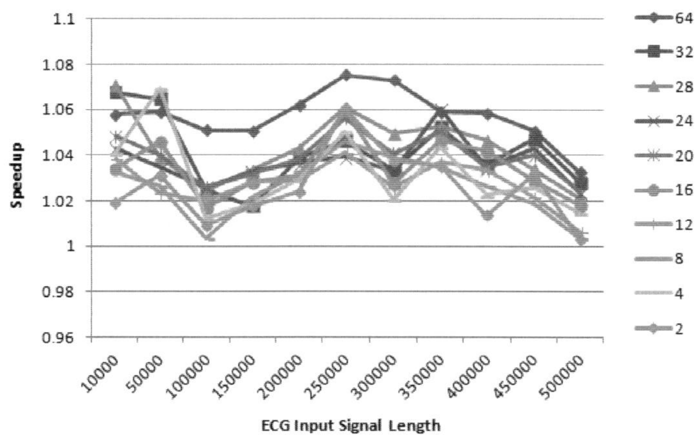

Fig. 11. Speedup of O3 optimization approach using loop unrolling for filter length of 1000.

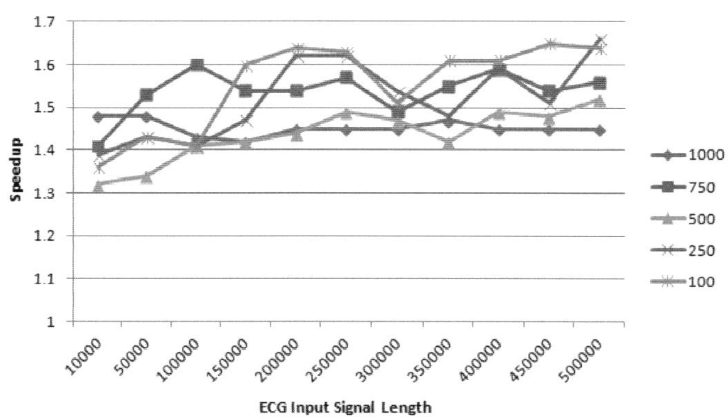

Fig. 12. Speedup obtained by the O4 optimization approach by decreasing the precision

5.5. Element version

The conducted experiments show that the element version actually decreases the performance. This happens due to the synchronization and bank conflicts that need to be performed in each step.

6. Discussion

Figure 13 combines the optimizations compared to the double precision version of the previous code. We observe that Shared Input and Output version speedups the code by 13% and 78%. Using Constant memory version is 15% faster compared to the previous code [2]. Combining these, the maximum speedup of 2.4 is obtained. Thus we can conclude that the hypothesis we set is confirmed.

We were also interested to investigate the effect of the loop unrolling. We observe that using a loop unrolling with a length of 64 introduces additional 4 % performance gain. Additionally, when the precision is decreased to 32 bit, the speedup increases up to 6. An interesting case was to see whether the element version performs better. We observed that, due to the internal synchronization and bank conflicts performed on each step, the performance of the element version is not attractive.

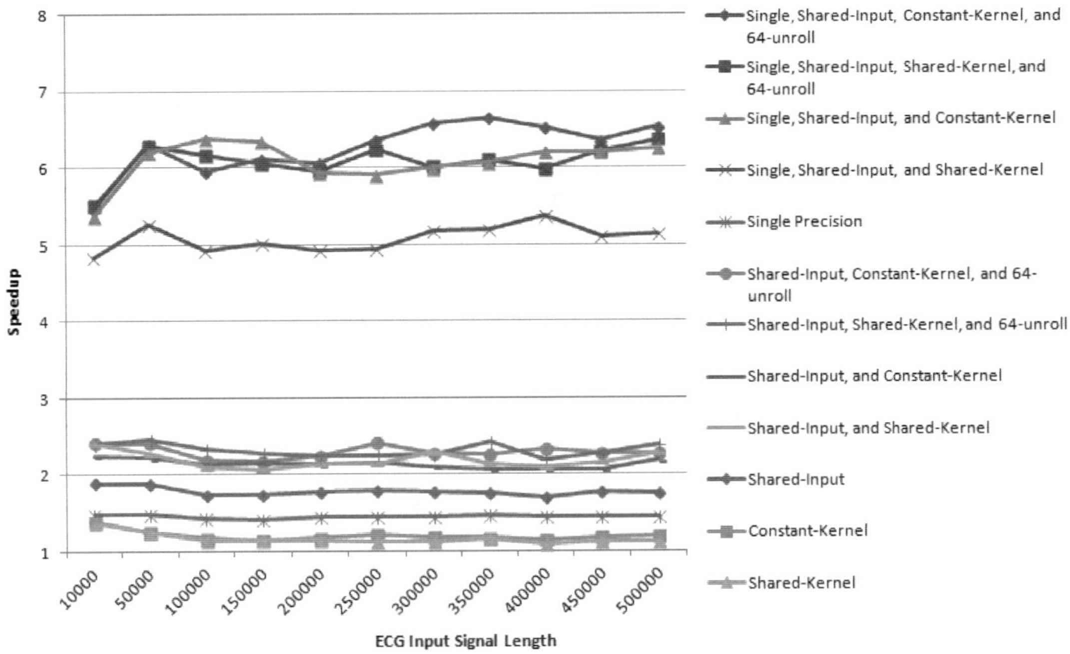

Fig. 13. Performance gain by a combination of optimization approaches.

Anwar et al. [9] have published a solution using shared memory, without analyzing the possibilities of the constant memory. Wefers et al. [12] have concentrated on implementing a real-time convolution on frequency domain. Herdeg et al. [13] have also concentrated on the frequency and time domain fast convolution. On time domain convolution, though, authors have not provided any information related to memory storage, access, and optimizations.

Previously we have considered parallelizing DSP filters on a Maxeler dataflow engine [1]. We obtained promising results, with linear speedups proportional to the kernel length. We calculated speedup by measuring a number of sequential steps needed to convolve the signal, whereas here we compared execution times.

Additionally, we have focused on parallelizing DSP filtering algorithm on CUDA [2], by utilizing thousands of GPU cores. The experiments have shown linear speedups, proportional to the kernel size. Obtained results serve as a basis for this research. We have also observed that threads per block (TPB) should be kept as high as possible by considering the upper limit being provided by the specification of the GPU card.

This research was concentrated on time-domain convolution. Generally, data intensive streams are convolved on the frequency domain, by actually multiplying two signals on the frequency domain. Gained results were much faster, but the results were not accurate. This is because ECG signal is not always periodic; since the heart beat rate changes due to physical activity, emotional state, etc. So, in the case of ECG, frequency-domain convolution is not recommended.

We also investigated the usability of NVIDIA's CUBLAS library to dispatch each of the required convolution calculations as a Level 1 BLAS (Basic Linear Algebra. Subprograms) caxpy (Scalar Alpha X Plus Y) operation. Caxpy computes a constant alpha times a vector x plus a vector y, and finally, overwrites the initial values of vector y. In the case of ECG signal, the impulse response has variable filter elements, making the use of this operation useless.

Moreover, we designed and implemented the element version of convolution for ECG signal context using explicit synchronization for reduction and summation. Even though the reduction operation has logarithmic complexity, our tests have shown that this approach decreases the performance. We plan to develop an optimized algorithm where multiplications will be distributed across different blocks, and the synchronization will be handled inside threads of a block. We believe that this will optimize the current version of the code.

Higher speedups are obtained when using single instead of double precision. This is directly related to the achieved GFLOPS (billions of floating point operations per second) of the device. On GPUs, double precision GFLOPS is smaller than the single precision [11], which results in faster executions when floating point numbers are used.

In this research, we have also considered using loop unrolling. Approximately 1-5 % performance gain is obtained with double precision. We have also tested to decrease the precision, and nearly doubled the performance of the algorithm.

7. Conclusion

This research contributes to the CUDA GPU optimization strategies for the noise elimination on ECG heart signals. The proposed optimizations, take advantage of intra-block shared memory and the general constant memory. By storing ECG segments on shared memories, we have optimized the memory accesses.

Additionally, we used storing of the filter coefficients on the constant memory, thus consecutive reads of the same address does not generate any additional memory traffic. We are aware of the limitations, where filter length should be smaller than the maximum number of threads per block due to the internal synchronization inside a block.

Results obtained by executing optimized algorithms show they are identical for each of the filter types. From the obtained results, we can conclude that proper usage of shared and constant memory has a positive impact on the performance. Our analysis showed that their combined effect yield 2.4 times faster executions compared to the previous code. We can, therefore, conclude that the hypothesis is confirmed.

Considering loop unrolling speeds up the code by 1-5%. Moreover, we tested the decreased precision effect on the performance and got nearly 1.5 faster code when on 1000 filter length. We observed that the element version is not effective when ported on GPU.

It is important to note that each of the proposed optimization techniques adds up to the combined speedup. We observed that the best-combined effect had a speedup of 6.

As future work, we plan to carry out more tests on multiple GPUs tied together. We will analyze shared memory parallelism by using OpenMP or similar solutions. We also consider developing an efficient algorithm for filter sizes greater than the available maximum number of threads per block. Additionally, we plan to investigate the element algorithm version on CUDA and Intel IPP. In this algorithm, threads are defined on the element level. We expect that the parallelized code will scale linearly with increasing filter size.

8. References

[1] E. Domazet, M. Gushev, and S. Ristov, "Dataflow DSP filter for ECG signals," in 13th International Conference on Informatics and Information Technologies, Bitola, Macedonia, 2016, in press

[2] E. Domazet, M. Gushev, and S. Ristov, "CUDA DSP filter for ECG signals," in 6th International Conference on Applied Internet and Information Technologies, Bitola, Macedonia, 2016, in press

[3] T. S. Lugovaya, "Biometric human identification based on ECG," http://www.physionet.org/pn3/ecgiddb/biometric.shtml, 2005, last visited on 28.05.2016

[4] "Cuda C best practices guide," http://docs.nvidia.com/cuda/pdf/CUDA_C_Best_Practices_Guide.pdf, licensed under the Creative Commons Attribution 3.0 Unsupported license, last visited on 28.05.2016.

[5] S. W. Smith, Digital signal processing: a practical guide for engineers and scientists. Newnes, 2003

[6] "Schematic diagram of normal sinus rhythm for a human heart as seen on ECG", Wikipedia, https://commons.wikimedia.org/wiki/File:SinusRhythmLabels.svg, released work into the public domain, last visited on 28.05.2016.

[7] Whitehead, Nathan, and Alex Fit-Florea. "Precision & performance: Floating point and IEEE 754 compliance for NVIDIA GPUs." rn (A+ B) 21 (2011): 1-1874919424.

[8] "Tuning CUDA Applications for Maxwell," http://docs.nvidia.com/cuda/maxwell-tuning-guide/, last visited on 28.05.2016.

[9] Anwar, Sajid, and Wonyong Sung. "Digital signal processing filtering with GPU." (2009).

[10] Sörnmo, Leif, and Pablo Laguna. "Electrocardiogram (ECG) signal processing." Wiley Encyclopedia of Biomedical Engineering (2006).

[11] Geeks3D, "AMD Radeon and NVIDIA GeForce FP32/FP64 GFLOPS Table", http://www.geeks3d.com/20140305/amd-radeon-and-nvidia-geforce-fp32-fp64-gflops-table-computing/, last visited on 28.05.2016

[12] Wefers, Frank, and Jan Berg. "High-performance real-time FIR-filtering using fast convolution on graphics hardware." Proc. of the 13th Conference on Digital Audio Effects, 2010

[13] McGraw-Herdeg, Michael P., Douglas P. Enright, and B. Scott Michel. "Benchmarking the NVIDIA 8800GTX with the CUDA Development Platform." HPEC 2007 Proceedings (2007)

DOI: 10.2507/27th.daaam.proceedings.092

DESIGN AND SIMULATION OF A FLEXIBLE CUTTING MACHINE FOR ROLL FORMING PRODUCT

Hong Seok Park, Duc Viet Dang, Gyu Bong Lee & Ivica Veža

This Publication has to be referred as: Park, H[ong] S[eok]; Dang, D[uc] V[iet]; Lee, G[yu] B[ong] & Veza, I[vica] (2016). Design and Simulation of a Flexible Cutting Machine for Roll Forming Product, Proceedings of the 27th DAAAM International Symposium, pp.0633-0644, B. Katalinic (Ed.), Published by DAAAM International, ISBN 978-3-902734-08-2, ISSN 1726-9679, Vienna, Austria
DOI: 10.2507/27th.daaam.proceedings.092

Abstract

Development of a virtual simulation for machine is the imperative task to resolve time issue, improve cost saving and iterative improvements of physical prototype. The objective of the present paper was to design a flexible cutting machine and analyze the interactions between the machine structure and control system via a multi-body simulation (MBS) model and the virtual controller. Firstly, a reliable model of the flexible cutting machine was developed with the help of SOLIDWORKS and Automatic Dynamic Analysis of Mechanical System (ADAMS). Subsequently, a virtual controller used adaptive siding mode PID (ASMP) based on MATLAB/Simulink was designed. Finally, the co-simulation was performed to derive the machine movements under the effects of manufacturing conditions. The simulation results showed that the virtual mechanical model could precisely described the dynamic behavior of the flexible cutting machine. Furthermore, the proposed ASMP controller is robust regardless of the system parameter and disturbance variations. Therefore, this work is expected as a contribution to improve the prototype's performance of the flexible cutting machine.

Keywords: Flexible cutting machine; Roll forming; Multi-body simulation; Adaptive sliding mode; PID controller; Modelling and controlling.

1. Introduction

Roll forming parts are frequently applied to many kinds of industry and automobile sector. Along with straight roll forming product, some of roll forming products require a cured path (usually in arc form). In this case, the cut-off process is more difficult than that of straight products. The roll forming process requires the cutting operation to separate a long formed strip to a desired length. The cutting devices for straight roll forming parts are developed at a high level as illustrated in Fig. 1. However, the cutting device for curved products is still underdeveloped. Consequently, design a flexible and multi-functional cutting devices that can cut different cross section and product with arc or curvature in the longitudinal direction of roll forming product is a significant research. In order to develop a new product, the physical prototype can no longer suitable due to the time -consuming and cost-intensive manufacturing and testing to detect weak spots and optimize design [1]. Fortunately, literature reviews have performed that approaches based upon virtual prototype simulation, using well-defined material properties,

Fig. 1 The cutting devices for straight product

numerical models, and controllers, can be considered as an intelligent solution to obtain reliable results. With the support of the virtual prototype, the designer can analysis and optimize the mechanical design or the control design to achieve a perfect system, therefore it saves time and cost and reduces the risk of device damage caused by the conflict between both systems. There has been a considerable amount of research in the virtual prototype simulation which demonstrated the advantages of this method, such as Catalin et al. [2] simulated the windshield wiper system with the support of the virtual prototyping platform. Zouhaier [3] has built the crank slider mechanism to illustrate the behavior of the whole mechatronic system under different control strategies. Hajicek [4] presented a methodology for establishing virtual CNC machine tool. Moreover, the virtual prototypes of industrial equipment have been proposed to investigate dynamic behaviour in the initial stage [5] [6]. The results of the aforementioned efforts indicated that virtual prototyping techniques could be considered as a powerful approach to simulation and optimization the physical prototype of the new product. Therefore, this works attempted to develop a flexible cutting machine with the virtual prototype simulation.

This paper is organized as follows: the research methodology is described briefly in section 2. The concept of the flexible cutting machine is performed in section 3 and the multi-body system model is shown in section 4. Section 5 will be explained the virtual controller. The results of co-simulation are showed section 6 and the conclusions are presented in this section 7.

2. Research methodology

The virtual prototype simulation platform includes CAD-Computer aided design, MBS-Multibody systems, FEA-Finite element analysis and DFC-Design for control programs [2], as depicted in Fig. 2. The CAD environment is used to create the geometric model of the system from the concept of the manufacturer, which contains information about the mass and inertia properties of the rigid parts. The MBS software, which represents the central component of the virtual prototyping platform, is used for analyzing, optimizing and simulating the kinematic and dynamic behavior of the mechanical system. The FEA software is used for modeling flexible components. DFC is a software product which is used to design the control system. This software exchange information with the MBS software. The exchange process creates a closed loop in which the outputs of the MBS model are the inputs for the control system and conversely. The combination between the MBS and the DFC system will be implemented the virtual prototype.

The virtual prototype simulation of the flexible cutting machine is established by the combination ADAMS software and MATLAB/Simulink software as described in Fig. 3. With the concept of the machine, the modeling will be established in SOLIDWORKS and then exported to ADAMS software to create the multi-body system, with the possibility of virtual measurement of any parameters of any components in the virtual model. The controller will be designed in MATLAB environments which is well-known for designing a control system. The co-simulation model of both softwares will generate the virtual prototype which is performed the real behavior of the machine.

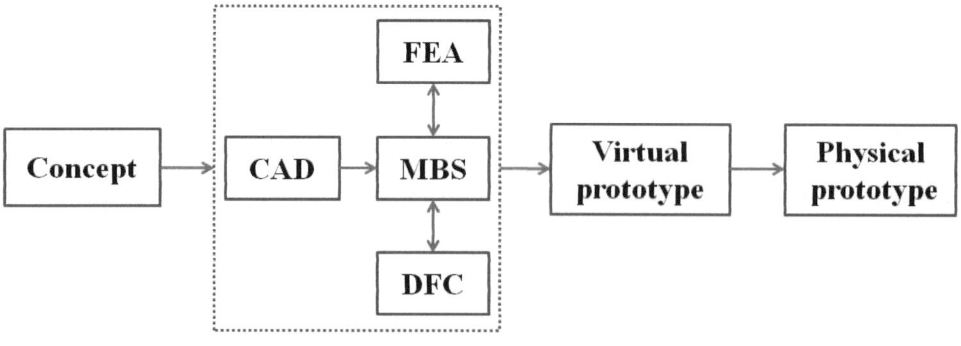

Fig.2. The virtual prototyping platform

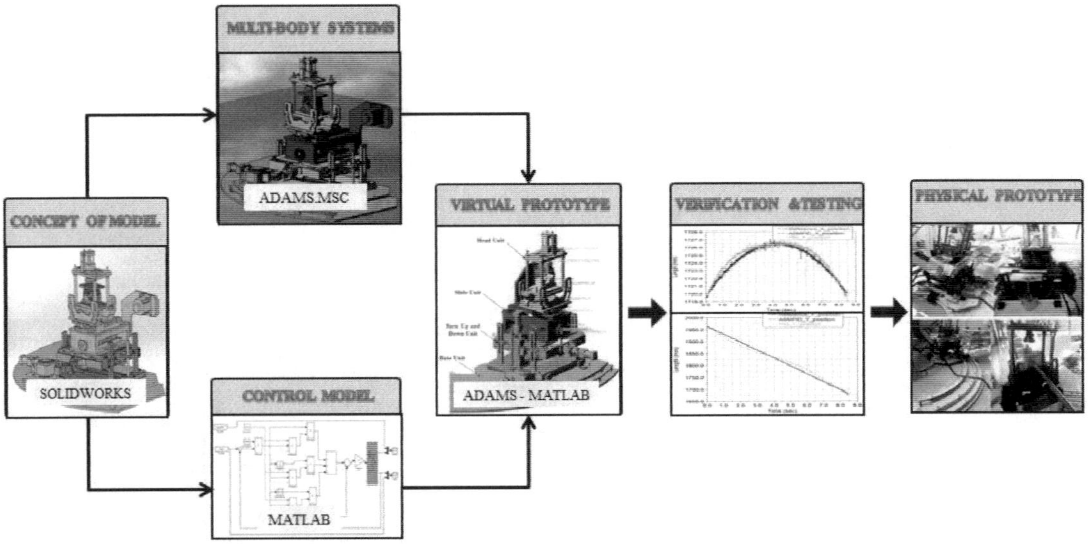

Fig. 3. The strategy for design and simulation of the flexible cutting machine.

3. The concept of the flexible cutting machine

Based on the straight forming machine, we attempt to develop the flexible cutting machine. This machine is used to manufacture the products require a cured path (usually in arc form), as shown in Fig. 4.

Fig. 4. The main types of work piece will be manufactured by the developed flexible cutting machine

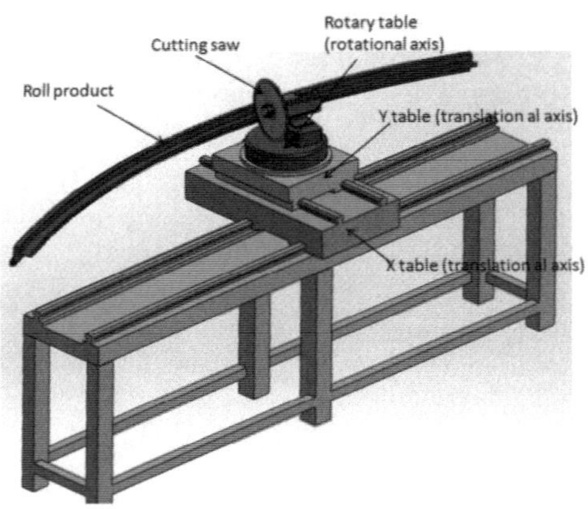

Fig. 5. The cutting machine with three degrees of freedom

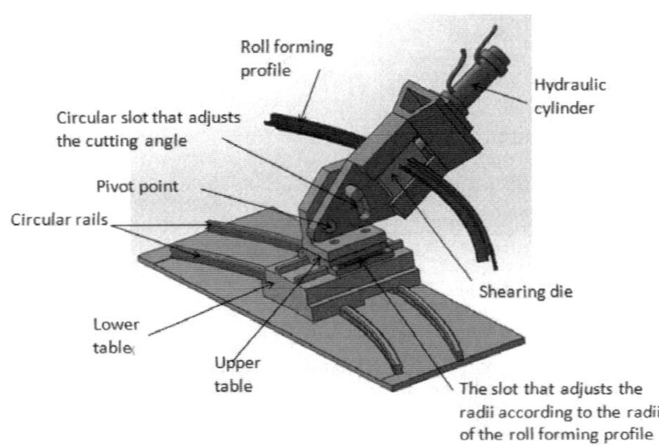

Fig. 6. The cutting machine with two degrees of freedom

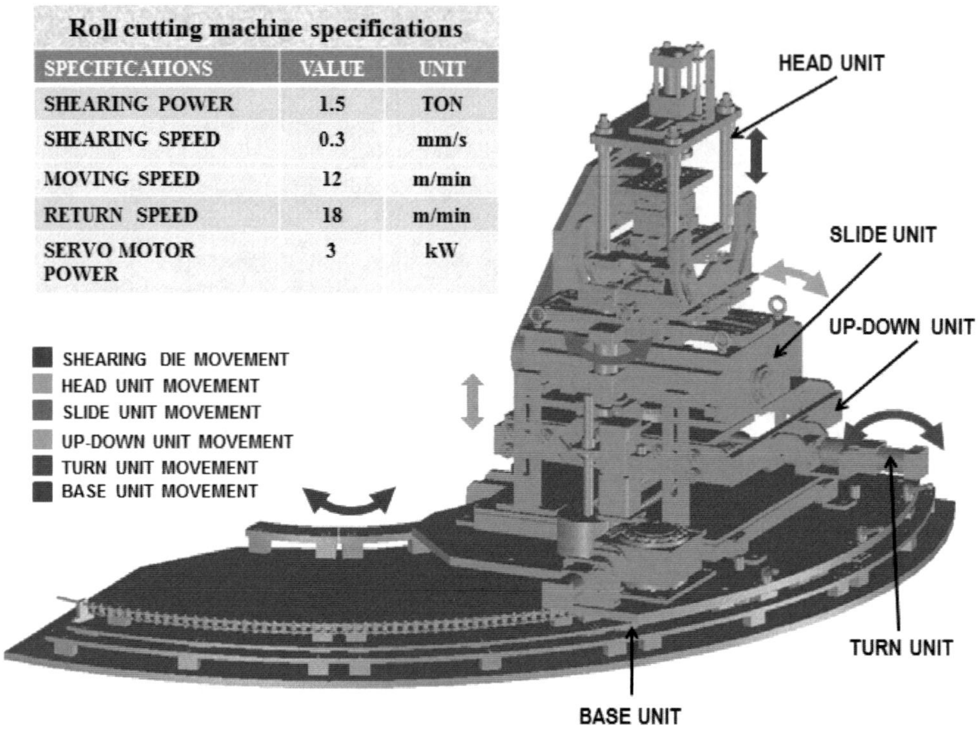

Roll cutting machine specifications

SPECIFICATIONS	VALUE	UNIT
SHEARING POWER	1.5	TON
SHEARING SPEED	0.3	mm/s
MOVING SPEED	12	m/min
RETURN SPEED	18	m/min
SERVO MOTOR POWER	3	kW

Fig. 7. The specifications and functionality of the flexible cutting machine

To satisfy the design requirements of the flexible cutting machine, two conceptual designs were proposed. The cutting machine using three degrees of freedoms and cutting saw was shown in Fig. 5, which includes two translations (longitudinal and traverse axis) and a rotational axis. The cutting movements are conducted based a combination of the machine bed and rotary table.

The conceptual machine using two degrees of freedoms is presented in Fig. 6. During forming process, the machine moves on two concentric circular rails. We can change the radius of the moving path by adjusting the position of the upper table. The cutting angle can be adjusted by means of a rotational pivot.

Based on the combination of the two proposed concepts, a concept design of the flexible cutting machine was proposed using five axes. The specifications and the construction of the roll cutting machine are shown in Fig. 7. During forming process, the base, turn, up-down, slide, and head unit are fixed. The head unit moves in the specific curved paths that is similar to the product profiles. The cutting movement at the end point of the contour is conducted by shearing die with the support of hydraulic cylinder.

The developed forming machine is a kind of moving cutting mechanism one. The cutting mechanism that not only generates the curved parts, but also cut off into accurate distance. The advantages of the flexible cutting machine are as follows compared to the conventional one:

- The machine can cut any kind of roll forming product regardless of straight or curve profile.

- Cutting distance is accurate because the mechanism is driven by servo motor.

- Hydraulic shearing mechanism with low noise.

4. Multi-body system for the flexible cutting machine

4.1. Analytical modeling of a constrained multi-body mechanical system

In this work, the flexible cutting machine was investigated as a multi-body mechanical system including of several rigid bodies. In the multi-body system [7], the generalized coordinates associated with rigid body *i* is denoted in what follows by

$$\boldsymbol{q}_i = \begin{bmatrix} p_i \\ \varepsilon_i \end{bmatrix} \tag{1}$$

with p_i, ε_i is the position and the orientation of a rigid body, respectively. For the system model containing nb bodies, the vector $\boldsymbol{q}$ is represented to describe the position and orientation of each body at a given time in the system.

$$\boldsymbol{q} = [\boldsymbol{q}_1^T \;\; \boldsymbol{q}_2^T \;\; \dots \;\; \boldsymbol{q}_{nb}^T]^T \tag{2}$$

Joints are regarded as constraints in the multi-body system which generally are obtained as

$$\Phi = \Phi(\boldsymbol{q}, t) = 0 \tag{3}$$

The entire constraint equations imposed by the joints are showed as

$$\Phi(\boldsymbol{q}, t) = [\Phi_1^T(\boldsymbol{q}, t) \;\; \Phi_2^T(\boldsymbol{q}, t) \;\; \dots . \;\; \Phi_{nj}^T(\boldsymbol{q}, t)]^T \tag{4}$$

where, nj is the number of joints in the system. By taking one time derivative of the position kinematic constraint equation (4), the velocity kinematic constraint equations are expressed as follows

$$\Phi_q(\boldsymbol{q}, t)\dot{\boldsymbol{q}} = -\Phi_t(\boldsymbol{q}, t) \tag{5}$$

By taking one time derivative of the kinematic constraint equations Eqn (5), the acceleration kinematic constraint equations are obtained as

$$\Phi_q(\boldsymbol{q}, t)\ddot{\boldsymbol{q}} = -(\Phi_q \dot{\boldsymbol{q}})_q \dot{\boldsymbol{q}} - 2\Phi_{qt}\dot{\boldsymbol{q}} - \Phi_{tt}(\boldsymbol{q}, t) = \gamma \tag{6}$$

The Lagrange multipliers form of the constraints acting on the multi body system can be written as

$$\boldsymbol{M}\ddot{\boldsymbol{q}} + \Phi_q^T \lambda = \boldsymbol{F} \tag{7}$$

where, $\boldsymbol{M}$ is the generalized mass matrix that equates $diag[\boldsymbol{M}_1 \;\; \boldsymbol{M}_2 \;\; \dots \;\; \boldsymbol{M}_n]$; $\boldsymbol{F}$ is the system force vector; and λ is the array of Lagrange multipliers. Combining Eqns (6) and (7), the complete set of constrained equations of motion can be written as follows

$$\begin{bmatrix} \boldsymbol{M} & \Phi_q^T \\ \Phi_q & 0 \end{bmatrix} \begin{bmatrix} \ddot{\boldsymbol{q}} \\ \lambda \end{bmatrix} = \begin{bmatrix} F \\ \gamma \end{bmatrix} \tag{8}$$

Theories of multi-body systems are embedded to modelling and analysis using computational method. It allows the designer to observe how the mechanism functions without having to build an expensive real world model and doing labour intensive measurements.

4.2. Multi-body system model

In order to generate the multi-body system of the flexible cutting machine, all components or elements which have the shape and dimensions of the physical model are modeled as 3D solid by SOLIDWORKS. The geometry was transferred to ADAMS environment using the standard file parasolid. The modeling process is constructed as the following sequence for an easy modification in the design phase. Firstly, the geometrical parameters of the parts such as material properties must be defined, then mass and inertial matrices are generated automatically with main parameters in Table 1. The flexible cutting machine model in ADAMS takes several aspects into account, such as gravity, contact constraints, joints, friction, motion and reference markers. All these definitions and setting must be defined properly for good approximation of real machine behavior in the virtual environment. The multi-bosy system of the flexible cutting machine in ADAMS software is depicted as Fig. 8.

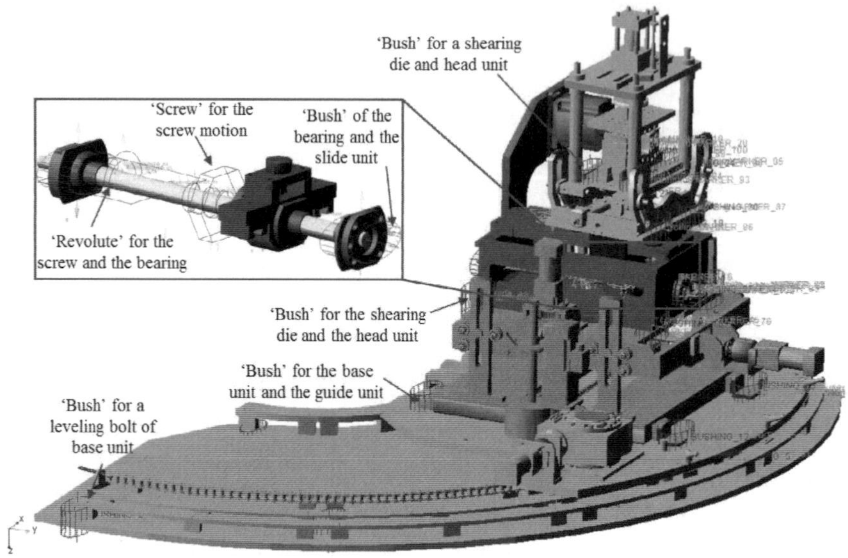

Fig. 8 . The multi-body system model of the flexible cutting machine in ADAMS. MSC

Body name	**Material**	**Mass (kg)**	**I_{xx} (kg.m^2)**	**I_{yy} (kg.m^2)**	**I_{zz} (kg.m^2)**
Base Unit	Cast steel	1424.6	1063.65	796.51	278.63
Turn Up and Down Unit	Cast steel	421.34	78.44	52.62	43.95
Slide Unit	Cast steel	280.76	42.24	35.73	13.38
Head Unit	Cast steel	172.69	857.78	699.35	1175.74
Shearing Die	Die steel	38.01	188.19	154.14	258.99

Table 1. Characteristics of main components of the flexible cutting machine in ADAMS

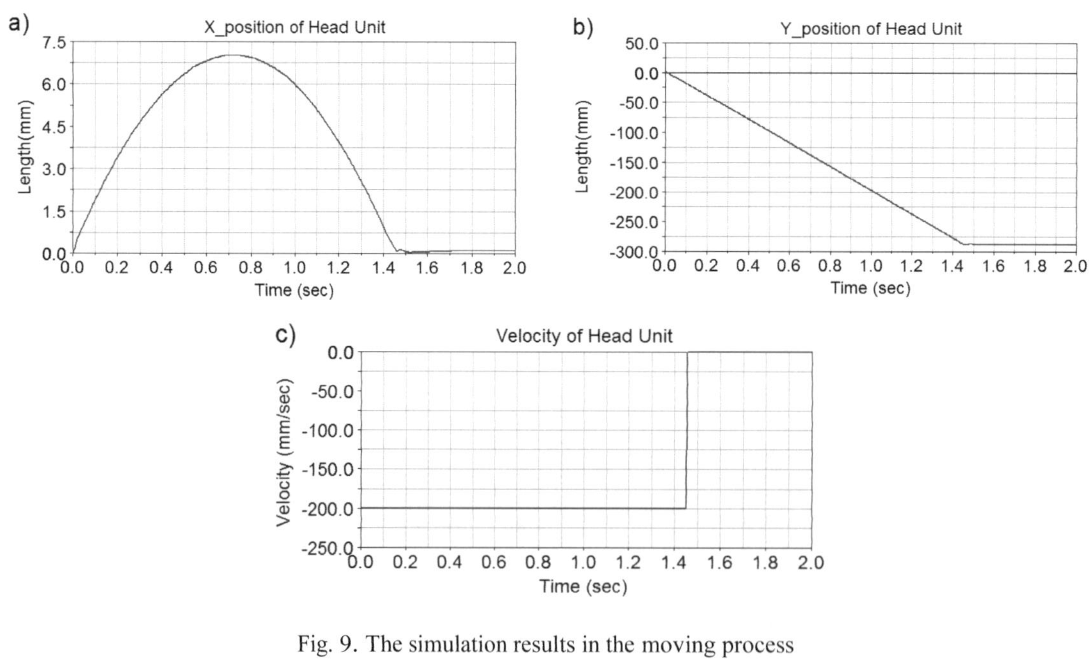

Fig. 9. The simulation results in the moving process

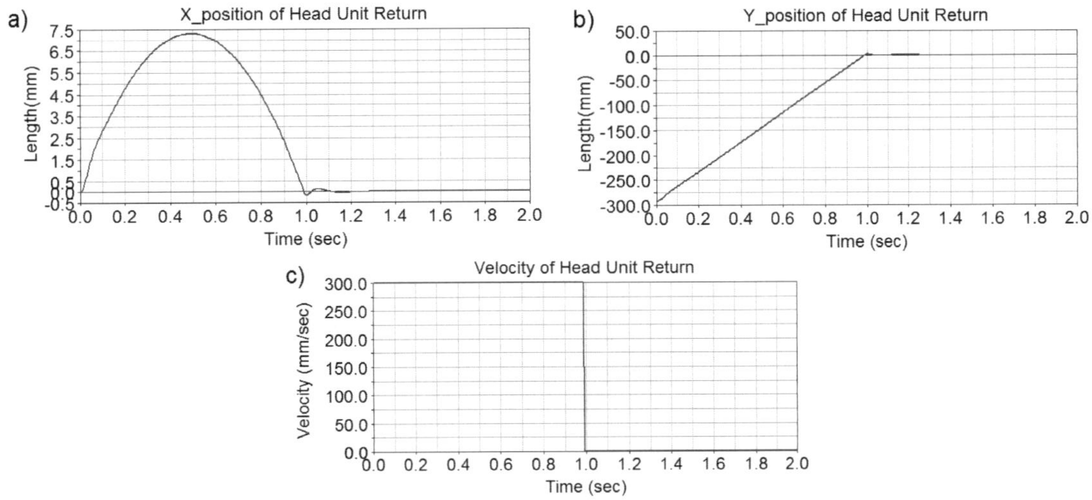

Fig. 10. The simulation results in the return process

The simulation results of the position and velocity of the head unit in ADAMS software as shown in Fig. 9 and Fig. 10 The trajectory of the head unit in X-axis and Y-axis are reflected correctly the shape of the product. Moreover, the velocity of the head unit is 200mm/s in the moving process and 300mm/s in the return process as the specifications of the machine when design. Therefore, the multi body simulation model can be used to imitate the characteristics of the flexible cutting machine.

5. Development of a control system for the flexible cutting machine

5.1. The interaction between the multi-body system and the control system

The principle of the connection between ADAMS and MATLAB/Simulink is depicted as Fig. 11. The input signal is a torque of the main servo motor and the output signals are the measured parameters of the position of the head unit. Subsequently, this model is exported to MATLAB/Simulink which have a *.m file and an adams_sys are created as shown in Fig. 12. It is also generates a command file (*.cmd) and a dataset file (*.adm), which will be used during the simulation process. The adams_sys block is created based on the information from the *.m file. With these file, the control system block will be created in MATLAB/Simulink.

5.2. Controller design

The PID control method can be applied easily and is widely known in many control applications because of its simplicity and effectiveness. However, this method is insensitive to parameter changes as proportional gain K_P, integral gain K_I, derivative gain K_D and only suitable for the linear system. Luckily, there has been extensive interest in self-tuning these three controller gains. For examples, the PID self-tuning methods based on the relay feedback technique were presented for a class of systems [8]. An adaptive PID control tuning was proposed to cope with the control problem for a class of uncertain chaotic systems with external disturbance [9]. Moreover, sliding mode control (SMC) is one of the popular strategies to deal with uncertain control systems [10], [11]. The main feature of SMC is the robustness against parameter variations and external disturbances. Consequently, the adaptive PID with SMC control system (ASMP) can be updated online the PID gains with an adequate adaptation mechanism that is adapted with variations of system parameters and external disturbances. In this work, the ASMP is proposed to control the rolling process of machine. Fig. 13. shows the block diagram of the adaptive SMC with PID control system for the flexible cutting machine.

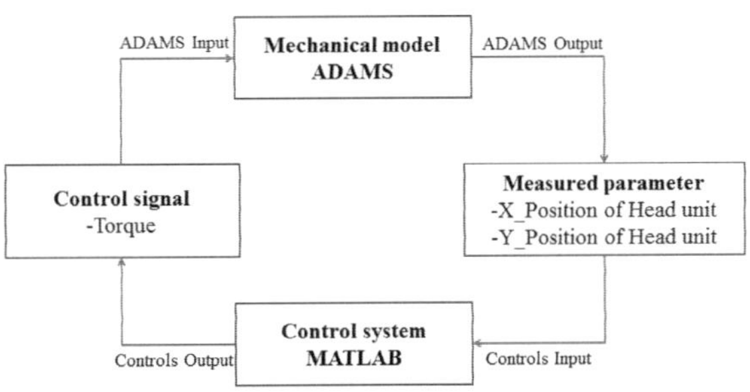

Fig.11. The principle of the connection between ADAMS and MATLAB/Simulink

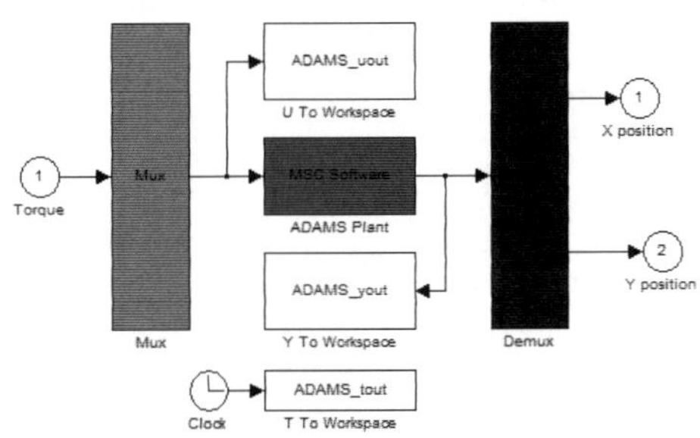

Fig.12. The adams_sub block

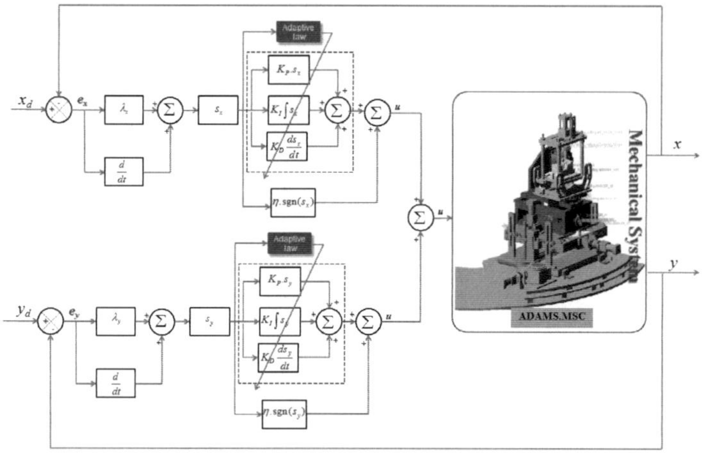

Fig. 13. Proposed ASMP control system scheme.

The flexible cutting machine is considered as a system with a single-input and multi-output. The state-space equation which expresses the head unit position can be expressed as follows:

$$\begin{aligned} \dot{x}_1(t) &= x_2(t) \\ \dot{x}_2(t) &= f_1(X) + b_1(X)u_x + d_1(t) \\ \dot{x}_3(t) &= x_4(t) \\ \dot{x}_4(t) &= f_2(X) + b_2(X)u_y + d_2(t) \end{aligned} \tag{9}$$

where, $X = (x_1, x_2, x_3, x_4)$ is a state variable vector that represents the position and velocity of the head unit in the X-axis and Y-axis, respectively. The $f_1(X)$, $b_1(X)$, $f_2(X)$ and $b_2(X)$ are the nonlinear functions. The $d_1(t)$ and $d_2(t)$ are presented for the bounded lumped disturbances which include the system parameter variations and external disturbances. The u_x and u_y are the control input to control the position of the head unit in X-axis and Y-axis, respectively.

From (9), the mechanical model has two subsystems: the position of the head unit in X-axis and Y-axis. Therefore, the sliding surfaces of the two subsystems are defined as:

$$s_x = \dot{e}_x + \lambda_x e_x \tag{10}$$

$$s_y = \dot{e}_y + \lambda_y e_y \tag{11}$$

In this paper, a control input u were designed to control simultaneously the position of the head unit in X-axis and Y-axis. The design methods for these two controllers are similar, therefore, we will show the algorithm to drive the head unit's position in X-axis. With the equation (10),
where $e_x = x_d - x$; x_d: desired position; x: measured position; and λ_x: positive constant.
Derivative (10) and substituting $\dot{x}_2 = \ddot{x}$

$$\dot{s}_x = \ddot{x}_d - f_1(X) - b_1(X)u_x - d_1(t) + \lambda_x \dot{e}_x \tag{12}$$

The control input of PID controller is designed based on (12):

$$u_{PID} = \frac{1}{b_1(X)}[\ddot{x}_d - f_1(X) - d_1(t) + \lambda_x \dot{e}_x] = AB + \varepsilon \tag{13}$$

where, $A = [K_P\ K_I\ K_D]$ is the vector of the gain of PID controller and $B = [s \int s \ \frac{ds}{dt}]^T$ is a basic vector of PID controller; ε is an appropriate error. The control signal u_x of the controller is determined as:

$$u_x = u_{PID} + u_h = \hat{A}B + u_h \tag{14}$$

where, $\hat{A} = [\hat{K}_P\ \hat{K}_I\ \hat{K}_D]$ is estimated value of vector A; u_h: control signal of the auxiliary controller.

Substituting (14) into (12) yields

$$\begin{aligned} \dot{s}_x &= \ddot{x}_d - f_1(X) - b_1(X)[\hat{A}B + u_h] - d_1(t) + \lambda_x \dot{e}_x \\ &= b_1 A \ B + b_1 \varepsilon - b_1 u_h \end{aligned} \tag{15}$$

where $A = A - \hat{A}$ is estimation error. In order to prove the stability, the Lyapuvov function can be used.

$$V = \frac{1}{2} s^2_x + \frac{1}{2\gamma} A^2 \tag{16}$$

Derivative (16) yields:

$$\begin{aligned} \dot{V} &= s_x \dot{s}_x + \frac{1}{\gamma} A \dot{A} \\ &= s_x(b_1 AB + b_1 \varepsilon - b_1 u_h) + \frac{1}{\gamma} A \dot{A} \\ &= (s_x b_1 B + \frac{1}{\gamma} \dot{A})A + s_x b_1 \varepsilon - b_1 u_h s_x \leq 0 \end{aligned} \tag{17}$$

From (17), we have $(s_x b_1 B + \frac{1}{\gamma} \dot{A}) = 0$

Hence, $\dot{A} = -\dot{\hat{A}} = -\gamma s_x b_1 B$

The three PID gains *(K_P, K_I, and K_d)* are on-line updated by the following adaptive laws:

$$\begin{aligned} \dot{\hat{K}}_P &= \gamma s_x b_1 s_x, \\ \dot{\hat{K}}_I &= \gamma s_x b_1 \int s_x, \\ \dot{\hat{K}}_D &= \gamma s_x b_1 \frac{ds}{dt} \end{aligned} \tag{18}$$

Considering (17) $\dot{V} = s_x b_1 \varepsilon - b_1 u_h s_x \leq 0$

With the auxiliary controller : $u_h = \eta \operatorname{sgn}(s)$; the sign function

$$\text{Where: } \operatorname{sgn}(s) = \begin{cases} 1 \ if \ s > 0 \\ 0 \ if \ s = 0 \\ -1 \ if \ s < 0 \end{cases} \tag{19}$$

We have :

$$\begin{aligned} \dot{V} &= s_x b_1 \varepsilon - b_1 \eta \operatorname{sgn}(s) s_x \\ &= s_x b_1 \varepsilon - b_1 \eta |s_x| < b_1 |s_x| (|\varepsilon| - \eta) < 0 \\ &\Rightarrow \eta > |\varepsilon|, \end{aligned} \tag{20}$$

Equation (17) proves that the sliding surface is stability. The control input for controlling the flexible cutting machine is composed of the position control input u_x and u_y as:

$$u = u_x + u_y \tag{21}$$

6. Simulation results

In this section, the co-simulation is implemented between the multi-body system and the ASMP controller. First of all, the proposed ASMP is applied for tracking control of the head unit of the roll forming machine, and the comparisons with PID control are presented in Fig. 14. From this figure, we can see that both PID and ASMP have the overshoot and settling time is nearly zero, and the steady state error has satisfied the stable control criteria, and disturbances are perfectly eliminated very fast responses to overcome the initial error conditions. However, the proposed ASMP control obtained much better tracking performance than PID control, indeed, the error tracking of ASMP control is smaller than PID control in the X-direction and Y-direction. Consequently, the ASMP controller approach is suitable for the roll forming machine which robust regardless of the system parameter and disturbance variations strategy.

Furthermore, the simulation results of the virtual prototype machine were conducted to derive the important parameters to optimize the machine prototype. Fig. 15 shows the dynamic simulation of the head unit, it was verified accurately with the specification of machine. Moreover, the value force and torque of the ball-screw and the head unit is shown in Fig. 16. The force of the ball screw impact to the head unit to implemented the rolling process is maximum to 750N, as illustrated in Fig. 16a. To make the head unit moving followed the curved part of the product, the largest contact force is 130N, as presented in Fig. 16b. Fig. 16c demonstrated the maximum torque to the ball screw is nearly 2000 N-mm. Therefore, these results were considered to support design engineers to decide the best parameters for the machine.

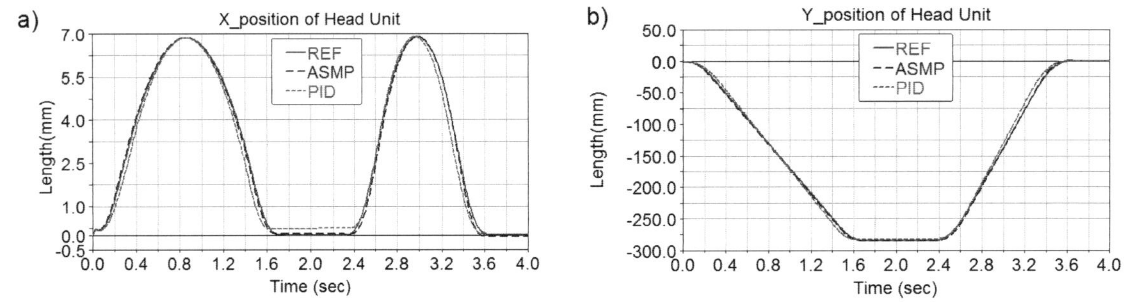

Fig. 14. Tracking the performance of the head unit of roll forming machine

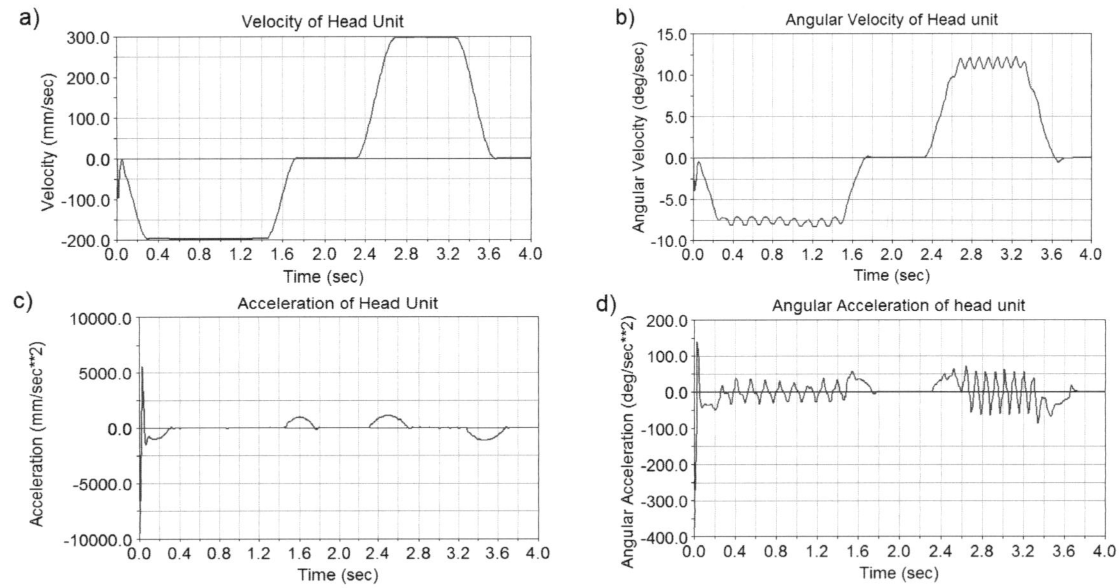

Fig. 15. The dynamic behavior of the head unit

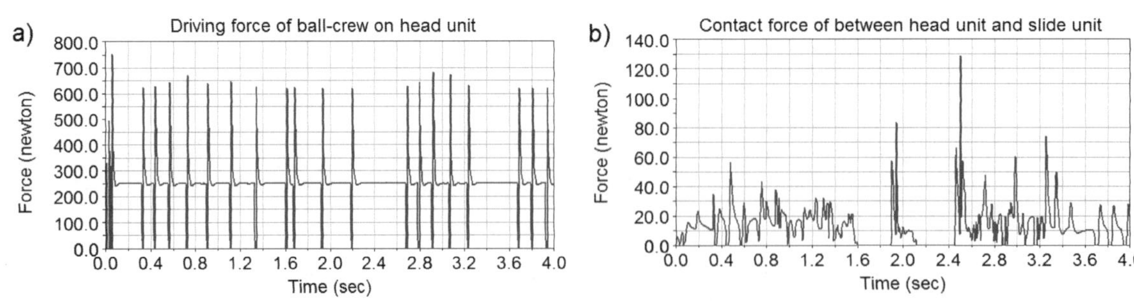

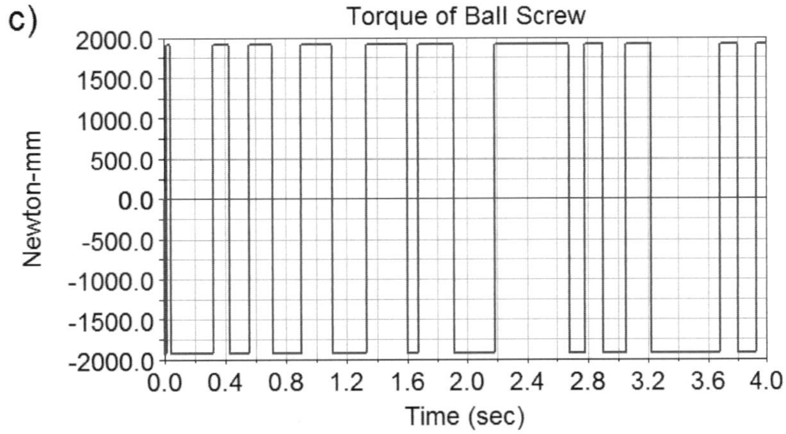

Fig. 16. The force of main components in the roll forming machine

Fig. 17. Physical prototype of the flexible cutting machine for roll forming product.

7. Conclusion

In this paper, the flexible cutting machine for roll forming product was designed and simulated, based on a multi-body system and the implementation of modern control technology. The mechanical model of the flexible cutting machine is established with the concept ideas to response the cutting requirements of roll forming product. Subsequently, the modelling of machine was transferred to ADAMS environment for analysis the dynamic behavior and the ASMP controller was designed for controlling the nonlinear process movement of the machine in MATLAB/Simulink. The virtual prototype simulation has already performed that the ASMP controller robustly against disturbance, increased productivity by reducing downtime, the virtual mechanical model could precisely described the dynamic behavior of the flexible cutting machine. Significantly, the approach can support the designers to modify the mechanical and control designs to eliminate errors and increase the accuracy. Therefore, this work is expected as a contribution to improving the prototype's performance of the flexible cutting machine.

In spite of the simulation models are useful, physical experiments to verify the simulation and optimization results are indispensable. Fig.17 shows the physical prototype is established in the company. The verification of the proposed ASMP method will be implemented in this test and the results of the experimental study will be presented in a future paper.

8. References

[1] C. Brecher, M. Weck, S. Witt Y. Altintas. (2005). Virtual Machine Tool CIRP Annals - Manufacturing Technology, vol. 54, no. 2, pp. 115-138

[2] Catalin Alexandru, Claudiu Pozna. (2009), Dynamic Modeling and Control of the Windshield Wiper Mechanisms, Wseas transactions on systems, vol. 8, no. 7

[3] Zouhaier Affi, Lotfi Romdhane. (2005), ADAMS/Simulink interface for Dynamic Modeling and Control of Closed Loop Mechanisms, Proceedings of the 7th WSEAS international conference on Automatic control, modeling and simulation, USA, ISBN:960-8457-12-2, pp. 353-356

[4] Zdenek Hajicek. (2015), Simulation of virtual machine tool during the development phase, Proceedings of the 26th DAAAM International Symposium, ISSN 1726-9679, ISBN 978-3-902734-07-5, pp. 0948-0954

[5] Park Hong Seok & Le Ngoc Tran. (2011), A 3d simulation system for mobile harbourcrane based on virtual prototyping technology, Proceedings of the 22nd International DAAAM Symposium, ISSN 1726-9679, ISBN 978-3-901509-83-4, pp. 0035-0036

[6] Zoran Pandilov, Andrzej Milecki, Amadeusz Nowak, Filip Górski, Damian Grajewski Damir Ciglar, Tihomir Mulc & Miho Klaić. (2016), Virtual Modelling and Simulation of a CNC Machine Feed Drive System, Transaction of FAMENA, vol. 39, no. 4, pp. 37-54

[7] Dan Negrut, Brett Harris. (2001), ADAMS Theory in a Nutshell

[8] A.Leva. (1993), PID autotuning algorithm based on relay feedback, IEE Porc-Control Theory Appl., vol. 140, pp. 328-337

[9] W. D. Chang & J.J.Yan. (2005), Adaptive robust PID controller design based on a sliding mode for uncertain chaotic systems, Chaos Solitions & Fractals, vol. 26, pp. 167-175

[10] K. D.Young, V. I. Utkin & Ü. Özgüner. (1999), A control engineer's guide to sliding mode control, IEEE Trans. Control Sys. Tech, vol. 7, pp. 328-342

[11] A. S.I.Zinober. (1994), Variable Structure and Lyapnuov Control, Berlin: Springer-Verlag, 1994

DOI: 10.2507/27th.daaam.proceedings.093

RESEARCH ON SIMILARITY VALUES RELEVANT FOR FUZZY MATRIX CONSISTENCY CHECK

Zedina Lavic, Nedzad Dukic, Mugdim Pasic & Branko Vucijak

This Publication has to be referred as: Lavic, Z[edina]; Dukic, N[edzad]; Pasic, M[ugdim] & Vucijak, B[ranko] (2016). Research on Similarity Values Relevant for Fuzzy Matrix Consistency Check, Proceedings of the 27th DAAAM International Symposium, pp.0645-0649, B. Katalinic (Ed.), Published by DAAAM International, ISBN 978-3-902734-08-2, ISSN 1726-9679, Vienna, Austria
DOI: 10.2507/27th.daaam.proceedings.093

Abstract

There is a problem of inherent inconsistency when forming pairwise comparison matrices in both classical and fuzzy Analytic Hierarchy Process. The smallest decision making element of fuzzy Analytic Hierarchy Process is square, positive, fuzzy reciprocal matrix 3x3, and this paper focus to the research on fuzzy numbers' similarities regarding the consistency check of fuzzy matrix 3x3. The research is performed using Java application created specifically for this purpose. The application, for the selected fuzzy scale, forms all 3x3 matrices, for each matrix calculates all similarities values relevant for consistency check, average similarity and determines the minimal similarity value. The application calculates CR (consistency ratio) values for corresponding classical matrices. Research results are presented and future research suggested.

Keywords: Pairwise Comparison Matrix; Consistency; Fuzzy Numbers, Fuzzy Scale; Similarity Measure; Fuzzy Analytic Hierarchy Process

1. Introduction

Numerous different methods are developed to solve multicriteria decision making problems, mostly related to the strict numerical valuation of selected criteria - examples of papers using such methods written by the authors relate to multicriteria evaluation of supervisory boards[1], selection of buildings' insulation [2] or selection of best alternative for highway tunnel doors [3].

One of frequently used multicriteria decision making aid methods is Analytic Hierarchy Process (classical, but also fuzzy). Limited research is done so far regarding evaluation of consistency in forming the pairwise comparison matrices, with few papers available dealing with the issue of consistency of fuzzy pairwise comparison matrices (examples are [4, 5, 6, 7]). According to the work of Buckley [8], fuzzy pairwise comparison matrix $\tilde{A} = \{\tilde{a}_{ij}\}$ is consistent if $\tilde{a}_{ik} \otimes \tilde{a}_{kj} \approx \tilde{a}_{ij}$ where $i, j, k = 1, 2, \dots, n$, and where "$\otimes$" represents multiplication of fuzzy numbers, while "$\approx$" means "fuzzy equal" [9].

The concept of fuzzy equality has the same meaning as the notion of similarity, since the similarity in essence is a generalization of the concept of equivalence [10]. Similarity is a relation which is an extension to the concept of functions belonging to the set of elements, where, abandoning the idea of exact equality, the elements of the domain are considered to have different degrees of similarity [11].

The application of fuzzy numbers in resolving problems requires the application of different similarity measures, calculating the degree of similarity between two fuzzy number; similarity measures appropriate for solving one type of problem often are not appropriate for solving the other problem or type of problems. Therefore, various similarity measures of fuzzy numbers exist [11].. The similarity measure of triangular fuzzy numbers proposed by Chen and Lin ([7]) is based on distances between the midpoints and the appropriate boundary points of the support interval: if $A = (a^l,\ a^m, a^u)$ and $B = (b^l,\ b^m, b^u)$ are triangular fuzzy numbers, similarity can be calculated with

$$S_{MB}(A,B) = 1 - \frac{|a^l - b^l| + |a^m - b^m| + |a^u - b^u|}{3} \quad (1)$$

Hsieh and Chen have presented the utility value concept of triangular fuzzy numbers, and based on these values have also defined the similarity [9]:

$$S_{UV}(A,B) = \frac{U_{UV}(A)xU_{UV}(B)}{\max((U_{UV}(A))^2,(U_{UV}(B))^2)}. \quad (2)$$

The similarity measure is function fulfilling the following two conditions [9]:

- Reflexivity, i. e. $S(A,A) = 1$ and
- Symmetry, i.e. $S(A,B) = S(B,A)$.

Similarity measure based on the distance is used in this paper (as defined in [9]):

$$S(A,B) = \frac{1}{1+d(A,B)}, \quad (3)$$

where $d\ (A,B)$ is a measure of the distance defined with:

$$d(A,B) = |a^l - b^l| + |a^m - b^m| + |a^u - b^u|, \quad (4)$$

so the similarity measure used is

$$S(A,B) = \frac{1}{1+|a^l-b^l|+|a^m-b^m|+|a^u-b^u|}. \quad (5)$$

This similarity measure fulfils the conditions of reflexivity and symmetry.

2. Research

Pairwise comparison matrices of the fuzzy Analytic Hierarchy Process (fAHP) based decision making models are positive and reciprocal. The smallest matrix in fAHP is 3x3 pairwise comparison matrix, square, positive and reciprocal:

$$\begin{bmatrix} (1\,1\,1) & a_{12} & a_{13} \\ {}^1/_{a_{12}} & (1\,1\,1) & a_{23} \\ {}^1/_{a_{13}} & {}^1/_{a_{23}} & (1\,1\,1) \end{bmatrix}. \quad (6)$$

The elements a_{ij} (i,j=1,2,3) are obtained using appropriate fuzzification of scale given by Saaty [13]. Original Saaty's scale and fuzzified Saaty's scale are presented with Tables 1 & 2 respectively.

Definition	Importance intensity	Reciprocals
Equal importance	1	1
Moderate importance	3	1/3
Strong importance	5	1/5
Very strong importance	7	1/7
Extreme importance	9	1/9
Intermediate values	2, 4, 6 and 8	1/2, 1/4, 1/6 and 1/8

Table 1. Scale given by Saaty

Definition	Importance intensity	Reciprocals
Equal importance	(1, 1, 1)	(1, 1, 1)
Moderate importance	(2, 3, 4)	(1/4, 1/3, 1/2)
Strong importance	(4, 5, 6)	(1/6, 1/5, 1/4)
Very strong importance	(6, 7, 8)	(1/8, 1/7, 1/6)
Extreme importance	(9, 9, 9)	(1/9, 1/9, 1/9)
Intermediate values	(1, 2, 3), (3, 4, 5), (5, 6, 7) and (7, 8, 9)	(1/3, 1/2, 1), (1/5, 1/4, 1/3), (1/7, 1/6, 1/5) and (1/9, 1/8, 1/7)

Table 2. Fuzzified Saaty's scale

In order to explore relationship between the similarity measures value and consistency of fuzzy matrices, Java application is created and used. The application creates $17^{n\,(n-1)\,/\,2}$ matrices, where n is dimension of the particular matrix explored. For each created matrix application calculates similarity of fuzzy numbers $\tilde{a}_{ik}\otimes\tilde{a}_{kj}$ and $\tilde{a}_{ij}$, for every $i, j, k = 1, 2, \ldots, n$. After that, for each matrix the application finds minimum, calculates the average similarity and forms MS Excel file where all the fuzzy matrices with the corresponding minimum and average similarities are given. The application also calculates and presents CR value (consistency ratio) [13] for classical matrices derived from the fuzzy matrices. Matrices are formed using classical matrices' elements a_{ij} representing the midpoint of the fuzzy numbers interval $\tilde{a}_{ij} = (a_{ij}^{l}, a_{ij}^{m}, a_{ij}^{u})$. Thus, the elements of the classical matrix are $a_{ij} = a_{ij}^{m}$ for all $i, j = 1, 2, \ldots, n$. Similarity values (minimal and average) and CR values for fuzzy and classical matrices 3x3 respectively are explored. It should be noted that these matrices, as the corresponding values of displayed indicators, would be different if fuzzy scales would be different than those presented in the Table 2.

3. Results and Discussion

For the case of selected fuzzy scale, all the values of minimal similarity of matrices 3x3 are rounded to 4 decimal places and they belong to the interval [0.0041,1]. There are total of 162 minimal similarity values, with the smallest pop-up value of 0.0041, for the selected fuzzy scale. Analysis of the output Excel file, containing all matrices so as corresponding minimal and average similarities and CR values of correspondent classic matrices, is performed. It was found that for the fuzzy matrices whose minimal similarities belong to the interval [0.0041,0.0294] all correspondent classical matrices have CR values greater than 10%, what means that they are not consistent. The total number of these matrices is 3576 and indicates the level of the problem of inconsistency check with matrices of dimension three. The total number of different minimal similarities within the range [0.0041,0.0294] is 122. For these matrices, the lowest average similarity is 0.6577, the largest average similarity is 0.8297, and the smallest CR is 10.1587%. The focus of further analysis of the minimal similarities values is at the values belonging to the interval [0.0303,1]. The values for these similarities (a total of 40 values) are presented with Table 3.

Ordinal Number	Minimal Similarity	Ordinal Number	Minimal Similarity	Ordinal Number	Minimal Similarity	Ordinal Number	Minimal Similarity
1.	0.0303	11.	0.0417	21.	0.0667	31.	0.2069
2.	0.0321	12.	0.0435	22.	0.0714	32.	0.2500
3.	0.0323	13.	0.0455	23.	0.0759	33.	0.2727
4.	0.0333	14.	0.0476	24.	0.0769	34.	0.4000
5.	0.0346	15.	0.0500	25.	0.0833	35.	0.4839
6.	0.0355	16.	0.0522	26.	0.0909	36.	0.5455
7.	0.0357	17.	0.0526	27.	0.1000	37.	0.5932
8.	0.0370	18.	0.0556	28.	0.1111	38.	0.6316
9.	0.0397	19.	0.0588	29.	0.125	39.	0.6632
10.	0.0400	20.	0.0625	30.	0.1429	40.	1

Table 3. Values of the minimal similarities from the interval [0.0303 to 1]

Analysis of the entire range of minimal values of similarities [0.0041. 1] has shown that:

- For the minimal similarities lower than 0.0303 there is no consistent classical matrix.
- For some minimal similarities within the range (0.0303 , 0.1429) there are corresponding average similarities for which both consistent and inconsistent classical matrices exist.
- For some minimal similarities within the range (0.0303 , 0.1] there are corresponding average similarities for which classical matrices are always consistent (Table 4).
- When the minimal similarities are higher than 0.1 (except if the minimal similarity is 0.1429 and corresponding average similarity is 0.7873, when CR of classical matrix is 12.7714%). classical matrices are consistent.,
- For minimal similarities higher than 0.1429 all the classic matrices are consistent.

Minimal Similarity	Average Similarity
0.0333	0.7919
0.0357	0.8154
	0.796
0.037	0.7545
	0.7627
	0.7641
0.0417	0.7583
	0.7652
	0.7962
	0.7969
0.0435	0.7673
	0.7985
0.0476	0.7494
	0.7513
0.05	0.7621
	0.7984
0.0526	0.7731
0.0556	0.7245
	0.7425
	0.7637
	0.8135
0.0588	0.7559
	0.7766
0.0625	0.7699
	0.8017
0.0667	0.7312

Minimal Similarity	Average Similarity
	0.7459
	0.7762
0.0714	0.7316
	0.7485
	0.7647
	0.7658
	0.7999
0.0769	0.7677
	0.7888
	0.7966
	0.8338
0.0833	0.7159
	0.7767
	0.7992
0.0909	0.7105
	0.7403
	0.7492
	0.7872
0.1	0.7126
	0.7586
	0.7734
	0.7808
	0.8015
	0.8166
	0.8509

Table 4. Minimal and average similarities in the range (0.0303 to 0.1]. CR <10%

4. Conclusion

Both for classical and for fuzzy Analytic Hierarchy Process, a problem of inconsistency of decision maker appears when forming the pairwise comparison matrices and this paper focuses on resolving it. Appropriate fuzzy scale, obtained by fuzzification of Saaty´s scale, is used in this paper. Research on similarities of fuzzy numbers, relevant in terms of consistency check, resulted with the conclusion that the inconsistency issues appear even with matrices of smaller dimensions. For the case of selected fuzzy scale, all the values of minimal similarity of matrices 3x3, rounded to 4 decimal places, belong to the interval [0.0041 , 1]. It is found that for the fuzzy matrices whose minimal similarities are within the interval [0.0041 , 0.0294] all correspondent classical matrices have CR values greater than 10%, what means they are not consistent. Total number of these matrices is 3576 (total number of fuzzy matrices is 4913). For some minimal similarities within the range (0.0303 , 0.1] there are corresponding average similarities for which classical matrices are always consistent. For minimal similarities higher than 0.1 (except if the minimal similarity is 0.1429 and corresponding average similarity is 0.7873, when CR of classical matrix is 12.7714%). classical matrices are consistent, and for minimal similarities higher than 0.1429 all the related classic matrices are consistent.

Since the matrix 3x3 is the basic element in fuzzy (and classical) Analytic Hierarchy Process, consistency check of fuzzy matrices of dimensions larger than 3 is feasible using decomposition of larger matrices into the matrices 3x3 , what will be the subject of the further research.

5. References

[1] Karic E., Vucijak B., Pasic M., Bijelonja I. (2011): „Multi-Criteria Evaluation of Supervisory Board Effectiveness", Proceedings of the 22nd DAAAM International Symposium on Intelligent Manufacturing and Automation, Volume 22, Vienna, 23-26 November 2011, 251 – 252

[2] Civic A. and Vucijak B. (2013): „Multi-Criteria Optimization of Insulation Options for Warmth of Buildings to Increase Energy Efficiency", 24th DAAAM International Symposium on Intelligent Manufacturing and Automation, Zadar, 23-26 October 2013, Procedia Engineering 69 (2014) 911 – 920

[3] Vucijak B., Pasic. M., Zorlak A. (2015): „Use of Multi-Criteria Decision Aid Methods for Selection of the Best Alternative for the Highway Tunnel Doors ", 25th DAAAM International Symposium on Intelligent Manufacturing and Automation, Zadar, 23-26 October 2013, Procedia Engineering 100 (2015) 656 – 665

[4] Ramik, J., Vlach, M. (2013). Measuring consistency and inconsistency of pair comparison systems, Kybernetika 49(3): 465-486 (2013)

[5] Ramik, J. (2014). Incomplete fuzzy preference matrix and its application to ranking of alternatives, Intelligent Systems 29, 8 (2014), 787-806

[6] Ramik, J., Korviny, P. (2010). Inconsistency of pair-wise comparison matrix with fuzzy elements based on geometric mean, Fuzzy Sets and Systems 161(11): 1604-1613 (2010)

[7] Ramik, J. (2009). Consistency of Pair-wise Comparison Matrix with Fuzzy Elements. IFSA/EUSFLAT Conf. 2009: 98-101

[8] Buckley, J. J. (1985), Fuzzy hierarchical analysis. Fuzzy Sets and Systems, 17(2), 233-247.

[9] Javanbarg, M. B., Scawthorn, C., Kiyono, J., Shahbodaghkhan, B. (2012), Fuzzy AHP-based multicriteria decision making systems using particle swarm optimization, Expert Systems vith Applications, 39 (2012), 960-966

[10] Zadeh, L. A. (1971). Similarity Relations and Fuzzy Orderings. Information Sciences 3, p. 177—200.

[11] Dukić, N. (2006), The Semantic Distance, Fuzzy Dependency and Fuzzy Formulas, Sarajevo Journal of Mathematics, Vol. 2 (15), 137-146.

[12] Zhang, X., Ma, W. and Chen, L. (2014), New Similarity of Triangular Fuzzy Number and Its Application, Hindawi Publishing Corporation, The Scientific World Journal, Volume 2014

[13] Saaty, T. L. (1980), *Analytic hierarchy process*, New York, McGraw-Hill.

DOI: 10.2507/27th.daaam.proceedings.094

FUZZY LOGISTICS MULTICRITERIA DECISION MAKING MODEL FOR FACILITIES LAYOUT

Zedina Lavic, Mugdim Pasic, Branko Vucijak & Nedzad Dukic

This Publication has to be referred as: Lavic, Z[edina]; Pasic, M[ugdim]; Vucijak, B[ranko] & Dukic, N[edzad] (2016). Fuzzy Logistics Multicriteria Decision Making Model for Facilities Layout, Proceedings of the 27th DAAAM International Symposium, pp.0650-0657, B. Katalinic (Ed.), Published by DAAAM International, ISBN 978-3-902734-08-2, ISSN 1726-9679, Vienna, Austria
DOI: 10.2507/27th.daaam.proceedings.094

Abstract

The paper is focused to the facilities layout problem formulated as a problem of fuzzy multicriteria (multi attributive) decision making. The goal is to develop a model for evaluation and selection of solution from Pareto set of solutions (facilities layout alternatives) applying theory for fuzzy sets and fuzzy logic in multicriteria logistics decision.
Decision criteria are selected and grouped according to the logistics objectives, meaning levels of logistics decision making (strategic, tactical and operational level). The decision making problem is structured to decision hierarchy and Fuzzy Analytic Hierarchy process is implemented, which by its essence best corresponds to the problem and the context of decision making, so as to the number and type of selected criteria. The new model is created and tested on a concrete example.

Keywords: facilities layout; multicriteria decision making; fuzzy analytic hierarchy process; logistics objectives

1. Introduction

The problem of optimizing the layout of facilities has been subject of many researches in the last sixty years. Several recent publications, conferences and researches confirmed the importance of the layout of facilities to meet the requirements set by the market and to achieve the goals of the company [1, 2, 3, 4]. The layout of facilities is an important issue not only for production but also for organizations providing services. Since the 1990s, an additional emphasis on this issue came from the strong requirement of the market for mass satisfying of customers in terms of the required product, quantity, variety, place and time of delivery, and this problem is further complicated because of an increase in the degree of uncertainty and ambiguity. Satisfying these demands requires quick reaction and rearrangement of facilities in the shortest possible time and best possible way, in order to reduce costs. The importance of the problem of layout is also reflected by the fact that the costs of material handling, and other costs which depend on the layout, constitute 20-50% of total operating costs [5]. Requirement to have an efficient process of supply for point of receipt from the point of supply implies the establishment of a logistics system that will allow supply of the right product, in the right condition, at the right time, right place, and at minimum cost.

Complexity of the logistics system imposes the need for multicriteria optimization and decision-making in order to achieve the basic logistical goal: a smooth flow of goods and information through the system. In order to achieve this goal, it is necessary for the facilities layout to be optimal. The facilities include physical entities such as machines, work centers, production cells, workshops, sections, warehouses and everything else that facilitates achievement of set or expected performances of work. Optimum design of a facility minimizes the costs of handling of the material, the production cycle duration, capital investment, inventories, increases efficiency of work and use of space, safety and comfort of employees, eliminates bottlenecks, enables visual control and maintains flexibility in changing conditions. "Multicriteria optimization can significantly facilitate and accelerate the decision making process"[6].

The facilities layout problem is a point of interest for employees at all levels in the organizational hierarchy. Top management is interested in this problem in terms of the size and efficiency of capital investment. Managers at lower levels are interested in this problem in terms of costs and efficiency of operations, and the workers have interest in terms of safety, comfort, their productivity and earnings.

There are many objectives to be achieved in the making of the facilities layout. They can be shown in the form of functions or in the form of evaluation criteria. According to Langevin and Riopel [7], the objectives of the facilities layout are divided into the following three groups:

- strategic objectives,
- tactical objectives,
- operational objectives.

Due to the complexity of the facilities layout problem, different models used just a few of a large number of layout objectives, and the most commonly used objectives are optimization of the flow (materials, information and personnel), minimization of material handling costs, optimization of equipment use and optimization of capital investment.

The development of fuzzy logic and its application in these models enabled better use of quality objectives (or criteria for the evaluation of the layout alternatives) that are difficult to measure or are difficult to quantify and measure without the use of fuzzy logic (classic).

The facilities layout problem solved in this paper is formulated as a problem of fuzzy multiple criteria (multiple attribute) decision making.

The main objective of this paper is to develop a model for the evaluation and selection of solutions from the Pareto set of solutions (alternatives for layout facilities) using the theory of fuzzy sets and fuzzy logic in multicriteria logistics decision making. The model should also include the criteria (quantitative and qualitative) for strategic, tactical and operational level, which are immanent to the real problems of logistics decision making on the facilities layout, when a decision maker is unable to accurately determine the value of individual criteria or when they are hardly measurable.

2. Fuzzy logistics multicriteria decision making model forfacilities layout

The development of the model is implemented through a series of steps:

1. Selection of the decision making criteria
2. Selection of methods for multicriteria decision aid
3. Structuring of the decision making hierarchy
4. Application of the Fuzzy Analytic Hierarchy Process method with checking the consistency of matrices
5. Testing of the model.

2.1 Decision criteria

The criteria are selected in accordance with the logistics objectives and they are immanent to real problems of logistics decision making on the facilities layout. It is possible to evaluate the extent to which a certain alternative contributes to achieving the objective referred to each other and to compare alternatives on the basis of criteria (attributes of alternatives). Analogous to division of logistical objectives, the criteria are also divided into 3 groups of criteria:

- Strategic level criteria (12 criteria given in Table 1)
- Tactical level criteria (9 criteria given in Table 2)
- Operational level criteria (9 criteria given in Table 3).

Criteria	Designation	Unit of Measure	Max/Min
Share of unrealized (unfulfilled) purchase orders (ordered quantity), caused by lack of capacity, in total ordered quantities per year	SC1	%	Min
Modularity (plan for future expansion)	SC2	Linguistic value	Max
Consistency with company image, promotional value, public or community relations	SC3	Linguistic value	Max
Net present value	SC4	BAM	Max
Internal rate of return	SC5	%	Max
Payback period	SC6	Number of years	Min
Installation period	SC7	h	Min
Noise emission level	SC8	dB	Min
Quantity of pollutants' emission to the air	SC9	t/year	Min
Quantity of energy used for conditioning, heating and cooling	SC10	J/year	Min
Distance of output / input from the place of loading / unloading	SC11	m	Min
Total distance of exit from the entrance to other facilities	SC12	m	Min

Table 1. Strategic level criteria

Criteria	Designation	Unit of Measure	Max/Min
Fits into the organizational structure	TC1	Linguistic value	Max
Facilitating the monitoring, control and communications	TC2	Linguistic value	Max
Optimizing the use of space	TC3	Linguistic value	Max
Number of requests for non-standard elements (equipment, tools, work surfaces, ...)	TC4	#	Min
Maintain flexibility in scheduling and operations	TC5	Linguistic value	Max
Average time from service request to service implementation	TC6	h	Min
Number of departments in which the intensity of daylight is not sufficient	TC7	#	Min
Facilitating the maintenance and household	TC8	Linguistic value	Max
Distance of departments staffed by employees with disabilities to the departments with which they are in close cooperation, or restaurant, toilets	TC9	m	Min

Table 2. Tactical level criteria

Criteria	Designation	Unit of Measure	Max/ Min
„Work in progress" turnover	OC1	#	Max
Cost of material flow	OC2	BAM	Min
Cost of the flow of information and personnel	OC3	BAM	Min
Handling optimization	OC4	Linguistic value	Max
Number of injuries of employees per year	OC5	#	Min
Number of thefts of materials and equipment per year	OC6	#	Min
To provide convenience for workers and improve job satisfaction	OC7	Linguistic value	Max
Percentage of equipment that performs the function in the observed period	OC8	%	Max
Daily effective engagement of employees	OC9	h/day	Max

Table 3. Operational level criteria

2.2 *Selection of multicriteria decision making method*

Due to the complexity of the logistics system and large interdependence of decisions on strategic, tactical and operational level as well as due to number and nature of the criteria imposed by the need to rely on the assessment, Fuzzy Analytic Hierarchy Process is selected and applied as a multicriteria decision making method, which, in its essence is the most appropriate for the problem and the context of decision making. The applied Fuzzy Analytic Hierarchy Process method is based on Chang's fuzzy extent analysis [8]. The basic concept, i.e. hierarchical structure of decision making, pairwise comparison and the bottom up synthesis of priorities through the hierarchy is the same as in classical Analytic Hierarchy Process. The key difference compared to the classical Analytic Hierarchy Process is that with the Fuzzy Analytic Hierarchy Process, decision makers are not required to give strict assessment, but instead, imprecision in the evaluation is allowed.

2.3 *Hierarchy of decision making and model parameters*

The problem of decision making is structured in a hierarchy that has 4 hierarchical levels:

1. Objective: selection of the best alternative from the Pareto set of alternatives for facilities layout
2. Criteria groups
3. Criteria (listed in the criteria groups)
4. Alternatives (Pareto set of alternatives).

The number of alternatives is a variable in the model and the specific parameters of the model are:

1. Number of levels in a hierarchical structure (4),
2. Number of criteria groups (3),
3. Total number of criteria (30),
4. Number of strategic level criteria (12),
5. Number of tactical level criteria (9),
6. Number of operational level criteria (9).

2.4 *Input/output*

Inputs for the model are expert assessments of criteria importance and priorities of alternatives that are entered into pairwise comparison matrices:

1. Pairwise comparison matrix of criteria groups with respect to the objective
2. Pairwise comparison matrix of strategic level criteria with respect to the corresponding group
3. Pairwise comparison matrix of tactical level criteria with respect to the corresponding group
4. Pairwise comparison matrix of operational level criteria with respect to the corresponding group
5. Pairwise comparison matrices of alternatives with respect to each of the criterion.

The outputs of the model are priority values for alternatives sorted from largest to smallest. The best alternative has the highest value of priority.

2.5 *Consistency check*

Pairwise comparison matrices are positive, squared and reciprocal. Elements of matrices are triangular fuzzy numbers from a fuzzy scale obtained through appropriate fuzzification of Saaty's scale [9] in order to derive corresponding classical matrices from the fuzzy matrices and to check the consistency. Consistency of a classical matrix implies consistency of the correspondent fuzzy matrix. Fuzzy scale is shown in Table 4.

Definition	**Importance intensity**	**Reciprocals**
Equal importance	(1, 1, 1)	(1, 1, 1)
Moderate importance	(2, 3, 4)	(1/4, 1/3, 1/2)
Strong importance	(4, 5, 6)	(1/6, 1/5, 1/4)
Very strong importance	(6, 7, 8)	(1/8, 1/7, 1/6)
Extreme importance	(9, 9, 9)	(1/9, 1/9, 1/9)
Intermediate values	(1, 2, 3), (3, 4, 5), (5, 6, 7) and (7, 8, 9)	(1/3, 1/2, 1), (1/5, 1/4, 1/3), (1/7, 1/6, 1/5) and (1/9, 1/8, 1/7)

Table 4. Fuzzified Saaty's scale

2.6 Algorithm model

The model can be realized through the algorithm whose flowchart is presented in Figure 1.

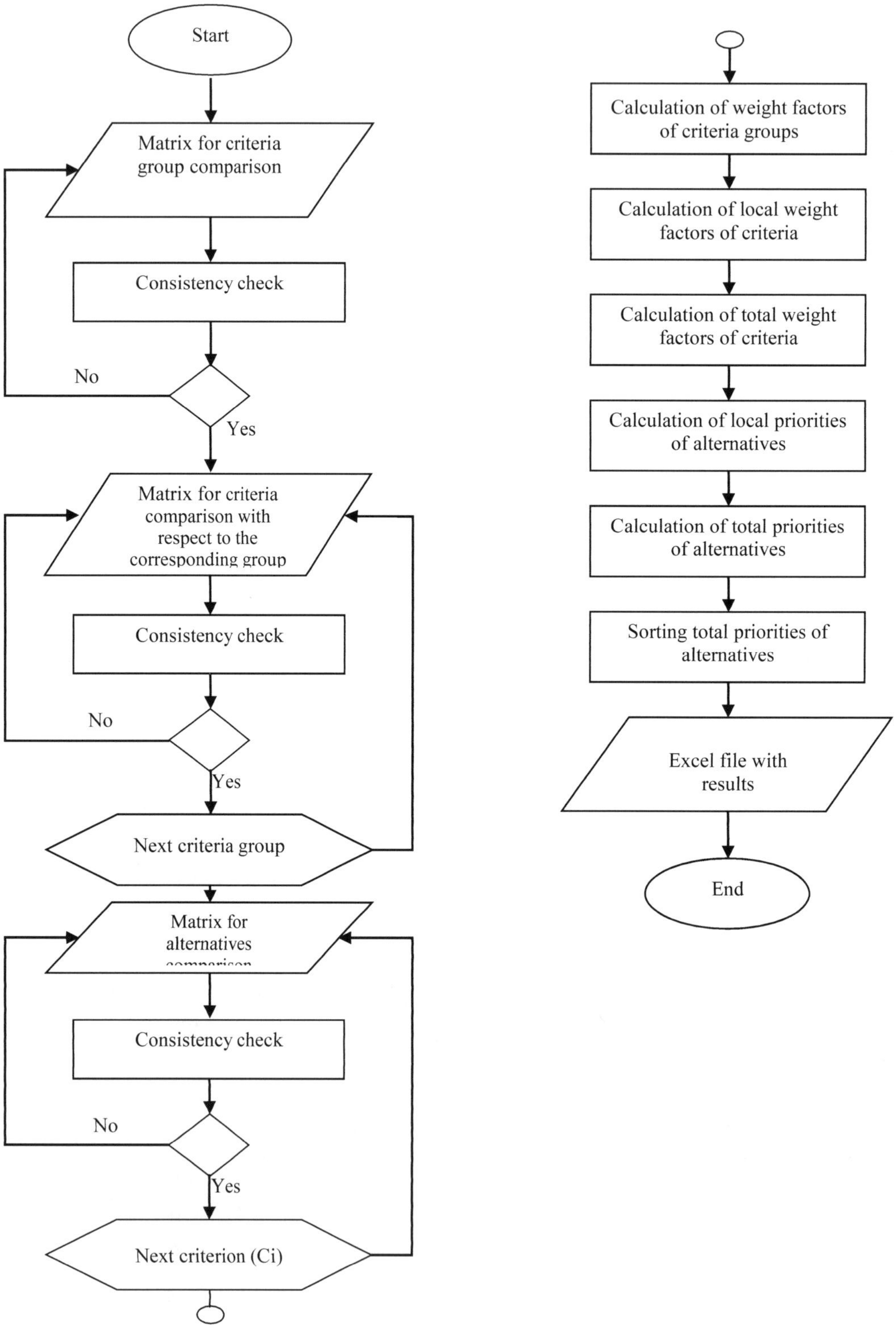

Fig. 1. Flowchart of algorithm model

3. Testing model on a concrete example

The developed model is tested on a concrete example of the selection problem for the best layout alternatives of 3 "workplaces" in a design office, using a Java application made on the basis of the algorithm of the developed model. The facilities here are "workplaces" (desk and chair on an area envisaged to accommodate the movement of the chair when sitting/getting up and working. Layouts of the "workplaces", work desk and floor space envisaged to accommodate furniture and movement of the chair have a form of a rectangle. Layout options (alternatives) from the Pareto set are presented in Figure 2 a-c.

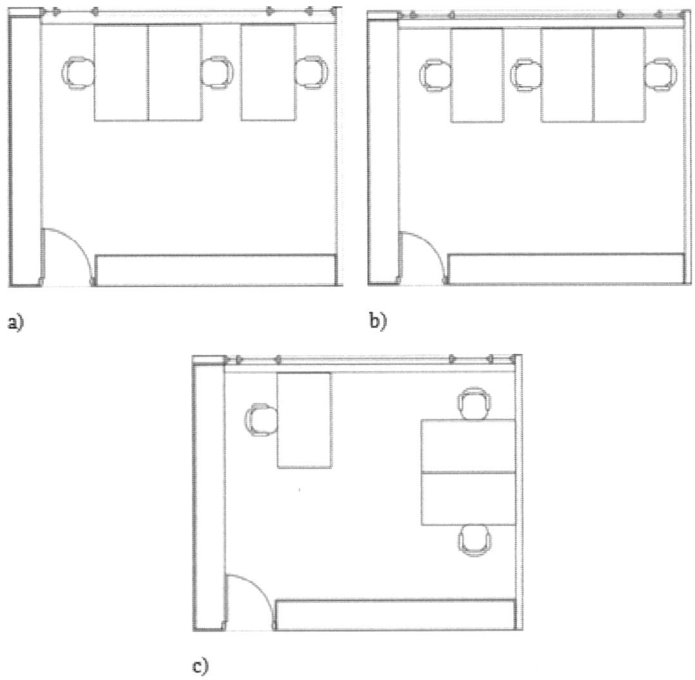

Fig. 2 a) A1, b) A2, c)A3

The decision maker is a sector manager (tactical level) and directly manages the work of employees who will be assigned to "workplaces". Along with the previous explanations of the hierarchical structure of decision making and the use of the Java application, the decision maker made the pairwise comparisons to give assessment of importance of criteria groups with respect to the objective (matrix 1 in Table 1), assessment of criteria importance with respect to the corresponding criteria group (matrices 2-4 in Table 1) and the assessment of the priorities of alternatives with respect to each of the criteria (matrices 5-34 in Table 2). The results are given in Table 7.

<table>
<tr><th>Matrix No</th><th>Matrix</th><th>CR (%)</th></tr>
<tr><td>1</td><td>[(1, 1, 1) (1/3, 1/2, 1) (1/3, 1/2, 1) | (1, 2, 3) (1, 1, 1) (1, 2, 3) | (1, 2, 3) (1/3, 1/2, 1) (1, 1, 1) |]</td><td>5,23</td></tr>
<tr><td>2</td><td>[(1, 1, 1) (1, 2, 3) (1, 2, 3) (1/4, 1/3, 1/2) (1/3, 1/2, 1) (1/3, 1/2, 1) (1/5, 1/4, 1/3) (1, 1, 1) (1, 1, 1) (1/5, 1/4, 1/3) (1/3, 1/2, 1) (1/3, 1/2, 1) | (1/3, 1/2, 1) (1, 1, 1) (1/3, 1/2, 1) (2, 3, 4) (1, 1, 1) (1, 1, 1) (1, 2, 3) (2, 3, 4) (1, 1, 1) (1, 1, 1) (1, 2, 3) (3, 4, 5) | (1/3, 1/2, 1) (1, 2, 3) (1, 1, 1) (1/3, 1/2, 1) (1/3, 1/2, 1) (1/3, 1/2, 1) (1, 1, 1) (1, 1, 1) (1, 1, 1) (1, 2, 3) (1, 2, 3) (1, 2, 3) | (2, 3, 4) (1/4, 1/3, 1/2) (1, 2, 3) (1, 1, 1) (1/4, 1/3, 1/2) (1/4, 1/3, 1/2) (1, 1, 1) (1, 1, 1) (1, 1, 1) (1, 1, 1) (1, 1, 1) (1, 2, 3) | (1, 2, 3) (1, 1, 1) (1, 2, 3) (2, 3, 4) (1, 1, 1) (1, 1, 1) (1, 2, 3) (2, 3, 4) (1, 1, 1) (1, 1, 1) (1, 2, 3) (2, 3, 4) | (1, 2, 3) (1, 1, 1) (1, 2, 3) (2, 3, 4) (1, 1, 1) (1, 1, 1) (1, 2, 3) (2, 3, 4) (1, 1, 1) (1, 1, 1) (1, 2, 3) (2, 3, 4) | (3, 4, 5) (1/3, 1/2, 1) (1, 1, 1) (1, 1, 1) (1/3, 1/2, 1) (1/3, 1/2, 1) (1, 1, 1) (1, 2, 3) (1, 1, 1) (1, 1, 1) (1, 2, 3) (2, 3, 4) | (1, 1, 1) (1/4, 1/3, 1/2) (1, 1, 1) (1, 1, 1) (1/4, 1/3, 1/2) (1/4, 1/3, 1/2) (1/3, 1/2, 1) (1, 1, 1) (1/4, 1/3, 1/2) (1, 1, 1) (1, 1, 1) (2, 3, 4) | (1, 1, 1) (1, 1, 1) (1, 1, 1) (1, 1, 1) (1, 1, 1) (1, 1, 1) (1, 1, 1) (2, 3, 4) (1, 1, 1) (1, 1, 1) (1, 2, 3) (2, 3, 4) | (3, 4, 5) (1, 1, 1) (1/3, 1/2, 1) (1, 1, 1) (1, 1, 1) (1, 1, 1) (1, 1, 1) (1, 1, 1) (1, 1, 1) (1, 1, 1) (1, 2, 3) (2, 3, 4) | (1, 2, 3) (1/3, 1/2, 1) (1/3, 1/2, 1) (1, 1, 1) (1/3, 1/2, 1) (1/3, 1/2, 1) (1/3, 1/2, 1) (1, 1, 1) (1/3, 1/2, 1) (1/3, 1/2, 1) (1, 1, 1) (1, 1, 1) | (1, 2, 3) (1/5, 1/4, 1/3) (1/3, 1/2, 1) (1/3, 1/2, 1) (1/4, 1/3, 1/2) (1/4, 1/3, 1/2) (1/4, 1/3, 1/2) (1/4, 1/3, 1/2) (1/4, 1/3, 1/2) (1/4, 1/3, 1/2) (1, 1, 1) (1, 1, 1) |]</td><td>7,96</td></tr>
</table>

<table>
<tr><th>Matrix No</th><th>Matrix</th><th>CR (%)</th></tr>
<tr><td>3</td><td>[(1, 1, 1) (1, 1, 1) (1, 2, 3) (2, 3, 4) (1, 2, 3) (1, 1, 1) (1, 1, 1) (1, 2, 3) (1, 1, 1)| (1, 1, 1) (1, 1, 1) (1, 1, 1) (1, 2, 3) (1, 2, 3) (1, 1, 1) (1, 1, 1) (1, 2, 3) (1, 1, 1)| (1/3, 1/2, 1) (1, 1, 1) (1, 1, 1) (1, 1, 1) (1, 1, 1) (1, 1, 1) (1, 1, 1) (1, 1, 1) (1, 1, 1)| (1/4, 1/3, 1/2) (1/3, 1/2, 1) (1, 1, 1) (1, 1, 1) (1, 1, 1) (1, 1, 1) (1, 1, 1) (1, 1, 1) (1, 1, 1)| (1/3, 1/2, 1) (1/3, 1/2, 1) (1, 1, 1) (1, 1, 1) (1, 1, 1) (1, 1, 1) (1, 1, 1) (1, 1, 1) (1, 1, 1)| (1, 1, 1) (1, 1, 1) (1, 1, 1) (1, 1, 1) (1, 1, 1) (1, 1, 1) (1, 1, 1) (1, 1, 1) (1, 1, 1)| (1, 1, 1) (1, 1, 1) (1, 1, 1) (1, 1, 1) (1, 1, 1) (1, 1, 1) (1, 1, 1) (1, 1, 1) (1, 1, 1)| (1/3, 1/2, 1) (1/3, 1/2, 1) (1, 1, 1) (1, 1, 1) (1, 1, 1) (1, 1, 1) (1, 1, 1) (1, 1, 1) (1, 1, 1)| (1, 1, 1) (1, 1, 1) (1, 1, 1) (1, 1, 1) (1, 1, 1) (1, 1, 1) (1, 1, 1) (1, 1, 1) (1, 1, 1)|]</td><td>1,67</td></tr>
<tr><td>4</td><td>[(1, 1, 1) (1/3, 1/2, 1) (1/3, 1/2, 1) (1/3, 1/2, 1) (1/3, 1/2, 1) (1/3, 1/2, 1) (1/4, 1/3, 1/2) (1, 1, 1) (1, 1, 1)| (1, 2, 3) (1, 1, 1) (1, 1, 1) (1, 1, 1) (1, 1, 1) (2, 3, 4) (2, 3, 4) (2, 3, 4) (2, 3, 4)| (1, 2, 3) (1, 1, 1) (1, 1, 1) (1, 1, 1) (1, 1, 1) (2, 3, 4) (2, 3, 4) (2, 3, 4) (2, 3, 4)| (1, 2, 3) (1, 1, 1) (1, 1, 1) (1, 1, 1) (1, 1, 1) (1, 2, 3) (2, 3, 4) (3, 4, 5) (1, 1, 1)| (1, 2, 3) (1, 1, 1) (1, 1, 1) (1, 1, 1) (1, 1, 1) (1, 1, 1) (1, 2, 3) (2, 3, 4) (3, 4, 5)| (1, 2, 3) (1/4, 1/3, 1/2) (1/4, 1/3, 1/2) (1/3, 1/2, 1) (1, 1, 1) (1, 1, 1) (1, 1, 1) (1, 1, 1) (2, 3, 4)| (2, 3, 4) (1/4, 1/3, 1/2) (1/4, 1/3, 1/2) (1/4, 1/3, 1/2) (1/3, 1/2, 1) (1, 1, 1) (1, 1, 1) (1, 1, 1) (1/4, 1/3, 1/2)| (1, 1, 1) (1/4, 1/3, 1/2) (1/4, 1/3, 1/2) (1/5, 1/4, 1/3) (1/4, 1/3, 1/2) (1, 1, 1) (1, 1, 1) (1, 1, 1) (1, 2, 3)| (1, 1, 1) (1/4, 1/3, 1/2) (1/4, 1/3, 1/2) (1, 1, 1) (1/5, 1/4, 1/3) (1/4, 1/3, 1/2) (2, 3, 4) (1/3, 1/2, 1) (1, 1, 1)|]</td><td>6,75</td></tr>
</table>

Table 5. Pairwise comparison matrices 1-4

<table>
<tr><th>Matrix No</th><th>Matrix</th><th>CR(%)</th></tr>
<tr><td>5</td><td>[(1, 1, 1) (1, 1, 1) (1/4, 1/3, 1/2)| (1, 1, 1) (1, 1, 1) (1/3, 1/2, 1)| (2, 3, 4) (1, 2, 3) (1, 1, 1)|]</td><td>1,92</td></tr>
<tr><td>6</td><td>[(1, 1, 1) (1, 1, 1) (1/5, 1/4, 1/3)| (1, 1, 1) (1, 1, 1) (1/5, 1/4, 1/3)| (3, 4, 5) (3, 4, 5) (1, 1, 1)|]</td><td>0,00</td></tr>
<tr><td>7</td><td>[(1, 1, 1) (1, 1, 1) (2, 3, 4)| (1, 1, 1) (1, 1, 1) (2, 3, 4)| (1/4, 1/3, 1/2) (1/4, 1/3, 1/2) (1, 1, 1)|]</td><td>0,00</td></tr>
<tr><td>8</td><td>[(1, 1, 1) (1, 1, 1) (1, 1, 1)| (1, 1, 1) (1, 1, 1) (1, 2, 3)| (1, 1, 1) (1/3, 1/2, 1) (1, 1, 1)|]</td><td>4,79</td></tr>
<tr><td>9</td><td>[(1, 1, 1) (1, 1, 1) (1, 1, 1)| (1, 1, 1) (1, 1, 1) (1, 2, 3)| (1, 1, 1) (1/3, 1/2, 1) (1, 1, 1)|]</td><td>4,79</td></tr>
<tr><td>10</td><td>[(1, 1, 1) (1, 1, 1) (1, 1, 1)| (1, 1, 1) (1, 1, 1) (1, 2, 3)| (1, 1, 1) (1/3, 1/2, 1) (1, 1, 1)|]</td><td>4,79</td></tr>
<tr><td>11</td><td>[(1, 1, 1) (1, 1, 1) (3, 4, 5)| (1, 1, 1) (1, 1, 1) (3, 4, 5)| (1/5, 1/4, 1/3) (1/5, 1/4, 1/3) (1, 1, 1)|]</td><td>0,00</td></tr>
<tr><td>12</td><td>[(1, 1, 1) (1, 1, 1) (1, 1, 1)| (1, 1, 1) (1, 1, 1) (1/3, 1/2, 1)| (1, 1, 1) (1, 2, 3) (1, 1, 1)|]</td><td>4,79</td></tr>
<tr><td>13</td><td>[(1, 1, 1) (1, 1, 1) (1, 1, 1)| (1, 1, 1) (1, 1, 1) (1/3, 1/2, 1)| (1, 1, 1) (1, 2, 3) (1, 1, 1)|]</td><td>4,79</td></tr>
<tr><td>14</td><td>[(1, 1, 1) (1, 1, 1) (1, 1, 1)| (1, 1, 1) (1, 1, 1) (1, 2, 3)| (1, 1, 1) (1/3, 1/2, 1) (1, 1, 1)|]</td><td>4,79</td></tr>
<tr><td>15</td><td>[(1, 1, 1) (1, 2, 3) (2, 3, 4)| (1/3, 1/2, 1) (1, 1, 1) (1, 2, 3)| (1/4, 1/3, 1/2) (1/3, 1/2, 1) (1, 1, 1)|]</td><td>0,96</td></tr>
<tr><td>16</td><td>[(1, 1, 1) (1, 2, 3) (2, 3, 4)| (1/3, 1/2, 1) (1, 1, 1) (1, 2, 3)| (1/4, 1/3, 1/2) (1/3, 1/2, 1) (1, 1, 1)|]</td><td>0,96</td></tr>
<tr><td>17</td><td>[(1, 1, 1) (1, 1, 1) (1, 2, 3)| (1, 1, 1) (1, 1, 1) (1, 2, 3)| (1/3, 1/2, 1) (1/3, 1/2, 1) (1, 1, 1)|]</td><td>0,00</td></tr>
<tr><td>18</td><td>[(1, 1, 1) (1, 1, 1) (2, 3, 4)| (1, 1, 1) (1, 1, 1) (1, 2, 3)| (1/4, 1/3, 1/2) (1/3, 1/2, 1) (1, 1, 1)|]</td><td>1,78</td></tr>
<tr><td>19</td><td>[(1, 1, 1) (1/3, 1/2, 1) (1, 2, 3)| (1, 2, 3) (1, 1, 1) (2, 3, 4)| (1/3, 1/2, 1) (1/4, 1/3, 1/2) (1, 1, 1)|]</td><td>0,96</td></tr>
<tr><td>20</td><td>[(1, 1, 1) (1, 1, 1) (3, 4, 5)| (1, 1, 1) (1, 1, 1) (3, 4, 5)| (1/5, 1/4, 1/3) (1/5, 1/4, 1/3) (1, 1, 1)|]</td><td>0,00</td></tr>
<tr><td>21</td><td>[(1, 1, 1) (1, 1, 1) (1/3, 1/2, 1)| (1, 1, 1) (1, 1, 1) (1/3, 1/2, 1)| (1, 2, 3) (1, 2, 3) (1, 1, 1)|]</td><td>0,00</td></tr>
<tr><td>22</td><td>[(1, 1, 1) (1/3, 1/2, 1) (2, 3, 4)| (1, 2, 3) (1, 1, 1) (2, 3, 4)| (1/4, 1/3, 1/2) (1/4, 1/3, 1/2) (1, 1, 1)|]</td><td>5,64</td></tr>
<tr><td>23</td><td>[(1, 1, 1) (1, 1, 1) (3, 4, 5)| (1, 1, 1) (1, 1, 1) (3, 4, 5)| (1/5, 1/4, 1/3) (1/5, 1/4, 1/3) (1, 1, 1)|]</td><td>0,00</td></tr>
<tr><td>24</td><td>[(1, 1, 1) (1, 1, 1) (2, 3, 4)| (1, 1, 1) (1, 1, 1) (2, 3, 4)| (1/4, 1/3, 1/2) (1/4, 1/3, 1/2) (1, 1, 1)|]</td><td>0,00</td></tr>
<tr><td>25</td><td>[(1, 1, 1) (1, 1, 1) (1, 2, 3)| (1, 1, 1) (1, 1, 1) (1, 2, 3)| (1/3, 1/2, 1) (1/3, 1/2, 1) (1, 1, 1)|]</td><td>0,00</td></tr>
<tr><td>26</td><td>[(1, 1, 1) (1, 1, 1) (1, 2, 3)| (1, 1, 1) (1, 1, 1) (1, 2, 3)| (1/3, 1/2, 1) (1/3, 1/2, 1) (1, 1, 1)|]</td><td>0,00</td></tr>
<tr><td>27</td><td>[(1, 1, 1) (1, 1, 1) (1, 2, 3)| (1, 1, 1) (1, 1, 1) (1, 2, 3)| (1/3, 1/2, 1) (1/3, 1/2, 1) (1, 1, 1)|]</td><td>0,00</td></tr>
<tr><td>28</td><td>[(1, 1, 1) (1, 1, 1) (2, 3, 4)| (1, 1, 1) (1, 1, 1) (2, 3, 4)| (1/4, 1/3, 1/2) (1/4, 1/3, 1/2) (1, 1, 1)|]</td><td>0,00</td></tr>
<tr><td>29</td><td>[(1, 1, 1) (1, 1, 1) (2, 3, 4)| (1, 1, 1) (1, 1, 1) (1, 2, 3)| (1/4, 1/3, 1/2) (1/3, 1/2, 1) (1, 1, 1)|]</td><td>1,78</td></tr>
<tr><td>30</td><td>[(1, 1, 1) (1, 1, 1) (3, 4, 5)| (1, 1, 1) (1, 1, 1) (3, 4, 5)| (1/5, 1/4, 1/3) (1/5, 1/4, 1/3) (1, 1, 1)|]</td><td>0,00</td></tr>
<tr><td>31</td><td>[(1, 1, 1) (1, 2, 3) (2, 3, 4)| (1/3, 1/2, 1) (1, 1, 1) (1, 2, 3)| (1/4, 1/3, 1/2) (1/3, 1/2, 1) (1, 1, 1)|]</td><td>0,96</td></tr>
<tr><td>32</td><td>[(1, 1, 1) (1, 1, 1) (1/5, 1/4, 1/3)| (1, 1, 1) (1, 1, 1) (1/6, 1/5, 1/4)| (3, 4, 5) (4, 5, 6) (1, 1, 1)|]</td><td>0,75</td></tr>
<tr><td>33</td><td>[(1, 1, 1) (1, 1, 1) (1, 2, 3)| (1, 1, 1) (1, 1, 1) (1, 2, 3)| (1/3, 1/2, 1) (1/3, 1/2, 1) (1, 1, 1)|]</td><td>0,00</td></tr>
<tr><td>34</td><td>[(1, 1, 1) (1, 1, 1) (4, 5, 6)| (1, 1, 1) (1, 1, 1) (4, 5, 6)| (1/6, 1/5, 1/4) (1/6, 1/5, 1/4) (1, 1, 1)|]</td><td>0,00</td></tr>
</table>

Table 6. Pairwise comparison matrices 5-34

Alternative	Fuzzy AHP	Clasical AHP
A 1	0,4262	0,3868
A 2	0,4244	0,3962
A 3	0,1492	0,2168

Table 7. Results

According to Fuzzy Analytic Hierarchy Process the best alternative is A1, while according to classical Analytic Hierarchy Process the best alternative is A2. Result (A1) is the expected.

4. Conclusion

In this paper facilities layout problem formulated as a problem of fuzzy multicriteria (multi attributive) decision making is solved and a model for evaluation and selection of solution from Pareto set of solutions (facilities layout alternatives) applying theory for fuzzy sets and fuzzy logic in multicriteria logistics decision is developed. The proposed fuzzy model for logistics multicriteria decision making on the facilities layout, developed on the basis of systemic thinking and a holistic approach to solving logistics problems, includes a relatively large number of criteria (quantitative and qualitative) is applicable in situations where it is difficult to measure criteria or it is difficult to quantify and measure them without the application of fuzzy logic.

Compared to models based on the classic Analytic Hierarchy Process method, the advantage of the proposed model based on the Fuzzy Analytic Hierarchy Process method is reflected in the fact that the expert assessments expressed verbally do not appear as strict assessments, but as to some extent overlapping normal, convex fuzzy sets (triangular fuzzy numbers) that are a more realistic presentation of expert assessments of the relative importance of the criteria and the relative priority of alternatives.

Future research can be carried out to the direction of integration of the proposed model for multicriteria decision making with models for the generation of facilities layout alternatives and to the direction of the modification of the model for team decision-making.

5. References

[1] Yang L., Deuse J. (2012): Multiple-attribute Decision Making for an Energy Efficient Facility Layout Design, Procedia CIRP, Vol 3, Pages 149-154, 45th CIRP Conference on Manufacturing Systems 2012

[2] Karande P., Chakraborty S.: A Facility Layout Selection Model using MACBETH Method, Proceedings of the 2014 International Conference on Industrial Engineering and Operations Management Bali, Indonesia, January 7-9, 2014.

[3] Bacudio L, Esmeria G. J., Promentilla M. A.: A Fuzzy Analytic Hierarchy Process Approach for Optimal Selection of Manufacturing Layout, Presented at the DLSU Research Congress 2016, De La Salle University, Manila, Philippines, March 7-9, 2016.

[4] Shokri H., Ashjari B., Saberi M., Yoon J. H (2013): An Integrated AHP-VIKOR Methodology for Facility Layout Design, Industrial Engineering & Management Systems, Vol 12, No 4, December 2013

[5] Tompkins J A., White J. A., Bozer Y. A. & Tanchoco, J. M. A.: *Facilities Planning*, 4th Edition, John Wiley & Sons, Inc, USA, 2010.,

[6] Vucijak, B., Pasic, M. & Zorlak, AJ. (2015). Use of Multi-criteria Decision Aid Methods for Selection of the Best Alternative for the Highway Tunnel Doors, 25th DAAAM International Symposium on Intelligent Manufacturing and Automation, DAAAM 2014, Procedia Engineering, Vol. 100, (2015), pp. 656-665

[7] Langevin A., Riopel D. (2005): *Logistics Systems: Design and Optimization*, Springer Science + Business Media, New York

[8] Kahraman, C. (2008) *Fuzzy Multi-Criteria Decision Making, Theory and Applications with Recent Developments*, New Jork, Springer Science+Business Media, LLC.

[9] Saaty, T. L. (1980), *Analytic hierarchy process*, New York, McGraw-Hill

DOI: 10.2507/27th.daaam.proceedings.095

MAPREDUCE-BASED FACE DETECTION IN IMAGES

Ana Pinjuh, Linda Vickovic & David Cavar

This Publication has to be referred as: Pinjuh, A[na]; Vickovic, L[inda] & Cavar, D[avid] (2016). Mapreduce-Based Face Detection in Images, Proceedings of the 27th DAAAM International Symposium, pp.0658-0663, B. Katalinic (Ed.), Published by DAAAM International, ISBN 978-3-902734-08-2, ISSN 1726-9679, Vienna, Austria
DOI: 10.2507/27th.daaam.proceedings.095

Abstract
Huge amounts of visual data have been generated daily on the Internet. Created data requires not only massive storage but also massive computational power for processing and analyzing data in efficient manner. Therefore, new distributed platforms like Hadoop are becoming more popular because of their capability to deal with huge amount of complex data in real time. MapReduce, a part of Hadoop, is frequently used programming model for writing distributed applications. This paper describes and presents results of MapReduce-based face detection process using Hadoop platform on BIOID image database.

Keywords: image processing; Hadoop; face detection.

1. Introduction

Today, in the information age computerization of data has led to the explosive increase in information generated on daily basis. Development of mobile technologies and social networks had a great influence on data explosion. Also, with the invention of digital cameras humans are given the possibility to store their every moment in picture form. Number of multimedia content is constantly growing at an increasing rate. Amount of the image data significantly increased not only due to the use of social networks but also because of the use of surveillance cameras and satellite images. Therefore, development of digital technologies made creating and storing visual data equally frequent as the text documents. Availability of such large amount of constantly growing visual data requires tools which can efficiently find and retrieve visual information on demand. The challenge is how to effectively manage all the computations and storage requirements put on by the influx of data.

Hadoop, at the moment is one of the most popular platforms for Big Data analysis and as such is useful in wide range of different applications. Some of them include: sorting, analyzing, machine learning, prediction models, risk assessment, pattern recognition and many other operations. Using Hadoop, users can write distributed applications and analyse huge data sets without understanding the underlying details of distributed computing and storage. This paper describes the role of Hadoop platform in image processing, specifically in face detection process.

The remainder of the document is organized as follows. Section 2 gives the survey of related work in the area of image processing and Hadoop. Hadoop, as one of the leading platforms is described in Section 3. Section 4 reports the process as well as results of face detection in experimental dataset. Finally, Section 5 concludes the paper and presents possible extensions.

2. Survey of the related work

Researches conducted in the field of image processing using Hadoop prove it as a valuable tool because of its ability to execute multiple parallel processes in real time. MapReduce, a part of Hadoop, is especially interesting as a large amount of problems are easily expressible as Hadoop-based MapReduce computations [1]. As a result, many researches are developing Hadoop-based MapReduce algorithms for sophisticated data analysis like is presented in [2], [3], [4], [5]. Current computer vision applications are not able to use big volumes of data because of the lack of resources for analyzing and storing large amount of data. Therefore, distributed platforms are becoming more popular as well as image processing features for those platforms. For example, design of Hadoop Image Processing Interface (HIPI) is described in [6]. HIPI is designed to enable: free libraries for image processing in MapReduce framework, easy image storing and filtering in MapReduce and to parallelise processes without any knowledge of parallelisation details. Also HIPI provides integration with OpenCV libraries which are very useful for computer vision algorithms.

Additionally, MapReduce implementations of several popular computer vision algorithms, such as clustering for image data and classifier training are described in [7]. Presented results show significant improvement compared to traditional implementations.

A distributed MapReduce algorithm for face matching based monitoring movement of individuals over large surveillance space is presented in [8]. The goal is to show benefits of using Hadoop for analyzing video data.

In [9] research have been conducted in using Hadoop for processing medical images. Authors have designed a MIFAS system for fast and efficient access to medical images using Hadoop. Furthermore, in [10] Hadoop is used to process ballistic images with a goal of recognizing samples from crime scenes when compared to image database.

Although, the data manipulation is one of the most explored Hadoop properties its performances are also under study. So in [11] different techniques for managing scheduling scheme are discussed, while in [12] shuffling phase of Hadoop's cluster have been researched.

3. Hadoop platform

Hadoop is an open source, batch data processing system for enormous amounts of data. This platform consists of two primary components: a reliable shared storage (Hadoop Distributed File System) and analysis system (MapReduce). However, these are not the only components of Hadoop. The Hadoop ecosystem provides a set of applications, libraries and systems for developing scalable Big Data applications. These numerous tools are based around Hadoop core (HDFS and MapReduce) for specific client's needs. For example, Mahout is a scalable machine learning and data mining library. Pig is high level data flow language while Hive provides ad hoc querying. Hbase is non - relational, distributed database written in Java. Zookeeper is a high performance coordination service for maintenance. Therefore, all these tools are in use for specific assignments and they represent upgrade on primary components of Hadoop.

3.1. Hadoop architecture

Machines controlled by Hadoop are organized in clusters, where a cluster is a group of connected computing nodes that work and act like a single system using local area network. One node, called *master*, receives request for particular job from client and it controls the job execution. All other nodes in cluster, responsible for computation are called *slaves*. The simple master – slave architecture of Hadoop cluster is shown in

Fig.**1**. First, it is important to notice that both, *master* and *slave* consist of two layers [13]: MapReduce and HDFS.

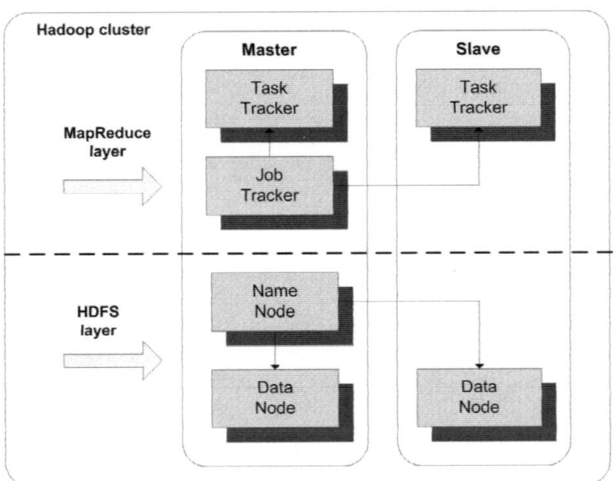

Fig.1. Hadoop architecture

MapReduce layer is responsible for job execution, so it contains two types of nodes a *jobtracker* located on *master* and a number of *tasktrackers* located on *slaves*. The *jobtracker* coordinates all the jobs run on the system by scheduling tasks to run on *tasktrackers*. *Tasktrackers* run tasks and send progress reports to the *jobtracker*, which keeps record of overall progress of each job. On the other hand, a HDFS layer is responsible for data storage. It also includes two types of nodes: a *namenode* (*master*) and a number of *datanodes* (*slaves*). The *namenode* keeps record of every file and block in the file system in memory. Without the *namenode* file system cannot be used. *Datanodes* are responsible for storing and retrieving blocks when they are told to by client or *namenode* and they report back to *namenode* periodically with list of blocks that they are storing. *Master* node keeps record of available *slave* nodes and lets *slave* node know what the next task is and where required data are located.

3.2. *MapReduce*

MapReduce, one of the core components of Hadoop, is linearly scalable programming model and execution environment for data processing. It runs on large clusters of commodity machines [13] and works by dividing the execution processing into two phases: map phase and reduce phase.

In map phase, the input dataset splits into chunks which are processed by the *mappers* in a parallel way. Each chunk can be processed by only one *mapper*. Once the data is processed by *mappers* the output of this phase are {*key*, *value*} pairs which are then forwarded to *reducers*. Hadoop sorts the outputs of the *mappers* according to the *key* values, which are then the input to be processed by the *reducers*. Each *reducer* gets the list of values related to a given *key* and the output of the reduce phase is saved into the distributed file system. Reduce phase cannot start until the map phase is completely finished.

Hadoop job not only consists of parallel map and reduce tasks but also of a shuffle and sort phase. Users write code for map and reduce phase while the Hadoop MapReduce framework automatically handles the shuffling, sorting and parallelizing the processes over multiple nodes [12]. MapReduce job execution is shown in

Fig.**2**.

Hadoop Distributed File System (HDFS) is designed to store data with high bandwidth across the cluster nodes and Hadoop job is design to read that data. Usually, file sizes in HDFS range from gigabytes to terabytes in size. They are broken into block-sized chunks, which are stored as independent units and distributed to hundreds of nodes in a single cluster. Default size of the block in HDFS is 64 MB. To insure fault tolerance and availability each block is replicated to a number of physically separate machines [13]. By default replication is set to three.

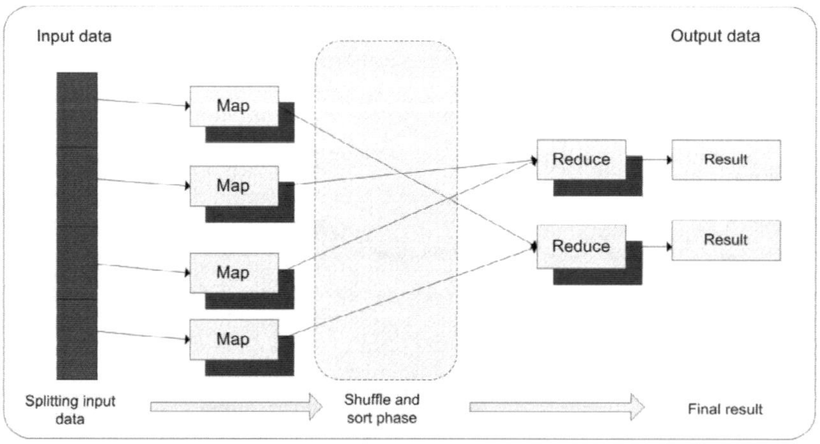

Fig.2. MapReduce job execution

4. Face detection experiment

The main goal of the experiment is to determine whether human faces appear in input image and its position. Dataset used for this experiment is composed of 1000 images (384 x 288 pixel, grayscale) of different persons downloaded from BIOID database [14]. Experiment is run on a single node Hadoop cluster with use of Hadoop version 2.6.4 and Hadoop Image Processing Interface [6]. HIPI is necessary feature for preserving the whole image as an input to mapper. By default HDFS is creating parts or splits the input data which are then replicated and stored in different computer nodes so without HIPI successful face detection would be difficult. For face detection process is used Haar-cascade classifier.

Face detection process in this experiment is conducted in several steps. First of all, Hadoop Image Processing Interface is storing data as a collection of images called HIPI Image Bundle which are then sent as an input to mapper. Input values are distributed among the available map tasks. The MAPPER class has a MAP method that is called once for each input. Map method is executing the following initial steps: loading OpenCV libraries to enable image processing, converting

HIPI Image Bundle for compatible OpenCV format and creating classifier for content-based image analyzing. Afterwards, face detection process is executing and the result is emitted to REDUCER in {*key*, *value*} form where key is metadata from image and value is coordinates of detected faces in images.

The REDUCER class has a REDUCE method that is called once for each unique key. Reduce method calculates number of detected faces and emits final result to HDFS. See Fig.3. where step-by-step Hadoop face detection process is presented, while its logic for a concrete implementation is given in Algorithm 1.

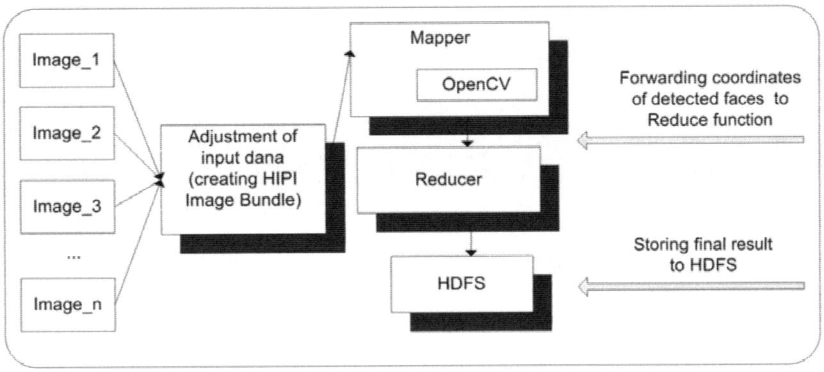

Fig.3 Hadoop face detection process

```
class MAPPER
    method MAP (metadatameta, image i)
    m ← convertToCompatibleFormat (image i)
    cascade_data ← load()
    c ← initClassifier(cascade cascade_data)
    detections ← detectFace(mat m, classifier c)
    for all detection d ∈ detections do
        key ← getSource(medatada meta)
        value ← formulateValue(detection d)
        EMIT(source key, frame value)
class REDUCER
    method REDUCE(source key, frames[r1, r2, r3, ... ])
    collection ← initCollection()
    for all frame f ∈ frames do
        appendToCollection(frame f, frame_collection collection)
    EMIT(Hash key, frame_collection collection)
```

Algorithm 1 MapReduce algorithm for face detection in images

Table **1**. shows the results of MapReduce-based face detection. From total 1000 faces in all images 976 faces are successfully detected. There is a 124 false positive results as well as the 24 false negative which gives overall accuracy of 86.83%. The total execution time for this process was 3 minutes 39 seconds. Results show that MapReduce-based face detection gives high accuracy of detection on great amount of images in real time. Therefore, it can be useful tool to solve different real-time problems which require face detection and face recognition.

Observed Items	**Values**
Faces in images	1000
Faces successfully detected	976
False positive detection	124
False negative detection	24
Time elapsed	3m 39s
Accuracy	86.83%

Table 1. Hadoop face detection experiment results

Some of the sample images used for face detection from BIOID database are presented in Fig. **4**. One of the original images from BIOID database before face detection process is presented in Fig.4. a), while Fig.4. b) shows detected face on the same image afterwards. Fig. 4. c) presents sample of image with false positive detection.

a) b) c)

Fig. 4. Sample images from BIOID database: a) original image, b) image after face detection process, c) image with false positive detection

5. Conclusion

Face detection is normal human psychological process by which humans easily locate faces in visual scene. Although, recognising content of image and especially face detection is natural process for humans, doing the same task for computers is much more complex process. Therefore it is actual and interesting research area because of the all spheres of real life where face detection can be put to use like in security systems and surveillance cameras.

Efficient usage of huge amount of visual data currently generated requires distributed storage and processing. One of the most popular software frameworks for distributed storage and distributed processing is Hadoop. It provides a simple MapReduce programming model, a distributed file system and job management as well as the cluster management.

Results presented in this paper show that MapReduce-based face detection is promising tool in terms of accuracy of detection as well as the time spent on execution. However, by increasing the number of nodes in Hadoop cluster the performances would improve linearly which can be studied in future work. Also, future research should consider implementing more pose variations as well as the greater amount of images. Additionally, different elements can partially cover the faces like glasses or beards which can affect performance of face detection. Therefore, different classifiers can be used and their performances can be compared.

However, it is shown that Hadoop can be successfully used in computer vision algorithms for real time processing because of its performances.

6. References

[1] Garcia, C. (2013). Demystifying MapReduce. *Procedia Computer Science*, Vol. 20., Complex Adaptive Systems Publication 3 (2013.), pp. 484-489.

[2] Shelly & Raghava, N.S. (2011). Iris recognition on Hadoop: A biometrics system implementation on cloud computing. *2011 IEEE International Conference on Cloud Computing Computing and Intelligence Systems,* (September 2011.), pp. 482-485., ISSN 2376-5933

[3] Almeer, M. (2012.). Cloud Hadoop map reduce for remote sensing image analysis, *International Journal of Emerging Technology and Advanced Engineering*, Vol. 2, No.4, 2012, ISSN 2250-2459

[4] Pavlech, M. (2011). Framework for development of distributed evolutionary algorithms based on mapreduce. *Annals of DAAAM for 2011 &Proceedings of the 22nd International DAAAM Symposium*, Vienna, Austria (2011) Vol.22, No.1, ISSN 1726-9679

[5] Leokhin, Y.; Myagkov, A. & Panfilov, P. (2016). GoMapReduce Parallel Computing Model Implementation on a Cluster of Plan9 Virtual Machines, *Proceedings of the 26th DAAAM International Symposium*, pp. 0656-0662, B.Katalinic (ed.), Published by DAAAM International, ISBN 978-3-902734-07-5, Vienna, Austria (2016), ISSN 1726-9679, DOI: 10.2507/26th.daaam.proceedings.089

[6] Sweeney, C.; Liu, L.; Arietta, S. & Lawrence, J. (2011). HIPI: A Hadoop image processing interface for image-based MapReduce tasks, B.S. Thesis, Department of Computer Science, University of Virginia, USA

[7] White, B; Yeh, T.; Lin, J. &Davis, L. (2010) Web-scale computer vision using MapReduce for multimedia data mining, *Proceedings of the Tenth International Workshop on Multimedia Data Mining*, New York, USA, ISBN 978-1-4503-0220-3, ACM, NY, DOI 10.1145/1814245.1814254

[8] Mishra, R.; Kumar, P; Chaudhury, S & S, I. (2013) Monitoring a large surveillance space through distributed face matching, *Fourth National Conference on Computer Vision, Pattern Recognition, Image Processing and Graphics (NCVPRIPG)*, 18-21 December 2013, Jodhpur, ISBN 978-1-4799-1586-6, pp.1-5, IEEE, DOI: 10.1109/NCVPRIPG.2013.6776185

[9] Yang, C.; Chen, L; Chou, W & Wang, K. (2010). Implementation of a medical image file accessing system on cloud computing, *IEEE 13th International Conference on Computational Science and Engineering (CSE)*, 11-13 December 2010, Hong Kong, ISBN 978-1-4244-9591-7, pp. 321-326, IEEE, DOI: 10.1109/CSE.2010.48

[10] Kocakulak, H & Temizel, T. (2011). A Hadoop solution for ballistic image analysis and recognition, *International Conference on High Performance Computing and Simulation (HPCS)*, 4-8 July 2011, Istanbul, ISBN 978-1-61284-380-3, pp. 836-842, IEEE, DOI: 10.1109/HPCSim.2011.5999917

[11] Xie, J.; Meng, F; Wang, H.; Pan, H.; Cheng, J.& Qin, X. (2013). Research on scheduling scheme for Hadoop clusters, *International Conference on Computational Science*, Procedia Computer Science, Vol. 18, pp. 2468-2471, Elsevier B.V., DOI: 10.1016/j.procs.2013.05.423

[12] Xie, J.; Tian, Y.; Yin, S.; Zhang, J.; Ruan, X. & Qin, X. (2013). Adaptive preshuffling in Hadoop clusters, *International Conference on Computational Science*, Procedia Computer Science, Vol. 18, pp. 2458-2467, Elsevier B.V., DOI: 10.1016/j.procs.2013.05.422

[13] White, T; (2012.). *Hadoop: The definitive guide, third edition*, OReilly Media, INC., ISBN 978-1-449-31152-0, CA, USA

[14] https://facedetection.com/datasets/ (2016). Face Detection & Recognition Homepage, Accessed on: 2016-04-04

DOI: 10.2507/27th.daaam.proceedings.096

EMISSIVITY FACTOR OF THERMOPLASTIC MATERIALS FOR A ROBOT BENDING PROCESS

Steiner Franz*, Bleicher Friedrich & Klingenbrunner Martin

This Publication has to be referred as: Steiner, F[ranz]; Bleicher, F[riedrich] & Klingenbrunner, M[artin] (2016). Emissivity Factor of Thermoplastic Materials for a Robot Bending Process, Proceedings of the 27th DAAAM International Symposium, pp.0664-0670, B. Katalinic (Ed.), Published by DAAAM International, ISBN 978-3-902734-08-2, ISSN 1726-9679, Vienna, Austria
DOI: 10.2507/27th.daaam.proceedings.096

Abstract

During thermal incremental forming of semi-finished products, the temperature, spring-back, and morphological structure of the material are interdependent. To determine the forming temperature of the semi-finished product, knowledge about the material-specific properties is required. The processing heat for amorphous polymers is near the glass transition temperature. For semi-crystalline materials, this heating temperature is near the melting point. In approximation to these temperatures, the material can be bent such that a hot core in the semi-finished product leads to low applied force at the robot end effector. Therefore, the semi-finished product is brought near the processing heat by the supply of heat. Subsequently, the core temperature is determined by infrared thermography to observe the heating process. This method provides a connection between the bending force and core temperature, as well as determines the quality of bending.

Keywords: infrared thermography; emissivity factor; semi-finished thermoplastic; robot bending

1. Introduction

The test field at the Institute of Production Engineering and Laser Technology (IFT) is used for detecting bending parameters with the influence of heat. Bending parameters for forming operations of synthetic materials are the material involved and temperature-dependent spring-back. To investigate material performance, knowledge on the temperature profile during the bending process must be provided. Infrared (IR) thermography, an imaging technique used to display temperatures on object surfaces and for managing production processes, can be used for this purpose. Its main advantage compared with common contact thermometers is contact-free, reactionless temperature measurement, which operates by absorbing thermal radiation from an object. Other advantages include better in-process performance based on computer compatibility and flexibility of thermographic devices. A thermal imaging camera detects and receives this IR radiation. This process is used to characterize materials [1] or measure the bending load of thermoplastic composites as described in [2]. Investigations on damaging impact [3] and linear thermal expansion coefficients of polymeric materials [4] have been reported. The use of IR thermography for controlling a temperature-dependent manufacturing process and rising material temperature and inhomogeneity has been presented [5].

Thermoplastic heating methods are used for plasticizing ceramic, quartz, halogen, or carbon radiators [6]. Advantages and disadvantages of different simulation models involving quartz tube heaters or the finite difference method to evaluate material parameters and the heating cycle are demonstrated in [7-8].

2. Bending robot process

For the forming of semi-finished plastics, a robot end effector can be used as a bending tool. Further advantages of this arrangement are the possibility of manipulation through the 3D bending space and low bending forces acting on the work piece. The bending line is divided into increments (inc), i.e., the section to be bent is heated, formed, and cooled. Cooling freezes the molecular chains to maintain the bent increment in the desired position in space. The test field at IFT consists of two parts: plastics processing and manipulation via a robot (Figure 1).

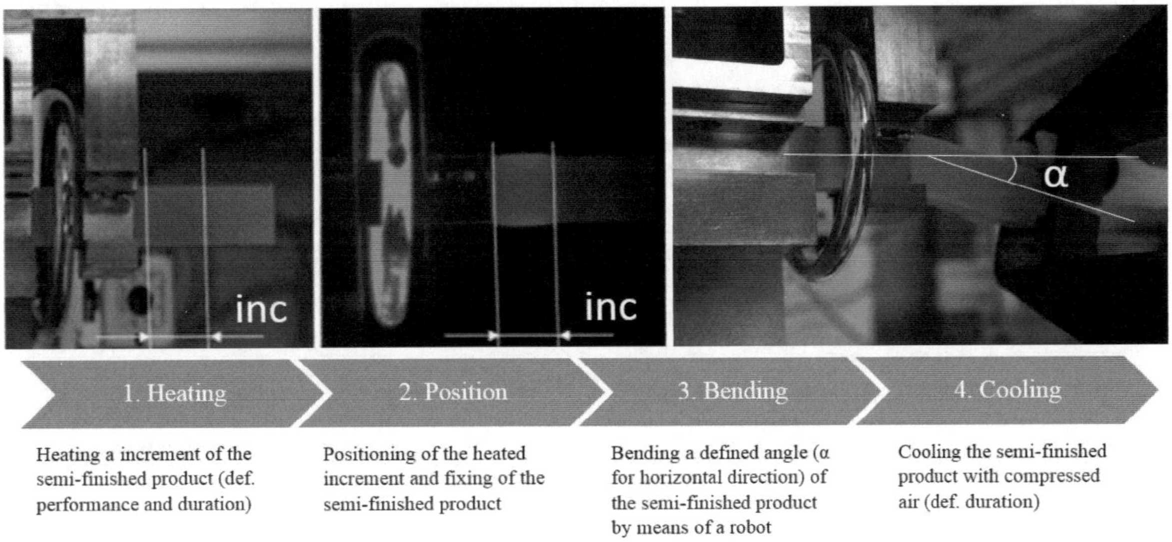

Fig. 1. Bending process: the semi-finished product is heated, positioned, bent (horizontal), and cooled

The process is initiated by the supply of the semi-finished product to the system. The product is then positioned in the heating zone along a linear axis. A circular pronounced IR heating element heats the bent increment until the material reaches the dependent forming temperature field. In this field, the molecular chains are highly flexible and semi-finished products can be plasticized and processed. This plasticized increment is brought into the area outside plastics processing and clamped pneumatically. The area between pneumatic clamping and the end effector is defined as the bending zone. At this position, the end effector engages the material and moves it in space (see Figure 1: α defines the angle of a horizontal bend). The positions in the x-, y-, and z-directions for movement in space are passed via the control. The bending process data are identified along a polyline and provided as a text file. The content of this text file is converted by machine control into bending parameters. The bending parameters act along the center of gravity of the semi-finished product and define the neutral phase of the bending line. The scanning or design of the CAD file for the front end is described in [9].

For process stability, the temperature curve must be known inside and around the work piece. A test environment is created by an IR camera, which provides information on the external and internal temperatures (core temperature) of the semi-finished product.

Important characteristic properties for the polymer are the glass transition temperature T_g and melting temperature T_m. Below T_g, polymers are very brittle, the structure is "frozen," and polymer chains have limited movement. However, if the temperature increases to T_g, the polymer chains start to vibrate. Above T_g, the material behaves rubbery and the chains can easily move. T_m, an important value for crystalline polymers, varies because of the percentage of crystallinity [10]. Thus, by reaching the forming temperature in the core, low forces for bending are applied to the semi-finished product.

To accurately detect an object's surface temperature, the material and surface-specific emissivity factor ε must be determined. According to (1), ε is defined as a factor that compares the specific radiation of a real object ("gray body") to a black body at the same temperature. If the evaluated object is a black body, ε is equal to 1. Otherwise, $\varepsilon < 1$ (gray body).

$$\varepsilon = \frac{M(T)}{\sigma * T^4} \quad (1)$$

Referring to (1), M(T) is the thermal emissivity of the observed object depending on its temperature (W/m²), σ is the Stefan–Boltzmann constant, and T is the temperature of the black body (K), which is equal to the observed temperature. Thermographic devices (e.g., IR cameras) detect M(T) and calculate the surface temperature with ε as the input [11].

3. Determination of the emissivity factors

To determine the emissivity factors, thermoplastic samples (compare Figure 2 data sheets in [12]) are measured using an IR camera (Optris PI400 properties in [13]) and contact thermometer (Testo 925) for reference purposes. The specimen's surface condition exerts remarkable effects on the emissivity factors. Given the lack of standardized descriptions, the determined emissivity factor must be re-examined every time the supervised specimen's surface varies, even if the material is the same.
The following surfaces supervised by the IR camera can be described as smooth finished after milling them plain.

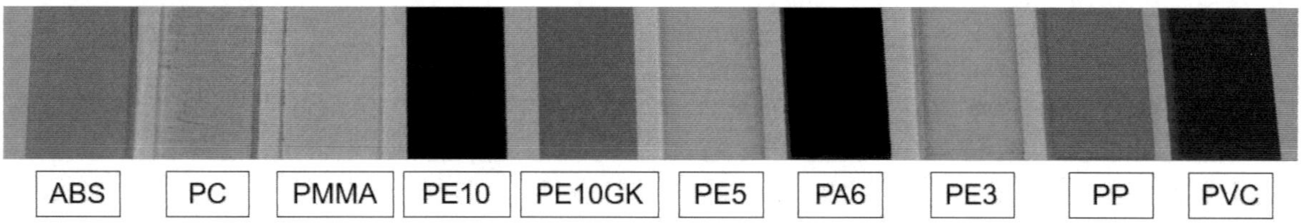

Fig. 2. Observed samples acrylonitrile-butadiene-styrene (ABS), polycarbonate (PC), poly(methyl methacrylate) (PMMA), ultra-high-molecular-weight (uhmw) polyethylene (PE10), uhmw polyethylene + glassballs, PE10 GB(PE10GK), high-molecular-weight (hmw) polyethylene (PE5), poly(hexano -6-lactam) polyamide 6 (PA6), high-density polyethylene (PE3), polypropylene (PP), and poly(vinyl chloride) (PVC)

First, the samples are prepared (Figure 3). A large hole (d_1 = 3 mm, $l_1 >> d_1$) is needed to realize a black body. According to [11], this size can be achieved by drilling a hole with a depth that is several times its diameter. The second hole is for measuring the core temperature with an ordinary contact thermometer for reference. A small difference between the diameters of the holes and thermometer is essential for precise measurement.

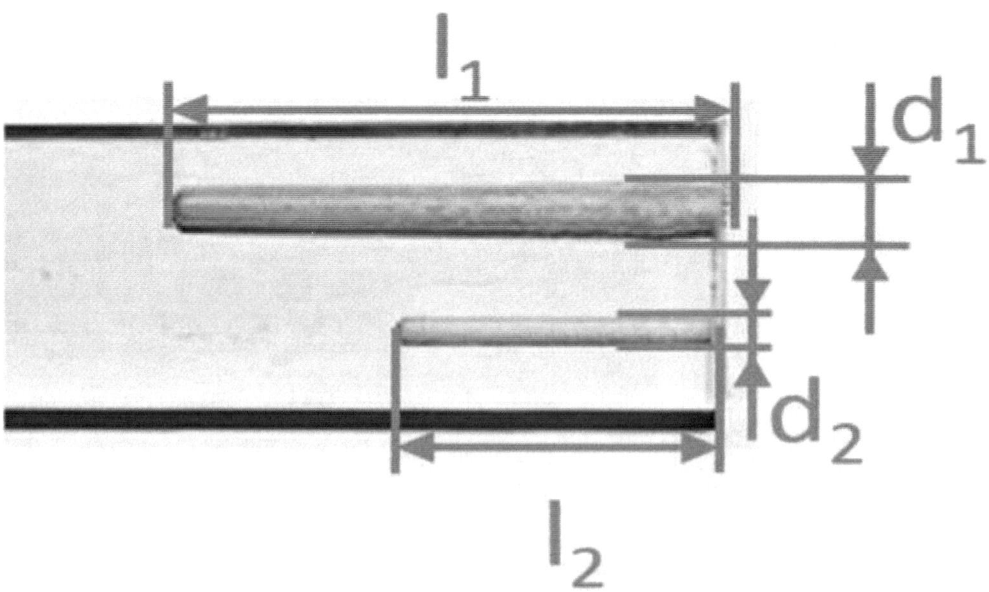

Fig. 3. Preparation of samples of an amorphous semi-finished product poly(methyl methacrylate)

To reduce external distortion radiation, the whole setup is placed in a black box (Figure 4). The inner surface is coated with a black and rough material to impede reflection. To ensure correct measurement, a distance of at least 500 mm is necessary between the supervised specimen and IR camera.

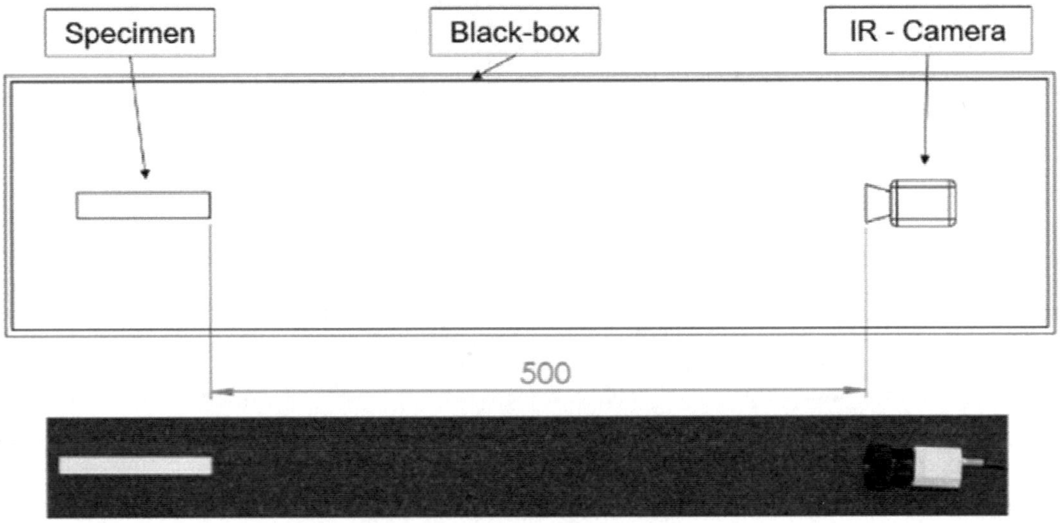

Fig. 4. Experimental setup: top - exemplary, down - real view

Figure 5 shows the surface of one sample, as detected by the IR camera. The two bright spots correspond to the holes in Figure 3. The right cavity shows that the black body has a higher thermal radiation than its surroundings. This feature is due to the ability of a black body to absorb incident thermal radiation completely, with $\varepsilon = 1$, because of Kirchhoff's law [11].

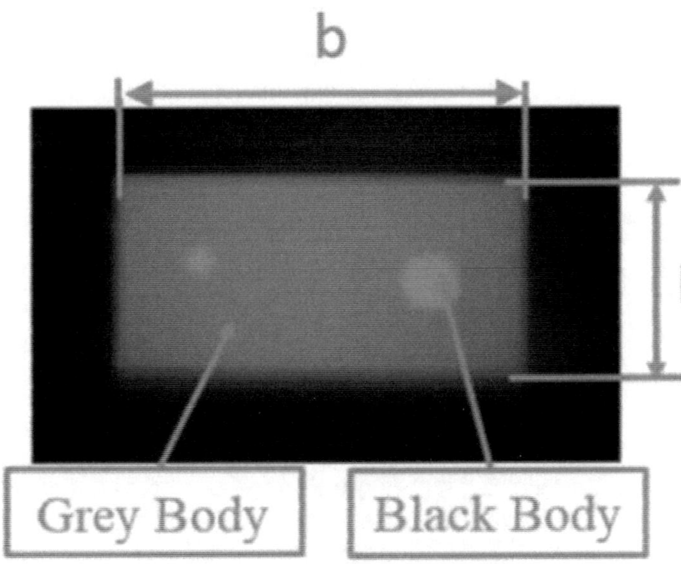

Fig. 5: Surface of the sample through infrared camera
$b = 20$ mm, $h = 10$ mm.

Using this knowledge, one measuring point of the processing program evaluates the temperature of the black body with a dedicated ε of 1 (Figure 6). The second measuring point evaluates the mean surface temperature via the displayed rectangle, excluding the two holes. To achieve the same temperature as the first measuring point (thermal equilibrium) and determine the specific emissivity factor, ε of the rectangle is modified until the two temperatures are equal [14].

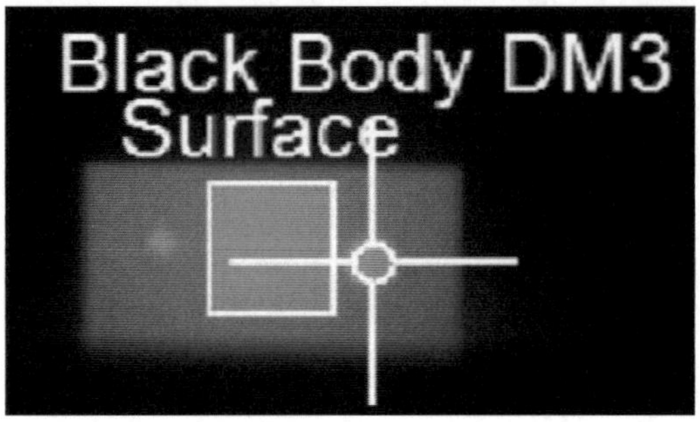

Fig. 6. Measuring surface temperature

4. Results

The following Matlab plot (Figure 7) displays the measured temperatures for PE5 and PMMA for amorphous and semi-crystalline materials, respectively. The black and blue lines indicate the black body and surface temperatures over time evaluated by the IR camera, respectively. The red and cyan lines represent the core and surface temperatures measured by an ordinary contact thermometer for reference purposes, respectively. The scale of the vertical axis, as well as the small deviation of the lines, shows suitable results.

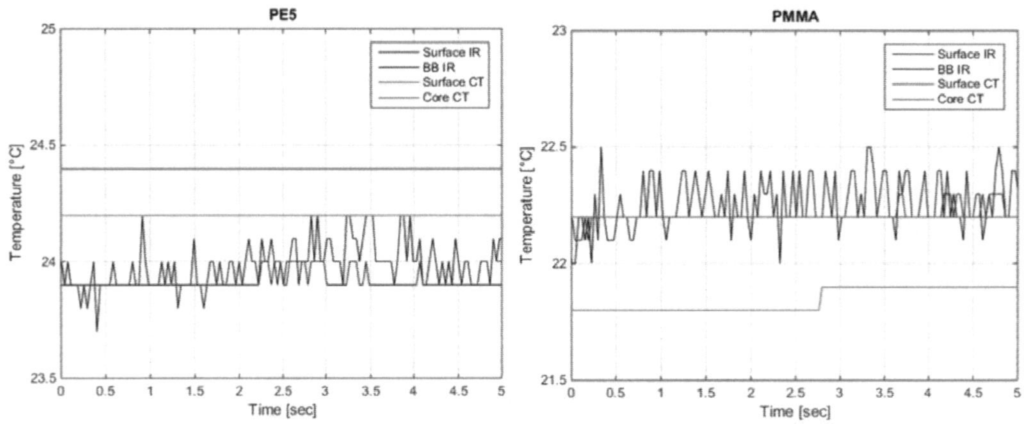

Fig. 7. Visualization of measurement results for PE5 (left) and PMMA (right).

Figure 8 provides an overview of the average measured temperatures over the supervised time on the example polyacrylic (PA6). The first two columns represent the surface and black body temperatures evaluated by the IR camera. The next two columns represent the core and surface reference temperatures measured by the contact thermometer.

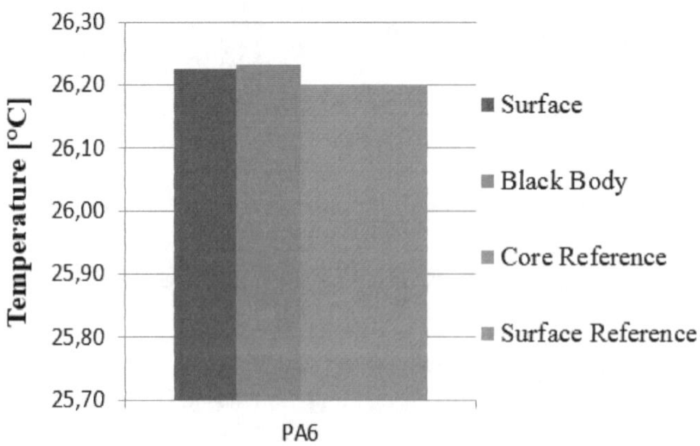

Fig. 8. Measurements of PA6.

Table 1 shows the determined emissivity factors. Column three describes the relative error (RE), which is evaluated from the mean surface temperature and absolute temperature of the black body (see Figure 8). A small variation of 0.015 in ε is found across the observed materials.

Sample	**ε**	**RE [%]**
ABS	0,965	0,73
PA6	0,965	0,02
PC	0,975	0,15
PE3	0,96	0,46
PE5	0,965	0,02
PE10	0,975	0,56
PE10GK	0,965	0,03
PMMA	0,96	0,26
PP	0,965	0,28
PVC	0,965	0,25

Table 1. Emissivity factor ε and relative error of the samples.

The experimental setup provides a simple method to determine the emissivity factor. The effectiveness of this method is displayed in the third column of Table 1 and its small relative error. According to the provided systematic measurement errors of the used devices in Table 2, a measured temperature of approximately 270 °C is necessary until the RE exceeds the systematic measurement error of the IR camera in the worst case (ABS).

Systematic measurement error IR	2 °C
Systematic measurement error CT	0.5 °C

Table 2. Systematic measurement error of infrared camera and contact thermometer (CT).

Conclusion

Thermography is a modern and highly useful technology for managing production processes via surface temperature. In addition to the accurate observation of the heating process using the IFT's prototype, ε may support the selection of materials intended for thermoforming because of Kirchhoff's law. This law states that the absorption of thermal radiation is equal to ε.

The emissivity factor ε plays a key role in determining radiation, and this relationship can be used as a parameter for process control. Through the detection of contact-free radiation, the temperature can be determined at the surface and in the core. Thus, plasticity in the core will allow application-oriented bending forces in the future.

The emissivity is evaluated by means of a test set-up consisting of different thermoplastic samples and an IR camera embedded in a black box. The lowest value of the tested samples is ε = 0,965 (ABS, PA6, PE5, PE10GK, PP, PVC) and the highest value is ε = 0,975 (PC, PE10). These values are needed to obtain a statement about the core temperature. This core temperature is crucial for the behavior of the bending line. Higher temperature in the core leads to lower spring-back effects.

The results of this study will benefit future studies, because approximation to glass transition temperature and melting temperature leads to low stress in the material. Stress in the working part during bending affects dimensional tolerance or size accuracy and results in another bending process parameter, the spring-back. The determination of the process parameters and their interrelationships allows reproducible and economic process control.

5. References

[1] Meola, C.; Carlomagno, G. M. & Giorleo, L. (2007) The use of infrared thermography for materials characterization, In: Journal of Materials Processing Technology, Vol. 155-156, (30 November 2004) pp. 1132–1137

[2] Boccardi, S.; Carlomagno, G. M.; Meola, C.; Simeoli, G. & Russo, P. (2015) Infrared thermography to evaluate thermoplastic composites under bending load, In: Composite Structures, Vol. 134, (15 December 2015) pp. 900–904.

[3] Meola, C.; Boccardi, S.; Boffa, N. D.; Ricci, F.; Simeoli, G.; Russo, P. & Carlomagnoa, G. M. (2016) New perspectives on impact damaging of thermoset- and thermoplastic-matrix composites from thermographic images, In: Composite Structures, Vol. 152, (15 September 2016) pp. 746–754

[4] Wang, M.; Liao, Y. & Chen, D. (2013) Determination of linear thermal expansion coefficient of polymeric materials by infrared thermography, In: Polymer Testing, Vol. 32, Issue 2, (April 2013) pp. 175–178

[5] Meola, C.; Giorleo, G.; Nele, L.; Squillace, A. & Carlomagno, G. M. (2002) Infrared thermography in the quality assurance of manufacturing systems, In: Nondestructive Testing and Evaluation, Vol. 18, Issue 2, (2002) pp. 83-90

[6] Schwarzmann, P. (2015) Heating of Intermediates, In: Kunststoffe 1/2015, Germany: Carl Hanser Verlag, pp. 48-51

[7] Brogan, M. T. & Monaghan, P. F. (1996) Thermal simulation of quartz tube infra-red heaters used in the processing of thermoplastic composites, In: Composites Part A: Applied Science and Manufacturing, Vol. 27, Issue 4, (1996) pp. 301-306.

[8] Fertschej, A.; Langecker, G. R. & Kertz, J. (2007) Finite Difference Method: Simulation of the Heating Cycle during Thermoforming, In: Kunststoffe 1/2007, Germany: Carl Hanser Verlag, pp. 62-65

[9] Steiner, F. & Bleicher, F. (2016) Freiformen im sphärischen Biegekopf, In: Kunststoffe 1/2016, Germany: Carl Hanser Verlag, pp. 56-59

[10] Stuart, B. H., Polymer Analysis. ISBN: 978-0-471-81363-7, England: John Wiley & Sons, Ltd, 2002

[11] Vollmer, M. & Möllmann, K. P., Infrared Thermal Imaging: Fundamentals, Research and Applications, ISBN: 978-3-527-40717-0, Germany: Wiley-VCH Verlag GmbH & Co. KGaA, 2010

[12] Faigle Industrieplast GmbH, faigle semi-finished products - plastic technical data, 2009

[13] Optris GmbH, Basic Principles of Non-Contact Temperature Measurement, Germany

[14] Raytek Corporation, The Principles of Noncontact Temperature, USA, 2016

DOI: 10.2507/27th.daaam.proceedings.097

STANDALONE APPLICATION USING JAVA AND ANFIS FOR PREDICTING ELECTRIC ENERGY CONSUMPTION BASED ON FORECASTED TEMPERATURE

Dragan Mlakić, Srete Nikolovski & Dean Vucinic

This Publication has to be referred as: Mlakic, D[ragan]; Nikolovski, S[rete] & Vucinic, D[ean] (2016). Standalone Application Using JAVA and ANFIS for Predicting Electric Energy Consumption Based on Forecasted Temperature, Proceedings of the 27th DAAAM International Symposium, pp.0671-0677, B. Katalinic (Ed.), Published by DAAAM International, ISBN 978-3-902734-08-2, ISSN 1726-9679, Vienna, Austria
DOI: 10.2507/27th.daaam.proceedings.097

Abstract

Predicting energy consumption is the ability, which is useful anywhere where the energy resource is the main product. In particular, the electric energy is a very expensive product to be wasted without a concrete need. The need for fast prediction of the electric energy consumption is based on various inputs covering the following aspects: any time, any sampling resolution, any depth based on well-known and available data. Today, the advancing of the technology allows the new generation of software applications to integrate different methods for the prediction of the electric energy consumption, which is based on various available data coming from anywhere. Focus of this paper is flexibility of AI methods in predicting energy consumption and availability of input data for predicament. In this paper, the method for predicting el. energy consumption is based on a daily temperature using the Adaptive Neuro Fuzzy Inference System (ANFIS). Comparison of results are presented in paper between ANN and ANFIS on same input data, and that puts ANFIS method as best for predicting energy consumption. The results analysis is done with the developed MatLab software and the data acquisition module is specifically developed in JAVA. The data source is the web service available as "weather underground" web service. All the components are integrated in a performing and compact lightweight application that supports mobility, easy to use and easy to change functionality. The modelling, testing and validation are made on the data acquired from the electric energy consumption provided by the distribution system of the town on Kreševo in Bosnia and Hercegovina.

Keywords: ANFIS; Fuzzy; JAVA; electric energy consumption; MatLab; Simulink; application; mobility.

1. Introduction

The use of the Artificial Intelligence (AI) methods for predicting the electric energy consumption is becoming an on-going trend, which has recently started. Many papers, researches and ideas have emerged and presented interesting work based on ANN, Fuzzy logic, GA, FA, and weather conditions gathered from many different places [1]. Today. The weather data are available from the web services, like wunderground.com, weather.com, intellicast.com, etc., which are accurate and diverse data to be reused. The capturing of those data and using them as input to the AI methods for predicting the energy consumption is not a new idea [2] [3] [4] [5] [6] [7]. Bringing them to a higher level of abstraction and make them available on any system and at any time, through a simple GUI, which nicely brings together highly modular and complex software tool is the focus of this research paper. Performing fast analysis, with mobility and easiness of use, are enabled through different set of tools. The importance of the energy consumption forecasting, in this case, is based on the weather forecast input, as a very important parameter for avoiding suboptimal management of an energy network, and for dispatcher to control in advance the load or unload of the network, and in such way to prevent the unnecessary outages or waste of the costly and valuable electrical energy.

In praxis, it is very important to make fast prediction of the consumption for the next custom period, as sampled in this use case for the distribution area of the town Kreševo located in Bosnia and Herzegovina. It is well known that every time is needed to perform finding, segmenting, preparing, loading, training, testing and presenting data. In order to reduce time and simplify such work it was essential to make further analysis, like management of line breakers in distribution network and power flow without risk of outage or energy waste [6].

Many research papers have been applying the AI methods for predicting the electric energy consumption based on variety of data input [2] [3] [4] [5]. Some papers compare the AI methods to find the most appropriate method to calculate such results. By considering the results of those papers, the Artificial Neural Network (ANN) is among the best and most accurate methods found for such forecasting and predicting [9] [10] [11].

In this paper, ANN is compared to the Adaptive Neuro Fuzzy Inference System (ANFIS), which is validated on the same input data according to the measured data from the Automatic Meter Reading (AMR) system, which is installed on the sampled electric energy network in the area of Kreševo. For the comparison of these results the best method was chosen between ANFIS and ANN, and used for the application implemented with the MatLab GUIDE. The JAVA programming language is used for gathering the data from the web service wunderground.com, based on the direct link giving access to the forecasting weather results.

The JAVA application reads the entire web page by recognising the temperature data needed for ANFIS in order to predict the consumption. The buffer for storing the input data is generated as *.txt file. The output data are also stored in the next generated *.txt file for further use, which file size is unlimited. The flexibility of and multiplatform ability of such JAVA application are the reasons for the development of the data gathering application in the selected JAVA language. The call-back for the JAVA application module is done within the GUIDE application, and based on the user choice.

The result of this applied approach is a fast, accurete and flexible application with an intuitive GUI and opening variety of choices for the end user

2. Electric energy distribution for the area of Kreševo

The geographic area, for which the forecasting is conducted is completely measurable on every giving/receiving point between the other areas, and data are already calculated in format of kWh, with resolution which is available for the daily or monthly forecasting. The Automatic Meter Reading (AMR) is conducted with the system Automatic Meter Management (AMM) based on information taken from meters in every substation within the area of Kreševo, having the secondary voltage level of 35/10 kV transformer in the energy receiving point located in the substation TS 35/10 kV Kreševo. One point is border between area Kreševo and area Kiseljak. Fig. 1. shows a single line diagram for the distribution area of Kreševo with the energy receiving points marked at the border point with area Kiseljak. The energy consumption of area Kreševo is calculated using (1):

$$E = E_R - E_{Kis} \quad (1)$$

Where is:
E_R – energy received to area Kreševo;
E_{Kis} – energy supplied to area Kiseljak;
E – energy we seek to forecast.

Most of the time the area of Kreševo is supplying energy to the area of Kiseljak, thus it is not taken into the consideration the situation, when energy is coming from Kiseljak to Kreševo, as those situations are rare. The gathered measured data from AMR/AMM, are input data for creating forecasting methods, ANFIS and ANN, developed in Matlab.

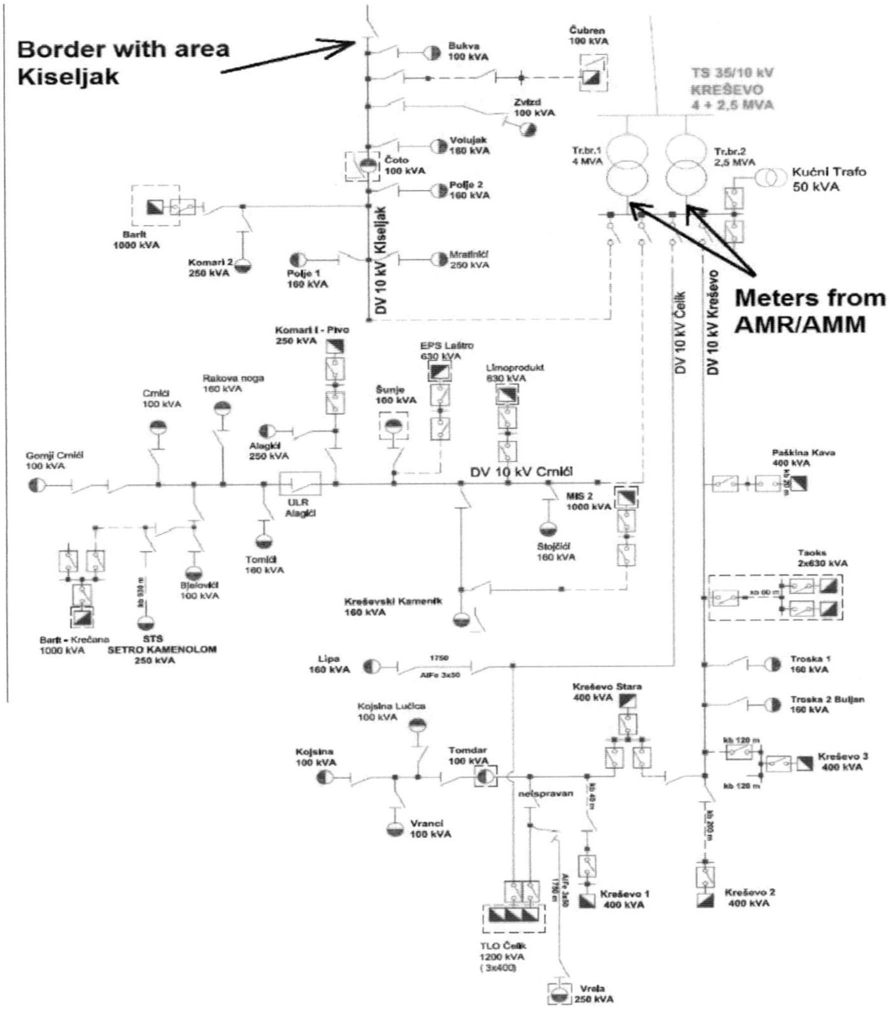

Fig. 1: Single line diagram of the distribution area of Kreševo.

3. Forecasting methods

Methods for forecasting the energy consumption were discussed through many conferences, research papers, and articles [7] [9] [10] [11]. Most of those papers discuss the accuracy and complexity of the proposed methods, which are based on their mathematical models, requirements for training, and checking results [8] [9] [12]. It was noticed that a few papers have in focus the elegancy and effectiveness of those methods applied to the everyday praxis. In this paper two methods were discussed and compared with same input data and tested according to measured values from AMR system. The input data are maximum, average and minimum temperature, acquired for one day within the targeted month. Those data are joined with the measured values of consumed electric energy during that day of the targeted month. By considering such data and combine it with, weather temperature and energy consumption, the method for forecasting the next month consumption is created and analysed.

3.1. ANFIS

The Adaptive Neuro Fuzzy Inference System is widely used in recent time and already introduced in these papers. In this case, the created ANFIS tool has 3 inputs and 1 output, with the Sugeno decision type, and the Gauss membership functions. Inputs for its training consists of 257 samples, and the average error of measured data is around 9%. The created rules are 64, defuzzy method is weighted as average since based on the Sugeno decision type. The tool for training, testing and validating ANFIS is Matlab Fuzzy module. The generated rules are shown in Fig. 2, and surface of created ANFIS is shown in Fig.3.

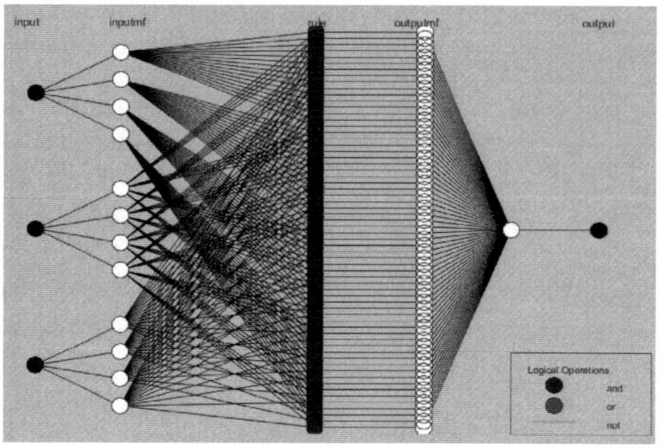

Fig. 2. Trained ANFIS for forecasting electric energy consumption based on weather online data.

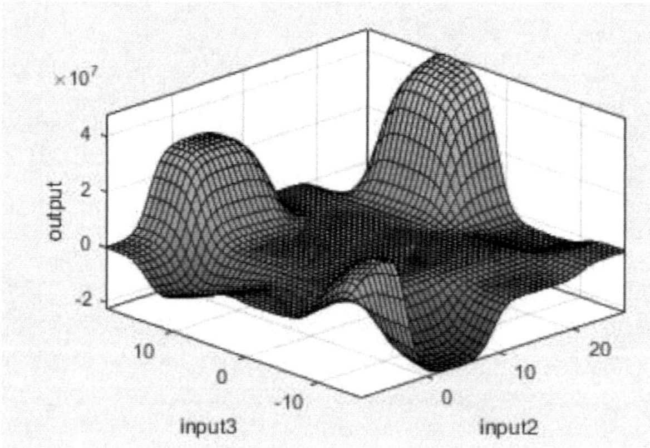

Fig. 3. Surface of created ANFIS.

3.2. *Artificial Neural Network*

The second method used is ANN in form of the Neural Network Fitting Tool from Matlab. The introduction of ANN is done in previous papers [11]. The Matlab nftool is used for creating neural fitting function with 365 samples, 3 inputs, 1 output, 10 hidden layers, training function is Bayesian Regularization method, performance measured by Mean Squared Error, and the calculations are done by using the MEX function. Fig. 4 shows Error Histogram for the creation process of ANN. Fig 5. Shows regression of created ANN function.

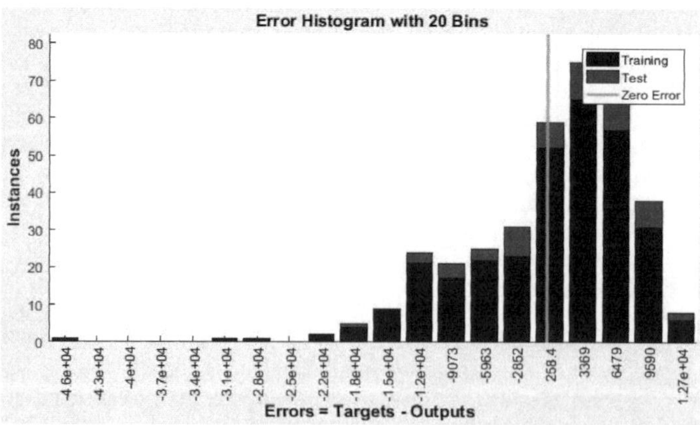

Fig. 4. Error histogram for created Neural Network.

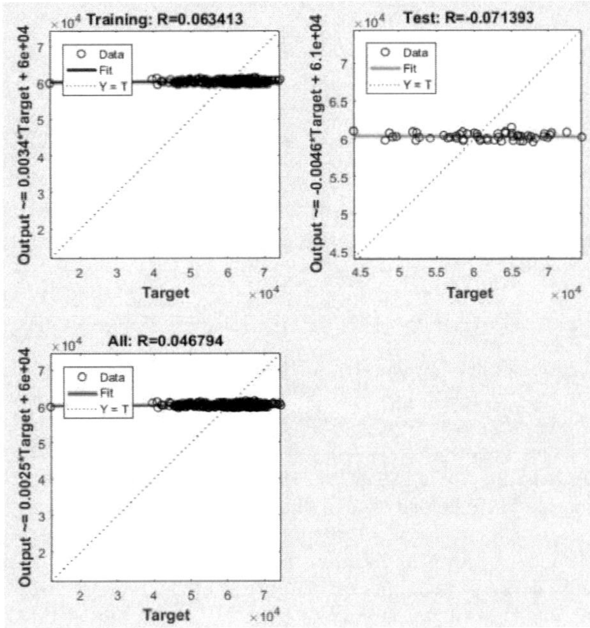

Fig. 5. Regression of Neural Network created function.

3.3. Comparison

Results from generated artificial intelligence methods are clearly showing ANFIS advantage for fitting the measured data [14] [15]. Fig. 6 shows graphical presentation of results from ANFIS, ANN and measured data from AMR/AMM.

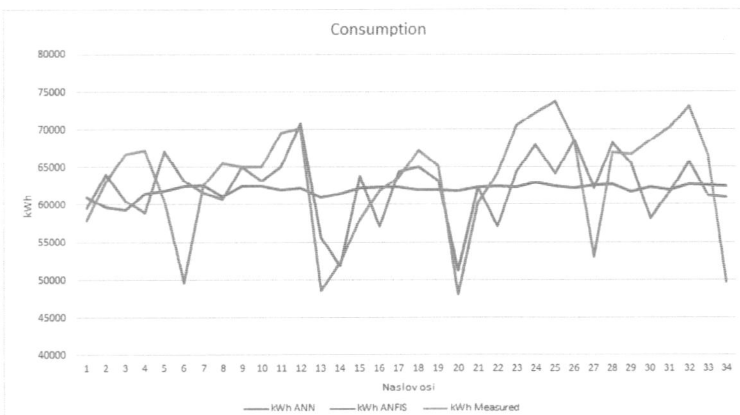

Fig. 6. Compared results from ANFIS and ANN for May 2016, for area Kreševo.

4. Application structure

The block scheme presents the structure of the forecasting procedure, as shown in Fig. 7. The diagram shown in Fig. 7 describes the way of process that starts with the web weather service, in this case www.wunderground.com, which gives the needed forecasting maximum, average and minimum temperature for the time period of 1 month in 1 day resolution. By using the developed JAVA application as a standalone runnable .jar, the, data from the web service are transferred to the editable and light textual file with extension .txt. In the Matlab application, the available input data are coming from the generated .txt file and transferred to the online memory of the application. When this is done, the application feedback is the "DONE!!!" note appearing in the application window. The next step is to use the input data for the forecasting consumption concern with the electric energy area, in this case the town of Kreševo, located in Bosnia and Hercegovina. This specific step is done by applying the artificial intelligence method, in this case ANFIS. If successfully done, the user receives the feedback message "DONE!!!". The generated data are stored in output textual file with extension .txt. The entire application is developed using Matlab 64 bit edition GUIDE tool and JAVA code in .jar file developed by using the Eclipse IDE.

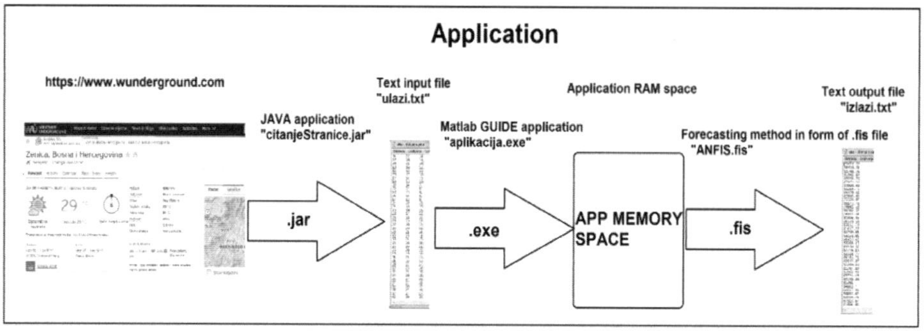

Fig. 7. Block diagram of application for forecasting electric energy consumption.

The purpose of the developed application was to make the method for forecasting the energy consumption mobile and agile, as much as possible, by using the known tools and data sources. It must be noted that today, the availability of the weather forecasting data is huge and the provided data are available and accurate, without the need to further filter or add information. Today the web services, like the one used in this paper, have the high variety of frames to present such data. Those frames are simple to be extracted in form, like textual file shown in this case. By taking into consideration the multiplatform ability of the developed JAVA application module and the strength of the operations available through JAVA, it was a natural choice to develop small application to segment and prepare weather data for Matlab application to integrate with. The only requirement of the application is to be run on the 64 bit OS Fig. 8 shows the developed application outputs by running it from beginning to the end of the entire process, as described before in a few lines.

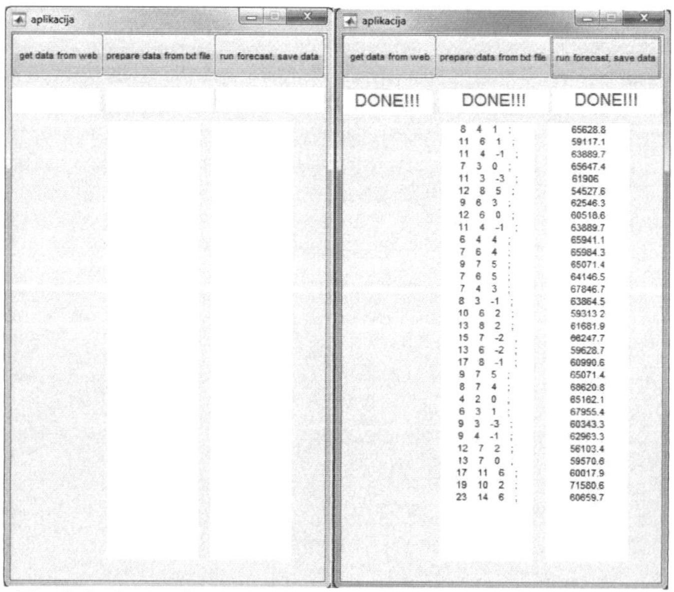

Fig. 8. Left - application at start, right - application at finished run.

5. Conclusion

Many times engineers in they work are presented with problem of predicting network consumption which is one of main characteristics of power grid management. Fast results are needed for engineers to work with deadlines [8]. The target of this research was the real life sample, and it is suitable for its ease of measurements, one place to gather all data from field to check error of research. The mobility of the JAVA application, the availability of the weather data and the expertise of the Artificial Intelligence methods are representing the main elements motivating the research described in this paper to give appropriate solution to be accessible, quick and accurate final product. The developed approach was done applying the above-mentioned tools and methods, which are enabled to work with all types of JAVA data, from integer and char to strings and matrix, data from web services, and with most of the AI methods. The data used in this research paper are available for everyone to use, and presented methods are found fast to be implemented with such data. The developed ANFIS algorithm that consist of ANN and with Fuzzy logic, integrated with the MatLab GUIDE application, and customized with the input information in JAVA. This compiled application solution is changeable to any new example from real life and can be extended with any number of input data. The described approach can be applied to the distribution network of any voltage level with restriction to the single weather zone. The ANFIS method results show very good accuracy compared to the measured data, thus making the developed mobile application an excellent tool for the fast distribution analysis. Further research plan is to extend the set of knowledge base of ANFIS, and enrich the JAVA application to be customized for the final user, with multiple years of weather reports available to be chosen. The next step of this research will make available to the user different weather areas, with more distribution networks, and more weather history data. All that will be incorporating the present features in a single application, which is the goal for future research.

6. Acknowledgments

Creating this paper would be impossible without help of Electric energy Distribution company EP HZ-HB d.d. Mostar, Bosnia and Herzegovina, lending data from their Automatic Meter Reading (AMR) system for purpose of this research.

7. References

[1] Abdoulaye Camara; Wang Feixing; Liu Xiuqin (2016). Energy Consumption Forecasting Using Seasonal ARIMA with Artificial Neural Networks Models. International Journal of Business and Management, April 2016, DOI: 10.5539/ijbm.v11n5p231

[2] Nicolás Marrugo; Jorge Ospina; Darío Amaya (2016). Estimated Energy Consumption for a Building Based on Weather and Time Conditions Using Neural Networks. March 2016, DOI: 10.14257/ijca.2016.9.3.02

[3] Qi Zeng; Ning Zhang; Yi Wang; Yuxiao Liu; Chongqing Kang; Zhijian Zeng; Wei Yang; Min Luo (2016). An Optimum Regression Approach for Analyzing Weather Influence on the Energy Consumption. 2016 International Conference on Probabilistic Methods Applied to Power Systems (2016 PMAPS), At Beijing, China, October 2016

[4] Kristopher T. Williams; Juan Gomez (2016). Predicting future monthly residential energy consumption using building characteristics and climate data: A statistical learning approach, Energy and Buildings, June 2016, DOI: 10.1016/j.enbuild.2016.06.076

[5] Deyin Zhao; Ming Zhong; Xu Zhang; Xing Su (2016). Energy consumption predicting model of VRV (Variable refrigerant volume) system in office buildings based on data mining, Energy 102:660-668, May 2016, DOI: 10.1016/j.energy.2016.02.134

[6] Mohammad Ahsan Habib; Md. Hasanuzzaman; M. Hosenuzzaman; Asif Salman; Md Riyad Mehadi (2016). Energy consumption, energy saving and emission reduction of a garment industrial building in Bangladesh, Energy 112:91-100, October 2016, DOI: 10.1016/j.energy.2016.06.062

[7] Sasan Barak; S. Saeedeh Sadegh (2016). Forecasting energy consumption using ensemble ARIMA–ANFIS hybrid algorithm. International Journal of Electrical Power & Energy Systems 82:92-104, November 2016, DOI: 10.1016/j.ijepes.2016.03.012

[8] H.H. Goh; Q.S. Chua; S.W. Lee; B.C. Kok; K.C. Goh; K.T.K. Teo (2015). Evaluation for Voltage Stability Indices in Power System Using Artificial Neural Network, Procedia Engineering Volume 118, September 2015, Pages 1127-1136, doi:10.1016/j.proeng.2015.08.454

[9] Aleksandr Fedorov; Sergey Zobnin;Viacheslav Shkodyrev; Egor Goloschchapov;Vyacheslav Potekhin (2015). Fuzzy Decision Support In Distributed V2v Systems. Proceedings of the 26th DAAAM International Symposium, pp.0097-0103, B. Katalinic (Ed.), Published by DAAAM International, ISBN 978-3-902734-07-5, ISSN 1726-9679, Vienna, Austria, DOI:10.2507/26th.daaam.proceedings.014

[10] Nyrkov, Anatoliy; Chernyi, Sergei; Zhilenkov, Anton, Sokolov, Sergey (2016). The use of Fuzzy Neural Structures to Increase the Reliability of Drilling Platforms, Proceedings of the 26th DAAAM International Symposium, pp.0672-0677, B. Katalinic (Ed.), Published by DAAAM International, ISBN 978-3-902734-07-5, ISSN 1726-9679, Vienna, Austria,DOI: 10.2507/26th.daaam.proceedings.091

[11] David Walker; Enrico Creaco; Lydia Vamvakeridou-Lyroudia; Raziyeh Farmani; Zoran Kapelan; Dragan Savić (2015). Forecasting Domestic Water Consumption from Smart Meter Readings Using Statistical Methods and Artificial Neural Networks. Procedia Engineering Volume 119, September 2015, Pages 1419-1428, doi:10.1016/j.proeng.2015.08.1002

[12] Xiaoqing Wei; Nianping Li; Wenjie Zhang (2015). Statistical Analyses of Energy Consumption Data in Urban Office Buildings of Changsha, China. Procedia Engineering, Volume 121, October 2015, Pages 1158-1163, doi:10.1016/j.proeng.2015.09.125

[13] Chirag Deb; Lee Siew Eang; Junjing Yang; Mattheos Santamouris (2015). Forecasting Energy Consumption of Institutional Buildings in Singapore. Procedia Engineering Volume 121, October 2015, Pages 1734-1740, doi:10.1016/j.proeng.2015.09.144

[14] Hyojoo Son; Changwan Kim (2015). Forecasting Short-term Electricity Demand in Residential Sector Based on Support Vector Regression and Fuzzy-rough Feature Selection with Particle Swarm Optimization. Procedia Engineering, Volume 118, September 2015, Pages 1162-1168, doi:10.1016/j.proeng.2015.08.459

[15] Fahimifard S.M; Homayounifar M; Mahmood Sabouhi; Moghaddamnia A.R (2009). Comparison of ANFIS, ANN, GARCH and ARIMA Techniques to Exchange Rate Forecasting. Journal of Applied Sciences 9(20), December 2009, DOI: 10.3923/jas.2009.3641.3651

DOI: 10.2507/27th.daaam.proceedings.098

NUMERICAL SIMULATION OF DELAMINATION IN LAMINATED COMPOSITE PLATES USING COHESIVE ELEMENTS

Michal Skovajsa, Frantisek Sedlacek

This Publication has to be referred as: Skovajsa, M[ichal] & Sedlacek, F[rantisek] (2016). Numerical Simulation of Delamination in Laminated Composite Plates Using Cohesive Elements, Proceedings of the 27th DAAAM International Symposium, pp.0678-0685, B. Katalinic (Ed.), Published by DAAAM International, ISBN 978-3-902734-08-2, ISSN 1726-9679, Vienna, Austria
DOI: 10.2507/27th.daaam.proceedings.098

Abstract

The paper deals with the comparison of numerical simulation and analytic computation of the "Double Cantilever Beam" (DCB) test for composite materials. For numerical simulation was used the software NX Nastran 11. Comparisons are used to understand the Mode-I interlaminar fracture toughness of undirectional fibre-reinforced polymer matrix composites. The numerical simulation and analytic computation was performed according to ASTM D5528 norm. The numerical simulation was performed different size and different type of 3D fine elements.

Keywords: DCB test; composite materials; numerical simulation; analytic solution.

1. Introduction

Composite materials are very often used, but strength with large deformation is still problem [11]. A key factor in the design of structures from composite materials is the prediction of failure of the fibre, matrix and interlaminar delamination. There are three independent fracture modes, the crack opening mode (Mode I) is regarded as more important than the in-plane shear mode (Mode II) and out-plane shear mode (Mode III) because the Mode I fracture toughness is usually smaller than those of Mode II and Mode III and so the fracture is easily initiated and propagated under the Mode I loading condition. There is also a combination of these modes called 'Mixed-mode'. The individual modes of crack opening are shown in Figure 1.

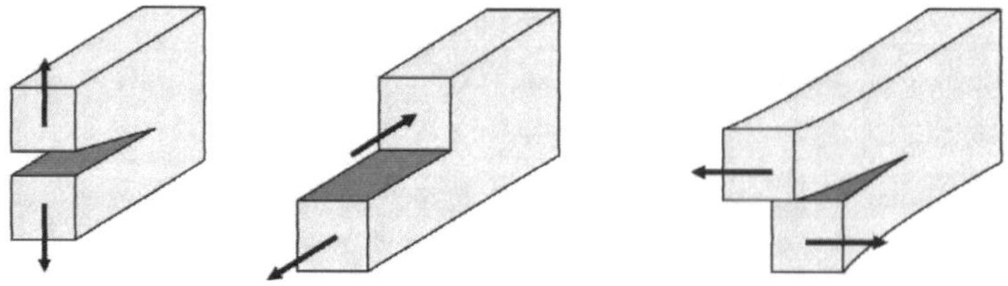

Fig. 1. – The crack opening modes (a) Mode I; (b) Mode II; (c) Mode III)

2. Specimen

The dimension of the specimen was defined by standard ASTM D5528 but it does not define the exact dimensions of the specimen. The shape of the specimen is given in Figure 2. These dimensions were selected for this test. The individual dimensions of the specimen are given in Table 1.

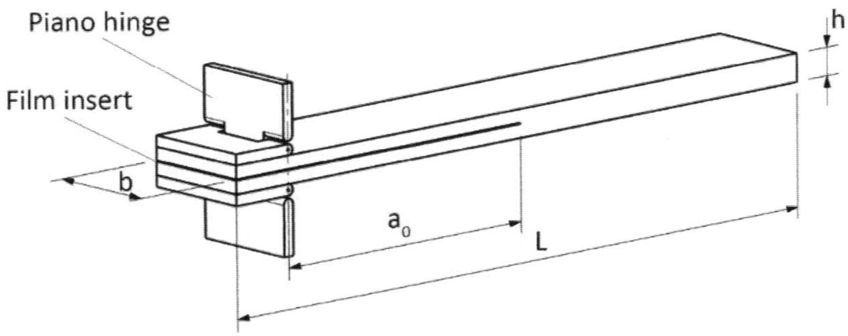

Fig. 2. Dimensions of specimen according to ASTM D5528

Shortcut	Name	Norm [mm]	Selected dimensions [mm]
L	Length of specimen	min. 125	130
a_0	Initial delamination length	approx. 50	50
b	Specimen width	20 - 25	20
h	Specimen thickness	3 - 5	4

Table 1. Dimensions of specimen according to ASTM D5528

The ratio a/h must be greater than 10. The ratio is 12,5 for selected dimensions. For materials with low-flexural modulus or high interlaminar fracture toughness, it may be necessary to increase the number of plies, that is, increase the laminate thickness or decrease the delamination length to avoid large deflections of the specimen arms. The specimen thickness and initial delamination length, a_0, shall be designed to satisfy the following criteria: [1]

$$a_0 \leq 0{,}0472\sqrt{\frac{h^3 E_{11}}{G_{Ic}}} \quad (1)$$

$$h \geq 8{,}28\left(\frac{G_{Ic} a_0^2}{E_{11}}\right)^{1/3} \quad (2)$$

Where a_0 is initial delamination length; h is specimen thickness, E_{11} is lamina modulus of elasticity in the fibre direction and G_{Ic}-Critical value of the strain energy release for Mode I conditions.

Carbon reinforcement material T300/976 with the carbon fibre oriented along the length of the specimen was chosen for this specimen. The individual parameters of the material are given in Table 2. The value of the material was assumed from [2].

Shortcut	Name	Value
E_{11}	Young's Modulus, axial direction	139.3 [GPa]
E_{22}	Young's Modulus, transverse direction	9.72 [GPa]
E_{33}	Young's Modulus, normal (out-of-plane) direction	9.72 [GPa]
G_{12}	Shear Modulus, , axial direction	5.58 [GPa]
G_{13}	Shear Modulus, transverse direction	5.58 [GPa]
G_{23}	Shear Modulus, normal (out-of-plane) direction	3.45 [GPa]
v	Poisson's Ratio in plane	0.29[-]
G_{Ic}	Opening Mode I interlaminar fracture toughness	0.0876 [N/mm]
C_{IIc}	Opening Mode II interlaminar fracture toughness	0.3152 [N/mm]
C_{IIIc}	Opening Mode III interlaminar fracture toughness	0.3152 [N/mm]

Table 2. Material description

3. Analytic computation

The analytic solution of DCB test was calculated based on Bernoulli beam theory and linear elastic fracture mechanics (3,4) [3,4,5]:

$$R_{el} = \frac{3}{2}\frac{E_1 I}{a_0^3}\Delta \tag{3}$$

$$R_{del} = \sqrt{\frac{3}{2}\frac{(bG_{IC}E_1 I)^{3/2}}{E_1 I \Delta}} \tag{4}$$

Where R_{el} is the linear part of the reaction force (before delamination); R_{del} is the non-linear part of the reaction force (during delamination); Δ is the prescribed displacement; I is the moment of inertia of the delaminated part; a_0 is the prescribed delamination length; E_1 is Young´s modulus; G_{Ic} is the critical value of the strain energy release for Mode I conditions.

4. Cohesive elements

In order to model the possible inter-laminar damage, a mesoscopic approach is used, where a thin layer is inserted between two plies of the laminate (Figure 3). A specific non-linear softening law is assigned to the material of this thin layer, and its stiffness and strength can decrease and become equal to zero over the loading, simulating a de-cohesion between the plies. [8]

The approach presented in [9] is available in NX Nastran. The related constitutive polynomial softening material law for the interface is depicted in Figure 4a. Without going into the details of [9], the constitutive equation in the interface takes the following form (5), where d is the damage variable, identical for all the directions, and k_i^0 the initial stiffness:[7,8,]

$$\begin{vmatrix}\sigma_{31}\\ \sigma_{32}\\ \sigma_{33}\end{vmatrix} = \begin{vmatrix}k_1^0(1-d) & & \\ & k_2^0(1-d) & \\ & & k_3^0(1-d)\end{vmatrix}\begin{vmatrix}\gamma_{31}\\ \gamma_{32}\\ \gamma_{33}\end{vmatrix} \tag{5}$$

Where d is equal to zero the interface is intact, while $d = 1$ corresponds to inter-laminar de-cohesion. The evolution of the damage d for the polynomial law is given by (6), [7,8].

$$d = \left(\frac{n}{n+1}\frac{\langle Y - Y_0\rangle_+}{Y_C - Y_0}\right)^n \qquad 0 \le d \le 1 \tag{6}$$

Where Y is an equivalent thermodynamic force and n, YC are material parameters. The damage appears after a threshold Y0 defined by the user (represented by the grey regions in Figure 4). The form of the associated fracture criterion is given in (7), [7,8].

$$\left\{\left(\frac{Y_I}{C_{Ic}}\right)^\alpha + \left(\frac{Y_{II}}{C_{IIc}}\right)^\alpha + \left(\frac{Y_{III}}{C_{IIIc}}\right)^\alpha\right\}^{1/\alpha} \le \frac{Y}{Y_C} \tag{7}$$

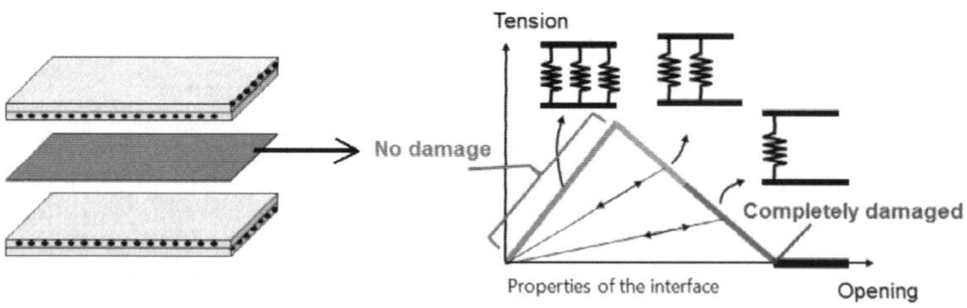

Fig. 3 Inserting an interface element between two plies of a laminate [8]

In NX Nastran, bi-triangular and exponential laws are also available [10]: they are illustrated in Figure 4. For the polynomial and bi-triangular cases, the damage appears after the threshold Y_0 defined by the user. For the exponential law, the damage directly appears when the interface is loaded. [7,8]

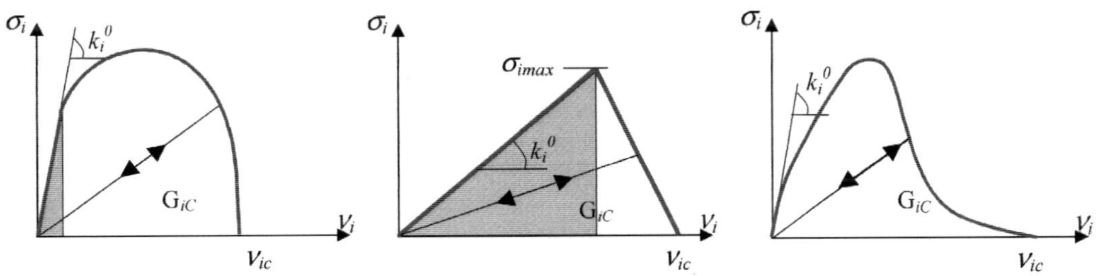

Fig. 4 The constitutive softening laws available in NX Nastran for the interface elements [8]

5. Numerical simulation

Numerical simulation of the DCB specimen was done using the non-linear solver NX Nastran 401. The 3D model of the specimen was created using 3D fine element mesh from linear and parabolic hexahedral elements. A cohesive layer was created between individual compounds of the DCB specimen. The bi-triangular law was used for the cohesive layer properties. The sizes (10 mm, 5 mm, 3 mm, 2 mm, 1 mm) of FE mesh were used for the numeric simulation. The 3D FE model with mesh and cohesive layer are given in Figure 5.

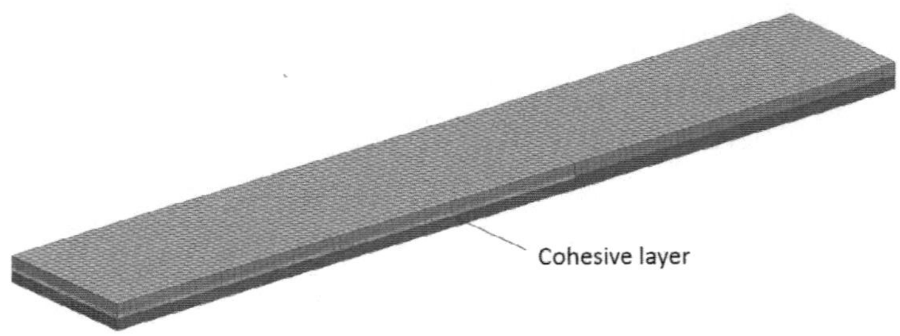

Fig. 5 3D fine element mesh model with cohesive layer

DCB Specimen was loaded by displacement on the end of the arm. The length of arms was a_0. The displacement of one arm was 2.5 mm with 50 steps. The two graphs below show the comparisons of all the different mesh sizes for linear and parabolic hexahedral elements. All the results are given in Figure 6.

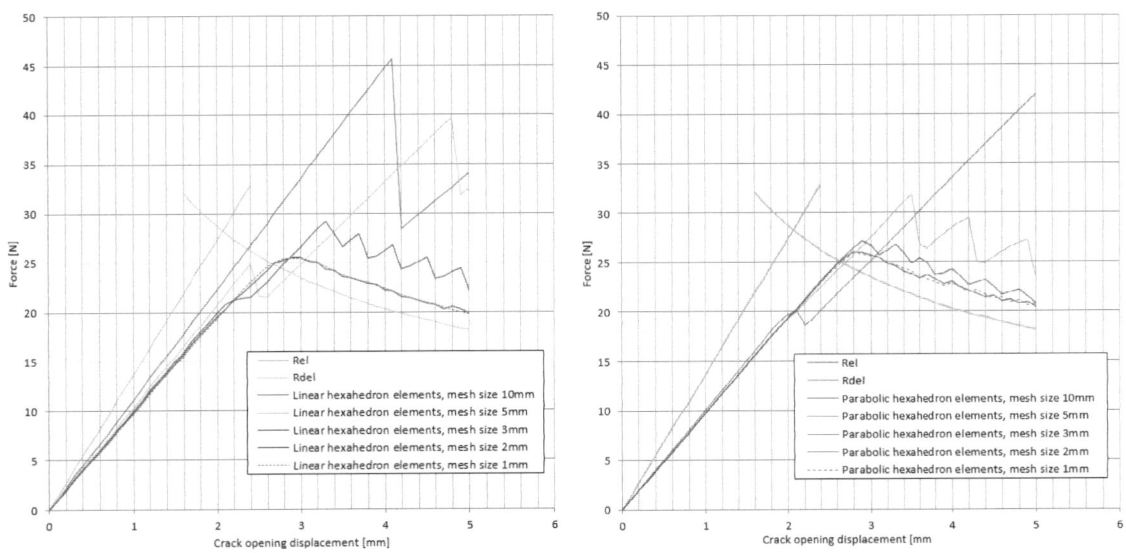

Fig. 6 Graphs with comparisons of all different sizes of mesh for linear and parabolic hexahedral elements of load displacement.

The numerical simulations for mesh sizes 10 mm, 5 mm and 3 mm were very accurate. The simulation with the best result used linear hexahedral elements with mesh size 1 mm. This simulation is shown in Figure 7b and Figure 8b. The different tilt of the first linear stage was created by a different calculation of the beam model. The analytic solution according to the Bernoulli beam model does not use the orthotropic properties of the material. The Bernoulli beam model used only single elasticity modules $E_{11} = E_{22} = E_{33}$ and the transverse shear deformation is not counted.[6] The results of the cohesive damage (BITR) are given in Figure 7a. The results of the pressure in cohesive layer are given in Figure 8a.

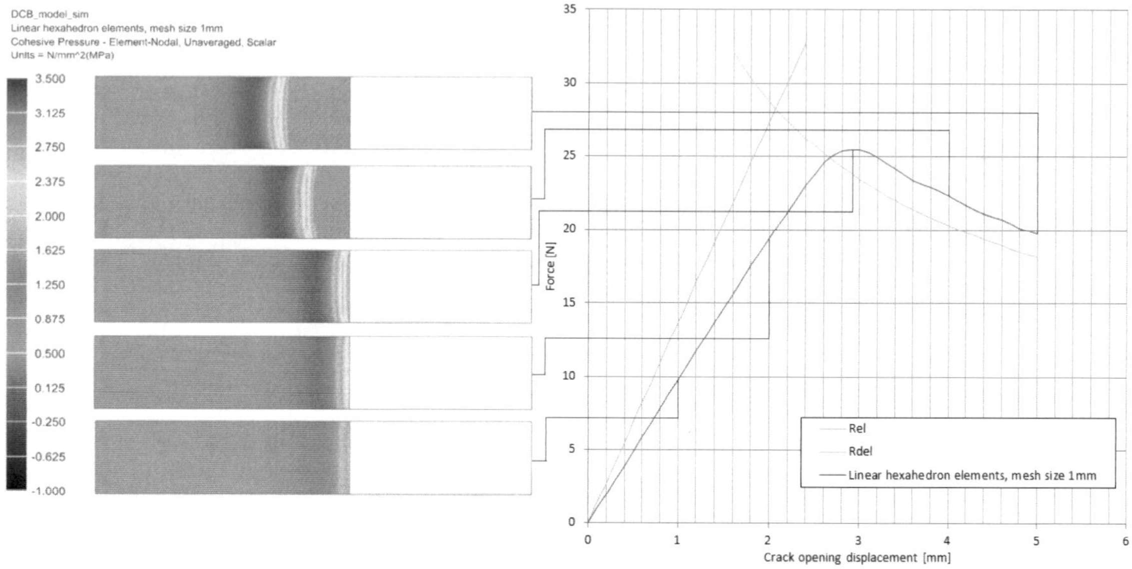

Figure 7 – (a) Results of cohesive damage (BITR); (b) load displacement trace of numerical simulation with linear hexahedral elements and with mesh size 1mm

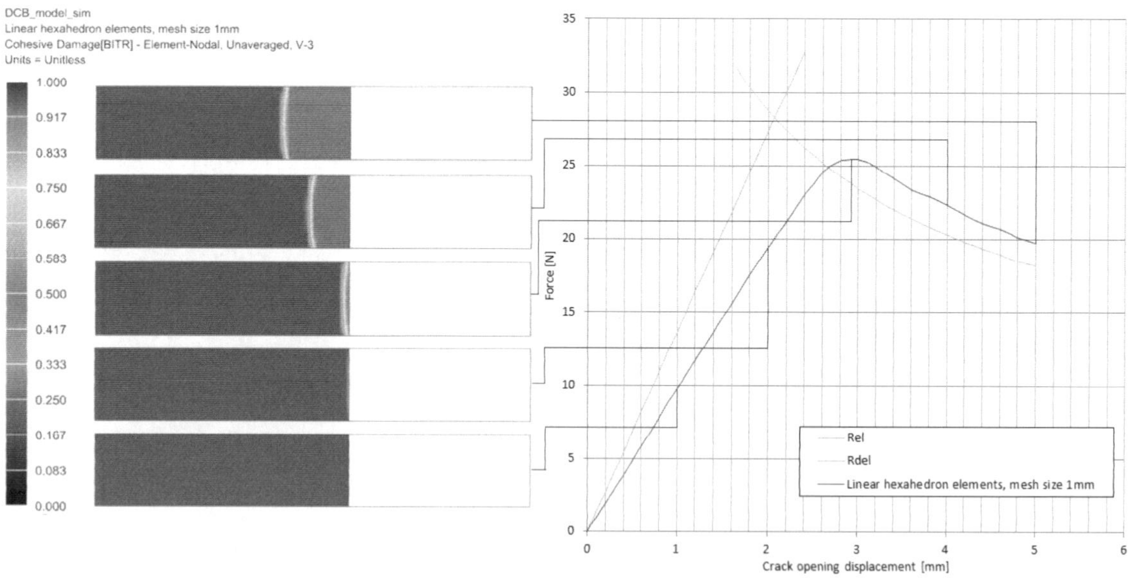

Figure 8 – (a) Results of the pressure in cohesive layer; (b) load displacement trace of numerical simulation with linear hexahedral elements and with mesh size 1mm

The applied load for analytic solution with crack opening displacement of 2.065 mm is 28.7 N. The applied load for numerical simulation with crack opening displacement 2.9 mm is 25.5 N. 11 % of the applied load was the difference between the numerical and analytical results. The maximum pressure in the cohesive layer was 3.45 MPa

Therefore, the graph was offset by the difference of the crack opening displacement (0.835mm). After offsetting the graph, the second part of the curve of the numerical simulation, which shows the damage to the cohesive layer was the same as the analytical solution. The offset result is given in Figure 9

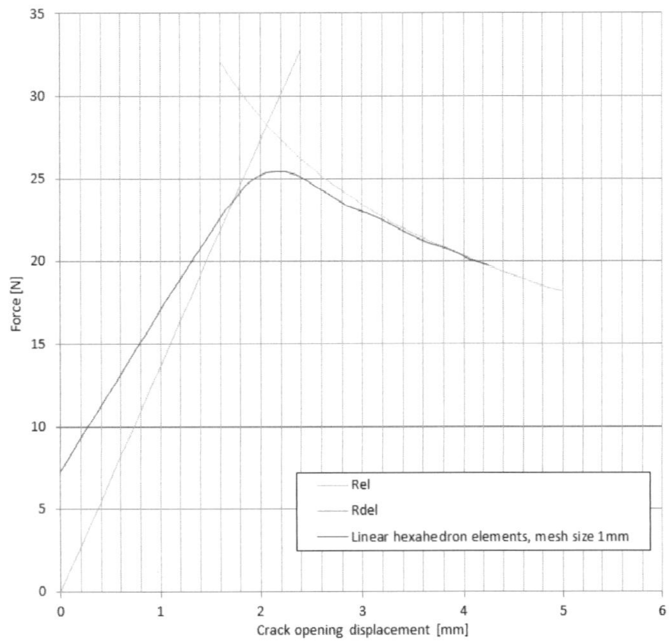

Figure 9 – Offset result of load displacement trace of numerical simulation with linear hexahedral elements and with mesh size 1 mm

The results of the stress analysis in direction 11 are shown in Figure 10, where the shortcut COD is crack opening displacement. The maximum stress of the numerical simulation with linear hexahedral elements and with mesh size 1 mm was 154.46 MPa.

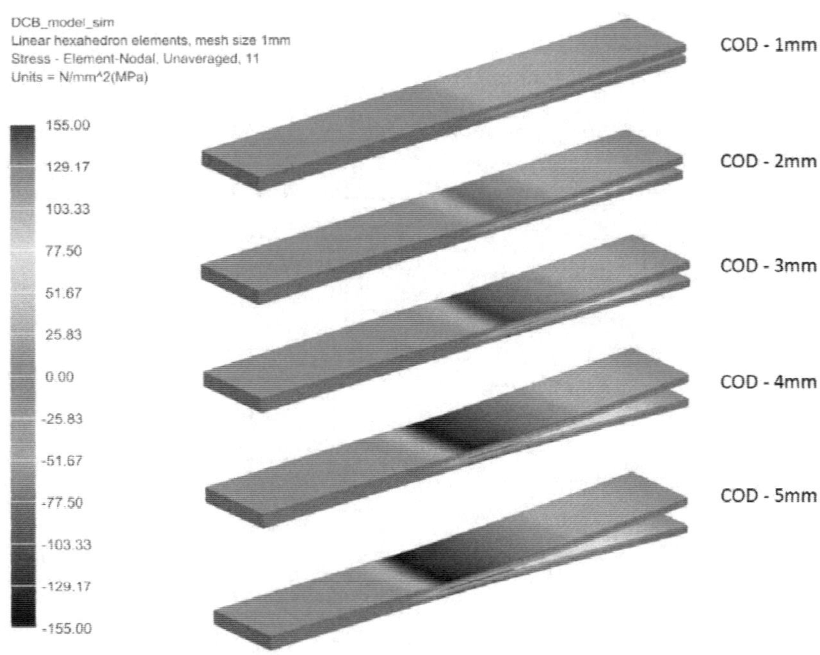

Figure 10 - The results of the stress analysis in direction 11

6. Comparison of the numerical simulation and the analytical solution

The comparison of the numerical simulation with analytical solution are given in Table 3.

Number of simulation	3D FE	Applied maximum load [N]	Shape of applied load curve	Comparison with analytic solution
1	Linear hexahedron elements, mesh size 10 mm	45.7	Very Inaccurate	159,5%
2	Linear hexahedron elements, mesh size 5 mm	39.6	Very Inaccurate	138,2%
3	Linear hexahedron elements, mesh size 3 mm	29.2	Inaccurate	101,9%
4	Linear hexahedron elements, mesh size 2 mm	25.4	Accurate	88,7%
5	Linear hexahedron elements, mesh size 1 mm	25.5	Very accurate	89,0%
6	Parabolic hexahedron elements, mesh size 10mm	unfailure	Incomparable	
7	Parabolic hexahedron elements, mesh size 5mm	31.8	Very Inaccurate	111,0%
8	Parabolic hexahedron elements, mesh size 3mm	27.1	Inaccurate	94,6%
9	Parabolic hexahedron elements, mesh size 2mm	26	Accurate	90,8%
10	Parabolic hexahedron elements, mesh size 1mm	26	Very accurate	90,8%

Table 3 - Comparison of the numerical simulation and analytical solution

The result with minimum difference of applied load between the numerical simulation and the analytical solution was achieved with result number 3, but the shape of this result curve is very inaccurate and has great amplitude. In this case the best result is number 5, where the shape of the result curve is very similar to the analytical solution

7. Conclusion

This paper deals with the comparison of numerical simulation and analytical computation according to Bernoulli beam theory. The question was: Which type of 3D fine elements is suitable to use for numerical simulation with new non-linear solver NX Nastran 401 in NX Nastran 11. The norm ASTM D 5528 was used to define the specimen dimensions. Linear and parabolic hexahedral elements with different mesh sizes were used for the numerical simulation. A cohesive layer was created between individual components of the DCB specimen. The best result of the numerical solution was achieved using linear hexahedral elements with mesh size 1mm.

The stiffness of the specimen was smaller when using the numerical simulation than with the analytical solution, because the analytical solutions only used single elasticity modules. The result with the minimum difference of applied load between the numerical simulation and analytical solution was result number 3, but the shape of this result curve was very inaccurate and has a great amplitude. In this case the best result is result number 5, where the shape of the result curve is very similar to the analytical solution.

The applied maximum load using numerical simulation with result number 5 (Table 3) was smaller than the analytical solution by about 11 %. The applied maximum load using the numerical simulation was 25.5 N. For numerical simulation of interlaminar fracture in Mode I should be used the linear hexahedron elements with mesh size 1 mm or 2mm. These elements provide sufficiently accurate results.

We are currently working on manufacturing a specimen for DCB for experimental testing and verification of the numerical simulations to best-fit the numerical simulation to a realistic value.

8. Acknowledgements

This paper is based on work sponsored by project SGS2016-012 (The complex support of designing of technical equipment III.).

9. References

[1] Standard Test Method for Mode I Interlaminar Fracture Toughness of Unidirectional Fiber-Reinforced Polymer Matrix Composites, Designation: D 5528 – 01 (Reapproved 2007)

[2] Hosseini-Toudeshky H, Hosseini S, Mohammadi B. Delamination buckling growth in laminated composites using layerwise-interface element. Compos Struct 2010;92(8):1846–56.

[3] Tenchev RT, Falzon BG. A pseudo-transient solution strategy for the analysis of delamination by means of interface elements. Finite Elem Anal Des 2006;42(8–9):698–708.

[4] Mi Y, Crisfield MA, Davies GAO. Progressive delamination using interface elements. J Compos Mater 1998;32(14):1246–72.

[5] Szekrényes A. Interlaminar stresses and energy release rates in delaminated orthotropic composite plates. Int J Solids Struct 2012;49(18):2460–70.

[6] Miroslav Marjanovic, Günther Meschke, Djordje Vuksanovic, A finite element model for propagating delamination in laminated composite plates based on the Virtual Crack Closure method., 2016

[7] m. Bruyneel , J.P. Delsemme , A.C. Goupil , P. Jetteur , C. Lequesne , T. Naito And Y. Urushiyama Damage modeling of laminated composites: validation of the inter-laminar damage model of samcef at the coupon level for ud plies

[8] M. Bruyneel1 , J.P. Delsemme1, P. Jetteur1 and F. Germain, A study of complex delaminations in laminated composite structures with SAMCEF, University of Liège, Belgium, 26-28 May 2008

[9] P. Ladevèze, O. Allix, B. Douchin and D. Lévêque. A computational method for damage intensity prediction in a laminated composite structure, Computational Mechanics – New Trends and Applications (Idelsohn S., Oñate E. and Dvorkin E., eds.), CIMNE, Barcelona, Spain, 1998.

[10] P. Jetteur. Model material for the interface element Report RS-246. 2006.

[11] Sedlacek, F.; Lasova, V. & Kottner, R.: Design of Flexible Composite Bellows Coupling, Proceedings of the 26th DAAAM International Symposium, pp.0930-0935, Published by DAAAM International, ISBN 978-3-902734-07-5, ISSN 1726-9679, Vienna, Austria

DOI: 10.2507/27th.daaam.proceedings.099

MONTE CARLO SIMULATION IN VALUATION OF INVESTMENT PROJECTS

Branislav Maric & Vanja Grozdic

University of Novi Sad, Faculty of Technical Sciences, Trg Dositeja Obradovića 6, 21000 Novi Sad

This Publication has to be referred as: Maric, B[ranislav] & Grozdic, V[anja] (2016). Monte Carlo Simulation in Valuation of Investment Projects, Proceedings of the 27th DAAAM International Symposium, pp.0686-0692, B. Katalinic (Ed.), Published by DAAAM International, ISBN 978-3-902734-08-2, ISSN 1726-9679, Vienna, Austria
DOI: 10.2507/27th.daaam.proceedings.099

Abstract

Valuation of investment projects is a crucial part of the investment decision making process. The most important part of any investment analysis are discounted cash flow methods: net present value (NPV) and internal rate of return (IRR). The aim of this paper is application of risk analysis technique – Monte Carlo simulation, including sensitivity and scenario analysis, in investment project valuation. For this purpose, we have used an investment project – the new plant for the combined production of electrical and thermal energy, as a case study. The input (key) variables used in this analysis, for performing a Monte Carlo simulation, are total revenue, variable costs, initial investment and discount rate, determined by the sensitivity analysis. As the output variables we have used net present value (NPV) and internal rate of return (IRR). Monte Carlo simulation results show that the probability of the project's success is 86.1%, according to the NPV rule, and 91.9%, according to the IRR rule of project's acceptance.

Keywords: investment project; sensitivity analysis; scenario analysis; Monte Carlo simulation

1. Introduction

This paper discusses the use of risk analysis techniques in investment decisions. Making a decision on the basis of assumptions, expectations, estimates and forecasts of future events involves taking risks. Risk and uncertainty characterize situations where the actual outcome for a particular event or activity is likely to deviate from the estimate or forecast value [1]. Risk can be quantified as folows:

$$\text{Risk} = \text{Event's probability x Magnitude of loss/gain} \quad (1)$$

Generaly, sensitivity analysis is about identification of critical factors and how their change affects behavior of investment project. Scenario analysis is a simple risk analysis technique which determines the „base case scenario", „best case scenario" and „worst case scenario" for the project. Monte Carlo simulation is a method in which we assighn probability distributions to the input variables (critical factors) and, on that basis, we calculate output variables and the probability of their occurence. According to PMI [2], Monte Carlo simulation is a technique that computes or iterates the project cost or schedule many times using input values selected at random from probability distributions of possible costs or durations, to calculate a distribution of possible total project cost or completion dates.

For the purpose of this paper we have used an investment project form the energy industry, as a case study. The most commonly used discounted cash flow methods are net present value (NPV) and internal rate of return (IRR), which were used as output variables in this analysis.

2. Brief literature review

Many researches have examined the use of risk analysis techniques in investment decision making process. Jovanovic [3] examines the aplicaton of sensitivity analysis, effect of discount rate and total investment value on the NPV, in investment project evaluation and concludes that, using this technique, we are able to obtain a great deal of information about the effect of input parameters on the examined criteria for investment project decision making. Acording to J.K. Kaldellis *et al.* [4], a remarkable growth of the energy production sector is expected based on the anaylsis of the influance of techno – economic parameters – such as capital cost, retutn on investment, electricity price etc. – on the economic viability and attractiveness of investment. E. Borgonovo *et al.* [5] applied sensitivity analysis to the project finance case study and analyzed investment criteria related to both the investors's and lenders' perspectives, where results showed that revenue assumptions and construction costs, are the most important drivers of economic performance from the perspectives of both sponsors and lenders. Monte Carlo simulation (MCS) is a widely used technique in the risk analysis of the energy projects. F. Kienzle [6] applies Monte Carlo simulation in a valuation of investments in multi – energy conversion, where he uses energy prices (gas, electricity and heat) as the input variables and calculate the present values (PVs) of an energy hub, and their distribution. L. Zhu [7] employs MCS in a nuclear power investment valuation where several technical and economic risk factors (investment cost, generating cost, electricity prices and nuclear accident) have been taken into account in the model. J. Munoz *et al.* [8] calculate NPV of an investment in a wind energy plant using MCS and input variables such as wind and production parameters, electricity price and investment cost, and compare this approach with the traditional NPV method. H. Laurikka [9] presents a simulation model which is used to explore the impact of the input variables, such as market prices of electricity, emission allowance and fuels on the investment's NPV, using two case studies: modifications of an existing condensing power plant and a new combined heat and power plant. F.A. Roques [10] illustrate, through a simple net present value model of three base load technologies (coal, gas, and nuclear plants), how Monte Carlo simulation can capture the impact of a wide range of uncertainties and risks, such as energy prices, fixed and variable costs, discount rate etc.

3. Investment project valuation – case study

Remer and Nieto [11] categorize methods and techniques used to evaluate economic desirability of projects into five basic types: net present value methods, rate of return methods, ratio methods, payback methods and accounting methods. A central feature of any investment analysis is discounted cash flow (DCF), which takes into consideration the time value of money, is regarded as theoretically correct, and includes at least four different discounting models: net present value (NPV), internal rate of return (IRR), modified internal rate of return (MIRR), and profitability index (PI) [12]. For the purpose of this study we will briefly explain the two most important discounted cash flow methods, i.e the net present value (NPV) and intetnal rate of return (IRR) method.

Net present value (NPV) is calculated by subtracting the initial investment from the sum of discounted values of future cash flows of the project. The NPV rule: If the NPV is positive – accept the project. The NPV formula:

$$NPV = \sum_{t=1}^{n} \frac{CF_t}{(1+r)^t} - CF_0 \tag{2}$$

Internal rate of return (IRR) is a discount rate that makes NPV equals zero. The IRR rule: If the IRR is greater than the cost of capital (discount rate) – accept the project. The IRR can be calculated using this formula:

$$\sum_{t=1}^{n} \frac{CF_t}{(1+r)^t} - CF_0 = 0 \tag{3}$$

For the purpose of this analysis, as a case study, we have used an investment project – the new plant for the combined production of electrical and thermal energy. The main objective of the project is the production of additional electrical and thermal energy. Duration time of the project is 20 years, and estimated construction time is 4 years. The calculation of the sales revenues from the electrical and thermal energy is based on the production of electrical and thermal energy in kWh and their selling prices. These revenues presents cash inflows of the project. Cash outflows of the project include initial investment, operational costs and income tax. Operational costs are divided into fixed costs (maintenance, earnings, insurance...) and variable costs (natural gas procurement costs and tax on emissions of NOx). Variable costs, i.e. natural gas procurement costs, have the dominant position in total costs structure (80% of the total costs). Future cash flows were discounted with the rate of 9%. The main results of the project's cash flow analysis are:

- Net present value (NPV) = 598.5 mil €,
- Internal rate of return (IRR) = 35%
- Payback period = 3 years (7 years, including construction time of 4 years)

4. Sensitivity and scenario analysis

Economic and financial analysis is based on the values which are subject to change over time and thus may affect the profitability of the investment project. This is the main reason for conducting sensisitivity analysis which has the goal to measure the effect of changes in a key variables on investment project's output (e.g. NPV).To perform a sensitivity analysis, all variables are fixed, except for the variable under consideration, which is allowed to fluctuate. As for this project, we have variables whose values are changing very offten, but beside that, we should bear in mind the fact that, regardless of the changing energy prices over time (e.g. electricity price is stable in one period, but natural gas price in that period is changing intensively), observed in the longer term, they grow and follow each other.

While sensitivity analysis measures sensitivity of project's NPV by varying one key variable at the time, scenario analysis constructs scenarios where several key variables change in each scenario. In this case, project is evaluated by looking his NPVs in all constructed scenarios. It is always possible to create a large number of alternative scenarios, but it is convenient to limit these scenarios to specific events and economic conditions that would be of great significance to the project. As mentioned before, this technique determines the „base case scenario", „best case scenario" and „worst case scenario" for the project. Table 1. and Fig. 1. show the sensitivity and scenario analysis, respectively.

No	Variable	Deviation (%)	NPV after change (in mil €)	Absolute change (in mil €)	Relative change (%)
1.	Revenue from electricity energy sale	-10	358.1	240.4	40
2.	Revenue from thermal energy sale	-10	573.2	25.3	4
3.	Initial investment	+10	570.8	27.7	5
4.	Maintaince cost	+10	588.9	9.6	2
5.	Earnings cost	+10	596.3	2.2	0.4
6.	Insurance premiums	+10	596.1	2.4	0.4
7.	Obligations to the state	+10	597.5	1	0.2
8.	Other costs	+10	598	0.5	0.08
9.	Natural gas procurement cost	+10	442.9	155.6	26
10.	Tax on emissions of NOx	+10	598.4	0.1	0.02
11.	Income tax	+10	591.8	6.7	1
12.	Discount rate	+10	545.5	53	9

Table 1. Sensitivity analysis of the project

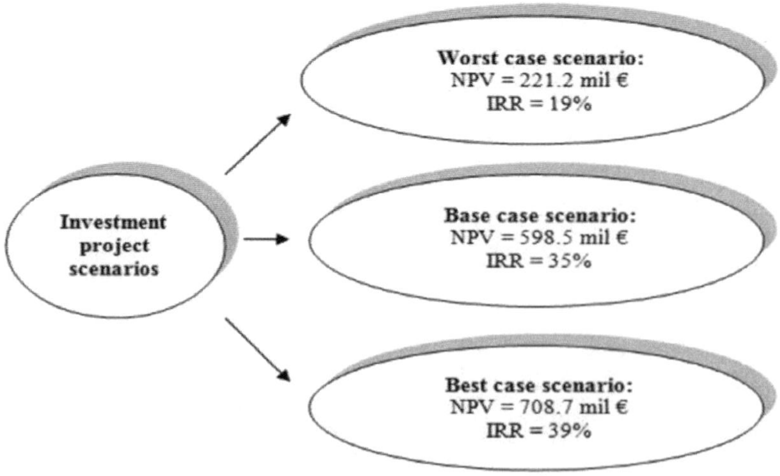

Fig. 1. Scenario analysis of the project

The results of the project's sensitivity analysis (Table 1.) show that the most influential variables on the investment project behavior are total revenue (production volume and the selling prices of electrical and thermal energy), varible costs (natural gas procurement cost), initial investment and discount rate. Generaly, the largest risk of investing in this project is related to the energy prices. If the price of natural gas increase, the production of electrical and thermal energy will become more expensive. On the other hand, increase of the electrical energy price, as well as the thermal energy price, will positively affect the project's profitability. As mentioned before, the price of electrical and thermal energy depends on the movement of the natural gas price. According to this, we can conclude that the project is most sensitive to changes in the natural gas price, which is logical, because this fuel is 80% of total costs of the production. In the case when only natural gas price is changing and other variables remain constant, the project is profitable as long as that increase does not reach 50%.

Unlike the sensitivity analysis, scenario analysis provides a likiely range of project's NPV and IRR (Fig. 1.). Base scenario assumes that the new plant will work during the whole year. In that case, the project's NPV is 598.5 mil € and IRR is 35%. However, engagement of this plant which uses expensive fuel such as natural gas, can drasticly decrese when the new capacity, which uses less expensive fuel, enters into operation for production of electrical energy. In this case, the plant will work only during the heating season, in cogeneration regime. This situation is considered as the worst case scenario, when both variables, total revenue and variable costs, are reduced by 40%. Best case scenario assumes increase of the energy (electrical, thermal and gas) prices, i.e. total revenue and variable costs by 10%. However, scenario analysis does not specify how likely are the occurrences of the scenarios.

5. Monte Carlo simulation

Since its conception in mid-20th century, Monte Carlo simulations have grown in popularity among the scientific community, becoming a mainstay of stochastic decision making tools, used in many fields including real options, air pollution, workspace-visualization, or hydro power plants availability, among a myriad of others [13]. The term Monte Carlo simulation refers to a broad spectrum of mathematical models and algorithms, whose most prominent feature is the use of random numbers in solving various problems [14]. The Monte Carlo method simulates the full system many times, each time randomly choosing a value for each variable from its probability distribution [15]. According to Hacura *et al.* [16], during the simulation process, random scenarios are built up using input values for the project's key uncertain variables, which are selected from appropriate probability distributions, and after which the results are collected and analyzed statistically so as to arrive at a probability distribution of the potential outcomes of the project, and to estimate various measures of project risk. Generaly, we need to take six steps to perform a Monte Carlo simulation:

1. Identification of the input variables
2. Identifaction of the output variables
3. Assigning probability distributions to the input variables
4. Establishing the relationship between the input variables
5. Running a simulation
6. Interpretation of the results

The first step in Monte Carlo simulation is to identify the input variables – key variables whose change has the greatest influence on the project . According to the sensitivity analysis, these variables are total revenue (the sum of electrical and thermal energy sale revenues), variable costs (natural gas procurement cost), initial investment and discount rate. The next step is to identify the output variables, which are net present value (NPV) and internal rate of return (IRR). The third step is related to defining of assumptions, i.e. assigning probability distributions to the input variables. Based on the opinions of experts and the historical data we have chosen the next probability distributions for the input variables:

- Total revenue – Triangular distribution – with the minimum, most likely and maximum values.
- Variable costs – Normal distribution – with the mean and standard deviation.
- Initial investment – Uniform distribution – with the minimum and maximum value.
- Discount rate – Triangular distribution – with the minimum, most likely and maximum values.

The next step is to define correlation between the input variables. Positive correlation coeficients, i.e. statistical dependency, between two input variables – total revenue and variable costs, are considered in the simulation. This means that when variable costs are growing, total revenues are also growing. This is related to the fact, as we mentioned before, that there is a statistical dependency between energy prices (electrical, thermal and gas) in the long term, i.e. they grow and follow each other. Between other key variables there is no statistical dependency, and they can be generated independently during the simulation.

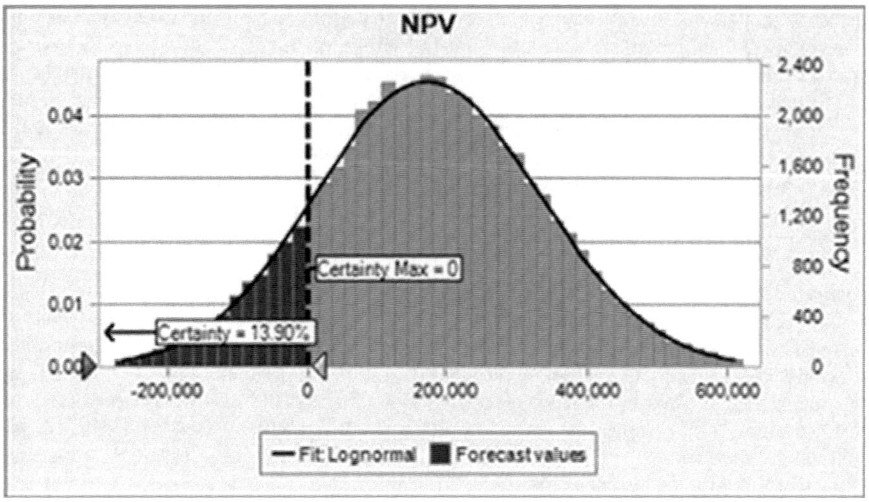

Fig. 2. Probability distribution of NPV

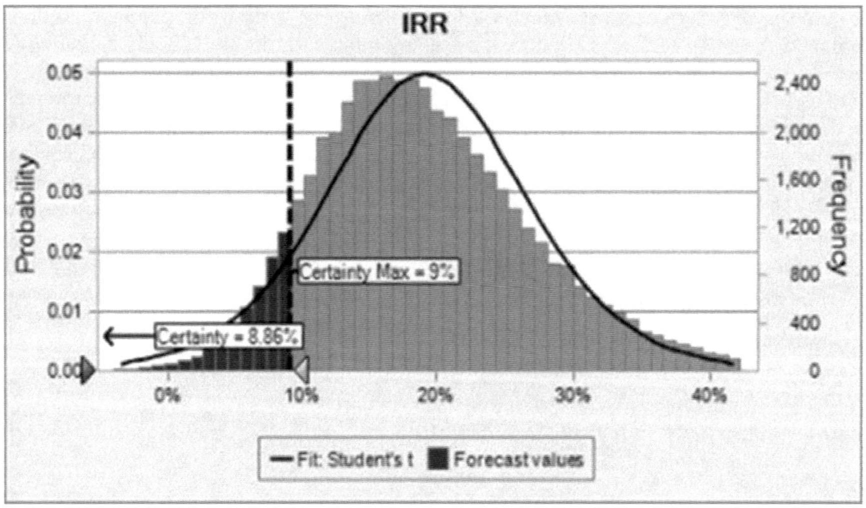

Fig. 3. Probability distribution of IRR

No	Statistics	Forecast values (NPV)	Forecast values (IRR)
1.	Trials	50,000	50,000
2.	Mean	172.9	19%
3.	Median	172.3	18%
4.	Mode	---	---
5.	Standard Deviation	160.2	8%
6.	Variance	25,669,832	1%
7.	Skewness	0.0089	0.6492
8.	Kurtosis	3.01	4.09
9.	Coeff. of Variability	0.9266	0.4337
10.	Minimum	-484	-56%
11.	Maximum	832.4	75%
12.	Range Width	1,316.4	131%
13.	Mean Std. Error	716.52	0%

Table 2. Statistical characteristics of the output variables

Monte Carlo simulation was performed with 50,000 trials and the confidence level of 95%, for the base case scenario of the investment project, using *Crystal Ball* risk analysis software application. The final step is related to the interpretation of the simulation results.

Fig. 2. presents the probability distribution of the NPV variable. The NPV variable has *lognormal* distribution, with the mean value of 172.9 mil € and the standard deviation of 160.2 mil € (Table 2.). The mean (172.9 mil €), actualy, represents the expected net present value (eNPV), which modifies standard NPV by multiplying each possible NPV with the probability of their occurrence. Because the difference between the mean (eNPV) and the standard deviation is not large, we can conclude that there is no high uncertainty of the project. In cases when the difference between eNPV and the standard deviation is drastic, uncertainty of the project grows, which means that the probability of loss is also high as a consequence of the high uncertainty [17]. Since the eNPV (172.9 mil €) is smaller than the base NPV (598.5 mil €), but still high and positive, we can expect that, in the reality, the project will bring lower profit than originaly calculated in the terms of determinism. We can also see (Fig. 2.) that there is 13.9% chance of negative NPV value, which means that the probability of the project's success is 86.1%. Table 2. also shows, among other statistical characteristics, coefficient of variabilty of 0.9266 for the NPV variable, which is the ratio between the standard deviation and the mean (eNPV). This coefficient is below 1, which means that the probability of loss is small.

Fig. 3. presents the probability distribution of the IRR variable. The IRR variable has *student's t* distribution, with the mean value of 19% and the standard deviation of 8% (Table 2.). Coefficient of variability for the IRR variable is 0.4337, also less than 1. As well for the NPV, we can see (Fig. 3.) that there is 8.9% chance for the IRR to be lower than discount rate, which is 9% in this case. This means that the probability of the project's success is 91.1%, according to the IRR rule of project's acceptance.

6. Conclusion

Planning of investment projects is always related to the future processes, which are never absolutely known, thus the fututre events can't be predicted with full certainty. This implies that the projections carry within themselves the risk and uncertainty, which must be incorporated into the project. So, we can't predict the future, but we can reduce the risk and uncertainty related to the future processes. This can be achieved by using certain tools, such as Monte Carlo simulation, which mathematically compute future scenarios and tell us how likely are the occurrences of the possible outcomes, their probability distributions and risks related to them. Actualy, we can decide which risk to take and which to avoid, in order to make the best investment decision.

In this paper, for performing Monte Carlo simulation, we have used an investment project from the energy sector, as a case study. As the input variables we have used total revenue, variable costs, initial investment and discount rate. The output variables were net present value (NPV) and internal rate of return (IRR). The results have shown that the probability of the project's success is 86.1%, according to the NPV rule, and 91.9%, according to the IRR rule of project's acceptance.

Small number of the project's risk factors can be considered as the limitation of this study. Future research will include several investment projects from different industries, and also, detailed analysis of larger number of input and output variables, i.e. risk factors related to those projects.

7. References

[1] J. Raftery, Risk Analysis in Project Management, E & FN Spon, London SE1 8HN, UK, 1994.

[2] PMBOK Guide, A guide to the project management body of knowledge: PMBOK guide, 3rd edition, Pennsylvania, USA, Project Management Institute, 2004.

[3] P. Jovanovic, Application of sensitivity analysis in investment project evaluation under uncertainty and risk, International Journal of Project Management, Vol. 17, No. 4, 1999, pp. 217–222.

[4] J.K. Kaldellis, Th.J. Gavras, The economic viability of commercial wind plants in Greece A complete sensitivity analysis, Energy Policy, Vol. 28, No. 8, 2000, pp. 509–517.

[5] E. Borgonovo, S. Gatti, and L. Peccati, What Drives Value Creation in Investment Projects? An Application of Sensitivity Analysis to Project Finance Transactions, European Journal of Operational Research, Vol. 205, No. 1, 2010, pp. 227-236.

[6] F. Kienzle, P. Ahcin, and G. Andersson, Valuing investments in multi-energy conversion, storage, and demand-side management systems under uncertainty, IEEE Transactions on Sustainable Energy, Vol. 2, No. 2, 2011, pp. 194–202.

[7] L. Zhu, A simulation based real options approach for the investment evaluation of nuclear power, Computers & Industrial Engineering, Vol. 6, No. 3, 2012, pp. 585–593.

[8] J. Munoz, J. Contreras, J. Caamano and P. Correia, Risk assessment of wind power generation project investments based on real options, Proceedings of IEEE PowerTech, 2009, pp. 1-8, Bucharest.

[9] H. Laurikka, Option value of gasification technology within an emissions trading scheme, Energy Policy, Vol. 34, No. 18, 2006, pp. 3916–3928.

[10] F.A. Roques, D.M. Newbery, W.J. Nuttall, Fuel mix diversification incentives in liberalized electricity markets: A Mean-Variance Portfolio theory approach, Energy Economics, Vol. 30, No. 4, 2008, pp. 1831-1849.

[11] D.S. Remer and A.P. Nieto, A compendium and comparison of 25 project evaluation techniques. Part 1: Net present value and rate of return methods, International Journal of Production Economics, Vol. 42, No. 1, 1995, pp. 79-96.

[12] K. Bennouna, G. Meredith and T. Marchant, Improved capital budgeting decision making: evidence from Canada, Management Decision, Vol. 48, No. 2, 2010, pp. 225-247.

[13] R. Špicar, M. Januška, Use of Monte Carlo modified Markov Chains in capacity planning, Procedia Engineering, Vol. 100, 2015, pp. 953–959.

[14] Lj.D. Milanovic, D.D. Milanovic, and M. Misita, The Evaluation of Risky Investment Projects, in FME Transactions 2, Faculty of Mechanical Engineering, Univ. Press, Belgrade, 2010, pp. 103-106.

[15] Y.H. Kwak, L. Ingall, Exploring Monte Carlo simulation applications for project management, Risk Management, Vol. 9, No. 1, 2007, pp. 44–57.

[16] A. Hacura, M. Jadamus-Hacura, A. Kocot, Risk analysis in investment appraisal based on the Monte Carlo simulation technique", The European Physical Journal B - Condensed Matter and Complex Systems, Vol. 20, No. 4, 2001, pp. 551-553.

[17] B. Marić, Managing Investments, Faculty of Technical Sciences, University of Novi Sad, Serbia, 2008.

DOI: 10.2507/27th.daaam.proceedings.100

DYNAMIC ANALYSIS AND MOTION CONTROL OF HYDRAULIC CRANE FOR MEN LIFTING USING MODELING AND SIMULATIONS

Ilir Doçi, Mirlind Bruqi* & Nexhat Qehaja

University of Prishtina, Faculty of Mechanical Engineering, 10000 Prishtina, Kosovo

This Publication has to be referred as: Doci, I[lir]; Bruqi, M[irlind] & Qehaja, N[exhat] (2016). Dynamic Analysis and Motion Control of Hydraulic Crane for Men Lifting Using Modeling and Simulations, Proceedings of the 27th DAAAM International Symposium, pp.0693-0700, B. Katalinic (Ed.), Published by DAAAM International, ISBN 978-3-902734-08-2, ISSN 1726-9679, Vienna, Austria
DOI: 10.2507/27th.daaam.proceedings.100

Abstract

This paper deals with modeling and simulation of Hydraulic crane for load lifting in order to determine its dynamics, methods for lifting control and minimization of oscillations. The proposed procedure is modelling of crane with method of schematic design with inerconnected elements that represents crane parts, and its 3-d visualization. Simulations will be planned and applied for lifting motion of Crane's Boom with hangind load. Dynamic analysis will be carried through simulations and solution of Euler differential equations of second order. Diagrams with results of main dynamic and kinematic parameters will be presented for main parts of crane as the solution results of the analyzed system. Based on these results conclusions will be presented. The aim is to find optimal lifting control process and dynamic behavior of crane which is important for motion planning, failure analysis and safety during work. Analysis will be done using modeling and simulations with computer application MapleSim.

Keywords: Hydraulic Crane; Men Lifting; Dynamics; Vibrations; Control; modeling; simulations;

1. Introduction

Many companies that work with Hydraulic cranes have difficulties dealing with load lifting due to oscillations and swinging of load, which can lead to safety problems. Hydraulic crane taken for study is also known as a Straight Boom Lift, Man Lift, Basket Crane. It is an elevated work platform that consists of a platform or Cabin at the end of a lifting system and is used to lift men and other load. It is important to find optimal procedure to lift the men and load in order to minimize swinging and oscillations. To do this, main kinematic and dynamic parameters of motion must be measured, which is difficult with instrumentation. Modelling and simulations helps determining crane's behavior and search for

* Corresponding author. Tel.: +377 44 503 577. *E-mail address:* mirlind.bruci@uni-pr.edu

main paramaters on its parts, like: wheels, chasis, boom, cabin, etc. Dynamic and kinematic parameters investigated are: motion length, velocity, acceleration, angular velocity, forces and torque that act in main parts of crane. Model of Hydraulic Crane is designed and modeled with software Maple Sim 6.1 [3]. Model is created based on manufacturer *JLG 660SJ* (Fig.1) [1]. Main technical data of crane are: Total weight of crane 12519 kg; Max work height 21.3 m; Max carrying load: Capacity: Q_{max} = 454 kg. Boom lifting speed is v_l = 0.25 m/s. (Fig.1).

Until now, authors have studied dynamics of hydraulic cranes, with suggestion of various crane model types [4], [6], [10], multibody dynamics [2], [4], [9], [13], design of crane hydraulics [4], motion control [4], [5], [12], simulation methods [6], [7], [8], [10], analysis approach [2], [5], [9], [12], and results representation [6], [7], [10], with the aim to find best crane models for analysis, search for dynamics of motion, and implement control. In this work, methodology of research is through Model Predictive Control Technology [5] with software, similar to Bond Graphs [6], with the design of schematic algorithm to create model, find differential equations and implement simulations procedure.

Research is in the field of crane dynamics and control and is limited to load lifting analysis and optimization of lifting motion through analysis of results gained through modeling and simulations. Authors approach is determination of crane's motion through dynamic analysis and control of motion in order to improve safety during crane's work.

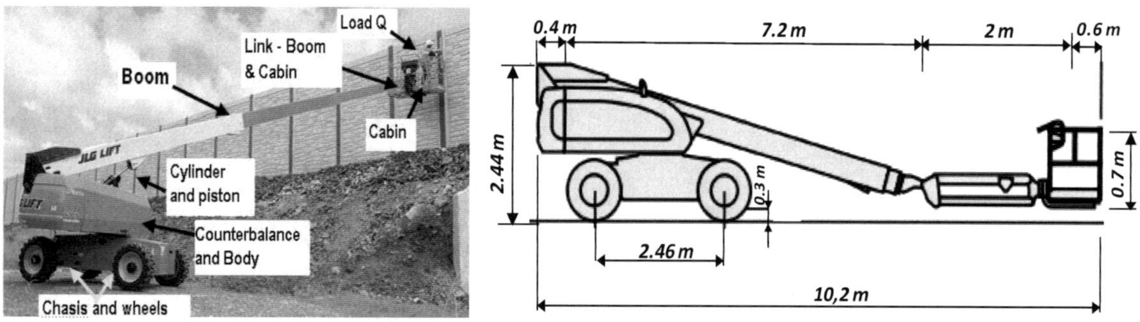

Fig.1. Hydraulic crane studied and its main dimensions [1]

2. Schematic design of crane model

In Fig.2 is presented schematic design and block diagram of Hydraulic crane [1], created with software that enables topological representation, and interconnects related components [3]. Schematic diagram is created for the purpose of analysis, generation of differential equations, applying simulations for control, and getting results.

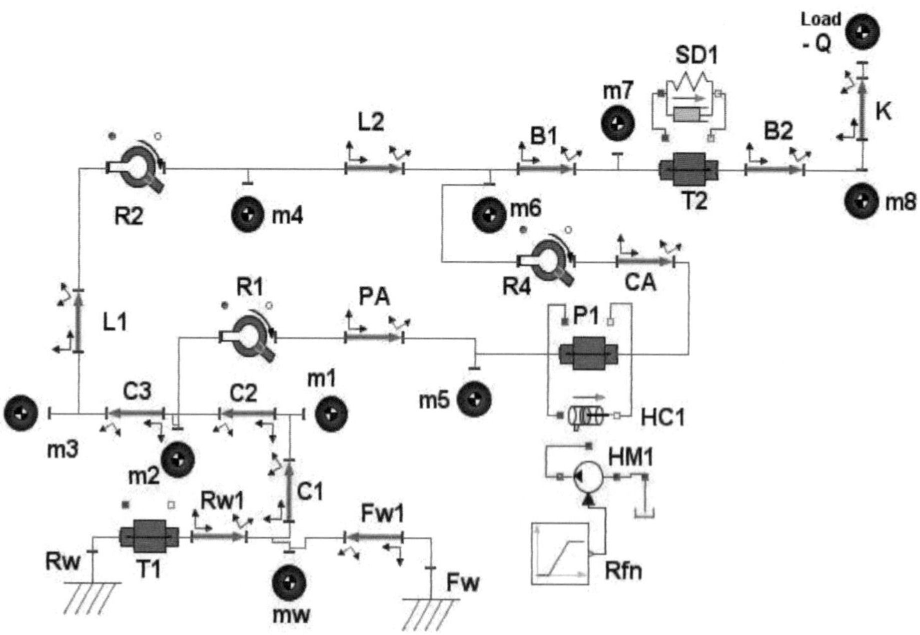

Fig.2. Schematic design (Block Diagram) of hydraulic crane with boom lifting motion

Block diagram starts from left, with basement of wheels, Chassis and Boom of crane, and continues to the right of scheme where Load Q and cabin K are connected. All crane parts are designed with these elements (Fig.2 & Fig.3):

- **Rigid body frames (bars):** Chassis- Rw1 & Fw1; Link between chassis and Body – C1; Body and Counterbalance – C2, C3; Cylinder frames – PA, CA; Link between Body and Boom – L1, Boom - L2 & B1; Link between Boom and Cabin – B2; Cabin – K. [9]
- **Concentrated masses** - Chassis mass – *mw.* Body & Counterbalance masses - *m1, m2, m3;* Boom masses- *m4, m6, m7;* Cylinder mass – *m5*, Cabin mass- *m8,* Load-*Q*;
- **Fixed Frames** – Front wheels - *Fw;* Rear Wheels – *Rw,* that represents wheels of truck and additional supports;
- **Revolute joints** - *R1, R2, R3, R4*;
- **Boom lifting piston**- P1;
- **Hydraulic cylinder for Boom lifting** – HC1;
- **Hydraulic motor for Boom Lifting** – HM1;
- **Translational Joints** – In chassis–T1; In Boom – T2;
- **Spring and damping element** - SD1 - represents oscillations of boom while lifting;
- **Ramp function -** Rfn – Function of fluid flow values in motor HM1 to implement pressure force in cylinder HC1

In Fig.3. is presented discrete-continuous model of crane used for model view and simulations. This model is 3-D visualization created by software recurring from Schematic design on Fig.2. On this model, simulations will be performed in time frame of 0< t < 15 s. During this simulation time, crane will lift up Boom (L2, B1, B2 elements), Load Q and cabin K.

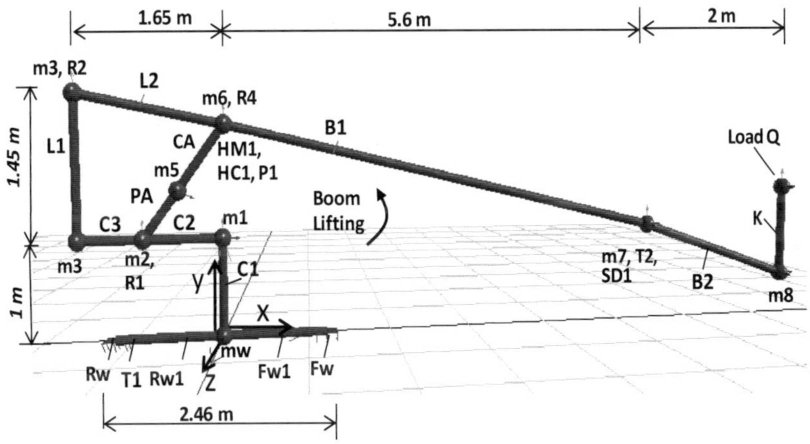

Fig.3. Discrete-continuous model of Hydraulic crane in form of 3-D visualization generated by software [3]

3. Differential equations of hydraulic crane

To formulate dynamics of this system, standard Euler-Lagrange methods are applied, by considering the crane as a multi-body system composed by concentrated masses, links and joints. For a controlled system with several degrees of freedom (DOF), the Euler-Lagrange equations are given as [2], [5], [11], [12]:

$$\frac{d}{dt}\left(\frac{\partial E_k}{\partial \dot{q}_i}\right)+\frac{\partial E_p}{\partial q_i}=Q_i\,,\ (\text{i=1 ,2...n}) \tag{1}$$

Where: q_i - are generalized coordinates for the system with n degrees of freedom, E_k is Kinetic Energy, E_p is Potential energy, Q is the n-vector of external non-conservative forces acting at joints. Kinetic energy for mechanical systems is in the form:

$$E_k\,(q,\dot{q}\,)=\frac{1}{2}\dot{q}^T\cdot M(q)\cdot\dot{q} \tag{2}$$

$E_p(q)$ – is potential energy that is a function of systems position.
M(q) - is a symmetric and positive matrix of inertias. [4]

Modern software calculates physical modeled systems through mathematical methods, numeric methods and Finite Elements Method [3],[10]. These calculations are based on Euler-Lagrange Equation (3.1), and forces applied for control of force/moments acting on crane. The modeling result is then an *n*-degree-of-freedom crane model whose position is described by generalized coordinates $\mathbf{q} = [\, q_1 \ldots q_l \,]^T$, and which is enforced, in addition to the applied forces, by m actuator forces/moments $\mathbf{u} = [u_1 \ldots u_m]^T$, where m<n [2]. The crane dynamic equations can be written in the following second order differential equation:

$$M(q)\cdot\ddot{q} + C(q,\cdot\dot{q})\dot{q} + \frac{\partial E_p}{\partial q} = Q(q,\cdot\dot{q}) - B^T u \tag{3}$$

where **M** is the $nx\,n$ generalized mass matrix, $C(q,\cdot\dot{q})$ is nxn matrix of Corriolis Forces, $\partial E_p/\partial E_q$ is the vector of gravity, **Q** is n-vector of generalized applied forces, and $\mathbf{B}^T$ is the nxm matrix of influence of control inputs **u** on the generalized actuating force vector $\mathbf{f_u} = -\mathbf{B}^T\mathbf{u}$. [2]

After design and testing of model, Software Maplesim has powerful module for symbolic generation of differential equations. There are 8 DOF from crane model (Fig.3), which gives 8 differential equations. Variables in differential equations are:

P1_F(t) – force in axes direction in piston P1 shown as translational joint ; ***P1_F2((t)*** – force in piston P1 in direction of y; ***T2_F(t)*** – force in translation Joint T2 in Boom B1; ***T2_s(t)*** – motion in axe of translation Joint T2 in Boom- B1; ***y(t)*** – variable of flow in cylinder HC1 implemented through ramp function RFn; ***HC1_s_rel(t)*** – Relative length of cylinder HC1; ***R1_θ(t)*** – Rotation of Revolute joint R1 around its axis (z), (Euler Angles); ***R2_θ(t)*** – Rotation of Revolute joint R2 around its axis (z).

3.1. Differential equations

8 Differential equations that represent boom lifting of crane are:

$$-\mathrm{T2\_F(t)} - 60000\cdot\mathrm{T2\_s(t)} - 2000\cdot\left(\frac{d}{dt}\mathrm{T2\_s(t)}\right) = 0 \tag{3}$$

$$-554\cdot\left(\frac{d}{dt}\mathrm{R2\_\theta(t)}\right)^2\cdot\mathrm{T2\_s(t)} + 654\cdot\left(\frac{d}{dt}\mathrm{R2\_\theta(t)}\right)^2 + 554\cdot\frac{d}{dt}\left(\frac{d}{dt}\mathrm{T2\_s(t)}\right) + \frac{50137}{10}\cdot\frac{d}{dt}\left(\frac{d}{dt}\mathrm{R2\_\theta(t)}\right) + \\ + \frac{271737}{10}\cdot\cos(\mathrm{R2\_\theta(t)}) - \mathrm{T2\_F(t)} = 0 \tag{4}$$

$$-\frac{41}{100} + \frac{33}{20}\cdot\cos(\mathrm{R2\_\theta(t)})\cdot\cos(\mathrm{R1\_\theta(t)}) - \frac{3}{10}\cdot\cos(\mathrm{R2\_\theta(t)})\cdot\sin(\mathrm{R1\_\theta(t)}) + \frac{3}{10}\cdot\sin(\mathrm{R2\_\theta(t)}) \\ \cdot\cos(\mathrm{R1\_\theta(t)}) + \\ + \frac{33}{20}\cdot\sin(\mathrm{R2\_\theta(t)})\cdot\sin(\mathrm{R1\_\theta(t)}) - \frac{18}{25}\cdot\cos(\mathrm{R1\_\theta(t)}) + \frac{29}{20}\cdot\sin(\mathrm{R1\_\theta(t)}) = 0 \tag{5}$$

$$\frac{8829}{25}\cdot\cos(\mathrm{R1\_\theta(t)}) - \frac{5886}{5}\cdot\sin(\mathrm{R1\_\theta(t)}) + \frac{784801}{10000}\cdot\frac{d}{dt}\left(\frac{d}{dt}\mathrm{R1\_\theta(t)}\right) + \frac{29}{20}\cdot\sin(\mathrm{R1\_\theta(t)}) - \frac{18}{25}\cdot\sin(\mathrm{R1\_\theta(t)}) \\ \cdot\mathrm{P1\_F2(t)} - \frac{18}{25}\cdot\cos(\mathrm{R1\_\theta(t)})\cdot\mathrm{P1\_F(t)} - \frac{29}{20}\cdot\cos(\mathrm{R1\_\theta(t)})\cdot\mathrm{P1\_F2(t)} + \frac{33}{20}\cdot\cos(\mathrm{R2\_\theta(t)})\cdot \\ \cos(\mathrm{R1\_\theta(t)})\cdot\mathrm{P1\_F(t)} + \frac{33}{20}\cdot\cos(\mathrm{R2\_\theta(t)})\cdot\sin(\mathrm{R1\_\theta(t)})\cdot\mathrm{P1\_F2(t)} + \frac{3}{10}\cdot\sin(\mathrm{R2\_\theta(t)})\cdot\cos(\mathrm{R1\_\theta(t)}) \\ \cdot\mathrm{P1\_F(t)} + \frac{3}{10}\cdot\sin(\mathrm{R2\_\theta(t)})\cdot\sin(\mathrm{R1\_\theta(t)})\cdot\mathrm{P1\_F2(t)} - \frac{33}{20}\cdot\sin(\mathrm{R2\_\theta(t)})\cdot\cos(\mathrm{R1\_\theta(t)})\cdot \\ \mathrm{P1\_F2(t)} + \frac{33}{20}\cdot\sin(\mathrm{R2\_\theta(t)})\cdot\sin(\mathrm{R1\_\theta(t)})\cdot\mathrm{P1\_F(t)} + \frac{3}{10}\cdot\cos(\mathrm{R2\_\theta(t)})\cdot\cos(\mathrm{R1\_\theta(t)})\cdot\mathrm{P1\_F2(t)} - \frac{3}{10}\cdot \\ \cos(\mathrm{R2\_\theta(t)})\cdot\sin(\mathrm{R1\_\theta(t)})\cdot\mathrm{P1\_F(t)} = 0 \tag{6}$$

$$\frac{142624647}{1000}\cdot\cos(\mathrm{R2\_\theta(t)}) + \frac{614106}{25}\cdot\sin(\mathrm{R2\_\theta(t)}) + 1108\cdot\mathrm{T2\_s(t)}\cdot\left(\frac{d}{dt}\mathrm{T2\_s(t)}\right) \\ \cdot\left(\frac{d}{dt}\mathrm{R2\_\theta(t)}\right) + \frac{50137}{10}\cdot\frac{d}{dt}\left(\frac{d}{dt}\mathrm{T2\_s(t)}\right) + \frac{204218471}{2000}\cdot\frac{d}{dt}\left(\frac{d}{dt}\mathrm{R2\_\theta(t)}\right) - 1308\cdot\left(\frac{d}{dt}\mathrm{T2\_s(t)}\right)\cdot\left(\frac{d}{dt}\mathrm{R2\_\theta(t)}\right) + \\ 554\cdot\mathrm{T2\_s(t)}^2\cdot\left(\frac{d}{dt}\left(\frac{d}{dt}\mathrm{R2\_\theta(t)}\right)\right) - 1308\cdot\frac{d}{dt}\left(\frac{d}{dt}\mathrm{R2\_\theta(t)}\right)\cdot\mathrm{T2\_s(t)} - \frac{271737}{50}\cdot\sin(\mathrm{R2\_\theta(t)})\cdot\mathrm{T2\_s(t)} - \frac{33}{20}\cdot \\ \cos(\mathrm{R2\_\theta(t)})\cdot\sin(\mathrm{R1\_\theta(t)})\cdot\mathrm{P1\_F(t)} - \frac{33}{20}\cdot\cos(\mathrm{R2\_\theta(t)})\cdot\sin(\mathrm{R1\_\theta(t)})\cdot\mathrm{P1\_F2(t)} - \frac{3}{10}\cdot\sin(\mathrm{R2\_\theta(t)}) \\ \cdot\cos(\mathrm{R1\_\theta(t)})\cdot\mathrm{P1\_F(t)} + \frac{33}{20}\cdot\sin(\mathrm{R2\_\theta(t)})\cdot\sin(\mathrm{R1\_\theta(t)})\cdot\mathrm{P1\_F2(t)} - \frac{33}{20}\cdot\sin(\mathrm{R2\_\theta(t)})\cdot \\ \cos(\mathrm{R1\_\theta(t)})\cdot\mathrm{P1\_F(t)} - \frac{3}{10}\cdot\cos(\mathrm{R2\_\theta(t)})\cdot\cos(\mathrm{R1\_\theta(t)})\cdot\mathrm{P1\_F2(t)} + \frac{3}{10}\cdot\cos(\mathrm{R2\_\theta(t)})\cdot \\ \sin(\mathrm{R1\_\theta(t)})\cdot\mathrm{P1\_F(t)} = 0 \tag{7}$$

$$y(t) = \frac{9}{10000} + \begin{cases} 0 & 0<t<12 \\ -\frac{9}{20000}\cdot t + \frac{27}{5000} & 12<t<14 \\ -\frac{9}{20000} & t>14 \end{cases} \tag{8}$$

$$\frac{d}{dt}HC1\_s\_rel(t) = 50\cdot y(t) \tag{9}$$

$$HC1\_s\_rel(t) = -\frac{3}{10}\cdot \cos(R1\_\theta(t))\cdot \cos(R2\_\theta(t)) - \frac{33}{20}\cdot \cos(R2\_\theta(t))\cdot \sin(R1\_\theta(t)) + \frac{33}{20}\cdot \cos(R1\_\theta(t)) \cdot \sin(R2\_\theta(t)) - \frac{3}{10}\cdot \sin(R1\_\theta(t))\cdot \sin(R2\_\theta(t)) - \frac{7}{5} + \frac{29}{20}\cdot \cos(R1\_\theta(t)) + \frac{18}{25}\cdot \sin(R1\_\theta(t)) \tag{10}$$

Solution of 8 differential equations will give results, which will be presented in graphical form.

4. Graphical results for main parts of crane

Results are achieved after simulations applied on designed system, Fig.2 & Fig.3. Simulations are planned to reflect real work of crane and boom lifting in order to achieve reliable results. Time of simulation is t = 15 s. Simulation has three phases [10],[13]:

First phase – Lifting of boom from its start position close to the ground (Fig.3). Lifting speed of boom v_l = 0.25 m/s. Time of simulation 0 s <t < 12 s. *Second phase* – Stopping phase of Boom Lifting that usually lasts few seconds, while there is no sudden stop of motion in reality. This phase start after first phase, and lasts 2 seconds, between simulation time 12 s <t < 14 s. *Third phase* – Boom is stopped at highest position and there is no motion, but there are oscillations after stopping. Starts after second phase, lasts between time 14 s < t < 15 s, which is end of simulation. It is implemented in order to monitor after motion oscillations.

Simulation of boom lifting is achieved with adjustment of flow rate of fluid in Hydraulic motor HM1 to give enough pressure force on piston P1 and cylinder HC1, with cross section A_c = 0.02 m^2 and regulate lifting speed. Flow rate is implemented with ramp function **Rfn,** shown in Fig.4. It starts with $\mathbf{q_{HM1}}$ **= 0.0009 $\mathbf{m^3/s}$** at first simulation phase (0 s <t < 12 s), and ends with $\mathbf{q_{HM1}}$ **= 0 $\mathbf{m^3/s}$** at the end of second simulation phase t = 14 s (Fig.4). Regulation of process of flow and its values on hydraulic motor HM1 is achieved through numerous tests to implement planned simulation and achieve lifting speed $v_l \approx 0.25$ m/s, in order to get best results with less oscillations [6], [10], [11]. This is the main process of regulation and control in this work. Higher values of $\mathbf{q_{HM1}}$ will give higher speeds which increase oscillations and safety risk, lower values of $\mathbf{q_{HM1}}$ will not lift the maximal load. Other parameters important for regulation and optimized results are for hoisting mechanism [7], [8]. This is for elements SD1 and T2, in order to minimize effect of vibrations which exist in boom during lifting. Spring constant for SD1 is determined as **k = 60 kN/m** and Damping constant is **d = 2 kNm/s.**

Based on model created, differential equations gained, and simulations, results are achieved for main dynamic parameters, shown with their symbols and units as follows: *Velocity v (m/s), Acceleration a (m/s²), Angular velocity ω (1/s), Angular acceleration aa (or α) (1/s²), Force F (N), Torque T (Nm), Motion* or *length s (m)*. [2], [4], [13]

Next will be presented graphical results for main parts of crane, where horizontal axis is time (t = 0…15 s) and vertical axes are corresponding values of dynamic and kinematic parameters. Only most significant graphs will be shown. On these graphs components towards **x** axis are shown with index 1, components towards **y** axis are shown with index 2 and components towards **z** axis are shown with index 3.

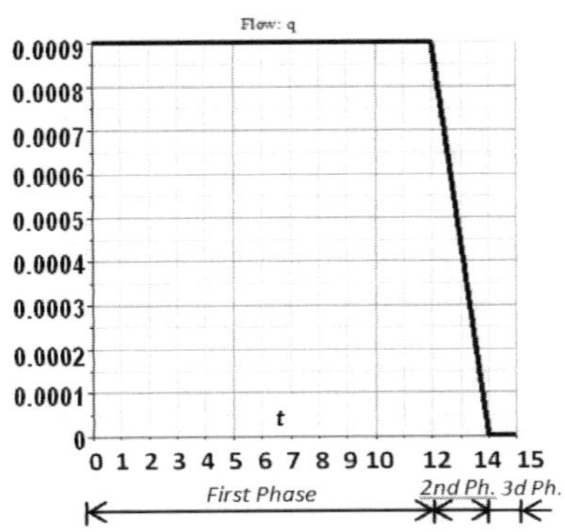

Fig.4. Ramp function **Rfn** of fluid flow rate $\mathbf{q_{HM1}}$ in motor HC1

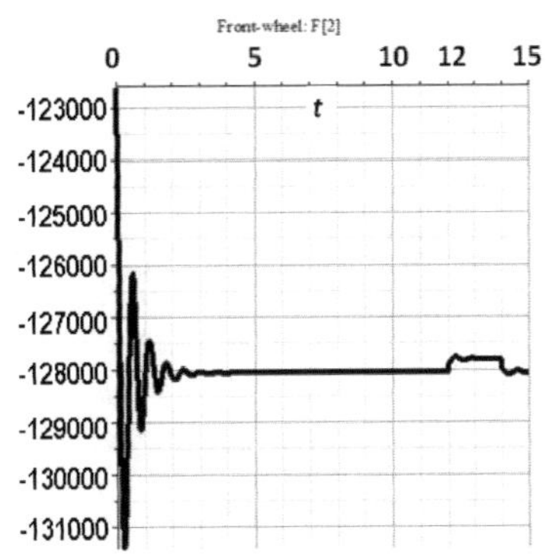

Fig.5. Force component F(y) or F(2) in Front wheels (Fw)

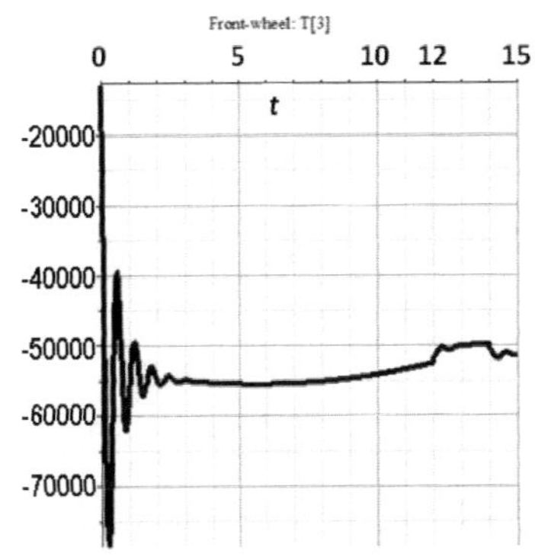

Fig.6. Torque T(z) or (T3) in Front Wheels (Fw)

4.1. Results of force and torque in Crane's wheels

Crane wheels are used for motion of crane and also to provide stability (Fig.1). In schematic diagram (Fig.2) wheels are represented with elements *Fw* and *Rw*. In Fig. 5 is shown graph of Componential Force *F(y)*, and in Fig. 6 graph of torque *T(z)* in Front wheels – *Fw*. In Fig.5, Force *F(y)* is Reaction Force of crane towards *y* axis. In this case values of other Reaction Forces F(1) or F(x) and F(3) or F(z) are very small and will not be shown in graphs.

Based on Fig.5, graph of *F(y)* shows change of curved line, with higher oscillations at the beginning of lifting process $0< t < 4$ s, and after lifting stoppage $t >12$. Oscillations at start are higher than near lifting stop. Max value of Force F(y) is: $F_{ymax} = |-1.32 \cdot 10^5|$ N at the start of process. Between time 4 s<t<12 s curve in diagram is less dynamic, almost constant, until $t >12$ when it again shows small oscillations. Values of *F(x)* component are small and negligible, up to $F_{xmax} = 450$ [N] and not shown in graphs. We can conclude that wheels undergo oscillations that are irregular, with amplitudes that are high at the beginning, up to $\Delta_F \approx 2000$ N, but drop fast after 2 seconds to almost 0, and are small at the stoppage of lifting, $\Delta_F \approx 200$ N, giving calm lifting process, starting from t = 2 s until the end of lifting. This is the intention of control of lifting motion. In Fig. 6 is graph of Torque T(z) in Front wheels. It has max value: $T_{zmax} = |-77000|$ Nm at the start of process, and drops to medium values of $T_{zm} = |-55000|$ Nm around time $t \approx 4$ s, which gives amplitudes of $\Delta_T \approx 20000$ Nm. Between time 4 s<t<12 graph drops to almost no oscillations and has curved line. After time $t >12$ oscillations appear again with smaller dynamic intensity than at the start, with $\Delta_T \approx 5000$ Nm. This graph concludes again that wheels undergo intensive and irregular oscillations at start, until t = 3 s, but after that have small oscillations, which shows god results of lifting and control. In Fig. 7 are shown graphical result for Force component *F(y)* in Rear wheels (Rw). Noticeable is that max value of $F_{ymax} = |-87000|$ N is smaller than in Front Wheels for about 52%, due to boom and load position on the side of Front wheels. In Fig.8 is shown value of Torque T(z) in Rear wheels. Conclusion is that dynamics in rear wheels is similar as in front wheels, but with different intensity.

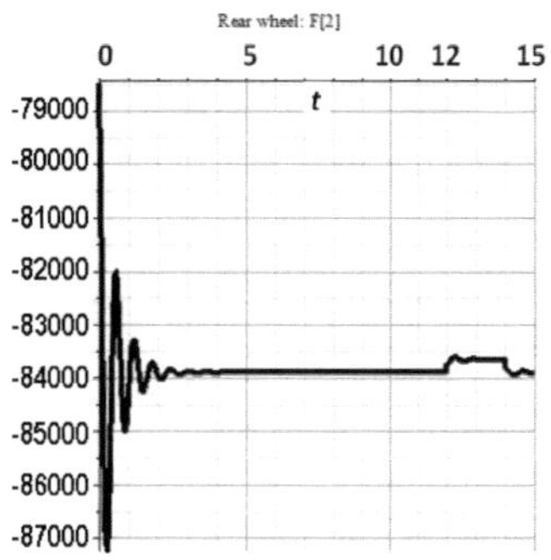

Fig.7. Force component F(y) in Rear wheels (Rw)

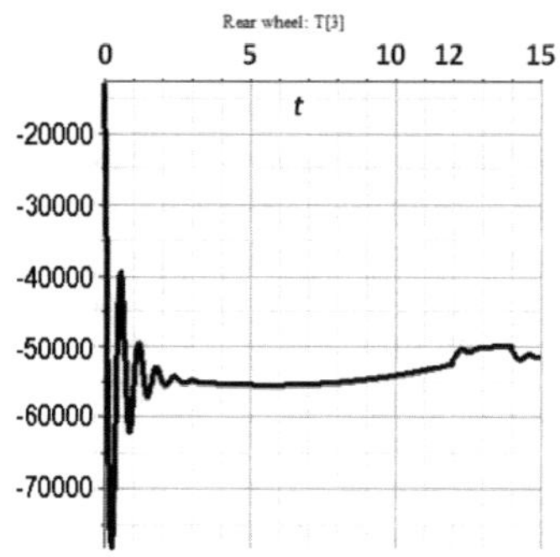

Fig.8. Torque T(z) or (T3) in Rear Wheels (Rw)

4.2. Results for Crane's Body and Counterbalance

Body and Counterbalance is rigid part of Crane that connects with Chassis of Crane in bottom, Boom, and Cylinder (Fig.1). In Fig.2 schematics it consists of elements *C2, C3, m2, m3*. In Fig. 9 to Fig.11 are shown graphs of body element **C3,** Componential Forces *F(x)* and *F(y)* and Torque *T(z)*. Between time 4 s< t <12s graphs shows inclined curve of forces and torque, due to change of distance of Load Q from Center of Body during Boom lifting. Conclusions are similar to those of wheels, at the start of lifting there are higher oscillations that drop fast after 3 seconds, and gives calm lifting process up to t = 12. Between 12s<t<14s there are some small oscillations due to stoppage, but have little effect on Cranes Body.

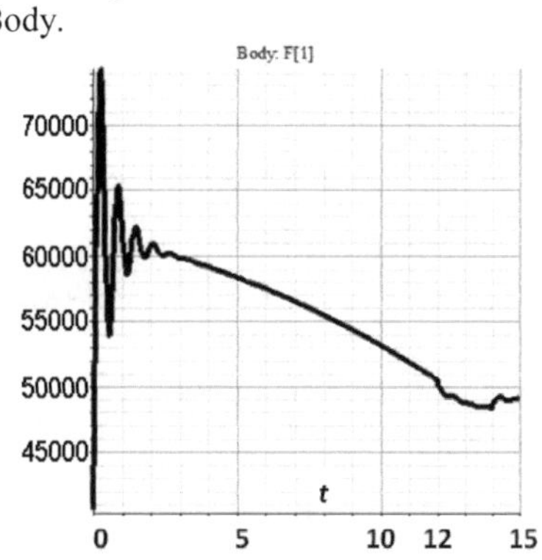

Fig.9. Force components F(x) in Body

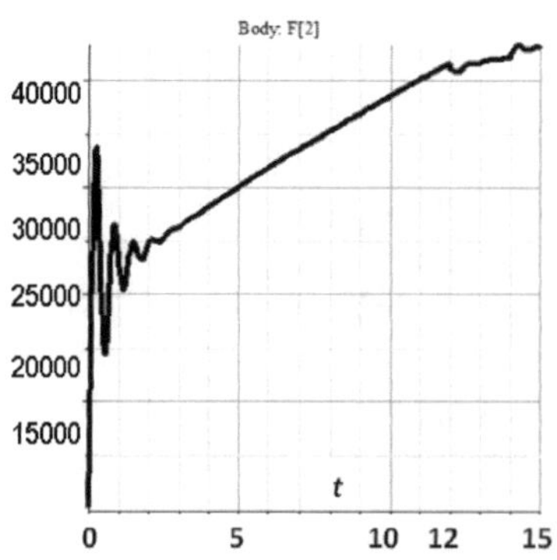

Fig.10. Force components F(y)

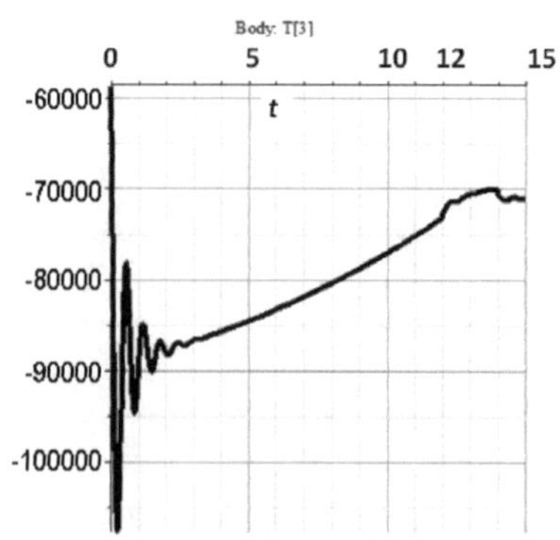

Fig.11. Torque T(z)

4.3. Results for Crane's Boom

In Fig.12 to Fig.14 are shown graphical results for Lifting Boom element **B1** for some main parameters. This is the part that passes the force from cylinder to lift the load **Q**, and is heavy loaded part. Load Component F(x) is increasing until t = 12 s, but has smaller intensity, with no oscillations, and has small effect on Boom. Force component F(y) and Torque T(z) have higher intensity, and are dynamic in nature, similar to Conclusions of motion and oscillations for wheels and Crane's Body.

Last Graph, Fig.14, is important kinematic parameter *Angular velocity w(z)* in Crane's Boom. Between time 1s<t<12s values are almost constant and don't change significantly, then drop down between time 12 s < t < 14 s, and have valuc 0 at time t=15 s due to motion stop. This change corresponds with simulations function, Fig.4.

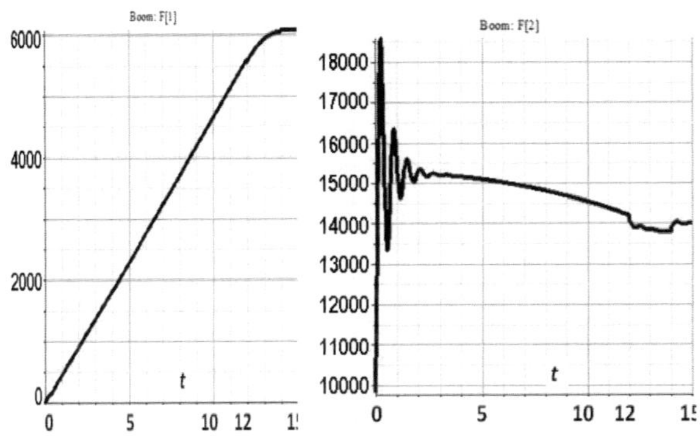

Fig.12. Force components F(x) and F(y) in Crane's Boom

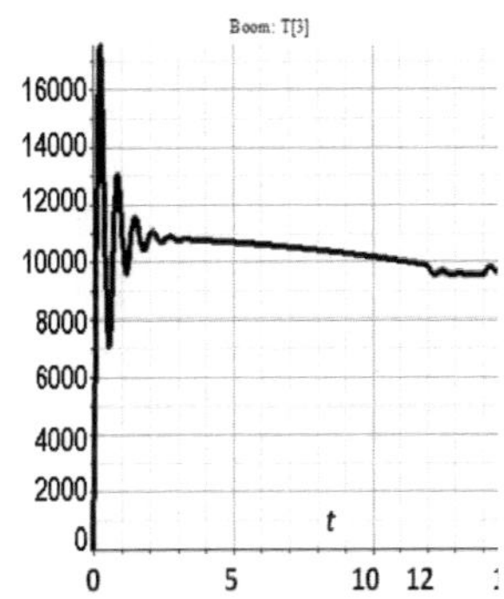

Fig.13. Torque T(z)

Fig.14. Angular velocity w(z) in Crane's Boom

4.4. Results for hydraulic cylinder and piston

In Fig. 15 to Fig.18 are shown graphical results for Hydraulic cylinder HC1 and piston P1. These are parts that give power and motion for boom lifting. Hydraulic motor HM1 transfers flow of fluid q_{HM1} to Hydraulic Cylinder HC1, which converts to pressure force to piston P1 and thereafter on Crane's Boom - L2 and B1. Based on Fig.15, the form of curve of Pressure Force in Cylinder HC1, as main power force of process, is passed to other parts of crane, similar in dynamic behavior and oscillations. Velocity of piston in Fig.17 corresponds to lifting Ramp function and fluid flow in Fig.4. In Fig.18, force components in piston have similar form as in Fig.15, and as in previous crane parts.

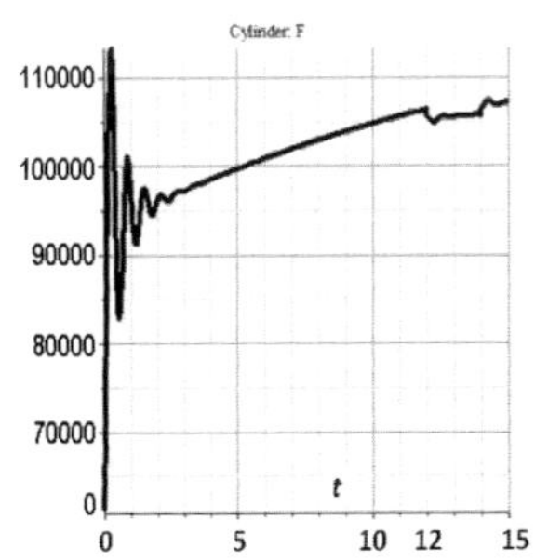

Fig.15. Pressure force in cylinder HC1

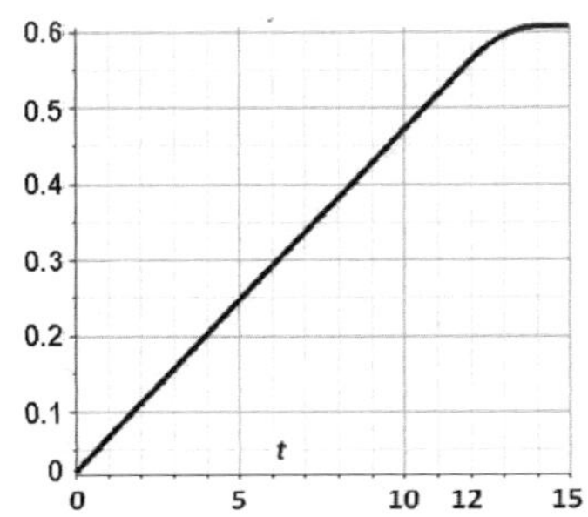

Fig.16. Motion length of piston P1

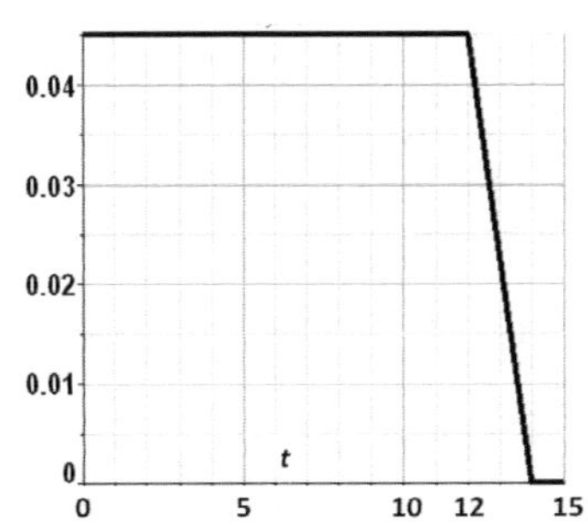

Fig.17. Velocity of piston P1 (m/s)

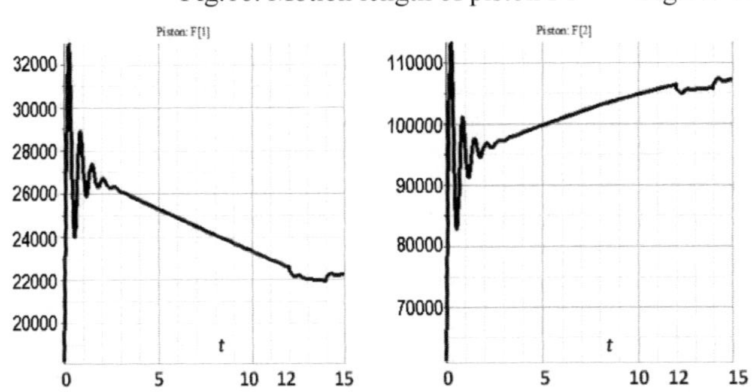

Fig.18. Force components F(x) and F(y) in piston P1

4.5. Results of oscillations for Load Q and cabin K

Load Q is the part being carried by crane, in this case men working and other type of load. Cabin K with load Q will make oscillations while being lifted. These oscillations influence directly and indirectly other parts of crane. It is important to identify dynamic behavior of load and cabin and minimize their oscillations. In Fig.19 is presented lifting of load from start to end, in unit of meters. The curve increases with very small oscillations. This is a sought result, while load has very little swinging due to regulation of Boom lifting speed. In Fig. 20, component *v(y)* has high oscillations until t =3 s, then is constant up to t = 12 s, with sought value of v ≈ 0.25 m/s. After this, they drop to 0 at t = 14 s, end of lifting. Conclusion is that load Q and Cabin K at the start of lifting have irregular motion with oscillations, which drops after 3 seconds, and remains low until the end of lifting. Fig.21 represents acceleration *a(y)* of load Q and Cabin K. Graph shows dynamic form with oscillations at the start of lifting 0s< t <3s with higher frequencies and amplitudes, which drop significantly at the end of process. Between time 4 s< t <12 s acceleration is close to 0, which concludes that lifting speed is properly regulated close to constant.

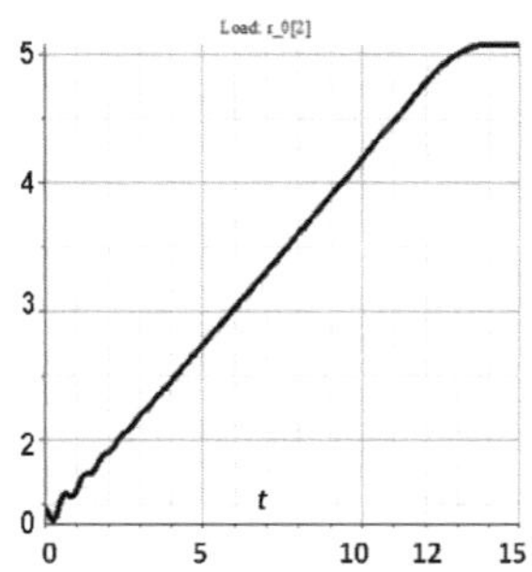

Fig.19. Q and K – Lifting r(y)

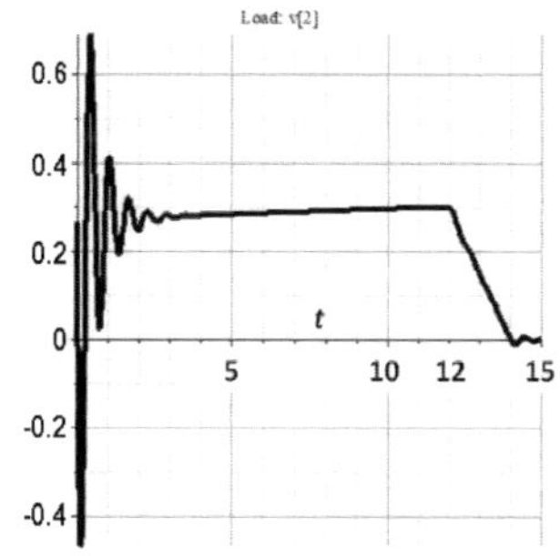

Fig.20. Q and K– Velocity v(y)

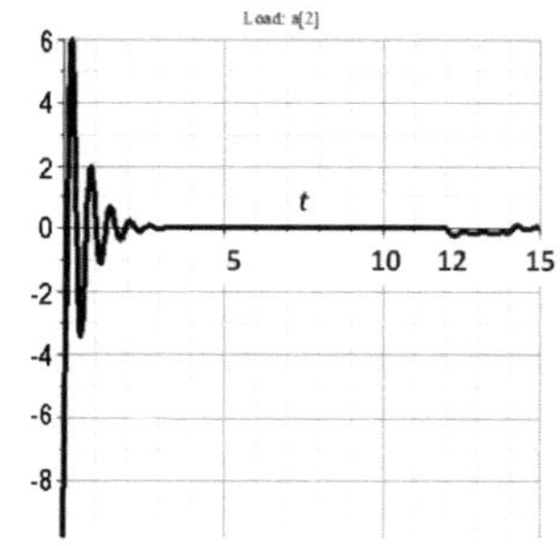

Fig.21. Q and K – Acceleration a(y)

5. Conclusion

The main problem during load lifting is oscillations in hydraulic crane as part of dynamic occurrences. In order to control them, it is important to identify them first. To do this we created crane model with schematic design and 3-d visualization, and implemented simulations. Important part of analysis is finding proper simulations plan that reflects real lifting of crane's boom, so that results are reliable. Results are gained for main dynamic parameters and presented in graphical form. From these diagrams can be noticed oscillations in some parts of crane, and mostly with irregular occurrence. They occur in different planes. Oscillations have high intensity at the start of lifting process, almost zero values in the middle of process, and small values at end of process [10], [13]. The intention of research was to minimize oscillations [4], [11]. This was achieved through numerous simulations, in order to find optimal boom lifting, fluid flow in hydraulic motor and accurate spring and dumping constants to minimize load swinging. All this is done to with the aim to optimize the speed of lifting which is main parameter for control and optimization. This work is also important for safety at work, while the load carried are usually humans. It can be used also for further analysis for crane's optimization, and in the future can be used for other work processes of crane, like travel with load.

6. References

[1] JLG 660SJ Lift, https://www.jlg.com/en/equipment/engine-powered-boom-lifts/telescopic/600-series.

[2] Garcíaorden, J. Carlos, Goicolea, José M., Cuadrado Javier. (2007). Multibody Dynamics, Computational methods and applications, p.91, 2007 Springer.

[3] MapleSim User Guide. (2014). Maplesoft, a division of Waterloo Maple Inc.

[4] La Hera, P. M. (2011). Dynamics modeling of an electro-hydraulically-actuated system, Umea University.

[5] S. Joe Qin, Badgwell Th.A. (2003). An overview of industrial model predictive control technology, Control Engineering Practice, A Journal of IFAC, ISSN: 0967-0661.

[6] Damica V., Cohodarb M., Kulenovic M. (2013). Modeling and Simulation of Hydraulic Actuated Multibody Systems by Bond Graphs, 24th DAAAM International Symposium on Intelligent Manufacturing and Automation.

[7] Ye F., Qiong W., b, Haiyang Y. (2010). Dynamic Simulation of Hydraulic Truck Crane Hoisting System Based on AMESim, Applied Mechanics and Materials Vols 29-32 (2010) pp 2031-2036.

[8] Yingguang C., Vilmar Æ. and Houxiang Zh., Modelling and simulation of an offshore hydraulic crane, (2012), Proceedings 28th European Conference on Modelling and Simulation ©ECMS, ISBN: 978-0-9564944-8-1.

[9] Elena S. Gebel, (2015), Mathematical Modeling of Dynamics of Multi-Lever Linkages, 26th DAAAM International Symposium, p. 470-477.

[10] Doçi I., Hamidi B. (2015). Studying rotational motion of Luffing Jib Cranes with maximum load using simulations, International Journal for science, technics and innovations for the industry-MTM 2015; 1(12):20-24.

[11] Vaughan J. (2008). Dynamics and control of mobile cranes, Georgia Institute of Technology.

[12] PARK, Hong Seok & LE, Ngoc Tran, (2011), A 3d simulation system for mobile harbor crane based on virtual prototyping technology, 22th DAAAM International Symposium, p.35-36.

[13] Doçi I, Buza Sh, Pajaziti A, Cakolli V, (2015), Studying dynamic effects of motion of telpher on console cranes using simulations. Journal of Fundamental Sciences and Applications. Tech-Sys 2015;21(2):337-342.

DOI: 10.2507/27th.daaam.proceedings.101

VACUUM INCREASE IN THE ELECTROLYTIC CELL WITH THE AIM OF REDUCING EMISSIONS INTO THE WORK ENVIRONMENT

Boris Vican, Vedrana Galić & Josip Ćorić

This Publication has to be referred as: Vican, B[oris]; Galic, V[edrana] & Coric, J[osip] (2016). Vacuum Increase in the Electrolytic Cell With the Aim of Reducing Emissions Into the Work Environment, Proceedings of the 27th DAAAM International Symposium, pp.0701-0707, B. Katalinic (Ed.), Published by DAAAM International, ISBN 978-3-902734-08-2, ISSN 1726-9679, Vienna, Austria
DOI: 10.2507/27th.daaam.proceedings.101

Abstract

In the production of the primary aluminum, waste gases occur as a byproduct in electrolytic cell. Such gases are drained to the Gas Treatment Centre by degassing system where, before discharging into the atmosphere, are purified by adsorbing to the active surface of alumina (Al_2O_3). Gases enriched alumina is, by pneumatic transportation, returned to electrolytic cell. When performing operations on the cell, such as anode replacing and aluminum tapping from the cells, these gases are emitted into the electrolysis hall, which directly affects the environment and the health of employees. Thus, the purpose of this work is to increase the vacuum in the cells exhaust system in order to reduce emissions into the electrolysis hall during performance of the aforementioned operations, with minimal investment and utilization of existing resources.

Keywords: electrolytic cell; the Gas Treatment Centre; waste gases; anode replacing; aluminum tapping; exhaust system

1. Introduction

In the process of aluminum reduction, there are number of factors that could adversely affect both the health of direct employees and their work environment. One of these factors includes emissions from the electrolytic cells that occur as a byproduct of the production of primary aluminum. Hydrogen fluoride (HF), as a byproduct of the electrolysis of alumina, is one of the most harmful gases when it comes to impact on human health and therefore place great emphasis on the reduction in the working environment [1]. The emergence of the HF in the electrolytic cell is attributed to two reactions:

The electrochemical oxidation of hydrogen at calcined carbon anode;
The reaction of moisture with present AlF_3 [2].

Mentioned reaction of forming HF is shown by the equation (1):

$$AlF_3 + H_2O \rightarrow Al_2O_3 + 6HF \qquad (1)\ [3]$$

High concentrations of hydrogen fluoride generated, leave the cell during the operation of anode replacing and aluminum tapping and thus directly endanger the health of employees.
In the last few decades, the aluminum industry has invested a great deal of knowledge and resources to the adverse impacts of aluminum production to the lowest possible level. For this purpose, the specific system for exhaust gases from an electrolytic cell and filtering before releasing to atmosphere has being developed.
Since the conditions, in the global aluminum market, are constantly changing, involved in the production process are daily faced with the need for increasing or decreasing the production of primary aluminum. Of course, with increased production of aluminum, there's an increase in the production of waste gases, which requires large investments in the field:

- Cell exhaust gas system
- Gas treatment centers capacity increase

It is important to emphasize that in addition to the above mentioned hydrogen fluoride, electrolytic aluminum production process, unfortunately, includes a range of waste gases generated by the oxidation of impurities present in the raw materials.
This paper will emphasize increase in exhaust gases from the electrolytic cell towards gas treatment center in during the following operations:

- Aluminum tapping
- Anode replacing

Considering that these activities include the longest stage of exposure to harmful impacts. These phases include the following operations:

- Removing anode butts
- Insertion of material removed under anode into molds
- Keeping the cell open during new anode replacing

The low price of aluminum on the world market and high electricity prices currently create a very unfavorable financial climate for the production of primary aluminum. This reduces investment of aluminum companies in the development of both existing and the construction of new systems to reduce emissions. Therefore, it is our goal to, with minimal investment and utilization of existing resources, increase the negative pressure in the cell when performing the above mentioned operations and thereby reduce emissions of gases in the electrolysis hall and the environment. The increase in negative pressure at the outlet suction tube of superstructure will be based on the installation of a system based on the principle of the ejector. Measurements will be made by a device for measuring the negative pressure, and the analysis of data will be used to evaluate the performance of the developed system.

2. Working principle of the ejector

Ejectors are devices that are used for fluid or mixture of fluids transfer. The working principle of the ejector is simple because does not require the driving engine but uses the energy of the input drive fluids for suction, compression and suppression of the vacuumed fluid. Operating fluid operates at a higher pressure, as opposed to the vacuumed fluid that streams under the influence of lower pressure. Ejectors can be used for high pressures and negative pressure for small and large flows, high and low temperatures [4].
The main parts of the ejector are:

- Nozzle with a connector for operating fluid
- Vacuum chamber
- Diffuser

The operating fluid, while passing through the nozzle, forms a stream jet that rapidly enters the vacuum chamber and creates a negative pressure for suction of the secondary fluid. At the entrance to the vacuum chamber particles of the operating fluid encounter particles of vacuumed fluid. Fluid particles collide and mix and form a homogenous mixture, which flows to the diffuser. On the way to the chamber, particles of the operating fluid lose some of their kinetic energy and at the same time slow down, and the particles of vacuumed fluid receive part of the kinetic energy and thereby accelerate. After the entrance of the diffuser, due to expansion of the streaming space, a stream flow of the mixture is slowing down, the speed decreases and the pressure increases until the exit of the ejector. The output pressure depends on the relationship between pressure and flow of the vacuumed and operating fluid and is always higher than the vacuumed and lower than the operating pressure [5].

It is worth mentioning that there are also ejectors without diffuser. In comparisons of ejectors with diffuser and without diffuser it has been observed that ejectors with a diffuser provide a higher injection coefficient for the same ejector diameter values and the same parameters of working fluid. For the same values of nozzle diameter, the injection coefficient is higher for lower values of mixing room diameter [12].
In our case the principle of ejector without the diffuser was used because if the diffuser was installed, the gas flow at normal vacuum from the electrolytic cells would be reduced.

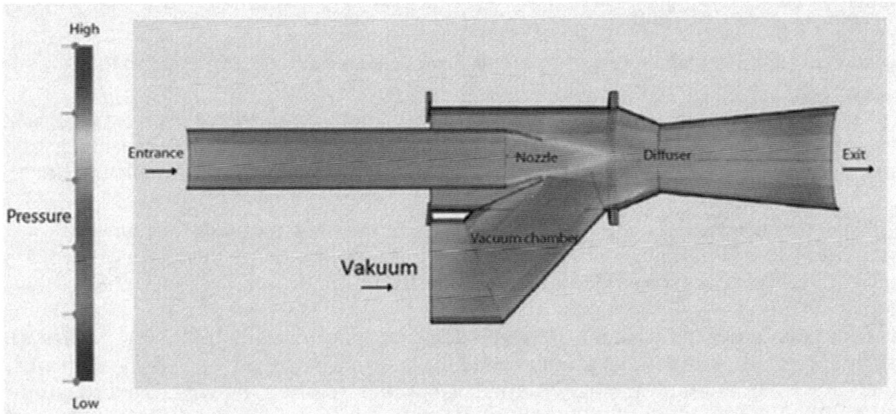

Figure 1: Working principle of the ejector [6]

3. Dry gas processing (Gas Treatment Centre)

Principle of the dry scrubbing system is simple [9]. Scrubbers operate by direct extraction of flue gases from electrolytic cells into reactors where gases react with active alumina. At first, poisonous fluoride and HF are adsorbed from the gas mixture onto active alumina surface. Gases are vacuumed into the Venturi tube shape like reactor located in front of the bag filters. In the opening of the reactor is injected fresh alumina and alumina from recirculation (enriched with fluorine). The alumina, enriched with gases and dust particles, stops on the filter bags and is being collected in the filter baskets. Effectiveness of the cleaning process depends on details of the system and the process. In terms of reaction mechanism, hydrogen fluoride, at purification temperatures up to 150 °C binds onto alumina by:

- Direct chemisorption of HF at reactive sites on the alumina surface with formation of Na-F and/or Al-F species and with additional monolayer being physically adsorbed on top of chemisorbed layer [10].
- Physical adsorption of HF on the alumina surface (hydroxyl groups bonded to surface and/or physisorbed by water to form alternating layers of HF and H2O) [11].

It is preferred to use such a structural form of alumina that has a larger specific surface area (**B.E.T. - m²/g**) and therefore the maximum capacity of chemisorption. By the implementation of the emitted gases through a layer of alumina it is possible to restore all the fluorine in cells. Under the normal operating conditions, all alumina required for the production of aluminum is used for catching fluorine i.e. goes through the gas treatment plant before dosing into cells.

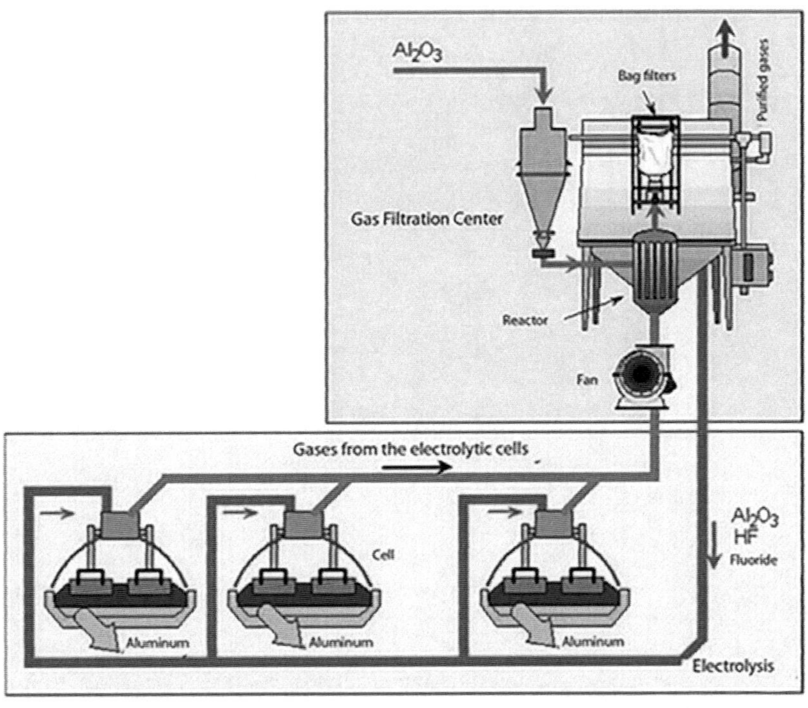

Figure 2: Schematic illustration of the filtration center operation

4. Negative pressure increase in electrolytic cell

For the operation of aluminum tapping and anode replacing in electrolytic cell, two systems for increasing the negative pressure were developed.

- The increase in negative pressure with a fan with an electric motor. This operation involves the operation of one fan at a group of cells, embedded in bypass tube. The system is quite expensive to install and maintain. This system called Process Gas Suction System, was developed by the company Norsk Hydro Aluminum (International Patent Application No. WO 01/36716) [7,13].
- The increase in negative pressure on the ejector principle. In these cases, the brand new line of a dividing pipeline is installed. A blower with a large amount of small pressure air is used as a source of air. This system was developed by the company AP Technology called JIBS (Jet Induced Boosted Suction) (U.S. Pat. No. 4,668,352) [8,13]

As already mentioned, the aim of this work is to develop a similar system with minimal investment and the use of its own resources. The system that we have developed is similar to the system developed by AP Technology. As the operational fluid, in this case, is used compressed air obtained from the compressor station through an existing line of the dividing pipeline that is used to start bath perforation cylinders and alumina dispenser cylinders and aluminum fluoride (AlF_3). Also, the type of the nozzle is different than one used in JISB system.

Compressor station has five compressors in which each compressor has a capacity of 37,7 m^3. Maximum operating pressure is 10 bars and the current pressure is 8.5 to 9 bars. Compressed air is supplied directly to superstructure of the cells. Connection to installation of the compressed air is made at the entrance of the pipeline in the superstructure of the cell.

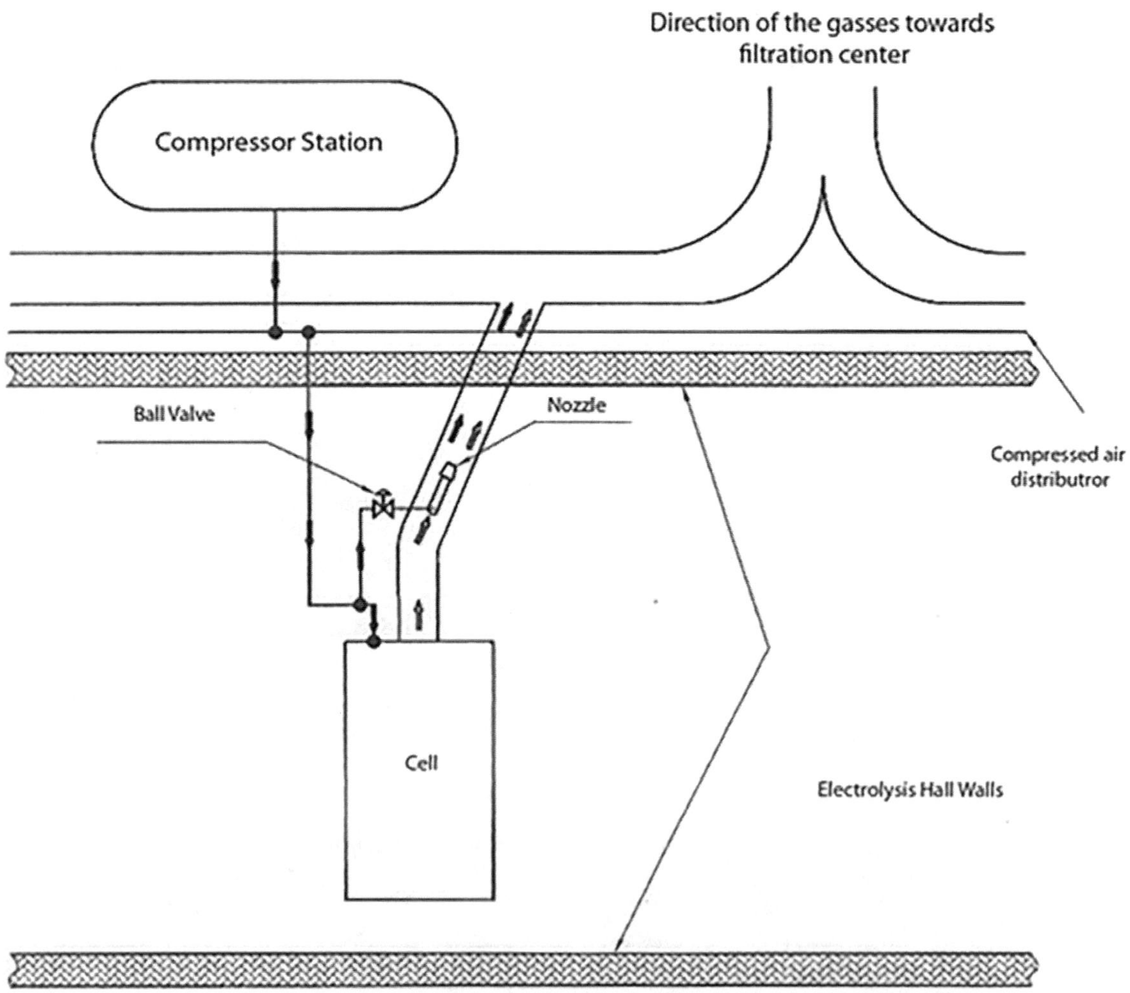

Figure 3: Schematic illustration of compressed air installation

The nozzle for the operating fluid is incorporated in gas discharge tube from electrolytic cells to the filtration center. During the installation of the system, the cell currently stays out of compressed air and, therefore, a very important factor is a system speed installation in order to avoid disturbances in the cell. Before the start of the operation of replacing the anodes and extraction of metals, the operational fluid, i.e. compressed air is passed by a ball valve.

Figure 4: Nozzle in the pipe

The system is installed to a group of 32 cells. Measurements shown in the *Table 1*, were done by differential pressure gauge with the range from -137 to + 137 bars.
The results of negative pressure on certain cells were below the average of other cells and, upon turning on the built-in system, errors of the bath perforator occurred at the same cells. The reason for this is, most likely, the system pressure drop. Subsequently, cartridge filters of compressed air are changed after which the results were similar to results from other cells.
As shown in the *Table 1*, a significant increase in negative pressure is obvious at all cells with installed system. In average, negative pressure is increased for 130% (from -2,7 mbar to -6,2 mbar)

Cell Number	System Off	System On
1	-1,8	-4,7
2	-2,1	-6,6
3	-1,9	-6,8
4	-2,0	-6,2
5	-2,2	-5,5
6	-2,3	-6,8
7	-2,3	-6,3
8	-2,7	-6,9
9	-1,9	-4,4
10	-2,7	-6,3

11	-2,1	-5,7
12	-2,3	-5,9
13	-2,5	-5,0
14	-2,1	-5,7
15	-3,6	-5,6
16	-3,9	-8,1
17	-2,8	-5,0
18	-3,0	-5,9
19	-3,5	-6,4
20	-3,3	-5,9
21	-2,0	-5,3
22	-3,1	-6,7
23	-3,7	-7,9
24	-3,7	-7,9
25	-3,2	-6,9
26	-3,3	-6,2
27	Off	
28	-3,5	-8,0
29	-3,2	-6,7
30	-2,0	-5,9
31	Off	
32	-2,0	-4,7
Average	-2,7	-6,2

Table 1: Measurement results

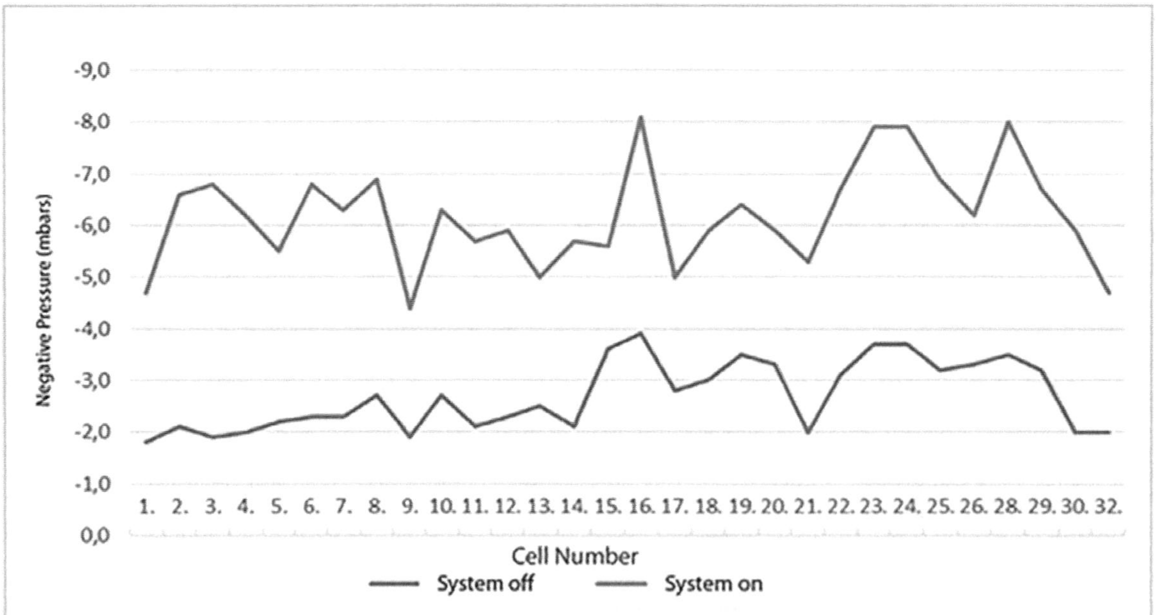

Figure 5: A diagram illustration of the measurement results

5. Conclusion

As mentioned in the introduction, the aim of this study was, with own resources and minimum investment, to develop a system which will help to increase the negative pressure in the electrolytic cell and thus reduce emissions of waste gases in the electrolysis hall and the environment.

The system that was installed at 32 electrolytic cells is completely developed with own resources. Measuring the negative pressure at these 32 cells, the increase for approx. 130% was registered. Installation and commissioning of the complete system takes about 10 minutes, and without obstruction of the technological process of production of primary aluminum. If further measurements of working conditions in cell environment prove that the installation of the system reduced emissions of HF in the hall of Electrolysis, considering investing in automating of the developed system will be intense. This means automatic start of the system for increase of the negative pressure on the cell control panel during aluminum tapping and anode replacing.

In addition please note that:

- Norsk Hydro Aluminium in its literature [7] with its system predicts increase of suction, i.e. gas flow from elecrolytic from 5000 Nm3/h to 15 000 Nm3/h
- AP Technology in its literature [8] with its system predicts icrease of suction, i.e., gas flow from the electrolytic cell as 2x the nominal base flow

6. Literature

[1] Planinić H.; Šarić M. & Šoše M, (1980). Technology of aluminium reduction-script for internal use in Aluminij Mostar, (Tehnologija zanimanja –interna skripta Aluminij Mostar), unpublished.

[2] Welch B. & Grjotheim K. (1988) *Aluminium smelting technology*, Aluminium-Verlag Dusseldorf

[3] Devadiga H.; Banjab A.; AlJallaf M.; Al Zarouni A.; Al Aswad K.; Kumar A.; Meintjes G.; Gowda S. & Khan M. (2013)., Reduction in HF emission through improvement in operational practices, Proceedings of Light Metal Symposia, March 3-7,2013, San Antonio-Texas, ISSN 1096-9586, ISBN 1118663071, 9781118663073, Sadler B. (Ed.), pp.763-767, John Wiley & Sons, 2013, DOI: 10.1002/9781118663189.ch130

[4] http://www.ejektor.co.rs/srpski/ejektori/oejektorima/oejektorima1.htm (1990). Ejektoinženjering, About Ejectors, Accessed on (2015-06-12)

[5] Jelić T. (2013). Air microturbine with ejector (Zračna mikroturbina sa ejektorom), Undergraduate thesis, University of Zagreb, Faculty of mechanical engineering and naval architecture, Zagreb, Croatia

[6] http://www.samson-agro.fr/media/1320/ejector_1.png Samson Agro A/S, Accessed on: (2015-06-12)

[7] Hydro Aluminium Primary Metal, Technology & Projects (2003). Feasibility study – Upgrade E2 at 185 kA in Aluminij-Mostar, unpublished

[8] https://www.ap-technology.com/SiteCollectionDocuments/AP_Factsheet_JIBS.pdf (2014). Rio Tinto Aluminium, Jet induced boosted suction (JIBS), Accessed on: (2016-01-12)

[9] Paulin I.; Donik Č. & Jenko M. (2009). Mechanism of HF bonding in dry scrubber in Aluminium Electrolysis, Materiali in Tehnologie 43, in press

[10] W. D. Lamb: A study of the equilibrium adsorption of hydrogen fluoride on smelter grade aluminas (1978). Light Metals 1978 proceedings of sessions, 107th AIME annual meeting, Denver, Colorado, Miller J.J. (Ed.), pp 425–444, American Institute of Mining, Metallurgical and Petroleum Engineers, 1978., New York

[11] Coyne J. F.; Wong P. J.; Wainwright M. S. & Brungs M.P. (1989). Factors influencing hydrogen fluoride adsorption on alumina, Light Metals 1989: Proceedings of the Technical Sessions by the TMS Light Metals Committee at the 118 TMS Annual Meeting, February 27-March 3, 1989, Las Vegas, Nevada, ISBN-10: 0873390822, Campbell P.G. (Ed.) pp. 113-119, The Minerals, Metals & Materials Society

[12] Krasniqi, F.; Bunjaku, J. & BUNJAKU, F. Analysis and optimisation of main components of ejectors in a distant heating system (2009). Proceedings of the 20th International DAAAM Symposium, Volume 20, No. 1, ISSN 1726-9679 ISBN 978-3-901509-70-4, Katalinic B. (Ed.), pp. 67-68, Published by DAAAM International, Vienna, Austria, EU

[13] Cantin P. & Girault G. (2009). System and process for collecting effluents from an electrolytic cell, Pub. No.: US 2009/0159434 A1, Avalilable from: http://www.google.com.na/patents/US20090159434, Accessed on : (2016-08-12)

DOI: 10.2507/27th.daaam.proceedings.102

TEMPERATURE INFLUENCE ON THE DRAWING PROCESS STABILITY WITH THE REDUCTION OF THE WALL THICKNESS

Himzo Đukić, Mirna Nožić & Darko Šunjić

This Publication has to be referred as: Dukic, H[imzo]; Nozic, M[irna] & Sunjic, D[arko] (2016). Temperature Influence on the Drawing Process Stability With the Reduction of the Wall Thickness, Proceedings of the 27th DAAAM International Symposium, pp.0708-0713, B. Katalinic (Ed.), Published by DAAAM International, ISBN 978-3-902734-08-2, ISSN 1726-9679, Vienna, Austria
DOI: 10.2507/27th.daaam.proceedings.102

Abstract

The cold drawing process takes places very quickly and during the conversion of mechanical energy into the heat energy the work peace and the tool heat up considerably. In order to achieve a stabile drawing process, it is necessary to keep tracks of the temperature growth of the work piece and the tool. The tool temperature has to be below the established limits, and it is regulated by the amount of the coolant, which is at the same time also a lubricant. The expressions for the temperature growth calculation in the conditions of adiabatic processes that can also be applied to the deep drawing process with the reduction of the wall thickness, due to the high deformation rate, have been presented in the work. The heat balance equation enables the determination of the coolant mass which will maintain the stabile tool temperature, what is confirmed by the experimental researches.

Keywords: Drawing with the reduction of wall thickness; heat balance; tool temperature.

1. Introduction

Deep drawing is the forming of smooth (sheet) blanks into hollow parts. It is a process which involves forming by tensile and compressive forces. [1]. Deep drawing process has to be surely considered as fully assessed since it was developed in the 1700s. Starting from that date, the process was extensively studied and it has become a strategic process in metalworking field [2]. Different cylindrical elements on the multi-tools are being made by the deep drawing process with the reduction of the wall thickness.

This process happens during mass production by the continuous passing of the work piece through several rings set one behind the other in the tool. The presentation of the process indicates a complex deformation mechanism, where during the passing of the work piece through the tool, the reduction of the wall at the diameter and the wall thickness occurs at the same time, as shown in Fig. 1 (a. 2D presentation, b. 3D presentation).

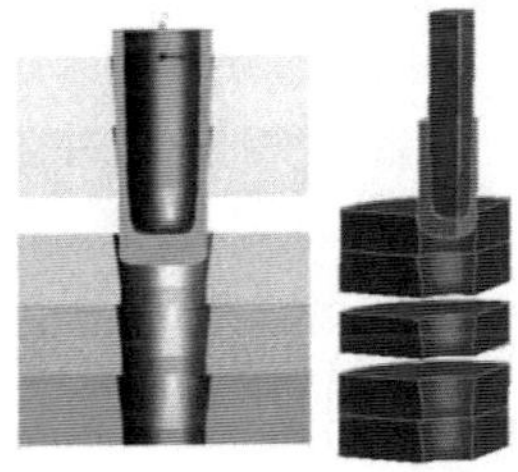

Fig. 1. Scheme of drawing on the multi-tools

When projecting a technological process of the deep drawing with the reduction of the wall thickness, special attention should be paid to the heat effect that has a significant role in the processes of cold deformation. In order to achieve a stabile drawing process, it is necessary to keep tracks of the temperature growth during the process and control it, i.e. to control the heating of the tool [3].

2. The heat effect during the deformation process

In order to keep tracks of the energy conversion during the deformation process it is needed to define the thermo-dynamic system. Since all the deformation process happens at high deformation rate, for the purpose of problem elaboration, this process can be observed as an idealized case within the closed system, isolated in relation to the environment. This means that the process takes place at approximately adiabatic conditions.

The expression for the temperature growth calculation during the drawing at high speed (for the conditions of the adiabatic process) has been given as following [4]:

$$\Delta t = \frac{\alpha \cdot x_1 \cdot F_1 \cdot h_1 \cdot 10^7}{V_0 \cdot \gamma \cdot c \cdot 4{,}184 \cdot 427} \quad (1)$$

Where:

$F_1 = d_1 \cdot \pi \cdot s \cdot k_1 \ [N]$ is the drawing force;

$V_0 \ [mm^3]$ is the work piece volume;

$h_1 \approx \frac{D_0^{\ 2} - d_1^{\ 2}}{4d_1}$ is the drawing depth $[mm]$;

$\alpha = 0{,}85 \div 0{,}90$ is the coefficient

$x_1 = 0{,}70 \div 0{,}85$ is the diagram fullness coefficient (F,h);

γ is the specific weight $[kg / dm^3]$ and

c is the specific heat $[J / kgK]$.

When reducing this expression (1), we obtain the following expression for the calculation of temperature growth during the deep drawing process:

$$\Delta t = \frac{\alpha \cdot x_1 \cdot \sigma_1 \cdot 10^7}{\gamma \cdot c \cdot 427 \cdot 4{,}184} \quad (2)$$

where:

$\sigma_1 [N / mm^2]$ represents the drawing voltage.

After including the recommended values for the heat transfer coefficient - $\alpha = 0{,}85$ and the diagram fullness coefficient we obtain:

$$\Delta t = \frac{3806 \cdot \sigma_1}{\gamma \cdot c} \quad (3)$$

where:

$\sigma_1 = \varphi_1 \cdot \sigma_m \left(\frac{1}{m_1} - 1 \right)$ is the deep drawing voltage.

for the relation $\frac{D_0}{s_0} \times 100 = 100 \div 200 \Rightarrow \varphi_1 \approx 1{,}6$

For the deep drawing process with the reduction of the wall thickness, the specific deformation stress of the shape change is calculated by the expression:

$$a_1 = k_1 \cdot \varphi_1 \tag{4}$$

where:

k_1 - specific deformation resistance of the material for the determined deformation stress in and

$[N/mm^2]$- value of logarithmic deformation.

φ_1 - value of logarithmic deformation.

Bearing in mind the (4), after the drawing process with the reduction of the wall thickness the temperature value is:

$$t = t_0 + \Delta t = 20^\circ + \frac{3806 \cdot k_1 \cdot \varphi_1}{\gamma \cdot c} \tag{5}$$

When it comes to deep drawing process without the reduction of the wall thickness, the calculation of temperature after the process is obtained by the following expression:

$$t = t_0 + \Delta t = 20^\circ + \frac{3806 \cdot \sigma_m \cdot \left(\frac{1}{m_1} - 1 \right)}{\gamma \cdot c} \tag{6}$$

Where m_1 stands for the drawing relation in the first operation.

Calculated temperature values according to the expression (6) for the deep drawing process without the reduction of the wall thickness for different materials have been given in the Table 1.

Material	Value of the drawing relation in the first operation			
	0,6	0,55	0,50	0,45
Low carbon steel	60	69	80	93
Stainless steel	94	112	132	156
Aluminium	38	42	47	53
Duralumin	65	76	88	103

Table 1. Calculated temperatures during the cold drawing process (adiabatic process)

3. Determining the energy of the intermediate form during the deformation process

The heat being released during the process causes the temperature growth of the work piece, which results in the reduction of its hardness, as well as in the reduced deformation resistance. The deformation process takes place at high deformation rates, leading to the softening of the workpiece material and increasing of its plasticity, as well as the reduction of the energy needed for deformation. According to the [5] during the deformation process around 80 – 90 % of mechanic energy spent for the deformation process is released in the metal in terms of heat, wheras the rest of 20 – 10 % remains in the metal in the form of residual stress. At higher deformation rates, the heat loss released in form of radiation to the environment is negligible. The energy conversion can be tracked if defining the thermo-dynamic system. Every thermo-dynamic system or his elements are characterized by a certain energy state. If we adopt the following as the system elements: the puller, work pieces, drawing rings and coolant, and if we adopt the supposition that the drawing process is seen as an idealized case within a closed system, isolated in relation to the environment it is possible to research, theoretically and experimentally, the energy of intermediate form and to determine the amount of heat and deformation stress during the deep drawing process with the reduction of the wall thickness. Exchanged amount of heat during the process from the initial state „l" (before drawing) to the final state „2" for the observed thermo-dynamic system is calculated according to the following expression [6]:

$$Q_{1,2} = m_i \cdot c_i \cdot (T_{i2} - T_{i1}) + N_i \cdot m \cdot c \cdot (T_2 - T_1) + + m_p \cdot c_p \cdot (T_{p2} - T_{p1}) + m_{rs} \cdot c_{rs} \cdot (T_{rs2} - T_{rs1}) \quad [J] \tag{7}$$

where:
m_i is the puller mass,
c_i is the specific heat of the puller's material for Tref=1/2(Ti1+Ti2),
T_{i1} is the puller's temperature in the state 1,
T_{i2} is the puller's temperature in the state 2,
N_i is number of work pieces drawn from the state 1 to state 2,
m is the mass of the work pieces,
c is the specific heat of the work piece material for Tref=1/2(To1+To2),
T_1 is the temperature of the work pieces in the state 1,
T_2 is the temperature of the work pieces in the state 2,
m_p is the mass of the drawing rings,
c_p is the specific heat of the ring materials for Tref=1/2(Tp1+Tp2),
T_{p1} is the rings temperature in the state 1,
T_{p2} is the rings temperature in the state 2,
m_{rs} is the coolant mass,
c_{rs} is the specific heat of the coolant for Tref=1/2(Trs1+Trs2),
T_{rs1} is the temperature of the coolant in the state 1,
T_{rs2} is the temperature of the coolant in the state 2.

Mechanical work (W) is another form of intermediate energy and it can be defined as a scale product of the force vector and the motion vector. If Fl= Fl (l), the mechanic work is calculated according to the following:

$$W_{1,2} = \int_1^2 \vec{F}(l)d\vec{l} = \int_1^2 F(l)\cos\varphi dl = \int_1^2 F_l(l)dl \quad [J] \tag{8}$$

In order to determine the deformation stress during the drawing process, we need to integrate the diagrams of the dependence of the deformation force on the drawing path and to multiply the given value with the number of the drawn work pieces between the referential states „1" and „2".

By comparing the heat value ($Q_{1,2}$) and the deformation stress ($W_{1,2}$) we get the answer on the percentual relation of the energy released for heat and the one spent during the deformation process.

The expression (7) enables the calculation of the needed amount of the coolant during the drawing process [7]:

$$m_{rs} = \frac{Q_{12} - m_i c_i (T_{i2} - T_{i1}) - N_i mc(T_2 - T_1) - m_p c_p (T_{p2} - T_{p1})}{c_{rs}(T_{rs2} - T_{rs1})} \tag{9}$$

By determing quantity of the revised heat and work of plastic deformation in deep drawing process with reduction of wall thickness it is possible to determine the amount of coolant necessary to ensure a stable drawing process.

4. Results of the experimental research of the intermediate form of energy during the drawing process

In order to determine the amount of energy released in the form of heat during the process of deep drawing with the reduction of the wall thickness, by an experimental research, it is necessary to measure and register the deformation force and to measure the temperature of all the system elements before and after the drawing process. The deformation force is determined by the usage of the tensometer device „Spider 8" with the appropriate software. Measuring chain sketch for registration of drawing force on multi-stage tools in given in Fig. 2.

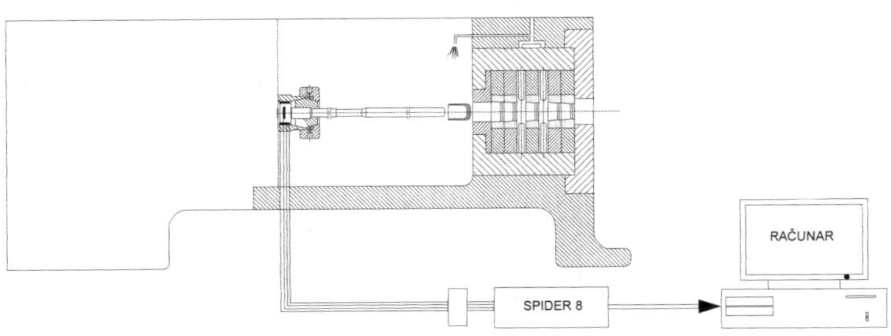

Fig. 2. Sketch of measuring chain in production condition

Using a measuring chain were obtained addiction diagrams of deformation force and draw stroke. Typical diagram from one measurement is given in Fig. 3.

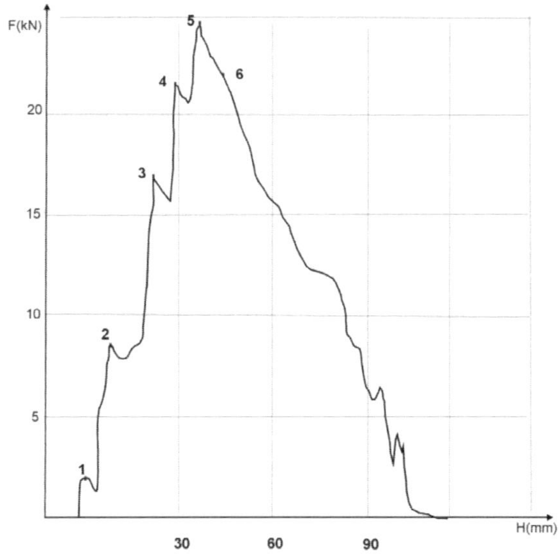

Fig. 3. Diagram of drawing force on multi-stage tool

By integrating the diagram of the dependence of the deformation force on the drawing path we obtain the value of the deformation stress or the energy spent on deformation. Registering of the temperatures before and after the drawing was done with laser measuring device and with thermography. Temperature measurement results before and after drawing are given in Table 2.

System elements temperature values before drawing						
No.	*Puller temperature T_{i1} (℃) Measuring spots*			*Blank temperature T_1(℃)*	*Rings temperature T_{p1}(℃)*	*Coolant temperature T_{rs1}(℃)*
	1	*2*	*3*			
1	*20*	*21*	*21*	*22*	*21*	*29*
2	*22*	*22*	*22*	*22*	*21*	*33*
3	*22*	*21*	*22*	*22*	*21*	*33*
System elements temperature values after drawing						
1	*40*	*35*	*34*	*39*	*29*	*35*
2	*40*	*35*	*34*	*40*	*30*	*35*
3	*40*	*39*	*35*	*39*	*30*	*36*

Table 2. System elements temperature values before and after drawing

Thermography of temperature fields on blanks and on drawing rings is given in Fig. 4.

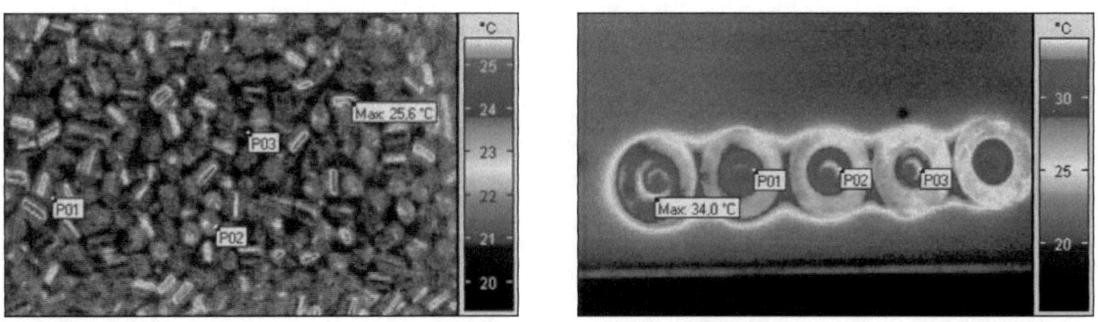

Pmax=25,6°C, P01=20,8°C, P02=21,7°C, P03=19,8°C   Pmax=34,0°C, P01=30,1°C, P02=30,6°C, P03=30,8°C

Fig. 4. Temperature fields on blanks and drawing rings

For the calculation of the exchanged heat amount between the two reference states of the observed system, the average values of the registered temperatures of the system elements, their masses and specific heats were used [8]:

- state "1"
  T_{i1}=21°C; T_1=22°C; T_{p1}=21°C; T_{rs1}=32°C
- state "2"
  T_{i2}=37°C; T_2=39°C; T_{p2}=30°C; T_{rs2}=35°C
- puller mass m_i = 0.127 kg
- work piece mass m= 0.0072 kg
- mass of the rings (5 rings) m_p=0.473 kg
- coolant mass m_{rs}= 69.87 kg
- specific heat of the puller material (steel) and temperature 20°C c_i=0.461 kJ/kg K
- specific heat of the work piece material (brass) and T_{ref}=31°C c=0.382 kJ/kg K
- specific heat of the material of rings (steel) and temperature 20°C c_p=0.461 kJ/kg K
- specific heat of the coolant (water) and T_{ref}=33.5°C c_{rs}=4.176 kJ/kg K
- number of work pieces N_i=1920.

Specific heat values of the puller materials and rings (steel) have been adopted as for the temperature of 20°C, due to the fact that charts on heats do not contain values for reference temperatures. By implementing the listed data into the expression (7), we get the value of the exchanged amount of heat for the observed system from state "1" to state "2" $Q_{1,2}$=967.79 kJ.

The value of deformation stress $W_{1,2}$ is obtained when multiplying the deformation stress for one work piece (W=0.638 kNm=0.638 kJ) with the amount of the drawn work pieces N_i=1920. The obtained value of the deformation stress is $W_{1,2}$=1224.96 kJ.

Measuring of the tool temperatures after the drawing process has shown that the puller has a slightly higher temperature (T_{i2}=37°C), and the temperature of the drawing rings is slightly lower (T_{p2}=30°C).

The expression (8) enables the determination of the coolant mass that allows a stabile drawing process. If the maximal temperature of the tool is restricted to T=40°C, in order to have a stabile drawing process, then it is, according to the expression (8), necessary to ensure that there is an amount of the coolant m_{rs}=69,67 kg circulating during the process.

5. Conclusion

Stability of drawing process with a reduction of wall thickness in production conditions at high speeds of drawing essentially depends on the correct certain amount of coolant and lubricant. In order to solve this problem experimental research of the energy transition forms (exchanged heat and work of plastic deformation) in the process of drawing with a reduction of wall thickness was done.

By recording drawing force on 1920 workpieces and simultaneous registration of the temperature on: workpieces, punch, puller, drawing rings and coolant, the results obtained, which allows calculation of mass of coolant that is required to calculate mass of coolant that which is required to limit the maximum temperature to T=40 ℃.

The calculated temperature value of the theoretical expression (6) in the drawing without reduction of the wall thickness, are shown in Table 1, indicate that in different materials occur high temperatures, which could impair the stability of the drawing process if the amount of coolant is not correctly determined.

Further researches should be expanded to other metal forming processes by deforming the blanks from various materials.

6. References

[1] Tschaetsch, Heinz: Metal Forming Practise Processes – Machines – Tools, Springer-Verlag Berlin Heidelberg, 2006. ISBN 978-3-540-33217-6

[2] G. Ambrogioa, L. Filice, G. Palumbo, S. Pinto: Prediction of formability extension in deep drawing when superimposing a thermal gradient, Journal of Materials Processing Technology 162–163 (2005) 454–460

[3] Đukić H., Nožić M. :„ Metal forming", Univerzitet „Džemal Bijedić" Mašinski fakultet Mostar, Mostar 2013

[4] L.A.Šofman :„ Theory and cold processes ", Mašinostroenie, Moskva, 1964

[5] Čaušević, M.: „ Metal plastic deformation", Svjetlost Sarajevo, 1979

[6] Nožić M., Đukić H. :„ Heat effect in the process of deep drawing with high speed ", MATRIB□12 International Conference MATERIALS,TRIBOLOGY, RECYCLING, 2012, Vela Luka, Croatia

[7] M.Nožić, H.Đukić, D.Denjo: „ Heat balance of drawing processes with a reduction of the wall thickness ", COMETa 2014, 2nd International Scientific Conference, decembar 2014, Jahorina, BiH

[8] Nožić, M. :„ Contribution to the study of individual and total loads on multistage tools for drawing with reduction of wall thickness", Doktorska disertacija, Mašinski fakultet, 2005

DOI: 10.2507/27th.daaam.proceedings.103

CONTROL SYSTEM FOR UNDERWATER VEHICLE WITH MULTILINK MANIPULATOR FOR AUTOMATIC MANIPULATION OPERATIONS

Vladimir Filaretov, Alexander Konoplin, Nikita Konoplin & Georgy Gorbachev

This Publication has to be referred as: Filaretov, V[ladimir] F[edorovich]; Konoplin, A[lexander]; Konoplin, N[ikita] & Gorbachev, G[eorgy] (2016). Control System for Underwater Vehicle With Multilink Manipulator for Automatic Manipulation Operations, Proceedings of the 27th DAAAM International Symposium, pp.0714-0720, B. Katalinic (Ed.), Published by DAAAM International, ISBN 978-3-902734-08-2, ISSN 1726-9679, Vienna, Austria
DOI: 10.2507/27th.daaam.proceedings.103

Abstract

This paper describes development and research of the new synthesis method of control system for autonomous and remotely controlled underwater vehicles and for multilink manipulators mounted on the vehicles. Such system admits to realize widely used underwater research manipulation operations in automatic mode. Some of them are: collecting bio-organisms, definition of composition and density of the soil with special probes and drills, taking precipitation samples with hermetically sealed soil tubes, measurements with thermistor sensors in different layers of sedimentary soil.

Keywords: underwater vehicles; multilink manipulators; Doppler lag; automatic mode; manipulation operations.

1. Introduction

The problem of automatic implementation of different manipulation operations types in the depths of the Global Ocean is very topical nowadays. Many researchers successfully solve problems of the underwater vehicles (UV) development. These UV are equipped with one or several multilink underwater manipulators (UM) [1-3]. Such systems are designed to carry out many technological and research operations. Field of its application is significantly expanding by means of realization of new kinds of operations.

The manipulator-equipped UV are the mostly used for research operations in hydrogeology, hydrobiology, geophysics, hydrochemistry, in oil and gaze extraction. The next operations are carried out in these cases: selective sampling of soil probes and geological rocks; collecting bio-organisms; set-up and servicing of different tools; definition of composition and density of the soil using special probes and drills; taking precipitation samples by means of hermetically sealed soil tubes; measurements with thermistor sensors in different layers of sedimentary soil etc.

In most cases, specially trained operators implement these operations in manual mode. However, it is very complicated to solve such tasks quickly and accurately without direct contact with work object (WO) and with knowing its location from the video picture only. Furthermore, there are only known single cases of using the autonomic UV for the simplest manipulation operations in automatic mode [3].

Nowadays, control systems for navigation and automatic movement of UV have been already developed and investigated [4, 5]. Methods and approaches that guarantee high precision holding of the UV near WO have been also developed [6-9]. It allows fixing the UV in the determined spatial point with desired spatial orientation and without grounding and roiling of near-bottom layers. Moreover, described in the paper [10] correction method of UM movement trajectories allows to carry out the manipulation operations even in conditions of unplanned displacements of the UV relative to WO.

The most important part of solving of manipulation tasks in automatic mode is the definition of the location and spatial orientation of the WO relative to the UV. In addition, it is the formation of special motion modes of UM and the UV, at which the task is to be solved with prescribed accuracy and without emergencies. Paper [3] offers the special markers fixation on WO that allows to define location of WO and its orientation relative to the UV. Also video picture processing algorithms are used in this paper to carry out the special operations of carbine capture of WO with previously known form and colour. In paper [11] the processing algorithms of video streams from several cameras are used in cases of complicated forms of WO. Previously installed light markers are also used in this paper.

However, it is impossible to install markers near the seabed surface during the implementation of underwater research operations. Form and colour of this surface do not allow to use video processing algorithms effectively. As a result, features and disadvantages of known approaches do not allow to automatize the implementation of widely used operations in the depths of the Global Ocean.

Creating of the effective approaches for solution of stated manipulation tasks in automatic mode allows to use the autonomic UV equipped with UM with expanding of vehicles functionality and reducing of the underwater works cost. In addition, automatization of UM allows to improve quality and increase the speed of operations, performing by the remote operated UV. The operator's psychological load, obviously, decreases.

2. Formulation of the problem

The task of development of the new approach for automatic performing of research manipulation operations by the UV equipped with UM near the seabed surface is set. Location of WO or seabed surface relative to the UV should be defined using onboard sensors during the realization of this approach. Also the UV movement control for the most convenient position assignment for performing of operation is used. Movement trajectories of working instruments of multilink UM should be formed considering the boundaries of the UM workspace and the information about model of WO. Along these trajectories, UM should affect the least possible force and torque effects on the UV.

3. Definition of the UV and seabed surface interposition

It is necessary to determine the location and spatial orientation of WO in real time scale to solve the manipulation tasks in automatic mode during entering of UV into the workspace. During implementation of research underwater operation, seabed surface serves as the WO. By deep-sea researches in hydrogeology, hydrobiology, geophysics and hydrochemistry using the UV in Sea of Japan, Sea of Okhotsk, Bering Sea, Chukchi Sea and the Pacific Ocean, it has been shown that in most cases seabed surface has only small roughness. That is why research manipulation operations, which are related to sampling of soil probes of sediment layers, are implementing in conditions of bottom low relief. In addition, seabed surface with sufficiently low relief can be tilted about the horizontal plane. This complicates entering of the UV into the workspace and implementation of manipulation tasks.

In condition of the UV nearness to seabed surface, this surface can be approximately described as a plane and its spatial orientation relative to the UV can be determined.

Advanced absolute Doppler lags allow high-precision determination of distance from log to bottom along axes of its hydro acoustic antennas, which usually not less than three. The obtained measurements are translated into right rectangular coordinate system (CS) that rigidly connected with the body of the UV. The origin of coordinates is located in the UV centre of buoyancy C, X^* axis coincides with the roll axis of the UV, Z^* axis coincides with the vertical axis of the UV, Y^* axis is right-handed. As a result, in CS $X^*Y^*Z^*$ we have an array of points, which belong to seabed surface.

It is necessary to approximate the obtained array of points by linear polynomial using least square method to derive the equation of plane. In the general case, equation of plane is written as next one [12]:

$$Ax + By + Cz + D = 0\ , \qquad (1)$$

where A, B, C, D are the constant coefficients, and A, B, C are not equal to zero simultaneously. After transformations, (1) is given by:

$$z = A^* x + B^* y + C^*\ , \qquad (2)$$

where $A^* = -\frac{A}{C}$, $B^* = -\frac{B}{C}$, $C^* = -\frac{D}{C}$.

According to least square method, it is necessary to determine parameters A^*, B^*, C^* such that the next conditions are implementing:

$$\sum_{i=1}^{n} e_i^2 = \sum_{i=1}^{n} [z_i - z(x_i, y_i)]^2 \rightarrow \min, \tag{3}$$

where x_i, y_i and z_i are the coordinates of points in the array of measurements; $z(x_i, y_i) = A^* x_i + B^* y_i + C^*$; $\sum_{i=1}^{n} e_i^2$ is the sum of squared deviations; $i = \overline{(1, n)}$, where n is the number of points in the array of measurements. To solve the problem of function (3) minimizing it is necessary to find its critical points, differentiating it with respect to unknown parameters A^*, B^*, C^*, setting derivatives to zero and solving obtained system of equations:

$$\begin{cases} \dfrac{\partial \sum_{i=1}^{n} e_i^2}{\partial A^*} = \dfrac{\partial \sum_{i=1}^{n} [z_i - A^* x_i - B^* y_i - C^*]^2}{\partial A^*} = 0, \\ \dfrac{\partial \sum_{i=1}^{n} e_i^2}{\partial B^*} = \dfrac{\partial \sum_{i=1}^{n} [z_i - A^* x_i - B^* y_i - C^*]^2}{\partial B^*} = 0, \\ \dfrac{\partial \sum_{i=1}^{n} e_i^2}{\partial C^*} = \dfrac{\partial \sum_{i=1}^{n} [z_i - A^* x_i - B^* y_i - C^*]^2}{\partial C^*} = 0. \end{cases} \tag{4}$$

The modified (4) is given by:

$$\begin{cases} A^* \sum_{i=1}^{n} x_i^2 + B^* \sum_{i=1}^{n} x_i y_i + C^* \sum_{i=1}^{n} x_i = \sum_{i=1}^{n} z_i x_i, \\ A^* \sum_{i=1}^{n} x_i y_i + B^* \sum_{i=1}^{n} y_i^2 + C^* \sum_{i=1}^{n} y_i = \sum_{i=1}^{n} z_i y_i, \\ A^* \sum_{i=1}^{n} x_i + B^* \sum_{i=1}^{n} y_i + nC^* = \sum_{i=1}^{n} z_i. \end{cases} \tag{5}$$

Solving the system using Cramer's rule [13], we find unknown A^*, B^*, C^*:

$$\Delta = \begin{vmatrix} \sum_{i=1}^{n} x_i^2 & \sum_{i=1}^{n} x_i y_i & \sum_{i=1}^{n} x_i \\ \sum_{i=1}^{n} x_i y_i & \sum_{i=1}^{n} y_i^2 & \sum_{i=1}^{n} y_i \\ \sum_{i=1}^{n} x_i & \sum_{i=1}^{n} y_i & n \end{vmatrix} \neq 0, \quad \Delta_1 = \begin{vmatrix} \sum_{i=1}^{n} z_i x_i & \sum_{i=1}^{n} x_i y_i & \sum_{i=1}^{n} x_i \\ \sum_{i=1}^{n} z_i y_i & \sum_{i=1}^{n} y_i^2 & \sum_{i=1}^{n} y_i \\ \sum_{i=1}^{n} z_i & \sum_{i=1}^{n} y_i & n \end{vmatrix} \neq 0,$$

$$\Delta_2 = \begin{vmatrix} \sum_{i=1}^{n} x_i^2 & \sum_{i=1}^{n} z_i x_i & \sum_{i=1}^{n} x_i \\ \sum_{i=1}^{n} x_i y_i & \sum_{i=1}^{n} z_i y_i & \sum_{i=1}^{n} y_i \\ \sum_{i=1}^{n} x_i & \sum_{i=1}^{n} z_i & n \end{vmatrix} \neq 0, \quad \Delta_3 = \begin{vmatrix} \sum_{i=1}^{n} x_i^2 & \sum_{i=1}^{n} x_i y_i & \sum_{i=1}^{n} z_i x_i \\ \sum_{i=1}^{n} x_i y_i & \sum_{i=1}^{n} y_i^2 & \sum_{i=1}^{n} z_i y_i \\ \sum_{i=1}^{n} x_i & \sum_{i=1}^{n} y_i & \sum_{i=1}^{n} z_i \end{vmatrix} \neq 0, \tag{6}$$

$$A^* = \frac{\Delta_1}{\Delta}, \; B^* = \frac{\Delta_2}{\Delta}, \; C^* = \frac{\Delta_3}{\Delta}.$$

As a result, (2) with parameters from (6) will be equation of plane, which characterizes the spatial orientation of seabed surface near the UV.

4. Approach of the UV with predetermined spatial orientation to WO

During automatic entering of the UV into workspace it is necessary that roll axis X^* of the UV is directing to the rise of the plane, because workspaces of UM, which are installed on the UV, are located in the front semisphere of these vehicles. Also sensors, cameras, lights and other equipment are concentrated at the front of the UV. For the UV attitude control it is necessary to control the UV roll and trim angles. But it will cause additional energy demands because stability of the well-ballasted UV tends to reset roll and trim angles to zero.

Besides, propulsion system of many UV of survey and working classes does not allow to control roll and trim angles, that is why such UV should be directed to the rise of seabed surface. It lets to carry out survey, searching and manipulation operations in the condition of bottom rugged relief.

It should be mentioned that UV direction to the rise of seabed surface allows to see the bottom in time with preventing of emergencies. Also this direction allows to move the UV up and down inclined with permanent visual contact with seabed surface.

To satisfy a mentioned condition it is necessary to direct the UV along the vector $\bar{n}^*$ direction. This vector is the projection of the plane normal vector $\bar{n} = [A, B, C]^T \in R^3$, which is described by (2), on plane CX^*Y^* of CS $X^*Y^*Z^*$ (see Fig. 1).

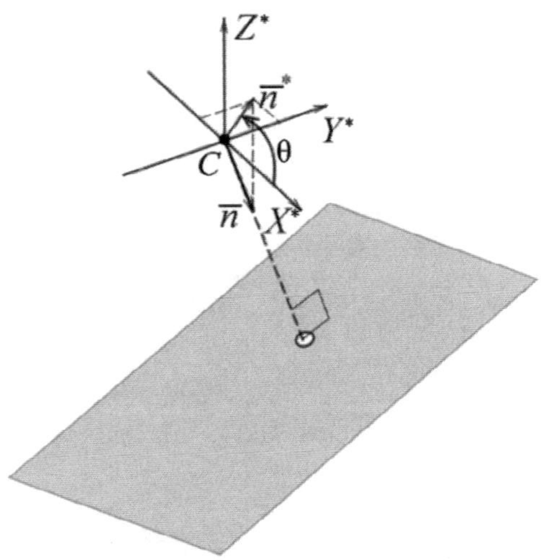

Fig. 1. Determination of the UV orientation

Using the vector $\bar{n}^* = [A, B, 0]^T \in R^3$ and the vector of roll axis of the UV $x^* = [1,0,0] \in R^3$, desired angle θ can be defined by the next equations:

$$\theta = \arccos\left(\frac{\bar{n}^* x^*}{\left|\bar{n}^*\right|\left|x^*\right|}\right) \text{ if } B>0;$$

$$\theta = -\arccos\left(\frac{\bar{n}^* x^*}{\left|\bar{n}^*\right|\left|x^*\right|}\right) \text{ if } B<0;$$

It is obvious that after rotation by angle θ relative to Z^* axis of CS $X^*Y^*Z^*$ roll axis Y^* of UV will be directed to the rise of seabed surface.

Doppler lags allow to update the array of points of seabed surface in CS $X^*Y^*Z^*$ several times per second. Therefore the plane equation (2) of seabed surface for current Doppler lag measurements should be defined during process of the UV movement and the UV stabilization in hang mode. It allows to use actual information about location and orientation of seabed surface for control of the UV and the UM movements.

5. UV trajectory formation

We consider approach for formation of UM working tool movement trajectory by the example of implementation of widely used underwater research operation of taking precipitation samples with hermetically sealed soil tubes. Sampler that configured as a tube with a reverse valve at the upper end is fixed in UM working tool for the sampling of a soil. Process of the soil sampling consists in immersing of tube perpendicular to seabed surface at least at 2/3 of length into soil, and then sampler is getting out and move to the transportation position.

The method for calculation of UM working tool movement trajectory for implementation of determined manipulation operation is further described.

For simplification of calculation of interposition of the UM and calculated seabed plane (2), workspace of UM is proposed to be represented as finite numbers s of segments inserted in it. These segments are parallelepiped-shape with equal bases. Vertical axis of described segments are parallel to the axis Z^* of CS $X^*Y^*Z^*$.

Divided into segments workspace of UM is shown at Fig. 2. In this figure the next labels are introduced: 1 is the seabed surface, 2 is the plane approximating the seabed surface, 3 is the approach point P_a of the UM working tool, 4 is the sampler, 5 is the point C_a of intersection of the segment vertical axis with the seabed surface 2. Number of segments is chosen because of workspace size, measures of sampler and required accuracy of manipulation operation implementation.

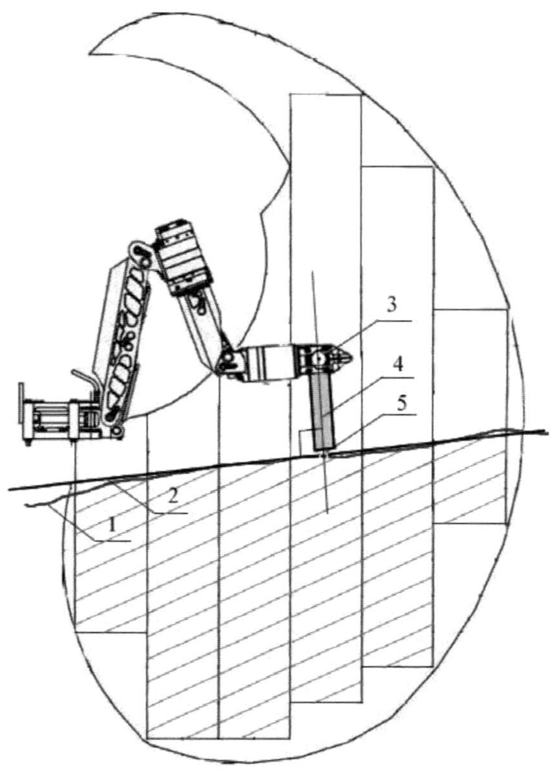

Fig. 2. Scheme of manipulation operation

For every segment $a = \overline{(1, s)}$ coordinates x_{ca} and y_{ca} of its vertical axis and X^*Y^* plane crossing are known. The point C_a is a crossing of vertical axis of segment a and seabed surface plane. Using (2), let's find coordinate z_{ca} of crossing point C_a using the formula:

$$z_{ca} = z(x_{ca}, y_{ca}) = A^* x_{ca} + B^* y_{ca} + C^*.$$

For the gained point C_a with coordinates (x_{ca}, y_{ca}, z_{ca}) let's find the approach point P_a on the perpendicular to plane (2) using the formula:

$$P_a = C_a - \frac{\overline{n}}{\|\overline{n}\|} l,$$

where l is a length of a sampler. Every calculated point P_a is checked for belonging to UM workspace.

For minimization of force and torque effects on UV from moving UM, lever from working tool to UV center of mass should have minimal length during the immersion of a tube into the soil. For that, from all belonging to UM workspace points P_a one point with the closest to point C of CS $X^*Y^*Z^*$ distance is chosen.

As a result, a trajectory of UM movement could be represented as a sequence of working tool transition from initial position to approach point P_a and then to C_a, in which sampler has the deepest soil immersing. After that the working tool firstly moves through P_a and then returns to initial position.

It should be mentioned that in case of no points C_a in workspace of UM, UV should keep immersion, and in case of no points P_a in workspace, UV should perform a re-entry into the work area.

6. Conclusion

In this paper the synthesis method of control system for automatic implementation of underwater research manipulation operations by means of UV with UM was developed. Proposed system controls linear and axial movements of the UV to define its best position for manipulation operations. Also this system forms spatial trajectories of the multilink UM working tools taking into account the information about continuously refined spatial location of the bottom surface. These trajectories are formed such that the UM acts the least possible force and torque effects on the UV during the process of operation, taking into account the borders of UM workspace.

The control system for remotely controlled UV and multilink UM mounted on the vehicle has been developed based on the proposed method. Sea tests in deep-sea expedition in Bering Sea with UV Sub-Atlantic Comanche 18 equipped with UM Schilling Orion 7P have been completed for research of operability and features of functioning of synthesized control system. Used UV and UM are shown on Fig. 3. In automatic mode this system defined the location of the bottom surface relative to the UV by means of the onboard Doppler lag. Implemented in software control system improved accuracy and speed of implementation of underwater research operations. Results of experiments approved efficiency and simplicity of practical realization of developed control system of UV and UM for implementation of manipulation operations in automatic mode.

Fig. 3. Sea tests with UV Comanche 18

7. Acknowledgments

This work was performed in Institute for Automation and Control Processes Far Eastern Branch of the Russian Academy of Sciences and also in Far Eastern Federal University (Vladivostok, Russia). This paper was supported by Russian Foundation for Basic Research (RFBR) grants 16-29-04195 ofi_m and 16-38-00488 mol_a. Experimental studies were performed in expedition financed by Russian Science Foundation (RSF) grant 14-14-00232.

8. References

[1] Hamed Farivarnejad, S. Ali A. Moosavian, "Multiple impedance control for object manipulation by a dual arm underwater vehicle–manipulator system" Ocean Engineering 89, pp. 82–98, 2014.

[2] E. Simetti, G. Casalino, "Whole body control of a dual arm underwater vehicle manipulator system" Annual Reviews in Control 40, pp. 191–200, 2015.

[3] G. Marani, S.K. Choi, and J. Yuh, "Underwater autonomous manipulation for intervention missions AUVs," Ocean Engineering, vol. 36, no. 1, pp. 15 – 23, 2009.

[4] A. Scherbatyuk, F. Dubrovin, "Development of Algorithms for an Autonomous Underwater Vehicle Navigation with a Single Mobile Beacon: The Results of Simulations and Marine Trials" Proc. of the XXIIth International Conference on Integrated Navigation Systems, ICINS 2015, Saint Petersburg, Russia. P. 144-152. ISBN 978–5-91995-023-3.

[5] A. Boreyko, A. Scherbatyuk, S. Moun, "Precise UUV positioning based on images processing for underwater construction inspection" Proceedings of the ISOPE Pacific/Asia Offshore Mechanics Symposium, 2008, Bangkok, Thailand. PP. 14 – 20.

[6] V.F. Filaretov, A.Yu Konoplin, "System of Automatic Stabilization of Underwater Vehicle in Hang Mode with Working Multilink Manipulator" 2015 International Conference on Computer, Control, Informatics and Its Applications (IC3INA), Bandung, Indonesia, International IEEE Conference, pp. 132-137, 2015.

[7] T.W. McLain, S.M. Rock, M.J. Lee, "Experiments in the coordinated control of an underwater arm/vehicle system", Autonomous Robots, vol. 3, 1996. no. 2-3, pp. 213–232.

[8] Santhakumar Mohan, "Investigation into the Dynamics and Control of an Underwater Vehicle-Manipulator System" Modelling and Simulation in Engineering. Vol. 2013, Article ID 839046, 13 pages, http://dx.doi.org/10.1155/2013/839046.

[9] V.F. Filaretov, A.Yu. Konoplin, "Experimental Definition of the Viscous Friction Coefficients for Moving Links of Multilink Underwater Manipulator", Proceedings of the 26th DAAAM International Symposium, pp.0762-0767, B. Katalinic (Ed.), Published by DAAAM International, ISBN 978-3-902734-07-5, ISSN 1726-9679, Vienna, Austria, DOI:10.2507/26th.daaam.proceedings.106.

[10] V.F. Filaretov, A.Yu. Konoplin, "System of Automatically Correction of Program Trajectory of Motion of Multilink Manipulator Installed on Underwater Vehicle" Procedia Engineering, Vol. 100, 2015. pp. 1441-1449.

[11] R. O. Faria, F. Kucharczak, G. M. Freitas, A. C. Leite, F. Lizarralde, M. Galassi, P. J. From "A Methodology for Autonomous Robotic Manipulation of Valves Using Visual Sensing" 2nd IFAC Workshop on Automatic Control in Offshore Oil and Gas Production OOGP 2015 – Florianópolis, Brazil, 27–29 May 2015, IFAC-Papers OnLine, Vol. 48, Issue 6, 2015, pp. 221-228.

[12] Yu. V. Linnik, " The method of least squares, and the foundations of mathematics and statistical theory of processing of observations" Vol. 2, 1962, 349 p. (Линник, Ю.В. Метод наименьших квадратов и основы математико-статистической теории обработки наблюдений. - Изд. 2-е, доп. и испр. - М. : Гос. изд-во физ.-мат. лит., 1962. - 349 с.)

[13] A. I. Maltsev, "Fundamentals of linear algebra" Vol. 3, 1970, 400 p. (Мальцев А. И. Основы линейной алгебры. — Изд. 3-е, перераб., М.: «Наука», 1970. — 400 с.)

DOI: 10.2507/27th.daaam.proceedings.104

INTER-ORGANIZATIONAL COLLABORATION FOR INNOVATION IN MANUFACTURING FIRMS

Bojan Lalić, Nemanja Tasić, Uglješa Marjanović, Milan Delić &Nela Cvetković

This Publication has to be referred as: Lalic, B[ojan]; Tasic, N[emanja]; Marjanovic, U[gljesa]; Delic, M[ilan] & Cvetkovic, N[ela] (2016). Inter-Organizational Collaboration for Innovation in Manufacturing Firms, Proceedings of the 27th DAAAM International Symposium, pp.0721-0729, B. Katalinic (Ed.), Published by DAAAM International, ISBN 978-3-902734-08-2, ISSN 1726-9679, Vienna, Austria
DOI: 10.2507/27th.daaam.proceedings.104

Abstract

From day to day, taking part in the collaboration process in different business aspects, especially in the innovation process continues to gain in importance. Innovation enables companies to strive for a more powerful market position, while inter-organizational cooperation is perceived as an efficient means of enhancement of firm innovativeness.
The purpose of this paper is to present the state of inter-organizational collaboration for innovation among manufacturing companies in Vojvodina, analyzing cooperation among manufacturing firms with different external parties, in different innovation areas. Furthermore, the aim is to examine the effects of inter-organizational collaboration on the innovativeness of manufacturing firms in Vojvodina.
The data used in this paper was gathered within the international project European Manufacturing Survey.
The results are presented using descriptive statistics and they demonstrate that the existence and frequency of inter-organizational collaboration, although quite present, depends on external parties as potential cooperation partners and on different innovation areas in which collaboration could occur. Also, the research shows there is a notable relation between inter-organizational collaboration and firm innovativeness.

Keywords: Inter-organizational collaboration; Cooperation; Manufacturing firms; Innovation; Innovativeness.

1. Introduction

In the present constant and dynamic changes in economic and technological conditions, inter-organizational collaboration appears as the industrial response [1]. In order to neutralize and cope with competition intensity and market heterogeneity, some companies decide to engage in joint cooperation with other firms and different organizations [2].

Various potential participants have different strategic objectives to take part in certain aspects of collaboration. Higher level of motivation in certain companies might be due to their perception of their capabilities and resources to perform various activities on their own. More specific, large companies can rely on their own resources, while small firms tend to acquire the necessary resources from the environment. However, albeit large organizations may have necessary resources,

SMEs are often a fruitful source of innovative ideas and fresh perspectives. In practice, collaboration has shown many benefits for companies, especially in terms of innovation capacity building. However, not all organizations recognize the benefits of inter-organizational collaboration and cooperation, although the percentage of companies seeing collaboration as desirable is relatively high. This can be explained by varying organizational cultures, environmental climates and similar, influencing those differences in the understanding of the collaboration concept by different parties.

We do not have a clear understanding of how inter-organizational collaboration affects innovation in small regions such as Autonomous Province of Vojvodina. Thus, this paper deals with the collaboration concept in manufacturing companies in the territory of Vojvodina. Related to that, three research questions are defined:

- The first research question of this paper is to present the level of collaboration of manufacturing firms with other different external parties, in different areas of innovation.
- The second research target is to investigate the perception of manufacturing firms regarding the importance of different innovation areas.
- The last research target is to examine the connection between inter-firm cooperation and firm innovativeness, more precisely, to analyze whether more intensive inter-organizational cooperation in new product development activities, as well as in R&D activities, influences the higher firm innovativeness in terms of innovative products.

For this purpose, we adopted Danneels' [3] explanation of "firm innovativeness as the outcome of the innovation process, defined as the capacity of the firm to develop and introduce a new product or service" and Dietz & Becker's [4] view on innovation output of firms as indicated by new products. Although innovativeness could be also observed from a wider perspective (including new processes, technological concepts etc.), in this paper we used the previously described definition to investigate the existing connections between collaboration practice and innovativeness. The assumption was that a greater degree of collaboration in new product development and R&D activities results in greater innovative development capacity, specifically, new products introduction success.

The data used were gathered through the European Manufacturing Survey (EMS), covering the results from 2015 in the territory of Vojvodina where the Faculty of Technical Sciences in Novi Sad had carried out the EMS survey. The results are presented using descriptive statistics. To the extent of our knowledge, so far there are no recent studies directly addressing this perspective on the topic in Vojvodina. The results show that there is a high degree of collaboration between manufacturing companies in Vojvodina and external parties. Nonetheless, the results demonstrate that manufacturing companies collaborating more often with other organizations (companies and research institutions) in product innovation activities or R&D have introduced a noticeably higher percentage of new products to the company and new products to the market than the companies not engaged in inter-organizational collaboration.

The paper is structured as follows. The theoretical part briefly introduces the concept of collaboration and innovation. The methodological section explains the concept of EMS and describes the process of data gathering. The empirical section analyses the level and forms of collaboration of manufacturing firms with external parties and potential connection between inter-firm cooperation and firm innovativeness. Finally, the conclusion summarizes the findings with certain implications.

2. Collaboration and Innovation

In their paper [5] the authors define the concept of inter-organizational collaboration as a feature of the innovation process referring to the extent to which other organizations form an important part in the innovation process, where organizations can stand for firms or for institutions.

According to [6], innovation represents 'the implementation of new or significantly improved product, process, new marketing method, or a new organizational method in business practice, workplace organization or external relation'.

Nowadays, as the highest priority in achieving competitive advantage is innovation [7], collaboration with partners, customers and competitors has become strategic imperative for companies [8].

So far, many incentives for inter-company and inter-organizational collaboration have been pointed out by different researchers and practitioners. In terms of positive effects on firm innovativeness, collaborative linkages enable all parts to exploit the benefits of knowledge sharing, to benefit from complementary skills merged from different firms [9], to take advantages of scale economies in the research process, cost sharing and risk spreading [10], [11], [12]. Furthermore, cooperation results in faster product/service development and faster speed to market [13], [14].

One of the main aims of any firm is to maximize profits by using and improving its resources and capabilities [15]. Consequently, knowledge flows and exchange are becoming essential priorities for firms. One of the most effective ways to accomplish this and improve the learning process is to cooperate with external partners [12].

SMEs are prompt for the establishment of cooperative relationships with other companies to achieve market strength, or to exploit new opportunities [12]. This could not be possible to that extent due to their limited resources, both human and financial. Although SMEs may be well innovative, they often do not have the commercial strength or professionalism required to successfully exploit their innovative ideas [16]. Related to that, the success of an organization is not guaranteed only by the ability to understand and generated advanced knowledge, but also to transfer it into marketable innovation as well [17]. On the other hand, large companies are less interested in cooperation for innovation, due to the fact they possess the needed technical and financial capability [11]. However, in practice, there is significant evidence that the global market leaders recognize the necessity to find cooperation partners to design an innovative value chain, combining their own core competencies with those of other leading firms [18]. Fostering mutual trust as a basis of cooperation, companies are able to exchange resources and key competences to achieve the goals unreachable when acting individually [19].

The decision to cooperate in innovation is significantly driven by the fact that cooperation is seen as an efficient way to enhance the probability of innovation project success [20], [4]. This is in direct relation to the belief that the implementation of additional external capacities raises the level of innovation output [4]. Nonetheless, analyses of the effect of cooperation demonstrate that, on average, cooperative firms have higher overall performance levels than non-cooperative ones [4] as well as higher R&D intensity [21].

3. Research methodology

The research methodology used was a survey research. Specifically, data was collected through The European Manufacturing Survey (EMS), coordinated by the Fraunhofer Institute for Systems and Innovation Research - ISI. EMS represents the largest European survey of manufacturing activities whose extensive questionnaire covers questions generally referring to organizational concepts, collaboration for innovation, cooperation issues, the application of advanced technologies and innovative concepts, project management, performance indicators and others. The main focus of the survey is the development and utilization of different types of innovations by companies in the manufacturing industries, and associated performance effects. In 2001, researchers at ISI took the initiative to make an international survey, aiming to gather internationally comparable data by means of a standardized method and core questionnaire that was carefully translated into each participating country's local language. Since then, the survey has been conducted every three years, constantly expanding the range of participating countries. So far, five multinational survey rounds have been carried out with about 3-year intervals while the sixth round is under way (2015/2016).

Our research was based on EMS data from the Serbian (territory of Vojvodina) subsample from the year 2015. The responding companies were from the manufacturing sector, having at least 20 employees. The total number of companies meeting the requirements in Vojvodina is 600. However, in order to ensure a representative sample of companies equally distributed by different industries and counties, 334 questionnaires had been sent, 123 of which were returned (36.8% response rate). Data gathering was performed by the Faculty of Technical Sciences in Novi Sad, firstly through the pre-test phase and two phases of mass distribution.

The information asked for and analyzed concerns the following:

- Existence of collaboration with external parties (suppliers, competitors, service organization and research centers and universities) in specific innovation areas (new products, new technical processes, new product related services and new organizational concepts).
- Cooperation in R&D with other organizations (suppliers, competitors, service organization and research centers and universities).
- New product development.
- Relevance of different innovation areas for manufacturing companies.

All questions were carefully translated from the original sample and adjusted to the specificities of the territory where the survey was conducted. An example of one question related to the main sources of innovative ideas for manufacturing companies in Vojvodina is presented in figure 1.

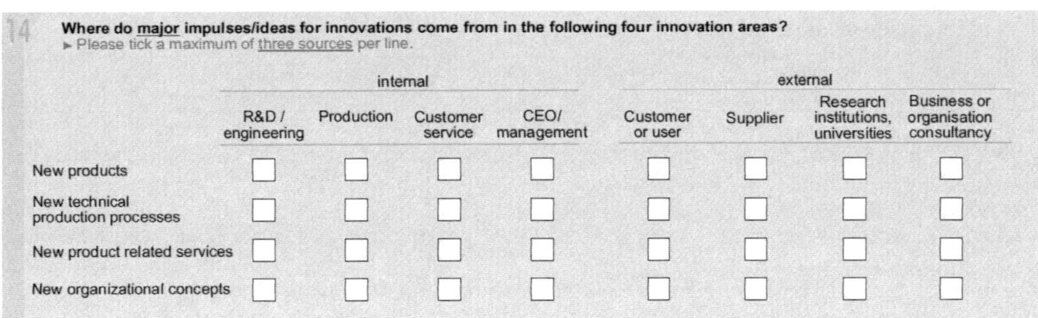
14 **Where do major impulses/ideas for innovations come from in the following four innovation areas?**
► Please tick a maximum of three sources per line.

	internal				external			
	R&D / engineering	Production	Customer service	CEO/ management	Customer or user	Supplier	Research institutions, universities	Business or organisation consultancy
New products	☐	☐	☐	☐	☐	☐	☐	☐
New technical production processes	☐	☐	☐	☐	☐	☐	☐	☐
New product related services	☐	☐	☐	☐	☐	☐	☐	☐
New organizational concepts	☐	☐	☐	☐	☐	☐	☐	☐

Fig.1. Question on innovation process in EMS 2015.

4. Results

For the purpose of testing research questions, descriptive statistics was conducted by IBM SPSS.

4.1. Inter-organizational cooperation in specific innovation areas

Firstly, the percentage of companies that collaborate with other organizations or parties in specific innovation areas is analyzed and presented. The defined areas of innovation in this survey are: new products, new technical production processes, new product related services and new organizational concepts. Following, figure 2 shows the percentage of manufacturing companies in Vojvodina according to the external party they cooperate with in specific innovation areas, those parties being: customers, suppliers, competitors, service organizations (business or organization consultancy) and research institutions and universities. Finally, the frequency of collaboration is presented in figure 3.

Table 1 presents the percentage of Vojvodina's manufacturing companies that cooperate with external organizations in specific innovation areas.

Innovation area	**Share**	**Rank**
New products	69.1%	1
New technical production process	58.5%	2
New product related services	50.4%	3
New organizational concepts	37.4%	4

Table 1. Share of manufacturing companies in Vojvodina that cooperate with external organizations in specific innovation areas.

The analysis demonstrates that organizations collaborate with other organizations and external parties mostly in the domain of product innovation (69.1% of respondents). Furthermore, in the case of innovation areas such as technical production process innovation and product related services innovation, the number of companies cooperating is also above 50% (58.5% and 50.4% respectively). The lowest share of manufacturing companies that collaborate with external parties is in the field of organizational concepts innovation (37.4%).

Following, figure 2 contains the percentage of manufacturing companies in Vojvodina collaborating with different external parties regarding each of the specific innovation areas.

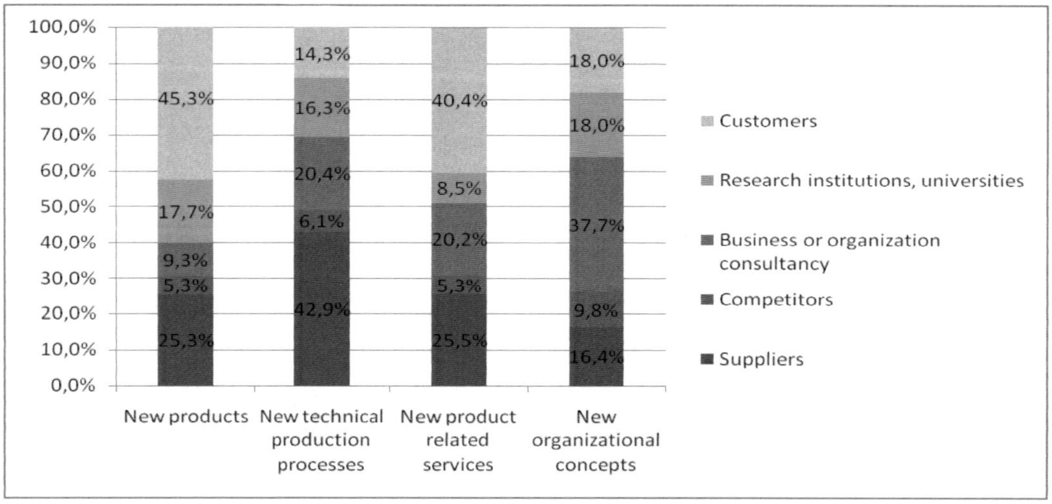

Figure 2.The percentage of manufacturing companies in Vojvodina collaborating with different external parties regarding each of the specific innovation areas.

Regarding new product development, manufacturing firms in Vojvodina mostly cooperate with their customers (45.3%), and with their suppliers in the second place (25.3%). When it comes to innovation in terms of technical production processes, companies cooperate by far most frequently with their suppliers (42.9%). In the area of new product related services the most frequent partner in cooperation activities are customers (40.4%) and following, suppliers (25.5%). Finally, most of the companies have marked business or organization consultancy as their collaboration partner in new organizational concepts development (37.7%).

In each of the specific areas of innovation, the lowest share of manufacturing companies has pointed out competitors as their collaborative partners. This may be explained by the insecurity of firms when it comes to sharing their specific knowledge, expertise or internal processes to the competitors and by the fact that every collaboration process requires intensive management of knowledge and information flows which increases in case of cooperation with competitors.

Figure 3 contains the frequency of collaboration of manufacturing companies in Vojvodina with different external parties regarding each of the specific innovation areas.

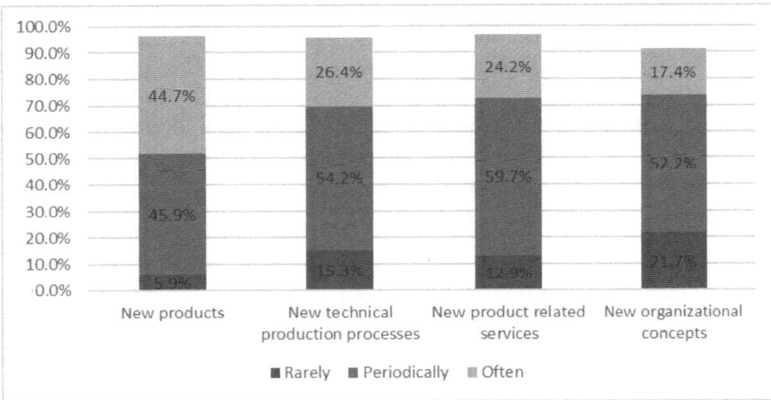

Figure 3. The frequency of collaboration of manufacturing companies in Vojvodina with different external parties regarding each of the specific innovation areas.

Figure 3 shows that the majority of the respondents reporting that collaboration of their firm with other companies or organizations exists, have also defined the frequency of their collaboration as periodical or frequent. However, it can be observed that the frequency of inter-organizational collaboration is the most intensive in the domain of new product development, mildly decreasing in the area of new technical processes and product related services. Finally, the frequency of collaboration is lowest when it comes to the innovation of organizational concepts.

Based on results presented in Table 1, Figure 1 and Figure 2, it can be discussed that most of the manufacturing firms are cooperating with other firms, suppliers, buyers, competitors and research institutions in the area on new product development (69.1%). Furthermore, the most common partners in collaboration for innovation of products or services are buyers (45.3%). Also, the results show that manufacturing firms are choosing different partners regarding different areas of innovation. Compared to the results from similar studies, there is notable difference. While in Vojvodina most manufacturing firms tend to develop stronger cooperation relationship with suppliers and buyers, in developed European countries most of the manufacturing firms are leaning towards more intensive cooperation with research institutions. Nonetheless, the same study shows that manufacturing firms in developed European countries are more often cooperating in new technical production processes development, new product related services development and new organizational concepts [22].

4.2. *Importance of different innovation areas for manufacturing companies in Vojvodina*

Figure 4 demonstrates the evaluation of importance of different innovation areas for manufacturing companies in Vojvodina. Manufacturing companies participating in EMS were asked to assess and rank 4 domains of innovation activities (improvement of product offer with related services, new organizational concepts, new technological production processes and new product development) in terms of their importance for their companies. Companies were evaluating using a scale 1-4, where 1 is the most important and 4 the least important innovation area for them.

Manufacturing companies have assessed new product development as the most important innovation domain for their firm and business. That is one of the reasons why more attention is dedicated to this innovation area in the next sections.

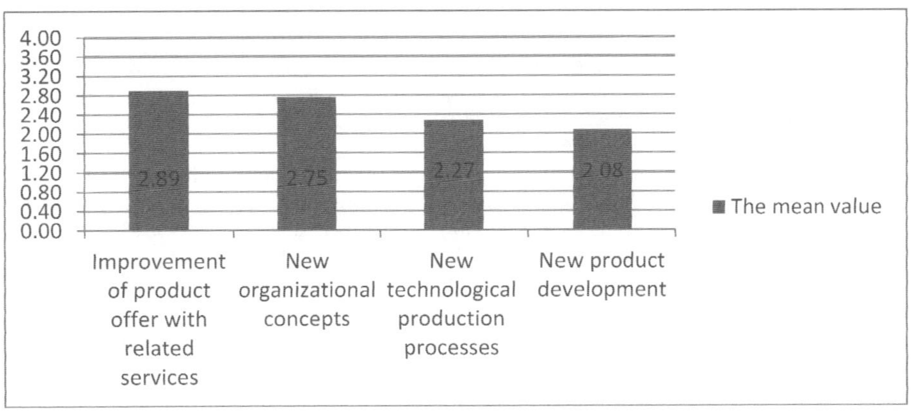

Figure 4. Innovation areas ranked by the importance for manufacturing firms

4.3. Inter-organizational collaboration in product innovation activities

The last research aim is to examine if there was a connection between the existence of inter-firm collaboration and the capability for new product development, as an aspect of innovativeness. In other words, the goal was to analyse if firms cooperating with other organizations in new product development activities have increased innovativeness in terms of new products development. In order to achieve that, the percentage of companies that have introduced innovative products to the firm and to the market in the last 4 years has been compared considering whether those firms are cooperating with other organizations in product innovation activities or not.

New product development requires interdisciplinary approach which can only be achieved by linking firms with other actors [23]. So far, firms where cooperation for innovation is common practice have been introducing innovations new not only to the firm, but also to the market 11] and in the same time with the opportunity to accelerate the process of product development through collaboration [24]. The main assumption is that the firms collaborating with external parties in the field of product innovation and in R&D innovate their products more often than the companies that do not cooperate or do so only rarely, both in the firm and on the market.

The analysis showed that 65% of manufacturing companies in Vojvodina claiming they are collaborating with other organizations in the new product development, have introduced new products since 2012, out of which almost 30% are innovations to the market. On the other hand, 25% of firms claiming they do not cooperate in the area of product innovation have introduced new products to the firm and only 5% new to the market. Considering this, the assumption that firms collaborating with external parties in the field of product innovation innovate their products more often than the companies that do not cooperate or do so rarely, both in the firm and market has been confirmed in the case of manufacturing companies in Vojvodina.

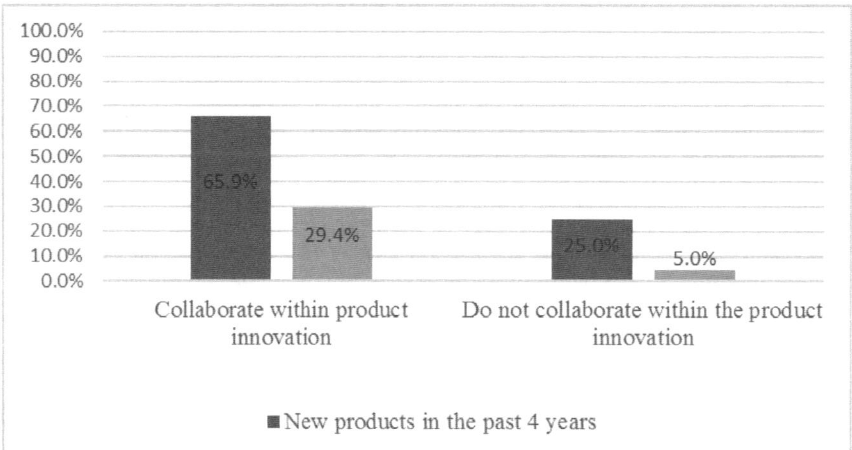

Figure 5. Connection between the existence of inter-organization collaboration and product innovation

4.4. Inter-organizational collaboration in R&D and firm innovativeness

Cooperation activities are considered as efficient means for the industrial organization of complex R&D and innovation process [11]. As companies are experiencing pressure in terms of shortened time from an innovative idea to its commercial success, they are realizing that R&D activities could not be performed on their own, but rather through acquiring knowledge and ideas from other companies and institutions [7]. Furthermore, it is perceived that cooperation increases the profitability of research and development [25].

One of the incentives for undertaking collaborative research and development with other firms or institutions is the fact that the capacity of a company to absorb and exploit the knowledge from the external environment has a positive effect on the probability of being a successful innovator [21]. As innovation activity is considered directly related to the amount of money companies spend on R&D activities [25] it is expected that firm collaboration and joint investment in R&D will result in higher innovation capacity. Generally, firms engaged in the innovation process are recognizing the necessity of taking part in R&D cooperation in order to obtain expertise which could not be generated in-house. Considering that, R&D collaboration with other companies and institutions is seen as the crucial way to make external resources available and usable, since it enables efficient knowledge transfer, resource exchange and organizational learning [4]. Finally, most commonly innovations are associated with research and development activities of the products [27].

Taking the previously mentioned into consideration, it seemed important to analyse the practice of R&D inter-organizational collaboration in Vojvodina, in order to present the degree of existence and frequency of such practice.

Furthermore, the assumption was that the firms claiming they have introduced a new innovative product in the last four years will mostly be those taking part in R&D cooperation. This hypothesis is based on the fact that the absorption of external resources through such cooperation leads to an extension of firms' capabilities of developing new products.

To analyze if firms cooperating with other organizations R&D activities are more innovative in terms of new products development, the percentage of companies that have introduced innovative products to the firm and to the market in the last 4 years has been compared considering whether those firms are cooperating with other organizations in research and development activities or not.

Figure 6 shows comparison between the percentage of manufacturing firms having introduced new products in the firm and to the market in the last four years depending on whether they collaborate in R&D or not.

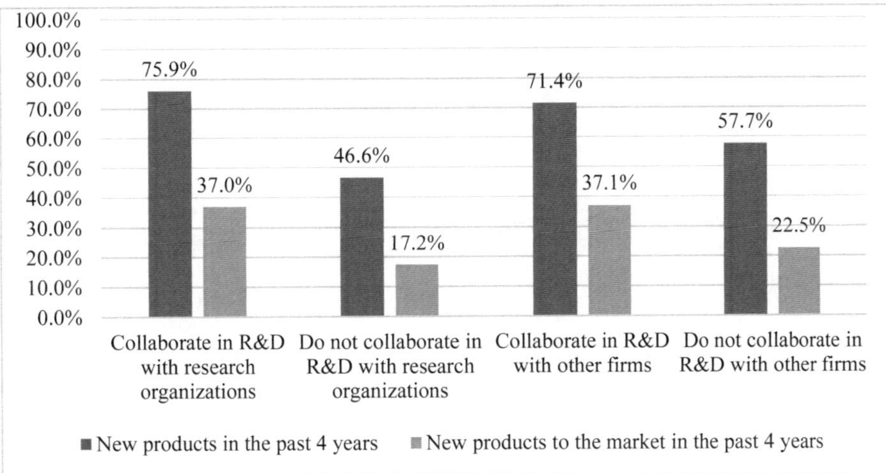

Figure 6. Connection between the existence of inter-organization R&D collaboration and product innovation

Regarding the firms collaborating in R&D, a high level of product innovations was observed. 75.9% of companies cooperating in R&D with research organizations confirmed they had developed new products in the last four years, out of which almost half (37%) were products completely new to the market. A very similar situation applies to the companies that choose to collaborate in R&D with other firms. However, a noticeably lower percentage of manufacturing firms in Vojvodina that do not cooperate in R&D with research organizations or other firms have introduced a new product, both to the firm and to the market, demonstrating that there is considerable difference in innovativeness between manufacturing firms collaborating in R&D and those not collaborating with other firms or research organizations.

Similar study covering European countries are aligned with the results presented in Figure 5 and Figure 6. Companies cooperating with other firms in R&D and product innovation activities demonstrate higher innovativeness in terms of new products commercialization. The European manufacturing companies show the increase of interest in collaboration in R&D and product innovation activities, since it had proved to be an efficient mean for the improvement of innovativeness [5]. Furthermore, the study shows that the percentage of European manufacturing companies engaging in collaboration for innovation increases more rapidly than in the past.

5. Conclusion

In this paper we have analyzed and presented the state of collaboration between manufacturing companies in Serbia and other firms or organizations in terms of innovation activities. The research questions are to investigate level and frequency of collaboration of manufacturing firms with other different external parties, in different areas of innovation, to investigate the perception of manufacturing firms regarding the importance of different innovation areas, and finally, to examine whether the more intensive inter-organizational cooperation in new product development and R&D activities influences the higher firm innovativeness in terms of innovative products.

To analyze previously mentioned issues, EMS (European Manufacturing Survey) was conducted and results were presented with descriptive statistics.

The results show that there is a high level of cooperation with customers, suppliers, business consultancy and research institutions. Collaboration with competitors in different areas of innovation exists, however, it is on a notably lower level than with the previously mentioned external parties. Out of four defined areas of innovation: new product development, new technological production processes, new organizational concepts and new product related services, inter-organizational cooperation most frequently occurs in product innovation activities. Additionally, product innovations have been emphasized as the most important area for manufacturing firms business.

Furthermore, the analysis demonstrates that manufacturing companies collaborating with other organizations in product development and R&D activities are notable more innovative, in terms of innovative products, than companies that are not collaborating. In case of manufacturing firms declared as participants in product innovation activities, a considerably higher percentage of them had introduced new products to the firm (65,9%) and to the market (29,4%), than firms not engaging in cooperation with other companies or institutions (25% and 5%, respectively). Moreover, when it comes to the companies cooperating in R&D activities, in average 73% of them had introduced new products to the firm, and 37% to the market, while in average 50% of companies that are not cooperating in R&D have introduced new product to the firm out of which only around 20% have introduced products new to the market. Considering that, it can be said that there is a strong bond between inter-organizational collaboration and innovativeness in the case of manufacturing firms in Vojvodina. Therefore, firms tending to strengthen their market position and competitiveness through improvement of their innovativeness should cooperate with other firms, suppliers, research organizations, buyers and competitors.

In this paper the location of collaboration partners was not taken in consideration. Related to that, as further research question we see investigation of influence of partners' location on collaboration forms, frequency, success and similar.

6. References

[1] Smith, H.L., Dickson, K. & Smith, S.L., (1991). " There are two sides to every story " : Innovation and collaboration within networks of large and small firms. Research Policy, Vol. 20., 1991, pp.457–468

[2] Lee, S.M., Olson, D.L. & Trimi, S., (2012). Co-innovation: convergenomics, collaboration, and co-creation for organizational values. Management decision, Vol. 50., No. 5., 2012, pp.817–831

[3] Danneels, E., (2002). The dynamics of product and firm competences. Strategic Management Journal, Vol. 23, 2002, pp.1095–1121

[4] Becker, W. & Dietz, J., (2004). R&D cooperation and innovation activities of firms-evidence for the German manufacturing industry. Research Policy, Vol. 33., No. 2, 2004, pp.209–223

[5] Alexiev, A.S., Volberda, H.W. & Van den Bosch, (2016). Interorganizational collaboration and firm innovativeness: Unpacking the role of the organizational environment. Journal of Business Research, Vol. 69., pp. 974-984

[6] Trifan, L., Guica R.-I. & Micu, C., (2012). Entrepreneurial creativity and innovation management in engineering: Proceedings of DAAAM International 2012, pp. 783-789

[7] Fajsi, A., Tekic, Z. & Moraca, S., (2016). Open innovation in manufacturing SMEs- integration into Value Networks: Proceedings of DAAAM International 2012, pp. 1076-1081

[8] Orcik, A., Anisic, Z. & Cosic, I., (2012). Collaboration and new product development in PLM environment: Annals and Proceedings of DAAAM International 2012, pp. 1091-1094

[9] Ahuja, G., (2000). Collaboration Networks, Structural Holes, and Innovation: A Longitudinal Study. Administrative Science Quarterly, Vol. 45., No. 3, 2000, pp.425–455

[10] Lopez, A., (2008). Determinants of R&D cooperation: Evidence from Spanish manufacturing firms. International Journal of Industrial Organization. Vol 26., 2008, pp. 113-136

[11] Cassiman, B. et al., (2002). R&D Cooperation and Spillovers: Some Empirical Evidence from Belgium. The American Economic Review, Vol. 92., No. 4., 2002, pp.1169–1190

[12] De Faria, P., Lima, F. & Santos, R., (2010). Cooperation in innovation activities: The importance of partners. Research Policy, Vol. 39, 2010, pp.1082–1092

[13] Hwee, S., (2008). Competitive intensity and collaboration: Impact on firm growth across technological environments. Strategic Management Journal, Vol. 29, 2008, pp.1057–1075

[14] Barringer, B.R. & Harrison, J.S., (2000). Walking a Tightrope: Creating Value Through Interorganizational Relationships. Journal of Management, Vol.26., No 3., 2000, pp. 367–403

[15] Tsang, E.W.K., (2000). Transaction Cost and Resource-Based Explanations of Joint Ventures: A Comparison and Synthesis. Organization Studies, Vol. 21., No. 1., 2000, pp.215–242

[16] Konsti-Laakso, S., Pihkala, T. & Kraus, S., (2012). Facilitating SME innovation capability through business networking. Creativity and Innovation Management, Vol. 21., No. 1., 2012, pp.93–105

[17] Tekić, Ž., Ćosić, I. & Katalinić, B., (2011). Innovation and knowledge creation mechanisms. In Annals & Proceedings of DAAAM International 2011. pp. 419–421

[18] Tapscott, D. & Williams, A., (2006). Wikinomics How Mass Collaboration Changes Everything, N. Y. Portfolio., New York

[19] Milovanović, B.M., Primorac, D. & Kozina, G., (2016). Two-dimensional analysis of the influence of strategic networking on entrepreneurial orientation and business performance among SMEs. Tehnicki vjesnik - Technical Gazette, Vol. 23., No. 1., pp.247–255

[20] Abramovsky, L. et al., (2005). Understanding co-operative R&D activity: Evidence from four European countries, The Institute for Fiscal Studies

[21] Sampson, R.C., (2007). R&D alliances and firm performance: the impact of technological diversity and alliance organization on innovation. Academy of Management Journal, Vol. 50., No. 2., pp.364–386.

[22] Robin, S. & Schubert T., (2010). Cooperation with public research institutions and success in innovation. Evidence from France and Germany. Fraunhofer ISI Discussion Papers Innovation System and Policy Analysis. Vol. 24.

[23] Palcic, I. et al., (2010). Manufacturing firms ' cooperation patterns – experiences from Slovenia and Spain. Advances in Production Engineering & Management, Vol. 5., No. 4, pp.253–261

[24] Wu, S. B., Gu, X., Wu, G.D., Zhou, Q., (2016). Cooperative R&D contract of supply chain considering the quality of product innovation. International Journal of Simulation Modelling. Vol. 15., No. 2., 341-351

[25] Belderbos, R., (2004). Heterogeneity in R&D cooperation strategies. International Journal of Industrial Organization, Vol. 22., No. 8-9, pp.1237–1263.

[26] Palcic, I., Buchmeister, B. & Polajnar, A., (2010). Analysis of innovation concepts in slovenian manufacturing companies. Strojniski Vestnik/Journal of Mechanical Engineering. Vol. 56., No. 12. pp. 803-810

[27] Koren, R. & Palcic, I., (2015). The impact of technical and organisational innovation concepts on product characteristics. Advances in Production Engineering & Management, Vol. 10., No. 1, pp. 27–39

DOI: 10.2507/27th.daaam.proceedings.105

YAGI-UDE ANTENNA AS WIFI ELECTROMAGNETIC SOURCE

Rene Hartansky, Martin Mierka & Marek Vimpel

This Publication has to be referred as: Hartansky, R[ene]; Mierka, M[artin] & Vimpel, M[arek] (2016). Yagi-Ude Antenna as WiFi Electromagnetic Source, Proceedings of the 27th DAAAM International Symposium, pp.0730-0733, B. Katalinic (Ed.), Published by DAAAM International, ISBN 978-3-902734-08-2, ISSN 1726-9679, Vienna, Austria
DOI: 10.2507/27th.daaam.proceedings.105

Abstract

There are a lot of applications that require signal transmission in high-frequency spectrum (e.g. data and sensor networks), because of that we need to build up a set of equipment that allows this transmission. In this article we will occupy with the design of electromagnetic (EM) field radiator, whose task is to create EM field and capture information from MEMS structure. In this case the EM field is influenced by the MEMS structure so, that the EM field is carrying information about processes happening in MEMS.

Keywords: Electromagnetic field; Antennas, Method of moment; Gain; VSWR.

1. Introduction

The article deals with Yagi's antenna design in range of 2.4GHz, which will have significant polarizing properties and gain minimally 12dBi in range from 2.4GHz to 2.5GHz. In the design process of the antenna has been taken to minimize the cross-polarization and maximize gain front-back ratio. In this case the transferring signals in sensor (sensor network) environment, the interference from another signal sources will be minimized [1]. Another reason is the fact that the useful signals are in range from 2.4GHz to 2.5GHz [2]. It means that the Yagi's antenna has to be relatively wide ranged (this must be valid especially for its radiation pattern) however the higher priority is, that the antennas input impedance has to be constant. By using a typical coaxial cable of 50Ω the input impedance has to be around 200Ω and its imaginary component has to be nearly zero. The computation and design of antenna was conducted by recomputing the tabular parameters from other Yagi and Ude works. However, this way designed antenna had not adequate parameters [3]. Because of this, a new computation based on moment method in FEKO software, has been done. The basic geometry of Yagi's antenna was put into the computer software. The more useful shape of radiation patter was found by moving the antenna elements. The result is antenna geometry, where the radiation patterns in above mentioned range are almost identical. The results presented in further are completed also by graphs. The next solved task is stabilizing the input impedance. This has been solved by changing the shape of scatter and also by changing size and distance of first director. More about this task you can find further in this article completed with a lot of graphs showing positive results.

2. Finding the geometry of Yagi antenna

We can say, that Yagi-Ude antenna seems to be very simple. In spite of this, the design of such an antenna is complicated and takes too long. There are two ways how to find a solution:

1. The Yagi and Ude theory gives us the possibility to design the structure and number of antenna components (directors). These components will ensure us quite enough antenna gain however, no radiation patterns and input impedance characteristics are obtained. Finally it is necessary to measure the antenna and correct the components sizes.
2. It is also possible to use an existing antenna with adequate gain and good shape of radiation pattern to adapt it to the given frequency range.

Comparing these two procedures, the first one is more complicated and time consuming than the second one. According to [4] the geometry of seven component antenna has been chosen and modified for frequency 2.5 GHz. The antenna components sizes are in Table 1.

	Reflector [mm]	Scatter [mm]	1. Dir. [mm]	2. Dir. [mm]	3. Dir. [mm]	4. Dir. [mm]	5. Dir. [mm]
Length	57.1	62.4	56.4	52.3	51.6	52.8	51.6
Distance	-30.0	0.0	6.0	40.6	89.4	128.1	178.8
Perimeter	0.4	0.4	0.4	0.4	0.4	0.4	0.4

Table 1. Antenna components dimensions

To simulate the radiation of our antenna, computer package FEKO has been used. More detailed description can be found in [3]. The results are radiation patterns in E and H plane (Fig. 1).

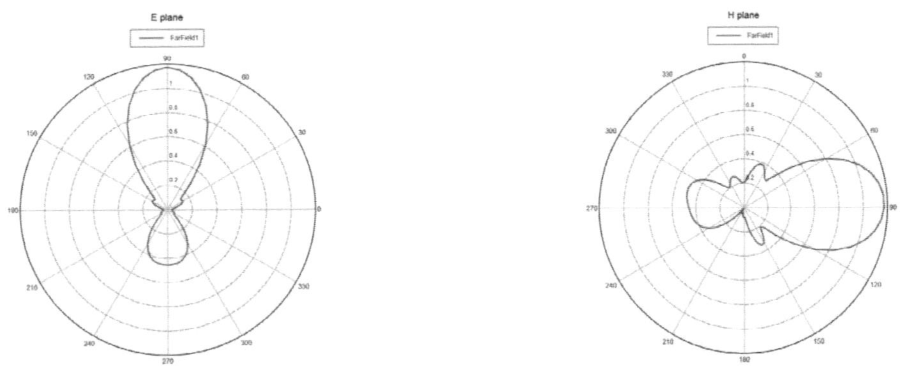

Fig. 1. Radiation pattern of designed antenna (E plane and H plane)

As can be seen from radiation pattern of Yagi antenna it's scattering into the back direction is very large. This means, that when we are transporting the signal, not only direct signal is being received, but also the reflected one, what leads to increased error rate of transmitted data. The designed antenna in this shape is unsatisfying for data signals transfer.

3. Antenna optimizing design

Because of the previous results, the antenna design has to be made by a different way as usual. The antenna must have an aligned input impedance and prescribed gain in wide frequency range. This leads to use large diameter directors. The used components have diameter equal to 6 mm. It´s also known, that the first director is used as matching one (matches the input impedance to characteristic impedance of feeder). This director does not influence neither the shape of radiation pattern nor gain of antenna [4]. So the beginning of Yagi antenna design is without matching director.

Firstly, the approximate length of half-wave dipole will be computed (this will be used to feed the antenna):

$$l \doteq \frac{\lambda}{2} = \frac{c}{2f} = \frac{3\ 10^8}{2\ 2.5\ 10^9} = 6cm \qquad (1)$$

According to [4], distance between director and reflector from radiator must be less than 0.4 l, length of reflector is greater than the length of radiator (scatter element) and length of director shorter than the length of radiator. These conditions had to be set at the beginning of the simulation process of Yagi antenna. Components have been set into the FEKO environment and the system had been enlarged step by step [5]. All sizes and distances have been chosen in such a way, that the gain of antenna should be as high as possible (for given number of components). Next component has been added only when the used were tuned to maximal gain. The last director was not added to get a higher gain, but to change the radiation pattern, especially to change (maximize) the front-back ratio.

You can find the dimension results in Table 2.

	Refl. [mm]	**Scat. [mm]**	**1. Dir. [mm]**	**2. Dir. [mm]**	**3. Dir. [mm]**	**4. Dir. [mm]**	**5. Dir. [mm]**	**6. Dir. [mm]**	**7. Dir. [mm]**
Length	61	58	50	47	45	43	42	42	42
Distance	-30.5	0.0	17.5	51.5	83.0	105.0	142.1	179.5	238.0
Perimeter	3	3	3	3	3	3	3	3	3

Table 2. Front-back optimizing antenna components dimensions

Radiation patterns of this antenna can be seen in Fig. 2.

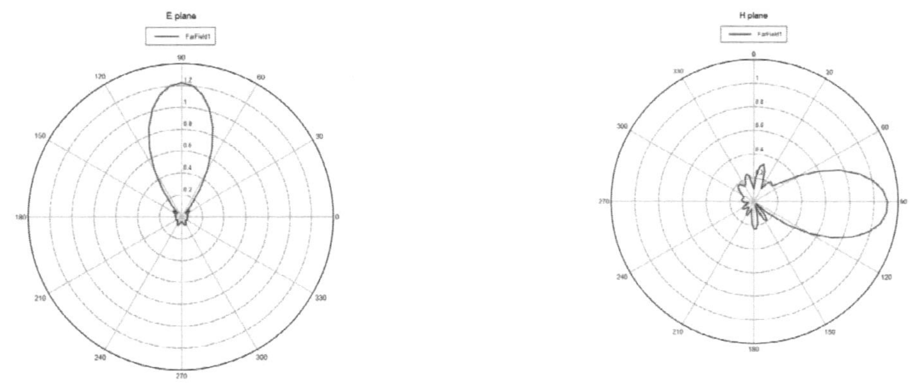

Fig. 2. Radiation pattern of optimized antenna (E plane and H plane)

4. Antenna optimizing input impedance

Two requests of Yagi antenna design were fulfilled till now (design for maximal gain and design for maximal front-back ratio of radiation pattern). The value of input impedance and its dependency on frequency (in frequency range from 2.4 to 2.5 GHz) has still to be made. According to [6], [7], the input impedance can be influenced by following three factors:

1. adding a matching director
2. changing the shape and size of scatter
3. changing the length of reflector.

These changes result using a nine component antenna can be found in Fig. 3.

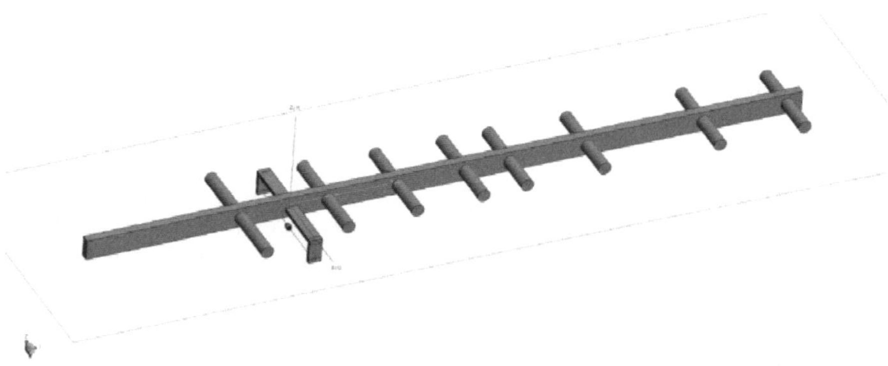

Fig. 3. Model of manufactured 12dB antenna

Feeding the antenna with 50Ω coaxial wire with an symmetrical loop at the end gives us a very good ratio of standing wave VSWR in wide frequency range. This can be seen in Fig. 4.

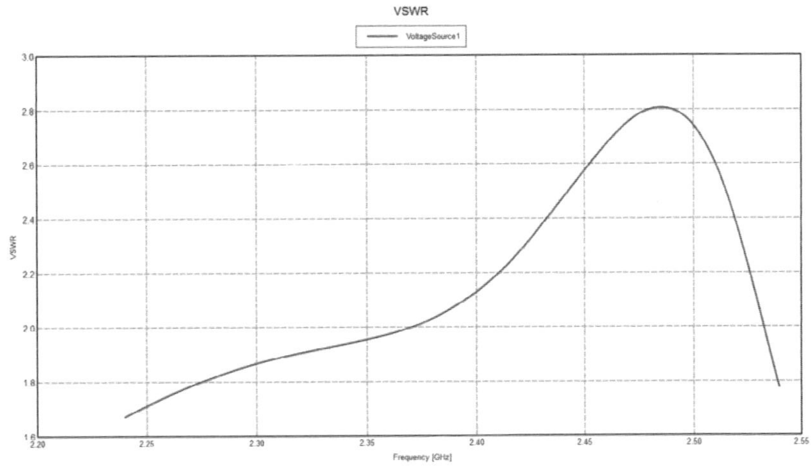

Fig. 4. Model of manufactured 12dB antenna

5. Conclusion

The point of this article is to give us instruction for analysis and synthesis more component Yagi-Ude antenna by fulfilling the input parameter requirements. We have focused on fulfilling the direction characteristics, better said maximum gain and maximum radiated power – minimalization VSWR [8]. Such proposed antenna will be used as source of electromagnetic field by solving project APVV-14-0076.

5. Acknowledgments

This work was supported by the project APVV-14-0076.

6. References

[1] Farkas, P., Michalovic, M., Majchrak, S. & Rakus, M. (2009) One Experiment with Wireless Sensor Network. TSP 2009 Telecommunications and Signal Processing : 32nd International Conference. Dunakiliti, Hungary, 26.-27.8.2009. Budapest : Asszisztencia Szervező Kft., 2009, pp. CD-Rom. ISBN: 978-963-06-7716-5.

[2] Kenyeres, J., Sajban, S., Farkas, P. & Rakus, M. (2010) Indoor Experiment with WSN Application. MIPRO 2010 : 33rd International Convention. Conference: Computers in Education. Opatija, Croatia, 24.-28.5.2010. Rijeka : Croatian Society for Information and Communication Technology, Electronics and Microelectronics - MIPRO, 2010, pp.863-866. ISBN: 978-953-233-054-0.

[3] Bittera, M. & Smiesko, V. (2012) Directivity Error of Broadband Antennas at EMI Measurement. Przeglad elektrotechniczny. Vol. 88, No. 6, 2012, pp.61-63., ISSN: 0033-2097.

[4] Kraus, D. J.: Antennas, McGRAW-HILL, ISBN: 0-07-100482-3.

[5] Bozek, P. & Turygin, Y. (2014) Measurement of the operating parameters and numerical analysis of the mechanical subsystem. Measurement Science Review. Vol. 14, No. 4, pp. 198-203, ISSN: 1335-8871.

[6] Hallon, J. & Kovac, K. (2013) Simplified generation of electromagnetic field within EMC immunity test area. Procedia Engineering: 24th DAAAM international symposium on intelligent manufacturing and automation. Zadar, Croatia, 23-26th October 2013, 69. pp. 27-34. ISSN: 1877-7058.

[7] Bittera, M., Smiesko, V., Kovac, K. & Hallon, J. (2010) Directional Properties of the Bilog Antenna as a Source of Radiated Electromagnetic Interference Measurement Uncertainty. IET Microwaves Antennas & Propagation. Vol. 4, Iss. 10, 2010, pp. 1469-1474. ISSN: 1751-8725.

[8] Hallon, J., Kovac, K. & Szolik, I. (2006) Geometrical Configuration of Cabling as Factor Influencing the Reproducibility of EMC Imunity Tests. Radioengineering. Vol. 15, No. 4, 2006, pp. 73-33. ISSN: 1210-2512.

DOI: 10.2507/27th.daaam.proceedings.106

GAS COMPOSITION INFLUENCE ON THE MICROSTRUCTURE AND GEOMETRY OF LASER-WELDED JOINTS IN DUPLEX STAINLESS STEEL

Angela Topić, Branko Bauer, Zoran Kožuh, Nikola Knezović

This Publication has to be referred as: Topic, A[ngela]; Bauer, B[ranko]; Kozuh, Z[oran] & Knezovic, N[ikola] (2016). Gas Composition Influence on the Microstructure and Geometry of Laser-Welded Joints in Duplex Stainless Steel, Proceedings of the 27th DAAAM International Symposium, pp.0734-0742, B. Katalinic (Ed.), Published by DAAAM International, ISBN 978-3-902734-08-2, ISSN 1726-9679, Vienna, Austria
DOI: 10.2507/27th.daaam.proceedings.106

Abstract

Shielding gases, i.e., argon or an argon-nitrogen mixture, are normally used during the laser-beam welding of duplex stainless steel. Helium is also often added to the argon. The effect of the type of shielding gas – argon, nitrogen, helium and their mixtures – on the geometrical characteristics of laser-welded joints of the duplex steel W.Nr. 1.4462 was studied. The welding was carried out according to the experimental model with mixtures. The effect of the welding speed and the gas flow rate were additionally explored using a factorial experiment. Obtained data were statistically processed and mathematical modeling, applying the method of response surfaces, was carried out. The analysis revealed that the impact of the shielding gas mixtures on the geometrical characteristics of the joint is significant, while it is not so significant for the microstructure. The effect of the gas flow rate on the geometrical characteristics was registered, while the effect of the heat input is the most significant. Special cubic, quadratic and reduced quadratic models, depicting effects of the shielding gas on geometrical characteristics and microstructure of welded joint, were obtained.

Keywords: laser; welding; duplex steel; shielding gas; weld geometry; microstructure

1. Introduction

Duplex and super-duplex steels are materials that have been increasingly used and applied in the last years. In many steel constructions, duplex steels are replacing the standard carbon steels and austenite stainless steels very effectively.

Duplex stainless steels are optimised with respect to mechanical properties and corrosion stability so that their structures contain equal proportions of austenite and ferrite. However, the weldability of duplex steel when the laser-welding process is applied is still under investigation and is particularly related to an increased rate of ferrite formation after welding, induced by the high cooling rates. A large ferrite content tends to produce a deterioration of the mechanical properties and in particular a reduction of the corrosion resistance. The ferrite content may be higher than 90 % in the microstructure of the weld metal and HAZ (Heat Affected Zone), while the target is to achieve at least 35 % of austenite. One of approaches to realize such a goal is to apply filler material that may increase the costs and the technical difficulties. Another approach includes heat treatment, applying a defocused laser beam, induction heating or treatment in a furnace.

Such methods also involve increased production costs. Heat treatment in a furnace additionally impedes the main advantage of laser welding, i.e., the production rate. Welding without filler material tends to produce a strong formation of ferrite and a large grain growth in the weld metal, but there is an advantage in the case of laser welding, for it much simplifies the actual execution of the welding [1–5].

The shielding gas has an important role in laser welding, fulfilling the following tasks: protection of molten pool and heat-affected zone from the effect of the surrounding atmosphere, affecting the shape of the weld and protection of the optics of the device against metal vapors and spatter droplets. The most used shielding gases for the laser welding of duplex steels are argon, helium and nitrogen. According to the reference data, argon as a shielding gas is the best selection in Nd:YAG laser welding. An acceptable shape of welded joint may be achieved with all shielding gases [6,7]. Argon and helium are inert gases, having no effects on the metallurgical processes during welding. Nitrogen is a reactive gas, affecting the metallurgical processes occurring in the welded joint [6]. When nitrogen shielding is used, a significant amount of nitrogen is absorbed into the welded joint. Doping the welded joint with nitrogen is particularly important in the welding of duplex steel, since nitrogen stimulates the formation of austenite [6,8].

The application of argon-based gas mixtures with the addition of nitrogen and/or helium for the Nd:YAG laser welding may reduce the ferritization rate. Current investigations indicated that the application of the mentioned mixtures ensures an acceptable appearance and quality of the welded joints [2,3]. In such investigations, testing with specific gas mixtures has been made without detecting the interaction between the components of gas mixtures and mathematical modeling. An experimental model for the mixtures appears to be suitable for the investigation and mathematical modeling of the effects of shielding-gas mixtures on the welding [9–13]. The mixture contains two or more constituents. The target of mixing certain components into the mixture is to investigate whether such a mixture has a more favourable effect upon the specific properties than a single component. The aim was to describe quantitatively the effect of the mixture composition on the geometrical characteristics and microstructure of the welded joint, applying a mathematical model, developed on the basis of the experimental data measured for a three-component mixture.

It is important to achieve the optimum weld shape and geometry to fulfill the quality requirements according to the standard EN ISO 131919-1. Also, it is important to see effects of input energy, type and flow rate of the shielding gas on achieving a two-phase microstructure with approximately equal portions of austenite and ferrite. Welding with maximum speed, without the use of filler material and with the use of the appropriate gas mixture that contains nitrogen would considerably increase the economy and productivity of laser welding.

The final objective of this research was to determine the influence of the shielding gas – argon, nitrogen, helium and their mixtures – on the geometrical characteristics and microstructure of the welded joint on a 2-mm-sheet of duplex stainless steel W.Nr. 1.4462 and additionally explore the influence of the welding speed and the gas flow rate for every shielding gas and mixture.

2. Experimental work

2.1. Equipment

The welding was performed using a Nd:YAG laser "ROFIN CW 020" with a continuous power of 2 kW. An optical fiber with a 600-µm core diameter was used for the beam transfer. Focusing optics of 120/120 mm were used. The beam diameter in the focus was equal to 0.6 mm. The focusing optics were attached to the robot arm, and the robot "IGM Limat RT 280" features 6 degrees of movement freedom.

2.2. Experimental

The welding was performed on 2-mm-thick sheets of duplex steel W.Nr. 1.4462. The sample dimensions were (250 × 130) mm and (230 × 130) mm. Prior to welding, the samples were machined and cleaned by applying emery paper and ethanol. The samples were fixed in the jigging tool and argon was used for the root shielding.
The tests were conducted according to a "simplex grid" planned experiment [9]. The simplex grid for a three-component mixture is represented by a triangle, within which the states of the experiments are distributed uniformly in a "grid" pattern. A uniform distribution of the states of the experiment in equal intervals of all the component portions in mixture is required to attain an acceptable definition of the response surface. Each response function can be associated with a polynomial and the appropriate coefficients for the equation can be calculated. Seven types for three-component mixtures for the simplex-grid model were selected. For every shielding gas – argon, nitrogen, helium and their mixtures – in the simplex grid, a factorial plan on two levels 2^n, with two factors for the welding speed and gas flow rate were additionally conducted. A complete plan of the experiment is shown in Figure 1.

The constant parameters in all states of the experiment were power (P=1800 W), optic focal length (f = 120 mm), tip diameter of the coaxial gas nozzle (5 mm) and distance of gas nozzle tip from the workpiece (8 mm).

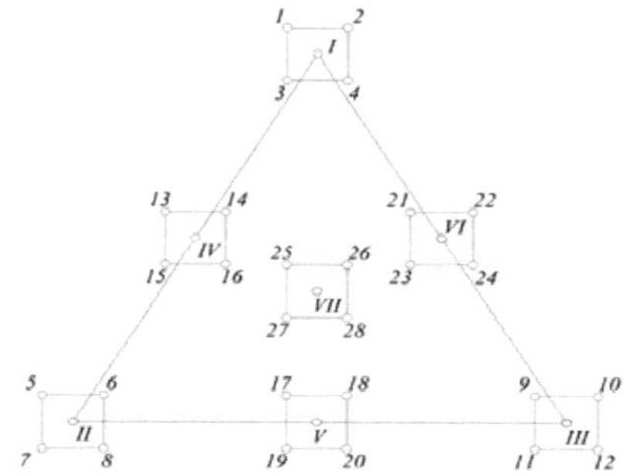

Fig. 1. Plan of experiment with 28 states

Shielding gas	State of exp.	Mixture components (%)			Process parameters		
		Ar	N_2	He	v (cm/min)	Q (l/min)	z (mm)
I	1	100	0	0	110	18	−1.0
	2				140	18	
	3				110	9	
	4				140	9	
II	5	0	100	0	110	18	−0.8
	6				140	18	
	7				110	9	
	8				140	9	
III	9	0	0	100	110	33	−1.2
	10				140	33	
	11				110	21	
	12				140	21	
IV	13	50	50	0	110	18	−0.9
	14				140	18	
	15				110	9	
	16				140	9	
V	17	0	50	50	110	25	−1.0
	18				140	25	
	19				110	15	
	20				140	15	
VI	21	50	0	50	110	25	−1.1
	22				140	25	
	23				110	15	
	24				140	15	
VII	25	33,33	33,33	33,33	110	23	−1.0
	26				140	23	
	27				110	13	
	28				140	13	

Table 1. States of experiment (v – welding speed, Q – gas flow rate, z – focus position relative to the workpiece surface)

2.3. Inspection and testing

The welded specimens were inspected visually and with radiographic control, followed by destructive testing methods. In order to determine the weld geometry, macro-etches of the joint cross-sections were prepared. The weld geometry measurement on the macro-etches was performed under 50-times magnification.

With the aim of determining the quality of the laser welded joint according to the standard EN ISO 13919-1, the weld and root reinforcement/underfill, linear misalignment of the sheets and the size of the undercut were measured. Measurements were also performed on the width of the weld b_zav, the width of the HAZ b_zut, the width of the weld root b_kor and the area of the weld cross-section A, in order to determine the influence of the parameters on the geometrical characteristics of the welded joint.

For microstructure inspecting the same specimens were used, but with 500-times magnification. Specimens were reviewed with optical microscope, connected with software for analysis obtained pictures. After erosion with special tool, it could be seen that austenite and ferrite are different coloured, and that was the basis for analysis. The results of the measurements of local mechanical properties of the welded joint were processed by applying the software package "Design Expert 6.0.6". A statistical analysis of measured data for all 7 states of the experiment revealed the appropriate model. The model could be linear, quadratic, special cubic, interactional or mean value if there are no significant effects of the tested factors. Subsequently, the significance of the model and the members of the response polynomial was tested by applying the variance analysis. For the linear model, only the members Ar, N2 and He appear. In the square model, members of second order, i.e., Ar*N2, Ar*He and N2*He appear, and in special cubic model, the member Ar*N2*He.

R-squared is a determination coefficient that represents the estimation of the total variation of the data described by the model. The adjusted "R-Squared" is R-squared adjusted to the number of members in the model in relation to the number of experiment states. The predicted "R-Squared" is a measure for the variation of the values within a new set of data described by the model. The values of both R-squares should be close to 1, and if they are equal to 1, then 100 % of the variations of the tested values are explained by the model. If the adjusted R-squared is higher than 75 %, the model can be considered as significant. A mathematical model established through such an approach provides very a distinctive graphic interpretation and the optimization of tested parameters [10,11].

3. Results and discussion

3.1. Geometrical characteristics of welded joints

The welds feature a slight reinforcement of the face and the root without undercuts. Cracks, open pores and other surface defects did not occur and welds without porosity were obtained. All the laser welds are classified into group B according to standard EN ISO 13919-1 i.e., they are of high quality.The appearance of the weld face is acceptable for all the experimental conditions, while increased spattering occurred on the root side on certain samples. Increased spattering is noted in all the samples welded in an argon shield. Increased spattering while applying a nitrogen shield is registered in the experimental conditions No. 5 and 7, where a lower welding speed was used. In samples No. 6 and 8, which were welded applying the same shielding gas, but a higher speed, spattering was not registered. A uniform root shape along the full weld length and full penetration were obtained for all the experimental conditions. All the applied shield gasses and mixtures produced faultless welds without defects and of acceptable geometric shape, this being the first target of the testing. Good results also showed up when nitrogen was used as the shielding gas. This is an important finding, because of its favourable effect upon ferritization [2,3].

3.1.1. Relationship between the geometric characteristics of welded joint and the welding parameters

For all shielding gasses and mixtures, samples welded with a lower speed have larger geometric characteristics, due to the higher energy input. With a lower welding speed the weld root is wider and therefore this compensates for the beam-guidance accuracy. A higher welding speed is economical but requires better edge preparation, while without any beam-guidance sensor a risk of a lack of fusion defects is higher. For welded samples, by applying a higher welding speed the root is significantly narrower than the joint face, except for the samples made in an argon shield. Thus, with respect to the root width, an argon shield is the best choice. The geometrical characteristics of the welds made applying the same welding speed and different shield flows do not differ significantly.

3.1.2. Relationship between geometric characteristics of the welded joint and the composition of the shielding gas

From the economic point of view, i.e., a reduction of costs, the best solution is to apply the highest welding speed and the lowest flow rate of the shielding gas, thus v_{max}, Q_{min}. Figure 2. shows cross-sections of the welded joint for experimental conditions applying all the shielding gases and mixtures at maximum welding speed and minimum flow rate of the shielding gas, i.e., condition a. Figure 3. shows a graphical representation of the geometric characteristics of the weld for the experimental condition a. The mixture Ar/N2 and pure nitrogen, no matter which welding parameters were used, produced welded joints with the largest cross-section and root width within the experiment. This fact may provide a higher content of austenite in the microstructure of the welded joint, thus improving the mechanical properties and the corrosion resistance of the joint.

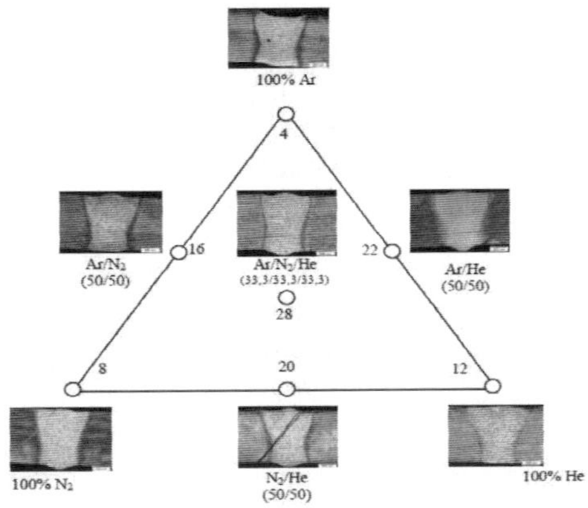

Fig. 2. Cross-sections of welded joints for all shielding gases and mixtures at maximum welding speed and minimum flow rate of the shielding gas – experimental condition a (v_{max}, Q_{min})

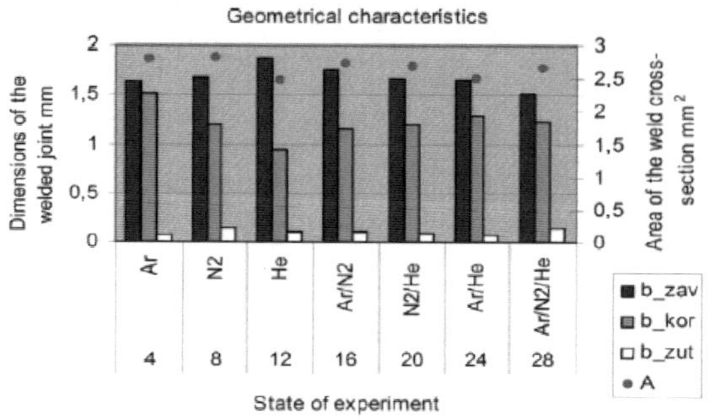

Fig. 3. Geometrical characteristics of the welded joint – conditions a(v_{max}, Q_{min}), (b_zav – weld width, b_kor – root width, b_zut – HAZ width, A – area of the weld cross-section)

3.1.3. Mathematical models depicting the relationship between the composition of the shielding gases and the geometrical characteristics of the welded joint

Mathematical models obtained for mixtures in states 1 (v_{min}, Q_{min}), a (v_{max}, Q_{min}), b (v_{min}, Q_{max}) and ab (v_{max}, Q_{max}) are listed in Table 2. "Outliers" did not occur in any model, indicating that none of the measured data has a significant deviation. An adjusted R-squared value for the weld joint width b_zav, Table 2, close to the value of one, and variance analysis of the model indicate that special cubic models are significant for all four combinations of parameters. From the graphic plot for state a (v_{max}, Q_{min}), Table 3, it could be noted that the widest face of the weld is obtained for pure He and that additions of N2 and Ar to helium cause a width reduction. The smallest width of the welded joint is obtained in the centre applying a (Ar/N2/He) mixture. It can be concluded that the type of shielding gas/mixture affects the weld width that can be easily predicted by applying a special cubic mathematic model.

The square (quadratic) model in state a and the special cubic model in state ab have been yielded for the root width b_kor, Table 2. For states 1 and b a negative Predicted R-squared was obtained, meaning that the mean value predicts the response better than the special cubic model. For low welding speeds, no significant model was obtained, while for high welding speeds different significant models are revealed. Different models may be explained by the effects of some other factors, unaccounted effects combined with the effects of mixtures. It can be concluded that the selection of the gas mixture has a significant effect on the root width in the case of high welding speeds, while in the case of low welding speeds it is not important which mixture is used. From the graphic plot for state a (v_{max}, Q_{min}, Table 3) it can be seen that the root width is the widest in the case of the application of pure Ar. Additions of N2 and He to argon reduce the root width.

An adjusted R-squared for the HAZ width for states 1 and ab has a value less than 0.75, while the predicted R-squared is negative, indicating that no one model is significant. Thus, it can be concluded that the mean value predicts the response of model more adequately. Also, it may be concluded that the type of gas mixture has no significant effect on the width of the joint HAZ. Significant reduced square models for states 1, a and b for the cross-section area of welded joint A were obtained, Table 2. A significant special cubic model for state ab has been developed. From graphic plots, relationships between the models produced for different conditions can not be detected. For state 1 the largest area is for pure argon and it is reduced with additions of nitrogen and helium. In state a (v_{max}, Q_{min}, Table 3), argon and nitrogen produce the largest area, which is reduced with additions of helium, attaining its minimum for pure helium. In state b, the largest area is obtained for pure nitrogen and is reduced with additions of argon and helium. In state ab, the largest surface is obtained for mixtures Ar/N2 and Ar/He and it is reduced for the center (Ar/N2/He) and pure gases. It can be concluded that the type of shielding gas/mixture affects the cross-section area of welded joint and it can be well predicted by the model. Also, it can be concluded that the model depends on a combination of welding speed and gas flow rate. Mixtures have a significant effect on the geometric characteristics of the welded joint. Special cubic, quadratic and reduced quadratic models were obtained. In state a (v_{max}, Q_{min}) the helium content increases the weld width and reduces the root width and the area of the cross-section. Nitrogen content reduces the root width.

Measured value	**State**			
	1 – v_{min}, Q_{min}	a – v_{max}, Q_{min}	b – v_{min}, Q_{max}	ab – v_{max}, Q_{max}
b_zav, mm	Special cubic	Special cubic	Special cubic	Special cubic
b_kor, mm	-	Quadratic	-	Special cubic
b_zut, mm	-	-	-	-
A, mm2	Reduced quadratic	Reduced quadratic	Reduced quadratic	Special cubic

Table 2. Mathematical models of weld width b_zav, root width b_kor, HAZ width b_zut, area of the weld cross-section A

No	Model	2D plot	3D plot X1=A=Ar; X2=B=N_2; X3=C=He
1	Special cubic "Prob>F"<0,0001 Adjusted R-squared=0,94 b_zav = +1.615 * Ar +1.670 * N2 +1.855 * He +0.430 * Ar * N2 -0.300 * Ar * He -0.450 * N2 * He -4.395*Ar*N2*He	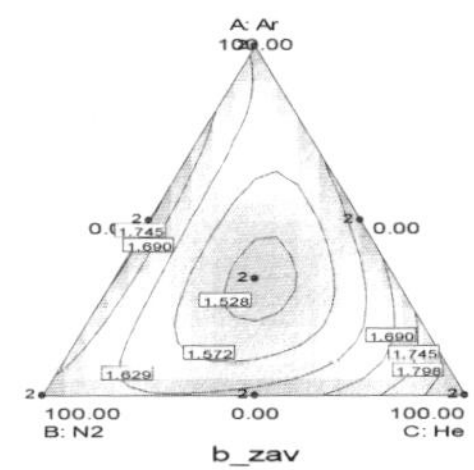	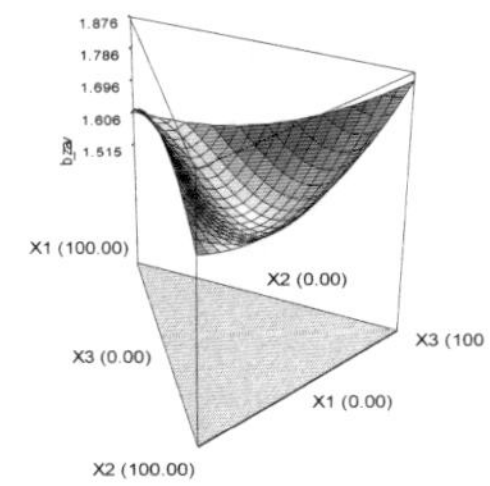
2	Quadratic "Prob>F"<0,0001 Adjusted R-squared=0,9920 b_kor = +1.49894 * Ar +1.21394 * N2 +0.94894 * He -0.76879 * Ar * N2 +0.32121 * Ar * He +0.49121 * N2 * He	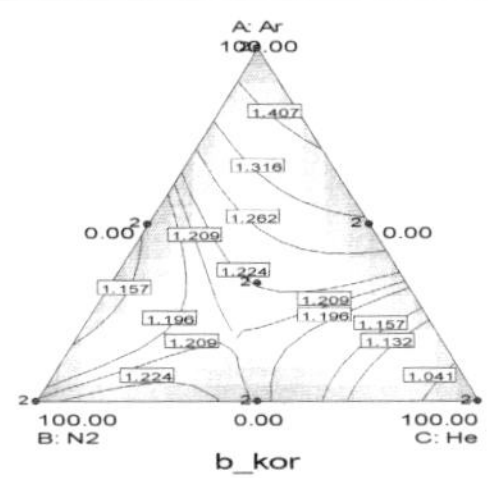	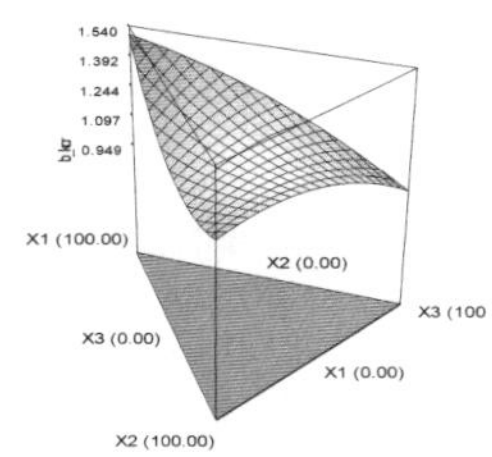
3	Reduced quadratic "Prob>F"<0,0001 Adjusted R-squared=0,9140 A = +2.750 * Ar +2.760 * N2 +2.437 * He -0.355 * Ar * He +0.385 * N2 * He	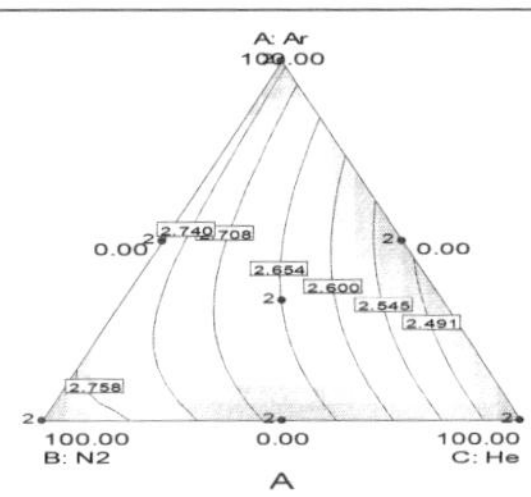	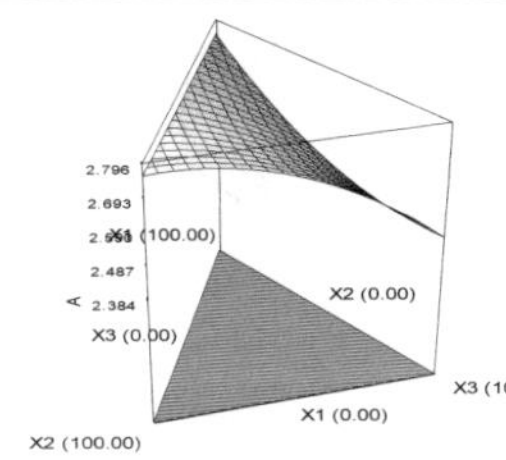

Table 3. Mathematical models and graphical plots for condition a (v_{max}, Q_{min})

3.2. *Microstructure of welded joints*

Structure of welded joint was made of austenite and ferrite. Ferritization happened in all states of experiment (ferrite portion was from 87% to 93%, what is much more than wanted 65%). In Table 4 portion of ferrite for every state of experiment is shown.

	I				II				III				IV			
State of ex.	1	2	3	4	5	6	7	8	9	10	11	12	13	14	15	16
% ferrite	89,3	90,3	91,1	91,3	87,8	91,8	91	89	92,1	89,8	91,6	87,4	90	88,2	91,7	91,4
	V				VI				VII							
State of ex.	17	18	19	20	21	22	23	24	25	26	27	28				
% ferrite	88,3	92,4	89,3	90	88,3	92	90,5	89	88,4	91,4	92,1	87,4				

Table 4. Ferrite portion in microstructure

Obtained models made possible to partially calculate ferrite portion that will be obtained with laser welding of duplex steels, using Ar, N_2, He and their mixtures as shielding gases, for all combinations of upper and lower level of welding parameters (maximum and minimum parameters).

3.2.1. *Relationship between microstructure of the welded joint and the composition of the shielding gas*

In Table 5. there is obtained mathematical model which shown gas mixture affect on ferrite portion in welded joint. An adjusted R-squared close to 1 and model variance analysis show there is significant special cubic model in state b, with all linear and quadratic members as significant, and one significant cubic member. In states 1, a and ab models are not significant, because adjusted R-squared is less than 0,75. In state b there i s an „Outlier“, which occurs using Ar/N_2/He mixture.

No	Model	2D plot	3D plot X1=A=Ar; X2=B=N_2; X3=C=He
b	Specijalni kubni "Prob>F"<0,0001 Prilagođeni R-kvadrat=0,9791 F = +88.300 * Ar +87.800 * N2 +92.100 * He +8.800 * Ar * N2 -7.600 * Ar * He -6.800 * N2 * He -14.250*Ar*N2*He	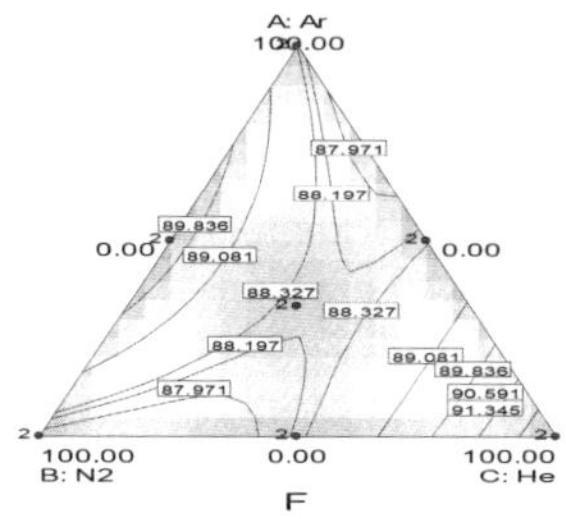	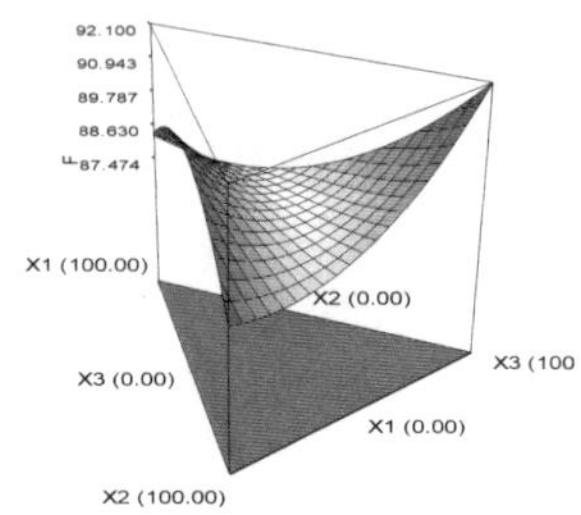

Table 5. Mathematical model and graphical plot for state b

From graphical plot of state b can be seen that helium and mixture Ar/N_2 give largest ferrite portion in welded joint. The lowest ferrite portion is obtained with nitrogen and argon, and mixing nitrogen and argon with helium, until helium portion is maximum 50%. Adding more than 50% of helium to nitrogen or argon causes ferrite portion starts to grow. Low ferrite portion is also obtained in center with mixture Ar/N_2/He. It can be concluded that type of shielding gas or mixture affects on portion of ferrite in welded joint only at minimum welding speed and maximum gas flow, what can be fine described with model. In other states of experiment mean value of measured ferrite portions better predicts response than the model.

3.2.2. *Relationship between the microstructure of welded joint and the welding parameters*

Significant models with Ar and N_2/He i Ar/N_2/He mixtures are obtained and they are shown in Table 6.

State	Model	2D plot	3D plot
Ar	Glavni efekti "Prob>F"=0,0066 Prilagođeni R- kvadrat=0,8119 F = +86.820 -0.180 * Q +0.045 * v	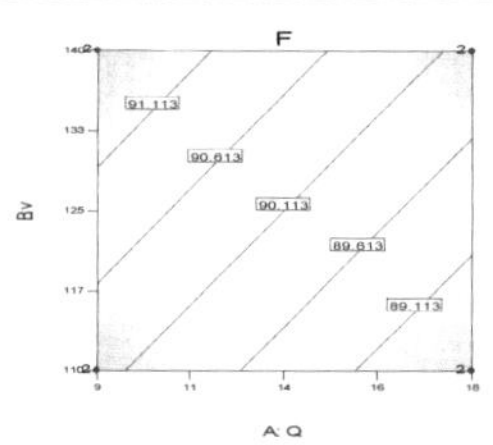	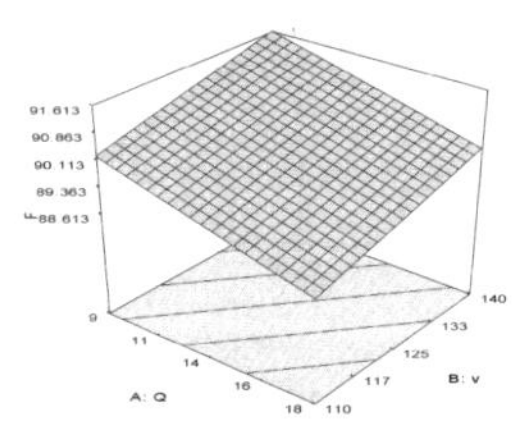
N_2/He	Interakcijski "Prob>F"=0,0006 Prilagođeni R- kvadrat=0,9677 F = +106.825 -1.351 * Q -0.145 * v +0.011 * Q * v	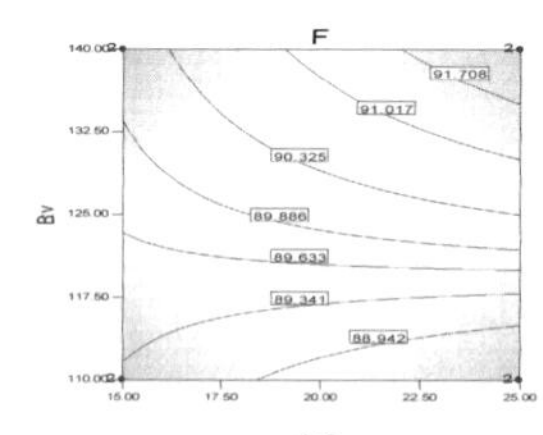	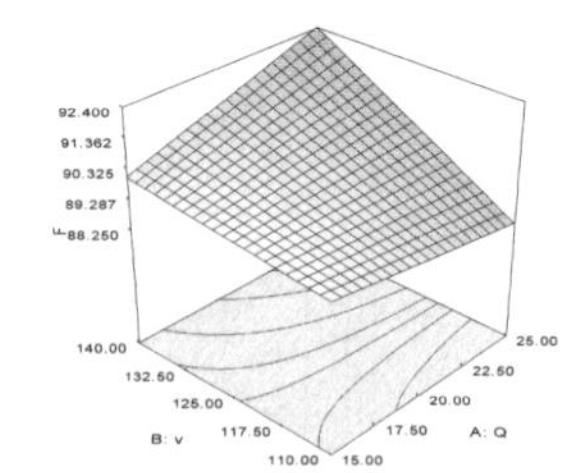
Ar/N_2/He	Interakcijski "Prob>F"=0,047 Prilagođeni R- kvadrat=0,9121 F = +123.803 -2.040 * Q **-0.241** * **v** +0.015 * Q * v	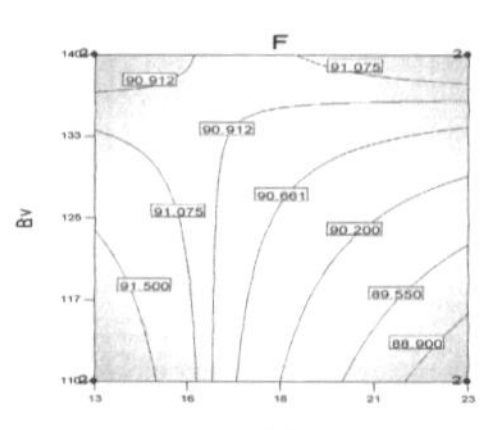	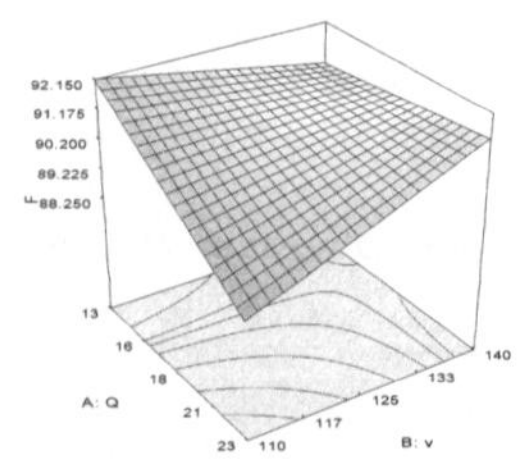

Table 6. Mathematical models depicting the impact of the welding parameters on portion of ferrite in welded joint

3.2.3. Mathematical models depicting the relationship between the composition of the shielding gases and the geometrical characteristics of the welded joint

In members of interaction models at N_2 and mixture $Ar/N_2/He$ gas flow is significant as linear member and as member in interaction with welding speed (Q*v). For mixture $Ar/N_2/He$ welding speed is not significant as linear member, but only in interaction with gas flow (Q*v). "Outliers" did not occur.
From graphical plot for mixture N_2/He it can be seen there is significant affect of welding speed and gas flow, and with decrease of welding speed and gas flow ferrite portion is reduced (for large flow more significant than for the low). Decrease of gas flow also affect reduction of ferrite portion (for high speed more significant than for the low). That is cause of interaction between welding speed and gas flow (Q*v). The lowest ferrite portion is obtained using minimum welding speed and minimum gas flow.

From graphical plot for mixture $Ar/N_2/He$ it can be seen that the lowest ferrite portion is obtained using minimum welding speed and maximum gas flow, and for mixture N_2/He only for minimum welding speed. For Ar there is significant linear impact of welding speed and gas flow. From graphical plot it can be seen that decreasing of welding speed and increasing of gas flow reduce ferrite portion. The lowest ferrite portion is obtained using minimum speed and maximum flow. Also, for argon there was an „Outlier“ in state 1 (v_{min}, Q_{min}).

For interaction and for linear models the lowest ferrite portion is obtained using minimum welding speed (that is accordant with literature), but impact of gas flow, so conclusion could be that impact of gas flow depends on type of mixture, but that should be subject for further researches.

4. Conclusion

Metal sheets of duplex steel W. Nr. 1.4462, 2-mm thick, were welded with a Nd:YAG laser. The shielding gases – argon, nitrogen, helium and their mixtures – were used and for root shielding argon was used. A coaxial shielding-gas nozzle was used. The welding was carried out according to the experimental model with mixtures. For the tested shielding gases and mixtures, high-quality welds were produced, meeting the requirements of the group B in EN ISO 13919-1 standard. No cracks occurred.

Statistical processing of the results and the developed mathematical models indicate that there is a significant effect of welding speed on the geometrical characteristics of the welded joint for the tested range of parameters, and not so much impact on microstructure. An increase of the welding speed reduces the geometrical characteristics, while the effect of the shielding gas flow rate is not significant.

Special cubic, quadratic and reduced quadratic mathematical models were obtained, which can accurately predict the effects of the shielding gases and their mixtures on the weld width, root width and the area of cross-section. The shielding gas/mixture has an effect on the geometric characteristics, while the model type depends on a combination of the welding speed and the flow rate of the shielding gas. No model is significant for the width of the HAZ, i.e., the mean value can better predict the response than the model. The type of mixture has no significant effect upon width of the HAZ. Using the obtained mathematical models it is possible to optimize the composition of the gas mixture with regard to the geometrical characteristics of the laser-beam welded joints.

In all states of experiment microstructure consisted of austenite and ferrite, and ferrite portion was from 87% to 93%. By analising mathematical models it was noticed that shielding gas type (or mixture) affects ferrite portion only using minimum welding speed and maximum gas flow. The lowest ferrite portion is obtained using nitrogen and argon as shielding gas, and mixing argon and nitrogen with helium, but until helium portion is maximum 50%.

Also, significant models which describe impact of welding speed and gas flow of ferrite portion in welded joint using argon and mixtures N_2/He i $Ar/N_2/He$ are obtained. The lowest ferrite portion is obtained using lower welding speeds, and gas flow impact on ferrite portion varies from model to model. That means that gas flow impact on ferrite portion in welded joint depends on mixture type, and to describe that more precisely further researches have to be done, and that also could be some proposal for anyone interested in that.

An analysis of the weld appearance quality and the geometrical characteristics of the welded joint, points towards the mixture (Ar/N2) as being optimal. Considering the root width, the best choice is argon. The best choice to prevent ferritization would be pure nitrogen, provided that an acceptable microstructure of the weld cross-section, while additions of helium cause a reduction. Duplex stainless steel W.Nr.1.4462 can be successfully welded by applying a laser using the shielding gases. The choice of shielding gas can affect the microstructure of the welded joint. An evaluation of the effects of the tested gas mixtures mechanical properties and corrosion resistance of duplex grade steels will be the topic of future research.

5. Acknowledgments

The materials and shielding gases were sponsored by Messer Croatia Plin d.o.o. We express our thanks to the Messer Croatia Plin and ROFIN-SINAR Laser, Hamburg for their technical and financial support.

6. References

[1] Šimunović V., Zavarivanje, Zagreb, 43 (2000), ¾, 75

[2] Steffens, H.-D., Honekamp E., Wilden J., DVS Berichte Band 194, Düsseldorf, (1999), 199

[3] Marquardt E., Sitte G., Plasma- und Laserstrahlschweißen von Feinblechen aus Duplexstahl 1.4462, Abschlußbericht, SLV Halle, 1996

[4] Dilthey U., Wieschemann A., DVS Berichte Band 205, Düsseldorf, (1999), 41

[5] Borggreen K., Klaestrup Kristensen J., Hansen L .E., Kocak M., dos Santos J. F., Laser welding of heavy section duplex stainless steel grade 2205, Proc. Of Stainless Steel World 99 Conference, (1999), 267-274

[6] Faerber M., Process gases for laser welding, Proc of CISFFEL 6, Toulon, (1998), 837-841

[7] Danzer W., Ammann T., Bauer B., Zavarivanje, Zagreb, 49 (2006), 1-2, 5

[8] Bauer B., Kralj S., Miculinić M., Welding in the world, 51 (2007), 835.

[9] Cornell J. A., Experiments with mixtures, John Willey & Sons, New York, 1981

[10] Panić V., Master Thesis, Faculty of Mechanical Engineering and Naval Architecture, Zagreb, Croatia, 2001.

[11] Šakić N., Panić V., Experimental design for investigating the influence of process variables and mixture components, Proc. Of 3rd International Conference Welding in Maritime Engineering, Hvar (2004), 495-500

[12] Bauer B., Ph. D. Thesis, Faculty of Mechanical Engineering and Naval Architecture, Zagreb, Croatia, 2006.

[13] Topić A., Ph. D. Thesis, Faculty of Mechanical Engineering and Computing, Mostar, Bosnia and Herzegovina, 2008.

DOI: 10.2507/27th.daaam.proceedings.107

MODEL OF FUZZY LOGIC FOR SELECTION INFRASTRUCTURAL INVESTMENT PROJECT OF WIND FARM LOCATIONS

Zajim Aljicevic, Aleksandra Kostic, Nedis Dautbasic & Gunay Karli

This Publication has to be referred as: Aljicevic, Z[ajim]; Kostic, A[leksandra]; Dautbasic, N[edis] & Karli, G[unay] (2016). Model of Fuzzy Logic for Selection Infrastructural Investment Project of Wind Farm Locations, Proceedings of the 27th DAAAM International Symposium, pp.0743-0748, B. Katalinic (Ed.), Published by DAAAM International, ISBN 978-3-902734-08-2, ISSN 1726-9679, Vienna, Austria
DOI: 10.2507/27th.daaam.proceedings.107

Abstract

The total energy consumption in world is exponentially increasing. Infrastructural investment in wind energy project is the best solution for resolving energy problem. Selecting the site for wind farm is a complex process involving not only technical requirement, but also physical, economical, social, environmental and political requirements that may result in conflicting objectives, and that's why, it is necessary to create model that provides most desirable location for investing.

The present paper uses fuzzy logic methodology and MATLAB to assess wind sites and decide which sites should be given the highest priority with respect to their benefits and costs. The criterion of evaluation using fuzzy logic is based on different parameters, i.e., wind resources, prevailing wind direction, above ground level, site capacity, soil conditions, site elevation, land cost, land roughness, temperature, cultural and environmental concerns, aviation/telecommunications conflicts, nearby resident's concerns, site environmental issue and distance to transmission line.

For verification of our solutions, we had deployed the available data on the wind farm in Podveležje and based on our model, we may conclude that the site selection is quite good.

Keywords: Fuzzy logic; Infrastructural project; Wind farm; Renewable resources; Wind power; Site selection

1. Introduction

Increasing environmental concern during the 21th Century has moved the research focus from conventional electricity sources to renewable and alternative energy solutions. In renewable power generation, wind energy has been noted as the fastest-growing energy technology in the world, due to the fact that world has enormous resources of wind energy. Bosnia and Herzegovina also has a big wind potential which need to be used.

At Global Wind Energy Council (GWEC, in April 2016, it is noted that wind power had another record-breaking year. After passing the 50 GW mark for the first time in a single year in 2014, it reached another milestone in 2015 as annual installations topped 63 GW, a 22% increase. By the end of last year, there were about 433 GW of wind power spinning around the globe, a cumulative 17% increase; and wind power supplied more new power generation than any other technology in 2015, according to the IEA[1] (Source: GWEC – Global Wind Energy Council)

Many studies have been proposed to optimize wind power generation. Shata and Hanitsch [2] have evaluated different wind sites in Egypt using classical statistical analysis to estimate the best wind site. Koçak [3] focused entirely on wind speed persistence during weather forecast, site selection for wind turbines and synthetic generation of the wind speed data. Herbert et al [4] developed models for wind resources assessment, site selection and aerodynamic including wake effect to improve the wind turbine performance and to increase its productivity. Lackner et al [5] utilized ground-based measurement devices instead of meteorological towers for wind resource assessment. They investigated the use of a monitoring strategy in more than one site to determine the best wind sites.

This paper utilizes fuzzy logic to evaluate various sites in the bases of their benefits and costs to adopt the higher priority sites in terms of different criteria. The paper is organized so that in Section 2 is given a general overview of the installed capacity of wind power. In Section 3 are given wind characteristics in Bosnia and Herzegovina. In Section 4 is shown fuzzy logic methodology. In Section 5 are presented fuzzy logic rules. In the sixth Section is explained defuzzification. The conclusion was made verification using data on existing wind farm in Podveležje, and plan future research.

2. Renewable resources

Today, new renewable resources provide only a small share of global energy production (Figure 1). However, the global market for wind power has been expanding faster than any other sources of renewable energy. The world wind power capacity has been duplicated around twenty five times from just 17.4 GW in 2000 to reach over than 432.8 GW at the end of 2015 (Figure 2).

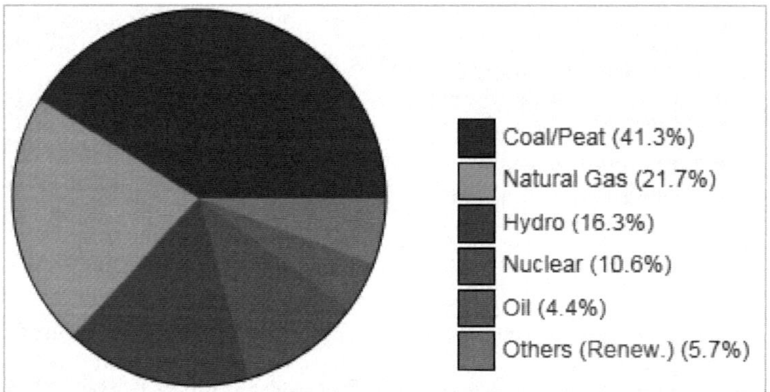

Fig. 1. Shares of world electricity generation *(EIA, 2015).*

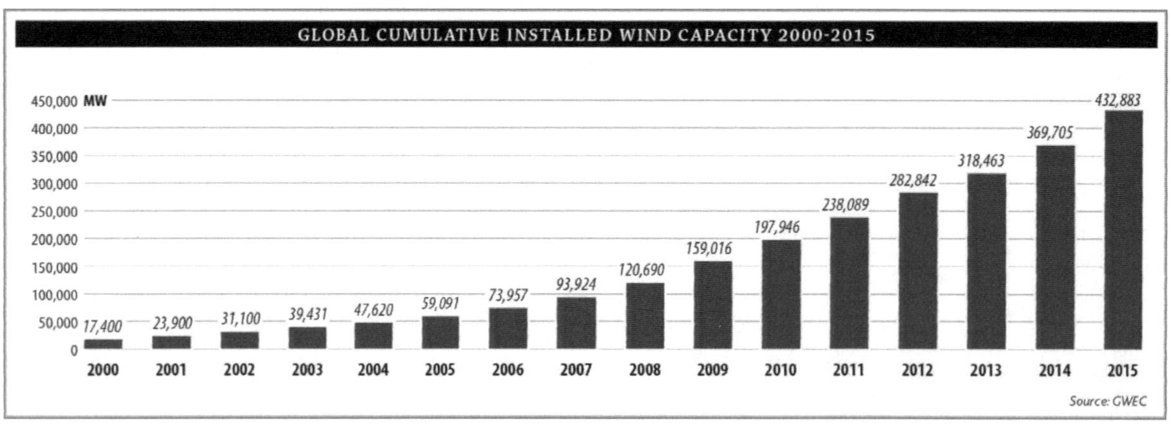

Fig. 2. Global cumulative wind power capacity (Source: GWEC 2015)

In 2015, increase in wind generation was equal to almost half of global electricity growth. This was surprising but welcome news. It became apparent from new IEA analysis revealing that, for the second successive year, global CO2 emissions remained stable despite growth in the world economy.

It is generally known that we need to invest in wind power. The main problem is how to choose the best location for investment in wind power plants.

This paper utilizes fuzzy logic to evaluate various sites in the bases of their benefits and costs to adopt the higher priority sites in terms of different criteria, i.e., same benefits factors are: wind resources, prevailing wind direction, above ground level, site capacity, site accessibility, soil conditions, site elevation. Cost factors are: land cost, land roughness, temperature, cultural and environmental concerns, aviation/telecommunications conflicts, nearby resident's concerns, site environmental issues (corrosion, humidity), and distance to transmission line, as is used in [6].

3. Wind characteristics in Bosnia and Herzegovina

In the previous period in Bosnia and Herzegovina wind characteristics have been measured at weather stations. Study of Energetic sector [7] shows average annual wind speed at height 50 m above ground for the period 1997-2006. (Figure 3)

Therefore, the first analysis, research and examinations of wind power potential are linked to the area of south Bosnia and Herzegovina. In April 2002 at the location Podveležje Mostar the first measuring station with appropriate equipment is installed, and then the other ones followed. Inclusive to 31.12.2007, at 13 areas the total number of locations where measuring equipment has been installed is 33. [8-10].

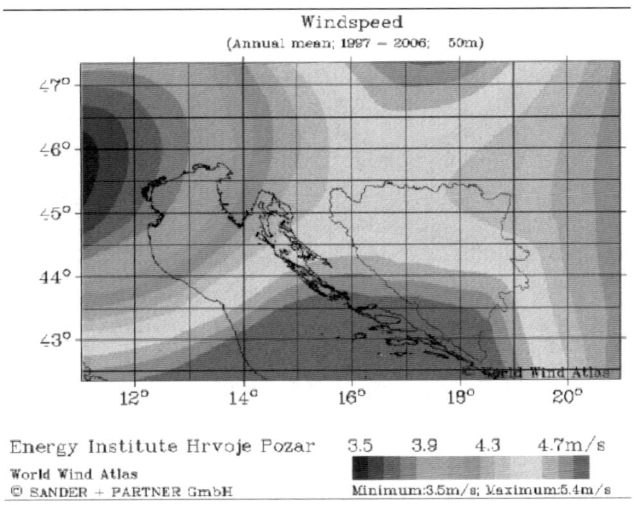

Fig. 3. Average annual wind speed at height 50 m above ground for the period 1997-2006

4. Fuzzy logic methodology for decision making

Fuzzy logic has a wide range of utilizations in decision making since it condenses a large amount of parameters into smaller fuzzy sets [11].

In this paper, we have selected Mamdami method because of its closeness to human understanding, since we had a large number of rules in decision-making.

The Fuzzy input/ output combination is shown in the Figure 4.

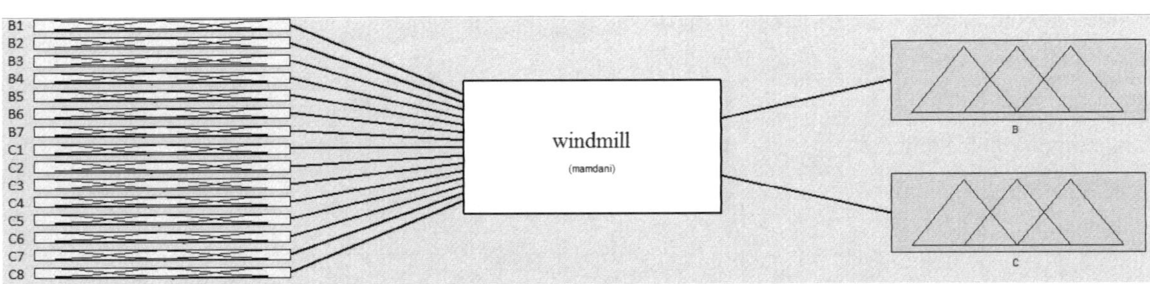

Fig. 4. Fuzzy input/output combination

The fuzzy logic decision selection of the sites options was applied according to benefits and costs:

Benefits:
B1= Wind resource,
B2= Prevailing wind direction,
B3= Above ground level (AGL) (m),
B4= Site capacity,
B5= Site accessibility,
B6= Soil conditions,
B7= Site elevation.

Costs:
C1= Land cost,
C2= Land roughness,
C3= Temperature (°C),
C4= Cultural and environmental concerns,
C5=Aviation/Telecommunications conflicts,
C6= Nearby resident's concerns,
C7= Site environmental issues,
C8= Distance to transmission line (m).

The fuzzy logic methodology is applied taking into account each site parameters for location Podveležje [8-10]. The inputs in tables 1 and 2 are considered to be the fuzzy variables, each variable can vary over a fixed range.

	B1	**B2**	**B3**	**B4**	**B5**	**B6**	**B7**	**Relative weight**	**Normalized relative weight**
Podveležje (BiH)	1	0.7	50	0.8	0.9	0.3	825	0.737	0.905

Table 1. Overall fuzzy weights for the selected sites based on benefits

	C1	**C2**	**C3**	**C4**	**C4**	**C6**	**C7**	**C8**	**Relative weight**	**Normalized relative weight**
Podveležje (BiH)	0.2	0.1	15	0.65	0.1	0.4	0.5	10	0.394	0.816

Table 2. Overall fuzzy weights for the selected sites based on costs

The linguistic variables used in the fuzzy methodology are: Very low (VL), Low (L), Normal (N), High (H), Very high (VH), Poor, Marginal, Satisfactory, Good, Excellent, Rock, Mostly rock, Rock/soil, Mostly soil, Soil, Moderate, None, Extensive, Negligible, Minor, Average, Very close, Close, Not far, Far, Very far. Each fuzzy set is addressed as listed in Table 3.

Parametres	**Symbol**	**Variable type**	**Linguistic variables**					**Range**
			1	**2**	**3**	**4**	**5**	
Wind resource	B1	INPUT	VL	L	N	H	VH	0 - 1
Prevailing wind direction	B2	INPUT	Poor	Marginal	Satisfactory	Good	Excellent	0 - 1
Above ground level	B3	INPUT	VL	L	N	H	VH	10-80 m
Site capacity	B4	INPUT	Poor	Marginal	Satisfactory	Good	Excellent	0 - 1
Site accessibility	B5	INPUT	Poor	Marginal	Satisfactory	Good	Excellent	0 - 1
Soil conditions	B6	INPUT	Rock	Mostly rock	Rock/Soil	Mostly soil	Siol	0 - 1
Site elevation	B7	INPUT	VL	L	N	H	VH	0-2100 m
Land cost	C1	INPUT	VL	L	N	H	VH	0 - 1
Land roughness	C2	INPUT	VL	L	N			0 - 1
Temperature	C3	INPUT	None	Moderate	Extensive			-40-40°C
Cultural and environmental concerns	C4	INPUT	None	Moderate	Extensive			0 - 1

Aviation/ Telecommunic ations conflicts	C5	INPUT	None	Moderat e	Extensive			0 - 1
Nearby resident's concerns	C6	INPUT	Negligible	Moderat e	Extensive			0 - 1
Site environmental issues	C7	INPUT	None	Minor	Average	Moderate	Extensiv e	0 - 1
Distance to transmission line	C8	INPUT	Very close	Close	Not far	Far	Very far	0-30 km
	B	**OUTPUT**	**VL**	**L**	**N**	**H**	**VH**	**0 - 1**
	C	**OUTPUT**	**VL**	**L**	**N**	**H**	**VH**	**0 - 1**

Table 3. Fuzzy sets

5. Constructing fuzzy rules

Sixty five rules were used in the current fuzzy method implementation to predict the most preferable options or option out of the Podveležje site.

Fuzzy logic enabled us to dense large amount of data, collected to compare between different sites, into a smaller set of variable rules, to make a decision in the basis of their merits and barriers to produce higher power output at low cost as well as to capture as maximum wind power as possible.

The benefit to cost ratio is shown in Table 4.

Site	Normalized benefit relative weight	Normalized cost relative weight	B/C	Normalized B/C
Podveležje (BiH)	0.905	0.816	1.109	0.671

Table 4. Benefit to cost ratio

6. Performing fuzzy interface

Finally, the mapping process takes place to provide the final decision as shown in Figure 5.

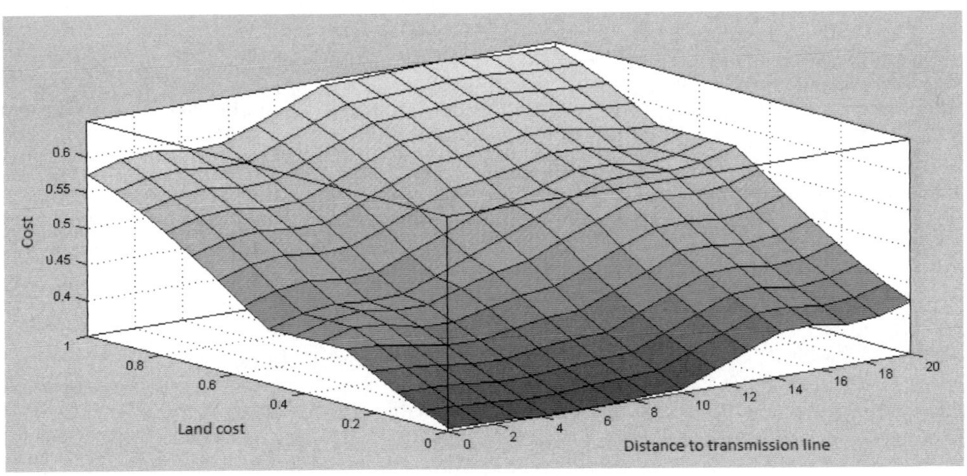

Fig. 5. Fuzzy implementation sequence

Defuzzification process assesses the outcome in the decision making process. The parameters are analyzed individually, and in this process it is possible to observe two parameters and decide which of them more influenced our results.

This is especially important and helpful in choosing a location for a wind farms. It helps in determining the optimal location. For example in the case that we have two locations with similar parameters then based on those parameters which are different we can make analysis and bring the correct conclusion that the location is favorable.

Also, on the basis of this method it is possible to perform a number of conclusions. For example analysis parameters, land cost and distance to transmission line can be concluded that the land cost is much more important in the decision-making process (Figure 5). The image is clear that the parameter 8 (distance to transmission line) has very little effect, and its only effect is that if the transmission line is over a distance of 10 km, or more than 10 km while any change in parameter C1 (land cost) leads to changes in the final results.

7. Conclusion

The purpose of this paper is to use fuzzy logic and MATLAB for predict the best location for wind farm.

In deciding the used 15 input parameters, where seven parameters presents benefits and eight parameters represents costs. Used a total of 65 rules for decision making.

This model includes all relevant parameters to determine the optimal location for the construction of a wind farm. The advantage of the model is the possibility of determining the advantages of a parameter to the second parameter to a certain value. It is notable that in some parameters higher changes their values do not lead to large differences in profitability of a wind farm until some parameters and their values minor changes lead to great changes in profitability.

For verification of our solutions, we used data on existing wind farm in Podveležje. Based on our model, we concluded that the choice of location in this case was quite good. Normalized benefit relative weight 0.905, normalized cost relative weight 0.816), or by defined outputs it is between high and very high.

In further research is necessary to collect all the relevant information for potential locations where it is possible to build wind parks. These data would be used in the current model presented in this paper to determine the viability of investments in the wind park.

8. References

[1] http://www.gwec.net, (2016). GWEC-Global Wind 2015 Report, Accessed on: 2016-09-27

[2] Shata, A.S. &Hanitsch, R. (2006). Evaluation of wind energy potential and electricity generation on the coast of Mediterranean Sea in Egypt, Renewable Energy, Vol.31, No.8, page numbers (1183-1202), 0960-1481

[3] Koçak, K. (2008). Practical ways of evaluating wind speed persistence, Energy, Vol.33, No.1, page numbers (65-70), 0360-5442

[4] Herbert, G.M.; Iniyan, S.; Sreevalsan, E. &Rajapandian, S.A. (2007). Review of wind energy technologies, Renewable and Sustainable Energy Reviews, Vol.11, No.6, page numbers (1117-1145), 1364-0321

[5] Lackner, M.A.; Rogers, A.L. &Manwell, J.F. (2008). The round robin site assessment method: A new approach to wind energy site assessment, Renewable Energy, Vol.33, No.9, page numbers (2019-2026), 0960-1481

[6] Badran, O.; Abdulhad, E. &El-Tous, Y. (2011). Fuzzy Logic Controller for Predicting Wind Turbine Power Generation, International Journal of Mechanical and Materials Engineering, Vol.6, No.1, page numbers (51-66), 2198-2791

[7] Study of Energetic sector in B&H, Energetski institut Hrvoje Požar - Croatia, Soluziona - Spain, Ekonomski institut Banjaluka, - BH, Rudarski institut Tuzla – BH, 2008

[8] Elvir, Z.; Mehmed, B.; Fuad, Ć. &Enes, S. (2008). Energy resources from wind energy in Bosnia & Herzegovina, current state and prospects, 12 International Research/Expert Conference "Trends in the Development of Machinery and Associated Technology" TMT 2008, Istanbul, Turkey

[9] Behmen, M.; Zlomušica, E. & Ćatović, F. (2004). The Influence of Electro-Supply Capacities on the State of the Environment in B&H, Status and Perspectives, 8th International Research/Expert Conference Trend in the Development of Machinery and Associated Technology-TMT 2004, Neum, Bosnia and Herzegovina

[10] Ćatović, F.; Behmen, M. & Zlomušica, E. (2004) Trends in the development of the electric power systems based on wind energy in world and in Bosnia and Herezegovina, Journal of Enviromental Protection and Ecology-Official Journal of the Balkan Enviromental Association (B.EN.A), Vol. 5, No.4, page numbers (836-840), 1311-5065

[11] Nyrkov, A; Chernyi, S.; Zhilenkov, A. & Sokolov, S. (2016). The use of Fuzzy Neural Structures to Increase the Reliability of Drilling Platforms, Proceedings of the 26th DAAAM International Symposium, pp.0672-0677, B. Katalinic (Ed.), Published by DAAAM International, ISBN 978- 3-902734-07-5, ISSN 1726-9679, Vienna, Austria

DOI: 10.2507/27th.daaam.proceedings.108

SIMULATION AND VISUALISATION OF A LABORATORY MODEL BY SUPPORT SOFTWARE TOOLS CONNECTION

Jiri Zatopek

Department of Automatic and Control Engineering, Faculty of Applied Informatics, Tomas Bata University in Zlin Nad Stranemi 4511, 760 05 Zlin, Czech Republic

This Publication has to be referred as: Zatopek, J[iri] (2016). Simulation and Visualisation of a Laboratory Model by Support Software Tools Connection, Proceedings of the 27th DAAAM International Symposium, pp.0749-0755, B. Katalinic (Ed.), Published by DAAAM International, ISBN 978-3-902734-08-2, ISSN 1726-9679, Vienna, Austria
DOI: 10.2507/27th.daaam.proceedings.108

Abstract

This article introduces the possibilities of the simulation and visualisation of the "Twin-Rotor MIMO System" laboratory model outputs by means of various support software tools. The 3D model of the system, (used for simulation and visualisation), is designed in SolidWorks 3D CAD software. Matlab/Simulink with extension libraries like Simscape and 3D Animation - (formerly Virtual Reality Toolbox), is used for 3D visualisation and simulation. The 3D Animation toolbox is only used for the visualisation of the mathematical and real models. The Simscape library - on the other hand, is used for the validation of the reverse control of the derived mathematical model´s correctness and for simulation and analysis purposes as a suitable substitution for real models. As a result of this, these supporting software tools streamline the overall suggested controls - from analysis to presentation of the results.

Keywords: Motion Control; Simulation; Visualisation; Matlab/Simulink; Simscape; 3D Animation

1. Introduction

Today, a wide range of support software tools are available. These tools facilitate work in analysing, modelling, simulating, and designing control laws for complex mechanical structures. By interconnecting these resources, a considerable simplification of the synthesis can be achieved - as well as clarity, by using visualisation in 3D space directly in the regulation process, or by creating a physical model in Simulink directly from CAD software (SolidWorks), without the need to derive the motion equations [3, 10].

The "Twin Rotor MIMO System" laboratory model is the subject of interest. It contains Teaching Toolboxes with a derived mathematical model, a Simulink model and 1DOF and 2DOF PID controllers [7, 11], which are used as a link between the laboratory model, the physical (Simscape) model, and the graphic (3D Animation) model.

First of all, the 3D model will be designed in the 3D CAD software – SolidWorks environment. The precision of the 3D CAD model should be a compromise between the complexity of the model and the actual impact on the overall dynamics of the system – i.e. it must adequately reflect the behaviour of a real model. The model contains masses of individual parts and the resulting dynamic properties of the entire assembly [9, 2]. The model, designed using 3D SolidWorks, is exported directly into a suitable format for the 3D Animation toolbox and the Simscape/SimMechanics library [10].

The 3D Animation toolbox - intended for visualisation, uses only the appearance properties of the SolidWorks model and is applied to the mathematical and laboratory model outputs´ visualisation. The mathematical model´s outputs or captured laboratory model variables (for virtual animation purposes) are used as 3D Animation model inputs [6, 9]. These parameters are converted into the appropriate format by a transformation matrix using Euler parameters [1, 3, 4].

The Simscape – i.e. the SimMechanics specifically, uses the SolidWorks model as a whole (e.g. its material properties, physical links, etc.) to build a physical model. The model can be used for analysis and simulation purposes, (the mathematical model does not need to know), as well as to validate the mathematical model´s correctness [10, 8, 5].

The results of this work could be used for: inductive identification, design of control laws, mathematical model correctness validation, or the presentation of results.

2. Laboratory model

The laboratory model is used for educational purposes as an example of a non-linear motion system. It represents a model similar to a tethered helicopter, with orthogonally disposed axes of rotation. The system is controlled by the voltage change of a pair of DC motors, in the 0 to 5V range. The laboratory model output is information about rotations in the vertical (i.e. elevation); and horizontal (azimuth) directions. Measurements of the rotation information is realised by incremental sensors [7, 11].

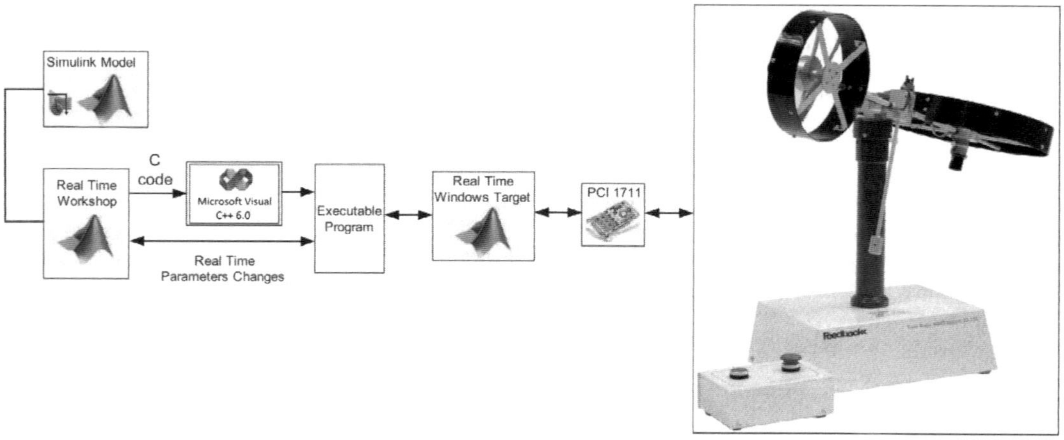

Fig. 1 Laboratory model in a block chart [7, 12]

3. SolidWorks model

The SolidWorks model consists of 9 parts - as shown in Fig. 2.b. Each part has an associated local coordinate system, mass, and material, and is bounded by structural bonds to the other parts of the assembly (Fig. 2.a). Working directly in SolidWorks, it is possible to determine the centres of mass positions, the inertia matrices or even to execute a physical analysis for example.

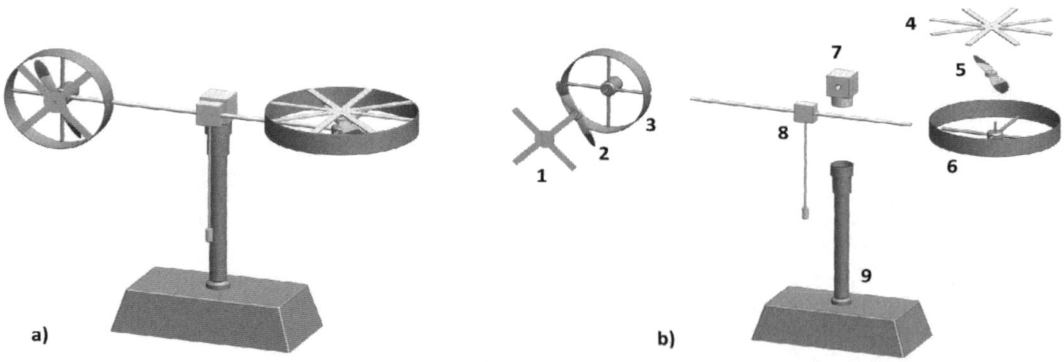

Fig. 2 SolidWorks model (a); and exploded view (b)

4. 3D Animation

Fig. 3 VR Sink block (a), and Web-browser visualisation (b), in 3D Animation

The 3D CAD model cannot be exported into a format for 3D Animation with linkages; thus all nine bodies are freed up. Their position in 3D space is determined by a direction vector with its origin in a global coordinate system, and its orientation is determined by a unit vector and the angle of rotation around this unit vector. However, the only parameters available are those of the azimuth and elevation - by means of which it is possible to calculate the desired parameters (transformation matrices and Euler's Theorem will be used for this) [9].

4.1. Transformation of coordinate systems

The transformation matrices are used to determine the position and orientation of a local coordinate system firmly connected with the component in the overall global coordinate system. In this case, we need to know only 3 transformation matrices since Part 9 in Fig. 2.b is fixed (i.e. connected to the ground); while Part 7 rotates azimuthally around a global axis (i.e. there is no need for a transformation matrix); and Parts 1, 3, 4, 6, 8 in Fig. 2.b are connected by mutual firm bonds - they are only relatively displaced (i.e. a mutual rotation transformation matrix).

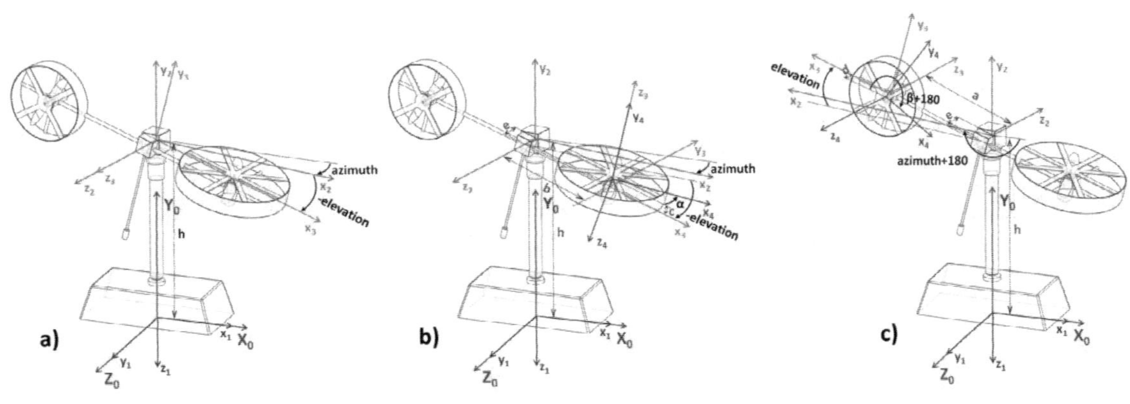

Fig. 4 Model with the selected coordinate system placement

The coordinate system placement for the determination of the transformation matrix is shown in Fig. 4. Fig. 4.a has the last local coordinate system linked with an arm-mounting point, and the resulting transformation matrix is used to determine the orientation of Parts 1, 3, 4, 6, 8 (shown in Fig. 2.b) in 3D space. Fig. 4.b has the last local coordinate system linked with the centre of rotation of the main propeller, and the resulting transformation matrix is used for determining the orientation of Part 5. Fig. 4.c has the same use as Fig. 4.b - but is for the secondary propeller.

The Denavit-Hartenberg notation/method of coordinate systems placement (DH) [1, 9] is used for the derivation of the transformation matrix. The interface for determining the general (i.e. symbolically specified) transformation matrix was created in Matlab. It has a graphical interface, and uses the Symbolic Math Toolbox (Fig. 5).

The resulting transformation matrices are transformations from the local coordinate system into the global coordinate system (${}^{G}T_{B}$) [1,9] and are listed in (1) for Fig. 4.a, (2) for Fig. 4.b and (3) for Fig. 4.c.

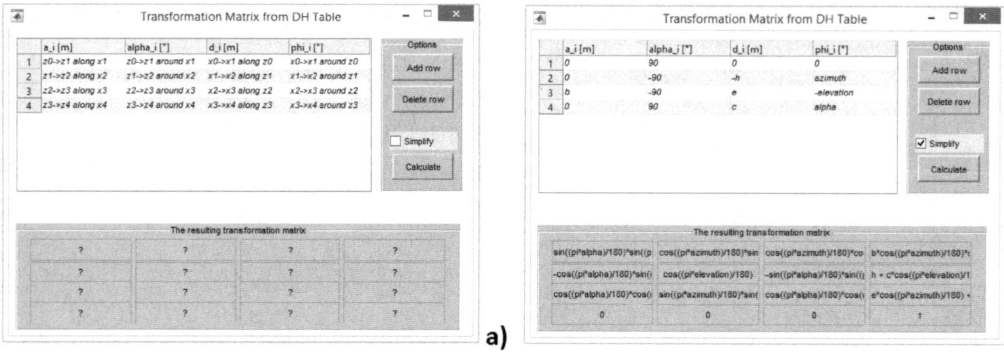

Fig. 5 Matlab program - designed for determining the transformation matrix

$$ {}^{G}T_{1\,B} = \begin{bmatrix} cos(\varepsilon)cos(\varphi) & cos(\varepsilon)sin(\varphi) & -sin(\varepsilon) & 0 \\ -sin(\varphi) & cos(\varphi) & 0 & h \\ sin(\varepsilon)cos(\varphi) & sin(\varepsilon)sin(\varphi) & cos(\varepsilon) & 0 \\ 0 & 0 & 0 & 1 \end{bmatrix} \tag{1} $$

$$ {}^{G}T_{2\,B} = \begin{bmatrix} sin(\alpha)sin(\varepsilon)+cos(\alpha)cos(\varepsilon)cos(\varphi) & cos(\varepsilon)sin(\varphi) & sin(\alpha)cos(\varepsilon)cos(\varphi)-cos(\alpha)sin(\varepsilon) & b\cdot cos(\varepsilon)cos(\varphi)-e\cdot sin(\varepsilon)+c\cdot cos(\varepsilon)sin(\varphi) \\ -cos(\alpha)sin(\varphi) & cos(\varphi) & -sin(\alpha)sin(\varphi) & h+c\cdot cos(\varphi)-b\cdot sin(\varphi) \\ cos(\alpha)sin(\varepsilon)cos(\varphi)-sin(\alpha)cos(\varepsilon) & sin(\varepsilon)sin(\varphi) & cos(\alpha)cos(\varepsilon)+sin(\alpha)sin(\varepsilon)cos(\varphi) & e\cdot cos(\varepsilon)+b\cdot sin(\varepsilon)cos(\varphi)+c\cdot sin(\varepsilon)sin(\varphi) \\ 0 & 0 & 0 & 1 \end{bmatrix} \tag{2} $$

$$ {}^{G}T_{3\,B} = \begin{bmatrix} -cos(\beta+\varphi)cos(\varepsilon) & -sin(\beta+\varphi)cos(\varepsilon) & -sin(\varepsilon) & -d\cdot sin(\varepsilon)-e\cdot sin(\varepsilon)-a\cdot cos(\varepsilon)cos(\varphi) \\ sin(\beta+\varphi) & -cos(\beta+\varphi) & 0 & h+a\cdot sin(\varphi) \\ -cos(\beta+\varphi)sin(\varepsilon) & -sin(\beta+\varphi)sin(\varepsilon) & cos(\varepsilon) & d\cdot cos(\varepsilon)+e\cdot cos(\varepsilon)-a\cdot sin(\varepsilon)cos(\varphi) \\ 0 & 0 & 0 & 1 \end{bmatrix} \tag{3} $$

In which:

ε ... azimuth
φ ... elevation

In order to determine the orientation of the individual parts of the system, the Axis-angle representation was predicted using Euler's Rotation Theorem [9].

4.2. Axis-angle representation - Euler's Rotation Theorem

Euler's Rotation Theorem dictates that any rotation or sequence of rotations of a rigid body in a three-dimensional space is equivalent to a pure rotation about a single, fixed, axis. Using the transformation matrix (${}^{G}R_{B}$) – i.e. Rodriguez' Rotation Formula, and Euler's Rotation Theorem, the following relations can be written [9]:

$$ \Phi = arccos\left(\frac{tr\left({}^{G}R_{B}\right)-1}{2}\right) \tag{4} $$

$$ \tilde{u} = \begin{pmatrix} 0 & -u_3 & u_2 \\ u_3 & 0 & -u_1 \\ -u_2 & u_1 & 0 \end{pmatrix} = \frac{1}{2\cdot sin(\Phi)}\cdot\left[{}^{G}R_{B}-\left({}^{G}R_{B}\right)^{T}\right] \tag{5} $$

Where:

(u_1, u_2, u_3) ... is the unit vector that determines the direction of the rotation axis
Φ ... is the rotation angle around the unit vector

After the substitution of the derived transformation matrix into (4) and (5), a transcription into the form of (u_1, u_2, u_3, Φ), we derive a vector which will be connected directly to a *VR Sink* block *rotation* parameter. The *centre* and *translation* parameters are also calculated using transformation matrices. The *centre* parameter is the origin of the unit vector, and the *translation* parameter is the position of the local coordinate system in the global coordinate system – a transformation matrix multiplied by the vector (0, 0, 0, 1) [9]. It is possible to finalize the animation at this moment by using Euler's Rotation Theorem and transformation matrices above-mentioned. All parts are virtually bounded over the transformations connected to VR Sink Block shown in Fig. 3.a. Final 3D animation of the laboratory model is shown in Web-browser in Fig. 3.b.

5. The Simscape/SimMechanics

The SolidWorks 3D model is then exported into the SimMechanics Second Generation scheme, along with all linkages and masses - and therefore contains the dynamics of the entire assembly, and is suitable for simulation purposes. Damping coefficients and spring stiffness can be added to the contact surfaces and joints, the parts can be modified (e.g. by changing the masses, moments of inertia, etc.), or the whole assembly´s kinematic structure can be changed. Also, it is possible to simulate the torques based on the generated kinematic movements, for each joint.

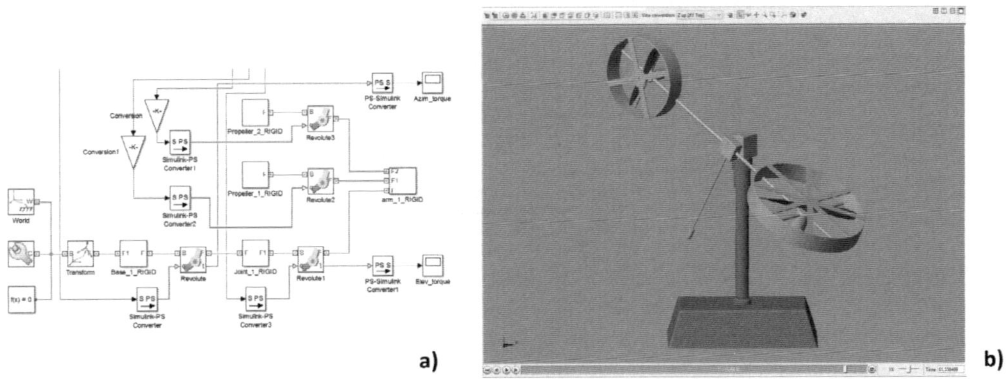

Fig. 6 SimMechanics Second Generation scheme (a) and its simulation (b)

The SimMechanics scheme is shown in Fig. 6.a. It includes 4 rotational kinematic pairs, which determine the types of linkages between the five independent parts (i.e. Base - Fig. 2.b.9; Joint - Fig. 2.b.7; Arm - Fig. 2.b.1, 3, 4, 6, 8; Propeller_1 - Fig. 2.b.2; Propeller_2 - Fig. 2.b.5). The gravitational field vector is in the negative direction of the global axis Y_0. Fig. 6.b shows a 3D simulation of a SimMechanics model. Thus, it is possible to observe dynamic processes occurring in the system in real time - with the possibility of slowing down, speeding up, or pausing the simulation.

6. The Matlab/Simulink interconnection

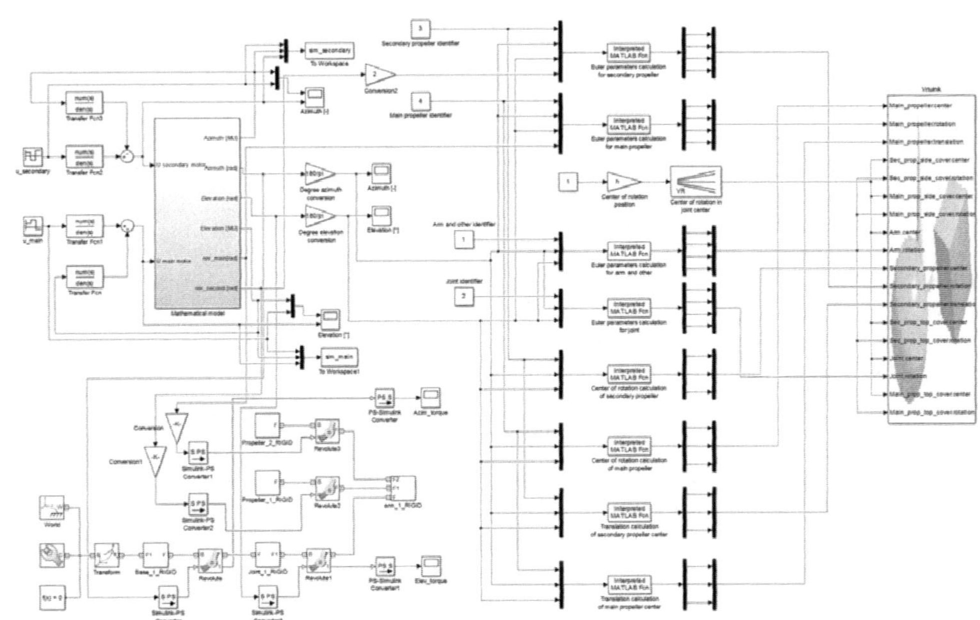

Fig. 7 Complete Simulink scheme

Fig. 7 depicts the complete Simulink scheme, which contains a Mathematical model part, the SimMechanics model, and the 3D Animation model with Euler Parameter Recalculations.

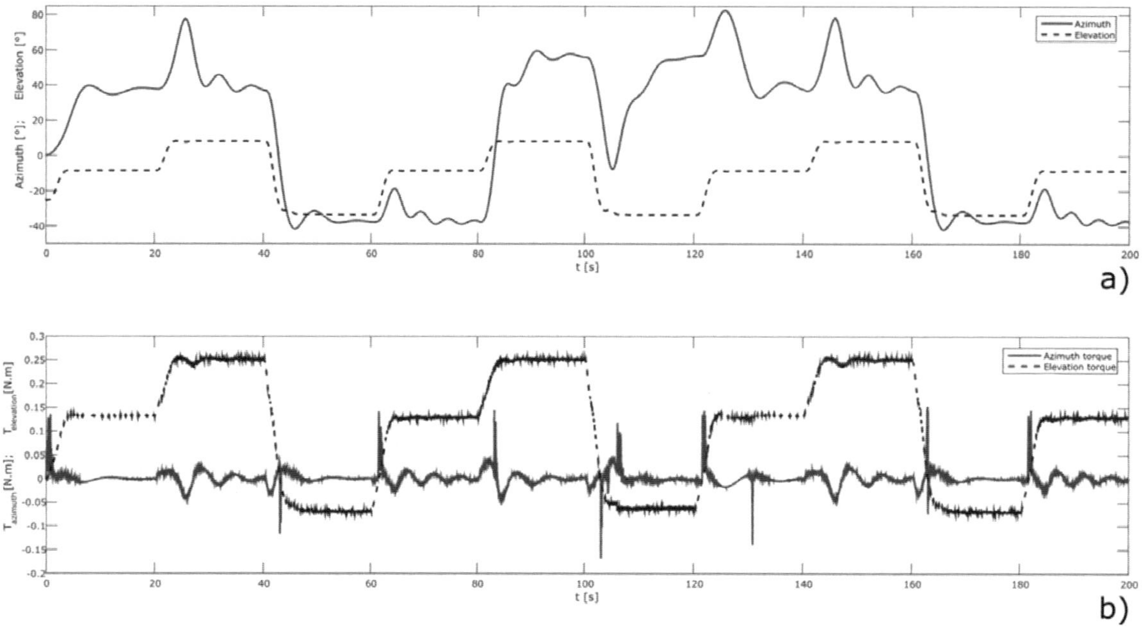

Fig. 8 Reference waveforms of laboratory model orientation (a) and required torque (b) needed to achieve them

The reference waveforms in Fig. 8.a are the controlled mathematical model outputs, which are then applied as required inputs into the SimMechanics model. The SimMechanics model is capable of generating the required moments (viz. Fig. 8.b), to perform the waveforms in Fig. 8.a. This is useful for instance, for the selection of suitable actuators for the system. If it is known - to what extent, and what dynamics the system requires, it is possible to use the SimMechanics model to determine what the requirements for the torque characteristics of the motor are.

In connection with the CAD model, the mathematical model, and the measurable state of the real model, the SimMechanics environment has many uses; this being only one of them.

7. Conclusion

Attention in this work was devoted to the simulation and visualisation of the "Twin Rotor MIMO System" laboratory model by using support software tool connections, which was the main intention of this paper. The whole assembly was designed in the SolidWorks 3D CAD system, which is used throughout this paper. The precision of the 3D CAD model should be a compromise between the complexity of the model and the actual impact on the overall dynamics of the system.

The first use of the 3D CAD model is for the visualisation of the mathematical model outputs using the 3D Animation toolbox. The position and orientation of each part is determined by The Denavit-Hartenberg Notation and Euler's Rotation Theorem. The Matlab interface was designed for the transformation matrix determined from the Denavit-Hartenberg table, and works with the Symbolic Math Toolbox. The 3D Animation toolbox is also applicable to visualise real model outputs and use of a web-browser could be helpful for the real-time presentation of a real model´s behaviour.

Another part relates to simulation by a physical model. This is also exported from the SolidWorks model, with all masses and links. The SimMechanics model could be used as a partial substitute for the real model - in the case where it is constructed with sufficient precision. The SimMechanics model was used as an aid for the selection of the appropriate actuators in this paper, by generating the required torque.

The significance of using the above-mentioned support software tools is their interconnection into one suitable software programme - Matlab. Simulation, visualisation, and other operations are solved by this interconnection; the Simulink model is prepared to further work with the "Twin Rotor MIMO System" laboratory model.

Further work will be focused on using the SimMechanics model for dynamic analysis purposes – e.g. the influence of the propellers´ rotation on the azimuth/elevation parameters from the perspective of centrifugal forces, the correction of friction coefficients by similarities observed between the real and physical models, or the online visualisation of real model outputs.

Support software tools connection is a popular theme today, especially 3D simulation/visualisation, not only for academics but also for students as a tool for the understanding of complex processes.

8- References

[1] Denavit, J. & Hartenberg, R. S. (1955). ASME Journal of Applied Mechanics", A kinematic notation for lower-pair mechanisms based on matrices, Vol. 77, No. 22, June 1955, pp. 215-221.

[2] R. N. Jazar. (2010). Theory of Applied Robotics, Springer US, ISBN: 978-1-4419-1750-8, Boston, MA

[3] Damic, V. & Cohodar, M. (2016). Dynamic Analysis and Visualization of Spatial Manipulators with Closed Structure, Proceedings of the 26th DAAAM International Symposium, pp.0109-0117, B. Katalinic (Ed.), Published by DAAAM International, ISBN 978-3-902734-07-5, ISSN 1726-9679, Vienna, Austria, DOI: 10.2507/26th.daaam.proceedings.016

[4] Shala, A., Likaj, R., Bruqi, M. & Bajrami, X. (2014). Propulsion Effect Analysis of 3Dof Robot under Gravity, Proceedings of the 25th DAAAM International Symposium, pp.206-212, B. Katalinic (Ed.), Published by Elsevier Ltd, ISSN 1877-7058, Vienna, Austria, DOI: 10.1016/j.proeng.2015.01.359

[5] Sudharsan, J. & Karunamoorthy, L. (2016). International Journal of Simulation Modelling", Path Planning and Co-Simulation Control of 8 DOF Anthropomorphic Robotic Arm, Vol. 15, No. 2, June 2016, pp. 302-312., ISSN: 1726-4529

[6] Flanders, M. & Kavanagh, R. C. (2015). Computer Applications In Engineering Education", Build-A-Robot: Using virtual reality to visualize the Denavit–Hartenberg parameters, Vol. 23, No. 6, Apr 2015, pp. 846-853., DOI: 10.1002/cae.21656

[7] http://www.feedback-instruments.com/products/education/control_instrumentation/twin_rotor_mimo, (2016). Feedback, Twin Rotor MIMO System, Accessed on: 2016-09-23

[8] Urednicek, Z. (2009). Electromechanical actuating devices, (In Czech) T. Bata university in Zlin, ISBN: 978-80-7318-835-1, Zlin

[9] Urednicek, Z. (2012). Robotics, (In Czech) T. Bata university in Zlin, ISBN: 978-80-7454-223-7, Zlin

[10] Zatopek, J. (2016). XLI. Seminar ASR '2016 "Instruments and Control", States of Nonlinear Unstable System Simulation, Connections between SolidWorks-Matlab/SimulinkSimMechanics-3D Animation, Apr 2016, pp. 163-172., ISBN: 978-80-248-3910-3, VSB-TUO, Ostrava

[11] Chalupa, P., Prikryl, J. & Novak, J. (2014). Modelling of Twin Rotor MIMO System, Proceedings of the 25th DAAAM International Symposium, pp.249-258, B. Katalinic (Ed.), Published by Elsevier Ltd, ISSN 1877-7058, Vienna, Austria, DOI: 10.1016/j.proeng.2015.01.365

[12] Belmonte, L. M., Morales, R., Fernandez-Caballero, A. & Somolinos, J. A. (2016). Sensors, Robust Decentralized Nonlinear Control for a Twin Rotor MIMO System, Vol. 16, No. 8, July 2016, pp. 1160-1181., DOI:10.3390/s16081160

DOI: 10.2507/27th.daaam.proceedings.109

MINING DATA STREAMS FOR THE ANALYSIS OF PARAMETER FLUCTUATIONS IN IoT-AIDED FRUIT COLD-CHAIN

Petar Juric, Marija Brkic Bakaric, Xiang Wang, Xiaoshuan Zhang & Maja Matetic

This Publication has to be referred as: Juric, P[etar]; Brkic Bakaric, M[arija]; Wang, X[iang]; Zhang, X[iaoshuan] & Matetic, M[aja] (2016). Mining Data Streams for the Analysis of Parameter Fluctuations in IoT-Aided Fruit Cold-Chain, Proceedings of the 27th DAAAM International Symposium, pp.0756-0761, B. Katalinic (Ed.), Published by DAAAM International, ISBN 978-3-902734-08-2, ISSN 1726-9679, Vienna, Austria
DOI: 10.2507/27th.daaam.proceedings.109

Abstract

The paper gives an overview of the state of the art methods and technologies in the field of stream data mining with applications in the Internet of Things systems for supporting fruit cold chain logistics. As the number of sensors used in on-line monitoring of the process is large, the amount of time series data is increasing rapidly. It is challenging to process such data in order to discover patterns, trends and outliers as a consequence of fluctuations of certain process parameters. In particular, the paper discusses methods for mining stream data collected in fruit cold chain aiming at real time control of fruit quality. A model of the centralized IoT system and the part responsible for monitoring fluctuations of temperature, humidity, and concentration of gases is proposed.

Keywords: Internet of Things, Mining data streams, Outlier detection, Parameter fluctuation, Cold-chain logistics

1. Introduction

Unlike the "classic" Internet which interconnects computer networks, the Internet of Things (IoT) refers to various areas with a common characteristic of connecting networks and devices from everyday life which can be uniquely identified on the Internet. IoT is used for monitoring and measuring parameters of interconnected physical world objects with the aim to improve business processes by data mining models and technics [1].

The first IoT prototype was devised by the scientists at the Auto-ID Centre of the Massachusetts Institute of Technology when they suggested designing a global network which would rely on the Radio Frequency Identification (RFID) technology for connecting devices. This network would enable identification and intelligent control over connected objects [2]. In the same year Kevin Ashtonn coined the term IoT to refer to an idea of using connected RFID devices within a distribution chain. Since data are generated by RFID and sensors, computers which have access to these data can observe, identify and understand the environment, and communicate (machine-machine) without limitations imposed by manual data input (time, concentration and precision) [3].

RFID technology in combination with wireless sensor networks enables monitoring environmental conditions important for quality preservation of temperature-sensitive food [4]. Beside temperature, it enables real-time humidity monitoring of agricultural products during transportation and storage in order to reduce loss and ensure high quality standards of these products [5]. State of the art devices also monitor other parameters that can affect food quality, such as shock, acceleration, light, temperature, sound, and combine these information with external information delivered over the Internet, such as rainfall, wind intensity, and route conditions [6].

IoT can help solving scalability and communication problems within sensor networks since it enables real time communication between sensor networks and cloud locations where data are analysed [10]. However, classic data mining methods cannot be applied since these systems generate Big Data [11], which is also stream data [15, 16]. This paper focuses on food cold chain data processing and proposes IoT cold chain model and its respective implementation with the support for simultaneous analysis of various environmental parameters and processing multitude of cold chains in both, online and offline modes. An example of processing data in time windows [17, 18] is provided. The purpose is to build an optimized system with quick autonomous reactions for food quality preservation.

2. Mining wireless sensor networks data

Applying wireless sensor networks (WSN) data mining algorithms can be seen as centralized (collection node) at the location of the computer centre and data warehouse or distributed across the network (sensor nodes) [7]. Centralized systems have more computing power at the expense of real-time data processing. Updating and sensor reactions are hence delayed or subsequent. Distributed systems process data real time and react quickly in case of deviation, but they do not enable more complex analysis due to low computing power [8].

Regardless of the system setup, the main goals of data mining entail frequent pattern detection, sequential pattern detection, clustering and classification. Data mining in distributed systems is mostly used for optimizing processing efficiency, energy consumption, and data and memory flow. Centralized systems are mostly used for prediction, raising result precision, and rationalization of operational costs [9].

Data mining within these systems faces scalability and communication problems when other sensor networks need to be incorporated.

IoT can help solving these problems since it enables real time communication between sensor networks and cloud locations where data are analysed. This meets the requirements for smart connecting devices and context aware observation activities [10].

3. IoT and Big Data

Since data generated within these systems has Big Data characteristics, classic data mining methods cannot be applied [11]. Observation, identification and understanding data generated by IoT systems is achievable by applying advanced machine learning algorithms adapted for real time Big Data stream processing.

Big Data is defined by three big Vs: Volume, Velocity at which data are generated and sent to a system, and Variety (structured, half-structured and unstructured dynamic data) [12]. Machine to machine (M2M) communication adds two more Vs to IoT Big Data complexity: Variablility, which refers to the oscillating speed of data generation and system input, and Value, which refers to the fact that not all data are equally important for analysis, i.e. those that can improve the process or model are more important. Data mining social networks can additionally raise complexity with other Vs like Veracity, which refers to data accuracy in expressing personal attitudes with regard to subjectivity [13].

4. Mining data streams

Big Data is mainly stream data. Stream data is every ordered pair of sequences (i, t) where *i* is an n-tuple and *t* is a positive time interval. Data size cannot be directly limited since stream data sequences are continuously generated in different intensities and spans. Thus it is potentially unlimited [14].

Classic data mining algorithms access file data or relational database data. On the other hand, stream data mining algorithms are applied at the very stream [15]. Stream data mining results are expected in real time. Single pass data processing is common for all stream data mining algorithms. It is usually done by using time windows, micro clustering, limited aggregation and approximation [16].

Windows can be defined physically by time spans or logically by the number of elements [17]. State of the art algorithms seek to determine optimal window size.

Data processing in time windows can handle data streams and important variables in the following manners:

- summative and continuously from the beginning of measurement with equal weights (landmark window)
- summative and continuously from the beginning of measurement with different weights – more obsolete data have lower weights (damped window)
- temporarily within defined time spans, e.g. only last 60 seconds (sliding window) [18].

The process of mining multiple data streams includes alignment, approximation, combination, building the model and adaptation to concept drift [19].

5. Stream data mining model within cold chain logistics

The analysis of wireless sensor network data produced by food distribution chains from the source to end-users can lead to process optimization which, in turn, can affect food quality preservation [20].

IoT solutions in food supply chain make parameters relevant for food preservation visible to all the stakeholders and enable cooperation between manufacturing, storage, transporting, selling and buying in order to perform automated activities in real-time with the aim to achieve maximum food quality, on-time delivery, and food preparation under optimal conditions [21].

Although cold chain and IoT in food transportation imply raising prices for end-users, due to implementation and maintenance costs and due to an increase in energy consumption, the acceptance of such systems facilitates process efficiency. Efficiency is reflected in raising standards for food safety and loss minimization [22].

IoT for supporting cold chain logistics is made possible by integrating the Internet and the existent system of wireless sensor networks. Data mining of these IoT systems should support M2M communication between the system for collecting data, namely sensors, and the system for real time knowledge extraction from multiple data streams. The goal is to build a system which would enable quick autonomous reactions and optimization in logistics of different cold chains for fruit quality preservation.

After careful observation of the IoT cold chain, we propose modelling data with approximate time windows. The window size should be defined by the time needed for detecting deviation of the combination of discretized values of elements that can be influenced upon (temperature, humidity, SO_2 concentration) from their limit values [23]. This model upgrades WSN and the respective model for managed traceability system (MTS) [24] in grape cold chain [25] with autonomous machine (computer) for detecting critical CO_2 values (values which indicate senescence), and for managing temperature, humidity and SO_2 concentration in order to preserve optimal grape quality. An example is shown in Figure 1.

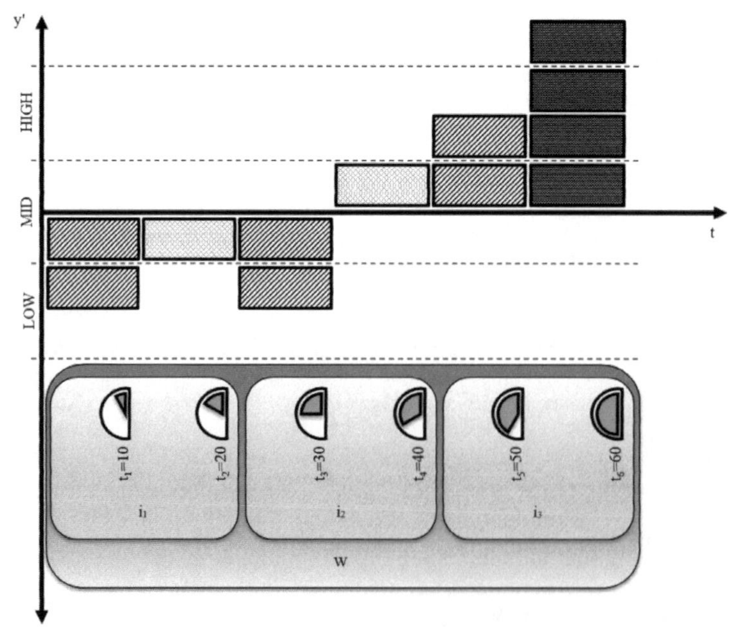

Fig. 1. Example of a discretized stream data mining of WSNMTS within grape cold chain

The window interval is made up of the following: the measured CO_2 concentration (dotted = optimal, diagonal = passing, trellis = critical), time *t*, time intervals *i* for data processing, window time span *w*, and the combination of temperature, humidity and SO_2 concentration *y'*. The example given in Fig. 1 has a window size of 60 seconds which is updated every 20 seconds. The change in the concentration of CO_2 from optimal to critical is detected after three intervals. The sensor data is sent twice per interval. The last detected CO_2 value in the interval is taken as the reference value for the interval. The window size should be adapted to real conditions in order to achieve timely and precise reactions based on M2M communication.

The problem with centralized WSN systems is reflected in their impossibility for timely reactions, i.e. real time reactions, since data are saved and loaded from the database instead of analysed as they are recorded [26].

The centralized IoT system capable of real time data mining and sending results back to sensors and computer devices within the cold chain in order to adjust to micro climate conditions can be achieved by using the proposed model and the newest technology for processing data streams. The system implementing such a model would need to deal with Big Data scalability, heterogeneity, and timeliness [27] considering that it would simultaneously analyse data of a number of cold chains, and potentially a number of groceries. From the IoT perspective, this system should be built as a cloud service [28].

The system could be based on open-source technologies for machine learning and for Big Data analysis of a third generation which enable real time data processing [29]. Systems suitable for cold chain data processing are Hadoop and Spark [30]. The proposed IoT cold chain model is given in Figure 2.

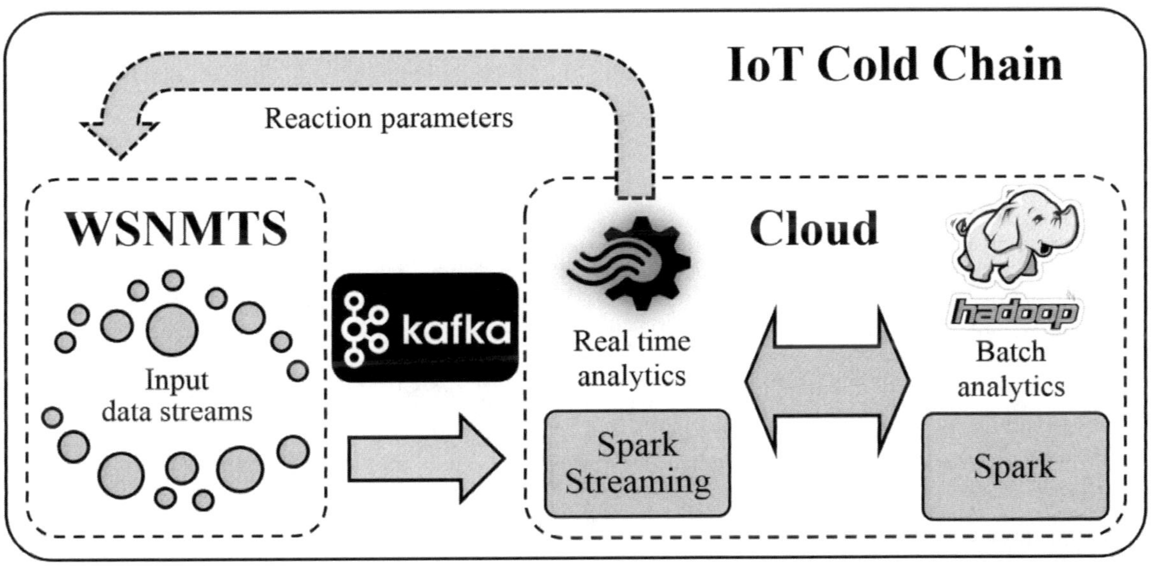

Fig. 2. IoT Cold-chain model

When designing such systems, problems and security threads which can affect data availability and data integrity need to be taken into account [31].

In order to ensure better data availability and data analysis, software technologies supporting clustering (e.g. Apache Kafka) could be used for input and bus architectures for data sent from multiple WSNMTS [32]. The real time stream data processing functionality of the system could be achieved with Spark Streaming [33]. A drawback of this approach is the failure to deliver data and real-time reactions when offline. If it would rely on the Spark architecture [34], the system would also be capable of on-demand batch data processing [35], i.e. Big Data mining over the entire dataset stored in the Hadoop file system [36] and supported data warehouses.

The combination of both data processing modes is a prerequisite for improving classification accuracy and evaluating stream data mining model. The system could be updated in online and offline (batch) modes. Such approach would lessen the problem with Internet connectivity because the system could base input reactions on the last version of the cumulatively trained offline model. The model could even be stored on computer equipment used in warehouses or vehicles, although this might negatively affect energy consumption.

6. Conclusion

The paper deals with the problem of real time communication between sensor networks and cloud locations in food cold chain. We propose a model for the centralized IoT system capable of real time data mining and sending results back to sensors and computer devices within the cold chain with a special focus on the part responsible for monitoring fluctuations of temperature, humidity, and concentration of gases. Lastly, we discuss and evaluate possible technological solutions for the proposed model.

In our future work we will deal with problems which relate to the failure to deliver certain variables important for analysis and prediction due to sensor failure or signal loss. Since in such cases real data are delivered with delay, the problem of resynchronizing multiple stream data occurs [37]. Therefore, system implementation would have to take advantage of techniques memorizing the last known state or compensating the stream with virtual data approximating the trend until then.

7. Acknowledgements

This work has been fully supported by the University of Rijeka under the project number 13.13.1.3.03.

8. References

[1] Winter, J. S. (2014). Privacy Challenges for the Internet of Things, In: Encyclopedia of Information Science and Technology, Third Edition, Khosrow-Pour, M. (Ed.), pp. 4373-4383, ISBN 978-1-4666-5888-2, IGI Global.

[2] Xiaoping, X.; Luoxian, L.; Mingyang, L. & Guobin, L. (2012). Perspectives on Internet of Things and Its Applications, Proceedings of the 2012 International Conference on Computer Application and System Modeling, Taiyuan Institute of Science and Technology, Taiyuan, Shanxi, China, ISSN 1951-6851, ISBN 978-94-91216-00-8, pp. 20-24, Atlantis Press, DOI:10.2991/iccasm.2012.6.

[3] Ashton, K. (2009). That 'Internet of Things' Thing, Available from: http://www.rfidjournal.com/articles/view?4986, Accessed: 2016-09-07.

[4] Ting, P. H. (2013). An Efficient and Guaranteed Cold-Chain Logistics for Temperature-Sensitive Foods: Applications of RFID and Sensor Networks, International Journal of Information Engineering and Electronic Business, Vol. 5, No. 6, pp. 1-5, ISSN 2074-9031.

[5] Li, Y.; Peng, Y.; Zhang, L.; Wei, J. & Li, D. (2015). Quality Monitoring Traceability Platform of Agriculture products Cold Chain Logistics Based on the Internet of Things, Chemical Engineering Transactions, Vol 46, pp. 517-522, ISSN 2283-9216.

[6] Capello, F.; Toja, M. & Trapani, N. (2016). A Real-Time Monitoring Service based on Industrial Internet of Things to manage agrifood logistics, Proceedings of the 6th International Conference on Information Systems, Logistics and Supply Chain, Bordeaux, France, Available from: http://ils2016conference.com/wp-content/uploads/2015/03/ILS2016_FB01_1.pdf, Accessed: 2016-10-21.

[7] Flammini, A. & Sisinni, E. (2014). Wireless Sensor Networking in the Internet of Things and Cloud Computing Era, Procedia Engineering, Vol. 87, pp. 672-679, ISSN 1877-7058.

[8] Lojka, T. & Zolotova, I. (2014). Distributed sensor network – data stream mining and architecture, Advances in Information Science and Applications, Vol. 1, Proceedings of the 18th International Conference on Computers (part of CSCC '14), Santorini Island, Greece, Recent Advances in Computer Engineering Series, ISSN 1790-5109, ISBN 978-1-61804-236-1, pp. 98-103.

[9] Mahmood, A.; Shi, K. & Khatoon, S. (2012). Mining Data Generated by Sensor Networks: A Survey, Information Technology Journal, Vol. 11, No. 11, pp. 1534-1543, ISSN 1812-5638.

[10] Gubbia, J.; Buyyab, R.; Marusic, S. & Palaniswami, M. (2013). Internet of Things (IoT): A vision, architectural elements, and future directions, Future Generation Computer Systems, Vol. 29, No. 7, pp. 1645-1660, ISSN 0167-739X.

[11] Fouad, M. M.; Oweis, N. E.; Gaber, T.; Ahmed, M. & Snasel, V. (2015). Data Mining and Fusion Techniques for WSNs as a Source of the Big Data. Procedia Computer Science, Vol. 65, pp. 778-786, ISSN 1877-0509.

[12] Gandomi, A. & Haider, M. (2015). Beyond the hype: Big data concepts, methods, and analytics, International Journal of Information Management, Vol. 35, No. 2, pp. 137-144, ISSN 0268-4012.

[13] Özkösea, H.; Ari, E. S. & Gencerb, C. (2015). Yesterday, Today and Tomorrow of Big Data, Procedia – Social and Behavioral Sciences, Vol. 195, pp. 1042-1050, ISSN 1877-0428.

[14] Namiot, D. (2015). On Big Data Stream Processing, International Journal of Open Information Technologies, Vol. 3, No. 8, pp. 48-51, ISSN 2307-8162.

[15] Hebrail, G. (2008). Data stream management and mining, In: Mining Massive Data Sets for Security, Fogelman-Soulie, F.; et al. (Eds.), pp. 89-102, IOS Press, ISBN 978-1-58603-898-4.

[16] Han, J. & Gao, J. (2009). Research Challenges for Data Mining in Science and Engineering, In: Next Generation of Data Mining, Kargupta, H.; et al. (Eds.), pp. 3-28, Chapman & Hall, ISBN 978-1-4200-8586-0.

[17] Joseph, S.; Jasmin, E. A. & Chandran, S. (2015). Stream Computing: Opportunities and Challenges in Smart Grid, Procedia Technology, Vol. 21, pp. 49-53, ISSN 2212-0173.

[18] Shah, H. M. & Kaur, N. (2014). Improve Frequent Pattern Mining in Data Stream, International Journal of Research in Engineering & Technology, Vol. 2, No. 5, pp. 143-152, ISSN 2321-8843.

[19] Spiliopoulou, M. (2012). Mining Multiple Interdependent Streams, Tutorial at European Conference on Machine Learning and Principles and Practice of Knowledge Discovery in Databases, 24-28 September 2012, Bristol, UK.

[20] Ruiz-Garcia, L.; Lunadei, L.; Barreiro, P. & Robla, J. I. (2009). A Review of Wireless Sensor Technologies and Applications in Agriculture and Food Industry: State of the Art and Current Trends, Sensors, Vol. 9, No. 6, pp. 4728-4750, ISSN 1424-8220.

[21] Xiaorong, Z.; Honghui, F.; Hongjin, Z.; Zhongjun, F. & Hanyu, F. (2015). The Design of the Internet of Things Solution for Food Supply Chain, Proceedings of the 5th International Conference on Education, Management, Information and Medicine (EMIM 2015), Shenyang, China, ISSN 2352-5428, ISBN 978-94-62520-68-4, pp. 314-318, Atlantis Press, DOI:10.2991/emim-15.2015.61.

[22] Liu, X.; Xu, M. & Yu, C. (2016). Food Cold Chain Logistics Based on Internet of Things Technology, Proceedings of the 6th International Conference on Applied Science, Engineering and Technology (ICASET 2016), Qingdao, China, ISSN 2352-5401, ISBN 978-94-6252-186-5, pp. 92-96, Atlantis Press, DOI:10.2991/icaset-16.2016.18.

[23] Kasetty, S.; Stafford, C.; Walker, G. P.; Wang, X. & Keogh, E. (2008). Real-Time Classification of Streaming Sensor Data, Proceedings of the 20th IEEE International Conference on Tools with Artificial Intelligence (ICTAI 2008), Vol. 1, Dayton, Ohio, USA, ISSN 1082-3409, ISBN 978-0-7695-3440-4, pp. 149-156, IEEE Computer Society, Los Alamitos, CA 90720-1314.

[24] Zhang, J.; Liu, L.; Mu, W.; Moga, L. M. & Zhang, X. (2009). Development of temperature-managed traceability system for frozen and chilled food during storage and transportation, Journal of Food, Agriculture & Environment, Vol. 7, No. 3, pp. 28-31, ISSN 1459-0255.

[25] Xiao, X.; Wang, X.; Zhang, X.; Chen, E. & Li, J. (2015). Effect of the Quality Property of Table Grapes in Cold Chain Logistics-Integrated WSN and AOW, Applied Sciences, Vol. 5, No. 4, pp. 747-760, ISSN 2076-3417.

[26] Duhaney, J.; Khoshgoftaar, T. M.; Agarwal, A. & Sloan, J. C. (2010). Mining and storing data streams for reliability analysis, Proceedings of the 16th International Society of Science and Applied Technologies on Reliability and Quality in Design, Washington D.C, USA, Pham, H. (Ed.), pp. 314-317, ISBN 978-0-9763486-6-5, International Society of Science and Applied Technologies, Piscataway, NJ 08855.

[27] Cortes, R.; Bonnaire, X.; Marin, O. & Sens, P. (2015). Stream processing of healthcare sensor data: studying user traces to identify challenges from a big data perspective, Procedia Computer Science, Vol. 52, pp. 1004-1009, ISSN 1877-0509.

[28] Gnimpieba, Z. D. R.; Nait-Sidi-Moh, A.; Durand, D. & Fortin, J. (2015). Using Internet of Things technologies for a collaborative supply chain: Application to tracking of pallets and containers, Procedia Computer Science, Vol. 56, pp. 550-557, ISSN 1877-0509.

[29] Agneeswara, V. S. (2014). Big Data Analytics Beyond Hadoop: Real-Time Applications with Storm, Spark, and More Hadoop Alternatives, Pearson, ISBN 978-0-13-383794-0, New Jersey 07458, USA.

[30] Ko, D.; Kwak, Y.; Choi, D. & Song, S. (2015). Design of Smart Cold Chain Application Framework Based on Hadoop and Spark, International Journal of Software Engineering and Its Applications, Vol. 9, No. 12, pp. 99-106, ISSN 1738-9984.

[31] Cvitic, I.; Vujic, M. & Husnjak, S. (2015). Classification of Security Risks in the IoT Environment, Proceedings of the 26th DAAAM International Symposium, pp. 0731-0740, B. Katalinic (Ed.), Published by DAAAM International, ISBN 978-3-902734-07-5, ISSN 1726-9679, Vienna, Austria, DOI:10.2507/26th.daaam.proceedings.102.

[32] Apache Kafka. Available: http://kafka.apache.org, Accessed: 2016-09-22.

[33] Spark Streaming. Available: http://spark.apache.org/streaming, Accessed: 2016-09-22.

[34] Apache Spark. Available: http://spark.apache.org, Accessed: 2016-09-22.

[35] Karau, H.; Konwinski, A.; Wendell, P. & Zaharia, M. (2015). Learning Spark, O'Reilly Media, ISBN 978-1-4493-5862-4, Sebastopol, CA 95472, USA.

[36] Apache Hadoop. Available: http://hadoop.apache.org, Accessed: 2016-09-20.

[37] Krempl, G.; et al. (2014). Open Challenges for Data Stream Mining Research, ACM SIGKDD Explorations Newsletter - Special issue on big data archive, Vol. 16, No. 1, pp. 1-10, ISSN 1931-0145.

DOI: 10.2507/27th.daaam.proceedings.110

REALIZATION OF THE ONTOLOGICALLY BASED METHOD FOR CHECKING STRUCTURAL INCONSISTENCES OF RELATIONAL DATABASES

Alexander A. Kropotin, Andrei V. Grigoryev, Yuliya V. Bidulya,
Alexander G. Ivashko & Nikita S. Durynin

This Publication has to be referred as: Kropotin, A[lexander]; Grigoryev V., A[ndrei]; Bidulya V., Y[uliya]; Ivashko G., A[lexander] & Durynin S., N[ikita] (2016). Realization of the Ontologically Based Method for Checking Structural Inconsistences of Relational Databases, Proceedings of the 27th DAAAM International Symposium, pp.0762-0767, B. Katalinic (Ed.), Published by DAAAM International, ISBN 978-3-902734-08-2, ISSN 1726-9679, Vienna, Austria
DOI: 10.2507/27th.daaam.proceedings.110

Abstract

In paper the problem of use of technologies of a semantic web for automatic detection of structural inconsistenses of relational databases in case of software integration is solved. The description of the conflicts of relational databases in case of their integration is provided in chapter 1, methods of their determination and reasons for development of own method of the solution of an objective. The description of a method of automatic detection of structural mismatches of relational databases with use of technologies of a semantic web is provided in chapter 2. The description of experimental evaluating of the developed method is provided in a chapter 3 and its results are represented. In the inference plans for future development in this area are defined.

Keywords: Ontology; description logics; database; ER-model.

1. Introduction

At present the integration of information systems and their databases is one of the popular research task [1, 2, 3]. Data integration may be formally defined like synchronization of data between databases of the software which can be provided by different data schemes in different sources.

A lot of scientific schools around the world take part in researches related to information systems integration, such like: Database Technology University of Zurich [1], Wright State University [2], Dublin City University School of Computing [3] and other. Besides, there are different methods and approaches to integration of data [4, 5, 6, 7], but all of them face with the problems of integration of data connected to a structural mismatch of data schemes [4], which may be defined as conflicts related to

a) schemes inconsistencies, it may arise in solving problems of integrating schemes;
b) data inconsistencies, it may arise in solving problems of integrating schemes.

The following data schemes differences may be associated with schemes inconsistencies[4, 5]:

a) a difference in terminologies: homonymy and synonymy of data in different sources;

b) a difference in data semantics: equivalent real-world entities are organized at various levels of abstraction in different sources;

c) a difference in data structure: equivalent real-world entities organized in different data structures in different sources.

In turn, semantic (b) and structure (c) differences in these schemes may be defined as:

d) a difference in the cardinality of relationships set: equivalent relationships set from different sources have a different cardinality;

e) a difference in the degree of participation of relationships set: equivalent relationships set from different sources have different degrees of participation;

f) an absence of elements of schemes: an absence of value sets, entities sets, relationships sets in different sources;

g) a difference in «entities sets ¬ value sets»: equivalent real-world entities represented by a entities sets in some sources and a value sets in the other;

h) a difference in the «value sets ¬ set value sets»: properties of equivalent real-world entities are represented in value sets of some sources and a set of value sets in the other;

i) a difference in the restrictions value sets: restrictions on permissible and existing values in a difference of sources;

j) a difference in the types of data: equivalent value sets of different sources presented different types of data;

k) a difference in the set of valid values: equivalent value sets of different sources identified different sets of constants;

The difference in «entities sets ¬ value sets» (a) as well as differences in relationships sets (c, d), the absence of value sets, entities sets and relationship sets may be associated with a different view on the logical level of conceptual designs as:

a) entities sets that come into connection 1:1;

b) entities sets that come into relationships sets superclass / subclass;

Also, the absence of value sets, entities sets and relationships sets (c, d) can be associated with a different view on the logical level of conceptual designs as unary (recursive) connections.

The following data differences may be associated with data inconsistencies [4, 5]:

a) difference of data format ¬ presentation equivalent values in different sources define different formats;

b) difference values ¬ equivalent values in different sources have different views.

From the assumption that by means of application of means of a semantic web [8] perhaps implementation of verification of several diagrams of relational databases regarding existence of their structural distinctions the task of their verification using a formalism of description logics from each other appeared.

In the course of search of the existing solution of an objective operations in which mathematical models and algorithms of ontological submission of diagrams of relational databases in terms of description logics are described [9, 10, 11, 12, 13, 14, 15, 16, 17, 18, 19, 20, 21] were found and probed. The study of these operations led to an output that submission of diagrams of relational databases in ontology is possible, however this representation doesn't allow to solve an objective.

All these facts prove actuality of this paper. And thus, the main idea of paper is a development and realization of the ontologically based method for checking structural inconsistences of relational databases [25] during software integration.

2. Method description

We propose the OWL DL implementation to detect problems during the process of data schemes integration. The developed method includes three steps:

a) at the first step, each conceptual schema of relational database must be mapped into ontology.

b) at the second step these ontologies have to be merged into one.

c) at the last step the reasoning for this ontology have to be implemented.

Special for realization of the method steps were developed three models:

a) the ontological model of a conceptual domain objects representation has been developed special for realization of the method first step. The model is is presented as a rules for describing database concepts into description logics terms. The model basedon the expressive description logic SROIQ.

b) the model of ontologies union of an conceptual domain objects has been developed special for realization of the method second step.

c) the model analysis of an united ontology has been developed special for realization of the method last step.

The detailed description of the developed models has been submitted in [22] therefore further, we will provide their short description in the form of rules of the mapping binary entity-relationship diagram elements into ontology elements.

2.1. The mapping conceptual schema of relational database into ontology

Following we present some rules for mapping of semantic network "binary entity-relationship" models to OWL DL ontology. The set of terminological axioms and assertional axioms is proposed to term the conceptual domain objects ontology (1).

$$\mathcal{K} = \mathcal{T} \cup \mathcal{A}, \tag{1}$$

where $\mathcal{K}$ is the conceptual domain objects ontology; $\mathcal{T}$ is a terminological axioms; $\mathcal{A}$ is an assertional axioms.

Entity set and value set. Both entities sets and values sets should be mapped into class. While transforming entities sets and values sets, it should be taken care of tagging classes names to correspond to entities sets names and values sets names. Each class should be made as disjoint with other.

Entity and value. Both entities and values would be mapped into individuals. While transforming entities and values, it should be taken care of making local unique names. Each individual should be maked as different with other in frame of one class, when it putted in line with entity.

Relationship set and attribute. Both relationships sets and attributes should be mapped into object properties. While transforming relationships sets and attributes, it should be taken care of tagging object properties names to correspond to relationships sets and attributes names.

While transforming attributes, in addition to object properties datatype properties should be created. Thus, each attribute should be mapped into composition of object property and datatype property. While transforming attributes, it should be taken care of making local unique names.

Each object property and datatype property should be made as functional, otherwise OWL DL allows object property and datatype property to take many values by default.

While transforming relationship sets cardinality of «one-to-many», in addition for object properties would be created and tagged as inverse object properties. While transforming relationship sets cardinality of «one-to-one», object properties would be tagged as symmetric. Thus relationship sets would be mapped into two inverse object property or one symmetric object property for each.

The domain and the range of each object property is the class. The domain of each data property is the class and range is the actual datatype (int, string, etc).

Primary key and unique constraint. Each datatype property and object property, when it putted in line with primary key attribute, should be tagged as hasKey of their class.

2.2. The merging rules of ontologies conceptual domain objects

The merging ontologies domain conceptual objects (1) is proposed to term ontology of set of domain conceptual object different representation (2).

$$\mathcal{K} = \bigcup_{1 \le l \le L} \mathcal{K}_l, \tag{2}$$

where $\mathcal{K}$ is the ontology of set of domain conceptual object different representation; L is the count of different ontology union $\mathcal{K}_l$: $L \ge 2$; $\mathcal{K}_l$ is the domain conceptual objects ontology.

Then in this case in integrated terminological axiom set $\mathcal{T} \subseteq \mathcal{K}$ should be defined equal atomic concepts (1) of different ontologies in terminological axiom terms of a concepts equivalence (3).

$$\{\{\{R_i^l \equiv R_i^h\}_{i=\overline{1,I^l}}\}_{l<h\le L}\}_{l=\overline{1,L}}, \tag{3}$$

where R_i^l, R_i^h are object properties and datatype properties defined on the ontology $\mathcal{K}_l$ and $\mathcal{K}_h$ respectively; $I^l$is a count of all object properties and datatype properties defined on the ontology $\mathcal{K}_l$; L is a count of all ontologies $\mathcal{K}_l$.

In addition, in integrated terminological axiom set $\mathcal{T} \subseteq \mathcal{K}$ equal classes (4) of different ontologies should be defined in terminological axiom terms of a concepts equivalence (4), (a) if the count of ontology $\mathcal{K}_l$ object properties is greater than count of ontology $\mathcal{K}_h$ or (b) otherwise.

$$\begin{cases} \{\{\{A_n^l \equiv A_n^h\}_{n=\overline{1,N^l}}\}_{l<h\le L}\}_{l=\overline{1,L}} & \text{(a)} \\ \{\{\left\{A_n^l \prod_{i=1}^{I} \exists S^l.C^i \equiv A_n^h\right\}_{n=\overline{1,N^l}}\}_{l<h\le L}\}_{l=\overline{1,L}} & \text{(b)} \end{cases}, \tag{4}$$

where A_n^l, A_n^h are classes of the ontologies $\mathcal{K}_l$ и $\mathcal{K}_h$ respectively; S^l is the transitiv object property of the ontology K_l; I is a count of all properties of the concept A_n^l are putted in line with properties of the concept A_n^h which have not equal roles from the ontology $\mathcal{K}_l$; N^l is a count of all classes of the ontology $\mathcal{K}_l$; L is a count of the all ontologies $\mathcal{K}_l$.

In case of integrated assertional axioms the set $\mathcal{A} \subsetneq \mathcal{K}$ contains at least two assertional axioms sets $\mathcal{A}_l \subsetneq \mathcal{A}$: $\mathcal{A}_l \neq \emptyset$, in which individuals were defined, in these assertional axioms set the individuals equivalence of different ontologies $\mathcal{K}_l$ should be defined in assertional axioms terms (5).

$$\{\{\{a_j^l \approx a_j^h\}_{j=\overline{1,J^l}}\}_{l<h\le L}\}_{l=\overline{1,L}}, \tag{5}$$

where a_j^l, a_j^h are individuals of the ontologies $\mathcal{K}_l$ и $\mathcal{K}_h$ respectively; J^l is a count of all individuals of the ontology $\mathcal{K}_l$; L is a count of all ontologies $\mathcal{K}_l$.

2.3. *The checking structural inconsistences of relational databases*

We propose to implement the description logics calculus [8, 23, 24] to ontology of set of domain conceptual object different representation for checking structural inconsistences of relational databases.

Thus it is necessary to resolve three algorithmic problems of description logics calculus, such that:

a) the inconsistence algorithmic problem of an ontology terminological (only Tbox);
b) the classification algorithmic problem of an ontology terminological (only Tbox);
c) the inconsistence algorithmic problem of an ontology (Tbox and Abox);

The resolving of inconsistence algorithmic problem of an ontology terminological could be able to follows detecting relational databases integration problems [4, 5, 6, 7, 8]:

a) difference in the cardinality of relationships set the domain and/or domain range do not overlap;
b) difference of the restrictions value sets range of roles (concrete domain) do not intersect;
c) difference of the types of data range of roles (concrete domain) do not intersect;
d) difference of the set of valid values range of roles (concrete domain) do not intersect.

The resolving of classification algorithmic problem of the terminological ontology could be able to follow detecting semantic network "binary entity-relationship" models integration problems:

a) an absence of elements of schemes 1) if there is no composition of atomic roles, then transitive role is not completed in the place of the absent, and one concept is embedded in the another, but about another we could allege this, so they are not equivalent; 2) absence of equivalent concepts from concepts of one of terminology; 3) absence of equivalent concepts from concepts of one of terminology;

b) difference of «entities sets ¬ value sets» equivalent concepts belong simultaneously to concepts EntitySet and ValueSet;

c) difference of the «value sets ¬ set value sets» 1) equivalent concepts has no concepts in a one terminology; 2) one concept in one terminology has many equivalent concepts in the other;

The resolving of inconsistence algorithmic problem of an ontology could be able to follows detecting relational databases integration problems:

a) difference in the modality of relationships set null value is defined as {nominal} implication for all Role with not a Concept;

b) difference values equivalent individuals belong to equivalent roles with different values.

3. Evaluating

Software implementing presented method was developed. It get database descriptions from Papyrus XML file and transforms it to knowledge bases. Software was developed using Java language, because of there is a library called OWL API that allows creating knowledge bases in OWL language. Besides, there are many open source reasoning software allowing solving task of logical reasoning. To check same names of the tables we use greedy algorithm with similarity checking. In the future researches we consider to use a method based on the Kun algorithm for max weighted matching in bipartite graph.

HermiT reasoner is using as reasoning tool, because of HermiT shows great results on past OWL Reasoning Evaluation Workshop [24].

Unfortunately, we test our software only during development on test databases with no more than 10 tables and tests shows 100% regular results (table 1).

Distinction title	Count of structural distinctions	The method result
distinction "entity set – value set"	5	5
distinction "value set – set of value set"	5	5
distinction in a set of admissible values	5	5
distinction in data type	5	5
distinction in semantics of communications	5	5

Tab. 1. The method evaluating result.

4. Conclusion

This paper describes a problem of automatic checking structural inconsistences of relational databases. This paper presents a framework for the detecting relational databases integration problems. We have provided rules to transform relational databases concepts into equivalent OWL ontology for semantic web.

This proposed framework helps software engineers in upgrading the structured analysis and design relational databases. Our ongoing research on this topic is to handle other cases of relationships that are not binary, which require reification.

Software implementing presented method was developed. It get database descriptions from Papyrus XML file and transforms it to knowledge bases. HermiT reasoner is using as reasoning tool. Software was developed using Java language and OWL API.

Unfortunately, we test our software only during development on test databases with no more than 10 tables and tests shows 100% regular results.

In the future, we will extend our researches, we plan to develop, and present ready software tested on enterprise databases. Moreover, there are few systems with opened database creating scripts, and to build knowledge bases we will use data description using data definition in web services. For example, WSDL-definition with a XSD schemas can be considered as a source of the database schema.

5. Acknowledgements

We would like to thank our colleagues who have helped in the development of independent parts of the software, but who wish to remain outside the scope of the co-authors of the article: Shaykhullina Irina, Michael Samoilov, Alexander Stupnikov and Dusso Bruno Jean-Marie.

6. References

[1] Patrick Ziegler, Klaus R. Dittrich, "Data Integration — Problems, Approaches, and Perspectives." In Conceptual Modelling in Information Systems Engineering, Springer, 2007. (Conference or Workshop Paper). p. 39-58.

[2] Adila Alfa Krisnadhi. "Ontology Pattern-Based Data Integration", PhD Thesis, Wright State University. December 18, 2015.

[3] Pahl, Claus and Zhu, Yaoling. "Data integration in mediated service compositions" In Computing and Informatics, Volume 31 Issue 6, 2001, pp. 1129-1149.

[4] Beloshitsky D. A. Data Integration in Information Systems - Electron. Dan. – [Electronic resource] / D. A. Beloshitsky. – Access: http://sntbul.bmstu.ru/doc/602635.html – (reference date: 20. 03. 2015).

[5] Ramon Lawrence, Ken Barker. "Integrating relational database schemas using a standardized dictionary." In Proceeding SAC '01 Proceedings of the 2001 ACM symposium on Applied computing. ACM New York, NY, USA, 2001, pp. 225-230.

[6] Stephen Mc Kearney, "Schema Integration.", 2002. Available at URL: http://www.smckearney.com/adb/notes/lecture.schema.integration.pdf.

[7] Batini C., Lenzerini M., Navathe S. B. "A comparative analysis of methodologies for database schema integration." In Journal ACM Computing Surveys, Volume 18 Issue 4, Dec. 1986, pp. 323-364.

[8] Gruber, T. A Translation Approach to Portable Ontology Specifications / T. Gruber // Knowledge Acquisition – 1993. – pp.199-220.

[9] Astrova, I. Mapping of SQL relational schemata to OWL ontologies / I. Astrova, A. Kalja // Proceedings of the 6th WSEAS International Conference on Applied Informatics and Communications. 18-20 August 2006, Elounda, Greece / Institute of Cybernetics Tallinn University of Technology Akadeemia tee 21, 12618 Tallinn Etonia. – P. 375-380.

[10] Barrasa Rodríguez, J. R2O, an Extensible and Semantically based Database-to-Ontology Mapping Language / J. Barrasa Rodríguez, O. Corcho, A. Gómez-Pérez // Proceedings of the Second Workshop on Semantic Web and Databases. August 2004, Toronto, Canada / Springer-Verlag, Berlín, Alemania. – P. 1069-1070

[11] Berners-Lee, T. The Semantic Web / T. Berners-Lee, J. Hendler, O. Lassila, // Scientific American – 2001. – Vol. 284 – pp. 34-43.

[12] Bumans, G. Mapping between Relational Databases and OwL Ontologies: an example / G. Bunmas // Computer science and information technologies – 2010. – Vol. 756 – pp. 99-117.

[13] Chujai, P. On Transforming the ER Model to Ontology Using Protégé OWL Tool / P. Chujai, N. Kerdprasop, K. Kerdprasop // International Journal of Computer Theory and Engineering – 2014. – Vol. 6 – pp. 484-489.

[14] Colomb, R. M. Issues in Mapping Metamodels in the Ontology Development Metamodel Using QVT / R. M. Colomb, A. Gerber, M. Lawley // Proceedings of the First International Workshop on the Model-Driven Semantic

Web. The First International Workshop on the Model-Driven Semantic Web. 20-24 September 2004, Monterey, California.

[15] Cullot, N. DB2OWL: A Tool for Automatic Databaseto-Ontology Mapping / N. Cullot, R. Ghawi, K. Yétongnon // Proceedings of the 15th Italian Symposium on Advanced Database Systems. 2007. – P. 491-494.

[16] Fahad, M. ER2OWL: Generating OWL Ontology from ER Diagram / M. Fahad // Intelligent Information Processing IV – 5th IFIP International Conference on Intelligent Information Processing. The International Federation for Information Processing Volume 288. 19-22 October 2008, Beijing, China / Mohammad Ali Jinnah University, Islamabad, Pakistan. – P. 28-37.

[17] Louhdi, M. R. C. Transformation Rules For Building Owl Ontologies From Relational Databases / M. R. C. Louhdi, H. Behja, S. O. El Alaoui // Second International Conference on Advanced Information Technologies and Applications. November 2013. P. 271-283.

[18] Shihan, Y. Semi-automatically building ontologies from relational databases / Y. Shihan, Z. Ying, Y. Xuehui // Proceedings of the 3rd IEEE International Conference on Computer Science and Information Technology. 2010. – P. 150 - 154.

[19] Telnarova, Z. Relational database as a source of ontology creation / Z. Telnarova // Proceedings of the International Multiconference on Computer Science and Information Technology. 2010. – P. 135-139.

[20] Natalya F. Noy. “Semantic integration: a survey of ontology-based approaches.” In Newsletter ACM SIGMOD Record Homepage archive, Volume 33 Issue 4, December 2004. pp. 65-70.

[21] AnHai Doan, Alon Y. Halevy. “Semantic-integration research in the database community.” In Journal AI Magazine - Special issue on semantic integration archive, Volume 26 Issue 1, March 2005 pp. 83-94.

[22] Andrey V. Grigoryev, Alexander A. Kropotin, Alexander G. Ivashko. Database Schema Method for Automatic Semantic Errors Resolving During Information Systems Integration. Proceedings of the 10th AICT2016 International Conference, pp.187-190, IEEE CFP1656H-PRT, 2016, Baku, Azerbaijan.

[23] Ivashko A.G., Grigoryev A.V., Grigoryev M.V. “Tableau algorithm modification based on complex concepts disjointness checking.” In Bulletin of the Tyumen State University. - 2012, №4. - P. 143-150.

[24] Birte Glimm, Ian Horrocks, Boris Motik, Giorgos Stoilos, Zhe Wang. “HermiT: An OWL 2 Reasoner” In Journal of Automated Reasoning October 2014, Volume 53, Issue 3, pp 245-269.

[25] Salibekyan Sergey & Panfilov Peter. Database Architecture for Specifying and Modeling Spatio-Temporal Relations, Proceedings of the 26th DAAAM International Symposium, pp.0589-0598, B. Katalinic (Ed.), Published by DAAAM International, ISBN 978-3-902734-07-5, ISSN 1726-9679, Vienna, Austria

DOI: 10.2507/27th.daaam.proceedings.111

DEVELOPMENT OF ALGORITHM MODEL FOR EXHAUST GASES SYSTEM OF DIESEL ENGINE WITH ELECTRONIC CONTROL DIAGNOSTICS

Miroslav Grubisic, Boris Crnokic

University of Mostar, Faculty of Mechanical Engineering and Computing,
Matice hrvatske bb, Mostar 88000, Bosnia and Herzegovina

This Publication has to be referred as: Grubisic, M[iroslav] & Crnokic, B[oris] (2016). Development of Algorithm Model for Exhaust Gases System of Diesel Engine with Electronic Control Diagnostics, Proceedings of the 27th DAAAM International Symposium, pp.0768-0774, B. Katalinic (Ed.), Published by DAAAM International, ISBN 978-3-902734-08-2, ISSN 1726-9679, Vienna, Austria
DOI: 10.2507/27th.daaam.proceedings.111

Abstract

This paper discusses the problem of increased emissions of harmful exhaust gases of the Diesel engine in situations where the engine control unit (ECU) has not memorized a single error that indicates a malfunction in the control system of the engine. In order to detect hidden problems in the Diesel engine performance, which directly affect the increase in its emissions, this paper has developed an optimal algorithmic model for system diagnostics of Diesel engines with electronic control exhaust gases after-treatment. Functionality check of developed aphorism model was conducted experimentally on Diesel motors with electronic control and built-in exhaust gas recirculation into the intake engine manifold (EGR), oxidation catalyst, diesel particulate filter (DPF) and selective catalytic reduction (SCR) system.

Keywords: Diesel engine; exhaust gases; diagnostic model; exhaust gas recirculation; EGR; diesel particulate filter; DPF; selective catalytic reduction; SCR

1. Introduction

Diesel engine is undoubtedly one of the most important inventions of the nineteenth century and the fact is that a large part of the world system of transportation of people and goods is based on it. On the other hand, in the last years of the twentieth century was definitely, by means of analysis and monitoring, found that motor vehicles driven by internal combustion engines, besides the industry, are the biggest air pollutants due to increased emissions of harmful combustion products and greenhouse gas emissions. This emission has a direct impact on human health, which is particularly acute in urban city environments. Diesel engine has its own harmful emissions, such as carbon monoxide (CO), unburned hydrocarbons (HC), nitrogen oxides (NOx) and particulates of soot (PM). As a product of combustion in the Diesel engine, there is the emission of carbon dioxide (CO_2) called greenhouse gas that leads to a warming of the earth due to the greenhouse effect.

In order to protect the environment, the European Union has, to all the manufacturers of vehicles with Diesel engines, imposed requirements (Euro norms) in terms of emissions. These legal requirements strictly determine how much harmful combustion products the Diesel engine exhaust gases can contain in various conditions of use [1]. Having in mind the Kyoto Protocol on reducing greenhouse gas emissions into the atmosphere [2], it is a matter of time when the emission of carbon dioxide (CO_2) from the Diesel engines will be limited.

With the introduction of legislation on limiting emissions of Diesel engine, a various studies have been conducted in order to reduce the emissions. Researches on optimizing the combustion process in the cylinder by improvements of the engine (combustion chamber shape, the flow in the cylinder, the fuel injection mode, etc.) have led to a reduction in the so-called "raw emissions" [3]. Studying the fuel, reducing sulfur content and constant improvement of fuel quality led to reducing emissions also. However, studies have shown that the harmful emission is most reduced by conducting the purification of exhaust gases after they came out of the engine, i.e by subsequent processing.

With the appearance of the Euro 1 the amount of carbon monoxide (CO) and unburned hydrocarbons (HC) is reduced by using an oxidation catalyst that is built into the Diesel engine [4]. Since the Diesel engine is running lean, due to the large excess air in the exhaust gases there was no possibility of supervising the catalyst lambda probe, i.e. was not included in the OBD program, and engine control unit (ECU) had no information on the condition of the exhaust gas. In order to reduce the content of nitrogen oxides (NOx), a system for exhaust gas recirculation into the intake manifold Diesel Engine EGR was investigated and successfully implemented, which lowers the temperature of combustion in the cylinder, and thus the creation of nitrogen oxides (NOx) [5, 6]. When the Euro 5 entered the force in 2009, the permitted amount of particulate matter (PM) emission in the Diesel engine was significantly reduced. Researches in order to solve this problem [7] led to the development of so-called Diesel Particulate Filter (DPF) that, by its shape keeps solids in it, and from time to time is emptied by the combustion of accumulated particles by means of exhaust gases high temperature, resulting from the late fuel injection during the expansion stroke. Euro 6 normative, that entered the force in 2014, significantly reduces and limits the amount of nitrogen oxides (NOx) emission in the Diesel engine. In recent times, in order to reduce NOx emissions, a selective catalytic reduction (SCR) is more and more researched, developed and operated [8, 9]. The SCR system for reduction of nitrogen oxide uses the liquid reducing agent on the basis of ammonia (NH_3) which is injected into the exhaust pipe before the reduction catalyst and in the catalyst leads to chemical reactions which transform harmful nitrogen oxides into nitrogen (N_2) and water (H_2O). Concern for the protection of the human environment has become one constant race between the applicable legal regulations and techniques that must follow.

This paper discusses the problem of increased emissions of harmful exhaust gases of the Diesel engine in situations where the engine control unit (ECU) has not memorized a single error that indicates a malfunction in the control system of the engine. This issue was investigated on the Diesel engines whose control units (ECU) are connected, via CAN-Bus connection [10], with the control devices of other systems in the vehicle.

2. Development of the algorithmic model

Just absolutely correct operation of the Diesel engine, with the all elements of the control system correctly functional, guarantees the emissions within the prescribed limits. Any disruption in the engine, as well as a malfunction of individual components, causes improper combustion which results in an increase in harmful emissions. If the emission of harmful exhaust gases of the Diesel engine is increased, for reasons of failure of one or more components of the control system, the problem is detected by commonly known diagnostic methods and by removing detected defection the engine gas emission returns to the acceptable range.

However, there are situations when the Diesel engine has increased emissions of harmful exhaust gases and no component of the system of electronic engine management has any visible deficiency and no electrical malfunction in the form of errors stored in the engine control unit, which is read by diagnostic device. The consequence of increased emissions in these circumstances may be the wrong signal of the sensor which is within the limits of the possible values so engine control unit doesn't recognize it as a mistake and a part of the regulatory process in the work of the Diesel engine is performed towards it. Another option, which can also lead to increased emissions, is the imprecise work of the actuator even though they are regularly triggered by the control unit pulse signal; hence error related to their electrical malfunction is not recorded.

In order to detect these hidden problems in the Diesel engine, which directly affect the increase in its emissions, this paper has developed an optimal algorithmic model for system diagnostics of Diesel engines with electronic control exhaust gases after-treatment and its abbreviated version is shown in Figure 1.

Through a developed comprehensive algorithmic model for system diagnostics of Diesel engines with electronic control exhaust gases after-treatment, individual system components of the engine management test are done step by step, which ultimately leads to detection of the problem that directly affects the increase of the soot particles in the exhaust. By the complete algorithm model, among other, testing of differential pressure sensor, pressure sensor in the intake manifold, fuel injectors, clutch pedal position sensor, speed sensor engine, Hall sensor, EGR valve, turbocharger control, throttle the suction pipe, temperature sensors on DPF filter, is planned as well as the testing of all other elements of the engine whose malfunction leads to increased formation of soot particles and an increase in the total emission of harmful exhaust gases.

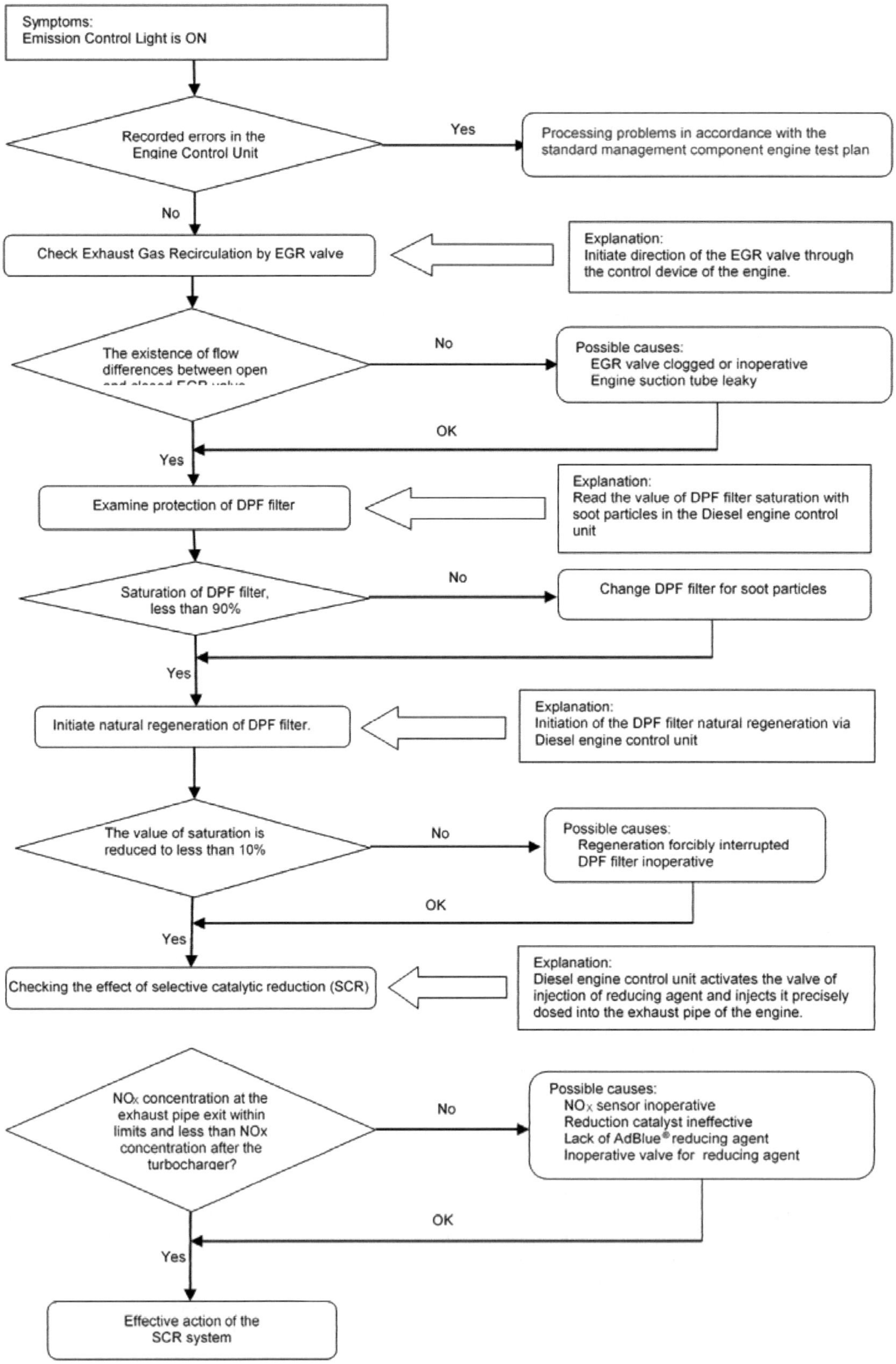

Fig. 1. Algorithm model for exhaust gases system diagnostics

3. Experimental setup

Functionality check of the developed algorithm model was done experimentally on 2.0 TDI and 1.6 TDI Diesel engines, type Volkswagen with electronic control, Common Rail fuel injection system, built-in system of exhaust gas recirculation in the intake manifold of the engine (EGR), oxidation catalyst, Diesel particulate filter (DPF) and selective catalytic reduction (SCR). In the process of measuring, testing and diagnosis during experimental research measurement and diagnostic equipment shown in Figure 2 was used.

1. Diagnostic device (computer with software)
2. Adapter cables for testing sensors and actuators
3. Multimeter (voltmeter, ammeter and ohmmeter)
4. OBD Bluetooth diagnostic head
5. Adapters for OBD diagnostic connector
6. Remap tool to modify the Engine control unit software (Chip-tuning)
7. OBD diagnostic cable for reading EEPROM of the Engine control unit
8. Diagnostic adapter for reading engine operation
9. OBD K+CAN device for reading codes
10. Dual-channel memory oscilloscope
11. EDC controller for Diesel motor
12. EDC controller for Diesel motor

Fig. 2. Measuring and diagnostic equipment used in the experiment

4. Research results

Experimental testing of the functionality of the developed algorithmic model for exhaust after-treatment system diagnostics is carried out in accordance with the anticipated test steps within the model. All the components (sensors and actuators) in the Diesel engine operation management system and in the exhaust after-treatment systems were tested. The deviation of the measured actual parameters of the engine in relation to the required parameters indicate malfunction of the Diesel engine, and thus the increased amount of harmful emissions in the exhaust gases.

In this paper, segments of the experimental testing of algorithmic models for exhaust after-treatment system diagnostics are stated as an example. Any disturbance in the regulation of the start and fuel injection quantity leads to incomplete combustion in the cylinder of the Diesel engine which results in an increase in emissions [11]. The amount of fuel injection is directly dependent on the duration of injection through the injectors. In order to test the algorithmic model presented in this paper, experimental testing process of control fuel injection was done. On the tested Diesel engine was established duration of the main and three subsequent injections during the running regeneration of DPF filter at intervals of about 20 to 30 seconds. Piezo fuel injectors that, with an electric pulse, trigger engine control unit via the piezo actuator were used on the tested engine. It has been found that the duration of the main injection was longer in the beginning of the regeneration process (about 300 μs) while the later has become shorter (about 290 μs). Experimentally determined duration of the main injection is expected, because the beginning of the regeneration process requires a greater amount of fuel to increase the exhaust temperature. Diagram of the main fuel injection during the regeneration is shown in Figure 3. The duration of subsequent injections at time intervals is shown in Figure 4. The diagram shows that the third post-injection had interruptions at certain points (injection duration value is 0), which had resulted in a slight disturbance in the exhaust after-treatment.

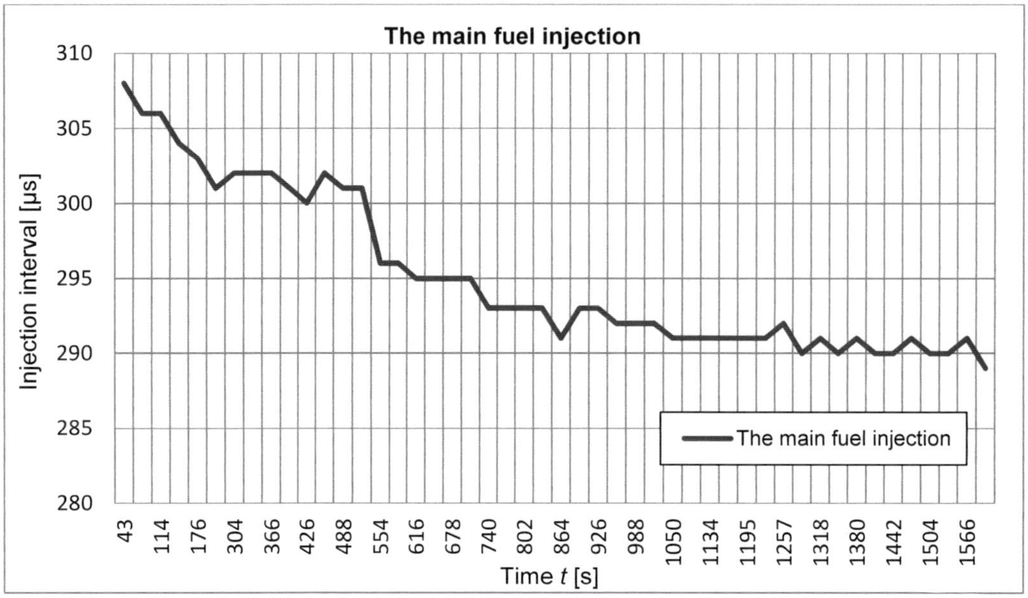

Fig. 3. The main fuel injection during DPF regeneration

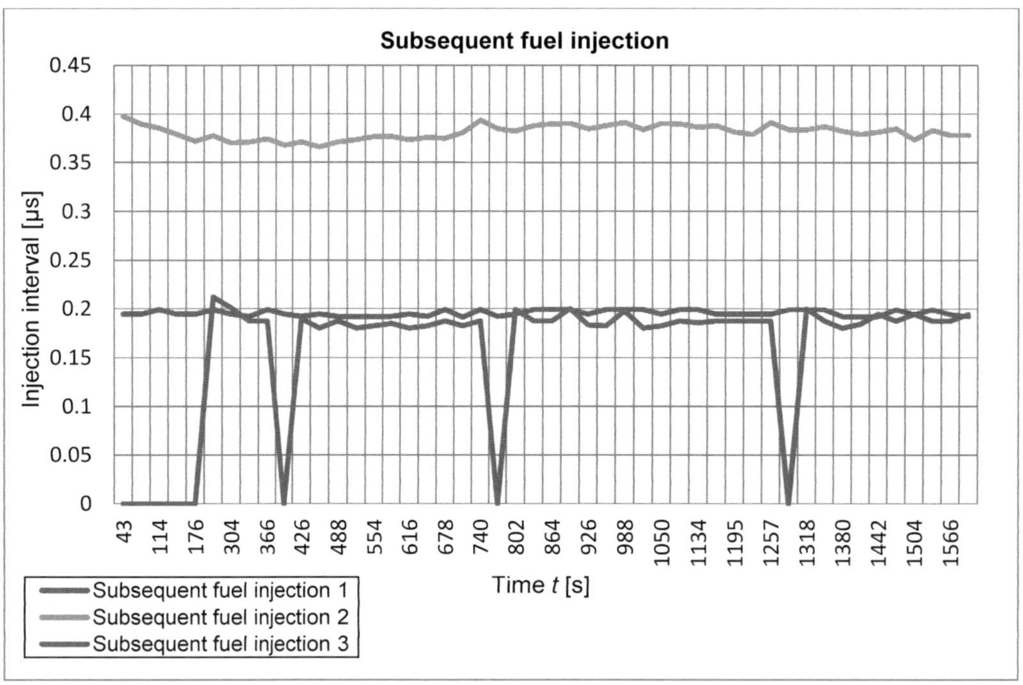

Fig. 4. The subsequent fuel injection during DPF regeneration

In the functionality check of the DPF filter, the element that most affects the reduction of soot particles, PTC temperature sensors in front of the turbocharger, before and after the DPF were tested. From the correctness of the sensor signal, depends the very process of regeneration in the DPF filter. In the process of forced regeneration, the soot particles deposited in the DPF filter burn due to an increase in the exhaust gas temperature conducted by the control device of the engine. In the experiment, during the process of regeneration, at intervals of about 20 seconds, temperature values of the exhaust gases were measured at three characteristic places (in front of the turbocharger, before and after the DPF Filter). The measured temperature values from the beginning of the process of regeneration (43 seconds after the warm-up phase) until the end of the regeneration process (1597 seconds after the start of the process) are shown in Figure 5. The continuous change in the value of temperature indicated the correctness of the tested temperature sensor. Each illogical abrupt change in temperature indicates a failure of any of the used temperature sensors which directly affects the process of regeneration and the creation of high emissions of soot.

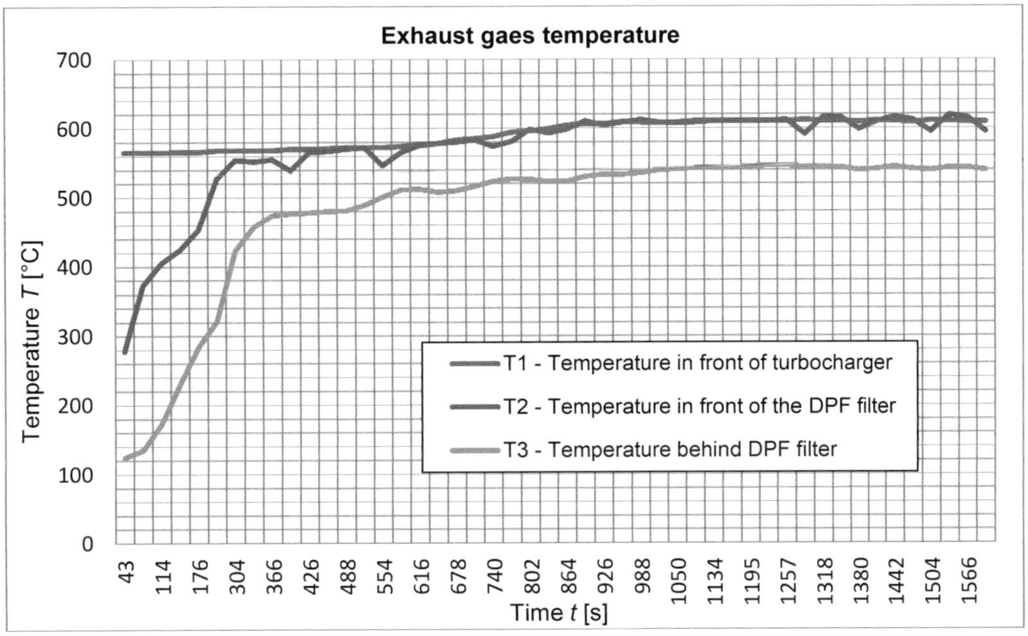

Fig. 5. Exhaust gases temperature change during DPF regeneration

In the process of experimental test of the developed diagnostics model, one of the steps was to monitor changes in the saturation of DPF filter with soot particulate. During the process of regeneration amount of soot particles must be continuously reduced as a result of combustion in the DPF filter. The mass of soot, on a tested Diesel engine, before the beginning of the regeneration was 24 grams, and the concentration is measured by the differential pressure sensor. Values of the mass of soot in the DPF filter, were measured at intervals during the regeneration and recorded in the Table 1.

Dimension	Value							
Time *t* [s]	43	83	114	145	176	204	304	335
Soot mass [g]	24,12	24,12	24,12	24,12	24,12	24,12	24,00	23,82
Time *t* [s]	366	395	426	457	488	516	554	585
Soot mass [g]	23,42	23,08	22,58	22,00	21,38	20,76	19,78	19,08
Time *t* [s]	616	647	678	709	740	771	802	833
Soot mass [g]	18,12	17,18	16,26	15,32	14,30	13,46	12,42	11,38
Time *t* [s]	864	895	926	957	988	1019	1050	1079
Soot mass [g]	10,44	9,42	8,36	7,50	6,70	6,00	5,44	4,96
Time *t* [s]	1134	1165	1195	1226	1257	1287	1318	1349
Soot mass [g]	4,22	3,90	3,64	3,40	3,20	3,02	2,88	2,74
Time *t* [s]	1380	1411	1442	1473	1504	1535	1566	1597
Soot mass [g]	2,60	2,48	2,36	2,24	2,14	2,04	1,92	1,84

Table 1. Measured values of soot mass during DPF regeneration

According to Table 1, each point of observation, in the process of regeneration, is associated with the value of the mass of soot at the time. The table clearly shows that during the experimental testing of the process of regeneration DPF passed without interference because the value of the mass of soot in the filter is continuously decreased without sudden changes. At the end of the regeneration process, combustion of soot particles was complete because the value of soot fell below 2 grams. This indicates the correctness of the DPF filter and overall system exhaust after-treatment. Also, the outcome of the process indicates the correctness of the entire Diesel engine management system.

The conducted experiments in this paper, in accordance with the steps of developed algorithm model for exhaust gases system of Diesel engine with electronic control diagnostics, has shown that any malfunction of the Diesel engine, that affects the harmful emissions, can be detected by this systematic approach.

5. Conclusion

After the research, it has been experimentally demonstrated and supported by diagnostic protocol, that the developed algorithmic model for exhaust gases system of diesel engine with electronic control diagnostics control is fully functional and effective. The results of experimental studies in this paper clearly and unequivocally indicate that using the developed algorithm model for exhaust gases system of Diesel engine with electronic control diagnostics, which includes interactive communication of engine control unit and engine diagnostic device, allows accurate diagnosis of faults and disturbances in the system exhaust after-treatment. Also, the developed model can specify the impact of failure of each individual component of the Diesel engine with electronic control in the process of after-treatment of exhaust gases of the engine. In the future, it is certainly expected emergence of new more sophisticated systems for Diesel engines exhaust gas after treatment, therefore the future goal of the research will be the adjustment of the algorithmic model to those systems.

6. References

[1] EUR-Lex: Regulation (EC) No 715/2007 of the European Parliament and of the Council on type approval of motor vehicles with respect to emissions from light passenger and commercial vehicles (Euro 5 and Euro 6) and on access to vehicle repair and maintenance information, Annex I. Emission Limits, Official Journal of the European Union, No. 284, 20.06.2007.

[2] United Nations: Kyoto Protocol to the United Nations Framework Convention on Climate Change, UNFCCC, Kyoto, 1998.

[3] Takeda, Y.; Keiichi, N.: Emission Characteristics of Premixed Lean Diesel Combustion With Extremely Early Staged Fuel Injection, SAE Technical Paper 961163, Dearborn, 1996.

[4] Bagly, S.; Gratz, L.; Johnson, J.; McDonald, J.: Effects of an Oxidation Catalytic Converter and a Biodiesel Fuel on the Chemical, Mutagenic, and Particle Size Characteristics of Emissions from a Diesel Engine, Environmental Science and Technology, Vol. 32, No. 9, pp 1183-1191, 1998.

[5] Dürnholz, M.; Eifler, G.; Endres, H.: Exhaust Gas Recirculation – A Measure to Reduce Exhaust Emissions of DI Diesel Engines, SAE Technical Paper 920725, Detroit, 1992.

[6] Sabau, A.; Barhalescu, M.; Oanta, E. & Dumitrache, C., Simplified Model for Combustion Reactions in Diesel Engine, Annals of DAAAM for 2009 & Proceedings of the 20th International DAAAM Symposium, pp. 0811-0812, B. Katalinic (Ed.), Published by DAAAM International Vienna, 2009.

[7] Adler, J.: Ceramic Diesel Particulate Filters, International Journal of Applied Ceramic Technology, Vol. 2, No. 6, pp 429-439, 2005.

[8] Brandenberger, S.; Krocher, O.; Tissler, A.; Althoff, R.: The State of the Art in Selective Catalytic Reduction of NO_x by Ammonia Using Metal-Exchanged Zeolite Catalysts, Catalysis Reviews: Science and Engineering, Vol.50, No. 2, pp 492-531, 2008.

[9] Kyu Yun, B.; Young Kim, M.: Modeling the selective catalytic reduction of NO_x by ammonia over a Vanadia-based catalyst from heavy duty diesel exhaust gases, Applied Thermal Engineering, Vol. 50, No. 1, pp 152-158, 2013.

[10] Dosek, R.; Janku, P.; Bliznak, M. & Varacha, P., Secure High Level Communication Protocol for CAN Bus, Proceedings of the 26th DAAAM International Symposium, pp.1009-1015, B. Katalinic (Ed.), Published by DAAAM International, Vienna, 2016.

[11] Blyankinshtein, I.; Voevodin, E. & Ruban, A., Techniques of Adjusting Opacity Estimates of the Diesel Exhaust Gases, Proceedings of the 26th DAAAM International Symposium, pp.0883-0891, B. Katalinic (Ed.), Published by DAAAM International, Vienna, 2016.

DOI: 10.2507/27th.daaam.proceedings.112

TECHNICAL DIAGNOSTICS IN THE PAPER INDUSTRY

Radoslav Krehel', Marek Kočiško & Peter Poór

This Publication has to be referred as: Krehel, R[adoslav]; Kocisko, M[arek] & Poor, P[eter] (2016). Technical Diagnostics in the Paper Industry, Proceedings of the 27th DAAAM International Symposium, pp.0775-0784, B. Katalinic (Ed.), Published by DAAAM International, ISBN 978-3-902734-08-2, ISSN 1726-9679, Vienna, Austria DOI: 10.2507/27th.daaam.proceedings.112

Abstract

Sample application article offers diagnostic sensor networks. It is shown how it is possible to diagnose production facilities that provide a network containing diagnostic sensors. Comprehensive and systematic view of the use of vibration diagnostics in the factory. Recovery of used vibration diagnostics and technical equipment was carried out on the printing machine Heidelberg Speedmaster directly in active operation of the particular stationery company. Based on the results of measurement of vibration (vibration) deposits are processed and the results are optimal solutions forward.

Keywords: Lathe processing; mechanical vibration; vibration acceleration amplitude; frequency; frequency spectrum.

1. Introduction

To detect impending defects or equipment failures, there are several methods and procedures. One of these methods is one of the main parts of the technical diagnostics and vibrodiagnostics, which deals with measuring and assessing vibration, or mechanical vibration devices diagnosed eg. electrical rotating machine. Based on the results obtained by measuring and analyzing vibration, we are able to prevent or at least delay the failure to diagnose the machine. This is a very impor-tant especially if it is an impor- tant in the manufacture of equipment whose failure could cause major losses and problems for the production. Some companies push the diagnostic data a step further and create forecasts – those opposed to the trapping law, pump failure, entire production operation and tweak it for longer uptime and greater effi ciency. Now we get more information about the network. Branches operate your business have come to this data only to determine the problem. However, currently operating more and more focus on the forecasts in order to eliminate problems. Machinery and equipment provide more information and utilization forecasts for possible defects detected and removed earlier. In the past it was considered a single task readiness and operational capability to produce at minimum cost optimal. Today, more and more manufacturing companies realize that maintenance has an impact on all areas of business risk and efficiency. Definitely affects the personal security, energy efficiency, product quality and meet customer needs [1, 2]. Vibrodiagnostiky main task is to determine the moment when the bearing is the end of their life. Bearings are final and life be destroyed, even if the subsequent installation and maintenance were in order. Thus, the diagnosis is not that we save the deposit, but to ensure reliable operation of equipment [3, 4]. In addition to natural wear of bearings such as find hidden errors, which then can cause rapid wear of bearings, or transported to a critical state, or sudden, unexpected failure. When the diagnosis is to determine the following errors: wear of bearings, release bearings, although people shaft, unbalance, incorrect setting transfer belt, bent shaft, free teeth disorders, resonance, and so on [5, 6, 7].

2. Measurement and analysis of technical condition of the paper printing machine brand He Delberg

The diagnostic measurements in a firm in which the measurement was carried out was to find technical condition and wear of moving parts printing machine He Delberg (figure 1).

2.1. Analysis of problems and selection of the optimal solution

The importance of technical diagnosis is high, despite its insufficient attention to even in many manufacturing operations. In tandem with adequate maintenance, but many times can extend the life of machinery and save money for frequent repairs of defects. Slowly it improves and the number of businesses that use a serious diagnosis is increasing. The technical diagnosis is an important part of maintenance of machinery and equipment. What types of diagnostics are used and what they have in common, or what are the differences? There are many types of diagnostics, but have a common goal which is to consider operating without dismantling the state apparatus, machinery or node. The challenge is to determine the functional status of the device based on the results vibrodiagnostics [8, 9, 10].

2.2. Conduct diagnostic measurements

The company was conducted to measure the printing machine He Delberg.

Fig. 1. Printing machine He Delberg, sample main propulsion printing machine and preview print one sheet

The measurement was designed to assess the vibration of the main printing machine operator (figure 1). The engine manufacturer, Siemens. Machine was diagnosed 13 years but still bearing the original [11, 12]. The measurement was necessary to define a revolution that is known as a rotary wheel of the machine, but takes into account the number of printed sheets (figure 1). One wheel of the machine is printing a single sheet [13, 14, 15].

2.3. Measurement procedure

As a diagnostic tool measuring points were selected in two places:

- measuring point is located on the main drive at the front
- measuring point is located on the main drive behind

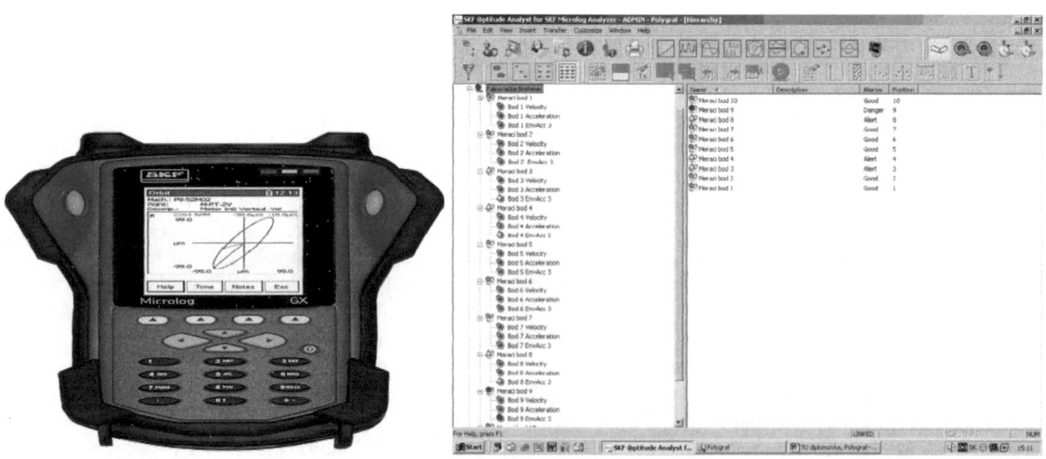

Fig. 2. Measuring device Microlog GX-S and create a database of measuring a computer program AptitudeAnalist for SKF microlog analiser

Using vibration diagnostic instrument was measuring device microlog GX-S (figure 2). Frequency spectra were analyzed using software AptitudeAnalist. The first task was to create a database of measuring a computer program AptitudeAnalist for SKF microlog analiser (figure 2), where the navolené number of measurement points, as well as signal processing method, or setting alarms. Subsequently, the data were uploaded into the measuring device. Then the vibration measurement itself took place [16, 17, 18].

2.4. Data collection

The printing machine was selected two measuring points (figure 3). In each of these points was measured vibration velocity (Velocity), vibration acceleration (Acceleration) and envelope acceleration (EnvAcc) depending on the speed. Measurements took place in the value of speed 3000 rpm 8000 rpm and 12,000 rpm [19, 20].

3. A graphic representation of the vibrations of certain points and analysis

Measurement of vibration diagnostic signals to the printing machine at 3000 rpm.

Fig. 3. Demonstration of the practical implementation of diagnostic measurements of the vibration signal in the first measuring point

Measurement results are displayed in graphs in figures (figure 4 to figure 15).

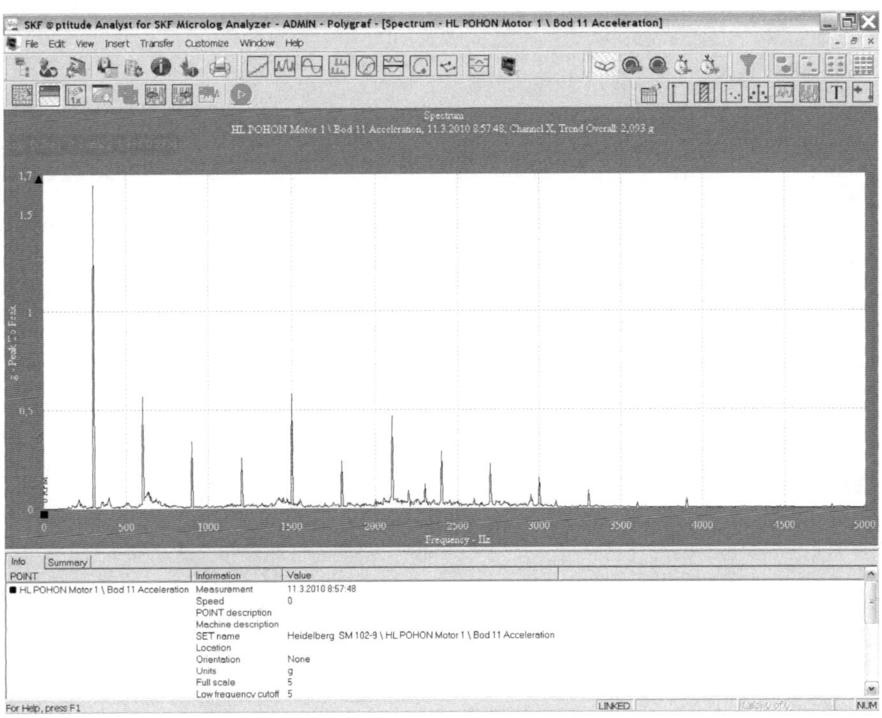

Fig. 4. Spectrum measured vibration diagnostic signals in the first measuring point at 3000 rpm

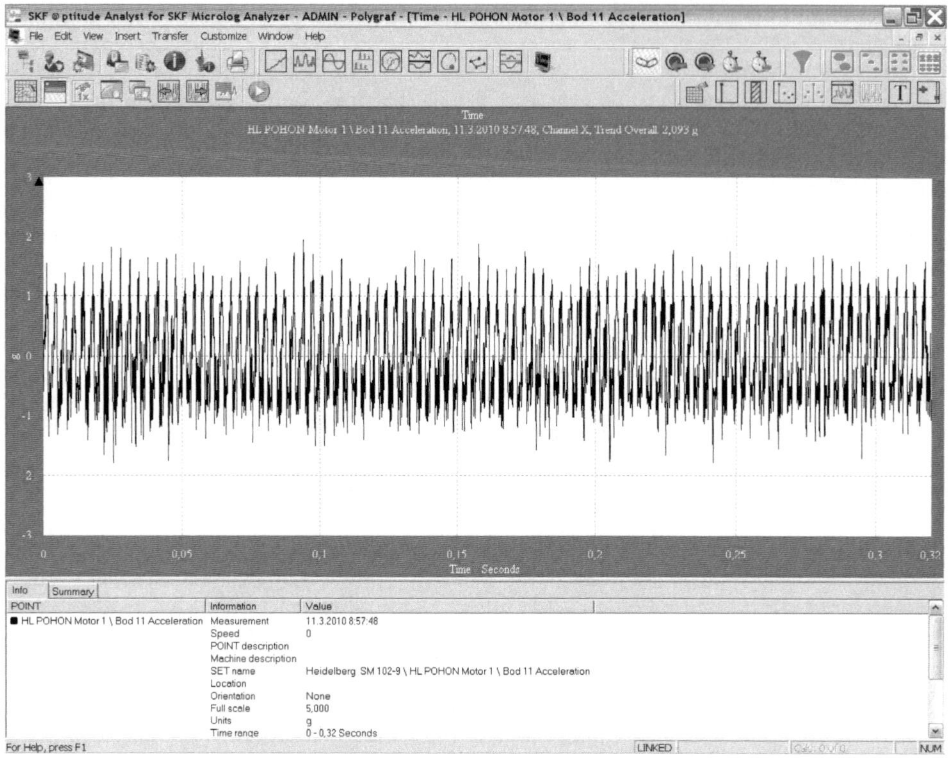

Fig. 5. Time record first measuring point at 3000 rpm.

Measurement of the first measuring point at 8000 rpm:

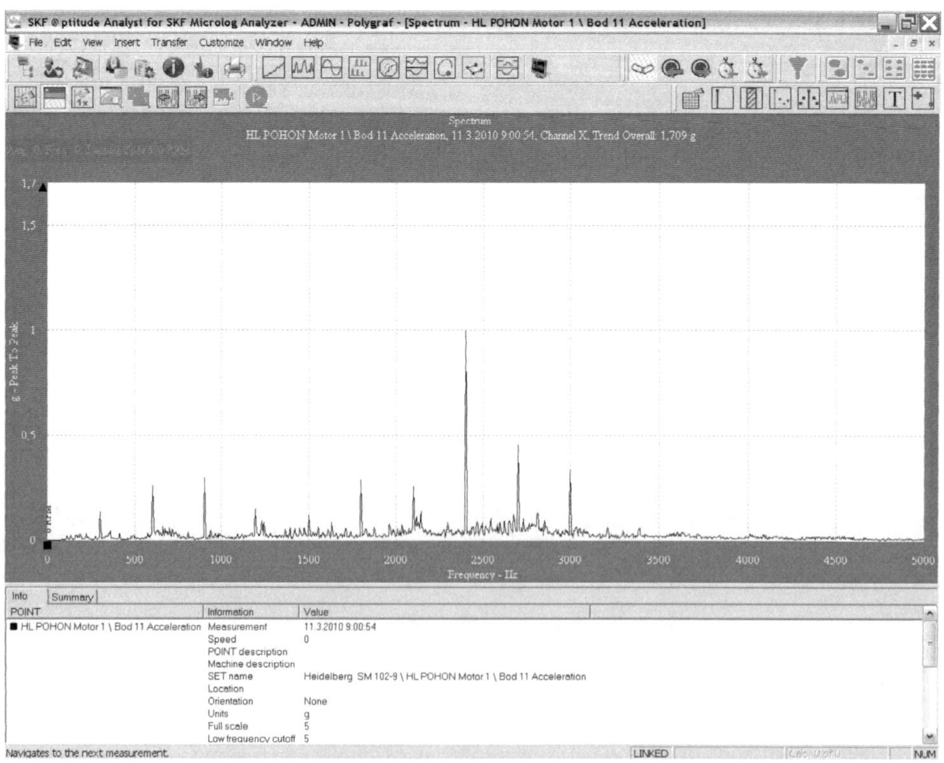

Fig. 6. Spectrum measured vibration diagnostic signals in the first measuring point at 8000 rpm

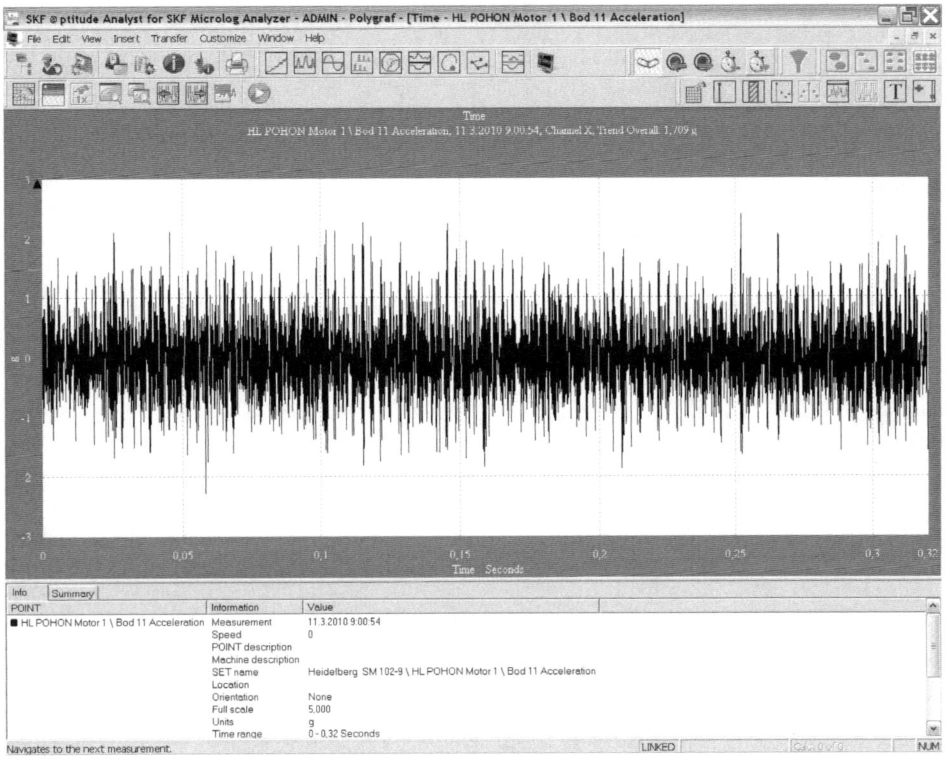

Fig. 7 Time record first measuring point at 8000 rpm.

Measurement of the first measuring point at 12000 rpm:

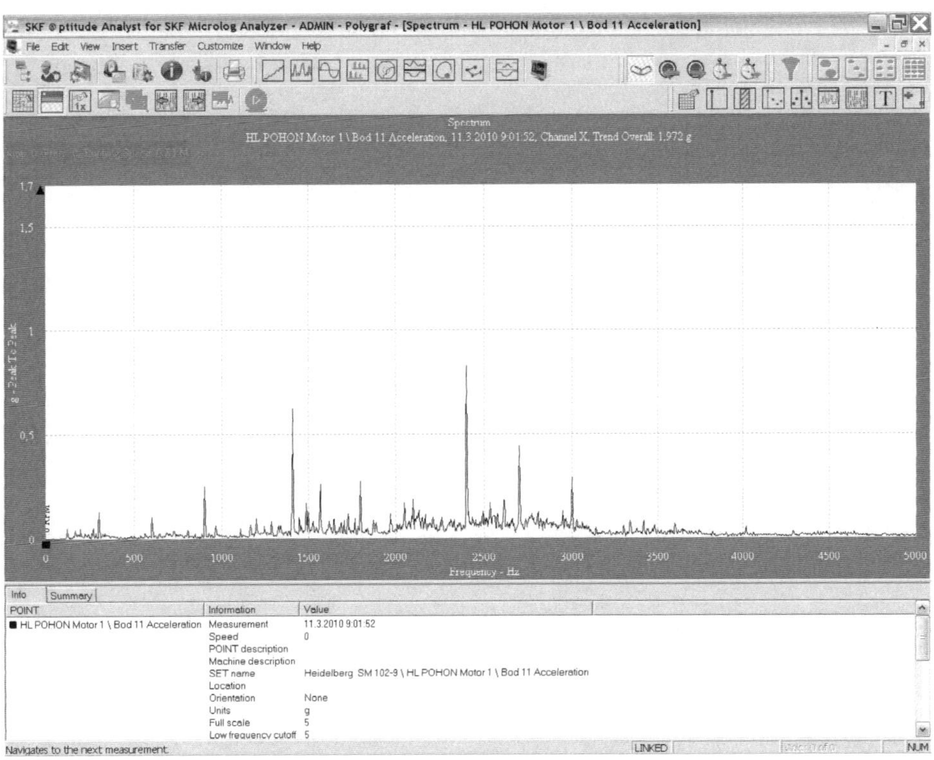

Fig. 8 Spectrum measured vibration diagnostic signals in the first measuring point at 12,000 rpm

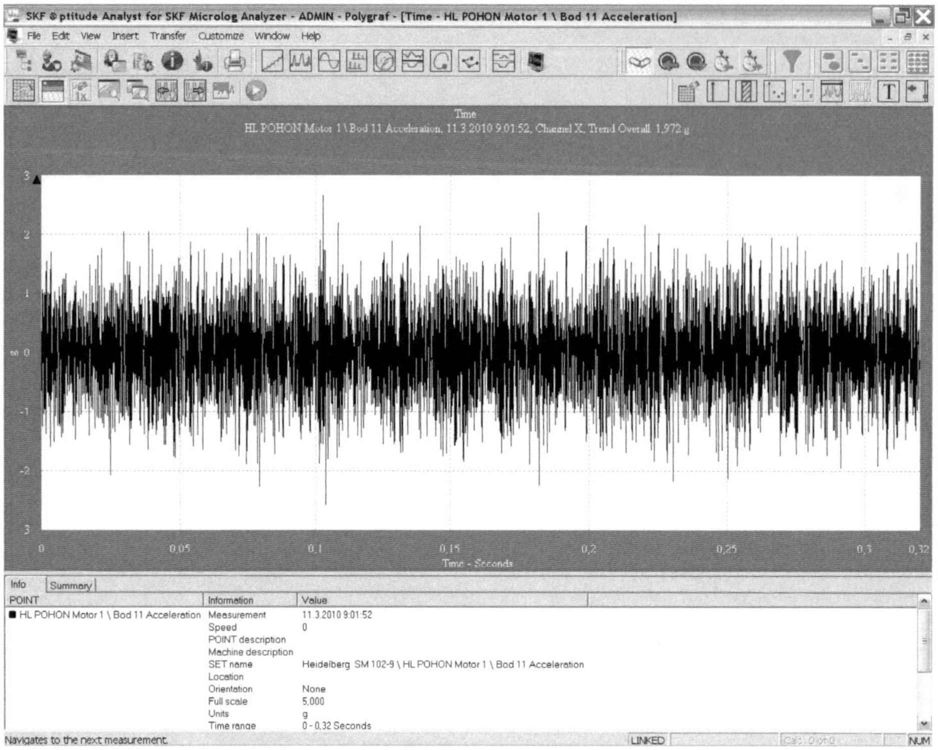

Fig. 9 Time record first measuring point at 12,000 rpm

Processed measurement results of the second measuring point printing machine at 3000 rpm:

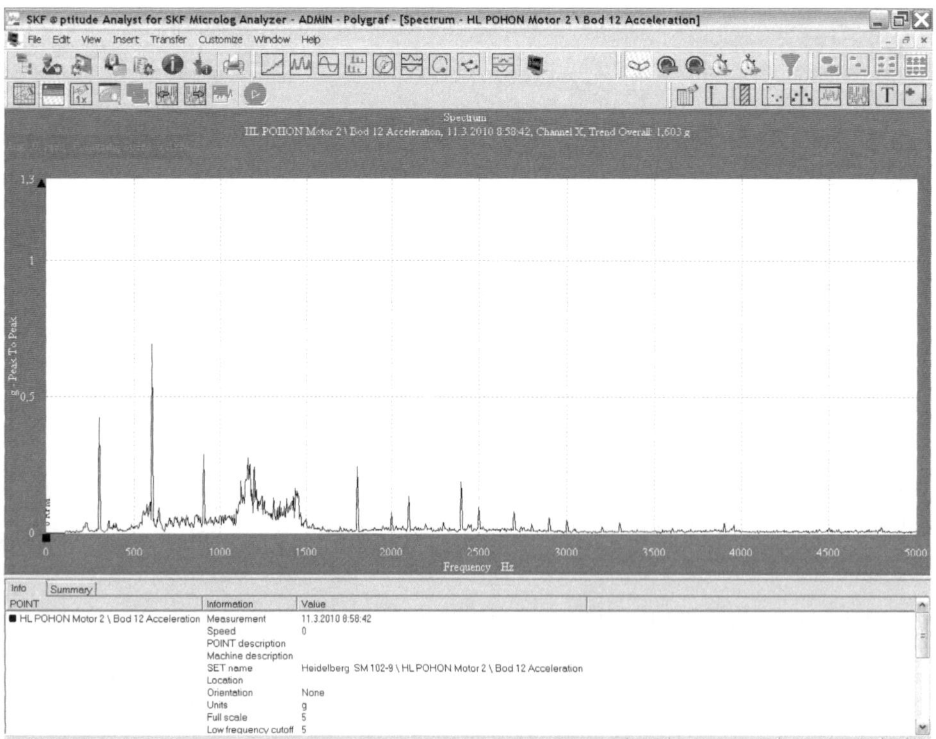

Fig. 10 Spectrum measured vibration diagnostic signals in the second measuring point at 3000 rpm

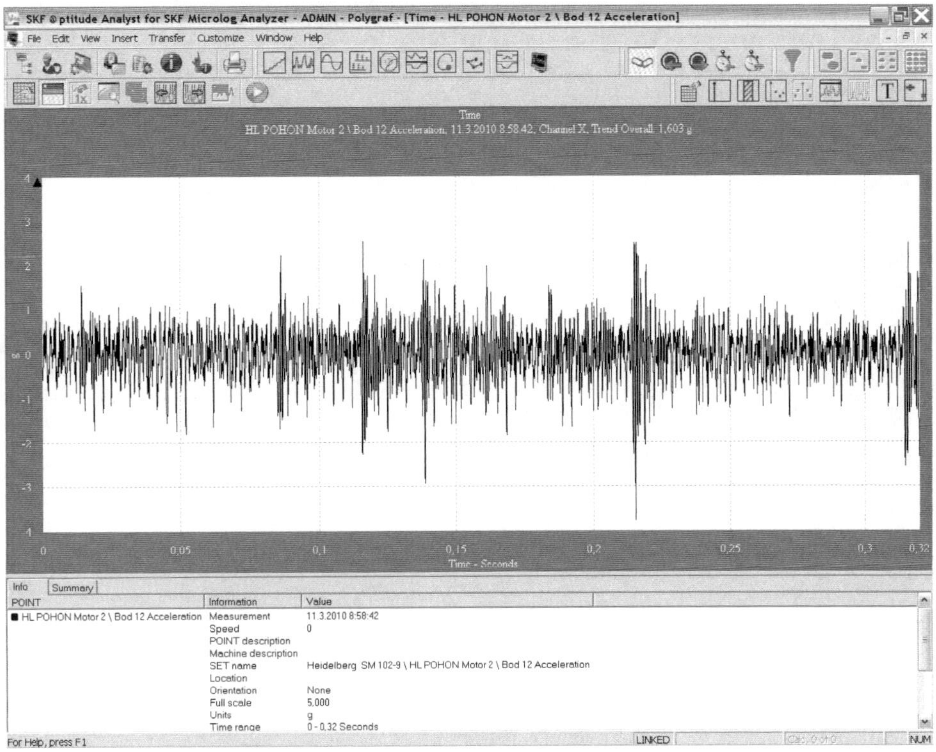

Fig. 11 Time recording the second measuring point at 3000 rpm

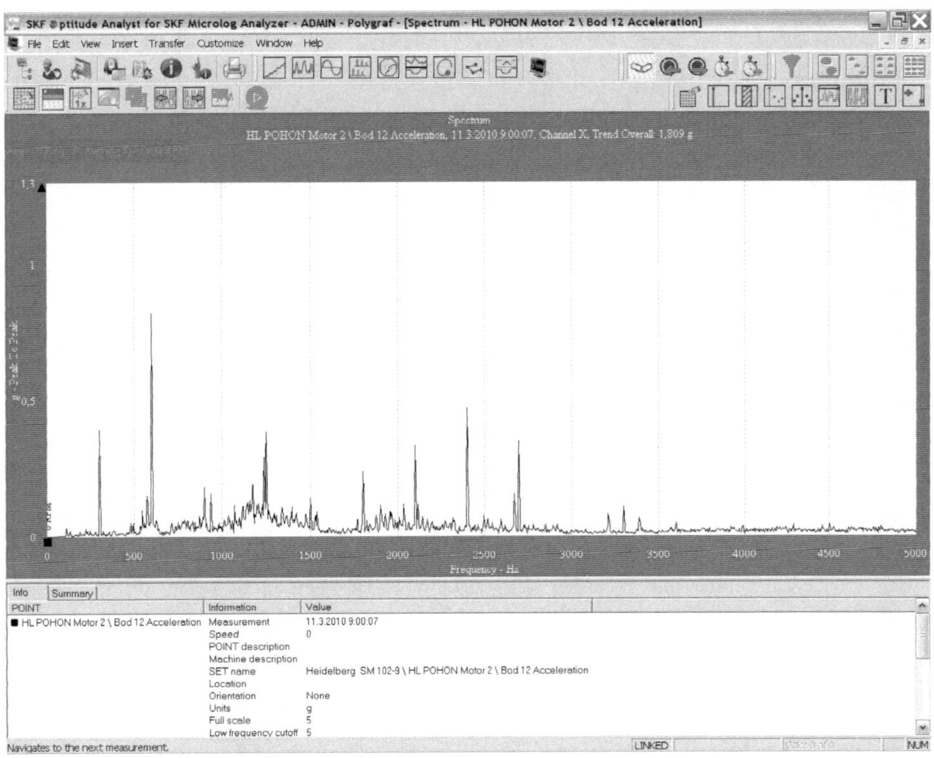

Fig. 12 Spectrum measured vibration diagnostic signals in the second measuring point at 8000 rpm

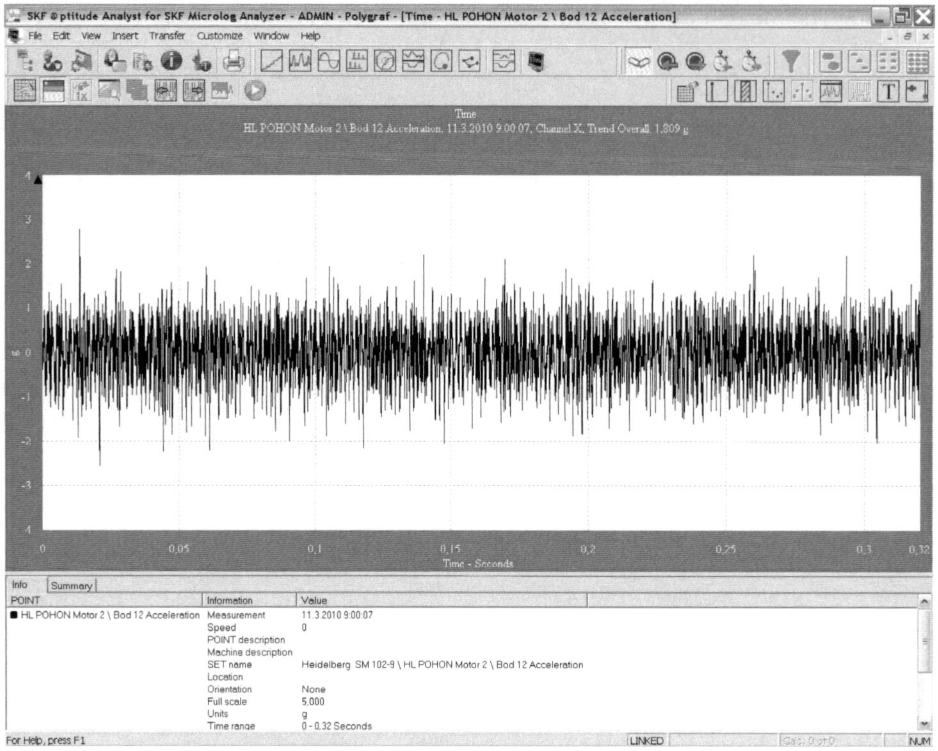

Fig. 13 Time recording the second measuring point at 8000 rpm

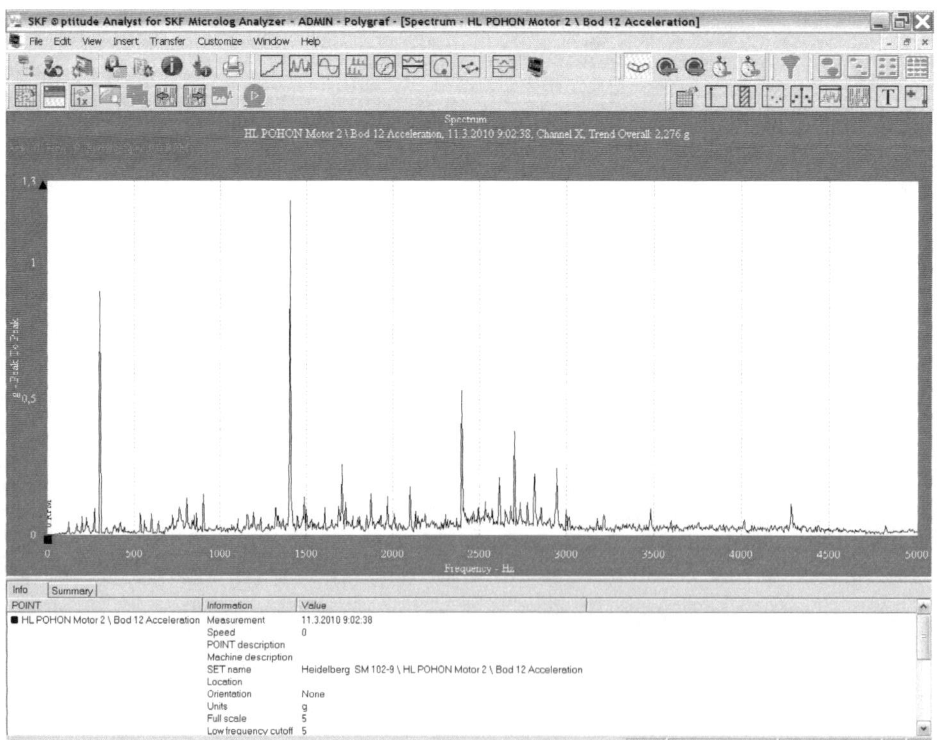

Fig. 14 Time recording the second measuring point at 12000 rpm

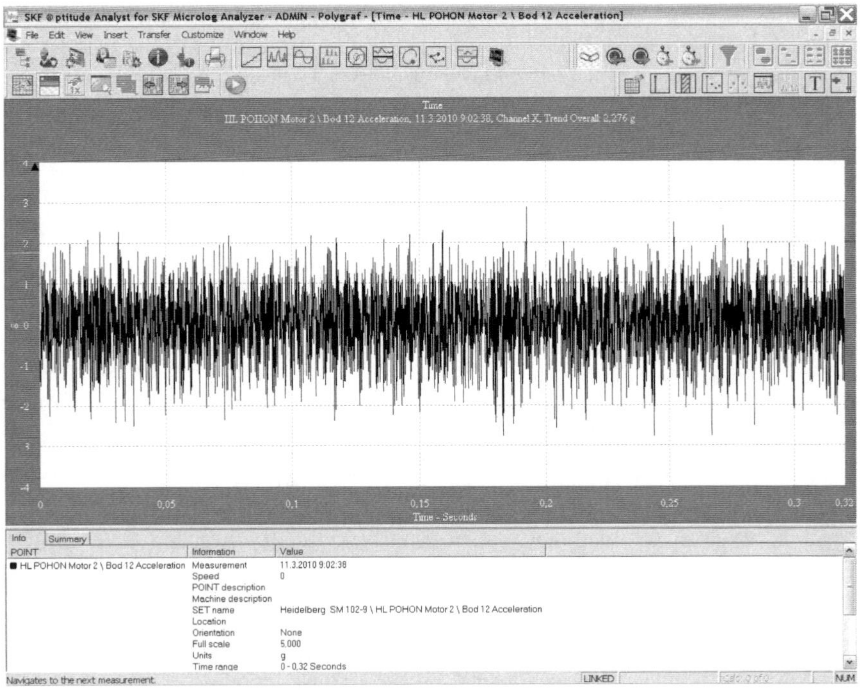

Fig. 15 Spectrum measured vibration diagnostic signals in the second measuring point at 12000 rpm

4. Conclusion

Vibration value measurements obtained had the following characteristics:

- The measurements were conducted for two measurement points and at 3000, 8000 and 12000 rpm.
- Acceleration value vibration is within the range of from zero to 1 g. What for the machine according to the norm does not exceed the value vibration acceleration of 1.8 g.
- Some acceleration value of the spectrum have a greater value in the lower frequency range, which gives information about possible impending failure and the impact of the release of structural parts of the machine.
- Although the record at the time showed values vibration acceleration greater than 1g it is not reflected on the spectrum.
- The machine of the overall results of operations satisfy the requirements and can work in operation.

Article provides a basic overview and introduction to vibration diagnostics, as a method for early diagnosis of emerging faults diagnosed by analyzing machine vibration measurement on that device, because a major impact on production efficiency interoperability has just emerging early diagnosis of faults in the use of machinery and its effective removal. Practical measurements confirmed the assumptions of conduct disorder has the final shape of the measured disturbance variables depending on the type of disorder. That article analyzed the vibrations measured on the printing machine He Delberg (Speedmaster), year 1997, the country of origin, Germany. The measurement device was used microlog GX-S. The measurements were performed on a motor bearings Siemens entry and exit of the main printing machine operator. Given the objective identification and subsequent processing of the measured values of vibration diagnostic measurements are measured at three different speeds for input and output section. Reported values are analyzed on the basis of spectra and time records. These records were evaluated using the diagram. Measurement results showed deposits of perfection, even after many years of continuous operation of the machine with minimal maintenance. The aim of this course is to describe the practical diagnosis of the technical problems of a particular printing machine using vibration diagnostic measurements. Despite the fact that the said vibration diagnostic measurements show no technical failure to measure the printing machine is shown that such a diagnosis can be scheduled to avoid serious disturbances in machinery, which are often expensive and difficult to repair. Therefore, I propose that similar measurements carried out regularly in a certain time interval and thus could prevent unwanted downtime due to failure.

The main part of the research, investigation and analysis of vibration of the operating reliability and operational conditions of the production system, building models to simulate the vibration of the operating reliability and lower operating costs of the production system. The importance of vibration diagnosis is large, but even in its manufacturing plants not given importance. In collaboration with adequate maintenance can prevent vibration diagnostics of machinery malfunctions. Can we save not only money but also expensive machines. The vibration diagnostic methods used much. In practice this means that for each type of machine there is a separate method. The main task is to determine the vibration diagnostics moment when the bearing is the end of their life. The analysis and diagnosis of vibration of the justification given in practice to prevent damage to machine parts such as bearings and the like. Vibration diagnostics is performed by measuring the vibration and vibration of moving parts of machinery. To measure in practice using a variety of devices that capture the vibration spectrum over time.

5. Acknowledgments

This article was written within the project: K-14-008-00. This article was written within the project: ITMS project code: 26220220125; Operational Programme Research and Development support research activities

References

[1] Jacko, P., Krenicky, T. , Salokyova, S. and Rimar, M. Vibration detection technology heads in the process of cutting water jet, Strojarstvo extra 2011, 46 (2013) 1-3

[2] Orlovsky, I., Hatala, M. and Duplak, J. Balance equation – an essential element of the definition of thy drying process. Advanced Materials Research 2014, 849 (2013) 220-4

[3] Jurko, J., Panda, A., and Pandova, I. Analysis of the tool wear of screw drill during the drilling of steel X04CR16Ni12MnTiN. Applied Mechanics and Materials 2014, 599-601 (2014) 32-5

[4] Jurko, J., Džupon, M., Panda. A., Zajac, J., Study influence of plastic deformation a new extra low carbon stainless steels XCr17Ni7MoTiN under the surface finish when drilling, Advanced Materials Research 2012, 541 (2012) 1312-1315

[5] Čačko, P., Krenický, T., Dobranský, J., Impact of an excessive wear of bearing on the mechatronic devices, Applied Mechanics and Materials, 460 (2014) 99-106

[6] Gaspar, S., Pasko,. Influence of technological factors of die casting on mechanical properties of castings from siluin, In: Lecure Notes in Electrical Engineering. Vol. 240 (2013) 713-722

[7] Mascenik. J., Pavlenko, S., Bicejova, L.: Component selected parameters geometrical tolerance value experimental specification, Applied Mechanis and Materials 2013, 389 (2013) 1096-1099

[8] Straka, L., Rimar, M,. Corny, I. and Mihalcova, J. Increasing oc operational reliability of technical system. Paper presented at the Annals of DAAAM and Proceedings of the International DAAAM Smposium, (2011) 1089-1090

[9] Šmeringaiová, A., Valíček, J., Hloch, S., Goban, J.: Three views on kinematic analysis of crank mechanism in educational process. International Scientific Geoconference and EXPO, 3 (2011) 1309-1314.

[10] Novak-Marcincin, J., Fecova, V., Novakova- Marcincinova, L., Janak, M., Barna, J., Kocisko, M.: Virtools and its application in MOCAP and creation of the scripts for animations of models. Engineering Review 2012, 32 (2012) 96-102.

[11] Góts, I., Zajac, J., Vojtko, I.: Vorrichtung zum Messen des Abnutzungsgrades von Schneidwerkzeugen, Technisches Messen /carent/, 62 (1995) 8-11.

[12] Gašpár, Š., Maščenik, J., Paško, J.: The effect of degassing pressure casting molds on the quality of pressure casting, Advanced Materials Research. 428 (2012) 43-46.

[13] Straka, L.: Operational reliability of mechatronic equipment based on pneumatic artificial muscle. Applied Mechanics and Materials. 460 (2014) 41-8.

[14] Monkova, K., Monka, P.: Vibrodiagnostics and its application in manufacturing practice. Applied Mechanics and Materials 2013, 390 (2013) 220-4.

[15] Monkova, K., Monka, P.: Vibrodiagnostics and its application in manufacturing practice. Applied Mechanics and Materials 2013, 390 (2013) 220-224.

[16] Vagaská, A.: Mathematical description and static characteristics of the spring actuator with pneumatic artificial muscle. Applied Mechanics and Materials 2014, 460 (2014) 65-72.

[17] Svetlík, J., Baron, P., Dobránsky, J., Kočiško, M.: Implementation of computer system for support of technological preparation of production for technologies of surface processing. Applied Mechanics and Materials 2014, 613 (2014) 418-25.

[18] Salokyová, Š, Jacko, P., Rimár, M.: Effect of grain size and type of abrasive spectrum of vibration acceleration in the process of water jet cutting, 5 (2011) 39/1-39/4.

[19] Fedak, M, Semanco, P, Micko, M.: Statistical process control method based on weight percent of al-si alloy for melting and holding process in die casting. Lecture Notes of the Institute for Computer Sciences, Social-Informatics and Telecommunications Engineering 2012, 62 (2012) 188-90.

[20] Vojtko, I., Simkulet, V., Baron, P., Orlovský, I.: Microstructural characteristics investigation of the chip-making process after machining. Applied Mechanics and Materials 2014, 616 (2014) 344-50.

DOI: 10.2507/27th.daaam.proceedings.113

DESIGN OF CONSTRUCTION FOR COMPOSITES AND POLYMERS TESTING

Josip Kožul

This Publication has to be referred as: Kozul, J[osip] (2016). Design of Construction for Composites and Polymers Testing, Proceedings of the 27th DAAAM International Symposium, pp.0785-0790, B. Katalinic (Ed.), Published by DAAAM International, ISBN 978-3-902734-08-2, ISSN 1726-9679, Vienna, Austria
DOI: 10.2507/27th.daaam.proceedings.113

Abstract

The aim of this scientific work is development of construction for measuring equipment ElectroForce BOSE, which should enable dinamical-mechanical testing of composites and polymers specimens by hydraulical electrical actuators. The claim of construction is to enable specimen testing for max. applied force of 10 kN. The construction will serve for specimen testing for aeronautical and automotiv industry, The construction claims biaxial material testing, and because of that the claim is also to enable as much as possible reliable simulation of working conditions. Also one of tha construction claims was to enable interchange for each of the actuators which could be in operation of work at the same time. It is necessary to give a conceptual solution for construction design, and than model it in NX Siemens software, give all dimensions, choose construction material by appling finite element methode, and compose tehnical documentation.

Keywords: Design of construction; composites and polymer testing; testing via. Actuators; NX Siemens; finite element methode

1. Introduction

Turbulent market of today is pressuring companies to respond on customer demands very fast, with right quality and acceptable price. On Institute of Polymer Product Engineering (IPPE JKU Linz) we have to give quick response on companies requirements based on research or testing of some specific part in automotive or aerospace industry. These three factors (Time, quality, and price) are in an interaction and interdependence, so ideal balance between them should be found with the aim to achieve company satisfactory. To make high quality testing results for companies is often hard job, with many risks. One of those risks can be failure appearance during testing process, it is always really important to simulate real conditions in laboratory. This problem can be solved with various desiging solutions, analysis, techniques, and methods, and one very famous methode is Finite element methode wider known as FEM.

In general this paper is focused on solving problem such as specific construction design for material testing. By usage FEM in NX Siemens construction has been tested in extreme conditions. The materials which are expose to tests are mostly different type of composites and polymers. Taking into the consideration that the work has been done at the Institute of Polymer Product Engineering at University of Linz when we had encountered after the arrival of new testing machine for dynamic mechanical testing. We should designed and adjust the construction that will enable dynamic testing of composites and polymers in laboratory conditions and multiaxial tensile conditions. The problem was how to come up with the preliminary solution, make construction model in NX Siemens, analyze it by finite element method in NASTRAN, and deliver all necessary technical documentation for it's manufacturing.

The finite element method (which is used in this test) obtained its real impetus in the 1960s and 1970s by the developments of J. H. Argyris with co-workers at the University of Stuttgart, R. W. Clough with co-workers at UC Berkeley, O. C. Zienkiewicz with co-workers Ernest Hinton, Bruce Irons and others at the University of Swansea, Philippe G. Ciarlet at the University of Paris and Richard Gallagher with co-workers at Cornell University. Further impetus was provided in these years by available open source finite element software programs. NASA sponsored the original version of NASTRAN, and UC Berkeley made the finite element program SAP IV widely available. [4]

Solution is found in the steel construction which is strong enough to hold out forces over 10 kN. Principle to achieve this is shown in Figure 3. So, in this paper I give design solution for this specific requirements. As result, I offer optimal model which satisfies all requiremets to achived precisious test results.

2. Defining the requirements in design, usage of experimental equipment and material selection

The process of progression from the initial specification (definition of the task or problem) to a solution, or complete set of information about each product can be viewed as a series of steps in which the alternate information processing and decision making. Each of the steps may indicate a decision which in any way can change the situation within the set of product information. A set of information about the product consists of all generated drawings, models, analyzes, calculations, technical instructions, notes and accumulated knowledge during the design process. Requirements in the design of construction were specifically define and given in a way:

- Strength of construction must allow performance tests of samples up to a maximum applied force of 10 kN;
- Dimensions of the construction must be compatible with implementation and adaptation of the experimental machine BOSE ElctroForce 3510;
- The design should provide the necessary interchangeability of all actuators that can be found in the construction;
- Design allow height adjustability and flexibility of the side plates;
- Adapt of the construction where actuators reinforce in the uniaxial mode withoit any complicated procedures, provide a replacement of the same, if necessary, in order to present valid test results.
- Dismantling of the upper plate

A detailed review and analysis of the properties of construction steel I came to the conclusion that the optimal solution of choice of materials for the construction is steel S235JRG2 / 1.0038 / Č0361, because it is very available on the market, affordable prices. [1]

Fig. 1: Taking measurements for construction on BOSE ElectroForce 3510 machine[1]

Test equipment ElectroForce® test instrument from BOSE offers outstanding precision and versatility for different types of testing samples. ElectroForce 3510 is designed for dynamic force up to 10 kN and it has a large range of frequency bands. Dynamics test of sample depends on the very characteristics of the sample, the method of fixing the machine, and the configuration of the sample. It's most commonly used for testing samples from the aerospace, and automotive industries, as well as for testing of composite and polymeric materials, and a variety of sporting goods and electronics. There are opportunities for testing the dynamic fatigue of material, work on testing fatigue is practically possible nonstop because the machine has a well-developed cooling system, and a lot of those. ElectroForce for linear actuator uses direct electromagnetic force, and quality design allows greater acceleration, more frequencies and high speed. [1,2]

3. Selection of computing tool - Siemens PLM NX

The system is Siemens PLM NX selected due to its comprehensive and posibility to upgrade modules for specific areas of design. This chapter presents modules Siemens PLM NX programs that allow companies efficient and effective innovation, and a control of whole range of services that are related to the individual phases of the product life cycle. The selection of this program is contributed due to the positive experiences of successful companies that use Siemens PLM NX, and numerous awards. NX (created by merging Unigraphics and I-DEAS program) is a complex CAD / CAM / CAE system that covers a large range of jobs in the construction and manufacturing systems: from the creation of the first preliminary proposal, simulation and analysis, modeling of individual components and assemblies, creating 2D documentation, programming CNC machines up to do quality control. NX is a system that allows data management and connection with a variety of information systems in the enterprise, such as Teamcenter system. NX is a fully associative with each other modules that unite the working environment and unique graphical objectively oriented database. Database that enables parallel operation of a engineer (Eng. Collaborative Engineering). In practice, this means that at some point develop a model can simultaneously perform the simulation of manufacturing parts on CNC machine, creating 2D documentation or analysis of the strength of the model. NX program consists of several modules:

- Modules Design (CAD)
- Modules for designing various kinds of tools
- Modules for the simulation of production on CNC machines (CAM)
- Modules for special areas of mechanical engineering
- Modules for the simulation of the finite element method (FEM, CAE)

Siemens vision of CAE simulations is to deliver engineering solutions that will decide on the management performance of the product throughout the production cycle. Siemens has improved and formed their products: Nastran, SDRC and I-DEAS NX ™ CAE program. Effective use of simulation can help project managers reduce many costs and risks. CAE simulation enables engineers to understand, predict and digital improve of the products performance. Like that concept of design can be more researched which in turn reduces direct costs associated with expensive real prototyping and enables more educated decisions. The key is to increase the value of simulation features in common use as early as possible, if possible, and further development of the process. CAE solutions: NX Scenario - for quick assessment of strength, currents, thermal status and trends; NX Master FEM - (formerly known as I-DEAS Master FEM); NX Nastran FEA [7,8]

4. Solution and construction goal

The problem was how to design a construction that will enable the dynamic-mechanical testing of composite and polymer samples by means of hydraulic and electrical actuators attached to the frame structure, while she herself was attached on ElektroForce instrument. The application of structure was to enable testing of the samples to the maximum applied force of 10 kN. Since such structure is not in everyday use for school performance tests specimens, but is used to test samples in everyday use that are in extreme conditions, such as parts for aircraft and automotive industry, and for an authentic simulation of real conditions, performance tests required a multi-axis voltage state, which this construction provided. Also one of the conditions was to provide structure for interchangeability of the three actuators, which can be found in operation at the same time. After preliminary design followed the modeling software in Siemens NX, material selection and application of the finite element method to the construction, and the construction has satisfied all the conditions for the implementation of the envisaged tests.

The solution was to design main part of the construction as on figure 3 which consists two plates, top and bottom, and four supporting pillars. Additional components of constructions are two side horizontal plate on which are mounted hydraulic or electric actuators, one auxiliary detachable vertical plate with the same purpose as the previous two, but this, if necessary, can be dismantled in the event that we do not need multi-axis testing, but only one axis, its stabilization allow two identical carriers. [1]

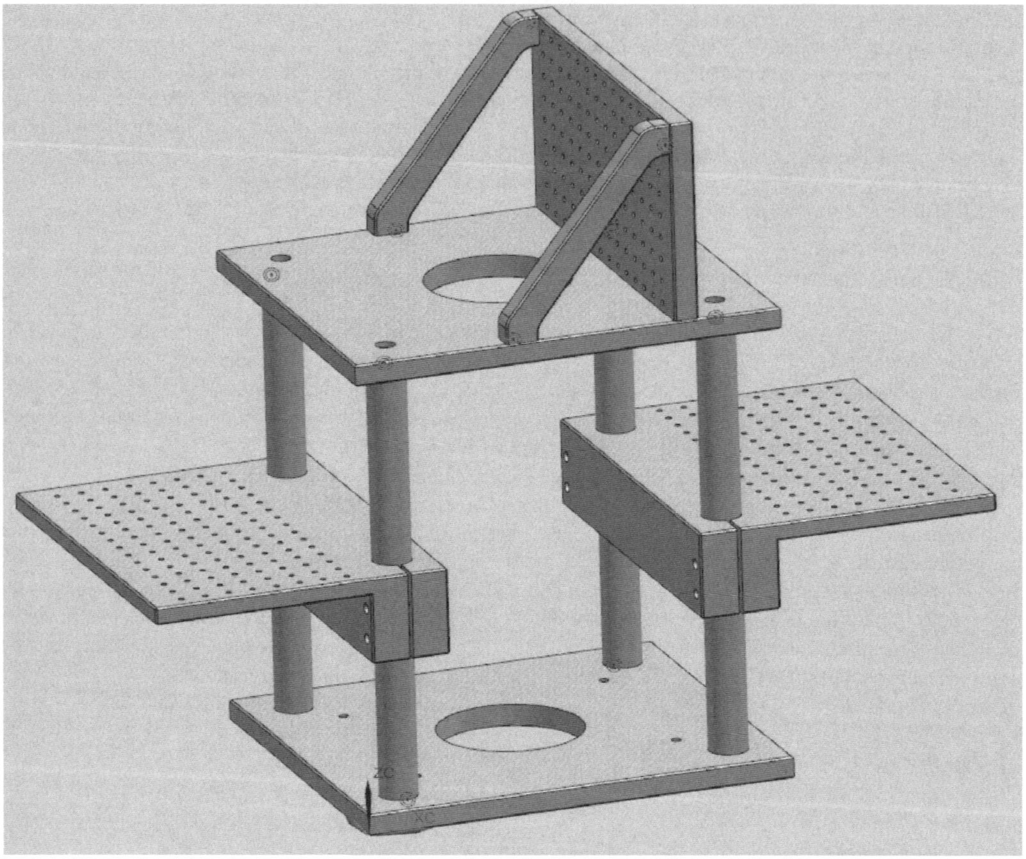

Fig. 2: Construction for composites and polymers testing [1]

5. Results of stress analysis by FEM

Analysis by finite element method was performed in the software NX Nastran, and the conditions under which analysis is perform are set in the way it is conducted under extreme conditions when the construction is weakest and that is when the side panels are on half of the construction height, or when they are at maximum distance from the console attachment, also analysis is carried out under conditions where the load (force) is the highest possible which can occur as a result of the test material, and it is 10 kN. Testing has been done in conditions set in the way that the ends of the pillars of the structure designed as a console, and a force that occurs as result of electric or hydraulic actuators that is acting in the horizontal direction in the middle of the construction at a height of 10 cm above the side panel, which corresponds to a distance shaft of electric actuators with side plates when they are mounted on the same. These conditions correspond to the real, but also to extreme conditions in which the construction can be found in exploitation. It is important to say that the structure is very rarely found in extreme conditions, but due to security measures, the analysis is carried out in a manner such that these are normal conditions. The aim of the construction is to be openable and detachable for easier assembly and disassembly, to withstand these conditions and that on it as such could perform a variety of tests. Strength is the ability of opposition appears undue damage that may arise due to the load. This load, because of stresses and strains that produce, causing two main types of illicit damage: fracture (or the occurrence of cracks, leading to breakage) and plastic deformation. Of course, the limit stresses are relevant characteristics of strength of materials. This means that they should be selected according to the one (characteristic) values of strength, which can not be achieved. If the stress time-varying (dynamic), the relevant characteristics of strength will be the dynamic strength (the fatigue limit). In the event of prolonged static load, especially at elevated temperatures, the relevant characteristics of strength will limit creep or long-term static strength, and so on. It is clear that the values of the boundary stresses are different for different types of load (tensile, compression, bending, shear, torsion). The ratio of the relevant characteristics of strength and work stress which shows how many times the relevant characteristics of the strength is greater than the operating stress σ is called a degree of safety. [5]

The degree of safety ν must be greater than or at least equal, very carefully and very responsibly chosen values so the required degree of safety ν needed. [6]

After this expression strength is controlled at the relevant place of the component and therefore it is a condition of strength. In doing so, the required level of security is determined on the basis of experience and knowledge, and limits to him by

estimating the extent of damage which could arise from unauthorized impairment (ceiling), and the least possible expenditure of materials, that is the price of the product (the lower limit).
The ratio of strength and degree of safety ν needed is the limit that the drive stress σ must never cross, and is called allowable stress.

After calculating the level of security of the structure which is 8.57, I carefully selected the required degree certainlity which is 3, the inclusion in the above formula have been given to maximum stress σ construction during the test is 42 N/mm², and it is not greater than the allowable stress σ which is 120 N/mm². So we came to the conclusion that the design meets the requirements of strength: $\sigma \leq \sigma$ allowable, 42 N / mm² ≤ 120 N / mm².
Comparing to other test results of similar construction this construction is strong and enough to hang out all possible loads in extreme conditions and give precisous results. [3]

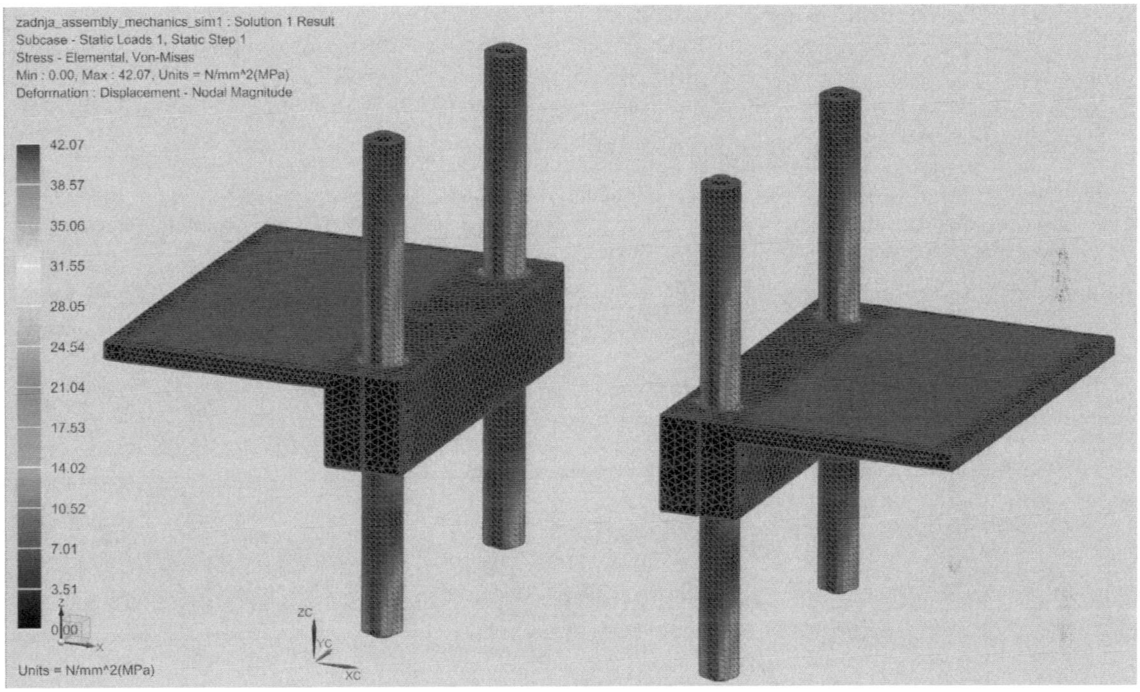

Fig. 3: Results of stress analysis by FEM [1]

6. Conclusion

In this paper, problem was to design and manufacture construction which enable composites and polymer testing in uniaxial and multiaxial conditions with max. applied force over 10 kN. This work gives solution by usage of principles of methodical design and development construction for testing composite and polymers materials in uniaxial and multiaxial stress conditions at the BOSE ElectroForce 3510. The procedure for making digital documentation for the construction and testing of composites and polymer materials to uniaxial and multiaxial stress is shown in the system Siemens PLM NX. In the practical part of the work is description of detail process of making a 3D model construction and its simulation by finite element method in NX Nastran, as well as the creation of technical drawings. Through the examples it's described development of construction and there are shown basic functions and features that are used in software Siemens PLM NX. It's very important that this work is based on new research. The results are showing that maximum stress σ in construction during the test is 42 N/mm², and it is not greater than the allowable stress σ which is 120 N/mm². So we came to the conclusion that the design meets the requirements of strength: $\sigma \leq \sigma$ allowable, 42 N / mm² ≤ 120 N / mm². Based on global growing production and manufacturing in future there will always be need for new researches and testing to implement our ideas in the real world and to protect ourselfs.

7. Acknowledgments

This work has been fully supported by Austrian federal ministry of science economy and research and Institute of Polymer Product Engineering at Johannes Kepler University in Linz.

8. References

[1] Kozul, J.,: Design of construction for testing composites and polymer materials in uniaxial and multiaxial stress state conditions at the BOSE ElectroForce® 3510 machine, Faculty of Mechanical Engineering and Computing, Mostar, 2015, Accessed: 2016-06-15

[2] Jelaska D.; Properties and application materials, Faculty of Mechanical Engineering and Naval Architecture, Zagreb, 2007, Accessed: 2016-08-05

[3] Filetin, T., Kovačiček, F., Indof, J .: Steel materials, Faculty of Mechanical Engineering and Naval Architecture, Zagreb, 2007, Accessed: 2016-07-10

[4] Biron, M .: Thermosets and Composites, Elsevier, Oxford, 2004, Accessed: 2016-07-12

[5] Filetin, T.,: Steel grades according to European standards; Faculty of Mechanical Engineering and Naval Architecture, Zagreb, 2008, Accessed: 2016-08-24

[6] Explore NX and discover your solution, Siemens Product Lifecycle Management Software Inc. 2013 URL: http://www.plm.automation.siemens.com/en_us/products/nx, Accessed: 2016-09-10

[7] Jelaska D .;Machine Elements; Faculty of Mechanical Engineering, Electrical Engineering and Naval Architecture; Split; 2005, Accessed: 2016-08-18

[8] Nx Siemens Product Lifecycle Management Software Inc. 2013 URL: http://www.plm.automation.sie Nx Siemens Product Lifecycle Management Software Inc. 2013 URL: http://www.plm.automation.siemens.com/en_us/products/nx, Accessed: 2016-09-10mens.com/en_us/products/nx, Accessed: 2016-09-10

DOI: 10.2507/27th.daaam.proceedings.114

CONTROLLING AND TESTING A FLAP SYSTEM FOR GENERATING GUSTY CROSSWINDS IN WIND TUNNEL EXPERIMENTS

Franz Leitner[a], Henning Wilhelmi[b], Klaus Ehrenfried[b], Markus Trenker[a]

[a] *University of Applied Sciences, FH Technikum Wien, 1200 Vienna, Austria*
[b] *Institute of Aerodynamics and Flow Technology, German Aerospace Center, 37073 Göttingen, Germany*

This Publication has to be referred as: Leitner, F[ranz]; Wilhelmi, H[enning]; Ehrenfried, K[laus] & Trenker, M[arkus] (2016). Controlling and Testing a Flap System for Generating Gusty Crosswinds in Wind Tunnel Experiments, Proceedings of the 27th DAAAM International Symposium, pp.0791-0796, B. Katalinic (Ed.), Published by DAAAM International, ISBN 978-3-902734-08-2, ISSN 1726-9679, Vienna, Austria
DOI: 10.2507/27th.daaam.proceedings.114

Abstract

A flap system for the generation of gusty crosswinds in a closed return wind tunnel with closed test section was developed. The system consists of four wings with oscillating flaps. This paper describes the construction of a first prototype of a single oscillating flap, which includes the programming of the drive and the mechanical testing of the flap movement. The flap is required to oscillate sinusoidal with a frequency of 50Hz. Therefore, the system is analysed by using a laser Doppler vibrometer. In the process the flap movement is examined with a frequency of a maximum of 50Hz and an amplitude of a maximum of 3°.
The results show an adequate movement of the flap concerning the programmed frequency, depicting that the amplitude of the oscillation seems to be frequency-dependent. The specified amplitude varies about 20% of the intended value. During this process, no relevant deformations of the mechanism were measured.

Keywords: Wind tunnel; prototype; flap system; oscillation; gusty crosswinds.

1. Introduction

Examining flow conditions for a certain application can require a combination of different approaches. This includes experiments in wind tunnels, field measurements and numerical simulations, which are important methods besides the analytical theory [1]. One issue of wind tunnel experiments is the difficulty of simulating unsteady flow conditions like gusts over a useful frequency range [2]. Gusts can occur for example at tunnel exits, passing trains or just randomly due to weather conditions. Taking unsteady flow in wind tunnel experiments into account helps to reproduce more realistic wind conditions. To define a repeatable and comparable basis for gust generation this research focuses on harmonic oscillations of the flow direction in the range of the test section of a wind tunnel. The harmonic oscillation is defined via its frequency and its amplitude which describes the maximum deflection of the flow direction. Increasing the frequency usually requires a reduction of the amplitude if the motor power and the moment of inertia are invariable. The only variable is the acceleration which is calculated as a product of the amplitude and the square of the angular velocity. Researching flap based systems for gust generation revealed a maximum frequency of about 12.5Hz along with an amplitude up to 2° [3] and a maximum amplitude of about 10° in combination with a frequency of about 5Hz [4]. Patel and Hancock [3] use for example an electro-hydraulic servo-mechanism for displacing movable plates at the nozzle outlet.

This contribution focuses on the development of a more dynamic system which should oscillate with a frequency of 50Hz and an amplitude of 3°. The system consists of four wings with oscillating flaps for generating gusty crosswinds in wind tunnel experiments. To fulfil the dynamic requirements the moveable flaps are built as small and light as possible. The wings of this system get vertically placed at the beginning of the test section of the crosswind test facility in Göttingen. This facility is used, among other things, to test the crosswind stability of trains [5].The moveable flaps of the wing system should induce oscillating flow conditions for the test model by deflecting the flow direction and cause a variation of the wind's angle of attack. For testing the feasibility of the system a prototype of a single flap is built. Thereby it is to be examined, whether the fast oscillating flaps can be driven directly with a common industrial servomotor. This includes the programming of the control unit and the mechanical testing of the flap movement with a laser Doppler vibrometer which is presented in this paper.

2. Theoretical background

The crosswind test facility at the German Aerospace Center, Göttingen is equipped with a rotary table which allows testing the crosswind stability of a train model under different yaw angles in the stationary flow of the wind tunnel. Thereby the direction of inflow is steady and no unsteady flow conditions are taken into account. Several articles have determined unsteady effects and higher forces and moments on cars compared to steady inflow conditions triggered by specific frequencies of gusty crosswinds [6–10]. If the induced wave length corresponds for example with the length of the train model, a higher yaw moment than with steady inflow will be measured as opposite lateral flow components appear in the front and the back part of the model (see fig. 1).

On the one hand, this describes the need for unsteady test conditions for certification procedures of high-speed trains which could, in turn, increase the safety standards in the future. On the other hand, the higher yaw moment is the reason for the frequency requirement of 50Hz. Train models are usually scaled by 1:25. Assuming that a train model with the length of 1m is placed in the test section and the flow velocity is 50m/s, a frequency of 50Hz is necessary in order to get a wave length which corresponds to the model length.

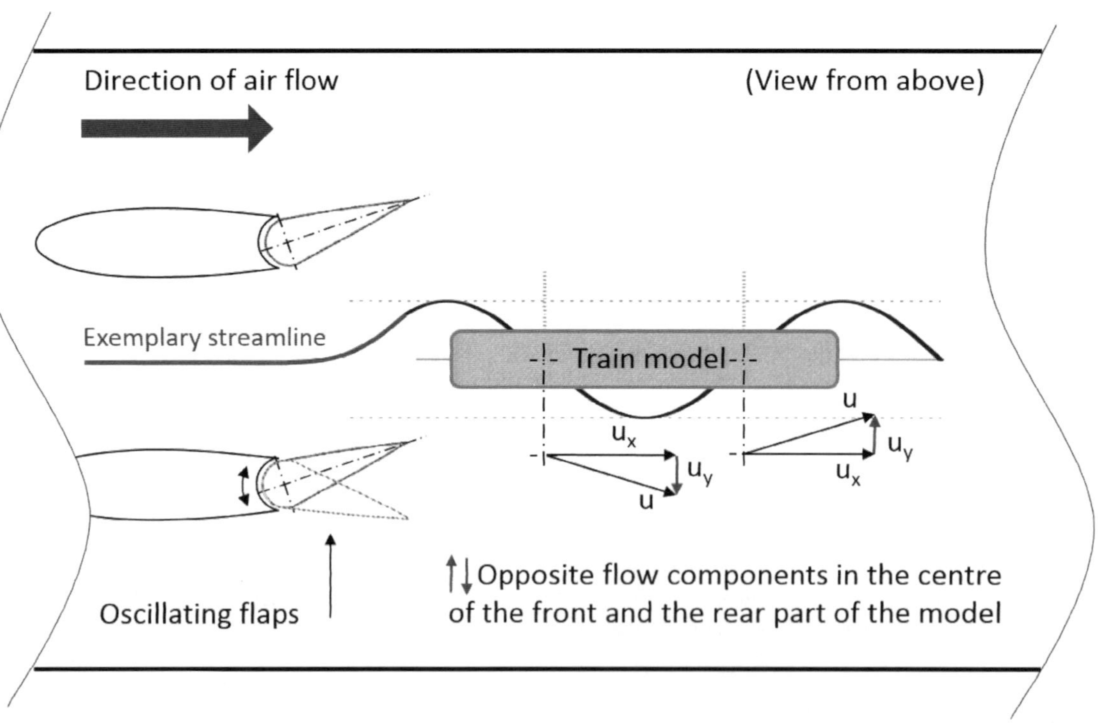

Fig. 1. Concept of oscillating flaps in the test section of the crosswind test facility for generating sinusoidal crosswinds in wind tunnel experiments (view from above)

Results from preceding simulations of the developed flap system with OpenFOAM show its potential. Fig. 2 illustrates the wake flow behind a two-wing system. The simulation was performed with oscillating flaps with a frequency of 100Hz and an amplitude of 5°. The velocity vectors of the flow are stretched ten times along the ordinate to make the induced variation of the flow direction visible. The depicted parameter α describes the current angle of deflection in degrees in relation to the direction of the main flow. The rotation axis of the flaps are located at a test section length of 0m.The deflection still exists in a sufficient extent up to 1.5m after the flaps. These results led to next step of building a prototype.

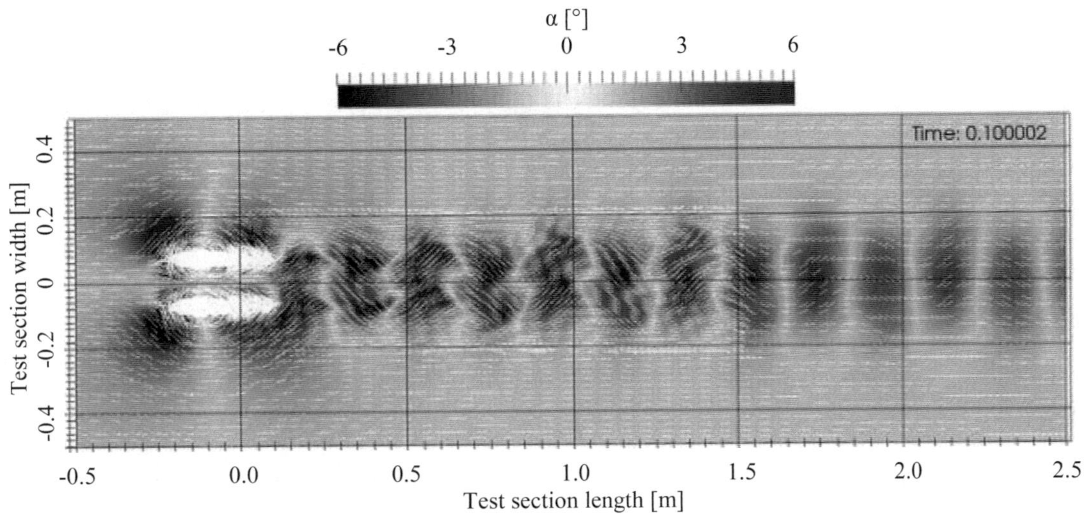

Fig. 2. Results from a flow calculation of a two-wing system by a general flow velocity of 40m/s

3. Materials and Methods

In the following section, the approach for building and testing the prototype is described. For the dimensioning of the drive chain, a sine function for describing the oscillation is used. The required motor torque is calculated via the necessary acceleration and the moment of inertia of the drive chain. A servomotor with supplementary equipment is chosen accordingly: A synchronous servomotor from Beckhoff (AM8036-0L0A) and an embedded PC (CX5130-0120), which is acquired for driving the flap system. For programming the industrial PC the engineering software TwinCAT3 is used. The industrial PC functions as a Programmable Logic Controller (PLC) which gives position command values to the servo module.

For measuring the flap movement, a laser Doppler vibrometer (LDV) from Polytec is used. The LDV consists of an OFV-5000 controller and an OFV-505 sensor head. The signals are analysed by using Matlab R2012b.

4. Practical Realization

The main structure of the prototype consists of aluminium profiles where the servomotor is attached to a mounting plate in the centre of the top of the structure. The flap system (see red arrow in fig. 3) is located on the top of the main structure and is connected to the servomotor (see green arrow in fig. 3) by a metal bellow coupling (see yellow arrow in fig. 3). It consists of a split NACA0018 airfoil where the front part (grey) is fixed and the rear part (black) acts as a moveable flap. The moveable flap makes up a third of the chord length. It is made out of carbon fibre and a steel shaft is pasted into it at each end and is guided by a slide bearing. The wooden plate makes up part of the bottom of the wind tunnel. The whole drive is placed beneath the wind tunnel and the vertical flap is about half the height of the test section (h_K = 800mm).

Furthermore, a control program and a user interface is developed for defining the flap movement. It is possible to specify the desired frequency and amplitude of the sinusoidal oscillation. The amplitude describes the maximum deflection of the flap in degrees. The user can start and stop the flap movement and define the oscillation parameters via a graphical user interface. The program is developed as a finite state machine with six main states. It is therefore not possible to change the settings while the system is running. It is possible to define a maximum frequency of 50Hz and maximum amplitude of 10°. However, the system was tested with a maximum frequency of 50Hz along with a maximum amplitude of 3° in a first approach.

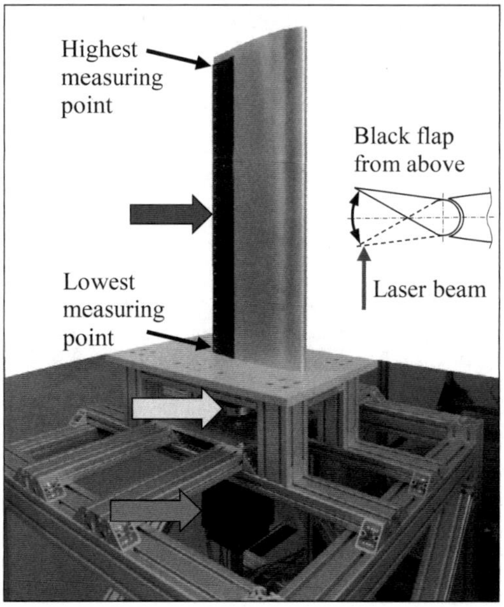

Fig. 3. Construction of the single flap system

For measuring the actual movement of the flap, it is equipped with 23 measuring points at a distance of 75mm from the axis of rotation (see white little markings close to the trailing edge of the flap in fig. 3). The velocity of the flap $v_K(t)$ is determined at each of these measuring points by using the laser Doppler vibrometer and logged synchronous with the actual position signal from the encoder system of the servomotor $\alpha_{act}(t)$. A resolver serves as an encoder system in the utilised servomotor and returns the actual position of the rotor shaft in degrees. Analysing the phase differences of these signals for different movement parameters makes it possible to form a statement about the rigidity of the drive chain as a possible torsion results in a measurable change in the phase differences.

5. Results

At first, a Fast Fourier Transformation (FFT) is applied on the velocity signals $v_K(t)$ from the laser Doppler vibrometer to verify the frequency f of the movement. The set frequency f is properly met as it constitutes the highest spectral component $|v_K(f)|$ of the underlying velocity signal (see fig. 4). The achieved frequency of f = 50Hz of the moveable flap exceeds the dynamic of all researched gust generating systems. The frequency is often a limiting factor in practical applications as it is usually connected to a reduction of the amplitude and reaches mechanical limits [11].

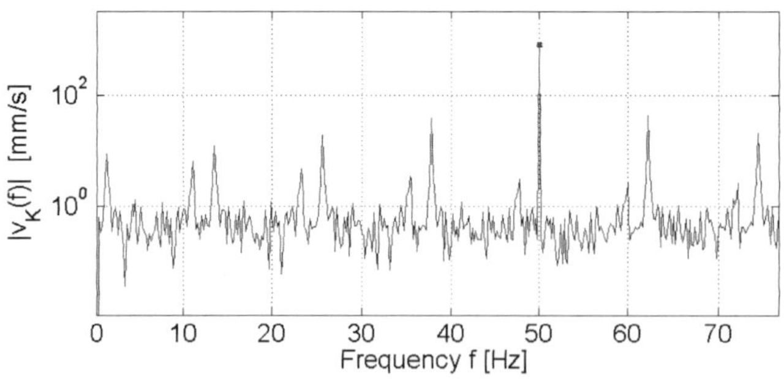

Fig. 4. Frequency spectrum $|v_K(f)|$ of the velocity signal $v_K(t)$ for an oscillation with a frequency of f = 50Hz and an amplitude of $\hat{\alpha}$ = 3°

The next step is to test the actual amplitude $\hat{\alpha}_{act}$ for different frequencies f. This is done by calculating the actual amplitude from the position signal $\alpha_{act}(t)$ and depicting its ratio to the set amplitude $(\hat{\alpha}_{act}/\hat{\alpha})$ over the frequency f. The results of this examinations show a strong frequency dependence of the amplitude $\hat{\alpha}_{act}$ (see fig. 5). The achieved amplitude of the system of over 2° at a frequency of 50Hz exceeds the dynamic characteristics of the presented systems in the introduction.

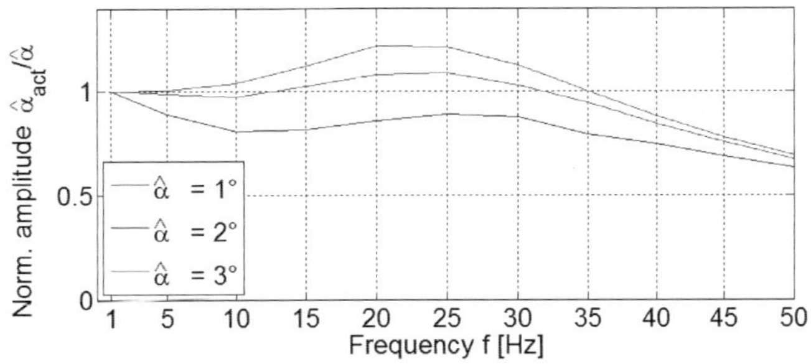

Fig. 5. Ratio of the actual amplitude of the resolver to the set amplitude ($\hat{\alpha}_{act}/\hat{\alpha}$) over the frequency f (measured at the lowest measuring point, see fig. 3)

The actual amplitude $\hat{\alpha}_{act}$ overshoots the via the user interface defined amplitude $\hat{\alpha}$ by more than 20% in a frequency range of f = 20-25Hz. As from f = 25Hz onwards the amplitude $\hat{\alpha}_{act}$ decreases continuously. The source of these deviations is not explicitly clear but may be found in the internal control system of the servo module. By checking the phase differences between the velocity signal of the lowest measuring point and the resolver signal the torsion in the coupling can be approximated. This analysis shows just very little twisting angles in the coupling which can be neglected.

The final step concerns the torsion of the flap. The variation of the amplitude along the flap span is determined by integrating the velocity signal of the laser Doppler vibrometer $v_K(t)$ of each measuring point. The resulting distance signal describes the travelled path of the related measuring point. Dividing the amplitude of this oscillating distance signal by the radius (75mm) approximates the amplitude of the flap $\hat{\alpha}_K$. A variation of $\hat{\alpha}_K$ along the flap is linked to a torsion of it. Fig. 6 shows the torsion along the flap for different frequencies. By a frequency of f = 50Hz a stationary wave with a node at a span of h_K = 750mm is depicted (see red line in fig. 6). The examined differences show a maximum value of 0.1°.

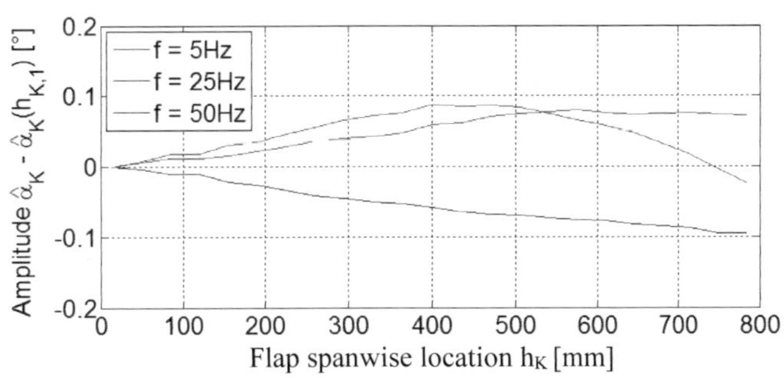

Fig. 6. Torsion along the flap span h_K indicated by the difference of the amplitude $\hat{\alpha}_K$ - $\hat{\alpha}_K(h_{K,1})$ at a set amplitude of $\hat{\alpha}$ = 3° and different frequencies

6. Conclusion

This paper addresses the problem of generating gusty crosswinds in wind tunnel experiments. Using oscillating flaps for simulating gusts demands a high dynamic drive chain. Gust generators described in previous literature do not fulfil the frequency requirements of the German Aerospace Center. Therefore, a new flap system was developed. The flaps are smaller and lighter in order to reduce their moment of inertia. The building and testing of a prototype with a single flap is described in this paper. The single flap is successfully driven by a common industrial servomotor and programmed via a PLC.

The prototype is able to carry out sinusoidal oscillations with an adjustable amplitude and frequency. It is able to oscillate with a frequency of up to 50Hz and an amplitude of up to 3° which exceeds the dynamic characteristics of the researched gust generators. The set frequency gets properly transformed whereas the amplitude shows significant variations which should be considered in future wind tunnel experiments. All measured torsions are insignificantly small. The rigidity of the coupling and the flap can therefore be seen as sufficient. It is assumed that the small twisting angles of the flap will not cause disruptive flow effects in future wind tunnel tests. Further research includes the validation of the system in the wind tunnel by measuring its flow effects for example with hot-wire anemometers.

7. References

[1] Böswirth, L. (2013). Technische Strömungslehre: Lehr- und Übungsbuch (Technical Fluid Mechanics), Vieweg+Teubner Verlag, ISBN 3322943674, Wiesbaden, Germany

[2] Tang D. M., Cizmas P. G. A. & Dowell, E. H. (1996). Experiments and analysis for a gust generator in a wind tunnel, Journal of Aircraft, Vol. 33, No. 1, pp. 139–148

[3] Patel, M. H. & Hancock G. J. (1977). A Gust tunnel facility. Reports and Memoranda No. 3802, Queen Mary College, University of London

[4] Michelbach, A. & Blumrich, R. (2015). Upgrade of the full-scale aeroacoustic wind tunnel of Stuttgart University by FKFS, In: 15. Internationales Stuttgarter Symposium: Automobil- und Motorentechnik (15th International Stuttgart Symposium: Automotive and Engine Technology), Bargende, M.; Reuss, H. C. & Wiedemann, J., pp. 497-523, Springer Fachmedien, ISBN 978-3-658-08844-6, Wiesbaden, Germany

[5] DLR, Institute of Aerodynamics and Flow Technology – Side wind stability of high-speed trains. [Online] Available: http://www.dlr.de/as/en/desktopdefault.aspx/tabid-4702/7791_read-26403/. Accessed on: 2016-03-01

[6] Theissen, P. (2012). Unsteady vehicle aerodynamics in gusty crosswind, Ph.D. Dissertation, Technical University, Munich, Germany

[7] Schröck, D. (2011). Eine Methode zur Bestimmung der aerodynamischen Eigenschaften eines Fahrzeugs unter böigem Seitenwind (A method for determining the aerodynamic characteristics of a vehicle during gusty crosswind), Ph.D. Dissertation, Technical University, Stuttgart, Germany

[8] Wiedemann, J.; Schröck, D. & Krantz W. (2010). Fahrzeugdynamik: Das Gesamtsystem Fahrer-Fahrzeug-Regelsysteme (Vehicle Dynamics: The Complete System Driver-Vehicle-Control System), Themenheft Forschung, No. 7, pp. 38–46

[9] Passmore, M. A.; Richardson, S. & Imam, A. (2001). An experimental study of unsteady vehicle aerodynamics, Proceedings of the Institution of Mechanical Engineers, Part D: Journal of Automobile Engineering, Vol. 215, No. 7, pp. 779–788

[10] Wojciak, J. D. (2012). Quantitative analysis of vehicle aerodynamics during crosswind gusts, Ph.D. Dissertation, Technical University, Munich, Germany

[11] Abdessemed, A. C. & Viba, J. (2016). Aerodynamic and Motion Study of Flapping Wings Vehicle, Proceedings of the 26th DAAAM International Symposium, pp.0834-0841, B. Katalinic (Ed.), Published by DAAAM International, ISBN 978-3-902734-07-5, ISSN 1726-9679, Vienna, Austria, DOI:10.2507/26th.daaam.proceedings.116

DOI: 10.2507/27th.daaam.proceedings.115

DEVELOPMENT OF A FULLY AUTOMATED TUNING SYSTEM FOR ORGAN PIPES

Clemens Sulz[a], Wendelin Eberle[b], Markus Trenker[a]

[a] *University of Applied Sciences, FH Technikum Wien, 1200 Vienna, Austria*
[b] *Rieger Orgelbau GmbH, 6858 Schwarzach, Austria*

This Publication has to be referred as: Sulz, C[lemens]; Eberle, W[endelin] & Trenker, M[arkus] (2016). Development of a Fully Automated Tuning System for Organ Pipes, Proceedings of the 27th DAAAM International Symposium, pp.0797-0803, B. Katalinic (Ed.), Published by DAAAM International, ISBN 978-3-902734-08-2, ISSN 1726-9679, Vienna, Austria
DOI: 10.2507/27th.daaam.proceedings.115

Abstract

Many pipe organs consist of thousands of pipes, divided basically into two different types: flue pipes and reed pipes. Because of the fact, that the principle of sound generation differs, reed pipes must be tuned by hand periodically, which is a time-consuming and thus expensive process. The aim of this project was to do a feasibility study, to determine if this tuning process can be automated and to build up several prototypes for extensive testing. Thereby different actuator technologies were examined and evaluated. Finally a very cheap and compact actuator solution was developed. Appropriate software for controlling the system was programmed and the required drive electronics were developed. Tests with the prototypes have shown that the system is able to perform the tuning process in much shorter time than a human being with satisfying precision.

Keywords: organ, pipe; reed pipe; autonomous tuning system; automatic tuning application

1. Introduction

The pipe organ, called the king of instruments, has fascinated people for hundreds of years. It is the only instrument, which is played by feet and hands simultaneously, produces a huge range of tone colours and covers the whole frequency spectrum of the human hearing. Pressing a key causes air to stream into specific pipes, whereby each pipe produces one tone with a determined tone pitch and timbre. There can be thousands of pipes in a single pipe organ, with each pipe producing a unique sound. Basically, two types of pipes are used in pipe organs: flue pipes and reed pipes (left side of Fig. 1). The sound of the flue pipes is generated in the same way as in a real flute. The air stream strikes against the lip and begins oscillating. The result is a standing wave or vibrating column of air inside the pipe body. These are the facade pipes a beholder can generally see in a church and which represent the majority of the pipe stock. The pipes of the other type, reed pipes, work in a completely different way and are hidden inside the organ. Within the pipe foot there is a metal tongue, which begins to oscillate, if air flows through the pipe (right side of Fig. 1). The tuning spring is used to adjust the pitch of the reed pipe, because it defines the oscillatory length of the tongue. The tone colour of reed pipes allows imitating trumpets, clarinets or other wind instruments.

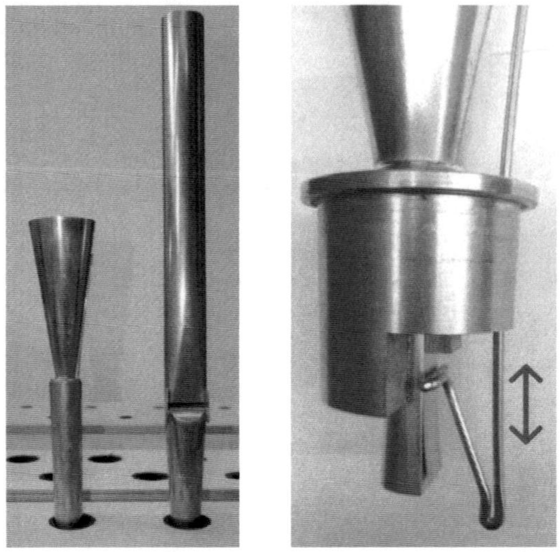

Fig. 1. Left picture: both types of organ pipes (reed pipe and flue pipe);
Right picture: inner parts of reed pipe (tuning spring is moved to tune the pipe)

2. Problem description

The pitch of flue pipes depends directly on the velocity of sound, which in turn depends on air temperature. So the pitch is lowered, if the temperature is decreased and vice versa. Because of the fact, that in reed pipes the tongue oscillates and not the air, the pitch of these pipes stays almost constant. A temperature change of just 1-2 °C causes an audible detuning of the organ. Not least because of the lower number of reed pipes and their easier tunability, the pitch of the reeds is tuned to the pitch of the flue pipes. To tune the pipes the tuning spring has to be moved up or down for each single pipe. Generally this tuning process requires two people (one sitting at the keyboard pressing down the keys) and takes between a few hours and several days for big organs. Because of the associated expense, the reed pipes often are not in tune and are not used by the organist. The aim of this project was to develop a system, which can tune reed pipes automatically. Refined, the aim was to determine whether a technical system is basically able to tune the reed pipes with satisfying precision in an acceptable amount of time. Furthermore, because of the number of reed pipes (usually a few hundred) the solution should be very cost-effective. Especially the little reed pipes were a challenging object of research due to the high sensibility of the tuning spring. Thereby movements of less than a micrometre are required to adjust the pitch exactly enough. The final stage of the project was to build a few prototypes to test and demonstrate the abilities of the system.

3. State of the art

At the beginning of this project an extensive market and patent review was done to find out, if any similar applications are already on the market. Thereby a few patents for automated organ tuning were found belonging to the German organ builder Voigt ([1], [2] and [3]). Furthermore, a project within the framework of a bachelor thesis by Fachhochschule Kiel [4] was found. But in contrast to the idea of automating the tuning process for reed pipes all these applications are developed to modify the pitch of flue pipes, whereby these projects are primarily concerned with conceptual studies. The company Rieger Orgelbau [5], which was the main cooperation partner for this project, has developed a system, which allows the tuning person to control the organ with a smartphone app. Specifically it is possible to play the keys of the organ via the smartphone, so the second person is no longer needed. This system represented the newest state of technology at the beginning of this project. If an actuator, which should be developed in the course of this project, would be combined with this system, a fully automated tuning application would be established.

4. Actuator research

Following the analysis of the mentioned system a research on actuators was performed. Thereby the most important criteria were cost efficiency and space requirements. Furthermore the components and structure of reed pipes should not be modified, or if it is unavoidable, as little as possible. This would make it feasible to upgrade already existing organs with the tuning system. Before the research took place, force investigations on various tuning springs on three different pipes of different size were conducted to determine, how much force an actuator should be able to apply. The highest value which was measured was 6,0N. Including an appropriate safety surcharge for the following research a minimum guide value of 10N was defined.

4.1. Piezoelectric drives

Because of the required precision, piezoelectric drives were examined as a first step. One possible new type of piezo drive is the motor X15G (Fig. 2) from Elliptec [6]. If the piezo crystal inside this actuator is driven by the natural resonant frequency of the whole actuator, the rotor begins to move forward. With a second specific frequency the motor could also be moved backwards. According to the datasheet the drive can also be used as a linear actuator, whereby the drive could be attached directly to the tuning spring to move it up or down (see right side of Fig. 2). Unfortunately, it became apparent that this drive can only raise 1.2N, which is much too little for this application.

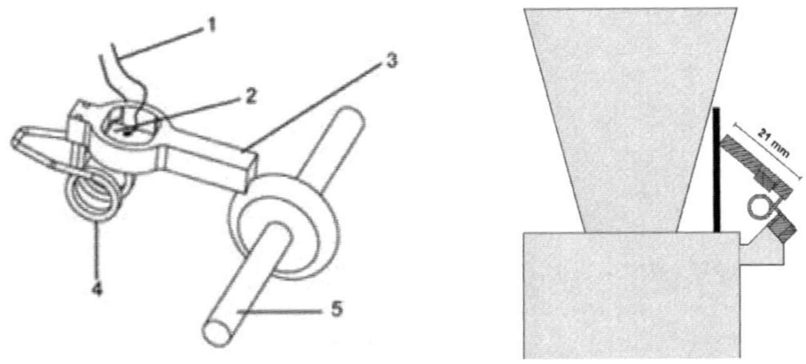

Fig. 2. Piezo motor Elliptec X15G; Left side: 1...wires, 2...piezo ceramic, 3...resonator, 4...spring, 5...rotor (Source: based on [6]); Right side: possible implementation on reed pipe

A second interesting piezo drive is the so-called Squiggle motor (Fig. 3) from New Scale Technologies [7]. Because of the coordinated movement of four piezo elements, the threaded rod inside the drive begins to move translationally without twist. If the tuning spring would be threaded, the Squiggle motor could be attached to it directly. In this case this drive would be very interesting (very low space requirement), but it also became obvious that the force of the motor is too low (max. 0.2N) for the tuning application.

Fig. 3. Squiggle motor (Source: [7])

A last piezoelectric drive, which was investigated, is the Piezomike (Fig. 4) from PI GmbH [8]. With 20N thrust, it was the first piezo device found, which would be strong enough for the tuning application. The piezo crystal inside the actuator is expanded slowly because of the controlled increase of voltage, whereby the gripper starts rotating the screw. If the final position is reached, the voltage is switched off and the gripper goes back to the starting position jerkily without moving the screw. Right side of Fig. 4 shows a schematic diagram for a possible implementation on a reed pipe with a spring, whereby the tuning spring is pulled against the Piezomike. A resulting advantage with this kind of drive would be the possibility to tune the pipe manually through rotation of the screw shaft without disassembling the tuning system. Due to the good applicability the manufacturer of this piezo drive was contacted to get the price of the system. The exorbitantly high price of about 500$/pcs. inhibits the application in this project.

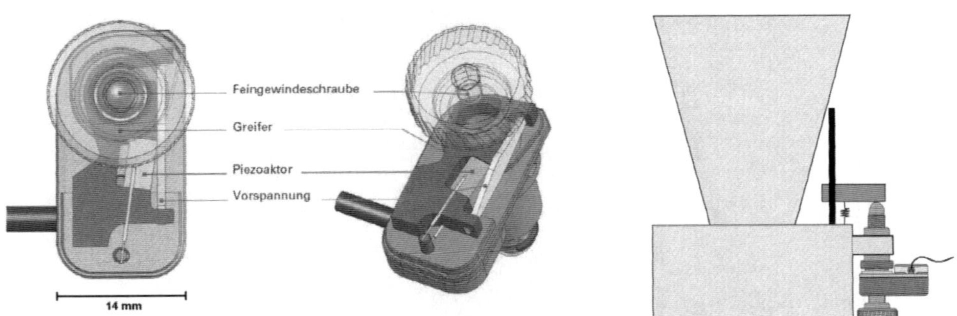

Fig. 4. Piezomike and possible implementation (Source: [8])

4.2. Stepper motors

Because of the possibility of fine positioning of stepper motors, these drives were investigated following the piezo drives. Thereby stepper motors with already mounted threaded-spindle shafts were examined in detail. A possible application is illustrated in Fig. 5.

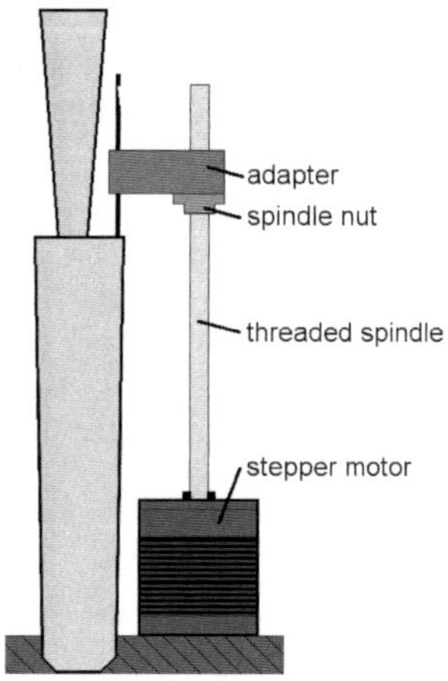

Fig. 5. Stepper motor with threaded shaft as tuning device

The spindle nut in combination with a connected adapter part could transform the rotation into translational movements for the tuning spring. This would certainly be a solution, which would generate the required forces. But because of the centric motor shaft and the frame size of a stepper motor this motor type would not fit between the pipes.

4.3. DC gearbox drives

As a smaller and cheaper alternative to stepper motors DC gearbox motors were explored. Thereby the structure with a threaded shaft, as in the last section on stepper motors, should be used. Due to the gear reduction such a motor could be significantly smaller. Also the drive electronics would be simpler to implement.

5. Practical realization

In the following section the implemented solution is described in detail.

5.1. Implemented drive technology

Owing to the fact, that the last described drive technology with a gearbox motor seems to be the best one, it was chosen to build a prototype and for further evaluation. An appropriate gearbox motor was available, wherefore this drive was used for a first prototype (Fig. 6).

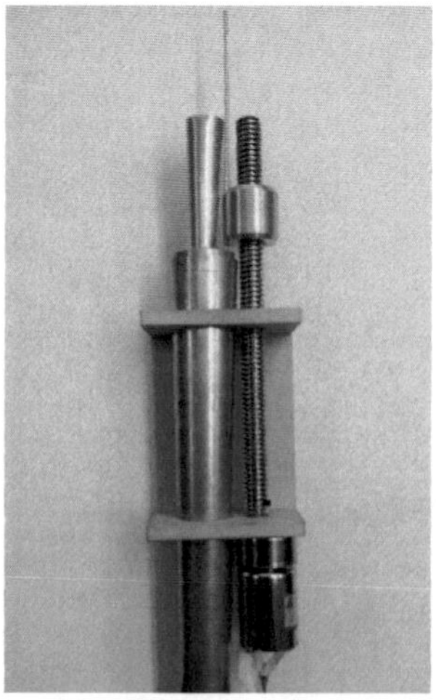

Fig. 6. First prototype with gearbox motor

For testing the prototype, software, which will be described in section 5.3, was developed in parallel. With the experimental setup, first successes in tuning the pipe were achieved. Nonetheless searching for alternative gearbox drives was continued, whereby a very compact and cheap gearbox motor was found, which is perfectly suited to the tuning application. This drive already has a threaded M3 or M4 (selectable) output flange. Furthermore, there is an alternative "`Flip-Type"' of this motor available. Thereby the input and output shaft are on the same side of the gearbox. This allows a more compact design and the reduction of the distance between the tuning spring and the threaded shaft. In low quantities this version is not available with a threaded shaft, but according to the manufacturer it could be modified for higher order quantities. For prototyping, an adapter for mounting a threaded rod was manufactured manually. In Fig. 7 prototypes with both kinds of motors are pictured. Note that the left prototype contains the same pipe as in Fig. 6 to enable one to see thc difference in size.

Fig. 7. Prototypes with compact gearbox motors (motor housing length with gearbox is 25mm)

To transform the rotating movement of the spindle into translation for the tuning spring, an adapter component of high-strength plastic was manufactured. In this component the tuning spring is fixed with a grub screw. If this single screw is loosened, the pipe can be tuned manually without disassembling the automatic tuning system. Thus an excellent, mechanical solution for the tuning system was found. Because of the low thread pitch, high precision positioning in micrometre range creates no problems for this actuator system. Additionally, the low thread pitch and the high gear reduction result in the very low drive torque needed for the motor. Performed force measurements showed, that the solution can generate about 60N before the drive blocks, which is 10-times more than required.

To move the tuning spring and correct the pitch of the pipes, a control loop is necessary. This loop must contain the final control element or actuator, which executes the calculated move, frequency detection and a logic unit or software, which processes the frequency measurements. In Figure 8 such a control loop is depicted.

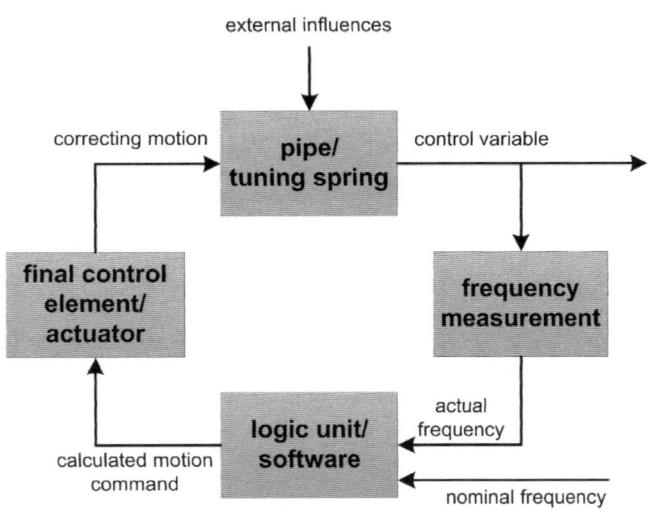

Fig.8. Control circuit for automated pipe tuning

5.2. Frequency detection

At the beginning of this project the bought-in tuning device TLA CTS-32-C [9] was used for pitch detection. It communicates with the software part over an USB-Interface and was especially developed for organ builders and their needs. Because of the high price of the tuning device, an own solution for detecting the frequency was developed. Using a variable bandpass filter, it is possible to extract a sinusoidal wave with the fundamental frequency of the pipe from a complex audio signal, which is recorded by a microphone. Through detecting the zero-crossing-rate of this sinus the pitch of the pipe can be calculated directly, using an Arduino platform for this purpose in prototype stage.

5.3. Software

For calculating the required movements of the motors from the frequency measurement, appropriate software was developed in C#. For controlling the tuning system in the prototyping phase a graphical user interface (GUI) was designed. The tuning device and the electronics (described in the following section) are connected via USB to the computer, on which the software is executed. On the GUI the current divergence to nominal frequency is charted in real-time. The motors are not driven continuously, but stepwise. The length of the switched-on pulses depends on the divergence to nominal frequency of the pipe, followed by a stop until the next pulse length is calculated. This stepwise mode is needed because of the very high sensibility of the reed. At the smallest pipes 1 micrometre motion of the tuning spring results in 0.5 cents deviation of pitch.

5.4. Electronics

To transform the calculated pulses from the software into voltage for the motors, drive electronics and an appropriate logic unit are needed. Therefore, an Arduino board with three motor shields (extension boards) was used. Each board can drive two motors, so six pipes can be connected simultaneously for prototyping. Furthermore, the motor shields support motor current measurement, so it can be detected without additional sensors, if the motor is stalling, e.g. if the tuning spring has reached its end position.

6. Results

After finishing the constructing phase, the prototyping setup was tested extensively. An endurance test was performed with one pipe to verify fatigue strength of the system. Thereby the motor moved the tuning spring for about 20 hours continuously (3715 tuning cycles), until one gear wheel was abraded. This number of tuning cycles would never be reached in a real organ, so the drive is applicable from this point of view.

The precision and the speed of the automated tuning process meet the requirements set for this project. A pipe can be tuned in less than ten seconds with satisfying precision (±0.5 cents), whereby the manual process takes about 30 seconds for each pipe. The system can perform tuning even more accurately, whereby the tuning time increases.

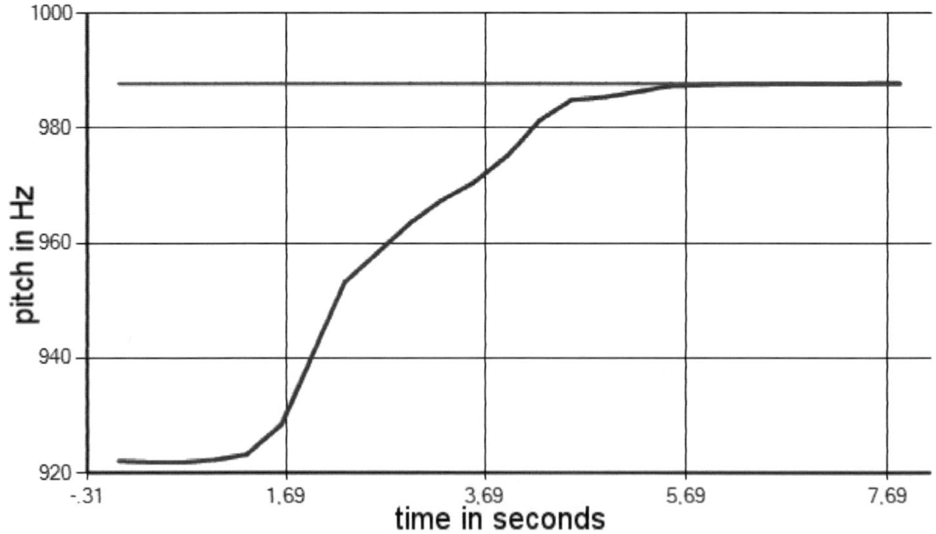

Fig. 9. Resulting tuning process of reed pipe (green…nominal value, red…actual value)

7. Conclusion and outlook

Overall, the main aim of this project, to evaluate the possibility of automated reed pipe tuning, was reached at an early stage and extensive additional developmental work was done. Because of the low price and the small size of the implemented actuator the results actually exceeded the author's own expectations by far. In future work the software should be transformed from PC to an embedded system and should be integrated into the real organ control system. Thereby the organ could be programmed to tune itself at specific dates or tuned by starting the process from a smartphone from anywhere. Using more than one bandpass filter would enable tuning several pipes simultaneously. That would be a significant advantage over to manual pipe tuning.

8. References

[1] Voigt, D. & Voigt, M. (2012). Pfeifenorgel mit selbstregulierender Stimmung, Patent DE102011013444

[2] Voigt, M. (2014). Stimmungseinrichtung für Orgelpfeifen, Patent DE102012021644

[3] Voigt, M. (2015). Stimmungseinrichtung für gedackte Orgelpfeifen, Patent DE102013012821

[4] Bothe, T. & Kablitz, J. (2014). Selbststimmende Orgelpfeife, Bachelorprojekt, FH Kiel

[5] Rieger Orgelbau GmbH (2016). Rieger Orgelbau, http://www.rieger-orgelbau.at/ Accessed: 2016-10-01

[6] Elliptec GmbH (2016). Elliptec Motor X15G, Available at: http://www.elliptec.com/de/produkte/ motor-x15g/, Accessed 2016-10-01

[7] New Scale Technologies (2016). Squiggle micro motor technology, Available at: http://www.newscaletech.com/technology/squiggle-motors.php, Accessed 2016-10-01

[8] Physik Instrumente (PI) GmbH (2016). Piezomike Linearaktoren, Available at: http://www.physikinstrumente.de/technologie/piezomike-linearmotoren.html, Accessed 2016-10-01

[9] Langbein, T. (2008). Tuning Set CTS-32-C,Bedienungsanleitung, Abtsgmünd, Available at: http://www.tuning-set.de/Beda32C_D_Internet.pdf, Accessed 2016-10-01

DOI: 10.2507/27th.daaam.proceedings.116

RESEARCH OF RENEWABLE ENERGY SOURCES BASED ON WIND IN AREA OF HERZEGOVINA

Božo Goluža, Ivana Matijević

This Publication has to be referred as: Goluza, B[ozo] & Matijevic, I[vana] (2016). Research of Renewable Energy Sources Based on Wind in Area of Herzegovina, Proceedings of the 27th DAAAM International Symposium, pp.0804-0810, B. Katalinic (Ed.), Published by DAAAM International, ISBN 978-3-902734-08-2, ISSN 1726-9679, Vienna, Austria

DOI: 10.2507/27th.daaam.proceedings.116

Abstract

Modern ways of developing product and techniques in its production are based on maximum optimization. To ensure this it is inevitable for function of development in company to be based on principles of innovation activities in technical and modern principles of organizing of company in every segment of its activity. On the example of researching meteorological characteristics in Herzegovina the model of optimization in phase of researching and designing of development of energy facilities based on wind energy is presented. For this part of paper, I used results of domestic companies' researches which adjusted this process to EU standards so investment process can be adjusted to criteria of European Investment Bank.

Keywords: renewable energy sources; wind energy; optimization; research and design

1. Introduction

Characteristics of Dinara system in climatological assessment when it comes to wind are classified into complex conditions and are similar as parts of the area in Ireland, Spain and the mountains of Austria. Based on research of characteristics of the wind on the mountain Čabulja slopes (Velika Vlajna) the methodology of necessary information research is shown so the investment process in the sector of wind energy could be realized. The meteorological station in the city of Mostar has a database on the characteristics of the wind back to the period of Austro-Hungary. In the period from 1956 it started with modern methods of measuring weather conditions where in the graphs are displayed in segments of hour defined meteorological condition of wind. [1] Appropriate methods of modelling respecting graphic forms in the period 2006-2011 through the project FP6 were obtained results on the characteristics of the wind every 10 minutes by which is a database for projects of wind power plants extended for a period of 55 years. [2] These databases through statistical analysis provide extremely high-quality results in forecasting of meteorological characteristics that there is not one country in the EU, which increases the interest of future investors because the risks of the investments with the technical aspects are minimized.

2. Selection of micro location for measuring station installation

Location Velika Vlajna is considered as a place for energy use of wind resource for analysis of the possibility of using wind energy in the region of Western Herzegovina.

In the analysis of the use of wind the first step for the production of electricity is the selection of locations for measuring poles and their installation, which is done by the following criteria:

- Wind Resources
- Topography
- Geology and Seismology
- Access to electricity grid
- Access roads
- Freezing
- Lightning
- The vegetation and other obstacles
- Legislation of the environment protection
- Private ownership
- High quality signal of GSM network [3]

Terrain of Velika belong to the extreme south-eastern part of the mountain Čabulja. Location Velika Vlajna (Jastrebinka) is located north-west of Mostar. The topography is suitable for the development of wind power station because of mild slope and any obstacles. On location Velika Vlajna are installed 10 m and 50 m high measuring stations.

Selected point (6477498, 4805321) for 10-meter pole at a height of 1050 meters due to favourable topography, flat lofty slope, and wide flat parts at the top and free access has good conditions for the development of wind farms. It is well exposed to all wind directions. It is possible to access by car.

Selected point (6478089, 4805625) for 50-meter pole is near a communication pole. It is well exposed to all wind directions. Communication pole has no effect on measurement due to braced structures and sufficient distance from measuring pole (approximately 200 m). On this monitoring station are set two levels of measurements of wind: 30 m and 50 m.

3. Review and analysis of wind potential data

The following graph shows the availability of data and the mean wind in months. The red line shows the mean wind speed for a one-year period 11/01/2004 to 10/31/2005 (7.22 m / s). This period was chosen because of its high data availability (96.98%).

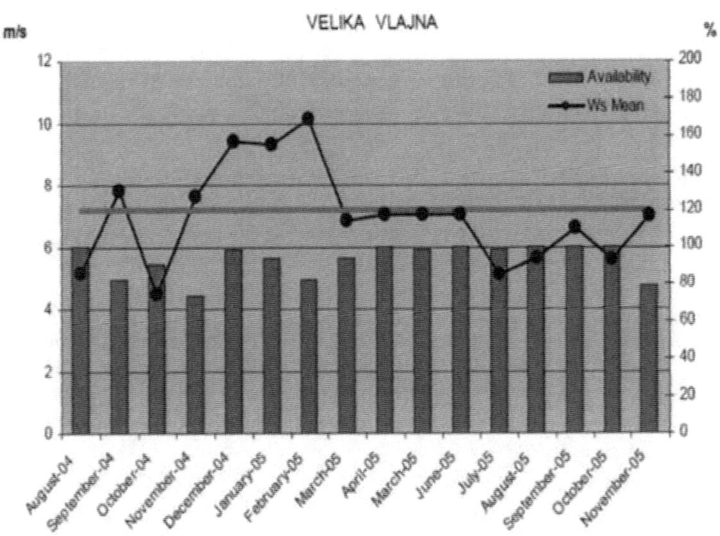

Fig. 1. Diagram of average wind speed and availability of data on the measuring station Velika Vlajna 10m [4]

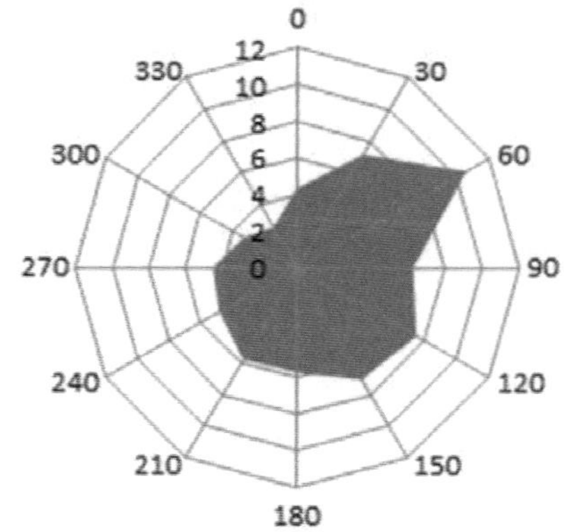

Fig. 2. Wind speed rose [4]

Wind rose describes wind speeds from 12 different directions, and it looks like a compass. Each of the 12 courses includes 30 ° horizon as the standard for the European Wind Atlas. The characteristics of the wind at the location can vary from year to year to 10%, so it is recommended to take measurements for more years as it is worked in this study.

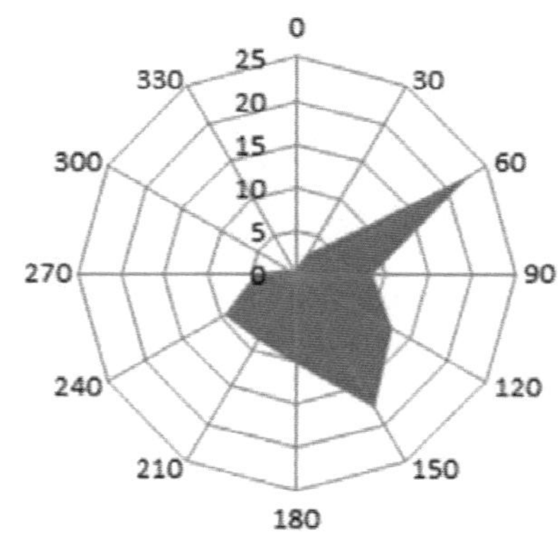

Fig. 3. Frequency rose [4]

Temperature (°C)	**Min**	**Mean**	**Max**
	-15,2	11,1	33,4
Speed (m/s)	**Min**	**Mean**	**Max**
	0,4	7,2	42,8
Dominant wind direction	NNI		
Frost days (%/god)	0,5		
Days of quite (%/god)	17,17		
Turbulence intensity (%)	11,1		
Air density (kg/m^3)	1,098		

Table 1. Meteorological data [4]

Dominant wind direction is north-northeast. Based on these results it was decided to install 50-meter pole. After installing the 50-meter pole it went to test wind speed and direction at both locations, and within three years the data for the wind were obtained:

- 10 m measuring stations data are available from 07.19.2004. to 10.13.2007. (8 days unrecorded data). After three-year measuring wind speed is an average of 7.1 m/s
- 50 m measuring stations data are available from 07.14.2005. to 01.23.2007. (large lack of data from 12.06.2005. to 01.07.2006). With this lack there is no larger measuring defects. If we examine the correspondence of wind courses of measuring stations 10 m and 50 m, according to Figure 4 can be concluded that they are not linear. Due to the different orography and distance of station there is a difference in signal.

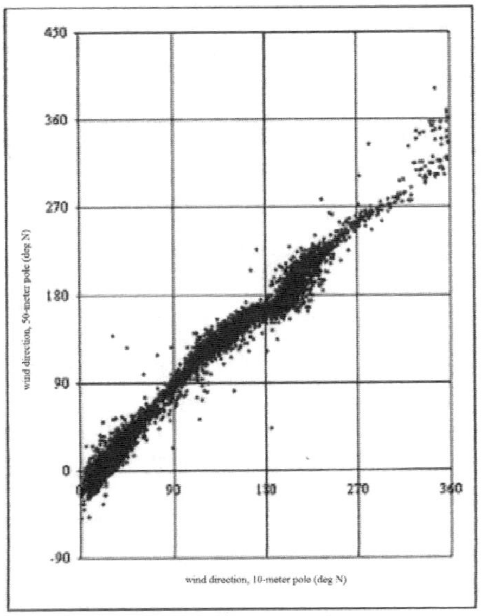

Fig. 4. Comparison of wind directions with both measuring stations [4]

Therefore, it is approached to the polynomial approximation and gets the following picture:

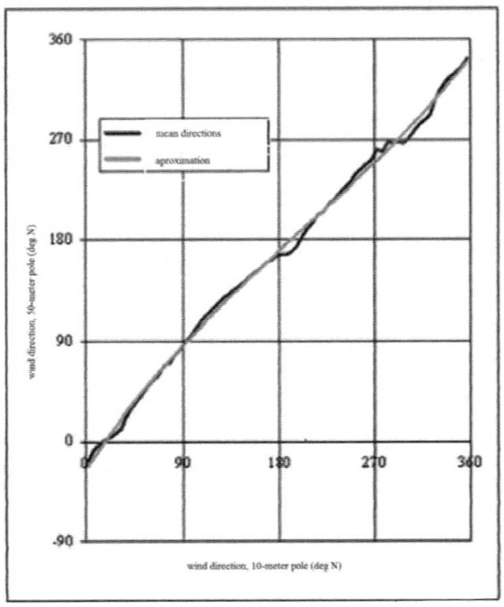

Fig. 5. The relationship between the wind direction signal and their approximation [4]

In this way are filled the deficiencies of data with 50 m pole. Gradient of wind speed varies moderately, but during the day obviously. From diagram it is evident that the influence of orography and roughness of the terrain on the wind directions at the measuring station is minimal. The relatively small oscillations indicate that there is a slightly variety in the angle of action for different wind speeds (north, south). Figure 6 shows average wind speeds during the day.

From the diagram it is evident that the winter winds (wind) energy-efficient (the wind speed). High intensity is more pronounced at night because of the greater temperature difference between Herzegovina - Bosnia (winter) and Herzegovina - Adriatic Sea (in summer). The following figure shows the daily turbulence intensity.

Turbulence in the day shows the influence of thermal effects caused by radiation, which is significantly more expressed during the day than at night.

Small wind speeds have high turbulence intensity and significantly affect the overall turbulence. The following illustration shows the overall turbulence intensity as a function of the wind direction.

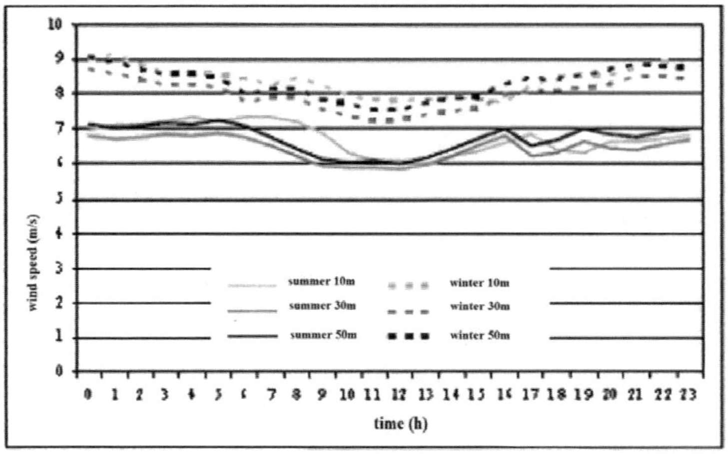

Fig. 6. Daily average wind speed [4]

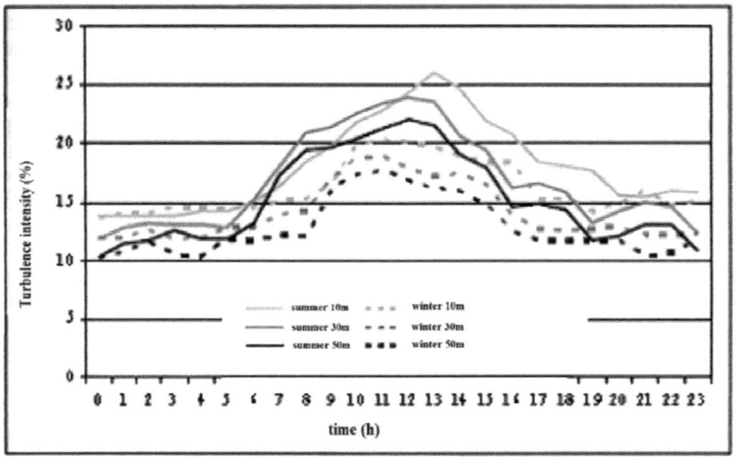

Fig. 7. Daily mean intensity of turbulence [4]

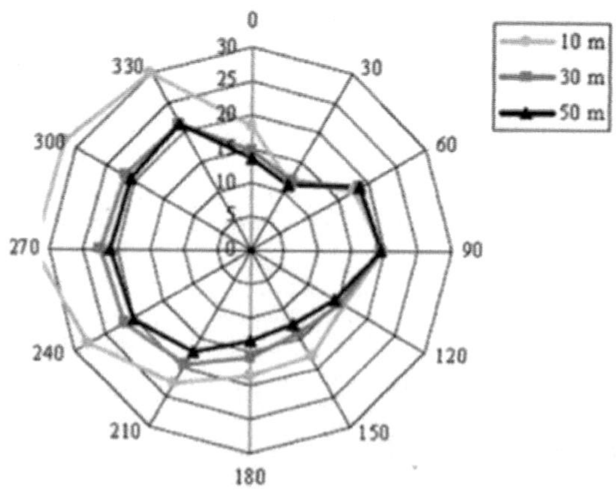

Fig. 8. Mean intensity of turbulence in percentage as a function of wind direction for all wind speeds [4]

In the part from 0 ° to 90 ° turbulence on both measuring poles are almost identical except for the north of 10 m measuring station, where it is visible the impact of a pinwheel that increases its turbulence in this sector. The fact that the level of turbulence between the measuring stations in this direction are comparable means that wind follows the surface of the ground and reaches the 10-meter pole from below the top without major disturbances.

On the other side, great turbulence on the 10 m measuring station on the southern side over the southwest to the northwest indicate the possible impact of higher roughness in that direction. There is a bigger number of shrubs and trees that obviously have an impact. It is noted that turbulence is significantly lower at higher altitudes above the earth in this direction. However, the influence of the smaller top to the northwest side causes moderately increased turbulence at higher altitudes also.

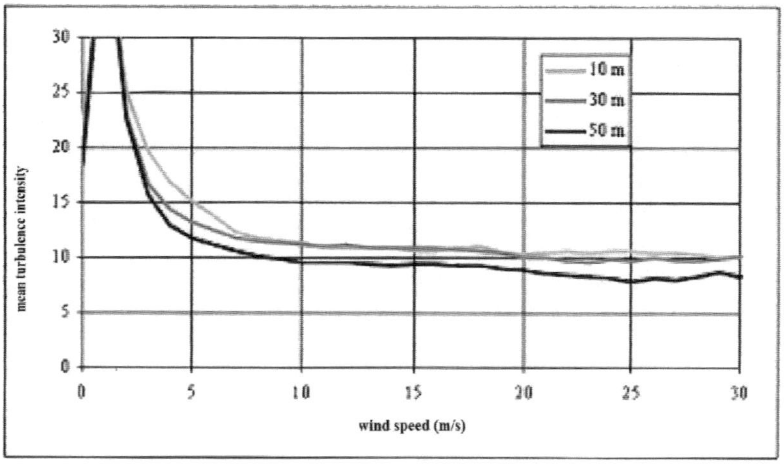

Fig. 9. Mean turbulence intensity as a function of wind speed [4]

Little impact of anemometer mounting at a height of 30 m can be recognized on the part of the 180 ° to 210 °. It is assumed that at the part of 90 ° is there the impact of a telecommunication pole. According to Figure 8., wind speed relevant to the operation of wind turbines causes pretty low turbulence, which meets the requirements for the selection of the turbine class II.

Figures 10 and 11 show the importance of the dominant direction:

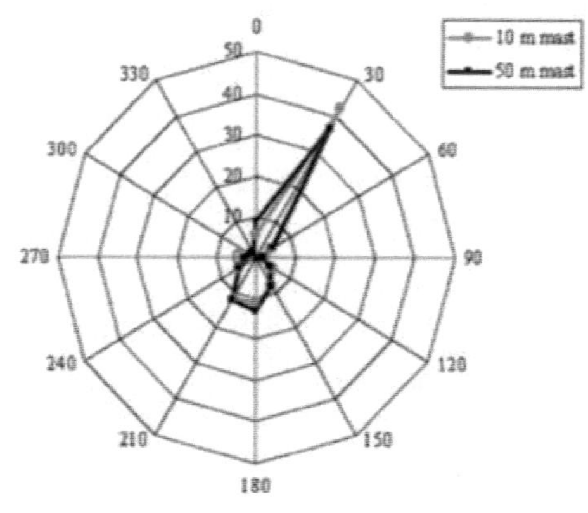

Fig. 10. A three-year wind frequency at both measuring stations (%) [4]

Figure 11 shows a lower wind speed in the north due to the position of a pinwheel. If you take into account Figure 10 this difference has no significance. The mean wind speed at height of 30 m is 30 °, and also at 60 ° and 90 ° is similar on both measurement station which is explained by the influence of turbulence as already. If it takes exception effect from one pole to the other difference in the mean wind speed at a height of 30 m and 50 m is 0.28 m / s. The uncertainty caused by measuring poles wind speed at 10 m and 50 m pole, difference in the mean wind speed is around 0.1-0.2 m / s at the top of the 50 m pole.

This paper clearly shows that the dominant wind direction is NNE and NE. Turbulence in the day shows the influence of thermal effects caused by radiation, which is significantly more expressed during the day than at night.

For extrapolation purposes of wind data from wind station, for the long term have been used wind speeds and directions from weather stations Mostar, Šibenik, Split-Marjan, Split-Resnik and Dubrovnik-airport. By all the data has been verified stability and errors. Some individual data is discarded.

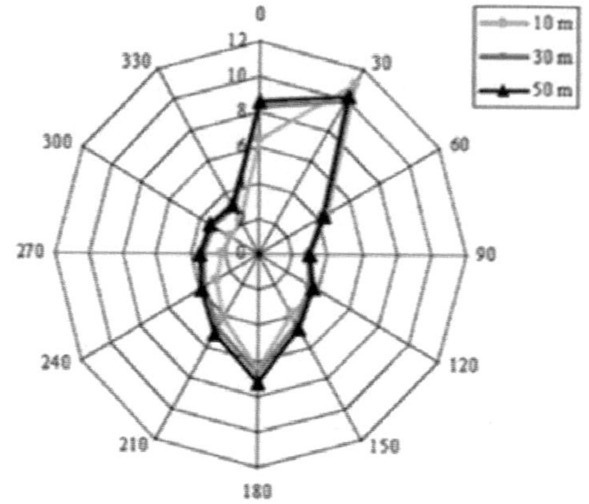

Fig. 11. Average wind speed in 3 years (m/s) [4]

By the extrapolation of wind data on a long-term period it seemed that the first 5 years of data available with a frequency of 6 hours and a midnight data partially missing. It was also found several deficiencies in the data in all sets. Recent data have partly hour resolution also. A number of data are often lacking, and sometimes at night and during the temperature below 0 ° C.

4. Conclusion

The aim of this paper was to represent a results of wind energy research at mountain of Čabulja (Velika Vlajna). This paper clearly shows that the dominant wind direction is NNE and NE. Turbulence in the day shows the influence of thermal effects caused by radiation, which is significantly more expressed during the day than at night.

For extrapolation purposes of wind data from wind station, for the long term have been used wind speeds and directions from weather stations Mostar, Šibenik, Split-Marjan, Split-Resnik and Dubrovnik-airport. By all the data has been verified stability and errors. Some individual data is discarded.

By the extrapolation of wind data on a long-term period it seemed that the first 5 years of data available with a frequency of 6 hours and a midnight data partially missing. It was also found several deficiencies in the data in all sets. Recent data have partly hour resolution also. A number of data are often lacking, and sometimes at night and during the temperature below 0 ° C.

The conclusion is that there is a huge potential for building wind farm on Velika Vlajna. The next step would be installation of wind farm, which is in the procedure. After installation the results would be constantly recorded, analysed and compared with research results.

5. References

[1] Zlomušica, E.; Behmen, M.; Ćatović, F.; Maričić, Đ. (2004). Perspektive BiH u instaliranju vjetroenergetskih kapaciteta, 3. Skup Industrijska energetika 2004. u organizaciji Društva tehničara SiCG, 2004, Donji Milanovac

[2] Behmen, M.; Zlomušica, E. (2008). Renevable energy sources- wind energy in BiH, The Network of Academies of Sciences in OIC Countries Nasic, 2008, Kazan, Tatarstan

[3] Behmen, M.; Čampara, M.; Zlomušica, E.; Samardžić, Ž; Domljan, V. (2009). The risks of building wind farms in Bosnia and Herzegovina, International economic conference, 2009, Tuzla

[4] Matijevic, I. et al. Contribution to the study of wind energy in complex terrain in the case of wind farms Velika Vlajna, Mostar, 2009

[5] Behmen, M.; Zlomušica, E.; Dedić, R.; Manjgo, M.; Badžak, I., Possibilities of cluster production of equipment for renewable energy sources in BiH, Mostar First International Scientific Conference on Engineering 'MAT 2010'; 2010, Mostar

[6] Zlomušica,, E.; Behmen, M; Ćatović, F.; Sarač, E. (2005). 'Wind farms Designing Principles Based on Maximum Energy and Investment Effect', 9th International Reserch/Expert Conference "Trends in the Development of Machinery and Associated Technology TMT 2005; 2005, Antalya

[7] Federal law about renewable energy sources (2013), Available from http://www.fmeri.gov.ba/zakon-o-obnovljivim-izvorima-energije-i-kogeneracije.aspx Acessed: 2016-09-25

[8] Analysis for the Use of Wind Power for Electricity Generation in Bosnia and Herzegovina, "NIP, SA", 2006, Madrid

DOI: 10.2507/27th.daaam.proceedings.117

APPLICATION OF GRAPH THEORY IN AUTOMATION

Ivana Zubac, Snjezana Rezic & Marijana Bandic Glavas

This Publication has to be referred as: Zubac, I[vana]; Rezic, S[njezana] & Bandic Glavas, M[arijana] (2016). Application of Graph Theory in Automation, Proceedings of the 27th DAAAM International Symposium, pp.0811-0815, B. Katalinic (Ed.), Published by DAAAM International, ISBN 978-3-902734-08-2, ISSN 1726-9679, Vienna, Austria DOI: 10.2507/27th.daaam.proceedings.117

Abstract

In many ways,engineering has always been dependent on mathematical background. This was usually mathematical analysis. In recent years, one of the areas of discrete mathematics - graph theory has been increasingly used in many fields, such as automation, programming or algorithms.In automatization process directed graphs are used for description and analysis of control system, where nodes represent states and arrows direct edges or transitions between states automaton that manages processes. In fact, as well as in other disciplines, automation for analysis and synthesis of control units are used in achievements of discrete mathematics. In this paper we show how to use graph to simplify the process of automation.We analyze the possibilities of using directed graphs for the purposes of analysis and synthesis of dynamic control system (elevator) with the aim of its simplification or minimization.

Keywords: graph theory; digraph; automation; matrix; Euler digraph

1. Introduction

Many problems in natural, technical and social science can be successfully formulated interms ofgraph theory. Today the graph theory is well developed, strongly stimulated by technical and chemical applications. For a general background on graph theory and terminology we refer the reader tothe classical book by West D.B[1][2]. For theory of digraphs not definedhere we also recommend [3,4].In paper [5] it was researchedhow a bond graph model of a robotic manipulator may be used to create a model-based observer to improve the control of manipulator by using estimated link angular velocities in a conventional proportional and derivative feedback controller.

Graph theory is study of points and lines. In particular, it involves the ways in which sets of points, called vertices, can be connected by lines or arcs, called edges. Graphs in this context differ from more familiar coordinate plots that portray mathematical relations and functions.

1.1. Basic definition of graph and digraph

GraphG consists of a finite nonempty set $V$of p points (vertex) together with a prescribed set X of q unordered pairs of distinct points of V. Each pair $x = \{u, v\}$ of points in X is a line (edge) of G, and x is said to join u and v. A graph with p points and q lines is called a (p, q) graph.

DigraphD consists of a finite set V of points (vertex) and a collection of ordered pairs of distinct points. Any such pair (u, v) is called an arc or directed line and it is usually denoted by uv. The arc uv goes from u to v and is incident to u and v. We also say that u is adjacent to v and v is adjacent from$u$. The *outdegree* $od(v)$ of a point v is the number of points adjacent from it, and the *indegree* $id(v)$ is the number adjacent to it.

In this paper all digraphs are finite and may have loops and multiple edges (edges with the same initial and final vertices). Digraph D is simple if D has no loops (edges e with $\delta(e) = (v, v)$) and there is at most one edge from x to y for any $x, y \in V(D)$. Note that oppositely oriented edges between two vertices are allowed in simple digraphs, and such a pair of edges is called a digon. The completedigraph of order t, denoted $\vec{K}_t$, is a digraph with t vertices and a digon between each pair of vertices.A *subgraph* is a subset of a digraph's edges (and associated vertices) that constitutes a digraph.

1.2. Representation graph by matrix

It is customary to represent a graph by means of a diagram and by matrix representation. A graph is completely determined by either its adjacencies or its incidences. This information can be conveniently stated in matrix form. The *adjancency matrix*$A = [a_{ij}]$ of a labelled graph G with p points is the $p \times p$ matrix in which $a_{ij} = 1$ if v_i is adjacent with v_j and $a_{ij} = 0$ otherwise. A second matrix, associated with graph G in which the points and lines are labelled, is incidence to matrix $B = [b_{ij}]$. This $p \times q$ matrix has $b_{ij} = 1$ if v_i and x_j are incident and $b_{ij} = 0$ otherwise.
The *adjancency matrix*$A(D)$ of a digraph D is the $p \times p$ matrix $[a_{ij}]$ with $a_{ij} = 1$ if $v_i v_j$ is an arc of D, and 0 otherwise (Fig.1.). It is often possible to make use of these matrices in order to identify certain properties of a graph. For a digraph, the degree of a vertex is the sum of entries in the column and the row corresponding to its adjacency matrix. The entries in the row corresponding to vertex v represent those edges with v as initial vertex and the entries in the column corresponding to v represent the edges with v as final vertex. The sum of all entries in the adjacency matrix is, of course, the total number of edges in the digraph.

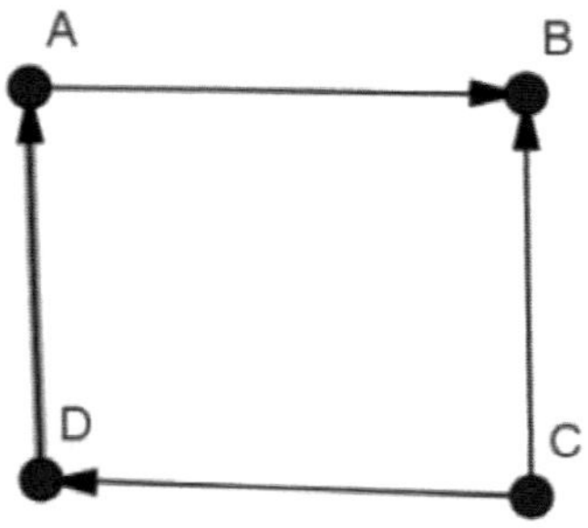

$$A(D) = \begin{bmatrix} 0 & 1 & 0 & 0 \\ 0 & 0 & 0 & 0 \\ 0 & 1 & 0 & 1 \\ 1 & 0 & 0 & 0 \end{bmatrix}$$

Fig. 1. Example of Digraph D and his adjancency matrix $A(D)$

A binary relation on a set can be represented by a digraph. Let ***R*** be a binary relation on a set ***A***, that is ***R*** is a subset of AxA. Then a digraph representing ***R*** can be constructed as follows:
Let the elements of ***A*** be the vertices of the digraph ***G***, and let $\langle x, y \rangle$ be an arc of ***G*** from vertex ***x*** to vertex ***y*** if and only if $\langle x, y \rangle$ is in ***R***.

2. Process of automation

Automation of the process is replacing human labour with machines, not only in terms of strength, but also intellectual work. Technically, an automatic machine (technical system) is formed of three groups of elements:

- the senses (sensors, cameras, microphones, etc.)
- controllers (processors which process information)
- executive elements.

For theory of automation of the process not defined here we also recommend [6][7][8], (Fig.2.).*The management process* in a system, in which one or more input variables, over the legality which characterizes this system, affect other variables as output values. *The process* is quantitative and/or qualitative change depending on the time, and it takes place in nature, society and technology.

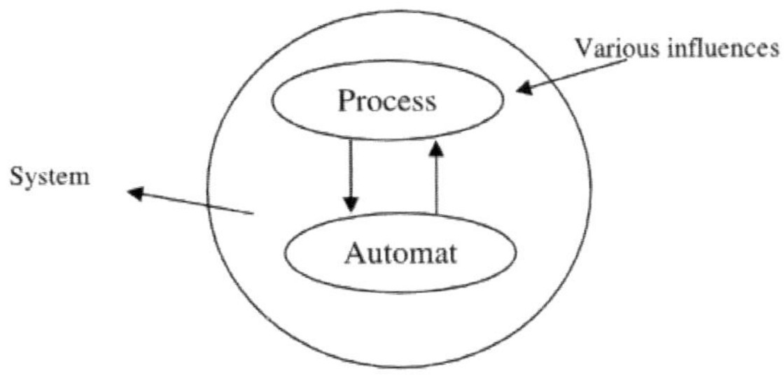

Fig. 2. Process of automation

Information is data about a particular phenomenon, concept or event. The holder of information is signal. The signal is changeable and measurable variable at the entry or exit of the system, and can take a variety of physical forms. Classification of signals primarilydepends on respective parameters of amplitude and time. We distinguish: continuous, discrete and binary signals.Digital automat is a universal sequential circuit, whose behaviour onlydepends on the current and previous input of data-events. Working machines can be explained by theory of systems management.

Automat measures state of the process and resolves all its essential condition. On the basis of current and previous events in the process, the optimal action or series of actions can be determined. These actions should result in automat process in the optimal mode.To make this possible, the automat must have sufficient set of actions, in order to compensate any predictable impact of the environment. We say that the process must be controllable, so that the management could work.If the value of output variable depends not only on the current values of the input variable, but also of the past values (total input-output) for the digital system is said to be a sequential system or automatic.

The properties of automata:

- *Define finality (there are a finite number of states, the final memory);*
- *Define discretion (working in discrete time);*
- *Define the digital mode (available digital inputs and outputs);*
- *Define determination (unambiguously perform its function);*
- *Define the specificity (completely - all possible sequences of input events - expects an arbitrary set of inputs; incomplete - only a series of input events are possible);*
- *Define synchronicity (discrete time defined by phased signal).*

3. Solution of automaton problem for elevator

Elevator serves two floors. A button for call the elevator is installed on each floor, and in the cabin there are two buttons for selecting the desired floor. Draw a graph and make a synthesis automaton describing control of the lift while the following should be taken into account

- *The pressure on each button is a special input*
- *An output signal represents work or inaction engine*
- *The state of the automat represents inaction lifts on a floor*

First we need to define the input, output and state automata.

Inputs for automation:

- *IK - button on the first floor*
- *IKL - button in the elevator on the first floor*
- *IIK - button on the second floor*
- *IIKL - button on the elevator on the second floor*

Conditions for automaton:

- *S1 - first floor*
- *S2 - second floor*

Outputs for automaton:

- *MR - the engine is running or Y1*
- *MM - the engine is at rest or Y2*

According to these labels (inputs, outputs and states), we can draw the following graph (Fig. 3). The states represent nodes, and input and outputarcs.

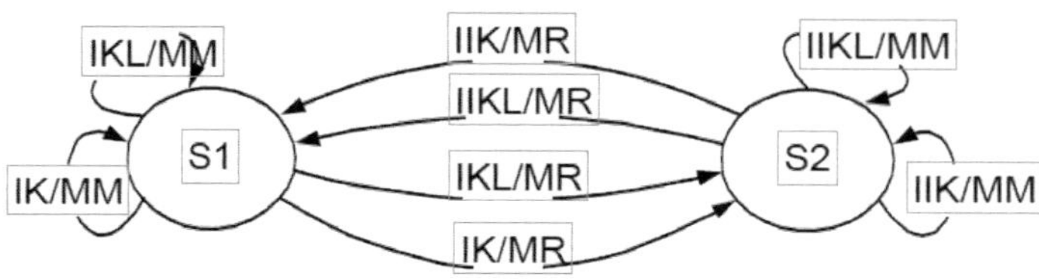

Fig. 3. Graph of transition state the elevator

Automat shownin previous graph can be displayed in a table format crossing / output (Table 1.).

Inputs/state	**S1**	**S2**
IK	S1/MM	S1/MR
IKL	S1/MM	S1/MR
IIK	S2/MR	S2/MM
IIKL	S2/MR	S2/MM

Table 1. Table of crossing/outputs

It may be noted that this is a completely defined automaton. The next step is the minimization of automata. First you need to make coding sheet of input and output (Table 2.). Based on set of coding modes, and using Grey code form transition / exit table is coded (Table 3.).

Inputs	X_1	X_0	**States**	**S**	**Outputs**	**Y**
P	0	0	S1	0	MM	0
PL	0	1	S2	1	MR	1
D	1	1				
DL	1	0				

Table 2. Table of code

	0	1
00	0/0	0/1
01	0/0	0/1
11	1/1	1/0
10	1/1	1/0

Table 3. The coded transition / output tables

The resulting graph of automata operation clearly shows the connection between all its internal states, and state of operand combination from which it originates, transitions, ie. The flow of information and signals. All this allows us to describe and analyze the entire control system, but also easier implementation and simulation of theoretical and actual system, ie. their synthesis.

Among the classical methods of mathematical descriptions of control systems and grapho-analytical methods there is a correlation of both models.Developed grapho-analytical methods can easily get to a complex mathematical model, which significantly reduces model setting time, synthesis of management and development of the control algorithm.

Def.1. A digraph is connected (or weakly connected) if its underlying undirected graph is connected. A digraph is *strongly connected* if, for every ordered pair of vertices (v,w), there is a directed path from v to w.

Th.1. A digraph which represents automata of elevator is strongly connected digraph.

There are notions of Eulerian digraph and Hamiltonian digraph defined in the obvious way; Eulerian digraph has a closed directed path containing every edge and Hamiltonian digraph has a directed cycle passing through every vertex.

Def.2. A connected digraph is *Eulerian* if and only if the in-degree equals the out-degree for every vertex.

Th.2. A digraph who represents automata is Eulerian digraph.

The problem of immersing a complete digraph on t vertices in a simple digraph [9] was considered in the paper. For Eulerian digraphs, it is shown that such an immersion always exists whenever minimum degree is at least $t(t-1)$, and for $t \leq 4$ minimum degree at least $t-1$ suffices.

Th.3. A digraph which represents automata is transitive.

Recal that digraph D is transitive if, for every pair xy and yz of arcs in D with $x \neq z$, the arc xz is also in D. Transitive digraphs from underlying graph-theoretical model in a number of applications. For example, transitive oriented graphs correspond very naturally to partial orders. Clearly, a strong digraph D is transitive if and only if D is complete. [10]

4. Conclusion

Unlike conventional methods of mathematical modelling and mathematical description of automatic control systems grapho-analytical methods took over the primacy not only in practice but also in research, because they lead to a simpler and understandable description of the system, and thus to its analysis and ultimately synthesis of optimal solutions.

The fact that the edge connecting node A to node B with no requirement for feedback is the reason for application of directed graph in developing automation for management processes, since in description, analysis and synthesis of automation for process management feedbackis not always necessary.

5. Acknowledgments

Our paper is a result of research made within the Faculty of Mechanical Engineering and Computing, University of Mostar. Faculty is a sponsor of this paper.

6. References

[1] West D.B. (2007). Introduction to Graph Theory, Prentice Hall, Inc.,978-0130144003, Upper Saddle River, NJ
[2] Harary F. (1972). Graph Theory, Addison Wesley Longman Publishing Co.,978-0201410334, Massachusetts
[3] Bang-Jensen, J.; Gutin, G. (2007). Digraphs Theory, Algorithms and Applications, Springer-Verlag, 978-1-84800-998-1, London
[4] Garnier R.; Taylor J. (2002). Discrete Mathematics for New Technology, Institute of Physics Publishing, 0 7503 06521, Bristoland Philadelphia
[5] Rezic, S.; Pehar, S. & Crnokic, B. (2009). Design of a Bond Graph Model for Robot Control, Annals of DAAAM for 2009 & Proceedings of the 20th International DAAAM Symposium, 25-28th November 2009, Vienna, Austria, ISSN 1726-9679, ISBN 978-3-901509-70-4, Katalinic, B. (Ed.), pp. 1887-1888, Published by DAAAM International Vienna, Vienna
[6] Balach M. (2003). Complete Digital Design, McGraw-Hill,978-0071737708, New York
[7] Hopcroft, J.& Ullman, J. (1979). Introduction to Automata Theory, Languages, and Computation, Addison-Wesley, 81-7808-347-7, Boston
[8] http://www.intechopen.com, (2009). Automation Control – Theory and Practice, Edited by Rodic, A.D. InTech,Accessed on: 2016-10-03
[9] DeVos, M.; McDonald, J.; Mohar, B.;&Scheide, D. (2012). Immersing complete digraphs. Europ. J. Combin., No.33, 1294-1302
[10] Lam C.W.H. Distance transitive digraphs, Discrete Mathematics, Vol. 29, 1980, 265-274

DOI: 10.2507/27th.daaam.proceedings.118

REDISTRIBUTION OF STOCK SECTIONS ACCORDING TO THE MATERIALS TURNOVER AND THE ABC METHOD

Milan Jemelka, Bronislav Chramcov & Pavel Kříž

This Publication has to be referred as: Jemelka, M[ilan]; Chramcov, B[ronislav] & Kriz, P[avel] (2016). Redistribution of Stock Sections According to the Materials Turnover and the ABC Method, Proceedings of the 27th DAAAM International Symposium, pp.0816-0821, B. Katalinic (Ed.), Published by DAAAM International, ISBN 978-3-902734-08-2, ISSN 1726-9679, Vienna, Austria
DOI: 10.2507/27th.daaam.proceedings.118

Abstract

The paper focuses on redistribution of stock sections and shortening forklift handling distances. By analysing the stock according to the turnover of the raw materials and redistributing the storage sections according to the ABC method based on the turnover leads to shortening tracks of forklifts. Moreover, a comparison of the original 12 section layout based on practical experience of warehouse operators is also added to the article and compared with the computed solution according to the turnover of the raw materials based on the ABC method.

Keywords: ABC classification; ABC analysis; putaway strategy; inventory planning.

1 Introduction

ABC analysis is a well-established categorization technique based on the Pareto Principle for determining which items should get priority in the management of a company's inventory. In discussing this topic, today's operations management and supply chain textbooks focus on dollar volume as the sole criterion for performing the categorization. The authors argue that today's businesses and supply chains operate in a World where the ability to deliver the right products rapidly to very specific markets is key to survival. With suppliers, intermediaries, and customers all over the globe, and products live decreasing rapidly, this focus on a single criterion. The large body of research was summarized based on multiple criteria ABC analysis that has accumulated since the 1980s and recommend that textbooks incorporate their key findings and methods into their discussions of this topic. Suggestions are offered on how this discussion might be structured.

Inventory control was a well-known problem in operations research. A large number of authors solved this issue [1, 2, 3, 21, 22] and several models have been developed to solve inventory problems [23]. Business companies have hundreds of different types of materials. Therefore, it is easy to lose control of managing the materials. Inventory classification using ABC analysis is one of the most widely used techniques in organizations. ABC classification allows an organization to separate stock keeping units (SKU) into three groups: A, the most important; B, important; and C, the least important. Most management efforts and oversights are expended on managing A items. C items get the least attention and B items are in-between. The purpose of classifying items into groups is to establish appropriate levels of control over each item [4, 5, 6]. The major advantage of ABC analysis is the easiness of use and simplicity to understand.

The items are classified according to the annual use value, which is the product of annual demand and the average unit price [7, 8]. The classification of items into A, B, and C groups has generally been implemented according to one criterion.

For inventory items, the criterion is frequently the annual dollar usage of the item. However, it has been generally recognized that the traditional ABC analysis has a serious drawback that may inhibit the effectiveness of the procedure in some situations. Using one criterion only may create problems of significant financial loss. For example, class C items with long lead time or class A items prone to obsolescence may incur financial losses due to a possible interruption of production or huge inventory levels. [9, 10, 11] Therefore, it has been proposed that multi-criteria ABC classification, such as lead time, criticality of a stock out of the item, the rate of obsolescence, the scarcity, substitutability, and order size requirement of the item, can provide a more comprehensive managerial control and to take other important criteria into consideration [12, 13, 14].

In this paper, we are concerned with the ABC classification method aimed to facilitate inventory management and to divide the inventory material into new sections [15, 16, 17]. This classification method will affect the distribution of the materials and we will perform efficiency measurement [18, 19,20].

2 Problem formulation

An inefficient layout of a warehouse can have a negative effect on business performance. The two main variables on which every warehouse's performance is based are the space required for allocation of items and the time needed for their handling [8]: space and time are obviously related to the selected storage policy. Without any loss of generality, it is possible to say that permanently assigning certain slots to each SKU ensures a notable ease in tracing items locations. This procedure, combined with the opportunity of properly placing fast/slow mover products in specific warehouse areas also enables a reduction of material handling times [9].

The original warehouse is divided into 12 sections and put away of raw materials into individual sections is carried out intuitively, based on the past experience of forklift operators. The aim is to enhance all the stock operation based on the ABC analysis [12]. The observed variable is the total distance of the warehouse forklift operations.

2.1 Description of current storage location

The storage hall has a total size of 23,520 square meters. The length of the hall is 210 meters and the width is 56 meters. This storage hall is designed for storage of raw materials. All the inbound deliveries and the outbound deliveries are processed in one point. All operations are performed by forklifts. Shelves are used for each of these sections. SKUs are stored in bins. Bin is a special location of a shelve to store multiple SKUs at one place.

The materials are originally placed randomly in these shelves, which are divided in 12 sections (see Fig.1a). These sections were designed by warehouse operators based on their experience. All operations are managed by forklifts and operation of this storage area is continual. 1063 types of raw materials are being stored in warehouse. The warehouse capacity comprises 71 442 storage positions (Bins). Maximum capacity and number of stored raw materials types are displayed down here in the Table 1.

Sections	Maximum number of bins	Number of material types
A01	2912	81
A02	75	10
A03	11474	141
A04	6004	77
A05	7543	98
A06	2785	37
A07	9881	99
A08	8462	113
A09	4429	61
A10	7958	100
A11	3580	51
A12	6339	195
SUM	**71442**	**1063**

Table 1. The existing structure and capacity of the warehouse

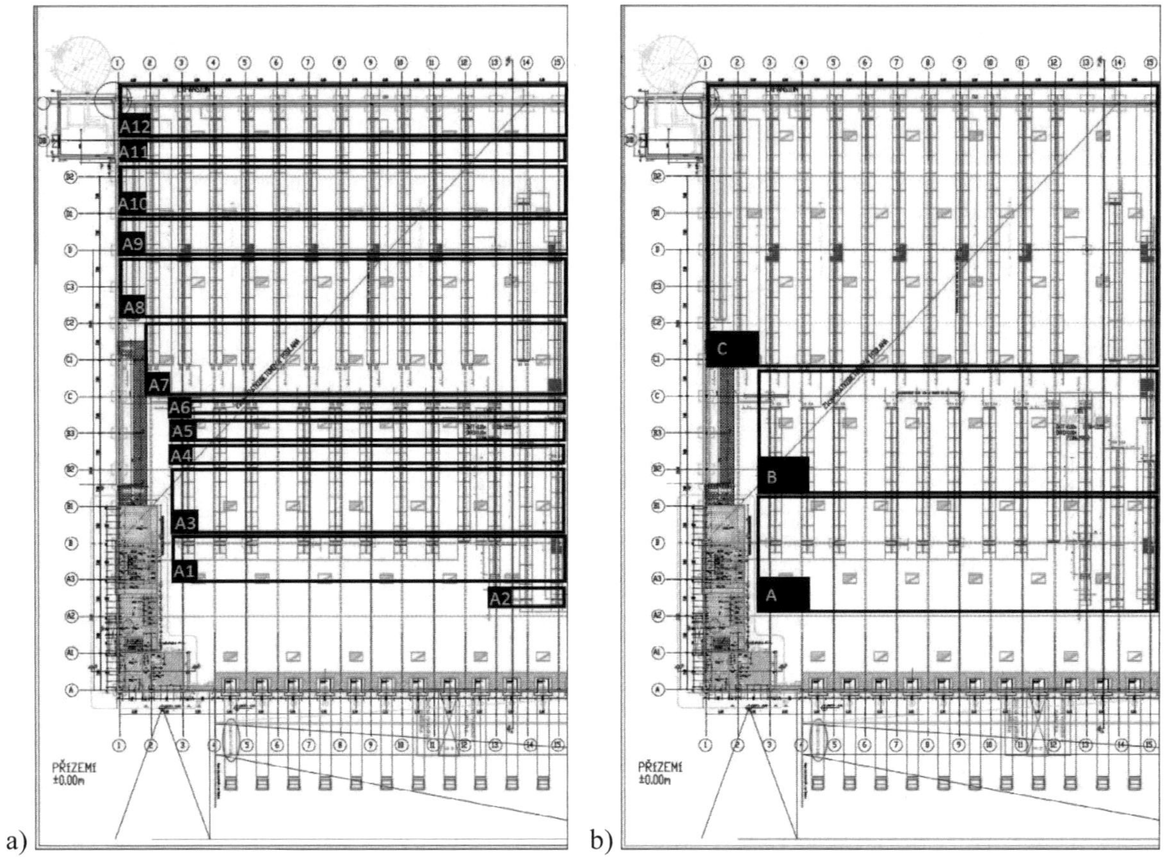

Fig. 1. Warehouse layout (a-current state, b-new design)

3 New warehouse layout

The new warehouse layout is based on the ABC analysis. Based on this analysis, the storage hall is divided into three managed areas (see Fig. 1b). The largest storage section is the area A, smaller section is the area B and the smallest section is the area C. Ratio 4:2:1 is determined for the individual sections allocation.

All the raw materials are divided into three types A, B, C (Pareto principle) according to their turnover. The A raw material is used the most frequently. Their turnover rate is less than one day. The consumption of these raw materials comprises approximately 80% of the total consumption of raw materials. The B raw materials have turnover less than one week. The consumption of these raw materials comprises approximately 15% of the total consumption. The C raw materials have turnover greater than one week. According to this classification, raw materials are put-away into individual sections of A, B, or C.

4 Analysis and results

4.1 Analysis of the current state

During the design and analysis phases were used real data obtained from the information system of the warehouse. Period of 90 days was observed. Data of this period was analysed and A, B, and C groups were determined based on the turnover information. The values are clearly presented in Table 2.

	A	B	C
Total number of raw material types	142	249	672
Turnover for 90 days	135	14	1.8

Table 2. The average turnover of each raw material type

Analysis of the current 12 sections state is outlined in details in Table 3. The table presents the distribution of each raw material type in sections 1 to 12. It is evident that the division into individual sections does not consider the raw materials turnover. Distribution is intuitive, more or less random. These sections were designed by warehouse workers based on their own experience. Furthermore, the table indicates the number of units that were withdrawn in the period of 90 days in all the sections and the distance needed to dispatch all these SKUs.

Sections	**A**	**B**	**C**	**The number of SKUs in 90 days**	**The track of a forklift (km)**
A01	26	9	46	3719	31.83
A02	3	1	6	430	7.45
A03	13	40	88	2473	126.85
A04	19	13	45	2828	290.33
A05	14	24	60	2334	332.56
A06	9	9	19	1375	237.69
A07	15	26	58	2493	523.80
A08	4	27	82	1066	281.31
A09	6	9	46	1019	307.56
A10	15	33	52	2581	873.00
A11	12	11	28	1824	679.06
A12	6	47	142	1724	691.80
SUM	**142**	**249**	**672**	**23866**	**4383.25**

Table 3. Results of the current state analysis

4.2 *Results for ABC classification*

All the ABC analysis results are clearly presented in Table 4. It is apparent from the table, that approximately 80% of SKUs, which are transported, are included in sector A. Contrary the least transported raw materials are located in sector C. The average length of the tracks in each section was used for calculating the overall track of forklifts. These lengths are also recorded in Table 4.

	A	**B**	**C**	**SUM**
The number of raw material types	142	249	672	1063
Turnover of one material type for 90 days (estimate)	135	14	1.8	
The number of SKUs in 90 days	19170	3486	1210	23866
Average forklift track into each section and back [m]	120	300	390	
The total track of forklift [km]	2300.4	1045.8	471.744	**3817.944**

Table 4. Results for the ABC analysis

4.3 *Overall conclusion*

Benefit of the ABC method is obvious from the evident Table 5. One line was completed with the results of the random manipulation of raw materials for interest reasons.

The average total length of forklift truck route is 160 m and this length includes both routes in one operation. Due to rearrangement of the warehouse area, the total route distance of the forklift decreased by 12.9% that means 565.3 km.

The 12.9% saving of the route distance enables the company to reduce staff, saving labour costs every month. Moreover, this new finding will have a positive impact on reducing the amount of labour intensity of forklift operators and will have positive effect on a greater well-being at work and to society by reducing the total distance.

	The total track of forklift	The average forklift length during manipulation of one SKU
The current status of the 12 sectors	4383.25 km	183.7 m
ABC analysis	3817.94 km	160 m
Random stock (12 sections)	5011.86 km	210 m

Table 5. Comparison of the results

5 Conclusion

Traditionally, ABC analysis has been used to classify various inventory items into three categories - A, B, and C. The design of a warehouse, ensuring its optimum layout configuration and effective material handling system, is strictly related to an accurate and structured allocation of products according to the ABC method. The main aim of the proposed formulation was to enhance the necessary storage space while determining a proper ABC material allocation to reduce handling times and distances. The proof is a real industrial case, showed the effectiveness of the formulation, depicting performances notably close to the best feasible case. Though using a dedicated storage policy approach, the outcomes obtained through the developed formulation resulted in being equal to the lower bound computed using a randomized policy which, distinct from the presented technique, should be unavoidably sustained by warehouse management system software. Thus, the proposed approach allows companies of all sizes to optimize their warehouse, reducing the required storage space to a minimum while adopting the ABC method within their ERP system.

The problem was to verify, if the current storage location is distributed correctly. Our calculation confirmed, that ABC analysis delivers the effective warehouse layout and saved 13% of forklift manipulation movements. The positive outcome is, that this method is fast and most of the world enterprise resource planning software can generate the ABC analysis by themselves based on various criteria.

This is a real industrial case, which can be implemented in any warehouse location.

The future plans are to use different ABC ratios to distribute the warehouse area and to find the most effective one.

6 Acknowledgments

This work was supported by the Ministry of Education, Youth and Sports of the Czech Republic within the National Sustainability Programme project No. LO1303 (MSMT-7778/2014) and also by the European Regional Development Fund under the project CEBIA-Tech No. CZ.1.05/2.1.00/03.0089 and also by the Internal Grant Agency of Tomas Bata University under the project No. IGA/FAI(CEBIA)/2016/033.

7 References

[1] Cecil-Wright, J. (1986). How the boardroom can influence warehousing costs. International Journal of Retail & Distribution Management, 14 (3), 67-69.

[2] Segerstedt, A., Pettersson, A.I. (2012). Measurements of excellence in a supply chain. International Journal of Logistics Systems and Management, 13 (1), 65-80.

[3] Hausman, W.H., Schwarz, L.B. and Graves, S.C. (1976). Optimal storage assignment in automatic warehousing system. Management Science, 22 (6), 629–638.

[4] Muppani, V.R., Adil, G.K. and Bandyopadhyay, A. (2010). A review of methodologies for class-based storage location assignment in a warehouse. International Journal of Advanced Operations Management, 2 (3-4), 274-291.

[5] Kovacs, A. (2011). Optimizing the Storage Assignment in a Warehouse Served by Milkrun Logistics. International Journal of Production Economics, 133 (1), 312-318.

[6] Glock, C.H. and Grosse, E.H. (2012). Storage policies and order picking strategies in U-shaped order picking systems with a movable base. International Journal of Production Research, 50 (16), 4344-4357.

[7] De Koster, R.B.M., Le-Duc, T. and Zaerpour, N. (2012). Determining the number of zones in a pick and-sort order picking system. International Journal of Production Research, 50 (3), 757-771.

[8] Bragg S. M. (2013) Inventory Management, Accounting Tools, Colorado.

[9] Waters D. (2003) Inventory Control and Management (Business), John Wiley and Sons, 2nd edition.

[10] Piasecki D. J. (2009)., Inventory Management Explained, Ops Publishing.

[11] Granville D. (2007). Excellence in Inventory Management: How to Minimise Costs and Maximise Services, Cambridge Academic; First edition.

[12] Milner C. (2015). Inventory Management: Advanced Methods for Managing Inventory within Business Systems, Kogan Page; 1 edition.

[13] Grinsted S. (2013). The Logistics and Supply Chain Toolkit: Over 90 Tools for Transport, Warehousing and Inventory Management, Kogan Page; 1 edition.

[14] Muller M. (2011). Essentials of Inventory Management, Amacom, 2nd edition.

[15] Bottani E., R. Montanari, M. Rinaldi and G. Vignali (2015). Intelligent algorithms for warehouse management. Intelligent Systems Reference Library, 87, pp. 645-667.

[16] López J. A, Mendoza A. and J. Masini (2013). A classic and effective approach to inventory management. International Journal of Industrial Engineering: Theory Applications and Practice, 20 (56), pp. 372-386.

[17] Xiao Y., Zhang R. and I. Kaku (2011). A new approach of inventory classification based on loss profit. Expert Systems with Applications, 38 (8), pp. 9382-9391.

[18] Paweł P., Marek F. and G. Paulina (2008). Using ABC Classification to determine production sequence in automotive industry, In Proceedings of 2008 World Automation Congress, WAC 2008, IEEE, pp. 1-6.

[19] Smith A.D. (2011). Inventory management and ABC analysis practices in competitive environments. International Journal of Procurement Management, 4 (4), pp. 433-454.

[20] Miculescu, M. N.; Lut, D. M. & Miculescu, C. (2011). Current Trends of Production Cost Accounting, Annals of DAAAM for 2011 & Proceedings of the 22nd International DAAAM Symposium, 23-26th November 2011, Vienna, Austria, Volume 22, No. 1, ISSN 1726-9679, ISBN 978-3-901509-83-4, Katalinic, B. (Ed.), pp. 0941-0942, Published by DAAAM International Vienna, Vienna

[21] Pasic, M.; Kadric, E. R. & Bajric, H. (2010). Relationship between Inventory Investment and Forecasting and Inventory Control, Annals of DAAAM for 2010 & Proceedings of the 21st International DAAAM Symposium, 20-23rd October 2010, Zadar, Croatia, ISSN 1726-9679, ISBN 978-3-901509-73-5, Katalinic, B. (Ed.), pp. 0511-0512, Published by DAAAM International Vienna, Vienna

[22] Knezevic, B. (2011). Usage of Simulation in Inventory Management Education, Annals of DAAAM for 2011 & Proceedings of the 22nd International DAAAM Symposium, 23-26th November 2011, Vienna, Austria, Volume 22, No. 1, ISSN 1726-9679, ISBN 978-3-901509-83-4, Katalinic, B. (Ed.), pp. 1197-1198, Published by DAAAM International Vienna, Vienna

[23] Gastermann, B. C.; Luftensteiner, F.; Stopper, M. & Katalinic, B.: Multiple Stage Production Planning in Plain Manufacturing Environments, Annals of DAAAM for 2012 & Proceedings of the 23rd International DAAAM Symposium, ISBN 978-3-901509-91-9, ISSN 2304-1382, pp 0883 - 0886, Editor B[ranko] Katalinic, Published by DAAAM International, Vienna, Austria, 2012

DOI: 10.2507/27th.daaam.proceedings.119

VIBRATION ASSISTED DEEP HOLE MICRO-DRILLING: A PRELIMINARY EXPERIMENTAL STUDY

Todić Rajko, Bartulović Ante

This Publication has to be referred as: Todic, R[ajko] & Bartulovic, A[nte] (2016). Vibration Assisted Deep Hole Micro-Drilling: a Preliminary Experimental Study, Proceedings of the 27th DAAAM International Symposium, pp.0822-0826, B. Katalinic (Ed.), Published by DAAAM International, ISBN 978-3-902734-08-2, ISSN 1726-9679, Vienna, Austria
DOI: 10.2507/27th.daaam.proceedings.119

Abstract

The process of deep hole micro-drilling is characterized with a higher depth to diameter ratio that leads to difficult evacuation of the chip from the hole and increased friction between the tool and the workpiece. This causes an increase of drill wear, drilling forces and torque. In combination with the small stiffness of the drill could often lead to the tool breakage, less productivity and sometimes to the loss of the workpiece. One option to increase the reliability of the process is superimposing vibrations to the standard kinematics of the drilling process. Preliminary experimental investigations on vibration assisted deep-hole micro-drilling of AlMg3 material, with hole aspect ratio of 1:20, are performed. Drill lifetime related with drill breakage is compared between vibration assisted and conventional micro drilling process.

Keywords: Micro-drilling; Deep hole drilling; Vibration assisted drilling, Drill breakage; Tool lifetime

1. Introduction

The increase of production of devices with micro features, mainly in medical equipment, electronic and optical devices, automotive parts, etc., demands efficient and reliable technologies of micro machining. Mechanical micro-drilling is one of the methods for producing micro holes among other modern technologies like electrical discharge machining, ultrasound machining, laser machining and electro-chemical machining. Advantages of using mechanical micro-drilling are process independence from workpiece material properties, reduced thermal deformations, acceptable material removal rates with satisfying shape and surface quality, less requirements for the final machining, etc. Deep hole mechanical micro-drilling produces micro-holes with a diameter less than 1 mm with the bigger hole aspect ratio (depth/diameter) more than 15. The main problems related with deep micro-drilling are low rigidity of the drill, difficult chip extraction and difficult lubrication supply to the deep micro hole due to big hole aspect ratio. This could often lead to the tool breakage, less productivity and sometimes to the loss of the workpiece. The application of vibrations at the micro-drilling, especially for deep holes, presents a possible progress in solving those issues to increase the reliability of micro drilling processes.

2. Micro drilling and vibration assisted drilling

Vibration assisted drilling occurs when the vibrations are superimposed to the standard kinematics of the drilling process in the drill axis direction or torsional, on the side of the drill or of the workpiece. Vibrations can vary from frequencies of 100 Hz to ultrasonic vibrations (more than 20 kHz). Vibrations are changing the mechanism of formation of the chip, a thinner and smaller chip is created, which reduces friction between the chip and the drill, as well as between the chip and the workpiece hole. The thinner chip can easily evacuate from the hole, thus reducing the forces acting on the drill bit. Cutting mechanism is shown on the Fig. 1, where the tool vibrates with displacement x, vibration amplitude A, angular frequency ω, tool vibration speed v_t and cuts workpiece in period t_c.

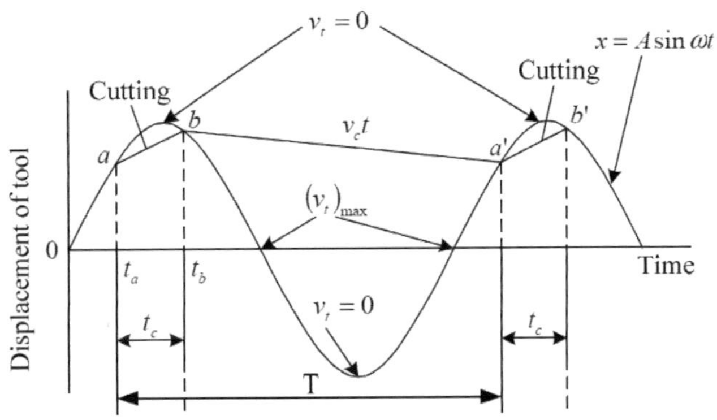

Fig. 1. Vibration cutting state in vibration assisted drilling

If $A\omega > v_c$ tool will separate from the workpiece and the condition for the vibration cutting is satisfied, otherwise it stays a conventional cutting process. The relationship of contact time between tool and workpiece $r = t_c/T$ has been introduced [1]. For vibration assisted cutting r should be as low as possible, which can be achieved either with increasing the frequency and amplitude of vibration or with lowering the cutting speed v_c.

The results of the application of vibration assisted drilling with conventional drilling show many positive effects such as reduced cutting forces and wear, longer tool lifetime and better surface quality, improved geometry and dimensions of the hole, better surface quality of the bore, smaller burr formation, etc. Evidence of the benefits of vibration in drilling were mainly focused on ultrasonic vibration assisted drilling of relatively large diameters or small depth vibration assisted micro drilling. Most of them were based on experimental results, selectively taking into account, in various combinations, characteristics like force of drilling process, chip characteristics, drill lifetime, material removal rate, quality of the hole and burr forming.

One main advantage of vibration for larger diameter drilling is the reduction of forces and moments acting on the drill bit. Reduction of forces and moments acting on the drill bit with vibration assistance is documented during drilling of various types of materials like aluminum [2], copper [3], nickel [4] and titanium [5] alloys, as well as cortical bone [6] and composite materials like carbon fiber-reinforced plastics [7]. Vibration assistance provides a mechanism to create shorter chip and their better extraction from the hole [8], thus avoiding the need for periodic drill retracting. Reduction of chip size is listed as one of the main reasons for the increase drills in life expectancy [9]. Vibration assisted micro-drilling investigations show significantly lower level of drilling force [10], longer drill lifetime [11] and smaller chip size [12].

Vibration assisted deep hole micro-drilling, after review of available literature, has not been sufficiently investigated, there is no sufficient number of experimental nor theoretical models of the process, which is understandable due to the unconventional mechanisms of cutting and complexity of the problem. There are many challenges facing the use of vibrations in the drilling: additional equipment or machine modifications, vibration drilling control systems are relatively undeveloped, vibration parameters depend on the application or require adjustment during drilling. This article presents preliminary experimental analysis of influence of vibration application on the drill breakage related with drill lifetime and productivity in the deep hole micro-drilling process. This is a part of experimental investigations with aim to establish a model of vibration assisted deep hole micro-drilling as an important step to the realization in industrial applications.

3. Experiment

The experimental setup developed in HSTec consists of the following components (Fig. 2):

- Motor spindle
- Z-axis dynamic linear motor
- Vibration generator
- Force sensors
- Control computer / data base

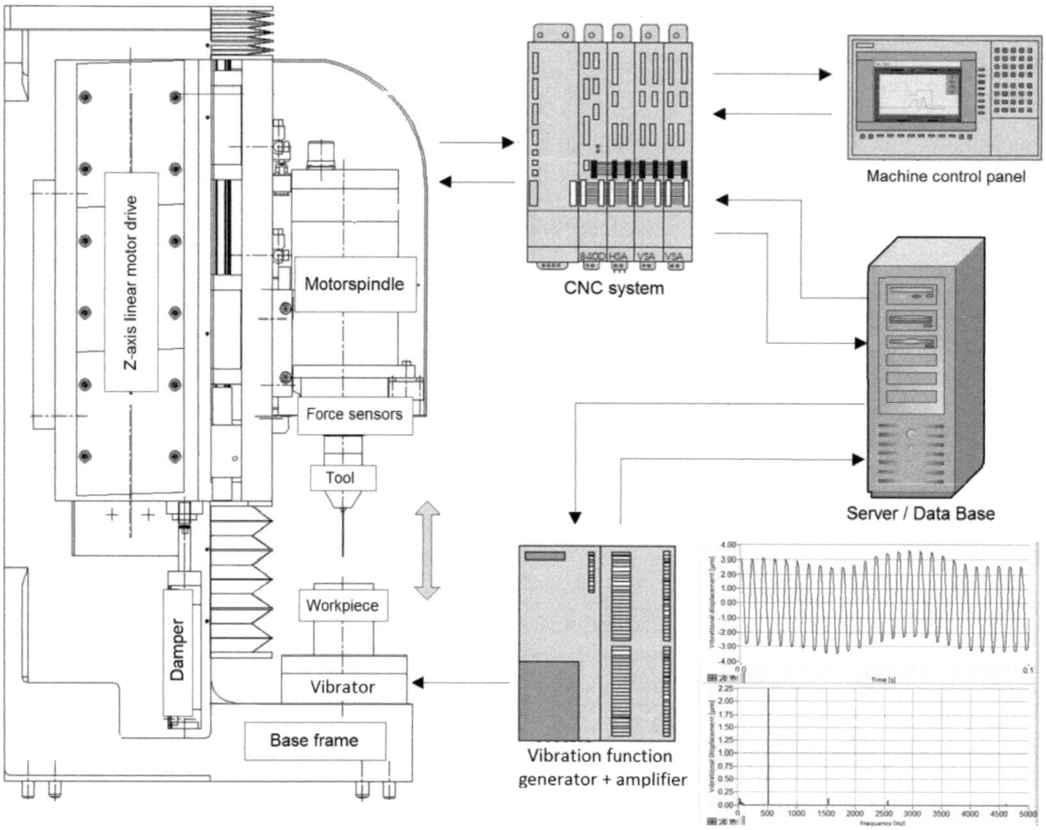

Fig. 2. Schematic diagram of drilling experimental setup

The drilling machine has a 2-pole asynchronous motor with rated power of 4,9 kW and rated speed of 39.400 min-1 installed. It has a manual clamping system and is equipped with HSK 32 tool interface. Highly dynamic vertical axis is performed by use of a permanent, magnet-triggered, synchronous linear motor with rated thrust of 625N. A pneumatic cylinder is installed in the Z-axis for balancing and damping purposes.

Fig. 3. Conventional micro drilling tests on the micro-drilling machine

The system for generating vibrations is installed below the clamping device, so the vibrations are generated on the side of the workpiece. Electromagnetic vibration generator [Fig. 4] can create vibration frequencies up to 3 kHz, but vibration frequency and amplitude are inversely proportional, which means that with higher frequency generator forms smaller amplitudes. Besides this limitation, magnetic fields created in the generator make an influence on the chip extraction during machining of steel and that was the reason that nonmagnetic material as aluminum was used for this stage of investigation.

Fig. 4. Vibration assisted micro-drilling tests with vibration generator

Deep micro drilling experiments are performed with conventional spiral solid carbide drills Gühring 706 HVM with diameter of 0.5 mm, depth of holes 10 mm, with and without the application of vibration, in order to define the impact of the application of vibration to the deep micro drilling. Drilling and vibration parameters are presented in Table 1.

Experimental Setup No.	**Rotational speed (*rpm*)**	**Feed rate (*mm/rev*)**	**Peck drilling feed (*mm*)**	**Vibration assist. drilling**	
				Frequency (*Hz*)	**Amplitude (*μm*)**
Setup 1	15.000	0,008	0,15	no vibrations	
Setup 2	15.000	0,008	0,15	250	4,5
Setup 3	15.000	0,07	0,15	250	4,5

Table 1. Drilling parameters and vibration parameters

4. Results

The results shows that the vibration assistance has extremely positive impact on a micro-drill lifetime. It can be concluded that vibration assisted micro-drilling (Setup 2) prolongs drill lifetime in comparison with conventional micro-drilling (Setup 1), when using the same parameters. It proved that the use of the vibration assistance on the deep hole micro drilling has a similar effect on the lifetime as documented in the literature for the vibration assisted drilling with larger diameter drills and for smaller depth holes micro-drilling.

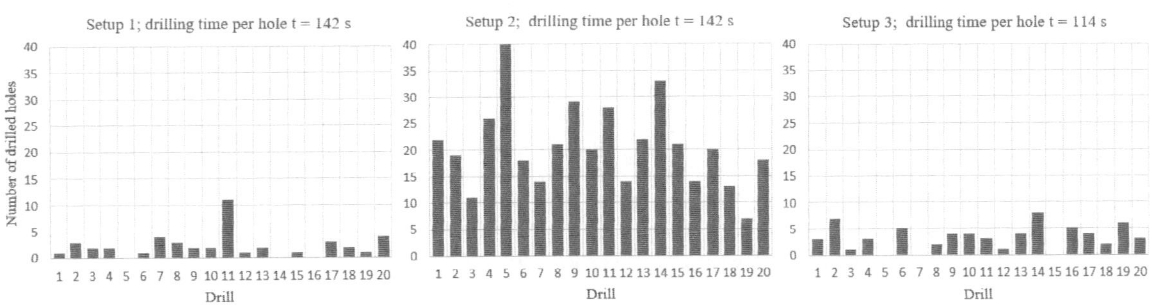

Fig. 5. Number of drilled holes related to the setup type

Another comparison shows that approximately the same number of drilled holes can be achieved when using vibration assisted micro-drilling with almost nine times higher feed rate (Setup 3) than the conventional micro-drilling (Setup 1). That enables shorter drilling time and higher productivity. Shorter and thinner chip is produced with the vibration assisted process [Fig 6] with the assumption that it leads to less chip clogging in the hole and to reducing of cutting forces. Reduced load on the micro-drill, which is characterized with small rigidity, could be a reason for higher drill durability.

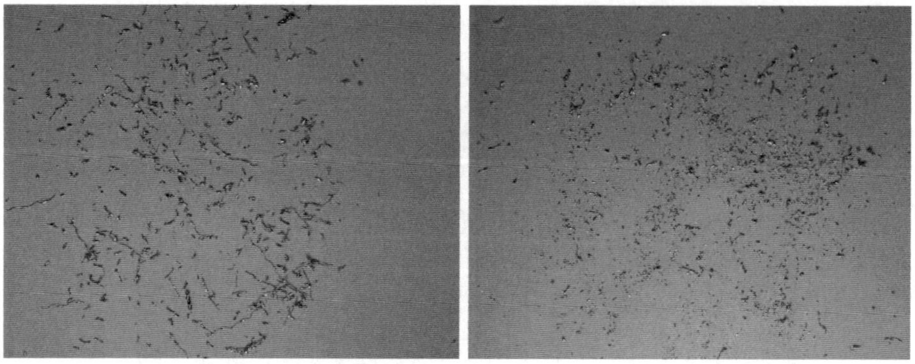

Fig. 6. Chip from conventional (left) and vibration assisted (right) deep hole micro-drilling

5. Conclusion and future work

Preliminary experimental study showed advantages of using vibration assistance which led to thinner and smaller chip which enabled bigger number of drilled holes per drill with the same drilling parameters. Another advantage is possibility of using increased feed which leads to shorter drilling time and higher productivity. In the future work emphasis will be placed on the analysis of the impact of drilling and vibration parameters, in order to better define the rules necessary for the development of this hybrid process. Development of a of deep micro-vibration assisted drilling model could allow selection of drilling parameters (speed, feed or retraction of the drill) and vibration parameters (frequency and amplitude), related with the characteristics of the workpiece material and dimensions of the hole. The assumption is that implementing selection of process parameters will prevent tool breakage, reduce wear and prolong the life of the drill, with the potential to improve the quality holes. The developed model could be used in the development of the adaptive control of vibration assisted micro drilling processes.

6. References

[1] Nath, C., & Rahman, M. (2008). Effect of machining parameters in ultrasonic vibration cutting. International Journal of Machine Tools and Manufacture,48(9), 965-974.

[2] Chang, S. S., & Bone, G. M. (2009). Thrust force model for vibration-assisted drilling of aluminum 6061-T6. International Journal of Machine Tools and Manufacture, 49(14), 1070-1076.

[3] Azarhoushang, B., & Akbari, J. (2007). Ultrasonic-assisted drilling of Inconel 738-LC. International Journal of Machine Tools and Manufacture, 47(7), 1027-1033.

[4] Heisel, U., Wallaschek, J., Eisseler, R., & Potthast, C. (2008). Ultrasonic deep hole drilling in electrolytic copper ECu 57. CIRP Annals-Manufacturing Technology, 57(1), 53-56.

[5] Pujana, J., Rivero, A., Celaya, A., & de Lacalle, L. L. (2009). Analysis of ultrasonic-assisted drilling of Ti6Al4V. International Journal of Machine Tools and Manufacture, 49(6), 500-508.

[6] Alam, K., Mitrofanov, A. V., & Silberschmidt, V. V. (2011). Experimental investigations of forces and torque in conventional and ultrasonically-assisted drilling of cortical bone. Medical engineering & physics, 33(2), 234-239.

[7] Makhdum, F., Jennings, L. T., Roy, A., & Silberschmidt, V. V. (2012). Cutting forces in ultrasonically assisted drilling of carbon fibre-reinforced plastics. InJournal of Physics: Conference Series (Vol. 382, No. 1, p. 012019). IOP Publishing.

[8] Brinksmeier, E., Pecat, O., & Rentsch, R. (2015). Quantitative analysis of chip extraction in drilling of Ti 6 Al 4 V. CIRP Annals-Manufacturing Technology,64(1), 93-96.

[9] Vakili Azghandi, B., & Razfar, M. R. (2011). An experimental study on the effects of ultrasonic assisted drilling on chip characteristics and tool life. InAdvanced Materials Research (Vol. 325, pp. 351-356). Trans Tech Publications.

[10] Zhang, Z., & Babitsky, V. I. (2011). Finite element modeling of a micro-drill and experiments on high speed ultrasonically assisted micro-drilling. Journal of Sound and Vibration, 330(10), 2124-2137.

[11] Chern, G. L., & Lee, H. J. (2006). Using workpiece vibration cutting for micro-drilling. The International Journal of Advanced Manufacturing Technology, 27(7-8), 688-692.

[12] Liu, Z. F., & Guo, T. (2014). Experimental Study on Ultrasonic Vibration Drilling of Stainless Steel 0Cr17Ni4Cu4Nb Micro-Deep-Hole. In Applied Mechanics and Materials (Vol. 692, pp. 381-386). Trans Tech Publications.

DOI: 10.2507/27th.daaam.proceedings.120

SIMULATING THE MAGNETRON SPUTTERING PROCESS FOR INDUSTRIAL APPLICATIONS

Ionuţ Spătaru, Adrian Nicolescu, Cezara Coman, Cozmin Cristoiu, Andrei Ivan

Faculty for Engineering and Management of Technological Systems, Politehnica University of Bucharest, Splaiul Independentei 313, Bucharest 060042, sector 6, Romania

This Publication has to be referred as: Spataru, I[onut]; Nicolescu, A[drian] F[lorin]; Coman, C[ezara]; Cristoiu, C[ozmin] & Andrei, I[van] (2016). Simulating the Magnetron Sputtering Process for Industrial Applications, Proceedings of the 27th DAAAM International Symposium, pp.0827-0837, B. Katalinic (Ed.), Published by DAAAM International, ISBN 978-3-902734-08-2, ISSN 1726-9679, Vienna, Austria
DOI: 10.2507/27th.daaam.proceedings.120

Abstract

This paper presents a new method of simulating the process of thin layer glass coating by magnetron sputtering on real industrial scale equipment and in real industrial application as well as necessary experimental results for the validation of the new developed mathematical formulation and software algorithm. Based on major differences between applications realized with laboratory equipment studied until now by other authors and the real scale industrial installations analyzed in present paper, development of a new mathematical model, simulation method and process control of magnetron sputtering systems by ionized physical vapor deposition (IPVD) was needed. Improvement of the deposition process by applying this new concept on industrial scale is discussed for different material types, specific deposition energy levels and the uniformity of the deposed layers based on simulation and experimental results analysis.

Keywords: magnetron, sputtering, industrial scale equipment, modeling, simulation, experimental tests.

1. Introduction

At the moment there are two known techniques for applying thin metallic layers on glass: by vapor deposition in chemical / physical phase (CVD - Chemical vapor deposition / PVD - Physical vapor deposition). These methods are also called "on-line method" respectively "off-line method". On-line method involves glass coating during its pyro-lytic production process (spreading of the melted material on the hot glass floating in the oven on a tin bath) and represents the process by which a thin layer is applied to the glass by CVD. In contrast to this, off-line method is applied after the glass has been produced, using a magnetron and by vacuum deposition in PVD methods, usually sputtering type.

In terms of a comparative analysis of the two deposition methods, in general, off-line / PVD method has as main advantages the widest range of options for covering materials and superior optical performance of the coated glass against on-line / CVD method. As main disadvantages offline method is characterized by increased lead times and color variability between batches.

For this reason most of the industrial research efforts are focused on improving operating performances of magnetron sputtering PVD type equipment, with special focus on analyzing the possibilities of optimization of PVD process parameters for industrial production, in accordance with the desired performance to be obtained for different product categories and their reproducibility for multiple manufacturing batches. How this research is marked by the existence of a serious competition both between companies producing magnetron sputtering PVD type equipment as well as between their users (manufacturers of coated glass products), in the related literature there is an extremely low number of information regarding such researches [1].

But unlike the absence of such studies, in the related literature are presented a big number of studies on the development of theoretical and experimental researches conducted on pilot production equipment and experimental research laboratory facilities. Unfortunately a large number of theoretical and experimental results presented in such studies, although contribute to founding of PVD processes, are not applicable in industrial production processes. because laboratory machines / research equipment have different characteristics than industrial scale production machines / equipment. [1]

Taking into account above aspects, this paper presents a part of theoretical and experimental research developed along first author's PhD thesis elaboration, by using a new approach for simulating the operation of magnetron sputtering PVD equipment for industrial applications. The major goal of this research was to create and experimentally validate a new software simulation environment for the magnetron sputtering PVD processes performed at industrial scale by means of flexible manufacturing lines for glass coating.

2. Current state of magnetron sputtering theoretical and experimental researches

Magnetron sputtering uses a PVD technique for deposition of metallic layers also called as "spraying" and consists in applying layers of metal oxide, in vacuum conditions on glass substrates. This process allow obtaining a wide range of products with different formulas / number of layers deposited on the glass substrate, and different thickness variations of the layers of deposited metal materials. As a result of this deposits one of the well-known properties of the so treated glass is the ability to reflect different types of solar radiation (ultraviolet, infrared etc.), as are the glass products from the "Low class", necessary for sun protection of enclosures used in warm climates as well as those that require improved heat insulation. For each product of this type should be used specific manufacturing "formulas" depending on a large number of process parameters. From this point of view existing studies in the related literature describe in detail almost exclusively only glass processing on laboratory magnetron sputtering equipment or pilot manufacturing facilities with a small number of production cells (usually 1 or maximum 2 cells). [1]

Two variants of such kind of specific systems for research developed in laboratory are presented below.

2.1. Research on laboratory equipment for experimental study of magnetron sputtering facilities

A synthesis of the related literature regarding existing studies for modeling the phenomena that occur in magnetron sputtering equipment is presented in table 1, highlighting the influence of main factors affecting this process. [2].

Table 1 [2] presents mainly interactions that occur on target surface in magnetron sputtering process (see also fig 1, [3]) included formulas highlighting the complexity of physical phenomena occurring in such facilities. To simulate the process unfolded in the industrial environment, the mathematical models based on simulation of such installations must include all these interactions, so becoming very complicated and impossible to be practically used.

Description	Formula	Unit
Surface binding energy	$U_{sb}=(\Delta_{fus}H + \Delta_{vap}H)\Delta N_a$	eV
Energy transfer mass factor	$\Gamma=4m_p m_r /(m_p + m_r)^2$	dimensionless
Average energy of recoil	$E_{ave} = U_{sb}\ln(\gamma E/U_{sb})$	eV
Thomas- Fermi screening length of projectile	$a_p = 0.8853a_0 /(Z_p^{\frac{2}{3}} + Z_r^{\frac{2}{3}})^{1/2}$	Ά
Thomas- Fermi screening length of recoil	$a_r = 0.8853a_0 /(2Z_r^{\frac{2}{3}})^{1/2}$	Ά
Reduce energy of the projectile	$e_p(E) = \left[\frac{a_p m_r}{Z_p Z_r q^2 (m_p + m_r)}\right] E$	dimensionless
Reduce average energy of recoil	$e_{ave}(E) = \left[\frac{a_r}{Z_p^2 q^2}\right] E_{ave}$	dimensionless
Nuclear stopping power of the projectile	$\frac{dE}{dR \uparrow_{np}} = S_{np} n_t$	eV/ Ά
Nuclear stopping power for recoil	$\frac{dE}{dR \uparrow_{nr}} = S_{nr} n_t$	eV/ Ά
Range of the projectile	$R_p(E) = E/(\frac{dE}{dR \uparrow_{np}})$	Ά
Range of the recoil	$R_r(E_{ave}) = (E_{ave} - U_{sb})/(\frac{dE}{dR \uparrow_{np}})$	Ά
Numbers of recoils	$N = E/E_{ave}$	dimensionless
Efficiency (Yield)	$Y=N(R_p/R_r)1/4$	dimensionless

Table 1. Equations that describe the collisions at the target surface [2]

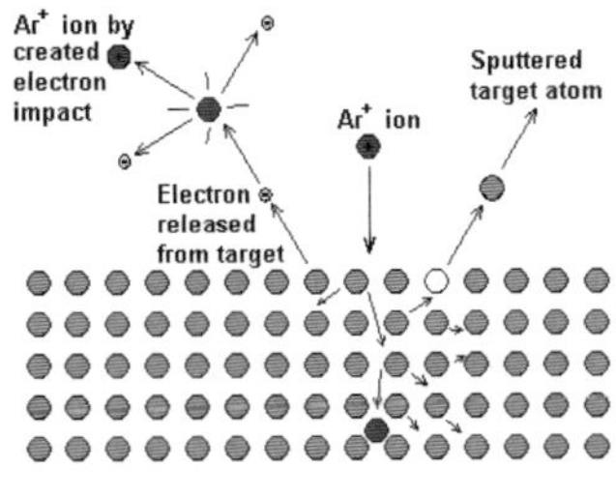

Fig. 1 Collision at target surface [3]

Fig. 2 Magnetron sputtering laboratory research equipment [4]

In fig. 2 the complexity of an experimental facility (an AJA Orion sputtering system) from a research laboratory (Cornell Nano-Scale Science and Technology Facility) can be observed too [4]. This equipment comes with a lot of sensors that can provide accurate values not only for the experiment results, but also for trajectory of all ions of sprayed elements. In addition the input parameters control is completely determined. A big plus that laboratory facilities have is that they transmit only one wave of energetic charged ions allowing to perform individual spraying, being able to identify precisely the full set of process parameters and offering the possibility to analyze carefully the results of developed experiments [3], [4].

Unlike these laboratory facilities, in industrial environment : spraying is continuous and is made with multiple cells for materials deposition in different process conditions; the outcome of multiple sprays on the same substrate can be observed only cumulatively by optically measuring the characteristics of the resulted deposed material layers [1].

2.2. Research on laboratory pilot installations for glass processing with magnetron sputtering

A laboratory pilot installation for glass processing with magnetron sputtering (for multilayer coating deposition on both sides of the sheet glass) is presented in fig 3. For oxide coating the method of magnetron reactive sputtering on AC current is used. For getting high economic benefits, rotary magnetrons with metal and ceramic targets are used. For quality monitoring, optical methods of measuring the spectral characteristics of transmission and reflection are applied. Modeling the coating structure and the subsequent reproduction of these parameters in vacuum machine allows receiving very good coating quality on large area glass sheets. [5]

Laboratory (pilot) manufacturing line, UV100 [5] (fig. 3), is a coating equipment consisting of three rooms: the load module, deposit module and coating module. The modules are distinguished by gates that serve as valves, each individual module having a pressure up to 2×10^{-5} torr. The pumping system is based on turbo-molecular pumps which removes the possibility of oil contamination. In addition, the use of turbo-molecular pumps is suitable for stable pumping rate in the range of pressures, necessary for the magnetron spray process. [5] The loading module is used to load and unload the glass sheet and for pumping before entering in the deposition area. The deposit module comprises two rotatable cathodes, placed on one side of the module and three planar cathodes, installed on the other side. This makes possible to install up to 5 different target materials for deposition. The maximum length of the objects that can be coated is 800 mm. The deposition module is maintained at a constant pressure of de 2×10^{-5}torr, which ensure the elimination of any atmospheric contamination. [5]. For the pre-treatment of the sub-layer, an ion source is used with an energy up to 1000 eV. On the machine presented in fig. 3 gas distributors are applied, allowing production of gas and controlled flowing along magnetrons length. The flow control is important as other studies have shown, the structure of thin films deposited by magnetron sputtering being characterized by a function of gas flow and film thickness. [10] The machine is equipped with a box that generates vacuum. The box has a variable speed travel. Sources of power on the vacuum machine may be DC or AC power supplies. [5]

Developing a working software environment for accurate simulations using such systems is however very difficult, because pilot installations have a limited number of cell deposition (usually 1 maximum 2) compared to industrial production facilities that have usually 7, 15 or 22 deposition cells, and because each manufacturing line type has a specific construction, characteristic functional parameters and different control possibilities. [1]

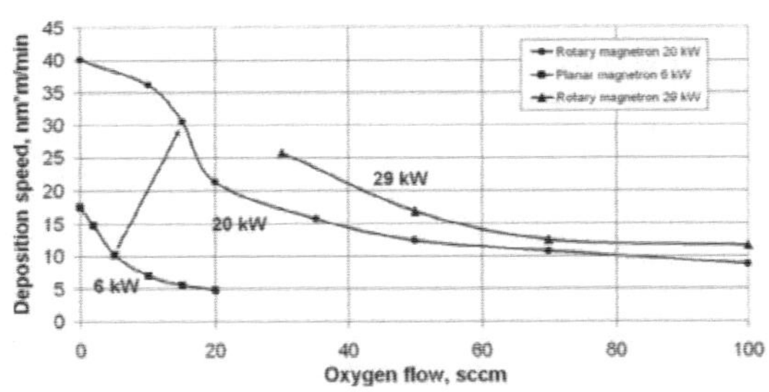

Fig. 3.Laboratory pilot line UV100 [5] Fig. 4. Deposition speed of TiO2 at different powers and oxygen flows [5]

As concerning the experimental results obtained by using this facility, in Figure 2 the dependences of deposition speed for titanium oxide in case of different powers and oxygen flows are presented. The experiments were conducted on planar and rotary magnetrons by the length 800 mm. From experimental data it may be seen that the rotary magnetrons makes it possible to increase the deposition rate by several times as the planar magnetrons, similar results being obtained for deposition of silicon oxide.[5]

2.3. Mathematical modeling of specific processes for magnetron sputtering

Most of the current theoretical models and simulations environments for magnetron sputtering are based on the formulas developed by P. Sigmund [6], the most used simulation model based on Sigmund's formulas being the Monte – Carlo simulation algorithm. Simulating the thickness profile of sputtered layers not only compels the simulation of the transport through the gas phase but also the knowledge of the initial properties when the particles leave the target. [9]

This algorithm / simulation model covers the process in terms of energy transfer in the system and the characteristics of used materials.

$$Y(E) = \frac{0.04}{U}\alpha\,(Mt/Mi)Sn(E) \tag{1}$$

Sn (E) - the energy of the surface collision (energy lost in the impact of molecules);
U - the energy of sublimation.

$$\text{Sn(E)= } 85\,\frac{ZiZt}{(Zi^{\frac{2}{3}}+Zt^{\frac{2}{3}})^{0.5}} * \frac{Mi}{Mt+Mi}Sn(\varepsilon) \tag{2}$$

Sn (ε) is a function of ion energy reduction and is constant for each combination of ion and target.

$$Sn(\varepsilon) = \frac{3.441\sqrt{\varepsilon}\ln(\varepsilon + 2.719)}{1 + 6.355\sqrt{\varepsilon} + \varepsilon\left(6.882\sqrt{\varepsilon} - 1.708\right)} \tag{3}$$

$$\varepsilon = \frac{0.03255}{\text{ZiZt}(\text{Zi}^{\frac{2}{3}} + \text{Zt}^{\frac{2}{3}})^{0.5}} * \frac{\text{Mt}}{\text{Mt} + \text{Mi}}\text{E} \tag{4}$$

The model is completed by total efficiency evaluation by formula:

$$\text{Y(E)} = \frac{0.04}{\text{U}}\alpha\left(\frac{\text{Mt}}{\text{Mi}}\right)85\,\frac{\text{ZiZt}}{\left(\text{Zi}^{\frac{2}{3}} + \text{Zt}^{\frac{2}{3}}\right)^{0.5}} * \frac{\text{Mi}}{\text{Mt} + \text{Mi}}\frac{3.441\sqrt{\varepsilon}\ln(\varepsilon + 2.719)}{1 + 6.355\sqrt{\varepsilon} + \varepsilon\left(6.882\sqrt{\varepsilon} - 1.708\right)} \tag{5}$$

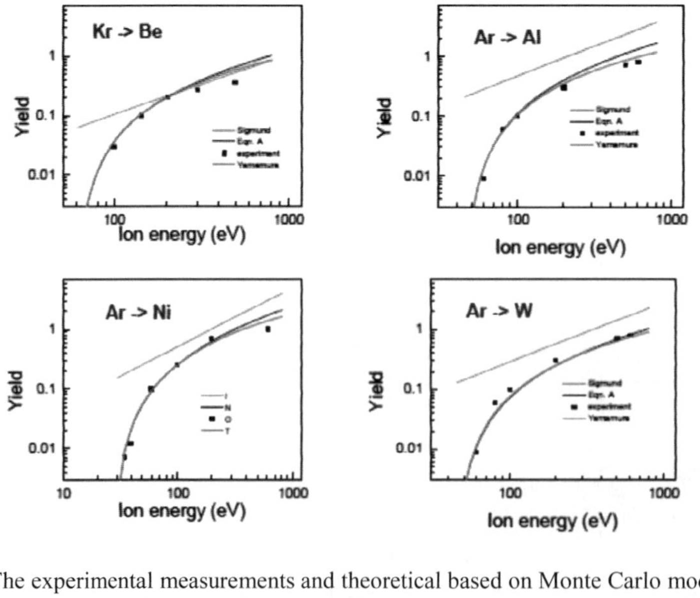

Fig. 5. The experimental measurements and theoretical based on Monte Carlo model [6]

However, Sigmund's equation and model based on that equation is treating the process like a single interaction between one particle and the target material like is presented in fig. 5 [6].In fig. 5 it can be seen that the results obtained by using this mathematical model are characterized by following issues: different low levels energy in the system (green line represents the results predicted by Sigmund's equation and blackheads are the real values obtained experimentally). Increasing system's energy, the values obtained from simulation are more accurate.[1] Thus, generally speaking, studies involving real industrial scale magnetron sputtering machines, characterized by energies levels much higher than those presented in fig.5, may be performed by using similar mathematical formulations and may be expected to conduct to an increasing accuracy of the simulation results.[1] That is why the model created by Sigmund underlies most simulation models further developed in the literature and is also experimentally verified. However, this model cannot be directly applied in the industrial environment for studying the phenomenon as the interaction of a single particle and individual spraying procedures because in the real industrial scale machines the magnetron interaction appears in different processing environments, the substrate deposition is exclusively performed by using multiple spraying procedures as well as the magnetron sputtering process being usual in continuing operation for several days. [1]

3. Conceptual and operational specificities of industrial installation for glass deposition by magnetron sputtering for which mathematical model was developed and software simulation for deposition process performed

3.1. Specific conceptual features of the flexible manufacturing line for thin layer glass deposition by magnetron sputtering

In fig 6 is shown the real industrial scale flexible manufacturing line for thin layer glass coating by magnetron sputtering process, for which has been developed the mathematical model and the software simulation environment of the deposition process. The flexible manufacturing line is an "Apollon" type production line [7] - designed and supplied by Leyboldoptics Company - for deposition of nanometric layers by magnetron sputtering process, being installed in Saint-Gobain Glass production facility located in Călăraşi, Romania (fig.6). [7]

Processing area of Apollon system [7] consists of flexible manufacturing cells allowing the production of different coating layers on glass. The divided modular design is focused on individual cells, each of them being a standardized spraying environment. Thus, each unit of the line become a flexible manufacturing cell integrated into the flexible manufacturing line by joining all flexible cells through the same transfer and transport integrated system (fig. 6) [7].

The flexibility of the system is very important to create all the necessary new coatings especially for architectural glass, used in building facades. Reconfiguration of the system is easy to be achieved, in order for both, to maximize flexible manufacturing line productivity and as well continue extend the products portfolio.

This technological process is very suitable for the study of interactions through a flexible manufacturing process due to the specific of magnetron deposition method. The industrial installation includes a series of manufacturing cells, (it can reach a number of 200 interconnected units) all of them making successive deposits on the same substrate represented by plane glass with dimensions that can vary from 2,1 m×3.21 m up to 6 m×3.21 m. In each flexible manufacturing cell the deposition can be made using the same materials and gases as in previous cells in order to get a very high productivity of the manufacturing line or respectively by using totally different materials and gases in order for obtaining products with a high complexity of deposited layers (fig 6 and fig.7) [1], [7].

a. General overview of the full flexible manufacturing line [7]

Fig 6 Real industrial scale flexible manufacturing line for thin layer glass coating by magnetron sputtering process [7]

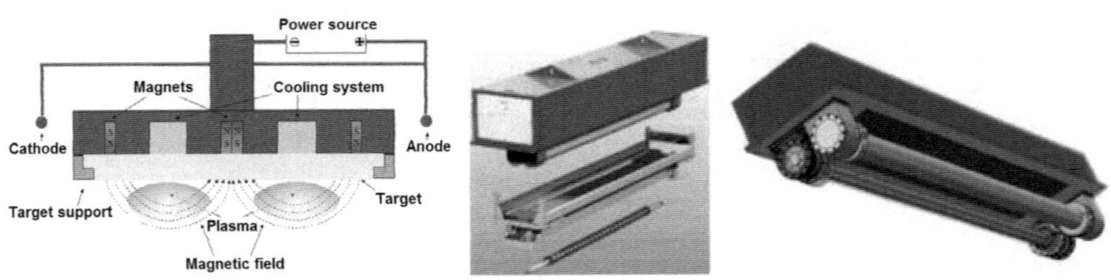

Fig 7 Functional principle, real industrial processing unit and virtual prototype of a flexible cell from the flexible manufacturing line for thin layer glass coating by magnetron sputtering process [1], [7]

3.2. Operating principle of the flexible manufacturing line for thin layer glass coating by magnetron sputtering process. Specific parameters in magnetron sputtering processes. Particular procedures in adjusting the thin layer's homogeneity by mean of process parameters' adjustment

Flexible manufacturing lines for thin layer deposition by magnetron sputtering consists in an assembly of flexible manufacturing cells. Each cell can have a deposition process that can be totally different from the previous due to: target type, reaction gases, input power, source type (fig.8) [1].
The sistem is composed by cells that performs succesive material depositions on a sublayer, in this way the layer deposited by the first cell becomes a sublayer for the second cell depositon. Each cell is a sandalone manufacturing system with it's own parameters and those are: G1 is the reaction gas type introduced in the system, P1 is the flow and in consequence the pressure of the gas injected on the main pipe, P2 is the flow from the secondary pipes, A1 is the input energy amperage, M1 is the material to be deposited in cell process (fig.8) [1].

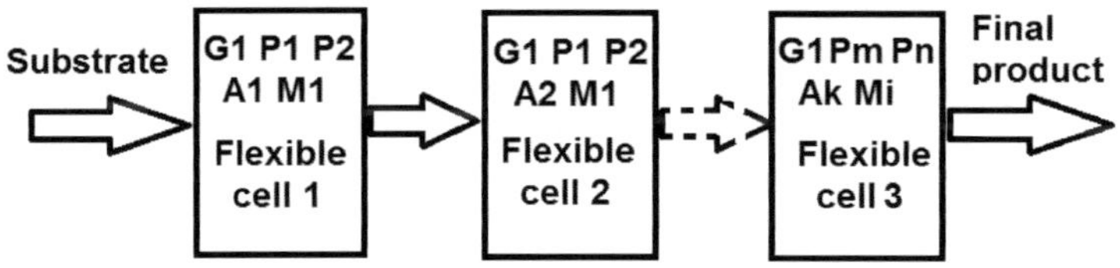

Fig.8 Operating principle and specific processing parameters setting for flexible manufacturing line / cells [1]

A table (Table 2) [1] for system influences was built to show how a material layer deposition can be controlled and manipulated. This table is valid only for a single cell; the process being repeated every time a glass sheet passes through the manufacturing cell.This table describes how the simulation that must be realized should be able to provide exact values for regulating and control of deposition rate of each manufacturing cell from the system. In table 2 are presented in the left side different typologies of the deposition profile that can result from the reaction. In the column called "corrective action to adjust thin layer homogeneity " reaction factors are presented and the way in which those must be modified to improve the profile homogeneity.

Deposition profile	Corrective action to adjust thin layer homogeneity
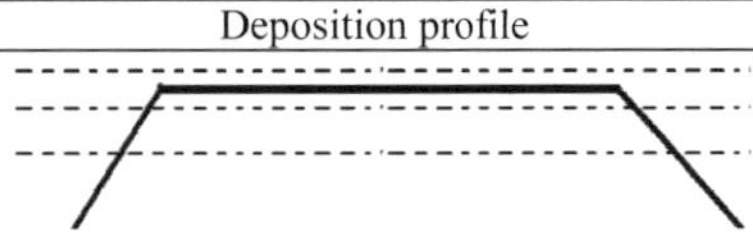	Increasing value of P2
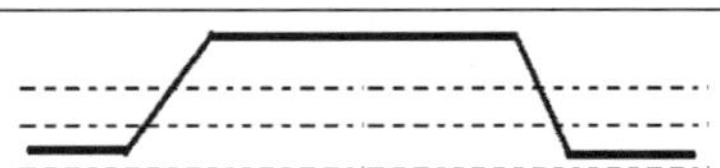	Decreasing value of P1
	Increasing value of P1
	Decreasing value of A1
	Increasing value of A1
	Check if P1 and P2 have reached saturation value
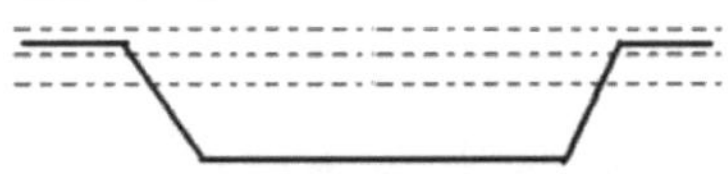	Check if P1 exceeded system saturation value
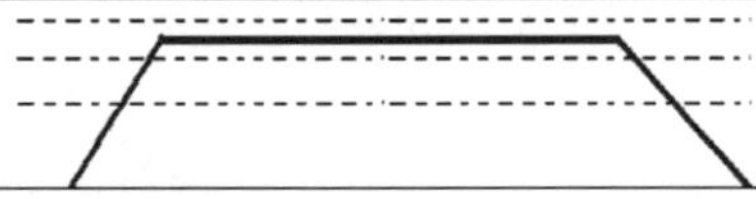	Check if P2 exceeded system saturation value

Table 2. Specific stages in adjusting thin layer homogeneity in magnetron sputtering deposition [1]

*3.3. Thin layer deposition uniformity control: the international system for color identification CIELAB ("CIE L*a*b*")*

The original mathematical model further presented in the paper is compatible with the international system for color identification CIELAB (L* a* b*) illustrated in fig. 9, this compatibility being necessary for experimental results comparison by mean of the specific investigation technique and characteristic values supplied by an electro photometer.

CIE L* a* b* (CIELAB) is a spherical color space specified by International Commission for Illumination (France - "CIE") [8]. This spatial color identification system describe all visible colors for the human eye that has been specially created to serve as an independent device (and in the meantime reference element) in color evaluation.

The three coordinates of the CIELAB color space represents the color luminosity as follows: L* = 0 for black şi L* = 100 indicate the white diffuse light; - a* characterize the color included between red and magenta: a* is corresponding to red color in case of positive values and to magenta color in case of negative values; b* characterize the color included yellow and blue: b* is corresponding to blue color in case of negative values and yellow in case of positive values.

However, as a 3D model CIE L* a* b* may be correctly represented only in spatial (3D) coordinate system, the 2D dimensional representation being able to include only chromatic diagrams [8]. The most used coordinate system to represent CIE L* a* b* is the spherical coordinate system, as an alternative mean, less used, being possible to appeal too on a cylindrical coordinate system.

3.4. The flow chart for thin layer glass coating by magnetron sputtering process simulation and experimental research

Based on the manufacturing flow presented in fig. 8, on the influences table (Table 2) and on the previously presented conclusions the logic scheme of magnetron manufacturing process was built in fig. 10. The parameters are introduced in the simulation and the program calculates de dimension of the deposition layer and if it is homogenous. Following experimental research was conducted on the same logic algorithm.

A first advantage of this working method with the simulation preceding the experiments on the manufacturing line is that different parameters sets can be tested without losses due to system stops or consumption of materials and energy, respectively rejection of some product batches as non-corresponding.

The second step is comparing the simulation results with the results desired to be obtained through experiment and analyze of performance parameters level characterizing coated material layer obtained through simulation / to be deposited effectively on glass sub-layer in experimental manufacturing processes. If from the simulation results that the coated layer is not homogenous, initial parameters adjusting must be done according to Table 2 procedures. If from the simulation results that the coated layer is uniform then the gases distribution is also uniform and can proceed to the next step of simulating the technological processes.

The next step is checking if the layer thickness matches the desired thickness. If the layer thickness is smaller than the one necessary for the product, correction procedures presented in Table 2 must be applied, process parameters values being adjusted iteratively until the desired effect is obtained.

This process is repeated for each parameter of each manufacturing cell, simulation time being influenced by de the complexity of the desired product (how many material types and gases are used in the process), and also by the number of cells used to obtain the final product.

After the possible materials with specified thicknesses have been identified, in the fourth stage the program is mixing this data. This way is obtained both necessary layer thickness and in layer position of material.

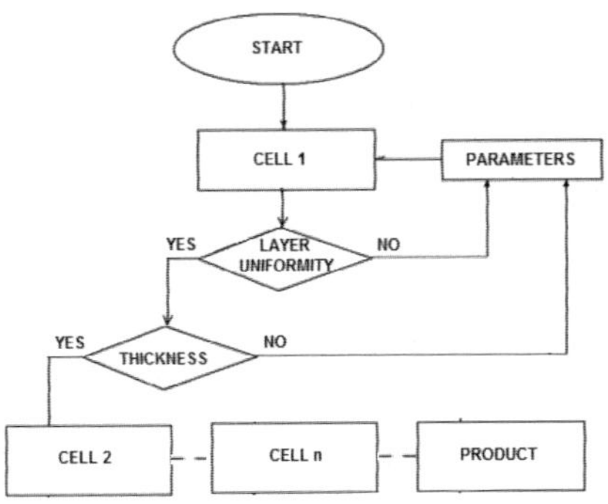

Fig. 10 Simulation flow chart

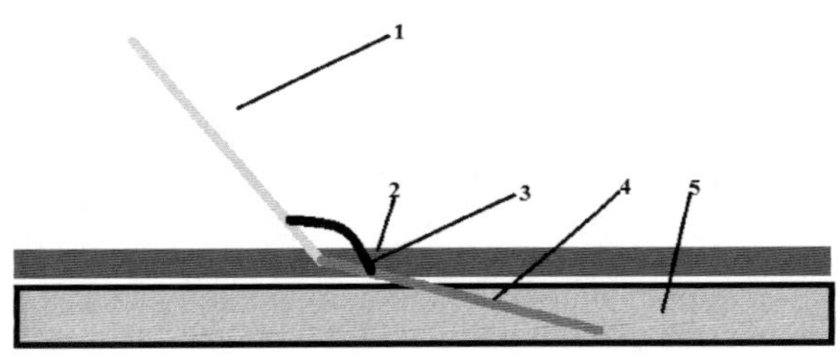

Fig. 11 Light beam path

3.5. The core original mathematical model used for thin layer glass coating by magnetron sputtering process simulation and experimental research

The model created by Sigmund [6] underlies the most simulation models in the specialty literature, experimental developed and checked. This model cannot be directly applied in the industrial environment because it studies the phenomenon as an interaction between a single particle and the sputtering environment being useful only in case of modeling some experiments on lab equipment.

Using the same principles developed in Sigmund equations but adapted to the specifics of industrial processes, equations (6), (7) and (8) [1] have been developed. Thus the equations, (6), (7) and (8) describe the core of the original mathematical model created for simulation of industrial system.

The core of the new original mathematical model consists primarily in viewing the process from the point of view of the layer deposited on the precedent sub layer and not from the point of view of a particle and its interactions with the magnetron sputtering process environment. The second and most important change that the mathematical model's equations include is the identification of color values of light passing through the deposited layers of glass based on CIELAB standard. These improvements are necessary both for proper implementation in simulation model of the real industrial scale processes parameters / procedures and evaluation standards, as well as for experimental validation of mathematical formulations by experimental results obtained by performing magnetron sputtering process in industrial environment.

$$L^*c=[116f(\frac{Y}{Yn})-16]^*\{n^*[P^*(\frac{\text{D plasma gas} * d1}{\text{D reaction gas} * d2})]\} \quad (6)$$

$$a^*c=[500[f(\frac{X}{Xn})-f(\frac{Y}{Yn})]^*\{n^*[P^*(\frac{\text{D plasma gas} * d1}{\text{D reaction gas} * d2})]\} \quad (7)$$

$$b^*c=[200[f(\frac{Y}{Yn})-f(\frac{Z}{Zn})]^*\{ n^*[P^*(\frac{\text{D gaz plasma} * d1}{\text{D reaction gas} * d2})]\} \quad (8)$$

In order to be able to simulate the layer homogeneity, in the simulation mathematical model, yield formula for coating was modified with the addition of a new factor "d". This is a dimensionless parameter and takes account of the gas dispersion in the room (the distance between the simulated point and the nozzles position through which the plasma and reaction gasses are dispersed).
In comparison with Sigmund original formula:

$$Y(E) = \frac{0.04}{U}\alpha\left(\frac{Mt}{Mi}\right)Sn(E) \quad (9)$$

the formula used in the program for regulating the homogeneity becomes:

$$Y(E) = \frac{0.04}{U}\alpha\left(\frac{Mt}{Mi * d}\right)Sn(E) \quad (10)$$

Complementary to the mathematical relations that form the core of the model other aspects must be considered too. Thus, a first element necessary to be taken in consideration is the sputtering yield. This is measured in nm of coating / m / min (the speed at which the sub-layer / glass sheet is transported in the experimental installation).
By convention the transport speed of the sub-layer in the industrial installations is about 1m / min. The speed of production line can however be adjusted depending on the desired performance parameters for the final product, because it influences directly the coated layer thickness in every manufacturing cell (due to the fact that the technological system realizes a constant sputtering both in terms of quality and quantity, the time spent by the glass sheet in the sputtering area influences the coating thickness).

$$Y(E) = \frac{Xnm}{m} = \frac{Xnm}{1} = Xnm \quad (11)$$

where X is the thickness of the coated layer.
In the same time, for correlating the results of experimental testing with the parameters for identifying the coated layers thickness CIELAB, in the mathematical model were included specific formalizations. Thus, if for example it is taken in consideration the coating with a single copper layer with a thickness of X nm, the light passed through this layer will have the coordinates:

$$X^*c_1Cu = L^*C$$
$$X^*c_2\,Cu = a^*C \qquad (11)$$
$$X^*c_3\,Cu = b^*C \qquad (12)$$

where c_1, c_2 ,c_3 are color characteristics of copper material.

If for example it is taken in consideration as a requirement getting a product with 3 layers of material, results that for obtaining the specific values of $L_3{}^*a_3{}^*b_3{}^*$ the characteristics of the all 3 material layers must be added up in each of the three coordinates and must look after an equilibrium between their value and the layer thickness.

$$L_3^*= (L_2^* c_1Cu^*X) +(L_1^* c_1Cu^*X)+ (L_0^* c_1Cu^*X)$$
$$a_3^*=(a_2^* c_1Cu^*X) + (a_1^* c_1Cu^*X)+ (a_0^* c_1Cu^*X) \quad (12) \quad (13)$$
$$b_3^*=(b_2^* c_1Cu^*X)+ (b_1^* c_1Cu^*X) + (b_0^* c_1Cu^*X)$$

In fig. 11 it is represented in a simplified way the base principle of the mathematical model. The light beam, marked with nr. 1 having color coordinates for natural white light as they are described in the international standard D 65, passes the coated layer, marked with nr. 2, on the sub-layer represented on the plane glass, marked with nr. 5. Starting from the initial color coordinates L*,a*,b* due to the refraction phenomenon, angle of refraction being marked with nr. 3, results the light beam, marked with 4, having color coordinates L*c,a*c,b*c.

Due to the fact that the developed formula is similar for the three coordinates, the only difference is the different initial value of the coordinates.

The coated layer of material by magnetron process is given by equation (14) where P is system input power measured in KW/h; D is plasma gas plasma gas flow introduced in system measured in s.s.c; d_{1is} dimensionless material constant for plasma gas; D is reactive gas flow introduced in system measured in s.s.c; d_2 is dimensionless material constant for reaction gas:

$$(P^*[(D\ \text{plasma gas}^*\ d1)/(D\ \text{reaction gas}\ ^*\ d2)] \quad (14)$$

3.6. Experimental works on industrial scale flexible manufacturing line for thin layer magnetron sputtering deposition

For each manufacturing cell there are already set some specific features of the flexible manufacturing lines by producer including: typical flexible cell dimensions: 80 ×15 × 5 m (L×W×H); minimum industrial room height: 6,0 m; maximum size of glass substrate according to industry standards PLF and DLF: 3,66 × 2,54 m2; 6,00 × 3,30 m^2; minimum dimensions of the glass layer: 0,8 × 0,30 m^2; usual glass substrate thickness: 2.0 - 12 mm; interior width of the processing compartment: 850 mm [7]. The main operating parameters to be set include: system input power applied on the cathodes expressed in Kw; (by increasing the power, material deposition rate increases in direct proportion to the introduced system power); plasma gas used, usually Helium, Neon, Argon, Krypton, Xenon being used. (Depending on the molar mass of this material, the target is bombarded with particles with higher energy resulting in greater impact, a higher spraying rate and a high deposition rate); reaction gas (in some products, in layers can be found beside metals also oxide or nitride layers. To obtain these layers oxygen or nitrogen is introduced in system which reacts with the sprayed on the surface of the targets, optical and electrical proprieties of coated films being dependent by the nature and flow of these gases [11]).

A comparison between simulation and experimental sets of results performed for four types of materials deposed M_1, M_2, M_3, M_4 (for experimentally works performed along four months, only partially sets of results being included in the paper) are presented in table 3 and table 4. Relative error between simulation results and experimentally results as well as appropriate deviations are also presented in table 4 and table 5 as well as fig. 12, fig 13 and fig.14.

date	a* simulation	a* error	a* experim.	calculus deviation a*	b* simulation	b* error	b* experim.	calculus deviation b*	M_1 kW	M_2 kW	M_3 kW	M_4 kW
24.03.2015	3.529394	0.029394	3.5	-0.029394	-8.970605	-0.470605	-8.5	0.470605	96	288	25.3	336
26.04.2015	3.136491	0.636491	2.5	-0.636491	-10.303750	-0.803750	-9.5	0.803750	92	279	24.5	336
13.05.2015	0.166666	1.333333	1.5	1.333333	-8.893617	-0.393617	-8.5	0.393617	92	279	19	375
15.06.2015	3.905788	1.105788	2.8	-1.105788	-9.572830	-0.572830	-9	0.572830	82	237	25.5	336
18.07.2015	4.973798	2.473798	2.5	-2.473798	-8.466442	1.033557	-9.5	-1.033557	100	252	24.5	336

Table 5. Simulation and experimental works set of results and relative error between them and simulation deviation

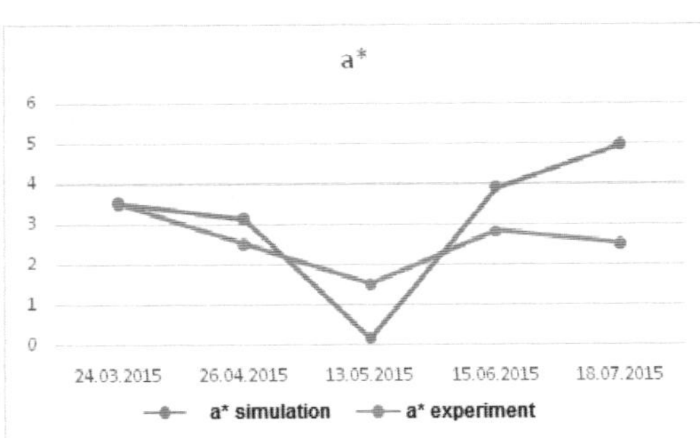

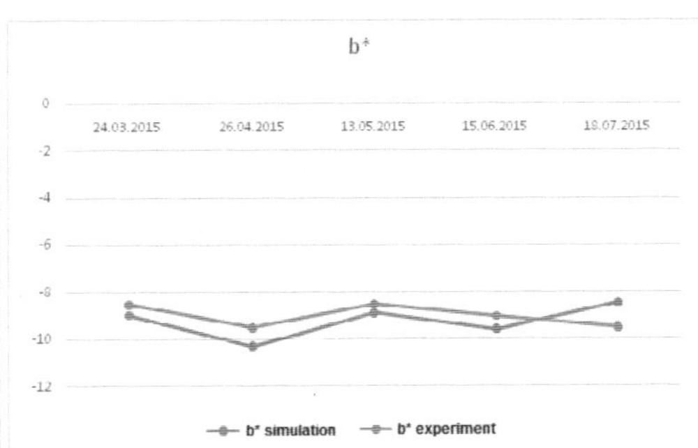

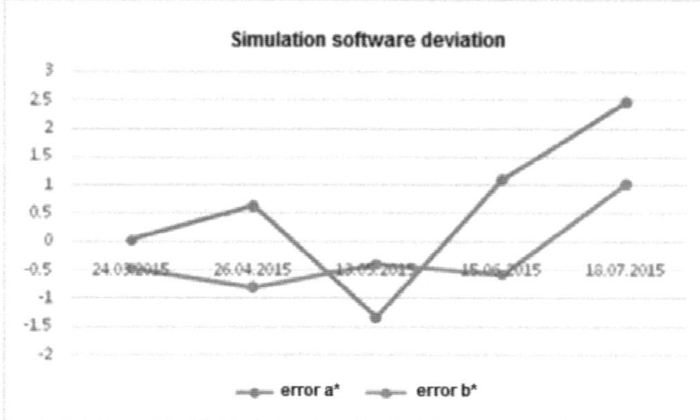

Fig. 12 a. Comparison between simulation and experimental results for a*, b* values. Simulation software deviation

4. Conclusions

Use of the simulation models presented in the literature until now is not possible in the industrial environment because the laboratory and the industrial experimental equipment / manufacturing facilities are different. The laboratory experimental equipment is usually used to completely characterize a singular experiment already performed by using a single pulverization of a single deposition material type. Besides, in the industrial environment the manufacturing lines are including a large number of magnetron sputtering cells / chambers that are interconnected, in order to allow successively depositing of different materials in specific conditions of processing gases' pressure and substrate speeds. Another main difference between ways of conducting experiments is due to the experiments duration difference: the experimental laboratory equipment is dedicated to the study of the reactions lasting only a few seconds, comparatively to the real industrial scale equipment where the reaction is continuous for several days. These issues have been solved by present paper specific approach. As result, present paper illustrates a new approach regarding the study of magnetron sputtering discharge process on an industrial scale flexible manufacturing line. For this purpose a new mathematical model and simulation software have been developed. Based on this new mathematical model and simulation software, the thin layer glass coating by magnetron sputtering processes performed in industrial environment has been theoretically studied and the results have been validated by experimental research conducted on the industrial site.

Developing a simulator for the whole process of facilities at industrial scale represents a way of studying the possibilities to optimize the technological processes, allowing time shortage of facility setup for realizing a new product, from few weeks to few hours or even less. Using the simulator presented in this paper a significant improvement of product performance parameters is possible, regarding the reducing of the quantity of consumed energy, coating materials and rate of non conform product batches. Major difference in experimental approach of vacuum deposition processes for laboratory research equipment (presented in the literature) and industrial processing equipment (presented in current paper) consist in focusing on investigating the interaction between the target material and particles as well as particle trajectory in vacuum environment for laboratory equipment, respectively on effective deposition process and the thickness and uniformity of the deposed layers on the substrate for industrial equipment.

A second major aspect of the new presented approach consist in mathematical formulation used to simulate the final result of the deposition process, which in present paper is illustrated using the specific industrial standards(CIELAB) to characterize the light passing through the deposited layer on the substrate, this approach being necessary in order to create first the possibility to validate the mathematical formalization by experimental research performed on real industrial scale production equipment and second to use specific industrial means to characterize the final product obtained in such type of equipment (optic photometer). For this purpose, the operating principle of simulation software is first to predict the thickness of the layer deposed in the magnetron process based on the international system D65 and, then evaluate the layer thickness starting from the measured white light coordinates, (computing the color of light resulted from passing through the deposed layer) by mean of an optic photometer used for measuring the value of the light passing through the substrate and deposed layer. All simulation results were validated on an industrial flexible manufacturing line. As result of applying this new concept on industrial scale processing system an improvement of the deposition process control has been obtained leading to an increasing of the uniformity of deposed layers from ±10% to ±2% and differences between theoretical and applicative vaporizing energies have been identified for two materials.

5. Acknowledgments

This work was partially supported by the strategic grant POSDRU/159/1.5/S/137070 (2014) of the Ministry of National Education, Romania, co-financed by the European Social Fund – Investing in People, within the Sector Operational Program Human Resources Development 2007-2013".

6. References

[1] I. Spataru - Contributions to modeling and simulation of flexible manufacturing systems" PhD Thesis, Politehnica University of Bucharest, public presentation september 2016, Bucharest

[2] N. Brenning, D. Lundin, M. A. Raadu, C. Huo, C. Vitelaru, G. D. Stancu, T. Minea , and U. Helmersson, Understanding deposition rate loss in high power impulse magnetron sputtering, Plasma sources science & technology (Print), (21), 2, 2012, ID no. 025005, http://dx.doi.org/10.1088/0963-0252/21/2/025005

[3] Murtaza Saleem, Overview of Magnetron Sputtering System (DaON 1000S), Lahore University of Management Sciences, http://physlab.org/wp-content/uploads/2016/04/Murtaza_presentation.pdf

[4] https://en.wikipedia.org/wiki/Sputtering

[5] V. Kozlov, E. Yadin, G. Taiminsh, and V. Fomin, New vacuum equipment for multilayer coating deposition on large area glass, Society of Vacuum Coaters 505/856-7188, 50th Annual Technical Conference Proceedings (2007) ISSN 0737-5921, 2007

[6] P.Sigmund, "Sputtering by ion bombardment: theoretical concepts", Chapter 2, in Sputtering by particle bombardment I, edited by R. Behrish, Volume 47 of the series Topics in Applied Physics pp 9-71, Springer-Verlag, 2005

[7] *** Solutions for large area coating, Leybold Optics documentation 2016, Bühler Alzenau GmbH, www.buhlergroup.com

[8] *** CIE 15 - Technical Report - International comission on illumination, ISBN 3-901-906-33-9, Washington, USA

[9] S. Mahieu, G. Buyle, D. Depla, S. Heirwegh, P. Ghekiere, R. De Gryse, "Monte Carlo simulation of the transport of atoms in DC magnetron sputtering", in Nuclear Instruments and Methods in Physics Research Section B: Beam Interactions with Materials and Atoms, Volume 243, Issue 2, 2006, Elsevier

[10] S. Van Steenberge, W.P. Leroy, D. Depla, "Influence of oxygen flow and film thickness on the texture and microstructure of sputtered ceria thin films", in Thin Solid Films, Volume 553, 2014, Elsevier

[11] M. Bender, W. Seelig, C. Daube, H. Frankenberger, B. Ocker, J. Stollenwerk, "Dependence of oxygen flow on optical and electrical properties of DC-magnetron sputtered ITO films", in Thin Solid Films, Volume 326, Issues 1–2, 1998, Elsevier

DOI: 10.2507/27th.daaam.proceedings.121

ATTENUATION OF WIRELESS COMMUNICATION UNDER IEEE 802.11AH

Pavel Tomasek

This Publication has to be referred as: Tomasek, P[avel] (2016). Attenuation of Wireless Communication Under IEEE 802.11ah, Proceedings of the 27th DAAAM International Symposium, pp.0838-0844, B. Katalinic (Ed.), Published by DAAAM International, ISBN 978-3-902734-08-2, ISSN 1726-9679, Vienna, Austria
DOI: 10.2507/27th.daaam.proceedings.121

Abstract

This contribution aims at shielding communication under a new standard for wireless communication, IEEE 802.11ah. This standard is very probably going to be applied in the rapidly growing area of the Internet of Things (IoT) due to its perfect battery saving system and range of operation. This work focuses on a design of a frequency selective filter shielding a very narrow band of frequencies used just under IEEE 802.11ah and thus passing all of the other frequency ranges used under earlier standards for wireless communication. The motivation for shielding of such a communication is primarily in elimination of the possibility of eavesdropping of delicate or private data.

Keywords: IEEE 802.11ah; Frequency Selective Filter; Internet of Things; Eavesdropping.

1. Introduction

The new standard from the Institute of Electrical and Electronics Engineers (IEEE) with the name of 802.11ah is a quite hot topic in the field of the Internet of Things (IoT). For the purpose of a wireless communication between very small electrical devices a very special energy saving way of transmission of information has been developed.

The world of IoT is full of electrical devices, sensors, wearables, security elements, various appliances utilizable in Smart Home (for instance lighting, cooking, heating) and also agriculture monitoring, industrial automation and smart metering. The number of mentioned devices is supposed to raise rapidly in the near future what is related with potentially significant security risks.

Therefore the main goal of this study is aimed at analysis of possible ways of shielding such a communication against eavesdropping outside a room or a building.

This study is based on the previous work [1] where the goal was to reflect wireless communication under older standard IEEE 802.11b,g.

This article contains the following content: section 2 contains brief description of the new standard of IEEE 802.11ah. In section 3, there are very simply described two simple ways of shielding of this type of wireless communication. Section 4 informs about a chosen way of optimization of a filter based on a Frequency Selective Surface. And finally Section 5 presents synthetic results and section 6 concludes the whole idea presented in this study.

2. IEEE 802.11ah

The IEEE standard of 802.11ah (sometimes called "Wi-Fi HaLow") has a great potential in the area of IoT because of two main reasons [2]:

- Low power consumption thanks to native power saving mechanisms including sleep modes should consume much less energy in comparison with Bluetooth or Wireless-Fidelity (Wi-Fi) of earlier standards b, g, a or n. This makes the electrical appliances operate longer on a battery.
- Long range (can penetrate walls more easily), the range of operation of various Wi-Fi standards is depicted in Fig. 1.

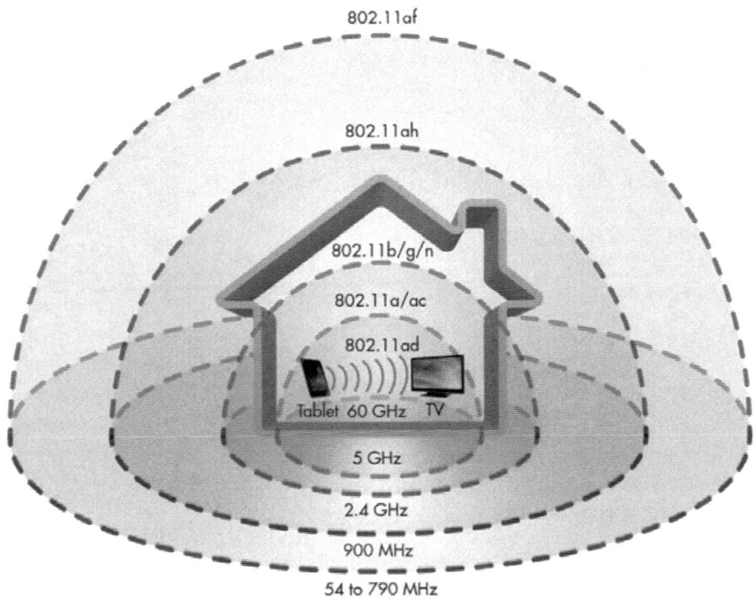

Fig. 1. Comparison of different Wi-Fi ranges [3].

Both key aspects mentioned above are based on the technological features of IEEE 802.11ah [4]:

- Sub 1 GHz frequency operation.
- Design of new Physical Layer (PHY layer) and Media Access Control Layer (MAC layer). These new layers include several modifications with respect to consolidated IEEE standards. The IEEE 802.11ah MAC layer incorporates most of the main IEEE 802.11 characteristics, adding some novel power management mechanisms.
- Typical range of IEEE 802.11ah is 100 – 1000 m.
- Transmission power is from <10 mW to <1 W (depending on the country's regulations).
- Battery operation should be from months to years (also thanks to long sleeping periods).

The mentioned standard is very new. It was standardized and introduced at the beginning of 2016 (in January 4 2016) [5] (the first IEEE 802.11 standard was released already in June 1997). The first certified devices should come soon (probably in 2017 or 2018). Due to these data, the topic of this study is potentially very important and interesting from the point of view of secure communication and transmission of sensitive data.

General security vulnerabilities together with basic introduction to the Internet of Things is also presented in other contemporary papers, for instance in [19].

3. Shielding

The first ideas of how to shield a communication in a room, in a small building or area considering wireless communication between electrical devices using IEEE 802.11ah is to use:

- A very simple Faraday cage or
- A very specific and hard to develop wallpaper reflecting only a desired frequency range.

3.1. Faraday cage

Faraday cages are named after the English scientist Michael Faraday. Faraday shield (cage) is an enclosure made from a conductive material or by a mesh of such material to block electric fields.

These shields – cages can be used to protect different kinds of electronic equipment from electrostatic discharges. They cannot block magnetic fields like Earth's magnetic field, but they can protect the interior from electromagnetic radiation coming from the outside. An external electrical field leads to rearrangement of the charges, and this cancels the field inside. Electric fields (applied externally) create forces on electrons in the conductor, creating a current, which will further result in charge rearrangement. The current will cease when the charges rearrange and the applied field inside is cancelled [6].

This approach is cheap but has several very negative side effects. First of all, whole frequencies coming to or from a cage are reflected, including:

- Global System for Mobile (GSM),
- Universal Mobile Telecommunications System (UMTS),
- Long-Term Evolution (LTE),
- All standards of Wi-Fi and its' frequency ranges (sub 1 GHz, 2.4 and 5 GHz),
- Bluetooth,
- And possibly also the visible light if not using a dense mesh.

This approach may go against the original aim to use IEEE 802.11ah in longer distances.

3.2. FSS

Frequency Selective Surfaces (FSSs) are important spatial filters, which can efficiently filter desired band of frequencies. Therefore these can play a significant role in electromagnetic related problems. Frequency selective surfaces can be used and adjusted to prepare a structure reflecting just a desired narrow range of a spectrum.

To briefly sketch the history, the beginning of FSS relates to Ben A. Munk who was the guru of this approach [7]. In the last decade, the idea of FSS has spread out into many applications. Example of a band-pass FSS is in [8] where the goal was to transmit GSM signals through energy efficient windows. One of the first FSS absorbers was presented by Salisbury and Jaumann [9]. Great research has been already done in the field of FSS including also the analysis of frequency characteristics of dielectric period structures [10] and another analysis of characteristics of dielectric grating of left-handed and right-handed materials [11]. FSS are also used in the antenna theory and experiments like analysis of ultra wide band planar monopole antenna and its design [12].

The second idea of how to shield the communication (using FSS) is to use a special pattern/wallpaper selectively attenuating just the frequency range used in IEEE 802.11ah.

With respect to the design, rules and law of various countries the frequency range for instance for Europe is 863 – 868 MHz, in USA it is 902 – 928 MHz and in China it is 755 – 787 MHz [13]. Fig. 2 presents the ranges in more detail.

The standard of IEEE 802.11ah is operating in sub-gigahertz frequencies in comparison with traditional IEEE 802.11b or IEEE 802.11g working at 2.4 GHz and IEEE 802.11a working at 5 GHz.

The schema of a typical FSS structure: simple cross and a Jerusalem-cross is presented in Fig. 3. Both models consist of simple rectangular elements. Theoretically, the second geometry may have better reflection. Moreover double-layer should provide a narrower band-stop filter. Several comparisons have been presented in [14].

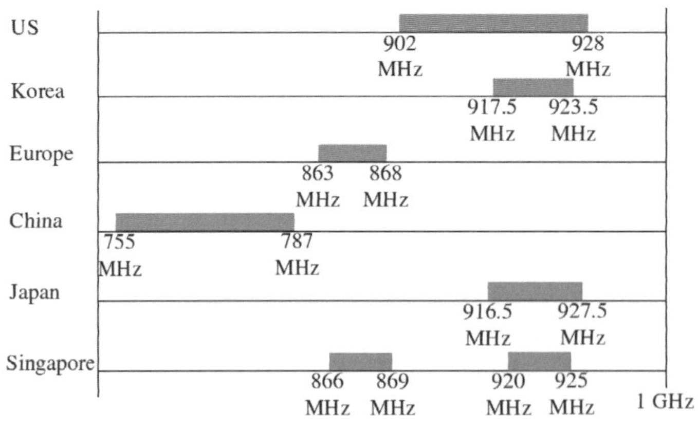

Fig. 2. Sub 1 GHz spectrum specified in the IEEE 802.11ah channelization [13].

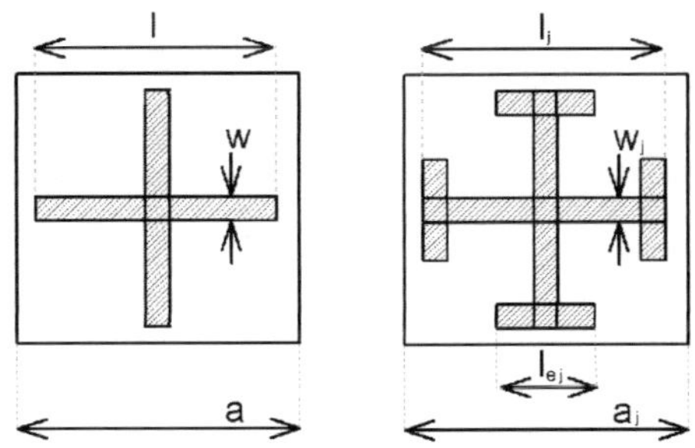

Fig. 3. Schema of a cell containing the simple cross (on the left) and the Jerusalem-cross (on the right).

In Fig. 3, a and a_j represent the width and height of a cell (a cell is just one square element of the whole structure of FSS; index j relates to the structure depicted on the right: the Jerusalem-cross), l and l_j is the total width and height of the cross, w and w_j is the width of an arm and l_{ej} represents the length of the bar connected to the end of an arm of the Jerusalem-cross.

There is a special software suitable for optimization of FSS elements. It is FSSMR software [15], which was developed at Tomas Bata University in Zlin and which analyses the planar periodic structures and tries to optimize them with respect to defined optimization goals. Therefore this software is suitable for estimation of proper values of design variables (a_j, l_j, w_j, and l_{ej}) to meet the optimization goals (and thus to reflect the desired frequency band in this case of IEEE 802.11ah).

There are also some shortages in this approach. One of the most questionable aspects of the FSS approach is the influence of the angle of incidence which must be also examined. Another problem is with windows when attempting to secure a room against transmitting sub 1 GHz frequencies outside the room.

4. Optimization

A frequency range, an initial geometry with design variables (e.g. width and height of the arms of the cross) and optimization goals must be set before performing the optimization of an FSS filter. The transmission coefficient depends on frequency and other parameters forming the parameter vector of the filter which specifies the geometry (defined by design variables). An optimization method searches for the set of parameters which satisfies the given objectives, at least approximately, being thus in a certain sense optimal. An optimization goal is defined by a frequency range where the transmission coefficient must be lower or greater than a defined threshold value.

This paper contains an experiment dedicated to shielding IoT communication in the United States of America where the dedicated range of frequencies is between 902 and 928 MHz. In our experiment three optimization goals were modelled (also graphically presented by green colour in Fig. 4 together with results of the initial configuration):

- To transmit frequencies from 0.1000 to 0.86924 GHz (threshold: -2.5 dB),
- To reflect frequencies from 0.86924 to 0.96075 GHz (this range relates to the Q factor equal to 10, $f_{stop\ range} = f_c / 10 = 91.5$ MHz; threshold: -20.0 dB),
- To transmit frequencies from 0.96075 to 3.0000 GHz (threshold: -2.5 dB).

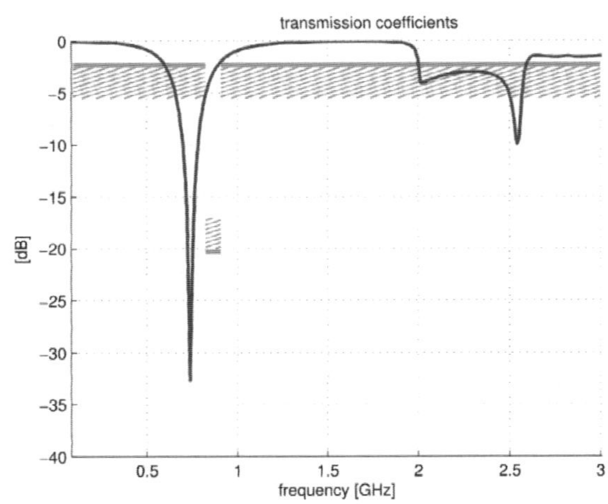

Fig. 4. Transmission coefficients of the optimized geometry

The initial values of design parameters with lower and upper bounds were guessed (by the earlier professional experiences with FSS filters) and they are mentioned in Table 1 where $l = ka$, $l_j = k_{1j}\, a_j$ and $l_{ej} = k_{2j}\, l_j$.

Variable	Description	Initial Value	Lower Bound	Upper Bound
a_j	The width and height of a cell [m]	0.15	0.1	0.2
w_j	The width of an arm [m]	0.0055	0.001	0.010
k_{1j}	The width parameter ($k_{1j} = l_j / a_j$)	0.85	0.7	1.0
k_{2j}	The length parameter ($k_{2j} = l_{ej} / l_j$)	0.35	0.2	0.5

Table 1. Table example

In this work, optimization was performed numerically using an implementation of local optimizer of Levenberg-Marquardt (a possible alternative is fmincon [16] or fminsearchbnd [17] which can be directly used in Matlab, evolutionary algorithms are another possible way).

The settings of the optimization process:

- Optimization technique: Levenberg-Marquardt,
- *FunTol* = 10^{-3}, this represents the threshold tolerance,
- *MaxIter* = 100, this constant defines the maximal number of iterations,
- *NormStep* = 0.06, this is the constraint on maximal norm of a step,
- Constraints on design variables a_j, w_j, k_{1j} and k_{2j} reflect the lower and upper bounds mentioned in Table 1,
- Cost function uses the method of moments [7, 18] to analyse and estimate the FSS transmission coefficients.

All computations by optimization were based on perpendicular angle of incidence only.

In this study, FSSMR software [15] was used. It was developed at Tomas Bata University in Zlin and it analyses the planar periodic structures and tries to optimize them with respect to the optimization goals.

5. Results

The optimization procedure results in a filter which suppresses well the whole range dedicated to the specific wireless communication. The attenuation is below -20 dB what means that the filter perform very well. The final transmission coefficients are presented in Fig. 5. The process of optimization in Matlab took at about 4 hours using an average computer (mainly depends on the desired amount of computed frequency points and the density of mesh cells in computation). The filter seems to reflect the part of the spectrum used in 802.11ah in the US region perfectly.

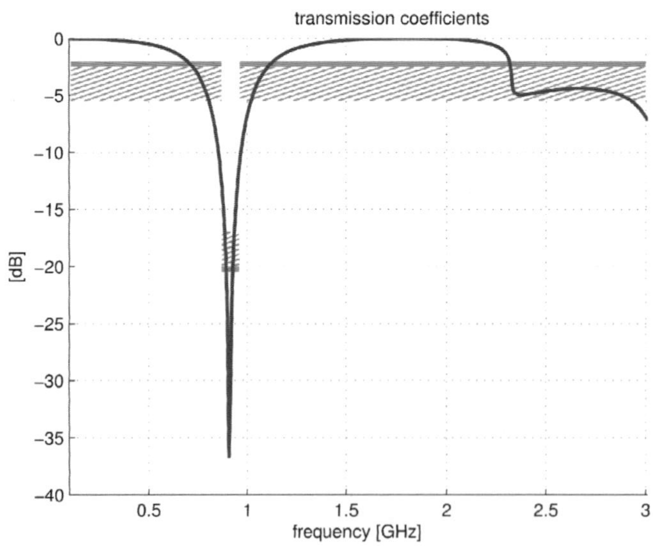

Fig. 5. Transmission coefficients of the optimized geometry of the FSS IEEE 802.11ah filter (US region).

There is also some kind of a subsidiary weak attenuation between 2.3 and 3.0 GHz. The optimized values of design parameters of Jerusalem-cross FSS filter are presented in the list below together with computed l_j and l_{ej}:

- $a_j = 0.129$ m,
- $w_j = 0.009$ m,
- $k_{1j} = 0.84$,
- $k_{2j} = 0.37$,
- $l_j = k_{1j}\, a_j$; $l_j = 0.1084$ m (the total width and height of the Jerusalem-cross),
- $l_{ej} = k_{2j}\, l_j$; $l_{ej} = 0.0401$ m (the total length of the outer arms).

6. Conclusion

A very new standard for wireless communication suitable for the Internet of Things, IEEE 802.11ah, has been introduced in this preliminary study together with analysis of possible ways of how to shield such a communication in the region of the United States of America. A theoretical concept of a wallpaper with a possibly deep practical impact has been revealed. A technique of how to generate an optimized geometry of an FSS filter with adaptation to the narrow specific range of frequencies used in IEEE 802.11ah has been presented. Also, some shortages of this approach have been described including unwanted subsidiary attenuation in higher frequencies and the limitation on perpendicular angle of incidence.

The boom of the Internet of Things is expectable. It can make life simpler (like other technologies in the recent history), but it also contains a great portion of a threat of abuse (also like other technologies in the recent history). This article tries to points this out.

Further analysis with practical experiments and measurements are proposed as a possible further work in this domain.

7. Acknowledgments

This work was partially supported by Internal Grant Agency of Tomas Bata University under the project No. IGA/CebiaTech/2016/004. This work was supported by the Ministry of Education, Youth and Sports of the Czech Republic within the National Sustainability Programme project No. LO1303 (MSMT-7778/2014).

8. References

[1] Tomasek, P. & Gona, S. (2013). Automated Design of Frequency Selective Surfaces with the Application to Wi-Fi Band-Stop Filter, Proceedings of PIERS Electromagnetics Academy, Cambridge, pp. 221-224, ISBN: 9781934142264

[2] Khorov, E.; Lyakhov, A.; Krotov, A. & Guschin, A. (2014). A survey on IEEE 802.11ah: An enabling networking technology for smart cities, Computer Communications, Vol. 58, pp. 53-69, ISSN: 0140-3664

[3] DeLisle, J.J. (2015). What's the Difference Between IEEE 802.11af and 802.11ah?, Microwaves and RF, Vol. 54, pp. 69-72, ISSN: 0745-2993

[4] Adame, T.; Bel, A.; Bellalta, B.; Barcelo, J. & Oliver, M. (2014). IEEE 802.11ah: The WiFi Approach For M2M Communications, IEEE Wireless Communications, Vol. 21, No. 6, pp. 144-152, ISSN: 1536-1284

[5] http://www.wifi.org/news-events/newsroom/wi-fi-allianceintroduces-low-power-long-range-wi-fi-halow, (2016). Wi-Fi Alliance(R) introduces low power, long range Wi-Fi HaLow(TM) (2016), Accessed on: 2016-03-30

[6] http://www.faradaycage.org/, (2016). Faraday Cage, Accessed on: 2016-03-30

[7] Munk, B. (2000). Frequency selective surfaces – theory and design, Willey & Sons, New York, USA, ISBN: 978-0-471-37047-5

[8] Kiani, G.; Olsson, L.; Karlsson, A.; Esselle, K. & Nilsson, M. (2011). Cross-Dipole Bandpass Frequency Selective Surface for Energy-Saving Glass Used in Buildings, IEEE Transactions on Antennas and Propagation, Vol. 59, No. 2, pp. 520-525, ISSN: 0018-926X

[9] Haupt, R. (2006). Scattering from Small Salisbury Screens, IEEE Transactions on Antennas and Propagation, Vol. 54, No. 6, pp. 1807-1810, ISSN: 0018-926X

[10] Mehrnejad, R. & Razmjoueian, R. (2012). Frequency characteristics of dielectric periodic structures, ACMIN'12 Proceedings of the 14th international conference on Automatic Control, Modelling & Simulation, and Proceedings of the 11th international conference on Microelectronics, Nanoelectronics, Optoelectronics, WSEAS, Stevens Point, Wisconsin, USA, pp. 187-190, ISBN: 978-1-61804-080-0

[11] Mehrnejad, R.; Razmjoueian, R. (2012). Characteristics of Dielectric Grating of Left Handed and Right Handed Materials, ACMIN'12 Proceedings of the 14th international conference on Automatic Control, Modelling & Simulation, and Proceedings of the 11th international conference on Microelectronics, Nanoelectronics, Optoelectronics, WSEAS, Stevens Point, Wisconsin, USA, pp. 205-208, ISBN: 978-1-61804-080-0

[12] Rahim, M.; Masri, T.; Majid, H.; Ayop, O. & Zubir, F. (2011). Design and analysis of ultra wide band planar monopole antenna, WSEAS Transactions on Communications, WSEAS, Vol. 10, No. 7, pp. 212-221, ISSN: 1109-2742

[13] Sun, W.; Choi, M. & Choi, S. (2013). IEEE 802.11ah: A Long Range 802.11 WLAN at Sub 1 GHz, Journal of ICT Standardization, Vol. 1, pp. 83-108, ISSN: 2246-0853

[14] Tomasek, P. (2014). Optimization of FSS Filters, International Journal of Circuits, Systems and Signal Processing, Vol. 8, No. 1, pp. 594-599, ISSN: 1998-4464

[15] Gona, S. & Kresalek, V. (2008). Development of a Versatile Planar Periodic Structure Simulator in MATLAB, COMITE, Vol. 14, No. 1

[16] http://www.mathworks.com/help/optim/ug/fmincon.html, (2016). MathWorks, Find minimum of constrained nonlinear multivariable function - matlab, Accessed on: 2016-03-30

[17] http://www.mathworks.com/matlabcentral/fileexchange/8277-fminsearchbnd--fminsearchcon, (2012). J. D'Errico, fminsearchbnd, fminsearchcon - File Exchange - MATLAB Central, Accessed on: 2016-03-30

[18] Wu, T.K. (1995). Frequency selective surfaces and grid arrays, Willey & Sons, New York, USA, ISBN: 0-471-31189-8

[19] Cvitić, I.; Vujic, M. & Husnjak, S. (2016). Classification of Security Risks in the IoT Environment, Proceedings of 26th DAAAM International Symposium, Vienna, Austria: DAAAM International, pp. 731-740, DOI: 10.2507/26th.daaam.proceedings.102

DOI: 10.2507/27th.daaam.proceedings.122

APPLICATION OF WIRELESS MULTIMEDIA SENSOR NETWORKS FOR GREEN BORDERLINE SURVEILLANCE

Astrit Hulaj, Adrian Shehu & Xhevahir Bajrami*

This Publication has to be referred as: Hulaj, A[strit]; Shehu, A[drian] & Bajrami, X[hevahir] (2016). Application of Wireless Multimedia Sensor Networks for Green Borderline Surveillance, Proceedings of the 27th DAAAM International Symposium, pp.0845-0853, B. Katalinic (Ed.), Published by DAAAM International, ISBN 978-3-902734-08-2, ISSN 1726-9679, Vienna, Austria
DOI: 10.2507/27th.daaam.proceedings.122

Abstract

The security of the state borderline is one of the global concerns and main goal of every country. Intensive human involvement in application of conventional systems for border patrol is one of the serious problematic issues.
In recent times, border security systems in their composition include high-tech equipment, such as: various unattended sensors, unmanned aerial vehicle, surveillance towers equipped with cameras, etc. However, these systems are faced with unsolvable problems, as regards their application along border areas with difficult terrain and covered with high forests. In this paper we will give ideas for a new system, which can be placed alongside the green borderline with a much cheaper cost and which could offer better results. In the new system for the green borderline surveillance, we propose the application of advanced sensors technology, respectively multimedia wireless sensor nodes (MWSN). This technology will have direct impact on reducing the involvement of human factor (police patrols) to the security of the green borderline and will impact on elimination of false alarms.

Keywords: WMSN; borderline; security; sensor; surveillance; architecture.

1. Introduction

Although today, in many countries of the world the trans border security has increased, yet in the world has enough conflicts and problems regarding protecting and securing of the state borders, such as Iraq, Syria, Afghanistan, etc. Problems with securing the borders have not only countries such as: Iraq, Syria, Afghanistan, but also developed countries of the world, such as: EU states, the USA, Turkey, etc., but these countries, unlike the first countries, their security border is threatened from: illegal border crossings of migrants, smugglers of drugs and weapons, etc. Therefore, each state depending on economic conditions and geopolitical factors creates strategies and policies for security and monitoring of its national borders. The security of national borders is a key element of the sovereignty of the country. Each state may have borders that are characterized by challenges that are different with those of other countries. Therefore, also the surveillance system located along the border of a country can't be appropriate for another country.

In this paper we shall analyze the challenges by which the green borderline surveillance is characterized. Today

* Corresponding author. E-mail address: xhevahirbajrami@gmail.com

still many countries, realize the green borderline surveillance through conventional systems. With other words, the border surveillance is carried out by police or military patrols. These patrols carry out borderline control physically in designated time intervals and places. This kind of surveillance, except that can't provide continuous control of the border, is also very troublesome. Border security, according to this method of surveillance depends directly on the human factor. The objective of this research is to identify possible ways of increased security and efficiency of border control along the green borderline with a much cheaper cost. In this paper, firstly we will define problems and analyse possible scenarios of the green borderline. Then we will analyze techniques and existing surveillance systems which are applicable. In the third section we will offer the concrete proposals for solutions, and technology architecture proposed for borderline surveillance. In the end we will analyze the challenges by which the application of proposed technology is characterized.

2. Problem formulation and analysis of possible cases of green borderline.

2.1. Problem formulation

Today, almost all countries of the world are being faced with illegal border crossings. Illegal border crossings can be for different purposes, such as: illegal immigration, terrorism, smuggling of arms, drugs and other things of value, guerrilla (destabilization of neighbouring country), etc. Depending on the nature of illegal border crossings and roads used for illegal border crossing, each country tries to undertake measures to prevent such illegal border crossings. Therefore, border control is vital to the security of a nation and its citizens [1]. One of the technologies that are being considered to be applicable for border surveillance is the sensors technology. However, the application of this technology for the green border surveillance continually is characterized by some significant challenges such as:

- Power supply of sensor nodes;
- False Alarms (reliability);
- Failure of any of the sensor nodes and the difficulties of its replacement, etc.

Regarding these challenges, today many authors have tried to provide solutions and different proposals for these challenges in order that the sensor technology can find application for green borderline surveillance. Although today, there are some papers that have addressed these problems, yet these are the main challenges that have affected the sensor technology not to find wide application.

2.2. Analysis of some of potential cases of green border.

With state borderline is implied the line that separates the territory of a state, with other neighbouring states. This is also the state borderline. State borderline of a country can be divided into: terrestrial, water and air. **Terrestrial border** - also known as "green border" and means any borderline between that country and neighbouring countries, with the exception of official border crossing points. Green border, in many countries constitutes the main part of the border. This part of the border in many states is more problematic border, regarding application of surveillance and security systems. Before we begin with the discussion of border surveillance techniques, initially we will analyze several potential scenarios, along the green borderline such as:

Scenario 1 - These areas are more difficult for surveillance because they are composed of mountains, rocks and grasses, and are characterized by steep terrains. In such a scenario presented in the Fig. 1, the leaves will create very difficult for surveillance of the borderline by air or through long-range cameras of operation but will be very good shield for illegal border crossings.

Fig. 1. Scenario with leaves, rocks, and grasses

Criminals consider the difficulties that have the security authorities for surveillance of these areas often can be the target of their illegal border crossing. Therefore, the borderline as in the case presented in Fig. 1, entering in one of the most difficult areas for surveillance.

Scenario 2 - In the scenario presented in Fig. 2 we have presented another case of possible along the green borderline. As can be seen in Figure 2, in this scenario we have to do deal with the case where most of the border is characterized by rocks and grasses. So, as shown in Fig. 2, compared with scenario 1 and 3, this border area is stripped from forests. However, the border areas are enough steep which includes holes that are enough difficult to be monitored from the ground surveillance systems, as well as by air systems.

Fig. 2. A possible scenario along the borderline with rocks and grasses

Scenario 3 - In this scenario is presented another case of possible illegal border crossing along the green borderline, where areas along the border are covered with high forests. Such a case is respectively shown in Fig. 3, 3a and 3b. In Figure 3b, is presented a case where along the borderline, the terrain is partially flat, but covered with high forests. While, in fig. 3a is presented approximately the same case, but in this case the border area is steeper.

For the scenario presented in Figure 3, the green borderline surveillance through systems or equipment with a long-range of the operation is almost impossible. Also, the surveillance by aircrafts is unsuccessful. Namely difficulties of monitoring these areas can make these areas very attractive for criminals. Should be noted that these areas are characterized by very dense forests and many parts are almost impassable. However these areas are inhabited by very different animals that live and which create many paths that can be easily used by criminals for entering inland the country, in order to achieve their objectives.

a) A scenario along the green borderline between Kosovo - Albania

b) A possible scenario along the green borderline with very dense forests

Fig. 3. Possible scenarios along the border with high forests

Scenario 4 - In Fig. 4 is presented another possible scenario along the borderline. From Figure 4 we can see that in this case we are dealing with a borderline even and without forests. So, this is the easiest scenario possible along the borderline. In this case compared with the other scenarios, observation and security the borderline, is much easier by the Border Security Authorities. Along such a borderline can be installed and applied various technology and security systems. In this case, in other words there are no restrictions concerning the environment, compared with other cases presented in Figure 1-3, where the deployment of technologies requires commitment and high costs, and in some cases application it is almost impossible.

Fig. 4. Another possible scenario along green borderline

In this section we tried to analyze some possible scenario along the green borderline from the easiest one for surveillance (Fig. 4) to the surveillance of most difficult ones (Fig. 3) from the Border Security Authorities.

3. Existing techniques and systems for surveillance of the borderline

The security of the state border has existed since the existence of the state itself. However, the techniques of state border security has changed with time. Few years ago, security and surveillance of the state borderline has been conducted by military and police patrols. Border security is based entirely on the human factor. In this case for a full coverage of the borderline through police or military patrols it needed a commitment of a considerable number of patrols and support staff. Increasing the number of police or military patrols, supported mainly on human factor directly affects the growth of the cost. Besides cost, border control is realized in certain periods of time. So that, full control of borderline in this way is very troublesome, problematic and not to the very high level of accuracy, because the criminals very easily can detect the movements of border patrols and when they have space for illegal border crossing of the borderline, they can easily realize it.

By technology development, also the border control techniques started to change. So the state border management concept started to change itself. Therefore in order to increase the efficiency of border patrols along the borderline the technology is included as part of the border patrol. Some equipment used as part of border patrols are: various communication equipment, unmanned aircraft for detection of the illegal border crossings, mobile vehicles equipped with advanced technology, etc. Some of the equipments used by border patrols are shown in fig. 5.

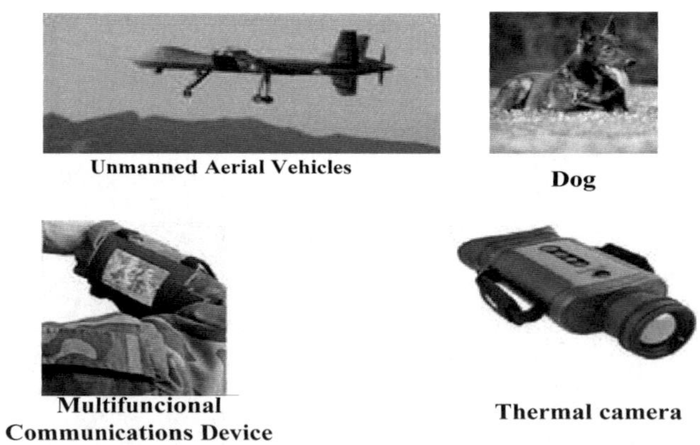

Fig. 5. Some equipment as part of the border patrol

A very big assistance for border patrols has given the use of unmanned aircraft. With the better coordination of the activities of border patrol and surveillance aircraft can prevent or minimize the illegal border crossing by criminals.
The use of unmanned aircraft for aerial surveillance of the border has affected in easier detection of illegal border crossings along the borderline and even after crossing the border enabled tracking of criminals. Due to the large coverage and high mobility that have these aircraft, intensive human involvement in surveillance activities, border patrols can be reduced, respectively. So the use of these aircraft directly affects in the reduction of border patrols. Thus, human resources may be engaged in other tasks, such as those of the management and analysis of information collected from equipment used for border surveillance.

However, similarly with the conventional border patrol systems, unmanned aircraft alone can't completely cover the borderline at any time. In this case, there may be times when certain parts of the border will not be monitored by such devices [2]. These aircrafts can't constantly keep under surveillance the borderline or must engage a large number of them. Also, the involvement of aircraft for surveillance of borderline for the case of the scenario 1 and 3, doesn't give good results because very easily criminals can cross the border without being detected by the aircraft. In other words, the engagement of aircraft for border surveillance, offers enough uncovered spaces that the criminals can use for border crossing. Moreover, unmanned aircraft have considerable cost and requires very specialized staff in monitoring the activities of these aircraft. In addition, bad weather conditions such as rain and snow may have the effect in limiting of their supervision skills.

Taking into consideration these issues that we have addressed, we could conclude that the use of these unmanned aircraft, nevertheless to the high cost, allow enough areas that can't be under surveillance and which can be used by criminals for illegal border crossing. In order to conduct the best possible surveillance of the border line the surveillance towers with long-range operation are used. However, these surveillance towers are suitable for areas where there is visibility. These towers are suitable for those areas that are approximately as in Scenario 4. Surveillance towers are not suitable to apply in those cases as in Scenario 1 and 3. Therefore in order to find a solution for those cases where existing technologies do not provide solutions, in last decade the use of WSN technology along the border line is meant to be used. This technology, from some authors, is intended as additional technology of existing technologies. Today, regarding the application of WSN technology along the borderline, we can find some research and concrete proposals.

In [2], hybrid architecture of wireless sensor networks is presented for application along the state borderline. This system architecture of BorderSense has three layers. The unattended ground sensors and the underground sensors constitute the lower layer of the architecture, which provide higher granularity for monitoring. At the second layer, surveillance towers improve the accuracy of the system through visual information. Finally, mobile ground robots and unmanned aerial vehicles constitute the higher layer that provides additional coverage and flexibility. This architecture is appropriate to be applied, mainly in those border areas, as in Scenario 4 or similar. Namely it is appropriate in those areas that aren't characterized by steep terrain, and covered with dense forests, as in Scenario 1, 2 and 3. The application of such a system for scenario 1, 2, 3 is a cost, almost unbearable and very equipment included in this system are unnecessarily.

In [1], authors have studied the Turkish border security system via simulation to identify possible ways of increasing border control and security along the land borders.

In [3] authors FemtoNode and the adaptable middleware platform for military surveillance have presented. In order to illustrate the use of the proposed platform infrastructure, i.e. the customizable FemtoNode and the adaptable middleware, from authors an area surveillance application is studied. In this application, low-end sensors nodes are scattered on the ground along a borderline. In case of an unauthorized vehicle crosses the borderline limit, the sensors issue an alarm which will trigger the use of Unmanned Aerial Vehicles (UAVs) equipped with a more sophisticated sensors, such as radars or visible light cameras, in order to perform the recognition of the vehicle. However, this studied platform isn't appropriate to apply for the borderline cases that are presented in scenario 1, 2, and 3.

In [4], authors have proposed a Energy-efficient routing algorithm, for the case when WSN are applied along the borderline. Routing algorithm will be extended to sleep mode and therefore a longer network lifetime can be achieved. Then the Degree of Aggregation algorithm DOA means the minimum number of reports about an event that a leader of a group waits to receive from its group members, before reporting the ship location to the base station.

In this section we analyzed some of the papers which dealt with the application of WSN along the borderline, and the problems of their application. However, it should be noted that today there are many other papers that dealt with different problems of WSN, but which we haven't analyzed in this paper.

4. Results and discussion

In this section we will propose a new method of application of sensor technology along the green borderline, which will provide an increase security performance along the state green borderline. As per discussion, existing technologies do not provide an appropriate solution for increased security along the green border line. Moreover the application of existing systems along scenarios 1, 2, and 3, are with a high cost or almost impossible to be applied.

If we analyze scenarios 1 and 3 we can see that we have to deal with the border areas which are characterized by high and very dense forests along which is very difficult to pass. However, through these forests may have multiple paths that can be created from wild animals which can be present in such forests. Also, these crossing paths can be created by human factor for various purposes of exploitation of these forests. These paths created by animals or human factor can be used easily by criminals for smuggling. On the other hand, these paths will be almost impossible to be monitored 24 hours from conventional border patrol or unmanned aircraft. Also application of surveillance towers is unsuccessful because through these towers is impossible to be realized surveillance of such paths as a result of high forests.

Unlike the scenario 1 and 3, scenario 2 is characterized by steep areas, but without high forests, where the visibility is better. In the areas that are of the type of scenario 2, we can also apply to existing technologies analyzed in this paper, such as surveillance towers, unmanned aircraft, etc.

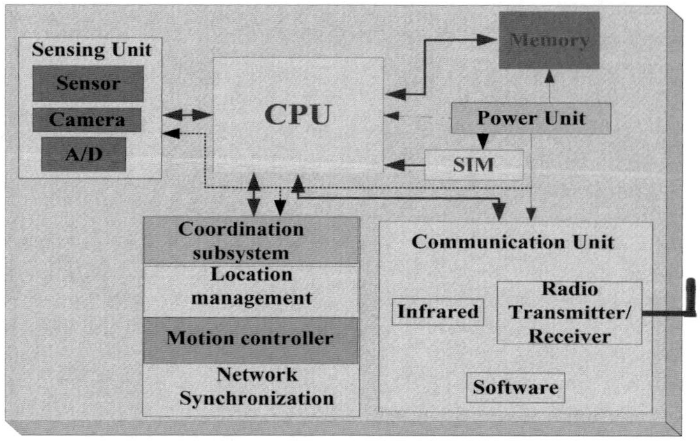

Fig. 6. The architecture of a multimedia sensor node.

In order to monitor these paths along the forests with very steep terrains and high forests or for monitoring the green borderline, in this paper we propose the application of a new method. In this new method for surveillance of these paths and green border line, WMSN is used. Communication WMSN based on IEEE 802.15.4 protocol [5, 6]. The architecture of a WMSN is presented in fig. 6. This node consists of: **Sensing unit** - this unit include: A / D, scalar sensor and thermal camera. The function of this unit is the capture of images in space observed by the camera; **Communication Unit** - This unit enables the communication of sensor node with other nodes in the network. Also, this unit allows the sensor node that information captured from the surroundings to send to the desired destination; **Power Unit - Battery -** This unit serves as an energy source for other units of the node; **Processing Unit** - This unit includes the CPU and memory. This unit enables information processing and storage of information; **SIM card** - it could be any of the operators of telephony, etc.

The images captured by the sensor nodes, routing through the network to the nearest police station, and then sent to The central security monitoring room [7, 8] for analysis of information. After analyzing the information collected, depending on the results of the analysis has been realized by sending information relevant to patrol units. The network architecture for WMSN along the green borderline is presented in the figure 7.

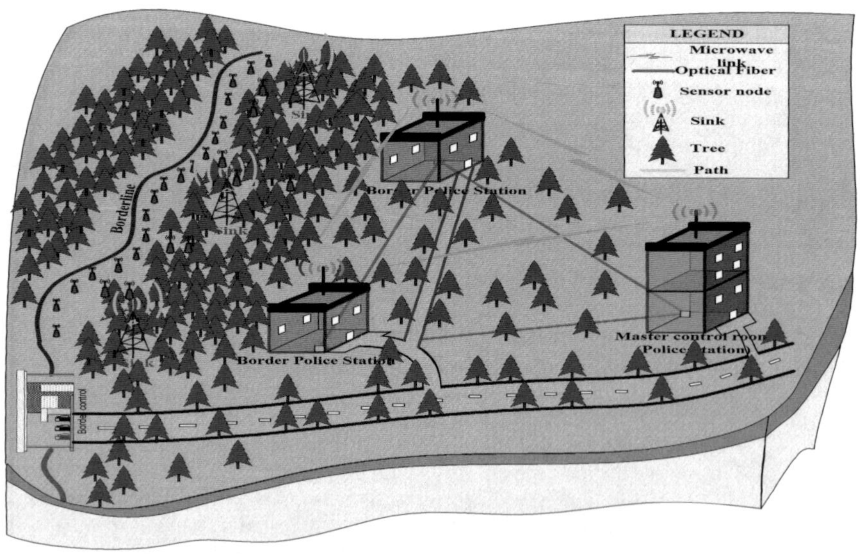

Fig. 7. The proposed network architecture for WMSN

From Fig. 7, we can see that the network architecture is divided into three parts. The first part includes the green borderline, or end network. The second part includes the end buildings of the police stations, which are nearest along the borderline. The third part is the central monitoring room. The communication Node-Sink-Police Station is realized through microwave network. Communication between Police stations can be realized through optical or microwave network.

Application of WMSN for the purpose of enhancing security along the borderline, affects the reduction of other sensor types. If we analyze, scenario 1, 2, and 3, we can see that in those areas, it is impossible for vehicles to cross. The opening of underground tunnels is also something unaffordable by interest groups. Therefore, it isn't necessary for the application of seismic sensors, along the green border. Since we propose the application of multimedia sensor, also it isn't necessary for the application of pressure sensors.

Placement of such a system, along the green borderline, can have direct impact in the efficiency of border police or military patrols through continuous information about the possible illegal border crossings. The intervention patrols will be realized only after the captured images are analyzed. This technology eliminates false alarms, which have been a major concern of sensor nodes. On the other hand, the cost of application of this technology will not be very high. WMSN could be placed along the borderline completely or only in those hot spot areas and paths where border authorities determine that can have illegal border crossings. A deployment of such technology directly affects reducing border patrols.

4.1. Area of coverage of a wireless multimedia sensor node

In this section we will analyse the coverage of the borderline through WMSN distribution. The radius of coverage of a multimedia sensor node depends on the type of sensor, the purpose and place of deployment. However, we will examine the potential coverage area of a WMS node which is characterized by a rotation angle of 360 degrees. These sensors offer a multidimensional coverage. The area which is covered directly by the sensor node is called as Field of View (FOV). The radius of coverage of a sensor node and FOV are shown in fig. 8. FOV is calculated by the formula:

$$FoV = 2 * \operatorname{atan}(\frac{h}{2r}) \quad (1)$$

Where: h - is the horizontal FOV; r - radius of coverage.

The proposed multimedia sensor node for application is considered to have a rotation angle of 360 degrees in the horizontal direction and an angle of rotation of 35-108 degrees in the vertical direction.

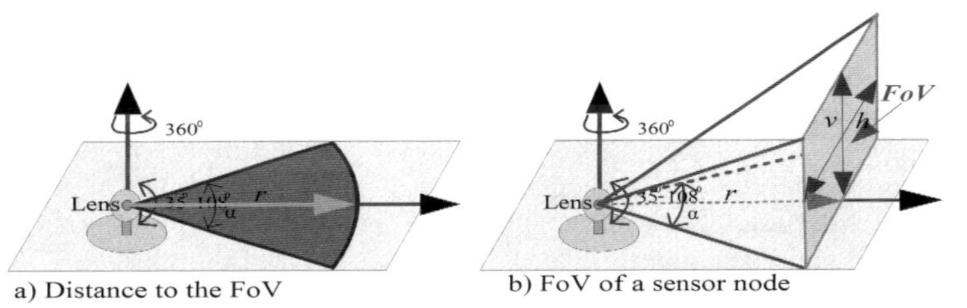

Fig. 8. Schematic presentation of the distance and FOV of a multimedia sensor node

For the monitoring of green border, through the application of multimedia sensor nodes, we propose the distributing sensor nodes, as in Figure 9. This case foresees that the sensor nodes should be deployed in a strip along the borderline, one after another. Each sensor node is responsible for detecting, analyzing and reporting on movements across the area covered. Formula to find the necessary number of sensor nodes to cover the border line is:

$$d = 2r * n \Rightarrow n = \frac{d}{2r} \quad (2)$$

Where: r-radius of the sensor node; n - the number of sensor nodes; d- length of the borderline.

If a sensor node has a coverage radius of 13 m. To cover a distance of 1 km of the border line with node sensor, the required number of sensor nodes is 39. This logic is not valid for coverage paths along the forest. Placement of sensor nodes, along paths is realized depending on the need and forecasting of border authorities.

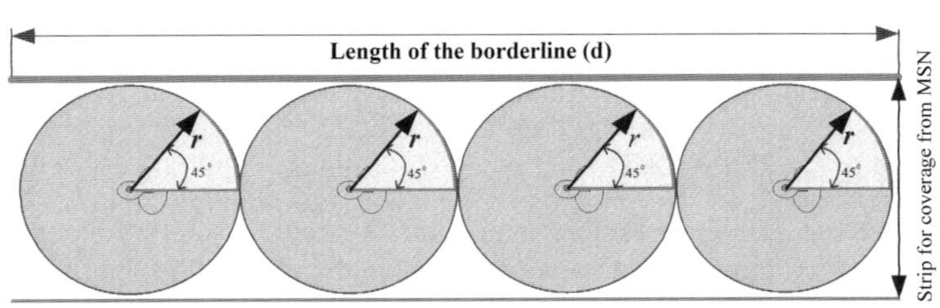

Fig. 9. Sensor nodes deployed in a strip along the borderline

4.2. Challenges by which the WMSN application is characterized

Placement and functioning of WMSN along the border line or in specific areas along the border line is characterized by several challenges, but we will discuss some among the most important, such as:

- Network infrastructure;
- Placement and maintenance of sensor nodes;
- Power supply of sensor nodes;
- Damage of sensor nodes from animals or human factor (criminals), etc.

Network infrastructure - for multimedia sensor node operation, initially border areas should be covered by the network infrastructure. For the case scenario 1, 2, and 3 of the border area coverage network, are easier through any of the wireless technologies.

Multimedia sensor nodes placement - placement of NMS should be realized by human factor, manually. Thus, placement and maintenance of sensor nodes, along steep areas is difficult. Placement the sensor nodes can be realized in these areas, along the borderline, specific areas along the borderline or in certain parts across paths. The place and position of the sensor node placement must be determined and realized from the experts of the border authorities.

Power supply of sensor nodes - lifetime a sensor node is directly dependent on the power supply (battery node). On the other hand, sensor nodes should be placed along the border areas that can't be accessed by military machinery. Therefore, manual battery replacement sensor node is one of the main challenges. With the aim of saving of energy, sensors should stay in inactive state during the time when there isn't movement and active state at the moment when are detected movement in the area covered by the sensor node. Also, captured images from the sensor nodes should be converted to black and white images "in press" [9] so that power for transmission of image and loads of transmission link to be minimal. In this case, we will have a minimal consumption of electrical energy from sensor nodes.

Damages to the sensor nodes from the animal or the human factor - can happen that the sensor nodes are damaged by the animals during their movement through paths from one side to the other side of the borderline. However, damage of the sensor nodes can also occur by the criminals so that they can realize crossing the border without detected of border security authorities. So should these sensor nodes to be replaced by border security authorities, but the replacement can't be done immediately. But usually the placement of sensor nodes should be realized in such a way that each sensor node should protect the other sensor node. So if a node is damaged, other node which is located in the same path, but in another position can replace the damaged node.

We should note that although the proposed system and technology is characterized by several challenges, the impact of this technology in increasing the efficiency of border patrols in preventing crossing of criminals along the borderline is considered to be satisfactory. Also, the design and implementation of such a system for surveillance of the green border can be realized in not such a high cost. On the other hand, this applicable proposed technology resolves the problem of false alarms (false alarms can be verified immediately) which has been one of the main factors that have prevented the application of sensor technology for the purposes of border security.

By comparing existing systems for surveillance the state border with the proposed system and method we can see that the proposed system compared to other systems and technologies offers a very suitable solution. Moreover, the proposed system, except that can be applied with a low-cost, offers choice for many existing concerns regarding application of sensor technology along the green border.

5. Conclusion

In this paper, we presented a new system and method of surveillance the green borderline for border patrol, to reduce intensive human involvement and to improve the detection accuracy of illegal border crossings. This proposed architecture includes application of WMSN having a direct impact on the elimination of the problems that have existing systems for green borderline surveillance.

Application of WMSN effects on increasing the efficiency of border patrols to the detection of illegal border crossings and impacts on reducing of intensive patrols. In other words, the technology proposed WMSN border surveillance affects the elimination of problems that sensor technology has regarding the large number of false alarms. Patrols will be activated and will react only after they have received and verified images captured from sensor nodes.

In the future we will design an algorithm which will enable multimedia sensor nodes to be able to make the difference between animals and humans. This will have direct affect in electricity saving to the sensor nodes and also will affect the reducing of load transmission link. Also, after the designing of the algorithm in order to test the system performance, the real tests will be conducted.

6. References

[1]. Çelik, G. and Sabuncuoğlu, İ. (2007). Simulation modelling and analysis of a border security system. European Journal of Operational Research, Vol. 180, No. 3, pp.1394-1410. Ankara, Turkey. Elsevier.

[2]. Sun, Z., Wang, P., Vuran, M.C., Al-Rodhaan, M.A., Al-Dhelaan, A.M. and Akyildiz, I.F. (2011). BorderSense: Border patrol through advanced wireless sensor networks. Ad Hoc Networks, Vol. 9, No. 3, pp.468-477. Elsevier.

[3]. Freitas, E. P., Allgayer, R. S., Wehrmeister, M. A., Pereira, C. E., & Larsson, T. (2009). Supporting platform for heterogeneous sensor network operation based on unmanned vehicles systems and wireless sensor nodes. In Intelligent Vehicles Symposium, pp. 786-791. IEEE.

[4]. Babu B. Swaminathan A, D. C. Joy D.C. Boarder Analysis with Ensora and Doa Using Wireless Sensor Networks. International Conference on Emerging trends in Engineering and Technology. ISSN: 2248-9622, pp.76-83.

[5]. Nedelcu, A. V., Duguleana, M., & Sandu, F. (2014). Evaluating the Energy Overhead Generated by Interferences within the 2.4 GHz Band for a Hybrid RFID Network. 24th DAAAM International Symposium on Intelligent Manufacturing and Automation, 2013. Procedia Engineering,Vol. 69, pp. 210-215. Elsevier.

[6]. Cvitić, I., Vujić, M. and Husnjak, S., 2016, January. Classification of Security Risks in the IoT Environment. DAAAM international symposium on intelligent manufacturing and automation. ISBN 978-3-902734-07-5. ISSN 1726-9679. pp.0731-0740. Vienna, Austria

[7]. Shehu, A. and Hulaj, A., (2013). The analysis of delays in the network for video and voice applications through OPNET software package", Advances in Circuits, Systems, Telecommunications and Control, ISBN: 978-960-474-341-4. France, Paris. WSEAS.

[8]. Hulaj, A., and Shehu, A., (2014). Optimization of network delays through implementation of EIGRP routing protocol, International Journal of Computers and Communications, Vol. 8. NAUN.

[9]. Shehu, A., Hulaj, A., and Bajrami, Xh., (2016). An Algorithm for Edge Detection of the Image for Application in WSN, Applied Physics, System Science and Computers (APSAC 2016), Dubrovnik, Croatia, September 28-30, "in press" Springer.

DOI: 10.2507/27th.daaam.proceedings.123

ANALYSIS OF SETTLEMENT OF FOUNDATION PLATES BY FINITE DIFFERENCE METHOD

Maja Prskalo & Anton Vrdoljak

This Publication has to be referred as: Prskalo, M[aja] & Vrdoljak, A[nton] (2016). Analysis of Settlement of Foundation Plates by Finite Difference Method, Proceedings of the 27th DAAAM International Symposium, pp.0854-0859, B. Katalinic (Ed.), Published by DAAAM International, ISBN 978-3-902734-08-2, ISSN 1726-9679, Vienna, Austria
DOI: 10.2507/27th.daaam.proceedings.123

Abstract

This paper provides an analysis of settlement of foundation plates using finite difference method, and is designed in the aim to prove how this method, implemented in our Matlab program code, will give results similar to those obtained with the finite element method, and to show that plate' settlements are less dependent on the change of modulus of soil reaction.

Keywords: foundation plate; finite difference method; settlement; modulus of soil reaction; Matlab.

1. Introduction

To do the high-quality calculation and reliable performance of the foundations of the building, it is necessary, among other things, to resolve the interaction of the object, foundations and soil during all phases of construction. In all these cases, if the foundation feet are too close, or if they overlap, or if we have to reduce the differential settlement, we have to determine for the foundation plates. The mechanical properties of the soil are very complex and the solution to the problem of interaction between foundation soil with the structural elements requires certain simplifications, because without them the problem will be unsolvable or economic unprofitable. With this in mind, the greatest number of solutions is developed for the soil as linearly elastic, homogenous and isotropic half-space [5]. These simplifications and the improvement of the computer technology allow us the use of the numerical methods for calculation of the stress as well as for settlement of foundation plates [4]. In this paper we will use finite difference method (FDM), the oldest numerical method for solving partial differential equations [3], [4].

2. Previous research

Various theoretical and numerical procedures are developed to obtain approximate solutions of real distribution of stresses in a structure-soil interface. A practically reasonable approximation of the actual state is achieved by developing numerical procedures and thoroughly investigating deformation properties of soil [7].

The problem of numerical resolution of partial differential equations, or finding and/or improving the algorithms for solving them is studied in numerical analysis - a branch of numerical mathematics. The most famous numerical method for solving such problems is the finite difference method (or mesh method) which idea is to approximate the derivatives of function (of one or more variables) with the divided differences. There are several other methods in use, such are collocation method, Galerkin method, Ritz method and finite element method.

Finite difference method replaces the main differential equation with the system of algebraic equations that links shifts of observed points relative to neighbouring points. These points are distributed in a grid/mesh of the surface of the plate. The method was first formulated (introduced) by Allen and Severn, and is described in different publications.

3. Numerical method - finite difference method

Finite difference method replaces the main differential equation with the system of algebraic equations that links shifts of observed points relative to neighbouring points. It is placed in every point of the grid/mesh on the surface of the plate, thus a system of algebraic equations is formed, the system based on a rectangular coordinate system, and on a load, as well as on the physical and elastic constants of the loaded plate. This load-deformation equation establishes a linear relationship between the load at the focal point and the unknown value of the deformation of the plate at that point, but in the other active points too, so by solving these equations the approximate values of the deformation of the plate in the active points of the mesh are obtained. Refinement of the mesh will result with increasing of the order for matrix of the system, and with growing of the complexity for the algorithm which would be almost impossible to solve without using a computer [1].

Figure bellow is showing a division/distribution of the surface of the plate with grid lines, what is resulting with rectangular elements of the grid/mesh. In direction, we are having a interval between grid lines h, and rh in x direction. Places of intersecting for two grid lines we will call grid/mesh points. Some of them are marked with letters (N, E, S, W,...) and are called active points, and a central/middle point of an active group of points is called the focal point (Z). To simplify the writing of equations, the marks of the active grid points are used also as the size of settlement w, and the middle plains are observed in specific grid point on the plate.

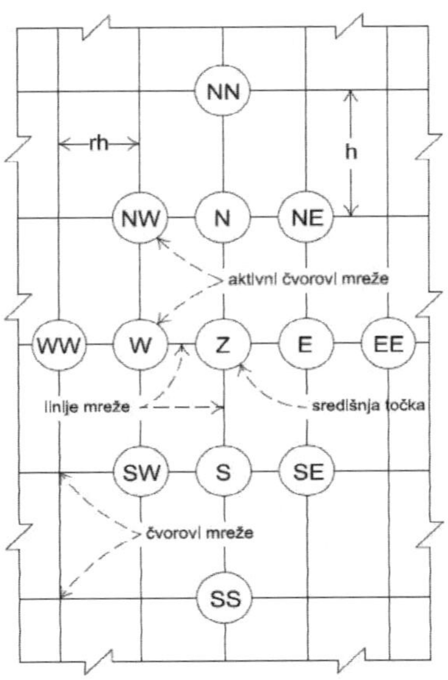

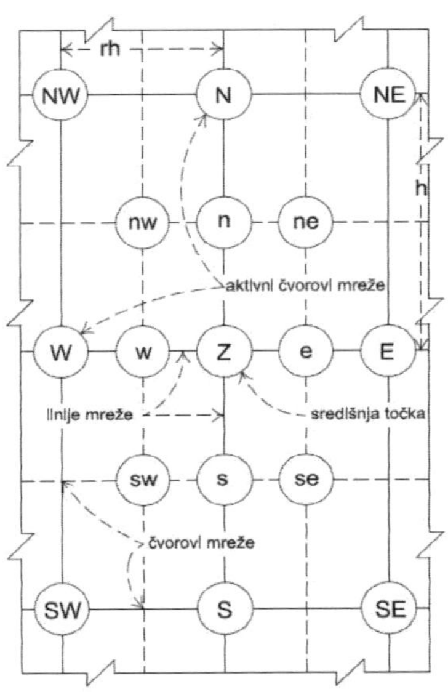

Fig. 1. The system of marking for the grid points

By using a standard procedure of finite difference method, then by including a value of deflection of the plate in the form of designation of active grid points and finally by arranging, the Lagrange differential equation will translate into the following equation:

$$\frac{1}{r^4}(WW - 4W + 6Z - 4E + EE) + \frac{2}{r^2}(NW - 2N + NE - 2W + 4Z - 2E + SW - 2S + SE) + (NN - 4N + 6Z - 4S + SS) = \frac{P_z h^2}{rD} \quad (1)$$

Equation (1) is the basic load-deformation relation to any internal active point. Next, this general form of the equation is being used for special cases which involving the boundary cases, as well as for all relations connecting the deformation of the plate with the loads, momentums, reactions and shear [2]. Essentially, this load-deformation equation establishes a linear relationship between the load at the focal point and the unknown value of the deformation of the plate at that point, but in the other active points too, so by solving these equations the approximate values of the deformation of the plate in the active points of the mesh are obtained.

4. Determination for the modulus of soil reaction

The foundation of the building/object is defined as an integral part of the structure that is in direct contact with soil or rock massif and which receives and transmits the load from the building to the soil/ground or rock massif. Due to object's load the soil is being deformed, and while deformed it returns the reactive load to the building structure. The soil and the structure of the building are being deformed together, which is why it is necessary to consider projecting/designing the foundations together with the construction of the building and vice versa. Their mutual consideration and design leads to more favourable technical and economic solutions for entire building/object [8]. During the analysis of the foundation plate, the soil can be observed in two ways:

- as a group of springs, according Winkler's theory;
- as a continuum, which is usually observed as elastic half-space.

The modulus of soil reaction can be determined in the following ways:

- by using circular plate;
- by using a table of typical values and correlations;
- by calculating of settlement.

The usual way of obtaining for modulus of soil reaction is by applying loads to the soil through a circular plate with a diameter of 30 cm. The disadvantage of this method is regarding the load, but only for the layer of lower thickness (lower in comparison to the layer loaded with the foundation plate). The modulus of soil reaction can also be estimated on the basis of predicted settlement for real foundation. In this case it can be assumed that the foundation is solid and under pressure corresponding to the total load on the foundation. The settlement is projected by standard methods of soil mechanics. In this case, the modulus of soil reaction we can calculate as follows:

$$k = \frac{\bar{q}}{\bar{w}} \tag{2}$$

where $\bar{q}$ is average pressure, and $\bar{w}$ is settlement for solid foundation.

This procedure can take into account the different layers exposed to different levels of stress, what was not possible in the previous known methods. In addition, thanks to short-term and long-term predictions for the settlement, we are able to get modules for short-term and long-term analysis of the behaviour of the foundation, taking into account both, the consolidation and creep of soil. It is difficult to find a relation between k and Young's modulus for elastic half-space because the answer depends on the stiffness of the foundation itself. Such a relation is obtained by equating the settlement of solid plate in homogeneous elastic half-space, with the settlement of the same plate on Winkler's springs. It follows:

$$k = \frac{1}{B \cdot I_B}\left(\frac{E}{1 - \nu^2}\right) \tag{3}$$

The modulus of soil reaction is not constant value, but depends on the quantity of the load (p), and on the shape and the size of the loaded test plane. Terzaghi is suggesting the values for the modulus of soil reaction (k_1), for test surface 30 cm × 30 cm and for different types of soil and its compaction (1955), which are given in table (1). In a case of larger test areas, Terzaghi is proposing corrections in the form of [6]:

- for incoherent soil:

$$k = k_1 \left(\frac{B' + 1}{2B'}\right)^2 \tag{4}$$

- for coherent soil:

$$k = \frac{k_1}{B} \tag{5}$$

where

$B' = B/30.3$,

and B is width of the foundation (cm).

Sand		Incoherent	Half-coherent	Coherent
1.	Dry or wet	0.6 - 1.9	1.9 - 9.6	9.6 – 32
2.	Partially sunken	1.3	4.2	16
3.	Sunken	0.8	2.6	9.6
Clay		Consistency		
		Stiff	Very stiff	Solid
1.	q_u(kN/m^2)	100 - 200	200 - 400	> 400
2.	Limits	1.6 - 3.2	3.2 - 6.4	6.4
3.	Partially sunken	2.4	4.8	9.6

Table 1. Examples of calculations for selected modulus of soil reaction

The following examples will demonstrate the results of calculations for foundation foot and foundation plate for different types of soil, with an accent on their maximum settlements. In these calculations we used the modified program code, which is initially developed by Salar Delaware Ghashghaei in Matlab [9].

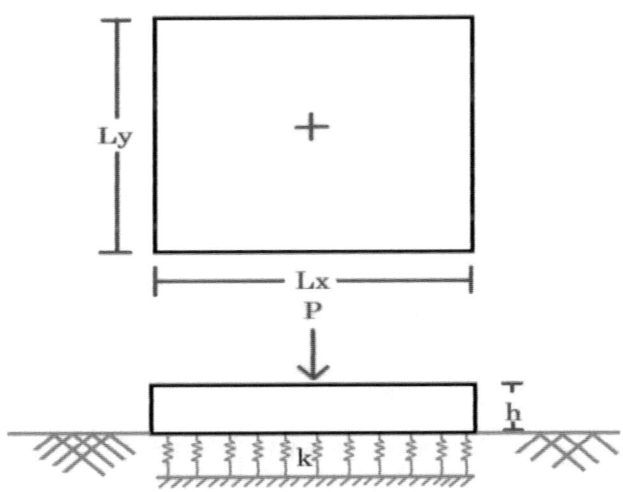

Fig. 2. Foundation foot on Winkler's springs with stiffness k

First example is an example of a foundation foot with the following characteristics:

$L_x = 120\ cm, L_y = 120\ cm, h = 100\ cm, E = 2.1 \cdot 10^5\ MPa, \nu = 0.20,$

and the results of calculations are represented with table 2 and figure 3.

Type of soil:	k_1 (10^4 kN/m^3)	k (10^4 kN/m^3)	δ_{max} (cm)
Incoherent sand (dry or wet)	1.9	0.47975	-3.353
Coherent sand (dry or wet)	32	8.08	-2.321
Clay with stiff consistency	2.4	0.606	-3.326
Clay with solid consistency	9.6	2.424	-2.986

Table 2. Results of settlement' calculations for foundation foot

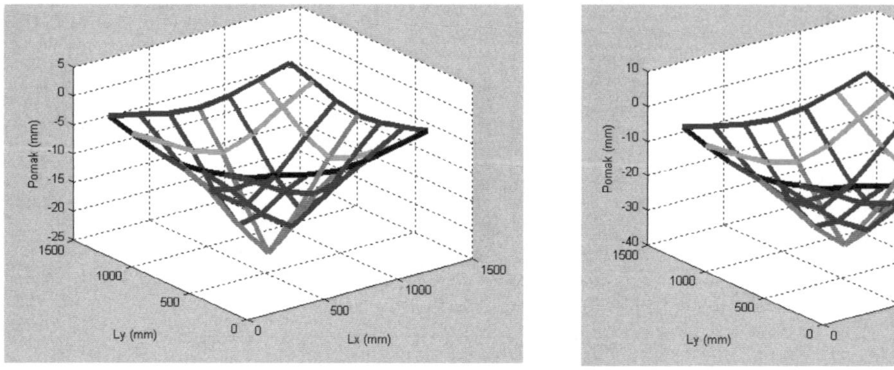

Fig. 3. Diagrams of deflections for foundation foot on elastic soil/base, obtained by finite difference method [9]

Second example is an example of a foundation plate, with dimension 500 cm × 500 cm, height $h = 100\ cm$, and with total load/force distributed symmetrically in four ultimate points of internal network. For this example, the results of calculations are represented with table 3 and figure 4.

Type of soil:	k_1 (10^4 kN/m^3)	k (10^4 kN/m^3)	δ_{max} (cm)
Incoherent sand (dry or wet)	1.9	1279.3	-2.06
Coherent sand (dry or wet)	32	19392	-1.55
Clay with stiff consistency	2.4	14544	-2.05
Clay with solid consistency	9.6	5817.6	-1.9

Table 3. Results of settlement' calculations for foundation plate

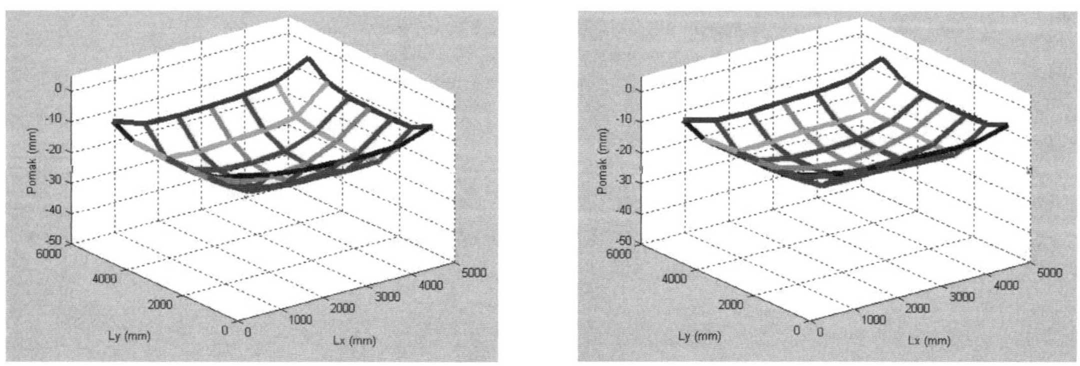

Fig. 4. Diagrams of deflections for foundation plate on elastic soil/base, obtained by finite difference method [9]

5. Conclusion

Examples of calculations for foundation plates, taken from practice and references, shows differences in accuracy for some numerical methods in the analysis of stresses and deformations/strains. In most of calculations the compressible soil is replaced with the system of elastic springs, which are deformed just below the loaded part of the plate. This model is not entirely satisfactory because deformations in real soil are spread beyond the loaded parts, but is still very widely used because the analytical solutions of differential equations is relatively simple, and the results are more realistic than those obtained with the assumption of linear distribution of soil reaction. Furthermore, in this model are not considered elongations (stresses) due to temperature.

By summarizing our results we obtain that finite difference method gives results similar to those obtained with the finite element method, but the contours of the bending moments obtained with finite difference method are less smooth than those obtained with finite element method. The density of the network has an impact on the results of each of the numerical methods. The higher density gives smoother diagrams of bending moments and deformation of the foundation plate. Our examples with calculations made by using modified Matlab program code, and their results, indicates that plate' settlement is less dependent on the change of modulus of soil reaction, as well as that the maximum values of those settlings are less than the values of settlings for foundation foot. These benefits and reduction of differential settlements are the most common reasons for selection of the foundation plate as a way of foundation of the building. The focus of further study has to be on including temperature strains, as well as dynamic load in calculations.

6. References

[1] Hemsley, J. A. (2000). Design applications of raft foundations, Thomas Telford Publishing, London

[2] Moody, W. T. (1990). Moments and reactions for rectangular plates, U.S. Government Printing Office, Washington

[3] Scitovski, R. (2004). Numerical mathematics, Department of Mathematics, University of Osijek, ISBN 953–6032–24–4, Osijek

[4] Bowles, J. E. (1974). Analytic and computer methods in foundation engineering, McGraw-Hill Inc.,US, ISBN 978–0070067530, New York

[5] Aron, C. & Jonas, E. (2012). Structural element approaches for soil-structure interaction, M.Sc. Thesis, Department of Civil and Environmental Engineering, Chalmers University of Technology, Göteborg, Sweden

[6] Nonveiller, E. (1990). Soil mechanics and foundations of buildings, Školska knjiga, ISBN 978–8603996456, Zagreb

[7] Verić, F. (2000). Savitljive temeljne konstrukcije (Flexible foundation structure), Authorized lectures, Faculty of Civil Engineering, University of Zagreb, Institute of Geotechnics, Zagreb

[8] Roje-Bonacci, T. (2007). Posebna poglavlja temeljenja (Special chapters of foundation), Faculty of Civil Engineering and Architecture, University of Split, Split

[9] Matlab, The language of technical computing, http://www.mathworks.com/products/matlab/

DOI: 10.2507/27th.daaam.proceedings.124

POLLUTION MANAGEMENT THROUGH ENVIRONMENTAL PERMIT – CASE STUDY

Tomislav Lukić, Nevenko Herceg & Svjetlana Stanić-Koštroman

This Publication has to be referred as: Lukic, T[omislav]; Herceg, N[evenko] & Stanic-Kostroman, S[vjetlana] (2016). Pollution Management Through Environmental Permit – Case Study, Proceedings of the 27th DAAAM International Symposium, pp.0860-0865, B. Katalinic (Ed.), Published by DAAAM International, ISBN 978-3-902734-08-2, ISSN 1726-9679, Vienna, Austria

DOI: 10.2507/27th.daaam.proceedings.124

Abstract

Processing industry in Bosnia and Herzegovina had been developed by 1991 in several centers. In the postwar period, the majority of industrial plants were extinguished with only the most productive sectors remaining as carriers of development, but their priority was not the environmental protection. The adoption of the Law on Environmental Protection ("Official Gazette of the F B&H", number 33/03 of 19 July 2003) involving the European Directive IPCC (*Integrated Pollution, Prevention and Control*) created the preconditions for an integrated approach to preventing and supervising of pollution and practical application of a new legal instrument – environmental permit. First environmental permits were issued in 2004, and up to date some 900 permits have been put into force in the Federation of Bosnia and Herzegovina. With this paper we want to show the effect of the administrative procedure of issuing environmental permit on the development of integral system of environmental protection of the Federation of Bosnia and Herzegovina, and to emphasize the need of changing the existing legislation in this sector.

Keywords: Federation of Bosnia and Herzegovina; environment; environmental permit; industry; pollution

1. Introduction

In the socialist Yugoslavia, Bosnia and Herzegovina was the second most underdeveloped republic only after Macedonia. The population occupied themselves with agriculture, especially in Posavina and in the part of Herzegovina, and this only on the farms which were small and unprofitable while food was mainly imported. What was favored was heavy, processing and military industry with especially developed special-purpose industry in the field of armament, military vehicles and equipment. Not legging behind were other industrial branches such as mining, ferrous metallurgy and steel production, aluminum production, machine building, paper, textile and wood industries. The care of environment was symbolic, resulting in quite a number of accumulated industrial refuse dumps while emissions into the air, water and soil were enormous. An additional damage to the environment was caused by the war in Bosnia and Herzegovina which was waged from 1992 to 1995. The war conflicts destroyed the economy and infrastructure of Bosnia and Herzegovina, and the production recorded an 80 % fall. In the postwar period, starting with 1995, the production started to recover but with a growth rather slowed down. The private initiative started to emerge with a part of state companies undergoing privatization. However, the consequences of centrally planned economy in the socialist Yugoslavia are felt even today, and one of the key problems was a very large number of employees in industry. Due to an unstable political situation and the division of the country into two entities and a district, the growth of the GDP was not at a satisfactory level in the next ten years, and B&H became one of the most polluted countries in the region.[1], [2].

In cooperation with EU, in 2003 legislature institutions of B&H enacted a set of laws which were to a greater extent adjusted to EU directives, whereby the preconditions were created for the interventions of the state into the environment. According to the F B&H Constitution, Chapter III, Article 2, Indent c), the environmental protection policy is in the joint jurisdiction of the Federation of Bosnia and Herzegovina and cantons. [3]. In the F B&H Constitution there is no explicit provision defining the competence for sustainable development, but Chapter III, Article 1, Indents d) and i) stipulates that the making of economic and energy policies is the exclusive responsibility of the Federation of Bosnia and Herzegovina. The set of laws from 2003 regulating the field of environmental protection in the Federation of Bosnia and Herzegovina includes: Law on Environmental Protection, Law on Water Protection, Law on Air Protection, Law on Waste Management, Law on Nature Protection and Law on Environmental Protection Fund. The water management and protection is covered by the Law on Waters ("Official Gazette of the F B&H", number 70/06). However, the sub-legislation prescribed by certain laws does not exist, and for example out of 23 by-laws, whose adoption is prescribed by the Law on Environmental Protection, only 10 have been adopted. One of the major activities on adjusting B&H legislation to the European Union legislation is the introduction of an integrated approach to preventing and supervising of pollution based on IPPC directive (*Integrated Pollution Prevention and Control Directive*). A new legal document called *Environmental Permit* with its legal instruments in the form of the provisions in the Law on Environmental Protection ("Official Gazette of the F B&H", number 33/03 and 38/09) and implementation regulations should act as prevention against excessive environmental protection. [4], [5].

Environmental Permit is based on the acquisition of the European Union legislation, in this case of the so called horizontal legislation and EU directives: EIA Directive 85/337/EEC, Supplementary Directives 97/11/EC and 2003/35/EC, SEA Directive 2001/42/EC, Directive on Public Access to Environmental Information 2003/4/EC as well as EU Directive on Industrial Pollution Control, IPPC Directive 96/61/EC with the supplement of Directive 2003/35/EC 2003/87/EC and Regulations EC/1882/2003 and EC/166/2006, then COMAH Directive 96/82/EC and LCP Directive 2001/80/EC [6].

Some of the implementation by-laws regulating the area of environmental permit issuance are:

- Ordinance on plants and facilities that require environmental impact assessment and on plants and facilities which can be constructed and put into operation only subject to environmental permit ("Official Gazette of the F B&H", number: 19/04),
- Ordinance on the conditions for the submission of the request for the issuance of environmental permit for plants and facilities which had been granted a permission before coming into force of the Law on Environmental Protection ("Official Gazette of the F B&H", number: 45/09 i 43/10),
- Ordinance on deadlines for submitting the request for the issuance of permit for plants and facilities which had been granted a permission before coming into force of the Law on Environmental Protection ("Official Gazette of the F B&H", number: 68/05 and 43/10),
- Ordinance on Registers on plants and pollutions ("Official Gazette of the F B&H", number: 82/07).

All plants which produce or may produce loads on the components of the environment are obliged to provide a working permit or environmental permit in a short time. This permit obliges them to adept their operations to the most effective and most advanced levels of development of the activity and related modes of production and maintenance of plants available to the operator, which adaption must be based on the demands of the best available technique – so called BAT (*Best Available Technique*) [4]. Pursuant to the provisions of Article 87 of the Law on Environmental Protection ("Official Gazette of the F B&H", number 33/03 and 38/09) and Ordinance on adopting the best available technique to achieve environmental quality standards, the Federal Ministry of Environment and Tourism passed a decision adopting Technical instructions for the best available techniques of BAT documents for six sub-sectors of food processing industry and agriculture:

- 1. Slaughterhouses for Bovine Cattle
- 2. Meet Processing,
- 3. Fish Processing,
- 4. Fruit and Vegetable Processing,
- 5. Milk Production and Processing
- 6. Beer Production and
- 7. Fish Growing.

Obligations that environmental legislation imposes on economy, related to the obligation of environmental permit acquisition, meeting the requirements for waste management, obligation of monitoring, reporting and compensation for the damage done, have already largely become a reality. Clear responsibility for the consequences on the environment prescribed by environmental permit imposes a number of new responsibilities for companies, financially demanding, but profitable in the long run. However, what should be additionally improved is certainly the change of the existing perception according to which socially responsible business operations are considered as the fulfillment of legal obligation. What will help in this matter is certainly the interest of the public in the environmental issues, especially through the engagement of the NGO sector which from the regulatory bodies and economy requires greater engagement and investments in environmental protection. [4], [7].

2. Environmental Permit

Environmental permit is integral administrative act prescribing measures for the protection of all environmental components: air, water, soil, biological diversity, waste management etc., for the purpose of achieving a high level of their protection. In the legal proceedings preceding the issuance of town-planning permit, the investor whose project is considered to have or may have a negative impact on the environment is obliged to obtain environmental permit beforehand.

In the Federation of B&H this permit should be obtained in the following cases:

- Construction and putting into operation new plants and facilities where the aim is for the investor to anticipate, in the earliest phase of the project development, the measures for preventing a negative effect the project may have on the environment, most often in agreement with the Study on the environmental impact assessment,
- The existing plants and facilities which had been granted permits before coming into force of the 2003 Law on Environmental Protection, where an adjustment to the environmental protection standards is carried out through the Plan of Activities with measures and deadlines for gradual reduction of emissions,
- Devastated plants as well as plants and facilities out of operation, which had been granted permits before coming into force of the 2003 Law on Environmental Protection, and
- In case of major changes in the operation of existing facilities subject to the issuance of environmental permit [4], [7].

By the implementation regulation and Ordinance on plants and facilities that require environmental impact assessment and on plants and facilities which can be constructed and put into operation only subject to environmental permit ("Official Gazette of the F B&H", number: 19/04) a special control regime and records of the activities as per branches of economy that endanger or may endanger environment have been regulated for new plants and facilities. Also, in its separate chapters this Ordinance defines the content of the study on environmental impact assessment as well as criteria assessing the need for the environmental impact assessment implementation. According to the said Ordinance and classification of plants and facilities by types, quantity, capacity and other criteria, environmental permit is issued by the Federal Ministry of Environment and Tourism, while for other plants with parameters below the said quantities and capacities and/or other criteria, for middle-sized and small plants and facilities, permit is issued by the competent cantonal ministries (Article 2 of the Ordinance). For very small plants and facilities environmental permit is not required, rather the conditions and environmental protection measures are determined within a town-planning consent and building permit issued by a municipality. Allowing for the need of companies to adjust their operations to the Law requirements, interim periods and the conditions for the environmental permit acquisition have been envisaged. One of the key conditions is the Plan of Activities for those plants to which the adjustment to the European standards causes considerable changes and adjustment deadlines for the permit to be obtained. [4], [7].

Environmental Permit contains:

- marginal values for the emissions of polluting substances,
- measures for the reduction of emissions in order to protect air, soil, water, plants and animals,
- measures for the management of the waste produced by plants and facilities,
- protection measures against noise,
- measures for the reduction or cessation of the cross-border pollution,
- monitoring system with the defining of methodology and measuring frequency.
- conditions and measures for the prevention of accidents on a larger scale and
- rehabilitation measures upon work completion, in other words measures for the restoration of the environment.

When considering the request for the issuance of environmental permit, the Ministry takes into account the following criteria:

- project characteristics: size, amassment with other structures, exploitation of natural resources, waste production, pollution and disturbances, accident risks etc.,
- project location and the sensitivity of the environment of the geographic areas which may also be affected by the project: the present use of land, availability, quality and regenerative capacity of natural resources, absorption capacity of natural ecosystem (wetlands, coastal zones, protected areas etc.),
- The characteristics of the possible effects: the scope of the effect, cross-border nature of the effect, the size and complexity of the effect, the frequency and reversibility.

Environmental permits are issued for a period of five years. The owner of the plant (operator) is obliged to perform the prescribed monitoring and send the results to the competent ministry. If in the set period the operator fails to fulfill the set obligations for environmental protection, the competent ministry will revoke the issued permit. Also, upon the expiry of a five-year period, all operators are obliged to apply for the issuance of a new environmental permit. This approach ensures:

- the possibility of survival and development of enterprises, preservation of jobs and fulfillment of obligations towards broader social community,
- the possibility of planning and implementing of world-wide accepted environmental protection measures in the interim period of five years under the supervision of planned measures by the competent ministries of environment and inspection,
- the information of the public about the impact of the enterprises on environment as prescribed in the Law on Environmental Protection, and
- the reduction of the harmful effects on the environment and the reaching of the levels of protection prescribed by EU Directives and Guidelines.

With legislation like this, B&H started to come nearer to the EU environmental policy faster than some other countries in the region. [4], [7]. This is corroborated by the latest records in the data basis which the Federal Ministry of Environment and Tourism updates and in a transparent way presents to the Government of the Federation of B&H and the public by means of the reports and on its web-site: www.fmoit.gov.ba.

3. Statistics about Issuing of Environmental Permits in the Federation of B&H

In the period from 2004 to 2016, the Federal Ministry of Environment and Tourism issued a total of 892 environmental permits in its competence (Fig 1). As we deal here with quite a new administrative procedure, at the very beginning, that is in 2004 only five permits were issued. However, by the end of 2015 their number rose up to more than a hundred permits a year. Considering the complexity of the administrative procedure, of acquisition of legally prescribed documents, and especially concerning harmonization with area-planning documents of the municipality on whose territory the plant-facility is situated, procurement of opinions of competent authorities, as well as preceding actions in the procedure which include the evaluation of the Study of environmental impact assessment for new installations or the evaluation of acceptability of Plans of Activity as adjustment plans for existing plants and facilities, [4], [7], we can conclude that the number of issued permits at the Federal level is very good. The said studies with waste management plans are evaluated by expert commissions. At the same time, in this complex procedure it is necessary to follow the procedure of procuring the prior water consent on meeting water conditions or acquisition of water consent.

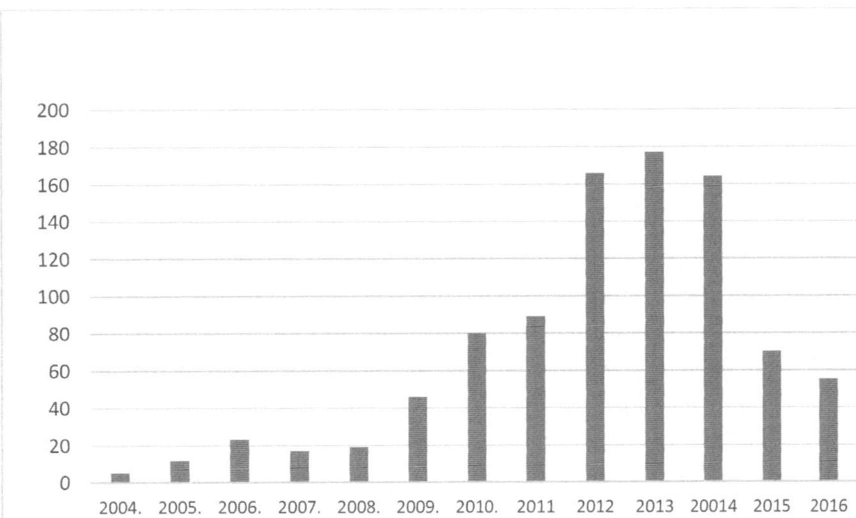

Fig 1.The number of issued environmental permits at the federal level of competence in the 2005-2016 period.

The Sector for the issuance of environmental permits within the Federal Ministry of Environment and Tourism is under staffed. Despite this, the number of issued environmental permits in the Federation of Bosnia and Herzegovina is very high (Fig 1, Table 1). European average for the issuance of environmental permits is four permits per an employee, and by comparing these standards with the number of employees in the federal and cantonal ministries of environment it can be concluded that the results are above average.

Canton	The number of issued environmental permits
Canton 1. Una-Sana (USC)	360
Canton 2. Posavina (PC)	49
Canton 3. Tuzla (TC)	441
Canton 4. Zenica-Doboj (ZDC)	331
Canton 5. Bosnia-Podrinje (BPC)	37
Canton 6. Central-Bosnia (SBC)	154
Canton 7. Herzegovina-Neretva (HNC)	254
Canton 8. Western Herzegovina (ZHC)	41
Canton 9. Sarajevo Canton (SC)	157
Canton 10. Hercegbosnia (HBC)	15
TOTAL	**1839**

Table 1 The number of environmental permits at the cantonal level of competence issued by 2015

Analyses have shown that up to date, as per branches of economy, the majority of environmental permits have been issued in the energy sector, the sector of excavating industries (mainly quarries and mines), then chemical, metal and food processing industries, waste management, infrastructure, textile, leather, wood and paper industries, tourism, water management etc. [4], [7], [8], [9].

However, many enterprises have not started the procedure of acquisition of environmental permits yet whether it be the federal or cantonal competence and in many cases for the objective reasons which can be summarized as follows:

- non-existence of sector strategies,
- non-compliance of sector/department policies with the new legal obligations regarding environmental permit acquisition,
- inadequate privatization procedure of enterprises from the environmental permit acquisition point of view,
- low level of consciousness and financial unpreparedness of enterprises for the implementation of legal procedure for environmental permit acquisition,
- the complexity of the administrative procedure for the issuance of environmental permits, and
- the lack of experience for the implementation of such a complex administrative procedure.

The advantages of environmental permit for the economy and the overall state of the environment in the Federation of B&H are great, and they are reflected in the following:

- It contributes to a complete observation of business processes and their impacts on the environment as well as to a transparent presentation of the data on such impacts,
- It works as a binding element, because representatives of all organizational units must participate in its preparation: from procurement and planning to production, maintenance and sale,
- It is good for the representatives of public authorities (local and regional self-governments, inspection bodies, legal persons with public authorities etc.) because of integrating, harmonizing and uniting of all the important data for the plant (farm, waste dump) in one document,
- The procedure in the case of a plant close-down is also described,
- An obviously interested public, which as a rule is included in the very procedure of the issuance of the environmental permit,
- Strengthened consciousness of a company management on the implementation of the conditions from the environmental permit,
- Improved quality condition of all the environmental components,
- The damage inflicted on the protected plant and/or animal species and their habitats as well as to landscape, water, sea, soil and Earth's crust has been significantly reduced, and
- The risk for the life and health of people living near these plants has been reduced [4], [7].

4. Conclusion

Judging from the experiences in the last 13 years, since the existence of a new legal frame and some 40 implementing regulations in the field of environmental protection, it can be concluded that B&H is entering the circle of European countries which implement systematic environmental protection. The Law on Environmental Protection contains regulations on integrated framework for the issuance of environmental permits based on the concept of integrated pollution prevention and control (IPPC). Ordinances, directives and instructions have been adopted to prescribe the method that plants and facilities are obliged to follow in order to fulfill all the requirements for the environmental permit acquisition.

The quality of this systematic approach is based on the creation of an obligatory environmental legal act – environmental permit, which instigates the prevention and control of pollution management procedure not only for new plants, but also for the existing ones that with their technologies must gradually adapt to new stricter requirements regarding the reduction of pollution levels and application of the new pollution thresholds prescribed by the law. In that way, the procedure of the remediation of the environment in B&H and work ambient will be gradually carried out, creating more favorable business space which encourages responsible management and technological processes control in order to reduce excessive environmental pollution.

5. References

[1] Herceg, N. (2009). The State of the Environment in the Federation B&H, the Federal Ministry of Environment and Tourism, Sarajevo

[2] The Ministry of Commerce and Economic Relations of Bosnia and Herzegovina (2012). The Report on the State of the Environment in Bosnia and Herzegovina, Sarajevo

[3] Official Gazette FB&H, nos. 1/94, 13/97, 16/02, 22/02, 52/02, 63/03, 9/04, 20/04, 33/04, 71/05, 72/05 and 88/08.

[4] Herceg, N.; Rudež, M.; Lukić, T. & Bandur, M. (2011). Environmental Permit – Systematic Approach to Pollution Management and Legal Guarantee of the Environmental Protection in Bosnia and Herzegovina, XII International Symposium on Quality "With Quality to Business Excellence", Osijek

[5] Herceg, N. & Cero, M. (2009). Environmental EU Acquis in Federation B&H, state and perspective, International scientific thematic conference "Soil Protection Activities and Soil Quality Monitoring in South Eastern Europe", Sarajevo

[6] Herceg, N. (2009). Environmental Policy through the Strategy of the Environmental Protection. *Modern issues*, Vol. 4, No. 7, (May, 2009)

[7] Herceg, N. (2010). The Challenges of the Environmental Permit, the Federal Ministry of Environment and Tourism, Sarajevo

[8] Herceg, N & Rudež, M. (2010). The Importance of the Environmental Permit and the Waste Management Plan in Prevention and Waste Disposal Control in B&H, International Conference "Management of Dangerous and Harmless Waste in the Region", Zenica

[9] Herceg N. & Cero, M. (2010). Waste in the Report on the State of the Environment in the Federation B&H, International Conference "Management of Dangerous and Harmless Waste in the Region", Zenica

[10] Herceg, N. (2013). Environment and Sustainable Development, University of Mostar/Synopsis Zagreb, Mostar-Zagreb

DOI: 10.2507/27th.daaam.proceedings.125

STRUCTURE AND WORKING MODES OF THE INTELLIGENT ADVISER MODULE

Damir Haskovic, Branko Katalinic, Ilija Zec,
Ilya Kukushkin & Alina Zavrazhina

This Publication has to be referred as: Haskovic, D[amir]; Katalinic, B[ranko]; Zec, I[lija]; Kukushkin, I[lya] & Zavrazhina, A[lina] (2016). Structure and Working Modes of the Intelligent Adviser Module, Proceedings of the 27th DAAAM International Symposium, pp.0866-0875, B. Katalinic (Ed.), Published by DAAAM International, ISBN 978-3-902734-08-2, ISSN 1726-9679, Vienna, Austria
DOI: 10.2507/27th.daaam.proceedings.125

Abstract

This paper describes the Intelligent Adviser Module (IAM) as a decision making support tool for the system operator in Bionic Assembly System (BAS). The efficiency of BAS depends on many parameters and also on the ability of the system operator to reach quality decisions in good time. The main target of the control system is to reach the highest machine utilization in the given working conditions. Working scenarios are realized with the execution of all activities which are needed to assemble a continuous stream of products. Continuous stream of products is made with assembly orders. One assembly order means to assemble one run of product. In a normal working mode there is always a small difference between planned and realized working scenarios. This difference is compensated by the automatic control system. If the difference increases beyond the automatic control system's ability to compensate it, the system operator has to make decisions which bring the system back to a normal working mode, efficiently. Efficiency depends on the relation between the quality of the decisions and time needed to make them. Higher efficiency means higher quality of decision made in shorter time. The best way is to make decisions while the machines are working. This is not always possible and in that case the machines must wait for the decision. Machine waiting is lost machine time. This has to be avoided or minimised. This time shortage brings stress to the system operator and forces him to make lower quality decisions. In this situation, the IAM offers a decision support in form of proposals. A working IAM prototype is developed, realized and analysed in predefined laboratory conditions. The analysis of the results show that the IAM is a promising operator decision support tool for complex assembly systems.

Keywords: Intelligent Adviser; Bionic Assembly System; hybrid control structure; expert systems; knowledge discovery

1. Introduction

Current product development is defined with increased product complexity, variety and shorter lifetime [1]. Modern assembly systems need to respond to these challenges through adaptability, efficiency and robustness. They need to be capable of assembling a large variety of products with high complexity within defined customer deadlines. As a result, the working scenarios are becoming more complex and the sophistication of the control structure is proportionally increasing [2].

As an answer, a Bionic Assembly System (BAS) concept based on task distribution, intelligence and self-organization is introduced [3]. It presents the next generation of hybrid assembly systems which combine two basic control structures and principles. First is centralized upper control system based on hierarchy and the second is the self-organizing control system based on heterarchy [4].

BAS is a human centric assembly system which promotes the integration of humans within the assembly process [5]. The main decision maker is the human system operator. The efficiency of the Bionic Assembly System depends on his limited ability to reach repetitive, high quality decisions in good time. Good time means that the machines are not waiting for his decision. Main goal is to achieve minimum lost machine time by avoiding standstills during assembly.

However, this introduces new challenges for the system operator. The system complexity, large amounts of data, distractions and disturbances have a high impact on his capabilities to run the entire system and to perform his tasks. In addition to that, as a human, he has limited physical and cognitive capacity to reach repetitive quality decisions over a long period of time [6]. He needs a decision support system which will support his decision making abilities. Such a system needs to be capable of replicating human intelligence, learning and thought processes. This is becoming possible thanks to the rapid development of information technologies (IT) and the introduction of Artificial Intelligence (AI) [7].

As an answer to these challenges, the Intelligent Manufacturing Systems (IMS) group from Vienna University of Technology has introduced the Intelligent Adviser Module (IAM) [8]. It is a decision support tool integrated as a part of the BAS control structure as shown in Fig. 1. On one side, the expert system platform of the IAM enables it to model and represent human knowledge. On the other side, it uses the recorded past system data for learning through knowledge extraction.

The knowledge extraction process was described in the previous paper [9]. It demonstrated how recorded past system data was used to identify a correlation between components from different suppliers and the quality of the final product. Once the results were evaluated, the new knowledge was specified in form of rules. This paper describes the IAM structure, its working modes, how new knowledge is imported into the KB and used to propose new solutions to the system operator.

2. Intelligent Adviser Module Structure

The final decisions have to be made exclusively by the system operator. The IAM needs to support him during the decision making process. The main concept is that the IAM learns and improves its accuracy over time. This is realized through the IAM structure and its integration within BAS. As a result of this integration, data can be collected and analysed and useful proposals can be presented to the system operator when he needs to make a decision. As shown in Fig. 2., the main IAM structure is composed from the following submodules:

- System Monitoring
- Knowledge Management
- Decision Support

2.1. System monitoring

Working scenarios are executed as a continuous stream of one or more parallel assembly orders. One assembly order means to assemble one run of product. The realization of working scenarios is tracked through real time system monitoring. The progress of this realization is determined from the analysis of the target and actual state of the system. If the difference starts to be critical, the IAM notifies the operator.

IAM is an integral part of the BAS control system and is connected to all other system components [8]. It serves as an information hub and as such it presents current system information to the system operator. This includes performance of the shop floor elements as well as the disturbances that can occur, resource levels and status of assembly order completion. If there is a critical error or a malfunction, an audio-visual notification is sent to the operator.

Notifications can be triggered via:

- Threshold values (minimum, average, maximum) – This kind of notifications are based on measurements. They can be sent if the stock of resources is getting too low, if the duration of an operation is above a maximum scheduled time, if a robot's battery is almost depleted, etc.
- Boolean logic (yes / no, on / off) – These notifications can be sent according to a specific sensor readout. This can include if a station is shut down or turned on, if a robot is moving, if the shop floor operator is present at a specific position, quality control is good or bad etc.
- Trend predictions (incline, stagnation, decline) – Trend prediction notifications are based on measurements, data analysis and correlation functions. These methods are being successfully implemented in the prediction of tool breakage [10]. This enables the IAM to proactively warn the system operator before a failure occurs. As a result, the system operator can make decisions with which disturbances can be avoided and consequently increase the efficiency of the assembly system.

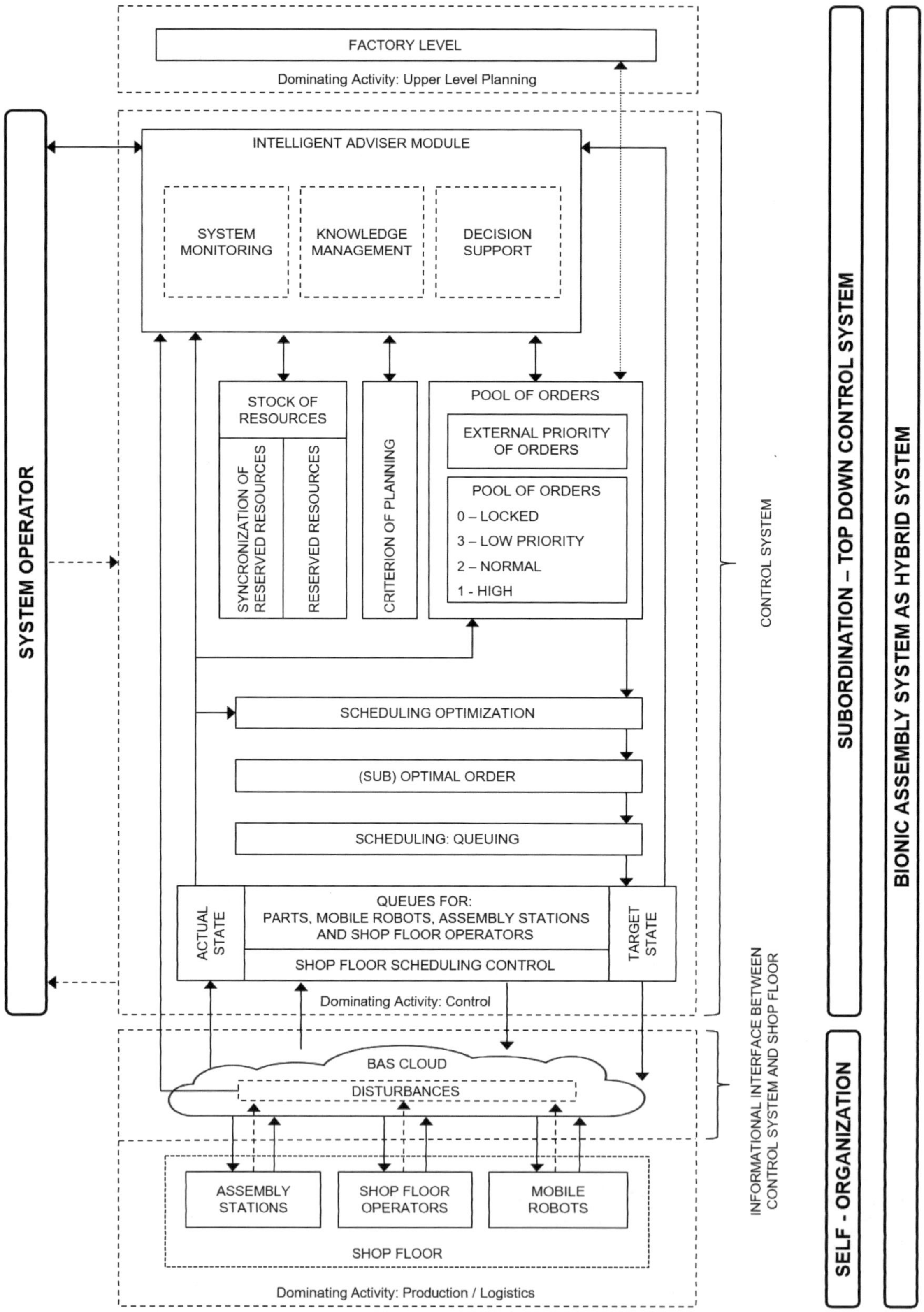

Fig. 1. Intelligent Adviser Module within BAS control structure

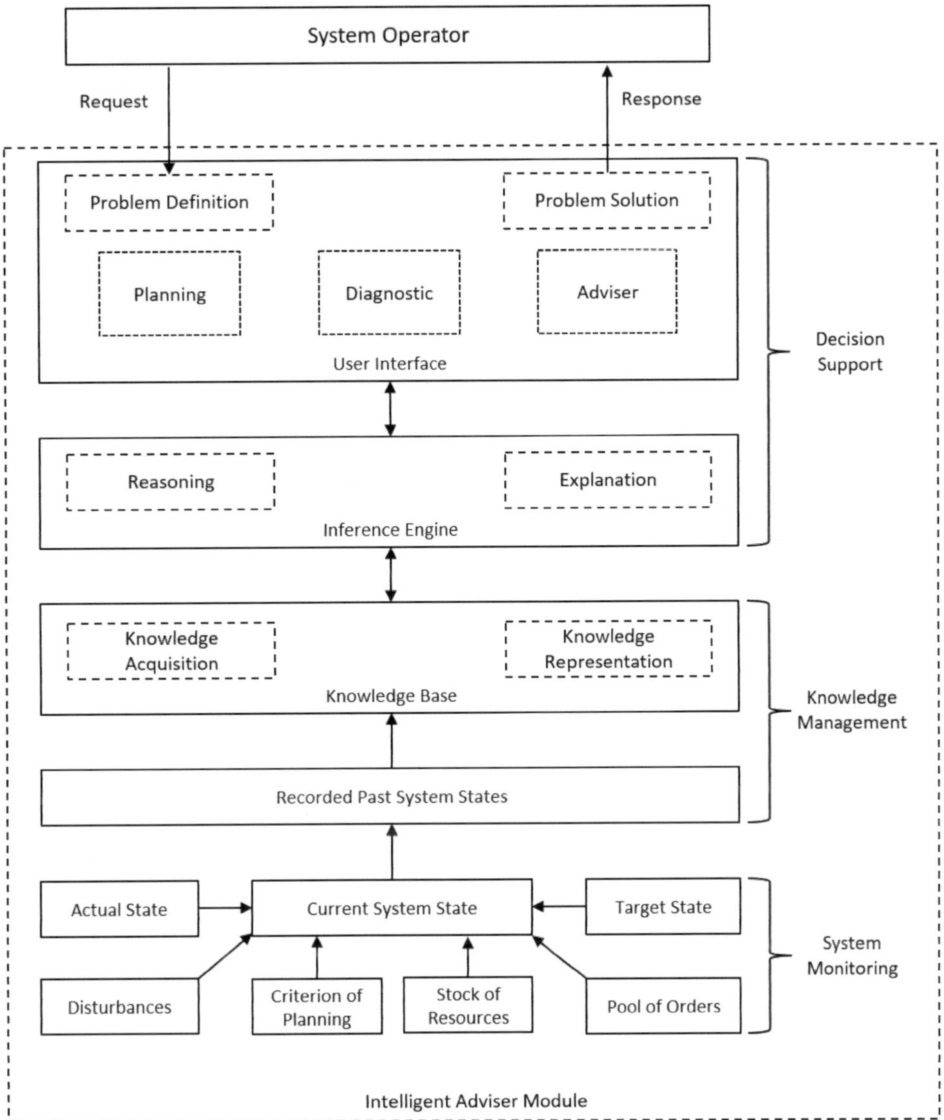

Fig. 2. Intelligent Adviser Module structure

2.2. *Knowledge Management*

The output from the IAM are proposals. The quality of these proposals have a direct influence on the system operator during his decision making process. He has to be able to reach quality decisions in good time. This can be achieved if a previously unknown correlation between predicting datasets (suppliers, hardware age or performance, process duration…) and target datasets (end product quality, breakdowns…) derived from past system data can be identified. In other words, the system needs to be able to acquire new knowledge based on data derived from the execution of past working scenarios.

Knowledge Management is based on:

- **Recorded Past System States** – System monitoring produces large amounts of data which is constantly recorded. This recorded data represents past system states and is the basis for the Knowledge Base (KB). The main information flow channel is between the control system and the shop floor. A real time exchange of information in both directions is taking place. Control system sends instructions, commands and data to people, machines and robots. They send feedback information about the realization and hardware status. The IAM receives all data from this communication channel and stores it within a database. This data consists out of measurements, quality reports, disturbance occurrences, robot travel times, operation times, external and internal resource tracking, status of orders, selection of orders based on specific criteria, past system operator decisions, etc.

- **Knowledge Base** –KB contains different knowledge types in a format which is understandable to a computer system. It uses Knowledge Representation (KR) methods to store facts, theories, rules, procedures, strategies as well as heuristics. The IAM is able to learn from past events through the Knowledge Acquisition (KA) process. This process consists of data preparation, exploration, modelling and evaluation. It is used to describe past events or to predict future system behaviour if a correlation between parameters is found. The quality of the KB improves over time as more and more data is accumulated. It is possible to identify the influence of particular causes on the overall results. The recorded data also contains the past system operator decisions in contrast to the past system states. This allows to extract his expert knowledge and to define expert decision rules which can be used in future. If there is a similar situation where a decision needs to be made, the IAM can quickly recognize it and make a proposal according to the past results. If the past decision from a system operator was satisfactory, it can be reused. On the other hand, if the decision he made had negative impact on the system performance, it can be shown as a warning to avoid similar negative decisions.

2.3. Decision Support

The IAM decision support is realized through the implementation of an expert system. It gives an ability to reason about the acquired knowledge from the KB. Once the KB has been defined, it is connected to the reasoning module also known as the Inference Engine or the Rule Interpreter. It is a computer algorithm which uses, acquires and manipulates knowledge in order to define a solution to a specific problem as well as asserting new knowledge back to the KB. It decides which rules to apply and resolves conflict situations when multiple rules are applicable to a particular case [11].

The inference engine in the IAM enables the following two capabilities:

a) Reasoning

Reasoning allows to reach conclusion based on facts and rules derived from the KB. It can be used for diagnostics or to identify problems. Reasoning knowledge can include [12]:

- Deduction: starts from a general theory based on wide range of facts, rules and principles and narrows it down to a specific conclusion
- Induction: starts from specific observations (measurements), identifies the existence of patterns or regularities and at the end defines a conclusion
- Classification: grouping concepts, ideas and objects into related categories
- Analogy: reaching conclusions after recognizing that two or more thing have specific characteristics in common. By analogy, if they have one, they can have more characteristics in common.

b) Explanation

The assembly process within BAS is very complex and it is hard to cover all scenarios and disturbances. One of IAM advantages is that it becomes more accurate as more data is accumulated. However, as with any other system, mistakes are possible. This can happen because its KB is incomplete or contains incorrect facts or rules. As a result, conclusions can be inaccurate.

This is why it is very important that the IAM can demonstrate the logic behind the reasoning. This is completed using the explanation capabilities of the inference engine.

During the decision making process, the system operator has several options when presented with a proposal:

- Acceptance – In this case the system operator accepts the IAM proposal without considering the its background reasoning logic. It can happen because of multiple reasons. This includes high stress situations, when he is not concentrated, when a decision is needed immediately, when his knowledge is not sufficient or simply when he agrees with the proposal.
- Inspection – If the system operator is not satisfied or suspects the validity of the proposal, he will inspect it using the explanation mechanism. It shows every step of the reasoning process through the graphical user interface. Presentation methods can include a text based report or a visual reasoning tree. The advantage of a rule based knowledge representation over procedural code is that it is much easier to present explanations to the system operator in a more natural and understandable way. This allows to clearly view which rules were fired and what new facts were asserted.
- Rejection – The system operator can reject proposals if he has enough knowledge about the specific problem or if he decides that the reasoning is flawed after performing an inspection. In either case, his decision is recorded for future reference. Based on his past decisions, new rules can be extracted from the recorded data and the inference engine reasoning improves.

3. Working modes

Functionality and efficiency are the two most important requirements of BAS. The main goal of the entire system is to achieve efficient processing of continuous stream of one or more parallel assembly orders. One assembly order means to assemble one run of product. Each product is assembled according to a defined sequence of operations. For every operation there is a group of machines which can complete them.

To ensure that BAS is working with the highest possible efficiency, it has to be organised so that the minimal sum of lost machine time is achieved. This can be accomplished through the organisation of the system according to the following secondary goals:

- Machines standstills are avoided by ensuring their uninterrupted work. To achieve this, every machine has to have the necessary NC programs, tools, workpieces and various other resources at their disposal at the right time.
- Machine setups should be as short as possible. Setup procedures should start without delay and all necessary resources have to be available at the right time.
- Removal of machine standstills. In case that a machine is not working, it has to be brought back to a working state as soon as possible or replaced if possible.

For this reason, target and the actual state of the system are monitored. There is always a difference between planned and realized working scenarios as shown in Fig. 3. System efficiency directly depends on this difference. Smaller difference means higher system efficiency and smaller number of minor disturbances. Depending on the size of the difference, there are three main working modes during the execution of BAS working scenarios: normal, fluctuating and emergency working mode.

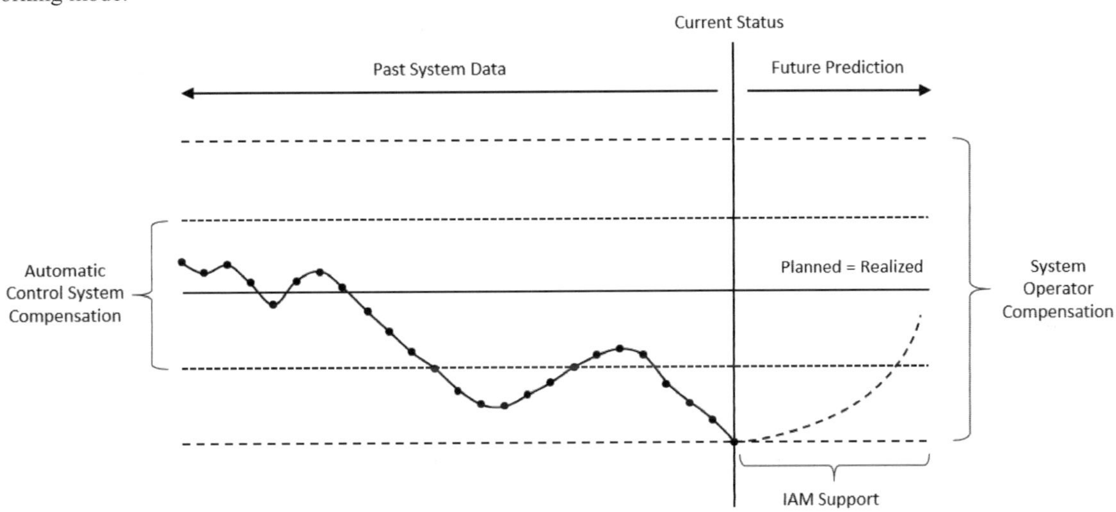

Fig. 3. Intelligent Adviser Module working modes

3.1 Normal working mode

Small differences between planned and realized working scenarios are compensated by the automatic control system and the self-organizing nature of the shop floor. In this mode, the entire assembly process is regulated without the intervention from the system operator. These small differences are caused by chance causes. They are negligible and inevitable variations which occur in random manner during assembly. Such variations cannot be anticipated, detected, identified or eliminated from practical or economic reasons [13].

Causes of small differences can include [14]:

- minor variations in operation start / end times
- mobile robot traveling times
- occasional failed quality control checks
- slight variations in raw material
- vibrations caused by operating hardware
- normal wear and tear
- computer latency
- minor imprecise shop floor operator actions
- working conditions (light, noise, temperature, humidity, dust, ventilation...)

For example, if a product during its assembly did not pass a quality control check, the system reroutes it to a repair station. However, a certain percentage of errors are non-reparable. In that case the automatic compensation system needs to ensure that the exact number of ordered products is completed. That means that the defective product is discarded and an automatic replacement order takes place. If there are multiple failed products it implies that there are more serious problems in the system. The automatic control system cannot compensate for it.

3.2 Fluctuating working mode

The automatic control system is limited. If the difference is increasing beyond its ability to compensate it, the system operator is notified to actively bring the system back to a normal working mode. The cause of these more serious differences are called assignable causes [15]. They result in a large amount of variations during assembly. They can be detected, identified and eliminated.

Fluctuating working mode can be caused by:
- Defective raw material
- Shop floor operator is absent or asleep
- Wrong equipment adjustment or calibration
- Defective mobile robots
- Machine malfunction
- Power fluctuation

In a fluctuating mode the system operator has enough time and ability to solve a simple problem. Simple problem means that it is relatively easy to diagnose and to determine a solution. Because assignable causes can be detected, the system operator uses the IAM to help him quickly determine what the problem could be and where it occurred. For example, an assembly station is waiting a long time because a mobile robot has a faulty wheel. When a new problem is detected, a Problem Record File (PRF) is created. It is registered with its ID which is composed from a timestamp (YY-MM-DD-HH:MM:SS) and an additional 4 digit identifier (####). PRF contains all the available data which includes times, status of involved hardware, sensor readouts, possible solutions (if applicable) etc. All this information is presented via menus, tables, graphs etc. to the system operator. He uses this information to investigate and make a decision. The applied solution as well as the final results are recorded in the PRF. In this case, after replacing the mobile robot with the faulty wheel, the waiting error was removed which means that the solution was successful. All this information can be useful to the system operator if a similar problem repeats.

3.3 Emergency working mode

Extreme differences between planned and realized working scenarios mean that there is a problem caused by multiple errors. In such an emergency working mode, the system operator does not have enough time or ability to make a decision. He uses the IAM as a decision support software which helps him to reach higher quality decisions in good time. Emergency working mode can be caused by previously described assignable causes. When these causes occur at the same time it can result with:
- Bad quality in an assembly run
- Inability to keep up with deadlines
- Multiple machine failures

Unlike in fluctuating working mode, the system operator does not have enough time to investigate and analyse all the options and navigate through multiple menus. He needs to be presented with the most important information which includes problem description, possible explanations and a solution. For this reason, he activates the adviser function. There are two ways how the IAM can output proposals: from KB history or through a dialog manager.

When a problem occurs, the KB history is searched in order to find any matches which could be used. There can be multiple matches. The system operator selects the option with the highest match percentage. He is then presented with the details which include the past problem description, causes, applied solution and the results. From this, the system operator decides if the proposal is applicable.

In case that there were no matches found in the KB history or if the system operator does not think that the presented options are applicable, he can activate the dialog manager. It serves as a problem solving interface between the IAM and the system operator. The exchange of information is performed like a conversation as shown in Fig. 4. This makes it more natural and intuitive for the system operator. The dialog manager was developed in CLIPS 6.3, which is an open source programming tool for building expert systems. The KB contains expert knowledge represented through complex rules functions and object [16]. Using the inference machine, KB and its expert knowledge the IAM performs reasoning in order to find a solution. The system operator answers the questions and follows the instructions. At the end, he is asked if the proposed solution solved the problem. If yes, it is recorded in the PRF. If not, other possibilities are tried until a solution is found or there are no more rules. This is the worst case scenario which is possible in young systems.

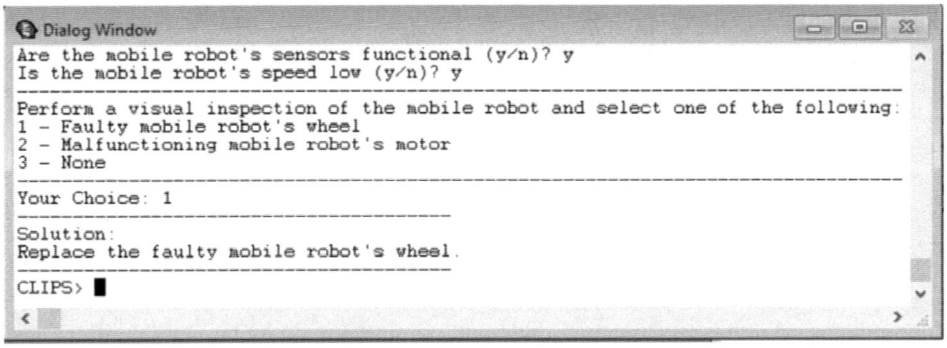

Fig. 4. IAM Dialog Manager – developed in CLIPS 6.3

4. Learning from past system data through Knowledge Record File

As stated, the KB contains knowledge from a specific domain. This knowledge is defined as rules which represent the expertise from human experts, troubleshooting manuals, domain specific facts etc. This is appropriate for hardware related problems, to solve malfunctions and for transferring general experiences of the domain expert into the KB.

The advantage of the IAM within BAS is that it is able to learn specific system related knowledge which is generated during the operation of the entire assembly system. This includes knowledge about events which have not occurred ever before and as such are very difficult or impossible to predict and include in the KB. These events represent the past BAS system states.

This example is a continuation of the knowledge acquisition procedure described in the previous paper [9]. The IAM has detected that there is a high number of specific products with unsatisfactory quality results. A new PRF has been created where all the facts are listed which includes involved assembly stations, shop floor operators, mobile robots, used components, their suppliers etc. The system operator is notified and he uses the dialog manager to solve the active product quality problem. He answers the questions and follows the instructions, but a solution is not found. The problem is very specific to the current execution of BAS working scenarios. The following message is shown to the system operator:

"Unable to determine the cause of problem.
The Knowledge Base needs to be updated."

To solve this, new knowledge needs to be imported. After the recorded past system data has been analysed, the best predictor has been identified with its matching set of new rules as shown in Table 1. This represents specific new knowledge based on past system data. The connection between the knowledge management and the decision support submodules is the Knowledge Record File (KRF) generator as shown in Fig .5. The main function of the KRF generator is to translate these new rules into a format which CLIPS can understand. A part of the KRF code for these new rules is shown in Table 2. Each time new knowledge is discovered and evaluated a new KRF is generated. Each KRF is also registered with its ID which is composed from a timestamp (YY-MM-DD-HH:MM:SS) and an additional 4 digit identifier (####).All these KR files are imported into the KB and the entire system becomes "smarter".

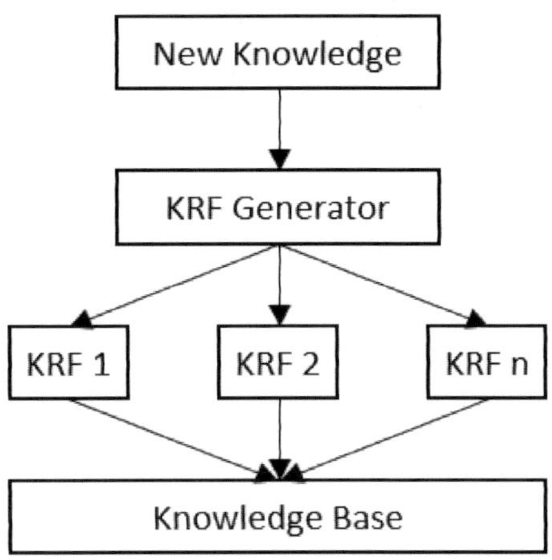

Fig. 5. Expanding the Knowledge Base with Knowledge Record Files

Timestamp_ID	2016-11-19-16-01-29_340		New Rules
Supplier ID	product_quality		
	OK	NO	
SUP_114	29	0	IF Supplier ID = SUP_114 THEN Quality = OK (29 OK > 0 NO)
SUP_442	72	4	IF Supplier ID = SUP_442 THEN Quality = OK (72 OK > 4 NO)
SUP_559	8	28	IF Supplier ID = SUP_559 THEN Quality = NO (8 OK < 28 NO)
SUP_746	57	2	IF Supplier ID = SUP_746 THEN Quality = OK (57 OK > 2 NO)

Table 1. Best predictor identified after Knowledge Acquisition (9)

Now, the new set of rules have been imported into the KB and is possible to go back to the example where the IAM reports that there is a problem with unsatisfactory quality results. The generated PRF among other data contains that the supplier of the used components was "SUP_559". In this case, if the system operator selects the option to solve the active problem again, the following solution is displayed:

"The components are delivered by the supplier SUP_559.
Final product quality is unsatisfactory because the supplier SUP_559 is bringing subpar components."

```
;===========================================================
; Intelligent Adviser Module for Bionic Assembly System
; Knowledge Record File: KRF_2016-11-19-16-01-29_340
;===========================================================

(defrule KRF_2016-11-19-16-01-29_340/SUP_559 "Supplier ID: SUP_559"
(SUP_559)
=>
(assert (solution ok))
(assert (product_quality_NO))
(printout t "The components are delivered by the supplier SUP_559." crlf)
(printout t "Final product quality is unsatisfactory because the supplier SUP_559 is bringing subpar components."
crlf
crlf))
...
```

Table 2. Part of the KRF code for CLIPS 6.3

5. Conclusion

A working prototype of the IAM was developed and analysed. The analysis results show that the IAM can be successfully integrated in the control system of BAS. The proposed IAM structure is designed to integrate system monitoring, data collection and knowledge discovery functions with a decision support expert system. Such integration takes advantage of large amounts of data generated during the execution of BAS working scenarios. The rule based method for knowledge representation gives promising results. The main advantage is the possibility to include predefined expert knowledge, facts and definitions with new system specific knowledge. This makes it possible for the quality of IAM proposals to constantly improve. When the IAM gives more precise proposals, the system operator can make better quality decisions in shorter time and avoid or minimise lost machine time. Further results demonstrate the IAM usability in all possible work flows which can occur in every day operation of BAS. It offers flexible levels of support according to varying difference between planned and realized working scenarios. This makes it possible for the system operator to select a working mode according to his wishes. Based on the results presented in this paper, future research will focus on the investigation of the IAM performance in real industry working scenarios. Main topics of research will include IAM stability, maintenance, optimization and general usability. Additionally, different methods of information exchange between the IAM and the system operator will be examined. One possible direction of further development is the implementation of speech recognition and voice assistant capabilities of the IAM. This could introduce new possibilities in the work flow of the system operator.

6. Acknowledgments

This paper would not be possible without the mentorship and encouragement from Univ. Prof. Dr.sc. Dr.mult.h.c. Prof.h.c. Branko Katalinic.

7. References

[1] Hobday, M. (1998). Product complexity, innovation and industrial organisation. Research policy, 26(6), 689-710

[2] Goldratt, Eliyahu M. (1988). Computerized shop floor scheduling, The International Journal of Production Research 26.3, Pages: 443-455

[3] Katalinic, B., Kukushkin, I., Pryanichnikov, V., & Haskovic, D. (2014). Cloud Communication Concept for Bionic Assembly System, Procedia Engineering, Volume 69, 2014, Pages 1562-1568, ISSN 1877-7058, doi:10.1016/j.proeng.2014.03.156

[4] Katalinic, B., Kukushkin, I. K., Cesarec, P., & Kettler, R. (2012). Hybrid control structure and scheduling of bionic assembly system. In Proc. 8th International Conference of DAAAM Baltic, Industrial Engineering (pp. 483-489). Tallinn, Estonia

[5] Post, P. (2014). "Smart Systems for Intelligent Manufacturing - Industry 4.0" Plenary Lecture, 25th DAAAM International Symposium, Vienna, Austria

[6] Saaty, T. L. (1990). Decision making for leaders: the analytic hierarchy process for decisions in a complex world. RWS publications

[7] National Research Council (US). Committee on Innovations in Computing, Communications, & Lessons from History. (1999). Funding a revolution: government support for computing research. National Academies Press

[8] Haskovic, D., Katalinic, B., & Kukushkin, I. (2015). Role of the Adviser Module in the Hybrid Assembly Subordinating Control Structure. Procedia Engineering, 100, 1706-1713

[9] Haskovic, D[amir]; Katalinic, B[ranko]; Kildibekov, A[skar] & Kukushkin, I[lya] (2016). Intelligent Adviser Module for Bionic Assembly Control System: Functions and Structure Concept, Proceedings of the 26th DAAAM International Symposium, pp.1158-1165, B. Katalinic (Ed.), Published by DAAAM International, ISBN 978-3-902734-07-5, ISSN 1726-9679, Vienna, Austria

[10] Hsueh, Y. W., & Yang, C. Y. (2008). Prediction of tool breakage in face milling using support vector machine. The International Journal of Advanced Manufacturing Technology, 37(9-10), 872-880

[11] Buchanan, B. G., & Shortliffe, E. H. (Eds.). (1984). Rule-based expert systems (Vol. 3). Reading, MA: Addison-Wesley

[12] Dunn, M. (2013). What other types of reasoning are there? theoryofknowledge.net. http://www.theoryofknowledge.net/ways-of-knowing/reason/what-other-types-of-reasoning-are-there/ Last accessed: 26th May 2016

[13] Shewhart, W. A. (1931). Economic control of quality of manufactured product. ASQ Quality Press

[14] Deming, W. E. & Edwards, D. W. (1982). Quality, productivity, and competitive position (Vol. 183). Cambridge, MA: Massachusetts Institute of Technology, Center for advanced engineering study

[15] Duncan, A. J. (1971). The economic design of-charts when there is a multiplicity of assignable causes. Journal of the American Statistical Association, 66(333), 107-121

[16] Giarratano, J. C. (1993). CLIPS User's guide. NASA Technical Report, Lyndon B Johnson Center

DOI: 10.2507/27th.daaam.proceedings.126

SIMULATION APPROACH FOR ANALYSIS OF THE INVENTORY CAPACITY FOR FILLING AND PACKAGING PLANTS

Ilya Kukushkin, Alina Zavrazhina, Johann Grabenweger,
Branko Katalinic & Damir Haskovic

This Publication has to be referred as: Kukushkin, I[lya]; Zavrazhina, A[lina]; Grabenweger, J[ohann]; Katalinic, B[ranko] & Haskovic, D[amir] (2016). Simulation Approach for Analysis of the Inventory Capacity for Filling and Packaging Plants, Proceedings of the 27th DAAAM International Symposium, pp.0876-0885, B. Katalinic (Ed.), Published by DAAAM International, ISBN 978-3-902734-08-2, ISSN 1726-9679, Vienna, Austria
DOI: 10.2507/27th.daaam.proceedings.126

Abstract

This paper shows a new approach for the logistics analysis for modern filling and packaging plants. It compares different internal logistic strategies, considering the inventory sizes. The goal is to analyse the overall equipment effectiveness depending on each strategy using the simulation modeling.
The first experiment is to analyze the Just-in-time concept. Just-in time implies a lack of inventory. However, in real conditions the complete absence of inventory is impossible. Therefore, in the case of Just-in-time we creates the one storage rack that allows placing a small amount of pallets. In every new experiment the amount of storage racks will be increased by one. Running of the new experiments will cease when OEE value growth stops.

Keywords: Overall Equipment Effectiveness; Inventory; Warehouse space; Inventory Control Systems; Simulation.

1. Introduction

The main objective of any commercial organization is increasing of their own cash profit. The profit can be expressed by the following formula:

$$\text{Profit} = (\text{ Market Price} - \text{Production Costs}) \cdot \text{Output} \tag{1}$$

Market Price is fixed for a manufacturer. However, the manufacturer can affect on the Production Costs and Output. The Formula 1 leads to the conclusion that the higher the Output value the higher the Profit. The Output depends on various factors. For instance, equipment breakdown, machines deterioration, production of defective products, etc. Analysis of all the factors influencing the Output is possible using the Overall Equipment Effectiveness (OEE). If OEE is 100% it means Output is maximum.

Any Production Costs result to a reduction in Profit. Production Costs consist of the Capital Expenditure (CAPEX) and Operative Expenditure (OPEX) [1].

OPEX includes day-to-day costs for the goods production. A list of these costs for filling and packaging plants are following:

• Energy and electricity costs
• Packaging costs (preforms, labels, caps, foil, boxes, pallets)
• Labour costs
• Maintenance costs

CAPEX is an investment in equipment and buildings.
Therefore, three ways to increase Profit are:

1. To reduce CAPEX
2. To reduce OPEX
3. To increase output

To reduce the CAPEX costs is possible due to the reduction of the building and equipment costs. Savings on equipment may decrease the Output since production capacity reduces. The building costs can be decreased due to space reducing, for instance, warehouse space. Insufficient size of raw materials warehouse could cause a stoppage in production because of lack of raw materials. Lack of space at finished products warehouse could also cause the production stoppage since goods have to be stored under specific conditions. Both situations lead to the Output losses. Excessively large warehouses result to unnecessary costs for their construction (CAPEX) and their further maintenance (OPEX). The problem is to find the golden mean between the warehouses size and OEE.

In our previous paper we described a development of a simulation model of the filling and packaging line [2]. The model allows to carry out the overall equipment effectiveness analysis in case of different situations emerging on the filling and packaging lines. In the present paper we supplement the current model with elements of logistics and carry out experiments to search for optimal inventory size for the particular packaging plant.

2. Inventory control system

Companies should be flexible since customer's desires can be very diverse and changeable. The company satisfying the customer's needs before all competitors gets a project and all profit. Effective inventory control system allows to manage customer demand and to reduce inventory and storage costs [3]. There are two systems of inventory control: push and pull.

Push system means a production of enough amount of product to meet the consumer's demand. Producers are forced to predict how many products should be produced. Inaccuracies of demand forecast can be the cause of excess in the warehouse. It results in additional inventory and storage costs [4].

Examples of push systems are the MRP and MRP II strategies [5]. They are implemented as an application package. MRP provides coverage of the enterprise resources and enables the operational, manufacturing and financial planning.

Pull system are characterized by production of enough product to carry out customer's orders. The advantage of pull system over the push system is minimization of the inventory size and it leads to a decrease of storage costs [4].

The most widespread logistical concept in the literature is the Just-in-Time» (JIT). It implies that the components from the previous operation are delivered to next operation at the precise moment when they are needed, but not before. This approach reduces costs by eliminating the warehouse space [6]. In other words JIT is example of pull system. On the other hand if a supplier of raw materials can't deliver the raw material at exactly specified time, it could result in losses or even completely stop of the production process.

The available today theoretical methods and computer systems such as ERP and MPR give only approximate recommendations to minimize inventory in case of warehouse design development.

To see how effectively inventory management functions and what changes is reasonable to accept became possible with simulation modeling. The capacity of modern computer systems and the presence of professional software makes it possible the development of the factory simulation model and conducting experiments on a variety of processes taking place at the plant.

The simulation doesn't suppose a rejection of the technologies listed above. It offers their joint use because many of useful features already built into these systems.

3. Description of the production line

The initial model is developed for soft drinks filling and packaging line. The line can be configured to work with 1.5 and 0.5-liter PET bottles. In both cases the bottles are blown from preforms directly in the line. The production line represents a number of machines connected with each other through conveyors. List of machines and description of their functions are shown in Table 1.

Machine name	Function
Blower	Intended for the production of plastic bottles by blowing them out preforms.
Mixer	Mixes the ingredients to get the finished beverage
Filler	Fills the beverage into the empty bottles and close them with a cap.
Labeling machine	Pastes the label on the bottle filled with beverage. The quality of gluing is checked.
Packing machine	Packs bottles in six-pack or different packaging.
Carry strip master	Attaches the handle to the package for easy carrying bottles.
Palletizer	Lays out received boxes on pallets.
Pallet wrapper	Tightens the pallet with a protective foil.

Table 1. Description of the machines functions

Line is formed in 2-block design, which means that the blowing and filling machines are interconnected in one block. The only accumulation table is located between filling and labeling machines, as shown in Fig.1. Other conveyors are not designed for additional accumulation.

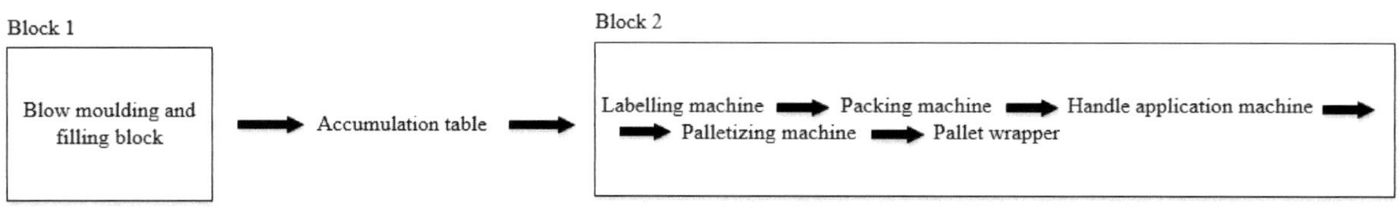

Fig. 1. 2-block design

The input information about orders contains following fields:
- Beverage Sort type (Cola, Cola light, carbonated water, etc.)
- Bottle volume, l
- Lot size, bottle
- Type of label
- CIP matrix including cleaning time before the order and cleaning time after the order, sec
- Changeover matrix including changeover time before and after the order, sec

4. The warehouse description

The warehouse area is divided into two parts:
- Raw material warehouse
- Finished products warehouse

It is necessary to pr
ovide the presence of loading zone and discharge zone for transport. The layout plan of warehouse is shown in Figure 2.

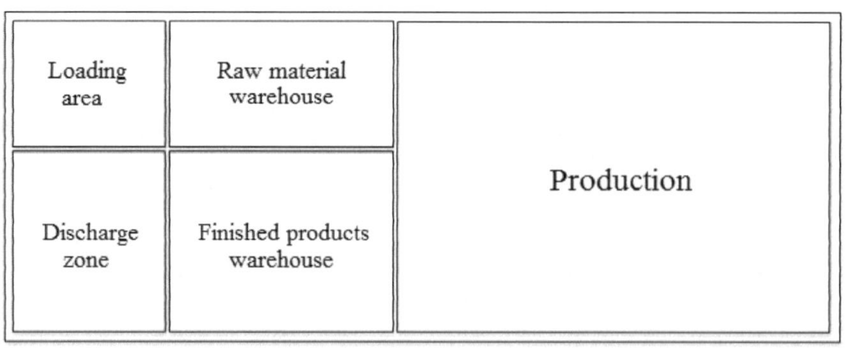

Fig.2. Layout plan of warehouse

There are two groups of forklifts working in warehouse. The first group is engaged in loading and unloading of coming trucks and the second group delivers raw materials to machines, unused materials back to the raw material warehouse and finished goods to finished products warehouse. Passages for forklifts are provided between storage racks.

5. Model development

Simulation model is designed with the use of professional tools Anylogic. This program allows to solve problems in various fields of modeling: discrete event, system dynamics and agent based modeling. The main difference of AnyLogic from other existing solutions is the ability to combine and simultaneously use different modeling paradigms in the development of one model. It allows to create complex models, which are most compatible with the real production. Developed model consist of two simulation methods – agent-based and discrete-event modeling. Discrete-event simulation allows to observe the processes occurring at the production line and agent-based modeling allows to set up machine behavior logic. Integration of Java compiler in AnyLogic opens up almost unlimited possibilities for the model development.

5.1 Warehouse development

Discrete event simulation is most often used in the simulation of queuing systems, production processes and logistics [7]. Discrete Event modeling approach is implemented in AnyLogic using Enterprise Library. Enterprise Library is full of a variety of objects that allows to perform various operations. For example, the capture of the resource, the delay, the creation and destruction of the agents, etc.

Distributors trucks have an access to the discharge area and trucks from suppliers have an access to the loading area. Trucks have different capacity, customizable via such variables as ShippingTruckCapacity and TruckCapacity respectively. Also maximum movement speed for trucks on the territory was specified via variable TruckSpeed.

Warehouse can be represented by the following elements of Enterprise Library (Fig. 3).

Fig. 3. Warehouse elements

Two groups of forklifts are connected with model (Fig. 5).

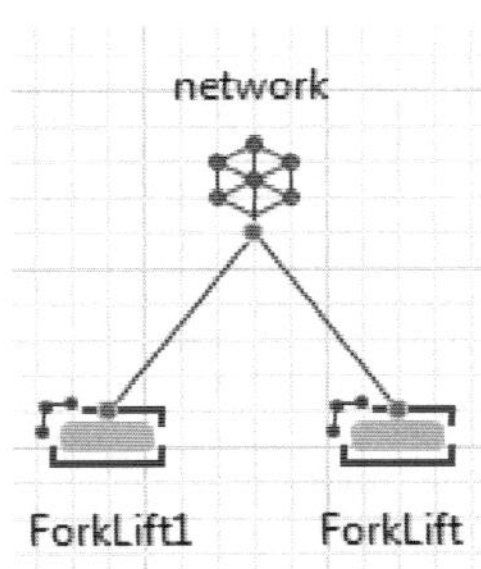

Fig. 4. Model network

Each storage rack can be developed according to individual design (Fig. 5).

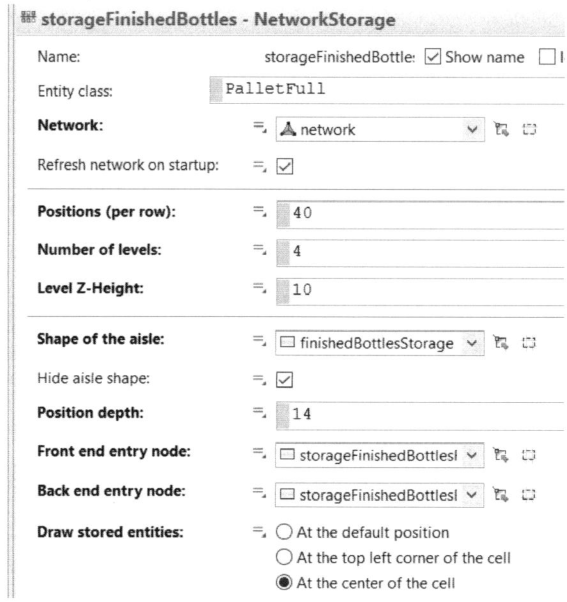

Fig. 5. Configuring of storage rack

The following processes were implemented by using the elements of standard library AnyLogic:

- Delivery of raw materials to the plant;
- Unloading of raw materials in the warehouse;
- Raw materials delivery from the warehouse to the machine;
- Transportation of the finished products to the warehouse;
- Shipment of finished products from the warehouse.

Figure 6 presents the process of delivery of raw materials to the plant, unloading of raw materials in the warehouse and delivery of raw materials from the warehouse to the machine. Truck advent causes the creation of preforms boxes and set a schedule for the delivery truck. Configuring the number of preforms into one box is easily adjusted by changing variable BoxsCapacity. Then the boxes with preforms are loaded into the truck and go to the loading area. After these operations there is branching: preforms are sent to the raw material warehouse and the truck is waiting in the loading until all boxes with preforms will be removed from it. RestrictedAreaStart RestrictedAreaStop elements set a limit on the number of preforms boxes that one forklift can carry. BoxAtMachine element lets forklift to know when it should stop the boxes delivery to machine. Then dropoffRawParts element opens the boxes and provides the preforms to Blower.

From time to time a Changeover happens on the line. In reference [8] the following changeover concept is gaven: changeover is the process of converting a line or machine from running one product to another. In this case we need to pick up the rest of PETs, labels and caps and bring them back to the raw material warehouse (Fig.7).

Almost every element requires a writing additional code. For example, Fig. 8 shows the code that calls a forklift with a box of caps for Filler.

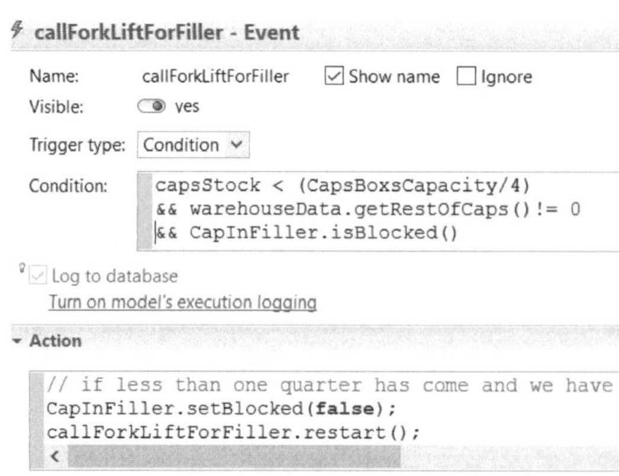

Fig. 8. Event of forklift call

Figure 9 shows the processes of transporting finished products to the warehouse and the shipment of finished products from the warehouse.

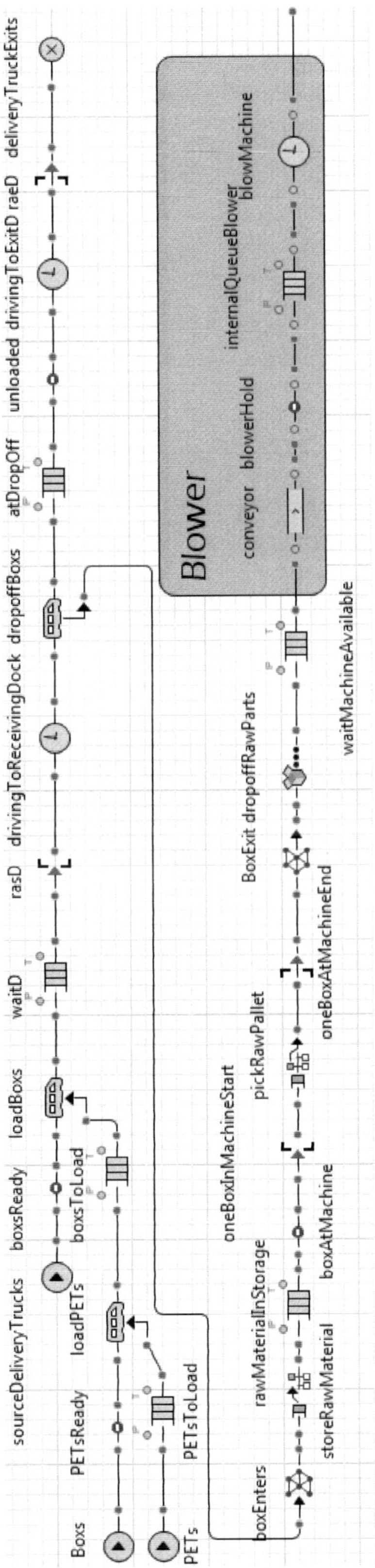

Fig. 6. Processes of delivery of raw materials to the plant, unloading of raw materials in the warehouse and delivery of raw materials from the warehouse to the machine represented via elements of standard library

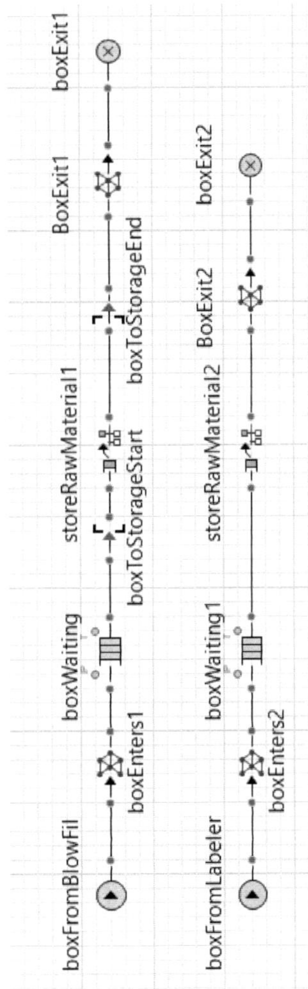

Fig. 7. Rest of raw material delivery back to the raw material warehouse

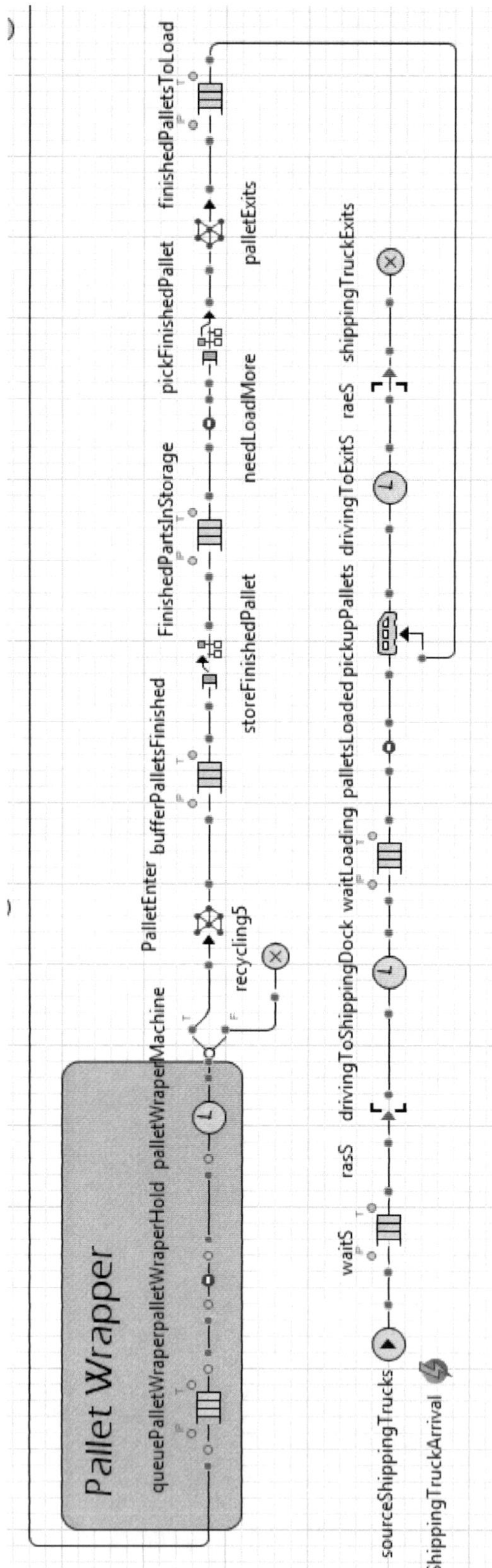

Fig. 9. Processes of transporting finished products to the warehouse and the shipment of finished products from the warehouse

Pallet Wraper wraps pallets by protective film. Than they includes in the network, where they are picked up by trucks and unloaded at the warehouse for further storage. When Shipping Truck Arrival event is active the trucks come for shipment of the finished products. Also truck stop was provided in the discharge zone as long as the tuck will be fully loaded.

The 2D graphics (Fig. 10) and Gantt chart (Fig. 11) were developed to observe the functioning of the system.

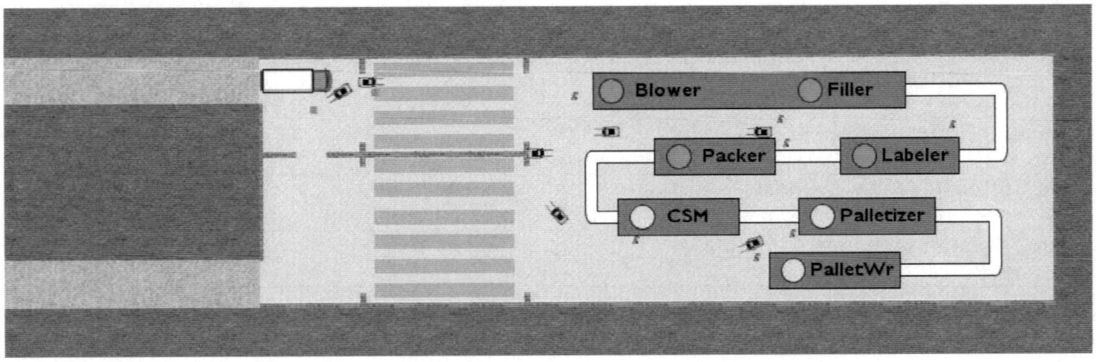

Fig. 10. 2D graphics of model

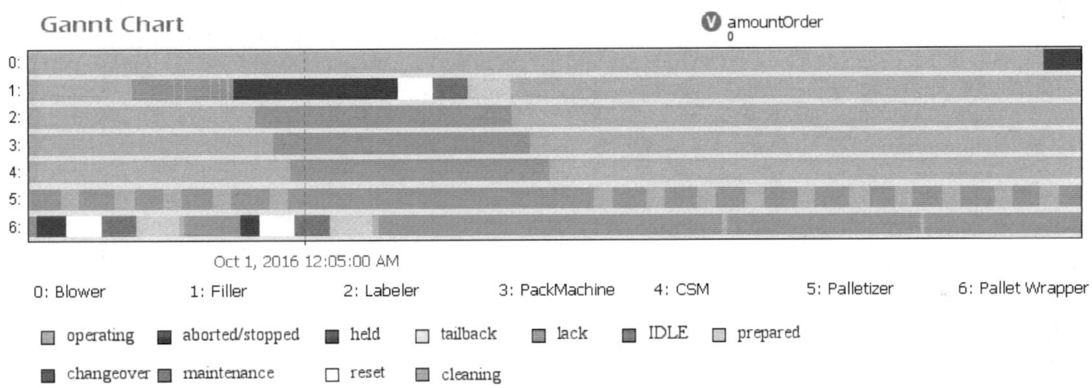

Fig. 11. Gantt chart

6. Experiments

The aim the experiment is to determine the optimal warehouse space for the particular plant. The first experiment is to analyze the Just-in-time concept. Just-in time implies a lack of inventory. However, in real conditions the complete absence of inventory is impossible. Therefore, in the case of Just-in-time we creates the one storage rack in raw material warehouse and one in finished products warehouse that allows placing a small amount of pallets. In every new experiment the amount of storage racks will be increased by one. Running of the new experiments will cease when OEE value growth stops.

The analysis is based on the value of overall equipment effectiveness. OEE is used for monitoring and improving production efficiency in the industry. The higher the OEE value the more effectively filler is used, accordingly the greater production output.

In case of raw materials warehouse workload, the call of truck will be canceled. Stop of production does not occur immediately because the other machines can continue working on conditions the free space in buffer. However, the lack of space in the finished products warehouse can lead to production stop. The resumption of production will begin after free up space at least for one pallet.

Table 2 gives the simulation results. The first experiment shows the Just-in-Time situation. In each subsequent experiment we add the extra storage racks.

In second and third experiments the drop sizes increase due to increasing of raw material warehouse space. Average utilization of raw material zone has the average performance because new supplies can come with a delay. Delays in the supply of raw materials are generated by asymmetric triangular distribution, where the minimum value is 0 hours, maximum value is 7 hours and most likely is 3 hours.

Further increase of raw material warehouse does not lead to a significant performance improvement, therefore the third experiment is taken in the following experiments. In fourth and fifth experiments we add extra storage rack in finished products warehouse (Table 3).

	Just-in-Time (1st Experiment)	**1 extra storage rack (2nd Experiment)**	**2 extra storage rack (3rd Experiment)**
		in raw material warehouse	
OEE	42.4	44.2	47.9
Maximum Utilization of Raw material zone, %	91.7	86.5	95.2
Maximum Utilization of finished bottles zone, %	97.8	97.8	97.8
Average utilization of Raw material zone, %	46.4	49.9	54.4
Average Utilization of finished bottles zone, %	81.8	82.9	84
Filler Utilization, %	43.7	43.9	45.6
Lead time, sec	1 039 804	1 038 822	993 967

Table 2. The simulation results

	1 extra storage rack (4th Experiment)	**3 extra storage rack (5th Experiment)**
	in finished products warehouse	
OEE	49.3	50
Maximum Utilization of Raw material zone, %	85.7	85.1
Maximum Utilization of finished bottles zone, %	98.3	98.9
Average utilization of Raw material zone, %	52.8	52.3
Average Utilization of finished bottles zone, %	78.7	81.2
Filler Utilization, %	46.8	49
Lead time, sec	977 968	907 101

Table 3. The simulation results

Increasing of the finished products warehouse space has a strong impact on the lead time. The pallet wrapper will be stopped when there is no enough space in finished products warehouse. If the buffer placed in front of it is full, the palletizer will be also stopped and so on.

7. Conclusion

The simulation results show that the economy of warehouse space can lead to frequent production stops and bring the losses for manufacturer. Simulation have led to increasing of OEE from 42.4% to 50% and reduction of lead time by 132 703 seconds. We can take into account that one bottle makes a profit equals 1-euro cent, filler succeeds to fill from 8 to 10 bottles in one second depending on the volume of the bottle, the simulated scenario is performed for two weeks, and saved lead is 132 703 seconds for two work weeks. By changing the warehouses sizes as it was shown in 5th experiment the production output of 0.5-liter bottles will increase by 1 327 030 (filler capacity is 10 bottles/sec). The company's profits will be increased more than 10 000 euros per every two work weeks or more than 25 000 euros per year.

Simulation provides an opportunity to develop the warehouse design for a particular product line. The results demonstrate that the inventory capacity can significantly affect the downtime in production and consequently the final profit.

8. References

[1] Benaskar F. Cost analysis for a continuously operated fine chemicals production plant at 10 kg/day using a combination of microprocessing and microwave heating //Journal of Flow Chemistry. – 2011. – T. 1. – №. 2. – C. 74-89.

[2] Kukushkin, I[lya]; Zavrazhina, A[lina]; Grabenweger, J[ohann]; Kildibekov, A[skar]; Katalinic, B[ranko] & Haskovic, D[amir] (2016). Model-based concept for scheduling analysis of packaging lines, Proceedings of the 26th DAAAM International Symposium, pp.1149-1157, B. Katalinic (Ed.), Published by DAAAM International, ISBN 978-3-902734-07-5, ISSN 1726-9679, Vienna, Austria

[3] Axsater S. Inventory control. – Springer Science & Business Media, 2007. – T. 90.

[4] Ghrayeb, Omar, Nipa Phojanamongkolkij, and Boon Aik Tan. "A hybrid push/pull system in assemble-to-order manufacturing environment." Journal of Intelligent Manufacturing 20.4 (2009): 379-387.

[5] Cochran, J. K., and H. A. Kaylani. "Optimal design of a hybrid push/pull serial manufacturing system with multiple part types." International Journal of Production Research 46.4 (2008): 949-965.

[6] Modak N. M., Panda S., Sana S. S. Optimal just-in-time buffer inventory for preventive maintenance with imperfect quality items //Tekhne. – 2015. – T. 13. – №. 2. – C. 135-144.

[7] Boev V.D. Computer simulation. Handbook for practical classes, course and diploma projects in AnyLogic 7/VAS, 2014. 432 p.

[8] Henry J.R. Achieving Lean Changeover: Putting SMED to Work/ Productivity Press, 2012. 205 p.

DOI: 10.2507/27th.daaam.proceedings.127

STUDY OF THE TRANSIENT PROCESSES IMPACT ON THE DIESEL FUEL CONSUMPTION WHILE IN SERVICE

Sergey Ovcharenko, Vitaliy Chetvergov

This Publication has to be referred as: Ovcharenko, S[ergey] & Chetvergov, V[italiy] (2016). Study of the Transient Processes Impact on the Diesel Fuel Consumption While in Service, Proceedings of the 27th DAAAM International Symposium, pp.0886-0891, B. Katalinic (Ed.), Published by DAAAM International, ISBN 978-3-902734-08-2, ISSN 1726-9679, Vienna, Austria
DOI: 10.2507/27th.daaam.proceedings.127

Abstract

This paper shows the study results of transient processes parameters of the diesel locomotive engine of the shunting locomotive. The paper discusses the transient processes impact on the diesel fuel consumption using results of experimental runs. The result is a theoretical model for calculating the coefficient of the transient processes impact on the fuel consumption.

Keywords: diesel; fuel consumption; transient process; locomotive; fuel consumption rate.

1. Introduction

Providing traffic by railway transport associated with need to perform an auxiliary shunting operations, that has a number of features complicating the formation of fuel consumption rates per trip. The complexity of consideration of this factor leads to an increase of the diesel fuel consumption and traffic costs.

Exploitation of shunting locomotives associated with frequent changes of diesel-generator set (DGS) operating mode. According to the author [1] the overall duration of the diesel shunting locomotive in a transient mode can reach 20-40% of the total operating time.

Solution to the problem of formation of the science-based standards of the fuel consumption rate per trip for shunting work is possible through the use of a mathematical model that takes into account the impact of transient processes on the increase in fuel consumption. Scientific papers of authors [2, 3, 4] and several other are dedicated to research of work process parameters when changing the operation mode of diesel. The main problem which is solved in these studies is the optimization of operating influences of the fuel equipment management system in order to improve economic and ecological characteristics of diesel.

The dependence of the increase in fuel consumption in the transition process as compared with the established operation mode of diesel was set as a result of experimental studies. Duration of the transition process was set when passing of the diesel operation mode from one to another control notch. Influence coefficient of transient processes number on the diesel fuel consumption during operation was proposed in the paper. Correlation dependence of the influence coefficient of transient processes number on the diesel fuel consumption on the number of transient processes per trip was obtained.

2. Analysis of experiments

The analysis of the experimental data obtained by several motive-power depot of industrial railway transport allows to state that within 10-12 operating hours of a shunting locomotive the switching number n of control notches (of transient processes) can reach 2500. Figure 1 shows the results of the studies presented in the form of distribution of the switching number n of control notches per trip.

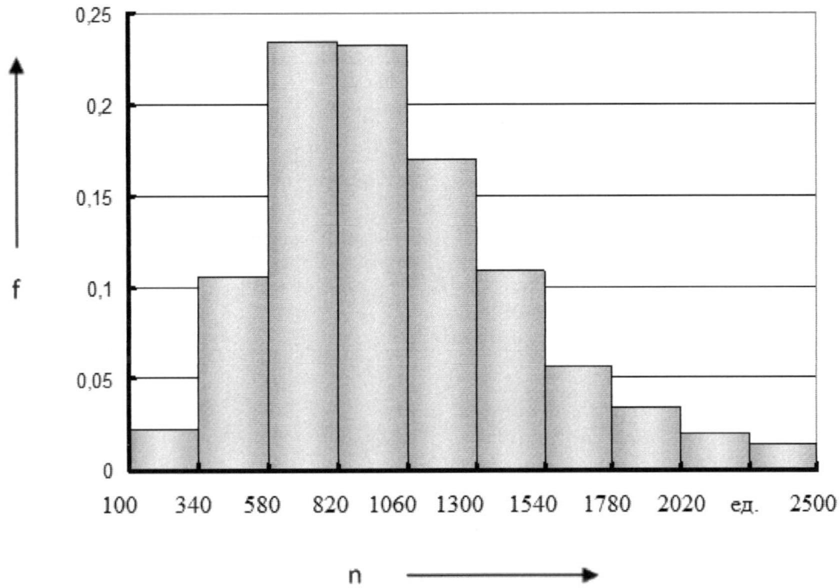

Fig. 1. The distribution of the switching number n of control notches per shift

When changing the controller handle from one position to another the speed governor of crankshaft and capacity controls the fuel supply by moving the fuel injection pump racks. The rack rotates a fuel pump plunger, thereby changing the positioning of edge 1 (Fig. 2) relative to the opening 2 through which the fuel fills a space above the plunger. The plunger explosion stroke increases (decreases) and a certain amount of diesel fuel supplied into the combustion chamber. Figure 3 presents the dependence of the cyclic fuel delivery (b, g) on the rack output of fuel pump (h, mm) for diesel PD1M (ПД1М – in original).

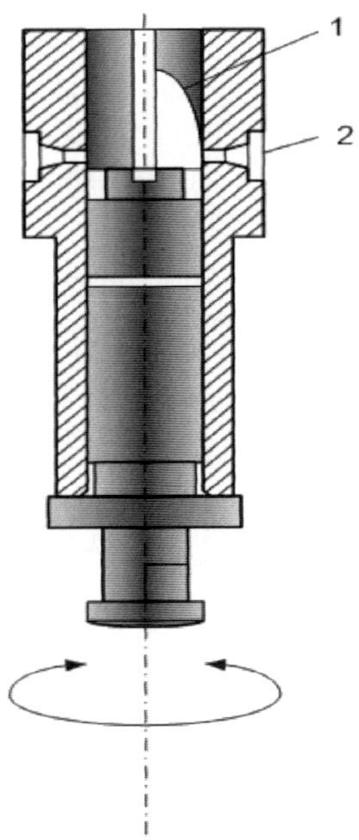

Fig. 2. Fuel injection pump plunger

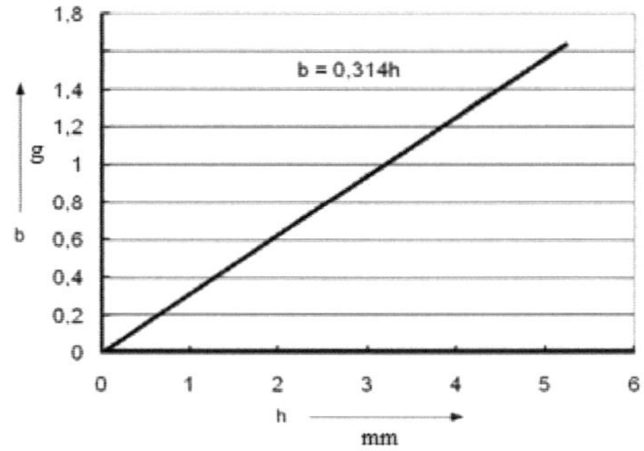

Fig. 3. The dependence of the cyclic fuel delivery on the effective stroke of fuel injection pump plunger

The frequency stabilization process of engine crankshaft rotation corresponding to the installed operation mode of diesel-generator set occurs for a while when the fuel consumption is increased. The equation for an approximate evaluation of the difference between the cycle fuel supply in the transient and steady operation mode of diesel-generator set is provided in [5]:

$$\Delta q_{\text{ő}} = J_{\Sigma} \frac{dP_{\text{Ã}}}{dt} \left(\frac{\partial M_{\text{Ã}}}{\partial \omega_{\text{Ã}}} \omega_{\text{Ã}} + M_{\text{Ã}} \right)^{-1} H_u^{-1} \eta_e^{-1} \omega_{\text{Ã}}^{-1} z \pi \tau \tag{1}$$

where J_{Σ} is a moment of inertia of the rotating diesel-generator masses and auxiliary equipment of locomotive with diesel drive;

$P_{\text{Ã}}$ is a traction generator capacity;

t is a time;

$M_{\text{Ã}}$ is a moment, being passed to generator shaft;

$\omega_{\text{Ã}}$ is aт angular velocity of diesel crankshaft;

H_u is a fuel calorific efficiency;

η_e is an effective efficiency of diesel in steady mode;

z is a number of diesel cylinders;

τ is a cycle coefficient of diesel.

Taking into account that the work process parameters change in transient mode, the use of equation (1) for calculating the influence coefficient of transition process on the diesel fuel consumption is difficult in practical calculations.

To evaluate the impact degree of a number of main controller switching on increasing the fuel consumption the experimental studies were conducted on shunting locomotives at carrying out of the rheostat tests. The inductive displacement sensor was mounted on the fuel pump rack. It allows to fix the rack output with 1 mm resolution and maximum displacement of 60 mm, provided that the maximum possible rack output of rails is 40 mm. The rack displacement was recorded through 0.1 s. Figures 4 and 5 presents the temporary implementation of fuel pump rack during transient processes of the diesel PD1M of the locomotive series TEM2 obtained from the results of the experiment in load conditions and idle running. The control notches set was carried out from minimum to maximum and back again with time delay in order to stabilize the fuel supply at every transient process.

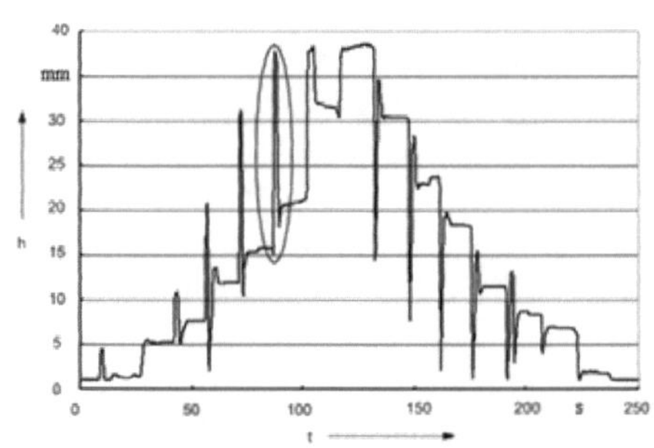

Fig. 4. Output of fuel injection pump racks during transient processes in load conditions

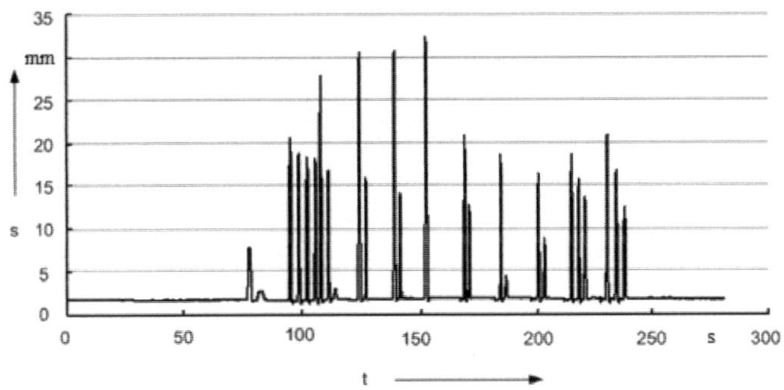

Fig. 5. Output of fuel injection pump racks during transient processes in idle running

The fuel pump rack output is proportional to the cyclic fuel supply. To evaluate the degree of transient processes influence over the diesel fuel consumption is enough to compare the area bounded by line 1 (Figure 6) in the transient process and the rectangle 2, characterizing the fuel consumption in the steady process.

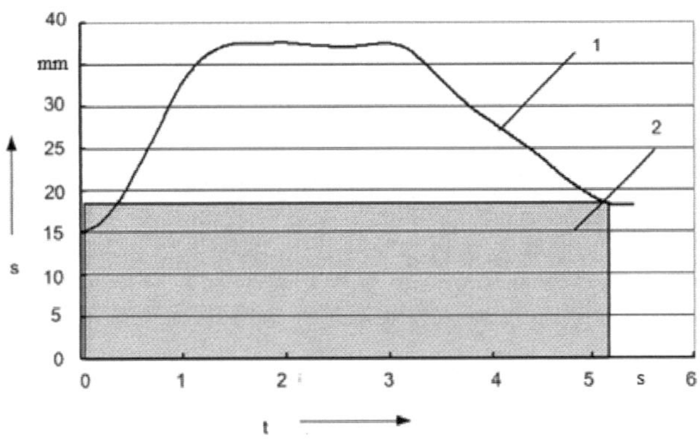

Fig. 6. Fragment of the transient process

As a result of the experimental data processing we found that the duration of the transient process for diesel PD1M is an average of 5 seconds. To calculate the area bounded by a line 1 (Fig. 6), the fuel pump rack output in the transient process was described by 5th degree Lagrange polynomial:

$$H(t)=\sum_{i=0}^{5} h_i \prod_{\substack{j=0\\ j\neq i}}^{5} \frac{t-t_j}{t_i-t_j} \tag{2}$$

The area obtained by polynomial integrating in the time interval 0 - 5 s:

$$S=\int_0^5 H(t)dt \tag{3}$$

According to analysis of the experiment results the mathematical expectation of exceeding the fuel consumption during the transient processes relative to the steady state is accepted as 43% in load conditions and 25% in idle running.

In practical calculations the transient processes impact on the fuel consumption is taken into account by coefficient $K_{\ddot{I}}$. Theoretical value of $K_{\ddot{I}}$ can be obtained based on the assumption of equality calculated diesel fuel consumption in all operation modes of the diesel-generator set and fuel consumption subject to the experimentally established impact degree of the transient processes in load conditions and idle running.

$$\left(\sum_{i=0}^{8} b_{xi}t_{xi}+\sum_{i=1}^{8} b_{ni}N_i t_{ni}\right)K_{\ddot{I}}=\left(\sum_{i=0}^{8} b_{xi}t_{xi}+\sum_{i=1}^{8} b_{ni}N_i t_{ni}\right)+\sum_{i=0}^{8} b_{xi}\left(d_x-1\right)nt_{\ddot{I}}K_x+\sum_{i=1}^{8} b_{ni}\left(d_n-1\right)nt_{\ddot{I}}K_n \tag{4}$$

where i is a number of control notches;

b_{xi} is a fuel consumption on control notches in idle running, kg/h;

b_{ni} is a fuel consumption on control notches in load conditions, kg/kW·h;

t_{xi}, t_{ni} are a work time of locomotive diesel-generator set on control notches in idle running and under load conditions respectively, h;

$t_{П}$ = 0,0013 is a set time of the transient process, h;

d_x, d_n are a transient processes share attributable to the idle running and load conditions respectively;

K_x, K_n are an impact coefficients of transient processes on increasing the fuel consumption in idle running and load conditions respectively;

N_i is a locomotive diesel-generator set capacity on control notches, kW;

n is a number of transient processes.

From the equation (4):

$$K_П = 1 + \frac{\sum_{i=0}^{8} b_{xi}(d_x - 1)nt_П K_x + \sum_{i=1}^{8} b_{ni}(d_n - 1)nt_П K_n}{\sum_{i=0}^{8} b_{xi}t_{xi} + \sum_{i=1}^{8} b_{ni}N_i t_{ni}} \qquad (5)$$

Figure 7 provides the results of numerical simulation of $K_П$ when changing the number of transient processes from 0 to 2500 and the ratio of the transient processes share in load conditions and idle running d_n, d_x ($d_n + d_x = 1$).

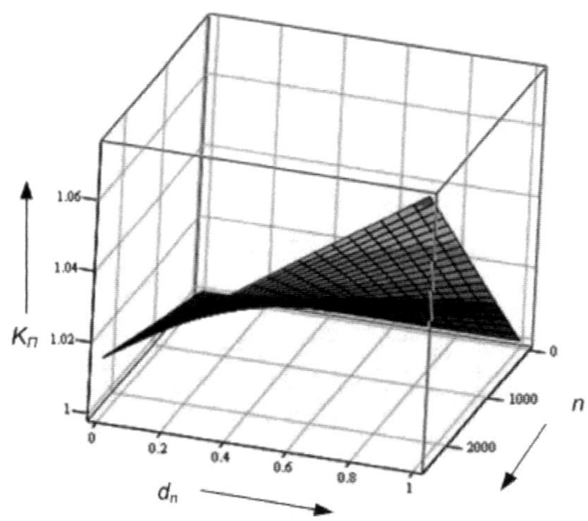

Fig. 7. The surface of the coefficient $K_П$ values on the number of transient processes and their redistribution on operating modes of diesel-generator set

Random factors influence the fuel consumption during the operation of locomotives. To take into account these factors is very problematic in the theoretical simulation. Experimental trip in which operation modes of locomotive diesel-generator set were recorded, a number of transient processes and actual fuel consumption were conducted to evaluate the impact of transient processes on increasing the fuel consumption. Fuel quantity consumed by diesel-generator set was assessed during the obtaining of the dependence $K'_П = f(n)$ from an actual fuel consumption using combustion characteristic. Remaining fuel share refers to the increase due to transient processes. The impact of number of diesel-generator set transient processes on fuel consumption was found as a result of experimental trips (Fig. 8). Average number of control notches switching in a minute by the trip results is indicated on the abscissa. Constituent value of the coefficient of fuel consumption increasing is located on the ordinate. Analysis of regression dependence allows to assert that the impact degree of transient processes established by experimental trip results on the fuel consumption agrees with the results of theoretical research and can reach 6-8%.

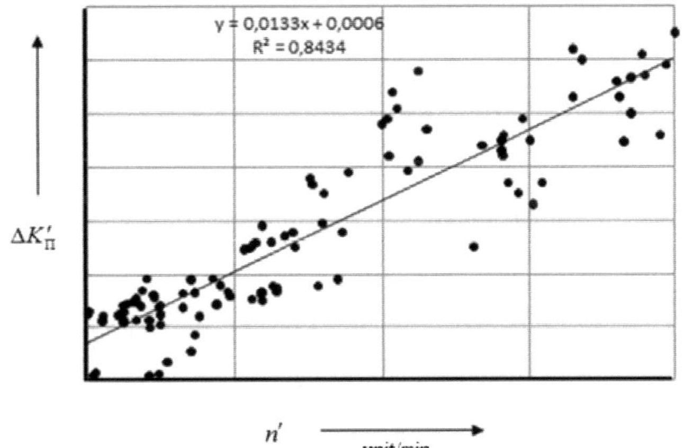

x

Fig. 8. Influence of transient processes on fuel consumption coefficient by experimental trip results

3. Conclusion

1) When diesel operates on transient process, a fuel consumption is increased by 43% in load conditions and 25% in idle running;
2) The results of theoretical calculations allow us to conclude that by increasing the operating time of diesel-generator set in the transient processes, the increase in fuel consumption per shift can reach 6-8%;
3) Equation for calculating K_{Π} was obtained for practical calculations. The equation allows taking into account the impact of transients processes on locomotive fuel consumption;
4) The results allow to move to the implementation of a science-based fuel consumption rate per trip using the onboard electronic systems for controlling the work parameters of locomotive diesel on shunting operations.

4. References

[1] Homich, A.; Tupitsyn, O.; Simson, A. (1975). Fuel Economy and Thermotechnical Modernization of Locomotives, Transport

[2] Kurapin, A.; Slavutskiy, V. (2014). Research of transient processes in fuel-injection equipment of autotractor diesel by experimental frequency method. Proceedings of the Volgograd Technical University, Vol. 145, No. 18, pp. 21-26

[3] Markov, V.; Polukhin, E. (2008). Diesel transient processes with angle control system of fuel injection timing advance. Proceedings of the Universities. Mechanical Engineering, No. 5, pp. 33-65

[4] Gorelik, G. (2008). Study of the relationship between fuel supply and mixture formation with the working process in diesels operating on sharing modes. Journal of Pacific State University, Vol. 8, No. 1, pp. 75-90

[5] Kosov, E.; Sukhoparov, S. (1999). Operating Modes Optimization of the Locomotive Diesel Generators, Intext

DOI: 10.2507/27th.daaam.proceedings.128

TREAD PROFILE OF WHEEL PAIR SUPPORT UNIT

Valeriy Shiler, Alexander Shiler, Alexander Lyubchenko

This Publication has to be referred as: Shiler, V[aleriy]; Shiler, A[lexander] & Lyubchenko, A[lexander] (2016). Tread Profile of Wheel Pair Support Unit, Proceedings of the 27th DAAAM International Symposium, pp.0892-0895, B. Katalinic (Ed.), Published by DAAAM International, ISBN 978-3-902734-08-2, ISSN 1726-9679, Vienna, Austria
DOI: 10.2507/27th.daaam.proceedings.128

Abstract

One of the reasons of contact-fatigue damages generation on the tread of the standard wheel pair is the spurious wheel slippage along rails while moving. The spurious slippage makes an additional stress in metal of the wheel in the area of contact region "wheel-rail" and increased wear of the tread. The objective of our research is the development of an engineering solution aimed at the reduction of the metal stressed condition in the area of contact region "wheel-rail". In order to reach the formulated objective, the following tasks were solved: a novel block construction of the wheel set was developed with the main feature, which is an independent rotation of the tread of support unit wheels (leading and supporting) contacting with surfaces of track rail; a novel lateral tread profile of supporting wheel was designed and justified. The research had been carried out using the methods of simulation and physical modeling.
We have got and developed new lateral tread profile of wheel pair support unit, which ensures constant single-point contact of wheel with rail and guarantees the lack of spurious wheel slippage along a rail. Therefore, the corresponding shear stress at the contact point of "wheel-rail" is also lacking. This approach resulted in significant reduction of the metal stressed condition of the supporting wheel in the area of contact region "wheel-rail".

Keywords: lateral profile; tread; wheel slippage; stress.

1. Introduction

Rail transport leans on the standard wheel pairs, which consist of two wheels immovably mounted on the axle. Possessing equal angular velocity, wheel contacting surfaces with rails (three surfaces) have different values of tread radius. Therefore, movement of the wheel set is accompanied with spurious slippage along the rails, which is a source of: auto-oscillation of wheel set (nosing motion) while moving, increased resistance to movement depending on the square of progressive motion velocity etc. As a result of increased axial load, weight of a train and speed of motion, the wear rate of wheels and rails increased greatly. For example, if 15-20 years ago metal technological exit of wheel bandage amounted to 30-35% of their total mass, then nowadays this measure reaches 75-80%. At present, greasing (lubrication) of wheels and rail contacting surfaces is widely utilized for wear reduction [1-3]. This approach decreases the rate of deterioration processes, but it does not eliminate their reasons. Therefore, it is considered that any even compromise solution towards tread lateral profile of wheel is positive [2, 3]. For a solution of the problem of deterioration in "wheel-rail" system, the authors went by means of the improvement of the whole wheel set construction.

2. Problem definition

For a solution of the problem of dynamic interaction of wheel pair and rail gauge, the authors have proposed a block construction of the wheel set, which main feature is an independent rotation all surfaces their wheels contacting with surfaces of track rail. Besides, new lateral tread profile for the supporting wheel was developed. These technical solutions allowed realization of block wheel pair movement along rails without spurious slippage and lack of auto-oscillation mode (nosing motion), which are inherent in the standard wheel set.

In the presented work, the task of the estimation of influence of new technological solutions on internal metal stress of the supporting wheel of block wheel pair in the area of its contact with surface of rail.

3. Theory

First, consider the wheel deterioration model of the standard wheel pair [3], which is presented in Fig. 1. Lateral tread profiles of wheel and rail functionally are divided into the following three areas: A, B and C (Fig. 1). In area (A), contact stresses and lateral slippage are the lowest of all possible values that creates favorable conditions for stability of equipage movement. Contact area (B) is small, and the contact is characterized by extremely difficult stressed condition of wheel material. There are three possible variants of the contact of wheel of the standard wheel pair with a rail: single-point; double-point and conformal contact [3].

Single-point contact inflicts the greatest damages on the wheel and rail. Double-point is inherent in intense slippage and wear, which results in occurrence of material plastic flow on the wheel tread. Conformal comber contact appears to the extent deterioration of working wheel fillet up to the general profile [3-5], and has a number of advantages: contact area rises and specific pressure decreases. It is significant that for the standard wheel pair in running-out mode for all contact areas (A, B, C) and variants of contacting, the slippage is spurious conditioned by peculiarities of the standard wheel pair construction.

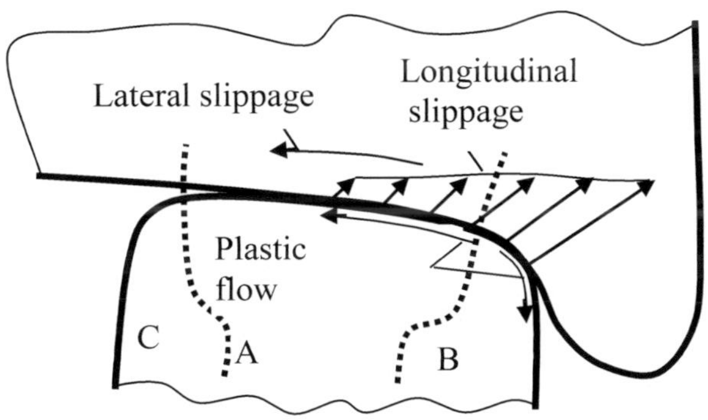

Fig. 1. Constant deterioration model of standard wheel

New construction of the wheel [6, 7], presented in Fig. 2, with working name "wheel unit" consists of leading wheel 1 and supporting wheel 2, which rotate independently of one another on wheel set axle

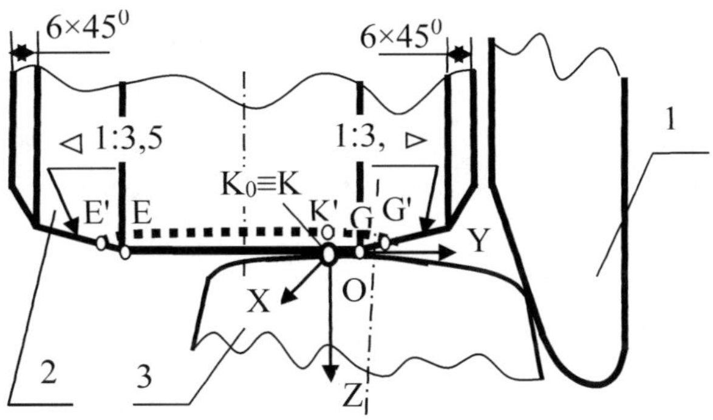

Fig. 2. Lateral profile of wheel treads of support unit

With accordance to the existed requirements for making lateral tread profiles and structural features of wheel unit, the lateral profile for supporting wheel is developed and is marked with "2" in Fig. 2. The proposed profile is made up of 5 surfaces. Central part of the lateral profile (segment EG) has cylindrical surface. Extreme sections of the profile have equal delineation in the form of two pairs of conic surfaces, which have the following dimensions: $\triangleleft 1:3,5$ and 6×45^0. These cones have the same purpose that the standard lateral profiles have. In exploitation, the central cylindrical part of the lateral tread profile (segment EK'$_0$G) is extended and comes out on adjacent conic surfaces (segment EE′ и GG′). Thus, delineation stability of the cylindrical part is maintained.

Models and experimental data devoted to the research the characteristics of "wheel-rail" system and presented in proceedings of foreign authors [8-10] are directed on the analysis of multibody systems. In this work, comparative calculations for the supporting and standard wheels were implemented for the objective of contact material stresses estimation in "wheel-rail" system. For each construction of the wheels, extreme conditions of contacting with rail head were accepted: for standard wheel – outworn tread profile and single-point contact with rail; for wheel unit – new linear profiles of the supporting and leading wheels and corresponding to them single-point contacts.

For normal problem solution the Hertz contact theory was applied [3]. Stresses at contact point lying on the pressure axle at a depth on O_KZ axis are equal to [3]:

$$\begin{aligned}
\sigma_x &= -\left[\frac{a^2+2z^2}{a\sqrt{a^2+z^2}} - \frac{2z}{a}\right] q_0; \\
\sigma_z &= -\frac{a}{\sqrt{a^2+z^2}} q_0; \\
\sigma_y &= \mu(\sigma_x+\sigma_z) = -2\mu\frac{\sqrt{a^2+z^2}-z}{a} q_0; \\
\tau_{xz} &= 0,
\end{aligned} \tag{1}$$

where σ_X and σ_Z are the main normal stresses operating at the areas, which are perpendicular to O_KX and O_KZ axis; are main stresses operating at the areas lying in plane of tread cross-section; μ is the Poisson ratio.

The largest shear stress from normal load in plane XOZ (Fig. 2) is defined as follows:

$$\tau_{max} = \frac{\sigma_x - \sigma_z}{2} = \frac{z}{a}\left[1 - \frac{z}{\sqrt{a^2+z^2}}\right] q_0 \tag{2}$$

Along with vertical load in horizontal plane of "wheel-rail" contact of the standard wheel, spurious friction forces of slippage have an effect, which make longitudinal and lateral stress components at the "wheel-rail" contact point. On the basis of the theory of independency of force effect, the contact stresses are summed up in equivalent stress [3].

4. Experimental Results

We have obtained that, according to the calculation results and the Hertz theory, maximum static stress of the grip occurs on the wheel tread, and maximum shear stress is at a depth of 2.0-4.0 mm. If normal deformation on the contact surface has elastic nature, then plastic deformations can appear under contact surface in the area of the maximum shear stresses.

When a wheel passes along a rail, repeated stress grip-strain occurs under contact area surface. The stress leads to accumulation of subsurface plastic deformation and origination of residual stresses in material [3], which are casual for various types of contact-fatigue damages on the tread of wheel and rail. When tangential force (for example, spurious slippage) is applied to the contact area, maximum shear stress rises and displaces closer to the tread. Further increase of the tangential force can result in appearance of alone region of potential damage of the material [3].

5. Discussion of Results

For demonstration of shear stress influence on contact bearing capacity (for example, due to spurious slippage), the so-called Johnson adaptability diagram (Fig. 3) is used. In this diagram, the border (line 1) between regions of different material behavior as a dependency of loading factor q0/[τ] on coefficient μ = T/N, where q0 is the normal contact pressure, [τ] is the yield point of material, T is the tangential force and N is the normal force. The largest level of plastic flow is reached on the contact surface if calculation results are above the curve 1 (Fig. 3).

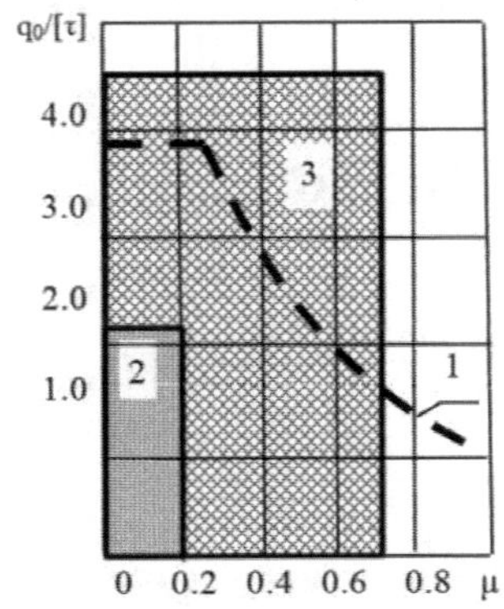

Fig. 3. Adaptability diagram of wheel material.

In accordance with the diagram in Fig. 3, in the interval of the accepted parameters of "wheel-rail" system, the stressed possible conditions of the material of the wheel pair support unit (area 2) are significantly less (in 2-3 times) in compassion with the standard wheel (area 3).

6. Conclusion

1. The proposed lateral cylindrical tread profile of wheel pair support unit ensures single-point of the supporting wheel with rail head in wheel unit position relative to track rail.
2. In process of forward movement, the supporting wheel rolls along a rail in the pure rolling mode without spurious slippage thanks to the independent rotation and cylindrical surface.
3. Lack of the spurious slippage reduces in 2-3 times the stressed wheel metal condition in the contact area of wheel and rail and brings to the minimum of their tread wear.

Thus, for the contact stresses, the block construction of wheel pair with cylindrical central part of the supporting wheel tread ensures conditions for the increase of axial load and motion speed of rolling stock for the existed railroad track architecture and improvement of operation safety.

7. References

[1] V.M. Bogdanov, "Reduction of deterioration rate of wheel flange and rail lateral deterioration," Railways of the world, vol. 12, 1992, pp. 30-34

[2] Zhiguo Li, Qing He, "Predicting failure times of railcar wheels and trucks by using wayside detector signals," IEEE International Conference on Mechatronics and Automation, 2014, IEEE, pp. 1113-1118. (DOI: 10.1109/ICMA.2014.6885854)

[3] W.J. Harris, S.M. Zakharov, J. Lundgren, H. Tournay, W. Ebersöhn, "Guidelines to best practices for heavy haul railway operations: wheel and rail interface issues," Virginia Beach, International Heavy Haul Association Publ., 2001. 408 p.

[4] I.V. Kragelsky, N.B. Demkin, "Contact area of rough surface," Wear, 1960, vol. 3, iss. 3, pp.170-187. (DOI: 10.1016/0043-1648(60)90136-8)

[5] N.B. Demkin, V.V. Izmailov, "Surface topography and properties of frictional contactsSurface topography and properties of frictional contactsSurface topography and properties of frictional contacts," Tribology International, 1991, vol. 24, Iss. 1, pp. 21-24. (DOI: 10.1016/0301-679X(91)90058-H)

[6] Patent Num.2011 154 293 (RF) MKI V60V 9/1 (2006.01) Tread lateral profile cushioned wheel bandage of wheel set, Shiler V.V., Shiler A.N. Application: 2011154293/11 from 28.12.2011. Applicant Shiler V.V. Published vBI, 10.07.2013 Bulletin Num. 18

[7] A. Shiler, "Analysis and simulation of new wheel pair construction," Procedia Engineering, 2015, vol. 100, Elsevier, pp. 1714-1723. (DOI: 10.1016/j.proeng.2015.01.547)

[8] A. Rovira, A. Roda, M.B. Marshallb, H. Brunskill, R. Lewis, "Experimental and numerical modelling of wheel–rail contact and wear," Wear, 2011, vol. 271, Elsevier, pp. 911-924. (DOI: 10.1016/j.wear.2011.03.024)

[9] S. Magheri, M. Malvezzi, E. Meli, A. Rindi, "An innovative wheel–rail contact model for multibody applications," Wear, 2011, vol. 271, Elsevier, pp. 462-471. (DOI: 10.1016/j.wear.2010.10.038)

[10] M. Mahdi Jalilia, H. Salehi, "Wheel/rail contact model for rail vehicle dynamics," ComptesRendusMecanique, 2011, vol. 339, Elsevier, pp. 700-707. (DOI: 10.1016/j.crme.2011.07.006)

DOI: 10.2507/27th.daaam.proceedings.129

NEW WHEELSET DESIGN FOR RAIL TRANSPORT

Valeriy Shiler, Alexander Shiler, Ilham Galiev

This Publication has to be referred as: Shiler, V[aleriy]; Shiler, A[lexander] & Galiev, I[lkham] (2016). New Wheelset Design for Rail Transport, Proceedings of the 27th DAAAM International Symposium, pp.0896-0900, B. Katalinic (Ed.), Published by DAAAM International, ISBN 978-3-902734-08-2, ISSN 1726-9679, Vienna, Austria
DOI: 10.2507/27th.daaam.proceedings.129

Abstract

Economic efficiency and competitiveness of rail transport depends mainly on the reducing the level of the dynamic interaction between wheel and rail. This paper presents a new technical solution for the wheelset in the form of a block design of the wheels. Analysis of the motion path of contact points was carried out in the system «wheelset – rail track». The paper evaluates the stress state of the block construction elements of the wheelset and shows the results of comparative theoretical and prototype tests of the block and standard wheelsets.

Keywords: block wheel; wheelset; independent rotation; motion path; contact; rail.

1. Introduction

The main disadvantages of a standard wheelset are significant unsprung weight (1500 ÷ 5000 kg) acting on the contact wheel-to-rail surfaces and the rigid connections concerning the coordinate rotation between points on the surface of wheels between the coordinate points of rotation of the wheel contacting with surfaces of rails [1, 2]. During the moving all points on the surfaces of wheels rotate at the same angular speed and wheels pass a different path length on the rails due to the presence of the inevitable difference of diameters of running surfaces. Such motion is accompanied by a «parasitic» wheel sliding on the rails and within the gap with rail track the wheelset moves in a winding path (continuous process of searching for equality of rolling circles diameters) – self-oscillation mode.

The winding motion occurs by energy of a forward motion, increasing resistance to its movement. But such a motion path for the standard wheelset is a blessing since the wheel flange contacts with the rail not constantly but only in 20 ÷ 30 m (wavelength of the winding motion).

Search of the wheelset design without disadvantages inherent in standard wheelset are conducted for a long time. Currently, it has accumulated more than 60 variants of technical solutions. The main features of these variants: independent rotation of the wheels in the wheelset, tire springing relative to wheel centre, wheels mounting at an angle relative to the vertical axis, division of the rigid axle of the wheelset on two parts connected by an elastic or freewheeling clutch [3].

2. Block design of the wheelset

Based on the analysis of numerous wheelsets designs the authors have proposed new technical solution of wheelset in which the wheels have a block design (Fig. 1) [4]. Block wheel consist of supporting and guiding 3 wheels. Rim of the guiding wheel has a standard shape of flange on the perimeter. The guiding wheel 3 is tightly fixed to the wheelset axle 8 and it perceives the horizontal guiding forces from side surface of rail. Supporting wheel consists of wheel centre 5, rubber pad 11 and flexible tire 1 and it is mounted on wheelset axle 8 by a pair of wheel bearing 13. Wheelset axle is mounted on the wheel bearing 9 of journal box and it transmits a vertical load through a wheel centre 5, elastic medium 11 and flexible tire 1, that rolls along a running surface of rail.

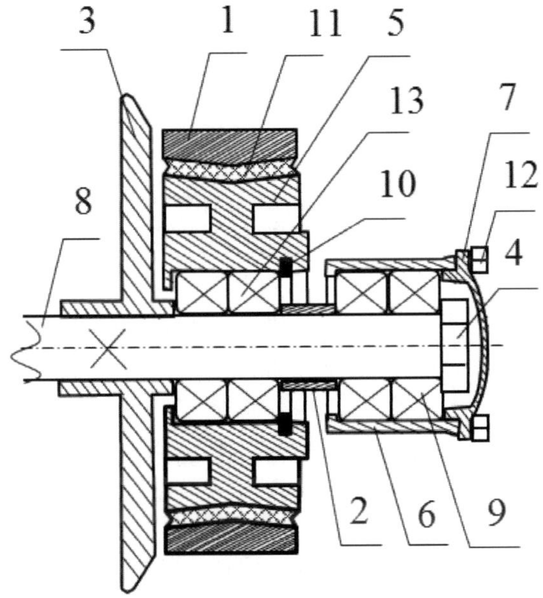

Fig. 1. Block wheel of a wheelset of a rail vehicle

Figure 2 shows the kinematic scheme of mechanical linkages on the rotation coordinate between the elements of the block design of the wheelset. In the mechanical system of «block wheelset – track», as in «standard wheelset – track», there are three mechanical loop with the same the basic sizes. In the system «block wheelset – track» the first (I) and second (II) mechanical loops are open (Fig. 2b.) due to the wheel centres mounting on bearings 13 (Fig. 1). The third loop (III) is formed by two guiding wheels 3 (fig. 1 and 2) tightly fixed to the wheelset axle. Only one of two guiding wheels can be in contact with the rail since the guaranteed open gap equal to 0.01 m (Fig. 2b and 2c) is provided between the wheel and rail gauges. Therefore, at any position of the wheelset in the rail gauge in the system «block wheelset – track» all mechanical loops (I, II and III) are open (Fig. 2b and 2c.). It allows to create the design of block wheelset without mechanism of the winding motion and eliminate such deficiencies as the impact of the difference in the wheel diameter on the position of the wheelset to the rail gauge, «parasitic» wheel slippage on rails during the forward motion in direct and curved parts of railway track, significantly reduce the unsprung weight in the system «wheelset – rail gauge».

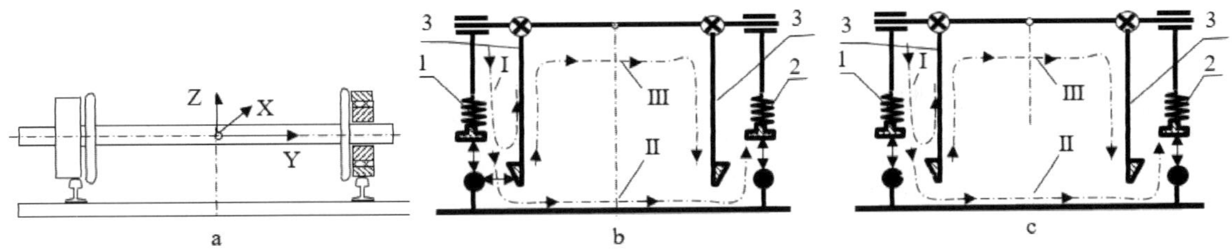

Fig. 2. Schemes of the mechanical loops of the system «block wheelset – rail gauge»
a) kinematic scheme of the block design of the wheelset;
b) mechanical loops when the flange of the guiding wheel contacts with a rail;
c) mechanical loops when the contact between flange of the guiding and rail is absent.

3. Analysis of motion trajectories of contacts points in the system «wheelset – rail gauge»

Figure 3 shows the motion trajectory of the contact points of the running surfaces of block and standard wheels in contact with the surfaces of a rail head, where K is contact point of running surface of flexible tire and standard wheels, B is flange of standard wheel and D is guiding wheel block design of wheelset. Trajectory of point (K) is an ordinary cycloid (Fig. 3, Line 1). The contact point (D) of the guiding wheel moves along the trajectory of ordinary cycloid (Fig. 3, Line 3), provided that the angle of attack on the rail is zero. As is known, the main feature of the ordinary cycloid is the equality of distance traversed by center of the generating circle (O_k) and its expanded length which corresponds to the wheel motion without slippage.

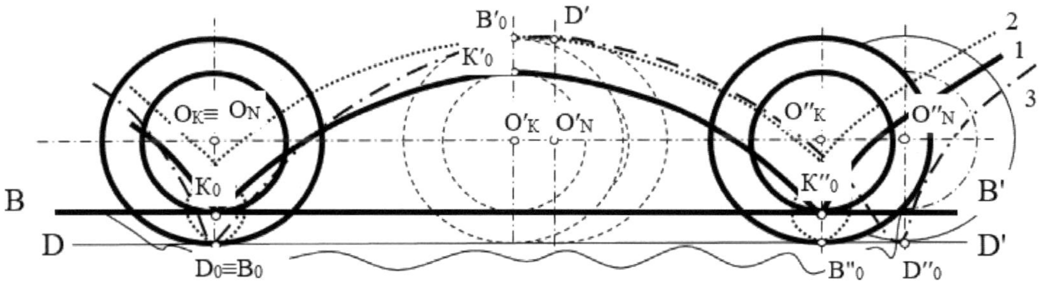

Fig. 3. The motion trajectory of the contact points of the running surfaces of block and standard wheels
Line 1 is trajectory of the point (K_0) of the rolling surface of wheels (the curve $K_0K_0'K_0''$ is an ordinary cycloid);
Line 2 is trajectory of the point (B_0) of the side surface of the standard wheel flange
(the curve $B_0B_0'B_0''$is an elongated cycloid);
Line 3 is trajectory of the contact point (D_0) of the side surface of the guiding wheel flange
(the curve $D_0D_0'D_0''$ is an ordinary cycloid).

The trajectory of the contact point (B) of the standard wheelset flange looks like an elongated cycloid (Fig. 3 and 4, line 2). The presence of a loop of elongated cycloid indicates a repetition of the traversed distance and consequently a «parasitic» slippage of standard wheel flange along the side surface of the rail head.

Instantaneous radius (r_{D_1}) of point of rotation D_1 relative to the instantaneous center (K_0) is determined by the equation:

$$r_{D_1} = \sqrt{t^2 + \lambda_p^2} \qquad (1)$$

where t is a height of the contact point location D_0 relative to the rolling surface of the rail (K_0), (t = 10 mm). λ_p^2 is a value of «race» of the contact point D_1 relative to the vertical axis of the wheels (D_0O_0) and determined by the formula [3]:

$$\lambda_p = (r_K + t) \cdot tg(\alpha + \gamma) \cdot tg(\tau) \qquad (2)$$

where α is a angle of attack of the wheel flange on the rail;
τ is an angle of slope of the side surface of the flange to the horizon;
γ is an angle of withdrawal of the rail gauge.

The length of sliding distance of the contact point of the standard flange (D_1) along the rail head in one wheel revolution with regard to the equation (1) has the following correspondence:

$$L_{D_1} = 2 \cdot \pi \cdot r_{D_1} = 2 \cdot \pi \cdot \sqrt{t^2 + \lambda_p^2} \qquad (3)$$

When the standard wheel flange steps on the rail at an angle the contact point (Fig. 4) is displaced relative to the vertical axis of the wheel on the value of the «race» (λ_p) and is located at a point (D_1). In this case the radius of elongated cycloid loop increases and thereby the length of sliding distance of the flange side surface on the side surface.

When the guiding wheel flange of block wheelset steps on the rail at an angle the contact point (Fig. 4) is also displaced relative to the vertical axis of the wheel to the same value of the «race» (λ_p) and is located at a point (D_1). But the end of the generatrix O_0D_1 (point D_1) describes the curved trajectory $D_1D_2D_1$, which corresponds to the determinant of the irregular curve, known as «Conchoid of Nicomedes». When the guiding wheel moves translationally the generatrix O_0D_1

slides in point D_1 (instantaneous center of rotation) on the side surface of the rail head in two directions: the rotation relative to the point D1 (spin-torque) and the linear movement along a generator O_0D_1. The rotation of the generatrix O_0D_1 in point D_1 was not considered in the calculation of the value of the length of sliding distance of guiding wheel flange on the rail head because the smallness of the area of contact (D_1 point).

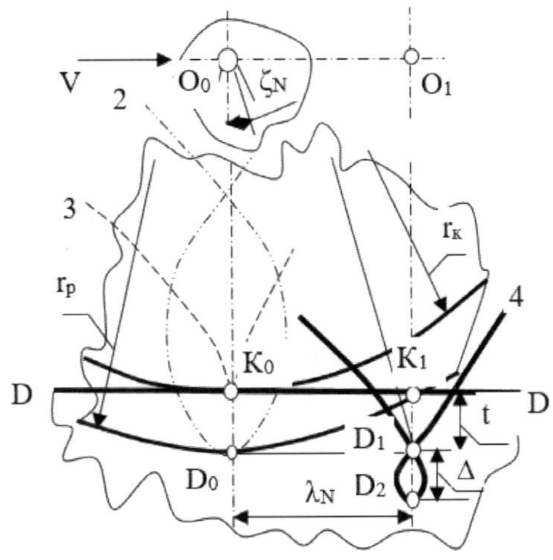

Fig. 4. Trajectory of the contact point of flange surface of guiding wheel with the rail head

The value of the length of sliding is determined from the geometric relationship in a right triangle:

$$\Delta = \sqrt{(r_k + t)^2 + \lambda_p^2} - (r_k + t) \quad (4)$$

where r_k is a radius of the running surface of the supporting wheel, mm;

Calculation results of the length of sliding distance of guiding wheel flange and standard wheel flange are fulfilled using equations 3 and 4, and finalized to one wheel rotation (Fig. 5, curves 1 and 2 respectively).

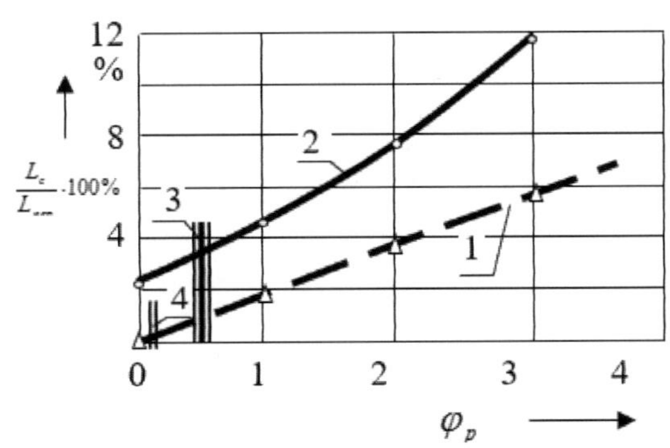

Fig. 5. Dependence of the sliding length of flanges along the rail head on the angle of attack of the wheels on the rails: 1– standard wheel; 2 – guiding wheel of block design wheelset. The maximum possible value of the angle of attack of the flanges on the rail: 3 – standard wheel; 4 – guiding wheel of the block design of wheelset.

According to the graph (Fig. 5), the main advantage of the block design of wheels is that at the zero value angle of attack of the guiding wheel on the rail the flange rolls along the side surface of the rail head without sliding in resistance to rolling regime. When the guiding wheel steps on the rail at an angle (α) the length of «parasitic» sliding distance of the flange on the side surface of trail significantly less (in 2 - 8 times) as compared to standard wheel flange. According to the results of full-scale study and theoretical calculations was determined that the at a straight section of the distance for the standard wheelset the maximum value of the angle of attack of flange is 0,5 ° [1, 5], the calculated value for the block design of wheelset is 0,2° (Fig. 5, lines 3 and 4 respectively).

4. Stress state of the elements of the block design of wheelset

If the independent rotation of the supporting and guiding wheels in block wheelset is absent the «parasitic» sliding of wheels on the rails which form the tangential stresses at the contact point of wheels and rails in the elements of standard design and the twisting moment of force in the wheel axle. The cylindrical surface of the flexible flange (Fig. 6) has a constant one-point contact ($K_0 \equiv K_p$) in the central part of the rail head surface at any position of the wheelset relative to the rail gauge and the level of excessive wear of the contacting surfaces. Therefore, the contact stress zone of normal reaction forces of the supporting and guiding wheels are located at a substantial distance from each other that provides their equal distribution across the perimeter of the rail head and wheel surfaces. It reduces significantly the levels of the contact stresses and, accordingly, a fatigue damage of metal of the block wheel and rail.

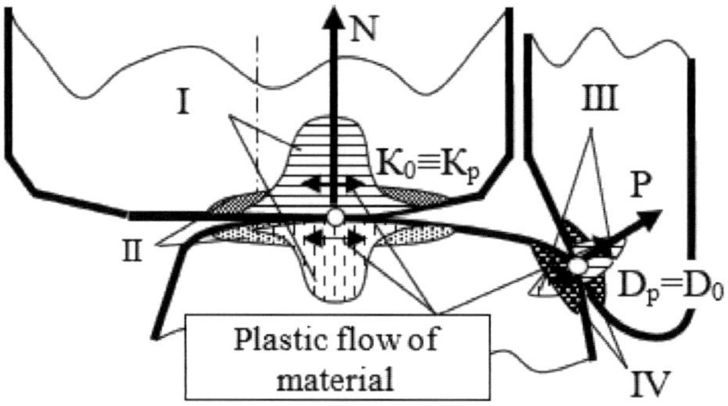

Fig. 6. Model of metal stress state of the rail and the wheel of the block design under normal (I and III) and worn-out (II and IV) profiles of running surfaces of the support and guiding wheels

5. The results of comparative tests

The results of comparative theoretical, prototype and bench tests of two wheelsets variants (block wheel and standard wheel) show the following:

- in a range of accommodate speeds up to 100 km/h the resistance to motion of block wheelset is less in 1.5-2 times as compared with the standard design (in direct and curved sections of distance);
- motion trajectory of the block wheelset in a horizontal plane depends only on a relative position of wheelset in a bogie. Irregularities of the rail gauge in the plan do not influence a horizontal dynamics of block wheelset;
- in a biaxial bogie when the rail flange touches the first block wheelset on the motion, the second block wheelset rolls on a tractrix trajectory (line of least resistance) in the direction of the guiding rail till establish contact with it. As a result, in the curved sections of the distance the bogie with the block wheelset automatically takes a chordal position in the whole speed range of the rolling stock motion;
- setting of block wheelset into existing bogies design of passenger and freight wagons allows to essentially increase an axial load and motion speed under the condition the permissible values of stresses in the elements of rails and wheelsets designs.

On the basis of executed researches the following conclusions were reached:

New technical solution of a wheelset design in which there are no mechanical closed loops, that allow to eliminate a parasitic slippage of wheels on rails in the system «wheelset - rail». It leads to a significant reduction in the intensity of the formation of such types of wear and defects as clearance under flange, side undercut of flange and slid flat. In turn, this allow to increase the run between the rolling stock maintenance up to 1 million kilometres.

6. References

[1] Pankin, I. (1992). The Nature of the Solids Friction Force, Railway Transport, Vol. 7, pp. 52-56

[2] Improving Communication between the Rolling Stock and Track, Railways of the World, No. 2, 2003, pp. 48-53

[3] W. J. Harris, S. M. Zakharov, J. Landgren, Ch. Turne, and V. Ebersen, Generalization of the World Experience of Heavy-Weight Railway Traffic: The Problems of Wheel-Rail Interaction (Intekst, Moscow, 2002) [in Russian].

[4] Galiev, I., Shiler, V., Gorbunov, P., Kychakov, K., Nikolaev, V., Talovsky, D. The Transverse Profile of Flange (patent), Patent №2378127 Russian Federation, №2008131365 at the request from 10.12.2009, pp. 1-10

[5] Kamaev, A.; Soroka, M.; Kolpakov, F. (1971). Impact on the Tracks into the Circular Curve Bogies with Freely Rotating Wheels, Dynamics of Railway Rolling Stock, Bryansk Institute of Transport Engineering, Vol. 23, pp. 156-159